College Physics

PhET Simulations
Available in the Pearson eText and in the Study Area of MasteringPhysics

*Indicates an associated tutorial available in the MasteringPhysics Item Library.

ActivPhysics OnLine™ Activities (MP) www.masteringphysics.com

About the Author   Hugh D. Young

Hugh D. Young is Emeritus Professor of Physics at Carnegie Mellon University. He earned both his undergraduate and graduate degrees from that university. He earned his Ph.D. in fundamental particle theory under the direction of the late Richard Cutkosky. He joined the faculty of Carnegie Mellon in 1956 and retired in 2004. He also had two visiting professorships at the University of California, Berkeley.

Dr. Young's career has centered entirely on undergraduate education. He has written several undergraduate-level textbooks, and in 1973 he became a coauthor with Francis Sears and Mark Zemansky for their well-known introductory texts. In addition to his authorship of Sears & Zemansky's *College Physics*, he is also coauthor, with Roger Freedman, of Sears & Zemansky's *University Physics*.

Dr. Young earned a bachelor's degree in organ performance from Carnegie Mellon in 1972 and spent several years as Associate Organist at St. Paul's Cathedral in Pittsburgh. He has played numerous organ recitals in the Pittsburgh area. Dr. Young and his wife, Alice, usually travel extensively in the summer, especially overseas and in the desert canyon country of southern Utah.

Sears & Zemansky's
College Physics

Hugh D. Young
Carnegie Mellon University

9th Edition

Addison-Wesley

Boston Columbus Indianapolis
New York San Francisco Upper Saddle River
Amsterdam Cape Town Dubai London
Madrid Milan Munich Paris Montréal Toronto
Delhi Mexico City São Paulo Sydney
Hong Kong Seoul Singapore Taipei Tokyo

Publisher:	Jim Smith
Executive Editor:	Nancy Whilton
Editorial Manager:	Laura Kenney
Director of Development:	Michael Gillespie
Senior Development Editor:	Margot Otway
Editorial Assistant:	Steven Le
Associate Media Producer:	Kelly Reed
Managing Editor:	Corinne Benson
Production Project Manager:	Beth Collins
Production Management and Composition:	PreMediaGlobal
Proofreaders:	Elka Block and Frank Purcell
Interior Designers:	Gary Hespenheide Design, Derek Bacchus
Cover Designer:	Derek Bacchus
Illustrators:	Rolin Graphics
Senior Art Editor:	Donna Kalal
Photo Researcher:	Eric Shrader
Manufacturing Buyer:	Jeff Sargent
Senior Marketing Manager:	Kerry Chapman
Cover Photo Credit:	Mike Kemp/Rubberball/Corbis

Credits and acknowledgments borrowed from other sources and reproduced, with permission, in this textbook appear on the appropriate page within the text or on p. C-1.

Library of Congress Cataloging-in-Publication Data

Young, Hugh D.

Sears & Zemansky's college physics.—9th ed. / Hugh D. Young.

p. cm.

Includes bibliographical references and index.

ISBN-13: 978-0-321-73317-7 (alk. paper)

ISBN-10: 0-321-73317-7 (alk. paper)

1. Physics—Textbooks. I. Sears, Francis Weston, 1898–1975. College physics. II. Title. III. Title: College physics. IV. Title: Sears and Zemansky's college physics.

QC23.2.Y68 2012

530—dc22

2010046658

College Physics—Complete Edition
ISBN 10: 0-321-73317-7; ISBN 13: 978-0-321-73317-7 (Student edition)
ISBN 10: 0-321-73315-0; ISBN 13: 978-0-321-73315-3 (Exam copy)

1 2 3 4 5 6 7 8 9 10—WBC—14 13 12 11 10

Addison-Wesley
is an imprint of

PEARSON

www.pearsonhighered.com

Brief Contents

Build Skills

Learn basic and advanced skills that help solve a broad range of physics problems.

Problem-Solving Strategies coach students in how to approach specific types of problems. ▶

This text's uniquely extensive set ▶
of **Examples** enables students to explore problem-solving challenges in exceptional detail.

Consistent
The **Set Up / Solve / Reflect** format, used in all Examples, encourages students to tackle problems thoughtfully rather than skipping to the math.

Visual
Most Examples employ a diagram— often a **pencil sketch** that shows what a student should draw.

PROBLEM-SOLVING STRATEGY 24.1 **Image formation by mirrors** (MP)

SET UP

1. The principal-ray diagram is to geometric optics what the free-body diagram is to mechanics! When you attack a problem involving image formation by a mirror, *always* draw a principal-ray diagram first if you have enough information. (And apply the same advice to lenses in the sections that follow.) It's usually best to orient your diagrams consistently, with the incoming rays traveling from left to right. Don't draw a lot of other rays at random; stick with the principal rays—the ones that you know something about.

2. If your principal rays don't converge at a real image point, you may have to extend them straight backward to locate a virtual image point. We recommend drawing the extensions with broken lines. Another useful aid is to color-code your principal rays consistently; for example, referring to the preceding definitions of principal rays, Figure 24.19 uses purple for 1, green for 2, orange for 3, and pink for 4.

SOLVE

3. Identify the known and unknown quanties, such as s, s', R, and f. Make lists of the known and unknown quantities, and identify the relationships among them; then substitute the known values ~~and solve for the~~ image distances, ~~s, and object and~~

EXAMPLE 24.1 **Image from a concave mirror**

A lamp is placed 10 cm in front of a concave spherical mirror that forms an image of the filament on a screen placed 3.0 m from the mirror. What is the radius of curvature of the mirror? If the lamp filament is 5.0 mm high, how tall is its image? What is the lateral magnification?

SOLUTION

SET UP Figure 24.14 shows our diagram.

SOLVE Both object distance and image distance are positive; we have $s = 10$ cm and $s' = 300$ cm. To find the radius of curvature, we use Equation 24.4:

$$\frac{1}{s} + \frac{1}{s'} = \frac{2}{R},$$
$$\frac{1}{10 \text{ cm}} + \frac{1}{300 \text{ cm}} = \frac{2}{R},$$

and $R = 19.4$ cm. To find the height of the image, we use Equation 24.7:

$$m = \frac{y'}{y} = -\frac{s'}{s},$$
$$\frac{y'}{5.0 \text{ mm}} = -\frac{300 \text{ cm}}{10 \text{ cm}},$$

and $y' = -150$ mm. The lateral magnification m is

$$m = \frac{y'}{y} = \frac{-150 \text{ mm}}{5 \text{ mm}} = -30.$$

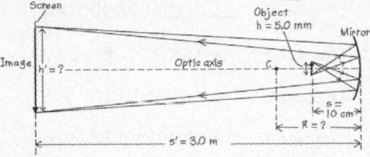

▲ **FIGURE 24.14** Our diagram for this problem (not to scale).

REFLECT The image is inverted (as we know because $m = -30$ is negative) and is 30 times taller than the object. Notice that the filament is *not* located at the mirror's focal point; the image is not formed by rays parallel to the optic axis. (The focal length of this mirror is $f = R/2 = 9.7$ cm.)

Practice Problem: A concave mirror has a radius of curvature $R = 25$ cm. An object of height 2 cm is placed 15 cm in front of the mirror. What is the image distance? What is the height of the image? *Answers:* $s' = 75$ cm, $y' = -10$ cm.

...work for all four ...plane and spherical ...ities mentioned in ...the sign rules care-

---

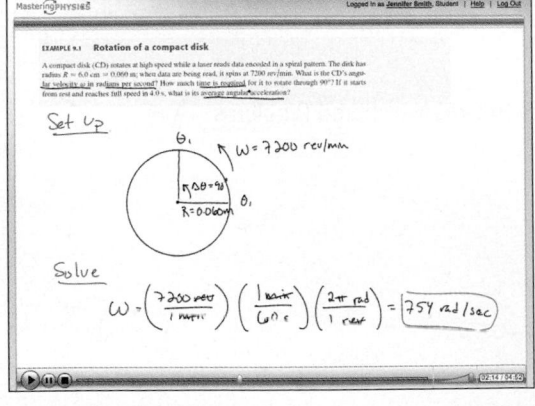

▲
NEW! Video Tutor Solution for Every Example
Each Example is explained and solved by an instructor in a Video Tutor solution provided in the Study Area of MasteringPhysics® and in the Pearson eText.

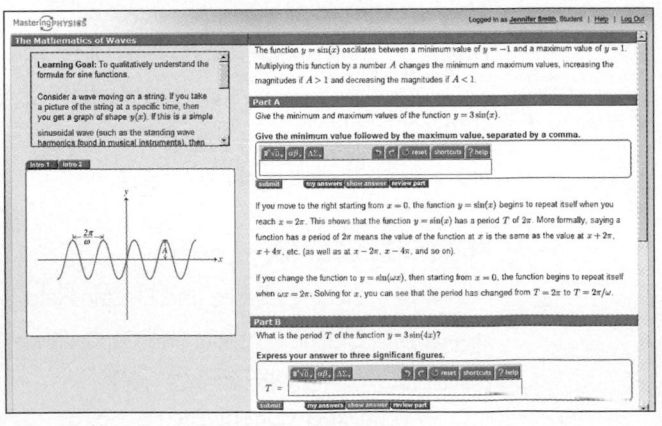

▲
NEW! Mathematics Review Tutorials
MasteringPhysics offers an extensive set of assignable mathematics review tutorials, covering all the areas in which students typically have trouble.

Build Confidence

Develop problem-solving confidence through a range of practice options—from guided to unguided.

NEW! **Passage Problems**, which use the same reading-passage format as most MCAT questions, develop students' ability to apply physics to a real-world situation (often biological or biomedical in nature). ▶

About 20% of the **End-of-Chapter Problems** are new or revised. These revisions are driven by detailed student-performance data gathered nationally through MasteringPhysics.®

Problem difficulty is indicated by a three-dot ranking system based on data from MasteringPhysics.

22. •• A grasshopper leaps into the air from the edge of a vertical cliff, as shown in Figure 3.38. Use information from the figure to find (a) the initial speed of the grasshopper and (b) the height of the cliff.

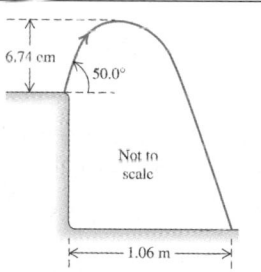

▲ FIGURE 3.38 Problem 22.

23. •• Firemen are shooting a stream of water at a burning building. A high-pressure hose shoots out the water with a speed of 25.0 m/s as it leaves the hose nozzle. Once it leaves the hose, the water moves in projectile motion. The firemen adjust the angle of elevation of the hose until the water takes 3.00 s to reach a building 45.0 m away. You can ignore air resistance; assume that the end of the hose is at ground level. (a) Find the angle of elevation of the hose. (b) Find the speed and acceleration of the water at the highest point in its trajectory. (c) How high above the ground does the water strike the building, and how fast is it moving just before it hits the building?

24. •• Show that a projectile achieves its maximum range when it is fired at 45° above the horizontal if $y = y_0$.

Passage Problems

BIO Stimulating the brain.

Communication in the nervous system is based on propagating electrical signals called action potentials that travel along the extended nerve cell processes, the axons. Action potentials are generated when the electrical potential difference across the membrane changes so that the inside of the cell becomes more positive. Researchers in clinical medicine and neurobiology want to stimulate nerves non-invasively at specific locations in conscious subjects. But using electrodes to apply current on the skin is painful and requires large currents.

Anthony Barker and colleagues at the University of Sheffield in England developed a technique that is now widely used called transcranial magnetic stimulation (TMS). In the TMS technique, a coil positioned near the skull produces a time-varying magnetic field, which induces electric currents in the conductive brain tissue sufficient enough to cause action potentials in nerve cells. For example, if the coil is placed near the motor cortex, the region of the brain that controls voluntary movement, scientists can monitor the contraction of muscles and assess the state of the connections between the brain and the muscle.

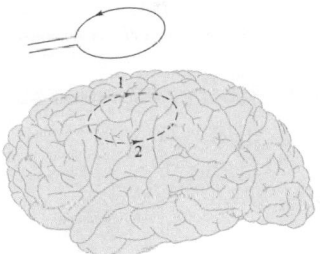

▲ FIGURE 21.70 Problems 64–66.

64. In the diagram of TMS shown in Figure 21.70, a current pulse increases to a peak and then decreases to zero in the direction shown in the stimulating coil. What will be the direction (1 or 2) of the induced current (dotted line) in the brain tissue?
   A. 1
   B. 2
   C. 1 while the current increases in the stimulating coil and 2 while the current decreases
   D. 2 while the current increases in the stimulating coil, 1 while the current decreases

NEW! **Enhanced End-of-Chapter Problems in MasteringPhysics** ▶
Select end-of-chapter problems will now offer additional support such as problem-solving strategy hints, relevant math review and practice, and links to the eText. These new enhanced problems bridge the gap between guided tutorials and traditional homework problems.

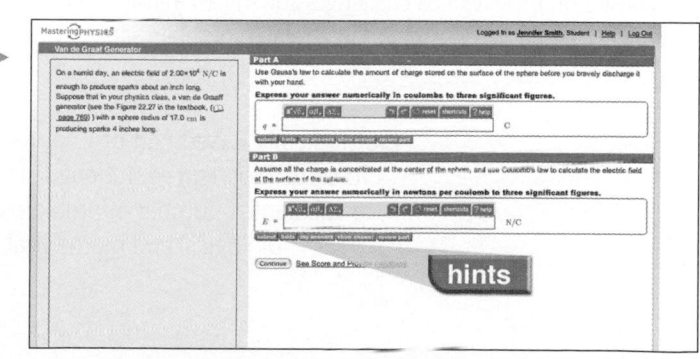

Bring Physics **to Life**

Deepen knowledge of physics by building connections to the real world.

ENHANCED! Applications of Physics ▶
Throughout the text, captioned photos apply physics to real situations, with particular emphasis on applications of biomedical and general interest.

▲ **BIO Application Real-time molecular biology.** The patch clamp technique is an ingenious way to investigate how cells work. To create a patch clamp, a polished microelectrode is carefully manipulated to the outer membrane of a cell to make a tight seal. Protein molecules called ion channel is isolated within the experimenter can then study the electric properties of the single channel, manipulate the voltage membrane (and thus the channel) we know that an electrical force across the membrane and thus the movement of membrane. At the appropriate channel flicks open for a (ms) and then closes. The opening of voltage-gated channels electrical signaling in most many aspects of cellular inventors of the patch clamp awarded the Nobel Prize in medicine in 1991.

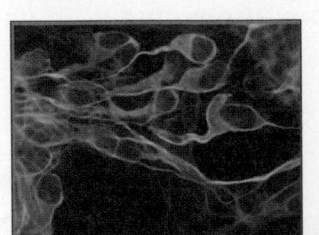

▲ **BIO Application A long time coming.** Proportional reasoning may not make sense for two variables with a non-linear relationship. Albert Einstein's seminal work on Brownian motion showed that the distance that molecules diffuse is related to the time diffusion by a quadratic relation $x = kt^2$, where k incorporates information about the diffusing molecule and the medium through which the molecules are moving. Proteins and informational molecules can diffuse from the center of a typical spherical biological cell (diameter 20 μm) to the cell's periphery in about 1 s. In the case of a motor neuron, which the cell nucleus (in red) is in the spinal cord, an extension of the cell called the axon is sent to the target muscle and may be 1 m long in humans. Proportional reasoning shows that the time required for molecules to diffuse from the nucleus to the end of the cell where they are used would be 10^{10} s or about 300 years. This is an impossibly long time, neurons have evolved sophisticated mechanisms for moving materials around much more rapidly.

▲ **BIO Application Knowing up from down.** Plant roots have exquisitely developed mechanisms for sensing gravity and growing downward. If the direction of a root is changed, say by running into a rock in the soil, the root is forced to grow horizontally; however, as soon as it can, it again turns downward. This ability involves a number of cellular signals, but the primary sensor detects acceleration. The sensing cells contain statoliths, specialized starch-containing granules that are denser than the fluid in the cells and, in response to changes in the direction of gravity, fall (are accelerated) to a different location within the cells. This triggers an active localized response of the cells and causes the growth direction of the root to re-orient so that the root once again grows down.

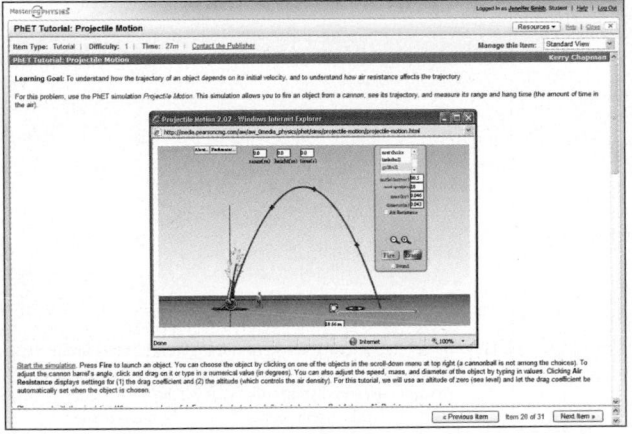

◀ **NEW! PhET Simulations and Tutorials**
76 PhET simulations are provided in the Study Area of the MasteringPhysics® website and in the Pearson eText. In addition, MasteringPhysics contains 16 new, assignable PhET-based tutorials.

NEW! Video Tutor Demonstrations and Tutorials
"Pause and predict" demonstration videos of key physics concepts engage students by asking them to submit a prediction before seeing the outcome. These videos are available through the Study Area of MasteringPhysics and in the Pearson eText. A set of assignable tutorials based on these videos challenge students to transfer their understanding of the demonstration to a related problem situation.

Biomedically Based End-of-Chapter Problems ▶
To serve biosciences students, the text offers a substantial number of problems based on biological and biomedical situations.

38. • **Chin brace.** A person with an injured jaw has a brace
BIO below his chin. The brace is held in place by two cables directed at 65° above the horizontal. (See Figure 1.23.) The cables produce forces of equal magnitude having a vertical resultant of 2.25 N upward. (a) Make a scale drawing showing both the forces produced by the cables and the resultant force. Estimate the angle carefully or measure it with a protractor. (b) Use your scale drawing to estimate the magnitude of the force due to each cable.

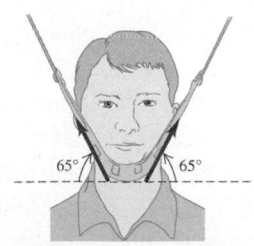

▲ **FIGURE 1.23** Problem 38.

Make a Difference with
MasteringPhysics®

www.masteringphysics.com

MasteringPhysics is the most effective and widely used online science tutorial, homework, and assessment system available.

NEW! Pre-Built Assignments ▶
For every chapter in the book, MasteringPhysics now provides pre-built assignments that cover the material with a tested mix of tutorials and end-of-chapter problems of graded difficulty. Professors may use these assignments as-is or take them as a starting point for modification.

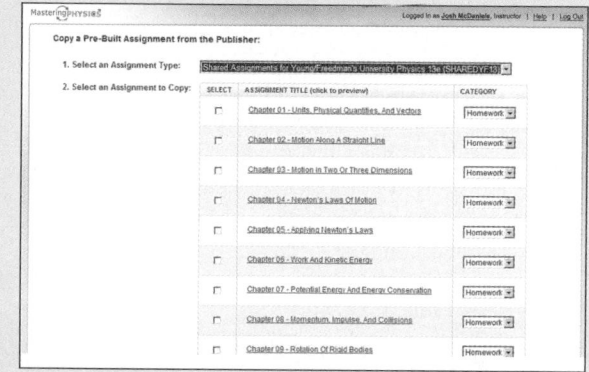

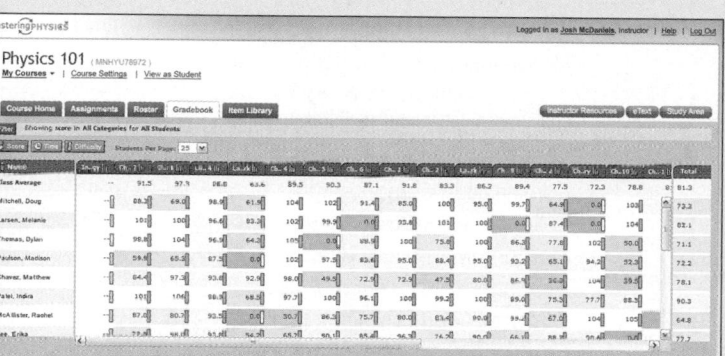

◀ **Gradebook**
- Every assignment is graded automatically.
- Shades of red highlight vulnerable students and challenging assignments.

Class Performance on Assignment ▶
Click on a problem to see which step your students struggled with most, and even their most common wrong answers. Compare results at every stage with the national average or with your previous class.

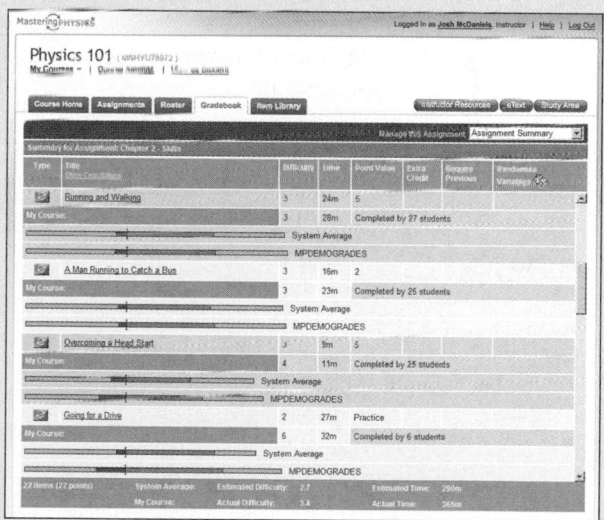

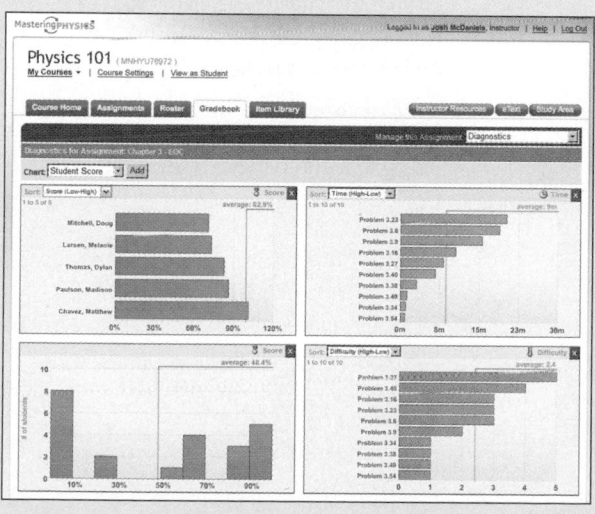

◀ **Gradebook Diagnostics**
This screen provides your favorite weekly diagnostics. With a single click, charts summarize the most difficult problems, vulnerable students, grade distribution, and even improvement in scores over the course.

Real-World Applications

BIO indicates bioscience applications.

xv

Real-World Applications (continued)

To the Student

How to Succeed in Physics

"Is physics hard? Is it too hard for me?" Many students are apprehensive about their physics course. However, while the course can be challenging, almost certainly it is *not* too hard for you. If you devote time to the course and use that time wisely, you can succeed.

Here's how to succeed in physics.

1. **Spend time studying.** The rule of thumb for college courses is that you should expect to study about 2 to 3 hours per week for each unit of credit, *in addition* to the time you spend in class. And budget your time: 3 hours every other day is much more effective than 33 hours right before the exam.

   The good news is that physics is consistent. Once you've learned how to tackle one topic, you'll use the same study techniques to tackle the rest of the course. So if you find you need to give the course extra time at first, do so and don't worry—it'll pay dividends as the course progresses.

2. **Don't miss class.** Yes, you could borrow a friend's notes, but listening and participating in class are far more effective. Of course, *participating* means paying active attention, and interacting when you have the chance!

3. **Make this book work for you.** This text is packed with decades of teaching experience—but to make it work for you, you must read and use it *actively*. *Think* about what the text is saying. *Use* the illustrations. Try to *solve* the Examples and the Quantitative and Conceptual Analysis problems on your own, before reading the solutions. If you *underline*, do so thoughtfully and not mechanically.

   A good practice is to skim the chapter before going to class to get a sense for the topic, and then read it carefully and work the examples after class.

4. **Approach physics problems systematically.** While it's important to attend class and use the book, your *real* learning will happen mostly as you work problems—*if* you approach them correctly. Physics problems aren't math problems. You need to approach them in a different way. (If you're "not good at math," this may be good news for you!) What you do before and after solving an equation is more important than the math itself. The worked examples in this book help you develop good habits by consistently following three steps—*Set Up*, *Solve*, and *Reflect*. (In fact, this global approach will help you with problem solving in all disciplines—chemistry, medicine, business, etc.)

5. **Use campus resources.** If you get stuck, get help. Your professor probably has office hours and email; use them. Use your TA or campus tutoring center if you have one. Partner with a friend to study together. But also try to get unstuck on your own *before* you go for help. That way, you'll benefit more from the help you get.

   So remember, you *can* succeed in physics. Just devote time to the job, work lots of problems, and get help when you need it. Your book is here to help. Have fun!

SET UP
Think about the physics involved in the situation the problem describes. What information are you given and what do you need to find out? Which physics principles do you need to apply? Almost always you should *draw a sketch* and label it with the relevant known and unknown information. (Many of the worked examples in this book include hand-drawn sketches to coach you on what to draw.)

SOLVE
Based on what you did in Set Up, identify the physics and appropriate equation or equations and do the algebra. Because you started by *thinking about the physics* (and *drawing a diagram*), you'll know which physics equations apply to the situation—you'll avoid the "plug and pray" trap of picking any equation that seems to have the right variables.

REFLECT
Once you have an answer, ask yourself whether it is plausible. If you calculated your weight on the Moon to be 10,423 kg—you must have made a mistake somewhere! Next, check that your answer has the right units. Finally, think about what *you* learned from the problem that will help you later.

Preface

College Physics has been a success for over half a century, placing equal emphasis on quantitative, qualitative, and conceptual understanding.. Guided by tested and proven innovations in education research, we have revised and enhanced previous material and added new features focusing on more explicit problem-solving steps and techniques, conceptual understanding, and visualization and modeling skills. Our main objectives are to teach a solid understanding of the fundamentals, help students develop critical thinking and quantitative reasoning, teach sound problem-solving skills, and spark the students' interest in physics with interesting and relevant applications.

This text provides a comprehensive introduction to physics at the beginning college level. It is intended for students whose mathematics preparation includes high-school algebra and trigonometry but no calculus. The complete text may be taught in a two-semester or three-quarter course, and the book is also adaptable to a wide variety of shorter courses.

New to This Edition

- **New Chapter 0 (Mathematics Review)** covers math concepts that students will need to use throughout the course: Exponents; scientific notation and powers of 10; algebra; direct, inverse, and inverse-square relationships; logarithmic and exponential functions; areas and volumes; and plane geometry and trigonometry. This review chapter includes worked examples and end-of-chapter problems.
- **New margin applications** include over 40 new biosciences-related applications with photos added to the text, including those focused on cutting-edge technology. BIO icons signify the bio-related applications.
- **Changes to the end-of-chapter problems** include the following:

  - **15–20% of the problems are new.**
  - **Many additional biosciences-related problems.**
  - **One set of MCAT-style passage problems** added at the end of most chapters, many of them bio-related.

  The addition of new biological and biomedical real-world applications and problems gives this edition more coverage in the biosciences than nearly every other book on the market.

- **Over 70 PhET simulations** are linked to the Pearson eText and are provided in the study area of the MasteringPhysics website (with icons in the print text). These powerful simulations allow students to interact productively with the physics concepts they are learning. PhET clicker questions are also included on the Instructor Resource DVD.
- **Video Tutors bring key content to life throughout the text:**

  - **Dozens of Video Tutors feature "pause-and-predict" demonstrations of key physics concepts** and incorporate assessment as the student progresses, to actively engage the student in understanding the key conceptual ideas underlying the physics principles.
  - **Every Worked Example in the book is accompanied by a Video Tutor Solution** that walks students through the problem-solving process, providing a virtual teaching assistant on a round-the-clock basis.
  - **All of these Video Tutors play directly through links within the Pearson eText.** Many also appear in the Study area within MasteringPhysics.

Complete and Two-Volume Editions

With MasteringPhysics:
- **Complete Edition:** Chapters 1–30
  (ISBN 978-0-321-74980-2)

Without MasteringPhysics:
- **Complete Edition:** Chapters 1–30
  (ISBN 978-0-321-73317-7)
- **Volume 1:** Chapters 1–16
  (ISBN 978-0-321-76624-3)
- **Volume 2:** Chapters 17–30
  (ISBN 978-0-321-76623-6)

- Assignable MasteringPhysics tutorials are based on the Video Tutor Demonstrations and PhET simulations.

  - **Video Tutor Demonstrations will be expanded to tutorials in Mastering** by requiring the student to transfer their understanding to a new problem situation so that these will be gradable and distinct from the "pause and predict" demonstrations alone.
  - Sixteen new **PhET tutorials** enable students to not only explore the PhET simulations but also answer questions, helping them make connections between real-life phenomena and the underlying physics that explain such phenomena.

Key Features of *College Physics*

- **A systematic approach to problem solving.** To solve problems with confidence, students must learn to approach problems effectively at a global level, must understand the physics in question, and must acquire the specific skills needed for particular types of problems. The Ninth Edition provides research-proven tools for students to tackle each goal.
- The **worked examples** all follow a consistent and explicit **global problem-solving strategy** drawn from educational research. This three-step approach puts special emphasis on how to **set-up** the problem before trying to **solve** it, and the importance of how to **reflect** on whether the answer is sensible.
- Worked example solutions emphasize the steps and decisions students often skip. In particular, many worked examples include **pencil diagrams:** hand-drawn diagrams that show exactly what a student should draw in the **set-up** step of solving the problem.
- **Conceptual Analysis** and **Quantitative Analysis** problems help the students practice their qualitative and quantitative understanding of the physics. The Quantitative Analysis problems focus on skills of quantitative and proportional reasoning—skills that are key to success on the MCATs. The CAs and QAs use a multiple-choice format to elicit specific common misconceptions.
- **Problem-solving strategies** teach the students tactics for particular types of problems—such as problems requiring Newton's second law, energy conservation, etc.—and follow the same 3-step global approach (set-up, solve, and reflect).
- **Unique, highly effective figures incorporate the latest ideas from educational research.** Extraneous detail has been removed and color used only for strict pedagogical purposes—for instance, in mechanics, **color is used to identify the object of interest,** while all other objects are grayscale. **Illustrations include helpful blue annotated comments** to guide students in 'reading' graphs and physics figures. Throughout, **figures, models, and graphs are placed side by side** to help students 'translate' between multiple representations. **Pencil sketches** are used consistently in worked examples to emphasize what students should draw.
- **Visual chapter summaries** show each concept in words, math, and figures to reinforce how to 'translate' between different representations and address different student learning styles.
- **Rich and diverse end-of-chapter problem sets.** The renowned Sears & Zemansky problems, refined over five decades of use, have been revised, expanded and enhanced for today's courses, based on data from MasteringPhysics.
- Each chapter includes a set of **multiple-choice problems** that test the skills developed by the Qualitative Analysis and Quantitative Analysis problems in the chapter text. The multiple-choice format elicits specific common misconceptions, enabling students to pinpoint their misunderstandings.

- The General Problems contain many **context-rich problems** (also known as **real-world problems**), which require students to simplify and model more complex real-world situations. Many problems relate to the field of biology and medicine; these are all labeled BIO.
- **Connections of physics to the student's world.** In-margin photos with explanatory captions provide diverse, interesting, and self-contained examples of physics at work in the world. Many of these real-world "applications" are also related to the fields of biology and medicine and are labeled BIO.
- **Writing that is easy to follow and rigorous.** The writing is friendly yet focused; it conveys an exact, careful, straightforward understanding of the physics, with an emphasis on the connections between concepts.

Instructor Supplements

Note: For convenience, all of the following instructor supplements (except for the Instructor Resource DVD) can be downloaded from the Instructor Area, accessed via the left-hand navigation bar of MasteringPhysics (www.masteringphysics.com).

Instructor Solutions, prepared by A. Lewis Ford (Texas A&M University) and Forrest Newman (Sacramento City College) contain complete and detailed solutions to all end-of-chapter problems. All solutions follow consistently the same Set Up/Solve/Reflect problem-solving framework used in the textbook. Download only from the MasteringPhysics Instructor Area or from the Instructor Resource Center (www.pearsonhighered.com/irc).

The cross-platform **Instructor Resource DVD** (ISBN 978-0-321-76570-3) provides a comprehensive library of more than 420 applets from ActivPhysics OnLine as well as all line figures from the textbook in JPEG format. In addition, all the key equations, problem-solving strategies, tables, and chapter summaries are provided in editable Word format. Lecture outlines in PowerPoint are also included along with over 70 PhET simulations as well as Pause and Predict Video Tutors and Video Tutor Solutions.

MasteringPhysics® (www.masteringphysics.com) is the most advanced, educationally effective, and widely used physics homework and tutorial system in the world. Eight years in development, it provides instructors with a library of extensively pre-tested end-of-chapter problems and rich, multipart, multistep tutorials that incorporate a wide variety of answer types, wrong answer feedback, individualized help (comprising hints or simpler sub-problems upon request), all driven by the largest metadatabase of student problem-solving in the world. NSF-sponsored published research (and subsequent studies) show that Mastering-Physics has dramatic educational results. MasteringPhysics allows instructors to build wide-ranging homework assignments of just the right difficulty and length and provides them with efficient tools to analyze both class trends, and the work of any student in unprecedented detail.

MasteringPhysics routinely provides instant and individualized feedback and guidance to more than 100,000 students every day. A wide range of tools and support make MasteringPhysics fast and easy for instructors and students to learn to use. Extensive class tests show that by the end of their course, an unprecedented eight of nine students recommend MasteringPhysics as their preferred way to study physics and do homework.

MasteringPhysics enables instructors to:

- Quickly build homework assignments that combine regular end-of-chapter problems and tutoring (through additional multi-step tutorial problems that offer wrong-answer feedback and simpler problems upon request).
- Expand homework to include the widest range of automatically graded activities available–from numerical problems with randomized values, through algebraic answers, to free-hand drawing.

- Choose from a wide range of nationally pre-tested problems that provide accurate estimates of time to complete and difficulty.
- After an assignment is completed, quickly identify not only the problems that were the trickiest for students but the individual problem types where students had trouble.
- Compare class results against the system's worldwide average for each problem assigned, to identify issues to be addressed with just-in-time teaching.
- Check the work of an individual student in detail, including time spent on each problem, what wrong answers they submitted at each step, how much help they asked for, and how many practice problems they worked.

ActivPhysics OnLine™ (accessed through the Study Area within www.masteringphysics.com) provides a comprehensive library of more than 420 tried and tested ActivPhysics applets updated for web delivery using the latest online technologies. In addition, it provides a suite of highly regarded applet-based tutorials developed by education pioneers Alan Van Heuvelen and Paul D'Alessandris. Margin icons throughout the text direct students to specific exercises that complement the textbook discussion.

The online exercises are designed to encourage students to confront misconceptions, reason qualitatively about physical processes, experiment quantitatively, and learn to think critically. The highly acclaimed ActivPhysics OnLine companion workbooks help students work through complex concepts and understand them more clearly. More than 420 applets from the ActivPhysics OnLine library are also available on the Instructor Resource DVD for this text.

The **Test Bank** contains more than 2,000 high-quality problems, with a range of multiple-choice, true/false, short-answer, and regular homework-type questions. Test files are provided both in TestGen (an easy-to-use, fully networkable program for creating and editing quizzes and exams) and Word format. Download only from the MasteringPhysics Instructor Area or from the Instructor Resource Center (www.pearsonhighered.com/irc).

Five Easy Lessons: Strategies for Successful Physics Teaching (ISBN 978-0-8053-8702-5) by Randall D. Knight (California Polytechnic State University, San Luis Obispo) is packed with creative ideas on how to enhance any physics course. It is an invaluable companion for both novice and veteran physics instructors.

Student Supplements

The **Student Solutions Manual,** (ISBN 978-0-321-74769-3), written by Lewis Ford (Texas A&M University) and Forrest Newman (Sacramento City College), contains detailed, step-by-step solutions to more than half of the odd-numbered end-of-chapter problems from the textbook. All solutions consistently follow the same Set Up/Solve/Reflect problem-solving framework used in the textbook, reinforcing good problem-solving behavior.

 **MasteringPhysics®** (www.masteringphysics.com) is a homework, tutorial, and assessment system based on years of research into how students work physics problems and precisely where they need help. Studies show that students who use MasteringPhysics significantly increase their scores compared to handwritten homework. MasteringPhysics achieves this improvement by providing students with instantaneous feedback specific to their wrong answers, simpler sub-problems upon request when they get stuck, and partial credit for their method(s). This individualized, 24/7 Socratic tutoring is recommended by 9 out of 10 students to their peers as the most effective and time-efficient way to study.

Pearson eText is available through MasteringPhysics, either automatically when MasteringPhysics is packaged with new books, or available as a purchased upgrade online. Allowing students access to the text wherever they have access to the Internet, Pearson eText comprises the full text, including figures that can be enlarged for better viewing. With eText, students are also able to pop up definitions and terms to help with vocabulary and the reading of the material. Students can also take notes in eText using the annotation feature at the top of each page.

Pearson Tutor Services (www.pearsontutorservices.com). Each student's subscription to MasteringPhysics also contains complimentary access to Pearson Tutor Services, powered by Smarthinking, Inc. By logging in with their MasteringPhysics ID and password, they will be connected to highly qualified e-instructors who provide additional interactive online tutoring on the major concepts of physics. Some restrictions apply; offer subject to change.

ActivPhysics OnLine (accessed through the Study Area within www. masteringphysics.com) provides students with a suite of highly regarded applet-based tutorials (see above). The following workbooks help students work through complex concepts and understand them more clearly.

ActivPhysics OnLine Workbook, Volume 1: Mechanics * Thermal Physics * Oscillations & Waves (978-0-8053-9060-5)

ActivPhysics OnLine Workbook, Volume 2: Electricity & Magnetism * Optics * Modern Physics (978-0-8053-9061-2)

Acknowledgments

I want to extend my heartfelt thanks to my colleagues at Carnegie Mellon for many stimulating discussions about physics pedagogy and for their support and encouragement during the writing of several successive editions of this book. I am equally indebted to the many generations of Carnegie Mellon students who have helped me learn what good teaching and good writing are, by showing me what works and what doesn't. I'm pleased to acknowledge also the contributors of problems, applications, and other essential elements for this new edition, including Ken Robinson, Charlie Hibbard, Forrest Newman, Larry Coleman, and Biman Das. Special thanks are due to the Addison-Wesley people, especially Laura Kenney and Margot Otway, who brought this all together, and to Nancy Whilton and Kerry Chapman. Thanks also to Jared Sterzer at PrcMediaGlobal. Finally, and most importantly, it is always a joy and a privilege to express my gratitude to my wife Alice and our children Gretchen and Rebecca for their love, support, and emotional sustenance during the writing of several successive editions of this book. May all men and women be blessed with love such as theirs.

—H.D.Y.

Reviewers and Classroom Testers

Susmita Acharya, *Cardinal Stritch University*

Hamid Aidinejad, *Florida Community College, Jacksonville*

Alice Hawthorne Allen, *Virginia Tech*

Jim Andrews, *Youngstown State University*

Charles Bacon, *Ferris State University*

Jennifer Blue, *Miami University*

Richard Bone, *Florida International University*

Phillip Broussard, *Covenant College*

Young Choi, *University of Pittsburgh*

Orion Ciftja, *Prairie View A&M University*

Dennis Collins, *Grossmont College*

Lloyd Davis, *Montreat College*

Diana Driscoll, *Case Western Reserve University*

Laurencin Dunbar, *St. Louis Community College, Florissant Valley*

Alexander Dzyubenko, *California State University, Bakersfield*

Robert Ehrlich, *George Mason University*

Mark Fair, *Grove City College*

Shamanthi Fernando, *Northern Kentucky University*

Len Feuerhelm, *Oklahoma Christian University*

Carl Frederickson, *University of Central Arkansas*

Mikhail Goloubev, *Bowie State University*

Alan Grafe, *University of Michigan, Flint*

William Gregg, *Louisiana State University*

John Gruber, *San Jose State University*

Robert Hagood, *Washtenaw Community College*

Scott Hildreth, *Chabot College*

Andy Hollerman, *University of Louisiana, Lafayette*

John Hubisz, *North Carolina State University*

Manuel Huerta, *University of Miami*

Todd Hurt, *Chatham College*

Adam Johnston, *Weber State University*

Roman Kezerashvili, *New York City College of Technology*

Ju Kim, *University of North Dakota*

Jeremy King, *Clemson University*

David Klassen, *Rowan University*

Ichishiro Konno, *University of Texas, San Antonio*

Ikka Koskelo, *San Francisco State University*

Jon Levin, *University of Kentucky*

David Lind, *Florida State University*

Dean Livelybrooks, *University of Oregon, Eugene*

Estella Llinas, *University of Pittsburgh, Greensburg*

Craig Loony, *Merrimack College*

Rafael Lopez-Mobilia, *University of Texas, San Antonio*

Barbra Maher, *Red Rocks Community College*

Dan Mazilu, *Virginia Tech*

Randy McKee, *Tallahassee Community College*

Larry McRae, *Berry College*

William Mendoza, *Jacksonville University*

Anatoli Mirochnitchenko, *University of Toledo*

Charles Myles, *Texas Tech University*

Austin Napier, *Tufts University*

Erin O' Connor, *Allan Hancock College*

Christine O'Leary, *Wallace State College*

Jason Overby, *College of Charleston*

James Pazun, *Pfeiffer University*

Unil Perera, *Georgia State University*

David Potter, *Austin Community College*

Michael Pravica, *University of Nevada, Las Vegas*

Sal Rodano, *Harford Community College*

Rob Salgado, *Dillard University*

Surajit Sen, *SUNY Buffalo*

Bart Sheinberg, *Houston Community College*

Natalia Sidorovskaia, *University of Louisiana*

Chandralekha Singh, *University of Pittsburgh*

Marlina Slamet, *Sacred Heart University*

Daniel Smith, *South Carolina State University*

Gordon Smith, *Western Kentucky University*

Kenneth Smith, *Pennsylvania State University*

Zhiyan Song, *Savannah State University*

Sharon Stephenson, *Gettysburg College*

Chuck Stone, *North Carolina A&T State University*

George Strobel, *University of Georgia*

Chun Fu Su, *Mississippi State University*

Brenda Subramaniam, *Cypress College*

Mike Summers, *George Mason University*

Eric Swanson, *University of Pittsburgh*

Colin Terry, *Ventura College*

Vladimir Tsifrinovich, *Polytechnic University*

Gajendra Tulsian, *Daytona Beach Community College*

Paige Uozts, *Lander University*

James Vesenka, *University of New England*

Walter Wales, *University of Pennsylvania*

John Wernegreen, *Eastern Kentucky University*

Dan Whitmire, *University of Louisiana, Lafayette*

Sue Willis, *Northern Illinois University*

Jaehoon Yu, *University of Texas, Arlington*

Nouredine Zettili, *Jacksonville State University*

Bin Zhang, *Arkansas State University*

Detailed Contents

18 Electric Potential and Capacitance   582

19 Current, Resistance, and Direct-Current Circuits   618

20 Magnetic Field and Magnetic Forces   658

21 Electromagnetic Induction   698

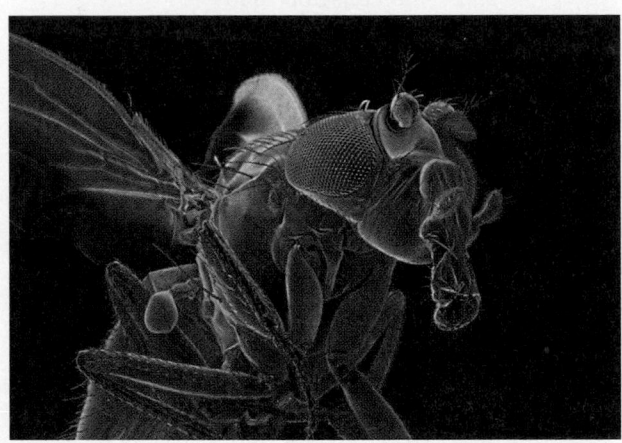

College Physics

0 Mathematics Review

A study of physics at the level of this textbook requires some basic math skills. The relevant math topics are summarized in this chapter. We strongly recommend that you review this material, practice end-of-chapter problems, and become comfortable with these as quickly as possible, so that during your physics course, you can focus on the physics concepts and procedures that are being introduced, without being distracted by unfamiliarity with the math being used. Note that the beauty of physics cannot be enjoyed if you do not have adequate mastery of basic mathematical skills.

The arrangement of seeds in a sunflower is a classic example of how natural processes give rise to patterns that can be expressed by means of fairly simple mathematics. In this chapter, we will review most of the mathematics you will need for this course.

0.1 Exponents

Exponents are used frequently in physics. When we write 3^4, the superscript 4 is called an **exponent** and the **base number** 3 is said to be raised to the fourth power. The quantity 3^4 is equal to $3 \times 3 \times 3 \times 3 = 81$. Algebraic symbols can also be raised to a power—for example, x^4. There are special names for the operation when the exponent is 2 or 3. When the exponent is 2, we say that the quantity is **squared**; thus, x^2 means x is squared. When the exponent is 3, the quantity is **cubed**; hence, x^3 means x is cubed.

Note that $x^1 = x$, and the exponent 1 is typically not written. Any quantity raised to the zero power is defined to be unity (that is, 1). Negative exponents are used for reciprocals: $x^{-4} = 1/x^4$. An exponent can also be a fraction, as in $x^{1/4}$. The exponent $\frac{1}{2}$ is called a **square root,** and the exponent $\frac{1}{3}$ is called a **cube root.**

For example, $\sqrt{6}$ can also be written as $6^{1/2}$. Most calculators have special keys for calculating numbers raised to a power—for example, a key labeled y^x or one labeled x^2.

Exponents obey several simple rules, which follow directly from the meaning of raising a quantity to a power:

1. The product rule: $(x^m)(x^n) = x^{m+n}$.
    For example, $(3^3)(3^2) = 3^5 = 243$. To verify this result, note that $3^3 = 27$, $3^2 = 9$, and $(27)(9) = 243$.

2. The quotient rule: $\dfrac{x^m}{x^n} = x^{m-n}$.

    For example, $\dfrac{3^3}{3^2} = 3^{3-2} = 3^1 = 3$. To verify this result, note that $\dfrac{3^3}{3^2} = \dfrac{27}{9} = 3$.

    A special case of this rule is, $\dfrac{x^m}{x^m} = x^{m-m} = x^0 = 1$.

3. The first power rule: $(x^m)^n = x^{mn}$.
    For example, $(2^2)^3 = 2^6 = 64$. To verify this result, note that $2^2 = 4$, so $(2^2)^3 = (4)^3 = 64$.

4. Other power rules:

    $$(xy)^m = (x^m)(y^m), \text{ and } \left(\frac{x}{y}\right)^m = \frac{x^m}{y^m}.$$

    For example, $(3 \times 2)^4 = 6^4 = 1296$. To verify the first result, note that $3^4 = 81$, $2^4 = 16$, and $(81)(16) = 1296$.
    If the base number is negative, it is helpful to know that $(-x)^n = (-1)^n x^n$, and $(-1)^n$ is $+1$ if n is even and -1 if n is odd. You can verify easily the other power rules for any x and y.

EXAMPLE 0.1   Simplifying an exponential expression

Simplify the expression $\dfrac{x^3 y^{-3} x y^{4/3}}{x^{-4} y^{1/3}(x^2)^3}$, and calculate its numerical value when $x = 6$ and $y = 3$.

SOLUTION

SET UP AND SOLVE We simplify the expression as follows:

$$\frac{x^3 x}{x^{-4}(x^2)^3} = x^3 x^1 x^4 x^{-6} = x^{3+1+4-6} = x^2;$$

$$\frac{y^{-3} y^{4/3}}{y^{1/3}} = y^{-3+\frac{4}{3}-\frac{1}{3}} = y^{-2}.$$

Therefore,

$$\frac{x^3 y^{-3} x y^{4/3}}{x^{-4} y^{1/3}(x^2)^3} = x^2 y^{-2} = x^2 \left(\frac{1}{y}\right)^2 = \left(\frac{x}{y}\right)^2.$$

For $x = 6$ and $y = 3$, $\left(\dfrac{x}{y}\right)^2 = \left(\dfrac{6}{3}\right)^2 = 4$.

If we evaluate the original expression directly, we obtain

$$\frac{x^3 y^{-3} x y^{4/3}}{x^{-4} y^{1/3}(x^2)^3} = \frac{(6^3)(3^{-3})(6)(3^{4/3})}{(6^{-4})(3^{1/3})([6^2]^3)}$$

$$= \frac{(216)(1/27)(6)(4.33)}{(1/1296)(1.44)(46,656)} = 4.00,$$

which checks.

REFLECT This example demonstrates the usefulness of the rules for manipulating exponents.

EXAMPLE 0.2   Solving an exponential expression for the base number

If $x^4 = 81$, what is x?

SOLUTION

SET UP AND SOLVE We raise each side of the equation to the $\frac{1}{4}$ power: $(x^4)^{1/4} = (81)^{1/4}$. $(x^4)^{1/4} = x^1 = x$, so $x = (81)^{1/4}$ and $x = +3$ or $x = -3$. Either of these values of x gives $x^4 = 81$.

REFLECT Notice that we raised *both sides* of the equation to the $\frac{1}{4}$ power. As explained in Section 0.3, an operation performed on both sides of an equation does not affect the equation's validity.

0.2 Scientific Notation and Powers of 10

In physics, we frequently encounter very large and very small numbers, and it is important to use the proper number of significant figures when expressing a physical quantity. Both these issues are addressed by using **scientific notation,** in which a quantity is expressed as a decimal number with one digit to the left of the decimal point, multiplied by the appropriate power of 10. If the power of 10 is positive, it is the number of places the decimal point is moved to the right to obtain the fully written-out number. For example, $6.3 \times 10^4 = 63,000$. If the power of 10 is negative, it is the number of places the decimal point is moved to the left to obtain the fully written-out number. For example, $6.56 \times 10^{-3} = 0.00656$. In going from 6.56 to 0.00656, the decimal point is moved three places to the left, so 10^{-3} is the correct power of 10 to use when the number is written in scientific notation. Most calculators have keys for expressing a number in either decimal (floating-point) or scientific notation.

When two numbers written in scientific notation are multiplied (or divided), multiply (or divide) the decimal parts to get the decimal part of the result, and multiply (or divide) the powers of 10 to get the power-of-10 portion of the result. You may have to adjust the location of the decimal point in the answer to express it in scientific notation. For example,

$$
\begin{aligned}
(8.43 \times 10^8)(2.21 \times 10^{-5}) &= (8.43 \times 2.21)(10^8 \times 10^{-5}) \\
&= (18.6) \times (10^{8-5}) = 18.6 \times 10^3 \\
&= 1.86 \times 10^4.
\end{aligned}
$$

Similarly,

$$
\frac{5.6 \times 10^{-3}}{2.8 \times 10^{-6}} = \left(\frac{5.6}{2.8}\right) \times \left(\frac{10^{-3}}{10^{-6}}\right) = 2.0 \times 10^{-3-(-6)} = 2.0 \times 10^3.
$$

Your calculator can handle these operations for you automatically, but it is important for you to develop good "number sense" for scientific notation manipulations.

When adding, subtracting, multiplying, or dividing numbers, keeping the proper number of significant figures is important. See Section 1.5 to review how to keep the proper number of significant figures in these cases.

0.3 Algebra

Solving Equations

Equations written in terms of symbols that represent quantities are frequently used in physics. An **equation** consists of an equal sign and quantities to its left and to its right. Every equation tells us that the combination of quantities on the left of the equals sign has the same value as (that is, equals) the combination on the right of the equals sign. For example, the equation $y + 4 = x^2 + 8$ tells us that $y + 4$ has the same value as $x^2 + 8$. If $x = 3$, then the equation $y + 4 = x^2 + 8$ says that $y = 13$.

Often, one of the symbols in an equation is considered to be the *unknown,* and we wish to solve for the unknown in terms of the other quantities. For example, we might wish to solve the equation $2x^2 + 4 = 22$ for the value of x. Or we might wish to solve the equation $x = v_0 t + \frac{1}{2} at^2$ for the unknown a in terms of x, t, and v_0. Use the following rule to solve an equation:

An equation remains true if any valid operation performed on one side of the equation is also performed on the other side. The operations could be (a) adding or subtracting a number or symbol, (b) multiplying or dividing by a number or symbol, or (c) raising each side of the equation to the same power.

EXAMPLE 0.3 **Solving a numerical equation**

Solve the equation $2x^2 + 4 = 22$ for x.

SOLUTION

SET UP AND SOLVE First we subtract 4 from both sides. This gives $2x^2 = 18$. Then we divide both sides by 2 to get $x^2 = 9$. Finally, we raise both sides of the equation to the $\frac{1}{2}$ power. (In other words, we take the square root of both sides of the equation.) This gives $x = \pm\sqrt{9} = \pm 3$. That is, $x = +3$ or $x = -3$. We can verify our answers by substituting our result back into the original equation: $2x^2 + 4 = 2(\pm 3)^2 + 4 = 2(9) + 4 = 18 + 4 = 22$, so $x = \pm 3$ does satisfy the equation.

REFLECT Notice that a square root always has *two* possible values, one positive and one negative. For instance, $\sqrt{4} = \pm 2$, because $(2)(2) = 4$ and $(-2)(-2) = 4$. Your calculator will give you only a positive root; it's up to you to remember that there are actually two. Both roots are correct mathematically, but in a physics problem only one may represent the answer. For instance, if you can get dressed in $\sqrt{4}$ minutes, the only physically meaningful root is 2 minutes!

EXAMPLE 0.4 **Solving a symbolic equation**

Solve the equation $x = v_0 t + \frac{1}{2}at^2$ for a.

SOLUTION

SET UP AND SOLVE We subtract $v_0 t$ from both sides. This gives $x - v_0 t = \frac{1}{2}at^2$.

Now we multiply both sides by 2 and divide both sides by t^2, giving

$$a = \frac{2(x - v_0 t)}{t^2}.$$

REFLECT As we've indicated, it makes no difference whether the quantities in an equation are represented by variables (such as x, v, and t) or by numerical values.

The Quadratic Formula

Using the methods of the previous subsection, we can easily solve the equation $ax^2 + c = 0$ for x:

$$x = \pm\sqrt{\frac{-c}{a}}.$$

For example, if $a = 2$ and $c = -8$, the equation is $2x^2 - 8 = 0$ and the solution

$$x = \pm\sqrt{\frac{-(-8)}{2}} = \pm\sqrt{4} = \pm 2.$$

The equation $ax^2 + bx = 0$ is also easily solved by factoring out an x on the left side of the equation, giving $x(ax + b) = 0$. (To *factor* out a quantity means to isolate it so that the rest of the expression is either multiplied or divided by the quantity.) The equation $x(ax + b) = 0$ is true (that is, the left side equals zero) if either $x = 0$ or $x = -\frac{b}{a}$. These are the two solutions of the equation. For example, if $a = 2$ and $b = 8$, the equation is $2x^2 + 8x = 0$ and the solutions are $x = 0$ and $x = -\frac{8}{2} = -4$.

But if the equation is in the form $ax^2 + bx + c = 0$, with a, b, and c nonzero, we cannot use the previous simple methods to solve for x. Such equation is called a **quadratic equation**, and its solutions are expressed by the **quadratic formula:**

Quadratic formula

For a quadratic equation in the form $ax^2 + bx + c = 0$, where a, b, and c are real numbers and $a \neq 0$, the solutions are given by the quadratic formula:

$$x = \frac{-b \pm \sqrt{b^2 - 4ac}}{2a}$$

In general, a quadratic equation has two roots (solutions), which may be real or complex numbers.

If $b^2 - 4ac = 0$, then the two roots are equal and real numbers.

If $b^2 > 4ac$, that is, $b^2 - 4ac$ is positive, then the two roots are unequal and real numbers.

By contrast, if $b^2 < 4ac$, that is, $b^2 - 4ac$ is negative, then the roots are unequal complex numbers and cannot represent physical quantities. In such a case, the quadratic equation has mathematical solutions, but no physical solutions.

EXAMPLE 0.5 **Solving a quadratic equation**

Find the values of x that satisfy the equation $2x^2 - 2x = 24$.

SOLUTION

SET UP AND SOLVE First we write the equation in the standard form $ax^2 + bx + c = 0$: $2x^2 - 2x - 24 = 0$. Then $a = 2$, $b = -2$, and $c = -24$. Next, the quadratic formula gives the two roots as

$$x = \frac{-(-2) \pm \sqrt{(-2)^2 - 4(2)(-24)}}{(2)(2)}$$

$$= \frac{+2 \pm \sqrt{4 + 192}}{4} = \frac{2 \pm 14}{4},$$

so $x = 4$ or $x = -3$. If x represents a physical quantity that takes only nonnegative values, then the negative root $x = -3$ is non physical and is discarded.

REFLECT As we've mentioned, when an equation has more than one mathematical solution or root, it's up to *you* to decide whether one or the other or both represent the true physical answer. (If neither solution seems physically plausible, you should review your work.)

Simultaneous Equations

If a problem has two unknowns—for example, x and y—then it takes two independent equations in x and y (that is, two equations for x and y, where one equation is not simply a multiple of the other) to determine their values uniquely. Such equations are called **simultaneous equations** because you solve them together. A typical procedure is to solve one equation for x in terms of y and then substitute the result into the second equation to obtain an equation in which y is the only unknown. You then solve this equation for y and use the value of y in either of the original equations in order to solve for x. A pair of equations in which all quantities are symbols can be combined to eliminate one of the common unknowns. In general, to solve for n unknowns, we must have n independent equations. Simultaneous equations can also be solved **graphically** by plotting both equations using the same scale on the same graph paper. The solutions are the coordinates of the points of intersection of the graphs.

EXAMPLE 0.6 **Solving two equations in two unknowns**

Solve the following pair of equations for x and y:

$$x + 4y = 14$$
$$3x - 5y = -9$$

SOLUTION

SET UP AND SOLVE The first equation gives $x = 14 - 4y$. Substituting this for x in the second equation yields, successively, $3(14 - 4y) - 5y = -9$, $42 - 12y - 5y = -9$, and $-17y = -51$. Thus, $y = \frac{-51}{-17} = 3$. Then $x = 14 - 4y = 14 - 12 = 2$. We can verify that $x = 2$, $y = 3$ satisfies both equations.

An alternative approach is to multiply the first equation by -3, yielding $-3x - 12y = -42$. Adding this to the second equation gives, successively, $3x - 5y + (-3x) + (-12y) =$ $-9 + (-42)$, $-17y = -51$, and $y = 3$, which agrees with our previous result.

REFLECT As shown by the alternative approach, simultaneous equations can be solved in more than one way. The basic methods we describe are easy to keep straight; other methods may be quicker, but may require more insight or forethought. Use the method you're comfortable with.

EXAMPLE 0.7 **Solving two symbolic equations in two unknowns**

Use the equations $v = v_0 + at$ and $x = v_0 t + \frac{1}{2}at^2$ to obtain an equation for x that does not contain a.

SOLUTION

SET UP AND SOLVE We solve the first equation for a:

$$a = \frac{v - v_0}{t}.$$

We substitute this expression into the second equation:

$$x = v_0 t + \frac{1}{2}\left(\frac{v - v_0}{t}\right)t^2 = v_0 t + \frac{1}{2}vt - \frac{1}{2}v_0 t$$

$$= \frac{1}{2}v_0 t + \frac{1}{2}vt = \left(\frac{v_0 + v}{2}\right)t.$$

REFLECT When you solve a physics problem, it's often best to work with symbols for all but the final step of the problem. Once you've arrived at the final equation, you can plug in numerical values and solve for an answer.

0.4 Direct, Inverse, and Inverse-Square Relationships

The essence of physics is to describe and verify the relationships among physical quantities. The relationships are often simple. For example, two quantities may be directly proportional to each other, inversely proportional to each other, or one quantity may be inversely proportional to the square of the other quantity.

Direct Relationship

Two quantities are said to be **directly proportional** to one another if an increase (or decrease) of the first quantity causes an increase (or decrease) of the second quantity by the same factor. If y is directly proportional to x, the direct proportionality is written as $y \propto x$. The ratio y/x is a constant, say, k. That is, $\frac{y_1}{x_1} = \frac{y_2}{x_2} = k$. For example, the ratio of the circumference C to the diameter d of a circle is always 3.14, known as π (pi). Therefore, the circumference of a circle is directly proportional to its diameter as $C = \pi d$, where π is the constant of proportionality. Another simple example of direct proportionality is the stretching or compression of an ordinary helical spring (discussed in Section 5.4). The spring has a certain length at rest, which increases when it is suspended vertically with weights attached to the bottom. If the amount of stretch is not too long, the amount of force F (measured in pounds or newtons) on the spring and the amount of stretch x are directly proportional to each other (Figure 0.1). Thus, $F = kx$ where k is the constant of proportionality.

In general, in a direct proportion, $\dfrac{a}{b} = \dfrac{c}{d}$. Multiplying both sides by bd, we find

$$\cancel{b}d \cdot \frac{a}{\cancel{b}} = b\cancel{d} \cdot \frac{c}{\cancel{d}} \quad \text{or} \quad a \cdot d = b \cdot c$$

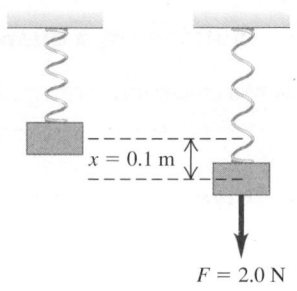

$F = 2.0\ \text{N}$

▲ **FIGURE 0.1**

Graph of Direct Proportionality Relationship

When y is directly proportional to x, $y = kx$, and the graph of y versus x is a straight line passing through the origin as shown in Figure 0.2. In the graph, the change of the quantity x is labeled as Δx (which is often called "run") and the corresponding change in y is labeled as Δy (which is often called "rise").

Here,

$$\Delta x = x_2 - x_1$$
$$\Delta y = y_2 - y_1,$$

where (x_1, y_1) and (x_2, y_2) are coordinates of the two points on the line. The constant of proportionality between Δy and Δx is also k. Thus,

$$\Delta y = k\,\Delta x$$

The steepness of the line is measured by the ratio $\Delta y / \Delta x$ and is called the *slope* of the line. Thus,

$$\text{Slope} = \frac{\Delta y}{\Delta x} = k$$

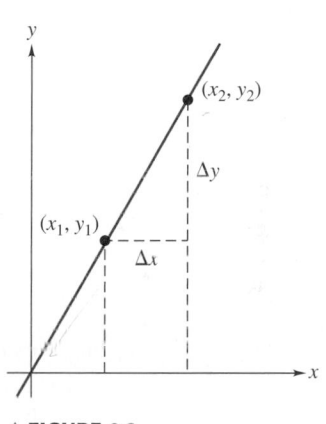

▲ **FIGURE 0.2**

The slope of a line can be *positive*, *negative*, *zero*, or *undefined*, as shown in Figure 0.3 (a–d).

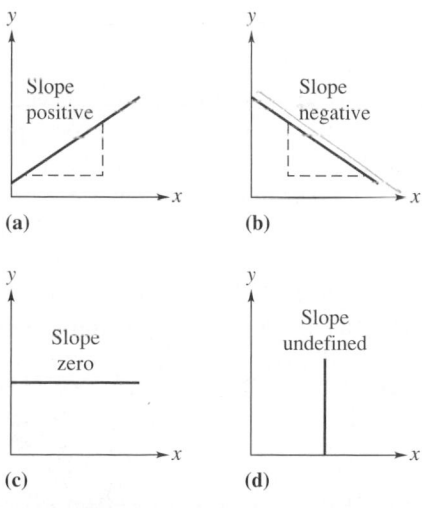

(a)   (b)

(c)   (d)

▲ **FIGURE 0.3**

Note that: slope *positive* means *y increases as x increases*; slope *negative* means *y decreases as x increases*; slope *zero* means *y* does not change—that is, the line is *parallel to the x-axis*; slope *undefined* means *x* does not change—that is, the line is *parallel to the y-axis*.

EXAMPLE 0.8 Solving for a quantity in direct proportion

If y is directly proportional to x, and $x = 2$ when $y = 8$, what is y when $x = 10$?

SOLUTION

SET UP AND SOLVE Since y is directly proportional to x,

$$\frac{y_1}{x_1} = \frac{y_2}{x_2} = k.$$

Substituting the values we get $\dfrac{y}{10} = \dfrac{8}{2}.$

Multiplying both sides by 2 and 10 to get rid of the fractions gives $2y = 10 \cdot 8 = 80$.

Divide by 2 to isolate y: $y = \dfrac{80}{2} = 40$.

REFLECT Although this is a simple problem, this gives you the strategy for how to solve problems in direct proportion. Note that x has increased by a factor of 5, so y must also increase by the same factor.

EXAMPLE 0.9 Solving for the stiffness constant of a spring

A spring is suspended vertically from a fixed support. When a weight of 2.0 newtons is attached to the bottom of the spring, the spring stretches by 0.1 m. (The newton, abbreviated N, is the SI unit for force; the hanging weight exerts a downward force on the spring.) Determine the spring constant of the spring.

SOLUTION

SET UP AND SOLVE We have force, $F = 2.0$ N and stretch, $x = 0.10$ m.

The sketch of the problem is shown in Figure 0.1. The applied weight and the amount of stretch are related by a direct proportion, as expressed by $F = kx$. Using this equation we can solve for the stiffness constant:

$$k = \frac{F}{x} = \frac{2.0 \text{ N}}{0.10 \text{ m}} = 20 \text{ N/m}$$

REFLECT The stiffness constant in this equation is the constant of proportionality in the direct proportion between the force and the amount of stretch. Its unit is the ratio of the units of F and x.

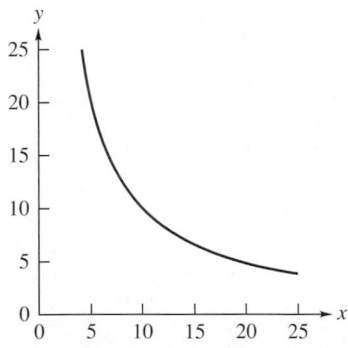

▲ FIGURE 0.4

Inverse Proportion

When one quantity increases and the other quantity decreases in such a way that their product stays the same, they are said to be in **inverse proportion**. In inverse proportion, when one quantity approaches zero, the other quantity becomes extremely large, so that the product remains the same. For example, the product of the pressure and volume of an ideal gas remains constant if the temperature of the gas is maintained constant, as you will find in Section 15.2. Mathematically, if y is in inverse proportion to x, $y \propto 1/x$. This gives $y = k/x$, or $xy = k$, where k is the constant of proportionality. That is, when x changes from x_1 to x_2, y changes from y_1 to y_2 so that $x_1 y_1 = x_2 y_2 = k$. This type of behavior is illustrated in Figure 0.4 (for an arbitrary choice of $k = 100$).

EXAMPLE 0.10 Solving for a quantity (volume of an ideal gas) in inverse proportion

According to Boyle's law, if the temperature of an ideal gas is kept constant, its pressure, P, is *inversely proportional* to its volume, V. A cylindrical flask is fitted with an airtight piston and contains an ideal gas. Initially, the pressure of the inside gas is 11×10^4 pascals (Pa) and the volume of the gas is 8.0×10^{-3} m^3. Assuming that the system is always at the temperature of 330 kelvins (K), determine the volume of the gas when its pressure increases to 24×10^4 Pa.

SOLUTION

SET UP AND SOLVE Since the pressure P is inversely proportional to the volume V according to Boyle's law, the product PV remains constant.

That is,

$$P_1 V_1 = P_2 V_2.$$

We divide by P_2 to solve for V_2. $V_2 = \dfrac{P_1 V_1}{P_2}$.

In this problem, $P_1 = 11 \times 10^4$ Pa, $V_1 = 8.0 \times 10^{-3}$ m³, and $P_2 = 24 \times 10^4$ Pa.

Substituting the values of P_1, V_1, and P_2, we solve for V_2.

$$V_2 = \frac{(11 \times 10^4 \,\text{Pa}) \times (8.0 \times 10^{-3}\,\text{m}^3)}{24 \times 10^4 \,\text{Pa}} = \frac{11 \times 8.0}{24} \times 10^{-3}\,\text{m}^3$$
$$= 3.7 \times 10^{-3}\,\text{m}^3$$

REFLECT Since the pressure increased, the final volume has decreased, as expected in an inverse proportion. Note that the pascal (Pa) and K/kelvin (K) are the SI units for pressure and temperature, respectively.

Inverse Square Proportion

Inverse square dependence is common in the laws of nature. For example, the force of gravity due to a body decreases as the inverse square of the distance from the body, as expressed by Newton's law of gravitation in Section 6.3. Similarly, the electrostatic force due to a point electric charge decreases as the square of the distance from the charge, as expressed by Coulomb's law in Section 17.4. The intensity of sound and of light also decreases as the inverse square of the distance from a point source, as you will find in Section 12.10. (Intensity in these cases is a measure of the power of the sound or light per unit area.) Mathematically, if y varies inversely with the square of x, then

$$y \propto \frac{1}{x^2} \quad \text{or} \quad y = \frac{k}{x^2} \quad \text{or} \quad x^2 y = k,$$

where k is the constant of proportionality. That is, when x changes from x_1 to x_2, y changes from y_1 to y_2 so that $x_1^2 y_1 = x_2^2 y_2 = k$. Or

$$\frac{y_1}{y_2} = \frac{x_2^2}{x_1^2}$$

This relationship is illustrated in Figure 0.5 (for an arbitrary choice of $k = 100$).

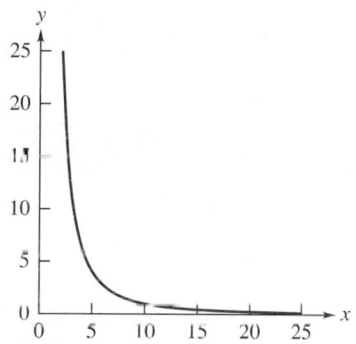

▲ **FIGURE 0.5**

EXAMPLE 0.11   Solving for a quantity (sound intensity) that varies as the inverse square

A small source of sound emits sound equally in all directions. The intensity of emitted sound at a point is given by the equation $I = k/r^2$, where k is a constant of proportionality and r is the distance of the point from the source. If the sound intensity is 0.05 watt/meter² (W/m²) at a distance 0.5 m from the source, find the sound intensity at a distance of 20 m from the source.

SOLUTION

SET UP AND SOLVE The intensity of sound is given by the equation $I = k/r^2$. We apply this equation to solve the problem. However, we do not need to know the value of the constant k. In this problem, we have an initial distance, $r_1 = 0.5$ m, sound intensity, $I_1 = 0.05$ W/m², and final distance, $r_2 = 20$ m, where intensity I_2 is to be determined.

From the inverse square relationship, we have (where the variables x and y have been replaced by r and I in the equation)

$$\frac{I_1}{I_2} = \frac{r_2^2}{r_1^2}$$

We take the reciprocal of both sides and multiply by I_1 to solve for I_2.

$$I_2 = I_1 \frac{r_1^2}{r_2^2} = (0.05\,\text{W/m}^2)\frac{(0.5\,\text{m})^2}{(20\,\text{m})^2}$$
$$= 3.1 \times 10^{-5}\,\text{W/m}^2$$

REFLECT As the distance increases, the intensity decreases. Note that because the intensity decreases as the square of the distance, the result for intensity is less than it would have been for the case of a simple inverse proportion. Thus, the result makes sense.

0.5 Logarithmic and Exponential Functions

The base-10 logarithm, or **common logarithm** (log), of a number y is the power to which 10 must be raised to obtain y: $y = 10^{\log y}$. For example, $1000 = 10^3$, so $\log(1000) = 3$; you must raise 10 to the power 3 to obtain 1000. Most calculators have a key for calculating the log of a number.

Sometimes we are given the log of a number and are asked to find the number. That is, if $\log y = x$ and x is given, what is y? To solve for y, write an equation in which 10 is raised to the power equal to either side of the original equation: $10^{\log y} = 10^x$. But $10^{\log y} = y$, so $y = 10^x$. In this case, y is called the **antilog** of x. For example, if $\log y = -2.0$, then $y = 10^{-2.0} = 1.0 \times 10^{-2} = 0.010$.

The log of a number is positive if the number is greater than 1. The log of a number is negative if the number is less than 1, but greater than zero. The log of zero or of a negative number is not defined, and $\log 1 = 0$.

Another base that occurs frequently in physics is the quantity $e = 2.718\ldots$. The **natural logarithm** (ln) of a number y is the power to which e must be raised to obtain y: $y = e^{\ln y}$. If $x = \ln y$, then $y = e^x$, which is called an **exponential function**, also written as exp (x). Most calculators have keys for $\ln x$ and for e^x. For example, $\ln 10.0 = 2.30$, and if $\ln x = 3.00$, then $x = e^{3.00} = 20.1$. Note that $\ln 1 = 0$. The plot of the function e^x is shown in Figure 0.6. The plot of e^{-x} will be a curve that is a reflection of Figure 0.6 about the y-axis.

The exponential function and natural logarithm occur in natural phenomena (such as, radioactive decay, Section 30.3) where the rate of increase or decrease of some quantity depends on the quantity. Exponential growth and decay of electric charges and electric current are also common in electric circuits as described in Sections 19.8 and 21.11.

Logarithms with any choice of base, including base 10 or base e, obey several simple and useful rules:

1. $\log(ab) = \log a + \log b$.

2. $\log\left(\dfrac{a}{b}\right) = \log a - \log b$.

3. $\log(a^n) = n \log a$.

A particular example of the second rule is

$$\log\left(\frac{1}{a}\right) = \log 1 - \log a = -\log a,$$

since $\log 1 = 0$.

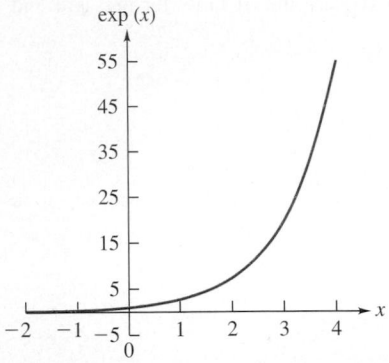

exp (x)

▲ **FIGURE 0.6**

EXAMPLE 0.12 **Solving a logarithmic equation**

If $\frac{1}{2} = e^{-\alpha T}$, solve for T in terms of α.

SOLUTION

SET UP AND SOLVE We take the natural logarithm of both sides of the equation: $\ln\left(\frac{1}{2}\right) = -\ln 2$ and $\ln\left(e^{-\alpha T}\right) = -\alpha T$. The equation thus becomes $-\alpha T = -\ln 2$, and it follows that $T = \frac{\ln 2}{\alpha}$.

REFLECT The equation $y = e^{\alpha x}$ expresses y in terms of the exponential function $e^{\alpha x}$. The general rules for exponents in Section 0.1 apply when the base is e, so $e^x e^y = e^{x+y}$, $e^x e^{-x} = e^{x+(-x)} = e^0 = 1$, and $(e^x)^2 = e^{2x}$.

0.6 Areas and Volumes

Figure 0.7 illustrates the formulas for the areas and volumes of common geometric shapes:

- A rectangle with length a and width b has area $A = ab$.
- A rectangular solid (a box) with length a, width b, and height c has volume $V = abc$.

- A circle with radius r has diameter $d = 2r$, circumference $C = 2\pi r = \pi d$, and area $A = \pi r^2 = \pi d^2/4$.
- A sphere with radius r has surface area $A = 4\pi r^2$ and volume $V = \frac{4}{3}\pi r^3$.
- A cylinder with radius r and height h has volume $V = \pi r^2 h$.

0.7 Plane Geometry and Trigonometry

Following are some useful results about angles:

1. Interior angles formed when two straight lines intersect are equal. For example, in Figure 0.8, the two angles θ and ϕ are equal.
2. When two parallel lines are intersected by a diagonal straight line, the alternate interior angles are equal. For example, in Figure 0.9, the two angles θ and ϕ are equal.
3. When the sides of one angle are each perpendicular to the corresponding sides of a second angle, then the two angles are equal. For example, in Figure 0.10, the two angles θ and ϕ are equal.
4. The sum of the angles on one side of a straight line is 180°. In Figure 0.11, $\theta + \phi = 180°$.
5. The sum of the angles in any triangle is 180°.

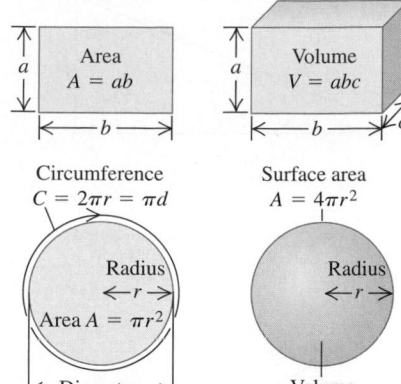

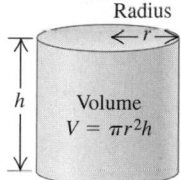

▲ **FIGURE 0.7**

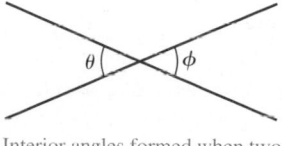

Interior angles formed when two straight lines intersect are equal:
$$\theta = \phi$$

▲ **FIGURE 0.8**

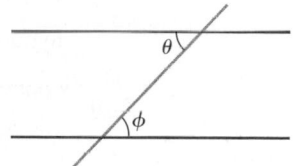

When two parallel lines are intersected by a diagonal straight line, the alternate interior angles are equal:
$$\theta = \phi$$

▲ **FIGURE 0.9**

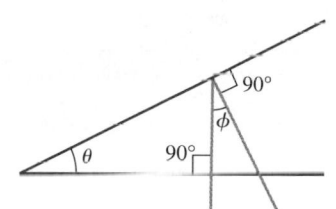

When the sides of one angle are each perpendicular to the corresponding sides of a second angle, then the two angles are equal:
$$\theta = \phi$$

▲ **FIGURE 0.10**

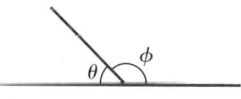

The sum of the angles on one side of a straight line is 180°:
$$\theta + \phi = 180°$$

▲ **FIGURE 0.11**

Similar Triangles

Triangles are **similar** if they have the same shape, but different sizes or orientations. Similar triangles have equal corresponding angles and equal ratios of corresponding sides. If the two triangles in Figure 0.12 are similar, then $\theta_1 = \theta_2$, $\phi_1 = \phi_2$, $\gamma_1 = \gamma_2$, and $\dfrac{a_1}{a_2} = \dfrac{b_1}{b_2} = \dfrac{c_1}{c_2}$.

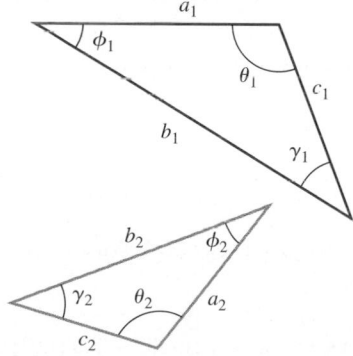

Two similar triangles: Same shape but not necessarily the same size.

▲ **FIGURE 0.12**

If two similar triangles have the same size, they are said to be **congruent.** If triangles are congruent, one can be flipped and rotated so that it can be placed precisely on top of the other.

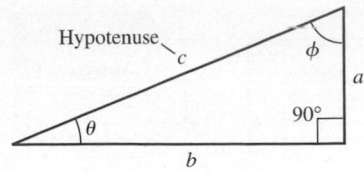

For a right triangle:
$\theta + \phi = 90°$
$c^2 = a^2 + b^2$ (Pythagorean theorem)

▲ **FIGURE 0.13**

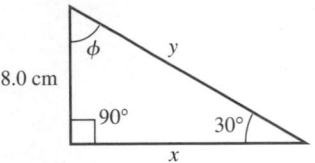

▲ **FIGURE 0.14**

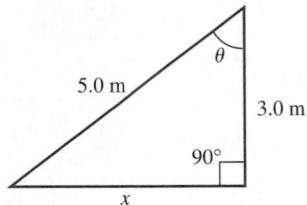

▲ **FIGURE 0.15**

Right Triangles and Trig Functions

In a **right triangle,** one angle is 90°. Therefore, the other two acute angles (*acute* means less than 90°) have a sum of 90°. In Figure 0.13, $\theta + \phi = 90°$. The side opposite the right angle is called the **hypotenuse** (side c in the figure). In a right triangle, the square of the length of the hypotenuse equals the sum of the squares of the lengths of the other two sides. For the triangle in Figure 0.13, $c^2 = a^2 + b^2$. This formula is called the **Pythagorean Theorem.**

If two right triangles have the same value for one acute angle, then the two triangles are similar and have the same ratio of corresponding sides. This true statement allows us to define the functions **sine, cosine,** and **tangent** that are ratios of a pair of sides. These functions, called **trigonometric** functions or **trig functions,** depend only on one of the angles in the right triangle. For an angle θ, these functions are written $\sin\theta$, $\cos\theta$, and $\tan\theta$.

In terms of the triangle in Figure 0.13, the sine, cosine, and tangent of the angle θ are as follows:

$$\sin\theta = \frac{\text{opposite side}}{\text{hypotenuse}} = \frac{a}{c},$$

$$\cos\theta = \frac{\text{adjacent side}}{\text{hypotenuse}} = \frac{b}{c}, \text{ and}$$

$$\tan\theta = \frac{\text{opposite side}}{\text{adjacent side}} = \frac{a}{b}.$$

Note that $\tan\theta = \dfrac{\sin\theta}{\cos\theta}$. For angle ϕ, $\sin\phi = \dfrac{b}{c}$, $\cos\phi = \dfrac{a}{c}$, and $\tan\phi = \dfrac{b}{a}$.

In physics, angles are expressed in either degrees or radians, where π radians $= 180°$. (For more on radians, see Section 9.1.) Most calculators have a key for switching between degrees and radians. Always be sure that your calculator is set to the appropriate angular measure.

Inverse trig functions, denoted, for example, by $\sin^{-1}x$ (or arcsin x) have a value equal to the angle that has the value x for the trig function. For example, $\sin30° = 0.500$, so $\sin^{-1}(0.500) = \arcsin(0.500) = 30°$. Note that $\sin^{-1}x$ does *not* mean $\dfrac{1}{\sin x}$. Also, note that when you determine an angle using inverse trigonometric functions, the calculator will always give you the smallest correct angle, which may or may not be the right answer. Use the knowledge of which quadrant you are working in to determine the correct angle in the situation.

EXAMPLE 0.13 **Using trigonometry I**

A right triangle has one angle of 30° and one side with length 8.0 cm, as shown in Figure 0.14. What is the angle ϕ and what are the lengths x and y of the other two sides of the triangle?

SOLUTION

SET UP AND SOLVE $\phi + 30° = 90°$, so $\phi = 60°$.

$$\tan30° = \frac{8.0 \text{ cm}}{x}, \text{ so } x = \frac{8.0 \text{ cm}}{\tan30°} = 13.9 \text{ cm.}$$

To find y, we use the Pythagorean Theorem: $y^2 = (8.0 \text{ cm})^2 + (13.9 \text{ cm})^2$, so $y = 16.0$ cm.

Or we can say $\sin30° = 8.0$ cm/y, so $y = 8.0$ cm/$\sin30° = 16$ cm, which agrees with the previous result.

REFLECT Notice how we used the Pythagorean Theorem in combination with a trig function. You will use these tools constantly in physics, so make sure that you can employ them with confidence.

EXAMPLE 0.14 **Using trigonometry II**

A right triangle has two sides with lengths as specified in Figure 0.15. What is the length x of the third side of the triangle, and what is the angle θ, in degrees?

SOLUTION

SET UP AND SOLVE The Pythagorean Theorem applied to this right triangle gives $(3.0 \text{ m})^2 + x^2 = (5.0 \text{ m})^2$, so $x = \sqrt{(5.0 \text{ m})^2 - (3.0 \text{ m})^2} = 4.0 \text{ m}$. (Since x is a length, we take the positive root of the equation.) We also have

$$\cos\theta = \frac{3.0 \text{ m}}{5.0 \text{ m}} = 0.60, \text{ so } \theta = \cos^{-1}(0.60) = 53°.$$

REFLECT In this case, we knew the lengths of two sides, but none of the acute angles, so we used the Pythagorean Theorem first and then an appropriate trig function.

In a right triangle, all angles are in the range from $0°$ to $90°$, and the sine, cosine, and tangent of the angles are all positive. This must be the case, since the trig functions are ratios of lengths. But for other applications, such as finding the components of vectors, calculating the oscillatory motion of a mass on a spring, or describing wave motion, it is useful to define the sine, cosine, and tangent for angles outside that range. Graphs of $\sin\theta$ and $\cos\theta$ are given in Figure 0.16. The values of $\sin\theta$ and $\cos\theta$ vary between $+1$ and -1. Each function is periodic, with a period of $360°$. Note the range of angles between $0°$ and $360°$ for which each function is positive and negative. The two functions $\sin\theta$ and $\cos\theta$ are $90°$ out of phase (that is, out of step). When one is zero, the other has its maximum magnitude (i.e., its maximum or minimum value).

For any triangle (see Figure 0.17)—in other words, not necessarily a right triangle—the following two relations apply:

1. $\dfrac{\sin\alpha}{a} = \dfrac{\sin\beta}{b} - \dfrac{\sin\gamma}{c}$ (law of sines).

2. $c^2 = a^2 + b^2 - 2ab\cos\gamma$ (law of cosines).

Some of the relations among trig functions are called trig identities. The following table lists only a few, those most useful in introductory physics:

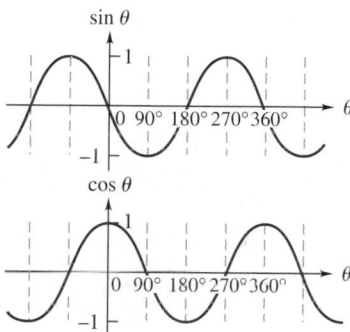

▲ FIGURE 0.16

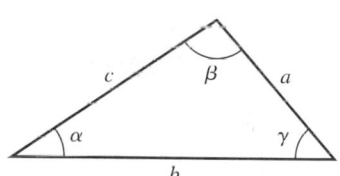

▲ FIGURE 0.17

Useful trigonometric identities

$$\sin(-\theta) = -\sin(\theta) \quad (\sin\theta \text{ is an odd function})$$
$$\cos(-\theta) = \cos(\theta) \quad (\cos\theta \text{ is an even function})$$
$$\sin 2\theta = 2\sin\theta\cos\theta$$
$$\cos 2\theta = \cos^2\theta - \sin^2\theta = 2\cos^2\theta - 1 = 1 - 2\sin^2\theta$$
$$\sin(\theta \pm \phi) = \sin\theta\cos\phi \pm \cos\theta\sin\phi$$
$$\cos(\theta \pm \phi) = \cos\theta\cos\phi \mp \sin\theta\sin\phi$$
$$\sin(180° - \theta) = \sin\theta$$
$$\cos(180° - \theta) = -\cos\theta$$
$$\sin(90° - \theta) = \cos\theta$$
$$\cos(90° - \theta) = \sin\theta$$

For small angle θ (in radians),

$$\cos\theta \approx 1 - \frac{\theta^2}{2} \approx 1$$

$$\sin\theta \approx \theta$$

Problems

0.1 Exponents

Use the exponent rules to simplify the following expressions:

1. $(-3x^4y^2)^2$

2. $\dfrac{(2^3 4^4)^2}{(8)^4}$

3. $\left(\dfrac{4x^2}{2y^3}\right)^2$

4. $\left(-\dfrac{x^2y^4}{xy^{-2}}\right)^3$

0.2 Scientific Notation and Powers of 10

Express the following expressions in scientific notation:

5. 475000

6. 0.00000472

7. 123×10^{-6}

8. $\dfrac{8.3 \times 10^5}{7.8 \times 10^2}$

0.3 Algebra

Solve the following equations using any method:

9. $4x + 6 = 9x - 14$
10. $F = 9/5\, C + 32$ (solve for C)
11. $4x^2 + 6 = 3x^2 + 18$
12. $-196 = -9.8t^2$
13. $x^2 - 5x + 6 = 0$
14. $x^2 + x - 1 = 0$
15. $4.9t^2 + 2t - 20 = 0$
16. $5x - 4y = 1, 6y = 10x - 4$
17. $\dfrac{x}{2} + \dfrac{y}{3} = 2, 2x - y = 1$

0.4 Direct, Inverse, and Inverse-Square Relationships

18. If x is proportional to y and $x = 2$ when $y = 10$, what is the value of x when $y = 8$?

19. A hand exerciser uses a coiled spring and the force needed to compress the spring is *directly proportional* to the amount of compression. If a force of 80 N is needed to compress the spring by 0.02 m, determine the force required to compress the spring by 0.05 m.

20. According to the ideal gas equation (Chapter 15.2), the volume of an ideal gas is directly proportional to its temperature in kelvins (K) if the pressure of the gas is constant. An ideal gas occupies a volume of 4.0 liters at 100 K. Determine its volume when it is heated to 300 K while held at a constant pressure.

21. For a sound coming from a point source the amplitude of sound is inversely proportional to the distance. If the displacement amplitude of an air molecule in a sound wave is 4.8×10^{-6} m at a point

1.0 m from the source, what would be the displacement amplitude of the same sound when the distance increases to 4.0 m?

22. For a small lamp that emits light uniformly in all directions, the light intensity is given by the inverse square law, $I = k/r^2$. If the light intensity reaching an object at a distance of 0.40 m is 60.0 lux, determine the light intensity 1.80 m from the lamp.

23. The force of gravity on an object (which we experience as the object's weight) varies inversely as the square of the distance from the center of the earth. Determine the force of gravity on an astronaut when he is at a height of 6000 km from the surface of the earth if he weighs 700 newtons (N) when on the surface of the earth. The radius of the earth is 6.38×10^6 m. (If the astronaut is in orbit, he will float "weightlessly," but gravity still acts on him – he and his spaceship appear weightless because they are falling freely in their orbit around the earth.)

0.5 Logarithmic and Exponential Functions

24. Use the properties of logarithms and write each expression in terms of logarithms of x, y, and z.
    (a) $\log(x^4y^2z^8)$ (b) $\log\sqrt{x^3y^7}$ (c) $\log\sqrt[3]{\dfrac{x^2y^6}{z^3}}$

25. Simplify the expression.
    (a) $4\log x + \log y - 3\log(x + y)$
    (b) $\log(xy + x^2) - \log(xz + yz) + 2\log z$

26. $\beta = 10\log\left(\dfrac{I}{10^{-12}}\right)$, find β when $I = 10^{-4}$.

27. $\beta = 10\log\left(\dfrac{I}{10^{-12}}\right)$, find I when $\beta = 60$.

28. $N = N_0\, e^{-(0.210)t}$. If $N_0 = 2.00 \times 10^6$, solve for t when $N = 2.50 \times 10^4$.

0.6 Areas and Volumes

29. (a) Compute the circumference and area of a circle of radius 0.12 m. (b) Compute the surface area and volume of a sphere of radius 0.21 m. (c) Compute the total surface area and volume of a rectangular solid of length 0.18 m, width 0.15 m, and height 0.8 m. (d) Compute the total surface area and volume of a cylinder of radius 0.18 m and height 0.33 m.

0.7 Plane Geometry and Trigonometry

30. A right triangle has a hypotenuse of length 20 cm and another side of length 16 cm. Determine the third side of the triangle and the other two angles of the triangle.

31. In a stairway, each step is set back 30 cm from the next lower step. If the stairway rises at an angle of 36° with the horizontal, what is the height of each step?

I Models, Measurements, and Vectors

The study of physics is an adventure. You'll find it challenging, sometimes frustrating, and often richly rewarding and satisfying. It will appeal to your sense of beauty as well as to your rational intelligence. Our present understanding of the physical world has been built on foundations laid by scientific giants such as Galileo, Newton, and Einstein. You can share some of the excitement of their discoveries when you learn to use physics to solve practical problems and to gain insight into everyday phenomena. Above all, you'll come to see physics as a towering achievement of the human intellect in its quest to understand the world we all live in.

In this opening chapter we go over some important preliminaries that we'll need throughout our study. We need to think about the philosophical framework in which we operate—the nature of physical theory and the role of idealized *models* in representing physical systems. We discuss systems of *units* that are used to describe physical quantities, such as length and time, and we examine the *precision* of a number, often described by means of *significant figures*. We look at examples of problems in which we can't or don't want to make precise calculations, but in which rough estimates can be interesting and useful. Finally, we study several aspects of *vector algebra*. We use vectors to describe and analyze many physical quantities, such as velocity and force, that have direction as well as magnitude.

Like these players, you already know a lot of physics. Your brain easily computes forces and motions in complex situations. In this course, your goal is simply to build a conscious model of the physical world you live in.

1.1 Introduction

Physics is an *experimental* science. Physicists observe the phenomena of nature and try to discover patterns and principles that relate these phenomena. These patterns are called *physical theories* or, when they are very broad and well established, *physical laws*. The development of physical theory requires creativ-

(a)   **(b)**

▶ **FIGURE 1.1** Two research laboratories. (a) The Leaning Tower of Pisa (Italy). According to legend, Galileo studied the motion of freely falling objects by dropping them from the tower. He is also said to have gained insights into pendulum motion by observing the swinging of the chandelier in the cathedral to the left of the tower. (b) The Hubble Space Telescope, the first major telescope to operate in space, free from the effects of earth's atmosphere.

ity at every stage. The physicist has to learn to ask appropriate questions, design experiments to try to answer the questions, and draw appropriate conclusions from the results. Figure 1.1 shows two famous experimental facilities.

According to legend, Galileo Galilei (1564–1642) dropped light and heavy objects from the top of the Leaning Tower of Pisa to find out whether they would fall at the same rate or at different rates. Galileo recognized that only an *experimental* investigation could answer this question. From the results of his experiments, he had to make the inductive leap to the principle, or theory, that the rate at which an object falls is independent of its weight.

The development of physical theory is always a two-way process that starts and ends with observations or experiments. Physics is not a collection of facts and principles; it is the *process* by which we arrive at general principles that describe the behavior of the physical universe. And there is always the possibility that new observations will require revision of a theory. We can *disprove* a theory by finding a phenomenon that is inconsistent with it, but we can never *prove* that a theory is *always* correct.

Getting back to Galileo, suppose we drop a feather and a cannonball. They certainly *do not* fall at the same rate. This doesn't mean that our statement of Galileo's theory is wrong, but that it is incomplete. One complicating feature is air resistance. If we drop the feather and the cannonball in vacuum, they *do* fall at the same rate. Our statement of Galileo's theory is valid for bodies that are heavy enough that air resistance has almost no effect on them. A feather or a parachute clearly does not have this property.

Every physical theory has a range of validity and a boundary outside of which it is not applicable. The range of Galileo's work with falling bodies was greatly extended half a century later by Newton's laws of motion and his law of gravitation. Nearly all the principles in this book are so solidly established by experimental evidence that they have earned the title **physical law.** Yet there are some areas of physics in which present-day research is continuing to broaden our understanding of the physical world. We'll discuss some examples of these areas in the final four chapters of the book.

An essential part of the interplay of theory and experiment is learning how to apply physical principles to a variety of practical problems. At various points in our study, we'll discuss systematic problem-solving procedures that will help you set up problems and solve them efficiently and accurately. Learning to solve problems is absolutely essential; you don't *know* physics unless you can *do* physics. This means not only learning the general principles, but also learning how to apply them in a variety of specific situations.

▲ **Application Where are the tunes coming from?** These dancers may not realize it, but they are enjoying the application of fundamental principles of physics that makes their music possible. In later chapters, we will learn about electric charge and current and about how batteries are used to store electrical potential energy. We will see how this energy can be harnessed to perform the mechanical work of rotating the CD within the CD player. We will learn about lasers and optics, which provide the means to "read" a CD. And we will learn about harmonic motion, sound waves, and how these waves travel through space to bring us these "good vibes."

1.2 Idealized Models

In everyday conversation, we often use the word **model** to mean either a small-scale replica, such as a model train or airplane, or a human figure that displays articles of clothing. **In physics, a *model* is a simplified version of a physical system that would be too complicated to analyze in full without the simplifications.**

Here's an example: We want to analyze the motion of a baseball thrown through the air. How complicated is this problem? The ball is neither perfectly spherical nor perfectly rigid; it flexes a little and spins as it moves through the air. Wind and air resistance influence its motion, the earth rotates beneath it, the ball's weight varies a little as its distance from the center of the earth changes, and so on. If we try to include all these things, the analysis gets pretty hopeless. Instead, we invent a simplified version of the problem. We neglect the size and shape of the ball and represent it as a *particle*—that is, as an object that has no size but is completely localized at a single point in space. We neglect air resistance, treating the ball as though it moves in vacuum; we forget about the earth's rotation, and we make the weight of the ball exactly constant. *Now* we have a problem that is simple enough to deal with: The ball is a particle moving along a simple parabolic path. We'll analyze this model in detail in Chapter 3.

The point is that we have to overlook quite a few minor effects in order to concentrate on the most important features of the motion. That's what we mean by making an idealized model of the system. Of course, we have to be careful not to neglect *too much*. If we ignore the effect of gravity completely, then when we throw the ball, it will travel in a straight line and disappear into space, never to be seen again! We need to use some judgment and creativity to construct a model that simplifies a system without throwing out its essential features.

When we analyze a system or predict its behavior on the basis of a model, we always need to remember that the validity of our predictions is limited by the validity of the model. If the model represents the real system quite precisely, then we expect predictions made from it to agree quite closely with the actual behavior of the system. Going back to Pisa with Galileo once more, we see that our statement of his theory of falling objects corresponds to an idealized model that does not include the effects of air resistance, the rotation of the earth, or the variation of weight with height. This model works well for a bullet and a cannonball, not so well for a feather.

The concept of an idealized model is of the utmost importance in all of physical science and technology. In applying physical principles to complex systems, we always make extensive use of idealized models, and it is important to be aware of the assumptions we are making. Indeed, the *principles* of physics themselves are stated in terms of idealized models; we speak about point masses, rigid bodies, ideal insulators, and so on. Idealized models will play a crucial role in our discussion of physical theories and their applications to specific problems throughout this book.

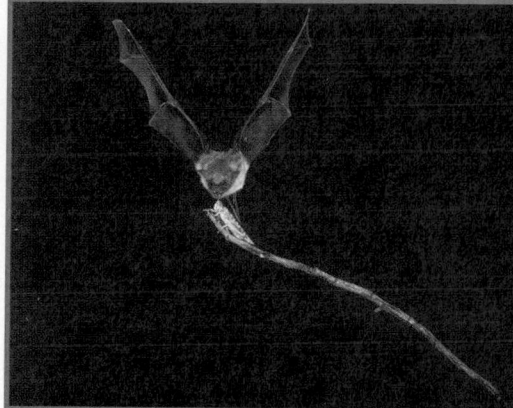

▲ **BIO Application How can I find lunch in the dark?** Insect-eating bats utilize the laws of physics in their quest for prey in a process known as echolocation. To locate and capture an insect, these flying mammals first emit an extremely rapid series of 20 to 200 bursts of high-frequency sound. Without using a stopwatch or calculator, they are able to determine the insect's position by sensing the direction of the reflected sound waves and the time it takes for these waves to reach their ears. Luckily for us, the sounds they emit are above the range of human hearing, as they can be over 100 decibels, or about as loud as a typical smoke alarm.

1.3 Standards and Units

Any number that is used to describe an observation of a physical phenomenon quantitatively is called a **physical quantity.** Some physical quantities are so fundamental that we can define them only by describing a procedure for measuring them. Such a definition is called an **operational definition.** In other cases, we define a physical quantity by describing a way to *calculate* the quantity from other quantities that we can measure. As an example of an operational definition, we might use a ruler to measure a length or a stopwatch to measure a time interval. As

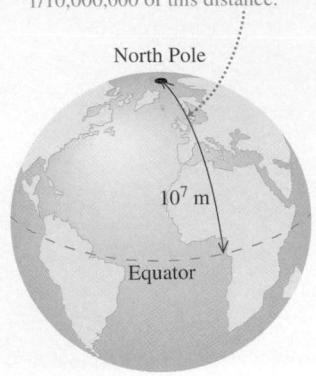

The meter was originally defined as 1/10,000,000 of this distance.

North Pole

10^7 m

Equator

▲ **FIGURE 1.2** The original definition of the meter, from 1791.

(a)

(b)

▲ **FIGURE 1.3** (a) NIST-F1, the atomic clock that serves as the primary time standard for the United States. This clock is accurate to one part in 10^{15}, meaning that it will neither gain nor lose more than 1 second in 30 million years. (b) This platinum–iridium cylinder, kept in a vault in Sèvres, France, is the standard for the unit of the kilogram. Duplicates are kept at various locations around the world.

an example of a quantity we calculate, we might define the average speed of a moving object as the distance traveled (measured with a ruler) divided by the time of travel (measured with a stopwatch).

When we measure a quantity, we always compare it with some reference standard. When we say that a rope is 30 meters long, we mean that it is 30 times as long as a meterstick, which we *define* to be 1 meter long. We call such a standard a **unit** of the quantity. The meter is a unit of length, and the second is a unit of time.

> **NOTE** ► When we use a number to describe a physical quantity, we *must* specify the unit we are using; to describe a distance simply as "30" wouldn't mean anything. ◄

To make precise measurements, we need definitions of units of measurement that do not change and that can be duplicated by observers in various locations. When the metric system was established in 1791 by the Paris Academy of Sciences, the **meter** was originally defined as one ten-millionth of the distance from the equator to the North Pole (Figure 1.2), and the **second** was defined as the time taken for a pendulum 1 meter long to swing from one side to the other.

These definitions were cumbersome and hard to duplicate precisely, and in more recent years they have been replaced by more refined definitions. Since 1889, the definitions of the basic units have been established by an international organization, the General Conference on Weights and Measures. The system of units defined by this organization is based on the metric system, and since 1960 it has been known officially as the **International System,** or **SI** (the abbreviation for the French equivalent, Système International).

Time

From 1889 until 1967, the unit of time was defined in terms of the length of the day, directly related to the time of the earth's rotation. The present standard, adopted in 1967, is much more precise. It is based on an atomic clock (Figure 1.3a) that uses the energy difference between the two lowest energy states of the cesium atom. Electromagnetic radiation (microwaves) of precisely the proper frequency causes transitions from one of these states to the other. One second is defined as the time required for 9,192,631,770 cycles of this radiation.

Length

In 1960 an atomic standard for the meter was established, using the orange-red light emitted by atoms of krypton in a glow discharge tube. In November 1983 the standard of length was changed again, in a more radical way. The new definition of the meter is the distance light travels (in vacuum) in $1/299,792,458$ second. The adoption of this definition has the effect, in turn, of *defining* the speed of light to be precisely 299,792,458 m/s; we then define the meter to be consistent with this number and with the atomic-clock definition of the second. This approach provides a much more precise standard of length than the one based on a wavelength of light.

Mass

The standard of *mass* is the mass of a particular cylinder of platinum-iridium alloy. Its mass is defined to be 1 **kilogram,** and it is kept at the International Bureau of Weights and Measures at Sèvres, near Paris (Figure 1.3b). An atomic standard of mass has not yet been adopted, because, at present, we cannot measure masses on an atomic scale with as great precision as on a macroscopic scale.

TABLE 1.1 **Prefixes for powers of ten**

Power of ten	Prefix	Abbreviation
10^{-18}	atto-	a
10^{-15}	femto-	f
10^{-12}	pico-	p
10^{-9}	nano-	n
10^{-6}	micro-	μ
10^{-3}	milli-	m
10^{-2}	centi-	c
10^{3}	kilo-	k
10^{6}	mega-	M
10^{9}	giga-	G
10^{12}	tera-	T
10^{15}	peta-	P
10^{18}	exa-	E

Prefixes

Once the fundamental units have been defined, it is easy to introduce larger and smaller units for the same physical quantities. One of the great strengths of the metric system is that these other units are always related to the fundamental units by multiples of 10 or 1/10. (By comparison, the English system bristles with inconvenient factors such as 12, 16, and 36.) Thus, 1 kilometer (1 km) is 1000 meters, 1 centimeter (1 cm) is 1/100 meter, and so on.

We usually express these multiplicative factors in exponential notation, sometimes called *powers-of-ten* notation or *scientific* notation. For example; $1000 = 10^3$ and $1/100 = 10^{-2}$. With this notation,

$$1 \text{ km} = 10^3 \text{ m}, \qquad 1 \text{ cm} = 10^{-2} \text{ m}.$$

The names of the additional units are always derived by adding a **prefix** to the name of the fundamental unit. For example, the prefix "kilo-," abbreviated k, always means a unit larger by a factor of 1000; thus,

$$1 \text{ kilometer} = 1 \text{ km} = 10^3 \text{ meters} = 10^3 \text{ m},$$
$$1 \text{ kilogram} = 1 \text{ kg} = 10^3 \text{ grams} = 10^3 \text{ g},$$
$$1 \text{ kilowatt} = 1 \text{ kW} = 10^3 \text{ watts} = 10^3 \text{ W}.$$

Table 1.1 lists the standard SI prefixes with their meanings and abbreviations. Note that most of these are multiples of 10^3.

When pronouncing unit names with prefixes, always accent the *first* syllable: KIL-o-gram, KIL-o-meter, CEN-ti-meter, and MIC-ro-meter. (Kilometer is sometimes pronounced kil-OM-eter.)

Here are a few examples of the use of multiples of 10 and their prefixes with the units of length, mass, and time. Some additional time units are also included. Figure 1.4 gives a partial idea of the range of length scales studied in physics.

Length

$$1 \text{ nanometer} = 1 \text{ nm} = 10^{-9} \text{ m} \quad \text{(a few times the size of an atom)}$$
$$1 \text{ micrometer} = 1 \text{ } \mu\text{m} = 10^{-6} \text{ m} \quad \text{(size of some bacteria and cells)}$$
$$1 \text{ millimeter} = 1 \text{ mm} = 10^{-3} \text{ m} \quad \text{(point of a ballpoint pen)}$$
$$1 \text{ centimeter} = 1 \text{ cm} = 10^{-2} \text{ m} \quad \text{(diameter of your little finger)}$$
$$1 \text{ kilometer} = 1 \text{ km} = 10^{3} \text{ m} \quad \text{(distance traveled in a 10 minute walk)}$$

10^{21} m
A galaxy

10^{-5} m
Red blood cells

10^9 m
Our sun

10^{-9} m
Individual atoms

10^0 m
You

10^7 m
The earth

10^{-14} m
An atomic nucleus

▲ **FIGURE 1.4** These images show part of the range of sizes with which physics deals.

Mass

1 microgram = 1 μg = 10^{-9} kg   (mass of a 1 mm length of hair)

1 milligram = 1 mg = 10^{-6} kg   (mass of a grain of salt)

1 gram   = 1 g   = 10^{-3} kg   (mass of a paper clip)

1 kilogram = 1 kg = 10^3 g   (mass of a 1 liter bottle of water)

Time

1 nanosecond = 1 ns = 10^{-9} s   (time required for a personal computer to add two numbers)

1 microsecond = 1 μs = 10^{-6} s   (time required for a 10-year-old personal computer to add two numbers)

1 millisecond = 1 ms = 10^{-3} s   (time required for sound to travel 0.35 m)

1 minute   = 1 min = 60 s

1 hour   = 1 h   = 3600 s

1 day   = 1 day = 86,400 s

Finally, we mention the British system of units. These units are used only in the United States and a few other countries, and in most of those countries they are being replaced by SI units. British units are now officially *defined* in terms of SI units, as follows:

Length: 1 inch = 2.54 cm (exactly).

Force: 1 pound = 4.448221615260 newtons (exactly).

The fundamental British unit of time is the second, defined the same way as in SI. British units are used only in mechanics and thermodynamics; there is no British system of electrical units.

In this book we use SI units for almost all examples and problems, but in the early chapters we occasionally give approximate equivalents in British units. As you solve problems using SI units, try to think of the approximate equivalents in British units, but also try to think in SI as much as you can. Even in the United States, we now use metric units in many contexts; we speak about 4 liter engines, 50 mm lenses, 10 km races, 2 L soda bottles, and so on. The use of SI in everyday life is not far off.

Quantitative Analysis 1.1

SI measurement of hand

With your hand flat and your fingers and thumb close together, the width of your hand is about

A. 50 cm
B. 10 cm
C. 10 mm

SOLUTION If you're not familiar with metric units of length, you can use your body to develop intuition for them. One meter (100 cm) is slightly more than a yard, so the distance from elbow to fingertips is about 50 cm; this eliminates answer A. Ten millimeters (1 cm) is about the width of your little finger. Ten centimeters is about 4 inches and is about the width of your hand.

1.4 Unit Consistency and Conversions

We use equations to express relations among physical quantities that are represented by algebraic symbols. Most algebraic symbols for physical quantities denote both a number and a unit. For example, d might represent a distance of 10 m, t a time of 5 s, and v (for velocity) a speed of 2 m/s.

An equation must always be **dimensionally consistent.** You can't add apples and pomegranates; two terms may be added or equated only if they have the same units. For example, if a body moving with constant speed v travels a distance d in a time t, these quantities are related by the equation

$$d = vt. \quad (\text{distance} = \text{speed} \times \text{time})$$

If d is measured in meters, then the product vt must also be expressed in meters. Using the above numbers as an example, we may write

$$10 \text{ m} = (2 \text{ m/s})(5 \text{ s}).$$

Because the unit 1/s cancels the unit s in the last factor, the product vt does have units of meters, as it must. In calculations, with respect to multiplication and division, units are treated just like algebraic symbols.

NOTE ▶ When a problem requires calculations using numbers with units, *always* write the numbers with the correct units and carry the units through the calculation, as in the preceding example. This provides a very useful check for calculations. If at some stage in a calculation you find that an equation or an expression has inconsistent units, you know you have made an error somewhere. In this book we will always carry units through all calculations; we strongly urge you to follow this practice when you solve problems. ◀

▲ **Application A $125 million unit consistency error.** In 1998, the Mars Climate Orbiter was sent by NASA to orbit Mars at an altitude above the Martian atmosphere; instead, it entered the atmosphere in 1999 and burned up! The key error that led to this disaster was a miscommunication between the Jet Propulsion Laboratory (JPL) and the spacecraft engineers who built it. The engineers specified the amount of thrust required to steer the craft's trajectory in units of British pounds, while the JPL assumed the numbers were in SI units of newtons. Thus, each correction of the trajectory applied a force 4.45 times larger than needed. While there were other compounding errors, this unit inconsistency was a major factor in the failure of the mission and the loss of the Orbiter.

PROBLEM-SOLVING STRATEGY 1.1 Unit conversions

Units are multiplied and divided just like ordinary algebraic symbols. This fact gives us an easy way to convert a quantity from one set of units to another. The key idea is that we can express the same physical quantity in two different units and form an equality. For example, when we say that 1 min = 60 s, we don't mean that the number 1 is equal to the number 60, but rather that 1 min represents the same physical time interval as 60 s. In a physical sense, multiplying by the quantity 1 min/60 s is really multiplying by unity, which doesn't change the physical meaning of the quantity. To find the number of seconds in 3 min, we write

$$3 \text{ min} = (3 \text{ min})\left(\frac{60 \text{ s}}{1 \text{ min}}\right) = 180 \text{ s}.$$

This expression makes sense. The second is a smaller unit than the minute, so there are more seconds than minutes in the same time interval.

A ratio of units, such as 1 min/60 s (or 1 min = 60 s), is called a **conversion factor**. A table of conversion factors is given in Appendix D and on the inside front cover. If you are used to English units, it may be useful to realize that 1 m is about 3 ft and that 1 gallon is about 4 liters $(1 \text{ gal} \approx 4 \text{ L})$.

Quantitative Analysis 1.2   Inches vs. centimeters

If you express your height both in inches and in centimeters, which unit will give the larger number for your height?

A. The result in centimeters is larger, since you are comparing your height to a unit of length smaller than the inch.
B. Your height in inches is larger, since an inch is larger than a centimeter.
C. Either approach will give the same number, since your height does not depend on the unit used to express it.

SOLUTION Suppose you had a ruler just 1 cm long. To measure the height of a 5 ft person, you would have to move the ruler end-to-end up the person's body about 150 times. However, if your ruler were 1 inch long, you would have to move it just 60 times $(5 \text{ ft} = 60 \text{ in})$. Thus, the smaller your unit of measure, the larger the number you will get when you measure something with it. The correct answer is A.

EXAMPLE 1.1   Driving the autobahn

Although there is no maximum speed limit on the German autobahn, signs in many areas recommend a top speed of 130 km/h. Express this speed in meters per second and miles per hour.

SOLUTION

SET UP We know that 1 km = 1000 m. From Appendix D or the inside front cover, 1 mi = 1.609 km. We also know that 1 h = 60 min = 60 × (60 s) = 3600 s.

SOLVE We use these conversion factors with the problem-solving strategy we've outlined above:

$$130 \text{ km/h} = \left(\frac{130 \text{ km}}{1 \text{ h}}\right)\left(\frac{1000 \text{ m}}{1 \text{ km}}\right)\left(\frac{1 \text{ h}}{3600 \text{ s}}\right) = 36.1 \text{ m/s},$$

$$130 \text{ km/h} = \left(\frac{130 \text{ km}}{1 \text{ h}}\right)\left(\frac{1 \text{ mi}}{1.609 \text{ km}}\right) = 80.8 \text{ mi/h}.$$

REFLECT In physics, speeds are often expressed in meters per second, but if you're American, you're probably used to thinking in miles per hour. These calculations show that the number of miles per hour is roughly twice the number of meters per second; that is, a speed of 36.1 m/s isn't very different from 80.8 mi/h.

Practice Problem: Suppose you leave the autobahn and reduce your speed to 55 km/h. What is this speed in m/s and mi/h? *Answer:* 55 km/h = 15 m/s = 34 mi/h.

EXAMPLE 1.2 **The age of the universe**

The universe is thought to be about 13.7 billion years old. **(a)** What is the age of the universe in seconds? **(b)** Our species evolved about 4.7×10^{12} seconds ago. How many years is that?

SOLUTION

SET UP AND SOLVE We calculate the answer by using the appropriate conversion factors, starting with $1 \text{ y} = 3.156 \times 10^7 \text{ s}$.

Part (a): $13.7 \times 10^9 \text{ y} = (13.7 \times 10^9 \text{ y})\left(\dfrac{3.156 \times 10^7 \text{ s}}{1 \text{ y}}\right)$

$$= 4.32 \times 10^{17} \text{ s}.$$

Part (b): $4.7 \times 10^{12} \text{ s} = (4.7 \times 10^{12} \text{ s})\left(\dfrac{1 \text{ y}}{3.156 \times 10^7 \text{ s}}\right)$

$$= 1.5 \times 10^5 \text{ y}.$$

REFLECT These numbers tell us that the universe is about 10^5 times older than the human species. That is, the universe has existed about 100,000 times as long as we have. A handy approximate conversion factor is $1 \text{ y} \approx \pi \times 10^7 \text{ s}$; this is within 0.5% of the correct value.

Practice Problem: The average human life span in the United States is about 76.1 years. Calculate this time in seconds. *Answer:* 2.40×10^9 s.

Quantitative Analysis 1.3 **Consistency of units**

When an object moves through air, the air exerts a force on it. (You notice this force when you put your hand out of the window of a moving car.) The strength of the force, which we denote by F, is related to the speed v of the object by the equation $F = bv^2$, where b is a constant. In SI units $(\text{kg}, \text{m}, \text{s})$, the units of force are $\text{kg} \cdot \text{m/s}^2$ and the units of speed are m/s. For the preceding equation to have consistent units (the same units on both sides of the equation), the units of b must be

A. kg/m
B. $\text{kg} \cdot \text{m/s}$
C. kg/m^3

SOLUTION To solve this problem, we start with the equation $F = bv^2$. For each symbol whose units we know (in this case, F and v), we replace the symbol with those units. For example, we replace v with (m/s). We obtain

$$\left(\frac{\text{kg} \cdot \text{m}}{\text{s} \cdot \text{s}}\right) = b\left(\frac{\text{m}}{\text{s}}\right)\left(\frac{\text{m}}{\text{s}}\right).$$

(Algebraically, units are treated like numbers.)

We now solve this equation for b. When we multiply the equation on both sides by s^2, the seconds (s) cancel, and one of the meter units (m) cancels. We obtain $b = \text{kg/m}$, so the correct answer is A.

1.5 Precision and Significant Figures

Measurements always have uncertainties. When we measure a distance with an ordinary ruler, it is usually reliable only to the nearest millimeter, but with a precision micrometer caliper, we can measure distances with an uncertainty of less than 0.01 mm. We often indicate the precision of a number by writing the number, the symbol $\pm$, and a second number indicating the likely uncertainty. If the diameter of a steel rod is given as 56.47 ± 0.02 mm, this expression means that the true value is unlikely to be less than 56.45 mm or greater than 56.49 mm.

When we use numbers having uncertainties or errors to compute other numbers, the computed numbers are also uncertain. It is especially important to understand this when comparing a number obtained from measurements with a value obtained from a theoretical prediction. Suppose you want to determine the value of π (pi), as in Figure 1.5, using the relation of the circumference C of a circle to its diameter D: $C = \pi D$. Then

$$\pi = \frac{C}{D} = \frac{\text{circumference}}{\text{diameter}}.$$

The true value of this ratio, to 10 digits, is 3.141592654. To make your own calculation, you cut a circular piece out of a sheet of plywood and measure its diameter and circumference to the nearest millimeter, using a flexible measuring tape. Suppose you obtain the values 135 mm and 424 mm, respectively. You

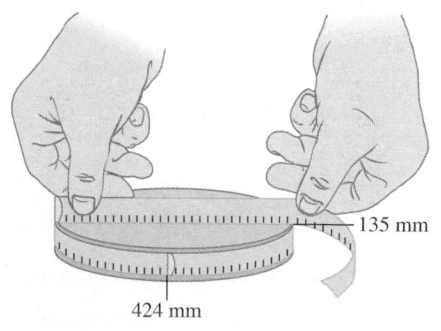

▲ **FIGURE 1.5** Finding π with a plywood disk and a tape measure.

▲ **Application What went wrong?** While a 0.01% error may sound insignificant, in this case it resulted in a spectacular mishap. In a journey of 500 kilometers, a 0.01% error represents only 50 meters. Yet going 0.01% too far had tragic results in this instance. In physics, measurements can be considerably more precise than ±0.01%. For example, according to the National Institute of Standards and Technology, the mass of the electron has been determined to be $9.1093826 \pm 0.0000016 \times 10^{-31}$ kg. This represents a relative uncertainty of 1.7×10^{-7}, or only ±0.000017%, in measuring the mass of the electron.

punch these numbers into your calculator and obtain the quotient 3.140740741. Does this quantity agree with the true value or not?

First, the last seven digits in this answer are meaningless, because they imply greater precision than is possible with your measurements. The number of meaningful digits in a number is called the number of **significant figures** or **significant digits.** ("Figure" in this context means "digit.") **When numbers are multiplied or divided, the number of significant figures in the result is no greater than in the factor with the fewest significant figures.** For example, $3.1416 \times 2.34 \times 0.58 = 4.3$. Thus your measured value of π using the plywood circle has only three significant figures and should be stated simply as 3.14. Your value *does* agree with the true value, although only to three significant figures.

NOTE ▶ Zero can be a significant figure. The number 0.580 has three significant figures. The initial zero doesn't matter, but the final zero implies that the number is probably between 0.579 and 0.581. Writing 0.58 implies only that the number is probably between 0.57 and 0.59. Leading zeroes aren't significant, but trailing zeroes are. ◀

When you add or subtract, the rule for handling significant figures is slightly different than when you multiply or divide: **When you add or subtract, the answer can have no more decimal places than the term with the fewest decimal places.** For example, $5.232 + 0.31 = 5.54$, not 5.542.

Here's one additional point: In symbolic relationships, when unitless numerical constants appear, usually they should be taken to have infinitely great accuracy, with no uncertainty at all. For example, in the relation $C = 2\pi R$ between the radius R and circumference C of a circle, the factor 2 is taken to be exact, with no uncertainty. It could appropriately be written as 2.000000000000000.

Most examples and problems in this book give data to two or three significant figures (occasionally more). In each problem, your answer should have no more significant figures than the preceding rules allow. You may do the arithmetic with a calculator having a display with 5 to 10 digits. But you should *not* give a 10 digit answer for a calculation using numbers with three significant figures. To do so is not only unnecessary; it is genuinely wrong, because it misrepresents the precision of the results. Always round your answer to keep only the correct number of significant figures.

When we calculate with very large or very small numbers, we can show significant figures much more easily by using powers-of-ten notation, sometimes called *scientific notation,* as we discussed in Section 1.3. For example, the distance from the earth to the sun is about 149,000,000,000 m, but writing the number in this form gives no indication of the number of significant figures. Certainly, not all 12 are significant! Instead, we move the decimal point 11 places to the left (corresponding to dividing by 10^{11}) and multiply by 10^{11}. That is,

$$149{,}000{,}000{,}000 \text{ m} = 1.49 \times 10^{11} \text{ m.} \quad \text{(three significant figures)}$$

In this form, it is clear that we have three significant figures. In scientific notation, a common practice is to express a quantity as a number between 1 and 10, multiplied by the appropriate power of 10.

EXAMPLE 1.3   **Mass is energy**

The energy E corresponding to the mass m of an electron is given by Einstein's famous equation

$$E = mc^2, \quad \text{(equivalence of mass and energy)}$$

where c is the speed of light. Calculate the value of E for an electron, using powers of ten and three significant figures.

Continued

SOLUTION

SET UP We find the needed numbers inside the front cover or in Appendix D: $m = 9.11 \times 10^{-31}$ kg and $c = 3.00 \times 10^8$ m/s.

SOLVE We use these numbers to calculate the value of E:

$$E = (9.11 \times 10^{-31} \text{ kg})(3.00 \times 10^8 \text{ m/s})^2$$
$$= (9.11)(3.00)^2 (10^{-31})(10^8)^2 (\text{kg} \cdot \text{m}^2)/\text{s}^2$$
$$= (82.0)(10^{[-31+(2\times8)]})(\text{kg} \cdot \text{m}^2)/\text{s}^2$$
$$= 8.20 \times 10^{-14} \text{ kg} \cdot \text{m}^2/\text{s}^2.$$

REFLECT Most pocket calculators can use scientific notation and do this addition of exponents automatically for you; but you should be able to do such calculations by hand when necessary. Incidentally, the value used for c in Example 1.3 has three significant figures, even though two of them are zeros. To greater precision, $c = 2.997925 \times 10^8$ m/s.

Conceptual Analysis 1.4

Significant figures

To seven significant figures, the speed of light is $c = 2.997925 \times 10^8$ m/s. Which of the following choices is or are correct?

A. $c = 3.0 \times 10^8$ m/s
B. $c = 3.00 \times 10^8$ m/s
C. $c = 3.000 \times 10^8$ m/s

SOLUTION Answer C is incorrect, because when we round to four significant figures, we get 2.998, not 3.000. But A and B are both correct; they represent different levels of precision, even though the numerical value is the same.

1.6 Estimates and Orders of Magnitude

We've talked about the importance of knowing the precision of numbers that represent physical quantities. But there are also situations in which even a very crude estimate of a quantity may give us useful information. Often we know how to calculate a certain quantity but have to guess at the data we need for the calculation. Or the calculation may be too complicated to carry out exactly, so we make some crude approximations. In either case our result is also a guess, but many times such a guess is useful, even if it is uncertain by a factor of 2, or 10, or more. Such calculations are often called **order-of-magnitude estimates.** The great nuclear physicist Enrico Fermi (1901–1954) called them "back-of-the-envelope calculations."

PhET: Estimation

EXAMPLE 1.4   Espionage

You are writing an international espionage novel in which the hero escapes across the border with a billion dollars worth of gold in his suitcase. Is this possible? Would that amount of gold fit in a suitcase? Would it be too heavy to carry?

SOLUTION

SET UP AND SOLVE We can guess at both the mass of gold needed and its density. From these numbers we can then find the volume of gold needed.

Gold sells for around $400 an ounce. On a particular day, the price might be $200 or $600, but never mind. An ounce is about 30 grams. Actually, an ordinary (avoirdupois) ounce is 28.35 g; an ounce of gold is a troy ounce, which is 9.45% more. Again, never mind. Ten dollars worth of gold has a mass somewhere around 1 gram, so a billion (10^9) dollars worth of gold is a hundred million (10^8) grams, or a hundred thousand (10^5) kilograms. This corresponds to a weight in British units of around 200,000 lb, or a hundred tons.

We can also estimate the *volume* of this gold. If its density were the same as that of water (1 g/cm^3), the volume would be 10^8 cm³, or 100 m³. But gold is a heavy metal; we might guess its density to be 10 times that of water. In fact, it is over 20 times as dense as water. But guessing 10, we find a volume of 10 m³. Visualize 10 cubical stacks of gold bricks, each 1 meter on a side, and ask whether they would fit in a suitcase!

REFLECT Whether the precise weight is 50 tons or 200 doesn't matter. Either way, the hero is not about to carry it across the border in a suitcase. And 10 cubic meters certainly won't fit in the suitcase!

Practice Problem: Suppose our hero needs only a million dollars, not a billion. Are the answers the same or different? Would he do better if he used cut diamonds instead of gold? *Answer:* Weight is a few hundred pounds; volume is comparable to a small carry-on suitcase. Diamonds would be better.

1.7 Vectors and Vector Addition

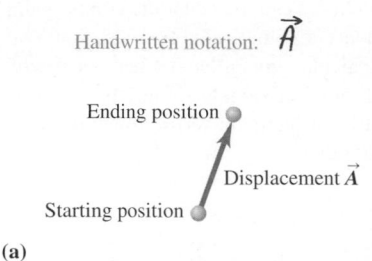

Handwritten notation: $\vec{A}$

Ending position

Displacement $\vec{A}$

Starting position

(a)

The displacement depends only on the starting and ending positions—not on the path taken.

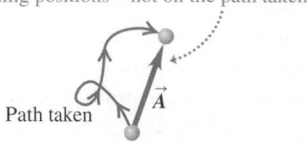

Path taken

$\vec{A}$

(b)

If the object makes a round-trip, the total displacement is 0, regardless of the distance traveled.

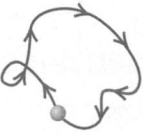

(c)

▲ **FIGURE 1.6** Displacement is a vector quantity. A displacement vector represents a change in position (distance and direction) of an object from one point in space to another.

Some physical quantities, such as time, temperature, mass, density, and electric charge, can be described completely by a single number with a unit. Many other quantities, however, have a *directional* quality and cannot be described by a single number. A familiar example is velocity. To describe the motion of an airplane, we have to say not only *how fast* it is moving, but also *in what direction*. To fly from Chicago to New York, the plane has to head east, not south. Force is another example. When we push or pull on a body, we exert a force on it. To describe a force, we need to describe the direction in which it acts, as well as its magnitude, or "how hard" the force pushes or pulls. A tow truck usually pulls a stalled car forward, not sideways.

When a physical quantity is described by a single number, we call it a **scalar quantity.** In contrast, a **vector quantity** has both a **magnitude** (the "how much" or "how big" part) and a **direction** in space. Calculations with scalar quantities use the operations of ordinary arithmetic. For example, 6 kg + 3 kg = 9 kg, or $(12\,\text{m})/(3\,\text{s}) = 4\,\text{m/s}$. However, combining vector quantities requires a different set of operations. Vector quantities play an essential role in all areas of physics, so let's talk next about what vectors are and how they combine.

We start with the simplest of all vector quantities: **displacement.** Displacement is simply a change in position of an object, which we will assume can be treated as a pointlike particle. In Figure 1.6a, we represent the object's change in position by an arrow that points from the starting position to the ending position. Displacement is a *vector quantity* because we must state not only *how far* the particle moves, but also *in what direction.* Walking 3 km north from your front door doesn't get you to the same place as walking 3 km southeast.

We usually represent a vector quantity, such as a displacement, by a single letter, such as $\vec{A}$ in Figure 1.6. In this book, we always print vector symbols **in boldface type with an arrow above them,** to remind you that vector quantities have properties different from those of scalar quantities. In handwriting, vector symbols are usually written with an arrow above, as shown in Figure 1.6a, to indicate that they represent vector quantities.

NOTE ▶ When you write a symbol for a vector quantity, *always* put an arrow over it. If you don't distinguish between scalar and vector quantities in your notation, you probably won't make the distinction in your thinking either, and hopeless confusion will result. ◀

A displacement is always a straight-line segment, directed from the starting point to the endpoint, even though the actual path of the object may be curved. In Figure 1.6b, the displacement is still $\vec{A}$, even though the object takes a roundabout path. Also, note that displacement is not related directly to the total *distance* traveled. If the particle makes a round trip, returning to its starting point as shown in Figure 1.6c, the displacement for the entire trip is *zero*.

The vector $\vec{A}'$ from point 3 to point 4 in Figure 1.7 has the same length and direction as the vector $\vec{A}$ from point 1 to point 2. These two displacements are equal, even though they start at different points. Thus we may write $\vec{A}' = \vec{A}$. By definition, two vector quantities are equal if they have the same magnitude (length) and direction, no matter where they are located in space. The vector $\vec{B}$, however, is *not* equal to $\vec{A}$, because its direction is *opposite* to that of $\vec{A}$. We define the *negative* of a vector as a vector having the same magnitude as the original vector, but opposite direction. The negative of the vector quantity $\vec{A}$ is denoted $-\vec{A}$. If $\vec{A}$ is 87 m south, then $-\vec{A}$ is 87 m north.

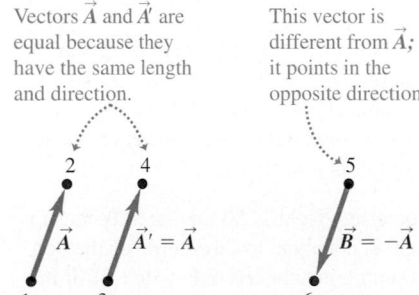

Vectors $\vec{A}$ and $\vec{A}'$ are equal because they have the same length and direction.

This vector is different from $\vec{A}$; it points in the opposite direction.

2   4

5

$\vec{A}$ $\vec{A}' = \vec{A}$ $\vec{B} = -\vec{A}$

1   3   6

▲ **FIGURE 1.7** Two displacement vectors are equal if they have the same length (magnitude) and the same (not opposite) direction.

Thus, the relation between $\vec{A}$ and $\vec{B}$ may be written as $\vec{A} = -\vec{B}$ or $\vec{B} = -\vec{A}$. When two vectors $\vec{A}$ and $\vec{B}$ have opposite directions, we say that they are *antiparallel*.

NOTE ▶ Remember that two vector quantities are equal only when they have the same magnitude *and* the same (not opposite) direction. ◀

We usually represent the magnitude of a vector quantity (its length, in the case of a displacement vector) by the same letter used for the vector, but in light italic type rather than boldface italic. An alternative notation is the vector symbol with vertical bars on both sides. Thus,

Magnitude of $\vec{A} = A = |\vec{A}|$.   (notation for magnitude of a vector quantity)

By definition, the magnitude of a vector quantity is a scalar quantity (a single number) and is always positive. We also note that a vector can never be equal to a scalar, because they are different kinds of quantities. The expression $\vec{A} = 6$ m is just as wrong as "2 oranges = 3 apples" or "6 lb = 7 km"!

Now suppose a particle undergoes a displacement $\vec{A}$, followed by a second displacement $\vec{B}$ (Figure 1.8a). The final result is the same as though the particle had started at the same initial point and undergone a single displacement $\vec{C}$, as shown. We call displacement $\vec{C}$ the **vector sum,** or **resultant,** of displacements $\vec{A}$ and $\vec{B}$; we express this relationship symbolically as

$$\vec{C} = \vec{A} + \vec{B}.\quad \text{(vector sum of two vectors)}$$

The plus sign in this equation denotes the geometrical operation of **vector addition** that we've described; it is *not* the same operation as adding two scalar quantities such as $2 + 3 = 5$.

If we make the displacements $\vec{A}$ and $\vec{B}$ in the reverse order, as in Figure 1.8b, with $\vec{B}$ first and $\vec{A}$ second, the result is the same, as the figure shows. Thus,

$$\vec{C} = \vec{B} + \vec{A} \quad \text{and} \quad \vec{A} + \vec{B} = \vec{B} + \vec{A} \quad \text{(Vector sum is commutative.)}$$

The last equation shows that the order of the terms in a vector sum doesn't matter; addition of vectors is **commutative.** We draw the tail of the second vector at the point of the first, and the results are the same with either order.

Figure 1.8b also suggests an alternative graphical representation of the vector sum. When vectors $\vec{A}$ and $\vec{B}$ are both drawn with their tails at the same point, vector $\vec{C}$ is the diagonal of a parallelogram constructed with $\vec{A}$ and $\vec{B}$ as two adjacent sides.

Figure 1.9 shows the special cases in which two vectors $\vec{A}$ and $\vec{B}$ are parallel (Figure 1.9a) or antiparallel (Figure 1.9b). When the vectors are parallel, the magnitude of the vector sum equals the *sum* of the magnitudes of $\vec{A}$ and $\vec{B}$; when they are antiparallel, it equals the *difference* of their magnitudes.

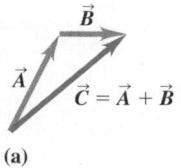

(a)

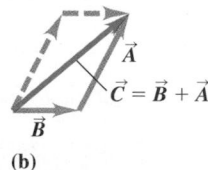

(b)

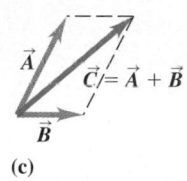

(c)

▲ FIGURE 1.8 Vector $\vec{C}$ is the vector sum of vectors $\vec{A}$ and $\vec{B}$. The order in vector addition doesn't matter. (Vector addition is commutative.)

$$\vec{A} \qquad \vec{B}$$
$$\vec{C} = \vec{A} + \vec{B}$$

(a) The sum of two parallel vectors

$$\vec{A}$$
$$\vec{C} = \vec{A} + \vec{B} \qquad \vec{B}$$

(b) The sum of two antiparallel vectors

▲ FIGURE 1.9 The vector sum of (a) two parallel vectors and (b) two antiparallel vectors. Note that the vectors $\vec{A}$, $\vec{B}$, and $\vec{C}$ in (a) are not the same as the vectors $\vec{A}$, $\vec{B}$, and $\vec{C}$ in (b).

Conceptual Analysis 1.5

Resultant vector

Two vectors, one with magnitude 3 m and the other with magnitude 4 m, are added together. The resultant vector could have a magnitude as small as

A. 1 m;
B. 3 m;
C. 4 m.

SOLUTION The resultant has the largest magnitude when the two vectors are parallel and the smallest when they are antiparallel—that is, pointing in opposite directions. In the latter case, the resultant will be the difference between their lengths, which is 1 m. Thus, answer A is correct.

To find the sum of these three vectors . . .

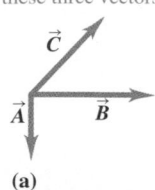

(a)

we could add $\vec{A}$ and $\vec{B}$ to get $\vec{D}$ and then add $\vec{C}$ to $\vec{D}$ to get the final sum (resultant) $\vec{R}$, . . .

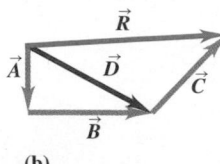

(b)

or we could add $\vec{B}$ and $\vec{C}$ to get $\vec{E}$ and then add $\vec{A}$ to $\vec{E}$ to get $\vec{R}$, . . .

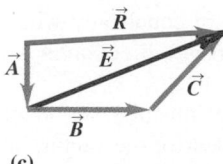

(c)

or we could add $\vec{A}$, $\vec{B}$, and $\vec{C}$ to get $\vec{R}$ directly, . . .

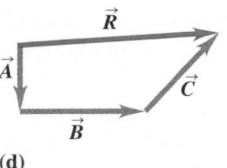

(d)

or we could add $\vec{A}$, $\vec{B}$, and $\vec{C}$ in any other order and still get $\vec{R}$.

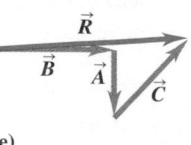

(e)

▲ **FIGURE 1.10** Several constructions for finding the vector sum $\vec{A} + \vec{B} + \vec{C}$.

When we need to add more than two vectors, we may first find the vector sum of any two, add it vectorially to the third, and so on. Figure 1.10a shows three vectors $\vec{A}$, $\vec{B}$, and $\vec{C}$. In Figure 1.10b, vectors $\vec{A}$ and $\vec{B}$ are first added, giving a vector sum $\vec{D}$; vectors $\vec{D}$ and $\vec{C}$ are then added by the same process to obtain the vector sum $\vec{R}$:

$$\vec{R} = (\vec{A} + \vec{B}) + \vec{C} = \vec{D} + \vec{C}.$$

Alternatively, we can first add $\vec{B}$ and $\vec{C}$ (Figure 1.10c) to obtain vector $\vec{E}$ and then add $\vec{A}$ and $\vec{E}$ to obtain $\vec{R}$:

$$\vec{R} = \vec{A} + (\vec{B} + \vec{C}) = \vec{A} + \vec{E}.$$

We don't even need to draw vectors $\vec{D}$ and $\vec{E}$; all we need to do is draw the given vectors in succession, with the tail of each at the head of the one preceding it, and complete the polygon by a vector $\vec{R}$ from the *tail* of the first to the *head* of the last, as shown in Figure 1.10d. The order makes no difference; Figure 1.10e shows a different order, and we invite you to try others.

NOTE ▶ When adding vectors graphically, always arrange them tip to tail, so that the tip of one points to the tail of the next. The vectors can be in any order. The resultant always points from the tail of the first vector to the tip of the final one. ◀

▲ **Application Where will it end up?**
Although this caribou is trying to swim straight across the river, it will not end up directly across from its starting point. Its final position on the opposite bank can be determined by the addition of two vectors. One represents its movement directly across the river, and a second represents the downstream movement of the water. If it paddles across at the same speed as the river flows, summing these two vectors shows us that it will actually cross the river at a 45° angle. At the other side, it will end up a distance downstream equal to the width of the river, having traveled a distance of $\sqrt{2}$ times the river's width.

Quantitative Analysis 1.6

Pythagorean vectors

If you walk 40 m east, then 30 m north, how far from the starting point are you?

A. 10 m
B. 50 m
C. 70 m

SOLUTION The two legs of your trip form the sides of a right triangle; your final displacement is the hypotenuse. Draw a diagram showing the two successive displacements and the vector sum. You can use the Pythagorean theorem to calculate the length of the vector sum, 50 m.

You can confirm the length of this resultant vector by using what you know about parallel and antiparallel vectors. If the two legs of your trip were antiparallel (rather than at right angles), the resultant would have a magnitude of 10 m. So A can't be right; the correct answer must be larger. If the two legs were parallel (i.e., if you walked 40 m and then continued 30 m in the same direction), the resultant would have a magnitude of 70 m, so C can't be right either; the correct answer must be smaller. That leaves B as the correct answer.

Diagrams for the addition of displacement vectors don't have to be drawn actual size. Often, we'll use a scale similar to those used for maps, in which the distance on the diagram is *proportional* to the actual distance, such as 1 cm for 5 km. When we work with other vector quantities with different units, such as force or velocity, we *must* use a scale. In a diagram for velocity vectors, we might use a scale in which a vector 1 cm long represents a velocity with magnitude 5 m/s. A velocity with magnitude 20 m/s would then be represented by a vector 4 cm long in the appropriate direction.

A vector quantity such as a displacement can be multiplied by a scalar quantity (an ordinary number). As shown in Figure 1.11, the displacement $2\vec{A}$ is a displacement (vector quantity) in the same direction as the vector $\vec{A}$, but twice as long. The scalar quantity used to multiply a vector may also be a physical quantity having units. For example, when an object moves in a straight line with constant velocity $\vec{v}$ for a time t, its total displacement $\vec{d}$ is equal to $\vec{v}t$. The magnitude d of the displacement is equal to the magnitude v of the velocity multiplied by the time t. The directions of $\vec{d}$ and $\vec{v}$ are the same (assuming that t is positive). The units of d are the product of the units of v and t.

We've already mentioned that $-\vec{A}$ is, by definition, a vector having the same magnitude as $\vec{A}$ but opposite direction. This definition provides the basis for defining vector *subtraction*. (See Figure 1.12.) We define the *difference* $\vec{A} - \vec{B}$ of two vectors $\vec{A}$ and $\vec{B}$ to be the vector sum of $\vec{A}$ and $-\vec{B}$:

$$\vec{A} - \vec{B} = \vec{A} + (-\vec{B}). \quad \text{(vector difference)}$$

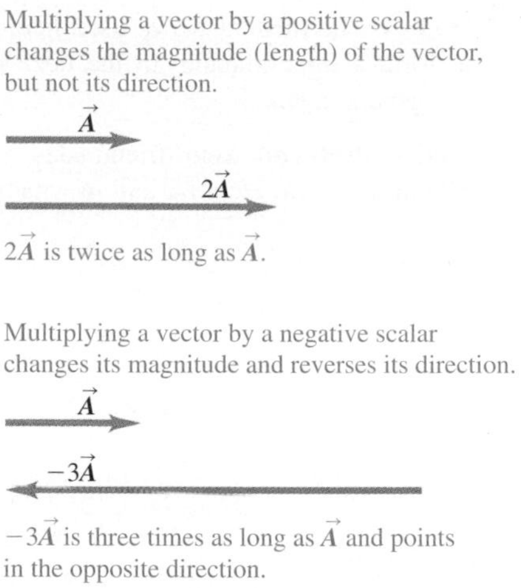

Multiplying a vector by a positive scalar changes the magnitude (length) of the vector, but not its direction.

$2\vec{A}$ is twice as long as $\vec{A}$.

Multiplying a vector by a negative scalar changes its magnitude and reverses its direction.

$-3\vec{A}$ is three times as long as $\vec{A}$ and points in the opposite direction.

▲ **FIGURE 1.11** The effect of multiplying a vector by a scalar.

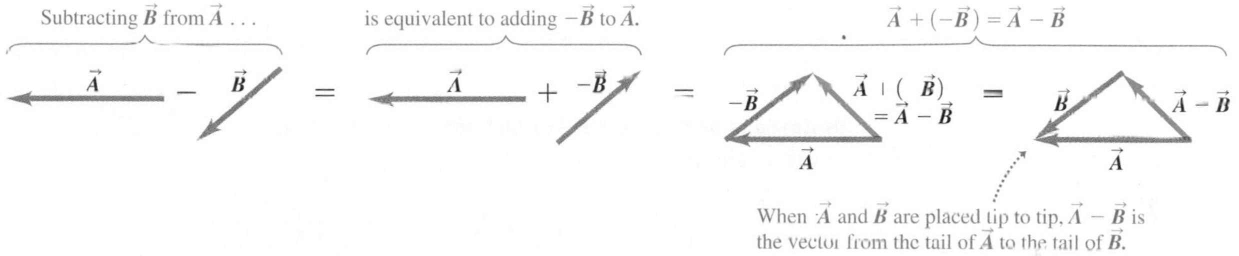

Subtracting $\vec{B}$ from $\vec{A}$ ...   is equivalent to adding $-\vec{B}$ to $\vec{A}$. $\vec{A} + (-\vec{B}) = \vec{A} - \vec{B}$

$$\vec{A} - \vec{B} = \vec{A} + -\vec{B} = \vec{A} + (-\vec{B}) = \vec{A} - \vec{B}$$

When $\vec{A}$ and $\vec{B}$ are placed tip to tip, $\vec{A} - \vec{B}$ is the vector from the tail of $\vec{A}$ to the tail of $\vec{B}$.

▲ **FIGURE 1.12** The definition of vector subtraction.

EXAMPLE 1.5   Displacement of a cross-country skier

On a cross-country ski trip, you travel 1.00 km north and then 2.00 km east. **(a)** How far and in what direction are you from your starting point? **(b)** What are the magnitude and direction of your resultant displacement?

SOLUTION

SET UP Figure 1.13 is a scale diagram. By careful measurement on the diagram, we find that the distance d from the starting point is about 2.2 km and the angle ϕ is about 63°. But it's much more accurate to *calculate* the result. The vectors in the diagram form a right triangle, and we can find the length of the hypotenuse by using the Pythagorean theorem. The angle can be found by simple trigonometry, from the definition of the tangent function.

SOLVE Part (a): Use the Pythagorean theorem to find the length d of the resultant displacement vector (the distance from the starting point):

$$d = \sqrt{(1.00 \text{ km})^2 + (2.00 \text{ km})^2}$$
$$= 2.24 \text{ km}. \quad \text{(Magnitude of displacement vector)}$$

From the definition of the tangent function,

$$\tan\phi = \frac{\text{opposite side}}{\text{adjacent side}} = \frac{2.00 \text{ km}}{1.00 \text{ km}},$$
$$\phi = 63.4°. \quad \text{(direction of displacement vector)}$$

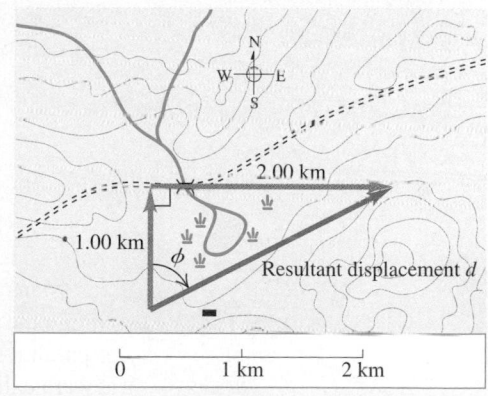

▲ **FIGURE 1.13** A vector diagram, drawn to scale, for a cross-country ski trip.

Part (b): The magnitude of the resultant displacement is just the distance d we found in part (a), 2.24 km. We can describe the direction as 63.4° east of north, or 26.6° north of east. Take your choice!

Continued

REFLECT The method we've used here works only for vectors that form a right triangle. In the next section, we'll present a more general method.

Practice Problem: Your friend skis 3.00 km east and then 1.50 km north. **(a)** How far and in what direction is he from the starting point? **(b)** What are the magnitude and direction of his resultant displacement? *Answers:* (a) 3.35 km, 26.6°; (b) 3.35 km, 26.6° north of east.

1.8 Components of Vectors

We need to develop efficient and general methods for adding vectors. In Example 1.5 in the preceding section, we added two vectors by drawing and measuring a scale diagram or by making use of the properties of right triangles. Measuring a diagram offers only very limited precision, and our example was a special case in which the two vectors were perpendicular. With an angle other than 90° or with more than two vectors, we would get into repeated trigonometric solutions of oblique triangles, which can get horribly complicated.

So we need a simple, but more general, method for adding vectors. The usual method makes use of *components* of vectors. As a prelude to explaining this method, let's review the definitions of the basic trigonometric functions.

For a right triangle, the functions $\sin\theta$, $\cos\theta$ and $\tan\theta$ are defined as follows:

Definitions of trigonometric functions
For the right triangle shown,

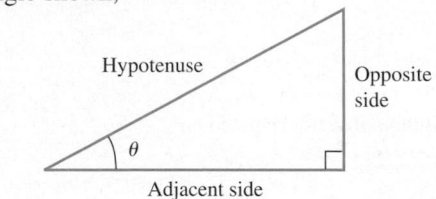

$$\sin\theta = \frac{\text{opposite side}}{\text{hypotenuse}}, \qquad \cos\theta = \frac{\text{adjacent side}}{\text{hypotenuse}},$$
$$\tan\theta = \frac{\text{opposite side}}{\text{adjacent side}}. \qquad (1.1)$$

From these definitions, we can prove that, for any angle θ,

$$\tan\theta = \frac{\sin\theta}{\cos\theta} \quad \text{and} \quad \sin^2\theta + \cos^2\theta = 1.$$

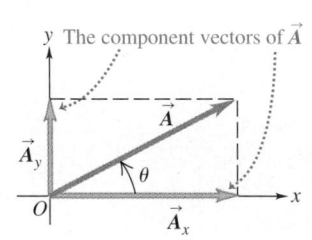

▲ FIGURE 1.14 (a) The component vectors of $\vec{A}$ in the direction of the x and y axes. (b) The x and y components of $\vec{A}$.

Now, what are components, and how are they used to add and subtract vectors? To define what we mean by components, let's start with a rectangular (Cartesian) coordinate axis system as in Figure 1.14. We can represent any vector lying in the x-y plane as the vector sum of a vector parallel to the x axis and a vector parallel to the y axis. These two vectors are labeled $\vec{A}_x$ and $\vec{A}_y$ in Figure 1.14a; they are called the **component vectors** of vector $\vec{A}$, and their vector sum is equal to $\vec{A}$. In symbols,

$$\vec{A} = \vec{A}_x + \vec{A}_y. \quad \text{(vector } \vec{A} \text{ as a sum of component vectors)}$$

By definition, each component vector lies along one of the two coordinate-axis directions. Thus, we need only a single number to describe each component. When the vector points in the positive axis direction, we define the number A_x to be the magnitude of $\vec{A}_x$. When $\vec{A}_x$ points in the $-x$ direction, we define the number A_x to be the *negative* of that magnitude, keeping in mind that the magnitude of a vector quantity is always positive. We define the number A_y the same way. The two numbers A_x and A_y are called the **components** of the vector $\vec{A}$. We haven't yet described what they're good for; we'll get to that soon!

NOTE ▶ Be sure that you understand the relationship between $\vec{A}$ (a vector), A or $|\vec{A}|$ (the vector's magnitude), $\vec{A}_x$ (a component vector) and A_x (a component). The following tips are useful:

- A subscript x or y indicates that the quantity is a component vector or component along the x or y axis. Watch for these subscripts!
- A *magnitude* is always positive.
- A *component* is positive if the corresponding component vector points in the positive direction of the axis, negative if the component vector points in the negative direction of the axis. ◀

If, for the vector $\vec{A}$ in Figure 1.14, we know both the magnitude A and the direction, given by the angle θ, we can calculate the components, as shown in Figure 1.14b. From the definitions of the trigonometric cosine and sine functions,

$$\frac{A_x}{A} = \cos\theta \qquad \text{and} \qquad \frac{A_y}{A} = \sin\theta;$$

$$A_x = A\cos\theta \qquad \text{and} \qquad A_y = A\sin\theta. \tag{1.2}$$

NOTE ▶ These equations are correct for *any* angle θ, from 0 to 360°, provided that the angle is measured counterclockwise from the positive x axis. This is the usual convention, but sometimes you'll find it more convenient to measure θ from a different axis; then you'll have to identify the appropriate right triangle and use the definitions of the trigonometric functions for that triangle. Be careful! ◀

In Figure 1.15, the component B_x is negative because its direction is opposite that of the positive x axis. This is consistent with Equations 1.2; the cosine of an angle in the second quadrant is negative. B_y is positive, but both C_x and C_y are negative.

We can describe a vector completely by giving either its magnitude and direction or its x and y components. Equations 1.2 show how to find the components if we know the magnitude and direction. Applying the Pythagorean theorem to Figure 1.14, we find

$$A = \sqrt{A_x^2 + A_y^2}. \quad \text{(magnitude of vector } \vec{A}\text{)} \tag{1.3}$$

Also, from the definition of the tangent of an angle,

$$\tan\theta = \frac{A_y}{A_x} \qquad \text{and} \qquad \theta = \tan^{-1}\frac{A_y}{A_x}. \quad \text{(direction of vector } \vec{A}\text{)} \tag{1.4}$$

NOTE ▶ There's a slight complication in using Equations 1.4 to find θ. Suppose $A_x = 2$ m and $A_y = -2$ m; then $\tan\theta = -1$. But, as shown in Figure 1.16, there are two angles having tangents of -1, namely, 135° and 315° (or -45°). To decide which is correct, we have to look at the individual components. Because A_x is positive and A_y is negative, the angle must be in the fourth quadrant; thus, $\theta = 315$° (or -45°) is the correct value. Most pocket calculators give $\tan^{-1}(-1) = -45$°. In this case, that is

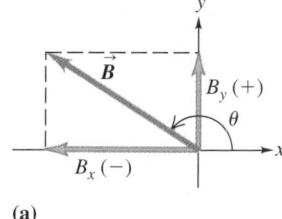

(a)

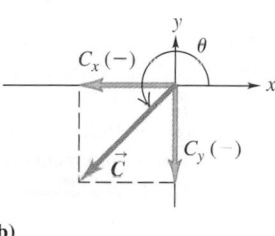

(b)

▲ FIGURE 1.15 The components of a vector are numbers; they may be positive or negative.

In this case, $\tan\theta = \dfrac{A_y}{A_x} = -1$.

Two angles have tangents of -1: 135° and 315°. Inspection of the diagram shows that θ must be 315°.

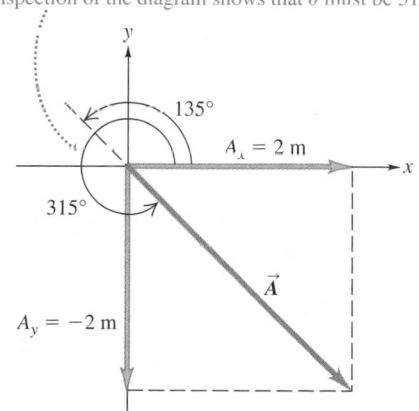

▲ FIGURE 1.16 Two angles may have the same tangent; sketching the vector shows which angle is correct.

correct; but if, instead, we have $A_x = -2$ m and $A_y = 2$ m, then the correct angle is 135°. You should always draw a sketch to check which of the two possibilities is the correct one. Similarly, when A_x and A_y are both negative, the tangent is positive, but the angle is in the third quadrant. ◄

The next example will give us some practice in finding components of individual vectors; then we'll talk about how the components are used to find vector sums and differences.

EXAMPLE 1.6 **Finding components of vectors**

Raoul and Maria set out walking from their aunt's house. Raoul walks a certain distance straight east and then a certain distance straight north. A crow passing overhead starts from the same point as Raoul, but flies in a straight line. Raoul observes that the crow flies a total distance of 500 m in a direction 35° north of east to reach the same final point as Raoul. Maria starts from the same point as Raoul, but she walks first straight west and then straight south. Her final point is 700 m from the starting point in a direction that is 55° south of west. How far did Raoul walk on the east leg of his trip? How far on the north leg? Similarly, how far did Maria walk on the west leg and how far on the south leg? Express these quantities in terms of displacement vectors and their components, using a coordinate system with the $+x$ axis pointing east and the $+y$ axis pointing north.

SOLUTION

SET UP First we sketch a diagram showing each displacement, as in Figure 1.17. We draw a rectangular coordinate system and label the axes, with the $+x$ axis east and the $+y$ axis north. Be sure to make your sketch large enough so that you will be able to draw and label the components of each vector. Note that the vector representing Maria's displacement lies in the third quadrant, so both the x and y components of this vector will be negative.

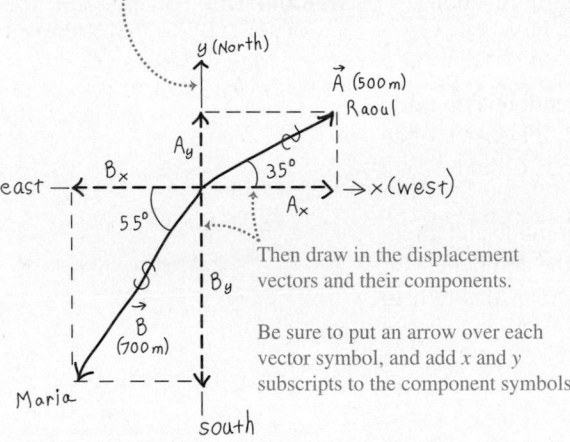

First draw your axes. Make them big enough.

Then draw in the displacement vectors and their components.

Be sure to put an arrow over each vector symbol, and add x and y subscripts to the component symbols.

▲ **FIGURE 1.17** The diagram we draw for this problem.

SOLVE We'll label Raoul's displacement vector $\vec{A}$ and Maria's $\vec{B}$. Then to find the components we use Equations 1.2:

$$A_x = A\cos\theta \qquad \text{and} \qquad B_x = B\cos\theta$$
$$A_y = A\sin\theta \qquad \qquad B_y = B\sin\theta$$

We measure the angles counterclockwise from the $+x$ axis. So the direction of Maria's displacement is $180° + 55° = 235°$. The components of the two displacement vectors are as follows:

Raoul	Maria
$A_x = 500$ m $(\cos 35°)$	$B_x = 700$ m $(\cos 235°)$
$\quad = 410$ m (east)	$\quad = -402$ m (west)
$A_y = 500$ m $(\sin 35°)$	$B_y = 700$ m $(\sin 235°)$
$\quad = 287$ m (north)	$\quad = -573$ m (south)

REFLECT Because Maria walked west and south, the x and y components of her displacement are both negative. To get the correct signs, we must measure the angles counterclockwise from the $+x$ axis.

Practice Problem: Raoul's friend Johnny sets out from his house in the city. His displacement is 600 m 40° north of west. What are the x and y components of his displacement vector? *Answers:* $A_x = -460$ m; $A_y = 386$ m.

Using components to add vectors

Now, at last, we're ready to talk about using components to calculate the vector sum of several vectors. The component method requires only right triangles and simple computations, and it can be carried out with great precision. Here's the basic idea: Figure 1.18a shows two vectors $\vec{A}$ and $\vec{B}$ and their vector sum (resultant) $\vec{R}$, along with the x and y components of all three vectors. You can see from the diagram that the x component R_x of the vector sum is simply the sum

$\vec{R}$ is the sum (resultant) of $\vec{A}$ and $\vec{B}$.

The y component of $\vec{R}$ equals the sum of the y components of $\vec{A}$ and $\vec{B}$.

The same goes for the x components.

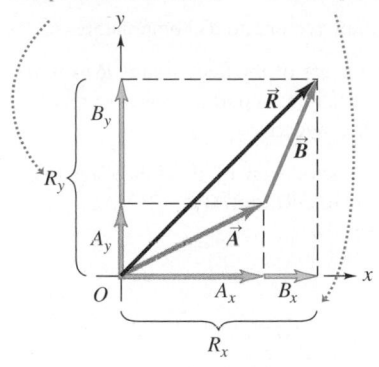

(a)

When we replace a vector with its components, we cross out the original vector with a squiggle for clarity.

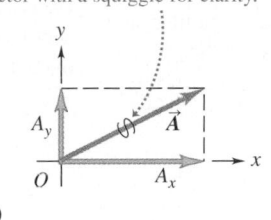

(b)

◀ **FIGURE 1.18** (a) When two vectors $\vec{A}$ and $\vec{B}$ are added to give a vector sum (resultant) $\vec{R}$, the sum of their x components is the x component of $\vec{R}$, and the sum of their y components is the y component of $\vec{R}$. (b) When we replace a vector with its components, we cross out the original vector to avoid confusion.

$(A_x + B_x)$ of the x components of the vectors being added. The same is true for the y components. In symbols,

$$R_x = A_x + B_x,$$
$$R_y = A_y + B_y.$$ (components of vector sum)   (1.5)

Once we know the components of $\vec{A}$ and $\vec{B}$, perhaps by using Equations 1.2, we can compute the components of the vector sum $\vec{R}$. Then if the magnitude and direction of $\vec{R}$ are needed, we can obtain them from Equations 1.3 and 1.4, with the A's replaced by R's.

Mastering PHYSICS

PhET: Vector Addition

NOTE ▶ As Figure 1.18b shows, when you represent a vector in terms of its components, it's always a good idea to use a squiggly line to cross out the original vector in your diagram. Otherwise there's danger that you'll count the same vector twice when you have to perform a vector addition or some other vector operation. ◀

This procedure can easily be extended to find the vector sum of any number of vectors. Let $\vec{R}$ be the vector sum of $\vec{A}$, $\vec{B}$, $\vec{C}$, $\vec{D}$, and $\vec{E}$. Then

$$R_x = A_x + B_x + C_x + D_x + E_x,$$
$$R_y = A_y + B_y + C_y + D_y + E_y.$$

NOTE ▶ In sums such as this, usually some of the components are positive and some are negative, so be very careful about signs when you evaluate such sums. And remember that x and y components are *never* added together in the same equation! ◀

EXAMPLE 1.7   Adding two vectors

Vector $\vec{A}$ has a magnitude of 50 cm and a direction of 30°, and vector $\vec{B}$ has a magnitude of 35 cm and a direction of 110°. Both angles are measured counterclockwise from the positive x axis. Use components to calculate the magnitude and direction of the vector sum (i.e., the resultant) $\vec{R} = \vec{A} + \vec{B}$.

SOLUTION

SET UP We set up a rectangular (Cartesian) coordinate system and place the tail of each vector at the origin, as shown in Figure 1.19a. We label the known magnitudes and angles on our diagram.

SOLVE We find the x and y components of each vector, using Equations 1.2, and record the results in the table shown next. We add the x components to obtain the x component of the vector sum $\vec{R}$, and we add the y components to obtain the y component

Continued

of $\vec{R}$. (But be very careful *not* to add the x and y components together!) The resulting table is as follows:

Magnitude	Angle	x component	y component
$A = 50$ cm	$30°$	43.3 cm	25.0 cm
$B = 35$ cm	$110°$	-12.0 cm	32.9 cm
		$R_x = 31.3$ cm	$R_y = 57.9$ cm

Now we use Equations 1.3 and 1.4 to find the magnitude and direction of the vector sum $\vec{R}$. From the Pythagorean theorem,

$$R = \sqrt{R_x^2 + R_y^2} = \sqrt{(31.3\text{ cm})^2 + (57.9\text{ cm})^2} = 66\text{ cm}.$$

From the definition of the inverse tangent,

$$\theta = \tan^{-1}\frac{R_y}{R_x} = \tan^{-1}\frac{57.9\text{ cm}}{31.3\text{ cm}} = 62°.$$

Figure 1.19b shows this resultant vector and its components.

REFLECT We used the components of the vector sum $\vec{R}$ to find its magnitude and direction. Note how important it was to recognize that the x component of $\vec{B}$ is negative.

Practice Problem: Find the vector sum of the following two vectors by using components: 45 m, 55°, and 70 m, 135°. *Answers:* magnitude = 90 m; direction = 105°.

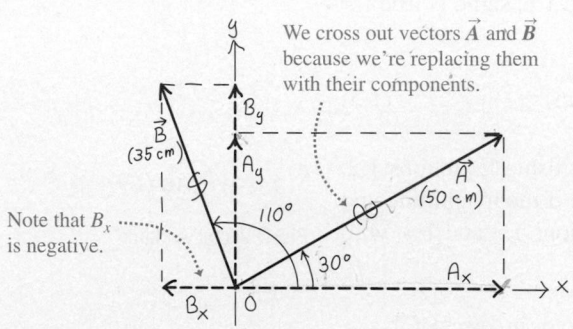

(a) Our diagram for this problem

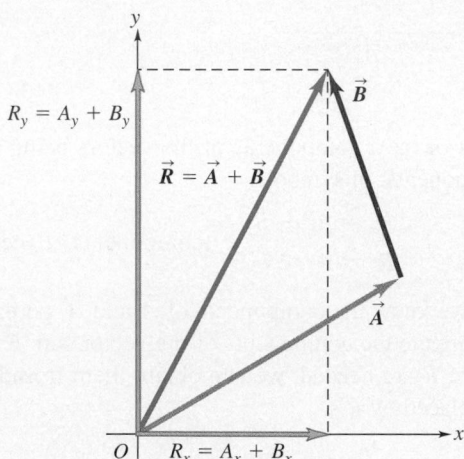

(b) The resultant $\vec{R}$ and its components

▲ FIGURE 1.19

EXAMPLE 1.8   Vector addition helps win a Porsche

The four finalists in a hog-calling contest are brought to the center of a large, flat field. Each is given a meterstick, a compass, a calculator, a shovel, and (in a different order for each contestant) the following three displacements:

    72.4 m, 32.0° east of north;
    57.3 m, 36.0° south of west;
    17.8 m straight south.

The three displacements lead to the point where the keys to a new Porsche are buried. The finder gets the Porsche. Three of the four contestants start measuring immediately, but the winner first *calculates* where to go. What does he calculate?

SOLUTION

SET UP The situation is shown in Figure 1.20. We have chosen the x axis as east and the y axis as north. This is the usual choice for maps. Let $\vec{A}$ be the first displacement, $\vec{B}$ the second, and $\vec{C}$ the third. We can estimate from the diagram that the vector sum $\vec{R}$ is about 10 m, 40° west of north. The angles of the vectors, measured counterclockwise from the $+x$ axis, are 58°, 216°, and 270°. We have to find the components of each.

SOLVE The components of $\vec{A}$ are

$$A_x = A\cos\theta = (72.4\text{ m})(\cos 58°) = 38.37\text{ m},$$
$$A_y = A\sin\theta = (72.4\text{ m})(\sin 58°) = 61.40\text{ m}.$$

The table below shows the components of all the displacements, the addition of components, and the other calculations. To find the magnitude R and direction θ, use the Pythagorean theorem $R = \sqrt{R_x^2 + R_y^2}$ and the definition of the inverse tangent,

Continued

$\theta = \tan^{-1}\dfrac{R_y}{R_x}$. Always arrange your component calculations systematically as in the table. Note that we have kept one too many significant figures in the components; we'll wait until the end to round to the correct number of significant figures. Here's the table:

Distance	Angle	x component	y component
A = 72.4 m	58°	38.37 m	61.40 m
B = 57.3 m	216°	−46.36 m	−33.68 m
C = 17.8 m	270°	0.00 m	−17.80 m
		R_x = −7.99 m	R_y = 9.92 m

$$R = \sqrt{(-7.99 \text{ m})^2 + (9.92 \text{ m})^2} = 12.7 \text{ m}$$
$$\theta = \tan^{-1}\frac{9.92 \text{ m}}{-7.99 \text{ m}} = 129° = 39° \text{ west of north.}$$

REFLECT The losers try to measure three angles and three distances totaling 147.5 m, 1 meter at a time. The winner measures only one angle and one much shorter distance.

Practice Problem: In a second contest, the three displacements given are 63.5 m, 52° west of north; 12.6 m straight north; and

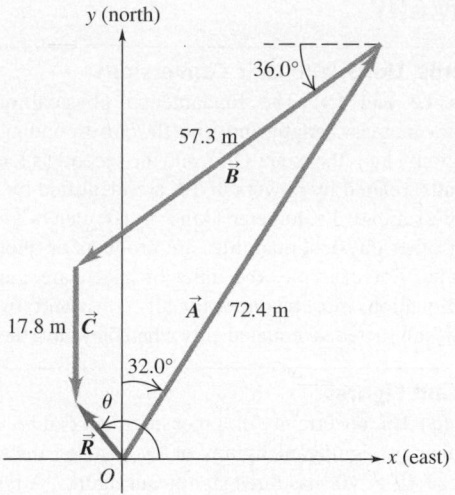

▲ **FIGURE 1.20** Three successive displacements $\vec{A}$, $\vec{B}$, and $\vec{C}$ and the resultant (vector sum) displacement $\vec{R} = \vec{A} + \vec{B} + \vec{C}$.

81.9 m, 41° south of east. What magnitude and direction does the winner calculate for the displacement to find the keys to the Porsche? *Answers:* R = 11.9 m and θ = 9.8° south of east.

Our discussion of vector addition has centered mostly on combining *displacement* vectors, but the method is applicable to all other vector quantities as well, including force, velocity, acceleration, and many other vector quantities. In Chapter 4, when we study the concept of *force* and its effects, we'll make extensive use of the fact that forces can be combined according to the same methods of vector addition that we've used with displacement.

SUMMARY

Standards, Units, and Unit Conversions

(Sections 1.3 and 1.4) The fundamental physical quantities of mechanics are mass, length, and time; the corresponding SI units are the kilogram (kg), the meter (m), and the second (s), respectively. Other units, related by powers of 10, are identified by adding prefixes: for example, 1 kilometer $(\text{km}) = 10^3$ meters (m). Derived units for other physical quantities are products or quotients of the basic units. For example, the units of speed are meters/second (m/s). Equations must be dimensionally consistent; two terms can be added, subtracted, or equated only when they have the same units.

Significant Figures

(Section 1.5) The uncertainty of a measurement can be indicated by the number of significant figures or by a stated uncertainty. For example, 1.40×10^6 has three significant figures. A typical metric ruler has an uncertainty of 1 mm. When numbers are multiplied or divided in a calculation, the number of significant figures in the result is no greater than in the factor with the least number of significant figures. Thus, the product $3.1416 \times 2.34 \times 0.58 = 4.3$ has only two significant figures, because 0.58 has only two significant figures.

Calculating pi from measured values of radius r and circumference C:

Three significant figures (the leading zero doesn't count)⋯⋯⋅

We are dividing, so the result can have only three significant figures.

$$\pi = \frac{C}{2r} = \frac{0.424 \text{ m}}{2(0.06750 \text{ m})} = 3.14 \leftarrow⋯⋯$$

This exact number has no uncertainty and thus doesn't affect the significant figures in the result.

⋯⋯ Four significant figures (the trailing zero counts)

Vector Addition

(Section 1.7) Vector quantities have direction as well as magnitude and are combined according to the rules of vector addition. To add vectors graphically, we place the tail of the second vector at the tip, or point, of the first vector. To subtract vectors, as in $\vec{A} - \vec{B}$, we first obtain the vector $(-\vec{B})$ by reversing the direction of $\vec{B}$. Then we write $\vec{A} - \vec{B} = \vec{A} + (-\vec{B})$ and apply the usual rule for vector addition.

$$\vec{A} + \vec{B} = \vec{A} + \vec{B}$$

To add vectors graphically, arrange them tip to tail in any order.

$$\left(\vec{A} - \vec{B}\right) = \left(\vec{A} + -\vec{B}\right) = \left(\begin{array}{c}\vec{A} - \vec{B} \\ = \vec{A} + (-\vec{B})\end{array}\right)$$

Subtracting $\vec{B}$ from $\vec{A}$ is equivalent to adding $-\vec{B}$ to $\vec{A}$.

Components of Vectors

(Section 1.8) We can represent any vector $\vec{A}$ lying in the x-y plane as the sum of a vector $\vec{A}_x$ parallel to the x axis and a vector $\vec{A}_y$ parallel to the y axis; these are called x and y **component vectors** of $\vec{A}$. If $\vec{A}_x$ points in the positive x direction, we define the number A_x to be its magnitude. If $\vec{A}_x$ points in the $-x$ direction, we define A_x to be the negative of its magnitude. (Remember that a magnitude is always positive.) We define the number A_y the same way. The two numbers A_x and A_y are called the **components** of the vector $\vec{A}$.

Vector addition can be carried out using components of vectors. If A_x and A_y are the components of vector $\vec{A}$, and B_x and B_y are the components of vector $\vec{B}$, the components R_x and R_y of the vector sum $\vec{R} = \vec{A} + \vec{B}$ are given by

$$R_x = A_x + B_x \quad \text{and} \quad R_y = A_y + B_y. \quad (1.5)$$

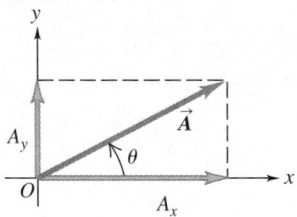

$$A_x = A \cos\theta; \quad A_y = A \sin\theta; \quad A = \sqrt{A_x^2 + A_y^2}; \quad \theta = \tan^{-1}\frac{A_y}{A_x}$$

Conceptual Questions

1. A guidebook describes the rate of climb of a mountain trail as 120 m per kilometer. How can you express this as a number with no units?
2. Suppose you are asked to compute the cosine of 3 meters. Is this possible?
3. A highway contractor stated that in building a bridge deck he poured 250 yards of concrete. What do you think he meant?
4. (a) Would 200 cm be a reasonable height for an adult male? (b) Is it unusual for a human to have an age of 6×10^8 s? (c) Would 10^6 g be a reasonable mass for an adult person?
5. What physical phenomena (other than those discussed in the text) could be used to define a time standard?
6. Atomic quantities are now used to define the second and the meter. Could some atomic quantity be used for a definition of the kilogram? What advantages (or disadvantages) might this definition have compared with that based on the 1 kg platinum cylinder kept at Sèvres?
7. Suppose your lab partner tells you to calculate the volume of a cone of radius r and height h using the formula $V = \pi r^3 h$. Could this formula be correct?
8. Why is an angle a dimensionless number when measured in radians?
9. How could you determine the average thickness of a page of this textbook with an ordinary ruler?
10. (a) The mass m of an object is proportional to its volume V: $m = kV$. What are the SI units of k? (b) If the distance d an object moves in time t is given by the equation $d = At^2$, find the SI units of A. (c) If the speed v of an object depends on time t according to the equation $v = A + Bt + Ct^4$, what are the SI units of A, B, and C?
11. Based *only* on consistency of units, which of the following formulas could *not* be correct? In each case, A is area, V is volume, C is circumference, and R is radius. (a) $A = 2\pi R$, (b) $V = \frac{4}{3}\pi R^2$, (c) $A = 2\pi R^2$, (d) $V = 6\pi R^3$, (e) $C = \pi R^2$, (f) $V = CA$.
12. Based *only* on consistency of units, which of the following formulas could *not* be correct? In each case, x is distance, v is speed, and t is time. (a) $v = x/t$, (b) $x = vt$, (c) $t = v/x$, (d) $vx = t$, (e) $xt = v$, (f) $v_2^2 = v_1^2 + 2x^2$.
13. (a) In adding scalars, is 1 m + 1 m necessarily equal to 2 m? (b) In adding two vectors, each of magnitude 1 m, does the resultant *necessarily* have a magnitude of 2 m? Explain your reasoning in each case.
14. Is it possible for the magnitude of a vector (a) to be smaller than the magnitude of any of its components or (b) to be equal to the magnitude of one of its components? How?
15. (a) Does it make sense to say that a vector is *negative*? Why? (b) Does it make sense to say that one vector is the negative of another? Why? Does your answer here contradict what you said in part (a)?

Multiple-Choice Problems

1. Andy Roddick can sometimes serve a tennis ball at 150 mi/h (about 75 m/s). The length of a tennis court is 78 ft. Approximately how long does his opponent have to react, from the instant he sees Roddick hit the ball until the ball reaches the opponent?
   A. 1/10 s   B. 1/3 s   C. 3/4 s   D. 1 s

2. Suppose that bacon is sold for \$2 per pound. If you were to buy 1 kilogram of bacon, what would you expect to pay for it?
   A. More than \$2
   B. Less than \$2
   C. \$2
3. The distance from an adult's elbow to her fingertips is closest to
   A. 5 cm   B. 10 cm   C. 50 cm   D. 1 m
4. A vector is directed at an angle θ above the $+x$ axis, pointing into the first quadrant. If the x component of this vector is 10 m, the magnitude of the vector is
   A. greater than 10 m
   B. equal to 10 m
   C. less than 10 m
   D. the same as the y component of the vector
5. A mass of 1.0×10^{-3} μkg is the same as
   A. 10^{-12} g   B. 10^{-6} g   C. 1 kg   D. 1 mg
6. If the vector $\vec{B}$ has a magnitude of 25 m and makes an angle of $30°$ with the $+x$ axis, the vector $-\vec{B}$
   A. has a magnitude of 25 m.
   B. has the same direction as $\vec{B}$, but a magnitude of -25 m.
   C. makes an angle of $-30°$ with the $+x$ axis.
   D. has the opposite direction of $\vec{B}$ and a magnitude of -25 m.
7. The mass of a dime is closest to
   A. 1 mg   B. 1 g   C. 100 g   D. 1 kg
8. The number of kilometers in 1 mile is closest to
   A. 0.001   B. 1   C. 100   D. 1000
9. If vector $\vec{A}$ has components A_x and A_y and makes an angle θ with the $+x$ axis, then
   A. $A = A_x + A_y$ (where A is the magnitude of $\vec{A}$).
   B. $\theta = A_y/A_x$.
   C. $\cos\theta = \dfrac{A_x}{\sqrt{A_x^2 + A_y^2}}$.
   D. $\tan\theta = A_x/A_y$.
10. If the distance d (in meters) traveled by an object in time t (in seconds) is given by the formula $d = A + Bt^2$, the SI units of A and B must be
    A. meters for both A and B.
    B. m/s for A and m/s^2 for B.
    C. meters for A and m/s^2 for B.
    D. m/s^2 for both A and B.
11. The surface area of a typical classroom floor is closest to
    A. 1 cm^2
    B. 1 m^2
    C. 10 m^2
    D. 100 m^2
12. If vector $\vec{A}$ has unity magnitude and makes an angle of $45°$ with the $+x$ axis, then the x and y components of $\vec{A}$ are
    A. $A_x = A_y = 1$.
    B. $A_x = A_y = \dfrac{1}{\sqrt{2}}$.
    C. $A_x = A_y = \frac{1}{2}$.
    D. $A_x = A_y = \sqrt{2}$.
13. If vector $\vec{A}$ has components $A_x = -3$ and $A_y = -4$, then the magnitude of $\vec{A}$ is
    A. 7
    B. -7
    C. $\sqrt{7}$
    D. 5
    E. -5

Problems

1.3 Standards and Units

1.4 Unit Consistency and Conversions

1. • (a) The maximum sodium intake for a person on a 2000 calo-
BIO rie diet should be 2400 mg/day. How many grams of sodium is
this per day? (b) The recommended daily allowance (RDA) of
the trace element chromium is 120 μg/day. Express this dose in
grams per day. (c) An intake of vitamin C up to 500 mg/day is
considered safe (although the safety of higher doses has not yet
been established). Express this intake in grams per day. (d) The
electrical resistance of the human body is approximately
1500 ohms when it is dry. Express this resistance in kilohms.
(e) An electrical current of about 0.020 amp can cause muscular
spasms so that a person cannot, for example, let go of a wire
with that amount of current. Express this current in milliamps.

2. • (a) How many ohms are there in a 7.85-megohm resistor?
(b) Typical laboratory capacitors are around 5 picofarads. How
many farads are they? (c) The speed of light in vacuum is
3.00×10^8 m/s. Express this speed in gigameters per second.
(d) The wavelength of visible light is between 400 nm and
700 nm. Express this wavelength in meters. (e) The diameter
of a typical atomic nucleus is about 2 femtometers. Express
this diameter in meters.

3. • (a) The recommended daily allowance (RDA) of the trace
BIO metal magnesium is 410 mg/day for males. Express this quantity in μg/day. (b) For adults, the RDA of the amino acid lysine
is 12 mg per kg of body weight. How many grams per day
should a 75 kg adult receive? (c) A typical multivitamin tablet
can contain 2.0 mg of vitamin B$_2$ (riboflavin), and the RDA is
0.0030 g/day. How many such tablets should a person take
each day to get the proper amount of this vitamin, assuming
that he gets none from any other sources? (d) The RDA for the
trace element selenium is 0.000070 g/day. Express this dose in
mg/day.

4. • (a) Starting with the definition 1.00 in. = 2.54 cm, find the
number of kilometers in 1.00 mile. (b) In medicine, volumes
are often expressed in milliliters (ml or mL). Show that a milli-
liter is the same as a cubic centimeter. (c) How many cubic
centimeters of water are there in a 1.00 L bottle of drinking
water?

5. •• (a) The density (mass divided by volume) of water is
1.00 g/cm³. What is this value in kilograms per cubic meter?
(b) The density of blood is 1050 kg/m³. What is this density in
g/cm³? (c) How many kilograms are there in a 1.00 L bottle of
drinking water? How many pounds?

6. • Calculate the earth's speed in its orbit around the sun, in
m/s, km/h and mi/h, using information from Appendices D
and E.

7. • How many nanoseconds does it take light to travel 1.00 ft in
vacuum? (This result is a useful quantity to remember.)

8. • **Metric wrenches.** (a) You have a new set of metric wrenches,
but need to loosen a $\frac{3}{8}$ inch bolt. To find out which size metric
wrench to use, convert the $\frac{3}{8}$ in. to millimeters, accurate to the
nearest tenth of a millimeter. (b) If you want to tighten a 12 mm
bolt, what size wrench should you use in inches, accurate to the
nearest hundredth of an inch? (c) English-unit wrenches often
come in $\frac{1}{8}$ inch intervals, not in decimal units. What size wrench
should you use in part (b)?

9. • **Gasoline mileage.** You are considering buying a European
car and want to see if its advertised fuel efficiency (expressed
in km/L) is better than that of your present car. If your car gets
37.5 miles per gallon, how many km/L is this? Consult
Appendix D.

10. •• While driving in an exotic foreign land, you see a speed-
limit sign on a highway that reads 180,000 furlongs per fortnight. How many miles per hour is this? (One furlong is $\frac{1}{8}$ mile,
and a fortnight is 14 days. A furlong originally referred to the
length of a plowed furrow.)

11. •• **Fill 'er up!** You fill up your gas tank in Europe when the
euro is worth \$1.25 and gasoline costs 1.35 euros per liter.
What is the cost of gasoline in dollars per gallon? How does
your answer compare with the cost of gasoline in the United
States? (Use conversion factors in Appendix D.)

12. •• **Bacteria.** Bacteria vary somewhat in size, but a diameter
BIO of 2.0 μm is not unusual. What would be the volume (in cubic
centimeters) and surface area (in square millimeters) of such a
bacterium, assuming that it is spherical? (Consult Chapter 0
for relevant formulas.)

13. • Compute the number of seconds in (a) an hour, (b) a 24 hour
day, and (c) a 365 day year.

14. • **Some commonly occurring quantities.** All of the quantities that follow will occur frequently in your study of physics.
(a) Express the speed of light $(3.00 \times 10^8$ m/s$)$ in mi/s and
mph. (b) Find the speed of sound in air at 0°C $(1100$ ft/s$)$ in
m/s and mph. (c) Show that 60 mph is the same as 88 ft/s.
(d) Convert the acceleration of a freely falling body $(9.8$ m/s²$)$
to ft/s².

1.5 Precision and Significant Figures

15. • Express each of the following numbers to three, five, and
eight significant figures: (a) $\pi = 3.141592654 \ldots$, (b) $e =
2.718281828 \ldots$, (c) $\sqrt{13} = 3.605551275. \ldots$

16. • Express each of the following approximations of π to six
significant figures: (a) 22/7, (b) 355/113. (c) Are these approximations accurate to that precision?

17. • An angle is given, to one significant figure, as 4°, meaning
that its value is between 3.5° and 4.5°. Find the corresponding
range of possible values of (a) the cosine of the angle, (b) the
sine of the angle, and (c) the tangent of the angle.

18. • **Blood is thicker than water.** The density (mass divided by
BIO volume) of pure water is 1.00 g/cm³, that of whole blood is
1.05 g/cm³, and the density of seawater is 1.03 g/cm³. What is
the mass (in grams) of 1.00 L of each of these substances?

19. •• **White dwarfs and neutron stars.** Recall that density is
mass divided by volume, and consult Chapter 0 and Appendix E
as needed. (a) Calculate the average density of the earth in
g/cm³, assuming our planet to be a perfect sphere. (b) In about
5 billion years, at the end of its lifetime, our sun will end up as a
white dwarf, having about the same mass as it does now, but
reduced to about 15,000 km in diameter. What will be its density
at that stage? (c) A neutron star is the remnant left after certain
supernovae (explosions of giant stars). Typically, neutron stars
are about 20 km in diameter and have around the same mass as
our sun. What is a typical neutron star density in g/cm³?

20. •• **Atoms and nuclei.** The atom helium (He) consists of two
protons, two neutrons, and two electrons. (Recall that density
is mass divided by volume, and consult Appendices A and E

and Table 1.1 as needed.) (a) The diameter of the He atom is approximately 0.10 nm. Calculate the density of the He atom in g/cm^3 (assuming that it is a sphere), and compare it with that of pure water, which is $1.0\ g/cm^3$. (b) The diameter of the He nucleus is about 2.0 fm. Assuming the nucleus to be a sphere, calculate its density in g/cm^3 and compare it with that of a neutron star in the previous exercise.

21. •• **Critical mass of neptunium.** In the fall of 2002, a group of scientists at Los Alamos National Laboratory determined that the critical mass of neptunium-237 is about 60.0 kg. (The critical mass of a fissionable material is the minimum amount that must be brought together to start a nuclear chain reaction.) Neptunium has a density of $19.5\ g/cm^3$. What would be the radius of a sphere made of this material that has a critical mass? (Recall that density is mass divided by volume.)

22. •• **Cell walls.** Although these quantities vary from one type of
BIO cell to another, a cell can be 2.0 μm in diameter with a cell wall 50.0 nm thick. If the density (mass divided by volume) of the wall material is the same as that of pure water, what is the mass (in mg) of the cell wall, assuming the cell to be spherical and the wall to be a very thin spherical shell?

23. •• A brass washer has an outside diameter of 4.50 cm with a hole of diameter 1.25 cm and is 1.50 mm thick. (See Figure 1.21.) The density of brass is $8600\ kg/m^3$. If you put this washer on a laboratory balance, what will it "weigh" in grams? (Recall that density is mass divided by volume and consult Chapter 0 as needed.)

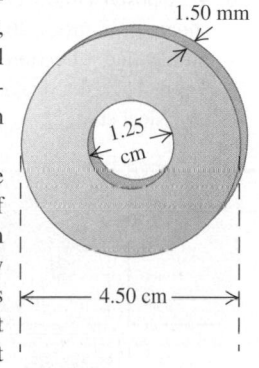

1.50 mm

1.25 cm

4.50 cm

▲ **FIGURE 1.21** Problem 23.

1.6 Estimates and Orders of Magnitude

24. • Estimate the total mass of all the humans presently living on earth.

25. •• How many words are there in this book?

26. •• **How many cells in the body?** Although their sizes vary,
BIO cells can be modeled as spheres 2.0 μm in diameter, on the average. About how many cells does a typical human contain, and approximately what is the mass of an average cell?

27. • How many times does a typical person blink her eyes in a
BIO lifetime?

28. • You are using water to dilute small amounts of chemicals in the laboratory, drop by drop. How many drops of water are in a 1.0 L bottle? (*Hint:* Start by estimating the diameter of a drop of water.)

29. • How many dollar bills would you have to stack to reach the moon? Would that be a cheaper way to get there than building and launching a spacecraft (which costs a few billion dollars)?

30. •• **Space station.** You are designing a space station and want
BIO to get some idea how large it should be to provide adequate air for the astronauts. Normally, the air is replenished, but for security, you decide that there should be enough to last for two weeks in case of a malfunction. (a) Estimate how many cubic meters of air an average person breathes in two weeks. A typical human breathes about 1/2 L per breath. (b) If the space

station is to be spherical, what should be its diameter to contain all the air you calculated in part (a)?

31. •• **A beating heart.** How many times does a human heart
BIO beat during a lifetime? How many gallons of blood does it pump in that time if, on the average, it pumps 50 cm^3 of blood with each beat?

32. •• How long would it take you to walk to the moon, and how many steps would you have to take, assuming that you could somehow walk normally in space?

33. •• How much would it cost to paper the entire United States (including Alaska and Hawaii) with dollar bills? What would be the cost to each person in the nation?

1.7 Vectors and Vector Addition

34. • On a single diagram, carefully sketch each force vector to scale and identify its magnitude and direction on your drawing: (a) 60 lb at 25° east of north. (b) 40 lb at $\pi/3$ south of west. (c) 100 lb at 40° north of west. (d) 50 lb at $\pi/6$ east of south.

35. • Hearing rattles from a snake, you make two rapid displacements of magnitude 1.8 m and 2.4 m. In sketches (roughly to scale), show how your two displacements might add up to give a resultant of magnitude (a) 4.2 m; (b) 0.6 m; (c) 3.0 m.

36. • A ladybug starts at the center of a 12-in.-diameter turntable and crawls in a straight radial line to the edge. While this is happening, the turntable turns through a 45° angle. (a) Draw a sketch showing the bug's path and the displacement vector for the bug's progress. (b) Find the magnitude and direction of the ladybug's displacement vector.

37. • For the vectors $\vec{A}$ and $\vec{B}$ shown in Figure 1.22, carefully sketch (a) the vector sum $\vec{A} + \vec{B}$; (b) the vector difference $\vec{A} - \vec{B}$; (c) the vector $-\vec{A} - \vec{B}$; (d) the vector difference $\vec{B} - \vec{A}$.

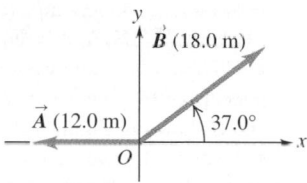

$\vec{B}$ (18.0 m)

$\vec{A}$ (12.0 m)

37.0°

▲ **FIGURE 1.22** Problem 37.

38. • **Chin brace.** A person with
BIO an injured jaw has a brace below his chin. The brace is held in place by two cables directed at 65° above the horizontal. (See Figure 1.23.) The cables produce forces of equal magnitude having a vertical resultant of 2.25 N upward. (a) Make a scale drawing showing both the forces produced by the cables and the resultant force. Estimate the angle carefully or measure it with a protractor. (b) Use your scale drawing to estimate the magnitude of the force due to each cable.

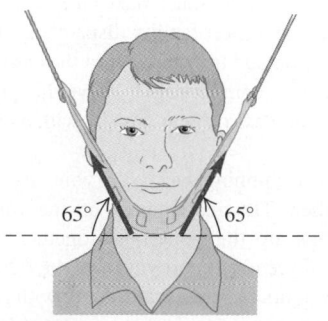

65°   65°

▲ **FIGURE 1.23** Problem 38.

1.8 Components of Vectors

39. • A rocket fires two engines simultaneously. One produces a thrust of 725 N directly forward, while the other gives a 513-N thrust at 32.4° above the forward direction. Find the magnitude and direction (relative to the forward direction) of the resultant force that these engines exert on the rocket.

40. • In each of the cases that follow, the magnitude of a vector is given along with the counterclockwise angle it makes with the $+x$ axis. Use trigonometry to find the x and y components of the vector. Also, sketch each vector approximately to scale to see if your calculated answers seem reasonable. (a) 50.0 N at 60.0°, (b) 75 m/s at $5\pi/6$ rad, (c) 254 lb at 325°, (d) 69 km at 1.1π rad.

41. • In each of the cases that follow, the components of a vector $\vec{A}$ are given. Use trigonometry to find the magnitude of that vector and the counterclockwise angle it makes with the $+x$ axis. Also, sketch each vector approximately to scale to see if your calculated answers seem reasonable. (a) $A_x = 4.0$ m, $A_y = 5.0$ m, (b) $A_x = -3.0$ km, $A_y = -6.0$ km, (c) $A_x = 9.0$ m/s, $A_y = -17$ m/s, (d) $A_x = -8.0$ N, $A_y = 12$ N.

42. • A woman takes her dog Rover for a walk on a leash. To get the little pooch moving forward, she pulls on the leash with a force of 20.0 N at an angle of 37° above the horizontal. (a) How much force is tending to pull Rover forward? (b) How much force is tending to lift Rover off the ground?

43. • If a vector $\vec{A}$ has the following components, use trigonometry to find its magnitude and the counterclockwise angle it makes with the $+x$ axis: (a) $A_x = 8.0$ lb, $A_y = 6.0$ lb, (b) $A_x = -24$ m/s, $A_y = -31$ m/s, (c) $A_x = -1500$ km, $A_y = 2000$ km, (d) $A_x = 71.3$ N, $A_y = -54.7$ N.

44. • Compute the x and y components of the vectors $\vec{A}$, $\vec{B}$, and $\vec{C}$ shown in Figure 1.24.

45. • Vector $\vec{A}$ has components $A_x = 1.30$ cm, $A_y = 2.25$ cm; vector $\vec{B}$ has components $B_x = 4.10$ cm, $B_y = -3.75$ cm. Find (a) the components of the vector sum $\vec{A} + \vec{B}$; (b) the magnitude and direction of $\vec{A} + \vec{B}$; (c) the components of the vector difference $\vec{B} - \vec{A}$; (d) the magnitude and direction of $\vec{B} - \vec{A}$.

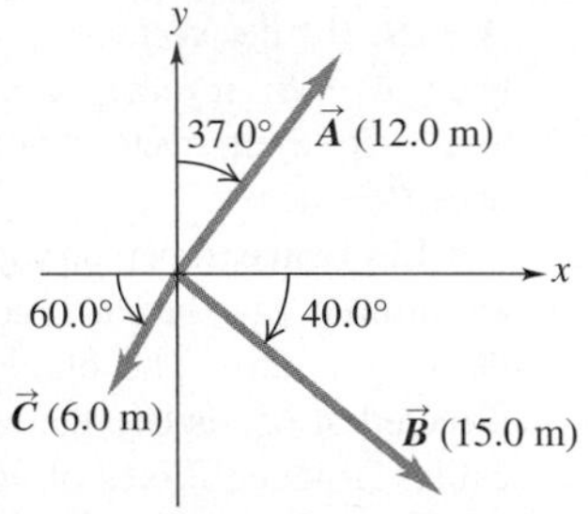

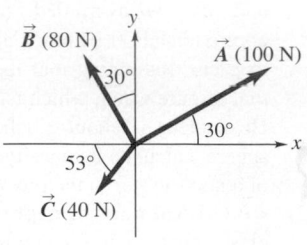

▲ **FIGURE 1.24** Problem 44.

46. •• A plane leaves Seattle, flies 85 mi at 22° north of east, and then changes direction to 48° south of east. After flying at 115 mi in this new direction, the pilot must make an emergency landing on a field. The Seattle airport facility dispatches a rescue crew. (a) In what direction and how far should the crew fly to go directly to the field? Use components to solve this problem. (b) Check the reasonableness of your answer with a careful graphical sum.

47. •• You're hanging from a chinning bar, with your arms at right angles to each other. The magnitudes of the forces exerted by both your arms are the same, and together they exert just enough upward force to support your weight, 620 N. (a) Sketch the two force vectors for your arms, along with their resultant, and (b) use components to find the magnitude of each of the two "arm" force vectors.

48. • Three horizontal ropes are attached to a boulder and produce the pulls shown in Figure 1.25. (a) Find the x and y components of each pull. (b) Find the components of the resultant of the three pulls. (c) Find the magnitude and direction (the counterclockwise angle with the $+x$ axis) of the resultant pull. (d) Sketch a clear *graphical* sum to check your answer in part (c).

▲ **FIGURE 1.25** Problem 48.

49. • A disoriented physics professor drives 3.25 km north, then 4.75 km west, and then 1.50 km south. (a) Use components to find the magnitude and direction of the resultant displacement of this professor. (b) Check the reasonableness of your answer with a graphical sum.

50. • A postal employee drives a delivery truck along the route shown in Figure 1.26. Use components to determine the magnitude and direction of the truck's resultant displacement. Then check the reasonableness of your answer by sketching a graphical sum.

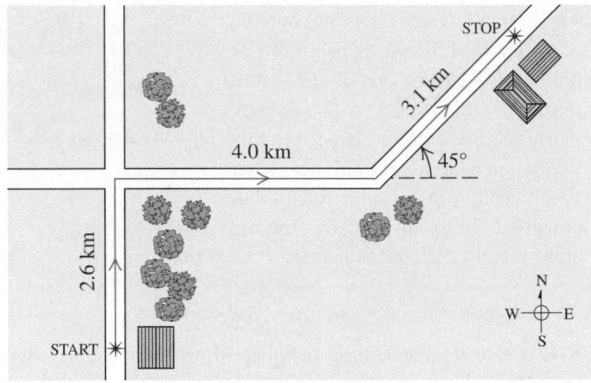

▲ **FIGURE 1.26** Problem 50.

General Problems

51. • **Baseball mass.** Baseball rules specify that a regulation ball shall weigh no less than 5.00 ounces nor more than $5\frac{1}{4}$ ounces. What are the acceptable limits, in grams, for a regulation ball? (See Appendix D and use the fact that 16 oz = 1 lb.)

52. • As you eat your way through a bag of chocolate chip cookies, you observe that each cookie is a circular disk with a diameter of 8.50 cm and a thickness of 0.050 cm. Find (a) the volume of a single cookie and (b) the ratio of the diameter to the thickness, and express both in the proper number of significant figures.

53. •• **Breathing oxygen.** The density of air under standard laboratory conditions is 1.29 kg/m³, and about 20% of that air consists of oxygen. Typically, people breathe about $\frac{1}{2}$ L of air per breath. (a) How many grams of oxygen does a person breathe in a day? (b) If this air is stored uncompressed in a cubical tank, how long is each side of the tank?

54. •• The total mass of Earth's atmosphere is about 5×10^{15} metric tonnes (1 metric tonne = 1000 kg). Suppose you

breathe in about 1/3 L of air with each breath, and the density of air at room temperature is about 1.2 kg/m^3. About how many breaths of air does the entire atmosphere contain? How does this compare to the number of atoms in one breath of air (about 1.2×10^{22})? It's sometimes said that every breath you take contains atoms that were also breathed by Albert Einstein, Confucius, and in fact anyone else who ever lived. Based on your calculation, could this be true?

55. •• **How much blood in a heartbeat?** A typical human con-
BIO tains 5.0 L of blood, and it takes 1.0 min for all of it to pass through the heart when the person is resting with a pulse rate of 75 heartbeats per minute. On the average, what volume of blood, in liters and cubic centimeters, does the heart pump during each beat?

56. •• **Muscle attachment.** When muscles attach to bones, they
BIO usually do so by a series of tendons, as shown in Figure 1.27. In the figure, five tendons attach to the bone. The uppermost tendon pulls at 20.0° from the axis of the bone, and each tendon is directed 10.0° from the one next to it. (a) If each tendon exerts a 2.75 N pull on the bone, use vector components to find the magnitude and direction of the resultant force on this bone due to all five tendons. Let the axis of the bone be the $+x$ axis. (b) Draw a graphical sum to check your results from part (a).

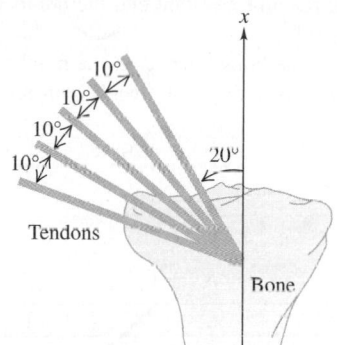

▲ **FIGURE 1.27** Problem 56.

57. •• **Hiking the Appalachian Trail.** The Appalachian Trail runs from Mt. Katahdin in Maine to Springer Mountain in Georgia, a total distance of 2160 mi. If you hiked for 8 h per day, estimate (a) how many steps it would take to hike this trail and b) how many days it would take to hike it.

58. •• Estimate the number of atoms in your body. (*Hint:* Based
BIO on what you know about biology and chemistry, what are the most common types of atom in your body? What is the mass of each type of atom? Appendix C gives the atomic masses for different elements, measured in atomic mass units; you can find the value of an atomic mass unit, or 1 u, in Appendix E.)

59. •• Biological tissues are typically made up of 98% water.
BIO Given that the density of water is $1.0 \times 10^3 \text{ kg/m}^3$, estimate the mass of (a) the heart of an adult human; (b) a cell with a diameter of 0.5 μm; (c) a honeybee.

60. •• **Density of the human body.** (a) Make simple measure-
BIO ments on your own body, and use them to calculate your body's average density (mass divided by volume) in kg/m^3. (b) How does this result compare with the density of water, which is 1000 kg/m^3? Is your result surprising?

61. •• While surveying a cave, a spelunker follows a passage 180 m straight west, then 210 m in a direction 45° east of south, and then 280 m at 30.0° east of north. After a fourth unmeasured displacement, she finds herself back where she started. Use vector components to find the magnitude and direction of the fourth displacement. Then check the reasonableness of your answer with a graphical sum.

62. •• A sailor in a small sailboat encounters shifting winds. She sails 2.00 km east, then 3.50 km southeast, and then an additional distance in an unknown direction. Her final position is 5.80 km directly east of her starting point. (See Figure 1.28.) Find the magnitude and direction of the third leg of the journey. Draw the vector addition diagram, and show that it is in qualitative agreement with your numerical solution.

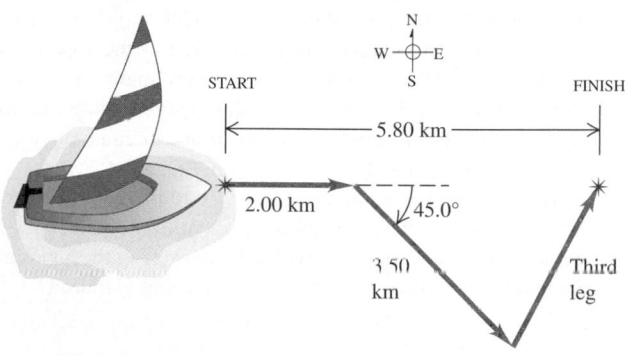

▲ **FIGURE 1.28** Problem 62.

63. •• **Dislocated shoulder.** A patient with a dislocated shoulder
BIO is put into a traction apparatus as shown in Figure 1.29. The pulls $\vec{A}$ and $\vec{B}$ have equal magnitudes and must combine to produce an outward traction force of 5.60 N on the patient's arm. How large should these pulls be?

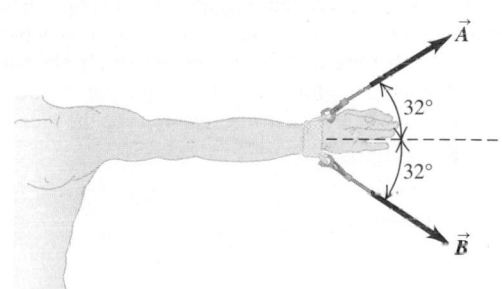

▲ **FIGURE 1.29** Problem 63.

64. •• On a training flight, a stu-
dent pilot flies from Lincoln, Nebraska to Clarinda, Iowa, then to St. Joseph, Missouri, and then to Manhattan, Kansas (Fig. 1.30). The directions are shown relative to north: 0° is north, 90° is east, 180° is south, and 270° is west. Use the method of components to find (a) the distance she has to fly from Manhattan

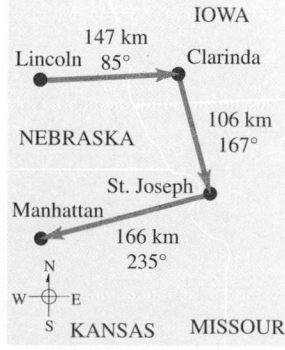

▲ **FIGURE 1.30** Problem 64.

to get back to Lincoln, and (b) the direction (relative to north) she must fly to get there. Illustrate your solutions with a vector diagram.

65. •• **Bones and muscles.** A patient in therapy has a forearm
BIO that weighs 20.5 N and lifts a 112.0 N weight. The only other significant forces on his forearm come from the biceps muscle (which acts perpendicularly to the forearm) and the force at the elbow. If the biceps produce a pull of 232 N when the forearm is raised 43° above the horizontal, find the magnitude and direction of the force that the elbow exerts on the forearm. (*Hint:* The elbow force and the biceps force together must balance the weight of the arm and the weight it is carrying, so their vector sum must be 132.5 N upward.)

66. •• **Googols and googolplexes!** When the mathematician Edward Kasner asked his young nephew to coin a name for the huge number 10^{100}, the boy said *googol*. (a) Express the googol in standard notation as a 1 followed by the appropriate number of zeroes. (b) Approximately how many googols of atoms does our sun contain? For simplicity, assume that the sun consists of only protons and electrons, in equal numbers, which is approximately true. (Consult Appendix E.) (c) The *googolplex* is an even larger number, 10 to the googol power: 10^{googol}. Express the googolplex in scientific notation. If you wrote it in standard notation with a 1 followed by the appropriate number of zeroes, how many zeroes would you need?

Passage Problems

BIO Calculating lung volume in humans. The portion of the human lung that is engaged in exchanging oxygen and carbon dioxide with the blood comprises many small sacs called alveoli. These alveoli are formed during embryonic development by repeated branching of the tubes of the airways and function to provide a large surface area for gas exchange. Recent careful measurements show that the total number of alveoli in a typical human lung pair is about 480×10^6 and that the average volume of a single alveolus is $4.2 \times 10^6 \ \mu m^3$. It may be useful to recall the equation for the volume of a sphere, $V = 4/3 \ \pi r^3$, and the equation for the area of a sphere, $A = 4\pi r^2$.

67. What is total volume of the gas-exchanging region of the lungs?
   A. 2000 μm^3
   B. 2 m³
   C. 2.0 L
   D. 120 L

68. Assuming the alveoli are spherical, what is the diameter of a typical alveolus?
   A. 0.20 mm
   B. 2 mm
   C. 20 mm
   D. 200 mm

69. Individuals vary considerably in total lung volume. Figure 1.31 shows the results of measurement of total lung volume (horizontal axis) and average alveolar volume (vertical axis) for six individuals. What can you infer about the relationship between alveolar size, total lung volume and number of alveoli per individual from these data?
   A. As the total volume of the lungs increases, the number and volume of individual alveoli increase.
   B. As the total volume of the lungs increases, the number of alveoli increases and the volume of individual alveoli decreases.
   C. As the total volume of the lungs increases, the volume of the individual alveoli remains constant and the number of alveoli increases.
   D. As the total volume of the lungs increases, the number of alveoli and the volume of individual alveoli both remain constant.

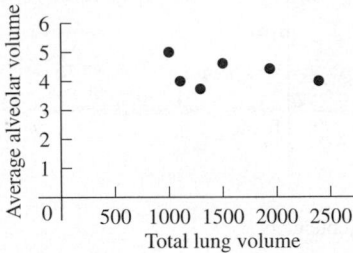

▲ **FIGURE 1.31** Problem 69.

2 Motion Along a Straight Line

How do you describe the motion of a racehorse coming down the homestretch, galloping toward the finish line? When you throw a baseball straight up in the air, how high does it go? How fast do you have to throw it to reach that height? When a glass slips from your hand, how much time do you have to catch it before it hits the floor? These are the kinds of questions you'll learn to answer in this chapter.

We're beginning our study of physics with **mechanics,** the study of the relationships among force, matter, and motion. Our first goal is to develop mathematical methods for *describing* motion; this discipline is called **kinematics.** Later we'll study the more general subject of **dynamics,** the relation of motion to its causes.

Motion is a continuous change of position with time. In this chapter we'll limit our discussion to the simplest possible situation, a single object moving along a straight line. We'll assume that the object can be treated as a particle, as described in Section 1.2. We'll always take the line to be one of the axes of a coordinate system. Then we can describe the position of the particle by its displacement from the origin of coordinates along this line. Displacement is a *vector* quantity, as we discussed in Chapter 1. But when a particle moves along a line that is a coordinate axis, only one component of displacement is different from zero, and the particle's position is described by a single **coordinate.** In Chapter 3 we'll consider more general motions in a plane, in which the particle has two coordinates.

A fall is a classic example of straight-line motion. If air resistance could be neglected, these skydivers would be accelerating toward the earth at a constant 9.8 m/s². Luckily for them, air resistance exists—and later, when they deploy their parachutes, it will give them an *upward* acceleration.

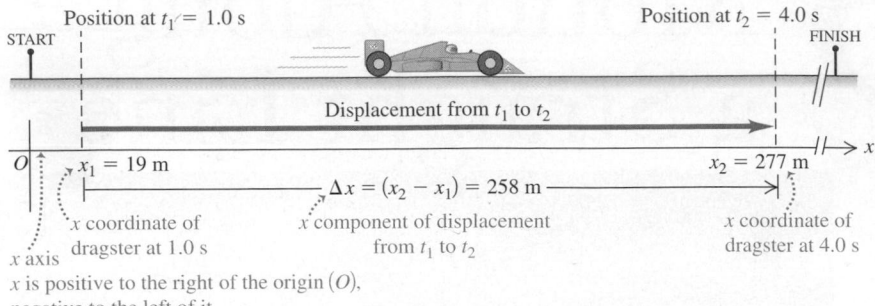

▲ **FIGURE 2.1** Positions of a AA-fuel dragster at two times during its run.

2.1 Displacement and Average Velocity

Suppose a drag racer drives his AA-fuel dragster in a straight line down the track. We'll use this line as the x axis of our coordinate system, as shown in Figure 2.1. We'll place the origin O at the starting line. The distance of the front of the dragster from the origin is given by the coordinate x, which varies with time. Note that x is positive when the car is to the right of O, negative when it is to the left. Suppose that, at a time 1 second (1.0 s) after the start of the race, the car is at x_1, 19 m to the right of the origin, and that, at 4.0 s after the start, it is at x_2, 277 m from the origin. Then it has traveled a distance of $(277\text{ m} - 19\text{ m}) = 258\text{ m}$ in the direction of increasing x in a time interval of $(4.0\text{ s} - 1.0\text{ s}) = 3.0$ s. We define the car's **average velocity** during this interval to be a vector quantity whose x component is the change in x (258 m), divided by the time interval (3.0 s); that is, $(258\text{ m})/(3.0\text{ s}) = 86\text{ m/s}$.

Now let's generalize this discussion. At time t_1 a point at the front of the car is at position x_1, and at time t_2 it is at position x_2. The displacement during the time interval from t_1 to t_2 is the vector from x_1 to x_2. The x component of this vector is $(x_2 - x_1)$; all other components are zero, because there is no motion in any direction other than along the x axis. For motion in the x direction, we define the average velocity as a vector in the direction from x_1 to x_2, with an x component of average velocity $v_{\text{av},x}$ given by

$$v_{\text{av},x} = \frac{x_2 - x_1}{t_2 - t_1}. \tag{2.1}$$

All other components of the average velocity vector are zero.

Changes in quantities, such as $(x_2 - x_1)$ and $(t_2 - t_1)$, occur so often throughout physics that it's worthwhile to use a special notation. From now on, we'll use the Greek letter Δ (capital delta) to represent a *change* in any quantity. Thus, we write

$$\Delta x = x_2 - x_1. \tag{2.2}$$

Be sure you understand that Δx is *not the product* of Δ and x. It is a *single symbol*, and it means "the change in the quantity x." Here, Δx is the x-component of the displacement of the particle. With this same notation, we denote the time interval from t_1 to t_2 as $\Delta t = t_2 - t_1$.

NOTE ▶ Values such as Δx or Δt always denote the final value minus the initial value, never the reverse. ◀

We can now define the x component of the average velocity.

Definition of average velocity

The x component of an object's average velocity is defined as the x component of displacement Δx, divided by the time interval Δt in which the displacement occurs. We represent this quantity by the letter v, with a subscript "av" to signify "average value" and a subscript "x" to signify "x component." Thus,

$$v_{av,x} = \frac{x_2 - x_1}{t_2 - t_1} = \frac{\Delta x}{\Delta t}. \qquad (2.3)$$

Unit: m/s

When we express distance in meters and time in seconds, a component of velocity is expressed in meters per second (m/s). Other common units of velocity are kilometers per hour (km/h), miles per hour (mi/h), and knots $(6080\ ft/h)$. In the preceding example, we have $x_1 = 19$ m, $x_2 = 277$ m, $t_1 = 1.0$ s, and $t_2 = 4.0$ s, and Equation 2.3 gives

$$v_{av,x} = \frac{277\ m - 19\ m}{4.0\ s - 1.0\ s} = 86\ m/s.$$

In some cases, $v_{av,x}$ is negative, as shown in Figure 2.2. Suppose an official's truck is moving *toward* the origin, with $x_1 = 277$ m and $x_2 = 19$ m. If $t_1 = 6.0$ s and $t_2 = 16.0$ s, then $\Delta x = -258$ m, $\Delta t = 10.0$ s, and the x component of average velocity is -26 m/s. Whenever x is positive and decreasing, or negative and becoming more negative, the object is moving in the negative x direction and $v_{av,x}$ is negative.

We stress again that average velocity is a *vector* quantity, and Equation 2.3 defines the x component of this vector. In this chapter, all vectors have *only one* nonzero component, frequently only an x component. We'll often call x the displacement and $v_{av,x}$ the average velocity, remembering that these are really the x components of vector quantities that, in straight-line motion, have *only* x components. In Chapter 3, displacement, velocity, and acceleration vectors will have two nonzero components.

When we deal with motion along a vertical line, we'll usually use a vertical y axis instead of an x axis, so our components will be called y components. Then Equation 2.3, with the x's replaced by y's, gives the y component of average velocity, $v_{av,y}$.

In Figures 2.1 and 2.2, it doesn't matter whether the car's velocity is or isn't constant during the time interval Δt. Another dragster may have started from rest,

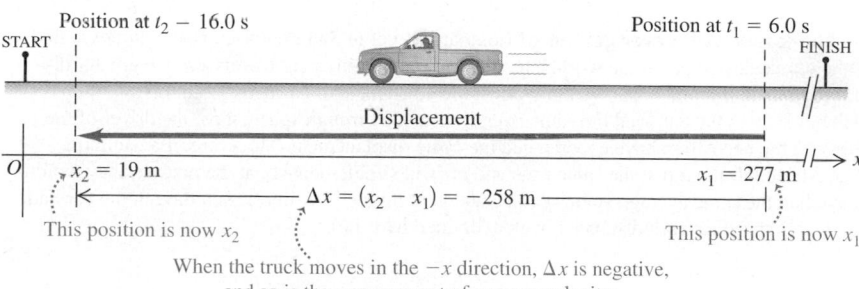

When the truck moves in the $-x$ direction, Δx is negative, and so is the x component of average velocity:

$$v_{av,x} = \frac{x_2 - x_1}{t_2 - t_1} = -26\ m/s$$

▲ **FIGURE 2.2** Positions of an official's truck at two times during its drive up the racetrack.

reached a maximum velocity, blown its engine, and then slowed down. To calculate the average velocity, we need only the total displacement $\Delta x = x_2 - x_1$ and the total time interval $\Delta t = t_2 - t_1$.

EXAMPLE 2.1 Swim competition

During a freestyle competition, a swimmer performs the crawl stroke in a pool 50.0 m long, as shown in Figure 2.3. She swims a length at racing speed, taking 24.0 s to cover the length of the pool. She then takes twice that time to swim casually back to her starting point. Find **(a)** her average velocity for each length and **(b)** her average velocity for the entire swim.

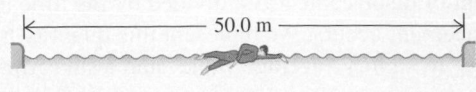

▲ **FIGURE 2.3**

SOLUTION

SET UP As shown in Figure 2.4, we choose a coordinate system with the origin at the starting point (often a convenient choice) and x increasing to the right. We add the information given to the diagram we sketch for the problem.

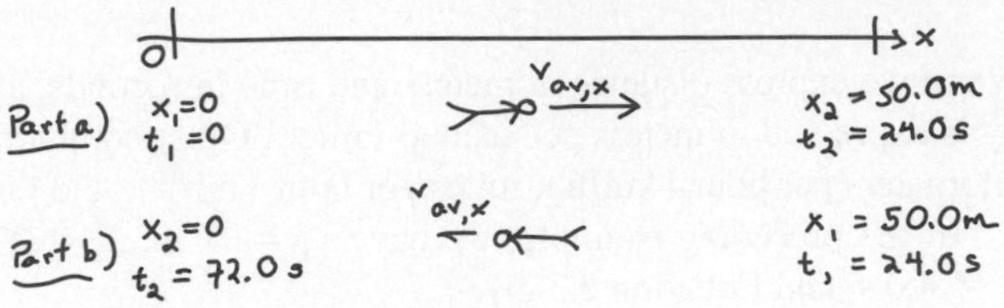

SOLVE Part (a): For the first length, we have $x_1 = 0$, $x_2 = 50.0$ m, $t_1 = 0$, and $t_2 = 24.0$ s. Using the definition of average velocity given by Equation 2.3, we find that

$$v_{av,x} = \frac{x_2 - x_1}{t_2 - t_1} = \frac{50.0 \text{ m} - 0 \text{ m}}{24.0 \text{ s} - 0 \text{ s}} = 2.08 \text{ m/s}.$$

For the return trip, we have $x_1 = 50.0$ m, $x_2 = 0$, $t_1 = 24.0$ s, and $t_2 = 24.0$ s $+ 48.0$ s $= 72.0$ s. Using the definition of average velocity again, we obtain

$$v_{av,x} = \frac{x_2 - x_1}{t_2 - t_1} = \frac{0 \text{ m} - 50.0 \text{ m}}{72.0 \text{ s} - 24.0 \text{ s}} = -1.04 \text{ m/s}.$$

Part (b): The starting and finishing points are the same: $x_1 = x_2 = 0$. The average velocity for a round-trip is zero!

▲ **FIGURE 2.4** Our sketch for this problem.

REFLECT The average x component of velocity in part (b) is negative because x_2 lies to the left of x_1 on the axis and x is decreasing during this part of the trip. Thus, the average velocity *vector* points to the right for part (a) and to the left for part (b). The average velocity for the total out-and-back trip is zero because the total displacement is zero.

Practice Problem: If the swimmer could cross the English Channel (32 km), how many hours would she need if she could maintain the same average velocity as for the first 50 m in the pool? (Actual times are around 15 hours.) *Answer:* 4.3 hr.

You may have noticed that we haven't used the term *speed* in our discussion of particle motion. The terms *speed* and *velocity* are often used interchangeably (and erroneously), but there is a very important distinction between them. The average **speed** of a point during any motion is a *scalar* quantity equal to the total *distance* traveled (disregarding the direction of motion) during a specified time interval, divided by the time interval. Average *velocity* is a *vector* quantity, equal to the vector *displacement* of the particle, divided by the time interval. In Example 2.1, the average velocity during the two laps is zero because the total

◄ **Application Will I ever get there?** Lombard Street in San Francisco is sometimes called "the crookedest street in the world." In descending the hill, a car travels a much greater distance than does a pedestrian walking down the adjacent stairs directly from top to bottom. However, both have moved the same overall distance through space; thus, the driver of the car and the pedestrian have experienced the same displacement. Therefore, if a car and a pedestrian left the top at the same time and arrived simultaneously at the bottom, they would have had the same average velocity (displacement divided by time), even though the car had a greater speed (overall distance traveled, divided by time).

displacement is zero. But the average speed is the total *distance* traveled (100.0 m), divided by the total time (72.0 s): $100.0 \text{ m}/72.0 \text{ s} = 1.39 \text{ m/s}$. The speedometer of a car is correctly named; it measures speed, irrespective of the direction of motion. To make a velocity meter, you would need a speedometer and a compass.

In Example 2.1, we could just as well have chosen the origin to be the other end of the pool and the positive direction for x to be to the left instead of to the right. Then the first x component of velocity would be negative and the second positive.

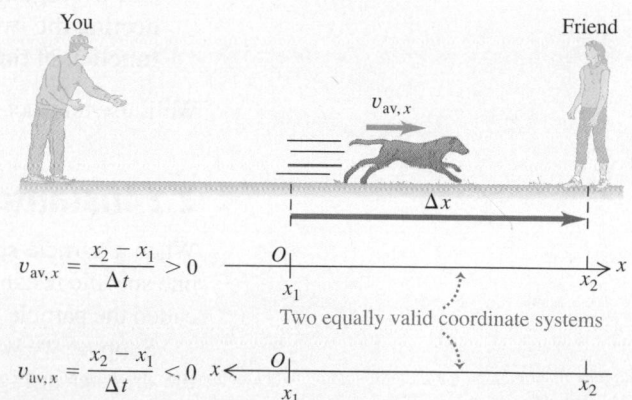

Conceptual Analysis 2.1

Signs and coordinate choices

You and a friend stand 10 m apart. Your dog, initially midway between you, runs in a straight line toward your friend. Each of you independently defines a coordinate system (position of origin and positive direction for the coordinate x) and calculates the dog's average x component of velocity. (Remember in this context, when we say "velocity," we really mean "x component of velocity.") What can you say with *certainty* about the average velocity?

A. The average velocity you calculate is positive.
B. The average velocity your friend calculates is negative.
C. Neither of you can obtain a negative velocity.
D. Both of you can obtain a negative velocity.

SOLUTION The question makes sense only if the coordinate axis is along the line joining you and your friend. But as Figure 2.5 shows, the axis can point to the left or to the right. If it points to the right, the dog moves in the positive x direction, so Δx is positive; if it points to the left, the dog moves in the negative x direction, so Δx is negative. Either person can choose either axis, so either person can obtain either sign for velocity. Thus, the correct

▲ **FIGURE 2.5** Your choice of coordinate system determines the sign of velocity.

answer is D. Notice that the sign of the average velocity does *not* depend on where you place the origin, but only on the direction you define as positive.

We've described motion of a particle along a straight line by stating the values of the coordinate x at various times t. In mathematical language, we say that x is a *function* of t. It's often useful to represent the relationship of x and t by means of a graph. We choose an axis system in which the coordinate x of the particle is plotted along the vertical axis and time t along the horizontal axis.

Figure 2.6 is a graph of the position of the dragster as a function of time. The curve in the figure *does not* represent the car's path in space; as Figure 2.1 shows,

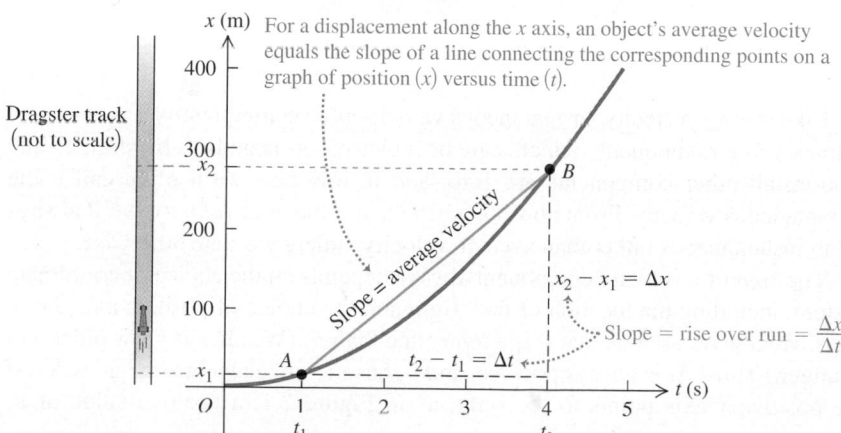

◀ **FIGURE 2.6** Graph of the position of the dragster as a function of time.

the path is a straight line. Rather, the graph is a pictorial way to represent the car's position at various times. Points A and B on the graph correspond to positions x_1 and x_2, respectively. The graph shows that, at time $t = 1.0$ s, the dragster's x coordinate (its displacement from the origin) is $x = 19$ m and that, at time $t = 4.0$ s, it is $x = 277$ m. During the 3.0-s interval from $t = 1.0$ s to $t = 4.0$ s, the x coordinate changes from 19 m to 277 m. The horizontal interval is $\Delta t = 3.0$ s, the vertical interval is $\Delta x = 258$ m. The average velocity $v_{av,x} = \Delta x/\Delta t$ is the vertical side of the triangle (Δx) divided by the horizontal side (Δt). This quotient is the **slope** of the line joining A and B (often described as the "rise" divided by the "run"). Thus, we obtain the following general principle:

> The average velocity between two positions is the slope of a line connecting the two corresponding points on a graph of position as a function of time.

We'll use this fact in the next section when we define instantaneous velocity.

2.2 Instantaneous Velocity

When a particle speeds up or slows down, we can still define its velocity at any one specific instant of time or at one specific point in the path. Such a velocity is called the particle's **instantaneous velocity,** and it needs to be defined carefully.

Suppose we want to find the instantaneous velocity (i.e., the x component) of the dragster in Figure 2.1 at the point x_1. We can take the second point x_2 to be closer and closer to the first point x_1, and we can compute the average velocity over these shorter and shorter displacements and time intervals. Although both Δx and Δt become very small, their quotient does not necessarily become small. We then define the instantaneous velocity at x_1 as the value that this series of average velocities *approaches* as the time interval becomes very small and x_2 comes closer and closer to x_1.

In mathematical language, the instantaneous velocity is called the **limit** of $\Delta x/\Delta t$ as Δt approaches zero or (in the language of calculus) the **derivative** of x with respect to t. Using the symbol v_x with no subscript for instantaneous velocity, we obtain the following definition:

Definition of instantaneous velocity
The x component of instantaneous velocity is the limit of $\Delta x/\Delta t$ as Δt approaches zero:

$$v_x = \lim_{\Delta t \to 0} \frac{\Delta x}{\Delta t}. \qquad (2.4)$$

Like average velocity, instantaneous velocity is a vector quantity. Equation 2.4 defines its x component, which can be positive or negative. In straight-line motion, all other components are zero, and in this case we'll often call v the instantaneous velocity. From now on, when we use the term *velocity,* we'll always mean instantaneous rather than average velocity, unless we state otherwise.

The *sign* of a velocity component always depends on the choice of coordinate system, including the location of the origin and the choice of positive axis directions. We always assume that t_2 is a *later* time than t_1. (We always grow older, not younger.) Thus, Δt is always positive, and v_x has the same algebraic sign as Δx. If the positive x axis points to the right, as in Figure 2.1, a positive value of v_x

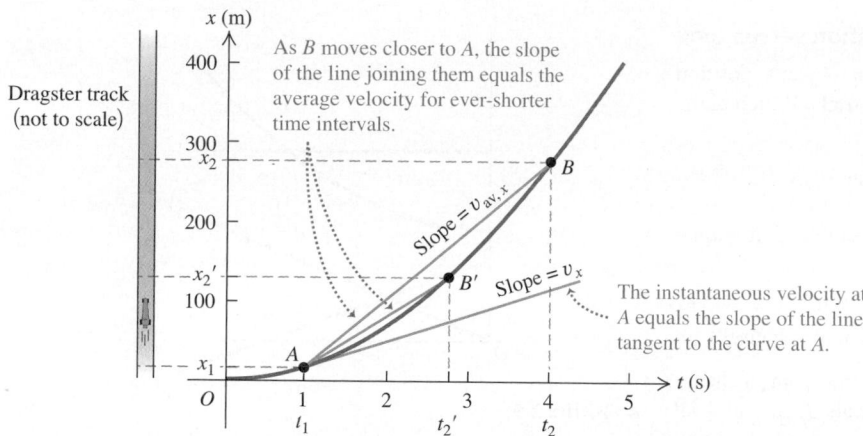

means motion toward the right, and a negative value means motion toward the left. But an object can have a positive x and a negative v_x, or the reverse; x tells us where the particle *is*, and v_x tells us how it is *moving*. In Figure 2.2 the object is to the right of O but is moving toward the left, so it has a positive x and a negative v_x. An object to the left of O moving toward the right has a negative x and a positive v_x. Can you see what the situation is when *both* x and v_x are negative?

Figure 2.7 shows the relation of the graph of x vs t to the average and instantaneous velocities. As we saw in Figure 2.6, the average velocity between any two points A and B on a graph of x versus t is the slope of the line connecting the points. To approach the instantaneous velocity, as we just learned, we move x_2 closer and closer to x_1 (and t_2 closer and closer to t_1). Graphically, that means finding the slope for ever smaller intervals as point B approaches point A. As this interval becomes very small, the slope equals that of a line tangent to the curve at point A. Thus,

> **On a graph of a coordinate x as a function of time t, the instantaneous velocity at any point is equal to the slope of the tangent line to the curve at that point.**

As shown in Figure 2.8, if the tangent line slopes upward to the right, its slope is positive, v_x is positive, and the motion of the particle is in the positive x direction. If the tangent slopes downward to the right, the velocity is negative, and the motion of the particle is in the negative x direction. Where the tangent is horizontal, the slope is zero and the velocity is zero.

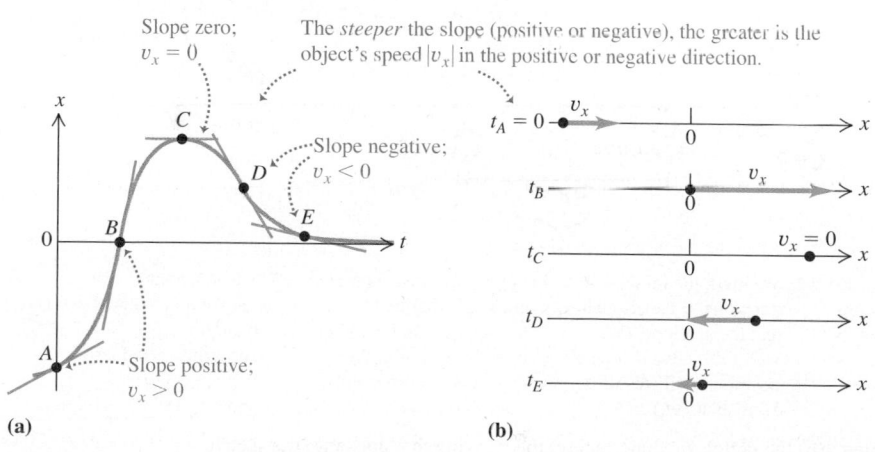

(a)

(b)

◀**FIGURE 2.8** On a graph of position as a function of time, the slope of the tangent line at any point is the velocity at that point.

Conceptual Analysis 2.2

Velocity from graphs of position versus time

Figure 2.9 shows graphs of an object's position along the x axis versus time for four separate trials. Which of the following statements is or are correct?

A. The velocity is greater in trial 2 than in trial 3.
B. The velocity is not constant during trial 1.
C. During trial 4, the object changes direction as it passes through the point $x = 0$.
D. The velocity is constant during trials 2, 3, and 4.

(Remember that "velocity" means "x component of velocity.")

SOLUTION The graph of the motion during trial 1 has a changing slope and therefore v_x isn't constant. Trials 2, 3, and 4 all have graphs with constant slope and thus correspond to motion with constant velocity. (The velocity is zero during trial 2, as indicated by the slope of zero, but this value is still constant.) Trial 3

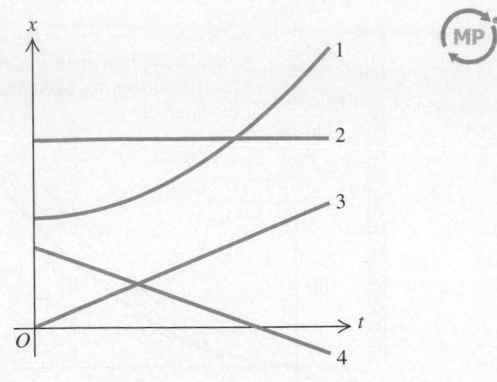

▲ **FIGURE 2.9**

has a positive velocity, while the velocity in trial 4 is negative. Thus, choices B and D are correct.

EXAMPLE 2.2 Average and instantaneous velocities

A cheetah is crouched in ambush 20.0 m to the east of an observer's vehicle, as shown in Figure 2.10a. At time $t = 0$, the cheetah charges an antelope in a clearing 50.0 m east of the observer. The cheetah runs along a straight line; the observer estimates that, during the first 2.00 s of the attack, the cheetah's coordinate x varies with time t according to the equation

$$x = 20.0 \text{ m} + (5.00 \text{ m/s}^2)t^2.$$

(a) Find the displacement of the cheetah during the interval between $t_1 = 1.00$ s and $t_2 = 2.00$ s. **(b)** Find the average velocity during this time interval. **(c)** Estimate the instantaneous velocity at time $t_1 = 1.00$ s by taking $\Delta t = 0.10$ s.

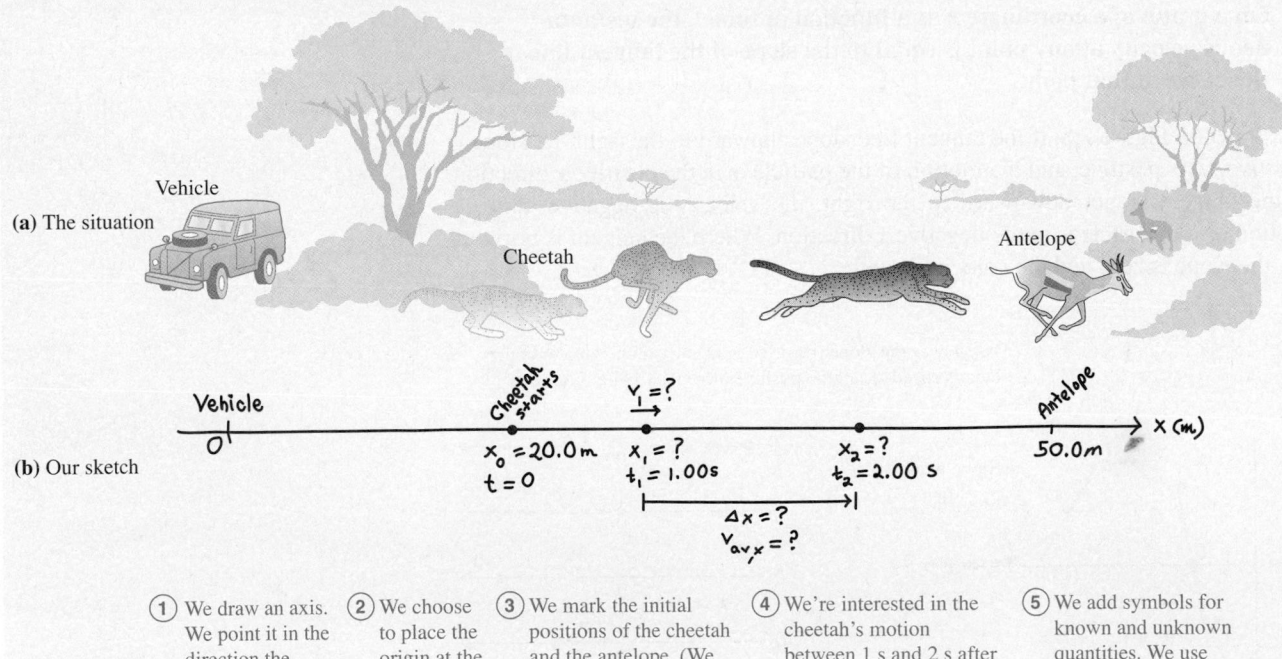

▲ **FIGURE 2.10** (a) The situation for this problem, (b) the sketch we draw, and (c) the thinking that goes into the sketch.

Continued

SOLUTION

SET UP Figure 2.10b shows the diagram we sketch. First we define a coordinate system, orienting it so that the cheetah runs in the $+x$ direction. We add the points we are interested in, the values we know, and the values we will need to find for parts (a) and (b).

SOLVE Part (a): To find the displacement Δx, we first find the cheetah's positions (the values of x) at time $t_1 = 1.00$ s and at time $t_2 = 2.00$ s by substituting the values of t into the given equation. At time $t_1 = 1.00$ s, the cheetah's position x_1 is

$$x_1 = 20.0 \text{ m} + (5.00 \text{ m/s}^2)(1.00 \text{ s})^2 = 25.0 \text{ m}.$$

At time $t_2 = 2.00$ s, the cheetah's position x_2 is

$$x_2 = 20.0 \text{ m} + (5.00 \text{ m/s}^2)(2.00 \text{ s})^2 = 40.0 \text{ m}.$$

The displacement during this interval is

$$\Delta x = x_2 - x_1 = 40.0 \text{ m} - 25.0 \text{ m} = 15.0 \text{ m}.$$

Part (b): Having the displacement from 1.00 s to 2.00 s, we can now find the average velocity for that interval:

$$v_{av,x} = \frac{\Delta x}{\Delta t} = \frac{40.0 \text{ m} - 25.0 \text{ m}}{2.00 \text{ s} - 1.00 \text{ s}} = \frac{15.0 \text{ m}}{1.00 \text{ s}} = 15.0 \text{ m/s}.$$

Part (c): The instantaneous velocity at 1.00 s is approximately (but not exactly) equal to the average velocity in the interval from $t_1 = 1.00$ s to $t_2 = 1.10$ s (i.e., $\Delta t = 0.10$ s). At $t_2 = 1.10$ s,

$$x_2 = 20.0 \text{ m} + (5.00 \text{ m/s})(1.10 \text{ s})^2 = 26.05 \text{ m},$$

so that

$$v_{av,x} = \frac{\Delta x}{\Delta t} = \frac{26.05 \text{ m} - 25.0 \text{ m}}{1.10 \text{ s} - 1.00 \text{ s}} = 10.5 \text{ m/s}.$$

If you use the same procedure to find the average velocities for time intervals of 0.01 s and 0.001 s, you will get 10.05 m/s and 10.005 m/s, respectively. As Δt gets smaller and smaller, the average velocity gets closer and closer to 10.0 m/s. We conclude that the instantaneous velocity at time $t = 1.0$ s is 10.0 m/s.

REFLECT As the time interval Δt approaches zero, the average velocity in the interval is closer and closer to the limiting value 10.0 m/s, which we call the instantaneous velocity at time $t = 1.00$ s. Note that when we calculate an average velocity, we need to specify two times—the beginning and end times of the interval—but for instantaneous velocity at a particular time, we specify only that one time.

NOTE ▶ Here's a reminder about the term *speed*. It is distance traveled, divided by time taken, on either an average or an instantaneous basis. Speed is a scalar quantity, not a vector; it has no direction. If a race car makes one lap around the Indianapolis race track (2.5 mi in circumference) in 1.00 min ($= 1/60$ h), its average speed is $(2.5 \text{ mi})/(1/60 \text{ h}) = 150 \text{ mi/h}$. Its instantaneous speed may or may not be constant, but its velocity is certainly not constant because the car changes direction at the turns. At the end of one lap, the race car is back where it started, so its average *velocity* for one lap is zero. ◀

2.3 Average and Instantaneous Acceleration

When the velocity of a moving object changes with time, we say that the object has *acceleration*. Just as velocity is a quantitative description of the rate of change of *position* with time, acceleration is a quantitative description of the rate of change of *velocity* with time. Like velocity, acceleration is a vector quantity. In straight-line motion its only nonzero component is along the coordinate axis.

To introduce the concept of acceleration, let's consider again the motion of an object (such as the dragster of Section 2.1) along the x axis. As usual, we assume that the object can be treated as a particle. Suppose that at time t_1 the particle is at point x_1 and has an x component of instantaneous velocity v_{1x} and that at a later time t_2 it is at point x_2 and has an x component of velocity v_{2x}.

Definition of average acceleration

The **average acceleration** a_{av} of an object as it moves from x_1 (at time t_1) to x_2 (at time t_2) is a vector quantity whose x component is the ratio of the change in the x component of velocity, $\Delta v_x = v_{2x} - v_{1x}$, to the time interval $\Delta t = t_2 - t_1$:

$$a_{av,x} = \frac{v_{2x} - v_{1x}}{t_2 - t_1} = \frac{\Delta v_x}{\Delta t}. \tag{2.5}$$

Unit: If we express velocity in meters per second and time in seconds, then acceleration is in meters per second per second $(\text{m/s})/\text{s}$. This unit is usually written as m/s^2 and is read "meters per second squared."

▲ **Application Haven't I been here before?** Gil de Ferran has just won the 2003 Indianapolis 500 auto race by completing 200 laps of the 2.5-mile oval course more rapidly than anyone else. He did this in under three hours, for an average speed of close to 200 miles per hour for the 500-mile race. However, since he finished the race right where he started, his total displacement during the race was zero. So don't tell him now, but despite his high average speed, his average velocity for the trip was 0 miles per hour!

In straight-line motion, we'll usually call $a_{\text{av},x}$ the average acceleration, although we recognize that it is actually the x component of the average acceleration.

EXAMPLE 2.3 Acceleration in a space walk

An astronaut has left the space shuttle on a tether to test a new personal maneuvering device. She moves along a straight line directly away from the shuttle. Her onboard partner measures her velocity before and after certain maneuvers, and obtains the following results:

(a) $v_{1x} = 0.8$ m/s, $v_{2x} = 1.2$ m/s; (speeding up)
(b) $v_{1x} = 1.6$ m/s, $v_{2x} = 1.2$ m/s; (slowing down)
(c) $v_{1x} = -0.4$ m/s, $v_{2x} = -1.0$ m/s; (speeding up)
(d) $v_{1x} = -1.6$ m/s, $v_{2x} = -0.8$ m/s. (slowing down)

If $t_1 = 2$ s and $t_2 = 4$ s in each case, find the average acceleration for each set of data.

SOLUTION

SET UP We use the diagram in Figure 2.11 to organize our data.

SOLVE To find the astronaut's average acceleration in each case, we will use the definition of average acceleration (Equation 2.5): $a_{\text{av},x} = \Delta v_x / \Delta t$. The time interval is $\Delta t = 2.0$ s in all cases; the change in velocity in each case is $\Delta v_x = v_{2x} - v_{1x}$.

Part (a): $a_{\text{av},x} = \dfrac{1.2 \text{ m/s} - 0.8 \text{ m/s}}{4 \text{ s} - 2 \text{ s}} = +0.2 \text{ m/s}^2$;

Part (b): $a_{\text{av},x} = \dfrac{1.2 \text{ m/s} - 1.6 \text{ m/s}}{4 \text{ s} - 2 \text{ s}} = -0.2 \text{ m/s}^2$;

Part (c): $a_{\text{av},x} = \dfrac{-1.0 \text{ m/s} - (-0.4 \text{ m/s})}{4 \text{ s} - 2 \text{ s}} = -0.3 \text{ m/s}^2$;

Part (d): $a_{\text{av},x} = \dfrac{-0.8 \text{ m/s} - (-1.6 \text{ m/s})}{4 \text{ s} - 2 \text{ s}} = +0.4 \text{ m/s}^2$.

REFLECT The astronaut speeds up in cases (a) and (c) and slows down in (b) and (d), but the average acceleration is positive in (a) and (d) and negative in (b) and (c). In other words, negative acceleration *does not* necessarily indicate a slowing down.

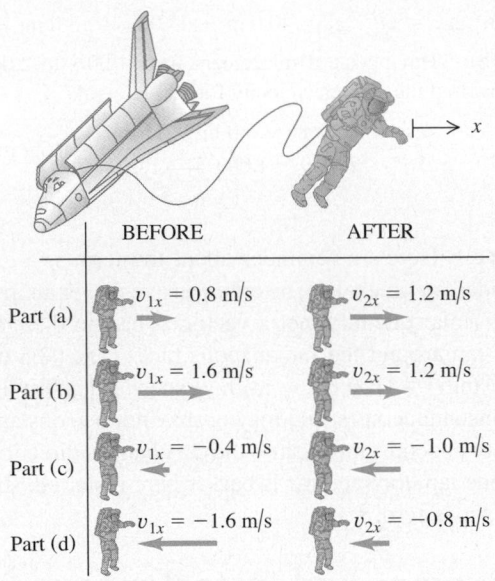

	BEFORE	AFTER
Part (a)	$v_{1x} = 0.8$ m/s	$v_{2x} = 1.2$ m/s
Part (b)	$v_{1x} = 1.6$ m/s	$v_{2x} = 1.2$ m/s
Part (c)	$v_{1x} = -0.4$ m/s	$v_{2x} = -1.0$ m/s
Part (d)	$v_{1x} = -1.6$ m/s	$v_{2x} = -0.8$ m/s

▲ **FIGURE 2.11** We can use a sketch and table to organize the information given in the problem.

The preceding example raises the question of what the sign of acceleration means and how it relates to the signs of displacement and velocity. Because average acceleration (i.e., the x component) is defined as $a_{\text{av},x} = \Delta v_x / \Delta t$, and Δt is always positive, the sign of acceleration is the same as the sign of Δv_x. Figure 2.12 shows the four possible cases; make sure you understand why the acceleration is positive or negative in each case.

NOTE ▶ As Figure 2.12 shows, the velocity and acceleration of an object do not necessarily have the same sign. For instance, an object that is moving in the positive direction, but slowing down, has positive velocity and negative acceleration. Velocity describes the motion of an object at a given instant of time; acceleration describes how the object's motion is *changing* with time. ◀

The term *deceleration* is sometimes used for a decrease in speed; such a decrease can correspond to either positive or negative acceleration. Because of

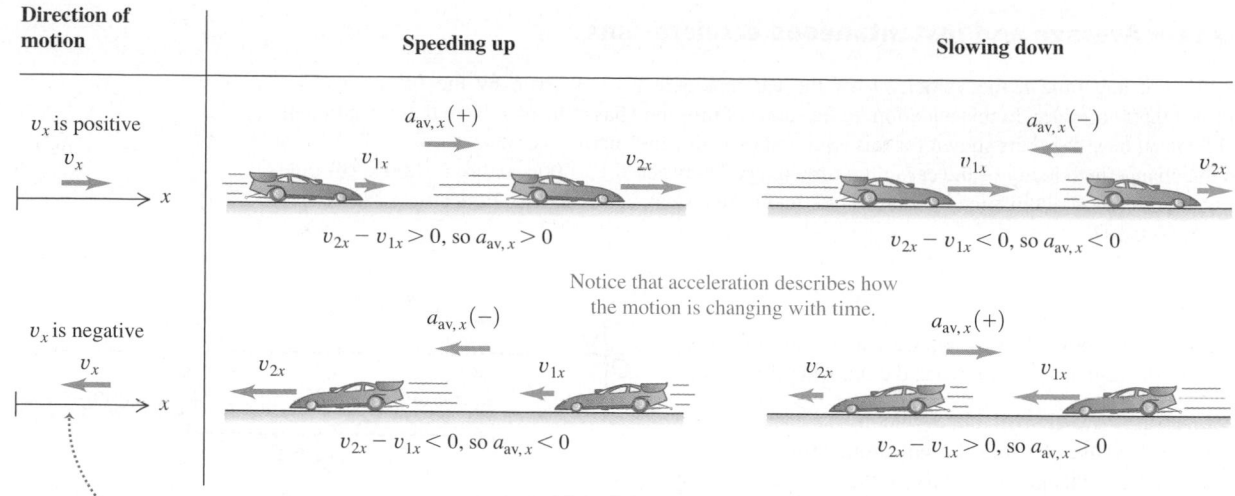

▲ FIGURE 2.12 The relationship between the sign of acceleration, the sign of Δv, and the direction chosen as positive.

this ambiguity, we won't use the term in what follows. Instead, we recommend careful attention to the interpretation of the algebraic sign of a_x in relation to that of v_x.

We can now define **instantaneous acceleration,** following the same procedure that we used for instantaneous velocity. Consider the situation shown in Figure 2.13: A sports car has just entered the final straightaway at the Grand Prix. It reaches point x_1 at time t_1, moving with velocity v_{1x}. It passes point x_2, closer to the finish line, at time t_2 with velocity v_{2x}.

To define instantaneous acceleration at point x_1, we take the second point x_2 in Figure 2.13 to be closer and closer to the first point x_1, so the average acceleration is computed over shorter and shorter time intervals. The instantaneous acceleration at point x_1 is defined as the limit approached by the average acceleration when point x_2 is taken closer and closer to point x_1 and t_2 is taken closer and closer to t_1.

Definition of instantaneous acceleration
The x component of instantaneous acceleration is defined as the limit of $\Delta v_x/\Delta t$ as Δt approaches zero:

$$a_x = \lim_{\Delta t \to 0} \frac{\Delta v_x}{\Delta t}. \qquad (2.6)$$

Instantaneous acceleration plays an essential role in the laws of mechanics. From now on, when we use the term *acceleration*, we will always mean instantaneous acceleration, not average acceleration.

▲ Application Is this car accelerating?
Shortly after this jet-propelled car reached its top speed, the parachute was deployed to help bring it to a stop. As you know from your study of physics, the car is indeed accelerating while it is slowing down. Because acceleration is defined as a change in velocity per unit time, in this case the car has a negative acceleration component in the forward direction.

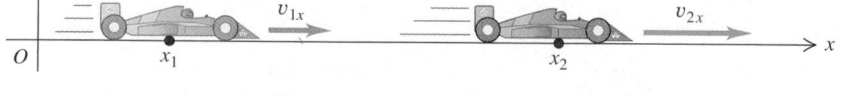

◄ FIGURE 2.13 A race car at two points on the straightaway.

EXAMPLE 2.4 **Average and instantaneous accelerations**

Suppose that, at any time t, the velocity v of the car in Figure 2.13 is given by the equation $v_x = 60.0 \text{ m/s} + (0.500 \text{ m/s}^3)t^2$. In this equation, v_x has units of m/s and t has units of s. Note that the numbers 60 and 0.50 *must* have the units shown for this equation to be dimensionally consistent.
(a) Find the change in velocity of the car in the time interval between $t_1 = 1.00$ s and $t_2 = 3.00$ s. **(b)** Find the average acceleration in this time interval. **(c)** Estimate the instantaneous acceleration at time $t_1 = 1.00$ s by taking $\Delta t = 0.10$ s.

$t \approx 2 \text{ s}$

SOLUTION

SET UP Figure 2.14 shows the diagram we use to establish a coordinate system and organize our known and unknown information.

SOLVE Part (a): We first find the velocity at each time by substituting each value of t into the equation. From these values, we can find the change in velocity in the interval. Thus, at time $t_1 = 1.00$ s,

$$v_{1x} = 60.0 \text{ m/s} + (0.500 \text{ m/s}^3)(1.00 \text{ s})^2 = 60.5 \text{ m/s}.$$

At time $t_2 = 3.00$ s,

$$v_{2x} = 60.0 \text{ m/s} + (0.500 \text{ m/s}^3)(3.00 \text{ s})^2 = 64.5 \text{ m/s}.$$

The change in velocity Δv_x is then

$$\Delta v_x = v_{2x} - v_{1x} = 64.5 \text{ m/s} - 60.5 \text{ m/s} = 4.00 \text{ m/s}.$$

Part (b): The change in velocity, divided by the time interval, gives the average acceleration in each interval. The time interval is $\Delta t = 3.00 \text{ s} - 1.00 \text{ s} = 2.00$ s, so

$$a_{\text{av},x} = \frac{v_{2x} - v_{1x}}{t_2 - t_1} = \frac{4.0 \text{ m/s}}{2.00 \text{ s}} = 2.00 \text{ m/s}^2.$$

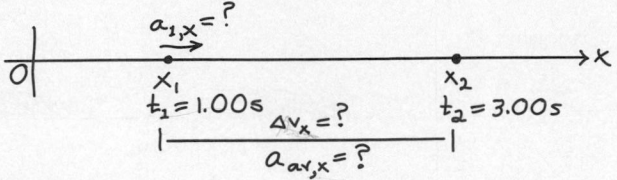

▲ **FIGURE 2.14** Our sketch for this problem.

Part (c): We use the same procedure to approximate the instantaneous acceleration at x_1 by calculating the average acceleration over 0.100 s. When $\Delta t = 0.100$ s, $t_2 = 1.10$ s, and it follows that

$$v_{2x} = 60.0 \text{ m/s} + (0.500 \text{ m/s}^3)(1.10 \text{ s})^2 = 60.605 \text{ m/s},$$
$$\Delta v_x = 0.105 \text{ m/s},$$
$$a_{\text{av},x} = \frac{\Delta v_x}{\Delta t} = \frac{0.105 \text{ m/s}}{0.100 \text{ s}} = 1.05 \text{ m/s}^2.$$

REFLECT If we repeat the calculation in part (c) for $\Delta t = 0.01$ s and $\Delta t = 0.001$ s, we get $a_{\text{av},x} = 1.005 \text{ m/s}^2$ and $a_{\text{av},x} = 1.0005 \text{ m/s}^2$, respectively (although we aren't entitled to this many significant figures). As Δt gets smaller and smaller, the average acceleration gets closer and closer to 1.00 m/s^2. We conclude that the instantaneous acceleration at $t = 1.00$ s is 1.00 m/s^2.

As with average and instantaneous velocity, we can gain added insight into the concepts of average and instantaneous acceleration by plotting a graph with velocity v_x on the vertical axis and time t on the horizontal axis. Figure 2.15 shows such a graph for the race car from Figure 2.13. Notice that now we are plotting *velocity* against time, not position against time as before. If we draw a line between any two points on the graph, such as A and B (corresponding to a displacement from x_1 to x_2 and to the time interval $\Delta t = t_2 - t_1$), the slope of that line equals the average acceleration over that interval. If we then take point B and move it closer and closer to point A, the slope of the line AB approaches the slope

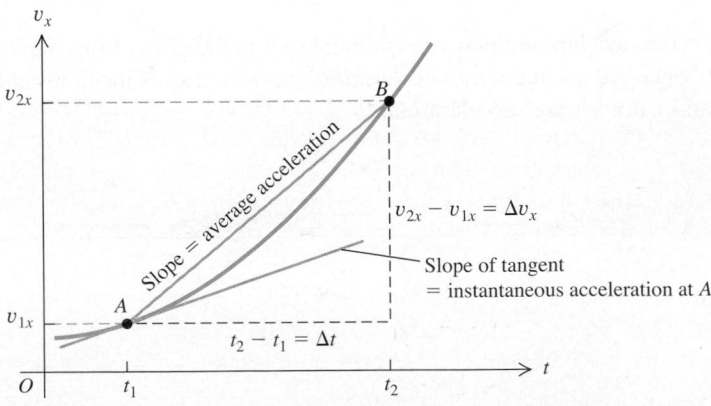

▶ **FIGURE 2.15** A graph of velocity versus time for the motion of the race car in Figure 2.13.

of the line that is tangent to the curve at point A. Accordingly, we can interpret a graph of velocity versus time as follows:

> **The average acceleration between any two points on a graph of velocity versus time equals the slope of a line connecting those points.**

> **The instantaneous acceleration at any point on the graph equals the slope of the line tangent to the curve at that point.**

Let's emphasize once again the significance of the *sign* of acceleration. First, the velocity v_x and acceleration a_x of an object don't necessarily have the same sign; they are independent quantities. When v_x and a_x *do* have the same sign, the object is speeding up. If both are positive, the object is moving in the positive direction with increasing speed. If both are negative, the object is moving in the negative direction with a velocity that becomes more and more negative with time, and again the object's speed increases. If v_x is positive and a_x is negative, the object is moving in the positive direction with decreasing speed. If v_x is negative and a_x is positive, the object is moving in the negative direction with a velocity that is becoming less negative, and again the object slows down.

 **Conceptual Analysis 2.3** | **Translating a graphical description to a verbal description**

Figure 2.16 shows a graph of velocity versus time for a moving object. (Notice that the vertical axis is velocity in this case, not position.) The object moves on an east-west axis, with west chosen as the positive direction. Which of the following verbal descriptions matches the graph?

A. The object speeds up continuously while moving westward.
B. The object moves toward the origin while slowing, then moves away from the origin while speeding up.
C. The object moves in the positive direction while slowing to a stop, then moves east while speeding up.
D. The object moves east while slowing to a stop, then moves west while speeding up.

SOLUTION The direction of motion and the sign of the velocity tell us the same thing; in this case, a positive velocity means that the object is going west and a negative velocity means that it is going east. A large negative velocity means that the object is going east at high speed, while a small negative velocity means that it is going east slowly. Since the vertical axis of this graph represents velocity, the velocity coordinate of any point on the plot tells you the object's direction and speed. The plot starts

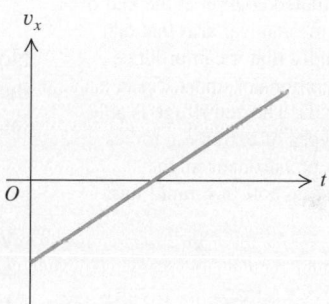

▲ **FIGURE 2.16**

with a large, negative value of velocity, indicating that the object is moving east fast. When the plot reaches $v_x = 0$ (the point at which it crosses the horizontal axis), the object has slowed to a halt. From there, the object has progressively higher positive values of v_x; it is moving to the west with increasing speed. Thus, A cannot be true, because the object starts with negative v_x. Nor can B be true, since knowing your speed and direction of motion says nothing about where you are. The answer D matches the reasoning given here.

Acceleration and the Human Body

Accelerations require forces; we'll study the relationship between them in Chapter 4. An object subjected to sufficiently large accelerations and the associated forces may suffer damage. The earth's gravitational attraction causes objects to fall with a downward acceleration with a magnitude of about 9.8 m/s^2 (assuming that effects such as air resistance can be neglected). This acceleration is often denoted by g; we'll study its effects in greater detail in Section 2.5. For now, we simply note that $g = 9.8 \text{ m/s}^2$.

Because earth's land life evolved in an environment characterized by this magnitude of acceleration, it isn't surprising that the human body can withstand accelerations on the order of g without damage. Sport cars can accelerate at about $0.5g$, or 4.9 m/s^2. Accelerations on the order of $3g$ in a jet fighter cause serious impairment of the pilot's blood circulation, and above about $5g$ the pilot loses

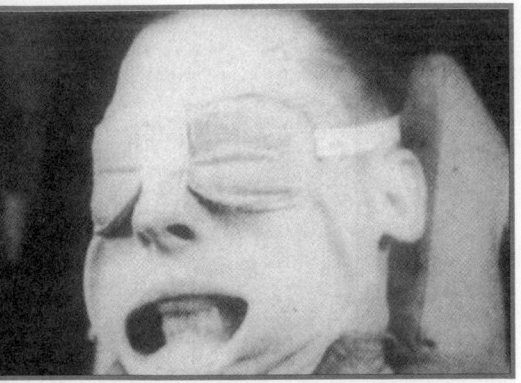

▲ BIO Application Are we there yet?
Fortunately for this test pilot, the extreme *g* force he is experiencing will last only a second or two. He is strapped into a human centrifuge, an instrument designed to test the limits of human endurance when exposed to extreme *g* forces. The device consists of a simulated cockpit at the end of an 8-meter-long mechanical arm that can be rotated so rapidly that it can produce temporary accelerations of up to 15*g* in a period of 3 seconds. The centrifuge is able to simulate the types of extreme *g* forces a fighter jet pilot or astronaut might experience during takeoff or a rapid turn.

Mastering**PHYSICS**

PhET: Forces in 1 Dimension
ActivPhysics 1.1: Analyzing Motion Using Diagrams
ActivPhysics 1.3: Predicting Motion from Graphs
ActivPhysics 1.4: Predicting Motion from Equations
ActivPhysics 1.5: Problem-Solving Strategies for Kinematics
ActivPhysics 1.6: Skier Races Downhill

consciousness. The primary function of air bags in cars is to reduce the maximum acceleration during a collision to a value that can be sustained by the body without catastrophic damage. The human body can withstand accelerations on the order of 50*g* for very short time intervals (about 0.05 s).

2.4 Motion with Constant Acceleration

The simplest accelerated motion is straight-line motion with *constant* acceleration. In this case, the velocity changes at the same rate throughout the motion. Figure 2.17 shows four ways to represent the motion of a race car undergoing constant acceleration. Notice that the plot of velocity versus time (Figure 2.17c) is a straight line; that is, its slope is constant. That tells you that the car's velocity changes by equal amounts Δv_x in equal time intervals Δt. This characteristic of motion with constant acceleration makes it easy for us to derive equations for the position x and velocity v_x as functions of time. Let's start with velocity. In the definition of average acceleration (Equation 2.5), we can replace the average acceleration $a_{\text{av},x}$ by the constant (instantaneous) acceleration a_x. We then have

$$a_x = \frac{v_{2x} - v_{1x}}{t_2 - t_1} \quad \text{(for an object moving with constant acceleration).} \quad (2.7)$$

Now we let $t_1 = 0$ and t_2 be any arbitrary later time t. We use the symbol v_{0x} for the velocity (again, the x component) at the initial time $t = 0$; the velocity at the later time t is v_x. Then Equation 2.7 becomes the following:

Velocity as a function of time for an object moving with constant acceleration
For an object that has x component of velocity v_{0x} at time $t = 0$ and moves with constant acceleration a_x, we find the velocity v_x at any later time t:

$$a_x = \frac{v_x - v_{0x}}{t - 0}, \quad \text{or}$$

$$v_x = v_{0x} + a_x t. \quad (2.8)$$

Unit: m/s

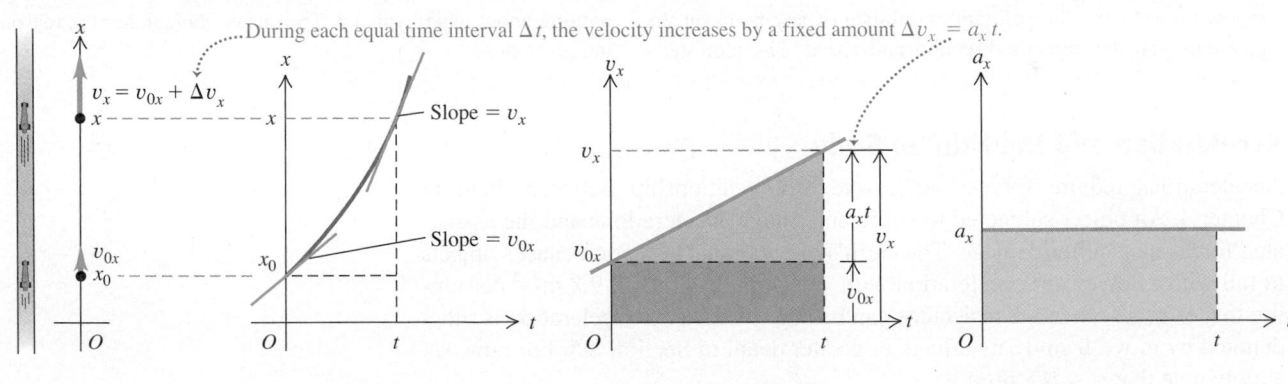

(a) A race car moves in the *x* direction with constant acceleration.

(b) The graph of *x* versus *t*.

(c) The graph of v_x versus *t*.

(d) The graph of a_x versus *t*.

▲ **FIGURE 2.17** Four ways to represent the motion of a race car moving with constant acceleration.

We can interpret this equation as follows: The acceleration a_x is the constant rate of change of velocity, that is, the change in velocity per unit time. The term $a_x t$ is the product of the change in velocity per unit time, a_x, and the time interval t. Therefore, it equals the *total* change in velocity from the initial time $t = 0$ to the later time t. The velocity v_x at any time t is the sum of the initial velocity v_{0x} (at $t = 0$) and the *change* in velocity $a_x t$. Figure 2.17c shows this analysis in graphical terms. The height v_x of the graph at any time t is the sum of two segments: one with length v_{0x} equal to the initial velocity, the other with length $a_x t$ equal to the change in velocity during time t.

MasteringPHYSICS

PhET: The Moving Man
ActivPhysics 1.8: Seat Belts Save Lives
ActivPhysics 1.9: Screeching to a Halt
ActivPhysics 1.11: Car Starts, Then Stops
ActivPhysics 1.12: Solving Two-Vehicle Problem
ActivPhysics 1.13: Car Catches Truck
ActivPhysics 1.14: Avoiding a Rear-End Collision

EXAMPLE 2.5   Passing speed

A car initially traveling along a straight stretch of highway at 15 m/s accelerates with a constant acceleration of 2.0 m/s^2 in order to pass a truck. What is the velocity of the car after 5.0 s?

SOLUTION

SET UP Figure 2.18 shows what we draw. We take the origin of coordinates to be at the initial position of the car, where $v_x = v_{0x}$ and $t = 0$, and we let the $+x$ direction be the direction of the car's initial velocity. With this coordinate system, $v_{0x} = +15 \text{ m/s}$ and $a_x = +2.0 \text{ m/s}^2$.

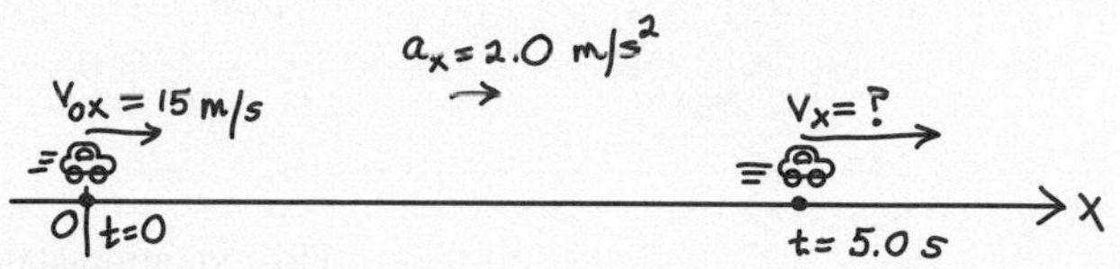

SOLVE The acceleration is constant during the 5.0 s time interval, so we can use Equation 2.8 to find v_x:

$$v_x = v_{0x} + a_x t = 15 \text{ m/s} + (2.0 \text{ m/s}^2)(5.0 \text{ s})$$
$$= 15 \text{ m/s} + 10 \text{ m/s} = 25 \text{ m/s}.$$

REFLECT The velocity and acceleration are in the same direction, so the speed increases. An acceleration of 2.0 m/s^2 means

▲ **FIGURE 2.18** The diagram we draw for this problem.

that the velocity increases by 2.0 m/s every second, so in 5.0 s the velocity increases by 10 m/s.

Practice Problem: If the car maintains its constant acceleration, how much additional time does it take the car to reach a velocity of 35 m/s? *Answer:* 5.0 s.

Next we want to derive an equation for the *position* x of a particle moving with constant acceleration. To do this, we make use of two different expressions for the average velocity $v_{\text{av},x}$ during the interval from $t = 0$ to any later time t. First, when the acceleration is constant and the graph of velocity vs. time is a straight line, as in Figure 2.17, the velocity changes at a uniform rate. In this particular case, the average velocity throughout any time interval is simply the average of the velocities at the beginning and the end of the interval. Thus, for the time interval from 0 to t,

$$v_{\text{av},x} = \frac{v_{0x} + v_x}{2} \quad \text{(for constant acceleration).} \tag{2.9}$$

(This equation does *not* hold if the acceleration varies with time; in that case, the graph of velocity vs. time is a curve, as in Figure 2.15.)

We also know that v_x, the velocity at any time t, is given by Equation 2.8. Substituting that expression for v_x into Equation 2.9, we find that

$$v_{\text{av},x} = \tfrac{1}{2}(v_{0x} + v_{0x} + a_x t)$$
$$= v_{0x} + \tfrac{1}{2} a_x t. \tag{2.10}$$

We can also get an expression for $v_{\text{av},x}$ from the general definition, Equation 2.1:

$$v_{\text{av},x} = \frac{x_2 - x_1}{t_2 - t_1}.$$

We call the position at time $t = 0$ the *initial position*, denoted by x_0. The position at the later time t is simply x. Thus, for the time interval $\Delta t = t - 0$ and the corresponding displacement $x - x_0$, Equation 2.1 gives

$$v_{av,x} = \frac{x - x_0}{t}. \qquad (2.11)$$

Finally, we equate Equations 2.10 and 2.11 and simplify the result to obtain the following equation:

Position as a function of time for an object moving with constant acceleration
For constant acceleration a_x, we solve

$$v_{0x} + \tfrac{1}{2}a_x t = \frac{x - x_0}{t} \qquad \text{to find}$$

$$x = x_0 + v_{0x}t + \tfrac{1}{2}a_x t^2. \qquad (2.12)$$

This equation states that if, at the initial time $t = 0$, a particle is at position x_0 and has velocity v_{0x}, its new position x at any later time t is the sum of three terms: its initial position x_0, plus the distance $v_{0x}t$ that it would have moved if its velocity had been constant, plus an additional distance $\tfrac{1}{2}a_x t^2$ caused by the changing velocity.

EXAMPLE 2.6   Passing distance

What distance does the car in Example 2.5 travel during its 5.0 seconds of acceleration?

SOLUTION

SET UP We use the same coordinates as in Example 2.5. As before, $v_{0x} = +15$ m/s and $a_x = +2.0$ m/s^2.

SOLVE We want to solve for $x - x_0$, the distance traveled by the car during the 5.0 s time interval. The acceleration is constant, so we can use Equation 2.12 to find

$$x - x_0 = v_{0x}t + \tfrac{1}{2}a_x t^2$$
$$= (15 \text{ m/s})(5.0 \text{ s}) + \tfrac{1}{2}(2.0 \text{ m/s}^2)(5.0 \text{ s})^2$$
$$= 75 \text{ m} + 25 \text{ m} = 100 \text{ m}.$$

REFLECT If the speed were constant and equal to the initial value $v_0 = 15$ m/s, the car would travel 75 m in 5.0 s. It actually travels *farther*, because the speed is increasing. From Example 2.5,

we know that the final velocity is $v_x = 25$ m/s, so the *average velocity* for the 5.0 s segment of motion is

$$v_{av,x} = \frac{v_{0x} + v_x}{2} = \frac{15 \text{ m/s} + 25 \text{ m/s}}{2} = 20 \text{ m/s}.$$

An alternative way to obtain the distance traveled is to multiply the average velocity by the time interval. When we do this, we get $x - x_0 = v_{av,x}t = (20 \text{ m/s})(5.0 \text{ s}) = 100$ m, the same result we obtained using Equation 2.12.

Practice Problem: If the car in Example 2.5 maintains its constant acceleration for a total time of 10 s, what total distance does it travel? *Answer:* 250 m.

For some problems, we aren't given the time interval for the motion, and we need to obtain a relation for x, v_x, and a_x that doesn't contain t. To obtain such a relation, we can combine Equations 2.8 and 2.12 to eliminate t. This involves a little algebra, but the result is worth the effort!

We first solve Equation 2.8 for t and then substitute the resulting expression into Equation 2.12 and simplify:

$$t = \frac{v_x - v_{0x}}{a_x},$$

$$x = x_0 + v_{0x}\left(\frac{v_x - v_{0x}}{a_x}\right) + \frac{1}{2}a_x\left(\frac{v_x - v_{0x}}{a_x}\right)^2.$$

We transfer the term x_0 to the left side and multiply through by $2a_x$:

$$2a_x(x - x_0) = 2v_{0x}v_x - 2v_{0x}^2 + v_x^2 - 2v_{0x}v_x + v_{0x}^2.$$

Finally, we combine terms and simplify to obtain the following equation:

Velocity as a function of position for an object moving with constant acceleration

For an object moving in a straight line with constant acceleration a_x,

$$v_x^2 = v_{0x}^2 + 2a_x(x - x_0). \qquad (2.13)$$

This equation gives us the particle's velocity v_x at any position x without needing to know the *time* when it is at that position.

EXAMPLE 2.7   Entering the freeway

A sports car is sitting at rest in a freeway entrance ramp. The driver sees a break in the traffic and floors the car's accelerator, so that the car accelerates at a constant 4.9 m/s² as it moves in a straight line onto the freeway. What distance does the car travel in reaching a freeway speed of 30 m/s?

SOLUTION

SET UP As shown in Figure 2.19, we place the origin at the initial position of the car and assume that the car travels in a straight line in the $+x$ direction. Then $v_{0x} = 0$, $v_x = 30$ m/s, and $a_x = 4.9$ m/s².

SOLVE The acceleration is constant; the problem makes no mention of time, so we can't use Equation 2.8 or Equation 2.12 by themselves. We need a relation between x and v_x, and this is provided by Equation 2.13:

$$v_x^2 = v_{0x}^2 + 2a_x(x - x_0).$$

Rearranging and substituting numerical values, we obtain

$$x - x_0 = \frac{v_x^2 - v_{0x}^2}{2a_x} = \frac{(30 \text{ m/s})^2 - 0}{2(4.9 \text{ m/s}^2)} = 92 \text{ m}.$$

Alternative Solution: We could have used Equation 2.8 to solve for the time t first:

$$t = \frac{v_x - v_{0x}}{a_x} = \frac{30 \text{ m/s} - 0}{4.9 \text{ m/s}^2} = 6.12 \text{ s}.$$

Then Equation 2.12 gives the distance traveled:

$$x - x_0 = v_{0x}t + \tfrac{1}{2}a_x t^2 = 0 + \tfrac{1}{2}(4.9 \text{ m/s}^2)(6.12 \text{ s})^2 = 92 \text{ m}.$$

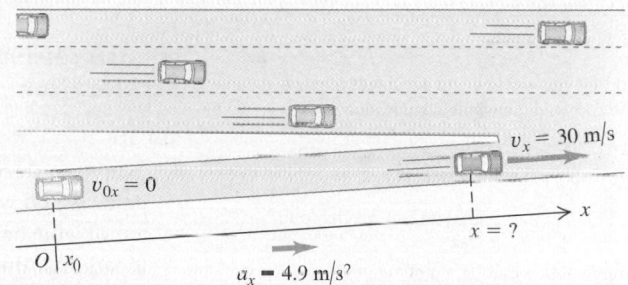

▲ **FIGURE 2.19** A car accelerating on a ramp merging onto a freeway.

REFLECT We obtained the same result in one step when we used Equation 2.13 and in two steps when we used Equations 2.8 and 2.12. When we use them correctly, Equations 2.8, 2.12, and 2.13 *always* give consistent results.

The final speed of 30 m/s is about 67 mph, and 92 m is about 100 yd. Does this distance correspond to your own driving experience?

Practice Problem: What distance has the car traveled when it has reached a speed of 20 m/s? *Answer:* 41 m.

Equations 2.8, 2.12, and 2.13, are the *equations of motion with constant acceleration.* Any kinematic problem involving motion of an object in a straight line with constant acceleration can be solved by applying these equations.

Here's a summary of the equations:

Equations of motion for constant acceleration

$v_x = v_{0x} + a_x t$	(Gives v_x if t is known.)	(2.8)
$x = x_0 + v_{0x}t + \tfrac{1}{2}a_x t^2$	(Gives x if t is known.)	(2.12)
$v_x^2 = v_{0x}^2 + 2a_x(x - x_0)$	(Gives v_x if x is known.)	(2.13)

We can get one more useful relationship by equating the two expressions for $v_{av,x}$ given by Equations 2.9 and 2.11, and multiplying through by t. Doing this, we obtain the following equation:

Position, velocity, and time for an object moving with constant acceleration

$$x - x_0 = \frac{v_{0x} + v_x}{2}t. \qquad (2.14)$$

This equation says that the total displacement $x - x_0$ from time $t = 0$ to a later time t is equal to the average velocity $(v_{0x} + v_x)/2$ during the interval, multiplied by the time t. Equation 2.14 doesn't contain the acceleration a_x, so it is sometimes useful when a_x is not given in the problem.

Figure 2.20 is a graphical representation of Equations 2.12 and 2.8, respectively, for the case of positive velocity and positive acceleration. In Figure 2.20a, the blue curve represents Equation 2.12. The straight dashed line shows the plot we would get if the acceleration is zero: If $a_x = 0$, then $x = x_0 + v_{0x}t$. For this line, the slope (velocity) is constant. The vertical separation between the two curves represents the term $\frac{1}{2}a_x t^2$, which is the effect of acceleration. In Figure 2.20b, the green curve represents Equation 2.8. Make sure you understand how this graph represents the individual terms of the equation.

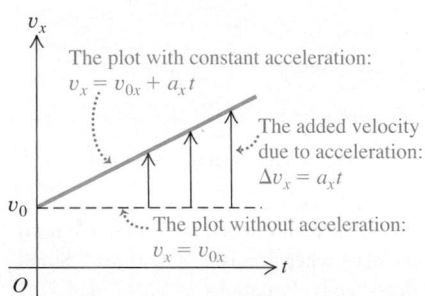

The plot with constant acceleration:
$x = x_0 + v_{0x}t + \frac{1}{2}a_x t^2$

The effect of acceleration:
$\frac{1}{2}a_x t^2$

The plot we would get with zero acceleration:
$x = x_0 + v_{0x}t$

(a) Plot of x versus t for an object moving with positive constant acceleration

The plot with constant acceleration:
$v_x = v_{0x} + a_x t$

The added velocity due to acceleration:
$\Delta v_x = a_x t$

The plot without acceleration:
$v_x = v_{0x}$

(b) Plot of v_x versus t for the same object

▲ **FIGURE 2.20** The graphical meaning of Equations 2.8 and 2.12 for an object moving with positive velocity and positive constant acceleration.

PROBLEM-SOLVING STRATEGY 2.1 Motion with constant acceleration

SET UP

1. You *must* decide at the beginning of a problem where the origin of coordinates is and which axis direction is positive. The choices are usually a matter of convenience; it is often easiest to place the origin at the object's location at time $t = 0$; then $x_0 = 0$. Draw a diagram showing your choices. Then sketch any relevant later positions of the object in the same diagram.

2. Once you have chosen the positive axis direction, the positive directions for velocity and acceleration are also determined. It would be wrong to define x as positive to the right of the origin and velocities as positive toward the left.

3. It often helps to restate the problem in prose first and then translate the prose description into symbols and equations. *When* (i.e, at what value of t) does the particle arrive at a certain point? *Where* is the particle when its velocity has a certain value? (That is, what is the value of x when v_x has the specified value?) The next example asks, "Where is the motorcyclist when his velocity is 25 m/s?" Translated into symbols, this becomes "What is the value of x when $v_x = 25$ m/s?"

4. Either on your diagram or in a list, write known and unknown quantities such as x, x_0, v_x, v_{0x}, and a_x. Write in the values for those that are known. Be on the lookout for implicit information; for example, "A car sits at a stoplight" usually means that $v_{0x} = 0$, and so on.

SOLVE

5. Once you've identified the unknowns, you may be able to choose an equation from among Equations 2.8, 2.12, 2.13, and 2.14 that contains only one of the unknowns. Solve for that unknown; then substitute the known values and compute the value of the unknown. Carry the units of the quantities along with your calculations as an added consistency check.

REFLECT

6. Take a hard look at your results to see whether they make sense. Are they within the general range of magnitudes you expected? If you change one of the given quantities, do the results change in a way you can predict?

EXAMPLE 2.8   Constant acceleration on a motorcycle

A motorcyclist heading east through a small Iowa town accelerates after he passes a signpost at $x = 0$ marking the city limits. His acceleration is constant: $a_x = 4.0 \text{ m/s}^2$. At time $t = 0$ he is 5.0 m east of the signpost and has a velocity of $v_x = 15 \text{ m/s}$. **(a)** Find his position and velocity at time $t = 2.0 \text{ s}$. **(b)** Where is the motorcyclist when his velocity is 25 m/s?

SOLUTION

SET UP Figure 2.21 shows our diagram. The problem tells us that $x = 0$ at the signpost, so that is the origin of coordinates. We point the x axis east, in the direction of motion. The (constant) acceleration is $a_x = 4.0 \text{ m/s}^2$. At the initial time $t = 0$ the position is $x_0 = 5.0 \text{ m}$, and the initial velocity is $v_{0x} = 15 \text{ m/s}$.

SOLVE Part (a): We want to know the position and velocity (the values of x and v_x, respectively) at the later time $t = 2.0 \text{ s}$. Equation 2.12 gives the position x as a function of time t:

$$x = x_0 + v_{0x}t + \tfrac{1}{2}a_x t^2$$
$$= 5.0 \text{ m} + (15 \text{ m/s})(2.0 \text{ s}) + \tfrac{1}{2}(4.0 \text{ m/s}^2)(2.0 \text{ s})^2 = 43 \text{ m}.$$

Equation 2.8 gives the velocity v_x as a function of time t:

$$v_x = v_{0x} + a_x t$$
$$= 15 \text{ m/s} + (4.0 \text{ m/s}^2)(2.0 \text{ s}) = 23 \text{ m/s}.$$

Part (b): We want to know the value of x when $v_x = 25 \text{ m/s}$. Note that this occurs at a time later than 2.0 s and at a point farther than 43 m from the signpost. From Equation 2.13, we have

$$v_x^2 = v_{0x}^2 + 2a_x(x - x_0),$$
$$(25 \text{ m/s})^2 = (15 \text{ m/s})^2 + 2(4.0 \text{ m/s}^2)(x - 5.0 \text{ m}),$$
$$x = 55 \text{ m}.$$

Alternatively, we may use Equation 2.8 to find first the time when $v_x = 25 \text{ m/s}$:

$$v_x = v_{0x} + a_x t,$$
$$25 \text{ m/s} = 15 \text{ m/s} + (4.0 \text{ m/s}^2)(t),$$
$$t = 2.5 \text{ s}.$$

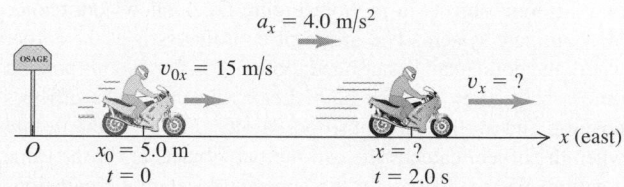

▲ **FIGURE 2.21**

Then, from Equation 2.12,

$$x = x_0 + v_{0x}t + \tfrac{1}{2}a_x t^2$$
$$= 5.0 \text{ m} + (15 \text{ m/s})(2.5 \text{ s}) + \tfrac{1}{2}(4.0 \text{ m/s}^2)(2.5 \text{ s})^2$$
$$= 55 \text{ m}.$$

REFLECT Do the results make sense? The cyclist accelerates from 15 m/s (about 34 mi/h) to 23 m/s (about 51 mi/h) in 2.0 s while traveling a distance of 38 m (about 125 ft). This is fairly brisk acceleration, but well within the realm of possibility for a high-performance bike.

Practice Problem: If the acceleration is 2.0 m/s^2 instead of 4.0 m/s^2, where is the cyclist, and how fast is he moving, 5.0 s after he passes the signpost if $x_0 = 5.0 \text{ m}$ and $v_{0x} = 15 \text{ m/s}$? *Answers:* 105 m, 25 m/s.

EXAMPLE 2.9   Pursuit!

A motorist traveling at a constant velocity of 15 m/s passes a school-crossing corner where the speed limit is 10 m/s (about 22 mi/h). A police officer on a motorcycle stopped at the corner starts off in pursuit with constant acceleration of 3.0 m/s^2 (Figure 2.22a). **(a)** How much time elapses before the officer catches up with the car? **(b)** What is the officer's speed at that point? **(c)** What is the total distance the officer has traveled at that point?

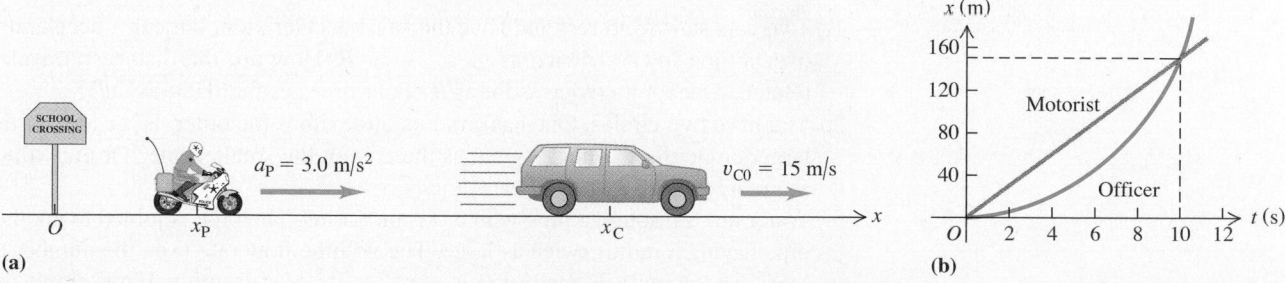

(a)

(b)

▲ **FIGURE 2.22** (a) A diagram of the problem. (b) Graphs of position as a function of time for the police officer and the car.

Continued

SOLUTION

SET UP Both objects move in a straight line, which we'll designate as the x axis. Then all positions, velocities, and accelerations have only x components. We'll omit the subscripts x in our solution; but always keep in mind that when we say "velocity," we really mean "x component of velocity," and so on.

The motorcycle and the car both move with constant acceleration, so we can use the formulas we have developed. We have two different objects in motion; Figure 2.22a shows our choice of coordinate system. The origin of coordinates is at the corner where the policeman is stationed; both objects are at this point at time $t = 0$, so $x_0 = 0$ for both. Let x_P be the police officer's position and x_C the car's position, at any time t. At the instant when the officer catches the car, the two objects are at the same position. We need to apply the constant-acceleration equations to each, and find the time when x_P and x_C are equal. We denote the initial x components of velocities (i.e., x components) as v_{P0} and v_{C0} and the accelerations (x components) as a_P and a_C. From the data given, we have $v_{P0} = 0$, $v_{C0} = 15$ m/s, $a_P = 3.0$ m/s^2, and $a_C = 0$.

SOLVE Applying Equation 2.12 to each object, we find that

$$x_P = x_0 + v_{P0}t + \tfrac{1}{2}a_Pt^2 = 0 + 0 + \tfrac{1}{2}(3.0 \text{ m/s}^2)t^2$$
$$= \tfrac{1}{2}(3.0 \text{ m/s}^2)t^2,$$
$$x_C = x_0 + v_{C0}t + \tfrac{1}{2}a_Ct^2 = 0 + (15 \text{ m/s})t + 0 = (15 \text{ m/s})t.$$

Part (a): At the time the officer catches the car, they must be at the same position, so at this time, $x_C = x_P$. Equating the preceding two expressions, we have

$$\tfrac{1}{2}(3.0 \text{ m/s}^2)t^2 = (15 \text{ m/s})t,$$

or

$$t = 0, 10 \text{ s}.$$

There are *two* times when the two vehicles have the same x coordinate; the first is the time ($t = 0$) when the car passes the parked motorcycle at the corner, and the second is the time when the officer catches up.

Part (b): From Equation 2.8, we know that the officer's velocity v_P at any time t is given by

$$v_P = v_{P0} + a_Pt = 0 + (3.0 \text{ m/s}^2)t,$$

so when $t = 10$ s, $v_P = 30$ m/s. When the officer overtakes the car, she is traveling twice as fast as the motorist is.

Part (c): When $t = 10$ s, the car's position is

$$x_C = (15 \text{ m/s})(10 \text{ s}) = 150 \text{ m},$$

and the officer's position is

$$x_P = \tfrac{1}{2}(3.0 \text{ m/s}^2)(10 \text{ s})^2 = 150 \text{ m}.$$

This result verifies that, at the time the officer catches the car, they have gone equal distances and are at the same position.

REFLECT A graphical description of the motion is helpful. Figure 2.22b shows graphs of x_P and x_C as functions of time. At first x_C is greater, because the car is ahead of the officer. But the car travels with constant velocity, while the officer accelerates, closing the gap between the two. At the point where the curves cross, the officer has caught up to the car. We see again that there are two times when the two positions are the same. Note that the two vehicles don't have the same speed at either of these times.

Practice Problem: If the officer's acceleration is 5.0 m/s^2, what distance does she travel before catching up with the car? What is her velocity when she has caught up? During what time interval is she moving more slowly than the car? *Answer:* 90 m, 30 m/s, 0 to 3 s.

2.5 Proportional Reasoning

The relationships for straight-line motion that we've developed in this chapter offer an ideal background for introducing an analytical technique that is useful in many areas of physics. We'll call this technique **proportional reasoning.** It's particularly useful when we want to describe general relationships between variables, without reference to specific details of numerical values.

Following are some examples of the sort of problems for which proportional reasoning is useful. Some are trivial, but they illustrate the methods we'll describe.

1. Two cars start from rest and have the same acceleration, but car *A* accelerates over a time interval twice as great as car *B*. How are the distances traveled related? Does *A* go twice as far as *B*? Four times as far? Half as far?
2. You have two circles. One has a radius three times the other. Is the area of the larger circle three times as great as the area of the smaller one? Or nine times as great? Or one-third as great?
3. Water flows through a pipe with a certain radius. The pipe is joined to another pipe having a radius twice as large. The volume flow rate (i.e., the number of liters per second) in each pipe is equal to its cross-sectional area times the

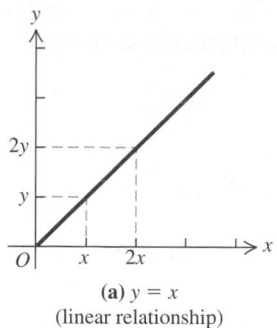

(a) $y = x$
(linear relationship)

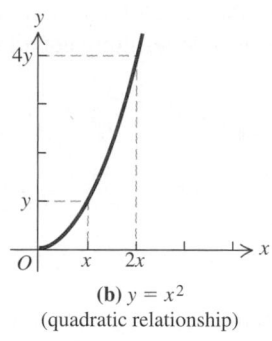

(b) $y = x^2$
(quadratic relationship)

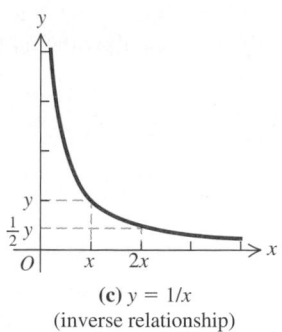

(c) $y = 1/x$
(inverse relationship)

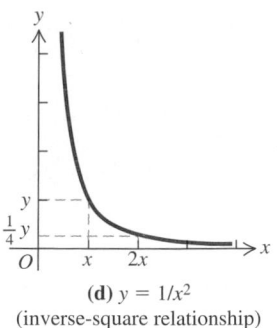

(d) $y = 1/x^2$
(inverse-square relationship)

▲ **FIGURE 2.23** Graphs of some simple relationships between two variables.

speed of the water's motion. If the volume flow rate is the same in both pipes, does the water move twice as fast in the larger pipe? Half as fast?

In all of these situations, we have two variables, such as distance and time or the radius and area of a circle, that are related in a simple way. If we call the two variables x and y, the relationship between them often takes one of the following simple forms:

$$y = kx \ (\text{linear relation});$$
$$y = kx^2 \ (\text{quadratic relation});$$
$$y = k/x \ (\text{inverse relation});$$
$$y = k/x^2 \ (\text{inverse-square relation}).$$

In all these forms, k is a constant or a product of constants, and the numerical value of k may or may not be known. Figure 2.23 shows graphs of these four forms of relationship; for simplicity, we have taken the constant k equal to unity in all of them.

Let's see how the preceding relational forms can be applied to Problem 1. From our analysis in this chapter, we know that when an object starts from rest at the point $x = 0$ on the x axis and moves along the axis with constant acceleration a_x, its position x at any time t is given by the simple relation

$$x = \tfrac{1}{2}a_x t^2.$$

We call the two cars A and B; they have the same acceleration a_x. If car A accelerates for a time t_A, it travels a distance x_A given by

$$x_A = \tfrac{1}{2}a_x t_A^2,$$

and if car B accelerates for a time t_B, it travels a distance x_B given by

$$x_B = \tfrac{1}{2}a_x t_B^2.$$

We'd like to get a general relationship between the two distances and the two times that doesn't contain the common acceleration, which may or may not be known. There are two ways to do this: We can divide one of the preceding equations by the other. When we do, the factor $a_x/2$ divides out, and we obtain

$$\frac{x_A}{x_B} = \frac{\tfrac{1}{2}a_x t_A^2}{\tfrac{1}{2}a_x t_B^2} = \frac{t_A^2}{t_B^2} = \left(\frac{t_A}{t_B}\right)^2.$$

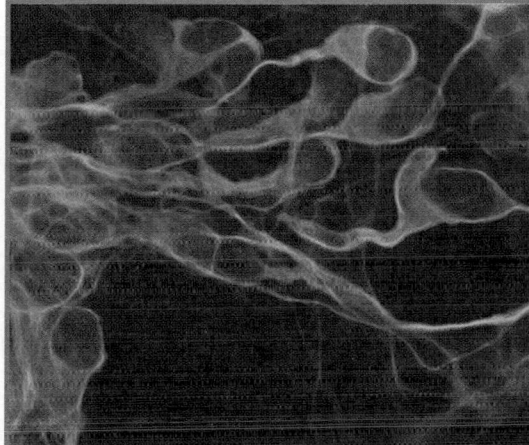

▲ **BIO Application A long time coming.**
Proportional reasoning may not make sense for two variables with a non-linear relationship. Albert Einstein's seminal work on Brownian motion showed that the distance that molecules diffuse is related to the time of diffusion by a quadratic relation $x = kt^2$, where k incorporates information about the diffusing molecule and the medium through which the molecules are moving. Proteins and informational molecules can diffuse from the center of a typical spherical biological cell (diameter 20 μm) to the cell's periphery in about 1 s. In the case of a motor neuron, in which the cell nucleus (in red) is in the spinal cord, an extension of the cell called the axon is sent to the target muscle and may be 1 m long in humans. Proportional reasoning shows that the time required for molecules to diffuse from the nucleus to the end of the cell where they are used would be 10^{10} s or about 300 years. This is an impossibly long time, so neurons have evolved sophisticated mechanisms for moving materials around much more rapidly.

Or we can put all the variables in each equation on one side and all the constants on the other side, equate the two resulting equations, and obtain the same result:

$$\frac{x_A}{t_A^2} = \frac{1}{2}a_x \quad \text{and} \quad \frac{x_B}{t_B^2} = \frac{1}{2}a_x, \quad \text{so}$$

$$\frac{x_A}{t_A^2} = \frac{x_B}{t_B^2}, \quad \text{or} \quad \frac{x_A}{x_B} = \left(\frac{t_A}{t_B}\right)^2 \quad (\text{as before}).$$

Either of these procedures can be used to yield a relation that contains only the values of x and t.

Now let's get back to the original problem: If car A travels for twice as much time as car B, then $t_A = 2t_B$, or $t_A/t_B = 2$. This equation becomes

$$\frac{x_A}{x_B} = (2)^2 = 4,$$

and we see that $x_A = 4x_B$; Car A, with twice as much time to accelerate as car B, travels four times as far.

Now let's summarize what we've done:

1. First, we choose the quantities we want to consider as our variables (in this case, x and t) and find an equation relating them (in this case, $x = a_x t^2/2$).
2. Next, we write this equation for two different values of each of the chosen variables (t_A, x_A and t_B, x_B in our example) and then use one of the procedures described above to eliminate the constants and obtain an equation containing only the two values of each variable.
3. Finally, we specify a relation between two values of one variable (in this case, t_A and t_B) and solve for the relation between the corresponding values of the other variable (in this case, x_A and x_B). We don't need the numerical value of a_x (or k in the generic examples in Figure 2.23); everything is done with ratios and proportions. Note, incidentally, that this relation is *quadratic* in form, the second example in the figure.

Here's another example: The same two cars, A and B, have just finished a speed run; they cross the finish line at the same time, with different speeds v_A and v_B. They both apply their brakes as they cross the line, giving them the same (negative) acceleration, and they come to a stop at distances x_A and x_B, respectively, beyond the finish line. If car A has $1\frac{1}{2}$ times the speed of car B at the finish line, by what factor does its stopping distance exceed that of car B?

For Step 1, we choose v and x as our variables; we need to find a relation between them. No times are mentioned in the statement of the problem; this suggests that we use Equation 2.13. Dropping the subscript x to simplify our notation, we write this equation as

$$v^2 = v_0^2 + 2a(x - x_0).$$

We place the origin of coordinates at the finish line (the starting point for our analysis), so that $x_0 = 0$ for both cars. The final velocity is zero (i.e., $v = 0$) for both cars. Thus, the relationship we need is simply

$$v_0^2 = -2ax.$$

That is, a car that crosses the finish line with speed v_0 and brakes with constant (negative) acceleration a travels a distance x before stopping.

For Step 2, we rewrite the preceding equation, placing the variables on the left side and the constants on the right:

$$\frac{v_0^2}{x} = -2a, \quad \text{so} \quad \frac{v_{0A}^2}{x_A} = -2a \quad \text{and} \quad \frac{v_{0B}^2}{x_B} = -2a.$$

Hence $\dfrac{x_A}{x_B} = \left(\dfrac{v_{0A}}{v_{0B}}\right)^2$.

The relationship is again quadratic, but now our variables are x and v instead of x and t.

Finally, for Step 3, if car A is initially moving $1\frac{1}{2}$ times as fast as car B, it travels a distance $\left(\frac{3}{2}\right)^2 = \frac{9}{4}$ as great as car B before stopping.

Here's an alternative language for the relationships we've discovered: In our first example, in which the variables were x and t, we say that x is proportional to t^2; in the second, x is proportional to v^2. The special symbol $\propto$ is often used to represent proportionality; thus, in the above examples, we would write $x \propto t^2$ and $x \propto v^2$, read as "x is proportional to t^2" and "x is proportional to v^2."

Quantitative Analysis 2.4

Brakes on a wet road

When you are driving your car on wet pavement and need to stop suddenly, the maximum (negative) acceleration you can apply to the car without skidding is about one-third as much as on dry pavement. For a given initial velocity, how does your stopping distance x_w on wet pavement differ from your stopping distance x_d on dry pavement?

A. $x_w = x_d$
B. $x_w = 3x_d$
C. $x_w = 9x_d$

SOLUTION As in the previous example, the variables are related by $v_0^2 = -2ax$. This time we consider the variables to be x and a, and the initial velocity v_0 is constant. Separating constants and variables, we find that

$$xa = -\tfrac{1}{2}v_0^2, \quad \text{so} \quad x_d a_d = -\tfrac{1}{2}v_0^2, \quad x_w a_w = -\tfrac{1}{2}v_0^2,$$

$$x_d a_d = x_w a_w, \quad \frac{x_w}{x_d} = \frac{a_d}{a_w}, \quad \text{and} \quad x \propto \frac{1}{a}.$$

This result confirms our intuition that x varies *inversely* with a: If $a_w = \frac{1}{3}a_d$, then $x_w = 3x_d$, and choice B is the correct one. The relation of x to a is shown by the graph in Figure 2.23c.

2.6 Freely Falling Objects

The most familiar example of motion with (nearly) constant acceleration is that of an object falling under the influence of the earth's gravitational attraction. Such motion has held the attention of philosophers and scientists since ancient times. Aristotle thought (erroneously) that heavy objects fall faster than light objects, in proportion to their weight. Galileo argued that an object should fall with an acceleration that is constant and *independent* of its weight. We mentioned in Section 1.1 that, according to legend, Galileo experimented by dropping musket balls and cannon balls from the Leaning Tower of Pisa.

The motion of falling bodies has since been studied with great precision. When air resistance is absent, as in a vacuum chamber or on the moon, Galileo is right: All objects at a particular location fall with the same acceleration, regardless of their size or weight. Figure 2.24 shows an apple and a feather falling in a vacuum chamber; they are released at the same time and fall at the same rate. If the distance of the fall is small compared with the radius of the earth (or of the moon, planet, or other astronomical body on which the fall occurs), the acceleration is constant. In the discussion that follows, we'll use an idealized model in which we neglect air resistance, the earth's rotation, and the decrease in an object's acceleration with increasing altitude. We call this idealized motion **free fall**, although it includes rising as well as falling motion.

The constant acceleration of a freely falling object is called the **acceleration due to gravity,** or the *acceleration of free fall;* we denote its magnitude with the letter g. At or near the earth's surface, the value of g is approximately 9.80 m/s^2, 980 cm/s^2, or 32.2 ft/s^2. Because g is the magnitude of a vector quantity, it is always a *positive* number. On the surface of the moon, the acceleration due to gravity is caused by the attractive force of the moon rather than the earth, and $g = 1.62 \text{ m/s}^2$. Near the surface of the sun, $g = 274 \text{ m/s}^2$.

Mastering**PHYSICS**

PhET: Lunar Lander
ActivPhysics 1.7: Balloonist Drops Lemonade
ActivPhysics 1.10: Pole-Vaulter Lands

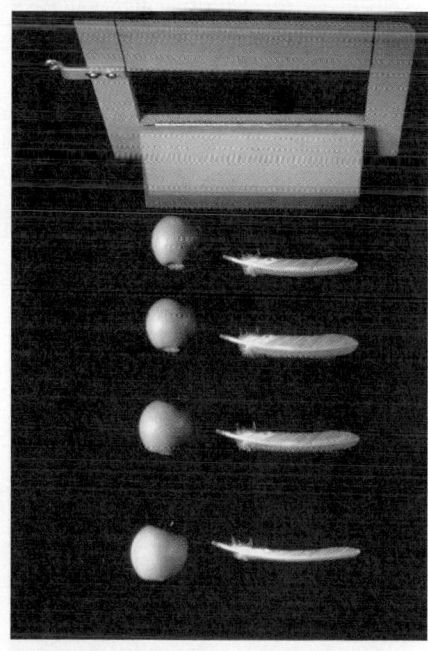

▲ **FIGURE 2.24** In vacuum, objects with different weights fall with the same acceleration and thus fall side by side.

In the examples that follow, we use the constant-acceleration equations developed in Section 2.4. We suggest that you review the problem-solving strategy discussed in that section before you study these examples.

EXAMPLE 2.10 Falling euro in Pisa

A 2-euro coin is dropped from the Leaning Tower of Pisa. It starts from rest and falls freely. Compute its position and velocity after 1.0, 2.0, and 3.0 s.

SOLUTION

SET UP Figure 2.25 shows the diagram we sketch. Because the motion is downward, we choose a vertical axis; by convention, we call it the y axis rather than the x axis. Choosing this axis means that we will replace all the x's in the constant-acceleration equations by y's. We choose the upward direction to be positive, and we place the origin at the starting point. The position, velocity, and acceleration have only y components. The initial value y_0 of the coordinate y and the initial y component of velocity v_{0y} are both zero. The acceleration is downward (in the negative y direction), so the constant y component of acceleration is $a_y = -g = -9.8 \text{ m/s}^2$. (Remember that, by definition, g itself is *always* positive.)

SOLVE From Equations 2.12 and 2.8, with x replaced by y, we get

$$y = v_0 t + \tfrac{1}{2} a_y t^2 = 0 + \tfrac{1}{2}(-g)t^2 = (-4.9 \text{ m/s}^2)t^2,$$

$$v_y = v_{0y} + a_y t = 0 + (-g)t = (-9.8 \text{ m/s}^2)t.$$

Using subscripts 1, 2, and 3 for the three times, we find that when $t = t_1 = 1.0 \text{ s}$, $y_1 = (-4.9 \text{ m/s}^2)(1.0 \text{ s})^2 = -4.9 \text{ m}$ and $v_{1y} = (-9.8 \text{ m/s}^2)(1.0 \text{ s}) = -9.8 \text{ m/s}$. The coin is therefore 4.9 m below the origin (because y is negative) and has a downward velocity (v_y is negative) with magnitude 9.8 m/s.

The position and velocity at 2.0 and 3.0 s are found in the same way. The results are -19.6 m and -19.6 m/s at $t = 2.0 \text{ s}$ and -44.1 m and -29.4 m/s at $t = 3.0 \text{ s}$.

REFLECT All of our values of y and v_y are negative because we have chosen the positive y axis to be upward. We could just as well have chosen the downward direction to be positive. In that case, the y component of acceleration would have been

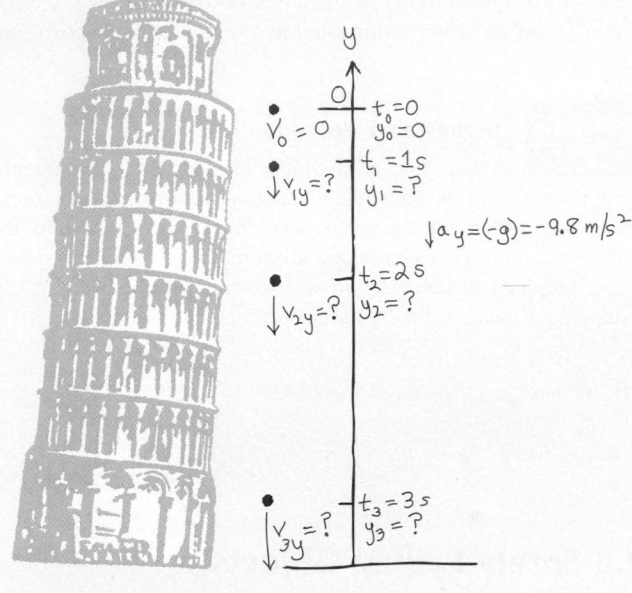

The Leaning Tower   Our sketch for the problem
▲ **FIGURE 2.25**

$a_y = g = +9.8 \text{ m/s}^2$. The crucial point is that you *must* decide at the start which direction is positive, and then stick with your decision.

Practice Problem: The coin is released 47 m above the ground. How much time is required for the coin to fall to the ground? What is its velocity just before it strikes the ground? *Answers:* 3.1 s, -30 m/s.

▲ **FIGURE 2.26** Strobe photo of a freely falling ball.

Figure 2.26 is a strobe photograph of a falling ball taken with a light source that produces a series of intense, brief flashes. The flashes occur at equal time intervals, so this photograph records the position of the ball at equal time intervals. Therefore, the average velocity of the ball between successive flashes is proportional to the distance between the corresponding images of the ball in the photograph. The increasing distances between images show that the ball's velocity increases continuously; in other words, the ball accelerates. Careful measurement shows that the change in velocity is the same in each time interval; thus the acceleration is constant.

EXAMPLE 2.11   **A ball on the roof**

Suppose you throw a ball vertically upward from the flat roof of a tall building. The ball leaves your hand at a point even with the roof railing, with an upward velocity of 15.0 m/s. On its way back down, it just misses the railing. Find **(a)** the position and velocity of the ball 1.00 s and 4.00 s after it leaves your hand; **(b)** the velocity of the ball when it is 5.00 m above the railing; and **(c)** the maximum height reached and the time at which it is reached. Ignore the effects of the air.

SOLUTION

SET UP As shown in Figure 2.27, we place the origin at the level of the roof railing, where the ball leaves your hand, and we take the positive direction to be upward. Here's what we know: The initial position y_0 is zero, the initial velocity v_{0y} (y component) is $+15.0$ m/s, and the acceleration (y component) is $a_y = -9.80$ m/s^2.

SOLVE What equations do we have to work with? The velocity v_y at any time t is

$$v_y = v_{0y} + a_y t = 15.0 \text{ m/s} + (-9.80 \text{ m/s}^2)t.$$

The position y at any time t is

$$y = v_{0y}t + \tfrac{1}{2}a_y t^2 = (15.0 \text{ m/s})t + \tfrac{1}{2}(-9.80 \text{ m/s}^2)t^2.$$

The velocity v_y at any position y is given by

$$v_y^2 = v_{0y}^2 + 2a_y(y - y_0)$$
$$= (15.0 \text{ m/s})^2 + 2(-9.80 \text{ m/s}^2)(y - 0).$$

Part (a): When $t = 1.00$ s, the first two equations give $y = +10.1$ m, $v_y = +5.20$ m/s.
The ball is 10.1 m above the origin (y is positive), and it has an upward velocity (v is positive) of 5.20 m/s (less than the initial velocity of 15.0 m/s, as expected).

When $t = 4.00$ s, the same equations give

$$y = -18.4 \text{ m}, \qquad v_y = -24.2 \text{ m/s}.$$

Thus, at time $t = 4.00$ s, the ball has passed its highest point and is 18.4 m *below* the origin (y is negative). It has a *downward* velocity (v_y is negative) with magnitude 24.2 m/s (greater than the initial velocity, as we should expect). Note that, to get these results, we don't need to find the highest point the ball reaches or the time at which it was reached. The equations of motion give the position and velocity at *any* time, whether the ball is on the way up or on the way down.

Part (b): When the ball is 5.00 m above the origin, $y = +5.00$ m. Now we use our third equation to find the velocity v_y at this point:

$$v_y^2 = (15.0 \text{ m/s})^2 + 2(-9.80 \text{ m/s}^2)(5.00 \text{ m}) = 127 \text{ m}^2/\text{s}^2,$$
$$v_y = \pm 11.3 \text{ m/s}.$$

We get two values of v_y, one positive and one negative. That is, the ball passes the point $y = +5.00$ m *twice*, once on the way up and again on the way down. The velocity on the way up is $+11.3$ m/s, and on the way down it is -11.3 m/s.

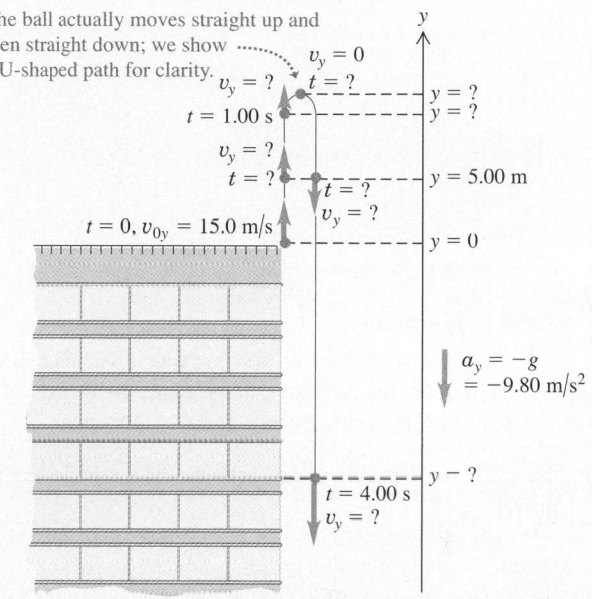

▲ **FIGURE 2.27**

Part (c): At the highest point, the ball's velocity is momentarily zero $(v_y = 0)$; it has been going up (positive v_y) and is about to start going down (negative v_y). From our third equation, we have

$$0 = (15.0 \text{ m/s})^2 - (19.6 \text{ m/s}^2)y,$$

and the maximum height (where $v_y = 0$) is $y = 11.5$ m.

We can now find the time t when the ball reaches its highest point from Equation 2.8, setting $v_y = 0$:

$$0 = 15.0 \text{ m/s} + (-9.80 \text{ m/s}^2)t,$$
$$t = 1.53 \text{ s}.$$

Alternative Solution: Alternatively, to find the maximum height, we may ask first *when* the maximum height is reached. That is, at what value of t is $v_y = 0$? As we just found, $v_y = 0$ when $t = 1.53$ s. Substituting this value of t back into the equation for y, we find that

$$y = (15 \text{ m/s})(1.53 \text{ s}) + \tfrac{1}{2}(-9.8 \text{ m/s}^2)(1.53 \text{ s})^2 = 11.5 \text{ m}.$$

REFLECT Although at the highest point the velocity is momentarily zero, the *acceleration* at that point is still -9.8 m/s^2. The ball stops for an instant, but its velocity is continuously changing, from positive values through zero to negative values.

Continued

The acceleration is constant throughout; it is *not* zero when $v_y = 0$!

Figure 2.28 shows graphs of position and velocity as functions of time for this problem. Note that the graph of v_y versus t has a constant negative slope. Thus, the acceleration is negative on the way up, at the highest point, and on the way down.

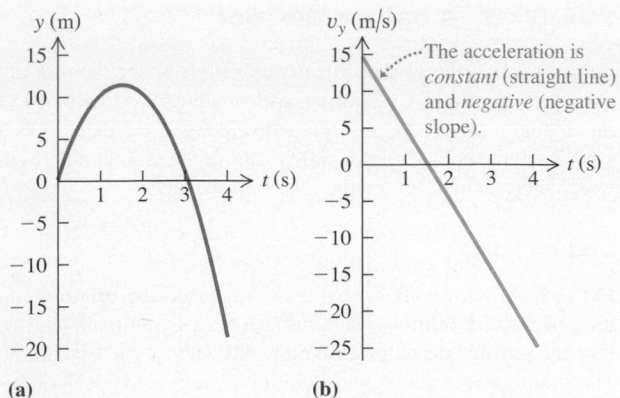

▲ **FIGURE 2.28** (a) Position and (b) velocity as functions of time for a ball thrown upward with an initial velocity of 15 m/s.

 **A pop fly**

In a baseball game, a batter hits a pitched ball that goes straight up in the air above home plate. At the top of its motion, the ball is momentarily motionless. At this point,

A. the acceleration is zero because the ball is motionless at that instant.
B. the acceleration is zero because the motion is changing from slowing down to speeding up.
C. the acceleration is zero because at that instant the force from the impact with the bat is balanced by the pull of the earth.
D. the acceleration is 9.8 m/s² downward.

SOLUTION A nonzero acceleration means that an object is either speeding up or slowing down. If an object has no acceleration, its speed does not change. Since a ball at the top of its motion is about to move downward, it must have acceleration, even though it is motionless at that instant. This argument rules out A, B, and C; the ball's acceleration cannot be zero. But is D correct? When velocity and acceleration are in opposite directions, the velocity is decreasing in magnitude. When they are in the same direction, the velocity is increasing in magnitude. At the top, the velocity is about to increase in the downward direction (i.e., becomes more and more negative), and its direction becomes the same as that of the acceleration. So D is correct.

 **Tossing two stones**

Suppose you throw two identical stones from the top of a tall building. You throw one directly upward with a speed of 10 m/s and the other directly downward with the same speed. When they hit the street below, how do their speeds compare? (Neglect air friction.)

A. The one thrown upward is traveling faster.
B. The one thrown downward is traveling faster.
C. Both are traveling at the same speed.

SOLUTION The stone thrown upward decreases its speed by 9.8 m/s every second as it goes upward and increases its speed by 9.8 m/s every second as it comes back down. Thus, when it returns to the point (at the top of the building) from which it was thrown, it has exactly its original speed; but it is moving downward instead of upward. Since it now has the same downward velocity as the stone that was originally thrown downward, the two stones must have the same velocity when they hit the ground.

2.7 Relative Velocity along a Straight Line

We always describe the position and velocity of an object with reference to a particular coordinate system. When we speak of the velocity of a moving car, we usually mean its velocity with respect to an observer who is stationary on the earth. But when two observers measure the velocity of a moving object, they get different results if one observer is moving relative to the other. The velocity seen by a particular observer is called the velocity *relative* to that observer, or simply the **relative velocity.**

Figure 2.29 shows an example of the concept of relative velocity in action. In this example we'll assume that all the motions are along the same straight line, the *x* axis. A woman walks in a straight line with a velocity (*x* component) of

1.0 m/s along the aisle of a train that is moving along the same line with a velocity of 3.0 m/s. What is the woman's velocity? This seems like a simple enough question, but in reality it has no single answer. As seen by a passenger sitting in the train, she is moving at 1.0 m/s. A person on a bicycle standing beside the train sees the woman moving at 1.0 m/s + 3.0 m/s = 4.0 m/s. An observer in another train going in the opposite direction would give still another answer. The velocity is different for different observers. We have to specify which observer we mean, and we speak of the velocity *relative* to a particular observer. The woman's velocity relative to the train is 1.0 m/s, her velocity relative to the cyclist is 4.0 m/s, and so on. Each observer, equipped in principle with a meterstick and a stopwatch, forms what we call a **frame of reference.** Thus, a frame of reference is a coordinate system plus a timer.

Let's generalize this analysis. In Figure 2.29, we call the cyclist's frame of reference (at rest with respect to the ground) C and we call the frame of reference of the moving train T (Figure 2.29b). The woman's position at any time, relative to a reference point O_T in the moving train (i.e, in frame of reference T), is given by the distance $x_{W/T}$. Her position at any time, relative to a reference point O_C that is stationary on the ground (i.e, in the cyclist's frame C), is given by the distance $x_{W/C}$, and the position of the train's reference point O_T relative to the stationary one O_C is $x_{T/C}$. We choose the reference points so that at time $t = 0$ all these x's are zero. Then we can see from the figure that at any later time they are related by the equation

$$x_{W/C} = x_{W/T} + x_{T/C}. \qquad (2.15)$$

That is, the total distance from the origin of C to the woman at point W is the sum of the distance from the origin O_C of the cyclist's frame to the origin O_T of the train's frame, plus the distance from O_T to the position of the woman.

Because Equation 2.15 is true at any instant, it must also be true that the *rate of change* of $x_{W/C}$ is equal to the sum of the rates of change of $x_{W/T}$ and $x_{T/C}$. But the rate of change of $x_{W/C}$ is just the velocity of W with respect to C, which we'll denote as $v_{W/C}$, and similarly for $x_{W/T}$ and $x_{T/C}$. Thus, we arrive at the velocity relation

$$v_{W/C} = v_{W/T} + v_{T/C}. \qquad (2.16)$$

Remember that C is the cyclist's frame of reference, T is the frame of reference of the train, and point W represents the woman. Using the preceding notation, we have

$$v_{W/T} = 1.0 \text{ m/s}, \qquad v_{T/C} = 3.0 \text{ m/s}.$$

From Equation 2.16, the woman's velocity $v_{W/C}$ relative to the cyclist on the ground is

$$v_{W/C} = 1.0 \text{ m/s} + 3.0 \text{ m/s} = 4.0 \text{ m/s},$$

as we already knew.

In this example both velocities are toward the right, and we have implicitly taken that as the positive direction. If the woman walks toward the *left* relative to the train, then $v_{W/T} = -1.0$ m/s, and her velocity relative to the cyclist is 2.0 m/s. The sum in Equation 2.16 is always an *algebraic* sum, and any or all of the velocities may be negative.

When the woman looks out the window, the stationary cyclist on the ground appears to her to be moving backward; we can call the cyclist's velocity relative to the woman on the train $v_{C/W}$. Clearly, this is just the negative of $v_{W/C}$. In general, if A and B are any two points or frames of reference,

$$v_{A/B} = -v_{B/A}. \qquad (2.17)$$

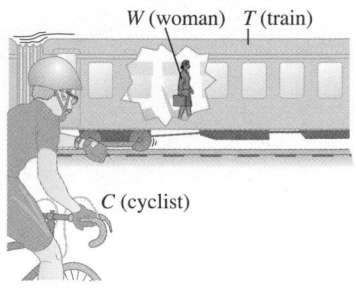

(a)

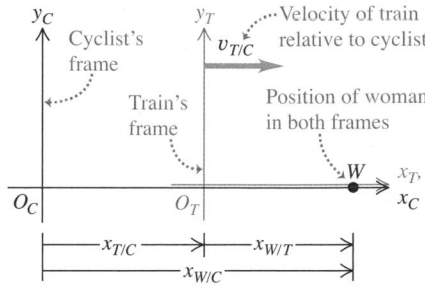

(b)

▲ **FIGURE 2.29** (a) A woman walks in a moving train while observed by a cyclist on the ground outside. (b) The woman's position W, at the instant shown, in two frames of reference: that of the train (T) and that of the cyclist (C).

Note the order of the double subscripts on the velocities in our discussion; $v_{A/B}$ always means "the velocity of A (first subscript) relative to B (second subscript)." These subscripts obey an interesting kind of algebra, as Equation 2.16 shows. If we regard each one as a fraction, then the fraction on the left side is the *product* of the fractions on the right sides: For a point P relative to frames A and B, $P/A = (P/B)(B/A)$. This is a handy rule to use when you apply Equation 2.16. If there are *three* different frames of reference, A, B, and C, we can write immediately

$$v_{P/A} = v_{P/C} + v_{C/B} + v_{B/A},$$

and so on.

EXAMPLE 2.12   Relative velocity on the highway

Suppose you are driving north on a straight two-lane road at a constant 88 km/h (Figure 2.30). A truck traveling at a constant 104 km/h approaches you (in the other lane, fortunately). **(a)** What is the truck's velocity relative to you? **(b)** What is your velocity with respect to the truck?

SOLUTION

SET UP Let you be Y, the truck be T, the earth be E, and let the positive direction be north (Figure 2.30). Then $v_{Y/E} = +88$ km/h.

SOLVE Part (a): The truck is approaching you, so it must be moving south, giving $v_{T/E} = -104$ km/h. We want to find $v_{T/Y}$. Modifying Equation 2.16, we have

$$v_{T/E} = v_{T/Y} + v_{Y/E},$$
$$v_{T/Y} = v_{T/E} - v_{Y/E}$$
$$= -104 \text{ km/h} - 88 \text{ km/h} = -192 \text{ km/h}.$$

The truck is moving 192 km/h south relative to you.

Part (b): From Equation 2.17,

$$v_{Y/T} = -v_{T/Y} = -(-192 \text{ km/h}) = +192 \text{ km/h}.$$

You are moving 192 km/h north relative to the truck.

REFLECT How do the relative velocities of Example 2.9 change after you and the truck have passed? They don't change at all! The relative positions of the objects don't matter. The velocity of

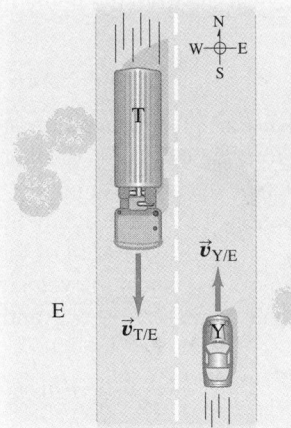

▲ **FIGURE 2.30** Velocities of you and the truck, relative to the ground.

the truck relative to you is still -192 km/h, but it is now moving away from you instead of toward you.

Theory of Relativity

When we derived the relative-velocity relations, we assumed that all the observers use the same time scale. This assumption may seem so obvious that it isn't even worth mentioning, but it is precisely the point at which Einstein's special theory of relativity departs from the physics of Galileo and Newton. When any of the speeds approach the speed of light in vacuum, denoted by c, the velocity-addition equation has to be modified. It turns out that if the woman could walk down the aisle at $0.30c$ and the train could move at $0.90c$, then her speed relative to the ground would be not $1.20c$ but $0.94c$. No material object can travel faster than c; we'll return to the special theory of relativity later in the book.

SUMMARY

Displacement and Velocity ~~Position~~

(Sections 2.1 and 2.2) When an object moves along a straight line, we describe its position with respect to an origin O by means of a coordinate such as x. If the object starts at position x_1 at time t_1 and arrives at position x_2 at time t_2, the object's **displacement** is a vector quantity whose x component is $\Delta x = x_2 - x_1$. The displacement doesn't depend on the details of how the object travels between x_1 and x_2.

The rate of change of position with respect to time is given by the **average velocity,** a vector quantity whose x component is

$$v_{\text{av},x} = \frac{x_2 - x_1}{t_2 - t_1} = \frac{\Delta x}{\Delta t}. \qquad (2.3)$$

On a graph of x versus t, $v_{\text{av},x}$ is the slope of the line connecting the starting point (x_1, t_1) and the ending point (x_2, t_2). As the average velocity is calculated for progressively smaller intervals of time Δt, the average velocity approaches the **instantaneous velocity** v_x: $v_x = \lim\limits_{\Delta t \to 0} (\Delta x / \Delta t)$ (Equation 2.4).

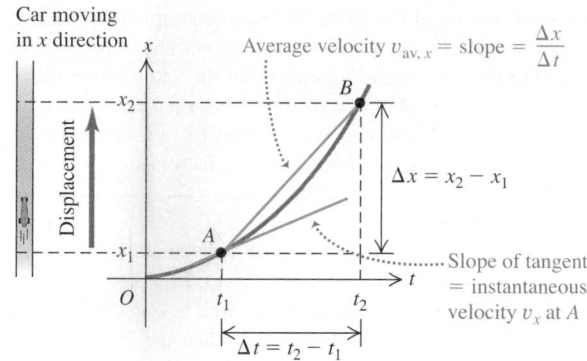

Acceleration

(Section 2.3) When the velocity of an object changes with time, we say that the object has an **acceleration.** Just as velocity describes the rate of change of *position* with time, acceleration is a vector quantity that describes rate of change of *velocity* with time. For an object with velocity v_{1x} at time t_1 and velocity v_{2x} at time t_2, the x component of average acceleration is

$$a_{\text{av},x} = \frac{v_{2x} - v_{1x}}{t_2 - t_1} = \frac{\Delta v}{\Delta t}. \qquad (2.5)$$

The average acceleration (x component) between two points can also be found on a graph of v_x versus t: $a_{\text{av},x}$ is the slope of the line connecting the first point (v_{1x} at time t_1) and the second point (v_{2x} at time t_2). As with velocity, when the average acceleration is calculated for smaller and smaller intervals of time Δt, the average acceleration approaches the **instantaneous acceleration** a_x.

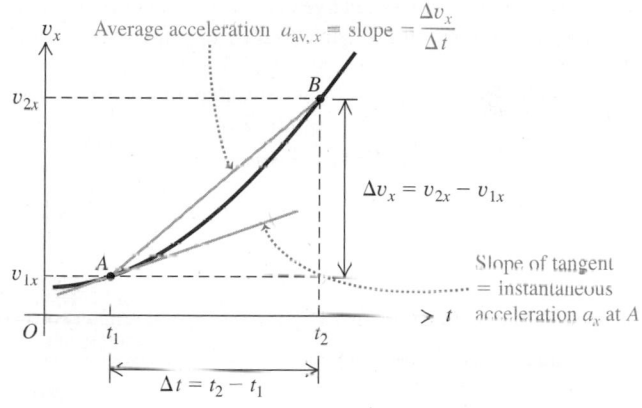

Motion with Constant Acceleration

(Section 2.4) When an object moves with constant acceleration in a straight line, the following two equations describe its position and velocity as functions of time: $x - x_0 = v_{0x}t + \frac{1}{2}a_x t^2$ (Equation 2.12), and $v_x = v_{0x} + a_x t$ (Equation 2.8).

In these equations, x_0 and v_{0x} are, respectively, the position and velocity at an initial time $t = 0$, and x and v_x are, respectively, the position and velocity at any later time t. The final position x is the sum of three terms: the initial position x_0, plus the distance $v_0 t$ that the body would move if its velocity were constant, plus an additional distance $\frac{1}{2}at^2$ caused by the constant acceleration.

The following equation relates velocity, acceleration, and position without explicit reference to time: $v_x^2 = v_{0x}^2 + 2a_x(x - x_0)$ (Equation 2.13).

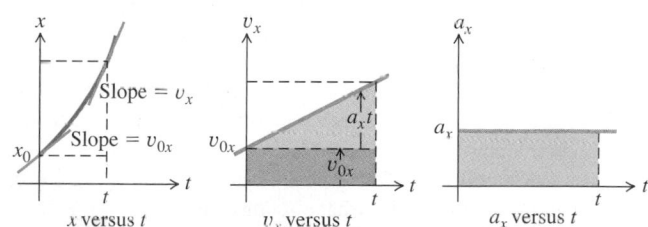

Three graphical views of constant-acceleration motion.

Continued

Proportional Reasoning

(Section 2.5) Many problems have two variables that are related in a simple way. One variable may be a constant times the other, or a constant times the square of the other, or a similar simple relationship. In such problems, a change in the value of one variable is related in a simple way to the corresponding change in the value of the other. This relation can be expressed as a relation between the quotient of two values of one variable and the quotient of two values of the other, even when the constant proportionality factors are not known.

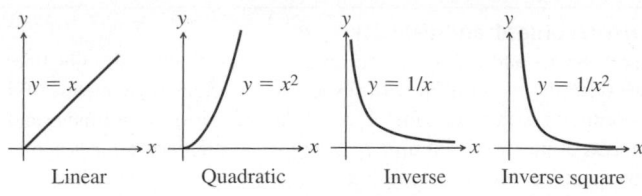

Four mathematical relationships common in physics

Freely Falling Objects

(Section 2.6) A freely falling object is an object that moves under the influence of a constant gravitational force. The term *free fall* includes objects that are initially at rest, as well as objects that have an initial upward or downward velocity. When the effects of air resistance are excluded, all bodies at a particular location fall with the same acceleration, regardless of their size or weight. The constant acceleration of freely falling objects is called the **acceleration due to gravity,** and we denote its magnitude with the letter g. At or near the earth's surface, the value of g is approximately 9.8 m/s^2.

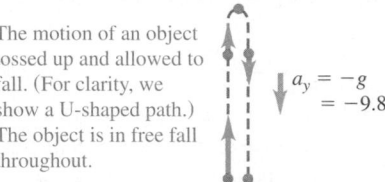

The motion of an object tossed up and allowed to fall. (For clarity, we show a U-shaped path.) The object is in free fall throughout.

Relative Velocity along a Straight Line

(Section 2.7) When an object P moves relative to an object (or reference frame) B, and B moves relative to a second reference frame A, we denote the velocity of P relative to B by $v_{P/B}$, the velocity of P relative to A by $v_{P/A}$, and the velocity of B relative to A by $v_{B/A}$. These velocities are related by this modification of Equation 2.16:

$$v_{P/A} = v_{P/B} + v_{B/A}.$$

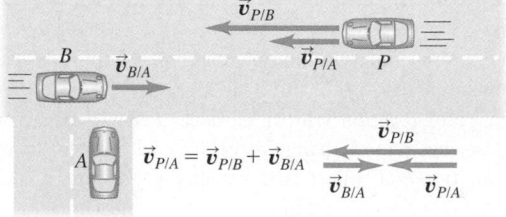

 For instructor-assigned homework, go to www.masteringphysics.com

Conceptual Questions

1. Give an example or two in which the magnitude of the displacement of a moving object is (a) equal to the distance the object travels and (b) less than the distance the object travels. (c) Can the magnitude of the displacement ever be greater than the distance the object travels?
2. Does the speedometer of a car measure speed or velocity?
3. Under what conditions is average velocity equal to instantaneous velocity?
4. If an automobile is traveling north, can it have an acceleration toward the south? Under what circumstances?
5. True or false? (a) If an object's average speed is zero, its average velocity must also be zero. (b) If an object's average velocity is zero, its average speed must also be zero. Explain the reasoning behind your answers. If the statement is false, give several examples to show that it is false.
6. Is it possible for an object to be accelerating even though it has stopped moving? How? Illustrate your answer with a simple example.
7. Can an object with constant acceleration reverse its direction of travel? Can it reverse its direction *twice*? In each case, explain your reasoning.

8. Under constant acceleration the average velocity of a particle is half the sum of its initial and final velocities. Is this still true if the acceleration is *not* constant? Explain your reasoning.
9. If the graph of the velocity of an object as a function of time is a straight line, what can you find out about the acceleration of this object? Consider cases in which the slope is positive, negative, and zero.
10. If the graph of the position of an object as a function of time is a straight line, what can you find out about the velocity of this object? Consider cases in which the slope is positive, negative, and zero.
11. The following table shows an object's position x as a function of time t:

x (in m)	6.50	6.75	7.00	7.25	7.50	7.75
t (in s)	2.00	4.00	6.00	8.00	10.0	12.0

(a) Just by looking at the table, what can you conclude about this object's velocity and acceleration? (b) To see if you are correct, calculate this object's average velocity and acceleration for a few time intervals. (c) How far did the object move between $t = 2.00$ s and $t = 12.0$ s?

12. The following table shows an object's velocity v as a function of time t:

v (in cm/s)	3.40	3.80	4.20	4.60	5.00	5.40
t (in s)	5.00	8.00	11.0	14.0	17.0	20.0

   (a) Just by looking at the table, what can you conclude about this object's acceleration? (b) To see if you are correct, calculate the object's average acceleration for several time intervals.

13. A dripping water faucet steadily releases drops 1.0 s apart. As these drops fall, will the distance between them increase, decrease, or stay the same? Prove your answer.

14. Figure 2.31 shows graphs of the positions of three different moving objects as a function of time. All three graphs pass through points A and B. (a) What can you conclude about the average velocities of these three objects between points A and $B?$ Why? b) At point $B,$ what characteristics of the motion do the three objects have in common? That is, are they in the same place, do they have the same velocity or speed, and do they have the same acceleration?

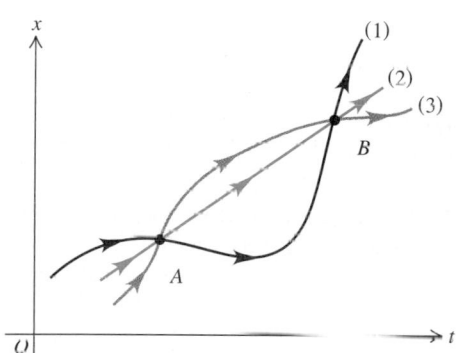

▲ **FIGURE 2.31** Question 14.

15. Figure 2.32 shows graphs of the velocities of three different moving objects as a function of time. All three graphs pass through points A and B. (a) What can you conclude about the average accelerations of these three objects between points A and $B?$ Why? (b) At point $A,$ what characteristics of the motion do the three objects have in common? That is, are they in the same place, do they have the same velocity or speed, and do they have the same acceleration?

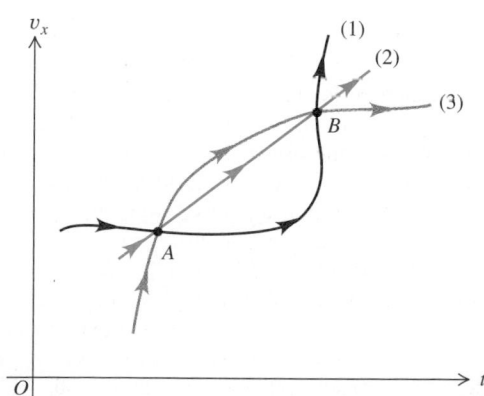

▲ **FIGURE 2.32** Question 15.

16. Figure 2.33 shows the graph of an object's position x as a function of time t. (a) Does this object ever reverse its direction of motion? If so, where? (b) Does the object ever return to its starting point? (c) Is the velocity of the object constant? (d) Is the object's speed ever zero? If so, where? (e) Does the object have any acceleration?

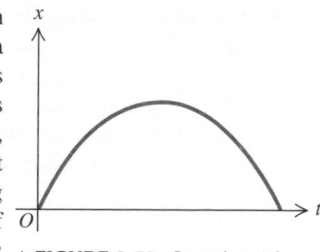

▲ **FIGURE 2.33** Question 16.

17. Figure 2.34 shows the graph of an object's velocity v_x as a function of time t. (a) Does this object ever reverse its direction of motion? b) Does the object ever return to its starting point? (c) Is the object's speed ever zero? If so, where? (d) Does the object have any acceleration?

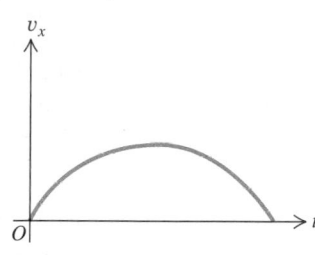

▲ **FIGURE 2.34** Question 17.

18. A ball is dropped from rest from the top of a building of height h. At the same instant, a second ball is projected vertically upward from the ground level, such that it has zero speed when it reaches the top of the building. When the two balls pass each other, which ball has the greater speed, or do they have the same speed? Explain. Where will the two balls be when they are alongside each other: at height $h/2$ above the ground, below this height, or above this height? Explain.

Multiple-Choice Problems

1. Which of the following statements about average speed is correct? (More than one statement may be correct.)
   A. The average speed is equal to the magnitude of the average velocity.
   B. The average speed can never be greater than the magnitude of the average velocity.
   C. The average speed can never be less than the magnitude of the average velocity.
   D. If the average speed is zero, then the average velocity must be zero.
   E. If the average velocity is zero, then the average speed must be zero.

2. A ball is thrown directly upwards with a velocity of $+20$ m/s. At the end of 4 s, its velocity will be closest to
   A. -50 m/s.   B.   30 m/s.   C. -20 m/s   D. -10 m/s.

3. Two objects start at the same place at the same time and move along the same straight line. Figure 2.35 shows the position x as a function of time t for each object. At point $A,$ what must be true about the motion of these objects? (More than one statement may be correct.)
   A. Both have the same speed.

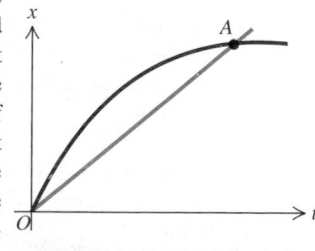

▲ **FIGURE 2.35** Multiple-choice problem 3.

B. Both have the same velocity.

C. Both are at the same position.

D. Both have traveled the same distance.

4. Based *only* on dimensional analysis, which formulas could *not* be correct? In each case, x is position, v is velocity, a is acceleration, and t is time. (More than one choice may be correct.)

   A. $v^2 = v_0^2 + 2at$

   B. $x = v_0 + 1/2\, at^2$

   C. $a = v^2/x$

   D. $v = v_0 + 2at$

5. An object starts from rest and accelerates uniformly. If it moves 2 m during the first second, then, during the first 3 seconds, it will move

   A. 6 m.   B. 9 m.   C. 10 m.   D. 18 m.

6. If a car moving at 80 mph takes 400 ft to stop with uniform acceleration after its brakes are applied, how far will it take to stop under the same conditions if its initial velocity is 40 mph?

   A. 20 ft   B. 50 ft   C. 100 ft   D. 200 ft

7. Figure 2.36 shows the velocity of a jogger as a function of time. What statements must be true about the jogger's motion? (More than one statement may be correct.)

   A. The jogger's speed is increasing.

   B. The jogger's speed is decreasing.

   C. The jogger's acceleration is increasing.

   D. The jogger's acceleration is decreasing.

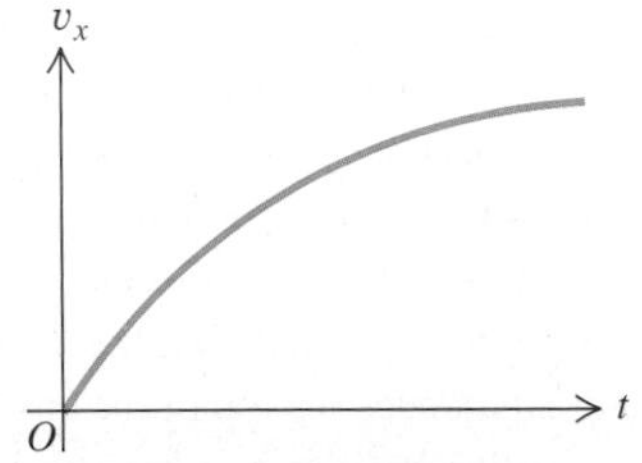

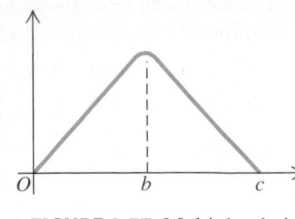

▲ **FIGURE 2.36** Multiple-choice problem 7.

8. A certain airport runway of length L allows planes to accelerate uniformly from rest to takeoff speed using the full length of the runway. Because of newly designed planes, the takeoff speed must be doubled, again using the full length of the runway and having the same acceleration as before. In terms of L, what must be the length of the new runway?

   A. $4L$ B. $2L$ C. $L/2$ D. L^2

9. A ball rolls off a horizontal shelf a height h above the floor and takes 0.5 s to hit. For the ball to take 1.0 s to reach the floor, the shelf's height above the floor would have to be

   A. $2h$ B. $3h$ C. $\sqrt{2}h$ D. $4h$

10. A frog leaps vertically into the air and encounters no appreciable air resistance. Which statement about the frog's motion is correct?

    A. On the way up its acceleration is $9.8\ \mathrm{m/s^2}$ upward, and on the way down its acceleration is $9.8\ \mathrm{m/s^2}$ downward.

    B. On the way up and on the way down its acceleration is $9.8\ \mathrm{m/s^2}$ downward, and at the highest point its acceleration is zero.

    C. On the way up, on the way down, and at the highest point its acceleration is $9.8\ \mathrm{m/s^2}$ downward.

    D. At the highest point, it reverses the direction of its acceleration.

11. You slam on the brakes of your car in a panic and skid a distance X on a straight, level road. If you had been traveling half as fast under the same road conditions, you would have skidded a distance

    A. $\dfrac{X}{4}$.   B. $\dfrac{X}{2}$.   C. X.   D. $\dfrac{X}{\sqrt{2}}$.

12. A cat runs in a straight line. Figure 2.37 shows a graph of the cat's position as a function of time. Which of the following statements about the cat's motion must be true? (There may be more than one correct choice.)

    A. At point c, the cat has returned to the place where it started.

    B. The cat's speed is zero at points a and c.

    C. The cat's velocity is zero at point b.

    D. The cat reverses the direction of its velocity at point b.

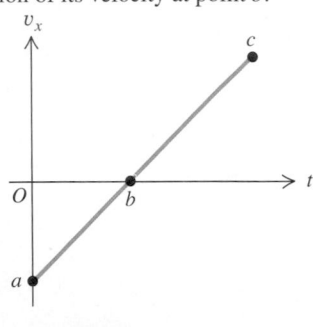

▲ **FIGURE 2.37** Multiple-choice problem 12.

13. A wildebeest is running in a straight line, which we shall call the x axis, with the positive direction to the right. Figure 2.38 shows this animal's velocity as a function of time. Which of the following statements about the animal's motion must be true? (There may be more than one correct choice.)

    A. Its acceleration is increasing.

    B. Its speed is decreasing from a to b and increasing from b to c.

    C. It is moving to the right between a and c.

    D. It is moving to the left between a and b and to the right between b and c.

▲ **FIGURE 2.38** Multiple-choice problem 13.

14. A brick falling freely from a helicopter drops 40 meters during a certain 1-second time interval. The distance it will fall in the *next* second is

    A. 45 m   B. 50 m   C. 56 m   D. 80 m

15. A bullet is dropped into a river from a very high bridge. At the same time, another bullet is fired from a gun straight down towards the water. If air resistance is negligible, the acceleration of the bullets just before they strike the water

    A. is greater for the dropped bullet.

    B. is greater for the fired bullet.

    C. is the same for both bullets.

    D. depends on how high they started.

Problems

2.1 Displacement and Average Velocity

1. • An ant is crawling along a straight wire, which we shall call the x axis, from A to B to C to D (which overlaps A), as shown in Figure 2.39. O is the origin. Suppose you take measurements and find that AB is 50 cm, BC is 30 cm, and AO is 5 cm. (a) What is the ant's position at points A, B, C, and D? (b) Find the displacement of the ant and the distance it has moved over each of the following intervals: (i) from A to B, (ii) from B to C, (iii) from C to D, and (iv) from A to D.

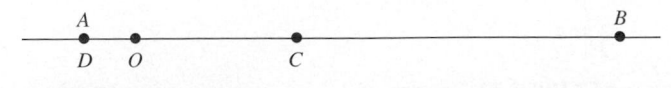

▲ **FIGURE 2.39** Problem 1.

2. •• A person is walking briskly in a straight line, which we shall call the *x* axis. Figure 2.40 shows a graph of the person's position *x* along this axis as a function of time *t*. (a) What is the person's displacement during each of the following time intervals: (i) between *t* = 1.0 s and *t* = 10.0 s, (ii) between *t* = 3.0 s and *t* = 10.0 s, (iii) between *t* = 2.0 s and *t* = 3.0 s, and (iv) between *t* = 2.0 s and *t* = 4.0 s? (b) What distance did the person move from (i) *t* = 0 s to *t* = 4.0 s, (ii) *t* = 2.0 s to *t* = 4.0 s, and (iii) *t* = 8.0 s to *t* = 10.0 s?

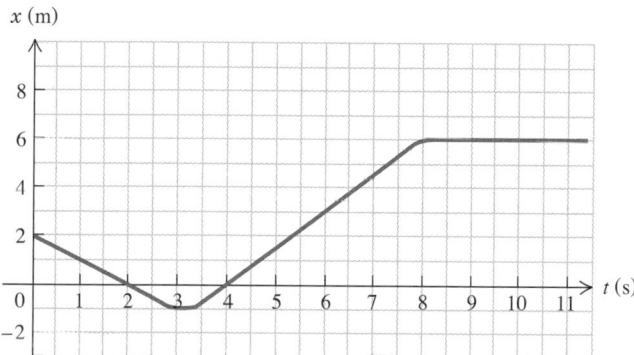

▲ **FIGURE 2.40** Problem 2.

3. • A dog runs from points *A* to *B* to *C* in 3.0 s. (See Figure 2.41.) Find the dog's average velocity and average speed over this 3-second interval.

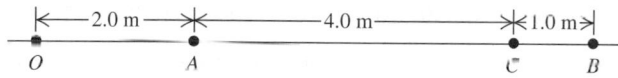

▲ **FIGURE 2.41** Problem 3.

4. • In an experiment, a shearwater (a seabird) was taken from its
BIO nest, flown 5150 km away, and released. The bird found its way back to its nest 13.5 days after release. If we place the origin in the nest and extend the +*x*-axis to the release point, what was the bird's average velocity in m/s (a) for the return flight, and (b) for the whole episode, from leaving the nest to returning?

5. •• Figure 2.42 shows the position of a moving object as a function of time. (a) Find the average velocity of this object from points *A* to *B*, *B* to *C*, and *A* to *C*. (b) For the intervals in part (a), would the average speed be less than, equal to, or greater than the values you found in that part? Explain your reasoning.

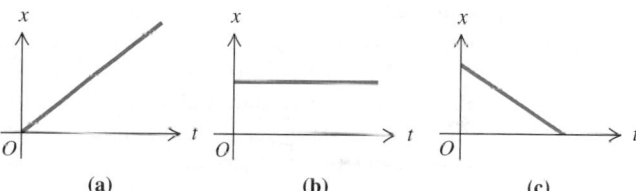

▲ **FIGURE 2.42** Problem 5.

6. •• An object moves along the *x* axis. Figure 2.43 shows a graph of its position *x* as a function of time. (a) Find the average velocity of the object from points *A* to *B*, *B* to *C*, and *A* to *C*. (b) For the intervals in part (a), would the average speed be less than, ▲ **FIGURE 2.43** Problem 6.

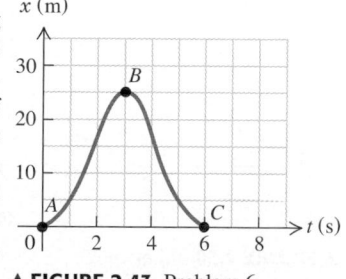

equal to, or greater than the values you found in that part? Explain your reasoning.

7. •• A boulder starting from rest rolls down a hill in a straight line, which we shall call the *x* axis. A graph of its position *x* as a function of time *t* is shown in Figure 2.44. Find (a) the distance the boulder rolled between the end of ▲ **FIGURE 2.44** Problem 7.
the first second and the end of the third second and (b) the boulder's average speed during (i) the first second, (ii) the second second, (iii) the third second, (iv) the fourth second, and (v) the first 4 seconds.

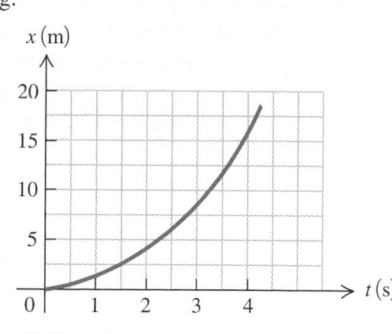

8. •• Each graph in Figure 2.45 shows the position of a running cat, called Mousie, as a function of time. In each case, sketch a clear *qualitative* (no numbers) graph of Mousie's velocity as a function of time.

▲ **FIGURE 2.45** Problem 8.

9. • In 1954, Roger Bannister became the first human to run a mile in less than 4 minutes. Suppose that a runner on a straight track covers a distance of 1.00 mi in exactly 4.00 min. What is his average speed in (a) mi/h, (b) ft/s, and (c) m/s?

10. • **Hypersonic scramjet.** On March 27, 2004, the United States successfully tested the hypersonic X-43A scramjet, which flew at Mach 7 (seven times the speed of sound) for 11 seconds. (A scramjet gets its oxygen directly from the air, rather than from fuel.) (a) At this rate, how many minutes would it take such a scramjet to carry passengers the approximately 5000 km from San Francisco to New York? (Use 331 m/s for the speed of sound in air.) (b) How many kilometers did the scramjet travel during its 11 second test?

11. • **Plate tectonics.** The earth's crust is broken up into a series of more-or-less rigid plates that slide around due to motion of material in the mantle below. Although the speeds of these plates vary somewhat, they are typically about 5 cm/yr. Assume that this rate remains constant over time. (a) If you and your neighbor live on opposite sides of a plate boundary at which one plate is moving northward at 5.0 cm/yr with respect to the other plate, how far apart do your houses move in a century? (b) Los Angeles is presently 550 km south of San Francisco, but is on a plate moving northward relative to San Francisco. If the 5.0 cm/yr velocity continues, how many years will it take before Los Angeles has moved up to San Francisco?

12. • A runner covers one lap of a circular track 40.0 m in diameter in 62.5 s. (a) For that lap, what were her average speed and average velocity? (b) If she covered the first half-lap in 28.7 s, what were her average speed and average velocity for that half-lap?

13. • Sound travels at a speed of about 344 m/s in air. You see a distant flash of lightning and hear the thunder arrive 7.5 seconds later. How many miles away was the lightning strike? (Assume the light takes essentially no time to reach you.)

14. • **Ouch!** Nerve impulses travel at different speeds, depending **BIO** on the type of fiber through which they move. The impulses for touch travel at 76.2 m/s, while those registering pain move at 0.610 m/s. If a person stubs his toe, find (a) the time for each type of impulse to reach his brain, and (b) the time *delay* between the pain and touch impulses. Assume that his brain is 1.85 m from his toe and that the impulses travel directly from toe to brain.

15. •• While driving on the freeway at 110 km/h, you pass a truck whose total length you estimate at 25 m. (a) If it takes you, in the driver's seat, 5.5 s to pass from the rear of the truck to its front, what is the truck's speed relative to the road? (b) How far does the truck travel while you're passing it?

16. •• A mouse travels along a straight line; its distance x from the origin at any time t is given by the equation $x = (8.5 \text{ cm} \cdot \text{s}^{-1})t - (2.5 \text{ cm} \cdot \text{s}^{-2})t^2$. Find the average velocity of the mouse in the interval from $t = 0$ to $t = 1.0$ s and in the interval from $t = 0$ to $t = 4.0$ s.

17. •• **The freeway blues!** When you normally drive the freeway between Sacramento and San Francisco at an average speed of 105 km/hr (65 mph), the trip takes 1.0 hr and 20 min. On a Friday afternoon, however, heavy traffic slows you down to an average of 70 km/hr (43 mph) for the same distance. How much longer does the trip take on Friday than on the other days?

18. •• Two runners start simultaneously at opposite ends of a 200.0 m track and run toward each other. Runner A runs at a steady 8.0 m/s and runner B runs at a constant 7.0 m/s. When and where will these runners meet?

2.2 Instantaneous Velocity

19. •• A physics professor leaves her house and walks along the sidewalk toward campus. After 5 min, she realizes that it is raining and returns home. The distance from her house as a function of time is shown in Figure 2.46. At which of the labeled points is her velocity (a) zero? (b) constant and positive? (c) constant and negative? (d) increasing in magnitude? and (e) decreasing in magnitude?

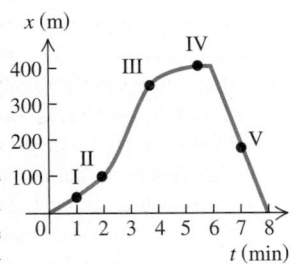
▲ **FIGURE 2.46** Problem 19.

20. •• A test car travels in a straight line along the x axis. The graph in Figure 2.47 shows the car's position x as a function of time. Find its instantaneous velocity at points A through G.

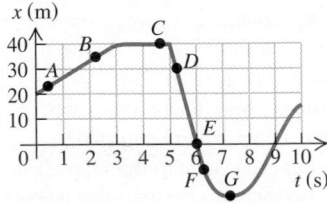
▲ **FIGURE 2.47** Problem 20.

21. •• Figure 2.48 shows the position x of a crawling spider as a function of time. Use this graph to draw a *numerical* graph of the spider's velocity as a function of time over the same time interval.

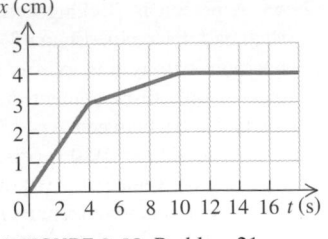
▲ **FIGURE 2.48** Problem 21.

2.3 Average and Instantaneous Acceleration

22. •• The graph in Figure 2.49 shows the velocity of a motorcycle police officer plotted as a function of time. Find the instantaneous acceleration at times $t = 3$ s, at $t = 7$ s, and at $t = 11$ s.

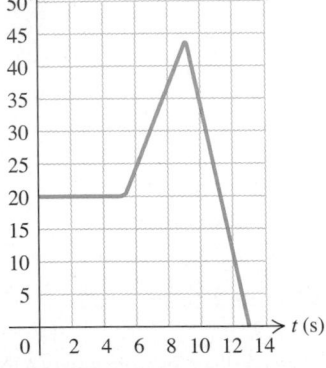
▲ **FIGURE 2.49** Problem 22.

23. •• A test driver at Incredible Motors, Inc., is testing a new model car having a speedometer calibrated to read m/s rather than mi/h. The following series of speedometer readings was obtained during a test run:

Time (s)	0	2	4	6	8	10	12	14	16
Velocity (m/s)	0	0	2	5	10	15	20	22	22

(a) Compute the average acceleration during each 2 s interval. Is the acceleration constant? Is it constant during any part of the test run? (b) Make a velocity–time graph of the data shown, using scales of 1 cm = 1 s horizontally and 1 cm = 2 m/s vertically. Draw a smooth curve through the plotted points. By measuring the slope of your curve, find the magnitude of the instantaneous acceleration at times $t = 9$ s, 13 s, and 15 s.

24. • (a) The pilot of a jet fighter will black out at an acceleration greater than approximately $5g$ if it lasts for more than a few seconds. Express this acceleration in m/s² and ft/s². (b) The acceleration of the passenger during a car crash with an air bag is about $60g$ for a very short time. What is this acceleration in m/s² and ft/s²? (c) The acceleration of a falling body on our moon is 1.67 m/s². How many g's is this? (d) If the acceleration of a test plane is 24.3 m/s², how many g's is it?

25. •• For each graph of velocity as a function of time in Figure 2.50, sketch a qualitative graph of the acceleration as a function of time.

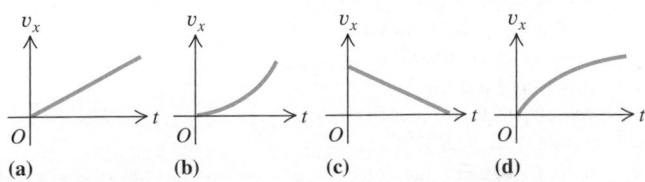
▲ **FIGURE 2.50** Problem 25.

26. •• A little cat, Bella, walks along a straight line, which we shall call the x axis, with the positive direction to the right. As an observant scientist, you make measurements of her motion and construct a graph of the little feline's velocity as a function of time. (See Figure 2.51.) (a) Find Bella's velocity at $t = 4.0$ s and at $t = 7.0$ s. (b) What is her acceleration at $t = 3.0$ s? at $t = 6.0$ s? at $t = 7.0$ s? (c) Sketch a clear graph of Bella's acceleration as a function of time.

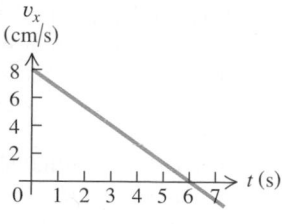

▲ **FIGURE 2.51** Problem 26.

2.4 Motion with Constant Acceleration

27. • A car driving on the turnpike accelerates uniformly in a straight line from 88 ft/s (60 mph) to 110 ft/s (75 mph) in 3.50 s. (a) What is the car's acceleration? (b) How far does the car travel while it accelerates?

28. • **Animal motion.** Cheetahs, the fastest of the great cats, can
BIO reach 45 mph in 2.0 s starting from rest. Assuming that they have constant acceleration throughout that time, find (a) their acceleration (in ft/s² and m/s²) and (b) the distance (in m and ft) they travel during that time.

29. •• A cat drops from a shelf 4.0 ft above the floor and lands on
BIO all four feet. His legs bring him to a stop in a distance of 12 cm. Calculate (a) his speed when he first touches the floor (ignore air resistance), (b) how long it takes him to stop, and (c) his acceleration (assumed constant) while he is stopping, in m/s² and g's.

30. •• **Blackout?** A jet fighter pilot wishes to accelerate from rest
BIO at $5g$ to reach Mach 3 (three times the speed of sound) as quickly as possible. Experimental tests reveal that he will black out if this acceleration lasts for more than 5.0 s. Use 331 m/s for the speed of sound. (a) Will the period of acceleration last long enough to cause him to black out? (b) What is the greatest speed he can reach with an acceleration of $5g$ before blacking out?

31. •• **A fast pitch.** The fastest measured pitched baseball left
BIO the pitcher's hand at a speed of 45.0 m/s. If the pitcher was in contact with the ball over a distance of 1.50 m and produced constant acceleration, (a) what acceleration did he give the ball, and (b) how much time did it take him to pitch it?

32. • If a pilot accelerates at more than $4g$, he begins to "gray
BIO out," but not completely lose consciousness. (a) What is the shortest time that a jet pilot starting from rest can take to reach Mach 4 (four times the speed of sound) without graying out? (b) How far would the plane travel during this period of acceleration? (Use 331 m/s for the speed of sound in cold air.)

33. • **Air-bag injuries.** During an auto accident, the vehicle's air
BIO bags deploy and slow down the passengers more gently than if they had hit the windshield or steering wheel. According to safety standards, the bags produce a maximum acceleration of $60\ g$, but lasting for only 36 ms (or less). How far (in meters) does a person travel in coming to a complete stop in 36 ms at a constant acceleration of $60\ g$?

34. •• Starting from rest, a boulder rolls down a hill with constant acceleration and travels 2.00 m during the first second. (a) How far does it travel during the second second? (b) How fast is it moving at the end of the first second? at the end of the second second?

35. • **Faster than a speeding bullet!** The Beretta Model 92S (the standard-issue U.S. army pistol) has a barrel 127 mm long. The bullets leave this barrel with a muzzle velocity of 335 m/s. (a) What is the acceleration of the bullet while it is in the barrel, assuming it to be constant? Express your answer in m/s² and in g's. (b) For how long is the bullet in the barrel?

36. •• An airplane travels 280 m down the runway before taking off. Assuming that it has constant acceleration, if it starts from rest and becomes airborne in 8.00 s, how fast (in m/s) is it moving at takeoff?

37. **Entering the freeway.** A car sits in an entrance ramp to a freeway, waiting for a break in the traffic. The driver accelerates with constant acceleration along the ramp and onto the freeway. The car starts from rest, moves in a straight line, and has a speed of 20 m/s (45 mi/h) when it reaches the end of the 120-m-long ramp. (a) What is the acceleration of the car? (b) How much time does it take the car to travel the length of the ramp? (c) The traffic on the freeway is moving at a constant speed of 20 m/s. What distance does the traffic travel while the car is moving the length of the ramp?

38. •• The "reaction time" of the average automobile driver is about 0.7 s. (The reaction time is the interval between the perception of a signal to stop and the application of the brakes.) If an automobile can slow down with an acceleration of 12.0 ft/s², compute the total distance covered in coming to a stop after a signal is observed (a) from an initial velocity of 15.0 mi/h (in a school zone) and (b) from an initial velocity of 55.0 mi/h.

39. •• According to recent typical test data, a Ford Focus travels 0.250 mi in 19.9 s, starting from rest. The same car, when braking from 60.0 mph on dry pavement, stops in 146 ft. Assume constant acceleration in each part of its motion, but not necessarily the same acceleration when slowing down as when speeding up. (a) Find this car's acceleration while braking and while speeding up. (b) If its acceleration is constant while speeding up, how fast (in mph) will the car be traveling after 0.250 mi of acceleration? (c) How long does it take the car to stop while braking from 60.0 mph?

40. •• A subway train starts from rest at a station and accelerates at a rate of 1.60 m/s² for 14.0 s. It runs at constant speed for 70.0 s and slows down at a rate of 3.50 m/s² until it stops at the next station. Find the *total* distance covered.

2.5 Proportional Reasoning

41. •• If the radius of a circle of area A and circumference C is doubled, find the new area and circumference of the circle in terms of A and C. (Consult Chapter 0 if necessary.)

42. •• In the redesign of a machine, a metal cubical part has each of its dimensions tripled. By what factor do its surface area and volume change?

43. • You have two cylindrical tanks. The tank with the greater volume is 1.20 times the height of the smaller tank. It takes 218 gallons of water to fill the larger tank and 150 gallons to

fill the other. What is the ratio of the radius of the larger tank to the radius of the smaller one?

44. • A speedy basketball point guard is 5 ft 10 inches tall; the
BIO center on the same team is 7 ft 2 inches tall. Assuming their bodies are similarly proportioned, if the point guard weighs 175 lb, what would you expect the center to weigh?

45. •• Two rockets having the same acceleration start from rest, but rocket *A* travels for twice as much time as rocket *B*. (a) If rocket *A* goes a distance of 250 km, how far will rocket *B* go? (b) If rocket *A* reaches a speed of 350 m/s, what speed will rocket *B* reach?

46. •• Two cars having equal speeds hit their brakes at the same time, but car *A* has three times the acceleration as car *B*. (a) If car *A* travels a distance *D* before stopping, how far (in terms of *D*) will car *B* go before stopping? (b) If car *B* stops in time *T*, how long (in terms of *T*) will it take for car *A* to stop?

47. •• Airplane *A*, starting from rest with constant acceleration, requires a runway 500 m long to become airborne. Airplane *B* requires a takeoff speed twice as great as that of airplane *A*, but has the same acceleration, and both planes start from rest. (a) How long must the runway be for airplane *B*? (b) If airplane *A* takes time *T* to travel the length of its runway, how long (in terms of *T*) will airplane *B* take to travel the length of its runway?

2.6 Freely Falling Bodies

48. • (a) If a flea can jump straight up to a height of 22.0 cm, what is its initial speed (in m/s) as it leaves the ground, neglecting air resistance? (b) How long is it in the air? (c) What are the magnitude and direction of its acceleration while it is (i) moving upward? (ii) moving downward? (iii) at the highest point?

49. • A brick is released with no initial speed from the roof of a building and strikes the ground in 2.50 s, encountering no appreciable air drag. (a) How tall, in meters, is the building? (b) How fast is the brick moving just before it reaches the ground? (c) Sketch graphs of this falling brick's acceleration, velocity, and vertical position as functions of time.

50. • **Loss of power!** In December of 1989, a KLM Boeing 747 airplane carrying 231 passengers entered a cloud of ejecta from an Alaskan volcanic eruption. All four engines went out, and the plane fell from 27,900 ft to 13,300 ft before the engines could be restarted. It then landed safely in Anchorage. Neglecting any air resistance and aerodynamic lift, and assuming that the plane had no vertical motion when it lost power, (a) for how long did it fall before the engines were restarted, and (b) how fast was it falling at that instant? c) In reality, why would the plane not be falling nearly as fast?

51. •• A tennis ball on Mars, where the acceleration due to gravity is 0.379g and air resistance is negligible, is hit directly upward and returns to the same level 8.5 s later. (a) How high above its original point did the ball go? (b) How fast was it moving just after being hit? (c) Sketch clear graphs of the ball's vertical position, vertical velocity, and vertical acceleration as functions of time while it's in the Martian air.

52. • **Measuring g.** One way to measure g on another planet or moon by remote sensing is to measure how long it takes an object to fall a given distance. A lander vehicle on a distant planet records the fact that it takes 3.17 s for a ball to fall freely 11.26 m, starting from rest. (a) What is the acceleration due to gravity on that planet? Express your answer in m/s² and in earth g's. b) How fast is the ball moving just as it lands?

53. •• **That's a lot of hot air!** A hot-air balloonist, rising vertically with a constant speed of 5.00 m/s, releases a sandbag at the instant the balloon is 40.0 m above the ground. (See Figure 2.52.) After it is released, the sandbag encounters no appreciable air drag. (a) Compute the position and velocity of the sandbag at 0.250 s and 1.00 s after its release. (b) How many seconds after its release will the bag strike the ground? (c) How fast is it moving as it strikes the ground? (d) What is the greatest height above the ground that the sandbag reaches? (e) Sketch graphs of this bag's acceleration, velocity, and vertical position as functions of time.

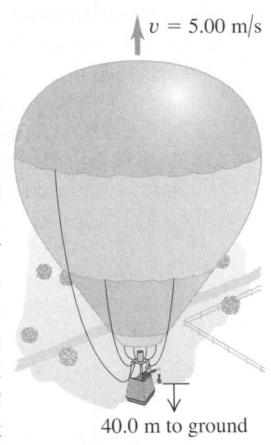

$v = 5.00$ m/s

40.0 m to ground

▲ **FIGURE 2.52**
Problem 53.

54. **Look out below.** Sam heaves a 16-lb shot straight upward, giving it a constant upward acceleration from rest of 45.0 m/s² for 64.0 cm. He releases it 2.20 m above the ground. You may ignore air resistance. (a) What is the speed of the shot when Sam releases it? (b) How high above the ground does it go? (c) How much time does he have to get out of its way before it returns to the height of the top of his head, 1.83 m above the ground?

55. •• **Astronauts on the moon.** Astronauts on our moon must function with an acceleration due to gravity of 0.170g. (a) If an astronaut can throw a certain wrench 12.0 m vertically upward on earth, how high could he throw it on our moon if he gives it the same starting speed in both places? (b) How much *longer* would it be in motion (going up and coming down) on the moon than on earth?

56. • A student throws a water balloon vertically downward from the top of a building. The balloon leaves the thrower's hand with a speed of 15.0 m/s. (a) What is its speed after falling freely for 2.00 s? (b) How far does it fall in 2.00 s? (c) What is the magnitude of its velocity after falling 10.0 m?

57. • A rock is thrown vertically upward with a speed of 12.0 m/s from the roof of a building that is 60.0 m above the ground. (a) In how many seconds after being thrown does the rock strike the ground? (b) What is the speed of the rock just before it strikes the ground? Assume free fall.

58. •• **Physiological effects of large acceleration.** The rocket-
BIO driven sled *Sonic Wind No. 2*, used for investigating the physiological effects of large accelerations, runs on a straight, level track that is 1080 m long. Starting from rest, it can reach a speed of 1610 km/h in 1.80 s. (a) Compute the acceleration in m/s² and in g's. (b) What is the distance covered in 1.80 s? (c) A magazine article states that, at the end of a certain run, the speed of the sled decreased from 1020 km/h to zero in 1.40 s and that, during this time, its passenger was subjected to more than 40g. Are these figures consistent?

59. •• Two stones are thrown vertically upward from the ground, one with three times the initial speed of the other. (a) If the faster stone takes 10 s to return to the ground, how long will it take the slower stone to return? (b) If the slower stone reaches a maximum height of *H,* how high (in terms of *H*) will the faster stone go? Assume free fall.

60. •• Two coconuts fall freely from rest at the same time, one from a tree twice as high as the other. (a) If the coconut from the taller tree reaches the ground with a speed *V,* what will be the speed (in terms of *V*) of the coconut from the other tree when it reaches the ground? (b) If the coconut from the shorter tree takes time *T* to reach the ground, how long (in terms of *T*) will it take the other coconut to reach the ground?

*2.7 Relative Velocity along a Straight Line

61. • A Toyota Prius driving north at 65 mph and a VW Passat driving south at 42 mph are on the same road heading toward each other (but in different lanes). What is the velocity of each car relative to the other (a) when they are 250 ft apart, just before they meet, and (b) when they are 525 ft apart, after they have passed each other?

62. • The wind is blowing from west to east at 35 mph, and an eagle in that wind is flying at 22 mph relative to the air. What is the velocity of this eagle relative to a person standing on the ground if the eagle is flying (a) from west to east relative to the air and (b) from east to west relative to the air?

63. •• A helicopter 8.50 m above the ground and descending at 3.50 m/s drops a package from rest (relative to the helicopter). Just as it hits the ground, find (a) the velocity of the package relative to the helicopter and (b) the velocity of the helicopter relative to the package. The package falls freely.

64. •• A jetliner has a cruising air speed of 600 mph relative to the air. How long does it take this plane to fly round-trip from San Francisco to Chicago, an east–west flight of 2,000 mi each way, (a) if there is no wind blowing and (b) if the wind is blowing at 150 mph from the west to the east?

General Problems

65. • **Life from space?** One rather controversial suggestion for
BIO the origin of life on the earth is that the seeds of life arrived here from outer space in tiny spores—a theory called *panspermia.* Of course, any life in such spores would have to survive a long journey through the frigid near vacuum of outer space. If these spores originated at the distance of our nearest star, Alpha Centauri, and traveled at a constant 1000 km/s, how many years would it take for them to make the journey here? (Alpha Centauri is approximately $4\frac{1}{4}$ light years away, and the light year is the distance light travels in 1 year.)

66. •• At the instant the traffic light turns green, an automobile that has been waiting at an intersection starts ahead with a constant acceleration of 2.50 m/s². At the same instant, a truck, traveling with a constant speed of 15.0 m/s, overtakes and passes the automobile. (a) How far beyond its starting point does the automobile overtake the truck? (b) How fast is the automobile traveling when it overtakes the truck?

67. •• On a 20-mile bike ride, you ride the first 10 miles at an average speed of 8 mi/h. What must your average speed over the next 10 miles be to have your average speed for the total 20 miles be (a) 4 mi/h? (b) 12 mi/h? (c) Given this average speed for the first 10 miles, can you possibly attain an average speed of 16 mi/h for the total 20-mile ride? Explain.

68. • You and a friend start out at the same time on a 10-km run. Your friend runs at a steady 2.5 m/s. How fast do you have to run if you want to finish the run 15 minutes before your friend?

69. •• Two rocks are thrown directly upward with the same initial speeds, one on earth and one on our moon, where the acceleration due to gravity is one-sixth what it is on earth. (a) If the rock on the moon rises to a height *H,* how high, in terms of *H,* will the rock rise on the earth? (b) If the earth rock takes 4.0 s to reach its highest point, how long will it take the moon rock to do so?

70. • **Prevention of hip fractures.** Falls resulting in hip fractures
BIO are a major cause of injury and even death to the elderly. Typically, the hip's speed at impact is about 2.0 m/s. If this can be reduced to 1.3 m/s or less, the hip will usually not fracture. One way to do this is by wearing elastic hip pads. (a) If a typical pad is 5.0 cm thick and compresses by 2.0 cm during the impact of a fall, what acceleration (in m/s² and in *g*'s) does the hip undergo to reduce its speed to 1.3 m/s? (b) The acceleration you found in part (a) may seem like a rather large acceleration, but to fully assess its effects on the hip, calculate how long it lasts.

71. • **Are we Martians?** It has been suggested, and not face-
BIO tiously, that life might have originated on Mars and been carried to Earth when a meteor hit Mars and blasted pieces of rock (perhaps containing primitive life) free of the surface. Astronomers know that many Martian rocks have come to Earth this way. (For information on one of these, search the Internet for "ALH 84001".) One objection to this idea is that microbes would have to undergo an enormous, lethal acceleration during the impact. Let us investigate how large such an acceleration might be. To escape Mars, rock fragments would have to reach its escape velocity of 5.0 km/s, and this would most likely happen over a distance of about 4.0 m during the impact. (a) What would be the acceleration, in m/s² and *g*'s, of such a rock fragment? (b) How long would this acceleration last? (c) In tests, scientists have found that over 40% of *Bacillus subtilis* bacteria survived after an acceleration of 450,000*g*. In light of your answer to part (a), can we rule out the hypothesis that life might have been blasted from Mars to Earth?

72. •• **Raindrops.** If the effects of the air acting on falling raindrops are ignored, then we can treat raindrops as freely falling objects. (a) Rain clouds are typically a few hundred meters above the ground. Estimate the speed with which raindrops would strike the ground if they were freely falling objects. Give your estimate in m/s, km/h, and mi/h. (b) Estimate (from your own personal observations of rain) the speed with which raindrops actually strike the ground. (c) Based on your answers to parts (a) and (b), is it a good approximation to neglect the effects of the air on falling raindrops? Explain.

73. •• **Egg drop.** You are on the roof of the physics building of your school, 46.0 m above the ground. (See Figure 2.53.) Your physics professor, who is 1.80 m tall, is walking alongside the building at a constant speed of 1.20 m/s. If you wish to drop

an egg on your professor's head, where should the professor be when you release the egg, assuming that the egg encounters no appreciable air drag.

74. •• A 0.525 kg ball starts from rest and rolls down a hill with uniform acceleration, traveling 150 m during the second 10.0 s of its motion. How far did it roll during the first 5.0 s of motion?

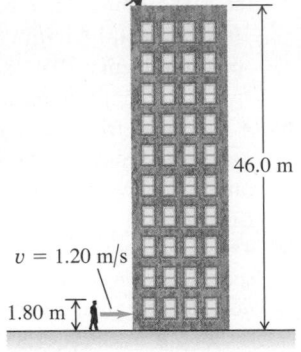

▲ FIGURE 2.53 Problem 73.

75. •• A large boulder is ejected vertically upward from a volcano with an initial speed of 40.0 m/s. Air resistance may be ignored. (a) At what time after being ejected is the boulder moving at 20.0 m/s upward? (b) At what time is it moving at 20.0 m/s downward? (c) When is the displacement of the boulder from its initial position zero? (d) When is the velocity of the boulder zero? (e) What are the magnitude and direction of the acceleration while the boulder is (i) moving upward? (ii) Moving downward? (iii) At the highest point? (f) Sketch a_y-t, v_y-t, and y-t graphs for the motion.

76. •• **Freeway ramps.** In designing a freeway on-ramp, you must make it long enough for cars to be able to accelerate and merge safely with traffic traveling at speeds of 70 mph. For an off-ramp, the cars must be able to come to a stop from 70 mph over the length of the ramp. Your traffic engineers provide the following data: Powerful cars can accelerate from rest to 60 mph in 5.0 s, while less powerful cars take twice as long to do this. When slowing down from 60 mph, a car with good tire treads will take 12.0 s to stop, whereas one with bald tires takes 20.0 s. In order to safely accommodate all the types of vehicles, what should be the minimum length (in meters) of on-ramps and off-ramps? Assume constant acceleration.

77. • A healthy heart pumping at a rate of 72 beats per minute
BIO increases the speed of blood flow from 0 to 425 cm/s with each beat. Calculate the acceleration of the blood during this process.

78. •• You decide to take high-speed strobe light photos of your little dog Holly as she runs along. Figure 2.54 shows some of these photos. The strobe flashes at a uniform rate, which means that the time interval between adjacent images is the same in all the photos. For each case, sketch clear *qualitative* (no numbers) graphs of Holly's position as a function of time and her velocity as a function of time.

▲ FIGURE 2.54 Problem 78.

79. •• A rocket blasts off vertically from rest on the launch pad with a constant upward acceleration of 2.50 m/s². At 20.0 s after blastoff, the engines suddenly fail, and the rocket begins free fall. (a) How high above the launch pad will the rocket eventually go? (b) Find the rocket's velocity and acceleration at its highest point. (c) How long after it was launched will the rocket fall back to the launch pad, and how fast will it be moving when it does so?

80. •• **Mouse at play, I.** A little mouse runs in a straight line, which we shall call the x axis. As an eager scientist, you record its position as a function of time and use the information you obtain to construct the graph plotted in Figure 2.55, which shows the mouse's position x as a function of time t. You have adopted the usual sign convention according to which quantities to the right are positive and those to the left are negative. Use the information on the graph to answer the following questions: (a) When is the mouse to the right of the origin? to the left of the origin? at the origin? (b) What are the magnitude and direction of the mouse's initial velocity? (c) Does the mouse move with constant acceleration throughout the 10 seconds? (d) What is the mouse's maximum speed? In which direction is it moving with that speed? (e) When is the mouse moving to the right? moving to the left? not moving? (f) Through what total distance has the mouse moved during the first 3 s? during the first 10 s? (g) When is the mouse speeding up? slowing down? (h) Do the kinematics formulas $v_x = v_{0x} + a_x t$ and $x = x_0 + v_{0x}t + \frac{1}{2}a_x t^2$ apply to this mouse's motion throughout the 10 s? Why or why not?

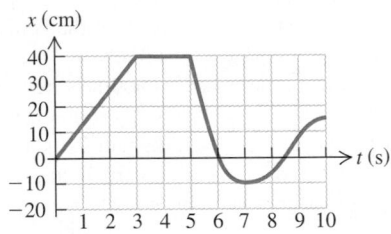

▲ FIGURE 2.55 Problem 80.

81. •• **Earthquake waves.** Earthquakes produce several types of shock waves. The best known are the P-waves (P for *primary* or *pressure*) and the S-waves (S for *secondary* or *shear*). In the earth's crust, P-waves travel at around 6.5 km/s while S-waves move at about 3.5 km/s. (The actual speeds vary with the type of material the waves are going through.) The time delay between the arrival of these two types of waves at a seismic recording station tells geologists how far away the earthquake that produced the waves occurred. (a) If the time delay at a seismic station is 33 s, how far from that station did the earthquake occur? (b) One form of earthquake warning system detects the faster (but less damaging) P-waves and sounds an alarm when they first arrive, giving people a short time to seek cover before the more dangerous S-waves arrive. If an earthquake occurs 375 km away from such a warning device, how much time would people have to take cover between the alarm and the arrival of the S-waves?

82. • An elite human sprinter reaches his top speed of 11.8 m/s at
BIO a time of 7.02 s after the starting gun. In the first 1.40 s, however, he reaches a speed of 8.00 m/s, with a nearly constant

acceleration. Calculate (a) his maximum acceleration during the starting phase and (b) his average acceleration to top speed. (c) Assuming constant acceleration for the first 1.40 s, how far does he travel during that time?

83. •• **How high is the cliff?** Suppose you are climbing in the High Sierra when you suddenly find yourself at the edge of a fog-shrouded cliff. To find the height of this cliff, you drop a rock from the top and, 10.0 s later, hear the sound of it hitting the ground at the foot of the cliff. (a) Ignoring air resistance, how high is the cliff if the speed of sound is 330 m/s? (b) Suppose you had ignored the time it takes the sound to reach you. In that case, would you have overestimated or underestimated the height of the cliff? Explain your reasoning.

Passage Problems

BIO Blood flow in the heart. The human circulatory system is closed; that is, the blood that is pumped out of the left ventricle of the heart into the arteries is constrained to a series of continuous, branching vessels as it passes through the capillaries and then into the veins as it returns to the heart. The blood in each of the four chambers of the heart comes briefly to rest after entering and before ejection by the contraction of the heart muscle.

84. If the contraction of the left ventricle of the heart lasts 250 ms and the rate of flow of the blood in the aorta (the large artery leaving the heart) is 1.0 m/s, what is the average acceleration of a red blood cell as it leaves the heart?
   A. 250 m/s^2
   B. 25 m/s^2
   C. 40 m/s^2
   D. 4.0 m/s^2

85. If the aorta (diameter of d_a) branches into two equal-sized arteries with a combined area equal to the aorta, what is the diameter of one of the branches?
   A. $\sqrt{d_a}$
   B. $d_a/\sqrt{2}$
   C. $2d_a$
   D. $d_a/2$

86. After its journey to the head, arms, and legs, a round trip distance of about 3 m, the red blood cell returns to the left ventricle after 1 minute. What is the average velocity for the trip?
   A. 0.05 m/s, downward
   B. 0.5 m/s, downward
   C. 0 m/s, round trip
   D. 0.05 m/s, upward

3 Motion in a Plane

A Navy F14 Tomcat jet fighter roars off from the deck of an aircraft carrier, moving 55 m/s relative to the deck. It quickly accelerates into the sky. The instruments in the cockpit tell the pilot the speed and altitude of the plane and warn of other airplanes nearby. Pilots must constantly be aware of the three-dimensional nature of their motion.

In the straight-line problems of Chapter 2, we could describe the position of a particle with a single coordinate. But to understand the curved flight of a baseball, the orbital motion of a satellite, or the path of the water from a fire hose, we need to extend our descriptions of motion to more than one dimension. We'll focus on motion in two dimensions—that is, in a plane—but the principles also apply to three-dimensional motion. For two-dimensional motion, the vector quantities displacement, velocity, and acceleration now have two components and lie in a plane rather than along a line. We'll also generalize the concept of relative velocity to motion in a plane, such as an airplane flying in a crosswind.

This chapter represents a merging of the vector language we learned in Chapter 1 with the kinematic language of Chapter 2. As before, we're concerned with *describing* motion, not with analyzing its causes. But the language you learn here will be an essential tool in later chapters when you use Newton's laws of motion to study the relation between force and motion.

3.1 Velocity in a Plane

To describe the *motion* of an object in a plane, we first need to be able to describe the object's *position*. (In this chapter, as in the preceding one, we assume that the objects we describe can be modeled as particles.) Often, it's useful to use a familiar *x-y* axis system (Figure 3.1a). For example, when a football player kicks a

When you see a cat jump, you probably don't think, "projectile!" However, except for the effects of air resistance and a couple of other minor factors, all objects that jump or are launched above the surface and then travel passively until they land move in exactly the same way and can be called projectiles. Projectile motion is a classic example of motion in a plane—the topic of this chapter.

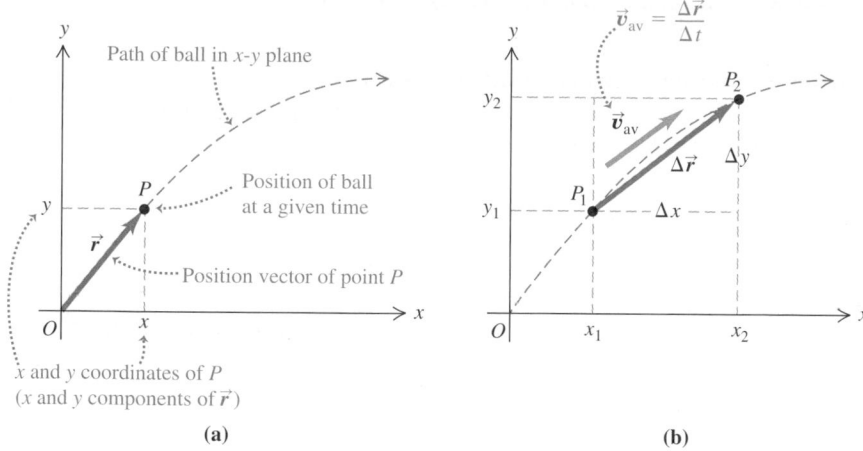

Path of ball in *x-y* plane

Position of ball
at a given time

Position vector of point *P*

$\vec{r}$

x and *y* coordinates of *P*
(*x* and *y* components of $\vec{r}$)

(a)

$\vec{v}_{\text{av}} = \dfrac{\Delta \vec{r}}{\Delta t}$

(b)

▲ **FIGURE 3.1** Position vectors can specify the location and displacement of a point in an *x-y* coordinate system. A position vector points from the origin to the point.

field goal, the ball (represented by point *P*) moves in a vertical plane. The ball's horizontal distance from the origin *O* at any time is *x*, and its vertical distance above the ground at any time is *y*. The numbers *x* and and *y* are called the *coordinates* of point *P*. The vector $\vec{r}$ from the origin *O* to point *P* is called the **position vector** of point *P*, and the Cartesian coordinates *x* and *y* of point *P* are the *x* and *y* components, respectively, of vector $\vec{r}$. (You may want to review Section 1.8, "Components of Vectors.") The *distance* of point *P* from the origin is the magnitude of vector $\vec{r}$:

$$r = |\vec{r}| = \sqrt{x^2 + y^2}.$$

Figure 3.1b shows the ball at two points in its curved path. At time t_1, it is at point P_1 with position vector $\vec{r}_1$; at the later time t_2, it is at point P_2 with position vector $\vec{r}_2$. The ball moves from P_1 to P_2 during the time interval $\Delta t = t_2 - t_1$. The change in position (the displacement) during this interval is the vector $\Delta \vec{r} = \vec{r}_2 - \vec{r}_1$. We define the **average velocity** $\vec{v}_{\text{av}}$ during the interval in the same way we did in Chapter 2 for straight-line motion:

Definition of average velocity

The average velocity of a particle is the displacement $\Delta \vec{r}$, divided by the time interval Δt:

$$\vec{v}_{\text{av}} = \frac{\vec{r}_2 - \vec{r}_1}{t_2 - t_1} = \frac{\Delta \vec{r}}{\Delta t}. \tag{3.1}$$

That is, the average velocity is a vector quantity having the same direction as $\Delta \vec{r}$ and a magnitude equal to the magnitude of $\Delta \vec{r}$, divided by the time interval Δt. The magnitude of $\Delta \vec{r}$ is always the straight-line distance from P_1 to P_2, regardless of the actual shape of the path taken by the object. Thus, the average velocity would be the same for *any* path that would take the particle from P_1 to P_2 in the same time interval Δt.

Figure 3.1b also shows that, during any displacement $\Delta \vec{r}$, Δx and Δy are the components of the vector $\Delta \vec{r}$; it follows that the components $v_{\text{av},x}$ and $v_{\text{av},y}$ of the average velocity vector are

$$v_{\text{av},x} = \frac{\Delta x}{\Delta t} \quad \text{and} \quad v_{\text{av},y} = \frac{\Delta y}{\Delta t}. \tag{3.2}$$

▲ **Application Not your parents' yo-yo.**
The yo-yo may seem like a child's toy, but looks can often be deceiving. Knowledge of basic principles of physics and engineering design has sparked the creation of a new generation of high-performance yo-yos. Some can be made to spin at the end of the string for up to 14 minutes, the current world record set in 2003. In the photo shown here, 1998 world champion Jennifer Baybrook is demonstrating a classic yo-yo trick. When the center of the spinning yo-yo remains at the same position, its displacement and average velocity are zero.

We define the **instantaneous velocity** $\vec{v}$ as follows:

Definition of instantaneous velocity $\vec{v}$ in a plane

The instantaneous velocity is the limit of the average velocity as the time interval Δt approaches zero:

$$\vec{v} = \lim_{\Delta t \to 0} \frac{\Delta \vec{r}}{\Delta t}. \tag{3.3}$$

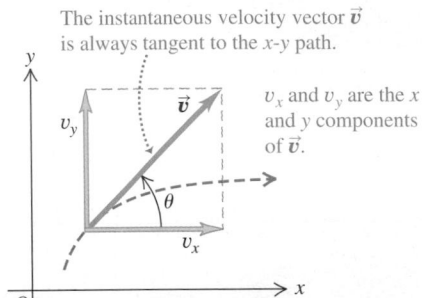

The instantaneous velocity vector $\vec{v}$ is always tangent to the x-y path.

v_x and v_y are the x and y components of $\vec{v}$.

▲ **FIGURE 3.2** The two velocity components for motion in the x-y plane.

As $\Delta t \to 0$, points P_1 and P_2 move closer and closer together. In the limit, the vector $\Delta \vec{r}$ becomes *tangent* to the curve, as shown in Figure 3.2. The direction of $\Delta \vec{r}$ in the limit is also the direction of the instantaneous velocity $\vec{v}$. This analysis leads to an important conclusion:

At every point along the path, the instantaneous velocity vector is tangent to the path.

During any displacement $\Delta \vec{r}$, the changes Δx and Δy in the coordinates x and y are the *components* of $\Delta \vec{r}$. It follows that the components v_x and v_y of the instantaneous velocity $\vec{v}$ are

$$v_x = \lim_{\Delta t \to 0} \frac{\Delta x}{\Delta t} \quad \text{and} \quad v_y = \lim_{\Delta t \to 0} \frac{\Delta y}{\Delta t}.$$

The instantaneous *speed* of the object is the *magnitude* v of the instantaneous velocity vector $\vec{v}$. This is given by the Pythagorean relation

$$|\vec{v}| = v = \sqrt{v_x^2 + v_y^2}. \quad \text{(speed of a particle in a plane)}$$

The direction of $\vec{v}$ is given by the angle θ in the figure. We see that when we measure θ in the usual way (counterclockwise from the $+x$ axis, as in Section 1.8),

$$\theta = \tan^{-1} \frac{v_y}{v_x}.$$

NOTE ▶ In this text, we always use Greek letters for angles; we'll use θ (theta) for the directions of vectors measured counterclockwise from the $+x$ axis and ϕ (phi) for most other angles. ◀

Because velocity is a vector quantity, we may represent it either in terms of its components or in terms of its magnitude and direction, as described in Chapter 1. The *direction* of an object's instantaneous velocity at any point is *always* tangent to the path at that point. But in general, the position vector $\vec{r}$ does *not* have the same direction as the instantaneous velocity $\vec{v}$. (The direction of the position vector depends on where you place the origin, while the direction of $\vec{v}$ is determined by the shape of the path.)

EXAMPLE 3.1   A model car

Suppose you are operating a radio-controlled model car on a vacant tennis court. The surface of the court represents the x-y plane, and you place the origin at your own location. At time $t_1 = 2.00$ s the car has x and y coordinates $(4.0 \text{ m}, 2.0 \text{ m})$, and at time $t_2 = 2.50$ s it has coordinates $(7.0 \text{ m}, 6.0 \text{ m})$. For the time interval from t_1 to t_2, find **(a)** the components of the average velocity of the car and **(b)** the magnitude and direction of the average velocity.

Continued

SOLUTION

SET UP Figure 3.3 shows the sketch we draw. We see that $\Delta x = 7.0\text{ m} - 4.0\text{ m} = 3.0\text{ m}$, $\Delta y = 6.0\text{ m} - 2.0\text{ m} = 4.0\text{ m}$, and $\Delta t = 0.50\text{ s}$.

SOLVE Part (a): To find the components of $\vec{v}_{av}$, we use their definitions (Equations 3.2):

$$v_{av,x} = \frac{\Delta x}{\Delta t} = \frac{3.0\text{ m}}{0.50\text{ s}} = 6.0\text{ m/s},$$

$$v_{av,y} = \frac{\Delta y}{\Delta t} = \frac{4.0\text{ m}}{0.50\text{ s}} = 8.0\text{ m/s}.$$

Part (b): The magnitude of $\vec{v}_{av}$ is obtained from the Pythagorean theorem:

$$|\vec{v}_{av}| = \sqrt{v_{av,x}{}^2 + v_{av,y}{}^2}$$
$$= \sqrt{(6.0\text{ m/s})^2 + (8.0\text{ m/s})^2} = 10.0\text{ m/s}.$$

The direction of $\vec{v}_{av}$ is most easily described by its angle, measured counterclockwise from the positive x axis. Calling this angle θ, we have

$$\theta = \tan^{-1}\frac{v_{av,y}}{v_{av,x}} = \tan^{-1}\frac{8.0\text{ m/s}}{6.0\text{ m/s}} = 53°.$$

Alternative Solution: Alternatively, the magnitude of $\vec{v}_{av}$ is the distance between the points $(4.0\text{ m}, 2.0\text{ m})$ and $(7.0\text{ m}, 6.0\text{ m})$ (i.e., 5.0 m, found using Pythagoras's theorem), divided by the time interval (0.5 s):

$$|\vec{v}_{av}| = \frac{5.0\text{ m}}{0.50\text{ s}} = 10.0\text{ m/s}.$$

$\vec{v}_{av}$ points from P_1 toward P_2. It doesn't matter how long you make it; its magnitude will be found mathematically.

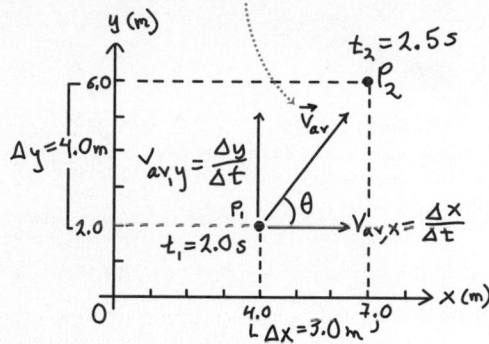

▲ **FIGURE 3.3** Our sketch for this problem.

Since the direction of $\vec{v}_{av}$ is the same as the direction of the displacement $\Delta\vec{r}$ between the two points, we can calculate it from Δx and Δy rather than $v_{av,x}$ and $v_{av,y}$:

$$\theta = \tan^{-1}\frac{\Delta y}{\Delta x} = \tan^{-1}\frac{4.0\text{ m}}{3.0\text{ m}} = 53°.$$

REFLECT Be sure you understand the relation between the two solutions to part (b). In the first, we calculated the magnitude and direction of $\vec{v}_{av}$ from the components of this vector quantity. In the alternative solution, we used the fact that the average velocity $\vec{v}_{av}$ and the displacement $\Delta\vec{r}$ have the same direction.

Practice Problem: Suppose you reverse the car's motion, so that it retraces its path in the opposite direction in the same time. Find the components of the average velocity of the car and the magnitude and direction of the average velocity. *Answers:* -6.0 m/s, -8.0 m/s, 10.0 m/s, 233°.

3.2 Acceleration in a Plane

Now let's consider the acceleration of an object moving on a curved path in a plane. In Figure 3.4, the vector $\vec{v}_1$ represents the particle's instantaneous velocity at point P_1 at time t_1, and the vector $\vec{v}_2$ represents the particle's instantaneous velocity at point P_2 at time t_2. In general, the two velocities differ in both magnitude and direction.

We define the **average acceleration $\vec{a}_{av}$** of the particle as follows:

Mastering PHYSICS

PhET: Maze Game

Definition of average acceleration $\vec{a}_{av}$

As an object undergoes a displacement during a time interval Δt, its average acceleration is its change in velocity, $\Delta\vec{v}$, divided by Δt:

$$\text{Average acceleration} = \vec{a}_{av} = \frac{\vec{v}_2 - \vec{v}_1}{t_2 - t_1} = \frac{\Delta\vec{v}}{\Delta t}. \tag{3.4}$$

Average acceleration is a vector quantity in the same direction as the vector $\Delta\vec{v}$. In Chapter 2, we stressed that acceleration is a quantitative description of the

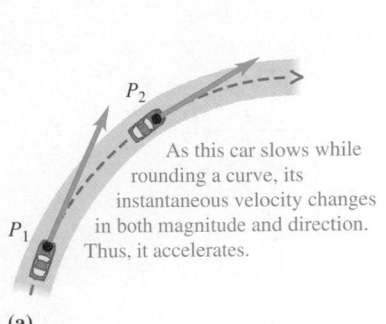

As this car slows while rounding a curve, its instantaneous velocity changes in both magnitude and direction. Thus, it accelerates.

(a)

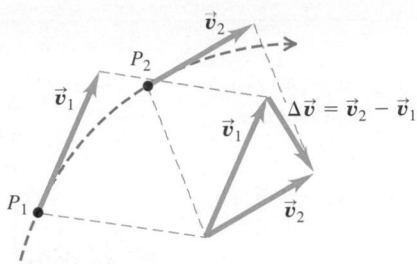

To find the car's average acceleration between P_1 and P_2, we first find the change in velocity $\Delta\vec{v}$ by subtracting $\vec{v}_1$ from $\vec{v}_2$. (Notice that $\vec{v}_1 + \Delta\vec{v} = \vec{v}_2$.)

(b)

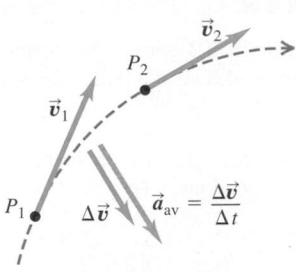

The average acceleration has the same direction as the change in velocity, $\Delta\vec{v}$.

(c)

▲ **FIGURE 3.4** Finding the average acceleration between two points in the x-y plane.

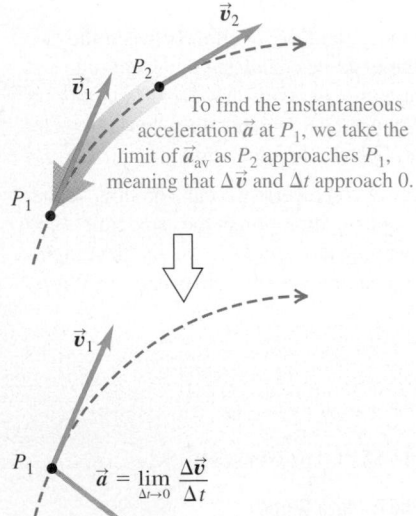

To find the instantaneous acceleration $\vec{a}$ at P_1, we take the limit of $\vec{a}_{av}$ as P_2 approaches P_1, meaning that $\Delta\vec{v}$ and Δt approach 0.

▲ **FIGURE 3.5** The instantaneous acceleration of a point in the x-y plane.

way an object's motion is *changing* with time. Figure 3.4b shows that the **final velocity $\vec{v}_2$ (at time t_2) is the vector sum of the original velocity $\vec{v}_1$ (at time t_1) and the change in velocity, $\Delta\vec{v}$, during the interval Δt.**

We define the **instantaneous acceleration** $\vec{a}$ at point P_1 as follows (Figure 3.5):

Definition of instantaneous acceleration $\vec{a}$

When the velocity of a particle changes by an amount $\Delta\vec{v}$ as the particle undergoes a displacement $\Delta\vec{r}$ during a time interval Δt, the instantaneous acceleration is the limit of the average acceleration as Δt approaches zero:

$$\text{Instantaneous acceleration} = \vec{a} = \lim_{\Delta t \to 0} \frac{\Delta\vec{v}}{\Delta t}. \tag{3.5}$$

The instantaneous acceleration vector at point P_1 in Figure 3.5 does *not* have the same direction as the instantaneous velocity vector $\vec{v}$ at that point; in general, there is no reason it should. (Recall from Chapter 2 that the velocity and acceleration components of a particle moving along a line could have opposite signs.) The construction in Figure 3.5 shows that the acceleration vector must always point toward the *concave* side of the curved path. **When a particle moves in a curved path, it *always* has nonzero acceleration, even when it moves with constant speed.** More generally, acceleration is associated with change of speed, change of direction of velocity, or both.

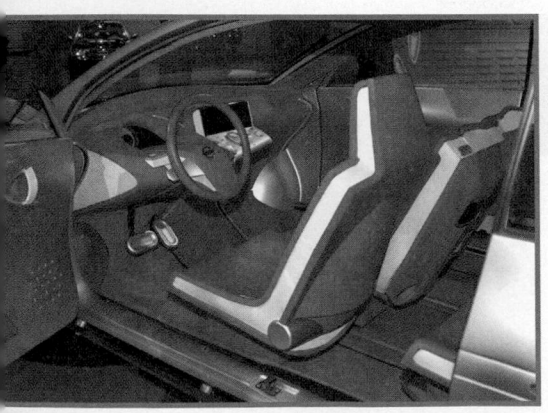

◄ **Application Can you spot the three accelerators?** Almost anyone will recognize the gas pedal as an accelerator—pressing it makes the car speed up. As a physics student, you probably also recognize the brake pedal as an accelerator—it slows the car. The third accelerator is the steering wheel, which changes the *direction* of the car's velocity.

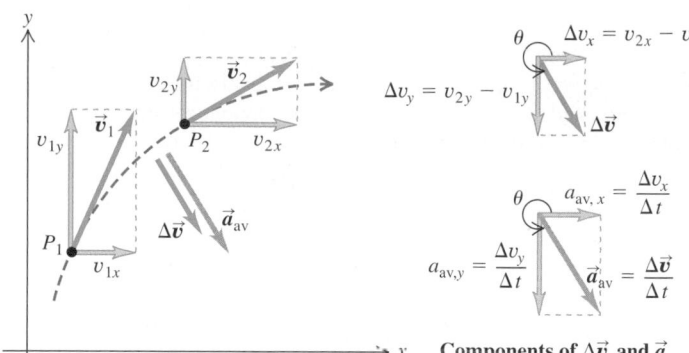

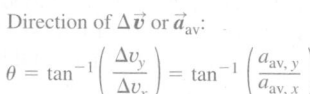

Components of $\Delta \vec{v}$ and $\vec{a}_{av}$.

Direction of $\Delta \vec{v}$ or $\vec{a}_{av}$:

$$\theta = \tan^{-1}\left(\frac{\Delta v_y}{\Delta v_x}\right) = \tan^{-1}\left(\frac{a_{av,\,y}}{a_{av,\,x}}\right)$$

Magnitude of $\vec{a}_{av}$:

$$\left|\vec{a}_{av}\right| = a_{av} = \sqrt{a_{av,\,x}^{\,2} + a_{av,\,y}^{\,2}}$$

(a) Components of average acceleration for the interval from P_1 to P_2.

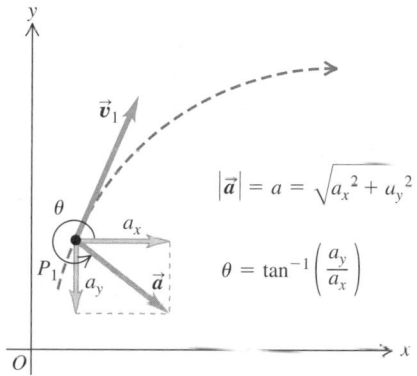

$$\left|\vec{a}\right| = a = \sqrt{a_x^{\,2} + a_y^{\,2}}$$

$$\theta = \tan^{-1}\left(\frac{a_y}{a_x}\right)$$

(b) Components of instantaneous acceleration at P_1.

▲ **FIGURE 3.6** Components of average and instantaneous acceleration.

We often represent the acceleration of a particle in terms of the components of this vector quantity. Like Figure 3.2, Figure 3.6 shows the motion of a particle as described in a rectangular coordinate system. During a time interval Δt, the velocity of the particle changes by an amount $\Delta \vec{v}$, with components Δv_x and Δv_y. So we can represent the average acceleration $\vec{a}_{av}$ in terms of its x and y components:

$$a_{av,x} = \frac{\Delta v_x}{\Delta t}, \qquad a_{av,y} = \frac{\Delta v_y}{\Delta t}. \tag{3.6}$$

Similarly, the x and y components of instantaneous acceleration, a_x and a_y, are

$$a_x = \lim_{\Delta t \to 0} \frac{\Delta v_x}{\Delta t}, \qquad a_y = \lim_{\Delta t \to 0} \frac{\Delta v_y}{\Delta t}.$$

If we know the components a_x and a_y, we can find the magnitude and direction of the acceleration vector $\vec{a}$, just as we did with velocity:

$$\left|\vec{a}\right| = a = \sqrt{a_x^2 + a_y^2}, \qquad \theta = \tan^{-1}\frac{a_y}{a_x}. \tag{3.7}$$

The angle θ gives the direction of $\vec{a}$, measured counterclockwise from the $+x$ axis.

▲ **BIO Application Knowing up from down.** Plant roots have exquisitely developed mechanisms for sensing gravity and growing downward. If the direction of a root is changed, say by running into a rock in the soil, the root is forced to grow horizontally; however, as soon as it can, it again turns downward. This ability involves a number of cellular signals, but the primary sensor detects acceleration. The sensing cells contain statoliths, specialized starch-containing granules that are denser than the fluid in the cells and, in response to changes in the direction of gravity, fall (are accelerated) to a different location within the cells. This triggers an active localized response of the cells and causes the growth direction of the root to re-orient so that the root once again grows down.

EXAMPLE 3.2 The model car again

Let's look again at the radio-controlled model car in Example 3.1. Suppose that at time $t_1 = 2.00$ s the car has components of velocity $v_x = 1.0$ m/s and $v_y = 3.0$ m/s and that at time $t_2 = 2.50$ s the components are $v_x = 4.0$ m/s and $v_y = 3.0$ m/s. Find **(a)** the components and **(b)** the magnitude and direction of the average acceleration during this interval.

SOLUTION

SET UP Figure 3.7 shows our sketch.

SOLVE Figure 3.6a outlines the relations we'll use.

Part (a): To find the components of average acceleration, we need the components of the change in velocity, Δv_x and Δv_y, and the time interval, which is $\Delta t = 0.50$ s. The change in v_x is $\Delta v_x = v_{2x} - v_{1x} = (4.0 \text{ m/s} - 1.0 \text{ m/s}) = 3.0 \text{ m/s}$, so the x component of average acceleration in the interval $\Delta t = 0.50$ s is

$$a_{av,x} = \frac{\Delta v_x}{\Delta t} = \frac{3.0 \text{ m/s}}{0.50 \text{ s}} = 6.0 \text{ m/s}^2.$$

The change in v_y is zero, so $a_{av,y}$ in this interval is also zero.

Part (b): The vector $\vec{a}_{av}$ has only an x component. The vector points in the $+x$ direction and has magnitude 6.0 m/s^2.

REFLECT We can always represent a vector quantity (such as displacement, velocity, or acceleration) either in terms of its components or in terms of its magnitude and direction.

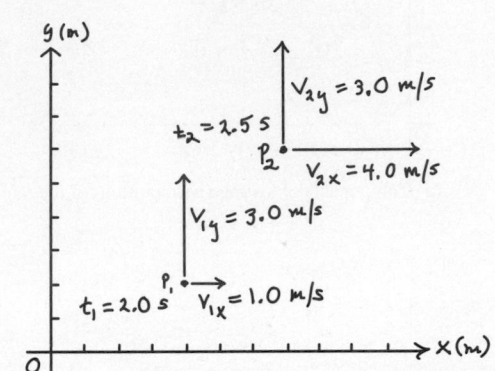

▲ **FIGURE 3.7** Our sketch for this problem.

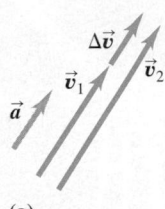

An acceleration parallel to an object's velocity changes the magnitude, but not the direction, of the velocity; the object moves in a straight line with changing speed.

(a)

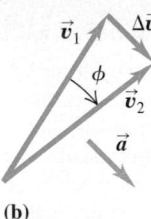

An acceleration perpendicular to an object's velocity changes the direction, but not the magnitude, of the velocity; the object follows a curved path at constant speed.

(b)

▲ **FIGURE 3.8** Two special cases: acceleration (a) parallel and (b) perpendicular to an object's velocity.

Figure 3.8 shows two special cases. In Figure 3.8a, the acceleration vector $\vec{a}$ is *parallel* to the velocity vector $\vec{v}_1$. (In Fig. 3.8a, $\vec{a}$ points in the same direction as $\vec{v}$; it could alternatively point in the opposite direction.) Then, because $\vec{a}$ gives the rate of change of velocity, the *change* in $\vec{v}$ during a small time interval Δt is a vector $\Delta \vec{v}$ that has the same direction as $\vec{a}$ and hence the same direction as $\vec{v}_1$. As a result, the velocity $\vec{v}_2$ at the end of Δt, given by $\vec{v}_2 = \vec{v}_1 + \Delta \vec{v}$, is a vector having the s*ame direction* as $\vec{v}_1$ but somewhat greater (or smaller) magnitude.

In Figure 3.8b, the acceleration $\vec{a}$ is *perpendicular* to the velocity $\vec{v}$. In a small time interval Δt, the change $\Delta \vec{v}$ in the velocity vector $\vec{v}$ is a vector very nearly perpendicular to $\vec{v}_1$, as shown. Again, $\vec{v}_2 = \vec{v}_1 + \Delta \vec{v}$, but in this case $\vec{v}_1$ and $\vec{v}_2$ have different directions. As the time interval Δt approaches zero, the angle ϕ in the figure also approaches zero. In this limit, $\Delta \vec{v}$ becomes perpendicular to *both* $\vec{v}_1$ and $\vec{v}_2$, and the two vectors have the same magnitude.

We conclude that when $\vec{a}$ is *parallel* (or antiparallel) to $\vec{v}$, its effect is to change the magnitude of $\vec{v}$, but not its direction. When $\vec{a}$ is *perpendicular* to $\vec{v}$, its effect is to change the direction of $\vec{v}$, but not its magnitude. In general, $\vec{a}$ may have components *both* parallel and perpendicular to $\vec{v}_1$, but the preceding statements are still valid for the individual components. In particular, when an object travels along a curved path with constant speed, its acceleration is *not* zero, even though the magnitude of $\vec{v}$ does not change. In this case, the acceleration is always perpendicular to $\vec{v}$ at each point. For example, when a particle moves in a circle with constant speed, the acceleration at each instant is directed toward the center of the circle. We'll consider this special case in detail in Section 3.4.

3.3 Projectile Motion

A **projectile** is any object that is given an initial velocity and then follows a path determined entirely by the effects of gravitational acceleration and air resistance. A batted baseball, a thrown football, and a package dropped from an airplane are all examples of projectiles. The path followed by a projectile is called its *trajectory.*

To analyze this common type of motion, we'll start with an idealized model. We represent the projectile as a single particle with an acceleration (due to the earth's gravitational pull) that is constant in both magnitude and direction. We'll neglect the effects of air resistance and the curvature and rotation of the earth. Like all models, this one has limitations. The curvature of the earth has to be considered in the flight of long-range ballistic missiles, and air resistance is of crucial importance to a skydiver. Nevertheless, we can learn a lot by analyzing this simple model.

We first notice that projectile motion is always confined to a vertical plane determined by the direction of the initial velocity. We'll call this plane the *x-y coordinate plane,* with the *x* axis horizontal and the *y* axis directed vertically upward. Figure 3.9 shows a view of this plane from the side, along with a typical trajectory.

The key to analyzing projectile motion is the fact that *we can treat the x and y coordinates separately.* Why is this so? Anticipating a relation that we'll study in detail in Chapter 4, we note that the instantaneous acceleration of an object is proportional to (and in the same direction as) the net *force* acting on the object. Because of the assumptions made in our model, the only force acting on the projectile is the earth's gravitational attraction; we assume that this is constant in magnitude and always vertically downward in direction. Thus the vertical component of acceleration is the same as if the projectile moved only in the *y* direction, as it did in Section 2.6. Figure 3.10 shows two trajectories; the vertical displacements of the two objects at any time are the same, even though their horizontal displacements are different.

We conclude that the *x* component of acceleration, a_x, is zero and the *y* component a_y is constant and equal in magnitude to the acceleration of free fall:

$$a_x = 0; \qquad a_y = -g = -9.80 \text{ m/s}^2.$$

NOTE ▶ Remember that, by definition, *g* is always positive (because it is the magnitude of the acceleration vector due to gravity), but with our choice of coordinate directions, a_y is negative. ◀

So we can think of projectile motion as a combination of horizontal motion with constant velocity and vertical motion with constant acceleration. We can then express all the vector relationships in terms of separate equations for the horizontal and vertical components. The actual motion is a combination of these separate motions. Figure 3.11 shows the horizontal and vertical components of motion for a projectile that starts at (or passes through) the origin of coordinates at time $t = 0$. As in Figure 3.10, the projectile is shown at equal time intervals.

The horizontal (x) and vertical (y) components of $\vec{a}$ for a projectile are

$$a_x = 0, \qquad a_y = -g.$$

We'll usually use $g = 9.80 \text{ m/s}^2$, but occasionally we'll use $g = 10 \text{ m/s}^2$ for approximate calculations.

MasteringPHYSICS

PhET: Projectile Motion
ActivPhysics 3.1: Solving Projectile Motion Problems
ActivPhysics 3.2: Two Balls Falling
ActivPhysics 3.3: Changing the *x*-Velocity
ActivPhysics 3.4: Projectile *x*- and *y*-Accelerations
ActivPhysics 3.5: Initial Velocity Components
ActivPhysics 3.6: Target Practice I
ActivPhysics 3.7: Target Practice II

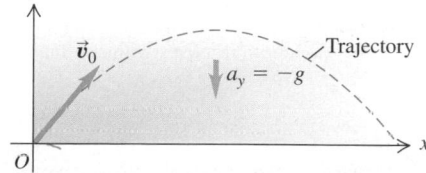

• A projectile moves in a vertical plane that contains the initial velocity vector $\vec{v}_0$.
• Its trajectory depends only on $\vec{v}_0$ and on the acceleration due to gravity.

▲ **FIGURE 3.9** The trajectory of a projectile.

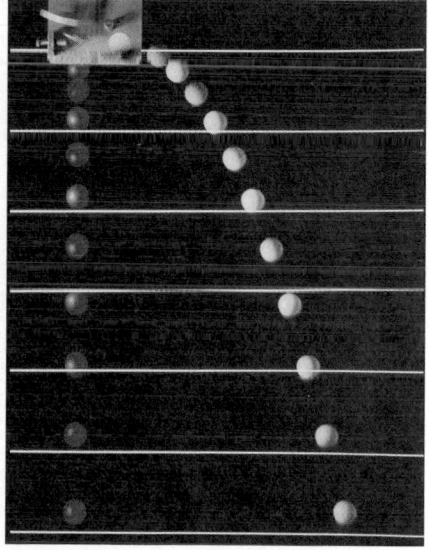

▲ **FIGURE 3.10** Independence of horizontal and vertical motion: At any given time, both balls have the same *y* position, velocity, and acceleration, despite having different *x* positions and velocities. Successive images are separated by equal time intervals.

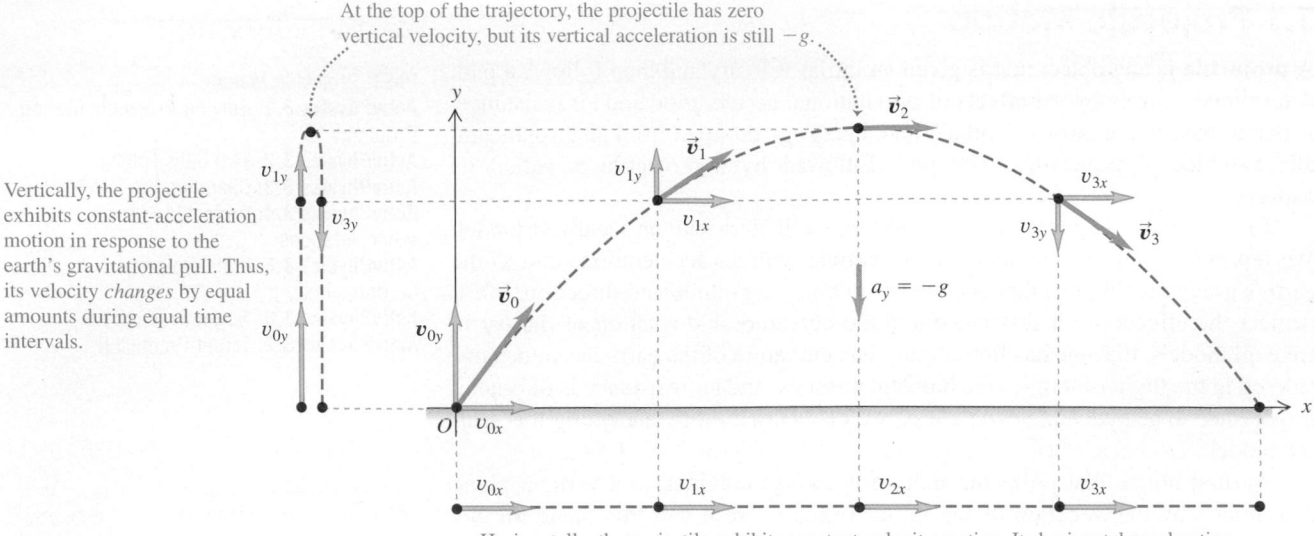

At the top of the trajectory, the projectile has zero vertical velocity, but its vertical acceleration is still $-g$.

Vertically, the projectile exhibits constant-acceleration motion in response to the earth's gravitational pull. Thus, its velocity *changes* by equal amounts during equal time intervals.

$a_y = -g$

Horizontally, the projectile exhibits constant-velocity motion: Its horizontal acceleration is zero, so it moves equal x distances in equal time intervals.

▲ **FIGURE 3.11** Independence of horizontal and vertical motion. In the vertical direction, a projectile behaves like a freely falling object; in the horizontal direction, it moves with constant velocity.

Conceptual Analysis 3.1

Horizontal stone throw

If you throw a stone horizontally out over the surface of a lake, the time it is in the air (i.e., before it hits the water) is determined *only* by

A. the height from which you throw it.
B. the height from which you throw it and its initial speed.
C. its initial speed and the horizontal distance to the point where it splashes down.

SOLUTION As we've just learned, the horizontal and vertical components of a projectile's motion are independent. Therefore, the stone reaches the water at the same time as a rock that was dropped vertically from the same starting height. The time required for the stone to reach the water depends only on the height from which it is thrown, so answer A is the correct one. The *distance* the stone travels before splashing down, however, depends on both the height from which it is thrown and the speed with which it is thrown.

In projectile motion, the vertical and horizontal coordinates both vary with constant-acceleration motion. (The horizontal component of acceleration is constant at zero.) Therefore, we can use the same equations we derived in Section 2.4 for constant-acceleration motion. Here's a reminder of these relationships:

$$v_x = v_{0x} + a_x t, \quad \text{(velocity as a function of time)} \tag{2.8}$$
$$x = x_0 + v_{0x}t + \tfrac{1}{2}a_x t^2, \quad \text{(position as a function of time)} \tag{2.12}$$
$$v_x^2 = v_{0x}^2 + 2a_x(x - x_0). \quad \text{(velocity as a function of position)} \tag{2.13}$$

Our procedure will be to use this set of equations separately for each coordinate. That is, we use this set of equations for the x coordinate, and then we use a second set with all the x's replaced by y's for the y coordinate. Next, we have to choose the appropriate components of the constant acceleration $\vec{a}$ and the initial velocity $\vec{v}_0$. At time $t = 0$, the particle is at the point (x_0, y_0), and its velocity components have the initial values v_{0x} and v_{0y}. The components of acceleration are constant: $a_x = 0$, $a_y = -g$. When we put all the pieces together, here's what we get:

Equations for projectile motion (assuming that $a_x = 0$, $a_y = -g$)

Considering the x motion, we substitute $a_x = 0$ in Equations 2.8 and 2.12, obtaining

$$v_x = v_{0x}, \tag{3.8}$$
$$x = x_0 + v_{0x}t. \tag{3.9}$$

For the y motion, we substitute y for x, v_y for v_x, v_{0y} for v_{0x}, and $-g$ for a to get

$$v_y = v_{0y} - gt \tag{3.10}$$
$$y = y_0 + v_{0y}t - \tfrac{1}{2}gt^2 \tag{3.11}$$

Usually it is simplest to take the initial position (at time $t = 0$) as the origin; in this case, $x_0 = y_0 = 0$. The initial position might be, for example, the position of a ball at the instant it leaves the thrower's hand or the position of a bullet at the instant it leaves the barrel of the gun.

As shown in Figure 3.11, the x component of acceleration for a projectile is zero, so v_x is constant, but v_y changes by equal amounts in equal times, corresponding to a constant y component of acceleration. At the highest point in the trajectory, $v_y = 0$. But a_y is still equal to $-g$ at this point. Make sure you understand why!

We can also represent the initial velocity $\vec{v}_0$ by its magnitude v_0 (the initial speed) and its angle θ_0 with the positive x axis, as shown in Figure 3.12. In terms of these quantities, the components v_{0x} and v_{0y} of initial velocity are

$$v_{0x} = v_0\cos\theta_0, \qquad v_{0y} = v_0\sin\theta_0.$$

Using these relations in Equations 3.8 through 3.11 and setting $x_0 = y_0 = 0$, we obtain the following alternative formulation for projectile motion:

Position and velocity of a projectile as functions of time t

$$x = (v_0\cos\theta_0)t \tag{3.12}$$
$$y = (v_0\sin\theta_0)t - \tfrac{1}{2}gt^2 \tag{3.13}$$
$$v_x = v_0\cos\theta_0 \tag{3.14}$$
$$v_y = v_0\sin\theta_0 - gt \tag{3.15}$$

We can get a lot of information from these equations. For example, the distance r of the projectile from the origin at any time (the magnitude of the position vector $\vec{r}$) is given by

$$r = \sqrt{x^2 + y^2}.$$

The projectile's speed v (the magnitude of its velocity) at any time is

$$v = \sqrt{v_x^2 + v_y^2}.$$

The *direction* of the velocity at any time, in terms of the angle θ it makes with the positive x axis, is given by

$$\theta = \tan^{-1}\frac{v_y}{v_x}.$$

The velocity vector $\vec{v}$ is tangent to the trajectory at each point. Note that v_x is constant, but that the *direction* of the velocity changes because v_y changes continuously.

The formulation just described gives us the coordinates and velocity components of a projectile as functions of time. The actual *shape* of the trajectory is a graph of y as a function of x. We can derive an equation for this relationship from Equations 3.12 and 3.13. First we solve Equation 3.12 for t, and then we

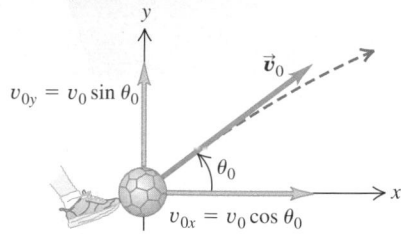

▲ **FIGURE 3.12** The initial velocity of a projectile, showing the components and the launch angle θ_0.

▲ **Application Which is accelerating the most?** If you answered "none of the above," you are correct. In the photo, one apple and one orange are moving up while the other two fruits are moving down. Despite their different positions and velocities, all four have the same rate of change of velocity. Like all freely falling objects near the earth's surface, they are accelerating downward at 9.8 m/s². The apple and orange moving up are losing upward velocity at exactly the same rate the other two fruits are gaining downward velocity. Even at the highest point of their path, where for an instant they move neither upward nor downward, they have the same acceleration, a rate of change in velocity of 9.8 m/s² toward the center of the earth.

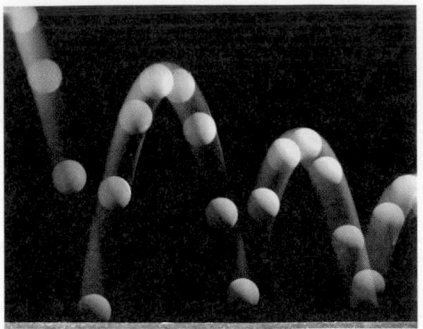

▲ **FIGURE 3.13** Strobe photo of a bouncing ball; the images are separated by equal time intervals. The ball follows a parabolic trajectory after each bounce. It rises a little less after each bounce because it loses energy during each collision with the floor.

substitute the resulting expression for t into Equation 3.13 and simplify the result. We find that $t = x/(v_0\cos\theta_0)$ and

$$y = v_0\sin\theta_0\left(\frac{x}{v_0\cos\theta_0}\right) - \tfrac{1}{2}g\left(\frac{x}{v_0\cos\theta_0}\right)^2,$$

$$y = (\tan\theta_0)x - \left(\frac{g}{2v_0{}^2\cos^2\theta_0}\right)x^2.$$

Don't worry about the details of this equation; the important point is its general form. The quantities v_0, $\tan\theta_0$, $\cos\theta_0$, and g are all constants, so the equation is a relation between the variables x and y, with the general form

$$y = bx - cx^2,$$

where b and c are positive constants. This is the equation of a *parabola*. With projectile motion, for our simple model, the trajectory is always a parabola (Figure 3.13).

Quantitative Analysis 3.2 **Horizontal paintball shot**

A paintball is shot horizontally in the positive x direction. At time Δt after the ball is shot, it is 3 cm to the right and 3 cm below its starting point. Over the next interval Δt, the *changes* in horizontal and vertical position are

A. $\Delta x = 3$ cm, $\Delta y = -3$ cm;
B. $\Delta x = 3$ cm, $\Delta y = -9$ cm;
C. $\Delta x = 6$ cm, $\Delta y = -6$ cm.

SOLUTION The ball is a projectile, so we assume that its horizontal component of velocity is constant. Therefore, the changes

Δx in horizontal position during equal time intervals are equal. This result rules out choice C. Because the ball speeds up as it falls, the second Δy must be larger than the first, eliminating answer A. However, is answer B correct? With constant acceleration, the distance of fall from rest is proportional to t^2. This means that, as time increases from 1 to 2 to 3 in units of t, the distance fallen will be y, $4y$, and $9y$, or -3 cm, -12 cm, and -27 cm, in vertical height. The change in y from one time interval to the next is -3 cm, -9 cm $[= -12\text{ cm} - (-3\text{ cm})]$, and -15 cm $[= -27\text{ cm} - (-12\text{ cm})]$. The ratio of these numbers is $1:3:5$, the sequence of odd integers. Answer B is correct.

EXAMPLE 3.3 **Paintball gun**

The contestant in Figure 3.14 shoots a paintball horizontally at a speed of 75.0 m/s from a point 1.50 m above the ground. The ball misses its target and hits the ground. **(a)** For how many seconds is the ball in the air? **(b)** Find the maximum horizontal displacement (which we'll call the *range* of the ball). Ignore air resistance.

SOLUTION

SET UP We choose to place the origin of the coordinate system at ground level, directly below the end of the gun barrel, as shown in Figure 3.15. (This choice of position of the origin avoids having to deal with negative values of y, a modest convenience.) Then $x_0 = 0$ and $y_0 = 1.50$ m. The gun is fired horizon-

tally, so $v_{0x} = 75.0$ m/s and $v_{0y} = 0$. The final position of the ball, at ground level, is $y = 0$.

SOLVE Part (a): We're asked to find the total time the paintball is in the air. This is equal to the time it would take the paintball to fall vertically from its initial height to the ground. In each case, the vertical position is given as a function of time by Equation 3.11,

▲ **FIGURE 3.14**

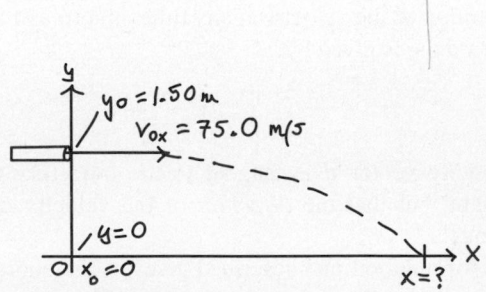

▲ **FIGURE 3.15** Our sketch for this problem.

Continued

$y = y_0 + v_{0y}t - \frac{1}{2}gt^2$. In this problem, $y_0 = 1.5$ m, $y = 0$, and $v_{0y} = 0$, so that equation becomes simply

$$0 = y_0 - \frac{1}{2}gt^2.$$

We need to find the time t. Solving for t, and substituting numerical values, we obtain

$$t = \sqrt{\frac{2y_0}{g}} = \sqrt{\frac{2(1.50 \text{ m})}{9.80 \text{ m/s}^2}} = 0.553 \text{ s}.$$

Part (b):   Now that we know the time t of the ball's flight through the air, we can find the range—that is, the horizontal dis-

tance x it travels during time t. We use Equation 3.9: $x = x_0 + v_{0x}t$. Setting $x_0 = 0$, we find that

$$x = x_0 + v_{0x}t = 0 + (75.0 \text{ m/s})(0.553 \text{ s}) = 41.5 \text{ m}.$$

REFLECT Actual ranges of paintballs are less than this, typically about 30 m. The difference is due primarily to air resistance, which decreases the horizontal component of velocity.

Practice Problem: If air resistance is neglected, what initial speed is required for a range of 20 m? *Answer:* 36.1 m/s.

PROBLEM-SOLVING STRATEGY 3.1 **Projectile motion**   (MP)

The strategies that we used in Sections 2.4 and 2.6 for solving straight-line, constant-acceleration motion problems are equally useful for projectile motion.

SET UP

1. Define your coordinate system and make a sketch showing your axes. Usually, it is easiest to place the origin at the initial ($t = 0$) position of the projectile, with the x axis horizontal and the y axis upward. Then $x_0 = 0$, $y_0 = 0$, $a_x = 0$, and $a_y = -g$.
2. List the known and unknown quantities. In some problems, the components (or magnitude and direction) of initial velocity will be given, and you can use Equations 3.12 through 3.15 to find the coordinates and velocity components at some later time. In other problems, you might be given two points on the trajectory and be asked to find the initial velocity. Be sure you know which quantities are given and which are to be found.
3. It often helps to state the problem in prose and then translate into symbols. For example, *when* does the particle arrive at a certain point (i.e, at what value of t)? *Where* is the particle when a velocity component has a certain value? (That is, what are the values of x and y when v_x or v_y has the specified value?)

SOLVE

4. At the highest point in a trajectory, $v_y = 0$. So the question "When does the projectile reach its highest point?" translates into "What is the value of t when $v_y = 0$?" Similarly, if $y_0 = 0$, then "When does the projectile return to its initial elevation?" translates into "What is the value of t when $y = 0$?" and so on.
5. Resist the temptation to break the trajectory into segments and analyze each one separately. You don't have to start all over, with a new axis system and a new time scale, when the projectile reaches its highest point. It is usually easier to set up Equations 3.8 through 3.11 (or Equations 3.12 through 3.15) at the start and use the same axes and time scale throughout the problem.

REFLECT

6. Try to make rough estimates as to what your answers should be, and then ask whether your calculations confirm your estimates. Ask, "Does this result make sense?"

▲ **BIO Application Ballistic spores.**
Animals generally depend on muscles to move about while other organisms such as plants and fungi are relatively immobile and require specialized mechanisms for seed and spore dispersal. Among the fungi, some of those living on animal dung have compartments called asci (shown) that contain spores (inset) and generate substantial hydrostatic pressure. The asci rupture and behave as "squirt guns," propelling the spores into the air. The initial acceleration of the spores has only recently been measured with specialized video cameras capturing up to 250,000 frames per second. These data reveal that the spores are ejected with accelerations of up to 1,800,000 m/s², or about 180,000 g. These are the largest natural accelerations ever measured.

EXAMPLE 3.4 **A home-run hit**

A home-run baseball is hit with an initial speed $v_0 = 37.0$ m/s at an initial angle $\theta_0 = 53.1°$. **(a)** Find the ball's position, and the magnitude and direction of its velocity, when $t = 2.00$ s. **(b)** Find the time when the ball reaches the highest point of its flight, and find its height h at that point. **(c)** Find the horizontal range R (the horizontal distance from the starting point to the point where the ball hits the ground).

SOLUTION

SET UP The ball is probably struck a meter or so above ground level, but we'll neglect this small distance and assume that it starts at ground level $(y_0 = 0)$. Figure 3.16 shows our sketch. We place the origin of coordinates at the starting point, so $x_0 = 0$. The components of initial velocity are

$$v_{0x} = v_0\cos\theta_0 = (37.0 \text{ m/s})(0.600) = 22.2 \text{ m/s},$$
$$v_{0y} = v_0\sin\theta_0 = (37.0 \text{ m/s})(0.800) = 29.6 \text{ m/s}.$$

SOLVE Part (a): We ignore the effects of the air. We want to find x, y, v_x, and v_y at time $t = 2.00$ s. We substitute these values into Equations 3.8 through 3.11, along with the value $t = 2.00$ s. The coordinates at $t = 2.00$ s are

$$x = v_{0x}t = (22.2 \text{ m/s})(2.00 \text{ s}) = 44.4 \text{ m},$$
$$y = v_{0y}t - \tfrac{1}{2}gt^2$$
$$= (29.6 \text{ m/s})(2.00 \text{ s}) - \tfrac{1}{2}(9.80 \text{ m/s}^2)(2.00 \text{ s})^2 = 39.6 \text{ m}.$$

The components v_x and v_y of the velocity vector $\vec{v}$ at time $t = 2.00$ s are

$$v_x = v_{0x} = 22.2 \text{ m/s},$$
$$v_y = v_{0y} - gt = 29.6 \text{ m/s} + (-9.80 \text{ m/s}^2)(2.00 \text{ s})$$
$$= 10.0 \text{ m/s}.$$

The magnitude and direction of the velocity vector $\vec{v}$ at time $t = 2.00$ s are, respectively,

$$v = \sqrt{v_x^2 + v_y^2} = \sqrt{(22.2 \text{ m/s})^2 + (10.0 \text{ m/s})^2} = 24.3 \text{ m/s},$$
$$\theta = \tan^{-1}\frac{10.0 \text{ m/s}}{22.2 \text{ m/s}} = \tan^{-1}0.450 = 24.2°.$$

Part (b): At the highest point in the ball's path, the vertical velocity component v_y is zero. When does this happen? Call that time t_1; then, using $v_y = v_{0y} - gt$, we find that

$$v_y = 0 = 29.6 \text{ m/s} - (9.80 \text{ m/s}^2)t_1,$$
$$t_1 = 3.02 \text{ s}.$$

The height h at this time is the value of y when $t = 3.02$ s. We use Equation 3.11,

$$y = y_0 + v_{0y}t + \tfrac{1}{2}(-g)t^2,$$

to obtain

$$h = 0 + (29.6 \text{ m/s})(3.02 \text{ s}) + \tfrac{1}{2}(-9.80 \text{ m/s}^2)(3.02 \text{ s})^2$$
$$= 44.7 \text{ m}.$$

Alternatively, we can use the constant-acceleration formula

$$v_y^2 = v_{0y}^2 + 2a(y - y_0) = v_{0y}^2 + 2(-g)(y - y_0).$$

At the highest point, $v_y = 0$ and $y = h$. Substituting these in, along with $y_0 = 0$, we find that

$$0 = (29.6 \text{ m/s})^2 - 2(9.80 \text{ m/s}^2)h,$$
$$h = 44.7 \text{ m}.$$

This is roughly half the height of the Toronto Skydome above the playing field.

Part (c): To find the range R, we start by asking *when* the ball hits the ground. This occurs when $y = 0$. Call that time t_2; then, from Equation 3.11,

$$y = 0 = (29.6 \text{ m/s})t_2 + \tfrac{1}{2}(-9.80 \text{ m/s}^2)t_2^2.$$

This is a quadratic equation for t_2; it has two roots:

$$t_2 = 0 \qquad \text{and} \qquad t_2 = 6.04 \text{ s}.$$

There are two times at which $y = 0$: $t_2 = 0$ is the time the ball *leaves* the ground, and $t_2 = 6.04$ s is the time of its return. The latter is exactly twice the time taken to reach the highest point, so the time of descent equals the time of ascent. (This is *always* true if the starting and ending points are at the same elevation and air resistance is neglected. We'll prove it in Example 3.5.)

The range R is the value of x when the ball returns to the ground—that is, when $t = 6.04$ s:

$$R = v_{0x}t_2 = (22.2 \text{ m/s})(6.04 \text{ s}) = 134 \text{ m}.$$

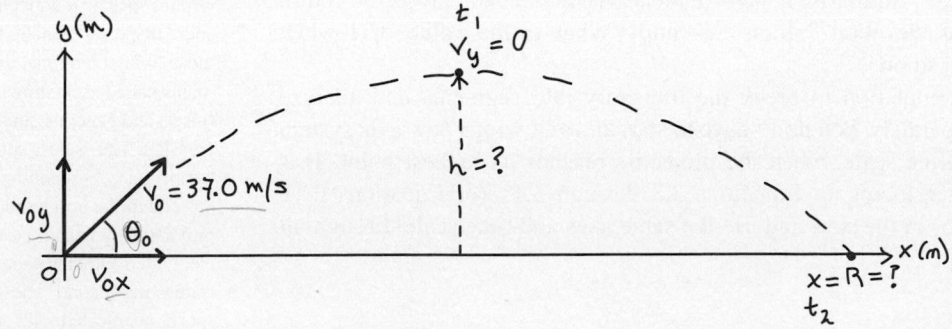

▲ **FIGURE 3.16** Our sketch for this problem.

Continued

For comparison, the distance from home plate to center field at Pittsburgh's PNC Park is 399 ft (about 122 m). If we could neglect air resistance, the ball really would be a home run.

At the instant the ball returns to $y = 0$, its vertical component of velocity is

$$v_y = 29.6 \text{ m/s} + (-9.80 \text{ m/s}^2)(6.04 \text{ s}) = -29.6 \text{ m/s}.$$

That is, v_y has the same magnitude as the initial vertical velocity v_{0y}, but the opposite sign. Since v_x is constant, the angle $\theta = -53.1°$ (below the horizontal) at this point is the negative of the initial angle $\theta_0 = 53.1°$.

REFLECT The actual values of the maximum height h and the range R are substantially less than the values we've found, because air resistance is *not* negligible. In fact, the range of a batted ball is substantially greater (on the order of 10 m for a home-run ball) in Denver than in Pittsburgh, because the density of air is almost 20% less in Denver.

Practice Problem: If the ball could continue to travel *below* its original level (through an appropriately shaped hole in the ground), then negative values of y corresponding to times greater than 6.04 s would be possible. Compute the ball's position and velocity 8.00 s after the start of its flight. *Answers:* $x = 178$ m, $y = -76.8$ m, $v_x = 22.2$ m/s, $v_y = -48.8$ m/s.

EXAMPLE 3.5 Range and maximum height of a home-run ball

For the situation of Example 3.4, derive general expressions for the maximum height h and the range R of a ball hit with an initial speed v_0 at an angle θ_0 above the horizontal (between 0 and 90°). For a given v_0, what value of θ_0 gives the maximum height? The maximum horizontal range?

SOLUTION

SET UP We can use the same coordinate system and diagram as for Example 3.4, so $y_0 = 0$.

SOLVE The basic relations are Equations 3.12 through 3.15. The solution will follow the same pattern as in Example 3.4, but now no numerical values are given. Thus the results won't be numbers, but rather symbolic expressions from which we can extract general relationships and proportionalities.

First, for a given θ_0, *when* does the projectile reach its highest point? At this point, $v_y = 0$, so, from Equation 3.15, the time t_1 at the highest point $(y = h)$ is given by

$$v_y = v_0 \sin\theta_0 - gt_1 = 0, \quad t_1 = \frac{v_0 \sin\theta_0}{g}.$$

Next, in terms of v_0 and θ_0, what is the value of y at this time? From Equations 3.10 and 3.13,

$$h = v_{0y}t - \tfrac{1}{2}gt^2 = v_0 \sin\theta_0 \left(\frac{v_0 \sin\theta_0}{g}\right) - \tfrac{1}{2}g\left(\frac{v_0 \sin\theta_0}{g}\right)^2,$$

$$h = \frac{v_0^2 \sin^2\theta_0}{2g}.$$

To derive a general expression for the range R, we again follow the procedure we used in Example 3.4. First we find an expression for the time t_2 when the projectile returns to its initial elevation. At that time, $y = 0$; from Equation 3.13,

$$t_2(v_0 \sin\theta_0 - \tfrac{1}{2}gt_2) = 0.$$

The two roots of this quadratic equation for t_2 are $t_2 = 0$ (the launch time) and $t_2 = 2v_0 \sin\theta_0/g$. The range R is the value of x at the *second* time. Substituting this expression for t_2 into Equation 3.12, we find that

$$R = (v_0 \cos\theta_0)\frac{2v_0 \sin\theta_0}{g}.$$

We now use the trigonometric identity $2\sin\theta_0 \cos\theta_0 = \sin 2\theta_0$ to rewrite this equation as

$$R = \frac{v_0^2 \sin 2\theta_0}{g}.$$

REFLECT If we vary θ_0, the *maximum* value of h occurs when $\sin\theta_0 = 1$ and $\theta_0 = 90°$, in other words, when the ball is hit straight up. That's what we should expect. In that case, $h = v_0^2/2g$. If the ball is launched horizontally, then $\theta_0 = 0$, and the maximum height is zero! The maximum range R occurs when $\sin 2\theta_0$ has its greatest value, namely, unity. That occurs when $2\theta_0 = 90°$, or $\theta_0 = 45°$. This angle gives the maximum range for a given initial speed.

Note that both h and R are proportional to the *square* of the initial speed v_0; this relationship may be a little surprising. If we double the initial speed, the range and maximum height increase by a factor of four! We can also use the proportional-reasoning methods of Section 2.5, along with the equations for h and R just derived, to confirm that both of these quantities are proportional to v_0^2 and inversely proportional to g.

Practice Problem: Show that the range is the same when the launch angle is 30° as when it is 60° and that it is the same for any launch angle θ_0 as for the complementary angle $(90° - \theta_0)$. *Answer:* $\sin\phi = \sin(180° - \phi)$, so $\sin 2\theta_0 = \sin 2(90° - \theta_0)$.

NOTE ▶ We don't recommend memorizing the equations for h and R. They are applicable only in the special circumstances that we have described. In particular, they are valid *only* when the launch and landing heights are equal. You may encounter problems to which these equations are *not* applicable. ◀

Figure 3.17 is based on a composite photograph of three trajectories of a ball projected from a spring gun at angles of $< 45°$, $45°$, and $> 45°$. The initial speed v_0 is approximately the same in all three cases. The horizontal ranges are nearly the same for the smallest and largest angles, and the range for $45°$ is greater than either.

Symmetry in Projectile Motion

Comparing the expressions for t_1 and t_2 in Example 3.5, we see that $t_2 = 2t_1$; that is, the total flight time is twice the time required to reach the highest point. It follows that the time required to reach the highest point equals the time required to fall from there back to the initial elevation, as we asserted in Example 3.4. More generally, Figure 3.11 shows that the path of the particle is symmetric about the highest point.

A 45° launch angle gives the greatest range; other angles fall shorter.

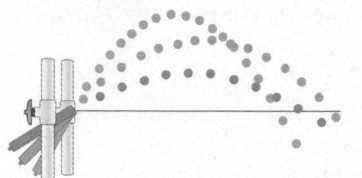

▲ **FIGURE 3.17** A firing angle of 45° gives the maximum horizontal range (based on a strobe photo).

Conceptual Analysis 3.3

Throwing stones

Two stones are launched from the top of a tall building. One stone is thrown in a direction 20° above the horizontal with a speed of 10 m/s; the other is thrown in a direction 20° below the horizontal with the same speed. How do their speeds compare just before they hit the ground below? (Neglect air friction.)

A. The one thrown upward is traveling faster.
B. The one thrown downward is traveling faster.
C. Both are traveling at the same speed.

SOLUTION We've just learned that the portion of a projectile's trajectory that lies above the initial height will be symmetric about its highest point. Thus, when the stone launched upward comes back to the level of the building's top, it is moving at 10 m/s in a direction 20° below the horizontal (symmetric to its launch velocity). At this point, it has the same speed and angle as the stone launched downward. The velocities of the two stones therefore match exactly at any position below the top of the building. So answer C is correct.

EXAMPLE 3.6 **Kicking a field goal**

For a field goal attempt, a football is kicked from a point on the ground that is a horizontal distance s from the goalpost. For the attempt to be successful, the ball must clear the crossbar, 10 ft (about 3.05 m) above the ground, as shown in Figure 3.18.

The ball leaves the kicker's foot with an initial speed of 20.0 m/s at an angle of 30° above the horizontal. What is the distance d between kicker and goalpost if the ball barely clears the crossbar?

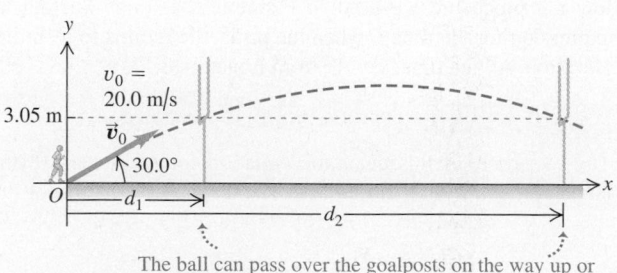

The ball can pass over the goalposts on the way up or on the way down, so there are two possible values of x.

▲ **FIGURE 3.18** Kicking a field goal.

Continued

SOLUTION

SET UP We use our idealized model of projectile motion, in which we assume level ground, neglect effects of air resistance, and treat the football as a point particle. We place the origin of coordinates at the point where the ball is kicked. Then $x_0 = y_0 = 0$,

$$v_{0x} = v_0 \cos 30.0° = (20.0 \text{ m/s})(0.866) = 17.3 \text{ m/s}, \quad \text{and}$$
$$v_{0y} = v_0 \sin 30.0° = (20.0 \text{ m/s})(0.500) = 10.0 \text{ m/s}.$$

SOLVE We first ask *when* (i.e., at what value of t) the ball is at a height of 3.05 m above the ground; then we find the value of x at that time. When that value of x is equal to the distance d, the ball is just barely passing over the crossbar.

To find the time t when $y = 3.05$ m, we use the equation $y = y_0 + v_{0y}t - \frac{1}{2}gt^2$. Substituting numerical values, we obtain

$$3.05 \text{ m} = (10.0 \text{ m/s})t - (4.90 \text{ m/s}^2)t^2.$$

This is a quadratic equation; to solve it, we first write it in standard form: $(4.90 \text{ m/s}^2)t^2 - (10.0 \text{ m/s})t + 3.05 \text{ m} = 0$. Then we use the quadratic formula. (See Chapter 0 if you need to review it.) We get

$$t = \frac{1}{2(4.9 \text{ m/s}^2)}$$
$$\times \left(10.0 \text{ m/s} \pm \sqrt{(10.0 \text{ m/s})^2 - 4(4.90 \text{ m/s}^2)(3.05 \text{ m})}\right)$$
$$= 0.373 \text{ s}, 1.67 \text{ s}.$$

There are two roots: $t = 0.373$ s and $t = 1.67$ s. The ball passes the height 3.05 m twice, once on the way up and once on the way down. We need to find the value of x at each of these times, using the equation for x as a function of time. Because $x_0 = 0$, we have simply $x = v_{0x}t$. For $t = 0.373$ s, we get $x = d_1 = (17.3 \text{ m/s})(0.373 \text{ s}) = 6.45$ m, and for $t = 1.67$ s, we get $x = d_2 = (17.3 \text{ m/s})(1.67 \text{ s}) = 28.9$ m. So if the goalpost is located between 6.45 m and 28.9 m from the initial point, the ball will pass over the crossbar; otherwise it will pass under it.

REFLECT The distance of 28.9 m is about 32 yards; field goal attempts are often successful at that distance. The ball passes the height $y = 10$ ft twice, once on the way up (when $t = 0.373$ s) and once on the way down (when $t = 1.67$ s). To verify this, we could calculate v_y at both times; when we do, we find that it is positive (upward) at $t = 0.373$ s and negative (downward) with the same magnitude at $t = 1.67$ s.

Practice Problem: If the kicker gives the ball the same initial speed and angle, but the ball is kicked from a point 25 m from the goalpost, what is the height of the ball above the crossbar as it crosses over the goalpost? *Answer:* 1.2 m.

EXAMPLE 3.7   Robin Hood shoots a pear

In a festival competition, the great archer Robin Hood is challenged to hit a falling pear with an arrow. At the sound of a horn, he is to shoot his arrow, and at the same instant the pear will be dropped from the top of a tall tower. As shown in Figure 3.19, Robin Hood aims *directly* at the initial position of the pear, seemingly making no allowance for the fact that the pear is dropping as the arrow moves toward it. His rivals assume that this is a mistake, but to their shock, he hits the pear. Show that if the arrow is aimed directly at the initial position of the pear, it will *always* hit the pear, regardless of the pear's initial location or initial speed (assuming that neither pear nor arrow hits the ground first).

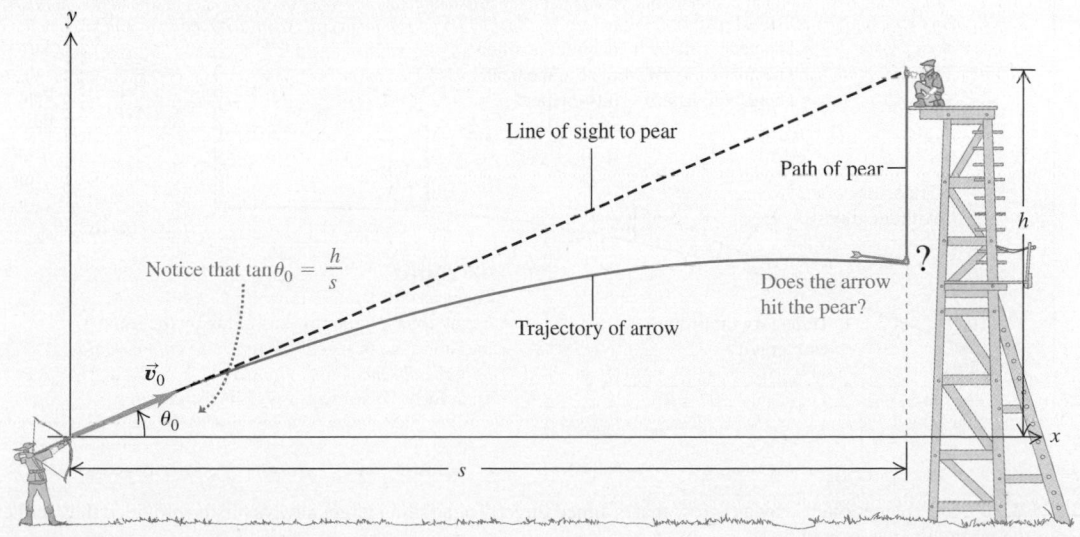

▲ **FIGURE 3.19** Robin Hood's shot.

Continued

SOLUTION

SET UP To show that the arrow hits the pear, we have to prove that there is some time when the arrow and the pear have the same x and y coordinates. We place the origin at the point from which Robin Hood shoots the arrow. Figure 3.19 shows that, initially, the pear is a horizontal distance s and a vertical distance $h = s \tan\theta_0$ from this point. First, we derive an expression for the time when the x coordinates are the same. Then we ask whether the y coordinates are *also* the same at this time; if they are, the arrow hits the pear.

SOLVE The pear drops straight down, so for it, $x_{\text{pear}} = s$ at *all* times. For the arrow, x_{arrow} is given by Equation 3.12: $x_{\text{arrow}} = (v_0 \cos\theta_0)t$. The time when the x coordinates of the arrow and pear are equal (i.e., $x_{\text{pear}} = x_{\text{arrow}}$) is given by $s = (v_0 \cos\theta_0)t$, or

$$t = \frac{s}{v_0 \cos\theta_0}.$$

Now we ask whether y_{arrow} and y_{pear} are also equal at this time; if they are, we have a hit. The pear is in one-dimensional free fall, and its position at *any* time t is given by Equation 3.11, that is, $y = y_0 + v_{0y}t - \frac{1}{2}gt^2$. The initial height is $y_0 = s \tan\theta_0$, the initial y component of velocity v_y is zero, and we find that

$$y_{\text{pear}} = s \tan\theta_0 - \tfrac{1}{2}gt^2.$$

For the arrow, we use Equation 3.13:

$$y_{\text{arrow}} = (v_0 \sin\theta_0)t - \tfrac{1}{2}gt^2.$$

If the y coordinates are equal ($y_{\text{pear}} = y_{\text{arrow}}$) at the same time that $t = s/v_0 \cos\theta_0$, the time when their x coordinates are equal, we have a hit. We see that this happens if $s \tan\theta_0 = (v_0 \sin\theta_0)t$ at time $t = s/v_0 \cos\theta_0$ (the time when the x coordinates are equal).

When we substitute this expression for t into the preceding equation, the right side becomes

$$(v_0 \sin\theta_0)t = (v_0 \sin\theta_0)\frac{s}{v_0 \cos\theta_0} = s \tan\theta_0.$$

We see that the right side is indeed equal to the left side, showing that, at the time when the x coordinates of pear and arrow are equal, their y coordinates are also equal. Robin Hood wins the contest!

REFLECT We have established the fact that, at the time the x coordinates are equal, the y coordinates are also equal. Thus, an arrow aimed at the initial position of the pear *always* hits it, no matter what v_0 is. With no gravity ($g = 0$), the pear would remain motionless, and the arrow would travel in a straight line to hit it. With gravity, both objects "fall" the same additional distance ($-\frac{1}{2}gt^2$) below their $g = 0$ positions, and the arrow still hits the pear. We can see this explicitly in the expressions for the y positions of the two objects, where the effects of gravity are highlighted:

Arrow: Pear:
$$y_{\text{arrow}} = v_{0y}\,t - \tfrac{1}{2}gt^2 \quad \text{and} \quad y_{\text{pear}} = y_{0,\text{pear}} - \tfrac{1}{2}gt^2$$
Effect of gravity is the same for both.

Figure 3.20 shows a comparison of the situations with and without gravity.

Practice Problem: Suppose the pear is released from a height of 6.00 m above Robin's arrow, the arrow is shot at a speed of 30.0 m/s, and the distance between Robin Hood and the base of the tower is 15.0 m. Find the time at which the arrow hits the pear, the distance the pear has fallen, and the height of the arrow above its release point. *Answers:* 0.54 s, 1.4 m, and 4.6 m.

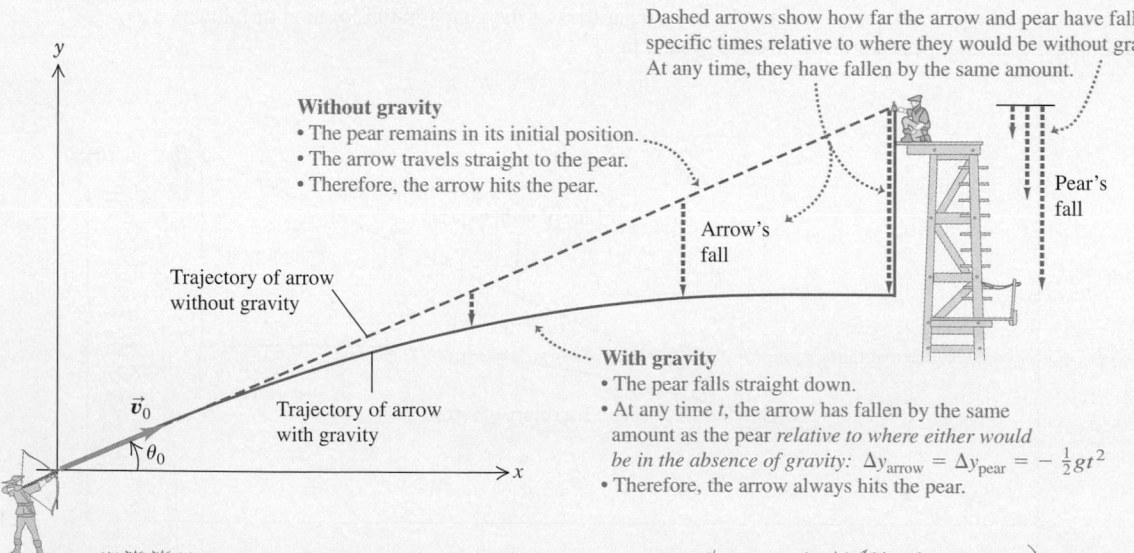

▲ FIGURE 3.20 An explanation of why an arrow that is aimed directly at a falling target always hits it (provided that neither the arrow nor the target hits the ground first).

We mentioned at the beginning of this section that air resistance isn't always negligible. When it has to be included, the calculations become a lot more complicated, because the air-resistance forces depend on velocity and the acceleration is no longer constant. Figure 3.21 shows a computer simulation of the trajectory of the baseball in Example 3.4 for 10 s of flight, with an air-resistance force proportional to the square of the particle's speed. We see that air resistance decreases the maximum height and range substantially, and the trajectory becomes asymmetric. In this case, the initial angle θ_0 that gives maximum range (for a given value of v_0) is less than 45°.

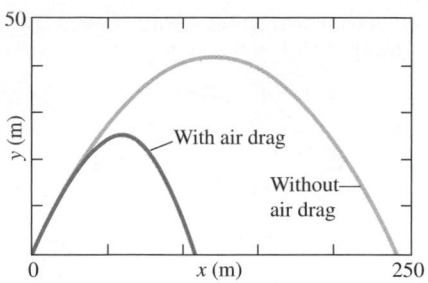

▲ **FIGURE 3.21** Computer-generated trajectories of a baseball with and without air resistance. Air resistance has a large cumulative effect on the flight of a baseball.

Quantitative Analysis 3.4

Shooting at a cliff

Suppose you shoot an arrow across a ravine at the vertical cliff opposite you, a distance d away, as shown in Figure 3.22. On your first try, you aim your arrow 15° above the horizontal, directly toward point A; it strikes the cliff a distance h below point A. On your next try, you aim the arrow 15° below the horizontal, toward point B, shooting it with the same speed. It will strike the cliff

A. the same distance h below point B.
B. less than distance h below point B.
C. more than distance h below point B.

SOLUTION If there were no gravitational force on the arrows, they would travel in straight lines, hitting the cliff at points A and B, respectively. The earth's gravitational attraction causes each arrow to fall below its straight-line path by an amount $\Delta y = -\frac{1}{2}gt^2$ for any time interval t. So, if the two arrows take the same time to reach the cliff, they will have fallen below their aiming points by the same amount $|\Delta y| = h$ when they strike, and answer A will be correct. Do they take the same time to reach the cliff? Remember that the time it takes a projectile to reach a given horizontal position x (the cliff, in this problem) depends on the projectile's horizontal component of velocity v_{0x}: $x - x_0 = v_{0x}t$ (where $x_0 = 0$ in this problem). Because the two arrows are shot at the same initial speed and at angles of 15° with respect to the horizontal, this component of velocity is the same for both. Therefore, the two arrows do in fact take the same time to reach the cliff and strike at equal distances below their aiming points.

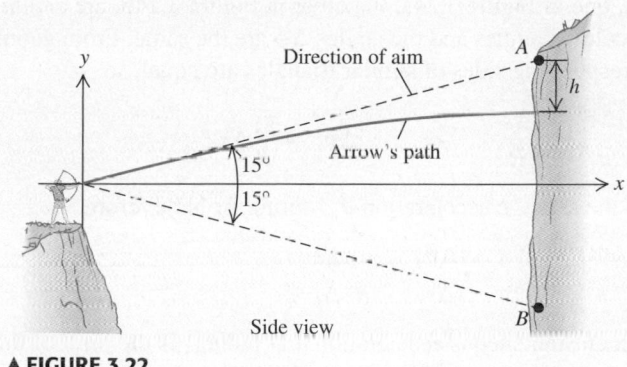

▲ **FIGURE 3.22**

3.4 Uniform Circular Motion

We discussed components of acceleration in Section 3.2. When a particle moves along a curved path, the direction of its velocity changes. Thus it *must* have a component of acceleration perpendicular to the path, even if its speed is constant.

When a particle moves in a circle with constant speed, the motion is called **uniform circular motion.** A car rounding a curve with a constant radius at constant speed, a satellite moving in a circular orbit, and an ice skater skating in a circle with constant speed are all examples of uniform circular motion. There is no component of acceleration parallel (tangent) to the path; otherwise, the speed would change (Figure 3.23). The component of acceleration perpendicular to the path, which causes the direction of the velocity to change, is related in a simple way to the speed v of the particle and the radius R of the circle. Our next project is to derive that relation.

First we note that this is a different problem from the projectile-motion situation in Section 3.3, in which the acceleration was always straight down and was constant in both magnitude and direction. Here the acceleration is perpendicular

MasteringPHYSICS

PhET: Ladybug Revolution
PhET: Motion in 2D
ActivPhysics 4.1: Magnitude of Centripetal Acceleration

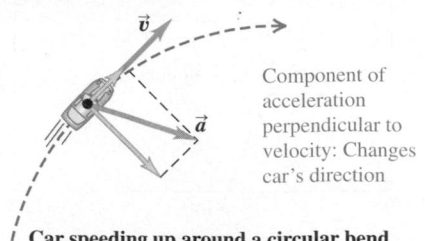

Component of acceleration parallel to velocity: Changes car's speed

Component of acceleration perpendicular to velocity: Changes car's direction

Car speeding up around a circular bend

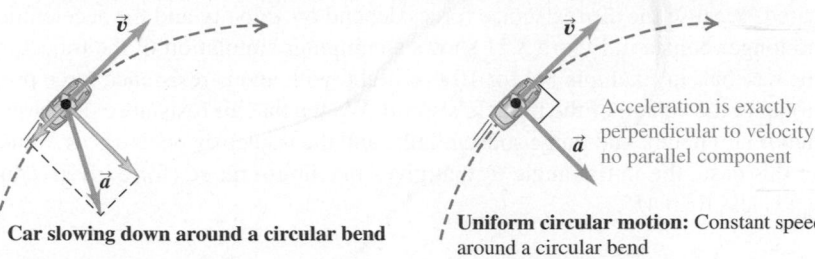

Car slowing down around a circular bend

Acceleration is exactly perpendicular to velocity; no parallel component

Uniform circular motion: Constant speed around a circular bend

▲ **FIGURE 3.23** In uniform circular motion, an object moves at constant speed in a circular path, and its acceleration is perpendicular to its velocity.

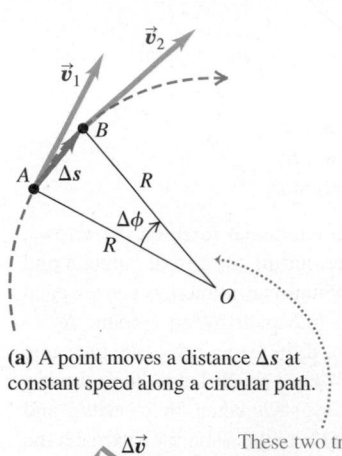

(a) A point moves a distance Δs at constant speed along a circular path.

These two triangles are similar.

(b) The corresponding change in velocity.

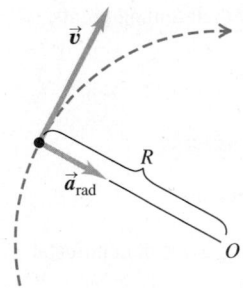

(c) The acceleration in uniform circular motion always points toward the center of the circle.

▲ **FIGURE 3.24** Finding the change in velocity $\vec{v}$ of an object moving in a circle at constant speed.

to the velocity at each instant; as the direction of the velocity changes, the direction of the acceleration also changes. As we will see, the acceleration vector at each point in the circular path is directed toward the *center* of the circle.

Figure 3.24a shows an object (represented by a dot) moving with constant speed in a circular path with radius R and center at O. The object moves from A to B, a distance Δs, in a time Δt. The vector change in velocity $\Delta \vec{v}$ during this time is shown in Figure 3.24b.

The two triangles, one in Figure 3.24a, the other in Figure 3.24b, are *similar*, because both are isosceles triangles and the angles $\Delta \phi$ are the same. From geometry, the ratios of corresponding sides of similar triangles are equal, so

$$\frac{|\Delta \vec{v}|}{v_1} = \frac{\Delta s}{R}, \qquad \text{or} \qquad |\Delta \vec{v}| = \frac{v_1}{R} \Delta s.$$

The magnitude a_{av} of the average acceleration $\vec{a}_{av}$ during Δt is therefore

$$a_{av} = \frac{|\Delta \vec{v}|}{\Delta t} = \frac{v_1}{R} \frac{\Delta s}{\Delta t}.$$

The magnitude a of the instantaneous acceleration $\vec{a}$ at point A is the limit of this expression as Δt approaches zero and point B gets closer and closer to point A:

$$a = \lim_{\Delta t \to 0} \frac{v_1}{R} \frac{\Delta s}{\Delta t} = \frac{v_1}{R} \lim_{\Delta t \to 0} \frac{\Delta s}{\Delta t}.$$

But the limit of $\Delta s / \Delta t$ is just the *speed* v_1 at point A. Also, A can be any point on the path, and the speed is the same at every point on the path. So we can drop the subscript and let v represent the speed at *any* point. Then we obtain the following relationship:

Acceleration in uniform circular motion

The acceleration of an object in uniform circular motion is *radial*, meaning that it always points toward the center of the circle and is perpendicular to the object's velocity $\vec{v}$. We denote it as $\vec{a}_{rad}$; its magnitude a_{rad} is given by

$$a_{rad} = \frac{v^2}{R}. \tag{3.16}$$

That is, the magnitude a_{rad} is proportional to the square of the speed $(a_{rad} \propto v^2)$ and inversely proportional to the radius $(a_{rad} \propto 1/R)$.

Because the acceleration of an object in uniform circular motion is always directed toward the center of the circle, it is sometimes called **centripetal acceleration.**

▶ **Application Where am I?** If you've ever used a global positioning system (GPS) unit for navigating or geocaching, you've used an application of uniform circular motion. This system uses a group of 24 satellites to pinpoint locations anywhere on the earth's surface or in the air, often to within as little as 3 meters. Although these satellites are moving at speeds of over 11,000 km/h, their orbits are precisely known, and their exact positions at any instant can be determined. In the field, distance readings taken simultaneously from several of the satellites to a GPS unit provide position vectors that are used to determine precise positions anywhere on the earth.

The word *centripetal* is derived from two Latin words meaning "seeking the center."

Figure 3.25 shows the directions of the velocity and acceleration vectors at several points for an object moving with uniform circular motion. Compare the motion shown in this figure with the projectile motion in Figure 3.11, in which the acceleration is always directed straight down and is *not* perpendicular to the path, except at the highest point in the trajectory.

It may seem odd that the centripetal acceleration is proportional to the *square* of the object's speed, rather simply proportional to the speed. Here's a way to make that relationship more plausible. Suppose that in Figure 3.24 we double the object's speed. This doubles the magnitude of the velocity vector, which would seem to double its rate of change when its direction changes. But note that the direction is also changing twice as rapidly, an effect that, by itself, would double the magnitude of the acceleration. So each of the two effects separately would double a_{rad}, and the combined effect is to increase a_{rad} by a factor of four when v is doubled. Hence we say that a_{rad} is proportional to v^2.

▲ FIGURE 3.25 For an object in uniform circular motion, the velocity at each point is tangent to the circle, and the acceleration is directed toward the center of the circle.

EXAMPLE 3.8 **Fast car, flat curve**

The 2005 Corvette claims a "lateral acceleration" of 0.92 g, which is $(0.92)(9.8 \text{ m/s}^2) = 8.9 \text{ m/s}^2$. If this represents the maximum centripetal acceleration that can be attained without skidding out of the circular path, and if the car is traveling at a constant 45 m/s (about 101 mi/h), what is the minimum radius of curve the car can negotiate? (Assume that the curve is unbanked.)

SOLUTION

SET UP AND SOLVE Figure 3.26 shows our diagram. The solution is a straightforward application of Equation 3.16. We find that

$$R = \frac{v^2}{a_{rad}} = \frac{(45 \text{ m/s})^2}{8.9 \text{ m/s}^2} = 230 \text{ m} \qquad \left(\text{about } \tfrac{1}{7} \text{ mi}\right).$$

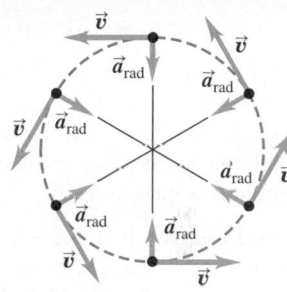

▲ FIGURE 3.26 Our diagram for this problem.

REFLECT Don't try this at home! The acceleration given represents an absolute maximum with smooth, dry pavement and very grippy tires. If the curve is banked, the radius can be smaller, as we'll see in Chapter 5.

Practice Problem: A more reasonable maximum acceleration for varying pavement conditions is 5.0 m/s². Under these condi-

tions, what is the maximum speed at which a car can negotiate a flat curve with radius 230 m? *Answer:* 34 m/s = 76 mi/h.

EXAMPLE 3.9 **A high-speed carnival ride**

The passengers in a carnival ride travel in a circle with radius 5.0 m (Figure 3.27). They make one complete circle in a time $T = 4.0$ s. What is their acceleration?

SOLUTION

SET UP Figure 3.28 shows our diagram.

SOLVE We again use Equation 3.16: $a = v^2/R$. To find the speed v, we use the fact that a passenger travels a distance equal to the circumference of the circle $(2\pi R)$ in the time T for one revolution:

$$v = \frac{2\pi R}{T} = \frac{2\pi (5.0 \text{ m})}{4.0 \text{ s}} = 7.9 \text{ m/s}.$$

The centripetal acceleration is

$$a_{\text{rad}} = \frac{v^2}{R} = \frac{(7.9 \text{ m/s})^2}{5.0 \text{ m}} = 12 \text{ m/s}^2.$$

REFLECT As in Example 3.8, the direction of $\vec{a}$ is always toward the center of the circle. The magnitude of $\vec{a}$ is greater than g, the acceleration due to gravity, so this is not a ride for the faint-hearted. (But some roller coasters subject their passengers to accelerations as great as $4g$.)

Practice Problem: If the ride increases in speed so that $T = 2.0$ s, what is a_{rad}? (This question can be answered by using proportional reasoning, without much arithmetic.) *Answer:* 49 m/s^2

▶ **FIGURE 3.27**

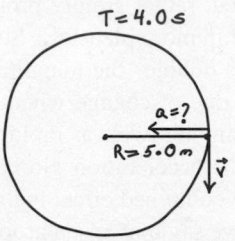

▲ **FIGURE 3.28** Our diagram for this problem.

3.5 Relative Velocity in a Plane

In Section 2.7, we introduced the concept of *relative velocity* for motion along a straight line. We can extend this concept to include motion in a plane by using vector addition to combine velocities. We suggest that you review Section 2.7 as a prelude to this discussion.

Suppose that the woman in Figure 2.29a is walking, not down the aisle of the railroad car, but from one side of the car to the other, with a speed of 1.0 m/s, as shown in Figure 3.29a. We can again describe the woman's position in two different frames of reference: that of the railroad car and that of the ground. Instead of coordinates x, we use position vectors $\vec{r}$. Let W represent the woman's position, C the frame of reference of the stationary ground observer (the cyclist), and T the frame of reference of the moving train. Then, as Figure 3.29b shows, the velocities are related by

$$\vec{v}_{W/C} = \vec{v}_{W/T} + \vec{v}_{T/C}.$$

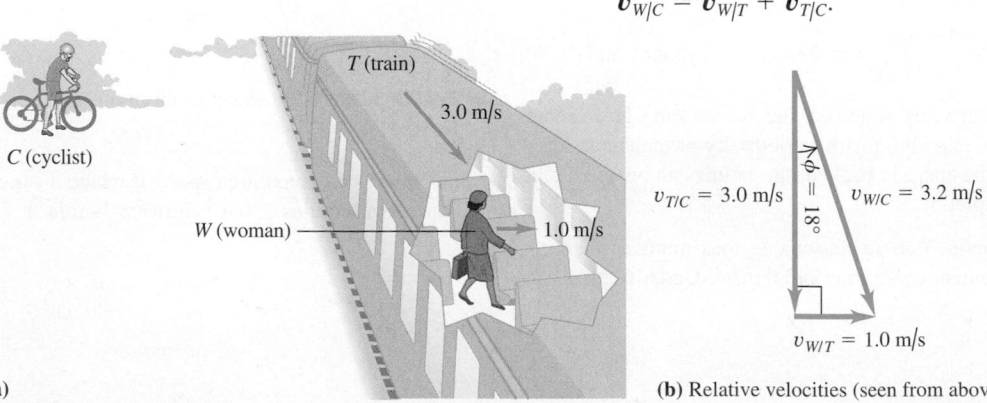

(a)

(b) Relative velocities (seen from above)

▲ **FIGURE 3.29** (a) A woman walking across a railroad car. (b) Vector diagram for the velocity of the woman relative to the cyclist. Recall that vector addition is commutative.

Relative motion in a plane

When an object W is moving with velocity $\vec{v}_{W/T}$ relative to an object (or observer) T, and T is moving with velocity $\vec{v}_{T/C}$ with respect to an object (or observer) C, the velocity $\vec{v}_{W/C}$ of W with respect to C is given by

$$\vec{v}_{W/C} = \vec{v}_{W/T} + \vec{v}_{T/C}. \tag{3.17}$$

If the train's velocity relative to the cyclist has magnitude $v_{T/C} = 3.0 \text{ m/s}$ and the woman's velocity relative to the train has magnitude $v_{W/T} = 1.0 \text{ m/s}$, then her velocity $\vec{v}_{W/C}$ relative to the cyclist is as shown in the vector diagram of Figure 3.29b. The Pythagorean theorem then gives us

$$v_{W/C} = \sqrt{(3.0 \text{ m/s})^2 + (1.0 \text{ m/s})^2} = \sqrt{10 \text{ m}^2/\text{s}^2} = 3.2 \text{ m/s}.$$

We can also see from the diagram that the *direction* of her velocity relative to the cyclist makes an angle ϕ with the train's velocity vector $\vec{v}_{T/C}$, where

$$\tan\phi = \frac{v_{W/T}}{v_{T/C}} = \frac{1.0 \text{ m/s}}{3.0 \text{ m/s}}, \qquad \phi = 18°.$$

As in Section 2.7, we have the general rule that if A and B are any two points or frames of reference, then

$$\vec{v}_{A/B} = -\vec{v}_{B/A}. \tag{3.18}$$

The velocity of the woman relative to the train is the negative of the velocity of the train relative to the woman, and so on.

PROBLEM-SOLVING STRATEGY 3.2   Relative velocity   (MP)

The strategy introduced in Section 2.7 is also useful here. The essential difference is that now the $\vec{v}$'s aren't all along the same line, so they have to be treated explicitly as vectors. For the double subscripts on the velocities, $\vec{v}_{A/B}$ always means "velocity of A relative to B." A useful rule for keeping the order of things straight is to regard each double subscript as a fraction. Then the fraction on the left side is the *product* of the fractions on the right sides: $P/A = (P/B)(B/A)$. This is helpful when you apply Equation 3.18 If there are *three* different frames of reference A, B, and C, you can write immediately

$$\vec{v}_{P/A} = \vec{v}_{P/C} + \vec{v}_{C/B} + \vec{v}_{B/A},$$

and so on. This is a *vector* equation, and you should always draw a vector diagram to show the addition of velocity vectors.

EXAMPLE 3.10   Flying in a crosswind

The compass of an airplane indicates that it is headed due north, and the airspeed indicator shows that the plane is moving through the air at 240 km/h. If there is a wind of 100 km/h from west to east, what is the velocity of the aircraft relative to the earth?

SOLUTION

SET UP Figure 3.30 shows the appropriate vector diagram. We choose subscript P to refer to the plane and subscript A to the moving air (which now plays the role of the railroad car in Figure 3.29). Subscript E refers to the earth. The information given is

$$\vec{v}_{P/A} = 240 \text{ km/h} \qquad \text{due north,}$$
$$\vec{v}_{A/E} = 100 \text{ km/h} \qquad \text{due east,}$$

and we want to find the magnitude and direction of $\vec{v}_{P/E}$.

Continued

SOLVE We adapt Equation 3.17 to the notation of this situation:

$$\vec{v}_{P/E} = \vec{v}_{P/A} + \vec{v}_{A/E}.$$

The three relative velocities and their relationship are shown in Figure 3.30. From this diagram, we find that

$$v_{P/E} = \sqrt{(240 \text{ km/h})^2 + (100 \text{ km/h})^2} = 260 \text{ km/h},$$

$$\alpha = \tan^{-1}\frac{100 \text{ km/h}}{240 \text{ km/h}} = 23° \text{ E of N.}$$

REFLECT The plane's velocity with respect to the air is straight north, but the air's motion relative to earth gives the plane's velocity with respect to earth a component toward the east.

Practice Problem: If the plane maintains its airspeed of 240 km/h, but the wind decreases, what is the wind speed if the plane's velocity with respect to earth is 15° east of north? *Answer:* 64 km/h

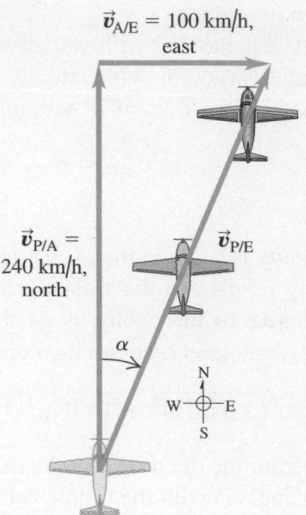

▲ **FIGURE 3.30** The plane is pointed north, but the wind blows east, giving the resultant velocity $\vec{v}_{P/E}$ relative to the earth.

EXAMPLE 3.11 **Compensating for a crosswind**

In what direction should the pilot in Example 3.10 head in order for the plane to travel due north? What will be her velocity relative to the earth then? (Assume that the wind velocity and the magnitude of her airspeed are the same as in Example 3.10.)

SOLUTION

SET UP Now the information given is

$$\vec{v}_{P/A} = 240 \text{ km/h} \qquad \text{direction unknown,}$$
$$\vec{v}_{A/E} = 100 \text{ km/h} \qquad \text{due east.}$$

Figure 3.31 shows the appropriate vector diagram. Be sure you understand why this is not the same diagram as that in Figure 3.30.

SOLVE We want to find $\vec{v}_{P/E}$; its magnitude is unknown, but we know that its direction is due north. Note that both this and the preceding example require us to determine two unknown quantities. In Example 3.10, these were the magnitude and direction of $\vec{v}_{P/E}$; in this example, the unknowns are the direction of $\vec{v}_{P/A}$ and the *magnitude* of $\vec{v}_{P/E}$.

The three relative velocities must still satisfy the vector equation

$$\vec{v}_{P/E} = \vec{v}_{P/A} + \vec{v}_{A/E}.$$

We find that

$$\beta = \sin^{-1}\frac{100 \text{ km/h}}{240 \text{ km/h}} = 25°,$$

$$v_{P/E} = \sqrt{(240 \text{ km/h})^2 - (100 \text{ km/h})^2} = 218 \text{ km/h.}$$

The pilot should head 25° west of north; her ground speed will then be 218 km/h.

REFLECT When a plane flies in a crosswind that is at right angles to the plane's velocity relative to the ground, its speed relative to

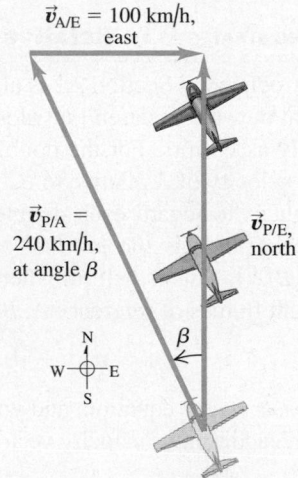

▲ **FIGURE 3.31** The pilot must point the plane in the direction of the vector $\vec{v}_{P/A}$ in order to travel due north relative to the earth.

the ground is always less than the airspeed; the trip takes longer than in calm air.

Practice Problem: Prove that if the wind speed is greater than or equal to 240 km/h, the plane can't fly straight north relative to the ground, no matter how hard it tries. Hurricanes often have wind speeds of greater than 240 km/h (about 150 mi/h).

SUMMARY

Position, Velocity, and Acceleration Vectors

(Sections 3.1 and 3.2) The position vector $\vec{r}$ of an object in a plane is the displacement vector from the origin to that object. Its components are the coordinates x and y. The average velocity $\vec{v}_{av}$ of the object during a time interval Δt is its displacement $\Delta \vec{r}$ (the change in the position vector $\vec{r}$), divided by Δt:

$$\vec{v}_{av} = \frac{\vec{r}_2 - \vec{r}_1}{t_2 - t_1} = \frac{\Delta \vec{r}}{\Delta t}. \tag{3.1}$$

Because Δt is a scalar quantity, the direction of average velocity vector $\vec{v}_{av}$ is determined entirely by the direction of the vector displacement $\Delta \vec{r}$.

The instantaneous velocity $\vec{v}$ is $\vec{v} = \lim\limits_{\Delta t \to 0} (\Delta \vec{r} / \Delta t)$; its components are $v_x = \lim\limits_{\Delta t \to 0} (\Delta x / \Delta t)$ and $v_y = \lim\limits_{\Delta t \to 0} (\Delta y / \Delta t)$.

The average acceleration $\vec{a}_{av}$ during the time interval Δt is the change in velocity, $\Delta \vec{v}$, divided by the time interval Δt:

$$\vec{a}_{av} = \frac{\vec{v}_2 - \vec{v}_1}{t_2 - t_1} = \frac{\Delta \vec{v}}{\Delta t}. \tag{3.4}$$

Instantaneous acceleration $\vec{a}$ is $\vec{a} = \lim\limits_{\Delta t \to 0} (\Delta \vec{v} / \Delta t)$; its components are $a_x = \lim\limits_{\Delta t \to 0} (\Delta v_x / \Delta t)$, $a_y = \lim\limits_{\Delta t \to 0} (\Delta v_y / \Delta t)$. An object has acceleration if *either* its speed or its direction of motion changes—that is, if either the magnitude or direction of its velocity changes.

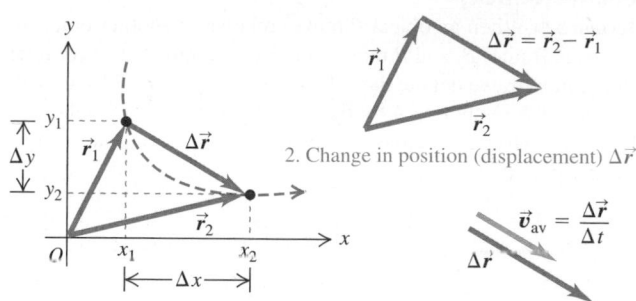

1. Position $\vec{r}$ in an x-y coordinate system
2. Change in position (displacement) $\Delta \vec{r}$
3. Average velocity $\vec{v}_{av}$ over a displacement

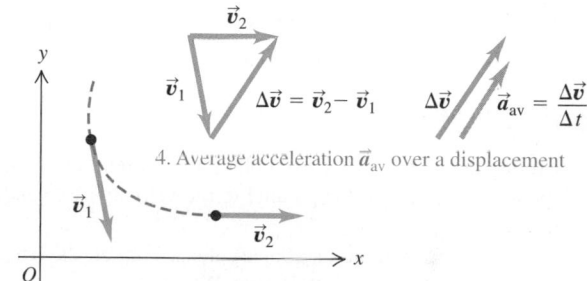

4. Average acceleration $\vec{a}_{av}$ over a displacement

Projectile Motion

(Section 3.3) Projectile motion occurs when an object is given an initial velocity and then follows a path determined entirely by the effect of a constant gravitational force. The path, called a **trajectory**, is a parabola in the x-y plane. The vertical motion of a projectile is independent of its horizontal motion.

In projectile motion, $a_x = 0$ (there is no horizontal component of acceleration) and $a_y = -g$ (a constant vertical component of acceleration due to a constant gravitational force). The coordinates and velocity components, as functions of time, are

$$x = (v_0 \cos\theta_0)t, \tag{3.12}$$

$$y = (v_0 \sin\theta_0)t - \frac{1}{2}gt^2, \tag{3.13}$$

$$v_x = v_0 \cos\theta_0, \tag{3.14}$$

$$v_y = v_0 \sin\theta_0 - gt. \tag{3.15}$$

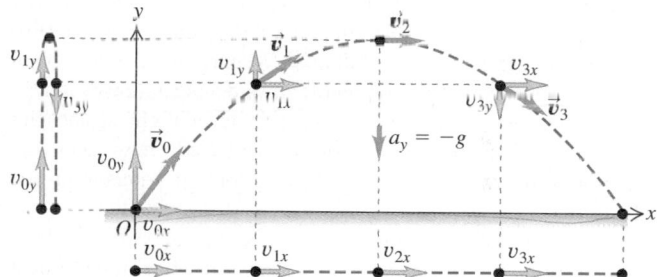

The vertical and horizontal components of a projectile's motion are independent

Uniform Circular Motion

(Section 3.4) When a particle moves in a circular path with radius R and with constant speed v, it has an acceleration with magnitude

$$a_{rad} = \frac{v^2}{R} \tag{3.16}$$

always directed toward the center of the circle and perpendicular to the instantaneous velocity v at each instant.

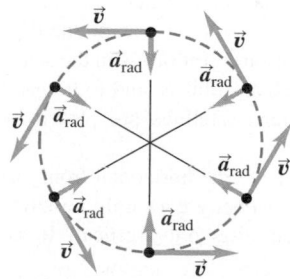

Continued

Relative Velocity

(Section 3.5) When an object P moves relative to another object (or reference frame) B, and B moves relative to a third object (or reference frame) A, we denote the velocity of P relative to B by $\vec{v}_{P/B}$, the velocity of P relative to A by $\vec{v}_{P/A}$, and the velocity of B relative to A by $\vec{v}_{B/A}$. These velocities are related by this variation of Equation 3.17:

$$\vec{v}_{P/A} = \vec{v}_{P/B} + \vec{v}_{B/A}.$$

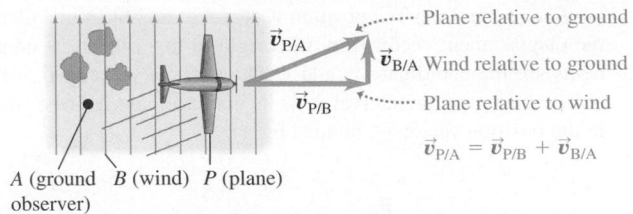

Plane relative to ground
$\vec{v}_{B/A}$ Wind relative to ground
Plane relative to wind

$$\vec{v}_{P/A} = \vec{v}_{P/B} + \vec{v}_{B/A}$$

A (ground B (wind) P (plane)
observer)

 For instructor-assigned homework, go to www.masteringphysics.com

Conceptual Questions

1. A football is thrown in a parabolic path. Is there any point at which the acceleration is parallel to the velocity? Perpendicular to the velocity?

2. When a rifle is fired at a stationary distant target, the barrel is not lined up exactly with the target. Why not?

3. A physicist measures the acceleration of a falling body in an elevator traveling at a constant speed of 9.8 m/s. What result does she obtain if the elevator is traveling (a) upward and (b) downward?

4. Intrigued by her result in the previous question, the physicist now measures the acceleration of a falling body in an elevator accelerating at a constant 9.8 m/s². What result does she obtain if the elevator is accelerating (a) upward and (b) downward?

5. A package falls from an airplane in level flight at constant speed. If air resistance can be neglected, how does the motion of the package look to the pilot? To a person on the ground?

6. If an athlete can give himself the same initial speed regardless of the direction in which he jumps, how is his maximum vertical jump (high jump) related to his maximum horizontal jump (long jump)?

7. The maximum range of a projectile occurs when it is aimed at a 45° angle if air resistance is neglected. At what angle should you launch it so that it will achieve the maximum time in the air? What would be its range in that case?

8. If an artificial earth satellite is in an orbit around the earth's equator with a period of exactly 1 day, how does its motion look to an observer on the rotating earth if it orbits in the same direction as the earth turns? (Such an orbit is said to be *geosynchronous;* most communications satellites are placed in geosynchronous orbits.)

9. A projectile is fired at an angle above the horizontal from the edge of a vertical cliff. (a) Is its velocity ever only horizontal? If so, when? (b) Is its velocity ever only vertical? If so, why?

10. An archer shoots an arrow from the top of a vertical cliff at an angle θ above the horizontal. When the arrow reaches the level ground at the bottom of the cliff, will its speed depend on the angle θ at which it was shot?

11. An observer draws the path of a stone thrown into the air, as shown in the Figure 3.32. What is wrong with the path shown? (There are *two* things wrong with it; can you spot both of them?)

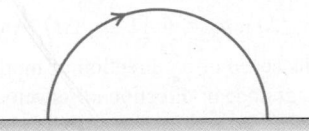

▲ **FIGURE 3.32** Problem 11.

12. A frisky grasshopper leaps into the air with an initial speed of 1.5 m/s at an angle of 60° above the level ground. The insect feels no air resistance. (a) Using the given information, what could you calculate regarding the grasshopper's motion? (Do not actually calculate it; just decide what things you *could* calculate.) (b) What information could you calculate if you knew only the grasshopper's initial speed, but not the angle at which it jumped?

13. In uniform circular motion, how does the acceleration change when the speed is increased by a factor of 3? When the radius is decreased by a factor of 2?

14. Suppose a space capsule with an astronaut inside is launched from the top of a tower at some angle above the horizontal. (a) While the capsule is in free fall, what are the acceleration of the astronaut and the capsule? (b) If the astronaut measures his acceleration compared with that of the capsule, what value will he get? (c) In what sense can we say that the astronaut is "weightless" in the capsule? Is he "weightless" because the gravity of the earth has gone away?

15. A hunter shoots a bullet from the top of a cliff. What is wrong with the drawing of the bullet's path in Figure 3.33?

16. According to what we have seen about circular motion, the earth is accelerating toward the sun, yet it is not getting any closer to the sun. To many people, this situation would seem like a contradiction. Explain why it really is *not* a contradiction.

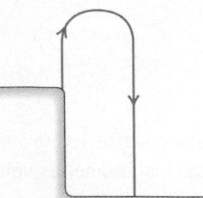

▲ **FIGURE 3.33** Problem 15.

Multiple-Choice Problems

1. A cannonball is fired toward a vertical building 400 m away with an initial velocity of 100 m/s at 36.9° above the horizontal. The ball will hit the building in
   A. 4.0 s.   B. 5.0 s.
   C. less than 4.0 s.   D. more than 5.0 s.

2. If the cannonball in the previous question is fired horizontally from a 150-m-high cliff, but still 400 m from the building, then the ball will hit the building in
   A. 4.0 s.   B. 5.0 s.
   C. less than 4.0 s.   D. more than 5.0 s.

3. A ball thrown horizontally from the top of a building hits the ground in 0.50 s. If it had been thrown with twice the speed in the same direction, it would have hit the ground in
   A. 4.0 s.   B. 1.0 s.   C. 0.50 s.
   D. 0.25 s.   E. 0.125 s.

4. Two balls are dropped from the top of the Leaning Tower of Pisa. The second ball is dropped a fraction of a second after the first ball. As they continue to accelerate to the ground, the distance between them will
   A. remain constant.
   B. decrease.
   C. increase.

5. Two balls are dropped at the same time from different heights. As they accelerate toward the ground, the distance between them will
   A. remain constant.
   B. decrease.
   C. increase.

6. A rock dropped from a small bridge encounters no air resistance and hits the pond below 1.0 s later. If it had been thrown horizontally with a speed of 10 m/s, then it would have hit the water
   A. more than 1.0 s later, since it has a greater distance to travel.
   B. 1.0 s later, since the vertical motion is not affected by the horizontal motion.
   C. less than 1.0 s later, since the higher speed would tend to get the ball to the ground more quickly.

7. An airplane flying at a constant horizontal velocity drops a package of supplies to a scientific mission in the Antarctic. If air resistance is negligibly small, the path of this package, *as observed by a person in the plane,* is
   A. a parabola.
   B. a straight line downward.
   C. a straight line pointing ahead of the plane.
   D. a straight line pointing behind the plane.

8. Your boat departs from the bank of a river that has a swift current parallel to its banks. If you want to cross this river in the shortest amount of time, you should direct your boat
   A. perpendicular to the current.
   B. upstream.
   C. downstream.
   D. so that it drifts with the current.

9. A car is driving toward a building at 10 mph. On the roof of the car, an insect is scurrying away from the building at 15 mph relative to the car's roof. To a person sitting in the building, the insect's velocity is
   A. 25 mph away from the building.
   B. 25 mph toward the building.
   C. 5 mph away from the building.
   D. 5 mph toward the building.

10. A ball is thrown horizontally from the top of a building and lands a distance d from the foot of the building after having been in the air for a time T and encountering no significant air resistance. If the building were twice as tall, the ball would have
    A. landed a distance $2d$ from the foot of the building.
    B. been in the air a time $2T$.
    C. been in the air a time $T\sqrt{2}$.
    D. reached the ground with twice the speed it did from the shorter building.

11. A child standing on a rotating carousel at a distance R from the center walks slowly inward until she's a distance $R/2$ from the center. Compared to her original centripetal acceleration, her acceleration is now
    A. half as great
    B. twice as great
    C. $\sqrt{2}$ times as great
    D. unchanged

12. An airplane whose air speed is 600 mi/h is flying perpendicular to a jet stream whose speed relative to the earth's surface is 100 mi/h. The airplane's speed relative to the earth's surface is
    A. 600 mi/h
    B. somewhat less than 600 mi/h
    C. somewhat more than 600 mi/h
    D. 100 mi/h

13. A golf ball is hit into the air, but not straight up, and encounters no significant air resistance. Which statements accurately describe its motion while it is in the air? (More than one choice may be correct.)
    A. On the way up it is accelerating upward, and on the way down it is accelerating downward.
    B. On the way up, both its horizontal and vertical velocity components are decreasing; on the way down, they are both increasing.
    C. Its vertical acceleration is zero at the highest point.
    D. Its horizontal velocity does not change once it is in the air, but its vertical velocity does change.

14. You shoot a bullet at a tasty apple high up in a tree some feet in front of you, hoping to knock it loose so you can catch it and eat it. In order to hit this apple, you should aim
    A. directly at it.
    B. above it.
    C. below it.

15. At the same time that rock A is dropped from rest from the top of a building, rock B is thrown horizontally away from the building, also starting at the top. Air resistance is not large enough to worry about. Which of the following statements are correct? (More than one choice may be correct.)
    A. Rock B hits the ground before rock A does.
    B. Both rocks have the same speed just as they reach the ground.
    C. Rock B has more acceleration than rock A.
    D. Both rocks reach the ground at the same time.

16. A stone is thrown horizontally with a speed of 15 m/s from the top of a vertical cliff at the edge of a lake. If the stone hits the water 2.0 s later, the height of the cliff is closest to
    A. 10 m.   B. 20 m.   C. 30 m.   D. 50 m.

17. An object traveling at constant speed V in a circle of radius R has an acceleration a. If *both* R and V are doubled, the new acceleration will be
    A. a.   B. $2a$.   C. $4a$.   D. $8a$.

Problems

3.1 Velocity in a Plane
3.2 Acceleration in a Plane

1. • A meteor streaking through the night sky is located with radar. At point A its coordinates are $(5.00 \text{ km}, 1.20 \text{ km})$, and 1.14 s later it has moved to point B with coordinates $(6.24 \text{ km}, 0.925 \text{ km})$. Find (a) the x and y components of its average velocity between A and B and (b) the magnitude and direction of its average velocity between these two points.

2. • At an air show, a jet plane has velocity components $v_x = 625 \text{ km/h}$ and $v_y = 415 \text{ km/h}$ at time 3.85 s and $v_x = 838 \text{ km/h}$ and $v_y = 365 \text{ km/h}$ at time 6.52 s. For this time interval, find (a) the x and y components of the plane's average acceleration and (b) the magnitude and direction of its average acceleration.

3. •• A dragonfly flies from point A to point B along the path shown in Figure 3.34 in 1.50 s. (a) Find the x and y components of its position vector at point A. (b) What are the magnitude and direction of its position vector at A? (c) Find the x and y components of the dragonfly's average velocity between A and B. (d) What are the magnitude and direction of its average velocity between these two points?

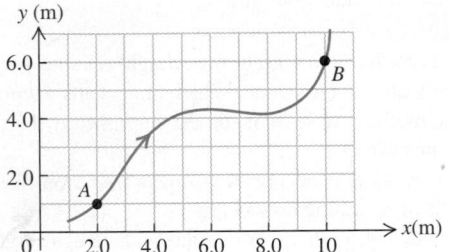

▲ **FIGURE 3.34** Problem 3.

4. • A coyote chasing a rabbit is moving 8.00 m/s due east at one moment and 8.80 m/s due south 4.00 s later. Find (a) the x and y components of the coyote's average acceleration during that time and (b) the magnitude and direction of the coyote's average acceleration during that time.

5. •• An athlete starts at point A and runs at a constant speed of 6.0 m/s around a round track 100 m in diameter, as shown in Figure 3.35. Find the x and y components of this runner's average velocity and average acceleration between points (a) A and B, (b) A and C, (c) C and D, and (d) A and A (a full lap). (e) Calculate the magnitude of the runner's average velocity between A and B. Is his average speed equal to the magnitude of his average velocity? Why or why not? (f) How can his velocity be changing if he is running at constant speed?

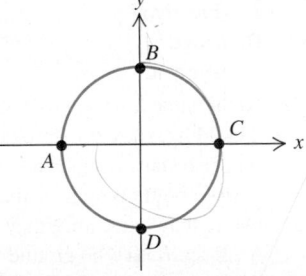

▲ **FIGURE 3.35** Problem 5.

3.3 Projectile Motion

6. • A stone is thrown horizontally at 30.0 m/s from the top of a very tall cliff. (a) Calculate its horizontal position and vertical position at 2-s intervals for the first 10.0 s. (b) Plot your positions from part (a) to scale. Then connect your points with a smooth curve to show the trajectory of the stone.

7. • A baseball pitcher throws a fastball horizontally at a speed of 42.0 m/s. Ignoring air resistance, how far does the ball drop between the pitcher's mound and home plate, 60 ft 6 in away?

8. • A physics book slides off a horizontal tabletop with a speed of 1.10 m/s. It strikes the floor in 0.350 s. Ignore air resistance. Find (a) the height of the tabletop above the floor, (b) the horizontal distance from the edge of the table to the point where the book strikes the floor, and (c) the horizontal and vertical components of the book's velocity, and the magnitude and direction of its velocity, just before the book reaches the floor.

9. • A tennis ball rolls off the edge of a tabletop 0.750 m above the floor and strikes the floor at a point 1.40 m horizontally from the edge of the table. (a) Find the time of flight of the ball. (b) Find the magnitude of the initial velocity of the ball. (c) Find the magnitude and direction of the velocity of the ball just before it strikes the floor.

10. • A military helicopter on a training mission is flying horizontally at a speed of 60.0 m/s when it accidentally drops a bomb (fortunately, not armed) at an elevation of 300 m. You can ignore air resistance. (a) How much time is required for the bomb to reach the earth? (b) How far does it travel horizontally while falling? (c) Find the horizontal and vertical components of the bomb's velocity just before it strikes the earth. (d) Draw graphs of the horizontal distance vs. time and the vertical distance vs. time for the bomb's motion. (e) If the velocity of the helicopter remains constant, where is the helicopter when the bomb hits the ground?

11. •• Inside a starship at rest on the earth, a ball rolls off the top of a horizontal table and lands a distance D from the foot of the table. This starship now lands on the unexplored Planet X. The commander, Captain Curious, rolls the same ball off the same table with the same initial speed as on earth and finds that it lands a distance $2.76D$ from the foot of the table. What is the acceleration due to gravity on Planet X?

12. •• A daring 510 N swimmer dives off a cliff with a running horizontal leap, as shown in Figure 3.36. What must her minimum speed be just as she leaves the top of the cliff so that she will miss the ledge at the bottom, which is 1.75 m wide and 9.00 m below the top of the cliff?

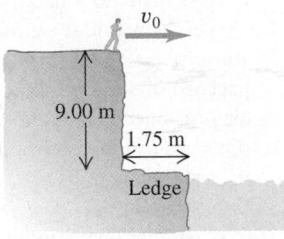

▲ **FIGURE 3.36** Problem 12.

13. •• **Leaping the river, I.** A 10,000 N car comes to a bridge during a storm and finds the bridge washed out. The 650 N driver must get to the other side, so he decides to try leaping it with his car. The side the car is on is 21.3 m above the river, while the opposite side is a mere 1.80 m above the river. The river itself is a raging torrent 61.0 m wide. (a) How fast should the car be traveling just as it leaves the cliff in order to clear the river and land safely on the opposite side? (b) What is the speed of the car just before it lands safely on the other side?

14. • A football is thrown with an initial upward velocity component of 15.0 m/s and a horizontal velocity component of 18.0 m/s. (a) How much time is required for the football to reach the highest point in its trajectory? (b) How high does it get above its release point? (c) How much time after it is

thrown does it take to return to its original height? How does this time compare with what you calculated in part (b)? Is your answer reasonable? (d) How far has the football traveled horizontally from its original position?

15. • A tennis player hits a ball at ground level, giving it an initial velocity of 24 m/s at 57° above the horizontal. (a) What are the horizontal and vertical components of the ball's initial velocity? (b) How high above the ground does the ball go? (c) How long does it take the ball to reach its maximum height? (d) What are the ball's velocity and acceleration at its highest point? (e) For how long a time is the ball in the air? (f) When this ball lands on the court, how far is it from the place where it was hit?

16. •• (a) A pistol that fires a signal flare gives it an initial velocity (muzzle velocity) of 125 m/s at an angle of 55.0° above the horizontal. You can ignore air resistance. Find the flare's maximum height and the distance from its firing point to its landing point if it is fired (a) on the level salt flats of Utah, and (b) over the flat Sea of Tranquility on the moon, where $g = 1.67$ m/s^2.

17. •• A major leaguer hits a baseball so that it leaves the bat at a speed of 30.0 m/s and at an angle of 36.9° above the horizontal. You can ignore air resistance. (a) At what two times is the baseball at a height of 10.0 m above the point at which it left the bat? (b) Calculate the horizontal and vertical components of the baseball's velocity at each of the two times you found in part (a). (c) What are the magnitude and direction of the baseball's velocity when it returns to the level at which it left the bat?

18. •• A balloon carrying a basket is descending at a constant velocity of 20.0 m/s. A person in the basket throws a stone with an initial velocity of 15.0 m/s horizontally perpendicular to the path of the descending balloon, and 4.00 s later this person sees the rock strike the ground. (See Figure 3.37.) (a) How high was the balloon when the rock was thrown out? (b) How far horizontally does the rock travel before it hits the ground? (c) At the instant the rock hits the ground, how far is it from the basket?

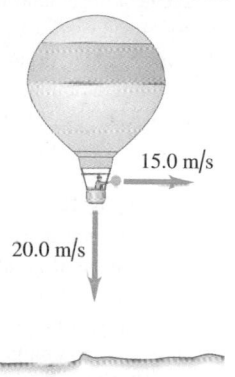

▲ **FIGURE 3.37**
Problem 18.

19. •• A batted baseball leaves the bat at an angle of 30.0° above the horizontal and is caught by an outfielder 375 ft from home plate at the same height from which it left the bat. (a) What was the initial speed of the ball? (b) How high does the ball rise above the point where it struck the bat?

20. •• A man stands on the roof of a 15.0-m-tall building and throws a rock with a velocity of magnitude 30.0 m/s at an angle of 33.0° above the horizontal. You can ignore air resistance. Calculate (a) the maximum height above the roof reached by the rock, (b) the magnitude of the velocity of the rock just before it strikes the ground, and (c) the horizontal distance from the base of the building to the point where the rock strikes the ground.

21. • **The champion jumper of the insect world.** The froghop-
BIO per, *Philaenus spumarius,* holds the world record for insect jumps. When leaping at an angle of 58.0° above the horizontal, some of the tiny critters have reached a maximum height of 58.7 cm above the level ground. (See *Nature,* Vol. 424, 31 July 2003, p. 509.) (a) What was the takeoff speed for such a leap? (b) What horizontal distance did the froghopper cover for this world-record leap?

22. •• A grasshopper leaps into the air from the edge of a vertical cliff, as shown in Figure 3.38. Use information from the figure to find (a) the initial speed of the grasshopper and (b) the height of the cliff.

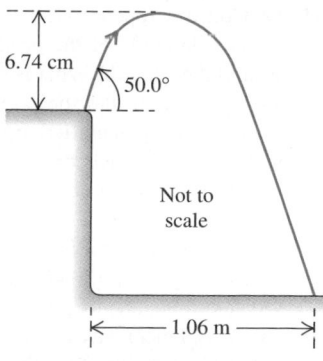

▲ **FIGURE 3.38** Problem 22.

23. •• Firemen are shooting a stream of water at a burning building. A high-pressure hose shoots out the water with a speed of 25.0 m/s as it leaves the hose nozzle. Once it leaves the hose, the water moves in projectile motion. The firemen adjust the angle of elevation of the hose until the water takes 3.00 s to reach a building 45.0 m away. You can ignore air resistance; assume that the end of the hose is at ground level. (a) Find the angle of elevation of the hose. (b) Find the speed and acceleration of the water at the highest point in its trajectory. (c) How high above the ground does the water strike the building, and how fast is it moving just before it hits the building?

24. •• Show that a projectile achieves its maximum range when it is fired at 45° above the horizontal if $y = y_0$.

25. •• A water balloon slingshot launches its projectiles essentially from ground level at a speed of 25.0 m/s. (a) At what angle should the slingshot be aimed to achieve its maximum range? (b) If shot at the angle you calculated in part (a), how far will a water balloon travel horizontally? (c) For how long will the balloon be in the air? (You can ignore air resistance.)

26. •• A certain cannon with a fixed angle of projection has a range of 1500 m. What will be its range if you add more powder so that the initial speed of the cannonball is tripled?

27. •• The nozzle of a fountain jet sits in the center of a circular pool of radius 3.50 m. If the nozzle shoots water at an angle of 65°, what is the maximum speed of the water at the nozzle that will allow it to land within the pool? (You can ignore air resistance.)

28. •• Two archers shoot arrows in the same direction from the same place with the same initial speeds but at different angles. One shoots at 45° above the horizontal, while the other shoots at 60.0°. If the arrow launched at 45° lands 225 m from the archer, how far apart are the two arrows when they land? (You can assume that the arrows start at essentially ground level.)

29. • A bottle rocket can shoot its projectile vertically to a height of 25.0 m. At what angle should the bottle rocket be fired to reach its maximum horizontal range, and what is that range? (You can ignore air resistance.)

30. •• An airplane is flying with a velocity of 90.0 m/s at an angle of 23.0° above the horizontal. When the plane is 114 m directly above a dog that is standing on level ground, a suitcase drops out of the luggage compartment. How far from the dog will the suitcase land? You can ignore air resistance.

3.4 Uniform Circular Motion

31. • You swing a 2.2 kg stone in a circle of radius 75 cm. At what speed should you swing it so its centripetal acceleration will be 9.8 m/s^2?

32. •• Consult Appendix E. Calculate the radial acceleration (in m/s^2 and g's) of an object (a) on the ground at the earth's equator and (b) at the equator of Jupiter (which takes 0.41 day to spin once), turning with the planet.

33. •• Consult Appendix E and assume circular orbits. (a) What is the magnitude of the orbital velocity, in m/s, of the earth around the sun? (b) What is the radial acceleration, in m/s, of the earth toward the sun? (c) Repeat parts (a) and (b) for the motion of the planet Mercury.

34. • A model of a helicopter rotor has four blades, each 3.40 m in length from the central shaft to the tip of the blade. The model is rotated in a wind tunnel at 550 rev/min. (a) What is the linear speed, in m/s, of the blade tip? (b) What is the radial acceleration of the blade tip, expressed as a multiple of the acceleration g due to gravity?

35. •• A wall clock has a second hand 15.0 cm long. What is the radial acceleration of the tip of this hand?

36. • A curving freeway exit has a radius of 50.0 m and a posted speed limit of 35 mi/h. What is your radial acceleration (in m/s²) if you take this exit at the posted speed? What if you take the exit at a speed of 50 mi/h?

37. •• **Dizziness.** Our balance is maintained, at least in part, by
BIO the endolymph fluid in the inner ear. Spinning displaces this fluid, causing dizziness. Suppose a dancer (or skater) is spinning at a very high 3.0 revolutions per second about a vertical axis through the center of his head. Although the distance varies from person to person, the inner ear is approximately 7.0 cm from the axis of spin. What is the radial acceleration (in m/s² and in g's) of the endolymph fluid?

38. • **Pilot blackout in a power dive.** A jet plane comes in for a
BIO downward dive as shown in Figure 3.39. The bottom part of the path is a quarter circle having a radius of curvature of 350 m. According to medical tests, pilots lose consciousness at an acceleration of 5.5g. At what speed (in m/s and mph) will the pilot black out for this dive?

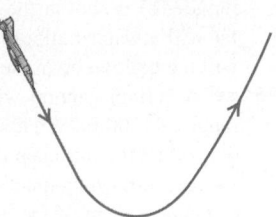

▲ **FIGURE 3.39** Problem 38.

3.5 Relative Velocity in a Plane

39. •• A canoe has a velocity of 0.40 m/s southeast relative to the earth. The canoe is on a river that is flowing 0.50 m/s east relative to the earth. Find the velocity (magnitude and direction) of the canoe relative to the river.

40. •• **Crossing the river, I.** A river flows due south with a speed of 2.0 m/s. A man steers a motorboat across the river; his velocity relative to the water is 4.2 m/s due east. The river is 800 m wide. (a) What is his velocity (magnitude and direction) relative to the earth? (b) How much time is required for the man to cross the river? (c) How far south of his starting point will he reach the opposite bank?

41. •• **Crossing the river, II.** (a) In which direction should the motorboat in the previous problem head in order to reach a point on the opposite bank directly east from the starting point? (The boat's speed relative to the water remains 4.2 m/s.) (b) What is the velocity of the boat relative to the earth? (c) How much time is required to cross the river?

42. • You're standing outside on a windless day when raindrops begin to fall straight down. You run for shelter at a speed of 5.0 m/s, and you notice while you're running that the raindrops appear to be falling at an angle of about 30° from the vertical. What's the vertical speed of the raindrops?

43. •• **Bird migration.** Canadian geese migrate essentially along
BIO a north–south direction for well over a thousand kilometers in some cases, traveling at speeds up to about 100 km/h. If one such bird is flying at 100 km/h relative to the air, but there is a 40 km/h wind blowing from west to east, (a) at what angle relative to the north–south direction should this bird head so that it will be traveling directly southward relative to the ground? (b) How long will it take the bird to cover a ground distance of 500 km from north to south? (*Note:* Even on cloudy nights, many birds can navigate using the earth's magnetic field to fix the north–south direction.)

General Problems

44. •• A test rocket is launched by accelerating it along a 200.0-m incline at 1.25 m/s² starting from rest at point A (Figure 3.40.) The incline rises at 35.0° above the horizontal, and at the instant the rocket leaves it, its engines turn off and it is subject only to gravity (air resistance can be ignored). Find (a) the maximum height above the ground that the rocket reaches, and (b) the greatest horizontal range of the rocket beyond point A.

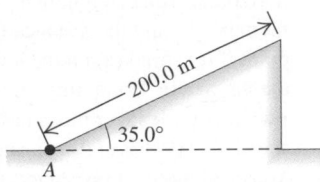

▲ **FIGURE 3.40** Problem 44.

45. •• A player kicks a football at an angle of 40.0° from the horizontal, with an initial speed of 12.0 m/s. A second player standing at a distance of 30.0 m from the first (in the direction of the kick) starts running to meet the ball at the instant it is kicked. How fast must he run in order to catch the ball just before it hits the ground?

46. •• **Dynamite!** A demolition crew uses dynamite to blow an old building apart. Debris from the explosion flies off in all directions and is later found at distances as far as 50 m from the explosion. Estimate the maximum speed at which debris was blown outward by the explosion. Describe any assumptions that you make.

47. •• **Fighting forest fires.** When fighting forest fires, airplanes work in support of ground crews by dropping water on the fires. A pilot is practicing by dropping a canister of red dye, hoping to hit a target on the ground below. If the plane is flying in a horizontal path 90.0 m above the ground and with a speed of 64.0 m/s (143 mi/h), at what horizontal distance from the target should the pilot release the canister? Ignore air resistance.

48. •• **An errand of mercy.** An airplane is dropping bales of hay to cattle stranded in a blizzard on the Great Plains. The pilot releases the bales at 150 m above the level ground when the plane is flying at 75 m/s 55° above the horizontal. How far in front of the cattle should the pilot release the hay so that the bales will land at the point where the cattle are stranded?

49. •• A cart carrying a vertical missile launcher moves horizontally at a constant velocity of 30.0 m/s to the right. It launches a rocket vertically upward. The missile has an initial vertical velocity of

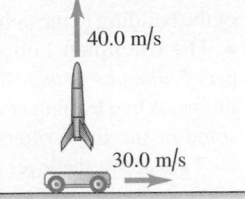

▲ **FIGURE 3.41** Problem 49.

40.0 m/s relative to the cart. (a) How high does the rocket go? (b) How far does the cart travel while the rocket is in the air? (c) Where does the rocket land relative to the cart?

50. •• **The longest home run.** According to the *Guinness Book of World Records,* the longest home run ever measured was hit by Roy "Dizzy" Carlyle in a minor-league game. The ball traveled 188 m (618 ft) before landing on the ground outside the ballpark. (a) Assuming that the ball's initial velocity was 45° above the horizontal, and ignoring air resistance, what did the initial speed of the ball need to be to produce such a home run if the ball was hit at a point 0.9 m (3.0 ft) above ground level? Assume that the ground was perfectly flat. (b) How far would the ball be above a fence 3.0 m (10 ft) in height if the fence were 116 m (380 ft) from home plate?

51. • A professional golfer can hit a ball with a speed of 70.0 m/s. What is the maximum distance a golf ball hit with this speed could travel on Mars, where the value of g is 3.71 m/s²? (The distances golf balls travel on earth are greatly shortened by air resistance and spin, as well as by the stronger force of gravity.)

52. •• A baseball thrown at an angle of 60.0° above the horizontal strikes a building 18.0 m away at a point 8.00 m above the point from which it is thrown. Ignore air resistance. (a) Find the magnitude of the initial velocity of the baseball (the velocity with which the baseball is thrown). (b) Find the magnitude and direction of the velocity of the baseball just before it strikes the building.

53. •• A boy 12.0 m above the ground in a tree throws a ball for his dog, who is standing right below the tree and starts running the instant the ball is thrown. If the boy throws the ball horizontally at 8.50 m/s, (a) how fast must the dog run to catch the ball just as it reaches the ground, and (b) how far from the tree will the dog catch the ball?

54. •• Suppose the boy in the previous problem throws the ball upward at 60.0° above the horizontal, but all else is the same. Repeat parts (a) and (b) of that problem

55. •• A firefighting crew uses a water cannon that shoots water at 25.0 m/s at a fixed angle of 53.0° above the horizontal. The firefighters want to direct the water at a blaze that is 10.0 m above ground level. How far from the building should they position their cannon? There are *two* possibilities; can you get them both? (*Hint:* Start with a sketch showing the trajectory of the water.)

56. • A gun shoots a shell into the air with an initial velocity of 100.0 m/s, 60.0° above the horizontal on level ground. Sketch quantitative graphs of the shell's horizontal and vertical velocity components as functions of time for the complete motion.

57. ••• **Look out!** A snowball rolls off a barn roof that slopes downward at an angle of 40.0°. (See Figure 3.42.) The edge of the roof is 14.0 m above the ground, and the snowball has a speed of 7.00 m/s as it rolls off the roof. Ignore air resistance. How far from the edge of the barn does the snowball strike the ground if it doesn't strike anything else while falling?

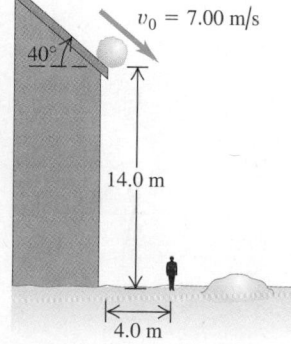

▲ **FIGURE 3.42** Problem 57.

58. •• **Spiraling up.** It is common to see birds of prey rising upward on thermals. The paths they take may be spiral-like. You can model the spiral motion as uniform circular motion combined with a constant upward velocity. Assume a bird completes a circle of radius 8.00 m every 5.00 s and rises vertically at a rate of 3.00 m/s. Determine: (a) the speed of the bird relative to the ground; (b) the bird's acceleration (magnitude and direction); and (c) the angle between the bird's velocity vector and the horizontal.

59. •• A water hose is used to fill a large cylindrical storage tank of diameter D and height $2D$ The hose shoots the water at 45° above the horizontal from the same level as the base of the tank and is a distance $6D$ away (Fig. 3.43). For what *range* of launch speeds (v_0) will the water enter the tank? Ignore air resistance, and express your answer in terms of D and g.

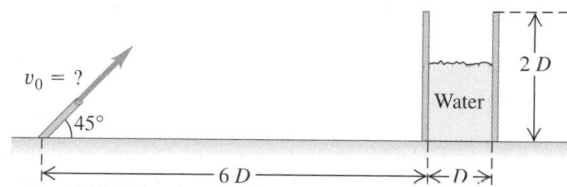

▲ **FIGURE 3.43** Problem 59.

60. •• **A world record.** In the shot put, a standard track-and-field event, a 7.3 kg object (the shot) is thrown by releasing it at approximately 40° over a straight left leg. The world record for distance, set by Randy Barnes in 1990, is 23.11 m. Assuming that Barnes released the shot put at 40.0° from a height of 2.00 m above the ground, with what speed, in m/s and mph, did he release it?

61. •• A Ferris wheel with radius 14.0 m is turning about a horizontal axis through its center, as shown in Figure 3.44. The linear speed of a passenger on the rim is constant and equal to 7.00 m/s. What are the magnitude and direction of the passenger's acceleration as she passes through (a) the lowest point in her circular motion and (b) the highest point in her circular motion? (c) How much time does it take the Ferris wheel to make one revolution?

▲ **FIGURE 3.44** Problem 61.

62. ••• **Leaping the river, II.** A physics professor did daredevil stunts in his spare time. His last stunt was an attempt to jump across a river on a motorcycle. (See Figure 3.45.) The takeoff ramp was inclined at 53.0°, the river was 40.0 m wide, and the far bank was 15.0 m lower than the top of the ramp. The river itself was 100 m below the ramp. You can ignore air resistance. (a) What should his speed have been at the top of the ramp for him to have just made it to the edge of the far bank? (b) If his speed was only half the value found in (a), where did he land?

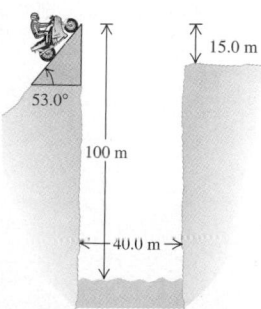

▲ **FIGURE 3.45** Problem 62.

63. •• A 76.0 kg boulder is rolling horizontally at the top of a vertical cliff that is 20.0 m above the surface of a lake, as shown in Figure 3.46. The top of the vertical face of a dam is located 100.0 m from the foot of the cliff, with the top of the dam level with the surface of the water in the lake. A level plain is 25.0 m below the top of the dam. (a) What must the minimum speed of the rock be just as it leaves the cliff so that it will travel to the plain without striking the dam? (b) How far from the foot of the dam does the rock hit the plain?

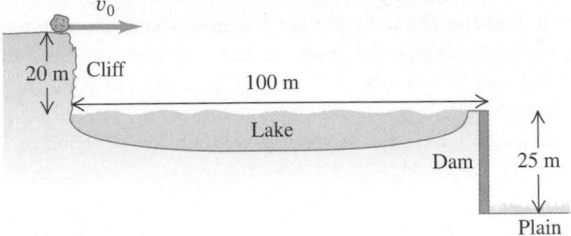

▲ **FIGURE 3.46** Problem 63.

64. •• A batter hits a baseball at a speed of 35.0 m/s and an angle of 65.0° above the horizontal. At the same instant, an outfielder 70.0 m away begins running away from the batter in the line of the ball's flight, hoping to catch it. How fast must the outfielder run to catch the ball? (Ignore air resistance, and assume the fielder catches the ball at the same height at which it left the bat.)

65. •• A shell is launched at 150 m/s 53° above the horizontal. When it has reached its highest point, it launches a projectile at a velocity of 100.0 m/s 30.0° above the horizontal relative to the *shell*. Find (a) the maximum height about the ground that the projectile reaches and (b) its distance from the place where the shell was fired to its landing place when it eventually falls back to the ground.

Passage Problems

BIO Throwing a ball on the moon. The acceleration due to gravity on the moon is approximately 1.6 m/s^2, or 1/6th that of the acceleration due to gravity on the earth. An astronaut who can throw a ball 5 m straight up on earth is now on the moon with the same ball. (Assume that air resistance on the earth is negligible.)

66. If the astronaut tosses the ball straight up on the moon, it will reach a height of
   A. 5 m
   B. 10 m
   C. 30 m
   D. 50 m

If the astronaut throws the ball at an angle of 45° (see figure 3.11) what can you say about the path of the ball compared to the path that it would have followed had it been thrown on the earth instead?

67. The ball will travel
   A. the same distance as on the earth
   B. farther
   C. not as far

68. The ball will stay aloft
   A. the same amount of time as on the earth
   B. a shorter amount of time
   C. a longer amount of time

69. The maximum height of the ball will be
   A. the same as on earth
   B. higher
   C. lower

70. If the time of flight of the ball is t seconds, at what point in time will the ball have zero vertical velocity?
   A. 0 s
   B. $t/4$ s
   C. $t/2$ s
   D. t s
   E. There is no place on the path where the vertical velocity is zero.

4 Newton's Laws of Motion

How can a tugboat push a cruise ship that's much heavier than the tug? Why is a long distance needed to stop the ship once it is in motion? Why does your foot hurt more when you kick a big rock than when you kick an empty cardboard box? Why is it harder to control a car on wet ice than on dry concrete? The answers to these and similar questions take us into the subject of **dynamics,** the relationship of motion to the forces associated with it. In the two preceding chapters we studied *kinematics,* the language for describing motion. Now we are ready to think about what makes objects move the way they do. In this chapter, we will use the kinematic quantities of displacement, velocity, and acceleration, along with two new concepts: *force* and *mass.*

All the principles of dynamics can be wrapped up in a neat package containing three statements called **Newton's laws of motion.** These laws, the cornerstone of mechanics, are based on experimental studies of moving objects. They are fundamental laws of nature; they cannot be deduced or proved from any other principles. They were clearly stated for the first time by Sir Isaac Newton (1642–1727), who published them in 1686 in his *Principia,* or *Mathematical Principles of Natural Philosophy.* Many other scientists before Newton contributed to the foundations of mechanics, especially Galileo Galilei (1564–1642), who died the same year Newton was born. Indeed, Newton himself said, "If I have been able to see a little farther than other men, it is because I have stood on the shoulders of giants."

Ordinarily we'd say that the elephant pushes the tree, not that the tree pushes the elephant. Yet the forces they exert on each other are equal in magnitude, so both statements are true. In this chapter we'll learn why.

4.1 Force

The concept of **force** gives us a quantitative description of the interaction between two objects or between an object and its environment (Figure 4.1). When you push on a car that is stuck in the snow, you exert a force on it. A locomotive exerts a force

- A force is a push or a pull.
- A force is an interaction between two objects or between an object and its environment.
- A force is a vector quantity, with magnitude and direction.

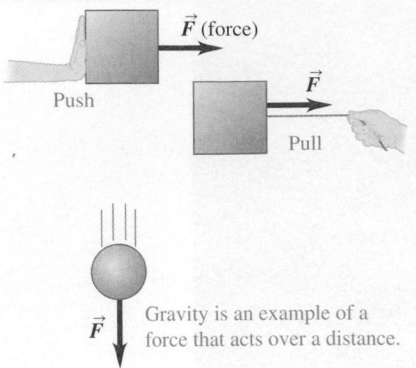

Gravity is an example of a force that acts over a distance.

▲ **FIGURE 4.1** The concept of a force.

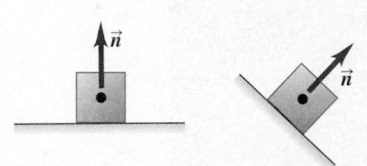

Normal force $\vec{n}$: When an object rests or pushes on a surface, the surface exerts a push on it that is directed perpendicular to the surface.

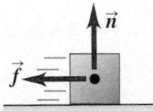

Friction force $\vec{f}$: In addition to the normal force, a surface may exert a frictional force on an object, directed parallel to the surface.

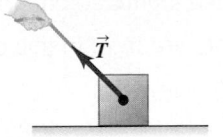

Tension force $\vec{T}$: A pulling force exerted on an object by a rope, cord, etc.

Weight $\vec{w}$: The pull of gravity on an object.

▲ **FIGURE 4.2** Some types of forces.

on the train it is pulling or pushing, a steel cable exerts a force on the beam it is lifting at a construction site, and so on. Each of these examples shows that force is a push or a pull acting on an object. In this chapter, we'll encounter several kinds of forces.

Types of Forces

When a force involves direct contact between two objects, we call it a **contact force.** When an object rests on a surface, there is always a component of force perpendicular to the surface; we call this component a **normal force,** denoted by $\vec{n}$ (Figure 4.2). There may also be a component of force parallel to the surface; we call this a **friction force,** denoted by $\vec{f}$. This force often (though not always) acts to resist sliding of the object on the surface.

When a rope or cord is attached to an object and pulled, the corresponding force applied to the object is referred to as a **tension,** denoted by $\vec{T}$.

A familiar force that we'll work with often is the gravitational attraction that the earth (or some other astronomical body) exerts on an object. This force is the object's **weight,** denoted by $\vec{w}$. You may be used to thinking of your weight as a property of your body, but actually it is a force exerted on you by the earth. The gravitational attraction of two objects acts even when the objects are not in contact. We'll discuss this concept at greater length in Section 4.4.

Measuring Force

Force is a vector quantity; to describe a force, we need to describe the direction in which it acts as well as its magnitude—the quantity that tells us "how much" or "how strongly" the force pushes or pulls. The SI unit of the magnitude of force is the *newton,* abbreviated N. (We haven't yet given a precise definition of the newton. The official definition is based on the standard kilogram; we'll get into that in Section 4.3.) The weight of a medium-sized apple is about 1 N; the pulling force of a diesel locomotive can be as much as 10^6 N.

When forces act on a solid object, they usually deform the object. For example, a coil spring stretches or compresses when forces act on its ends. This property forms the basis for a common instrument for *measuring* forces, called a *spring balance.* The instrument consists of a coil spring, enclosed in a case for protection, with a pointer attached to one end. When forces are applied to the ends of the spring, it stretches; the amount of stretch depends on the force. We can make a scale for the pointer and calibrate it by using a number of identical objects with weights of exactly 1 N each. When two, three, or more of these are suspended simultaneously from the balance, the total stretching force on each end of the spring is 2 N, 3 N, and so on, and we can label the corresponding positions of the pointer 2 N, 3 N, and so on. Then we can use our spring balance to measure the magnitude of an unknown force. We can also make a similar instrument that measures pushes instead of pulls.

Suppose we pull or push a box, as shown in Figure 4.3. As with other vector quantities, we can represent the force we exert on the box by a vector arrow. The arrow points in the direction of the force, and its length represents the magnitude

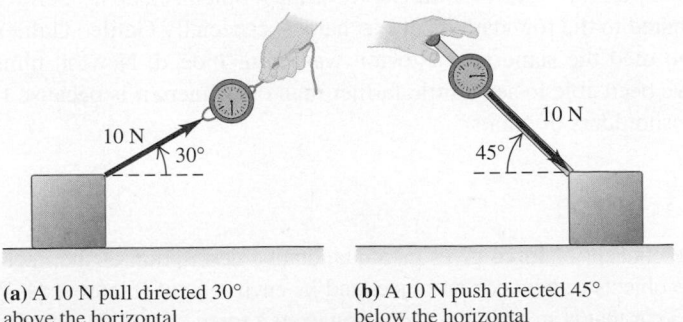

(a) A 10 N pull directed 30° above the horizontal

(b) A 10 N push directed 45° below the horizontal

▲ **FIGURE 4.3** Forces can be represented with vector arrows.

of the force. For example, we might scale the diagram so that an arrow 1 cm long represents a 5 N force.

Resultant of Forces

Experiment shows that when two forces $\vec{F}_1$ and $\vec{F}_2$ act at the same time on the same point of an object (Figure 4.4), the effect is the same as the effect of a single force equal to the vector sum of the original forces. This vector sum is often called the **resultant** of the forces or the *net force,* denoted by $\vec{R}$. That is, $\vec{R} = \vec{F}_1 + \vec{F}_2$. More generally, the effect of *any number* of forces applied at a point on an object is the same as the effect of a single force equal to the vector sum of the original forces. This important principle goes by the name **superposition of forces.**

The discovery that forces combine according to vector addition is of the utmost importance, and we'll use this fact many times throughout our study of physics. In particular, it allows us to represent a force by means of components, as we've done with displacement, velocity, and acceleration. (See Section 1.8 if you need a review.) In Figure 4.5a, for example, a force $\vec{F}$ acts on an object at a point that we'll designate as the origin of coordinates, O. The component vectors of $\vec{F}$ in the x and y directions are $\vec{F}_x$ and $\vec{F}_y$, and the corresponding components are F_x and F_y. As Figure 4.5b shows, if we apply two forces $\vec{F}_x$ and $\vec{F}_y$ to the object, the effect is exactly the same as applying the original force. **Any force can be replaced by its components, acting at the same point.**

In general, a force can have components in all three (x, y, and z) directions. It will usually be clear how to extend this discussion to three-dimensional situations, but we'll limit our discussion here to situations in which only x and y components are needed.

The x and y coordinate axes don't have to be vertical and horizontal. Figure 4.6 shows a stone block being pulled up a ramp by a force $\vec{F}$, represented by its components F_x and F_y with respect to axes parallel and perpendicular, respectively, to the sloping surface of the ramp. We draw a wiggly line through the force vector $\vec{F}$ to show that we have replaced it by its x and y components. Otherwise, the diagram would include the same force twice.

We'll often need to find the vector sum (resultant) of several forces acting on an object. We'll use the Greek letter Σ (capital sigma, equivalent to an S) as a shorthand notation for a sum. If the forces are labeled $\vec{F}_1$, $\vec{F}_2$, $\vec{F}_3$, and so on, we abbreviate the vector sum operation as

$$\vec{R} = \vec{F}_1 + \vec{F}_2 + \vec{F}_3 + \cdots = \Sigma\vec{F}, \quad \text{(resultant, or vector sum, of forces)} \quad (4.1)$$

where $\Sigma\vec{F}$ is read as "the vector sum of the forces" or "the net force." The component version of Equation 4.1 is the pair of component equations

$$R_x = \Sigma F_x, \qquad R_y = \Sigma F_y, \qquad \text{(components of vector sum of forces)} \quad (4.2)$$

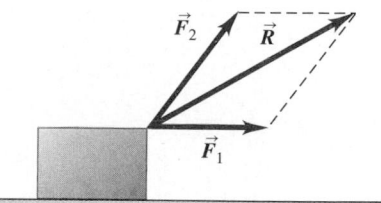

Two forces $\vec{F}_1$ and $\vec{F}_2$ acting on a point have the same effect as a single force $\vec{R}$ equal to their vector sum (resultant).

▲ **FIGURE 4.4** Superposition of forces.

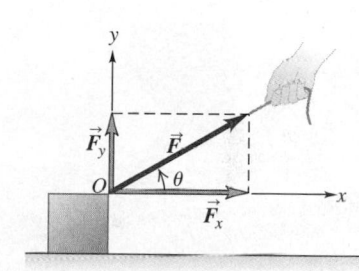

(a) Component vectors: $\vec{F}_x$ and $\vec{F}_y$
Components: $F_x = F \cos\theta$ and $F_y = F \sin\theta$

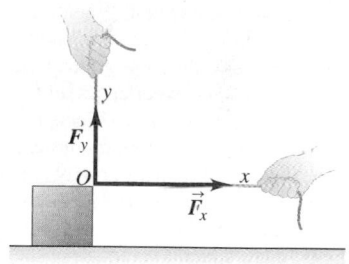

(b) Component vectors $\vec{F}_x$ and $\vec{F}_y$ together have the same effect as original force $\vec{F}$.

▲ **FIGURE 4.5** A force can be represented by its components.

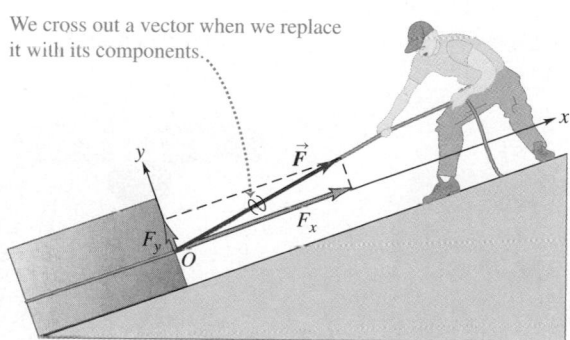

We cross out a vector when we replace it with its components.

▲ **FIGURE 4.6** The pull exerted by the mason can be replaced by components parallel and perpendicular to the direction of motion.

where $\sum F_x$ is the sum of the x components, and so on. Each component may be positive or negative; be careful with signs when you evaluate the sums in Equations 4.2.

Once we have the components R_x and R_y, we can find the magnitude and direction of the vector $\vec{R}$. The magnitude is

$$R = \sqrt{R_x^2 + R_y^2}, \qquad \text{(magnitude of vector sum of forces)}$$

and the angle θ between $\vec{R}$ and the x axis can be found from the relation $\tan\theta = R_y/R_x$. The components R_x and R_y may be positive or negative, and the angle θ may be in any of the four quadrants. As usual, we measure θ counterclockwise from the x axis. We've used this procedure in earlier chapters; in particular, you may want to review Section 1.8.

4.2 Newton's First Law

The fundamental role of a force is to change the state of motion of the object on which the force acts. The key word is "change." Newton's first law of motion, translated from the Latin of the *Principia* and put into modern language, is as follows:

Newton's first law
Every object continues either at rest or in constant motion in a straight line, unless it is forced to change that state by forces acting on it.

When *no* force acts on an object, or when the vector sum of forces on it is zero, the object either remains at rest or moves with constant velocity in a straight line. Once an object has been set in motion, no net force is needed to keep it moving. In other words, **an object acted on by no net force moves with constant velocity (which may be zero) and thus with zero acceleration.**

Everyday experience may seem to contradict this statement. Suppose you give a push to a hockey puck on a table, as shown in Figure 4.7a. After you stop pushing and take your hand away, the puck *does not* continue to move indefinitely; it slows down and stops. To keep it moving, you have to keep pushing. This is because, as the puck slides, the tabletop applies a frictional force to it in a direction *opposite* that of its motion.

But now imagine pushing the puck across the smooth ice of a skating rink (Figure 4.7b). Here, it will move a lot farther after you quit pushing before it stops. Put it on an air-hockey table, where it floats on a thin cushion of air, and it slides still farther (Figure 4.7c). The more slippery the surface, the less friction there is. The first law states that if we could eliminate friction completely, we would need *no forward force at all* to keep the puck moving with constant velocity once it had been started moving.

▲ **Application An object in motion.**
Catapults are founded on Newton's first law. When the catapult arm is released from its cocked position, it accelerates rapidly, carrying the projectile with it. When the arm hits the crossbar, it comes to a sudden stop—but the projectile keeps going. Once in flight, the projectile experiences no forward force; it flies on because objects in motion tend to stay in motion. It *does* experience the downward force of gravity and a retarding force due to air resistance; these forces will end its flight if it doesn't hit a wall first. Today, a modern version of the catapult is used on aircraft carriers to help launch airplanes, which can reach speeds of 250 kilometers per hour in less than 3 seconds.

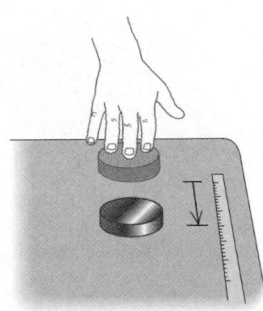

(a) Table: puck stops short

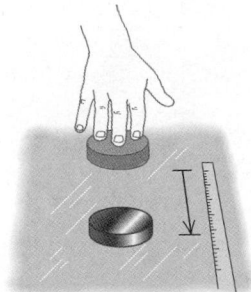

(b) Ice: puck slides farther

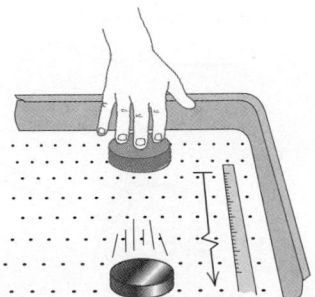

(c) Air-hockey table: puck slides even farther

▲ **FIGURE 4.7** An experimental approach to Newton's first law. As we reduce friction, the puck's motion comes closer and closer to constant velocity and the net force acting on the puck approaches zero, as predicted by Newton's first law.

▶ **Application Objects at rest?** This trick photo was taken an instant after a super-smooth table was very rapidly yanked out from underneath the dinner setting. The table was removed so rapidly that it exerted a force on the place setting only for a very short time. We can visualize the concept of inertia as described by Newton's first law. The objects at rest tend to stay at rest—but the force of gravity causes them to accelerate rapidly downward!

Inertia

The tendency of an object to remain at rest, or to keep moving once it is set in motion, results from a property called *inertia*. That's what you feel when you're behind home plate trying to catch a fastball pitch. That baseball really "wants" to keep moving. The tendency of an object at rest to remain at rest is also due to inertia. You may have seen a tablecloth yanked out from under the china without anything being broken. As the tablecloth slides, the force it exerts on the china isn't great enough to make the china move appreciably.

The quantitative measure of inertia is the physical quantity called **mass,** which we'll discuss in detail in the next section.

Conceptual Analysis 4.1

Science or fiction?

In a 1950s TV science fiction show, the hero is cruising the vacuum of outer space when the engine of his spaceship dies and his ship drifts to a stop. Will rescue arrive before his air runs out? What do you think about the physics of this plot?

SOLUTION The engine creates a force that accelerates the ship by ejecting mass out of its stern. When the engine dies and this process stops, there is no force on the ship (assuming that the gravitational forces exerted by planets and stars are negligible). The ship doesn't come to a stop, but continues to move in a straight line with constant speed. Science fiction sometimes contains more fiction than science!

When a single force acts on an object, it changes the object's state of motion. An object that is initially at rest starts to move. If the object is initially moving, a force in the direction opposite that of the motion causes the object to slow down. (If the force is maintained, eventually the object stops and reverses direction.) Suppose a hockey puck rests on a horizontal surface with negligible friction, such as an air-hockey table or a slab of ice. If the puck is initially at rest and a single horizontal force $\vec{F}_1$ acts on it (Figure 4.8a), the puck starts to move. If the puck is in motion at the start, the force makes it speed up, slow down, or change direction, depending on the direction of the force.

Now suppose we apply a second force $\vec{F}_2$ (Figure 4.8b), equal in magnitude to $\vec{F}_1$, but opposite in direction. The two forces are negatives of each other $(\vec{F}_2 = -\vec{F}_1)$, and their vector sum $\vec{R}$ is zero:

$$\vec{R} = \vec{F}_1 + \vec{F}_2 = 0.$$

We find that if the puck is at rest at the start, it remains at rest. If it is initially moving, it continues to move in the same direction with constant speed. These results show that, in Newton's first law, **zero resultant force is equivalent to no force at all.**

When an object is acted on by no forces or by several forces whose vector sum (resultant) is zero, we say that the object is in **equilibrium.** For an object in equilibrium,

$$\vec{R} = \Sigma\vec{F} = 0. \qquad \text{(equilibrium under zero resultant force)} \qquad (4.3)$$

For Equation 4.3 to be true, each component of $\vec{R}$ must be zero, so

$$\Sigma F_x = 0, \qquad \Sigma F_y = 0. \qquad \text{(object in equilibrium)} \qquad (4.4)$$

A puck on a frictionless surface accelerates when acted on by a single horizontal force.

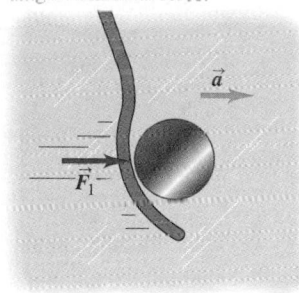

(a)

An object acted on by forces whose vector sum is zero behaves as though no forces act on it.

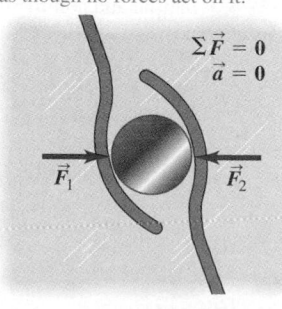

(b)

▲ **FIGURE 4.8** A puck responding to a nonzero net force (a) and to two forces whose vector sum is zero (b).

When Equations 4.4 are satisfied, the object is in equilibrium; that is, it has no acceleration. (We are assuming that the object can be modeled adequately as a particle. For an object with finite size, we also have to consider *where* on the object the forces are applied. We'll return to this point in Chapter 10.)

Conceptual Analysis 4.2

Air hockey

Your team decides to play leaf-blower ice hockey. You use your blower to apply a constant force that accelerates the puck in a straight line. An opponent skates up and uses her blower to apply an equally strong, constant force directed opposite to yours. Her force, acting together with yours, will cause the puck to

A. slow down and eventually stop.
B. stop immediately.
C. slow somewhat and then move with constant velocity.
D. continue to move, but with constant velocity.

SOLUTION The force from your blower alone causes the puck to accelerate. The force from your opponent's blower is equal in magnitude and opposite in direction to yours, so the vector sum of the two forces is zero. A puck that is acted upon by zero net force has no acceleration. Thus, the puck does not slow down or halt (which would be acceleration), but instead moves with constant velocity in accordance with Newton's first law. Thus, the answer is D.

Inertial Frames of Reference

In our discussions of relative velocity at the ends of Chapters 2 and 3, we stressed the concept of a *frame of reference.* This concept also plays a central role in Newton's laws of motion. Suppose you are sitting in an airplane as it accelerates down the runway during takeoff. You feel a forward force pushing on your back, but you don't start moving forward relative to the airplane. If you could stand in the aisle on roller skates, you would accelerate *backward* relative to the plane. In either case, it looks as though Newton's first law is not obeyed. Forward net force but no acceleration, or zero net force and backward acceleration. What's wrong?

The point is that the plane, accelerating with respect to the earth, is not a suitable frame of reference for Newton's first law. This law is valid in some frames of reference and not in others. A frame of reference in which Newton's first law *is* valid is called an **inertial frame of reference.** The earth is approximately an inertial frame of reference, but the airplane is not.

This may sound as though there's only one inertial frame of reference in the whole universe. On the contrary, if Newton's first law is obeyed in one particular reference frame, it is also valid in every other reference frame that moves with constant velocity relative to the first. All such frames are therefore inertial. For instance, Figure 4.9 shows three frames of reference: that of a person standing beside the runway, that of a truck driving at constant speed in a straight line, and that of the accelerating airplane. The stationary person's frame of reference is inertial. The truck moves with constant velocity relative to the person, so its frame of reference is also inertial. In both of these frames, Newton's first law is obeyed. However, the airplane, which is accelerating with respect to both of these observers, is in a non-inertial frame.

Thus, there is no single inertial frame of reference that is preferred over all others for formulating Newton's laws. If one frame is inertial, then every other frame moving relative to it with constant velocity is also inertial. Both the state of rest and the state of uniform motion (with constant velocity) can occur when the vector sum of forces acting on the object is zero. Because Newton's first law can be used to define what we mean by an inertial frame of reference, it is sometimes called the *law of inertia.*

The truck moves with constant velocity relative to the person on the ground. Newton's first law is obeyed in both frames, so they are **inertial**.

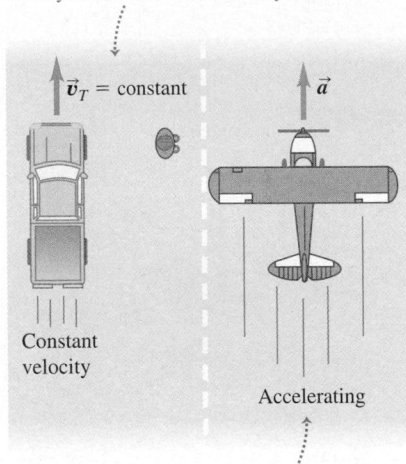

$\vec{v}_T$ = constant $\vec{a}$

Constant velocity

Accelerating

In the accelerating airplane, Newton's first law is not obeyed, so the frame of the airplane is **not inertial**.

▲ **FIGURE 4.9** Inertial and non-inertial frames of reference.

4.3 Mass and Newton's Second Law

Qualitatively, we've learned that when an object is acted on by a nonzero net force, it accelerates. We now want to know the relation of the acceleration to the force; this is what Newton's second law of motion is all about.

Let's look at several fundamental experiments. Consider a small object moving on a flat, level, frictionless surface. The object could be a puck on an air-hockey table, as we described in Section 4.2. Initially, it is moving to the right along the x axis of a coordinate system (Figure 4.10a). We apply a constant horizontal force $\vec{F}$ to the object, using the spring balance we described in Section 4.1, with the spring stretched a constant amount. We find that during the time the force is acting, the velocity of the object changes at a constant rate; that is, the object moves with *constant acceleration.* If we change the magnitude of the force, the acceleration changes in the same proportion. Doubling the force doubles the acceleration, halving the force halves the acceleration, and so on (Figures 4.10b and 4.10c). When we take away the force, the acceleration becomes zero and the object moves with constant velocity. We conclude that, for any given object, the acceleration is *directly proportional* to the force acting on it.

In another experiment (Figure 4.11), we give the object the same initial velocity as before. With no force, the velocity is constant (Figure 4.11a). With a constant force directed to the right (Figure 4.11b), the acceleration is constant and to the right. But when we reverse the direction of the force (Figure 4.11c), we find that the object moves more and more slowly to the right. (If we continue the experiment, the object stops and then moves toward the left with increasing speed.) The direction of the acceleration is toward the *left,* in the same direction as the force $\vec{F}$. We conclude that the *magnitude* of the acceleration is proportional to that of the force, and the *direction* of the acceleration is the same as that of the force, regardless of the direction of the velocity. If there is *no* force, the object has zero acceleration; no force is required to maintain motion with constant velocity.

A constant force $\vec{F}$ causes a constant acceleration $\vec{a}$.

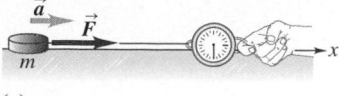

(a)

Doubling the force doubles the acceleration.

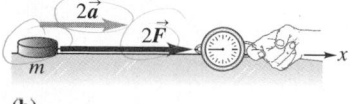

(b)

Halving the force halves the acceleration.

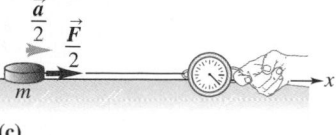

(c)

▲ **FIGURE 4.10** Experiments demonstrating the direct proportion between an applied force and the resulting acceleration.

A puck moving with constant velocity: $\Sigma\vec{F} = 0$, $\vec{a} = 0$

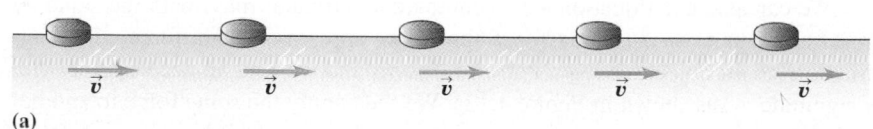

(a)

A constant force in the direction of motion causes a constant acceleration in the same direction as the force.

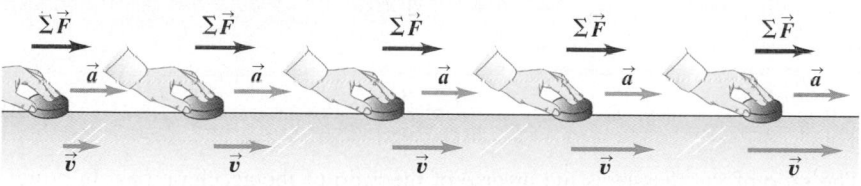

(b)

A constant force opposite to the direction of motion causes a constant acceleration in the same direction as the force.

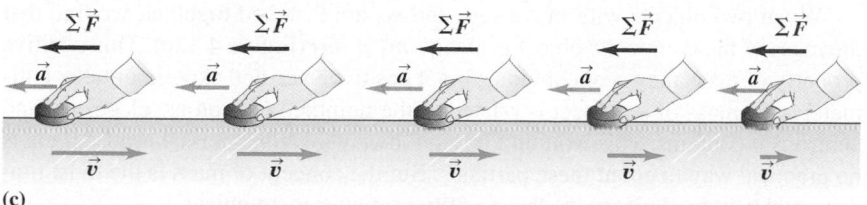

(c)

▲ **FIGURE 4.11** Experiments showing that an object's acceleration has the same direction as the net force acting on the object.

For a given object, the ratio of the magnitude F of the force to the magnitude a of the acceleration is constant. We call this ratio the inertial mass, or simply the **mass,** of the object and denote it as m. That is, $m = F/a$.

An equivalent statement is that when a force with magnitude F acts on an object with mass m, the magnitude a of the object's acceleration is

$$a = \frac{F}{m}. \qquad (4.5)$$

The concept of mass is a familiar one. If you hit a table-tennis ball with a paddle and then hit a basketball with the same force, the basketball has a much smaller acceleration because it has a much greater mass. When a large force is needed to give an object a certain acceleration, the mass of the object is large. When only a small force is needed for the same acceleration, the mass is small. The greater the mass, the more the object "resists" being accelerated. Thus, mass is a quantitative measure of inertia, which we discussed in Section 4.2.

The SI unit of mass is the **kilogram.** We mentioned in Section 1.3 that the kilogram is officially defined to be the mass of a chunk of platinum–iridium alloy kept in a vault near Paris. We can use this standard kilogram, along with Equation 4.5, to define the fundamental SI unit of force, the **newton,** abbreviated N:

Definition of the newton

One newton is the amount of force that gives an acceleration of 1 meter per second squared to an object with a mass of 1 kilogram. That is,

$$1 \text{ N} = (1 \text{ kg})(1 \text{ m/s}^2).$$

We can use this definition to calibrate the spring balances and other instruments that we use to measure forces.

We can also use Equation 4.5 to compare a particular mass with the standard mass and thus to *measure* masses. Suppose we apply a constant force with magnitude F to an object having a known mass m_1 and we find an acceleration with magnitude a_1, as shown in Figure 4.12a. We then apply the *same* force to another object having an unknown mass m_2, and we find an acceleration with magnitude a_2 (Figure 4.12b). Then, according to Equation 4.5,

$$m_1 a_1 = m_2 a_2,$$

or

$$\frac{m_2}{m_1} = \frac{a_1}{a_2}. \qquad (4.6)$$

The ratio of the masses is the inverse of the ratio of the accelerations. In principle, we could use this relation to measure an unknown mass m_2, but it is usually easier to determine an object's mass indirectly by measuring the object's weight. We'll return to this point in Section 4.4.

When two objects with masses m_1 and m_2 are fastened together, we find that the mass of the composite object is always $m_1 + m_2$ (Figure 4.12c). This additive property of mass may seem obvious, but it has to be verified experimentally. Ultimately, the mass of an object is related to the numbers of protons, electrons, and neutrons it contains. This wouldn't be a good way to *define* mass, because there is no practical way to count these particles. But the concept of mass is the most fundamental way to characterize the quantity of matter in an object.

A known force $\Sigma \vec{F}$ causes an object with mass m_1 to have an acceleration $\vec{a}_1$.

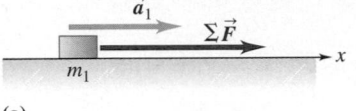

(a)

Applying the same force $\Sigma \vec{F}$ to a second object and noting the acceleration allows us to measure the mass.

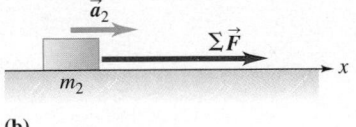

(b)

When the two objects are fastened together, the same method shows that their composite mass is the sum of their individual masses.

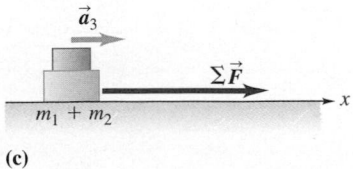

(c)

▲ **FIGURE 4.12** For a constant force, the magnitude of acceleration is inversely proportional to the mass of an object. The mass of a composite object is the sum of the masses of its parts.

We need to generalize Equation 4.5 in two ways. First, the object doesn't necessarily have to move along a straight line; its path may be a curve. Its velocity, its acceleration, and the force acting on it can then all change directions, so they have to be treated as vector quantities. Second, there may be several forces $(\vec{F}_1, \vec{F}_2, \cdots)$ acting on the object. As discussed earlier, we can combine the forces by using vector addition; the effect of the forces acting together is the same as the effect of a single force equal to their vector sum, which we usually write as $\sum \vec{F}$ or $\vec{R}$.

Newton wrapped up all these relationships and experimental results in a single concise statement that we now call **Newton's second law of motion:**

Newton's second law of motion (vector form)

The vector sum (resultant) of all the forces acting on an object equals the object's mass times its acceleration (the rate of change of its velocity):

$$\sum \vec{F} = m\vec{a}. \tag{4.7}$$

The acceleration $\vec{a}$ has the same direction as the resultant force $\sum \vec{F}$.

Newton's second law is a fundamental law of nature, the basic relation between force and motion. Most of the remainder of this chapter and all of the next are devoted to learning how to apply this principle in various situations.

Equation 4.7 is a *vector* equation. We'll usually use it in component form:

Newton's second law of motion (component form)

For an object moving in a plane, each component of the total force equals the mass times the corresponding component of acceleration:

$$\sum F_x = ma_x, \qquad \sum F_y = ma_y. \tag{4.8}$$

Equations 4.7 and 4.8 are valid only when the mass m is *constant*. It's easy to think of systems whose masses change, such as a leaking tank truck, a rocket ship, or a moving railroad car being loaded with coal. Such systems are better handled by using the concept of momentum; we'll get to that in Chapter 8.

Like the first law, Newton's second law is valid only in inertial frames of reference. We'll usually assume that the earth is an adequate approximation to an inertial frame, even though it is not precisely inertial because of its rotation and orbital motion.

Remember that we defined the newton as $1 \text{ N} = 1 \text{ kg} \cdot \text{m/s}^2$. Notice that the newton *must* be defined this way for $\sum \vec{F} = m\vec{a}$ to be dimensionally consistent.

In this chapter, as we learn how to apply Newton's second law, we'll begin with problems involving only motion along a straight line, and we'll always take this line to be a coordinate axis, often the x axis. In these problems, $\vec{v}$ and $\vec{a}$ have components only along this axis. Individual forces may have components along directions other than the x axis, but the *sum* of those components will always be zero.

In Chapter 5, we'll consider more general problems in which v_y, a_y, and $\sum F_y$ need not be zero. We'll derive more general problem-solving strategies for applying Newton's laws of motion to these situations.

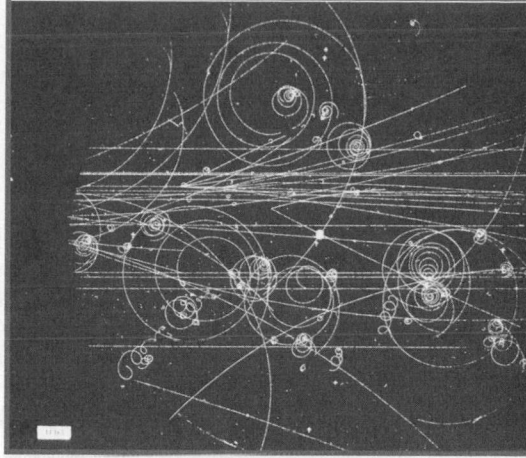

▲ **Application Too small to measure?**
Bubble chamber images like the one shown here are used both to study characteristics of known subatomic particles and to search for new types of matter and antimatter particles. For example, the linked spiral tracks indicate the simultaneous creation of an electron–positron pair from a photon. Using such images, we can apply Newton's second law $(F = ma)$ or its relativistic generalization to determine the masses of particles so minute that they cannot be measured directly.

EXAMPLE 4.1 **A box on ice**

A worker with spikes on his shoes pulls with a constant horizontal force of magnitude 20 N on a box with mass 40 kg resting on the flat, frictionless surface of a frozen lake. What is the acceleration of the box?

SOLUTION

SET UP Figure 4.13 shows our sketch. We take the $+x$ axis in the direction of the horizontal force. "Frictionless" means that the ice doesn't supply any force that opposes the motion (such as the friction force mentioned in Section 4.1). Thus, the only horizontal force is the one due to the rope that the worker is pulling. Because the box has no vertical acceleration, we know that the sum of the vertical components of force is zero. However, two vertical forces do act on the box: the downward force of the earth's gravitational attraction and the upward normal force exerted by the ice on the box. For completeness, we show these two vertical forces in our sketch; their vector sum must be zero because $a_y = 0$.

SOLVE The acceleration is given by Newton's second law. There is only one horizontal component of force, so we have

$$\sum F_x = ma_x,$$
$$a_x = \frac{F_x}{m} = \frac{20 \text{ N}}{40 \text{ kg}} = \frac{20 \text{ kg} \cdot \text{m/s}^2}{40 \text{ kg}} = 0.5 \text{ m/s}^2.$$

The force is constant, so the acceleration is also constant. If we are given the initial position and velocity of the box, we can find

The box has no vertical acceleration, so the vertical components of the net force sum to zero. Nevertheless, for completeness, we show the vertical forces acting on the box.

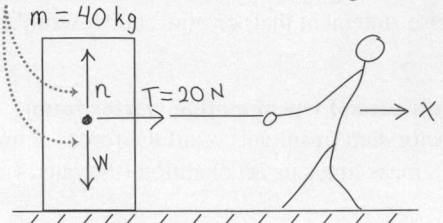

▲ **FIGURE 4.13** Our sketch for this problem.

the position and velocity at any later time from the equations of motion with constant acceleration.

REFLECT Two vertical (y) components of force act on the box—its weight $\vec{w}$ and an upward normal force $\vec{n}$ exerted on the bottom of the box by the ice—but their vector sum is zero.

Practice Problem: If the box starts from rest, what is its speed after it has been pulled a distance of 0.36 m? *Answer: 0.6 m/s.*

EXAMPLE 4.2 **The ketchup slide**

A counter attendant in a diner shoves a ketchup bottle with mass 0.20 kg along a smooth, level lunch counter. The bottle leaves her hand with an initial velocity of 2.8 m/s. As it slides, it slows down because of the horizontal friction force exerted on it by the countertop. The bottle slides a distance of 1.0 m before coming to rest. What are the magnitude and direction of the friction force acting on it?

SOLUTION

SET UP Figure 4.14 shows our diagrams—one for the bottle's motion and one for the forces on the bottle. We place the origin at the point where the waitress releases the bottle and point the $+x$ axis in the direction the bottle moves. Because the bottle has no vertical acceleration, we know that the vertical forces on it (the weight $\vec{w}$ and the upward normal force $\vec{n}$) must sum to zero. The only horizontal force on the bottle is the friction force $\vec{f}$. Because the bottle is slowing down (a_x is negative in our coordinate system), we know that this force points in the $-x$ direction.

SOLVE We need to find the relation between v_x and a_x; to do this, we can use the constant-acceleration equation $v_x^2 - v_{0x}^2 =$

$2a_x(x - x_0)$ (Equation 2.13). The initial x component of velocity is $v_{0x} = 2.8$ m/s, and the final value is $v_x = 0$. Also, $x_0 = 0$ and $x = 1.0$ m. Once a_x is known, we can use the component form of Newton's second law, Equation 4.8 $(\sum F_x = ma_x)$, to find the magnitude f of the friction force. Because f is the only horizontal component of force, $\sum F_x = -f$. Using the numerical values in Equation 2.13, we first find the bottle's acceleration:

$$v_x^2 - v_0^2 = 2a_x(x - x_0),$$
$$(0) - (2.8 \text{ m/s})^2 = 2a_x(1.0 \text{ m} - 0),$$
$$a_x = -3.9 \text{ m/s}^2.$$

We draw one diagram for the bottle's motion and one showing the forces on the bottle.

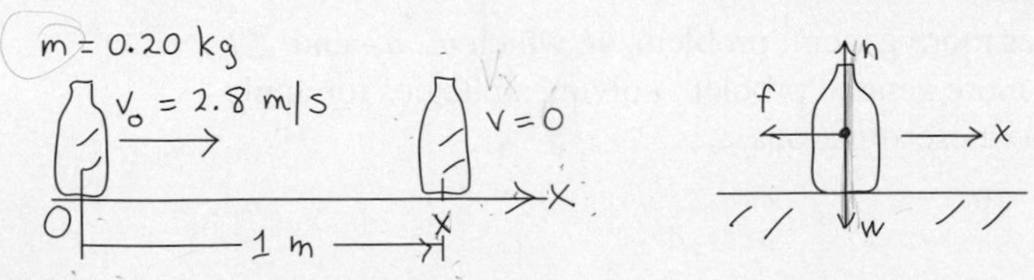

▲ **FIGURE 4.14** Our sketch for this problem.

Continued

We can now find the net force acting on the bottle.

$$\Sigma F_x = -f = ma_x$$
$$-f = (0.20 \text{ kg})(-3.9 \text{ m/s}^2) = -0.78 \text{ kg} \cdot \text{m/s}^2 = -0.78 \text{ N}.$$

Since the frictional force is the *only* horizontal force acting on the bottle, this answer gives us its magnitude (0.78 N) and also the fact that it acts in the $-x$ direction (which we already knew).

REFLECT Notice that we did most of the physics before doing any math. We knew that the vertical forces on the bottle summed to zero, and we knew the direction of the friction force. We did math only to find the magnitude of that force.

Practice Problem: Suppose the bottle travels 1.5 m before stopping, instead of 1.0 m. What is its initial velocity? *Answer:* 3.4 m/s.

EXAMPLE 4.3   A TV picture tube

In a color TV picture tube, an electric field exerts a net force with magnitude 1.60×10^{-13} N on an electron $(m = 9.11 \times 10^{-31} \text{ kg})$. What is the electron's acceleration?

SOLUTION

SET UP Figure 4.15 shows our simple diagram. The statement of the problem tells us that the *net force* on the electron is due to the electric field, so we consider only the electric-field force. (If other forces act on the electron, they are negligible or sum to zero.) The direction of the acceleration must be the same as that of the force; we designate this direction as the $+x$ direction.

SOLVE We use Newton's second law in component form, rearranging it slightly to solve for acceleration:

$$a_x = \frac{\Sigma F_x}{m} = \frac{1.60 \times 10^{-13} \text{ N}}{9.11 \times 10^{-31} \text{ kg}} = \frac{1.60 \times 10^{-13} \text{ kg} \cdot \text{m/s}^2}{9.11 \times 10^{-31} \text{ kg}}$$
$$= 1.76 \times 10^{17} \text{ m/s}^2.$$

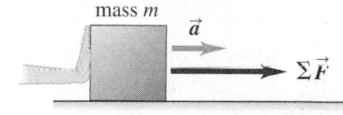

▲ **FIGURE 4.15** Even for a simple problem, a quick sketch helps us to see the situation and avoid mistakes.

REFLECT Notice that the magnitudes of these quantities are far outside the range of everyday experience. Because the mass of the electron is miniscule, a tiny force produces a huge acceleration.

Practice Problem: If the electron starts from rest, what time is required for it to travel 0.20 m from its source to the screen of the picture tube? *Answer:* 1.51×10^{-9} s.

Quantitative Analysis 4.3

Raising a toolbox

Suppose you use a rope to hoist a box of tools vertically at a constant speed v. The rope exerts a constant upward force of magnitude F_{up} on the box, and gravity exerts a constant downward force (the weight of the box). No other forces act on the box. To raise the box twice as fast, the force of the rope on the box would have to have

A. the same magnitude F_{up}.
B. a magnitude of $2F_{up}$.

SOLUTION The key is that the box moves at constant velocity in both cases; it is in equilibrium. Thus, by Newton's first law, no *net force* acts on it; the vector sum of the upward and downward forces is zero. Since the downward force (the box's weight) is always the same, the upward force must be the same, too, regardless of the box's speed. Note that this is true only while the box is moving at constant speed, not during the initial acceleration from rest.

4.4 Mass and Weight

We've mentioned that the weight of an object is a force—the force of gravitational attraction of the earth (or whatever astronomical body the object is near, such as the moon or Mars). We'll study gravitational interactions in detail in Chapter 6, but we need some preliminary discussion now. The terms *mass* and *weight* are often misused and interchanged in everyday conversation. It's absolutely essential for you to understand clearly the distinctions between these two physical quantities.

Mass characterizes the *inertial* properties of an object. Mass is what keeps the china on the table when you yank the tablecloth out from under it. The greater the mass, the greater is the force needed to cause a given acceleration; this meaning is reflected in Newton's second law, $\Sigma \vec{F} = m\vec{a}$. Weight, on the other hand, is a *force* exerted on an object by the gravitational pull of the earth or some other astronomical body (Figure 4.16). Everyday experience shows us that objects that have large mass also have large weight. A cart loaded with bricks is hard to get started rolling because of its large *mass,* and it is also hard to lift off the ground

The relation of mass to net force: $\Sigma \vec{F} = m\vec{a}$

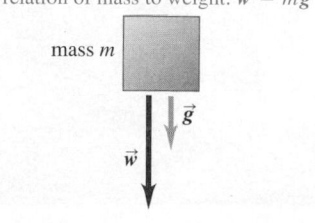

▲ **FIGURE 4.16** The relation of mass to net force and to weight.

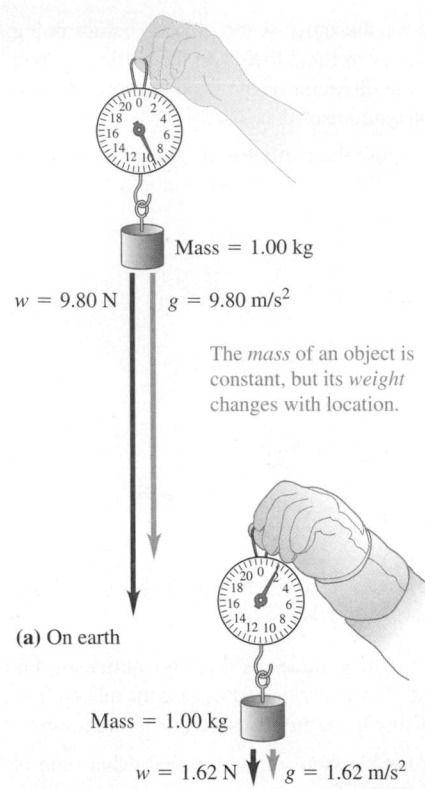

Mass = 1.00 kg

w = 9.80 N

g = 9.80 m/s²

The *mass* of an object is constant, but its *weight* changes with location.

(a) On earth

Mass = 1.00 kg

w = 1.62 N

g = 1.62 m/s²

(b) On the moon

▲ **FIGURE 4.17** Mass and weight for a 1 kg object on the earth and on the moon.

▲ **FIGURE 4.18** Astronaut Buzz Aldrin on the surface of the moon, trying to cope with his own inertia.

because of its large *weight*. On the moon, the cart would be just as hard to get rolling, but it would be easier to lift. So what exactly *is* the relationship between mass and weight?

The answer to this question, according to legend, came to Newton as he sat under an apple tree watching the apples fall. A freely falling object has an acceleration $\vec{a}$ with magnitude equal to g—that is, the acceleration due to gravity (also called the acceleration of free fall). Because of Newton's second law, this acceleration requires a force. If a 1 kg object falls with an acceleration of magnitude $g = 9.8$ m/s², the required force $\vec{F}$ must have magnitude

$$F = ma = (1 \text{ kg})(9.8 \text{ m/s}^2) = 9.8 \text{ kg} \cdot \text{m/s}^2 = 9.8 \text{ N}.$$

But the force that makes the object accelerate downward is the gravitational pull of the earth—that is, the *weight* of the object. Any object near the surface of the earth that has a mass of 1 kg *must* have a weight with magnitude 9.8 N in order for it to have the acceleration we observe when it is in free fall. More generally, the relation of mass to weight must be as follows:

Relation of mass to weight

The weight of an object with mass m must have a magnitude w equal to the mass times the magnitude of acceleration due to gravity, g:

$$w = mg. \tag{4.9}$$

Because weight is a force, we can write this equation as a vector relation:

$$\vec{w} = m\vec{g}. \tag{4.10}$$

NOTE ▶ Remember that g is the *magnitude* of $\vec{g}$, the acceleration due to gravity, so, by definition, g is always a positive number, Thus, w, given by Equation 4.9, is the *magnitude* of the weight, and it, too, is always positive. In the discussion that follows, we'll sometimes call w "the weight" and F "the force," even though, strictly speaking, we should call w "the magnitude of the weight" and F "the magnitude of the force." ◀

We'll use the value $g = 9.80$ m/s² for problems dealing with the earth. The value of g actually varies somewhat from point to point on the earth's surface, from about 9.78 to 9.82 m/s², because the earth is not perfectly spherical and because of effects due to its rotation, orbital motion, and composition. At a point where $g = 9.80$ m/s², the weight of a standard kilogram is $w = 9.80$ N. At a different point, where $g = 9.78$ m/s², the weight is $w = 9.78$ N, but the mass is still 1 kg. The weight of an object varies from one location to another; the mass does not. If we take a standard kilogram to the surface of the moon, where the acceleration of free fall is 1.62 m/s², the weight of the standard kilogram is 1.62 N, but its mass is still 1 kg (Figure 4.17). An 80 kg man has a weight on earth of $(80.0 \text{ kg})(9.80 \text{ m/s}^2) = 784$ N, but on the moon his weight is only $(80.0 \text{ kg})(1.62 \text{ m/s}^2) = 130$ N.

The following brief excerpt from Astronaut Buzz Aldrin's book *Men from Earth* offers some interesting insights into the distinction between mass and weight (Figure 4.18):

Our portable life-support system backpacks looked simple, but they were hard to put on and tricky to operate. On earth the portable life-support system and space suit combination weighed 190 pounds, but on the moon it was only 30. Combined with my own body weight, that brought me to a total lunar-gravity weight of around 60 pounds.

▶ **BIO Application No brakes needed.** When considering the deceleration of massive objects such as a car, air friction doesn't matter because inertial considerations dominate. However, small objects living in more viscous fluids such as water have vastly different experiences. Under those circumstances, inertia is unimportant and the force of viscous drag dominates. Astonishingly, if a bacterium propelling itself through water stops beating its flagellum (the oar that propels it), it will come to rest in about 0.01 nm, a dimension that is small on the scale of atoms. It lives in a very different world than we do.

One of my tests was to jog away from the lunar module to see how maneuverable an astronaut was on the surface. I remembered what Isaac Newton had taught us two centuries before: Mass and weight are not the same. I weighed only 60 pounds, but my *mass* was the same as it was on Earth. Inertia was a problem. I had to plan ahead several steps to bring myself to a stop or to turn without falling.

NOTE ▶ As indicated by Buzz Aldrin's words, the pound is actually a unit of force, not mass. In fact, 1 pound is defined officially as exactly 4.448221615260 newtons. It's handy to remember that a pound is about 4.4 N and a newton is about 0.22 pound. Next time you want to order a "quarter-pounder," try asking for a "one-newtoner" and see what happens. ◀

Here's another useful fact: An object with a mass of 1 kg has a weight of about 2.2 lb at the earth's surface. In the British system, the unit of force is the *pound* (or pound-force), and the unit of mass is the *slug*. The unit of acceleration is 1 foot per second squared, so 1 pound = 1 slug · ft/s². In the cgs metric system (not used in this book), the unit of mass is the gram, equal to 10^{-3} kg, and the unit of distance is the centimeter, equal to 10^{-2} m. The corresponding unit of force is called the *dyne:* 1 dyne = 1 g · cm/s² = 10^{-5} N.

It's important to understand that the weight of an object, as given by Equation 4.10, acts on the object *all the time*, whether it is in free fall or not. When a 10-kg flowerpot hangs suspended from a chain, it is in equilibrium, and its acceleration is zero. But its 98 N weight is still pulling down on it. In this case the chain pulls up on the pot, applying a 98 N upward force. The *vector sum* of the two 98 N forces is zero, and the pot is in equilibrium.

Mastering**PHYSICS**

ActivPhysics 2.9: Pole-Vaulter Vaults

EXAMPLE 4.4 **Quick stop of a heavy car**

A big luxury car weighing 1.96×10^4 N (about 4400 lb), traveling in the $+x$ direction, makes a fast stop; the x component of the net force acting on it is -1.50×10^4 N. What is its acceleration?

SOLUTION

SET UP Figure 4.19 shows our diagram. We draw the weight as a force with magnitude w. Because the car does not accelerate in the y direction, the road exerts an upward normal force of equal magnitude on the car; for completeness, we include that force in our diagram.

SOLVE To find the acceleration, we'll use Newton's second law, $\sum F_x = ma_x$. First, however, we need the car's mass. Since we know the car's weight, we can find its mass m from Equation 4.9, $w = mg$. We obtain

$$m = \frac{w}{g} = \frac{1.96 \times 10^4 \text{ N}}{9.80 \text{ m/s}^2} = \frac{1.96 \times 10^4 \text{ kg} \cdot \text{m/s}^2}{9.80 \text{ m/s}^2} = 2000 \text{ kg}.$$

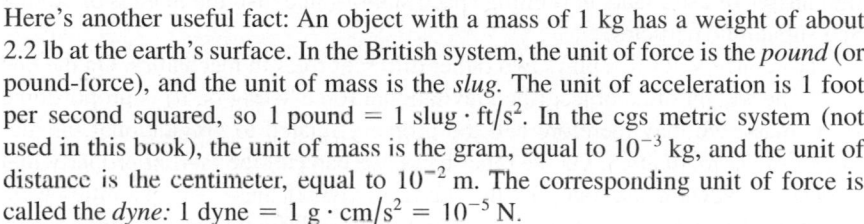

▲ **FIGURE 4.19** Our diagram for this problem.

Continued

Then, $\sum F_x = ma_x$ gives

$$a_x = \frac{F_x}{m} = \frac{-1.50 \times 10^4 \text{ N}}{2000 \text{ kg}} = \frac{-1.50 \times 10^4 \text{ kg} \cdot \text{m/s}^2}{2000 \text{ kg}}$$
$$= -7.5 \text{ m/s}^2.$$

REFLECT This acceleration can be written as $-0.77g$. Note that -0.77 is also the ratio of -1.50×10^4 N to 1.96×10^4 N. An

acceleration of this magnitude is possible only on a dry paved road. Don't expect it on a wet or icy road!

Practice Problem: If passenger comfort requires that the acceleration should be no greater in magnitude than $0.10g$, what distance is required to stop the car if its speed is initially 30 m/s (roughly 60 mi/h)? What force is required? *Answers:* 459 m, 1960 N.

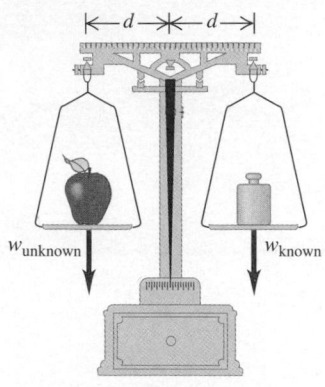

▲ **FIGURE 4.20** An equal-arm balance determines the mass of an object by comparing its weight to a known weight.

Usually, the easiest way to measure the mass of an object is to measure its weight, often by comparison with a standard weight. In accordance with Equation 4.10, two objects that have the same weight at a particular location also have the same mass. We can compare weights very precisely; the familiar equal-arm balance (Figure 4.20) can determine with great precision (to one part in 10^6) when the weights of two objects are equal and hence when their masses are equal. This method doesn't work, however, in the "zero-gravity" environment of outer space. Instead, we have to use Newton's second law directly: We apply a known force to the object, measure its acceleration, and compute the mass as the ratio of force to acceleration. This method, or a variation of it, is used to measure the masses of astronauts in orbiting space stations and also the masses of atomic and subatomic particles.

The concept of mass plays two rather different roles in mechanics. On the one hand, the weight of an object (the gravitational force acting on it) is proportional to its mass; we may therefore call the property related to gravitational interactions *gravitational mass*. On the other hand, we can call the inertial property that appears in Newton's second law the *inertial mass*. If these two quantities were different, the acceleration of free fall might well be different for different objects. However, extraordinarily precise experiments have established that, in fact, the two *are* the same, to a precision of better than one part in 10^{12}. Recent efforts to find departures from this equivalence, possibly showing evidence of a previously unknown force, have been inconclusive.

Finally, we remark that the SI units for mass and weight are often misused in everyday life. Incorrect expressions such as "This box weighs 6 kg" are nearly universal. What is meant is that the *mass* of the box, probably determined indirectly by *weighing,* is 6 kg. This usage is so common that there's probably no hope of straightening things out, but be sure you recognize that the term *weight* is often used when *mass* is meant. To keep your own thinking clear, be careful to avoid this kind of mistake! In SI units, weight (a force) is always measured in newtons, mass always in kilograms. (However, we'll often refer to the weight of an object when, strictly speaking, we mean the *magnitude* of its weight.)

4.5 Newton's Third Law

At the beginning of this chapter, we noted that force is a quantitative description of the interaction between two objects. A force acting on an object is always the result of that object's interaction with another object, so forces always come in pairs. As my hand pushes your shoulder, your shoulder pushes back on my hand. When you kick a football, the force your foot exerts on the ball launches it into its trajectory, but you also feel the force the ball exerts on your foot.

Experiments show that whenever two objects interact, the two forces they exert on each other are equal in magnitude and opposite in direction. This fact is called *Newton's third law.* In Figure 4.21, $\vec{F}_{\text{foot on ball}}$ is the force applied by the foot to the ball, and $\vec{F}_{\text{ball on foot}}$ is the force applied by the ball to the foot. The directions of the forces in this example correspond to a *repulsive* interaction,

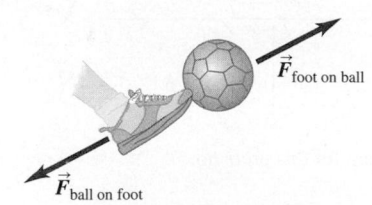

▲ **FIGURE 4.21** An interacting ball and foot exert forces on each other that are equal in magnitude and opposite in direction.

▶ **Application An action shot.** This Olympic biathlon competitor knows her physics: Every action has an equal and opposite reaction. When she fires her rifle, the force propelling the bullet forward is matched by an equal and opposite force pushing against the rifle and her. If she had her skis pointed in the direction in which she is firing, she would be pushed backward on the slippery mat and maybe end up flat on her back. By keeping her back ski firmly planted sideways on the mat, she can avoid the possible consequences of ignoring Newton's third law.

tending to push the objects apart. However, interactions can also be *attractive,* such as the gravitational attraction of two masses or the electrical attraction of two particles with opposite charges.

Newton's third law

For two interacting objects A and B, the formal statement of Newton's third law is

$$\vec{F}_{A \text{ on } B} = -\vec{F}_{B \text{ on } A}. \quad (4.11)$$

Newton's own statement, translated from the Latin of the *Principia,* is

> To every action there is always opposed an equal reaction; or, the mutual actions of two objects upon each other are always equal, and directed to contrary parts.

NOTE ▶ In the preceding statement, "action" and "reaction" are the two opposite forces. This terminology is not meant to imply any cause-and-effect relationship: We can consider either force as the action and the other the reaction. We often say simply that the forces are "equal and opposite" meaning that they have equal magnitudes and opposite directions. ◀

EXAMPLE 4.5 Newton's apple on a table

An apple sits in equilibrium on a table. What forces act on the apple? What is the reaction force to each of the forces acting on it? What are the action–reaction pairs?

SOLUTION

SET UP First, in Figure 4.22a, we diagram the situation and the two forces that act on the apple. $\vec{F}_{\text{earth on apple}}$ is the weight of the apple—the downward gravitational force exerted *by* the earth *on* the apple. Similarly, $\vec{F}_{\text{table on apple}}$ is the upward normal force exerted *by* the table *on* the apple.

Action–reaction pairs always represent a mutual interaction of two different objects.

The two forces acting on the apple are not a mutual interaction between objects, so they are not an action–reaction pair.

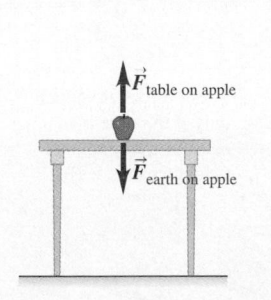

$\vec{F}_{\text{apple on earth}} = -\vec{F}_{\text{earth on apple}}$

$\vec{F}_{\text{apple on table}} = -\vec{F}_{\text{table on apple}}$

(a) The forces acting on the apple

(b) The action–reaction pair for the interaction between the apple and the earth

(c) The action–reaction pair for the interaction between the apple and the table

(d) Why the two forces acting on the apple don't form an action–reaction pair

▲ **FIGURE 4.22** The two forces in an action–reaction pair always act on different objects.

Continued

SOLVE The two forces acting on the apple are $\vec{F}_{\text{earth on apple}}$ and $\vec{F}_{\text{table on apple}}$. Because the apple is in equilibrium, it follows from Newton's first law that the vector sum of these two forces on the apple must be zero: $\vec{F}_{\text{earth on apple}} + \vec{F}_{\text{table on apple}} = 0$.

What is the reaction force for $\vec{F}_{\text{earth on apple}}$? As the earth pulls down on the apple, the apple exerts an equally strong upward pull $\vec{F}_{\text{apple on earth}}$ on the earth, as shown in Figure 4.22b. The two forces are

$$\vec{F}_{\text{apple on earth}} = -\vec{F}_{\text{earth on apple}}. \qquad \text{(action and reaction}$$
$$\text{for earth and apple)}$$

These forces represent the mutual interaction between the apple and the earth; one force cannot exist without the other.

Also, as the table pushes up on the apple with force $\vec{F}_{\text{table on apple}}$, the corresponding reaction is the downward force $\vec{F}_{\text{apple on table}}$ exerted on the table by the apple (Figure 4.22c), and we have

$$\vec{F}_{\text{table on apple}} = -\vec{F}_{\text{apple on table}}. \qquad \text{(action and reaction}$$
$$\text{for apple and table)}$$

The two forces acting on the apple are $\vec{F}_{\text{table on apple}}$ and $\vec{F}_{\text{earth on apple}}$. Are they an action–reaction pair? No, they aren't, despite the fact that they are "equal and opposite" (a consequence of Newton's *first* law, not the third). They do not represent the mutual interaction between two objects; they are two different forces acting on the same object. **The two forces in an action–reaction pair *never* act on the same object.** Figure 4.22d shows another way to look at this situation. Suppose we suddenly yank the table out from under the apple. The two forces $\vec{F}_{\text{table on apple}}$ and $\vec{F}_{\text{apple on table}}$ then become zero. But $\vec{F}_{\text{earth on apple}}$ and $\vec{F}_{\text{apple on earth}}$ are still there, and they *do* form an action–reaction pair. Thus, $\vec{F}_{\text{table on apple}}$ can't possibly be the negative of $\vec{F}_{\text{earth on apple}}$ with the table removed.

REFLECT Interaction forces always come in pairs; if you can't find both members of the pair, your diagrams aren't complete. Also, remember that the two forces in an action–reaction pair *never* act on the same object.

EXAMPLE 4.6 **Sliding a stone across the floor**

A stonemason drags a marble block across a floor by pulling on a rope attached to the block (Figure 4.23a). The block has just started to move, and it may not be in equilibrium. How are the various forces related? What are the action–reaction pairs?

SOLUTION

SET UP We have three interacting objects: the block, the rope, and the mason. First we draw separate diagrams showing the horizontal forces acting on each object (Figure 4.23b). Instead of writing out "block on rope," etc., we designate the block as B, the rope R, and the mason M.

SOLVE Vector $\vec{F}_{\text{M on R}}$ represents the force exerted *by* the mason *on* the rope. The reaction to $\vec{F}_{\text{M on R}}$ is the "equal and opposite" force $\vec{F}_{\text{R on M}}$ exerted *on* the mason *by* the rope. Vector $\vec{F}_{\text{R on B}}$ represents the force exerted on the block by the rope. The reaction to it is the equal and opposite force $\vec{F}_{\text{B on R}}$ exerted on the rope by the block. From Newton's third law,

$$\vec{F}_{\text{R on M}} = -\vec{F}_{\text{M on R}}$$

and

$$\vec{F}_{\text{B on R}} = -\vec{F}_{\text{R on B}}. \qquad \text{(action–reaction pairs)}$$

The forces $\vec{F}_{\text{M on R}}$ and $\vec{F}_{\text{B on R}}$ are *not* an action–reaction pair (Figure 4.23c). Both of these forces act on the *same* object (the rope); an action and its reaction *must* always act on *different* objects. Furthermore, the forces $\vec{F}_{\text{M on R}}$ and $\vec{F}_{\text{B on R}}$ are not necessarily equal in magnitude (Figure 4.23d). Applying Newton's second law to the rope (mass m_R, acceleration $\vec{a}_R$), we get

$$\vec{F}_{\text{M on R}} + \vec{F}_{\text{B on R}} = m_R\vec{a}_R.$$

If block and rope are accelerating to the right, the rope is not in equilibrium, and $\vec{F}_{\text{M on R}}$ must have greater magnitude than

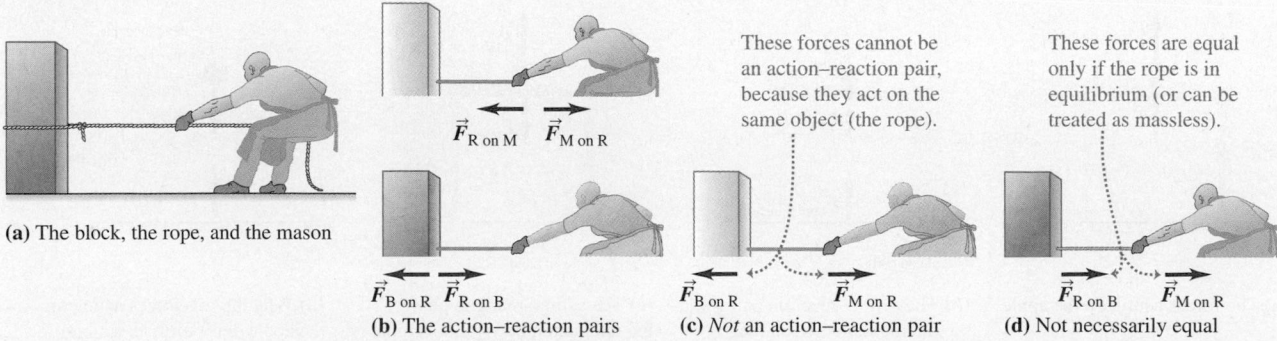

(a) The block, the rope, and the mason

$\vec{F}_{\text{R on M}}$ $\vec{F}_{\text{M on R}}$

$\vec{F}_{\text{B on R}}$ $\vec{F}_{\text{R on B}}$
(b) The action–reaction pairs

These forces cannot be an action–reaction pair, because they act on the same object (the rope).

$\vec{F}_{\text{B on R}}$ $\vec{F}_{\text{M on R}}$
(c) *Not* an action–reaction pair

These forces are equal only if the rope is in equilibrium (or can be treated as massless).

$\vec{F}_{\text{R on B}}$ $\vec{F}_{\text{M on R}}$
(d) Not necessarily equal

▲ **FIGURE 4.23** Analyzing the forces involved when a stonemason pulls on a block.

Continued

$\vec{F}_{\text{B on R}}$. In the special case in which the rope is in equilibrium, the forces $\vec{F}_{\text{M on R}}$ and $\vec{F}_{\text{B on R}}$ are equal in magnitude, but this is an example of Newton's *first* law, not his *third*. Newton's third law holds whether the rope is accelerating or not. The action–reaction forces $\vec{F}_{\text{M on R}}$ and $\vec{F}_{\text{R on M}}$ are always equal in magnitude to *each other*, as are $\vec{F}_{\text{R on B}}$ and $\vec{F}_{\text{B on R}}$. But $\vec{F}_{\text{M on R}}$ is *not* equal in magnitude to $\vec{F}_{\text{B on R}}$ when the rope is accelerating.

REFLECT If you feel as though you're drowning in subscripts at this point, take heart. Go over this discussion again, comparing the symbols with the vector diagrams, until you're sure you see what's going on.

In the special case in which the rope is in equilibrium, or when we can consider it as massless, then the total force on the rope is zero, and $\vec{F}_{\text{B on R}}$ equals $-\vec{F}_{\text{M on R}}$ because of Newton's *first* or *second* law. Also, $\vec{F}_{\text{B on R}}$ always equals $-\vec{F}_{\text{R on B}}$ by Newton's *third* law, so in this special case, $\vec{F}_{\text{R on B}}$ also equals $\vec{F}_{\text{M on R}}$. We can then think of the rope as "transmitting" to the block, without change, the force that the stonemason exerts on it (Figure 4.23d). This is a useful point of view, but you have to remember that it is valid only when the rope has negligibly small mass or is in equilibrium.

▶ **Application** **Out of this world.** This photo shows the liftoff of a Delta Two rocket carrying the *Pathfinder* probe bound for the surface of Mars. A common misconception is that a rocket is propelled by the exhaust gases pushing against the earth. As we know from Newton's third law, the explosive force on the gases being rapidly expelled from the tail of the rocket is matched by an equal and opposite forward force on the rocket itself. It is this second force that pushes the rocket upward and allows it to continue accelerating toward Mars long after it is away from the ground. If the rocket's motion did depend on its gases pressing against the earth, it would be unable to maneuver in outer space and the *Pathfinder* could never reach the Red Planet.

An object, such as the rope in Figure 4.23, that has pulling forces applied at its ends is said to be in **tension,** as we mentioned in Section 4.1. The tension at any point is the magnitude of force acting at that point. In Figure 4.23c, the tension at the right-hand end of the rope is the magnitude of $\vec{F}_{\text{M on R}}$ (or of $\vec{F}_{\text{R on M}}$), and the tension at the left-hand end equals the magnitude of $\vec{F}_{\text{B on R}}$ (or of $\vec{F}_{\text{R on B}}$). If the rope is in equilibrium and if no forces act except at its ends, the tension is the same at both ends and throughout the rope. If the magnitudes of $\vec{F}_{\text{B on R}}$ and $\vec{F}_{\text{M on R}}$ are 50 N each, the tension in the rope is 50 N (*not* 100 N). Resist the temptation to add the two forces; remember that the total force $\vec{F}_{\text{B on R}} + \vec{F}_{\text{M on R}}$ on the rope in this case is zero!

NOTE ▶ Finally, we emphasize once more a fundamental truth: The two forces in an action–reaction pair *never* act on the same object. Remembering this simple fact can often help you avoid confusion about action–reaction pairs and Newton's third law. ◀

Conceptual Analysis 4.4

Giving a push

You use your car to push a friend's pickup truck that has broken down. To bring both of you to a stop, your friend applies the brakes in the truck. Then, in magnitude,

A. the force of the car on the truck is greater than the force of the truck on the car.
B. the force of the car on the truck is equal to the force of the truck on the car.
C. the force of the car on the truck is less than the force of the truck on the car.

SOLUTION Whenever two objects interact with each other, the interaction is described by an action–reaction pair. From Newton's third law, the two forces in this pair are always equal in magnitude and opposite in direction (even though in this case the car and truck may have different masses). So the answer is B. Also, Newton's third law is still obeyed when the objects are accelerating, as they are in this example. Unequal masses and accelerations of the objects don't alter the validity of the third law.

4.6 Free-Body Diagrams

Newton's three laws of motion are a beautifully wrapped package containing the basic principles we need to solve a wide variety of problems in mechanics. They are very simple in form, but the process of applying them to specific situations can pose real challenges.

Let's talk about some useful techniques. When you apply Newton's first or second law ($\sum \vec{F} = 0$ for an equilibrium situation or $\sum \vec{F} = m\vec{a}$ for a non-equilibrium situation), you *must* apply it to some specific object. It is absolutely essential to decide at the beginning what object you're talking about. This statement may sound trivial, but it isn't. Once you have chosen an object, then you have to identify all the forces acting on it. These are the forces that are included in $\sum \vec{F}$. As you may have noticed in Section 4.5, it is easy to get confused between the forces acting *on* an object and the forces exerted *by* that object *on* some other object. Only the forces acting *on* the object go into $\sum \vec{F}$.

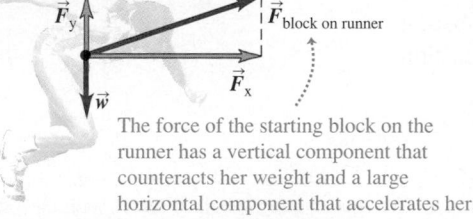

The force of the starting block on the runner has a vertical component that counteracts her weight and a large horizontal component that accelerates her.

(a)

(b)

To jump up, this player will push down against the floor, increasing the upward reaction force of the floor on him.

This player is a freely falling object.

(c)

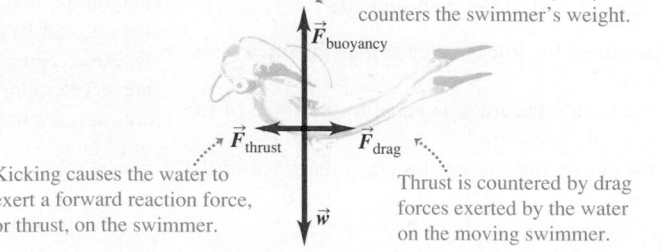

The water exerts a buoyancy force that counters the swimmer's weight.

Kicking causes the water to exert a forward reaction force, or thrust, on the swimmer.

Thrust is countered by drag forces exerted by the water on the moving swimmer.

▲ **FIGURE 4.24** Some examples of free-body diagrams.

To help identify the relevant forces, draw a **free-body diagram.** What's that? It is a diagram showing the chosen object by itself, "free" of its surroundings, with vectors drawn to show the forces applied to it by the various other objects that interact with it. We have already shown free-body diagrams in some previously worked-out examples. Be careful to include *all* the forces acting on the object, but be equally careful *not* to include any forces that the object exerts on any other object. In particular, the two forces in an action–reaction pair must *never* appear in the same free-body diagram, because they never act on the same object. Figure 4.24 shows some examples of free-body diagrams.

When you have a complete free-body diagram, you should be able to answer, for each force shown, the question "What other object is applying this force?" If you can't answer that question, you may be dealing with a nonexistent force. Sometimes you will have to take the problem apart and draw a separate free-body diagram for each part.

Conceptual Analysis 4.5   Dribble

A boy bounces a basketball straight down against the floor of the court. To answer the following question, it's helpful to draw a free-body diagram and consider the acceleration of the ball.

During the time the ball is in contact with the floor,

A. the force of the floor on the ball becomes much larger than the force of the ball on the floor.
B. the force of the ball on the floor plus the force of the floor on the ball equals the force of acceleration of the ball.
C. the force of the floor on the ball is equal to the force of gravity on the ball plus the force of inertia of the ball.
D. the force of the floor on the ball becomes much larger than the force of gravity on the ball.

SOLUTION First, there is no such thing as "force of acceleration" or "force of inertia." These are erroneous and misleading concepts, so answers B and C should be eliminated immediately from consideration. During the very short time the ball is touching the floor, the upward force exerted by the floor quickly becomes greater in magnitude than the ball's weight. The ball continues to move downward, but with decreasing speed (and upward acceleration), stops for an instant, then begins to move upward with increasing speed (again upward acceleration). For upward acceleration, the upward force exerted by the floor on the ball must be greater in magnitude than the weight of the ball. Thus, answer D is correct. But Newton's third law requires the force of the floor on the ball and the force of the ball on the floor to be equal and opposite; therefore, answer A cannot be correct. Forces by the ball on other objects have no effect on the acceleration of the ball. The floor pushes up on the ball, and the earth pulls down on it. Because the direction of acceleration is upward as the ball slows, stops, and then accelerates upward, the larger force must be directed upward to produce this upward acceleration. The answer is D.

Quantitative Analysis 4.6   Relating graphs and free-body diagrams

An object moves along the *x* axis. Assuming that the positive coordinate is to the right and that the forces are constant, which of the following graphs of v_x as a function of time is consistent with the forces shown in the free-body diagram in Figure 4.25?

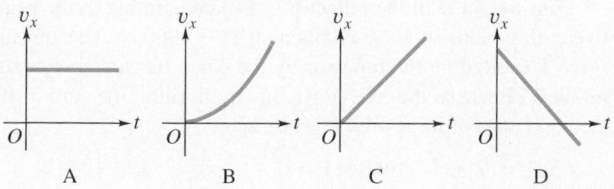

▶ **FIGURE 4.25**
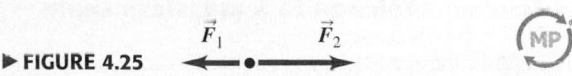

SOLUTION The force diagram shows the larger force to the right, the positive direction. Hence, the acceleration must be to the right also; a_x is positive. This means that the object is either (a) moving toward the left (the negative direction) and slowing down or (b) moving toward the right and speeding up. None of the graphs show leftward, slowing motion. D shows rightward, slowing motion, while A shows a constant velocity. Both B and C show rightward and speeding-up motion, but only C has a constant slope (corresponding to constant acceleration). Since the forces are constant, the acceleration is constant also, as indicated by the straight line in answer C.

PROBLEM-SOLVING STRATEGY 4.1 **Newton's laws**

SET UP

1. Always define your coordinate system. A diagram showing the positive axis direction is essential. If you know the direction of the acceleration, it's often convenient to take that as your positive direction. If you will need to calculate displacements of the moving object, you also need to specify the location of the origin of your coordinate system.

2. Be consistent with signs. Once you define the *x* axis and its positive direction, velocity, acceleration, and force components in that direction are also positive.

3. In applying Newton's laws, always concentrate on a specific object. Draw a free-body diagram showing all the forces (magnitudes and directions) acting *on* this object, but *do not* include forces that the object exerts on any other object. The acceleration of the object is determined by the forces acting on it, not by the forces it exerts on something else. Represent the object as a dot or by a simple sketch; you don't have to be an artist.

SOLVE

4. Identify the known and unknown quantities, and give each unknown quantity an algebraic symbol. If you know the direction of a force at the start, use a symbol to represent the *magnitude* of the force (always a positive quantity). Keep in mind that the *component* of this force along a particular axis direction may still be either positive or negative.

5. Always check units for consistency. When appropriate, use the conversion $1 \text{ N} = 1 \text{ kg} \cdot \text{m/s}^2$.

REFLECT

6. Check to make sure that your forces obey Newton's third law. When you can guess the direction of the net force, make sure that its direction is the same as that of the acceleration, as required by Newton's second law.

In Chapter 5 we'll expand this strategy to deal with more complex problems, but it's important for you to use it consistently from the very start, to develop good habits in the systematic analysis of problems.

EXAMPLE 4.7 **Tension in a massless chain**

To improve the acoustics in an auditorium, a sound reflector with a mass of 200 kg is suspended by a chain from the ceiling. What is its weight? What force (magnitude and direction) does the chain exert on it? What is the tension in the chain? Assume that the mass of the chain itself is negligible.

SOLUTION

SET UP The reflector is in equilibrium, so we use Newton's first law, $\Sigma \vec{F} = 0$. We draw separate free-body diagrams for the reflector and the chain (Figure 4.26). We take the positive *y* axis to be upward, as shown. Each force has only a *y* component. We give symbolic labels to the magnitudes of the unknown forces, using *T* to represent the tension in the chain.

SOLVE The magnitude of the weight $\vec{w}$ of the reflector is given by Equation 4.9:

$$w = mg = (200 \text{ kg})(9.80 \text{ m/s}^2) = 1960 \text{ N}.$$

Armed with this number and Newton's first law, we can compute the unknown force magnitudes.

The weight $\vec{w}$ of the reflector is a force pointing in the negative *y* direction, so its *y* component is -1960 N. The upward force $\vec{T}$ exerted on the reflector by the chain has unknown magnitude *T*. Because the reflector is in equilibrium, the sum of the *y* components of force on it must be zero:

$$\Sigma F_y = T + (-1960 \text{ N}) = 0, \quad \text{so} \quad T = 1960 \text{ N}.$$

The chain pulls *up* on the reflector with a force $\vec{T}$ of magnitude 1960 N. By Newton's third law, the reflector pulls *down* on the chain with a force of magnitude 1960 N.

Continued

The chain is also in equilibrium, so the vector sum of forces on it must equal zero. For this to be true, an upward force with magnitude 1960 N must act on it at its top end. The tension in the chain is $T = 1960$ N.

REFLECT Note that we have defined T to be the *magnitude* of a force, so it is always positive. But the y component of force acting on the chain at its lower end is $-T = -1960$ N. The tension is the same at every point in the chain; imagine cutting a link and then pulling the cut ends together. To hold them together requires a force with magnitude T on each cut end. If the chain isn't massless, the situation is different, as the next example shows.

Practice Problem: The weight of the reflector (the earth's gravitational pull) and the force the reflector exerts on the earth form an action–reaction pair. Determine the magnitude and direction of each of these forces. *Answers:* 1960 N down, 1960 N up.

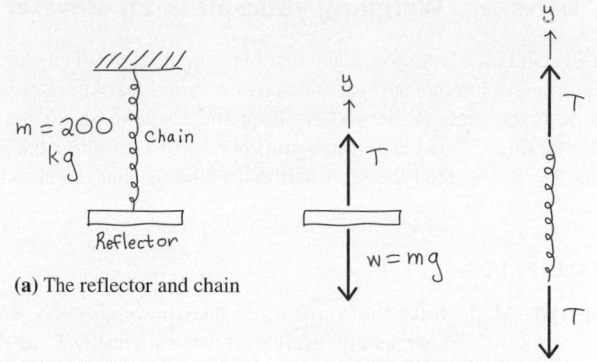

(a) The reflector and chain

(b) Free-body diagrams for the reflector and the chain

▲ **FIGURE 4.26** The diagrams we draw for this problem.

EXAMPLE 4.8   Tension in a chain with mass

Suppose the mass of the chain in Example 4.7 is not negligible, but is 10.0 kg. Find the forces at the ends of the chain.

SOLUTION

SET UP Again we draw separate free-body diagrams for the reflector and the chain (Figure 4.27); each is in equilibrium. We take the y axis to be vertically upward. The weight w_C of the chain is $w_C = m_C g = (10.0 \text{ kg})(9.8 \text{ m/s}^2) = 98$ N. The free-body diagrams are different from those in Figure 4.26 because the magnitudes of the forces acting on the two ends of the chain are no longer equal. (Otherwise the chain couldn't be in equilibrium.) We label these two forces T_1 and T_2.

SOLVE Note that the two forces labeled T_2 form an action–reaction pair; that's how we know that they have the same magnitude. The equilibrium condition $\Sigma F_y = 0$ for the reflector is

$$T_2 + (-1960 \text{ N}) = 0, \quad \text{so} \quad T_2 = 1960 \text{ N}.$$

There are now three forces acting on the chain: its weight and the forces at the two ends. The equilibrium condition $\Sigma F_y = 0$ for the chain is

$$T_1 + (-T_2) + (-98 \text{ N}) = 0.$$

Note that the y component of T_1 is positive because it points in the $+y$ direction, but the y components of both T_2 and 98 N are negative. When we substitute the value $T_2 = 1960$ N and solve for T_1, we find that

$$T_1 = 2058 \text{ N}.$$

Alternative Solution: An alternative procedure is to draw a free-body diagram for the composite object consisting of the reflector and the chain together (Figure 4.27b). The two forces on this composite object are the upward force T_1 at the top of the chain and the total weight, with magnitude 1960 N + 98 N =

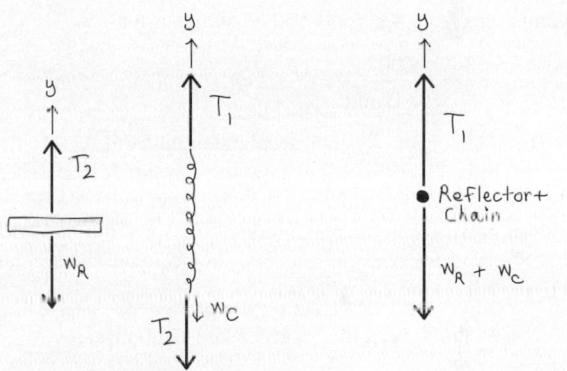

(a) Free-body diagrams for reflector and chain.

(b) Free-body diagram for reflector and chain treated as a unit.

▲ **FIGURE 4.27** Two alternative diagrams we could draw for this problem.

2058 N. Again, we find that $T_1 = 2058$ N. Note that we cannot find T_2 by this method; we still need a separate free-body diagram for one of the objects.

REFLECT When a composite object can be broken into two or more component parts, there are usually several alternative solutions. Often, the calculations can be simplified by a clever choice of subsystems. Also, solving a problem in two or more alternative ways provides a consistency check that's useful in finding errors.

Practice Problem: Solve the problem described in this example by using two free-body diagrams, one for the composite reflector–chain object, and one for the reflector alone.

EXAMPLE 4.9   **Weighing yourself in an elevator**

You stand on a bathroom scale that rests on the floor of an elevator. (Don't ask why!) Standing on the scale compresses internal springs and activates a dial that indicates your weight in newtons. When the elevator is at rest, the scale reads 600 N. Then the elevator begins to move upward with a constant acceleration $a_y = 2.00$ m/s². **(a)** Determine your mass. **(b)** Determine the scale reading while the elevator is accelerating. **(c)** If you read the scale without realizing that the elevator is accelerating upward, what might you *think* your mass is?

SOLUTION

SET UP All the forces are vertical; we take the y axis to be vertically upward. We neglect the weight of the scale itself. When the scale is in equilibrium (not accelerating), the total force acting on it must be zero. When it reads 600 N, forces of magnitude 600 N (with opposite directions) are applied to its top and bottom surfaces. Using this analysis, we draw separate free-body diagrams for you and the scale (Figure 4.28).

SOLVE Part (a): At first, you, the scale, and the elevator are at rest in an inertial system; all three objects are in equilibrium. You push down on the scale with a force of magnitude 600 N, and (from Newton's third law) the scale applies an equal and opposite upward force to your feet. The only other force on you is your weight; you are in equilibrium, so the *total y* component of force on you is zero. Your weight is 600 N, and your mass is

$$m = \frac{w}{g} = \frac{600 \text{ N}}{9.80 \text{ m/s}^2} = 61.2 \text{ kg} \qquad \text{(about 135 lb).}$$

Part (b): When the system accelerates upward at 2.00 m/s², your weight is still 600 N, but now the upward force the scale applies to your feet is different because you are no longer in equilibrium. We'll call the magnitude of the upward force F (Figure 4.28b). We use Newton's second law to relate the acceleration to the net force:

$$F - 600 \text{ N} = ma = (61.2 \text{ kg})(2.00 \text{ m/s}^2)$$

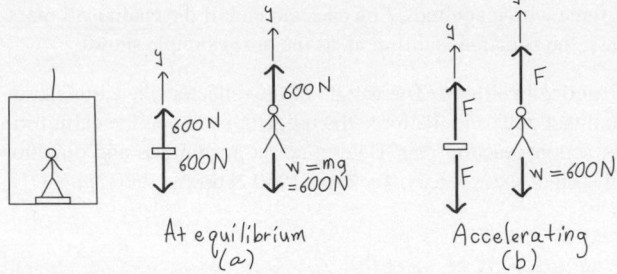

▲ **FIGURE 4.28** Our diagrams for this problem.

and

$$F = 722 \text{ N.}$$

Part (c): If you think your weight is 722 N, then you are likely to conclude that your mass is $m = w/g = (722 \text{ N})/(9.8 \text{ m/s}^2) = 73.7$ kg, instead of the actual value of 61.2 kg.

REFLECT Relate what we've said here to your own feeling of being lighter or heavier in an elevator. Do these feelings occur only when the elevator is accelerating?

Practice Problem: Suppose the elevator is accelerating *downward* at 2.00 m/s². What does the scale read during the acceleration? *Answer:* 478 N.

SUMMARY

Force

(Section 4.1) Force, a vector quantity, is a quantitative measure of the mechanical interaction between two objects. When several forces act on an object, the effect is the same as when a single force, equal to the vector sum, or resultant, of the forces, acts on the object.

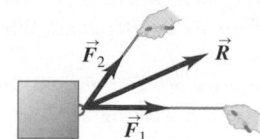

Two forces acting on a point have the same effect as a single force $\vec{R}$ equal to their vector sum.

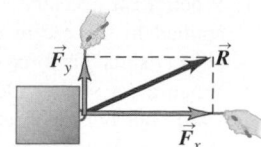

A force may be treated as the vector sum of its component vectors.

Newton's First Law

(Section 4.2) Newton's first law states that when no force acts on an object or when the vector sum of all forces acting on it is zero, the object is in equilibrium. If the object is initially at rest, it remains at rest; if it is initially in motion, it continues to move with constant velocity. This law is valid only in inertial frames of reference.

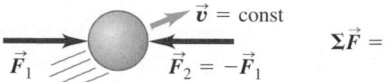

If the vector sum of the forces acting on an object is zero, the object remains at rest or (as here) in motion with constant velocity.

Mass and Newton's Second Law

(Section 4.3) The inertial properties of an object are characterized by its *mass*. The acceleration of an object under the action of a given set of forces is directly proportional to the vector sum of the forces and inversely proportional to the mass of the object. This relationship is Newton's second law:

$$\sum \vec{F} = m\vec{a}, \qquad (4.7)$$

or, in component form, $\sum F_x = ma_x$ and $\sum F_y = ma_y$ (Equations 4.8).

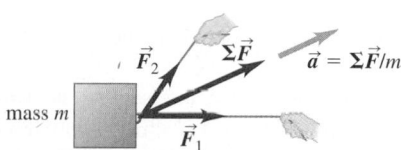

An object's acceleration depends on its mass and on the net force acting on it.

Mass and Weight

(Section 4.4) The weight of an object is the gravitational force exerted on it by the earth (or whatever other object exerts the gravitational force). Weight is a force and is therefore a vector quantity. The magnitude of the weight of an object at any specific location is equal to the product of the mass m of the object and the magnitude of the acceleration due to gravity, g, at that location:

$$w = mg. \qquad (4.9)$$

The weight of an object depends on its location, but the mass is independent of location.

An object's weight depends on its mass and on the acceleration due to gravity at its location.

Newton's Third Law

(Section 4.5) Newton's third law states that "action equals reaction"; when two bodies interact, they exert forces on each other that, at each instant, are equal in magnitude and opposite in direction. The two forces in an action–reaction pair always act on two different bodies; they never act on the same object.

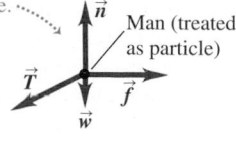

An action–reaction pair: $\vec{F}_{\text{ball on foot}} = -\vec{F}_{\text{foot on ball}}$. The two forces represent a *mutual interaction* of two objects, and each acts on a *different* object.

Free-Body Diagrams

(Section 4.6) In Newton's second law, only the forces acting on an object go into $\sum \vec{F}$. To help identify the relevant forces, draw a free-body diagram. Such a diagram shows the chosen object by itself, "free" of its surroundings, with vectors drawn to show the forces applied to it by the various other objects that interact with it. Be careful to include *all* the forces acting on the object, but be equally careful *not* to include any forces that the object exerts on any other object. In particular, the two forces in an action–reaction pair must *never* appear in the same free-body diagram, because they never act on the same object.

A free-body diagram of a man dragging a crate. The diagram shows *all* the forces acting on the man, and *only* forces acting on the man.

Man (treated as particle)

Conceptual Questions

1. When a car accelerates starting from rest, where is the force applied to the car in order to cause its acceleration? What object exerts this force on the car?

2. When a car stops suddenly, the passengers are "thrown forward" out of their seats (if they are not wearing seat belts). What causes this to happen? Are they really *thrown* forward?

3. For medical reasons, it is important for astronauts in outer space to determine their body mass at regular intervals. Devise a scheme for measuring body mass in a zero-gravity environment.

4. Although in everyday life the earth seems to be an inertial frame of reference, the paths of objects moving freely across the earth's surface, such as air masses, artillery shells, and even thrown baseballs, have a tendency to curve slightly to their right (in the Northern Hemisphere) as seen by an observer on the surface. Can you explain why this is?

5. A passenger in a bus notices that a ball which has been at rest in the aisle suddenly starts to roll toward the front of the bus. What can the passenger conclude about the motion of the bus from this observation?

6. If you hit the sidewalk with a hammer, the hammer bounces back at you. How does this process involve Newton's third law?

7. In a head-on collision between a compact hybrid car and a gas-guzzling SUV, which vehicle is acted upon by the greater force? So why are people in the smaller car more likely to get injured than those in the large one, assuming that both cars are equally strong?

8. Suppose you are in a car in a very dense fog and cannot see the meters on your dashboard or anything outside the car. (Maybe you should not be driving under these conditions!) (a) How can you tell if your car is speeding up or slowing down? (b) Can you tell if your car is at rest or is moving with uniform velocity?

9. It would be much easier to lift a bowling ball on the moon than on the earth. Would it be similarly easier to *catch* the bowling ball on the moon, if someone threw it to you? Explain.

10. You are in a spaceship far from any stars or planets when you notice that if you release an object, it accelerates toward your feet at 9.8 m/s². (a) What can you conclude about the motion of the spaceship? (b) What force, if any, do you feel on your feet from the floor? (c) Explain how your observations in parts (a) and (b) lead you to believe that gravity is present. (d) Is gravity *really* present in the spaceship? How has the ship generated artificial gravity?

11. When a car is hit from behind, people in that car can receive a whiplash injury to their necks. Use Newton's laws of motion to explain why this occurs.

12. If your hands are wet and no towel is handy, you can remove excess water by shaking them. Use Newton's laws to explain why doing this gets rid of the water.

13. It is possible to play catch with a softball in an airplane in level flight just as though the plane were at rest. Is this still possible when the plane is making a turn?

14. Newton's third law tells us that if you push a box with a 15 N force, it pushes back on you with a 15 N force. How can you ever accelerate this box if it always pushes back with the same force you exert on it?

15. If you drop a 10 lb rock and a 10 lb pillow from the same height with no air resistance, which one is moving faster when they reach the ground? So why would you rather be hit by the pillow than the rock?

16. Which "feels" a greater pull due to gravity, a heavy object or a light object? So why do heavy objects *not* accelerate faster than light objects?

17. If you step hard on the accelerator pedal of your car, your body feels "pushed back" in the seat. (a) Why do you feel this? Are you really pushed back? (b) Make a free-body diagram of your body under these circumstances.

Multiple-Choice Problems

1. A force F is required to push a crate along a rough horizontal floor at a constant speed V with friction present. What force is needed to push this crate along the same floor at a constant speed $3V$ if friction is the same as before?
   A. A constant force $3F$ is needed.
   B. A force that gradually increases from F to $3F$ is needed.
   C. A constant force F is needed.
   D. No force is needed, since the crate has no acceleration.

2. When you're driving on the freeway it's necessary to keep your foot on the accelerator to keep the car moving at a constant speed. In this situation
   A. the net force on the car is in the forward direction.
   B. the net force on the car is toward the rear.
   C. the net force on the car is zero.
   D. the net force on the car depends on your speed.

3. Assuming you like chocolate cake, which would you rather have:
   A. a piece of cake that weighs 1 N on earth.
   B. a piece of cake that weighs 1 N on the moon.
   C. a piece of cake that weighs 1 N on Jupiter.
   D. The weight doesn't matter; the amount of cake is the same in all three cases.

4. A golfer tees off and hits the ball with a mighty swing. During the brief time the golf club is in contact with the ball,
   A. the ball pushes just as hard on the club as the club pushes on the ball.
   B. the push of the club on the ball is much greater than the push of the ball on the club, since the club makes the ball move.
   C. the push of the ball on the club would be equal to the push of the club on the ball only if the ball did not move.
   D. the club pushes harder on the ball than the ball does on the club because the club has a greater mass than the ball.

5. Three books are at rest on a horizontal table, as shown in Figure 4.29. The *net* force on the middle book is
   A. 5 N downward.
   B. 15 N upward.
   C. 15 N downward.
   D. 0 N.

6. In outer space, where there is no gravity or air, an astronaut pushes with an equal force of 12 N on a 2 N moon rock and on a 4 N moon rock.

▲ **FIGURE 4.29** Multiple-choice problem 5.

A. Since both rocks are weightless, they will have the same acceleration.

B. The 4 N rock pushes back on the astronaut twice as hard as the 2 N rock.

C. Both rocks push back on the astronaut with 12 N.

D. Since both rocks are weightless, they do not push back on the astronaut.

7. A rocket firing its engine and accelerating in outer space (no gravity, no air resistance) suddenly runs out of fuel. Which of the following best describes its motion after burnout?

A. It continues to accelerate at the same rate.

B. It continues to accelerate but at a gradually decreasing rate, until it reaches a constant velocity.

C. It immediately stops accelerating, and continues moving at the velocity it had when burnout occurred.

D. It immediately begins slowing down and gradually approaches zero velocity.

8. A 3 lb physics book rests on an ordinary scale that is placed on a horizontal table. The reaction force to the downward 3 lb force on the book is

A. an upward 3 lb force on the scale due to the table.

B. an upward 3 lb force on the book due to the scale.

C. an upward 3 lb force on the table due to the floor.

D. an upward 3 lb force on the earth due to the book.

9. A person pushes horizontally with constant force P on a 250 N box resting on a frictionless horizontal floor. Which of the following statements about this box is correct?

A. The box will accelerate no matter how small P is.

B. The box will not accelerate unless $P > 250$ N.

C. The box will move with constant velocity because P is constant.

D. Once the box is set moving, it will come to rest after P is removed.

10. Suppose the sun, including its gravity, suddenly disappeared. Which of the following statements best describes the subsequent motion of the earth?

A. The earth would continue moving in a straight line tangent to its original direction, but would gradually slow down.

B. The earth would speed up because the sun's gravity would not be able to slow it down.

C. The earth would continue moving in a straight line tangent to its original direction and would not change its speed.

D. The earth would move directly outward with constant speed away from the sun's original position.

11. Three weights hang by very light wires as shown in Figure 4.30. What must be true about the tensions in these wires? (There may be more than one correct choice.)

A. The tension in A is the greatest.

B. The tension in C is the greatest.

C. All three wires have the same tension because the system is in equilibrium.

D. The tension in C is the least.

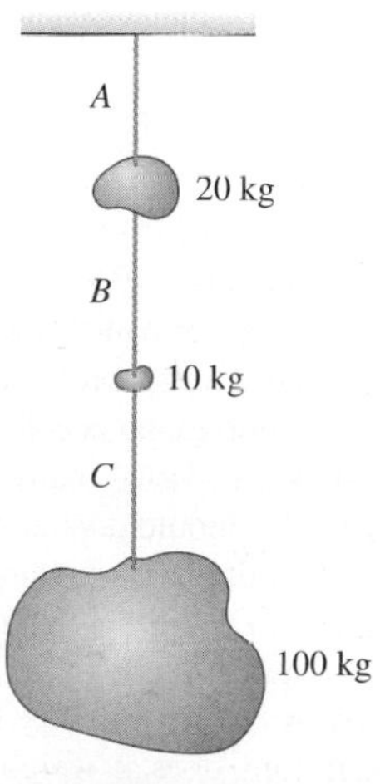

12. A woman normally weighs 125 lb. If she is standing on a spring scale in

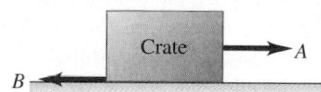

▲ **FIGURE 4.30** Multiple-choice problem 11.

an elevator that is traveling *downward*, but slowing up, the scale will read

A. more than 125 lb.

B. 125 lb.

C. less than 125 lb.

D. It is impossible to answer this question without knowing the acceleration of the elevator.

13. A worker pushes horizontally on a 2000 N refrigerator on a kitchen floor with a force of 25 N.

A. If the refrigerator moves forward, this person feels it pushing back with less than 25 N.

B. The person always feels a 25 N push from the refrigerator.

C. The person feels a 25 N push from the refrigerator only if it does not move.

D. The person feels a 2000 N push from the refrigerator.

14. Far from any gravity, an astronaut accidentally releases a metal wrench inside a spaceship that is traveling at a constant velocity of 250 km/s away from earth in outer space. What will be the subsequent behavior of that wrench as observed by the astronaut?

A. It will fall to the floor with constant velocity.

B. It will fall to the floor with constant acceleration.

C. It will fall to the back of the spaceship with constant velocity.

D. It will fall to the back of the spaceship with constant acceleration.

E. It will remain right where the astronaut released it.

15. A worker pulls horizontally on a crate on a rough horizontal floor, causing it to move forward with constant velocity. In Figure 4.31, force A is the pull of the worker and force B is the force of friction due to the floor. Which one of the following statements about these forces is correct?

A. $A < B$ B. $A = B$ C. $A > B$

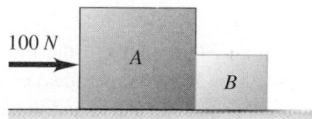

▲ **FIGURE 4.31** Multiple-choice problem 15.

16. A person pushes two boxes with a horizontal 100 N force on a frictionless floor, as shown in Figure 4.32. Box A is heavier than box B. Which of the following statements about these boxes is correct?

A. Box A pushes on box B with a force of 100 N, and box B pushes on box A with a force of 100 N.

B. Box A pushes on box B harder than box B pushes on box A.

C. Boxes A and B push on each other with equal forces of less than 100 N.

D. The boxes will not begin to move unless the total weight of the two boxes is less than 100 N.

▲ **FIGURE 4.32** Multiple-choice problem 16.

Problems

4.1 Force

1. • A warehouse worker pushes a crate along the floor, as shown in Figure 4.33, by a force of 10 N that points downward at an angle of 45° below the horizontal. Find the horizontal and vertical components of the push.

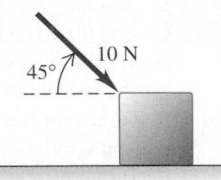

▲ **FIGURE 4.33** Problem 1.

2. • Two dogs pull horizontally on ropes attached to a post; the angle between the ropes is 60.0°. If dog A exerts a force of 270 N and dog B exerts a force of 300 N, find the magnitude of the resultant force and the angle it makes with dog A's rope.

3. • A man is dragging a trunk up the loading ramp of a mover's truck. The ramp has a slope angle of 20.0°, and the man pulls upward with a force $\vec{F}$ of magnitude 375 N whose direction makes an angle of 30.0° with the ramp. (See Figure 4.34.) Find the horizontal and vertical components of the force $\vec{F}$.

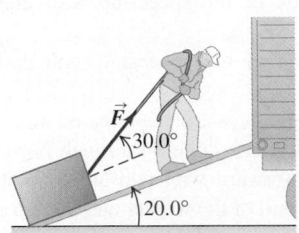

▲ **FIGURE 4.34** Problem 3.

4. • **Jaw injury.** Due to a jaw injury, a patient must wear a strap (see Figure 4.35) that produces a net upward force of 5.00 N on his chin. The tension is the same throughout the strap. To what tension must the strap be adjusted to provide the necessary upward force?

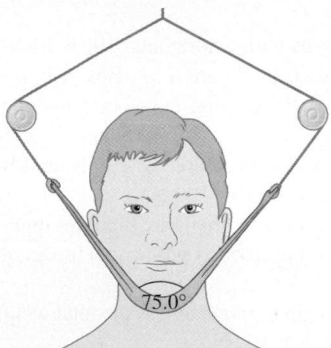

▲ **FIGURE 4.35** Problem 4.

5. • Workmen are trying to free an SUV stuck in the mud. To extricate the vehicle, they use three horizontal ropes, producing the force vectors shown in Figure 4.36. (a) Find the x and y components of each of the three pulls. (b) Use the components to find the magnitude and direction of the resultant of the three pulls.

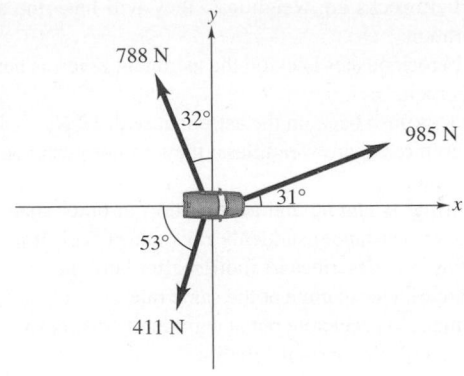

▲ **FIGURE 4.36** Problem 5.

4.3 Mass and Newton's Second Law

6. • A box rests on a frozen pond, which serves as a frictionless horizontal surface. If a fisherman applies a horizontal force with magnitude 48.0 N to the box and produces an acceleration of magnitude 3.00 m/s², what is the mass of the box?

7. • In outer space, a constant net force of magnitude 140 N is exerted on a 32.5 kg probe initially at rest. (a) What acceleration does this force produce? (b) How far does the probe travel in 10.0 s?

8. •• A 68.5 kg skater moving initially at 2.40 m/s on rough horizontal ice comes to rest uniformly in 3.52 s due to friction from the ice. What force does friction exert on the skater?

9. •• **Animal dynamics.** An adult 68 kg cheetah can accelerate
BIO from rest to 20.1 m/s (45 mph) in 2.0 s. Assuming constant acceleration, (a) find the net external force causing this acceleration. (b) Where does the force come from? That is, what exerts the force on the cheetah?

10. •• A hockey puck with mass 0.160 kg is at rest on the horizontal, frictionless surface of a rink. A player applies a force of 0.250 N to the puck, parallel to the surface of the ice, and continues to apply this force for 2.00 s. What are the position and speed of the puck at the end of that time?

11. •• A dock worker applies a constant horizontal force of 80.0 N to a block of ice on a smooth horizontal floor. The frictional force is negligible. The block starts from rest and moves 11.0 m in the first 5.00 s. What is the mass of the block of ice?

4.4 Mass and Weight

12. • (a) What is the mass of a book that weighs 3.20 N in the laboratory? (b) In the same lab, what is the weight of a dog whose mass is 14.0 kg?

13. • Superman throws a 2400-N boulder at an adversary. What horizontal force must Superman apply to the boulder to give it a horizontal acceleration of 12.0 m/s²?

14. • (a) How many newtons does a 150 lb person weigh?
BIO (b) Should a veterinarian be skeptical if someone said that her adult collie weighed 40 N? (c) Should a nurse question a medical chart which showed that an average-looking patient had a mass of 200 kg?

15. • (a) An ordinary flea has a mass of 210 μg. How many new-
BIO tons does it weigh? (b) The mass of a typical froghopper is 12.3 mg. How many newtons does it weigh? (c) A house cat typically weighs 45 N. How many pounds does it weigh and what is its mass in kilograms?

16. • Calculate the mass (in SI units) of (a) a 160-lb human being;
BIO (b) a 1.9-lb cockatoo. Calculate the weight (in English units) of (c) a 2300-kg rhinoceros; (d) a 22-g song sparrow.

17. •• An astronaut's pack weighs 17.5 N when she is on earth but only 3.24 N when she is at the surface of an asteroid. (a) What is the acceleration due to gravity on this asteroid? (b) What is the mass of the pack on the asteroid?

18. • **Interpreting a medical chart.** You, a resident physician,
BIO are reading the medical chart of a normal adult female patient. Carelessly, one of the nurses has entered this woman's weight as a number without units. Another nurse has offered a suggestion for what the units might be. In each of the following cases, decide whether this nurse's suggestion is physically reasonable: (a) The number is 150, and the nurse suggests that the units are kilograms. (b) The number is 4.25, and the nurse suggests that the units are slugs. (c) The number is 65,000, and the nurse suggests that the units are grams.

19. • What does a 138 N rock weigh if it is accelerating (a) upward at $12 \ m/s^2$, (b) downward at $3.5 \ m/s^2$? (c) What would be the answers to parts (a) and (b) if the rock had a mass of 138 kg? (d) What would be the answers to parts (a) and (b) if the rock were moving with a constant upward velocity of 23 m/s?

20. •• At the surface of Jupiter's moon Io, the acceleration due to gravity is $1.81 \ m/s^2$. If a piece of ice weighs 44.0 N at the surface of the earth, (a) what is its mass on the earth's surface? (b) What are its mass and weight on the surface of Io?

21. •• A scientific instrument that weighs 85.2 N on the earth weighs 32.2 N at the surface of Mercury. (a) What is the acceleration due to gravity on Mercury? (b) What is the instrument's mass on earth and on Mercury?

22. •• **Planet X!** When venturing forth on Planet X, you throw a 5.24 kg rock upward at 13.0 m/s and find that it returns to the same level 1.51 s later. What does the rock weigh on Planet X?

23. •• The driver of a 1750 kg car traveling on a horizontal road at 110 km/h suddenly applies the brakes. Due to a slippery pavement, the friction of the road on the tires of the car, which is what slows down the car, is 25% of the weight of the car. (a) What is the acceleration of the car? (b) How many meters does it travel before stopping under these conditions?

4.5 Newton's Third Law

24. • You drag a heavy box along a rough horizontal floor by a horizontal rope. Identify the reaction force to each of the following forces: (a) the pull of the rope on the box, (b) the friction force on the box, (c) the normal force on the box, and (d) the weight of the box.

25. • Imagine that you are holding a book weighing 4 N at rest on the palm of your hand. Complete the following sentences: (a) A downward force of magnitude 4 N is exerted on the book by _____. (b) An upward force of magnitude _____ is exerted on _____ by the hand. (c) Is the upward force in part (b) the reaction to the downward force in part (a)? d) The reaction to the force in part (a) is a force of magnitude _____, exerted on _____ by _____. Its direction is _____. (e) The reaction to the force in part (b) is a force of magnitude _____, exerted on _____ by _____. Its direction is _____. (f) The forces in parts (a) and (b) are "equal and

opposite" because of Newton's _____ law. (g) The forces in parts (b) and (e) are "equal and opposite" because of Newton's _____ law.

26. • Suppose now that you exert an upward force of magnitude 5 N on the book in the previous problem. (a) Does the book remain in equilibrium? (b) Is the force exerted on the book by your hand equal and opposite to the force exerted on the book by the earth? (c) Is the force exerted on the book by the earth equal and opposite to the force exerted on the earth by the book? (d) Is the force exerted on the book by your hand equal and opposite to the force exerted on your hand by the book? Finally, suppose that you snatch your hand away while the book is moving upward. (e) How many forces then act on the book? (f) Is the book in equilibrium?

27. • The upward normal force exerted by the floor is 620 N on an elevator passenger who weighs 650 N. What are the reaction forces to these two forces? Is the passenger accelerating? If so, what are the magnitude and direction of the acceleration?

4.6 Free-Body Diagrams

28. • A person throws a 2.5 lb stone into the air with an initial upward speed of 15 ft/s. Make a free-body diagram for this stone (a) after it is free of the person's hand and is traveling upward, (b) at its highest point, (c) when it is traveling downward, and (d) while it is being thrown upward, but is still in contact with the person's hand.

29. • The driver of a car traveling at 65 mph suddenly hits his brakes on a horizontal highway. (a) Make a free-body diagram of the car while it is slowing down. (b) Make a free-body diagram of a passenger in a car that is accelerating on a freeway entrance ramp.

30. • A tennis ball traveling horizontally at 22 m/s suddenly hits a vertical brick wall and bounces back with a horizontal velocity of 18 m/s. Make a free-body diagram of this ball (a) just before it hits the wall, (b) just after it has bounced free of the wall, and (c) while it is in contact with the wall.

31. •• Two crates, A and B, sit at rest side by side on a frictionless horizontal surface. The crates have masses m_A and m_B. A horizontal force $\vec{F}$ is applied to crate A and the two crates move off to the right. (a) Draw clearly labeled free-body diagrams for crate A and for crate B. Indicate which pairs of forces, if any, are third-law action–reaction pairs. (b) If the magnitude of force $\vec{F}$ is less than the total weight of the two crates, will it cause the crates to move? Explain.

32. •• A ball is hanging from a long string that is tied to the ceiling of a train car traveling eastward on horizontal tracks. An observer inside the train car sees the ball hang motionless. Draw a clearly labeled free-body diagram for the ball if (a) the train has a uniform velocity, and (b) the train is speeding up uniformly. Is the net force on the ball zero in either case? Explain.

33. • A person drags her 65 N suitcase along the rough horizontal floor by pulling upward at 30° above the horizontal with a 50 N force. Make a free-body diagram of this suitcase.

34. •• A factory worker pushes horizontally on a 250 N crate with a force of 75 N on a horizontal rough floor. A 135 N crate rests on top of the one being pushed and moves along with it. Make a free-body diagram of *each* crate if the friction force exerted by the floor is less than the worker's push.

35. ●● A dock worker pulls two boxes connected by a rope on a horizontal floor, as shown in Figure 4.37. All the ropes are horizontal, and there is some friction with the floor. Make a free-body diagram of each box.

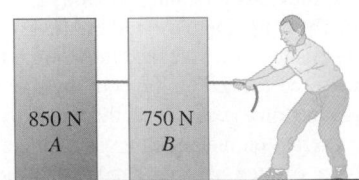

▲ **FIGURE 4.37** Problem 35.

36. ●● A hospital orderly pushes horizontally on two boxes of equipment on a rough horizontal floor, as shown in Figure 4.38. Make a free-body diagram of each box.

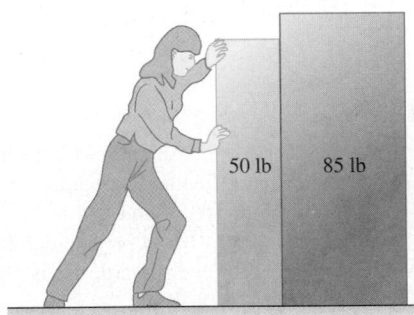

▲ **FIGURE 4.38** Problem 36.

37. ● A uniform 25.0 kg chain 2.00 m long supports a 50.0 kg chandelier in a large public building. Find the tension in (a) the bottom link of the chain, (b) the top link of the chain, and (c) the middle link of the chain.

38. ● An acrobat is hanging by his feet from a trapeze, while supporting with his hands a second acrobat who hangs below him. Draw separate free-body diagrams for the two acrobats.

39. ● A 275 N bucket is lifted with an acceleration of 2.50 m/s² by a 125 N uniform vertical chain. Start each of the following parts with a free-body diagram. Find the tension in (a) the top link of the chain, (b) the bottom link of the chain, and (c) the middle link of the chain.

40. ●● **Human biomechanics.** World-class sprinters can spring
BIO out of the starting blocks with an acceleration that is essentially horizontal and of magnitude 15 m/s². (a) How much horizontal force must a 55-kg sprinter exert on the starting blocks during a start to produce this acceleration? (b) What exerts the force that propels the sprinter, the blocks or the sprinter himself?

41. ●● A chair of mass 12.0 kg is sitting on the horizontal floor; the floor is not frictionless. You push on the chair with a force $F = 40.0$ N that is directed at an angle of 37.0° below the horizontal, and the chair slides along the floor. (a) Draw a clearly labeled free-body diagram for the chair. (b) Use your diagram and Newton's laws to calculate the normal force that the floor exerts on the chair.

42. ●● **Human biomechanics.** The fastest pitched baseball was
BIO measured at 46 m/s. Typically, a baseball has a mass of 145 g. If the pitcher exerted his force (assumed to be horizontal and

constant) over a distance of 1.0 m, (a) what force did he produce on the ball during this record-setting pitch? (b) Make free-body diagrams of the ball during the pitch and just *after* it has left the pitcher's hand.

43. ●● You walk into an elevator, step onto a scale, and push the "up" button. You also recall that your normal weight is 625 N. Start each of the following parts with a free-body diagram. (a) If the elevator has an acceleration of magnitude 2.50 m/s², what does the scale read? (b) If you start holding a 3.85 kg package by a light vertical string, what will be the tension in this string once the elevator begins accelerating?

44. ●● A truck is pulling a car on a horizontal highway using a horizontal rope. The car is in neutral gear, so we can assume that there is no appreciable friction between its tires and the highway. As the truck is accelerating to highway speeds, draw a free-body diagram of (a) the car and (b) the truck. (c) What force accelerates this system forward? Explain how this force originates.

45. ●● **The space shuttle.** During the first stage of its launch, a space shuttle goes from rest to 4973 km/h while rising a vertical distance of 45 km. Assume constant acceleration and no variation in g over this distance. (a) What is the acceleration of the shuttle? (b) If a 55.0 kg astronaut is standing on a scale inside the shuttle during this launch, how hard will the scale push on her? Start with a free-body diagram of the astronaut. (c) If this astronaut did not realize that the shuttle had left the launch pad, what would she think were her weight and mass?

46. ●● A woman is standing in an elevator holding her 2.5-kg briefcase by its handles. Draw a free-body diagram for the briefcase if the elevator is accelerating downward at 1.50 m/s², and calculate the downward pull of the briefcase on the woman's arm while the elevator is accelerating.

General Problems

47. ●● An advertisement claims that a particular automobile can "stop on a dime." What net force would actually be necessary to stop a 850-kg automobile traveling initially at 45.0 km/h in a distance equal to the diameter of a dime, which is 1.8 cm?

48. ●● A rifle shoots a 4.20 g bullet out of its barrel. The bullet has a muzzle velocity of 965 m/s just as it leaves the barrel. Assuming a constant horizontal acceleration over a distance of 45.0 cm starting from rest, with no friction between the bullet and the barrel, (a) what force does the rifle exert on the bullet while it is in the barrel? (b) Draw a free-body diagram of the bullet (i) while it is in the barrel and (ii) just *after* it has left the barrel. (c) How many g's of acceleration does the rifle give this bullet? (d) For how long a time is the bullet in the barrel?

49. ●● A parachutist relies on air resistance (mainly on her parachute) to decrease her downward velocity. She and her parachute have a mass of 55.0 kg, and at a particular moment air resistance exerts a total upward force of 620 N on her and her parachute. (a) What is the weight of the parachutist? (b) Draw a free-body diagram for the parachutist (see Section 4.6). Use that diagram to calculate the net force on the parachutist. Is the net force upward or downward? (c) What is the acceleration (magnitude and direction) of the parachutist?

50. ●● A spacecraft descends vertically near the surface of Planet X. An upward thrust of 25.0 kN from its engines slows it down at

a rate of 1.20 m/s^2, but it speeds up at a rate of 0.80 m/s^2 with an upward thrust of 10.0 kN. (a) In each case, what is the direction of the acceleration of the spacecraft? (b) Draw a free-body diagram for the spacecraft. In each case, speeding up or slowing down, what is the direction of the net force on the spacecraft? (c) Apply Newton's second law to each case, slowing down or speeding up, and use this to find the spacecraft's weight near the surface of Planet X.

51. •• **A standing vertical jump.** Basketball player Darrell Griffith is on record as attaining a standing vertical jump of 1.2 m (4 ft). (This means that he moved upward by 1.2 m after his feet left the floor.) Griffith weighed 890 N (200 lb). (a) What was his speed as he left the floor? (b) If the time of the part of the jump before his feet left the floor was 0.300 s, what were the magnitude and direction of his acceleration (assuming it to be constant) while he was pushing against the floor? (c) Draw a free-body diagram of Griffith during the jump. (d) Use Newton's laws and the results of part (b) to calculate the force he applied to the ground during his jump.

52. •• You leave the doctor's office after your annual checkup and recall that you weighed 683 N in her office. You then get into an elevator that, conveniently, has a scale. Find the magnitude and direction of the elevator's acceleration if the scale reads (a) 725 N, (b) 595 N.

53. •• **Human biomechanics.** The fastest served tennis ball, served by "Big Bill" Tilden in 1931, was measured at 73.14 m/s. The mass of a tennis ball is 57 g, and the ball is typically in contact with the tennis racquet for 30.0 ms, with the ball starting from rest. Assuming constant acceleration, (a) what force did Big Bill's tennis racquet exert on the tennis ball if he hit it essentially horizontally? (b) Make free-body diagrams of the tennis ball during the serve and just after it has moved free of the racquet.

54. •• **Extraterrestrial physics.** You have landed on an unknown planet, Newtonia, and want to know what objects will weigh there. You find that when a certain tool is pushed on a frictionless horizontal surface by a 12.0 N force, it moves 16.0 m in the first 2.00 s, starting from rest. You next observe that if you release this tool from rest at 10.0 m above the ground, it takes 2.58 s to reach the ground. What does the tool weigh on Newtonia and what would it weigh on Earth?

55. •• An athlete whose mass is 90.0 kg is performing weight-lifting exercises. Starting from the rest position, he lifts, with constant acceleration, a barbell that weighs 490 N. He lifts the barbell a distance of 0.60 m in 1.6 s. (a) Draw a clearly labeled free body force diagram for the barbell and for the athlete. (b) Use the diagrams in part (a) and Newton's laws to find the total force that the ground exerts on the athlete's feet as he lifts the barbell.

56. •• **Jumping to the ground.** A 75.0 kg man steps off a platform 3.10 m above the ground. He keeps his legs straight as he falls, but at the moment his feet touch the ground his knees begin to bend, and, treated as a particle, he moves an additional 0.60 m before coming to rest. (a) What is his speed at the instant his feet touch the ground? (b) Treating him as a particle, what are the magnitude and direction of his acceleration as he slows down if the acceleration is constant? (c) Draw a free-body diagram of this man as he is slowing down. (d) Use Newton's laws and the results of part (b) to calculate the force the ground exerts on him while he is slowing down. Express this force in newtons and also as a multiple of the man's weight. (e) What are the magnitude and direction of the reaction force to the force you found in part (c)?

57. •• An electron (mass $= 9.11 \times 10^{-31}$ kg) leaves one end of a TV picture tube with zero initial speed and travels in a straight line to the accelerating grid, which is 1.80 cm away. It reaches the grid with a speed of 3.00×10^6 m/s. If the accelerating force is constant, compute (a) the acceleration of the electron, (b) the time it takes the electron to reach the grid, and (c) the net force that is accelerating the electron, in newtons. (You can ignore the gravitational force on the electron.)

Passage Problems

BIO Bacterial motion. A bacterium using its flagellum as propulsion can move through liquids at a rate of 0.003 m/s. For a 50 μm-long bacterium, that is the equivalent of 60 cell lengths per second. Bacteria of that size have a mass of approximately 1×10^{-12} g.

The viscous drag on a swimming bacterium is so great that if it stops beating its flagellum it will stop within a distance of 0.01 nm.

58. What is the acceleration that stops the bacterium?
    A. 1.2×10^4 m/s^2
    B. 5×10^5 m/s^2
    C. 6×10^5 m/s^2
    D. 9×10^5 m/s^2

59. What is average magnitude of this viscous force?
    A. 4×10^{-7} N
    B. 1.7×10^{-8} N
    C. 9×10^{-9} N
    D. 5×10^{-10} N

60. What amount of force must the flagellum generate to propel the bacterium at a constant velocity of 0.003 m/s?
    A. 1.5×10^{-10} N
    B. 5×10^{-10} N
    C. 9×10^{-7} N
    D. 1.8×10^{-7} N

61. If the bacterium wished to accelerate at a rate of 0.001 m/s^2, how much additional force would be necessary?
    A. 1×10^{-18} N
    B. 3×10^{-18} N
    C. 1×10^{-15} N
    D. 4×10^{-15} N

5 Applications of Newton's Laws

Newton's three laws of motion, the foundation of classical mechanics, can be stated very simply, as we have seen. But applying these laws to situations such as a locomotive, a suspension bridge, a car rounding a banked curve, or a toboggan sliding down a hill requires some analytical skills and some problem-solving technique. In this chapter we introduce no new principles, but we'll try to help you develop some of the problem-solving skills you'll need to analyze such situations.

The soles of hiking boots are designed to stick, not slip, on rocky surfaces. In this chapter we'll learn about the interactions that give good traction.

We begin with equilibrium problems, concentrating on systems at rest. Then we generalize our problem-solving techniques to include systems that are not in equilibrium, for which we need to deal precisely with the relationships between forces and motion. We'll learn how to describe and analyze the contact force that acts on an object when it rests or slides on a surface, as well as the elastic forces that are present when a solid object is deformed. Finally, we take a brief look at the fundamental nature of force and the kinds of forces found in the physical universe.

5.1 Equilibrium of a Particle

We learned in Chapter 4 that an object is in **equilibrium** when it is at rest or moving with constant velocity in an inertial frame of reference. A hanging lamp, a rope-and-pulley setup for hoisting heavy loads, a suspension bridge—all these are examples of equilibrium situations. In this section, we consider only the equilibrium of an object that can be modeled as a particle. (Later, in Chapter 10, we'll consider the additional principles needed when an object can't be represented adequately as a particle.) The essential physical principle is Newton's first law: **When an object is at rest or is moving with constant velocity in an inertial**

frame of reference, the vector sum of all the forces acting on it must be zero.
That is, as discussed in Chapter 4, we have the following principle:

Necessary condition for equilibrium of an object
For an object to be in equilibrium, the vector sum of the forces acting on it
must be zero:

$$\sum \vec{F} = 0. \qquad (4.3)$$

This condition is sufficient only if the object can be treated as a particle,
which we assume in the next principle and throughout the remainder of the
chapter.

We'll usually use Equation 4.3 in component form:

Equilibrium conditions in component form
An object is in equilibrium if the sum of the components of force in each axis
direction is zero:

$$\sum F_x = 0, \qquad \sum F_y = 0. \qquad (4.4)$$

This chapter is about solving problems. We strongly recommend that you
study carefully the strategies that follow, look for their applications in the
worked-out examples, and then try to apply them when you solve problems.

PROBLEM-SOLVING STRATEGY 5.1 **Solving problems involving an**
object in equilibrium

SET UP
1. Draw a simple sketch of the apparatus or structure, showing all relevant
   dimensions and angles.
2. Identify the object or objects in equilibrium that you will consider.
3. Draw a free-body diagram for each object identified in step 2.
   a. Assuming that the object can be modeled as a particle, you can represent
      it by a large dot. Do not include other objects (such as a surface the object
      may be resting on or a rope pulling on it) in your free-body diagram.
   b. Identify all the ways in which other things interact with the object,
      either by touching it or via a noncontact force such as gravity. Draw a
      force vector to represent each force *acting on the object*. Do not include
      forces exerted *by* the object. For each force, make sure that you can
      answer the question "What other object interacts with the chosen object
      to cause that force?" If you can't answer that question, you may be
      imagining a force that isn't there.
   c. Label each force with a symbol representing the *magnitude* of the force
      (*w* for the object's weight; *n* for a normal force; *T* for a tension force,
      etc.); indicate the directions with appropriate angles. (If the object's
      mass is given, use $w = mg$ to find its weight.)
4. Choose a set of coordinate axes, and represent each force acting on the
   object in terms of its components along those axes. Cross out each force
   that has been replaced by its components, so that you don't count it twice.
   You can often simplify the problem by using a particular choice of coordinate axes. For example, when an object rests or slides on a plane surface, it
   is usually simplest to take the axes in the directions parallel and perpendicular to this surface, even when the plane is tilted.

Continued

▲ **Application** **Easy does it!** This crane is
unloading a 50 ton, 8.2 meter mirror for the
European Southern Observatory in Antofa-
gasta, Chile. It is one of four such mirrors
that make up an instrument known as the
Very Large Telescope, one of the world's
biggest, most advanced optical telescopes.
The crane moves the mirror steadily, so that
it remains nearly in equilibrium—the vector
sum of the forces on it is almost zero. However, that does not ensure the mirror's safety,
because it is not a particle. The forces acting
on different *parts* of it are not the same.
Warping or deforming would be disastrous.
The mirror is heavy in part because it is
heavily reinforced.

SOLVE

5. For each object, set the algebraic sum of all x components of force equal to zero. In a separate equation, set the algebraic sum of all y components equal to zero. (Never add x and y components in a single equation.) You can then solve these equations for up to two unknown quantities, which may be force magnitudes, components, or angles.

6. If you are dealing with two or more objects that interact with each other, use Newton's third law to relate the forces they exert on each other. You need to find as many independent equations as the number of unknown quantities. Then solve these equations to obtain the values of the unknowns. This part is algebra, not physics, but it's an essential step.

REFLECT

7. Whenever possible, look at your results and ask whether they make sense. When the result is a symbolic expression or formula, try to think of special cases (particular values or extreme cases for the various quantities) for which you can guess what the results ought to be. Check to see that your formula works in these particular cases. Think about what the problem has taught you that you can apply to other problems in the future.

EXAMPLE 5.1 **One-dimensional equilibrium**

A gymnast has just begun climbing up a rope hanging from a gymnasium ceiling. She stops, suspended from the lower end of the rope by her hands. Her weight is 500 N, and the weight of the rope is 100 N. Analyze the forces on the gymnast and on the rope.

SOLUTION

SET UP First we sketch the situation (Figure 5.1a). Then we draw two free-body diagrams, one for the gymnast and the other for the rope (Figures 5.1b and c). (We'll explain Figure 5.1d below.) The forces acting on the gymnast are her weight (magnitude 500 N) and the upward force (magnitude T_1) exerted on her by the rope. We *don't* include the downward force she exerts on the rope, because it isn't a force that acts *on* her. We take the y axis to be directed vertically upward, the x axis horizontally. There are no x components of force; that's why we call this a one-dimensional problem.

SOLVE The gymnast is motionless, so we know that she is in equilibrium. Since we know her weight, we can use Equation 4.4, $\Sigma F_y = 0$, to find the magnitude of the upward tension T_1 on her. This force pulls in the positive y direction. Her weight acts in the negative y direction, so its y component is the *negative* of the magnitude—that is, -500 N. Thus,

$$\Sigma F_y = 0,$$
$$T_1 + (-500\,\text{N}) = 0, \quad \text{(equilibrium of gymnast)}$$
$$T_1 = 500\,\text{N}.$$

The two forces acting on the gymnast are *not* an action–reaction pair, because they act on the same object.

Next we need to consider the forces acting on the rope (Figure 5.1c). Newton's third law tells us that the gymnast exerts a force on the rope that is equal and opposite to the force it exerts on her. In other words, she pulls down on the rope with a force

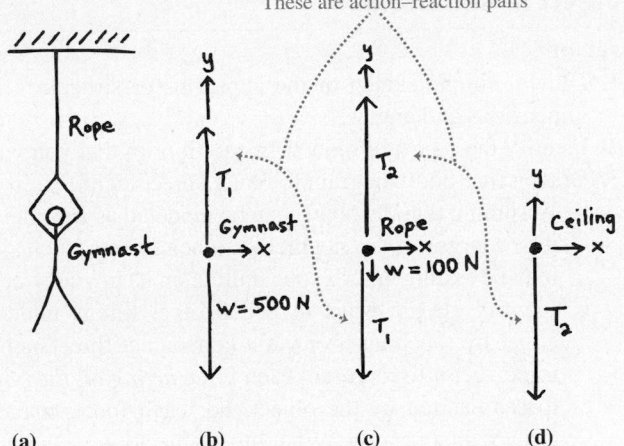

These are action–reaction pairs

▲ FIGURE 5.1 Our diagrams for this problem. We sketch the situation (a), plus free-body diagrams for the gymnast (b) and rope (c). (d) A partial free-body diagram for the ceiling. (See "Reflect.")

whose magnitude T_1 is 500 N. As you probably expect, this force equals the gymnast's weight.

The other forces on the rope are its own weight (magnitude 100 N) and the upward force (magnitude T_2) exerted on its upper end by the ceiling. The equilibrium condition $\Sigma F_y = 0$ for the rope gives

$$\Sigma F_y = T_2 + (-100\,\text{N}) + (-500\,\text{N}) = 0, \quad \text{(equilibrium}$$
$$T_2 = 600\,\text{N}. \qquad\qquad\qquad\qquad \text{of rope)}$$

Continued

REFLECT The tension is 100 N greater at the top of the rope (where it must support the weights of both the rope and the gymnast) than at the bottom (where it supports only the gymnast).

Figure 5.1d is a partial free-body diagram of the ceiling, showing that the rope exerts a downward force with magnitude T_2 on the ceiling. From Newton's third law, the magnitude of this force is also 600 N.

Practice Problem: Determine the tension at the midpoint of the rope. *Answer: 550 N.*

The strength of the rope in Example 5.1 probably isn't a major concern. The breaking strength of a string, rope, or cable is described by the maximum tension it can withstand without breaking. A few typical breaking strengths are shown in Table 5.1.

TABLE 5.1 Approximate breaking strengths

Thin white string	50 N
$\frac{1}{4}''$ nylon clothesline rope	4000 N
11 mm Perlon mountaineering rope	3×10^4 N
$1\frac{1}{4}''$ manila climbing rope	6×10^4 N
$\frac{1}{4}''$ steel cable	6×10^4 N

EXAMPLE 5.2 Two-dimensional equilibrium

In Figure 5.2a, a car engine with weight w hangs from a chain that is linked at point O to two other chains, one fastened to the ceiling and the other to the wall. Find the tension in each of the three chains, assuming that w is given and the weights of the chains themselves are negligible.

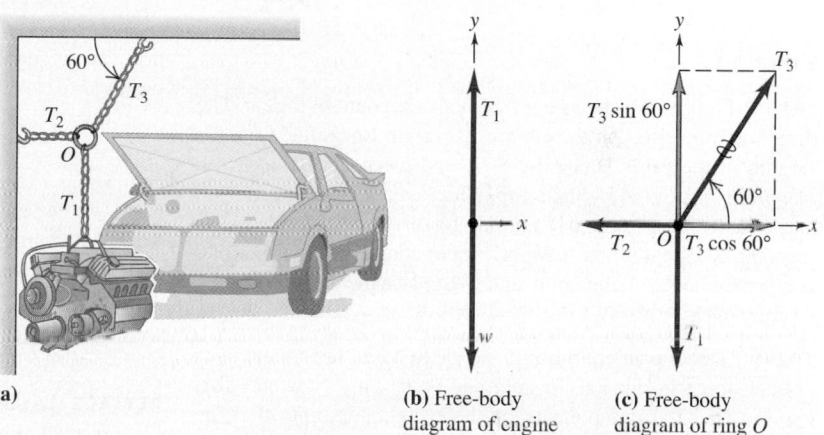

(a)

▶ **FIGURE 5.2**

(b) Free-body diagram of engine

(c) Free-body diagram of ring O

SOLUTION

SET UP Figure 5.2b is our free-body diagram for the engine. Without further ceremony, we can conclude that $T_1 = w$. (Because we neglect the weights of the chains, the tension is the same throughout the length of each chain.) The horizontal and slanted chains do not exert forces on the engine itself, because they are not attached to it, but they do exert forces on the ring (point O), where three chains join. So let's consider the *ring* as a particle in equilibrium; the weight of the ring itself is negligible.

In our free-body diagram for the ring (Figure 5.2c), T_1, T_2, and T_3 are the *magnitudes* of the forces; their directions are shown by the vectors in the diagram. We add an *x-y* coordinate system and resolve the force with magnitude T_3 into its x and y components. Note that the downward force with magnitude T_1 of the chain acting on the ring and the upward force with magnitude T_1 of the chain acting on the engine are not an action–reaction pair.

SOLVE We now apply the equilibrium conditions *for the ring*, writing separate equations for the x and y components. (Note that

x and y components are *never* added together in a single equation.) We find that

$$\Sigma F_x = 0, \qquad T_3 \cos 60° + (-T_2) = 0,$$
$$\Sigma F_y = 0, \qquad T_3 \sin 60° + (-T_1) = 0.$$

Because $T_1 = w$, we can rewrite the second equation as

$$T_3 = \frac{T_1}{\sin 60°} = \frac{w}{\sin 60°} = 1.155w.$$

We can now use this result in the first equation:

$$T_2 = T_3 \cos 60° = (1.155w) \cos 60° = 0.577w.$$

So we can express all three tensions as multiples of the weight w of the engine, which we assume is known. To summarize, we have

$$T_1 = w,$$
$$T_2 = 0.577w,$$
$$T_3 = 1.155w.$$

If the engine's weight is $w = 2200$ N (about 500 lb), then

$$T_1 = 2200 \text{ N},$$
$$T_2 = (0.577)(2200 \text{ N}) = 1270 \text{ N},$$
$$T_3 = (1.155)(2200 \text{ N}) = 2540 \text{ N}.$$

Continued

REFLECT Each of the three tensions is proportional to the weight w; if we double w, all three tensions double. Note that T_3 is greater than the weight of the engine. If this seems strange, observe that T_3 must be large enough for its vertical component to be equal in magnitude to w, so T_3 itself must have a magnitude *larger* than w.

Practice Problem: If we change the angle of the upper chain from 60° to 45°, determine the new expressions for the three tensions. *Answers:* $T_1 = w$, $T_2 = w\cos 45°/\sin 45° = w$, $T_3 = w/\sin 45° = \sqrt{2}\,w$.

EXAMPLE 5.3   Car on a ramp

A car with a weight of 1.76×10^4 N rests on the ramp of a trailer (Figure 5.3a). The car's brakes and transmission lock are released; only the cable prevents the car from rolling backward off the trailer. The ramp makes an angle of 26.0° with the horizontal. Find the tension in the cable and the force with which the ramp pushes on the car's tires.

We replace the weight by its components.

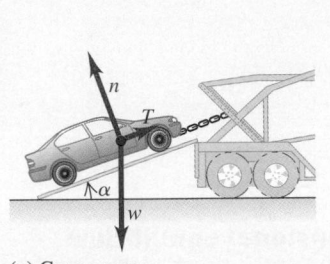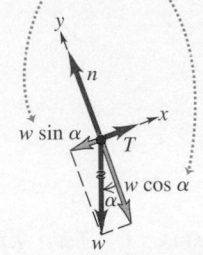

▶ **FIGURE 5.3**   (a) Car on ramp   (b) Free-body diagram for car

---

SOLUTION

SET UP Figure 5.3b shows our free-body diagram for the car. The three forces exerted on the car are its weight (magnitude w), the tension in the cable (magnitude T), and the normal force with magnitude n exerted by the ramp. (Because we treat the car as a particle, we can lump the normal forces on the four wheels together as a single force.) We orient our coordinate axes parallel and perpendicular to the ramp, and we replace the weight force by its components.

SOLVE The car is in equilibrium, so we first find the components of each force in our axis system and then apply Newton's first law. To find the components of the weight, we note that the angle α between the ramp and the horizontal is equal to the angle α between the weight vector and the normal to the ramp, as shown. The angle α is *not* measured in the usual way, counterclockwise from the $+x$ axis. To find the components of the weight (w_x and w_y), we use the right triangles in Figure 5.3b. We find that $w_x = -w\sin\alpha$ and $w_y = -w\cos\alpha$. The equilibrium conditions then give us

$$\sum F_x = 0, \qquad T + (-w\sin\alpha) = 0,$$
$$\sum F_y = 0, \qquad n + (-w\cos\alpha) = 0.$$

Be sure you understand how the signs are related to our choice of coordinate axis directions. Remember that, by defini-

tion, T, w, and n are *magnitudes* of vectors and are therefore positive.

Solving these equations for T and n, we find that

$$T = w\sin\alpha,$$
$$n = w\cos\alpha.$$

Finally, inserting the numerical values $w = 1.76 \times 10^4$ N and $\alpha = 26°$, we obtain

$$T = (1.76 \times 10^4 \text{ N})(\sin 26°) = 7.72 \times 10^3 \text{ N},$$
$$n = (1.76 \times 10^4 \text{ N})(\cos 26°) = 1.58 \times 10^3 \text{ N}.$$

REFLECT To check some special cases, note that if the angle α is zero, then $\sin\alpha = 0$ and $\cos\alpha = 1$. In this case, the ramp is horizontal; no cable tension T is needed to hold the car, and the magnitude of the total normal force n is equal to the car's weight. If the angle is 90° (the ramp is vertical), then $\sin\alpha = 1$ and $\cos\alpha = 0$. In that case, the cable tension T equals the weight w and the normal force n is zero.

We also note that these results would still be correct if the car were on a ramp on a car transport trailer traveling down a straight highway at a constant speed of 65 mi/h. Do you see why?

Practice Problem: What ramp angle would be needed in order for the cable tension to equal one-half of the car's weight? *Answer:* 30°.

EXAMPLE 5.4   Lifting granite and dumping dirt

Blocks of granite, each with weight w_1, are being hauled up a 15° slope out of a quarry (Figure 5.4). For environmental reasons, dirt is also being dumped into the quarry to fill up old holes. You have been asked to find a way to use this dirt to move the granite out more easily. You design a system that lets the dirt (weight w_2, including the weight of the bucket) that drops vertically into the quarry pull out a granite block on a cart with steel wheels (total weight w_1), rolling on steel rails. Ignoring the weight of the cable and friction in the pulley and wheels, determine how the weights w_1 and w_2 must be related for the system to move with constant speed.

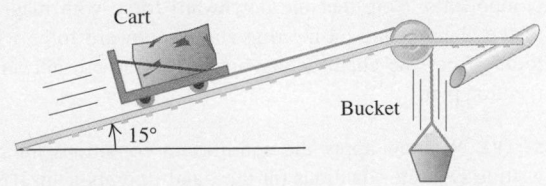

▲ **FIGURE 5.4**

Continued

SOLUTION

SET UP Figure 5.5a shows an idealized model for the system. We draw two free-body diagrams, one for the dirt and bucket (Figure 5.5b) and one for the granite block on its cart (Figure 5.5c). In drawing the coordinate axes for each object, we're at liberty to orient them differently for the two objects; the choices shown are the most convenient ones. (But we must *not* change the orientation of the objects themselves.) We represent the weight of the granite block in terms of its components in the chosen axis system. The tension T in the cable is the same throughout, because we are assuming that the cable's weight is negligible. (The pulley changes the *directions* of the cable forces, but if it is frictionless and massless, it can't change their *magnitudes*.)

SOLVE Each object is in equilibrium. Applying $\Sigma F_y = 0$ to the bucket (Figure 5.5b), we find that

$$T + (-w_2) = 0, \qquad T = w_2.$$

Applying ΣF_x to the cart (Figure 5.5c), we get

$$T + (-w_1 \sin 15°) = 0, \qquad T = w_1 \sin 15°.$$

Equating the two expressions for T yields

$$w_2 = w_1 \sin 15° = 0.26 w_1.$$

REFLECT If the weight w_2 of the dirt and bucket totals 26% of the weight w_1 of the granite block and cart, the system can move in either direction with constant speed. In the real world, where friction tends to hold back the motion, would w_2 need to be more or less than $0.26w_1$ for the granite block to move up with constant speed?

Practice Problem: Determine the total normal force n that the tracks apply to the wheels of the granite cart, as a fraction of w_1. *Answer:* $n = w_1 \cos 15° = 0.97 w_1$.

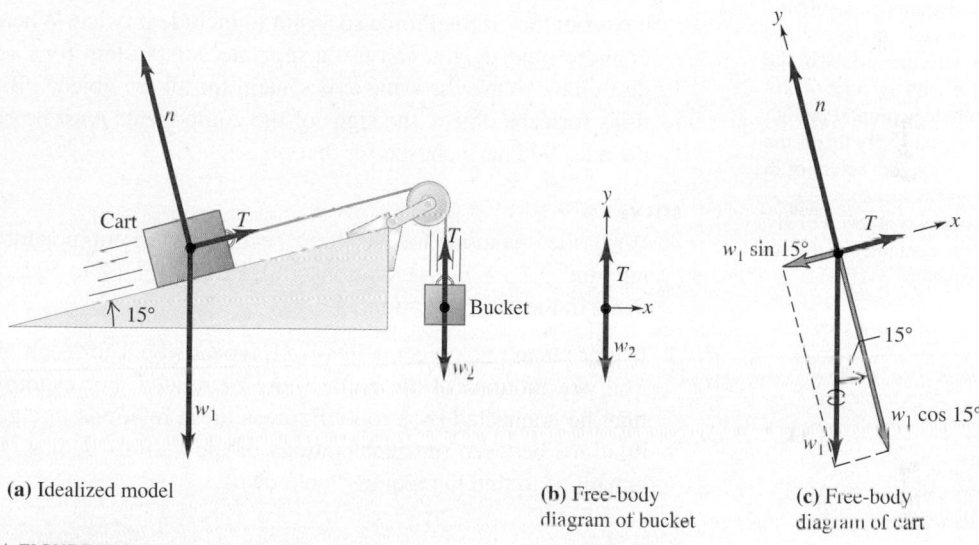

(a) Idealized model

(b) Free-body diagram of bucket

(c) Free-body diagram of cart

▲ **FIGURE 5.5**

5.2 Applications of Newton's Second Law

We're now ready to discuss problems in **dynamics,** showing applications of Newton's second law to systems that are *not* in equilibrium. Here's a restatement of that law:

Newton's second law

An object's acceleration equals the vector sum of the forces acting on it, divided by its mass. In vector form, we rewrite this statement as

$$\Sigma \vec{F} = m\vec{a}. \tag{4.7}$$

However, we'll usually use this relation in its component form:

$$\Sigma F_x = ma_x, \qquad \Sigma F_y = ma_y. \tag{4.8}$$

Mastering**PHYSICS**

PhET: Lunar Lander
ActivPhysics 2.1.5: Car Race
ActivPhysics 2.2: Lifting a Crate
ActivPhysics 2.3: Lowering a Crate
ActivPhysics 2.4: Rocket Blasts Off
ActivPhysics 2.11: Modified Atwood Machine

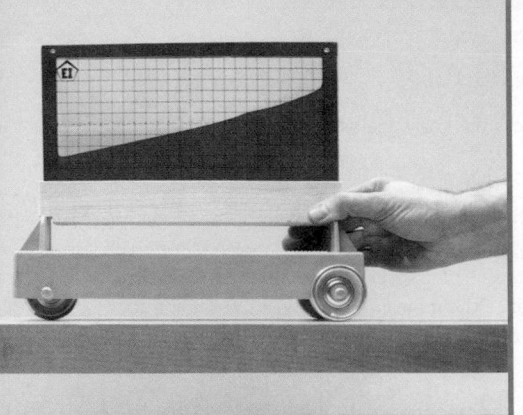

▲ Application Gravity-defying liquid?
Although this container is on a level table-top, the liquid inside forms a slant. How can that be? You may have guessed that the photo shows a demonstration of a simple liquid-filled accelerometer. As the container is accelerated to the left, the surface of the liquid forms an angle with the horizontal. The tangent of the angle is proportional to the acceleration. Can you verify that if the container is 19.6 cm wide, the height of the liquid at the right end (in cm) above its level when the apparatus is at rest will give the acceleration of the container itself, in meters per second squared?

The following problem-solving strategy is similar to our strategy for solving equilibrium problems, presented in Section 5.1:

PROBLEM-SOLVING STRATEGY 5.2 Using Newton's second law

SET UP

1. Draw a sketch of the physical situation, and identify the moving object or objects to which you will apply Newton's second law.
2. Draw a free-body diagram for each chosen object, showing all the forces acting *on* that object, as described in the strategy for Newton's first law (Problem-Solving Strategy 5.1). Label the magnitudes of unknown forces with algebraic symbols. Usually, one of the forces will be the object's weight; it is generally best to label this as mg. If a numerical value of mass is given, you can compute the corresponding weight.
3. Show your coordinate axes explicitly in each free-body diagram, and then determine components of forces with reference to these axes. If you know the direction of the acceleration, it is usually best to take that direction as one of the axes. When you represent a force in terms of its components, cross out the original force so as not to include it twice. When there are two or more objects, you can use a separate axis system for each object; you don't have to use the same axis system for all the objects. But in the equations for each object, the signs of the components *must* be consistent with the axes you have chosen for that object.

SOLVE

4. Write the equations for Newton's second law in component form: $\sum F_x = ma_x$ and $\sum F_y = ma_y$ (Equations 4.8). Be careful not to add x and y components in the same equation.

5. If more than one object is involved, repeat step 4 for each object. In addition, the motions of the bodies may be related. For example, the objects may be connected by a rope. Express these relations in algebraic form as relations between the accelerations of the various bodies. Then solve the equations to find the required unknowns.

REFLECT

6. Check particular cases or extreme values of quantities, when possible, and compare the results for these particular cases with your intuitive expectations. Ask yourself, "Does this result make sense?" Think about what the problem has taught you that you can apply to other problems in the future.

Quantitative Analysis 5.1 **A two-cart train**

The carts in Figure 5.6 are speeding up as they are pulled to the right with increasing speed across a frictionless surface. The ropes have negligible mass. We can conclude that

A. The pull of rope 1 on cart A has greater magnitude than the pull of rope 1 on cart B.
B. The pull of rope 2 on cart B has greater magnitude than the pull of rope 1 on cart B.
C. The pull of rope 1 on cart A has greater magnitude than the pull of rope 2 on cart B.

SOLUTION Answer B is correct: For cart B to accelerate, there must be a net force to the right. Answer A cannot be right; if the

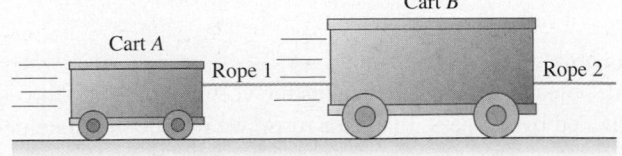

▲ FIGURE 5.6

rope is massless, the forces acting on its two ends must add to zero. Nor can answer C be right: The two carts have the same acceleration; therefore, the tension in rope 2 must be great enough to give this acceleration to both carts, while the tension in rope 1 is just great enough to give cart A the same acceleration.

EXAMPLE 5.5 A simple accelerometer

You tape one end of a piece of string to the ceiling light of your car and hang a key with mass m to the other end (Figure 5.7). A protractor taped to the light allows you to measure the angle the string makes with the vertical. Your friend drives the car while you make measurements. When the car has a constant acceleration with magnitude a toward the right, the string hangs at rest (relative to the car), making an angle β with the vertical. **(a)** Derive an expression for the acceleration a in terms of the mass m and the measured angle β. **(b)** In particular, what is a when $\beta = 45°$? When $\beta = 0$?

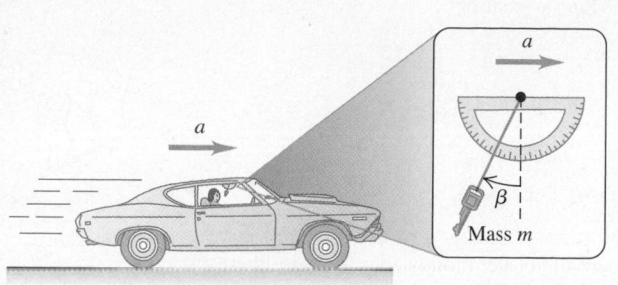

(a) Low-tech accelerometer

(b) Free-body diagram for the key

▲ **FIGURE 5.7**

SOLUTION

SET UP Our free-body diagram is shown in Figure 5.7b. The forces acting on the key are its weight $w = mg$ and the string tension T. We direct the x axis to the right (in the direction of the acceleration) and the y axis vertically upward.

SOLVE Part (a): This problem may look like an equilibrium problem, but it isn't. The string and the key are at rest with respect to the car, but car, string, and key are all accelerating in the $+x$ direction. Thus, there must be a horizontal component of force acting on the key. Therefore, we use Newton's second law in component form: $\Sigma F_x = ma_x$ and $\Sigma F_y = ma_y$.

We find the components of the string tension, as shown in Figure 5.7b. The sum of the horizontal components of force is

$$\Sigma F_x = T\sin\beta,$$

and the sum of the vertical components is

$$\Sigma F_y = T\cos\beta + (-mg).$$

The x component of acceleration is the acceleration a_x of the car, string, and key, and the y component of acceleration is zero, so

$$\Sigma F_x = T\sin\beta = ma_x,$$
$$\Sigma F_y = T\cos\beta + (-mg) = ma_y = 0.$$

Rearranging terms, we obtain

$$T\sin\beta = ma_x, \qquad T\cos\beta = mg.$$

When we divide the first equation by the second, we get

$$a_x = g\tan\beta.$$

The acceleration a_x is thus proportional to the tangent of the angle β.

Part (b): When $\beta = 0$, the key hangs vertically and the acceleration is zero; when $\beta = 45°$, $a_x = g$, and so on.

REFLECT We note that β can never be 90°, because that would require an infinite acceleration. We note also that the relation between a_x and β doesn't depend on the mass of the key, but it *does* depend on the acceleration due to gravity.

Practice Problem: What is the angle β if the acceleration is $g/2$? *Answer:* 26.6°.

EXAMPLE 5.6 Acceleration down a hill

A toboggan loaded with vacationing students (total weight w) slides down a long, snow-covered slope. The hill slopes at a constant angle α, and the toboggan is so well waxed that there is virtually no friction at all. Find the toboggan's acceleration and the magnitude n of the normal force the hill exerts on the toboggan.

SOLUTION

SET UP As a reminder of the importance of drawing sketches, we'll show the diagrams we draw in pencil, rather than in printed style (Figure 5.8). The only forces acting on the toboggan are its weight w and the normal force n (Figure 5.8b). The direction of

the weight is straight down, but the direction of the normal force is perpendicular to the surface of the hill, at an angle α with the vertical. We take axes parallel and perpendicular to the surface of the hill and resolve the weight into x and y components.

Continued

SOLVE There is only one *x* component of force, so

$$\sum F_x = w \sin\alpha.$$

From $\sum F_x = ma_x$, we have

$$w \sin\alpha = ma_x,$$

and since $w = mg$, the acceleration is

$$a_x = g \sin\alpha.$$

The *y*-component equation gives $\sum F_y = n + (-mg\cos\alpha)$. We know that the *y* component of acceleration is zero, because there is no motion in the *y* direction. So $\sum F_y = 0$ and $n = mg\cos\alpha$.

REFLECT The mass *m* does not appear in the expression for a_x; this means that *any* toboggan, regardless of its mass or number of passengers, slides down a frictionless hill with an acceleration of $g \sin\alpha$. In particular, when $\alpha = 0$ (a flat surface with no slope at all), the acceleration is $a_x = 0$, as we should expect. When the surface is vertical, $\alpha = 90°$ and $a_x = g$ (free fall).

Note that the magnitude *n* of the normal force exerted on the toboggan by the surface of the hill $(n = mg\cos\alpha)$ is *proportional* to the magnitude *mg* of the toboggan's weight; the two are *not* equal, except in the special case where $\alpha = 0$.

Practice Problem: At what angle does the hill slope if the acceleration is $g/2$? *Answer:* 30°.

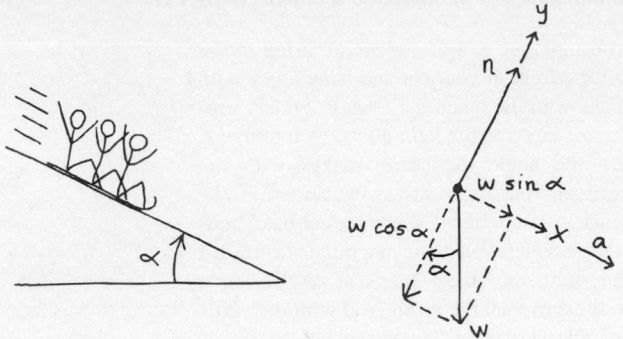

(a) The situation **(b)** Free-body diagram for toboggan
▲ **FIGURE 5.8** Our diagrams for this problem.

EXAMPLE 5.7 **An air track in a physics lab**

Figure 5.9 shows a glider with mass m_1 that moves on a level, frictionless air track in a physics lab. It is connected by a string passing over a small frictionless pulley to a hanging weight with total mass m_2. The string is light and flexible, and it doesn't stretch. Find the acceleration of each object and the tension in the string.

SOLUTION

SET UP The two objects have different motions, so we need to draw a separate free-body diagram and coordinate system for each, as shown in Figure 5.9b and c. We are free to use different coordinate axes for the two objects; in this case, it's convenient to take the $+x$ direction to the right for the glider and the $+y$ direction downward for the hanging weight. Then the glider has only an *x* component of acceleration a_{1x} (i.e., $a_{1y} = 0$), and the weight hanger has only a *y* component a_{2y} (i.e., $a_{2x} = 0$). There is no fric-

tion in the pulley, and we consider the string to be massless, so the tension *T* in the string is the same throughout; it applies a force with magnitude *T* to each object. The weights are m_1g and m_2g.

SOLVE We apply Newton's second law, in component form, to each object in turn.

For the glider on the track, Newton's second law gives

$$\sum F_x = T = m_1 a_{1x}$$
$$\sum F_y = n + (-m_1 g) = m_1 a_{1y} = 0,$$

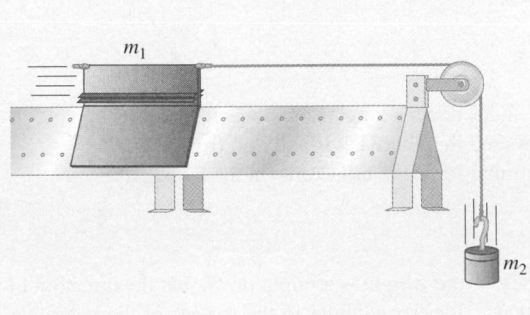

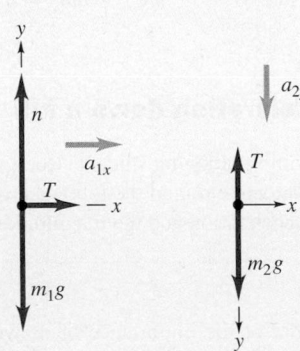

(a) Apparatus

▲ **FIGURE 5.9**

(b) Free-body diagram for glider

(c) Free-body diagram for weight

Continued

and for the hanging weight,

$$\Sigma F_y = m_2 g + (-T) = m_2 a_{2y}.$$

Now, because the string doesn't stretch, the two objects must move equal distances in equal times, and their *speeds* at any instant must be equal. When the speeds change, they change by equal amounts in a given time, so the accelerations of the two bodies must have the same magnitude a. We can express this relation as

$$a_{1x} = a_{2y} = a.$$

(The directions of the two accelerations are different, of course.) The two Newton's-second-law equations are then

$$T = m_1 a,$$
$$m_2 g + (-T) = m_2 a.$$

We'd like to get separate expressions for T and a, in terms of the masses and g.

An easy way to get an expression for a is to replace $-T$ in the second equation by $-m_1 a$ and rearrange the result; when we do so, we obtain $m_2 g + (-m_1 a) = m_2 a$, $m_2 g = (m_1 + m_2)a$, and, finally,

$$a = \frac{m_2}{m_1 + m_2} g.$$

Then, to get an expression for T, we substitute this expression for a back into the first equation. The result is

$$T = \frac{m_1 m_2}{m_1 + m_2} g = \frac{m_1}{m_1 + m_2} m_2 g.$$

REFLECT It's always a good idea to check general symbolic results such as this for particular cases where we can guess what the answer ought to be. For example, if the mass of the glider is zero ($m_1 = 0$), we expect that the hanging weight (mass m_2) would fall freely with acceleration g and there would be no tension in the string. When we substitute $m_1 = 0$ into the preceding expressions for T and a, they do give $T = 0$ and $a = g$, as expected. Also, if $m_2 = 0$, there is nothing to create tension in the string or accelerate either mass. For this case, the equations give $T = 0$ and $a = 0$. Thus, in these two special cases, the results agree with our intuitive expectations.

We also note that, in general, the tension T is *not* equal to the weight $m_2 g$ of the hanging mass m_2, but is *less* by a factor of $m_1/(m_1 + m_2)$. If T *were* equal to $m_2 g$, then m_2 would be in equilibrium, but it isn't.

5.3 Contact Forces and Friction

We've seen several problems in which an object rests or slides on a surface that exerts forces on the object. At the beginning of Chapter 4, we introduced the terms *normal force* and *friction force* to describe these forces. Whenever two objects interact by direct contact (touching) of their surfaces, we call the interaction forces *contact forces*. Normal and friction forces are both contact forces. Friction is important in many aspects of everyday life, for both facilitating and preventing the slipping of an object in contact with a surface. For example, the oil in a car engine facilitates sliding between moving parts, but friction between the tires and the road prevents slipping. Similarly, air drag decreases automotive fuel economy, but makes parachutes work. Without friction, nails would pull out, light bulbs and bottle caps would unscrew effortlessly, and riding a bicycle would be hopeless. A "frictionless surface" is a useful concept in idealized models of mechanical systems in which the friction forces are negligibly small. In the real world, such an idealization is unattainable, although driving a car on wet ice comes fairly close!

Let's consider an object sliding across a surface. When you try to slide a heavy box of books across the floor, the box doesn't move at all until you reach a certain critical force. Then the box starts moving, and you can usually keep it moving with less force than you needed to get it started. If you take some of the books out, you need less force than before to get it started or keep it moving. What general statements can we make about this behavior?

First, when an object rests or slides on a surface, we can always represent the contact force exerted by the surface on the object in terms of components of force perpendicular and parallel to the surface (Figure 5.10). We call the perpendicular component the *normal force,* denoted by $\vec{n}$. (*Normal* is a synonym for *perpendicular.*) The component parallel to the surface is the *friction force,* denoted

MasteringPHYSICS

PhET: Forces in 1 Dimension
PhET: Friction
PhET: The Ramp
ActivPhysics 2.1.1: Force Magnitudes
ActivPhysics 2.1.4: Sliding on an Incline
ActivPhysics 2.5: Truck Pulls Crate
ActivPhysics 2.6: Pushing a Crate Up a Wall
ActivPhysics 2.7: Skier Goes Down a Slope
ActivPhysics 2.8: Skier and Rope Tow
ActivPhysics 2.10: Truck Pulls Two Crate

The frictional and normal forces are really components of a single contact force.

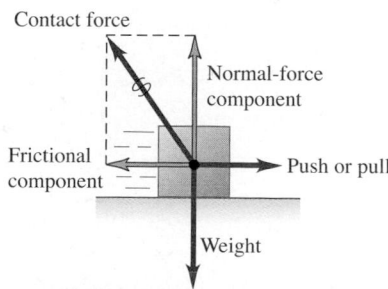

▲ **FIGURE 5.10** The frictional and normal forces for a block sliding on a rough surface.

▲ **Application Friction can be fun.** As this rock climber shows, frictional forces can be important. Without them, a climber could not move or maintain an equilibrium position. Notice how the climber is demonstrating one of the key concepts of frictional forces: The frictional force is proportional to the normal force. Therefore, even in a narrow crack, a good climber tries to contact the rock at an angle as close as feasible to 90°. This maximizes the normal force and therefore also maximizes the frictional force.

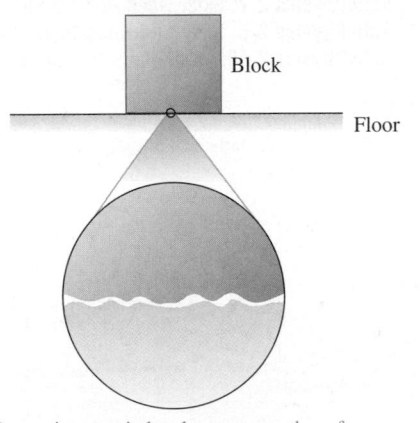

On a microscopic level, even smooth surfaces are rough; they tend to catch and cling.

▲ **FIGURE 5.11** The origin of frictional forces.

by $\vec{f}$. By definition, $\vec{n}$ and $\vec{f}$ are always perpendicular to each other. When an object is sliding with respect to a surface, the associated friction force is called a *kinetic-friction* force, with magnitude f_k. When there is no relative motion, we speak of a *static-friction* force, with magnitude f_s. We'll discuss these two possibilities separately.

When one object is moving relative to a surface, the *direction* of the kinetic-friction force on the object is always opposite to the direction of motion of that object relative to the surface. For example, when a book slides from left to right along a tabletop, the friction force on it acts to the left, as in Figure 5.10. According to Newton's third law, the book simultaneously applies a force to the table that is equal in magnitude, but directed to the right; in effect, the book tries to drag the table along with it.

The *magnitude* f_k of a kinetic-friction force usually increases when the normal force magnitude n increases. Thus, more force is needed to slide a box full of books across the floor than to slide the same box when it is empty. This principle is also used in automotive braking systems: The harder the brake pads are squeezed against the rotating brake disks, the greater is the braking effect. In some cases, the magnitude of the sliding friction force f_k is found to be approximately *proportional* to the magnitude n of the normal force. In such cases, we call the ratio f_k/n the **coefficient of kinetic friction,** denoted as μ_k:

Relation between kinetic-friction force and normal force

When the magnitude of the sliding friction force f_k is roughly proportional to the magnitude n of the normal force, the two are related by a constant μ_k called the coefficient of kinetic friction:

$$f_k = \mu_k n. \tag{5.1}$$

Because μ_k is the ratio of two force magnitudes, it has no units.

The more slippery the surface, the smaller the coefficient of friction is. Because it is a quotient of two force magnitudes, μ_k is a pure number, without units.

NOTE ▶ The friction force and the normal force are always perpendicular, so Equation 5.1 is *not* a vector equation, but a relation between the *magnitudes* of the two forces. ◀

The numerical value of the coefficient of kinetic friction for any two surfaces depends on the materials and the surfaces. For Teflon on steel, μ_k is about 0.04. For rubber tires on rough dry concrete (or sneakers on a dry gym floor), μ_k can be as large as 1.0 to 1.2. For two smooth metal surfaces, it is typically 0.4 to 0.8.

Although these numbers are given to two significant figures, they are only approximate values; indeed, Equation 5.1 is just an *approximate* representation of a complex phenomenon. On a microscopic level, friction and normal forces result from the intermolecular forces (fundamentally electrical in nature) between two rough surfaces at high points where they come into contact. Figure 5.11 suggests the nature of the interaction. The actual area of contact is usually much smaller than the total surface area. When two smooth surfaces of the same metal are brought together, these forces can cause a "cold weld"; a similar effect occurs with aluminum pistons in an unlubricated aluminum cylinder. Indeed, the success of lubricating oils depends on the fact that they maintain a film between the surfaces that prevents them from coming into actual contact. Friction forces can also depend on the *velocity* of the object relative to the surface. We'll ignore that complication for now in order to concentrate on the simplest cases.

Friction forces may also act when there is *no* relative motion between the surfaces of contact. If you try to slide that box of books across the floor, the box may not move at all if you don't push hard enough, because the floor exerts an equal and opposite friction force on the box. This force is called a **static-friction force** $\vec{f}_s$. In Figure 5.12a, the box is at rest, in equilibrium, under the action of its weight $\vec{w}$ and the upward normal force $\vec{n}$, which is equal in magnitude to the weight and exerted on the box by the floor. Now we tie a rope to the box (Figure 5.12b) and gradually increase the tension T in the rope. At first the box remains at rest, because, as T increases, the force of static friction f_s also increases (staying equal in magnitude to T).

At some point, however, T becomes greater than the maximum friction force f_s that the surface can exert; the box then "breaks loose" and starts to slide. Figure 5.12c shows the force diagram when T is at this critical value, and Figure 5.12d is the case when T exceeds this value and the box begins to move. Figure 5.12e shows how the friction force varies if we start with no applied force ($T = 0$) and gradually increase the force until the object starts to slide. Note that the force f_k of kinetic friction is somewhat less than the maximum force $f_{s,\text{max}}$ of static friction: Less force is needed to keep the box sliding with constant speed than to start it moving initially.

For a given pair of surfaces, the maximum value of f_s depends on the normal force. In some cases, the maximum value of f_s is approximately *proportional* to n;

▲ Application **Friction can be a problem.** If you're trying to saw stone, friction is not necessarily a good thing. If this saw blade were not lubricated, friction would make it hard to draw the saw through the cut, and the heat from friction could warp or melt the blade. Therefore, the blade of this "wet saw" in a commercial stone quarry is continuously showered with water to minimize frictional forces and carry away heat. Even so, a cut like this will take time to complete.

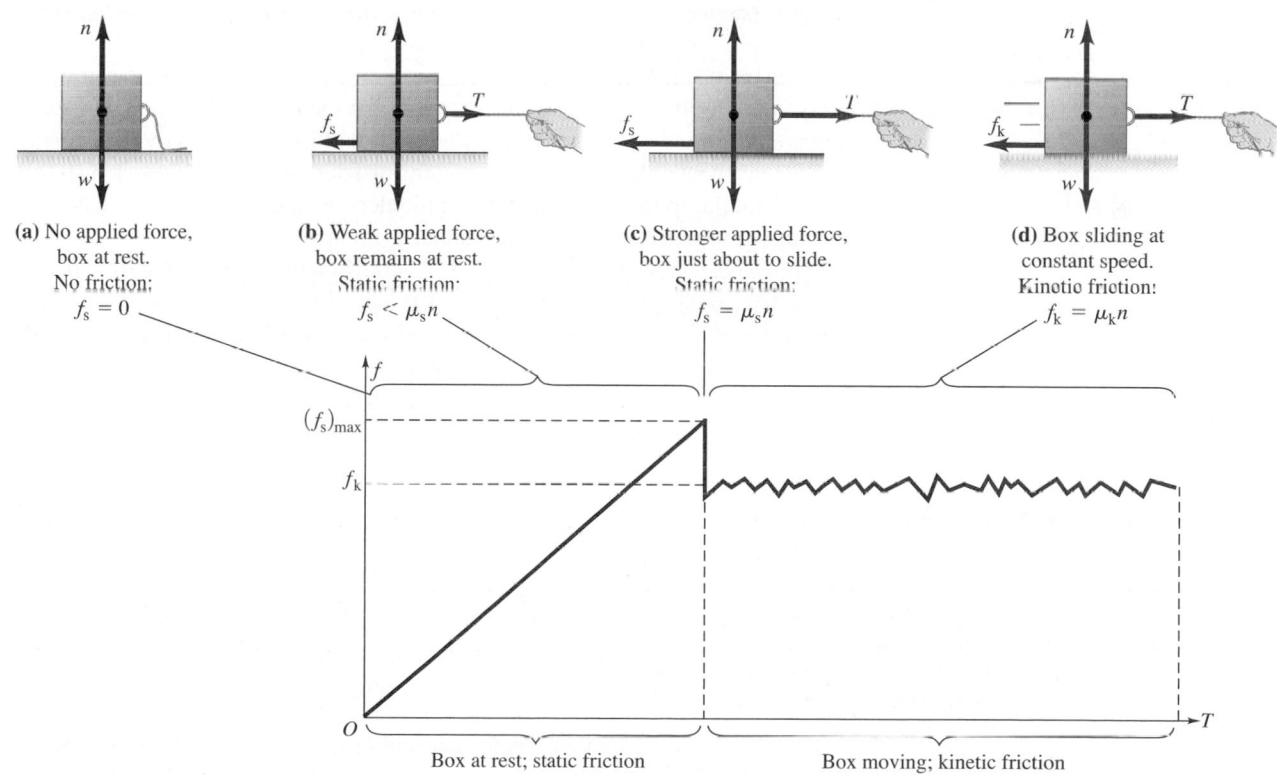

(a) No applied force, box at rest. No friction: $f_s = 0$

(b) Weak applied force, box remains at rest. Static friction: $f_s < \mu_s n$

(c) Stronger applied force, box just about to slide. Static friction: $f_s = \mu_s n$

(d) Box sliding at constant speed. Kinetic friction: $f_k = \mu_k n$

$(f_s)_{\text{max}}$

f_k

Box at rest; static friction

Box moving; kinetic friction

(e)

▲ **FIGURE 5.12** Static and kinetic friction for a box subjected to a gradually increasing pulling force.

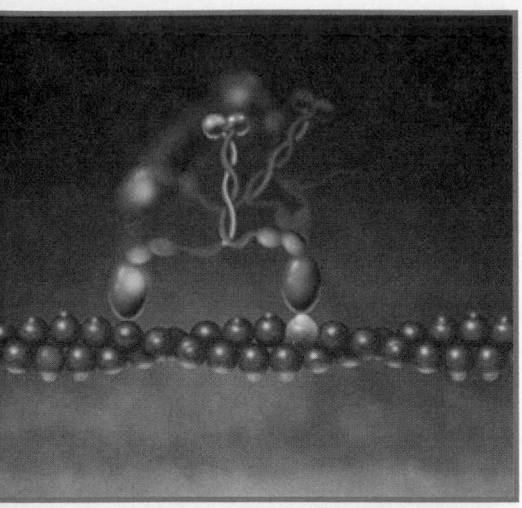

▲ BIO Application Molecular motors.
Nature has engineered a small number of molecular motors that convert the chemical energy of an ATP molecule into mechanical force. These nanomachines rely on a kind of friction to direct this force to their loads. A two-headed motor protein called myosin (in blue) binds to a structural track called actin (in red), along which it applies force through electrical interactions that are the small-scale version of friction. During the power stroke, when the myosin is applying force to the actin, each myosin head applies 1-10 pN of force to the actin, depending on the type of myosin. The aggregate force of many such molecules working in concert is the basis of the forces generated by animal skeletal muscles.

we call the proportionality factor μ_s the **coefficient of static friction.** In a particular situation, the actual force of static friction can have any magnitude between zero (when there is no other force parallel to the surface) and a maximum value $f_{s,max}$ given by $f_{s,max} = \mu_s n$. Like μ_k, μ_s has no unit.

Relation between normal force and maximum static-friction force
When the maximum magnitude of the static-friction force can be represented as proportional to the magnitude of the normal force, the two are related by a constant μ_s called the coefficient of static friction:

$$f_s \leq \mu_s n. \tag{5.2}$$

The equality sign holds only when the applied force T, parallel to the surface, has reached the critical value at which motion is about to start (Figure 5.12c). When T is *less* than this value (Figure 5.12b), the inequality sign holds. In that case, we have to use the equilibrium conditions $\sum \vec{F} = 0$ to find f_s. For any given pair of surfaces, the coefficient of kinetic friction is usually *less* than the coefficient of static friction. As a result, when sliding starts, the friction force usually *decreases,* as Figure 5.12e shows.

In some situations, the surfaces will alternately stick (static friction is operative) and slip (kinetic friction arises); this alternation is what causes the horrible squeak made by chalk held at the wrong angle while you're writing on the blackboard. Another slip-and-stick phenomenon is the squeaky noise your windshield-wiper blades make when the glass is nearly dry; still another is the outraged shriek of tires sliding on asphalt pavement. A more positive example is the sound produced by the motion of a violin bow against the string.

Liquids and gases also show frictional effects. Friction between two solid surfaces separated by a layer of liquid or gas is determined primarily by the *viscosity* of the fluid. In a car engine, the pistons are separated from the cylinder walls by a thin layer of oil. When an object slides on a layer of gas, friction can be made very small. In the familiar linear air track (like the one in Example 5.7), the gliders are supported on a layer of air. The frictional force is velocity dependent, but at typical speeds the effective coefficient of friction is on the order of 0.001. A device similar to the air track is the frictionless air table, on which pucks are supported by an array of small air jets about 2 cm apart.

For a wheeled vehicle, we can define a **coefficient of rolling friction,** μ_r, which is the horizontal force needed for constant speed on a flat surface, divided by the upward normal force exerted by the surface. Typical values of μ_r are 0.002 to 0.003 for steel wheels on steel rails and 0.01 to 0.02 for rubber tires on concrete. These values are one reason that railroad trains are, in general, much more fuel efficient than highway trucks.

Quantitative Analysis 5.2   **A friction experiment**

A block is pulled along a horizontal surface (with friction) by a constant force with magnitude F_x, and the acceleration a_x is measured. The experiment is repeated several times, with different values of F_x, and a graph of acceleration as a function of force is plotted. For the various trials, which of the graphs of acceleration versus force shown at right is most nearly correct?

SOLUTION No motion occurs unless the force F_x exceeds the maximum static-friction force; when it does, the block starts to

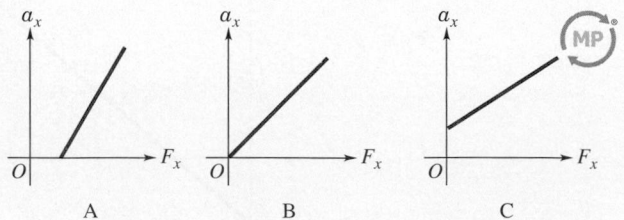

move under the action of F_x and the smaller kinetic-friction force. Only answer A shows no acceleration at small values of the applied force.

Turning things over

You are pushing the heavy plastic block in Figure 5.13 across the room with constant velocity. You decide to turn the block on end, reducing the surface area in contact with the floor by half. In this new orientation, to push this same block across the same floor with the same speed as before, the magnitude of force that you must apply is

A. twice as great as before.
B. the same as before.
C. half as great as before.

SOLUTION For many surfaces, dry sliding friction depends only on the normal force with which the object presses against the sur-

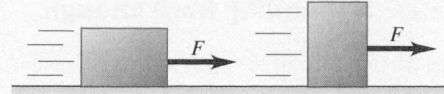

▲ **FIGURE 5.13**

face, in this case a force equal to the weight of the block. The frictional force does not depend on the area of contact. Since the frictional force is the same in both orientations, the applied force needed is the same. The correct answer is B.

EXAMPLE 5.8 Delivering the goods

A delivery company has just unloaded a 500 N crate full of home exercise equipment in your level driveway. You find that to get it started moving toward your garage, you have to pull with a horizontal force of magnitude 230 N. Once it "breaks loose" and starts to move, you can keep it moving at constant velocity with only 200 N of force. What are the coefficients of static and kinetic friction?

SOLUTION

SET UP As shown in Figure 5.14, the forces acting on the crate are its weight (magnitude w), the force applied by the rope (magnitude T), and the normal and frictional components (magnitudes n and f, respectively) of the contact force the driveway surface exerts on the crate. We draw free-body diagrams for the crate at the instant it starts to move (when the static frictional force is maximum) and while it is in motion.

SOLVE Whether the crate is moving or not, the friction force opposes your pull because it tends to prevent the crate from sliding relative to the surface. An instant before the crate starts to move, the static-friction force has its maximum possible value, $f_{s,max} = \mu_s n$. The state of rest and the state of motion with constant velocity are both equilibrium conditions. Remember that w, n, and f are the *magnitudes* of the forces; some of the components have negative signs. For example, the magnitude of the weight is 500 N, but its y component is -500 N. We write Newton's second law in component form, $\Sigma F_x = 0$ and $\Sigma F_y = 0$, for the crate. Then

$$\Sigma F_y = n + (-w) = n - 500\text{ N} = 0, \qquad n = 500\text{ N},$$

$$\Sigma F_x = T + (-f_s) = 230\text{ N} - f_s = 0, \qquad f_s = 230\text{ N},$$

$$f_{s,max} = \mu_s n \qquad \text{(motion about to start)},$$

$$\mu_s = \frac{f_{s,max}}{n} = \frac{230\text{ N}}{500\text{ N}} = 0.46.$$

After the crate starts to move, the friction force becomes that of kinetic friction (Figure 5.14c). The crate moves with constant velocity, so it is still in equilibrium. Thus, the vector sum of the forces is still zero, and we have

$$\Sigma F_y = n + (-w) = n - 500\text{ N} = 0, \qquad n - 500\text{ N},$$

$$\Sigma F_x = T + (-f_k) = 200\text{ N} - f_k = 0, \qquad f_k - 200\text{ N},$$

$$f_k = \mu_k n \qquad \text{(moving)}.$$

The coefficient of kinetic friction is

$$\mu_k - \frac{f_k}{n} - \frac{200\text{ N}}{500\text{ N}} = 0.40.$$

REFLECT There is no vertical motion, so $a_y = 0$ and $n = w$. It is easier to keep the crate moving than to start it moving from rest, because $\mu_k < \mu_s$. Note, however, that just as the crate begins to move, the acceleration is nonzero while the velocity changes from zero to its final, constant value.

Practice Problem: On a different part of the driveway, $\mu_s = 0.30$ and $\mu_k = 0.25$. What horizontal force is required to start the crate moving? What force is required to keep it sliding at constant velocity? *Answers:* 150 N, 125 N.

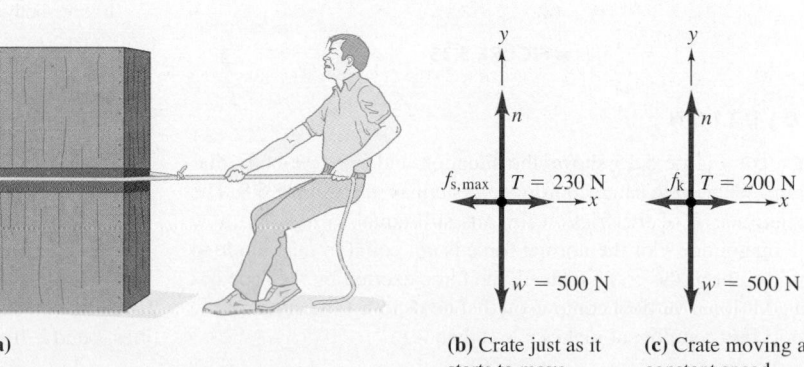

(a)

▶ **FIGURE 5.14**

(b) Crate just as it starts to move

(c) Crate moving at constant speed

EXAMPLE 5.9 **Not pulling hard enough**

What is the friction force if the crate in Example 5.8 is at rest on the surface and a horizontal force of 50 N is applied to it?

SOLUTION

SET UP AND SOLVE The setup is the same as for Example 5.8. From the equilibrium conditions, we have

$$\sum F_x = T + (-f_s) = 50\text{ N} - f_s = 0,$$
$$f_s = 50\text{ N}.$$

REFLECT In this case, $f_s < \mu_s n$. The frictional force can prevent motion for any horizontal force less than 230 N.

Quantitative Analysis 5.4 **Pushing too hard**

A baggage handler applies a constant horizontal force with magnitude F to push a box across a rough horizontal surface with a very small constant acceleration, starting from rest. Then he pushes a second box, identical to the first, with a force twice as great in magnitude. How does the acceleration of the second box compare with that of the first?

A. It is more than double the first acceleration.
B. It is exactly double the first acceleration.
C. It is less than double the first acceleration.

SOLUTION From Newton's second law, $F - f_k = ma_x$. In the first case, the acceleration is small, so F is only a little greater than f_k, and the net horizontal component of force $(F - f_k)$ is much smaller than F. Therefore, if F is doubled, the net horizontal force is much larger than before. Thus the acceleration is also much larger than before. Answer A is correct.

EXAMPLE 5.10 **Moving the exercise equipment again**

Suppose you try to move the crate full of exercise equipment in Example 5.8 by pulling upward on the rope at an angle of 30° above the horizontal. How hard do you have to pull to keep the crate moving with constant velocity? Is this easier or harder than pulling horizontally? Recall that $w = 500$ N and $\mu_k = 0.40$ for this crate.

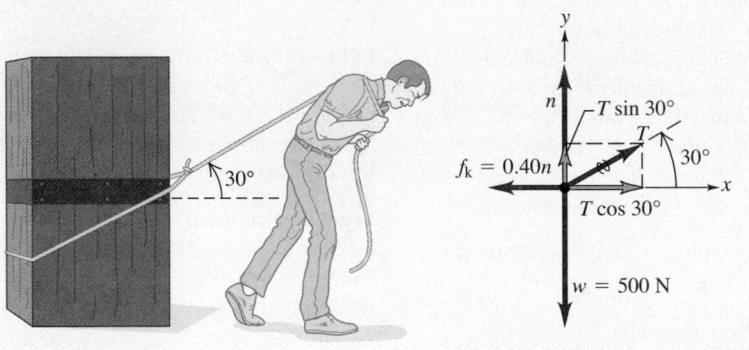

(a)

(b) Free-body diagram for moving crate

▲ FIGURE 5.15

SOLUTION

SET UP Figure 5.15 shows the situation and our free-body diagram. We use the same coordinate system as in Example 5.8. The magnitude f_k of the friction force is still equal to $\mu_k n$, but now the magnitude n of the normal force is *not* equal in magnitude to the weight of the crate. Instead, the force exerted by the rope has an additional vertical component that tends to lift the crate off the floor; this component makes n less than w.

SOLVE The crate is moving with constant velocity, so it is still in equilibrium. Applying $\sum \vec{F} = 0$ in component form, we find that

$$\sum F_x = T\cos 30° + (-f_k) = T\cos 30° - 0.40n = 0,$$
$$\sum F_y = T\sin 30° + n + (-500\text{ N}) = 0.$$

These are two simultaneous equations for the two unknown quantities T and n. To solve them, we can eliminate one unknown and

Continued

solve for the other. There are many ways to do this. Here is one way: Rearrange the second equation to the form

$$n = 500 \text{ N} - T\sin 30°.$$

Then substitute this expression for n back into the first equation:

$$T\cos 30° - 0.40(500 \text{ N} - T\sin 30°) = 0.$$

Finally, solve this equation for T, and then substitute the result back into either of the original equations to obtain n. The results are

$$T = 188 \text{ N}, \qquad n = 406 \text{ N}.$$

REFLECT The normal-force magnitude n is *less* than the weight ($w = 500$ N) of the box, and the tension required is a little less than the force needed (200 N) when you pulled horizontally.

Practice Problem: Try pulling at 22°; you'll find that you need even less force. (In fact, this angle gives the smallest possible value of T.) *Answers:* $T = 186$ N, $n = 430$ N.

EXAMPLE 5.11 Toboggan ride with friction

Let's go back to the toboggan we studied in Example 5.6. The wax has worn off, and there is now a coefficient of kinetic friction μ_k. The slope has just the right angle to make the toboggan slide with constant velocity. Derive an expression for the slope angle in terms of w and μ_k.

SOLUTION

SET UP Figure 5.16 shows our sketch and free-body diagram. The slope angle is α. The forces on the toboggan are identified by their magnitudes: its weight (w) and the normal (n) and frictional (f_k) components of the contact force exerted on it by the sloping surface. We take axes perpendicular and parallel to the surface and represent the weight in terms of its components in these two directions as shown.

SOLVE The toboggan is moving with constant velocity and is therefore in equilibrium. The equilibrium conditions are

$$\sum F_x = w\sin\alpha + (-f_k) = w\sin\alpha - \mu_k n = 0,$$
$$\sum F_y = n + (-w\cos\alpha) = 0.$$

Rearranging terms, we get

$$\mu_k n = w\sin\alpha \qquad n = w\cos\alpha$$

Note that the normal force n is less than the weight w. When we divide the first of these equations by the second, we find that

$$\mu_k = \frac{\sin\alpha}{\cos\alpha} = \tan\alpha.$$

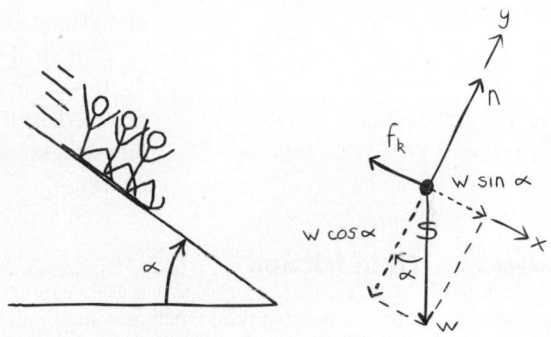

▲ **FIGURE 5.16** Our diagrams for this problem.

REFLECT The normal force is *not* equal in magnitude to the weight; it is always smaller. The weight w doesn't appear in the expression for μ_k; *any* toboggan, regardless of its weight, slides down an incline with constant speed if the tangent of the slope angle of the hill equals the coefficient of kinetic friction. The greater the coefficient of friction, the steeper the slope has to be for the toboggan to slide with constant velocity. This behavior is just what we should expect.

EXAMPLE 5.12 A steeper hill

Suppose the toboggan in Example 5.11 goes down a steeper hill. The coefficient of friction is the same, but this time the toboggan accelerates. Derive an expression for the acceleration in terms of g, α, μ_k, and w.

SOLUTION

SET UP Our free-body diagram (Figure 5.17) is almost the same as for Example 5.11, but the toboggan is no longer in equilibrium: a_y is still zero, but a_x is not.

SOLVE We apply Newton's second law in component form. From $\sum F_x = ma_x$ and $\sum F_y = ma_y$, we get the two equations

$$\sum F_x = mg\sin\alpha + (-f_k) = ma_x,$$
$$\sum F_y = n + (-mg\cos\alpha) = 0.$$

From the second equation, we find that $n = mg\cos\alpha$; then, from Equation 5.1 ($f_k = \mu_k n$), we get $f_k = \mu_k n = \mu_k mg\cos\alpha$.

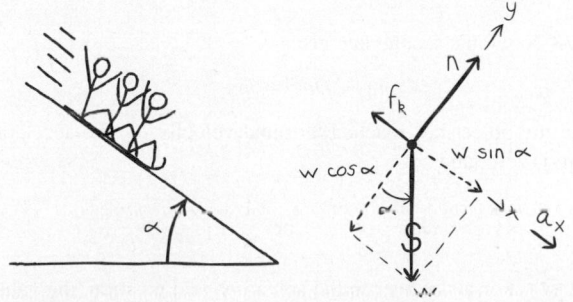

▲ **FIGURE 5.17** Our diagrams for the toboggan on a steeper hill.

Continued

That is, the magnitude of the friction force equals μ_k times the component of the weight normal to the surface. We substitute this expression for f_k back into the x-component equation. The result is

$$mg\sin\alpha + (-\mu_k mg\cos\alpha) = ma_x,$$
$$a_x = g(\sin\alpha - \mu_k\cos\alpha).$$

REFLECT Does this result make sense? Let's check some special cases. First, if the hill is *vertical,* $\alpha = 90°$; then $\sin\alpha = 1$, $\cos\alpha = 0$, and $a_x = g$. This is free fall, just what we would expect. Second, on a hill with angle α with *no* friction, $\mu_k = 0$. Then $a_x = g\sin\alpha$. The situation is the same as in Example 5.6,

and we get the same result; that's encouraging! Next, suppose that $a_x = 0$. Then there is just enough friction to make the toboggan move with constant velocity. In that case, $a_x = 0$, and it follows that

$$\sin\alpha = \mu_k\cos\alpha \qquad \text{and} \qquad \mu_k = \tan\alpha.$$

This agrees with our result from Example 5.11; good! Finally, note that there may be *so much* friction that $\mu_k\cos\alpha$ is actually greater than $\sin\alpha$. In that case, a_x is negative. If we give the toboggan an initial downhill push, it will slow down and eventually stop.

We have pretty much beaten the toboggan problem to death, but there is an important lesson to be learned. We started out with a simple problem and then extended it to more and more general situations. Our most general result includes all the previous ones as special cases, and that's a nice, neat package! Don't memorize this package; it is useful only for this one set of problems. But do try to understand how we obtained it and what it means.

One final variation that you may want to try out is the case in which we give the toboggan an initial push *up* the hill. The direction of the friction force is now reversed, so the acceleration is different from the downhill value. It turns out that the expression for a_x is the same as before, except that the minus sign becomes a plus sign.

EXAMPLE 5.13 Fluid friction

When an object moves through a fluid (such as water or air), it exerts a force on the fluid to push it out of the way. By Newton's third law, the fluid pushes back on the object, in a direction opposite to the object's velocity relative to the fluid, always opposing the object's motion and usually increasing with speed. In high-speed motion through air, the resisting force is approximately proportional to the square of the object's speed v; it's called a *drag force,* or simply *drag.* We can represent its magnitude F_D by

$$F_D = Dv^2,$$

where D is a proportionality constant that depends on the shape and size of the object and the density of air.

When an object falls vertically through air, the drag force opposing the object's motion increases, and the downward acceleration decreases. Eventually, the object reaches a *terminal velocity:* its acceleration approaches zero and the velocity becomes nearly constant. Derive an expression for the terminal velocity magnitude v_T in terms of D and the weight mg of the object.

SOLUTION

SET UP As shown in Figure 5.18, we take the positive y direction to be downward, and we neglect any force associated with buoyancy in the fluid. The net vertical component of force is $mg - Dv^2$.

SOLVE Newton's second law gives

$$mg - Dv_y^2 = ma_y.$$

When the object has reached terminal velocity v_T, the acceleration a_y is zero, and

$$mg - Dv_T^2 = 0, \qquad \text{or} \qquad v_T = \sqrt{\frac{mg}{D}}.$$

REFLECT For a skydiver in the spread-eagled position, the value of the constant D is found experimentally to be about 0.25 kg/m.

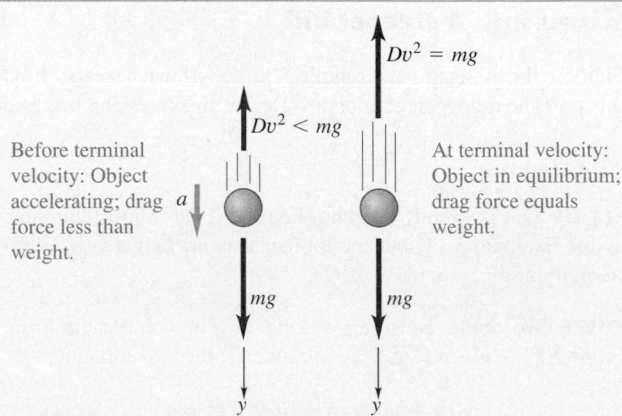

Before terminal velocity: Object accelerating; drag force less than weight.

$Dv^2 < mg$

$Dv^2 = mg$

At terminal velocity: Object in equilibrium; drag force equals weight.

▲ **FIGURE 5.18** An object falling through air, before and after reaching its terminal velocity.

Continued

(Does this number have the correct units?) If the skydiver's mass is 80 kg, the terminal velocity magnitude is

$$v_T = \sqrt{\frac{mg}{D}} = \sqrt{\frac{(80 \text{ kg})(9.8 \text{ m/s}^2)}{0.25 \text{ kg/m}}} = 56 \text{ m/s}.$$

This terminal velocity magnitude is about 125 mi/h. Does this seem reasonable?

Practice Problem: For an average-sized cat, the constant D is approximately one-fourth that for a human, or about $D = 0.062 \text{ kg/m}$. Find the terminal velocity magnitude of a 4.5 kg cat. *Answer:* 27 m/s.

Quantitative Analysis 5.5

Skydivers everywhere

As we just saw, objects falling through the air at high speeds are acted upon by a drag force proportional in magnitude to v^2. A falling object speeds up until the magnitude of the force from air drag equals the magnitude of the object's weight. If a 120 lb woman using a parachute falls with terminal velocity v, a 240 lb man using an identical chute will fall with terminal velocity

A. v.   B. $\sqrt{2}v$.   C. $2v$.

SOLUTION Since both people use the same chute, the constant D is the same for both. The mass of the man is twice the mass of the woman, so he requires twice the drag force to produce zero acceleration. The drag force is proportional to v^2; hence, it doubles if the velocity increases by a factor of $\sqrt{2}$. The correct answer is B.

Rolling vehicles also experience air drag, which usually increases proportionally to the square of the speed. Air drag is often negligible at low speeds, but at highway speeds it is comparable to or greater than rolling friction.

Mastering**PHYSICS**

ActivPhysics 2.1.2: Skydiver

5.4 Elastic Forces

In Section 4.1, when we introduced the concept of *force,* we mentioned that when forces are applied to a solid object, they usually deform the object. In particular, we discussed the use of a coil spring in a device called a spring balance that can be used to measure the magnitudes of forces.

Let's look at the relation of force to deformation in a little more detail. Figure 5.19 shows a coil spring with one end attached to an anchor point and the other end attached to an object that can slide without friction along the x axis. The coordinate x describes the position of the right end of the spring. When $x = 0$, the spring is neither stretched nor compressed. When the object is moved a distance x to the right, the spring is stretched by an amount x. It then exerts a force with magnitude F_{spr} on the object and an "equal and opposite" force on the anchor point. When the string is stretched more, greater forces are required. For many common materials, such as steel coil springs, experiments show that stretching the spring an amount $2x$ requires forces with magnitude $2F_{spr}$; that is, the magnitude of force required is *directly proportional* to the amount of stretch.

If the spring has spaces between its coils, as in the figure, then it can be compressed as well as stretched (Figure 5.19d). Again, experiments show that the amount of compression is directly proportional to the magnitude of the force. In both stretching and compressing, the sign of the x coordinate of the right end of the spring is always opposite the sign of the x component of force the spring applies to the object at its right end. When x is positive, the spring is stretched and applies a force that pulls to the left (the negative x direction) on the object. When x is negative, the spring is compressed and pushes to the right (the positive x direction) on the object.

This proportionality of force to stretching and compressing can be summarized in a simple equation called *Hooke's law* (after Robert Hooke, a contemporary of Newton):

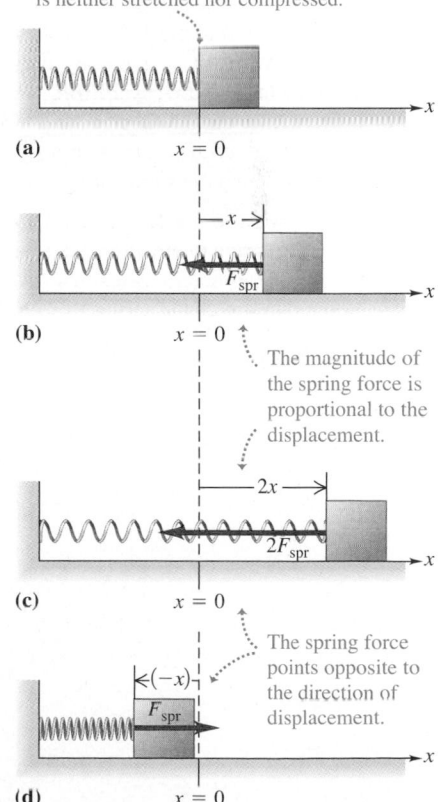

▲ FIGURE 5.19 Some experiments exploring spring forces.

Elastic behavior of springs (Hooke's law)

For springs, the spring force F_{spr} is approximately proportional to the distance x by which the spring is stretched or compressed:

$$F_{spr} = -kx. \tag{5.3}$$

In this equation, k is a positive proportionality constant called the **force constant,** or sometimes the *spring constant,* of the spring. If force is measured in newtons and x in meters, the units of k are N/m. The negative sign indicates that the spring force on the object is always opposite to the displacement.

A very stiff spring requires a large force for a little deformation, corresponding to a large value of the constant k; a weak spring, made with thin wire, requires only a small force for the same deformation, corresponding to a smaller value of k.

Although this relationship is called Hooke's *law,* it is only an approximate relation. Most metallic springs obey it quite well for deformations that are small compared with the overall length of the spring, but rubber bands don't obey it even for small deformations. So it would be safer to call it "Hooke's rule of thumb" (without intending any disparagement of Hooke, one of the great pioneers of late 17th-century science).

EXAMPLE 5.14 **Fishy business**

A spring balance used to weigh fish is built with a spring that stretches 1.00 cm when a 12.0 N weight is placed in the pan. When the 12.0 N weight is replaced with a 1.50 kg fish, what distance does the spring stretch?

SOLUTION

SET UP Figure 5.20a and b show how the spring responds to the 12.0 N weight and to the 1.50 kg fish. The spring and fish are in equilibrium. We draw a free-body diagram for the fish (Figure 5.20c), showing the downward force of gravity and the upward force exerted by the stretched spring. We point the x axis vertically downward, so x is zero with no fish and positive when the fish is attached and the spring stretches.

SOLVE We want to use Hooke's law $(F_{spr} = -kx)$ to relate the stretch of the spring to the force, but first we need to find the force constant k. The problem states that forces of magnitude 12.0 N are required to stretch the spring 1.00 cm $(= 1.00 \times 10^{-2} \text{ m})$, so

$$k = -\frac{-12.0 \text{ N}}{1.00 \times 10^{-2} \text{ m}} = 1200 \text{ N/m}.$$

The weight of the 1.50 kg fish is $w = mg = (1.50 \text{ kg}) \cdot (9.80 \text{ m/s}^2) = 14.7$ N. The equilibrium condition for the fish is

$$\Sigma F_x = 0, \qquad mg + F_{spr} = 0,$$
$$14.7 \text{ N} - (1200 \text{ N/m})x = 0,$$

and finally, $x = 0.0123 \text{ m} = 1.23$ cm.

REFLECT The weight of the fish is greater than the amount of force needed to stretch the spring 1.00 cm, so we expect the fish to stretch the spring more than 1.00 cm. We've used the fact that x can stand for any length change of a Hooke's law spring and F_{spr} for the corresponding change in the component of the spring force.

Practice Problem: If, instead of the fish, we place a 3.00 kg rock in the pan, how much does the spring stretch? *Answer:* Twice as much, 2.45 cm.

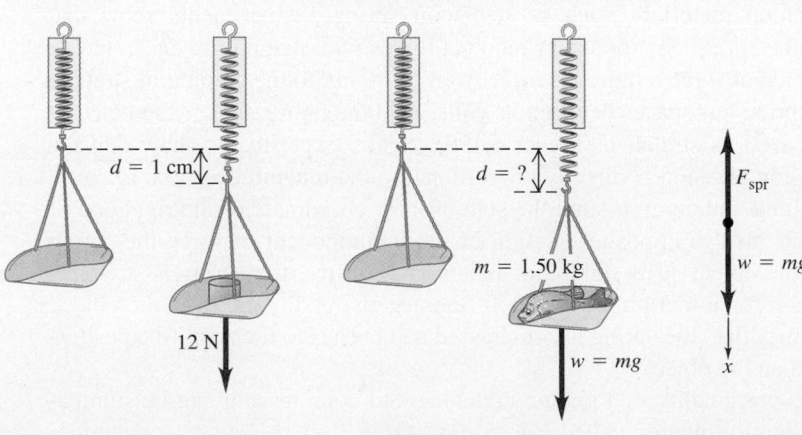

(a) The scale stretched by a known weight

(b) The scale stretched by a known mass

(c) Free-body diagram for the fish

▶ **FIGURE 5.20** Weighing a fish.

EXAMPLE 5.15 **An innerspring mattress**

You have just bought an innerspring mattress that contains coil springs in a rectangular array 20 coils wide and 40 coils long. You estimate that when you lie on the mattress, your weight is supported by about 200 springs (about one-fourth of the total number of springs in the mattress). Lacking further funds for a foundation spring, you lay the mattress directly on the floor. You observe that the springs compress about 2.6 cm when you lie on the mattress. Assuming that your weight of 800 N is supported equally by 200 springs, find the force constant of each spring.

SOLUTION

SET UP You are the object in equilibrium, so start with a free-body diagram of yourself (Figure 5.21). Observe that the spring forces are *additive,* in the sense that the total upward force they exert on your body is just 200 times the force of each individual spring. Your body is in equilibrium under the action of the total spring force and your weight.

SOLVE For equilibrium, the total upward force of all the supporting springs must be 800 N, so $(200)k(2.6 \times 10^{-2}\,\mathrm{m}) = 800\,\mathrm{N}$, and

$$k = 154\,\mathrm{N/m} = 1.54\,\mathrm{N/cm}.$$

REFLECT To borrow a term from electric circuit analysis, the springs are *in parallel;* the deformation is the same for each, and the total force is the sum of the individual spring forces. Of course, the assumption that all the deformations are the same is probably not realistic.

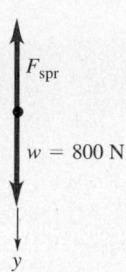

▲ **FIGURE 5.21** Free-body diagram of yourself on the mattress.

Practice Problem: Suppose you lay a thin, lightweight sheet of plywood on top of the mattress, so that your weight is spread over all 800 springs. How much does each spring compress? *Answer:* 0.65 cm.

5.5 Forces in Nature

The historical development of our understanding of the forces (or interactions) found in nature has traditionally placed them into four distinct classes (Figure 5.22). Two are familiar in everyday experience. The other two involve fundamental particle interactions that we cannot observe with the unaided senses.

Of the two familiar classes, **gravitational interactions** were the first to be studied in detail. The *weight* of an object results from the earth's gravitational pull acting on it. The sun's gravitational attraction for the earth keeps the earth in its nearly circular orbit around the sun. Newton recognized that the motions of the planets around the sun and the free fall of objects on earth are both the result of gravitational forces. In the next chapter, we'll study gravitational interactions in greater detail and we'll analyze their vital role in the motions of planets and satellites.

The second familiar class of forces, **electromagnetic interactions,** includes electric and magnetic forces. When you run a comb through your hair, you can then use the comb to pick up bits of paper or fluff; this interaction is the result of an *electric* charge on the comb, which also causes the infamous "static cling." *Magnetic* forces occur in interactions between magnets or between a magnet and a piece of iron. These forces may seem to form a different category, separate from electric forces, but magnetic interactions are actually the result of electric charges in motion. In an electromagnet, an electric current in a coil of wire causes magnetic interactions. One of the great achievements of 19th-century physics was the *unification* of theories of electric and magnetic interactions into a single theoretical framework. We'll study these interactions in detail in the second half of this book.

Gravitational and electromagnetic interactions differ enormously in their strength. The electrical repulsion between two protons at a given distance is

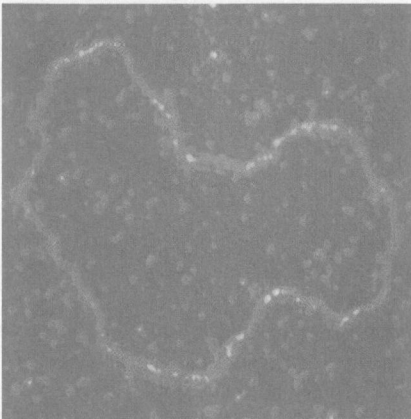

◀**FIGURE 5.22** Instances of the four fundamental forces. (a) Gravitational interactions hold the earth and moon in orbit around each other. (b) Electromagnetic interactions hold together the atoms of a DNA molecule. (c) The strong nuclear interaction is involved in the nuclear reactions that release energy in stars, such as our sun. (d) The weak nuclear interaction plays a role in the release of energy in supernovae, which partly involves the decay of unstable nuclei. (The arrow in the left image indicates the star that has become a supernova in the right image.)

stronger than their gravitational attraction by a factor on the order of 10^{35}. Gravitational forces play no significant role in atomic or molecular structure. But in bodies of astronomical size, positive and negative charge are usually present in nearly equal amounts, and the resulting electrical interactions nearly cancel out. Gravitational interactions are the dominant influence in the motion of planets and also in the internal structure of stars.

The other two classes of interactions are less familiar. One, the **strong interaction,** is responsible for holding the nucleus of an atom together. Nuclei contain electrically neutral and positively charged particles. The charged particles repel each other, and a nucleus could not be stable if it were not for the presence of an *attractive* force of a different kind that counteracts the repulsive electrical interactions. In this context, the strong interaction is also called the *nuclear force*. It has a much shorter range than electrical interactions, but within its range it is much stronger. The strong interaction is also responsible for the creation of unstable particles in high-energy particle collisions.

Finally, there is the **weak interaction.** This force plays no direct role in the behavior of ordinary matter, but it is of vital importance in the behavior of fundamental particles. The weak interaction is responsible for beta decay (one kind of nuclear radiation), as well as the decay of many unstable particles produced in high-energy collisions of fundamental particles. It also plays a significant role in the evolution of stars.

During the 1970s, a unified theory of the electromagnetic and weak interactions was developed. This interaction is now called the *electroweak interaction.* In a sense, it reduces the number of classes of forces from four to three. Similar attempts have been made to understand all strong, electromagnetic, and weak interactions on the basis of a single unified theory called *quantum chromodynamics* (QCD), or a *grand unified theory* (GUT), in which only two forces are postulated. The most promising of the GUT theories make use of vibrating entities called *strings,* whose description requires a space with many more dimensions than our familiar three-dimensional space.

Finally, the ultimate goal is to include gravitational forces in the unification, to produce a *theory of everything* (TOE). Theories of this nature are intimately related to the very early history of the universe, following the Big Bang. Such theories are still highly speculative and raise many unanswered questions. The entire field is in much ferment today. Recent research has had the remarkable effect of connecting our understanding of the *smallest* objects in the universe—that is, fundamental particles—with our understanding of the *largest* objects—the universe itself and its evolution in time since the Big Bang.

SUMMARY

Equilibrium of a Particle

(Section 5.1) When an object is in equilibrium, the vector sum of the forces acting on it must be zero: $\sum \vec{F} = 0$. (Equation 4.3). In component form,

$$\sum F_x = 0, \qquad \sum F_y = 0. \qquad (4.4)$$

Free-body diagrams are useful in identifying the forces acting on the object being considered. Newton's third law is also frequently needed in equilibrium problems. The two forces in an action–reaction pair never act on the same object.

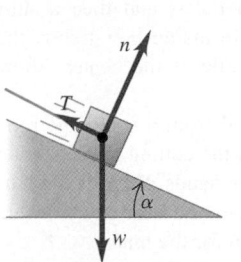

An object moving at constant velocity down a frictionless ramp is in equilibrium: $\sum \vec{F} = 0$.

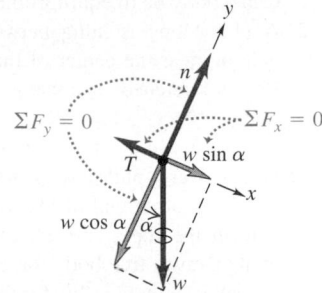

A free-body diagram and coordinate system for the object. The weight vector is replaced by its components.

Applications of Newton's Second Law

(Section 5.2) When the vector sum of the forces on an object is not zero, the object has an acceleration determined by Newton's second law, $\sum \vec{F} = m\vec{a}$ (Equation 4.7). In component form,

$$\sum F_x = ma_x, \qquad \sum F_y = ma_y. \qquad (4.8)$$

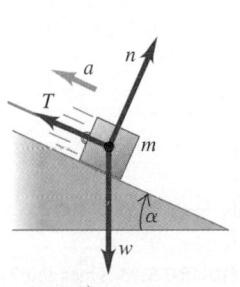

$\sum \vec{F} = m\vec{a}$.

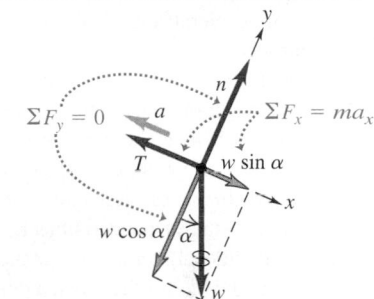

A free-body diagram for the object.

Contact Forces and Friction

(Section 5.3) The contact force between two objects can always be represented in terms of a normal component n perpendicular to the surface of interaction and a frictional component f parallel to the surface. When sliding occurs, the kinetic-friction force f_k is often approximately proportional to n. Then the proportionality constant is μ_k, the coefficient of kinetic friction: $f_k = \mu_k n$ (Equation 5.1).

When there is no relative motion, the maximum possible friction force is approximately proportional to the normal force, and the proportionality constant is μ_s, the coefficient of static friction. The governing equation is $f_s \leq \mu_s n$ (Equation 5.2).

The actual static-friction force may be anything from zero to the maximum value given by the equality in Equation 5.2, depending on the situation. Usually, μ_k is less than μ_s for a given pair of surfaces.

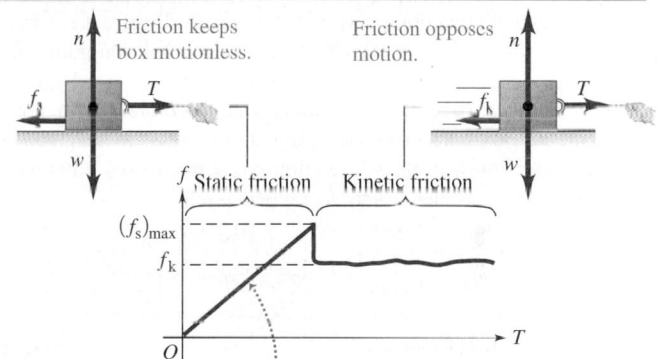

The static-friction force remains equal in magnitude to the tension force until its maximum value $(f_s)_{max}$ is exceeded.

Elastic Forces

(Section 5.4) When forces act on a solid object, the object usually deforms. In some cases, such as a stretched or compressed spring, the deformation is approximately proportional to the magnitude of the applied force. This proportionality is called Hooke's law:

$$F_{spr} = -kx. \qquad (5.3)$$

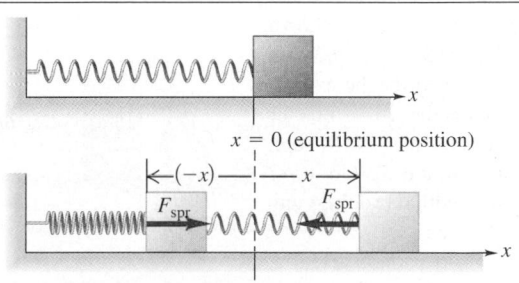

When stretched or compressed from equilibrium, the spring exerts a force $\vec{F}_{spr}$ whose magnitude is approximately proportional to the displacement.

Forces in Nature

(Section 5.5) Historically, forces have been classified as strong, electromagnetic, weak, and gravitational. Present-day research includes intensive efforts to create a unified description of forces in all these categories.

Conceptual Questions

1. Can a body be in equilibrium when only one force acts on it?
2. A clothesline is hung between two poles, and then a shirt is hung near the center of the line. No matter how tightly the line is stretched, it always sags a little at the center. Show why.
3. A man sits in a seat that is suspended from a rope. The rope passes over a pulley suspended from the ceiling, and the man holds the other end of the rope in his hands. What is the tension in the rope, and what force does the seat exert on the man? Draw a free-body force diagram for the man.
4. Is it physically possible for the coefficient of friction ever to be greater than one? What would this mean?
5. Why is it so much more difficult to walk on icy pavement than dry pavement?
6. A car accelerates gradually to the right with power on the two rear wheels. What is the direction of the friction force on these wheels? Show why. Is it static or kinetic friction?
7. Without doing any calculations, decide whether the tension T in Figure 5.23 will be equal to, greater than, or less than the weight of the suspended object.
8. A box slides up an incline, comes to rest, and then slides down again,

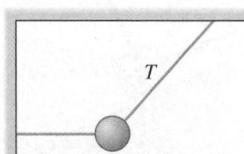

▲ **FIGURE 5.23** Question 7.

accelerating in both directions. Answer the following questions, using only a free-body diagram and *without* doing any calculations: (a) If there is no friction on the incline, how will the box's acceleration going up compare in magnitude *and* direction with its acceleration going down? (b) If there is kinetic friction on the incline, how will the box's acceleration going up compare in magnitude *and* direction with its acceleration going down? (c) What will be the box's acceleration at its highest point?
9. For the objects shown in Figure 5.24, will the tension in the wire be greater than, equal to, or less than the weight W? Decide *without* doing any calculations.
10. A woman is pushing horizontally on two boxes on a factory floor, as shown in Figure 5.25. Which is greater, the force the woman exerts on the 10 lb box or the force the 10 lb box exerts on the 250 lb box? Decide without doing any calculations.

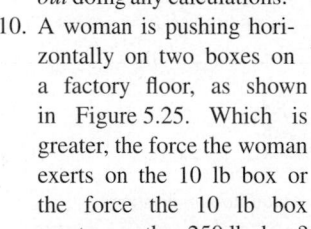

▲ **FIGURE 5.24** Question 9.

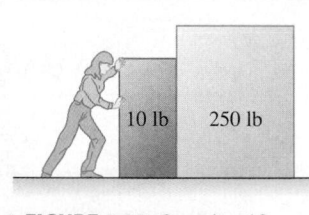

▲ **FIGURE 5.25** Question 10.

11. In a world *without* friction, could you (a) walk on a horizontal sidewalk, (b) climb a ladder, (c) climb a vertical pole, (d) jump into the air, (e) ride your bike, and (f) drive around a curve on a flat roadway? Explain your reasoning.
12. How could you measure the acceleration of a spaceship by using only simple instruments (such as springs, strings, and rulers) located *within* the ship?
13. You can classify scales for weighing objects as those that use springs and those that use standard masses to balance unknown masses. Which group would be more accurate when you use it in an accelerating spaceship? When you use it on the moon?
14. (a) Show that, in terms of the fundamental quantities, the units of the force constant k for a spring reduce to kg/s^2. (b) Why is it physically more sensible to express k in N/m instead of in its fundamental units of kg/s^2?
15. When you stand with bare feet in a wet bathtub, the grip feels fairly secure, and yet a catastrophic slip is quite possible. Explain this in terms of the two coefficients of friction.
16. True or false? The minus sign in Hooke's law $(F_{spr} = -kx)$ means that the spring's force always points in the negative direction. If the statement is false, what *does* the minus sign mean?

Multiple-Choice Problems

1. A horizontal force accelerates a box across a rough horizontal floor with friction present, producing an acceleration a. If the force is now tripled, but all other conditions remain the same, the acceleration will become
   A. greater than $3a$.
   B. equal to $3a$.
   C. less than $3a$.
2. You slide an 800-N table across the kitchen floor by pushing with a force of 100 N. If the table moves at a constant speed, the friction force with the floor must be
   A. 100 N   B. more than 100 N
   C. 800 N   D. more than 800 N

3. An artist wearing spiked shoes pushes two crates across her frictionless horizontal studio floor. (See Figure 5.26.) If she exerts a horizontal 36 N force on the smaller crate, then the smaller crate exerts a force on the larger crate that is closest to

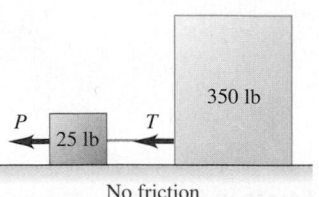

▲ **FIGURE 5.26** Multiple-choice problem 3.

   A. 36 N.   B. 30 N.   C. 200 N.   D. 240 N.
4. A horizontal pull P pulls two wagons over a horizontal frictionless floor,

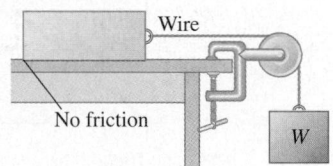

▲ **FIGURE 5.27** Multiple-choice problem 4.

   as shown in Figure 5.27. The tension in the light horizontal rope connecting the wagons is
   A. equal to P, by Newton's third law.
   B. equal to 2000 N.
   C. greater than P.
   D. less than P.
5. A horizontal pull P drags two boxes connected by a horizontal rope having tension T, as shown in Figure 5.28. The floor is horizontal and frictionless. Decide which of the following statements is true *without* doing any calculations:

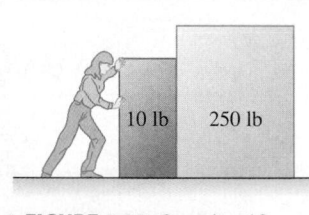

▲ **FIGURE 5.28** Multiple-choice problem 5.

   A. $P > T$ B. $P = T$ C. $P < T$
   D. We need more information to decide which is greater.

6. A crate slides up an inclined ramp and then slides down the ramp after momentarily stopping near the top. This crate is acted upon by friction on the ramp and accelerates both ways. Which statement about this crate's acceleration is correct?
   A. The acceleration going up the ramp is greater than the acceleration going down.
   B. The acceleration going down the ramp is greater than the acceleration going up.
   C. The acceleration is the same in both directions.

7. A bungee cord with a normal length of 2 ft requires a force of 30 lb to stretch it to a total length of 5 ft. The force constant of the bungee cord is
   A. 6 lb/ft
   B. 10 lb/ft
   C. 15 lb/ft
   D. 30 lb/ft

8. A weightless spring scale is attached to two equal weights as shown in Figure 5.29. The reading in the scale will be
   A. 0.
   B. W.
   C. 2W.

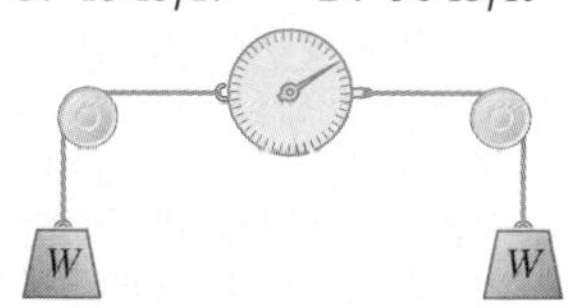

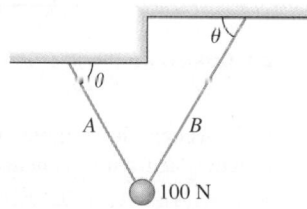

▲ **FIGURE 5.29** Multiple-choice problem 8.

9. Two objects are connected by a light wire as shown in Figure 5.30, with the wire pulling horizontally on the 400 N object. After this system is released from rest, the tension in the wire will be
   A. less than 300 N.
   B. 300 N.
   C. 200 N.
   D. 100 N.

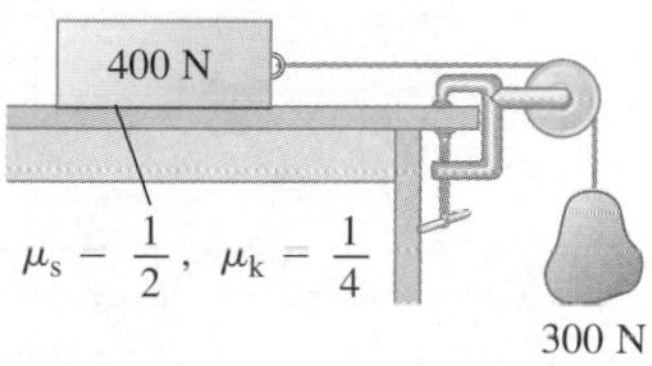

$\mu_s = \frac{1}{2}, \ \mu_k = \frac{1}{4}$

400 N

300 N

▲ **FIGURE 5.30** Multiple-choice problem 9.

10. A 100 N weight is supported by two weightless wires *A* and *B* as shown in Figure 5.31. What can you conclude about the tensions in these wires?
    A. The tensions are equal to 50 N each.
    B. The tensions are equal, but less than 50 N each.
    C. The tensions are equal, but greater than 50 N each.
    D. The tensions are equal to 100 N each.

θ

A   B

100 N

▲ **FIGURE 5.31** Multiple-choice problem 10.

11. The two blocks in Figure 5.32 are in balance, and there is no friction on the surface of the incline. If a slight vibration occurs, what will *M* do after that?
    A. It will accelerate up the ramp.
    B. It will accelerate down the ramp.
    C. It will not accelerate.

▲ **FIGURE 5.32** Multiple-choice problem 11.

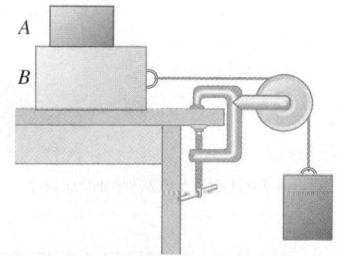

▲ **FIGURE 5.33** Multiple-choice problem 12.

12. The system shown in Figure 5.33 is released from rest, there is no friction between *B* and the tabletop, and all of the objects move together. What must be true about the friction force on *A*?
    A. It is zero.
    B. It acts to the right on *A*.
    C. It acts to the left on *A*.
    D. We cannot tell whether there is any friction force on *A* because we do not know the coefficients of friction between *A* and *B*.

13. In the system shown in Figure 5.34, $M > m$, the surface of the bench is horizontal and frictionless, and the connecting string pulls horizontally on *m*. As more and more weight is gradually added to *m*, which of the following statements best describes the behavior of the system after it is released?

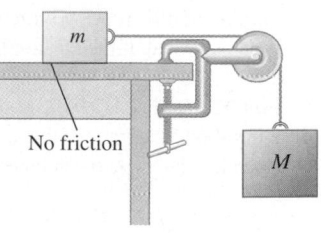

m

No friction

M

▲ **FIGURE 5.34** Multiple-choice problem 13.

   A. The acceleration remains the same in all cases, since there is no friction and the pull of gravity on *M* is the same.
   B. The acceleration becomes zero when enough weight is added so that $m = M$.
   C. The velocity becomes zero when $m = M$.
   D. None of the preceding statements is correct.

Problems

5.1 Equilibrium of a Particle

1. • A 15.0 N bucket is to be raised at a constant speed of 50.0 cm/s by a rope. According to the information in Table 5.1, how many kilograms of cement can be put into this bucket without breaking the rope if it is made of (a) thin white string, (b) $\frac{1}{4}$ in. nylon clothesline, (c) $1\frac{1}{4}$ in. manila climbing rope?

2. • In a museum exhibit, three equal weights are hung with identical wires, as shown in Figure 5.35. Each wire can support a tension of no more than 75.0 N without breaking. Start each of the following parts with an appropriate free-body diagram. (a) What is the maximum value that *W* can be without breaking any wires? (b) Under these conditions, what is the tension in each wire?

W

W

W

▲ **FIGURE 5.35** Problem 2.

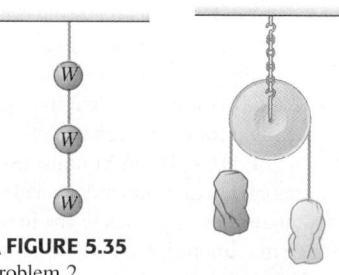

▲ **FIGURE 5.36** Problem 3.

3. • Two 25.0 N weights are suspended at opposite ends of a rope that passes over a light, frictionless pulley. The pulley is attached to a chain that is fastened to the ceiling. (See Figure 5.36.) Start solving this problem by making a free-body diagram of each weight. (a) What is the tension in the rope? (b) What is the tension in the chain?

4. • Two weights are hanging as shown in Figure 5.37. (a) Draw a free-body diagram of each weight. (b) Find the tension in cable *A*. (c) Find the tension in cables *B* and *C*.

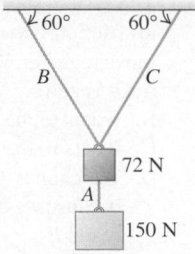

▲ FIGURE 5.37 Problem 4.

5. • An adventurous archaeologist crosses between two rock cliffs by slowly going hand over hand along a rope stretched between the cliffs. He stops to rest at the middle of the rope (Figure 5.38). The rope will break if the tension in it exceeds 2.50×10^4 N. Our hero's mass is 90.0 kg. (a) If the angle θ is 10.0°, find the tension in the rope. Start with a free-body diagram of the archaeologist. (b) What is the smallest value the angle θ can have if the rope is not to break?

▲ FIGURE 5.38 Problem 5.

6. •• A 1130-kg car is held in place by a light cable on a very smooth (frictionless) ramp, as shown in Fig. 5.39. The cable makes an angle of 31.0° above the surface of the ramp, and the ramp itself rises at 25.0° above the horizontal. (a) Draw a free-body diagram for the car. (b) Find the tension in the cable. (c) How hard does the surface of the ramp push on the car?

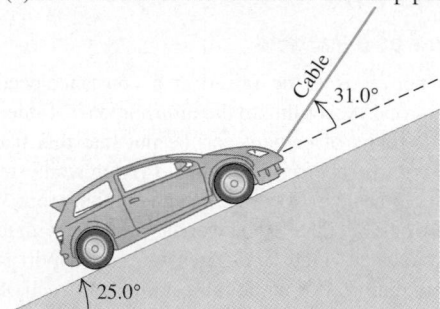

▲ FIGURE 5.39 Problem 6.

7. •• **Tension in a muscle.** Muscles are attached to bones by **BIO** means of tendons. The maximum force that a muscle can exert is directly proportional to its cross-sectional area A at the widest point. We can express this relationship mathematically as $F_{max} = \sigma A$, where σ (sigma) is a proportionality constant. Surprisingly, σ is about the same for the muscles of all animals and has the numerical value of 3.0×10^5 in SI units. (a) What are the SI units of σ in terms of newtons and meters and also in terms of the fundamental quantities (kg, m, s)? (b) In one set of experiments, the average maximum force that the gastrocnemius muscle in the back of the lower leg could exert was measured to be 755 N for healthy males in their midtwenties. What does this result tell us was the average cross-sectional area, in cm², of that muscle for the people in the study?

8. •• **Muscles and tendons.** The gastrocnemius muscle, in the **BIO** back of the leg, has two portions, known as the medial and lateral heads. Assume that they attach to the Achilles tendon as shown in Figure 5.40. The cross-sectional area of each of these two muscles is typically 30 cm² for many adults. What is the maximum tension they can produce in the Achilles tendon? (See the previous problem for some useful information.)

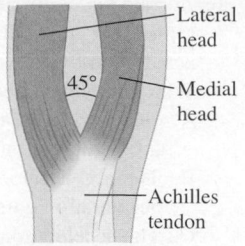

▲ FIGURE 5.40 Problem 8.

9. • **Traction apparatus.** In order to **BIO** prevent muscle contraction from misaligning bones during healing (which can cause a permanent limp), injured or broken legs must be supported horizontally and at the same time kept under tension (traction) directed along the leg. One version of a device to accomplish this aim, the Russell traction apparatus, is shown in Figure 5.41. This system allows the apparatus to support the full weight of the injured leg and at the same time provide the traction along the leg. If the leg to be supported weighs 47.0 N, (a) what must be the weight of W and (b) what traction force does this system produce along the leg?

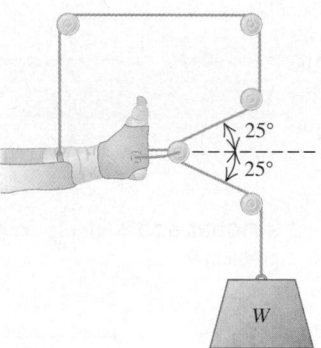

▲ FIGURE 5.41 Problem 9.

10. • **A broken thigh bone.** When the thigh is fractured, the **BIO** patient's leg must be kept under traction. One method of doing so is a variation on the Russell traction apparatus. (See Figure 5.42.) If the physical therapist specifies that the traction force directed along the leg must be 25 N, what must W be?

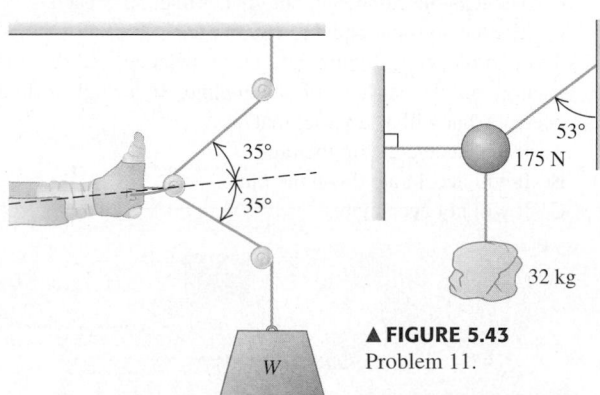

▲ FIGURE 5.42 Problem 10.

▲ FIGURE 5.43 Problem 11.

11. •• Two artifacts in a museum display are hung from vertical walls by very light wires, as shown in Figure 5.43. (a) Draw a

free-body diagram of each artifact. (b) Find the tension in each of the three wires. (c) Would the answers be different if each wire were twice as long, but the angles were unchanged? Why or why not?

12. •• In a rescue, the 73 kg police officer is suspended by two cables, as shown in Figure 5.44. (a) Sketch a free-body diagram of him. (b) Find the tension in each cable.

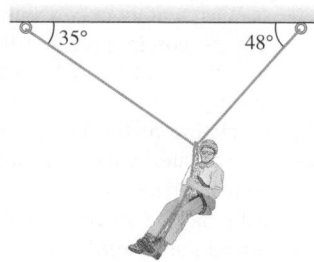

▲ **FIGURE 5.44** Problem 12.

13. •• A tetherball leans against the smooth, frictionless post to which it is attached. (See Figure 5.45.) The string is attached to the ball such that a line along the string passes through the center of the ball. If the string is 1.40 m long and the ball has a radius of 0.110 m with mass 0.270 kg, (a) make a free-body diagram of the ball. (b) What is the tension in the rope? (c) What is the force the pole exerts on the ball?

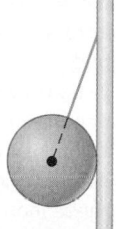

▲ **FIGURE 5.45**
Problem 13.

14. •• Find the tension in each cord in Figure 5.46 if the weight of the suspended object is 250 N.

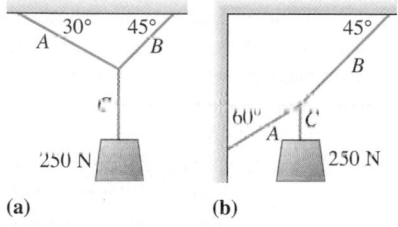

(a)   (b)

▲ **FIGURE 5.46** Problem 14.

15. •• Two blocks, each with weight w, are held in place on a frictionless incline as shown in Figure 5.47. In terms of w and the angle α of the incline, calculate the tension in (a) the rope connecting the blocks and (b) the rope that connects block A to the wall. (c) Calculate the magnitude of the force that the incline exerts on each block. (d) Interpret your answers for the cases $\alpha = 0$ and $\alpha = 90°$.

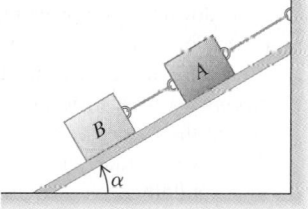

▲ **FIGURE 5.47** Problem 15.

16. •• A man pushes on a piano of mass 180 kg so that it slides at a constant velocity of 12.0 cm/s *down* a ramp that is inclined at 11.0° above the horizontal. No appreciable friction is acting on the piano. Calculate the magnitude and direction of this push (a) if the man pushes parallel to the incline, (b) if the man pushes the piano *up* the plane instead, also at 12.0 cm/s parallel to the incline, and (c) if the man pushes horizontally, but still with a speed of 12.0 cm/s.

5.2 Applications of Newton's Second Law

17. •• **Air-bag safety.** According to safety standards for air bags,
BIO the maximum acceleration during a car crash should not exceed 60g and should last for no more than 36 ms. (a) In such a case, what force does the air bag exert on a 75 kg person? Start with a free-body diagram. (b) Express the force in part (a) in terms of the person's weight.

18. •• **Forces during chin-ups.** People who do chin-ups raise
BIO their chin just over a bar (the chinning bar), supporting themselves only by their arms. Typically, the body below the arms is raised by about 30 cm in a time of 1.0 s, starting from rest. Assume that the entire body of a 680 N person who is chinning is raised this distance and that half the 1.0 s is spent accelerating upward and the other half accelerating downward, uniformly in both cases. Make a free-body diagram of the person's body, and then apply it to find the force his arms must exert on him during the accelerating part of the chin-up.

19. •• **Force on a tennis ball.** The record speed for a tennis ball that is served is 73.14 m/s. During a serve, the ball typically starts from rest and is in contact with the tennis racquet for 30 ms. Assuming constant acceleration, what was the average force exerted on the tennis ball during this record serve, expressed in terms of the ball's weight W?

20. •• A 75,600 N spaceship comes in for a vertical landing. From an initial speed of 1.00 km/s, it comes to rest in 2.00 min with uniform acceleration. (a) Make a free-body diagram of this ship as it is coming in. (b) What braking force must its rockets provide? Ignore air resistance.

21. •• **Force during a jump.** An average person can reach a
BIO maximum height of about 60 cm when jumping straight up from a crouched position. During the jump itself, the person's body from the knees up typically rises a distance of around 50 cm. To keep the calculations simple and yet get a reasonable result, assume that the *entire body* rises this much during the jump. (a) With what initial speed does the person leave the ground to reach a height of 60 cm? (b) Make a free-body diagram of the person during the jump. (c) In terms of this jumper's weight W, what force does the ground exert on him or her during the jump?

22. •• A short train (an engine plus four cars) is accelerating at 1.10 m/s². If the mass of each car is 38,000 kg, and if each car has negligible frictional forces acting on it, what are (a) the force of the engine on the first car, (b) the force of the first car on the second car, (c) the force of the second car on the third car, and (d) the force of the third car on the fourth car? In solving this problem, note the importance of selecting the correct set of cars to isolate as your object.

23. •• A large fish hangs from a spring balance supported from the roof of an elevator. (a) If the elevator has an upward acceleration of 2.45 m/s² and the balance reads 60.0 N, what is the true weight of the fish? (b) Under what circumstances will the balance read 35.0 N? (c) What will the balance read if the elevator cable breaks?

24. •• A 750.0-kg boulder is raised from a quarry 125 m deep by a long uniform chain having a mass of 575 kg. This chain is of uniform strength, but at any point it can support a maximum tension no greater than 2.50 times its weight without breaking. (a) What is the maximum acceleration the boulder can have

and still get out of the quarry, and (b) how long does it take to be lifted out at maximum acceleration if it started from rest?

25. •• The TGV, France's high-speed train, pulls out of the Lyons station and is accelerating uniformly to its cruising speed. Inside one of the cars, a 3.00 N digital camera is hanging from the luggage compartment by a light, flexible strap that makes a 12.0° angle with the vertical. (a) Make a free-body diagram of this camera. (b) Apply Newton's second law to the camera, and find the acceleration of the train and the tension in the strap.

26. •• Which way and by what angle does the accelerometer in Figure 5.48 deflect under the following conditions? (a) The cart is moving toward the right with speed increasing at 3.0 m/s². (b) The cart is moving toward the left with speed decreasing at 4.5 m/s². (c) The cart is moving toward the left with a constant speed of 4.0 m/s.

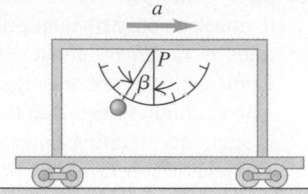

▲ **FIGURE 5.48** Problem 26.

27. • A skier approaches the base of an icy hill with a speed of 12.5 m/s. The hill slopes upward at 24° above the horizontal. Ignoring all friction forces, find the acceleration of this skier (a) when she is going up the hill, (b) when she has reached her highest point, and (c) after she has started sliding down the hill. In each case, start with a free-body diagram of the skier.

28. • At a construction site, a 22.0 kg bucket of concrete is connected over a very light frictionless pulley to a 375 N box on the roof of a building. (See Figure 5.49.) There is no appreciable friction on the box, since it is on roller bearings. The box starts from rest. (a) Make free-body diagrams of the bucket and the box. (b) Find the acceleration of the bucket. (c) How fast is the bucket moving after it has fallen 1.50 m (assuming that the box has not yet reached the edge of the roof)?

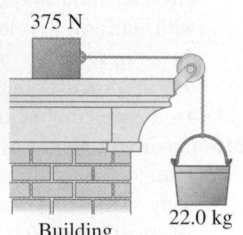

▲ **FIGURE 5.49** Problem 28.

29. •• Two boxes are connected by a light string that passes over a light, frictionless pulley. One box rests on a frictionless ramp that rises at 30.0° above the horizontal (see Figure 5.50), and the system is released from rest. (a) Make free-body diagrams of each box. (b) Which way will the 50.0 kg box move, up the plane or down the plane? Or will it even move at all? Show why or why not. (c) Find the acceleration of each box.

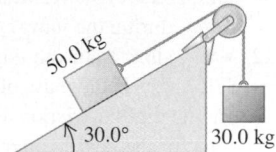

▲ **FIGURE 5.50** Problem 29.

5.3 Frictional Forces

30. • An 80 N box initially at rest is pulled by a horizontal rope on a horizontal table. The coefficients of kinetic and static friction between the box and the table are $\frac{1}{4}$ and $\frac{1}{2}$, respectively. What is the friction force on this box if the pull is (a) 0 N, (b) 25 N, (c) 39 N, (d) 41 N, (e) 150 N?

31. • A box of bananas weighing 40.0 N rests on a horizontal surface. The coefficient of static friction between the box and the surface is 0.40, and the coefficient of kinetic friction is 0.20.

(a) If no horizontal force is applied to the box and the box is at rest, how large is the friction force exerted on the box? (b) What is the magnitude of the friction force if a monkey applies a horizontal force of 6.0 N to the box and the box is initially at rest? (c) What minimum horizontal force must the monkey apply to start the box in motion? (d) What minimum horizontal force must the monkey apply to keep the box moving at constant velocity once it has been started?

32. • **Friction at the hip joint.** Fig-
BIO ure 5.51 shows the bone structure at the hip joint. The bones are normally not in direct contact, but instead are covered with cartilage to reduce friction. The space between them is filled with water-like synovial fluid, which further decreases friction. Due to this fluid, the coefficient of kinetic friction between the bones can range from 0.0050 to 0.020. (The wide range of values is due to the fact that motion such as running causes more fluid to squirt between the bones, thereby reducing friction when they strike each other.) Typically, approximately 65% of a person's weight is above the hip—we'll call this the *upper weight*. (a) Show that when a person is simply standing upright, each hip supports half of his upper weight. (b) When a person is walking, each hip now supports up to 2.5 times his upper weight, depending on how fast he is walking. (Recall that when you walk, your weight shifts from one leg to the other and your body comes down fairly hard on each leg.) For a 65 kg person, what is the maximum kinetic friction force at the hip joint if μ_k has its minimum value of 0.0050? (c) As a person gets older, the aging process, as well as osteoarthritis, can alter the composition of the synovial fluid. In the worst case, this fluid could disappear, leaving bone-on-bone contact with a coefficient of kinetic friction of 0.30. What would be the greatest friction force for the walking person in part (b)? The increased friction causes pain and, in addition, wears down the joint even more.

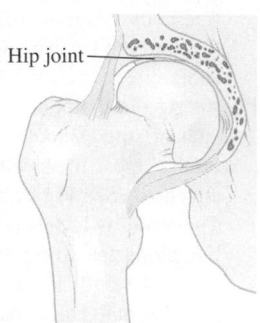

▲ **FIGURE 5.51** Problem 32.

33. • At a construction site, a pallet of bricks is to be suspended by attaching a rope to it and connecting the other end to a couple of heavy crates on the roof of a building, as shown in Figure 5.52. The rope pulls horizontally on the lower crate, and the coefficient of static friction between the lower crate and the roof is 0.666. (a) What is the weight of the heaviest pallet of bricks that can be supported this way? Start with appropriate free-body diagrams. (b) What is the friction force on the upper crate under the conditions of part (a)?

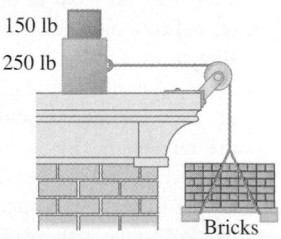

▲ **FIGURE 5.52** Problem 33.

34. •• Two crates connected by a rope of negligible mass lie on a horizontal surface. (See Figure 5.53.) Crate A has mass m_A and crate B has mass m_B. The coefficient of kinetic friction between each crate and the surface is μ_k. The crates are pulled to the right at a constant velocity of 3.20 cm/s by a horizontal force $\vec{F}$. In terms of m_A, m_B, and μ_k, calculate

(a) the magnitude of the force $\vec{F}$ and (b) the tension in the rope connecting the blocks. Include the free-body diagram or diagrams you used to determine each answer.

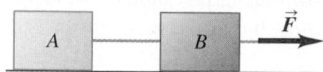

▲ FIGURE 5.53 Problem 34.

35. •• A hockey puck leaves a player's stick with a speed of 9.9 m/s and slides 32.0 m before coming to rest. Find the coefficient of friction between the puck and the ice.

36. •• **Stopping distance of a car.** (a) If the coefficient of kinetic friction between tires and dry pavement is 0.80, what is the shortest distance in which you can stop an automobile by locking the brakes when traveling at 29.1 m/s (about 65 mi/h)? (b) On wet pavement, the coefficient of kinetic friction may be only 0.25. How fast should you drive on wet pavement in order to be able to stop in the same distance as in part (a)? (*Note:* Locking the brakes is *not* the safest way to stop.)

37. •• An 85-N box of oranges is being pushed across a horizontal floor. As it moves, it is slowing at a constant rate of 0.90 m/s each second. The push force has a horizontal component of 20 N and a vertical component of 25 N downward. Calculate the coefficient of kinetic friction between the box and floor.

38. •• **Rolling friction.** Two bicycle tires are set rolling with the same initial speed of 3.50 m/s on a long, straight road, and the distance each travels before its speed is reduced by half is measured. One tire is inflated to a pressure of 40 psi and goes 18.1 m; the other is at 105 psi and goes 92.9 m. What is the coefficient of rolling friction μ_r for each? Assume that the net horizontal force is due to rolling friction only.

39. •• A stockroom worker pushes a box with mass 11.2 kg on a horizontal surface with a constant speed of 3.50 m/s. The coefficients of kinetic and static friction between the box and the surface are 0.200 and 0.450, respectively. (a) What horizontal force must the worker apply to maintain the motion of the box? (b) If the worker stops pushing, what will be the acceleration of the box?

40. •• The coefficients of static and kinetic friction between a 476 N crate and the warehouse floor are 0.615 and 0.420, respectively. A worker gradually increases his horizontal push against this crate until it just begins to move and from then on maintains that same maximum push. What is the acceleration of the crate after it has begun to move? Start with a free-body diagram of the crate.

41. •• **Measuring the coefficients of friction.** One straightforward way to measure the coefficients of friction between a box and a wooden surface is illustrated in Figure 5.54. The sheet of wood can be raised by pivoting it about one edge. It is first raised to an angle θ_1 (which is measured) for which the box just begins to slide downward. The sheet is then immediately

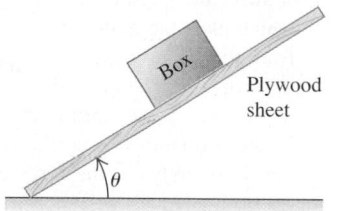

▲ FIGURE 5.54 Problem 41.

lowered to an angle θ_2 (which is also measured) for which the box slides with constant speed down the sheet. Apply Newton's second law to the box in both cases to find the coefficients of kinetic and static friction between it and the wooden sheet in terms of the measured angles θ_1 and θ_2.

42. With its wheels locked, a van slides down a hill inclined at 40.0° to the horizontal. Find the acceleration of this van a) if the hill is icy and frictionless, and b) if the coefficient of kinetic friction is 0.20.

43. • **The Trendelberg position.** In emergencies with major
BIO blood loss, the doctor will order the patient placed in the Trendelberg position, which is to raise the foot of the bed to get maximum blood flow to the brain. If the coefficient of static friction between the typical patient and the bedsheets is 1.2, what is the maximum angle at which the bed can be tilted with respect to the floor before the patient begins to slide?

44. •• **Injuries to the spinal column.** In treating spinal injuries,
BIO it is often necessary to provide some tension along the spinal column to stretch the backbone. One device for doing this is the Stryker frame, illustrated in part (a) of Figure 5.55. A weight W is attached to the patient (sometimes around a neck collar, as shown in part (b) of the figure), and friction between the person's body and the bed prevents sliding. (a) If the coefficient of static friction between a 78.5 kg patient's body and the bed is 0.75, what is the maximum traction force along the spinal column that W can provide without causing the patient to slide? (b) Under the conditions of maximum traction, what is the tension in each cable attached to the neck collar?

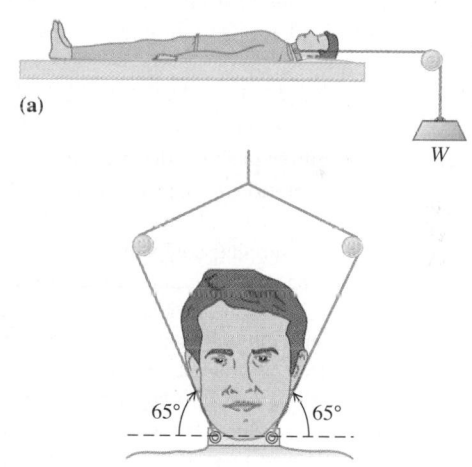

▲ FIGURE 5.55 Problem 44.

45. •• A winch is used to drag a 375 N crate up a ramp at a constant speed of 75 cm/s by means of a rope that pulls parallel to the surface of the ramp. The rope slopes upward at 33° above the horizontal, and the coefficient of kinetic friction between the ramp and the crate is 0.25. (a) What is the tension in the rope? (b) If the rope were suddenly to snap, what would be the acceleration of the crate immediately after the rope broke?

46. •• A toboggan approaches a snowy hill moving at 11.0 m/s. The coefficients of static and kinetic friction between the snow and the toboggan are 0.40 and 0.30, respectively, and the hill slopes upward at 40.0° above the horizontal. Find the acceleration of the toboggan (a) as it is going up the hill and

(b) after it has reached its highest point and is sliding down the hill.

47. •• A 25.0-kg box of textbooks rests on a loading ramp that makes an angle α with the horizontal. The coefficient of kinetic friction is 0.25, and the coefficient of static friction is 0.35. (a) As the angle α is increased, find the minimum angle at which the box starts to slip. (b) At this angle, find the acceleration once the box has begun to move. (c) At this angle, how fast will the box be moving after it has slid 5.0 m along the loading ramp?

48. •• A person pushes on a stationary 125 N box with 75 N at 30° below the horizontal, as shown in Figure 5.56. The coefficient of static friction between the box and the horizontal floor is 0.80. (a) Make a free-body diagram of the box. (b) What is the normal force on the box? (c) What is the friction force on the box? (d) What is the largest the friction force could be? (e) The person now replaces his push with a 75 N pull at 30° above the horizontal. Find the normal force on the box in this case.

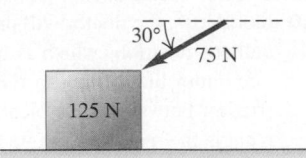

▲ FIGURE 5.56 Problem 48.

49. •• A crate of 45.0-kg tools rests on a horizontal floor. You exert a gradually increasing horizontal push on it and observe that the crate just begins to move when your force exceeds 313 N. After that you must reduce your push to 208 N to keep it moving at a steady 25.0 cm/s. (a) What are the coefficients of static and kinetic friction between the crate and the floor? (b) What push must you exert to give it an acceleration of 1.10 m/s²? (c) Suppose you were performing the same experiment on this crate but were doing it on the moon instead, where the acceleration due to gravity is 1.62 m/s². (i) What magnitude push would cause it to move? (ii) What would its acceleration be if you maintained the push in part (b)?

50. •• You are working for a shipping company. Your job is to stand at the bottom of an 8.0-m-long ramp that is inclined at 37° above the horizontal. You grab packages off a conveyor belt and propel them up the ramp. The coefficient of kinetic friction between the packages and the ramp is $\mu_k = 0.30$. (a) What speed do you need to give a package at the bottom of the ramp so that it has zero speed at the top of the ramp? (b) Your coworker is supposed to grab the packages as they arrive at the top of the ramp, but she misses one and it slides back down. What is its speed when it returns to you?

51. •• The drag coefficient for a spherical raindrop with a radius of 0.415 cm falling at its terminal velocity is 2.43×10^{-5} kg/m. Calculate the raindrop's terminal velocity in m/s.

52. What is the acceleration of a raindrop that has reached half of its terminal velocity? Give your answer in terms of g.

53. •• An object is dropped from rest and encounters air resistance that is proportional to the square of its speed. Sketch qualitative graphs (no numbers) showing (a) the air resistance on this object as a function of its speed, (b) the *net* force on the object as a function of its speed, (c) the *net* force on the object as a function of time, (d) the speed of the object as a function of time, and (e) the acceleration of the object as a function of time.

5.4 Elastic Forces

54. • You find that if you hang a 1.25 kg weight from a vertical spring, it stretches 3.75 cm. (a) What is the force constant of this spring in N/m? (b) How much mass should you hang from the spring so it will stretch by 8.13 cm from its original, unstretched length?

55. • An unstretched spring is 12.00 cm long. When you hang an 875 g weight from it, it stretches to a length of 14.40 cm. (a) What is the force constant (in N/m) of this spring? (b) What total mass must you hang from the spring to stretch it to a total length of 17.72 cm?

56. • **Heart repair.** A surgeon is using material from a donated **BIO** heart to repair a patient's damaged aorta and needs to know the elastic characteristics of this aortal material. Tests performed on a 16.0 cm strip of the donated aorta reveal that it stretches 3.75 cm when a 1.50 N pull is exerted on it. (a) What is the force constant of this strip of aortal material? (b) If the maximum distance it will be able to stretch when it replaces the aorta in the damaged heart is 1.14 cm, what is the greatest force it will be able to exert there?

57. • An "extreme" pogo stick utilizes a spring whose uncompressed length is 46 cm and whose force constant is 1.4×10^4 N/m. A 60-kg enthusiast is jumping on the pogo stick, compressing the spring to a length of only 5.0 cm at the bottom of her jump. Calculate (a) the net upward force on her at the moment the spring reaches its greatest compression and (b) her upward acceleration, in m/s² and g's at that moment.

58. •• A student measures the force required to stretch a spring by various amounts and makes the graph shown in Figure 5.57, which plots this force as a function of the distance the spring has stretched. (a) Does this spring obey Hooke's law? How do you know? (b) What is the force constant of the spring, in N/m? (c) What force would be needed to stretch the spring a distance of 17 cm from its unstretched length, assuming that it continues to obey Hooke's law?

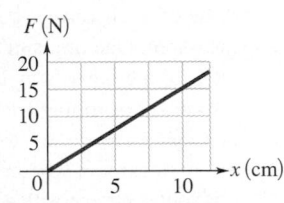

▲ FIGURE 5.57 Problem 58.

59. •• A student hangs various masses from a spring and records the resulting distance the spring has stretched. She then uses her data to construct the graph shown in Figure 5.58, which plots the mass hung from the spring as a function of the distance the mass caused the spring to stretch beyond its unstretched length. (a) Does this spring obey Hooke's law? How do you know? (b) What is the force constant of the spring, in N/m? (c) By how much will the spring have stretched if you hang a 2.35 kg mass from it, assuming that it continues to obey Hooke's law?

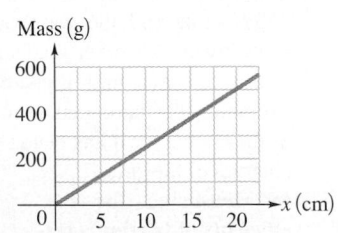

▲ FIGURE 5.58 Problem 59.

60. • Three identical 6.40 kg masses are hung by three identical springs, as shown in Figure 5.59. Each spring has a force

constant of 7.80 kN/m and was 12.0 cm long before any masses were attached to it. (a) Make a free-body diagram of each mass. (b) How long is each spring when hanging as shown? (*Hint:* First isolate only the bottom mass. Then treat the bottom two masses as a system. Finally, treat all three masses as a system.)

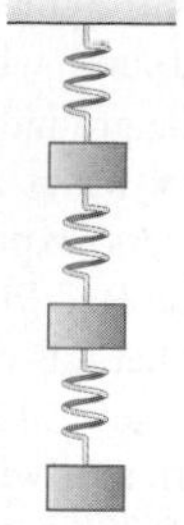

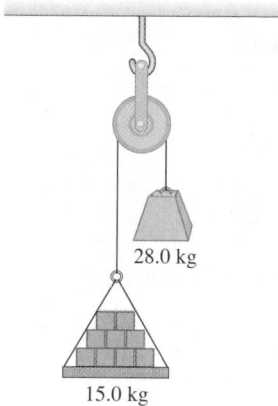

▲ **FIGURE 5.59** Problem 60.

61. • A light spring having a force constant of 125 N/m is used to pull a 9.50 kg sled on a horizontal frictionless ice rink. If the sled has an acceleration of 2.00 m/s², by how much does the spring stretch if it pulls on the sled (a) horizontally, (b) at 30.0° above the horizontal?

62. •• In the previous problem, what would the answers in both cases be if there were friction and the coefficient of kinetic friction between the sled and the ice were 0.200?

General Problems

63. •• **Prevention of hip injuries.** People (especially the elderly)
BIO who are prone to falling can wear hip pads to cushion the impact on their hip from a fall. Experiments have shown that if the speed at impact can be reduced to 1.3 m/s or less, the hip will usually not fracture. Let us investigate the worst-case scenario, in which a 55 kg person completely loses her footing (such as on icy pavement) and falls a distance of 1.0 m, the distance from her hip to the ground. We shall assume that the person's entire body has the same acceleration, which, in reality, would not quite be true. (a) With what speed does her hip reach the ground? (b) A typical hip pad can reduce the person's speed to 1.3 m/s over a distance of 2.0 cm. Find the acceleration (assumed to be constant) of this person's hip while she is slowing down and the force the pad exerts on it. (c) The force in part (b) is very large. To see if it is likely to cause injury, calculate how long it lasts.

64. •• **Modeling elastic hip pads.** In the previous problem, we
BIO assumed constant acceleration (and hence a constant force) due to the hip pad. But of course, such a pad would most likely be somewhat elastic. Assuming that the force you found in the previous problem was the maximum force over the period of impact with the ground, and that the hip pad is elastic enough to obey Hooke's law, calculate the force constant of this pad, using appropriate data from the previous problem.

65. •• You've attached a bungee cord to a wagon and are using it to pull your little sister while you take her for a jaunt. The bungee's unstretched length is 1.3 m, and you happen to know that your little sister weighs 220 N and the wagon weighs 75 N. Crossing a street, you accelerate from rest to your normal walking speed of 1.5 m/s in 2.0 s, and you notice that while you're accelerating, the bungee's length increases to about 2.0 m. What's the force constant of the bungee cord, assuming it obeys Hooke's law?

66. •• **Atwood's Machine.** A 15.0-kg load of bricks hangs from one end of a rope that passes over a small, frictionless pulley. A 28.0-kg counterweight is suspended from the other end of the rope, as shown in Fig. 5.60. The system is released from rest. (a) Draw two free-body diagrams, one for the load of bricks and one for the counterweight. (b) What is the magnitude of the upward acceleration of the load of bricks? (c) What is the tension in the rope while the load is moving? How does the tension compare to the weight of the load of bricks? To the weight of the counterweight?

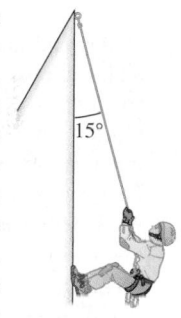

▲ **FIGURE 5.60** Problem 66.

67. • **Mountaineering.** Figure 5.61 shows a technique called *rappelling,* used by mountaineers for descending vertical rock faces. The climber sits in a rope seat, and the rope slides through a friction device attached to the seat. Suppose that the rock is perfectly smooth (i.e., there is no friction) and that the climber's feet push horizontally onto the rock. If the climber's weight is 600.0 N, find (a) the tension in the rope and (b) the force the climber's feet exert on the rock face. Start with a free-body diagram of the climber.

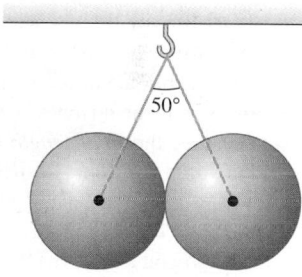

▲ **FIGURE 5.61** Problem 67.

68. •• Two identical, perfectly smooth 71.2 N bowling balls 21.7 cm in diameter are hung together from the same hook in the ceiling by means of two thin, light wires, as shown in Figure 5.62, find (a) the tension in each wire and (b) the force the balls exert on each other.

▲ **FIGURE 5.62** Problem 68.

69. •• **Stay awake!** An astro-
BIO naut is inside a 2.25×10^6 kg rocket that is blasting off vertically from the launch pad. You want this rocket to reach the speed of sound (331 m/s) as quickly as possible, but you also do not want the astronaut to black out. Medical tests have shown that astronauts are in danger of blacking out for an acceleration greater than $4g$. (a) What is the maximum thrust the engines of the rocket can have to just barely avoid blackout? Start with a free-body diagram of the rocket. (b) What force, in terms of her weight W, does the rocket exert on the astronaut? Start with a free-body diagram of the astronaut. (c) What is the shortest time it can take the rocket to reach the speed of sound?

70. •• The stretchy silk of a certain species of spider has a force
BIO constant of 1.10 mN/cm. The spider, whose mass is 15 mg,

has attached herself to a branch as shown in Figure 5.63. Calculate (a) the tension in each of the three strands of silk and (b) the distance each strand is stretched beyond its normal length.

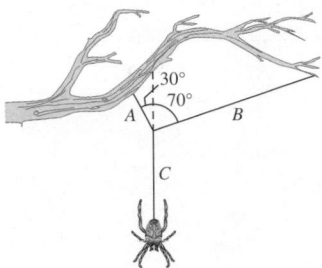

▲ **FIGURE 5.63** Problem 70.

71. •• Block *A* in Figure 5.64 weighs 60.0 N. The coefficient of static friction between the block and the surface on which it rests is 0.25. The weight *w* is 12.0 N and the system remains at rest. (a) Find the friction force exerted on block *A*. (b) Find the maximum weight *w* for which the system will remain at rest.

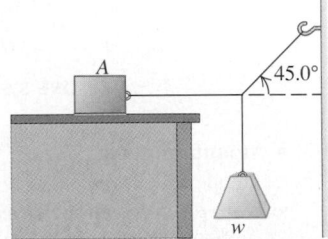

▲ **FIGURE 5.64** Problem 71.

72. •• **Friction in an elevator.** You are riding in an elevator on the way to the 18th floor of your dormitory. The elevator is accelerating upward with $a = 1.90 \text{ m/s}^2$. Beside you is the box containing your new computer; the box and its contents have a total mass of 28.0 kg. While the elevator is accelerating upward, you push horizontally on the box to slide it at constant speed toward the elevator door. If the coefficient of kinetic friction between the box and the elevator floor is $\mu_k = 0.32$, what magnitude of force must you apply?

73. •• A student attaches a series of weights to a tendon and
BIO measures the *total length* of the tendon for each weight. He then uses the data he has gathered to construct the graph shown in Figure 5.65, giving the weight as a function of the length of the tendon. (a) Does this tendon obey Hooke's law? How do you know? (b) What is the force constant (in N/m) for the tendon? (c) What weight should you hang from the tendon to make it stretch by 8.0 cm from its unstretched length?

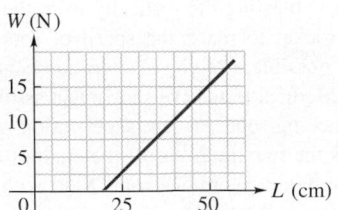

▲ **FIGURE 5.65** Problem 73.

74. •• A 65.0-kg parachutist falling vertically at a speed of
BIO 6.30 m/s impacts the ground, which brings him to a complete stop in a distance of 0.92 m (roughly half of his height).

Assuming constant acceleration after his feet first touch the ground, what is the average force exerted on the parachutist by the ground?

75. •• **Mars Exploration Rover landings.** In January 2004 the Mars Exploration Rover spacecraft landed on the surface of the Red Planet, where the acceleration due to gravity is 0.379 what it is on earth. The descent of this 827 kg vehicle occurred in several stages, three of which are outlined here. In Stage I, friction with the Martian atmosphere reduced the speed from 19,300 km/h to 1600 km/h in a 4.0 min interval. In Stage II, a parachute reduced the speed from 1600 km/h to 321 km/h in 94 s, and in Stage III, which lasted 2.5 s, retrorockets fired to reduce the speed from 321 km/h to zero. As part of your solution to this problem, make a free-body diagram of the rocket during each stage. Assuming constant acceleration, find the force exerted on the spacecraft (a) by the atmosphere during Stage I, (b) by the parachute during Stage II, and (c) by the retrorockets during Stage III.

76. •• Block *A* in Figure 5.66 weighs 1.20 N and block *B* weighs 3.60 N. The coefficient of kinetic friction between all surfaces is 0.300. Find the magnitude of the horizontal force $\vec{F}$ necessary to drag block *B* to the left at a constant speed of 2.50 cm/s (a) if *A* rests on *B* and moves with it (Figure 5.66a); (b) if *A* is held at rest by a string (Figure 5.66b). (c) In part (a), what is the friction force on block *A*?

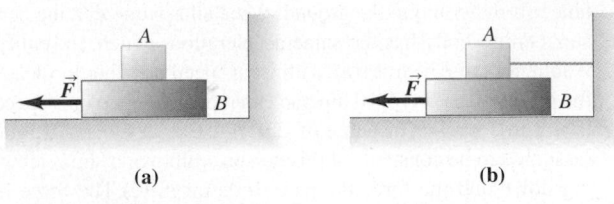

(a)   (b)

▲ **FIGURE 5.66** Problem 76.

77. •• **Crash of the Genesis Mission.** You are assigned the task of designing an accelerometer to be used inside a rocket ship in outer space. Your equipment consists of a very light spring that is 15.0 cm long when no forces act to stretch or compress it, plus a 1.10 kg weight. One end of this spring will be attached to a friction-free tabletop, while the 1.10 kg weight is attached to the other end, as shown in Figure 5.67. (Such a spring-type accelerometer system was actually used in the ill-fated Genesis Mission, which collected particles of the solar wind. Unfortunately, because it was installed *backward,* it did not measure the acceleration correctly during the craft's descent to earth. As a result, the parachute failed to open and the capsule crashed on Sept. 8, 2004.) (a) What should be the force constant of the spring so that it will stretch by 1.10 cm when the rocket accelerates forward at 2.50 m/s²? Start with a free-body diagram of the weight. (b) What is the acceleration (magnitude *and* direction) of the rocket if the spring is *compressed* by 2.30 cm?

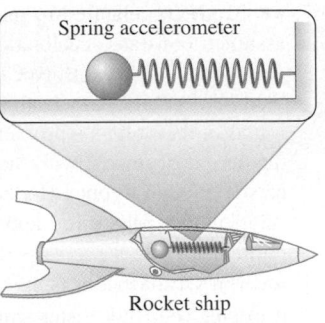

▲ **FIGURE 5.67** Problem 77.

78. •• A block with mass m_1 is placed on an inclined plane with slope angle α and is connected to a second hanging block with mass m_2 by a cord passing over a small, frictionless pulley (Fig. 5.68). The coefficient of static friction is μ_s and the coefficient of kinetic friction is μ_k. (a) Find the mass m_2 for which block m_1 moves up the plane at constant speed once it is set in motion. (b) Find the mass m_2 for which block m_1 moves down the plane at constant speed once it is set in motion. (c) For what range of values of m_2 will the blocks remain at rest if they are released from rest?

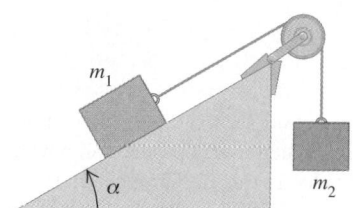

▲ **FIGURE 5.68** Problem 78.

79. •• A pickup truck is carrying a toolbox, but the rear gate of the truck is missing, so the box will slide out if it is set moving. The coefficients of kinetic and static friction between the box and the bed of the truck are 0.355 and 0.650, respectively. Starting from rest, what is the shortest time this truck could accelerate uniformly to 30.0 m/s ($\approx$ 60 mph) without causing the box to slide. Include a free-body diagram of the toolbox as part of your solution. (*Hint:* First use Newton's second law to find the maximum acceleration that static friction can give the box, and then solve for the time required to reach 30.0 m/s.)

80. •• **Accident analysis.** You have been called to testify as an expert witness in a trial involving an automobile accident. The speed limit on the highway where the accident occurred was 40 mph. The driver of the car slammed on his brakes, locking his wheels, and left skid marks as the car skidded to a halt. You measure the length of these skid marks to be 219 ft, 9 in., and determine that the coefficient of kinetic friction between the wheels and the pavement at the time of the accident was 0.40. How fast was this car traveling (to the nearest number of mph) just before the driver hit his brakes? Was he guilty of speeding?

81. •• A window washer pushes his scrub brush up a vertical window at constant speed by applying a force $\vec{F}$ as shown in Fig. 5.69. The brush weighs 12.0 N and the coefficient of kinetic friction is $\mu_k = 0.150$. Calculate (a) the magnitude of the force $\vec{F}$ and (b) the normal force exerted by the window on the brush.

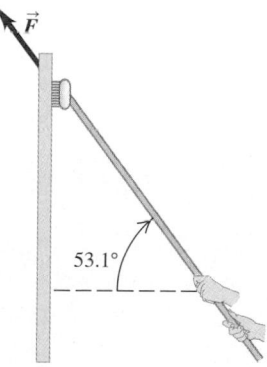

▲ **FIGURE 5.69** Problem 81.

82. • **A fractured tibia.** While a
BIO fractured tibia (the larger of the two major lower leg bones in mammals) is healing, it must be held horizontal and kept under some tension so that the bones will heal properly to prevent a permanent limp. One way to do this is to support the leg by using a variation of the Russell traction apparatus. (See Figure 5.70.) The lower leg (including the foot) of a particular patient weighs 51.5 N, all of which must be supported by the traction apparatus. (a) What must be the mass of W, shown in the figure? (b) What traction force does the apparatus provide along the direction of the leg?

83. • You push with a horizontal force of 50 N against a 20 N box, pressing it against a rough vertical wall to hold it in place. The coefficients of kinetic and static friction between this box and the wall are 0.20 and 0.50, respectively. (a) Make a free-body diagram of this box. (b) What is the friction force on the box? (c) How hard would you have to press for the box to slide downward with a uniform speed of 10.5 cm/s?

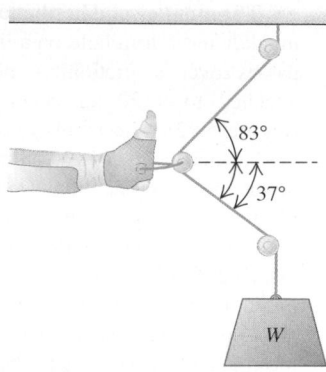

▲ **FIGURE 5.70** Problem 82.

84. •• Some sliding rocks approach the base of a hill with a speed of 12 m/s. The hill rises at 36° above the horizontal and has coefficients of kinetic and static friction of 0.45 and 0.65, respectively, with these rocks. Start each part of your solution to this problem with a free-body diagram. (a) Find the acceleration of the rocks as they slide up the hill. (b) Once a rock reaches its highest point, will it stay there or slide down the hill? If it stays there, show why. If it slides down, find its acceleration on the way down.

85. •• **Elevator design.** You are designing an elevator for a hospital. The force exerted on a passenger by the floor of the elevator is not to exceed 1.60 times the passenger's weight. The elevator accelerates upward with constant acceleration for a distance of 3.0 m and then starts to slow down. What is the maximum speed of the elevator?

86. •• At night while it is dark, a driver inadvertently parks his car on a drawbridge. Some time later, the bridge must be raised to allow a boat to pass through. The coefficients of friction between the bridge and the car's tires are $\mu_s = 0.750$ and $\mu_k = 0.550$. Start each part of your solution to this problem with a free-body diagram of the car. (a) At what angle will the car just start to slide? (b) If the bridge attendant sees the car suddenly start to slide and immediately turns off the bridge's motor, what will be the car's acceleration after it has begun to move?

87. •• A block is placed against the vertical front of a cart as shown in Figure 5.71. What acceleration must the cart have in order that block A does not fall? The coefficient of static friction between the block and the cart is μ_s.

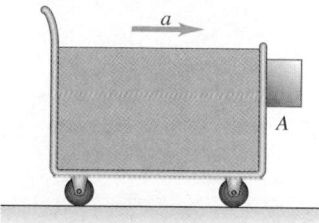

▲ **FIGURE 5.71** Problem 87.

88. •• **The monkey and her bananas.** A 20 kg monkey has a firm hold on a light rope that passes over a frictionless pulley and is attached to a 20 kg bunch of bananas (Figure 5.72). The monkey looks up, sees the bananas, and starts to climb the rope to get them. (a) As the monkey climbs, do the bananas move up, move down, or remain at rest? (b) As the monkey climbs, does the distance between the monkey and the bananas decrease, increase, or remain constant? (c) The monkey releases her hold on the rope. What happens to the distance between the monkey and the bananas while she is falling? (d) Before reaching the ground, the monkey grabs the rope to stop her fall. What do the bananas do?

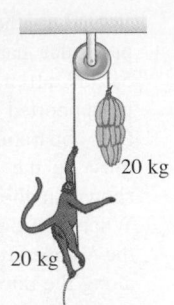

20 kg

20 kg

▲ **FIGURE 5.72**
Problem 88.

Passage Problems

BIO Back to the toboggan! Examples 5.6, 5.11, and 5.12 explored the motion of a toboggan filled with students flying down a hill. Our clever physics students have now designed a sail system that they attach to the toboggan. The sail uses the force of the wind, which blows up the hill at a constant speed to allow them to ride up the hill as well as down. The slope of the hill is the angle α; the combined weight of the students, the toboggan, and the sail is w.

89. If the toboggan is well waxed so that there is no friction, what force does the sail have to provide to move the student-filled toboggan up the hill at a constant velocity (once it gets started)?
    A. $g\sin\alpha$
    B. $w\sin\alpha$
    C. w
    D. $wg\sin\alpha$

90. If the wind were to change direction and blow downhill, what would be the acceleration of the toboggan?
    A. $2w\sin\alpha$
    B. $w\sin\alpha$
    C. $2g\sin\alpha$
    D. $g\sin\alpha$

91. The wax wears off so that the coefficient of kinetic friction is now μ_k. How much force does the sail now have to provide for a constant velocity ride back up the hill (once it gets started)?
    A. $\mu_k g\sin\alpha$
    B. $\mu_k w\sin\alpha$
    C. $\mu_k w\cos\alpha$
    D. $w\sin\alpha + \mu_k w\cos\alpha$
    E. $w\sin\alpha - \mu_k w\cos\alpha$

6 Circular Motion and Gravitation

In Chapter 5, we learned how to apply Newton's laws to a variety of mechanics problems. In this chapter, we'll continue our study of dynamics, concentrating especially on *circular motion*. Objects move in circular paths in a wide variety of situations, such as cars on racetracks, rotating machine parts, and the motions of planets and satellites, so it's worthwhile to study this special class of motions in detail. We also return to *gravitational* interactions for a detailed study of Newton's law of gravitation. This law plays a central role in planet and satellite motion, and it also provides a fuller understanding of the concept of weight than does the formula $w = mg$ by itself.

6.1 Force in Circular Motion

We explored the *kinematics* of circular motion in Section 3.4; we suggest that you review that section now to help you understand what is to come. Here's a brief summary of what we learned. When a particle moves along any curved path, the direction of its velocity must change. Thus, it *must* have a component of acceleration perpendicular to its path, even if its speed is constant.

In particular, when a particle moves in a circular path with constant speed, there is *no* component of acceleration parallel to the instantaneous velocity (tangent to the path); otherwise the speed would change. Motion with constant speed in a circular path (either a complete circle or a fraction of a circle) is called *uniform circular motion*. In this motion, the particle's acceleration at each instant is directed toward the center of the circle (perpendicular to the instantaneous velocity), as shown in Figure 6.1. As we found in Chapter 3, the magnitude a_{rad} of

Why do these skaters lean deeply into their turn? They do it so that the force of the ice on their skates has a large component directed inward toward the center of the turn. In this chapter, we will learn why turning motions require inward-directed forces.

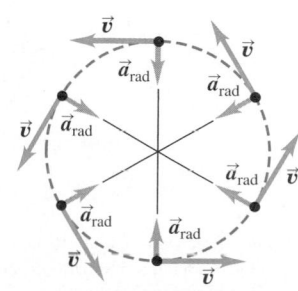

▲ **FIGURE 6.1** The kinematics of uniform circular motion.

the acceleration is constant and is given in terms of the speed v and the radius R of the circle by the following equation:

Acceleration in uniform circular motion

When an object moves at constant speed around a circle (or an arc of a circle) the magnitude of its acceleration (a_{rad}) is proportional to the square of its speed v and inversely proportional to the radius R of the circle:

$$a_{\text{rad}} = \frac{v^2}{R}. \tag{3.16}$$

MasteringPHYSICS®

ActivPhysics 4.1: Magnitude of Centripetal Acceleration
ActivPhysics 4.5: Car Circles a Track

The subscript "rad" is a reminder that at each point the direction of the acceleration is radially inward toward the center of the circle, perpendicular to the instantaneous velocity. We explained in Section 3.4 why this acceleration is sometimes called *centripetal acceleration*.

We can also express the centripetal acceleration a_{rad} in terms of the **period** T, the time for one revolution. The speed v is equal to the distance $(2\pi R)$ the particle travels in one revolution (i.e., the circumference of the circle), divided by the time T, so

$$v = \frac{2\pi R}{T}. \qquad \text{(relation of period to speed)} \tag{6.1}$$

Expressed in terms of the period,

$$a_{\text{rad}} = \frac{4\pi^2 R}{T^2}. \tag{6.2}$$

NOTE ▶ In earlier chapters, we've used T to denote tension in a rope or string. Here we use T to denote the period of a periodic motion, including uniform circular motion. From here on, we'll usually denote tension by F_T. ◄

A ball whirls on a string attached to a rotating peg in the center of a frictionless table (viewed from above). Suddenly, the string breaks, releasing the radial force on the ball.

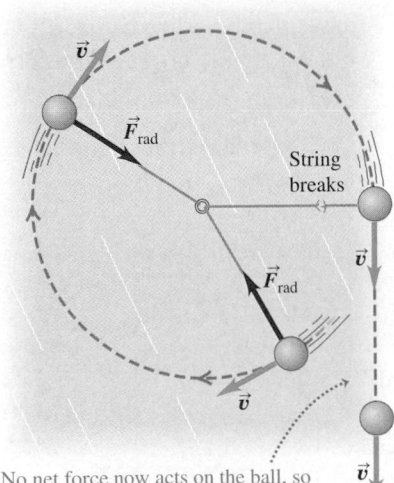

No net force now acts on the ball, so it obeys Newton's first law—it moves straight at constant velocity.

▲ **FIGURE 6.2** An object moves in a circle because of a centripetal net force.

Like all other motions of an object that can be modeled as a particle, circular motion is governed by Newton's second law. The object's acceleration toward the center of the circle must be caused by a force, or several forces, such that their vector sum $\Sigma\vec{F}$ is a vector that is always directed toward the center, with constant magnitude. The magnitude of the radial acceleration is given by $a_{\text{rad}} = v^2/R$, so the magnitude of the net inward radial force F_{net} on an object with mass m must be equal to the object's mass m, multiplied by this acceleration:

$$F_{\text{net}} = m\frac{v^2}{R}. \qquad \text{(relation of net force to acceleration)} \tag{6.3}$$

Here's a familiar example: We fasten a small object, such as a ball, to a string and whirl it in a horizontal circle (Figure 6.2). To keep the ball moving in a circle,

▶ **Application Slip slidin' away.** When a car rounds a curve on a level surface, the centripetal force that keeps it moving in a circle is provided by the friction force of the road on the tires. For a bobsledder racing on a frozen track, friction is minimal, and level turns are impossible. Therefore, the curves are steeply and precisely banked. The normal force of the track presses on the sled, providing the necessary horizontal centripetal force. This force allows sharp, nearly frictionless turns at speeds of over 120 km/hr and helped Jill Bakken and Vonetta Flowers win the gold medal at the 2002 Winter Olympics—the first USA bobsledding medal since 1956.

the string has to pull constantly toward the center. If the string breaks, then the inward force no longer acts, and the ball moves off in a direction tangent to the circle—not radially outward.

When several forces act on an object in uniform circular motion, their *vector sum* must be a vector with magnitude given by Equation 6.3 and direction always toward the center of the circle. The net force magnitude in Equation 6.3 is sometimes called *centripetal force*. This term tends to imply that such a force is somehow different from ordinary forces, and it really isn't. The term *centripetal* refers to the fact that, in uniform circular motion, the direction of the net force that causes the acceleration toward the center must also point toward the center. In the equation $\sum \vec{F} = m\vec{a}$, the sum of the forces must include only the real physical forces—pushes or pulls exerted by strings, rods, gravitation, and so on. The quantity $m(v^2/R)$ does *not* appear in $\sum \vec{F}$, but rather in the $m\vec{a}$ side of the equation.

Two cars on a curving road

A small car with mass m and a large car with mass $2m$ drive around a highway curve of radius R with the same speed v. As they travel around the curve, their accelerations are

A. equal.
B. along the direction of motion.
C. in the ratio of 2 to 1.
D. zero.

SOLUTION Equation 3.16 gives the acceleration of an object moving with uniform speed v in a circular path of radius R. The acceleration depends only on the speed and the radius. Both cars have the *same* speed and radius, so both must have the same acceleration. The correct answer is A. Because their masses are different, different forces are required for the two cars, even though their accelerations are the same. Answers B and D are certainly not right: The acceleration is not zero (otherwise the cars would move in a straight line), nor is it along the direction of motion (otherwise the speed would increase or decrease).

Skateboarder takes the hills

The following pictures show a skateboarder (seen from the side) moving along a straight or curved track.

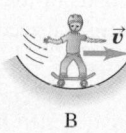
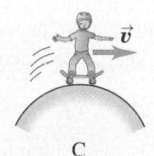

Which picture is consistent with the free-body diagram shown in Figure 6.3?

SOLUTION Force $\vec{F}$ is the net normal force of the track on the skateboard wheels. Force $\vec{w}$ has a larger magnitude than force $\vec{F}$. The net force is directed downward, so the acceleration must also

▲ **FIGURE 6.3**

be downward. In A, there is no vertical component of acceleration, so this choice cannot be correct. In B, the acceleration is upward, toward the center of the circular path. This would require a net upward force, not downward. The center of the circular path in C is downward from the ball. The acceleration is downward (toward the center), requiring a net downward force, as in Figure 6.3. The correct picture is C.

PROBLEM-SOLVING STRATEGY 6.1 **Circular motion**

The strategies for dynamics problems outlined at the beginning of Section 5.2 are equally applicable here, and we suggest that you reread them before studying the examples that follow.

NOTE ▶ A serious peril in circular-motion problems is the temptation to regard $m(v^2/R)$ as a *force*, as though the object's circular motion somehow generates an extra force in addition to the real physical forces exerted by strings, contact with other bodies, gravitation, or whatever. *Resist this temptation!* The quantity $m(v^2/R)$ is *not* a force; it corresponds to the $m\vec{a}$ side of $\sum \vec{F} = m\vec{a}$ and *must not* appear in $\sum \vec{F}$. It may help to draw the free-body diagram with colored pencils, using one color for the forces and a different color for the acceleration. Then, in the $\sum \vec{F} = m\vec{a}$ equations, write the force terms in the force color and the $m(v^2/R)$ term in the acceleration color. This may sound silly, but if it helps, do it! ◀

EXAMPLE 6.1 **Model airplane on a string**

Suppose you fly a propeller-driven model airplane on a 5.00 m string in a horizontal circle. The airplane, which has a mass of 0.500 kg, flies level and at constant speed and makes one revolution every 4.00 seconds. How hard must you pull on the string to keep the plane flying in a circle?

SOLUTION

SET UP Figure 6.4 shows our sketches. For the free-body diagram, we assume that the string has negligible mass, so that the force you exert on it equals the tension force it exerts on the plane (which keeps the plane moving in a circle). The plane flies horizontally, so its weight must be countered by an upward lift force from its wings. Thus, the net force is the tension from the string. (The plane also encounters drag from the air and thrust from its propeller, neither of which is shown in the plane of our free-body diagram. However, since the plane travels at constant speed, these forces also sum to zero.)

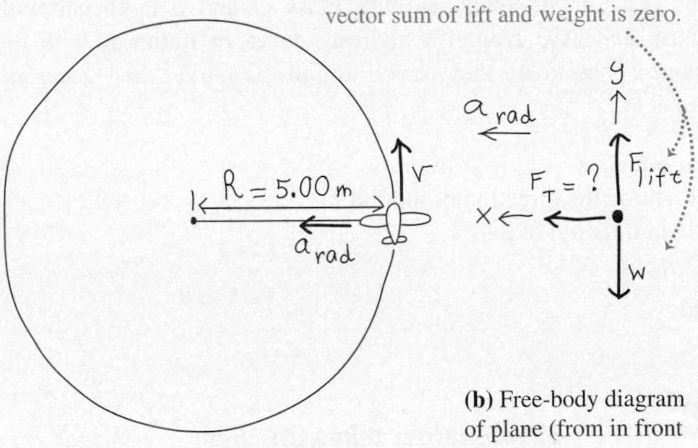

SOLVE We first find the airplane's centripetal acceleration and then use Newton's second law to find the tension force F_T. We begin by calculating the plane's speed v, using Equation 6.1 and the period T and radius R given:

$$v = \frac{2\pi R}{T} = \frac{2(3.14)(5.00\ \text{m})}{(4.00\ \text{s})}$$
$$= 7.85\ \text{m/s}, \quad \text{(speed of airplane in circular path)}$$

We now use this value with Equation 3.16 to calculate the centripetal acceleration:

$$a_{\text{rad}} = \frac{v^2}{R} = \frac{(7.85\ \text{m/s})^2}{(5.00\ \text{m})} = 12.3\ \text{m/s}^2. \quad \begin{array}{l}\text{(acceleration in} \\ \text{circular motion)}\end{array}$$

Alternatively, we could use Equation 6.2 to find a_{rad}:

$$a_{\text{rad}} = \frac{4\pi^2 R}{T^2} = \frac{4\pi^2(5.00\ \text{m})}{(4.00\ \text{s})^2} = 12.3\ \text{m/s}^2.$$

The plane flies horizontally, so the vector sum of lift and weight is zero.

(a) Sketch (from above)

(b) Free-body diagram of plane (from in front or behind)

▲ **FIGURE 6.4** Our diagrams for this problem.

Finally, we use Newton's second law to calculate the radial force necessary to produce this centripetal acceleration:

$$\sum F_x = F_T = ma_{\text{rad}} = (0.500\ \text{kg})(12.3\ \text{m/s}^2) = 6.15\ \text{N}.$$

This is the magnitude of force exerted by the string on the airplane; it is also the magnitude of force you have to exert on the string to keep the airplane flying in its circular path.

REFLECT The free-body diagram shows that the force on an object moving in a circle at constant speed is inward (toward the center of the circle), *not* outward.

Practice Problem: What if the string you are using is rated for no more than 50 N (approx 10 lb). How fast could you fly your model plane before the string breaks? *Answer:* 22.4 m/s.

EXAMPLE 6.2   At the end of the tether

A tetherball is attached to a swivel in the ceiling by a light cord of length L. When the ball is hit by a paddle, it swings in a horizontal circle with constant speed v, and the cord makes a constant angle β with the vertical direction. The ball goes through one revolution in time T. Assuming that T, the mass m, and the length L of rope are known, derive algebraic expressions for the tension F_T in the cord and the angle β.

SOLUTION

SET UP Figure 6.5 shows our diagrams. Two forces act on the ball: its weight $\vec{w}$ and the tension force $\vec{F}_T$ exerted by the cord. In our free-body diagram, we point the x axis toward the center of the circle (in the direction of the centripetal acceleration a_{rad}), and we replace the tension force by its x and y components: $F_x = F_T\sin\beta$ and $F_y = F_T\cos\beta$, respectively. The y components of force sum to zero, so the net force on the ball is the x component of tension, which acts toward the center of the ball's circular path.

SOLVE Because the period T is known, we express a_{rad} in terms of T and the radius R of the circle, using Equation 6.2: $a_{\text{rad}} = 4\pi^2 R/T^2$. (Note that the radius of the circular path is $L\sin\beta$, not L.)

To find F_T, we use Newton's second law for the x component of net force:

$$\Sigma F_x = ma_x = ma_{\text{rad}}, \qquad F_T\sin\beta = m\frac{4\pi^2 R}{T^2}.$$

Next, we substitute the expression for the radius R of the circle ($R = L\sin\beta$) into the preceding equation:

$$F_T\sin\beta = m\frac{4\pi^2(L\sin\beta)}{T^2}.$$

The factor $\sin\beta$ divides out, and we get

$$F_T = m\frac{4\pi^2 L}{T^2}.$$

We're halfway home; the foregoing equation gives the rope tension F_T in terms of known quantities. To obtain an expression for the angle β, we use Newton's second law for the y component of net force. (Remember that this component must be zero ($\Sigma F_y = 0$), because there is no y component of acceleration.) Newton's second law yields

$$\Sigma F_y = F_T\cos\beta + (-mg) = 0.$$

We substitute our expression for F_T into this equation to get

$$m\frac{4\pi^2 L}{T^2}\cos\beta + (-mg) = 0, \qquad \cos\beta = \frac{gT^2}{4\pi^2 L}.$$

So we've succeeded in expressing the tension F_T and the cosine of the angle β in terms of known quantities.

REFLECT For a given length L, $\cos\beta$ decreases and β increases as T becomes smaller and the ball makes more revolutions per second. The angle can never be 90°, however; that would require that $T = 0$, $F_T = \infty$, and $v = \infty$. The relationship between T and β doesn't contain the *mass* of the ball, but the force magnitude F_T is directly proportional to m. When β is very small, F_T is approximately equal to the ball's weight, $\cos\beta$ is approximately unity, and the expressions we found above give the approximate expression $T = 2\pi\sqrt{(L/g)}$.

Practice Problem: What is the speed v of the ball if $T = 2.0$ s and $L = 2.0$ m? *Answer:* 5.5 m/s.

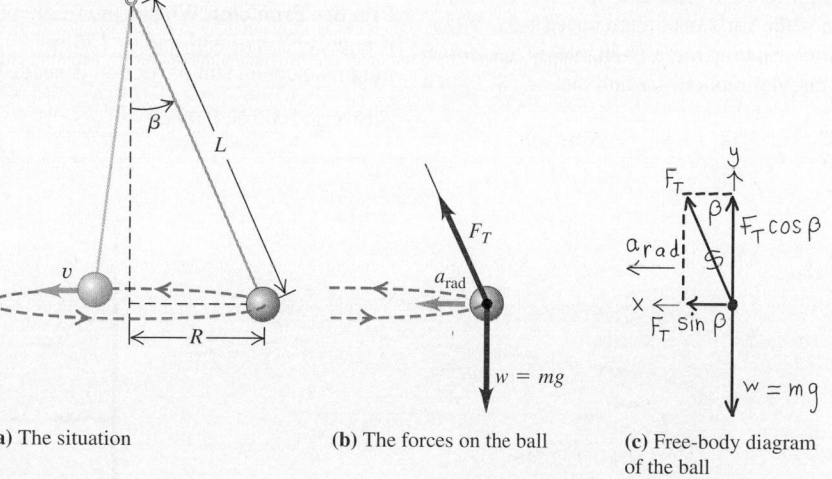

(a) The situation   (b) The forces on the ball   (c) Free-body diagram of the ball

▲ **FIGURE 6.5**

NOTE ▶ You may be tempted to add an extra outward force to the forces in Figure 6.5b, to "keep the object out there" at angle β or to "keep it in equilibrium." You may have heard the term *centrifugal force;* "centrifugal means" "fleeing from the center." Resist that temptation! First, the object *doesn't* stay "out there"; it is in constant motion around its circular path. Its velocity is constantly changing in direction—it accelerates—so it is *not* in equilibrium. Second, if there *were* an additional outward ("centrifugal") force to balance the inward force, there would then be *no* net inward force to cause the circular motion, and the object would move in a straight line, not a circle. In an inertial frame of reference, there is no such thing as centrifugal force, and we promise not to mention this term again. ◄

EXAMPLE 6.3 Rounding a flat curve

A car rounds a flat, unbanked curve with radius R. **(a)** If the coefficient of static friction between tires and road is μ_s, derive an expression for the maximum speed v_{max} at which the driver can take the curve without sliding. **(b)** What is the maximum speed if $R = 250$ m and $\mu_s = 0.90$?

SOLUTION

SET UP Figure 6.6 shows the situation, plus a free-body diagram for the car. As in the preceding example, we point the x axis toward the center. Note that the friction between a rolling wheel and the pavement is *static* friction, because the portion of the wheel in contact with the pavement doesn't slide relative to the pavement. (If the wheel slides or skids, however, the friction force is kinetic.) This friction force f_s is responsible for the car's centripetal acceleration $a_{rad} = v^2/R$; it points toward the center of the circle, in the x direction. In the y direction, the car's weight and the normal force on the car sum to zero.

SOLVE Part (a): The component equations from Newton's second law are

$$f_s = m\frac{v^2}{R}, \qquad n + (-mg) = 0.$$

In the first equation, f_s is the force *needed* to keep the car moving in its circular path at a given speed v; this force increases with v. But the maximum friction force *available* is $f_s = \mu_s n = \mu_s mg$; this limitation determines the car's maximum speed v_{max}. When we equate the maximum friction force available to the force mv^2/R required for the circular motion, we find that

$$\mu_s mg = m\frac{v_{max}^2}{R} \qquad \text{and} \qquad v_{max} = \sqrt{\mu_s gR}.$$

Part (b): If $\mu_s = 0.90$ and $R = 250$ m, then

$$v_{max} = \sqrt{(0.90)(9.8 \text{ m/s}^2)(250 \text{ m})} = 47 \text{ m/s},$$

or about 105 mi/h.

REFLECT If the car travels around the curve at less than v_{max}, f_s is less than $\mu_s n$, but it is still true that $f_s = mv^2/R$. If the car is moving at more than v_{max}, not enough friction is available to provide the needed centripetal acceleration. Then the car skids. Because the kinetic friction acting on skidding tires is less than the static friction acting before the car skids, a car always skids toward the outside of the curve.

It's possible to bank a curve so that at one certain speed no friction at all is needed. Then a car moving at just the right speed can round the curve even on wet ice with Teflon® tires. Airplanes *always* bank at this angle when making turns, and bobsled racing depends on the same idea. We'll explore the banking of curves in Example 6.4.

Practice Problem: What is the friction force on a 1200 kg car if it rounds a curve with radius 150 m at a speed of 25 m/s? What minimum coefficient of friction is needed?

Answers: 5000 N, 0.43.

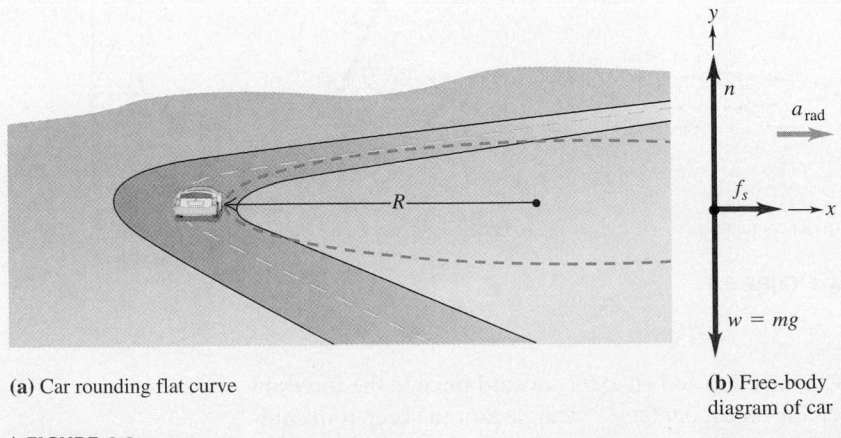

(a) Car rounding flat curve

(b) Free-body diagram of car

▲ **FIGURE 6.6**

Quantitative Analysis 6.3

Speeding up on the curves

A frictional force f_s provides the centripetal force as a car goes around an unbanked curve of radius R at speed v. If the driver speeds up on this curve until the speed of the car is tripled to $3v$, and if the car still follows the curved path, the frictional force must now be

A. $3f_s$. B. $\frac{1}{3}f_s$. C. $9f_s$. D. $\frac{1}{9}f_s$.

SOLUTION The centripetal acceleration, and therefore the frictional force needed, is proportional to v^2. When the speed increases by a factor of 3, the force needed increases by a factor of $3^2 = 9$. Thus, C is the correct answer. Note that both m and R are constant. If this isn't clear, you may want to review Example 6.3.

EXAMPLE 6.4   **Rounding a banked curve**

An engineer proposes to rebuild the curve in Example 6.3, banking it so that, at a certain speed v, no friction at all is needed for the car to make the curve. At what angle β should it be banked if $v = 25$ m/s (56 mi/h)?

SOLUTION

SET UP Figure 6.7 shows the free-body diagram for a car following a horizontal, circular path around the curve at the designated speed. We use the same coordinate system as in Example 6.3. The car's centripetal acceleration is now due to the horizontal component of the normal force exerted by the banked pavement. No friction force is needed.

SOLVE The horizontal component of $\vec{n}$ must now cause the acceleration v^2/R; there is no vertical component of acceleration. The component equations from Newton's second law are

$$n\sin\beta = \frac{mv^2}{R}, \qquad n\cos\beta + (-mg) = 0.$$

Dividing $n\sin\beta$ by $n\cos\beta$, we find that

$$\tan\beta = \frac{v^2}{gR}.$$

As in Example 6.3, $R = 250$ m. For the specified speed $v = 25$ m/s,

$$\beta = \tan^{-1}\frac{(25 \text{ m/s})^2}{(9.8 \text{ m/s}^2)(250 \text{ m})} = 14°$$

REFLECT The banking angle depends on the speed as well as the radius, but *not* on the mass of the vehicle. (Mass does not appear in the equation.) That's why it's possible to make banked curves that work for both cars and trucks, provided that the vehicles do not drive too fast. Highway designers assume friction will be available under good driving conditions and bank curves at angles less than β.

Practice Problem: In this example, what is the magnitude of the normal force on the 1200 kg car as it rounds the curve? *Answer:* 12,000 N.

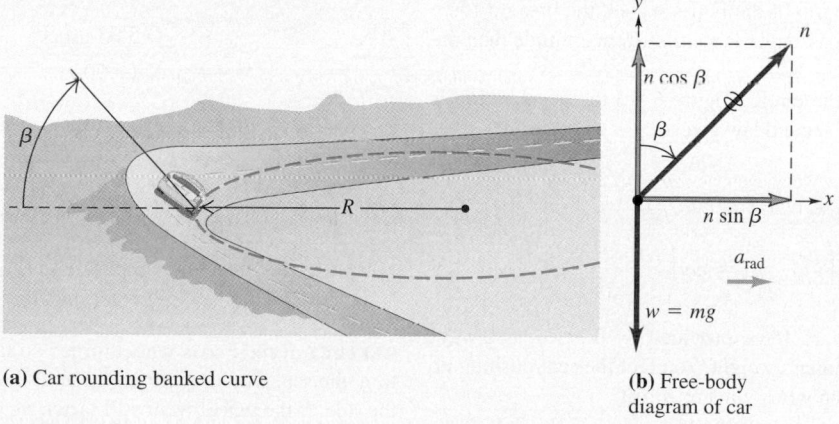

(a) Car rounding banked curve

(b) Free-body diagram of car

▲ **FIGURE 6.7**

6.2 Motion in a Vertical Circle

In all the examples in Section 6.1, the object moved in a horizontal circle. Motion in a *vertical* circle is no different in principle, but the weight of the object has to be treated carefully. The next example shows what we mean.

EXAMPLE 6.5 **Dynamics of a Ferris wheel ride**

A passenger on a Ferris wheel moves in a vertical circle of radius R with constant speed v. **(a)** Assuming that the seat remains upright during the motion, derive expressions for the magnitude of the upward force the seat exerts on the passenger at the top and bottom of the circle if the passenger's mass is m. **(b)** What are these forces if the passenger's mass is 60.0 kg, the radius of the circle is $R = 8.00$ m, and the wheel makes one revolution in 10.0 s? How do they compare with the passenger's actual weight?

SOLUTION

SET UP Figure 6.8 shows the Ferris wheel, with free-body diagrams for the passenger at the top and at the bottom. All the vector quantities (forces and accelerations) have only vertical components. We take the positive y direction as upward in both cases. We use n_T and n_B to denote the magnitude of force the seat applies to the passenger at the top and bottom of the wheel, respectively.

SOLVE Part (a): To derive expressions for the magnitudes n_T and n_B, we use Newton's second law for the y component of net force: $\sum F_y = ma_y$. Notice that the sign of a_y is negative at the top of the wheel, but positive at the bottom, although its magnitude is v^2/R at both places. Also, the y component of weight (magnitude mg) is negative, but the y components of both n_T and n_B are positive.

Thus, at the top of the Ferris wheel,

$$n_{T,y} + w_y = ma_y, \quad n_T + (-mg) = m\left(-\frac{v^2}{R}\right),$$

$$\text{or} \quad n_T = m\left(g - \frac{v^2}{R}\right) = mg\left(1 - \frac{v^2}{gR}\right).$$

This means that, at the top of the Ferris wheel, the upward force the seat applies to the passenger is *smaller* in magnitude than the passenger's weight.

At the bottom of the circle (Figure 6.8c) the acceleration is upward, and Newton's second law gives

$$n_B + (-mg) = +m\frac{v^2}{R}, \quad \text{or}$$

$$n_B = m\left(g + \frac{v^2}{R}\right) = mg\left(1 + \frac{v^2}{gR}\right).$$

At this point, the upward force provided by the seat is always *greater than* the passenger's weight. You feel the seat pushing up on you more firmly than when you are at rest.

Part (b): We want to calculate n_T and n_B for a 60.0 kg passenger riding a Ferris wheel with radius $R = 8.00$ m and a period of 10.0 s. The speed v is the circumference $2\pi R$, divided by the time T for one revolution:

$$v = \frac{2\pi R}{T} = \frac{2\pi(8.00 \text{ m})}{10.0 \text{ s}} = 5.03 \text{ m/s}.$$

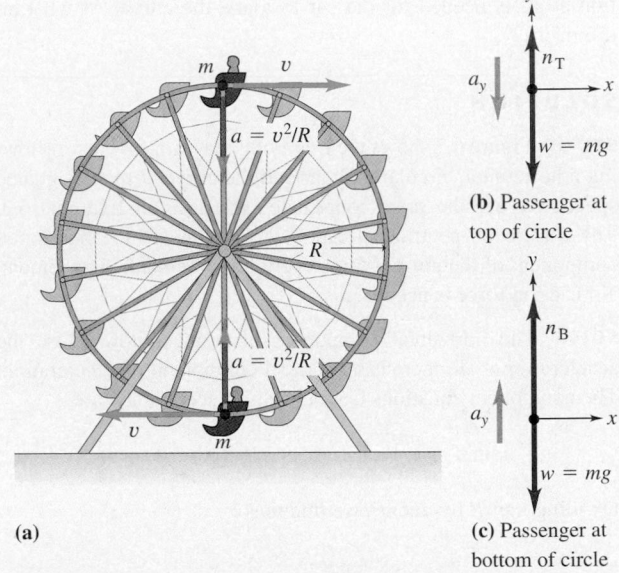

(b) Passenger at top of circle

(c) Passenger at bottom of circle

(a)

▲ **FIGURE 6.8** A passenger on a Ferris wheel.

The magnitude of the radial acceleration is

$$a_{rad} = \frac{v^2}{R} = \frac{(5.03 \text{ m/s})^2}{(8.00 \text{ m})} = 3.16 \text{ m/s}^2.$$

The two forces (the apparent weights at top and bottom) are

$$n_T = (60.0 \text{ kg})(9.80 \text{ m/s}^2 - 3.16 \text{ m/s}^2) = 398 \text{ N}$$
$$n_B = (60.0 \text{ kg})(9.80 \text{ m/s}^2 + 3.16 \text{ m/s}^2) = 778 \text{ N}.$$

The passenger's weight is 588 N, and n_T and n_B are, respectively, about 30% less than and greater than the weight.

REFLECT If the Ferris wheel turned so fast that v^2/R were equal to g, the seat would apply *no* force to the passenger at the top of the ride. If the wheel went still faster, n_T would become negative, so that a *downward* force would be needed to keep the passenger in the seat. We could supply a seat belt, or we could glue the passenger to the seat.

Some military aircraft are capable of an apparent gravitational acceleration of $9g$ when pulling out of a dive. (That is, $g + v^2/R = 9g$.) However, even wearing special flight suits, human pilots black out quickly at such accelerations.

NOTE ▶ When we tie a string to an object and whirl it in a vertical circle, the preceding analysis isn't directly applicable. The speed v isn't constant in this case, because, at every point except the top and bottom, there is a component of force (and therefore of acceleration) tangent to the circle. Even worse, we can't use the constant-acceleration formulas to relate the speeds at various points, since *neither* the magnitude nor the direction of the acceleration is constant. The speed relations we need are best obtained by using energy relations. We'll consider such problems in Chapter 7. ◀

Quantitative Analysis 6.4

A stone-age toy

A stone of mass m is attached to a strong string and whirled in a vertical circle of radius R. At the lowest point of the path, the tension in the string is six times the stone's weight. The stone's speed at this point is given by

A. $5\sqrt{gR}$ B. $\sqrt{gR}$ C. $\sqrt{5gR}$

SOLUTION At the lowest point of the circle, the forces acting on the stone are its weight (downward, with magnitude mg) and the string tension force (upward, with magnitude $6mg$). The net upward force has magnitude $5mg$ and must equal the mass m times the magnitude of the centripetal acceleration v^2/R. Thus, $5mg = mv^2/R$, and it follows that $v = \sqrt{5gR}$. The correct answer is C.

Conceptual Analysis 6.5

A very slippery ramp

A box slides down a circular, frictionless ramp, as shown in Figure 6.9. The forces on the box at the instant it reaches point A, and their relative magnitudes, are

A. the upward push of the ramp, which is greater than the inertial force ma of the box.
B. the upward push of the ramp, which equals the downward pull of the earth.
C. the upward push of the ramp, which is greater than the downward pull of the earth.

SOLUTION The quantity ma is not a force; there is no such thing as an inertial force. Answer A is wrong both quantitatively and

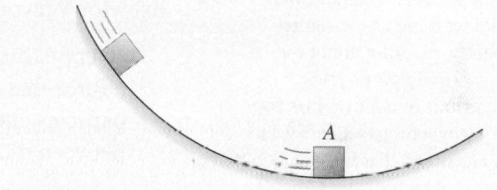

▲ **FIGURE 6.9**

conceptually. If B were correct, the net force on the box would be zero and it would have to move in a straight line, not a circle. For circular motion at point A, there must be a net upward force; the correct answer is C.

Walking and Circular Motion

The swinging motion of your legs when you walk can be modeled approximately as circular motion, and this model can be used to find the maximum speed a person of a given leg length can walk. Humans keep the weight-bearing leg straight while walking. Consequently, as shown in Figure 6.10, each hip in turn moves in a circular arc with a radius R equal to the length of the leg and with the weight-bearing foot at the center of the circle. In the simplest model, we imagine the person's entire mass M to be concentrated near the hip. In this model, when a person walks with speed v, the force required to keep the hips in circular motion is

$$F_{rad} = \frac{Mv^2}{R}.$$

Gravity provides this force, so the maximum available force is the person's weight Mg. Thus, in this simple model, the maximum speed v_{max} a person can walk is given by

$$Mg = \frac{M(v_{max})^2}{R}, \quad \text{or} \quad v_{max} = \sqrt{gR}.$$

The average adult leg length is about 1 m long; assuming this length, we get

$$v_{max} = \sqrt{(9.8\,\text{m/s}^2)(1.0\,\text{m})} = 3.1 \text{ m/s, or } 7.0 \text{ mi/h}.$$

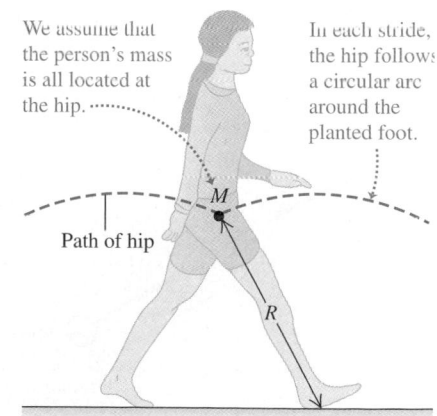

We assume that the person's mass is all located at the hip.

In each stride, the hip follows a circular arc around the planted foot.

Path of hip

M

R

▲ **FIGURE 6.10** A simple model of human walking.

This number is quite realistic. Adults can walk at speeds up to about 2.7 m/s, although usual walking speeds are below 2.0 m/s. Race walkers use a special motion that flattens the arc followed by the hip and allows them to walk at up to 4.0 m/s.

6.3 Newton's Law of Gravitation

About 300 years ago, Newton discovered that gravitation—the force that makes an apple fall out of a tree—also keeps the planets in their nearly circular orbits around the sun. As we remarked in Section 5.5, gravitation is one of the four classes of forces found in nature, and it was the earliest one of the four to be studied extensively.

Our most familiar encounter with the gravitational force is the *weight* of an object—the force that attracts it toward the earth (or toward whatever astronomical body it is near, such as a moon, planet, or star). Newton discovered the **law of gravitation** during his study of the motions of planets around the sun, and he published it in 1686.

Newton's law of gravitation

Newton's law of gravitation may be stated as follows:

Every particle of matter in the universe attracts every other particle with a force that is directly proportional to the product of the masses of the particles and inversely proportional to the square of the distance between them.

For two particles with masses m_1 and m_2, separated by a distance r (Figure 6.11), Newton's law of gravitation can be expressed as

$$F_g = G\frac{m_1 m_2}{r^2}, \qquad (6.4)$$

where F_g is the magnitude of the gravitational force on either particle and G is a fundamental physical constant called the **gravitational constant.** Gravitational forces are always attractive; gravitation always pulls objects toward each other.

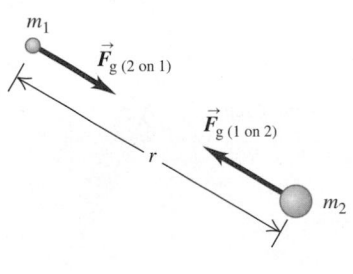

$$F_{g\,(1\text{ on }2)} = F_{g\,(2\text{ on }1)}$$

▲ **FIGURE 6.11** Two particles exert attractive gravitational forces of equal magnitude on each other, even if their masses are quite different.

Gravitational forces between two particles always obey Newton's third law; they always act along the line joining the two particles, and they form an action–reaction pair. Even when the masses of the particles are different, the two interaction forces have equal magnitude. The attractive force that your body exerts on the earth has the same magnitude as the force it exerts on you.

We've stated the law of gravitation in terms of the interaction forces between two *particles*. It turns out that the interaction between any two objects having *spherically symmetric* mass distributions (such as solid spheres or spherical shells) is the same as though all the mass of each were concentrated at its center, as in Figure 6.12. In effect, we can replace each spherically symmetric object with a particle at its center. Thus, if we model the earth as a solid sphere with mass m_E, then the force exerted by it on a particle or a spherically symmetric object with mass m at a distance r between the two centers is

$$F_g = G\frac{m m_E}{r^2}, \qquad (6.5)$$

provided that the object lies outside the earth. A force of the same magnitude is exerted *on* the earth by the object.

NOTE ▶ Notice that r is the distance between the *centers* of the two objects, not between their surfaces. ◀

At points *inside* the earth, the situation is different. If we could drill a hole to the center of the earth and measure the gravitational force on a particle at various depths, we would find that the force *decreases* as we approach the center, rather than increasing as $1/r^2$. As the object entered the interior of the earth (or another spherical object), some of the earth's mass would be on the side of the object opposite from the center and would pull in the opposite direction. Exactly at the center, the gravitational force on the object would be zero.

To determine the value of the gravitational constant G, we have to *measure* the gravitational force between two objects with known masses m_1 and m_2 at a known distance r. For bodies of reasonable size, the force is extremely small, but it can be measured with an instrument called a *torsion balance,* used by Sir Henry Cavendish in 1798 to determine G.

The Cavendish balance consists of a light, rigid rod shaped like an inverted T (Figure 6.13) and supported by a very thin vertical quartz fiber. Two small spheres, each of mass m_1, are mounted at the ends of the horizontal arms of the T. When we bring two large spheres, each of mass m_2, to the positions shown, the attractive gravitational forces twist the T through a small angle. To measure this angle, we shine a beam of light on a mirror fastened to the T; the reflected beam strikes a scale, and as the T twists, the beam moves along the scale.

After calibrating the instrument, we can measure gravitational forces and thus determine G. The currently accepted value (in SI units) is

$$G = 6.674 \times 10^{-11}\ \text{N} \cdot \text{m}^2/\text{kg}^2.$$

Because $1\ \text{N} = 1\ \text{kg} \cdot \text{m}/\text{s}^2$, the units of G can also be expressed (in fundamental SI units) as $\text{m}^3/(\text{kg} \cdot \text{s}^2)$.

Gravitational forces combine vectorially. If each of two masses exerts a force on a third, the *total* force on the third mass is the vector sum of the individual forces due to the first two. That is, gravitational forces obey the principle of superposition, introduced in Section 4.1.

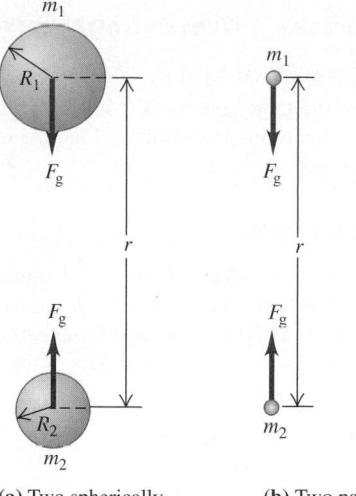

(a) Two spherically symmetric masses m_1 and m_2

(b) Two particles with masses m_1 and m_2

▲ **FIGURE 6.12** Spherically symmetric objects interact gravitationally as though all the mass of each were concentrated at its center.

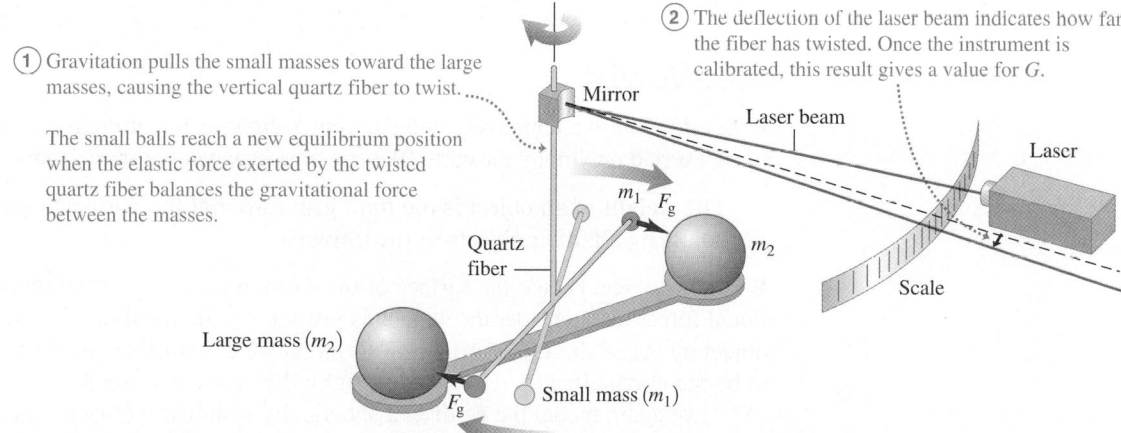

① Gravitation pulls the small masses toward the large masses, causing the vertical quartz fiber to twist.

The small balls reach a new equilibrium position when the elastic force exerted by the twisted quartz fiber balances the gravitational force between the masses.

② The deflection of the laser beam indicates how far the fiber has twisted. Once the instrument is calibrated, this result gives a value for G.

Mirror

Laser beam

Laser

m_1 F_g

m_2

Quartz fiber

Scale

Large mass (m_2)

F_g Small mass (m_1)

▲ **FIGURE 6.13** The principle of the Cavendish balance. The angle of deflection is exaggerated for clarity.

EXAMPLE 6.6 **Gravitational forces in a Cavendish balance**

The mass m_1 of one of the small spheres of a Cavendish balance is 10.0 g ($=0.0100$ kg), the mass m_2 of one of the large spheres is 0.500 kg, and the center-to-center distance between each large sphere and the nearer small one is 0.0500 m. Find the magnitude of the gravitational force F_g on each sphere due to the nearest other sphere.

SOLUTION

SET UP AND SOLVE Figure 6.14 shows our sketch. Assuming that the spheres have spherically symmetric mass distributions, we can apply Newton's law of gravitation, taking the distance r to be the distance between centers of the spheres. The magnitude of each force is

$$F_g = G\frac{m_1 m_2}{r^2}$$

$$= (6.67 \times 10^{-11} \text{ N} \cdot \text{m}^2/\text{kg}^2)\frac{(0.0100 \text{ kg})(0.500 \text{ kg})}{(0.0500 \text{ m})^2}$$

$$= 1.33 \times 10^{-10} \text{ N}.$$

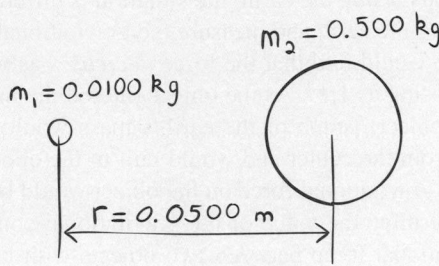

▲ FIGURE 6.14 Our sketch for this problem.

REFLECT Gravitational forces between objects of familiar sizes are extremely small. Also, gravitational forces obey Newton's third law: The force that mass m_1 exerts on mass m_2 is equal in magnitude to the force that m_2 exerts on m_1, even though their masses are very different.

Practice Problem: What is the force magnitude F_g if the separation between centers of the spheres is doubled to 0.100 m? *Answer:* One-fourth as much: 3.34×10^{-11} N.

EXAMPLE 6.7 **Accelerations of Cavendish spheres**

Suppose one large and one small sphere in Example 6.6 are detached from the apparatus and placed 0.0500 m (between centers) from each other at a point in space far removed from all other objects. What is the magnitude of the acceleration of each sphere relative to an inertial system?

SOLUTION

SET UP AND SOLVE The force on each sphere has the same magnitude that we found in Example 6.6. From Newton's second law, the acceleration magnitude a_1 of the smaller sphere is

$$a_1 = \frac{F_g}{m_1} = \frac{1.33 \times 10^{-10} \text{ N}}{0.0100 \text{ kg}} = 1.33 \times 10^{-8} \text{ m/s}^2.$$

The acceleration magnitude a_2 of the larger sphere is

$$a_2 = \frac{F_g}{m_2} = \frac{1.33 \times 10^{-10} \text{ N}}{0.500 \text{ kg}} = 2.66 \times 10^{-10} \text{ m/s}^2.$$

REFLECT Although the forces on the objects are equal in magnitude, the two accelerations are *not* equal in magnitude. Also, they are not constant, because the gravitational forces increase in magnitude as the spheres start to move toward each other.

6.4 Weight

In Section 4.4, we tentatively defined your *weight* as the attractive gravitational force exerted on you by the earth. We can now broaden our definition:

> The weight of an object is the *total* gravitational force exerted on the object by all other objects in the universe.

When the object is near the surface of the earth, we can neglect all other gravitational forces and consider the weight as just the gravitational force exerted on the object by the earth. At the surface of the *moon* we can consider an object's weight to be the gravitational force exerted on it by the moon, and so on.

If we again model the earth as a spherically symmetric object with radius R_E and mass m_E, then the weight w of a small object of mass m at the earth's surface (a distance R_E from its center) is

$$w = F_g = \frac{Gmm_E}{R_E^2}. \tag{6.6}$$

But we also know from Section 4.4 that the weight w of an object is its mass m times the acceleration g of free fall at the location of the object: $w = mg$; weight is the force that causes the acceleration of free fall. Thus, Equation 6.6 must be equal to mg:

$$mg = \frac{Gmm_E}{R_E^2}.$$

Dividing by m, we find

$$g = \frac{Gm_E}{R_E^2}. \tag{6.7}$$

Because this result does not contain m, the acceleration due to gravity g is independent of the mass m of the object. We already knew that, but we can now see how it follows from the law of gravitation.

Furthermore, we can *measure* all the quantities in Equation 6.7 except for m_E, so this relation allows us to compute the mass of the earth. We first solve Equation 6.7 for m_E, obtaining

$$m_E = \frac{gR_E^2}{G}.$$

Taking $R_E = 6380$ km $= 6.38 \times 10^6$ m and $g = 9.80$ m/s^2, we find that

$$m_E = 5.98 \times 10^{24} \text{ kg}.$$

Once Cavendish had measured G, he could compute the mass of the earth! He described his measurement of G with the grandiose phrase "weighing the earth." That's not really what he did; he certainly didn't hang our planet from a spring balance. But after he had determined G, he carried out the preceding calculation and determined the mass (not the weight) of the earth.

The weight of an object decreases inversely with the square of its distance from the earth's center. At a radial distance $r = 2R_E$ from earth's center, the weight is one-quarter of its value at the earth's surface. Figure 6.15 shows how weight varies with height above the earth for an astronaut who weighs 700 N on earth.

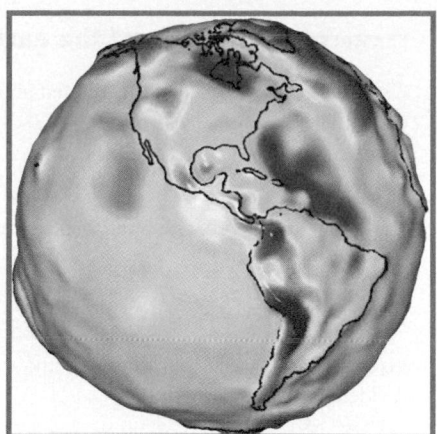

▲ Application **Falling with style.** This is a gravity map of the earth. Protrusions represent areas of stronger gravity (because of nonuniform mass distribution), and dips show areas of weaker gravity. For example, the Andes represent an area of strong gravity because of the extra mass of the mountain range. The map was made by the Gravity Recovery and Climate Experiment (GRACE) project. To measure gravity, GRACE employs a pair of satellites that orbit the earth in formation while precisely monitoring the distance between them. For instance, as the pair approaches a mountain range, the lead satellite will speed up in response to the mountains' gravitational pull, increasing the distance between the two satellites. As the satellites straddle the mountains, the pull of the extra mass then shrinks the distance between them, and as they pass the mountains, the distance increases again.

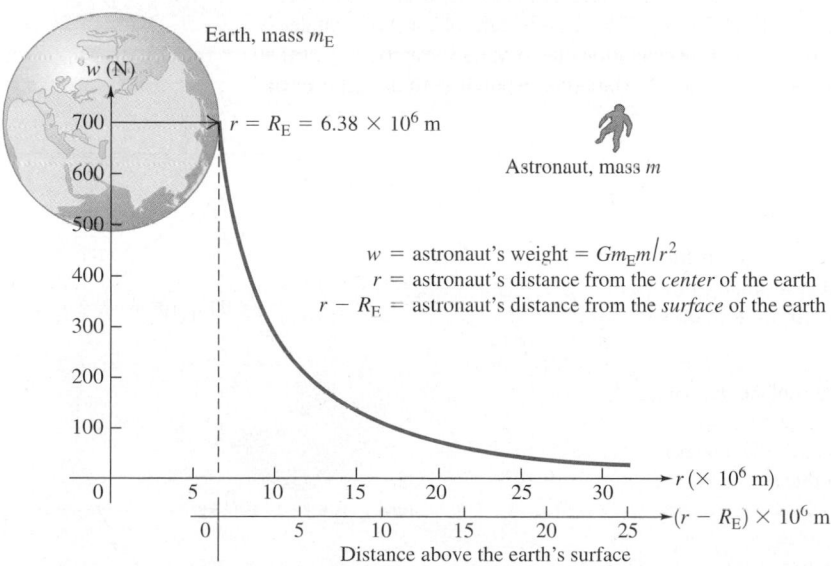

▲ **FIGURE 6.15** A graph of weight (gravitational interaction) as a function of distance from the earth's surface.

EXAMPLE 6.8 **Density of the earth**

Using the values of m_E and R_E quoted above, determine the average *density* (mass per unit volume) of the earth.

SOLUTION

SET UP The volume, assuming a spherical earth is

$$V = \frac{4}{3}\pi R_E^3 = \frac{4}{3}\pi (6.38 \times 10^6 \text{ m})^3 = 1.09 \times 10^{21} \text{ m}^3.$$

SOLVE The average density ρ of the earth is the total mass divided by the total volume:

$$\rho = \frac{M}{V} = \frac{5.98 \times 10^{24} \text{ kg}}{1.09 \times 10^{21} \text{ m}^3} = 5500 \text{ kg/m}^3 = 5.50 \text{ g/cm}^3.$$

REFLECT For comparison, the density of water is $1.00 \text{ g/cm}^3 = 1000 \text{ kg/m}^3$. The density of igneous rock near the earth's surface, such as basalt, is about $3 \text{ g/cm}^3 = 3000 \text{ kg/m}^3$. So the interior of the earth must be much more dense than the surface. In fact, the maximum density at the center, where the pressure is enormous, is thought to be nearly 15 g/cm^3. Figure 6.16 is a graph of density as a function of distance from the earth's center.

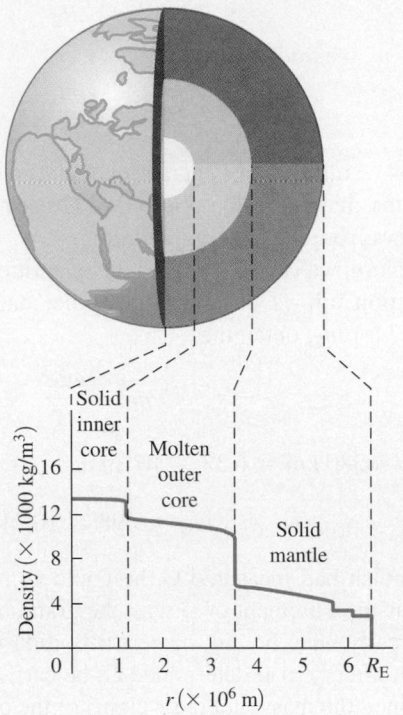

▶ **FIGURE 6.16** The density of the earth decreases with increasing distance from its center. The "steps" in the graph represent major changes in composition or in phase (solid versus liquid).

EXAMPLE 6.9 **Gravity on Mars**

You're involved in the design of a mission carrying humans to the surface of the planet Mars, which has a radius $R_{Mars} = 3.38 \times 10^6$ m and a mass $m_{Mars} = 6.42 \times 10^{23}$ kg. The earth weight of the Mars lander is 39,200 N. Calculate the weight F_g of the Mars lander and the acceleration due to Mars's gravity, g_{Mars}, **(a)** at the surface of Mars, and **(b)** at 6.00×10^6 m above the surface of Mars (corresponding to the orbit of the Martian moon Phobos).

SOLUTION

SET UP Figure 6.17 diagrams the problem. Notice that the altitude given is 6.00×10^6 m above the *surface* of Mars. To find the gravitational forces, we need to translate that altitude to the distance from the *center* of Mars: $r = 6.00 \times 10^6 \text{ m} + 3.38 \times 10^6 \text{ m} = 9.38 \times 10^6 \text{ m}$.

SOLVE Part (a): In Equations 6.6 and 6.7 we replace m_E with m_{Mars} and R_E with R_{Mars}. The value of G is the same everywhere; it is a fundamental physical constant. The mass m of the lander is its *earth* weight w, divided by the acceleration due to gravity, g, on the earth:

$$m = \frac{w}{g} = \frac{3.92 \times 10^4 \text{ N}}{9.80 \text{ m/s}^2} = 4.00 \times 10^3 \text{ kg}.$$

The mass is the same whether the lander is on the earth, on Mars, or in between. From Equation 6.6,

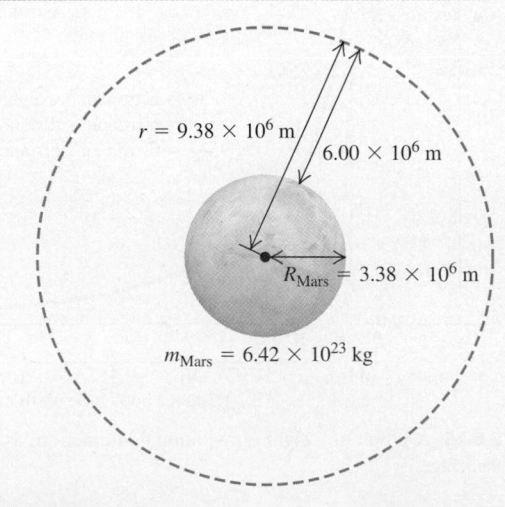

▲ **FIGURE 6.17**

Continued

$$F_g = \frac{Gmm_{Mars}}{r^2}$$
$$= \frac{(6.67 \times 10^{-11}\,\text{N} \cdot \text{m}^2/\text{kg}^2)(4.00 \times 10^3\,\text{kg})(6.42 \times 10^{23}\,\text{kg})}{(3.38 \times 10^6\,\text{m})^2}$$
$$= 1.50 \times 10^4\,\text{N}.$$

The acceleration due to the gravity of Mars at its surface is

$$g_{Mars} = \frac{F_g}{m} = \frac{1.50 \times 10^4\,\text{N}}{4.00 \times 10^3\,\text{kg}} = 3.75\,\text{m/s}^2.$$

We can also obtain g_{Mars} directly from Equation 6.7:

$$g_{Mars} = \frac{Gm_{Mars}}{R_{Mars}{}^2}$$
$$= \frac{(6.67 \times 10^{-11}\,\text{N} \cdot \text{m}^2/\text{kg}^2)(6.42 \times 10^{23}\,\text{kg})}{(3.38 \times 10^6\,\text{m})^2}$$
$$= 3.75\,\text{m/s}^2.$$

Part (b): To find F_g and g at a point 6.0×10^6 m above the surface of Mars, we repeat the preceding calculations, replacing $R_{Mars} = 3.38 \times 10^6$ m with $r = 9.38 \times 10^6$ m. At that altitude, both F_g and g are smaller than they are at the surface, by a factor of

$$\left(\frac{3.38 \times 10^6\,\text{m}}{9.38 \times 10^6\,\text{m}}\right)^2 = 0.130,$$

and we find that $F_g = 1.95 \times 10^3$ N and $g = 0.487$ m/s^2. The latter is also the acceleration of Phobos in its orbit due to Mars's gravity, even though Phobos has far more mass than our lander.

REFLECT On the surface of Mars, F_g and g are roughly 40% as large as they are on the surface of the earth. At a point in the orbit of Phobos, they are 13% as large as they are on the surface of Mars.

6.5 Satellite Motion

Satellites orbiting the earth are a familiar fact of contemporary life. We know that they differ only in scale, not in principle, from the motion of our moon around the earth or the motion of the moons of Jupiter. But we still have to deal with questions such as "What holds that thing up there anyway?" So let's see how we can use Newton's laws and the law of gravitation for a detailed analysis of satellite motion.

To begin, think back to the discussion of projectile motion in Section 3.3. In Example 3.3, we fired a paintball horizontally from some distance above the ground; it followed a parabolic path until it struck the ground. In a similar situation, a motorcycle rider rides horizontally off the edge of a cliff, launching himself into a parabolic path that ends on the flat ground at the base of the cliff (Figure 6.18).

If he survives and repeats the experiment with an increased launch speed, he will land farther from the starting point. We can imagine him launching himself with great enough speed that the earth's curvature becomes significant. As he falls, the earth curves away beneath him. If he is going fast enough, and if his launch point is high enough that he clears the mountaintops, he may be able to go right on around the earth without ever landing.

Figure 6.19 shows a variation on this theme. We launch a projectile from point A in a direction tangent to the earth's surface (i.e., horizontally). Trajectories

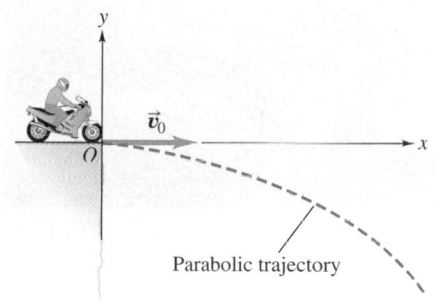

▲ **FIGURE 6.18** The trajectory of a biker launching himself horizontally from a cliff.

PhET: My Solar System
ActivPhysics 4.6: Satellites Orbit

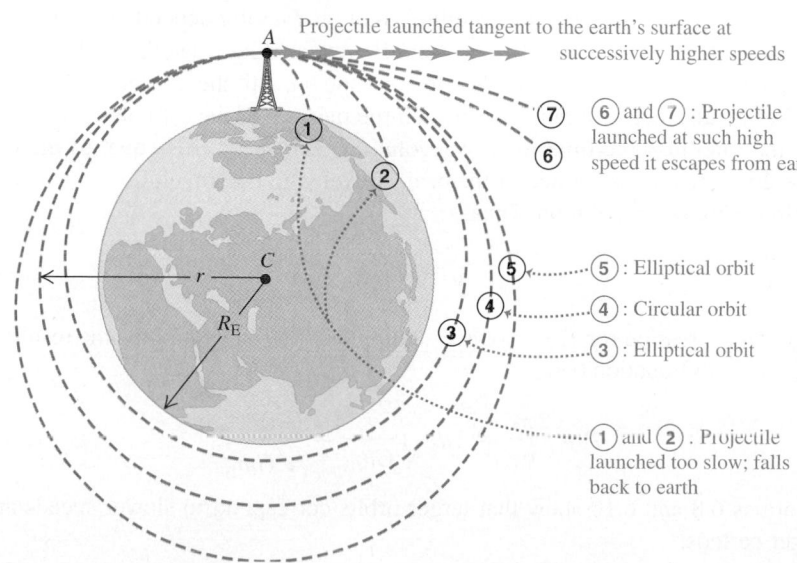

Projectile launched tangent to the earth's surface at successively higher speeds

⑥ and ⑦ : Projectile launched at such high speed it escapes from earth

⑤ : Elliptical orbit

④ : Circular orbit

③ : Elliptical orbit

① and ② : Projectile launched too slow; falls back to earth

◀ **FIGURE 6.19** An orbit is not fundamentally different from familiar trajectories on earth. If you launch a projectile slowly, it falls back to earth. If you launch it fast enough, the earth curves away from it as it falls, and it goes into orbit.

1 through 7 show the effect of increasing the initial speed. Trajectory 3 just misses the earth, and the projectile has become an earth satellite. If there is no retarding force (such as air resistance), its speed when it returns to point A is the same as its initial speed, and it repeats its motion indefinitely.

Trajectories 1 through 5 are ellipses or segments of ellipses; trajectory 4 is a circle (a special case of an ellipse). Trajectories 6 and 7 are not closed orbits; in these trajectories, the projectile never returns to its starting point, but travels farther and farther away from the earth.

A circular orbit is the simplest case and is the only one we'll analyze in detail. Artificial satellites often have nearly circular orbits, and the orbits of most of the planets around the sun are also nearly circular. As we learned in Section 3.4, a particle in uniform circular motion with speed v and radius r has an acceleration with magnitude $a_{rad} = v^2/r$, always directed toward the center of the circle. For a satellite with mass m, the *force* that provides this acceleration is the gravitational attraction of the earth (mass m_E), as shown in Figure 6.20. If the radius of the circular orbit, measured from the *center* of the earth, is r, then the gravitational force is given by the law of gravitation: $F_g = Gmm_E/r^2$.

The principle governing the motion of the satellite is Newton's second law; the force is F_g, and the acceleration is v^2/r, so the equation $\sum \vec{F} = m\vec{a}$ becomes

$$\frac{Gmm_E}{r^2} = \frac{mv^2}{r}.$$

Solving this equation for v, we find that

$$v = \sqrt{\frac{Gm_E}{r}}. \tag{6.8}$$

Equation 6.8 shows that we can't choose the radius r of the orbit and the speed v of the satellite independently; if we choose a value of r, then v is determined. Equation 6.8 also shows that the satellite's motion *does not* depend on its mass m, because m doesn't appear in the equation. If we could cut a satellite in half without changing its speed, each half would continue on with the original motion.

We can also derive a relation between the radius r of the orbit and the period T of the satellite—the time for one revolution. The speed v is equal to the distance $2\pi r$ (the circumference of the orbit) traveled in one revolution, divided by the time T for one revolution. Thus,

$$v = \frac{2\pi r}{T}. \tag{6.9}$$

To get an expression for T, we solve Equation 6.9 for T and combine the resulting equation with Equation 6.8:

$$T = \frac{2\pi r}{v} = 2\pi r \sqrt{\frac{r}{Gm_E}} = \frac{2\pi r^{3/2}}{\sqrt{Gm_E}}. \tag{6.10}$$

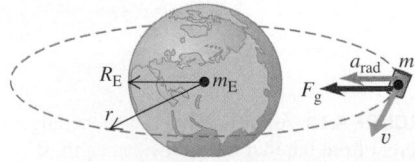

▲ **FIGURE 6.20** The force F_g due to the earth's gravitational attraction provides the centripetal force that keeps a satellite in orbit.

Equations 6.8 and 6.10 show that larger orbits correspond to slower speeds and longer periods.

Conceptual Analysis 6.6

Orbital speed of a satellite

A satellite in a circular orbit travels at constant speed because

A. the net force acting on it is zero.
B. the pull of gravity is balanced by the centrifugal force.
C. there is no component of force along its direction of motion.

SOLUTION The net force is not zero; if it were, the satellite would travel in a straight line, not a circle. Thus, A can't be right. There is no such thing as a balancing centrifugal force; if the pull of gravity were balanced by some mythical outward force, the satellite would move in a mythical straight line! So B is wrong, too. The net force on the satellite is the pull of the earth's gravity on the satellite. This pull is always directed toward the center of the orbit, and it provides the centripetal acceleration to maintain the satellite in circular motion. The net force is perpendicular to the satellite's velocity and therefore acts only to change the direction of the velocity. It has no component along the direction of the satellite's motion; if it, did, the speed of the satellite would change. The correct answer is C.

Conceptual Analysis 6.7

Why doesn't the moon fall?

The moon is pulled directly toward the earth by the earth's gravity, yet it does not fall to the earth because

A. the net force on it is zero.
B. the pull of earth's gravity is balanced by the centrifugal force.
C. the moon pulls back on the earth with an equal and opposite force.
D. . . . but it *is* constantly falling!

SOLUTION The correct answer is D. Recall our earlier discussion about the daredevil motorcyclist going off the cliff. If he goes fast enough, the earth curves away beneath him. At one very high speed, he falls at just the right rate to stay on a circular path. This same argument applies to the moon. Without gravity, it would shoot away from earth in a straight line. The moon does fall, but just enough to keep it on a circular path around the earth. Concerning answer A, the net force is certainly *not* zero. With respect to answer B, a balancing centrifugal force is a mythical "beast," like the unicorn and the free lunch. Answer C tries to bring in an additional force that doesn't act *on* the moon. It is a misuse of Newton's second law!

EXAMPLE 6.10 A weather satellite

Suppose we want to place a weather satellite into a circular orbit 300 km above the earth's surface. What speed, period, and radial acceleration must it have? The earth's radius is 6380 km = 6.38×10^6 m, and its mass is 5.98×10^{24} kg. (See Appendix F.)

SOLUTION

SET UP AND SOLVE Figure 6.21 shows our sketch. The radius of the satellite's orbit is

$$r = 6380 \text{ km} + 300 \text{ km} = 6680 \text{ km} = 6.68 \times 10^6 \text{ m}.$$

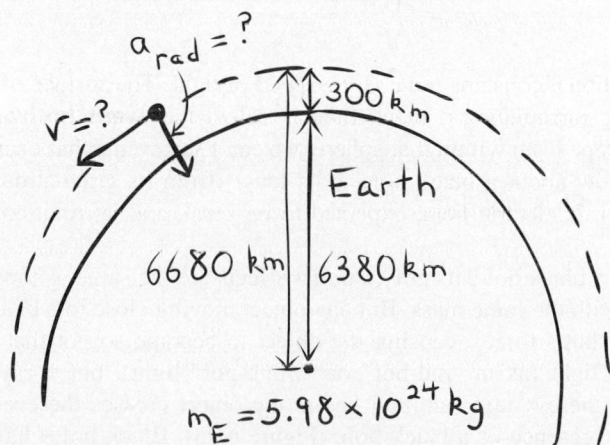

▲ **FIGURE 6.21** Our sketch for this problem.

From Equation 6.8,

$$\frac{Gmm_E}{r^2} = \frac{mv^2}{r},$$

$$v = \sqrt{\frac{Gm_E}{r}} = \sqrt{\frac{(6.67 \times 10^{-11} \text{ N} \cdot \text{m}^2/\text{kg}^2)(5.98 \times 10^{24} \text{ kg})}{6.68 \times 10^6 \text{ m}}}$$
$$= 7730 \text{ m/s}.$$

From Equation 6.10,

$$T = \frac{2\pi r}{v} = \frac{2\pi (6.68 \times 10^6 \text{ m})}{7730 \text{ m/s}}$$
$$= 5430 \text{ s} = 90.5 \text{ min}.$$

The radial acceleration is

$$a_{rad} = \frac{v^2}{r} = \frac{(7730 \text{ m/s})^2}{6.68 \times 10^6 \text{ m}}$$
$$= 8.95 \text{ m/s}^2.$$

Continued

REFLECT The radial acceleration of the satellite is somewhat less than the value of the free-fall acceleration $g = 9.8$ m/s^2 at the earth's surface; it is equal to the value of g at a height of 300 km above the earth's surface. The value of T we obtained is only a little larger than the smallest possible period for an earth satellite (corresponding to the smallest possible orbit radius, i.e., the radius of earth itself).

Practice Problem: Find the speed and orbit radius for an earth satellite with a period of 1 day (86,400 s). *Answer:* 3.07 × 10^3 m/s, 4.23 × 10^7 m (from the center of the earth).

▲ **FIGURE 6.22** Saturn's rings consist of myriad separate particles orbiting the planet independently.

Masses of Astronomical Objects

We've talked mostly about artificial earth satellites, but we can apply the same analysis to the circular motion of *any* object acted upon by a gravitational force from a stationary object. Other familiar examples are our moon, the moons of other planets, and the planets orbiting the sun in nearly circular paths. The rings of Saturn (Figure 6.22), Jupiter, and Neptune are composed of small pieces of matter traveling in circular orbits around those planets.

Observations of satellite motions have been used to determine the masses of the planets in our solar system and of many other astronomical objects. For example, Jupiter has over 30 moons; the orbit radii and periods of many of them have been measured by astronomical observations. The appropriate relationship is Equation 6.10; we replace m_E with $m_{Jupiter}$ and then solve for $m_{Jupiter}$:

$$T^2 = \frac{4\pi^2 r^3}{G m_{Jupiter}}, \qquad m_{Jupiter} = \frac{4\pi^2 r^3}{G T^2}.$$

If the orbit radius r and period T of a satellite of Jupiter can be measured, we can calculate the mass of Jupiter.

Black Holes

A more dramatic example of the sort of analysis just described is the discovery of **black holes**—aggregations of matter that are so massive and so dense that nothing (not even light) can escape from their gravitational field. In 1916, Karl Schwarzschild predicted, on the basis of Einstein's general theory of relativity, that if an aggregation of matter with total mass M has a radius less than a critical value R_S, called the **Schwarzschild radius,** then nothing (not even light) can escape from its gravitational field. The value of R_S is given by

$$R_S = \frac{2GM}{c^2},$$

where G is the gravitational constant and c is the speed of light. The surface of a sphere with radius R_S surrounding a black hole is called the **event horizon;** because light can't escape from within that sphere, we can't see events that occur inside. All we can know about a black hole is its mass (from its gravitational field), its charge (from its electric field, expected to be zero), and its rotational characteristics.

At points far from a black hole, its gravitational effects are the same as those of any normal object with the same mass. But any object moving close to a black hole is acted upon by huge forces, causing the object to become so hot that it emits not just visible light (as in "red-hot" or "white-hot" light), but x rays. Astronomers look for these x rays (emitted *before* the object crosses the event horizon) to signal the presence of a black hole (Figure 6.23). Black holes have also been observed indirectly by the effects their gravitational fields have on the motions of nearby astronomical objects.

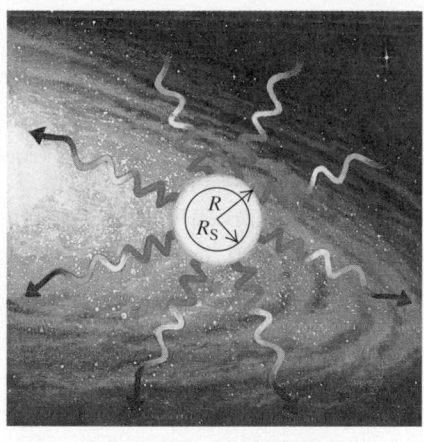 

(a)   (b)

▲ **FIGURE 6.23** (a) When a spherical object with mass M has a radius R greater than the Schwarzschild radius R_S, matter and energy can escape from its gravitational attraction. (b) When the radius is less than R_S, the object is a black hole. No matter or energy at a distance less than R_S from its center (the event horizon) can escape, because the escape speed would need to be greater than the speed of light.

The concept of a black hole is one of the most interesting and startling products of modern gravitational theory. Experimental and theoretical studies of black holes are a vital and exciting area of research in contemporary physics.

Weightlessness

What does it mean to say that an astronaut orbiting the earth in a spacecraft is *weightless?* The earth's gravitation still acts on the spacecraft and its occupants, as it must to provide the centripetal acceleration that keeps the craft in orbit. Yet, to the occupants, the environment *seems* to have no gravity. How can we understand this apparent paradox?

To answer this question, let's return to Example 4.10, a person standing in an accelerating elevator. We learned there that if the person stands on a spring scale, the scale doesn't read the true weight of the person during the acceleration. Now let's consider an implausible, though instructive, extreme. Suppose the elevator cables have all been broken and all the safety brakes have failed, so that both the elevator and its contents are in free fall, with a downward acceleration of magnitude $g = 9.8 \text{ m/s}^2$. What does the spring scale read now? The scale can't exert any upward force on the person; otherwise her acceleration would be less than g. So the scale must read zero, and the passenger *seems* to be weightless, even though the earth's gravitation still acts on her and her actual weight hasn't changed. If, in the excitement of the moment, she lets go of her purse, it doesn't fall to the floor, because it is already accelerating downward with the same acceleration as the elevator!

The same discussion is applicable to astronauts in a spacecraft orbiting the earth (Figure 6.24). The earth's gravity acts on both the spacecraft and its occupants, giving them the same acceleration. Thus, when an object is released inside the spacecraft, it doesn't fall relative to the craft, so it *appears* to be weightless.

The physiological effects of prolonged apparent weightlessness afford an interesting medical problem that is being actively explored. Gravity plays an important role in distributing blood in the body; one reaction to apparent weightlessness is a decrease in blood volume through increased excretion of water. In some cases, astronauts returning to earth have experienced temporary impairment of their sense of balance, a greater tendency toward motion sickness, and decalcification of their bones.

▲ **FIGURE 6.24** Astronauts appear to be weightless as they orbit the earth, because their acceleration is the same as that of their spacecraft. They still interact gravitationally with the earth, however, and thus have weight in the true sense; this weight is the force that holds them in orbit.

Historical Footnote

We close this chapter with a bit of historical perspective about the role of gravitation and orbital motion in the evolution of scientific thought. The study of planetary motion played a pivotal role in the early development of physics. Johannes Kepler (1571–1630) spent several painstaking years analyzing the motions of the planets, basing his work on remarkably precise measurements made by the Danish astronomer Tycho Brahe (1546–1601) *without the aid of a telescope.* (The telescope was invented in 1608.) Kepler discovered that the orbits of the planets are (nearly circular) ellipses and that the period of a planet in its orbit is proportional to the three-halves power of the radius of the orbit, as Equation 6.10 shows.

But it remained for Isaac Newton (1642–1727) to show, with his laws of motion and law of gravitation, that this celestial behavior of the planets could be understood on the basis of the very same physical principles he had developed to analyze *terrestrial* motion. Newton recognized that the falling apple (Section 4.4) was pulled to the earth by gravitational attraction. But to arrive at the $1/r^2$ form of the law of gravitation, he had to do something much subtler than this qualitative observation: He had to take Brahe's observations of planetary motions, as systematized by Kepler, and show that Kepler's rules demanded a $1/r^2$ force law.

From our historical perspective more than 300 years later, there is absolutely no doubt that this **Newtonian synthesis,** as it has come to be called, is one of the greatest achievements in the entire history of science, certainly comparable in significance to the development of quantum mechanics, the theory of relativity, and the understanding of DNA in the 20th century. It was an astonishing leap in understanding, made by a giant intellect!

SUMMARY

Circular Motion Dynamics

(Sections 6.1 and 6.2) For an object in uniform circular motion, the acceleration vector $\vec{a}_{\text{rad}}$ is directed toward the center of the circle and has constant magnitude $a_{\text{rad}} = v^2/R$ (Equation 3.16). The period T is the time required for the object to make one complete circle; the magnitude of the acceleration vector can also be written $a_{\text{rad}} = 4\pi^2 R/T^2$ (Equation 6.2). Provided that the object can be treated as a particle, circular motion (like other motions) is governed by Newton's second law, $\sum \vec{F} = m\vec{a}$. Because the net force vector points in the same direction as the acceleration vector, an object in uniform circular motion is acted upon by a net force directed toward the center of the circle, with magnitude $F_{\text{net}} = m(v^2/R)$ (Equation 6.3).

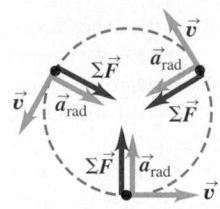

Newton's Law of Gravitation

(Section 6.3) Two particles with masses m_1 and m_2, a distance r apart, attract each other gravitationally with forces of magnitude

$$F_{\text{g}} = G\frac{m_1 m_2}{r^2}. \qquad (6.4)$$

These forces form an action–reaction pair in accordance with Newton's third law. If the objects cannot be treated as particles, but are spherically symmetric, this law is still valid; then, r is the distance between their centers. The gravitational interaction of a spherically symmetric object, from any point outside the object, is just the same as though all its mass were concentrated at the object's center.

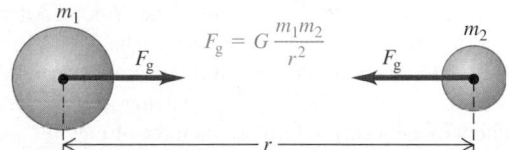

Weight

(Section 6.4) The weight w of a body is the total gravitational force exerted on it by all other bodies in the universe. Near the surface of the earth, an object's weight is very nearly equal to the gravitational force exerted by the earth alone. If the earth's mass is m_{E}, the weight of an object of mass m is

$$w = F_{\text{g}} = \frac{Gmm_{\text{E}}}{R_{\text{E}}^2} \qquad (6.6)$$

and the acceleration due to gravity g is

$$g = \frac{Gm_{\text{E}}}{R_{\text{E}}^2}. \qquad (6.7)$$

Satellite Motion

(Section 6.5) When a satellite moves in a circular orbit, the centripetal acceleration is provided by the gravitational attraction of the astronomical body it orbits. If the satellite orbits the earth in an orbit of radius r, its speed v and period T are, respectively,

$$v = \sqrt{\frac{Gm_{\text{E}}}{r}}, \qquad (6.8)$$

$$T = \frac{2\pi r}{v} = 2\pi r\sqrt{\frac{r}{Gm_{\text{E}}}} = \frac{2\pi r^{3/2}}{\sqrt{Gm_{\text{E}}}}. \qquad (6.10)$$

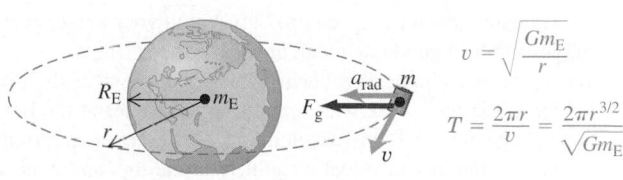

A black hole is an aggregation of matter that is so dense that nothing can escape its gravitational attraction. A nonrotating spherical distribution of mass is a black hole if its total mass M is contained within a radius of less than $R_{\text{s}} = 2GM/c^2$ (the Schwarzschild radius), where c is the speed of light.

Conceptual Questions

1. If there is a net force on a particle in uniform circular motion, why doesn't the particle's speed change?

2. As a car rounds a banked circular curve at constant speed, several forces are acting on it: for example, air resistance toward the rear, friction from the pavement in the forward direction, gravity, and the normal force from the tilted road surface. In what direction does the *net* force point? Explain.

3. A student wrote, "The reason an apple falls downward to meet the earth instead of the earth falling upward to meet the apple is that the earth is much more massive than the apple and therefore exerts a much greater pull than the apple does." Is this explanation correct? If not, what is the correct one?

4. In discussions of satellites by laymen, one often hears questions such as "What keeps the satellite moving in its orbit?" and "What keeps the satellite up?" How do you answer these questions? Are your answers also applicable to the moon?

5. Suppose you are sealed inside a lab on the planet Vulcan. After some time, you notice that objects do not fall to the floor when released. Without looking outside the lab, could you determine whether you and the lab are in free fall or whether the Vulcans have somehow found a way to turn off the force of gravity?

6. During an actual interview for a college teaching position in physics, several candidates said that the earth does not fall into the sun because the gravitational pull from the outer planets prevents it from doing so. If you were on the interview committee, would you hire any of these candidates? Explain your reasoning.

7. If two planets have the same mass, will they *necessarily* produce the same gravitational pull on 1.0 kg objects that are (a) at their surfaces; (b) the same center-to-center distance from both planets (but above their surfaces)? Explain.

8. True or false? Astronauts in satellites orbiting around the earth are weightless because the earth's gravity is so weak up there that it is negligible. Explain.

9. True or false? If a rock is acted upon by a gravitational force F from the earth when it is a distance d above the surface of our planet, it will be acted upon by a force $F/4$ if it is raised to twice the height $(2d)$ above the surface. Explain your reasoning.

10. On an icy road, you approach a curve that has the banking angle calculated for 55 mph. Your passenger suggests you slow down below 55 mph, just to be on the safe side, but you say that you should maintain your speed at 55 mph. (a) Who is correct, you or your passenger? (b) What would happen if you were to slow down (or speed up)? (c) Would your passenger's suggestion be a good one on an *unbanked* road?

11. When you're riding a slow Ferris wheel, you may feel slightly lighter at the top of the wheel, and a little heavier at the bottom. If the wheel keeps accelerating, however, at a certain speed of rotation you'll feel weightless at the top, and double your normal weight at the bottom. What's special about this particular rotation speed?

12. The moon is accelerating toward the earth. Does this mean that it is getting closer to our planet? Explain your reasoning.

13. A passenger in a car rounding a sharp curve feels "thrown" toward the outside of the curve. (a) What causes this to happen? Is the person really *thrown* away from the center of the curve? (b) Make a free-body diagram of the person.

14. How does a lettuce spinner remove water from lettuce?

15. What is *wrong* with this statement? "A satellite stays in orbit because the outward centrifugal force just balances the inward centripetal force." Make a free-body diagram of the satellite.

Multiple-Choice Problems

1. Objects in orbiting satellites are apparently weightless because
   A. they are too far from earth to feel its gravity.
   B. earth's gravitational pull is balanced by the outward centrifugal force.
   C. earth's gravitational pull is balanced by the centripetal force.
   D. both they and the satellite are in free fall toward the earth.

2. If the earth had twice its present mass, its orbital period around the sun (at our present distance from the sun) would be
   A. $\sqrt{2}$ years.
   B. 1 year.
   C. $\dfrac{1}{\sqrt{2}}$ year.
   D. $\frac{1}{2}$ year.

3. An astronaut is floating happily outside her spaceship, which is orbiting the earth at a distance above the earth's surface equal to 1 earth radius. The astronaut's weight is
   A. zero
   B. equal to her normal weight on earth
   C. half her normal weight on earth
   D. one-fourth her normal weight on earth

4. A stone of mass m is attached to a strong string and whirled in a vertical circle of radius R. At the exact top of the path, the tension in the string is three times the stone's weight. At this point, the stone's speed is
   A. $2\sqrt{gR}$.
   B. $\sqrt{2gR}$.
   C. $3\sqrt{gR}$.
   D. $\sqrt{3gR}$.

5. A frictional force f provides the centripetal force as a car goes around an unbanked curve of radius R at speed V. Later, the car encounters a similar curve, except of radius $2R$, and the driver continues around this curve at the same speed V. In order to make this second curve, the frictional force on the car must be equal to
   A. $2f$.
   B. f.
   C. $\frac{1}{2}f$.
   D. $\frac{1}{4}f$.

6. In the 1960s, during the Cold War, the Soviet Union put a rocket into a nearly circular orbit around the earth. One U.S. senator expressed concern that a nuclear bomb could be dropped on our nation if it were released when the rocket was over the United States. If such a bomb were, in fact, released when the rocket was directly over the United States, where would it land?
   A. Directly below, on the United States.
   B. It would follow a curved path and most likely land somewhere in the ocean.
   C. It would not hit the United States, because it would remain in orbit with the rocket.
   D. It would move in a straight line out into space.

7. Two masses m and $2m$ are each forced to go around a curve of radius R at the same constant speed. If, as they move around this curve, the smaller mass is acted upon by a net force F, then the larger one is acted upon by a net force of
   A. $\frac{1}{2}F$.
   B. F.
   C. $2F$.
   D. $4F$.

8. A weight W is swung in a vertical circle at constant speed at the end of a rigid bar that pivots about the opposite end (like the spoke of a wheel). If the tension in this bar is equal to $2W$ when the weight is at its lowest point, then the tension when the weight is at the highest point will be
   A. 0.   B. W.
   C. $2W$.   D. $3W$.

9. If a planet had twice the earth's radius, but only half its mass, the acceleration due to gravity at its surface would be
   A. G.   B. $\frac{1}{2}g$.
   C. $\frac{1}{4}g$.   D. $\frac{1}{8}g$.

10. When a mass goes in a horizontal circle with speed V at the end of a string of length L on a frictionless table, the tension in the string is T. If the speed of this mass were doubled, but all else remained the same, the tension would be
    A. $2T$.   B. $4T$.
    C. $T\sqrt{2}$.   D. $\dfrac{T}{\sqrt{2}}$.

11. In the previous problem, if *both* the speed and the length were doubled, the tension in the string would be
    A. T.   B. $2T$.
    C. $4T$.   D. $T\sqrt{2}$.

12. Two 1.0 kg point masses a distance D apart each exert a gravitational attraction F on the other one. If 1.0 kg is now added to each mass, the gravitational attraction exerted by each would be
    A. $2F$.   B. $3F$.   C. $4F$.

13. Two massless bags contain identical bricks, each brick having a mass M. Initially, each bag contains four bricks, and the bags mutually exert a gravitational attraction F_1 on each other. You now take two bricks from one bag and add them to the other bag, causing the bags to attract each other with a force F_2. What is the closest expression for F_2 in terms of F_1?
    A. $F_2 = \frac{3}{4}F_1$.
    B. $F_2 = \frac{1}{2}F_1$.
    C. $F_2 = \frac{1}{4}F_1$.
    D. $F_2 = F_1$ because neither the total mass nor the distance between the bags has changed.

14. When two point masses are a distance D apart, each exerts a gravitational attraction F on the other mass. To reduce this force to $\frac{1}{3}F$, you would have to separate the masses to a distance of
    A. $D\sqrt{3}$.   B. $3D$.
    C. $9D$.   D. $\frac{1}{3}D$.

15. If human beings ever travel to a planet whose mass and radius are both twice that of the earth, their weight on that planet will be
    A. equal to their weight on earth
    B. twice their weight on earth
    C. half their weight on earth
    D. one-fourth their weight on earth

Problems

6.1 Force in Circular Motion

1. ● A racing car drives at *constant speed* around the horizontal track shown in Figure 6.25. At points A, B, and C, draw a vector showing the magnitude and direction of the net force on this car. Make sure that the lengths of your arrows represent the relative magnitudes of the three forces.

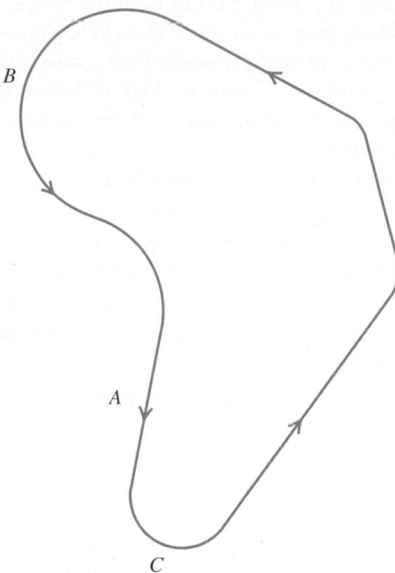

▲ **FIGURE 6.25** Problem 1.

2. ● A stone with a mass of 0.80 kg is attached to one end of a string 0.90 m long. The string will break if its tension exceeds 60.0 N. The stone is whirled in a horizontal circle on a frictionless tabletop; the other end of the string remains fixed. (a) Make a free-body diagram of the stone. (b) Find the maximum speed the stone can attain without breaking the string.

3. ●● **Force on a skater's wrist.** A 52 kg ice skater spins about a
BIO   vertical axis through her body with her arms horizontally outstretched, making 2.0 turns each second. The distance from one hand to the other is 1.50 m. Biometric measurements indicate that each hand typically makes up about 1.25% of body weight. (a) Draw a free body diagram of one of her hands. (b) What horizontal force must her wrist exert on her hand? (c) Express the force in part (b) as a multiple of the weight of her hand.

4. ● A flat (unbanked) curve on a highway has a radius of 220 m. A car rounds the curve at a speed of 25.0 m/s. (a) Make a free-body diagram of the car as it rounds this curve. (b) What is the minimum coefficient of friction that will prevent sliding?

5. ●● The "Giant Swing" at a county fair consists of a vertical central shaft with a number of horizontal arms attached at its upper end. (See Figure 6.26.) Each arm supports a seat suspended from a 5.00-m-long rod, the upper end of which is

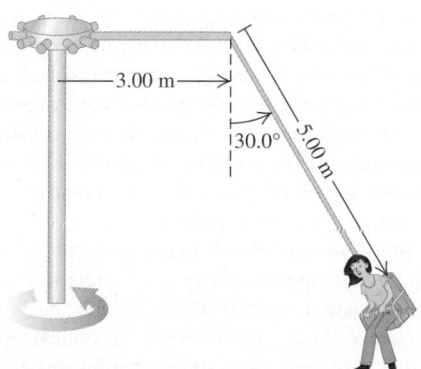

▲ **FIGURE 6.26** Problem 5.

fastened to the arm at a point 3.00 m from the central shaft. (a) Make a free-body diagram of the seat, including the person in it. (b) Find the time of one revolution of the swing if the rod supporting the seat makes an angle of 30.0° with the vertical. (c) Does the angle depend on the weight of the passenger for a given rate of revolution?

6. • A small button placed on a horizontal rotating platform with diameter 0.320 m will revolve with the platform when it is brought up to a speed of 40.0 rev/min, provided the button is no more than 0.150 m from the axis. (a) What is the coefficient of static friction between the button and the platform? (b) How far from the axis can the button be placed, without slipping, if the platform rotates at 60.0 rev/min?

7. • Using only astronomical data from Appendix E, calculate (a) the speed of the planet Venus in its essentially circular orbit around the sun and (b) the gravitational force that the sun must be exerting on Venus. Start with a free-body diagram of Venus.

8. • A highway curve with radius 900.0 ft is to be banked so that a car traveling 55.0 mph will not skid sideways even in the absence of friction. (a) Make a free-body diagram of this car. (b) At what angle should the curve be banked?

9. •• **The Indy 500.** The Indianapolis Speedway (home of the Indy 500) consists of a 2.5 mile track having four turns, each 0.25 mile long and banked at 9°12′. What is the no-friction-needed speed (in m/s and mph) for these turns? (Do you think drivers actually take the turns at that speed?)

6.2 Motion in a Vertical Circle

10. • A bowling ball weighing 71.2 N is attached to the ceiling by a 3.80 m rope. The ball is pulled to one side and released; it then swings back and forth like a pendulum. As the rope swings through its lowest point, the speed of the bowling ball is measured at 4.20 m/s. At that instant, find (a) the magnitude and direction of the acceleration of the bowling ball and (b) the tension in the rope. Be sure to start with a free-body diagram.

11. •• The Cosmoclock 21 Ferris wheel in Yokohama City, Japan, has a diameter of 100 m. Its name comes from its 60 arms, each of which can function as a second hand (so that it makes one revolution every 60.0 s). (a) Find the speed of the passengers when the Ferris wheel is rotating at this rate. (b) A passenger weighs 882 N at the weight-guessing booth on the ground. What is his apparent weight at the highest and at the lowest point on the Ferris wheel? (c) What would be the time for one revolution if the passenger's apparent weight at the highest point were zero? (d) What then would be the passenger's apparent weight at the lowest point?

12. •• A 50.0 kg stunt pilot who has been diving her airplane vertically pulls out of the dive by changing her course to a circle in a vertical plane. (a) If the plane's speed at the lowest point of the circle is 95.0 m/s, what should the minimum radius of the circle be in order for the centripetal acceleration at this point not to exceed 4.00g? (b) What is the apparent weight of the pilot at the lowest point of the pullout?

13. •• **Effect on blood of walking.** While a person is walking, **BIO** his arms swing through approximately a 45° angle in $\frac{1}{2}$ s. As a reasonable approximation, we can assume that the arm moves with constant speed during each swing. A typical arm is 70.0 cm long, measured from the shoulder joint. (a) What is the acceleration of a 1.0 gram drop of blood in the fingertips at the bottom of the swing? (b) Make a free-body diagram of the drop of blood in part (a). (c) Find the force that the blood vessel must exert on the drop of blood in part (b). Which way does this force point? (d) What force would the blood vessel exert if the arm were not swinging?

14. • **Stay dry!** You tie a cord to a pail of water, and you swing the pail in a vertical circle of radius 0.600 m. What minimum speed must you give the pail at the highest point of the circle if no water is to spill from it? Start with a free-body diagram of the water at its highest point.

15. • Stunt pilots and fighter pilots who fly at high speeds in a **BIO** downward-curving arc may experience a "red out," in which blood is forced upward into the flier's head, potentially swelling or breaking capillaries in the eyes and leading to a reddening of vision and even loss of consciousness. This effect can occur at centripetal accelerations of about 2.5g's. For a stunt plane flying at a speed of 320 km/h, what is the minimum radius of downward curve a pilot can achieve without experiencing a red out at the top of the arc? (*Hint:* Remember that gravity provides part of the centripetal acceleration at the top of the arc; it's the acceleration required *in excess of gravity* that causes this problem.)

6.3 Newton's Law of Gravitation

16. • If two tiny identical spheres attract each other with a force of 3.0 nN when they are 25 cm apart, what is the mass of each sphere?

17. • What gravitational force do the two protons in the helium nucleus exert on each other? Their separation is approximately 1.0 fm. (Consult Appendix E and Table 1-1.)

18. • **Rendezvous in space!** A couple of astronauts agree to rendezvous in space after hours. Their plan is to let gravity bring them together. One of them has a mass of 65 kg and the other a mass of 72 kg, and they start from rest 20.0 m apart. (a) Make a free-body diagram of each astronaut, and use it to find his or her initial acceleration. As a rough approximation, we can model the astronauts as uniform spheres. (b) If the astronauts' acceleration remained constant, how many days would they have to wait before reaching each other? (Careful! They *both* have acceleration toward each other.) (c) Would their acceleration, in fact, remain constant? If not, would it increase or decrease? Why?

19. • What is the ratio of the gravitational pull of the sun on the moon to that of the earth on the moon? (Assume that the distance of the moon from the sun is approximately the same as the distance of the earth from the sun.) Use the data in Appendix E. Is it more accurate to say that the moon orbits the earth or that the moon orbits the sun? Why?

20. • A 2150 kg satellite used in a cellular telephone network is in a circular orbit at a height of 780 km above the *surface* of the earth. What is the gravitational force on the satellite? What fraction is this force of the satellite's weight at the surface of the earth?

21. • An interplanetary spaceship passes through the point in space where the gravitational forces from the sun and the earth on the ship exactly cancel. (a) How far from the center of the earth is it? Use the data in Appendix E. (b) Once it reached the point found in part (a), could the spaceship turn off its engines and just hover there indefinitely? Explain.

22. •• Find the magnitude and direction of the net gravitational force on mass *A* due to masses *B* and *C* in Figure 6.27. Each mass is 2.00 kg.

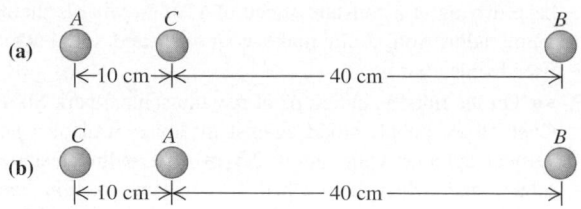

▲ FIGURE 6.27 Problem 22.

23. • How far from a very small 100-kg ball would a particle have to be placed so that the ball pulled on the particle just as hard as the earth does? Is it reasonable that you could actually set up this as an experiment? Why?

24. •• Each mass in Figure 6.28 is 3.00 kg. Find the magnitude and direction of the net gravitational force on mass A due to the other masses.

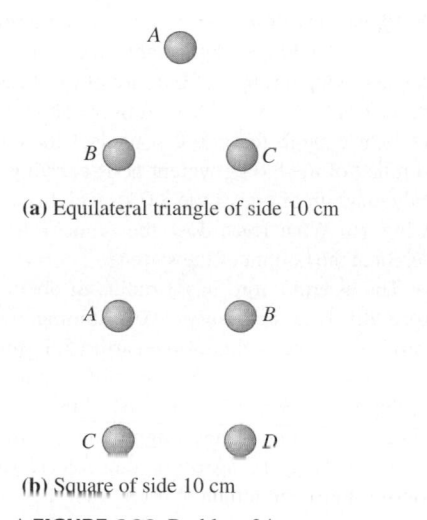

(a) Equilateral triangle of side 10 cm

(b) Square of side 10 cm

▲ FIGURE 6.28 Problem 24.

25. •• An 8.00-kg point mass and a 15.0-kg point mass are held in place 50.0 cm apart. A particle of mass m is released from a point between the two masses 20.0 cm from the 8.00-kg mass along the line connecting the two fixed masses. Find the magnitude and direction of the acceleration of the particle.

6.4 Weight

26. • How many kilometers would you have to go above the *surface* of the earth for your weight to decrease to half of what it was at the surface?

27. •• Your spaceship lands on an unknown planet. To determine the characteristics of this planet, you drop a 1.30 kg wrench from 5.00 m above the ground and measure that it hits the ground 0.811 s later. You also do enough surveying to determine that the circumference of the planet is 62,400 km. (a) What is the mass of the planet, in kilograms? (b) Express the planet's mass in terms of the earth's mass.

28. •• If an object's weight is W on the earth, what would be its weight (in terms of W) if the earth had (a) twice its present mass, but was the same size, (b) half its present radius, but the same mass, (c) half its present radius and half its present mass, (d) twice its present radius and twice its present mass?

29. •• **Huygens probe on Titan.** In January 2005 the *Huygens* probe landed on Saturn's moon Titan, the only satellite in the solar system having a thick atmosphere. Titan's diameter is 5150 km, and its mass is 1.35×10^{23} kg. The probe weighed 3120 N on the earth. What did it weigh on the surface of Titan?

30. •• The mass of the moon is about 1/81 the mass of the earth, its radius is $\frac{1}{4}$ that of the earth, and the acceleration due to gravity at the earth's surface is 9.80 m/s². Without looking up either body's mass, use this information to compute the acceleration due to gravity on the moon's surface.

31. •• Neutron stars, such as the one at the center of the Crab Nebula, have about the same mass as our sun, but a *much* smaller diameter. If you weigh 675 N on the earth, what would you weigh on the surface of a neutron star that has the same mass as our sun and a diameter of 20.0 km?

32. •• The asteroid 234 Ida has a mass of about 4.0×10^{16} kg and an average radius of about 16 km (it's not spherical, but you can assume it is). (a) Calculate the acceleration of gravity on 234 Ida. (b) What would an astronaut whose earth weight is 650 N weigh on 234 Ida? (c) If you dropped a rock from a height of 1.0 m on 234 Ida, how long would it take to reach the ground? (d) If you can jump 60 cm straight up on earth, how high could you jump on 234 Ida? (Assume the asteroid's gravity doesn't weaken significantly over the distance of your jump.)

6.5 Satellite Motion

33. • An earth satellite moves in a circular orbit with an orbital speed of 6200 m/s. (a) Find the time of one revolution of the satellite. (b) Find the radial acceleration of the satellite in its orbit. (c) Make a free-body diagram of the satellite.

34. • What is the period of revolution of a satellite with mass m that orbits the earth in a circular path of radius 7880 km (about 1500 km above the surface of the earth)?

35. • What must be the orbital speed of a satellite in a circular orbit 780 km above the surface of the earth?

36. •• **Planets beyond the solar system.** On October 15, 2001, a planet was discovered orbiting around the star HD68988. Its orbital distance was measured to be 10.5 million kilometers from the center of the star, and its orbital period was estimated at 6.3 days. What is the mass of HD68988? Express your answer in kilograms and in terms of our sun's mass. (Consult Appendix E.)

37. •• **Communications satellites.** Communications satellites are used to bounce radio waves from the earth's surface to send messages around the curvature of the earth. In order to be available all the time, they must remain above the same point on the earth's surface and must move in a circle above the equator. (a) How long must it take for a communications satellite to make one complete orbit around the earth? (Such an orbit is said to be *geosynchronous*.) (b) Make a free-body diagram of the satellite in orbit. (c) Apply Newton's second law to the satellite and find its altitude above the earth's *surface*. (d) Draw the orbit of a communications satellite to scale on a sketch of the earth.

38. • (a) Calculate the speed with which you would have to throw a rock to put it into orbit around the asteroid 234 Ida, near the surface (see Problem 32; assume 234 Ida is spherical, with the given radius and mass). (b) How long would it take for your rock to return and hit you in the back of the head?

39. •• In March 2006, two small satellites were discovered orbiting Pluto, one at a distance of 48,000 km and the other at

64,000 km. Pluto already was known to have a large satellite Charon, orbiting at 19,600 km with an orbital period of 6.39 days. Assuming that the satellites do not affect each other, find the orbital periods of the two small satellites *without* using the mass of Pluto.

40. •• **Apparent weightlessness in a satellite.** You have probably seen films of astronauts floating freely in orbiting satellites. People often think the astronauts are weightless because they are free of the gravity of the earth. Let us see if that explanation is correct. (a) Typically, such satellites orbit around 400 km above the *surface* of the earth. If an astronaut weighs 750 N on the ground, what will he weigh if he is 400 km above the surface? (b) Draw the orbit of the satellite in part (a) to scale on a sketch of the earth. (c) In light of your answers to parts (a) and (b), are the astronauts weightless because gravity is so weak? Why are they apparently weightless?

41. •• **Baseball on Deimos!** Deimos, a moon of Mars, is about 12 km in diameter, with a mass of 2.0×10^{15} kg. Suppose you are stranded alone on Deimos and want to play a one-person game of baseball. You would be the pitcher, and you would be the batter! With what speed would you have to throw a baseball so that it would go into orbit and return to you so you could hit it? Do you think you could actually throw it at that speed?

General Problems

42. •• **International Space Station.** The International Space Station, launched in November 1998, makes 15.65 revolutions around the earth each day in a circular orbit. (a) Draw a free-body diagram of this satellite. (b) Apply Newton's second law to the space station and find its height (in kilometers) above the earth's surface. (Consult Appendix E.)

43. •• **Artificial gravity.** One way to create artificial gravity in a space station is to spin it. If a cylindrical space station 275 m in diameter is to spin about its central axis, at how many revolutions per minute (rpm) must it turn so that the outermost points have an acceleration equal to g?

44. •• **Hip wear on the moon.** (a) Use data from Appendix E to **BIO** calculate the acceleration due to gravity on the moon. (b) Calculate the friction force on a walking 65 kg astronaut carrying a 43 kg instrument pack on the moon if the coefficient of kinetic friction at her hip joint is 0.0050. (If necessary, see problem 32 in Chapter 5 and recall that approximately 65% of the body weight is above the hip.) (c) What would be the friction force on earth for this astronaut?

45. •• **Volcanoes on Io.** Jupiter's moon Io has active volcanoes (in fact, it is the most volcanically active body in the solar system) that eject material as high as 500 km (or even higher) above the surface. Io has a mass of 8.94×10^{22} kg and a radius of 1815 km. Ignore any variation in gravity over the 500 km range of the debris. How high would this material go on earth if it were ejected with the same speed as on Io?

46. •• You are driving with a friend who is sitting to your right on the passenger side of the front seat of your car. You would like to be closer to your friend, so you decide to use physics to achieve your romantic goal by making a quick turn. (a) Which way (to the left or to the right) should you turn the car to get your friend to slide toward you? (b) If the coefficient of static friction between your friend and the car seat is 0.55 and you

keep driving at a constant speed of 15 m/s, what is the maximum radius you could make your turn and still have your friend slide your way?

47. •• On the ride "Spindletop" at the amusement park Six Flags Over Texas, people stood against the inner wall of a hollow vertical cylinder with radius 2.5 m. The cylinder started to rotate, and when it reached a constant rotation rate of 0.60 rev/s, the floor on which the people were standing dropped about 0.5 m. The people remained pinned against the wall. (a) Draw a free-body diagram for a person on this ride after the floor has dropped. (b) What minimum coefficient of static friction is required if the person on the ride is not to slide downward to the new position of the floor? (c) Does your answer in part (b) depend on the mass of the passenger? (*Note:* When the ride is over, the cylinder is slowly brought to rest. As it slows down, people slide down the walls to the floor.)

48. •• **Physical training.** As part of a training program, an athlete **BIO** runs while holding 8.0 kg weights in each hand. As he runs, the weights swing through a 30.0° arc in $\frac{1}{3}$ s at essentially constant speed. His hands are 72 cm from his shoulder joint, and they are light enough that we can neglect their weight compared with that of the 8.0 kg weight he is carrying. (a) Make a free-body diagram of one of the 8.0 kg weights at the bottom of its swing. (b) What force does the runner's hand exert on each weight at the bottom of the swing?

49. •• The asteroid Toro has a radius of about 5.0 km. Consult Appendix E as necessary. (a) Assuming that the density of Toro is the same as that of the earth (5.5 g/cm^3), find its total mass and find the acceleration due to gravity at its surface. (b) Suppose an object is to be placed in a circular orbit around Toro, with a radius just slightly larger than the asteroid's radius. What is the speed of the object? Could you launch yourself into orbit around Toro by running?

50. •• A 1125-kg car and a 2250-kg pickup truck approach a curve on the expressway that has a radius of 225 m. (a) At what angle should the highway engineer bank this curve so that vehicles traveling at 65.0 mi/h can safely round it regardless of the condition of their tires? Should the heavy truck go slower than the lighter car? (b) As the car and truck round the curve at 65.0 mi/h, find the normal force on each one due to the highway surface.

51. •• **Exploring Europa.**
BIO Europa, a satellite of Jupiter, is believed to have an ocean of liquid water (with the possibility of life) beneath its icy surface. (See Figure 6.29.) Europa is 3130 km in diameter and has a mass of 4.78×10^{22} kg. In the future, we will surely want to send astronauts to investigate Europa. In planning such a future mission, what is the fastest that such an astronaut could walk on the surface of Europa if her legs are 1.0 m long?

▲ **FIGURE 6.29** Problem 51.

52. • The star Rho[1] Cancri is 57 light-years from the earth and has a mass 0.85 times that of our sun. A planet has been

detected in a circular orbit around Rho¹ Cancri with an orbital radius equal to 0.11 times the radius of the earth's orbit around the sun. What are (a) the orbital speed and (b) the orbital period of the planet of Rho¹ Cancri?

53. •• **Catch a piece of a comet.** The *Stardust* spacecraft was designed to visit a comet and bring samples of its material back to the earth. The craft is 1.7 m across and has a mass of 385 kg. In January 2004, from a distance of 237 km, it took the photograph of the nucleus of comet Wild2, shown in Figure 6.30. The distance across this nucleus

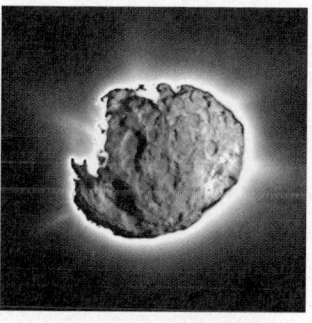

▲ **FIGURE 6.30** Problem 53.

is 5 km, and we can model it as an approximate sphere. Since astronomers think that comets have a composition similar to that of Pluto, we can assume a density the same as Pluto's, which is 2.1 g/cm³. The samples taken were returned to earth on Jan. 15, 2006. (a) What gravitational force did the comet exert on the spacecraft when it took the photo in the figure? (b) Compare this force with the force that the earth's gravity exerted on it when the spacecraft was on the launch pad.

54. •• The 4.00 kg block in the figure is attached to a vertical rod by means of two strings. When the system rotates about the axis of the rod, the strings are extended as shown in Figure 6.31 and the tension in the upper string is 80.0 N. (a) What is the tension in the lower cord? Start with a free-body diagram of the block. (b) What is the speed of the block?

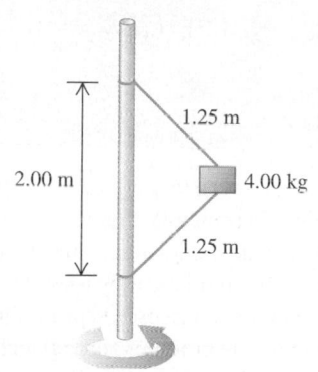

▲ **FIGURE 6.31** Problem 54

55. •• As your bus rounds a flat curve at constant speed, a package with mass 0.500 kg, suspended from the luggage compartment of the bus by a string 45.0 cm long, is found to hang at rest relative to the bus, with the string making an angle of 30.0° with the vertical. In this position, the package is 50.0 m from the center of curvature of the curve. What is the speed of the bus?

56. •• **Artificial gravity in space stations.** One problem for humans living in outer space is that they are apparently weightless. One way around this problem is to design a cylindrical space station that spins about an axis through its center at a constant rate. (See Figure 6.32.) This spin creates "artificial gravity" at the outside rim of the station. (a) If the diameter of the space station is 800.0 m, how fast must the rim be moving in order for the "artificial gravity" acceleration to be *g* at the outer rim? (b) If the space station is a waiting area for travelers going to Mars, it might be desirable to simulate the acceleration due to gravity on the Martian surface. How fast must the rim move in this case? (c) Make a free-body diagram of an astronaut at the outer rim.

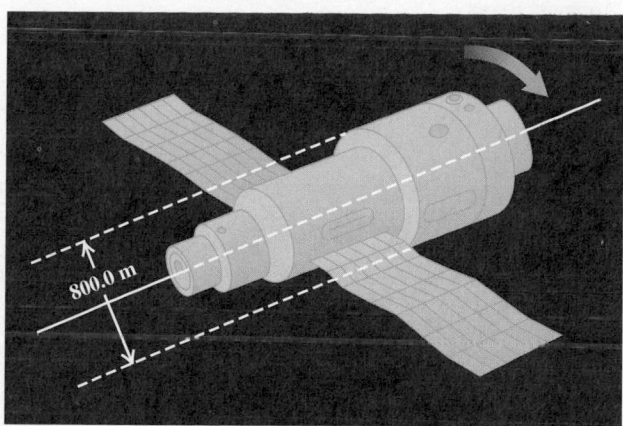

▲ **FIGURE 6.32** Problem 56.

Passage Problems

BIO Weightlessness and artificial gravity. Astronauts who live under weightless (zero gravity) conditions for a prolonged time can experience health risks as a result. One way to avoid these adverse physiological effects is to provide an artificial gravity to simulate what is naturally experienced on the earth.

In one design a space station is constructed in the shape of a long cylinder that spins at a constant rate about its longitudinal axis. Astronauts standing on the inside lateral surface of the cylinder experience a centripetal acceleration (due to their circular motion about the axis of the cylinder) that simulates the effect of gravity. The magnitude of the simulated gravity can be increased or decreased to the desired value by changing the rotation rate of the cylinder.

57. If the diameter of the space station is 1000 m, how fast must the outer edge of the space station move to give an astronaut the experience of a reduced "gravity" of 5 m/s² (roughly ½ earth normal)? You may assume that the astronaut is standing on the inner wall at a distance of nearly 1000 m from the axis of the space station.
   A. 5000 m/s
   B. 2500 m/s
   C. 71 m/s
   D. 50 m/s

58. What would be the resulting period of rotation of the space station? You may assume that the space station has a circumference of approximately 3000 m.
   A. 0.6 s
   B. 3 s
   C. 42 s
   D. 60 s
   E. 380 s

59. Under these same conditions, how much force would an astronaut need to exert on a 2-kg space helmet that she is holding to keep it moving on the same circular path that she follows? *Note*: This force is known as the "effective weight" of the space helmet.
   A. 20 N
   B. 10 N
   C. 5 N
   D. 0 N

7 Work and Energy

Sculptures like the one above are activated by the energy of a rolling ball. The ball acquires energy when it is raised to its starting position, and as it rolls downward, its energy is transferred to other elements in the sculpture, setting them in motion. Sculptures like this are based on the principles discussed in this chapter.

In the preceding three chapters, we've developed an understanding of mechanics as embodied in Newton's three laws of motion. In principle, these laws provide a complete description of the relation between the motion of an object and the forces that act on it. But some problems aren't as simple as they look. Suppose you try to find the speed of a toy glider launched by a compressed spring that applies a varying force. You apply Newton's laws and all the problem-solving techniques that we've learned, but you find that the acceleration isn't constant and the simple methods we've learned can't be implemented directly. As we'll see, there are powerful alternative methods for dealing with such problems.

In this chapter, we'll introduce a new concept—*energy*—that is one of the most important unifying concepts in all of physical science. Its importance stems from the principle of *conservation of energy*. We'll concentrate on mechanical energy, which is associated with the motions and positions of mechanical systems. In the opening section of the chapter, we'll tour the main ideas associated with energy on a qualitative level. Then we'll develop the concepts of work and kinetic energy (energy of motion) and the relationship between them. In the remainder of the chapter, we'll learn how to apply these concepts to the analysis of problems.

7.1 An Overview of Energy

The concept of **energy** appears throughout every area of physics, yet it's not easy to define just what energy is. This concept is the cornerstone of a fundamental law of nature called **conservation of energy,** which states that the total energy in any isolated system is constant, no matter what happens within the system. (Later in the chapter, we'll explain what an *isolated system* is.) Energy can be converted from one form to another, but it cannot be created or destroyed.

Conservation laws, including conservation of mass, energy, momentum, angular momentum, electric charge, and other quantities, play vital roles in every area of physics and provide unifying threads that run through the whole fabric of the subject. Every conservation law says that the total amount of some physical quantity in every isolated system is constant.

An example is the conservation of mass in chemical reactions. The total mass of the material participating in a chemical reaction is always the same (within the limits of experimental precision) before and after the reaction. This generalization is called the principle of *conservation of mass,* and it is obeyed in all chemical reactions.

Something analogous happens in collisions between objects. For a particle with mass m moving with speed v, we define a quantity $\frac{1}{2}mv^2$ called the **kinetic energy** of the particle. When two hard, springy objects, such as steel ball bearings, collide, we find that the individual speeds change, but that the total kinetic energy (the sum of the $\frac{1}{2}mv^2$ quantities for the two balls) is very nearly the same after the collision as before. When it is *exactly* the same, we say that kinetic energy is *conserved.* We haven't really said what kinetic energy *is,* only that this product of half the mass and the square of the speed is useful in representing a conservation principle in a particular class of interactions.

When two soft, deformable objects (such as two balls of putty or chewing gum) collide, we find that kinetic energy is *not* conserved, but the bodies become warmer. It turns out that there is a definite relationship between the rise in temperature and the loss of kinetic energy. We can define a new quantity, *internal energy,* that increases with temperature in a definite way such that the *sum* of the kinetic energy and the internal energy *is* conserved in these collisions. We can also define additional forms of energy that enable us to extend the principle of conservation of energy to broader classes of phenomena.

Energy is transformed from one form to another, but it is not created or destroyed. There is energy associated with heat, elastic deformations, electric and magnetic fields, and, according to the theory of relativity, even with mass itself. Newtonian mechanics has its limitations; it doesn't work for very fast motion, for which we have to use the theory of relativity, or for atomic or subatomic systems, for which we need quantum mechanics. But conservation of energy holds even in these realms. It is a *universal* conservation principle; no exception has ever been found. Studying conservation of energy is like studying a snowflake under a microscope, or the finer and finer details of a fractal pattern, or the structure of a Bach fugue. The more closely you look, the richer the subject becomes.

Figure 7.1 conveys a sense of energy transfer and transformation. A little of the sun's nuclear energy is successively transformed into other forms and eventually appears as energy you spend when exercising. This energy is not created spontaneously, and it is not "used up"; instead, it is transformed from one form to another.

Forms of Mechanical Energy

In this chapter, we'll be concerned only with **mechanical energy**—the energy associated with motion, position, and deformation of objects. We'll find that, in some interactions, mechanical energy is conserved; in others, there is a conversion of mechanical energy to or from other forms. In Chapters 15 and 16, we'll study in detail the relation of mechanical energy to heat; in later chapters, we'll define still other forms of energy.

As we mentioned at the beginning of this chapter, the energy associated with motion of an object is called kinetic energy. For an object with constant mass, kinetic energy depends only on its speed v and mass m. When a moving object

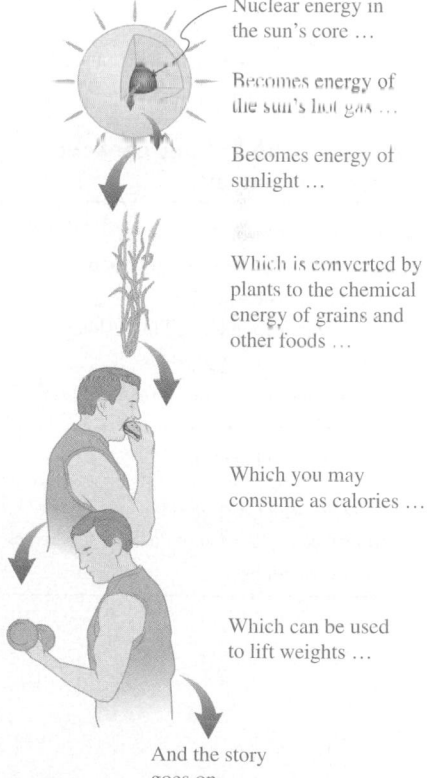

Nuclear energy in the sun's core …

Becomes energy of the sun's hot gas …

Becomes energy of sunlight …

Which is converted by plants to the chemical energy of grains and other foods …

Which you may consume as calories …

Which can be used to lift weights …

And the story goes on.

▲ **FIGURE 7.1** A typical sequence of energy transfers and transformations—one that sustains life on earth.

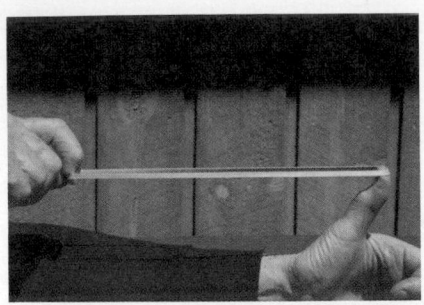

▲ **FIGURE 7.2** Elastic potential energy stored in a stretched rubber band.

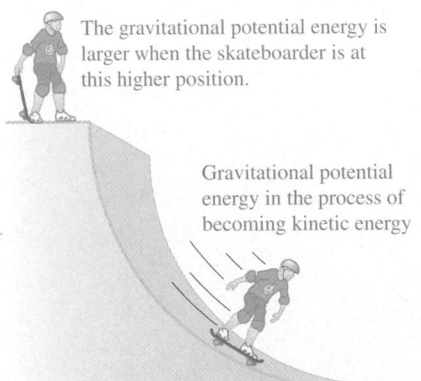

The gravitational potential energy is larger when the skateboarder is at this higher position.

Gravitational potential energy in the process of becoming kinetic energy

▲**FIGURE 7.3** Transformation of gravitational potential energy to kinetic energy.

TABLE 7.1 The analogy between energy and money

How money is like energy:
- It can take multiple forms (coins, bills, checks, bank accounts).
- You can transform it (e.g., by cashing a check) or transfer it to others.
- These transfers and transformations do not change its total amount.
- It can be stored or spent.

How money is not like energy:
- Money can be created or destroyed, whereas energy cannot be.

such as a kicked football or a speeding car interacts with another object, obvious changes (such as broken windows or bent fenders) result, and associated with these changes are transfers of kinetic energy to other forms of energy.

The other class of mechanical energy we'll consider is energy that is not directly associated with motion, but rather a quantity that is *stored* as a result of the position or elastic deformation of an object. Consider a stretched rubber band (Figure 7.2). You have to expend energy to stretch it, and as soon as it is released, physical changes occur; for instance, it may fly across the room. An unstretched rubber band lacks this capability. The act of stretching the rubber band stores energy in it, and releasing the rubber band allows this stored energy to be transformed to other types of energy, such as kinetic energy. We call this stored energy **potential energy.** Energy that is stored in an elastic object when you stretch, compress, twist, or otherwise deform it is called **elastic potential energy.**

Another example of potential energy is provided by the skateboarder in Figure 7.3. When he descends from the top edge of the quarter-pipe, he speeds up and gains kinetic energy. While he momentarily stops at the top of the opposite edge, he has no kinetic energy. As soon as he begins his descent, he gains kinetic energy as the gravitational force (his weight) pulls him downward. Thus, his initial elevated position represents a form of *stored energy;* we call it **gravitational potential energy.** Whenever an object moves vertically while a gravitational force (its weight) acts on it, the gravitational potential energy changes. The greater the change in height, the greater is the change in gravitational potential energy.

Kinetic energy, elastic potential energy, and gravitational potential energy are all forms of mechanical energy.

Energy and Money

Earlier, we remarked that even physicists have a hard time defining what energy *is*. You may now feel that same difficulty. What *is* this entity that goes from object to object and from one form to another? An analogy that you're probably familiar with is *money*. Money takes many forms, such as a coin, a bill, and an electronic number in a bank's computer. You can store money, or you can spend it to buy things. Even though money is real, it is also in a sense a pure abstraction. Energy, similarly, is both an abstract idea and a very real quantity that is involved in changes and transformations of physical systems. Table 7.1 sums up this analogy between money and energy.

Conservation of Energy

In discussing energy relations, we'll often use the concept of a **system.** A system usually consists of one or more objects that can interact, move, and undergo deformations. In general, a system can interact with its surroundings. In the special case where a system has *no* interaction with its surroundings, we call it an **isolated system.** As we mentioned earlier, the principle of conservation of energy states that **the total energy in any isolated system is constant, no matter what happens within the system.** Energy cannot be created or destroyed. This statement is believed to be an *absolute* conservation law; no exception has ever been observed. It is one of the core principles of physics.

Here's another example of energy relations: Suppose we use a slingshot to shoot a rock straight upward (Figure 7.4). For the present, we'll ignore air resistance. Initially, the rock sits in the pouch of the stretched sling (Figure 7.4a). At this time, the sling's stretched elastic bands hold elastic potential energy. When the pouch is released, the sling propels the rock upward. The sequence in Figure 7.4b begins where the rock has just lost contact with the pouch. Most of the elastic potential energy that was stored in the sling has now become kinetic energy of the rock. As the rock rises, it slows; the gravitational potential energy of the system increases as the rock's kinetic energy decreases. The total energy of the system is constant, but there is a transformation from kinetic energy to gravitational potential energy. At the highest point in the rock's path, it comes to rest momentarily and has zero kinetic energy. At this point, all of the energy is gravitational potential energy. As the rock falls, the gravitational potential energy again becomes kinetic energy. Every change in the system represents a transformation or transfer of energy, but *the total* mechanical energy (kinetic plus potential) remains the same.

Dissipation of Mechanical Energy

Real-world systems often include interactions that allow mechanical energy to be transformed into *non*mechanical types of energy. In the case of the slingshot, for instance, we ignored air drag, which would remove kinetic energy from the system without any compensating increase in potential energy. And when the rock hits the ground, all its kinetic energy disappears, and we have a rock sitting on the ground. What became of this lost mechanical energy?

Most of the mechanical energy of this system goes into slightly warming the rock, the air through which it passes, and the ground the rock strikes. To understand this transformation, think about a brick kicked across the floor. As the brick slides, friction eventually brings it to a halt. On a microscopic level, the brick consists of molecules vibrating randomly, held together by a web of springlike bonds (Figure 7.5). *The hotter an object, the more strongly its molecules vibrate.* This random, atomic-scale vibration is a form of energy called **internal energy.** On a microscopic scale, internal energy consists of the kinetic and potential energies of individual molecules, but it is useful to distinguish it from energy of position and motion of macroscopic (large-scale) objects.

When two objects slide against each other, they catch and tug on each other. These interactions cause the surface molecules of each object to vibrate more strongly. As shown in Figure 7.6, when friction slows a sliding block, the block's kinetic energy is transformed into internal energy of random vibrations of molecules in the block, the floor, and the surrounding air.

In our slingshot example, the rock may lose kinetic energy to air drag (a form of friction). Figure 7.7 shows the effect of air drag on a tennis ball. As the ball passes through air, some of its kinetic energy is transferred to the air, increasing the random motion of the air molecules. Thus, air drag converts some of an object's kinetic energy to internal energy of both the air and the surface of the moving object.

Any force, such as friction, that turns mechanical energy into nonmechanical forms of energy, is called a **dissipative** force, because it dissipates (reduces) the mechanical energy in a system, converting it to other forms of energy.

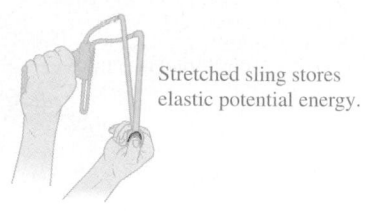

Stretched sling stores elastic potential energy.

(a) Before release

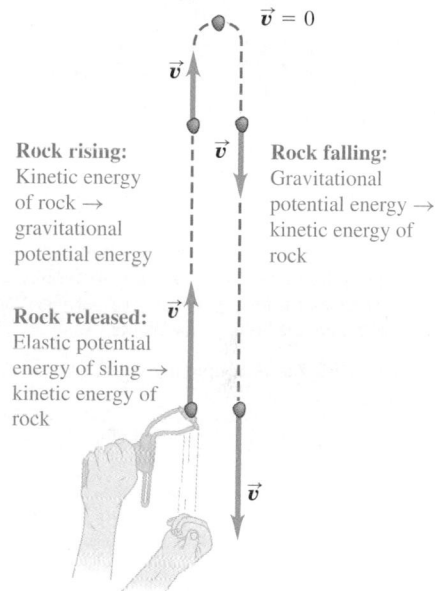

Top of trajectory:
Gravitational potential energy only

$\vec{v} = 0$

$\vec{v}$

Rock rising:
Kinetic energy of rock →
gravitational potential energy

$\vec{v}$

Rock falling:
Gravitational potential energy →
kinetic energy of rock

Rock released: $\vec{v}$
Elastic potential energy of sling →
kinetic energy of rock

$\vec{v}$

(b) After release

▲ **FIGURE 7.4** Energy is conserved as it is first transferred from a slingshot to a rock and then transformed between kinetic energy and gravitational potential energy during the rock's flight.

▶ **BIO Application Efficient locomotion through springs.** The hopping gait of kangaroos is exceptionally energy efficient. Much of this efficiency comes from the fact that kangaroo's hind legs are, in effect, big pogo sticks. The massive tendons you see in the photo are elastic, like stiff rubber bands. At the end of each hop, the impact stretches these tendons, transforming energy of motion into elastic potential energy. This stored energy helps to launch the animal on its next hop, thus transforming back to kinetic energy. Because most of the kinetic energy from each hop is saved and reused, the kangaroo's muscles need to add only a little kinetic energy at the start of each hop.

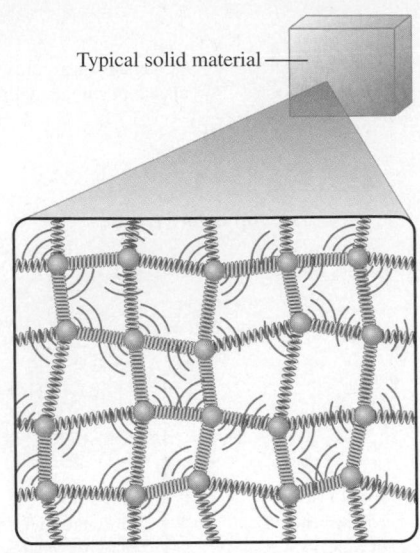

Typical solid material

The atoms and molecules of a solid can be thought of as particles vibrating randomly on springlike bonds. This random vibration is an example of **internal energy**. The stronger the vibration, the hotter is the object.

▲ **FIGURE 7.5** A simple model of internal energy in a solid.

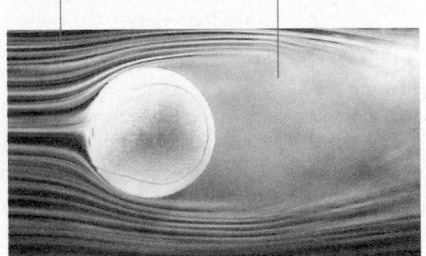

Undisturbed air ahead of the ball. The lines are thin smoke streams.

In the ball's wake, the air's motion is stronger and more random.

▲ **FIGURE 7.7** The motion of air in the wake of a tennis ball. This photo is from a wind-tunnel study of the aerodynamics of tennis balls at typical court speeds of over 100 mi/hr. The wind tunnel blows air past a fixed ball, but the effects are exactly the same as for a ball moving through still air. The lines in the air are due to thin smoke streams that are used to make the air's motion visible.

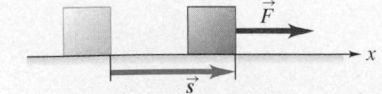

▲ **FIGURE 7.8** Work is the product of the magnitudes of a displacement $\vec{s}$ and a constant force $\vec{F}$ in the direction of $\vec{s}$.

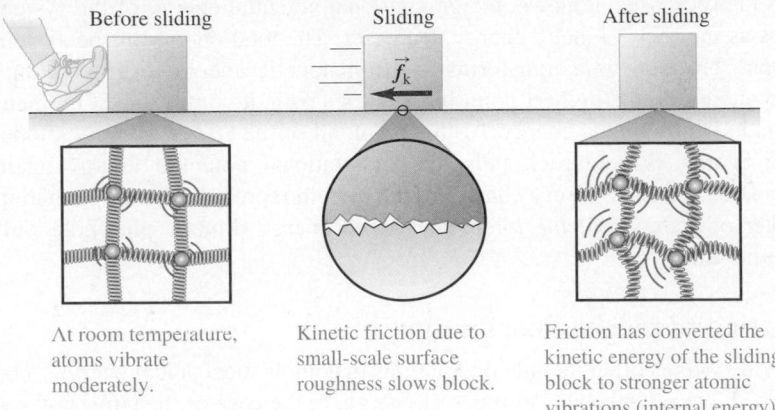

Before sliding

Sliding

After sliding

At room temperature, atoms vibrate moderately.

Kinetic friction due to small-scale surface roughness slows block.

Friction has converted the kinetic energy of the sliding block to stronger atomic vibrations (internal energy).

▲ **FIGURE 7.6** How friction converts kinetic energy to internal energy. The "atomic-scale" view in the square insets represents a far higher level of magnification than the small-scale roughness in the middle inset.

NOTE ▶ When we speak of "dissipation of mechanical energy," we really mean conversion of mechanical energy into other forms. Physicists and engineers sometimes say that a system "loses" mechanical energy. The energy isn't really lost; mechanical energy is transformed into nonmechanical energy, but the total energy (mechanical and nonmechanical) doesn't change. ◀

Let's summarize the key points we've established in this section:

- Energy can take various forms. In this chapter, we'll concentrate on kinetic energy, gravitational potential energy, and elastic potential energy, all of which are forms of mechanical energy.
- The total energy (mechanical plus nonmechanical) of an isolated system is *always* conserved. It can be transferred between objects or transformed into different forms, but its total amount remains the same.
- Mechanical energy can be dissipated to nonmechanical forms of energy. For instance, friction dissipates kinetic energy, converting it to internal energy.

7.2 Work

In everyday usage, *work* is any activity that requires muscular or mental exertion. Physicists have a much more specific definition, involving a force acting on an object while the object undergoes a displacement. The importance of the concept of work stems from the fact that, in any motion of a particle, the change in *kinetic energy* of the particle equals the total work done by the forces acting on it. We'll develop this relationship in Section 7.3; but first, let's learn how to calculate work in a variety of situations.

The simplest case is straight-line motion of an object acted on by a constant force $\vec{F}$. When the displacement $\vec{s}$ and force $\vec{F}$ have the same direction (Figure 7.8), we define the work W done by the force as the product of the magnitudes of these two quantities ($W = Fs$). More generally, if the force makes an angle ϕ with the displacement (Figure 7.9), then the component of the force along the direction of the displacement is $F_{\parallel} = F\cos\phi$ (where the subscript $\parallel$ means *parallel,* indicating that this component is parallel to the direction of displacement). We can now define work as the product of the *component* of force in the direction of displacement and the magnitude s of the displacement.

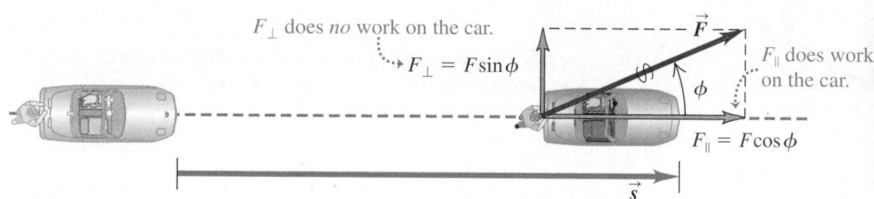

▲ **FIGURE 7.9** The work done by a constant force $\vec{F}$ during a straight-line displacement $\vec{s}$ is $(F\cos\phi)s$.

Definition of work

When an object undergoes a displacement $\vec{s}$ with magnitude s along a straight line, while a constant force $\vec{F}$ with magnitude F, making an angle ϕ with $\vec{s}$ (Figure 7.9), acts on the object, the **work** done by the force on the object is

$$W = F_{\parallel}s = (F\cos\phi)s. \qquad (7.1)$$

Unit: joule (J). In SI units, the unit of force is the newton, the unit of distance is the meter, and their product is the *newton-meter* $(\text{N} \cdot \text{m})$. This combination appears so often in mechanics that we give it a special name, the **joule** (abbreviated J and pronounced "jewel"). It was named in honor of the English physicist James Prescott Joule:

$$1 \text{ joule} = 1 \text{ newton-meter} \qquad \text{or} \qquad 1 \text{ J} = 1 \text{ N} \cdot \text{m}.$$

In the British system, the unit of force is the pound (lb), the unit of distance is the foot, and the unit of work is the *foot-pound* $(\text{ft} \cdot \text{lb})$. The following conversions are useful:

$$1 \text{ J} = 0.7376 \text{ ft} \cdot \text{lb}, \qquad 1 \text{ ft} \cdot \text{lb} = 1.356 \text{ J}.$$

▲ **Application Low and sleek.** Many engineering problems boil down to reducing dissipative forces. For instance, a standard bicycle places the rider in a position that catches a lot of wind, creating significant air drag. On a recumbent bicycle, the rider's low profile causes much less air drag. The real-life difference is so great that, in 1933, a second-string racer rode a recumbent at nearly 30 miles an hour, smashing by 10% a record almost 20 years old. After that, recumbent bicycles were banned from racing. A *fairing*—an aerodynamic shell—further reduces air drag and thus increases speed even more; faired recumbent bicycles can travel 36 mph and can break 65 mph for short distances.

EXAMPLE 7.1 Pushing a stalled car

Steve is trying to impress Elaine with his new car, but the engine dies in the middle of an intersection. While Elaine steers, Steve pushes the car 19 m to clear the intersection. If he pushes with a constant force with magnitude 210 N (about 47 lb), how much work does he do on the car **(a)** if he pushes in the direction the car is heading and **(b)** if he pushes at 30° to that direction?

SOLUTION

SET UP AND SOLVE From Equation 7.1, $W = F\cos\phi \, s$.

Part (a): When Steve pushes in the direction the car is headed, $\phi = 0$, $\cos\phi = 1$, and

$$W = (F\cos\phi)s = (210 \text{ N})(1)(19 \text{ m}) = 4.0 \times 10^3 \text{ J}.$$

Part (b): In this case, $\phi = 30°$, $\cos\phi = 0.866$, and

$$W = (F\cos\phi)s = (210 \text{ N})(0.866)(19 \text{ m}) = 3.5 \times 10^3 \text{ J}.$$

REFLECT In part (a), Steve pushed the car in the direction he wanted it to go. In (b), he pushed at an angle of 30° to that direction. Only the component of force in the direction of the car's motion, $F\cos\phi$ is effective in moving the car.

Practice Problem: What force would be required for Steve to do as much work when he pushes at $\phi = 30°$ as when he pushed in the $\phi = 0$ direction? *Answer:* 240 N.

It's important to understand that work is a *scalar* quantity, even though it is calculated by using two vector quantities (force and displacement). A 5 N force toward the east acting on an object that moves 6 m to the east does exactly the same 30 J of work as a 5 N force toward the north acting on an object that moves 6 m to the north. Work can be positive or negative. When the force has a component in the *same direction* as the displacement (ϕ between zero and 90°), $\cos\phi$ in

ActivPhysics 5.1: Work Calculations

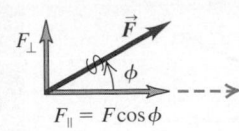
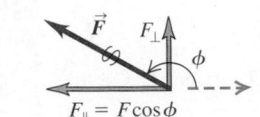

(a)

The force has a component in the direction of displacement:
• The work on the object is positive. (The object speeds up.)
• $W = F_{\parallel}s = (F\cos\phi)s$

$F_{\parallel} = F\cos\phi$

(b)

The force has a component opposite to the direction of displacement:
• The work on the object is negative. (The object slows down.)
• $W = F_{\parallel}s = (F\cos\phi)s$
• Mathematically, $W < 0$ because $F\cos\phi$ is negative for $90° < \phi < 270°$.

$F_{\parallel} = F\cos\phi$

(c)

The force is perpendicular to the direction of displacement:
• The force does *no* work on the object.
• More generally, if a force acting on an object has a component $F_{\perp}$ perpendicular to the object's displacement, that component does no work on the object.

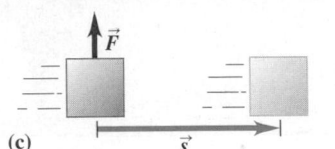
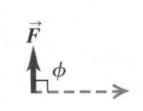

▲ **FIGURE 7.10** Three cases in which a single constant force does work on an object.

Equation 7.1 is positive, and the work W is *positive* (Figure 7.10a). When the force has a component *opposite* to the displacement (ϕ between 90° and 270°), $\cos\phi$ is negative, and the work is *negative* (Figure 7.10b). When the force is *perpendicular* to the displacement, $\phi = 90°$, and the work done by the force is *zero* (Figure 7.10c).

Work and orbital motion

Conceptual Analysis 7.1

A communications satellite moves in a circular orbit at constant speed in response to gravity, as shown in Figure 7.11. Which of the following statements is correct?

A. The earth does positive work on the satellite.
B. The earth does negative work on the satellite.
C. The earth does no work on the satellite.
D. Once a coordinate system is specified, the work changes sign every half orbit, so that the average work is zero.

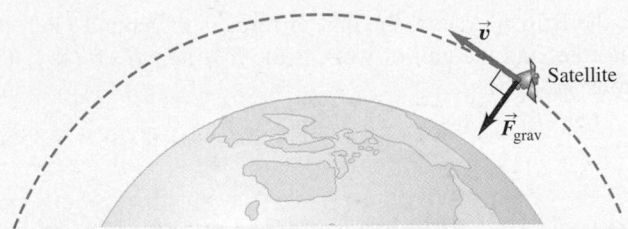

▲ **FIGURE 7.11** A satellite in a circular orbit.

SOLUTION In circular gravitational orbits, the force on the orbiting object is always perpendicular to the object's velocity, as indicated in the figure. The displacement of the object is always along the direction of the velocity, so the force is always perpendicular to the displacement. Therefore, $F_{\parallel} = 0$, and the work is zero for any displacement along the orbit. This result still might seem confusing, since your intuition may correctly guess that orbiting objects have energy. But remember that *work is the transfer of energy that changes an object's speed.* Since the speed of an object is constant in uniform circular motion, $W = 0$, and all answers other than choice C can be rejected. As we'll see later, energy is required to set an object in orbital motion, and the orbital system holds onto this energy, but no additional energy needs to be supplied to maintain the object in its orbit. Our solar system has been executing orbital motion for about 5 billion years and does not require an energy source to maintain its orbits.

We always speak of work done *by* a specific force *on* a particular object. Always be sure to specify exactly what force is doing the work you are talking about. When a spring is stretched, the work done on the spring by the stretching force is positive. When an object is lifted, the work done on the object by the lifting force is positive. However, the work done by the *gravitational* force on an object being lifted is *negative* because the (downward) gravitational force is opposite to the (upward) displacement. You might think it is "hard work" to hold this book out at arm's length for five minutes, but you aren't actually doing any work at all on the book, because there is no displacement. Even when you walk with constant velocity on a level floor while carrying the book, you do no work on it, because the (vertical) supporting force you exert on the book has no component in the direction of the (horizontal) motion. In this case, in Equation 7.1, $\phi - 90°$ and $\cos\phi = 0$. When an object slides along a surface, the work done by the normal force acting on the object is zero; and when an object moves in a circle, the work done by the centripetal force on the object is also zero. (See Figure 7.11.).

How do we calculate work when several forces act on an object? One way is to use Equation 7.1 to compute the work done by each separate force. Then, because work is a scalar quantity, the *total* work W_{total} done on the object by all the forces is the algebraic sum of the quantities of work done by the individual forces. An alternative route to finding the total work is to compute the vector sum (resultant) of the forces and then use this vector sum as $\vec{F}$ in Equation 7.1.

EXAMPLE 7.2 **Sliding down a ramp**

A package with mass m is unloaded from a truck with an inclined ramp, as shown in Figure 7.12. The ramp has rollers that eliminate friction, and the truck unloads the package from an initial height h. The ramp is inclined at an angle β. Find an algebraic expression in terms of these quantities for the work done on the package during its trip down the ramp.

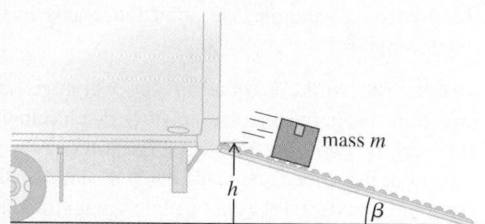

▲ **FIGURE 7.12** A package sliding down a friction-less ramp.

SOLUTION

SET UP Figure 7.13 shows our sketches. We diagram the situation, indicating the package's displacement $\vec{s}$ and the height h. Then we draw a free-body diagram for the package, pointing the x axis in the direction of the displacement. Only the weight has a component parallel to the direction of displacement, so only this force does work on the package. The normal force $\vec{n}$ is perpendicular to the displacement, so it does no work on the package.

SOLVE We solve for the work done on the package by the x (parallel) component of weight. Notice, however, that the angle we are given, β, is *not* the angle between the direction of displacement and the weight force (ϕ in Figure 7.13b). To use Equation 7.1 directly, we could define $\phi = 90 - \beta$. Instead, we choose to use β directly, modifying the trigonometry to fit. Because weight is the only force doing work on the package, the work done on the package is

$$W = (F\sin\beta)s$$
$$= (mg\sin\beta)s.$$

The statement of the problem gives the height h rather than the distance s measured along the ramp, but we can solve for h by recognizing from the geometry in Figure 7.13 that $s\sin\beta = h$. Therefore,

$$W - (mg\sin\beta)s = mgh.$$

REFLECT Because we are neglecting friction, the only force that does work on the package is its weight mg. We've found that the work W is the product of the weight mg and the vertical distance h through which the package falls. In Section 7.5, we'll restate this result in terms of potential energy, which we haven't yet defined precisely. We'll find that the work done on the package is the negative of the change in the potential energy. When the force of gravity does a positive amount of work $W = mgh$ on the package, the potential energy must decrease by the same amount. The work done on the package by gravity depends only on the change in the package's vertical position h.

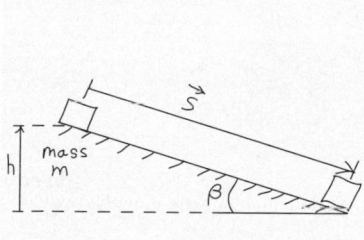

(a) Sketch of situation

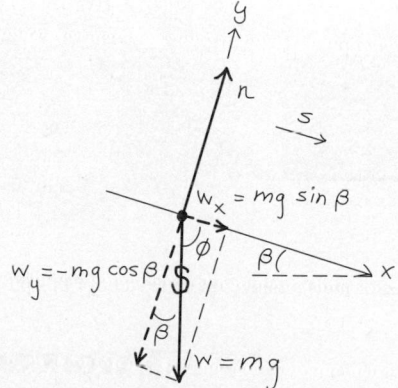

(b) Free-body diagram of package

▲ **FIGURE 7.13** Our sketches for this problem.

EXAMPLE 7.3   **Work done by several forces**

Farmer Johnson hitches his tractor to a sled loaded with firewood and pulls it a distance of 20.0 m along level frozen ground (Figure 7.14a). The total weight of sled and load is 14,700 N. The tractor exerts a constant force $\vec{F}_T$ with magnitude 5000 N at an angle of $\phi = 36.9°$ above the horizontal, as shown. A constant 3500 N friction force opposes the motion. Find the work done on the sled by each force individually and the total work done on the sled by all the forces.

SOLUTION

SET UP Figure 7.14b shows a free-body diagram and a coordinate system, identifying all the forces acting on the sled. As in the preceding example, we point the x axis in the direction of displacement.

SOLVE The work W_w done by the weight is zero because its direction is perpendicular to the displacement. (The angle between the two directions is 90°, and the cosine of the angle is zero.) For the same reason, the work W_n done by the normal force $\vec{n}$ (which, incidentally, is *not* equal in magnitude to the weight) is also zero. So $W_w = W_n = 0$.

That leaves F_T and f. From Equation 7.1, the work W_T done by the tractor is (with $\cos \phi = \cos 36.9° = 0.800$)

$$W_T = (F_T \cos \phi)s$$
$$= (5000 \text{ N})(0.800)(20.0 \text{ m}) = 80,000 \text{ N} \cdot \text{m} = 80.0 \text{ kJ}.$$

The friction force f is opposite to the displacement, so, for this force, $\phi = 180°$ and $\cos \phi = -1$. The work W_f done by the friction force is

$$W_f = fs \cos 180° = -(3500 \text{ N})(20.0 \text{ m})$$
$$= -70,000 \text{ N} \cdot \text{m} = -70.0 \text{ kJ}.$$

The total work W_{total} done by all of the forces on the sled is the algebraic sum (*not* the vector sum) of the work done by the individual forces:

$$W_{total} = W_T + W_w + W_n + W_f$$
$$= 80.0 \text{ kJ} + 0 + 0 + (-70.0 \text{ kJ}) = 10.0 \text{ kJ}.$$

Alternative Solution: In the alternative approach, we first find the vector sum (resultant) of the forces and then use it to compute the total work. The vector sum is best found by using components. From Figure 7.14b,

$$\Sigma F_x = (5000 \text{ N}) \cos 36.9° - 3500 \text{ N} = 500 \text{ N},$$
$$\Sigma F_y = (5000 \text{ N}) \sin 36.9° + n + (-14,700 \text{ N}).$$

We don't really need the second equation; we know that the y component of force is perpendicular to the displacement, so it does no work. Besides, there is no y component of acceleration, so ΣF_y has to be zero anyway. The work done by the total x component is therefore the total work:

$$W_{total} = (500 \text{ N})(20.0 \text{ m}) = 10,000 \text{ J} = 10.0 \text{ kJ}.$$

This is the same result that we found by computing the work of each force separately.

REFLECT Be sure you understand why the two solution methods are equivalent and give the same result.

Practice Problem: Suppose the tractor pulls horizontally on the sled instead of at an angle of 36.9°. As a result, the magnitude of the friction force increases to 4400 N. What is the total work done on the sled? *Answer:* 1.20×10^4 J.

(a)

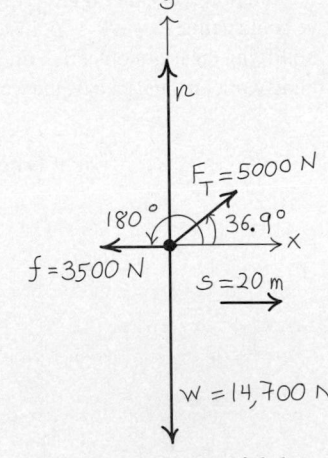

(b) Free-body diagram of sled

▲ **FIGURE 7.14** (a) A tractor pulls a sled. (b) Our sketch for this problem.

7.3 Work and Kinetic Energy

Now we're ready to develop the relation between the work done on an object and that object's change in kinetic energy. This important relationship is called the *work–energy theorem* (more precisely, the work–kinetic energy theorem).

Let's consider a particle with mass m moving along the x axis under the action of a constant net force with magnitude F_{total} directed along the x axis. The particle's acceleration is constant and is given by Newton's second law, $F_{\text{total}} = ma$. Suppose the particle's speed increases from v_i to v_f while the particle undergoes a displacement s from point x_i to x_f. Using the constant-acceleration equation (Equation 2.13) and replacing v_{0x} by v_i, v_x by v_f, and $(x - x_0)$ by s, we have

$$v_f^2 = v_i^2 + 2as,$$

$$a = \frac{v_f^2 - v_i^2}{2s}.$$

When we multiply the last equation by m and replace ma with F_{total}, we obtain

$$F_{\text{total}} = ma = m\frac{v_f^2 - v_i^2}{2s}$$

and

$$F_{\text{total}}s = \tfrac{1}{2}mv_f^2 - \tfrac{1}{2}mv_i^2. \tag{7.2}$$

The product $F_{\text{total}}s$ is the work done by the net force F_{total} and thus is equal to the total work W_{total} done by all the forces. In Section 7.1, we defined the quantity $\tfrac{1}{2}mv^2$ as kinetic energy. We'll denote this quantity as K. Then

$$K = \tfrac{1}{2}mv^2. \tag{7.3}$$

The first term on the right side of Equation 7.2 is the final kinetic energy of the particle, $K_f = \tfrac{1}{2}mv_f^2$ (after the displacement). The second term is the initial kinetic energy, $K_i = \tfrac{1}{2}mv_i^2$, and the difference between these terms is the *change* in the particle's kinetic energy during the displacement s. So Equation 7.2 relates the particle's change in kinetic energy to the work done by the forces that act on the particle.

Work–energy theorem

The kinetic energy K of a particle with mass m moving with speed v is $K - \tfrac{1}{2}mv^2$. During any displacement of the particle, the work done by the net external force on it is equal to its change in kinetic energy, or

$$W_{\text{total}} = K_f - K_i = \Delta K. \tag{7.4}$$

This theorem is the foundation for most of what follows in this chapter.

Kinetic energy, like work, is a scalar quantity. It can never be negative, although work can be either positive or negative. The kinetic energy of a moving particle depends only on its speed (the magnitude of its velocity), not on the direction of its motion. A car (viewed as a particle) has the same kinetic energy when going north at 5 m/s as when going east at 5 m/s. The *change* in kinetic energy during any displacement is determined by the total work W_{total} done by all the forces acting on the particle. When W_{total} is *positive,* the particle speeds up during the displacement, K_f is greater than K_i, and the kinetic energy *increases.* When W_{total} is *negative,* the kinetic energy and speed *decrease;* and when $W_{\text{total}} = 0$, K is *constant.*

In SI units, m is measured in kilograms and v in meters per second, so the quantity $\tfrac{1}{2}mv^2$ (kinetic energy) has the units $\text{kg} \cdot \text{m}^2/\text{s}^2$. From Equation 7.2 or Equation 7.4, kinetic energy must have the same units as work, which we saw is measured in joules (newton-meters). To confirm that these units are the same, we recall from Newton's second law that $1 \text{ N} = 1 \text{ kg} \cdot \text{m}/\text{s}^2$, so

$$1 \text{ J} = 1 \text{ N} \cdot \text{m} = 1 \left(\text{kg} \cdot \text{m}/\text{s}^2\right) \cdot \text{m} = 1 \text{ kg} \cdot \text{m}^2/\text{s}^2.$$

The joule is the SI unit of both work and kinetic energy and, indeed, as we will see later, of all kinds of energy. In the British system,

$$1 \, \text{ft} \cdot \text{lb} = 1 \, \text{ft} \cdot \text{slug} \cdot \text{ft}/\text{s}^2 = 1 \, \text{slug} \cdot \text{ft}^2/\text{s}^2.$$

NOTE ▶ In calculating W_{total} for Equation 7.4, remember what we learned in Section 7.2: First, only the component of net force parallel to the displacement does work on the particle; second, W_{total} can be calculated either as the work done by the net force or as the algebraic sum of the amounts of work done by the individual forces. ◀

Conceptual Analysis 7.2

Work and kinetic energy

Two blocks of ice, one twice as heavy as the other, are at rest on a frozen lake. A person pushes each block a distance of 5 m with a constant force (the same magnitude of force for each block). Assume that friction may be neglected. The kinetic energy of the light block after the push is

A. smaller than that of the heavy block.
B. equal to that of the heavy block.
C. larger than that of the heavy block.

SOLUTION Three forces act on each block: its weight, the normal force exerted by the surface, and the push applied by the person. The weight and the normal force are perpendicular to the displacement, so they do no work on the block. Only the person's push does work on each block. Both this force and the displacement are the same for the two blocks, so the amounts of work done on them are equal. The work–energy theorem, $W_{\text{total}} = K_f - K_i = \Delta K$, says that the change in each block's kinetic energy equals the net work done on it. Both blocks have the same initial kinetic energy ($K_i = 0$) and have equal amounts of work done on them. Therefore, they must have equal final kinetic energies K_f. Answer B is correct.

PROBLEM-SOLVING STRATEGY 7.1   **Work and kinetic energy**

SET UP

1. Draw a free-body diagram; make sure you show all the forces that act on the object. List the forces; represent any unknown forces by algebraic symbols.
2. List the initial and final kinetic energies K_i and K_f. If a quantity such as v_i or v_f is unknown, express it in terms of an algebraic symbol. When you calculate kinetic energies, make sure you use the *mass* of the object, not its *weight*.

SOLVE

3. Calculate the total work done on the object. You can do this in either of two ways: by calculating the work done by each individual force and taking the algebraic sum (usually the easiest method) or by finding the net force and then calculating the work it does on the object. Be sure to check signs. When a force has a component in the same direction as the displacement, its work is positive; when the direction is opposite to the displacement, the work is negative. When force and displacement are perpendicular, the work is zero.
4. Use the relationship $W_{\text{total}} = K_f - K_i = \Delta K$; insert the results from the preceding steps and solve for whatever unknown is required.

REFLECT

5. Remember that kinetic energy can never be negative. If you come up with a negative K, you've made a mistake. Maybe you interchanged subscripts i and f or made a sign error in one of the calculations of work.

EXAMPLE 7.4   **Using work and energy to calculate speed**

Let's revisit the sled from Example 7.3. The free-body diagram is shown again in Figure 7.15. We found that the total work done on the sled by all the forces is 10,000 J = 10.0 kJ, so the kinetic energy of the sled must increase by 10.0 kJ. The mass of the sled is $m = (14,700 \, \text{N})/(9.80 \, \text{m/s}^2) = 1500 \, \text{kg}$. Suppose the sled's initial speed v_i is 2.00 m/s. What is its final speed?

Continued

SOLUTION

SET UP Steps 1 and 2 of the problem-solving strategy were done in Example 7.3, where we found that $W_{total} = 10.0$ kJ. The initial kinetic energy K_i is

$$K_i = \tfrac{1}{2}mv_i^2$$
$$= \tfrac{1}{2}(1500 \text{ kg})(2.00 \text{ m/s})^2 = 3000 \text{ kg} \cdot \text{m}^2/\text{s}^2 = 3000 \text{ J}.$$

The final kinetic energy K_f is

$$K_f = \tfrac{1}{2}(1500 \text{ kg})v_f^2,$$

where v_f is the unknown final speed that we want to find.

SOLVE Equation 7.4 gives

$$K_f = K_i + W_{total},$$
$$\tfrac{1}{2}(1500 \text{ kg})v_f^2 = 3000 \text{ J} + 10{,}000 \text{ J} = 13{,}000 \text{ J}.$$

Solving for v_f, we find

$$v_f = 4.16 \text{ m/s}.$$

REFLECT This problem can also be done without the work–energy theorem. We can find the acceleration from $F = ma$ and then use the equations of motion with constant acceleration to find v_f. (Recall from Example 7.3 that the sled moves through a displacement of 20.0 m.) We have

$$a = \frac{F}{m} = \frac{4000 \text{ N} - 3500 \text{ N}}{1500 \text{ kg}} = 0.333 \text{ m/s}^2,$$

and

$$v_f^2 = v_i^2 + 2as = (2.0 \text{ m/s})^2 + 2(0.333 \text{ m/s}^2)(20 \text{ m})$$
$$= 17.3 \text{ m}^2/\text{s}^2,$$
$$v_f = 4.16 \text{ m/s}$$

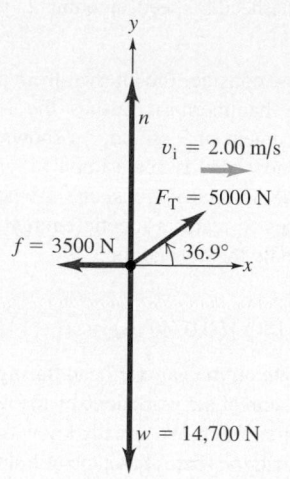

▲ **FIGURE 7.15** The free-body diagram from Example 7.3.

The work–energy approach lets us avoid the intermediate step of finding the acceleration. You'll find several other examples and problems in this chapter and the next that *can* be done without using energy considerations, but that are easier when energy methods are used. Also, when a problem can be done by two different methods, doing it both ways is always a good way to check your work.

Practice Problem: If the tractor pulls horizontally and the friction force again has magnitude 4400 N, find the magnitude of the force the tractor must exert to make the sled move with constant speed. Verify that in this case the kinetic energy of the sled doesn't change. *Answer:* 4400 N.

EXAMPLE 7.5 Forces on a hammerhead

In a pile driver, a steel hammerhead with mass 200 kg is lifted 3.00 m above the top of a vertical I-beam being driven into the ground (Figure 7.16a). The hammer is then dropped, driving the I-beam 7.40 cm farther into the ground. The vertical rails that guide the hammerhead exert a constant 60.0 N friction force on it. Use the work–energy theorem to find **(a)** the speed of the hammerhead just as it hits the I-beam and **(b)** the average force the hammerhead exerts on the I-beam.

SOLUTION

SET UP Figures 7.16b and 7.16c are free-body diagrams for the hammerhead during two intervals: first, while it falls (from point 1 to point 2 in Figure 7.16a) and second, while it pushes the I-beam (from point 2 to point 3). Because the displacement is vertical, any horizontal forces that may be present do no work. Therefore, our free-body diagrams show only the vertical forces on the hammerhead.

SOLVE The forces on the hammerhead during the 3.00 m drop are its weight, with magnitude $mg = (200 \text{ kg})(9.80 \text{ m/s}^2) = 1960 \text{ N}$ and the upward friction force, with magnitude 60.0 N. The net

downward force on the hammerhead has magnitude 1900 N, and the total work done on it during the 3.00 m drop is

$$W_{total} = (1900 \text{ N})(3.00 \text{ m}) = 5700 \text{ J}.$$

Part (a): In place of our usual subscripts i and f, we'll use 1 and 2, referring to points 1 and 2 in Figure 7.16. The hammerhead is initially at rest (point 1), so its initial kinetic energy K_1 is zero. Equation 7.4 gives

$$W_{total} = K_2 - K_1,$$
$$5700 \text{ J} = \tfrac{1}{2}(200 \text{ kg})v_2^2 - 0,$$
$$v_2 = 7.55 \text{ m/s}.$$

Continued

This is the hammerhead's speed at point 2, just as it hits the beam.

Part (b):   We now consider the interval from point 2 to point 3, during which the hammerhead pushes the I-beam through a downward displacement of 7.40 cm. As shown in Figure 7.16c, the hammerhead now also is acted upon by an upward normal force $\vec{n}$ (which we assume is constant). At point 3, where the hammerhead comes to rest, its kinetic energy K_3 is again zero. The work W_n done by the normal force is

$$W_n = (F\cos\phi)s$$
$$= (n\cos 180°)(0.0740 \text{ m}) = n(-1)(0.0740 \text{ m}).$$

The total work done on the hammerhead during its displacement of 0.0740 m is the sum of the work done by the weight, the friction force, and the normal force. We already know that the vector sum of the weight and friction forces has a magnitude of 1900 N, so we sum the work done by that resultant force and by the normal force:

$$W_{\text{total}} = (1900 \text{ N})(0.0740 \text{ m}) + n(0.0740 \text{ m})(-1).$$

The total work is equal to the change in kinetic energy, $K_3 - K_2 = 0 - 5700$ J, so we have

$$(1900 \text{ N} - n)(0.0740 \text{ m}) = 0 - 5700 \text{ J},$$
$$n = 79,000 \text{ N}.$$

The force *on* the I-beam is the equal, but opposite, *downward* reaction force of 79,000 N (about 9 tons).

REFLECT The total change in the hammerhead's kinetic energy during the whole process is zero; a relatively small force does positive work over a large distance, and then a much larger force does negative work over a much smaller distance. The same thing happens if you speed your car up gradually and then drive it into a brick wall. The very large force needed to reduce the kinetic energy to zero over a short distance is what does the damage to your car—and possibly to you!

Practice Problem: Suppose the pile is driven 15.0 cm instead of 7.4 cm. What force (assumed constant) is exerted on it? *Answer:* 39,900 N.

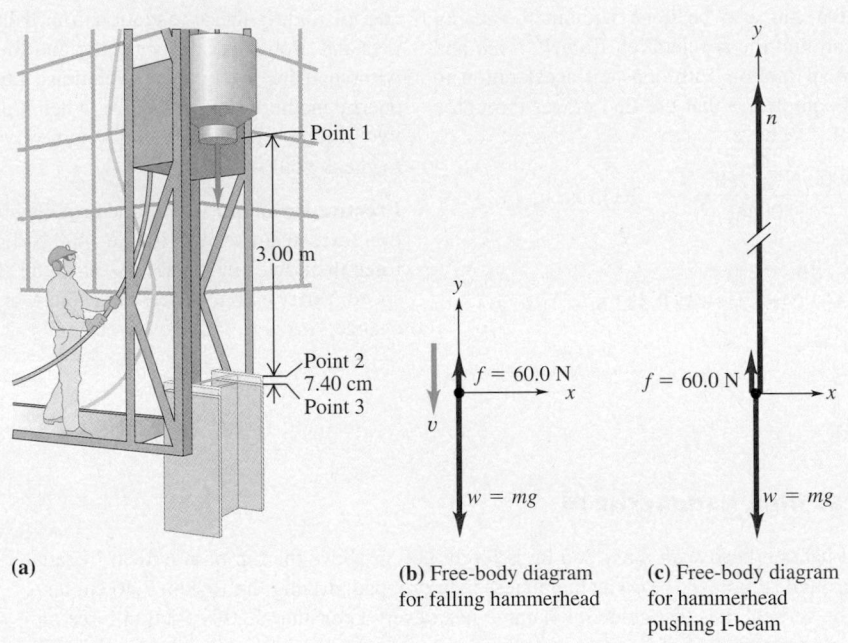

(a)

(b) Free-body diagram for falling hammerhead

(c) Free-body diagram for hammerhead pushing I-beam

▲ **FIGURE 7.16**

NOTE ▶ Because we used Newton's laws in deriving the work–kinetic energy theorem, we must use the theorem only in an inertial frame of reference. The speeds we use to compute the kinetic energies and the distances we use to compute work *must* be measured in an inertial frame. But the work–energy theorem is valid in *all* inertial frames. For a given situation, the work and kinetic energies will be different in different frames of reference because the speed of an object is different in different frames. ◀

7.4 Work Done by a Varying Force

In Section 7.2, we defined work done by a *constant* force. But what happens when you stretch a spring? The more you stretch it, the harder you have to pull, so the force is *not* constant as the spring is stretched. You can think of many other situations in which a force that varies, in both magnitude and direction, acts on an

object moving along a curved path. We need to be able to compute the work done by the force in these more general situations.

To add only one complication at a time, let's consider straight-line motion under the influence of a force that is directed along the line, but that may change in magnitude as the object moves. For example, imagine a train on a straight track, with the engineer constantly changing the locomotive's throttle setting or applying the brakes. Suppose a particle moves along the x axis from point x_i to point x_f, as shown in Figure 7.17a. Figure 7.17b is a graph of the x component of force as a function of the particle's coordinate x. To find the work done by this force, we divide the total displacement into small segments Δx_1, Δx_2, and so on (Figure 7.17c). We approximate the work done by the force during segment Δx_1 as the average force F_{1x} in that segment, multiplied by its length Δx_1. We do this for each segment and then add the results for all the segments. The work done by the force in the total displacement from x_i to x_f is approximately

$$W = F_{1x}\,\Delta x_1 + F_{2x}\,\Delta x_2 + F_{3x}\,\Delta x_3 + \cdots.$$

We see that $F_{1x}\,\Delta x_1$ represents the *area* of the first vertical strip in Figure 7.17c, $F_{2x}\,\Delta x_2$ the area of the second, and so on. The sum represents the area under the curve of Figure 7.17b between x_i and x_f. **On a graph of force as a function of position, the total work done by the force is represented by the area under the curve between the initial and final positions.**

Now let's apply this analysis to a stretched spring. To keep a spring stretched an amount x beyond its unstretched length, we have to apply a force with magnitude F at each end (Figure 7.18). As we learned in Chapter 5, if the elongation is not too great, the spring obeys Hooke's law; that is, F is directly proportional to x, or

$$F = kx, \tag{7.5}$$

where k is a constant called the force constant (or spring constant) of the spring. Equation 7.5 shows that the units of k are force divided by distance. In SI, the units are N/m.

Suppose we apply forces with equal magnitude and opposite direction to the ends of the spring and gradually increase the forces, starting from zero. We hold the left end stationary; the force at this end does no work on the spring. The force at the moving end *does* do work. Figure 7.19 is a graph of the x component of force, F_x, as a function of x (the elongation of the spring). The total work done on the spring by the force F_x when the elongation goes from zero to a maximum value X is represented by the area under the graph (the shaded triangle in the figure). The area is equal to half the product of the base (X) and altitude (kX), or

$$W = \tfrac{1}{2}(X)(kX) = \tfrac{1}{2}kX^2. \tag{7.6}$$

This result also says that the work is equal to the *average* force $kX/2$, multiplied by the total displacement X. Thus, the total work is proportional to the *square* of the final elongation X. When the elongation is doubled, the total work needed increases by a factor of four.

The spring also exerts a force on the hand, which moves during the stretching process. The displacement of the hand is the same as that of the

Particle moving from x_i to x_f in response to a changing force in the x direction

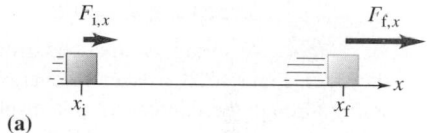

(a)

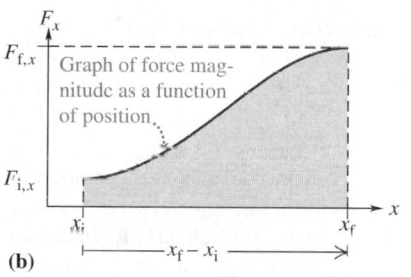

(b)

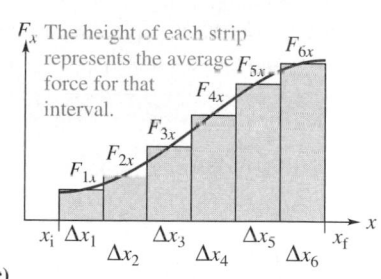

(c)

▲ **FIGURE 7.17** Graphs of force as a function of displacement for a particle moving in response to a varying force.

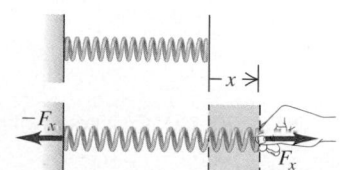

▲ **FIGURE 7.18** The force needed to stretch an ideal spring is proportional to the elongation: $F_x = kx$ (Hooke's law).

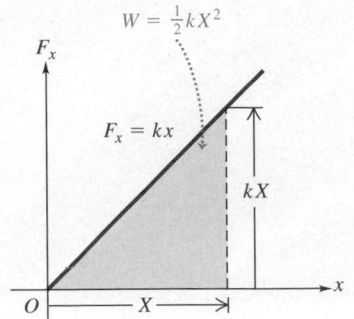

The area under the curve represents the work done on the spring as the spring is stretched from $x = 0$ to a maximum value X:

$$W = \tfrac{1}{2}kX^2$$

$F_x = kx$

kX

▲ **FIGURE 7.19** The work done in stretching a spring is equal to the area of the shaded triangle.

moving end of the spring, but the force on it is opposite in direction to the force on the spring, because the two forces form an action–reaction pair. Thus, the work done *on* the hand *by* the spring is the negative of the work done on the spring, namely, $-\tfrac{1}{2}kX^2$. In problems involving work and its relation to kinetic energy, we will nearly always want to find the work done *by* a force *on* the object under study, and we have to be careful to write work quantities with the correct sign.

Now, suppose the spring is stretched a distance x_i at the start. Then the work we have to do on it to stretch it to a greater elongation x_f is

$$W = \tfrac{1}{2}kx_f^2 - \tfrac{1}{2}kx_i^2. \tag{7.7}$$

If the spring has spaces between the coils when it is unstretched, then it can be compressed as well as stretched, and Hooke's law holds for compression as well as stretching. In this case, F and x in Equation 7.5 are both negative, the force again has the same direction as the displacement, and the work done by F is again positive. So the total work is still given by Equation 7.6 or 7.7, even when X or either or both of x_i and x_f are negative.

Quantitative Analysis 7.3

Stretching a spring

A spring is stretched from $x = 0$ to $x = 2a$, as shown in Figure 7.20a. Is more energy required to stretch the spring through the first half of this displacement (from $x = 0$ to $x = a$) or through the second half (from $x = a$ to $x = 2a$)?

SOLUTION Hooke's law $(F_x = kx)$ tells us that the force required to stretch a spring increases with increasing elongation. Thus, more energy is required to stretch the spring through the second half of its displacement. The graph in Figure 7.20b shows how much more. As we've seen, the area under a graph of force versus displacement for an object represents the work done by the force on the object. In this case, the area under the curve for the second half of the displacement can be divided into three triangles, each of which has an area equal to that under the curve for the first half. Thus, three times as much energy is needed to stretch the spring through the second half of the displacement than through the first half.

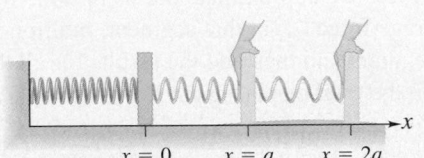

(a) Stretching a spring through two equal halves of a total displacement $2a$

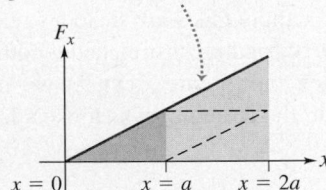

This half of the area can be divided into three triangles, each equal in area to the triangle under the first half of the curve.

F_x

(b) Force-versus-distance curve for the two halves of the displacement

▲ **FIGURE 7.20**

EXAMPLE 7.6 **Work done on a spring scale**

A woman weighing 600 N steps on a bathroom scale containing a heavy spring (Figure 7.21). The spring compresses by 1.0 cm under her weight. Find the force constant of the spring and the total work done on it during the compression.

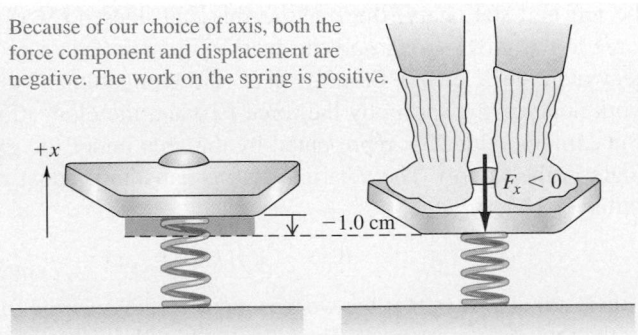

Because of our choice of axis, both the force component and displacement are negative. The work on the spring is positive.

$+x$

$F_x < 0$

-1.0 cm

▶ **FIGURE 7.21** Compressing the spring of a bathroom scale.

Continued

SOLUTION

SET UP AND SOLVE If positive values of x correspond to elongation, then $x = -0.010$ m when $F_x = -600$ N. We need to use Equation 7.5 (Hooke's law) to find the force constant k and then use Equation 7.6 to find the work W done on the spring. We have

$$k = \frac{F_x}{x} = \frac{-600 \text{ N}}{-0.010 \text{ m}} = 6.0 \times 10^4 \text{ N/m},$$

$$W = \tfrac{1}{2}kX^2 = \tfrac{1}{2}(6.0 \times 10^4 \text{ N/m})(-0.010 \text{ m})^2$$
$$= 3.0 \text{ N} \cdot \text{m} = 3.0 \text{ J}.$$

REFLECT The force of gravity is the woman's weight, a constant 600 N. As she steps on the scale, the force pushing down on the top of the spring varies from 0 to 600 N, with an average value of 300 N, as it is compressed 0.010 m. The total work done on the spring is the average force (300 N) multiplied by the displacement (0.010 m), or 3.0 J.

Practice Problem: Suppose we want to modify the scale to permit a weight of 900 N when the spring is compressed 1.00 cm. What should the force constant of the spring be? *Answer:* 9.0×10^4 N/m.

7.5 Potential Energy

An object gains or loses kinetic energy because it interacts with other objects that exert forces on it. We've learned that, during any interaction, the change in a particle's kinetic energy is equal to the total work done on the particle by the forces that act on it during the interaction.

But there are many situations in which it seems as though kinetic energy has been *stored* in a system, to be recovered later. It's like a savings account in a bank: You deposit money, and then you can draw it out later. When we raise the steel hammerhead of Example 7.5 into the air, we're storing energy in the system, energy that is later converted into kinetic energy as the hammer is released and falls back down. Or consider the humble pogo stick, in which a spring is compressed as you jump on the stick and then rebounds, launching you into the air. Or a mousetrap, in which the spring stores energy that is released when the trigger is tripped. And so on.

All these examples point to the idea of an energy associated with the *position* of bodies in a system. In some cases, changes in this energy may accompany opposite changes in kinetic energy in such a way that the *total* energy remains constant, or is *conserved*. Energy associated with position is called potential energy, and forces that can be associated with a potential energy are called **conservative forces.** A system in which the total mechanical energy, kinetic and potential, is constant, is called a **conservative system.** As we'll see, there is potential energy associated with an object's weight and its height above the ground. There is also potential energy associated with *elastic* deformations of an object, such as stretching or compressing a spring (Figure 7.21).

Another way to look at potential energy is that "potential" refers to the potential for a force to do work on an object. When the pile-driver hammer is raised, the earth's gravitational force (the hammer's weight) gains the *opportunity* to do work on the hammer as it falls. When we set a mousetrap, we give it the opportunity to do work on whatever is in its way when the spring snaps back. So potential energy is associated with the opportunity, or potential, to do work.

The kinetic energy of a falling object increases as it drops because a force (the earth's gravity, or, in other words, the object's weight) is doing positive work on it. Remember, $W_{\text{total}} = \Delta K$. Another way to say this is that the kinetic energy increases because the potential energy decreases. When the force we are considering (such as the pile-driver's weight) does positive work, the potential energy must decrease. So we can think of potential energy as a shorthand way to calculate the work done by some of the forces in a system.

Gravitational Potential Energy

To begin our detailed study of potential energy, let's apply the work–energy theorem to the work done on an object when it moves under the action of the earth's

▲ **BIO Application Energy-efficient transport.** In Indonesia and other countries, some women are known to carry loads of up to 70% of their body weight in baskets on their heads. Westerners who attempt to carry a load in this way expend almost twice as much energy as people who do it normally. This puzzling difference prompted an additional study, which found that, when Westerners walk, they have more up-and-down motion than do women accustomed to carrying baskets. For larger vertical motion, more work is done by gravity on the load; thus, more energy is required to raise the load a higher distance against the force of gravity.

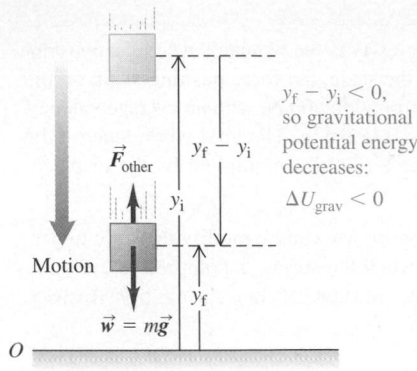

$y_f - y_i < 0,$
so gravitational potential energy decreases:

$\Delta U_{grav} < 0$

(a)

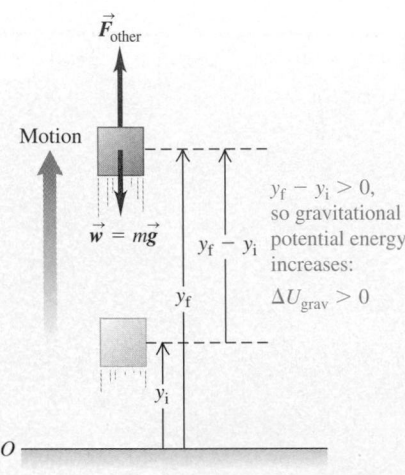

$y_f - y_i > 0,$
so gravitational potential energy increases:

$\Delta U_{grav} > 0$

(b)

▲ **FIGURE 7.22** The work done by the gravitational force w as an object moves downward or upward changes the gravitational potential energy.

gravitational force (again, the object's weight). This will lead directly to the concept of *gravitational potential energy*—energy associated with the *position* of the object relative to the earth.

Let's consider an object with mass m that moves along the (vertical) y axis (Figure 7.22). The forces acting on the object are its weight, with magnitude $w = mg$, and possibly some other forces; we call the resultant of all the other forces $\vec{F}_{other}$. We'll assume that the object stays close enough to earth's surface that the weight is constant. We want to find the work done by the weight when the object drops from a height y_i above the origin to a smaller height y_f (Figure 7.22a). The weight and displacement are in the same direction, so the work W_{grav} done on the object by its weight is positive and is given by

$$W_{grav} = Fs = mg(y_i - y_f) = mgy_i - mgy_f. \tag{7.8}$$

This expression also gives the correct work when the object moves upward and y_f is greater than y_i (Figure 7.22b). In that case, the quantity $(y_i - y_f)$ is negative, and W_{grav} is negative because the weight and displacement are opposite in direction.

Equation 7.8 shows that we can express W_{grav} in terms of the values of the quantity mgy at the beginning and at the end of the displacement. This quantity, the product of the weight mg and the height y above the origin of coordinates, is called the gravitational potential energy and is denoted by U_{grav}.

Definition of gravitational potential energy

When an object with mass m is a vertical distance y above the origin of coordinates, in a uniform gravitational field g, the gravitational potential energy U_{grav} of the system is

$$U_{grav} = mgy. \tag{7.9}$$

Like work and all forms of energy, gravitational potential energy is measured in joules.

In Figure 7.22, the initial value of U_{grav} is $U_{grav,i} = mgy_i$, and its final value is $U_{grav,f} = mgy_f$. We can express the work W_{grav} done by the gravitational force during the displacement from y_i to y_f as

$$W_{grav} = U_{grav,i} - U_{grav,f} = -\Delta U_{grav}. \tag{7.10}$$

The negative sign in front of ΔU_{grav} is essential. Remember that ΔU_{grav} always means the final value minus the initial value. When the object moves down, y decreases, the gravitational force does *positive* work, and the potential energy *decreases* (Figure 7.22a). It's like drawing some money out of the bank (decreasing U) and spending it (doing positive work). When the object moves up (Figure 7.22b), the work done by the gravitational force is *negative,* and the potential energy *increases*—you put money into the bank.

> **NOTE** ▶ We use the subscript "grav" to distinguish gravitational potential energy from other forms of potential energy we'll introduce later. But in problems where the *only* potential energy is gravitational, we'll sometimes drop this subscript to simplify the notation. ◀

Here's a simple example of the usefulness of the concept of potential energy. An object moves along a straight vertical line. Suppose the object's weight is the *only* force acting on it, so that $\vec{F}_{other} = 0$. The object may be in free fall, or perhaps we throw it upward. Let its speed at point y_i be v_i and its speed at y_f be v_f. The work–energy theorem, Equation 7.4, says that $W_{total} = K_f - K_i$, and in this case, $W_{total} = W_{grav} = U_i - U_f$. Putting these equations together, we get

$$U_i - U_f = K_f - K_i,$$

which we can rewrite as

$$K_i + U_i = K_f + U_f,$$

or

$$\tfrac{1}{2}mv_i^2 + mgy_i = \tfrac{1}{2}mv_f^2 + mgy_f. \qquad (7.11)$$

We now define $K + U$ to be the **total mechanical energy** (kinetic plus potential) of the system; let's call it E. Then $E_i = (K_i + U_i)$ is the total energy when the object is at position y_i, and $E_f = (K_f + U_f)$ is the total energy at y_f. Equation 7.11 says that when the object's weight is the only force doing work on it, $E_i = E_f$. That is, E is constant; it has the same value at y_i and at all points during the motion. **When $\vec{F}_{other} = 0$, the total mechanical energy is constant or conserved.** This is our first example of the principle of **conservation of energy**.

When we throw a ball straight upward, it slows down on the way up as kinetic energy is converted to potential energy. On the way back down, potential energy is converted back to kinetic energy, and the ball speeds up. But the total energy, kinetic plus potential, is the same at every point in the motion. (We're assuming, of course, that air resistance and other frictional effects are negligible.)

EXAMPLE 7.7 Height of a baseball from energy conservation

You throw a 0.150 kg baseball straight up in the air, giving it an initial upward velocity with magnitude 20.0 m/s. Use conservation of energy to find how high it goes, ignoring air resistance.

SOLUTION

SET UP The only force doing work on the ball after it leaves your hand is its weight, and we can use Equation 7.11. We place the origin at the starting (initial) point, (point i), where the ball leaves your hand; then $y_i = 0$ (Figure 7.23). At this point, $v_i = 20.0$ m/s. We want to find the height at the final point (point f), where the ball stops and begins to fall back to earth. At this point, $v_f = 0$ and y_f is unknown.

SOLVE Equation 7.11 says that $K_i + U_i = K_f + U_f$, or

$$\tfrac{1}{2}mv_i^2 + mgy_i = \tfrac{1}{2}mv_f^2 + mgy_f,$$
$$\tfrac{1}{2}(0.150 \text{ kg})(20.0 \text{ m/s})^2 + (0.150 \text{ kg})(9.80 \text{ m/s}^2)(0)$$
$$= \tfrac{1}{2}(0.150 \text{ kg})(0)^2 + (0.150 \text{ kg})(9.80 \text{ m/s}^2)y_f,$$
$$y_f = 20.4 \text{ m}.$$

The mass divides out, as we should expect; we learned in Chapter 2 that the motion of an object in free fall doesn't depend on its mass.

REFLECT We could have substituted the values $y_i = 0$ and $v_f = 0$ in Equation 7.11 and then solved algebraically for y_f to get

$$y_f = \frac{v_i^2}{2g}.$$

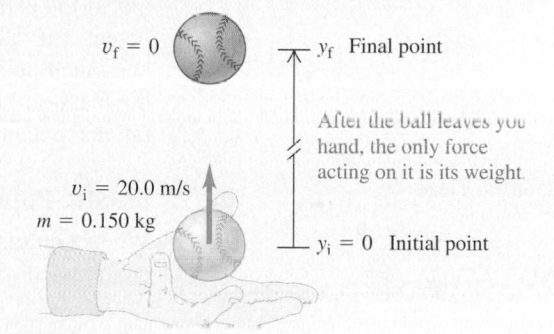

$v_f = 0$

y_f Final point

After the ball leaves your hand, the only force acting on it is its weight.

$v_i = 20.0$ m/s
$m = 0.150$ kg

$y_i = 0$ Initial point

▲ **FIGURE 7.23**

We could also have derived this result by using the methods of Section 2.6, without energy considerations.

Practice Problem: You throw the same baseball up from the surface of the moon (where the acceleration due to gravity is 1.62 m/s^2) with the same speed as before. Find how high the ball goes. *Answer:* 123 m.

An important point about gravitational potential energy is that it doesn't matter where we put the origin of coordinates. If we shift the origin for y, both y_i and y_f change, but the *difference* $(y_f - y_i)$ does not. It follows that, although the potential energies U_i and U_f depend on where we place the origin, their difference $(U_f - U_i)$ does not. The physically significant quantity is not the value of U at a particular point, but only the *difference* in U between two points. We can define U

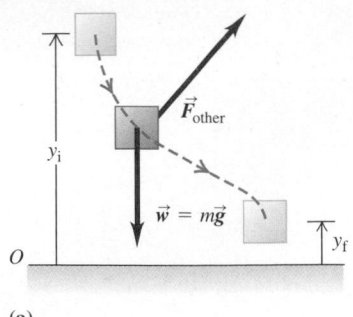

(a)

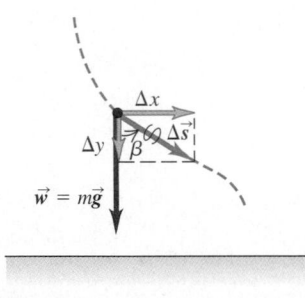

(b)

▲ **FIGURE 7.24** (a) A displacement along a curved path. (b) The work done by the gravitational force w depends only on the vertical component of the displacement Δy.

to be zero at any point we choose (in this case, at the origin); the difference in the value of U between any two points is independent of this choice.

In the preceding examples, the object moved along a straight vertical line. What happens when the path is slanted or curved, as in Figure 7.24a? The forces include the gravitational force $w = mg$ and possibly an additional force $\vec{F}_{other}$. The work done by the gravitational force during the displacement of the object is still given by Equation 7.10:

$$W_{grav} = U_i - U_f = mgy_i - mgy_f. \tag{7.12}$$

To prove that this equation is valid, we divide the path into small segments Δs; a typical segment is shown in Figure 7.24b. The work done in this segment by the gravitational force is the magnitude of the force, multiplied by the vertical component of the displacement, $\Delta y = -\Delta s \cos\beta$. (The negative sign is needed because Δs is positive, while Δy is negative.) As the figure shows, the angle between the force and displacement vectors is β. The work done during the displacement is therefore $mg\,\Delta s \cos\beta$. But $\Delta s \cos\beta$ is equal to Δy, so the work done by w is

$$mg\,\Delta s \cos\beta = -mg\,\Delta y.$$

The work is the same as though the object had been displaced vertically a distance Δy. This is true for every segment, so the *total* work done by the gravitational force depends only on the *total* vertical displacement $(y_f - y_i)$. The total work is $-mg(y_f - y_i)$; this work is independent of any horizontal motion that may occur.

We haven't spoken about the gravitational potential energy *of an object* moving in the earth's gravitational field because the energy belongs not just to the object, but rather to the *system* consisting of the object and the earth. The earth has much more mass than the objects we've been considering. However, when the masses of the objects are more comparable, it is more obvious that the gravitational potential energy cannot be given to one of them.

Elastic Potential Energy

When a railroad car runs into a spring bumper at the end of the track, the spring is compressed as the car is brought to a stop. If there is no friction, the bumper springs back, and the car rolls away in the opposite direction. During the interaction with the spring, the car's kinetic energy is "stored" in the elastic deformation of the spring. Something similar happens in a spring gun. Work is done on the spring by the force that compresses it; that work is stored in the spring until you pull the trigger and the spring gives kinetic energy to the projectile.

This is the same pattern that we saw with the pile driver: Do work on the system to store energy; then this energy can later be converted to kinetic energy. We can apply the concept of elastic potential energy to the storage process. We proceed just as we did for gravitational potential energy. We begin with the work done by the force that the spring exerts on an object (which we will call the **spring force**) and then combine this work with the work–energy theorem.

In Section 7.4, we discussed work associated with stretching or compressing a spring. Figure 7.25 shows the spring from Figure 7.18, with its left end held stationary and its right end attached to an object with mass m that can move along the x axis. The object is at $x = 0$ when the spring is neither stretched nor compressed (Figure 7.25a). An external force $\vec{F}_{ext}$ acts on the object to make it undergo a displacement (Figure 7.25b), while the spring force $\vec{F}_{spring}$ pulls on it in the other direction. How much work does the spring force do on the object?

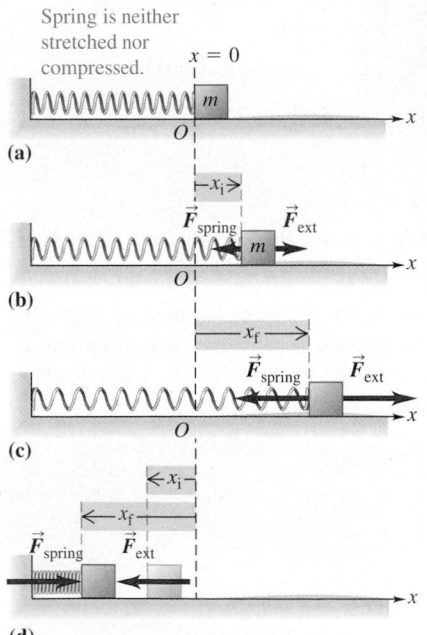

▲ **FIGURE 7.25** A block and spring on a horizontal surface. Both the spring force $\vec{F}_{spring}$ and the external force $\vec{F}_{ext}$ do work on the block.

We found in Section 7.4 that the work we must do *on* the spring to stretch it from an elongation x_i to a greater elongation x_f is $\frac{1}{2}kx_f^2 - \frac{1}{2}kx_i^2$, where k is the force constant of the spring. Now we need to find the work done *by* the spring *on* the object. From Newton's third law, the two quantities of work are just negatives of each other. Changing the signs in the expression on the right-hand side of Equation 7.7, we find that, in a displacement from x_i to x_f, the spring does an amount of work W_{el} given by

$$W_{el} = \tfrac{1}{2}kx_i^2 - \tfrac{1}{2}kx_f^2,$$

where the subscript "el" stands for *elastic*. When x_f is greater than x_i (Figure 7.25c), this quantity is negative because the object moves in the $+x$ direction while the spring pulls on it in the $-x$ direction. If the spring can be compressed as well as stretched, then x_i, x_f, or both may be negative (Figure 7.25d), and the expression for W_{el} is still valid.

Following the same procedure as for gravitational work, we define **elastic potential energy** as follows:

Definition of elastic potential energy
When a spring that obeys Hooke's law, $F = kx$, is stretched or compressed a distance x from its undistorted state, the associated potential energy U_{el} is given by

$$U_{el} = \tfrac{1}{2}kx^2. \tag{7.13}$$

The units of k are N/m, so the units of U_{el} are J. (Remember that $1\text{ J} = 1\text{ N} \cdot \text{m}$.) The joule is the unit used for *all* energy and work quantities.

NOTE ▶ In the expression $U_{el} = \frac{1}{2}kx^2$, x *must* be the displacement of the spring from its *unstretched* length (because we have used $F = kx$ to derive the expression). ◀

We can use Equation 7.13 to express the work W_{el} done on the object by the spring force in terms of the change in potential energy:

$$W_{el} = \tfrac{1}{2}kx_i^2 - \tfrac{1}{2}kx_f^2 = U_{el,i} - U_{el,f} = -\Delta U_{el}. \tag{7.14}$$

When x increases, W_{el} is *negative* and the stored energy U_{el} *increases;* when x decreases, W_{el} is *positive* and U_{el} *decreases.* If the spring can be compressed as well as stretched, then x is negative for compression. But, as Equation 7.13 shows, U_{el} is positive for both positive and negative x, and Equations 7.13 and 7.14 are valid in both cases. When $x = 0$, $U_{el} = 0$, and the spring is neither stretched nor compressed.

The work–energy theorem says that $W_{total} = K_f - K_i$, no matter what kind of forces are acting. If the spring force is the *only* force that does work on the object, then

$$W_{total} = W_{el} = U_i - U_f.$$

The work–energy theorem then gives us

$$K_i + U_{el,i} = K_f + U_{el,f} \quad \text{and} \quad \tfrac{1}{2}mv_i^2 + \tfrac{1}{2}kx_i^2 = \tfrac{1}{2}mv_f^2 + \tfrac{1}{2}kx_f^2 \tag{7.15}$$

In this case, the total energy $E = K + U_{el}$ is *conserved.*

▲ **BIO Application Climbing El Capitan.** One of the world's most famous climbing rocks is the 884 m El Capitan in Yosemite National Park. The red line shows a common climbing route, which involves an overnight stay on a ledge. For a typical 75 kg male climber, the change in gravitational potential energy from the bottom to the top of El Capitan is about 6.5×10^5 J. That sounds like a lot, but in fact, it's less than the energy you get from eating a typical candy bar! The work that results in the net change in gravitational potential energy represents just a small part of the total energy expended on such a climb.

▲ **BIO Application Miniature machines.** The inner ears of vertebrates contain "hair cells" which convert the mechanical energy of fluid motion to electrical signals that are sent to the brain. Each hair cell has a bundle of cross-linked cilia that flex in response to fluid movement. The stiffness of these bundles is due to the mechanical properties of the molecule actin. Direct microscopic measurement of bundle stiffness reveals a force constant of about 600 μN/m. From that measurement, we can estimate the structural rigidity of actin molecules, because the number of cilia per hair cell is easily determined and each cilium has about 30 actin molecules. The physical principles derived from the study of the macroscopic world apply with remarkably little modification to the micro- and nano-scale world.

EXAMPLE 7.8 **Potential energy on an air track**

A glider with mass $m = 0.200$ kg sits on a frictionless, horizontal air track, connected to a spring of negligible mass with force constant $k = 5.00$ N/m. You pull on the glider, stretching the spring 0.100 m, and then release it with no initial velocity. The glider begins to move back toward its equilibrium position $(x = 0)$. What is its speed when $x = 0.0800$ m?

SOLUTION

SET UP Figure 7.26 shows our diagram. We mark the $x = 0$ position at which the spring is relaxed, and we show the initial and final positions we will use. As the glider starts to move, potential energy is converted into kinetic energy. The spring force is the only force doing work on the glider, so $W_{other} = 0$, the total mechanical energy is constant, and we may use Equation 7.15. The energy quantities are

$$K_i = \tfrac{1}{2}(0.200 \text{ kg})(0)^2 = 0,$$
$$U_i = \tfrac{1}{2}(5.00 \text{ N/m})(0.100 \text{ m})^2 = 0.0250 \text{ J},$$
$$K_f = \tfrac{1}{2}(0.200 \text{ kg})v_f^2,$$
$$U_f = \tfrac{1}{2}(5.00 \text{ N/m})(0.0800 \text{ m})^2 = 0.0160 \text{ J}.$$

SOLVE From Equation 7.15,

$$\tfrac{1}{2}mv_i^2 + \tfrac{1}{2}kx_i^2 = \tfrac{1}{2}mv_f^2 + \tfrac{1}{2}kx_f^2,$$
$$0 + 0.0250 \text{ J} = \tfrac{1}{2}(0.200 \text{ kg})v_f^2 + 0.0160 \text{ J},$$
$$v_f = \pm 0.300 \text{ m/s}.$$

REFLECT The quadratic equation for v_f has two roots. The object doesn't stop at $x = 0$, but overshoots; when it returns to $x = 0$, the direction of its motion is reversed. Note that this problem

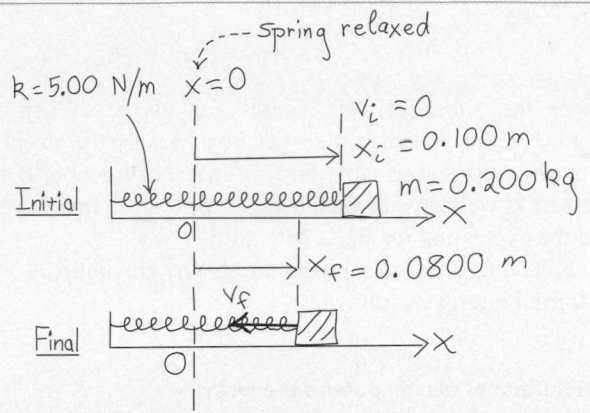

▲ **FIGURE 7.26** Our sketch for this problem. We show the initial and final states we will use in solving the problem.

cannot be solved by using the equations of motion with constant acceleration, because the spring force varies with position. The energy method, in contrast, offers a simple and elegant solution.

Practice Problem: What is the value of x when the object's speed is 0.15 m/s? *Answer:* ± 0.095 m.

▲ **BIO Application Catapult legs.** To jump, a grasshopper must straighten its hind legs faster than its leg muscles can contract. It does so by using a catapult mechanism. The black swelling on the first joint of each jumping leg marks a spot where the insect's exoskeleton has been modified into a pair of stiff springs shaped like flattened C's. Before a jump, the limb muscles build up elastic potential energy by compressing these springs while the joint is flexed. To jump, the animal releases the springs, which cause the legs to snap straight. Such mechanisms rely on conservation of mechanical energy.

7.6 Conservation of Energy

In the examples of the preceding section, we classified the forces acting on an object as gravitational, elastic, and "other." The general form of the work–energy theorem is

$$\Delta K = K_f - K_i = W_{total} = W_{grav} + W_{el} + W_{other}.$$

We've learned that the work done by the gravitational and elastic forces can be represented in terms of changes in their associated potential energy U_{grav} and U_{el}:

$$W_{grav} = U_{grav,i} - U_{grav,f} = -\Delta U_{grav} \quad \text{and} \quad W_{el} = U_{el,i} - U_{el,f} = -\Delta U_{el}.$$

When we substitute these expressions into the general work–energy theorem, we find that

$$K_f - K_i = U_{grav,i} - U_{grav,f} + U_{el,i} - U_{el,f} + W_{other},$$

or

$$K_f + U_{grav,f} + U_{el,f} = K_i + U_{grav,i} + U_{el,i} + W_{other}. \tag{7.16}$$

This equation bristles with subscripts and may seem complicated. But when we look at the individual terms, we recognize the quantity $K + U_{grav} + U_{el}$, with subscript i (initial) or f (final), as representing the total mechanical energy of the system. Therefore, Equation 7.16 states that the total mechanical energy of a system at the end (f) of any process equals the total mechanical energy at the beginning (i), plus the work W_{other} done by the forces *other than* gravitational and elastic forces. If we abbreviate the total potential energy as $U = U_{grav} + U_{el}$, we can say, more simply, that

$$K_f + U_f = K_i + U_i + W_{other}.$$

If there are no forces other than gravitational and elastic forces, or if the vector sum of all other forces is zero, then $\vec{F}_{other} = 0$, $W_{other} = 0$, and *all* the work can be represented in terms of changes in potential energy. In this special case, the *total* mechanical energy, $E = K + U_{grav} + U_{el}$, is constant, or *conserved*. Earlier, we mentioned that such a system is called a *conservative system* because mechanical energy is conserved. In this section, we'll discuss several examples of conservative systems.

MasteringPHYSICS

ActivPhysics 5.2: Upward-Moving Elevator Stops
ActivPhysics 5.3: Stopping a Downward-Moving Elevator
ActivPhysics 5.4: Inverse Bungee Jumper

Conceptual Analysis 7.4

Mass-and-pulley system

Two unequal masses are connected by a massless cord passing over a frictionless pulley as shown in Figure 7.27. Considering the two masses as a system, which of the following statements is true about the gravitational potential energy U_{grav} and kinetic energy K after the masses are released from rest?

A. U_{grav} increases and K increases.
B. U_{grav} decreases and K increases.
C. Both U_{grav} and K remain constant.

SOLUTION The total kinetic energy K of the system is the sum of the two objects' individual kinetic energies, and the total gravitational potential energy U_{grav} is the sum of the individual gravitational potential energies.

Because both objects start from rest and end up moving, the total kinetic energy of the system increases. (Remember that kinetic energy depends on speed, not velocity, so the *direction* of motion has no effect on the kinetic energy.)

Now let's consider the change in gravitational potential energy. The downward motion of the heavy block decreases the

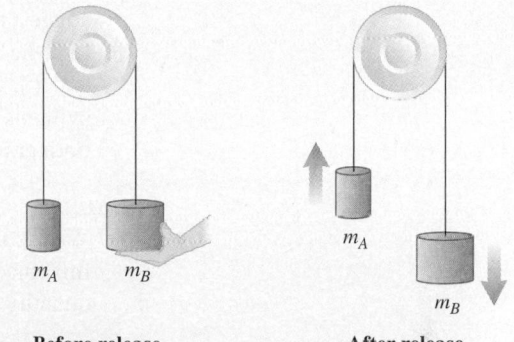

Before release   **After release**

▲ **FIGURE 7.27**

U_{grav} associated with this block. The upward motion of the light block increases the associated U_{grav}. However, since U_{grav} depends directly on mass, the decrease due to the heavy block dominates, so the total gravitational potential energy decreases. Thus, choice B is correct.

Quantitative Analysis 7.5

Total mechanical energy

In Figure 7.28, two blocks with masses m_A and m_B, where $m_A > m_B$, are attached to a thin string that passes over a frictionless pulley. The larger mass is resting on a spring that is compressed a distance s from its equilibrium position. The spring's compression is maintained by a trigger mechanism. What is the total mechanical energy for this system?

A. $m_A g h_A + m_B g h_B + \frac{1}{2}kh_A^2$
B. $m_A g h_A + m_B g h_B + \frac{1}{2}ks^2$
C. $m_A g h_A - m_B g h_B + \frac{1}{2}ks^2$

SOLUTION The total mechanical energy E is the sum of the energies for the individual parts of the system and is given by $E = K + U$, where $K = \frac{1}{2}m_B v_B^2 + \frac{1}{2}m_A v_A^2$ and $U = U_{grav} + U_{el}$. Using the tabletop for the reference level as shown in Figure 7.28, we find that $U_{grav} = m_A g h_A + m_B g h_B$. The spring is compressed a distance s, so $U_{el} = \frac{1}{2}ks^2$. With the trigger in place,

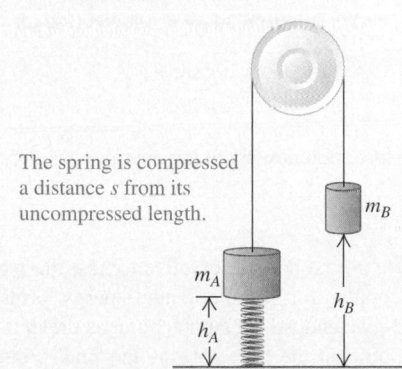

The spring is compressed a distance s from its uncompressed length.

▲ **FIGURE 7.28**

neither block is moving, and $K = 0$. Thus, $E = U_{grav} + U_{el} = m_A g h_A + m_B g h_B + \frac{1}{2}ks^2$ (choice B).

PROBLEM-SOLVING STRATEGY 7.2 **Conservation of energy with conservative forces**

SET UP

1. Identify the system you will analyze, and decide on the initial and final states (positions and velocities) you will use in solving the problem. Use the subscript i for the initial state and f for the final state. Draw one or more sketches showing the initial and final states.

2. Define your coordinate system, particularly the zero points for gravitational and elastic potential energies (the point at which $y = 0$ in the case of gravitational potential energy or at which the spring or other elastic object is relaxed in the case of elastic potential energy). For gravitational potential energy, Equation 7.9 assumes that the positive direction for y is upward; we suggest that you use this choice consistently.

3. List the initial and final kinetic and potential energies—that is, K_i, K_f, U_i, and U_f. Some of these will be known and some unknown. Use algebraic symbols for any unknown coordinates or velocities. In general, U includes both gravitational and elastic potential energy.

SOLVE

4. Write expressions for the total initial mechanical energy and for the total final mechanical energy, equate them, and solve to find whatever unknown quantity is required.

REFLECT

5. Check to make sure there are no forces other than gravitational and elastic forces that do work on the system. If there are additional forces, W_{other} is not zero, and you need to use the method we'll discuss in the next section (Problem-Solving Strategy 7.3).

EXAMPLE 7.9 **Maximum height of a home-run ball**

In Example 3.5 (Section 3.3) we derived an expression for the maximum height h of a projectile launched with initial speed v_0 at initial angle θ_0 (neglecting air resistance):

$$h = \frac{v_0^2 \sin^2 \theta_0}{2g}.$$

Derive this expression now by using energy considerations.

SOLUTION

SET UP There are no nonconservative forces; the work done by gravity is accounted for in the potential-energy terms. As shown in Figure 7.29, we choose the launch point as the initial point and the highest point on the trajectory as the final point. We place $y = 0$ at the launch level; then $U_i = 0$ and $U_f = mgh$, where h is unknown. We can express the kinetic energy at each point in terms of the components of velocity, using the fact that, at each point, $v^2 = v_x^2 + v_y^2$:

$$K_i = \tfrac{1}{2}m(v_{i,x}^2 + v_{i,y}^2)$$
$$K_f = \tfrac{1}{2}m(v_{f,x}^2 + v_{f,y}^2)$$

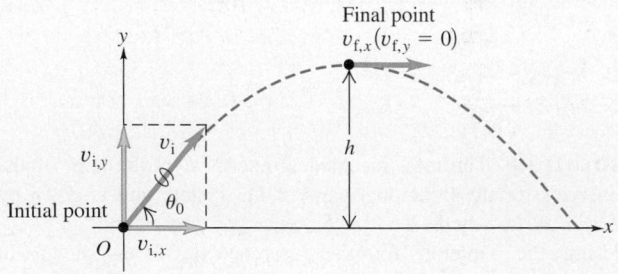

▲ **FIGURE 7.29**

Continued

SOLVE Conservation of energy gives $K_i + U_i = K_f + U_f$, so

$$\tfrac{1}{2}m(v_{i,x}^2 + v_{i,y}^2) + 0 = \tfrac{1}{2}m(v_{f,x}^2 + v_{f,y}^2) + mgh.$$

To simplify this equation, we multiply through by $2/m$ to obtain

$$v_{i,x}^2 + v_{i,y}^2 = v_{f,x}^2 + v_{f,y}^2 + 2gh.$$

Now for the *coup de grace:* We recall that, in projectile motion, the x component of acceleration is zero, so the x component of velocity is constant, and $v_{i,x} = v_{f,x}$. Also, because the final point is the highest point, the vertical component of velocity is zero at that point: $v_{f,y} = 0$. Subtracting the v_x^2 terms from both sides, we get

$$v_{i,y}^2 = 2gh.$$

But $v_{i,y}$ is just the y component of initial velocity, which is equal to $v_0 \sin\theta_0$. Making this substitution and solving for h, we find that

$$h = \frac{v_0^2 \sin^2\theta_0}{2g}.$$

REFLECT This result agrees with our result from Example 3.5. The energy approach has the advantage that we don't have to solve for the time when the maximum height is reached.

Practice Problem: Use the energy method to show that, for a given initial speed, the maximum height is greatest when $\theta_0 = 90°$.

EXAMPLE 7.10 Calculating speed along a vertical circle

Your cousin Throckmorton skateboards down a quarter-pipe with radius $R = 3.0$ m (Figure 7.30). The total mass of Throcky and the skateboard is 25.0 kg. If he starts from rest and there is no friction, derive an algebraic expression for his speed at the bottom of the ramp. Evaluate this expression with the values given. (Throcky's center of mass moves in a circle with radius somewhat smaller than R; neglect this small difference.)

SOLUTION

SET UP This is a conservative system. We can't use the equations of motion with constant acceleration; the acceleration isn't constant, because the slope decreases as Throcky descends. However, if there is no friction, the only force other than his weight is the normal force $\vec{n}$ exerted by the ramp (Figure 7.30b). The work done by this force is zero because, at each point, it is perpendicular to his velocity at that point. Thus, $W_{other} = 0$, and mechanical energy is conserved. We take his starting point as the initial point and the lowest point in the pipe as the final point. Then $y_i = R$ and $y_f = 0$. Throcky starts from rest at the top, so $v_i = 0$. The various energy quantities are

$$K_i = 0, \qquad K_f = \tfrac{1}{2}mv_f^2,$$
$$U_i = mgR, \quad U_f = 0.$$

SOLVE From conservation of energy,

$$K_i + U_i = K_f + U_f,$$
$$0 + mgR = \tfrac{1}{2}mv_f^2 + 0,$$
$$v_f = \sqrt{2gR} = \sqrt{2(9.80 \text{ m/s}^2)(3.00 \text{ m})} = 7.67 \text{ m/s}.$$

REFLECT Throcky's speed at the bottom is the same as if he had fallen vertically through a height R, and it is independent of his mass.

Practice Problem: Find Throcky's speed at the bottom of the pipe if he is given a push at the top edge, so that he has an initial downward speed of 2.00 m/s. *Answer:* 7.93 m/s.

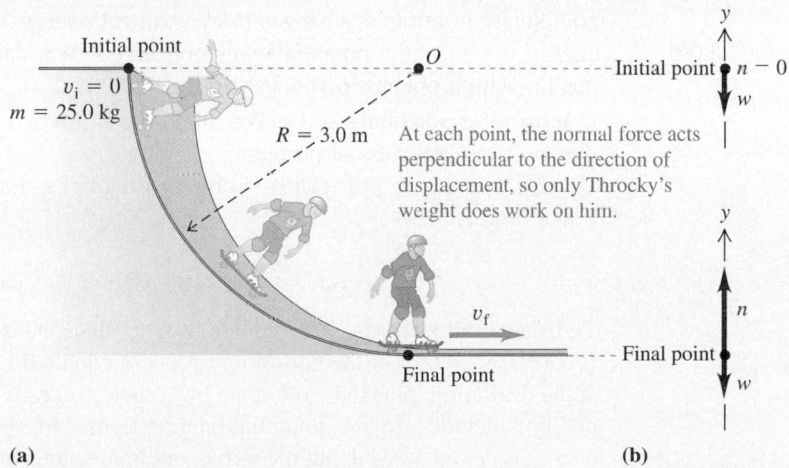

(a) **(b)**

▲ **FIGURE 7.30** (a) Motion showing initial and final points. (b) Free-body diagrams for initial and final points.

EXAMPLE 7.11 **Elevator safety (or not)**

In a "worst-case" design scenario, a 2000 kg elevator with broken cables is falling at 8.00 m/s when it first contacts a cushioning spring at the bottom of the shaft. The spring is supposed to stop the elevator, compressing 3.00 m as it does so (Figure 7.31). As an energy consultant, you are asked to determine what the force constant of the spring should be. Ignore air resistance and friction in the elevator guides.

SOLUTION

SET UP AND SOLVE We take the initial point as the elevator's position when it first contacts the spring $(y_i = 3.00 \text{ m})$ and the final point as its position when the spring is fully compressed $(y_f = 0)$. The elevator's initial speed is $v_i = 8.00 \text{ m/s}$, so

$$K_i = \tfrac{1}{2}mv_i^2 = \tfrac{1}{2}(2000 \text{ kg})(8.00 \text{ m/s})^2 = 64{,}000 \text{ J}.$$

The elevator stops at the final point; thus, $K_f = 0$. The elastic potential energy at the initial point is zero because the spring isn't yet compressed. The initial gravitational potential energy is

$$U_i = mgy_i = (2000 \text{ kg})(9.80 \text{ m/s}^2)(3.00 \text{ m}) = 58{,}800 \text{ J}.$$

At the final point, the gravitational potential energy is zero (because $y_f = 0$) and the elastic potential energy is

$$U_f = \tfrac{1}{2}kx^2 = \tfrac{1}{2}k(-3.00 \text{ m})^2,$$

where the force constant k is to be determined.

Inserting all these values in Equation 7.16, we find that

$$K_i + U_i = K_f + U_f,$$
$$64{,}000 \text{ J} + 58{,}800 \text{ J} = 0 + \tfrac{1}{2}k(-3.00 \text{ m})^2,$$
$$k = 2.73 \times 10^4 \text{ N/m}.$$

REFLECT You still have to explain to the client that the elevator won't stay at the bottom of the shaft, but will bounce back up and

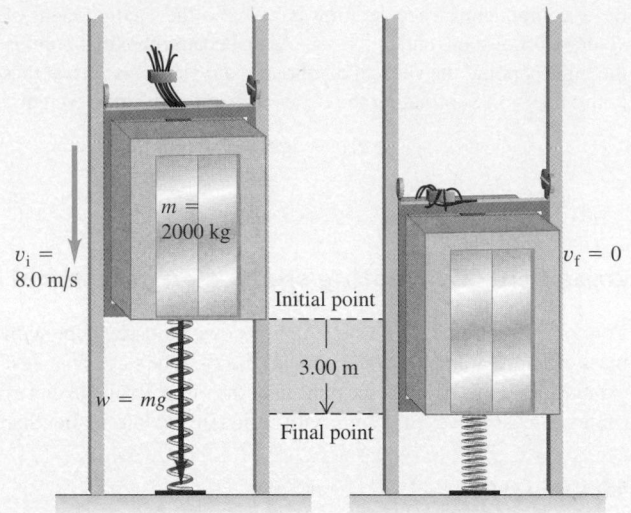

▲ **FIGURE 7.31**

then return to hit the spring again and again until enough energy has been removed by friction for it to stop.

Practice Problem: If the mass of the elevator (with four fewer passengers than before) is 1600 kg, what should the spring constant be for a maximum compression of 3.00 m? *Answer:* $2.18 \times 10^4 \text{ N/m}$.

7.7 Conservative and Nonconservative Forces

In the preceding section, we discussed several problems involving work done on an object by conservative forces. In these problems, the work could be described completely in terms of changes in mechanical energy, and the *total* mechanical energy (kinetic plus potential) was conserved. We also hinted at methods for dealing with problems involving forces that do work W_{other} that can't be described in terms of a potential energy. We called these forces $\vec{F}_{\text{other}}$. Let's now return to this more general class of problem.

The key to dealing with problems involving nonconservative forces is Equation 7.16:

$$K_f + U_{\text{grav,f}} + U_{\text{el,f}} = K_i + U_{\text{grav,i}} + U_{\text{el,i}} + W_{\text{other}}. \qquad (7.16)$$

Translated into words, this equation says that the total mechanical energy $(K + U_{\text{grav}} + U_{\text{el}})$ at the end of any process equals the total mechanical energy at the beginning, plus the work done by "other" forces whose work is not (or cannot be) included in the potential-energy terms. In such systems, mechanical energy is *not* conserved, but the initial and final values of total mechanical energy are related simply to the work done by the "other" forces.

Following are several examples of this more general class of problems and of the application of Equation 7.16. Then we'll close this section with a more general discussion of the nature of conservative and nonconservative forces.

PROBLEM-SOLVING STRATEGY 7.3 **Conservation of energy with nonconservative forces**

The strategy outlined in Section 7.6 is equally useful here. In the list of kinetic and potential energies in Step 3, you should include both gravitational and elastic potential energies when appropriate. Remember that every force that does work must be represented *either* in U or in W_{other}, but *never* in both places. The work done by the gravitational and elastic forces is accounted for by the potential energies; the work of the other forces, W_{other}, has to be included separately. In some cases there will be "other" forces that do no work, as we mentioned in Section 7.2. Remember that, in the expression $U = \frac{1}{2}kx^2$, the coordinate x must be the displacement of the spring from its *unstretched* length (because we have used $F = -kx$ to derive U).

EXAMPLE 7.12 **Work and energy on an air track**

For the system of Example 7.8, suppose the glider is initially at rest at $x = 0$, with the spring unstretched. Then you apply a constant force $\vec{F}$ with magnitude 0.610 N to the glider. What is the glider's speed when it has moved to $x = 0.100$ m?

SOLUTION

SET UP Figure 7.32 shows our sketch. Mechanical energy is not conserved, because of the work W_{other} done by the force $\vec{F}$, but we can still use the energy relation of Equation 7.16. Let the initial point be $x = 0$ and the final point be $x = 0.100$ m. The energy quantities are

$K_i = 0, \qquad K_f = \frac{1}{2}(0.200\ \text{kg})v_f^2,$
$U_i = 0, \qquad U_f = \frac{1}{2}(5.00\ \text{N/m})(0.100\ \text{m})^2 = 0.0250\ \text{J},$
$W_{other} = (0.610\ \text{N})(0.100\ \text{m}) = 0.0610\ \text{J}.$

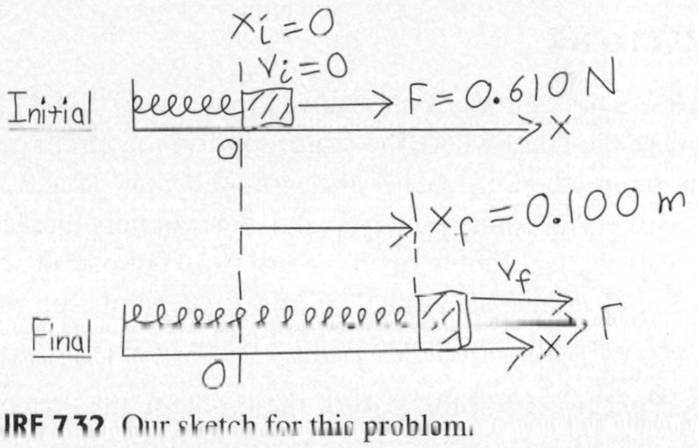

SOLVE Putting the pieces into Equation 7.16, we obtain

$$K_f + U_f = K_i + U_i + W_{other},$$
$$\tfrac{1}{2}(0.200\ \text{kg})v_f^2 + 0.0250\ \text{J} = 0 + 0 + 0.0610\ \text{J},$$
$$v_f = 0.600\ \text{m/s}.$$

REFLECT The total mechanical energy of the system changes by an amount equal to the work W_{other} done by the applied force. Even

▲ **FIGURE 7.32** Our sketch for this problem.

though mechanical energy isn't conserved, energy considerations simplify this problem greatly.

Practice Problem: If the 0.610 N force is removed when the glider reaches the 0.100 m point, at what distance from the starting point does the glider come to rest? *Answer:* 0.156 m.

EXAMPLE 7.13 **Throcky on the half-pipe again**

Suppose the ramp in Example 7.10 is not frictionless and that Throcky's speed at the bottom is only 7.00 m/s. What work was done by the frictional force acting on him?

SOLUTION

SET UP We choose the same initial and final points and reference level as in Example 7.10 (Figure 7.30). In this case, $W_{other} = W_f$. The energy quantities are

$K_i = 0,$
$U_i = mgR = (25.0\ \text{kg})(9.80\ \text{m/s}^2)(3.00\ \text{m}) = 735\ \text{J},$
$K_f = \frac{1}{2}mv_f^2 = \frac{1}{2}(25.0\ \text{kg})(7.00\ \text{m/s})^2 = 613\ \text{J},$
$U_f = 0.$

SOLVE From Equation 7.16,

$$W_{other} = K_f + U_f - K_i - U_i$$
$$= 613\ \text{J} + 0 - 0 - 735\ \text{J} = -122\ \text{J}.$$

The work done by the friction force is -122 J, and the total mechanical energy *decreases* by 122 J.

REFLECT W_{other} has to be negative because, at each point, the friction force is opposite in direction to Throcky's velocity. In this problem, as in several others in this chapter, Throcky's motion is

Continued

determined by Newton's second law, $\sum \vec{F} = m\vec{a}$. But it would be very difficult to apply that law here because the acceleration is continuously changing, in both magnitude and direction, as Throcky skates down. The energy approach, in contrast, relates the motions at the top and bottom of the ramp without involving the details of what happens in between. This simplification is typical of many problems that are easy if we use energy considerations, but very complex if we try to use Newton's laws directly.

The work W_f done by the nonconservative friction forces is -122 J. As Throcky rolls down, the wheels, the bearings, and the ramp all get a little warmer. The same temperature changes could have been produced by adding 122 J of heat to these objects. Their *internal* energy increases by 122 J; the sum of this energy and the final mechanical energy equals the initial mechanical energy, and the total energy of the system (including nonmechanical forms of energy) is conserved.

Practice Problem: If Throcky's speed at the bottom of the quarter-pipe is exactly half of what it would be without friction, how much work does the friction force do? *Answer:* -551 J.

EXAMPLE 7.14   **Loading a crate onto a truck**

A crate full of machine parts sits on the floor; the total mass is 8.0 kg. The crate must be raised to the floor of a truck by sliding it up a ramp 2.5 m long, inclined at 30°. The shop foreman, giving no thought to the force of friction, calculates that he can get the crate up the ramp by giving it an initial speed of 5.0 m/s at the bottom and letting it go. Unfortunately, friction is *not* negligible; the crate slides 1.6 m up the ramp, stops, and slides back down. Figure 7.33 shows the situation. **(a)** Assuming that the friction force acting on the crate is constant, find its magnitude. **(b)** How fast is the crate moving when it reaches the bottom of the ramp?

SOLUTION

SET UP We have three critical points: ground level (point 1 in Figure 7.33a), the point where the crate stops on the ramp (point 2), and the point where it reaches the ground again (point 3 in Figure 7.33b). (We count points 1 and 3 separately because the crate's velocity is different on the return trip.) For part (a), point 1 is the initial point and point 2 is the final point; for part (b), point 1 is the initial point and point 3 is the final point.

SOLVE Part (a): For this part of the problem, the final point is $(1.6\text{ m})(\sin 30°) = 0.80$ m above the floor. If we take $U = 0$ at floor level, we have $y_1 = 0$, $y_2 = 0.80$ m. The energy quantities are

$$K_1 = \tfrac{1}{2}(8.0\text{ kg})(5.0\text{ m/s})^2 = 100\text{ J}, \qquad K_2 = 0,$$
$$U_1 = 0, \qquad U_2 = (8.0\text{ kg})(9.8\text{ m/s}^2)(0.80\text{ m}) = 62.7\text{ J},$$
$$W_{\text{other}} = -fs = -f(1.6\text{ m}),$$

where f is the magnitude of the unknown friction force. Using Equation 7.16, we find that

$$K_1 + U_1 + W_{\text{other}} = K_2 + U_2,$$
$$100\text{ J} + 0 - f(1.6\text{ m}) = 0 + 62.7\text{ J},$$
$$f = 23\text{ N}.$$

Part (b): This time we use points 1 and 3 as our initial and final points, respectively. The friction force reverses direction for the return trip, so the frictional work is negative for both halves of the trip, and the total work done by friction is twice the previous amount. The total gravitational work for the round-trip is zero, so the total work is

$$W_{\text{other}} = W_f = -2(1.6\text{ m})(23.3\text{ N}) = -74.6\text{ J}.$$

The other energy quantities are

$$K_1 = 100\text{ J}, \qquad U_1 = U_3 = 0, \qquad K_3 = \tfrac{1}{2}(8.0\text{ kg})v_3^2,$$

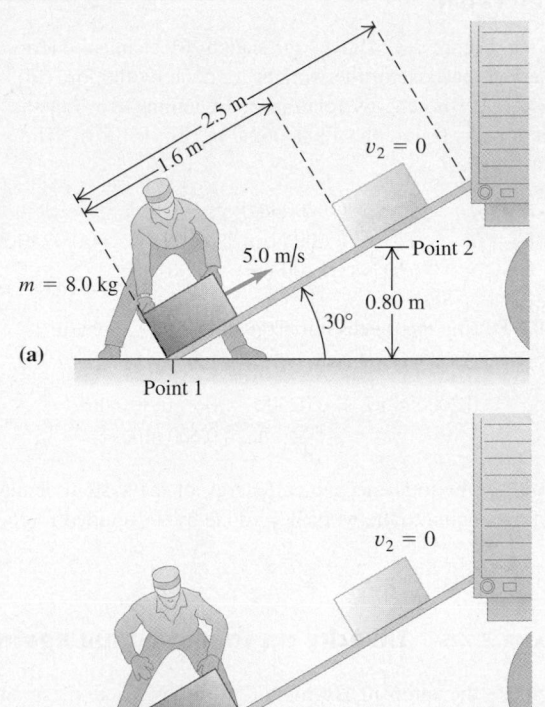

▲ FIGURE 7.33

Continued

where v_3 is the unknown final speed. Substituting these values into Equation 7.16, we get

$$100 \text{ J} + 0 + (-74.6 \text{ J}) = \tfrac{1}{2}(8.0 \text{ kg})v_3^2 + 0,$$
$$v_3 = 2.5 \text{ m/s}.$$

REFLECT The crate's speed and kinetic energy when it returns to the bottom of the ramp are less than when it left that point;

mechanical energy is lost to friction. We could have applied Equation 7.16 to points 2 and 3, considering the second half of the trip by itself.

Practice Problem: Follow the suggestion to apply Equation 7.16 to points 2 and 3 to verify that you get the same result for v_3 as before.

Nonconservative Forces

In our discussions of potential energy, we've talked about "storing" kinetic energy by converting it to potential energy. When you throw a ball up in the air, it slows down as kinetic energy is converted into potential energy. On the way down, the conversion is reversed, and potential energy is converted completely back to kinetic energy (if air resistance is neglected.).

Similarly, when a glider moving on a frictionless, horizontal air track runs into a spring bumper at the end of the track, the spring compresses and the glider stops. But then the glider bounces back with the same speed and kinetic energy that it had before the collision. Again, there is a two-way conversion from kinetic to potential energy and back.

A force that permits two-way conversion between kinetic and potential energies is called a **conservative force.** The work of a conservative force is always *reversible:* Anything that we deposit in the energy "bank" can later be withdrawn without loss. In some cases (including gravitational forces), an object may move from an initial to a final point by various paths (Figure 7.34), but if the force is conservative, the work done by the force is the same for all of these paths. If an object moves around a *closed* path, ending at the same point where it started, the *total* work done by the gravitational force is always zero.

The work done by a conservative force has these properties:

1. It can always be expressed as the difference between the initial and final values of a *potential-energy* function.
2. It is reversible.
3. It is independent of the path of the object and depends only on the starting and ending points.
4. When the starting and ending points are the same, the total work is zero.

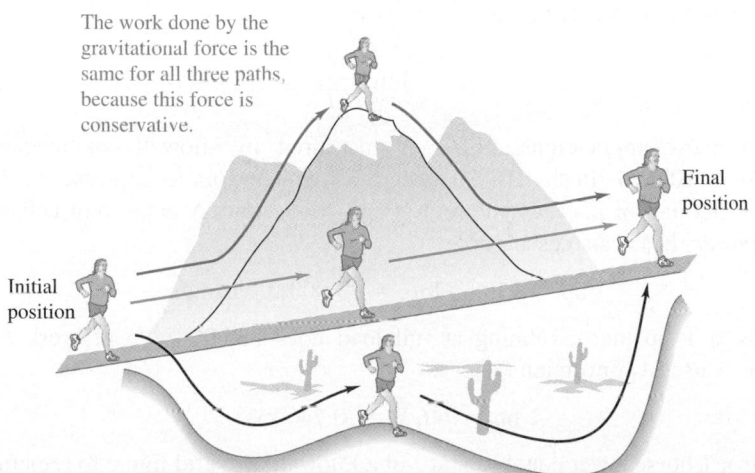

The work done by the gravitational force is the same for all three paths, because this force is conservative.

Initial position

Final position

◄ **FIGURE 7.34** The work done by a conservative force is independent of the path taken.

When all the work done on an object is done by conservative forces, the total mechanical energy $E = K + U$ is constant.

Not all forces are conservative. Consider the friction force acting on the crate sliding on a ramp in Example 7.14. There is no potential-energy function for this force. When the object slides up and then back down to the starting point, the total work done on it by the friction force is *not* zero. When the direction of motion reverses, so does the friction force, and it does *negative* work in *both* directions. When a car with its brakes locked skids across the pavement with decreasing speed (and decreasing kinetic energy), the lost kinetic energy cannot be recovered by reversing the motion (or, for that matter, in any other way), and mechanical energy is *not* conserved.

Such a force is called a **nonconservative force** or a **dissipative force.** To describe the associated energy relations, we have to introduce additional kinds of energy and a more general energy-conservation principle. When an object slides on a rough surface, such as tires on pavement or a box dropped on a conveyer belt, the surfaces become hotter; the energy associated with this change in the state of the materials is called *internal energy*. In later chapters, we'll study the relation of internal energy to temperature changes, heat, and work. This is the heart of the area of physics called *thermodynamics*.

7.8 Power

Time considerations aren't involved directly in the definition of work. If you lift a barbell weighing 400 N through a vertical distance of 0.5 m at constant velocity, you do 200 J of work on it, whether it takes you 1 second, 1 hour, or 1 year to do it. Often, though, we need to know how quickly work is done. The time rate at which work is done or energy is transferred is called **power**. Like work and energy, power is a scalar quantity. We define **average power** as follows:

Definition of average power P_{av}

When a quantity of work ΔW is done during a time interval Δt, the average power P_{av}, or work per unit time, is defined as

$$P_{av} = \frac{\Delta W}{\Delta t}. \tag{7.17}$$

Unit: watt (W), where 1 watt is defined as 1 joule per second (J/s).

The rate at which work is done isn't necessarily constant. Even when it varies, however, we can define the **instantaneous power** P as the limit of this quotient as Δt approaches zero:

$$P = \lim_{\Delta t \to 0} \frac{\Delta W}{\Delta t}.$$

For many applications, power is measured in kilowatts or megawatts $(1 \text{ MW} = 10^6 \text{ W})$. In the British system, in which work is expressed in foot-pounds, the unit of power is the foot-pound per second. A larger unit called the *horsepower* (hp) is also used:

$$1 \text{ hp} = 550 \text{ ft} \cdot \text{lb/s} = 33,000 \text{ ft} \cdot \text{lb/min.}$$

That is, a 1 hp motor running at full load does 33,000 ft · lb of work every minute. A useful conversion factor is

$$1 \text{ hp} = 746 \text{ W} = 0.746 \text{ kW.}$$

Note that 1 horsepower equals about $\frac{3}{4}$ of a kilowatt, a useful figure to remember.

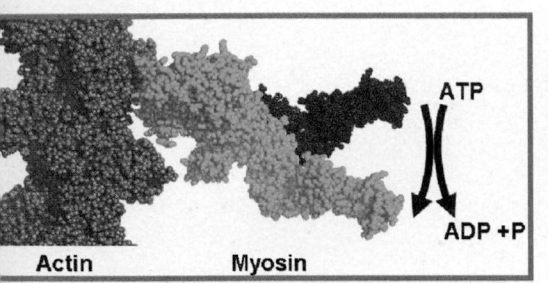

Actin   Myosin
ATP
ADP +P

▲ **BIO Application Work at the molecular scale.** Ultimately, everything a living organism does, including doing work, depends on the aggregate actions of organized molecules. High-resolution pictures of molecules, such as those provided by x-ray crystallography, provide insight into the workings of molecular machines. When a muscle contracts, the force generated is due to the interaction of a motor protein called myosin, and a structural track called actin. The head of the myosin molecule attaches to the actin and rotates about a pivot point, as shown. The force applied by the myosin to the actin in a skeletal muscle is in the range of 1–10 pN and the distance moved in a single power stroke is perhaps 5 nm. The work done in a single power stroke ($\frac{1}{2} kx^2$) derives its energy from a single ATP molecule, the body's energy currency. One ATP molecule can do up to 60 pN · nm of work, consistent with the force constant of the elastic element of the myosin of 5 pN · nm. This work represents about half the energy available from one ATP molecule.

The watt is a familiar unit of *electrical* power; a 100 W lightbulb converts 100 J of electrical energy into light and heat each second. But there is nothing inherently electrical about the watt; a lightbulb could be rated in horsepower, and some automobile manufacturers rate their engines in kilowatts as well as horsepower.

Power units can be used to define new units of work or energy. The *kilowatt-hour* (kWh) is the usual commercial unit of electrical energy. One kilowatt-hour is the total work done in 1 hour (3600 s) when the power is 1 kilowatt (10^3 J/s), so

$$1 \text{ kWh} = (10^3 \text{ J/s})(3600 \text{ s}) = 3.6 \times 10^6 \text{ J} = 3.6 \text{ MJ}.$$

The kilowatt-hour is a unit of *work* or *energy*, not power.

When a force acts on a moving object, it does work on the object (unless the force and velocity are always perpendicular). The corresponding power can be expressed in terms of force and velocity. Suppose a force $\vec{F}$ acts on an object while it undergoes a vector displacement $\Delta\vec{s}$. If $F_\parallel$ is the component of $\vec{F}$ tangent to the path (parallel to $\Delta\vec{s}$), then the work is $\Delta W = F_\parallel \Delta s$, where Δs is the magnitude of the displacement $\Delta\vec{s}$, and the average power is

$$P_{\text{av}} = \frac{F_\parallel \Delta s}{\Delta t} = F_\parallel \frac{\Delta s}{\Delta t} = F_\parallel v_{\text{av}}.$$

Instantaneous power P is the limit of P_{av} as $\Delta t \to 0$:

$$P = F_\parallel v, \tag{7.18}$$

where v is the magnitude of the instantaneous velocity.

It's a curious fact of modern life that although energy is an abstract physical quantity, it is bought and sold. We don't buy a newton of force or a meter per second of velocity, but a kilowatt-hour of electrical energy usually costs from 2 to 10 cents, depending on one's location and the amount purchased.

EXAMPLE 7.15 Power in a jet engine

A jet airplane engine develops a thrust (a forward force on the plane) of 15,000 N (roughly 3000 lb). When the plane is flying at 300 m/s (roughly 670 mi/h), what horsepower does the engine develop?

SOLUTION

SET UP AND SOLVE From Equation 7.18,

$$P = Fv = (1.50 \times 10^4 \text{ N})(300 \text{ m/s}) = 4.50 \times 10^6 \text{ W}$$
$$= (4.50 \times 10^6 \text{ W})\left(\frac{1 \text{ hp}}{746 \text{ W}}\right) = 6030 \text{ hp}.$$

REFLECT This power is comparable to the power developed by a Diesel railroad locomotive. The airplane travels 10 times as fast as the locomotive, with a much smaller cargo weight.

Practice Problem: Suppose the airplane carries 150 passengers with an average mass of 55 kg each. After disembarking, all these passengers get on a freight elevator, which itself has a mass of 750 kg. How fast can the elevator's 6000 hp motor lift the loaded elevator? (That is, at what speed is 6000 hp just enough to keep the elevator rising?) *Answer:* 51 m/s.

EXAMPLE 7.16 A marathon stair climb

As part of a charity fund-raising drive, a Chicago marathon runner with mass 50.0 kg runs up the stairs to the top of the Willis Tower, the tallest building in Chicago (443 m), in 15.0 minutes (Figure 7.35). What is her average power output in watts? In kilowatts? In horsepower?

Continued

▲ **FIGURE 7.35** Climbing the Willis Tower.

SOLUTION

SET UP AND SOLVE The runner's total work W is her weight mg, multiplied by the height h she climbs:

$$W = mgh = (50.0 \text{ kg})(9.80 \text{ m/s}^2)(443 \text{ m}) = 2.17 \times 10^5 \text{ J}.$$

The time is 15 min = 900 s, so the average power is

$$P_{\text{av}} = \frac{2.17 \times 10^5 \text{ J}}{900 \text{ s}} = 241 \text{ W} = 0.241 \text{ kW} = 0.323 \text{ hp}.$$

Alternative Solution: The average vertical component of velocity is $(443 \text{ m})/(900 \text{ s}) = 0.492$ m/s, so the average power is

$$\begin{aligned} P_{\text{av}} &= Fv_{\text{av}} = (mg)v_{\text{av}} \\ &= (50.0 \text{ kg})(9.80 \text{ m/s}^2)(0.492 \text{ m/s}) = 241 \text{ W}. \end{aligned}$$

REFLECT A horse can supply considerably more than 1 hp for short periods of time.

Practice Problem: If all the runner's work output could be collected and sold at the rate of 10 cents per kilowatt-hour, what would be the cash value of the work? *Answer:* 0.603 cents.

SUMMARY

Overview of Energy

(Section 7.1) Energy is one of the most important unifying concepts in all of physical science. Its importance stems from the principle of conservation of energy. Energy is exchanged and transformed during interactions of systems, but the total energy in a closed system is constant. This chapter is concerned with mechanical energy of three types: kinetic energy, associated with the motion of objects that have mass; gravitational potential energy, associated with gravitational interactions; and elastic potential energy, associated with elastic deformations of objects. Potential energy can be viewed as a stored quantity that represents the potential for doing work.

Work

(Section 7.2) When a force acts on an object that undergoes a displacement, the force does work on the object and transfers energy to it. For a constant force, $W = F_\| s$ (Equation 7.1), where $F_\|$ is the component of force parallel to the object's displacement (of magnitude s). This component can be positive, negative, or zero. Work is a scalar quantity, not a vector. If $F_\|$ points opposite to the displacement, then $W < 0$. When several forces act on an object, the total work done by all the forces is the sum of the amounts of work done by the individual forces.

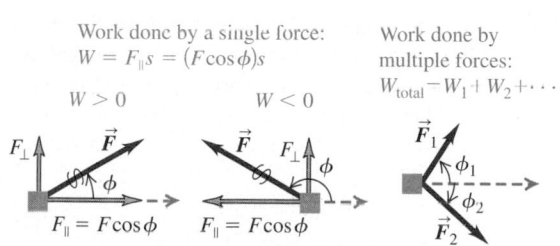

Work done by a single force:
$$W = F_\| s = (F\cos\phi)s$$

$W > 0$ $W < 0$

$F_\| = F\cos\phi$ $F_\| = F\cos\phi$

Work done by multiple forces:
$$W_{total} = W_1 + W_2 + \cdots$$

Work and Kinetic Energy

(Section 7.3) The kinetic energy K of an object is defined in terms of the object's mass m and speed v as $K = \frac{1}{2}mv^2$. When forces act on an object, the change in its kinetic energy is equal to the total work done by all the forces acting on it: $W_{total} = K_f - K_i = \Delta K$ (Equation 7.4). When W is positive, the object speeds up; when W is negative, it slows down.

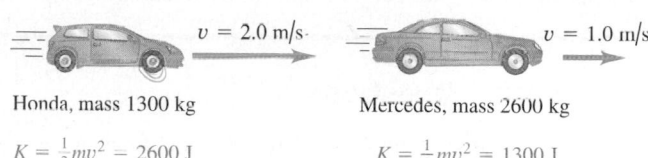

$v = 2.0$ m/s $v = 1.0$ m/s

Honda, mass 1300 kg   Mercedes, mass 2600 kg

$K = \frac{1}{2}mv^2 = 2600$ J $K = \frac{1}{2}mv^2 = 1300$ J

Work Done by a Varying Force

(Section 7.4) When an object moves under the action of a varying force parallel to the displacement, the work done by the force is represented graphically as the area under a graph of the force as a function of displacement. If the magnitude of force F required to stretch a spring a distance x beyond its natural length is proportional to x, then $F = kx$, where k is a constant for the spring, called its force constant or spring constant. The total work W needed to stretch the spring from $x = 0$ to $x = X$ is $W = \frac{1}{2}kX^2$.

Potential Energy

(Section 7.5) Potential energy can be thought of as *stored* energy. For an object with mass m at a height y above a chosen origin (where $y = 0$), the gravitational potential energy is $U_{grav} = mgy$ (Equation 7.9). When a spring is stretched or compressed a distance x from its uncompressed length, the spring stores elastic potential energy $U_{el} = \frac{1}{2}kx^2$ (Equation 7.13). Potential energy is associated only with conservative forces, not with nonconservative forces such as friction.

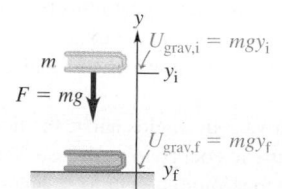

Gravitational potential energy: For an object at vertical position y,

$$U_{grav} = mgy$$

For a change in vertical position,

$$\Delta U_{grav} = U_{grav,f} - U_{grav,i}$$

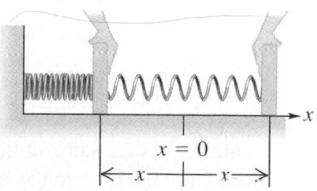

Elastic potential energy: For a spring stretched or compressed by a distance x from equilibrium,

$$U_{el} = \frac{1}{2}kx^2$$

Continued

Conservation of Energy

(Sections 7.6 and 7.7) When only conservative forces act on an object, the total mechanical energy (kinetic plus potential) is constant; that is, $K_i + U_i = K_f + U_f$, where U may include both gravitational and elastic potential energies. If some of the forces are nonconservative, we label their work as W_{other}. The change in total energy (kinetic plus potential) of an object during any motion is equal to the work W_{other} done by the nonconservative forces: $K_i + U_i + W_{other} = K_f + U_f$ (Equation 7.16). Nonconservative forces include friction forces, which usually act to decrease the total mechanical energy of a system.

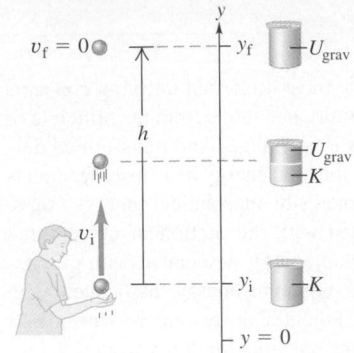

Power

(Section 7.8) Power is the time rate of doing work or the rate at which energy is transferred or transformed. For mechanical systems, power is the time rate at which work is done by or on an object or a system. When an amount of work ΔW is done during a time interval Δt, the average power is $P_{av} = \frac{\Delta W}{\Delta t}$ (Equation 7.17).

 For instructor-assigned homework, go to www.masteringphysics.com

Conceptual Questions

1. How does friction produce internal energy? Explain what is happening at the molecular level to produce this form of energy.
2. True or false? If hydrogen molecules and oxygen molecules have the same kinetic energy, then they have the same speed. Explain your answer.
3. Can the *total* work done on an object during a displacement be negative? Explain. If the total work is negative, can its magnitude be greater than the initial kinetic energy of the object? Explain.
4. An elevator is hoisted by its cables at constant speed. Is the *total* work done on the elevator positive, negative, or zero? Explain your reasoning.
5. A rope tied to an object is pulled, causing the object to accelerate. According to Newton's third law, the object pulls back on the rope with an equal and opposite force. Is the total net work done on the object then zero? If so, how can the object's kinetic energy change?
6. If a projectile is fired upward at various angles above the horizontal, it has the same initial kinetic energy in each case. Why does it *not* then rise to the same maximum height in each case?
7. When you use a jack to lift a car, the force you exert on the jack is much less than the weight of the car. Does this mean that less work is done on the car than if the car were lifted directly?
8. An advertisement for a portable electrical generating unit claims that the unit's diesel engine produces 28,000 hp to drive an electrical generator that produces 30 MW of electrical power. Is this possible? Explain.
9. A compressed spring is clamped in its compressed position and is then dissolved in acid. What becomes of the spring's potential energy?

10. You bounce on a trampoline, going a little higher with each bounce. Explain how you increase your total mechanical energy.
11. When you jump from the ground into the air, where does your kinetic energy come from? What force does work on you to lift you into the air? Could you jump if you were floating in outer space?
12. Hydroelectric energy comes from gravity pulling down water through dams in rivers. Explain how such energy is really just a form of solar energy by tracing how the sun's energy is able to get the water from the ocean to the reservoir behind the dam.
13. Does the kinetic energy of a car change more when the car speeds up from 10 mph to 15 mph or from 15 mph to 20 mph?
14. Does an object's speed at the bottom of a frictionless ramp depend on the shape of the ramp or just on its height? Explain. What if the ramp is *not* frictionless?
15. You often hear that the energy we use on the earth comes from the sun. Explain how this is true. For example, how does the kinetic energy of a running person come from the sun?
16. On your electrical "power" bill, you are charged for kilowatt-hours (kWh). Is this really a power bill? What kind of quantity is the kilowatt-hour?

Multiple-Choice Problems

1. A spiral spring is compressed so as to add U units of potential energy to it. When this spring is instead stretched two-thirds of the distance it was compressed, its remaining potential energy in the same units will be (see Section 2.5 for a review of proportional reasoning)
   A. $2U/3$ B. $4U/9$ C. $U/3$ D. $U/9$
2. A block slides a distance d down a frictionless plane and then comes to a stop after sliding a distance s across a rough

horizontal plane, as shown in the accompanying figure. What fraction of the distance s does the block slide before its speed is reduced to one-third of the maximum speed it had at the bottom of the ramp?

A. $s/3$ B. $2s/3$ C. $s/9$ D. $8s/9$

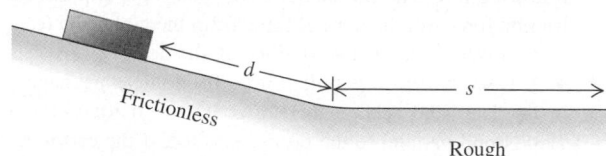

▲ **FIGURE 7.36** Multiple-choice problem 2.

3. You slam on the brakes of your car in a panic and skid a distance d on a straight and level road. If you had been traveling twice as fast, what distance would the car have skidded under the same conditions?

A. $4d$ B. $2d$ C. $\sqrt{2}d$ D. $d/2$

4. You wish to accelerate your car from rest at a constant acceleration. Assume that there is negligible air drag. To create a constant acceleration, the car's engine must
A. maintain a constant power output.
B. develop ever-decreasing power.
C. develop ever-increasing power.

5. Consider two frictionless inclined planes with the same vertical height. Plane 1 makes an angle of 30° with the horizontal, and plane 2 makes an angle of 60° with the horizontal. Mass m_1 is placed at the top of plane 1, and mass m_2 is placed at the top of plane 2. Both masses are released at the same time. At the bottom, which mass is going faster?
A. m_1
B. m_2
C. Neither; they both have the same speed at the bottom.

6. A brick is dropped from the top of a building through the air (friction is present) to the ground below. How does the brick's kinetic energy (K) just before striking the ground compare with the gravitational potential energy (U_{grav}) at the top of the building? Set $y = 0$ at the ground level.
A. K is equal to U_{grav}.
B. K is greater than U_{grav}.
C. K is less than U_{grav}.

7. Which of the following statements about work is or are true? (More than one statement may be true.)
A. Negative net work done on an object always reduces the object's kinetic energy.
B. If the work done on an object by a force is zero, then either the force or the displacement must have zero magnitude.
C. If a force acts downward, it does negative work.
D. The formula $W = Fd\cos\theta$ can be used only if the force is constant over the distance d.

8. A 10 kg stone and a 100 kg stone are released from rest at the same height above the ground. There is no appreciable air drag. Which of the following statements is or are true? (More than one statement may be true.)
A. Both stones have the same initial gravitational potential energy.
B. Both stones will have the same acceleration as they fall.
C. Both stones will have the same speed when they reach the ground.

D. Both stones will have the same kinetic energy when they reach the ground.
E. Both stones will reach the ground at the same time.

9. Two *identical* objects are pressed against two different springs so that each spring stores 50 J of potential energy. The objects are then released from rest. One spring is quite stiff (hard to compress), while the other one is quite flexible (easy to compress). Which of the following statements is or are true? (More than one statement may be true.)
A. Both objects will have the same maximum speed after being released.
B. The object pressed against the stiff spring will gain more kinetic energy than the other object.
C. Both springs are initially compressed by the same amount.
D. The stiff spring has a larger spring constant than the flexible spring.
E. The flexible spring must have been compressed more than the stiff spring.

10. For each of two objects with *different* masses, the gravitational potential energy is 100 J. They are released from rest and fall to the ground. Which of the following statements is or are true? (More than one statement may be true.)
A. Both objects are released from the same height.
B. Both objects will have the same kinetic energy when they reach the ground.
C. Both objects will have the same speed when they reach the ground.
D. Both objects will accelerate toward the ground at the same rate.
E. Both objects will reach the ground at the same time.

11. Two objects with *different* masses are launched vertically into the air by identical springs. The two springs are compressed by the same amount before launching. Which of the following statements is or are true? (More than one statement may be true.)
A. Both masses reach the same maximum height.
B. Both masses leave the springs with the same energy.
C. Both masses leave the springs with the same speed.
D. Both masses leave the springs with the same kinetic energy.
E. The lighter mass will gain more gravitational potential energy than the heavier mass.

12. Two objects with *unequal* masses are released from rest from the *same height*. They slide without friction down a slope and then encounter a rough horizontal region, as shown in the accompanying figure. The coefficient of kinetic friction in the rough region is the same for both masses. Which of the following statements is or are true? (More than one statement may be true.)
A. Both masses start out with the same gravitational potential energy.
B. Both objects have the same speed when they reach the base of the slope.

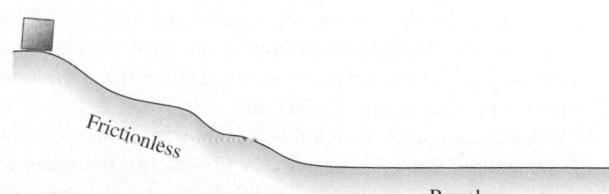

▲ **FIGURE 7.37** Multiple-choice problem 12.

C. Both masses have the same kinetic energy at the bottom of the slope.

D. Both masses travel the same distance on the rough horizontal surface before stopping.

E. Both masses will generate the same amount of thermal energy due to friction on the rough surface.

13. Spring #1 has a force constant of k, and spring #2 has a force constant of $2k$. Both springs are attached to the ceiling, identical weights are hooked to their ends, and the weights are allowed to stretch the springs. The ratio of the energy stored by spring #1 to that stored by spring #2 is

  A. 1:1   B. 1:2   C. 2:1

  D. $1:\sqrt{2}$ E. $\sqrt{2}:1$

14. Two balls having *different* masses reach the same height when shot into the air from the ground. If there is no air drag, which of the following statements must be true? (More than one statement may be true.)

A. Both balls left the ground with the same speed.

B. Both balls left the ground with the same kinetic energy.

C. Both balls will have the same gravitational potential energy at the highest point.

D. The heavier ball must have left the ground with a greater speed than the lighter ball.

E. Both balls have no acceleration at their highest point.

15. The stone in the accompanying figure can be carried from the bottom to the top of a cliff by various paths. Which path requires more work?

A. *AC*

B. *ABC*

C. The work is the same for both paths.

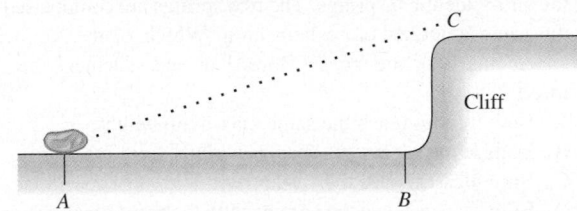

▲ **FIGURE 7.38** Multiple-choice problem 15.

Problems

7.2 Work

1. • A fisherman reels in 12.0 m of line while landing a fish, using a constant forward pull of 25.0 N. How much work does the tension in the line do on the fish?

2. • A tennis player hits a 58.0 g tennis ball so that it goes straight up and reaches a maximum height of 6.17 m. How much work does gravity do on the ball on the way up? On the way down?

3. • A boat with a horizontal tow rope pulls a water skier. She skis off to the side, so the rope makes an angle of 15.0° with the forward direction of motion. If the tension in the rope is 180 N, how much work does the rope do on the skier during a forward displacement of 300.0 m?

4. • A constant horizontal pull of 8.50 N drags a box along a horizontal floor through a distance of 17.4 m. (a) How much work does the pull do on the box? (b) Suppose that the same pull is exerted at an angle above the horizontal. If this pull now does 65.0 J of work on the box while pulling it through the same distance, what angle does the force make with the horizontal?

5. • You push your physics book 1.50 m along a horizontal tabletop with a horizontal push of 2.40 N while the opposing force of friction is 0.600 N. How much work does each of the following forces do on the book? (a) your 2.40 N push, (b) the friction force, (c) the normal force from the table, and (d) gravity? (e) What is the net work done on the book?

6. • A 128.0 N carton is pulled up a frictionless baggage ramp inclined at 30.0° above the horizontal by a rope exerting a 72.0 N pull parallel to the ramp's surface. If the carton travels 5.20 m along the surface of the ramp, calculate the work done on it by (a) the rope, (b) gravity, and (c) the normal force of the ramp. (d) What is the net work done on the carton? (e) Suppose that the rope is angled at 50.0° above the horizontal, instead of being parallel to the ramp's surface. How much work does the rope do on the carton in this case?

7. • A factory worker moves a 30.0 kg crate a distance of 4.5 m along a level floor at constant velocity by pushing horizontally on it. The coefficient of kinetic friction between the crate and the floor is 0.25. (a) What magnitude of force must the worker apply? (b) How much work is done on the crate by the worker's push? (c) How much work is done on the crate by friction? (d) How much work is done by the normal force? By gravity? (e) What is the net work done on the crate?

8. •• An 8.00 kg package in a mail-sorting room slides 2.00 m down a chute that is inclined at 53.0° below the horizontal. The coefficient of kinetic friction between the package and the chute's surface is 0.40. Calculate the work done on the package by (a) friction (b) gravity, and (c) the normal force. (d) What is the net work done on the package?

9. •• Two tugboats pull a disabled supertanker. Each tug exerts a constant force of 1.80×10^6 N, one 14° west of north and the other 14° east of north, as they pull the tanker 0.75 km toward the north. What is the total work they do on the supertanker?

10. •• A tow truck pulls a car 5.00 km along a horizontal roadway using a cable having a tension of 850 N. (a) How much work does the cable do on the car if it pulls horizontally? If it pulls at 35.0° above the horizontal? (b) How much work does the cable do on the tow truck in both cases of part (a)? (c) How much work does gravity do on the car in part (a)?

11. •• A boxed 10.0 kg computer monitor is dragged by friction 5.50 m up along the moving surface of a conveyor belt inclined at an angle of 36.9° above the horizontal. If the monitor's speed is a constant 2.10 cm/s, how much work is done on the monitor by (a) friction, (b) gravity, and (c) the normal force of the conveyor belt?

7.3 Work and Kinetic Energy

12. • It takes 4.186 J of energy to raise the temperature of 1.0 g of water by 1.0°C. (a) How fast would a 2.0 g cricket have to jump to have that much kinetic energy? (b) How fast would a 4.0 g cricket have to jump to have the same amount of kinetic energy?

13. • A bullet is fired into a large stationary absorber and comes to rest. Temperature measurements of the absorber show that the bullet lost 1960 J of kinetic energy, and high-speed photos of the bullet show that it was moving at 965 m/s just as it struck the absorber. What is the mass of the bullet?

14. •• **Animal energy.** Adult cheetahs, the fastest of the great **BIO** cats, have a mass of about 70 kg and have been clocked at up

to 72 mph (32 m/s). (a) How many joules of kinetic energy does such a swift cheetah have? (b) By what factor would its kinetic energy change if its speed were doubled?

15. •• A racing dog is initially running at 10.0 m/s, but is slowing down. (a) How fast is the dog moving when its kinetic energy has been reduced by half? (b) By what fraction has its kinetic energy been reduced when its speed has been reduced by half?

16. •• If a running house cat has 10.0 J of kinetic energy at speed v, (a) At what speed (in terms of v) will she have 20.0 J of kinetic energy? (b) What would her kinetic energy be if she ran half as fast as the speed in part (a)?

17. • A 0.145 kg baseball leaves a pitcher's hand at a speed of 32.0 m/s. If air drag is negligible, how much work has the pitcher done on the ball by throwing it?

18. ••• A 1.50 kg book is sliding along a rough horizontal surface. At point A it is moving at 3.21 m/s, and at point B it has slowed to 1.25 m/s. (a) How much work was done on the book between A and B? (b) If -0.750 J of work is done on the book from B to C, how fast is it moving at point C? (c) How fast would it be moving at C if $+0.750$ J of work were done on it from B to C?

19. •• **Stopping distance of a car.** The driver of an 1800 kg car (including passengers) traveling at 23.0 m/s slams on the brakes, locking the wheels on the dry pavement. The coefficient of kinetic friction between rubber and dry concrete is typically 0.700. (a) Use the work–energy principle to calculate how far the car will travel before stopping. (b) How far would the car travel if it were going twice as fast? (c) What happened to the car's original kinetic energy?

20. • **Meteor crater.** About 50,000 years ago, a meteor crashed into the earth near present-day Flagstaff, Arizona. Recent (2005) measurements estimate that this meteor had a mass of about 1.4×10^8 kg (around 150,000 tons) and hit the ground at 12 km/s. (a) How much kinetic energy did this meteor deliver to the ground? (b) How does this energy compare to the energy produced in one day by a standard coal-fired power plant, which generates about 1 billion joules per second?

21. •• You throw a 20 N rock into the air from ground level and observe that, when it is 15.0 m high, it is traveling upward at 25.0 m/s. Use the work–energy principle to find (a) the rock's speed just as it left the ground and (b) the maximum height the rock will reach.

22. • A 0.420 kg soccer ball is initially moving at 2.00 m/s. A soccer player kicks the ball, exerting a constant 40.0 N force in the same direction as the ball's motion. Over what distance must her foot be in contact with the ball to increase the ball's speed to 6.00 m/s?

23. •• A 61 kg skier on level snow coasts 184 m to a stop from a speed of 12.0 m/s. (a) Use the work–energy principle to find the coefficient of kinetic friction between the skis and the snow. (b) Suppose a 75 kg skier with twice the starting speed coasted the same distance before stopping. Find the coefficient of kinetic friction between that skier's skis and the snow.

24. •• A block of ice with mass 2.00 kg slides 0.750 m down an inclined plane that slopes downward at an angle of $36.9°$ below the horizontal. If the block of ice starts from rest, what is its final speed? You can ignore friction.

7.4 Work Done by a Varying Force

25. • To stretch a certain spring by 2.5 cm from its equilibrium position requires 8.0 J of work. (a) What is the force constant of this spring? (b) What was the maximum force required to stretch it by that distance?

26. •• A spring is 17.0 cm long when it is lying on a table. One end is then attached to a hook and the other end is pulled by a force that increases to 25.0 N, causing the spring to stretch to a length of 19.2 cm. (a) What is the force constant of this spring? (b) How much work was required to stretch the spring from 17.0 cm to 19.2 cm? (c) How long will the spring be if the 25 N force is replaced by a 50 N force?

27. •• A spring of force constant 300.0 N/m and unstretched length 0.240 m is stretched by two forces, pulling in opposite directions at opposite ends of the spring, that increase to 15.0 N. How long will the spring now be, and how much work was required to stretch it that distance?

28. •• An unstretched spring has a force constant of 1200 N/m. How large a force and how much work are required to stretch the spring: (a) by 1.0 m from its unstretched length, and (b) by 1.0 m beyond the length reached in part (a)?

29. •• The graph in the accompanying figure shows the magnitude of the force exerted by a given spring as a function of the distance x the spring is stretched. How much work is needed to stretch this spring: (a) a distance of 5.0 cm, starting with it unstretched, and (b) from $x = 2.0$ cm to $x = 7.0$ cm?

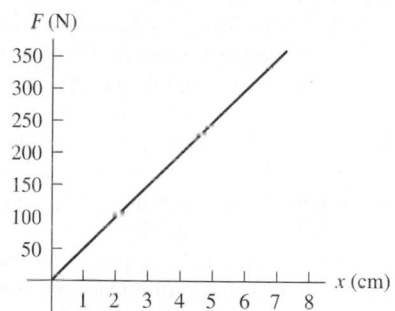

▲ **FIGURE 7.39** Problem 29.

7.5 Potential Energy

30. • A 575 N woman climbs a staircase that rises at 53° above
BIO the horizontal and is 4.75 m long. Her speed is a constant 45 cm/s. (a) Is the given weight a reasonable one for an adult woman? (b) How much has the gravitational potential energy increased by her climbing the stairs? (c) How much work has gravity done on her as she climbed the stairs?

31. • **How high can we jump?** The maximum height a typical
BIO human can jump from a crouched start is about 60 cm. By how much does the gravitational potential energy increase for a 72 kg person in such a jump? Where does this energy come from?

32. • A 72.0-kg swimmer jumps into the old swimming hole from a diving board 3.25 m above the water. Use energy conservation to find his speed just he hits the water (a) if he just holds his nose and drops in, (b) if he bravely jumps straight up (but just beyond the board!) at 2.50 m/s, and (c) if he manages to jump downward at 2.50 m/s.

33. •• A 2.50-kg mass is pushed against a horizontal spring of force constant 25.0 N/cm on a frictionless air table. The spring

is attached to the tabletop, and the mass is not attached to the spring in any way. When the spring has been compressed enough to store 11.5 J of potential energy in it, the mass is suddenly released from rest. (a) Find the greatest speed the mass reaches. When does this occur? (b) What is the greatest acceleration of the mass, and when does it occur?

34. • A force of magnitude 800.0 N stretches a certain spring by 0.200 m from its equilibrium position. (a) What is the force constant of this spring? (b) How much elastic potential energy is stored in the spring when it is: (i) stretched 0.300 m from its equilibrium position and (ii) compressed by 0.300 m from its equilibrium position? (c) How much work was done in stretching the spring by the original 0.200 m?

35. • **Tendons.** Tendons are strong elastic fibers that attach muscles to bones. To a reasonable approximation, they obey **BIO** Hooke's law. In laboratory tests on a particular tendon, it was found that, when a 250 g object was hung from it, the tendon stretched 1.23 cm. (a) Find the force constant of this tendon in N/m. (b) Because of its thickness, the maximum tension this tendon can support without rupturing is 138 N. By how much can the tendon stretch without rupturing, and how much energy is stored in it at that point?

36. ••• A certain spring stores 10.0 J of potential energy when it is stretched by 2.00 cm from its equilibrium position. (a) How much potential energy would the spring store if it were stretched an additional 2.00 cm? (b) How much potential energy would it store if it were compressed by 2.00 cm from its equilibrium position? (c) How far from the equilibrium position would you have to stretch the string to store 20.0 J of potential energy? (d) What is the force constant of this spring?

37. • In designing a machine part, you need a spring that is 8.50 cm long when no forces act on it and that will store 15.0 J of energy when it is compressed by 1.20 cm from its equilibrium position. (a) What should be the force constant of this spring? (b) Can the spring store 850 J by compression?

38. •• **Leg presses.** As part of your daily workout, you lie on your back and push with your feet against a platform attached to two stiff springs arranged side by side so that they are parallel to each other. When you push the platform, you compress the springs. You do 80.0 J of work when you compress the springs 0.200 m from their uncompressed length. (a) What magnitude of force must you apply to hold the platform in this position? (b) How much *additional* work must you do to move the platform 0.200 m *farther*, and what maximum force must you apply?

39. •• **Food calories.** The *food calorie,* equal to 4186 J, is a **BIO** measure of how much energy is released when food is metabolized by the body. A certain brand of fruit-and-cereal bar contains 140 food calories per bar. (a) If a 65 kg hiker eats one of these bars, how high a mountain must he climb to "work off" the calories, assuming that all the food energy goes only into increasing gravitational potential energy? (b) If, as is typical, only 20% of the food calories go into mechanical energy, what would be the answer to part (a)? *(Note: In this and all other problems, we are assuming that 100% of the food calories that are eaten are absorbed and used by the body. This is actually not true. A person's "metabolic efficiency" is the percentage of calories eaten that are actually used; the rest are eliminated by the body. Metabolic efficiency varies considerably from person to person.)*

40. •• **A good workout.** You overindulged on a delicious dessert, **BIO** so you plan to work off the extra calories at the gym. To accomplish this, you decide to do a series of arm raises holding a 5.0 kg weight in one hand. The distance from your elbow to the weight is 35 cm, and in each arm raise you start with your arm horizontal and pivot it until it is vertical. Assume that the weight of your arm is small enough compared with the weight you are lifting that you can ignore it. As is typical, your muscles are 20% efficient in converting the food energy they use up into mechanical energy, with the rest going into heat. If your dessert contained 350 food calories, how many arm raises must you do to work off these calories? Is it realistic to do them all in one session?

41. •• **An exercise program.** A 75 kg person is put on an exer- **BIO** cise program by a physical therapist, the goal being to burn up 500 food calories in each daily session. Recall that human muscles are about 20% efficient in converting the energy they use up into mechanical energy. The exercise program consists of a set of consecutive high jumps, each one 50 cm into the air (which is pretty good for a human) and lasting 2.0 s, on the average. How many jumps should the person do per session, and how much time should be set aside for each session? Do you think that this is a physically reasonable exercise session?

7.6 Conservation of Energy

42. • Tall Pacific Coast redwood trees (*Sequoia sempervirens*) can reach heights of about 100 m. If air drag is negligibly small, how fast is a sequoia cone moving when it reaches the ground if it dropped from the top of a 100 m tree?

43. • The total height of Yosemite Falls is 2425 ft. (a) How many more joules of gravitational potential energy are there for each kilogram of water at the top of this waterfall compared with each kilogram of water at the foot of the falls? (b) Find the kinetic energy and speed of each kilogram of water as it reaches the base of the waterfall, assuming that there are no losses due to friction with the air or rocks and that the mass of water had negligible vertical speed at the top. How fast (in m/s and mph) would a 70 kg person have to run to have that much kinetic energy? (c) How high would Yosemite Falls have to be so that each kilogram of water at the base had twice the kinetic energy you found in part (b); twice the speed you found in part (b)?

44. • **The speed of hailstones.** Although the altitude may vary considerably, hailstones sometimes originate around 500 m (about 1500 ft) above the ground. (a) Neglecting air drag, how fast will these hailstones be moving when they reach the ground, assuming that they started from rest? Express your answer in m/s and in mph. (b) From your own experience, are hailstones actually falling that fast when they reach the ground? Why not? What has happened to most of the initial potential energy?

45. •• Pebbles of weight w are launched from the edge of a vertical cliff of height h at speed v_0. How fast (in terms of the quantities just given) will these pebbles be moving when they reach the ground if they are launched (a) straight up, (b) straight down, (c) horizontally away from the cliff, and (d) at an angle θ above the horizontal? (e) How would the answers to the previous parts change if the pebbles weighed twice as much?

46. • **Volcanoes on Io.** Io, a satellite of Jupiter, is the most volcanically active moon or planet in the solar system. It has

volcanoes that send plumes of matter over 500 km high (see the accompanying figure). Due to the satellite's small mass, the acceleration due to gravity on Io is only 1.81 m/s², and Io has no appreciable atmosphere. Assume that there is no variation in gravity over the distance traveled. (a) What

▲ **FIGURE 7.40** Problem 46.

must be the speed of material just as it leaves the volcano to reach an altitude of 500 km? (b) If the gravitational potential energy is zero at the surface, what is the potential energy for a 25 kg fragment at its maximum height on Io? How much would this gravitational potential energy be if it were at the same height above earth?

47. •• **Human energy vs. insect energy.** For its size, the com-
BIO mon flea is one of the most accomplished jumpers in the animal world. A 2.0-mm-long, 0.50 mg critter can reach a height of 20 cm in a single leap. (a) Neglecting air drag, what is the take-off speed of such a flea? (b) Calculate the kinetic energy of this flea at takeoff and its kinetic energy per kilogram of mass. (c) If a 65 kg, 2.0-m-tall human could jump to the same height compared with his length as the flea jumps compared with its length, how high could he jump, and what takeoff speed would he need? (d) In fact, most humans can jump no more than 60 cm from a crouched start. What is the kinetic energy per kilogram of mass at takeoff for such a 65 kg person? (e) Where does the flea store the energy that allows it to make such a sudden leap?

48. •• A 25 kg child plays on a swing having support ropes that are 2.20 m long. A friend pulls her back until the ropes are 42° from the vertical and releases her from rest. (a) What is the potential energy for the child just as she is released, compared with the potential energy at the bottom of the swing? (b) How fast will she be moving at the bottom of the swing? (c) How much work does the tension in the ropes do as the child swings from the initial position to the bottom?

49. • **Tarzan and Jane.** Tarzan, in one tree, sights Jane in another tree. He grabs the end of a vine with length 20 m that makes an angle of 45° with the vertical, steps off his tree limb, and swings down and then up to Jane's open arms. When he arrives, his vine makes an angle of 30° with the vertical. Determine whether he gives her a tender embrace or knocks her off her limb by calculating Tarzan's speed just before he reaches Jane. You can ignore air resistance and the mass of the vine.

50. •• A slingshot obeying Hooke's law is used to launch pebbles vertically into the air. You observe that if you pull a pebble back 20.0 cm against the elastic band, the pebble goes 6.0 m high. (a) Assuming that air drag is negligible, how high will the pebble go if you pull it back 40.0 cm instead? (b) How far must you pull it back so it will reach 12.0 m? (c) If you pull a pebble that is twice as heavy back 20.0 cm, how high will it go?

51. •• When a piece of wood is pressed against a spring and compresses the spring by 5.0 cm, the wood gains a maximum kinetic energy K when it is released. How much kinetic energy (in terms of K) would the piece of wood gain if the spring were compressed 10.0 cm instead?

52. • A 1.5 kg box moves back and forth on a horizontal friction-less surface between two different springs, as shown in the accompanying figure. The box is initially pressed against the stronger spring, compressing it 4.0 cm, and then is released from rest. (a) By how much will the box compress the weaker spring? (b) What is the maximum speed the box will reach?

Frictionless surface

▲ **FIGURE 7.41** Problem 52.

53. •• A 12.0 N package of whole wheat flour is suddenly placed on the pan of a scale such as you find in grocery stores. The pan is supported from below by a vertical spring of force constant 325 N/m. If the pan has negligible weight, find the maximum distance the spring will be compressed if no energy is dissipated by friction.

54. •• A spring of negligible mass has force constant $k = 1600$ N/m. (a) How far must the spring be compressed for 3.20 J of potential energy to be stored in it? (b) You place the spring vertically with one end on the floor. You then drop a 1.20-kg book onto it from a height of 0.80 m above the top of the spring. Find the maximum distance the spring will be compressed.

7.7 Conservative and Nonconservative Forces

55. • A 1.50 kg brick is sliding along on a rough horizontal surface at 13.0 m/s. If the brick stops in 4.80 s, how much mechanical energy is lost, and what happens to this energy?

56. • A fun-loving 11.4 kg otter slides up a hill and then back down to the same place. If she starts up at 5.75 m/s and returns at 3.75 m/s, how much mechanical energy did she lose on the hill, and what happened to that energy?

57. • A 12.0 g plastic ball is dropped from a height of 2.50 m and is moving at 3.20 m/s just before it hits the floor. How much mechanical energy was lost during the ball's fall?

58. • You and three friends stand at the corners of a square whose sides are 8.0 m long in the middle of the gym floor, as shown in the accompanying figure. You take your physics book and push it from one person to the other. The book has a mass of 1.5 kg, and the coefficient of kinetic friction between the

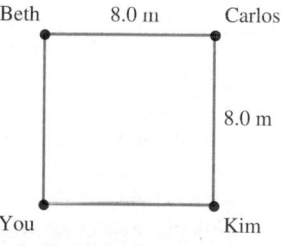

▲ **FIGURE 7.42** Problem 58.

book and the floor is $\mu_k = 0.25$. (a) The book slides from you to Beth and then from Beth to Carlos, along the lines connecting these people. What is the work done by friction during this displacement? (b) You slide the book from you to Carlos along the diagonal of the square. What is the work done by friction during this displacement? (c) You slide the book to Kim who then slides it back to you. What is the total work done by friction during this motion of the book? (d) Is the friction force on the book conservative or nonconservative? Explain.

59. • While a roofer is working on a roof that slants at 36° above the horizontal, he accidentally nudges his 85.0 N toolbox, causing it to start sliding downward, starting from rest. If it

starts 4.25 m from the lower edge of the roof, how fast will the toolbox be moving just as it reaches the edge of the roof if the kinetic friction force on it is 22.0 N?

60. •• A block with mass 0.50 kg is forced against a horizontal spring of negligible mass, compressing the spring a distance of 0.20 m, as shown in the accompanying figure. When released, the block moves on a horizontal tabletop for 1.00 m before coming to rest. The spring constant k is 100 N/m. What is the coefficient of kinetic friction μ_k between the block and the tabletop?

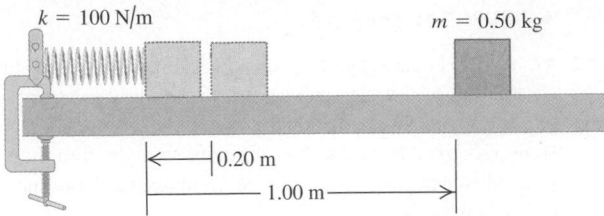

▲ **FIGURE 7.43** Problem 60.

61. •• A loaded 375 kg toboggan is traveling on smooth horizontal snow at 4.5 m/s when it suddenly comes to a rough region. The region is 7.0 m long and reduces the toboggan's speed by 1.5 m/s. (a) What average friction force did the rough region exert on the toboggan? (b) By what percent did the rough region reduce the toboggan's (i) kinetic energy and (ii) speed?

62. •• A 62.0 kg skier is moving at 6.50 m/s on a frictionless, horizontal snow-covered plateau when she encounters a rough patch 3.50 m long. The coefficient of kinetic friction between this patch and her skis is 0.300. After crossing the rough patch and returning to friction-free snow, she skis down an icy, frictionless hill 2.50 m high. (a) How fast is the skier moving when she gets to the bottom of the hill? (b) How much internal energy was generated in crossing the rough patch?

7.8 Power

63. • (a) How many joules of energy does a 100 watt lightbulb use every hour? (b) How fast would a 70 kg person have to run to have that amount of kinetic energy? Is it possible for a person to run that fast? (c) How high a tree would a 70 kg person have to climb to increase his gravitational potential energy relative to the ground by that amount? Are there any trees that tall?

64. • The engine of a motorboat delivers 30.0 kW to the propeller while the boat is moving at 15.0 m/s. What would be the tension in the towline if the boat were being towed at the same speed?

65. •• At 7.35 cents per kilowatt-hour, (a) what does it cost to operate a 10.0 hp motor for 8.00 hr? (b) What does it cost to leave a 75 W light burning 24 hours a day?

66. • A tandem (two-person) bicycle team must overcome a force of 165 N to maintain a speed of 9.00 m/s. Find the power required per rider, assuming that each contributes equally.

67. •• An elevator has mass 600 kg, not including passengers. The elevator is designed to ascend, at constant speed, a vertical distance of 20.0 m (five floors) in 16.0 s, and it is driven by a motor that can provide up to 40 hp to the elevator. What is the maximum number of passengers that can ride in the elevator? Assume that an average passenger has mass 65.0 kg.

68. •• **U.S. power use.** The total consumption of electrical energy in the United States is about 1.0×10^{19} joules per year. (a) Express this rate in watts and kilowatts. (b) If the U.S. population is about 310 million people, what is the average rate of electrical energy consumption per person?

69. •• **Solar energy.** The sun transfers energy to the earth by radiation at a rate of approximately 1.0 kW per square meter of surface. (a) If this energy could be collected and converted to electrical energy with 25% efficiency, how large an area (in square kilometers) would be required to collect the electrical energy used by the United States? (See the previous problem.) (b) If the solar collectors were arranged in a square array, what would be the length of its sides in kilometers and in miles? Does an array of these dimensions seem technologically feasible?

70. •• A 20.0 kg rock slides on a rough horizontal surface at 8.00 m/s and eventually stops due to friction. The coefficient of kinetic friction between the rock and the surface is 0.200. What average thermal power is produced as the rock stops?

71. •• **Horsepower.** In the English system of units, power is expressed as *horsepower* (hp) instead of watts, where 1.00 hp = 746 W. Horsepower is often used for motors and automobiles. (a) What should be the power rating in horsepower of a 100 W lightbulb? (b) How many watts does a 75 hp motor produce? (c) Electrical resistors are rated in watts to indicate how much power they can tolerate without burning up. If a resistor is rated at 2.00 W, what should be its rating in horsepower? (d) Our sun radiates 3.92×10^{26} J of energy each second. What is its horsepower rating? (e) How many kilowatt-hours of energy is produced when you run a 25 hp motor for 90 minutes?

72. •• **Maximum sustainable human power.** The maximum *sustainable* mechanical power a human can produce is about $\frac{1}{3}$ hp. How many food calories can a human burn up in an hour by exercising at this rate? (Remember that only 20% of the food energy used goes into mechanical energy.)

73. •• A typical flying insect applies an average force equal to twice its weight during each downward stroke while hovering. Take the mass of the insect to be 10 g, and assume the wings move an average downward distance of 1.0 cm during each stroke. Assuming 100 downward strokes per second, estimate the average power output of the insect.

74. •• When its 75-kW (100-hp) engine is generating full power, a small single-engine airplane with mass 700 kg gains altitude at a rate of 2.5 m/s (150 m/min, or 500 ft/min). What fraction of the engine power is being used to make the airplane climb? (The remainder is used to overcome the effects of air resistance and of inefficiencies in the propeller and engine.)

75. •• **The power of the human heart.** The human heart is a powerful and extremely reliable pump. Each day it takes in and discharges about 7500 L of blood. Assume that the work done by the heart is equal to the work required to lift that amount of blood a height equal to that of the average American female, approximately 1.63 m. The density of blood is 1050 kg/m³. (a) How much work does the heart do in a day? (b) What is the heart's power output in watts? (c) In fact, the heart puts out more power than you found in part (b). Why? What other forms of energy does it give the blood?

76. •• At the site of a wind farm in North Dakota, the average wind speed is 9.3 m/s, and the average density of air is 1.2 kg/m³. (a) Calculate how much kinetic energy the wind

contains, per cubic meter, at this location. (b) No wind turbine can capture all of the energy contained in the wind, the main reason being that capturing all the energy would require stopping the wind completely, meaning that air would stop flowing through the turbine. Suppose a particular turbine has blades with a radius of 41 m and is able to capture 35% of the available wind energy. What would be the power output of this turbine, under average wind conditions?

General Problems

77. • **Bumper guards.** You are asked to design spring bumpers for the walls of a parking garage. A freely rolling 1200 kg car moving at 0.65 m/s is to compress the spring no more than 7.0 cm before stopping. (a) What should be the force constant of the spring, and what is the maximum amount of energy that gets stored in it? (b) If the springs that are actually delivered have the proper force constant but can become compressed by only 5.0 cm, what is the maximum speed of the given car for which they will provide adequate protection?

78. •• **Human terminal velocity.** By landing properly and on soft ground (and by being lucky!), humans have survived falls from airplanes when, for example, a parachute failed to open, with astonishingly little injury. Without a parachute, a typical human eventually reaches a terminal velocity of about 62 m/s. Suppose the fall is from an airplane 1000 m high. (a) How fast would a person be falling when he reached the ground if there were no air drag? (b) If a 70 kg person reaches the ground traveling at the terminal velocity of 62 m/s, how much mechanical energy was lost during the fall? What happened to that energy?

79. •• A wooden rod of negligible mass and length 80.0 cm is pivoted about a horizontal axis through its center. A white rat with mass 0.500 kg clings to one end of the stick, and a mouse with mass 0.200 kg clings to the other end. The system is released from rest with the rod horizontal. If the animals can manage to hold on, what are their speeds as the rod swings through a vertical position?

80. •• **Mountain climbing!** A 75.0 kg mountain climber is holding his 60.0 kg partner over a cliff when he suddenly steps on frictionless ice at the horizontal top of the cliff, as shown in the accompanying figure. The rope has negligible mass and is held horizontally by the climber. There is no appreciable friction at the icy edge of the cliff. Use energy methods to calculate the speed of the climbers after the lower one has descended 1.50 m starting from rest.

75.0 kg

60.0 kg

▲ **FIGURE 7.44** Problem 80.

81. •• **More mountain climbing!** What would be the speed of the climbers in the previous problem after they had moved 1.50 m if there were friction between the upper climber and the ice with a coefficient of kinetic friction of 0.250?

82. •• **Ski jump ramp.** You are designing a ski jump ramp for the next Winter Olympics. You need to calculate the vertical height h from the starting gate to the bottom of the ramp. The skiers push off hard with their ski poles at the start, just above the starting gate, so they typically have a speed of 2.0 m/s as they reach the gate. For safety, the skiers should have a speed of no more than 30.0 m/s when they reach the bottom of the ramp. You determine that for a 85.0-kg skier with good form, friction and air resistance will do total work of magnitude 4000 J on him during his run down the slope. What is the maximum height h for which the maximum safe speed will not be exceeded?

83. • **Rescue.** Your friend (mass 65.0 kg) is standing on the ice in the middle of a frozen pond. There is very little friction between her feet and the ice, so she is unable to walk. Fortunately, a light rope is tied around her waist and you stand on the bank holding the other end. You pull on the rope for 3.00 s and accelerate your friend from rest to a speed of 6.00 m/s while you remain at rest. What is the average power supplied by the force you applied?

84. •• On an essentially frictionless horizontal ice-skating rink, a skater moving at 3.0 m/s encounters a rough patch that reduces her speed by 45% due to a friction force that is 25% of her weight. Use the work-energy principle to find the length of the rough patch.

85. ••• **Pendulum.** A small 0.12 kg metal ball is tied to a very light (essentially massless) string 0.80 m long to form a pendulum that is then set swinging by releasing the ball from rest when the string makes a 45° angle with the vertical. Air drag and other forms of friction are negligible. What is the speed of the ball when the string passes through its vertical position, and what is the tension in the string at that instant?

86. •• A pump is required to lift 750 liters of water per minute from a well 14.0 m deep and eject it with a speed of 18.0 m/s. How much work per minute does the pump do?

87. •• A 350 kg roller coaster starts from rest at point A and slides down the frictionless loop-the-loop shown in the accompanying figure. (a) How fast is this roller coaster moving at point B? (b) How hard does it press against the track at point B?

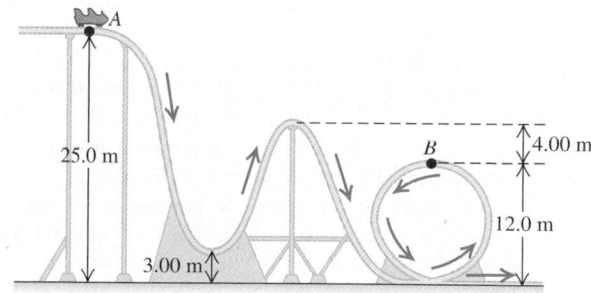

A
25.0 m
3.00 m
B
4.00 m
12.0 m

▲ **FIGURE 7.45** Problem 87.

88. ••• **Automobile air-bag safety.** An automobile air bag cushions the force on the driver in a head-on collision by absorbing her energy before she hits the steering wheel. Such a bag can be modeled as an elastic force, similar to that produced by a

spring. (a) Use energy conservation to show that the effective force constant k of the air bag is $k = mv_0^2/x_{max}^2$, where m is the mass of the driver, v_0 is the speed of the car (and driver) at the instant of the accident, and x_{max} is the maximum distance the bag gets compressed, which, in a severe accident, would be the distance from the driver's body to

▲ FIGURE 7.46 Problem 88.

the steering wheel. (b) Show that the maximum force the air bag would exert on the driver is $F_{max} = mv_0^2/x_{max}$. (c) Now let's put in some realistic numbers. Experimental tests have shown that injury occurs when a force *density* greater than 5.0×10^5 N/m^2 acts on human tissue. (The *total* force is this force density times the area over which it acts.) As the accompanying figure shows, the force of the air bag acts mostly on the upper front half of the driver's body, over an area of about 2500 cm^2. (Check your own body to see if this is reasonable.) Use this value to calculate the total force on the driver's body at the threshold of injury. (d) Use your results to calculate the effective force constant k of the air bag and the maximum speed for which the bag will prevent injury to a 65 kg driver if she is 30 cm from the steering wheel at the instant of impact. Express the speed in m/s and mph. (e) How could you design a safer air bag for higher speed collisions? What things could you alter to do this? Would it be safe to make a stiffer air bag by inflating it more? Explain your reasoning.

89. •• In creating his definition of horsepower, James Watt, the
BIO inventor of the steam engine, calculated the power output of a horse operating a mill to grind grain or cut wood. The horse walked in a 24-ft diameter circle, making, according to Watt, 144 trips around the circle in an hour. (a) Using the currently accepted value of 746 watts for 1 horsepower, calculate the force (in pounds) with which Mr. Watt's horse must have been pulling. (See Appendix D for useful conversion factors.) (b) Calculate the power output, in hp, of a 70-kg human being who climbs a 3.0-m-high set of stairs in 5.0 seconds.

90. •• All birds, independent of their size, must maintain a power
BIO output of 10–25 watts per kilogram of body mass in order to fly by flapping their wings. (a) The Andean giant hummingbird (*Patagona gigas*) has mass 70 g and flaps its wings 10 times per second while hovering. Estimate the amount of work done by such a hummingbird in each wingbeat. (b) A 70-kg athlete can maintain a power output of 1.4 kW for no more than a few seconds; the *steady* power output of a typical athlete is only 500 W or so. Is it possible for a human-powered aircraft to fly for extended periods by flapping its wings? Explain.

91. •• A 250 g object on a frictionless, horizontal lab table is pushed against a spring of force constant 35 N/cm and then released. Just before the object is released, the spring is compressed 12.0 cm. How fast is the object moving when it has gained half of the spring's original stored energy?

92. •• **Automobile accident analysis.** In an auto accident, a car hit a pedestrian and the driver then slammed on the brakes to stop the car. During the subsequent trial, the driver's lawyer claimed that the driver was obeying the posted 35 mph speed limit, but that the limit was too high to enable him to see and

react to the pedestrian in time. You have been called as the state's expert witness. In your investigation of the accident site, you make the following measurements: The skid marks made while the brakes were applied were 280 ft long, and the tread on the tires produced a coefficient of kinetic friction of 0.30 with the road. (a) In your testimony in court, will you say that the driver was obeying the posted speed limit? You must be able to back up your answer with clear numerical reasoning during cross-examination. (b) If the driver's speeding ticket is $10 for each mile per hour he was driving above the posted speed limit, would he have to pay a ticket, and if so, how much would it be?

93. •• **Bungee jump.** A bungee cord is 30.0 m long and, when stretched a distance x, it exerts a restoring force of magnitude kx. Your father-in-law (mass 95.0 kg) stands on a platform 45.0 m above the ground, and one end of the cord is tied securely to his ankle and the other end to the platform. You have promised him that when he steps off the platform he will fall a maximum distance of only 41.0 m before the cord stops him. You had several bungee cords to select from, and you tested them by stretching them out, tying one end to a tree, and pulling on the other end with a force of 380.0 N. When you do this, what distance will the bungee cord that you should select have stretched?

94. •• **Riding a loop-the-loop.** A car in an amusement park ride travels without friction along the track shown in the accompanying figure, starting from rest at point A. If the loop the car is currently on has a radius of 20.0 m, find the minimum height h so that the car will not fall off the track at the top of the circular part of the loop.

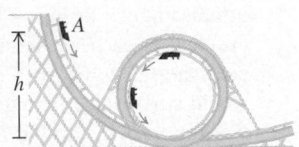

▲ FIGURE 7.47 Problem 94.

95. •• A 2.0 kg piece of wood slides on the surface shown in the accompanying figure. All parts of the surface are frictionless, except for a 30-m-long rough segment at the bottom, where the coefficient of kinetic friction with the wood is 0.20. The wood starts from rest 4.0 m above the bottom. (a) Where will the wood eventually come to rest? (b) How much work is done by friction by the time the wood stops?

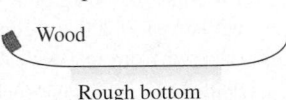

▲ FIGURE 7.48 Problem 95.

96. •• A 68 kg skier approaches the foot of a hill with a speed of 15 m/s. The surface of this hill slopes up at 40.0° above the horizontal and has coefficients of static and kinetic friction of 0.75 and 0.25, respectively, with the skis. (a) Use energy conservation to find the maximum height above the foot of the hill that the skier will reach. (b) Will the skier remain at rest once she stops, or will she begin to slide down the hill? Prove your answer.

97. •• **Energy requirements of the body.** A 70 kg human uses
BIO energy at the rate of 80 J/s, on average, for just resting and sleeping. When the person is engaged in more strenuous activities, the rate can be much higher. (a) If the individual did nothing but rest, how many food calories per day would she or he have to eat to make up for those used up? (b) In what forms is energy used when a person is resting or sleeping? In other words, what happens to those 80 J/s? (*Hint: What kinds of energy, mechanical and otherwise, do our body*

components have?) (c) If an average person rested and did other low-level activity for 16 hours (which consumes 80 J/s) and did light activity on the job for 8 hours (which consumes 200 J/s), how many calories would she or he have to consume per day to make up for the energy used up?

98. •• The aircraft carrier *USS George Washington* has mass 1.0×10^8 kg. When its engines are developing their full power of 260,000 hp, the *George Washington* travels at its top speed of 35 knots (65 km/h). If 70% of the power output of the engines is applied to pushing the ship through the water, what is the magnitude of the force of water resistance that opposes the carrier's motion at this speed?

99. •• Two paint buckets are connected by a lightweight rope passing over a pulley of negligible mass and friction. (a) As shown in the accompanying figure, the system is released from rest with the 12.0 kg bucket 2.00 m above the floor. Use energy conservation to find the speed with which this bucket strikes the floor. (b) Suppose the pulley had appreciable mass but no bearing friction. Would the bucket's speed be greater than, less than, or the same as you found in part (a)? Explain your reasoning in terms of energy.

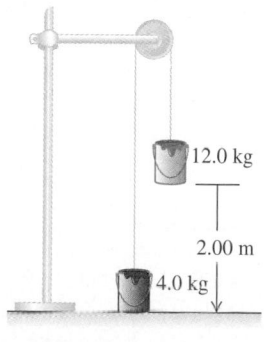

▲ **FIGURE 7.49** Problem 99.

100. ••• A ball is thrown upward with an initial velocity of 15 m/s at an angle of 60.0° above the horizontal. Use energy conservation to find the ball's greatest height above the ground.

101. ••• **Automotive power.** A truck engine transmits 28.0 kW (37.5 hp) to the driving wheels when the truck is traveling at a constant velocity of magnitude 60.0 km/h (37.7 mi/h) on a level road. (a) What is the resisting force acting on the truck? (b) Assume that 65% of the resisting force is due to rolling friction and the remainder is due to air resistance. If the force of rolling friction is independent of speed, and the force of air resistance is proportional to the square of the speed, what power will drive the truck at 30.0 km/h? At 120.0 km/h? Give your answers in kilowatts and in horsepower.

102. ••• **Mass extinctions.** One of the greatest mass extinctions occurred about 65 million years ago, when, along with many other life-forms, the dinosaurs went extinct. Most geologists and paleontologists agree that this event was caused when a large asteroid hit the earth. Scientists estimate that this asteroid was about 10 km in diameter and that it would have been traveling at least as fast as 11 km/s. The density of asteroid material is about 3.5 g/cm³, on the average. (a) What would be the approximate mass of the asteroid, assuming it to be spherical? (b) How much kinetic energy would the asteroid have delivered to the earth? (c) In order to put the amount of energy you found in part (b) in perspective, consider the following: the total amount of energy used in one year by the human race is roughly 500 exajoules (see Appendix E). If this rate of energy use

remained constant, how many years would it take the human species to use an amount of energy equal to the amount delivered by this asteroid?

103. •• **Avoiding mass extinctions.** It has been suggested that we can protect the earth from devastating asteroidal impacts such as the one discussed in the previous problem by using nuclear devices to alter the orbits of such asteroids around the sun so that they will miss our planet. If this is done very far from earth, it is necessary to move them only a few centimeters to spare the earth a mass extinction. How much energy would it take to move the asteroid that has been implicated in the dinosaur extinction by a few centimeters? (See the previous problem.) To make the calculation reasonable, assume that we need to exert a force on the asteroid that will accelerate it uniformly from rest through a distance of 5.0 cm in 0.50 s. The energy we must give to the asteroid is the added kinetic energy from this motion. To see if it is feasible to do this, how many 1.0 megaton bombs would it take to accomplish the task?

104. •• The spring of a spring gun has force constant $k = 400$ N/m and negligible mass. The spring is compressed 6.00 cm, and a ball with mass 0.0300 kg is placed in the horizontal barrel against the compressed spring. The spring is then released, and the ball is propelled out the barrel of the gun. The barrel is 6.00 cm long, so the ball leaves the barrel at the same point that it loses contact with the spring. The gun is held so the barrel is horizontal. (a) Calculate the speed with which the ball leaves the barrel if you can ignore friction. (b) Calculate the speed of the ball as it leaves the barrel if a constant resisting force of 6.00 N acts on the ball as it moves along the barrel. (c) For the situation in part (b), at what position along the barrel does the ball have the greatest speed, and what is that speed? (In this case, the maximum speed does not occur at the end of the barrel.)

105. •• A graph of the potential energy stored in a tendon as a **BIO** function of the *square* of the distance x it has stretched from its equilibrium position is shown in the accompanying figure. (a) Does the tendon obey Hooke's law? Explain your reasoning. (b) What is the force constant of the tendon? (c) How far must the tendon be stretched from its equilibrium position to store 10.0 J of energy? (d) What force is necessary to hold the tendon in place in part (c)? (e) Sketch a clear graph of the potential energy stored in the tendon as a function of x.

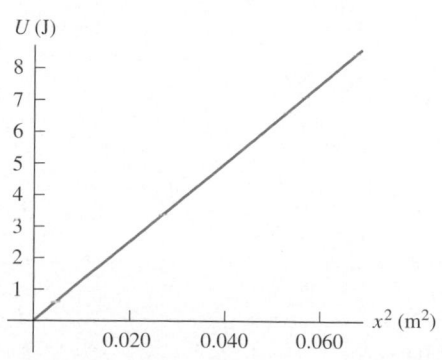

▲ **FIGURE 7.50** Problem 105.

106. •• A sled with rider having a combined mass of 125 kg travels over the perfectly smooth icy hill shown in the accompanying figure. How far does the sled land from the foot of the cliff?

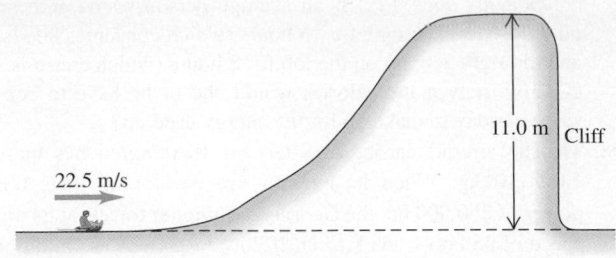

▲ **FIGURE 7.51** Problem 106.

$\boldsymbol{8}$ Momentum

When an 18-wheeler collides head-on with a compact car, why are the occupants of the car much more likely to be injured than those of the truck? How do you decide how to aim the cue ball in pool so as to knock the eight ball into the pocket? How can a rocket engine accelerate a space shuttle in outer space, where there's nothing to push against?

To answer these and similar questions, we need two new concepts—momentum and impulse—and a new conservation law: conservation of momentum. This law is every bit as important as conservation of energy, and its validity extends far beyond the bounds of classical mechanics, to include relativistic mechanics (the mechanics of the very fast) and quantum mechanics (the mechanics of the very small). Within classical mechanics, it enables us to analyze many situations that would be difficult to handle if we used Newton's laws directly. Among these situations are collision problems, in which two colliding objects exert large forces on each other for a short time.

The tugboat and the container ship have the same velocity, but if they accidentally collided with a dock, the container ship would do far more damage because of its greater mass. In this chapter we consider the combination of mass multiplied by velocity, which we call *momentum*.

8.1 Momentum

When a particle with mass m moves with velocity $\vec{v}$, we define its **momentum** $\vec{p}$ as follows:

Definition of momentum

The momentum of a particle, denoted by $\vec{p}$, is the product of its mass m and its velocity $\vec{v}$:

$$\vec{p} = m\vec{v}. \tag{8.1}$$

Unit: kg · m/s (This combination has no special name.)

Momentum $\vec{p}$ is a vector quantity; a particle's momentum has the same direction as its velocity $\vec{v}$.

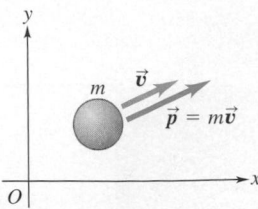

▲ **FIGURE 8.1** Momentum is a vector quantity.

Because velocity is a vector quantity, momentum is also a vector quantity. (See Figure 8.1.) Momentum has a magnitude (mv, mass times speed) and a direction (the same as that of the velocity vector). The momentum of a car driving north at 20 m/s is different from the momentum of the same car driving east at the same speed (even though both cars have the same kinetic energy). A fastball thrown by a major-league pitcher has greater momentum than the same ball thrown by a child, because the speed is greater. An 18-wheeler going 55 mi/h has greater momentum than a Saturn automobile going the same speed, because the truck's mass is greater.

Conceptual Analysis 8.1

Momentum and kinetic energy

Two objects have the same momentum but different masses. Which of the following statements about them is correct?

A. The one with less mass has more kinetic energy.
B. Both objects have the same kinetic energy.
C. The one with more mass has more kinetic energy.

SOLUTION Momentum is defined as $\vec{p} = m\vec{v}$, so if two objects have the same magnitude of momentum $p = mv$, the one with

the smaller mass m must have the greater speed v. However, kinetic energy depends on the *square* of speed: $K = \frac{1}{2}mv^2$. The kinetic energy can be expressed as

$$K = \frac{1}{2}mv^2 = \frac{1}{2}m\left(\frac{p}{m}\right)^2 = \frac{p^2}{2m}.$$

Solving for p, we get $p = \sqrt{2mK}$. Thus, if two objects have the same magnitude of momentum, p, the one with smaller mass m must have greater kinetic energy K. So choice A is right.

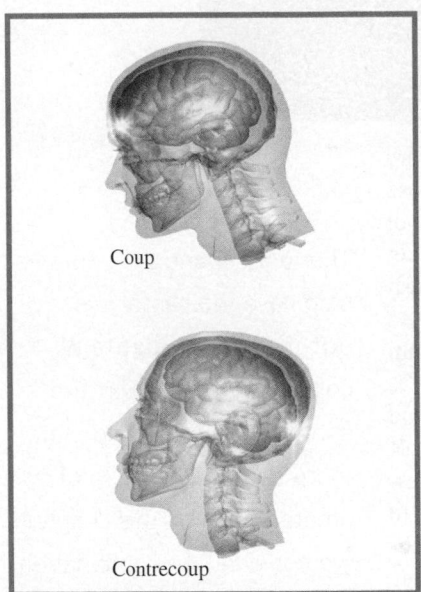

Coup

Contrecoup

▲ **BIO Application Contrecoup injury.**
The large mass of the human brain makes us susceptible to a type of head trauma known as a contrecoup brain injury (*contrecoup* is French for "across from the blow"). A blow to the head accelerates both the skull and brain. The brain's large mass gives it considerable momentum. When the skull stops moving, the brain runs into it. Depending on how suddenly the skull slows down, this second, contrecoup blow to the brain can cause as much or more injury as the initial blow. Animals with less massive brains rarely suffer this type of injury.

We'll often express the momentum of a particle in terms of its components. It follows from Equation 8.1 that if the particle has velocity components v_x, v_y, and v_z, its components of momentum, p_x, p_y, and p_z, are given by

$$p_x = mv_x$$
$$p_y = mv_y \qquad \text{(components of momentum)} \qquad (8.2)$$
$$p_z = mv_z$$

These three component equations are equivalent to Equation 8.1. In most of the problems in this chapter, the velocities will lie in the x-y plane and $\vec{v}$ will have only x and y components, but all the component relations are easily generalized to three-dimensional situations in which three components of velocity are needed.

The importance of momentum can be traced to its close relation to Newton's second law, $\sum\vec{F} = m\vec{a}$. We assume for now that the mass m of the particle is constant. (Later we'll learn how to deal with systems in which the mass changes.) When m is constant, we can express Newton's second law for a particle in terms of momentum. First, the instantaneous acceleration $\vec{a}$ of the particle is defined as

$$\vec{a} = \lim_{\Delta t \to 0} \frac{\Delta \vec{v}}{\Delta t},$$

so Newton's second law can be written as

$$\sum\vec{F} = m \lim_{\Delta t \to 0} \frac{\Delta \vec{v}}{\Delta t}.$$

Now if m is constant, the right side of this expression is just the instantaneous rate change of momentum of the particle. To show this, we note that if m is constant, $m\,\Delta\vec{v}$ is the change of momentum during the interval Δt:

$$m\,\Delta\vec{v} = m(\vec{v}_2 - \vec{v}_1) = m\vec{v}_2 - m\vec{v}_1 = \vec{p}_2 - \vec{p}_1 = \Delta\vec{p},$$

so we obtain

Newton's second law in terms of momentum

The vector sum of forces acting on a particle equals the rate of change of momentum of the particle with respect to time:

$$\Sigma\vec{F} = \lim_{\Delta t \to 0} \frac{\Delta\vec{p}}{\Delta t}. \tag{8.3}$$

This, not $\Sigma\vec{F} = m\vec{a}$, is the modern equivalent of Newton's original statement of his second law. It is actually more general than our original form of the second law, because it includes the possibility of a change in momentum resulting from a change in mass as well as from a change in velocity. Note that both forms of the second law are valid only in inertial frames of reference.

Total Momentum

We define the *total momentum* of two or more particles as the vector sum of the momenta (plural of momentum) of the particles. Thus, if two particles A and B have momenta $\vec{p}_A$ and $\vec{p}_B$, their total momentum $\vec{P}$ is the vector sum

$$\vec{P} = \vec{p}_A + \vec{p}_B.$$

We can extend this definition directly to any number of particles.

Definition of total momentum

The total momentum $\vec{P}$ of any number of particles is equal to the vector sum of the momenta of the individual particles:

$$\vec{P} = \vec{p}_A + \vec{p}_B + \vec{p}_C + \cdots.$$
$$\text{(total momentum of a system of particles)} \tag{8.4}$$

We'll often use Equation 8.4 in component form. If $p_{A,x}$, $p_{A,y}$, and $p_{A,z}$ are the components of momentum of particle A, $p_{B,x}$, $p_{B,y}$, and $p_{B,z}$ the components for particle B, and so on, then Equation 8.4 is equivalent to the component equations

$$P_x = p_{A,x} + p_{B,x} + \cdots,$$
$$P_y = p_{A,y} + p_{B,y} + \cdots, \qquad \text{(components of total momentum)} \tag{8.5}$$
$$P_z = p_{A,z} + p_{B,z} + \cdots.$$

Momentum is sometimes called *linear momentum* to distinguish it from *angular momentum,* which plays an important role in rotational motion. We won't usually use the prefix *linear,* except where there is danger of confusion with angular momentum.

> **NOTE** ▶ It's absolutely essential to remember that momentum is a vector quantity; Equations 8.1 through 8.5 are *vector* equations. (Equations 8.2 and 8.5 express a vector relationship in terms of components of the vectors.) We'll often add momentum vectors by the use of components, so everything that you've learned about components of vectors will be useful with momentum. ◀

EXAMPLE 8.1 **Preliminary analysis of a collision**

A small compact car with a mass of 1000 kg is traveling north on Morewood Avenue with a speed of 15 m/s. At the intersection of Morewood and Fifth Avenues, it collides with a truck with a mass of 2000 kg that is traveling east on Fifth Avenue at 10 m/s. Treating each vehicle as a particle, find the total momentum (magnitude and direction) just before the collision.

SOLUTION

SET UP Because momentum is a vector quantity, we need coordinate axes. We draw a sketch (Figure 8.2a), labeling the car A and the truck B.

SOLVE We need to find (1) the components of momentum of each vehicle, (2) the components of total momentum, and then (3) the magnitude and direction of the total momentum vector (as sketched in Figure 8.2b). The components of momentum of the two vehicles are

$$p_{A,x} = m_A v_{A,x} = (1000 \text{ kg})(0) = 0,$$
$$p_{A,y} = m_A v_{A,y} = (1000 \text{ kg})(15 \text{ m/s}) = 1.5 \times 10^4 \text{ kg} \cdot \text{m/s},$$
$$p_{B,x} = m_B v_{B,x} = (2000 \text{ kg})(10 \text{ m/s}) = 2.0 \times 10^4 \text{ kg} \cdot \text{m/s},$$
$$p_{B,y} = m_B v_{B,y} = (1000 \text{ kg})(0) = 0$$

There are no z components of velocity or momentum.

From Equations 8.5, the components of the total momentum $\vec{P}$ are

$$P_x = m_A v_{A,x} + m_B v_{B,x} = 0 + 2.0 \times 10^4 \text{ kg} \cdot \text{m/s}$$
$$= 2.0 \times 10^4 \text{ kg} \cdot \text{m/s},$$
$$P_y = m_A v_{A,y} + m_B v_{B,y} = 1.5 \times 10^4 \text{ kg} \cdot \text{m/s} + 0$$
$$= 1.5 \times 10^4 \text{ kg} \cdot \text{m/s}.$$

The total momentum $\vec{P}$ is a vector quantity with these components. Its magnitude is

$$P = \sqrt{(2.0 \times 10^4 \text{ kg} \cdot \text{m/s})^2 + (1.5 \times 10^4 \text{ kg} \cdot \text{m/s})^2}$$
$$= 2.5 \times 10^4 \text{ kg} \cdot \text{m/s}.$$

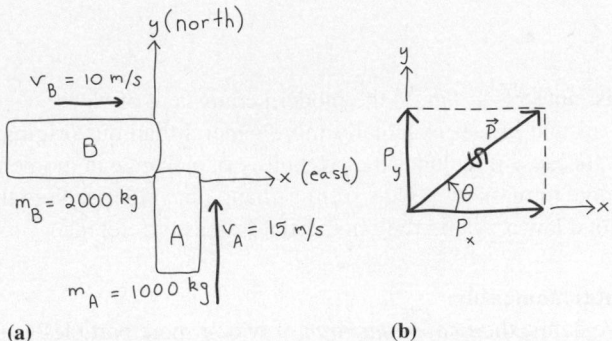

(a)   **(b)**

▲ **FIGURE 8.2** Our sketches for this problem.

Its direction is given by the angle θ in Figure 8.2b, where

$$\tan\theta = \frac{1.5 \times 10^4 \text{ kg} \cdot \text{m/s}}{2.0 \times 10^4 \text{ kg} \cdot \text{m/s}} = \frac{3}{4}, \qquad \theta = 36.9°.$$

REFLECT It is essential to treat momentum as a vector. When we add the momenta of the two objects we must do vector addition, using the method of components. Each vehicle has momentum in only one coordinate direction so has only one nonzero component of momentum, but the total momentum of the system has both x and y components.

Practice Problem: If the car is moving at 15 m/s, how fast must the truck move for the total momentum vector to make a 45° angle with the $+x$ and $+y$ axes? *Answer:* 7.5 m/s.

8.2 Conservation of Momentum

The concept of momentum is particularly important in situations in which two or more objects interact. Let's consider first an idealized system consisting of two objects that interact with each other, but not with anything else—for example, two astronauts who touch each other as they float freely in the zero-gravity environment of outer space (Figure 8.3). Think of the astronauts as particles. Each particle exerts a force on the other; according to Newton's third law, the two forces are always equal in magnitude and opposite in direction. From Newton's second law, as reformulated in Equation 8.3, the vector sum of forces acting on each particle equals the rate of change of that particle's momentum. This means that, at any instant, the two rates of change of momentum are equal in magnitude and opposite in direction, so the rate of change of the *total* momentum is zero, and the total momentum is constant. We say that total momentum is *conserved*.

Let's focus our attention on a definite collection of particles that we'll call a **system.** For instance, the two astronauts in Figure 8.3 could be taken as a system. For any system, the various particles exert forces on each other. We'll call these **internal forces.** In addition, forces may be exerted on any part of the system by

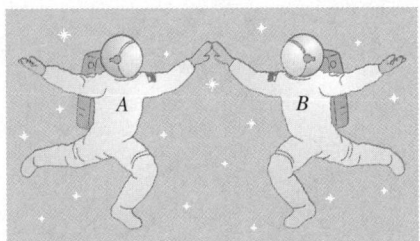

The forces exerted by the astronauts on each other form an action–reaction pair.

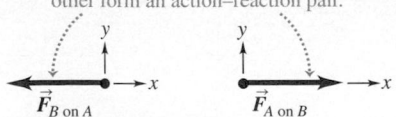

▲ **FIGURE 8.3** The internal forces the astronauts exert on each other don't change the total momentum of the two astronauts.

▶ **Application Strike!** A strike in bowling demonstrates conservation of momentum. The system consists of the ball and pins. Assuming that friction is negligible, the external forces acting on it—weight and normal forces—sum to zero, so the system can be regarded as isolated. Before the strike, the system's momentum is all in the moving ball (because the pins are stationary). Afterward, it is divided among the ball and the scattering pins—but the vector sum of the momenta of the ball and pins equals the momentum of the ball before impact.

objects *outside* the system; we call these **external forces.** A system that is acted upon by *no* external forces is called an **isolated system.** We have defined the total momentum of the system, P, as the vector sum of the momenta of the individual particles:

$$\vec{P} = \vec{p}_A + \vec{p}_B + \vec{p}_C + \cdots. \qquad \text{(total momentum of a system)}$$

The essential point to be made is that **internal forces cannot change the total momentum of a system.**

Here's why: Suppose particle A exerts a force $\vec{F}_{A\,\text{on}\,B}$ on particle B. Then, from Newton's third law, particle B must simultaneously exert a force $\vec{F}_{B\,\text{on}\,A}$ on particle A that is equal in magnitude and opposite in direction: $\vec{F}_{A\,\text{on}\,B} = -\vec{F}_{B\,\text{on}\,A}$. Each force by itself would cause the total momentum of the system to change, according to Equation 8.3. But because the two forces are negatives of each other, the two rates of change of momentum are also negatives of each other, and the *total* rate of change of momentum from this interaction is zero. Similarly, the changes in momentum caused by *all* the internal forces cancel out in pairs, and the rate of change of the total momentum $\vec{P}$ with respect to time due to internal forces is always zero. Internal forces can change the momenta of individual particles, but not the total momentum of the system.

If the internal forces cannot change the total momentum, and if there are no external forces, then the total momentum cannot change. We conclude that **the total momentum of an isolated system is constant.** Furthermore, if there are external forces, but their vector sum is zero, the total momentum is again constant.

Conservation of momentum

The total momentum of a system is constant whenever the vector sum of the external forces on the system is zero. In particular, the total momentum of an isolated system is constant.

NOTE ▶ We have used Newton's second law to derive the principle of conservation of momentum, so we have to be careful to use this principle only in inertial frames of reference. ◀

In some ways, the principle of conservation of momentum is more general than the principle of conservation of mechanical energy. For example, it is valid *even when the internal forces are not conservative;* by contrast, mechanical energy is conserved *only when the internal forces are conservative.* In this chapter, we'll analyze some situations in which both momentum and mechanical energy are conserved and others in which only momentum is conserved. These two principles play a fundamental role in all areas of physics, and we'll encounter them throughout our study.

PROBLEM-SOLVING STRATEGY 8.1 **Conservation of momentum**

Momentum is a vector quantity; you *must* use vector addition to compute the total momentum of a system. Using components is usually the simplest method.

SET UP

1. Define the system you are analyzing. Choose a coordinate system, specifying the positive direction for each axis. Often, it is easiest to choose the x axis as having the direction of one of the initial velocities. Make sure that you are using an inertial frame of reference.

2. Treat each object as a particle. Sketch "before" and "after" diagrams, including your coordinate system. Add vectors on each diagram to represent all known velocities. Label the vectors with magnitudes, angles, components, or whatever information is given, and give each unknown magnitude, angle, or component an algebraic symbol. It's often helpful to use multiple subscripts on velocity and momentum symbols to ensure complete identification. For example, the initial value (i) of the x component of velocity of object A can be written as $v_{A,i,x}$.

SOLVE

3. Compute the x and y components of momentum of each particle, both before and after the interaction, using the relations $p_x = mv_x$ and $p_y = mv_y$. Some of the components will be expressed in terms of symbols representing unknown quantities. Even when all the velocities lie along a line (such as the x axis), the components of velocity and momentum along this line can be positive or negative, so be careful with signs!

4. Determine whether both the x and the y components of total momentum are conserved, and why or why not. If they are conserved, write a relation that equates the total *initial* x component of momentum to the total *final* x component of momentum. Write another equation for the y components. These two equations express conservation of momentum in component form. In some problems, only one component of momentum is conserved. If the sum of the x components of the external forces is zero, then the x component of total momentum is conserved, and so on.

5. Solve the equations you wrote in step 4 to determine whatever quantities the problem asks for. In some problems, you'll have to convert from the x and y components of a velocity to its magnitude and direction, and in others you'll need to make the opposite conversion from magnitude and direction to x and y components.

   Remember that the x and y components of velocity or momentum are *never* added together in the same equation.

REFLECT

6. As with all problem solving, especially when the results are expressed in symbols rather than numbers, try to think of particular cases in which you can guess what the results of your analysis should be. What happens when two masses are equal? When one mass is zero? Is the result what you would expect?

EXAMPLE 8.2 **Astronaut rescue**

An astronaut finds herself floating in space 100 m from her ship when her safety cable becomes unlatched. She and the ship are motionless relative to each other. The astronaut's mass (including space suit) is 100 kg; she has a 1.0 kg wrench and only a 20 minute air supply. Thinking back to her physics classes, she devises

Continued

a plan to use conservation of momentum to get back to the ship safely by throwing the wrench away from her. In what direction should she throw the wrench? What is the magnitude of her recoil velocity if she throws the wrench at 10 m/s? Will her recoil velocity be enough to get her back to the spacecraft before she runs out of air?

SOLUTION

SET UP We choose a system consisting of the astronaut plus the wrench. No external forces act on this system, so the total momentum is conserved. We draw diagrams showing the situation just before and just after she throws the wrench; we point the *x* axis to the right (Figure 8.4). We use the subscripts i and f (initial and final) to label the situations before and after the wrench is thrown, and we use subscripts *A* and *W* (astronaut and wrench) to identify the parts of the system.

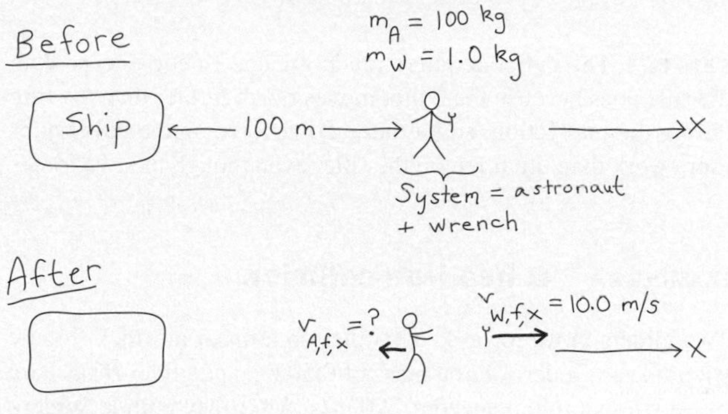

SOLVE The astronaut needs to acquire a velocity *toward* the spaceship. Because the total momentum of herself and the wrench is zero, she should throw the wrench directly *away* from the ship.

All the vector quantities lie along the *x* axis, so we're concerned only with *x* components. We write an equation expressing the equality of the initial and final values of the total *x* component of momentum:

$$m_A(v_{A,i,x}) + m_W(v_{W,i,x}) = m_A(v_{A,f,x}) + m_W(v_{W,f,x})$$

In this case, the initial velocity of each object is zero, so the left side of the equation is zero. That is, $m_A(v_{A,f,x}) + m_W(v_{W,f,x}) = 0$. Solving for $v_{A,f,x}$, the astronaut's *x* component of velocity after she throws the wrench, we get

$$v_{A,f,x} = \frac{-m_W(v_{W,f,x})}{m_A} = \frac{-(1.0 \text{ kg})(10.0 \text{ m/s})}{(100 \text{ kg})} = -0.10 \text{ m/s}.$$

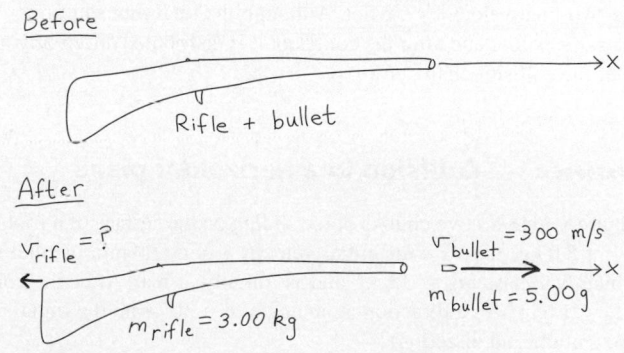

Before
Ship ← 100 m →
$m_A = 100$ kg
$m_W = 1.0$ kg
System = astronaut + wrench

After
$v_{A,f,x} = ?$ ←
$v_{W,f,x} = 10.0$ m/s →

▲ **FIGURE 8.4** Our sketches for this problem.

The negative sign indicates that the astronaut is moving toward the ship, opposite to our chosen +*x* direction.

To find the total time required for the astronaut to travel 100 m (at constant velocity) to reach the ship, we use $x = vt$, or

$$t = \frac{100 \text{ m}}{0.10 \text{ m/s}} = 1.00 \times 10^3 \text{ s} = 16 \text{ min } 40 \text{ s}.$$

REFLECT With a 20 minute air supply, she makes it back to the ship safely, with 3 min 20 s, to spare.

Practice Problem: If the astronaut has only a 10 minute air supply left, how fast must she throw the wrench so that she makes it back to the spaceship in time? *Answer:* 17 m/s.

EXAMPLE 8.3 **Recoil of a rifle**

A marksman holds a 3.00 kg rifle loosely, allowing it to recoil freely when fired, and fires a bullet of mass 5.00 g horizontally with a speed $v_B = 300$ m/s. What is the recoil speed v_R of the rifle? What are the final kinetic energies of the bullet and the rifle?

SOLUTION

SET UP We consider an idealized model in which the horizontal forces the marksman exerts on the rifle are negligible. We also neglect the momentum of the expanding gases. Then we can consider the rifle and bullet as an isolated system. (The net vertical force is zero.) The total momentum $\vec{P}$ of the system is zero both before and after the marksman fires the rifle. We take the positive *x* axis to be the direction in which the rifle is aimed; Figure 8.5 shows our sketch.

SOLVE Before the bullet is fired, the *x* component of the total momentum is zero. After the bullet is fired, the *x* component of its momentum is $(5.00 \times 10^{-3} \text{ kg})(300 \text{ m/s})$ and that of the rifle is $(3.00 \text{ kg})v_{R,f,x}$. Conservation of the *x* component of total momentum, P_x, before and after the firing, gives us

$$P_{i,x} = P_{f,x} = 0 = (5.00 \times 10^{-3} \text{ kg})(300 \text{ m/s})$$
$$+ (3.00 \text{ kg})(v_{R,f,x}) \text{ (zero total momentum)},$$
$$v_{R,f,x} = -0.500 \text{ m/s}.$$

Before

Rifle + bullet

After
$v_{rifle} = ?$ ←
$m_{rifle} = 3.00$ kg
$v_{bullet} = 300$ m/s →
$m_{bullet} = 5.00$ g

▲ **FIGURE 8.5** Our sketches for this problem.

The negative sign means that the recoil of the rifle is in the direction opposite that of the bullet. If the butt of a rifle were to hit your shoulder traveling at this speed, you would probably feel the "kick."

Continued

The final kinetic energy of the bullet is

$$K_{bullet} = \tfrac{1}{2}(5.00 \times 10^{-3} \text{ kg})(300 \text{ m/s})^2 = 225 \text{ J},$$

and the final kinetic energy of the rifle is

$$K_{rifle} = \tfrac{1}{2}(3.00 \text{ kg})(0.500 \text{ m/s})^2 = 0.375 \text{ J}.$$

REFLECT The bullet acquires much greater kinetic energy than the rifle does because the bullet moves much farther than the rifle during the interaction, so the interaction force on the bullet does more work than the force on the rifle (even though the two forces are equal and opposite). In fact, the ratio of the two kinetic energies, 600 : 1, is equal to the inverse ratio of the masses. Note that because the bullet has much less mass than the rifle, it has much greater acceleration during the firing.

Practice Problem: The same rifle fires a bullet with mass 10.0 g with the same speed as before. For the same idealized model, find the ratio of the final kinetic energies of the bullet and rifle. *Answer:* 450 J/1.5 J, or a ratio of 300 : 1, independent of the muzzle velocity of the bullet.

EXAMPLE 8.4   A head-on collision

Two gliders move toward each other on a linear air track (Figure 8.6), which we assume is frictionless. Glider A has a mass of 0.50 kg, and glider B has a mass of 0.30 kg; both gliders move with an initial speed of 2.0 m/s. After they collide, glider B moves away with a final velocity whose x component is +2.0 m/s (Figure 8.6c). What is the final velocity of A?

SOLUTION

SET UP We take the x axis as lying along the air track, with the positive direction to the right. All the velocities and momenta have only x components. Our system consists of the two gliders. There are no external horizontal forces, so the x component of total momentum is the same before and after the collision.

SOLVE Let the final x component of velocity of A be $v_{A,f,x}$. Then we write an expression for the total x component of momentum before the collision (subscript i) and one for after the collision (subscript f):

$$P_{i,x} = (0.50 \text{ kg})(2.0 \text{ m/s}) + (0.30 \text{ kg})(-2.0 \text{ m/s}),$$
$$P_{f,x} = (0.50 \text{ kg})(v_{A,f,x}) + (0.30 \text{ kg})(+2.0 \text{ m/s}).$$

From conservation of the x component of momentum, these two quantities must be equal. When we solve the resulting equation for $v_{A,f,x}$, we get

$$v_{A,f,x} = -0.40 \text{ m/s}. \qquad \text{(final x velocity of glider A)}$$

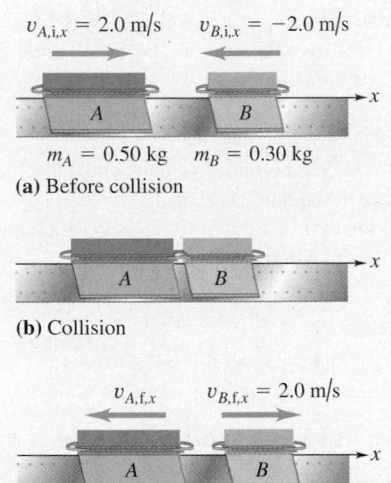

► **FIGURE 8.6**   (c) After collision

REFLECT The two initial velocities are equal in magnitude, but the two final velocities are not. Although the total momentum is the same before and after the collision, it is distributed differently after the collision than before.

Practice Problem: Suppose the initial x component of velocity of glider A is changed to 3.0 m/s, with all other values remaining the same as before. Find the velocity of glider A after the collision. *Answer:* 0.60 m/s.

EXAMPLE 8.5   Collision in a horizontal plane

Figure 8.7 shows two chunks of ice sliding on the surface of a frictionless frozen pond. Chunk A, with mass $m_A = 5.0$ kg, moves with initial velocity $v_{A,i} = 2.0$ m/s parallel to the x axis. It collides with chunk B, which has mass $m_B = 3.0$ kg and is initially at rest. After the collision, the velocity of A is found to be $v_{A,f} = 1.0$ m/s in a direction at an angle $\alpha = 30°$ with the initial direction. What is the final velocity of B (magnitude and direction)?

SOLUTION

SET UP The velocities are not all along a single line, so we have to use both x and y components of momentum for each chunk of ice. There are no horizontal external forces, so the total horizontal momentum of the system is the same before and after the collision. Conservation of momentum requires that the sum of the x components before the collision must equal their sum after the collision, and similarly for the y components. We must

Continued

write a separate equation for each component. We draw in the coordinate axes as shown in Figure 8.7. We also express the final velocities in terms of their components.

SOLVE We start by writing expressions for the total x component of momentum before and after the collision. For the x components, we have

$$m_A(v_{A,i,x}) + m_B(v_{B,i,x})$$
$$= (5.0\,\text{kg})(2.0\,\text{m/s}) + (3.0\,\text{kg})(0), \quad (\text{before})$$
$$m_A(v_{A,f,x}) + m_B(v_{B,f,x})$$
$$= (5.0\,\text{kg})(1.0\,\text{m/s})(\cos 30°) + (3.0\,\text{kg})(v_{B,f,x}). \quad (\text{after})$$

Equating these two expressions and solving for $v_{B,f,x}$, we find that

$$v_{B,f,x} = 1.89\,\text{m/s}. \quad (x \text{ component of final velocity of } B)$$

Conservation of the y component of total momentum gives

$$m_A(v_{A,i,y}) + m_B(v_{B,i,y})$$
$$= (5.0\,\text{kg})(0) + (3.0\,\text{kg})(0), \quad (\text{before})$$
$$m_A(v_{A,f,y}) + m_B(v_{B,f,y})$$
$$= (5.0\,\text{kg})(1.0\,\text{m/s})(\sin 30°) + (3.0\,\text{kg})(v_{B,f,y}). \quad (\text{after})$$

Equating these two expressions and solving for $v_{B,f,y}$, we obtain

$$v_{B,f,y} = -0.83\,\text{m/s}. \quad (y \text{ component of final velocity of } B)$$

We now have the x and y components of the final velocity $\vec{v}_{B,f}$ of chunk B. The magnitude of $\vec{v}_{B,f}$ is

$$|\vec{v}_{B,f}| = \sqrt{(1.89\,\text{m/s})^2 + (-0.83\,\text{m/s})^2}$$
$$= 2.1\,\text{m/s}, \quad (\text{final } speed \text{ of } B)$$

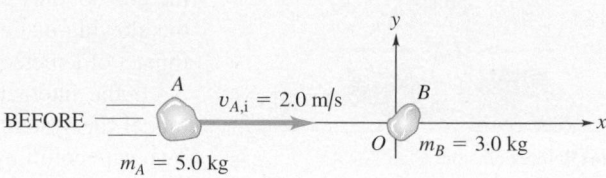

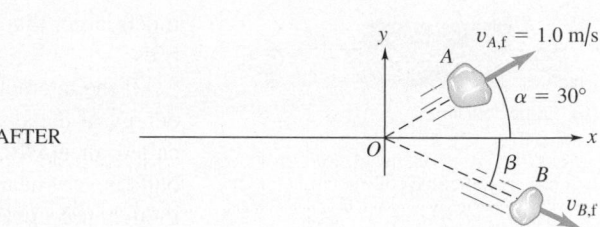

▲ **FIGURE 8.7**

and the angle β of its direction from the positive x axis is

$$\beta = \tan^{-1}\frac{-0.83\,\text{m/s}}{1.89\,\text{m/s}} = -24°.$$

REFLECT It's essential to treat velocity and momentum as vector quantities, representing them in terms of components. You can then use these components to find the magnitudes and directions of unknown vector quantities.

Practice Problem: If chunk B has an initial velocity of magnitude 2.0 m/s in the $+y$ direction instead of being initially at rest, find its final velocity (magnitude and direction). *Answer:* 2.2 m/s, 32°.

Conceptual Analysis 8.2

Exploding projectile

A model rocket travels as a projectile in a parabolic path after its first stage burns out. At the top of its trajectory, where its velocity points horizontally to the right, a small explosion separates it into two sections with equal masses. One section falls straight down, with no horizontal motion. What is the direction of the motion of the other part just after the explosion?

A. Up and to the left
B. Straight up
C. Up and to the right

SOLUTION We consider the two rocket sections as a system. During the explosion, internal forces dominate, and the external forces of gravity and air drag can be neglected. Therefore, we can assume that, during the short time of the explosion, the total momentum is conserved. Figure 8.8 illustrates our reasoning. Before the explosion, the rocket's momentum points horizontally to the right. The *total* momentum of the system must be the same just after the explosion as just before. We're told that one section

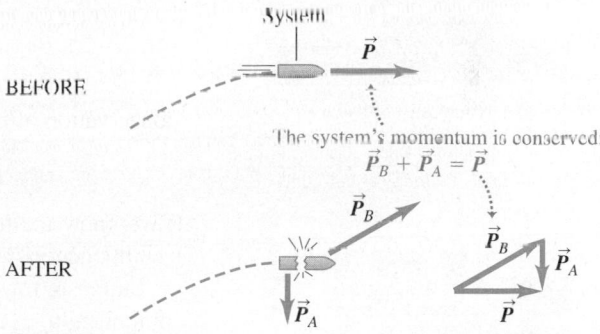

▲ **FIGURE 8.8**

moves straight down. Therefore, the other section must move up in order for the system as a whole to still have zero vertical momentum. This section must also move to the right, so that the system retains its original horizontal momentum. Thus, the correct choice is C.

8.3 Inelastic Collisions

In everyday speech, *collision* is likely to mean some sort of automotive disaster. We'll use it in that sense, but we'll also broaden the meaning to include any strong interaction between two objects that lasts a relatively short time. So we

Mastering**PHYSICS**

ActivPhysics 6.8: Skier and Cart

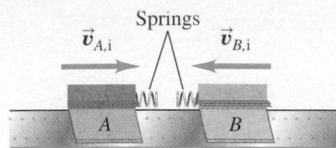

(a) Before collision

Kinetic energy is stored as potential energy in compressed springs.

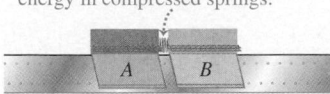

(b) Elastic collision

The system of the two gliders has the same kinetic energy after the collision as before it.

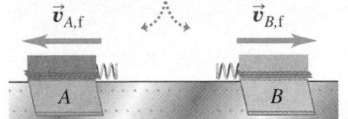

(c) After collision

▲ **FIGURE 8.9** An elastic collision.

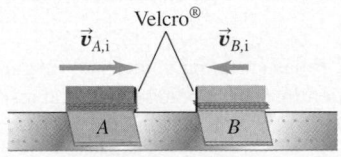

(a) Before collision

The gliders stick together

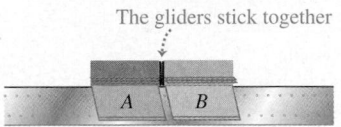

(b) Completely inelastic collision

The system of the two gliders has less kinetic energy after the collision than before it.

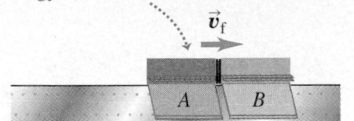

(c) After collision

▲ **FIGURE 8.10** A completely inelastic collision.

include not only car accidents, but also balls hitting each other on a billiard table, the slowing of neutrons in a nuclear reactor by encounters with nuclei, and the impact of a meteor on the Arizona desert.

If the interaction forces are much larger than any external forces, we can model the system as an *isolated* system, neglecting the external forces entirely. Two cars colliding at an icy intersection provide a good example. Even two cars colliding on dry pavement can be treated as an isolated system during the collision if, as happens all too often, the interaction forces between the cars are much larger than the external forces, such as friction forces of pavement against tires.

If the interaction forces between the objects are conservative, the total kinetic energy of the system is the same after the collision as before. Such a collision is called an **elastic collision.** A collision between two steel balls or two billiard balls is very nearly elastic, and collisions between subatomic particles are often, though not always, elastic. Figure 8.9 shows a model of an elastic collision. When the objects collide, the springs are momentarily compressed, and some of the original kinetic energy is converted to elastic potential energy. Then the objects bounce apart, the springs expand, and the potential energy is reconverted to kinetic energy.

A collision in which the total kinetic energy after the collision is *less* than that before the collision is called an **inelastic collision.** In one kind of inelastic collision, the colliding objects stick together and move as one object after the collision. In Figure 8.10, we've replaced the springs in Figure 8.9 with strips of Velcro®, causing the gliders to have a completely inelastic collision. A paintball striking a window shade, a bullet embedding itself in a block of wood, and two cars colliding and locking their fenders are additional examples of completely inelastic collisions.

Let's do a general analysis of a completely inelastic collision. Imagine two objects (A and B) that move along a straight line, which we'll designate as the x axis. Because they stick together after the collision, their final velocities must be equal, and the final total momentum is $\vec{P}_f = (m_A + m_B)\vec{v}_f$. We'll use $v_{f,x}$ to denote the x component of the common final velocity $\vec{v}_f$. We have

$$\vec{v}_{A,f} = \vec{v}_{B,f} = \vec{v}_f, \qquad v_{A,f,x} = v_{B,f,x} = v_{f,x}. \qquad \text{(equal final velocities)}$$

Conservation of momentum gives the relation

$$m_A(v_{A,i,x}) + m_B(v_{B,i,x}) = (m_A + m_B)(v_{f,x}). \qquad (8.6)$$

If we know the masses and initial velocities of the objects, we can compute the x component $v_{f,x}$ of the common final velocity $\vec{v}_f$.

Suppose, for example, that an object with mass m_A and initial velocity with component $v_{A,i,x}$ along the $+x$ axis collides inelastically with an object with mass m_B that is initially at rest $(v_{B,i,x} = 0)$. From Equation 8.6, the common x component of velocity $v_{f,x}$ of the two objects after the collision is

$$v_{f,x} = \frac{m_A}{m_A + m_B}(v_{A,i,x}).$$

(common final velocity in completely inelastic collision)   (8.7)

Let's verify that the total kinetic energy after this completely inelastic collision is less than before the collision. Using the above expression for $v_{f,x}$, we find that the kinetic energies K_i and K_f before and after the collision are, respectively,

$$K_i = \tfrac{1}{2}m_A(v_{A,i,x})^2,$$

$$K_f = \tfrac{1}{2}(m_A + m_B)(v_{f,x})^2 = \tfrac{1}{2}(m_A + m_B)\left(\frac{m_A}{m_A + m_B}\right)^2(v_{A,i,x})^2.$$

Dividing the second of these equations by the first, we find that the ratio of final to initial kinetic energy is

$$\frac{K_f}{K_i} = \frac{m_A}{m_A + m_B}. \qquad (8.8)$$

Because $m_A + m_B$ is always greater than m_A, the right side of this equation is always less than 1. Therefore, as expected, the final kinetic energy of the system is less than the initial kinetic energy.

When two objects with equal mass have a completely inelastic collision, the final kinetic energy is one-half the initial value. Even when the initial velocity of m_B is not zero, it is not hard to verify that the kinetic energy after a completely inelastic collision is always less than before.

NOTE ▶ We don't recommend memorizing the preceding equations. We derived them only to prove that kinetic energy is always lost in a completely inelastic collision. ◀

▲ **Application Get outta my way!** Since momentum is the product of mass and velocity, even slow-moving objects can have substantial momentum if they are massive enough.

Quantitative Analysis 8.3   Energy in an inelastic collision

An iron block B with mass m_B slides with speed v_B across a frictionless horizontal plane (Figure 8.11). It collides with and sticks to a magnet M with mass m_M. The magnet is attached to a spring with spring constant k. We can find the speed V of the two blocks immediately after the collision by applying conservation of momentum and solving: $(m_B + m_M)V = m_B v_B$. The masses stop when the spring is compressed an amount x_{max}. Which of the following equations would you use to find x_{max}?

A. $\frac{1}{2}m_B V^2 = \frac{1}{2}kx_{max}^2$
B. $\frac{1}{2}m_B v_B^2 = \frac{1}{2}kx_{max}^2$
C. $\frac{1}{2}(m_M + m_B)V^2 = \frac{1}{2}kx_{max}^2$

SOLUTION Because the two masses stick together, mechanical energy is not conserved. Therefore, choice B, which equates the

initial kinetic energy of the iron block and the final potential energy of the compressed spring, can't be correct. However, once the masses have collided and stuck, mechanical energy is conserved while the masses compress the spring. Thus, the correct choice is C, which equates the kinetic energy of the blocks just after the collision with the final elastic potential energy of the spring. We have to find the speed V of the combined masses $(m_M + m_B)$ and then apply conservation of energy to the compression of the spring.

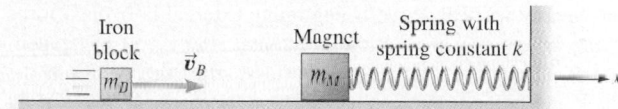

▲ **FIGURE 8.11**

EXAMPLE 8.6   Inelastic collision on an air track

We perform a collision experiment using the Velcro® equipped gliders of Figure 8.10. As in Example 8.4, glider A has a mass of 0.50 kg and glider B has a mass of 0.30 kg; both move with initial speeds of 2.0 m/s. Find the final velocity of the joined gliders, and compare the initial and final kinetic energies.

SOLUTION

SET UP Figure 8.12 shows our sketches for this problem. As in Example 8.4, we point the x axis in the direction of motion. All the velocity and momentum vectors lie along the x axis.

SOLVE We need to write expressions for the total x component of momentum before and after the collision and equate them. From conservation of the x component of momentum, we have

$$(0.50 \text{ kg})(2.0 \text{ m/s}) + (0.30 \text{ kg})(-2.0 \text{ m/s})$$
$$= (0.50 \text{ kg} + 0.30 \text{ kg})(v_{f,x})$$
$$v_{f,x} = 0.50 \text{ m/s}.$$

Because $v_{f,x}$ is positive, the gliders move together to the right (the $+x$ direction) after the collision. Before the collision, the kinetic energy of glider A is

$$K_{A,i} = \frac{1}{2}m_A(v_{A,i,x})^2 = \frac{1}{2}(0.50 \text{ kg})(2.0 \text{ m/s})^2 = 1.0 \text{ J},$$

Before $A \xrightarrow{v_{A,i,x} = 2.0 \text{ m/s}}$ $\xleftarrow{v_{B,i,x} = -2.0 \text{ m/s}} B$ $\longrightarrow x$
$m_A = 0.50 \text{ kg}$ $m_B = 0.30 \text{ kg}$

After $\boxed{A\,B} \xrightarrow{v_{f,x} = ?}$ $\longrightarrow x$

▲ **FIGURE 8.12** Our sketches for this problem.

and that of glider B is

$$K_{B,i} = \frac{1}{2}m_B(v_{B,i,x})^2 = \frac{1}{2}(0.30 \text{ kg})(-2.0 \text{ m/s})^2 = 0.60 \text{ J}.$$

Note that the initial kinetic energy of glider B is positive, even though the x components of its initial velocity, $v_{B,i,x}$, and momentum, $m_B(v_{B,i,x})$, are both negative. (Remember that

Continued

kinetic energy is a scalar, not a vector component, and that it can never be negative.)

The total kinetic energy before the collision is 1.6 J. The kinetic energy after the collision is

$$\tfrac{1}{2}(m_A + m_B)(v_{\text{f},x})^2 = \tfrac{1}{2}(0.50 \text{ kg} + 0.30 \text{ kg})(0.50 \text{ m/s})^2$$
$$= 0.10 \text{ J}.$$

REFLECT The final kinetic energy is only $\frac{1}{16}$ of the original, and $\frac{15}{16}$ is "lost" in the collision. Of course, it isn't really lost; it is converted from mechanical energy to various other forms of energy. For instance, if there were a wad of chewing gum between the gliders, it would squash irreversibly on impact and become warmer. If the gliders coupled together like two freight cars, the energy would go into elastic waves that would eventually dissipate. If there were a spring between the gliders that compressed as they locked together, then the energy would be stored as potential energy in the spring. In all of these cases, the *total* energy of the system *is* conserved, even though *kinetic* energy is not. However, in an isolated system, momentum is *always* conserved, whether the collision is elastic or not.

Practice Problem: If glider A has twice the mass of glider B, determine the ratio of final to initial kinetic energy. *Answer:* $\frac{1}{9}$.

EXAMPLE 8.7 **The ballistic pendulum**

How fast is a speeding bullet? Figure 8.13 shows a simple form of ballistic pendulum, a system for measuring the speed of a bullet. The bullet, with mass m_B, is fired into a block of wood with mass m_W suspended like a pendulum. The bullet makes a completely inelastic collision with the block, becoming embedded in it. After the impact of the bullet, the block swings up to a maximum height h. Given values of h, m_B, and m_W, how can we find the initial speed v of the bullet? What becomes of its initial kinetic energy?

SOLUTION

SET UP We analyze this event in two stages: first, the embedding of the bullet in the block, and second, the subsequent swinging of the block on its strings. During the first stage, the bullet embeds itself in the block so quickly that the block doesn't have time to swing appreciably away from its initial position. So, during the impact, the supporting strings remain very nearly vertical, there is negligible external horizontal force acting on the system, and the horizontal component of momentum is conserved. In the second stage, after the collision, the block and bullet move as a unit. The only forces are the weight (a conservative force) and the string tensions (which do no work). As the pendulum swings upward and to the right, mechanical energy is conserved.

To keep the notation simple, we'll denote the x component of the bullet's velocity just before impact as v and the x component of the velocity of block plus bullet just after the impact as V.

SOLVE Figure 8.13 shows the situation just before impact, the situation just after impact, and the block (with the embedded bullet) at the highest point of its path. Conservation of momentum just before impact and just after gives us

$$m_B v = (m_B + m_W)V,$$
$$v = \frac{m_B + m_W}{m_B}V.$$

The kinetic energy of the system just after the collision is $K = \tfrac{1}{2}(m_B + m_W)V^2$. The pendulum comes to rest (for an instant) at a height h, where its kinetic energy $\tfrac{1}{2}(m_B + m_W)V^2$ has all become potential energy $(m_B + m_W)gh$; then it swings back down. Energy conservation gives

$$\tfrac{1}{2}(m_B + m_W)V^2 = (m_B + m_W)gh.$$

Assuming that h can be measured, we solve this equation for V:

$$V = \sqrt{2gh}. \quad \text{(velocity of block and bullet just after impact)}$$

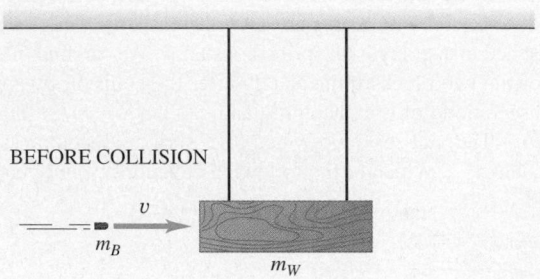

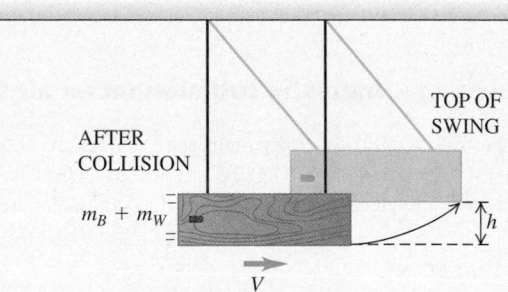

▲ **FIGURE 8.13** Using a ballistic pendulum to find a bullet's speed.

Now we substitute this result into the momentum equation to find v:

$$v = \frac{m_B + m_W}{m_B}\sqrt{2gh}. \quad \text{(speed of bullet just before impact)}$$

By measuring m_B, m_W, and h, we can compute the original velocity v of the bullet. For example, if $m_B = 5.00 \text{ g} = 0.00500 \text{ kg}$, $m_W = 2.00 \text{ kg}$, and $h = 3.00 \text{ cm} = 0.0300 \text{ m}$, then

$$v = \frac{2.00 \text{ kg} + 0.00500 \text{ kg}}{0.00500 \text{ kg}}\sqrt{2(9.80 \text{ m/s}^2)(0.0300 \text{ m})}$$
$$= 307 \text{ m/s}.$$

Continued

The x component V of velocity of the block just after impact is

$$V = \sqrt{2gh} = \sqrt{2(9.80 \text{ m/s}^2)(0.0300 \text{ m})}$$
$$= 0.767 \text{ m/s}.$$

Once we have the needed velocities, we can compute the kinetic energies just before and just after impact. The total kinetic energy just before impact is $K_i = \frac{1}{2}m_B v^2 = \frac{1}{2}(0.00500 \text{ kg}) \cdot (307 \text{ m/s})^2 = 236 \text{ J}$. We find that just after impact, it is $K_f = \frac{1}{2}(m_B + m_W)V^2 = \frac{1}{2}(2.005 \text{ kg}) \cdot (0.767 \text{ m/s})^2 = 0.590 \text{ J}$. Only a small fraction of the initial kinetic energy remains.

REFLECT When an object collides inelastically with a stationary object having much greater mass, nearly all of the first object's kinetic energy is lost. In this problem, the wood splinters, and the bullet and wood become hotter as mechanical energy is converted to internal energy.

Practice Problem: Suppose the mass of the bullet, the mass of the block, and the height of the block's swing have the same values as above. If the bullet goes all the way through the block and emerges with half its initial velocity, what was its initial speed?

Answer: $v = \dfrac{2m_W}{m_B}\sqrt{2gh} = 615 \text{ m/s}$.

EXAMPLE 8.8 Collision analysis, continued

In Example 8.1, we considered the impending collision of a small car with a larger truck. It's now 2 seconds later, and the collision has occurred. Fortunately, all occupants were wearing seat belts, and there were no injuries; but the two vehicles became thoroughly tangled and moved away from the point of impact as one mass. The insurance adjuster has asked you to help find the velocity of the wreckage just after impact.

SOLUTION

SET UP Figure 8.14 shows our sketches. The "before" diagram is the same as the one for Example 8.1. We worked out the initial momentum components in that example. We suggest you review that calculation before proceeding. To find out what happens *after* the collision, we'll assume that we can treat the vehicles as an isolated system during the collision. This approach may seem implausible; the pavement certainly exerts substantial friction forces on the tires. But the interaction forces between the vehicles during the collision are much larger—so much larger, in fact, that they crumple the vehicles. So it's reasonable to neglect the friction forces and consider the two vehicles together as an isolated system during the impact.

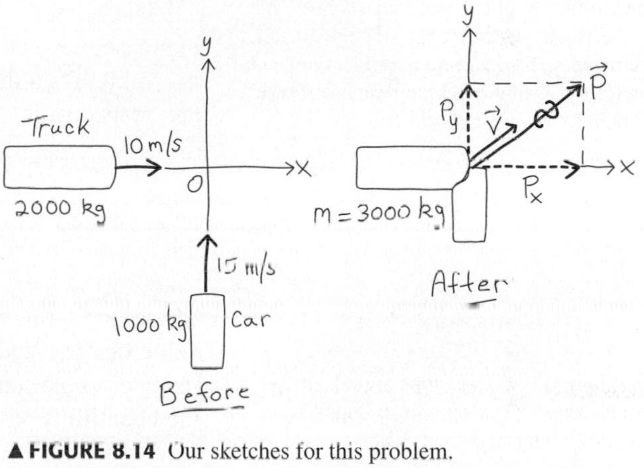

SOLVE The total momentum $\vec{P}$ of the vehicles just after the collision is the same as we calculated in Example 8.1: $2.5 \times 10^4 \text{ kg} \cdot \text{m/s}$ in a direction 36.9° north from straight east. Assuming that no parts fall off the wreckage, the total mass of wreckage is $M = 3000 \text{ kg}$. Calling the final velocity $\vec{V}$, we have $\vec{P} = M\vec{V}$. The direction of the velocity $\vec{V}$ is the same as that of the momentum $\vec{P}$, and its magnitude is

$$V = \frac{P}{M} = \frac{2.5 \times 10^4 \text{ kg} \cdot \text{m/s}}{3000 \text{ kg}} = 8.3 \text{ m/s}. \quad \text{(final speed)}$$

This is an inelastic collision, so we expect the total kinetic energy to be less after the collision than before. When you make the calculations, you'll find that the initial kinetic energy is $2.1 \times 10^5 \text{ J}$ and the final kinetic energy is $1.0 \times 10^5 \text{ J}$. More than half of the initial kinetic energy is converted to other forms of energy.

REFLECT If you were tempted to find the final velocity by taking the vector sum of the initial velocities, go back two squares and ask yourself why you would expect that to work. Is there a law of conservation of velocities? *Absolutely not;* the conserved quantity is the total *momentum* of the system.

▲ **FIGURE 8.14** Our sketches for this problem.

Let's look again at the matter of neglecting the friction of tires against pavement and treating the two vehicles as an isolated system. Here are some numbers: The mass of the truck is 2000 kg, so its weight is about 20,000 N. If the coefficient of friction is 0.5, the friction force is somewhere around 10,000 N. That sounds like a large force. But suppose the truck runs into a brick wall while going 10 m/s. Its kinetic energy just before impact is $\frac{1}{2}(2000 \text{ kg})(10 \text{ m/s})^2 = 1.0 \times 10^5 \text{ J}$, so the force applied by the wall must do $-1.0 \times 10^5 \text{ J}$ of work on the truck to stop (and crumple) it. The truck may crumple 0.2 m or so; for the stopping force to do $-1.0 \times 10^5 \text{ J}$ of work in 0.2 m, the force must have a magnitude of $5.0 \times 10^5 \text{ N}$, or 50 times as great as the friction force. You'll see now why it's not so unreasonable to neglect the friction forces on both vehicles and assume that during the collision they form an isolated system.

Practice Problem: Suppose the vehicles are both moving at 10 m/s, the 2000 kg truck heading east and the 1000 kg car heading north. Find the velocity (magnitude and direction) of the wreckage. *Answer:* 7.5 m/s, 27° N of E.

▲ **Application Light sail.** All current space-ships use rockets of one sort or another. A disadvantage of a rocket is that the craft must carry its own fuel. One alternative is to use a light sail—an enormous sheet of reflective material. In 1873, James Clerk Maxwell showed that light reflected by a mirror applies a slight pressure to the mirror. Thus, in principle, light bouncing off a reflective sail could propel a spacecraft. The more perfectly reflective the sail, the greater its efficiency; for maximum momentum transfer we want the light to bounce off the sail rather than being absorbed.

Mastering**PHYSICS**

ActivPhysics 6.2: Collisions and Elasticity
ActivPhysics 6.3: Momentum Conservation and Collisions
ActivPhysics 6.4: Collision Problems
ActivPhysics 6.5: Car Collision: Two Dimensions
ActivPhysics 6.9: Pendulum Bashes Box
ActivPhysics 6.10: Pendulum Person—Projectile Bowling

8.4 Elastic Collisions

In Section 8.3, we defined an elastic collision in an isolated system to be a collision in which kinetic energy (as well as momentum) is conserved. Elastic collisions occur when the interaction forces between the objects are conservative. When two steel balls collide, they squash a little near the surface of contact, but then they spring back. Some of the kinetic energy is stored temporarily as elastic potential energy, but by the end of the collision it is reconverted to kinetic energy.

Let's look at an elastic collision between two objects A and B. We start with a head-on collision, in which all the velocities lie along the same line; we choose this line to be the x axis. Each momentum and velocity then has only an x component. (Later we'll consider collisions in two dimensions.) We'll simplify our notation, calling the x components of velocity before the collision $v_{A,i}$ and $v_{B,i}$ and those after the collision $v_{A,f}$ and $v_{B,f}$. From conservation of kinetic energy, we have

$$\tfrac{1}{2}m_A v_{A,i}^2 + \tfrac{1}{2}m_B v_{B,i}^2 = \tfrac{1}{2}m_A v_{A,f}^2 + \tfrac{1}{2}m_B v_{B,f}^2,$$

and conservation of momentum gives

$$m_A v_{A,i} + m_B v_{B,i} = m_A v_{A,f} + m_B v_{B,f}.$$

If the masses m_A and m_B and the initial velocities (i.e., x components) $v_{A,i}$ and $v_{B,i}$ are known, these two equations can be solved simultaneously to find the two final velocities $v_{A,f}$ and $v_{B,f}$.

The general solution is a little complicated, so we'll concentrate on the particular case in which object B is at rest before the collision. Think of it as a target for m_A to hit. Then the equations for conservation of kinetic energy and momentum are, respectively,

$$\tfrac{1}{2}m_A v_{A,i}^2 = \tfrac{1}{2}m_A v_{A,f}^2 + \tfrac{1}{2}m_B v_{B,f}^2 \tag{8.9}$$

and

$$m_A v_{A,i} = m_A v_{A,f} + m_B v_{B,f}. \tag{8.10}$$

These are a pair of simultaneous equations that can be solved for the two final velocities $v_{A,f}$ and $v_{B,f}$ in terms of the masses and the initial velocity $v_{A,i}$. This involves some fairly strenuous algebra; we'll omit the details and go straight to the beautifully simple final results. We find that

$$v_{A,f} = \frac{m_A - m_B}{m_A + m_B} v_{A,i}, \tag{8.11}$$

$$v_{B,f} = \frac{2m_A}{m_A + m_B} v_{A,i}. \tag{8.12}$$

Let's interpret these results by looking at several particular cases. Suppose object A is a Ping-Pong ball and B is a bowling ball (Figure 8.15a). We expect A to bounce back after the collision with a velocity nearly equal to its original

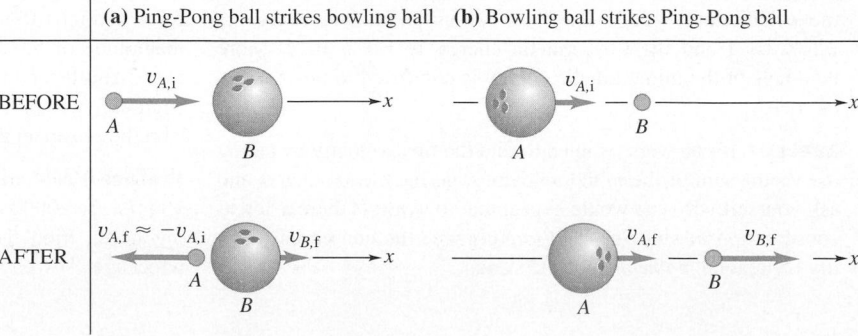

▶ **FIGURE 8.15** Collisions between a Ping-Pong ball and a bowling ball, one or the other of which is initially at rest.

(a) Ping-Pong ball strikes bowling ball   **(b)** Bowling ball strikes Ping-Pong ball

value, but in the opposite direction, and we expect B's final velocity to be much smaller in magnitude. That's just what the equations predict: When m_A is much smaller than m_B, the fraction in the expression for $v_{A,f}$ in Equation 8.11 is approximately equal to -1, so $v_{A,f}$ is approximately equal to $-v_{A,i}$. Also, when m_A is much smaller than m_B, the fraction in Equation 8.12 is much smaller than 1, so $v_{B,f}$ (the final velocity of the bowling ball) is much less than $v_{A,i}$.

Figure 8.15b shows the opposite case, in which A is the bowling ball, B is the Ping-Pong ball, and m_A is much larger than m_B. What do you expect to happen in this case? Check your predictions against Equations 8.11 and 8.12.

Another interesting case occurs when the masses are equal, as in the case of the billiard balls in Figure 8.16. If $m_A = m_B$, then Equations 8.11 and 8.12 give $v_{A,f} = 0$ and $v_{B,f} = v_{A,i}$. That is, the object that was moving stops dead; it gives all of its momentum and kinetic energy to the object that was initially at rest. This behavior is familiar to all pool players and marble shooters.

Now comes a surprise bonus. From Equations 8.11 and 8.12, we can find the x component of the velocity of B *relative to* A after the collision (that is, $v_{B,f} - v_{A,f}$), by simply subtracting Equation 8.12 from Equation 8.11. The result is astonishingly simple:

$$v_{B,f} - v_{A,f} = v_{A,i}. \quad \text{(relative velocity before and after collision)} \quad (8.13)$$

Now, $v_{B,f} - v_{A,f}$ is the velocity of B relative to A *after* the collision, and $v_{A,i}$ is the negative of the velocity of B relative to A *before* the collision. Thus, **the relative velocity has the same magnitude, but opposite sign, before and after the collision.**

We have proved this result only for straight-line collisions in which one object is initially at rest, but it turns out that a similar *vector* relationship is a general property of *all* elastic collisions, even when both objects are moving initially and the velocities do not all lie along the same line. That is,

$$\vec{v}_{B,f} - \vec{v}_{A,f} = -(\vec{v}_{B,i} - \vec{v}_{A,i}).$$

This result provides an alternative and equivalent definition of an elastic collision: **In an elastic collision, the relative velocity of the two objects has the same magnitude before and after the collision, and the two relative velocities have opposite directions.** Whenever this condition is satisfied, the total kinetic energy is also conserved.

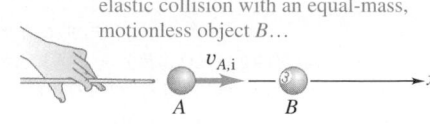

When a moving object A has a 1-D elastic collision with an equal-mass, motionless object B…

$v_{A,i}$

A B

…all of A's momentum and kinetic energy are transferred to B.

$v_{A,f} = 0$ $v_{B,f} = v_{A,i}$

A B

▲ FIGURE 8.16 An elastic collision between two objects of equal mass, one of which is initially at rest.

EXAMPLE 8.9 Elastic collision on an air track

Let's carry out another collision between the air-track gliders of Examples 8.4 and 8.6. This time, we equip the gliders with spring bumpers so that the collision will be elastic. As before, glider A has a mass of 0.50 kg, glider B has a mass of 0.30 kg, and each moves with an initial speed of 2.0 m/s as they approach each other. What are the velocities of A and B after the collision?

SOLUTION

SET UP We draw "before" and "after" sketches, using the same coordinate system as in Example 8.4 (Figure 8.17). The masses and initial velocities of the gliders are the same as in Examples 8.4 and 8.6, so the initial total kinetic energy (1.6 J) is also the same as in those examples.

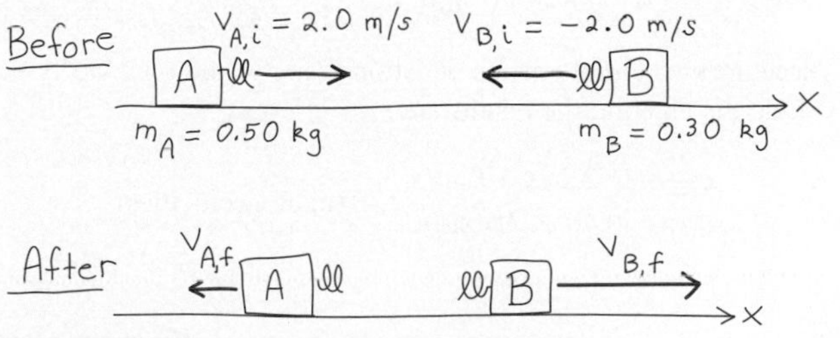

SOLVE We can't use Equations 8.11 and 8.12 because in this instance neither object is at rest before the collision. But we can use conservation of momentum, along with the relative-velocity relation we've just discussed, to obtain two simultaneous equations for the two final velocities. Because we're dealing with a one-dimensional problem, we'll drop the x subscripts on the

Before $v_{A,i} = 2.0$ m/s $v_{B,i} = -2.0$ m/s

A B

$m_A = 0.50$ kg $m_B = 0.30$ kg

After $v_{A,f}$ A B $v_{B,f}$

▲ FIGURE 8.17 Our sketches for this problem.

velocities, while keeping in mind that they are all x components of velocities.

Continued

From conservation of momentum,

$$m_A v_{A,i} + m_B v_{B,i} = m_A v_{A,f} + m_B v_{B,f},$$
$$(0.50 \text{ kg})(2.0 \text{ m/s}) + (0.30 \text{ kg})(-2.0 \text{ m/s})$$
$$= (0.50 \text{ kg})v_{A,f} + (0.30 \text{ kg})v_{B,f},$$
$$0.50 v_{A,f} + 0.30 v_{B,f} = 0.40 \text{ m/s}.$$

(In the last equation we've divided through by the unit "kg.") From the relative-velocity relation for an elastic collision,

$$v_{B,f} - v_{A,f} = -(v_{B,i} - v_{A,i})$$
$$= -(-2.0 \text{ m/s} - 2.0 \text{ m/s}) = 4.0 \text{ m/s}.$$

Solving these equations simultaneously, we obtain

$$v_{A,f} = -1.0 \text{ m/s}, \qquad v_{B,f} = 3.0 \text{ m/s}.$$
(final x components of velocity)

REFLECT Both gliders reverse their directions of motion; A moves to the left at 1.0 m/s and B moves to the right at 3.0 m/s. This result is different from the result of Example 8.4; but that collision was *not* an elastic one, so we shouldn't expect the results to be the same. In this case, the total kinetic energy after the collision is

$$K_f = \frac{1}{2}(0.50 \text{ kg})(-1.0 \text{ m/s})^2 + \frac{1}{2}(0.30 \text{ kg})(3.0 \text{ m/s})^2$$
$$= 1.6 \text{ J}.$$

This equals the total kinetic energy before the collision, as expected.

Practice Problem: Suppose we interchange the two gliders, making $m_A = 0.30$ kg and $m_B = 0.50$ kg. (We also turn them around, so the springs still meet.) If the initial speeds are the same as before, find the two final velocities. *Answers:* -3.0 m/s, 1.0 m/s.

EXAMPLE 8.10 Moderator in a nuclear reactor

High-speed neutrons are produced in a nuclear reactor during nuclear fission processes. Before a neutron can trigger additional fissions, it has to be slowed down by collisions with nuclei of a material called the *moderator.* In some reactors (including the one involved in the Chernobyl accident), the moderator consists of carbon in the form of graphite. The masses of nuclei and subatomic particles are measured in units called *atomic mass units,* abbreviated u, where 1 u = 1.66×10^{-27} kg. Suppose a neutron (mass 1.0 u) traveling at 2.6×10^7 m/s makes an elastic head-on collision with a carbon nucleus (mass 12 u) that is initially at rest. What are the velocities after the collision?

SOLUTION

SET UP Figure 8.18 shows our sketches. We use subscripts n and C to denote the neutron and carbon nucleus, respectively.

SOLVE Because we have a head-on elastic collision, all velocity and momentum vectors lie along the x axis, and we can use Equations 8.10 and 8.13, with $m_n = 1.0$ u, $m_C = 12$ u, and $v_{n,i} = 2.6 \times 10^7$ m/s. We'll get two simultaneous equations that we'll need to solve for $v_{n,f}$ and $v_{C,f}$.

First, conservation of the x component of momentum gives

$$m_n(v_{n,i}) + m_C(v_{C,i}) = m_n(v_{n,f}) + m_C(v_{C,f}),$$
$$(1.0 \text{ u})(2.6 \times 10^7 \text{ m/s}) + (12.0 \text{ u})(0)$$
$$= (1.0 \text{ u})v_{n,f} + (12.0 \text{ u})v_{C,f}.$$

Second, the relative-velocity relation yields

$$v_{n,i} - v_{C,i} = -(v_{n,f} - v_{C,f}),$$
$$2.6 \times 10^7 \text{ m/s} = v_{C,f} - v_{n,f}.$$

These are two simultaneous equations for $v_{n,f}$ and $v_{C,f}$. We'll let you do the algebra; the results are

$$v_{n,f} = -2.2 \times 10^7 \text{ m/s},$$
$$v_{C,f} = 0.40 \times 10^7 \text{ m/s}.$$
(final x velocities)

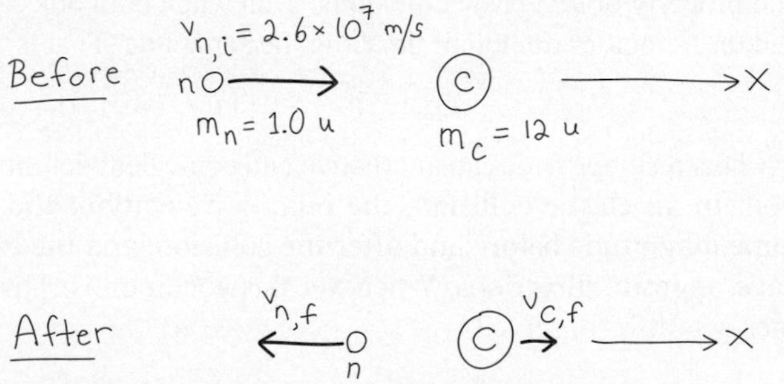

▲ **FIGURE 8.18** Our sketches for this problem.

REFLECT The neutron ends up with $\frac{11}{13}$ of its original speed, and the speed of the recoiling carbon nucleus is $\frac{2}{13}$ of the neutron's original speed. Kinetic energy is proportional to speed squared, so the neutron's final kinetic energy is $\left(\frac{11}{13}\right)^2$, or about 0.72 of its original value. If the neutron makes a second such collision, its kinetic energy is $(0.72)^2$, or about half its original value, and so on. After many such collisions, the neutron's kinetic energy is reduced to a small fraction (1/100 or less) of its initial value.

Practice Problem: If the neutron's kinetic energy is reduced to $\frac{9}{16}$ of its initial value in a single collision, what is the mass of the moderator nucleus? *Answer:* 7.0 u.

When an elastic collision between two objects isn't head-on, the velocities don't all lie along a single line. If they all lie in a plane, each final velocity has two unknown components, and there are four unknowns in all. Conservation of energy and conservation of the x and y components of momentum give only three equations. To determine the final velocities, we need additional information, such as the direction or magnitude of one of the final velocities.

EXAMPLE 8.11 An off-center elastic collision on an air table

Two pucks collide on a frictionless air-hockey table. The pucks have equal masses $m = 0.20$ kg. Puck A has an initial velocity of 4.0 m/s in the positive x direction and a final velocity of 2.0 m/s at an angle of $60°$ to the $+x$ axis. Puck B is initially at rest. Find the final velocity of puck B (magnitude and direction). Is the collision elastic?

SOLUTION

SET UP Figure 8.19 shows "before" and "after" diagrams. We take the x axis as the horizontal and the positive direction to the right.

SOLVE We need to write two equations, one for conservation of the x component of momentum and one for the y component. We denote the unknown components of the final velocity of B as $v_{B,f,x}$ and $v_{B,f,y}$. The x and y components of momentum of A after the collision are, respectively,

$$m_A(v_{A,f,x}) = (0.20 \text{ kg})(2.0 \text{ m/s})\cos 60°$$

and

$$m_A(v_{A,f,y}) = (0.20 \text{ kg})(2.0 \text{ m/s})\sin 60°.$$

The x and y components of momentum of B after the collision are, respectively,

$$m_B(v_{B,f,x}) = (0.20 \text{ kg})(v_{B,f,x})$$

and

$$m_B(v_{B,f,y}) = (0.20 \text{ kg})(v_{B,f,y}).$$

Conservation of the x component of total momentum gives

$$m_A(v_{A,i,x}) + m_B(v_{B,i,x}) = m_A(v_{A,f,x}) + m_B(v_{B,f,x}),$$
$$(0.20 \text{ kg})(4.0 \text{ m/s}) + 0 = (0.20 \text{ kg})(2.0 \text{ m/s})\cos 60°$$
$$+ (0.20 \text{ kg})v_{B,f,x},$$

and

$$v_{B,f,x} = 3.0 \text{ m/s}. \quad \text{(final } x \text{ component of velocity of } B)$$

Conservation of the y component of total momentum gives the equations

$$m_A(v_{A,i,y}) + m_B(v_{B,i,y}) = m_A(v_{A,f,y}) + m_B(v_{B,f,y}),$$
$$0 = (0.20 \text{ kg})(2.0 \text{ m/s})\sin 60° + (0.20 \text{ kg})v_{B,f,y},$$

and

$$v_{B,f,y} = -\sqrt{3} \text{ m/s} = -1.73 \text{ m/s}.$$
$$\text{(final } y \text{ component of velocity of } B)$$

The magnitude $v_{B,f}$ of the final velocity of B is

$$v_{B,f} = \sqrt{v_{B,f,x}^2 + v_{B,f,y}^2} = \sqrt{(3.0 \text{ m/s})^2 + (-\sqrt{3} \text{ m/s})^2}$$
$$= 2\sqrt{3} \text{ m/s} = 3.5 \text{ m/s}.$$

Its direction (an angle θ measured from the $+x$ axis) is

$$\theta = \tan^{-1}\frac{v_{B,f,y}}{v_{B,f,x}} = \tan^{-1}\left(\frac{-\sqrt{3} \text{ m/s}}{3.0 \text{ m/s}}\right) = -30°.$$

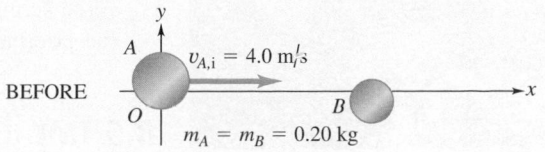

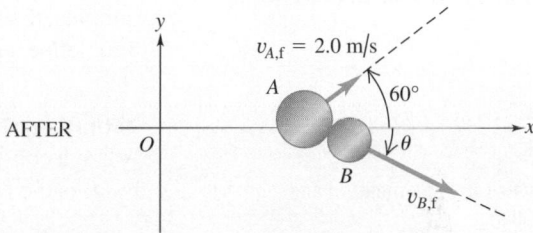

▲ **FIGURE 8.19**

Is kinetic energy conserved? The initial kinetic energy of A (equal to the initial *total* kinetic energy) is

$$K_{A,i} = K_i = \tfrac{1}{2}m_A(v_{A,i})^2 = \tfrac{1}{2}(0.20 \text{ kg})(4.0 \text{ m/s})^2 = 1.6 \text{ J}.$$

The final total kinetic energy of A and B together is

$$K_f = \tfrac{1}{2}m_A(v_{A,f})^2 + \tfrac{1}{2}m_B(v_{B,f})^2$$
$$= \tfrac{1}{2}(0.20 \text{ kg})(2.0 \text{ m/s})^2 + \tfrac{1}{2}(0.20 \text{ kg})(2\sqrt{3} \text{ m/s})^2$$
$$= 1.6 \text{ J}.$$

The total initial and final kinetic energies are equal, so the collision is elastic.

REFLECT We used a separate equation for conservation of each component of momentum. The x and y components of momentum are *never* added together in a conservation-of-momentum equation. Because the masses are equal, we could have simplified the calculations by dividing the factor m out of each momentum equation. The two final velocities are perpendicular. In fact, it can be shown that in *every* elastic collision between two objects with equal mass, when one object is initially at rest, the two final velocities are *always* perpendicular (except when the collision is head-on and one final velocity is zero). Every good billiards player is familiar with this fact.

Practice Problem: Suppose the final direction of puck A's velocity is at $30°$ to the $+x$ axis instead of $60°$. Find the final velocity (magnitude and direction) of puck B. *Answers:* 2.5 m/s, $-24°$.

We've classified collisions according to energy considerations. A collision in which kinetic energy is conserved is called *elastic*. A collision in which the total kinetic energy decreases is called *inelastic*. There are also cases in which the final

▲ **FIGURE 8.20** A racquet hitting a tennis ball. Typically, the ball is in contact with the racket for approximately 0.01 s. The ball flattens noticeably on both sides, and the frame of the racket vibrates during and after the impact.

ActivPhysics 6.1: Momentum and Energy Change

▲ **BIO Application Long legs for leaping.** Why are this frog's legs so long? Short legs could supply an equal magnitude of force. The answer hinges on the fact that muscles are limited in the force they can develop and the speed with which they can contract. The longer the frog's legs, the greater is the time interval during which the legs can deliver their force, and thus the greater is the total impulse delivered. Since impulse represents a change in momentum, longer legs result in a larger change in momentum, imparting a greater velocity to a frog of a given mass.

kinetic energy is *greater* than the initial kinetic energy. A recoiling rifle, discussed in Example 8.3, is an example.

> **NOTE** ▶ We emphasize again that we can sometimes use conservation of momentum even when external forces act on the system. If the vector sum of the external forces is zero, momentum is conserved. Also, if the internal forces between colliding objects are much stronger than the net external force acting on the system, we may use an idealized model that neglects the external forces during the actual collision. ◀

8.5 Impulse

When a force acts on an object, the object's change in momentum depends on the force and on the time interval during which it acts. We've seen several examples of interactions in which large forces act during a short time of impact. A recoiling rifle, exploding projectiles, a tennis ball struck by a racket (Figure 8.20)—all are examples of such interactions. To analyze these events in detail, we'll find it useful to define a quantity called **impulse:**

Definition of impulse
When a constant force $\vec{F}$ acts on an object, the impulse of the force, denoted by $\vec{J}$, is the force multiplied by the time interval during which it acts:

$$\vec{J} = \vec{F}(t_f - t_i) = \vec{F}\,\Delta t. \qquad (8.14)$$

Notice that impulse is a vector quantity; its direction is the same as that of the force $\vec{F}$.

Unit: Force times time $(N \cdot s)$. Because $1\,N = 1\,kg \cdot m/s^2$, an alternative set of units for impulse is $kg \cdot m/s$; thus, the units of impulse are the same as those of momentum.

To see what impulse is good for, let's go back to Newton's second law, as we restated it in terms of momentum in Equation 8.4, namely,

$$\vec{F} = \lim_{\Delta t \to 0} \frac{\Delta \vec{p}}{\Delta t},$$

where $\vec{F}$ represents the vector sum (resultant) of forces acting on an object. If $\vec{F}$ doesn't vary with time, we can simplify this equation to

$$\vec{F} = \frac{\Delta \vec{p}}{\Delta t}, \quad \text{or} \quad \Delta \vec{p} = \vec{F}\,\Delta t.$$

Combining this result with Equation 8.14, we get the following relation, called the **impulse–momentum theorem:**

Relation of impulse to change in momentum:
The impulse–momentum theorem
When a constant force $\vec{F}$ acts on an object during a time interval $\Delta t = t_f - t_i$, the change in the object's momentum is equal to the impulse of the force acting on the object, or

$$\Delta \vec{p} = \vec{F}\,\Delta t = \vec{F}(t_f - t_i) = \vec{J}. \qquad (8.15)$$

Unit: mass times velocity $(kg \cdot m/s)$

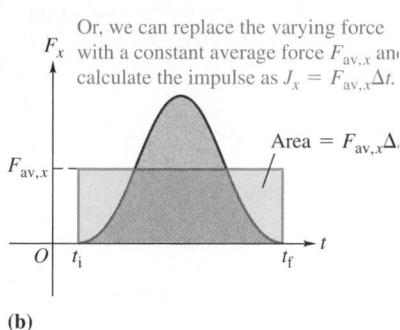

To approximate the impulse for a force that varies with time, we can divide the total interval $t_f - t_i$ into subintervals and calculate the impulse as $F_x\Delta t$ for each.

Area of strip $= F_{5x}\Delta t_5$

Or, we can replace the varying force F_x with a constant average force $F_{av,x}$ and calculate the impulse as $J_x = F_{av,x}\Delta t$.

Area $= F_{av,x}\Delta$

(a)

(b)

▲ **FIGURE 8.21** Determining the impulse of a force that varies with time.

As noted, impulse and momentum are both vector quantities, and Equations 8.14 and 8.15 are vector equations. In specific problems, it's often easiest to use these equations in component form:

$$J_x = F_x\,\Delta t = p_{f,x} - p_{i,x} = mv_{f,x} - mv_{i,x},$$
$$J_y = F_y\,\Delta t = p_{f,y}\;\;\; p_{i,y} = mv_{f,y} - mv_{i,y}. \qquad (8.16)$$

Now, what if the force isn't constant? Suppose, for example, that the x component of force varies, as shown in Figure 8.21. Then we can divide the total time interval $t_f - t_i$ into small subintervals, as in Figure 8.21a, apply Equation 8.15 to each, and add all the $\vec{F}\Delta t$ quantities. Thus, the impulse–momentum relation $\vec{J} = \vec{p}_f - \vec{p}_i$ is valid even when the force $\vec{F}$ varies with time. Figure 8.21a also shows that the impulse of the force is the area under the graph of force versus time.

As shown in Figure 8.21b, we can also define an average force $\vec{F}_{av}$ such that even when $\vec{F}$ isn't constant, its impulse $\vec{J}$ is given by

$$\vec{J} = \vec{F}_{av}\,(t_f - t_i) \qquad (8.17)$$

Conceptual Analysis 8.4

Soften the blow

A baseball player catching a fastball lets his hand move backward (in the direction of the ball's motion) during the catch. This maneuver reduces the force of impact on his hand principally because:

A. the speed of impact is lessened.
B. the time of impact is increased.
C. the impulse is reduced.

SOLUTION Whether the hand moves or not, $\Delta\vec{p}$, and therefore the impulse $\vec{J}$, is roughly the same, because the speed of the ball is much larger than any motion of the hand. However, for the same impulse, the relation $\vec{J} = \vec{F}_{av}\,\Delta t$ shows that if we increase the time interval Δt of the interaction, we decrease the average force. The correct answer is B.

EXAMPLE 8.12　A ball hits a wall

Suppose you throw a ball with mass 0.40 kg against a brick wall (Figure 8.22). It hits the wall moving horizontally to the left at 30 m/s and rebounds horizontally to the right at 20 m/s. Find the impulse of the force exerted on the ball by the wall. If the ball is in contact with the wall for 0.010 s, find the average force on the ball during the impact.

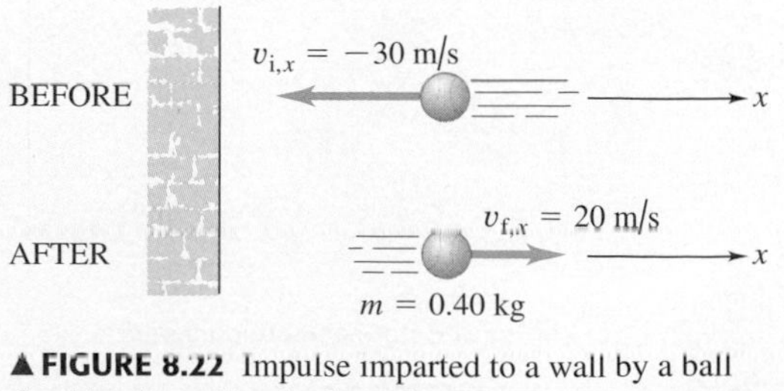

BEFORE $v_{i,x} = -30$ m/s

AFTER $v_{f,x} = 20$ m/s

$m = 0.40$ kg

▲ **FIGURE 8.22** Impulse imparted to a wall by a ball that hits the wall and rebounds.

Continued

SOLUTION

SET UP We take the x axis as horizontal and the positive direction to the right.

SOLVE We need to find the x component of momentum just before and just after impact, find the *change* in momentum, and then equate this change to the impulse. Then we divide by the time interval to find the average force on the ball.

The initial x component of momentum of the ball is

$$p_{i,x} = mv_{i,x} = (0.40\text{ kg})(-30\text{ m/s}) = -12\text{ kg} \cdot \text{m/s}.$$

The final x component of momentum is

$$p_{f,x} = mv_{f,x} = +(0.40\text{ kg})(20\text{ m/s}) = 8.0\text{ kg} \cdot \text{m/s}.$$

The change in the x component of momentum is

$$\Delta p_x = p_{f,x} - p_{i,x} = mv_{f,x} - mv_{i,x}$$
$$= 8.0\text{ kg} \cdot \text{m/s} - (-12\text{ kg} \cdot \text{m/s}) = 20\text{ kg} \cdot \text{m/s}.$$

(Note the signs carefully. The initial x component of momentum is negative, and we are subtracting it from the final value to find the change Δp_x.)

According to Equation 8.17, the preceding result equals the x component of impulse of the force exerted on the ball by the wall, so $J_x = 20\text{ kg} \cdot \text{m/s} = 20\text{ N} \cdot \text{s}$.

If the force acts for 0.010 s, then, from $J_x = F_{av,x}\,\Delta t$,

$$F_{av,x} = \frac{J_x}{\Delta t} = \frac{20\text{ N} \cdot \text{s}}{0.010\text{ s}} = 2000\text{ N}.$$

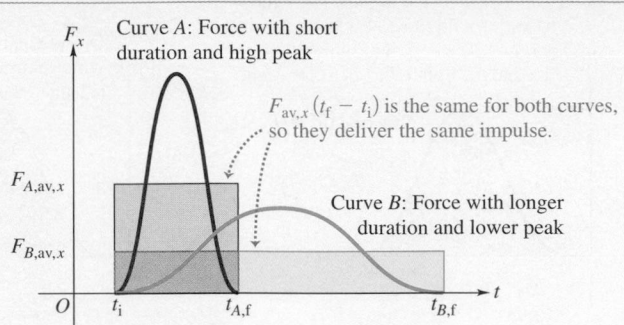

▲ **FIGURE 8.23** Forces that differ in duration and peak magnitude may deliver the same impulse.

REFLECT The force exerted by the wall on the ball isn't constant; its variation with time may be similar to a varying force described by one of the curves in Figure 8.23. The force is zero before impact, rises to a maximum, and then decreases to zero when the ball loses contact with the wall. If the ball is relatively rigid, like a baseball or a golf ball, the time of collision is small and the maximum force is large, as in curve A of the figure. If the ball is softer, like a tennis ball, the collision time is larger and the maximum force is less, as in curve B. Either way, the area under the curve represents the same impulse: $J_x = F_{av,x}\,\Delta t$.

Practice Problem: Suppose you have a pitching machine that throws five balls per second, with the same mass and initial and final velocities as given in this example. Find the average force exerted on the wall by this stream of balls. *Answer:* 100 N.

EXAMPLE 8.13 Kicking a soccer ball

Let's consider the soccer ball in Figure 8.24a. The ball has mass 0.40 kg. Initially it moves horizontally to the left at 20 m/s, but then it is kicked and given a velocity with magnitude 30 m/s and direction 45° upward and to the right. Find the impulse of the force and the average force on the ball, assuming a collision time $\Delta t = 0.010$ s.

SOLUTION

SET UP The velocities aren't along the same line, so we must treat momentum and impulse as vector quantities, using their x and y components. As shown in Figure 8.24a, we point the x axis to the right and the y axis upward. We find the following initial and final velocity components:

$$v_{i,x} = -20\text{ m/s}, \qquad v_{i,y} = 0,$$
$$v_{f,x} = v_{f,y} = \cos 45°(v_f) = (0.707)(30\text{ m/s}) = 21\text{ m/s}.$$

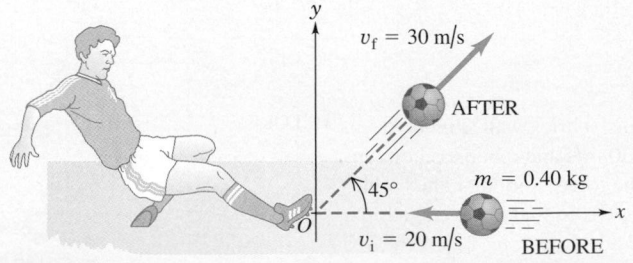

(a) Before-and-after diagram

(b) Average force on the ball

▲ **FIGURE 8.24**

Continued

SOLVE The x component of the impulse is equal to the x component of the change in momentum, and the same is true for the y components:

$$J_x = m(v_{f,x} - v_{i,x}) = (0.40\text{ kg})[21\text{ m/s} - (-20\text{ m/s})]$$
$$= 16.4\text{ kg}\cdot\text{m/s},$$
$$J_y = m(v_{f,y} - v_{i,y}) = (0.40\text{ kg})(21\text{ m/s} - 0) = 8.4\text{ kg}\cdot\text{m/s}.$$

The components of average force on the ball (Figure 8.24b) are

$$F_{av,x} = \frac{J_x}{\Delta t} = 1640\text{ N}, \qquad F_{av,y} = \frac{J_y}{\Delta t} = 840\text{ N}.$$

The magnitude and direction of the average force are, respectively,

$$F_{av} = \sqrt{(1640\text{ N})^2 + (840\text{ N})^2} = 1.8 \times 10^3\text{ N},$$
$$\theta = \tan^{-1}\frac{840\text{ N}}{1640\text{ N}} = 27°,$$

where θ is measured counterclockwise from the $+x$ axis.

REFLECT The direction of the ball's final velocity is not the same as that of the average force acting on it. Indeed, there's no reason it should be. The average magnitude of force on the ball is about 400 lb; that's roughly three times as much force as though you just stood on the ball!

Practice Problem: Suppose the ball is moving at 30° downward from the horizontal before it is kicked. If all the numerical values are the same as before, find the magnitude and direction of the average force on the ball. *Answer:* 2.0×10^3 N, 39°.

NOTE ▶ The relation between impulse and change in momentum, $\Delta\vec{p} = \vec{F}\Delta t = \vec{J}$ (Equation 8.15), has a superficial resemblance to the work–kinetic energy theorem that we developed in Chapter 7, namely, $W_{total} = K_f - K_i$ (Equation 7.4). However, there are important differences. First, impulse is a product of a force and a *time* interval, while work is a product of a force and a *distance* and depends on the angle between the force and displacement vectors. Second, impulse and momentum are vector quantities, and work and kinetic energy are scalars. Even in straight-line motion, in which only one component of a vector is involved, force, velocity, and momentum may have components along this line that are either positive or negative. **◀**

8.6 Center of Mass

We can restate the principle of conservation of momentum in a useful way with the help of the concept of **center of mass,** which we define as follows:

Definition of center of mass
Suppose we have several particles A, B, etc., with masses m_A, m_B, ... Let the coordinates of A be (x_A, y_A), let those of B be (x_B, y_B), and so on. We define the center of mass of the system as the point having coordinates (x_{cm}, y_{cm}) given by

$$x_{cm} = \frac{m_A x_A + m_B x_B + m_C x_C + \cdots}{m_A + m_B + m_C + \cdots},$$
$$y_{cm} = \frac{m_A y_A + m_B y_B + m_C y_C + \cdots}{m_A + m_B + m_C + \cdots}. \qquad (8.18)$$

In statistical language, the center of mass is a *mass-weighted average* position of the particles.

For a solid object, which we often model as having a continuous distribution of matter, locating the center of mass often frequently requires calculus. However, symmetry considerations can be helpful. For example, whenever a homogeneous solid object has a geometric center, such as a billiard ball, an ice cube, or a can of frozen orange juice, the center of mass is at the geometric center (Figure 8.25). Also, whenever an object has an axis of symmetry, such as a wheel or a pulley, the center of mass always lies on that axis. Note, however, that the center of mass isn't always within the object. For example, the center of mass of a doughnut is right in the middle of the hole.

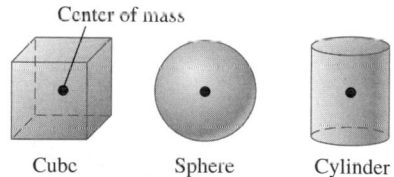

If a homogeneous object has a geometric center, that is where the center of mass is located.

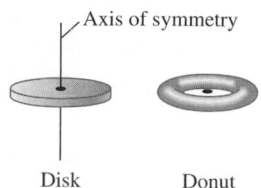

If an object has an axis of symmetry, the center of mass lies along it. As in the case of the donut, the center of mass may not be within the object.

▲ FIGURE 8.25 Location of the center of mass for some symmetric objects.

When the particles in a system move, the center of mass may also move. Equations 8.18 give the coordinates (x_{cm}, y_{cm}) of the center of mass in terms of the coordinates of the particles. The rates of change of these quantities (the components of the velocity of the center of mass) are related to the components of the velocity of the particles in the same way. Therefore, we immediately find the following relation:

Velocity of center of mass

The velocity $\vec{v}_{cm}$ of the center of mass of a collection of particles is the mass-weighted average of the velocities of the individual particles:

$$\vec{v}_{cm} = \frac{m_A\vec{v}_A + m_B\vec{v}_B + m_C\vec{v}_C + \cdots}{m_A + m_B + m_C + \cdots}. \tag{8.19}$$

In terms of components,

$$v_{cm,x} = \frac{m_A v_{A,x} + m_B v_{B,x} + m_C v_{C,x} + \cdots}{m_A + m_B + m_C + \cdots},$$

$$v_{cm,y} = \frac{m_A v_{A,y} + m_B v_{B,y} + m_C v_{C,y} + \cdots}{m_A + m_B + m_C + \cdots}. \tag{8.20}$$

We denote the total mass $m_A + m_B + \cdots$ by M; we can then rewrite Equation 8.19 as

$$M\vec{v}_{cm} = m_A\vec{v}_A + m_B\vec{v}_B + m_C\vec{v}_C + \cdots.$$

The right side is the total momentum of the system, so we have $\vec{P} = M\vec{v}_{cm}$.

Total momentum $\vec{P}$ in terms of center of mass

For a system of particles, the total momentum $\vec{P}$ is the total mass $M = m_A + m_B + \cdots$ times the velocity $\vec{v}_{cm}$ of the center of mass:

$$M\vec{v}_{cm} = m_A\vec{v}_A + m_B\vec{v}_B + m_C\vec{v}_C + \cdots = \vec{P} \tag{8.21}$$

It follows that, for an *isolated* system, in which the total momentum is constant, the velocity of the center of mass is also constant.

EXAMPLE 8.14 Quarreling pets

A 2.0 kg cat and a 3.0 kg dog are moving toward each other along the x axis, heading for a fight. At a particular instant, the cat is 1.0 m to the right of the origin and is moving in the $+x$ direction with speed 3.0 m/s, and the dog is 2.0 m to the right of the origin, moving in the $-x$ direction with speed 1.0 m/s. Find the position and velocity of the center of mass of the two-pet system, and also find the total momentum of the system.

SOLUTION

SET UP Figure 8.26 shows our sketch. We represent the pets' motion as x components of velocity, being careful with signs.

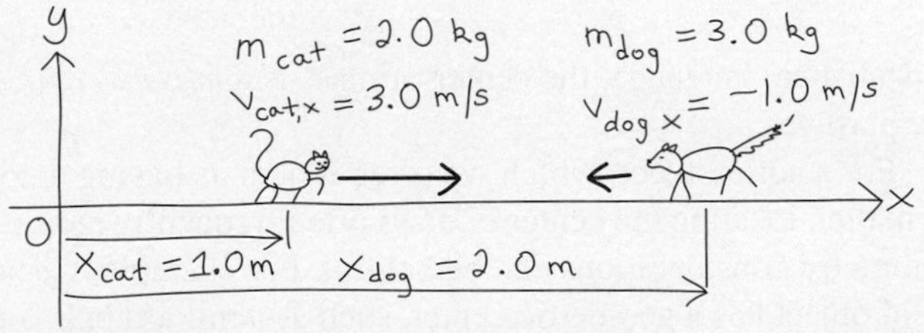

SOLVE To find the center of mass of the two-pet system, we use Equation 8.18:

$$x_{cm} = \frac{m_{cat}(x_{cat}) + m_{dog}(x_{dog})}{m_{cat} + m_{dog}}$$
$$= \frac{(2.0\,\text{kg})(1.0\,\text{m}) + (3.0\,\text{kg})(2.0\,\text{m})}{2.0\,\text{kg} + 3.0\,\text{kg}} = 1.6\,\text{m}.$$

▲ FIGURE 8.26 Our sketch for this problem.

Continued

To find the velocity of the center of mass, we use Equation 8.20:

$$v_{cm,x} = \frac{m_{cat}v_{cat,x} + m_{dog}v_{dog,x}}{m_{cat} + m_{dog}}$$

$$= \frac{(2.0 \text{ kg})(3.0 \text{ m/s}) + (3.0 \text{ kg})(-1.0 \text{ m/s})}{2.0 \text{ kg} + 3.0 \text{ kg}}$$

$$= 0.60 \text{ m/s}.$$

The total x component of momentum is the sum of the x components of momenta of the two animals:

$$P_x = m_{cat}(v_{cat,x}) + m_{dog}(v_{dog,x})$$

$$= (2.0 \text{ kg})(3.0 \text{ m/s}) + (3.0 \text{ kg})(-1.0 \text{ m/s})$$

$$= 3.0 \text{ kg} \cdot \text{m/s}.$$

Alternatively, the total momentum is the total mass M, times the velocity of the center of mass:

$$P_x = Mv_{cm,x} = (5.0 \text{ kg})(0.60 \text{ m/s}) = 3.0 \text{ kg} \cdot \text{m/s}.$$

REFLECT As always, remember that an object moving in the $-x$ direction has a negative x component of velocity. The total momentum of the system is equal to the momentum of a single particle with mass equal to the total mass of the system and with velocity equal to the velocity of the center of mass of the system.

Practice Problem: At the instant described, the cat decides to avoid combat. It quickly turns around and runs in the $-x$ direction with speed 2.0 m/s. Find the position and velocity of the center of mass, and the total momentum of the system, at this instant. *Answers:* $x_{cm} = 1.6$ m, $v_{cm} = -1.4$ m/s, $P_x = -7.0$ kg · m/s.

8.7 Motion of the Center of Mass

In Section 8.6, we defined the center of mass of a system of particles as an average position of the particles, weighted according to their masses. We found that the total momentum of the system is related simply to the total mass and the velocity of the center of mass. Now our final step is to look at the relationship between the acceleration of the center of mass and the forces acting on the system. Again we'll find a simple and elegant relation.

Equations 8.20 and 8.21 give the velocity of the center of mass in terms of the velocities of the individual particles. Proceeding one additional step, we take the rate of change of each term in these equations to show that the accelerations are related in the same way. Let $\vec{a}_{cm}$ be the acceleration of the center of mass (the rate of change of $\vec{v}_{cm}$ with respect to time); then

$$M\vec{a}_{cm} = m_A\vec{a}_A + m_B\vec{a}_B + m_C\vec{a}_C + \cdots \qquad (8.22)$$

Now, $m_A\vec{a}_A$ is equal to the vector sum of forces on particle A, and so on, so the right side of Equation 8.22 is equal to the vector sum $\Sigma\vec{F}$ of *all* the forces acting on *all* the particles. Just as we did in Section 8.2, we may classify each force as *internal* or *external*. The sum of forces on all the particles is then

$$\Sigma\vec{F} = \Sigma\vec{F}_{ext} + \Sigma\vec{F}_{int} = M\vec{a}_{cm}.$$

Because of Newton's third law, the internal forces all cancel in pairs, and $\Sigma\vec{F}_{int} = 0$. Only the sum of the external forces is left, and we have the following result:

▲ Application **Please don't sneeze!** To the physics students in the audience, these acrobats are demonstrating the concept of the center of mass of a system. Individually, each of them has his or her own separate mass. However, while moving together, they can be considered as one large mass centered at the mass-weighted average position of all the acrobats. As long as this center of mass is positioned directly above the bicycle, they can keep up their precarious balancing act.

Acceleration of center of mass

When an object or a collection of particles is acted on by external forces, the center of mass moves just as though all the mass were concentrated at that point and were acted on by a resultant force equal to the sum of the external forces on the system. Stated symbolically, this relationship is

$$\Sigma\vec{F}_{ext} = M\vec{a}_{cm}. \qquad (8.23)$$

Let's look at some applications of this result. Suppose the center of mass of an adjustable wrench lies partway down the handle. Figure 8.27 shows what happens when we send the wrench spinning across a smooth tabletop. The overall motion appears complicated, but the center of mass (marked by a white dot) follows a straight line, as though all the mass were concentrated at that point. As

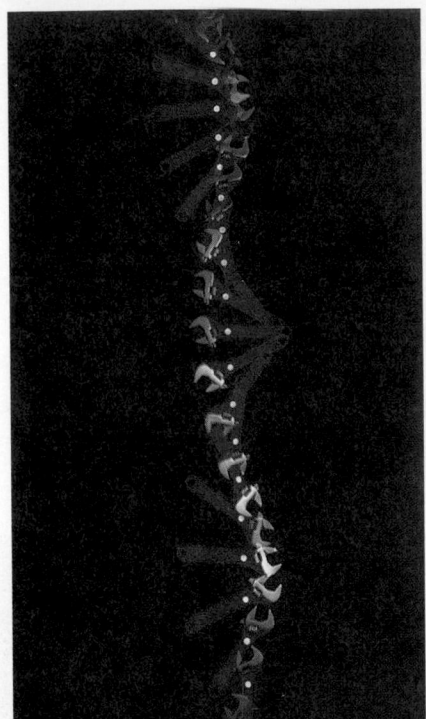

▲ **FIGURE 8.27** The center of mass of this wrench is marked with a white dot. The total external force acting on the wrench is small. As the wrench spins on a smooth horizontal surface, the center of mass moves in a straight line.

another example, suppose an explosive shell traveling in a parabolic trajectory (neglecting air resistance) explodes in flight, splitting into two fragments with equal mass (Figure 8.28a). The fragments follow new parabolic paths, but the center of mass of the shell continues on the original parabolic trajectory, just as though all the mass were still concentrated at that point. Exploding fireworks are a spectacular example of this principle (Figure 8.28b).

The property of the center of mass we've just discussed will be important when we analyze the motion of extended rigid objects in Chapter 10. The center of mass also plays an important role in the motion of astronomical objects. The earth and the moon revolve in orbits centered on their center of mass, as do the two stars in a binary system.

Finally, we note again that, for an isolated system, $\Sigma \vec{F}_{\text{ext}} = 0$. In this case, Equation 8.23 shows that the acceleration $\vec{a}_{\text{cm}}$ of the center of mass is zero, so its velocity $\vec{v}_{\text{cm}}$ is constant, and from Equation 8.21, the total momentum $\vec{P}$ is also constant. This conclusion reaffirms our statement in Section 8.2 that the total momentum of an isolated system is constant.

8.8 Rocket Propulsion

Momentum considerations are particularly useful when we have to analyze a system in which the masses of parts of the system change with time. In such cases, we can't use Newton's second law $(\Sigma \vec{F} = m\vec{a})$ directly, because m changes. Rocket propulsion offers a typical and interesting example of this kind of analysis. A rocket is propelled forward by the rearward ejection of burned fuel that initially was in the rocket. The forward force on the rocket is the reaction to the backward force on the ejected material. The total mass of the system is constant, but the mass of the rocket itself decreases as material is ejected. As a simple example, we consider a rocket fired in outer space, where there is no gravitational field and no air resistance (so that the system is isolated). We choose our x axis to be along the rocket's direction of motion.

NOTE ▶ Be especially careful with signs in this discussion. Negative signs may appear where you don't expect them; be sure you understand why they have to be where they are. ◀

Figure 8.29 shows the rocket at a time t after firing, when its mass is m and the x component of its velocity relative to our coordinate system is v. The x component of the total momentum P at this instant is $P = mv$. In a short time interval

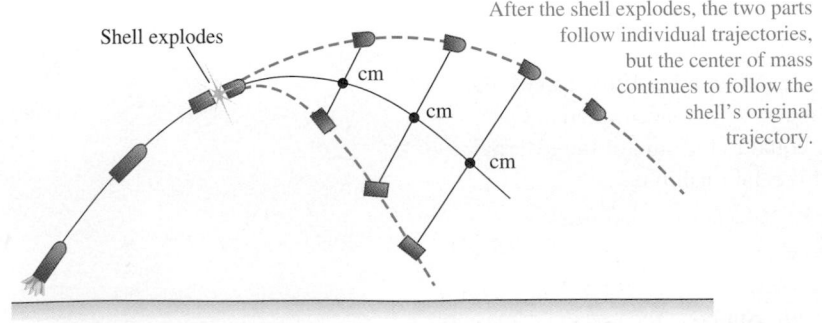

Shell explodes

After the shell explodes, the two parts follow individual trajectories, but the center of mass continues to follow the shell's original trajectory.

cm
cm
cm

(a)

(b)

▲ **FIGURE 8.28** (a) A projectile explodes in flight. If air resistance can be ignored, the fragments follow individual parabolic trajectories, but the center of mass continues on the same trajectory the projectile was following before it exploded. (b) A firework represents an exploding projectile.

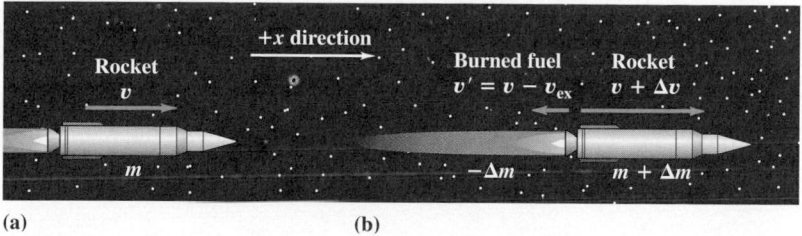

At time t, the rocket has mass m and x component of velocity v.

At time $t + \Delta t$, the rocket has mass $m + \Delta m$ (where Δm is inherently *negative*) and x component of velocity $v + \Delta v$. The burned fuel has x component of velocity $v' = v - v_{ex}$ and mass $-\Delta m$. (The minus sign is needed to make $-\Delta m$ *positive*, because Δm is negative.)

+x direction

Rocket
v

Burned fuel
$v' = v - v_{ex}$

Rocket
$v + \Delta v$

m

$-\Delta m$

$m + \Delta m$

(a)

(b)

▲ **FIGURE 8.29** A rocket moving in gravity-free outer space.

Δt, the mass of the rocket changes by an amount Δm. This is an inherently negative quantity, because the rocket's mass m *decreases* with time as fuel is burned. During Δt, a positive mass $(-\Delta m)$ of burned fuel is ejected from the rocket. Let v_{ex} be the exhaust speed of this material relative to the rocket. Then the x component of the velocity v' of the fuel relative to our coordinate system is

$$v' = v - v_{ex},$$

and the x component of momentum of the mass $-\Delta m$ is

$$(-\Delta m)v' = (-\Delta m)(v - v_{ex}).$$

At the end of the time interval Δt, the mass of the rocket (including the unburned fuel) has decreased to $m + \Delta m$, and the rocket's velocity has increased to $v + \Delta v$ (Figure 8.29). (Remember that Δm is negative.) The rocket's momentum at this time is

$$(m + \Delta m)(v + \Delta v).$$

Thus, the total x component of the momentum of the rocket plus the ejected fuel at time $t + \Delta t$ is

$$P = (m + \Delta m)(v + \Delta v) + (-\Delta m)(v - v_{ex}).$$

According to our initial assumption, the rocket and fuel are an isolated system. Hence, momentum is conserved; the total momentum P of the system must be the same at time t and at time $t + \Delta t$:

$$mv = (m + \Delta m)(v + \Delta v) + (-\Delta m)(v - v_{ex}).$$

This equation can be simplified to

$$m\,\Delta v = -\Delta m v_{ex} - \Delta m\,\Delta v.$$

We can neglect the term $\Delta m\,\Delta v$ because it is a product of two small quantities and thus is much smaller than the other terms. Dropping that term, dividing by Δt, and rearranging terms, we find that

$$m\frac{\Delta v}{\Delta t} = -v_{ex}\frac{\Delta m}{\Delta t}. \tag{8.24}$$

Because $\Delta v / \Delta t$ is the x component of acceleration of the rocket, the left side of this equation (mass times acceleration) equals the x component of the resultant force F, or *thrust,* on the rocket, so

$$F = -v_{ex}\frac{\Delta m}{\Delta t}. \quad \text{(thrust of rocket)} \tag{8.25}$$

▲ **BIO Application A jet-propelled . . . jellyfish?** These jellyfish move by using the same physical principles of propulsion that we see in interplanetary rockets. By forcing water out of their bells in one direction, they are able to move in the opposite direction.

The thrust is proportional both to the relative speed v_{ex} of the ejected fuel and to the (positive) mass of fuel ejected per unit time, $-\Delta m/\Delta t$. (Remember that $\Delta m/\Delta t$ is negative because it is the rate of change of the rocket's mass.)

The x component of acceleration a of the rocket is

$$a = \frac{\Delta v}{\Delta t} = -\frac{v_{ex}}{m}\frac{\Delta m}{\Delta t}. \qquad \text{(acceleration of rocket)} \qquad (8.26)$$

The rocket's mass m decreases continuously while the fuel is being consumed. If v_{ex} and $\Delta m/\Delta t$ are constant, the acceleration increases until all the fuel is gone.

EXAMPLE 8.15 Launch of a rocket

A rocket floats next to an interplanetary space station far from any planet. It ignites its engine. In the first second of its flight, it ejects $\frac{1}{60}$ of its mass with a relative velocity magnitude of 2400 m/s. Find its acceleration.

SOLUTION

SET UP AND SOLVE We are given that $\Delta m = -m/60$ and $\Delta t = 1.0$ s. We use the same coordinate system as in Figure 8.29, with the $+x$ axis to the right.

The acceleration is given by Equation 8.26:

$$a = -\frac{v_{ex}}{m}\frac{\Delta m}{\Delta t}$$
$$= -\frac{2400 \text{ m/s}}{m}\left(\frac{-m/60}{1.0 \text{ s}}\right) = 40 \text{ m/s}^2.$$

REFLECT At the start of the flight, when the velocity of the rocket is zero, the ejected exhaust is moving to the left, relative to our coordinate system, at 2400 m/s. At the end of the first sec-

ond, the rocket is moving at 40 m/s. The speed of the exhaust relative to our system is 2360 m/s. We could now compute the acceleration at $t = 1.0$ s (using the decreased rocket mass), find the velocity at $t = 2.0$ s, and so on, stepping through the calculation one second at a time until all the fuel is used up. As the mass decreases, the acceleration in successive 1 s intervals increases.

Detailed calculation shows that, after about 22 s, the rocket's velocity in our coordinate system passes 2400 m/s. The exhaust ejected after this time therefore moves *forward*, not backward, in our system. The final velocity acquired by the rocket can be greater in magnitude (and is often *much* greater) than the relative speed v_{ex}.

Practice Problem: Find the rocket's velocities 2.0 s and 3.0 s after launch. *Answers:* 81 m/s, 122 m/s.

In the early days of rocket propulsion, people who didn't understand conservation of momentum thought that a rocket couldn't function in outer space because it wouldn't have anything to "push against." On the contrary, rockets work *best* in outer space! Figure 8.30 shows a dramatic example of rocket propulsion. The rocket is *not* "pushing against the ground" to get into the air.

▲ **FIGURE 8.30** Launch of the Titan 4B Centaur, a dramatic example of rocket propulsion.

SUMMARY

Momentum

(Section 8.1) The momentum $\vec{p}$ of a particle with mass m moving with velocity $\vec{v}$ is defined as the vector quantity $\vec{p} = m\vec{v}$ (Equation 8.1). Newton's second law can be restated as follows: The rate of change of momentum of a particle equals the vector sum of the forces acting on it:

$$\sum \vec{F} = \lim_{\Delta t \to 0} \frac{\Delta \vec{p}}{\Delta t}. \tag{8.3}$$

The total momentum $\vec{P}$ of a system of particles is the vector sum of the momenta of the individual particles:

$$\vec{P} = \vec{p}_A + \vec{p}_B + \cdots = m_A\vec{v}_A + m_B\vec{v}_B + \cdots. \tag{8.4, 8.5}$$

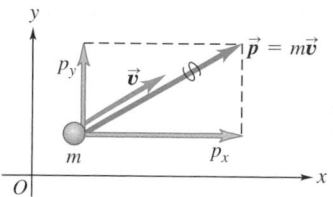

Momentum is a vector quantity; it has the same direction as the object's velocity and can be represented by components.

Conservation of Momentum

(Section 8.2) An internal force is a force exerted by one part of a system on another part of the same system; an external force is a force exerted on part or all of a system by something outside the system. An isolated system is a system that experiences either no external forces or external forces whose vector sum is zero. The total momentum of an isolated system is constant: $\vec{P}_i = \vec{P}_f$. Each component of momentum is conserved separately.

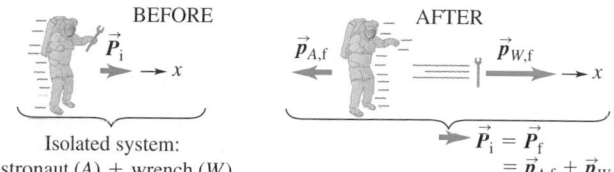

When the astronaut throws the wrench, the total momentum $\vec{P}$ of the system is conserved, although the individual momenta $\vec{p}_A$ and $\vec{p}_W$ are not.

Collisions

(Sections 8.3 and 8.4) Collisions can be classified according to energy relations and final velocities. In an elastic collision between two objects, kinetic energy is conserved and the initial and final relative velocities have the same magnitude. In an inelastic two-object collision, the final kinetic energy is less than the initial kinetic energy; if the two objects have the same final velocity (for instance, if they stick together), the collision is completely inelastic.

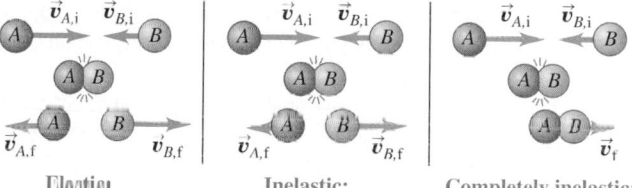

Elastic:
K conserved

Inelastic:
Some K lost

Completely inelastic:
Objects have same $\vec{v}_f$

Impulse

(Section 8.5) The impulse $\vec{J}$ of a constant force $\vec{F}$ acting over a time interval Δt from t_i to t_f is the vector quantity $\vec{J} = \vec{F}(t_f - t_i) = \vec{F}\,\Delta t$ (Equation 8.14). The change in momentum of a particle during any time interval equals the impulse of the total force acting on the particle: $\Delta \vec{p} = \vec{F}\,\Delta t = \vec{F}(t_f - t_i) = \vec{J}$ (Equation 8.15).

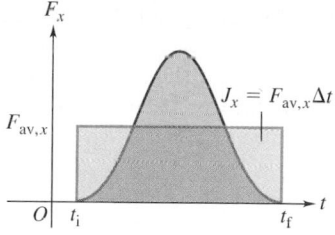

The impulse $\vec{J}$ of a force over a certain time interval is the area under the graph of force versus time. If the force varies with time, the impulse may be calculated by using the average force.

Center of Mass

(Sections 8.6 and 8.7) The coordinates x_{cm} and y_{cm} of the center of mass of a system of particles are defined as

$$x_{cm} = \frac{m_A x_A + m_B x_B + m_C x_C + \cdots}{m_A + m_B + m_C + \cdots},$$

$$y_{cm} = \frac{m_A y_A + m_B y_B + m_C y_C + \cdots}{m_A + m_B + m_C + \cdots}. \tag{8.18}$$

The total momentum $\vec{P}$ of the system equals the total mass M, multiplied by the velocity $\vec{v}_{cm}$ of the center of mass:

$$M\vec{v}_{cm} = m_A\vec{v}_A + m_B\vec{v}_B + m_C\vec{v}_C + \cdots = \vec{P} \tag{8.21}$$

The center of mass of a system moves as though all the mass were concentrated at the center of mass: $\sum \vec{F}_{ext} = M\vec{a}_{cm}$ (Equation 8.23).

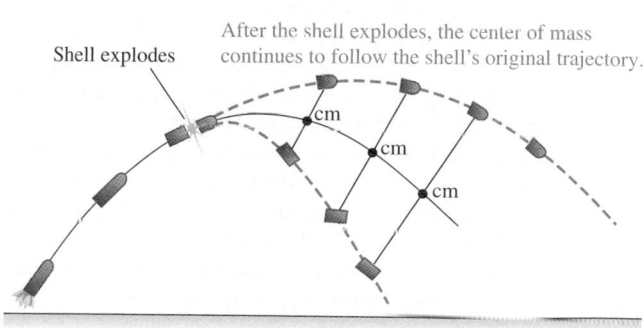

Shell explodes

After the shell explodes, the center of mass continues to follow the shell's original trajectory.

Continued

Rocket Propulsion

(Section 8.8) In rocket propulsion, the mass of the rocket changes as the fuel is burned. Analysis of the momentum relations must include the momentum carried away by the fuel, as well as the momentum of the rocket itself.

 For instructor-assigned homework, go to www.masteringphysics.com

Conceptual Questions

1. The objects shown in Figure 8.31 move together. Identify internal and external forces for each of the following systems: (a) The system consists of only block *A*. (b) The system consists of only block *B*. (c) The system consists of both blocks *A* and *B*.

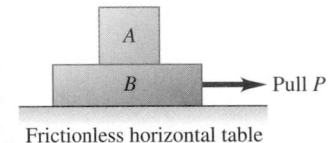

Frictionless horizontal table

▲ **FIGURE 8.31** Question 1.

2. Gliders *A* and *B* are headed directly toward each other on an air track and collide head-on. Identify the internal and external forces on each glider during the collision. What are the internal and external forces if the system consists of *both* gliders?

3. Example 8.3, as well as some of the problems in this chapter, describe how, when an object explodes into two fragments, the lighter one gets more kinetic energy than the heavier one. How can this be so, since both fragments are acted upon by the same force for the same amount of time during the explosion? (*Hint:* Does the force act over the same *distance* on each fragment?)

4. Is the momentum of a satellite in a circular orbit conserved? What about its kinetic energy? Explain your reasoning.

5. Golfers, tennis players, and baseball batters are told to "follow through" with their swing. Why? What does a follow-through enable them to do?

6. When a catcher in a baseball game catches a fast ball, he does not hold his arms rigid, but relaxes them so that the mitt moves several inches while the ball is being caught. Why is this important?

7. A woman stands in the middle of a perfectly smooth, frictionless, frozen lake. She can set herself in motion by throwing things, but suppose she has nothing to throw. Can she propel herself to shore *without* throwing anything? Explain.

8. When rockets were first suggested, some people scoffed that they would not work in space because there is nothing out there for them to push against. So why *do* they work in space, since there really is nothing out there for them to push against?

9. (a) If the momentum of a *single* object is equal to zero, must its kinetic energy also be zero? (b) If the momentum of a *pair* of objects is equal to zero, must the kinetic energy of those objects also be zero? (c) If the kinetic energy of a pair of objects is equal to zero, must the momentum of those objects also be zero? Explain your reasoning in each case.

10. (a) When a large car collides with a small car, which one undergoes the greater change in momentum, the large one or the small one? Or is it the same for both? (b) In light of your answer to part (a), why are the occupants of the small car more likely to be hurt than those of the large car, assuming that both cars are equally sturdy?

11. When rain falls from the sky, what becomes of the momentum of the raindrops as they hit the ground? Is your answer also valid for Newton's famous apple?

12. An egg is released from the roof of a building and falls to the ground. As the egg falls, what happens to the momentum of the system of the egg plus the earth?

13. A machine gun is fired at a steel plate. Is the impulse imparted to the plate from the impact of the bullets greater if the bullets bounce off or if they are squashed and stick to the plate?

14. In a zero-gravity environment, can a rocket-propelled spaceship ever attain a speed greater than the relative speed with which the burnt fuel is exhausted? Explain.

15. At the highest point in its parabolic trajectory, a shell explodes into two fragments. Is it possible for *both* fragments to fall straight down after the explosion? Why or why not?

16. If the movie "supergorilla" King Kong beats his chest on a frozen frictionless lake, will he go backward? Why?

Multiple-Choice Problems

1. A small car collides head-on with a large SUV. Which of the following statements concerning this collision are correct? (There may be more than one correct choice.)
   A. Both vehicles are acted upon by the same magnitude of average force during the collision.
   B. The small car is acted upon by a greater magnitude of average force than the SUV.
   C. The small car undergoes a greater change in momentum than the SUV.
   D. Both vehicles undergo the same change in magnitude of momentum.

2. A ball of mass 0.18 kg moving with speed 11.3 m/s collides head-on with an identical stationary ball. (Notice that we do not know the type of collision.) Which of the following quantities can be calculated from this information alone?
   A. The force each ball exerts on the other.
   B. The velocity of each ball after the collision.
   C. Total kinetic energy of both balls after the collision.
   D. Total momentum of both balls after the collision.

3. A proton with a speed of 50,000 km/s makes an *elastic* head-on collision with a stationary carbon nucleus. We can look up the masses of both particles. Which of the following quantities can be calculated from only the known information? (There may be more than one correct choice.)

A. The velocity of the proton and carbon nucleus after the collision.

B. The kinetic energy of each of the particles after the collision.

C. The momentum of each of the particles after the collision.

4. In which of the following collisions would you expect the kinetic energy to be conserved? (There may be more than one correct choice.)

A. A bullet passes through a block of wood.

B. Two bull elk charge each other and lock horns.

C. Two asteroids collide by a glancing blow, but do not actually hit each other, and their only interaction is through gravity.

D. Two cars with springlike bumpers collide at fairly low speeds.

5. A rifle of mass M is initially at rest, but is free to recoil. It fires a bullet of mass m with a velocity $+v$ relative to the ground. After the rifle is fired, its velocity relative to the ground is

A. $-\sqrt{m/M}\,v$. B. $-mv/(m + M)$.

C. $-mv/M$. D. $-v$.

6. Two carts, one twice as heavy as the other, are at rest on a horizontal frictionless track. A person pushes each cart with the same force for 5 s. If the kinetic energy of the *lighter* cart after the push is K, the kinetic energy of the heavier cart is

A. $\frac{1}{4}K$. B. $\frac{1}{2}K$. C. K.

D. $2K$. E. $4K$.

7. A 70-kg wide receiver running west at 8 m/s collides head-on with, and is seized by, a 140-kg lineman lumbering eastward at 4 m/s. In this collision, the greater change in kinetic energy is experienced by

A. the wide receiver. B. the lineman.

C. neither; they experience the same change in kinetic energy.

8. Two masses, M and $5M$, are at rest on a horizontal frictionless table with a compressed spring of negligible mass between them. When the spring is released (there may be more than one correct choice),

A. the two masses receive equal magnitudes of momentum.

B. the two masses receive equal amounts of kinetic energy from the spring.

C. the heavier mass gains more kinetic energy than the lighter mass.

D. the lighter mass gains more kinetic energy than the heavier mass.

9. Cart A, of mass 1 kg, approaches and collides with cart B, which has a mass of 4 kg and is initially at rest. (See Figure 8.32.) When the springs have reached their maximum compression,

A. cart A has come to rest relative to the ground.

B. both carts have the same velocity.

C. both carts have the same momentum.

D. all the initial kinetic energy of cart A has been converted to elastic potential energy.

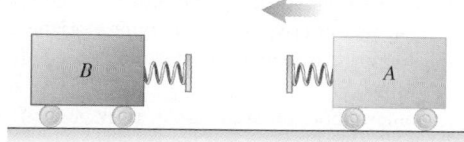

▲ **FIGURE 8.32** Multiple-choice problem 9.

10. A glider airplane is coasting horizontally when a very heavy object suddenly falls out of it. As a result of dropping this object, the glider's speed will

A. increase.

B. decrease.

C. remain the same as it was.

11. Which of the following statements is true for an *inelastic* collision? (There may be more than one correct choice.)

A. Both momentum and kinetic energy are conserved.

B. Momentum is conserved, but kinetic energy is not conserved.

C. Kinetic energy is conserved, but momentum is not conserved.

D. The amount of momentum lost by one object is the same as the amount gained by the other object.

E. The amount of kinetic energy lost by one object is the same as the amount gained by the other object.

12. Which of the following statements is true for an *elastic* collision? (There may be more than one correct choice.)

A. Both momentum and kinetic energy are conserved.

B. Momentum is conserved, but kinetic energy is not conserved.

C. Kinetic energy is conserved, but momentum is not conserved.

D. The amount of momentum lost by one object is the same as the amount gained by the other object.

E. The amount of kinetic energy lost by one object is the same as the amount gained by the other object.

13. Two lumps of clay having equal masses and speeds, but traveling in opposite directions on a frictionless horizontal surface, collide and stick together. Which of the following statements about this system of lumps must be true? (There may be more than one correct choice.)

A. The momentum of the system is conserved during the collision.

B. The kinetic energy of the system is conserved during the collision.

C. The two masses lose all their kinetic energy during the collision.

D. The velocity of the center of mass of the system is the same after the collision as it was before the collision.

14. A heavy rifle initially at rest fires a light bullet. Which of the following statements about these objects is true? (There may be more than one correct choice.)

A. The bullet and rifle both gain the same magnitude of momentum.

B. The bullet and rifle are both acted upon by the same average force during the firing.

C. The bullet and rifle both have the same acceleration during the firing.

D. The bullet and the rifle gain the same amount of kinetic energy.

15. You drop an egg from rest with no air resistance. As it falls,

A. only its momentum is conserved.

B. only its kinetic energy is conserved.

C. both its momentum and its mechanical energy are conserved.

D. its mechanical energy is conserved, but its momentum is not conserved.

Problems

8.1 Momentum

1. • For each case in Figure 8.33, the system consists of the masses shown with the indicated velocities. Find the net momentum of each system.

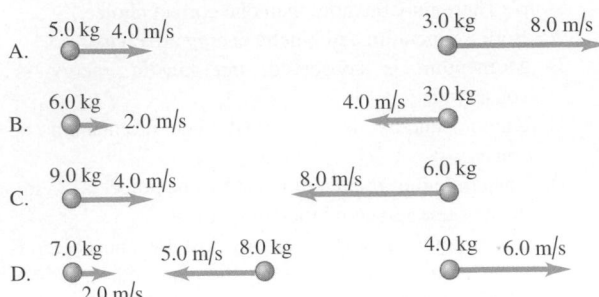

▲ **FIGURE 8.33** Problem 1.

2. • For each case in Figure 8.34, the system consists of the masses shown with the indicated velocities. Find the x and y components of the net momentum of each system.

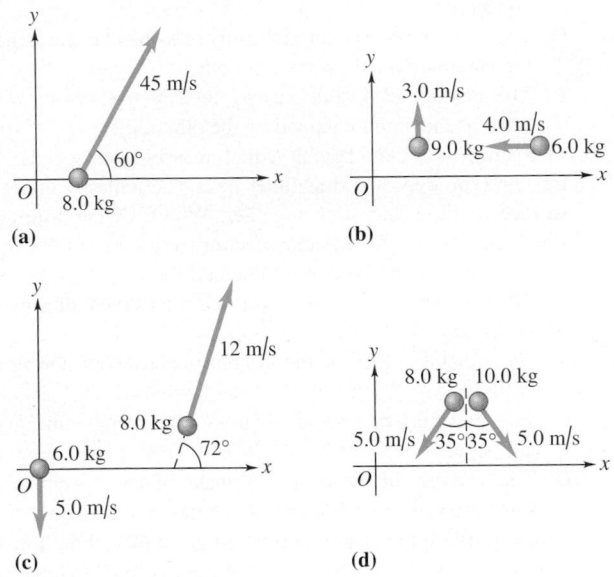

▲ **FIGURE 8.34** Problem 2.

3. • Three objects A, B, and C are moving as shown in Figure 8.35. Find the x and y components of the net momentum of the particles if we define the system to consist of (a) A and C, (b) B and C, (c) all three objects.

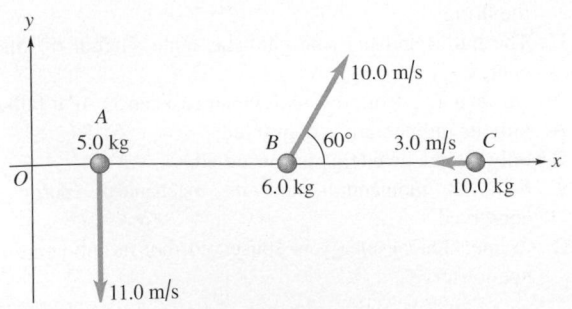

▲ **FIGURE 8.35** Problem 3.

4. • A 1200 kg car is moving on the freeway at 65 mph. (a) Find the magnitude of its momentum and its kinetic energy in SI units. (b) If a 2400 kg SUV has the same speed as the 1200 kg car, how much momentum and kinetic energy does it have?

5. • The speed of the fastest-pitched baseball was 45 m/s, and the ball's mass was 145 g. (a) What was the magnitude of the momentum of this ball, and how many joules of kinetic energy did it have? (b) How fast would a 57 gram ball have to travel to have the same amount of (i) kinetic energy, and (ii) momentum?

6. •• **Some useful relationships.** The following relationships between the momentum and kinetic energy of an object can be very useful for calculations: If an object of mass m has momentum of magnitude p and kinetic energy K, show that (a) $K = (p^2/2m)$, and (b) $p = \sqrt{2mK}$. (c) Find the momentum of a 1.15 kg ball that has 15.0 J of kinetic energy. (d) Find the kinetic energy of a 3.50 kg cat that has 0.220 kg · m/s of momentum.

7. •• The magnitude of the momentum of a cat is p. What would be the magnitude of the momentum (in terms of p) of a dog having three times the mass of the cat if it had (a) the same speed as the cat, and (b) the same kinetic energy as the cat?

8.2 Conservation of Momentum

8. • Two figure skaters, one weighing 625 N and the other 725 N, push off against each other on frictionless ice. (a) If the heavier skater travels at 1.50 m/s, how fast will the lighter one travel? (b) How much kinetic energy is "created" during the skaters' maneuver, and where does this energy come from?

9. • **Recoil speed of the earth.** In principle, any time someone jumps up, the earth moves in the opposite direction. To see why we are unaware of this motion, calculate the recoil speed of the earth when a 75 kg person jumps upward at a speed of 2.0 m/s. Consult Appendix E as needed.

10. • On a frictionless air track, a 0.150 kg glider moving at 1.20 m/s to the right collides with and sticks to a stationary 0.250 kg glider. (a) What is the net momentum of this two-glider system before the collision? (b) What must be the net momentum of this system after the collision? Why? (c) Use your answers in parts (a) and (b) to find the speed of the gliders after the collision. (d) Is kinetic energy conserved during the collision?

11. • **Baseball.** A regulation 145 g baseball can be hit at speeds of 100 mph. If a line drive is hit essentially horizontally at this speed and is caught by a 65 kg player who has leapt directly upward into the air, what horizontal speed (in cm/s) does he acquire by catching the ball?

12. • You are standing on a sheet of ice that covers the football stadium parking lot in Buffalo; there is negligible friction between your feet and the ice. A friend throws you a 0.400 kg ball that is traveling horizontally at 10.0 m/s. Your mass is 70.0 kg. (a) If you catch the ball, with what speed do you and the ball move afterwards? (b) If the ball hits you and bounces off your chest, so that afterwards it is moving horizontally at 8.00 m/s in the opposite direction, what is your speed after the collision?

13. •• On a frictionless, horizontal air table, puck A (with mass 0.250 kg) is moving to the right toward puck B (with mass 0.350 kg), which is initially at rest. After the collision, puck A has a velocity of 0.120 m/s to the left, and puck B has a velocity of 0.650 m/s to the right. (a) What was the speed of puck A

before the collision? (b) Calculate the change in the total kinetic energy of the system that occurs during the collision.

14. •• Block *A* in Figure 8.36 has mass 1.00 kg, and block *B* has mass 3.00 kg. The blocks are forced together, compressing a spring *S* between them; then the system is released from rest on a level, frictionless surface. The spring, which has negligible mass, is not fastened to either block and drops to the surface after it has expanded. Block *B* acquires a speed of 1.20 m/s. (a) What is the final speed of block *A*? (b) How much potential energy was stored in the compressed spring?

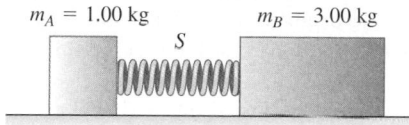

▲ **FIGURE 8.36** Problem 14.

15. •• Two ice skaters, Daniel (mass 65.0 kg) and Rebecca (mass 45.0 kg), are practicing. Daniel stops to tie his shoelace and, while at rest, is struck by Rebecca, who is moving at 13.0 m/s before she collides with him. After the collision, Rebecca has a velocity of magnitude 8.00 m/s at an angle of 53.1° from her initial direction. Both skaters move on the frictionless, horizontal surface of the rink. (a) What are the magnitude and direction of Daniel's velocity after the collision? (b) What is the change in total kinetic energy of the two skaters as a result of the collision?

16. •• You (mass 55 kg) are riding your frictionless skateboard (mass 5.0 kg) in a straight line at a speed of 4.5 m/s when a friend standing on a balcony above you drops a 2.5 kg sack of flour straight down into your arms. (a) What is your new speed, while holding the flour sack? (b) Since the sack was dropped vertically, how can it affect your *horizontal* motion? Explain. (c) Suppose you now try to rid yourself of the extra weight by throwing the flour sack straight up. What will be your speed while the sack is in the air? Explain.

17. •• A 4.25 g bullet traveling horizontally with a velocity of magnitude 375 m/s is fired into a wooden block with mass 1.12 kg, initially at rest on a level frictionless surface. The bullet passes *through* the block and emerges with its speed reduced to 122 m/s. How fast is the block moving just after the bullet emerges from it?

18. •• A ball with a mass of 0.600 kg is initially at rest. It is struck by a second ball having a mass of 0.400 kg, initially moving with a velocity of 0.250 m/s toward the right along the *x* axis. After the collision, the 0.400 kg ball has a velocity of 0.200 m/s at an angle of 36.9° above the *x* axis in the first quadrant. Both balls move on a frictionless, horizontal surface. (a) What are the magnitude and direction of the velocity of the 0.600 kg ball after the collision? (b) What is the change in the total kinetic energy of the two balls as a result of the collision?

19. •• **Combining conservation laws.** A 5.00 kg chunk of ice is sliding at 12.0 m/s on the floor of an ice-covered valley when it collides with and sticks to another 5.00 kg chunk of ice that is initially at rest. (See Figure 8.37.) Since the valley is icy, there is no friction. After the collision, how high above the valley floor will the combined chunks go? (*Hint:* Break this problem into two parts—the collision and the behavior after the collision—and apply the appropriate conservation law to each part.)

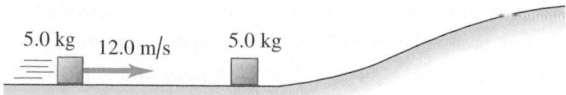

▲ **FIGURE 8.37** Problem 19.

20. •• **Combining conservation laws.** A 15.0 kg block is attached to a very light horizontal spring of force constant 500.0 N/m and is resting on a frictionless horizontal table. (See Figure 8.38.) Suddenly it is struck by a 3.00 kg stone traveling horizontally at 8.00 m/s to the right, whereupon the stone rebounds at 2.00 m/s horizontally to the left. Find the maximum distance that the block will compress the spring after the collision. (*Hint:* Break this problem into two parts—the collision and the behavior after the collision—and apply the appropriate conservation law to each part.)

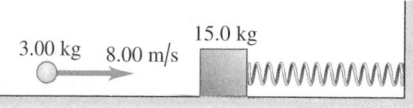

▲ **FIGURE 8.38** Problem 20.

8.3 Inelastic Collisions

21. • Three identical boxcars are coupled together and are moving at a constant speed of 20.0 m/s on a level track. They collide with another identical boxcar that is initially at rest and couple to it, so that the four cars roll on as a unit. Friction is small enough to be neglected. (a) What is the speed of the four cars? (b) What percentage of the kinetic energy of the boxcars is dissipated in the collision? What happened to this energy?

22. • On a highly polished, essentially frictionless lunch counter, a 0.500 kg submarine sandwich moving 3.00 m/s to the left collides with a 0.250 kg grilled cheese sandwich moving 1.20 m/s to the right. (a) If the two sandwiches stick together, what is their final velocity? (b) How much mechanical energy dissipates in the collision? Where did this energy go?

23. •• An astronaut in space cannot use a scale or balance to weigh objects because there is no gravity. But she does have devices to measure distance and time accurately. She knows her own mass is 78.4 kg, but she is unsure of the mass of a large gas canister in the airless rocket. When this canister is approaching her at 3.50 m/s, she pushes against it, which slows it down to 1.20 m/s (but does not reverse it) and gives her a speed of 2.40 m/s. (a) What is the mass of this canister? (b) How much kinetic energy is "lost" in this collision, and what happens to that energy?

24. •• On a very muddy football field, a 110-kg linebacker tackles an 85-kg halfback. Immediately before the collision, the linebacker is slipping with a velocity of 8.8 m/s north and the halfback is sliding with a velocity of 7.2 m/s east. What is the velocity (magnitude and direction) at which the two players move together immediately after the collision?

25. •• A 5.00 g bullet is fired horizontally into a 1.20 kg wooden block resting on a horizontal surface. The coefficient of kinetic friction between block and surface is 0.20. The bullet remains embedded in the block, which is observed to slide 0.230 m along the surface before stopping. What was the initial speed of the bullet?

26. •• You and your friends are doing physics experiments on a frozen pond that serves as a frictionless, horizontal surface. Sam, with mass 80.0 kg, is given a push and slides eastward. Abigail, with mass 50.0 kg, is sent sliding northward. They collide, and after the collision Sam is moving at 37.0° north of east with a speed of 6.00 m/s and Abigail is moving at 23.0° south of east with a speed of 9.00 m/s. (a) What was the speed of each person before the collision? (b) By how much did the total kinetic energy of the two people decrease during the collision?

27. •• A hungry 11.5 kg predator fish is coasting from west to east at 75.0 cm/s when it suddenly swallows a 1.25 kg fish swimming from north to south at 3.60 m/s. Find the magnitude and direction of the velocity of the large fish just after it snapped up this meal. Neglect any effects due to the drag of the water.

28. •• **Bird defense.** To protect their young in the nest, peregrine
BIO falcons will fly into birds of prey (such as ravens) at high speed. In one such episode, a 600 gram falcon flying at 20.0 m/s ran into a 1.5 kg raven flying at 9.0 m/s. The falcon hit the raven at right angles to its original path and bounced back with a speed of 5.0 m/s. (These figures were estimated by one of the authors (WRA) as he watched this attack occur in northern New Mexico.) By what angle did the falcon change the raven's direction of motion?

29. •• **Accident analysis.** Two cars collide at an intersection. Car A, with a mass of 2000 kg, is going from west to east, while car B, of mass 1500 kg, is going from north to south at 15 m/s. As a result of this collision, the two cars become enmeshed and move as one afterwards. In your role as an expert witness, you inspect the scene and determine that, after the collision, the enmeshed cars moved at an angle of 65° south of east from the point of impact. (a) How fast were the enmeshed cars moving just after the collision? (b) How fast was car A going just before the collision?

30. •• A hockey puck B rests on frictionless, level ice and is struck by a second puck A, which was originally traveling at 40.0 m/s and which is deflected 30.0° from its original direction. (See Figure 8.39.) Puck B acquires a velocity at a 45.0° angle to the original direction of A. The pucks have the same mass. (a) Compute the speed of each puck after the collision. (b) What fraction of the original kinetic energy of puck A dissipates during the collision?

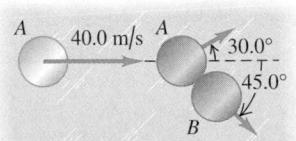

▲ **FIGURE 8.39** Problem 30.

8.4 Elastic Collisions

31. • A 0.300 kg glider is moving to the right on a frictionless, horizontal air track with a speed of 0.80 m/s when it makes a head-on collision with a stationary 0.150 kg glider. (a) Find the magnitude and direction of the final velocity of each glider if the collision is elastic. (b) Find the final kinetic energy of each glider.

32. • On a cold winter day, a penny (mass 2.50 g) and a nickel (mass 5.00 g) are lying on the smooth (frictionless) surface of a frozen lake. With your finger, you flick the penny toward the nickel with a speed of 2.20 m/s. The coins collide elastically; calculate both their final velocities (speed and direction).

33. •• **Nuclear collisions.** Collisions between atomic and subatomic particles are often perfectly elastic. In one such collision, a proton traveling to the right at 258 km/s collides head-on and elastically with a stationary alpha particle (a helium nucleus, having mass 6.65×10^{-27} kg). Consult Appendix E as needed. (a) Find the magnitude and direction of the velocity of each particle after the collision. (b) How much kinetic energy does the proton lose during the collision? (c) How can the collision be elastic if the proton loses kinetic energy?

34. •• On an air track, a 400.0 g glider moving to the right at 2.00 m/s collides elastically with a 500.0 g glider moving in the opposite direction at 3.00 m/s. Find the velocity of each glider after the collision.

35. •• Blocks A (mass 2.00 kg) and B (mass 10.00 kg) move on a frictionless, horizontal surface. Initially, block B is at rest and block A is moving toward it at 2.00 m/s. The blocks are equipped with ideal spring bumpers, as in Example 8.9. The collision is head-on, so all motion before and after the collision is along a straight line. (a) Find the maximum energy stored in the spring bumpers and the velocity of each block at that time. (b) Find the velocity of each block after they have moved apart.

36. •• Two identical objects traveling in opposite directions with the same speed V make a head-on collision. Find the speed of each object after the collision if (a) they stick together and (b) if the collision is perfectly elastic.

8.5 Impulse

37. • A catcher catches a 145 g baseball traveling horizontally at 36.0 m/s. (a) How large an impulse does the ball give to the catcher? (b) If the ball takes 20 ms to stop once it is in contact with the catcher's glove, what average force did the ball exert on the catcher?

38. • A block of ice with a mass of 2.50 kg is moving on a frictionless, horizontal surface. At $t = 0$, the block is moving to the right with a velocity of magnitude 8.00 m/s. Calculate the magnitude and direction of the velocity of the block after each of the following forces has been applied for 5.00 s: (a) a force of 5.00 N directed to the right; (b) a force of 7.00 N directed to the left.

39. •• **Biomechanics.** The mass of a regulation tennis ball is
BIO 57 g (although it can vary slightly), and tests have shown that the ball is in contact with the tennis racket for 30 ms. (This number can also vary, depending on the racket and swing.) We shall assume a 30.0 ms contact time throughout this problem. The fastest-known served tennis ball was served by "Big Bill" Tilden in 1931, and its speed was measured to be 73.14 m/s. (a) What impulse and what force did Big Bill exert on the tennis ball in his record serve? (b) If Big Bill's opponent returned his serve with a speed of 55 m/s, what force and what impulse did he exert on the ball, assuming only horizontal motion?

40. •• To warm up for a match, a tennis player hits the 57.0 g ball vertically with her racket. If the ball is stationary just before it is hit and goes 5.50 m high, what impulse did she impart to it?

41. •• A 150 gram baseball is hit toward the left by a bat. The magnitude of the force the bat exerts on the ball as a function of time is shown in Figure 8.40. (a) Find the magnitude and direction of the impulse that the bat imparts to the ball. (b) Find the magnitude and direction of the ball's velocity just

after it is hit by the bat if the ball is initially (i) at rest, and (ii) moving to the right at 30.0 m/s.

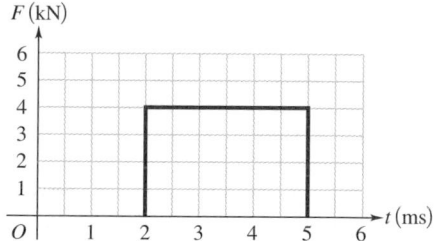

▲ **FIGURE 8.40** Problem 41.

42. •• Your little sister (mass 25.0 kg) is sitting in her little red wagon (mass 8.50 kg) at rest. You begin pulling her forward and continue accelerating her with a constant force for 2.35 s, at the end of which time she's moving at a speed of 1.80 m/s. (a) Calculate the impulse you imparted to the wagon and its passenger. (b) With what force did you pull on the wagon?

43. •• **Bone fracture.** Experimental tests have shown that bone
BIO will rupture if it is subjected to a force density of $1.0 \times 10^8 \, \text{N/m}^2$. Suppose a 70.0 kg person carelessly roller-skates into an overhead metal beam that hits his forehead and completely stops his forward motion. If the area of contact with the person's forehead is 1.5 cm², what is the greatest speed with which he can hit the wall without breaking any bone if his head is in contact with the beam for 10.0 ms?

44. •• A bat strikes a 0.145 kg baseball. Just before impact, the ball is traveling horizontally to the right at 50.0 m/s, and it leaves the bat traveling to the left at an angle of 30° above horizontal with a speed of 65.0 m/s. (a) What are the horizontal and vertical components of the impulse the bat imparts to the ball? (b) If the ball and bat are in contact for 1.75 ms, find the horizontal and vertical components of the average force on the ball.

8.7 Motion of the Center of Mass

45. • Calculate the location of the center of mass of the earth-moon system (that is, find the distance of the center of mass from the earth's center). Use data from Appendix E and assume the orbital radius of the moon is equal to the distance between the centers of the earth and the moon. What can you say about the position of the center of mass with respect to the earth's surface?

46. • Two small-sized objects are placed on a uniform 9.00 kg plastic beam 3.00 m long, as shown in Figure 8.41. Find the location of the center of mass of this system by setting $x = 0$ at (a) the left end of the beam and (b) the right end of the beam. (c) Do you get the same result both ways?

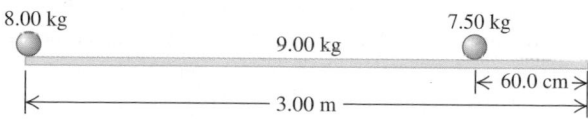

▲ **FIGURE 8.41** Problem 46.

47. •• **Detecting planets around other stars.** Roughly 500 planets have so far been detected beyond our solar system. This is accomplished by looking for the effect the planet has on the star. The star is not truly stationary; instead, it and its planets orbit around the center of mass of the system. Astronomers can measure this wobble in the position of a star. (a) For a star with the mass and size of our sun and having a planet with five times the mass of Jupiter, where would the center of mass of this system be located, relative to the center of the star, if the distance from the star to the planet was the same as the distance from Jupiter to our sun? (Consult Appendix E.) (b) If the planet had earth's mass, where would the center of mass of the system be located if the planet was just as far from the star as the earth is from the sun? (c) In view of your results in parts (a) and (b), why is it much easier to detect stars having large planets rather than small ones?

48. • Three odd-shaped blocks of chocolate have the following masses and center-of-mass coordinates: (1) 0.300 kg, (0.200 m, 0.300 m); (2) 0.400 kg, (0.100 m, −0.400 m); (3) 0.200 kg, (−0.300 m, 0.600 m). Find the coordinates of the center of mass of the system of three chocolate blocks.

49. •• A machine part consists of a thin, uniform 4.00-kg bar that is 1.50 m long, hinged perpendicular to a similar vertical bar of mass 3.00 kg and length 1.80 m. The longer bar has a small but dense 2.00-kg ball at one end (Fig. 8.42). By what distance will the center of mass of this part move horizontally and vertically if the vertical bar is pivoted counterclockwise through 90° to make the entire part horizontal?

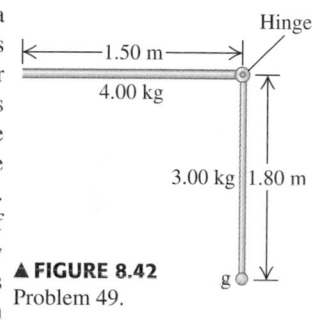

▲ **FIGURE 8.42** Problem 49.

50. •• **Changing your center of mass.** To keep the calculations
BIO fairly simple, but still reasonable, we shall model a human leg that is 92.0 cm long (measured from the hip joint) by assuming that the upper leg and the lower leg (which includes the foot) have equal lengths and that each of them is uniform. For a 70.0 kg person, the mass of the upper leg would be 8.60 kg, while that of the lower leg (including the foot) would be 5.25 kg. Find the location of the center of mass of this leg, relative to the hip joint, if it is (a) stretched out horizontally and (b) bent at the knee to form a right angle with the upper leg remaining horizontal.

51. • A 1200 kg station wagon is moving along a straight highway at 12.0 m/s. Another car, with mass 1800 kg and speed 20.0 m/s, has its center of mass 40.0 m ahead of the center of mass of the station wagon. (See Figure 8.43.) (a) Find the position of the center of mass of the system consisting of the two automobiles. (b) Find the magnitude of the total momentum of the system from the given data. (c) Find the speed of the center of mass of the system. (d) Find the total momentum of the system, using the speed of the center of mass. Compare your result with that of part (b).

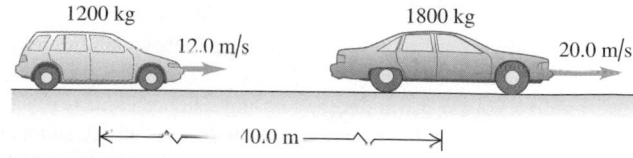

▲ **FIGURE 8.43** Problem 51.

52. •• **Walking in a boat.** A 45.0 kg woman stands up in a 60.0 kg canoe of length 5.00 m. She walks from a point 1.00 m from one

end to a point 1.00 m from the other end. (See Figure 8.44.) If the resistance of the water is negligible, how far does the canoe move during this process?

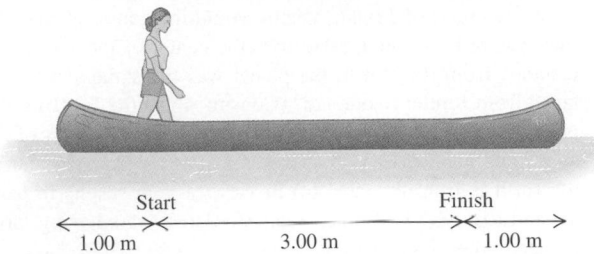

▲ **FIGURE 8.44** Problem 52.

8.8 Rocket Propulsion

53. • A small rocket burns 0.0500 kg of fuel per second, ejecting it as a gas with a velocity of magnitude 1600 m/s relative to the rocket. (a) What is the thrust of the rocket? (b) Would the rocket operate in outer space, where there is no atmosphere? If so, how would you steer it? Could you brake it?

54. • A rocket is fired in deep space, where gravity is negligible. If the rocket has an initial mass of 6000 kg and ejects gas at a relative velocity of magnitude 2000 m/s, how much gas must it eject in the first second to have an initial acceleration of 25.0 m/s²?

55. • A rocket is fired in deep space, where gravity is negligible. In the first second, it ejects $1/160$ of its mass as exhaust gas and has an acceleration of 15.0 m/s². What is the speed of the exhaust gas relative to the rocket?

56. •• In outer space, where gravity is negligible, a 75,000 kg rocket (including 50,000 kg of fuel) expels this fuel at a steady rate of 135 kg/s with a speed of 1200 m/s relative to the rocket. (a) Find the thrust of the rocket. (b) What are the initial acceleration and the maximum acceleration of the rocket? (c) After the fuel runs out, what happens to this rocket's acceleration? Does it (i) remain the same as it was just as the fuel ran out, (ii) suddenly become zero, or (iii) gradually drop to zero? Explain your reasoning. (d) After the fuel runs out, what happens to the rocket's speed? Does it (i) remain the same as it was just as the fuel ran out, (ii) suddenly become zero, or (iii) gradually drop to zero? Explain your reasoning.

57. •• A 70-kg astronaut floating in space in a 110-kg MMU (manned maneuvering unit) experiences an acceleration of 0.029 m/s² when he fires one of the MMU's thrusters. (a) If the speed of the escaping N₂ gas relative to the astronaut is 490 m/s, how much gas is used by the thruster in 5.0 s? (b) What is the thrust of the thruster?

General Problems

58. •• In 1.00 second an automatic paintball gun can fire 15 balls, each with a mass of 0.113 g, at a muzzle velocity of 88.5 m/s. Calculate the average recoil force experienced by the player who's holding the gun.

59. •• In a volcanic eruption, a 2400-kg boulder is thrown vertically upward into the air. At its highest point, it suddenly explodes (due to trapped gases) into two fragments, one being three times the mass of the other. The lighter fragment starts out with only horizontal velocity and lands 274 m directly north of the point of the explosion. Where will the other fragment land? Neglect any air resistance.

60. •• A 5.00 kg ornament is hanging by a 1.50 m wire when it is suddenly hit by a 3.00 kg missile traveling horizontally at 12.0 m/s. The missile embeds itself in the ornament during the collision. What is the tension in the wire immediately after the collision?

61. •• A stone with a mass of 0.100 kg rests on a frictionless, horizontal surface. A bullet of mass 2.50 g traveling horizontally at 500 m/s strikes the stone and rebounds horizontally at right angles to its original direction with a speed of 300 m/s. (a) Compute the magnitude and direction of the velocity of the stone after it is struck. (b) Is the collision perfectly elastic?

62. •• A steel ball with a mass of 40.0 g is dropped from a height of 2.00 m onto a horizontal steel slab. The ball rebounds to a height of 1.60 m. (a) Calculate the impulse delivered to the ball during the impact. (b) If the ball is in contact with the slab for 2.00 ms, find the average force on the ball during the impact.

63. •• A movie stuntman (mass 80.0 kg) stands on a window ledge 5.0 m above the floor (Fig. 8.45). Grabbing a rope attached to a chandelier, he swings down to grapple with the movie's villain (mass 70.0 kg), who is standing directly under the chandelier. (Assume that the stuntman's center of mass moves downward 5.0 m. He releases the rope just as he reaches the villain.) (a) With what speed do the entwined foes start to slide across the floor? (b) If the coefficient of kinetic friction of their bodies with the floor is $\mu_k = 0.250$, how far do they slide?

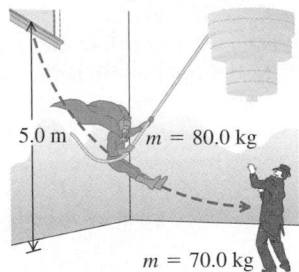

▲ **FIGURE 8.45** Problem 63.

64. •• **Tennis, anyone?** Tennis players sometimes leap into the air to return a volley. (a) If a 57 g tennis ball is traveling horizontally at 72 m/s (which does occur), and a 61 kg tennis player leaps vertically upward and hits the ball, causing it to travel at 45 m/s in the reverse direction, how fast will her center of mass be moving horizontally just after hitting the ball? (b) If, as is reasonable, her racket is in contact with the ball for 30.0 ms, what force does her racket exert on the ball? What force does the ball exert on the racket?

65. •• Two identical masses are released from rest in a smooth hemispherical bowl of radius R, from the positions shown in Fig. 8.46. You can ignore friction between the masses and the surface of the bowl. If they stick together when they collide, how high above the bottom of the bowl will the masses go after colliding?

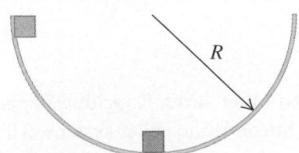

▲ **FIGURE 8.46** Problem 65.

66. ●● Two identical 1.50 kg masses are pressed against opposite ends of a light spring of force constant 1.75 N/cm, compressing the spring by 20.0 cm from its normal length. Find the speed of each mass when it has moved free of the spring on a frictionless horizontal lab table.

67. ●● A rifle bullet with mass 8.00 g strikes and embeds itself in a block with a mass of 0.992 kg that rests on a frictionless, horizontal surface and is attached to a coil spring. (See Figure 8.47.) The impact compresses the spring 15.0 cm. Calibration of the spring shows that a force of 0.750 N is required to compress the spring 0.250 cm. (a) Find the magnitude of the block's velocity just after impact. (b) What was the initial speed of the bullet?

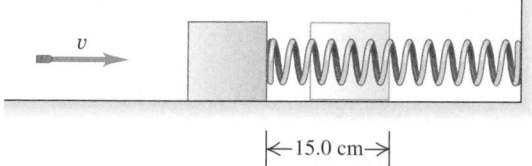

▲ FIGURE 8.47 Problem 67.

68. ●● A 5.00 g bullet traveling horizontally at 450 m/s is shot *through* a 1.00 kg wood block suspended on a string 2.00 m long. If the center of mass of the block rises a distance of 0.450 cm, find the speed of the bullet as it emerges from the block.

69. ●● Jonathan and Jane are sitting in a sleigh that is at rest on frictionless ice. Jonathan's weight is 800 N, Jane's weight is 600 N, and that of the sleigh is 1000 N. They see a poisonous spider on the floor of the sleigh and immediately jump off. Jonathan jumps to the left with a velocity (relative to the ice) of 5.00 m/s at 30.0° above the horizontal, and Jane jumps to the right at 7.00 m/s at 36.9° above the horizontal (relative to the ice). Calculate the sleigh's horizontal velocity (magnitude and direction) after they jump out.

70. ●● **Animal propulsion.** Squids and octopuses propel themBIO selves by expelling water. They do this by taking the water into a cavity and then suddenly contracting the cavity, forcing the water to shoot out of an opening. A 6.50 kg squid (including the water in its cavity) that is at rest suddenly sees a dangerous predator. (a) If this squid has 1.75 kg of water in its cavity, at what speed must it expel the water to suddenly achieve a speed of 2.50 m/s to escape the predator? Neglect any drag effects of the surrounding water. (b) How much kinetic energy does the squid create for this escape maneuver?

71. ●● The objects in Figure 8.48 are constructed of uniform wire bent into the shapes shown. Find the position of the center of mass of each one.

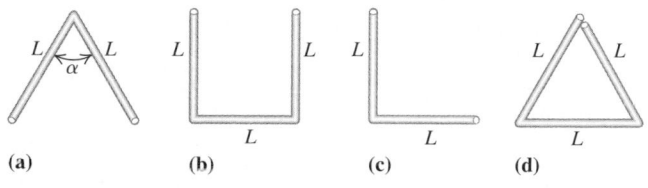

▲ FIGURE 8.48 Problem 71.

72. ●● **Origin of the moon.** Astronomers believe that our moon originated when a Mars-sized body collided with the earth over 4.5 billion years ago, knocking off matter that condensed to form the moon. To get some idea of how such a collision would affect the earth, assume that the collision occurred head-on (with the two bodies traveling in opposite directions), that the Mars-sized body merged completely with the earth and was originally the same mass as Mars, and that the earth's orbital speed was the same as it is now. Assume further that the speed of the colliding body was equal to the earth's escape velocity of 11 km/s (although it could have been greater). (a) By how much (in m/s) did such a collision change the earth's speed? Consult Appendix E. (b) How much thermal energy would such a collision create? (c) To how many megaton bombs is the energy in part (b) equivalent? (One ton of TNT releases 4.184×10^9 J of energy.)

73. ●● **Changing mass.** A railroad hopper car filled with sand is rolling with an initial speed of 15.0 m/s on straight, horizontal tracks. You can ignore frictional forces on the railroad car. The total mass of the car plus sand is 85,000 kg. The hopper door is not fully closed so sand leaks out the bottom. After 20 min, 13,000 kg of sand has leaked out. Then what is the speed of the railroad car? (Compare your analysis with that used to solve Problem 16.)

74. ●● **Forensic science.** Forensic scientists can measure the muzzle velocity of a gun by firing a bullet horizontally into a large hanging block that absorbs the bullet and swings upward. (See Figure 8.49.) The measured maximum angle of swing can be used to calculate the speed of the bullet. In one such test, a rifle fired a 4.20 g bullet into a 2.50 kg block hanging by a thin wire 75.0 cm long, causing the block to swing upward to a maximum angle of 34.7° from the vertical. What was the original speed of this bullet?

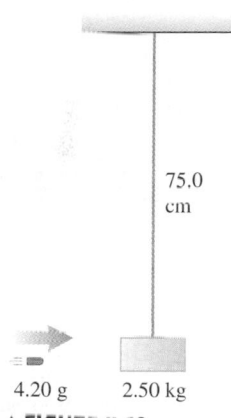

▲ FIGURE 8.49 Problem 74.

75. ●● A 20.0-kg lead sphere is hanging from a hook by a thin wire 3.50 m long, and is free to swing in a complete circle. Suddenly it is struck horizontally by a 5.00-kg steel dart that embeds itself in the lead sphere. What must be the minimum initial speed of the dart so that the combination makes a complete circular loop after the collision?

76. ●● A blue puck with mass 0.0400 kg, sliding with a velocity of magnitude 0.200 m/s on a frictionless, horizontal air table, makes a perfectly elastic, head-on collision with a red puck with mass m, initially at rest. After the collision, the velocity of the blue puck is 0.050 m/s in the same direction as its initial velocity. Find (a) the velocity (magnitude and direction) of the red puck after the collision; and (b) the mass m of the red puck.

77. ●● **The structure of the atom.** During 1910–1911, Sir Ernest Rutherford performed a series of experiments to determine the structure of the atom. He aimed a beam of alpha particles (helium nuclei, of mass 6.65×10^{-27} kg) at an extremely thin sheet of gold foil. Most of the alphas went right through with little deflection, but a small percentage bounced directly back. These results told him that the atom must be mostly empty space with an extremely small nucleus. The alpha particles that bounced back must have made a head-on collision with this nucleus. A typical speed for the alpha particles before the collision was 1.25×10^7 m/s, and the gold atom has a mass of 3.27×10^{-25} kg. Assuming (quite reasonably) elastic

collisions, what would be the speed after the collision of a gold atom if an alpha particle makes a direct hit on the nucleus?

78. •• **Rocket failure!** Just as it has reached an upward speed of 5.0 m/s during a vertical launch, a rocket explodes into two pieces. Photographs of the explosion reveal that the lower piece, with a mass one-fourth that of the upper piece, was moving downward at 3.0 m/s the instant after the explosion. (a) Find the speed of the upper piece just after the explosion. (b) How high does the upper piece go above the point where the explosion occurred?

79. •• In a common physics demonstration, two identical carts having rigid metal surfaces and equal speeds collide with each other. Each cart has a piece of Velcro® at one end and a spring at the other end. For each collision shown in the figure, find the magnitude and direction of the velocity of each cart after the collision. Are both collisions elastic? What causes one of them to be inelastic, and what happens to the "lost" kinetic energy?

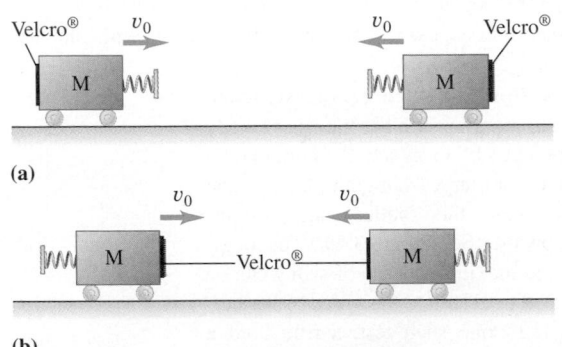

▲ **FIGURE 8.50** Problem 79.

80. •• A 7.0 kg shell at rest explodes into two fragments, one with a mass of 2.0 kg and the other with a mass of 5.0 kg. If the heavier fragment gains 100 J of kinetic energy from the explosion, how much kinetic energy does the lighter one gain?

81. •• **Radioactive beta decay.** *Beta decay* is a radioactive decay in which a neutron in the nucleus of an atom breaks apart (decays) to form a proton, an electron, and an antineutrino. The electron is also known as a *beta particle.* The proton remains in the nucleus, while the electron and antineutrino shoot out. Before the existence of the antineutrino was suspected, it was assumed that only the electron was emitted. Assuming (incorrectly) that there is no antineutrino, and that the neutron is initially at rest inside the nucleus, find (a) the ratio of the speed of the electron to the speed of the proton just after the decay and (b) the ratio of the kinetic energy of the electron to that of the proton just after the decay. Look up the necessary masses in Appendix E. (c) Use the results from parts (a) and (b) to explain why it is the electron, and not the proton, that shoots out of the nucleus. (*Note:* Electrons actually emerge with a range of speeds; this observation provided the first experimental evidence that an additional particle (the antineutrino) must also be emitted.)

82. •• A 15.0 g acorn falls from rest from the top of a 35.0-m-high oak tree. When it is halfway to the ground, a 135 g bird gliding horizontally at 75.0 cm/s scoops it up with its beak. Find (a) the horizontal and vertical components of the bird's velocity, and (b) the speed of the bird and the angle its velocity makes with the vertical, just after the bird scoops up the acorn.

83. •• **Accident analysis.** A 1500 kg sedan goes through a wide intersection traveling from north to south when it is hit by a 2200 kg SUV traveling from east to west. The two cars become enmeshed due to the impact and slide as one thereafter. On-the-scene measurements show that the coefficient of kinetic friction between the tires of these cars and the pavement is 0.75, and the cars slide to a halt at a point 5.39 m west and 6.43 m south of the impact point. How fast was each car traveling just before the collision?

Passage Problems

BIO Momentum and the squirting squid. An interesting use of "rocket power" is that used by cephalopods such as octopi and squids. These animals take in seawater and then squirt it out at high speed. A 2.5-kg squid can expel 0.25 kg of seawater (in a short burst of 0.20 s) with a speed of 600 cm/s.

84. What is the momentum of one squirt of water?
    A. 1.2 kg · m/s in the direction of the squirt
    B. 1.5 kg · m/s in the direction of the squirt
    C. 12 kg · m/s in the direction of the squirt
    D. 15 kg · m/s in the direction of the squirt

85. How much momentum does this give the squid?
    A. 1.2 kg · m/s in the direction of the squirt
    B. 1.5 kg · m/s in the direction of the squirt
    C. 15 kg · m/s in the direction of the squirt
    D. 1.2 kg · m/s in the direction opposite to the squirt
    E. 1.5 kg · m/s in the direction opposite to the squirt
    F. 15 kg · m/s in the direction opposite to the squirt

86. What would be the speed of the squid immediately after one squirt?
    A. 10 cm/s
    B. 20 cm/s
    C. 40 cm/s
    D. 60 cm/s
    E. 120 cm/s

87. After three quick squirts (in the same direction) what is the speed of the squid? Ignore the drag on the squid from the ocean water.
    A. 60 cm/s
    B. 120 cm/s
    C. 180 cm/s
    D. 200 cm/s
    E. 240 cm/s

9 Rotational Motion

In previous chapters, we would have considered these cyclists as so many particles in motion up the road. However, while each bicycle moves along the road, its wheels move in circles. In this chapter we start to look at objects that cannot be treated as particles. In particular, we will look at the motion of objects that *rotate*.

What do a Ferris wheel, a circular saw blade, and a ceiling fan have in common? None of these objects can be represented adequately as a moving *point;* each is a rigid object that *rotates* about an axis that is stationary in some inertial frame of reference. Rotation occurs on all size scales, from the motion of electrons in atoms to the motion of hurricanes, to the motions of entire galaxies.

In this chapter and the next, we'll consider the motion of objects that have a definite and unchanging size and shape; we'll neglect any stretching, twisting, and squeezing caused by the forces acting on the object. We call this idealized model a **rigid body.** (We could call it a rigid *object,* but physicists usually call it a rigid body.) We'll begin with kinematic language for describing rotational motion. Next, we'll look at the kinetic energy of rotation—the key to applying energy methods to rotational motion. Then we develop principles of rotational dynamics, analogous to Newton's laws for particles, to relate the forces on an object to its rotational motion.

9.1 Angular Velocity and Angular Acceleration

Let's think first about a rigid body that rotates about a stationary axis. (We'll also use the term *fixed axis.*) Both terms refer to an axis that is at rest in some inertial frame of reference and doesn't change direction relative to that frame. The body might be the shaft of a motor, a chunk of roast beef on a barbecue skewer, or a merry-go-round. Figure 9.1 shows a rigid body rotating about a stationary axis passing through point O (which may be the origin of a coordinate system) and perpendicular to the plane of the diagram. Line OP is fixed in the body and

Generic rigid body rotating counterclockwise around an axis at the origin

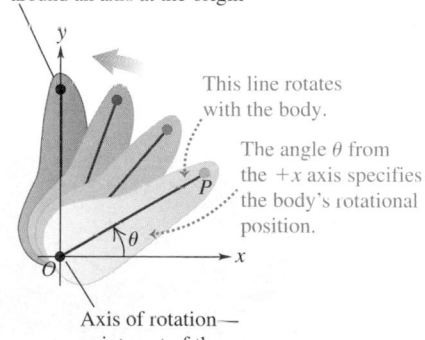

This line rotates with the body.

The angle θ from the $+x$ axis specifies the body's rotational position.

Axis of rotation— points out of the page

▲ **FIGURE 9.1** A rigid body rotating about an axis pointing out of the paper.

One radian is the angle at which the arc *s* has the same length as the radius *r*.

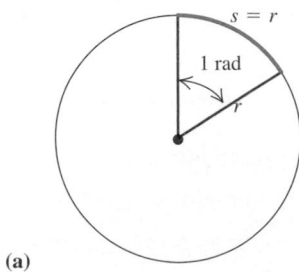

(a)

An angle θ in radians is the ratio of the arc length *s* to the radius *r*.

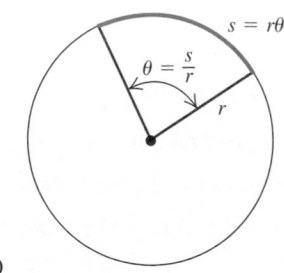

(b)

▲ **FIGURE 9.2** An angle θ in radians is defined as the ratio of the arc length *s* to the radius *r*.

Angular displacement $\Delta\theta$ of a rotating rigid body over a time interval Δt:

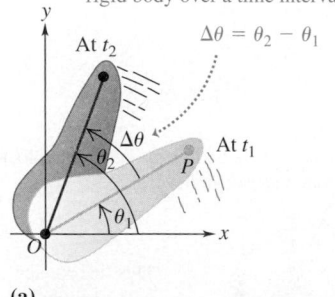

$\Delta\theta = \theta_2 - \theta_1$

(a)

(b)

▲ **FIGURE 9.3** (a) Angular displacement $\Delta\theta$ of a rotating body. (b) Every part of a rigid body has the same angular velocity.

rotates with it. The angle θ that this line makes with the $+x$ axis describes the rotational position of the body and serves as a *coordinate* for the body's rotational position. Thus, θ is analogous to x, which we used as the coordinate to specify position in our treatment of straight-line motion of a particle.

We usually measure the angle θ in **radians.** As is shown in Figure 9.2a, 1 radian (denoted 1 rad) is defined as the angle subtended at the center of a circle by an arc with a length equal to the radius of the circle. In Figure 9.2b, an angle θ is subtended by an arc of length s on a circle of radius r; θ (in radians) is equal to s divided by r:

$$\theta = \frac{s}{r}, \quad \text{or} \quad s = r\theta. \tag{9.1}$$

An angle measured in radians is the quotient of two lengths, so it is a pure number, without dimensions. If $s = 1.5$ m and $r = 1.0$ m, then $\theta = 1.5$, but we'll often write this as 1.5 rad to distinguish it from an angle measured in degrees or revolutions.

The circumference of a circle is 2π times the radius, so there are 2π (about 6.283) radians in one complete revolution $(360°)$. Therefore,

$$1 \text{ rad} = \frac{360°}{2\pi} = 57.3°.$$

Similarly, $180° = \pi$ rad, $90° = \pi/2$ rad, and so on.

We'll often encounter angles greater than $360°$; for a body in constant rotation, the angle θ may increase or decrease continuously.

Angular Velocity

Just as we described straight-line motion in terms of the rate of change of the coordinate x, we can describe rotational motion of a rigid body in terms of the rate of change of θ. In Figure 9.3a, a reference line OP in a rotating body makes an angle θ_1 with the $+x$ axis at time t_1. At a later time t_2, the angle has changed to θ_2. We define the **average angular velocity** ω_{av} of the body as follows:

Definition of average angular velocity
When a rigid body rotates through an angular displacement $\Delta\theta = \theta_2 - \theta_1$ in a time interval $\Delta t = t_2 - t_1$, the average angular velocity ω_{av} is defined as the ratio of the angular displacement to the time interval:

$$\omega_{av} = \frac{\theta_2 - \theta_1}{t_2 - t_1} = \frac{\Delta\theta}{\Delta t}. \tag{9.2}$$

Unit: rad/s

We can then define the **instantaneous angular velocity** ω as the limit of ω_{av} as Δt approaches zero:

$$\omega = \lim_{\Delta t \to 0} \frac{\Delta\theta}{\Delta t}. \tag{9.3}$$

Because the body is rigid, all lines in it rotate through the same angle in the same time (Figure 9.3b). Thus, **at any given instant, every part of a rigid body has the same angular velocity.** If the angle θ is measured in radians, the unit of angular velocity is 1 radian per second (1 rad/s). Other units, such as the revolution per minute (rev/min or rpm), are often used. Two useful conversions are

$$1 \text{ rev/s} = 2\pi \text{ rad/s} \quad \text{and} \quad 1 \text{ rev/min} = 1 \text{ rpm} = \frac{2\pi}{60} \text{ rad/s}. \quad (9.4)$$

Conceptual Analysis 9.1

Merry-go-round

A boy and a girl ride a turning merry-go-round. The boy is at the outer edge; the girl is near the axis of rotation. How do their angular velocities compare?

A. The boy has the greater angular velocity.
B. The girl has the greater angular velocity.
C. Both children have the same angular velocity.

SOLUTION An object's *angular* velocity, as opposed to its (straight-line) velocity, is determined by the change in angle of a line drawn from the axis of rotation to the object during any given unit of time (Figure 9.4). Both children have fixed posi-

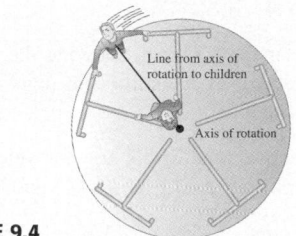

▶ **FIGURE 9.4**

tions on the same rotating object; they rotate through the same angle $\Delta\theta$ in the same time interval Δt. Their angular velocity is the same as that of the merry-go-round. Answer C is correct.

Angular Acceleration

When the angular velocity of a rigid body changes, the body has an angular acceleration. If ω_1 and ω_2 are the instantaneous angular velocities at times t_1 and t_2, respectively (Figure 9.5), we define the **average angular acceleration** α_{av} as follows:

Definition of average angular acceleration

The average angular acceleration α_{av} of a rotating rigid body is the change in angular velocity, $\Delta\omega$, divided by the time interval Δt:

$$\alpha_{av} = \frac{\omega_2 - \omega_1}{t_2 - t_1} = \frac{\Delta\omega}{\Delta t}. \quad (9.5)$$

Unit: rad/s^2

The average angular acceleration is the change in angular velocity divided by the time interval:

$$\alpha_{av} = \frac{\omega_2 - \omega_1}{t_2 - t_1} = \frac{\Delta\omega}{\Delta t}$$

ω_1 $\qquad$ ω_2

At t_1 $\qquad$ At t_2

▲ **FIGURE 9.5** The meaning of angular acceleration.

The **instantaneous angular acceleration** α is the limit of α_{av} as $\Delta t \to 0$:

$$\alpha = \lim_{\Delta t \to 0} \frac{\Delta\omega}{\Delta t}. \quad (9.6)$$

▶ **Application A boring tool?** The old-fashioned hand drill uses the principle of relative angular velocity to allow a carpenter to bore holes in very hard woods. Each part of the large wheel connected to the drill's handle has the same angular velocity. However, the large size of this wheel accommodates a large number of gear teeth compared with the few teeth that fit on the small wheel turning the drill bit. When these wheels are coupled together, a relatively small angular velocity of the handle can cause a much greater angular velocity of the drill bit, allowing it to penetrate dense woods with speed and precision.

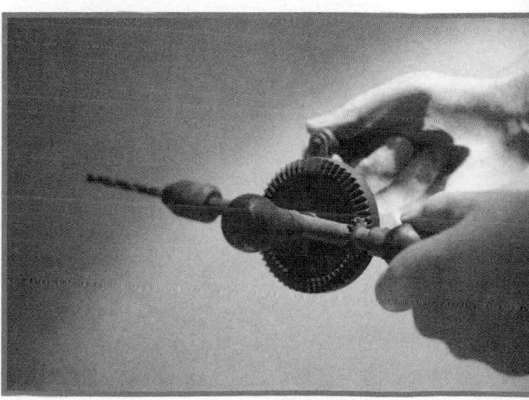

EXAMPLE 9.1 **Rotation of a compact disc**

A compact disc (CD) rotates at high speed while a laser reads data encoded in a spiral pattern. The disc has radius $R = 6.0$ cm $= 0.060$ m; when data are being read, it spins at 7200 rev/min. What is the CD's angular velocity ω in radians per second? How much time is required for it to rotate through 90°? If it starts from rest and reaches full speed in 4.0 s, what is its average angular acceleration?

SOLUTION

SET UP We make a sketch, showing the data (Figure 9.6). We need to be careful to use consistent units; we'll choose radians, meters, and seconds.

SOLVE To convert the disc's angular velocity ω to radians per second, we multiply by $(1\ \text{min}/60\ \text{s})$ and by $(2\pi\ \text{rad}/1\ \text{rev})$:

$$\omega = \frac{7200\ \text{rev}}{1\ \text{min}} = \frac{7200\ \text{rev}}{1\ \text{min}}\left(\frac{1\ \text{min}}{60\ \text{s}}\right)\left(\frac{2\pi\ \text{rad}}{1\ \text{rev}}\right) = 754\ \text{rad/s}.$$

From the definition of angular velocity $(\omega = \Delta\theta/\Delta t)$, we find the time Δt for the disc to turn through 90° $(\pi/2\ \text{rad})$:

$$\Delta t = \frac{\Delta\theta}{\omega} = \frac{\pi/2\ \text{rad}}{754\ \text{rad/s}} = 0.0021\ \text{s} = 2.1\ \text{ms}.$$

From the definition of angular acceleration $(\alpha_{av} = \Delta\omega/\Delta t)$, if the disc reaches a final angular velocity of 754 rad/s in 4.0 s (starting from rest), the average angular acceleration is

$$\alpha_{av} = \frac{\Delta\omega}{\Delta t} = \frac{754\ \text{rad/s}}{4.0\ \text{s}} = 189\ \text{rad/s}^2.$$

REFLECT A typical circular (table or radial-arm) saw found in woodworking shops runs at 3600 rev/min and reaches full

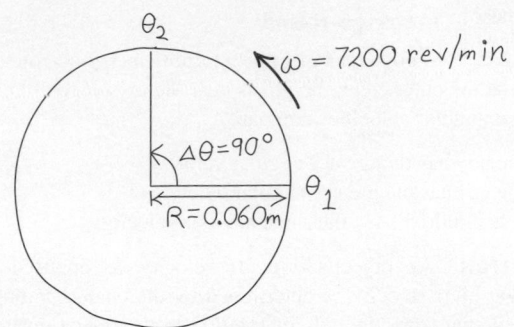

▲ **FIGURE 9.6** Our sketch for this problem.

speed about 1 s after it is turned on. The radius of a typical blade is 12 cm. Note, however, that the radius is not needed for the angular acceleration calculation. The angular acceleration is about twice that for the CD.

Practice Problem: An old-fashioned vinyl record is designed to turn at 33 rev/min. Find the angular velocity and period (the time for one revolution) of this record. *Answers:* $\omega = 3.5\ \text{rad/s}$, $T = 1.8\ \text{s}$.

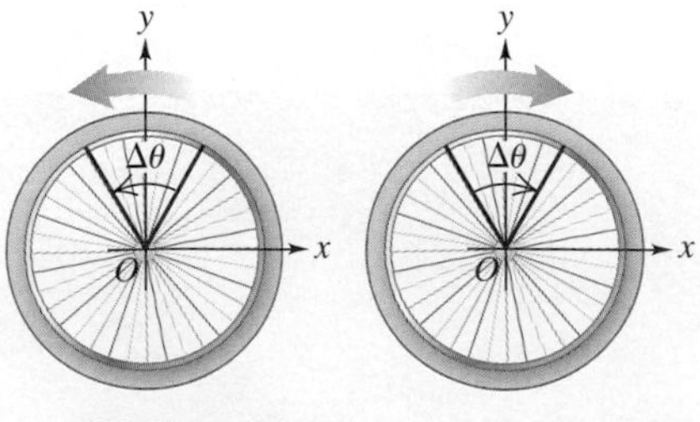

You've probably noticed that we're using Greek letters for angular kinematic quantities: θ for angular position, ω for angular velocity, and α for angular acceleration. These are analogous to x for position, v_x for the x component of velocity, and a_x for the x component of acceleration, respectively, in straight-line motion. In each case, velocity is the rate of change of position, and acceleration is the rate of change of velocity.

We've used the word *velocity* rather loosely in our discussion of angular quantities; we haven't said anything about vector aspects of these quantities. We'll return to this point in Chapter 10. But in the preceding discussion, the angular quantities *do* have algebraic signs. When an object rotates about an axis that is perpendicular to the plane of the drawing, it's customary to take the counterclockwise sense of rotation as positive and the clockwise sense as negative (Figure 9.7). For example, if the compact disk in Example 9.1 were to reverse its sense of rotation from 800 rad/s counterclockwise (positive) to 600 rad/s clockwise (negative) in 4.0 s, the average angular acceleration would be

$$\alpha_{av} = \frac{\omega_2 - \omega_1}{t_2 - t_1} = \frac{(-600\ \text{rad/s}) - (800\ \text{rad/s})}{4.0\ \text{s}} = -350\ \text{rad/s}^2.$$

Counterclockwise rotation positive:
$\Delta\theta > 0$, so
$\omega = \Delta\theta/\Delta t > 0$

Clockwise rotation negative:
$\Delta\theta < 0$, so
$\omega = \Delta\theta/\Delta t < 0$

▲ **FIGURE 9.7** The convention for the sign of angular velocity.

MasteringPHYSICS

ActivPhysics 7.7: Rotational Kinematics

9.2 Rotation with Constant Angular Acceleration

In Chapter 2, we found that analyzing straight-line motion is particularly simple when the acceleration is constant. This is also true of rotational motion: When the angular acceleration is constant, we can derive equations for angular velocity and angular position using exactly the same procedure that we used in Section 2.4. In fact, the equations that we are about to derive are identical to Equations 2.8, 2.12, and 2.13 if we replace x with θ, v_x with ω, and a_x with α. We suggest that you review Section 2.4 before continuing.

Let's proceed to derive relationships among the various angular quantities, assuming a constant angular acceleration α. We define θ_0 to be the angular position at time $t = 0$, θ the angular position at any later time t, ω_0 the angular velocity at time $t = 0$, and ω the angular velocity at any later time t. The angular acceleration α is constant and equal to the average value for any interval. Using Equation 9.5 with the interval from 0 to t, we find that

$$\alpha = \frac{\omega - \omega_0}{t - 0}, \quad \text{or}$$

$$\omega = \omega_0 + \alpha t. \quad \text{(angular velocity as function of time)} \tag{9.7}$$

The product αt is the total change in ω between $t = 0$ and the later time t; the angular velocity ω at time t is the sum of the initial value ω_0 and this total change.

With constant angular acceleration, the angular velocity changes at a uniform rate, so its average value ω_{av} between 0 and t is the average of the initial and final values:

$$\omega_{av} = \frac{\omega_0 + \omega}{2}. \tag{9.8}$$

We also know that ω_{av} is the total angular displacement $(\theta - \theta_0)$ divided by the time interval $(t - 0)$:

$$\omega_{av} = \frac{\theta - \theta_0}{t - 0}. \tag{9.9}$$

When we equate Equations 9.8 and 9.9 and multiply the result by t, we get

$$\theta - \theta_0 = \tfrac{1}{2}(\omega_0 + \omega)t. \tag{9.10}$$

Note that this relation doesn't contain the angular acceleration α.

To obtain a relation between θ and t that doesn't contain ω, we substitute Equation 9.7 into Equation 9.10:

$$\theta - \theta_0 = \tfrac{1}{2}(\omega_0 + \omega_0 + \alpha t)t,$$

$$\theta = \theta_0 + \omega_0 t + \tfrac{1}{2}\alpha t^2. \tag{9.11}$$

(angular position with constant angular acceleration)

That is, if at the initial time $t = 0$, the body is at angular position θ_0 and has angular velocity ω_0, then its angular position θ at any later time t is the sum of three terms: its initial angular position θ_0, plus the rotation $\omega_0 t$ it would have if the angular velocity were constant, plus an additional rotation $\tfrac{1}{2}\alpha t^2$ caused by the changing angular velocity.

Following the same procedure as in Section 2.4, we can combine Equations 9.7 and 9.11 to obtain a relation between θ and ω that doesn't contain t. We invite you to work out the details, following the same procedure we used to get Equation 2.13. In fact, because of the perfect analogy between straight-line and rotational quantities, we can simply take Equation 2.13 and replace each straight-line quantity by its rotational analog; we get

$$\omega^2 = \omega_0^2 + 2\alpha(\theta - \theta_0). \tag{9.12}$$

(angular velocity in terms of angular displacement)

Keep in mind that all of these results are valid *only* when the angular acceleration α is constant; be careful not to try to apply them to problems in which α is

TABLE 9.1 Comparison of linear and angular motion with constant acceleration

Straight-line motion with constant linear acceleration	Fixed-axis rotation with constant angular acceleration
$a = $ constant	$\alpha = $ constant
$v = v_0 + at$ (2.8)	$\omega = \omega_0 + \alpha t$ (9.7)
$x = x_0 + v_0 t + \frac{1}{2}at^2$ (2.12)	$\theta = \theta_0 + \omega_0 t + \frac{1}{2}\alpha t^2$ (9.11)
$v^2 = v_0^2 + 2a(x - x_0)$ (2.13)	$\omega^2 = \omega_0^2 + 2\alpha(\theta - \theta_0)$ (9.12)
$x - x_0 = \frac{1}{2}(v + v_0)t$ (2.14)	$\theta - \theta_0 = \frac{1}{2}(\omega + \omega_0)t$ (9.10)

not constant. Table 9.1 compares Equations 9.7, 9.10, 9.11, and 9.12, for fixed-axis rotation with constant angular acceleration, to the analogous equations for straight-line motion with constant acceleration.

EXAMPLE 9.2 Rotation of a bicycle wheel

The angular velocity of the rear wheel of a stationary exercise bike is 4.00 rad/s at time $t = 0$, and its angular acceleration is constant and equal to 2.00 rad/s^2. A particular spoke coincides with the $+x$ axis at time $t = 0$ (Figure 9.8). What angle does this spoke make with the $+x$ axis at time $t = 3.00$ s? What is the wheel's angular velocity at this time?

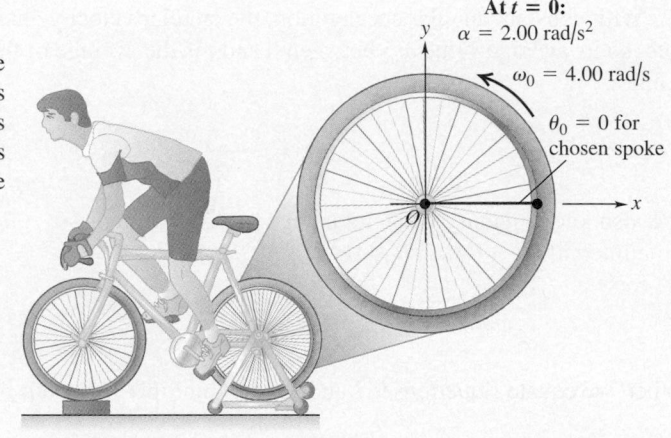

At $t = 0$:
$\alpha = 2.00$ rad/s^2
$\omega_0 = 4.00$ rad/s
$\theta_0 = 0$ for chosen spoke

▲ FIGURE 9.8

SOLUTION

SET UP We take $\theta = 0$ at the position where the chosen spoke is horizontal, and we take the counterclockwise sense of rotation to be positive. The initial conditions are $\theta_0 = 0$, $\omega_0 = 4.00$ rad/s, and $\alpha = 2.00$ rad/s^2.

SOLVE We can use Equations 9.11 and 9.7 to find θ and ω at any time. The angle θ is given as a function of time by Equation 9.11:

$$\theta = \theta_0 + \omega_0 t + \frac{1}{2}\alpha t^2$$
$$= 0 + (4.00 \text{ rad/s})(3.00 \text{ s}) + \frac{1}{2}(2.00 \text{ rad/s}^2)(3.00 \text{ s})^2$$
$$= 21.0 \text{ rad} = 21.0 \text{ rad}\left(\frac{1 \text{ rev}}{2\pi \text{ rad}}\right) = 3.34 \text{ rev}.$$

The wheel has turned through three complete revolutions plus an additional 0.34 rev, or $(0.34 \text{ rev})(2\pi \text{ rad/rev}) = 2.14 \text{ rad} = 123°$. Thus, the chosen spoke is at an angle of 123° with the $+x$ axis.

The wheel's angular velocity ω is given as a function of time by Equation 9.7, $\omega = \omega_0 + \alpha t$. At time $t = 3.0$ s,

$$\omega = 4.00 \text{ rad/s} + (2.00 \text{ rad/s}^2)(3.00 \text{ s}) = 10.0 \text{ rad/s}.$$

Alternatively, from Equation 9.12, we get

$$\omega^2 = \omega_0^2 + 2\alpha(\theta - \theta_0)$$
$$= (4.00 \text{ rad/s})^2 + 2(2.00 \text{ rad/s}^2)(21.0 \text{ rad}) = 100 \text{ rad}^2/\text{s}^2,$$
$$\omega = 10.0 \text{ rad/s} = 1.59 \text{ rev/s}.$$

REFLECT Note the step-by-step similarity between these equations and the kinematic relations for straight-line motion.

Practice Problem: Suppose we double the angular acceleration. What would be the new values for θ and ω? *Answers:* $\theta = 30.0$ rad, $\omega = 16.0$ rad/s.

MasteringPHYSICS

PhET: Ladybug Revolution

9.3 Relationship between Linear and Angular Quantities

In order to develop the principles of dynamics for rigid bodies, we need to be able to find the velocity and acceleration of a particular point in a rotating rigid body. For example, to find the kinetic energy of a rotating object, we have to start from $K = \frac{1}{2}mv^2$ for a particle. That requires knowing v for a particle in a rigid body. And in Chapter 10 we'll need to be able to find the acceleration of a point in a rigid body in order to work out general relations between force and motion for rotation, starting from $\Sigma\vec{F} = m\vec{a}$.

So it's worthwhile to develop general relations between the angular velocity and acceleration of a rigid body rotating about a fixed axis, on the one hand, and the velocity and acceleration of a specific point in the body, on the other. We'll sometimes use the terms *linear velocity* and *linear acceleration* for the familiar quantities that we defined in Chapters 2 and 3, to make sure we distinguish clearly between these and the *angular* quantities introduced in this chapter.

When a rigid body rotates about a fixed axis, every particle in the body moves in a circular path that is centered on the axis and lies in a plane perpendicular to the axis. The speed of a particle is directly proportional to the body's angular velocity; the faster the body rotates, the greater is the speed of each particle. In Figure 9.9, point P is a constant distance r away from the axis of rotation, and it moves in a circle of radius r. At any time, the angle θ (in radians) and the arc length s are related by the formula

$$s = r\theta.$$

When the angle θ increases by a small amount $\Delta\theta$ during a time interval Δt, the particle moves through an arc length $\Delta s = r\,\Delta\theta$. Thus the average speed of the particle is given by

$$v_{av} = \frac{\Delta s}{\Delta t} = r\frac{\Delta\theta}{\Delta t} = r\omega_{av}.$$

In the limit as $\Delta t \to 0$, this equation becomes

$$v = r\omega. \tag{9.13}$$

We'll continue to use the term *speed* to mean the magnitude of a particle's linear velocity; the *direction* of the particle's linear velocity is tangent to its circular path at each point.

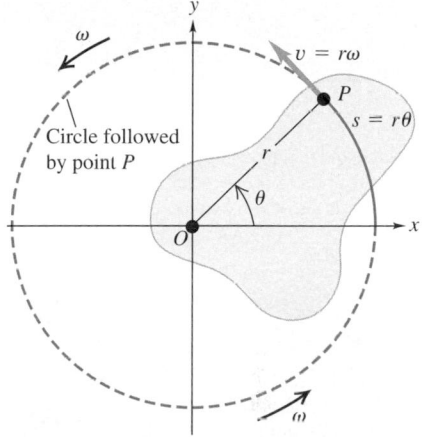

▲ **FIGURE 9.9** The distance s traversed by a point P on a rigid body equals $r\theta$ if θ is measured in radians.

Quantitative Analysis 9.2

Rotating disk

A disk rotates about an axis perpendicular to its plane. Point B on the disk is three times farther from the axis than point A. If the speed of point B is v, what is the speed of point A?

A. $3v$ B. $\frac{1}{3}v$ C. v

SOLUTION Speed in this case means the magnitude of *linear* velocity, not the magnitude of *angular* velocity. Because the disk is a rigid object, all points on the disk rotate with the same angular velocity ω. The speed of a point is directly proportional to its distance r from the axis of rotation: $v = r\omega$. Because point A is one-third the distance from the axis as point B, its speed is $v/3$ (choice B).

We can represent the acceleration $\vec{a}$ of a particle in terms of its radial and tangential components (Figure 9.10). By definition, the radial component a_{rad} at each point is along a radial line from the particle to the center of its circular path. The tangential component a_{tan} is tangent to the path at each point (and thus parallel or antiparallel to the velocity at each point). Because the particles in a rotating rigid body move in circular paths, every particle *must* have a radial component of acceleration. When the body undergoes angular acceleration, every particle must also have a tangential component of acceleration.

To find the **tangential component of acceleration**, a_{tan}, we note that if the angular velocity ω changes by an amount $\Delta\omega$, the particle's speed changes by an amount Δv given by

$$\Delta v = r\,\Delta\omega.$$

◀ **Application** **Not so fast.** This is a velocipede, one of the earliest bicycles produced. When invented in the 1800s, it was more of a curiosity than a practical means of transportation. Because the pedals attach directly to the front wheel, the wheel turns only once for each revolution of the pedals. Therefore, the only way to achieve a reasonable forward speed at a feasible rate of pedaling is to use a huge front wheel. These bicycles are difficult to mount and quite dangerous. Bicycles did not become widely used for transportation until the development of the chain-and-sprocket mechanism, which allowed each turn of the pedals to rotate a smaller wheel more than once.

This corresponds to a component of acceleration a_{tan} tangent to the particle's circular path. If these changes take place in a small time interval Δt, then

$$(a_{tan})_{av} = \frac{\Delta v}{\Delta t} = r\frac{\Delta \omega}{\Delta t},$$

and in the limit as $\Delta t \to 0$,

$$a_{tan} = r\alpha, \qquad (9.14)$$

where a_{tan} is the **tangential component of instantaneous acceleration.** This component is parallel to the instantaneous velocity at each point; it acts to change the *magnitude* of velocity of the particle (i.e., its speed). It is equal to the rate of change of speed. This component of a particle's acceleration is always tangent to the circular path of the particle.

The **radial component of acceleration,** a_{rad}, of the particle is associated with change in *direction* of the particle's velocity. In Section 3.4, we worked out the relation $a_{rad} = v^2/r$ for a particle moving in a circle with radius r. We can express this in terms of ω by using Equation 9.13:

$$a_{rad} = \frac{v^2}{r} = \omega^2 r. \qquad (9.15)$$

This is true at each instant, even when ω and v are not constant.

The vector sum of the radial and tangential components of acceleration of a particle in a rotating body is the linear acceleration $\vec{a}$ (Figure 9.10). We use the modifier *linear* to emphasize that $\vec{a}$ is the rate of change of velocity $\vec{v}$, *not* the rate of change of angular velocity. The radial and tangential components are always perpendicular, so the magnitude of $\vec{a}$ is given by

$$|\vec{a}| = a = \sqrt{a_{rad}^2 + a_{tan}^2}.$$

NOTE ▶ Remember that Equation 9.1 $(s = r\theta)$ is valid *only* when θ is measured in radians. The same is true of any equation derived from that relation, including Equations 9.13, 9.14, and 9.15. When you use these equations, you *must* express the angular quantities in radians, not revolutions or degrees. ◀

Radial and tangential acceleration components:
• a_{rad} is point P's centripetal acceleration.
• a_{tan} means that P's rotation is speeding up (the body has angular acceleration).

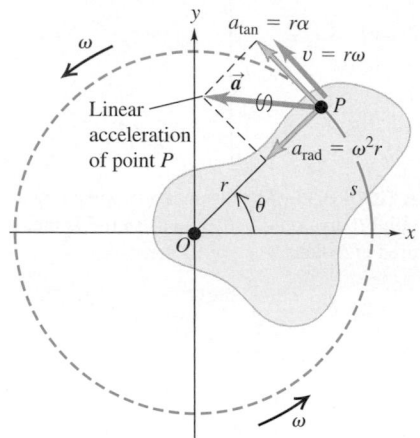

▲ **FIGURE 9.10** The components of acceleration for a point P on a rigid body whose angular rotation is speeding up.

Conceptual Analysis 9.3

Angular and centripetal acceleration

A disk spins with increasing angular velocity. Point B on the disk is twice as far from the axis of rotation as point A is. Which of the following statements is or are correct?

A. The angular acceleration of A is greater than that of B.
B. The radial acceleration of A is greater than that of B.
C. A and B have the same angular acceleration.
D. A and B have the same radial acceleration.

SOLUTION The disk's angular velocity is increasing, so the disk has an angular acceleration. All points on the disk rotate through the same angle in a given time interval and thus have the same angular velocity and angular acceleration. Therefore, choice A is incorrect and C is correct. The radial acceleration of each point is $a_{rad} = \omega^2 r$. Point B is farther from the axis than point A is, so its radial acceleration is greater than that of A, and answers B and D are incorrect. The only correct choice is C.

EXAMPLE 9.3 **Throwing a discus**

A discus thrower turns with angular acceleration $\alpha = 50 \text{ rad/s}^2$, moving the discus in a circle of radius 0.80 m (Figure 9.11). Find the radial and tangential components of acceleration of the discus (modeled as a point) and the magnitude of its acceleration at the instant when the angular velocity is 10 rad/s.

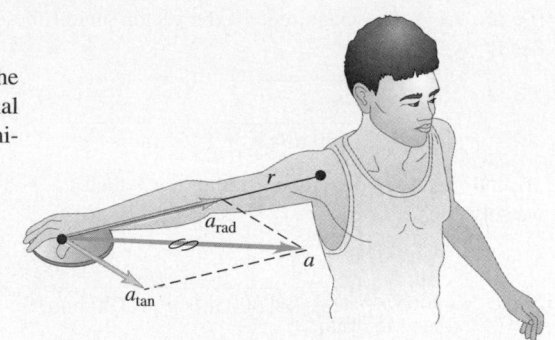

▲ **FIGURE 9.11**

SOLUTION

SET UP As shown in Figure 9.12, we model the thrower's arm as a rigid body, so r is constant. (This may not be completely realistic.) We model the discus as a particle moving in a circular path.

SOLVE The components of acceleration are given by Equations 9.14 and 9.15:

$$a_{\text{rad}} = \omega^2 r = (10 \text{ rad/s})^2 (0.80 \text{ m}) = 80 \text{ m/s}^2,$$
$$a_{\text{tan}} = r\alpha = (0.80 \text{ m})(50 \text{ rad/s}^2) = 40 \text{ m/s}^2.$$

The magnitude of the acceleration vector is

$$a = \sqrt{a_{\text{rad}}^2 + a_{\text{tan}}^2} = 89 \text{ m/s}^2.$$

REFLECT The magnitude of the acceleration is about nine times the acceleration due to gravity; the corresponding force supplied by the thrower's arm must be about nine times the weight of the discus. Note that we have dropped the unit *radian* from our final results; we can do this because a radian is a dimensionless quantity.

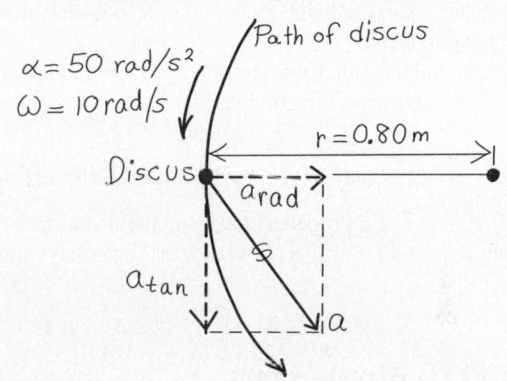

▲ **FIGURE 9.12** Our sketch for this problem.

Practice Problem: What is the direction of the acceleration vector of the discus if the angular acceleration is 25 rad/s²? *Answer:* 14° from a radial line toward the center of the circle.

EXAMPLE 9.4 **Designing a propeller**

You are asked to design an airplane propeller to turn at 1500 rev/min. The forward airspeed of the plane is to be 90 m/s, and the speed of the tips of the propeller blades through the air must not exceed 300 m/s, the speed of sound in very cold air (Figure 9.13). Determine the maximum radius of the propeller.

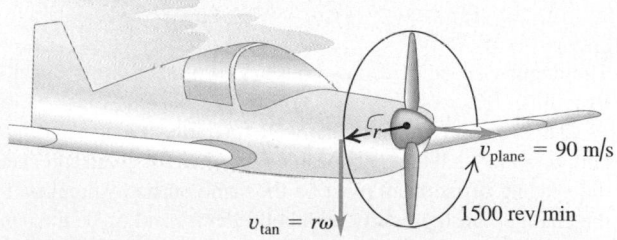

▲ **FIGURE 9.13**

SOLUTION

SET UP Figure 9.14 shows how we sketch this problem. The velocity of each propeller tip is the vector sum of the velocity $\vec{v}_{\text{plane}}$ of the plane through the air (magnitude 90 m/s) and the tangential velocity $\vec{v}_{\text{tan}}$ of the tip relative to the plane (magnitude $v_{\text{tan}} = r\omega$, where r is to be determined).

SOLVE First we convert the angular velocity ω to rad/s:

$$\omega = 1500 \text{ rpm} = \left(\frac{1500 \text{ rev}}{1 \text{ min}}\right)\left(\frac{2\pi \text{ rad}}{1 \text{ rev}}\right)\left(\frac{1 \text{ min}}{60 \text{ s}}\right) = 157 \text{ rad/s}.$$

The tangential velocity $v_{\text{tan}} = r\omega$ of the propeller tip is perpendicular to the forward velocity v_{plane} of the plane through the air

Continued

(Figure 9.14). The magnitude of the vector sum of these velocities is

$$v_{tip} = \sqrt{v_{plane}^2 + v_{tan}^2} = \sqrt{v_{plane}^2 + (r\omega)^2}$$
$$= \sqrt{(90 \text{ m/s})^2 + (157 \text{ rad/s})^2 r^2}.$$

To find the desired propeller radius (at which $v_{tip} = 300$ m/s), we solve this equation for r:

$$(90 \text{ m/s})^2 + (157 \text{ rad/s})^2 r^2 = (300 \text{ m/s})^2,$$
$$r = \frac{1}{157 \text{ rad/s}} \sqrt{(300 \text{ m/s})^2 - (90 \text{ m/s})^2}$$
$$= 1.82 \text{ m}.$$

REFLECT If the radius r is greater than 1.82 m, the airspeed at the propeller tips is greater than 300 m/s, meaning that it is supersonic in cold air.

We can find the radial acceleration of a propeller tip with our calculated radius, using Equation 9.15:

$$a_{rad} = \omega^2 r$$
$$= (157 \text{ rad/s})^2 (1.82 \text{ m}) = 4.49 \times 10^4 \text{ m/s}^2.$$

From $\sum \vec{F} = m\vec{a}$, the propeller must provide a centripetal force of magnitude 4.49×10^4 N (about 10,000 pounds) for each

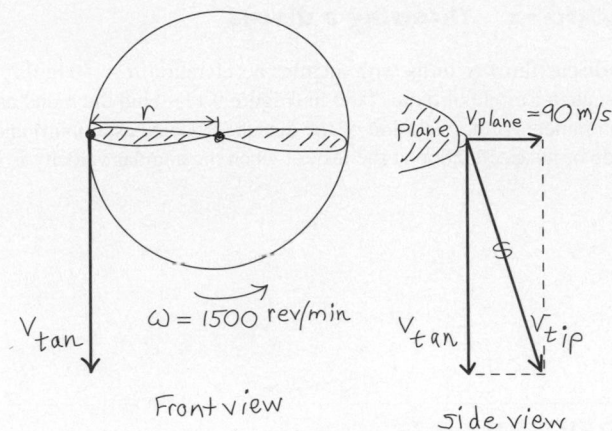

▲ **FIGURE 9.14** Our sketches for this problem.

kilogram of tip! The propeller has to be made out of tough material.

Practice Problem: If the plane's forward airspeed is 180 m/s and $\omega = 3000$ rev/min, what is the maximum propeller radius? *Answer: 0.764 m.*

EXAMPLE 9.5   Bicycle gears

How are the angular speeds of the two bicycle sprockets in Figure 9.15 related to their respective numbers of teeth, N_A and N_B?

SOLUTION

SET UP AND SOLVE The chain does not slip or stretch; therefore, it moves at the same tangential speed v on both sprockets. We use the relation $v = r\omega$ for each sprocket and equate the two tangential speeds:

$$v = r_A \omega_A = r_B \omega_B, \quad \text{or} \quad \frac{\omega_B}{\omega_A} = \frac{r_A}{r_B}.$$

The angular speed of each sprocket is inversely proportional to its radius.

This relationship holds as well for pulleys connected by a belt, provided that the belt doesn't slip. Also, for chain sprockets, the spacing of the teeth must be the same on both sprockets for the chain to mesh properly with both. Let N_A and N_B be the numbers of teeth on the two sprockets; then

$$\frac{2\pi r_A}{N_A} = \frac{2\pi r_B}{N_B}, \quad \text{or} \quad \frac{r_A}{r_B} = \frac{N_A}{N_B}.$$

Combining this equation with the previous one, we get

$$\frac{\omega_B}{\omega_A} = \frac{N_A}{N_B}.$$

REFLECT The angular speed of each sprocket is inversely proportional to its number of teeth. On an 18-speed bike, you get the highest angular speed ω_B of the rear wheel for a given pedaling

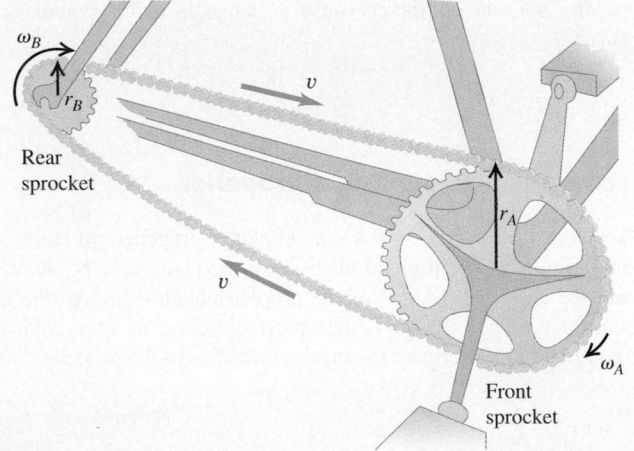

▲ **FIGURE 9.15**

rate ω_A using the larger-radius front sprocket and the smallest-radius rear sprocket, giving the maximum front-to-back-teeth ratio N_A/N_B.

Practice Problem: If the front sprocket has four times as many teeth as the rear sprocket and the rear wheel radius is 0.33 m, how fast is the bike moving when the pedaling rate is $\omega_A = 0.80$ rev/s? *Answer: 6.6 m/s.*

9.4 Kinetic Energy of Rotation and Moment of Inertia

A rotating rigid body consists of mass in motion, so it has kinetic energy. We can express this kinetic energy in terms of the body's angular velocity and a new quantity called moment of inertia, that we'll define shortly. To develop this relationship, we think of the body as made up of a lot of particles, with masses m_A, m_B, m_C, ..., at distances r_A, r_B, r_C, ..., from the axis of rotation. The particles don't necessarily all lie in the same plane, so we specify that each r is the perpendicular distance from the particle to the axis.

When a rigid body rotates about a fixed axis, the speed v of a typical particle is given by Equation 9.13, $v = r\omega$, where ω is the body's angular velocity in radians per second. Particle A has a perpendicular distance r_A from the axis of rotation and has a speed v_A given by $v_A = r_A\omega$. Different particles have different values of r and v, but ω is the same for all. (Otherwise the body wouldn't be rigid.) The kinetic energy K_A of particle A can be expressed as

$$K_A = \tfrac{1}{2}m_A v_A^2 = \tfrac{1}{2}m_A\left(r_A^2\omega^2\right).$$

The total kinetic energy K of the body is the sum of the kinetic energies of all its particles:

$$\begin{aligned}
K &= \tfrac{1}{2}m_A v_A^2 + \tfrac{1}{2}m_B v_B^2 + \tfrac{1}{2}m_C v_C^2 + \cdots \\
&= \tfrac{1}{2}m_A\left(r_A^2\omega^2\right) + \tfrac{1}{2}m_B\left(r_B^2\omega^2\right) + \tfrac{1}{2}m_C\left(r_C^2\omega^2\right) + \cdots \\
&= \tfrac{1}{2}\left(m_A r_A^2 + m_B r_B^2 + m_C r_C^2 + \cdots\right)\omega^2.
\end{aligned}$$

The quantity in parentheses in the last line, obtained by multiplying the mass of each particle by the square of its distance from the axis of rotation and adding all the products together, is called the **moment of inertia** of the body, denoted by I:

Definition of moment of inertia
A body's moment of inertia, I, describes how its mass is distributed in relation to an axis of rotation:

$$I = m_A r_A^2 + m_B r_B^2 + m_C r_C^2 + \cdots. \tag{9.16}$$

Unit: $\text{kg} \cdot \text{m}^2$

In terms of moment of inertia I, the kinetic energy K of a rotating rigid body is

$$K = \tfrac{1}{2}I\omega^2. \tag{9.17}$$

This is analogous to the expression $K = \tfrac{1}{2}mv^2$ for the kinetic energy of a particle; thus moment of inertia is analogous to mass, and angular velocity ω is analogous to speed v. This kinetic energy is not a new form of energy; it's the same physical quantity that we expressed as $\tfrac{1}{2}mv^2$ for a single particle. But Equation 9.17 is much easier to use when we have to find the kinetic energy of a rotating body.

MasteringPHYSICS

ActivPhysics 7.7: Rotational Kinematics

▶ **Application Faster, not heavier.** One way to store energy is in a rotating disk, called a flywheel. Traditional flywheels, like the one shown here, are massive, carefully machined steel disks. The danger of such flywheels is that if the disk breaks, the parts fly off like cannonballs. As we've learned, however, the flywheel's kinetic energy is proportional to the *square* of its angular velocity: $K = \tfrac{1}{2}I\omega^2$. Therefore, a flywheel's angular velocity is more important than its mass (or moment of inertia) for storing energy. Engineers are now developing lightweight composite flywheels that rotate at very high speed. These flywheels break by unraveling, not by fragmenting into projectiles, so they pose less danger than traditional flywheels. Future generations of hybrid autos may use such flywheels in place of batteries to store energy.

NOTE ▶ As in our previous discussion, ω *must* be measured in radians per second, not revolutions or degrees per second. ◀

EXAMPLE 9.6 **An abstract sculpture**

An artist is designing a part for a kinetic sculpture. The part consists of three massive disks connected by light supporting rods (Figure 9.16) that can be considered massless. Find the moment of inertia about an axis passing through disks B and C and the moment of inertia about an axis that passes through disk A and is perpendicular to the plane of that disk. If the object rotates about the axis through disk A with angular velocity 4.0 rad/s, find its kinetic energy.

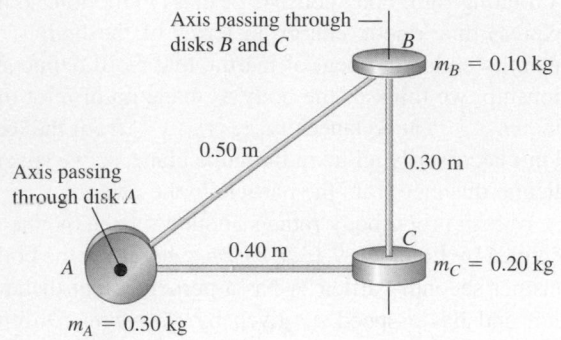

▲ **FIGURE 9.16**

SOLUTION

SET UP For each calculation, we need to find the perpendicular distance from each disk to the axis of rotation. We'll model each disk as a point. For the axis passing through disks B and C, their distance from that axis is zero. Similarly, the distance of disk A from the axis passing through it is zero.

SOLVE The moment of inertia is defined as $I = m_A r_A^2 + m_B r_B^2 + m_C r_C^2$, where each r is measured from the chosen axis of rotation. For axis BC, disks B and C both lie *on* the axis $(r_B = r_C = 0)$, so neither contributes to the moment of inertia. Only disk A contributes; $m_A = 0.30$ kg, $r_A = 0.40$ m, and we find that

$$I_{BC} = m_A r_A^2 = (0.30 \text{ kg})(0.40 \text{ m})^2 = 0.048 \text{ kg} \cdot \text{m}^2.$$

For the axis through point A perpendicular to the plane of the diagram, disk A lies on the axis, so $r_A = 0$. The masses are the same as before, but now

$$r_B = 0.50 \text{ m} \qquad \text{and} \qquad r_C = 0.40 \text{ m}.$$

The moment of inertia for this axis is

$$\begin{aligned} I_A &= m_B r_B^2 + m_C r_C^2 \\ &= (0.10 \text{ kg})(0.50 \text{ m})^2 + (0.20 \text{ kg})(0.40 \text{ m})^2 \\ &= 0.057 \text{ kg} \cdot \text{m}^2. \end{aligned}$$

If the object rotates about this axis with angular velocity $\omega = 4.0$ rad/s, its kinetic energy is

$$K = \tfrac{1}{2} I_A \omega^2 = \tfrac{1}{2}(0.057 \text{ kg} \cdot \text{m}^2)(4.0 \text{ rad/s})^2 = 0.46 \text{ J}.$$

REFLECT The moment of inertia of an object depends on the location of the axis. We can't just say, "The moment of inertia of this body is 0.048 kg · m²." We have to say, "The moment of inertia of this object *with respect to axis BC* is 0.048 kg · m²." Note that whenever a particle lies on the axis of rotation, $r = 0$, so that particle does not contribute to the moment of inertia.

Practice Problem: Find the moment of inertia and the corresponding rotational kinetic energy if the sculpture rotates with angular velocity $\omega = 4.0$ rad/s about an axis through point B, perpendicular to the plane of the diagram. *Answers:* 0.093 kg · m², 0.74 J.

In Example 9.6, we represented the body as several point masses and we evaluated the sum in Equation 9.16 directly. When the body is a continuous distribution of matter, such as a solid cylinder or plate, we need methods of integral calculus to evaluate the sum. Table 9.2 gives moments of inertia for several familiar shapes in terms of the masses and dimensions. In the formulas in the table, note that dimensions measured parallel to the axis of rotation never appear in the expressions. For example, two solid cylinders with the same mass and radius have equal moments of inertia about their axes, even if their lengths are different.

NOTE ▶ You may be tempted to try to compute the moment of inertia of a body by assuming that all of its mass is concentrated at the center of mass and then multiplying the total mass by the square of the distance from the center of mass to the axis. Resist that temptation; it doesn't work! For example, when a uniform thin rod with length L and mass M rotates about an axis through one end and perpendicular to the rod, the moment of inertia is $I = ML^2/3$. (Table 9.2b). If we took the mass as concentrated at the center of the rod, a distance $L/2$ from the axis, we would obtain the *incorrect* result $I = M(L/2)^2 = ML^2/4$. ◀

TABLE 9.2   Moments of inertia for various bodies

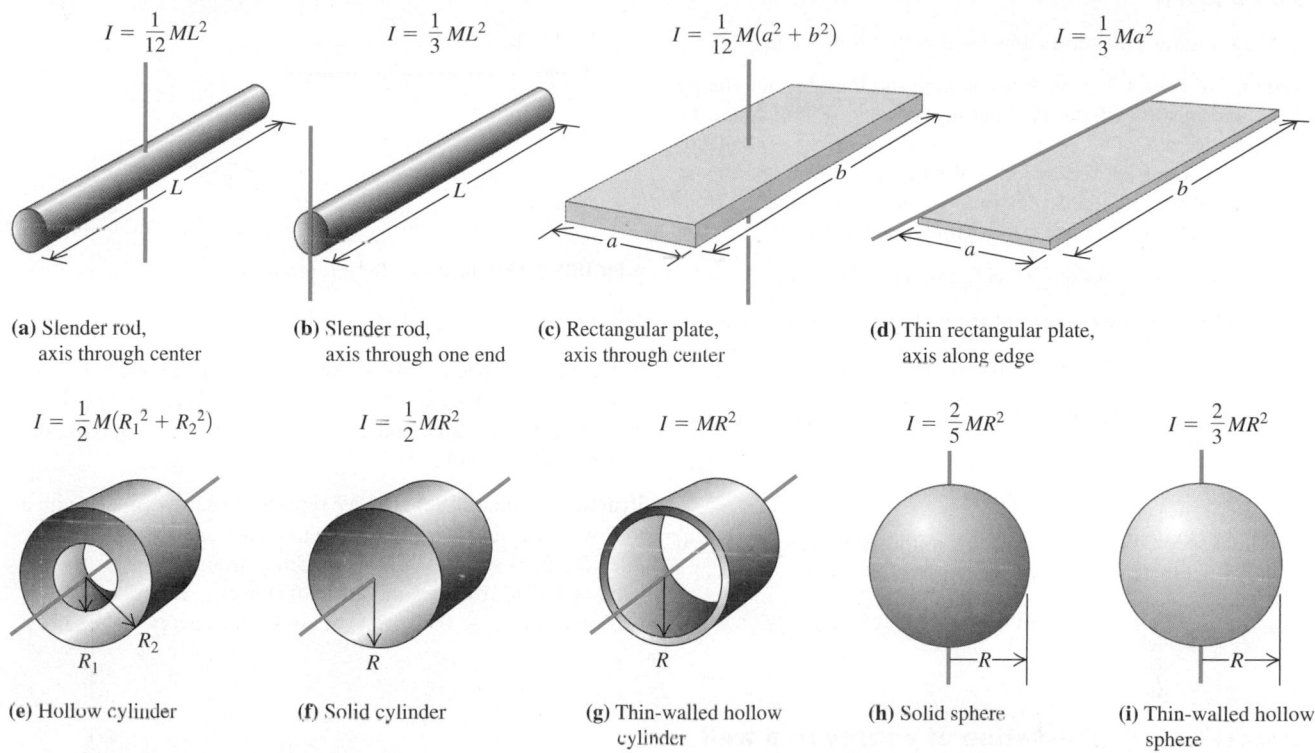

$I = \frac{1}{12}ML^2$

(a) Slender rod, axis through center

$I = \frac{1}{3}ML^2$

(b) Slender rod, axis through one end

$I = \frac{1}{12}M(a^2 + b^2)$

(c) Rectangular plate, axis through center

$I = \frac{1}{3}Ma^2$

(d) Thin rectangular plate, axis along edge

$I = \frac{1}{2}M(R_1{}^2 + R_2{}^2)$

(e) Hollow cylinder

$I = \frac{1}{2}MR^2$

(f) Solid cylinder

$I = MR^2$

(g) Thin-walled hollow cylinder

$I = \frac{2}{5}MR^2$

(h) Solid sphere

$I = \frac{2}{3}MR^2$

(i) Thin-walled hollow sphere

Now that we know how to calculate the kinetic energy of a rotating rigid body, we can apply the energy principles of Chapter 7 to rotational motion. Here are some points of strategy and some examples.

PROBLEM-SOLVING STRATEGY 9.1   Rotational energy   (MP)

We suggest that you review the problem-solving strategies outlined in Sections 7.3, 7.6, and 7.7; they are equally useful here. The only new idea is that the kinetic energy K is expressed in terms of the moment of inertia I and angular velocity ω of the body instead of its mass m and speed v. You can use work–energy relations and conservation of energy to find relations involving the position and motion of a rotating body.

The kinematic relations of Section 9.3, particularly Equations 9.13 and 9.14, are often useful, especially when a rotating cylindrical body functions as any sort of pulley. Examples 9.7 and 9.8 illustrate this point; in both, the angular velocity ω and the tangential component of velocity v_{tan} at the surface of the cylinder are related by $v_{\text{tan}} = r\omega$.

EXAMPLE 9.7   A cable unwinding from a winch

A light, flexible, nonstretching cable is wrapped several times around a winch drum—a solid cylinder with mass 50 kg and diameter 0.12 m that rotates about a stationary horizontal axis that turns in frictionless bearings (Figure 9.17). The free end of the cable is pulled with a constant force of magnitude 9.0 N for a distance of 2.0 m. It unwinds without slipping, turning the cylinder as it does so. If the cylinder is initially at rest, find its final angular velocity ω and the final speed v of the cable.

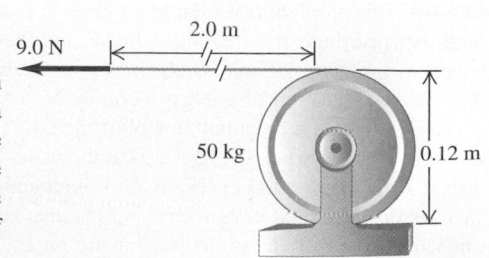

▲ **FIGURE 9.17**

Continued

SOLUTION

SET UP Figure 9.18 shows how we sketch this problem.

SOLVE No energy is lost in friction, so the final kinetic energy $K = \frac{1}{2}I\omega^2$ of the cylinder is equal to the work $W = Fs$ done by the force (of magnitude F, pulling through a distance s parallel to the force) acting on the cylinder. We need to use the mass and diameter of the cylinder to determine its moment of inertia, from Table 9.2. The work done on the cylinder by the force is

$$W = Fs = (9.0 \text{ N})(2.0 \text{ m}) = 18 \text{ J}.$$

From Table 9.2f, the moment of inertia for a solid cylinder is

$$I = \frac{1}{2}MR^2 = \frac{1}{2}(50 \text{ kg})(0.060 \text{ m})^2 = 0.090 \text{ kg} \cdot \text{m}^2.$$

The work–energy theorem $(W = \Delta K)$ from Chapter 7 then gives

$$18 \text{ J} = \frac{1}{2}(0.090 \text{ kg} \cdot \text{m}^2)\omega^2,$$
$$\omega = 20 \text{ rad/s}.$$

The final speed of the cable is equal to the final tangential speed v_{tan} of the cylinder, given by Equation 9.13:

$$v_{\text{tan}} = R\omega = (0.060 \text{ m})(20 \text{ rad/s}) = 1.2 \text{ m/s}.$$

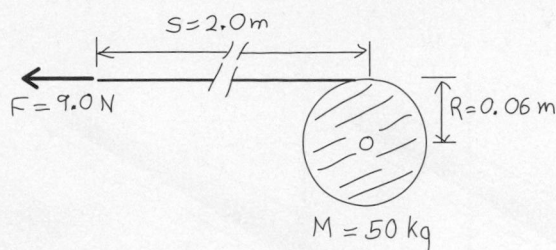

▲ **FIGURE 9.18** Our sketch for this problem.

REFLECT The speed of the cable at each instant equals the tangential component of velocity of the cylinder because there is no slipping between cable and cylinder; if there were slipping, the winch wouldn't work.

Practice Problem: Suppose we replace the solid cylinder with a thin-walled cylinder having the same mass and radius. (a) How does the final speed of the cable differ from our result for the solid cylinder? (b) Does this mean that energy is not conserved? *Answers:* (a) $v_{\text{tan}} = 0.85$ m/s; (b) no; kinetic energy is the same.

EXAMPLE 9.8 Conservation of energy in a well

In an old-fashioned well, a bucket is suspended over the well shaft by a winch and rope (Figure 9.19). The winch includes a solid cylinder with mass M and radius R and rotates without friction about a horizontal axis. The bucket (mass m) must descend a height h to reach the water; it is suspended by a rope of negligible mass that wraps around the winch. If the winch handle falls off, releasing the bucket to fall to the water, rotating the cylinder as it falls, find the speed v of the bucket and the angular velocity ω of the cylinder just before the bucket hits the water.

▶ **FIGURE 9.19**

SOLUTION

SET UP This problem is similar to Example 9.7, but it isn't identical, because both the rotating cylinder and the falling bucket have kinetic energy. Figure 9.20 shows our sketches. Since we will use conservation of energy, we draw before and after diagrams. We take the potential energy of the bucket–earth system as zero $(U = 0)$ when the bucket is at the water level. Then the bucket's initial potential energy is $U_i = mgh$ and the final value U_f is zero. The initial kinetic energy K_i is also zero. Just before impact, the bucket (mass m) has kinetic energy $\frac{1}{2}mv^2$ and the

rotating cylinder has kinetic energy $\frac{1}{2}I\omega^2$. The speed v of the dropping bucket at any time is proportional to the angular velocity ω of the cylinder at that time: $v = r\omega$ (assuming that the rope unwinds without slipping). There are no nonconservative forces, so the initial and final total energies are equal.

SOLVE Conservation of energy tells us that $K_i + U_i = K_f + U_f$. (The cable does no net work; at the bucket end the force and displacement are in the same direction, and at the winch end they are in opposite directions. So the *total* work done by the two ends of the cable is zero.)

Continued

Just before the bucket hits the water, both it and the cylinder have kinetic energy. The total kinetic energy K_f at that time is

$$K_f = \tfrac{1}{2}mv^2 + \tfrac{1}{2}I\omega^2.$$

From Table 9.2, we find that the moment of inertia of the cylinder is $I = \tfrac{1}{2}MR^2$. Also, v and ω are related by $v = R\omega$, because the speed of mass m must equal the tangential speed of the outer surface of the cylinder. Using these relations and equating the initial and final total energies, we find that

$$K_i + U_i = K_f + U_f,$$

$$0 + mgh = \tfrac{1}{2}mv^2 + \tfrac{1}{2}\left(\tfrac{1}{2}MR^2\right)\left(\frac{v}{R}\right)^2 = \tfrac{1}{2}\left(m + \tfrac{1}{2}M\right)v^2,$$

$$v = \sqrt{\frac{2gh}{1 + M/2m}}.$$

The final angular velocity ω is obtained from $\omega = v/R$.

REFLECT We can check some particular cases. When M is much larger than m (i.e., a very heavy cylinder and a tiny bucket), v is very small, as we would expect. When M is much *smaller* than m (i.e., a very light cylinder and a heavy bucket), there isn't much to hold the bucket back, and its final speed is nearly equal to the speed of a body in free fall with initial height h, namely, $\sqrt{2gh}$. Does it surprise you that v doesn't depend on the radius of the cylinder?

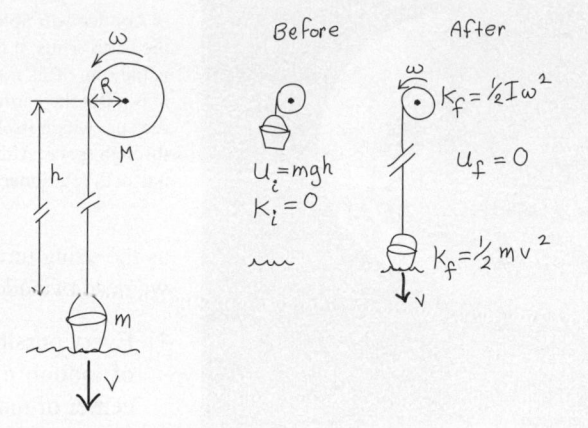

▲ **FIGURE 9.20** Our sketches for this problem.

Practice Problem: If the winch has the form of a thin cylindrical shell with mass M and radius R, derive expressions for the speed of the bucket and the angular velocity of the cylinder just before the bucket hits the water. *Answer:*

$$v = \sqrt{\frac{2gh}{1 + M/m}}.$$

Potential Energy of a Rigid Body

Later we'll need to know how to calculate the potential energy of a rigid body. In a uniform gravitational field, the gravitational potential energy is the same as though all the mass were concentrated at the center of mass of the body. Suppose we take the y axis vertically upward and define the potential energy to be zero at $y = 0$. Then, for an object with total mass M, the potential energy U is simply

$$U = Mgy_{cm}, \qquad (9.18)$$

where y_{cm} is the vertical coordinate of the center of mass. As always, we need to define the origin of coordinates, where $y_{cm} = 0$ and $U = 0$.

To prove Equation 9.18, we again represent the body as a collection of mass elements m_A, m_B, m_C, $\cdots$. The weight of element m_A is $m_A g$, its potential energy is $m_A g y_A$, and the *total* potential energy is

$$U = m_A g y_A + m_B g y_B + m_C g y_C + \cdots.$$

But from Equations 8.18, which define the coordinates of the center of mass,

$$m_A y_A + m_B y_B + m_C y_C + \cdots = (m_A + m_B + m_C + \cdots)y_{cm} = My_{cm},$$

where $M = m_A + m_B + m_C + \cdots$ is the total mass. Combining this with the earlier expression for U, we obtain

$$U = Mgy_{cm},$$

in agreement with Equation 9.18.

MasteringPHYSICS

ActivPhysics 7.12: Woman and Flywheel Elevator—Dynamics Approach
ActivPhysics 7.13: Rotoride—Energy Approach

9.5 Rotation about a Moving Axis

In the examples of rigid-body motion we've talked about so far, the axis of rotation has been stationary. Now let's consider some situations in which the axis moves. Familiar examples include a ball rolling down a hill or a yo-yo dropping

as the string unwinds. We can apply energy considerations to such problems, but we need two additional principles:

1. **Every possible motion of a rigid body can be represented as a combination of motion of the center of mass and rotation about an axis through the center of mass.**
2. **The total kinetic energy can always be represented as the sum of a part associated with motion of the center of mass, treated as a point, plus a part associated with rotation about an axis through the center of mass.**

Figure 9.21 illustrates the first of these principles. We'll often use the terms *translational* and *rotational* motion to refer to motion of the center of mass and rotational motion, respectively.

Suppose the rigid body has mass M and moment of inertia I_{cm} about the axis through the center of mass. If the center of mass has translational motion with velocity of magnitude v_{cm}, and the object rotates with angular velocity ω about this axis, the total kinetic energy K_{total} is given by

$$K_{total} = \tfrac{1}{2}Mv_{cm}^2 + \tfrac{1}{2}I_{cm}\omega^2. \qquad (9.19)$$

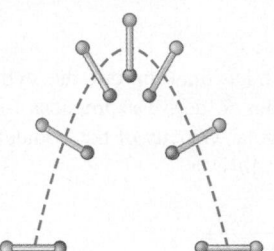

This simple baton toss can be represented as a combination of **rotation** and **translation**:

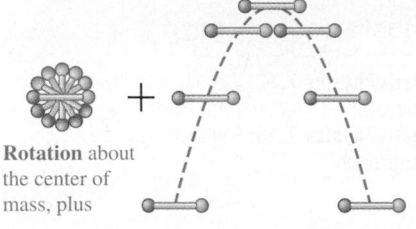

Rotation about the center of mass, plus

Translation of the center of mass

▲ **FIGURE 9.21** The motion of a rigid body can always be represented as a combination of rotation about the center of mass and translation of the center of mass.

We'll omit detailed derivation of this relationship and simply remark that it looks reasonable and that it can be proved rigorously. We'll often refer to the two terms in Equation 9.19 as *translational kinetic energy* ($K_{trans} = \tfrac{1}{2}Mv_{cm}^2$, associated with motion of the center of mass) and *rotational kinetic energy* ($K_{rot} = \tfrac{1}{2}I_{cm}\omega^2$, associated with rotation about an axis through the center of mass). With the help of Equation 9.19, we can apply conservation of energy to many problems involving rigid bodies whose axis of rotation moves. We'll continue to assume, however, that the *direction* of the axis of rotation doesn't change.

NOTE ▶ When applying Equation 9.19, we *must* use the moment of inertia about the axis through the center of mass; otherwise this relationship isn't applicable.

Conceptual Analysis 9.4

Racing canned food

A can of dog food (with fairly solid contents) and a can of chicken broth are released simultaneously to roll down an inclined plane. The two cans have the same diameter and the same mass. Which can (if either) reaches the bottom of the plane first?

A. The dog food.
B. The chicken broth.
C. They arrive simultaneously.

SOLUTION Because the equal masses roll down from the same height, the decrease in gravitational potential energy U_{grav} is the same for both cans. This energy is converted to kinetic energy;

some of it becomes *translational* kinetic energy K_{trans} (energy of motion of the center of mass), and some becomes *rotational* kinetic energy K_{rot}. Conservation of energy gives $U_{grav,i} = K_{trans,f} + K_{rot,f}$. Because both cans have the same $U_{grav,i}$, the can with the smaller $K_{rot,f}$ will have the larger $K_{trans,f}$, and therefore the larger value of $\tfrac{1}{2}mv^2$ and of v. The can with the larger v wins the race. Both cans have the same mass, but much of the fluid chicken broth does not rotate—mostly just the can rotates—and this results in a smaller K_{rot} for the chicken broth, a larger K_{trans}, and therefore a larger v. Choice B is correct; the chicken broth wins the race to the bottom of the slope. (We've neglected the fact that some mechanical energy may be dissipated by the internal motion of the viscous chicken broth.)

EXAMPLE 9.9 A primitive yo-yo

Figure 9.22 shows a primitive (and not very practical) yo-yo in the form of a solid disk with radius R and total mass M. It is released from rest, the supporting hand is stationary, and the string unwinds without slipping. Find an expression for the speed v_{cm} of the center of mass of the solid cylinder after it has dropped a distance h.

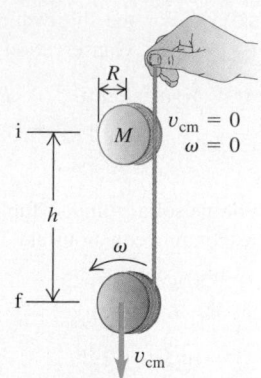

▲ FIGURE 9.22

SOLUTION

SET UP We take the potential energy to be zero at a distance h below the starting point. Then the initial and final potential energies are $U_i = Mgh$ and $U_f = 0$. The initial kinetic energy K_i is zero, and the final kinetic energy K_f is given by Equation 9.19:

$$K_f = \tfrac{1}{2}Mv_{cm}^2 + \tfrac{1}{2}I_{cm}\omega^2.$$

There are no nonconservative forces, so the total energy is conserved:

$$K_i + U_i = K_f + U_f.$$

SOLVE From conservation of energy,

$$0 + Mgh = \tfrac{1}{2}Mv_{cm}^2 + \tfrac{1}{2}I_{cm}\omega^2.$$

Both v_{cm} and ω are unknown; but if we assume that the string unwinds without slipping, these two quantities are *proportional*. When the cylinder has rotated through one complete revolution (2π radians), the string has unwound a distance equal to the circumference of the cylinder ($2\pi R$). Thus the distance traveled during any time interval Δt is R times the *angular* displace-

ment during that interval. It follows that $v_{cm} = R\omega$. Also, from Table 9.2, $I_{cm} = \tfrac{1}{2}MR^2$, so conservation of energy gives

$$Mgh = \tfrac{1}{2}Mv_{cm}^2 + \tfrac{1}{2}\left(\tfrac{1}{2}MR^2\right)\left(\frac{v_{cm}}{R}\right)^2 = \tfrac{3}{4}Mv_{cm}^2$$

and

$$v_{cm} = \sqrt{\tfrac{4}{3}gh}.$$

REFLECT If the object had fallen freely from a height h, its final speed would be $v_{cm} = \sqrt{2gh}$. The actual speed is less than this, because the same total energy has to be shared by the translational (center-of-mass) motion and the rotational motion. Note that both the mass and radius of the object cancel out, so the results would be the same for a long, skinny solid cylinder and a thin solid disk (but not for a cylindrical shell or a doughnut-shaped disk).

Practice Problem: Carry out the same analysis for a thin-walled cylinder. *Answer:* $v_{cm} = \sqrt{gh}$.

EXAMPLE 9.10 Race of the rolling objects

In a lecture demonstration, an instructor "races" various round objects by releasing them from rest at the top of an inclined plane and then rolling them down the plane (Figure 9.23). The students guess (and sometimes bet on) which object will win. Suppose a thin-walled hollow cylinder (Table 9.2g) and a solid cylinder (Table 9.2f) race. Which one would you bet on to win the race?

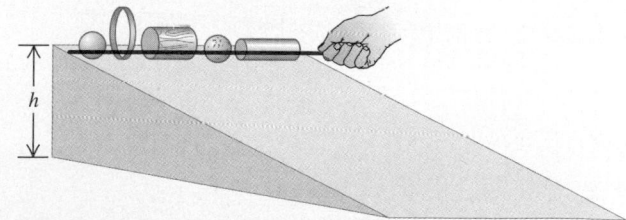

▲ FIGURE 9.23

SOLUTION

SET UP We use conservation of energy, ignoring rolling friction and air drag. If the objects roll without slipping, then no work is done by friction and the total energy is conserved. Each object starts from rest at the top of an incline with height h, so $K_i = 0$, $U_i = mgh$, and $U_f = 0$ for each. From Equation 9.19,

$$K_f = \tfrac{1}{2}mv_{cm}^2 + \tfrac{1}{2}I_{cm}\omega^2.$$

Both v_{cm} and ω are unknown; but if we assume that the objects roll without slipping, these two quantities are *proportional*. When an object with radius R has rotated through one complete revolution (2π radians), it has rolled a distance equal to its circumference ($2\pi R$). Thus the distance traveled during any time interval Δt is R times the *angular* displacement during that interval, and it follows that $v_{cm} = R\omega$. (The geometry of this situation is identical to that in Example 9.9, in which the speed of the center-of-mass of the yo-yo is proportional to its angular velocity.)

Continued

SOLVE For the thin-walled cylindrical shell, Table 9.2g gives $I_{shell} = MR^2$. Conservation of energy then results in

$$0 + Mgh = \tfrac{1}{2}Mv_{cm}^2 + \tfrac{1}{2}I_{shell}\omega^2 = \tfrac{1}{2}Mv_{cm}^2 + \tfrac{1}{2}(MR^2)(v_{cm}/R)^2$$
$$= \tfrac{1}{2}Mv_{cm}^2 + \tfrac{1}{2}Mv_{cm}^2 = Mv_{cm}^2,$$
$$v_{cm} = \sqrt{gh}.$$

For the solid cylinder, Table 9.2f gives $I_{solid} = \tfrac{1}{2}MR^2$, and the corresponding equations are

$$0 + Mgh = \tfrac{1}{2}Mv_{cm}^2 + \tfrac{1}{2}I_{solid}\omega^2 = \tfrac{1}{2}Mv_{cm}^2 + \tfrac{1}{2}(\tfrac{1}{2}MR^2)(v_{cm}/R)^2$$
$$= \tfrac{1}{2}Mv_{cm}^2 + \tfrac{1}{4}Mv_{cm}^2 = \tfrac{3}{4}Mv_{cm}^2,$$
$$v_{cm} = \sqrt{\tfrac{4}{3}gh}.$$

We see that the solid cylinder's speed at the bottom of the hill is greater than that of the hollow cylinder by a factor of $\sqrt{\tfrac{4}{3}}$, so the solid cylinder wins the race.

We can generalize this result in an elegant way. We note that the moments of inertia of all the round objects in Table 9.2 (about axes through their centers of mass) can be expressed as $I_{cm} = \beta MR^2$, where β is a pure number between 0 and 1 that depends on the shape of the body. For a thin-walled hollow cylinder, $\beta = 1$; for a solid cylinder, $\beta = \tfrac{1}{2}$; and so on. From conservation of energy,

$$0 + Mgh = \tfrac{1}{2}Mv_{cm}^2 + \tfrac{1}{2}I\omega^2 = \tfrac{1}{2}Mv_{cm}^2 + \tfrac{1}{2}(\beta MR^2)(v_{cm}/R)^2$$
$$= \tfrac{1}{2}(1 + \beta)Mv_{cm}^2,$$
$$v_{cm} = \sqrt{\frac{2gh}{1 + \beta}}.$$

REFLECT This is a fairly amazing result; the final speed of the center of mass doesn't depend on either the mass M of the body or its radius R. All uniform solid cylinders have the same speed at the bottom, even if their masses and radii are different, because they have the same β. All solid spheres have the same speed, and so on. The smaller the value of β, the faster the body is moving at the bottom (and at any point on the way down). Small-β bodies always beat large-β bodies because they have less kinetic energy tied up in rotation and have more available for translation. For the hollow cylinder $(\beta = 1)$, the translational and rotational energies at any point are equal, but for the solid cylinder $(\beta = \tfrac{1}{2})$, the rotational energy at any point is half the translational energy. Reading the values of β from Table 9.2, we see that the order of finish is as follows: any solid sphere, any solid cylinder, any thin spherical shell, and any thin cylindrical shell.

Practice Problem: (a) Find the ratio of the final speeds for a solid sphere and a solid cylinder. (b) Why aren't the expressions for moments of inertia of these two objects the same? *Answers:* (a) 1.035; (b) the sphere has a greater fraction of its mass close to the axis than the cylinder has.

SUMMARY

Angular Velocity and Angular Acceleration

(Section 9.1) When a rigid body rotates about a stationary axis, its position is described by an angular coordinate θ. When an object turns through an angle $\Delta\theta$ in a time interval Δt, the average angular velocity ω_{av} is defined as $\omega_{av} = \Delta\theta/\Delta t$ (Equation 9.2). The instantaneous angular velocity ω is

$$\omega = \lim_{\Delta t \to 0} \frac{\Delta\theta}{\Delta t}. \tag{9.3}$$

If the angular velocity changes by an amount $\Delta\omega$ in a time interval Δt, the average angular acceleration α_{av} is defined as $\alpha_{av} = \Delta\omega/\Delta t$ (Equation 9.5), and the instantaneous angular velocity α is

$$\alpha = \lim_{\Delta t \to 0} \frac{\Delta\omega}{\Delta t}. \tag{9.6}$$

Angular velocity ω: Rate of change of angular position:

$$\omega_{av} = \frac{\Delta\theta}{\Delta t}$$

Angular acceleration α: Rate of change of angular velocity:

$$\alpha_{av} = \frac{\Delta\omega}{\Delta t}$$

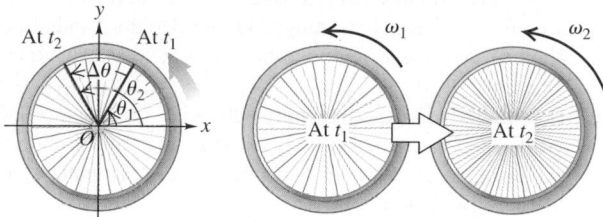

Rotation with Constant Angular Acceleration

(Section 9.2) When a body rotates with constant angular acceleration, the angular position, velocity, and acceleration are related by the equations: $\omega = \omega_0 + \alpha t$ (Equation 9.7), $\theta - \theta_0 = \frac{1}{2}(\omega_0 + \omega)t$ (Equation 9.10), $\theta = \theta_0 + \omega_0 t + \frac{1}{2}\alpha t^2$ (Equation 9.11), and $\omega^2 = \omega_0^2 + 2\alpha(\theta - \theta_0)$ (Equation 9.12), where θ_0 and ω_0 are the initial values of the angular position and angular velocity, respectively, of the body. These expressions are analogous to the equations for linear motion found in Chapter 2.

Relationship between Linear and Angular Quantities

(Section 9.3) A particle in a rotating rigid body at a distance r from the axis of rotation has a tangential speed v given by $v = r\omega$ (Equation 9.13). The particle's acceleration $\vec{a}$ has a tangential component $a_{tan} = r\alpha$ (Equation 9.14) and a radial component

$$a_{rad} = \frac{v^2}{r} = \omega^2 r. \tag{9.15}$$

Kinetic Energy of Rotation and Moment of Inertia

(Section 9.4) The moment of inertia I of a body with respect to a given axis is defined as $I = m_A r_A^2 + m_B r_B^2 + m_C r_C^2 + \cdots$ (Equation 9.16) where r_A is the distance of mass element m_A from the axis of rotation, and so on. The kinetic energy of a rigid body rotating about a stationary axis is given by $K = \frac{1}{2}I\omega^2$ (Equation 9.17). In a uniform gravitational field, the gravitational potential energy of a rigid body with center of mass at a distance y_{cm} above the reference level (where $U = 0$) is $U = Mgy_{cm}$ (Equation 9.18).

Rotation about a Moving Axis

(Section 9.5) When a rigid body undergoes both motion of its center of mass and rotation about an axis through the center of mass, the total kinetic energy is given by

$$K_{total} = \frac{1}{2}Mv_{cm}^2 + \frac{1}{2}I_{cm}\omega^2. \tag{9.19}$$

$$K_{total} \quad = \quad \frac{1}{2}Mv_{cm}^2 \quad + \quad \frac{1}{2}I_{cm}\omega^2$$

MP For instructor-assigned homework, go to www.masteringphysics.com

Conceptual Questions

1. What is the difference between the tangential acceleration and the radial acceleration of a point on a rotating object?
2. A flywheel rotates with constant angular velocity. Does a point on its rim have tangential acceleration? radial acceleration? Are these accelerations constant in magnitude? in direction?
3. A flywheel rotates with constant angular acceleration. Does a point on its rim have tangential acceleration? radial acceleration? Are these accelerations constant in magnitude? in direction?
4. How might you experimentally determine a moment of inertia of an *irregularly* shaped body by using materials from a typical introductory physics laboratory (such as pulleys, turntables, timers, etc.)?
5. A uniform ring of mass M and radius R and a point mass M a distance R from the axis of the ring both rotate about that axis, which is perpendicular to the ring at its center. Both objects have the same formula for their moment of inertia, $I = MR^2$, about this axis. (a) Is there a good physical reason that they have the same formula? (b) If the ring were replaced by a solid uniform disk of mass M and radius R, would the disk and point mass then have the same moment-of-inertia formula? Why or why not?
6. What is the purpose of the spin cycle of a washing machine? Explain in terms of acceleration components.
7. According to experienced riders, you make a bike less tiring to ride if you reduce its weight in the wheels rather than in the frame. Why is this so?
8. A solid ball, a solid cylinder, and a hollow cylinder roll down a slope. Which one reaches the bottom first? last? Does it matter whether the radii are the same? What about the masses?
9. Experienced cooks can tell whether an egg is raw or hard boiled by rolling it down a slope (and taking care to catch it at the bottom). How is this possible? Which type of egg should reach the bottom of the slope first?
10. Part of the kinetic energy of a moving automobile is in the rotational motion of its wheels. When the brakes are applied hard on an icy street, the wheels lock and the car starts to slide. What becomes of the rotational kinetic energy?
11. Can you think of a body that has the same moment of inertia for all possible axes? If so, give an example, and if not, explain why this is not possible. Can you think of a body that has the same moment of inertia for all axes passing through a certain point? If so, give an example and indicate where the point is located.
12. A client has come to you with two metal balls of the same size and mass. He thinks that one might be solid and the other might be hollow inside. By rolling both balls down a slope, how could you tell which one is hollow and which is solid?
13. In what ways does a jogger's body have *rotational* kinetic energy?
14. As a car of mass M travels with speed V, its kinetic energy is $\frac{1}{2}MV^2$. But this is not the whole story. What additional contributions are there to this car's kinetic energy from *internal* motion, both translational and rotational?
15. If a ball rolls down an *irregularly* shaped hill without slipping, it is very easy to find its speed at the bottom by using conservation of energy. Why could we not just as easily use Newton's second law to find the speed?

Multiple-Choice Problems

1. When a wheel turns through one complete rotation, the angle (in *radians*) that it has turned through is closest to
   A. 57.   B. 6.   C. 360.
2. Two points are on a disk that rotates about an axis perpendicular to the plane of the disk at its center. Point B is 3 times as far from the axis as point A. If the linear speed of point B is V, then the linear speed of point A is
   A. $9V$.   B. $3V$.   C. V.
   D. $\dfrac{V}{3}$.   E. $\dfrac{V}{9}$.
3. A bicycle wheel rotating at a rate of 12 rad/s begins to accelerate at a rate of 2 rad/s². After 5 seconds the rate of rotation will be
   A. 2 rad/s   B. 10 rad/s   C. 17 rad/s   D. 22 rad/s
4. You are designing a flywheel to store large amounts of kinetic energy. Which one of the following uniform shapes will be the *most* effective for storing the greatest amount of kinetic energy if all the objects have the same mass and same angular velocity?
   A. A solid sphere of diameter D rotating about a diameter.
   B. A solid cylinder of diameter D rotating about an axis perpendicular to each face through its center.
   C. A thin-walled hollow cylinder of diameter D rotating about an axis perpendicular to the plane of the circular face at its center.
   D. A solid thin bar of length D rotating about an axis perpendicular to it at its center.
5. Four uniform objects having the same mass and diameter are released simultaneously from rest at the same distance above the bottom of a hill and roll down without slipping. The objects are a solid sphere, a solid cylinder, a hollow cylinder, and a thin-walled hollow cylinder, as illustrated in Table 9.2. Which of these objects will be the first one to reach the bottom of the hill?
   A. The solid sphere
   B. The solid cylinder
   C. The hollow cylinder
   D. The thin-walled hollow cylinder
6. Which of the objects in question 5 will be the last one to reach the bottom of the hill?
   A. The solid sphere
   B. The solid cylinder
   C. The hollow cylinder
   D. The thin-walled hollow cylinder
7. For the objects in question 5, which of the following statements are correct? (There may be more than one correct choice.)
   A. All these objects have the same forward speed at the bottom of the hill.
   B. All these objects have the same kinetic energy at the bottom of the hill.
   C. All these objects have the same rotational kinetic energy at the bottom of the hill.
   D. All these objects have the same angular velocity at the bottom of the hill.
8. Two uniform solid spheres of the same size, but different mass, are released from rest simultaneously at the same height on a hill and roll to the bottom without slipping. Which of the

following statements about these spheres are true? (There may be more than one correct choice.)
A. Both spheres reach the bottom at the same time.
B. The heavier sphere reaches the bottom ahead of the lighter one.
C. Both spheres arrive at the bottom with the same forward speed.
D. Both spheres arrive at the bottom with the same total kinetic energy.

9. A disk starts from rest and has a constant angular acceleration. If it takes time T to make its first revolution, in time $2T$ (starting from rest) the disk will make
A. $\sqrt{2}$ revolutions.
B. 2 revolutions.
C. 4 revolutions.
D. 8 revolutions.

10. Two unequal masses m and $2m$ are attached to a thin bar of negligible mass that rotates about an axis perpendicular to the bar. When m is a distance $2d$ from the axis and $2m$ is a distance d from the axis, the moment of inertia of this combination is I. If the masses are now interchanged, the moment of inertia will be
A. $\frac{2}{3}I$. B. I. C. $\frac{3}{2}I$.
D. $2I$. E. $4I$.

11. A thin uniform bar has a moment of inertia I about an axis perpendicular to it through its center. If *both* the mass and length of this bar are doubled, its moment of inertia about the same axis will be
A. $2I$. B. $4I$.
C. $8I$. D. $16I$.

12. Two small objects of equal weight are attached to the ends of a thin weightless bar that spins about an axis perpendicular to the bar at its center. Each mass is a distance d from the axis of rotation. The system spins at a rate so that its kinetic energy is K. If *both* masses are now moved closer in so that each is $d/2$ from the rotation axis, while the angular velocity does not change, the kinetic energy of the system will be
A. $\frac{1}{4}K$. B. $\frac{1}{2}K$.
C. $2K$. D. $4K$.

13. A disk starts from rest and rotates with constant angular acceleration. If the angular velocity is ω at the end of the first two revolutions, then at the end of the first eight revolutions it will be
A. $\sqrt{2}\omega$. B. 2ω.
C. 4ω. D. 16ω.

14. Two identical merry-go-rounds are rotating at the same speed. One is crowded with riding children; the other is nearly empty. If both merry-go-rounds cut off their motors at the same time and coast to a stop, slowed only by friction (which you can assume is the same for both merry-go-rounds), which will take longer to stop?
A. The crowded merry-go-round
B. The empty merry-go-round
C. The same time for both.

15. A solid sphere and a hollow sphere, both uniform and having the same mass and radius, are released from rest at the same height on a ramp. If they roll down the ramp without slipping, which one will have the *greater* forward velocity at the bottom of the ramp?
A. The solid sphere
B. The hollow sphere
C. Both will have the same forward velocity.

16. A uniform ball rolls without slipping toward a hill with a forward speed V. In which case will this ball go farther up the hill?
A. There is enough friction on the hill to prevent slipping of the ball.
B. The hill is frictionless.
C. The ball will reach the same height in both cases, since it has the same initial kinetic energy in each case.

Problems

9.1 Angular Velocity and Angular Acceleration

1. • A flexible straight wire 75.0 cm long is bent into the arc of a circle of radius 2.50 m. What angle (in radians and degrees) will this arc subtend at the center of the circle?

2. • (a) What angle in radians is subtended by an arc 1.50 m in length on the circumference of a circle of radius 2.50 m? What is this angle in degrees? (b) An arc 14.0 cm in length on the circumference of a circle subtends an angle of 128°. What is the radius of the circle? (c) The angle between two radii of a circle with radius 1.50 m is 0.700 rad. What length of arc is intercepted on the circumference of the circle by the two radii?

3. • (a) Calculate the angular velocity (in rad/s) of the second, minute, and hour hands on a wall clock. (b) What is the period of each of these hands?

4. • The once-popular LP (long-play) records were 12 in. in diameter and turned at a constant $33\frac{1}{3}$ rpm. Find (a) the angular speed of the LP in rad/s and (b) its period in seconds.

5. • If a wheel 212 cm in diameter takes 2.25 s for each revolution, find its (a) period and (b) angular speed in rad/s.

6. • Find the angular velocity, in rad/s, of (a) the earth due to its daily spin on its axis, (b) the earth due to its yearly motion around the sun, and (c) our moon due to its monthly motion around the earth. Consult Appendix E as needed.

7. • A laser beam aimed from the earth is swept across the face of the moon. (a) If the beam is rotated at an angular velocity of 1.50×10^{-3} rad/s, at what speed does the laser light move across the moon's surface? (See Appendix E for the moon's orbital radius.) (b) If the diameter of the laser spot on the moon is 6.00 km, what is the angle of divergence of the laser beam?

8. • **Communications satellites.** Communications satellites are placed in orbits so that they always remain above the same point of the earth's surface. (a) What must be the period of such a satellite? (b) What is its angular velocity in rad/s?

9. • An airplane propeller is rotating at 1900 rpm. (a) Compute the propeller's angular velocity in rad/s. (b) How many seconds does it take for the propeller to turn through 35°? (c) If the propeller were turning at 18 rad/s, at how many rpm would it be turning? (d) What is the period (in seconds) of this propeller?

10. •• A wall clock on Planet X has two hands that are aligned at midnight and turn in the same direction at uniform rates, one at 0.0425 rad/s and the other at 0.0163 rad/s. At how many seconds after midnight are these hands (a) first aligned and (b) next aligned?

9.2 Rotation with Constant Angular Acceleration

11. • A turntable that spins at a constant 78.0 rpm takes 3.50 s to reach this angular speed after it is turned on. Find (a) its angular

acceleration (in rad/s²), assuming it to be constant, and (b) the number of degrees it turns through while speeding up.

12. • When the power is turned off on a turntable spinning at 78.0 rpm, you find that it takes 10.5 revolutions for it to stop while slowing down at a uniform rate. (a) What is the angular acceleration (in rad/s²) of this turntable? (b) How long does it take to stop after the power is turned off?

13. • **DVDs.** The angular speed of digital video discs (DVDs) varies with whether the inner or outer part of the disc is being read. (CDs function in the same way.) Over a 133 min playing time, the angular speed varies from 570 rpm to 1600 rpm. Assuming it to be constant, what is the angular acceleration (in rad/s²) of such a DVD?

14. • A circular saw blade 0.200 m in diameter starts from rest. In 6.00 s, it reaches an angular velocity of 140 rad/s with constant angular acceleration. Find the angular acceleration and the angle through which the blade has turned in this time.

15. • A wheel turns with a constant angular acceleration of 0.640 rad/s². (a) How much time does it take to reach an angular velocity of 8.00 rad/s, starting from rest? (b) Through how many revolutions does the wheel turn in this interval?

16. • An electric fan is turned off, and its angular velocity decreases uniformly from 500.0 rev/min to 200.0 rev/min in 4.00 s. (a) Find the angular acceleration in rev/s² and the number of revolutions made by the motor in the 4.00 s interval. (b) How many more seconds are required for the fan to come to rest if the angular acceleration remains constant at the value calculated in part (a)?

17. •• A flywheel in a motor is spinning at 500.0 rpm when a power failure suddenly occurs. The flywheel has mass 40.0 kg and diameter 75.0 cm. The power is off for 30.0 s, and during this time the flywheel slows down uniformly due to friction in its axle bearings. During the time the power is off, the flywheel makes 200.0 complete revolutions. (a) At what rate is the flywheel spinning when the power comes back on? (b) How long after the beginning of the power failure would it have taken the flywheel to stop if the power had not come back on, and how many revolutions would the wheel have made during this time?

18. •• A flywheel having constant angular acceleration requires 4.00 s to rotate through 162 rad. Its angular velocity at the end of this time is 108 rad/s. Find (a) the angular velocity at the beginning of the 4.00 s interval; (b) the angular acceleration of the flywheel.

19. •• Emilie's potter's wheel rotates with a constant 2.25 rad/s² angular acceleration. After 4.00 s, the wheel has rotated through an angle of 60.0 rad. What was the angular velocity of the wheel at the beginning of the 4.00 s interval?

20. •• Derive Eq. 9.12 by combining Eqs. 9.7 and 9.11 to eliminate t.

9.3 Relationship between Linear and Angular Quantities

21. • A car is traveling at a speed of 63 mi/h on a freeway. If its tires have diameter 24.0 in and are rolling without sliding or slipping, what is their angular velocity?

22. • (a) A cylinder 0.150 m in diameter rotates in a lathe at 620 rpm. What is the tangential speed of the surface of the cylinder? (b) The proper tangential speed for machining cast iron is about 0.600 m/s. At how many revolutions per minute should a piece of stock 0.0800 m in diameter be rotated in a lathe to produce this tangential speed?

23. • A wheel rotates with a constant angular velocity of 6.00 rad/s. (a) Compute the radial acceleration of a point 0.500 m from the axis, using the relation $a_{rad} = \omega^2 r$. (b) Find the tangential speed of the point, and compute its radial acceleration from the relation $a_{rad} = v^2/r$.

24. • **Ultracentrifuge.** Find the required angular speed (in rpm) of an ultracentrifuge for the radial acceleration of a point 2.50 cm from the axis to equal 400,000g.

25. •• **Exercise!** An exercise bike that you pedal in place has a bicycle chain connecting the wheel you pedal to the large wheel in front, as shown in Figure 9.24. For the wheel *diameters* shown, how many rpm must you produce to turn the large wheel at 75 rpm?

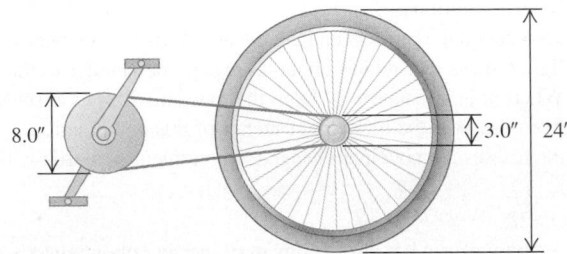

▲ **FIGURE 9.24** Problem 25.

26. •• A flywheel with a radius of 0.300 m starts from rest and accelerates with a constant angular acceleration of 0.600 rad/s². Compute the magnitude of the tangential acceleration, the radial acceleration, and the resultant acceleration of a point on its rim (a) at the start, (b) after it has turned through 60.0°, and (c) after it has turned through 120.0°.

27. • **Electric drill.** According to the shop manual, when drilling a 12.7-mm-diameter hole in wood, plastic, or aluminum, a drill should have a speed of 1250 rev/min. For a 12.7-mm-diameter drill bit turning at a constant 1250 rev/min, find (a) the maximum linear speed of any part of the bit and (b) the maximum radial acceleration of any part of the bit.

28. •• **Dental hygiene.** Electric toothbrushes can be effective in removing dental plaque. One model consists of a head 1.1 cm in diameter that rotates back and forth through a 70.0° angle 7600 times per minute. The rim of the head contains a thin row of bristles. (See Figure 9.25.) (a) What is the average angular speed in each direction of the rotating head, in rad/s? (b) What is the average linear speed in each direction of the bristles against the teeth? (c) Using your own observations, what is the approximate speed of the bristles against your teeth when you brush by hand with an *ordinary* toothbrush?

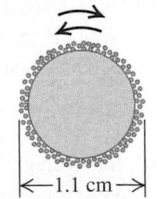

▲ **FIGURE 9.25** Problem 28.

29. •• The spin cycles of a washing machine have two angular speeds, 423 rev/min and 640 rev/min. The internal diameter of the drum is 0.470 m. (a) What is the ratio of the maximum radial force on the laundry for the higher angular speed to that for the lower speed? (b) What is the ratio of the maximum tangential speed of the laundry for the higher angular speed to that for the lower speed? (c) Find the laundry's maximum tangential speed and the maximum radial acceleration, in terms of g.

9.4 Kinetic Energy of Rotation and Moment of Inertia

30. • A twirler's baton is made of a slender metal cylinder of mass *M* and length *L*. Each end has a rubber cap of mass *m*, and you can accurately treat each cap as a particle in this problem. Find the total moment of inertia of the baton about the usual twirling axis (perpendicular to the baton through its center).

31. •• A thin uniform bar has two small balls glued to its ends. The bar is 2.00 m long and has mass 4.00 kg, while the balls each have mass 0.500 kg and can be treated as point masses. Find the moment of inertia of this combination about each of the following axes: (a) an axis perpendicular to the bar through its center; (b) an axis perpendicular to the bar through one of the balls; (c) an axis parallel to the bar through both balls.

32. • Use the formulas of Table 9.2 to find the moment of inertia about each of the following axes for a rod that is 0.300 cm in diameter and 1.50 m long, with a mass of 0.0420 kg: (a) about an axis perpendicular to the rod and passing through its center; (b) about an axis perpendicular to the rod and passing through one end; and (c) about an axis along the length of the rod.

33. •• Four small 0.200 kg spheres, each of which you can regard as a point mass, are arranged in a square 0.400 m on a side and connected by light rods. (See Figure 9.26.) Find the moment of inertia of the system about an axis (a) through the center of the square, perpendicular to its plane at point *O*; (b) along the line *AB*; and (c) along the line *CD*.

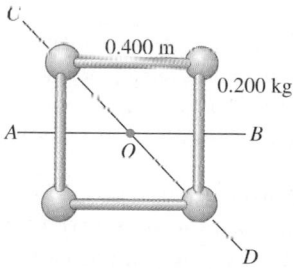

▲ FIGURE 9.26 Problem 33.

34. •• **Compound objects.** Moment of inertia is a scalar. Therefore, if several objects are connected together, the moment of inertia of this compound object is simply the scalar (algebraic) sum of the moments of inertia of each of the component objects. Use this principle to answer each of the following questions about the moment of inertia of compound objects: (a) A thin uniform 2.50 kg bar 1.50 m long has a small 1.25 kg mass glued to each end. What is the moment of inertia of this object about an axis perpendicular to the bar through its center? (b) What is the moment of inertia of the object in part (a) about an axis perpendicular to the bar at one end? (c) A 725 g metal wire is bent into the shape of a hoop 60.0 cm in diameter. Six wire spokes, each of mass 112 g, are added from the center of the hoop to the rim. What is the moment of inertia of this object about an axis perpendicular to it through its center?

35. •• **Energy from the Moon?** Suppose that sometime in the future we decide to tap the moon's rotational energy for use on earth. In additional to the astronomical data in Appendix E, you may need to know that the moon spins on its axis once every 27.3 days. Assume that the moon is uniform throughout. (a) How much total energy could we get from the moon's rotation? (b) The world presently uses about 5.0×10^{20} J of energy per year. If in the future the world uses five times as much energy yearly, for how many years would the moon's rotation provide us energy? In light of your answer, does this seem like a cost-effective energy source in which to invest?

36. •• A wagon wheel is constructed as shown in Figure 9.27. The radius of the wheel is 0.300 m, and the rim has a mass of 1.40 kg. Each of the wheel's eight spokes, which come out from the center and are 0.300 m long, has a mass of 0.280 kg. What is the moment of inertia of the wheel about an axis through its center and perpendicular to the plane of the wheel?

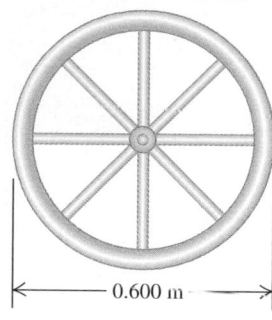

▲ FIGURE 9.27 Problem 36.

37. •• You need to design an industrial turntable that is 60.0 cm in diameter and has a kinetic energy of 0.250 J when turning at 45.0 rpm (rev/min). (a) What must be the moment of inertia of the turntable about the rotation axis? (b) If your workshop makes this turntable in the shape of a uniform solid disk, what must be its mass?

38. • A grinding wheel in the shape of a solid disk is 0.200 m in diameter and has a mass of 3.00 kg. The wheel is rotating at 2200 rpm about an axis through its center. (a) What is its kinetic energy? (b) How far would it have to drop in free fall to acquire the same amount of kinetic energy?

39. •• The flywheel of a gasoline engine is required to give up 500 J of kinetic energy while its angular velocity decreases from 650 rev/min to 520 rev/min. What moment of inertia is required?

40. •• An airplane propeller is 2.08 m in length (from tip to tip) with mass 117 kg and is rotating at 2400 rpm (rev/min) about an axis through its center. You can model the propeller as a slender rod. (a) What is its rotational kinetic energy? (b) Suppose that, due to weight constraints, you had to reduce the propeller's mass to 75.0% of its original mass, but you still needed to keep the same size and kinetic energy. What would its angular speed have to be, in rpm?

41. •• **Storing energy in flywheels.** It has been suggested that we should use our power plants to generate energy in the off-hours (such as late at night) and store it for use during the day. One idea put forward is to store the energy in large flywheels. Suppose we want to build such a flywheel in the shape of a hollow cylinder of inner radius 0.500 m and outer radius 1.50 m, using concrete of density 2.20×10^3 kg/m³. (a) If, for stability, such a heavy flywheel is limited to 1.75 second for each revolution and has negligible friction at its axle, what must be its length to store 2.5 MJ of energy in its rotational motion? (b) Suppose that by strengthening the frame you could safely double the flywheel's rate of spin. What length of flywheel would you need in that case? (Solve this part *without* reworking the entire problem!)

42. •• A light string is wrapped around the outer rim of a solid uniform cylinder of diameter 75.0 cm that can rotate without friction about an axle through its center. A 3.00 kg stone is tied

to the free end of the string, as shown in Figure 9.28. When the system is released from rest, you determine that the stone reaches a speed of 3.50 m/s after having fallen 2.50 m. What is the mass of the cylinder?

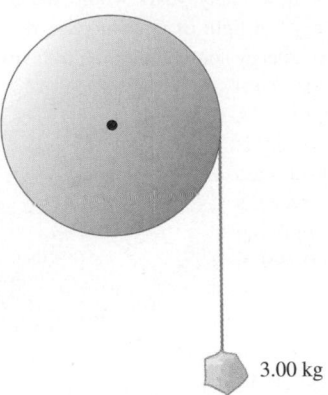

3.00 kg

▲ **FIGURE 9.28** Problem 42.

43. ●● A solid uniform 3.25 kg cylinder, 65.0 cm in diameter and 12.4 cm long, is connected to a 1.50 kg weight over two massless frictionless pulleys as shown in Figure 9.29. The cylinder is free to rotate about an axle through its center perpendicular to its circular faces, and the system is released from rest. (a) How far must the 1.50 kg weight fall before it reaches a speed of 2.50 m/s? (b) How fast is the cylinder turning at this instant?

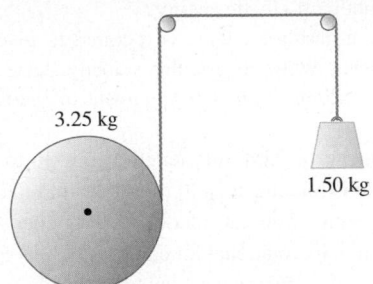

3.25 kg

1.50 kg

▲ **FIGURE 9.29** Problem 43.

44. ●● A compound disk of outside diameter 140.0 cm is made up of a uniform solid disk of radius 50.0 cm and area density 3.00 g/cm² surrounded by a concentric ring of inner radius 50.0 cm, outer radius 70.0 cm, and area density 2.00 g/cm². Find the moment of inertia of this object about an axis perpendicular to the plane of the object and passing through its center.

9.5 Rotation about a Moving Axis

45. ● **Gymnastics.** We can roughly model a gymnastic tumbler
BIO as a uniform solid cylinder of mass 75 kg and diameter 1.0 m. If this tumbler rolls forward at 0.50 rev/s, (a) how much total kinetic energy does he have and (b) what percent of his total kinetic energy is rotational?

46. ●● A bicycle racer is going downhill at 11.0 m/s when, to his horror, one of his 2.25 kg wheels comes off when he is 75.0 m above the foot of the hill. We can model the wheel as a thin-walled cylinder 85.0 cm in diameter and neglect the small mass of the spokes. (a) How fast is the wheel moving when it reaches the foot of the hill if it rolled without slipping all the way down? (b) How much total kinetic energy does the wheel have when it reaches the bottom of the hill?

47. ● A 2.20-kg hoop 1.20 m in diameter is rolling to the right without slipping on a horizontal floor at a steady 3.00 rad/s. (a) How fast is its center moving? (b) What is the total kinetic energy of the hoop?

48. ●● A solid uniform sphere and a uniform spherical shell, both having the same mass and radius, roll without slipping down a hill that rises at an angle θ above the horizontal. Both spheres start from rest at the same vertical height h. (a) How fast is each sphere moving when it reaches the bottom of the hill? (b) Which sphere will reach the bottom first, the hollow one or the solid one?

49. ●● A size-5 soccer ball of diameter 22.6 cm and mass 426 g rolls up a hill without slipping, reaching a maximum height of 5.00 m above the base of the hill. We can model this ball as a thin-walled hollow sphere. (a) At what rate was it rotating at the base of the hill? (b) How much rotational kinetic energy did it then have?

50. ●● A solid uniform marble and a block of ice, each with the same mass, start from rest at the same height H above the bottom of a hill and move down it. The marble rolls without slipping, but the ice slides without friction. (a) Find the speed of each of these objects when it reaches the bottom of the hill. (b) Which object is moving faster at the bottom, the ice or the marble? (c) Which object has more kinetic energy at the bottom, the ice or the marble?

51. ●● What fraction of the total kinetic energy is rotational for the following objects rolling without slipping on a horizontal surface? (a) a uniform solid cylinder; (b) a uniform sphere; (c) a thin-walled, hollow sphere; (d) a hollow cylinder with outer radius R and inner radius $R/2$.

52. ●● A string is wrapped several times around the rim of a small hoop with a radius of 0.0800 m and a mass of 0.180 kg. If the free end of the string is held in place and the hoop is released from rest (see Figure 9.30), calculate the angular speed of the rotating hoop after it has descended 0.750 m.

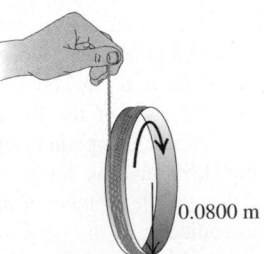

0.0800 m

▲ **FIGURE 9.30** Problem 52.

53. ●● An apparatus for launching a small boat consists of a 150.0 kg cart that rides down a set of tracks on four solid steel wheels, each with radius 20.0 cm and mass 45.0 kg. The tracks slope at an angle of 7.50° to the horizontal, and the boat's mass is 750.0 kg. If the boat is released from rest a distance of 16.0 m from the water (measured along the slope), how fast will it be moving when it reaches the water? Assume the wheels roll without slipping, and that there is no energy loss due to friction.

54. •• A uniform marble rolls down a symmetric bowl, starting from rest at the top of the left side. The top of each side is a distance h above the bottom of the bowl. The left half of the bowl is rough enough to cause the marble to roll without slipping, but the right half has no friction because it is coated with oil. (a) How far up the smooth side will the marble go, measured vertically from the bottom? (b) How high would the marble go if both sides were as rough as the left side? (c) How do you account for the fact that the marble goes *higher* with friction on the right side than without friction?

General Problems

55. •• A 7300 N elevator is to be given an acceleration of $0.150g$ by connecting it to a cable of negligible weight wrapped around a turning cylindrical shaft. If the shaft's diameter can be no larger than 16.0 cm due to space limitations, what must be its minimum angular acceleration to provide the required acceleration of the elevator?

56. •• A 392-N wheel comes off a moving truck and rolls without slipping along a highway. At the bottom of a hill it is rotating at 25.0 rad/s. The radius of the wheel is 0.600 m, and its moment of inertia about its rotation axis is $0.800\,MR^2$. Friction does work on the wheel as it rolls up the hill to a stop, a height h above the bottom of the hill; this work has absolute value 3500 J. Calculate h.

57. •• **Odometer.** The odometer (mileage gauge) of a car tells you the number of miles you have driven, but it doesn't count the miles directly. Instead, it counts the number of revolutions of your car's wheels and converts this quantity to mileage, assuming a standard size tire and that your tires do not slip on the pavement. (a) A typical midsize car has tires 24 inches in diameter. How many revolutions of the wheels must the odometer count in order to show a mileage of 0.10 mile? (b) What will the odometer read when the tires have made 5,000 revolutions? (c) Suppose you put oversize 28-inch-diameter tires on your car. How many miles will you really have driven when your odometer reads 500 miles?

58. •• **Speedometer.** Your car's speedometer works in much the same way as its odometer (see the previous problem), except that it converts the angular speed of the wheels to a linear speed of the car, assuming standard-size tires and no slipping on the pavement. (a) If your car has standard 24-inch-diameter tires, how fast are your wheels turning when you are driving at a freeway speed of 55 mph? (b) How fast are you going when your wheels are turning at 500 rpm? (c) If you put on undersize 20-inch-diameter tires, what will the speedometer read when you are actually traveling at 50 mph?

59. •• When a toy car is rapidly scooted across the floor, it stores energy in a flywheel. The car has mass 0.180 kg, and its flywheel has moment of inertia 4.00×10^{-5} kg·m². The car is 15.0 cm long. An advertisement claims that the car can travel at a scale speed of up to 700 km/h (440 mi/h). The scale speed is the speed of the toy car multiplied by the ratio of the length of an actual car to the length of the toy. Assume a length of 3.0 m for a real car. (a) For a scale speed of 700 km/h, what is the actual translational speed of the car? (b) If all the kinetic energy that is initially in the flywheel is converted to the translational kinetic energy of the toy, how

much energy is originally stored in the flywheel? (c) What initial angular velocity of the flywheel was needed to store the amount of energy calculated in part (b)?

60. •• A passenger bus in Zurich, Switzerland, derived its motive power from the energy stored in a large flywheel. Whenever the bus was stopped at a station, the wheel was brought up to speed with the use of an electric motor that could then be attached to the electric power lines. The flywheel was a solid cylinder with a mass of 1000 kg and a diameter of 1.80 m; its top angular speed was 3000 rev/min. At this angular speed, what was the kinetic energy of the flywheel?

61. •• **Kinetic energy of the earth.** Consult Appendix E as necessary. Assuming that the earth is of uniform density inside, find the kinetic energy of the earth (a) due to its yearly motion around the sun and (b) due to its daily spin on its axis. (c) What percent of the earth's total kinetic energy is contained in its axial spin motion?

62. •• **Compact discs.** When a compact disc (CD) is playing, the angular speed of the turntable is adjusted so that the laser beam, which reads the digital information encoded in the surface of the disc, maintains a constant *tangential* speed. The laser begins tracking at the inside of the disc and spirals outward as the disc plays, while the angular

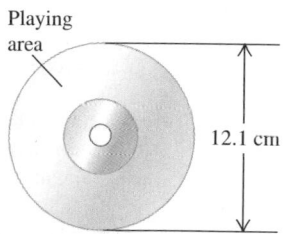

▲ **FIGURE 9.31** Problem 62.

speed of the disc varies between 500 rpm and 200 rpm. A certain CD (see Figure 9.31) plays for 74 minutes and has a playing area of outer diameter 12.1 cm (which is also essentially the outside diameter of the disc). (a) What is the tangential speed of the laser beam? (b) What is the diameter of the inside of the playing area of this CD? (c) What is the angular acceleration of the CD while it is playing, assuming it to be constant?

63. •• A vacuum cleaner belt is looped over a shaft of radius 0.45 cm and a wheel of radius 2.00 cm. The motor turns the shaft at 60.0 rev/s and the moving belt turns the wheel, which in turn is connected by another shaft to the roller that beats the dirt out of the rug being vacuumed (see Figure 9.32). Assume that the belt doesn't slip on either the shaft or the wheel. (a) What is the speed of a point on the belt? (b) What is the angular velocity of the wheel, in rad/s?

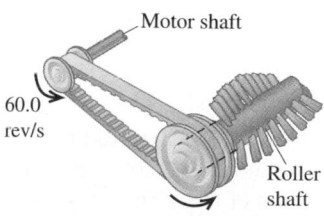

▲ **FIGURE 9.32** Problem 63.

64. •• A basketball (which can be closely modeled as a hollow spherical shell) rolls down a mountainside into a valley and then up the opposite side, starting from rest at a height H_0 above the bottom. In Figure 9.33, the rough part of the terrain prevents slipping while the smooth part has no friction. (a) How high, in terms of H_0, will it go up the other side?

(b) Why doesn't the ball return to height H_0? Has it lost any of its original potential energy?

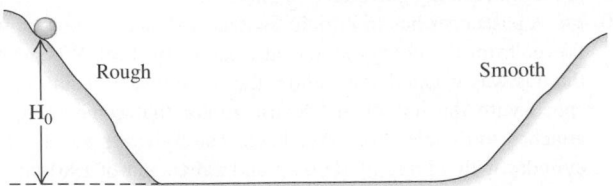

▲ **FIGURE 9.33** Problem 64.

65. •• **Human rotational energy.** A dancer is spinning at 72 rpm
BIO about an axis through her center with her arms outstretched, as shown in Figure 9.34. From biomedical measurements, the typical distribution of mass in a human body is as follows:

Head: 7.0%
Arms: 13% (for both)
Trunk and legs: 80.0%

Suppose you are this dancer. Using this information plus length measurements on your own body, calculate (a) your moment of inertia about your spin axis and (b) your rotational kinetic energy. Use the figures in Table 9.2 to model reasonable approximations for the pertinent parts of your body.

▲ **FIGURE 9.34** Problem 65.

66. •• A solid uniform spherical boulder rolls down the hill as shown in Figure 9.35, starting from rest when it is 50.0 m above the bottom. The upper half of the hill is free of ice, so the boulder rolls without slipping. But the lower half is covered with perfectly smooth ice. How fast is the boulder moving when it reaches the bottom of the hill?

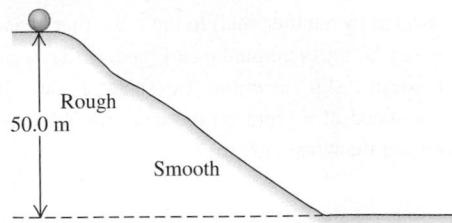

▲ **FIGURE 9.35** Problem 66.

67. •• A thin uniform rod 50.0 cm long with mass 0.320 kg is bent at its center into a V shape, with a 70.0° angle at its vertex. Find the moment of inertia of this V-shaped object about an axis perpendicular to the plane of the V at its vertex.

68. •• In redesigning a piece of equipment, you need to replace a solid spherical part of mass M with a hollow spherical shell of the same size. If both parts must spin at the same rate about an axis through their center, and the new part must have the same kinetic energy as the old one, what must be the mass of the new part in terms of M?

69. •• A solid uniform spherical stone starts moving from rest at the top of a hill. At the bottom of the hill the ground curves upward, launching the stone vertically a distance H below its start. How high will the stone go (a) if there is no friction on the hill and (b) if there is enough friction on the hill for the stone to roll without slipping? (c) Why do you get two different answers even though the stone starts with the same gravitational potential energy in both cases?

70. •• A solid, uniform ball rolls without slipping up a hill, as shown in Figure 9.36. At the top of the hill, it is moving horizontally; then it goes over the vertical cliff. (a) How far from the foot of the cliff does the ball land, and how fast is it moving just before it lands? (b) Notice that when the ball lands, it has a larger translational speed than it had at the bottom of the hill. Does this mean that the ball somehow *gained* energy by going up the hill? Explain!

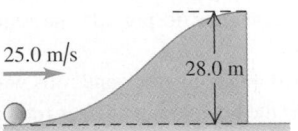

25.0 m/s

28.0 m

▲ **FIGURE 9.36** Problem 70.

71. •• **The kinetic energy of walking.** If a person of mass M
BIO simply moved forward with speed V, his kinetic energy would be $\frac{1}{2}MV^2$. However, in addition to possessing a forward motion, various parts of his body (such as the arms and legs) undergo rotation. Therefore, his total kinetic energy is the sum of the energy from his forward motion plus the rotational kinetic energy of his arms and legs. The purpose of this problem is to see how much this rotational motion contributes to the person's kinetic energy. Biomedical measurements show that the arms and hands together typically make up 13% of a person's mass, while the legs and feet together account for 37%. For a rough (but reasonable) calculation, we can model the arms and legs as thin uniform bars pivoting about the shoulder and hip, respectively. In a brisk walk, the arms and legs each move through an angle of about ±30° (a total of 60°) from the vertical in approximately 1 second. We shall assume that they are held straight, rather than being bent, which is not quite true. Let us consider a 75 kg person walking at 5.0 km/h, having arms 70 cm long and legs 90 cm long. (a) What is the average angular velocity of his arms and legs? (b) Using the average angular velocity from part (a), calculate the amount of rotational kinetic energy in this person's arms and legs as he walks. (c) What is the total kinetic energy due to both his forward motion and his rotation? (d) What percentage of his kinetic energy is due to the rotation of his legs and arms?

72. •• **The kinetic energy of running.** Using the previous prob-
BIO lem as a guide, apply it to a person running at 12 km/h, with his arms and legs each swinging through ±30° in $\frac{1}{2}$ s. As before, assume that the arms and legs are kept straight.

73. •• The pulley in Fig. 9.37 has radius R and a moment of inertia I. The rope does not slip over the pulley, and the pulley spins on a frictionless axle. The coefficient of kinetic friction between block A and the tabletop is μ_k The system is released

from rest, and block B descends. Block A has mass m_A and block B has mass m_B. Use energy methods to calculate the speed of block B as a function of the distance d that it has descended.

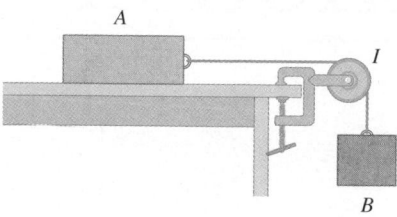

▲ **FIGURE 9.37** Problem 73.

Passage Problems

BIO The spinning wheel. When we first observe a wheel it is spinning *clockwise* at 30 rad/s, but when we observe it 10 seconds later it is spinning *counterclockwise* at 30 rad/s. (See Figure 9.38.) Assume that the wheel's angular acceleration is constant during the 10-second interval and that our point of view does not change.

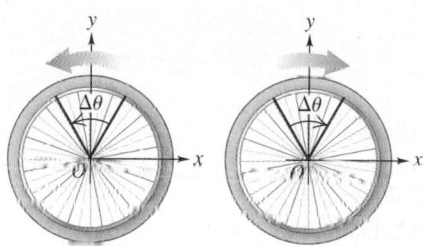

▲ **FIGURE 9.38** Problems 74–77.

74. What is the wheel's average angular velocity during the 10-s interval?
   A. 0 rad/s
   B. 15 rad/s
   C. 30 rad/s
   D. 60 rad/s

75. What is the wheel's angular acceleration during the 10-s interval?
   A. 0 rad/s²
   B. 1.5 rad/s²
   C. 3 rad/s²
   D. 6 rad/s²
   E. 60 rad/s²

76. What is the wheel's angular displacement over the 10-s interval?
   A. 0 rad
   B. 300 rad
   C. 600 rad
   D. 1500 rad

77. Assuming that the wheel continues with the same constant angular acceleration, what would its angular displacement be over the next 10-s interval (i.e., between 10 s and 20 s)?
   A. 0 rad
   B. 300 rad
   C. 450 rad
   D. 600 rad

10 Dynamics of Rotational Motion

This disk galaxy exemplifies conservation of *angular momentum*—the rotational analog of linear momentum. As this air moves inward, its angular velocity increases . . . as you see.

When you use a wrench to loosen lug nuts when you change a flat tire, the whole wheel starts to turn unless you find a way to hold it still. What is it about the force you apply to the handle of the wrench that makes the wheel turn? More generally, what is it that gives a rotating body an *angular* acceleration? A force is a push or a pull, but to produce rotational motion, we need a twisting or turning action.

In this chapter we'll define a physical quantity, *torque*, that describes the twisting or turning effort of a force. We'll incorporate this quantity into a fundamental dynamic relationship for rotational motion. We also need to look at work and power in rotational motion to understand such problems as how energy is transmitted by a rotating shaft (e.g., a drive shaft or an axle in a car or the shaft of a rotating turbine in a jet engine). Next, we'll develop a new conservation principle, *conservation of angular momentum*, that is useful in analyzing rotational motion, but that has much broader significance as well. We'll explore the relevance of torque for problems involving *equilibrium* of a rigid body. Then we'll conclude this chapter by studying briefly how gyroscopes work. Rotational motion can have some interesting surprises, as you may know if you've ever played with a toy gyroscope.

10.1 Torque

What is it about a force that determines how effective it is in causing or changing a rotational motion? The magnitude and direction of the force certainly play a role, but the point at which the force is applied is also important. When you try to swing a heavy door open, it's a lot easier if you push near the knob than if you push close to the hinges. The farther from the axis of rotation you push, the more effective the push is. In Figure 10.1, a wrench is being used to loosen a tight bolt. Note that force $\vec{F}_b$, applied at the end of the handle, is more effective than force $\vec{F}_a$, applied

MasteringPHYSICS

ActivPhysics 7.1: Calculating Torques

near the bolt, even though the two forces have equal magnitude. Force $\vec{F}_c$, directed along the length of the handle, does no good at all.

The quantitative measure of the tendency of a force to cause or change a body's rotational motion is called **torque.** Figure 10.2 shows a body that can rotate about an axis that passes through point O and is perpendicular to the plane of the figure. The body is acted on by a force $\vec{F}_1$ in the plane of the figure. The tendency of $\vec{F}_1$ to cause a rotation about point O depends on the magnitude F_1 of the force. It also depends on the perpendicular distance l_1 between the line of action of the force (i.e., the line along which the force vector lies) and O. The distance l_1 plays the role of the wrench handle, and we call it the **moment arm** (or *lever arm*) of force $\vec{F}_1$ about O. The twisting effort is proportional to both F_1 and l_1. We define the **torque,** or **moment,** of the force $\vec{F}_1$ with respect to point O as the product $F_1 l_1$. We use the Greek letter τ ("tau") to denote torque:

Definition of torque

Torque is a quantitative measure of the tendency of a force to cause or change rotational motion around a chosen axis. Torque is the product of the magnitude of the force and the moment arm, which is the perpendicular distance between the axis and the line of action of the force:

$$\tau = Fl \tag{10.1}$$

Unit: newton-meter (N · m)

NOTE ▶ The SI unit of torque is the newton-meter. In our discussion of work and energy, we called this combination the *joule.* But torque is *not* work or energy, and torque should be expressed in newton-meters, not joules. ◀

Figure 10.3 shows two additional forces acting on the body shown in Figure 10.2. The moment arm of $\vec{F}_2$ is the perpendicular distance l_2, just as the moment arm of $\vec{F}_1$ is the perpendicular distance l_1. Force $\vec{F}_3$ has a line of action that passes *through* the reference point O, so the moment arm for $\vec{F}_3$ is zero, and its torque with respect to point O is zero. Note that torque is always defined with reference to a specific point that could serve as an axis of rotation. Often (but not always), we take this point as the origin of our coordinate system. If we change the position of that point, the torque of each force usually also changes.

Force $\vec{F}_1$ in Figure 10.3 tends to cause *counterclockwise* rotation about O, and $\vec{F}_2$ tends to cause *clockwise* rotation. To distinguish between these two possibilities, we'll use the convention that **counterclockwise torques are positive and clockwise torques are negative.** With this convention, the torque τ_1 of $\vec{F}_1$ about O is

$$\tau_1 = +F_1 l_1$$

and the torque τ_2 of $\vec{F}_2$ is

$$\tau_2 = -F_2 l_2.$$

Note that this choice of signs is consistent with our choice for angular position, velocity, and acceleration in Section 9.1. Thus, we'll sometimes refer to counterclockwise rotation as the **positive sense of rotation.**

Figure 10.4 shows a force $\vec{F}$ applied at a point P. The vector $\vec{r}$ represents the position of P with respect to the chosen axis through O. There are several alternative ways to calculate the torque of this force. We can find the moment arm l and use $\tau = Fl$. Or we can determine the angle ϕ between the directions of vectors $\vec{F}$ and $\vec{r}$; the moment arm is $r\sin\phi$, so $\tau = rF\sin\phi$. Finally, we can represent $\vec{F}$ in

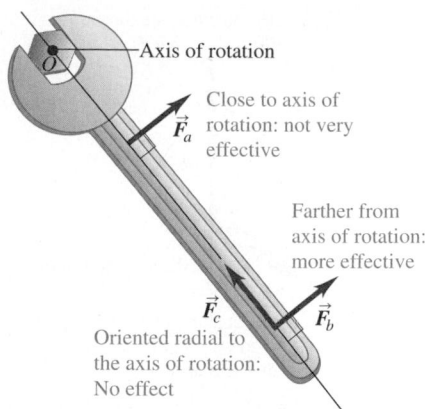

▲ **FIGURE 10.1** These three equal forces are *not* equally effective at loosening the nut.

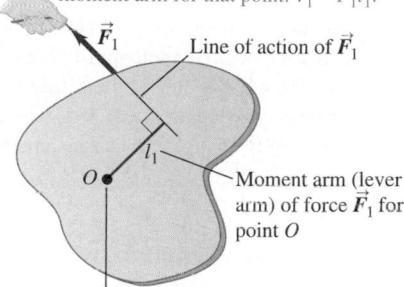

The torque τ exerted by $\vec{F}_1$ about point O is the product of the force magnitude and the moment arm for that point: $\tau_1 = F_1 l_1$.

▲ **FIGURE 10.2** The torque of a force about a point is the product of the magnitude of the force and the moment arm.

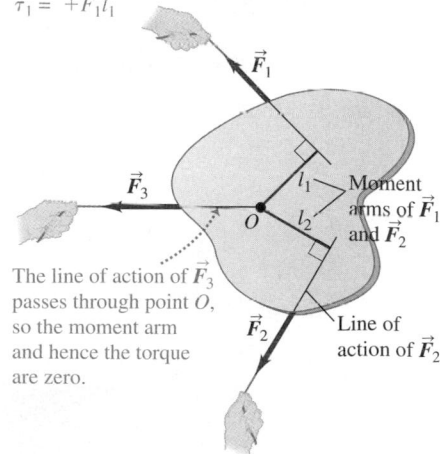

$\vec{F}_1$ tends to cause *counterclockwise* rotation about point O, so its torque is *positive:* $\tau_1 = +F_1 l_1$

The line of action of $\vec{F}_3$ passes through point O, so the moment arm and hence the torque are zero.

$\vec{F}_2$ tends to cause *clockwise* rotation about point O, so its torque is *negative:* $\tau_2 = -F_2 l_2$

▲ **FIGURE 10.3** The sign of a torque depends on whether it tends to cause clockwise or counterclockwise rotation about the chosen point. A force whose line of action passes through the chosen point has no torque.

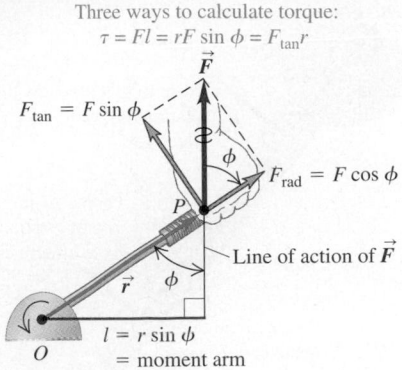

Three ways to calculate torque:
$\tau = Fl = rF\sin\phi = F_{tan}r$

$F_{tan} = F\sin\phi$

$F_{rad} = F\cos\phi$

Line of action of $\vec{F}$

$l = r\sin\phi$
= moment arm

▲ **FIGURE 10.4** The magnitude of the torque caused by a force $\vec{F}$ depends only on the tangential component of the force: $\tau = rF\sin\phi$.

terms of a radial component F_{rad} along the direction of $\vec{r}$ (the radial line from O to P) and a tangential component F_{tan} at right angles, perpendicular to $\vec{r}$. (We call F_{tan} a tangential component because when the object rotates, the point P at which the force acts moves in a circle, and this component is tangent to that circle.) Then $F_{tan} = F\sin\phi$, and

$$\tau = F_{tan}r = r(F\sin\phi) = Fl. \tag{10.2}$$

The component F_{rad} has no torque with respect to O, because its moment arm with respect to that point is zero.

NOTE ▶ Equations 10.1 and 10.2 give the *magnitude* of torque. When we use the sign conventions we've described (counterclockwise positive, clockwise negative), we have to be sure to attach the correct sign to each torque. ◀

EXAMPLE 10.1 A weekend plumber

An amateur plumber, unable to loosen a pipe fitting, slips a piece of scrap pipe (a "cheater") over the handle of his wrench. He then applies his full weight of 900 N to the end of the cheater by standing on it. The distance from the center of the fitting to the point where the weight acts is 0.80 m, and the wrench handle and cheater make an angle of 19° with the horizontal (Figure 10.5a). Find the magnitude and direction of the torque of his weight about the center of the pipe fitting.

SOLUTION

SET UP Figure 10.5b diagrams the position of the force exerted by the plumber in relation to the axis of rotation at point O.

SOLVE We have a choice of several methods for solving this problem. From Figure 10.5b, the angle ϕ between the vectors $\vec{r}$ and $\vec{F}$ is 109°, and the moment arm l is

$$l = (0.80\text{ m})(\sin 109°) = (0.80\text{ m})(\sin 71°)$$
$$= 0.76\text{ m}.$$

We can find the magnitude of the torque from either Equation 10.1 or Equation 10.2. From Equation 10.1,

$$\tau = Fl = (900\text{ N})(0.76\text{ m}) = 680\text{ N}\cdot\text{m}.$$

Or, from Equation 10.2,

$$\tau = rF\sin\phi = (0.80\text{ m})(900\text{ N})(\sin 109°) = 680\text{ N}\cdot\text{m}.$$

ALTERNATIVE SOLUTION Alternatively, we can start with the components of $\vec{F}$. From Figure 10.6b, we see that the component of force $\vec{F}$ perpendicular to $\vec{r}$ (which we call F_{tan}, as explained earlier) is

$$F_{tan} = F(\cos 19°) = (900\text{ N})(\cos 19°) = 850\text{ N}.$$

Then the torque is

$$\tau = F_{tan}r = (850\text{ N})(0.80\text{ m}) = 680\text{ N}\cdot\text{m}.$$

REFLECT The force tends to produce a counterclockwise rotation about O, so with the convention described above, this torque is $+680\text{ N}\cdot\text{m}$, *not* $-680\text{ N}\cdot\text{m}$. Also, please don't try to use this plumbing technique at home; you're very likely to break a pipe.

Practice Problem: If the pipe fitting under consideration can withstand a maximum torque of $1000\text{ N}\cdot\text{m}$ without breaking, what is the maximum moment arm at which our amateur plumber can safely apply his full weight? *Answer:* 1.12 m.

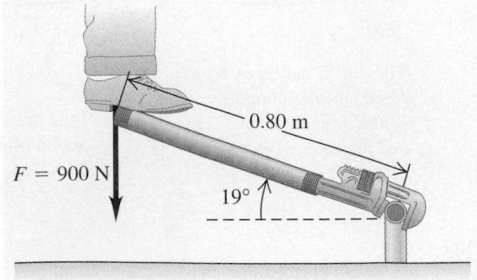

$F = 900\text{ N}$

0.80 m

19°

(a) Diagram of situation

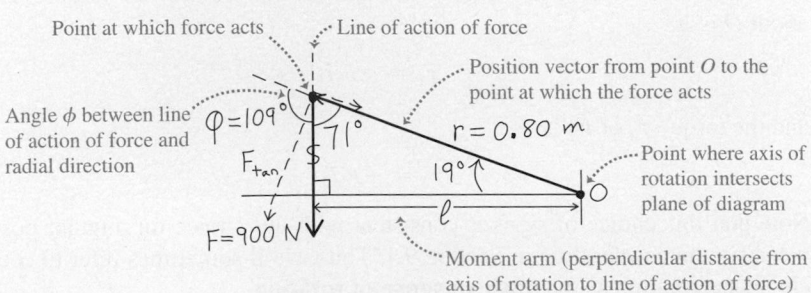

Point at which force acts

Line of action of force

Angle ϕ between line of action of force and radial direction

$\phi = 109°$

71°

F_{tan}

$F = 900\text{ N}$

Position vector from point O to the point at which the force acts

$r = 0.80\text{ m}$

19°

ℓ

Point where axis of rotation intersects plane of diagram

Moment arm (perpendicular distance from axis of rotation to line of action of force)

(b) Free-body diagram

▲ **FIGURE 10.5**

▶ **Application Don't fall for it.** To stay balanced while walking the tightrope, this acrobat must keep her center of mass directly above the rope. If the center of mass shifts to one side, gravity will exert a torque on the acrobat, tending to cause a rotation about the rope—a fall. To stay balanced, the acrobat must create a countertorque having equal magnitude and opposite sign. That is why tightrope walkers carry a long pole. Because the torque due to a force is the product of the magnitude of the force and the moment arm, the gravitational forces on the ends of the pole exert significant torques. Thus, the acrobat controls the net torque acting on her center of gravity by manipulating the pole. Also, the acrobat holds the pole low, moving her center of gravity closer to the rope.

Often, one of the important forces acting on a body is its weight. This force is not concentrated at a single point, but is distributed over the entire body. Nevertheless, it turns out that **in a uniform gravitational field, the total torque on a body due to its weight is the same as though all the weight were concentrated at the *center of mass* of the body.** We'll derive this relationship later in this chapter and use it for several problems.

10.2 Torque and Angular Acceleration

We're now ready to develop the fundamental dynamical relation for rotational motion of a rigid body. We'll show that the angular acceleration of a rotating body is directly proportional to the sum of the torques with respect to the axis of rotation. The proportionality factor is the moment of inertia.

To develop this relationship, we again imagine a body as being made up of a large number of particles. Figure 10.6 shows a particular particle, labeled A, that has mass m_A and is located a perpendicular distance r_A from the axis of rotation of the body. The *total force* $\vec{F}_A$ acting on this particle has a component $F_{A,\text{rad}}$ along the radial direction and a component $F_{A,\text{tan}}$ tangent to the circle of radius r_A in which the particle moves as the body rotates. (The force $\vec{F}_A$ may also have a component parallel to the axis of rotation, but this component has no torque with respect to the axis and therefore can be ignored in the analysis that follows.) Newton's second law for the tangential components is

$$F_{A,\text{tan}} = m_A a_{A,\text{tan}}. \tag{10.3}$$

We can express $a_{A,\text{tan}}$ in terms of the angular acceleration α, using Equation 9.14: $a_{A,\text{tan}} = r_A \alpha$. Using this relation and multiplying both sides of Equation 10.3 by r_A we obtain

$$F_{A,\text{tan}} r_A = m_A r_A (r_A \alpha) = (m_A r_A^2)\alpha. \tag{10.4}$$

Now, $F_{A,\text{tan}} r_A$ is just the torque τ_A of the force F_A with respect to the axis through point O. Also, $m_A r_A^2$ is the moment of inertia of particle A about this axis. The component $F_{A,\text{rad}}$ acts along a line passing through the axis of rotation and so has no torque with respect to the axis. Hence, we can rewrite Equation 10.4 as

$$\tau_A = I_A \alpha = (m_A r_A^2)\alpha.$$

We can write an equation like this for every particle in the body and then add all the equations together:

$$\tau_A + \tau_B + \tau_C + \cdots = (m_A r_A^2 + m_B r_B^2 + m_C r_C^2 + \cdots)\alpha$$

The left side of this equation is the sum of *all* the torques acting on *all* the particles. We'll call this sum simply $\Sigma\tau$. The right side of the equation is the total moment of inertia (Equation 9.16), namely,

$$I = m_A r_A^2 + m_B r_B^2 + m_C r_C^2 + \cdots,$$

multiplied by the angular acceleration α, which is the same for every particle in the body. Thus, for the entire body, we have the following relationship:

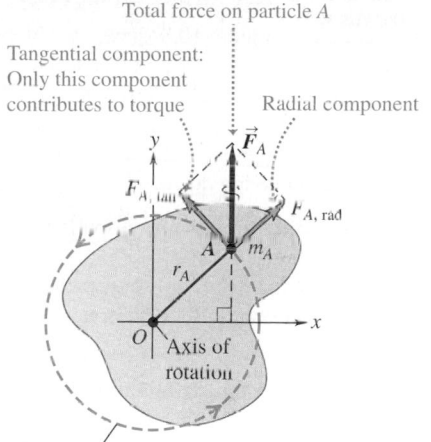

Total force on particle A

Tangential component: Only this component contributes to torque

Radial component

Path of particle A as body rotates

▲ **FIGURE 10.6** The components of the net force acting on one of the particles of a rigid body.

MasteringPHYSICS

ActivPhysics 7.8: Rotoride—Dynamics Approach
ActivPhysics 7.9: Falling Ladder
ActivPhysics 7.10: Woman and Flywheel Elevator—Dynamics Approach

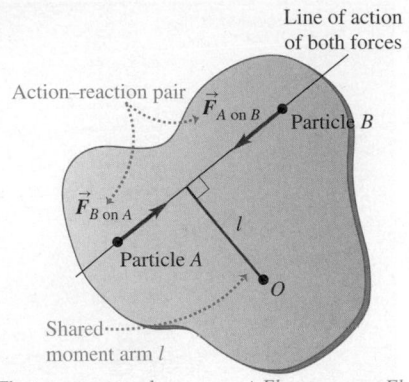

Line of action
of both forces

Action–reaction pair

Particle B

Particle A

Shared
moment arm l

The torques cancel: $\tau_{A \text{ on } B} = +Fl$; $\tau_{B \text{ on } A} = -Fl$

▲ **FIGURE 10.7** Why internal forces do not contribute to net torque.

Relation of torque to angular acceleration

The net torque (about a chosen axis) of all the forces acting on a rigid body equals the body's moment of inertia (about that axis), multiplied by its angular acceleration:

$$\sum \tau = I\alpha. \qquad (10.5)$$

This equation is the rotational analog of Newton's second law, $\sum \vec{F} = m\vec{a}$. It is the dynamic relationship we need to relate the rotational motion of a rigid body to the forces acting on it.

Now let's think a little more about the forces. The force $\vec{F}$ on each particle is the vector sum of the external and internal forces acting on it, as we defined these in Section 8.2. The internal forces that any pair of particles exert on each other are an action–reaction pair. According to Newton's third law, they have equal magnitude and opposite direction (Figure 10.7). Assuming that they act along the line joining the two particles, their *moment arms* with respect to any axis are also equal, so the *torques* for each such pair add to zero. Indeed, **the torques due to all the internal forces add to zero,** so the sum of the torques on a body includes only the torques of the *external* forces acting on the body.

Conceptual Analysis 10.1

Net torque on a pulley

Figure 10.8 shows two blocks with equal mass suspended by a cord over a pulley whose mass is half the mass of each block. Block B is given an initial push downward, after which it moves downward with constant velocity. There is no friction in the pulley axle, and the cord's weight can be ignored. While B is moving down,

A. the left cord pulls on the pulley with greater force than the right cord.
B. the left and right cords pull with equal force on the pulley.
C. the right cord pulls with greater force on the pulley than the left cord.

SOLUTION After the initial push, each block moves with constant velocity and so is in equilibrium. The tension in each side of the cord is equal to the weight of one block, so the tensions on the two sides of the cord are equal. The torques applied to the pulley

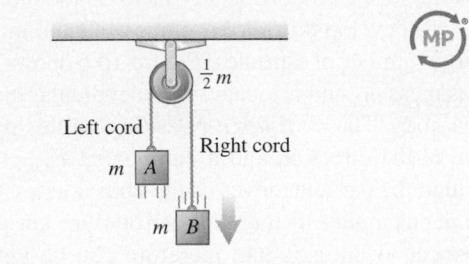

$\frac{1}{2}m$

Left cord

Right cord

m A

m B

▲ **FIGURE 10.8**

by the two sides of the cord are equal in magnitude but opposite in sign. Therefore, the net torque on the pulley is zero, and its angular acceleration is zero. The two blocks move with constant velocity, and the pulley turns with constant angular velocity. If either answer A or answer C were correct, the pulley would have a nonzero angular acceleration. Answer B is correct.

Quantitative Analysis 10.2

Rotating masses

Two heavy masses are mounted on a light rod that can be rotated by a string wrapped around a central cylinder of negligible mass, forming a winch (Figure 10.9). A force F is applied to the string to turn the system. With respect to the variables given in the figure, the equation for the angular acceleration α is

A. $rF/2mR^2$.
B. $RF/2mr^2$.
C. $rF/3mR^2$.
D. $RF/3mr^2$.

SOLUTION From Equation 10.5, $\alpha = (\sum \tau)/I$. The only torque with respect to the axis of rotation is supplied by the string force

(magnitude F), which is always tangent to the surface of the small cylinder. The moment arm of the force is the radius r of the cylinder, so the torque is $\tau = rF$. The moment of inertia I is the sum of the mR^2 terms for each mass. Because both m and $2m$ are at distance R from the axis of rotation, the total moment of inertia is $3mR^2$. Combining the results for torque and moment of inertia, we get $rF/3mR^2$ (answer C).

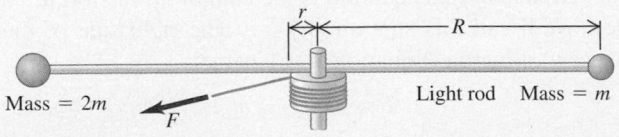

r

R

Mass = $2m$

F

Light rod   Mass = m

▲ **FIGURE 10.9**

PROBLEM-SOLVING STRATEGY 10.1 **Rotational dynamics**

Our strategy for solving problems in rotational dynamics is similar to
Problem-Solving Strategy 5.2 on the use of Newton's second law.

SET UP

1. Sketch the situation and select a body or bodies to analyze. You will apply
    $\Sigma\vec{F} = m\vec{a}$ or $\Sigma\tau = I\alpha$, or sometimes both, to each object.
2. Draw a free-body diagram for each body. Because we're now dealing with
    extended bodies rather than objects that can be treated as particles, your
    diagram must show the shape of the body accurately, with all dimensions
    and angles that you'll need for calculations of torque. A body may have
    translational motion, rotational motion, or both. As always, include all the
    forces that act on each body (and no others), including the body's weight
    (taken to act at the center of mass).
3. Choose coordinate axes for each body. Remember that, by convention, the
    positive sense of rotation is counterclockwise; this is the positive sense for
    angular position, angular velocity, angular acceleration, and torque.
4. If more than one object is involved, carry out the preceding steps for each
    object. There may be geometrical relations between the motions of two or
    more objects or between the translational and rotational motions of the
    same object. Express these in algebraic form, usually as relations between
    two linear accelerations or between a linear acceleration and an angular
    acceleration.

SOLVE

5. Write a separate equation of motion ($\Sigma\vec{F} = m\vec{a}$ or $\Sigma\tau = I\alpha$, or both) for
    each body, including separate equations for the x and y components of
    forces if necessary. Then solve the equations to find the unknown quanti-
    ties. This may involve solving a set of simultaneous equations.

REFLECT

6. Check particular cases or extreme values of quantities when possible, and
    compare the results for these particular cases or values with your intuitive
    expectations. Ask yourself whether the result makes sense.

EXAMPLE 10.2 **Unwinding a winch (again)**

Figure 10.10a shows the same situation that we analyzed in Example 9.7. A cable is wrapped several times
around a uniform solid cylinder with diameter 0.12 m and mass 50 kg that can rotate freely about its axis.
The cable is pulled by a force with magnitude 9.0 N. Assuming that the cable unwinds without stretching or
slipping, find the magnitude of its acceleration.

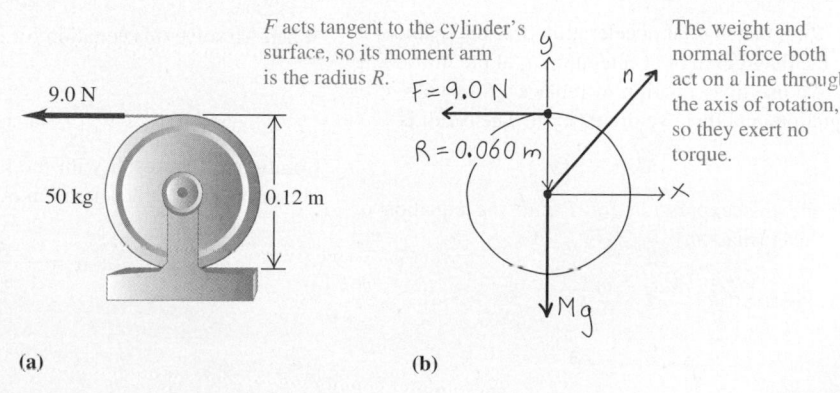

▶ **FIGURE 10.10** (a)   (b)

Continued

SOLUTION

SET UP We draw a free-body diagram for the cylinder (Figure 10.10b). Because the force acts tangent to the outside of the cylinder, its moment arm relative to the axis of rotation is the cylinder's radius.

SOLVE The torque is $\tau = Fl = (9.0 \text{ N})(0.060 \text{ m}) = 0.54 \text{ N} \cdot \text{m}$. The moment of inertia is

$$I = \tfrac{1}{2}MR^2 = \tfrac{1}{2}(50 \text{ kg})(0.060 \text{ m})^2 = 0.090 \text{ kg} \cdot \text{m}^2.$$

The angular acceleration α is given by Equation 10.5:

$$\alpha = \frac{\tau}{I} = \frac{0.54 \text{ N} \cdot \text{m}}{0.090 \text{ kg} \cdot \text{m}^2} = 6.0 \text{ rad/s}^2.$$

The magnitude a of the acceleration of the cable is given by Equation 9.14:

$$a = R\alpha = (0.060 \text{ m})(6.0 \text{ rad/s}^2)$$
$$= 0.36 \text{ m/s}^2.$$

REFLECT In the equation $\alpha = \tau/I$, check the units to make sure they are consistent.

Practice Problem: Use the preceding results, together with an equation from Chapter 2, to determine the speed of the cable after it has been pulled 2.0 m. Compare your result with Example 9.7, in which we found this speed by using work and energy considerations. *Answer:* 1.2 m/s.

EXAMPLE 10.3 Dynamics of a bucket in a well

Let's go back to the old-fashioned well in Example 9.8. Find the acceleration of the bucket (mass m) and the angular acceleration of the winch cylinder.

SOLUTION

SET UP We start by sketching the situation, using the information from Example 9.8 (Figure 10.11a). As before, the winch cylinder has mass M and radius R; the bucket has mass m and falls a distance h. We must treat the two objects separately, so we draw a free-body diagram for each (Figure 10.11b). We take the positive direction of the y coordinate for the bucket to be downward; the positive sense of rotation for the winch is counterclockwise. We assume that the rope is massless and that it unwinds without slipping.

SOLVE Newton's second law applied to the bucket (mass m) gives

$$\Sigma F_y = mg - T = ma_y.$$

The forces on the cylinder are its weight Mg, the upward normal force n exerted by the axle, and the rope tension T. The first two forces act along lines through the axis of rotation and thus have no torque with respect to that axis. The moment arm of the force T on the cylinder is R, and the torque is $\tau = TR$. Applying Equation 10.5 and the expression for the moment of inertia of a solid cylinder (Table 9.2) to the winch cylinder, we find that

$$\Sigma\tau = I\alpha \quad \text{and} \quad TR = \tfrac{1}{2}MR^2\alpha.$$

As in Example 10.2, the angular acceleration α of the cylinder is proportional to the magnitude of acceleration a_y of the unwinding rope: $a_y = R\alpha$. We use this equation to replace $(R\alpha)$ with a_y in the cylinder equation, and then we divide by R. The result is

$$T = \tfrac{1}{2}Ma_y.$$

Now we substitute this expression for T into the equation of motion for the bucket (mass m):

$$mg - \tfrac{1}{2}Ma_y = ma_y.$$

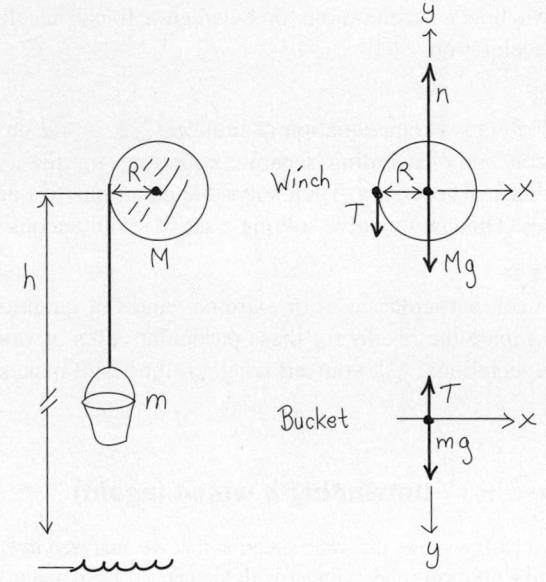

(a) Diagram of situation   (b) Free-body diagrams

▲ **FIGURE 10.11** Our sketches for this problem.

When we solve this equation for a_y, we get

$$a_y = \frac{g}{1 + M/2m}.$$

Combining this result with the kinematic equation $a_y = R\alpha$, we obtain the angular acceleration α:

$$\alpha = \frac{g/R}{1 + M/2m}.$$

Continued

Finally, we can substitute the equation for a_y back into the equation $\sum \vec{F} = m\vec{a}$ for m to get an expression for the tension T in the rope:

$$T = mg - ma_y = mg - m\frac{g}{1 + M/2m} \quad \text{and} \quad T = \frac{mg}{1 + 2m/M}.$$

REFLECT First, note that the tension in the rope is *less than* the weight mg of the bucket; if the two forces were equal (in magnitude), the bucket wouldn't accelerate downward.

We can check two particular cases. When M is much larger than m (a massive winch and a little bucket), the ratio m/M is much smaller than unity. Then the tension is nearly equal to mg, and the acceleration a_y is correspondingly much *less* than g. When M is zero, $T = 0$ and $a = g$; the bucket then falls freely. If it starts from rest at a height h above the water, its speed v when it strikes the water is given by $v^2 = 2a_yh$; thus,

$$v = \sqrt{2a_yh} = \sqrt{\frac{2gh}{1 + M/2m}}.$$

This is the same result that we obtained from energy considerations in Example 9.8. But note that there we couldn't find the tension in the rope or the accelerations of the winch cylinder and bucket using energy considerations.

Practice Problem: (a) What is the relation between m and M if the bucket's acceleration is half the acceleration of free fall? (b) In this case, what is the rope tension T? *Answer:* (a) $m = M/2$; (b) $T = mg/2 = Mg/4$.

NOTE ▶ In problems such as Example 10.3, in which an object is suspended by a rope, you may be tempted to assume at the start that the tension in the rope is equal to the object's weight. If the object is in equilibrium, this may be true, but when the object accelerates, the tension is usually *not* equal to the object's weight. Be careful! ◀

Rotation about a Moving Axis

In Section 9.5, we discussed the kinetic energy of a rigid body rotating about a moving axis. We can extend our analysis of the dynamics of rotational motion to some cases in which the axis of rotation moves. In these cases, the body has both translational and rotational motion. As we noted in Section 9.5, every possible motion of a rigid body can be represented as a combination of translational motion of the center of mass and rotation about an axis through the center of mass. A rolling wheel or ball and a yo-yo unwinding at the end of a string are familiar examples of such motion.

The key to this more general analysis, which we won't derive in detail, is that Equation 10.5 ($\sum \tau = I\alpha$) is valid *even when the axis of rotation moves* if the following two conditions are met: (1) The axis must be an axis of symmetry and must pass through the center of mass of the body; (2) the axis must not change direction. Note that the moving axis is not necessarily at rest in an inertial frame of reference. The moment of inertia must be computed with respect to the moving axis through the center of mass; to emphasize this point, we'll denote it as I_{cm}.

MasteringPHYSICS

ActivPhysics 7.11: Race Between a Block and a Disk

PROBLEM-SOLVING STRATEGY 10.2 **Rotation about a moving axis** (MP)

The problem-solving strategy outlined in Section 10.2 is equally useful here. There is one new wrinkle: When a rigid body undergoes translational and rotational motion at the same time, we need two separate equations of motion *for the same body*. (The situation is reminiscent of our analysis of projectile motion in Chapter 3, where we used separate equations for the x and y coordinates of a projectile.)

For a rigid body, one of the equations of motion is based on $\sum \vec{F} = m\vec{a}$ for the translational motion of the center of mass of the body. We showed in Equation 8.23 that for a body with total mass M, the acceleration $\vec{a}_{cm}$ of the center of mass is the same as that of a point mass M acted on by all the forces that act on the actual body. The other equation of motion is based on $\sum \tau = I_{cm}\alpha$ for the rotational motion of the body about the axis through the center of mass with moment of inertia I_{cm}. In addition, there is often a geometric relation between the linear and angular accelerations, such as when a wheel rolls without slipping or a string rotates a pulley while passing over it.

EXAMPLE 10.4 Dynamics of a primitive yo-yo

Let's consider again the yo-yo we analyzed in Example 9.9. As shown in Figure 10.12a, the yo-yo consists of a solid disk with radius R and mass M. In our previous analysis, we used energy considerations to find the yo-yo's speed after it had dropped a certain distance. Now let's find the acceleration of the yo-yo and the tension in the string (which we can't do by using energy considerations alone).

SOLUTION

SET UP We start with a free-body diagram, including a coordinate system (Figure 10.12b). If the string unwinds without slipping, the linear displacement of the center of mass of the cylindrical yo-yo in any time interval equals R times its angular displacement. This gives us the kinematic relations $v_{cm,y} = R\omega$ and $a_{cm,y} = R\alpha$. (If this scenario isn't obvious, imagine moving along with the center of mass of the cylinder and watching the string unwind. From that point of view, the kinematic situation is just the same as in Examples 10.2 and 10.3.)

SOLVE The equation for the translational motion of the center of mass is

$$\sum F_y = Mg - T = Ma_{cm,y}.$$

The moment of inertia for an axis through the center of mass is $I_{cm} = \frac{1}{2}MR^2$, and the equation for rotational motion about the axis through the center of mass is

$$\sum \tau = TR = I_{cm}\alpha = \frac{1}{2}MR^2\alpha.$$

We can now combine this equation with the relation $a_{cm,y} = R\alpha$ to eliminate α, obtaining $TR = \frac{1}{2}MRa_{cm,y}$, or $T = \frac{1}{2}Ma_{cm,y}$. Finally, we combine this result with the equation for the center-of-mass motion ($\sum F_y = Mg - T = Ma_{cm,y}$) to obtain expressions for T and $a_{cm,y}$. The results are amazingly simple:

$$a_{cm,y} = \frac{2}{3}g, \qquad T = \frac{1}{3}Mg.$$

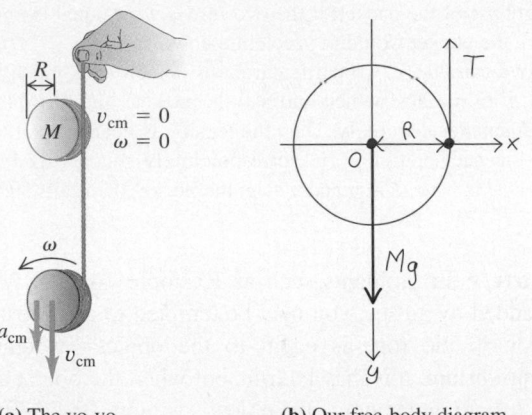

(a) The yo-yo   (b) Our free-body diagram

▲ **FIGURE 10.12**

REFLECT The acceleration and tension are both proportional to g, as we should expect. The acceleration is independent of M and R, and the mass cancels out of the dynamic equations. For a yo-yo in the form of a solid disk, the acceleration is exactly two-thirds the acceleration of free fall.

Practice Problem: If the solid cylinder is replaced by a thin cylindrical shell with the same mass and radius as before, find $a_{cm,y}$ and T. *Answer:* $a_{cm,y} = g/2$, $T = Mg/2$.

EXAMPLE 10.5 A rolling bowling ball

A bowling ball rolls without slipping down the return ramp at the side of the alley (Figure 10.13a). The ramp is inclined at an angle β to the horizontal. What is the ball's acceleration? What is the friction force acting on the ball? Treat the ball as a uniform solid sphere, ignoring the finger holes.

SOLUTION

SET UP Again, let's start with a free-body diagram, with the positive coordinate directions indicated (Figure 10.13b). Referring back to Table 9.2, we see that the moment of inertia of a solid sphere about an axis through its center of mass is $I_{cm} = \frac{2}{5}MR^2$. Because the ball rolls without slipping, the acceleration $a_{cm,x}$ of the center of mass is proportional to the ball's angular acceleration α ($a_{cm,x} = R\alpha$). We've represented the forces in terms of their components; it's most convenient to take the axes as parallel and perpendicular to the sloping ramp. The contact force of the ramp on the ball is represented in terms of its normal (n) and frictional (f_s) components. Note that f_s is a *static* friction force, because the ball doesn't slip on the ramp.

SOLVE The equations of motion for translational motion and for rotation about the axis through the center of mass are:

Translation: $\sum F_x = Mg\sin\beta - f_s = Ma_{cm,x},$
Rotation: $\sum \tau = f_s R = I_{cm}\alpha = \left(\frac{2}{5}MR^2\right)\alpha.$

We eliminate α from the second equation by using the kinematic relation $a_{cm} = R\alpha$. Then we express I_{cm} in terms of M and R and divide through by R:

$$f_s R = I_{cm}\alpha = \left(\frac{2}{5}MR^2\right)\left(\frac{a_{cm,x}}{R}\right) \quad \text{and} \quad f_s = \frac{2}{5}Ma_{cm,x}.$$

Next, we insert this expression for f_s into the first equation above and solve for $a_{cm,x}$:

$$Mg\sin\beta - \frac{2}{5}Ma_{cm,x} = Ma_{cm,x},$$
$$a_{cm,x} = \frac{5}{7}g\sin\beta.$$

Finally, we substitute the right-hand side of this result back into the expression for f_s to obtain

$$f_s = \frac{2}{7}Mg\sin\beta.$$

Continued

REFLECT The acceleration is $\frac{5}{7}$ as large as it would be if the ball could *slide* without friction down the slope, like the toboggan in Example 5.12 (Section 5.2). That is why the friction force f_s is a *static* friction force; it is needed to prevent slipping and to give the ball its angular acceleration. We can derive an expression for the minimum coefficient of static friction μ_s needed to prevent slipping. The magnitude of the normal force is $n = Mg\cos\beta$. To prevent slipping, the coefficient of friction must be at least as great as

$$\mu_s = \frac{\frac{2}{7}Mg\sin\beta}{Mg\cos\beta} = \frac{2}{7}\tan\beta.$$

If the plane is tilted only a little, β and $\tan\beta$ are small, and only a small value of μ_s is needed to prevent slipping. But as the angle increases, the required value of μ_s also increases, as we would expect intuitively.

Finally, suppose μ_s is *not* large enough to prevent slipping. Then we have a whole new ball game, so to speak. Our basic dynamic equations are still valid, but now the ball slides and rotates at the same time. There is no longer a definite relation between $a_{cm,x}$ and α, and the friction force is now a *kinetic* friction force, given by $f_k = \mu_k n = \mu_k Mg\cos\beta$. The acceleration of the center of mass is then

$$a_{cm,x} = g(\sin\beta - \mu_k\cos\beta).$$

This is the same result that we found in Example 5.12 (Section 5.3), in which we hoped that the toboggan would slide, and not roll, down the hill.

Practice Problem: Suppose we replace the bowling ball with a solid cylinder. Then the minimum coefficient of static friction needed to prevent slipping is $\mu_s = k\tan\beta$, where k is a dimensionless constant (equal to $\frac{2}{7}$ for the bowling ball). Determine the value of k for the cylinder. *Answer:* $k = \frac{1}{3}$.

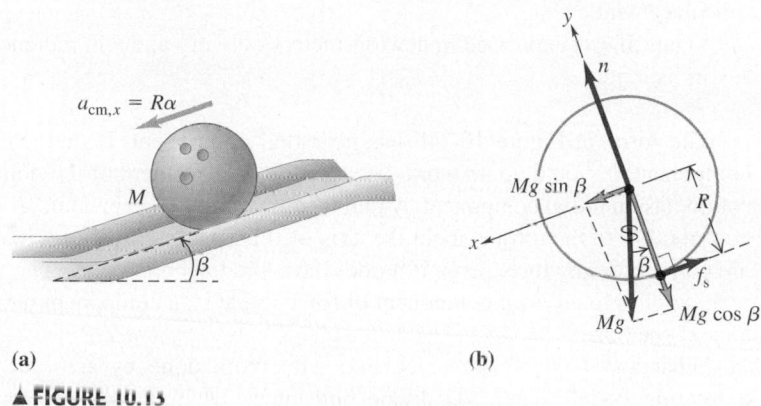

(a)

(b)

▲ **FIGURE 10.13**

10.3 Work and Power in Rotational Motion

When you pedal a bicycle, you apply forces to a rotating body and do work on it. Similar things happen in many real-life situations, such as a rotating motor shaft driving a power tool or a car engine propelling the vehicle. We can express this work in terms of torque and angular displacement.

Suppose a tangential force with constant magnitude F_{tan} acts at the rim of a merry-go-round platform with radius R (Figure 10.14) while the platform rotates through an angle $\Delta\theta$ about a fixed axis during a small time interval Δt. The displacement of the point at which the force acts is Δs. By definition, the work ΔW done by the force F_{tan} is $\Delta W = F_{tan}\Delta s$. If θ is measured in radians, $\Delta s = R\Delta\theta$, and it follows that

$$\Delta W = F_{tan}R \, \Delta\theta.$$

Now, $F_{tan}R$ is the *torque* τ due to the force F_{tan}, so

$$\Delta W = \tau \, \Delta\theta. \tag{10.6}$$

If the torque is *constant* while the angle changes by a finite amount $\Delta\theta = \theta_2 - \theta_1$, then the work done on the body by the torque is as follows:

▲ **BIO Application Sensing angular acceleration.** How do we detect angular movement so that we can perform complex activities such as tumbling and diving? The mammalian inner ear has three semicircular canals filled with a viscous fluid. The canals are oriented such that they represent each of the three spatial axes. In response to angular acceleration, the canals and the sensory cells they contain rotate relative to the internal fluid, and this motion is detected by the cells and converted to electrical signals that are sent to the brain. Long before humans understood them, physical laws shaped evolution's design of life.

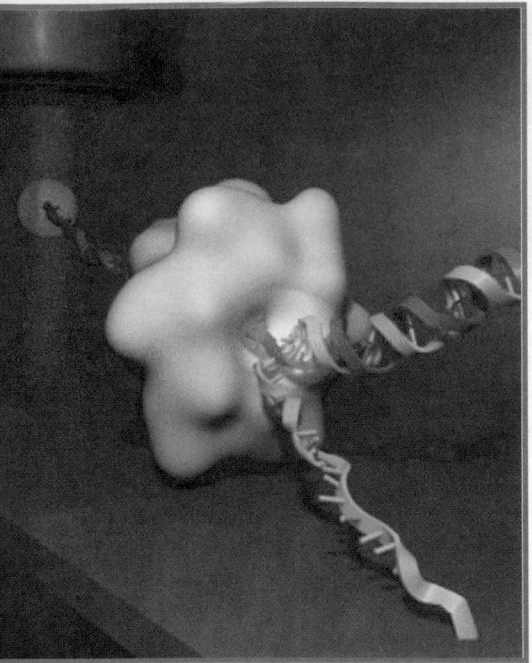

◄ **BIO** Application **Physics showed the way.** When Watson and Crick proposed the famous DNA double helix as the repository of genetic information, some scientists argued that an insurmountable amount of work would be required to unwind the helix during its replication, a necessary step in duplicating genes during cell division. The very long DNA molecule must rotate as it unwinds, and the friction between the rotating DNA and the fluid requires a continuous application of torque. Some thought that the torque required might be beyond the ability of biological molecular mechanisms to supply. A straightforward physics calculation showed that the energy required to overcome resistance to rotation was insignificant. The molecular motor responsible for unwinding DNA is called helicase (in yellow), which uses the energy stored in ATP to do its work, as is typical with molecular motors.

Work done by a constant torque

The work done on a body by a constant torque equals the product of the torque and the angular displacement of the body:

$$W = \tau(\theta_2 - \theta_1) = \tau \, \Delta\theta. \qquad (10.7)$$

This equation is the rotational analog of Equation 7.1, $W = F_{\parallel}s$, for the work done by a constant force with component $F_{\parallel}$ in the direction of displacement.

Unit: If τ is expressed in newton-meters $(\text{N} \cdot \text{m})$ and θ in radians, the work is in joules.

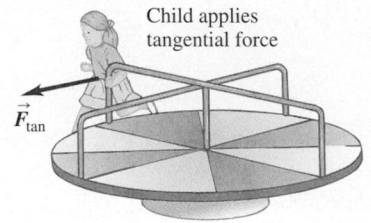

The force in Figure 10.14b has no radial component. If there were such a component, it would do no work, because the displacement of the point of application has no radial component. A radial component of force would also make no contribution to the torque about the axis of rotation, so Equations 10.6 and 10.7 are correct for any force, even if it does have a radial component. The same argument applies to an *axial* component of force—that is, a component parallel to the axis of rotation.

What about the *power* associated with work done by a torque acting on a rotating body? When we divide both sides of Equation 10.7 by the time interval Δt during which the angular displacement occurs, we find that

$$\frac{\Delta W}{\Delta t} = \tau \frac{\Delta\theta}{\Delta t}.$$

But $\Delta W / \Delta t$ is the rate of doing work, or *power P*, and $\Delta\theta / \Delta t$ is the body's angular velocity ω, so we obtain the following relation for power in rotational motion:

$$P = \tau\omega. \qquad (10.8)$$

When a torque τ (with respect to the axis of rotation) acts on a body that rotates with angular velocity ω, its power (rate of doing work) is the product of torque and angular velocity. This is the analog of the relation $P = Fv$ that we developed in Chapter 7 for particle motion.

NOTE ► In the relations just derived, angular displacements *must* be expressed in radians and angular velocities in radians per second. ◄

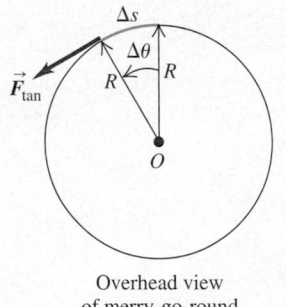

(b)

▲ **FIGURE 10.14** The work done by a tangential force acting on a body.

EXAMPLE 10.6 **Power of an electric motor**

An electric motor exerts a constant torque of $\tau = 10 \, \text{N} \cdot \text{m}$ on a grindstone mounted on its shaft; the moment of inertia of the grindstone is $I = 2.0 \, \text{kg} \cdot \text{m}^2$. If the system starts from rest, find the work done by the motor in 8.0 s and the kinetic energy at the end of this time. What was the average power delivered by the motor?

Continued

SOLUTION

SET UP AND SOLVE We want to use Equation 10.7 ($W = \tau\,\Delta\theta$) to find the total work done and then divide that by the time interval $\Delta t = 8.0$ s to find the average power (work per unit time). But first we need to find the total angle $\Delta\theta$ (in radians) through which the grindstone turns. The angular acceleration α is constant, so we can use $\tau = I\alpha$ to find α and then apply the formulas developed in Chapter 9 for constant angular acceleration.

The angular acceleration is

$$\alpha = \frac{\tau}{I} = \frac{10\,\text{N}\cdot\text{m}}{2.0\,\text{kg}\cdot\text{m}^2} = 5.0\,\text{rad/s}^2.$$

For constant angular acceleration,

$$\theta = \theta_0 + \omega_0 t + \tfrac{1}{2}\alpha t^2 = 0 + 0 + \tfrac{1}{2}(5.0\,\text{rad/s}^2)(8\,\text{s})^2 = 160\,\text{rad}.$$

The total work is

$$W - \tau\,\Delta\theta = (10\,\text{N}\cdot\text{m})(160\,\text{rad}) = 1600\,\text{J}.$$

Finally, the average power is the total work divided by the time interval:

$$P_{av} = \frac{W}{\Delta t} = \frac{1600\,\text{J}}{8.0\,\text{s}} = 200\,\text{J/s} = 200\,\text{W}.$$

REFLECT We can check the energy relations. The angular velocity of the grindstone after 8.0 s is

$$\omega = \alpha t = (5.0\,\text{rad/s}^2)(8.0\,\text{s}) = 40\,\text{rad/s}.$$

The grindstone's kinetic energy at this time is

$$K = \tfrac{1}{2}I\omega^2 = \tfrac{1}{2}(2.0\,\text{kg}\cdot\text{m}^2)(40\,\text{rad/s})^2 = 1600\,\text{J}.$$

This amount equals the total work done, as of course it must. Note that the power is *not* constant during this process. The torque is constant, but as the angular velocity increases, so does the rate of doing work. The instantaneous power, $\tau\omega$, increases linearly with time, from zero, when $\omega = 0$, to 400 W, with an average value of 200 W.

Practice Problem: In this problem, suppose we use a larger motor that applies a constant torque of 20 N · m. Find the total work done in 8.0 s and the average power during this period. *Answer:* 6400 J, 800 W.

10.4 Angular Momentum

An Olympic ice skater with outstretched arms sets herself spinning about a vertical axis. When she pulls her arms in to her sides, she spins faster. When an Olympic diver goes off the high board in a "layout" position and then pulls his knees in toward his chest into a "tuck" position, he spins faster, even though there is no appreciable external torque acting on him. Both of these familiar phenomena, and many others (Figure 10.15), can be understood on the basis of a quantity called *angular momentum*, which we'll introduce in this section.

Every rotational quantity that we've seen so far in this chapter and the previous one is the analog of some quantity related to the motion of a particle. Angular momentum is the analog of the linear momentum ($\vec{p} = m\vec{v}$) of a particle. We define angular momentum as follows:

Definition of angular momentum

When a rigid body with moment of inertia I (with respect to a specified symmetry axis) rotates with angular velocity ω about that axis, the **angular momentum** of the body with respect to the axis is the product of the moment of inertia I about the axis and the angular velocity ω. We denote angular momentum by L:

$$L = I\omega. \tag{10.9}$$

The sign of angular momentum depends on the sign of ω; thus, according to our usual convention, it is positive for counterclockwise rotation and negative for clockwise rotation.

Unit: $\text{kg}\cdot\text{m}^2/\text{s}$

▲ **FIGURE 10.15** As this diver goes into a tuck, her rotation speeds up, because angular momentum is conserved.

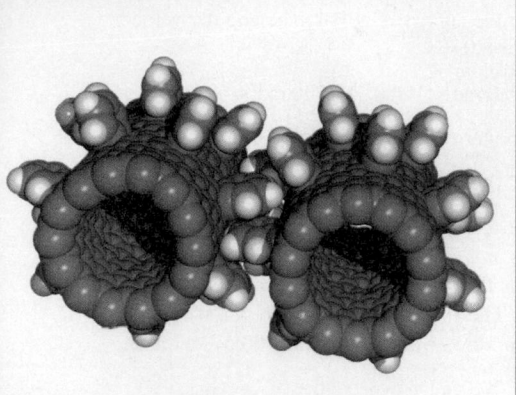

▲ **Application Gearing down.**
Nanoengineering involves the construction of submicroscopic machines, often built through the manipulation of individual atoms. Engineers have recently made tiny gears from carbon "nanotubes" with interlocking teeth made of organic molecules precisely placed on the tube surfaces. The operation of these gears is based on the same physical principles of torque and angular momentum as are applied to their macroscopic counterparts. To turn the gears, a laser shines tangentially onto the atoms at the end of the cylinder, creating a torque. This torque can cause a gear to rotate up to several thousand times per nanosecond.

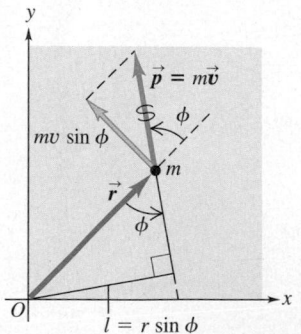

▲ **FIGURE 10.16** A particle with mass m rotating in the x-y plane. The angular momentum of the particle has magnitude $L = mvl$.

The basic dynamic principle for rotational motion (Equation 10.5, $\sum \tau = I\alpha$) can be restated in terms of angular momentum. If the axis doesn't move within the rotating object, then I is constant. In that case,

$$I\alpha = I \frac{\Delta \omega}{\Delta t} = \frac{\Delta (I\omega)}{\Delta t} = \frac{\Delta L}{\Delta t},$$

and Equation 10.5 can be rewritten as follows:

Torque and rate of change of angular momentum
The rate of change of angular momentum of a rigid body with respect to any axis equals the sum of the torques of the forces acting on it with respect to that axis:

$$\sum \tau = \frac{\Delta L}{\Delta t}. \tag{10.10}$$

We'll use this result soon when we discuss a conservation principle for angular momentum.

We can also define angular momentum for a single particle. In Figure 10.16, a particle moves in the x-y plane; its radial and tangential components of velocity are shown. The instantaneous velocity and linear momentum of the particle lie along a line at a perpendicular distance l from the axis through point O. We define the angular momentum of this particle as

$$L = mvl. \tag{10.11}$$

In the figure, the directions of $\vec{v}$ and $\vec{p}$ correspond to a counterclockwise (positive) sense of rotation. The distance l serves as a moment arm for the velocity, analogous to the moment arm (or lever arm) of a force in the definition of torque.

Equation 10.11 is consistent with our original definition of angular momentum of a rigid body. Suppose the object consists of a collection of particles with masses $m_A, m_B, m_C, \ldots$ at distances $r_A, r_B, r_C, \ldots$ from the axis through O. For each particle, the angle ϕ in Figure 10.16 is 90°, so the moment arm l for particle A is its distance r_A from the axis of rotation. When the object rotates with angular velocity ω, the magnitude v_A of the velocity of particle A is $v_A = \omega r_A$, and the angular momentum L_A of this particle is

$$L_A = m_A v_A r_A = m_A(\omega r_A) r_A = m_A r_A^2 \omega.$$

If the object is rigid, *all* the particles must have the same angular velocity ω; the total angular momentum of all the particles is then

$$L = (m_A r_A^2 + m_B r_B^2 + m_C r_C^2 + \cdots)\omega = I\omega,$$

in agreement with our previous definition of the angular momentum of a rigid body.

NOTE ▶ As with all angular quantities, angular momentum is defined with reference to a particular axis of rotation. It is essential to specify this axis and use it consistently when applying the principles we've just discussed. ◀

Equation 10.10 will be particularly useful when we develop the principle of conservation of angular momentum in the next section. We've discussed changes in angular momentum that occur when the angular velocity ω of a body changes, but there are also cases in which ω is constant and I changes because of a change of shape or rearrangement of component parts of a body. In cases where I isn't constant, Equation 10.5 ($\sum \tau = I\alpha$) isn't valid. But the angular momentum is still given at each instant by $L = I\omega$, and the rate of change of angular momentum is still given by Equation 10.10. The next section includes some examples in which we have to take this more general view.

We've discussed angular momentum in the context of macroscopic bodies, but it is vitally important in physics at all scales, from nuclear to galactic. Angular momentum plays an essential role in classifying the quantum states of nuclei, atoms, and molecules and in describing their interactions. Angular momentum considerations are also an important tool in analyzing the motions of stars, galaxies, and nebulae. Along with mass, energy, and momentum, angular momentum is one of the unifying threads woven through the entire fabric of physical science.

EXAMPLE 10.7   A kinetic sculpture

A part of a mobile suspended from the ceiling of an airport terminal building consists of two metal spheres, each with mass 2.0 kg, connected by a uniform metal rod with mass 3.0 kg and length $s = 4.0$ m. The assembly is suspended at its midpoint by a wire and rotates in a horizontal plane, making 3.0 revolutions per minute. Find the angular momentum and kinetic energy of the assembly.

SOLUTION

SET UP Figure 10.17 shows our sketch. We draw the assembly as viewed from above (or below), in its plane of rotation.

SOLVE First, we need to find the moment of inertia of the assembly about an axis through its midpoint; we'll assume that each sphere can be treated as a point. Then the angular momentum L is given by $L = I\omega$, and the kinetic energy is $K = \frac{1}{2}I\omega^2$.

The moment of inertia of each sphere (treated as a point) is

$$I_{sphere} = m\left(\frac{s}{2}\right)^2 = (2.0 \text{ kg})(2.0 \text{ m})^2 = 8.0 \text{ kg} \cdot \text{m}^2.$$

From Table 9.2, the moment of inertia of a rod about an axis through its midpoint and perpendicular to its length is

$$I_{rod} = \frac{1}{12}Ms^2 = \frac{1}{12}(3.0 \text{ kg})(4.0 \text{ m})^2 = 4.0 \text{ kg} \cdot \text{m}^2.$$

So the total moment of inertia of the assembly is

$$I_{total} = 2(8.0 \text{ kg} \cdot \text{m}^2) + 4.0 \text{ kg} \cdot \text{m}^2 = 20 \text{ kg} \cdot \text{m}^2.$$

The angular velocity is given in revolutions per minute; we must convert it to radians per second:

$$\omega = 3.0 \text{ rev/min} = \left(\frac{3.0 \text{ rev}}{1 \text{ min}}\right)\left(\frac{1 \text{ min}}{60 \text{ s}}\right)\left(\frac{2\pi \text{ rad}}{1 \text{ rev}}\right) = 0.31 \text{ rad/s}.$$

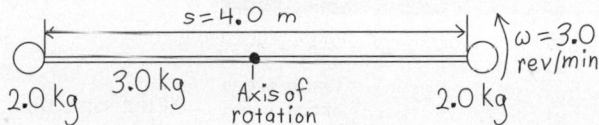

▲ **FIGURE 10.17**

Now we can calculate the angular momentum of the assembly:

$$L = I\omega = (20 \text{ kg} \cdot \text{m}^2)(0.31 \text{ rad/s}) = 6.2 \text{ kg} \cdot \text{m}^2/\text{s}.$$

The kinetic energy is

$$K = \frac{1}{2}I\omega^2 = \frac{1}{2}(20 \text{ kg} \cdot \text{m}^2)(0.31 \text{ rad/s})^2$$
$$= 0.96 \text{ kg} \cdot \text{m}^2/\text{s}^2 = 0.96 \text{ J}.$$

REFLECT The mass of the rod is comparable to that of the spheres, but it contributes less to the total moment of inertia because most of its mass is closer to the axis of rotation than are the masses of the spheres.

Practice Problem: How long would an equal-mass rod have to be in order for its contribution to the total moment of inertia to equal that of the two spheres together? *Answer:* 8.0 m.

10.5 Conservation of Angular Momentum

We've seen that angular momentum can be used in an alternative statement of the basic dynamic principle for rotational motion. Angular momentum also forms the basis for a very important conservation principle: the principle of **conservation of angular momentum.** Like conservation of energy and conservation of linear momentum, this principle appears to be a universal conservation law, valid at all scales, from atomic and nuclear systems to the motions of galaxies.

The principle is based on the concept of an **isolated system,** introduced in Section 8.2. We generalize this term to mean a system in which the total external force is zero *and* the total external torque is zero. Then our new conservation principle states simply that **the total angular momentum of an isolated system is constant.**

PhET: Torque
ActivPhysics 7.14: Ball Hits Bat

The proof of this principle follows directly from Equation 10.10 ($\sum \tau = \Delta L/\Delta t$). When a system has several parts, the internal forces that the parts exert on each other cause changes in the angular momenta of the parts, but the *total* angular momentum of the system doesn't change. Here's an example: When two bodies A and B interact with each other but not with anything else, A and B together form an *isolated system*. Suppose body A exerts a force $\vec{F}_{A \text{ on } B}$ on body B; the corresponding torque (with respect to whatever axis we choose) is $\tau_{A \text{ on } B}$. According to Equation 10.10, this torque is equal to the rate of change of angular momentum of B:

$$\tau_{A \text{ on } B} = \frac{\Delta L_B}{\Delta t}.$$

At the same time, body B exerts a force $\vec{F}_{B \text{ on } A}$ on body A, with a corresponding torque $\tau_{B \text{ on } A}$, and

$$\tau_{B \text{ on } A} = \frac{\Delta L_A}{\Delta t}.$$

From Newton's third law, $\vec{F}_{B \text{ on } A} = -\vec{F}_{A \text{ on } B}$. Furthermore, if the forces act along the same line, their moment arms with respect to the chosen axis are equal. Figure 10.7 demonstrated this point; we repeat it here as Figure 10.18. Thus, the torques of these two forces are equal and opposite, and $\tau_{B \text{ on } A} = -\tau_{A \text{ on } B}$. So when we add the preceding two equations, we find that

$$\frac{\Delta L_A}{\Delta t} + \frac{\Delta L_B}{\Delta t} = 0,$$

or, because $L_A + L_B$ is the *total* angular momentum L of the system,

$$\frac{\Delta L}{\Delta t} = 0. \tag{10.12}$$

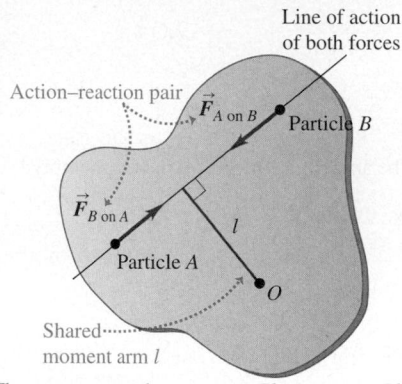

Line of action of both forces

Action–reaction pair $\vec{F}_{A \text{ on } B}$ Particle B

$\vec{F}_{B \text{ on } A}$

l

Particle A

O

Shared moment arm l

The torques cancel: $\tau_{A \text{ on } B} = +Fl$; $\tau_{B \text{ on } A} = -Fl$

▲ **FIGURE 10.18** Why internal forces cannot change the total angular momentum of a system.

That is, the total angular momentum of the system is constant; we also say that it is *conserved*. The torques of the internal forces can transfer angular momentum from one interacting object to another, but they can't change the *total* angular momentum of the system.

Conservation of angular momentum
When the sum of the torques of all the external forces acting on a system is zero, the total angular momentum of the system is constant (conserved).

An ice skater performing a pirouette on the toe of one skate takes advantage of this principle. Figure 10.19 shows the skater at the start of her pirouette, with her arms extended and with a counterclockwise angular momentum. As she pulls in her arms, she reduces her moment of inertia I. Because her angular momentum $L = I\omega$ must remain constant as I decreases, her angular velocity ω increases; she spins faster. That is,

$$I_i \omega_i = I_f \omega_f. \tag{10.13}$$

▲ **FIGURE 10.19** As this skater reduces her moment of inertia by pulling in her arms, she spins faster, conserving angular momentum.

A diver going into a tuck uses the same principle. Only by reducing his moment of inertia is he able to rotate multiple times before hitting the water. Notice that the skater and the diver can both be treated as isolated systems. The net external forces and torques on the skater are slight. Gravity has a major effect on the diver's translational motion, but once he is in the air it exerts no net torque on him. Thus, in terms of angular momentum, he is isolated.

EXAMPLE 10.8 Two rotating disks interacting

Figure 10.20 shows two disks, one an engine flywheel, the other a clutch plate attached to a transmission shaft. Their moments of inertia are I_A and I_B; initially, they are rotating with constant angular velocities ω_A and ω_B, respectively. We then push the disks together with forces acting along the axis, so as not to apply any torque on either disk. The disks rub against each other and eventually reach a common final angular velocity ω_f. Derive an expression for ω_f.

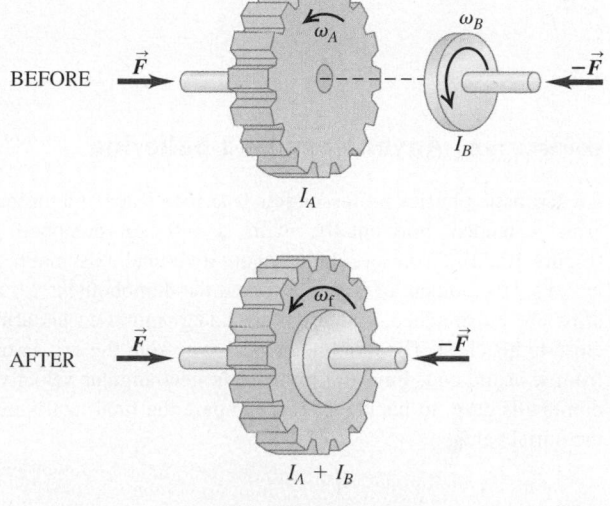

▶ **FIGURE 10.20**

SOLUTION

SET UP The two disks are an isolated system. The only torque acting on either disk is the torque applied by the other disk; there are no external torques. Thus the total angular momentum of the system is the same before and after they are pushed together. At the end, they rotate together as one body with total moment of inertia $I_f = I_A + I_B$ and angular velocity ω_f.

SOLVE Conservation of angular momentum gives

$$I_A\omega_A + I_B\omega_B = (I_A + I_B)\omega_f,$$
$$\omega_f = \frac{I_A\omega_A + I_B\omega_B}{I_A + I_B}.$$

REFLECT This is the analog of a completely inelastic collision of two particles, which we analyzed in Section 8.3. The final kinetic energy is always less than the initial value. This can be proved in general, using the above expression for ω_f.

Practice Problem: Suppose $I_B = 2I_A$. How must ω_A and ω_B be related if the entire system comes to rest after the interaction? *Answer:* $\omega_A = -2\omega_B$.

Quantitative Analysis 10.3

Mass on a string: I

In Figure 10.21, a block slides in a circular path on a horizontal frictionless plane under the action of a string that passes through a hole in the plane and is held vertically underneath it. By pulling on the lower end of the string, we reduce the radius to half of its original value. If the initial speed of the block is v_i, then the speed of the block after the string is shortened is

A. $\frac{1}{2}v_i$. B. $2v_i$. C. $4v_i$.

SOLUTION From Equation 10.11 the angular momentum of the block about a vertical axis through the hole is mvr. The force from the string pulls directly toward the axis. Therefore, it exerts no torque with respect to this axis, and it cannot change the block's angular momentum. Thus the angular momentum mvr of

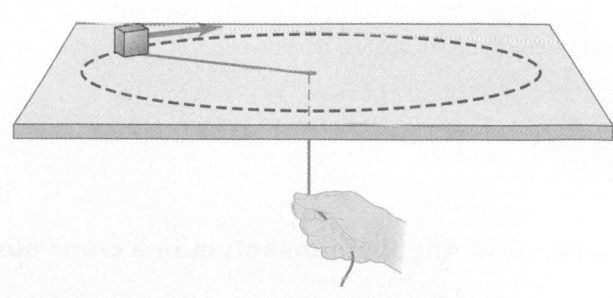

▲ **FIGURE 10.21**

the block must be constant. If r is reduced by half, then v must double to keep the angular momentum constant (answer B).

Quantitative Analysis 10.4

Mass on a string: II

Let's consider again the whirling block shown in Figure 10.21. By letting additional string pass through the hole in the plane, we increase the radius to twice its original value. If the initial tension in the string is T, then the tension in the string after the radius is increased is

A. $\frac{1}{2}T$.
B. $\frac{1}{4}T$.
C. $\frac{1}{8}T$.

SOLUTION From Quantitative Analysis 10.3, we see that in this situation the angular momentum of the block doesn't change: if r is doubled, then v must decrease by half. The tension T provides the needed centripetal force, according to the relationship $T = mv^2/r$. In the situation presented here, r doubles and v is reduced to half its original value. Thus, in the expression for T, the denominator doubles and the numerator is reduced to one-fourth its initial value, for an overall reduction of T to one-eighth of its initial value (answer C).

EXAMPLE 10.9   Anyone can be a ballerina

An acrobatic physics professor stands at the center of a turntable, holding his arms extended horizontally, with a 5.0 kg dumbbell in each hand (Figure 10.22). He is set rotating about a vertical axis, making one revolution in 2.0 s. His moment of inertia (without the dumbbells) is $3.0 \text{ kg} \cdot \text{m}^2$ when his arms are outstretched, and drops to $2.2 \text{ kg} \cdot \text{m}^2$ when his arms are pulled in close to his chest. The dumbbells are 1.0 m from the axis initially and 0.20 m from it at the end. Find the professor's new angular velocity if he pulls the dumbbells close to his chest, and compare the final total kinetic energy with the initial value.

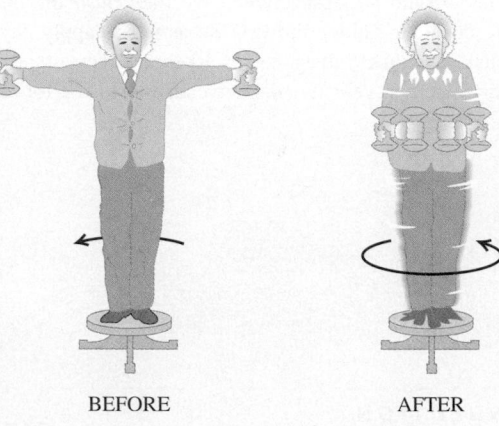

▶ **FIGURE 10.22**   BEFORE   AFTER

SOLUTION

SET UP If we neglect friction in the turntable, there are no external torques with respect to the vertical axis, and the angular momentum about this axis is constant. That is,

$$I_i\omega_i = I_f\omega_f,$$

where I_i and ω_i are the initial total moment of inertia and angular velocity, respectively, and I_f and ω_f are the final values.

SOLVE In each case, $I = I_{prof} + I_{dumb}$; thus,

$$I_i = 3.0 \text{ kg} \cdot \text{m}^2 + 2(5.0 \text{ kg})(1.0 \text{ m})^2 = 13 \text{ kg} \cdot \text{m}^2$$
$$I_f = 2.2 \text{ kg} \cdot \text{m}^2 + 2(5.0 \text{ kg})(0.20 \text{ m})^2 = 2.6 \text{ kg} \cdot \text{m}^2$$
$$\omega_i = 2\pi\frac{1 \text{ rev}}{2.0 \text{ s}} = \pi \text{ rad/s}.$$

From conservation of angular momentum,

$$(13 \text{ kg} \cdot \text{m}^2)(\pi \text{ rad/s}) = (2.6 \text{ kg} \cdot \text{m}^2)\omega_f,$$
$$\omega_f = 5.0\pi \text{ rad/s} = 2.5 \text{ rev/s}.$$

That is, the angular velocity increases by a factor of five.

The initial kinetic energy is

$$K_i = \tfrac{1}{2}I_i\omega_i^2 = \tfrac{1}{2}(13 \text{ kg} \cdot \text{m}^2)(\pi \text{ rad/s})^2$$
$$= 64 \text{ J}.$$

The final kinetic energy is

$$K_f = \tfrac{1}{2}I_f\omega_f^2 = \tfrac{1}{2}(2.6 \text{ kg} \cdot \text{m}^2)(5.0\pi \text{ rad/s})^2$$
$$= 320 \text{ J}.$$

REFLECT Where did the "extra" 256 J of kinetic energy come from?

Practice Problem: Suppose the professor drops the dumbbells and then pulls his arms close to his chest. What is his final angular velocity? *Answer:* 0.68 rev/s.

EXAMPLE 10.10   Angular momentum in a crime bust

A uniform door 1.0 m wide with a mass of 15 kg is hinged at one side so that it can rotate without friction about a vertical axis. The door is unlatched. A police detective fires a bullet with a mass of 10 g and a speed of 400 m/s into the exact center of the door in a direction perpendicular to the plane of the door. Find the angular velocity of the door just after the bullet embeds itself in it. Is kinetic energy conserved?

Continued

SOLUTION

SET UP We consider the door and bullet together as a system. We denote the mass of the bullet as m, the mass of the door as M, and the door's width as s. Figure 10.23 shows our sketch. There is no external torque about the axis defined by the hinges, so angular momentum about this axis is conserved. We find the initial angular momentum of the bullet from Equation 10.11 ($L = mvl$). This quantity is equal to the final total angular momentum $I_{\text{total}}\omega_f$, where $I_{\text{total}} = I_{\text{door}} + I_{\text{bullet}}$. From Table 9.2, $I_{\text{door}} = Ms^2/3$. (Table 9.2 uses L for the length of the body, measured perpendicular to the axis; we substitute s here to avoid confusion with angular momentum L.)

SOLVE We compute all moments of inertia and angular momenta with respect to the axis defined by the door's hinges. The initial angular momentum of the bullet is

$$L_{\text{bullet}} = m_{\text{bullet}}v_{\text{bullet}}l = (0.010\text{ kg})(400\text{ m/s})(0.50\text{ m})$$
$$= 2.0\text{ kg} \cdot \text{m}^2/\text{s}.$$

The moment of inertia of the door is

$$I_{\text{door}} = \frac{Ms^2}{3} = \frac{(15\text{ kg})(1.0\text{ m})^2}{3} = 5.0\text{ kg} \cdot \text{m}^2.$$

The moment of inertia of the bullet (with respect to the axis along the hinges) is

$$I_{\text{bullet}} = ml^2 = (0.010\text{ kg})(0.50\text{ m})^2 = 0.0025\text{ kg} \cdot \text{m}^2.$$

Conservation of angular momentum requires that $m_{\text{bullet}}v_{\text{bullet}}l = I_{\text{total}}\omega_f$, so we can solve for the final angular velocity ω_f:

$$2.0\text{ kg} \cdot \text{m}^2/\text{s} = (5.0\text{ kg} \cdot \text{m}^2 + 0.0025\text{ kg} \cdot \text{m}^2)\omega_f,$$
$$\omega_f = 0.40\text{ rad/s}.$$

The initial kinetic energy of the bullet is

$$K_i = \tfrac{1}{2}mv^2 = \tfrac{1}{2}(0.010\text{ kg})(400\text{ m/s})^2$$
$$= 800\text{ J}.$$

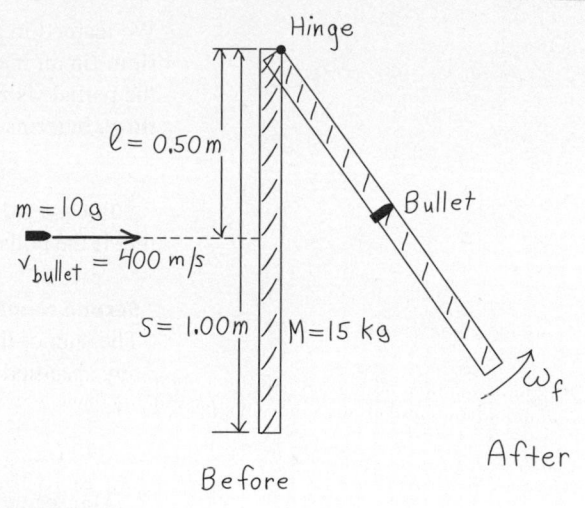

▲ **FIGURE 10.23**

The final total kinetic energy of the door and bullet is

$$K_f = \tfrac{1}{2}I\omega^2 = \tfrac{1}{2}(5.0025\text{ kg} \cdot \text{m}^2)(0.40\text{ rad/s})^2$$
$$= 0.40\text{ J}.$$

REFLECT Kinetic energy is *not* conserved. The final kinetic energy is only $\frac{1}{2000}$ of the initial value. The collision of bullet and door is inelastic because nonconservative forces act during the impact. Thus, we do not expect kinetic energy to be conserved.

Practice Problem: Suppose the bullet goes through the door and emerges with a speed of 100 m/s. Find the final angular velocity of the door just after the bullet goes through it. *Answer:* 0.30 rad/s.

10.6 Equilibrium of a Rigid Body

Here's an experiment to try: Face a wall, with your toes just touching the wall. Now try standing on tiptoe. What happens? You start falling over backward. Why? There is a net torque on you, and you are no longer in equilibrium. The need to consider torque occurs with a suspension bridge, a ladder leaning against a wall, or a crane hoisting a bucket full of concrete.

A body that can be modeled as a particle is in equilibrium whenever the vector sum of the forces acting on it is zero. But for the problems that we've just described, that condition isn't enough. Many bodies can't be adequately modeled as particles. The vector sum of the forces on such a body must still be zero, but an additional requirement must be satisfied to ensure that the body has no tendency to *rotate*. This requirement is based on the principles of rotational dynamics developed in this chapter. For a rigid body in equilibrium, the sum of the *torques* about any axis must be zero. We can compute the torque due to the weight of a body using the concept of center of mass from Section 8.6 and the related concept of center of gravity, which we'll introduce in this section.

Conditions for Equilibrium

We learned in Section 5.1 that a particle acted on by several forces is in equilibrium (in an inertial frame of reference) if the vector sum of the forces acting on the particle is zero ($\sum \vec{F} = 0$). This is often called the **first condition for equilibrium.** In terms of x and y components,

$$\sum F_x = 0, \qquad \sum F_y = 0. \qquad (10.14)$$

For a rigid body, there is a **second condition for equilibrium** that must be met if the body is not to rotate:

Second condition for equilibrium

The sum of the torques due to all forces acting on the body, with respect to any specified axis, must be zero:

$$\sum \tau = 0. \qquad \text{(about any axis)} \qquad (10.15)$$

The second condition for equilibrium is based on the dynamics of rotational motion in exactly the same way that the first condition is based on Newton's first law. A rigid body at rest has zero angular momentum ($L = 0$). For it to *remain* at rest, the *rate of change* of angular momentum $\Delta L/\Delta t$ must also be zero. From Equation 10.10, this means that the sum of the torques ($\sum \tau$) of all the forces acting on the object must be zero. Furthermore, because every point in the body is at rest in an inertial frame, any axis can be used as a reference line. Therefore, the sum of the torques about *any* axis must be zero. The object doesn't actually have to be pivoted about an axis through the chosen point. If a rigid body is in equilibrium, it can't have any tendency to begin to rotate about *any* axis. Thus, the sum of the torques must be zero, *no matter what axis is chosen.* (But don't compute some of the torques with respect to one axis and others with respect to a different axis; choose one axis and use it consistently.)

Although the choice of axis is arbitrary, once we choose it we must use the same axis to calculate all of the torques. An important element of the strategy for problems involving torques is to pick the axis or axes so as to simplify the calculations as much as possible.

Torque Due to Gravitational Force

In most equilibrium problems, one of the forces acting on the object is its weight. We need to be able to calculate the torque of this force. The weight doesn't act at a single point; rather, it is distributed over the entire body. But when the acceleration due to gravity is the same at every point in the body, we can always calculate the torque due to the body's weight by assuming that the entire force of gravity is concentrated at the center of mass of the object, which we defined in Equation 8.18. We stated this principle, without detailed proof, in Section 10.1. Because of it, the center of mass of an object is sometimes called its *center of gravity* (cg). Note, however, that the center of mass is defined independently of any gravitational effect.

We can often use symmetry considerations to locate the center of gravity of a body. **The center of gravity of a homogeneous sphere, cube, circular disk, or rectangular plate is at its geometric center. The center of gravity of a right circular cylinder or cone is on the axis of symmetry, and so on.**

The center of gravity has several other important properties. First, when a body is in equilibrium in a gravitational field, supported or suspended at a single point, the center of gravity is always at or directly above or below the point of suspension. If it were anywhere else, the weight would have a torque with respect to that point, and the body could not be in rotational equilibrium.

This fact can be used to determine experimentally the location of the center of gravity of an irregular body (Figure 10.24). Try the following experiment: Cut an

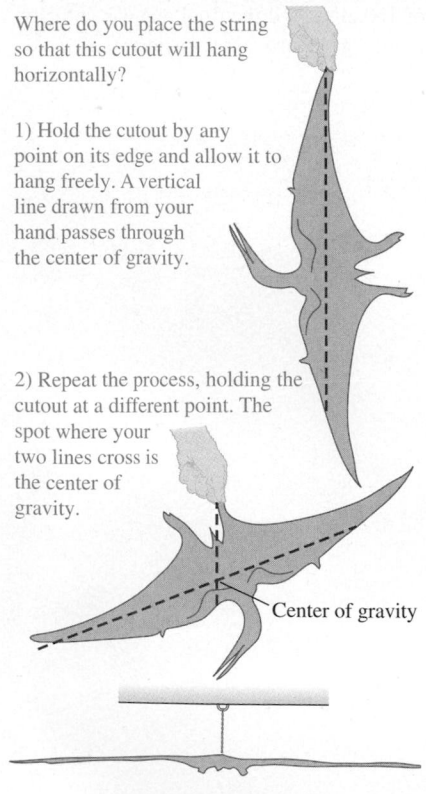

Where do you place the string so that this cutout will hang horizontally?

1) Hold the cutout by any point on its edge and allow it to hang freely. A vertical line drawn from your hand passes through the center of gravity.

2) Repeat the process, holding the cutout at a different point. The spot where your two lines cross is the center of gravity.

Center of gravity

When suspended from the center of gravity, the cutout hangs level.

▲ **FIGURE 10.24** How to find the center of gravity of a flat object.

irregular shape out of cardboard, and then grasp it gently between your thumb and forefinger. The body rotates until its center of gravity is directly below the point you're grasping. Draw a vertical line through this point; the center of gravity must lie along this line. Repeat this procedure for a different suspension point; the intersection of the two lines must be the center of gravity. Pick a third point; the three lines must intersect at a single point. Once you've located the intersection point, poke a small hole in the body at that point and insert a pointed pen or pencil to confirm that the body balances there.

A force applied to a rigid body at its center of gravity doesn't tend to cause the body to rotate about an axis through that point. A force applied at any other point can cause both rotational and translational motion. We used this fact implicitly in the examples of Section 10.2.

PROBLEM-SOLVING STRATEGY 10.3 **Equilibrium of a rigid body**

The principles of rigid-body equilibrium are few and simple: The vector sum of the forces on the object must be zero, and the sum of the torques about any axis must be zero. The more challenging part is applying these principles to specific problems. Careful and systematic problem-solving methods always pay off. The following strategy is similar to the suggestions in Section 5.1 for equilibrium of a particle.

SET UP

1. Sketch the physical situation, including dimensions.
2. Choose some appropriate body as the body in equilibrium and draw a free-body diagram, showing the forces acting *on* this body and no others. Follow all the rules and sign conventions we've discussed previously. *Do not include forces exerted by this body on other bodies.* Be careful to show correctly the point at which each force acts; this is crucial for correct torque calculations.
3. Draw coordinate axes, and represent the forces in terms of their components with respect to these axes.

SOLVE

4. Write equations expressing the equilibrium conditions. Remember that $\sum F_x = 0$, $\sum F_y = 0$, and $\sum \tau = 0$ are always separate equations; *never* add x and y components in a single equation.
5. In choosing an axis of rotation for computing torques, note that if a force has a line of action that goes *through* a particular rotation axis, the torque of the force with respect to that axis is zero. You can often eliminate unknown forces or components from the torque equation by a clever choice of axis. Also remember that when a force is represented in terms of its components, you can compute the torque of that force by finding the torque of each component separately, each with its appropriate moment arm and sign, and adding the results. This approach is often easier than determining the moment arm of the original force.

REFLECT

6. You always need as many equations as you have unknowns. Depending on the number of unknowns, you may need to compute torques with respect to two or more rotation axes to obtain enough equations. There often are several equally good sets of force and torque equations for a particular problem; there is usually no single "right" combination of equations. When you have as many independent equations as unknowns, you can solve the equations simultaneously. We'll illustrate this point in some of the examples that follow by using various sets of equations.

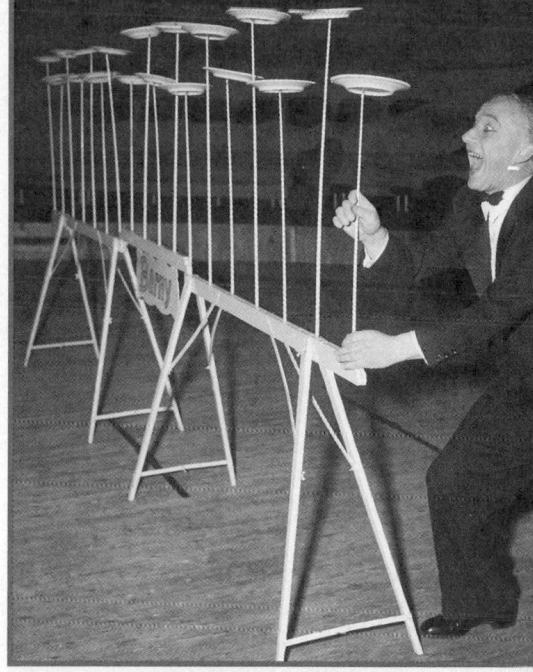

▲ **Application An accident waiting to happen?** If these plates were not spinning rapidly, it would be nearly impossible to balance them on the ends of their sticks. When they are spinning, each stick exerts frictional forces on the underside of its plate; these forces tend to move the point of contact of the stick toward the center of the plate. Once centered, each spinning plate behaves like a gyroscope flywheel with its axis vertical and its center of mass directly above the point of support. Although the plates aren't in equilibrium (because they are slowing down), they can maintain their spinning motion for a considerable time.

EXAMPLE 10.11 **Playing on a seesaw**

You and a friend play on a seesaw. Your mass is 90 kg, and your friend's mass is 60 kg. The seesaw board is 3.0 m long and has negligible mass. Where should the pivot be placed so that the seesaw will balance when you sit on the left end and your friend sits on the right end?

SOLUTION

SET UP Figure 10.25 shows our sketches. First we sketch the physical situation, and then we make a free-body diagram, placing the pivot at a distance x from the left end. Note that each weight is the corresponding mass multiplied by g, so g will divide out of the final result. Remember that counterclockwise torques are positive and clockwise torques are negative. For the seesaw to be in equilibrium, the sum of the torques about any axis must be zero. We choose to take torques about the pivot axis; then the normal force acting at the pivot doesn't enter into the analysis. The moment arms are x for you and $(3.0 \text{ m} - x)$ for your friend.

SOLVE Your torque is $\tau_{\text{you}} = +(90 \text{ kg})gx$. Your friend is $3.0 \text{ m} - x$ from the pivot, and the corresponding torque is $\tau_{\text{friend}} = -(60 \text{ kg})g(3.0 \text{ m} - x)$. For equilibrium, the sum of these torques must be zero:

$$\tau_{\text{you}} + \tau_{\text{friend}} = 0$$
$$+(90 \text{ kg})gx - (60 \text{ kg})g(3.0 \text{ m} - x) = 0.$$

We divide through by g and solve for x, obtaining $x = 1.2$ m.

REFLECT If we put the pivot at the center of the board $(x = 1.5 \text{ m})$, we would have equal moment arms but unequal weights, so it's clear that the pivot must be to the *left* of center.

(a) Sketch of physical situation

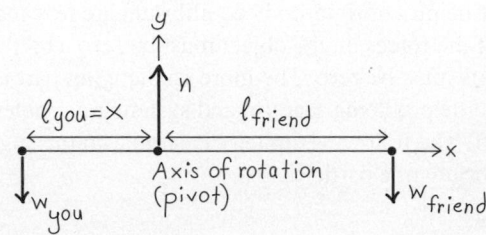

(b) Free-body diagram

▲ **FIGURE 10.25**

Practice Problem: If the board has a mass of 20 kg, where should the pivot be placed for balance? *Answer:* 1.24 m from the left end.

EXAMPLE 10.12 **A heroic rescue**

Sir Lancelot is trying to rescue the Lady Elayne from the Black Castle by climbing a uniform ladder that is 5.0 m long and weighs 180 N. Lancelot, who weighs 800 N, stops a third of the way up the ladder (Figure 10.26a). The bottom of the ladder rests on a horizontal stone ledge and leans across the castle's moat in equilibrium against a vertical wall that is frictionless because of a thick layer of moss. The ladder makes an angle of 53° with the horizontal, conveniently forming a 3–4–5 right triangle. (a) Find the normal and friction forces on the ladder at its base. (b) Find the minimum coefficient of static friction needed to prevent slipping. (c) Find the magnitude and direction of the contact force on the ladder at the base.

SOLUTION

SET UP Figure 10.26b shows a free-body diagram for the ladder. Since it is described as "uniform," we can assume that its center of gravity is at its center, halfway between the base and the wall. Lancelot's weight pushes down on the ladder at a point one-third of the way from the base toward the wall. The contact force n_1 at the top of the ladder is horizontal and to the left, because the wall is frictionless. The components of the contact force at the base are the upward normal force n_2 and the friction force f_s, which must point to the right to prevent slipping.

SOLVE Part (a) The first condition for equilibrium, in component form, gives

$$\Sigma F_x = f_s + (-n_1) = 0,$$
$$\Sigma F_y = n_2 + (-800 \text{ N}) + (-180 \text{ N}) = 0.$$

These are two equations for the three unknowns n_1, n_2, and f_s. The first equation tells us that the two horizontal forces must be equal and opposite, and the second equation gives $n_2 = 980$ N. That is, the ledge pushes up with a force of 980 N to balance the total (downward) weight $(800 \text{ N} + 180 \text{ N})$.

Continued

We don't yet have enough equations, but now we can use the second condition for equilibrium. We can take torques about any point we choose. The smart choice is the point that will give us the fewest terms and fewest unknowns in the torque equation. With this thought in mind, we choose point B, at the base of the ladder. The two forces n_2 and f_s have no torque about that point. From Figure 10.26b, we see that the moment arm for the ladder's weight is 1.5 m, the moment arm for Lancelot's weight is 1.0 m, and the moment arm for n is 4.0 m. The torque equation for point B is thus

$$\sum \tau_B = n_1 (4.0 \text{ m}) - (180 \text{ N})(1.5 \text{ m})$$
$$- (800 \text{ N})(1.0 \text{ m}) + n_2(0) + f_s(0) = 0.$$

Solving for n_1, we get

$$n_1 = 268 \text{ N}.$$

Now we substitute this result back into the equation $\sum F_x = 0$ to get

$$f_s = 268 \text{ N}.$$

Part (b) The static friction force f_s cannot exceed $\mu_s n_2$, so the *minimum* coefficient of static friction to prevent slipping is

$$(\mu_s)_{\min} = \frac{f_s}{n_2} = \frac{268 \text{ N}}{980 \text{ N}} = 0.27.$$

Part (c) The components of the contact force at the base are the static friction force $f_s = 286$ N and the normal force $n_2 = 980$ N; the magnitude of $\vec{F}_B$ is

$$F_B = \sqrt{(268 \text{ N})^2 + (980 \text{ N})^2} = 1020 \text{ N},$$

and its direction (Figure 10.26c) is

$$\theta = \tan^{-1} \frac{980 \text{ N}}{268 \text{ N}} = 75°.$$

REFLECT First, the contact force $\vec{F}_B$ at an angle of 75° is *not* directed along the length of the ladder, which is at an angle of 53° If it were, the sum of torques with respect to the point where Lancelot stands couldn't be zero.

Second, as Lancelot climbs higher on the ladder, the moment arm and torque of his weight about B increase, thereby increasing the values of n_1, f_s, and $(\mu_s)_{\min}$. At the top, his moment arm would be nearly 3 m, giving a minimum coefficient of static friction of nearly 0.7. An aluminum ladder on a wood floor wouldn't have a value of μ_s this large, so present-day aluminum ladders are usually equipped with nonslip rubber feet.

Third, a larger ladder angle (i.e., a more upright ladder) would decrease the moment arms of the weights of the ladder and Lancelot with respect to B and would increase the moment arm of n_1, all of which would decrease the required friction force. Some manufacturers recommend that their ladders be used at an angle of 75°. (Why not 90°?)

Practice Problem: Carry out an alternative procedure for obtaining the results asked for in Parts (a), (b), and (c) by taking torques about an axis at the top of the ladder; check to see that your results agree with the previous ones.

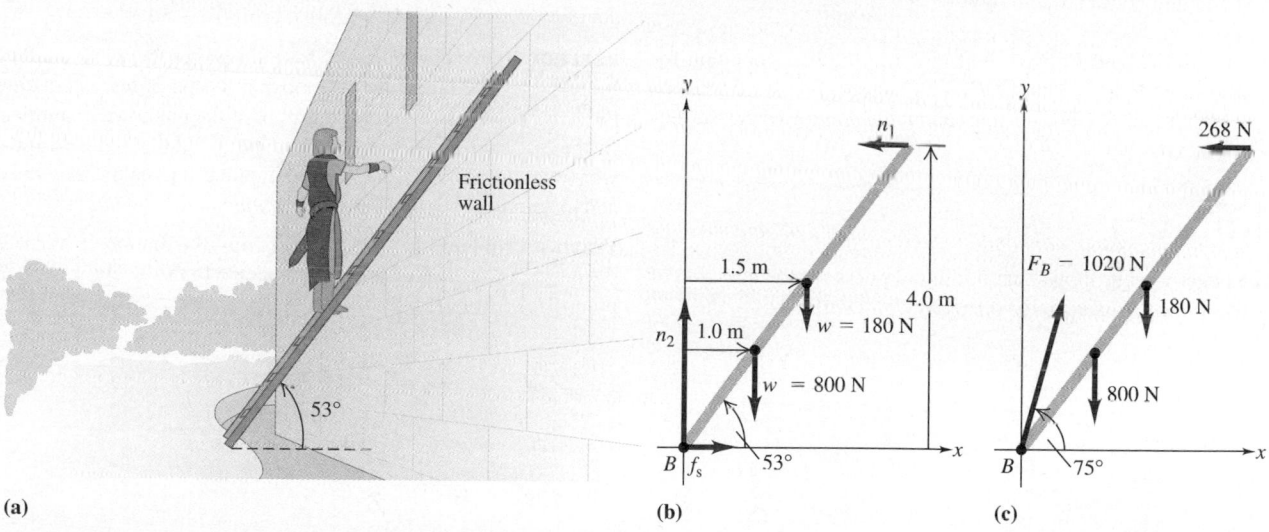

(a)   **(b)**   **(c)**

▲ **FIGURE 10.26**

EXAMPLE 10.13 **Equilibrium and pumping iron**

Figure 10.27a shows a human arm lifting a dumbbell. The forearm is in equilibrium under the action of the weight w of the dumbbell, the tension T in the tendon connected to the biceps muscle at point A, and the forces exerted on the forearm by the upper arm at the elbow joint. For clarity, the tendon force has been displaced away from the elbow farther than its actual position. The weight w and the angle θ are given. Find the tension T in the tendon and the two components of force (E_x and E_y) at the elbow (three unknown quantities in all). Neglect the weight of the forearm itself. Evaluate your results for $w = 50$ N, $d = 0.10$ m, $l = 0.50$ m, and $\theta = 80°$.

Continued

SOLUTION

SET UP As the body in equilibrium, we choose the forearm plus dumbbell (the colored elements in Figure 10.27a). Figure 10.27b shows our free-body diagram. We apply a coordinate system and draw the three forces on the forearm, representing the tension force and the forces on the elbow in terms of components along the x and y directions. Note that we don't yet know the sign of the elbow force component E_y (although we drew it as positive for convenience); it could come out either way. Study this free-body diagram; be sure you understand how these forces allow you to lift things.

SOLVE First we use the components T_x and T_y of the tension force, along with the given angle θ, to represent the unknown magnitude:

$$T_x = T\cos\theta \qquad T_y = T\sin\theta.$$

Next we note that if we take torques about the elbow joint (point E), the resulting torque equation does not contain E_x, E_y, or T_x because the lines of action of all these forces pass through this point. The torque equation is then simply

$$\sum\tau_E = lw - dT_y = 0.$$

From this equation, we find that

$$T_y = \frac{lw}{d} \qquad \text{and} \qquad T = \frac{lw}{d\sin\theta}.$$

To find E_x and E_y, we could now use the first condition for equilibrium: $\sum F_x = 0$ and $\sum F_y = 0$. Instead, for added practice in using torques, we take torques about the point A where the tendon is attached:

$$(l - d)w + dE_y = 0 \qquad \text{and} \qquad E_y = -\frac{(l - d)w}{d}.$$

The negative sign shows that our initial guess for the direction of E_y was wrong; it is actually vertically *downward*.

Finally, we take torques about point B in the figure and remember to use the shortest perpendicular distance to the vector $\vec{w}$ to simplify our calculations.

$$\sum\tau_B = lw - hE_x = 0 \qquad \text{and} \qquad E_x = \frac{lw}{h}.$$

Evaluating our expressions for $w = 50$ N, $d = 0.10$ m, $l = 0.50$ m, and $\theta = 80°$, we get $\tan\theta = h/d$, and we find $h = d\tan\theta = (0.10\text{ m})(5.67) = 0.57$ m. We then have

$$T = \frac{(0.50\text{ m})(50\text{ N})}{(0.10\text{ m})(0.98)} = 250\text{ N},$$

$$E_y = -\frac{(0.50\text{ m} - 0.10\text{ m})(50\text{ N})}{0.10\text{ m}} = -200\text{ N},$$

$$E_x = \frac{(0.50\text{ m})(50\text{ N})}{0.57\text{ m}} = 44\text{ N}.$$

The magnitude of the force at the elbow is

$$E = \sqrt{E_x^2 + E_y^2} = 200\text{ N}.$$

As we mentioned earlier, we have not explicitly used the first condition for equilibrium, that the vector sum of the forces be zero. To check our answer, we can compute $\sum F_x$ and $\sum F_y$ to verify that they really *are* zero. Such checks help verify internal consistency. Checking, we obtain

$$\sum F_x = E_x - T_x = 44\text{ N} - (250\text{ N})\cos 80° = 0,$$
$$\sum F_y = E_y + T_y - w$$
$$= -200\text{ N} + (250\text{ N})\sin 80° - 50\text{ N} = 0.$$

REFLECT Notice how much we have simplified these calculations by using a little ingenuity in choosing the point for calculating torques so as to eliminate one or more of the unknown quantities. In the last step, the force T has no torque about point B; thus, when the torques of T_x and T_y are computed separately, they must add to zero.

Practice Problem: Double the weight of the dumbbell. What is the new tension T in the tendon connected to the biceps muscle at point B? *Answer:* $T = 510$ N.

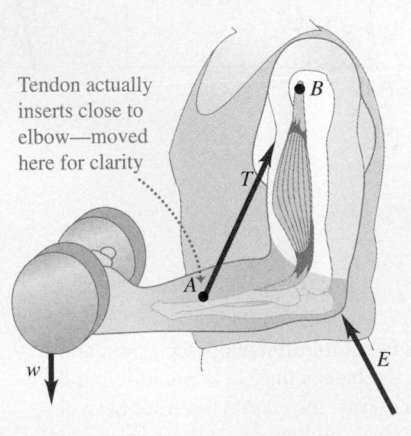

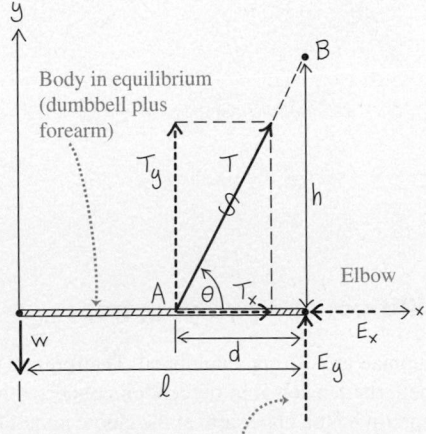

We don't know the sign of this component; we draw it positive for convenience.

(a)

(b)

▲ **FIGURE 10.27**

10.7 Vector Nature of Angular Quantities

Rotational quantities associated with an axis of rotation, such as angular velocity, angular momentum, and torque, can be represented as *vector* quantities. For a body rotating about a fixed axis that is an axis of symmetry, the directions of these vector quantities are along the axis of rotation, in a direction given by the **right-hand rule,** illustrated in Figure 10.28. If you curl the fingers of your *right* hand in the direction the object rotates, then your thumb points in the direction of the vector angular velocity $\vec{\omega}$ and angular momentum $\vec{L}$. In the case of a torque, you curl the fingers of your right hand in the direction the torque would turn the body if acting on its own; then your thumb points in the torque's direction.

So far, we've dealt with rotating bodies whose axis of rotation, whether stationary or moving, has a fixed direction that is perpendicular to the planes in which the particles of the body revolve. (We've oriented our coordinate systems so that the particles of the body move parallel to the *x-y* plane and the axis of rotation is perpendicular to that plane.) When the axis of rotation can change direction, the situation becomes much richer in the variety of physical phenomena (some quite unexpected) that can occur. For example, if you try to balance a stationary top on its point, it falls over; but when it's spinning, it balances easily. Why? And why are bicycles "tippy" when moving slowly but stable when moving fast? General answers to questions such as these are quite complex, but we can go a little way into understanding this behavior.

For a symmetric body turning about its axis of symmetry, the magnitude of the angular momentum is $L = I\omega$ (Equation 10.9). The corresponding relationship between *vector* angular momentum and *vector* angular velocity is

$$\vec{L} = I\vec{\omega}, \tag{10.16}$$

where I is the moment of inertia about the axis of rotation. Thus, for rotation about an axis of symmetry, the angular velocity and angular momentum vectors have the same direction. Furthermore, we can define a vector torque that is related to the rate of change of vector angular momentum with respect to time by a vector equation analogous to Equation 10.10:

$$\sum\vec{\tau} = \frac{\Delta\vec{L}}{\Delta t}, \quad \text{or} \quad \Delta\vec{L} = \sum\vec{\tau}\,\Delta t. \tag{10.17}$$

In the problems we've worked out so far, $\sum\vec{\tau}$ and $\vec{L}$ have always had the same direction, so only one component of each vector quantity is different from zero. But Equation 10.17 is valid even when $\sum\vec{\tau}$ and $\vec{L}$ have different directions. Then the *direction* of the axis of rotation may change, as with a spinning top, and we have to consider the vector nature of the various angular quantities carefully.

An example of a three-dimensional application of Equation 10.17 is the gyroscope. Figure 10.29 shows a typical toy gyroscope. We set the flywheel spinning by wrapping a string around its shaft and pulling. When the shaft is supported at only one end, as shown in the figure, one possible motion is a steady circular motion of the axis in a horizontal plane, combined with the spinning motion of the flywheel about the axis. This phenomenon is quite unexpected when you see it for the first time; intuition suggests that the free end of the axis should simply drop if it isn't supported. Indeed, if the flywheel isn't spinning, that *is* what happens. The vector nature of angular momentum is the key to understanding the horizontal motion of the axis when the wheel is spinning.

Suppose we take a stationary gyroscope and lift the flywheel until its axis is horizontal, as shown in Figure 10.30. Then when we let it go, two forces act on the flywheel: its weight $\vec{w}$, acting downward at the center of mass, and the normal force $\vec{n}$ at the pivot point O. The weight has a moment arm r; the resulting torque, whose direction is given by the right-hand rule, causes the free end of the flywheel

Angular velocity and angular momentum: Curl the fingers of your right hand in the direction of rotation. Your thumb then points in the direction of angular velocity and momentum.

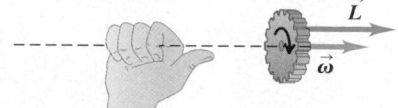

You must use your right hand!

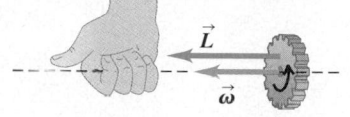

Torque: Curl the fingers of your right hand in the direction the torque would cause the body to rotate. Your thumb points in the torque's direction.

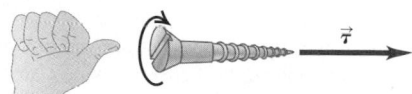

Right-hand screws are threaded so that they move in the direction of the torque applied to them.

▲ **FIGURE 10.28** The right-hand rule for angular quantities.

When the flywheel and its axis are stationary, they will fall to the table surface. When the flywheel spins, it and its axis "float" in the air while moving in a circle about the pivot.

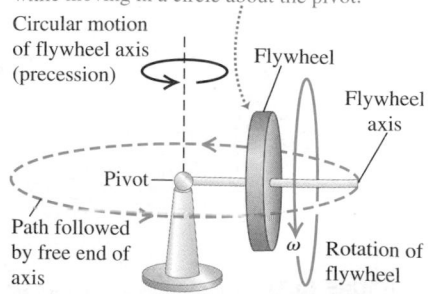

▲ **FIGURE 10.29** A typical toy gyroscope.

When the flywheel is rotating, the system starts with an angular momentum $\vec{L}_i$ parallel to the flywheel's axis of rotation.

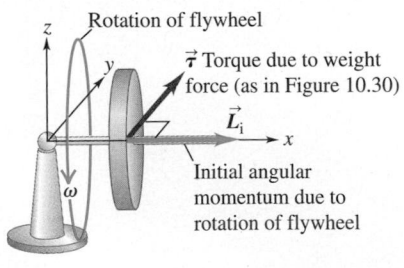

(a)

Now the effect of the torque is to cause the angular momentum to precess around the pivot. The gyroscope circles around its pivot without falling.

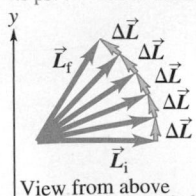

(b)

▲ **FIGURE 10.31** What happens when we set a gyroscope rotating before releasing it.

When the flywheel is not rotating, its weight creates a torque around the pivot, causing it to fall along a circular path until its axis rests on the table surface.

In falling, the flywheel rotates about the pivot and thus has an angular momentum $\vec{L}$. The *direction* of $\vec{L}$ stays constant.

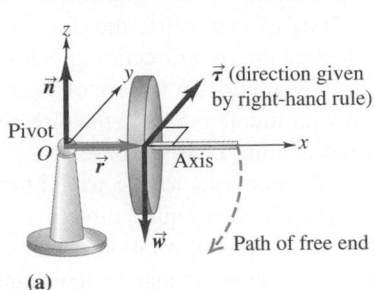

(a)

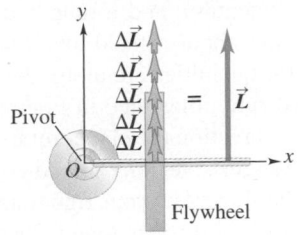

View from above as flywheel falls

(b)

▲ **FIGURE 10.30** What happens when we lift and release a gyroscope that is not rotating.

axis to fall along a path forming a circle in a vertical plane, centered on the pivot, until it hits the table. In doing so, the torque gives the flywheel angular momentum according to Equation 10.17: $\Delta\vec{L} = \sum\vec{\tau}\,\Delta t$. Because the axis of rotation is stationary, the directions of both the torque and the angular momentum are constant. Indeed, if we divide the motion into time intervals Δt (Figure 10.30b), we see that the successive $\Delta\vec{L}$'s are all parallel. These $\Delta\vec{L}$'s add to give a horizontal angular momentum that is constant in *direction* but continuously increases in *magnitude*.

Now let's start over; this time we lift the flywheel and start it rotating before we let it go (Figure 10.31). Now the initial angular momentum $\vec{L}$ of the flywheel (before we let it go) is *along* the flywheel's axis of rotation. Once we let go, the flywheel is acted upon by a torque due to the weight and normal forces, as before. Again, we divide the motion into time intervals Δt (Figure 10.31b). Each $\Delta\vec{L}$ is perpendicular to the flywheel's axis. This causes the *direction* of $\vec{L}$ to change, but not its *magnitude*. The center of mass of the flywheel moves in a horizontal circle. The normal force magnitude n is then equal to the flywheel's weight w, and the corresponding torque is $\tau = rw$. (The situation is analogous to a particle moving in uniform circular motion, where the centripetal force causes changes $\Delta\vec{p}$ in linear momentum that are always perpendicular to the linear momentum $\vec{p}$ that already exists, changing the direction of $\vec{p}$, but not its magnitude.)

In Figure 10.31a, at the instant shown, the gyroscope has a definite nonzero angular momentum described by the vector $\vec{L}_i$. The *change* $\Delta\vec{L}$ in angular momentum in a short time interval Δt following this instant is given by Equation 10.17:

$$\Delta\vec{L} = \vec{\tau}\,\Delta t.$$

The direction of $\Delta\vec{L}$ is the same as that of $\vec{\tau}$. After a time Δt, the angular momentum is $\vec{L}_i + \Delta\vec{L}$. As the vector diagram shows, this means that the gyroscope axis has turned through a small angle $\Delta\phi$ given by $\Delta\phi = |\Delta\vec{L}|/|\vec{L}|$. We see that such a motion of the axis, called **precession,** is required by the torque–angular momentum relationship.

The rate at which the axis moves, $\Delta\phi/\Delta t$, is called the *precession angular velocity*. Denoting this quantity by Ω, we find that

$$\Omega = \frac{\Delta\phi}{\Delta t} = \frac{|\Delta\vec{L}|/|\vec{L}|}{\Delta t} = \frac{\tau}{L} = \frac{wr}{I\omega}. \tag{10.18}$$

Thus, the precession angular velocity is *inversely* proportional to the angular velocity of spin about the axis. A rapidly spinning gyroscope precesses slowly; as the flywheel slows down, the precession angular velocity *increases*.

EXAMPLE 10.14 A laboratory gyroscope

Figure 10.32 shows a top view of a gyroscope wheel in the form of a solid cylinder, driven by an electric motor. The pivot
is at O, and the mass of the axle and
motor are negligible. Is the precession
clockwise or counterclockwise, as seen
from above? If the gyroscope takes 4.0 s
for one revolution of precession, what is
the wheel's angular velocity?

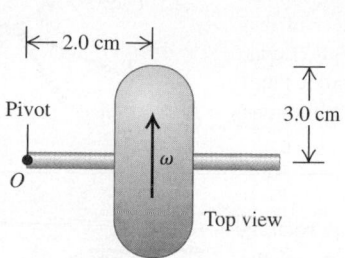

(a) Top view of spinning cylindrical
gyroscope wheel

(b) Vector diagram

▲ **FIGURE 10.32**

SOLUTION

SET UP The right-hand rule shows that $\vec{\omega}$ and $\vec{L}$ are to the left
(Figure 10.32b). The weight $\vec{w}$ points into the page, the torque $\vec{\tau}$
is toward the top of the page, with magnitude rw (where r is the
distance from the pivot point O to the center of mass of the cylinder), and $\Delta\vec{L}/\Delta t$ is also toward the top of the page. Adding a
small $\Delta\vec{L}$ to the $\vec{L}$ that we have at the start changes the direction
of $\vec{L}$ as shown, so the precession is clockwise as seen from above.

Be careful not to mix up ω and Ω. The precession angular
velocity Ω is given as $\Omega = (1\text{ rev})/(4.0\text{ s}) = (2\pi\text{ rad})/
(4.0\text{ s}) = \pi/2\text{ rad/s}$. The weight is equal to mg, and the moment
of inertia of a solid cylinder with radius R is $I = \frac{1}{2}mR^2$.

SOLVE As shown in the "Set Up" discussion, the precession is
clockwise as seen from above. Solving Equation 10.18 for ω, we
find that

$$\omega = \frac{wr}{I\Omega} = \frac{mgr}{(mR^2/2)\Omega}$$

$$= \frac{2(9.8\text{ m/s}^2)(2.0 \times 10^{-2}\text{ m})}{(3.0 \times 10^{-2}\text{ m})^2(\pi/2\text{ rad/s})} = 280\text{ rad/s} = 2700\text{ rev/min}.$$

REFLECT The mass of the flywheel divides out and does not
appear in the final result. If we double the mass, the torque causing the precession doubles, but the moment of inertia also doubles, and the two effects cancel.

As the gyroscope precesses, the center of mass moves in a
circle with radius r in a horizontal plane. Its vertical component
of acceleration is zero, so the pivot must exert an upward normal
force $\vec{n}$ just equal in magnitude to the weight. The circular
motion of the center of mass with angular velocity Ω requires a
force $\vec{F}$ directed toward the center of the circle, with magnitude
$F = m\Omega^2 r$. This force must also be supplied by the pivot.

The preceding analysis of the gyroscope gives us a glimpse into the richness of the dynamics of rotational motion, which can involve some complex
phenomena.

SUMMARY

Torque

(Section 10.1) When a force F acts on a body, the torque τ of that force with respect to a point O is given by $\tau = Fl = rF\sin\phi$ (Equations 10.1 and 10.2). We use the sign convention that a torque tending to cause counterclockwise rotation is positive, while one tending to cause clockwise rotation is negative.

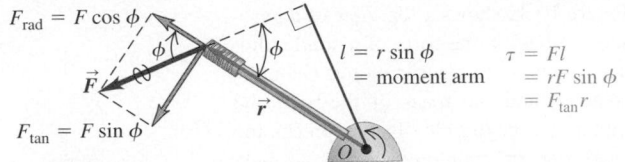

Torque and Angular Acceleration

(Section 10.2) The angular acceleration α of a rigid body rotating about a stationary axis is related to the total torque $\Sigma\tau$ and the moment of inertia I of the body by $\Sigma\tau = I\alpha$ (Equation 10.5). This relation is also valid for a body with a moving axis of rotation, provided that the rotation axis is an axis of symmetry that passes through the center of mass and does not change direction.

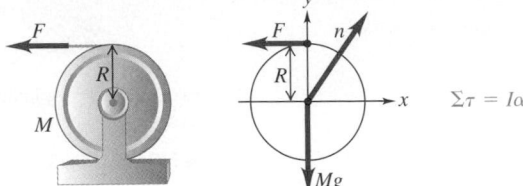

$\Sigma\tau = I\alpha$

Work and Power

(Section 10.3) When a constant torque τ acts on a rigid body that undergoes an angular displacement $\Delta\theta$ from θ_1 to θ_2, the work W done by the torque is $W = \tau(\theta_2 - \theta_1) = \tau\,\Delta\theta$ (Equation 10.7). When the object rotates with angular velocity ω, the power P (rate at which the torque does work) is $P = \tau\omega$ (Equation 10.8).

Angular Momentum

(Section 10.4) The angular momentum L, with respect to an axis through a point O, of a particle with mass m and velocity $\vec{v}$ is $L = mvl$ (Equation 10.11). When a rigid body with moment of inertia I rotates with angular velocity ω about a stationary axis, the angular momentum with respect to that axis is $L = I\omega$ (Equation 10.9).

In terms of torque τ and angular momentum L, the basic dynamic relation for rotational motion can be restated as

$$\Sigma\tau = \frac{\Delta L}{\Delta t}. \qquad (10.10)$$

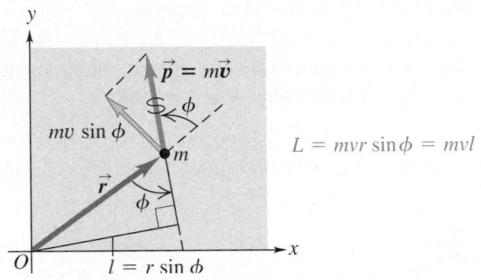

$L = mvr\sin\phi = mvl$

Conservation of Angular Momentum

(Section 10.5) If a system consists of bodies that interact with each other but not with anything else, or if the total torque associated with the external forces is zero, then the total angular momentum of the system is constant (conserved): $\vec{L}_i = \vec{L}_f$.

Equilibrium of a Rigid Body

(Section 10.6) The two conditions for equilibrium of a rigid body are $\Sigma\vec{F} = 0$ (Equation 10.14) and $\Sigma\tau = 0$ (Equation 10.15). The former ensures that the center of mass of the body has no linear acceleration a, and the latter ensures that it has no angular acceleration α about any axis.

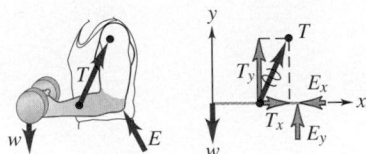

Continued

Vector Nature of Angular Quantities

(Section 10.7) Vector angular velocity $\vec{\omega}$ is a vector quantity whose direction is that of the axis of rotation and whose magnitude is the (scalar) angular velocity. The sense of the vector $\vec{\omega}$ is given by the right-hand rule. When a body rotates about an axis of symmetry, the vector angular momentum $\vec{L}$ is given by $\vec{L} = I\vec{\omega}$.

In terms of vector torque $\vec{\tau}$ and angular momentum $\vec{L}$, the basic dynamic relation for rotational motion can be restated as

$$\sum \vec{\tau} = \frac{\Delta \vec{L}}{\Delta t}. \qquad (10.17)$$

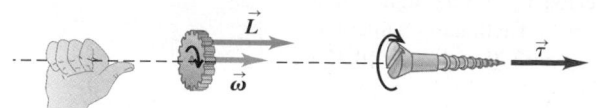

 For instructor-assigned homework, go to www.masteringphysics.com

Conceptual Questions

1. When tightening a bolt, mechanics sometimes extend the length of a wrench handle by slipping a section of pipe over the handle. Why could this procedure easily damage the bolt?

2. Must there necessarily be matter at the center of mass of an object? Illustrate your answer with several common examples.

3. Two identical uniform 12 inch bricks are placed one atop the other and pushed past the edge of a table. (See Figure 10.33.) What is the greatest that x can be?

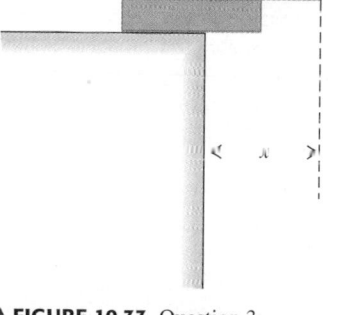

▲ **FIGURE 10.33** Question 3.

4. A uniform 10 cm × 20 cm box is placed on a ramp that rises at an angle θ above the horizontal. (See Figure 10.34.) Assuming that there is enough friction to prevent this box from sliding, what is the largest that θ can be without tipping it over?

5. (a) Can you change the location of your body's center of mass? How? (b) Can you change your body's moment of inertia? How?

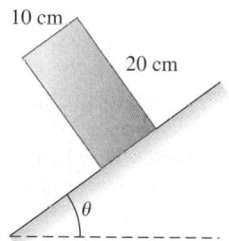

▲ **FIGURE 10.34** Question 4.

6. Serious bicyclists say that if you reduce the weight of a bike, it is more effective if you do so in the wheels rather than in the frame. Why would reducing weight in the wheels make it easier on the bicyclist than reducing the same amount in the frame?

7. When a tall, heavy object, such as a refrigerator, is pushed across a rough floor, what things determine whether it slides or tips?

8. (a) If the forces on an object balance, do the torques necessarily balance? (b) If the torques on an object balance, do the forces necessarily balance? Illustrate your answers with clear examples.

9. Why is a tapered water glass with a narrow base easier to tip over than one with straight sides? Does it matter whether the glass is full or empty?

10. True or false? In picking an axis about which to compute torques in rotational equilibrium, it is nearly always best to choose an axis through the center of mass of the object. Explain your reasoning.

11. People sometimes prop open a door by wedging an object in the space between the vertical hinged side of the door and the frame. Explain why this often results in the hinge screws being ripped out.

12. **Global warming.** As the earth's climate continues to warm, ice near the poles will melt and be added to the oceans. What effect will this have on the length of the day? (*Hint:* Consult a map to see where the oceans lie.)

13. In terms of torques, discuss the action of a claw hammer in pulling out nails.

14. You make two versions of the same object out of the same material having uniform density. For one version all the dimensions are exactly twice as great as for the other one. If the same torque acts on both versions, giving the smaller version angular acceleration α, what will be the angular acceleration of the larger version in terms of α?

15. If two spinning objects have the same angular momentum, will they necessarily have the same rotational kinetic energy? If they have the same rotational kinetic energy, will they necessarily have the same angular momentum?

Multiple-Choice Problems

1. You are designing a wheel that must have a fixed mass and diameter, but that can have its mass distributed in various uniform ways. If the torque you exert on it is also fixed, which of the wheels shown will have the *smallest* angular acceleration about an axis perpendicular to it at its center?

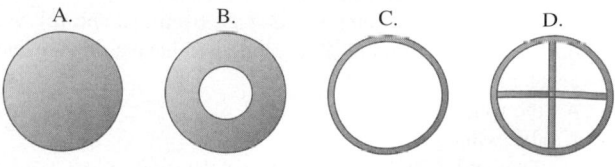

A.   B.   C.   D.

2. Two equal masses *m* are connected by a very light string over a frictionless pulley of mass *m*/2. (See Figure 10.35.) The system has been given a push to get it moving as shown, but that push is no longer acting. In which segment of the string is the tension greater?
   A. The tension in *A* is greater.
   B. The tension in *B* is greater.
   C. The two tensions are the same.

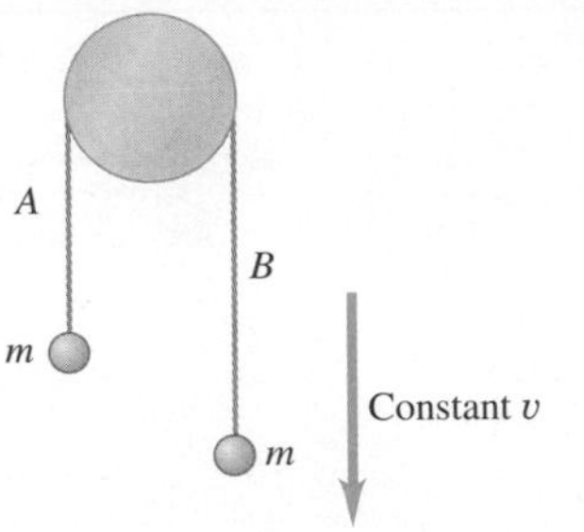

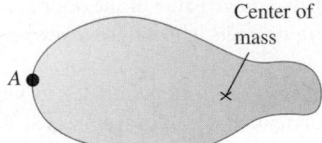

▲ **FIGURE 10.35** Multiple-choice problem 2.

3. The irregular object shown in Figure 10.36 is dropped from rest on the moon, where there is no air. Its center of mass is as shown in the figure. What will it do after it is dropped?
   A. It will start rotating clockwise about its center of mass.
   B. It will start rotating counterclockwise about its center of mass.
   C. It will rotate until point *A* is directly below its center of mass, and then it will stop turning.
   D. It will rotate until point *A* is directly above its center of mass, and then it will stop turning.
   E. It will not rotate.

▲ **FIGURE 10.36** Multiple-choice problem 3.

4. A student is sitting on a frictionless rotating stool with her arms outstretched holding equal heavy weights in each hand. If she suddenly lets go of the weights, her angular speed will
   A. increase
   B. stay the same
   C. decrease.

5. If the torques on an object balance, then it follows that this object (there could be more than one correct choice)
   A. cannot be rotating.
   B. cannot have any angular acceleration.
   C. cannot be moving.
   D. cannot be accelerating.

6. If the forces on an object balance, then it follows that this object (there could be more than one correct choice)
   A. cannot be rotating.
   B. cannot have any angular acceleration.
   C. cannot be moving.
   D. cannot be accelerating.

7. A solid uniform ball and a solid uniform cylinder with the same mass and diameter are released from rest and roll without slipping down a hill. Which one will have a *greater linear* acceleration on the hill?
   A. the sphere.
   B. the cylinder.
   C. They will both have the same acceleration.

8. A uniform beam is suspended horizontally as shown in Figure 10.37. It is attached to the wall by a small hinge. The force that the hinge exerts *on the beam* has components
   A. upward and to the right.
   B. upward and to the left.
   C. downward and to the right.
   D. downward and to the left.

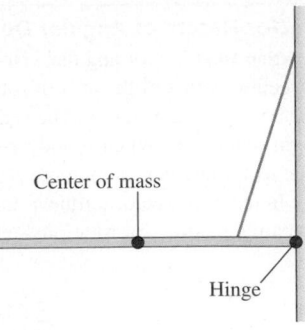

▲ **FIGURE 10.37** Multiple-choice problem 8.

9. Two identical cars are traveling in the same lane 100 m apart on a freeway. The lead car is moving at 88 km/h and the trailing car is moving at 82 km/h. From the point of view of a stationary observer by the road, what is happening to the center of mass of the system consisting of the two cars?
   A. It is moving backward at 3 km/h.
   B. It is moving forward at 3 km/h.
   C. It is moving backward at 85 km/h.
   D. It is moving forward at 85 km/h.

10. A person pushes vertically downward with force *P* on a lever of length *L* that is inclined at an angle θ above the horizontal as shown in Figure 10.38. The torque that the person's push produces about point *A* is
    A. $PL \sin\theta$.
    B. $PL \cos\theta$.
    C. PL.

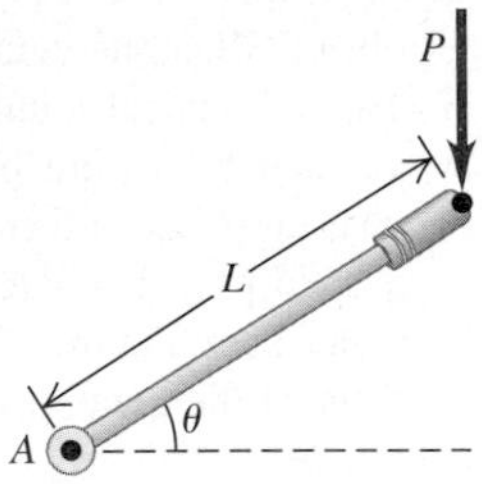

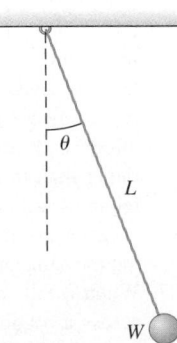

▲ **FIGURE 10.38** Multiple-choice problem 10.

11. String is wrapped around the outer rim of a solid uniform cylinder that is free to rotate about a frictionless axle through its center. When the string is pulled with a force *P* tangent to the rim, it gives the cylinder an angular acceleration α. If the cylinder had *twice* the radius, but everything else was the same, the angular acceleration would be
    A. 4α.   B. 2α.
    C. $\alpha/2$.   D. $\alpha/4$.

12. A weight *W* swings from a hook in the ceiling by a light string of length *L*, as shown in Figure 10.39. *T* is the tension in the string. When the string makes an angle θ with the vertical, the *net* torque about the hook is
    A. WL.
    B. $(W - T)L$.
    C. TL.
    D. $WL \sin\theta$.
    E. $WL \cos\theta$.

▲ **FIGURE 10.39** Multiple-choice problem 12.

13. A ball of mass 0.20 kg is whirled in a horizontal circle at the end of a light string 75 cm long at a speed of 3.0 m/s. If the string is lengthened to 1.5 m while the ball is being twirled, then the speed of the ball will now be
    A. 12 m/s.   B. 6.0 m/s.   C. 1.5 m/s.   D. 0.75 m/s.

14. A heavy solid disk rotating freely and slowed only by friction applied at its outer edge takes 100 seconds to come to a stop. If the disk had twice the radius and twice the mass, but the frictional force remained the same, the time it would it take the wheel to come to a stop from the same initial rotational speed is
    A. 25 s   B. 100 s   C. 400 s   D. 800 s

15. A uniform metal meterstick is balanced as shown in Figure 10.40 with a 1.0 kg rock attached to the left end of the stick. (Pay attention to the scale of the diagram.) What is the mass of the meterstick?
    A. 0.60 kg.   B. 1.0 kg.   C. 2.0 kg.   D. 3.0 kg.

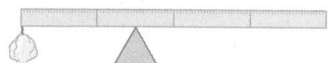

▲ **FIGURE 10.40** Multiple-choice problem 15.

Problems

10.1 Torque

1. • Calculate the torque (magnitude and direction) about point O due to the force $\vec{F}$ in each of the situations sketched in Figure 10.41. In each case, the force $\vec{F}$ and the rod both lie in the plane of the page, the rod has length 4.00 m, and the force has magnitude 10.0 N.

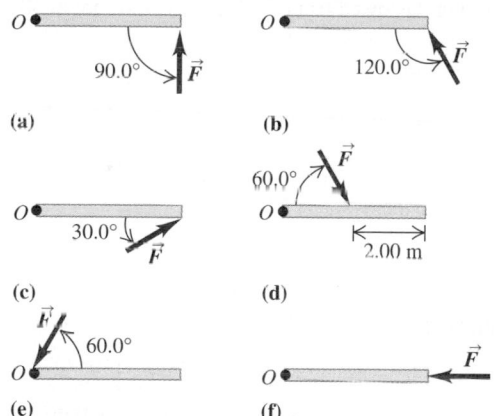

▲ **FIGURE 10.41** Problem 1.

2. • Calculate the net torque about point O for the two forces applied as in Figure 10.42. The rod and both forces are in the plane of the page.

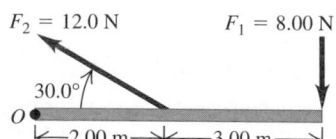

▲ **FIGURE 10.42** Problem 2.

3. •• Three forces are applied to a wheel of radius 0.350 m, as shown in Fig. 10.43. One force is perpendicular to the rim, one is tangent to it, and the other one makes a 40.0° angle with the radius. What is the net torque on the wheel due to these three forces for an axis perpendicular to the wheel and passing through its center?

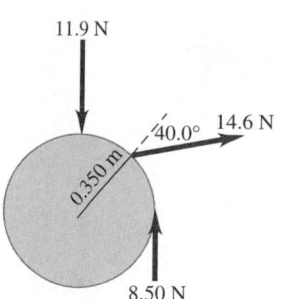

▲ **FIGURE 10.43** Problem 3.

▲ **FIGURE 10.44** Problem 4.

4. • In Figure 10.44, forces $\vec{A}$, $\vec{B}$, $\vec{C}$, and $\vec{D}$, each have magnitude 50 N and act at the same point on the object. (a) What torque (magnitude and direction) does each of these forces exert on the object about point P? (b) What is the total torque about point P?

5. •• A square metal plate 0.180 m on each side is pivoted about an axis through point O at its center and perpendicular to the plate. (See Figure 10.45.) Calculate the net torque about this axis due to the three forces shown in the figure if the magnitudes of the forces are $F_1 = 18.0$ N, $F_2 = 26.0$ N, and $F_3 = 14.0$ N. The plate and all forces are in the plane of the page.

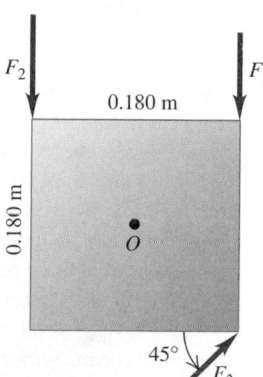

▲ **FIGURE 10.45** Problem 5.

10.2 Torque and Angular Acceleration

6. • A cord is wrapped around the rim of a wheel 0.250 m in radius, and a steady pull of 40.0 N is exerted on the cord. The wheel is mounted on frictionless bearings on a horizontal shaft through its center. The moment of inertia of the wheel about this shaft is $5.00 \text{ kg} \cdot \text{m}^2$. Compute the angular acceleration of the wheel.

7. • A certain type of propeller blade can be modeled as a thin uniform bar 2.50 m long and of mass 24.0 kg that is free to rotate about a frictionless axle perpendicular to the bar at its midpoint. If a technician strikes this blade with a mallet 1.15 m from the center with a 35.0 N force perpendicular to the blade, find the maximum angular acceleration the blade could achieve.

8. •• A 750 gram grinding wheel 25.0 cm in diameter is in the shape of a uniform solid disk. (We can ignore the small hole at the center.) When it is in use, it turns at a constant 220 rpm about an axle perpendicular to its face through its center. When the power switch is turned off, you observe that the wheel stops in 45.0 s with constant angular acceleration due to friction at the axle. What torque does friction exert while this wheel is slowing down?

9. •• A grindstone in the shape of a solid disk with diameter 0.520 m and a mass 50.0 kg is rotating at 850 rev/min. You press an ax against the rim with a normal force of 160 N (see Figure 10.46), and the grindstone comes to rest in 7.50 s. Find the coefficient of kinetic friction between the ax and the grindstone. There is negligible friction in the bearings.

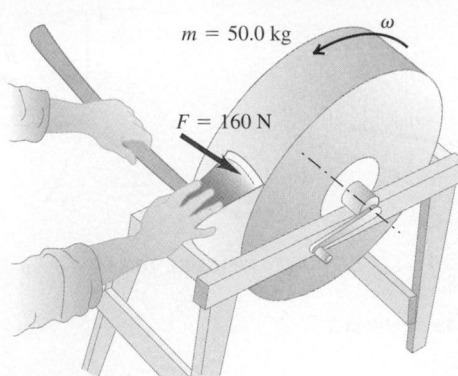

$m = 50.0$ kg
ω
$F = 160$ N

▲ **FIGURE 10.46** Problem 9.

10. •• A solid, uniform cylinder with mass 8.25 kg and diameter 15.0 cm is spinning at 220 rpm on a thin, frictionless axle that passes along the cylinder axis. You design a simple friction brake to stop the cylinder by pressing the brake against the outer rim with a normal force. The coefficient of kinetic friction between the brake and rim is 0.333. What must the applied normal force be to bring the cylinder to rest after it has turned through 5.25 revolutions?

11. •• A 2.00 kg stone is tied to a thin, light wire wrapped around the outer edge of the uniform 10.0 kg cylindrical pulley shown in Figure 10.47. The inner diameter of the pulley is 60.0 cm, while the outer diameter is 1.00 m. The system is released from rest, and there is no friction at the axle of the pulley. Find (a) the acceleration of the stone, (b) the tension in the wire, and (c) the angular acceleration of the pulley.

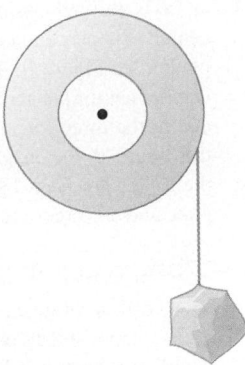

▲ **FIGURE 10.47**
Problem 11.

12. •• A light rope is wrapped several times around a large wheel with a radius of 0.400 m. The wheel rotates in frictionless bearings about a stationary horizontal axis, as shown in Figure 10.48. The free end of the rope is tied to a suitcase with a mass of 15.0 kg. The suitcase is released from rest at a height of 4.00 m above the ground. The suitcase has a speed of 3.50 m/s when it reaches the ground. Calculate (a) the angular velocity of the wheel when the suitcase reaches the ground and (b) the moment of inertia of the wheel.

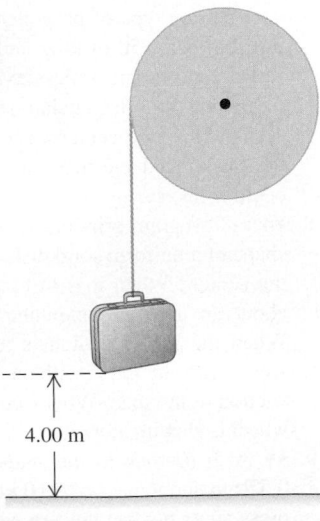

4.00 m

▲ **FIGURE 10.48** Problem 12.

13. •• A 22,500 N elevator is to be accelerated upward by connecting it to a counterweight using a light (but strong!) cable passing over a solid uniform disk-shaped pulley. There is no appreciable friction at the axle of the pulley, but its mass is 875 kg and it is 1.50 m in diameter. (a) How heavy should the counterweight be so that it will accelerate the elevator upward through 6.75 m in the first 3.00 s, starting from rest? (b) Under these conditions, what is the tension in the cable on each side of the pulley?

14. •• A thin, light string is wrapped around the rim of a 4.00 kg solid uniform disk that is 30.0 cm in diameter. A person pulls on the string with a constant force of 100.0 N tangent to the disk, as shown in Figure 10.49. *The disk is not attached to anything and is free to move and turn.* (a) Find the angular acceleration of the disk about its center of mass and the linear acceleration of its center of mass. (b) If the disk is replaced by a hollow thin-walled cylinder of the same mass and diameter, what will be the accelerations in part (a)?

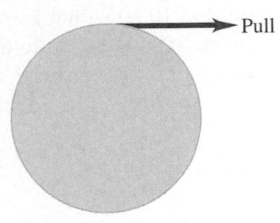

Pull

▲ **FIGURE 10.49**
Problem 14.

15. •• A uniform, 8.40-kg, spherical shell 50.0 cm in diameter has four small 2.00-kg masses attached to its outer surface and equally spaced around it. This combination is spinning about an axis running through the center of the sphere and two of the small masses (Fig. 10.50). What friction torque is needed to reduce its angular speed from 75.0 rpm to 50.0 rpm in 30.0 s?

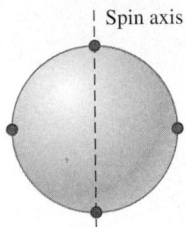

Spin axis

▲ **FIGURE 10.50** Problem 15.

16. •• A hollow spherical shell with mass 2.00 kg rolls without slipping down a 38.0° slope. (a) Find the acceleration of the shell and the friction force on it. Is the friction kinetic or static friction? Why? (b) How would your answers to part (a) change if the mass were doubled to 4.00 kg?

17. •• A solid disk of radius 8.50 cm and mass 1.25 kg, which is rolling at a speed of 2.50 m/s, begins rolling without slipping up a 10.0° slope. How long will it take for the disk to come to a stop?

10.3 Work and Power in Rotational Motion

18. • What is the power output in horsepower of an electric motor turning at 4800 rev/min and developing a torque of 4.30 N · m?

19. • A playground merry-go-round has a radius of 4.40 m and a moment of inertia of 245 kg · m² and turns with negligible friction about a vertical axle through its center. (a) A child applies a 25.0 N force tangentially to the edge of the merry-go-round for 20.0 s. If the merry-go-round is initially at rest, what is its angular velocity after this 20.0 s interval? (b) How much work did the child do on the merry-go-round? (c) What is the average power supplied by the child?

20. • The flywheel of a motor has a mass of 300.0 kg and a moment of inertia of 580 kg · m². The motor develops a constant torque of 2000.0 N · m, and the flywheel starts from rest. (a) What is the angular acceleration of the flywheel? (b) What is its angular velocity after it makes 4.00 revolutions? (c) How much work is done by the motor during the first 4.00 revolutions?

21. •• (a) Compute the torque developed by an industrial motor whose output is 150 kW at an angular speed of 4000.0 rev/min. (b) A drum with negligible mass and 0.400 m in diameter is attached to the motor shaft, and the power output of the motor is used to raise a weight hanging from a rope wrapped around the drum. How heavy a weight can the motor lift at constant speed? (c) At what constant speed will the weight rise?

10.4 Angular Momentum

22. • Calculate the angular momentum and kinetic energy of a solid uniform sphere with a radius of 0.120 m and a mass of 14.0 kg if it is rotating at 6.00 rad/s about an axis through its center.

23. •• (a) Calculate the magnitude of the angular momentum of the earth in a circular orbit around the sun. Is it reasonable to model it as a particle? (b) Calculate the magnitude of the angular momentum of the earth due to its rotation around an axis through the north and south poles, modeling it as a uniform sphere. Consult Appendix D and the astronomical data in Appendix E.

24. • A small 0.300 kg bird is flying horizontally at 3.50 m/s toward a 0.750 kg thin bar hanging vertically from a hook at its upper end, as shown in Figure 10.51. (a) When the bird is far from the bar, what are the magnitude and direction (clockwise or counterclockwise) of its angular momentum about a horizontal axis perpendicular to the plane of the figure and passing through (i) point A, (ii) point B, and (iii) point C? (b) Repeat part (a) when the bird is just ready to hit the bar but is still flying horizontally.

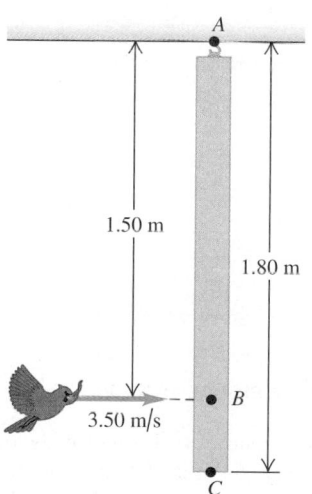

▲ **FIGURE 10.51** Problem 24.

1.50 m

1.80 m

3.50 m/s

A

B

C

25. • A small 4.0 kg brick is released from rest 2.5 m above a horizontal seesaw on a fulcrum at its center, as shown in Figure 10.52. Find the angular momentum of this brick about a horizontal axis through the fulcrum and perpendicular to the plane of the figure (a) the instant the brick is released and (b) the instant before it strikes the seesaw.

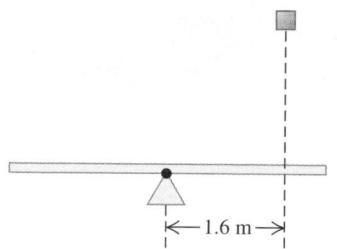

|← 1.6 m →|

▲ **FIGURE 10.52** Problem 25.

26. • A woman with mass 50.0 kg is standing on the rim of a large disk that is rotating at 0.50 rev/s about an axis perpendicular to it through its center. The disk has a mass of 110 kg and a radius of 4.0 m. Calculate the magnitude of the total angular momentum of the woman-plus-disk system, assuming that you can treat the woman as a point.

27. •• A certain drawbridge can be modeled as a uniform 15,000 N bar, 12.0 m long, pivoted about its lower end. When this bridge is raised to an angle of 60.0° above the horizontal, the cable holding it suddenly breaks, allowing the bridge to fall. At the instant after the cable breaks, (a) what is the torque on this bridge about the pivot and (b) at what rate is its angular momentum changing?

10.5 Conservation of Angular Momentum

28. • On an old-fashioned rotating piano stool, a woman sits holding a pair of dumbbells at a distance of 0.60 m from the axis of rotation of the stool. She is given an angular velocity of 3.00 rad/s, after which she pulls the dumbbells in until they are only 0.20 m distant from the axis. The woman's moment of inertia about the axis of rotation is 5.00 kg · m² and may be considered constant. Each dumbbell has a mass of 5.00 kg and may be considered a point mass. Neglect friction. (a) What is the initial angular momentum of the system? (b) What is the angular velocity of the system after the dumbbells are pulled in toward the axis? (c) Compute the kinetic energy of the system before and after the dumbbells are pulled in. Account for the difference, if any.

29. • **The spinning figure skater.** The outstretched hands and arms of a figure skater preparing for a spin can be considered a slender rod pivoting about an axis through its center. (See Figure 10.53.) When the skater's hands and arms are brought in and wrapped around his body to execute the spin, the hands and

▲ **FIGURE 10.53** Problem 29.

arms can be considered a thin-walled hollow cylinder. His hands and arms have a combined mass of 8.0 kg. When outstretched, they span 1.8 m; when wrapped, they form a cylinder of radius 25 cm. The moment of inertia about the axis of rotation of the remainder of his body is constant and equal to 0.40 kg · m². If the skater's original angular speed is 0.40 rev/s, what is his final angular speed?

30. •• A small block on a frictionless horizontal surface
has a mass of 0.0250 kg.
It is attached to a mass-
less cord passing through a
hole in the surface. (See
Figure 10.54.) The block is
originally revolving at a
distance of 0.300 m from the hole with an angular speed of
1.75 rad/s. The cord is then pulled from below, shortening the
radius of the circle in which the block revolves to 0.150 m. You
may treat the block as a particle. (a) Is angular momentum
conserved? Why or why not? (b) What is the new angular
speed? (c) Find the change in kinetic energy of the block.
(d) How much work was done in pulling the cord?

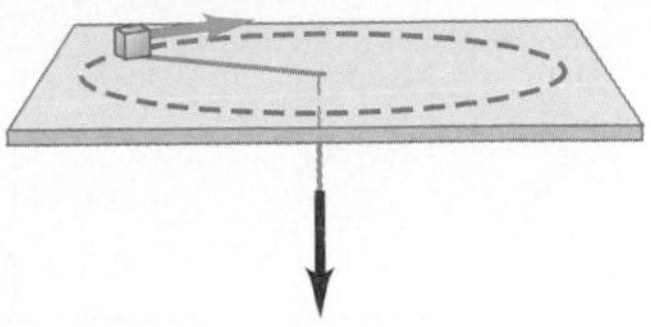

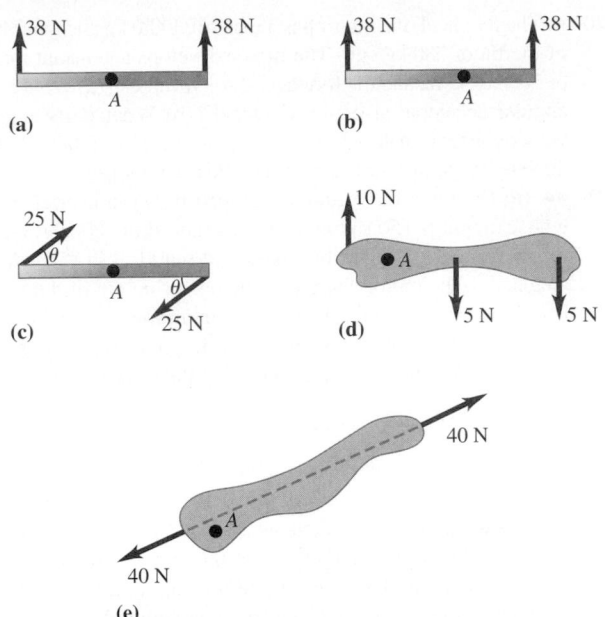
▲ **FIGURE 10.54** Problem 30.

31. •• A uniform 4.5 kg square solid wooden gate 1.5 m on each
side hangs vertically from a frictionless pivot at the center of
its upper edge. A 1.1 kg raven flying horizontally at 5.0 m/s
flies into this gate at its center and bounces back at 2.0 m/s in
the opposite direction. (a) What is the angular speed of the
gate just after it is struck by the unfortunate raven? (b) During
the collision, why is the angular momentum conserved, but not
the linear momentum?

32. •• A diver comes off a board with arms straight up and legs
straight down, giving her a moment of inertia about her rotation axis of 18 kg·m². She then tucks into a small ball,
decreasing this moment of inertia to 3.6 kg·m². While tucked,
she makes two complete revolutions in 1.0 s. If she hadn't
tucked at all, how many revolutions would she have made in
the 1.5 s from board to water?

33. •• A large turntable rotates about a fixed vertical axis, making
one revolution in 6.00 s. The moment of inertia of the turntable
about this axis is 1200 kg·m². A child of mass 40.0 kg, initially standing at the center of the turntable, runs out along a
radius. What is the angular speed of the turntable when the
child is 2.00 m from the center, assuming that you can treat the
child as a particle?

34. •• A large wooden turntable in the shape of a flat disk has a
radius of 2.00 m and a total mass of 120 kg. The turntable is
initially rotating at 3.00 rad/s about a vertical axis through its
center. Suddenly, a 70.0 kg parachutist makes a soft landing on
the turntable at a point on its outer edge. Find the angular
speed of the turntable after the parachutist lands. (Assume that
you can treat the parachutist as a particle.)

10.6 Equilibrium of a Rigid Body

35. • Which of the objects shown in Figure 10.55 are in only
translational equilibrium, only rotational equilibrium (about
the axis A), both translational and rotational equilibrium
(about A), or neither equilibrium?

36. • (a) In each of the objects indicated in Figure 10.56, what
magnitude of force F (if any) is needed to put the object into
rotational equilibrium about the axis A shown? (b) After you
have found the F required to put the object into rotational equilibrium, find out which (if any) of these objects is also in translational equilibrium.

37. •• **Supporting a broken leg.** A therapist tells a 74 kg patient
BIO with a broken leg that he must have his leg in a cast suspended
horizontally. For minimum discomfort, the leg should be sup-

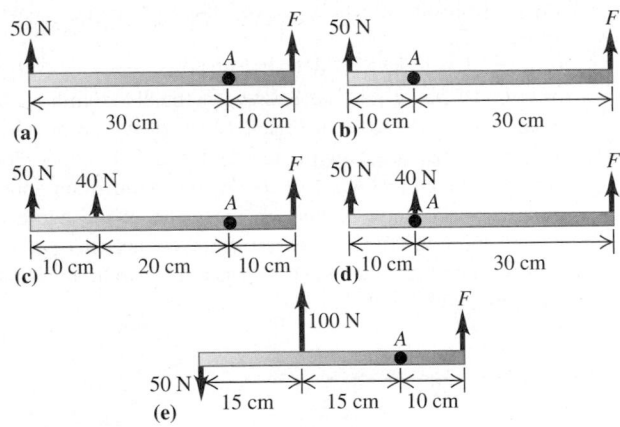
▲ **FIGURE 10.55** Problem 35.

▲ **FIGURE 10.56** Problem 36.

ported by a vertical strap
attached at the center of
mass of the leg–cast sys-
tem. (See Figure 10.57.) In
order to comply with these
instructions, the patient
consults a table of typical
mass distributions and finds
that both upper legs

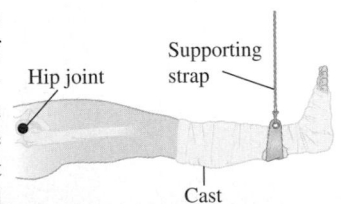
▲ **FIGURE 10.57** Problem 37.

(thighs) together typically account for 21.5% of body weight
and the center of mass of each thigh is 18.0 cm from the hip
joint. The patient also reads that the two lower legs (including
the feet) are 14.0% of body weight, with a center of mass
69.0 cm from the hip joint. The cast has a mass of 5.50 kg, and
its center of mass is 78.0 cm from the hip joint. How far from
the hip joint should the supporting strap be attached to the cast?

38. •• Two people are carrying a uniform wooden board that
is 3.00 m long and weighs 160 N. If one person applies an
upward force equal to 60 N at one end, at what point and with

what force does the other person lift? Start with a free-body diagram of the board.

39. •• **Push-ups.** To strengthen his arm and chest muscles, an
BIO 82 kg athlete 2.0 m tall is doing a series of push-ups as shown in Figure 10.58. His center of mass is 1.15 m from the bottom of his feet, and the centers of his palms are 30.0 cm from the top of his head. Find the force that the floor exerts on each of his feet and on each hand, assuming that both feet exert the same force and both palms do likewise. Begin with a free-body diagram of the athlete.

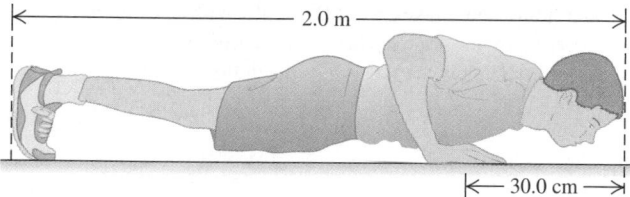

▲ **FIGURE 10.58** Problem 39.

40. •• Two people carry a heavy electric motor by placing it on a light board 2.00 m long. One person lifts at one end with a force of 400.0 N, and the other lifts at the opposite end with a force of 600.0 N. (a) Start by making a free-body diagram of the motor. (b) What is the weight of the motor? (c) Where along the board is its center of gravity located?

41. •• A 60.0-cm, uniform, 50.0-N shelf is supported horizontally by two vertical wires attached to the sloping ceiling (Fig. 10.59). A very small 25.0-N tool is placed on the shelf midway between the points where the wires are attached to it. Find the tension in each wire. Begin by making a free-body diagram of the shelf.

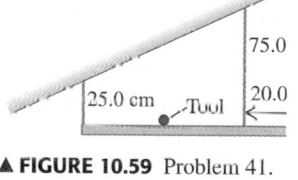

▲ **FIGURE 10.59** Problem 41.

42. •• The horizontal beam in Figure 10.60 weighs 150 N, and its center of gravity is at its center. First make a free-body diagram of the beam. Then find (a) the tension in the cable and (b) the horizontal and vertical components of the force exerted on the beam at the wall.

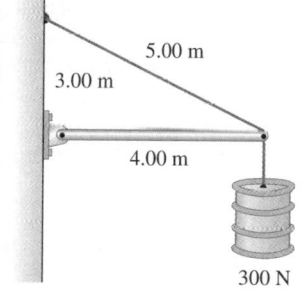

▲ **FIGURE 10.60** Problem 42.

43. •• The boom in Figure 10.61 weighs 2600 N and is attached to a frictionless pivot at its lower end. It is not uniform; the distance of its center of gravity from the pivot is 35% of its length. Find (a) the tension in the guy wire and (b) the horizontal and vertical components of the force exerted on the boom at its lower end. Start with a free-body diagram of the boom.

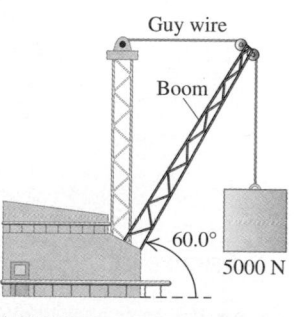

▲ **FIGURE 10.61** Problem 43.

44. • A uniform 250 N ladder rests against a perfectly smooth wall, making a 35° angle with the wall. (a) Draw a free-body diagram of the ladder. (b) Find the normal forces that the wall and the floor exert on the ladder. (c) What is the friction force on the ladder at the floor?

45. •• A uniform ladder 7.0 m long weighing 450 N rests with one end on the ground and the other end against a perfectly smooth vertical wall. The ladder rises at 60.0° above the horizontal floor. A 750 N painter finds that she can climb 2.75 m up the ladder, measured along its length, before it begins to slip. (a) Make a free-body diagram of the ladder. (b) What force does the wall exert on the ladder? (c) Find the friction force and normal force that the floor exerts on the ladder.

46. •• A 9.0 m uniform beam is hinged to a vertical wall and held horizontally by a 5.0 m cable attached to the wall 4.0 m above the hinge, as shown in Figure 10.62. The metal of this cable has a test strength of 1.00 kN, which means that it will break if the tension in it exceeds that amount. (a) Draw a free-body diagram of the beam. (b) What is the heaviest beam that the cable can support with the given configuration? (c) Find the horizontal and vertical components of the force the hinge exerts on the beam.

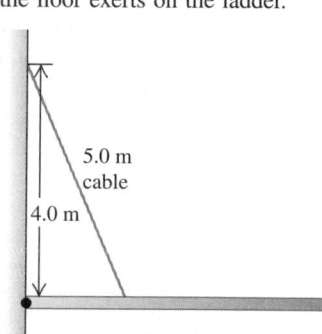

▲ **FIGURE 10.62** Problem 46.

47. •• A uniform beam 4.0 m long and weighing 2500 N carries a 3500 N weight 1.50 m from the far end, as shown in Figure 10.63. It is supported horizontally by a hinge at the wall and a metal wire at the far end. (a) Make a free-body diagram of the beam. (b) How strong does the wire have to be? That is, what is the minimum tension it must be able to support without breaking? (c) What are the horizontal and vertical components of the force that the hinge exerts on the beam?

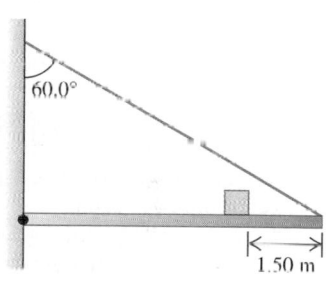

▲ **FIGURE 10.63** Problem 47.

48. •• **Leg raises.** In a simpli-
BIO fied version of the musculature action in leg raises, the abdominal muscles pull on the femur (thigh bone) to raise the leg by pivoting it about one end, as shown in Figure 10.64. When you are lying horizontally, these muscles make an angle of approximately 5° with the femur, and if you raise your legs, the muscles remain approximately horizontal, so the angle θ increases. We shall assume for simplicity that these muscles attach to the femur in only one place, 10 cm from the hip joint (although, in reality, the situation is more complicated). For a certain 80 kg person having a

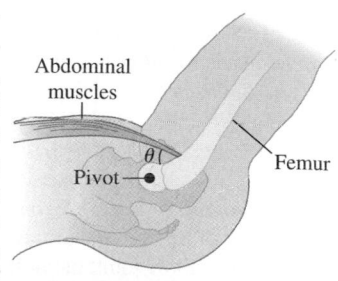

▲ **FIGURE 10.64** Problem 48.

leg 90 cm long, the mass of the leg is 15 kg and its center of mass is 44 cm from his hip joint as measured along the leg. If the person raises his leg to 60° above the horizontal, the angle between the abdominal muscles and his femur would also be about 60°. (a) With his leg raised to 60°, find the tension in the abdominal muscle on each leg. As usual, begin your solution with a free-body diagram. (b) When is the tension in this muscle greater, when the leg is raised to 60° or when the person just starts to raise it off the ground? Why? (Try this yourself to check your answer.) (c) If the abdominal muscles attached to the femur were perfectly horizontal when a person was lying down, could the person raise his leg? Why or why not?

49. ●● A diving board 3.00 m long is supported at a point 1.00 m from the end, and a diver weighing 500 N stands at the free end (Fig. 10.65). The diving board is of uniform cross section and weighs 280 N. Find (a) the force at the support point and (b) the force at the left-hand end.

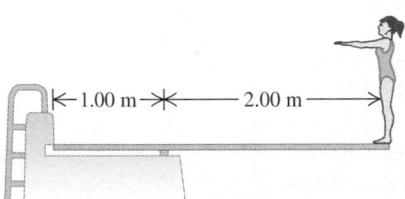

▲ **FIGURE 10.65** Problem 49.

50. ●● Two people carry a heavy electric motor by placing it on a light board 2.00 m long. One person lifts at one end with a force of 400 N, and the other lifts the opposite end with a force of 600 N. (a) What is the weight of the motor, and where along the board is its center of gravity located? (b) Suppose the board is not light but weighs 200 N, with its center of gravity at its center, and the two people each exert the same forces as before. What is the weight of the motor in this case, and where is its center of gravity located?

51. ●● **Pumping iron.** A
BIO 72.0 kg weight lifter is doing arm raises using a 7.50 kg weight in her hand. Her arm pivots around the elbow joint, starting 40.0° below the horizontal. (See Figure 10.66.) Biometric measurements have shown that both forearms and the hands together account for 6.00% of a person's weight. Since the upper arm is held vertically, the

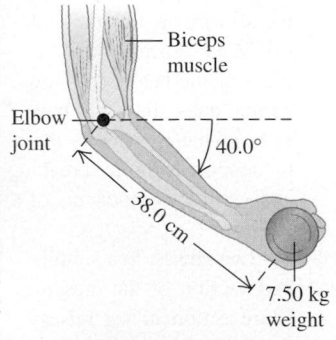

▲ **FIGURE 10.66** Problem 51.

biceps muscle always acts vertically and is attached to the bones of the forearm 5.50 cm from the elbow joint. The center of mass of this person's forearm–hand combination is 16.0 cm from the elbow joint, along the bones of the forearm, and the weight is held 38.0 cm from the elbow joint. (a) Make a free-body diagram of the forearm. (b) What force does the biceps muscle exert on the forearm? (c) Find the magnitude and direction of the force that the elbow joint exerts on the forearm. (d) As the weight lifter raises her arm toward a horizontal position, will the force in the biceps muscle increase, decrease, or stay the same? Why?

52. ●● **The deltoid muscle.** The deltoid muscle is the main mus-
BIO cle that allows you to raise your arm or even hold it out. It is connected to the humerus of the upper arm. (See part (a) of Figure 10.67.) The person shown is holding his arm out horizontally with a 2.50 kg weight in his hand. This weight is 60.0 cm from the shoulder joint. His forearm (including his hand) has a mass of 2.44 kg and is 34.0 cm long; its center of mass is 43 cm from the shoulder joint, measured along the arm. His upper arm is 26.0 cm long and has a mass of 2.63 kg; its center of mass is 13.0 cm from the shoulder joint. The deltoid muscle is attached to the humerus 15.0 cm from the shoulder joint and makes a 14.0° angle with the humerus. (See part (b) of the figure.) (a) Make a free-body diagram of the arm. (b) What is the tension in the deltoid muscle?

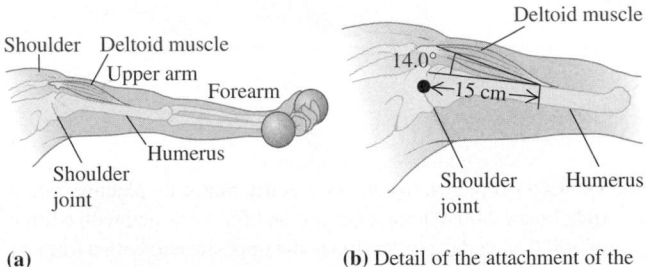

(a)

(b) Detail of the attachment of the deltoid muscle.

▲ **FIGURE 10.67** Problem 52.

53. ●● A uniform, 90.0-N table is 3.6 m long, 1.0 m high, and
BIO 1.2 m wide. A 1500-N weight is placed 0.50 m from one end of the table, a distance of 0.60 m from each of the two legs at that end. Draw a free-body diagram for the table and find the force that each of the four legs exerts on the floor.

10.7 Vector Nature of Angular Quantities

54. ● The rotor (flywheel) of a toy gyroscope has a mass of 0.140 kg. Its moment of inertia about its axis is $1.20 \times 10^{-4} \text{ kg} \cdot \text{m}^2$. The mass of the frame is 0.0250 kg. The gyroscope is supported on a single pivot (see Figure 10.68) with its center of mass a horizontal distance of 4.00 cm from the pivot. The gyroscope is precessing in a horizontal plane at the rate of 1 revolution in 2.20 s. (a) Find the upward force exerted by the pivot. (b) Find the angular speed with which the rotor is spinning about its axis, expressed in rev/min. (c) Copy the diagram and draw vectors to show the angular momentum of the rotor and the torque acting on it.

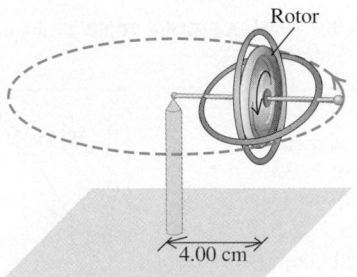

▲ **FIGURE 10.68** Problem 54.

55. ●● For each of the following rotating objects, describe the direction of the angular momentum vector: (a) the minute

hand of a clock; (b) the right front tire of a car moving backwards; (c) an ice skater spinning clockwise; (d) the earth, rotating on its axis.

General Problems

56. •• **Back pains during pregnancy.** Women often suffer from
BIO back pains during pregnancy. Let us investigate the cause of these pains, assuming that the woman's mass is 60 kg before pregnancy. Typically, women gain about 10 kg during pregnancy, due to the weight of the fetus, placenta, amniotic fluid, etc. To make the calculations easy, but still realistic, we shall model the unpregnant woman as a uniform cylinder of diameter 30 cm. We can model the added mass due to the fetus as a 10 kg sphere 25 cm in diameter and centered about 5 cm *outside* the woman's original front surface. (a) By how much does her pregnancy change the horizontal location of the woman's center of mass? (b) How does the change in part (a) affect the way the pregnant woman must stand and walk? In other words, what must she do to her posture to make up for her shifted center of mass? (c) Can you now explain why she might have backaches?

57. •• You are asked to design the decorative mobile shown in Figure 10.69. The strings and rods have negligible weight, and the rods are to hang horizontally. (a) Draw a free-body diagram for each rod. (b) Find the weights of the balls A, B, and C. Find the tensions S_1, S_2, and S_3 in the strings. (c) What can you say about the horizontal location of the mobile's center of gravity? Explain.

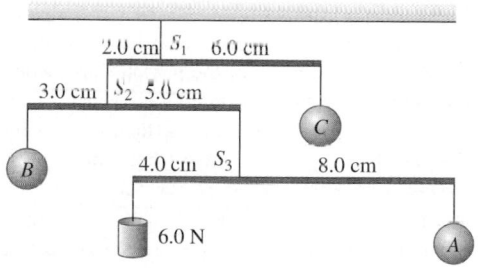

▲ **FIGURE 10.69** Problem 57.

58. •• **A good workout.** You
BIO are doing exercises on a Nautilus machine in a gym to strengthen your deltoid (shoulder) muscles. Your arms are raised vertically and can pivot around the shoulder joint, and you grasp the cable of the machine in your hand 64.0 cm from your shoulder joint. The deltoid muscle is attached to the humerus 15.0 cm from the shoulder joint and makes a 12.0° angle with that bone. (See Figure 10.70.) If you have set the tension in the cable of the machine

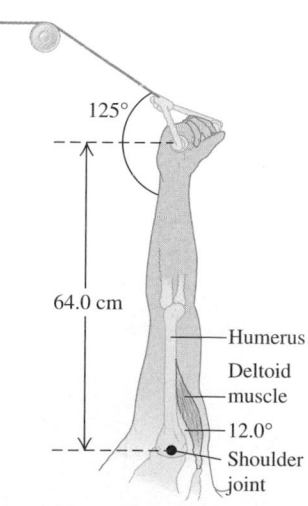

▲ **FIGURE 10.70** Problem 58.

to 36.0 N on each arm, what is the tension in each deltoid muscle if you simply hold your outstretched arms in place? (*Hint:* Start by making a clear free-body diagram of your arm.)

59. •• Prior to being placed in its hole, a 5700-N, 9.0-m-long, uniform utility pole makes some nonzero angle with the vertical. A vertical cable attached 2.0 m below its upper end holds it in place while its lower end rests on the ground. (a) Find the tension in the cable and the magnitude and direction of the force exerted by the ground on the pole. (b) Why don't we need to know the angle the pole makes with the vertical, as long as it is not zero?

60. •• A uniform drawbridge must be held at a 37° angle above the horizontal to allow ships to pass underneath. The drawbridge weighs 45,000 N, is 14.0 m long, and pivots about a hinge at its lower end. A cable is connected 3.5 m from the hinge, as measured along the bridge, and pulls horizontally on the bridge to hold it in place. (a) What is the tension in the cable? (b) Find the magnitude and direction of the force the hinge exerts on the bridge. (c) If the cable suddenly breaks, what is the initial angular acceleration of the bridge?

61. •• **Pyramid builders.** Ancient pyramid builders are balancing a uniform rectangular slab of stone tipped at an angle θ above the horizontal using a rope (Fig. 10.71). The rope is held by five workers who share the force equally. (a) If $\theta = 20.0°$, what force does each worker exert on the rope? (b) As θ increases, does each worker have to exert more or less force than in part (a), assuming they do not change the angle of the rope? Why? (c) At what angle do the workers need to exert *no force* to balance the slab? What happens if θ exceeds this value?

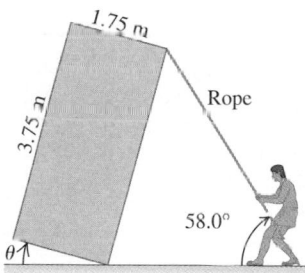

▲ **FIGURE 10.71** Problem 61.

62. •• **The farmyard gate.** A gate 4.00 m wide and 2.00 m high weighs 500 N. Its center of gravity is at its center, and it is hinged at A and B. To relieve the strain on the top hinge, a wire CD is connected as shown in Fig. 10.72. The tension in CD is increased until the horizontal force at hinge A is zero. (a) What is the tension in the wire CD? (b) What is the magnitude of the horizontal component of the force at hinge B? (c) What is the combined vertical force exerted by hinges A and B?

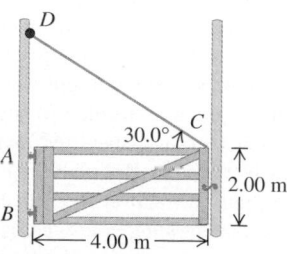

▲ **FIGURE 10.72** Problem 62.

63. •• **Atwood's machine.** Figure 10.73 illustrates an Atwood's machine. Find the linear accelerations of blocks A and B, the angular acceleration of the wheel C, and the tension in each side of the cord if there is no slipping between the cord and the surface of the wheel. Let the masses of blocks A and B be 4.00 kg and 2.00 kg, respectively, the moment of inertia of the wheel about its axis be $0.300 \text{ kg} \cdot \text{m}^2$, and the radius of the wheel be 0.120 m.

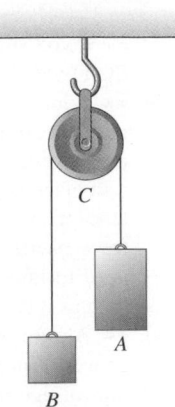

▲ **FIGURE 10.73** Problem 63.

64. •• **Neck muscles.** A student bends her
BIO head at 40.0° from the vertical while intently reading her physics book, pivoting the head around the upper vertebra (point P in Figure 10.74). Her head has a mass of 4.50 kg (which is typical), and its center of mass is 11.0 cm from the pivot point P. Her neck muscles are 1.50 cm from point P, as measured *perpendicular* to these muscles. The neck itself and the vertebrae are held vertical. (a) Draw a free-body diagram of the student's head. (b) Find the tension in her neck muscles.

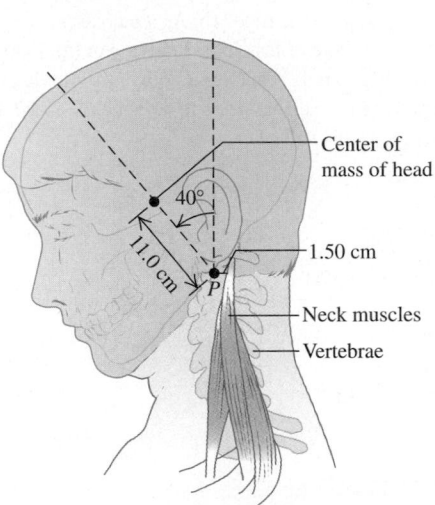

▲ **FIGURE 10.74** Problem 64.

65. •• **Russell traction apparatus.** The device shown in
BIO Figure 10.75 is one version of a Russell traction apparatus. It has two functions: to support the injured leg horizontally and at the same time provide a horizontal traction force on it. This can be done by adjusting the weight W and the angle θ. For this patient, his leg (including his foot) is 95.0 cm long (measured from the hip joint) and has a mass of 14.2 kg. Its center of mass is 41.0 cm from the hip joint. A support strap is attached to the patient's ankle 13.0 cm from the bottom of his foot. (a) What weight W is needed to support the leg horizontally? (b) If the therapist specifies that the traction force must be 12.0 N horizontally, what must be the angle θ? (c) What is the greatest traction force that this apparatus could supply to this patient's leg, and what is θ in that case?

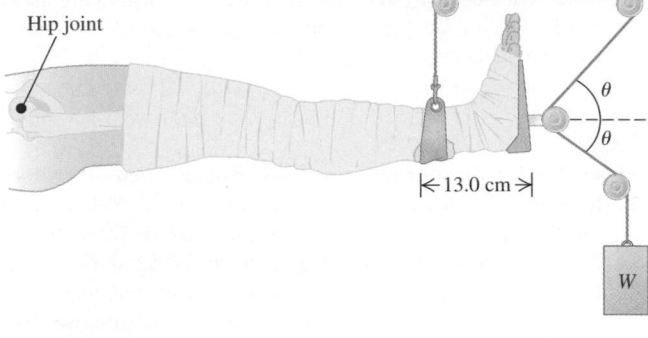

▲ **FIGURE 10.75** Problem 65.

66. •• **Supporting an injured**
BIO **arm: I.** A 650 N person must have her injured arm supported, with the upper arm horizontal and the forearm vertical. (See Figure 10.76.) According to biomedical tables and direct measurements, her upper arm is 26 cm long (measured from the shoulder joint), accounts for 3.50% of her body weight, and has a center of mass 13.0 cm from her shoulder joint. Her forearm (including the hand) is 34.0 cm long, makes up 3.25% of her body weight, and has a center of mass 43.0 cm from her shoulder joint. (a) Where is the center of mass of the person's arm when it is supported as shown? (b) What weight W is needed to support her arm? (c) Find the horizontal and vertical components of the force that the shoulder joint exerts on her arm.

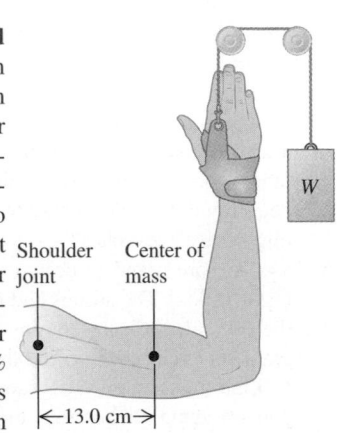

▲ **FIGURE 10.76** Problem 66.

67. •• **Supporting an injured arm: II.** As part of therapy, the
BIO person's arm in the previous problem is later to be supported with the upper arm horizontal, but with the forearm at an angle of 55° above the horizontal, as shown in Figure 10.77. (a) Where is the center of mass of the person's arm? (b) What weight W is needed to support her arm this way? (c) Find the horizontal and vertical components of the force that the shoulder joint exerts on her arm.

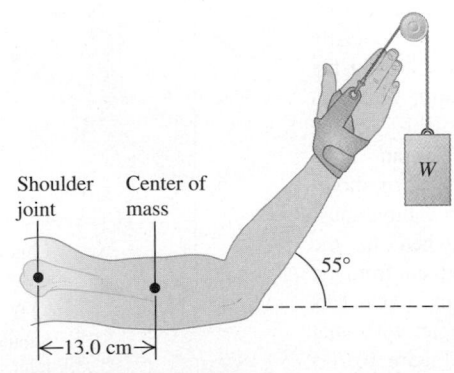

▲ **FIGURE 10.77** Problem 67.

68. •• **The forces on the foot.** A 750 N athlete standing on *one*
BIO foot on a very smooth gym floor lifts his body by pivoting
his foot upward through a 30.0° angle, balancing all of his
weight on the ball of the foot. The forces on the foot bones
from the rest of his body are due to the Achilles tendon and
the ankle joint. (See Figure 10.78.) The Achilles tendon acts
perpendicular to a line through the heel and toes; it is this
line that has rotated upward through 30.0°. Assume that the
weight of the foot is negligible compared with that of the rest
of the body, and begin by making a free-body diagram of the
foot. (a) What are the magnitude and direction of the force
that the floor exerts on the athlete's foot? (b) What is the tension in the Achilles tendon? (c) Find the horizontal and vertical components of the force that the ankle joint exerts on the
foot. Then use these components to find the magnitude of
this force.

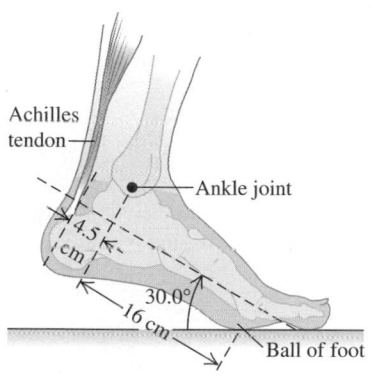

▲ **FIGURE 10.78** Problem 68.

69. •• A uniform solid cylinder of
mass M is supported on a ramp that
rises at an angle θ above the horizontal by a wire that is wrapped
around its rim and pulls on it tangentially parallel to the ramp (Fig.
10.79). (a) Show that there *must* be
friction on the surface for the cylinder to balance this way. (b) Show
that the tension in the wire must be
equal to the friction force, and find
this tension.

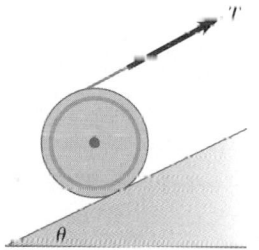

▲ **FIGURE 10.79**
Problem 69.

70. •• A uniform 8.0 m, 1500 kg
beam is hinged to a wall and
supported by a thin cable
attached 2.0 m from the free
end of the beam, as shown in
Figure 10.80. The beam is
supported at an angle of 30.0°
above the horizontal. (a) Make
a free-body diagram of the
beam. (b) Find the tension in
the cable. (c) How hard does
the beam push inward on the wall?

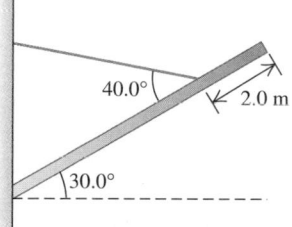

▲ **FIGURE 10.80** Problem 70.

71. •• You are trying to raise a bicycle wheel of mass m and
radius R up over a curb of height h. To do this, you apply a
horizontal force $\vec{F}$ (Fig. 10.81). What is the smallest magnitude of the force $\vec{F}$ that will succeed in raising the wheel onto

the curb when the force is applied (a) at the center of the
wheel, and (b) at the top of the wheel? (c) In which case is less
force required?

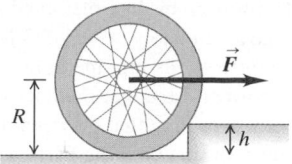

▲ **FIGURE 10.81** Problem 71.

72. •• An experimental bicycle wheel is placed on a test stand so
that it is free to turn on its axle. If a constant net torque of
5.00 N · m is applied to the tire for 2.00 s, the angular speed of
the tire increases from zero to 100 rev/min. The external
torque is then removed, and the wheel is brought to rest in
125 s by friction in its bearings. Compute (a) the moment of
inertia of the wheel about the axis of rotation, (b) the friction
torque, and (c) the total number of revolutions made by the
wheel in the 125 s time interval.

73. •• Under some circumstances, a star can collapse into an
extremely dense object made mostly of neutrons and called a
neutron star. The density of a neutron star is roughly 10^{14}
times as great as that of ordinary solid matter. Suppose we represent the star as a uniform, solid, rigid sphere, both before and
after the collapse. The star's initial radius was 7.0×10^5 km
(comparable to our sun); its final radius is 16 km. If the original star rotated once in 30 days, find the angular speed of the
neutron star.

74. •• Disks A and B are
mounted on shaft SS and
may be connected or dis-
connected by clutch C.
(See Figure 10.82.) Disk A
is made of a lighter material than disk B, so the
moment of inertia of disk A

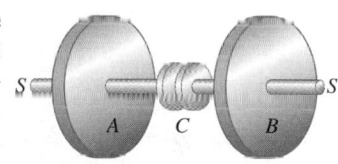

▲ **FIGURE 10.82** Problem 74.

about the shaft is one-third that of disk B. The moments of
inertia of the shaft and clutch are negligible. With the clutch
disconnected, A is brought up to an angular speed ω_0. The
accelerating torque is then removed from A, and A is coupled
to disk B by the clutch. (You can ignore bearing friction.) It is
found that 2400 J of thermal energy is developed in the clutch
when the connection is made. What was the original kinetic
energy of disk A?

75. •• While exploring a castle, Exena the Exterminator is spot-
ted by a dragon who chases her down a hallway. Exena runs
into a room and attempts to swing the heavy door shut before
the dragon gets her. The door is initially perpendicular to the
wall, so it must be turned through 90° to close. The door is
3.00 m tall and 1.25 m wide, and it weighs 750 N. You can
ignore the friction at the hinges. If Exena applies a force of
220 N at the edge of the door and perpendicular to it, how
much time does it take her to close the door?

76. •• **Downward-Facing Dog.** One yoga exercise, known as the
BIO "Downward-Facing Dog," requires stretching your hands
straight out above your head and bending down to lean against
the floor. This exercise is performed by a certain 750 N person,

as shown in the simplified model in Figure 10.83. When he bends his body at the hip to a 90° angle between his legs and trunk, his legs, trunk, head and arms have the dimensions indicated. Furthermore, his legs and feet weigh a total of 277 N, and their center of mass is 41 cm from his hip, measured along his legs. The person's trunk, head, and arms weigh 473 N, and their center of gravity is 65 cm from his hip, measured along the upper body. (a) Find the normal force that the floor exerts on each foot and on each hand, assuming that the person does not favor either hand or either foot. (b) Find the friction force on each foot and on each hand, assuming that it is the same on both feet and on both hands (but not necessarily the same on the feet as on the hands). [*Hint:* First treat his entire body as a system; then isolate his legs (or his upper body).]

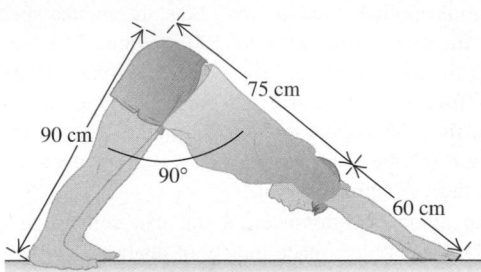

▲ **FIGURE 10.83** Problem 76.

77. •• A uniform, 7.5-m-long beam weighing 9000 N is hinged to a wall and supported by a thin cable attached 1.5 m from the free end of the beam. The cable runs between the beam and the wall and makes a 40° angle with the beam. What is the tension in the cable when the beam is at an angle of 30° above the horizontal?

Passage Problems

BIO Rotating space station. A space station, in the form of a rotating hollow cylinder, is designed to provide a comfortable living environment, complete with artificial gravity, for astronauts in orbit around the earth. The space station has a length of 1000 m, an outer radius of 250 m, an inner radius of 240 m, and is fabricated using an ultra light aluminum and graphite fiber composite. Forty small, adjustable rockets attached to the outer shell of the space station will provide the angular acceleration needed to spin the space station up to the required final angular velocity of 0.20 rad/s. According to design specifications, the moment of inertia of the space station about its longitudinal axis is 10^{12} kg · m².

78. What would be the radial (centripetal) acceleration of an astronaut standing on the inner surface of the space station, which is at a distance of 240 m from the axis of rotation, once the space station reaches its final angular velocity of 0.20 rad/s?
    A. 0 m/s²
    B. 4.8 m/s²
    C. 9.6 m/s²
    D. 48 m/s²

79. What would be the total torque produced by all 40 rockets? Assume that each rocket is adjusted so that its thrust is 1000 newtons in a direction that will provide maximum torque about the longitudinal axis of the space station.
    A. 2.40×10^5 N · m
    B. 2.50×10^5 N · m
    C. 1.00×10^7 N · m
    D. 1.0×10^{10} N · m

80. What would be the resulting angular acceleration of the space station when all 40 rockets provide maximum torque?
    A. 10^{-5} rad/s²
    B. 10^{-2} rad/s²
    C. 10^2 rad/s²
    D. 10^{22} rad/s²

81. How long will the rockets need to operate in order to bring the rotational velocity of the space station up from 0 rad/s to the final required value of 0.20 rad/s?
    A. 2×10^4 s
    B. 20 s
    C. 0.002 s
    D. 2×10^{-6} s

II Elasticity and Periodic Motion

In preceding chapters, we've often used idealized models to represent objects as particles or rigid bodies. A rigid body doesn't bend, stretch, or squash when forces act on it. But the rigid body is an idealized model; all real materials do deform to some extent. In this chapter we'll introduce concepts and principles that help us predict the deformations that occur when forces are applied to a real (not perfectly rigid) body. Elastic properties of materials are tremendously important. You want the wings of an airplane to be able to bend a little, but you don't want them to break off. The steel frame of an earthquake-resistant building has to be able to flex, but not too much. Many of the necessities of everyday life, from rubber bands to suspension bridges, depend on the elastic properties of materials.

Next we'll study a class of problems, including some that involve elastic behavior, in which an object repeats the same motion over and over again in a cyclic pattern. Any such motion is called a *periodic motion* or an *oscillation*. Many familiar examples of oscillating objects come to mind: the pendulum of a grandfather's clock, the vibrating strings of a musical instrument, the pistons in a gasoline engine, a child playing on a swing. Less obvious oscillatory phenomena are the vibrations of the quartz crystal in a watch, the electromagnetic waves of radio and television transmission, and the vibrations of atoms in the crystal structure of a solid material. Periodic motion plays a vital role in many areas of physics, and studying it will give us an important foundation for later work in several areas.

One topic we'll cover in this chapter is the motion of pendulums such as these swings. We'll learn that the frequency of a swing—how often it moves back and forth per unit of time—depends on the length of the ropes or chains suspending it but not on the mass of the rider. Whether one child or several use a swing, its frequency will be the same.

11.1 Stress, Strain, and Elastic Deformations

The rigid body is a useful idealized model, but the stretching, squeezing, and twisting of real objects when forces are applied are often too important to ignore. Figure 11.1 shows examples of these types of forces. For each case, we'll study

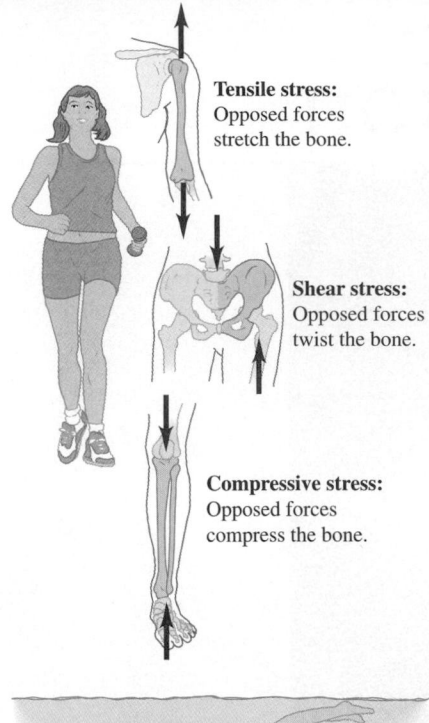

Tensile stress: Opposed forces stretch the bone.

Shear stress: Opposed forces twist the bone.

Compressive stress: Opposed forces compress the bone.

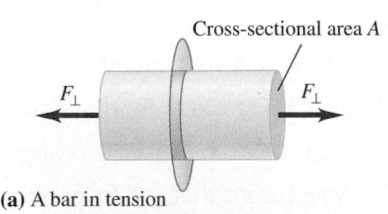

Volume stress: Water pressure squeezes the swimmer.

▲ **FIGURE 11.1** Four types of stress that we will study.

the relation between forces and deformations. We'll introduce a quantity called **stress** that characterizes the strength of the forces associated with the stretch, squeeze, or twist, usually on a "force per unit area" basis. Another quantity, **strain,** describes the deformation that occurs. When the stress and strain are small enough, we often find that the two are directly proportional. The harder you pull on something, the more it stretches; and the more you squeeze it, the more it compresses. The general pattern that emerges can be formulated as

$$\frac{\text{Stress}}{\text{Strain}} = \text{Constant}. \tag{11.1}$$

The proportionality between stress and strain (under certain conditions) is called **Hooke's law,** after Robert Hooke (1635–1703), a contemporary of Newton. We saw a special case of Hooke's law in Sections 5.4 and 7.4: The elongation of a spring is approximately proportional to the stretching forces involved. The constant in Equation 11.1 is often called an *elastic modulus.*

Tensile and Compressive Stress and Strain

The simplest elastic behavior to understand is the stretching of a bar, rod, or wire when its ends are pulled. Figure 11.2a shows a bar with uniform cross-sectional area A, with equal and opposite forces with magnitude $F_\perp$ pulling at its ends. We say that the bar is in **tension.** We've already talked a lot about tensions in ropes and strings; it's the same concept here. Figure 11.2b shows a cross section through the bar; the part to the right of the section pulls on the part to the left with a force with magnitude $F_\perp$ and vice versa. The force is distributed uniformly over the section, as shown by the short arrows in the figure. (This is always the case if the forces at the *ends* are uniformly distributed.)

We define the **tensile stress** at the cross section as the ratio of the force magnitude $F_\perp$ to the cross-sectional area A:

$$\text{Tensile stress} = \frac{F_\perp}{A}. \tag{11.2}$$

The SI unit of stress is the newton per square meter (N/m^2). This unit is also called the **pascal** (abbreviated Pa):

$$1 \text{ pascal} = 1 \text{ Pa} = 1 \text{ N/m}^2.$$

In the British system, the logical unit of stress would be the pound per square foot, but the pound per square inch (lb/in.², or psi) is more commonly used.

The units of stress are the same as those of *pressure,* which we'll encounter often in later chapters. Steel cables are commonly required to withstand tensile stresses on the order of 10^8 Pa (about 150,000 psi).

Cross-sectional area A

$F_\perp \qquad\qquad F_\perp$

(a) A bar in tension

We assume that the tension force is distributed evenly over any cross section through the bar.

$F_\perp \qquad\qquad F_\perp$

(b) Force on a cross section through the bar

▲ **FIGURE 11.2** Tensile stress.

Cross-sectional area A

$F_\perp \qquad\qquad F_\perp$

(a) A bar in compression

$F_\perp \qquad\qquad F_\perp$

(b) Force on a cross section through the bar

▲ **FIGURE 11.3** Compressive stress.

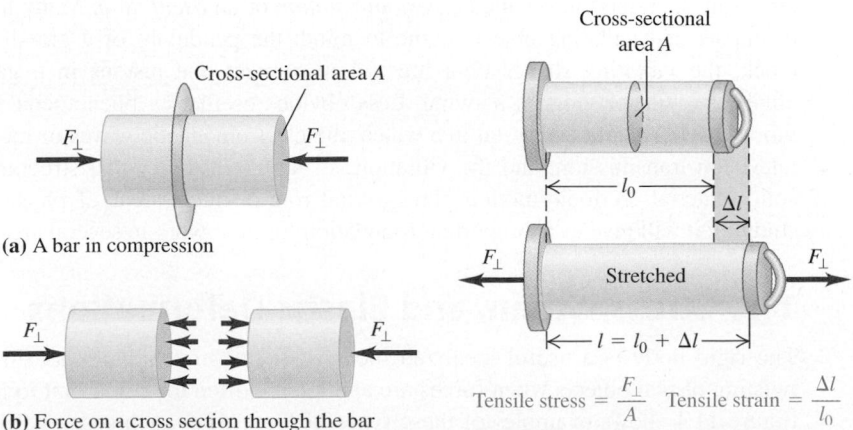

Cross-sectional area A

l_0

Δl

$F_\perp \qquad\qquad$ Stretched $\qquad\qquad F_\perp$

$l = l_0 + \Delta l$

$\text{Tensile stress} = \dfrac{F_\perp}{A} \qquad \text{Tensile strain} = \dfrac{\Delta l}{l_0}$

▲ **FIGURE 11.4** Tensile strain.

▶ **Application Bridging the gap.** The modern cable-stayed bridge is not only beautiful, but also a study in tension and compression. In this type of bridge, the weight of the roadway is carried by sets of diagonal cables attached to tall towers. The cables are in tension, while the roadway and the towers are in compression. Because the cables are oriented at angles to the vertical, they experience very high tensile stress and are thus quite stiff, making this a desirable type of bridge for windy conditions. In addition, the cables act as guy wires to stabilize the towers in the direction parallel to the roadway. This photo shows the Millau Viaduct in France while it was under construction. It is currently the world's tallest bridge.

When the forces on the ends of a bar are pushes rather than pulls (Figure 11.3a), the bar is in **compression,** and the stress is a **compressive stress.** At the cross section shown, each side pushes, rather than pulls, on the other.

The fractional change in length (the stretch) of an object under a tensile stress is called the **tensile strain.** Figure 11.4 shows a bar with unstretched length l_0 that stretches to a length $l = l_0 + \Delta l$ when equal and opposite forces with magnitude $F_\perp$ are applied to its ends. The elongation Δl, measured along the same line as l_0, doesn't occur only at the ends; every part of the bar stretches in the same proportion. The tensile strain is defined as the ratio of the elongation Δl to the original length l_0:

$$\text{Tensile strain} = \frac{l - l_0}{l_0} = \frac{\Delta l}{l_0}. \tag{11.3}$$

Tensile strain is the amount of stretch per unit length. It is a ratio of two lengths, always measured in the same units, so strain is a pure (dimensionless) number with no units. The **compressive strain** of a bar in compression is defined in the same way as tensile strain, but Δl has the opposite direction (Figure 11.5). In this case, it is often convenient to treat Δl as a negative quantity.

Experiments have shown that, for a sufficiently small tensile or compressive stress, stress and strain are proportional, as stated by Hooke's law, Equation 11.1. The corresponding proportionality constant, called **Young's modulus** (denoted by Y), is given by

$$Y = \frac{\text{Tensile stress}}{\text{Tensile strain}} \quad \text{or} \quad \frac{\text{Compressive stress}}{\text{Compressive strain}},$$

or

$$Y = \frac{F_\perp / A}{\Delta l / l_0} = \frac{l_0}{A} \frac{F_\perp}{\Delta l}. \tag{11.4}$$

Strain is a pure number, so the units of Young's modulus are the same as those of stress: force per unit area. Young's modulus is a property of a specific material, rather than of any particular object made of that material. A material with a large value of Y is relatively *unstretchable;* a large stress is required for a given strain. For example, steel has a much larger value of Y than rubber. Table 11.1 lists values of Y for a few representative materials.

When you stretch a wire or a rubber band, it gets *thinner* as well as longer. When Hooke's law holds, the fractional decrease in width is found to be proportional to the tensile strain, typically 0.2 to 0.4 times the tensile strain. Similarly, a material under compressive stress bulges at the sides, and again the fractional change in width is proportional to the compressive strain. For many materials, though not all, Young's modulus has the same values for both tensile and compressive stress.

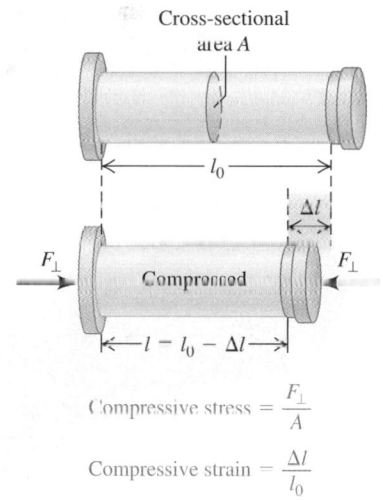

Cross-sectional area A

Compressed

$\text{Compressive stress} = \dfrac{F_\perp}{A}$

$\text{Compressive strain} = \dfrac{\Delta l}{l_0}$

▲ **FIGURE 11.5** Compressive strain.

TABLE 11.1 Young's modulus

Material	Y (Pa)
Aluminum	0.70×10^{11}
Brass	0.91×10^{11}
Copper	1.1×10^{11}
Glass	0.55×10^{11}
Iron	1.9×10^{11}
Steel	2.0×10^{11}
Tungsten	3.6×10^{11}

EXAMPLE 11.1 **A stretching elevator cable**

A small elevator with a mass of 550 kg hangs from a steel cable that is 3.0 m long when not loaded. The wires making up the cable have a total cross-sectional area of 0.20 cm², and with a 550 kg load, the cable stretches 0.40 cm beyond its unloaded length. Determine the cable's stress and strain. Assuming that the cable is equivalent to a rod with the same cross-sectional area, determine the value of Young's modulus for the cable's steel.

SOLUTION

SET UP AND SOLVE Figure 11.6 shows our diagram. The tension in the cable is $F_\perp = mg = (550 \text{ kg})(9.8 \text{ m/s}^2)$. We use the definitions of tensile stress and strain, and Young's modulus, given by Equations 11.2, 11.3, and 11.4:

$$\text{Stress} = \frac{F_\perp}{A} = \frac{(550 \text{ kg})(9.8 \text{ m/s}^2)}{2.0 \times 10^{-5} \text{ m}^2} = 2.7 \times 10^8 \text{ Pa},$$

$$\text{Strain} = \frac{\Delta l}{l_0} = \frac{0.0040 \text{ m}}{3.0 \text{ m}} = 0.00133,$$

$$Y = \frac{\text{Stress}}{\text{Strain}} = \frac{2.7 \times 10^8 \text{ Pa}}{0.00133} = 2.0 \times 10^{11} \text{ Pa}.$$

REFLECT The cable stretches by about 0.1% of its original length.

Practice Problem: If we want the elongation of the cable to be no greater than 0.001 m, what cross-sectional area must the cable have? *Answer:* 0.80 cm².

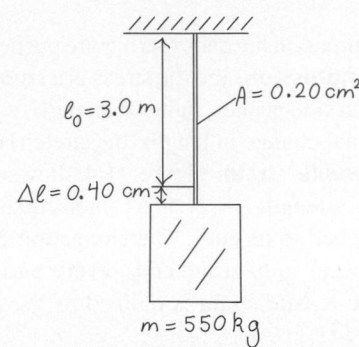

▲ **FIGURE 11.6** Our sketch for this problem.

▲ **Application Hooke, line, and sinker.** In addition to weighing a northern pike, this biologist is also demonstrating the relationship between stress and strain. Within reasonable limits, the spring inside the scale obeys Hooke's law: The displacement of the end of the spring (strain) is proportional to the force applied at the end by the weight of the fish (stress). Because the spring constant (force divided by elongation) does not vary, you can measure the displacement of the end of the spring and determine the mass of the fish.

Modeling an Elastic Material as a Spring

Young's modulus characterizes the elastic properties of a material under tension or compression in a way that is independent of the size or shape of the particular object. To find the relation of force to elongation of any specific rod, cable, or spring, we solve Equation 11.4 for the magnitude $F_\perp$ of the force at each end causing the elongation; we obtain

$$F_\perp = \frac{YA}{l_0} \Delta l.$$

Now let $k = YA/l_0$, call the elongation x instead of Δl, and call the force magnitude F_x instead of $F_\perp$; then we have

$$F_x = kx.$$

We recognize k as the force constant (or spring constant) for a spring; we encountered it in Sections 5.4 and 7.4. When Hooke's law is obeyed, the force F_x needed (on each end) to stretch a spring a distance x beyond its undistorted length is directly proportional to x, and the ratio of F_x to x is the force constant k. Similarly, the shortening of an object in compression is directly proportional to the compressing force. Hooke's law was originally stated in this form; it was reformulated in terms of stress and strain much later by other physicists.

Volume Stress and Strain

When you dive under water, the water exerts nearly uniform pressure everywhere on your surface and squeezes you to a slightly smaller volume (Figure 11.1). This situation is different from tensile and compressive stresses and strains. The stress is now a uniform pressure on all sides, and the resulting deformation is a change in volume. We use the terms **volume stress** and **volume strain** to describe these quantities. Another familiar example is the compression of a gas under pressure, such as the air in a car's tires. We'll use the terms *volume stress* and *pressure* interchangeably.

The **pressure** in a fluid, denoted by p, is the force $F_\perp$ per unit area A transmitted across any cross section of the fluid, against a wall of its container, or against a surface of an immersed object:

$$p = \frac{F_\perp}{A}. \tag{11.5}$$

When a solid object is immersed in a fluid and both are at rest, **the forces that the fluid exerts on the surface of the object are always perpendicular to the surface at each point.** If they were not, the fluid would move relative to the object.

When we apply pressure to the surface of a fluid in a container, such as the cylinder and piston shown in Figure 11.7, the pressure is transmitted through the fluid and also acts on the surface of any object immersed in the fluid. This principle is called **Pascal's law.** If pressure differences due to differences in depth within the fluid can be neglected, the pressure is the same at every point in the fluid and at every point on the surface of any submerged object.

Pressure has the same units as stress; commonly used units include 1 Pa (or 1 N/m^2) and 1 lb/in.^2 (or 1 psi). Also in common use is the **atmosphere,** abbreviated atm. One atmosphere is defined to be the average pressure of the earth's atmosphere at sea level:

$$1 \text{ atmosphere} = 1 \text{ atm} = 1.013 \times 10^5 \text{ Pa} = 14.7 \text{ lb/in}^2.$$

Pressure is a scalar quantity, not a vector quantity; it has no direction.

We've noted that pressure plays the role of volume stress in a volume deformation. The corresponding strain, called **volume strain,** is defined as the fractional change in volume (Figure 11.8)—that is, the ratio of the change in volume, ΔV, to the original volume V_0:

$$\text{Volume strain} = \frac{\Delta V}{V_0}. \tag{11.6}$$

Volume strain is change in volume per unit volume. Like tensile or compressive strain, it is a pure number, without units.

When Hooke's law is obeyed, the volume strain is *proportional* to the volume stress (change in pressure). The corresponding constant ratio of stress to strain is called the **bulk modulus,** denoted by B. When the pressure on an object changes by a small amount Δp, from p_0 to $p_0 + \Delta p$, and the resulting volume strain is $\Delta V/V_0$, Hooke's law takes the form

$$B = -\frac{\Delta p}{\Delta V/V_0}. \tag{11.7}$$

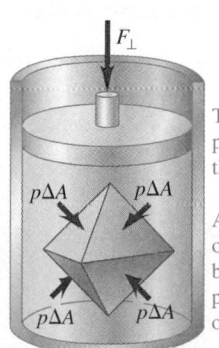

The force applied to the piston is distributed over the surface of the object.

At each point on the object, the force applied by the fluid is perpendicular to the object's surface.

▲ **FIGURE 11.7** Volume stress.

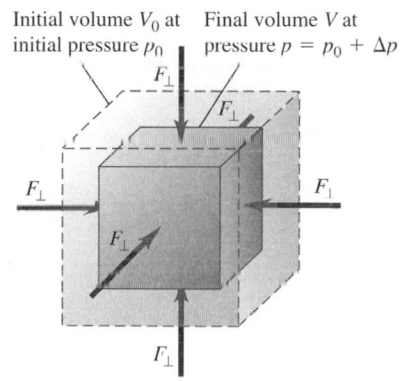

Initial volume V_0 at initial pressure p_0 Final volume V at pressure $p = p_0 + \Delta p$

Volume stress = Δp Volume strain = $\dfrac{\Delta V}{V_0}$

▲ **FIGURE 11.8** Volume strain.

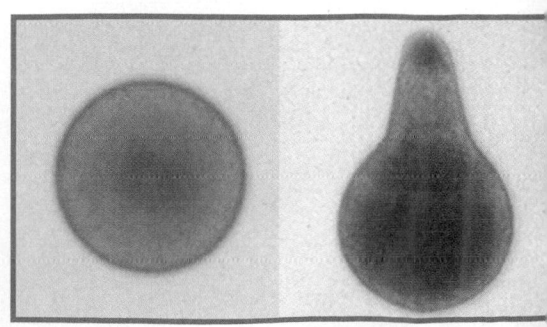

TABLE 11.2 Bulk modulus

Material	B (Pa)
Aluminum	0.70×10^{11}
Brass	0.61×10^{11}
Copper	1.4×10^{11}
Glass	0.37×10^{11}
Iron	1.0×10^{11}
Steel	1.6×10^{11}
Tungsten	2.0×10^{11}

We include a minus sign in this equation because an *increase* in pressure always causes a *decrease* in volume. In other words, when Δp is positive, ΔV is negative. The negative sign in Equation 11.7 makes B itself a positive quantity.

Table 11.2 lists values of the bulk modulus for several solid materials. Values for liquids are usually considerably smaller, typically between 1×10^9 Pa and 5×10^9 Pa. (I.e., in general, liquids are more easily compressed than solids). For small pressure changes in a solid or a liquid, we consider B to be constant. The bulk modulus of a *gas*, however, depends on the initial pressure p_0. The units of B—force per unit area, or pascals—are the same as those of pressure (and of tensile or compressive stress).

EXAMPLE 11.2 Squeezing water

The bulk modulus of water is 2.2×10^9 Pa $(= 2.2 \times 10^9 \text{ N/m}^2)$. By how much does a cubic meter of water decrease in volume when it is taken from the surface of the ocean down to a depth of 1.0 km, where the pressure is 9.8×10^6 Pa greater than at the surface?

SOLUTION

SET UP AND SOLVE From Equation 11.7, the bulk modulus is

$$B = -\frac{\Delta p}{\Delta V / V_0}.$$

We solve this equation for ΔV:

$$\Delta V = -\frac{\Delta p V_0}{B} = -\frac{(9.8 \times 10^6 \text{ Pa})(1.0 \text{ m}^3)}{(2.2 \times 10^9 \text{ N/m}^2)}$$
$$= -4.5 \times 10^{-3} \text{ m}^3 = -4500 \text{ cm}^3.$$

REFLECT The negative sign indicates that the volume has decreased while the pressure has increased. Although the pressure at a 1 km depth is roughly 100 times that at the surface, the fractional change in volume is only about 0.5%.

Practice Problem: A fluid with an initial volume of 0.35 m³ is subjected to a pressure decrease of 3.2×10^3 Pa. The volume is then found to have increased by 0.20 cm³. What is the bulk modulus of the fluid? *Answer:* $B = 5.6 \times 10^9$ Pa.

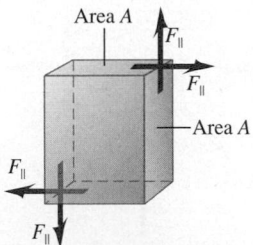

A block subjected to shear stress from two directions

▲ **FIGURE 11.9** Shear stress.

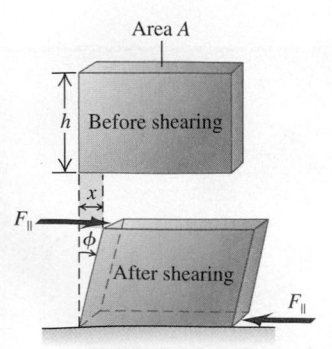

Shear stress $= \dfrac{F_{\parallel}}{A}$ Shear strain $= \dfrac{x}{h}$

▲ **FIGURE 11.10** Shear strain.

Shear Stress and Strain

Another kind of stress–strain situation shown in Figure 11.1 is called shear. The block in Figure 11.9 is under **shear stress,** which we define as the force $F_{\parallel}$ tangent to a material surface, divided by the area A of the surface on which the force acts:

$$\text{Shear stress} = \frac{F_{\parallel}}{A}. \tag{11.8}$$

Shear stress, like the other two types of stress, is a force per unit area. For systems in equilibrium, shear stress can exist only in *solid* materials; a fluid would simply flow in response to the tangential force.

Figure 11.10 shows an object deformed by shear. We define **shear strain** as the ratio of the displacement x to the transverse dimension h; that is,

$$\text{Shear strain} = \frac{x}{h} = \tan\phi, \tag{11.9}$$

with x and h defined as in Figure 11.10. In real-life situations, x is nearly always much smaller than h, $\tan\phi$ is very nearly equal to ϕ, and the strain is simply the angle ϕ, measured in radians. Like all strains, shear strain is a dimensionless number because it is a ratio of two lengths.

If the forces are small enough so that Hooke's law is obeyed, the shear strain is *proportional* to the shear stress. The corresponding proportionality constant (ratio of shear stress to shear strain), is called the **shear modulus,** denoted by S:

$$S = \frac{\text{Shear stress}}{\text{Shear strain}} = \frac{F_{\parallel}/A}{x/h} = \frac{h}{A}\frac{F_{\parallel}}{x} = \frac{F_{\parallel}/A}{\phi}. \tag{11.10}$$

For a given material, S is usually a third to a half as large as Y. A few representative values of S for common materials are listed in Table 11.3.

NOTE ▶ Keep in mind that only a *solid* material has a shear modulus. A liquid or gas flows freely under the action of a shear stress, so a fluid at rest cannot sustain such a stress. ◀

TABLE 11.3 Shear modulus

Material	S (Pa)
Aluminum	0.30×10^{11}
Brass	0.36×10^{11}
Copper	0.42×10^{11}
Glass	0.23×10^{11}
Iron	0.70×10^{11}
Steel	0.84×10^{11}
Tungsten	1.5×10^{11}

EXAMPLE 11.3 Minor earthquake damage

Suppose the brass baseplate of an outdoor sculpture experiences shear forces as a result of a mild earthquake. The plate is a square, 0.80 m on a side and 0.50 cm thick. How large a force $F_{\parallel}$ must be exerted on each of its edges if the displacement x in Figure 11.10b is 0.016 cm? The shear modulus of brass is listed in Table 11.3.

SOLUTION

SET UP Figure 11.11 shows our sketch. The shear stress at each edge of the plate is the force divided by the area of the edge:

$$\text{Shear stress} = \frac{F_{\parallel}}{A} = \frac{F_{\parallel}}{(0.80 \text{ m})(0.0050 \text{ m})} = (250 \text{ m}^{-2})F_{\parallel}.$$

The shear strain is

$$\text{Shear strain} = \frac{x}{h} = \frac{1.6 \times 10^{-4} \text{ m}}{0.80 \text{ m}} = 2.0 \times 10^{-4}.$$

SOLVE From Table 11.3, the shear modulus S is

$$S = \frac{\text{Shear stress}}{\text{Shear strain}} = 0.36 \times 10^{11} \text{ Pa} = \frac{(250 \text{ m}^{-2})F_{\parallel}}{2.0 \times 10^{-4}}.$$

Solving for $F_{\parallel}$, we obtain

$$F_{\parallel} = 2.9 \times 10^4 \text{ N}.$$

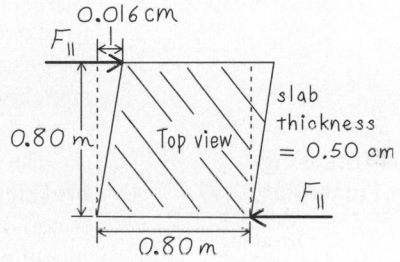

▲ **FIGURE 11.11** Our sketch for this problem.

REFLECT This force is nearly 7000 lb, or roughly twice the weight of a medium-size car.

Practice Problem: We replace the plate with another brass plate that has the same horizontal dimensions (0.80 m on a side), but a different thickness. If the shear forces have magnitude 1.2×10^5 N and the shear strain is the same as before, what is the plate's thickness? *Answer:* 2.1 cm.

Elasticity and Plasticity

Hooke's law, the proportionality between stress and strain in elastic deformations, has a limited range of validity. We know that if you pull, squeeze, or twist *anything* hard enough, it will bend or break. Just what *are* the limitations of Hooke's law?

Let's look at tensile stress and strain again. Suppose we gradually increase the tensile stress on an object, measure the corresponding strain, and plot a graph of stress as a function of strain. Figure 11.12 shows a typical stress–strain graph for a metal such as copper or soft iron subjected to tensile stress. The strain is shown as the *percent* elongation ($\Delta l/l_0 \times 100\%$).

The first part of the graph (to point A) is a straight line. During this phase, the object (say, a piece of copper wire) obeys Hooke's law: stress is proportional to strain, and the proportionality constant is Young's modulus. At point A, the plot ceases to be a straight line; strain is no longer exactly proportional to stress. However, up to point B, the wire still returns to its original, unstrained length if you release the stress. We call this reversible behavior an **elastic deformation.** (Note that an object can behave elastically without following Hooke's law.) This is the behavior you see when you pluck a guitar string; when it finishes vibrating, its length is unchanged.

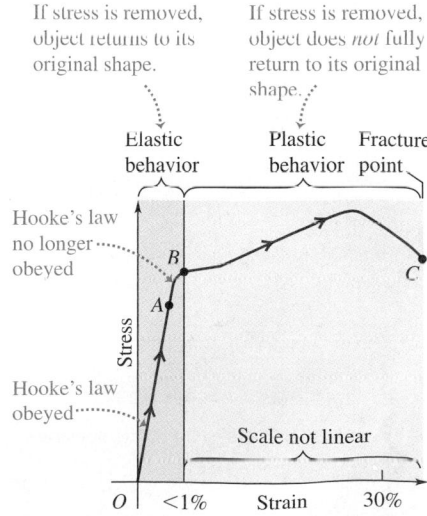

▲ **FIGURE 11.12** Typical stress–strain diagram for a ductile metal under tension.

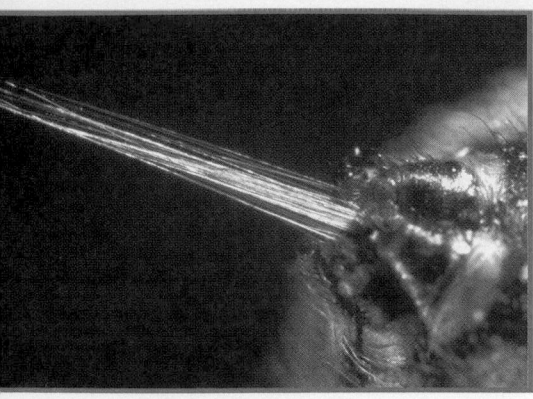

◀ **BIO Application As strong as silk.** This photograph shows silk fibers emerging from the spinning glands on a spider's abdomen. Spider silk has a unique combination of high strength and high elasticity. It can have a tensile strength of up to 17×10^8 Pa, comparable to some of the strongest high-tensile steels we can produce. Considering that its density is only about 20% that of steel, spider's silk has sometimes been called "five times stronger" than steel. It is also quite elastic, having a Young's elastic modulus of only 3×10^8 Pa, less than 1% that of steel. Therefore, in addition to being very strong, spider silk can stretch to at least twice its length before breaking, allowing it to absorb large amounts of energy from flying insects. It has even been proposed that a web made from pencil-thick spider's silk would be able to stop a jet airplane in flight!

At greater stresses, the object does *not* fully return to its original shape when the stress is removed; it has undergone some degree of permanent deformation. If you subject a copper wire to a stress in this range, then, when you release the stress, the wire springs back somewhat, but not all the way back to its original length. We say that the wire undergoes **plastic deformation.** Finally, at point *C*, the wire breaks.

For some materials, a large amount of plastic deformation takes place between the elastic limit and the fracture (breaking) point. Such a material is said to be *ductile*. But if fracture occurs soon after the elastic limit is passed, the material is said to be *brittle*. A soft iron wire that can undergo considerable permanent stretch without breaking is ductile; a steel piano string that breaks soon after its elastic limit is reached is brittle.

The stress required to cause actual fracture of a material is called the **breaking stress** or (in the case of tensile stress) the **tensile strength.** Two materials, such as two types of steel, may have very similar values of Young's modulus but vastly different breaking stresses. Table 11.4 gives a few typical values of breaking stress for several materials in tension. The conversion factor 6.9×10^8 Pa = 100,000 psi will help put these numbers in perspective. For example, if the breaking stress of a particular steel is 6.9×10^8 Pa, then a bar or cable with a 1 in.^{2} cross section has a breaking strength of 100,000 lb.

TABLE 11.4 Breaking stresses of materials

Material	Breaking stress (N/m^2 or Pa)
Aluminum	2.2×10^8
Arterial wall	2.0×10^6
Bone	$1-2 \times 10^8$
Brass	4.7×10^8
Collagen	1.0×10^8
Glass	10×10^8
Iron	3.0×10^8
Phosphor bronze	5.6×10^8
Steel	11.0×10^8
Wood	$1-2 \times 10^8$

11.2 Periodic Motion

An object such as a vibrating guitar string or the swinging pendulum of a grandfather clock repeats the same motion over and over. We refer to such motion as **periodic motion** or **oscillation.** A mechanical system that undergoes periodic motion always has a stable equilibrium position. When it is moved away from this position and released, a force comes into play to pull it back toward equilibrium. But by the time it gets there, it has picked up some kinetic energy, so it overshoots, stopping somewhere on the other side, and is again pulled back toward equilibrium. If this cycle repeats over and over, we say that the motion is periodic.

Simple Harmonic Motion

One of the simplest systems that can have periodic motion is shown in Figure 11.13. A glider with mass *m* moves along the *x* axis on a frictionless horizontal air track. It is attached to a spring that can be either stretched or compressed. The spring obeys Hooke's law and has force constant *k*. (You may want to review the definition of the force constant in Sections 5.4 and 11.1.) The left end of the spring is held stationary. The spring force is the only horizontal force acting on the object. (We neglect the very small friction force on the air track.) The two vertical forces (the normal and weight forces) always add to zero.

It is simplest to define our coordinate system so that the origin *O* is at the equilibrium position, where the spring is neither stretched nor compressed. Then *x* is the displacement of the object from equilibrium and is also the change in length of the spring. The *x* component of acceleration is given by $a_x = F_x/m$.

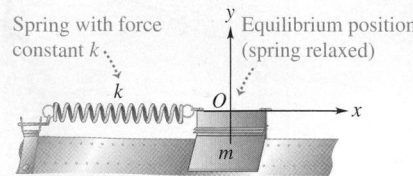

▲ **FIGURE 11.13** A system for demonstrating periodic motion.

Figure 11.14 shows complete free-body diagrams for three different positions of the object:

1. **x is negative** (Figure 11.14a). The object is to the left of O, so the spring is compressed. The spring pushes toward the right (toward the equilibrium point) on the object, so the acceleration is also toward the right. Thus, F_x and a_x are both positive. (The object may be *moving* toward either the left or right, however.)
2. **$x = 0$** (Figure 11.14b). The object is at O, the equilibrium position. The spring is neither stretched nor compressed; it exerts no force on the object. The acceleration is zero, and the object is instantaneously moving with constant velocity toward either the right or left.
3. **x is positive** (Figure 11.14c). The object is to the right of O, so the spring is stretched. The spring pulls toward the left (toward the equilibrium point) on the object, so the acceleration is also toward the left. Thus, F_x and a_x are both negative. (The object may be *moving* toward either the left or right, however.)

The force acting on the object tends always to restore it to the equilibrium position, and we call it a **restoring force. On either side of the equilibrium position, F_x and x always have opposite signs.** In Section 5.4, we represented the force acting *on* a stretched spring as $F_x = kx$. The force the spring exerts *on the object* is the negative of this, so the x component of force F_x acting on the body is

$$F_x = -kx. \tag{11.11}$$

This equation gives the correct magnitude and sign of the force, whether x is positive, negative, or zero. The force constant k is always positive and has units of N/m. A useful alternative set of units for k is kg/s^2.

The x component of acceleration, a_x, of the body at any point is determined by $\Sigma \vec{F} = m\vec{a}$, with the x component of force given by Equation 11.11:

$$a_x = \frac{F_x}{m} = -\frac{k}{m}x. \tag{11.12}$$

The acceleration is *not* constant, so don't even think of using the constant-acceleration equations from Chapter 2! Figure 11.15 shows a complete cycle of simple harmonic motion. If we displace the object to the right a distance A (so that $x = A$) and release it, its initial acceleration, from Equation 11.12, is $-kA/m$. As it approaches the equilibrium position, its acceleration decreases in magnitude, but its speed increases. It overshoots the equilibrium position and comes to a stop on the other side; we'll show later that the stopping point is at $x = -A$. Then it accelerates to the right, overshoots equilibrium again, and stops at the starting point $x = A$, ready to repeat the whole process. If there is no friction or other force to remove mechanical energy from the system, this motion repeats forever.

The motion that we've just described—periodic motion under the action of a restoring force that is *directly proportional* to the displacement from equilibrium, as given by Equation 11.11—is called **simple harmonic motion,** abbreviated **SHM.** Many periodic motions, such as the vibration of the quartz crystal in a watch, vibrations of molecules, the motion of a tuning fork, and the electric current in an alternating-current circuit, are approximately simple harmonic motion. However, not *all* periodic motions are simple harmonic.

Here are several terms that we'll use in discussing periodic motions:

Definitions of quantities that describe periodic motion

- The **amplitude** of the motion, denoted by A, is the maximum magnitude of displacement from equilibrium—that is, the maximum value of $|x|$. It is always positive. The total overall range of the motion is $2A$.

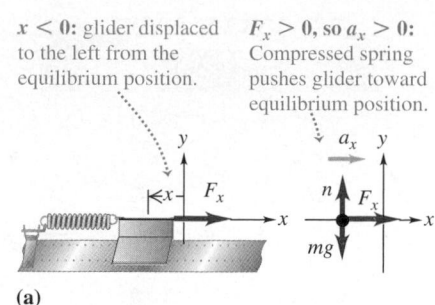

$x < 0$: glider displaced to the left from the equilibrium position.

$F_x > 0$, so $a_x > 0$: Compressed spring pushes glider toward equilibrium position.

(a)

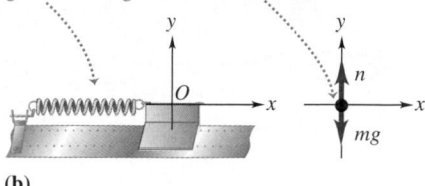

$x = 0$: The relaxed spring exerts no force on the glider, so the glider has zero acceleration.

(b)

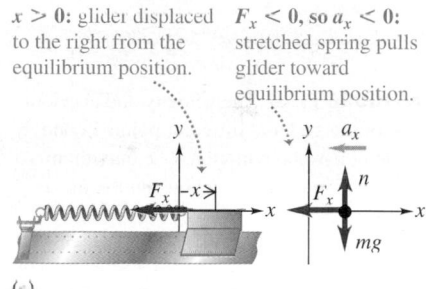

$x > 0$: glider displaced to the right from the equilibrium position.

$F_x < 0$, so $a_x < 0$: stretched spring pulls glider toward equilibrium position.

(c)

▲ **FIGURE 11.14** A model for periodic motion. If the spring obeys Hooke's law, the motion of the glider is simple harmonic motion.

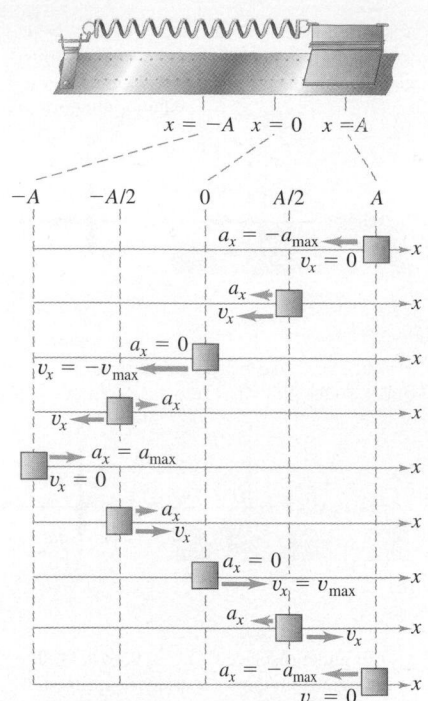

▲ FIGURE 11.15 The velocity and acceleration of the glider at different points in one cycle of harmonic motion. The maximum acceleration is $a_{max} = kA/m$ and the maximum speed is v_{max}.

- A **cycle** is one complete round trip—for example, from A to $-A$ and back to A, or from O to A, back through O to $-A$, and back to O. Note that motion from one side to the other (say, from $-A$ to A) is a half-cycle, not a whole cycle.
- The **period,** T, is the time for one whole cycle. It is always positive; the SI unit is the second, but it is sometimes expressed as "seconds per cycle."
- The **frequency,** f, is the number of cycles in a unit of time. It is always positive; the SI unit of frequency is the hertz, defined as follows:

$$1 \text{ hertz} = 1 \text{ Hz} = 1 \text{ cycle/s} = 1 \text{ s}^{-1}.$$

This unit is named in honor of Heinrich Hertz, one of the pioneers investigating electromagnetic waves during the late 19th century.

- The **angular frequency,** ω, is 2π times the frequency:

$$\omega = 2\pi f.$$

We'll learn later why ω is a useful quantity. It represents the rate of change of an angular quantity (not necessarily related to a rotational motion) that is always measured in radians, so its units are rad/s.

From the definitions of period T and frequency f, we see that each is the reciprocal of the other:

$$f = \frac{1}{T}, \qquad T = \frac{1}{f}. \tag{11.13}$$

Also, from the definition of ω,

$$\omega = 2\pi f = \frac{2\pi}{T}. \tag{11.14}$$

EXAMPLE 11.4 **Ultrasound for medical imaging**

An ultrasonic transducer (a kind of ultrasonic loudspeaker) used for medical diagnosis is oscillating at a frequency of 6.7 MHz (6.7×10^6 Hz) (Figure 11.16). How much time does each oscillation take, and what is the angular frequency?

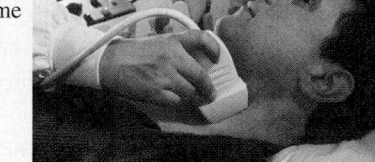

▶ FIGURE 11.16 Ultrasonic vibrations for medical diagnosis are produced by an electronic transducer.

SOLUTION

SET UP We first identify the frequency: $f = 6.7 \times 10^6$ Hz = 6.7×10^6 s^{-1}.

SOLVE The period T is given by Equation 11.13 and the angular frequency by Equation 11.14:

$$T = \frac{1}{f} = \frac{1}{6.7 \times 10^6 \text{ Hz}} = 1.5 \times 10^{-7} \text{ s} = 0.15 \text{ } \mu\text{s},$$
$$\omega = 2\pi f = 2\pi (6.7 \times 10^6 \text{ Hz}) = 4.2 \times 10^7 \text{ rad/s}.$$

REFLECT The transducer's rapid vibration corresponds to large values of f and ω and small values of T. The reverse would be true for a slow oscillator, such as the pendulum of a large grandfather clock.

Not all periodic motion is simple harmonic. **Only in SHM is the restoring force *directly proportional* to the displacement.** In more complex types of periodic motion, the force may depend on the displacement in a more complicated way. However, many periodic motions are *approximately* simple harmonic if the amplitude is small enough that the force is approximately proportional to the displacement. Thus, we can use SHM as an approximate model for many different periodic motions, such as the motion of a pendulum and the vibrations of atoms in molecules and solids.

11.3 Energy in Simple Harmonic Motion

In this section and the next, we'll analyze the spring–mass model of SHM that we've just described. What questions shall we ask? Given the mass m of the object and the force constant k of the spring, we'd like to be able to predict the period T and the frequency f. We also want to know how fast the object is moving at each point in its motion. That is, we want to know its x component of velocity, v_x, and its x component of acceleration a_x, as functions of position x (the x component of displacement from the equilibrium position). Best of all would be to have expressions for x, v_x, and a_x as functions of *time*, so that we can find where the object is and how it is moving at any specified time. In the next few pages we'll work all these things out.

PhET: Motion in 2D
PhET: Masses & Springs
ActivPhysics 9.3: Vibrational Energy
ActivPhysics 9.5: Ape Drops Tarzan

We'll approach these questions first by using energy considerations and then by using $\sum \vec{F} = m\vec{a}$ along with a geometric construction called the circle of reference. Let's start with energy. We've already noted that the spring force is the only horizontal force on the object. It is a *conservative* force, and the vertical forces do no work, so the total mechanical energy of the system is conserved. We'll also assume that the mass of the spring itself is negligible.

The kinetic energy of the object is $K = \frac{1}{2}mv_x^2$, and the potential energy of the spring is $U = \frac{1}{2}kx^2$, just as we found in Section 7.4. (You may want to review that section.) The equation for the conservation of mechanical energy is $K + U = E -$ constant, or

$$E = \tfrac{1}{2}mv_x^2 + \tfrac{1}{2}kx^2 - \text{constant}. \tag{11.15}$$

The total mechanical energy E is also directly related to the amplitude A of the motion. When the object reaches the point $x = A$ or $x = -A$ (its maximum displacement from equilibrium), it stops momentarily as it turns back toward the equilibrium position. That is, when $x = \pm A$, $v_x = 0$. At this point, the energy is entirely potential energy and $E = \frac{1}{2}kA^2$. Because E is constant, this quantity also equals E at any other point; combining the preceding expression for E with Equation 11.15, we get

$$E = \tfrac{1}{2}kA^2 = \tfrac{1}{2}mv_x^2 + \tfrac{1}{2}kx^2. \tag{11.16}$$

Solving Equation 11.16 for v_x, we find the following relationship:

Velocity of an object in SHM as a function of position
From the basic energy relation in Equation 11.16,

$$v_x = \pm\sqrt{\frac{k}{m}}\sqrt{A^2 - x^2}. \tag{11.17}$$

We can use this equation to find the magnitude of the velocity (but not its sign) for any given position x. For example, when $x = \pm A/2$,

$$v_x = \pm\sqrt{\frac{k}{m}}\sqrt{A^2 - \left(\pm\frac{A}{2}\right)^2} = \pm\sqrt{\frac{3}{4}}\sqrt{\frac{k}{m}}A.$$

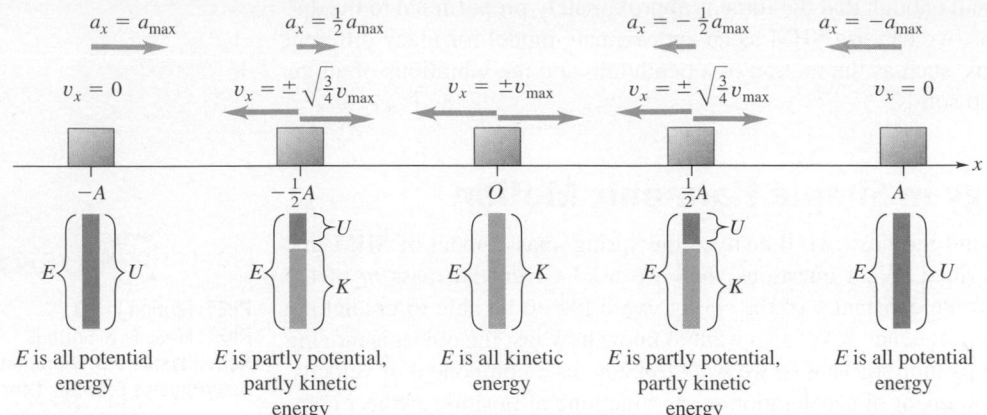

▲ **FIGURE 11.17** In simple harmonic motion, the total energy E is constant, continually transforming from potential energy U to kinetic energy K and back again.

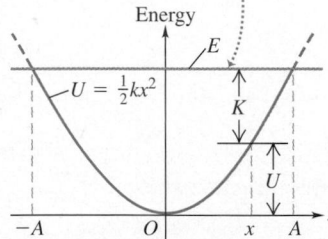

(a) The potential energy U and total energy E of an object in SHM as a function of x position.

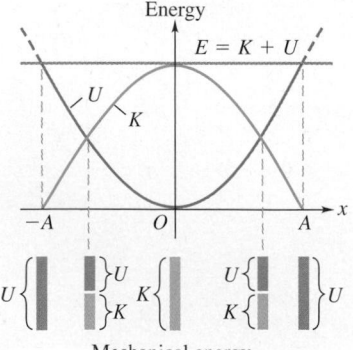

(b) The same graph as in (a), showing kinetic energy K as well.

▲ **FIGURE 11.18** Kinetic, potential, and total energies as functions of position for SHM. At each value of x, the sum of K and U equals the constant total energy E.

Equation 11.17 also shows that the *maximum* speed v_{max} occurs at $x = 0$ and is given by

$$v_{max} = \sqrt{\frac{k}{m}}A. \qquad (11.18)$$

Figure 11.17 is a pictorial display of the energy quantities at several x positions, and Figure 11.18 is a graphical display of Equation 11.15. Mechanical energy (kinetic, potential, and total) is plotted vertically, and the coordinate x is plotted horizontally. The parabolic curve in Figure 11.18a represents the potential energy $U = \frac{1}{2}kx^2$. The horizontal line represents the constant total energy E (which doesn't vary with x). This line intersects the potential-energy curve at $x = -A$ and $x = A$, where the energy is entirely potential.

Suppose we start on the horizontal axis at some value of x between $-A$ and A. The vertical distance from the x axis to the parabola is U, and since $E = K + U$, the remaining vertical distance up to the horizontal line is K. Figure 11.18b shows both K and U as functions of x. As the object oscillates between $-A$ and A, the mechanical energy is continuously transformed from potential to kinetic to potential and back. At $x = \pm A$, $E = U$ and $K = 0$.

We see again why A is the maximum displacement. If we tried to make x greater than A (or less than $-A$), U would be greater than E, and K would have to be negative. As we learned in Chapter 7, K can never be negative, so x can't be greater than A or less than $-A$. (In quantum mechanics, it *is* possible in some circumstances for a particle to traverse regions where classically it would have negative kinetic energy. This concept is important in alpha decay of nuclei and in the tunneling electron microscope, but it plays no role in the mechanics of macroscopic objects.)

Quantitative Analysis 11.1

Energy of a spring

A mass attached to a spring oscillates with SHM. If the maximum displacement from equilibrium is A and the total mechanical energy of the system is E, what is the system's kinetic energy when the position of the mass is half its maximum displacement?

A. $\frac{1}{2}E$. B. $\frac{3}{4}E$. C. E.

SOLUTION The total mechanical energy E of an oscillator is the sum of its kinetic and potential energies: $E = K + U$. But $U = \frac{1}{2}kx^2$. If $x = A$, then $U = E$ and $K = 0$. At $x = \frac{1}{2}A$, $U = \frac{1}{4}E$ and $K = E - U = \frac{3}{4}E$.

EXAMPLE 11.5 **Simple harmonic motion on an air track**

A spring is mounted horizontally on an air track (Figure 11.19a), with the left end held stationary. We attach a spring balance to the free end of the spring, pull toward the right, and measure the elongation. We determine that the stretching force is proportional to the displacement and that a force of 6.0 N causes an elongation of 0.030 m. We remove the spring balance and attach a 0.50 kg object to the end, pull it a distance of 0.040 m, release it, and watch it oscillate in SHM (Figure 11.19b). Find the following quantities:

(a) the force constant of the spring.
(b) the maximum and minimum velocities attained by the vibrating object.
(c) the maximum and minimum acceleration.
(d) the velocity and acceleration when the object has moved halfway to the center from its initial position.
(e) the kinetic energy, the potential energy, and the total energy in the halfway position.

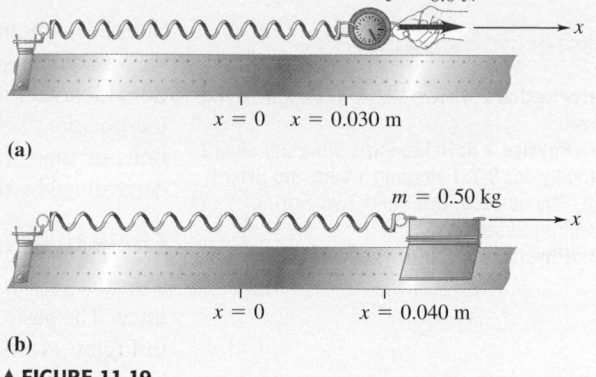

(a)

(b)

▲ **FIGURE 11.19**

SOLUTION

SET UP AND SOLVE For this glider, the amplitude is $A = 0.040$ m.

Part (a): When $x = 0.030$ m, the force the spring exerts on the object is $F_x = -6.0$ N. From Equation 11.11, the force constant k is

$$k = \frac{F_x}{x} = -\frac{-6.0 \text{ N}}{0.030 \text{ m}} = 200 \text{ N/m}.$$

Part (b): From Equation 11.18, the maximum velocity occurs at the equilibrium position, $x = 0$:

$$v_{max} = \sqrt{\frac{k}{m}} A = \sqrt{\frac{200 \text{ N/m}}{0.50 \text{ kg}}} (0.040 \text{ m}) = 0.80 \text{ m/s}.$$

The minimum (most negative) velocity also occurs at $x = 0$ and is the negative of the maximum value. The minimum *magnitude* of velocity, occurring at $x = \pm A$, is zero.

Part (c): The acceleration for any x is given by Equation 11.12:

$$a_x = -\frac{k}{m} x.$$

The maximum (most positive) acceleration a_{max} occurs at the most negative value of x, that is, $x = -A$; therefore,

$$a_{max} = -\frac{k}{m} (-A) = -\frac{200 \text{ N/m}}{0.50 \text{ kg}} (-0.040 \text{ m}) = 16.0 \text{ m/s}^2.$$

The minimum (most negative) acceleration is -16.0 m/s^2, occurring at $x = +A = +0.040$ m.

Part (d): The velocity v_x at any position x is given by Equation 11.17:

$$v_x = \pm \sqrt{\frac{k}{m}} \sqrt{A^2 - x^2}.$$

At the halfway point, $x = A/2 = 0.020$ m. We find that

$$v_x = -\sqrt{\frac{200 \text{ N/m}}{0.50 \text{ kg}}} \sqrt{(0.040 \text{ m})^2 - (0.020 \text{ m})^2} = -0.69 \text{ m/s}.$$

We've chosen the negative square root because the object is moving in the negative x direction, from $x = A$ toward $x = 0$. From Equation 11.12,

$$a_x = -\frac{k}{m} x = -\frac{200 \text{ N/m}}{0.50 \text{ kg}} (0.020 \text{ m}) = -8.0 \text{ m/s}^2.$$

Part (e): The energy quantities are the terms in Equation 11.16. From (d), when $x = A/2 = (0.040 \text{ m})/2$, $v_x = \pm 0.69$ m/s. At this point, the kinetic energy is

$$K = \tfrac{1}{2} m v_x^2 = \tfrac{1}{2} (0.50 \text{ kg}) (0.69 \text{ m/s})^2 = 0.12 \text{ J},$$

the potential energy is

$$U = \tfrac{1}{2} k x^2 = \tfrac{1}{2} (200 \text{ N/m}) (0.020 \text{ m})^2 = 0.040 \text{ J},$$

and the total energy is $E = K + U = 0.12 \text{ J} + 0.040 \text{ J} = 0.16 \text{ J}$. Alternatively,

$$E = \tfrac{1}{2} k A^2 = \tfrac{1}{2} (200 \text{ N/m}) (0.040 \text{ m})^2 = 0.16 \text{ J}.$$

The conditions at $x = 0$, $\pm A/2$, and $\pm A$ are shown in Figure 11.15.

REFLECT The equations we've used in this solution enable us to find relations among position, velocity, acceleration, and force. But they don't help us to find directly how these quantities vary with time. In the next section, we'll work out relations that enable us to do just that and also to relate the period and frequency of the motion to the mass and force constant of the system.

11.4 Equations of Simple Harmonic Motion

Simple harmonic motion is closely related to uniform circular motion, which we've studied several times. By examining this relationship, we can gain additional insight into both kinds of motion. In this section we'll derive equations for the position, velocity, and acceleration of an object undergoing SHM, as functions of time. We'll also show how the period and frequency of the motion are determined by the mass m and force constant k of the system.

Circle of Reference

Our analysis is based on a geometric representation called the **circle of reference.** The basic idea is shown in Figure 11.20 First, imagine the apparatus shown in Figure 11.20a. We glue a small ball to the rim of a rotating turntable on a flat horizontal tabletop (viewed from above in the figure). Behind the turntable, we set up a vertical screen. Then we illuminate the turntable from in front by a flat, horizontal beam of light placed at the level of the turntable. As the ball rotates, its shadow appears on the screen. And, while the ball moves in uniform circular motion, we find that **the shadow moves from side to side in SHM!**

Figure 11.20b shows the circle of reference—an abstract representation of the phenomenon in Figure 11.20a. Point Q (representing the ball) moves counterclockwise around a circle whose radius is equal to the amplitude A of the shadow's simple harmonic motion. The constant angular velocity ω of the ball (measured in rad/s) is equal to the angular frequency of the shadow's SHM. Thus, ω is the rate of change of the angle ϕ with time: $\omega = \Delta\phi/\Delta t$.

The vector from the origin O to point Q (which we can denote as OQ) is the position vector of point Q relative to O. This vector has constant magnitude A, and at time t it is at an angle ϕ (measured counterclockwise from the positive x axis). As Q moves, vector OQ rotates counterclockwise with constant angular velocity $\omega = \Delta\phi/\Delta t$. As we'll soon see, the x component of this vector represents the actual motion of the object under study.

Now let's prove that the horizontal component of the vector OQ represents the actual motion of the shadow. In Figure 11.20b, point P represents the ball's

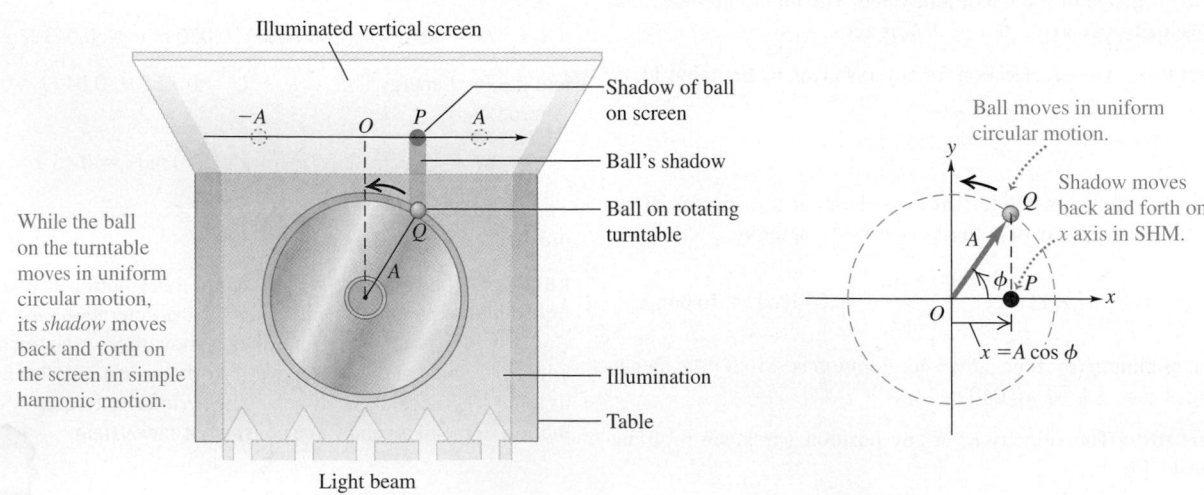

(a) Apparatus for creating the reference circle

(b) An abstract representation of the motion in **(a)**.

▲ **FIGURE 11.20** The circle of reference relates simple harmonic motion to uniform circular motion.

shadow; it lies on the diameter of the circle that coincides with the x axis, directly below point Q, just as the shadow in Figure 11.20a is always directly in line with the ball. We call Q the *reference point,* the circle itself the *circle of reference,* and P the *projection* of Q onto the diameter. As Q revolves, P moves back and forth along the diameter, staying always directly below (or above) Q.

The position of P with respect to the origin O at any time t is the coordinate x; from Figure 11.20a,

$$x = A\cos\phi.$$

If point Q is at the extreme right end of the diameter at time $t = 0$, then $\phi = 0$ when $t = 0$. Because ω is constant, ϕ increases uniformly with time; that is,

$$\phi = \omega t,$$

and

$$x = A\cos\omega t. \tag{11.19}$$

Now ω, the angular velocity of Q (in radians per second) is related to f, the number of complete revolutions of Q per second. There are 2π radians in one complete revolution, so $\omega = 2\pi f$. Furthermore, the point P makes one complete back-and-forth vibration (one cycle) for each revolution of Q. Hence f is also the number of vibrations per second or the *frequency* of vibration of point P. Thus we can also write Equation 11.19 as

$$x = A\cos 2\pi ft. \tag{11.20}$$

We can find the instantaneous velocity of P with the help of Fig. 11.21a. The reference point Q moves with a tangential speed $v_Q = v_\parallel$ given by Equation 9.13:

$$v_\parallel = \omega A = 2\pi fA.$$

Point P is always directly below or above the reference point, so the velocity v_x of P at each instant must equal the x component of the velocity of Q. That is, from Figure 11.21a,

$$v_x = -v_\parallel\sin\phi = \quad \omega A\sin\phi,$$

and it follows that

$$v_x = -\omega A\sin\omega t = -2\pi fA\sin 2\pi ft. \tag{11.21}$$

The minus sign is needed because, at the instant shown, the direction of the velocity is toward the left. When Q is below the horizontal diameter, the velocity of P is toward the right, but $\sin\phi$ is negative at such points. So the minus sign is needed in both cases, and Equation 11.21 gives the velocity of point P at any time.

We can also find the acceleration of point P. Because P is always directly below or above Q, its x component of acceleration must equal the x component of the acceleration of Q. Point Q moves in a circular path with constant angular velocity ω, and its acceleration is toward the center of the circle, with magnitude $a_Q = a_{\text{rad}}$ given by Equation 9.15:

$$a_{\text{rad}} = \omega^2 A = 4\pi^2 f^2 A.$$

From Figure 11.21b, the x component of this acceleration is

$$a_x = -a_{\text{rad}}\cos\phi;$$

or

$$a_x = -\omega^2 A\cos\omega t = -4\pi^2 f^2 A\cos 2\pi ft. \tag{11.22}$$

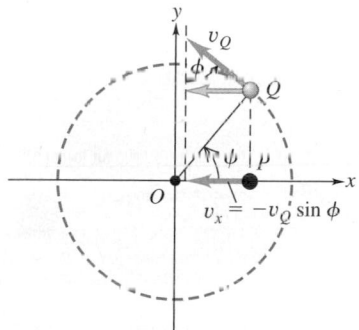

(a) Using the reference circle to determine the velocity of point P.

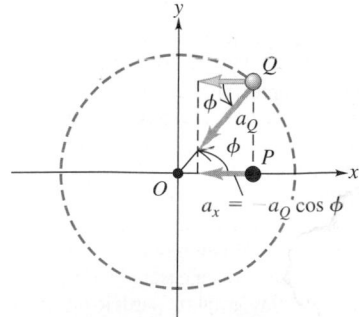

(b) Using the reference circle to determine the acceleration of point P.

▲ **FIGURE 11.21** The velocity and acceleration of point P are the horizontal components of the velocity and acceleration, respectively, of point Q.

The minus sign is needed because, at the instant shown, the acceleration is toward the left. When Q is to the left of the center, the acceleration of P is toward the right, but since $\cos \phi$ is negative at such points, the minus sign is still needed. Therefore, Equation 11.22 gives the acceleration of P at any time.

Now comes the reward for our lengthy analysis—the crucial step in showing that the motion of P is simple harmonic. We combine Equations 11.19 and 11.22, obtaining

$$a_x = -\omega^2 x. \tag{11.23}$$

Because ω is constant, **the acceleration a_x at each instant equals a negative constant times the displacement x at that instant.** But this is just the essential feature of simple harmonic motion, as given by Equation 11.12: Force and acceleration are proportional to the negative of the displacement from equilibrium. This shows that **the motion of P is indeed simple harmonic.**

Quantitative Analysis 11.2

Using equations of motion

An object undergoing simple harmonic motion along the x axis obeys the equation $x = 12.0\cos 0.25t$, where x is in cm and t is in seconds. Which of the following choices correctly gives the position, velocity, and acceleration of the object at the moment it is released (i.e., at time $t = 0$)?

A. $x = +12$ cm, $v_x = 0.00$ cm/s, $a_x = -0.75$ cm/s^2.
B. $x = 0.00$ cm, $v_x = 3.0$ cm/s, $a_x = 0.00$ cm/s^2.
C. $x = -12$ cm, $v_x = 0.00$ cm/s, $a_x = -0.75$ cm/s^2.
D. $x = 0.00$ cm, $v_x = -3.0$ cm/s, $a_x = 0.00$ cm/s^2.

SOLUTION At the initial time $t = 0$, the object has its maximum positive displacement, so at that instant $x = 12$ cm. The object is released from rest (i.e., with zero initial velocity), so $v_x = 0$. Only answer A is consistent with these two results. To confirm this choice, we can find the value of a_x at time $t = 0$. From Equation 11.19, $\omega = 0.25$ s^{-1}, and from Equation 11.22 at time $t = 0$, $a_x = -\omega^2 A = -(0.25 \text{ s}^{-1})^2(12 \text{ cm}) = -0.75$ cm/s^2 (also consistent with answer A). So answer A is correct.

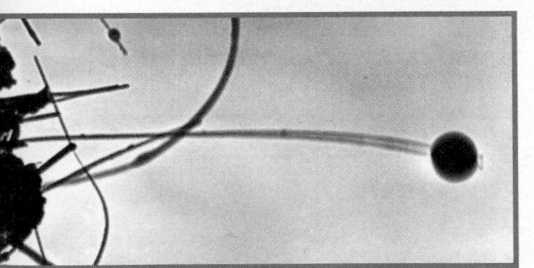

▲ **Application Hanging in the balance.**
Scientists recently used a single carbon nanotube to create a device to measure the mass of incredibly tiny objects, such as a tiny graphite particle with a mass of only 22 femtograms (22×10^{-15} g). Although submicroscopic, this "nanobalance" was shown to obey the same physical laws as a macroscopic mass oscillating on the end of an elastic beam—in particular, the period and frequency of oscillation depend on the mass at the end. The nanotube's resonant frequency of oscillation in response to an applied voltage was measured with and without the tiny graphite particle attached at the end. This procedure allowed the scientists to determine the particle's mass, and the value agreed well with the calculated mass for a graphite particle of that size. Thus, someday soon it may be possible to determine directly the mass of individual cells and even virus particles.

Frequency and Period of an Object in Simple Harmonic Motion

We're now ready to find how the frequency f and angular frequency ω of simple harmonic motion are related to the physical properties of the system (the spring constant k and the mass m). To make Equations 11.12 and 11.23 agree completely, we must choose an angular velocity ω for the reference point Q such that

$$\omega^2 = \frac{k}{m}, \quad \text{or} \quad \omega = \sqrt{\frac{k}{m}}. \tag{11.24}$$

With this choice, the frequency of motion of Q is equal to that of the simple harmonic motion of the actual particle P:

Frequency and period of SHM in terms of mass m and force constant k

$$f = \frac{\omega}{2\pi} = \frac{1}{2\pi}\sqrt{\frac{k}{m}}. \tag{11.25}$$

The period T of the motion (the time for one cycle) is the reciprocal of the frequency: $T = 1/f$. From Equation 11.25,

$$T = \frac{1}{f} = 2\pi\sqrt{\frac{m}{k}}. \tag{11.26}$$

The general form of Equations 11.25 and 11.26 can be understood intuitively. When m is large, a given displacement (and force) causes a relatively small acceleration, and we expect the motion to be slow and ponderous, corresponding to small f and large T. A large value of k means a very stiff spring, corresponding to a relatively large restoring force for a given displacement, resulting in large f and small T.

It may be surprising that these equations *do not* contain the amplitude A of the motion. Suppose we give our spring–mass system some initial displacement, release it, and measure its frequency. Then we stop it, give it a *different* initial displacement, and release it again. We find that the two frequencies are the same. To be sure, the maximum displacement, maximum speed, and maximum acceleration are all different in the two cases, but *not* the frequency. When the amplitude is increased, the mass has a greater distance to travel during each cycle, but the force and acceleration are also greater, on the average. In SHM, these two effects exactly cancel.

NOTE ▶ The preceding paragraph sums up one of the most important characteristics of simple harmonic motion: **The frequency of the motion does not depend on its amplitude.** For instance, a tuning fork vibrates with the same frequency, regardless of how hard you hit it (i.e., regardless of the amplitude of its vibration). As we'll learn in the next chapter, the pitch of a musical tone is determined primarily by the frequency of vibration. Thus, tuning forks always produce the same note. Indeed, if it were not for this property of simple harmonic motion, it would be impossible to play most musical instruments in tune. ◀

Quantitative Analysis 11.3

Frequency of a spring–mass system

An object with mass $m = 4\ kg$ is attached to a spring and vibrates in simple harmonic motion with frequency f. When this object is replaced by one with mass 1 kg, the new object vibrates at a frequency of

A. f.
B. $2f$.
C. $4f$.
D. $\frac{1}{2}f$.

SOLUTION Equation 11.25 states that the frequency is related to the inverse square root of the mass. If the mass is reduced, then the frequency must increase, so B and C are possible answers. (This also makes sense physically, since the same force applied to a lighter mass produces a greater acceleration at any given point.) When the mass changes from 4 kg to 1 kg, the square root of the mass changes from 2 to 1 and its reciprocal from 1 to 2. So B is the correct answer.

We can simplify many of the relationships in simple harmonic motion by using the angular frequency $\omega = 2\pi f$ rather than the frequency f. In terms of ω, we can rewrite Equations 11.12 and 11.17 respectively as

$$a_x = -\frac{k}{m}x = -\omega^2 x,$$
$$v_x = \pm\omega\sqrt{A^2 - x^2}. \tag{11.27}$$

In these and most of the other equations of simple harmonic motion, using ω instead of f lets us avoid a lot of factors of 2π. By convention, the hertz is *not* used as a unit of angular frequency; the usual unit for ω is the radian per second, or simply s^{-1}.

Equations for position, velocity, and acceleration of a simple harmonic oscillator as functions of time

Here's a summary of our results for the position x, velocity v_x, and acceleration a_x of a simple harmonic oscillator as functions of time:

$$x = A\cos\omega t = A\cos(2\pi ft), \tag{11.19, 11.20}$$
$$v_x = -\omega A\sin\omega t = -2\pi fA\sin(2\pi ft), \tag{11.21}$$
$$a_x = -\omega^2 A\cos\omega t = -(2\pi f)^2 A\cos(2\pi ft). \tag{11.22}$$

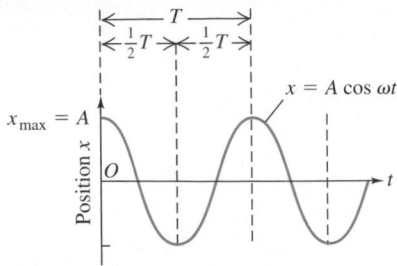

$x_{max} = A$

$x = A \cos \omega t$

(a) Position as a function of time

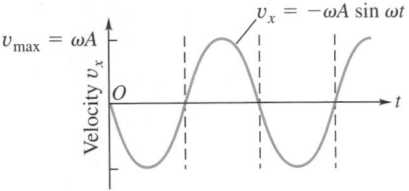

$v_{max} = \omega A$

$v_x = -\omega A \sin \omega t$

(b) Velocity as a function of time

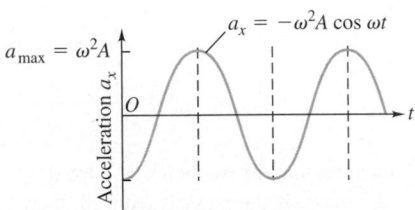

$a_{max} = \omega^2 A$

$a_x = -\omega^2 A \cos \omega t$

(c) Acceleration as a function of time

▲ **FIGURE 11.22** Graphs of position, velocity, and acceleration as functions of time for a particle in simple harmonic motion.

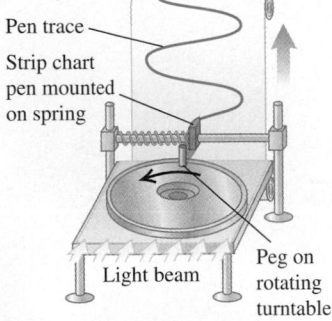

The strip chart pen moves in SHM on the spring while writing on the strip chart. The shadow of the peg follows the pen's motion exactly.

Pen trace

Strip chart pen mounted on spring

Light beam

Peg on rotating turntable

▲ **FIGURE 11.23** The strip chart pen, moving in simple harmonic motion, draws a curve of position as a function of time (the same curve as in Figure 11.22a). Meanwhile, the shadow of the peg on the turntable matches the motion of the pen.

Figure 11.22 shows graphs of these equations—for the position, velocity, and acceleration—as functions of time. We invite you to check that v_x at any time is the slope of the curve of x versus t and that a_x at any time is the slope of the curve of v_x versus t. We can plot a graph of x as a function of t with the help of an apparatus shown schematically in Figure 11.23.

Throughout this discussion, we've assumed that the initial position x_0 of the particle (at time $t = 0$) is its maximum positive displacement A (i.e., $x_0 = A$), but our analysis can be adapted to different starting conditions. Different initial positions correspond to different initial positions of the reference point Q. For example, if at time $t = 0$ the radial line OQ makes an angle ϕ_0 with the positive x axis, then the angle ϕ at time t is given, not by $\phi = \omega t$ as before, but by

$$\phi = \phi_0 + \omega t.$$

The object's position is then given as a function of time by

$$x = A \cos(\omega t + \phi_0).$$

With appropriate choices of the two quantities A and ϕ_0, this more general expression can be used to describe motion with any initial position and initial velocity. As before, the period and frequency are independent of the amplitude.

PROBLEM-SOLVING STRATEGY 11.1 **Simple harmonic motion**

SET UP

1. Be careful to distinguish between quantities that represent basic physical properties of the system and quantities that describe a particular motion that occurs when the system is set in motion in a specific way. The physical properties include the mass m, the force constant k, and the quantities derived from these, including the period T, the frequency f, and the angular frequency $\omega = 2\pi f$. In some problems, m or k, or both, can be determined from other information given about the system. Quantities that describe a particular motion include the amplitude A, the maximum velocity v_{max}, and any quantity that represents the position, velocity, or acceleration at a particular time.

SOLVE

2. If the problem involves a relation among position, velocity, and acceleration without reference to time, it is usually easier to use Equations 11.12 or 11.17 than to use the general expression for position as a function of time, Equation 11.19 or 11.20.
3. When detailed information about positions, velocities, and accelerations at various times is required, then Equations 11.20 through 11.22 must be used.
4. The energy equation, Equation 11.16, which includes the relation $E = \frac{1}{2}kA^2$, sometimes provides a convenient alternative to relations between velocity and position, especially when energy quantities are also required.

REFLECT

5. Try changing a value of one of the given quantities, and check whether the resulting changes in your answers make sense. Are they in the direction you would expect? Remember always to check for consistency of units in your results.

EXAMPLE 11.6 SHM on an air track, II: physical properties

Consider the oscillating air-track glider from Example 11.5 (Figure 11.24). As in that example, the force constant of the spring is $k = 200$ N/m and the glider mass is $m = 0.50$ kg. Find the angular frequency, frequency, and period of the glider's motion.

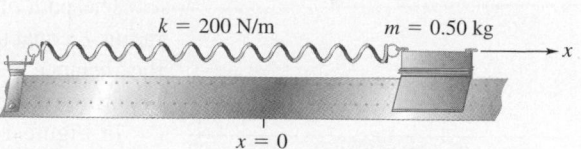

▲ FIGURE 11.24

SOLUTION

SET UP AND SOLVE All the necessary relations are contained in Equations 11.24 through 11.26. From Equation 11.24,

$$\omega = \sqrt{\frac{k}{m}} = \sqrt{\frac{200\ \text{N/m}}{0.50\ \text{kg}}} = \sqrt{\frac{200\ (\text{kg} \cdot \text{m/s}^2)/\text{m}}{0.50\ \text{kg}}} = 20\ \text{s}^{-1},$$

and from Equation 11.25,

$$f = \frac{\omega}{2\pi} = 3.2\ \text{Hz}, \qquad T = \frac{1}{f} = 0.31\ \text{s}.$$

REFLECT The angular frequency is not directly proportional to the force constant, as one might naively guess. To double the value of ω, we would have to increase k by a factor of four.

Practice Problem: Suppose we could "tune" this system by varying the mass; what mass would be required for a period of 1.0 s? *Answer:* 5.1 kg.

EXAMPLE 11.7 SHM on an air track, III: position vs. time

Suppose the object in Example 11.6 is released from rest at $x = 0.040$ m. How much time is required for it to move halfway to the center from the initial position?

SOLUTION

SET UP We can use the circle of reference, as shown in Figure 11.25. We need to find the angle ϕ of the radial line OQ when the glider has moved from the initial position $x = A$ to the later position $x = A/2$. Then we need to find the time required for ϕ to change by the corresponding amount.

SOLVE From Figure 11.25, we see that $x = A/2$ when $\cos\phi = 1/2$ and $\phi = 60° = \pi/3$. Also, $\phi = \omega t$, so we find that

$$t = \frac{\phi}{\omega} = \frac{\pi/3}{20\ \text{s}^{-1}} = 0.052\ \text{s}.$$

REFLECT The radial line OQ rotates through an angle of $\pi/3$ (1/6 revolution) in 0.052 s, corresponding to a time equal to the period, $T = 0.31$ s, needed for a full revolution. Note that we *must* express the angular quantity ϕ in radians.

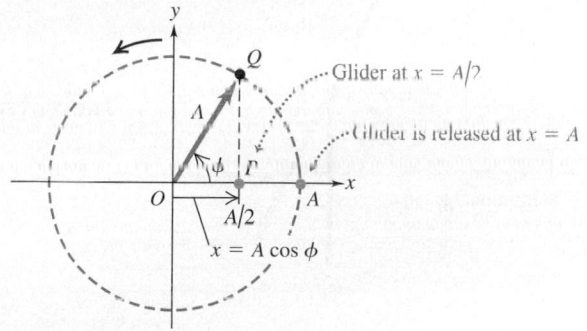

▲ FIGURE 11.25

Practice Problem: How much time is required for the glider to move from $x = A/2$ to $x = 0$? *Answer:* 0.026 s.

11.5 The Simple Pendulum

A **simple pendulum** is an idealized model consisting of a point mass suspended by a weightless, unstretchable string in a uniform gravitational field. When the mass is pulled to one side of its straight-down equilibrium position and released, it oscillates about the equilibrium position. Familiar situations such as a wrecking ball on a crane's cable, the plumb bob on a surveyor's transit, and a child on a swing can be modeled as simple pendulums. We can now analyze the motion of such systems, asking in particular whether it is simple harmonic.

PhET: Pendulum Lab
ActivPhysics 9.10: Pendulum Frequency
ActivPhysics 9.11: Risky Pendulum Walk

The path of the object (modeled as a point mass) is an arc of a circle with radius L equal to the length of the string (Figure 11.26). We use as our coordinate the distance x measured along the arc. If the motion is simple harmonic, the restoring force must be directly proportional to x, or (because $x = L\theta$) to θ. Is it?

In Figure 11.26, we represent the forces on the mass in terms of tangential and radial components. The restoring force F at each point is the component of force tangent to the circular path at that point:

$$F = -mg\sin\theta. \tag{11.28}$$

The restoring force is therefore proportional *not* to θ, but to $\sin\theta$, so the motion is *not* simple harmonic. However, *if the angle θ is small*, $\sin\theta$ is very nearly equal to θ (in radians). For example, when $\theta = 0.1$ rad (about $6°$), $\sin\theta = 0.0998$, a difference of only 0.2%. With this approximation, Equation 11.28 becomes

$$F = -mg\theta = -mg\frac{x}{L}, \quad \text{or}$$
$$F = -\frac{mg}{L}x. \tag{11.29}$$

The restoring force F is then proportional to the coordinate x *for small displacements,* and the constant mg/L represents the force constant k. From Equation 11.24, the angular frequency ω of a simple pendulum with small amplitude is given by the following expressions:

Frequency and period of simple pendulum with length L:

$$\omega = \sqrt{\frac{k}{m}} = \sqrt{\frac{mg/L}{m}} = \sqrt{\frac{g}{L}}. \tag{11.30}$$

The corresponding frequency and period relations are

$$f = \frac{\omega}{2\pi} = \frac{1}{2\pi}\sqrt{\frac{g}{L}}, \tag{11.31}$$

$$T = \frac{2\pi}{\omega} = \frac{1}{f} = 2\pi\sqrt{\frac{L}{g}}. \tag{11.32}$$

(a) A real pendulum

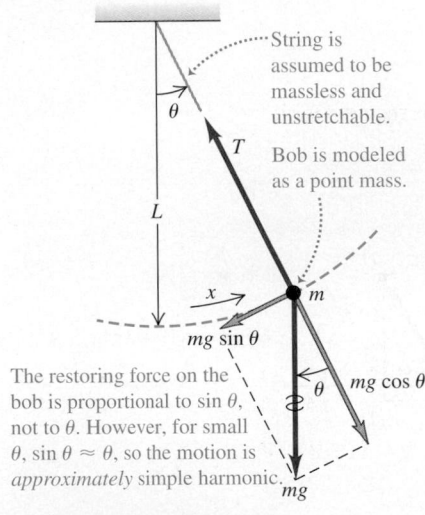

String is assumed to be massless and unstretchable.

Bob is modeled as a point mass.

The restoring force on the bob is proportional to $\sin\theta$, not to θ. However, for small θ, $\sin\theta \approx \theta$, so the motion is *approximately* simple harmonic.

(b) An idealized simple pendulum

▲ **FIGURE 11.26** The simple pendulum, an idealized model of a real pendulum.

Note that these expressions don't contain the *mass* of the particle. This is because the restoring force, a component of the particle's weight, is proportional to m. Thus, the mass appears on *both* sides of $\sum\vec{F} = m\vec{a}$ and cancels out. **For small oscillations, the period of a pendulum for a given value of g is independent of its mass and is determined entirely by its length.**

Galileo invented an elegant argument four centuries ago to show that the period should be independent of the mass. Make a simple pendulum, said Galileo, measure its period, and then split the pendulum down the middle, string and all. Because of the symmetry, neither half could have been pushing or pulling on the other before it was split. So splitting it can't change the motion, and each half must swing with the same period as the original. Therefore, the period cannot depend on the mass!

The dependence of the angular frequency, the frequency, and the period on L and g in Equations 11.30 through 11.32, respectively, is just what we should expect on the basis of everyday experience. A long pendulum has a longer period than a shorter one. Increasing g increases the restoring force, causing the frequency to increase and the period to decrease.

▶ Application **As the world turns.** In 1848, Jean-Bernard-Léon Foucault used a very long pendulum, now known as a Foucault pendulum, to demonstrate the rotation of the Earth itself. Newton's first law tells us that a pendulum tends to oscillate in a single plane in space unless a force deflects it. However, the plane of rotation of a Foucault pendulum seems to *rotate* in space: If the pendulum starts out swinging north–south, later in the day it will be swinging east–west. Is Newton wrong? No; the pendulum's plane of motion does not change, but the surface of the *earth* rotates beneath it. We humans see the pendulum as rotating because we mistake the earth's surface for an inertial frame of reference.

The simple pendulum is also a precise and convenient method for measuring the acceleration due to gravity (*g*), since *L* and *T* can be measured easily. Such measurements are often used in geophysics. Local deposits of ore or oil affect the local value of *g* because their density differs from that of their surroundings. Precise measurements of *g* over an area being surveyed often furnish valuable information about the nature of the underlying deposits.

NOTE ▶ We emphasize again that **the motion of a pendulum is only *approximately* simple harmonic.** For very small displacements, the approximation is quite precise. For example, when the angular amplitude is 15°, the actual period differs from the prediction of our approximate analysis by less than 0.5%. Indeed, the usefulness of the pendulum in clocks depends on the fact that, for small amplitudes, the motion is very nearly independent of amplitude. ◀

Quantitative Analysis 11.4

Changing the length of a pendulum

A pendulum with a string of length *L* swings with a period *T*. If the string is made twice as long, the pendulum's period will be about

A. *T*.
B. $\sqrt{2}\,T$.
C. 2*T*.

SOLUTION By now you should be able to work problems like this in your head quickly with thinking like the following: Since *T* is proportional to $\sqrt{L}$ (see Equation 11.32), if *L* is changed to 2*L*, then *T* changes to $\sqrt{2}\,T$. Thus, the answer is B. In more detail, Equation 11.23 can also be written as

$$\frac{T_{new}}{T_{old}} = \frac{\sqrt{L_{new}}}{\sqrt{L_{old}}} = \frac{\sqrt{2L_{old}}}{\sqrt{L_{old}}} = \sqrt{2}.$$

EXAMPLE 11.8 Period and frequency of a simple pendulum

Find the period and frequency of a simple pendulum 1.000 m long in Pittsburgh, where $g = 9.801$ m/s^2.

SOLUTION

SET UP AND SOLVE From Equation 11.32,

$$T = 2\pi \sqrt{\frac{1.000 \text{ m}}{9.801 \text{ m/s}^2}} = 2.007 \text{ s}$$

and

$$f = \frac{1}{T} = 0.4983 \text{ Hz.}$$

REFLECT The period is almost exactly 2 s; at one time there was a proposal to define the second as half the period of a 1-meter pendulum.

EXAMPLE 11.9 The lost Martian

A Martian who frequently gets lost in the solar system keeps a ball on a string so that he can always figure out what planet he happens to be on. On Mars, where the acceleration due to gravity is $g_{Mars} = 0.38 g_{earth}$, the ball oscillates with a period of 1.5 s when it is swung like a pendulum bob (Figure 11.27). During one journey, the Martian finds himself on a planet where the ball oscillates with a period of 0.92 s. What planet is he on?

Continued

SOLUTION

SET UP Each planet has a different value of the gravitational acceleration g near its surface. The Martian can measure g at his location, and from this he can determine what planet he's on. First we use the information about Mars to find the length L of the string that the Martian is swinging. Then we use that length to find the acceleration due to gravity on the unknown planet.

SOLVE First we use the Mars data to determine L. We solve Equation 11.32 for L:

$$T_{\text{Mars}} = 2\pi \sqrt{\frac{L}{g_{\text{Mars}}}} \quad \text{and}$$

$$L = \frac{T^2 g_{\text{Mars}}}{4\pi^2} = \frac{T^2(0.38 g_{\text{earth}})}{4\pi^2} = \frac{(1.5\text{ s})^2(0.38 \times 9.8\text{ m/s}^2)}{4\pi^2}$$
$$= 0.21\text{ m}.$$

Now we use this value of L with $T = 0.92$ s to solve for the acceleration due to gravity on the unknown planet, g_{planet}:

$$g_{\text{planet}} = \frac{4\pi^2 L}{T^2} = \frac{4\pi^2(0.21\text{ m})}{(0.92\text{ s})^2} = 9.8\text{ m/s}^2.$$

The Martian is surprised to find that he is on planet earth!

REFLECT In principle, the Martian can use this method to measure the acceleration due to gravity of any planet or satellite while traveling around the universe.

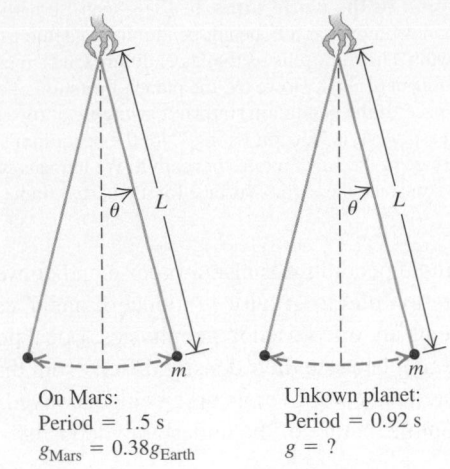

On Mars:
Period = 1.5 s
$g_{\text{Mars}} = 0.38 g_{\text{Earth}}$

Unkown planet:
Period = 0.92 s
$g = ?$

▲ **FIGURE 11.27**

Practice Problem: The acceleration due to gravity near the surface of the moon is about 1.6 m/s². If the Martian arrives there and swings the same pendulum bob, what will be the period of the pendulum? *Answer: T = 2.3 s.*

11.6 Damped and Forced Oscillations

The idealized oscillating systems that we've discussed thus far are frictionless. There are no nonconservative forces, the total mechanical energy is constant, and a system set into motion continues oscillating forever with no decrease in amplitude.

Real-world systems always have some friction, however, and oscillations do die out with time unless we provide some means for replacing the mechanical energy lost to friction. A pendulum clock continues to run because the potential energy stored in the spring or a hanging weight system replaces the mechanical energy lost because of friction in the pendulum and the gears. But the spring eventually runs down, or the weights reach the bottom of their travel. Then no more energy is available, and the pendulum swings decrease in amplitude and stop.

◄ **BIO Application Don't get hammered.** The oscillation of this construction worker's jackhammer is very effective at breaking up hard surfaces, such as concrete and asphalt. However, the repetitive motion can also damage the nerves and blood vessels in the hands, leading to the pain, tingling, and numbness characteristic of Raynaud's syndrome, or "white finger." It is not just the mechanical pounding on the tissue that causes the damage, but also the frequency of the vibration itself, which can set up damaging resonances in the human body. As little as five minutes of exposure to vibrations of 60 Hz can damage small arteries, and slower vibrations of 2 to 4 Hz can cause whole-body resonance as they are transmitted through the human skeleton. Occupational safety experts are designing new equipment and protective gear to minimize vibration-induced injury for workers using oscillating equipment.

The decrease in amplitude caused by dissipative forces is called **damping,** and the corresponding motion is called **damped oscillation.** The suspension system of an automobile is a familiar example of damped oscillations. The shock absorbers (Figure 11.28) provide a velocity-dependent damping force. The direction of the force always opposes the car's up-and-down motion, so that when the car goes over a bump, it doesn't continue bouncing forever. For optimal passenger comfort, the system should have enough damping that it bounces only once or twice after each bump. As the shocks get old and worn, the damping decreases and the bouncing is more persistent. Not only is this nauseating, but it is bad for steering, because the front wheels have less positive contact with the ground. Thus, damping is an advantage in this kind of system. But in a system such as a clock or an electrical oscillating system in a radio transmitter, it is often desirable to have as little damping as possible.

Figure 11.29 shows graphs of position x as a function of time for a damped oscillator—one graph for a relatively small damping force, the other for greater damping.

What happens when an additional periodically varying force is applied to an oscillating system? A factory floor supported by a slightly flexible steel framework may begin to vibrate when an oscillating machine rests on it. Or consider your little brother on a playground swing. You can keep him swinging with constant amplitude by giving him a little push once each cycle. More generally, we can maintain a constant-amplitude oscillation in a damped harmonic oscillator by applying a force that varies with time in a periodic or cyclic way, with a definite period and frequency. We call this additional force a **driving force.**

We denote the angular frequency of the driving force by ω_d. This angular frequency doesn't have to equal the angular frequency ω with which the system would naturally oscillate without a driving force. When a periodically varying driving force is present, the mass can undergo a periodic motion *with the same angular frequency ω_d as that of the driving force.* We call this motion a **forced oscillation,** or a *driven oscillation;* it is different from the motion that occurs when the system is simply set into motion and then left alone to oscillate with its natural frequency ω. The frequency of a forced oscillation is that of the driving force, not the system's natural frequency.

When the angular frequency of the driving force is nearly *equal* to the natural angular frequency $\omega = \sqrt{k/m}$ of the system, we are forcing the system to vibrate with a frequency that is close to the frequency it would have even with no driving force. In this case, we expect the amplitude of the resulting oscillation to be larger than it is when the two frequencies ω and ω_d are very different. This expectation is borne out by more detailed analysis and by experimental observation.

If we vary the frequency ω_d of the driving force, the amplitude of the resulting forced oscillation varies in an interesting way (Figure 11.30). When there is very little damping, the amplitude goes through a sharp peak as the driving angular frequency ω_d nears the natural oscillation angular frequency ω. When the damping is increased, the peak becomes broader and smaller in height and shifts toward lower frequencies.

The fact that there is an amplitude peak at driving frequencies close to the natural frequency of the system is called **resonance.** Physics is full of examples of resonance; building up the oscillations of a child on a swing by pushing with a frequency equal to the swing's natural frequency is one. A vibrating rattle in a car that occurs only at a certain engine speed or wheel-rotation speed is an all-too-familiar example. You may have heard of the dangers of a band marching in step

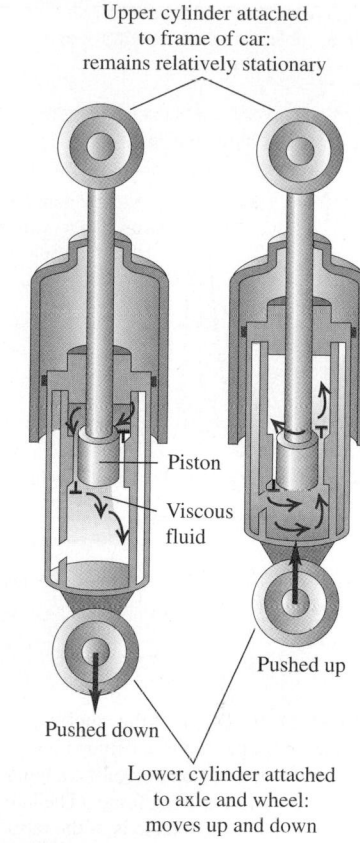

▲ **FIGURE 11.28** A shock absorber used in a car. The top part, connected to the piston, is attached to the car's frame; the bottom part, connected to the lower cylinder, is connected to the axle. The viscous fluid causes a damping force that depends on the relative velocity of the two ends of the unit. The device helps control wheel bounce and jounce.

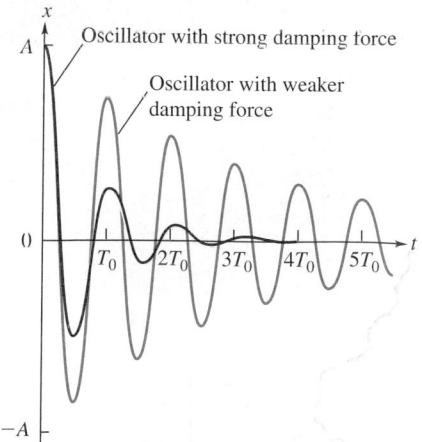

▲ **FIGURE 11.29** Graph showing damped harmonic oscillation for two oscillators, one strongly damped and one more weakly damped.

Each curve shows the amplitude for an oscillator subjected to a driving force at various angular frequencies.

Successive curves from blue to gold represent successively greater damping forces.

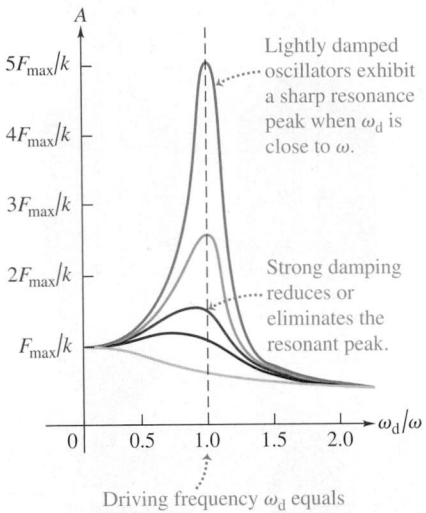

Lightly damped oscillators exhibit a sharp resonance peak when ω_d is close to ω.

Strong damping reduces or eliminates the resonant peak.

Driving frequency ω_d equals angular frequency ω of oscillator.

▲ **FIGURE 11.30** Graph of the amplitude A of the forced oscillation of a damped harmonic oscillator as a function of the angular frequency ω_d of the driving force. (The latter is plotted on the horizontal axis as the ratio of ω_d to the angular frequency $\omega = \sqrt{k/m}$ of an undamped oscillator.)

across a bridge. If the frequency of their steps is close to a natural vibration frequency of the bridge, dangerously large oscillations can build up. Inexpensive loudspeakers often have an annoying buzz or boom when a musical note happens to coincide with the resonant frequency of the speaker cone.

Finally, we mention the famous collapse of the Tacoma Narrows suspension bridge in 1940; nearly everyone has seen the film of this catastrophe (Figure 11.31). This is usually cited as an example of resonance, but there's some doubt as to whether it should be called that. The wind didn't have to vary *periodically* with a frequency close to a natural frequency of the bridge. The air flow past the bridge is turbulent, and vortices are formed with a regular frequency that depends on the flow speed. It is conceivable that this frequency may have resonated with a natural frequency of the bridge. But the cause may well have been something more subtle called a *self-excited oscillation,* in which the aerodynamic forces caused by a *steady* wind blowing on the bridge tended to displace it farther from equilibrium at times when it was already moving away from equilibrium.

It is as though we had a velocity-dependent damping force, but with the direction reversed. In that case, the force always has the *same* direction as the velocity and becomes an *anti-damping* force. Instead of draining mechanical energy away from the system, this force pumps energy into the system, building up the oscillations to destructive amplitudes. A similar phenomenon is the flapping of a flag in a wind.

The Tacoma Narrows bridge has been rebuilt, and engineers have now learned how to stabilize suspension bridges, both structurally and aerodynamically, to prevent such disasters.

▲ **FIGURE 11.31** The Tacoma Narrows Bridge collapsed four months and six days after it was opened for traffic. The main span was 2800 ft long and 39 ft wide, with 8-ft-high steel stiffening girders on both sides. The maximum amplitude of the twisting vibrations was 35°; the frequency was about 0.2 Hz.

SUMMARY

Stress, Strain, and Elastic Deformation

(Section 11.1) Forces that tend to stretch, squeeze, or twist an object constitute **stress.** The resulting deformation is called **strain.** Particularly for small deformations, stress and strain may be directly proportional: Stress/Strain = constant (Equation 11.1). We call this relationship **Hooke's law.**

Tensile and compressive stresses stretch and compress an object, respectively. Tensile stress is tensile force per unit area, $F_\perp/A$ (Equation 11.2). Tensile strain is the fractional change in length, $\Delta l/l_0$ (Equation 11.3).

Pressure or **volume stress** is force per unit area: $p = (F_\perp/A)$ (Equation 11.5). The force is a uniform force around an object and acts to increase or decrease its volume. The resulting deformation is volume **strain,** $\Delta V/V_0$ (Equation 11.6).

Shear stress is tangent force per unit area: $F_\parallel/A$ (Equation 11.8). The resulting deformation is **shear strain.**

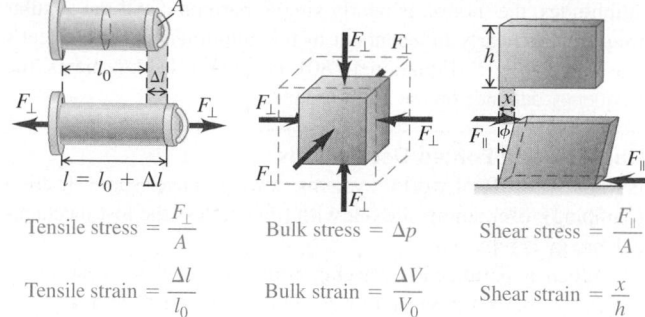

Tensile stress = $\dfrac{F_\perp}{A}$ Bulk stress = Δp Shear stress = $\dfrac{F_\parallel}{A}$

Tensile strain = $\dfrac{\Delta l}{l_0}$ Bulk strain = $\dfrac{\Delta V}{V_0}$ Shear strain = $\dfrac{x}{h}$

Elasticity and Plasticity

(Section 11.1) For small stresses, the strain of many materials is elastic; when the stress is removed, an object returns to its original size and shape. For stresses beyond a particular value, the deformation becomes **plastic:** when the stress is removed, the object does not return to its original shape. At a high enough stress, the object breaks.

Periodic Motion

(Section 11.2) Periodic motion is motion that repeats itself in a definite cycle. Periodic motion occurs whenever an object has a stable equilibrium position and is subject to a restoring force that acts when the object is displaced from equilibrium. When the restoring force is directly proportional to the displacement ($F_x = -kx$), which is often the case for small displacements, the resulting motion is called **simple harmonic motion.**

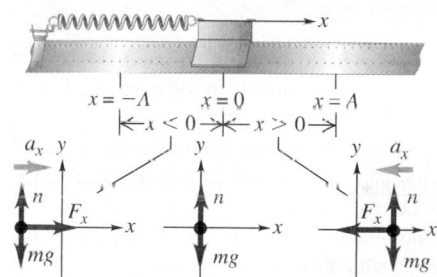

Energy in Simple Harmonic Motion

(Section 11.3) In simple harmonic motion, the total mechanical energy is constant: $E = \frac{1}{2}mv_x^2 + \frac{1}{2}kx^2 = \text{constant}$ (Equation 11.15). At points of maximum displacement, where $x = A$ and $v_x = 0$, all of the energy is in the form of elastic potential energy, so the (constant) total mechanical energy can be expressed as $E = \frac{1}{2}kA^2$ (Equation 11.16).

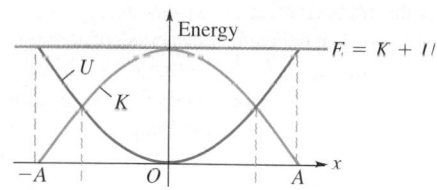

Equations of Simple Harmonic Motion

(Section 11.4) We obtain equations describing simple harmonic motion by using the **circle of reference.** We consider a point that moves counterclockwise with constant angular velocity ω around a circle of radius A. By evaluating the x components of the point's position, velocity, and acceleration, we find that

$$x = A\cos\omega t = A\cos(2\pi ft), \quad (11.19, 11.20)$$
$$v_x = -\omega A\sin\omega t = -2\pi fA\sin(2\pi ft), \quad (11.21)$$
$$a_x = -\omega^2 A\cos\omega t = -(2\pi f)^2 A\cos(2\pi ft). \quad (11.22)$$

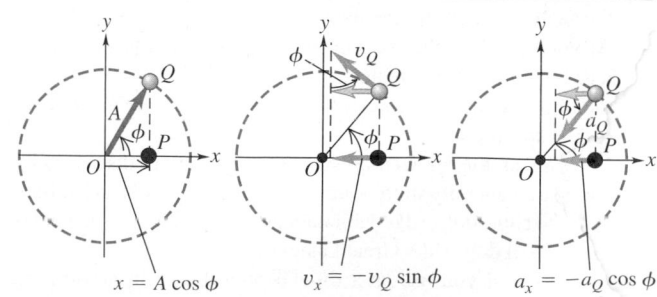

$x = A\cos\phi$ $v_x = -v_Q\sin\phi$ $a_x = -a_Q\cos\phi$

Continued

The Simple Pendulum

(Section 11.5) A **simple pendulum** is an idealized model consisting of a point mass suspended from a string with length L. For small amplitudes, the motion is nearly simple harmonic and the angular frequency is nearly independent of the amplitude and the object's mass: $\omega = \sqrt{g/L}$ (Equation 11.30). For larger displacements, the frequency depends on the amplitude.

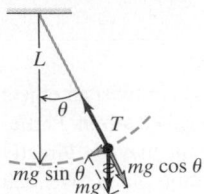

Damped and Forced Oscillations

(Section 11.6) Real-world systems always have some friction (damping); oscillations die out with time, unless the lost mechanical energy is replaced.

When a periodically varying force is added to a harmonic oscillator, the resulting motion is called a **forced oscillation.** The system can then oscillate with a frequency equal to that of the driving force ω_d, not the system's natural oscillation frequency $\omega = \sqrt{k/m}$. The amplitude peaks as ω_d nears the natural frequency ω; this phenomenon is called **resonance.**

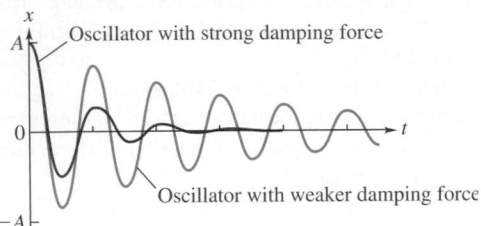

 For instructor-assigned homework, go to www.masteringphysics.com

Conceptual Questions

1. Think of several examples in everyday life of motion that is at least approximately simple harmonic. In what respects does each differ from SHM?
2. The analysis of simple harmonic motion in this chapter neglected the mass of the spring. How would the spring's mass affect the period and frequency of the motion? Explain your reasoning.
3. In any periodic motion, unavoidable friction always causes the amplitude to decrease with time. Does friction also affect the *period* of the motion? Give a qualitative argument to support your answer. (*Hint:* Does the friction affect the kinetic energy? If so, how does this affect the speed, and therefore the period, of a cycle?)
4. A mass attached by a light spring to the ceiling of an elevator oscillates vertically while the elevator ascends with constant acceleration. Is the period greater than, less than, or the same as when the elevator is at rest? Why?
5. A pendulum is mounted in an elevator that moves upward with constant acceleration. Is the period greater than, less than, or the same as when the elevator is at rest? Why?
6. At what point in the motion of a simple pendulum is (a) the tension in the string greatest, (b) the string tension least, (c) the radial acceleration greatest, (d) the angular acceleration least, (e) the speed greatest?
7. Looking at the molecular structure of matter, discuss why gases are generally more compressible than liquids and solids.
8. Why do short dogs (like Chihuahuas) walk with quicker strides than do tall dogs (like Great Danes)?
9. Which could you use as a timekeeping device in an orbiting satellite, a pendulum or a mass on a spring? Or could you use them both? Why?
10. What happens to the original energy as a damped pendulum (or spring) "runs down"? Explain *how* this energy gets dissipated.

11. Distinguish clearly between the angular frequency (ω) and the angular velocity (also ω) of a pendulum. Which of the two quantities is constant?
12. It is easy to get mixed up and think that the angular frequency ω of a pendulum is equal to $\sqrt{L/g}$ instead of $\sqrt{g/L}$. How can you use dimensional analysis to find out which version is the correct formula? Would such an analysis help to identify missing (or extra) factors of 2π? Why?
13. If a metal wire has its length doubled and its diameter tripled, by what factor does its Young's modulus change?
14. Would you expect a rubber band to have a larger or a smaller force constant than that of an iron wire? Why? How would you think that Young's modulus for these two materials would compare?
15. In designing structures in an earthquake-prone region, how should the natural frequencies of oscillation of a structure relate to typical earthquake frequencies? Why? Should the structures have a large or a small amount of damping?

Multiple-Choice Problems

1. A spring–mass system is undergoing simple harmonic motion of amplitude 2.0 cm and angular frequency 10.0 s^{-1}. The acceleration of the mass as it passes through the equilibrium position is
   A. 0.00 m/s^2.   B. 0.20 m/s^2.
   C. 2.0 m/s^2.   D. 31.4 m/s^2.
2. In a design for a piece of medical apparatus, you need a material that is easily compressed when a pressure is applied to it.
   A. This material should have a large bulk modulus.
   B. This material should have a small bulk modulus.
   C. The bulk modulus is not relevant to this situation.

3. If the force F in Figure 11.32 is constant over the area A, the pressure on that area is
   A. 2.50 Pa.
   B. 4.33 Pa.
   C. 5.00 Pa.

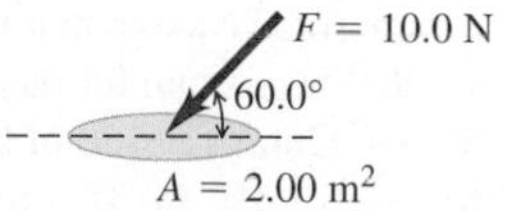

▲ **FIGURE 11.32** Multiple-choice problem 3.

4. A box with a mass of 5 kg, whose bottom measures 10×10 cm, sits on a table. The pressure the box exerts on the tabletop is approximately
   A. 0.5 Pa   B. 50 Pa   C. 500 Pa   D. 5000 Pa

5. When a mass attached to a spring is released from rest 3.0 cm from its equilibrium position, it oscillates with frequency f. If this mass were instead released from rest 6.0 cm from its equilibrium position, it would oscillate with frequency
   A. $2f$.   B. $\sqrt{2}f$ C. f.   D. $f/2$

6. As the bob on a pendulum swings down toward its lowest point, its angular frequency ω
   A. increases.
   B. decreases.
   C. does not change.

7. Suppose you increase the amplitude of oscillation of a mass vibrating on a spring. Which of the following statements about this mass are correct? (There may be more than one correct choice.)
   A. Its maximum speed increases.
   B. Its period of oscillation increases.
   C. Its maximum acceleration increases.
   D. Its maximum kinetic energy increases.

8. An object of mass M suspended by a spring vibrates with period T. If this object is replaced by one of mass $4M$, the new object vibrates with a period
   A. T.   B. $2T$.   C. $4T$.   D. $16T$.

9. When two wires of identical dimensions are used to hang 25 kg weights, wire A is observed to stretch twice as much as wire B. From this observation, you can make the following conclusions about the Young's moduli of these wires:
   A. $Y_A = 4Y_B$.   B. $Y_A = 2Y_B$.
   C. $Y_A = \frac{1}{2}Y_B$.   D. $Y_A = \frac{1}{4}Y_B$.

10. A mass on a spring oscillates with a period T. If both the mass and the force constant of the spring are doubled, the new period will be
    A. $4T$.   B. $\sqrt{2}T$.   C. T.
    D. $T/\sqrt{2}$.   E. $T/4$.

11. A pendulum oscillates with a period T. If both the mass of the bob and the length of the pendulum are doubled, the new period will be
    A. $4T$.   B. $\sqrt{2}T$.   C. T.
    D. $T/\sqrt{2}$.   E. $T/4$.

12. When a 100-kg mass is hung from a cable made of a certain material, the cable is observed to lengthen by 1%. If the cable is now replaced by one with twice the cross-sectional area and made of a material with twice the Young's modulus of the original one, and the mass hung from it is also doubled, how much stretching will be observed?
    A. 1%   B. 0.5%   C. 4%   D. 8%

13. An object with mass M suspended by a spring vibrates with frequency f. When a second object is attached to the first, the system now vibrates with frequency $f/2$. The mass of the second object is
    A. $4M$.   B. $3M$.   C. $2M$.   D. M.

14. A pendulum on earth swings with angular frequency ω. On an unknown planet, it swings with angular frequency $\omega/2$. The acceleration due to gravity on this planet is
    A. $4g$.   B. $2g$.   C. $g/2$.   D. $g/4$.

15. A mass oscillates with simple harmonic motion of amplitude A. The kinetic energy of the mass will equal the potential energy of the spring when the position is
    A. $x = 0$.   B. $x = A/2$.
    C. $x = A/\sqrt{2}$.   D. $x = A/4$.

Problems

11.1 Stress, Strain, and Elastic Deformations

1. • A thin, light wire 75.0 cm long having a circular cross section 0.550 mm in diameter has a 25.0 kg weight attached to it, causing it to stretch by 1.10 mm. (a) What is the stress in this wire? (b) What is the strain of the wire? (c) Find Young's modulus for the material of the wire.

2. • A petite young woman distributes her 500 N weight equally over the heels of her high-heeled shoes. Each heel has an area of 0.750 cm². (a) What pressure is exerted on the floor by each heel? (b) With the same pressure, how much weight could be supported by two flat-bottomed sandals, each of area 200 cm²?

3. •• Two circular rods, one steel and the other copper, are joined end to end. Each rod is 0.750 m long and 1.50 cm in diameter. The combination is subjected to a tensile force with magnitude 4000 N. For each rod, what are (a) the strain and (b) the elongation?

4. • A 5.0 kg mass is hung by a vertical steel wire 0.500 m long and 6.0×10^{-3} cm² in cross-sectional area. Hanging from the bottom of this mass is a similar steel wire, from which in turn hangs a 10.0 kg mass. For each wire, compute (a) the tensile strain and (b) the elongation.

5. • **Biceps muscle.** A relaxed biceps muscle requires a force of **BIO** 25.0 N for an elongation of 3.0 cm; under maximum tension, the same muscle requires a force of 500 N for the same elongation. Find Young's modulus for the muscle tissue under each of these conditions if the muscle can be modeled as a uniform cylinder with an initial length of 0.200 m and a cross-sectional area of 50.0 cm².

6. • **Stress on a mountaineer's rope.** A nylon rope used by mountaineers elongates 1.10 m under the weight of a 65.0 kg climber. If the rope is 45.0 m in length and 7.0 mm in diameter, what is Young's modulus for this nylon?

7. • A steel wire 2.00 m long with circular cross section must stretch no more than 0.25 cm when a 400.0 N weight is hung from one of its ends. What minimum diameter must this wire have?

8. • **Achilles tendon.** The Achilles tendon, which connects the **BIO** calf muscles to the heel, is the thickest and strongest tendon in the body. In extreme activities, such as sprinting, it can be subjected to forces as high as 13 times a person's weight. According to one set of experiments, the average area of the Achilles tendon is 78.1 mm², its average length is 25 cm, and its average Young's modulus is 1474 MPa. (a) How much tensile stress is required to stretch this muscle by 5.0% of its length? (b) If we model the tendon as a spring, what is its force constant? (c) If a 75 kg sprinter exerts a force of 13 times his weight on his Achilles tendon, by how much will it stretch?

9. •• **Artificial tendons.** The largest Young's modulus measured
BIO for any biological material is 30 GPa, for the S-layer of cell
walls. Suppose that in the future we are able to use this material to construct an artificial Achilles tendon (see previous
problem) for an athlete who has damaged his. The artificial
Achilles tendon should be just as long as the original one so
that it can connect at the proper places. (a) If the artificial tendon has a circular cross section, what should be its diameter be
so that it will have the same force constant as the natural
Achilles tendon? (b) How does this diameter compare with
that of a natural Achilles tendon?

10. •• **Human hair.** According to one set of measurements, the
BIO tensile strength of hair is 196 MPa, which produces a maximum strain of 0.40 in the hair. The thickness of hair varies
considerably, but let's use a diameter of 50 μm. (a) What is
the magnitude of the force giving this tensile stress? (b) If the
length of a strand of the hair is 12 cm at its breaking point,
what was its unstressed length?

11. •• **The effect of jogging on the knees.** High-impact activities
BIO such as jogging can cause considerable damage to the cartilage
at the knee joints. Peak loads on each knee can be eight times
body weight during jogging. The bones at the knee are separated by cartilage called the medial and lateral meniscus.
Although it varies considerably, the force at impact acts over
approximately 10 cm^2 of this cartilage. Human cartilage has a
Young's modulus of about 24 MPa (although that also varies).
(a) By what percent does the peak load impact of jogging compress the knee cartilage of a 75 kg person? (b) What would be
the percentage for a lower-impact activity, such as power walking, for which the peak load is about four times body weight?

12. • A solid gold bar is pulled up from the hold of the sunken
RMS *Titanic.* (a) What happens to its volume as it goes from
the pressure at the ship to the lower pressure at the ocean's surface? (b) The pressure difference is proportional to the depth.
How many times greater would the volume change have been
had the ship been twice as deep? (c) The bulk modulus of lead
is one-fourth that of gold. Find the ratio of the volume change
of a solid lead bar to that of a gold bar of equal volume for the
same pressure change.

13. • In the Challenger Deep of the Marianas Trench, the depth of
seawater is 10.9 km and the pressure is 1.16×10^8 Pa (about
1150 atmospheres). (a) If a cubic meter of water is taken to
this depth from the surface (where the normal atmospheric
pressure is about 1.0×10^5 Pa), what is the change in its volume? Assume that the bulk modulus for seawater is the same
as for freshwater $(2.2 \times 10^9 \text{ Pa})$. (b) At the surface, seawater
has a density of $1.03 \times 10^3 \text{ kg/m}^3$. What is the density of seawater at the depth of the Challenger Deep?

14. • **Effect of diving on blood.** It is reasonable to assume that
BIO the bulk modulus of blood is about the same as that of water
(2.2 GPa). As one goes deeper and deeper in the ocean, the pressure *increases* by 1.0×10^4 Pa for every meter below the surface. (a) If a diver goes down 33 m (a bit over 100 ft) in the
ocean, by how much does each cubic centimeter of her blood
change in volume? (b) How deep must a diver go so that each
drop of blood compresses to half its volume at the surface? Is
the ocean deep enough to have this effect on the diver?

15. • Shear forces are applied to a rectangular solid. The same
forces are applied to another rectangular solid of the same material, but with three times each edge length. In each case the forces

are small enough that Hooke's law is obeyed. What is the ratio of
the shear strain for the larger object to that of the smaller object?

16. •• **Compression of human bone.** The bulk modulus for bone
BIO is 15 GPa. (a) If a diver-in-training is put into a pressurized
suit, by how much would the pressure have to be raised (in
atmospheres) above atmospheric pressure to compress her
bones by 0.10% of their original volume? (b) Given that the
pressure in the ocean increases by 1.0×10^4 Pa for every
meter of depth below the surface, how deep would this diver
have to go for her bones to compress by 0.10%? Does it seem
that bone compression is a problem she needs to be concerned
with when diving?

17. • In Figure 11.33, suppose the
object is a square steel plate, 10.0 cm
on a side and 1.00 cm thick. Find the
magnitude of force required on each
of the four sides to cause a shear
strain of 0.0400.

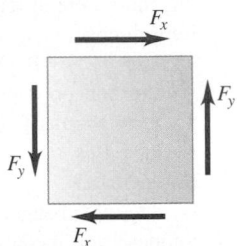

▲ **FIGURE 11.33**
Problem 17.

18. • In lab tests on a 9.25 cm cube of a
certain material, a force of 1375 N
directed at 8.50° to the cube, as
shown in Figure 11.34, causes the
cube to deform through an angle of
1.24°. What is the shear modulus of
the material?

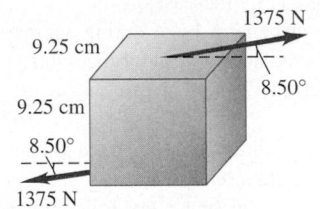

▲ **FIGURE 11.34** Problem 18.

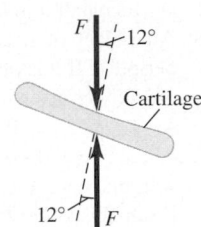

▲ **FIGURE 11.35**
Problem 19.

19. • **Downhill hiking.** During vigorous downhill hiking, the
BIO force on the knee cartilage (the medial and lateral meniscus)
can be up to eight times body weight. Depending on the angle
of descent, this force can cause a large shear force on the cartilage and deform it. The cartilage has an area of about 10 cm^2
and a shear modulus of 12 MPa. If the hiker plus his pack have
a combined mass of 110 kg (not unreasonable), and if the maximum force at impact is 8 times his body weight (which, of
course, includes the weight of his pack) at an angle of 12° with
the cartilage (see Figure 11.35), through what angle (in
degrees) will his knee cartilage be deformed? (Recall that the
bone below the cartilage pushes upward with the same force as
the downward force.)

20. • A steel wire has the following properties:

Length = 5.00 m
Cross-sectional area = 0.040 cm^2
Young's modulus = 2.0×10^{11} Pa
Shear modulus = 0.84×10^{11} Pa
Proportional limit = 3.60×10^8 Pa
Breaking stress = 11.0×10^8 Pa

The wire is fastened at its upper end and hangs vertically. (a) How great a weight can be hung from the wire without exceeding the proportional limit? (b) How much does the wire stretch under this load? (c) What is the maximum weight that can be supported?

21. •• A steel cable with cross-sectional area of 3.00 cm² has an elastic limit of 2.40×10^8 Pa. Find the maximum upward acceleration that can be given to a 1200 kg elevator supported by the cable if the stress is not to exceed one-third of the elastic limit.

22. •• **Weight lifting.** The legs of a weight lifter must ultimately **BIO** support the weights he has lifted. A human tibia (shinbone) has a circular cross section of approximately 3.6 cm outer diameter and 2.5 cm inner diameter. (The hollow portion contains marrow.) If a 90 kg lifter stands on both legs, what is the heaviest weight he can lift without breaking his legs, assuming that the breaking stress of the bone is 200 MPa?

11.2 Periodic Motion

23. • (a) **Music.** When a person sings, his or her vocal cords **BIO** vibrate in a repetitive pattern having the same frequency of the note that is sung. If someone sings the note B flat that has a frequency of 466 Hz, how much time does it take the person's vocal cords to vibrate through one complete cycle, and what is the angular frequency of the cords? (b) **Hearing.** When sound waves strike the eardrum, this membrane vibrates with the same frequency as the sound. The highest pitch that typical humans can hear has a period of 50.0 μs. What are the frequency and angular frequency of the vibrating eardrum for this sound? (c) **Vision.** When light having vibrations with angular frequency ranging from 2.7×10^{15} rad/s to 4.7×10^{15} rad/s strikes the retina of the eye, it stimulates the receptor cells there and is perceived as visible light. What are the limits of the period and frequency of this light? (d) **Ultrasound.** High-frequency sound waves (ultrasound) are used to probe the interior of the body, much as x rays do. To detect a small objects such as tumors, a frequency of around 5.0 MHz is used. What are the period and angular frequency of the molecular vibrations caused by this pulse of sound?

24. • Find the period, frequency, and angular frequency of (a) the second hand and (b) the minute hand of a wall clock.

25. • If an object on a horizontal frictionless surface is attached to a spring, displaced, and then released, it oscillates. Suppose it is displaced 0.120 m from its equilibrium position and released with zero initial speed. After 0.800 s, its displacement is found to be 0.120 m on the opposite side and it has passed the equilibrium position once during this interval. Find (a) the amplitude, (b) the period, and (c) the frequency of the motion.

26. • The graph shown in Figure 11.36 closely approximates the displacement x of a tuning fork as a function of time t as it is playing a single note. What are (a) the amplitude, (b) period, (c) frequency, and (d) angular frequency of this fork's motion?

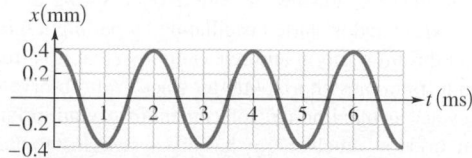

▲ **FIGURE 11.36** Problem 26.

27. • The wings of the Blue-throated Hummingbird (*Lampornis* **BIO** *clemenciae*), which inhabits Mexico and the southwestern United States, beat at a rate of up to 900 times per minute. Calculate (a) the period of vibration of the bird's wings, (b) the frequency of the wings' vibration, and (c) the angular frequency of the bird's wingbeats.

11.3 Energy in Simple Harmonic Motion

28. • A 0.500 kg glider on an air track is attached to the end of an ideal spring with force constant 450 N/m; it undergoes simple harmonic motion with an amplitude of 0.040 m. Compute (a) the maximum speed of the glider, (b) the speed of the glider when it is at $x = -0.015$ m, (c) the magnitude of the maximum acceleration of the glider, (d) the acceleration of the glider at $x = -0.015$ m, and (e) the total mechanical energy of the glider at any point in its motion.

29. • A 0.150 kg toy is undergoing SHM on the end of a horizontal spring with force constant 300.0 N/m. When the object is 0.0120 m from its equilibrium position, it is observed to have a speed of 0.300 m/s. Find (a) the total energy of the object at any point in its motion, (b) the amplitude of the motion, and (c) the maximum speed attained by the object during its motion.

30. • A 2.00 kg frictionless block is attached to an ideal spring with force constant 315 N/m. Initially the spring is neither stretched nor compressed, but the block is moving in the negative direction at 12.0 m/s. Find (a) the amplitude of the motion, (b) the maximum acceleration of the block, and (c) the maximum force the spring exerts on the block.

31. • Repeat the previous problem, but assume that initially the block has velocity −4.00 m/s and displacement +0.200 m.

32. •• You are watching an object that is moving in SHM. When the object is displaced 0.600 m to the right of its equilibrium position, it has a velocity of 2.20 m/s to the right and an acceleration of 8.40 m/s² to the left. How much farther from this point will the object move before it stops momentarily and then starts to move back to the left?

33. •• A mass is oscillating with amplitude A at the end of a spring. How far (in terms of A) is this mass from the equilibrium position of the spring when the elastic potential energy equals the kinetic energy?

34. •• (a) If a vibrating system has total energy E_0, what will its total energy be (in terms of E_0) if you double the amplitude of vibration? (b) If you want to triple the total energy of a vibrating system with amplitude A_0, what should its new amplitude be (in terms of A_0)?

11.4 Equations of Simple Harmonic Motion

35. • A 2.40 kg ball is attached to an unknown spring and allowed to oscillate. Figure 11.37 shows a graph of the ball's position x as a function of time t. For this motion, what are (a) the period, (b) the frequency, (c) the angular frequency, and (d) the amplitude? (e) What is the force constant of the spring?

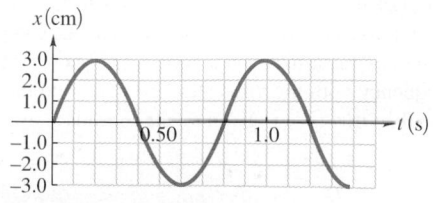

▲ **FIGURE 11.37** Problem 35.

36. • A proud deep-sea fisherman hangs a 65.0 kg fish from an ideal spring having negligible mass. The fish stretches the spring 0.120 m. (a) What is the force constant of the spring? (b) What is the period of oscillation of the fish if it is pulled down 3.50 cm and released?

37. •• One end of a stretched ideal spring is attached to an airtrack and the other is attached to a glider with a mass of 0.355 kg. The glider is released and allowed to oscillate in SHM. If the distance of the glider from the fixed end of the spring varies between 1.80 m and 1.06 m, and the period of the oscillation is 2.15 s, find (a) the force constant of the spring, (b) the maximum speed of the glider, and (c) the magnitude of the maximum acceleration of the glider.

38. • A mass of 0.20 kg on the end of a spring oscillates with a period of 0.45 s and an amplitude of 0.15 m. Find (a) the velocity when it passes the equilibrium point, (b) the total energy of the system, and (c) the equation describing the motion of the mass, assuming that x was a maximum at time $t = 0$.

39. •• A harmonic oscillator is made by using a 0.600 kg frictionless block and an ideal spring of unknown force constant. The oscillator is found to have a period of 0.150 s. Find the force constant of the spring.

40. •• **Weighing astronauts.** In order to study the long-term effects
BIO of weightlessness, astronauts in space must be weighed (or at least "*massed*"). One way in which this is done is to seat them in a chair of known mass attached to a spring of known force constant and measure the period of the oscillations of this system. If the 35.4 kg chair alone oscillates with a period of 1.25 s, and the period with the astronaut sitting in the chair is 2.23 s, find (a) the force constant of the spring and (b) the mass of the astronaut.

41. •• A mass m is attached to a spring of force constant 75 N/m and allowed to oscillate. Figure 11.38 shows a graph of its velocity v_x as a function of time t. Find (a) the period, (b) the frequency, and (c) the angular frequency of this motion. (d) What is the amplitude (in cm), and at what times does the mass reach this position? (e) Find the maximum acceleration of the mass and the times at which it occurs. (f) What is the mass m?

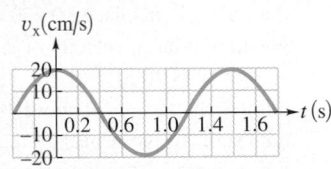

▲ **FIGURE 11.38** Problem 41.

42. •• An object of unknown mass is attached to an ideal spring with force constant 120 N/m and is found to vibrate with a frequency of 6.00 Hz. Find (a) the period, (b) the angular frequency, and (c) the mass of this object.

43. •• **Weighing a virus.** In February 2004, scientists at Purdue
BIO University used a highly sensitive technique to measure the mass of a vaccinia virus (the kind used in smallpox vaccine). The procedure involved measuring the frequency of oscillation of a tiny sliver of silicon (just 30 nm long) with a laser, first without the virus and then after the virus had attached itself to the silicon. The difference in mass caused a change in the frequency. We can model such a process as a mass on a spring. (a) Show that the ratio of the frequency with the virus attached (f_{S+V}) to the frequency without the virus (f_S) is given by the formula

$$\frac{f_{S+V}}{f_S} = \frac{1}{\sqrt{1 + \dfrac{m_V}{m_S}}},$$

where m_V is the mass of the virus and m_S is the mass of the silicon sliver. Notice that it is *not* necessary to know or measure the force constant of the spring. (b) In some data, the silicon sliver has a mass of 2.10×10^{-16} g and a frequency of 2.00×10^{15} Hz without the virus and 2.87×10^{14} Hz with the virus. What is the mass of the virus, in grams and femtograms?

11.5 The Simple Pendulum

44. • A science museum has asked you to design a simple pendulum that will make 25.0 complete swings in 85.0 s. What length should you specify for this pendulum?

45. • A simple pendulum in a science museum entry hall is 3.50 m long, has a 1.25 kg bob at its lower end, and swings with an amplitude of 11.0°. How much time does the pendulum take to swing from its extreme right side to its extreme left side?

46. • You've made a simple pendulum with a length of 1.55 m, and you also have a (very light) spring with force constant 2.45 N/m. What mass should you add to the spring so that its period will be the same as that of your pendulum?

47. •• **A pendulum on Mars.** A certain simple pendulum has a period on earth of 1.60 s. What is its period on the surface of Mars, where the acceleration due to gravity is 3.71 m/s²?

48. •• In the laboratory, a student studies a pendulum by graphing the angle θ that the string makes with the vertical as a function of time t, obtaining the graph shown in Figure 11.39. (a) What are the period, frequency, angular frequency, and amplitude of the pendulum's motion? (b) How long is the pendulum? (c) Is it possible to determine the mass of the bob?

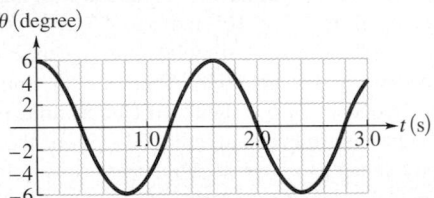

▲ **FIGURE 11.39** Problem 48.

49. •• (a) If a pendulum has period T and you double its length, what is its new period in terms of T? (b) If a pendulum has a length L and you want to triple its frequency, what should be its length in terms of L? (c) Suppose a pendulum has a length L and period T on earth. If you take it to a planet where the acceleration of freely falling objects is ten times what it is on earth, what should you do to the length to keep the period the same as on earth? (d) If you do *not* change the pendulum's length in part (c), what is its period on that planet in terms of T? (e) If a pendulum has a period T and you triple the mass of its bob, what happens to the period (in terms of T)?

11.6 Damped and Forced Oscillations

50. • A 1.35 kg object is attached to a horizontal spring of force constant 2.5 N/cm and is started oscillating by pulling it 6.0 cm from its equilibrium position and releasing it so that it is free to oscillate on a frictionless horizontal air track. You observe that after eight cycles its maximum displacement from equilibrium is only 3.5 cm. (a) How much energy has this system lost to damping during these eight cycles? (b) Where did the "lost" energy go? Explain physically how the system could have lost energy.

51. • A 2.50 kg rock is attached at the end of a thin, very light rope 1.45 m long and is started swinging by releasing it when the rope makes an 11° angle with the vertical. You record the observation that it rises only to an angle of 4.5° with the vertical after $10\frac{1}{2}$ swings. (a) How much energy has this system lost during that time? (b) What happened to the "lost" energy? Explain *how* it could have been "lost."

52. •• A mass is vibrating at the end of a spring of force constant 225 N/m. Figure 11.40 shows a graph of its position x as a function of time t. (a) At what times is the mass not moving? (b) How much energy did this system originally contain? (c) How much energy did the system lose between $t = 1.0$ s and $t = 4.0$ s? Where did this energy go?

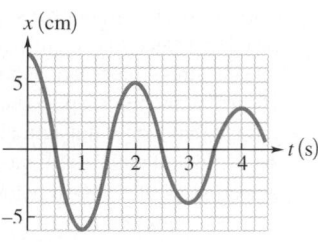

▲ **FIGURE 11.40** Problem 52.

General Problems

53. • What is the maximum kinetic energy of the vibrating ball in Problem 35, and when does it occur?

54. •• Inside a NASA test vehicle, a 3.50-kg ball is pulled along by a horizontal ideal spring fixed to a friction-free table. The force constant of the spring is 225 N/m. The vehicle has a steady acceleration of 5.00 m/s², and the ball is not oscillating. Suddenly, when the vehicle's speed has reached 45.0 m/s, its engines turn off, thus eliminating its acceleration but not its velocity. Find (a) the amplitude and (b) the frequency of the resulting oscillations of the ball. (c) What will be the ball's maximum speed relative to the vehicle?

55. •• Four passengers with a combined mass of 250 kg compress the springs of a car with worn-out shock absorbers by 4.00 cm when they enter it. Model the car and passengers as a single body on a single ideal spring. If the loaded car has a period of vibration of 1.08 s, what is the period of vibration of the empty car?

56. •• An astronaut notices that a pendulum which took 2.50 s for a complete cycle of swing when the rocket was waiting on the launch pad takes 1.25 s for the same cycle of swing during liftoff. What is the acceleration of the rocket? (*Hint:* Inside the rocket, it appears that g has increased.)

57. •• An object suspended from a spring vibrates with simple harmonic motion. At an instant when the displacement of the object is equal to one-half the amplitude, what fraction of the total energy of the system is kinetic and what fraction is potential?

58. •• On the planet Newtonia, a simple pendulum having a bob with a mass of 1.25 kg and a length of 185.0 cm takes 1.42 s, when released from rest, to swing through an angle of 12.5°, where it again has zero speed. The circumference of Newtonia is measured to be 51,400 km. What is the mass of the planet Newtonia?

59. •• An apple weighs 1.00 N. When you hang it from the end of a long spring of force constant 1.50 N/m and negligible mass, it bounces up and down in SHM. If you stop the bouncing and let the apple swing from side to side through a small angle, the frequency of this simple pendulum is half the bounce frequency. (Because the angle is small, the back-and-forth swings do not cause any appreciable change in the length of the spring.) What is the unstretched length of the spring (with the apple removed)?

60. •• A block with mass M rests on a frictionless surface and is connected to a horizontal spring of force constant k, the other end of which is attached to a wall (Figure 11.41). A second block with mass m rests on top of the first block. The coefficient of static friction between the blocks is μ_s. Find the maximum amplitude of oscillation such that the top block will not slip on the bottom block.

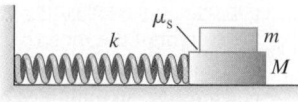

▲ **FIGURE 11.41** Problem 60.

61. •• In Fig. 11.42 the upper ball is released from rest, collides with the stationary lower ball, and sticks to it. The strings are both 50.0 cm long. The upper ball has mass 2.00 kg, and it is initially 10.0 cm higher than the lower ball, which has mass 3.00 kg. Find the frequency and maximum angular displacement of the motion after the collision.

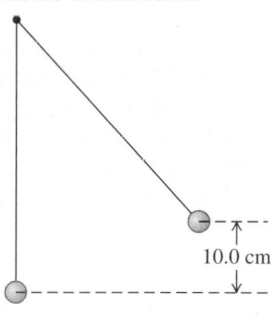

▲ **FIGURE 11.42** Problem 61.

62. •• A 15.0 kg mass fastened to the end of a steel wire with an unstretched length of 0.50 m is whirled in a vertical circle with angular velocity 2.00 rev/s at the bottom of the circle. The cross-sectional area of the wire is 0.010 cm². Calculate the elongation of the wire when the mass is at the lowest point of the path. (See Table 11.1.)

63. •• **Stress on the shinbone.** The compressive strength of our **BIO** bones is important in everyday life. Young's modulus for bone is approximately 14 GPa. Bone can take only about a 1.0% change in its length before fracturing. If Hooke's law were to hold up to fracture: (a) What is the maximum force that can be applied to a bone whose minimum cross-sectional area is 3.0 cm²? (This is approximately the cross-sectional area of a tibia, or shinbone, at its narrowest point.) (b) Estimate the maximum height from which a 70 kg man can jump and not fracture the tibia. Take the time between when he first touches the floor and when he has stopped to be 0.030 s, and assume that the stress is distributed equally between his legs.

64. •• You hang a floodlamp from the end of a vertical steel wire. The floodlamp stretches the wire 0.18 mm and the stress is proportional to the strain. How much would it have stretched (a) if the wire were twice as long? (b) If the wire had the same length but twice the diameter? (c) For a copper wire of the original length and diameter?

65. •• **Tendon-stretching exer-** **BIO** **cises.** As part of an exercise program, a 75 kg person does toe raises in which he raises his entire body weight on the ball of one foot, as shown in Figure 11.43. The Achilles tendon pulls straight upward on the heel bone of his foot. This tendon is 25 cm long and has a cross-sectional area of 78 mm² and a Young's modulus of 1470 MPa. (a) Make a free-body diagram of the

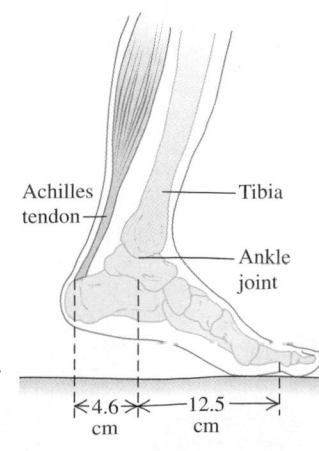

▲ **FIGURE 11.43** Problem 68.

person's foot (everything below the ankle joint). You can neglect the weight of the foot. (b) What force does the Achilles tendon exert on the heel during this exercise? Express your answer in newtons and in multiples of his weight. (c) By how many millimeters does the exercise stretch his Achilles tendon?

66. •• A 100 kg mass suspended from a wire whose unstretched length is 4.00 m is found to stretch the wire by 6.0 mm. The wire has a uniform cross-sectional area of 0.10 cm². (a) If the load is pulled down a small additional distance and released, find the frequency at which it vibrates. (b) Compute Young's modulus for the wire.

67. •• A brass rod with a length of 1.40 m and a cross-sectional area of 2.00 cm² is fastened end to end to a nickel rod with length L and cross-sectional area 1.00 cm². The compound rod is subjected to equal and opposite pulls of magnitude 4.00×10^4 N at its ends. (a) Find the length L of the nickel rod if the elongations of the two rods are equal. (b) What is the stress in each rod? (c) What is the strain in each rod?

68. •• **Rapunzel, Rapunzel, let down your golden hair.** In the Grimms' fairy tale *Rapunzel,* she lets down her golden hair to a length of 20 yards (we'll use 20 m, which is not much different) so that the prince can climb up to her room. Human hair has a Young's modulus of about 490 MPa, and we can assume that Rapunzel's hair can be squeezed into a rope about 2.0 cm in cross-sectional diameter. The prince is described as young and handsome, so we can estimate a mass of 60 kg for him. (a) Just after the prince has started to climb at constant speed, while he is still near the bottom of the hair, by how many centimeters does he stretch Rapunzel's hair? (b) What is the mass of the heaviest prince that could climb up, given that the maximum tensile stress hair can support is 196 MPa? (Assume that Hooke's law holds up to the breaking point of the hair, even though that would not actually be the case.)

69. •• Crude oil with a bulk modulus of 2.35 GPa is leaking from a deep-sea well 2250 m below the surface of the ocean, where the water pressure is 2.27×10^7 Pa. Suppose 35,600 barrels of oil leak from the wellhead; assuming all that oil reaches the surface, how many barrels will it be on the surface?

12 Mechanical Waves and Sound

Why do male frogs inflate their throats? These pouches serve as resonating chambers, amplifying the frog's voice.

When you go to the beach to enjoy the ocean surf, you're experiencing a wave motion. Ripples on a pond, musical sounds, sounds we *can't* hear, the wiggles of a Slinky® stretched out on the floor—all these are *wave* phenomena. Waves can occur whenever a system has an equilibrium position and when a disturbance away from that position can travel, or *propagate*, from one region of the system to another. Sound, light, ocean waves, radio and television transmission, and earthquakes are all wave phenomena. Waves occur in all branches of physical and biological science, and the concept of waves is one of the most important unifying threads running through the entire fabric of the natural sciences.

12.1 Mechanical Waves

This chapter is about mechanical waves. Every type of mechanical wave is associated with some material or substance called the **medium** for that type. (In Chapter 23, we'll study electromagnetic waves, such as radio and light waves, which do not require a medium and thus are not mechanical waves.) As the wave travels through the medium, the particles that make up the medium undergo displacements of various kinds, depending on the nature of the wave. Figure 12.1 shows waves produced by drops falling on a water surface. The speed of travel of a wave depends on the mechanical properties of the medium. Some waves are *periodic*, meaning that the particles of the medium undergo periodic motion during wave propagation. If the motion of each particle is simple harmonic (i.e., a sinusoidal function of time), the wave is called a *sinusoidal* wave.

Figure 12.2 shows three examples of mechanical waves. In Figure 12.2a, the medium is a stretched rope. (It could also be a spring or Slinky®.) If we give the left end a small up-and-down wiggle, the resulting pulse travels down the

▲ **FIGURE 12.1** A series of drops falling vertically into water produces a wave pattern that moves radially outward from its source. The wave crests and troughs are concentric circles.

length of the rope. Successive sections of rope undergo the same up-and-down motion that we gave to the end, but at successively later times. Because the displacements of particles of the medium are perpendicular (transverse) to the direction of travel of the wave along the medium, this type of wave is called a **transverse wave.**

In Figure 12.2b, the medium is a liquid or gas in a tube with a rigid wall at the right end and a movable piston at the left end. If we give the piston a back-and-forth motion, a displacement and a pressure fluctuation travel down the length of the medium. This time, the particles of the medium move back and forth parallel to the wave's direction of travel; we call this type of wave a **longitudinal wave.**

Longitudinal waves are familiar to you as sound. Your ears are designed to detect longitudinal waves carried by air, but sound waves can also be carried by liquids such as water (you can hear under water) and by solid objects such as walls.

In Figure 12.2c, the medium is water in a trough. When we move the flat board at the left end back and forth, a disturbance travels down the length of the trough. In this case, the displacements of the water particles are found to have *both* longitudinal and transverse components. The familiar waves of oceans and lakes are of this type; they are most often generated by wind moving over the water surface.

Earthquake waves have both longitudinal and transverse displacements. The longitudinal components travel faster and so give advance warning of the arrival of the more destructive transverse components—as much as 60 seconds at a distance of 200 miles from the source.

All the waves in Figure 12.2 have several characteristics in common. For each, there is an equilibrium state. For the stretched rope, it is the state in which the rope is at rest, stretched out along a straight line (if we neglect the sag due to gravity). For the fluid in a tube, it is a state of rest with uniform pressure, and for the water in a trough it is a smooth, level water surface. In each case, the wave motion is a disturbance from the equilibrium state that travels from one region of the medium to another. In each case, there are forces that tend to restore the system to its equilibrium position when it is displaced, just as the force of gravity tends to pull a pendulum toward its straight-down equilibrium position when it is displaced.

In each of these situations, the medium as a whole does not travel through space; its individual particles undergo back-and-forth motions around their equi-

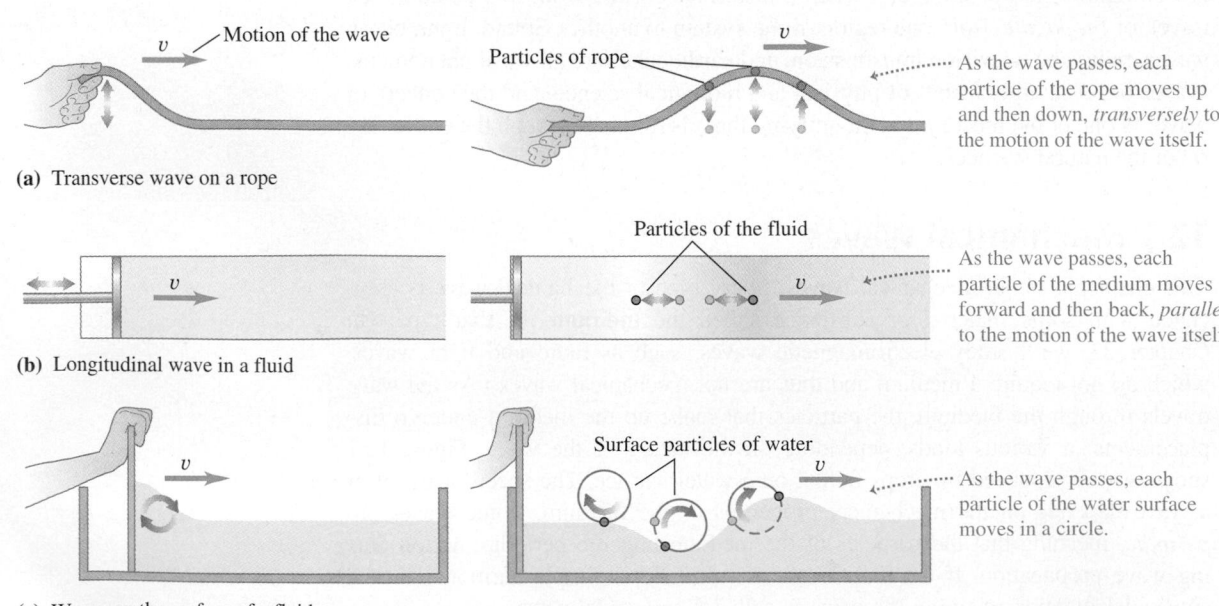

(a) Transverse wave on a rope

(b) Longitudinal wave in a fluid

(c) Waves on the surface of a fluid

▲ **FIGURE 12.2** Examples of (a) a transverse wave on a rope, (b) a longitudinal wave in a cylinder containing a gas or liquid, and (c) a surface wave with transverse and longitudinal components.

▶ Application **Two-minute warning.** Earthquakes are caused by large-scale movements of the Earth's crust; these movements generate both transverse and longitudinal waves in the crust. Using monitoring instruments known as seismographs, scientists can record these waves and detect the presence and location of distant earthquakes. The low-amplitude longitudinal waves travel faster than the more destructive transverse waves and arrive at a seismograph station sooner—several minutes sooner for an earthquake 1000 km away. Thus, prompt detection of the longitudinal waves from an earthquake can provide an early warning of the arrival of the destructive transverse waves. Often, even a minute or two of advance notice can be critical to safeguard life and property.

librium positions. What *does* travel is the pattern of the wave disturbance. To set any of these systems in motion, we have to put in energy by doing mechanical work on the system. The wave motion transports this energy from one region of the medium to another. **Waves transport energy, but not matter, from one region to another.** In each case, the disturbance is found to travel, or *propagate,* with a definite speed through the medium. This speed is called the speed of propagation or simply the **wave speed.** It is determined by the mechanical properties of the medium. We'll use the symbol v for wave speed.

Conceptual Analysis 12.1

Making waves

You and a friend hold the ends of a Slinky® that lies straight between you on a table. You create a wave by suddenly moving your hand toward and then away from your friend. The resulting wave

A. is purely transverse.
B. is purely longitudinal.
C. has both transverse and longitudinal components.

SOLUTION As the wave travels toward your friend, successive particles in the Slinky® undergo the same to-and-fro motion that you gave to the end. Since this motion is parallel to the direction in which the wave itself travels, the wave is longitudinal. (However, you could also cause the Slinky® to carry transverse waves by moving its end transversely.)

12.2 Periodic Mechanical Waves

The easiest kind of wave to demonstrate is a transverse wave on a stretched rope or flexible cable. Suppose we tie one end of a long, flexible rope to a stationary object and pull on the other end, stretching the rope out horizontally. In Figure 12.2a, we saw what happens when we give the end a single up-and-down shake. The resulting wiggle, or *wave pulse,* travels down the length of the rope. The tension in the rope restores the rope's straight-line shape once the wiggle has passed. (We are neglecting any sag in the rope due to the rope's weight.) Later we'll discuss what happens when the wiggle reaches the far end of the rope.

A more interesting situation develops when we give the free end of the rope a repetitive, or *periodic,* motion. In particular, suppose we use the setup in Figure 12.3 to move the end up and down with simple harmonic motion having

ActivPhysics 10.1: Properties of Mechanical Waves

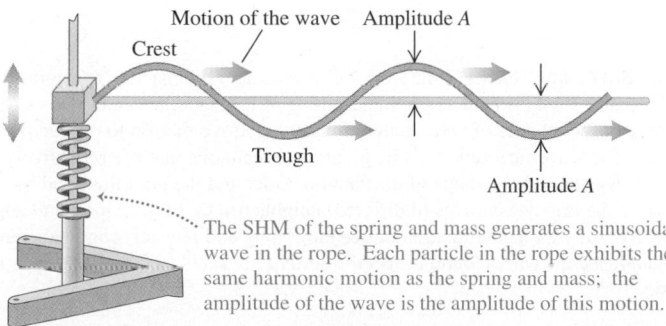

▲ **FIGURE 12.3** The simple harmonic motion of the spring–mass system produces a sinusoidal transverse wave on the rope.

The rope is shown at time intervals of $\frac{1}{8}$ period for a total of one period T. The highlighting shows the motion of one wave.

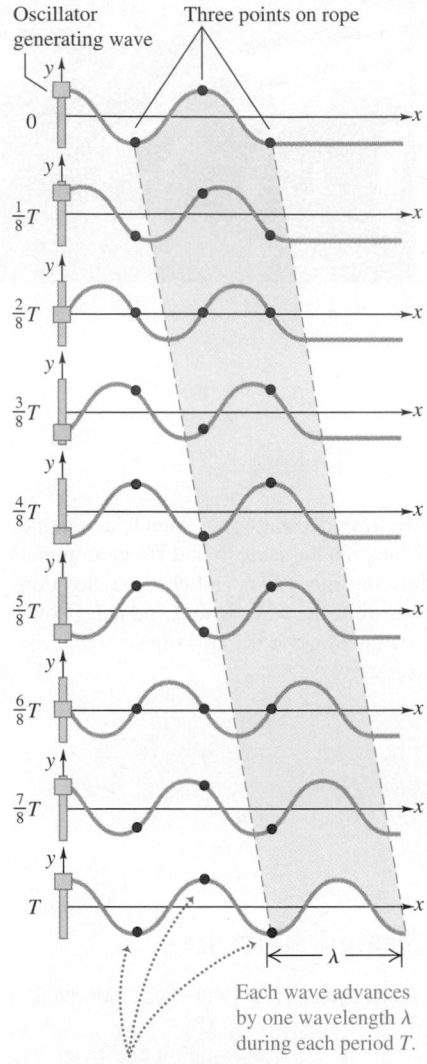

Each wave advances by one wavelength λ during each period T.

Each point moves up and down in place. Particles one wavelength apart move in phase with each other.

▲ **FIGURE 12.4** A sinusoidal transverse wave shown at intervals of one-eighth of a period.

amplitude A, frequency f, and period T. As usual, $f = 1/T = \omega/2\pi$, where ω is the angular frequency. A *continuous succession* of transverse sinusoidal waves then advances along the rope.

Figure 12.4 shows the shape of a part of the rope near the left end, at intervals of one-eighth of a period, for a total time of one period. The waveform advances steadily toward the right, while any one point on the rope oscillates up and down about its equilibrium position with simple harmonic motion. Be very careful to distinguish between the motion of a *waveform* (the shape of the displaced rope at any time), which moves with constant speed v *along* the rope, and the motion of *a particle of the rope,* which is simple harmonic and transverse (perpendicular) to the length of the rope. The **amplitude** A of the wave is defined as the amplitude of the transverse simple harmonic (sinusoidal) motion of a particle of the medium; it corresponds to the amplitude of the harmonic oscillation that creates the wave. As Figure 12.3 shows, for a periodic transverse wave, the amplitude is the distance from the equilibrium position to a peak or trough.

The shape of the rope at any instant is a repeating pattern—a series of identical shapes. For a periodic wave, the length of one complete wave pattern is the distance between any two points at corresponding positions on successive repetitions in the wave shape. We call this the **wavelength** of the wave, denoted by λ. The waveform travels with constant speed v and advances a distance of one wavelength λ in a time interval of one period T. We can relate these quantities as follows:

Relation among the speed, wavelength, period, and frequency of a transverse wave

The wave advances a distance of one wavelength λ during one period T, so the wave speed v is given by $v = \lambda/T$. Because frequency $f = 1/T$, the wave speed also equals the product of wavelength and frequency:

$$v = \lambda f = \frac{\lambda}{T}. \tag{12.1}$$

Note the familiar pattern: Speed equals distance divided by time.

To understand the mechanics of a *longitudinal* wave such as a sound wave in air, we consider a long tube filled with air, with a plunger at the left end, as shown in Figure 12.5. The cone of a loudspeaker is simply a sophisticated version of a plunger like this. If we push the plunger in, we compress the air near it, increasing the pressure in this region. The air in the region then pushes against the neighboring region of air, and so on, forming a wave pulse that moves along the tube.

◄ **Application Surf's up!** Thanks to the study of physics and the marvels of engineering, this surfer can enjoy a wild ride in a pool hundreds of miles from the ocean. This surf pool takes advantage of the physics of wave generation by repetitive motion to produce realistic "ocean" waves. The waves are generated by giant water cannons that fire repetitively at one end of the pool. By altering the shape of the cannon outlet and the pool floor and by changing the timing of the impulses, waves of different heights (up to 3 m), shapes, and angles can be generated. By studying a combination of oceanography and physics, engineers have even been able to duplicate the waves found in specific surfing hot spots such as California, Hawaii, and Indonesia.

The pressure fluctuations play the role of the restoring force, tending to restore the air to equilibrium and uniform pressure.

Now, suppose we move the plunger back and forth with simple harmonic motion along a line parallel to the direction of the tube. This motion forms regions where the pressure and density are alternately greater and less than the equilibrium values. We call a region of increased pressure a *compression*. A region of reduced pressure is an *expansion* (or *rarefaction*).

Figure 12.6 shows the waves in our apparatus at intervals of one-eighth of a period. The compressions and expansions move to the right with constant speed v. The motion of the individual particles of the medium (air) is simple harmonic, parallel to the direction of propagation. As we found for transverse waves, the amplitude A of a longitudinal wave is the amplitude of the periodic motion of the particles of the medium. That is, A represents the maximum displacement of a particle of the medium from its equilibrium position.

Earlier we mentioned that your ears detect longitudinal waves as sound. For a musical tone, you detect the amplitude of sound waves as loudness and their frequency as pitch—the higher pitched a sound, the greater its frequency. The wavelength is the distance between two successive compressions or two successive expansions. The same fundamental equation that we found for periodic transverse waves, $v = \lambda f$, also holds for periodic longitudinal waves and, indeed, for *all* types of periodic waves.

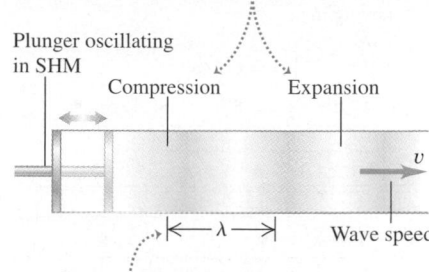

Forward motion of the plunger creates a compression (a zone of high pressure); backward motion creates an expansion (a zone of low pressure).

The wavelength λ is the distance between corresponding points on successive cycles.

▲ **FIGURE 12.5** A piston creating a sinusoidal longitudinal wave in a tube.

Quantitative Analysis 12.2

Distance from a sound source

A World War I infantryman sees the flash of a fired artillery gun; 20 s later he hears the gun's sound. The speed of sound in air is about a fifth of a mile per second. Which of the following is a reasonable estimate for the distance to the gun?

A. 1 mi.
B. 2 mi.
C. 4 mi.

SOLUTION The speed of *light* in air is roughly a million times as great as the speed of *sound* in air. Thus, in this example, the time between the creation of the flash and seeing it from a distance of several miles is negligible. In air, sound travels a fifth of a mile each second, or one mile in five seconds. It would take 20 seconds to travel 4 miles. A useful rule of thumb is that each mile gives a time delay of 5 seconds. If you see a lightning flash, and if you hear the accompanying thunder 10 seconds later, the flash was about 2 miles away.

EXAMPLE 12.1 **Frequencies of musical notes**

What is the wavelength of a sound wave in air at 20°C if the frequency is $f = 262$ Hz (the approximate frequency of the note "middle C" on the piano)? Assume that the speed of sound in air is 344 m/s.

SOLUTION

SET UP AND SOLVE The relation of wavelength λ to frequency f is given by Equation 12.1:

$$\lambda = \frac{v}{f} = \frac{344 \text{ m/s}}{262 \text{ s}^{-1}} = 1.31 \text{ m}.$$

REFLECT The "high C" sung by coloratura sopranos is two octaves above middle C (a factor of four in frequency), or about

1050 Hz. The corresponding wavelength is $(1.31 \text{ m})/4 = 0.327$ m.

Practice Problem: The frequency of the lowest note on large pipe organs is about 16 Hz. What is the corresponding wavelength? *Answer:* 21 m.

12.3 Wave Speeds

The speed of propagation of a wave depends on the mechanical properties of the medium. For transverse waves on a rope, these are the rope's tension and mass. Intuition suggests that increasing the tension should increase the restoring forces

ActivPhysics 10.2: Speed of Waves on a String

Longitudinal waves shown at intervals of $\frac{1}{8}T$ for one period T.

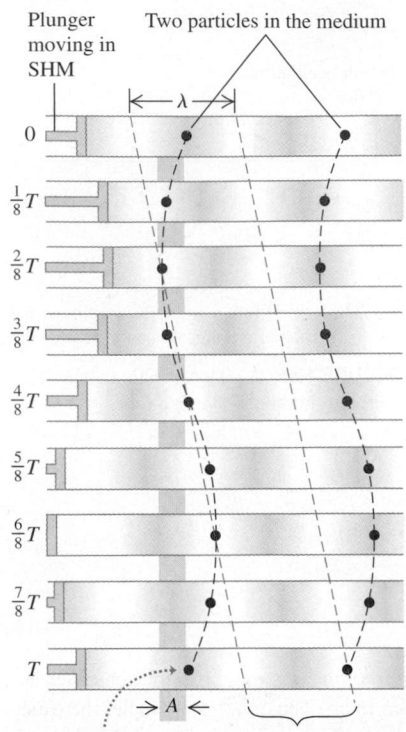

Particles oscillate with amplitude A. Waves advance by one wavelength each period.

▲ **FIGURE 12.6** A sinusoidal longitudinal wave shown at intervals of one-eighth of a period.

that tend to straighten the rope when it is disturbed and thus increase the wave speed. We might also guess that increasing the rope's mass should make the motion more sluggish and decrease the speed.

Both of these expectations are confirmed by more detailed analysis. The inertial property of the rope is best described not just by its total mass m (which depends on its length L), but on the mass *per unit length,* which we'll denote by μ. That is, $\mu = m/L$. We denote the tension in the rope by F_T. Analysis then shows that the wave speed v is given by the following expression:

Speed of a transverse wave

The speed of a transverse wave in a rope under tension is proportional to the square root of the tension, divided by the mass per unit length:

$$v = \sqrt{\frac{F_T}{\mu}}. \tag{12.2}$$

This result confirms our prediction that the wave speed v should increase when the tension F_T increases, but decrease when the mass per unit length μ increases.

NOTE ▶ The wave speed v is *directly* proportional, *not* to the tension F_T, but rather to the *square root* of F_T. Similarly, it is inversely proportional, not to the mass per unit length, μ, but to the square root of that quantity. ◀

Conceptual Analysis 12.3

Waves on a string

An oscillator creates periodic waves on a stretched rope. Which of the following parameters *cannot* be changed by altering the motion of the oscillator?

A. Wave frequency.
B. Wave speed.
C. Wavelength.

SOLUTION As we've just seen, the speed of a wave depends on the properties of the medium alone. Thus, at least choice B is correct. The frequency of the wave is determined only by the frequency of the oscillator, so choice A is *not* correct. What about wavelength (choice C)? Remember that wavelength equals wave speed divided by frequency: $\lambda = v/f$. Thus, wavelength depends partly on the medium and partly on the oscillator, so choice C, too, is incorrect.

Quantitative Analysis 12.4

A vibrating string

An oscillator creates periodic waves on a stretched string. If the oscillator frequency doubles, what happens to the wavelength and wave speed?

A. Wavelength doubles; wave speed is unchanged.
B. Wavelength is unchanged; wave speed doubles.
C. Wavelength is halved; wave speed is unchanged.

SOLUTION The speed of a wave depends only on the properties of the medium, so changing the frequency of the oscillator does not change the wave speed. Wavelength depends on frequency and wave speed: $\lambda = v/f$. As this equation shows, doubling the frequency will *halve* the wavelength, so C is the right answer. This conclusion makes intuitive sense: If you double the frequency without changing the wave speed, you must fit twice as many waves into a given length of string, so the waves must be half as long.

EXAMPLE 12.2 Mineral samples in a mine shaft

One end of a nylon rope is tied to a stationary support at the top of a vertical mine shaft that is 80.0 m deep (Figure 12.7). The rope is stretched taut by a box of mineral samples with mass 20.0 kg suspended from the lower end. The mass of the rope is 2.00 kg. The geologist at the bottom of the mine signals to his colleague at the top by jerking the rope sideways. What is the speed of a transverse wave on the rope? If a point on the rope is given a transverse simple harmonic motion with a frequency of 20 Hz, what is the wavelength of the wave?

SOLUTION

SET UP We can calculate the wave speed v from the tension F_T and the mass per unit length, $\mu = m/L$. The tension at the bottom of the rope is equal to the weight of the 20.0 kg load:

$$F_T = (20.0 \text{ kg})(9.80 \text{ m/s}^2)$$
$$= 196 \text{ N},$$

and the mass per unit length is

$$\mu = \frac{m}{L} = \frac{2.00 \text{ kg}}{80.0 \text{ m}} = 0.0250 \text{ kg/m}.$$

SOLVE The wave speed is given by Equation 12.2:

$$v = \sqrt{\frac{F_T}{\mu}} = \sqrt{\frac{196 \text{ N}}{0.0250 \text{ kg/m}}} = 88.5 \text{ m/s}.$$

From Equation 12.1,

$$\lambda = \frac{v}{f} = \frac{88.5 \text{ m/s}}{20.0 \text{ s}^{-1}} = 4.43 \text{ m}.$$

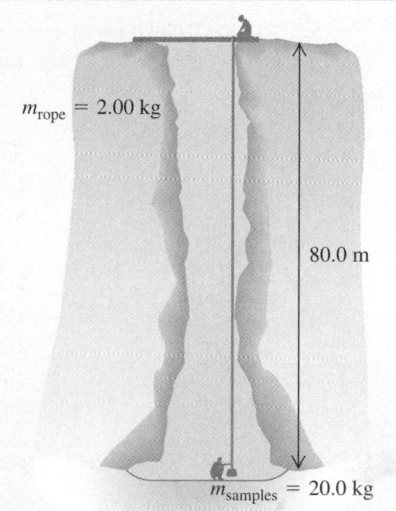

▶ **FIGURE 12.7**

REFLECT We have neglected the 10% increase in tension in the rope between bottom and top due to the rope's own weight. Can you verify that the wave speed at the top is 92.9 m/s?

Practice Problem: Another box of samples is hoisted by the same rope. If the rope is shaken with the same frequency as before, and the wavelength is found to be 8.85 m, what is the mass of this box of samples? *Answer:* 80.0 kg.

A transverse wave traveling along a rope carries *energy* from one region of the rope to another. If the wave is traveling from left to right, then the part of the rope to the left of a given point exerts a transverse force on the part to the right of the point as the point undergoes vertical displacements. Thus, the left part does *work* on the right part and transfers energy to it. A detailed calculation of the rate of doing work (power) shows that, for a sinusoidal wave with amplitude A and frequency f, the rate of energy transfer is proportional to the *square* of the amplitude and to the square of the frequency. (We'll return to this discussion later, in the context of sound waves.)

Longitudinal waves can occur in solid, liquid, and gaseous materials. In all cases, the wave speed is determined by the mechanical characteristics of the material. Ordinarily, the wave speed is proportional to the square root of an elastic factor (such as the bulk or shear modulus) and inversely proportional to the square root of the density of the material. Table 12.1 lists the speed of sound in several familiar materials. Notice that the speed of sound in a medium depends on temperature; in both air and water, the speed increases with increasing temperature.

12.4 Mathematical Description of a Wave

We've described several characteristics of periodic waves, using the concepts of wave speed, period, frequency, and wavelength. Sometimes, though, we need a more detailed description of the positions and motions of individual particles of the medium at particular times during wave propagation. As a specific example, let's look again at waves on a stretched string. If we ignore the sag of the string

TABLE 12.1 Speed of sound in various materials

Material	Speed of sound (m/s)
Gases	
Air (20°C)	344
(25°C)	347
(30°C)	350
Helium (20°C)	999
Hydrogen (20°C)	1330
Liquids	
Liquid helium (4 K)	211
Water (0°C)	1402
Water (100°C)	1543
Mercury (20°C)	1451
Solids	
Polystyrene	1840
Bone	3445
Brass	3480
Pyrex™ glass	5170
Steel	5000
Beryllium	12,870

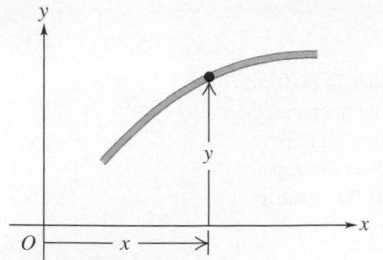

▲ **FIGURE 12.8** Coordinate system for a mathematical description of transverse waves on a string.

due to gravity, the equilibrium position of the string is along a straight line. We take this to be the *x* axis of a coordinate system. Waves on a string are *transverse:* During wave motion, a particle with equilibrium position *x* is displaced some distance *y* in the direction perpendicular to the *x* axis (Figure 12.8). The value of *y* depends on which particle we are talking about (i.e, on *x*) and also on the time *t* when we look at it. Thus, *y* is a *function* of *x* and *t*: $y = f(x, t)$. If we know this function for a particular wave motion, we can use it to find the displacement (from equilibrium) of any particle at any time.

Let's consider the application of the concept we've just described to a sinusoidal wave on a string. Suppose the left end of the string is at $x = 0$. We give this end a transverse displacement *y* that varies sinusoidally with time:

$$y = A\sin\omega t = A\sin 2\pi ft. \tag{12.3}$$

The resulting wave disturbance travels with speed *v* from $x = 0$ to all the points to the right of the origin (with positive values of *x*). Every such point undergoes simple harmonic motion with the same amplitude and frequency as the endpoint $(x = 0)$, but the motion of each point has a time delay x/v equal to the time required for a point on the wave shape to travel from $x = 0$ to the point *x*. Thus, the motion of point *x* at time *t* is the same as the motion of point $x = 0$ at the earlier time $t - x/v$. Therefore, we can find the displacement of any point *x* at any time *t* simply by replacing *t* in Equation 12.3 by $(t - x/v)$. We obtain

$$y(x, t) = A\sin\omega\left(t - \frac{x}{v}\right) = A\sin 2\pi f\left(t - \frac{x}{v}\right). \tag{12.4}$$

The notation $y(x, t)$ is a reminder that the displacement *y* is a function of both the equilibrium location *x* of the point and the time *t*. If we know the amplitude, frequency, and wave speed, we can use this equation to find the displacement *y* of *any* point on the string (i.e., any value of *x*) by substituting the appropriate values.

We can rewrite Equation 12.4 in several useful ways, presenting the same relationship in different forms. Here's one of the most useful forms expressing the wave properties in terms of the period $T = 1/f$ and the wavelength $\lambda = v/f$:

$$y(x, t) = A\sin 2\pi\left(\frac{t}{T} - \frac{x}{\lambda}\right). \tag{12.5}$$

At any specific instant of time *t*, Equation 12.4 or 12.5 gives the displacement *y* of a particle from its equilibrium position as a function of the *coordinate x* of the particle. If the wave is a transverse wave on a string, the equation represents the *shape* of the string at that instant, as if we had taken a stop-action photograph of the string. In particular, at time $t = 0$, Equation 12.5 becomes

$$y(x, 0) = A\sin 2\pi\left(0 - \frac{x}{\lambda}\right) = A\sin 2\pi\left(-\frac{x}{\lambda}\right).$$

This equation, which gives the shape of the string at time $t = 0$, is plotted in Figure 12.9a.

One period later, at time $t = T$, Equation 12.5 becomes

$$y(x, T) = A\sin 2\pi\left(1 - \frac{x}{\lambda}\right) = A\sin 2\pi\left(-\frac{x}{\lambda}\right).$$

(We've used the formula for the sine of the difference of two quantities.) This is the same expression as that for time $t = 0$, confirming the fact that the shape of the string varies periodically with period *T*, as we should expect. After a time *T* (and therefore also after 2*T*, 3*T*, and so on), the string instantaneously has the same shape that it had at time $t = 0$.

We can show that Equation 12.5 represents a periodic wave having a repeating pattern with wavelength λ. To do this, we replace *x* by $x + \lambda$ in Equation 12.5

If we use Equation 12.5 to plot *y* as a function of *x* for time $t = 0$, the curve shows the *shape* of the string at $t = 0$.

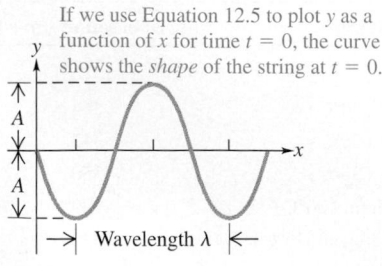

(a)

If we use Equation 12.5 to plot *y* as a function of *t* for position $x = 0$, the curve shows the *displacement y* of the particle at that coordinate as a function of time.

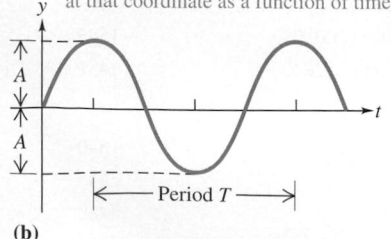

(b)

▲ **FIGURE 12.9** Two ways to graph Equation 12.5.

and again use the formula for the sine of the difference of two quantities. The result is that, at any time, point x has the same displacement as point $x + \lambda$, confirming that the wave shape is a repeating pattern with wavelength λ.

At any specific *coordinate x*, Equation 12.4 or Equation 12.5 gives the displacement y of the particle at that coordinate as a function of *time*. That is, it describes the motion of that particle. In particular, at the position $x = 0$,

$$y = A\sin\omega t = A\sin 2\pi\frac{t}{T}.$$

This equation is consistent with our original assumption that the motion at $x = 0$ is simple harmonic. The curve of the equation is plotted in Figure 12.9b; it is *not* a picture of the shape of the string, but rather a graph of the position y of a particle as a function of time.

12.5 Reflections and Superposition

In our study of mechanical waves thus far, we haven't been concerned with what happens when a wave arrives at an end or boundary of its medium. But there are many wave phenomena in which boundaries play a significant role. When you yell at a flat wall some distance away, the sound wave is reflected from the rigid surface, and an echo comes back. When you send a wave pulse down the length of a rope whose far end is tied to a rigid support, a reflected pulse comes back to you. In both cases, the initial and reflected waves overlap in the same region of the medium. Figure 12.10 is a multiflash photograph showing the reflection of a pulse at a stationary end of a stretched spring. Both the displacement and the direction of propagation of the reflected pattern are opposite to those of the initial pulse.

Figure 12.11 shows reflection of a wave pulse from an end that is held stationary and from one that is free to move transversely. In Figure 12.11a, a wave

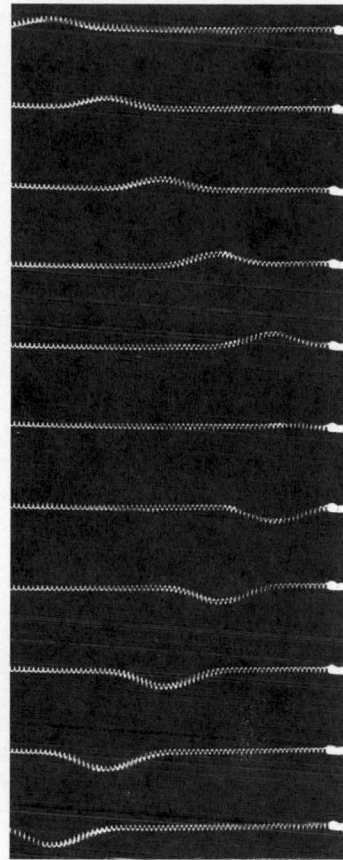

▲ **FIGURE 12.10** Successive strobe images of a stretched spring. The photos show a wave pulse traveling to the right, reflecting at the fixed end of the spring, and traveling back to the left.

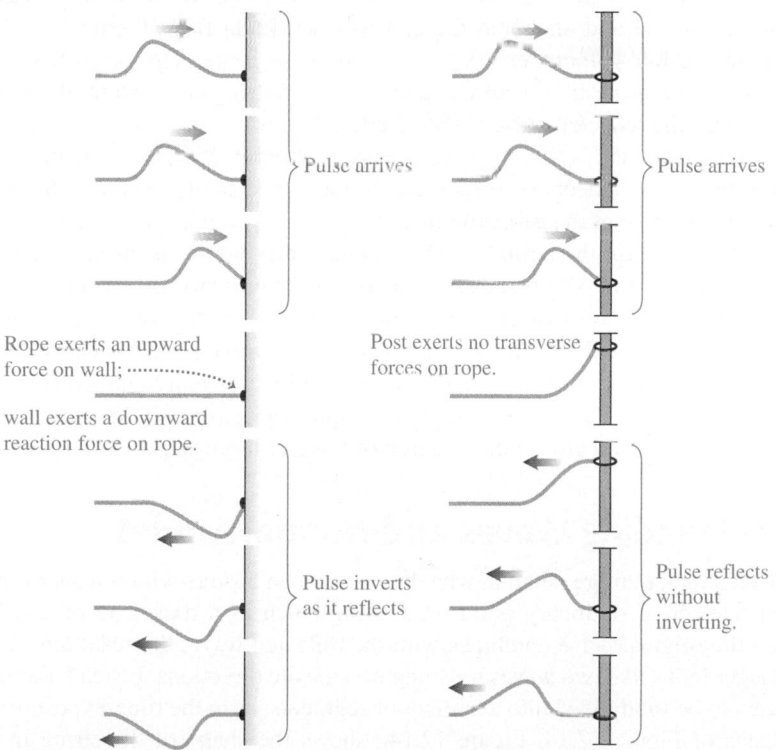

(a) Wave reflects from a fixed end   (b) Wave reflects from a free end

▲ **FIGURE 12.11** Reflection of a wave pulse (a) at a fixed end of a string and (b) at a free end.

As the pulses overlap, the displacement of the string at any point is the vector sum of the displacements due to the individual pulses.

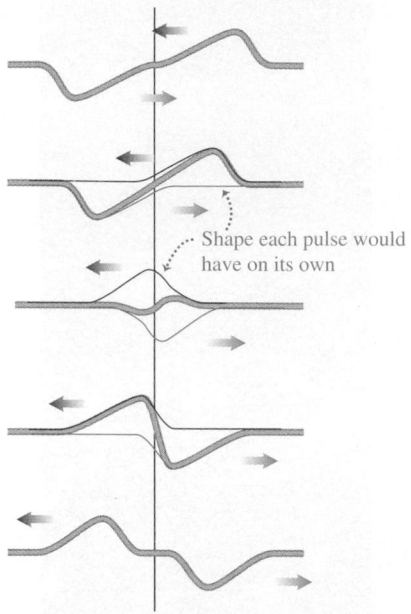

Shape each pulse would have on its own

O

▲ **FIGURE 12.12** An encounter between two waves traveling in opposite directions, one inverted with respect to the other.

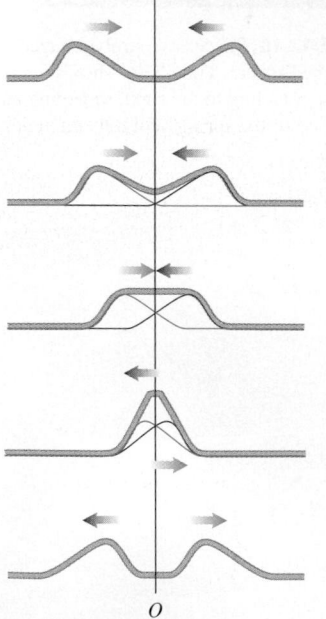

O

▲ **FIGURE 12.13** An encounter between two identical waves traveling in opposite directions.

Mastering**PHYSICS**

PhET: Fourier: Making Waves
PhET: Waves on a String
ActivPhysics 10.4: Standing Waves on Strings
ActivPhysics 10.5: Tuning a Stringed Instrument: Standing Waves
ActivPhysics 10.6: String Mass and Standing Waves

pulse arrives at an end of the rope that is fastened to a rigid support so that the end point of the rope can't move. The arriving wave exerts a force on the support; the reaction to this force, exerted *by* the support *on* the rope, "kicks back" on the rope and sets up a *reflected* pulse or wave traveling in the reverse direction, inverted with respect to the incoming pulse. We would see the same result for a sinusoidal wave arriving at a fixed end.

The opposite extreme (Figure 12.11b) is an end that is perfectly free to move in the direction transverse to the length of the rope. For example, the rope might be tied to a light ring that slides on a smooth rod perpendicular to the length of the rope. The ring and rod maintain the tension, but exert no transverse force. When a wave arrives at a free end, where there is no transverse force, the end "overshoots," and again a reflected wave is set up, this time *not* inverted. The conditions at the end of the rope, such as the presence of a rigid support or the complete absence of a transverse force, are called **boundary conditions.**

The formation of the reflected pulse is similar to the overlap of two pulses traveling in opposite directions. Figure 12.12 shows two pulses with the same shape, but one inverted with respect to the other, traveling in opposite directions. As the pulses overlap and pass each other, the total displacement of any point of the string is the *vector sum* of the displacements at that point in the individual pulses. At points along the vertical line in the middle of the figure, the total displacement is zero at all times. Thus, the motion of the left half of the string would be the same if we cut the string at point *O*, threw away the right side, and held the end at *O* stationary. The two pulses on the left side then correspond to the incident and reflected pulses, combining so that the total displacement at the end of the string is *always* zero. For this to occur, the reflected pulse must be inverted relative to the incident pulse.

Figure 12.13 shows two pulses with the same shape, traveling in opposite directions, but *not* inverted relative to each other. The displacement at the midpoint is not zero, but the slope of the string at that point is always zero. In this case, the motion of the left half of the string would be the same if we cut the string at point *O* and anchored the end with a sliding ring (Figure 12.11b) to maintain tension without exerting any transverse force. This situation corresponds to the reflection of a pulse at a free end of a string at the point in question. In this case, the reflected pulse is *not* inverted.

In each case, the actual displacement of any point on the rope is the vector sum of the displacements corresponding to the two separate pulses at that point. This is an example of the **principle of superposition,** which states that **whenever two waves overlap, the actual displacement of any point on the string, at any time, is obtained by vector addition of the following two displacements: the displacement the point would have if only the first wave were present and the displacement it would have with only the second wave present.** This principle holds for the overlap of sinusoidal waves as well as for individual wave pulses. Later in the chapter, we'll see that the principle applies to both longitudinal and transverse waves and also to the addition of three or more waves.

12.6 Standing Waves and Normal Modes

Now let's look in more detail at what happens when a *sinusoidal* transverse wave is reflected by a stationary point on a string (such as a fixed end of a string). When the original wave combines with the reflected wave, the resulting motion no longer looks like two waves traveling in opposite directions. Instead, the string appears to be subdivided into a number of segments, as in the time-exposure photographs of Figure 12.14. Figure 12.14e shows the shape of the string in Figure 12.14b at two different instants. Let's compare this behavior with that of the waves we studied earlier in the chapter. (We'll refer to those as *traveling waves,* because when there are no reflections, they travel in one direction through the

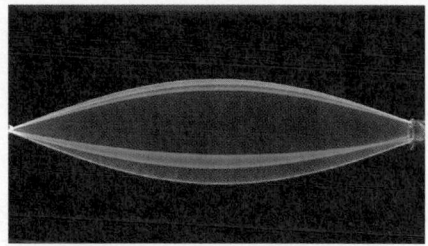

(a) String is one-half wavelength long.

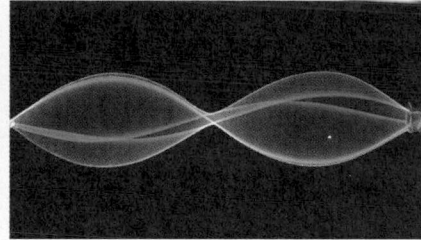

(b) String is one wavelength long.

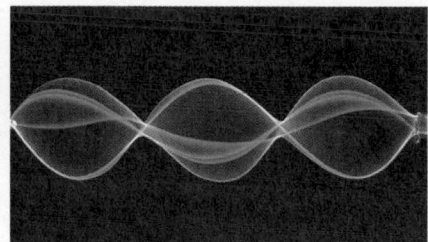

(c) String is one and a half wavelengths long.

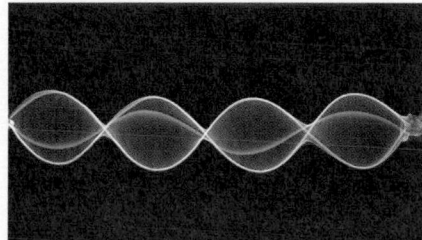

(d) String is two wavelengths long.

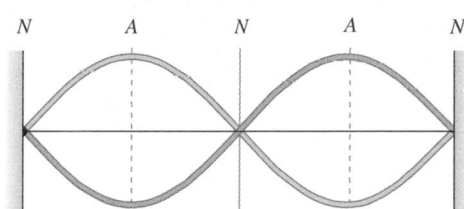

$N = $ **nodes:** points at which the string never moves.

$A = $ **antinodes:** points at which the amplitude of string motion is greatest.

(e) The shape of the string in **(b)** at two different instants.

▲ **FIGURE 12.14** (a–d) Standing waves in a stretched string (time exposure). (e) A standing wave, with nodes at the center and at the ends.

medium.) In a traveling wave, the amplitude is constant and the waveform moves with a speed equal to the wave speed. In the kind of wave we are examining here, the waveform remains in the same position along the string, and its amplitude fluctuates. There are particular points called **nodes** (labeled N in Figure 12.14c) that never move at all. Midway between the nodes are points called **antinodes** (labeled A in the figure), where the amplitude of motion is greatest. Because the wave pattern doesn't appear to be moving in either direction along the string, it is called a **standing wave.**

A standing wave, unlike a traveling wave, *does not* transfer energy from one end to the other. The two waves that form a standing wave would individually carry equal amounts of power in opposite directions. There is a local flow of energy from each node to the adjacent antinodes and back, but the *average* rate of energy transfer is zero at every point.

Figure 12.15 shows how we can use the superposition principle to understand the formation of a sinusoidal standing wave. The figure shows the waveform at instants one-sixteenth of a period apart. The thin red curves show a wave traveling to the left. The thin blue curves show a wave with the same propagation speed, wavelength, and amplitude, traveling to the right. (It could be a reflection of the red wave.) The brown curves show the resultant waveform—the actual shape a string would have at each given instant—obtained by applying the principle of superposition. At each point, the actual displacement of the string (the position of the brown curve) represents the vector sum of the displacements due to the two separate waves (i.e., the values of y for the two separate waves). The resultant (total) displacement at each node (marked N) is always zero. Midway between the nodes, the antinodes (marked A) have the greatest amplitude of displacement. We can see from the figure that **the distance between successive nodes or between successive antinodes is one half-wavelength, or $\lambda/2$.**

Normal Modes

We've mentioned that an echo of a sound wave can be produced by reflection at a rigid wall. Now suppose we have two parallel walls. If we stand between the walls and produce a sharp sound pulse (for example, by clapping our hands), the result is a series of regularly spaced echoes caused by repeated back-and-forth

▲ **Application Just passing through.** These two sets of ripples on water demonstrate the principle of superposition of mechanical waves. Notice how the two set of waves pass through each other without changing each other's overall shape or speed. At each point of intersection of individual waves, the vertical motion of the water is the vector sum of the motions of the two waves. Where the waves intersect exactly in step, they reinforce each other, resulting in a crest or trough height that is the sum of the heights of the individual waves. Where they intersect exactly out ! of step, they cancel each other and there is no vertical displacement of the water. This principle also holds true for a series of waves that are reflected from a boundary and return to interact with themselves.

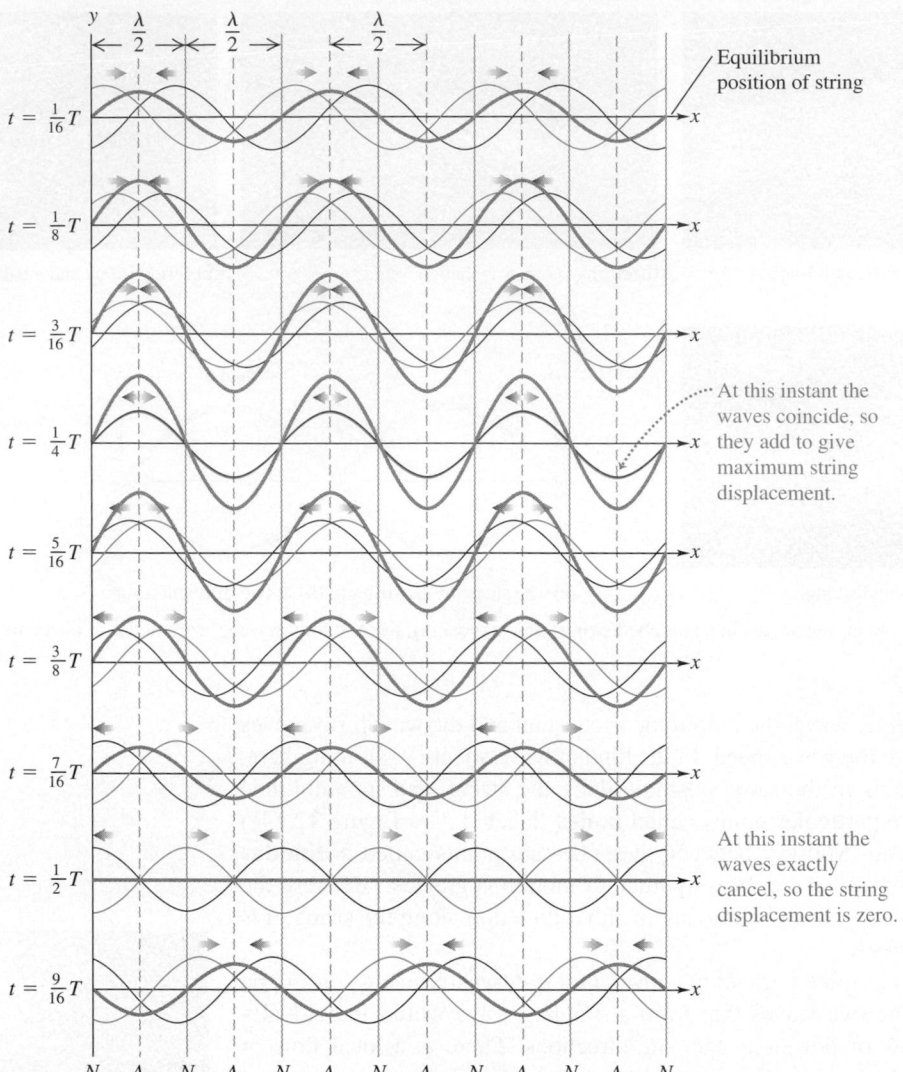

At this instant the waves coincide, so they add to give maximum string displacement.

At this instant the waves exactly cancel, so the string displacement is zero.

▲ **FIGURE 12.15** Formation of a standing wave. A wave traveling to the left (red) combines with an identical wave traveling to the right (blue) to form a standing wave on a string (brown). We see successive instants at intervals of $1/16$ of a period.

reflection between the walls. In room acoustics, this phenomenon is called "flutter echo"; it is the bane of acoustical engineers.

The analogous situation with transverse waves on a string is a string with some definite length L, rigidly held at *both* ends. If we produce a sinusoidal wave on a guitar string, the wave is reflected and re-reflected from the ends. The waves combine to form a standing wave like one of the waves shown in Figure 12.14. Both ends must be nodes, and adjacent nodes are one half-wavelength $(\lambda/2)$ apart. The length L of the string, therefore, has to be some integer multiple of $\lambda/2$ (i.e., $(\lambda/2), 2(\lambda/2), 3(\lambda/2), \dots$), that is, an integer number of half-wavelengths:

$$L = n\frac{\lambda}{2} \qquad (n = 1, 2, 3, \dots).$$

A standing wave can exist on a string with length L, held stationary at both ends, only when its wavelength satisfies the preceding equation. When we solve the equation for λ and label the possible values of λ as λ_n, we obtain the following expression:

Wavelengths for standing waves on a string

On a string that is fixed at both ends, the only possible standing waves have wavelengths λ_n that are related to the string length L by the formula

$$\lambda_n = \frac{2L}{n} \qquad (n = 1, 2, 3, \dots). \qquad (12.6)$$

When the wavelength is *not* equal to one of these values, no standing wave is possible.

Corresponding to this series of possible wavelengths is a series of possible frequencies f_n, each related to its corresponding wavelength by $f_n = v/\lambda_n$. The smallest frequency, f_1, corresponds to the largest wavelength (when $n = 1$), $\lambda = 2L$:

$$f_1 = \frac{v}{2L} \qquad \text{(string, held at both ends)}.$$

This frequency is called the **fundamental frequency.** The other frequencies are $f_2 = 2v/2L$, $f_3 = 3v/2L$, and so on. These are all integer multiples of f_1, such as $2f_1$, $3f_1$, $4f_1$, and so on, and we can express *all* the frequencies as follows:

Frequencies for standing waves on a string

On a string of length L that is held stationary at both ends, standing waves can have only frequencies f_n that are related to L and the wave speed v by the formula

$$f_n = n\frac{v}{2L} = nf_1, \qquad (n = 1, 2, 3, \dots). \qquad (12.7)$$

These frequencies, all integer multiples of f_1, are called **harmonics,** and the series is called a **harmonic series.** Musicians sometimes call f_2, f_3, and so on **overtones;** f_2 is the second harmonic or the first overtone, f_3 is the third harmonic or the second overtone, and so on. (You can think of overtones as the first, second, third, etc., "tones over" the fundamental frequency.)

A **normal mode** is a motion in which all particles of the string move sinusoidally (i.e., with simple harmonic motion) with the same frequency. Each of the frequencies given by Equation 12.7 corresponds to a possible normal-mode pattern. There are infinitely many normal modes, each with its characteristic frequency and vibration pattern. Figure 12.16 shows the four lowest-frequency

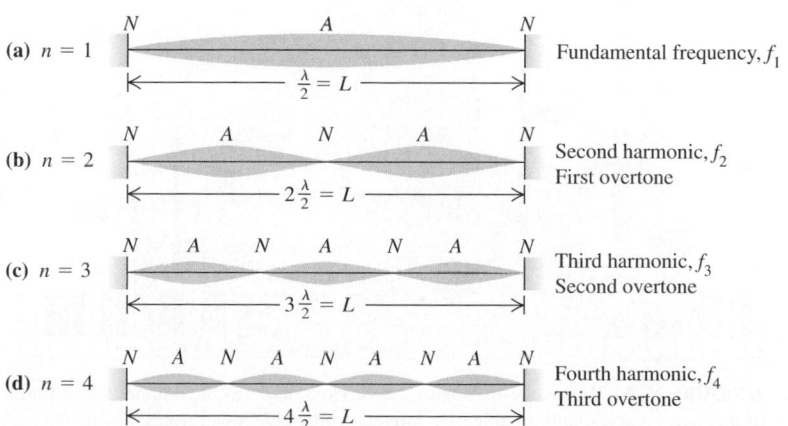

(a) $n = 1$ — N, A, N — Fundamental frequency, f_1 — $\frac{\lambda}{2} = L$

(b) $n = 2$ — N, A, N, A, N — Second harmonic, f_2 / First overtone — $2\frac{\lambda}{2} = L$

(c) $n = 3$ — N, A, N, A, N, A, N — Third harmonic, f_3 / Second overtone — $3\frac{\lambda}{2} = L$

(d) $n = 4$ — N, A, N, A, N, A, N, A, N — Fourth harmonic, f_4 / Third overtone — $4\frac{\lambda}{2} = L$

◀ **FIGURE 12.16** The first four normal modes of a string fixed at both ends.

normal-mode patterns and their associated frequencies and wavelengths. We can contrast this situation with a simpler vibrating system, the harmonic oscillator, which has only one normal mode and one characteristic frequency.

If we displace a string so that its shape is the same as one of the normal-mode patterns and then release it, it vibrates with the frequency of that mode. But when a piano string is struck or a guitar string is plucked, not only is the fundamental frequency present in the resulting vibration, but also many overtones. This motion is therefore a combination, or *superposition,* of many normal modes. Several frequencies and motions are present simultaneously, and the displacement of any point on the string is the vector sum (or superposition) of displacements associated with the individual modes.

As we've seen, the fundamental frequency of a vibrating string is $f_1 = v/2L$. The wave speed v is determined by Equation 12.2, $v = \sqrt{F_T/\mu}$. Combining these equations, we obtain the following expression:

Fundamental frequency of a vibrating string

The fundamental frequency f_1 of a vibrating string with length L, held at both ends, is

$$f_1 = \frac{1}{2L}\sqrt{\frac{F_T}{\mu}}, \tag{12.8}$$

where the string tension is F_T and the mass per unit length is μ.

All string instruments are "tuned" to the correct frequencies by varying the tension F_T. An increase in tension increases the wave speed v and thus also increases the frequency (and the pitch) for a string of any fixed length L. The inverse dependence of frequency on length L is illustrated by the long strings of the bass (low-frequency) section of the piano or the string bass, compared with the shorter strings of the piano treble or the violin (Figure 12.17). One reason for winding the bass strings of a piano with wire is to increase the mass per unit length μ so as to obtain the desired low frequency without resorting to a string that is inconveniently long or inflexible. In playing the violin or guitar, the usual means of varying the pitch is to press the strings against the fingerboard with the fingers to change the length L of the vibrating portion of the string. As Equation 12.8 shows, decreasing L increases f_1.

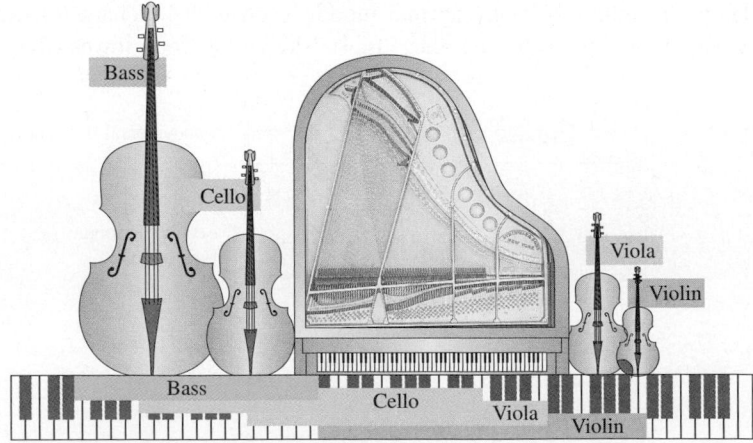

▲ **FIGURE 12.17** The ranges of a bass, cello, viola, and violin compared with that of a concert grand piano. In all cases, longer strings give lower notes.

PROBLEM-SOLVING STRATEGY 12.1 **Standing waves and normal modes**

In all wave problems, it's useful to distinguish between the purely kinematic quantities, such as the wave speed v, wavelength λ, and frequency f, and the dynamic quantities involving the properties of the medium, including F_T and μ. You can compute the wave speed if you know either λ and f or F_T and μ. Try to determine whether the properties of the medium are involved in the problem at hand or whether the problem is only kinematic in nature.

To visualize nodes and antinodes in standing waves, it's always helpful to draw a diagram. For a string, you can draw the shape at one instant and label the nodes N and the antinodes A. For longitudinal waves (discussed in the next section), it's not so easy to draw the shape, but you can still label the nodes and antinodes. The distance between two adjacent nodes or two adjacent antinodes is always $\lambda/2$, and the distance between a node and the adjacent antinode is always $\lambda/4$. For a string held at both ends, the fundamental mode has nodes only at the ends and one antinode in the middle.

Quantitative Analysis 12.5 **Changing the fundamental frequency**

In the apparatus shown in Figure 12.18, the wire is tensioned by the weight of the hanging block. If the vibrating portion of the wire has a fundamental frequency f when the block has mass m, what should the block's mass be in order to give a fundamental frequency of $2f$?

A. $\sqrt{2}m$
B. $2m$
C. $4m$

SOLUTION The frequency is related to the tension in the wire by Equation 12.8:

$$f_1 = \frac{1}{2L}\sqrt{\frac{F_T}{\mu}}.$$

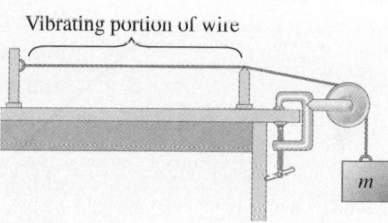
Vibrating portion of wire

▲ FIGURE 12.18

This equation says that the fundamental frequency is directly proportional to the square root of the tension F_T in the wire. We can assume that the tension is equal to the block's weight and thus is directly proportional to its mass. Thus, f_1 is directly proportional to the square root of the block's mass. In that case, to double the frequency, you must increase the mass by a factor of four—so C is the correct answer.

EXAMPLE 12.3 **A giant bass viol**

In an effort to get your name in the *Guinness Book of World Records,* you set out to build a bass viol with strings that have a length of 5.00 m between fixed points. One of the strings has a mass per unit length of 40.0 g/m and a fundamental frequency of 20.0 Hz. **(a)** Calculate the tension in this string. **(b)** Calculate the frequency and wavelength of the second harmonic. **(c)** Calculate the frequency and wavelength of the second overtone.

SOLUTION

SET UP Figure 12.19 shows our diagram. Remember that the second overtone is the third harmonic. (Refer to Figure 12.16 if you're confused.)

SOLVE Part (a): To find the tension F_T in the string, we first solve Equation 12.8 for F_T:

$$F_T = 4\mu L^2 f_1^2.$$

We are given that $\mu = 4.00 \times 10^{-2}$ kg/m, $L = 5.00$ m, and $f_1 = 20.0$ s^{-1}. Substituting these values, we find that

$$F_T = 4(4.00 \times 10^{-2}\,\text{kg/m})(5.00\,\text{m})^2(20.0\,\text{s}^{-1})^2 = 1600\,\text{N}.$$

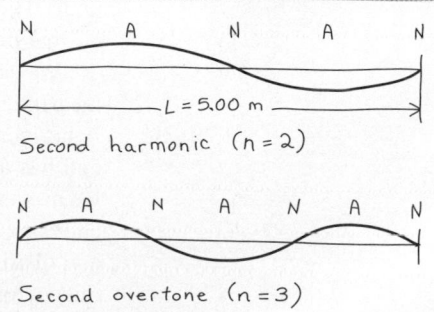

Second harmonic $(n=2)$

Second overtone $(n=3)$

▲ FIGURE 12.19 Our sketch for this problem.

Continued

(In a real bass viol, the tension in each string is typically a few hundred newtons.)

Part (b): From Equation 12.7, the second harmonic frequency $(n = 2)$ is

$$f_2 = 2f_1 = 2(20.0 \text{ Hz}) = 40.0 \text{ Hz}.$$

From Equation 12.6, the wavelength of the second harmonic is

$$\lambda_2 = \frac{2L}{2} = 5.00 \text{ m}.$$

Part (c): The second overtone is the "second tone over" (above) the fundamental; that is, $n = 3$, and

$$f_3 = 3f_1 = 3(20.0 \text{ Hz}) = 60.0 \text{ Hz}.$$

The wavelength is

$$\lambda_3 = \frac{2L}{3} = 3.33 \text{ m}.$$

REFLECT The 20.0 Hz fundamental frequency for this string is approximately the frequency of the E below the lowest C on the piano and is roughly at the threshold for human hearing. The second and third harmonics have higher frequencies and would thus be easier to hear.

Practice Problem: With the same values of μ and F_T, what length of string would be needed for a fundamental frequency of 33.0 Hz (the lowest C on the piano)? *Answer:* 3.03 m.

12.7 Longitudinal Standing Waves

When longitudinal waves propagate in a fluid (such as air or water) contained in a pipe with finite length, the waves are reflected from the ends in the same way that transverse waves on a string are reflected at its ends. The superposition of the waves traveling in opposite directions again forms a standing wave.

When reflection takes place at a *closed* end (an end with a rigid barrier or plug), the displacement of the particles at this end must always be zero. Just as with a stationary end of a string, there is no displacement at the end, and the end is a *node*. If the pipe is closed at both ends, then each end is a node; there may also be nodes at intermediate positions within the pipe. If an end of a pipe is *open* and the pipe is narrow in comparison with the wavelength (which is true for most wind instruments, including some organ pipes), the open end is an *antinode*—a point at which the displacement of the medium is maximal. (If this doesn't seem right, recall that a free end of a stretched string, like that in Figure 12.11b, is also a displacement antinode.) Thus, longitudinal waves in a column of fluid are reflected at the closed and open ends of a pipe in the same way that transverse waves in a string are reflected at stationary and free ends, respectively. A closed end is always a node, and an open end is always an antinode.

A standing longitudinal wave in a pipe always has alternating nodes and antinodes. The pressure variations in the pipe also have an alternating pattern. It turns out that the antinodes (points of maximum displacement) correspond to points where there is *no* pressure variation. We can call these *pressure nodes*. To distinguish the two kinds of nodes, we sometimes speak of *displacement nodes* and *pressure nodes,* and similarly for the antinodes. A displacement node is always a pressure antinode, and a displacement antinode is always a pressure node. The relationship will become clearer in the next section, where we discuss normal modes for longitudinal waves.

We can demonstrate longitudinal standing waves in a column of gas, and also measure the wave speed, with an apparatus called a Kundt's tube (Figure 12.20). A horizontal glass tube a meter or so long is closed at one end and at the other end has a flexible diaphragm that can transmit vibrations. As our sound source, we use a small loudspeaker, driven by a sinusoidal oscillator whose frequency we can vary. We place a small amount of light powder or cork dust inside the pipe and distribute it uniformly along the bottom side of the pipe. As we vary the frequency of the sound, we pass through frequencies at which the amplitude of the standing waves becomes large enough for the moving gas to sweep the cork dust along the pipe at all points where the gas is in motion. The powder therefore

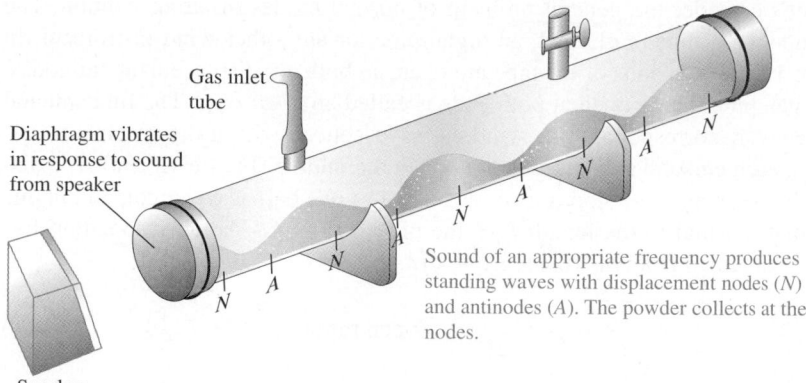

Gas inlet tube

Diaphragm vibrates in response to sound from speaker

Speaker

Sound of an appropriate frequency produces standing waves with displacement nodes (*N*) and antinodes (*A*). The powder collects at the nodes.

▲ **FIGURE 12.20** A Kundt's tube used to demonstrate standing longitudinal waves in a gas. The dust collects at the displacement nodes, where the displacement of gas due to the standing wave is minimal.

collects at the displacement nodes (where the gas is not moving). Adjacent nodes are separated by a distance equal to $\lambda/2$, and we can measure this distance. We read the frequency f from the oscillator dial, and we can then calculate the speed v of the waves from the relation $v = \lambda f$.

EXAMPLE 12.4 **Speed of sound in hydrogen**

At a frequency of 25 kHz, the distance from a closed end of a tube of hydrogen gas to the nearest displacement node of a standing wave is 0.026 m. Calculate the wave speed.

SOLUTION

SET UP AND SOLVE Figure 12.21 shows our sketch. This is a common way to represent a longitudinal standing wave in a tube. While the curves *look* like the real shape of a transverse standing wave on a string, for a longitudinal wave they don't show directly what the medium is doing. Rather, you can think of them as a graph of displacement (on the vertical axis) versus position (on the horizontal axis). As our sketch shows, the closed end of the tube is a displacement node. The statement of the problem tells us that the next displacement node is 0.026 m away. Adjacent nodes are always a half-wavelength apart, so we have $\lambda/2 = 0.026$ m and $\lambda = 0.052$ m. Then the wave speed is

$$v = \lambda f = (0.052 \text{ m})(25{,}000 \text{ s}^{-1}) = 1300 \text{ m/s}.$$

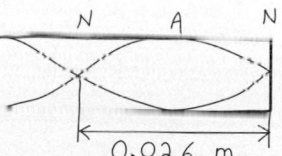

▲ **FIGURE 12.21** Our sketch for this problem.

REFLECT This result is close to the value given in Table 12.1 for the speed of sound in hydrogen at 20°C (1330 m/s). Either value is about four times the speed of sound in air at 20°C.

Practice Problem: If the hydrogen in the tube is replaced by air at 20°C, what frequency would be needed to produce standing waves with the same wavelength? *Answer:* 6600 Hz.

Normal Modes

Many musical instruments, including organ pipes and all the woodwind and brass instruments, use longitudinal standing waves (normal modes) in vibrating air columns to produce musical tones. Organ pipes are one of the simplest examples. Air is supplied by a blower, at a pressure typically on the order of 10^3 Pa, or 10^{-2} atm (above normal atmospheric pressure), to the bottom end of the pipe (Figure 12.22). A stream of air is directed against the mouth of each pipe, as shown in the figure. The column of air in the pipe is set into vibration, and there is a series of possible normal modes, just as with a stretched string. The mouth always acts as an open end; thus, it is a displacement antinode. The top of a pipe may be either open or closed. Figure 12.23 shows three normal modes in a schematic organ pipe open at the top end. (The pipes are shown on their sides.)

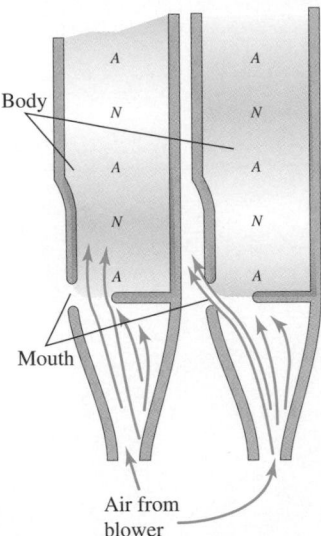

Vibrations from turbulent airflow set up standing waves in the pipe.

Body

Mouth

Air from blower

▲ **FIGURE 12.22** A cross section of an organ pipe. Vibrations from the turbulent air flow set up standing waves in the pipe.

Let's consider the general problem of normal modes in an air column. The pipe may be a flute, a clarinet, an organ pipe, or any other wind instrument. In Figure 12.23, both ends of the pipe are open, so both are displacement antinodes. An organ pipe that is open at both ends is called an *open pipe*. The fundamental frequency f_1 corresponds to a standing-wave pattern with a displacement antinode at each end and a displacement node in the middle (Figure 12.23a). The distance between adjacent nodes is always equal to one half-wavelength, and in this case that is equal to the length L of the pipe: $\lambda/2 = L$. The corresponding frequency, obtained from the relation $f = v/\lambda$, is

$$f_1 = \frac{v}{2L} \qquad \text{(open pipe).} \qquad (12.9)$$

The other two parts of Figure 12.23 show the second and third harmonics (first and second overtones); their vibration patterns have two and three displacement nodes, respectively. For these patterns, the length L of the pipe is equal to two or three half-wavelengths, so $\lambda/2$ is equal to $L/2$ and $L/3$, respectively, and the frequencies are twice and three times the fundamental frequency, respectively (i.e., $f_2 = 2f_1$ and $f_3 = 3f_1$). For *every* normal mode, the length L must be an integer number of half-wavelengths, and the possible wavelengths λ_n are given by

$$L = n\frac{\lambda_n}{2}, \qquad \text{or} \qquad \lambda_n = \frac{2L}{n} \qquad (n = 1, 2, 3, \dots).$$

The corresponding frequencies f_n are given by $f_n = v/\lambda_n$, so all the normal-mode frequencies for a pipe that is open at both ends are given by

$$f_n = n\frac{v}{2L} = nf_1 \qquad (n = 1, 2, 3, \dots). \qquad \text{(open pipe).} \qquad (12.10)$$

The value $n = 1$ corresponds to the fundamental frequency, $f_1 = v/2L$, $n = 2$ to the second harmonic (or first overtone), and so on.

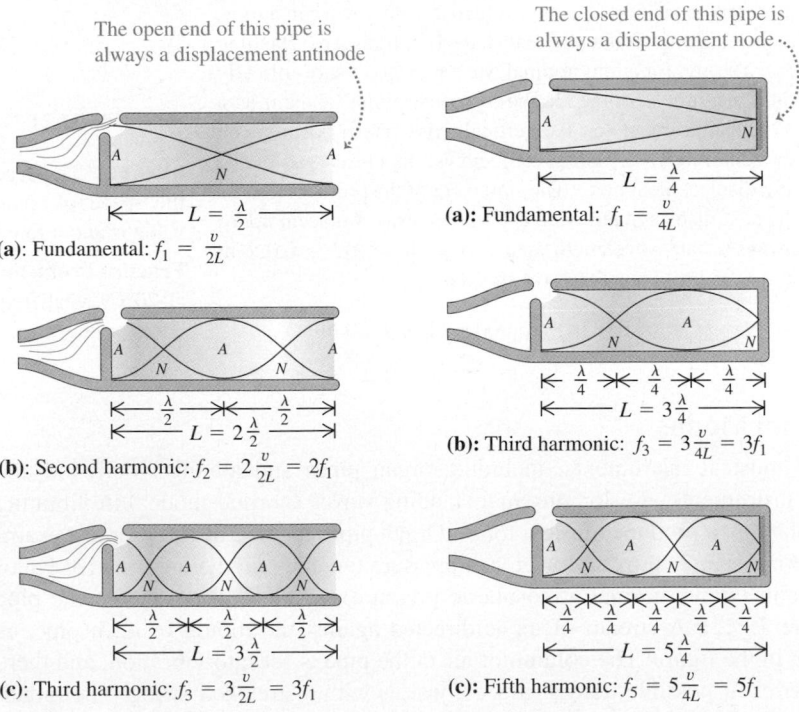

The open end of this pipe is always a displacement antinode

(a): Fundamental: $f_1 = \frac{v}{2L}$ $L = \frac{\lambda}{2}$

(b): Second harmonic: $f_2 = 2\frac{v}{2L} = 2f_1$ $\frac{\lambda}{2}$ $\frac{\lambda}{2}$ $L = 2\frac{\lambda}{2}$

(c): Third harmonic: $f_3 = 3\frac{v}{2L} = 3f_1$ $\frac{\lambda}{2}$ $\frac{\lambda}{2}$ $\frac{\lambda}{2}$ $L = 3\frac{\lambda}{2}$

The closed end of this pipe is always a displacement node

(a): Fundamental: $f_1 = \frac{v}{4L}$ $L = \frac{\lambda}{4}$

(b): Third harmonic: $f_3 = 3\frac{v}{4L} = 3f_1$ $\frac{\lambda}{4}$ $\frac{\lambda}{4}$ $\frac{\lambda}{4}$ $L = 3\frac{\lambda}{4}$

(c): Fifth harmonic: $f_5 = 5\frac{v}{4L} = 5f_1$ $\frac{\lambda}{4}$ $\frac{\lambda}{4}$ $\frac{\lambda}{4}$ $\frac{\lambda}{4}$ $\frac{\lambda}{4}$ $L = 5\frac{\lambda}{4}$

▲ **FIGURE 12.23** A cross section of an open pipe, showing the first three normal modes as well as the displacement nodes and antinodes.

▲ **FIGURE 12.24** A cross section of a stopped pipe, showing the first three normal modes as well as the displacement nodes and antinodes. Only odd harmonics are possible.

Figure 12.24 shows a pipe open at the left (mouth) end but closed at the right end. We call this a *stopped pipe.* The left (open) end is an antinode, but the right (closed) end is a node. The distance between a node and the adjacent antinode is always one quarter-wavelength. Figure 12.24a shows the fundamental frequency; the length of the pipe is a quarter-wavelength $(L = \lambda_1/4)$. The fundamental frequency is $f_1 = v/\lambda_1$, or

$$f_1 = \frac{v}{4L} \qquad \text{(stopped pipe).} \qquad (12.11)$$

This is one-half the fundamental frequency for an *open* pipe of the same length. In musical language, the *pitch* of a stopped pipe is one octave lower (a factor of two in frequency) than that of an open pipe of the same length. Figure 12.24b shows the next mode, for which the length of the pipe is *three-quarters* of a wavelength, corresponding to a frequency $3f_1$. In Figure 12.24c, $L = 5\lambda/4$ and the frequency is $5f_1$. The possible wavelengths are given by

$$L = n\frac{\lambda_n}{4} \qquad \text{or} \qquad \lambda_n = \frac{4L}{n} \qquad (n = 1, 3, 5, \dots).$$

The normal-mode frequencies are given by $f_n = v/\lambda_n$, or

$$f_n = n\frac{v}{4L} = nf_1 \qquad (n = 1, 3, 5, \dots), \qquad \text{(stopped pipe),} \quad (12.12)$$

with the fundamental frequency f_1 given by Equation 12.11. We see that the second, fourth, and, indeed, all *even* harmonics are missing. In a pipe that is closed at one end, the fundamental frequency is $f_1 = v/4L$, and only the odd harmonics in the series $(3f_1, 5f_1, 7f_1, \dots)$ are possible.

EXAMPLE 12.5 Harmonics of an organ pipe

On a day when the speed of sound is 345 m/s, the fundamental frequency of an open organ pipe is 690 Hz. If the second *harmonic* of this pipe has the same wavelength as the second *overtone* of a stopped pipe, what is the length of each pipe?

SOLUTION

SET UP Figure 12.25 shows our sketch. Note that the frequency of the second harmonic of the open pipe is twice its fundamental frequency, but the second overtone of the stopped pipe has a frequency five times that of the fundamental. Do you see why?

SOLVE For an open pipe, $f_1 = v/2L$, so the length of the open pipe is

$$L_{\text{open}} = \frac{v}{2f_1} = \frac{345 \text{ m/s}}{2(690 \text{ s}^{-1})} = 0.250 \text{ m.}$$

The second harmonic of the open pipe has a frequency of

$$f_2 = 2f_1 = 2(690 \text{ Hz}) = 1380 \text{ Hz.}$$

If the wavelengths are the same, the frequencies are the same, so the frequency of the second overtone of the stopped pipe is also 1380 Hz. The first overtone of a stopped pipe is at $3f_1$, and the second at $5f_1 = 5(v/4L)$. If this equals 1380 Hz, then

$$1380 \text{ Hz} = 5\left(\frac{345 \text{ m/s}}{4L_{\text{stopped}}}\right) \quad \text{and} \quad L_{\text{stopped}} = 0.313 \text{ m.}$$

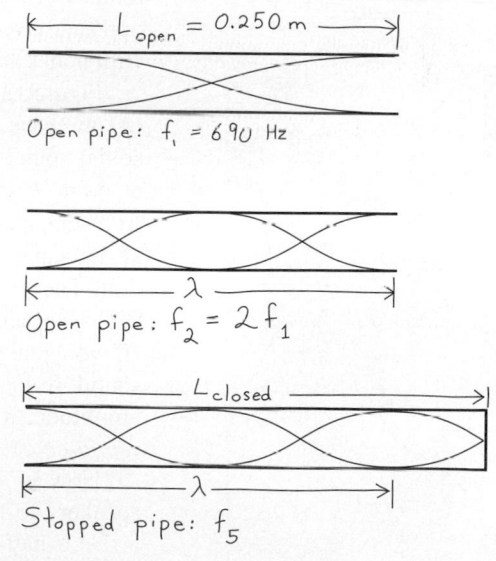

▲ **FIGURE 12.25** Our sketch for this problem.

REFLECT A final possibility is a pipe that is closed at *both* ends and therefore has nodes at both ends. This pipe wouldn't be of much use as a musical instrument, because there would be no way for the vibrations to get out of it.

Practice Problem: An open pipe and a stopped pipe have the same length. Does any harmonic of the open pipe have the same frequency as any harmonic of the stopped pipe? *Answer:* No.

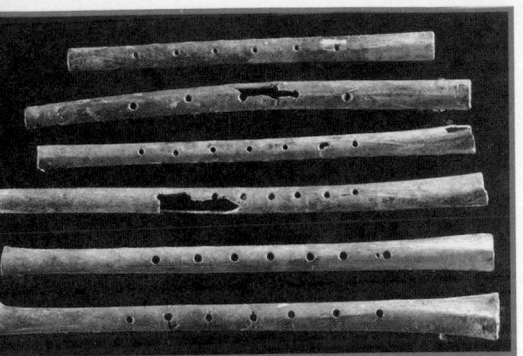

▲ **Application Boning up on physics.** You don't need a physicist's knowledge of standing waves to learn how to make a flute. This photo shows 9000-year-old bone flutes from Jiahu, China. The bone flute is one of the earliest known musical instruments. Blowing into the hollow bone causes a longitudinal wave to form in the internal air space. The wave produces a note with a fundamental frequency corresponding to a standing wave with a displacement antinode at each end of the flute. By placing their fingers over holes carved into the bone, these early musicians could change the effective length of the flute, thus varying the frequency of the waves produced and allowing them to play notes of different pitch.

In an actual organ pipe, several modes are always present at once. The situation is analogous to a string that is struck or plucked, producing several modes at the same time. In each case, the motion is a *superposition* of various modes. The extent to which modes higher than the fundamental mode are present depends on the cross section of the pipe, the proportion of the length to width, the shape of the mouth, and other, more subtle factors. The harmonic content of the tone is an important factor in determining its tone quality, or timbre. A very narrow pipe produces a tone that is rich in higher harmonics, which we hear as a thin and "stringy" tone; a fatter pipe produces mostly the fundamental mode, heard as a softer, more flutelike tone.

In this section we've talked mostly about organ pipes, but our discussion is also applicable to other wind instruments. The flute and the recorder are directly analogous. The most significant difference is that those instruments have holes along the pipe. Opening and closing the holes with the fingers changes the effective length L of the air column and thus changes the pitch. Any individual organ pipe, by comparison, plays only a single note. The flute and recorder behave as *open* pipes, but the clarinet acts as a *stopped* pipe (closed at the reed end, open at the bell).

Equations 12.9 and 12.11 show that the frequencies of any such instrument are proportional to the speed of sound v in the air column inside the instrument. In all gases (including air), v increases when the temperature increases. Thus, the pitch of all wind instruments rises with increasing temperature. An organ that has some of its pipes at one temperature and others at a different temperature is bound to sound out of tune.

12.8 Interference

Wave phenomena that occur when two or more periodic waves with the same frequency overlap in the same region of space are grouped under the heading of *interference*. As we have seen, standing waves are a simple example of an interference effect. Two waves with the same frequency and amplitude, traveling in opposite direction in a medium, combine to produce a standing-wave pattern, with nodes and antinodes that do not move.

Figure 12.26 shows an example of interference involving waves that spread out in space. Two speakers, driven by the same amplifier, emit identical sinusoidal sound waves with the same constant frequency. We place a microphone at point P in the figure, equidistant from the speakers. Wave crests emitted from the speakers at the same time travel equal distances and arrive at point P at the same time. The amplitudes add according to the principle of superposition. The total wave amplitude at P is twice the amplitude from each individual wave, and we can measure this combined amplitude with the microphone. (If you put your ear where the microphone is, the sound is *louder* than the sound from either speaker alone.) When interference causes an increase in amplitude, we call it **constructive interference,** and we say that the two waves *reinforce* each other.

Now let's move the microphone to point Q, where the distances from the two speakers to the microphone differ by a half-wavelength. Then the two waves arrive a half-cycle out of step; a positive crest from one speaker arrives at the same time as a negative crest from the other, and the net amplitude measured by the microphone at Q is much *smaller* than when only one speaker is present. When interference causes a decrease in amplitude, we call it **destructive interference** or *cancellation*.

Two speakers emit waves in step

Amplifier

$d_2 + \frac{\lambda}{2}$ d_2

d_1 d_1

Q

P

The path length from the speakers is the same; sounds from the two speakers arrive at P in step.

The path length from the speakers differs by $\frac{\lambda}{2}$; sounds from the two speakers arrive at Q out of step by $\frac{1}{2}$ cycle.

▲ **FIGURE 12.26** Two speakers driven by the same amplifier produce sound waves that interfere. The waves are in step as they emerge from the speakers; they interfere constructively at point P, destructively at point Q.

EXAMPLE 12.6 **Interference from two loudspeakers**

Two small loudspeakers A and B (Figure 12.27) are driven by the same amplifier and emit pure sinusoidal waves; crests from the speakers are emitted simultaneously ("in step"). If the speed of sound is 350 m/s, for what frequencies does maximum constructive interference occur at point P? For what frequencies does destructive interference (cancellation) occur at point P?

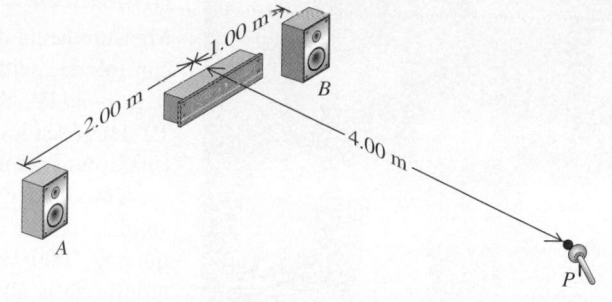

▶ **FIGURE 12.27**

SOLUTION

SET UP First we need to find the difference in path lengths from points A and B to point P. Maximum constructive interference occurs when the path difference d is an integer number of wavelengths; maximum destructive interference occurs when the path difference is a half-integer number of wavelengths. Once the wavelengths are known, the corresponding frequencies can be found from $f = v/\lambda$.

SOLVE The distance from speaker A to point P is $[(2.00\text{ m})^2 + (4.00\text{ m})^2]^{1/2} = 4.47$ m, and the distance from speaker B to point P is $[(1.00\text{ m})^2 + (4.00\text{ m})^2]^{1/2} = 4.12$ m. The path difference is $d = 4.47\text{ m} - 4.12\text{ m} = 0.35$ m.

Maximum constructive interference occurs when the path difference d is an integer multiple of λ: $d = 0, v/f, 2v/f, \ldots = nv/f$ $(n = 0, 1, 2, \ldots)$. So the possible frequencies for maximum constructive interference are

$$f_n = n\frac{v}{d} = n\frac{350\text{ m/s}}{0.35\text{ m}} \qquad (n = 1, 2, 3, \ldots)$$
$$= 1000\text{ Hz, }2000\text{ Hz, }3000\text{ Hz, }\ldots.$$

Maximum destructive interference (cancellation) occurs when the path difference for the two waves is a half-integer number of wavelengths:

$$d = \lambda/2, 3\lambda/2, 5\lambda/2, \ldots, \quad \text{or}$$
$$d = v/2f, 3v/2f, 5v/2f, \ldots.$$

The possible frequencies for destructive interference are

$$f_n = n\frac{v}{2d} = n\frac{350\text{ m/s}}{2(0.35\text{ m})} \qquad (n = 1, 3, 5, \ldots)$$
$$= 500\text{ Hz, }1500\text{ Hz, }2500\text{ Hz, }\ldots.$$

REFLECT As we increase the frequency, the sound at point P alternates between large and small amplitudes; the maxima and minima occur at the frequencies we have found. It would be hard to notice this effect in an ordinary room because of multiple reflections from the walls, floor, and ceiling. Such an experiment is best done in an anechoic chamber (Figure 12.28)—a room with walls that are almost completely nonreflecting for sound.

▲ **FIGURE 12.28** An anechoic chamber.

12.9 Sound and Hearing

Sound can travel in gases (such as air), liquids (such as water), and solid materials. The simplest sound waves are sinusoidal waves, which have definite frequency, amplitude, and wavelength. The human ear is sensitive to waves in the frequency range from about 20 to 20,000 Hz, but we also use the term *sound* for similar waves with frequencies above (**ultrasonic**) and below (**infrasonic**) the range of human hearing.

Sound waves may also be described in terms of variations of *pressure* at various points. In a sinusoidal sound wave in air, the pressure fluctuates above and below atmospheric pressure p_a with a sinusoidal variation having the same frequency as the motions of the air particles. The *maximum* variation in pressure above and below atmospheric pressure is called the *pressure amplitude,* denoted by p_{max}. Microphones and similar devices usually sense pressure variations, not displacements.

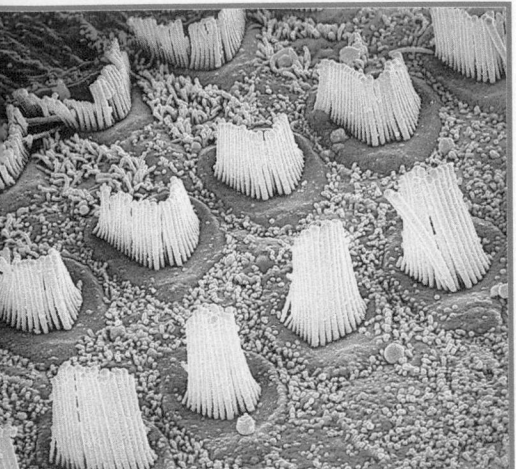

Amplitude

Measurements of sound waves show that, in the loudest sounds the human ear can tolerate without pain, the maximum pressure variations are on the order of $p_{max} = 30$ Pa above and below atmospheric pressure p_a (nominally, 1.013×10^5 Pa at sea level). If the frequency is 1000 Hz, the corresponding amplitude A (maximum displacement) is about $A = 0.012$ mm.

Thus, the displacement amplitude of even the loudest sound is *extremely* small. The maximum pressure variation in the *faintest* audible sound, of frequency 1000 Hz, is only about 3×10^{-5} Pa. The corresponding displacement amplitude is about 10^{-11} m. For comparison, the wavelength of yellow light is 6×10^{-7} m, and the diameter of a molecule is about 10^{-10} m. The ear is an extremely sensitive organ!

The Human Ear

Figure 12.29 shows the anatomy of the human ear. When a sound wave enters the auditory canal, it causes the eardrum to vibrate. The vibration is transmitted to the three small bones called the incus, the malleus, and the stapes. This motion is in turn transmitted to the fluid-filled *cochlea* of the inner ear. The motion of the cochlear fluid is then imparted to specialized hair cells in the cochlea, and the resulting nerve impulses are transmitted to the brain.

12.10 Sound Intensity

Like all other waves, sound waves transfer energy from one region to another. We define the **intensity** of a wave, denoted by I, as *the time-averaged rate at which energy is transported by the wave, per unit area,* across a surface perpendicular to the direction of propagation. That is, the intensity I is the average *power* transported per unit area. The intensity of the faintest sound wave that can be heard by a person with normal hearing is about 10^{-12} W/m^2, or 10^{-16} W/cm^2.

The average *total* power (energy per unit time) carried across a surface by a sound wave equals the product of the intensity (power per unit area) at the surface and the surface area (assuming that the intensity over the surface is uniform). The average total sound power emitted by a person speaking in an ordinary conversational tone is about 10^{-5} W, and a loud shout corresponds to about 3×10^{-2} W. If all the residents of New York City were to talk at the same time, the total sound power would be about 100 W, equivalent to the electric power requirement of a medium-sized light bulb! Yet the power required to fill a large auditorium with loud sound is considerable. The intensity at an individual point is proportional to the square of the amplitude of the sound wave and also proportional to the frequency of the wave.

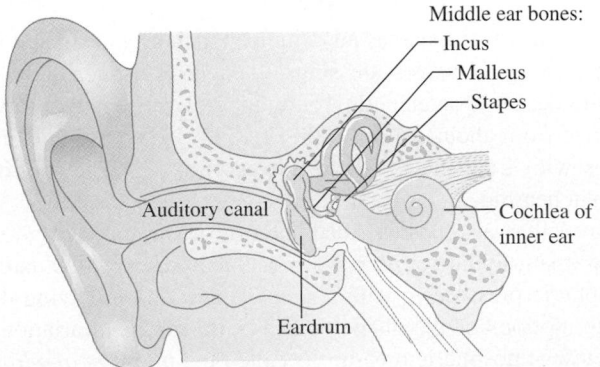

▲ **FIGURE 12.29** The structure of the human ear. The eardrum and the bones of the middle ear transmit sound waves into the cochlea, where the information is coded into nerve impulses.

EXAMPLE 12.7 **An auditorium sound system**

For a sound system for a 2000-seat auditorium, suppose we want the sound intensity over the surface of a hemisphere 20 m in radius to be 1.0 W/m². What acoustic power output would be needed from an array of speakers at the center of the sphere?

SOLUTION

SET UP AND SOLVE The total power needed is the intensity (power per unit area) times the area of the hemisphere. The area of the hemispherical surface is $\frac{1}{2}(4\pi R^2) = \frac{1}{2}(4\pi)(20\text{ m})^2$, or about 2500 m². The total acoustic power needed is

$$(1.0\text{ W/m}^2)(2500\text{ m}^2) = 2500\text{ W} = 2.5\text{ kW}.$$

REFLECT The efficiency of loudspeakers in converting electrical energy into sound is not very high (typically a few percent for ordinary speakers and up to 25% for horn-type speakers). The electrical power input to the speakers would need to be considerably larger than 2.5 kW.

However, 1.0 W/m² is *very* loud sound, at the threshold of pain and definitely in the range where permanent hearing damage is likely.

Practice Problem: In a smaller auditorium, we want the sound intensity over the surface of a hemisphere 10 m in radius to be 1.0 W/m². If the speakers have an efficiency of 20% (i.e., 20% of the electrical energy is converted into sound), what total electrical power input is required? *Answer:* 3.1 kW.

If a source of sound can be considered as a point, the intensity at a distance r from the source is inversely proportional to r^2. This relationship follows directly from energy conservation: If the power output of the source is P, then the average intensity I_1 through a sphere with radius r_1 and surface area $4\pi r_1^2$ is

$$I_1 = \frac{P}{4\pi r_1^2}.$$

The average intensity I_2 through a sphere with a different radius r_2 is given by a similar expression. If no energy is absorbed between the two spheres, the power P must be the same for both, and we have

$$4\pi r_1^2 I_1 = 4\pi r_2^2 I_2$$
$$\frac{I_1}{I_2} = \frac{r_2^2}{r_1^2}. \tag{12.13}$$

The intensity I at any distance r is therefore inversely proportional to r^2. This inverse-square relationship also holds for various other energy-flow situations involving a point source, such as light emitted by a point source.

Decibels

Because the ear is sensitive over such a broad range of intensities, a *logarithmic* intensity scale is usually used. The **intensity level** (or *sound level*) β of a sound wave is defined by the equation

$$\beta = (10\text{ dB})\log\frac{I}{I_0}. \tag{12.14}$$

In this equation, I_0 is a reference intensity chosen to be 10^{-12} W/m², and "log" denotes the logarithm to the base 10, not the natural logarithm (ln). The units of β are **decibels,** abbreviated dB. A sound wave with an intensity I equal to I_0 ($=10^{-12}$ W/m²) has an intensity level of 0 dB. This level corresponds roughly to the faintest sound that can be heard by a person with normal hearing, although that depends on the frequency of the sound. The intensity at the pain threshold, about 1 W/m², corresponds to an intensity level of 120 dB.

Table 12.2 gives the intensity levels in decibels of several familiar noises. It is taken from a survey made by the New York City Noise Abatement Commission.

TABLE 12.2 Noise levels due to various sources (representative values)

Type of noise	Sound level, dB	Intensity, W/m²
Rock concert	140	100
Threshold of pain	120	1
Riveter	95	3.2×10^{-3}
Elevated train	90	10^{-3}
Busy street traffic	70	10^{-5}
Ordinary conversation	65	3.2×10^{-6}
Quiet automobile	50	10^{-7}
Quiet radio in home	40	10^{-8}
Average whisper	20	10^{-10}
Rustle of leaves	10	10^{-11}
Threshold of hearing	0	10^{-12}

PROBLEM-SOLVING STRATEGY 12.2 Sound intensity

Quite a few quantities are involved in characterizing the amplitude and intensity of a sound wave, and it's easy to get lost in the maze of relationships. It helps to put them into categories: The amplitude is described by A or p_{max}; the frequency f is related to the wavelength λ and the wave speed v, which in turn is determined by the properties of the medium. Take a hard look at the problem at hand, identifying which of these quantities are given and which you have to find; then start looking for relationships that take you where you want to go.

In using Equation 12.14 for the sound intensity level, remember that I and I_0 must be in the same units, usually W/m². If they aren't, convert!

EXAMPLE 12.8 Temporary deafness

A 10-minute exposure to 120 dB sound typically shifts your threshold of hearing temporarily from 0 dB up to 28 dB. Studies have shown that, on average, 10 years of exposure to 92 dB sound causes a *permanent* shift of up to 28 dB. What intensities correspond to 28 dB and 92 dB?

SOLUTION

SET UP AND SOLVE We rearrange Equation 12.14, dividing both sides by 10 dB and then taking inverse logarithms of both sides:

$$I = I_0 10^{(\beta/10\,\mathrm{dB})}.$$

When $\beta = 28$ dB,

$$I = (10^{-12}\,\mathrm{W/m^2})10^{(28\,\mathrm{dB}/10\,\mathrm{dB})} = (10^{-12}\,\mathrm{W/m^2})(10^{2.8})$$
$$= 6.3 \times 10^{-10}\,\mathrm{W/m^2}.$$

When $\beta = 92$ dB,

$$I = (10^{-12}\,\mathrm{W/m^2})10^{(92\,\mathrm{dB}/10\,\mathrm{dB})} = (10^{-12}\,\mathrm{W/m^2})(10^{9.2})$$
$$= 1.6 \times 10^{-3}\,\mathrm{W/m^2}.$$

REFLECT If your answers are a factor of 10 too large, you may have entered 10×10^{-12} in your calculator instead of 1×10^{-12}. Be careful! Also, note that the human ear is sensitive to an extremely broad range of intensities.

EXAMPLE 12.9 A bird sings in a meadow

Consider an idealized model with a point source emitting a sound with constant power with an intensity inversely proportional to the square of the distance from the bird (Figure 12.30). By how many dB does the sound intensity level drop when you move to a point twice as far away from the bird?

Continued

SOLUTION

SET UP AND SOLVE We label the two points 1 and 2 (Figure 12.30), and we use Equation 12.14 twice. The difference in sound intensity level, $\beta_2 - \beta_1$, is given by

$$\beta_2 - \beta_1 = (10 \text{ dB})\left(\log\frac{I_2}{I_0} - \log\frac{I_1}{I_0}\right)$$
$$= (10 \text{ dB})[(\log I_2 - \log I_0) - (\log I_1 - \log I_0)]$$
$$= (10 \text{ dB})\log\frac{I_2}{I_1}.$$

Now we use Equation 12.13, inverted: $I_2/I_1 = r_1^2/r_2^2$. Then

$$\beta_2 - \beta_1 = (10 \text{ dB})\log\frac{r_1^2}{r_2^2} = (10 \text{ dB})\log\frac{r_1^2}{(2r_1)^2}$$
$$= (10 \text{ dB})\log\frac{1}{4} = -6.0 \text{ dB}.$$

A decrease in intensity of a factor of four corresponds to a 6 dB decrease in sound intensity level.

REFLECT The decibel scale is logarithmic; a factor of two in intensity corresponds to a 3 dB difference in sound intensity level,

▲ **FIGURE 12.30** When you double your distance from a point source of sound, how much does the sound intensity level drop?

a factor of eight to a 9 dB difference, and so on. In this problem, if two birds are singing, the sound intensity level increases by 3 dB.

Practice Problem: If the listener moves to a point four times the original distance from the bird, how many birds would have to sing in order to produce the same intensity level at this point as in the initial problem? *Answer:* 16 birds.

We invite you to prove that an increase in intensity of a factor of two corresponds to about a 3 dB increase in intensity level. This change is barely perceptible to the human ear. An increase of 8 to 10 dB in intensity level is usually interpreted by the ear as a doubling of loudness.

12.11 Beats

In several earlier sections, we've talked about *interference* effects that occur when two different waves with the same frequency overlap in the same region of space. Now let's look at the superposition of two waves with equal amplitude but slightly different frequencies. This phenomenon occurs, for example, when two guitar strings with slightly different frequencies are sounded together or when two organ pipes that are supposed to sound exactly the same pitch are slightly "out of tune." What you hear from two sound waves with slightly different frequencies is a slow pulsing or beating sound called **beats.**

Consider a particular point in space where the two waves overlap. The displacements of the individual waves at this point are plotted as functions of time in Figure 12.31a. The total length of the time axis represents about 1 s, and the frequencies are 16 Hz and 18 Hz. Applying the principle of superposition, we add the two displacements at each instant of time to find the total displacement at that time, obtaining the graph in Figure 12.31b. At certain times, the two waves are "in step": Their maxima coincide, and their amplitudes add. But as time goes on, they become more and more out of step because of their slightly different frequencies. Eventually, a maximum of one wave coincides with a maximum in the opposite direction for the other wave. The two waves then cancel each other, and the total amplitude is zero.

The resulting wave looks like a single sinusoidal wave with a varying amplitude that goes from a maximum to zero and back, as Figure 12.31b shows. In this

ActivPhysics 10.7: Beats and Beat Frequency

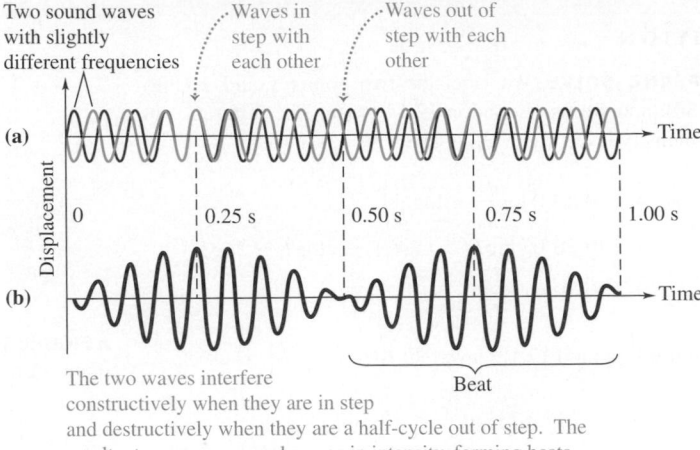

▲ FIGURE 12.31 When two waves with slightly different frequency interfere, the resultant wave fluctuates in amplitude. We hear these fluctuations as beats. (a) Two interacting waves; (b) the resultant wave formed by superposition of the two original waves.

▲ BIO Application Prey location by barn owls. Barn owls locate their prey by making use of the fact that each ear receives sounds at slightly different times if the owl's head is turned away from the source, due to the small difference in the distances traveled by the sound to reach each ear. This small difference is processed by the neural circuitry of the brain; the owl swivels its head until the waves reach both ears at the same time. At that null point, the head and eyes are facing directly toward the sound source. Owls are capable of detecting time differences in the sound waves received by each ear as small as 10 μs.

example, the amplitude goes through two maxima and two minima in 1 s; thus, the frequency of this amplitude variation is 2 Hz. The resulting variations in amplitude (loudness) are beats, and the frequency with which the amplitude varies is called the **beat frequency.** In this example the beat frequency is the *difference* of the two frequencies. If the beat frequency is a few hertz, we hear it as a waver or pulsation in the tone.

We can show that the beat frequency is *always* the difference of the two frequencies f_1 and f_2. Suppose f_1 is larger than f_2; then the corresponding periods are T_1 and T_2, with $T_1 < T_2$. If the two waves start out in step at time $t = 0$, they will be in step again at a time T_{beat} such that the first wave has gone through exactly one more cycle than the second. Let n be the number of cycles of the first wave in time T_{beat}; then the number of cycles of the second wave in the same time is $(n - 1)$, and we have the relations

$$T_{\text{beat}} = nT_1 \quad \text{and} \quad T_{\text{beat}} = (n - 1)T_2.$$

We solve the second equation for n and substitute the result back into the first equation (to eliminate n), obtaining

$$T_{\text{beat}} = \frac{T_1 T_2}{T_2 - T_1}.$$

Now, T_{beat} is just the *period* of the beat, and its reciprocal is the beat *frequency* $f_{\text{beat}} = 1/T_{\text{beat}}$. So

$$f_{\text{beat}} = \frac{T_2 - T_1}{T_1 T_2} = \frac{1}{T_1} - \frac{1}{T_2},$$

and finally,

$$f_{\text{beat}} = f_1 - f_2. \tag{12.15}$$

Beats between two tones can be heard up to a beat frequency of 6 or 7 Hz. Two piano strings or two organ pipes differing in frequency by 2 or 3 Hz sound wavery and "out of tune," although some organ stops contain two sets of pipes deliberately tuned to beat frequencies of about 1 to 2 Hz for a gently undulating effect. Listening for beats (or their absence) is an important technique in tuning all musical instruments.

At higher frequency differences, we no longer hear individual beats, and the sensation merges into one of *consonance* or *dissonance*, depending on the frequency ratio of the two tones. In some cases the ear perceives a tone called a *difference tone,* with a pitch equal to the beat frequency of the two tones.

In multiengine aircraft, the engines have to be synchronized so that the sounds don't cause annoying beats, which are heard as loud throbbing sounds. On some planes this is done electronically; on others the pilot does it by ear, just like tuning a piano.

12.12 The Doppler Effect

You've probably noticed that when a car approaches you with its horn sounding, the pitch seems to drop as the car passes. This phenomenon is called the **Doppler effect.** When a source of sound and a listener are in motion relative to each other, the frequency of the sound heard by the listener is not the same as the source frequency. We can work out a relation between the frequency shift and the source and listener velocities.

To keep things simple, we'll consider only the special case in which the velocities of both source and listener lie along the line joining them. We'll take this line to be the x axis, with the positive direction to the right, and we'll assume that the source S is to the right (at greater x) of the listener L. We assume that this coordinate system is at rest with respect to the air. Let v_S and v_L be the x components of velocity of source and listener, respectively, with respect to this coordinate system. Thus, the direction from the listener L to the source S is the positive direction for both v_S and v_L The speed of sound v is always considered positive, but v_S and v_L can be either positive or negative because they are components of velocity, a vector quantity. If $v_L > 0$, the listener is moving to the right; if $v_L < 0$, the listener is moving to the left. The same rules apply to v_S.

Moving Listener

Let's think first about a listener L moving with velocity v_L toward a stationary source S (Figure 12.32). The source emits a sound wave with frequency f_S and wavelength $\lambda = v/f_S$. The figure shows several wave crests, separated by equal

▲ Application **Found at sea.** Another important application of the physics of wave formation and propagation is sonar, which was used to find the long-lost steam ship Portland shown above, which sank in 1898. Sonar (SOund Navigation And Ranging) works by transmitting an ultrasonic pulse of sound waves through air or water and sensing the waves reflected from objects in the path of these waves. The frequency of the sound transmitted is about 50 kHz, well above the upper limit of the human ear, which is about 20 kHz. When the sound waves reach an object, they are reflected in various ways, depending on the position, shape, orientation, and surface characteristics of the object. The images from this shipwreck clearly show the side-by-side smokestacks and the diamond-shaped metal walking beam that provided power to the ship's side paddle wheels.

MasteringPHYSICS

ActivPhysics 10.8: Doppler Effect: Conceptual Introduction
ActivPhysics 10.9: Doppler Effect: Problems

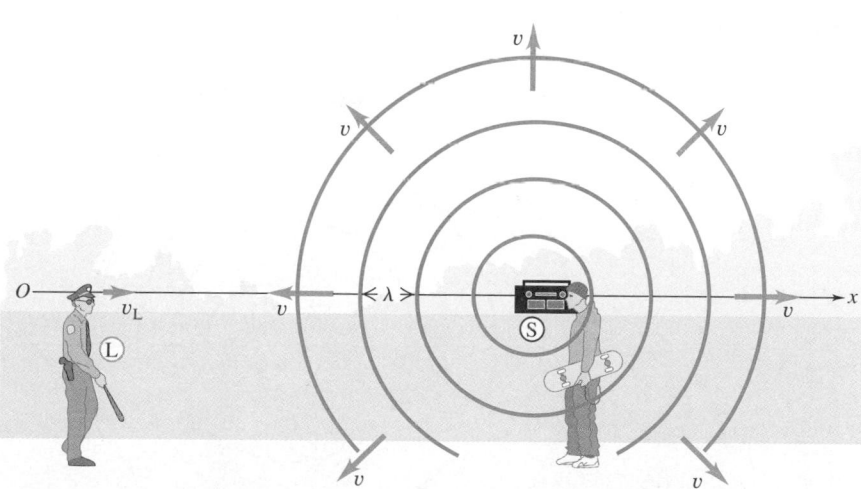

▲ **FIGURE 12.32** A listener moving toward a stationary sound source hears a frequency that is higher than the source frequency because the relative velocity of listener and wave is greater than the wave speed v.

distances λ. The waves approaching the moving listener have a speed of propagation of $v + v_L$ *relative to the listener,* so the frequency f_L with which the wave crests arrive at the listener's position (the frequency the listener hears) is

$$f_L = \frac{v + v_L}{\lambda} = \frac{v + v_L}{v/f_S} = \frac{v + v_L}{v}f_S. \tag{12.16}$$

So a listener moving toward a source $(v_L > 0)$ hears a sound that has a greater frequency (higher pitch) than a stationary listener hears. A listener moving away from the source $(v_L < 0)$ hears a sound that has a lesser frequency (lower pitch).

Moving Source

Now suppose the source is also moving, with velocity v_S (Figure 12.33). The wave speed relative to the air is still v; it is determined by the properties of the wave medium and is not changed by the motion of the source. But the wavelength is no longer equal to v/f_S. Here's why: The time for the emission of one cycle of the wave is the period $T = 1/f_S$. During this time, the wave travels a distance $vT = v/f_S$ and the displacement of the source is $v_ST = v_S/f_S$ (where v_s may be either positive or negative). The wavelength is the distance between successive wave crests and is determined by the *relative* displacement of source and wave. As Figure 12.33 shows, this is different in front of and behind the source. In the region to the right of the source in the figure, the wavelength is

$$\lambda = \frac{v}{f_S} - \frac{v_S}{f_S} = \frac{v - v_S}{f_S}. \tag{12.17}$$

In the region to the left of the source, the wavelength is

$$\lambda = \frac{v + v_S}{f_S}. \tag{12.18}$$

When v_S is positive (as it is in Figure 12.33), the waves are compressed in front of the source and stretched out behind it.

To find the frequency heard by the listener, we substitute Equation 12.18 into the first form of Equation 12.16 to obtain

$$f_L = \frac{v + v_L}{\lambda} = (v + v_L)\frac{f_S}{v + v_S},$$

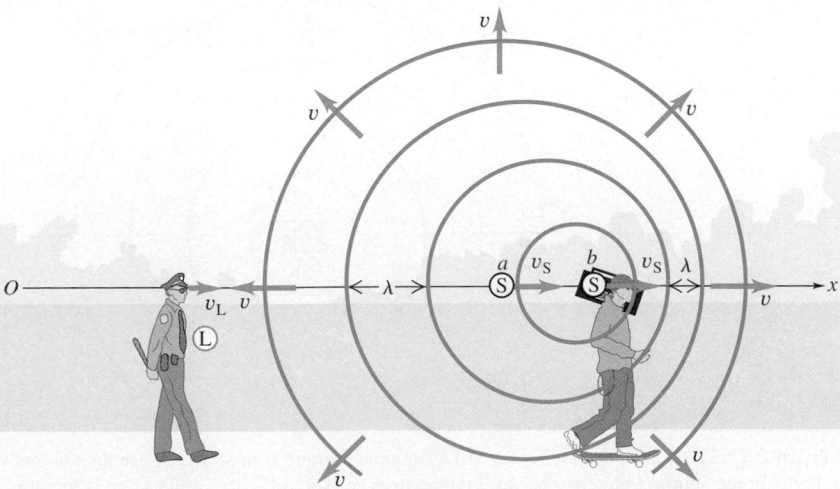

▲ **FIGURE 12.33** Wave crests emitted by a moving source are crowded together in front of the source and stretched out behind it.

which we can rewrite as follows:

Doppler effect when both listener and source may be moving

For a listener L and source S moving along the line joining them, the frequency f_L heard by the listener is related to the frequency f_S produced by the source as follows:

$$f_L = \frac{v + v_L}{v + v_S} f_S. \tag{12.19}$$

In this equation, v is the speed of sound and v_L and v_S are the x components of velocity of the listener and source, respectively.

Equation 12.19 includes all possibilities for the motion of source and listener (relative to the medium) along the line joining them. If the source happens to be at rest in the medium, v_S is zero; if the listener is at rest in the medium, v_L is zero. When both source and listener are at rest or have the same velocity relative to the medium, then $v_L = v_S$ and $f_L = f_S$. Whenever the source's or listener's velocity is in the negative x direction (opposite to the direction from listener toward source), the corresponding component of velocity $(v_L$ or $v_S)$ in Equation 12.19 is negative. But the wave speed v is always positive.

PROBLEM-SOLVING STRATEGY 12.3 Doppler effect

SET UP

1. Establish a coordinate system. Define the positive direction to be the direction from listener to source, and make sure that you know the signs of all the relevant velocities. A velocity in the direction from listener to source has a positive x component; a velocity in the opposite direction has a negative x component. Also, the velocities must all be measured relative to the air in which the sound is traveling.
2. Use consistent notation to identify the various quantities: subscript S for source, L for listener.

SOLVE

3. When a wave is reflected from a surface, either stationary or moving, the analysis can be carried out in two steps. In the first, the surface plays the role of listener; the frequency with which the wave crests arrive at the surface is f_L. Then think of the surface as a new source, emitting waves with this same frequency f_L. Finally, determine what frequency is heard by a listener detecting these new waves.

REFLECT

4. Ask whether your final result makes sense. If the source and listener are moving toward each other, f_L is always greater than f_S; if they are moving apart, f_L is always less than f_S. If source and listener have zero relative velocity, then $f_S = f_L$.

EXAMPLE 12.10 Police car chasing a speeder

A police siren emits a sinusoidal wave with frequency $f_S = 300$ Hz. The speed of sound is 340 m/s. **(a)** Find the wavelength of the waves if the siren is at rest in the air. **(b)** If the siren is moving with velocity $v_S = 30.0$ m/s, find the wavelengths of the waves ahead of and behind the source.

Continued

SOLUTION

SET UP (a) When the siren is at rest, $\lambda = v/f_s$. Figure 12.34 shows the situation for part (b). For the moving siren, we use Equations 12.17 and 12.18 with $v_S = 30.0$ m/s.

SOLVE Part (a): When the source is at rest,

$$\lambda = \frac{v}{f_s} = \frac{340 \text{ m/s}}{300 \text{ Hz}} = 1.13 \text{ m.}$$

Part (b): We use Equations 12.17 and 12.18. In front of the siren,

$$\lambda = \frac{v - v_S}{f_s} = \frac{340 \text{ m/s} - 30.0 \text{ m/s}}{300 \text{ Hz}} = 1.03 \text{ m.}$$

Behind the siren,

$$\lambda = \frac{v + v_S}{f_s} = \frac{340 \text{ m/s} + 30.0 \text{ m/s}}{300 \text{ Hz}} = 1.23 \text{ m.}$$

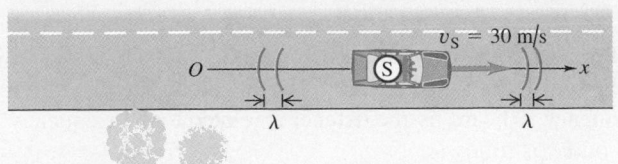

▲ **FIGURE 12.34** Wavelengths ahead of and behind the police siren when the siren is moving relative to the air.

REFLECT The wavelength is smaller in front of the moving siren, and greater behind it, than when the siren is at rest.

Practice Problem: If the siren speeds up until the wavelength in front of it is 1.00 m, what is its speed? *Answer:* 40.0 m/s.

EXAMPLE 12.11 A bystander observes the chase

If the listener is at rest and the siren is moving away from the listener at 30.0 m/s, what frequency does the listener hear?

SOLUTION

SET UP Figure 12.35 shows our diagram. We use Equation 12.19, with $v_L = 0$ and $v_S = 30.0$ m/s.:

SOLVE From Equation 12.19,

$$f_L = \frac{v + v_L}{v + v_S}f_S = \frac{340 \text{ m/s} + 0}{340 \text{ m/s} + 30.0 \text{ m/s}}(300 \text{ Hz}) = 276 \text{ Hz.}$$

REFLECT As the siren recedes, the listener hears sound of smaller frequency than when the siren is at rest.

$v_L = 0$ $v_S = 30$ m/s

O—(L)————————(S)—→×

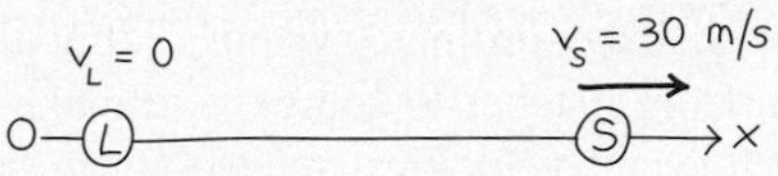

▲ **FIGURE 12.35** Our sketch for this problem.

Practice Problem: What frequency does the policeman hear? Answer: 300 Hz.

EXAMPLE 12.12 Driving past the stationary police car

If the siren is at rest and the listener is in a car moving away from the siren at 30 m/s, what frequency does the listener hear?

SOLUTION

SET UP Figure 12.36 shows our sketch. The positive direction is still from the listener toward the source, so this time $v_L = -30.0$ m/s and $v_S = 0$.

SOLVE From Equation 12.19,

$$f_L = \frac{v + v_L}{v + v_S}f_S$$

$$= \frac{340 \text{ m/s} + (-30.0 \text{ m/s})}{340 \text{ m/s} + 0}(300 \text{ Hz}) = 274 \text{ Hz.}$$

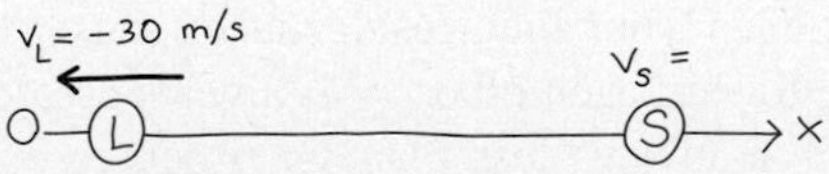

$v_L = -30$ m/s $v_S =$

O—(L)————————(S)—→×

▲ **FIGURE 12.36** Our sketch for this problem.

REFLECT The listener again hears sound of smaller frequency than when he and the siren are at rest, but not by the same amount as when the siren is moving.

EXAMPLE 12.13 **Following the police car**

If the siren is moving away from the listener with a speed of 45.0 m/s relative to the air, and the listener is moving in the same direction as the siren with a speed of 15.0 m/s relative to the air, what frequency does the listener hear?

SOLUTION

SET UP Figure 12.37 shows our sketch. In this case, $v_L = 15.0$ m/s and $v_s = 45.0$ m/s. (Both velocities are positive because both velocity vectors point in the $+x$ direction, from listener to source.)

$v_L = 15$ m/s $v_s = 45$ m/s

O — (L) —————————— (S) —→ ×

▲ **FIGURE 12.37** Our sketch for this problem.

SOLVE From Equation 12.17,

$$f_L = \frac{v + v_L}{v + v_S} f_S = \frac{340 \text{ m/s} + 15.0 \text{ m/s}}{340 \text{ m/s} + 45.0 \text{ m/s}} (300 \text{ Hz}) = 277 \text{ Hz}.$$

REFLECT When the source and listener are moving apart, the frequency f_L heard by the listener is always *less than* the frequency f_S emitted by the source. This is the case in all of the last three examples. Note that the *relative velocity* of source and listener (30.0 m/s) is the same in all three, but the Doppler shifts are all different because the velocities relative to the air are different.

Doppler Effect for Light

In our discussion thus far, the velocities v_L and v_S have always been measured *relative to the air* or whatever medium we are considering. There is also a Doppler effect for electromagnetic waves in empty space, such as light waves or radio waves. According to the special theory of relativity, the speed of light c is the same in all inertial frames of reference. When a source emits radiation with frequency f_S, a listener moving away from the source with speed v observes a smaller frequency f_L given by

$$f_L = \sqrt{\frac{c - v}{c + v}} f_S.$$

When the listener is moving directly toward the source, v is negative and f_L is *greater* than f_S.

A familiar application of the Doppler effect with radio waves is the radar device used in police cars to check other cars' speeds. The electromagnetic wave emitted by the device is reflected from a moving car, which then acts as a moving source, and the wave reflected back to the device is Doppler shifted in frequency. The transmitted and reflected signals are combined to produce beats, and the speed can be computed from the frequency of the beats.

The Doppler effect for light is of fundamental importance in astronomy. Light emitted by elements in stars of distant galaxies shows shifts toward longer wavelength and smaller frequency, compared with light from the same elements on earth. These shifts are Doppler shifts caused by the receding motion of the galaxies containing the stars. Because the shift is nearly always toward the longer wavelength (red end) of the visible spectrum, it is called a *red shift*. These observations provide solid evidence for the "big bang" theory, which describes a universe that has been expanding for about 14 billion years.

12.13 Applications of Acoustics

The importance of acoustic principles goes far beyond human hearing. Several animals use sound for navigation. Bats depend primarily on sound rather than sight for guidance during flight. That's how they can fly in the total darkness of a cave. They emit short pulses of ultrasonic sound, typically 30 to 150 kHz; the returning echoes give the bats information about the location and size of obstacles and potential prey such as flying insects.

Dolphins use an analogous system for underwater navigation. The frequencies are again ultrasonic, about 100 kHz. With such a system, the animal can

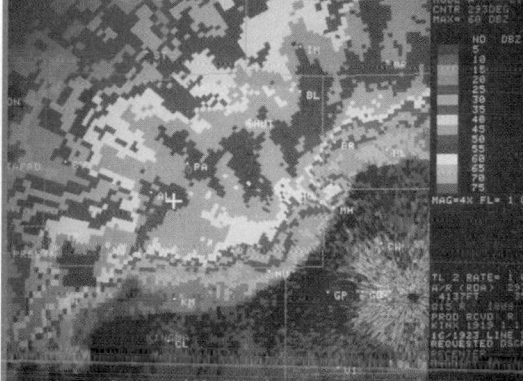

▲ **Application Take cover!** Meteorologists studying severe weather use radar (RAdio Detection And Ranging) and take advantage of the Doppler effect to monitor the occurrence and movement of thunderstorms, hurricanes, and tornadoes. Doppler radar towers transmit radio-frequency waves and can detect rain, hail, and snow by recording the waves reflected by these types of atmospheric moisture. Doppler radar also compares the frequencies of the transmitted and reflected waves to determine the movement of the moisture. Waves reflected from moisture moving toward the tower have a higher frequency than that of the transmitted wavelength, and the frequency of waves reflected from receding moisture is lower. Thus, knowledge of the physics of wave propagation allows us to receive advance warning of severe weather and can prevent unnecessary damage and injury.

sense objects of about the size of the wavelength of the sound (1.4 cm) or larger. The corresponding human-designed systems, used for submarine navigation, depth measurements, finding the locations of fish or wrecked ships, and so on, are called **sonar.** Sonar systems using the Doppler effect can measure the motion, as well as position, of submerged objects.

Analysis of elastic waves in the earth provides important information about its structure. The interior of the earth may be pictured crudely as being made of concentric spherical shells around a solid inner core. Mechanical properties such as density and elastic moduli are different in different shells. Waves produced by explosions or earthquakes are reflected and refracted at the interfaces between these shells, and analysis of these waves helps geologists measure the dimensions and properties of the shells. Local anomalies such as oil deposits can also be detected by studying wave propagation.

Acoustic phenomena have many important medical applications. Shock waves are used to break up kidney stones and gallstones without invasive surgery, by means of a technique with the impressive name *extracorporeal shock-wave lithotripsy.* A shock wave is produced outside the body and is then focused by a reflector or acoustic lens so that as much of its energy as possible converges on the stone. When the resulting stresses in the stone exceed its tensile strength, it breaks into small pieces and can be eliminated. This technique requires accurate determination of the location of the stone.

Ultrasonic imaging (the reflection of ultrasonic waves from regions in the interior of the body) is used for prenatal examinations, the detection of anomalous conditions such as tumors, and the study of heart-valve action, to name a few applications. Ultrasound is more sensitive than x rays in distinguishing various kinds of tissues, and it does not have the radiation hazards associated with x rays. At much higher power levels, ultrasound appears to have promise as a selective destroyer of pathological tissue in the treatment of arthritis and certain cancers. The sound is always produced by first generating an electrical wave and then using this wave to drive a loudspeakerlike device called a **transducer** that converts electrical waves to sound. Techniques have been developed in which transducers move over, or *scan,* the region of interest and a computer-reconstructed image is produced. An example of such an image is shown in Figure 12.38.

Acoustic principles have many important applications to environmental problems, such as noise control. The design of quiet mass-transit vehicles, for example, involves detailed study of the generation and propagation of sound in the motors, wheels, and supporting structures. Excessive noise levels often lead to permanent hearing impairment; studies have shown that many young rock musicians have hearing that is typical of persons 65 years of age. Prolonged listening to music at high intensity levels (100 to 120 dB) can lead to permanent hearing loss. Stereo headsets used at high volume levels pose similar threats to hearing. Be careful!

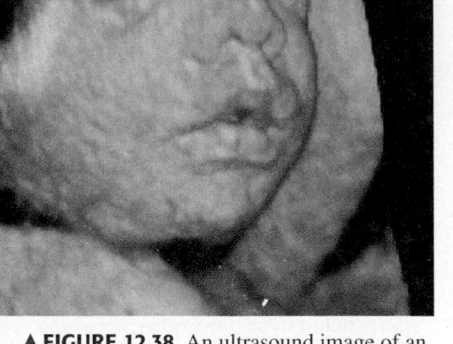

▲ **FIGURE 12.38** An ultrasound image of an unborn fetus.

12.14 Musical Tones

Several aspects of musical sound are directly related to the physical characteristics of sound waves. The **pitch** of a musical tone is the quality that lets us classify it as "high" or "low." Middle C on the piano has a frequency of 262 Hz. The musical interval of an *octave* corresponds to a factor of two in frequency. The C note an octave above middle C has a frequency of 2×262 Hz $= 524$ Hz, and the "high C" sung by coloratura sopranos, two octaves above middle C, has a frequency of 4×262 Hz, or 1048 Hz. The ear can be fooled, however: When a listener hears two sinusoidal tones with the same frequency but different intensities, the louder one usually seems to be slightly lower in pitch.

Musical tones usually contain many frequencies. A plucked, bowed, or struck string or the column of air in a wind instrument vibrates with a fundamental frequency and many harmonics at the same time. Two tones may have the same fundamental frequency (and thus the same pitch) but may sound different because of the presence of different intensities of the various harmonics. The difference is called *tone color, quality,* or **timbre,** often described in subjective terms such as reedy, golden, round, mellow, or tinny. A tone that is rich in harmonics usually sounds thin and "stringy" or "reedy," and a tone containing mostly a fundamental is more mellow and flutelike.

Another factor in determining tone quality is the behavior at the beginning and end of a tone. A piano tone begins percussively with a thump and then dies away gradually. A harpsichord tone, in addition to having different harmonic content, begins much more quickly and incisively with a click, and the higher harmonics begin before the lower ones. The ending of the tone, when the key is released, is also much more incisive on the harpsichord than the piano. Similar effects are present with other musical instruments; with wind and string instruments, the player has considerable control over the attack and decay of the tone, and these characteristics help to define the unique characteristics of each instrument.

When combinations of musical tones are heard together or in succession, even a listener with no musical training hears a relationship among them; they sound comfortable together, or they clash. Musicians use the terms *consonant* and *dissonant* to describe these effects. A *consonant interval* is a combination of two tones that sound comfortable or pleasing together. The most consonant interval is the *octave,* consisting of two tones with a frequency ratio of 2 : 1. An example is middle C on the piano and the next C above it. Another set of consonant tones is obtained by playing C, E, and G. These form what is called a *major triad,* and the frequencies are in the ratio 4 : 5 : 6. The interval from C to G, with a frequency ratio of 3 : 2, is a perfect fifth; C to E, with 5 : 4, is a major third; and so on. These combinations sound good together because they have many harmonics in common. For example, for the interval of the octave, every harmonic in the harmonic series of the lower-frequency tone is also present in the harmonic series of the upper tone. For other intervals, the overlap of harmonics is only partial, but it is there.

If we want to be able to play major triads starting on various tones on a keyboard instrument such as the piano or organ, complications arise. Either we add a lot of extra keys in each octave to get the exact frequencies we need, or else we compromise a little on the frequencies. The most common compromise in piano tuning is called *equal temperament,* in which every pair of adjacent keys (white and black) is tuned to the same frequency ratio of $2^{1/12}$. This interval is called a *semitone* or a *halftone;* a succession of 12 of these intervals then forms an exact 2 : 1 octave. The perfect fifth is seven semitones; the corresponding frequency ratio is $2^{7/12} = 1.4983$. This ratio is close to the ideal one of 3 : 2, or 1.5000, but a sensitive ear can hear the difference.

Thus, a piano tuned in equal temperament is not quite in tune, in terms of the ideal ratios, in *any* key, but it is equally good (or bad) in all keys. An alternative would be to tune the white keys to form ideal intervals. Doing this would make the instrument sound better for music in the key of C, but music in some other keys would sound worse in comparison with music played in equal temperament. So an instrument that is intended to be used for compositions in all keys is usually tuned to equal temperament.

In the Baroque period, however, keys with more than three sharps or flats were seldom used, and various compromise temperaments were invented to favor the commonly used keys. Organs and harpsichords that are intended primarily for music of this period are often tuned to one of these unequal temperaments. The great J. S. Bach favored such temperaments rather than equal temperament. His composition "The Well-Tempered Clavier" contains preludes and fugues in all the major and minor keys, but it is important not to misconstrue *well tempered* as meaning *equally tempered!*

SUMMARY

Mechanical Waves

(Sections 12.1–12.4) A mechanical wave is any disturbance from an equilibrium condition that propagates from one region of space to another through some material called the medium. In a periodic wave, the motion of each particle of the medium is periodic; the period T, frequency f, wavelength λ, and wave speed v are related by $v = \lambda f = \lambda/T$. For a string with mass per unit length μ and tension F_T, the speed of transverse waves is

$$v = \sqrt{\frac{F_T}{\mu}}. \qquad (12.2)$$

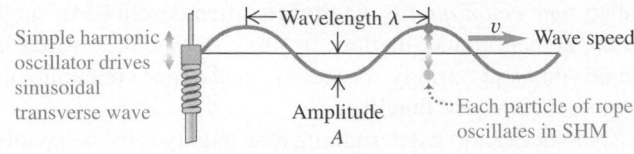

Simple harmonic oscillator drives sinusoidal transverse wave

Wavelength λ

v Wave speed

Amplitude A

Each particle of rope oscillates in SHM

Superposition, Standing Waves, and Normal Modes

(Sections 12.5–12.6) The **principle of superposition** states that when two waves overlap, the resulting displacement at any point is obtained by vector addition of the displacements that would be caused by the two individual waves. When a sinusoidal wave is reflected at a stationary or free end, the original and reflected wave combine to make a **standing wave.** At the **nodes** of a standing wave, the displacement is always zero; the **antinodes** are the points of maximum displacement. Fixed string ends are nodes; ends that are free to move transversely are antinodes.

Standing waves on a string of length L can have only certain specific frequencies. When both ends are held stationary, the allowed frequencies f_n are

$$f_n = n\frac{v}{2L} = nf_1, \qquad (n = 1, 2, 3, \dots) \qquad (12.7)$$

A **normal mode** is a motion in which all particles on the string move sinusoidally with one of the frequencies f_n. Each value of f_n corresponds to a different normal mode.

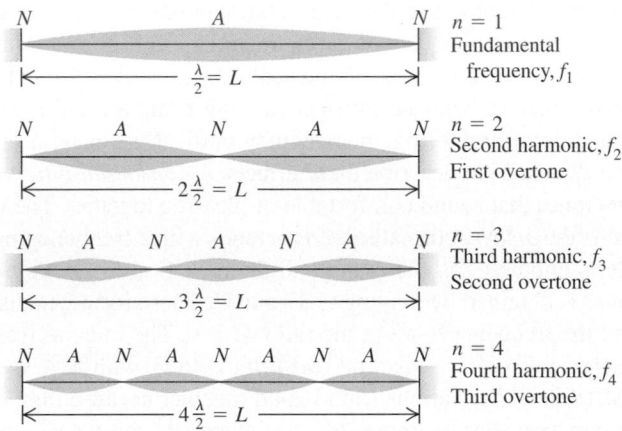

N A N $n = 1$ Fundamental frequency, f_1 — $\frac{\lambda}{2} = L$

N A N A N $n = 2$ Second harmonic, f_2 First overtone — $2\frac{\lambda}{2} = L$

N A N A N A N $n = 3$ Third harmonic, f_3 Second overtone — $3\frac{\lambda}{2} = L$

N A N A N A N A N $n = 4$ Fourth harmonic, f_4 Third overtone — $4\frac{\lambda}{2} = L$

Longitudinal Standing Waves

(Section 12.7) When a longitudinal wave propagates in a fluid in a pipe with a finite length, the waves reflect and form a standing wave. The closed end of a pipe is a *node,* and the open end is an *antinode.* For an open pipe of length L, the lowest frequency with antinodes at both ends is the fundamental frequency $f_1 = v/2L$ (Equation 12.9), and the harmonics are integer multiples of this fundamental frequency: $f_n = nf_1$. If a pipe is closed at one end, the fundamental frequency is $f_1 = v/4L$ (Equation 12.11) and the harmonics are

$$f_n = n\frac{v}{4L} = nf_1, \qquad (n = 1, 3, 5, \dots) \qquad (12.12)$$

Open pipe

$\frac{\lambda}{2}$

Fundamental: $f_1 = \frac{v}{2L}$

Second harmonic: $f_2 = 2\frac{v}{2L} = 2f_1$

Stopped pipe

$\frac{\lambda}{4}$

Fundamental: $f_1 = \frac{v}{4L}$

Third harmonic: $f_3 = 3\frac{v}{4L} = 3f_1$

Interference

(Section 12.8) Wave phenomena that occur when two or more waves overlap in the same region of space are grouped under the heading *interference.* According to the principle of superposition, if two periodic waves are *in step* at a point, then their amplitudes add together. This phenomenon is called **constructive interference.** If the waves are a half-cycle *out of step* at a point, the resulting amplitude is smaller and the phenomenon is called **destructive interference.**

Continued

Sound, Intensity, and Beats

(Sections 12.9–12.11) A *logarithmic* intensity scale is usually used to measure the **intensity level** β of a sound wave:

$$\beta = (10\ \text{dB})\log\frac{I}{I_0}. \qquad (12.14)$$

In this equation, I_0 is a reference intensity chosen to be $10^{-12}\ \text{W/m}^2$, and the units of β are **decibels,** abbreviated dB. When two sound waves of slightly different frequency interfere, alternating constructive and destructive interference produces **beats.** The beat frequency is $f_{\text{beat}} = f_1 - f_2$ (Equation 12.15).

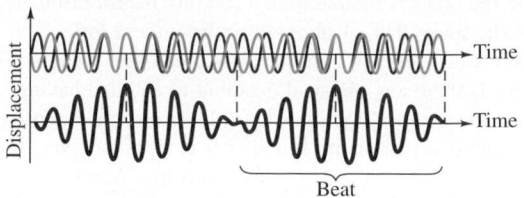

The Doppler Effect

(Section 12.12) The **Doppler effect** is the frequency shift that occurs when a listener is in motion relative to a source of sound. The source and listener frequencies f_S and f_L and their velocity components v_S and v_L are related by

$$f_L = \frac{v + v_L}{v + v_S}f_S. \qquad (12.19)$$

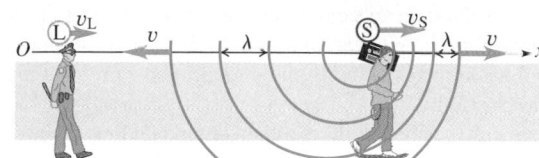

Applications of Acoustics

(Section 12.13) The importance of acoustic principles goes far beyond human hearing. Bats locate insects, and dolphins navigate, using ultrasonic waves. We study properties of the earth by measuring the reflected and refracted elastic waves passing through its interior. Ultrasonic shock waves are used to break up kidney stones or gallstones without invasive surgery. Ultrasonic imaging is used to view a fetus or tumor without the dangerous radiation associated with x-ray imaging.

Musical Tones

(Section 12.14) Several aspects of musical sound are directly related to the physical characteristics of sound waves. Musical tones usually contain many frequencies. A plucked, bowed, or struck string or the column of air in a wind instrument vibrates with a fundamental frequency and many harmonics at the same time. Two tones may have the same fundamental frequency (and thus the same pitch), but sound different because of the presence of different intensities of the various harmonics.

 For instructor-assigned homework, go to www.masteringphysics.com

Conceptual Questions

1. What kinds of energy are associated with waves on a stretched string? How could such energy be detected experimentally?
2. In most modern wind instruments, the pitch is changed by using keys or valves to change the length of the vibrating air column. The bugle, however, has no valves or keys, yet it can play many notes. How might this be possible? Are there restrictions on what notes a bugle can play?
3. Sci-fi movies sometimes show the explosion of a star accompanied by a great surge of noise. What is wrong with that picture?
4. Which of the characteristics of a sound wave (amplitude or frequency) is most closely related to musical pitch? To loudness?
5. Musical notes produced by different instruments (such as a flute and an oboe) may have the same pitch and loudness and yet sound different. What is the difference, in physical terms?
6. Energy can be transferred along a string by wave motion. However, in a standing wave, no energy can ever be transferred past a node. Why not?
7. Some opera singers, such as Enrico Caruso, have been reputed to be able to break a glass by singing the appropriate note. What physical phenomenon could account for such a feat? (In fact, this is a legend and is not known ever to have occurred.)
8. Lane dividers on highways sometimes have regularly spaced ridges or ripples. When the tires of a moving car roll along such a divider, a musical note is produced. Why? Could this phenomenon be used to measure the car's speed? How?

9. Two tuning forks have identical frequencies, but one is stationary while the other is mounted on a rotating record turntable. Describe the sound that a listener hears from these forks.

10. (a) Does a sound level of zero decibels mean that there is no sound? (b) Is there any physical meaning to a sound having a negative intensity level? What is it? (c) Does a sound intensity of zero mean that there is no sound? (d) Is there any physical meaning to a sound having a negative intensity? Why?

11. Players of stringed instruments tune them by tightening the string. Why does this work? How does a piano tuner tune the instrument?

12. You are standing several meters from the railroad tracks with a train approaching, sounding its whistle. (a) As the train gets closer and closer, what happens to the loudness of the sound? Why? (b) As the train gets closer and closer, the pitch you hear from the whistle gets lower and lower. Why? (c) Would both observations in parts (a) and (b) hold true if you were standing *on* the tracks? Why?

13. An organist in a cathedral plays a loud chord and then releases the keys. The sound persists for a few seconds and gradually dies away. Why does it persist? What happens to the sound energy when the sound dies away?

14. TV weather forecasters often refer to *Doppler radar* in predicting advancing storms. How do you think Doppler radar works?

15. A wire under tension and vibrating in its first overtone produces sound of wavelength λ. What is the new wavelength of the sound (in terms of λ) if the tension is doubled?

Multiple-Choice Problems

1. A hiker sees a lightning flash; 15 s later he hears the sound of the thunder. Recalling from his study of physics that the speed of sound in air is approximately $\frac{1}{3}$ km/s, he estimates that the distance to where the lightning flash occurred is approximately
   A. 5 km.　　B. 10 km.　　C. 15 km.　　D. 45 km.

2. A segment A of wire stretched tightly between two posts a distance L apart vibrates in its fundamental mode with frequency f. A segment B of an identical wire is stretched with the same tension, but between two different posts. You observe that the frequency of the second harmonic of wire B is the same as the fundamental frequency of wire A. The length of wire B must be
   A. L.　　B. $2L$.　　C. $4L$.

3. Two pulses of exactly the same size and shape are traveling toward each other along a stretched rope. They differ only in that one is upright while the other is inverted. Superposition tells us that when the pulses meet each other, they will cancel each other exactly at that instant and the rope will show no evidence of a pulse. What happens *afterwards?*
   A. Each pulse continues as though it had never met the other one.
   B. The rope remains straight, since the pulses have cancelled each other.
   C. The pulses rebound from each other, each going back in the direction from which it came.

4. An organ pipe open at one end, but closed at the other, is vibrating in its fundamental mode, producing sound of frequency 1000 Hz. If you now open the closed end, the new fundamental frequency will be
   A. 2000 Hz.　　B. 1000 Hz.　　C. 500 Hz.　　D. 250 Hz.

5. A person listening to a siren from a stationary police car observes the frequency and wavelength of that sound. This person now moves rapidly toward the police car.
   A. The wavelength of the sound the person observes is shorter than it was, but the frequency does not change.
   B. The frequency of the sound the person observes is higher than it was, but the wavelength does not change.
   C. The wavelength of the sound the person observes is shorter than it was, and the frequency is higher than it was.

6. A person listening to a siren from a stationary police car observes the frequency and wavelength of that sound. The car now drives rapidly toward the person.
   A. The wavelength of the sound the person observes will be shorter than it was, but the frequency will not be changed.
   B. The frequency of the sound the person observes will be higher than it was, but the wavelength will not be changed.
   C. The wavelength of the sound the person observes will be shorter than it was, and the frequency will also be higher than it was.

7. A string of length 0.600 m is vibrating at 100.0 Hz in its second harmonic and producing sound that moves at 340 m/s. What is true about the frequency and wavelength of this sound? (There may be more than one correct choice.)
   A. The wavelength of the sound is 0.600 m.
   B. The wavelength of the sound is 3.40 m.
   C. The frequency of the sound is 100.0 Hz.
   D. The frequency of the sound is 567 Hz.

8. When a 15 kg mass is hung vertically from a thin, light wire, pulses take time t to travel the full length of the wire. If an additional 15 kg mass is added to the first one without changing the length of the wire, the time taken for pulses to travel the length of the wire will be
   A. $2t$.　　　B. $\sqrt{2}t$.　　　C. $t/2$.　　　D. $t/\sqrt{2}$.

9. An omnidirectional loudspeaker produces sound waves uniformly in all directions. The total power received by a sphere of radius 2.0 m centered on the speaker is 100 W. A sphere of radius 4.0 m will receive a total power of
   A. 100 W.　　B. 50 W.　　C. 25 W.

10. An organ pipe open at both ends is resonating in its fundamental mode at frequency f. If you close both ends, the fundamental frequency will now be
   A. $2f$.　　B. f.　　C. $f/2$.　　D. $f/4$.

11. You are standing between two stereo speakers that are emitting sound of wavelength 10.0 cm in step with each other. They are very far apart compared with that wavelength. If you start in the middle and walk a distance x directly toward one speaker, you observe that the sound from these speakers first cancels when x is equal to
   A. 20.0 cm.　　B. 10.0 cm.　　C. 5.0 cm.　　D. 2.5 cm.

12. On a cold day, a siren emits sound waves with a wavelength of 17 cm. On a hot day, the wavelength of the sound produced by the same siren oscillating at the same frequency will be
   A. greater than 17 cm.
   B. less than 17 cm.
   C. 17 cm.

13. Traffic noise on Beethoven Boulevard has an intensity level of 80 dB; the traffic noise on Mozart Alley is only 60 dB. Compared

to the sound intensity on Beethoven Boulevard, the sound intensity on Mozart Alley is

A. 25% lower   B. 20 times lower
C. 100 times lower   D. 20 W/m² lower

14. A thin, light string supports a weight W hanging from the ceiling. In this situation, the string produces a note of frequency f when vibrating in its fundamental mode. In order to cause this string to produce a note one octave higher in its fundamental mode without stretching it, we must change the weight to

A. $4W$.   B. $2W$.   C. $\sqrt{2}W$.
D. $W/2$.   E. $W/4$.

15. String A weighs twice as much as string B. Both strings are thin and light and have the same length. If you hang equal weights at the bottom of each of these strings, the ratio of the speed of waves on string A to the speed of waves on string B will be

A. $v_A/v_B = 2$.   B. $v_A/v_B = \sqrt{2}$.
C. $v_A/v_B = 1/\sqrt{2}$.   D. $v_A/v_B = \frac{1}{2}$.

Problems

Unless otherwise indicated, assume that the speed of sound in air is 344 m/s.

12.2 Periodic Mechanical Waves

1. • (a) **Audible wavelengths.** The range of audible frequencies
BIO is from about 20 Hz to 20,000 Hz. What is range of the wavelengths of audible sound in air? (b) **Visible light.** The range of visible light extends from 400 nm to 700 nm. What is the range of visible frequencies of light? (c) **Brain surgery.** Surgeons can remove brain tumors by using a cavitron ultrasonic surgical aspirator, which produces sound waves of frequency 23 kHz. What is the wavelength of these waves in air? (d) **Sound in the body.** What would be the wavelength of the sound in part (c) in bodily fluids in which the speed of sound is 1480 m/s, but the frequency is unchanged?

2. • **The electromagnetic spectrum.** Electromagnetic waves, which include light, consist of vibrations of electric and magnetic fields, and they all travel at the speed of light. (a) **FM radio.** Find the wavelength of an FM radio station signal broadcasting at a frequency of 104.5 MHz. (b) **X rays.** X rays have a wavelength of about 0.10 nm. What is their frequency? (c) **The Big Bang.** Microwaves with a wavelength of 1.1 mm, left over from soon after the Big Bang, have been detected. What is their frequency? (d) **Sunburn.** Sunburn (and skin cancer) are caused by ultraviolet light waves having a frequency of around 10^{16} Hz. What is their wavelength? (e) **SETI.** It has been suggested that extraterrestrial civilizations (if they exist) might try to communicate by using electromagnetic waves having the same frequency as that given off by the spin flip of the electron in hydrogen, which is 1.43 GHz. To what wavelength should we tune our telescopes in order to search for such signals? (f) **Microwave ovens.** Microwave ovens cook food with electromagnetic waves of frequency around 2.45 GHz. What wavelength do these waves have?

3. • If an earthquake wave having a wavelength of 13 km causes the ground to vibrate 10.0 times each minute, what is the speed of the wave?

4. •• A fisherman notices that his boat is moving up and down periodically, owing to waves on the surface of the water. It takes 2.5 s for the boat to travel from its highest point to its lowest, a total distance of 0.62 m. The fisherman sees that the wave crests are spaced 6.0 m apart. (a) How fast are the waves traveling? (b) What is the amplitude of each wave? (c) If the total vertical distance traveled by the boat were 0.30 m, but the other data remained the same, how would the answers to parts (a) and (b) be affected?

12.3 Wave Speeds

5. • A steel wire 4.00 m long has a mass of 0.0600 kg and is stretched with a tension of 1000 N. What is the speed of propagation of a transverse wave on the wire?

6. • With what tension must a rope with length 2.50 m and mass 0.120 kg be stretched for transverse waves of frequency 40.0 Hz to have a wavelength of 0.750 m?

7. • One end of a horizontal rope is attached to a prong of an electrically driven tuning fork that vibrates at 120 Hz. The other end passes over a pulley and supports a 1.50 kg mass. The linear mass density of the rope is 0.0550 kg/m. (a) What is the speed of a transverse wave on the rope? (b) What is the wavelength? (c) How would your answers to parts (a) and (b) change if the mass were increased to 3.00 kg?

8. •• (a) If the amplitude in a sound wave is doubled, by what factor does the intensity of the wave increase? (b) By what factor must the amplitude of a sound wave be increased in order to increase the intensity by a factor of 9?

9. •• When a mass M hangs from a vertical wire of length L, waves travel on this wire with a speed V. What will be the speed of these waves (in terms of V) if (a) we double M without stretching the wire? (b) we replace the wire with an identical one, except twice as long? (c) we replace the wire with one of the same length, but three times as heavy? (d) we stretch the wire to twice its original length? (e) we increase M by a factor of 10, which stretches the wire to double its original length?

12.4 Mathematical Description of a Wave

10. • A certain transverse wave is described by the equation

$$y(x, t) = (6.50 \text{ mm}) \sin 2\pi \left(\frac{t}{0.0360 \text{ s}} - \frac{x}{0.280 \text{ m}} \right).$$

Determine this wave's (a) amplitude, (b) wavelength, (c) frequency, (d) speed of propagation, and (e) direction of propagation.

11. • Transverse waves on a string have wave speed 8.00 m/s, amplitude 0.0700 m, and wavelength 0.320 m. These waves travel in the x direction, and at $t = 0$ the $x = 0$ end of the string is at $y = 0$ and moving downward. (a) Find the frequency, period, and wave number of these waves. (b) Write the equation for $y(x, t)$ describing these waves. (c) Find the transverse displacement of a point on the string at $x = 0.360$ m at time $t = 0.150$ s.

12. • The equation describing a transverse wave on a string is

$$y(x, t) = (1.50 \text{ mm}) \sin[(157 \text{ s}^{-1})t - (41.9 \text{ m}^{-1})x].$$

Find (a) the wavelength, frequency, and amplitude of this wave, (b) the speed and direction of motion of the wave, and (c) the transverse displacement of a point on the string when $t = 0.100$ s and at a position $x = 0.135$ m.

13. •• Transverse waves are traveling on a long string that is under a tension of 4.00 N. The equation describing these waves is

$$y(x, t) = (1.25 \text{ cm}) \sin\left[(415 \text{ s}^{-1})t - (44.9 \text{ m}^{-1})x\right].$$

Find the linear mass density of this string.

12.5 Reflections and Superposition

12.6 Standing Waves and Normal Modes

14. • **Mapping the ocean floor.** The ocean floor is mapped by sending sound waves (sonar) downward and measuring the time it takes for their echo to return. From this information, the ocean depth can be calculated if one knows that sound travels at 1531 m/s in seawater. If a ship sends out sonar pulses and records their echo 3.27 s later, how deep is the ocean floor at that point, assuming that the speed of sound is the same at all depths?

15. • In Figure 12.39, each pulse is traveling on a string at 1 cm/s, and each square represents 1 cm. Draw the shape of the string at the end of 6 s, 7 s, and 8 s.

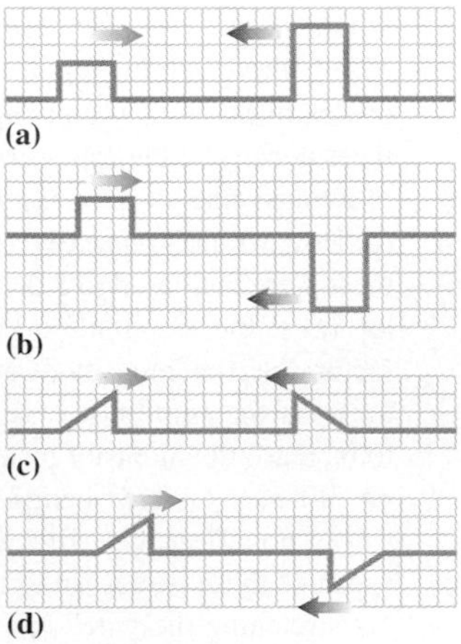

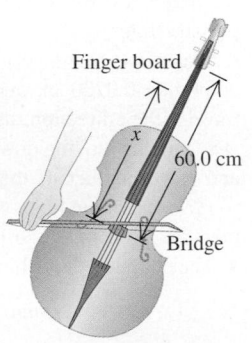

▲ **FIGURE 12.39** Problem 15.

16. • A 1.50-m-long rope is stretched between two supports with a tension that makes the speed of transverse waves 48.0 m/s. What are the wavelength and frequency of (a) the fundamental tone? (b) the second overtone? (c) the fourth harmonic?

17. • A piano tuner stretches a steel piano wire with a tension of 800 N. The wire is 0.400 m long and has a mass of 3.00 g. (a) What is the frequency of its fundamental mode of vibration? (b) What is the number of the highest harmonic that could be heard by a person who is capable of hearing frequencies up to 10,000 Hz?

18. • A wire with mass 40.0 g is stretched so that its ends are tied down at points 80.0 cm apart. The wire vibrates in its fundamental mode with frequency 60.0 Hz and with an amplitude of 0.300 cm at the antinodes. (a) What is the speed of propagation of transverse waves in the wire? (b) Compute the tension in the wire.

19. •• The portion of string between the bridge and upper end of the fingerboard (the part of the string that is free to vibrate) of a certain musical instrument is 60.0 cm long and has a mass of 2.00 g. The string sounds an A₄ note (440 Hz) when played. (a) Where must the player put a finger (at what distance x from the bridge) to play a D₅ note (587 Hz)? (See Figure 12.40.) For both notes, the string vibrates in its fundamental mode. (b) Without retuning, is it possible to play a G₄ note (392 Hz) on this string? Why or why not?

Finger board

x

60.0 cm

Bridge

▲ **FIGURE 12.40** Problem 19.

20. •• **Voiceprints.** In this chapter, we have been concentrating on sinusoidal waves. But most waves in the real world are far more complicated. However, many complicated waves can be created by adding together sine waves of varying amplitude and frequency. When a singer, for example, sings a note, the pitch we hear is the fundamental frequency at which his or her larynx is vibrating. But the larynx also vibrates in other frequencies (the overtones) at the same time. So the sound we hear is a superposition of the fundamental frequency plus all the overtones. This set of all the frequencies (with their respective amplitudes) is called the person's *voiceprint.* (a) To see how this works, carefully graph a sine wave of frequency 440 Hz (concert A), with time on the horizontal axis and displacement on the vertical axis. Let the amplitude be 1 unit. On the same set of axes, graph the first overtone of 880 Hz, but with an amplitude of $\frac{1}{2}$ unit. (b) Now add the two waves to find their superposition. Notice that the shape is no longer a sine wave.

21. •• **Voiceprints.** Suppose a singer singing F# (370 Hz, the fundamental frequency) has one overtone of frequency 740 Hz with half the amplitude of the fundamental and another overtone of frequency 1110 Hz having one-third the amplitude of the fundamental. Using the previous problem as a guide, graph the superposition of these three waves to show the complex sound wave produced by this singer.

22. •• **Guitar string.** One of the 63.5-cm-long strings of an ordinary guitar is tuned to produce the note B₃ (frequency 245 Hz) when vibrating in its fundamental mode. (a) Find the speed of transverse waves on this string. (b) If the tension in this string is increased by 1.0%, what will be the new fundamental frequency of the string? (c) If the speed of sound in the surrounding air is 344 m/s, find the frequency and wavelength of the sound wave produced in the air by the vibration of the B₃ string. How do these compare to the frequency and wavelength of the standing wave on the string?

12.7 Longitudinal Standing Waves

23. • Standing sound waves are produced in a pipe that is 1.20 m long. For the fundamental frequency and the first two overtones, determine the locations along the pipe (measured from the left end) of the displacement nodes if (a) the pipe is open at both ends; (b) the pipe is closed at the left end and open at the right end.

24. • Find the fundamental frequency and the frequency of the first three overtones of a pipe 45.0 cm long (a) if the pipe is open at both ends; (b) if the pipe is closed at one end. (c) For each of the preceding cases, what is the number of the highest harmonic that may be heard by a person who can hear frequencies from 20 Hz to 20,000 Hz?

25. • The longest pipe found in most medium-size pipe organs is 4.88 m (16 ft) long. What is the frequency of the note corresponding to the fundamental mode if the pipe is (a) open at both ends, (b) open at one end and closed at the other?

26. •• The fundamental frequency of a pipe that is open at both ends is 594 Hz. (a) How long is this pipe? If one end is now closed, find (b) the wavelength and (c) the frequency of the new fundamental.

27. • **The role of the mouth in sound.** The production of sound during speech or singing is a complicated process. Let's concentrate on the mouth. A typical depth for the human mouth is

about 8.0 cm, although this number can vary. (Check it against your own mouth.) We can model the mouth as an organ pipe that is open at the back of the throat. What are the wavelengths and frequencies of the first four harmonics you can produce if your mouth is (a) open, (b) closed? Use $v = 354$ m/s.

28. •• A certain pipe produces a fundamental frequency of 262 Hz in air at 20°C. (a) If the pipe is filled with helium at the same temperature, what fundamental frequency does it produce? (b) Does your answer to part (a) depend on whether the pipe is open or stopped? Why or why not?

29. •• **The vocal tract.** Many opera singers (and some pop
BIO singers) have a range of about $2\frac{1}{2}$ octaves or even greater. Suppose a soprano's range extends from A below middle C (frequency 220 Hz) up to E^b above high C (frequency 1244 Hz). Although the vocal tract is quite complicated, we can model it as a resonating air column, like an organ pipe, that is open at the top and closed at the bottom. The column extends from the mouth down to the diaphragm in the chest cavity, and we can also assume that the lowest note is the fundamental. How long is this column of air if $v = 354$ m/s? Does your result seem reasonable, on the basis of observations of your own body?

30. •• **Singing in the shower!** We all sound like great singers in the shower, due to standing waves. Assume that your shower is 2.45 m (about 8 ft) tall and can be modeled as an organ pipe. (a) What will we have at the floor and ceiling, displacement nodes or antinodes? (b) What are the wavelength and frequency of the fundamental harmonic for standing waves in this shower? (See the answer to Problem 28.) (c) What are the wavelength and frequency of the first two overtones for this shower?

31. •• **French horn.** The French horn, one of the most beautiful-sounding instruments in the orchestra, consists of about 3.7 m (roughly 12 ft) of thin tubing, rolled into a spiral shape (although sizes do vary). The player blows into the mouthpiece, which can be treated as a closed end, and places his hand in the opposite end, which has a large flared opening. In brass instruments, the fundamental note is not normally playable. Instead the *first overtone* is the lowest playable note. (a) If the player's hand keeps the large end open, what is the frequency of the lowest *playable* note? (b) If the player now closes the large end with his hand, what is the frequency of the lowest playable note? See the answer to Problem 28. (*Note:* The physics of the French horn is much more complex than is indicated here. The player's hand in the open end also changes the effective length of the tube, which then affects the frequency of the sound.)

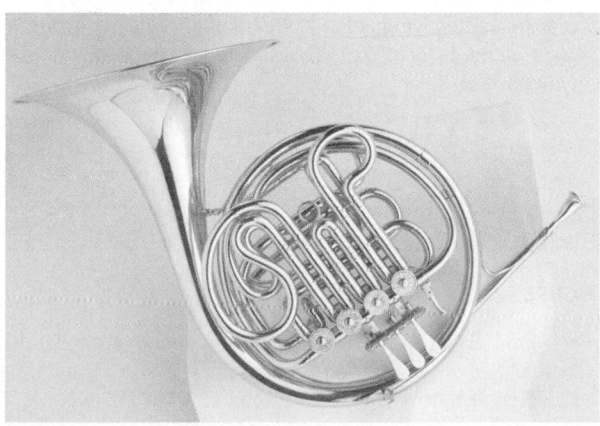

▲ **FIGURE 12.41** Problem 31.

32. •• You blow across the open mouth of an empty test tube and produce the fundamental standing wave of the air column inside the test tube. The speed of sound in air is 344 m/s and the test tube acts as a stopped pipe. (a) If the length of the air column in the test tube is 14.0 cm, what is the frequency of this standing wave? (b) What is the frequency of the fundamental standing wave in the air column if the test tube is half filled with water?

12.8 Interference

33. • Two small speakers A and B are driven in step at 725 Hz by the same audio oscillator. These speakers both start out 4.50 m from the listener, but speaker A is slowly moved away. (See Figure 12.42.) (a) At what distance d will the sound from the speakers first produce destructive interference at the location of the listener? (b) If A keeps moving, at what distance d will the speakers next produce destructive interference at the listener? (c) After A starts moving away, at what distance will the speakers first produce constructive interference at the listener?

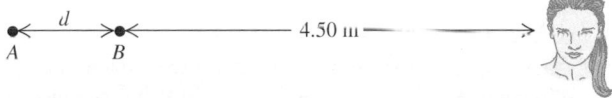

▲ **FIGURE 12.42** Problem 33.

34. •• In a certain home sound system, two small speakers are located so that one is 45.0 cm closer to the listener than the other. For what frequencies of audible sound will these speakers produce (a) destructive interference at the listener, (b) constructive interference at the listener? In each case, find only the three lowest audible frequencies.

35. •• Two small stereo speakers are driven in step by the same variable-frequency oscillator. Their sound is picked up by a microphone arranged as shown in Figure 12.43. For what frequencies does their sound at the speakers produce (a) constructive interference, (b) destructive interference?

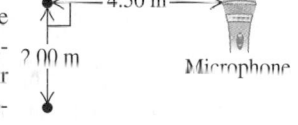

▲ **FIGURE 12.43** Problem 35.

12.9 Sound and Hearing

36. • **Human hearing.** The human
BIO outer ear contains a more-or-less cylindrical cavity called the *auditory canal* that behaves like a resonant tube to aid in the hearing process. One end terminates at the eardrum (tympanic membrane), while the other opens to the outside. (See Figure 12.44.) Typically, this canal is approximately 2.4 cm long. (a) At what frequencies would it resonate in its first two harmonics? (b) What are the corresponding sound wavelengths in part (a)?

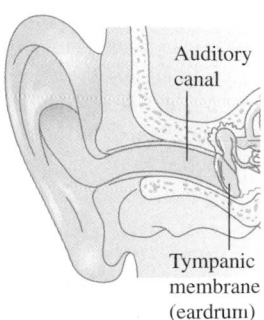

▲ **FIGURE 12.44** Problem 36.

37. • **Ultrasound and infrasound.** (a) **Whale communication.**
BIO Blue whales apparently communicate with each other using sound of frequency 17 Hz, which can be heard nearly 1000 km away in the ocean. What is the wavelength of such a sound in

seawater, where the speed of sound is 1531 m/s? (b) **Dolphin clicks.** One type of sound that dolphins emit is a sharp click of wavelength 1.5 cm in the ocean. What is the frequency of such clicks? (c) **Dog whistles.** One brand of dog whistles claims a frequency of 25 kHz for its product. What is the wavelength of this sound? (d) **Bats.** While bats emit a wide variety of sounds, one type emits pulses of sound having a frequency between 39 kHz and 78 kHz. What is the range of wavelengths of this sound? (e) **Sonograms.** Ultrasound is used to view the interior of the body, much as x rays are utilized. For sharp imagery, the wavelength of the sound should be around one-fourth (or less) the size of the objects to be viewed. Approximately what frequency of sound is needed to produce a clear image of a tumor that is 1.0 mm across if the speed of sound in the tissue is 1550 m/s?

38. •• A 75.0 cm wire of mass 5.625 g is tied at both ends and adjusted to a tension of 35.0 N. When it is vibrating in its second overtone, find (a) the frequency and wavelength at which it is vibrating and (b) the frequency and wavelength of the sound waves it is producing.

12.10 Sound Intensity

39. • A small omnidirectional stereo speaker produces waves in all directions that have an intensity of 6.50 W/m² at a distance of 2.50 m from the speaker. (a) At what rate does this speaker produce energy? (b) What is the intensity of this sound 7.00 m from the speaker? (c) What is the *total* amount of energy received each second by the walls (including windows and doors) of the room in which this speaker is located?

40. • Find the intensity (in W/m²) of (a) a 55.0 dB sound, (b) a 92.0 dB sound, (c) a −2.0 dB sound.

41. • Find the noise level (in dB) of a sound having an intensity of (a) 0.000127 W/m², (b) 6.53 × 10⁻¹⁰ W/cm², (c) 1.5 × 10⁻¹⁴ W/m².

42. • (a) By what factor must the sound intensity be increased to raise the sound intensity level by 13.0 dB? (b) Explain why you don't need to know the original sound intensity.

43. •• **Eavesdropping!** You are trying to overhear a juicy conversation, but from your distance of 15.0 m, it sounds like only an average whisper of 20.0 dB. So you decide to move closer to give the conversation a sound level of 60.0 dB instead. How close should you come?

44. •• **Energy delivered to the ear.** Sound is detected when a
BIO sound wave causes the tympanic membrane (the eardrum) to vibrate. (See Figure 12.29). Typically, the diameter of this membrane is about 8.4 mm in humans. (a) How much energy is delivered to the eardrum each second when someone whispers (20 dB) a secret in your ear? (b) To comprehend how sensitive the ear is to very small amounts of energy, calculate how fast a typical 2.0 mg mosquito would have to fly (in mm/s) to have this amount of kinetic energy.

45. •• **Human hearing.** A fan at a rock concert is 30 m from the
BIO stage, and at this point the sound intensity level is 110 dB. (a) How much energy is transferred to her eardrums each second? (See the previous problem.) (b) How fast would a 2.0 mg mosquito have to fly to have this much kinetic energy? Compare the mosquito's speed with that found for the whisper in part (a) of the previous problem.

46. •• The intensity due to a number of independent sound sources is the sum of the individual intensities. (a) When four quadruplets cry simultaneously, how many decibels greater is the sound intensity level than when a single one cries? (b) To increase the sound intensity level again by the same number of decibels as in part (a), how many more crying babies are required?

47. •• (a) What is the sound intensity level in a car when the sound intensity is 0.500 μW/m²? (b) What is the sound intensity in the air near a jackhammer when the sound intensity level is 103 dB?

12.11 Beats

48. • A trumpet player is tuning his instrument by playing an A note simultaneously with the first-chair trumpeter, who has perfect pitch. The first-chair player's note is exactly 440 Hz, and 2.8 beats per second are heard. What are the two possible frequencies of the other player's note?

49. • Two tuning forks are producing sounds of wavelength 34.40 cm and 33.94 cm simultaneously. How many beats do you hear each second?

50. • Two guitarists attempt to play the same note of wavelength 6.50 cm at the same time, but one of the instruments is slightly out of tune and plays a note of wavelength 6.52 cm instead. What is the frequency of the beat these musicians hear when they play together?

51. •• **Tuning a violin.** A violinist is tuning her instrument to concert A (440 Hz). She plays the note while listening to an electronically generated tone of exactly that frequency and hears a beat of frequency 3 Hz, which increases to 4 Hz when she tightens her violin string slightly. (a) What was the frequency of her violin when she heard the 3-Hz beat? (b) To get her violin perfectly tuned to concert A, should she tighten or loosen her string from what it was when she heard the 3-Hz beat?

12.12 The Doppler Effect

52. • A railroad train is traveling at 25.0 m/s in still air. The frequency of the note emitted by the locomotive whistle is 400 Hz. What is the wavelength of the sound waves (a) in front of the locomotive? (b) behind the locomotive? What is the frequency of the sound heard by a stationary listener (c) in front of the locomotive? (d) behind the locomotive?

53. • Two train whistles, A and B, each have a frequency of 392 Hz. A is stationary and B is moving toward the right (away from A) at a speed of 35.0 m/s. A listener is between the two whistles and is moving toward the right with a speed of 15.0 m/s. (See Figure 12.45.) (a) What is the frequency from A as heard by the listener? (b) What is the frequency from B as heard by the listener? (c) What is the beat frequency detected by the listener?

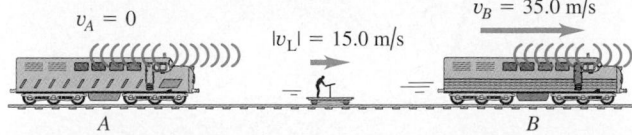

▲ **FIGURE 12.45** Problem 53.

54. • On the planet Arrakis, a male ornithoid is flying toward his stationary mate at 25.0 m/s while singing at a frequency of 1200 Hz. If the female hears a tone of 1240 Hz, what is the speed of sound in the atmosphere of Arrakis?

55. • A car alarm is emitting sound waves of frequency 520 Hz. You are on a motorcycle, traveling directly away from the car. How fast must you be traveling if you detect a frequency of 490 Hz?

56. • A railroad train is traveling at 30.0 m/s in still air. The frequency of the note emitted by the train whistle is 262 Hz. What frequency is heard by a passenger on a train moving in the opposite direction to the first at 18.0 m/s and (a) approaching the first; and (b) receding from the first?

57. •• The siren of a fire engine that is driving northward at 30.0 m/s emits a sound of frequency 2000 Hz. A truck in front of this fire engine is moving northward at 20.0 m/s. (a) What is the frequency of the siren's sound that the fire engine's driver hears reflected from the back of the truck? (b) What wavelength would this driver measure for these reflected sound waves?

58. •• A stationary police car emits a sound of frequency 1200 Hz that bounces off of a car on the highway and returns with a frequency of 1250 Hz. The police car is right next to the highway, so the moving car is traveling directly toward or away from it. (a) How fast was the moving car going? Was it moving towards or away from the police car? (b) What frequency would the police car have received if it had been traveling toward the other car at 20.0 m/s?

59. •• A container ship is traveling westward at a speed of 5.00 m/s. The waves on the surface of the ocean have a wavelength of 40.0 m and are traveling eastward at a speed of 16.5 m/s. (a) At what time intervals does the ship encounter the crest of a wave? (b) At what time intervals will the ship encounter wave crests if it turns around and heads eastward?

60. •• While sitting in your car by the side of a country road, you see your friend, who happens to have an identical car with an identical horn, approaching you. You blow your horn, which has a frequency of 260 Hz; your friend begins to blow his horn as well, and you hear a beat frequency of 6.0 Hz. How fast is your friend approaching you?

61. •• **Moving source vs. moving listener.** (a) A sound source producing 1.00 kHz waves moves toward a stationary listener at one-half the speed of sound. What frequency will the listener hear? (b) Suppose instead that the source is stationary and the listener moves toward the source at one-half the speed of sound. What frequency does the listener hear? How does your answer compare with that in part (a)? Did you expect to get the same answer in both cases? Explain on physical grounds why the two answers differ.

62. •• How fast (as a percentage of light speed) would a star have to be moving so that the frequency of the light we receive from it is 10.0% higher than the frequency of the light it is emitting? Would it be moving away from us or toward us? (Assume it is moving either directly away from us or directly toward us.)

General Problems

63. •• One end of a 14.0-m-long wire having a total mass of 0.800 kg is fastened to a fixed support in the ceiling, and a 7.50 kg object is hung from the other end. If the wire is struck a transverse blow at one end, how much time does the pulse take to reach the other end? Neglect the variation in tension along the length of the wire.

64. •• **Ultrasound in medicine.** A 2.00 MHz sound wave travels **BIO** through a pregnant woman's abdomen and is reflected from the fetal heart wall of her unborn baby. The heart wall is moving toward the sound receiver as the heart beats. The reflected sound is then mixed with the transmitted sound, and 85 beats per second are detected. The speed of sound in body tissue is 1500 m/s. Calculate the speed of the fetal heart wall at the instant this measurement is made.

65. •• A very noisy chain saw operated by a tree surgeon emits a total acoustic power of 20.0 W uniformly in all directions. At what distance from the source is the sound level equal to (a) 100 dB, (b) 60 dB?

66. •• **Tuning a cello.** A cellist tunes the C-string of her instrument to a fundamental frequency of 65.4 Hz. The vibrating portion of the string is 0.600 m long and has a mass of 14.4 g. (a) With what tension must she stretch that portion of the string? (b) What percentage increase in tension is needed to increase the frequency from 65.4 Hz to 73.4 Hz, corresponding to a rise in pitch from C to D?

67. •• A person is playing a small flute 10.75 cm long, open at one end and closed at the other, near a taut string having a fundamental frequency of 600.0 Hz. If the speed of sound is 344.0 m/s, for which harmonics of the flute will the string resonate? In each case, which harmonic of the string is in resonance?

68. •• A bat flies toward a wall, emitting a steady sound of frequency 2000 Hz. The bat hears its own sound, plus the sound reflected by the wall. How fast should the bat fly in order to hear a beat frequency of 10.0 Hz? (*Hint:* Break this problem into two parts, first with the bat as the source and the wall as the listener and then with the wall as the source and the bat as the listener.)

69. •• You're standing between two speakers that are driven by the same amplifier and are emitting sound waves with frequency 229 Hz. The two speakers are facing each other, 15 meters apart. (a) You begin walking away from one speaker toward the other one, and as you walk, you hear what sounds like beats, with a frequency of 2.50 Hz. How fast are you walking? (b) If the frequency of the sound emitted by the speakers increases to 573 Hz and you continue to walk at the same speed, what frequency of beats will you hear? [*Hint:* You can model this situation as a tube open at both ends; alternatively, you can treat it as a Doppler effect problem.]

70. •• The sound source of a ship's sonar system operates at a frequency of 22.0 kHz. The speed of sound in water (assumed to be at a uniform 20° C) is 1482 m/s. (a) What is the wavelength of the waves emitted by the source? (b) What is the difference in frequency between the directly radiated waves and the waves reflected from a whale traveling straight toward the ship at 4.95 m/s? The ship is at rest in the water.

71. •• **The range of human hearing.** A young person with nor-**BIO** mal hearing can hear sounds ranging from 20 Hz to 20 kHz. How many octaves can such a person hear? (Recall that if two tones differ by an octave, the higher frequency is twice the lower frequency.)

72. •• A person leaning over a 125-m-deep well accidentally drops a siren emitting sound of frequency 2500 Hz. Just before this siren hits the bottom of the well, find the frequency and wavelength of the sound the person hears (a) coming directly from the siren, (b) reflected off the bottom of the well. (c) What beat frequency does this person perceive?

73. •• A police siren of frequency f_{siren} is attached to a vibrating platform. The platform and siren oscillate up and down in simple

harmonic motion with amplitude A_P and frequency f_P. (a) Find the maximum and minimum sound frequencies that you would hear at a position directly above the siren. (b) At what point in the motion of the platform is the maximum frequency heard? The minimum frequency? Explain.

74. •• A flexible stick can oscillate in a standing-wave pattern, much as a string does. (a) Are the ends of the stick displacement nodes or antinodes? Why? (If in doubt, use a handy stick at home or in the laboratory and try it out.) (b) A flexible stick 3.00 m long is free to wiggle. Using the conditions from part (a), find the wavelengths of the first five harmonics for this stick.

75. •• A turntable 1.50 m in diameter rotates at 75 rpm. Two speakers, each giving off sound of wavelength 31.3 cm, are attached to the rim of the table at opposite ends of a diameter. A listener stands in front of the turntable. (a) What is the greatest beat frequency the listener will receive from this system? (b) Will the listener be able to distinguish individual beats?

76. •• **Musical scale.** The frequency ratio of a semitone interval on the equally tempered scale is $2^{1/12}$. (a) Show that this ratio is 1.059. (b) Find the speed of an automobile passing a listener at rest in still air if the pitch of the car's horn drops a semitone between the times when the car is coming directly toward him and when it is moving directly away from him.

77. •• Two organ pipes, open at one end but closed at the other, are each 1.14 m long. One is now lengthened by 2.00 cm. Find the frequency of the beat they produce when playing together in their fundamental.

Passage Problems

BIO Temperature and instrument tuning. The speed of a longitudinal sound wave in a gas depends on both the temperature and composition of the gas. For a given gas the speed of sound typically increases with increasing temperature: for instance, the speed of sound in air increases from about 330 m/s at 0°C to 350 m/s at 30°C. In addition, the speed of sound tends to be higher in gases composed of lightweight monatomic molecules (such as helium) and lower in gases composed of heavy polyatomic molecules (such as carbon dioxide)—for example, at 20°C, the speed of sound in helium is 999 m/s and the speed of sound in carbon dioxide is 260 m/s.

The flute is a musical instrument that behaves as a pipe that is open at both ends, while the violin is a musical instrument based on a tuned string. Initially a flute and a violin are tuned so that they both play middle C at 262 Hz, when the speed of sound in the air in the room is 350 m/s.

After the initial tuning, the temperature in the room drops so that the speed of sound in the air decreases by 2%. The instruments are not retuned to compensate.

78. By how much does middle C on the flute change?
   A. There is no change.
   B. It increases by 2%.
   C. It decreases by 2%.
   D. It increases by 4%.
   E. It decreases by 4%.

79. By how much does middle C on the violin change?
   A. There is no change.
   B. It increases by 2%.
   C. It decreases by 2%.
   D. It increases by 4%.
   E. It decreases by 4%.

80. Roughly, what would be the resulting beat frequency between the two instruments when played at the reduced temperature?
   A. 0 Hz
   B. 2 Hz
   C. 5 Hz
   D. 7 Hz
   E. 10.5 Hz

$I3$ Fluid Mechanics

Fluids play a vital role in many aspects of everyday life. We drink them, breathe them, and swim in them; they circulate through our bodies, they control our weather, airplanes fly through them, and ships float in them. A fluid is any substance that can flow; we use the term for both liquids and gases. Usually, gases are easily compressed while liquids are quite incompressible. We'll see some of the consequences of this difference.

We begin our study with *fluid statics,* the study of fluids at rest (in equilibrium). Like the analysis of other equilibrium situations, the study of fluid statics is based on Newton's first and third laws. The key concepts include density, pressure, buoyancy, and surface tension. *Fluid dynamics* is the study of fluids in motion. Fluid dynamics is much more complex and indeed is one of the most complex branches of mechanics. Fortunately, we can analyze many important situations with simple idealized models and familiar principles such as Newton's laws and conservation of energy. Even so, we will barely scratch the surface of this broad and interesting topic.

To sail well, the crew must understand how their boat interacts with two fluids, the water and the air.

13.1 Density

A block of Styrofoam™ can have the same mass as a steel nail, but a much greater volume. To express this difference, we need the concept of *density.* The **density** of a material is defined as its mass per unit volume (Figure 13.1). Because density is defined as *mass divided by volume,* all objects made of a given material have the same density, irrespective of their size.

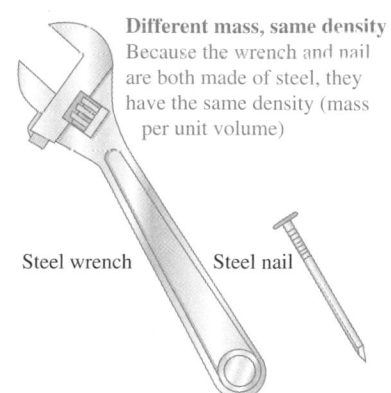

Different mass, same density Because the wrench and nail are both made of steel, they have the same density (mass per unit volume)

Steel wrench   Steel nail

▲ **FIGURE 13.1** Density is mass per unit volume.

▲ Application Layers of liquids. This coffee concoction was made by using liquids of differing densities and a steady hand. The liquid with the highest specific gravity (density relative to water) was poured first and the lighter liquid was carefully poured on top of the previous, more dense layer. As you know, any liquid will float on top of another liquid with a greater density, as long as no mixing occurs.

Definition of density

We use the Greek letter ρ (rho) for density. If a mass m of material has volume V, its density is

$$\rho = \frac{m}{V}. \tag{13.1}$$

Units: The SI unit of density is the kilogram per cubic meter (1 kg/m^3). The cgs unit, the gram per cubic centimeter (1 g/cm^3), is also widely used. (Notice that $1 \text{ g/cm}^3 = 1000 \text{ kg/m}^3$.)

A homogeneous material such as ice or iron has the same density throughout. Table 13.1 gives the densities of several common solids and liquids at ordinary temperatures. Note the wide range of magnitudes. The densest material found on earth is the metal osmium $(22,500 \text{ kg/m}^3, \text{ or } 22.5 \text{ g/cm}^3)$. The density of air at sea level is about 1.2 kg/m^3 (0.0012 g/cm^3), but the density of white-dwarf stars is on the order of 10^{10} kg/m^3, and that of neutron stars is on the order of 10^{18} kg/m^3!

The **specific gravity** of a material is the ratio of its density to the density of water; it is a pure (unitless) number. For example, the specific gravity of aluminum is 2.7; this means that the density of aluminum is 2.7 times as great as that of water. ("Specific gravity" is a poor term, since it has nothing to do with gravity; "relative density" would be better.)

Density measurements are an important analytical technique. For example, an auto mechanic can determine the freezing point of antifreeze by measuring its density. Antifreeze used in car engines is usually a solution of ethylene glycol (density $1.12 \times 10^3 \text{ kg/m}^3$) in water. The glycol concentration determines the freezing point of the solution; it can be found from a simple density measurement, performed routinely in service stations with the aid of a *hydrometer*. The auto mechanic measures density with this instrument by observing the level at which a calibrated body floats in a sample of the solution. (The hydrometer is discussed in Section 13.3.)

TABLE 13.1   Densities of Some Common Substances

Material	Density (kg/m^3)*	Material	Density (kg/m^3)*
Gases		Concrete	2×10^3
Air (1 atm, 20° C)	1.20	Aluminum	2.7×10^3
		Iron, steel	7.8×10^3
Liquids		Brass	8.6×10^3
Benzene	0.90×10^3	Copper	8.9×10^3
Ethanol	0.81×10^3	Silver	10.5×10^3
Water	1.00×10^3	Lead	11.3×10^3
Seawater	1.03×10^3	Gold	19.3×10^3
Blood	1.06×10^3	Platinum	21.4×10^3
Solids		Mercury	13.6×10^3
Glycerin	1.26×10^3	White-dwarf star	10^{10}
Ice	0.92×10^3	Neutron star	10^{18}

*To obtain the densities in grams per cubic centimeter, simply divide by 10^3.

Conceptual Analysis 13.1

Changing density?

If you cut a block of modeling clay in half, the density of each half is _____ the density of the original block.

A. twice   B. the same as   C. half

SOLUTION Because the clay in the block is a homogeneous material, its density (mass per unit volume) is the same for pieces of any size. Thus, B is the right answer.

EXAMPLE 13.1 **The weight of a roomful of air**

Find the mass of air, and its weight, in a living room with a 4.0 m × 5.0 m floor and a ceiling 3.0 m high. What would be the mass and weight of an equal volume of water?

SOLUTION

SET UP AND SOLVE Even for a problem like this, a simple sketch such as the one in Figure 13.2 helps to prevent mistakes. First we'll find the volume of the room; then we'll use the density of air to find the mass of air it contains. The room's volume is $V = l \times w \times h = (4.0 \text{ m})(5.0 \text{ m})(3.0 \text{ m}) = 60 \text{ m}^3$.

From Table 13.1, air has a density of $\rho = 1.2 \text{ kg/m}^3$. From Equation 13.1, $\rho = m/V$, so the mass of air in the room is $m = \rho V = (1.2 \text{ kg/m}^3)(60 \text{ m}^3) = 72 \text{ kg}$. The weight of the air is $w - mg - (72 \text{ kg})(9.8 \text{ m/s}^2) = 710 \text{ N}$.

From Table 13.1, the density of water is $\rho = 1.0 \times 10^3 \text{ kg/m}^3$. Thus, the mass of a volume of water that would fill the room is $m = \rho V = (1000 \text{ kg/m}^3)(60 \text{ m}^3) = 6.0 \times 10^4 \text{ kg}$. The weight is $w = (6.0 \times 10^4 \text{ kg})(9.8 \text{ m/s}^2) = 5.9 \times 10^5 \text{ N}$.

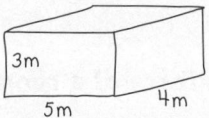

▲ **FIGURE 13.2** Our sketch for this problem.

REFLECT The weight of the air is about 160 lb. Does it surprise you that a roomful of air weighs this much? But the weight of the same volume of water is about 1.3×10^5 lb, or 66 tons! This much weight would certainly collapse the floor of an ordinary house. (We'll learn later why the weight of the room's air doesn't stress the floor.)

Practice Problem: What volume of water would have a mass equal to the mass of air in the room? *Answer:* $7.2 \times 10^{-2} \text{ m}^3$, or 72 L.

13.2 Pressure in a Fluid

When a fluid (either a liquid or a gas) is at rest, it exerts a force perpendicular to any surface in contact with it, such as the wall of a container or the surface of a body immersed in it. To see that this must be so, imagine a surface located *within* the fluid (Figure 13.3a). The fluid on either side of this surface exerts forces on it. Because the fluid is at rest, we know that the forces acting on the two sides must be equal in magnitude and opposite in direction (otherwise the surface would accelerate and the fluid would not remain at rest), and they must be oriented perpendicular to the surface. Similarly, any surface of an object immersed in the fluid is acted upon by a force perpendicular to the surface.

Pressure is force per unit area. We define the **pressure** p at a point in a fluid as the ratio of the normal force $F_\perp$ on a small area A around that point to the area.

The force exerted by the fluid on any surface must be perpendicular to the surface, otherwise it would have a component of shear that would cause the surface to accelerate.

Arbitrary surface in fluid

(a) Why forces due to fluid pressure must act normal to a surface.

Definition of pressure

The pressure p on a plane surface of area A is the magnitude $F_\perp$ of the force exerted on that surface, divided by the area:

$$p = \frac{F_\perp}{A}, \qquad F_\perp = pA. \qquad \text{(if } p \text{ is the same at all points on the surface)} \qquad (13.2)$$

Units: The SI unit of pressure is N/m^2. As we noted in Chapter 11, this unit is given a special name: the pascal (Pa). That is, $1 \text{ N/m}^2 = 1 \text{ Pa}$. Two related units, used principally in meteorology, are the *bar,* equal to 10^5 Pa, and the *millibar,* equal to 100 Pa. The British unit pounds per square inch (lb/in.², or psi) is still in common use in the United States. Another unit of pressure is the *atmosphere,* defined next. Later in this chapter, we'll encounter additional units used in specific situations.

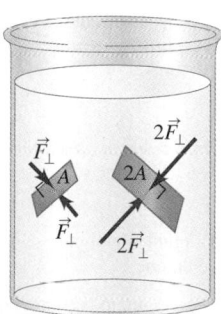

Although these two surfaces differ in area and orientation, the pressure on them (force divided by area) is the same.

Note that pressure is a scalar—it has no direction.

(b) Pressure equals force divided by area.

▲ **FIGURE 13.3** The forces due to pressure on arbitrary surfaces within a fluid.

Atmospheric pressure p_{atm} is the pressure of the earth's atmosphere—the pressure in the sea of air in which we live. This pressure varies with weather changes and with elevation. The average atmospheric pressure at sea level is used as a unit of pressure called the **atmosphere** (atm), defined as

$$1 \text{ atm} = 1.013 \times 10^5 \text{ Pa} = 14.7 \text{ psi} = 1.013 \text{ bars} = 1013 \text{ millibars}.$$

EXAMPLE 13.2 The force of a roomful of air

In the room described in Example 13.1, what is the total force on the floor due to the air above the surface if the air pressure is 1.00 atm?

SOLUTION

SET UP The total force on the floor has magnitude equal to the pressure multiplied by the area. We're given 1.00 atm of pressure—that is, $p = 1.013 \times 10^5$ Pa.

SOLVE The floor area is $A = (4.0 \text{ m})(5.0 \text{ m}) = 20 \text{ m}^2$. From Equation 13.3, the magnitude $F_\perp$ of the total force is

$$F_\perp = pA = (1.013 \times 10^5 \text{ Pa})(20 \text{ m}^2)$$
$$= 2.0 \times 10^6 \text{ N} \ (= 4.5 \times 10^5 \text{ lb} = 225 \text{ tons}).$$

REFLECT This force is equal to the total weight of all the air in a column directly above the floor. It is more than enough force to collapse the floor. So why doesn't it collapse? Because there is an upward force on the underside of the floor. If we neglect the thickness of the floor, this upward force is exactly equal in magnitude to the downward force on the top surface, and the total force due to air pressure is zero.

Practice Problem: If the room has a window with dimensions 0.50 m × 1.0 m, what is the total force on the window due to the air inside? Why doesn't this force push the window out? *Answer:* 5.1×10^4 N; there is equal pressure on the outside surface.

Pascal's Law

If the weight of the fluid can be neglected, the pressure in a fluid at rest is the same throughout. We used that assumption in our discussion of bulk stress and strain in Section 11.3. But often the fluid's weight is *not* negligible. The reason atmospheric pressure is greater at sea level than on a high mountain is that there is more air above you at sea level and greater weight pushing down on you. And when you dive into deep water, your ears tell you that the pressure increases rapidly with increasing depth below the surface.

We can derive a general relation between the pressure p at any point in a fluid in a gravitational field and the elevation h of that point relative to some reference level (where $h = 0$ and the pressure is p_0). We define h to be positive for points above the reference level. If the fluid is in equilibrium, every portion of it is in equilibrium. Consider the rectangular volume element shown in Figure 13.4. The bottom surface of the element is at $h = 0$, and the top surface is a distance h above that. The area of each of these surfaces is A. The height of the element is h, its volume is $V = hA$, its mass is $m = \rho V = \rho hA$, and its weight has magnitude $w = mg = \rho ghA$.

Let the pressure at the bottom surface of the volume element be p_0; then the total upward force on this surface (due to the fluid underneath pushing up) has magnitude $p_0 A$. The pressure at the top surface is p, and the total downward force on the top surface (due to the fluid above pushing down) has magnitude pA. The fluid in this volume is in equilibrium, so the total vertical component of force, including the weight of the fluid in the volume element and the forces at the bottom and top surfaces, must be zero. This condition gives

$$p_0 A - pA - \rho ghA = 0.$$

We can divide out the area A and rearrange terms, obtaining the following relation:

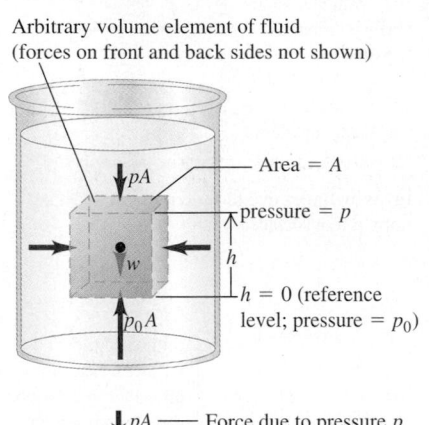

Arbitrary volume element of fluid (forces on front and back sides not shown)

Area = A

pressure = p

$h = 0$ (reference level; pressure = p_0)

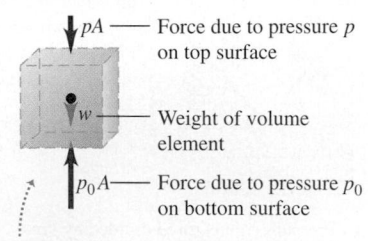

pA —— Force due to pressure p on top surface

w —— Weight of volume element

$p_0 A$ —— Force due to pressure p_0 on bottom surface

Because the fluid is in equilibrium, the vector sum of the vertical forces on the volume element must be zero:
$$p_0 A - pA - w = 0$$

▲ **FIGURE 13.4** The forces on an arbitrary volume element in a fluid.

Variation of pressure with height in a fluid

If the pressure at a point in a fluid is p_0, then, at a point in the fluid at a distance h above this point, the pressure is $p_0 - \rho g h$; that is,

$$p = p_0 - \rho g h. \qquad \text{(variation of pressure with height)} \qquad (13.3)$$

When h is positive, p is less than p_0; when h is negative (i.e., a point deeper in the fluid than the reference level $h = 0$), p is greater than p_0, as we would expect. Pressure increases with depth: When we decrease h, we increase p.

Let's apply this equation to a liquid in an open container, as shown in Figure 13.5. Again, we let p_0 be the pressure at $h = 0$ (a distance h below the surface). At the surface of the liquid, the pressure is atmospheric pressure, so $p - p_{\text{atm}}$. We're interested in the pressure p_0 at a depth h below the surface. In Equation 13.3, we substitute p_{atm} for p; we obtain $p_{\text{atm}} - p_0 = -\rho g h$, or

$$p_0 = p_{\text{atm}} + \rho g h. \qquad (13.4)$$

The pressure p_0 at a depth h below the surface is greater than the pressure p_{atm} at the surface by an amount $\rho g h$. It also follows from Equation 13.4 that if the pressure at the top surface is increased in any way—say, by inserting a piston on the top surface and pressing down on it—the pressure p_0 at any depth must increase by exactly the same amount. This fact was recognized in 1653 by the French scientist Blaise Pascal (1623–1662) and is called **Pascal's law:**

Pascal's law

Pressure applied to an enclosed fluid is transmitted undiminished to every portion of the fluid and the walls of the containing vessel. The pressure depends only on depth; the shape of the container does not matter.

For instance, notice that the liquid in Figure 13.6 rises to the same height in all of the arms, regardless of their shape. This is the level at which $p = p_{\text{atm}}$. At any level below the surface, the pressure is the same in all four arms.

The hydraulic jack shown schematically in Figure 13.7 illustrates Pascal's law. A piston with a small cross-sectional area A_1 exerts a force with magnitude F_1 on the surface of a liquid such as oil. The pressure $p = F_1/A_1$ is transmitted through the connecting pipe to a larger piston with area A_2. In this case, the two pistons are at the same height, so they experience the same pressure p:

$$p = \frac{F_1}{A_1} = \frac{F_2}{A_2} \qquad \text{and} \qquad F_2 = \frac{A_2}{A_1}F_1. \qquad (13.5)$$

Therefore, the hydraulic jack is a force-multiplying device with a multiplication factor equal to the ratio (A_2/A_1) of the areas of the two pistons. Dentists' chairs, car lifts and jacks, some elevators, and hydraulic brakes all use this principle.

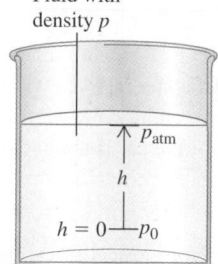

Fluid with density p

At a given level, the pressure p equals the external pressure (here, p_{atm}) plus the pressure due to the weight of the overlying liquid ($\rho g h$, where h is the distance below the surface): $p_0 = p_{\text{atm}} + \rho g h$

▲ **FIGURE 13.5** The pressure at a depth h in a liquid is greater than the surface pressure by $\rho g h$.

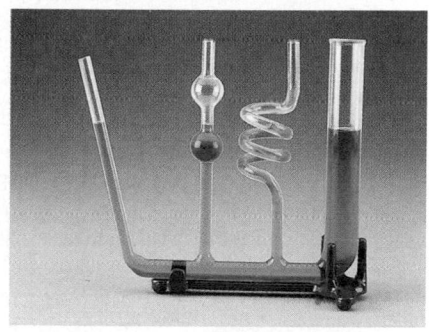

▲ **FIGURE 13.6** The pressure in a fluid is the same at all points having the same elevation, regardless of the container's shape.

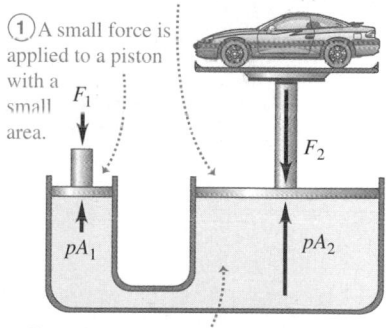

③ Acting on a piston with a large area, the pressure creates a force that can support a car.

① A small force is applied to a piston with a small area. F_1

F_2

pA_1 pA_2

② At any given height, the pressure p is the same everywhere in the fluid (Pascal's law).

▲ **FIGURE 13.7** The principle of the hydraulic lift, an application of Pascal's law.

EXAMPLE 13.3 The hydraulic lift: an application of Pascal's law

Suppose the hydraulic lift shown in Figure 13.7 has a small cylindrical piston with radius 5.0 cm and a larger piston with radius 20 cm. The mass of a car placed on the larger piston platform is 1000 kg. **(a)** Assuming that the two pistons are at the same height, what force must be applied to the small piston to lift the car? **(b)** How far must the small piston move down to lift the car through a height of 0.10 m?

Continued

SOLUTION

SET UP A downward external force with magnitude F_1 is applied on the smaller piston, and a downward force with magnitude F_2 equal to the weight of the car is applied on the larger piston. The weight of the car is $F_2 = w = mg = (1000 \text{ kg})(9.8 \text{ m/s}^2) = 9800 \text{ N}$.

SOLVE Part (a): From Pascal's law, the pressure at any given level is the same throughout the hydraulic fluid. Since the pistons are at the same level, we can use Equation 13.5. The downward force on the smaller piston needed to lift the car is

$$F_1 = \frac{A_1}{A_2} F_2 = \frac{\pi r_1^2}{\pi r_2^2} F_2 = \frac{\pi (0.050 \text{ m})^2}{\pi (0.20 \text{ m})^2}(9800 \text{ N}) = 610 \text{ N}.$$

So a force of 610 N can lift a car weighing 9800 N.

Part (b): If the small piston moves a distance d_1, the volume of fluid displaced is $V_1 = d_1 A_1$. Because the total volume of fluid is constant, V_1 must equal the volume V_2 displaced by the large piston. It follows that $d_1 A_1 = d_2 A_2$. If the car is hoisted a distance $d_2 = 0.10 \text{ m}$, the small piston moves a distance d_1 given by

$$d_1 = \frac{\pi r_2^2}{\pi r_1^2} d_2 = \frac{\pi (0.20 \text{ m})^2}{\pi (0.05 \text{ m})^2}(0.10 \text{ m}) = 1.6 \text{ m}.$$

ALTERNATIVE SOLUTION Part (b) may also be solved by using conservation of energy. If we neglect friction, the amount of work done by the applied force must equal the work done to lift the car. Thus,

$$W_1 = W_2 \quad \text{or} \quad F_1 d_1 = F_2 d_2,$$
$$(610 \text{ N})(d_1) = (9800 \text{ N})(0.10 \text{ m}),$$
$$d_1 = 1.6 \text{ m}.$$

REFLECT The hydraulic lift is a force-multiplying device. The multiplication factor is the ratio of the areas of the pistons. It seems as though we get something for nothing. But the much larger weight is lifted through a much smaller distance, and the two quantities of work are the same in magnitude. The device multiplies force, but it can't multiply *energy*.

Practice Problem: Engineers are designing a hydraulic lift that can be used to repair trucks with a mass of 5000 kg. What ratio of piston areas would you suggest for the design if the maximum available external force is to be 3000 N? *Answer:* $A_2/A_1 = 16.3$.

 **Conceptual Analysis 13.2**

Atmospheric pressure

Atmospheric pressure is about 1.013×10^5 Pa at sea level. This means that

A. the weight of a column of air 1 square meter in cross section extending up to the top of the atmosphere is about 1.013×10^5 N.

B. the weight of 1 cubic meter of air at sea level is about 1.013×10^5 N.

C. the density of air at sea level is about 1.013×10^5 times the density of water at sea level.

SOLUTION As we've just learned, the pressure at a given level in a fluid (such as air) is caused by the weight of the overlying fluid, plus any external pressure applied to the fluid. In the case of earth's atmosphere, there is no external pressure, because the atmosphere is open to outer space at the top. Remember also that pressure is force divided by area. In the case of the atmosphere, the force is the weight of the overlying air. If atmospheric pressure is 1.013×10^5 Pa, then the air exerts a weight force of 1.013×10^5 N per square meter. Thus, choice A is correct and choices B and C are not.

The assumption that the density ρ of a fluid is constant is often reasonable for liquids, which are relatively incompressible, but it is realistic for gases only over short vertical distances. When we calculated the weight of air in a room of known volume, it was reasonable to assume that the density of the air was the same throughout the room. To justify this assumption, we note that, in a room with a ceiling height of 3.0 m filled with air of density 1.2 kg/m³, the difference in pressure between floor and ceiling, given by Equation 13.3, is

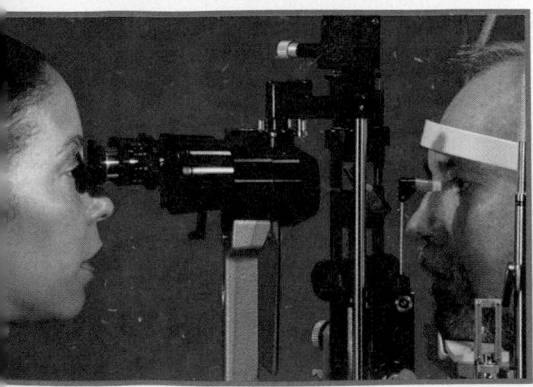

◄ **BIO Application A close look at pressure.** Fluid pressure in the human body is highly regulated, not just in the veins and arteries, but within the eye as well. This instrument, called a *tonometer*, is being used to measure the pressure of the fluid inside the eyeball as a means of diagnosing glaucoma, an eye disease characterized by a loss of vision from damage to the retina and the optic nerve. Glaucoma is often caused by excessive pressure within the eyeball. As Pascal's law states, if the weight of a fluid at rest can be neglected, the pressure in the fluid at rest is the same at all points. Therefore, using a tonometer to measure the pressure in the front of the eyeball, doctors can determine the pressure against the optic nerve at the back and diagnose whether excess pressure is the cause of vision problems.

$$\rho g h = (1.2 \text{ kg/m}^3)(9.8 \text{ m/s}^2)(3.0 \text{ m}) = 35 \text{ Pa},$$

or about 0.00035 atm, a very small difference. The resulting difference in density between floor and ceiling is correspondingly very small. But between sea level and the summit of Mount Everest (8850 m), the density of air changes by a factor of nearly three. In such situations, it wouldn't be correct to use Equation 13.3 $(p - p_0 = -\rho g h)$.

At great enough pressures, even liquids are compressed appreciably. The deepest point in the oceans is the Marianas Trench, which is 10,920 m deep. If the density of seawater were constant, the pressure at this depth would be 1.10×10^8 Pa (about 1000 atm or 16,000 lb/in.^{2}). But because seawater is compressed to greater density under high pressures, the actual pressure at this depth is 1.16×10^8 Pa (nearly 17,000 lb/in.^{2}).

If the pressure inside a car tire is equal to atmospheric pressure, the tire is flat. The pressure has to be *greater than* atmospheric to support the car, and the significant quantity is the *difference* between the inside and outside pressures. When we say that the pressure in a car tire is "32 pounds" (actually, 32 lb/in.^{2} = 32 psi, equal to 220 kPa, or 2.2×10^5 Pa), we mean that it is *greater* than atmospheric pressure $(14.7 \text{ lb/in.}^2, \text{ or } 1.01 \times 10^5 \text{ Pa})$ by that amount.

The *total* pressure in the tire is 47 lb/in.^{2}, or 320 kPa. The excess pressure above atmospheric pressure is usually called the **gauge pressure,** and the total pressure is called the **absolute pressure.** Engineers use the abbreviations psig and psia for "pounds per square inch gauge" and "pounds per square inch absolute," respectively. If the pressure is *less than* atmospheric, as in a partial vacuum, the gauge pressure is negative.

▲ Application **On top of the situation.** Snowshoes have been used for centuries to allow travel on foot in deep snow, long before we had defined pressure as force divided by area. If you try to walk in fresh snow without snowshoes, your feet sink in because you exert too much pressure for the snow to support. Snowshoes reduce the pressure you exert by distributing your weight over a larger area, allowing you to walk atop the snow without sinking in significantly.

Conceptual Analysis 13.3 An open U-tube

The equilibrium state of two nonmixing liquids in an open U-tube is shown in Figure 13.8. At which level(s) must the pressure be the same in both sides of this tube?

A. h_1.
B. h_2.
C. h_3.
D. The pressure at h_1 must equal the pressure at h_2.

SOLUTION The two liquids rise to different heights in the tube because the brown liquid is denser than the blue liquid. At level h_1, if the pressure in the two sides of the tube were *different,* the liquids would respond by moving. Since the liquids are in equilibrium (motionless), the pressure at h_1 is the same on both sides of the tube—so A is correct. (This argument also tells us that D is not correct.) At h_3, the pressure on both sides of the tube is atmospheric, so C is also correct. At h_2, the pressure on the left

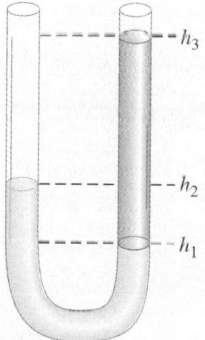

▲ **FIGURE 13.8**

side is atmospheric, but the pressure on the right side equals the atmospheric pressure *plus* the pressure due to the weight of the blue liquid above h_2. Thus, B is not correct.

EXAMPLE 13.4 **Finding absolute and gauge pressures**

A solar water-heating system uses solar panels on the roof, 12.0 m above the storage tank. The pressure at the level of the panels is 1 atmosphere. What is the absolute pressure at the top of the tank? The gauge pressure?

SOLUTION

SET UP Figure 13.9 shows our sketch.

SOLVE To find the absolute pressure p_{tank} at the tank, 12 m below the panels, we use Equation 13.4, with $p_0 = p_{\text{tank}}$:

$$p_{\text{tank}} = p_{\text{atm}} + \rho g h$$
$$= (1.01 \times 10^5 \text{ Pa})$$
$$\quad + (1.00 \times 10^3 \text{ kg/m}^3)(9.80 \text{ m/s}^2)(12.0 \text{ m})$$
$$= 2.19 \times 10^5 \text{ Pa}.$$

Continued

The gauge pressure is the amount by which p_{tank} exceeds atmospheric pressure—that is, $p_{\text{tank}} - p_{\text{atm}}$:

$$p_{\text{tank}} - p_{\text{atm}} = (2.19 - 1.01) \times 10^5 \, \text{Pa} = 1.18 \times 10^5 \, \text{Pa}$$
$$= 1.16 \, \text{atm} = 17.1 \, \text{lb/in}^2 = 17.1 \, \text{psig.}$$

REFLECT If the tank has a pressure gauge, it is usually calibrated to read gauge pressure rather than absolute pressure. As we have mentioned, the variation in *atmospheric* pressure over this height is negligibly small.

Practice Problem: At what distance below the panels is the gauge pressure equal to 1 atmosphere? *Answer:* 10.3 m.

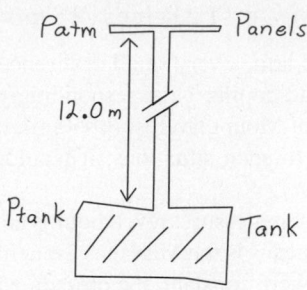

▲ **FIGURE 13.9** Our sketch for this problem.

Pressure Gauges

The simplest pressure gauge is the **open-tube manometer,** shown in Figure 13.10a. The U-shaped tube contains a liquid; one end of the tube is connected to the container in which the pressure is to be measured, and the other end is open to the atmosphere at pressure p_{atm}. According to Pascal's law, the pressures on the two sides of the tube at the level labeled $h = 0$ must be equal. On the left side, the pressure at this height is p_0, and on the right side (from Equation 13.4), it is $p_{\text{atm}} + \rho g h$. Equating these two expressions, we find that

$$p_0 = p_{\text{atm}} + \rho g h \qquad \text{and} \qquad p_{\text{gauge}} = \rho g h. \qquad (13.6)$$

In Equation 13.6, the pressure p_0 is the absolute pressure; the difference $p - p_{\text{atm}}$ between absolute and atmospheric pressure is the gauge pressure. Thus, the gauge pressure is proportional to the difference in height h of the liquid columns on the two sides.

Another common pressure gauge is the **mercury barometer;** it consists of a long glass tube, closed at one end, that has been filled with mercury and then inverted in a dish of mercury, as shown in Figure 13.10b. The space above the mercury column contains only mercury vapor; the pressure of the vapor is negligibly small, so the pressure above the mercury column is practically zero. From Equation 13.3, $(p = p_0 - \rho g h)$, with $p = 0$ and $p_0 = p_{\text{atm}}$,

$$p_{\text{atm}} = \rho g h.$$

NOTE ▶ The *shape* and *diameter* of the tubes in the open-tube manometer and mercury barometer don't matter. As we found in deriving Pascal's law, the fact that pressure is defined as force *per unit area* means that it is independent of area. ◀

Pressures are sometimes described in terms of the height of the corresponding mercury column as so many "inches of mercury" or "millimeters of mercury" (abbreviated mm Hg). Thus, the mercury barometer reads atmospheric pressure in mm Hg directly from the height of the mercury column. A pressure of 1 mm Hg is sometimes called 1 torr, after Evangelista Torricelli, inventor of the mercury barometer. The mm Hg and the torr depend on the density of mercury, which varies with temperature, and on the value of g, which varies with location, so the pascal is the preferred unit of pressure. As we noted earlier in this section, two related units are the *bar,* defined as 10^5 Pa, and the *millibar,* defined as 10^{-3} bar, or 10^2 Pa. Atmospheric pressures are on the order of 1000 millibars; the National Weather Service usually reports atmospheric pressures in this unit. The conversions are as follows:

$$1.013 \times 10^5 \, \text{Pa} = 1.013 \, \text{bars} = 1013 \, \text{millibars} = 1 \, \text{atm.}$$

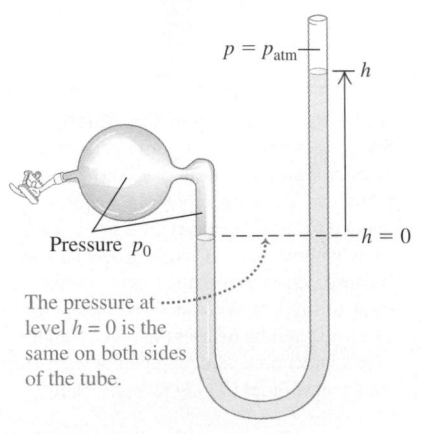

(a) Open-tube manometer

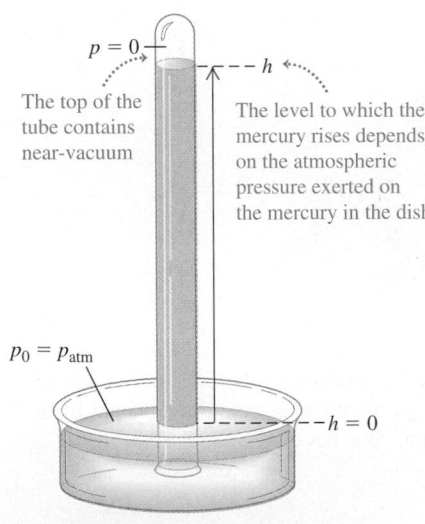

(b) Mercury barometer

▲ **FIGURE 13.10** Two types of pressure gauges.

Quantitative Analysis 13.4 **Two-liquid barometer**

The barometer in Figure 13.11 contains two liquids, A and B, differing in density. According to this device, the barometric pressure is

A. $\rho_A g h_A$.
B. $\rho_B g h_B$.
C. $\rho_A g h_A + \rho_B g h_B$.
D. $g(h_A + h_B)(\rho_A + \rho_B)/2$.

SOLUTION In the open dish, the surface of the liquid is at atmospheric pressure. Therefore, by Pascal's law, the liquid at that level inside the tube must also be at atmospheric pressure. But the pressure at this level in the tube *also* equals the pressure due to the weights of the overlying liquids. This pressure is given by the sum in choice C. Therefore, C gives the atmospheric (barometric) pressure.

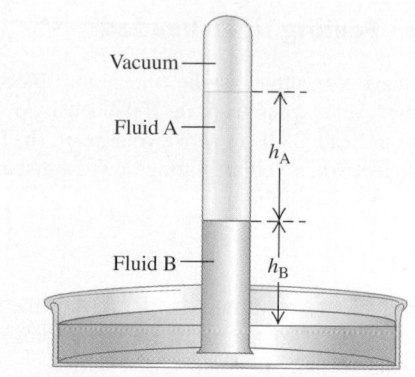

▲ **FIGURE 13.11**

EXAMPLE 13.5 **Atmospheric pressure**

Compute the atmospheric pressure p_{atm} on a day when the height of mercury in a barometer is 76.0 cm.

SOLUTION

SET UP Figure 13.12 shows our sketch; from Table 13.1, the density of mercury is $\rho = 13.6 \times 10^3$ kg/m^3. We assume that the pressure in the volume above the mercury column is zero.

SOLVE We use Equation 13.3 with $p = 0, p_0 = p_{atm}$:

$$p_{atm} = \rho g h = (13.6 \times 10^3 \text{ kg/m}^3)(9.80 \text{ m/s}^2)(0.760 \text{ m})$$
$$= 101,300 \text{ N/m}^2 = 1.013 \times 10^5 \text{ Pa} = 1.013 \text{ bars}.$$

REFLECT This is about 14.7 lb/in.^{2}. With weather variations, atmospheric pressure typically varies from about 0.97 bar to 1.03 bars; in the eye of a hurricane, it can be as little as 0.90 bar.

Practice Problem: On a very stormy day, the pressure is 0.950 bar. What is the height of mercury in a barometer? *Answer:* 71.3 cm (pressure 713 torr).

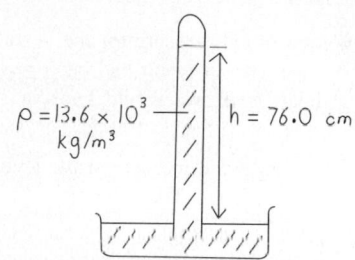

▲ **FIGURE 13.12** Our sketch for this problem.

Measuring Blood Pressure

Blood-pressure measurements are an important diagnostic tool in medicine. Blood-pressure readings, such as 130/80, refer to the maximum and minimum gauge pressures in the arteries, measured in mm Hg or torr. Blood pressure varies with height; the standard reference point is the upper arm, level with the heart.

In a commonly used blood-pressure gauge, known as a *sphygmomanometer,* an inflatable band, called a *cuff,* is placed around the upper arm, at heart level, as shown in Figure 13.13. The band is inflated by squeezing a flexible bulb, increasing the pressure on the arm. A stethoscope is used to detect the pulse at the elbow. When the pressure in the cuff exceeds the blood pressure, an artery under the cuff collapses and the pulse is no longer detectable. The pressure itself is measured with a mercury manometer or a mechanical pressure gauge.

To measure blood pressure, you listen through the stethoscope for arterial flow while letting the cuff deflate. Flow starts when the pressure in the deflating cuff reaches the arterial pressure.

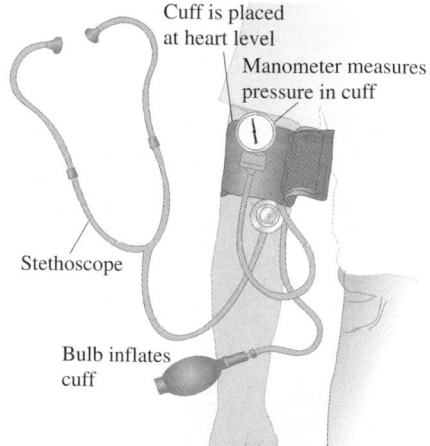

Cuff is placed at heart level

Manometer measures pressure in cuff

Stethoscope

Bulb inflates cuff

▶ **FIGURE 13.13** Measuring blood pressure with a pressure cuff and a stethoscope. (In practice, one measures two pressures: the *systolic* pressure, at which blood first starts to spurt discontinuously through the compressed vessel, and the lower *diastolic* pressure, at which blood flows continuously.)

EXAMPLE 13.6 **Feeling light-headed?**

Suppose you have a healthy systolic (maximum) blood pressure of 1.30×10^4 Pa (about 100 mm Hg), measured at the level of your heart. **(a)** If the density of blood is 1.06×10^3 kg/m³, find the blood pressure at a point in your head, 35.0 cm above your heart. **(b)** Find the difference in blood pressure between your head and your feet when you are sitting down, a distance of 1.10 m. (Ignore the fact that the blood is in motion and is not in equilibrium.)

SOLUTION

SET UP Figure 13.14 shows our sketch. We put the reference level $h = 0$ at the level of the heart, where the pressure is p_0. Notice that h_2, the distance to the foot, is *negative*.

SOLVE We use Equation 13.3, $p - p_0 = -\rho g h$, for the variation in pressure with height.

Part (a): Let p_1 be the pressure at the level of the head; p_0 is the pressure at the level of the heart, and the height h_1 is 35.0 cm = 0.350 m. From Equation 13.3,

$$p_1 = p_0 - \rho g h_1$$
$$= 1.30 \times 10^4 \text{ Pa}$$
$$\quad - (1.06 \times 10^3 \text{ kg/m}^3)(9.80 \text{ m/s}^2)(0.350 \text{ m})$$
$$= 9.4 \times 10^3 \text{ Pa}.$$

As expected, we get a blood pressure for the head that is lower than that at heart level. (If our result had been *greater* than the blood pressure at heart level, we would know we had made a mistake.)

Part (b): This time, p_2 is the pressure at the level of the foot, and the height is $h_2 = -1.10$ m. Then

$$p_2 = p_0 - \rho g h_2$$
$$= 1.30 \times 10^4 \text{ Pa}$$
$$\quad - (1.06 \times 10^3 \text{ kg/m}^3)(9.80 \text{ m/s}^2)(-1.10 \text{ m})$$
$$= 2.44 \times 10^4 \text{ Pa}.$$

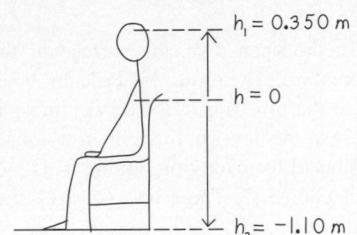

▲ **FIGURE 13.14** Our sketch for this problem.

As expected, the blood pressure at foot level is greater than that at heart level.

REFLECT Notice that we didn't need to know anything about the shape or volume of the blood vessels—just the relative heights of head, heart, and feet. If you lie down, then your head and feet are at the same level as your heart, and the differences we calculated disappear. Also, our results are somewhat simplistic because Pascal's law is valid only for fluids in equilibrium. As we will see later in the chapter, the viscosity of circulating blood has effects on its local pressure, in addition to the pressure differences due to relative height.

Practice Problem: When a person 1.70 m tall stands up, what is the difference in blood pressure between head and feet? *Answer:* 1.8×10^4 Pa.

13.3 Archimedes's Principle: Buoyancy

Buoyancy is a familiar phenomenon: An object immersed in water seems to weigh less than when it is in air. When the object is less dense than the fluid, it floats. Rafts inflated with air float in water, and a helium-filled balloon floats in air. Buoyancy is described by Archimedes's principle:

Archimedes's principle

When an object is completely or partially immersed in a fluid, the fluid exerts an upward force on the object equal to the weight of the fluid that is displaced by the object.

To derive this principle, we consider an arbitrary volume element of fluid at rest (in equilibrium), as shown in Figure 13.15a. The forces labeled $\vec{F}_\perp$ represent the forces exerted on the surface of the element by the surrounding fluid—that is, the forces due to the pressure of the fluid.

Because the entire beaker of fluid is in equilibrium, the sum of all the y components of force acting on the chosen element of fluid must be zero. Therefore, the sum of the y components of the surface forces must be an upward force equal in magnitude to the weight mg of the fluid inside the surface. Also, the sum of

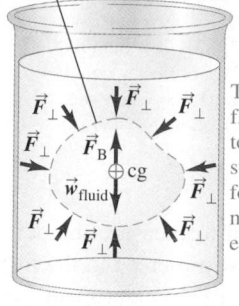

Arbitrary element of fluid in equilibrium

The forces on the fluid element due to pressure must sum to a buoyancy force equal in magnitude to the element's weight.

(a)

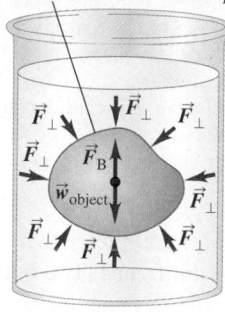

Fluid element replaced with solid object of the same size and shape.

The forces due to pressure are the same, so the object must be acted upon by the same buoyancy force as the fluid element, *regardless of the object's weight.*

(b)

▲ **FIGURE 13.15** The origin of the buoyancy force. (a) The forces on a volume element of fluid in equilibrium. (b) The fluid element is replaced with a solid object having the same shape and size.

the torques of these forces must be zero, so the line of action of the resultant *y* component of surface forces must pass through the center of gravity of this portion of fluid.

Now, suppose that we remove the chosen element of fluid and replace it with a solid object having exactly the same shape (Figure 13.15b). The pressure at each point on the surface of this object is exactly the same as before, so the total upward force exerted on the object by the surrounding fluid is also the same, again equal in magnitude to the weight *mg* of the fluid that formerly occupied this spot. We call this upward force the **buoyant force** on the solid object.

NOTE ▶ Notice that the object *need not have the same weight as the fluid it replaces.* If the object's weight is greater than that of the displaced fluid, the object sinks; if less, then it rises. Thus, objects that are denser than the surrounding fluid sink; those that are less dense rise. ◀

Notice also that the center of gravity of the object need not coincide with that of the fluid that's replaced. As we just explained, the line of action of the buoyant force passes through the center of gravity of the replaced portion of fluid. If the *object's* center of gravity does not lie along this line (as in Figure 13.16), the buoyant force exerts a torque on the object.

As a helium balloon loses helium, it passes through a condition in which it neither floats nor sinks, but is in equilibrium with the surrounding air. At this point, the balloon's weight must be the same as the weight of the air it displaces. That is, the average density of a balloon floating in equilibrium must be the same as that of the surrounding air. Similarly, when a submerged submarine is in equilibrium, its weight must equal the weight of the water it displaces.

An object whose average density is less than that of a liquid can float partially submerged at the free upper surface of the liquid. The greater the density of the liquid, the less far the object is submerged. When you swim in seawater (density $1030 \ \text{kg/m}^3$), your body floats higher than in fresh water ($1000 \ \text{kg/m}^3$). Unlikely as it may seem, lead floats in mercury. Very flat surfaced "float glass" is made by floating glass on molten tin and then letting it cool.

Another familiar example is the hydrometer, shown in Figure 13.17a. The calibrated float sinks into the fluid until the weight of the fluid it displaces is exactly equal to its own weight. The hydrometer floats *higher* in denser liquids than in less dense liquids. It is weighted at its bottom end so that the upright position is stable, and a scale in the top stem permits direct density readings.

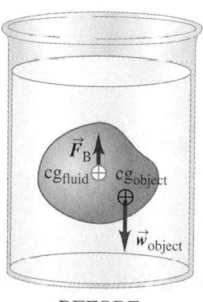

The cg of this object does not coincide with the cg of the displaced fluid.

BEFORE

Thus, the buoyant force exerts a torque about the object's cg, causing the object to rotate.

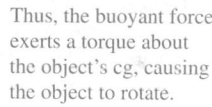
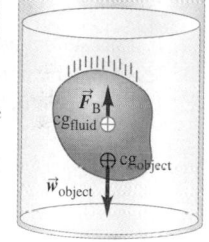

Because the object's weight is greater in magnitude than the buoyant force, the object also sinks.

AFTER

▲ **FIGURE 13.16** A case in which the buoyancy force does not act through the center of gravity of an immersed object.

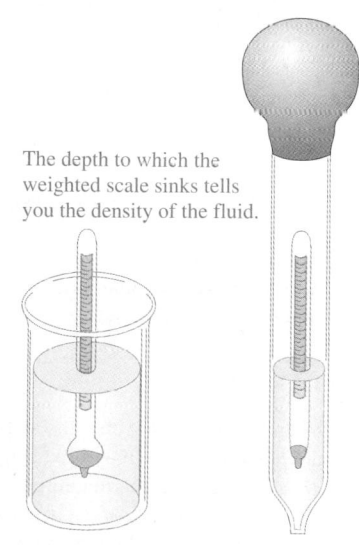

The depth to which the weighted scale sinks tells you the density of the fluid.

(a) A simple hydrometer

(b) Using a hydrometer to measure the density of battery acid or antifreeze

▲ **FIGURE 13.17** Hydrometers, for measuring the density of a fluid.

Figure 13.17b shows a hydrometer that is commonly used to measure the density of battery acid or antifreeze. The bottom of the tube is immersed in the liquid; the bulb is squeezed to expel air and is then released, like a giant medicine dropper. The liquid rises into the outer tube, and the hydrometer floats in this sample of the liquid.

Conceptual Analysis 13.5

Floating lead

A lead ball floats in liquid mercury and sinks in water. Which statement is true about the buoyant force acting on the lead ball?

A. It is greater when the ball is floating in the mercury.
B. It is greater when the ball is totally submerged in the water.
C. It is the same in both cases.

SOLUTION The ball's behavior depends on the difference in magnitude between its weight and the buoyant force. The ball's weight mg is the same in both experiments. The ball floats partially submerged in mercury, so the buoyant force equals the magnitude of the ball's weight. (It would be greater if the ball were completely submerged.) In water the ball sinks, so the buoyant force is less in magnitude than the ball's weight. Thus, A is correct: Mercury exerts a greater buoyant force on the ball than water does.

Conceptual Analysis 13.6

Buoyant Ping-Pong™ ball

A Ping-Pong™ ball is held submerged in a bucket of water by a string attached to the bucket's bottom. What happens to the tension in the string as water is added to the bucket, increasing the distance between the ball and the surface?

A. The tension does not change.
B. The tension increases.
C. The tension decreases.

SOLUTION The buoyant force is equal in magnitude to the weight of the water displaced by the Ping-Pong™ ball. As water is added to the bucket, the *pressure* at the level of the ball increases. We'll assume that water is nearly incompressible, so this pressure change is too small to cause a significant change in the *density* of water. Therefore, the weight of the displaced water remains the same, and so does the buoyant force. Thus, the tension doesn't change. (If the fluid were a compressible gas rather than water, changing the pressure might indeed have a significant effect on the buoyancy force.)

EXAMPLE 13.7 **Raising sunken treasure**

A 15.0 kg solid-gold statue is being raised from a sunken treasure ship, as shown in Figure 13.18a. **(a)** Find the tension in the hoisting cable when the statue is completely immersed. **(b)** Find the tension when the statue is completely out of the water.

SOLUTION

SET UP When immersed in water, the statue is acted upon by an upward buoyant force with magnitude F_B. From Archimedes's principle, this force is equal in magnitude to the weight of water displaced. Figure 13.18b shows a free-body diagram for the statue, including its weight, the tension T, and the upward buoyant force F_B.

SOLVE Part (a): To find F_B, we first find the volume of the statue, using its mass and the density of gold from Table 13.1:

$$V = \frac{m_{\text{statue}}}{\rho_{\text{gold}}} = \frac{15.0 \text{ kg}}{19.3 \times 10^3 \text{ kg/m}^3} = 7.77 \times 10^{-4} \text{ m}^3.$$

Continued

The weight of an equal volume of water is

$$m_{water}g = \rho_{water}Vg$$
$$= (1.00 \times 10^3 \text{ kg/m}^3)(7.77 \times 10^{-4} \text{ m}^3)(9.80 \text{ m/s}^2)$$
$$= 7.61 \text{ N}.$$

This weight is equal to the magnitude F_B of the buoyant force. In equilibrium, the tension in the cable equals the weight of the statue minus the magnitude of the buoyant force. Taking the positive y direction as upward, we find that

$$T + F_B - w = 0,$$
$$T + 7.61 \text{ N} \quad 147 \text{ N} = 0,$$
$$T = 139 \text{ N}.$$

We note that this tension is less than the weight of the statue.

Part (b): To find the tension when the statue is in air, we neglect the very small buoyant force of *air* (as we have done in earlier chapters). The tension when the block is in air is just equal to its weight:

$$m_{statue}g = (15.0 \text{ kg})(9.80 \text{ m/s}^2) = 147 \text{ N}.$$

REFLECT The tensions calculated here are for a statue at rest or being hoisted at *constant* speed. Here is a shortcut leading to the same result for the buoyant force: The density of water is less than that of gold by a factor of

$$\frac{1.00 \times 10^3 \text{ kg/m}^3}{19.3 \times 10^3 \text{ kg/m}^3} = \frac{1}{19.3} = 0.0518.$$

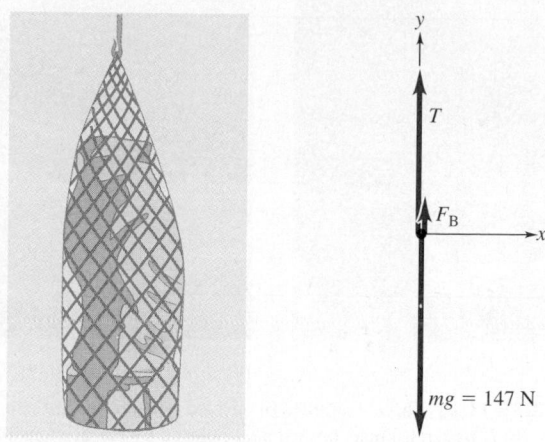

(a) Immersed statue in equilibrium   (b) Free-body diagram of statue

▲ **FIGURE 13.18**

That is, the weight of a given volume of water is 0.0518 times that of an equal volume of gold, and the weight of the displaced water is $(1.00/19.3)(147 \text{ N}) = 7.62 \text{ N}$.

Practice Problem: A brass statue with a mass of 0.500 kg and a density of $8.00 \times 10^3 \text{ kg/m}^3$ is suspended from a string. What is the tension in the string when the statue is completely immersed in water? *Answer:* $T = 4.29 \text{ N}$.

Conceptual Analysis 13.7

Weighing in water

Suppose you place a beaker of water on a scale, weigh it, and then immerse the statue from Example 13.7 in the water, suspending it so that it is under water, but does not touch the bottom of the beaker (Figure 13.19). Does the scale reading increase, decrease, or stay the same?

A. Increase.
B. Decrease.
C. Stay the same.

SOLUTION Consider the water and the statue together as a system; its total weight does not change when the statue is immersed. Thus, the total supporting force—the vector sum of the tension $\vec{T}$ and the upward force $\vec{F}$ of the scale on the container (equal to the reading on the scale)—is the same in both cases. But T decreases when the statue is immersed, so the scale reading F must *increase* by the same amount. Thus, A is correct. Alternatively, we can see that the scale force *must* increase because, when the statue is immersed, the water level

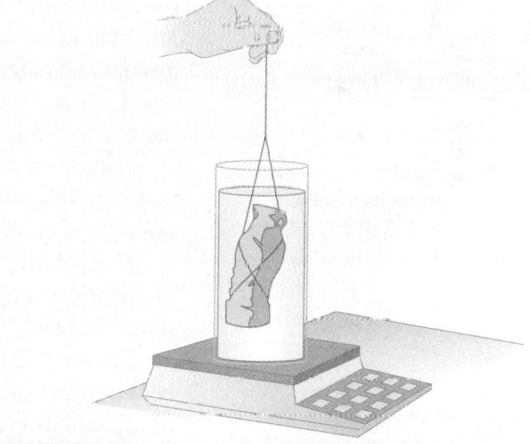

▲ **FIGURE 13.19**

in the container rises, increasing the pressure of water on the bottom of the container.

13.4 Surface Tension and Capillarity

What do the following observations (Figure 13.20) have in common? A liquid emerges from the tip of a pipette or medicine dropper as a succession of drops, not as a continuous stream. A paper clip, if placed carefully on a water surface, makes a small depression in the surface and rests there without sinking, even

(a)

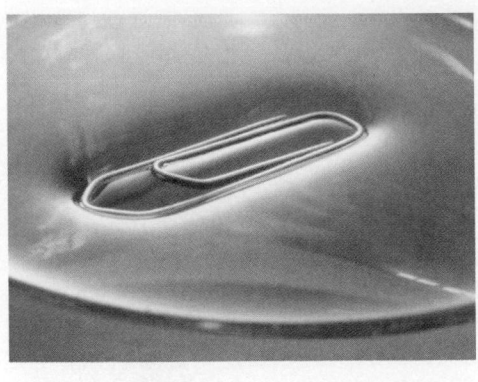
(b)

(c)

▲ **FIGURE 13.20** Three examples of surface tension. (a) Surface tension causes the drops at the ends of these pipettes to adopt a near-spherical shape. (b) A paperclip rests on water, supported by surface tension. (c) water striders spend their life walking on water.

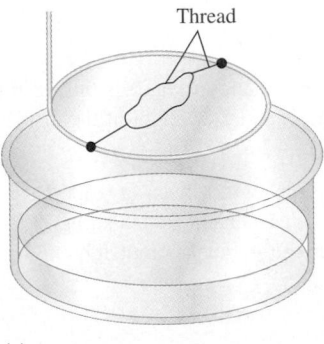

Thread

(a)

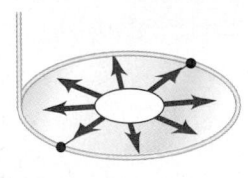

(b)

▲ **FIGURE 13.21** A flexible loop of thread in a wire ring that has been dipped in a soap solution (a) before and (b) after puncturing the film inside the loop.

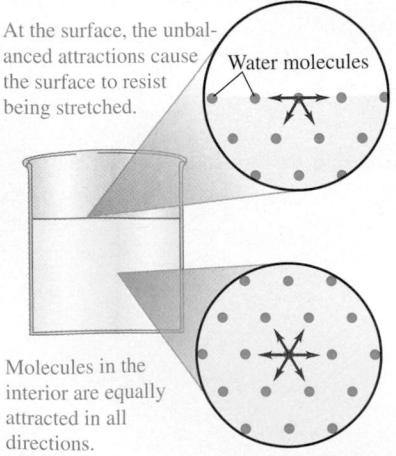

Molecules in a liquid are attracted by neighboring molecules.

At the surface, the unbalanced attractions cause the surface to resist being stretched.

Water molecules

Molecules in the interior are equally attracted in all directions.

▲ **FIGURE 13.22** The physical origin of surface tension.

though its density is several times that of water. Some insects can walk on the surface of water; their feet make indentations in the surface, but do not penetrate it. When a small, clean glass tube is dipped into water, the water rises in the tube, but when the tube is dipped in mercury, the mercury is depressed.

In all these situations, the surface of the liquid seems to be under *tension*. Here's another example: We attach a loop of thread to a wire ring, as shown in Figure 13.21a. We dip the ring and thread into a soap solution and remove them, forming a thin film of liquid in which the thread "floats" freely. When we puncture the film inside the loop, the thread springs out into a circular shape (Figure 13.21b) as the tension of the liquid surface pulls radially outward on it.

Why does the surface of a liquid behave this way? The key is that water molecules attract each other. (That's what makes liquid water cohere.) Figure 13.22 compares the forces of attraction on a water molecule on the surface with those on a molecule below the surface. Below the surface, a molecule is equally attracted in all directions. At the surface, a molecule is attracted to the molecules around and below it, but not to the overlying air. If you place a paper clip on the water surface, these forces cause the surface to resist being stretched; the resulting tension supports the clip.

A drop of liquid in free fall in vacuum is always spherical in shape (*not* teardrop shaped), because a surface under tension tends to have the minimum possible area. A sphere has a smaller surface area for a given volume than any other geometric shape has. Figure 13.23 is a beautiful example of the formation of spherical droplets in a very complex phenomenon: the impact of a drop on a liquid surface.

Figure 13.24 shows another example of surface tension. A piece of wire is bent into a U shape, and a second piece of wire slides on the arms of the U. When the apparatus is dipped into a soap solution and removed, the upward surface tension force may be greater in magnitude than the weight of the wire. If it is, the slider is pulled up toward the top of the inverted U by the upward surface tension force. To hold the slider in equilibrium, there must be a total downward force with a magnitude greater than the weight of the wire. When we apply a large enough downward force, we can pull the slider down, increasing the area of the film. Molecules move from the interior of the liquid (which is many molecules thick, even in a thin film) into the surface layers, which are not stretched like a rubber sheet. Instead, more surface is created by molecules moving from the bulk liquid.

Let l be the length of the wire slider. The film has two surfaces, so the surface force acts along a total length $2l$. The **surface tension** γ in the film is defined as *the ratio of the magnitude F of the surface force to the length d* (perpendicular to the force) *along which the force acts:*

Definition of surface tension (γ)
Surface tension is a *force per unit length:*

$$\gamma = \frac{F}{d}. \qquad (13.7)$$

Units: The SI unit of surface tension is the newton per meter (N/m), but the cgs unit, the dyne per centimeter (dyn/cm), is more commonly used: $1 \text{ dyn/cm} = 10^{-3} \text{ N/m}$.

In the case of the sliding wire, $d = 2l$ and

$$\gamma = \frac{F}{2l}.$$

Typical values of γ for familiar liquids are in the range from 20 to 100 dyn/cm. Surface tension for a particular liquid usually decreases as temperature increases.

Surface tension tends to pull the surface of a soap bubble inward, collapsing it. Excess pressure inside the bubble is needed to maintain its spherical shape. Detailed analysis shows that the excess pressure $p - p_{atm}$ inside a soap bubble with radius R, due to surface tension, is

$$p - p_{atm} = \frac{4\gamma}{R}. \qquad \text{(bubble)} \qquad (13.8)$$

For a liquid drop, which has only *one* surface film, the difference between the pressure of the liquid and that of the outside air is half that for a soap bubble:

$$p - p_{atm} = \frac{2\gamma}{R} \qquad \text{(liquid drop)}. \qquad (13.9)$$

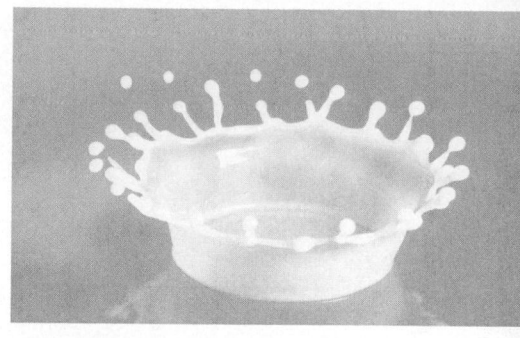

▲ **FIGURE 13.23** Aftermath of a drop splashing on a liquid surface.

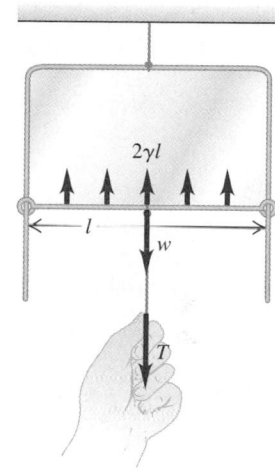

▲ **FIGURE 13.24** The horizontal slide wire is in equilibrium under the action of the upward surface-tension force $2\gamma l$ and downward pull $w + T$.

EXAMPLE 13.8 **A drop of mercury**

Mercury has an unusually large surface tension, about 465 dyn/cm. Calculate the excess pressure inside a drop of mercury 4.00 mm in diameter.

SOLUTION

SET UP AND SOLVE We first convert the surface tension γ to SI units:

$$\gamma = 465 \text{ dyn/cm} = 465 \times 10^{-3} \text{ N/m}.$$

Now we have, from Equation 13.9,

$$p - p_{atm} = \frac{2\gamma}{R} = \frac{(2)(465 \times 10^{-3} \text{ N/m})}{0.00200 \text{ m}}$$
$$= 465 \text{ Pa}$$
$$= 0.00459 \text{ atm}.$$

REFLECT The excess pressure (the difference between the inside and outside pressures) is a very small fraction of atmospheric pressure.

Practice Problem: What would the diameter of the drop be if the inside pressure were 2 atm? *Answer:* 0.009 mm.

When a gas–liquid interface meets a solid surface, such as the wall of a container (Figure 13.25), surface tension causes the interface to curve upward in a concave shape or curve downward in a convex shape, near the line where it meets the solid surface. The interface is concave if the fluid *wets* the solid surface—that is, if the surface attracts the fluid at a molecular level. It is convex if the solid does not attract the fluid (allowing surface tension to pull the fluid toward a spherical shape). As you can see from Figure 13.25a, water wets glass and forms a concave interface; mercury does not wet glass, and it forms a convex interface.

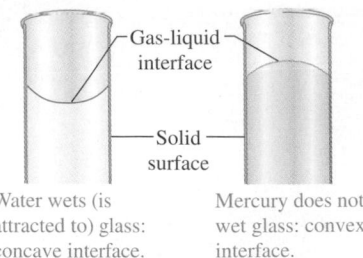

Water wets (is attracted to) glass: concave interface.

Mercury does not wet glass: convex interface.

(a) Gas–liquid interface in contact with a solid

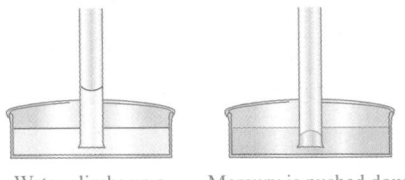

Water climbs up a glass tube because glass attracts it.

Mercury is pushed down by a glass tube because glass does not attract it.

(b) Capillarity

▲ **FIGURE 13.25** Two effects of surface tension when a gas–liquid interface meets a solid surface that the liquid either wets or does not wet.

If you push a thin glass tube into water, the attraction of the glass actually causes the water to climb a certain distance up into the tube, as shown in Figure 13.25b. Conversely, if you push a glass tube into mercury, surface tension makes it difficult for the mercury to enter the tube, so the tube pushes the mercury down. These two phenomena, called **capillarity,** are responsible for the absorption of water by paper towels, the rise of melted wax in a candle wick, and many other everyday phenomena.

Related to surface tension is the phenomenon of *negative pressure.* The stress in a liquid is ordinarily *compressive,* but in some circumstances liquids can sustain *tensile* stresses. Consider a cylindrical tube that is closed at one end and has a tight-fitting piston at the other. We fill the tube completely with a liquid that wets and adheres to both the inner surface of the tube and the piston face; the molecules of liquid adhere to all the surfaces. If the surfaces are very clean and the liquid very pure, then, when we pull the piston face, we observe a *tensile* stress and a slight *increase* in volume; we are *stretching* the liquid! Adhesive forces prevent it from pulling away from the walls of the container.

With water, tensile stresses as large as 300 atm have been observed in the laboratory. This situation is highly unstable: A liquid under tension tends to break up into many small droplets. In tall trees, however, negative pressures are a regular occurrence. Negative pressure is believed to be an important mechanism for the transport of water and nutrients from the roots to the leaves in the small xylem tubes (diameter on the order of 0.1 mm) in the tree's outer wood.

13.5 Fluid Flow

We're now ready to consider *motion* of a fluid. Fluid flow can be extremely complex, as shown by the flow of a river in flood or the swirling flames of a campfire. Despite this complexity, some situations can be represented by relatively simple idealized models. An **ideal fluid** is a fluid that is *incompressible* (i.e., that has constant density) and has no internal friction or viscosity. The assumption of incompressibility is usually a good approximation for liquids, and we may also treat a gas as incompressible whenever the pressure differences from one region to another are not too great. Internal friction in a fluid causes shear stresses when two adjacent layers of fluid move relative to each other, such as when the fluid flows inside a tube or around an obstacle. In some cases, these shear forces are negligible in comparison with forces arising from gravitation and pressure differences.

The path of an individual element of a moving fluid is called a **flow line.** If the overall flow pattern does not change with time, the flow is called **steady flow.** In steady flow, every element passing through a given point follows the same flow line. In this case the "map" of the fluid velocities at various points in space remains constant, although the velocity of a particular particle of the fluid changes in both magnitude and direction during its motion.

Figure 13.26 shows groups of flow lines for motion of a fluid flowing from left to right around a number of obstacles. The photographs were made by injecting dye into water flowing between two closely spaced glass plates. These patterns are typical of

◀ **BIO Application How does water get up there?** How do plants supply water all the way from the soil to the leaves at their highest points? It is especially mysterious how trees like the giant redwoods can lift water as high as 100 m, which requires a pressure of 10 atm. The answer is a kind of micro-capillarity in the leaves. As the conducting pipes of the tree (xylem) terminate in the leaves, the water flows into the spaces around the cells and evaporates through pores in the leaves, producing a thin layer of water on the surface of the cells. The irregular fibers outside the cell walls force the water to have very sharp radii of curvature of less than 100 nm as it conforms to the fibers. This generates pressure in the leaves as much as 30 atm less than that in the roots, sufficient to pull water to the tops of the highest trees and overcome the resistance to flow through the xylem. This process works only if the pores in the leaves are open so that the water evaporates continuously, thus maintaining the thin layer and the sharp curvature of the remaining water. At night, the pores in the leaves close and the flow of water stops.

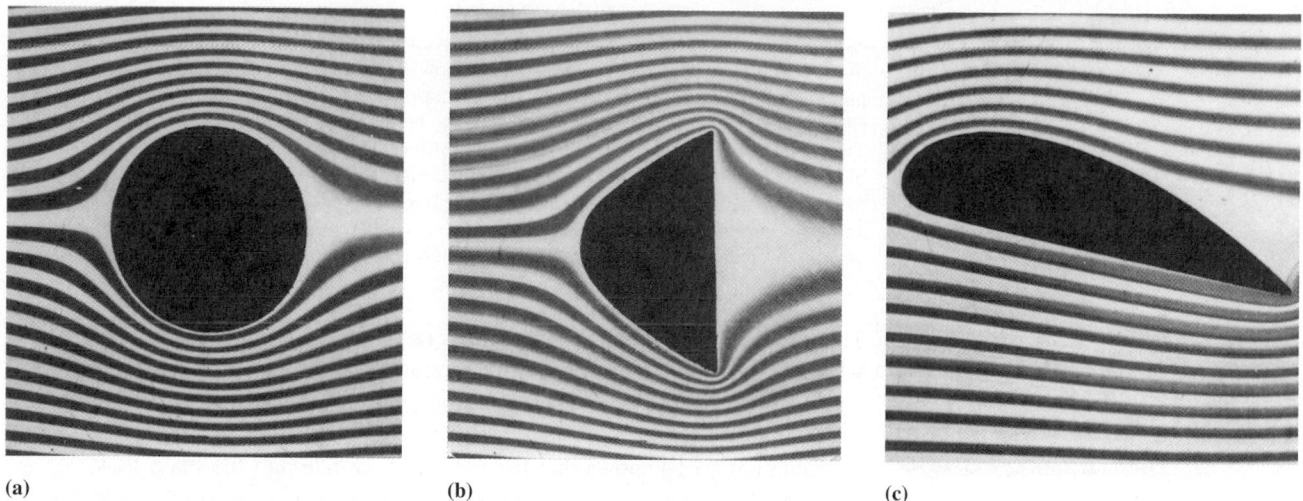

(a)   (b)   (c)

▲ **FIGURE 13.26** (a)–(c) Laminar flow around obstacles of different shapes.

laminar flow, in which adjacent layers of fluid slide smoothly past each other. At sufficiently high flow rates or when boundary surfaces cause abrupt changes in velocity, the flow becomes irregular and chaotic; this is called **turbulent flow.** Smoke rising from a candle or an incense stick typically exhibits laminar flow near the bottom and then breaks up into chaotic swirls as the flow becomes turbulent. (Figure 13.27). In turbulent flow, *there is no steady-state pattern;* the flow pattern continuously changes.

Conservation of mass in fluid flow gives us an important relationship called the **continuity equation.** The essence of this relation is contained in the familiar maxim "Still waters run deep." If the flow rate of a river with constant width is the same everywhere along a certain length, the water must run faster where it is shallow than where it is deep. As another example, the stream of water from a faucet tapers down in width because the water speeds up as it falls and a smaller diameter is needed to carry the same volume flow rate as at the top.

We can turn these observations into a quantitative relation. We'll consider only incompressible fluids, for which the density is constant. Figure 13.28 shows a fluid flowing through a gradually narrowing tube. We'll look at two points along the tube at which the cross-sectional areas are, respectively, A_1 and A_2 and the speeds of fluid flow are v_1 and v_2. During a small time interval Δt, the fluid at A_1 moves a distance $v_1 \Delta t$. The volume of fluid ΔV_1 that flows into the tube across A_1 during this interval is the fluid in the cylindrical element with base A_1 and height $v_1 \Delta t$; that is, $\Delta V_1 = A_1 v_1 \Delta t$.

If the density of the fluid is ρ, the mass Δm_1 that flows into the tube across A_1 during time Δt is the density ρ multiplied by the volume ΔV_1; that is, $\Delta m_1 = \rho \Delta V_1 = \rho A_1 v_1 \Delta t$. Similarly, the mass Δm_2 that flows out across A_2 in the same time is $\Delta m_2 = \rho A_2 v_2 \Delta t$. In steady flow the total mass in the tube is constant, so these two masses must be equal: $\Delta m_1 = \Delta m_2$, and

$$\rho A_1 v_1 \Delta t = \rho A_2 v_2 \Delta t, \quad \text{or} \quad A_1 v_1 = A_2 v_2.$$

Thus, we find the following:

Continuity equation

This equation expresses conservation of mass for an incompressible fluid flowing in a tube or some other channel. It says that the amount (either mass or volume) of fluid flowing through a cross section of the tube in a given time interval must be the same for all cross sections, or

$$\frac{\Delta V}{\Delta t} = A_1 v_1 = A_2 v_2. \tag{13.10}$$

▲ **FIGURE 13.27** Smoke rising from incense shows laminar flow near the bottom and turbulent flow farther up.

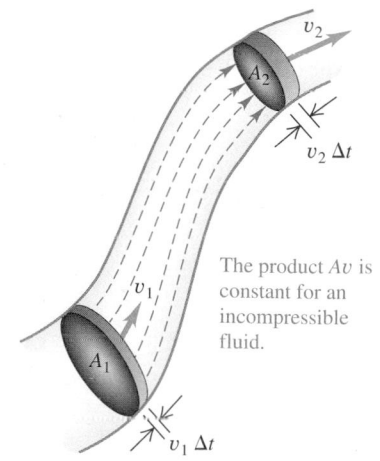

The product Av is constant for an incompressible fluid.

▲ **FIGURE 13.28** Flow in a tube with changing cross-sectional area.

◀ **Application Meandering along.** In a relatively straight, slow-moving waterway, the movement of the water approximates laminar flow; adjacent layers of water slide smoothly past each other with little mixing. As the flow approaches a bend in the river, it becomes slightly turbulent, and water on the outside of the bend travels faster than water on the inside. The turbulence causes erosion of the outside riverbank due to the higher flow rate and results in the deposition of suspended sediment on the inside riverbank as the water there slows. These combined processes lead to an ever-widening series of bends and turns, a phenomenon known as *meandering* that is characteristic of rivers on flat plains. So, if you're traveling by boat on this mature river, the shortest distance between two points may not be a straight line!

The product Av is the **volume flow rate** $(\Delta V / \Delta t = Av)$, the time rate at which a given volume crosses any section of the tube. Similarly, the product $\Delta m / \Delta t = \rho Av$ is the **mass flow rate,** the time rate at which a given mass crosses a section of the tube.

Equation 13.10 shows that the volume flow rate and the mass flow rate are constant along any given tube. When the cross section of the tube decreases, the speed of flow increases. The shallow part of a river has a smaller cross section and a faster current than the deep part, but the volume flow rates are the same in both. The stream from the faucet narrows as it gains speed during its fall, but $\Delta V / \Delta t$ is the same everywhere along the stream. If a water pipe with a 2 cm diameter is connected to a pipe with a 1 cm diameter, the flow speed is four times as great in the 1 cm part as in the 2 cm part (because the cross-sectional area is proportional to the square of the diameter).

We derived the continuity equation for a steady flow in a physically distinct tube, pipe, or channel. However, a larger steady flow pattern can be thought of as a bundle of side-by-side tubes with imaginary boundaries separating adjacent tubes. If there is no mixing across boundaries of the individual tubes, then the fluid in each tube behaves according to Equation 13.10. For instance, if a river has slowly moving water near the banks and swifter water in the middle, with little mixing between the two, the continuity equation can be applied separately to the faster-moving central portion and to a more slowly moving channel near the bank. In the discussion that follows, we'll use the term **flow tube** to describe the flow patterns in a fluid, even when the tube isn't bounded by a real physical wall.

Conceptual Analysis 13.8

Flow through a tube

Water flows steadily through a horizontal pipe that has a narrowed section. What happens to the volume flow rate as the water passes through the narrow section?

A. It decreases because the speed of the fluid decreases.
B. It remains constant.
C. It increases because the speed of the fluid increases.

SOLUTION As we've just seen, when an incompressible fluid such as water flows steadily through a pipe, the *same amount* of water must pass through each portion of the pipe in any given amount of time. This is equivalent to saying that the volume flow rate, $\Delta V / \Delta t$, is constant (choice B). For the product Av to be constant, the flow speed v must increase when the cross-sectional area A decreases. Thus, narrowing the pipe increases the flow speed.

13.6 Bernoulli's Equation

According to the continuity equation, the speed of fluid flow can vary along the paths of the fluid particles. The pressure can also vary; it depends on height, as in the static situation, and it also depends on the speed of flow. We can now use the concepts introduced in Section 13.5, along with the work–energy theorem, to derive an important relationship called **Bernoulli's equation** that relates the pressure, flow speed, and height for flow in an ideal fluid.

The dependence of pressure on fluid speed follows from Equation 13.10. When an incompressible fluid flows along a flow tube with varying cross section, the flow speed *must* change. Therefore, each element of fluid must have an acceleration, and the force that causes this acceleration has to be applied by the surrounding fluid. When an element of the fluid speeds up, we know that it is moving from a region of higher pressure to one of lower pressure; the pressure difference provides the net forward force needed to accelerate it. So when the cross section of a flow tube changes, the pressure *must* vary along the tube, even when there is no difference in elevation. If the elevation also changes, this causes an additional pressure difference.

To derive the Bernoulli equation, we use the work–energy theorem, which we studied in Chapter 7. We'll use the general form stated at the beginning of Section 7.6. Let's review that relationship; it is

$$\Delta K + \Delta U = W_{other},$$

where ΔK is the change in kinetic energy of an object during the motion we're considering, ΔU is the change in its potential energy (in this case, gravitational potential energy ΔU_{grav}), and W_{other} is the work done by the forces whose work is not included in ΔU_{grav}. We'll compute each of these three pieces for a system consisting of a segment of fluid in a flow tube; then we'll combine the pieces according to the work–energy theorem.

In Figure 13.29, we consider as our system the segment of fluid that lies between the two cross sections a and c at some initial time. The speed at the upstream end is v_1, and the pressure there is p_1. In a small time interval Δt, the fluid initially at a moves to b, a distance $\Delta s_1 = v_1 \Delta t$, where v_1 is the speed at this end of the tube. In the same time interval, the fluid initially at c moves to d, a distance $\Delta s_2 = v_2 \Delta t$. The cross-sectional areas at the two ends are A_1 and A_2, as shown. Because of the continuity relation, the volume ΔV of fluid passing *any* cross section during time Δt is the same at both ends: $\Delta V = A_1 \Delta s_1 = A_2 \Delta s_2$.

Note that the fluid between cross sections a and b is part of our system at the beginning of Δt, but not at the end, and that the fluid between cross sections c and d are part of our system at the end of Δt, but not at the beginning.

To apply the work–energy theorem, let's first compute W_{other}, which is the work done on this segment of fluid during Δt by the fluid adjacent to it. The pressure on the segment at cross section a is p_1; the total force exerted on this cross section by the adjacent fluid has magnitude p_1A_1, and the displacement of the segment during the interval Δt is Δs_1. Thus, the work done on the segment by the force at cross section a during time Δt (force times displacement) has magnitude $p_1A_1 \Delta s_1$. Similarly, the magnitude of the work done on the segment at cross section c is $p_2A_2 \Delta s_2$. Therefore, the net work W_{other} done on the segment by the adjacent fluid during this displacement is

$$W_{other} = p_1A_1 \Delta s_1 - p_2A_2 \Delta s_2$$
$$= (p_1 - p_2) \Delta V.$$

We need the negative sign in the second term because the force at c is opposite in direction to the displacement.

Next, let's find the change in gravitational potential energy during time Δt. The gravitational potential energy of the mass between a and b at the beginning of Δt is $\Delta m \, gy_1 = \rho \, \Delta V \, gy_1$, where y is the height of the mass above some reference level at which $y = 0$. At the end of the interval Δt, this mass has left our system, and there's a corresponding decrease in the potential energy of the system. But during the same interval, the mass between c and d has been added to the system, bringing with it a potential-energy increase $\Delta m \, gy_2 = \rho \, \Delta V \, gy_2$. Thus, the *net* change in potential energy ΔU_{grav} during Δt is

$$\Delta U_{grav} = \rho \, \Delta V \, g(y_2 - y_1).$$

Finally, we compute the change ΔK in kinetic energy of our system during Δt. At the beginning of Δt, the fluid between a and b has volume $A_1 \Delta s_1$, mass $\rho A_1 \Delta s_1$, and kinetic energy $\frac{1}{2}mv_1^2 = \frac{1}{2}\rho(A_1 \Delta s_1)v_1^2 = \frac{1}{2}\rho \, \Delta V \, v_1^2$. At the end of Δt, this fluid is no longer in the system, but the fluid between c and d is now included; it brings in kinetic energy $\frac{1}{2}\rho(A_2 \Delta s_2)v_2^2 = \frac{1}{2}\rho \, \Delta V \, v_2^2$. The net *change* in kinetic energy ΔK during time Δt is the difference of these quantities:

$$\Delta K = \tfrac{1}{2}\rho \, \Delta V(v_2^2 - v_1^2).$$

▲ **Application Raising the roof.** In severe weather accompanied by high winds, the roofs of houses may be blown completely off. At first you might think that the direct force of the wind pushed this roof off. In fact, the wind *pulled* the roof off—or, alternatively, the pressure inside the house blew it off. The wind that flowed up and over the house was forced to take a longer path than wind that did not meet the house. Therefore, it flowed faster. Bernoulli's principle tells us that the pressure in this fast-moving air was lower than the pressure elsewhere—such as inside the house. After a severe storm, you will often see that the windows of houses are blown out rather than in.

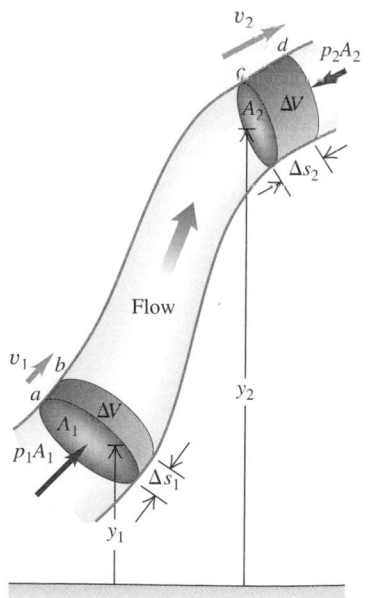

▲ **FIGURE 13.29** The net work done by the pressure equals the change in the kinetic energy plus the change in the gravitational potential energy.

(Note that there is no net change in kinetic energy of the fluid between cross sections b and c: There is just as much of our system contained between these sections at the end of the interval as at the beginning, and the flow velocity at any point doesn't change.)

Now we combine all these pieces in the work–energy theorem:

$$\Delta K + \Delta U_{\text{grav}} = W_{\text{other}},$$

$$\tfrac{1}{2}\rho\,\Delta V(v_2^2 - v_1^2) + \rho\,\Delta V g(y_2 - y_1) = (p_1 - p_2)\,\Delta V,$$

or

$$p_1 - p_2 = \tfrac{1}{2}\rho(v_2^2 - v_1^2) + \rho g(y_2 - y_1). \tag{13.11}$$

This is **Bernoulli's equation.** We may interpret it in terms of pressures. The second term on the right is the pressure difference between any two points in the fluid, caused by the weight of the fluid and the difference in elevation of the two ends. The first term on the right is the additional pressure difference associated with the change in speed of the fluid.

Equation 13.11 can also be written as follows:

Bernoulli's equation: pressure, flow speed, and height in a fluid

$$p_1 + \rho g y_1 + \tfrac{1}{2}\rho v_1^2 = p_2 + \rho g y_2 + \tfrac{1}{2}\rho v_2^2. \tag{13.12}$$

The subscripts 1 and 2 refer to *any* two points along the flow tube, so Bernoulli's equation may also be written as

$$p + \rho g y + \tfrac{1}{2}\rho v^2 = \text{constant}. \tag{13.13}$$

PROBLEM-SOLVING STRATEGY 13.1 **Bernoulli's equation** (MP)

Bernoulli's equation is derived from the work–energy relationship, so it isn't surprising that much of the problem-solving strategy suggested in Section 7.6 is equally applicable here.

SET UP

1. Always begin by identifying points 1 and 2 clearly, with reference to Bernoulli's equation.
2. Write down (or show on a sketch) the known and unknown quantities in Equation 13.12. The variables are p_1, p_2, v_1, v_2, y_1, and y_2, and the constants are ρ and g. What is given? What do you need to determine? Be sure that your pressures are either all gauge pressures or all absolute pressures; don't mix them. Be careful about consistency of units. In SI units, pressure is expressed in pascals, density in kilograms per cubic meter, and speed in meters per second.

SOLVE

3. In some problems you'll need to use the continuity equation, Equation 13.10, to get a relation between the two speeds in terms of cross-sectional areas of pipes or containers. Or perhaps you will know both speeds and will need to determine one of the areas.
4. The volume flow rate $\Delta V/\Delta t$ across any area A is given by $\Delta V/\Delta t = Av$, and the corresponding mass flow rate $\Delta m/\Delta t$ is $\Delta m/\Delta t = \rho Av$. These relations sometimes come in handy.

REFLECT

5. Always try to estimate comparative magnitudes of your results. Are they what you expected? Do they agree with commonsense predictions or estimates?

EXAMPLE 13.9 **Water pressure in the home**

Water enters a house through a pipe with an inside diameter of 2.0 cm at a gauge pressure of 4.0×10^5 Pa (about 4 atm, or 60 lb/in²). The cold-water pipe leading to the second-floor bathroom 5.0 m above is 1.0 cm in diameter (Figure 13.30). Find the flow speed and gauge pressure in the bathroom when the flow speed at the inlet pipe is 2.0 m/s. How much time would be required to fill a 100 L bathtub?

SOLUTION

SET UP Let point 1 be at the inlet pipe and point 2 at the bathroom. The speed v_2 of the water at the bathroom is obtained from the continuity equation, Equation 13.10. We take $y_1 = 0$ at the inlet and $y_2 = 5.0$ m at the bathroom. We are given p_1 and v_1; we can find p_2 from Bernoulli's equation, Equation 13.12. To find the time to fill the bathtub, we use the volume flow-rate relation $\Delta V / \Delta t = Av$. The bathtub's volume is 100 L = 100×10^{-3} m³.

SOLVE From the continuity equation, the flow speed v_2 in the bathroom is

$$v_2 = \frac{A_1}{A_2} v_1 = \frac{\pi (1.0 \text{ cm})^2}{\pi (0.50 \text{ cm})^2} (2.0 \text{ m/s}) = 8.0 \text{ m/s}.$$

From Bernoulli's equation, the gauge pressure p_2 in the bathroom is

$$\begin{aligned} p_2 &= p_1 - \tfrac{1}{2}\rho(v_2^2 - v_1^2) - \rho g(y_2 - y_1) \\ &= 4.0 \times 10^5 \text{ Pa} \\ &\quad - \tfrac{1}{2}(1.0 \times 10^3 \text{ kg/m}^3)(64 \text{ m}^2/\text{s}^2 - 4.0 \text{ m}^2/\text{s}^2) \\ &\quad - (1.0 \times 10^3 \text{ kg/m}^3)(9.8 \text{ m/s}^2)(5.0 \text{ m}) \\ &= 3.2 \times 10^5 \text{ Pa} \\ &= 3.2 \text{ atm} = 47 \text{ lb/in}^2. \quad \text{(gauge pressure)}. \end{aligned}$$

The volume flow rate is

$$\begin{aligned} \Delta V / \Delta t &= A_2 v_2 = \pi (0.50 \times 10^{-2} \text{ m})^2 (8.0 \text{ m/s}) \\ &= 6.3 \times 10^{-4} \text{ m}^3/\text{s} = 0.63 \text{ L/s}. \end{aligned}$$

The time needed to fill the tub is 100 L/(0.63 L/s) = 160 s.

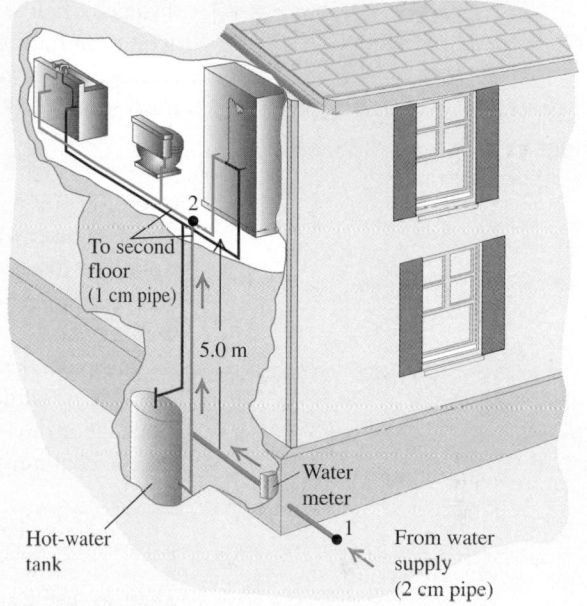

▲ **FIGURE 13.30** What is the water pressure in the second-story bathroom of this house?

REFLECT Note that when the water is turned off, the second term on the right of the pressure equation vanishes, and the pressure rises to 3.5×10^5 Pa. In fact, when the fluid is not moving, Bernoulli's equation (Equation 13.12) reduces to the pressure relation that we derived for a fluid at rest, Equation 13.3.

13.7 Applications of Bernoulli's Equation

The sections that follow discuss several practical applications of Bernoulli's equation.

Speed of Efflux; Torricelli's Theorem

Figure 13.31 shows a gasoline storage tank with cross-sectional area A_1 filled to a depth h with a liquid of density ρ. The space above the gasoline is vented to the atmosphere, so it contains air at atmospheric pressure p_{atm}, and the gasoline flows out through a short pipe with cross-sectional area A_2. How fast does the gasoline run out?

To answer this question, we consider the entire volume of moving liquid as a single flow tube; v_1 and v_2 are the speeds at points 1 and 2, respectively. The pressure at both points is atmospheric pressure p_{atm}, so the pressure differential is zero. We apply Bernoulli's equation to points 1 and 2 and take $y_2 = 0$ at the bottom of the tank.

When we set $y_1 = h$, Bernoulli's equation becomes

$$p_{\text{atm}} + \tfrac{1}{2}\rho v_1^2 + \rho gh = p_{\text{atm}} + \tfrac{1}{2}\rho v_2^2.$$

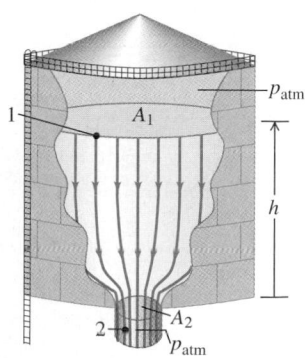

▲ **FIGURE 13.31** How fast does the gasoline run out of this storage tank?

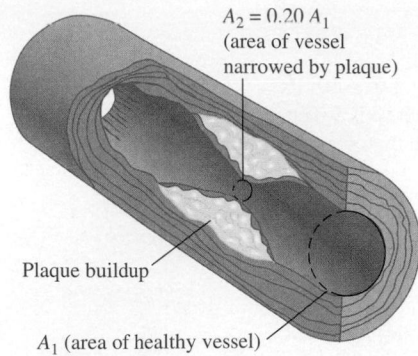

$A_2 = 0.20 A_1$
(area of vessel narrowed by plaque)

Plaque buildup

A_1 (area of healthy vessel)

▲ **FIGURE 13.32** An artery narrowed by plaque in arteriosclerosis.

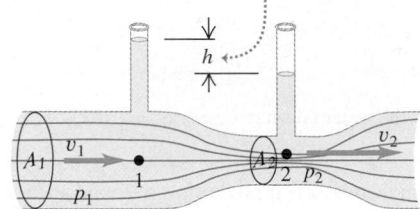

Difference in height results from reduced pressure in throat.

h

A_1 v_1 p_1 1 A_2 2 p_2 v_2

▲ **FIGURE 13.33** The Venturi meter.

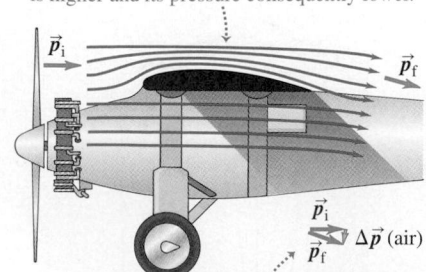

The air going over the wing takes a longer path than the air going under the wing, so its speed is higher and its pressure consequently lower.

$\vec{p}_i$

$\vec{p}_f$

$\vec{p}_i$

$\vec{p}_f$ $\Delta\vec{p}$ (air)

Alternatively, the wing's shape imparts a net downward momentum to the air; the reaction force on the airplane is upward.

▲ **FIGURE 13.34** How the flow of air around an airplane wing generates lift.

But A_2 is much smaller than A_1; therefore, v_1^2 is much smaller than v_2^2 and can be neglected. What remains is

$$\rho gh = \tfrac{1}{2}\rho v_2^2, \quad \text{or} \quad v_2 = \sqrt{2gh}.$$

This expression should look familiar. Recall from Chapter 7 that when an object of mass m falls through a height h, conservation of mechanical energy requires that $mgh = \tfrac{1}{2}mv^2$, or $v = \sqrt{2gh}$. Thus, the speed of efflux from an opening at a distance h below the top surface of a liquid is the same as the speed a body would acquire in falling freely through a height h. This equivalence should not be surprising, since we used conservation of energy to derive Bernoulli's equation. Each small portion of fluid acts like a particle with mass m.

Collapse of a Blood Vessel

Over 200,000 people die each year from arteriosclerosis. One aspect of this disease is a narrowing of blood vessels due to the buildup of plaque on the vessel wall. The disease is considered critical when a portion of vessel becomes 80% blocked. We can find the drop in pressure when blood flowing at 50 cm/s enters such a region.

In Figure 13.32, the cross-sectional area of the healthy portion of the blood vessel is A_1 while the cross-sectional area of the narrowing is A_2. Eighty percent blockage means that 20% of the original area remains, so $A_2 = 0.20A_1$. We first use the continuity equation to find the speed v_2 in the narrow part of the vessel:

$$A_1v_1 = A_2v_2 \quad \text{and} \quad v_2 = \frac{A_1}{A_2}v_1 = \frac{A_1}{0.2A_1}(0.50 \text{ m/s}) = 2.5 \text{ m/s}.$$

Then we use Bernoulli's equation, taking $y_2 = y_1$, to find the drop in blood pressure in the narrow region. The density of blood is about $1.06 \times 10^3 \text{ kg/m}^3$.

For the numerical values given, the drop in blood pressure is

$$p_1 - p_2 = \tfrac{1}{2}(1.06 \times 10^3 \text{ kg/m}^3)[(2.5 \text{ m/s})^2 - (0.50 \text{ m/s})^2]$$
$$= 3.2 \times 10^3 \text{ Pa}.$$

A pressure drop of this magnitude inside the narrow region can act to collapse the blood vessel. Thus, blood flow can stop abruptly even before plaque has entirely closed off the vessel.

The Venturi Meter

The Venturi meter, shown in Figure 13.33, is used to measure flow speed in a pipe. The narrow part of the pipe is called the *throat*. Because A_1 is greater than A_2, v_2 is greater than v_1, and the pressure p_2 in the throat is less than p_1. Thus, the difference in height h of the fluid in the two vertical tubes is directly related to v_1. By applying Bernoulli's equation to the wide (point 1) and narrow (point 2) parts, we can derive a relation between h and v_1. We'll omit the details of the calculation; the result is

$$v_1 = \sqrt{\frac{2gh}{(A_1/A_2)^2 - 1}}.$$

This same concept is used in some gasoline engines, in which gasoline is sprayed from a nozzle or an injector into a low-pressure region produced in a Venturi throat.

Lift on an Airplane Wing

Figure 13.34 shows flow lines around a cross section of an airplane wing. The lines crowd together above the wing, corresponding to increased flow speed and reduced pressure in this region, just as in the Venturi throat. Because the upward force on the underside of the wing is greater than the downward force on the top side, there is a net upward force, or *lift*. (This highly simplified discussion ignores the effect of turbulent flow and the formation of vortices; a more complete discussion would take these into account.)

We can also understand the lift force on the basis of momentum changes. Figure 13.34 shows that there is a net *downward* change in the vertical component of momentum of the air flowing past the wing, corresponding to the downward force the wing exerts on it. The reaction force *on* the wing is *upward,* as we concluded in the previous paragraph.

The Curveball

Does a curveball *really* curve? Yes, it certainly does. Figure 13.35 shows a ball moving through air from left to right. To an observer moving with the center of the ball, the air stream appears to move from right to left, as shown by the flow lines in the figure. Because of the large speeds that are ordinarily involved (up to 100 mi/h for a baseball or a served tennis ball), there is a region of turbulent flow behind the ball.

Figure 13.35b shows a spinning ball with "top spin," viewed from the side. When the ball is spinning, the viscosity of air causes layers of air near the ball's surface to be pulled around in the direction of spin. Therefore, the speed of the air relative to the ball's surface is smaller at the top of the ball than at the bottom (Also, the region of turbulence becomes asymmetric; turbulence occurs farther forward on the top side than on the bottom.) The difference in the speed of the air between the top and bottom of the ball causes a pressure difference: The average pressure at the top of the ball becomes greater than that at the bottom. The corresponding net downward force deflects the ball downward, as shown. This could be a side view of a tennis ball (Figure 13.35d) moving from left to right with "top spin." In a baseball curve pitch, the ball spins about a nearly vertical axis, and the actual deflection is mostly sideways. In that case, Figure 13.35c is a *top* view of the situation. Note that a curveball thrown by a left-handed pitcher curves *toward* a right-handed batter, making it harder to hit than the same ball thrown to a left-handed batter (in which case the ball would curve *away from* the batter).

A similar effect occurs with golf balls, which always have "backspin" from impact with the slanted face of the golf club. The resulting pressure difference

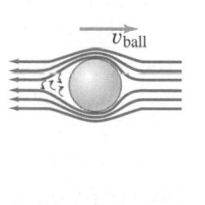

$\overrightarrow{v}_{\text{ball}}$

(a) Motion of air relative to a nonspinning ball

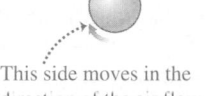

This side of the ball moves opposite to the air flow.

This side moves in the direction of the air flow.

(b) Motion of a spinning ball

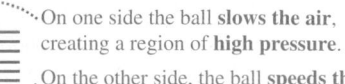

A moving ball drags the adjacent air with it. So, when air moves past a spinning ball:

On one side the ball **slows the air**, creating a region of **high pressure**.

On the other side, the ball **speeds the air**, creating a region of **low pressure**.

The resultant force points in the direction of the low-pressure side.

(c) Force generated when a spinning ball moves through air

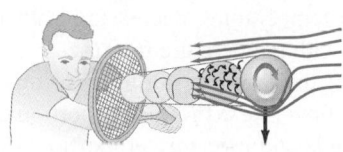

(d) Spin pushing a tennis ball downward

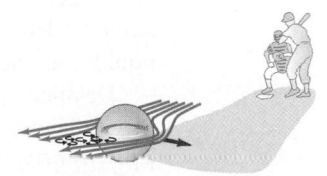

(e) Spin causing a curveball to be deflected sideways

▲ **FIGURE 13.35** How spin on a ball creates forces that cause the ball to follow a nonparabolic trajectory.

▲ **FIGURE 13.36** Strobe photo of a golf ball being struck by a club. The picture was taken at 1000 flashes per second. The ball rotates about once in eight pictures, corresponding to an angular velocity of 125 rev/s, or 7500 rpm.

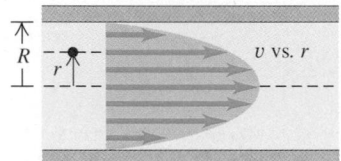

▲ **FIGURE 13.37** Velocity profile for a viscous fluid in a cylindrical pipe.

between the top and bottom of the ball causes a "lift" force that keeps the ball in the air considerably longer than would be possible without spin. From the tee, a well-hit drive appears to float, or even curve *upward,* during the initial portion of its flight. This is a real effect, not an illusion. The dimples on the ball play an essential role: Because of effects associated with the viscosity of air, a ball without dimples has a much shorter trajectory than a dimpled ball, given the same initial velocity and spin. One manufacturer even claims that the *shape* of the dimples is significant: Polygonal dimples are alleged to be more effective than round ones. Figure 13.36 shows the backspin of a golf ball just after it is struck by a club.

13.8 Real Fluids: Viscosity and Turbulence

Viscosity is internal friction in a fluid; viscous forces oppose the motion of one portion of a fluid relative to another. Viscosity is the reason that paddling a canoe through calm water requires effort, but it is also the reason the paddle works. Viscosity plays a vital role in the flow of fluids in pipes, the flow of blood, the lubrication of engine parts, and many other areas of practical importance. An ideal fluid, the model we used to derive the Bernoulli equation, has no viscosity, but in real-life fluids, especially liquids, effects associated with viscosity are often significant.

A viscous fluid always tends to cling to a solid surface that is in contact with it. There is always a thin *boundary layer* of fluid near the surface in which the fluid is nearly at rest with respect to the surface. That's why dust particles can cling to a fan blade even when it is rotating rapidly and why you can't get all the mud off your car by just squirting a hose at it.

Viscosity plays an important role in the flow of fluids through pipes. Let's think first about a fluid with zero viscosity, so that we can apply Bernoulli's equation, Equation 13.12. For a pipe of uniform cross section, if the two ends are at the same height (so that $y_1 = y_2$) and have the same flow speed ($v_1 = v_2$) and the fluid has zero acceleration, then the Bernoulli equation states that the pressure is the same at the two ends. But if viscosity is taken into account, the pressure must be greater at the upstream end than at the downstream end. To see why, let's consider the velocity profile for flow of a viscous fluid in a cylindrical pipe (Figure 13.37). The flow velocity is different at different points of a cross section. The outermost layer of fluid clings to the walls, so the speed is zero at the pipe walls. The walls exert a backward drag on this layer, which in turn drags backward on the next layer inward, and so on. The flow speed is greatest along the axis of the pipe.

The motion is like that of many coaxial telescoping tubes sliding relative to one another, with the central tube moving fastest and the outermost tube at rest. The viscous forces between the tubes resist this motion, so, for the fluid to keep moving, the pressure p_1 at the back end has to be greater than the pressure p_2 at the front end. The strength of the viscous resisting forces is characterized by a quantity called the *coefficient of viscosity,* or simply the *viscosity,* denoted by η (Greek eta). Very viscous fluids, such as molasses, have large values of η; gases have much smaller values, and water is in between. The coefficient of viscosity is strongly dependent on temperature, decreasing with increasing temperature for liquids, but increasing with temperature for gases.

Detailed analysis of the flow pattern of a viscous fluid in a cylindrical pipe shows that the volume flow rate $\Delta V/\Delta t$ is proportional to the pressure difference $p_1 - p_2$ between the ends and inversely proportional to the length L of the pipe. That is, the volume flow rate is proportional to the pressure change *per unit length,* which is called the *pressure gradient.* The volume flow rate is also proportional to the *fourth power* of the pipe's radius R. The specific relationship, called **Poiseuille's law,** is

$$\frac{\Delta V}{\Delta t} = \frac{\pi R^4}{8\eta} \frac{p_1 - p_2}{L}.$$

If the radius R is doubled, the volume flow rate increases by a factor of 16. This relation is familiar to physicians in connection with hypodermic needles. Needle size is much more important than thumb pressure in determining the flow rate from the needle; doubling the needle diameter has the same effect as increasing the thumb force by a factor of 16. Similarly, blood flow in arteries and veins can be controlled over a wide range by relatively small changes in diameter, an important temperature-control mechanism in warm-blooded animals. Relatively slight narrowing of arteries due to arteriosclerosis can result in elevated blood pressure and added strain on the heart muscle.

One more useful relation in viscous fluid flow is the expression for the magnitude F of force exerted on a sphere of radius r that moves with speed v through a fluid with viscosity η. When v is small enough that there is no turbulence, the relationship is simple:

$$F = 6\pi\eta rv.$$

This relation is called **Stokes's law.** We've encountered a similar velocity-dependent force before, in Example 5.13 (Section 5.3).

A sphere falling in a viscous fluid reaches a *terminal speed* v_t for which the total force, including the weight of the sphere and the viscous retarding force, is zero, and the sphere no longer accelerates. (We neglect the effect of buoyancy of the sphere in the fluid.) Let ρ be the density of the sphere; then the weight of the sphere is $\frac{4}{3}\pi r^3 \rho g$. When the sphere reaches terminal speed, the total force on it is zero:

$$6\pi\eta rv_t - \tfrac{4}{3}\pi r^3 \rho g = 0,$$

or

$$v_t = \frac{2}{9} \frac{r^2 g \rho}{\eta}.$$

The Stokes's-law force is what keeps clouds in the air. The terminal speed for a water droplet of radius 10^{-5} m is on the order of 1 mm/s. At higher speeds the flow often becomes turbulent, and Stokes's law is no longer valid. In air, the drag force at highway speeds is approximately proportional to v^2. We've discussed the applications of Stokes's law to skydiving (Example 5.13, Section 5.3) and to the air drag of a moving automobile (Section 7.9). Raindrops have terminal speeds of a few meters per second because of air drag; otherwise, they would smash everything in their path!

As we've mentioned previously, flow can be either laminar, with layers of fluid slipping smoothly past each other, or turbulent—irregular, complex, and ever changing. Figure 13.27 showed rising smoke changing from laminar to turbulent flow. Figure 13.38 compares laminar and turbulent flow in streams of water.

The transition from laminar to turbulent flow is often very sudden. A flow pattern that is stable at low speeds suddenly becomes unstable when a critical speed is reached. Irregularities in the flow pattern can be caused by roughness in a pipe wall, variations in density of the fluid, and many other factors. At small flow speeds, these disturbances are damped out and the flow pattern tends to

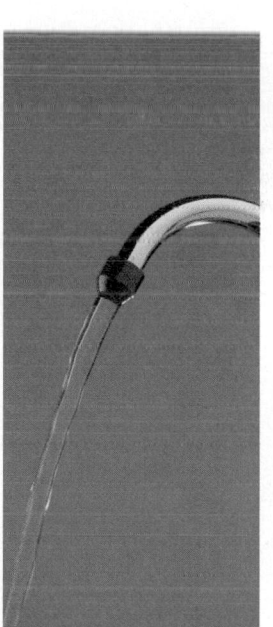

Laminar flow   Turbulent flow

▲ **FIGURE 13.38** Laminar and turbulent flow in a stream of water.

maintain its laminar nature. But when the critical speed is reached, the disturbances no longer damp out; instead, they grow until they destroy the entire laminar-flow pattern.

Turbulence poses some profound questions for theoretical physics. The motion of any particular particle in a fluid is presumably determined by Newton's laws. If we know the motion of the fluid at some initial time, shouldn't we be able to predict the motion at any later time? After all, when we launch a projectile, we can compute the entire trajectory if we know the initial position and velocity. How is it that a system that obeys well-defined and supposedly deterministic physical laws can have unpredictable, chaotic behavior?

There is no simple answer to these questions. Indeed, the study of chaotic behavior in deterministic systems is a highly active field of research in theoretical physics. The development of supercomputers in recent years has enabled scientists to carry out computer simulations of the behavior of physical systems on a scale that would have been hopeless only a few years ago. The results of such simulations have often guided new developments in the theory of chaotic behavior. A particularly significant result of these studies has been the discovery of common characteristics in the behavior of widely divergent kinds of systems, including turbulent fluid flow, population fluctuations in ecosystems, the growth of crystals, the shapes of coastlines, and many others.

SUMMARY

Density and Pressure in a Fluid

(Sections 13.1 and 13.2) Density is mass per unit volume. If a mass m of material has volume V, its density ρ is $\rho = m/V$ (Equation 13.1). **Specific gravity** is the ratio of the density of a material to the density of water.

   Pressure is force per unit area. Assuming that the pressure is constant over the area in question, we express pressure as $p = F_\perp/A$ (Equation 13.2). **Pascal's law** states that the pressure applied to the surface of an enclosed fluid is transmitted undiminished to every portion of the fluid. The total pressure at any level in a fluid is the external pressure plus the pressure due to the weight of the overlying fluid. The pressure p at a height h above or below a reference level where $h = 0$, relative to the pressure p_0 at the reference level, is $p - p_0 = -\rho gh$ (where ρ is the density of the fluid) (Equation 13.3).

   Absolute pressure is the total pressure in a fluid; **gauge pressure** is the difference between absolute pressure and atmospheric pressure. The SI unit of pressure is the pascal (Pa), where $1\ \text{Pa} = 1\ \text{N/m}^2$. Additional units of pressure are the bar ($1\ \text{bar} = 10^5\ \text{Pa}$), the torr $[1\ \text{torr} = 1\ \text{millimeter of mercury } (\text{mm Hg})]$, and the atmosphere (atm) ($1\ \text{atm} = 1.013 \times 10^5\ \text{Pa} = 760\ \text{torrs} = 1.013\ \text{bar}$)

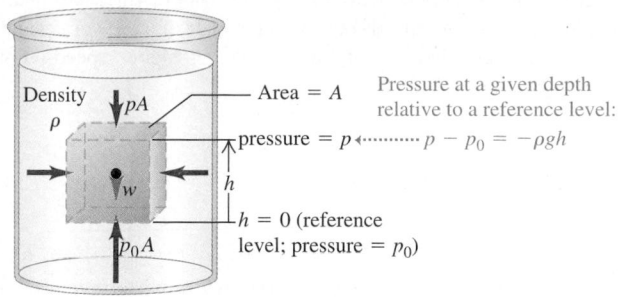

Density ρ

Area $= A$

pA

w

$p_0 A$

h

$h = 0$ (reference level; pressure $= p_0$)

Pressure at a given depth relative to a reference level:

pressure $= p$ $\cdots\cdots\cdots$ $p - p_0 = -\rho gh$

Archimedes's Principle: Buoyancy

(Section 13.3) Archimedes's principle states that when an object is immersed in a fluid, the fluid exerts an upward buoyant force on the object equal in magnitude to the weight of the fluid the object displaces. Thus, an object denser than the surrounding fluid sinks; one less dense than the surrounding fluid floats.

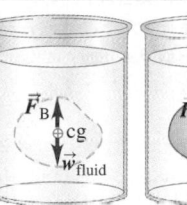

The pressure on any volume element of a fluid produces a buoyancy force $\vec{F}_B$ that exactly cancels the element's weight.

$\vec{F}_B$

cg

$\vec{w}_{\text{fluid}}$

$\vec{F}_B$

$\vec{w}_{\text{object}}$

The buoyancy force on a solid object is equal in magnitude to the weight of the displaced fluid.

Surface Tension and Capillarity

(Section 13.4) A boundary surface between a liquid and a gas behaves like a surface under tension (**surface tension**). Where a liquid–gas interface meets a solid surface, the interface is concave if the liquid is attracted to (wets) the solid and convex if not. In the related phenomenon of **capillarity**, a liquid climbs up a narrow tube if it wets the tube's material, but is pushed down by the tube if it does not.

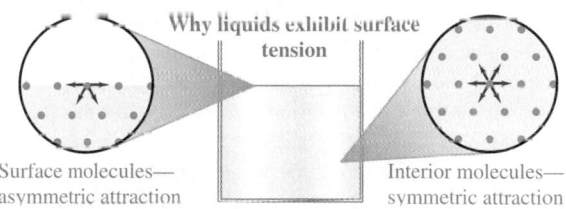

Why liquids exhibit surface tension

Surface molecules—asymmetric attraction

Interior molecules—symmetric attraction

Fluid Flow, Bernoulli's Equation

(Sections 13.5–13.7) An ideal fluid is an incompressible fluid with no viscosity. A flow line is the path of a fluid particle. In **laminar flow**, layers of fluid slide smoothly past each other. In **turbulent flow**, there is disorder and a constantly changing flow pattern.

   Conservation of mass for an incompressible fluid is expressed by the **equation of continuity:** For a liquid flowing in a tube with varying cross section, the flow speeds v_1 and v_2 at two cross sections with areas A_1 and A_2, respectively, are related by $A_1 v_1 = A_2 v_2$ (Equation 13.10). The product Av is the **volume flow rate, $\Delta V/\Delta t$,** the rate at which volume crosses a section of the tube: $\Delta V/\Delta t = Av$.

   Bernoulli's equation relates the pressure p, flow speed v, and elevation y in an ideal fluid. For any two points, indicated by subscripts 1 and 2, $p_1 + \rho gy_1 + \frac{1}{2}\rho v_1^2 = p_2 + \rho gy_2 + \frac{1}{2}\rho v_2^2$ (Equation 13.12). This equation states that faster flow is associated with lower pressure. That fact underlies the working of Venturi meters and is also responsible for the lift on airplane wings and for the nonparabolic trajectories of spinning baseballs and golf balls.

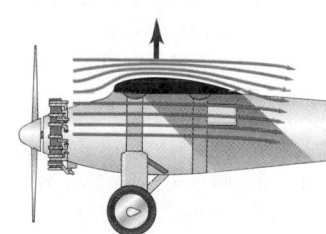

Lift due to Bernoulli's principle: Air travels farther and therefore faster over the top than the bottom of the wing; the result is lower pressure over the top of the wing.

Continued

Real Fluids: Viscosity and Turbulence

(Section 13.8) Real fluids have viscosity. Consequently, the boundary layer fluid in contact with a stationary wall or surface is nearly at rest, whereas flow speeds peak in the center of a pipe or channel. Flow is typically laminar at low speeds; at higher speeds, flow tends to become turbulent.

 For instructor-assigned homework, go to www.masteringphysics.com

Conceptual Questions

1. A clear plastic hose is attached to the narrow end of a funnel, and its free end is bent around to point upward (Figure 13.39). When water is poured into the funnel, it rises in the hose to the same level as in the funnel, despite the fact that the funnel has a lot more water in it than the hose. Why does it do this? Where does the force come from to support the weight of the extra water in the funnel?

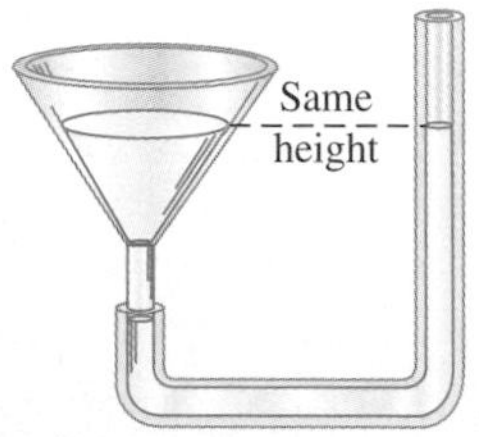

▲ **FIGURE 13.39** Question 1.

2. In describing the size of a large ship, one uses the expression "It displaces 20,000 tons." What does this expression mean? Can the weight of the ship be obtained from this information?

3. A popular, though expensive, sport is hot-air ballooning, in which a large nylon balloon is filled with air heated by a gas burner at the bottom. Why must the air be heated? How does the balloonist control the balloon's ascent and descent?

4. Equation 13.5 shows that an area ratio of 100 to 1 can give 100 times more output force than input force. Doesn't this violate conservation of energy? Explain.

5. Suppose the door of a room makes an airtight, but frictionless, fit in its frame. Do you think you could open the door if the air pressure on one side were standard atmospheric pressure and that on the other side differed from standard by 1%? Back up your answer with a simple but reasonable calculation.

6. When a smooth-flowing stream of water comes out of a faucet, it narrows as it falls. Why does it do this?

7. You push an empty glass jar into a tank of water with the open mouth downward, so that the air inside the jar is trapped and cannot get out. If you push the jar deeper into the water, does the buoyant force on the jar stay the same? If not, does it increase or decrease? Explain your answer.

8. A very smooth wooden block is pressed against the bottom of a container of water so that no water remains under the block. Will this block rise to the surface when it is released? Why? (*Hint:* Can the water exert a force on the bottom of the block?)

9. A marble is in a little box that is floating in a beaker of water. The marble is then removed from the box and dropped into the water, sinking to the bottom. What happens to the water level in the beaker when you do this? Does it go up, go down, or stay the same? Why? (It might help to draw a figure.)

10. A beaker of water rests on a scale. You slowly lower a piece of metal into the water, using a string attached to the metal, but you do not allow the metal to touch the walls or bottom of the beaker. (a) What happens to the reading on the scale when you do this? (b) How does the scale "know" that the metal is in the water, since the metal never touches either the beaker or the scale? (It might help to draw a figure.)

11. If a rocketship traveling through the vacuum of space is hit by a small asteroid, it could lose its air through the hole caused by the collision. Many people would describe this situation by saying that the air was *sucked out* by the vacuum. Is this a good *physical* explanation of what is happening? Does the vacuum actually suck out the air? What does make the air go out?

12. If you walk along a coastal seaport, you often see recently caught fish that are enormously bloated. These are fish that live very deep in the sea. What caused them to be so bloated? Were they that way deep down where they normally live?

13. Submarines can remain at equilibrium at various depths in the ocean, and they are made of rigid metal to withstand great pressures. (a) What must be the average density of a submarine? (b) Devise a simple way that a submarine could cause itself to go up (or down) without using its propellers.

14. You are told, "Bernoulli's equation tells us that where there is higher fluid speed, there is lower fluid pressure, and vice versa." Is this statement always true, even for an idealized fluid?

15. A helium-filled balloon is tied to a light string inside of a car at rest. The other end of the string is attached to the floor of the car, so the balloon pulls the string vertical. You now accelerate forward. Which way does the balloon move, forward or backward, or does it not move? Justify your reasoning with reference to buoyancy. (If you have a chance, try this experiment yourself—but with someone else doing the driving!)

Multiple-Choice Problems

1. Which has a greater buoyant force on it, a 25 cm³ piece of wood floating with part of its volume above water or a 25 cm³ piece of submerged iron?
   A. The floating wood.
   B. The submerged iron.
   C. They have the same buoyant force on them.
   D. It is impossible to tell without knowing their weights.

2. As a rigid (i.e., constant volume), lighter-than-air balloon leaves the ground and rises, the buoyant force on it
   A. increases.   B. decreases.   C. stays the same.

3. A mass of sunken lead is resting against the bottom in a glass of water. You take this lead, put it in a small boat of negligible mass, and float the boat in the water. Which of the following statements are true? (There may be more than one correct choice.)
   A. The sunken lead displaces a volume of water equal to the lead's own volume.
   B. The floating lead displaces a volume of water equal to the lead's own volume.
   C. The sunken lead displaces a volume of water whose weight equals the lead's weight.
   D. The floating lead displaces a volume of water whose weight equals the lead's weight.
4. Two equal-sized buckets are filled to the brim with water, but one of them has a piece of wood floating in it. Which bucket of water weighs more?
   A. The bucket with the wood.
   B. The bucket without the wood.
   C. They weigh the same amount.
5. Two equal *mass* pieces of metal are sitting side by side at the bottom of a deep lake. One piece is aluminum and the other is lead. Which piece has the greater buoyant force acting on it?
   A. The aluminum.
   B. The lead.
   C. They both have the same buoyant force acting on them.
6. If a 5 lb force is required to keep a block of wood 1 ft beneath the surface of water, the force needed to keep it 2 ft beneath the surface is
   A. 2.5 lb.   B. 5 lb.   C. 10 lb.
7. A horizontal pipe with water flowing through it has a circular cross section that varies in diameter. The diameter at the wide section is 3 times that of the diameter at the narrow section. If the rate of flow of the water in the narrow section is 9.0 L/min, the rate of flow of the water in the wide section is
   A. 1.0 L/min.   B. 3.0 L/min.   C. 9.0 L/min.
   D. 18 L/min.   E. 36 L/min.
8. A horizontal cylindrical pipe has a part with a diameter half that of the rest of the pipe. If V is the speed of the fluid in the wider section of pipe, then the speed of the fluid in the narrower section is
   A. $V/4$.   B. $V/2$.   C. $2V$.   D. $4V$.
9. If the *absolute* pressure at a depth d in a lake is P, the absolute pressure at a depth $2d$ will be
   A. $2P$.   B. P.
   C. Greater than $2P$.   D. Less than $2P$.
10. If the *gauge* pressure at a depth d in a lake is P, the gauge pressure at a depth $2d$ will be
    A. $2P$.   B. P.
    C. Greater than $2P$.   D. Less than $2P$.
11. There is a great deal of ice floating on the oceans near the North Pole. If this ice were to melt due to global warming, what would happen to the level of the oceans?
    A. The level would rise.
    B. The level would fall.
    C. The level would stay the same.
12. A rigid metal object is dropped into a lake and sinks to the bottom. The density of the water in this lake is the same everywhere. As the object sinks deeper and deeper below the surface, the buoyant force on it
    A. gets greater and greater.
    B. gets less and less.

C. does not change.
D. is equal to the weight of this object.
13. A spherical object has a density ρ. If it is compressed under high pressure to half of its original diameter, its density will now be
    A. $\rho/8$.   B. $\rho/4$.   C. 2ρ.
    D. 4ρ.   E. 8ρ.
14. Identical-size cubes of lead and aluminum are suspended at different depths by two wires in a tank of water. Which wire will have a greater tension?
    A. The wire holding the lead cube
    B. The wire holding the aluminum cube
    C. The tensions in the two wires will be equal.
15. Two small holes are drilled in the side of a barrel filled with water. One hole is twice as far below the surface as the other. If the speed of the water flowing from the upper hole is V, the speed of the water flowing from the lower hole is
    A. V B. $V/2$ C. $2V$ D. $\sqrt{2}V$

Problems

13.1 Density

1. • You purchase a rectangular piece of metal that has dimensions 5.0 mm × 15.0 mm × 30.0 mm and mass 0.0158 kg. The seller tells you that the metal is gold. To check this, you compute the average density of the piece. What value do you get? Were you cheated?
2. • A kidnapper demands a 40.0 kg cube of platinum as a ransom. What is the length of a side?
3. • Calculate the weight of air at 20°C in a room that measures 5.00 × 4.50 × 3.25 m. Give your answer in newtons and in pounds.
4. • By how many newtons do you increase the weight of your car when you fill up your 11.5 gal gas tank with gasoline? A gallon is equal to 3.788 L and the density of gasoline is 737 kg/m³.
5. •• **How big is a million dollars?** At the time this problem was written, the price of gold was about $1239 per ounce, while that of platinum was about $1508 an ounce. The "ounce" in this case is the *troy ounce*, which is equal to 31.1035 g. (The more familiar *avoirdupois ounce* is equal to 28.35 g.) The density of gold is 19.3 g/cm³ and that of platinum is 21.4 g/cm³. (a) If you find a spherical gold nugget worth 1.00 million dollars, what would be its *diameter*? (b) How much would a platinum nugget of this size be worth?
6. •• A cube 5.0 cm on each side is made of a metal alloy. After you drill a cylindrical hole 2.0 cm in diameter all the way through and perpendicular to one face, you find that the cube weighs 7.50 N. (a) What is the density of this metal? (b) What did the cube weigh before you drilled the hole in it?
7. •• A cube of compressible material (such as Styrofoam™ or balsa wood) has a density ρ and sides of length L. (a) If you keep its mass the same, but compress each side to half its length, what will be its new density, in terms of ρ? (b) If you keep the mass and shape the same, what would the length of each side have to be (in terms of L) so that the density of the cube was three times its original value?
8. • A hollow cylindrical copper pipe is 1.50 m long and has an outside diameter of 3.50 cm and an inside diameter of 2.50 cm. How much does it weigh?

9. • A uniform lead sphere and a uniform aluminum sphere have the same mass. What is the ratio of the radius of the aluminum sphere to the radius of the lead sphere?

13.2 Pressure in a Fluid

10. • **Blood pressure.** Systemic blood pressure is defined as the
BIO ratio of two pressures, both expressed in millimeters of mercury. Normal blood pressure is about $\frac{120 \text{ mm}}{80 \text{ mm}}$, which is usually just stated as $\frac{120}{80}$. (See also Problem 24.) What would normal systemic blood pressure be if, instead of millimeters of mercury, we expressed pressure in each of the following units, but continued to use the same ratio format? (a) atmospheres, (b) torr, (c) Pa, (d) N/m², (e) psi.

11. • **Blood.** (a) **Mass of blood.** The human body typically con-
BIO tains 5 L of blood of density 1060 kg/m³. How many kilograms of blood are in the body? (b) The average blood pressure is 13,000 Pa at the heart. What average force does the blood exert on each square centimeter of the heart? (c) **Red blood cells.** Red blood cells have a specific gravity of 5.0 and a diameter of about 7.5 μm. If they are spherical in shape (which is not quite true), what is the mass of such a cell?

12. • **Landing on Venus.** One of the great difficulties in landing on Venus is dealing with the crushing pressure of the atmosphere, which is 92 times the earth's atmospheric pressure. (a) If you are designing a lander for Venus in the shape of a hemisphere 2.5 m in diameter, how many newtons of inward force must it be prepared to withstand due to the Venusian atmosphere? (Don't forget about the bottom!) (b) How much force would the lander have to withstand on the earth?

13. • You are designing a diving bell to withstand the pressure of seawater at a depth of 250 m. (a) What is the gauge pressure at this depth? (You can ignore the small changes in the density of the water with depth.) (b) At the 250 m depth, what is the net force due to the water outside and the air inside the bell on a circular glass window 30.0 cm in diameter if the pressure inside the diving bell equals the pressure at the surface of the water? (You may ignore the small variation in pressure over the surface of the window.)

14. • **Glaucoma.** Under normal circumstances, the vitreous
BIO humor, a jelly-like substance in the main part of the eye, exerts a pressure of up to 24 mm of mercury that maintains the shape of the eye. If blockage of the drainage duct for aqueous humor causes this pressure to increase to about 50 mm of mercury, the condition is called *glaucoma*. What is the *increase* in the total force (in newtons) on the walls of the eye if the pressure increases from 24 mm to 50 mm of mercury? We can quite accurately model the eye as a sphere 2.5 cm in diameter.

15. • By means of physiological adaptations that are still not very well understood, sperm whales are thought to be able to hunt for their food at depths of between 400 m and 3000 m. (a) What range of gauge pressures (in Pa and atm) do the whales withstand at these depths? (b) Estimate the total inward force of water pressure on the surface of a sperm whale at a depth of 3000 m, modeling the whale as a cylinder 16 m long and 4 m in diameter.

16. • What gauge pressure must a pump produce to pump water from the bottom of the Grand Canyon (elevation 730 m) to Indian Gardens (elevation 1370 m)? Express your result in pascals and in atmospheres.

17. • **Intravenous feeding.** A hospital
BIO patient is being fed intravenously with a liquid of density 1060 kg/m³. (See Figure 13.40.) The container of liquid is raised 1.20 m above the patient's arm where the fluid enters his veins. What is the pressure this fluid exerts on his veins, expressed in millimeters of mercury?

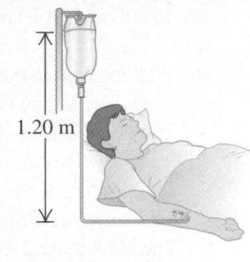

▲ **FIGURE 13.40**
Problem 17.

18. •• A 975-kg car has its tires each inflated to "32.0 pounds." (a) What are the absolute and gauge pressures in these tires in lb/in.², Pa, and atm? (b) If the tires were perfectly round, could the tire pressure exert any force on the pavement? (Assume that the tire walls are flexible so that the pressure exerted by the tire on the pavement equals the air pressure inside the tire.) (c) If you examine a car's tires, it is obvious that there is some flattening at the bottom. What is the total contact area for all four tires of the flattened part of the tires at the pavement?

19. •• An electrical short cuts off all power to a submersible diving vehicle when it is 30 m below the surface of the ocean. The crew must push out a hatch of area 0.75 m² and weight 300 N on the bottom to escape. If the pressure inside is 1.0 atm, what downward force must the crew exert on the hatch to open it?

20. • **Standing on your head.** (a) When you stand on your head,
BIO what is the *difference* in pressure of the blood in your brain compared with the pressure when you stand on your feet if you are 1.85 m tall? The density of blood is 1060 kg/m³. (b) What effect does the increased pressure have on the blood vessels in your brain?

21. •• You are designing a machine for a space exploration vehicle. It contains an enclosed column of oil that is 1.50 m tall, and you need the pressure difference between the top and the bottom of this column to be 0.125 atm. (a) What must be the density of the oil? (b) If the vehicle is taken to Mars, where the acceleration due to gravity is 0.379g, what will be the pressure difference (in earth atmospheres) between the top and bottom of the oil column?

22. •• **Ear damage from diving.** If the force on the tympanic
BIO membrane (eardrum) increases by about 1.5 N above the force from atmospheric pressure, the membrane can be damaged. When you go scuba diving in the ocean, below what depth could damage to your eardrum start to occur? The eardrum is typically 8.2 mm in diameter. (Consult Table 13.1.)

23. •• A barrel contains a 0.120 m layer of oil of density 600 kg/m³ floating on water that is 0.250 m deep. (a) What is the gauge pressure at the oil–water interface? (b) What is the gauge pressure at the bottom of the barrel?

24. •• **Blood pressure.** Systemic blood pressure is expressed as
BIO the ratio of the *systolic* pressure (when the heart first ejects blood into the arteries) to the *diastolic* pressure (when the heart is relaxed):

$$\text{systemic blood pressure} = \frac{\text{systolic pressure}}{\text{diastolic pressure}}$$

Both pressures are measured at the level of the heart and are expressed in millimeters of mercury (or torr), although the units are not written. Normal systemic blood pressure is $\frac{120}{80}$. (a) What are the maximum and minimum forces (in newtons) that the blood exerts against each square centimeter of the heart for a person with normal blood pressure? (b) As pointed out in the text, blood pressure is normally measured on the upper arm at the same height as the heart. Due to therapy for an injury, a patient's upper arm is extended 30.0 cm above his heart. In that position, what should be his systemic blood pressure reading, expressed in the standard way, if he has normal blood pressure? The density of blood is 1060 kg/m^3.

25. •• **Blood pressure on the moon.** When we eventually estab-
BIO lish lunar colonies, people living there will need to have their blood pressure taken. Assume that we continue to express the systemic blood pressure as we now do on earth (see previous problem) and that the density of blood does not change. Suppose also that normal blood pressure on the moon is still $\frac{120}{80}$ (which may not actually be true). If a lunar colonizer has her blood pressure taken at her upper arm when it is raised 25 cm above her heart, what will be her systemic blood pressure reading, expressed in the standard way, if she has normal blood pressure? The acceleration due to gravity on the moon is 1.67 m/s^2.

26. • The piston of a hydraulic automobile lift is 0.30 m in diameter. What gauge pressure, in pascals, is required to lift a car with a mass of 1200 kg? Now express this pressure in atmospheres.

27. •• **Hydraulic lift.** You are designing a hydraulic lift for an automobile garage. It will consist of two oil-filled cylindrical pipes of different diameters. A worker pushes down on a piston at one end, raising the car on a platform at the other end. (See Figure 13.41.) To handle a full range of jobs, you must be able to lift cars up to 3000 kg, plus the 500 kg platform on which they are parked. To avoid injury to your workers, the maximum amount of force a worker should need to exert is 100 N. (a) What should be the diameter of the pipe under the platform? (b) If the worker pushes down with a stroke 50 cm long, by how much will he raise the car at the other end?

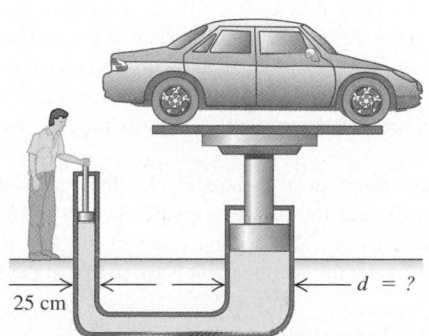

▲ **FIGURE 13.41** Problem 27.

28. •• There is a maximum depth at which a diver can breathe through a snorkel tube (Fig. 13.42), because as the depth increases, so does the pressure difference, which tends to collapse the diver's lungs. Since the snorkel connects the air in the lungs to the atmosphere at the surface, the pressure inside the lungs is atmospheric pressure. What is the external–internal pressure difference when the diver's lungs are at a depth of 6.1 m (about 20 ft)? Assume that the diver is in freshwater. (A scuba diver breathing from compressed air tanks can operate at greater depths than can a snorkeler, since the pressure of the air inside the scuba diver's lungs increases to match the external pressure of the water.)

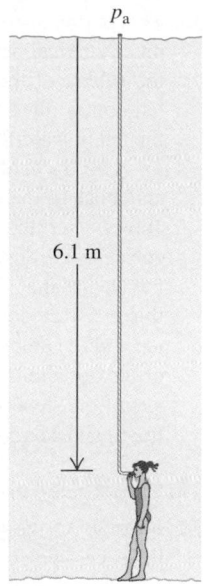

▲ **FIGURE 13.42** Problem 28.

13.3 Archimedes's Principle: Buoyancy

29. • A solid aluminum ingot weighs 89 N in air. (a) What is its volume? (See Table 13.1.) (b) The ingot is suspended from a rope and totally immersed in water. What is the tension in the rope (the *apparent* weight of the ingot in water)?

30. •• **Fish navigation.** (a) As you can tell by watching them in
BIO an aquarium, fish are able to remain at any depth in water with no effort. What does this ability tell you about their density? (b) Fish are able to inflate themselves using a sac (called the *swim bladder*) located under their spinal column. These sacs can be filled with an oxygen–nitrogen mixture that comes from the blood. If a 2.75 kg fish in fresh water inflates itself and increases its volume by 10%, find the *net* force that the *water* exerts on it. (c) What is the net *external* force on it? Does the fish go up or down when it inflates itself?

31. •• When an open-faced boat has a mass of 5750 kg, including its cargo and passengers, it floats with the water just up to the top of its gunwales (sides) on a freshwater lake. (a) What is the volume of this boat? (b) The captain decides that it is too dangerous to float with his boat on the verge of sinking, so he decides to throw some cargo overboard so that 20% of the boat's volume will be above water. How much mass should he throw out?

32. •• An ore sample weighs 17.50 N in air. When the sample is suspended by a light cord and totally immersed in water, the tension in the cord is 11.20 N. Find the total volume and the density of the sample.

33. •• A slab of ice floats on a freshwater lake. What minimum volume must the slab have for a 45.0 kg woman to be able to stand on it without getting her feet wet?

34. • Using data from Appendix E, calculate the average density of the planet Saturn. How does your answer compare to the density of water, and what does this imply about the buoyancy of Saturn, if you could find an ocean big enough to drop it into?

35. •• A hollow plastic sphere is held below the surface of a freshwater lake by a cord anchored to the bottom of the lake. The sphere has a volume of 0.650 m^3 and the tension in the cord is 900 N. (a) Calculate the buoyant force exerted by the water on the sphere. (b) What is the mass of the sphere? (c) The cord breaks and the sphere rises to the surface. When the sphere comes to rest, what fraction of its volume will be submerged?

36. •• (a) Calculate the buoyant force of air (density 1.20 kg/m^3) on a spherical party balloon that has a radius of 15.0 cm. (b) If the rubber of the balloon itself has a mass of 2.00 g and the balloon is filled with helium (density 0.166 kg/m^3), calculate the net upward force (the "lift") that acts on it in air.

37. •• **The tip of the iceberg.** Icebergs consist of freshwater ice and float in the ocean with only about 10% of their volume above water (the "tip of the iceberg," so to speak). This percentage can vary, depending on the condition of the ice. Assume that the ice has the density given in Table 13.1, although, in reality, this can vary considerably, depending on the condition of the ice and the amount of impurities in it. (a) What does this 10% observation tell us is the density of seawater? (b) What percentage of the icebergs' volume would be above water if they were floating in a large freshwater lake such as Lake Superior?

13.4 Surface Tension and Capillarity

38. • At 20° C, the surface tension of water is 72.8 dynes/cm. Find the excess pressure inside of (a) an ordinary-size water drop of radius 1.50 mm and (b) a fog droplet of radius 0.0100 mm.

39. • Find the gauge pressure in pascals inside a soap bubble 7.00 cm in diameter. The surface tension of this soap is 25.0 dynes/cm.

40. • What radius must a water drop have for the difference between the inside and outside pressures to be 0.0200 atm? The surface tension of water is 72.8 dynes/cm.

41. •• At 20° C, the surface tension of water is 72.8 dynes/cm and that of carbon tetrachloride (CCl_4) is 26.8 dynes/cm. If the gauge pressure is the same in two drops of these liquids, what is the ratio of the *volume* of the water drop to that of the CCl_4 drop?

13.5 Fluid Flow

42. • At a point where an irrigation canal having a rectangular cross section is 18.5 m wide and 3.75 m deep, the water flows at 2.50 cm/s. At a point downstream, but on the same level, the canal is 16.5 m wide, but the water flows at 11.0 cm/s. How deep is the canal at this point?

43. • Water is flowing in a pipe with a varying cross-sectional area, and at all points the water completely fills the pipe. At point 1, the cross-sectional area of the pipe is 0.070 m^2 and the magnitude of the fluid velocity is 3.50 m/s. What is the fluid speed at points in the pipe where the cross-sectional area is (a) 0.105 m^2, (b) 0.047 m^2?

44. • Water is flowing in a cylindrical pipe of varying circular cross-sectional area, and at all points the water completely fills the pipe. (a) At one point in the pipe, the radius is 0.150 m. What is the speed of the water at this point if the volume flow rate in the pipe is 1.20 m^3/s? (b) At a second point in the pipe, the water speed is 3.80 m/s. What is the radius of the pipe at this point?

45. •• A shower head has 20 circular openings, each with radius 1.0 mm. The shower head is connected to a pipe with radius 0.80 cm. If the speed of water in the pipe is 3.0 m/s, what is its speed as it exits the shower-head openings?

46. •• You're holding a hose at waist height and spraying water horizontally with it. The hose nozzle has a diameter of 1.80 cm, and the water splashes on the ground a distance of 0.950 m horizontally from the nozzle. Suppose you now constrict the nozzle to a diameter of 0.750 cm; how far horizontally from the nozzle will the water travel before hitting the ground? (Ignore air resistance.)

13.6 Bernoulli's Equation
13.7 Applications of Bernoulli's Equation

47. • A small circular hole 6.00 mm in diameter is cut in the side of a large water tank, 14.0 m below the water level in the tank. The top of the tank is open to the air. Find the speed at which the water shoots out of the tank.

48. • A sealed tank containing seawater to a height of 11.0 m also contains air above the water at a gauge pressure of 3.00 atm. Water flows out from the bottom through a small hole. Calculate the speed with which the water comes out of the tank.

49. •• What gauge pressure is required in the city water mains for a stream from a fire hose connected to the mains to reach a vertical height of 15.0 m? (Assume that the mains have a much larger diameter than the fire hose.)

50. • At one point in a pipeline, the water's speed is 3.00 m/s and the gauge pressure is 4.00×10^4 Pa. Find the gauge pressure at a second point in the line 11.0 m lower than the first if the pipe diameter at the second point is twice that at the first.

51. •• **Lift on an airplane.** Air streams horizontally past a small airplane's wings such that the speed is 70.0 m/s over the top surface and 60.0 m/s past the bottom surface. If the plane has a mass of 1340 kg and a wing area of 16.2 m^2, what is the net vertical force (including the effects of gravity) on the airplane? The density of the air is 1.20 kg/m^3.

52. •• A golf course sprinkler system discharges water from a horizontal pipe at the rate of 7200 cm^3/s. At one point in the pipe, where the radius is 4.00 cm, the water's absolute pressure is 2.40×10^5 Pa. At a second point in the pipe, the water passes through a constriction where the radius is 2.00 cm. What is the water's absolute pressure as it flows through this constriction?

53. •• Water discharges from a horizontal cylindrical pipe at the rate of 465 cm^3/s. At a point in the pipe where the radius is 2.05 cm, the absolute pressure is 1.60×10^5 Pa. What is the pipe's radius at a constriction if the pressure there is reduced to 1.20×10^5 Pa?

54. •• **Artery blockage.** A medical technician is trying to determine what percentage of a patient's artery is blocked by plaque. To do this, she measures the blood pressure just before the region of blockage and finds that it is 1.20×10^4 Pa, while in the region of blockage it is 1.15×10^4 Pa. Furthermore, she knows that blood flowing through the normal artery just before the point of blockage is traveling at 30.0 cm/s, and the specific gravity of this patient's blood is 1.06. What percentage of the cross-sectional area of the patient's artery is blocked by the plaque?

BIO

55. •• At a certain point in a horizontal pipeline, the water's speed is 2.50 m/s and the gauge pressure is 1.80×10^4 Pa. Find the gauge pressure at a second point in the line if the cross-sectional area at the second point is twice that at the first.

13.8 Real Fluids: Viscosity and Turbulence

56. • With what terminal speed would a steel ball bearing 2.00 mm in diameter fall in a liquid of viscosity 0.150 N · s/m^2 if we could neglect buoyancy?

57. •• What speed must a gold sphere of radius 3.00 mm have in castor oil for the viscous drag force to be one-fourth of the weight of the sphere? The density of gold is 19,300 kg/m^3 and the viscosity of the oil is 0.986 N · s/m^2.

58. •• A copper sphere with a mass of 0.20 g and a density of 8900 kg/m³ is observed to fall with a terminal speed of 6.0 cm/s in an unknown liquid. Find the viscosity of the unknown liquid if its buoyancy can be neglected.

59. •• **Clogged artery.** Viscous blood is flowing through an **BIO** artery partially clogged by cholesterol. A surgeon wants to remove enough of the cholesterol to double the flow rate of blood through this artery. If the original diameter of the artery is D, what should be the new diameter (in terms of D) to accomplish this for the same pressure gradient?

General Problems

60. •• Advertisements for a certain small car claim that it floats in water. (a) If the car's mass is 900 kg and its interior volume is 3.0 m³, what fraction of the car is immersed when it floats? You can ignore the volume of steel and other materials. (b) Water gradually leaks in and displaces the air in the car. What fraction of the interior volume is filled with water when the car sinks?

61. •• A U-shaped tube open to the air at both ends contains some mercury. A quantity of water is carefully poured into the left arm of the U-shaped tube until the vertical height of the water column is 15.0 cm (Figure 13.43). (a) What is the gauge pressure at the water–mercury interface? (b) Calculate the vertical distance h from the top of the mercury in the right-hand arm of the tube to the top of the water in the left-hand arm.

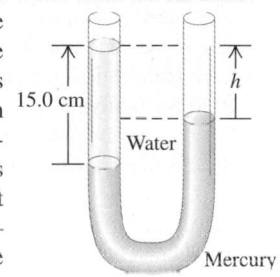

▲ **FIGURE 13.43** Problem 61.

62. •• An open barge has the dimensions shown in Figure 13.44. If the barge is made out of 4.0-cm-thick steel plate on each of its four sides and its bottom, what mass of coal can the barge carry in fresh water without sinking? Is there enough room in the barge to hold this amount of coal? (The density of coal is about 1500 kg/m³.)

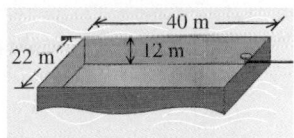

▲ **FIGURE 13.44** Problem 62.

63. •• A piece of wood is 0.600 m long, 0.250 m wide, and 0.080 m thick. Its density is 600 kg/m³. What volume of lead must be fastened underneath it to sink the wood in calm water so that its top is just even with the water level? What is the mass of this volume of lead?

64. •• A hot-air balloon has a volume of 2200 m³. The balloon fabric (the envelope) weighs 900 N. The basket with gear and full propane tanks weighs 1700 N. If the balloon can barely lift an additional 3200 N of passengers, breakfast, and champagne when the outside air density is 1.23 kg/m³, what is the average density of the heated gases in the envelope?

65. •• In seawater, a life preserver with a volume of 0.0400 m³ will support a 75.0 kg person (average density 980 kg/m³), with 20% of the person's volume above water when the life preserver is fully submerged. What is the density of the material composing the life preserver?

66. •• Block A in Figure 13.45 hangs by a cord from spring balance D and is submerged in a liquid C contained in beaker B. The mass of the beaker is 1.00 kg; the mass of the liquid is 1.80 kg. Balance D reads 3.50 kg and balance E reads 7.50 kg. The volume of block A is 3.80×10^{-3} m³. (a) What is the density of the liquid? (b) What will each balance read if block A is pulled up out of the liquid?

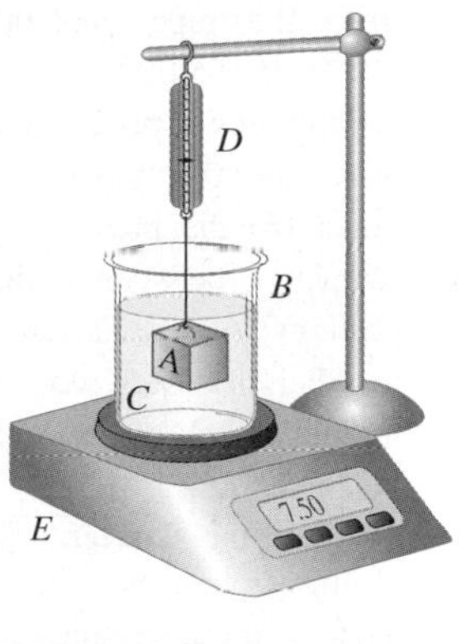

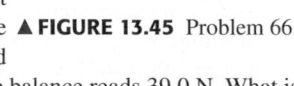

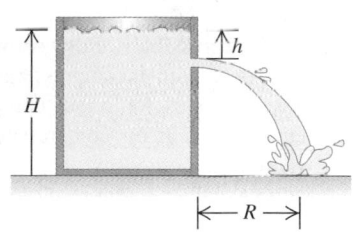

▲ **FIGURE 13.45** Problem 66.

67. •• A hunk of aluminum is completely covered with a gold shell to form an ingot of weight 45.0 N. When you suspend the ingot from a spring balance and submerge the ingot in water, the balance reads 39.0 N. What is the weight of the gold in the shell?

68. •• A liquid is used to make a mercury-type barometer, as described in Section 13.2. The barometer is intended for spacefaring astronauts. At the surface of the earth, the column of liquid rises to a height of 2185 mm, but on the surface of Planet X, where the acceleration due to gravity is one-fourth of its value on earth, the column rises to only 725 mm. Find (a) the density of the liquid and (b) the atmospheric pressure at the surface of Planet X.

69. •• An open cylindrical tank of acid rests at the edge of a table 1.4 m above the floor of the chemistry lab. If this tank springs a small hole in the side at its base, how far from the foot of the table will the acid hit the floor if the acid in the tank is 75 cm deep?

70. •• Water stands at a depth H in a large, open tank whose side walls are vertical (Fig. 13.46). A hole is made in one of the walls at a depth h below the water surface. (a) At what distance R from the foot of the wall does the emerging stream strike the floor? (b) How far above the bottom of the tank could a second hole be cut so that the stream emerging from it could have the same range as for the first hole?

▲ **FIGURE 13.46** Problem 70.

71. •• **Exploring Europa's oceans.** Europa, a satellite of Jupiter, appears to have an ocean beneath its icy surface. Proposals have been made to send a robotic submarine to Europa to see if there might be life there. There is no atmosphere on Europa, and we shall assume that the surface ice is thin enough that we can neglect its weight and that the oceans are fresh water having the same density as on the earth. The mass and diameter of Europa have been measured to be 4.78×10^{22} kg and 3130 km, respectively. (a) If the submarine intends to submerge to a depth of 100 m, what pressure must it be designed to withstand? (b) If you wanted to test this submarine before sending it to Europa, how deep would it have to go in our oceans to experience the same pressure as the pressure at a depth of 100 m on Europa?

72. •• The horizontal pipe shown in Figure 13.47 has a cross-sectional area of 40.0 cm² at the wider portions and 10.0 cm² at the constriction. Water is flowing in the pipe, and the discharge

from the pipe is 6.00 × 10^{-3} m³/s (6.00 L/s). Find (a) the flow speeds at the wide and the narrow portions; (b) the pressure difference between these portions; (c) the difference in height between the mercury columns in the U-shaped tube.

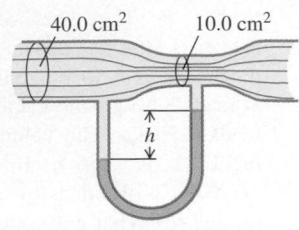

▲ **FIGURE 13.47** Problem 72.

73. •• **Venturi meter.** The Venturi meter is a device used to measure the speed of a fluid traveling through a pipe. Two cylinders are inserted in small holes in the pipes, as shown in Figure 13.48. Since the

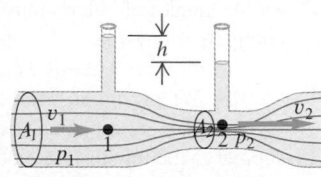

▲ **FIGURE 13.48** Problem 73.

cross-sectional area is different at the two places, the speed and pressure will be different there also. The difference (h) in the heights of the two columns can easily be measured, as can the cross-sectional areas A_1 and A_2. Notice that points 1 and 2 in the figure are both at the same vertical height. (a) Show that $\Delta p = \rho g h$, where ρ is the density of the fluid and Δp is the pressure difference between points 1 and 2. (b) Apply Bernoulli's equation and the continuity condition to show that the speed at point 1 is given by the equation

$$v_1 = \sqrt{\frac{2gh}{(A_1/A_2)^2 - 1}}.$$ (c) How would you find the speed at point 2?

74. •• **Compressible fluids.** Throughout this chapter, we have dealt only with incompressible fluids. But under very high pressure, fluids do, in fact, compress. (a) Show that the continuity condition for compressible fluids is $\rho_1 A_1 v_1 = \rho_2 A_2 v_2$, where ρ is the density of the fluid. (b) Show that your result reduces to the familiar result for incompressible fluids.

Passage Problems

BIO **Measuring blood pressure.** The standard procedure for measuring blood pressure determines both the systolic (maximum) and diastolic (minimum) pressures in the brachial artery during the

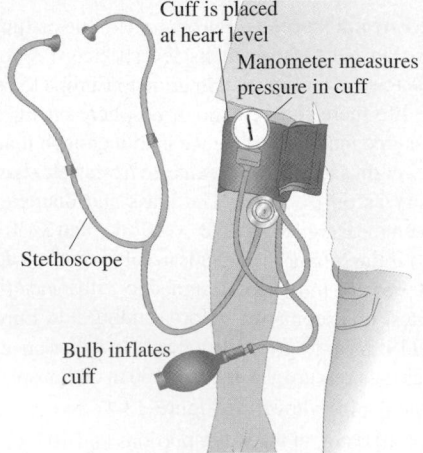

Cuff is placed at heart level

Manometer measures pressure in cuff

Stethoscope

Bulb inflates cuff

▲ **FIGURE 13.49** Problems 75–78.

cardiac cycle. (The brachial artery is the major artery that runs down the arm and provides blood to both the hand and arm.) See Figure 13.49.

First, the cuff is inflated to stop the flow of blood. As the pressure in the cuff is reduced to the patient's systolic blood pressure, some blood will begin to flow past the cuff, but in a turbulent fashion because the artery is still not fully open. This turbulence generates characteristic Korotkoff sounds that allow the physician to determine when to make the systolic pressure reading using the manometer. When the pressure is further reduced to the patient's diastolic blood pressure, the cuff no longer restricts arterial blood flow and the Korotkoff sounds disappear (as the blood flow becomes laminar again). A healthy patient typically has a systolic pressure that is between 90 and 120 mm Hg and a diastolic pressure of 60 to 80 mm. Note that all blood-pressure measurements are gauge pressures (i.e., they are the pressure difference between the arterial pressure and the ambient air pressure).

75. Roughly speaking, arteries carry blood from the heart and veins return blood to the heart. From your experience with small cuts or skinned knees, what can you say about the pressure in the venous system (veins)?
 A. It is much less than atmospheric pressure.
 B. It is slightly more than atmospheric pressure.
 C. It is approximately the same as atmospheric pressure.
 D. It is much greater than atmospheric pressure.
 E. It is slightly greater than atmospheric pressure.

76. Blood pressure is normally measured on the arm of a patient at a point that is nearly at the same height as the heart. If we were to measure the blood pressure on the lower leg of a standing patient, we would expect the measurement to be
 A. lower than
 B. higher than
 C. the same as the measurement on the arm.

77. On the earth, the pressure generated by the heart is sufficient to pump blood to a height of 1.3 m. (The density of blood is 1.04 g/cm³.) If the top of your head were 0.5 m above your heart, what would be the strongest gravitational acceleration that you could endure on another planet before your heart would be unable to pump blood to your brain?
 A. 2g
 B. 3g
 C. 5g
 D. 10g

78. The cross-sectional area of the aorta is 3 cm², and the average velocity of blood leaving the heart into the aorta is 30 cm/s. If the combined effective cross sectional area of the body's capillaries is 600 cm², what is the average flow rate in a capillary?
 A. 1 cm/s
 B. 2 cm/s
 C. 0.01 cm/s
 D. 0.15 cm/s
 E. 0.2 cm/s

14 Temperature and Heat

Which season do you like better, summer or winter? Whatever your choice, you use a lot of physics to keep comfortable in changing weather conditions. Your body's core needs to be kept at constant temperature. The body has effective temperature-control mechanisms, but sometimes it needs help. On a hot day, you wear less clothing, to improve heat transfer from your body to the air and to allow better cooling by evaporation of perspiration. You probably drink cold beverages, possibly with ice in them, and sit near a fan or in an air-conditioned room. On a cold day, you wear more clothes or stay indoors where it's warm. When you're outside, you keep active and drink hot liquids to stay warm. The concepts in this chapter will help you understand the basic physics of keeping warm or cool.

First we need to define temperature, including temperature scales and ways to measure temperature. Next we discuss changes in dimensions and in the volumes of materials caused by temperature changes. We'll encounter the concept of *heat*, which describes energy transfer caused by temperature differences, and we'll learn to calculate the *rates* of such energy transfers. This chapter lays the groundwork for our later study of *thermodynamics*, the study of energy transformations and their relationships to the properties of matter. Thermodynamics forms an indispensable part of the foundation of physics, chemistry, and the life sciences, and its applications turn up in such diverse places as car engines, refrigerators, biochemical processes, and the structure of stars.

Large animals have a relatively small ratio of heat-dissipating surface area to heat-producing internal volume, so they often have special systems to help them get rid of unwanted heat. People sweat, dogs pant, and antelope jackrabbits send blood into their large, thin-skinned, highly vascularized ears.

14.1 Temperature and Thermal Equilibrium

The concept of **temperature** is rooted in qualitative ideas of "hot" and "cold" based on our sense of touch. An object that feels hot usually has a higher temperature than when it feels cold. That's pretty vague, of course, but many properties of matter that we can measure quantitatively depend on temperature. The length of a metal rod increases as the rod becomes hotter. The pressure of gas in a container increases with temperature; a steam boiler may explode if it gets too hot. When a material is extremely hot, it glows "red hot" or "white hot," the color depending on its temperature.

Temperature is also related to the kinetic energies of the molecules of a material. In the next chapter, we'll look at that relationship in detail. It's important to understand, however, that temperature and heat can and must be defined independently of any detailed molecular picture.

Thermometers

Temperature is a measure of "hotness" or "coldness" on some scale. But *what* scale? We can use any measurable property of a system that varies with its hotness or coldness to assign numbers to various states of hotness and coldness. Figure 14.1a shows a familiar type of **thermometer** (a device used to measure temperature). When the system becomes hotter, the liquid (usually mercury or ethanol) expands and rises in the tube, and the value of L increases. Another simple system is a quantity of gas in a constant-volume container (Figure 14.1b). The pressure p, measured by the gauge, increases or decreases as the gas becomes hotter or colder. A third example is the electrical resistance R of a conducting wire, which also varies when the wire becomes hotter or colder. Each of these properties gives us a number (L, p, or R) that varies with hotness and coldness, so each property can be used to make a thermometer.

Thermal Equilibrium

To measure the temperature of an object, you place the thermometer in contact with the object. If you want to know the temperature of a cup of hot coffee, you stick the thermometer in the coffee. As the two interact, the thermometer becomes hotter and the coffee cools off a little. After the thermometer settles down to a steady value, you read the temperature. The system has reached an *equilibrium* condition, in which the interaction between the thermometer and the coffee causes no further change in the temperature of any part of the system. We call this a state of **thermal equilibrium.**

In a state of thermal equilibrium, the thermometer and the coffee have the same temperature. How do we know this? Consider the interaction of any two systems A and B. System A might be a hot potato and B a pan of cold water. We may use any kind of thermometer to monitor their temperatures; for example, we could use the tube-and-liquid system (Figure 14.1a) for system A and the container of gas and pressure gauge (Figure 14.1b) for system B. In general, when the two systems come into contact, the values of L and p change as the systems move toward thermal equilibrium. Initially, one system is hotter than the other; during the interaction, each system changes the temperature of the other.

If the systems are separated by an insulating material, or **insulator,** such as wood, plastic foam, or fiberglass, they influence each other more slowly. An *ideal* insulator is a material that permits *no* interactions at all between the two systems. It *prevents* the systems from attaining thermal equilibrium if they aren't in thermal equilibrium at the start. This is why camping coolers are made with insulating materials. We don't want the ice and cold food inside to warm up and attain thermal equilibrium with the hot summer air outside; the insulation delays this process, although not forever.

Zeroth Law of Thermodynamics

Now consider three systems A, B, and C that are initially *not* in thermal equilibrium (Figure 14.2). We surround them with an ideal insulating box so that they cannot interact with anything except each other. We separate A and B with an ideal insulating wall (the green slab in Figure 14.2a), but we let system C interact

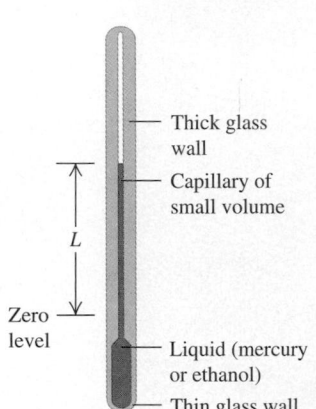

Thick glass wall

Capillary of small volume

L

Zero level

Liquid (mercury or ethanol)

Thin glass wall

(a) Changes in temperature cause the liquid's volume to change.

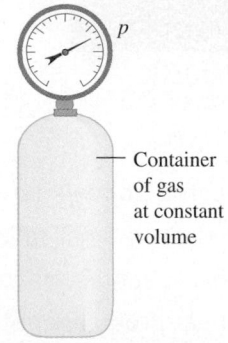

p

Container of gas at constant volume

(b) Changes in temperature cause the pressure of the gas to change.

▲ **FIGURE 14.1** Two types of thermometer.

with both *A* and *B*. This interaction is shown in the figure by a gold slab representing a thermal **conductor**—a material that *permits* thermal interactions through it. We wait until thermal equilibrium is attained; then *A* and *B* are each in thermal equilibrium with *C*. But are they in thermal equilibrium *with each other?*

To find out, we separate *C* from *A* and *B* with an ideal insulating wall (see Figure 14.2b), and then we replace the insulating wall between *A* and *B* with a *conducting* wall that lets *A* and *B* interact. What happens? Experiment shows that *nothing* happens: The states of *A* and *B* do not change. We thus conclude that if *C* is initially in thermal equilibrium with both *A* and *B*, then *A* and *B* are also in thermal equilibrium with each other. This principle is called the **zeroth law of thermodynamics.** It may seem trivial and obvious, but even so, it needs to be verified by experiment.

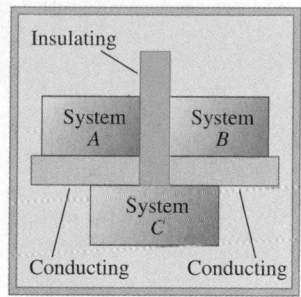

(a) If systems *A* and *B* are each in thermal equilibrium with system *C* …

Zeroth law of thermodynamics

Two systems that are each in thermal equilibrium with a third system are in thermal equilibrium with each other.

The importance of this law was recognized only after the first, second, and third laws of thermodynamics had been named. Since it is fundamental to all of them, the label "zeroth" is appropriate.

The temperature of a system determines whether the system is in thermal equilibrium with another system. **Two systems are in thermal equilibrium if and only if they have the same temperature.** When the temperatures of two systems are different, the systems *cannot* be in thermal equilibrium. A thermometer actually measures *its own* temperature, but when a thermometer is in thermal equilibrium with another object, the temperatures of the two must be equal. A temperature scale, such as the Celsius or Fahrenheit scale, is just a particular scheme for assigning numbers to temperatures.

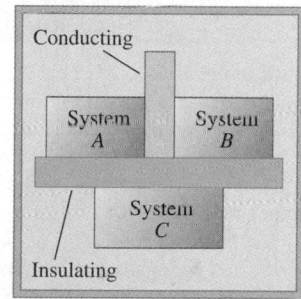

(b) … then systems *A* and *B* are in thermal equilibrium with each other.

▲ **FIGURE 14.2** The zeroth law of thermodynamics.

14.2 Temperature Scales

To make the thermometer in Figure 14.1a useful, we need to give it a scale with numbers on it. These numbers are arbitrary; historically, many different schemes have been used. Suppose we label the thermometer's liquid level when it is at the freezing temperature of pure water "zero" and the level at the boiling temperature "100"; then we divide the distance between these two points into 100 equal intervals called *degrees*. The result is the **Celsius temperature scale** (formerly called the *centigrade scale* in English-speaking countries). We interpolate between these reference temperatures 0°C and 100°C, or extrapolate beyond them, by using the liquid-in-tube thermometer or one of the other thermometers mentioned in Section 14.1. The temperature for a state colder than freezing water is a negative number. The Celsius scale, shown on the left side of Figure 14.3, is used, both in everyday life and in science and industry, everywhere in the world except the United States and a few other English-speaking countries.

The right side of Figure 14.3 shows the **Fahrenheit temperature scale,** still used in everyday life in the United States. The freezing temperature of water is assigned the value 32°F (32 degrees Fahrenheit), and the boiling temperature of water is 212°F, both at normal atmospheric pressure. Thus, there are 180 degrees between freezing and boiling, compared with 100 on the Celsius scale. One Fahrenheit degree represents only $\frac{100}{180}$, or $\frac{5}{9}$ as large a temperature change as one Celsius degree.

To convert temperatures from Celsius to Fahrenheit, we note that a Celsius temperature T_C is the number of Celsius degrees above freezing; to obtain the number of Fahrenheit degrees above freezing, we must multiply T_C by $\frac{9}{5}$. But freezing on the Fahrenheit scale is 32°F, so to obtain the actual Fahrenheit

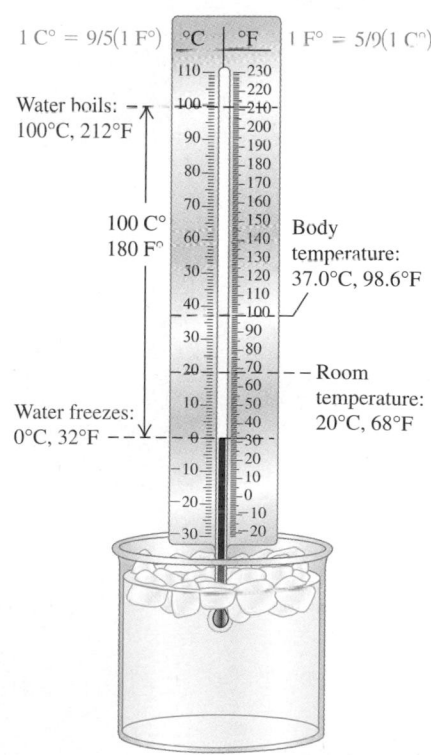

▲ **FIGURE 14.3** The Celsius and Fahrenheit temperature scales compared.

▲ Applications **Some like it hot.** We usually think of the boiling point of water as being 100°C, but remember that this is true only at normal atmospheric pressure. At higher pressures, water remains liquid at temperatures above 100°C. The photo shows hydrothermal vents on the ocean floor. The water emerging from such vents may be as hot as 400°C, but it remains liquid because of the great pressure due to the overlying water. Amazingly, the mineral "chimneys" built by these "smokers" host a rich population of microorganisms that can survive at temperatures higher than 110°C. Indeed, some *grow best* at temperatures above 100°C and find temperatures below 80°C (176°F) too cold for growth.

temperature T_F, we must first multiply the Celsius value by $\frac{9}{5}$ and then add 32°. Symbolically,

$$T_F = \frac{9}{5}T_C + 32°. \tag{14.1}$$

To convert Fahrenheit to Celsius, we first subtract 32° to obtain the number of Fahrenheit degrees above freezing and then multiply by $\frac{5}{9}$ to obtain the number of Celsius degrees above freezing—that is, the Celsius temperature. In symbols,

$$T_C = \frac{5}{9}(T_F - 32°). \tag{14.2}$$

We don't recommend memorizing Equations 14.1 and 14.2. Instead, try to understand the reasoning that led to them well enough that you can derive them on the spot when you need them. Check your reasoning with the relation 100°C = 212°F.

It's useful to distinguish between an actual temperature on a certain scale and a temperature *interval,* representing a *difference,* or change, in temperature. To make this distinction, we state an actual temperature of 20° as 20°C (20 degrees Celsius) and a temperature *interval* of 10° as 10 C° (10 Celsius degrees). Thus, a beaker of water heated from 20°C to 30°C undergoes a temperature change of 10 C°. We'll use this notation consistently throughout this book.

When we calibrate two thermometers, such as a liquid-in-tube system and a gas thermometer, so that they agree at 0°C and 100°C, they may not agree exactly at intermediate temperatures. The expansion properties of a gas may not match exactly the properties of a liquid. Ideally, we'd like to define a temperature scale that doesn't depend on the properties of a particular material. The gas thermometer, which we'll discuss next, comes close. In Chapter 16 we'll learn how to establish a scale that is truly independent of the material used.

The Kelvin Scale

In Section 14.1, we mentioned a thermometer that makes use of the variation in pressure of a gas with changing temperature. The gas, often helium, is placed in a constant-volume container (Figure 14.4a), and its pressure is measured by one of the devices described in Section 13.2. To calibrate a constant-volume gas ther-

(a) A constant-volume gas thermometer

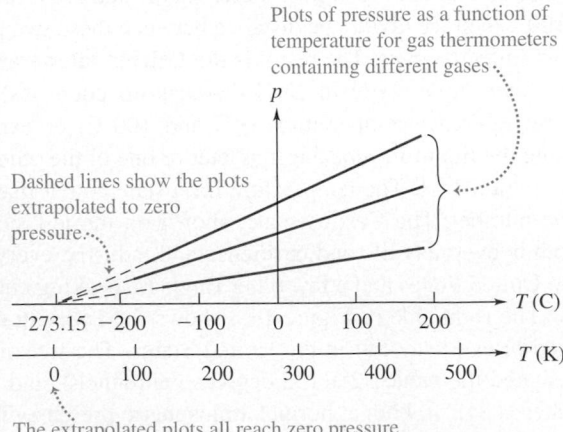

Plots of pressure as a function of temperature for gas thermometers containing different gases

Dashed lines show the plots extrapolated to zero pressure.

The extrapolated plots all reach zero pressure at the same temperature: −273.15C.

(b) Graphs of pressure versus temperature at constant volume for three gases

▲ **FIGURE 14.4** Use of a constant-volume gas thermometer to find the temperature at which a gas would have zero pressure (if it were still gaseous). This is the zero temperature on the Kelvin scale.

mometer, we measure the pressure at two temperatures, say, 0°C and 100°C, plot these points on a graph, and draw a straight line between them (Figure 14.4b). Then, from the graph, we can read the temperature corresponding to any other pressure.

By extrapolating this graph to lower and lower pressures, we see that there is a hypothetical temperature at which the gas pressure would become zero. We might expect that this temperature would be different for different gases, but it actually turns out to be the same for many different gases (at least in the limit of very low gas concentrations), namely, −273.15°C. We'll call this temperature *absolute zero*. We can't actually observe this zero-pressure condition, because gases liquefy and even solidify at very low temperatures; then the proportional relationship between pressure and temperature no longer holds. However, we can use this extrapolated zero-pressure temperature as the basis for a new temperature scale with its zero at that temperature.

This is the **Kelvin temperature scale,** named for Lord Kelvin (1824–1907). The units are the same size as those of the Celsius scale, but the zero is shifted so that the zero point is at absolute zero: 0 K = −273.15°C and 273.15 K = 0°C; thus,

$$T_K = T_C + 273.15° \tag{14.3}$$

A common room temperature, 20°C (= 68°F), is 20 + 273.15, or about 293 K. In SI nomenclature, "degree" is not used with the Kelvin scale; the room temperature 20°C is read "293 kelvins," not "293 degrees Kelvin." We capitalize Kelvin when it refers to the temperature scale, but the *unit* of temperature is the *kelvin*, not capitalized, but abbreviated K.

Figure 14.5 summarizes the relations among the Celsius, Fahrenheit, and Kelvin scales. The Kelvin scale is called an **absolute temperature scale,** and its zero point is called **absolute zero.** To define more completely what we mean by absolute zero, we need to use thermodynamic principles that we'll develop in Chapter 16.

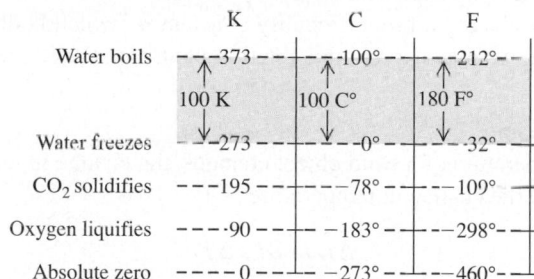

▲ **FIGURE 14.5** Relationships among Kelvin, Celsius, and Fahrenheit temperature scales. Temperatures have been rounded off to the nearest degree.

EXAMPLE 14.1 **Body temperature**

Normal internal body temperature for an average healthy human is 98.6°F. **(a)** Find this temperature in degrees Celsius and in kelvins. **(b)** When a human is relaxing on a couch, the skin temperature is about 2 C° lower than the internal temperature. How much lower is the skin temperature, in Fahrenheit degrees and in kelvins?

SOLUTION

SET UP AND SOLVE Part (a): To convert from Fahrenheit to Celsius temperatures, we first subtract 32° to obtain the number of Fahrenheit degrees above freezing and then multiply by $\frac{5}{9}$ to find the

number of Celsius degrees above freezing (i.e., the Celsius temperature). Thus, in degrees Celsius, normal body temperature is

$$T_C = \tfrac{5}{9}(T_F - 32°) = \tfrac{5}{9}(98.6° - 32°) = 37.0°C.$$

Continued

To find this temperature in kelvins, we add 273.15 K:

$$T_K = T_C + 273.15° = 37.0° + 273.15°$$
$$= 310.1 \text{ K} \ (= 310.1 \text{ kelvins}).$$

Part (b): To convert the number of Celsius degrees to the number of Fahrenheit degrees in a temperature interval, we multiply by 1 in the form (9 F°/5 C°). But one Celsius degree equals one kelvin, so to find the same interval on the Kelvin scale, we just multiply by 1:

$$2 \text{ Celsius degrees} = 2 \text{ C°} = (9 \text{ F°}/5 \text{ C°})2 \text{ C°} = 3.6 \text{ F°}$$

and

$$2 \text{ C°} = 2 \text{ K}.$$

REFLECT Note again that the degree symbol and the word *degree* are not used for the Kelvin scale. Also, for any calculations of temperature *changes*, the numbers of degrees on the Celsius and Kelvin scales are equal, but the number of degrees on the Fahrenheit scale is $\frac{9}{5}$ as great as either of these.

Practice Problem: On a stormy day in Orlando, Florida, the temperature dropped by 31.5 F°. What is this temperature change in Celsius degrees and in kelvins? *Answer:* $\Delta T = -17.5$ C° $= -17.5$ K.

14.3 Thermal Expansion

Most materials expand when their temperatures increase. Bridge decks need special joints and supports to allow for expansion. A completely filled and tightly capped bottle of water cracks when it is heated, but you can loosen a metal jar lid by running hot water on it. These are all examples of *thermal expansion*.

Linear Expansion

Suppose a rod of material has a length L_0 at some initial temperature T_0. When the temperature changes by a small amount ΔT, the length changes by ΔL (Figure 14.6). Experiment shows that if ΔT is not too large (say, less than 100 C° or so), ΔL is *directly proportional* to ΔT. If two rods made of the same material have the same temperature change, but one is twice as long as the other, then the *change* in its length is also twice as great. Therefore, ΔL must also be proportional to L_0. Introducing a proportionality constant α (which is different for different materials), we may express this relation in an equation:

Thermal expansion
When the temperature of a solid object changes, the change in length, ΔL, is proportional to the change in temperature, ΔT:

$$\Delta L = \alpha L_0 \, \Delta T. \tag{14.4}$$

The proportionality constant α is called the **coefficient of linear expansion.** If an object has length L_0 at temperature T_0, then its length L at a temperature $T = T_0 + \Delta T$ is

$$L = L_0 + \Delta L = L_0(1 + \alpha \, \Delta T). \tag{14.5}$$

Units: The units of α are K^{-1} or $(C°)^{-1}$. (Remember that a temperature *interval* is the same in Celsius and Kelvin.) In the discussion that follows, we'll usually use K^{-1}.

We can think of α as the fractional change in length of an object during a temperature change of 1 C°. For many materials, every linear dimension changes according to Equation 14.4 or 14.5. Thus, L could be the thickness of a rod, the length of the side of a square sheet, or (as we'll see in a minute) the diameter of a hole. Some materials, such as wood or single crystals, expand differently in different directions. We won't consider this complication.

We can understand thermal expansion qualitatively on a molecular basis. Picture the interatomic forces in a solid as springs (Figure 14.7). Each atom vibrates

For moderate temperature changes, ΔL is directly proportional to ΔT:

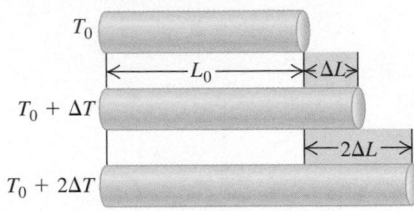

ΔL is also directly proportional to L_0:

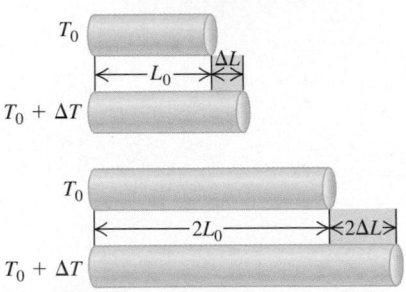

▲ **FIGURE 14.6** The dependence of linear thermal expansion on the change in temperature and the initial length.

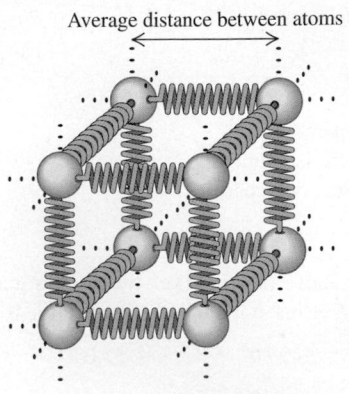

▲ **FIGURE 14.7** We can envision the forces between neighboring atoms in a solid by imagining the atoms to be connected by springs.

▶ **Application Bimaterial strips.** While coefficient of linear expansion may seem like a textbook concept, we depend on it daily. Many instruments, including mechanical thermostats and some thermometers and circuit breakers, employ sensors made of a *bimaterial strip*—that is, a strip made by welding together two thin strips of different materials so that the two materials form the two faces of the strip. The photo shows a sensor that uses such strips to measure temperature and humidity. In the temperature-sensing coil, the two faces of the strip consist of metals with different coefficients of linear expansion; changes in temperature cause the strip to either coil up or uncoil. In the humidity sensor, one face consists of paper or plastic that expands with increasing humidity, causing the strip to uncoil. In a mechanical thermostat, the coiling and uncoiling of a temperature-sensitive bimetallic strip makes and breaks electrical connections.

about its equilibrium position. When the temperature increases, the average distance between molecules also increases. As the atoms get farther apart, every dimension increases.

When you heat an object that has a hole in it, the hole gets bigger. Figure 14.8 shows what happens. If you heat a flat plate of material, its area increases. If you now cut a section from the center of the plate, both the cutout and the remaining plate (with a hole in it) behave as they did previously: The cut-out piece and the hole it came from both get bigger. When you heat an object with a hole in it, the *hole gets bigger.*

The direct proportionality between ΔL and ΔT expressed by Equation 14.4 isn't exact; it's *approximately* correct for sufficiently small temperature changes. For a given material, α varies somewhat with the initial temperature T_0 and the size of the temperature interval. We'll ignore this complication. Average values of α for several materials are listed in Table 14.1. Within the precision of these values, we don't need to worry about whether the initial temperature T_0 is 0°C, 20°C, or some other temperature. We also note that α is usually a very small number; even for a temperature change of 100 C°, the fractional change in length $\Delta L/L_0$ is on the order of $\frac{1}{1000}$. But don't be misled by this small number: The expansion of a 100 m bridge by 0.1 m has to be taken seriously!

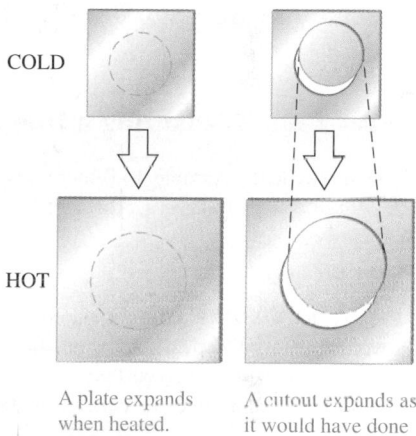

| A plate expands when heated. (Expansion exaggerated!) | A cutout expands as it would have done if still part of the plate—so the *hole must expand, too.* |

▲ **FIGURE 14.8** One way to understand why a hole in a solid object expands (rather than shrinks) when the object is heated.

PROBLEM-SOLVING STRATEGY 14.1 **Thermal expansion**

This strategy also applies to problems involving an expansion in *volume*, which we'll consider later in this section.

SET UP

1. Identify which of the quantities in Equation 14.4 or 14.5 are known and which are unknown. Often, you will be given two temperatures and will have to compute ΔT. Or you may be given an initial temperature T_0 and may have to find a final temperature corresponding to a given change in length or volume. In this case, find ΔT first; then the final temperature is $T_0 + \Delta T$.

SOLVE

2. Consistency of units is crucial, as always. L_0 and ΔL (or V_0 and ΔV) must have the same units, and if you use a value of α in K^{-1}, then ΔT must be in kelvins.

REFLECT

3. Remember that dimensions of holes in a material expand with temperature just the same way as any other linear dimensions do.

TABLE 14.1 Coefficients of linear expansion for selected materials

Material	α (K^{-1} or $C°^{-1}$)
Quartz (fused)	0.04×10^{-5}
Invar (nickel–iron alloy)	0.09×10^{-5}
Glass	$0.4-0.9 \times 10^{-5}$
Steel	1.2×10^{-5}
Copper	1.7×10^{-5}
Brass	2.0×10^{-5}
Aluminum	2.4×10^{-5}

EXAMPLE 14.2 **Change in length of a measuring tape**

A surveyor uses a steel measuring tape that is exactly 50.000 m long at a temperature of 20°C. What is its length on a hot summer day when the temperature is 35°C?

SOLUTION

SET UP AND SOLVE We have $L_0 = 50.000$ m, $T_0 = 20°C$, and $T = 35°C$. We use Equation 14.4 to find ΔL (Note that because ΔL is usually much smaller than L_0, we need to keep more significant figures in the lengths than in the temperatures.) From Table 14.1, the coefficient of linear expansion for steel is 1.2×10^{-5} K^{-1}. From Equation 14.4,

$$\Delta L = \alpha L_0 \Delta T$$
$$= (1.2 \times 10^{-5} \text{ K}^{-1})(50 \text{ m})(35°C - 20°C)$$
$$= 9.0 \times 10^{-3} \text{ m} = 9.0 \text{ mm}.$$

The new length is

$$L = L_0 + \Delta L = 50.000 \text{ m} + 0.0090 \text{ m} = 50.009 \text{ m}.$$

REFLECT The measuring tape lengthens by 9 mm, or 0.018%. In surveying a baseball field, this amount is negligible, but not when measuring the length of a steel bridge. Note that, although L_0 is given to five significant figures, we need only two of them to compute ΔL.

Practice Problem: A concrete slab forming part of a driveway is 20.000 m long when the temperature is 20°C. How much does the driveway increase in length when the temperature increases to 30°C? The coefficient of linear expansion for concrete is 1.2×10^{-5} K^{-1}. *Answer:* $\Delta L = 2.4$ mm.

EXAMPLE 14.3 **Measuring a true distance**

The surveyor in Example 14.2 measures a distance when the temperature is 35°C and obtains the result 35.794 m. What is the actual distance?

SOLUTION

SET UP AND SOLVE From Equation 14.5, the ratio L/L_0 for any specific material depends only on the temperature change. At 35°C, the distance between two successive meter marks on the tape is a little more than a meter, by the ratio

$$\frac{L}{L_0} = \frac{L_0(1 + \alpha T)}{L_0} = \frac{50.009 \text{ m}}{50.000 \text{ m}}.$$

Thus, the true distance is

$$\frac{50.009 \text{ m}}{50.000 \text{ m}}(35.794 \text{ m}) = 35.800 \text{ m}.$$

REFLECT The term $(1 + \alpha \Delta T)$ for a particular material depends only on the temperature change, within the limits of precision of Equation 14.5.

Volume Expansion

Increases in temperature usually cause increases in *volume* for both solid and liquid materials. Experiments show that if the temperature change ΔT is not too great (less than 100 C° or so), the increase in volume, ΔV, is approximately proportional to the temperature change. The change in volume is also proportional to the initial volume V_0, just as with linear expansion, so we obtain the following relationship:

TABLE 14.2 Coefficients of volume expansion

Material	β (K^{-1})
Solids	
Quartz (fused)	0.12×10^{-5}
Invar	0.27×10^{-5}
Glass	$1.2-2.7 \times 10^{-5}$
Steel	3.6×10^{-5}
Copper	5.1×10^{-5}
Brass	6.0×10^{-5}
Aluminum	7.2×10^{-5}
Liquids	
Mercury	18×10^{-5}
Glycerin	49×10^{-5}
Ethanol	75×10^{-5}
Carbon disulfide	115×10^{-5}

Volume expansion

When the temperature of an object changes, the change ΔV in its volume is proportional to the temperature change ΔT. That is,

$$\Delta V = \beta V_0 \Delta T \tag{14.6}$$

and

$$V = V_0(1 + \beta \Delta T). \tag{14.7}$$

The constant β characterizes the volume expansion properties of a particular material; it is called the **coefficient of volume expansion.** The units of β are

K^{-1}. As with linear expansion, β varies somewhat with temperature, and Equation 14.6 is an approximate relation that works well for small temperature changes. For many substances, β decreases at low temperatures. Table 14.2 lists values of β in the neighborhood of room temperature for several substances. Note that the values for liquids are much larger than those for solids.

For solid materials that expand equally in all directions, there is a simple relation between the volume expansion coefficient β and the linear expansion coefficient α. Expansion in each of the three directions contributes to changes in volume. Detailed analysis shows that the two coefficients are related simply by $\beta = 3\alpha$. We invite you to check this relationship for a few materials listed in Tables 14.1 and 14.2.

EXAMPLE 14.4 **Expansion of mercury**

A glass flask filled with a volume of 200 cm^3 is filled to the brim with mercury at 20°C. When the temperature of the system is raised to 100°C, does the mercury overflow? If so, by how much? The coefficient of volume expansion of the glass is 1.2×10^{-5} K^{-1}.

SOLUTION

SET UP Both the flask and the mercury expand. The mercury overflows because its coefficient of volume expansion (from Table 14.2) is much greater than that of the glass: 18×10^{-5} K^{-1} versus 1.2×10^{-5} K^{-1}.

SOLVE The increase in volume of the flask is $\Delta V_{flask} = \beta_{flask} V_0 \Delta T$, and the increase in the volume of mercury is $\Delta V_{mercury} = \beta_{mercury} V_0 \Delta T$:

$$\Delta V_{mercury} = (18 \times 10^{-5} \text{ K}^{-1})(200 \text{ cm}^3)(100°C - 20°C)$$
$$= 2.9 \text{ cm}^3;$$
$$\Delta V_{flask} = (1.2 \times 10^{-5} \text{ K}^{-1})(200 \text{ cm}^3)(100°C - 20°C)$$
$$= 0.19 \text{ cm}^3.$$

The volume of mercury that overflows is the difference between these values:

$$\Delta V_{mercury} - \Delta V_{flask} = 2.9 \text{ cm}^3 - 0.19 \text{ cm}^3 = 2.7 \text{ cm}^3.$$

REFLECT This is basically the way a mercury-in-glass thermometer works, except that instead of letting the mercury overflow and run all over the place, the thermometer has it rise inside a scaled tube as T increases.

Practice Problem: Suppose we replace the mercury with ethanol (less poisonous than mercury and more suitable for very cold weather). How much ethanol runs over? *Answer:* 11.8 cm^3

Thermal Expansion of Water

Above 4°C, liquid water expands with increasing temperature. However, between 0°C (the freezing point) and 4°C, water *decreases* in volume with increasing temperature; in this range, its coefficient of volume expansion is *negative*. Thus, water has its greatest density at 4°C. Water also expands when it freezes; that's why ice humps up in the middle of the compartments in an ice cube tray and why pipes full of water burst if allowed to freeze. Most other materials contract when they freeze.

This anomalous behavior of water has an important effect on plant and animal life in lakes. When a lake cools, the cooled water at the surface flows to the bottom because of its greater density. But when the temperature goes below 4°C, this flow ceases, and the water near the surface remains colder (and less dense) than that at the bottom. As the surface freezes, the ice floats because it is less dense than water. The water at the bottom remains at 4°C until nearly the entire lake is frozen. If water behaved like most substances, contracting continuously on cooling and freezing, lakes would freeze from the bottom up. Circulation due to density differences would continuously carry warmer (less dense) water to the surface for efficient cooling, and lakes would freeze solid much more easily. This would destroy all plant and animal life that can withstand cold water, but not freezing. All life forms known on earth depend on chemical systems based on aqueous solutions; if water did not have this special property, the evolution of various life-forms would have taken a very different course.

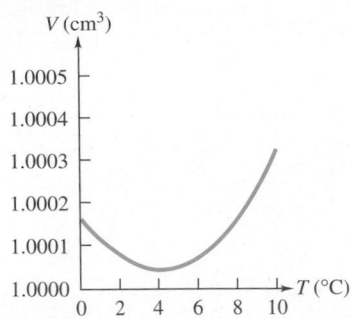

FIGURE 14.9 The volume of 1 g of water in the temperature range from 0°C to 10°C. The fact that water is denser at about 4°C than at the freezing point is vitally important for the ecology of lakes and streams.

FIGURE 14.10 The interlocking teeth of an expansion joint on a bridge. These joints are needed to accommodate changes in length that result from thermal expansion.

The anomalous expansion of water in the temperature range from 0°C to 10°C is shown in Figure 14.9.

Thermal Stress

If we clamp the ends of a rod rigidly to prevent expansion or contraction and then raise or lower the temperature, a compressive or tensile stress called **thermal stress** develops. The rod would like to expand or contract, but the clamps won't let it. The resulting stress may become large enough to strain the rod irreversibly or even break it. Concrete highways and bridge decks usually have gaps between sections, filled with a flexible material or bridged by interlocking teeth (Figure 14.10), to permit expansion and contraction of the concrete. If one end of a steel bridge is rigidly fastened to its abutment, the other end usually rests on rollers.

To calculate the thermal stress in a clamped rod, we compute the amount the rod *would* expand (or contract) if it were not held and then find the stress needed to compress (or stretch) it back to its original length. Suppose a rod with length L_0 and cross-sectional area A is held at constant length while the temperature is reduced (negative ΔT), causing a tensile stress. Then the fractional change in length if the rod were free to contract would be

$$\frac{\Delta L}{L_0} = \alpha \, \Delta T.$$

Both ΔL and ΔT are negative. The tension F must increase enough to produce an equal and opposite fractional change in length. From the definition of Young's modulus Y (Equation 11.4),

$$Y = \frac{F/A}{\Delta L/L_0} \quad \text{and} \quad \frac{\Delta L}{L_0} = \frac{F}{AY}.$$

If the *total* fractional change in length is zero, then

$$\alpha \, \Delta T + \frac{F}{AY} = 0, \quad \text{or} \quad F = -AY\alpha \, \Delta T. \tag{14.8}$$

For a decrease in temperature, ΔT is negative, so F is positive. The tensile *stress* F/A in the rod is

$$\frac{F}{A} = -Y\alpha \, \Delta T. \tag{14.9}$$

If ΔT is positive, F and F/A are negative, corresponding to *compressive* force and stress.

EXAMPLE 14.5 **Stress on a spacer**

An aluminum cylinder 10 cm long, with a cross-sectional area of 20 cm², is to be used as a spacer between two steel walls. At 17.2°C, it just slips in between the walls. Then it warms to 22.3°C. Calculate the stress in the cylinder and the total force it exerts on each wall, assuming that the walls are perfectly rigid and a constant distance apart.

SOLUTION

SET UP Figure 14.11 shows our sketch. We use Equation 14.9, which relates the stress F/A (compressive in this case) to the temperature change. From Table 11.1, $Y = 0.70 \times 10^{11}$ Pa, and from Table 14.1, $\alpha = 2.4 \times 10^{-5}$ K^{-1}. The temperature change is $\Delta T = 5.1$ K, and the cross-sectional area of the cylinder is 20 cm² = 20×10^{-4} m².

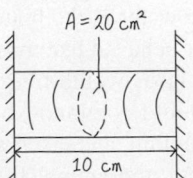

FIGURE 14.11 Our sketch for this problem.

Continued

SOLVE From Equation 14.9, the stress F/A is

$$\frac{F}{A} = -Y\alpha\,\Delta T$$
$$= -(0.70 \times 10^{11}\,\text{Pa})(2.4 \times 10^{-5}\,\text{K}^{-1})(5.1\,\text{K})$$
$$= -8.6 \times 10^{6}\,\text{Pa}\ (\text{about} -1240\,\text{lb/in.}^2).$$

The total force F exerted by the cylinder on a wall is

$$F = -AY\alpha\,\Delta T = (20 \times 10^{-4}\,\text{m}^2)(-8.6 \times 10^{6}\,\text{Pa})$$
$$= -1.7 \times 10^{4}\,\text{N}\ (\text{about 3800 lb}).$$

REFLECT The negative sign indicates compressive rather than tensile stress. The stress is independent of the length and cross-sectional area of the cylinder. The total force is nearly 2 tons.

Thermal stresses can be induced by non-uniform expansion due to temperature differences. Have you ever broken a glass jar by pouring boiling water into it? Thermal stress caused by temperature differences exceeded the breaking stress of the material, and it cracked. The same phenomenon makes ice cubes crack when they're dropped into a hot liquid. Heat-resistant glasses such as Pyrex™ have exceptionally low expansion coefficients and high strength.

14.4 Quantity of Heat

When you put a cold spoon into a cup of hot coffee, the spoon warms up and the coffee cools off as they approach thermal equilibrium. What kind of interaction is taking place to cause these temperature changes? During this interaction, *something* is transferred from the coffee to the spoon. We'll call this interaction a **heat transfer** or a *heat flow* from the coffee to the spoon. But what *is* this "something" that flows from the coffee to the spoon?

The answer to this question emerged gradually during the 18th and 19th centuries. Sir James Joule (1818–1889) discovered that water can be warmed by vigorous stirring with a paddle wheel (Figure 14.12a) and that *the rise in temperature is directly proportional to the work done* (and energy added) in turning the wheel. Since the same temperature change could have been caused by putting the water in contact with some hotter object (Figure 14.12b), that interaction must also involve an energy exchange. This experiment and many similar ones, have established the fact that heat flow is *energy transfer* and that there is an equivalence between heat and work: The same change of state of a system (such as a change in temperature or volume) can be produced by heat flow, work, or a combination of the two.

Energy transfer that takes place solely because of a temperature difference is called *heat flow,* or *heat transfer,* and energy transferred in this way is called **heat.** For example, cold water in a steam boiler is converted to steam by contact with a metal pipe kept at a high temperature by a hot flame from burning coal or gas. Having received energy from the flame, the water (as steam) has a greater ability to do *work* (by pushing against a turbine blade or the piston in a steam engine) than does an equal mass of cold liquid water.

NOTE ▶ The term *heat* always refers to the *transfer* of energy from one object or system to another due to a temperature difference, never to the amount of energy *contained within* a particular system. There is no such thing as "the heat *in* an object"; **the concept of heat has meaning only as energy in transit.** A detailed study of the relationship between heat and work leads to the *first law of thermodynamics,* which we'll study in detail in the next chapter. ◀

Because heat is energy transferred from one system to another, the SI unit of quantity of heat is the joule. The International Committee on Weights and Measures recommends use of the joule as the basic unit of energy in all forms, including heat. We'll follow that recommendation in this book. However, several older,

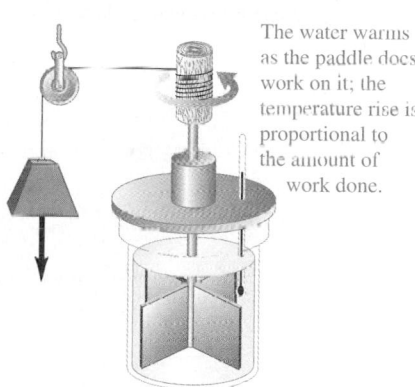

The water warms as the paddle does work on it; the temperature rise is proportional to the amount of work done.

(a) Raising the temperature of water by doing work on it

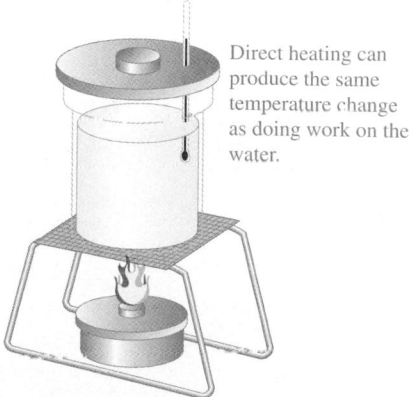

Direct heating can produce the same temperature change as doing work on the water.

(b) Raising the temperature of water by direct heating

▲ FIGURE 14.12 The temperature of a system may be changed by either (a) doing work on the system or (b) adding heat to it.

non-SI units are also in common use for heat. The most familiar of these is the calorie.

Definition of the calorie

The **calorie** (abbreviated cal) is defined as *the amount of heat required to raise the temperature of 1 gram of water from* 14.5°C *to* 15.5°C.

The familiar "Calorie" found on food labels is actually a kilocalorie (kcal), equal to 1000 cal. The kilocalorie, also called the food calorie, is sometimes written with a capital C (Calorie) to distinguish it from the calorie just defined. But as Figure 14.13 shows, much of the world worries about the joule content of their food, not the calorie (or Calorie) content.

A corresponding unit of quantity of heat that uses Fahrenheit degrees and British units is the **British thermal unit,** or Btu. One Btu is defined as the quantity of heat required to raise the temperature of 1 standard pound (weight) of water 1 F° from 63°F to 64°F.

Because heat is energy in transit, there must be a definite relation between the preceding units and the familiar energy unit, the joule. Experiments similar in concept to Joule's have shown that

$$1 \text{ cal} = 4.186 \text{ J},$$
$$1 \text{ kcal} = 1000 \text{ cal} = 4186 \text{ J},$$
$$1 \text{ Btu} = 778 \text{ ft} \cdot \text{lb} = 252 \text{ cal} = 1055 \text{ J}.$$

▲ **FIGURE 14.13** Packaged foods sold in many countries outside the United States, such as these sugar cubes from Germany, list their energy content in joules as well as calories.

Specific Heat Capacity

We use the symbol Q for quantity of heat (amount of energy transferred). The quantity of heat Q required to increase the temperature T of a mass m of a certain material by a small amount from T_1 to T_2 is found to be approximately proportional to the temperature change $\Delta T = T_2 - T_1$. To raise the temperature of a certain mass of substance from 20°C to 40°C ($\Delta T = 20$ K) requires approximately twice as much heat as the amount needed to raise it from 20°C to 30°C ($\Delta T = 10$ K). The heat required for a given temperature change is also proportional to the mass m of substance. Thus, when you heat water to make tea, you need twice as much heat for two cups as for one if the temperature interval is the same. Finally, the quantity of heat needed also depends on the nature of the material; to raise the temperature of 1 kilogram of water by 1 C° requires almost five times as much heat as for 1 kilogram of aluminum and the same temperature change.

Putting all these relationships together, we have the definition of specific heat capacity:

Specific heat capacity

The amount of heat Q needed for a certain temperature change ΔT is proportional to the temperature change and to the mass m of substance being heated; that is,

$$Q = mc \, \Delta T, \tag{14.10}$$

where c is a quantity, different for different materials, called the **specific heat capacity** (or sometimes *specific heat*) of the material. The SI unit of specific heat capacity is the joule per kilogram per kelvin $[\text{J}/(\text{kg} \cdot \text{K})]$. Also, 1 K = 1 C°.

In Equation 14.10, Q and ΔT can be either positive or negative. When they are positive, heat enters the object and its temperature increases; when they are negative, heat leaves the object and its temperature decreases.

NOTE ▶ Remember that Q does not represent a change in the amount of heat *contained* in an object; that is a meaningless concept. Heat is always energy *in transit* as a result of a temperature difference. There is no such thing as "the amount of heat in an object." The term *heat capacity* is unfortunate because it tends to suggest the erroneous idea that an object *contains* a certain amount of heat. ◀

The specific heat capacity of water is approximately

$$c_{water} = 4190 \text{ J}/(\text{kg} \cdot \text{K}), 1 \text{ cal}/(\text{g} \cdot \text{K}), \text{ or } 1 \text{ Btu}/(\text{lb} \cdot \text{F}°).$$

Specific heat capacities vary somewhat with the initial temperature and the temperature interval. For example, the specific heat capacity of water varies by almost 1% in the temperature range from 0°C to 100°C. In the problems and examples in this chapter, we'll usually ignore this small variation. Table 14.3 lists specific heat capacities for several common materials.

TABLE 14.3 Mean specific heat capacities (constant pressure, temperature range 0°C to 100°C)

Material	Specific heat capacity (c)	
	J/(kg·K)	cal/(g·K)
Solids		
Lead	0.13×10^3	0.031
Mercury	0.14×10^3	0.033
Silver	0.23×10^3	0.056
Copper	0.39×10^3	0.093
Iron	0.47×10^3	0.112
Marble ($CaCO_3$)	0.88×10^3	0.21
Salt	0.88×10^3	0.21
Aluminum	0.91×10^3	0.217
Beryllium	1.97×10^3	0.471
Ice (-25°C to 0°C)	2.01×10^3	0.48
Liquids		
Ethylene glycol	2.39×10^3	0.57
Ethanol	2.43×10^3	0.58
Water	4.19×10^3	1.00

Quantitative Analysis 14.1

The meaning of c

In the equation $Q = mc \, \Delta T$, m represents the mass in kilograms of the sample and c represents

A. the amount of heat required to raise the temperature of m kilograms of sample by 1 C°.
B. the amount of heat required to raise the temperature of 1 kilogram of the sample by 1 C°.
C. the amount of heat required to raise the temperature of m kilograms of the sample by ΔT.

SOLUTION The specific heat capacity c is a property of a particular material; it doesn't depend on *how much* of the material is present. It's also independent of the temperature interval: if we double ΔT, the total amount of heat doubles, but c doesn't change. Thus, neither A nor C can be right. The correct answer is B: c is the amount of heat needed *per unit mass* and *per unit temperature change*.

EXAMPLE 14.6 Running a fever

During a bout with the flu, an 80 kg man ran a fever of 2.0 C°; that is, his body temperature was 39.0°C (or 102.2°F) instead of the normal 37.0°C (or 98.6°F). Assuming that the human body is mostly water, how much heat was required to raise his temperature by that amount?

SOLUTION

SET UP AND SOLVE The temperature change is $\Delta T = 39.0°C - 37.0°C = 2.0 \, C° = 2.0 \, K$. From Table 14.3, the specific heat capacity of water is $4190 \text{ J}/(\text{kg} \cdot \text{K})$. From Equation 14.10,

$$Q = mc \, \Delta T = (80 \text{ kg})[4190 \text{ J}/(\text{kg} \cdot \text{K})](2.0 \text{ K}) =$$
$$= 6.7 \times 10^5 \text{ J}.$$

REFLECT This amount of heat is equal to 1.6×10^5 cal, or 160 kcal (160 food calories). This is roughly half the calorie intake from eating 1 ounce of milk chocolate.

Practice Problem: If a 35 kg child has a temperature of 104°F = 40°C, how much heat was needed to raise her temperature to that value from an initial value of 37°C? *Answer:* 4.4×10^5 J.

EXAMPLE 14.7 Circuit meltdown

You are designing an electronic circuit element made of 23 mg of silicon. The electric current through it adds energy at the rate of 7.4 mW $= 7.4 \times 10^{-3}$ J/s. If your design doesn't allow any heat transfer out of the element, at what rate does its temperature increase? The specific heat capacity of silicon is $705 \text{ J}/(\text{kg} \cdot \text{K})$.

Continued

SOLUTION

SET UP AND SOLVE In 1 second, $Q = (7.4 \times 10^{-3} \, \text{J/s})(1 \, \text{s}) = 7.4 \times 10^{-3} \, \text{J}$. From Equation 14.10, the temperature change in 1 second is

$$\Delta T = \frac{Q}{mc} = \frac{7.4 \times 10^{-3} \, \text{J}}{(23 \times 10^{-6} \, \text{kg})[705 \, \text{J}/(\text{kg} \cdot \text{K})]} = 0.46 \, \text{K}.$$

Alternatively, we can divide both sides of this equation by Δt and rearrange factors:

$$\frac{\Delta T}{\Delta t} = \frac{Q/\Delta t}{mc}$$

$$= \frac{7.4 \times 10^{-3} \, \text{J/s}}{(23 \times 10^{-6} \, \text{kg})[705 \, \text{J}/(\text{kg} \cdot \text{K})]} = 0.46 \, \text{K/s}.$$

REFLECT At this rate of temperature rise (27 K every minute), the circuit element would quickly self-destruct. Heat transfer is an important design consideration in electronic circuit elements.

NOTE ▶ We conclude this section with a short sermon. It is absolutely essential for you to keep clearly in mind the distinction between *temperature* and *heat*. Temperature depends on the physical state of a material and is a quantitative description of its hotness or coldness. Heat is energy transferred from one object to another because of a temperature difference. We can change the temperature of an object by adding heat to it or taking heat away from it or by adding or subtracting energy in other ways, such as doing mechanical work. If we cut an object in half, each half has the same temperature as the whole; but to raise the temperature of each half by a given interval, we add *half* as much heat as for the whole. ◀

Conceptual Analysis 14.2

Heat versus temperature

As Table 14.3 shows, the specific heat capacity of water is about 10 times that of iron. If you take equal masses of water and iron, initially in thermal equilibrium, and add 50 J of heat to each, which of the following statements is true?

A. They remain in thermal equilibrium.
B. They are no longer in thermal equilibrium; the iron is warmer.
C. They are no longer in thermal equilibrium; the water is warmer.

SOLUTION Because water has a larger specific heat capacity than iron, a greater amount of heat is needed to cause a given temperature change for a given mass of water than for an equal mass of iron. If the heat added is the same for both, it follows that the temperature change of the water is less than that of the iron. If they were initially at the same temperature, then, after the additions of heat, the iron is warmer than the water. The correct answer is B.

14.5 Phase Changes

We've discussed the energy transfer (heat) involved in changing an object's temperature. Heat is also involved in *phase changes,* such as the melting of ice or boiling of water. Once we understand these additional heat relationships, we can analyze a variety of problems involving quantity of heat Q.

We use the term **phase** to describe a specific state of matter, such as solid, liquid, or gas. The compound H_2O exists in the *solid phase* as ice, in the *liquid phase* as water, and in the *gaseous phase* as steam. A transition from one phase to another is called a **phase change** or phase transition. For any given pressure, a phase change usually takes place at a definite temperature and is usually accompanied by the absorption or emission of heat and a change in volume and density.

A familiar example of a phase change is the melting of ice. When we add heat to ice at 0°C and normal atmospheric pressure, the temperature of the ice *does not* increase. Instead, some of the ice melts to form liquid water. If we add the heat slowly, to maintain the system very close to thermal equilibrium, the temperature remains at 0°C until all the ice is melted. The effect of adding heat to this system is not to raise its temperature, but to change its *phase* from solid to liquid.

▶ **Application Flaking out?** Snow and ice are two familiar forms of the solid phase of water. Why do they look so different? Air always contains water vapor. When air above the freezing temperature of water cools to the point that it can't retain all its water vapor (i.e., the relative humidity exceeds 100%), some vapor condenses to liquid in the form of rain, fog, or dew. If the air cools further, the liquid water freezes to form hail or sleet. But if the temperature is below the freezing point of water, the water vapor goes directly from the gaseous to the solid phase. (This is the reverse of *sublimation*, where a solid material goes directly into the vapor phase.) Suspended in air, the ice crystals grow slowly, preserving their natural six-fold symmetry and forming snowflakes. A similar process causes frost to form on the cooling coils of a refrigerator.

To change 1 kg of ice at 0°C to 1 kg of liquid water at 0°C and normal atmospheric pressure requires 3.34×10^5 J of heat. The heat required, per unit mass is called the **heat of fusion** (or sometimes *latent* heat of fusion), denoted by L_f. Thus, the heat of fusion L_f of water, at normal atmospheric pressure, is

$$L_f = 3.34 \times 10^5 \text{ J/kg} = 79.6 \text{ cal/g} = 143 \text{ Btu/lb}.$$

More generally, to melt a mass m of material that has a heat of fusion L_f requires a quantity of heat Q given by

$$Q = mL_f.$$

This process is *reversible:* To freeze liquid water to ice at 0°C, we have to *remove* heat. The magnitude is the same, but in this case Q is negative because heat is removed rather than added. To cover both possibilities and to include other kinds of phase changes, we write

$$Q = \pm mL. \tag{14.11}$$

The plus sign (heat entering) is used when the material melts, the minus sign (heat leaving) when it freezes. The heat of fusion is different for different materials, and it also varies somewhat with pressure.

At any given pressure, there is a unique temperature at which liquid water and ice can coexist in a condition called **phase equilibrium.** The freezing temperature is always the same as the melting temperature.

We can tell this whole story again, changing the names of the characters, for *boiling*, or *evaporation*, a phase transition between the liquid and gaseous phases. The corresponding heat (per unit mass) is called the **heat of vaporization,** L_v. At normal atmospheric pressure, the heat of vaporization L_v for water is

$$L_v = 2.26 \times 10^6 \text{ J/kg} = 539 \text{ cal/g} = 970 \text{ Btu/lb}.$$

Note that over five times as much heat is required to vaporize a quantity of water at 100°C as is required to raise its temperature from 0°C to 100°C.

Both L_v and the boiling temperature of a material depend on pressure. Water boils at a lower temperature (about 95°C) in Denver than in Pittsburgh because Denver is at a higher elevation and the average atmospheric pressure is less. The heat of vaporization is somewhat greater at this lower pressure, about 2.27×10^6 J/kg.

Like melting, boiling is a reversible transition. When heat is removed from a gas at the boiling temperature, the gas returns to the liquid phase, or *condenses*, giving up to its surroundings the same quantity of heat (heat of vaporization) as was needed to vaporize it. At a given pressure, the boiling and condensation temperatures are the same.

Table 14.4 lists heats of fusion and vaporization for several materials, with the corresponding melting and boiling temperatures at normal atmospheric pressure. Very few *elements* have melting temperatures in the vicinity of ordinary room temperatures; one of the few is the metal gallium, shown in Figure 14.14.

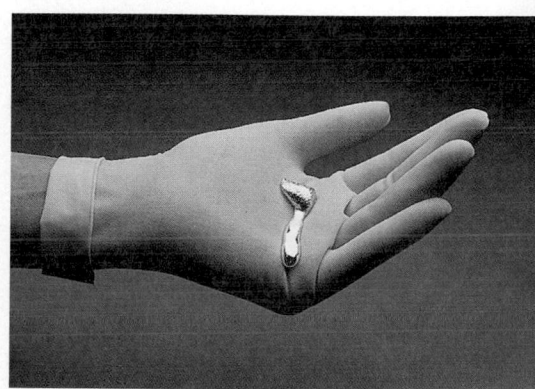

▲ **FIGURE 14.14** The metal gallium melts in your hand. This element is one of the few having a melting temperature near room temperature.

TABLE 14.4 **Heats of fusion and vaporization**

Substance	Normal melting point*		Heat of fusion, L_f, J/kg	Normal boiling point*		Heat of vaporization, L_v, J/kg
	K	**°C**		**K**	**°C**	
Helium	†	†	†	4.216	−268.93	20.9×10^3
Hydrogen	13.84	−259.31	58.6×10^3	20.26	−252.89	452×10^3
Nitrogen	63.18	−209.97	25.5×10^3	77.34	−195.81	201×10^3
Oxygen	54.36	−218.79	13.8×10^3	90.18	−182.97	213×10^3
Ethyl alcohol	159	−114	104.2×10^3	351	78	854×10^3
Mercury	234	−39	11.8×10^3	630	357	272×10^3
Water	273.15	0.00	334×10^3	373.15	100.00	2256×10^3
Sulfur	392	119	38.1×10^3	717.75	444.60	326×10^3
Lead	600.5	327.3	24.5×10^3	2023	1750	871×10^3
Antimony	903.65	630.50	165×10^3	1713	1440	561×10^3
Silver	1233.95	960.80	88.3×10^3	2466	2193	2336×10^3
Gold	1336.15	1063.00	64.5×10^3	2933	2660	1578×10^3
Copper	1356	1083	134×10^3	1460	1187	5069×10^3

* At normal atmospheric pressure.

†A pressure in excess of 25 atmospheres is required to make helium solidify. At 1 atmosphere pressure, helium remains a liquid down to absolute zero.

Conceptual Analysis 14.3

Getting burned

As a theoretical possibility, which (if any) of the following choices would transfer more heat to your skin and thus cause a worse burn?

A. Allowing 1 g of boiling water (at 100°C) to cool to body temperature in your hand.
B. Allowing 1 g of steam at 100°C to condense to liquid water at 100°C in your hand.
C. The heat transfer is the same in both cases.

SOLUTION When 1 g of liquid water cools from 100°C to 20°C, it gives off 4.186 J for each C° drop in temperature, for a total of 335 J. When 1 g of steam at 100°C condenses to liquid water at the same temperature, it gives off an amount of heat equal to the heat of vaporization, about 2260 J. The total amount of heat given off by the condensing steam is over six times as great as the amount given off by the cooling water. Answer B is correct.

Figure 14.15 shows how temperature varies with time when heat is added continuously to a specimen of ice with an initial temperature below 0°C. The temperature rises (from *a* to *b*) until the melting point is reached. Then, as more

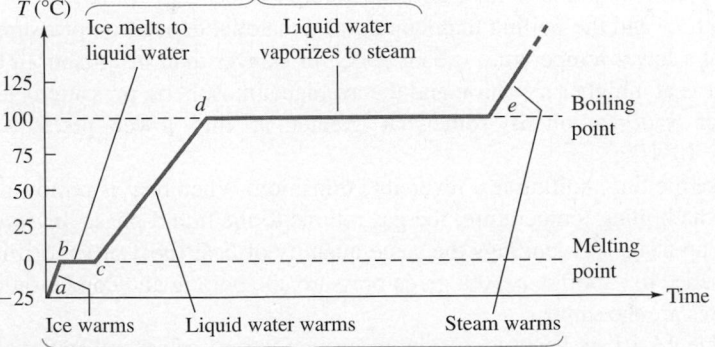

Phase change. As heat is added, temperature stays constant while phase change proceeds: $Q = \pm mL$.

Temperature change. Temperature rises as heat is added: $Q = mc\Delta T$.

▲ **FIGURE 14.15** A plot of temperature versus time for a sample of ice to which heat is added continuously.

heat is added (from b to c), the temperature remains constant until all the ice has melted. Next, the temperature starts to rise again (from c to d) until the boiling temperature is reached. At that point, the temperature is again constant (from d to e) until all the water is transformed into the vapor phase. If the rate of heat input is constant, the slope of the line for the liquid phase is smaller than that of the solid phase because, as Table 14.3 indicates, the specific heat capacity of ice is only about half that of liquid water.

Conceptual Analysis 14.4

Water and ice

What happens when a lump of ice at 0°C is placed in a beaker of water at 0°C in a room at 0°C?

A. Nothing.
B. All the ice melts.
C. Only some of the ice melts.
D. Only some of the water freezes.

SOLUTION The system is in thermal equilibrium; all parts are at the same temperature, and there is no heat transfer when the ice is placed in the beaker of water. If there were freezing or melting, there would have to be heat transfer. Therefore, nothing happens; the correct answer is A.

A substance can sometimes evaporate directly from the solid to the gaseous phase; this process is called **sublimation**, and the solid is said to *sublime*. The corresponding heat is called the **heat of sublimation**, L_s. Liquid carbon dioxide cannot exist at a pressure lower than about 5×10^5 Pa (about 5 atm), so "dry ice" (solid carbon dioxide) sublimes at atmospheric pressure. Sublimation of water from frozen food causes freezer burn, and it is the reason clothes can dry on a clothesline even at subfreezing temperatures. The reverse process, a phase change from gas to solid, occurs when frost forms on cold objects such as the cooling coils in a refrigerator.

The temperature-control mechanisms of many warm-blooded animals make use of the heat of vaporization, removing heat from the body by sweating or panting, both of which depend on evaporating water from the body. Evaporative cooling enables humans to maintain normal body temperature in hot, dry desert climates, where the air temperature may reach 55°C (about 130°F). The skin temperature may be as much as 30 C° cooler than the surrounding air. Under these conditions, a normal person may perspire several liters per day. Unless this lost water is replaced, dehydration, heat stroke, and death result. Old-time desert rats (such as the author) state that in the desert, any canteen that holds less than a gallon should be viewed as a toy!

Evaporative cooling is also used to cool buildings in hot, dry climates and to condense and recirculate "used" steam in coal-fired or nuclear-powered electric-generating plants. That's what is going on in the large cone-shaped concrete towers that you see at such plants.

Chemical reactions such as combustion involve definite quantities of heat. Complete combustion of 1 gram of gasoline produces about 46,000 J, or 11,000 cal, so the **heat of combustion** L_c of gasoline is

$$L_c = 46,000 \text{ J/g} = 46 \times 10^6 \text{ J/kg}.$$

Earlier, we mentioned that the energy value of food is measured in kilocalories (or Calories with a capital C), where 1 kcal = 1000 cal = 4186 J. When we say that a gram of peanut butter "contains 6 Calories," we mean that if we burned it completely—allowed it to undergo complete combustion with oxygen—it would release 6 kcal (25,000 J) of heat. In the body, essentially the same combustion takes place, although slowly and by means of enzyme-mediated reactions in cells. Not all of the energy liberated in the body's "burning" of food is available for mechanical work. We'll study the *efficiency* of energy conversion in Chapter 16.

▲ **BIO Application Heat stroke and hot tubs** At hot ambient temperatures above 30°C, the body loses heat more by evaporation of water than by radiation and convection. When body temperature rises above normal, feedback systems activate blood vessel dilation and increased blood flow to the skin. This leads to a drop in blood pressure and decreased blood flow to the brain and internal organs. In a hot tub or a sauna, where evaporative cooling cannot occur, the body's response fails to lower the core temperature and the body responds by further increasing surface blood flow. Because the water temperature is above normal body temperature, this leads to further warming of the blood and the body. Numerous deaths have been reported as a result of overheated hot tubs, often exacerbated by alcohol, which also increases blood flow to the skin. The body's feedback mechanisms for maintaining core temperature are complex and powerful, but they can be overcome in extreme situations, sometimes with disastrous results.

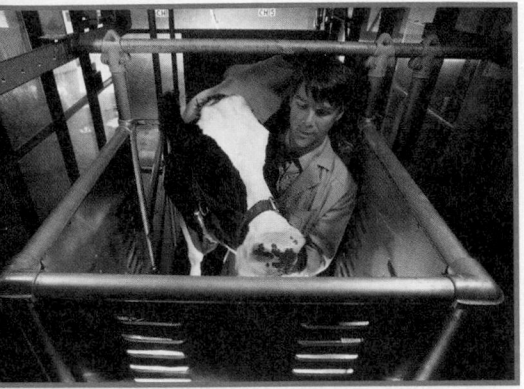

▲ BIO Application Meat, not heat.
A calorimeter is a device used to measure heat production or consumption by an isolated system. Typically, the calorimeter is a benchtop piece of equipment used in an analytical laboratory to study, for example, the amount of heat generated by chemical reactions. However, the calorimeter can also be used to measure heat production in a much larger isolated system, such as the farm animal shown here. The cow is fed a strictly controlled diet in this room-sized calorimeter for several days, and the amount of heat produced by the animal's metabolism is carefully measured. In this way, agricultural scientists have been able to determine which types of animal feed are most efficient at promoting weight gain without excessive loss of energy as heat.

14.6 Calorimetry

"Calorimetry" means "measuring heat." We've discussed the energy transfer (heat) involved in temperature changes and in phases changes. Now we're ready to look at some examples of calorimetry calculations (calculations with heat). The basic principle is simple: When heat flow occurs between two objects that are isolated from their surroundings, the amount of heat lost by one object must equal the amount of heat gained by the other. Heat is energy in transit, so this principle is really just another form of conservation of energy. We take each quantity of heat *added to* an object as *positive* and each quantity *leaving* an object as *negative*. When several objects interact, the *algebraic sum* of the quantities of heat transferred to *all* of them must be zero. The basic principle can be stated very simply as

$$\Sigma Q = 0.$$

Calorimetry, dealing entirely with one conserved scalar quantity, is in many ways the simplest of all physical theories!

PROBLEM-SOLVING STRATEGY 14.2 **Calorimetry problems**

SET UP

1. To avoid confusion about algebraic signs when calculating quantities of heat, use Equations 14.10 and 14.11 consistently for each object. Each Q is positive when heat enters an object and negative when it leaves. The algebraic sum of all the Q's must be zero.

SOLVE

2. Often you will need to find an unknown temperature. Represent it by an algebraic symbol such as T. Then, if an object has an initial temperature of 20°C and an unknown final temperature T, the temperature change for the object is $\Delta T = T_{final} - T_{initial} = T - 20°C$ (*not* 20°C − T) . . . and so on.
3. Compute the heat added to each interacting object, and then equate the algebraic sum of all the quantities of heat to zero (representing conservation of energy). When heat leaves an object, consider it as negative heat added to the object.

REFLECT

4. In problems in which a phase change takes place, as when ice melts, you may not know in advance whether *all* of the material undergoes a phase change or only part of it does. You can always assume one or the other, and if the resulting calculation gives an absurd result (such as a final temperature that is higher or lower than *all* of the initial temperatures), you know that the initial assumption was wrong. Back up and try again!

EXAMPLE 14.8 **Coffee in a metal cup**

A Turkish restaurant serves coffee in copper mugs (with the inside tin plated to avoid copper poisoning). A waiter fills a cup having a mass of 0.100 kg and initially at 20.0°C with 0.300 kg of coffee that is initially at 70.0°C. What is the final temperature after the coffee and the cup attain thermal equilibrium? (Assume that coffee has the same specific heat capacity as water and that there is no heat exchange with the surroundings.)

Continued

SOLUTION

SET UP AND SOLVE Call the final temperature T. The (negative) change in temperature of the coffee is $\Delta T_{coffee} = T - 70°C$, and the (positive) change in temperature of the cup is $\Delta T_{copper} = T - 20°C$. We need to find the heat lost by the coffee and gained by the copper cup and then set the algebraic sum equal to zero. Because the coffee loses heat, it has a negative value of Q:

$$Q_{coffee} = mc_{water}\Delta T_{coffee}$$
$$= (0.300 \text{ kg})[4190 \text{ J}/(\text{kg} \cdot \text{K})](T - 70.0°C).$$

(Note that ΔT_{coffee} is negative in this expression.) The heat gained by the copper cup is

$$Q_{copper} = mc_{copper}\Delta T_{copper}$$
$$= (0.100 \text{ kg})[390 \text{ J}/(\text{kg} \cdot \text{K})](T - 20.0°C).$$

(Note that ΔT_{copper} is positive in this expression.) We equate the sum of these two quantities of heat to zero, obtaining an algebraic equation for T:

$$Q_{coffee} + Q_{copper} = 0,$$
$$\text{or} \quad (0.300 \text{ kg})[4190 \text{ J}/(\text{kg} \cdot \text{K})](T - 70.0°C)$$
$$+ (0.100 \text{ kg})[390 \text{ J}/(\text{kg} \cdot \text{K})](T - 20.0°C) = 0.$$

Solving this equation for T, we find $T = 68.5°C$.

REFLECT The final temperature is much closer to the initial temperature of the coffee than to that of the cup; water has a much larger specific heat capacity than copper, and we have three times as much mass of water. We can also find the quantities of heat by substituting this value for T back into the original equations. When we do this, we find that $Q_{coffee} = -1890 \text{ J}$ and $Q_{copper} = +1890 \text{ J}$; Q_{coffee} is negative, as expected.

Practice Problem: Find the final temperature if the cup is a heavy ceramic cup with a mass of 0.200 kg and specific heat capacity $c = 800 \text{ J}/(\text{kg} \cdot \text{K})$. *Answer:* 64.3°C.

EXAMPLE 14.9 Chilling your soda

A physics student wants to cool 0.25 kg of Diet Omni-Cola (mostly water) initially at 20°C by adding ice initially at −20°C. How much ice should she add so that the final temperature will be 0°C with all the ice melted? Assume that the heat capacity of the paper container may be neglected.

SOLUTION

SET UP AND SOLVE The ice *gains* the heat (positive Q) required to warm it from −20°C to 0°C and to melt all of it. The cola *loses* (negative Q) the same amount of heat. Let the unknown mass of ice be m. The (negative) heat transfer to the Omni-Cola is

$$Q_{Omni} = m_{Omni}c_{water}\Delta T_{Omni}$$
$$= (0.25 \text{ kg})[4190 \text{ J}/(\text{kg} \cdot \text{K})](0°C - 20°C)$$
$$= -21,000 \text{ J}.$$

The specific heat capacity of ice (not the same as that for liquid water) is about $2.0 \times 10^3 \text{ J}/\text{kg} \cdot \text{K}$. The heat needed to warm a mass m of ice from −20°C to 0°C is

$$Q_{ice} = mc_{ice}\Delta T_{ice}$$
$$= m[2.0 \times 10^3 \text{ J}/(\text{kg} \cdot \text{K})][0°C - (-20°C)]$$
$$= m(4.0 \times 10^4 \text{ J}/\text{kg}).$$

The additional heat needed to melt the ice is the mass of the ice times the heat of fusion:

$$Q_{melt} = mL_f$$
$$= m(3.34 \times 10^5 \text{ J}/\text{kg}.)$$

The sum of these three quantities must equal zero:

$$Q_{Omni} + Q_{ice} + Q_{melt} = -21,000 \text{ J} + m(40,000 \text{ J}/\text{kg})$$
$$+ m(334,000 \text{ J}/\text{kg}) = 0.$$

Solving this equation for m, we get $m = 0.056 \text{ kg} = 56 \text{ g}$ (two or three medium-size ice cubes).

REFLECT A final state with all the ice melted isn't a very satisfying outcome. We probably want to add a couple of extra ice cubes so that the drink will stay cold for a while even though the temperature of the surrounding air may be considerably higher than 0°C.

14.7 Heat Transfer

In Section 14.1, we spoke qualitatively about *conductors* and *insulators*—materials that permit or prevent heat transfer between objects. We weren't concerned with how quickly or slowly the transfer takes place. But if you're boiling water to make tea or trying to build an energy-efficient house, you need to know how quickly or slowly heat is transferred. In this section, we'll study the three mechanisms of heat transfer—conduction, convection, and radiation—and the relation of the rate of heat transfer to the properties of the system. *Conduction* occurs within an object or between two objects that are in actual contact with each other.

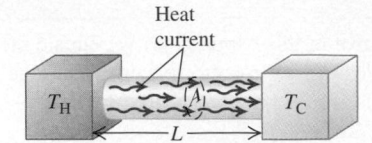

▲ **FIGURE 14.16** Steady-state heat flow due to conduction in a uniform rod.

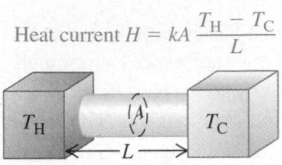

Heat current $H = kA\dfrac{T_H - T_C}{L}$

Doubling the cross-sectional area of the conductor doubles the heat current ($H \propto A$):

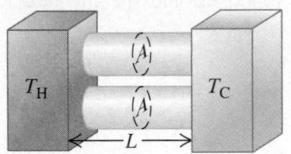

Doubling the length of the conductor halves the heat current ($H \propto 1/L$):

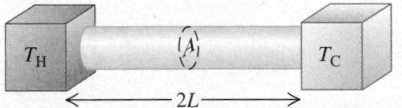

▲ **FIGURE 14.17** The relationship of heat current to the area and length of a conductor.

TABLE 14.5 Thermal conductivities

Material	k (W/(m·K))
Metals	
Lead	34.7
Steel	50.2
Brass	109
Aluminum	205
Copper	385
Silver	406
Other solids (representative values)	
Styrofoam™	0.01
Fiberglass	0.04
Wood	0.12–0.04
Insulating brick	0.15
Red brick	0.6
Concrete	0.8
Glass	0.8
Ice	1.6
Gases	
Air	0.024
Helium	0.14

Convection depends on the motion of mass (such as air or water) from one region of space to another, and *radiation* is heat transfer by electromagnetic radiation, such as sunshine, with no need for matter to be present in the space between objects. We'll consider these three kinds of processes in turn.

Conduction

If you hold one end of a copper rod and place the other end in a flame, the end you are holding gets hotter and hotter, even though it isn't in direct contact with the flame. Heat reaches the cooler end by **conduction** through the material. On the atomic level, the atoms in the hotter regions have more kinetic energy, on the average, than their cooler neighbors. They jostle their neighbors, giving them some of their energy. The neighbors then jostle *their* neighbors, and so it goes through the material. The atoms themselves do not move from one region of the material to another, but their energy does.

In metals, electron motion provides another mechanism for heat transfer. Most metals are good conductors of electricity because some electrons can leave their parent atoms and wander through the crystal lattice. These free electrons can carry energy from the hotter to the cooler regions of the metal. Good thermal conductors such as silver, copper, aluminum, and gold are also good electrical conductors.

Heat transfer occurs only between regions at different temperatures, and the direction of heat flow is always from higher to lower temperature. Figure 14.16 shows a rod of conducting material with cross-sectional area A and length L. The left end of the rod is kept at a temperature T_H and the right end at a lower temperature T_C, and heat flows from left to right. We assume that the rod is insulated, so that heat flows only from one end to the other, not in or out at the sides.

In discussing the rate of heat transfer with time, we'll change our notation slightly: We'll use ΔQ (rather than just Q) for the quantity of heat transferred in a time Δt. Then the rate of heat flow with time (heat transferred per unit time) is $\Delta Q/\Delta t$. We call this rate the **heat current** and denote it by H. That is, $H = \Delta Q/\Delta t$. Experiments show that the heat current is proportional to the cross-sectional area A. If we place two identical rods side by side, the total heat flow is twice as great as for each individual rod (Figure 14.17). It is also proportional to the temperature difference $(T_H - T_C)$. Finally, it is *inversely* proportional to the length L: Roughly speaking, if the heat has twice as far to go, it takes twice as long to get there.

Introducing a proportionality constant k called the **thermal conductivity** of the material, we have

$$H = \frac{\Delta Q}{\Delta t} = kA\frac{T_H - T_C}{L}. \qquad (14.12)$$

The quantity $(T_H - T_C)/L$ is the temperature difference *per unit length*; it is called the **temperature gradient.** The numerical value of k depends on the material of the rod. Materials with large k are good conductors of heat; materials with small k are poor conductors (and thus good insulators). Equation 14.12 also gives the heat current through a slab, or *any* homogeneous object with uniform cross-sectional area A perpendicular to the direction of flow; L is the length of the heat-flow path.

The units of heat current H are units of energy per unit time, or power; the SI unit of heat current is the joule per second, or watt (W). We can find the units of k by solving Equation 14.12 for k. We invite you to verify that the SI units of k are W/(m·K). Some numerical values of k are given in Table 14.5. This table makes it clear why copper and aluminum are often used in the bottoms of cooking pans.

The thermal conductivity of "dead" (nonmoving) air is very small. A wool sweater is warm because it traps air between the fibers. In fact, many insulating

▶ **BIO Application** **Hot-and-cold fish.** Most fish are the same temperature as the surrounding water because the heat their bodies generate is lost by conduction to the water as their blood circulates through the gills. However, some large predatory fishes, such as these bluefin tuna, use a special technique to warm key body parts—their central swimming muscles, their brains, and the retinas of their eyes, for example. Warming these organs allows the fishes to hunt effectively in cold water. To keep a body part warm, the arteries entering it run in intimate contact with the veins leaving it, allowing rapid transfer of heat by conduction between the vessels. The heat in the venous blood leaving the body part passes to the cold arterial blood approaching it; by the time the arterial blood reaches the part in question, it is warm. Thus, the "countercurrent" flow of blood in the adjoining vessels traps heat in a particular region.

materials, such as Styrofoam™ and fiberglass, are mostly dead air. Figure 14.18 shows a "space-age" ceramic material with highly unusual thermal properties, including very small conductivity.

Quantitative Analysis 14.5

Big house, little house

Two neighboring houses are built of the same materials, but one is twice as long, high, and wide as the other and also has walls twice as thick. To keep the two houses equally warm on a cold day, at what rate must the heater in the big house supply heat, compared with the heater in the small house?

A. at twice the rate.
B. at four times the rate.
C. at 16 times the rate.

SOLUTION We assume that the walls, roofs, and basement floors (if any) participate equally in the heat loss from the houses. The larger house is twice as large in each dimension, so it has four times the surface area as the smaller one. If the walls of the two houses were the same thickness, the larger one would lose heat at four times the rate of the small one. But its walls are twice as thick; therefore, the rate of heat flow for a given area is only half as great. We conclude that the larger house loses heat at a rate two times as great as the smaller one. The correct answer is A.

PROBLEM-SOLVING STRATEGY 14.3 **Heat conduction**

SET UP

1. Identify the direction of heat flow in the problem (from hot to cold). In Equation 14.12, L is always measured along that direction and A is always an area perpendicular to that direction. Often, when a box or other container has an irregular shape, but uniform wall thickness, you can approximate it as a flat slab with the same thickness and total wall area.

SOLVE

2. In some problems, the heat flows through two different materials in succession. The temperature at the interface between the two materials is then intermediate between T_H and T_C; represent it by a symbol such as T. The temperature differences for the two materials are then $(T_H - T)$ and $(T - T_C)$. In steady-state heat flow, the same heat has to pass through both materials in succession, so the heat current H must be *the same* in both materials.

3. If there are two *parallel* heat flow paths, so that some heat flows through each, then the total H is the sum of the quantities H_1 and H_2 for the separate paths. An example is heat flow from inside to outside a house, both through the glass in a window and through the surrounding frame. In this case, the temperature difference is the same for the two paths, but L, A, and k may be different.

4. As always, it is essential to use a consistent set of units. If you use a value of k expressed in $W/(m \cdot K)$, don't use distances in cm, heat in calories, or T in °F!

REFLECT

5. Ask whether your results make sense. If you find that Styrofoam™ conducts heat better than copper, you've made a mistake!

▲ **FIGURE 14.18** This protective tile, developed for use in the space shuttle, has extraordinary thermal properties. The extremely small thermal conductivity and small heat capacity of the material make it possible to hold the tile by its edges, even though it is hot enough to emit the light needed for this photograph.

EXAMPLE 14.10 **Conduction through a Styrofoam™ cooler**

The Styrofoam™ box in Figure 14.19 is used to keep drinks cold at a picnic. The total wall area (including the lid) is 0.80 m², and the wall thickness is 2.0 cm. The box is filled with ice and cans of Omni-Cola, keeping the inner surface at 0°C. What is the rate of heat flow into the box if the temperature of the outside surface is 30°C? How much ice melts in one day (24 h)?

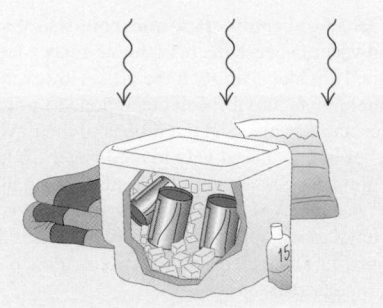

▶ **FIGURE 14.19**

SOLUTION

SET UP Figure 14.20 shows our sketch. We assume that the total heat flow is approximately the same as it would be through a flat slab of area 0.80 m² and thickness 2.0 cm = 0.020 m. We find k from Table 14.5. The heat of fusion of ice is 3.34×10^5 J/kg. There are 86,400 s in one day.

SOLVE From Equation 14.12, the heat current H (rate of heat flow) is

$$H = kA\frac{T_2 - T_1}{L}$$

$$= [0.010 \text{ W}/(\text{m} \cdot \text{K})](0.80 \text{ m}^2)\frac{30°C - 0°C}{0.020 \text{ m}}$$

$$= 12 \text{ W}.$$

The total heat flow Q in one day is

$$Q = Ht = (12 \text{ J/s})(86,400 \text{ s}) = 1.04 \times 10^6 \text{ J}.$$

The mass of ice melted by this quantity of heat is

$$m = \frac{Q}{L_f} = \frac{1.04 \times 10^6 \text{ J}}{3.34 \times 10^5 \text{ J/kg}} = 3.1 \text{ kg}.$$

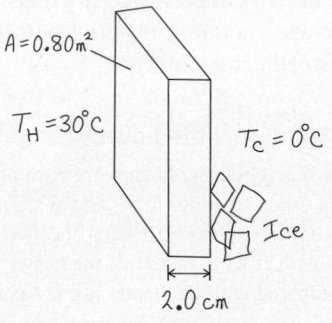

▲ **FIGURE 14.20** Our sketch for this problem.

REFLECT A typical bag of ice you buy at the market is 8 lb = 3.6 kg. So this amount is roughly equal to one 8 lb bag of ice.

Practice Problem: This same Styrofoam™ cooler is sitting in the sun, and the outside temperature reaches 40°C. How much time passes before the 8 lb of ice is completely melted? *Answer:* $t = 21$ h.

EXAMPLE 14.11 **Conduction in two bars in series**

A steel bar 10.0 cm long is welded end to end to a copper bar 20.0 cm long. Each bar has a square cross section 2.00 cm on a side. The free end of the steel bar is in contact with steam at 100°C, and the free end of the copper bar is in contact with ice at 0°C. Find the temperature at the junction of the two bars and the total rate of heat flow.

SOLUTION

SET UP Figure 14.21 shows our sketch; we use the subscripts s and c for steel and copper, respectively. The key to the solution is the fact that the heat currents in the two bars must be equal; otherwise, some sections would have more heat flowing in than out, or the reverse, and steady-state conditions could not exist.

SOLVE Let T be the unknown junction temperature; we use Equation 14.12 for each bar and equate the two expressions:

$$\frac{k_s A(100°C - T)}{L_s} = \frac{k_c A(T - 0°C)}{L_c}.$$

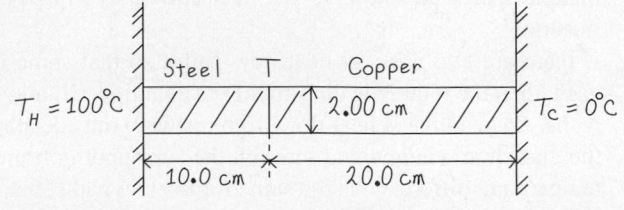

▲ **FIGURE 14.21** Our sketch for this problem.

Continued

The areas A are equal and may be divided out. Substituting numerical values, we find that

$$\frac{[50.2 \text{ W}/(\text{m} \cdot \text{K})](100°C - T)}{0.100 \text{ m}}$$
$$= \frac{[385 \text{ W}/(\text{m} \cdot \text{K})](T - 0°C)}{0.200 \text{ m}}.$$

Rearranging and solving for T, we obtain

$$T = 20.7°C.$$

To find the total heat current H, we substitute this value for T back into either of the previous expressions:

$$H = \frac{[50.2 \text{ W}/(\text{m} \cdot \text{K})](0.0200 \text{ m})^2(100°C - 20.7°C)}{0.100 \text{ m}}$$
$$= 15.9 \text{ W}$$

or

$$H = \frac{[385 \text{ W}/(\text{m} \cdot \text{K})](0.0200 \text{ m})^2(20.7°C - 0°C)}{0.200 \text{ m}}$$
$$= 15.9 \text{ W}.$$

REFLECT Even though the steel bar is shorter, the temperature drop across it is much greater than that across the copper bar because steel is a much poorer conductor of heat.

Practice Problem: If we change the length of the steel bar, what length is needed in order to make the temperature at the junction 50.0°C? *Answer:* 2.60 cm.

EXAMPLE 14.12 **Conduction in two bars in parallel**

Suppose the two bars in Example 14.11 are separated. One end of each bar is placed in contact with steam at 100°C, and the other end of each bar contacts ice at 0°C. What is the *total* rate of heat flow in the two bars?

SOLUTION

SET UP Figure 14.22 shows our sketch. In this case the bars are in parallel rather than in series. As before, we use the subscripts s and c for steel and copper, respectively.

SOLVE The total heat current H is now the *sum* of the currents in the two bars. The temperature difference between ends is the same for both bars: $T_H - T_C = 100°C - 0°C = 100$ K. We find that

$$H = \frac{\Delta Q}{\Delta t} = k_c A_c \frac{T_{H,c} - T_{C,c}}{L_c} + k_s A_s \frac{T_{H,s} - T_{C,s}}{L_s}$$
$$= \frac{[385 \text{ W}/(\text{m} \cdot \text{K})](0.0200 \text{ m})^2(100 \text{ K})}{0.200 \text{ m}}$$
$$+ \frac{[50.2 \text{ W}/(\text{m} \cdot \text{K})](0.0200\text{m})^2(100 \text{ K})}{0.100 \text{ m}}$$
$$= 77.0 \text{ W} + 20.1 \text{ W} = 97.1 \text{ W}.$$

REFLECT The heat flow in the copper bar is much greater than that in the steel bar, even though the copper bar is longer, because its thermal conductivity is much larger. The total heat flow is

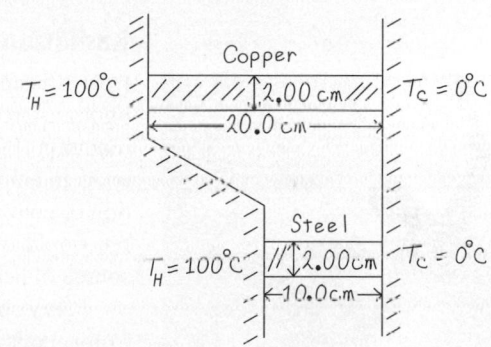

▲ **FIGURE 14.22** Our sketch for this problem.

much greater than that in Example 14.11, partly because the total cross section for heat flow is greater and partly because the full 100 C° temperature difference acts across each bar.

Practice Problem: Suppose we increase the cross-sectional area of the steel bar. What cross-sectional area would be needed to make the heat currents in the two bars equal? *Answer:* 15.3 cm².

Convection

Convection is the transfer of heat by the motion of a mass of fluid from one region of space to another. Familiar examples include hot-air and hot-water home heating systems, the cooling system of an automobile engine, and the heating and cooling of the body by the flow of blood. If the fluid is circulated by a blower or pump, the process is called *forced convection;* if the flow is caused by differences in density due to thermal expansion, such as hot air rising, the process is called *natural convection* or *free convection*.

Day: The land is warmer than the water; convection draws a sea breeze onshore.

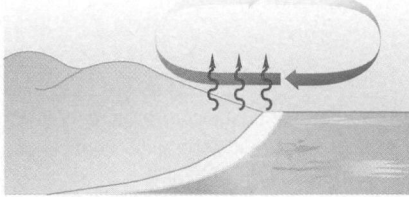

Night: The land is colder than the water; convection sends a land breeze offshore.

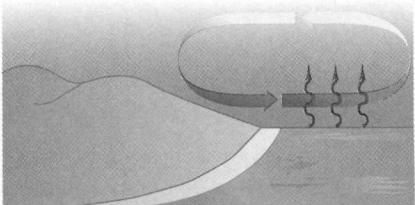

▲ **FIGURE 14.23** Convective currents in the air determine daytime sea breezes and nighttime land breezes near shore.

▲ **FIGURE 14.24** Heat transfer by radiation from the sun. The heat is not dangerous to human skin, but the ultraviolet radiation is.

Convection in the atmosphere plays a dominant role in determining weather patterns (Figure 14.23), and convection in the oceans is an important global heat-transfer mechanism. On a smaller scale, glider pilots make use of thermal updrafts from the warm earth. But sometimes the air cools enough as it rises to form thunderheads, which contain convection currents that can tear a glider to pieces.

The most important mechanism for heat transfer within the human body is forced convection of blood, with the heart serving as the pump. The total rate of heat loss from the body is on the order of 100 to 200 W (2000 to 4000 kcal per day). A dry, unclothed body in still air loses about 75 W by radiation, but during vigorous exercise and plentiful perspiration, evaporative cooling with convection is the dominant mechanism.

Convective heat transfer is a complex process, and there is no simple equation to describe it. Here are a few experimental facts: The heat current due to convection is directly proportional to the surface area involved. This is the reason for the large surface areas of radiators and cooling fins. In some cases, it is approximately proportional to the $\frac{5}{4}$ power of the temperature difference between the surface and the main body of fluid. The viscosity of fluids slows natural convection near a stationary surface. In air, this effect produces a surface film; forced convection decreases the thickness of this film, increasing the rate of heat transfer. This is the reason for the "wind-chill factor": You get cold faster in a cold wind than in still air with the same temperature.

Radiation

Heat transfer by **radiation** depends on electromagnetic waves such as visible light, infrared, and ultraviolet radiation. Everyone has felt the warmth of the sun's radiation (Figure 14.24) and the intense heat from a charcoal grill or the glowing coals in a fireplace. Heat from these very hot objects reaches you not by conduction or convection in the intervening air, but by *radiation*. This kind of heat transfer would occur even if there were nothing but vacuum between you and the source of heat.

Every object, even at ordinary temperatures, emits energy in the form of electromagnetic radiation. At ordinary temperatures, say, 20°C, nearly all the energy is carried by infrared waves with wavelengths much longer than those of visible light. As the temperature rises, the wavelengths shift to shorter values. At 800°C, an object emits enough visible radiation to be self-luminous and appears "red hot," although even at this temperature most of the energy is carried by infrared waves. At 3000°C, the temperature of an incandescent lamp filament, the radiation contains enough visible light that the object appears "white hot."

The rate of energy radiation from a surface is proportional to the surface area A. The rate increases very rapidly with temperature, in proportion to the fourth power of the absolute (Kelvin) temperature. The rate also depends on the nature of the surface; this dependence is described by a quantity e called the **emissivity,** a dimensionless number between 0 and 1. The heat current $H = \Delta Q / \Delta t$ due to radiation from a surface area A with emissivity e at absolute (Kelvin) temperature T can be expressed as

$$H = Ae\sigma T^4, \tag{14.13}$$

where σ is a fundamental physical constant called the **Stefan–Boltzmann constant.** This relation is called the **Stefan–Boltzmann law,** in honor of its late-19th-century discoverers. The numerical value of σ is found experimentally to be

$$\sigma = 5.6705 \times 10^{-8} \ \text{W}/(\text{m}^2 \cdot \text{K}^4).$$

▶ **Application Baked in a convection oven?** The volcanic Hawaiian Islands are a by-product of the earth's convective cooling system. Earth's deep interior generates heat, mainly by the decay of radioactive elements and the slow solidification of the liquid core. The temperature difference between the hot interior and the cold crust sets up density differences that cause the solid rock of the earth's crust and mantle to flow in (very) slow convective motions, carrying heat from the interior to the surface. The Hawaiian islands are built by a dramatic type of convective flow called a *mantle plume*—a vertical jet of hot rock that may rise from near the core itself. As this plume approaches the surface, the reduction in pressure causes some of the plume's rock to melt, creating the lava that builds the islands. In the foreground we see a cinder cone on the peak of Mauna Kea; in the background is the silhouette of Mauna Loa.

We invite you to check for consistency of units in Equation 14.13. Emissivity e is often larger for dark surfaces than for light ones. The emissivity of a smooth copper surface is about 0.3, but e for a dull black surface is nearly unity.

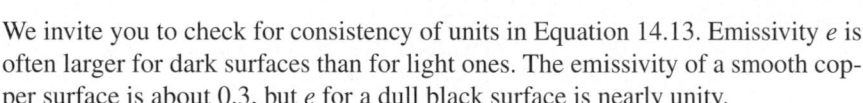

Quantitative Analysis 14.6

Raising the temperature

If the Kelvin temperature of a spherical object were increased by 20%, by approximately what factor would the rate of radiated energy emitted from its surface be changed?

A. 1.2.
B. 1.4.
C. 1.7.
D. 2.0.

SOLUTION The rate of radiation of energy is proportional to the fourth power of the absolute temperature. If the temperature increases by a factor of 1.2, the rate of radiation increases by a factor of $(1.2)^4$, or 2.07 (to three significant figures). The closest answer to this is D, the correct answer.

EXAMPLE 14.13 Radiation from a hot plate

A thin, square steel plate 10 cm on a side is heated in a blacksmith's forge to a temperature of 800°C. If the emissivity is 0.60, what is the plate's rate of radiation of energy?

SOLUTION

SET UP The total surface area of the plate, including both sides, is $2(0.10 \text{ m})^2 = 0.020 \text{ m}^2$. We must convert the temperature to the Kelvin scale: $800°C - 1073 \text{ K}$.

SOLVE Equation 14.13 gives

$$H = Ae\sigma T^4$$
$$- (0.020 \text{ m}^2)(0.60)[5.67 \times 10^{-8} \text{ W}/(\text{m}^2 \cdot \text{K}^4)](1073 \text{ K})^4$$
$$= 900 \text{ W}.$$

REFLECT This rate of emission of energy is comparable to the output of a small electric space heater.

Practice Problem: What temperature would be required for this plate to radiate at the rate of 1500 W? *Answer:* 950°C.

While an object at absolute (Kelvin) temperature T is radiating, its surroundings, at temperature T_s, are also radiating, and the object *absorbs* some of this radiation. If it is in thermal equilibrium with its surroundings, the rates of radiation and absorption must be equal. For this to be true, the rate of absorption must be given in general by $H = Ae\sigma T_s^4$. Then the *net* rate of radiation from an object at temperature T with surroundings at temperature T_s is

$$H_{\text{net}} = Ae\sigma T^4 - Ae\sigma T_s^4 = Ae\sigma(T^4 - T_s^4). \qquad (14.14)$$

In this equation, a positive value of H means a net heat flow *out of* the object. Equation 14.14 shows that, for radiation, as for conduction and convection, the heat current depends on the temperature *difference* between two objects.

EXAMPLE 14.14 Radiation from the human body

If the total surface area of a woman's body is 1.2 m² and her skin temperature is 30°C, find the total rate of radiation of energy from her body. She also *absorbs* radiation from her surroundings. If the surroundings are at a temperature of 20°C, what is her *net* rate of heat loss by radiation? The emissivity of the body is very close to unity, irrespective of skin pigmentation.

SOLUTION

SET UP We need to convert the temperatures to absolute (Kelvin) temperatures: 30°C = 303 K and 20°C = 293 K. The rate of radiation of energy per unit area is given by Equation 14.13.

SOLVE Taking $e = 1$, we find that

$$H = Ae\sigma T^4$$
$$= (1.2 \text{ m}^2)(1)[5.67 \times 10^{-8} \text{ W}/(\text{m}^2 \cdot \text{K}^4)](303 \text{ K})^4$$
$$= 574 \text{ W}.$$

This loss is partly offset by the *absorption* of radiation, which depends on the temperature of the surroundings. The *net* rate of radiative energy transfer is given by Equation 14.14:

$$H = Ae\sigma(T^4 - T_s^4)$$
$$= (1.2 \text{ m}^2)(1)[(5.67 \times 10^{-8} \text{ W}/(\text{m}^2 \cdot \text{K}^4)]$$
$$\cdot [(303 \text{ K})^4 - (293 \text{ K})^4]$$
$$= 72 \text{ W}.$$

REFLECT The heat produced by the human body at rest is about 75 W. Thus, this net rate of radiation is sufficient to prevent overheating. But when additional heat is produced by even mild exertion, heat must be removed by other means, such as evaporative cooling.

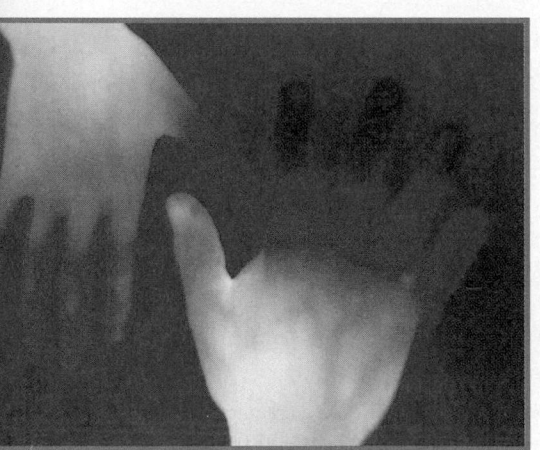

▲ **BIO Application See my heat.** Just as an ordinary photograph captures visible light from an object, an infrared image captures heat released in the form of long-wavelength infrared radiation (infrared light). We cannot see infrared light without the help of special cameras or goggles. However, the venomous snakes known as pit vipers, which include the rattlesnakes and water moccasins, have special "pit organs" on their snout that are sensitive to infrared radiation. These snakes hunt mostly at night; their pit organs help them to "see" their warm-blooded prey in the cool darkness.

An object that is a good absorber must also be a good emitter. An ideal radiator, with an emissivity of unity, is also an ideal absorber, absorbing *all* of the radiation that strikes it. Such an ideal surface is called an ideal blackbody, or simply a **blackbody.** Conversely, an ideal *reflector,* which absorbs *no* radiation at all, is a highly ineffective radiator. This is the reason for the silver coatings on vacuum, or thermos, bottles, which were invented by Sir James Dewar (1842–1923). A vacuum bottle has double glass walls, and the air is pumped out of the spaces between the walls, eliminating nearly all heat transfer by conduction and convection. The silver coating on the walls reflects most of the radiation from the contents back into the container, and the wall itself is a very poor emitter. Thus, a vacuum bottle can keep coffee or soup hot for several hours. The Dewar flask, used to store very cold liquefied gases, is exactly the same in principle.

Heat transfer by radiation is important in some surprising places. A premature baby in an incubator can be cooled dangerously by radiation if the walls of the incubator happen to be cold, even when the *air* in the incubator is warm. Some incubators regulate the air temperature by measuring the baby's skin temperature.

We can study the infrared radiation emitted by an object by using a camera with infrared-sensitive film or by means of a device, similar to a television camera, that is sensitive to infrared radiation. The resulting picture is called a *thermograph.* The rate of emission of energy depends strongly on temperature, so thermography permits the detailed study of temperature distributions, with a precision as great as 0.1 C°. Thermography is used to study energy losses in heated buildings, and it has several important medical applications. Various tumors, such as breast cancer, cause local temperature variations, and growths as small as 1 cm can be detected. Circulatory disorders that cause local temperature anomalies can be studied, and thermography has many other applications.

14.8 Solar Energy and Resource Conservation

The principles of heat transfer that we've studied in this chapter have many practical applications in a civilization such as ours, with growing energy consumption and dwindling energy resources. A substantial fraction of all the energy consumption in the United States is used to heat and cool homes and other buildings, to maintain a comfortable temperature and humidity inside when the outside temperature is much hotter or much colder.

For space heating, the objective is to prevent as much heat flow as possible from inside to outside; heating units replace the inevitable loss of heat. Walls insulated with material of low thermal conductivity, storm windows, and multiple-layer glass windows all help to reduce heat loss. It has been estimated that if all buildings used such materials, the total energy needed for space heating would be reduced by at least one-third.

Air-conditioning in summer poses the reverse problem: Heat flows from outside to inside, and energy must be expended in refrigeration units to remove it. Again, appropriate insulation can decrease this energy cost considerably.

Direct conversion of solar energy is a promising development in energy technology. Near the earth, the rate of energy transfer due to solar radiation is about 1400 W for each square meter of surface area. Not all of this energy reaches the earth's surface; even on a clear day, about one-fourth of the energy is absorbed by the earth's atmosphere. In a typical household solar heating system, large black plates facing the sun are backed with pipes through which water circulates. The black surface absorbs most of the sun's radiation; the heat is transferred by conduction to the water and then by forced convection to radiators inside the house. Heat loss by convection of air near the solar collecting panels is reduced by covering them with glass, with a thin air space between the glass and the panel. An insulated heat reservoir stores the energy collected for use at night and on cloudy days. Many larger-scale solar-energy conversion systems, including windmills, are also currently under study; a few examples are discussed in Section 16.9.

From an environmental standpoint, solar energy has multiple advantages over the use of fossil fuels (burning of coal or oil) or nuclear power. Fossil fuels are being used up; solar power continues indefinitely. Obtaining fossil fuels may involve strip mining, with its associated destruction of landscape and elimination of other useful land functions, such as farming or timber. Offshore oil drilling is a source of ocean water pollution. Air pollution from combustion products is a familiar problem, as is acid rain, which is directly attributable to coal smoke in many cases. Excess atmospheric carbon dioxide produced by the combustion of hydrocarbons is a major contributor to the so-called greenhouse effect. Many responsible scientists believe that the long-range effects on our climate will be serious or even catastrophic unless greenhouse gas emission is substantially reduced.

Nuclear power poses radiation hazards and the problem of disposing of radioactive waste material. Solar power avoids all these problems. Some practical problems still need to be worked out, however, including high installation costs and the need for energy storage facilities for nighttime and cloudy days.

Intelligent consideration of the effects of solar radiation in the design of buildings can also reduce energy consumption in heating and cooling. In recent years, there has been considerable growth in awareness of the need to design buildings with energy conservation in mind. The need to conserve is pressing and will undoubtedly become more so in years to come.

▲ **Application Driving off using the sunset.** Solar energy represents an almost limitless source of energy if we can figure out how to convert it efficiently into usable forms such as mechanical, chemical, or electrical energy. Thanks to ultralight materials and efficient solar panels and energy converters, the Nuna 2, pictured here, can reach speeds of over 100 mph (160 km/hr)! This car, designed and built in the Netherlands by physics and engineering students, shows how limitless human imagination, creativity, and innovation can be with enough study and hard work.

SUMMARY

Temperature and Temperature Scales

(Sections 14.1 and 14.2) Two objects in **thermal equilibrium** must have the same temperature. A conducting material between two objects permits interaction leading to thermal equilibrium; an insulating material prevents or impedes this interaction. The **Celsius** and **Fahrenheit** temperature scales are based on the freezing ($0°C = 32°F$) and boiling ($100°C = 212°F$) temperatures of water, related by the formulas $T_F = \frac{9}{5}T_C + 32°$ and $T_C = \frac{5}{9}(T_F - 32°)$ (Equations 14.1 and 14.2).

The **Kelvin** scale has its zero at the extrapolated zero-pressure temperature for a gas thermometer, which is $-273.15°C$. Thus, $0\ K = -273.15°C$, and $T_K = T_C + 273.15$ (Equation 14.3).

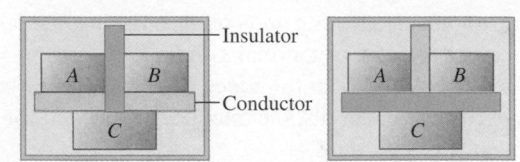

If A and B are each in thermal equilibrium with C … … then A and B are in thermal equilibrium with each other.

Thermal Expansion

(Section 14.3) The change ΔL in any linear dimension L_0 of a solid object with a temperature change ΔT is approximately $\Delta L = \alpha L_0 \Delta T$ (Equation 14.4), where α is the **coefficient of linear expansion.** The change ΔV in the volume V_0 of any solid or liquid material with a temperature change ΔT is approximately $\Delta V = \beta V_0 \Delta T$ (Equation 14.6), where β is the **coefficient of volume expansion.** For many solids, $\beta = 3\alpha$.

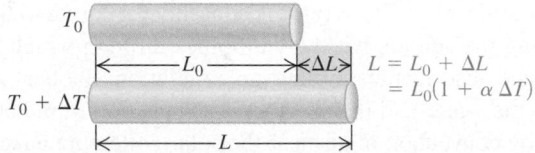

$$L = L_0 + \Delta L$$
$$= L_0(1 + \alpha \Delta T)$$

Heat, Phase Changes, and Calorimetry

(Sections 14.4–14.6) Heat is energy transferred from one object to another as a result of a temperature difference. The quantity of heat Q required to raise the temperature of a mass m of material by a small amount ΔT is $Q = mc\Delta T$ (Equation 14.10), where c is the **specific heat capacity** of the material.

To change a mass m of a material to a different phase (such as liquid to solid or liquid to vapor) at the same temperature requires the addition or subtraction of a quantity of heat Q given by $Q = \pm mL$ (Equation 14.11), where L is the **heat of fusion, heat of vaporization,** or **heat of sublimation.**

Calculations with heat are called **calorimetry calculations:** In an isolated system whose parts interact by heat exchange, the algebraic sum of the Q's for all parts of the system must be zero.

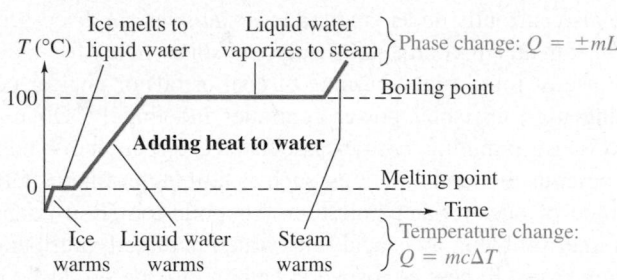

Heat Transfer

(Section 14.7) The three mechanisms of heat transfer are conduction, convection, and radiation. **Conduction** is transfer of energy of molecular motion within materials without bulk motion of the materials. **Convection** involves the motion of a mass from one region to another. **Radiation** is energy transfer through electromagnetic radiation.

The **heat current** H for conduction depends on the area A through which the heat flows, the length L of the heat path, the temperature difference $(T_H - T_C)$, and the thermal conductivity k of the material, according to the equation

$$H = kA\frac{T_H - T_C}{L}. \qquad (14.12)$$

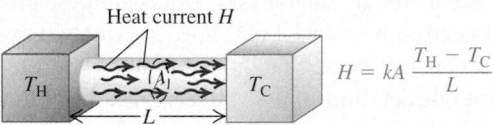

$$H = kA\frac{T_H - T_C}{L}$$

The heat current H due to radiation is given by $H = Ae\sigma T^4$ (Equation 14.13), where A is the surface area, e is the **emissivity** of the surface (a pure number between 0 and 1), T is the absolute (Kelvin) temperature, and σ is the Stefan–Boltzmann constant.

Continued

Solar Energy and Resource Conservation

(Section 14.8) A substantial fraction of all energy consumption in the United States is used to heat and cool homes and other buildings. Insulation in walls and multiple-layer glass windows can reduce heat flow by at least a third.

In a typical household solar heating system, large black plates facing the sun are backed with pipes through which water circulates. The black surface absorbs most of the sun's radiation; the heat is transferred by conduction to the water and then by forced convection to radiators inside the house.

 For instructor-assigned homework, go to www.masteringphysics.com

Conceptual Questions

1. When a block with a hole in it is heated, why doesn't the material around the hole expand into the hole and make it smaller? Explain this behavior by looking at the forces operating at the molecular level.

2. You have a drink that you want to cool off. You could put in 100 g of ice or 100 g of water, both at 0°C. Which one would be more effective in cooling your drink? Why?

3. A thermostat for controlling household heating systems often contains a bimetallic element consisting of two parallel strips of different metals welded together along their length. When the temperature changes, this composite strip bends one way or the other. Why?

4. Why is it sometimes possible to loosen caps on screw top bottles by dipping the capped bottle briefly into hot water?

5. Galileo supposedly designed a thermometer consisting of hollow metal spheres suspended in a liquid inside a vertical glass cylinder. At a low temperature, all the spheres were at the top of the liquid. As the temperature of the room (and hence of the liquid and the spheres) rose, the spheres, one by one, would descend from the top. Each sphere was marked with the temperature at which it descended. (Replicas of this thermometer are available today, one of which sits on the author's fireplace mantel.) Explain the physical principles behind the operation of this device. What do you suppose is different about the properties of the spheres?

6. Builders usually put steel bars (called *rebar*) inside of concrete slabs for reinforcement. Since the concrete and rebar are made of different materials, it might seem that they should expand at different rates when heated and that the rebar should actually *weaken* the concrete. So why does adding rebar strengthen concrete? What might happen if a metal other than steel were used?

7. To raise the temperature of an object, must you add heat to it? If you add heat to an object, must you raise its temperature? Explain.

8. You've probably seen what happens to lettuce or other vegetables if they freeze accidentally in the refrigerator. Freezing cells, as opposed to just making them cold, can kill them by causing them to burst. Why does a cell burst when frozen?

9. True or false? When two objects at different temperatures are placed in contact, the rise in temperature of the cooler object must equal the drop in temperature of the warmer one.

10. If you have wet hands and pick up a piece of metal that is below the freezing point of water, you may stick to it. This doesn't happen with wood. Why this difference in behavior?

11. If you add heat slowly to ice at 0°C, why doesn't the temperature of the ice increase? What becomes of this heat energy? Likewise, if you add heat to boiling water, the temperature remains at 100°C. What is happening to the heat you add?

12. In some household air conditioners used in dry climates, air is cooled by blowing it through a water-soaked filter, evaporating some of the water. How does this cool the air? Would such a system work well in a high-humidity climate? Why or why not?

13. A person pours a cup of hot coffee, intending to drink it 5 minutes later. To have it as hot as possible for drinking, should she put the cold milk into it now or wait until just before she drinks it? Explain your reasoning.

14. If you put your hand into boiling water at 212°F, you will immediately get a serious burn. Yet you readily reach into a much hotter 400°F oven without danger. Why the difference? (*Hint:* What are the significant differences between the properties of the air in the oven and the boiling water?)

15. You are going away for the weekend and plan to turn down your home thermostat to conserve energy. Your neighbor suggests that you not turn it down, because the walls and floors of your home will get cold and have to be reheated when you get back, which will take more energy than you save. Should you follow your neighbor's advice? Consider all three forms of heat transfer (conduction, convection, and radiation).

16. Why is snow, which is made up of ice crystals, a much better insulator than ice?

17. A cold block of metal feels colder than a block of wood at the same temperature. Why? A *hot* block of metal feels hotter than a block of wood at the same temperature. Again, why? Is there any temperature at which the two blocks feel equally hot or cold? What temperature is this?

18. Give some good reasons why the equation $H = e\sigma A T^4$ is valid only if the temperature is expressed in kelvins (rather than degrees Fahrenheit or Celsius).

Multiple-Choice Problems

1. If heat Q is required to increase the temperature of a metal object from 4°C to 6°C, the amount of heat necessary to increase its temperature from 6°C to 12°C is most likely
   A. Q. B. $2Q$. C. $3Q$. D. $4Q$.

2. A metal bar expands by 1.0 mm when its temperature is increased by 1.0 C°. Which of the following statements about this bar must be true? (There may be more than one correct choice.)
   A. It will expand by more than 1.0 mm if its temperature is increased by 1.0 K.
   B. It will expand by less than 1.0 mm if its temperature is increased by 1.0 F°.
   C. It will expand by 2.00 mm if its temperature is increased by 2.0 K.
   D. It will expand by more than 2.0 mm if its temperature is increased by 2.0 F°.

3. If an amount of heat Q is needed to increase the temperature of a solid metal sphere of diameter D from 4°C to 7°C, the amount of heat needed to increase the temperature of a solid sphere of diameter $2D$ of the same metal from 4°C to 7°C is
   A. Q. B. $2Q$. C. $4Q$. D. $8Q$.

4. If you mix 100 g of ice at 0°C with 100 g of boiling water at 100°C in a perfectly insulated container, the final stabilized temperature will be
   A. 0°C.   B. between 0°C and 50°C.
   C. 50°C.   D. between 50°C and 100°C.

5. A sphere radiates energy at a rate of 1.00 J/s when its temperature is 100°C. At what rate will it radiate energy if its temperature is increased to 200°C? (Neglect any heat transferred back into the sphere.)
   A. 1.27 J/s. B. 2.00 J/s. C. 2.59 J/s. D. 16.0 J/s.

6. Two identical closed, insulated containers hold equal amounts of water at a temperature of 50°C. One kilogram of ice at 0°C is added to the water in container A; one kilogram of steam at 100°C is added to container B. After all the ice has melted and all the steam has condensed, which container's temperature will have changed more?
   A. Container A   B. Container B
   C. Their temperatures change by the same amount.

7. The wall of a furnace conducts 1000 cal/min of heat out of the furnace and into the factory. If the wall's thickness is doubled, but nothing else changes (including the inner and outer surface temperatures), the amount of heat it conducts into the factory will be
   A. 4000 cal/min.   B. 2000 cal/min.
   C. 500 cal/min.   D. 250 cal/min.

8. A thin metal rod expands 1.5 mm when its temperature is increased by 2.0 C°. If an identical rod of the same material, except *twice as long*, also has its temperature increased by 2.0 C°, it will expand
   A. 0.75 mm. B. 1.0 mm. C. 3.0 mm. D. 6.0 mm.

9. Water has a specific heat capacity nearly 9 times that of iron. Suppose a 50-g pellet of iron at a temperature of 200°C is dropped into 50 g of water at a temperature of 20°C. When the system reaches thermal equilibrium, its temperature will be
   A. closer to 20°C.
   B. closer to 200°C.
   C. halfway between the two initial temperatures.

10. A cylindrical metal bar conducts heat at a rate R from a hot reservoir to a cold reservoir. If both its *length* and *diameter* are doubled, it will conduct heat at a rate
    A. R. B. $2R$. C. $4R$. D. $8R$.

11. Two rods P and Q of identical shape and size, but made of different metals, are joined end to end. The left end of rod P is kept at 100°C while the right end of rod Q is at 0°C, causing heat to flow down the rods. The graph in Figure 14.25 shows the temperature distribution along the rods from the hot end to the cold end. Which rod has the higher thermal conductivity?
    A. rod P.
    B. rod Q.
    C. We cannot tell from the available data.

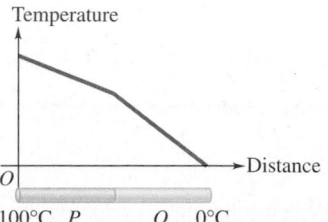

▲ **FIGURE 14.25** Multiple-choice problem 11.

12. The thermal conductivity of concrete is $0.80 \text{ W}/(\text{m} \cdot \text{K})$, and the thermal conductivity of a certain wood is $0.10 \text{ W}/(\text{m} \cdot \text{K})$. How thick would a solid concrete wall have to be to have the same rate of heat flow as an 8.0-cm-thick solid wall of this wood? Both walls have the same surface area and the same interior and exterior temperatures.
    A. 1.0 cm. B. 16 cm. C. 32 cm. D. 64 cm.

13. The graph in Figure 14.26 shows the temperature as a function of time for a sample of material being heated at a constant rate. What segment or segments of this graph show the sample existing in two phases (or states)?
    A. OA.
    B. AB.
    C. OA and BC.

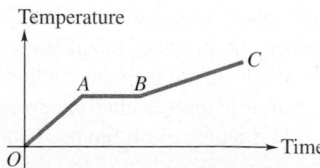

▲ **FIGURE 14.26** Multiple-choice problem 13.

14. For the sample in the previous question, what segment of the graph shows the specific heat capacity of the sample to be *greatest*?
    A. OA. B. AB. C. BC.

15. You enter a cold room containing a wooden table and a steel table; both have been there for a long time. Why does the steel table feel much colder than the wooden one?
    A. Steel is a much better conductor of heat than wood is.
    B. The steel stores much more cold than the wood does.
    C. Steel has a higher specific heat capacity than wood has.
    D. The temperature of the steel table is lower than that of the wooden table.

Problems

14.2 Temperature Scales

1. • (a) While vacationing in Europe, you feel sick and are told that you have a temperature of 40.2°C. Should you be concerned? What is your temperature in °F? (b) The morning weather report in Sydney predicts a high temperature of 12°C. Will you need to bring a jacket? What is this temperature in °F? (c) A friend has suggested that you go swimming in a pool having water of temperature 350 K. Is this safe to do? What would this temperature be on the Fahrenheit and Celsius scales?

2. • **Temperatures in biomedicine.** (a) **Normal body temperature.** The average normal body temperature measured in the mouth is 310 K. What would Celsius and Fahrenheit thermometers read for this temperature? (b) **Elevated body temperature.** During very vigorous exercise, the body's temperature can go as high as 40°C. What would Kelvin and Fahrenheit thermometers read for this temperature? (c) **Temperature difference in the body.** The surface temperature of the body is normally about 7 C° lower than the internal temperature. Express this temperature difference in kelvins and in Fahrenheit degrees. (d) **Blood storage.** Blood stored at 4.0°C lasts safely for about 3 weeks, whereas blood stored at −160°C lasts for 5 years. Express both temperatures on the Fahrenheit and Kelvin scales. (e) **Heat stroke.** If the body's temperature is above 105°F for a prolonged period, heat stroke can result. Express this temperature on the Celsius and Kelvin scales.

3. • (a) On January 22, 1943, the temperature in Spearfish, South Dakota, rose from −4.0°F to 45.0°F in just 2 minutes. What was the temperature change in Celsius degrees and in kelvins? (b) The temperature in Browning, Montana, was 44.0°F on January 23, 1916, and the next day it plummeted to −56.0°F. What was the temperature change in Celsius degrees and in kelvins?

4. • **Inside the earth and the sun.** (a) Geophysicists have estimated that the temperature at the center of the earth's core is 5000°C (or more), while the temperature of the sun's core is about 15 million K. Express both of these temperatures in Fahrenheit degrees.

5. •• (a) At what temperature do the Fahrenheit and Celsius scales coincide? (b) Is there any temperature at which the Kelvin and Celsius scales coincide?

6. • Convert the following Kelvin temperatures to the Celsius and Fahrenheit scales: (a) the midday temperature at the surface of the moon (400 K); (b) the temperature at the tops of the clouds in the atmosphere of Saturn (95 K); (c) the temperature at the center of the sun $(1.55 \times 10^7 \text{ K})$.

14.3 Thermal Expansion

7. • The Eiffel Tower in Paris is 984 ft tall and is made mostly of steel. If this is its height in winter when its temperature is −8.00°C, how much additional vertical distance must you cover if you decide to climb it during a summer heat wave when its temperature is 40.0°C? (b) Express the coefficient of linear expansion of steel in terms of Fahrenheit degrees.

8. • A steel bridge is built in the summer when its temperature is 35.0°C. At the time of construction, its length is 80.00 m. What is the length of the bridge on a cold winter day when its temperature is −12.0°C?

9. • A metal rod is 40.125 cm long at 20.0°C and 40.148 cm long at 45.0°C. Calculate the average coefficient of linear expansion of the rod's material for this temperature range.

10. • (a) Steel train rails are laid in 12.0-m-long segments placed end to end. The rails are laid on a winter day when their temperature is −2.00°C. How much space must be left between adjacent rails if they are just to touch on a summer day when their temperature is 33.0°C? (b) If the rails are mistakenly laid in contact with each other, what is the stress in them on a summer day when their temperature is 33.0°C?

11. • An underground tank with a capacity of 1700 L (1.70 m³) is completely filled with ethanol that has an initial temperature of 19.0°C. After the ethanol has cooled off to the temperature of the tank and ground, which is 10.0°C, how much air space will there be above the ethanol in the tank? (See Table 14.2, and assume that the volume of the tank doesn't change appreciably.)

12. • A copper cylinder is initially at 20.0°C. At what temperature will its volume be 0.150% larger than it is at 20.0°C?

13. •• A geodesic dome constructed with an aluminum framework is a nearly perfect hemisphere; its diameter measures 55.0 m on a winter day at a temperature of −15°C. How much more interior space does the dome have in the summer, when the temperature is 35°C?

14. •• The outer diameter of a glass jar and the inner diameter of its iron lid are both 725 mm at room temperature (20.0°C). What will be the size of the mismatch between the lid and the jar if the lid is briefly held under hot water until its temperature rises to 50.0°C, without changing the temperature of the glass?

15. •• A glass flask whose volume is 1000.00 cm³ at 0.0°C is completely filled with mercury at this temperature. When flask and mercury are warmed to 55.0°C, 8.95 cm³ of mercury overflow. Compute the coefficient of volume expansion of the glass. (Consult Table 14.2.)

16. •• **Ensuring a tight fit.** Aluminum rivets used in airplane construction are made slightly larger than the rivet holes and cooled by "dry ice" (solid CO_2) before being driven. If the diameter of a hole is 4.500 mm, what should be the diameter of a rivet at 23.0°C, if its diameter is to equal that of the hole when the rivet is cooled to −78.0°C, the temperature of dry ice? Assume that the expansion coefficient remains constant at the value given in Table 14.1.

17. •• The markings on an aluminum ruler and a brass ruler begin at the left end; when the rulers are at 0.00°C, they are perfectly aligned. How far apart will the 20.0 cm marks be on the two rulers at 100.0°C if the left-hand ends are kept precisely aligned?

14.4 Quantity of Heat

18. • (a) How much heat is required to raise the temperature of 0.250 kg of water from 20.0°C to 30.0°C? (b) If this amount of heat is added to an equal mass of mercury that is initially at 20.0°C, what is its final temperature?

19. • One of the moving parts of an engine contains 1.60 kg of aluminum and 0.300 kg of iron and is designed to operate at 210°C. How much heat is required to raise its temperature from 20.0° to 210°C?

20. • In an effort to stay awake for an all-night study session, a student makes a cup of coffee by first placing a 200.0 W electric immersion heater in 0.320 kg of water. (a) How much heat must be added to the water to raise its temperature from 20.0°C to 80.0°C? (b) How much time is required if all of the heater's power goes into heating the water?

21. • **Heat loss during breathing.** In very cold weather, a signif-
BIO icant mechanism for heat loss by the human body is energy
expended in warming the air taken into the lungs with each
breath. (a) On a cold winter day when the temperature is
$-20°C$, what is the amount of heat needed to warm to internal
body temperature $(37°C)$ the 0.50 L of air exchanged with
each breath? Assume that the specific heat capacity of air is
$1020\,\mathrm{J/(kg \cdot K)}$ and that 1.0 L of air has a mass of 1.3 g.
(b) How much heat is lost per hour if the respiration rate is
20 breaths per minute?

22. • A nail driven into a board increases in temperature. If 60%
of the kinetic energy delivered by a 1.80 kg hammer with a
speed of 7.80 m/s is transformed into heat that flows into the
nail and does not flow out, what is the increase in temperature
of an 8.00 g aluminum nail after it is struck 10 times?

23. • You are given a sample of metal and asked to determine its
specific heat. You weigh the sample and find that its weight is
28.4 N. You carefully add $1.25 \times 10^4\,$J of heat energy to the
sample and find that its temperature rises 18.0 C°. What is the
sample's specific heat?

24. •• A 25,000-kg subway train initially traveling at 15.5 m/s
slows to a stop in a station and then stays there long enough
for its brakes to cool. The station's dimensions are 65.0 m long
by 20.0 m wide by 12.0 m high. Assuming all the work done
by the brakes in stopping the train is transferred as heat uniformly to all the air in the station, by how much does the air
temperature in the station rise? Take the density of the air to be
$1.20\,\mathrm{kg/m^3}$ and its specific heat to be $1020\,\mathrm{J/(kg \cdot K)}$.

25. •• You add 8950 J of heat to 3.00 mol of iron. (a) What is the
temperature increase of the iron? (b) If this same amount of heat is
added to 3.00 kg of iron, what is the iron's temperature increase?
(c) Explain the difference in your results for parts (a) and (b).

26. • From a height of 35.0 m, a 1.25 kg bird dives (from rest)
into a small fish tank containing 50.0 kg of water. What is the
maximum rise in temperature of the water if the bird gives it
all of its mechanical energy?

27. •• A 15.0 g bullet traveling horizontally at 865 m/s passes
through a tank containing 13.5 kg of water and emerges with a
speed of 534 m/s. What is the maximum temperature increase
that the water could have as a result of this event?

28. •• **Maintaining body temperature.** While running, a 70 kg
BIO student generates thermal energy at a rate of 1200 W. To maintain a constant body temperature of 37°C, this energy must be
removed by perspiration or other mechanisms. If these mechanisms failed and the heat could not flow out of the student's
body, for what amount of time could a student run before irreversible body damage occurred? (Protein structures in the
body are damaged irreversibly if the body temperature rises to
44°C or above. The specific heat capacity of a typical human
body is $3480\,\mathrm{J/(kg \cdot K)}$, slightly less than that of water. The
difference is due to the presence of protein, fat, and minerals,
which have lower specific heat capacities.)

29. •• A technician measures the specific heat capacity of an unidentified liquid by immersing an electrical resistor in it. Electrical
energy is converted to heat, which is then transferred to the liquid
for 120 s at a constant rate of 65.0 W. The mass of the liquid is
0.780 kg, and its temperature increases from 18.55°C to 22.54°C.
(a) Find the average specific heat capacity of the liquid in this temperature range. Assume that negligible heat is transferred to the

container that holds the liquid and that no heat is lost to the surroundings. (b) Suppose that in this experiment heat transfer from
the liquid to the container or its surroundings cannot be ignored.
Is the result calculated in part (a) an overestimate or an underesti-
mate of the average specific heat capacity? Explain.

30. •• Much of the energy of falling water in a waterfall is converted into heat. If all the mechanical energy is converted into
heat that stays in the water, how much of a rise in temperature
occurs in a 100 m waterfall?

14.5 Phase Changes

31. • Consult Table 14.4. (a) How much heat is required to melt
0.150 kg of lead at 327.3°C? (b) How much heat would be
needed to evaporate this lead at 1750°C? (c) If the total heat
added from parts (a) and (b) were put into ice at 0.00°C, how
many grams of the ice would it melt?

32. •• A blacksmith cools a 1.20-kg chunk of iron, initially at a
temperature of 650.0°C, by trickling 15.0°C water over it. All
the water boils away, and the iron ends up at a temperature of
120.0°C. How much water did the blacksmith trickle over the iron?

33. • **Treatment for a stroke.** One suggested treatment for a
BIO person who has suffered a stroke is to immerse the patient in
an ice-water bath at 0°C to lower the body temperature,
which prevents damage to the brain. In one set of tests,
patients were cooled until their internal temperature reached
32.0°C. To treat a 70.0 kg patient, what is the minimum
amount of ice (at 0°C) that you need in the bath so that its
temperature remains at 0°C? The specific heat capacity of the
human body is $3480\,\mathrm{J/(kg \cdot C°)}$, and recall that normal body
temperature is 37.0°C.

34. • A container holds 0.550 kg of ice at $-15.0°C$. The mass of
the container can be ignored. Heat is supplied to the container
at the constant rate of 800.0 J/min for 500.0 min. (a) After how
many minutes does the ice *start* to melt? (b) After how many
minutes, from the time when the heating is first started, does
the temperature begin to rise above 0.00°C? (c) Plot a curve
showing the temperature as a function of the time elapsed.

35. •• An asteroid with a diameter of 10 km and a mass of
$2.60 \times 10^{15}\,$kg impacts the earth at a speed of 32.0 km/s,
landing in the Pacific Ocean. If 1.00% of the asteroid's kinetic
energy goes to boiling the ocean water (assume an initial water
temperature of 10.0°C), what mass of water will be boiled
away by the collision? (For comparison, the mass of water
contained in Lake Superior is about $2 \times 10^{15}\,$kg.)

36. • **Evaporative cooling.** The evaporation of sweat is an
BIO important mechanism for temperature control in some warm-
blooded animals. (a) What mass of water must evaporate
from the skin of a 70.0 kg man to cool his body 1.00 C°? The
heat of vaporization of water at body temperature (37°C) is
$2.42 \times 10^6\,$J/kg. The specific heat capacity of a typical human
body is $3480\,\mathrm{J/(kg \cdot K)}$. (b) What volume of water must the
man drink to replenish the evaporated water? Compare this
result with the volume of a soft-drink can, which is 355 cm³.

37. •• An ice-cube tray contains 0.350 kg of water at 18.0°C. How
much heat must be removed from the water to cool it to 0.00°C
and freeze it? Express your answer in joules and in calories.

38. •• How much heat is required to convert 12.0 g of ice at
$-10.0°C$ to steam at 100.0°C? Express your answer in joules
and in calories.

39. •• **Steam burns vs. water burns.** What is the amount of
BIO heat entering your skin when it receives the heat released
(a) by 25.0 g of steam initially at 100.0°C that cools to
34.0°C? (b) by 25.0 g of water initially at 100.0°C that cools
to 34.0°C? (c) What do these results tell you about the relative
severity of steam and hot-water burns?

40. •• **Bicycling on a warm day.** If the air temperature is the
BIO same as the temperature of your skin (about 30°C), your body
cannot get rid of heat by transferring it to the air. In that case, it
gets rid of the heat by evaporating water (sweat). During bicycling, a typical 70 kg person's body produces energy at a rate
of about 500 W due to metabolism, 80% of which is converted
to heat. (a) How many kilograms of water must the person's
body evaporate in an hour to get rid of this heat? The heat of
vaporization of water at body temperature is 2.42×10^6 J/kg.
(b) The evaporated water must, of course, be replenished, or
the person will dehydrate. How many 750 mL bottles of water
must the bicyclist drink per hour to replenish the lost water?
(Recall that the mass of a liter of water is 1.0 kg.)

41. •• **Overheating.** (a) By how much would the body tempera-
BIO ture of the bicyclist in the previous problem increase in an
hour if he were unable to get rid of the excess heat? (b) Is this
temperature increase large enough to be serious? To find out,
how high a fever would it be equivalent to, in °F? [Recall that
the normal internal body temperature is 98.6°F and the specific
heat capacity of the body is 3480 J/(kg · C°).]

14.6 Calorimetry

42. • You have 750 g of water at 10.0°C in a large insulated beaker.
How much boiling water at 100.0°C must you add to this beaker
so that the final temperature of the mixture will be 75°C?

43. •• A 0.500 kg chunk of an unknown metal that has been in
boiling water for several minutes is quickly dropped into an
insulating Styrofoam™ beaker containing 1.00 kg of water at
room temperature (20.0°C). After waiting and gently stirring
for 5.00 minutes, you observe that the water's temperature has
reached a constant value of 22.0°C. (a) Assuming that the Sty-
rofoam™ absorbs a negligibly small amount of heat and that
no heat was lost to the surroundings, what is the specific heat
capacity of the metal? (b) Which is more useful for storing
energy from heat, this metal or an equal weight of water?
Explain. (c) What if the heat absorbed by the Styrofoam™
actually is not negligible. How would the specific heat capacity you calculated in part (a) be in error? Would it be too large,
too small, or still correct? Explain your reasoning.

44. •• A copper pot with a mass of 0.500 kg contains 0.170 kg of
water, and both are at a temperature of 20.0°C. A 0.250 kg block
of iron at 85.0°C is dropped into the pot. Find the final temperature of the system, assuming no heat loss to the surroundings.

45. •• In a physics lab experiment, a student immersed 200 one-cent
coins (each having a mass of 3.00 g) in boiling water. After they
reached thermal equilibrium, she quickly fished them out and
dropped them into 0.240 kg of water at 20.0°C in an insulated
container of negligible mass. What was the final temperature of
the coins? [One-cent coins are made of a metal alloy—mostly
zinc—with a specific heat capacity of 390 J/(kg · K).]

46. •• A laboratory technician drops an 85.0 g solid sample of
unknown material at a temperature of 100.0°C into a calorimeter. The calorimeter can is made of 0.150 kg of copper and contains 0.200 kg of water, and both the can and water are initially

at 19.0°C. The final temperature of the system is measured to be
26.1°C. Compute the specific heat capacity of the sample.
(Assume no heat loss to the surroundings.)

47. •• A 4.00 kg silver ingot is taken from a furnace, where its
temperature is 750°C, and placed on a very large block of ice
at 0.00°C. Assuming that all the heat given up by the silver is
used to melt the ice and that not all the ice melts, how much
ice is melted?

48. •• An insulated beaker with negligible mass contains 0.250 kg
of water at a temperature of 75.0°C. How many kilograms of
ice at a temperature of −20.0°C must be dropped in the water
so that the final temperature of the system will be 30.0°C?

49. •• A Styrofoam™ bucket of negligible mass contains 1.75 kg
of water and 0.450 kg of ice. More ice, from a refrigerator at
−15.0°C, is added to the mixture in the bucket, and when thermal equilibrium has been reached, the total mass of ice in the
bucket is 0.778 kg. Assuming no heat exchange with the surroundings, what mass of ice was added?

14.7 Heat Transfer

50. • A slab of a thermal insulator with a cross-sectional area of
100 cm^2 is 3.00 cm thick. Its thermal conductivity is
0.075 W/(m · K). If the temperature difference between opposite faces is 80 C°, how much heat flows through the slab in
1 day?

51. •• You are asked to design a cylindrical steel rod 50.0 cm long,
with a circular cross section, that will conduct 150.0 J/s from
a furnace at 400.0°C to a container of boiling water under
1 atmosphere of pressure. What must the rod's diameter be?

52. • **Conduction through the skin.** The blood plays an impor-
BIO tant role in removing heat from the body by bringing this heat
directly to the surface where it can radiate away. Nevertheless,
this heat must still travel through the skin before it can radiate
away. We shall assume that the blood is brought to the bottom
layer of skin at a temperature of 37°C and that the outer surface
of the skin is at 30.0°C. Skin varies in thickness from 0.50 mm
to a few millimeters on the palms and soles, so we shall assume
an average thickness of 0.75 mm. A 165 lb, 6 ft person has a
surface area of about 2.0 m^2 and loses heat at a net rate of 75 W
while resting. On the basis of our assumptions, what is the
thermal conductivity of this person's skin?

53. •• A pot with a steel bottom 8.50 mm thick rests on a hot
stove. The area of the bottom of the pot is 0.150 m^2. The water
inside the pot is at 100.0°C, and 0.390 kg are evaporated every
3.00 min. Find the temperature of the lower surface of the pot,
which is in contact with the stove.

54. •• A carpenter builds an exterior house wall with a layer of
wood 3.0 cm thick on the outside and a layer of Styrofoam™
insulation 2.2 cm thick on the inside wall surface. The wood
has a thermal conductivity of 0.080 W/(m · K), and the Sty-
rofoam™ has a thermal conductivity of 0.010 W/(m · K).
The interior surface temperature is 19.0°C, and the exterior
surface temperature is −10.0°C. (a) What is the temperature at
the plane where the wood meets the Styrofoam™? (b) What is
the rate of heat flow per square meter through this wall?

55. •• A picture window has dimensions of 1.40 m × 2.50 m and
is made of glass 5.20 mm thick. On a winter day, the outside
temperature is −20.0°C, while the inside temperature is a comfortable 19.56°C. (a) At what rate is heat being lost through the

window by conduction? (b) At what rate would heat be lost through the window if you covered it with a 0.750-mm-thick layer of paper (thermal conductivity 0.0500 W/(m · K))?

56. •• One end of an insulated metal rod is maintained at 100°C, while the other end is maintained at 0°C by an ice–water mixture. The rod is 60.0 cm long and has a cross-sectional area of 1.25 cm². The heat conducted by the rod melts 8.50 g of ice in 10.0 min. Find the thermal conductivity k of the metal.

57. •• **Mammal insulation.** Animals in cold climates often
BIO depend on *two* layers of insulation: a layer of body fat [of thermal conductivity 0.20 W/(m · K)] surrounded by a layer of air trapped inside fur or down. We can model a black bear (*Ursus americanus*) as a sphere 1.5 m in diameter having a layer of fat 4.0 cm thick. (Actually, the thickness varies with the season, but we are interested in hibernation, when the fat layer is thickest.) In studies of bear hibernation, it was found that the outer surface layer of the fur is at 2.7°C during hibernation, while the inner surface of the fat layer is at 31.0°C. (a) What is the temperature at the fat–inner fur boundary, and (b) how thick should the air layer (contained within the fur) be so that the bear loses heat at a rate of 50.0 W?

58. • A box-shaped wood stove has dimensions of 0.75 m × 1.2 m × 0.40 m, an emissivity of 0.85, and a surface temperature of 205°C. Calculate its rate of radiation into the surrounding space.

59. • **Radiation by the body.** The amount of heat radiated by the
BIO body depends on its surface temperature and area. Typically, this temperature is about 30°C (although it can vary). The surface area depends on the person's height and weight. An empirical formula for the surface area of a person's body is

$$A(\text{in m}^2) = (0.202)M^{0.425} h^{0.725},$$

where M is the person's mass (in kilograms) and h is his or her height (in meters). (a) What would be the surface area of a 165 lb (75 kg), 6 ft (1.83 m) person? (This is a good chance to use the y^x key of your calculator.) (b) How much heat would the person radiate away per second at a skin temperature of 30°C? (At the low temperatures of room-temperature objects, nearly all the heat radiated is infrared radiation, for which the emissivity is essentially 1, regardless of the amount of skin pigment.) (c) How much *net* heat would radiation remove from the person's body if the air temperature is 20°C? (d) Take measurements on your own body to test the validity of the area formula. Treat yourself as a sphere and several cylinders.

60. • **How large is the sun?** By measuring the spectrum of wavelengths of light from our sun, we know that its surface temperature is 5800 K. By measuring the rate at which we receive its energy on earth, we know that it is radiating a total of 3.92×10^{26} J/s and behaves nearly like an ideal blackbody. Use this information to calculate the diameter of our sun.

61. •• **Basal metabolic rate.** The basal metabolic rate is the rate
BIO at which energy is produced in the body when a person is at rest. A 75 kg (165 lb) person of height 1.83 m (6 ft) would have a body surface area of approximately 2.0 m². (a) What is the net amount of heat this person could radiate per second into a room at 18°C (about 65°F) if his skin's surface temperature is 30°C? (At such temperatures, nearly all the heat is infrared radiation, for which the body's emissivity is 1.0, regardless of the amount of pigment.) (b) Normally, 80% of

the energy produced by metabolism goes into heat, while the rest goes into things like pumping blood and repairing cells. Also normally, a person at rest can get rid of this excess heat just through radiation. Use your answer to part (a) to find this person's basal metabolic rate.

62. • The emissivity of tungsten is 0.35. A tungsten sphere with a radius of 1.50 cm is suspended within a large evacuated enclosure whose walls are at 290 K. What power input is required to maintain the sphere at a temperature of 3000 K if heat conduction along the supports is negligible?

63. • **Size of a lightbulb filament.** The operating temperature of a tungsten filament in an incandescent lightbulb is 2450 K, and its emissivity is 0.35. Find the surface area of the filament of a 150 W bulb if all the electrical energy consumed by the bulb is radiated by the filament as light. (In reality, only a small fraction of the radiation appears as visible light.)

64. •• A spherical pot of hot coffee contains 0.75 L of liquid (essentially water) at an initial temperature of 95°C. The pot has an emissivity of 0.60, and the surroundings are at a temperature of 20.0°C. Calculate the coffee's rate of heat loss by radiation.

General Problems

65. •• An 8.50 kg block of ice at 0°C is sliding on a rough horizontal icehouse floor (also at 0°C) at 15.0 m/s. Assume that half of any heat generated goes into the floor and the rest goes into the ice. (a) How much ice melts after the speed of the ice has been reduced to 10.0 m/s? (b) What is the maximum amount of ice that will melt?

66. •• Use Fig. 14.9 to find the approximate coefficient of volume expansion of water at 2.0°C and at 8.0°C.

67. •• **Global warming.** As the earth warms, sea level will rise due to melting of the polar ice and thermal expansion of the oceans. Estimates of the expected temperature increase vary, but 3.5 C° by the end of the century has been plausibly suggested. If we assume that the temperature of the oceans also increases by this amount, how much will sea level rise by the year 2100 due only to the thermal expansion of the water? Assume, reasonably, that the ocean basins do not expand appreciably. The average depth of the ocean is 4000 m, and the coefficient of volume expansion of water at 20°C is $0.207 \times 10^{-3} (\text{C}°)^{-1}$.

68. •• A Foucault pendulum consists of a brass sphere with a diameter of 35.0 cm suspended from a steel cable 10.5 m long (both measurements made at 20.0°C). Due to a design oversight, the swinging sphere clears the floor by a distance of only 2.00 mm when the temperature is 20.0°C. At what temperature will the sphere begin to brush the floor?

69. •• **On-demand water heaters.** Conventional hot-water heaters consist of a tank of water maintained at a fixed temperature. The hot water is to be used when needed. The drawback is that energy is wasted because the tank loses heat when it is not in use, and you can run out of hot water if you use too much. Some utility companies are encouraging the use of *on-demand* water heaters (also known as *flash heaters*), which consist of heating units to heat the water as you use it. No water tank is involved, so no heat is wasted. A typical household shower flow rate is 2.5 gal/min (9.46 L/min) with the tap water

being heated from 50°F (10°C) to 120°F (49°C) by the on-demand heater. What rate of heat input (either electrical or from gas) is required to operate such a unit, assuming that all the heat goes into the water?

70. •• **Burning fat by exercise.** Each pound of fat contains
BIO 3500 *food* calories. When the body metabolizes food, 80% of this energy goes to heat. Suppose you decide to run without stopping, an activity that produces 1290 W of metabolic power for a typical person. (a) For how many hours must you run to burn up 1 lb of fat? Is this a realistic exercise plan? (b) If you followed your planned exercise program, how much heat would your body produce when you burn up a pound of fat? (c) If you needed to get rid of all of this excess heat by evaporating water (i.e., sweating), how many liters would you need to evaporate? The heat of vaporization of water at body temperature is 2.42×10^6 J/kg.

71. •• **Shivering.** You have no doubt noticed that you usually
BIO shiver when you get out of the shower. Shivering is the body's way of generating heat to restore its internal temperature to the normal 37°C, and it produces approximately 290 W of heat power per square meter of body area. A 68 kg (150 lb), 1.78 m (5 foot, 10 inch) person has approximately 1.8 m² of surface area. How long would this person have to shiver to raise his or her body temperature by 1.0 C°, assuming that none of this heat is lost by the body? The specific heat capacity of the body is about 3500 J/(kg · K).

72. •• A steel ring with a 2.5000-in. inside diameter at 20.0°C is to be warmed and slipped over a brass shaft with a 2.5020-in. outside diameter at 20.0°C. (a) To what temperature should the ring be warmed? (b) If the ring and the shaft together are cooled by some means such as liquid air, at what temperature will the ring just slip off the shaft?

73. •• **Pasta time!** You are making pesto for your pasta and have a cylindrical measuring cup 10.0 cm high made of ordinary glass $(\beta = 2.7 \times 10^{-5}(C°)^{-1})$ and that is filled with olive oil $(\beta = 6.8 \times 10^{-4}(C°)^{-1})$ to a height of 1.00 mm below the top of the cup. Initially, the cup and oil are at a kitchen temperature of 22.0°C. You get a phone call and forget about the olive oil, which you inadvertently leave on the hot stove. The cup and oil heat up slowly and have a common temperature. At what temperature will the olive oil start to spill out of the cup?

74. •• A copper calorimeter can with mass 0.100 kg contains 0.160 kg of water and 0.018 kg of ice in thermal equilibrium at atmospheric pressure. If 0.750 kg of lead at a temperature of 255°C is dropped into the can, what is the final temperature of the system if no heat is lost to the surroundings?

75. •• A piece of ice at 0°C falls from rest into a lake whose temperature is 0°C, and 1.00% of the ice melts. Compute the minimum height from which the ice has fallen.

76. •• **Hot air in a physics lecture.** (a) A typical student listening attentively to a physics lecture has a heat output of 100 W. How much heat energy does a class of 90 physics students release into a lecture hall over the course of a 50 min lecture? (b) Assume that all the heat energy in part (a) is transferred to the 3200 m³ of air in the room. The air has a specific heat capacity of 1020 J/(kg · K) and a density of 1.20 kg/m³. If none of the heat escapes and the air-conditioning system is off, how much will the temperature of the air in the room rise during the 50 min lecture? (c) If the class is taking an exam, the heat output per student rises to 280 W. What is the temperature rise during 50 min in this case?

77. •• **"The Ship of the Desert."** Camels require very little water because they are able to tolerate relatively large changes in their body temperature. While humans keep their body temperatures constant to within one or two Celsius degrees, a dehydrated camel permits its body temperature to drop to 34.0°C overnight and rise to 40.0°C during the day. To see how effective this mechanism is for saving water, calculate how many liters of water a 400-kg camel would have to drink if it attempted to keep its body temperature at a constant 34.0°C by evaporation of sweat during the day (12 hours) instead of letting it rise to 40.0°C. (*Note:* The specific heat of a camel or other mammal is about the same as that of a typical human, 3480 J/(kg · K). The heat of vaporization of water at 34°C is 2.42×10^6 J/kg.)

78. •• A worker pours 1.250 kg of molten lead at a temperature of 327.3°C into 0.5000 kg of water at a temperature of 75.00°C in an insulated bucket of negligible mass. Assuming no heat loss to the surroundings, calculate the mass of lead and water remaining in the bucket when the materials have reached thermal equilibrium.

79. •• **Time for a lake to freeze over.** When the air temperature is below 0°C, the water at the surface of a lake freezes to form a sheet of ice. If the upper surface of an ice sheet 25.0 cm thick is at -10.0°C and the bottom surface is at 0.00°C, calculate the time it will take to add 2.0 mm to the thickness of this sheet.

80. •• **Jogging in the heat of the day.** You have probably seen
BIO people jogging in extremely hot weather and wondered "Why?" As we shall see, there are good reasons not to do this! When jogging strenuously, an average runner of mass 68 kg and surface area 1.85 m² produces energy at a rate of up to 1300 W, 80% of which is converted to heat. The jogger radiates heat, but actually absorbs more from the hot air than he radiates away. At such high levels of activity, the skin's temperature can be elevated to around 33°C instead of the usual 30°C. (We shall neglect conduction, which would bring even more heat into his body.) The only way for the body to get rid of this extra heat is by evaporating water (sweating). (a) How much heat per second is produced just by the act of jogging? (b) How much *net* heat per second does the runner gain just from radiation if the air temperature is 40.0°C (104°F)? (Remember that he radiates out, but the environment radiates back in.) (c) What is the *total* amount of excess heat this runner's body must get rid of per second? (d) How much water must the jogger's body evaporate every minute due to his activity? The heat of vaporization of water at body temperature is 2.42×10^6 J/kg. (e) How many 750 mL bottles of water must he drink after (or preferably before!) jogging for a half hour? Recall that a liter of water has a mass of 1.0 kg.

81. •• **Overheating while jogging.** (a) If the jogger in the previ-
BIO ous problem were not able to get rid of the excess heat, by how much would his body temperature increase above the normal 37°C in a half hour of jogging? The specific heat capacity for a human is about 3500 J/(kg · K). (b) How high a fever (in °F) would this temperature increase be equivalent to? Is the increase large enough to be of concern? (Recall that normal body temperature is 98.6°F.)

82. •• A thirsty nurse cools a 2.00 L bottle of a soft drink (mostly water) by pouring it into a large aluminum mug of mass 0.257 kg and adding 0.120 kg of ice initially at -15.0°C. If the

soft drink and mug are initially at 20.0°C, what is the final temperature of the system, assuming no heat losses?

83. •• One experimental method of measuring an insulating material's thermal conductivity is to construct a box of the material and measure the power input to an electric heater inside the box that maintains the interior at a measured temperature above the outside surface. Suppose that in such an apparatus a power input of 180 W is required to keep the interior surface of the box 65.0 C° (about 120 F°) above the temperature of the outer surface. The total area of the box is 2.18 m², and the wall thickness is 3.90 cm. Find the thermal conductivity of the material in SI units.

84. •• The icecaps of Greenland and Antarctica contain about 1.75% of the total water (by mass) on the earth's surface; the oceans contain about 97.5%, and the other 0.75% is mainly groundwater. Suppose the icecaps, currently at an average temperature of about −30°C, somehow slid into the ocean and melted. What would be the resulting temperature decrease of the ocean? Assume that the average temperature of ocean water is currently 5.00°C.

85. •• **The effect of urbanization on plant growth.** A study
BIO published in July 2004 indicated that temperature increases in urban areas in the eastern United States are causing plants to bud up to 7 days early compared with plants in rural areas just a few miles away, thereby disrupting biological cycles. Average temperatures in the urban areas were up to 3.5 C° higher than in the rural areas. By what percent will the radiated heat per square meter increase due to such a temperature difference if the rural temperature was 0°C on the average?

86. •• **Basal metabolic rate.** The energy output of an animal
BIO engaged in an activity is called the basal metabolic rate (BMR) and is a measure of the conversion of food energy into other forms of energy. A simple calorimeter to measure the BMR consists of an insulated box with a thermometer to measure the temperature of the air. The air has a density of 1.29 kg/m³ and a specific heat capacity of 1020 J/(kg·K). A 50.0 g hamster is placed in a calorimeter that contains 0.0500 m³ of air at room temperature. (a) When the hamster is running in a wheel, the temperature of the air in the calorimeter rises 1.8 C° per hour. How much heat does the running hamster generate in an hour? (Assume that all this heat goes into the air in the calorimeter. Neglect the heat that goes into the walls of the box and into the thermometer, and assume that no heat is lost to the surroundings.) (b) Assuming that the hamster converts seed into heat with an efficiency of 10% and that hamster seed has a food energy value of 24 J/g, how many grams of seed must the hamster eat per hour to supply the energy found in part (a)?

87. •• **A thermos for liquid helium.** A physicist uses a cylindrical metal can 0.250 m high and 0.090 m in diameter to store liquid helium at 4.22 K; at that temperature the heat of vaporization of helium is 2.09×10^4 J/kg. Completely surrounding the metal can are walls maintained at the temperature of liquid nitrogen, 77.3 K, with vacuum between the can and the surrounding walls. How much helium is lost per hour? The emissivity of the metal can is 0.200. The only heat transfer between the metal can and the surrounding walls is by radiation.

$I5$ Thermal Properties of Matter

Y ou can't see air molecules, but they're moving all around you. Did you know that in one day you are hit 10^{32} times by air molecules? And did you know that the average speed of an air molecule is about 1000 mi/h? Or that the impacts of these molecules on a surface are responsible for the pressure that air exerts on that surface?

To understand and relate such facts, we need to be able to use both *macroscopic* (bulk) and *microscopic* viewpoints. The **macroscopic** perspective deals with large-scale variables such as pressure, volume, temperature, and mass of substance. In contrast, the **microscopic** point of view investigates small-scale quantities such as the speeds, kinetic energies, momenta, and masses of the individual molecules that make up a substance. Where these two perspectives overlap, they must be consistent. For example, the (microscopic) collision forces that occur when air molecules strike a surface (such as your skin) cause (macroscopic) atmospheric pressure.

In this chapter, we'll use both macroscopic and microscopic approaches to gain an understanding of the thermal properties and behavior of matter. One of the simplest kinds of matter to understand is the *ideal gas*. For this class of materials, we'll relate pressure, volume, temperature, and amount of substance to one another and to the speeds and masses of individual molecules. We will explore the molecular basis of the heat capacities of both gases and solids, and we will take a look at the various *phases* of matter—gas, liquid, and solid—and the conditions under which each occurs.

As we'll learn in this chapter, unconfined gases expand when heated. When hot ground heats the overlying air, the air expands and rises, forming a *thermal*. Thermals can lift soaring birds to great heights.

15.1 The Mole and Avogadro's Number

In our study of mechanics, we've usually described a quantity of a material or substance in terms of the *mass* of material. In this chapter, where we explore the relation between mechanics and thermal phenomena, we'll often find it useful to

describe a quantity of material in terms of the number of *moles* of the material rather than the mass. So we'll begin by discussing what we mean by a mole.

One **mole** of any pure chemical element or compound contains **a definite number of molecules**—the same number for all elements and compounds. The official SI definition of the mole is as follows:

Definition of the mole
One mole (1 mol) is the amount of substance that contains as many elementary entities as there are atoms in 0.012 kilogram of carbon 12.

In our discussion, the "elementary entities" are molecules. (In a monatomic substance such as carbon or helium, each molecule is a single atom, but in this discussion we'll still call it a molecule.)

The number of molecules in a mole is called **Avogadro's number,** denoted by N_A. The numerical value is

$$N_A = 6.022 \times 10^{23} \text{ molecules/mol.}$$

Figure 15.1 shows 1 mole of each of several familiar substances.

The **molar mass** M of a substance is the mass of 1 mole; this is equal to the mass m of a single molecule, multiplied by Avogadro's number N_A:

$$M = N_A m. \tag{15.1}$$

Molar mass is sometimes called "molecular mass" or "molecular weight." Both these terms are unfortunate, the first because M is *not* the mass of a molecule, the second because we are talking about mass, not weight. In this chapter, we'll use "molar mass" consistently.

We'll denote the number of moles of substance by n. In this chapter, we'll usually denote the total mass of substance by m_{total} (because later we'll want to use m as the mass of a single molecule). The total mass is equal to the number of moles multiplied by the mass per mole:

$$m_{total} = nM. \tag{15.2}$$

▲ **FIGURE 15.1** One mole each of several familiar substances.

EXAMPLE 15.1 Mass of a hydrogen atom

Find the mass of a single hydrogen atom and the mass of an oxygen molecule (consisting of two oxygen atoms and written O_2).

SOLUTION

SET UP From the periodic table of the elements (Appendix D), we find that the molar mass of hydrogen (i.e., the mass per mole of hydrogen atoms) is 1.008 g/mol. The molar mass of oxygen is 16.0 g/mol, so the molar mass of O_2 is 32.0 g/mol.

SOLVE We use Equation 15.1. The mass m_H of a single hydrogen atom is

$$m_H = \frac{M_H}{N_A} = \frac{1.008 \text{ g/mol}}{6.022 \times 10^{23} \text{ molecules/mol}}$$
$$= 1.674 \times 10^{-24} \text{ g/atom.}$$

The mass of a single molecule of O_2 is

$$m_{O_2} = \frac{32.0 \text{ g/mol}}{6.022 \times 10^{23} \text{ molecules/mol}}$$
$$= 53.1 \times 10^{-24} \text{ g/molecule.}$$

REFLECT A water molecule (H_2O) consists of one oxygen atom and two hydrogen atoms, but we can now see that the oxygen atom accounts for nearly 90% of its mass.

Practice Problem: Find the total mass of a molecule of nitrogen dioxide (NO_2), a component of smog. What fraction of this total mass is the mass of the nitrogen atom? *Answers:* 7.64×10^{-23} g/molecule, 0.30.

15.2 Equations of State

The conditions in which a particular material exists are described by physical quantities such as pressure, volume, and temperature. Usually we also need to specify the amount of substance in a system. For example, a tank of oxygen in a welding outfit has a pressure gauge and a label stating its volume. We could add a thermometer and place the tank on a scale. These variables (pressure, volume, and temperature) describe the **state** of the material, and they are called **state variables** or **state coordinates.** The amount of substance can be described by its total mass or the number of moles.

The volume V of a substance usually depends on its pressure p, temperature T, and amount of substance, measured either as mass m_{total} or number of moles n. Ordinarily, we can't change one of these variables without causing a change in another. When the tank of oxygen gets hotter, the pressure increases. When the tank gets too hot, the pressure causes it to explode; this happens occasionally with overheated steam boilers (Figure 15.2).

In a few cases, the relationship among p, V, T, and m_{total} (or n) is simple enough that we can express it as an equation called the **equation of state.** When it's too complicated for that, we can use graphs or numerical tables. Even then, the relation among the variables still exists; we call it an equation of state even when we don't know the actual equation.

In this discussion, we'll consider only **equilibrium states**—that is, states in which the temperature and pressure are uniform throughout the system. In principle, when a system changes from one state to another, *non-equilibrium* states must occur during the transition. For example, when a material expands, there must be mass in motion; this requires acceleration and non-uniform pressure. Similarly, when heat conduction occurs, different regions must have different temperatures. We'll assume that such changes occur so slowly that the system is always very nearly in an equilibrium state.

▲ **FIGURE 15.2** An overheated steam boiler can explode with great violence. This photo shows the rusted wreckage of a steam locomotive destroyed by a boiler explosion.

The Ideal-Gas Equation

Gases at low pressures have particularly simple equations of state. Figure 15.3 shows an experimental setup to study the behavior of a gas. The cylinder has a movable piston and is equipped with a pressure gauge and a thermometer. We can vary the pressure, volume, and temperature, and we can pump any desired mass of any gas into the cylinder. Measurements of the behavior of various gases at low pressures lead to several conclusions:

- First, **the volume V is proportional to the number of moles n and thus to the number of molecules.** Thus, if we double the number of moles, keeping the pressure and temperature constant, the volume doubles.

- Second, **the volume varies *inversely* with the pressure p.** Hence, if we double the pressure, holding the temperature T and amount of substance n constant, we compress the gas to one-half of its initial volume. This relation is called **Boyle's law,** after Robert Boyle (1627–1691), a contemporary of Newton. It states that pV = constant when n and T are constant.

- Third, **the pressure is directly proportional to the absolute (Kelvin) temperature.** Thus, if we double the absolute temperature, keeping the volume and quantity of material constant, the pressure doubles. This is called **Charles's law,** after Jacques Charles (1746–1823). It states that $p = (\text{constant})T$ when n and V are constant.

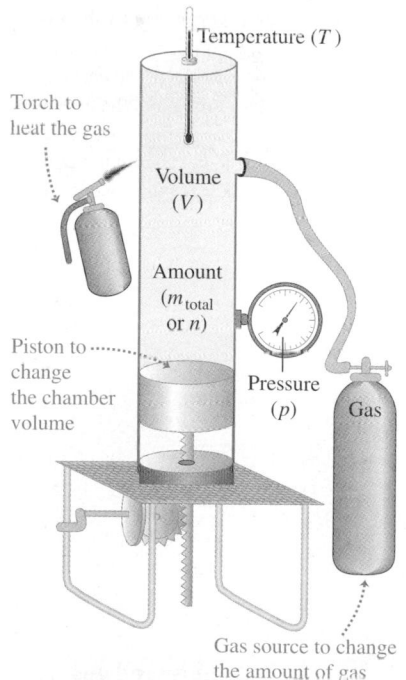

▲ **FIGURE 15.3** A hypothetical setup for studying the behavior of gases.

▲ **Application Martian dust devil.** This photo shows a dust devil that was caught on film while passing the *Mars Spirit Rover.* Although Mars is bitterly cold and has an atmosphere much thinner than earth's, the physics of dust devils is the same on both planets. First, a patch of sun-warmed ground transfers heat to the overlying air. By the ideal-gas equation ($pV = nRT$), this increase in the temperature of the air must increase the product pV. The air is not confined, so its volume V increases, lowering its density and causing it to rise buoyantly. As it rises, surrounding air flows in to take its place. If this surrounding air has a net rotation (as it often does), conservation of angular momentum causes the air to spin faster as it moves inward, creating the dust devil. On earth, where heating can be strong and the atmosphere is dense and energetic, this basic process results not just in dust devils, but in tornadoes and in great storms such as hurricanes and typhoons.

These three relationships can be combined neatly into a single equation:

Ideal-gas equation

For an ideal gas (defined in the following paragraphs), and as an approximation of real gas behavior, the product of pressure p and volume V is proportional to the product of the number n of moles of gas and the absolute temperature T:

$$pV = nRT. \tag{15.3}$$

The proportionality constant R is called the **ideal-gas constant.** Its numerical value depends on the units used for p, V, and T. In SI units, in which the unit of p is the Pa $(= N/m^2)$ and the unit of V is the cubic meter (m^3), $R = 8.3145 \, J/(mol \cdot K)$.

We call Equation (15.3) the **ideal-gas equation.** We might expect that the constant R in this equation would have different values for different gases, but it turns out to have the same value for *all* gases, at least at sufficiently high temperatures and low pressures. Note that the units of pressure times volume are the same as the units of work or energy (e.g., N/m^2 times m^3); that's why R has units of energy per mole per unit of absolute temperature.

In chemical calculations, volumes are sometimes expressed in liters (L) and pressures in atmospheres (atm). In this system,

$$R = 0.08206 \, L \cdot atm/(mol \cdot K).$$

We can express Equation 15.3 in terms of the mass m_{total} of material, using Equation 15.2:

$$pV = \frac{m_{total}}{M} RT.$$

From this relationship, we can get an expression for the density (mass per unit volume, $\rho = m_{total}/V$) of the gas:

$$\rho = \frac{pM}{RT}. \tag{15.4}$$

We can now define an ideal gas as **a gas for which Equation 15.3 holds precisely for *all* pressures and temperatures.** This is an idealized model; no real gas behaves exactly like an ideal gas. However, most real gases show nearly ideal behavior at very low pressures and high temperatures, when the gas molecules are far apart and in rapid motion. The match is reasonably good (within a few percent) at moderate pressures (such as a few atmospheres) and at temperatures well above those at which the gas liquefies.

For a *constant mass* (or constant number of moles) of an ideal gas, the product nR is constant, so the quantity pV/T is also constant. Thus, using the subscripts 1 and 2 to refer to any two states of the same mass of a gas, we have

$$\frac{p_1 V_1}{T_1} = \frac{p_2 V_2}{T_2} = \text{constant.} \tag{15.5}$$

Quantitative Analysis 15.1

A compressed gas

If a gas in a 1.0 L box at 800 K is compressed to a volume of 0.50 L while its pressure changes from 1.0 atm to 0.5 atm, what is the resulting temperature in kelvins? (Assume no phase change.)

A. 800 K.   B. 600 K.   C. 400 K.   D. 200 K.

The amount of gas doesn't change, so nR is constant, and we can use Equation 15.5, $p_1 V_1/T_1 = p_2 V_2/T_2$. Now we notice that the change of state involves *halving* both the volume and the pressure. If we rearrange the equation to solve for T_2/T_1, we see that $T_2/T_1 = p_2 V_2/p_1 V_1 = \frac{1}{4}$. Thus, the final absolute temperature is one-quarter of the original temperature, or 200 K. (Because our equation did not use the ideal-gas constant, the units used for pressure and volume did not matter, as long as they were the same for both states. However, for any ideal-gas calculation, temperature *must* be in kelvins.)

PROBLEM-SOLVING STRATEGY 15.1 **Using the ideal-gas law** (MP)

SET UP

1. List your known and unknown quantities. If the problem involves two different states of the same quantity of gas, designate one as state 1 and the other as state 2.
2. As always, be sure to use a consistent set of units. First decide which value of the gas constant R you are going to use, and then convert the units of the other quantities accordingly. You may have to convert atmospheres to pascals or liters to cubic meters. Sometimes the statement of the problem will make one system of units clearly more convenient than others. Decide on your system, and stick to it!
3. Don't forget that T must always be an *absolute* temperature. If you are given temperatures in degrees Celsius, be sure to add 273 to convert to kelvins.
4. You may sometimes have to convert between total mass m_{total} and number of moles n. The relationship is $m_{total} = Mn$, where M is the molar mass. Here's a tricky point: If you replace n in Equation 15.3 by (m_{total}/M), you *must* use the same mass units for m_{total} and M. So if M is in grams per mole (the usual units for molar mass), then m must also be in grams. If you want to use m_{total} in kilograms, then you must convert M to kilograms per mole. For example, the molar mass of oxygen is 32 g/mol or 32×10^{-3} kg/mol. Be careful!

SOLVE

5. If you are dealing with only one state of the system, you will use Equation 15.3, the ideal-gas equation. If you are dealing with two states of the same quantity of gas, *and* you know all but one of the p, V, and T values, you can use Equation 15.5; otherwise you have to use Equation 15.3.

REFLECT

6. The mathematical relations are simple, but it's easy to make mistakes with units and conversions of units. Be sure to carry units through your calculations; be especially careful to distinguish between grams, kilograms, and moles.

▲ **BIO Application Why do aircraft cabins have to be pressurized?** When we breathe in, air is warmed to 37° C and becomes saturated with water vapor, which has a "partial pressure" of 47 torr. In comparison, the partial pressure of O_2 in air is about 150 torr. Because the gases behave independently and the total pressure of the inhaled air does not change, the water displaces on a mole-to-mole basis an equal amount of air. The partial pressure of inhaled O_2 becomes $0.2P_B - 47$ torr, where P_B is barometric or ambient air pressure. As one goes up in altitude, P_B declines, but the partial pressure of the water vapor stays the same. Thus, at an altitude of between 50,000 and 60,000 ft, when P_B falls to 47 torr or less, a person can breathe in only water vapor even if breathing from an unpressurized source of 100% O_2. The amount of O_2 inspired becomes intolerably small even at lower altitudes. For this reason, commercial aircraft cabins require pressurization at altitudes higher than 9800 ft, and airlines must provide oxygen through drop-down masks in case of loss of pressurization.

EXAMPLE 15.2 **Volume of a gas at STP**

The condition called **standard temperature and pressure** (STP) for a gas is defined to be a temperature of 0°C = 273 K and a pressure of 1 atm = 1.013×10^5 Pa. If you want to keep a mole of an ideal gas in your room at STP, how big a container do you need?

SOLUTION

SET UP If we use the ideal-gas equation with SI units, we need to express R in units of $J/(mol \cdot K)$. The temperature in the ideal-gas equation must be an absolute temperature, expressed in kelvins.

SOLVE We rearrange the ideal-gas equation, solving for volume:

$$pV = nRT,$$
$$V = \frac{nRT}{p} = \frac{(1 \text{ mol})[8.314 \text{ J}/(mol \cdot K)](273 \text{ K})}{1.013 \times 10^5 \text{ Pa}}$$
$$= 0.0224 \text{ m}^3 = 22.4 \text{ L}.$$

REFLECT This is about the size of a 5-gallon gasoline can.

Practice Problem: How many moles would your entire room hold at STP if the room had dimensions of 3 m × 3 m × 2 m? *Answer: n = 800 mol.*

EXAMPLE 15.3 **Mass of air in a scuba tank**

A typical tank used for scuba diving (Figure 15.4) has a volume of 11.0 L (about $0.4 \, \text{ft}^3$) and a maximum gauge pressure of $2.10 \times 10^7 \, \text{Pa}$ (about 3000 lb/in² gauge). The "empty" tank contains 11.0 L of air at 21°C and 1 atm ($1.013 \times 10^5 \, \text{Pa}$). The air is hot when it comes out of the compressor; when the tank is filled, the temperature is 42°C and the gauge pressure is $2.10 \times 10^7 \, \text{Pa}$. What mass of air was added? (Air is a mixture of gases: about 78% nitrogen, 21% oxygen, and 1% miscellaneous gases; its average molar mass is 28.8 g/mol = 28.8×10^{-3} kg/mol.)

▶ **FIGURE 15.4**

SOLUTION

SET UP First we find the number of moles in the tank, at the beginning and at the end. We must remember to convert the temperatures from Celsius to kelvin and to convert the pressures from gauge to absolute by adding $1.013 \times 10^5 \, \text{Pa}$.

SOLVE From Equation 15.3, the number of moles n_1 in the "empty" tank is

$$n_1 = \frac{pV}{RT} = \frac{(1.013 \times 10^5 \, \text{Pa})(11.0 \times 10^{-3} \, \text{m}^3)}{[8.314 \, \text{J}/(\text{mol} \cdot \text{K})](294 \, \text{K})} = 0.456 \, \text{mol}.$$

The number of moles n_2 in the "full" tank is

$$n_2 = \frac{(2.11 \times 10^7 \, \text{Pa})(11.0 \times 10^{-3} \, \text{m}^3)}{[8.314 \, \text{J}/(\text{mol} \cdot \text{K})](315 \, \text{K})} = 88.6 \, \text{mol}.$$

We added 88.6 mol − 0.456 mol = 88.1 mol to the tank. The average molar mass of air is 28.8×10^{-3} kg/mol, so the added mass is $(88.1 \, \text{mol})(28.8 \times 10^{-3} \, \text{kg/mol}) = 2.54 \, \text{kg}$.

REFLECT Because the amount of gas was not the same for the two states of the system, Equation 15.5 isn't applicable; we have to use the ideal-gas equation, Equation 15.3.

EXAMPLE 15.4 **Density of air**

Find the density of air at 20°C and normal atmospheric pressure.

SOLUTION

SET UP AND SOLVE We use Equation 15.4, taking care to convert T to kelvins. As in the previous example, the average molar mass of air is 28.8 g/mol = 28.8×10^{-3} kg/mol. From Equation 15.4,

$$\rho = \frac{pM}{RT} = \frac{(1.013 \times 10^5 \, \text{Pa})(28.8 \times 10^{-3} \, \text{kg/mol})}{[8.314 \, \text{J}/(\text{mol} \cdot \text{K})](293 \, \text{K})}$$
$$= 1.20 \, \text{kg/m}^3.$$

REFLECT Remember that the density of water is an easy-to-remember $1.00 \times 10^3 \, \text{kg/m}^3$. Thus, the density of air at normal atmospheric pressure and ordinary temperatures is roughly a thousandth that of water.

When we deal with *mixtures* of ideal gases, the concept of **partial pressure** is useful. In a mixture, the partial pressure of each gas is the pressure that gas would exert if it occupied the entire volume by itself. The actual total pressure of the mixture is the sum of the partial pressures of the components. In air at normal atmospheric pressure, the partial pressure of nitrogen is about 0.8 atm, or $0.8 \times 10^5 \, \text{Pa}$, and the partial pressure of oxygen is about 0.2 atm, or $0.2 \times 10^5 \, \text{Pa}$.

The ability of the human body to absorb oxygen from the atmosphere depends critically on the partial pressure of oxygen. Absorption drops sharply when the partial pressure of oxygen is less than about $0.13 \times 10^5 \, \text{Pa}$, corresponding to an elevation above sea level of approximately 4700 m (15,000 ft). At partial pressures less than $0.11 \times 10^5 \, \text{Pa}$, oxygen absorption is not sufficient to maintain life. There is no permanently maintained human habitation on earth above 16,000 ft, although survival for short periods of time is possible at higher elevations. At the summit of Mount Everest (elevation 8882 m, or 29,141 ft) the partial pressure of

▶ **BIO Application Scuba hazards.** Scuba divers must breathe gas that is at the same pressure as the surrounding water; otherwise they couldn't inflate their lungs once they dove more than a short distance below the surface. However, that means that the partial pressures of the gases in the mix rise as the diver descends. Certain gases become toxic at elevated partial pressures. Nitrogen, which is harmless at its normal atmospheric partial pressure of 0.80 atm, becomes a narcotic at elevated pressures. A diver breathing ordinary compressed air begins to act drunk below about 30 m because of nitrogen narcosis. Oxygen also has toxic effects at elevated partial pressures, depending on the partial pressure and the length of exposure. To counteract these problems, technical divers sometimes use special gas mixtures (typically nitrogen–oxygen or helium–oxygen mixtures) chosen to fit the depth and duration of the intended dive.

oxygen is only about 0.07×10^5 Pa, and climbers nearly always carry oxygen tanks. For similar reasons, jet airplanes, which typically fly at altitudes of 8000 to 12,000 m, *must* have pressurized cabins for passenger comfort and health.

pV Diagrams

For a given quantity of a material, the equation of state is a relation among the three state coordinates: pressure p, volume V, and temperature T. A useful graphical representation of this relation is a set of graphs of pressure as a function of volume, each for a particular constant temperature. Such a diagram is called a ***pV diagram.*** Each curve, representing behavior at a specific temperature, is called an **isotherm,** or a *pV isotherm.* ("Isotherm" means "same temperature.")

Figure 15.5 shows pV isotherms for a constant amount of an ideal gas. The highest temperature is T_4; the lowest is T_1. The figure is a graphical representation of the ideal-gas equation of state. We can read off the volume V corresponding to any given pressure p and temperature T in the range shown.

Figure 15.6 shows a pV diagram for a material that *does not* obey the ideal-gas equation. At high temperatures such as T_4, the curves resemble the ideal-gas curves of Figure 15.5. However, at temperatures below T_c, the isotherms develop flat parts—regions where the material can be compressed without an increase in pressure. Observation of the gas shows that it is *condensing* from the vapor (gas) to the liquid phase. The flat parts of the isotherms in the shaded area of Figure 15.6 represent conditions of liquid–vapor *phase equilibrium.* As the volume decreases, more and more material goes from vapor to liquid, but the pressure does not change. To keep the temperature constant, we have to remove heat—the heat of vaporization we discussed in Section 14.5.

When we compress a gas such as that in Figure 15.6 at a constant temperature T_2, it is vapor until point a is reached. There it begins to liquefy; as the volume decreases further and heat of vaporization is removed, more material liquefies, with *both* pressure and temperature remaining constant. At point b, *all* the material is in the liquid state. After this, any further compression results in a very rapid rise in pressure because, in general, liquids are much less compressible than gases. At a lower constant temperature T_1, the gas behaves similarly, but the onset of condensation occurs at lower pressure and greater volume than at the constant temperature T_2.

We'll use pV diagrams often in this chapter and the next. We'll show that the area under a pV curve (regardless of whether it is or is not an isotherm) represents the *work* done by the system during a change in volume. This work, in turn, is directly related to heat transfer and changes in the *internal energy* of the system, which we'll study later in this chapter.

Another diagram that's useful, especially for describing phases of matter and phase transitions, is a graph with axes p and T, called a **phase diagram.** We learned in Section 14.5 that each phase is stable only in certain ranges of temperature and pressure. Ordinarily, a transition from one phase to another takes place under conditions of **phase equilibrium** between the two phases. For a given

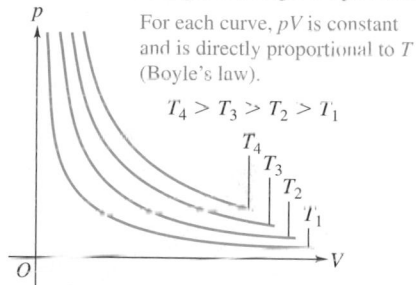

Each curve represents pressure as a function of volume for an ideal gas at a single temperature.

For each curve, pV is constant and is directly proportional to T (Boyle's law).

$T_4 > T_3 > T_2 > T_1$

▲ **FIGURE 15.5** Isotherms, or constant-temperature curves, for a constant amount of an ideal gas.

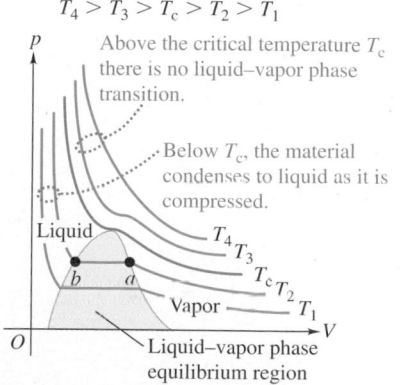

$T_4 > T_3 > T_c > T_2 > T_1$

Above the critical temperature T_c there is no liquid–vapor phase transition.

Below T_c, the material condenses to liquid as it is compressed.

▲ **FIGURE 15.6** A pV diagram for a non-ideal gas, showing pV isotherms for temperatures above and below the critical temperature T_c.

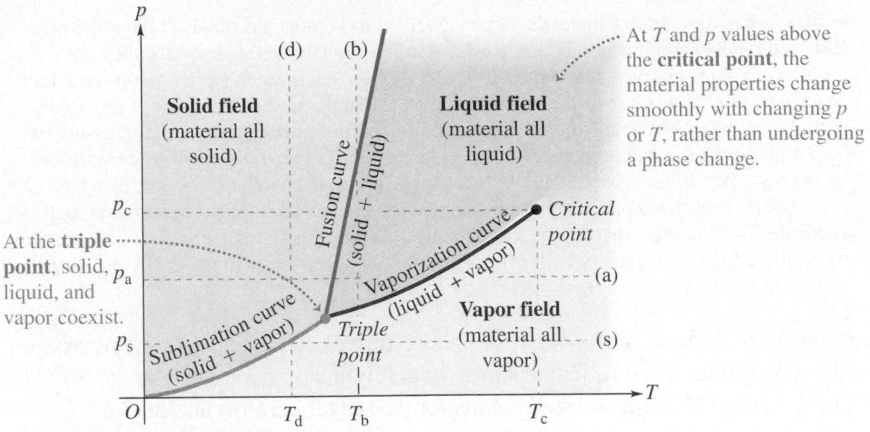

▲ **FIGURE 15.7** A typical pT phase diagram, showing regions of temperature and pressure in which the various phases exist and where phase changes occur.

pressure, this occurs at only one specific temperature. A phase diagram shows what phase occurs for each possible combination of temperature and pressure.

A typical example is shown in Figure 15.7. Each point on the diagram represents a pair of values of p and T. The three colored areas labeled "solid field," "liquid field," and "vapor field" represent points at which the material consists of a single phase: solid, liquid, and vapor, respectively. At any point along the three colored lines separating these fields, the material consists of two coexisting phases. On the sublimation curve, solid and vapor coexist; on the fusion curve, solid and liquid coexist; and on the vaporization curve, liquid and vapor coexist. When a material is at a point on one of these lines, a phase change can proceed.

If we increase the temperature of a substance, keeping the pressure constant, it passes through a sequence of states represented by points on a horizontal line such as the dashed gray line (a) in the figure. The melting and boiling temperatures at this pressure are the temperatures at which this line intersects the fusion and vaporization curves, respectively. If we *compress* a material by increasing the pressure while we hold the temperature constant, as represented by a vertical dashed line such as line (b), the material passes from vapor to liquid and then to solid at the points where the line crosses the vaporization curve and fusion curve, respectively.

When the pressure is low enough, constant-pressure heating can transform a substance from solid directly to vapor, as shown by line (s). As we learned in Chapter 14, this process is called *sublimation;* it occurs at a pressure and temperature corresponding to the intersection of the line and the sublimation curve. At atmospheric pressure, solid carbon dioxide (dry ice) undergoes sublimation; no liquid phase can exist at this pressure. When a substance with an initial temperature lower than the temperature at the point labeled "triple point" is compressed by increasing its pressure, it goes directly from the vapor to the solid phase, as shown by line (d). The melting of ice under pressure is what makes ice skating possible.

TABLE 15.1   Triple-point data

Substance	Temperature, K	Pressure, Pa
Hydrogen	13.8	0.0704×10^5
Neon	24.57	0.432×10^5
Nitrogen	63.18	0.125×10^5
Oxygen	54.36	0.00152×10^5
Ammonia	195.40	0.0607×10^5
Carbon dioxide	216.55	5.17×10^5
Water	273.16	0.00610×10^5

Triple Point

The intersection point of the equilibrium curves in Figure 15.7 is called the **triple point.** For any substance, this point represents the unique pressure–temperature combination at which *all three* phases can coexist. We call these the triple-point pressure and temperature. Triple-point data for a few substances are given in Table 15.1.

Critical Point

Figure 15.6 shows that a liquid–vapor phase transition occurs only when the temperature and pressure are less than those at the point lying at the top of the tongue-shaped area labeled "liquid–vapor equilibrium region." This point corresponds to the endpoint at the top of the vaporization curve in Figure 15.7. It is called the **critical point,** and the corresponding values of p and T are called the critical pressure p_c and critical temperature T_c. A gas at a temperature *above* the critical temperature doesn't separate into two phases when it is compressed isothermally (along a vertical line to the right of the critical point in Figure 15.7). Instead, its properties change gradually and continuously from those we ordinarily associate with a gas (low density and high compressibility) to those of a liquid (high density and low compressibility) *without a phase transition.* Similarly, when a material is heated at a constant pressure greater than the critical-point pressure (a horizontal line above the critical point in Figure 15.7), its properties change gradually from liquid-like to gas-like.

If the preceding discussion stretches your credulity, here's another point of view: Look at liquid–vapor phase transitions at successively higher points on the vaporization curve. As we approach the critical point, the *differences* in physical properties, such as density, bulk modulus, optical properties, and viscosity, between the liquid and vapor phases become smaller and smaller. Exactly *at* the critical point they all become zero, and at this point the distinction between liquid and vapor disappears. The heat of vaporization also grows smaller and smaller as we approach the critical point, and it too becomes zero at the critical point.

Table 15.2 lists critical constants for a few substances. The very low critical temperatures of hydrogen and helium show why these gases defied attempts to liquefy them for many years. For carbon dioxide, the triple-point temperature is $-56.6°C$ and the triple-point pressure is 5.11 atm. At atmospheric pressure, CO_2 can exist only as a solid or vapor—hence the name "dry ice." Liquid CO_2 can exist only at a pressure greater than 5.11 atm.

Many substances can exist in more than one solid phase. A familiar example is carbon, which exists as noncrystalline lampblack (roughly, soot) and crystalline graphite and diamond. (Graphite is the main constituent of pencil leads; it is also used in dry lubricants.) Water is another example: At least eight types of ice, differing in crystal structure and physical properties, have been observed at very high pressures.

▲ **Application How hot is hot?** For industrial applications, we often need to measure very hot and very cold temperatures, and the measurements involved require special thermometers such as the platinum resistance thermometer. Of course, these thermometers must be calibrated with the use of reproducible temperatures other than the freezing and boiling points of pure water used by the Fahrenheit and Celsius scales. The photo shows the calibration of a platinum resistance thermometer for high temperatures, using the freezing point of molten silver $(961°C = 1234 K)$. Calibration of thermometers for low-temperature use can be done at the triple point of argon, where solid, liquid, and gaseous phases coexist $(-189.3°C = 83.8 K)$. Some of these platinum resistance thermometers are sensitive enough to measure temperatures with an accuracy of $±0.05$ C° over a range of temperatures from $-190°C$ to over 1000°C.

TABLE 15.2   Critical-point data

Substance	Critical temperature, K	Critical pressure, Pa
Helium ($^4_2$He)	5.3	2.29×10^5
Hydrogen	33.3	13.0×10^5
Nitrogen	126.2	33.9×10^5
Oxygen	154.8	50.8×10^5
Ammonia	405.5	112.8×10^5
Carbon dioxide	304.2	73.9×10^5
Water	647.4	221.2×10^5

15.3 Kinetic Theory of an Ideal Gas

We've studied several properties of matter in bulk, including elasticity, density, surface tension, equations of state, and heat capacities, with only passing references to molecular structure. Now we want to look in more detail at the relation of bulk (macroscopic) behavior to microscopic structure. We begin with a general discussion of the molecular structure of matter. Then, in the next two sections, we'll develop the *kinetic-molecular model* of an ideal gas and derive the equation of state and an expression for molar heat capacity from this molecular model.

Molecular Properties of Matter

All familiar matter is made up of atoms or groups of connected atoms. For convenience, we'll use the term **molecule** to refer to the smallest unit of a substance, even in cases where this unit is a single atom. Individual atoms are on the order of 10^{-10} m in size; the largest molecules contain many atoms and are 10,000 or more times that large. In gases, the molecules move nearly independently; in liquids and solids, they are held together by intermolecular forces that are electrical in nature, arising from interactions of the electrically charged particles that make up the molecules. Gravitational forces between molecules are negligible in comparison with electrical forces.

When molecules of a gas are far apart, the electrical forces between them are very small and usually attractive. As a gas is compressed and its molecules are brought closer together, the attractive forces increase. At sufficiently small separations, the intermolecular attractions are large enough to make the material condense into a liquid or solid.

Molecules are always in motion; their kinetic energies usually increase with temperature. In *solids,* molecules vibrate about more or less fixed centers. In a crystalline solid, these centers are arranged in a recurring *crystal lattice.* Figure 15.8 shows the cubic crystal structure of sodium chloride (ordinary table salt). A scanning tunneling microscope photograph of the surface of a silicon crystal is shown in Figure 15.9. Each "bead" is a silicon atom.

In a *liquid,* the intermolecular distances are usually only slightly greater than in the solid phase of the same substance, but the molecules have much greater freedom of movement. Liquids show regularity of structure only in the immediate neighborhood of a few molecules.

The molecules of a *gas* are usually widely separated and exert only very small attractive forces on one another. A gas molecule moves in a straight line until it collides with another molecule or with a wall of the container. In molecular terms, an *ideal gas* is a gas whose molecules exert *no* attractive forces on one another and therefore have no *potential* energy.

At low temperatures, most common substances are in the solid phase. As the temperature rises, a substance melts and then vaporizes. From a *molecular* point of view, these transitions are in the direction of increasing molecular kinetic energy. Thus, temperature and molecular kinetic energy are closely related.

Kinetic Molecular Theory of an Ideal Gas

The goal of any molecular theory of matter is to understand the *macroscopic* properties of matter in terms of its atomic or molecular structure and behavior. Such theories are of tremendous practical importance; once we have this understanding, we can *design* materials to have specific desired properties. Such analysis has led to the development of high-strength steels, glasses with special optical properties, semiconductor materials for electronic devices, and countless other materials that are essential to contemporary technology.

One of the simplest examples of a molecular theory is the kinetic-molecular model of an ideal gas. With the help of this model, we can understand the relation

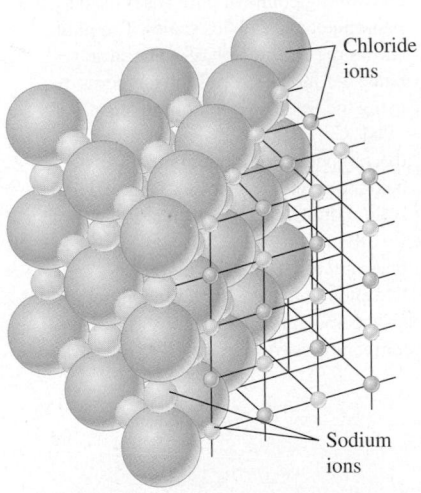

▲ **FIGURE 15.8** Schematic representation of the cubic crystal structure of sodium chloride (table salt).

Chloride ions

Sodium ions

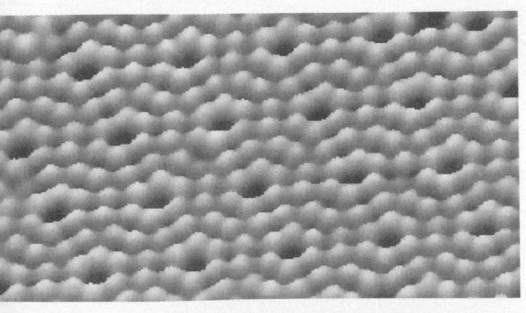

▲ **FIGURE 15.9** A scanning tunneling microscope image of silicon atoms on the surface of a silicon crystal. The regularity of the crystal structure is easy to see.

of ideal-gas behavior, including the ideal-gas equation of state (Equation 15.3) and the molar heat capacities of gases, to Newton's laws. The development that follows has several steps, and you may need to go over it several times to grasp how it all goes together. Don't get discouraged!

Here are the assumptions of the kinetic-molecular model:

1. A container with volume V contains a very large number N of identical molecules, each with mass m. The container has perfectly rigid walls that do not move.
2. The molecules behave as point particles; their size is small in comparison to the average distance between particles and to the dimensions of the container.
3. The molecules are in constant random motion; they obey Newton's laws. Each molecule occasionally makes a perfectly elastic (energy-conserving) collision with a wall of the container.
4. During collisions, the molecules exert *forces* on the walls of the container; these forces create the *pressure* that the gas exerts. In a typical collision (Figure 15.10), the velocity component v_y parallel to the wall is unchanged, and the component v_x perpendicular to the wall changes direction, but not magnitude.

Here's our program, one step at a time:

Step 1: We find the total change in momentum associated with each collision.
Step 2: We find the *number* of collisions per unit time for a certain wall area A.
Step 3: We find the total change in momentum per unit time due to these collisions and, from that, the force on area A needed to cause that change in momentum.
Step 4: Finally, we can obtain an expression for the pressure, which is force per unit area, and compare it with the ideal-gas equation.

We'll take these steps in order. Here we go:

Step 1. Let $|v_x|$ be the magnitude of the x component of velocity of a molecule of the gas. For now, we assume that all molecules have the same $|v_x|$. That isn't right, but making this temporary assumption helps to clarify the basic ideas. We will show later that the assumption isn't really necessary. For each collision, the x component of momentum changes from $-m|v_x|$ to $+m|v_x|$, so the *change* in the x component of momentum is $m|v_x| - (-m|v_x|) = 2m|v_x|$.

Step 2. If a molecule is going to collide with a given wall area A during a small time interval Δt, then, at the beginning of Δt, it must be within a distance $|v_x| \Delta t$ from the wall (Figure 15.11) and it must be headed toward the wall. So the number of molecules that collide with A during Δt is equal to the number of molecules that are within a cylinder with base area A and length $|v_x| \Delta t$ and that have their v_x aimed toward the wall. The volume of such a cylinder is $A|v_x| \Delta t$.

Assuming that the number of molecules per unit volume (N/V) is uniform, the *number* of molecules in this cylinder is $(N/V)(A|v_x| \Delta t)$. On the average, half of these molecules are moving toward the wall and half away from it. So the number of collisions with wall area A during Δt is

$$\tfrac{1}{2}\left(\frac{N}{V}\right)(A|v_x| \Delta t).$$

Step 3. The total change in momentum, ΔP_x, due to all these collisions during time Δt is the *number* of collisions multiplied by the change in momentum in one collision, $2m|v_x|$:

$$\Delta P_x = \tfrac{1}{2}\left(\frac{N}{V}\right)(A|v_x| \Delta t)(2m|v_x|) = \frac{NAmv_x^2 \Delta t}{V}.$$

(We're using capital P for momentum and small p for pressure. Also, capital V denotes volume, and small v represents a velocity component. Be careful!) The

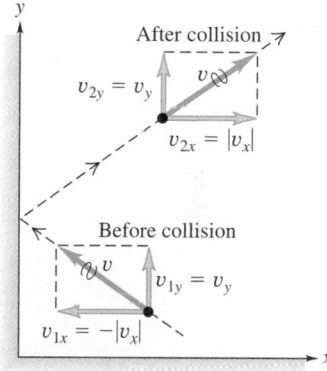

▲ FIGURE 15.10 Elastic collision of a molecule with an idealized container wall. The component v_y parallel to the wall does not change; the component v_x perpendicular to the wall reverses direction. The speed v does not change.

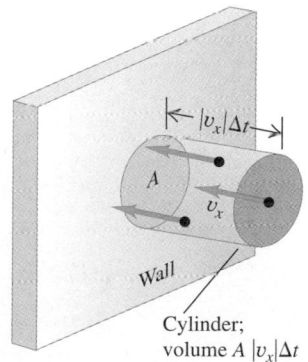

▲ FIGURE 15.11 Molecules moving toward the wall with speed $|v_x|$ collide with area A during the time interval Δt only if they are within a distance $|v_x| \Delta t$ of the wall at the beginning of the interval. All such molecules are contained within a volume $A|v_x| \Delta t$.

rate of change of the momentum component P_x with time is the preceding expression divided by Δt:

$$\frac{\Delta P_x}{\Delta t} = \frac{NAmv_x^2}{V}.$$

According to Newton's second law, this rate of change of momentum equals the force exerted by the wall area A on the molecules. From Newton's *third* law, this force is equal and opposite to the force exerted *on* the wall *by* the molecules.

Step 4. The pressure p of the gas is the magnitude of the force per unit area, and we obtain

$$p = \frac{F}{A} = \frac{Nmv_x^2}{V}, \qquad pV = Nmv_x^2. \tag{15.6}$$

Now we need one additional step: We need to express v_x^2 in this equation in terms of the average value of v^2 for all the molecules. First, the square of the speed of an individual molecule (v^2) is related to the sum of the squares of the components $(v_x, v_y$ and $v_z)$: $v^2 = v_x^2 + v_y^2 + v_z^2$. Second, we can take the *average* of this expression for all the molecules in the gas: When we do, we get $(v^2)_{av} = (v_x^2)_{av} + (v_y^2)_{av} + (v_z^2)_{av}$. Third, there is no distinction among the x, y, and z axis directions, so it must be true that $(v_x^2)_{av} = (v_y^2)_{av} = (v_z^2)_{av}$, and it follows that $(v^2)_{av} = 3(v_x^2)_{av}$, or $(v_x^2)_{av} = \left(\frac{1}{3}\right)(v^2)_{av}$.

We see that we didn't really need the initial assumption that all molecules had the same value of $|v_x|$, because in this step we work with *average values* of the velocity components. When we replace v_x^2 with $(v_x^2)_{av}$ in Equation 15.6, we get

$$pV = \tfrac{1}{3}Nm(v^2)_{av} = \tfrac{2}{3}N\left[\tfrac{1}{2}m(v^2)_{av}\right].$$

We notice that $\frac{1}{2}m(v^2)_{av}$ is the average *translational kinetic energy* of a single molecule. The product of this quantity and the total number N of molecules equals the *total* random kinetic energy K_{tr} of translational motion of *all* the molecules. (The notation K_{tr} reminds us that this energy is associated with *translational* motion and anticipates the possibility of additional energies associated with rotational and vibrational motion of molecules.) Hence, the product pV equals two-thirds of the total translational kinetic energy:

$$pV = \tfrac{2}{3}K_{tr}.$$

Now here comes the *coup de grace*: We compare this result with the ideal-gas equation, $pV = nRT$, which is based on experimental studies of gas behavior. For the two equations to agree, we must have the following relationship:

Kinetic energy of gas molecules
For an ideal gas, the total translational kinetic energy K_{tr} of all the molecules is proportional to the absolute (Kelvin) temperature T and to the quantity of gas (as measured by the number of moles, n):

$$K_{tr} = \tfrac{3}{2}nRT. \tag{15.7}$$

We'll use this important result several times in the discussion that follows.

The average translational kinetic energy of a *single* molecule is the total translational kinetic energy K_{tr} of *all* the molecules, divided by the number N of molecules:

$$K_{av} = \frac{K_{tr}}{N} = \tfrac{1}{2}m(v^2)_{av} = \frac{3nRT}{2N}.$$

► **Application Hold your hydrogen.** The cloud of gas and dust from which our solar system formed consisted mostly of diatomic hydrogen gas (H_2). However, earth and its sister rocky planets have practically no hydrogen gas in their atmospheres, whereas the gas giants, such as Jupiter, have a lot. Why? The main reason concerns mass—of hydrogen and of planets. Hydrogen is the least massive element; H_2 has a molar mass of 2.01 g/mol. Therefore, H_2 molecules have a higher average speed at a given temperature than more massive molecules. On a small planet like earth, a significant fraction of any H_2 molecules in the atmosphere are moving faster than the planet's escape speed and can simply fly off into outer space. Thus, small planets leak hydrogen until essentially none remains in their atmospheres. The gas giants, being far more massive, have enough gravity to keep their hydrogen.

Earth and
Jupiter to scale

Also, the total number of molecules, N, is the number of moles, n, multiplied by Avogadro's number N_A (i.e., $N = nN_A$), so the average translational kinetic energy per molecule is

$$K_{av} = \frac{3nRT}{2(nN_A)} = \tfrac{3}{2}\frac{R}{N_A}T.$$

The ratio R/N_A occurs frequently in molecular theory. It is called the **Boltzmann constant,** denoted by k and defined as

$$k = \frac{R}{N_A} = \frac{8.314 \text{ J}/(\text{mol} \cdot \text{K})}{6.022 \times 10^{23} \text{ molecules}/\text{mol}}$$
$$= 1.381 \times 10^{-23} \text{ J}/(\text{molecule} \cdot \text{K}).$$

We can rewrite the expression for K_{av} (the average translational kinetic energy of a single molecule) in terms of k as

$$K_{av} = \tfrac{1}{2}m(v^2)_{av} = \tfrac{3}{2}kT. \tag{15.8}$$

This result shows that, for an ideal gas, the average translational kinetic energy *per molecule* depends only on the temperature, not on the pressure, volume, or kind of molecule involved. We can obtain an equivalent statement by using the relation $M = N_A m$. The average translational kinetic energy *per mole* (which also depends only on T) is

$$N_A \tfrac{1}{2}m(v^2)_{av} = N_A(\tfrac{3}{2}kT) = \tfrac{3}{2}RT.$$

Finally, it's often useful to rewrite the ideal-gas equation on a molecular basis. We use $N = N_A n$ and $R = N_A k$ to obtain the following alternative form:

$$pV = NkT. \tag{15.9}$$

This equation shows that we can think of k as a gas constant on a "per molecule" basis instead of the usual "per-mole" basis with the gas constant R.

Quantitative Analysis 15.2

A cooled gas

An ideal gas is cooled at constant volume until the pressure is half the original pressure. When this happens, the average translational kinetic energy of the gas molecules is

A. twice the original.
B. unchanged.
C. half the original.

SOLUTION According to the ideal-gas equation, $pV = nRT$, if V is constant and p is reduced by a factor of one-half, the absolute temperature must also decrease by a factor of one-half. According to Equation 15.8, the average translational kinetic energy is proportional to the absolute temperature, so it too decreases by a factor of one-half. The correct answer is C.

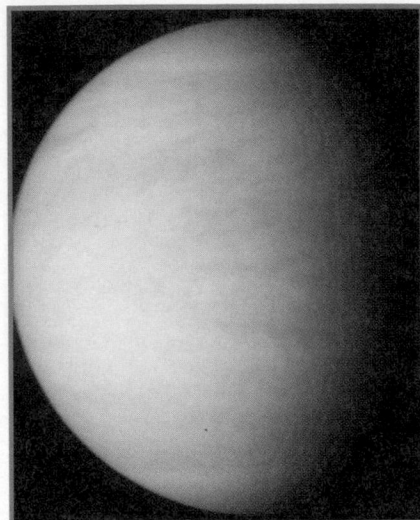

◀ **Application It could have been so nice there!** Science fiction authors once envisioned Venus as a cloud-shrouded tropical swamp planet—and indeed, when it formed, Venus probably had at least as much water as our own world. But Venus is now utterly dry, with a dense carbon dioxide atmosphere that keeps it hotter than a pizza oven. Where did the water go? Being closer to the sun, Venus was always warmer than earth. If it ever had oceans, they soon vaporized. With no oceans, carbon dioxide could not form limestone (as it does on earth), but instead stayed in the atmosphere. The thick, hot atmosphere let water vapor rise to altitudes at which UV light could split water molecules, releasing hydrogen atoms. Because of their low mass, these atoms easily reached escape speed and disappeared into outer space. Eventually, the water was gone. That is how our sister planet became an arid furnace. We're so lucky . . . .

Molecular Speeds in an Ideal Gas

From Equations 15.8 and 15.9, we can obtain an expression for the square root of $(v^2)_{av}$, called the **root-mean-square speed** v_{rms} of molecules in an ideal gas:

$$v_{rms} = \sqrt{(v^2)_{av}} = \sqrt{\frac{3kT}{m}} = \sqrt{\frac{3RT}{M}}. \tag{15.10}$$

This quantity increases with temperature, as we should expect; it also varies inversely with the square root of the molecular or molar mass.

To compute the rms speed, we square each molecular speed, add, divide by the number of molecules, and take the square root. That is, v_{rms} is the *root* of the *mean* of the *squares*. Example 15.6 illustrates this procedure.

We've mentioned that the molecules in a gas don't all have the same speed. Figure 15.12 is a graph of the relative numbers of molecules with various speeds for three different temperatures. As the temperature increases, the curve flattens and the peak moves to higher speed. This speed distribution curve is called the *Maxwell–Boltzmann* distribution.

Number of molecules

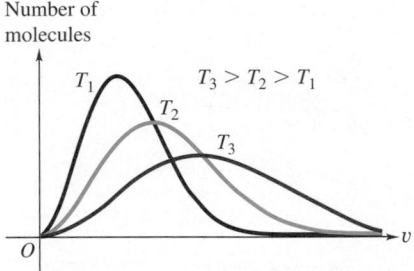

▲ **FIGURE 15.12** The distribution of speeds for the molecules of a gas at three different temperatures.

PROBLEM-SOLVING STRATEGY 15.2 **Kinetic-molecular theory**

As usual, using a consistent set of units is essential. Following are several places where caution is needed:

1. The usual units for molar mass M are grams per mole; the molar mass of O_2 is 32 g/mol, for example. These units are often omitted in tables. When you use SI units in equations, such as Equation 15.10, you *must* express M in kilograms per mole by multiplying the value shown in the table by $(1 \text{ kg}/10^3 \text{ g})$. Thus, in SI units, M for O_2 is 32×10^{-3} kg/mol.

2. Are you working on a "per molecule" basis or a "per mole" basis? Remember that m is the mass of a molecule and M is the mass of a mole. N is the number of molecules, and n is the number of moles; k is the gas constant per molecule, and R is the gas constant per mole. Although N, the number of molecules, is in one sense a dimensionless number, you can do a complete unit check if you think of N as having the unit "molecules;" then m has the unit "mass per molecule" and k has the unit "joules per molecule per kelvin."

3. Remember that T is always *absolute* (Kelvin) temperature.

EXAMPLE 15.5 **Kinetic energy of a molecule**

(a) What is the average translational kinetic energy of a molecule of oxygen (O_2) at a temperature of 27°C, assuming that oxygen can be treated as an ideal gas? **(b)** What is the *total* translational kinetic energy of the molecules in 1 mole of oxygen at this temperature? **(c)** Compare the root-mean-square speeds of oxygen and nitrogen (N_2) molecules at this temperature (assuming that they can be treated as ideal gases). The mass of an oxygen molecule is $m_{O_2} = 5.31 \times 10^{-26}$ kg, and the mass of a molecule of nitrogen is $m_{N_2} = 4.65 \times 10^{-26}$ kg.

Continued

SOLUTION

SET UP AND SOLVE Part (a):   First we convert the temperature to kelvins: 27°C = 300 K. Then we use Equation 15.8: For one molecule,

$$K_{av} = \tfrac{1}{2} m (v^2)_{av} = \tfrac{3}{2} kT$$
$$= \tfrac{3}{2}(1.38 \times 10^{-23} \text{ J/K})(300 \text{ K}) = 6.21 \times 10^{-21} \text{ J}.$$

Part (b):   Using Equation 15.7 (for 1 mole), we obtain

$$K_{tr} = \tfrac{3}{2} nRT = \tfrac{3}{2}(1 \text{ mol})[8.315 \text{ J/(mol} \cdot \text{K})](300 \text{ K}) = 3740 \text{ J}.$$

Part (c):   From Equation 15.10, we find that

$$v_{rms,O_2} = \sqrt{\frac{3kT}{m_{O_2}}} = \sqrt{\frac{3(1.38 \times 10^{-23} \text{ J/K})(300 \text{ K})}{5.31 \times 10^{-26} \text{ kg}}}$$
$$= 484 \text{ m/s,}$$

$$v_{rms,N_2} = \sqrt{\frac{3kT}{m_{N_2}}} = \sqrt{\frac{3(1.38 \times 10^{-23} \text{ J/K})(300 \text{ K})}{4.65 \times 10^{-26} \text{ kg}}}$$
$$= 517 \text{ m/s.}$$

Thus, the ratio of the rms speed of N_2 molecules to that of O_2 molecules is 1.07.

ALTERNATIVE SOLUTION

$$v_{rms,O_2} = \sqrt{\frac{3RT}{M_{O_2}}} = \sqrt{\frac{3(8.314 \text{ J/mol} \cdot \text{K})(300 \text{ K})}{32 \times 10^{-3} \text{ kg/mol}}} = 484 \text{ m/s,}$$

$$v_{rms,N_2} = \sqrt{\frac{3RT}{M_{N_2}}} = \sqrt{\frac{3(8.314 \text{ J/mol} \cdot \text{K})(300 \text{ K})}{28 \times 10^{-3} \text{ kg/mol}}} = 517 \text{ m/s.}$$

REFLECT These speeds are roughly 1000 mi/h. At any temperature, rms molecular speeds are inversely proportional to the square root of molar mass. Note that we have to be careful throughout to distinguish between "per molecule" and "per mole" quantities.

Practice Problem: What is the root-mean-square speed of a diatomic hydrogen molecule (H_2) at 300 K, assuming that it can be treated as an ideal gas? The mass of a hydrogen molecule is 3.348×10^{-27} kg. *Answer:* $v_{rms} = 1930$ m/s.

EXAMPLE 15.6   Gang of five

Five ideal-gas molecules chosen at random are found to have speeds of 500, 600, 700, 800, and 900 m/s, respectively. Find the rms speed for this collection. Is it the same as the *average* speed of these molecules?

SOLUTION

SET UP AND SOLVE The average value of v^2 for the molecules is

$$(v^2)_{av} = [(500 \text{ m/s})^2 + (600 \text{ m/s})^2 + (700 \text{ m/s})^2$$
$$+ (800 \text{ m/s})^2 + (900 \text{ m/s})^2]/5$$
$$= 5.10 \times 10^5 \text{ m}^2/\text{s}^2.$$

The square root of this is v_{rms}:

$$v_{rms} = 714 \text{ m/s.}$$

The *average* speed v_{av} is given by

$$v_{av} = \frac{500 \text{ m/s} + 600 \text{ m/s} + 700 \text{ m/s} + 800 \text{ m/s} + 900 \text{ m/s}}{5}$$
$$= 700 \text{ m/s.}$$

REFLECT In general, v_{rms} and v_{av} are not the same. Roughly speaking, v_{rms} gives greater weight to the larger speeds than does v_{av}.

Practice Problem: Suppose we double all the speeds in this example. How do v_{av} and v_{rms} change? *Answers:* $v_{av} = 1400$ m/s, $v_{rms} = 1428$ m/s.

EXAMPLE 15.7   Volume of a gas at STP, revisited

Find the number of molecules and the number of moles in 1 cubic meter of air at atmospheric pressure and 0°C.

SOLUTION

SET UP AND SOLVE We use the alternative form of the ideal-gas equation, $pV = NkT$ (Equation 15.9):

$$N = \frac{pV}{kT} = \frac{(1.013 \times 10^5 \text{ Pa})(1 \text{ m}^3)}{(1.38 \times 10^{-23} \text{ J/K})(273 \text{ K})}$$
$$= 2.69 \times 10^{25}.$$

The number of moles n is

$$n = \frac{N}{N_A} = \frac{2.69 \times 10^{25} \text{ molecules}}{6.022 \times 10^{23} \text{ molecules/mol}} = 44.7 \text{ mol.}$$

Continued

The total volume is 1.00 m³, so the volume of 1 mole is

$$\frac{1.00 \text{ m}^3}{44.7 \text{ mol}} = 0.0224 \text{ m}^3/\text{mol} = 22.4 \text{ L}/\text{mol}.$$

REFLECT This is the same result that we obtained in Example 15.2 (Section 15.2), using only macroscopic quantities.

15.4 Heat Capacities

In Section 14.4, we defined specific heat capacities in terms of the measured amount of heat needed to change the temperature of a specified quantity of material by a particular amount (in SI units, 1 K and 1 kg). We didn't attempt to *predict* heat capacities on the basis of more fundamental considerations. In some simple cases, these numbers *can* be predicted on theoretical grounds, using the relation between heat and molecular energy.

Especially when we deal with heat capacities of gases, it's often more convenient to describe the quantity of substance in terms of the number of *moles* rather than the mass m_{total} of material. We define the **molar heat capacity** (also called molar specific heat) of a substance as follows:

Molar heat capacity

The amount of heat Q needed for a certain temperature change ΔT is proportional to the temperature change and to the number of moles of substance; that is,

$$Q = nC\,\Delta T, \tag{15.11}$$

where C is a quantity, different for different materials, called the *molar heat capacity* for the material. The SI units of C are $\text{J}/(\text{mol} \cdot \text{K})$.

Comparing Equations 14.10 ($Q = m_{\text{total}}c\,\Delta T$) and 15.11 ($Q = nC\,\Delta T$), and recalling that $m_{\text{total}} = Mn$, we see that the molar heat capacity C of a material is related to its specific heat capacity c by the equation

$$C = Mc, \qquad \text{(Relation of specific and molar heat capacities)} \quad (15.12)$$

where M is the molar mass of the material. That is, heat per mole equals heat per unit mass times amount of mass per mole. For example, the molar heat capacity of liquid water is about $(4186 \text{ J}/(\text{kg} \cdot \text{K}))(18 \times 10^{-3} \text{ kg}/\text{mol}) = 75 \text{ J}/(\text{mol} \cdot \text{K})$.

In our discussion of heat capacities in this chapter, we'll keep the volume of material constant so that there is no energy transfer through mechanical work. If we were to let a gas expand, it would do work (force times distance) on the moving walls of its container. We'll return to this more general case in Section 15.7. For now, with the volume held constant, we'll denote the molar heat capacity at constant volume as C_V.

Molar Heat Capacities of Gases

In this section, we present a brief introduction to the analysis of heat capacities of gases on the basis of the simple kinetic-molecular model we discussed in the last section. In that model, the molecular energy consists only of the kinetic energy K_{tr} associated with the translational motion of pointlike molecules. This energy is directly proportional to the absolute temperature T, as shown by Equation 15.7: $K_{\text{tr}} = \frac{3}{2}nRT$. When the temperature changes by a small amount ΔT, the corresponding change in kinetic energy is

$$\Delta K_{\text{tr}} = \tfrac{3}{2}nR\,\Delta T.$$

From the definition of molar heat capacity at constant volume, C_V, we also have

$$Q = nC_V \Delta T,$$

where Q is the heat input needed for a temperature change ΔT. Now if K_{tr} represents the total molecular energy, as we have assumed, then Q and ΔK_{tr} must be *equal*. Equating the last two equations, we get

$$nC_V \Delta T = \tfrac{3}{2}nR\,\Delta T, \qquad (15.13)$$

$$C_V = \tfrac{3}{2}R \qquad (\text{monatomic gas})$$

This surprisingly simple result says that the molar heat capacity (at constant volume) of *every* gas whose molecules can be represented as points is equal to $3R/2$. In SI units,

$$C_V = \tfrac{3}{2}[8.314\ \text{J}/(\text{mol} \cdot \text{K})] = 12.47\ \text{J}/(\text{mol} \cdot \text{K}).$$

For comparison, Table 15.3 gives *measured* values of C_V for several gases. We see that, for monatomic gases, our prediction is right on the money, but that it is way off for diatomic and polyatomic gases.

Thus, our point-molecule model works for monatomic gases, but for diatomic and polyatomic molecules we need something more sophisticated. For example, we can picture a diatomic molecule as *two* point masses, like a little elastic dumbbell. Such a molecule can have additional kinetic energy associated with *rotation* about axes through its center of mass, and the atoms may also have a back-and-forth *vibrating* motion along the line joining them, with additional kinetic and potential energies. Hence, the molar heat capacities for diatomic gases should be greater than those for monatomic gases. These possibilities are shown in Figure 15.13.

Equation 15.7 $\left(K_{tr} = \tfrac{3}{2}nRT\right)$ shows that the average random *translational* kinetic energy of the molecules of an ideal gas is proportional to the absolute temperature T of the gas. It turns out that the additional energy of rotational and vibrational motion is also proportional to T. A diatomic molecule has an additional average *rotational* kinetic energy of kT per molecule or RT per mole. Thus, we expect a diatomic molecule to have an average kinetic energy of $K = \tfrac{3}{2}kT + kT = \tfrac{5}{2}kT$ per molecule or $\tfrac{5}{2}RT$ per mole, and the molar heat capacity should be $\tfrac{5}{2}R$. In SI units,

$$C_V = \tfrac{5}{2}R = \tfrac{5}{2}[8.314\ \text{J}/(\text{mol}\cdot\text{K})] = 20.8\ \text{J}/(\text{mol}\cdot\text{K}) \quad (\text{diatomic gas}). \quad (15.14)$$

We note that this value agrees, within a few percent, with the measured molar heat capacities of the diatomic gases listed in Table 15.3.

Vibrational motion can also contribute to the heat capacities of gases. Molecular bonds are not rigid; they can stretch and bend, and the resulting vibrations lead to additional degrees of freedom and additional energies. Because of quantum effects, including these energies in our formulation would be more complex than for the rotational energies of diatomic molecules, and we won't attempt a quantitative discussion. In Table 15.3, the large values of C_V for some polyatomic molecules show the contributions of vibrational energy. In addition, a molecule with three or more atoms that are not in a straight line has an average rotational kinetic energy of $\tfrac{3}{2}kT$ rather than just kT.

15.5 The First Law of Thermodynamics

Energy relationships are of central importance in the study of energy-conversion devices such as engines, batteries, and refrigerators and also in the functioning of living organisms. We've studied energy transfer through mechanical work

TABLE 15.3 Molar heat capacities of gases

Type of gas	Gas	$C_V\,(\text{J}/(\text{mol}\cdot\text{K}))$
Monatomic	He	12.47
	Ar	12.47
Diatomic	H_2	20.42
	N_2	20.76
	O_2	21.10
	CO	20.85
Polyatomic	CO_2	28.46
	SO_2	31.39
	H_2S	25.95

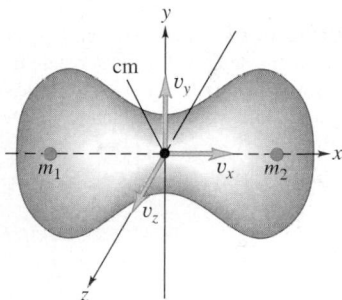

(a) Translational motion

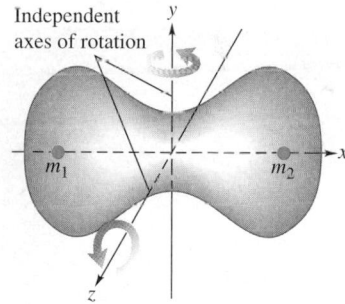

(b) Rotational motion

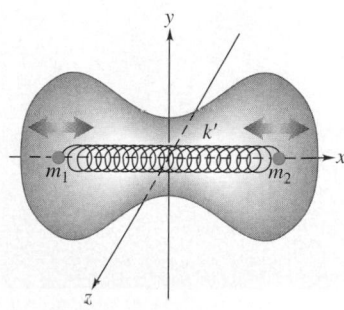

(c) Vibrational motion

▲ **FIGURE 15.13** A diatomic molecule can move in three ways: (a) by translational motion of the whole molecule through space, (b) by rotation of the molecule about an axis, and (c) by vibration of the two atoms on their springlike mutual bond.

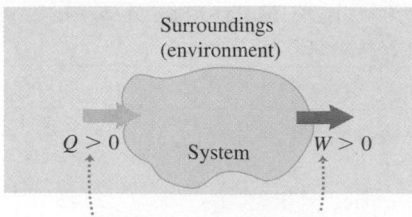

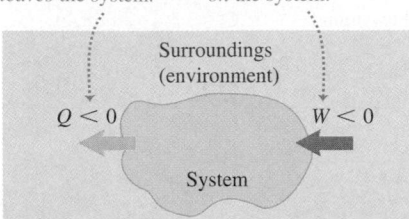

▲ **FIGURE 15.14** A thermodynamic system may exchange energy with its surroundings by means of heat (which may enter or leave the system) and work (which may be done by or on the system).

▲ **FIGURE 15.15** The catastrophic eruption of Mt. St. Helens on May 18, 1980, a dramatic illustration of the enormous pressure developed in a confined hot gas. The eruption column is driven in part by this pressure and in part by the buoyancy of the hot gas, which is much less dense than the surrounding air.

(Chapter 7) and through heat transfer (Chapters 14 and 15). Now we're ready to combine and generalize those principles. *Thermodynamics* is the study of energy relationships that involve heat, mechanical work, and other aspects of energy and energy transfer. The first law of thermodynamics extends the principle of conservation of energy to include heat as well as mechanical energy.

Energy, Heat, and Work

To state energy relationships precisely, we'll always use the concept of a *thermodynamic system,* and we'll discuss *heat* and *work* as two means of transferring energy into or out of such a system. The system might be a quantity of expanding steam in a turbine, the refrigerant in an air conditioner, some other specific quantity of material, or, sometimes, a particular device or an organism. A **thermodynamic system** (Figure 15.14) is a system that can interact (and exchange energy) with its surroundings, or environment, in at least two ways, one of which is transfer. A familiar example of a thermodynamic system is a quantity of gas confined in a cylinder with a piston, similar to the cylinders in internal-combustion engines. Energy can be added to the system by conduction of heat, and the system can also do *work* as the gas exerts a force on the piston and moves it through a displacement. With thermodynamic systems, as with all others, it's essential to define clearly at the start exactly what is and is not included in the system. Only then can we describe unambiguously the energy transfers into and out of that system.

Thermodynamics has its roots in practical problems. The engine in an automobile and the jet engines in an airplane use the heat of combustion of their fuel to perform mechanical work in propelling the vehicle. Muscle tissue in living organisms metabolizes chemical energy in food and performs mechanical work on the organism's surroundings. A steam engine or steam turbine uses the heat of combustion of coal or some other fuel to perform mechanical work such as driving an electric generator or pulling a train.

In all these situations, we describe the energy relations in terms of the quantity of heat Q added *to* the system and the work W done *by* the system. Both Q and W may be positive or negative. Figure 15.14 shows the sign conventions we'll use. As in the preceding chapter, a positive value of Q represents heat flow *into* the system, with a corresponding input of energy to it. Negative Q represents heat flow *out of* the system. A positive value of W represents work done *by* the system as it pushes outward against its surroundings, such as work done by an expanding gas. Thus, positive work corresponds to energy *leaving* the system. Negative W, such as work done *on a gas* by its surroundings while the gas is being compressed, represents energy *entering* the system. We'll use these rules consistently in the examples in this chapter and the next.

NOTE ▶ There is no uniformity in the literature concerning the signs of Q and W. In particular, W is sometimes defined with the opposite sign, as work done *on* the system rather than *by* it. The moral is, when you read other references, be careful! ◀

In this discussion, we'll be concerned mostly with *equilibrium* processes. In such processes, heat transfer takes place so slowly that the system is always very nearly in thermal equilibrium. (That is, the temperature is uniform throughout the system.). Also, changes in volume take place slowly enough so that the system is always very nearly in *mechanical equilibrium.* (That is, the pressure is uniform throughout the system.) Including both conditions, we speak of states of **thermodynamic equilibrium.** In non-equilibrium processes, there is no such thing as a single temperature or pressure for the system as a whole.

▶ **Application Your typical steamship.** It's a truism that the industrial revolution was powered by steam. In a steam engine, water is boiled to create steam, which later cools and condenses. The resulting changes in pressure and volume do work, such as moving a piston. We've all seen historical images of steam locomotives and steamships. What you may not realize is that today's nuclear submarines are . . . steamships. A reactor core is simply a source of heat. To do work with that heat—and, not incidentally, to prevent the core from overheating and melting down—the reactor uses steam. And what better method? Steam power is, after all, a mature technology.

Work Done during Volume Changes

A gas in a cylinder with a movable piston is a simple example of a thermodynamic system. Internal-combustion engines, steam engines, and compressors in refrigerators and air conditioners all use some version of such a system. In the next several sections, we'll use the gas-in-cylinder system to explore several kinds of processes that involve energy transformations. We'll look first at the *work* done by the system during a change in volume. When a gas expands, it pushes on its boundary surfaces as they move outward; an expanding gas always does positive work. The same thing is true of any solid or fluid material confined under pressure. Figure 15.15 shows an impressive example of the work done by expanding gases.

Figure 15.16 shows a solid or fluid in a cylinder with a movable piston. Suppose that the cylinder has cross-sectional area A and that the pressure exerted by the system at the piston face is p. Then the magnitude F of the total force exerted on the piston by the system is $F = pA$. When the piston moves out a small distance Δx, the work W done by the force is

$$W = F \, \Delta x = pA \, \Delta x. \tag{15.15}$$

The change in volume is $\Delta V = A \, \Delta x$, so we can express the work done by the system as

$$W = p \, \Delta V.$$

(We assume that the process proceeds so slowly that mechanical equilibrium is maintained; then, at any instant, the pressure is the same throughout the system.) If the pressure remains constant while the volume changes by a finite amount— say, from V_1 to V_2—the total work done by the system is

$$W = p(V_2 - V_1) \qquad \text{(constant pressure only)}. \tag{15.16}$$

But what if the pressure is *not* constant during the expansion? Suppose, for example, that we let an ideal gas expand, keeping its temperature constant. Then the product pV is constant, so as the volume V increases, the pressure p must decrease. This situation is similar to one in which work is done by a varying force, a phenomenon we studied in Chapter 7, and we can handle it in the same way. We divide the total change in volume into many small changes ΔV_1, ΔV_2, . . . , each so small that we can consider the pressure to be constant within each one. Calling the various pressures $p_1, p_2, \ldots$, we represent the total work as

$$W = p_1 \, \Delta V_1 + p_2 \, \Delta V_2 + \cdots.$$

As we take the ΔV's smaller and smaller, this approximation becomes more and more precise.

We can also draw a graph showing the variation of p with V, as in Figure 15.17a, with pressure p on the vertical axis and volume V on the horizontal axis. Then each product $p_1 \, \Delta V_1$, $p_2 \, \Delta V_2$, and so on, corresponds to the area of one of the rectangles in Figure 15.17b, and the sum of these areas is approximately equal to the total area under the curve between V_1 and V_2. As the ΔV's become smaller and smaller, the approximation gets better and better. So we have the following general result:

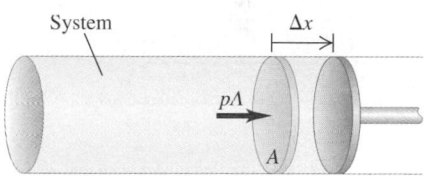

▲ **FIGURE 15.16** The work done by the system during the small expansion Δx is $\Delta W = F \, \Delta x = pA \, \Delta x$.

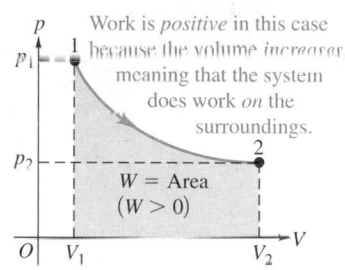

(a) pV diagram for a system undergoing a change in volume with varying pressure

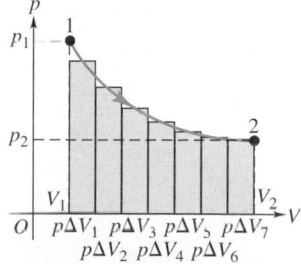

(b) The curve in **(a)** treated as a series of small constant-pressure intervals

▲ **FIGURE 15.17** The work done by a system equals the area under the curve on a pV diagram.

The volume *decreases* ($V_2 < V_1$), meaning that the surroundings do work on the system. Thus, work is *negative*.

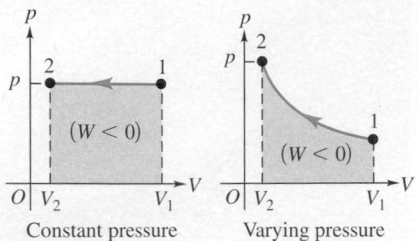

▲ **FIGURE 15.18** Two cases in which the work is negative (meaning that work is done *on* the system, reducing its volume).

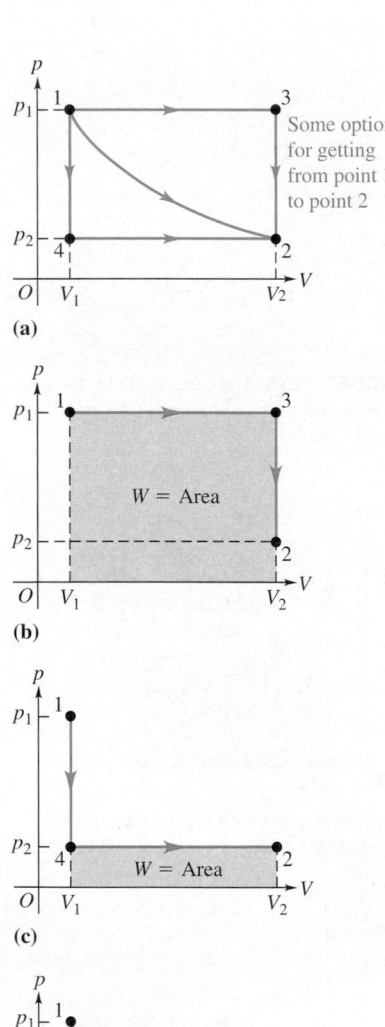

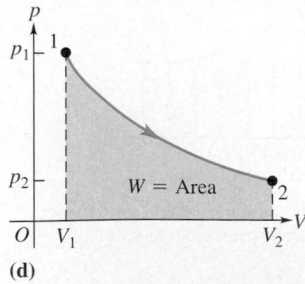

▲ **FIGURE 15.19** (a) Three different paths between state 1 and state 2. (b)–(d) The work done by the system during a transition between two states depends on the path chosen.

Graphical representation of work during a volume change

The work W done by a system is equal to the area under the curve on a pV diagram.

According to the sign rule that we stated earlier in this section, W is always the work done *by* the system, and it is always *positive* when a system *expands*. When a system expands from state 1 to state 2 as shown in Figure 15.17, the area is positive. In a *compression* from state 1 to state 2 (Figure 15.18), the change in volume is negative, and the total work done by the system on the piston by the system is negative. This corresponds to a *negative* area; thus, when a system is compressed, its volume decreases, and it does *negative* work on its surroundings.

A particular case that occurs often in problems is the *isothermal* (constant-temperature) expansion of an ideal gas. As the gas expands, the pressure decreases, so we can't use Equation 15.16. Methods of calculus can be used to show that when n moles of an ideal gas expand from volume V_1 to volume V_2 at constant temperature T, the total work W done by the gas is

$$W = nRT \ln \frac{V_2}{V_1}, \tag{15.17}$$

where "ln" stands for the natural logarithm (i.e., the logarithm to the base e).

When a system changes from an initial state to a final state, it passes through a series of intermediate states; we call this series of states a **path.** There are always infinitely many different possibilities for these intermediate states. When they are all equilibrium states, the path can be plotted on a pV diagram (Figure 15.19). Point 1 represents an initial state (1) with pressure p_1 and volume V_1, and point 2 represents a final state (2) with pressure p_2 and volume V_2. For example (Figure 15.19b), we could keep the pressure constant at p_1 while the system expands to volume V_2 (point 3 on the diagram) and then reduce the pressure to p_2 (probably by decreasing the temperature) while keeping the volume constant at V_2 (to point 2 on the diagram). The work done by the system during this process is the area under the line $1 \rightarrow 3$; the system does no work during the constant-volume process $3 \rightarrow 2$. Or the system might traverse the path $1 \rightarrow 4 \rightarrow 2$ (Figure 15.19c); in that case, the work is the area under the line $4 \rightarrow 2$. The smooth curve from 1 to 2 is another possibility (Figure 15.19d), and the work done in that case is different from that done in each of the other paths. If the system passes through non-equilibrium states during a change of state, we may not be able to speak of a pressure or temperature for the system as a whole. There is still a path for such a change of state, but it can't be represented in terms of points and curves on a pV diagram.

We conclude that **the work done by the system depends not only on the initial and final states, but also on the intermediate states—that is, on the path.** Furthermore, we can take the system through a series of states forming a closed loop, such as $1 \rightarrow 3 \rightarrow 2 \rightarrow 4 \rightarrow 1$. In this case, the final state is the same as the initial state, but the total work done by the system is *not* zero. (In fact, it is represented on the graph by the *area* enclosed by the loop; can you prove that?) It follows that it doesn't make sense to talk about the amount of work contained in a system. In a particular state, a system may have definite values of the state coordinates p, V, and T, but it wouldn't make sense to say that it has a definite value of W.

Heat Transfer during Volume Changes

Heat transfer and work are two means by which the energy of a thermodynamic system may increase or decrease. We've just seen that when a thermodynamic system undergoes a change of state, the *work* done by the system depends not

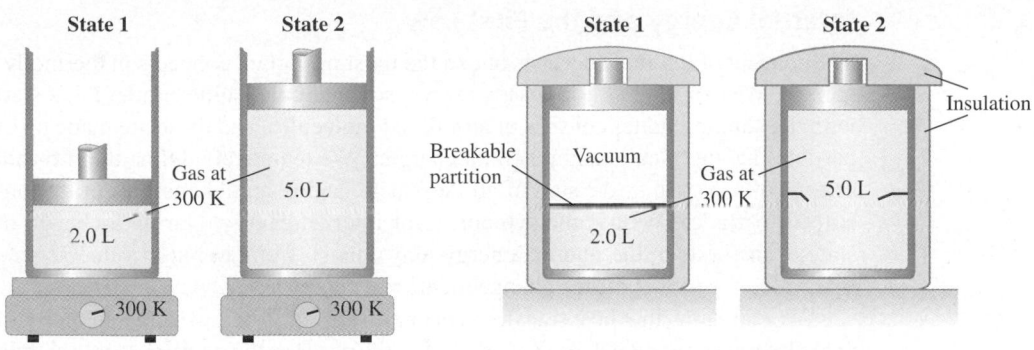

State 1 | State 2 | State 1 | State 2

(a) System does work on piston; hot plate adds heat to system $(W > 0$ and $Q > 0)$.

(b) System does no work; no heat enters or leaves system $(W = 0$ and $Q = 0)$.

▲ **FIGURE 15.20** (a) Slow, controlled isothermal expansion of a gas from an initial state 1 to a final state 2 with the same temperature, but lower pressure. (b) Rapid, uncontrolled expansion of the same gas starting at the same state 1 and ending at the same state 2.

only on the initial and final states, but also on the series of intermediate states through which it passes—that is, on the *path* from the initial state to the final state. Next we'll show that this is also true for the *heat* added to a system.

Here's an example: We want to change the volume of a certain quantity of an ideal gas from 2.0 L to 5.0 L while keeping the temperature constant at $T = 300$ K. Figure 15.20 shows two different ways we can do this. In Figure 15.20a, the gas is contained in a cylinder with a piston and has an initial volume of 2.0 L. We let the gas expand slowly, supplying heat from the electric heater to keep the temperature at 300 K. After expanding in this slow, controlled, isothermal manner, the gas reaches its final volume of 5.0 L; it absorbs a definite amount of heat in the process.

Figure 15.20b shows a different process leading to the same final state. The container is surrounded by insulating walls and is divided by a thin, breakable partition into two compartments. The lower part has a volume of 2.0 L, and the upper part has a volume of 3.0 L. In the lower compartment, we place the same amount of the same gas as in Figure 15.20a, again at $T - 300$ K. The initial state is the same as before. Now we break the partition; the gas undergoes a rapid, uncontrolled expansion, with no heat passing through the insulating walls. The final volume is 5.0 L, the same as in Figure 15.20a. This uncontrolled expansion of a gas into vacuum is called a **free expansion.** During the process, the gas is *not* in thermodynamic equilibrium.

Experiments have shown that when an ideal gas undergoes a free expansion, there is no temperature change. Therefore, the final equilibrium state of the gas is the same as that in Figure 15.20a. The intermediate states (pressures and volumes) during the transition from state 1 to state 2 are entirely different in the two cases; Figures 15.20a and 15.20b represent *two different paths* connecting the *same states* 1 and 2. For path (b), *no* heat is transferred into the system, and it does no work. **Like work, heat depends not only on the initial and final states, but also on the path.** The free expansion involves non-equilibrium states that can't be described by a single temperature or pressure for the system as a whole.

Because of this dependence on the path, it doesn't make sense to say that a system "contains" a certain quantity of heat. Suppose we were to assign an arbitrary value to "the heat in an object" in some standard reference state. Then, presumably, the heat in the object in some other state would equal the heat in the object in the reference state plus the heat added when the object goes into the second state. But that notion is ambiguous: The heat added depends on the *path* we take from the reference state to the second state. Thus, there is no consistent way to define "heat in an object;" it isn't a useful concept. It *does* make sense, however, to speak of the amount of *internal energy* in a body; this important concept is our next topic.

MasteringPHYSICS

ActivPhysics 8.5: Work Done by a Gas

Internal Energy and the First Law

The concept of internal energy is one of the most important concepts in thermodynamics. We can look at it in various ways, some simple, some subtle. Let's start with the simple. Matter consists of atoms and molecules, and these are made up of particles having kinetic and potential energies. We tentatively define the **internal energy** of a system as the sum of all the kinetic and potential energies of its constituent particles. We use the symbol U for internal energy. During a change of state of the system, the internal energy may change from an initial value U_1 to a final value U_2; we denote the change in internal energy as $\Delta U = U_2 - U_1$.

We also know that heat transfer is energy transfer. When we add a quantity of heat Q to a system and it does no work during the process, the internal energy should increase by an amount equal to Q; that is, $\Delta U = Q$. When a system does work W by expanding against its surroundings and no heat is added during the process, energy leaves the system and the internal energy decreases. That is, when W is positive, ΔU is negative, and conversely, so $\Delta U = -W$. When *both* heat transfer *and* work occur, the *total* change in internal energy is $U_2 - U_1 = \Delta U = Q - W$. We can rearrange this set of equations into the following form:

First law of thermodynamics

When heat Q is added to a system, some of this added energy remains within the system, changing its internal energy by an amount ΔU. The remainder leaves the system as the system does work W against its surroundings. Because W and Q may be positive, negative, or zero, we also expect ΔU to be positive, negative, or zero, depending on the process. We thus have

$$Q = \Delta U + W, \quad \text{or} \quad U_2 - U_1 = \Delta U = Q - W. \quad (15.18)$$

The first law of thermodynamics represents a generalization of the principle of conservation of energy to include energy transfer through heat as well as mechanical work. Actually *calculating* this energy for any real system, however, would be hopelessly complicated. By contrast, if we use Equation 15.18 to define changes in internal energy for thermodynamic processes, then we have to ask whether these changes may depend on the path taken as the system moves from one state to another. We've already seen that both work and heat exchange depend on the path taken; might this also be true for internal energy? If so, it wouldn't make sense to speak of the internal energy of an object, and it couldn't be the basis for an enlarged conservation law.

The only way to resolve such questions is through *experiment*. We study the properties of various materials; in particular, we measure Q and W for various changes of state and various paths in order to learn whether ΔU is or is not the same for different paths. The results of many such investigations are clear and unambiguous: ΔU *is independent of the path*. The change in internal energy of a system during any thermodynamic process depends only on the initial and final states, *not* on the path leading from one to the other. An equivalent statement is that the internal energy U of a system is a function of the state coordinates p, V, and T alone (actually, of any two of these, since the three variables are related by the equation of state).

Now let's return to the first law of thermodynamics. To say that the first law, given by Equation 15.18, represents conservation of energy for thermodynamic processes is correct—as far as it goes. But an important *additional* part of the content of the first law is the fact that internal energy depends only on the state of a system. In changes of state, the change in internal energy is path-independent.

If we take a system through a process that eventually returns it to its initial state (a *cyclic* process), the *total* internal energy change must be zero. Then

$$U_2 = U_1 \quad \text{and} \quad Q = W.$$

If a net quantity of work W is done by the system during this process, an equal amount of energy must have flowed into the system as heat Q. But there is no reason why either Q or W individually has to be zero.

An *isolated* system is a system that does no work on its surroundings and has no heat flow. For any process taking place in an isolated system,

$$W = Q = 0 \quad \text{and} \quad U_2 - U_1 = \Delta U = 0.$$

In other words, *the internal energy of an isolated system is constant.*

Conceptual Analysis 15.3

An expanding gas

When an ideal gas in a well-insulated container expands slowly by pushing a piston, the temperature of the gas drops because

A. the gas must do work, so its internal energy must decrease, and temperature is related to internal energy.

B. the molecules have more room to move around, which means that they can travel at a lower speed, and temperature is related to molecular speed.

C. pushing the piston robs the gas of heat, and loss of heat means that the temperature of the gas will drop.

SOLUTION The first law of thermodynamics says that $\Delta U = Q - W$. In this case the container is insulated, so $Q = 0$. The gas expands, so W is positive; therefore, ΔU must be negative. A is a correct answer. B is not correct: It's true that the molecules have room to move, but their slower speed is a consequence of doing work against the moving piston. C is not correct: The problem states that the container is well insulated, so there is no heat transfer.

PROBLEM-SOLVING STRATEGY 15.3 **First law of thermodynamics**

SET UP

1. As usual, consistent units are essential. If p is in pascals and V is in cubic meters, then W is in joules. If p and V are not in pascals and cubic meters, respectively, you may want to convert them into those units. If a heat capacity is given in terms of calories, the simplest procedure is usually to convert it to joules. Be especially careful with moles. When you use $n - m_{total}/M$ to convert between mass and number of moles, remember that if m_{total} is in kilograms, M must be in *kilograms* per mole. The usual units for M are *grams* per mole; be careful!

SOLVE

2. The internal energy change ΔU in any thermodynamic process or series of processes is independent of the path, no matter whether the substance is an ideal gas or not. This is of the utmost importance in the problems in this chapter and the next. Sometimes you will be given enough information about one path between given initial and final states to calculate ΔU for that path. Then you can use the fact that ΔU is the same for every other path between the same two states to relate the various energy quantities for other paths.

3. When a process consists of several distinct steps, it often helps to make a chart showing Q, W, and ΔU for each step. Put these quantities for each step on a different line, and arrange them so that the Q's, W's, and ΔU's form columns. Then you can apply the first law to each line; in addition, you can add each column and apply the first law to the sums. Do you see why?

REFLECT

4. Ask yourself whether the entries in the table in point 3 are consistent. For example, if you find that Q for a certain process is positive, but W and ΔU are both negative, you know that there's something wrong.

EXAMPLE 15.8 **Thermodynamics of boiling water**

One gram of water (1 cm^3) becomes 1671 cm^3 of steam when boiled at a constant pressure of 1 atm $(1.013 \times 10^5 \text{ Pa})$. The heat of vaporization at this pressure is $L_v = 2.256 \times 10^6 \text{ J/kg} = 2256 \text{ J/g}$. Compute **(a)** the work done by the water when it vaporizes; **(b)** its increase in internal energy.

SOLUTION

SET UP AND SOLVE Part (a): For a constant-pressure process, we may use Equation 15.16, $W = p(V_2 - V_1)$, to compute the work done by the vaporizing water.

The work done by the vaporizing water is

$$W = p(V_2 - V_1)$$
$$= (1.013 \times 10^5 \text{ Pa})(1671 \times 10^{-6} \text{ m}^3 - 1 \times 10^{-6} \text{ m}^3)$$
$$= 169 \text{ J}.$$

Part (b): We find the heat added to the water and then use the first law of thermodynamics. The heat added to the water is the heat of vaporization, $Q = mL_v$:

$$Q = mL_v$$
$$= (1 \text{ g})(2256 \text{ J/g})$$
$$= 2256 \text{ J}.$$

From the first law of thermodynamics, Equation 15.18, the change in internal energy is

$$\Delta U = Q - W$$
$$= 2256 \text{ J} - 169 \text{ J}$$
$$= 2087 \text{ J}.$$

REFLECT To vaporize 1 gram of water, we have to add 2256 J of heat. Most (2087 J) of this added energy remains in the system as an increase in its internal energy. The remaining 169 J leaves the system as the system does work against its surroundings while expanding from liquid to vapor. The increase in internal energy is associated mostly with the intermolecular forces that hold the molecules together in the liquid state. Because these forces are attractive, the associated potential energies are greater after work has been done on the molecules to pull them apart in forming the vapor state. It's like increasing the potential energy by pulling an elevator farther from the center of the earth.

Practice Problem: Suppose water is boiled in a pressure chamber in which the pressure is 2.00 atm, the heat of vaporization is 2200 J/g, and 1.00 kg of steam has a volume of 0.824 m^3. Determine the work done and the increase in internal energy. *Answers:* 167 J, 2033 J.

EXAMPLE 15.9 **An isothermal expansion**

Figure 15.21 shows a slow, controlled isothermal expansion of 3.0 moles of an ideal gas. The initial volume of the gas is 4.0 L and the final volume is 5.0 L. While the gas expands, heat is supplied from a flame to maintain the temperature at 300 K. How much work does the gas do during the expansion? What is the change in the internal energy of the gas?

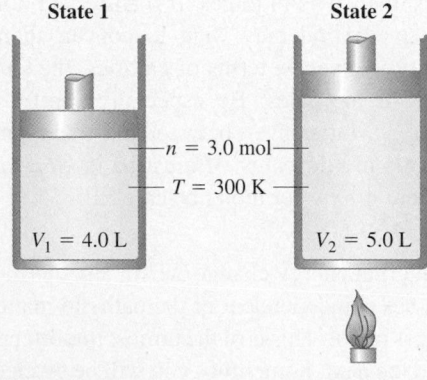

State 1   State 2

$n = 3.0 \text{ mol}$
$T = 300 \text{ K}$
$V_1 = 4.0 \text{ L}$ $V_2 = 5.0 \text{ L}$

▶ **FIGURE 15.21**

SOLUTION

SET UP AND SOLVE From Equation 15.17,

$$W = nRT \ln\left(\frac{V_2}{V_1}\right)$$

$$= (3.0 \text{ mol})[8.314 \text{ J/(mol} \cdot \text{K)}](300 \text{ K}) \ln\left(\frac{5.0 \text{ L}}{4.0 \text{ L}}\right)$$

$$= 1700 \text{ J}.$$

Because the temperature is constant and the substance is an ideal gas, the change in internal energy is zero.

REFLECT Because the internal energy does not change, all of the heat entering the system goes into work.

Practice Problem: If 2000 J of heat is added isothermally to 3.5 moles of an ideal gas with an initial volume of 3.0 L and a constant temperature of 300 K, what is the final volume of the gas? *Answer:* $V_2 = 3.8$ L.

EXAMPLE 15.10 **A series of thermodynamic processes**

A series of thermodynamic processes is shown in the pV diagram of Figure 15.22. In process ab, 150 J of heat are added to the system, and in process bd, 600 J of heat are added. Find **(a)** the internal energy change in process ab; **(b)** the internal energy change in process abd; **(c)** the total heat added in process acd.

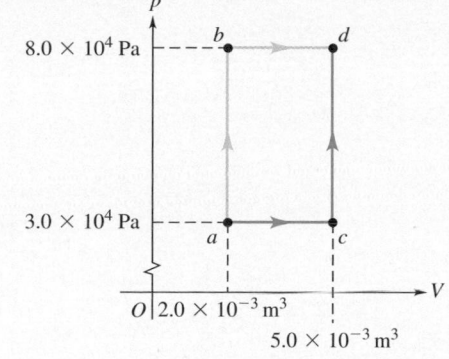

▶ **FIGURE 15.22** pV diagram showing various thermodynamic processes.

SOLUTION

SET UP AND SOLVE Part (a): No change in volume occurs during process ab, so $W_{ab} = 0$ and $\Delta U_{ab} = Q_{ab} = 150$ J.

Part (b): Process bd occurs at constant pressure, so the work done by the system during this expansion is

$$W_{bd} = p(V_2 - V_1)$$
$$= (8.0 \times 10^4 \text{ Pa})(5.0 \times 10^{-3} \text{ m}^3 - 2.0 \times 10^{-3} \text{ m}^3)$$
$$= 240 \text{ J}.$$

The total work for process abd is

$$W_{abd} = W_{ab} + W_{bd} = 0 + 240 \text{ J} = 240 \text{ J},$$

and the total heat added is

$$Q_{abd} = Q_{ab} + Q_{bd} = 150 \text{ J} + 600 \text{ J} = 750 \text{ J}.$$

Applying Equation 15.18 to process abd, we find that

$$\Delta U_{abd} = Q_{abd} - W_{abd} = 750 \text{ J} - 240 \text{ J} = 510 \text{ J}.$$

Part (c): Because ΔU is independent of the path, the internal energy change is the same for path acd as for abd; that is,

$$\Delta U_{acd} = \Delta U_{abd} = 510 \text{ J}.$$

The total work for the path acd is

$$W_{acd} = W_{ac} + W_{cd} = p(V_2 - V_1) + 0$$
$$= (3.0 \times 10^4 \text{ Pa})(5.0 \times 10^{-3} \text{ m}^3 - 2.0 \times 10^{-3} \text{ m}^3)$$
$$= 90 \text{ J}.$$

Now we apply Equation 15.18 to process acd:

$$Q_{acd} = \Delta U_{acd} + W_{acd} = 510 \text{ J} + 90 \text{ J} = 600 \text{ J}.$$

We see that although ΔU is the same (510 J) for abd and acd, W (240 J versus 90 J) and Q (750 J versus 600 J) are quite different for the two processes.

Here is a tabulation of the various quantities:

Step	Q	W	$\Delta U = Q - W$
ab	150 J	0	150 J
bd	600 J	240 J	360 J
abd	750 J	240 J	510 J

Step	Q	W	$\Delta U = Q - W$
ac	?	90 J	?
cd	?	0	?
acd	600 J	90 J	510 J

REFLECT The fact that ΔU is the same for all paths between given initial and final states plays a crucial role in this solution

15.6 Thermodynamic Processes

Here are four thermodynamic processes that occur often enough in practical problems to be worth some discussion. Their characteristics can be summarized briefly as "no heat transfer," "constant volume," "constant pressure," and "constant temperature," respectively. These four processes for an ideal gas are shown on the pV diagram in Figure 15.23.

Adiabatic Process

An **adiabatic process** is a process in which there is no heat transfer into or out of a system; in other words, $Q = 0$. We can prevent heat flow either by surrounding the system with thermally insulating material or by carrying out the process so quickly that there is not enough time for any appreciable heat flow. From the first law, we find that, for every adiabatic process,

$$U_2 - U_1 = \Delta U = -W \quad \text{(adiabatic process).} \quad (15.19)$$

MasteringPHYSICS

ActivPhysics 8.4: State Variables and Ideal Gas Law
ActivPhysics 8.6: Heat, Internal Energy, and First Law of Thermodynamics
ActivPhysics 8.8: Isochoric Process
ActivPhysics 8.9: Isobaric Process
ActivPhysics 8.10: Isothermal Process
ActivPhysics 8.11: Adiabatic Process

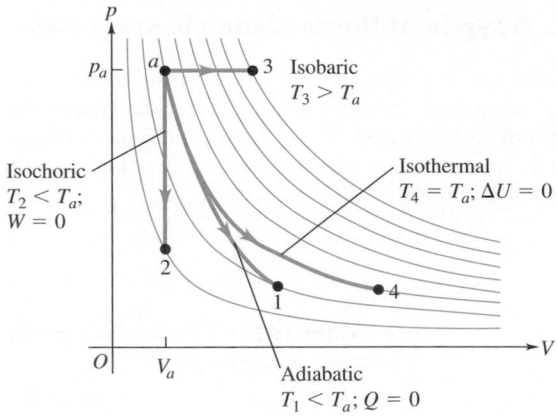

▲ **FIGURE 15.23** Four different processes for a constant amount of an ideal gas, all starting at state a. The temperature increases only during the isobaric expansion.

When a system expands under adiabatic conditions, W is positive, ΔU is negative, and the internal energy decreases. When a system is *compressed* adiabatically, W is negative and U increases. An increase in internal energy is often, though not always, accompanied by a rise in temperature.

The compression stroke in an internal-combustion engine is an example of a process that is approximately adiabatic. The temperature rises as the air–fuel mixture in the cylinder is compressed. Similarly, the expansion of the burned fuel during the power stroke is an approximately adiabatic expansion with a drop in temperature.

Isochoric Process

An **isochoric process** is a *constant-volume* process. When the volume of a thermodynamic system is constant, the system does no work on its surroundings. Then $W = 0$, and

$$U_2 - U_1 = \Delta U = Q \qquad \text{(isochoric process)}. \qquad (15.20)$$

In an isochoric process, all the energy added as heat remains in the system as an increase in its internal energy. Heating a gas in a closed constant-volume container is an example of an isochoric process.

Isobaric Process

An **isobaric process** is a *constant-pressure* process. In general, for an isobaric process, *none* of the three quantities ΔU, Q, and W in the first law is zero, but calculating W is easy, as we saw in Section 15.5. Figure 15.24 shows how an isobaric process looks on a pV diagram. For an isobaric process,

$$W = p(V_2 - V_1) \qquad \text{(isobaric process)}. \qquad (15.16)$$

Example 15.8 is an example of an isobaric process.

Isothermal Process

An **isothermal process** is a *constant-temperature* process. For a process to be isothermal, the system must remain in thermal equilibrium. This means that any heat flow into or out of the system must occur slowly enough to maintain thermal equilibrium. In general, *none* of the quantities ΔU, Q, and W is zero. Example 15.9 is an example of an isothermal process.

In some special cases (including ideal gases), the internal energy of a system depends *only* on its temperature, not on its pressure or volume. For such systems, if the temperature is constant, the internal energy is also constant; then $\Delta U = 0$

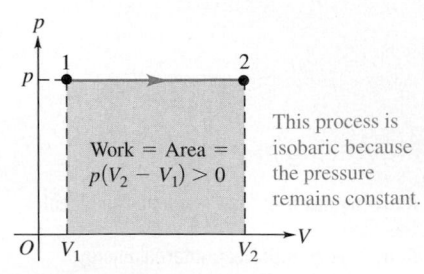

▲ **FIGURE 15.24** A pV diagram for an isobaric process.

and $Q = W$. That is, any energy entering the system as heat Q must leave it again as work W done by the system. Example 15.9, involving an ideal gas, is an example of an isothermal process in which U is also constant. For most systems other than ideal gases, the internal energy depends on pressure as well as temperature, so U may vary even when T is constant.

15.7 Properties of an Ideal Gas

We stated in Section 15.5 that, for an ideal gas, the internal energy U depends only on temperature, not on pressure or volume. How do we know this? Let's think again about the free-expansion experiment described in Section 15.5. A thermally insulated container with rigid walls is divided into two compartments by a partition (Figure 15.25). One compartment contains a quantity of an ideal gas, and the other is evacuated.

When the partition is removed or broken, the gas expands to fill both parts of the container. This process is called a *free expansion.* The gas does no work because the walls of the container don't move. Both Q and W are zero, so the internal energy U is constant. This is true of any substance, whether it is an ideal gas or not.

Does the temperature of a gas change during a free expansion? Suppose it *does* change, while the internal energy stays the same. In that case, the internal energy must depend on both temperature and volume, or both temperature and pressure, but certainly not on temperature alone. But if T is the *same* after the free expansion as before, a process for which we know that U is constant even though both p and V change, then we have to conclude that U depends *only* on T, not on p or V.

Many experiments have shown that when an ideal gas undergoes a free expansion, its temperature *does not* change. The conclusion is that **the internal energy of an ideal gas depends only on its temperature, not on its pressure or volume.** This property, in addition to the ideal-gas equation of state, is part of the ideal-gas model. In the next chapter, we'll use the preceding property several times, so make sure that you understand it.

For non-ideal gases, some temperature change occurs during free expansions, even though the internal energy is constant. This shows that the internal energy cannot depend *only* on temperature; it must depend on pressure as well. From a microscopic viewpoint, this dependence on pressure is not surprising. Non-ideal gases usually have attractive intermolecular forces. When molecules move farther apart, the associated potential energies increase. If the total internal energy is constant, the kinetic energies must decrease. Temperature is directly related to molecular *kinetic* energy, so, for a non-ideal gas, a free expansion is usually accompanied by a *drop* in temperature.

Heat Capacities of an Ideal Gas

We defined molar heat capacity in Section 15.4. The molar heat capacity of a substance depends on the conditions under which the heat is added. It is usually easiest to measure the heat capacity of a gas in a closed container under constant-volume conditions. The corresponding heat capacity is the **molar heat capacity at constant volume,** denoted by C_V. Heat capacity measurements for solids and liquids are usually carried out in the atmosphere under constant atmospheric pressure, and we call the resulting heat capacity the **molar heat capacity at constant pressure,** denoted by C_p. If neither p nor V is constant, we have an infinite number of possible heat capacities.

To measure C_V for a gas, we add heat to raise the temperature of the gas in a rigid container with constant volume. To measure C_p, we add heat while letting the gas expand just enough to keep the pressure constant as the temperature rises.

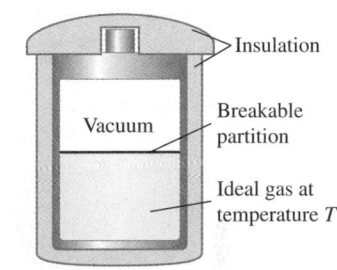

▲ **FIGURE 15.25** When the partition is broken, the gas expands freely into the vacuum region.

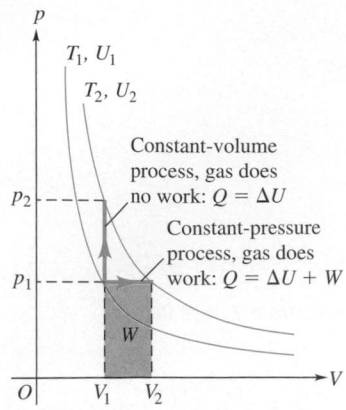

FIGURE 15.26 Constant-volume and constant-pressure processes for a sample of an ideal gas. For an ideal gas, U depends only on T.

Why should these two molar heat capacities be different? The answer lies in the first law of thermodynamics. In a constant-volume temperature change, the system does no work, and the change in internal energy, ΔU, equals the heat added, Q. In a constant-pressure temperature change, by contrast, the volume *must* increase; otherwise, the pressure could not remain constant. As the material expands, it does an amount of work W. According to the first law,

$$Q = \Delta U + W.$$

For a given temperature change, the heat input for a constant-pressure process must be *greater* than that for a constant-volume process because additional energy is needed to account for the work W done during the expansion, so C_p is greater than C_V. The pV diagram in Figure 15.26 shows this relationship. For air, C_p is 40% greater than C_V.

In fact, there's a simple relation between C_V and C_p for an ideal gas. We'll omit the details of the derivation; the result is

$$C_p = C_V + R. \tag{15.21}$$

For an ideal gas, C_p is always greater than C_V; the difference is the gas constant R. (Of course, R must be expressed in the same units as C_V and C_p, such as $J/(mol \cdot K)$.)

Conceptual Analysis 15.4

C_p versus C_V

The main reason more heat is needed to raise the temperature of an ideal gas by a given amount at constant pressure than at constant volume is that

A. at constant volume the molecules remain the same mean distance apart.
B. at constant pressure some energy must be provided to increase the volume occupied by the gas.

C. at constant pressure some energy must be provided to separate the molecules, which are subject to cohesive forces of attraction.

SOLUTION For an ideal gas, the molecules have no appreciable interaction, so answers A and C are eliminated. As the gas expands during the constant-pressure process, it has to do work against the walls of its container as the volume increases. This work represents an additional energy requirement; B is correct.

Measured values of C_p and C_V are given in Table 15.4 for several real gases at low pressures; the difference in most cases is approximately 8.31 $J/(mol \cdot K)$, as Equation 15.21 predicts. The table also shows that the molar heat capacity of a gas is related to its molecular structure, as we discussed in Section 15.4. In fact, the first two columns of Table 15.4 are the same as those in Table 15.3.

NOTE ▶ Here's a final reminder: For an ideal gas, the change in internal energy for *any* infinitesimal process is given by $\Delta U = nC_V \Delta T$, *whether the volume is constant or not*. This relation, which comes in handy in the next conceptual analysis example, holds for other substances *only* when the volume is constant. ◀

TABLE 15.4   Molar heat capacities of gases at low pressure

Type of gas	Gas	$C_V(J/(mol \cdot K))$	$C_p(J/(mol \cdot K))$	$C_p - C_V(J/(mol \cdot K))$	γ
Monatomic	He	12.47	20.78	8.31	1.67
	Ar	12.47	20.78	8.31	1.67
Diatomic	H_2	20.42	28.74	8.32	1.41
	N_2	20.76	29.07	8.31	1.40
	O_2	20.85	29.17	8.31	1.40
	CO	20.85	29.16	8.31	1.40
Polyatomic	CO_2	28.46	36.94	8.48	1.30
	SO_2	31.39	40.37	8.98	1.29
	H_2S	25.95	34.60	8.65	1.33

Conceptual Analysis 15.5

A compressed gas

An ideal gas in a container is compressed to half its original volume with no change in temperature. The internal energy of the gas is then

A. half its original value, because of the heat released.
B. unchanged, because the temperature is unchanged.
C. twice its original value, because the energy comes from work on the gas.
D. unchanged, because no heat is absorbed or released.

SOLUTION The internal energy of an ideal gas depends only on its temperature. In this case the temperature doesn't change, so the internal energy is also unchanged (answer B). Heat is released, and work is done on the gas, but these two quantities exactly balance each other. Thus, answers A, C, and D can't be correct.

Adiabatic Process for an Ideal Gas

An adiabatic process, defined in Section 15.5, is a process in which no heat transfer takes place between a system and its surroundings. In an adiabatic process, $Q = 0$ and (from the first law) $\Delta U = -W$. Zero heat transfer is an idealization, but a process is approximately adiabatic if the system is well insulated or if the process takes place so quickly that there is not enough time for any appreciable heat flow to occur.

An adiabatic process for an ideal gas is shown on the pV diagram of Figure 15.27. As the gas expands from volume V_a to V_b, its temperature drops because it does positive work, decreasing its internal energy. If point a, representing the initial state, lies on an isotherm at temperature $T + \Delta T$, then point b for the final state is on a different isotherm at a lower temperature T. An adiabatic curve at any point is always *steeper* than the isotherm passing through the same point. For an adiabatic *compression* from V_b to V_a, the situation is reversed and the temperature rises.

The air in the output pipes of air compressors used in gasoline stations and paint-spraying equipment and to fill scuba tanks is always warmer than the air entering the compressor because of the approximately adiabatic compression. When air is compressed in the cylinders of a diesel engine during the compression stroke, it gets so hot that fuel injected into the cylinders ignites spontaneously.

The first law of thermodynamics, together with the ideal-gas equation of state, can be used to derive a relation between pressure and volume for an adiabatic process for an ideal gas. This relation depends on the ratio of the two molar heat capacities at constant pressure and at constant volume. We'll denote this ratio by γ; that is, $\gamma = C_p/C_V$. For diatomic gases, γ is approximately 1.40, as Table 15.4 shows.

We recall that for an *isothermal* (constant-temperature) process, the product pV is constant, so p is inversely proportional to V. It turns out that for an *adiabatic* process, the product pV^γ is constant, and p is proportional to $1/V^\gamma$. So when V increases, the pressure p decreases more rapidly for an adiabatic process than an isothermal process. This is to be expected, because the temperature of an ideal gas always decreases when it expands isothermally.

When an ideal gas expands from state (p_1, V_1) to state (p_2, V_2), the state coordinates are related by

$$p_1V_1^\gamma = p_2V_2^\gamma = \text{constant.} \tag{15.22}$$

We can combine this result with the ideal-gas equation of state to obtain the following relation between volume and temperature:

$$T_1V_1^{\gamma-1} = T_2V_2^{\gamma-1}. \tag{15.23}$$

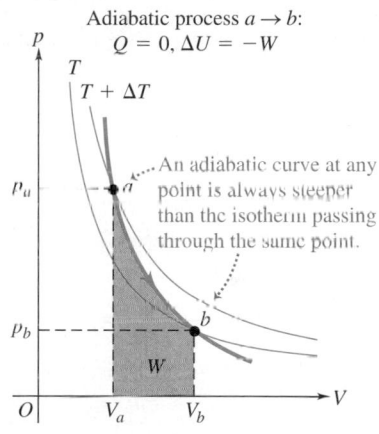

Adiabatic process $a \rightarrow b$:
$Q = 0, \Delta U = -W$

An adiabatic curve at any point is always steeper than the isotherm passing through the same point.

▲ **FIGURE 15.27** A pV diagram of an adiabatic process for an ideal gas. As the gas expands from V_a to V_b, its temperature drops from $T + \Delta T$ to T, corresponding to the decrease in internal energy due to the work W done by the gas (represented by the shaded area).

EXAMPLE 15.11 Adiabatic compression in a diesel engine

The compression ratio of a certain diesel engine is 15 to 1; this means that air in the cylinders is compressed to one-fifteenth of its initial volume (Figure 15.28). If the initial pressure is 1.01×10^5 Pa and the initial temperature is 27°C (300 K), find the final pressure and the temperature after compression. Air is mostly a mixture of diatomic oxygen and nitrogen; treat it as an ideal gas with $\gamma = 1.40$.

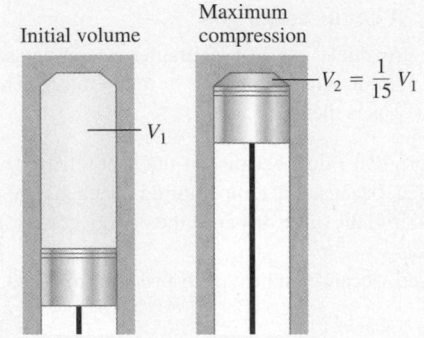

▶ FIGURE 15.28

SOLUTION

SET UP We solve Equation 15.22 for p_2. We don't need to know V_1 or V_2, but only their ratio. Similarly, we solve Equation 15.23 for T_2 and use the ratio of the initial and final volumes.

SOLVE We have $p_1 = 1.01 \times 10^5$ Pa, $T_1 = 300$ K, and $V_1/V_2 = 15$. From Equation 15.22,

$$p_2 = p_1 \left(\frac{V_1}{V_2} \right)^{\gamma} = (1.01 \times 10^5 \text{ Pa})(15)^{1.40}$$
$$= 44.8 \times 10^5 \text{ Pa} = 44 \text{ atm}.$$

From Equation 15.23,

$$T_2 = T_1 \left(\frac{V_1}{V_2} \right)^{\gamma - 1} = (300 \text{ K})(15)^{0.40} = 886 \text{ K} = 613° \text{ C}.$$

REFLECT If the compression had been isothermal, the final pressure would have been 15 atm, but because the temperature also increases during an adiabatic compression, the final pressure is much greater, about 650 psi. The high temperature attained during compression causes the fuel to ignite spontaneously, without the need for spark plugs, when it is injected into the cylinders near the end of the compression stroke.

SUMMARY

The Mole and Avogadro's Number

(Section 15.1) **Avogadro's number** $(N_A = 6.022 \times 10^{23}$ molecules/mol$)$ is the number of molecules in a mole. The **molar mass** of a substance is the mass of the substance in 1 mole.

Equations of State

(Section 15.2) The pressure, volume, and temperature of a given quantity of a substance are called state variables; they are related by an equation of state. The equation of state for an ideal gas is the **ideal-gas equation:** $pV = nRT$ (Equation 15.3); R is the **ideal-gas constant.**

Kinetic Theory of an Ideal Gas

(Section 15.3) The average translational kinetic energy of n moles of an ideal gas is directly proportional to the absolute temperature of the gas: $K_{tr} = \frac{3}{2}nRT$ (Equation 15.7). This relation can be expressed for a single particle in the gas by using Boltzmann's constant $k = R/N_A$: $K_{av} = \frac{1}{2}m(v^2)_{av} = \frac{3}{2}kT$ (Equation 15.8).

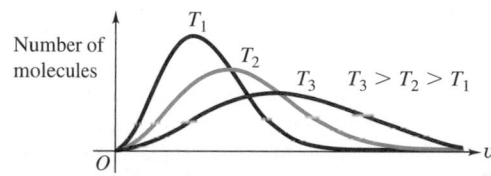

Heat Capacities

(Section 15.4) The **molar heat capacity** is the amount of heat required to change the temperature of 1 mole of material by 1 kelvin. The molar heat capacity at constant volume is the same for *every* ideal monatomic gas: $C_V = \frac{3}{2}R$ (Equation 15.13) For an ideal diatomic gas, including rotational, but not vibrational, kinetic energy, $C_V = \frac{5}{2}R$ (Equation 15.14).

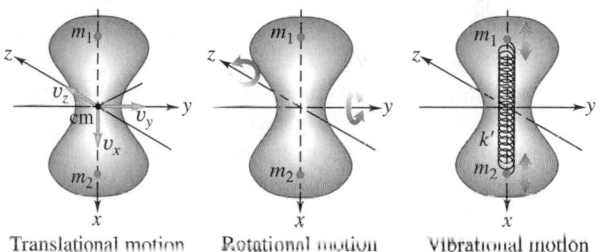

Translational motion   Rotational motion   Vibrational motion

The First Law of Thermodynamics

(Section 15.5) The **first law of thermodynamics** states that when heat Q is added to a system while work W is done by the system, the internal energy U changes by an amount $\Delta U = Q - W$ (Equation 15.18). The internal energy of any thermodynamic system depends only on its state. The change in internal energy in any process depends only on the initial and final states, not on the path taken.

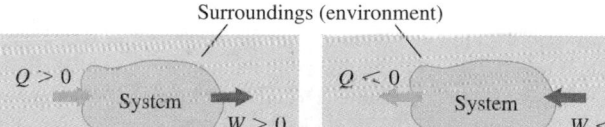

Thermodynamic Processes

(Section 15.6) Four common kinds of thermodynamic processes are as follows:

- **Adiabatic process:** No heat transfer in or out of a system; $Q = 0$.
- **Isochoric process:** Constant volume; $W = 0$.
- **Isobaric process:** Constant pressure; $W = p(V_2 - V_1)$.
- **Isothermal process:** Constant temperature.

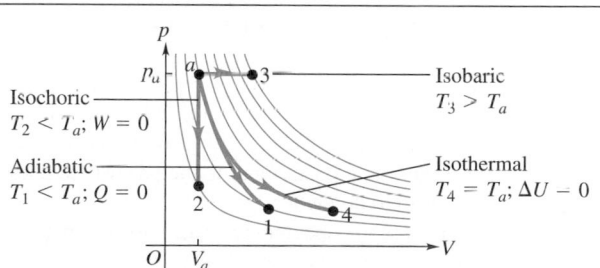

Properties of an Ideal Gas

(Section 15.7) The internal energy of an ideal gas depends only on its temperature, not its pressure or volume. For non-ideal gases, and for solids and liquids, the internal energy generally depends on both pressure and temperature. The molar heat capacities C_V and C_p of an ideal gas are related by $C_p = C_V + R$ (Equation 15.25). The ratio of C_p to C_V is denoted by γ: $\gamma = C_p/C_V$ (Equation 15.26). For an adiabatic process taking place in an ideal gas, the quantities $TV^{\gamma-1}$ and pV^γ are constant.

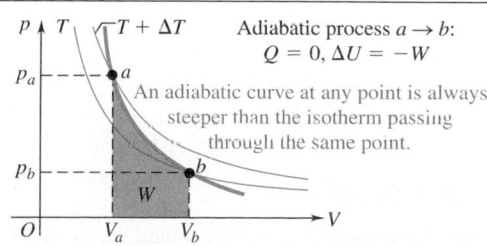

Conceptual Questions

1. In the ideal-gas equation, could you give the temperature in degrees Celsius rather than in kelvins if you used an appropriate numerical value of the gas constant R?

2. True or false? Equal masses of two different gases placed in containers of equal volume at equal temperature must exert equal pressures.

3. How does evaporation of perspiration from your skin cool your body?

4. The ideal-gas law is sometimes written in the form $p_1V_1/T_1 = p_2V_2/T_2$. Are there any extra limitations on this form of the law, or is it just as general as the form $pV = nRT$?

5. By thinking about matter on the microscopic level, explain why heat flows from a hot to a cold object, but not the other way around. (*Hint:* Look at the transfer of energy during molecular collisions.)

6. (a) If you double the absolute temperature of an ideal gas, what happens to the average kinetic energy of its molecules? To the root-mean-square speed of its molecules? (b) If you double the speeds of the molecules of an ideal gas, what happens to the absolute temperature of this gas?

7. Chemical reaction rates slow down as the temperature is decreased. (This is why we put food in the refrigerator to preserve it.) By thinking about matter on the microscopic (molecular) level, explain why this happens.

8. The *mean free path* of a gas molecule is the average distance the molecule travels before colliding with another molecule. Does this mean free path increase, decrease, or stay the same if the gas is compressed?

9. True or false? When two ideal gases are mixed, they must have the same average molecular speed when they have reached thermal equilibrium.

10. You hold an inflated balloon over a hot air vent in your house and watch it slowly expand. You then remove it and let it cool back to room temperature. During the expansion, which was larger: the heat added to the balloon or the work done by the air inside it? Explain. (Assume that air is an ideal gas.) Once the balloon has returned to room temperature, how does the net heat gained or lost by the air inside it compare to the net work done on it by the surrounding air?

11. In a constant-volume process, $\Delta U = nC_V \Delta T$; in a constant-pressure process, it is *not* true that $\Delta U = nC_p \Delta T$. Why not?

12. When a gas expands adiabatically, it does work on its surroundings. But if there is no heat input to the gas, where does the energy come from to do the work?

13. Since C_V is defined with specific reference to a constant-volume process, how can it be correct that, for an ideal gas, $\Delta U = nC_V \Delta T$ even when the volume is not constant?

14. The ratio γ found in Equations 15.22 and 15.23 must always be greater than 1. Why?

15. Do Equations 15.22 and 15.23 hold for adiabatic processes for liquids and solids? Why or why not?

Multiple-Choice Problems

1. To double both the pressure and the volume of a fixed amount of an ideal gas, you would multiply its absolute temperature by
   A. 1 (i.e., keep the temperature the same).
   B. 2.   C. 4.   D. 16.

2. Oxygen molecules are 16 times more massive than hydrogen molecules. If samples of these two gases are at the *same temperature,* what must be true about the motion of the molecules? (There may be more than one correct choice.)
   A. The rms molecular speed is the same for both gases.
   B. The average kinetic energy is the same for both gases.
   C. The rms speed of the hydrogen molecules is 4 times greater than that of the oxygen molecules.
   D. The rms speed of the hydrogen molecules is 16 times greater than that of the oxygen molecules.

3. An ideal gas in a cubical box having sides of length L exerts a pressure p on the walls of the box. If all of this gas is put into a box having sides of length $2L$ without changing its temperature, the pressure it exerts on the walls of the larger box will be
   A. $4p$.   B. $p/2$.   C. $p/4$.   D. $p/8$.

4. If you mix different amounts of two ideal gases that are originally at different temperatures, what must be true of the final state after the temperature stabilizes? (There may be more than one correct choice.)
   A. Both gases will reach the same final temperature.
   B. The final rms molecular speed will be the same for both gases.
   C. The final average kinetic energy of a molecule will be the same for both gases.

5. If you double the rms speed of the molecules of an ideal gas, which of the following statements is or are true about the gas?
   A. Its absolute temperature is doubled.
   B. Its Celsius temperature is doubled.
   C. Its absolute temperature is quadrupled.
   D. Its Celsius temperature is quadrupled.

6. In an ideal gas, which of the following quantities can be determined by measuring *just* the temperature of the gas?
   A. The average kinetic energy of the molecules.
   B. The total kinetic energy of the molecules.
   C. The pressure of the gas.

7. You add equal amounts of heat to two identical cylinders containing equal amounts of the same ideal gas. Cylinder A is allowed to expand, while cylinder B is not. How do the temperature changes of the two cylinders compare?
   A. The two cylinders will experience the same temperature change.
   B. Cylinder A will experience a greater temperature change.
   C. Cylinder B will experience a greater temperature change.

8. When ice melts at 0°C, its volume decreases. Compared to the amount of heat added, the change in internal energy is
   A. greater.   B. less.   C. the same.

9. The formula $\Delta U = nC_V \Delta T$ for the change in the internal energy of a fixed amount of an ideal gas is valid
   A. only for constant-volume processes.
   B. only for adiabatic processes.
   C. only for isobaric processes.
   D. for *any* process involving that fixed amount of the gas.

10. For the process shown in the pV diagram in Figure 15.29, the total work in going from a to d along the path shown is
    A. 15×10^5 J.   B. 9×10^5 J.   C. 6×10^5 J.   D. 1×10^5 J.

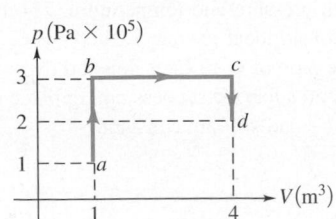

▶ **FIGURE 15.29** Multiple-choice problem 10.

11. You have two boxes, one containing some hot gas and the other containing some cold gas. *This is all you know about these boxes.* What can you validly conclude about the characteristics of the gas in the boxes? (There may be more than one correct choice.)
   A. The molecules in the hot gas are moving faster, on average, than those in the cold gas.
   B. The molecules of the hot gas have greater average kinetic energy than those of the cold gas.
   C. The molecules of the hot gas have more *total* kinetic energy than those of the cold gas.
   D. The pressure of the hot gas is greater than that of the cold gas.

12. The gas shown in Figure 15.30 is in a completely insulated rigid container. Weight is added to the frictionless piston, compressing the gas. As this is done,
   A. the temperature of the gas stays the same because the container is insulated.
   B. the temperature of the gas increases because heat is added to the gas.
   C. the temperature of the gas increases because work is done on the gas.
   D. the pressure of the gas stays the same because the temperature of the gas is constant.

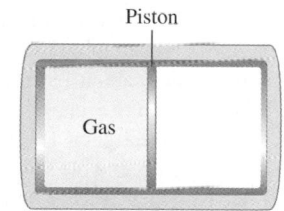

▲ **FIGURE 15.30** Multiple-choice problem 12.

13. Which of the following must be true about an ideal gas that undergoes an isothermal *expansion?* (There may be more than one correct choice.)
   A. No heat enters the gas.
   B. The pressure of the gas decreases.
   C. The internal energy of the gas does not change.
   D. The gas does positive work.

14. An ideal gas is initially confined to one side of a perfectly insulated rigid chamber by a movable frictionless piston, as shown in Figure 15.31. (The other side of the chamber is evacuated.) What is true of this gas after the piston is suddenly pulled to the right end of the chamber? (There may be more than one correct choice.)
   A. The expansion causes the temperature of the gas to decrease.
   B. The pressure of the gas decreases.
   C. The temperature of the gas does not change.
   D. The expansion causes the internal energy of the gas to decrease.

15. Suppose that, in the previous problem, the piston is not pulled but instead is allowed to move slowly to the right as it is hit by the gas molecules. What is true about this gas just as the piston reaches the right end of the chamber? (There may be more than one correct choice.)
   A. The expansion causes the temperature of the gas to decrease.
   B. The pressure of the gas decreases.

C. The temperature of the gas does not change.
D. The expansion causes the internal energy of the gas to decrease.

Problems

15.1 Equations of State

1. • A cylindrical tank has a tight-fitting piston that allows the volume of the tank to be changed. The tank originally contains 0.110 m^3 of air at a pressure of 3.40 atm. The piston is slowly pulled out until the volume of the gas is increased to 0.390 m^3. If the temperature remains constant, what is the final value of the pressure?

2. • Helium gas with a volume of 2.60 L under a pressure of 1.30 atm and at a temperature of 41.0°C is warmed until both the pressure and volume of the gas are doubled. (a) What is the final temperature? (b) How many grams of helium are there? The molar mass of helium is 4.00 g/mol.

3. • A 3.00 L tank contains air at 3.00 atm and 20.0°C. The tank is sealed and cooled until the pressure is 1.00 atm. (a) What is the temperature then in degrees Celsius, assuming that the volume of the tank is constant? (b) If the temperature is kept at the value found in part (a) and the gas is compressed, what is the volume when the pressure again becomes 3.00 atm?

4. • A 20.0 L tank contains 0.225 kg of helium at 18.0°C. The molar mass of helium is 4.00 g/mol. (a) How many moles of helium are in the tank? (b) What is the pressure in the tank, in pascals and in atmospheres?

5. • A room with dimensions 7.00 m by 8.00 m by 2.50 m is filled with pure oxygen at 22.0°C and 1.00 atm. The molar mass of oxygen is 32.0 g/mol. (a) How many moles of oxygen are required? (b) What is the mass of this oxygen, in kilograms?

6. • Three moles of an ideal gas are in a rigid cubical box with sides of length 0.200 m. (a) What is the force that the gas exerts on each of the six sides of the box when the gas temperature is 20.0°C? (b) What is the force when the temperature of the gas is increased to 100.0°C?

7. • A large cylindrical tank contains 0.750 m^3 of nitrogen gas at 27°C and 1.50×10^5 Pa (absolute pressure). The tank has a tight-fitting piston that allows the volume to be changed. What will be the pressure if the volume is decreased to 0.480 m^3 and the temperature is increased to 157°C?

8. • **Planetary atmospheres.** (a) Calculate the density of the atmosphere at the surface of Mars (where the pressure is 650 Pa and the temperature is typically 253 K, with a CO_2 atmosphere), Venus (with an average temperature of 730 K and pressure of 92 atm, with a CO_2 atmosphere), and Saturn's moon Titan (where the pressure is 1.5 atm and the temperature is −178°C, with a N_2 atmosphere). (b) Compare each of these densities with that of the earth's atmosphere, as determined in Example 15.4. Consult the periodic chart in Appendix C to determine molar masses.

9. • The gas inside a balloon will always have a pressure nearly equal to atmospheric pressure, since that is the pressure applied to the outside of the balloon. You fill a balloon with helium (a nearly ideal gas) to a volume of 0.600 L at a temperature of 19.0°C. What is the volume of the balloon if you cool it to the boiling point of liquid nitrogen (77.3 K)?

10. • **Lung volume.** The total lung volume for a typical person is
BIO 6.00 L. A person fills her lungs with air at an absolute pressure of 1.00 atm. Then, holding her breath, she compresses her chest cavity, decreasing her lung volume to 5.70 L. What is the pressure of the air in her compressed lungs, assuming that the temperature of the air remains constant?

11. • A Jaguar XK8 convertible has an eight-cylinder engine. At the beginning of its compression stroke, one of the cylinders contains 499 cm^3 of air at atmospheric pressure $(1.01 \times 10^5$ Pa$)$ and a temperature of 27.0°C. At the end of the stroke, the air has been compressed to a volume of 46.2 cm^3 and the gauge pressure has increased to 2.72×10^6 Pa. Compute the final temperature.

12. •• A diver observes a bubble of air rising from the bottom of a lake (where the absolute pressure is 3.50 atm) to the surface (where the pressure is 1.00 atm). The temperature at the bottom is 4.0°C and the temperature at the surface is 23.0°C. (a) What is the ratio of the volume of the bubble as it reaches the surface to its volume at the bottom? (b) Would it be safe for the diver to hold his breath while ascending from the bottom of the lake to the surface? Why or why not?

13. •• At an altitude of 11,000 m (a typical cruising altitude for a jet airliner), the air temperature is −56.5°C and the air density is 0.364 kg/m^3. What is the pressure of the atmosphere at that altitude? The molar mass of air is 28.8 g/mol.

14. •• If a certain amount of ideal gas occupies a volume V at STP on earth, what would be its volume (in terms of V) on Venus, where the temperature is 1003°C and the pressure is 92 atm?

15.2 Phases of Matter

15. • Calculate the volume of 1.00 mol of liquid water at a temperature of 20°C (at which its density is 998 kg/m^3), and compare this volume with the volume occupied by 1.00 mol of water at the critical point, which is 56×10^{-6} m^3. Water has a molar mass of 18.0 g/mol.

16. •• Solid water (ice) is slowly warmed from a very low temperature. (a) What minimum external pressure p_1 must be applied to the solid if a melting phase transition is to be observed? Describe the sequence of phase transitions that occur if the applied pressure p is such that $p < p_1$. (b) Above a certain maximum pressure p_2, no boiling transition is observed. What is this pressure? Describe the sequence of phase transitions that occur if $p_1 < p < p_2$.

17. •• The atmosphere of the planet Mars is 95.3% carbon dioxide (CO_2) and about 0.03% water vapor. The atmospheric pressure is only about 600 Pa, and the surface temperature varies from −30°C to −100°C. The polar ice caps contain both CO_2 ice and water ice. Could there be *liquid* CO_2 on the surface of Mars? Could there be liquid water? Why or why not?

15.3 Kinetic Theory of an Ideal Gas

18. • Find the mass of a single sulfur (S) atom and an ammonia (NH_3) molecule. Use the periodic table in Appendix C to find the molar masses.

19. • How many water molecules are there in a 1.00 L bottle of water? The molar mass of water is 18.0 g/mol.

20. • In the air we breathe at 72°F and 1.0 atm pressure, how many molecules does a typical cubic centimeter contain, assuming that the air is all N_2?

21. •• We have two equal-size boxes, A and B. Each box contains gas that behaves as an ideal gas. We insert a thermometer into each box and find that the gas in box A is at a temperature of 50°C while the gas in box B is at 10°C. This is all we know about the gas in the boxes. Which of the following statements *must* be true? Which *could* be true? (a) The pressure in A is higher than in B. (b) There are more molecules in A than in B. (c) A and B cannot contain the same type of gas. (d) The molecules in A have more average kinetic energy per molecule than those in B. (e) The molecules in A are moving faster than those in B. Explain the reasoning behind your answers.

22. •• (a) A deuteron, $^2_1$H, is the nucleus of a hydrogen isotope and consists of one proton and one neutron. The plasma of deuterons in a nuclear fusion reactor must be heated to about 300 million K. What is the rms speed of the deuterons? Is this a significant fraction of the speed of light $(c = 3.0 \times 10^8$ m/s$)$? (b) What would the temperature of the plasma be if the deuterons had an rms speed equal to $0.10c$?

23. • Oxygen (O_2) has a molar mass of 32.0 g/mol. (a) What is the root-mean-square speed of an oxygen molecule at a temperature of 300 K? (b) What is its average translational kinetic energy at that speed?

24. • Suppose some insects have speeds of 10.00 m/s, 8.00 m/s, 7.00 m/s, and 2.00 m/s. Find (a) the rms speed of these critters and (b) their average speed.

25. •• In a gas at standard temperature and pressure, what is the length of the side of a cube that contains a number of molecules equal to the population of the earth (about 6 billion people at present)?

26. •• At what temperature is the root-mean-square speed of nitrogen molecules equal to the root-mean-square speed of hydrogen molecules at 20.0°C? (*Hint:* The periodic table in Appendix C shows the molar mass (in g/mol) of each element under the chemical symbol for that element. The molar mass of H_2 is twice the molar mass of hydrogen atoms, and similarly for N_2.)

27. •• **Where is the hydrogen?** The average temperature of the atmosphere near the surface of the earth is about 20°C. (a) What is the root-mean-square speed of hydrogen molecules, H_2, at this temperature? (b) The escape speed from the earth is about 11 km/s. Is the average H_2 molecule moving fast enough to escape? (c) Compare the rms speeds of oxygen (O_2) and nitrogen (N_2) with that of H_2. (d) So why has the hydrogen been able to escape the earth's gravity, but the heavier gases (such as O_2 and N_2) have not, even though none of these gases has an rms speed equal to the escape speed of the earth? (*Hint:* Do *all* the molecules have the rms speed, or are some moving faster? Since the rms speed for H_2 is greater than that of O_2 and N_2, which gas would have a higher percentage of its molecules moving fast enough to escape?)

28. •• A flask contains a mixture of neon (Ne), krypton (Kr), and radon (Rn) gases. Compare (a) the average kinetic energies of the three types of atoms; (b) their root-mean-square speeds. (*Hint:* The periodic table in Appendix C shows the molar mass (in g/mol) of each element.)

29. •• **STP.** The conditions of standard temperature and pressure (STP) are a temperature of 0.00°C, and a pressure of 1.00 atm. (a) How many liters does 1.00 mol of any ideal gas occupy at STP? (b) For a scientist on Venus, an absolute pressure of 1 Venusian-atmosphere is 92 Earth-atmospheres. Of course she would use the Venusian-atmosphere to define STP. Assuming she kept the same temperature, how many liters would 1 mole of ideal gas occupy on Venus?

30. •• **Breathing at high altitudes.** If you have ever hiked or
BIO climbed to high altitudes in the mountains, you surely have
noticed how short of breath you get. This occurs because the
air is thinner, so each breath contains fewer O_2 molecules
than at sea level. At the top of Mt. Everest, the pressure is
only $\frac{1}{3}$ atm. Air contains 21% O_2 and 78% N_2, and an average
human breath is 0.50 L of air. At the top of Mt. Everest,
(a) how many O_2 molecules does each breath contain when
the temperature is $-15°F$, and (b) what percent is this of the
number of O_2 molecules you would get from a breath at sea
level at $-15°F$?

31. •• **How often do we need to breathe?** A resting person
BIO requires 14.5 L of O_2 per hour to maintain metabolic activities.
Such a person breathes in 0.50 L of air at approximately 20°C
with each breath. The inhaled air is 20.9% O_2, while the
exhaled air is 16.3% O_2. (a) How many breaths per minute
does a resting person need to take to provide the necessary
oxygen? (b) How many O_2 molecules does a resting person
inhale per breath?

15.4 Heat Capacities

32. • (a) How much heat does it take to increase the temperature
of 2.50 moles of an ideal monatomic gas from 25.0°C to
55.0°C if the gas is held at constant volume? (b) How much
heat is needed if the gas is diatomic rather than monatomic?
(c) Sketch a pV diagram for these processes.

33. • (a) If you supply 1850 J of heat to 2.25 moles of an ideal
diatomic gas initially at 10.0°C in a perfectly rigid container, what will be the final temperature of the gas?
(b) Suppose the gas in the container were an ideal
monatomic gas instead. How much heat would you need to
add to produce the same temperature change? (c) Sketch a
pV diagram of these processes.

34. • Compute the specific heat capacity (in $J/(kg \cdot K)$) at constant volume of nitrogen (N_2) gas, and compare it with the
specific heat capacity of liquid water. Use Appendix C to
determine the molar mass of N_2.

35. • Perfectly rigid containers each hold n moles of ideal gas,
one being hydrogen (H_2) and other being neon (Ne). If it
takes 100 J of heat to increase the temperature of the hydrogen
by 2.50 C°, by how many degrees will the same amount of heat
raise the temperature of the neon?

36. •• Assume that the gases in this problem can be treated as ideal
over the temperature ranges involved, and consult Appendix C
to determine the necessary molar masses. (a) How much heat is
needed to raise the temperature of 75.0 g of N_2 from 12.1°C to
49.5°C at constant volume? (b) If instead you want to produce
the same temperature change in 75.0 g of O_2 at constant volume,
how much heat do you need?

15.5 The First Law of Thermodynamics

37. • A metal cylinder with rigid walls contains 2.50 mol of oxygen gas. The gas is cooled until the pressure decreases to
30.0% of its original value. You can ignore the thermal contraction of the cylinder. (a) Draw a pV diagram of this process.
(b) Calculate the work done by the gas.

38. • A gas under a constant pressure of 1.50×10^5 Pa and with
an initial volume of 0.0900 m³ is cooled until its volume
becomes 0.0600 m³. (a) Draw a pV diagram of this process.
(b) Calculate the work done by the gas.

39. • Two moles of an ideal gas are heated at constant pressure
from $T = 27°C$ to $T = 107°C$. (a) Draw a pV diagram for this
process. (b) Calculate the work done by the gas.

40. •• Three moles of an ideal monatomic gas expands at a constant pressure of 2.50 atm; the volume of the gas changes from
3.20×10^{-2} m³ to 4.50×10^{-2} m³. (a) Calculate the initial
and final temperatures of the gas. (b) Calculate the amount of
work the gas does in expanding. (c) Calculate the amount of
heat added to the gas. (d) Calculate the change in internal
energy of the gas.

41. •• **Work done in a
cyclic process.** In Fig-
ure 15.32, consider the
closed loop $1 \rightarrow 2 \rightarrow
3 \rightarrow 4 \rightarrow 1$. This is a
cyclic process in which
the initial and final states
are the same. (a) Find
the total work done by
the system in this pro-
cess, and show that it is

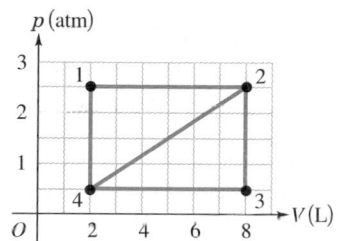

▲ **FIGURE 15.32** Problem 41.

equal to the area enclosed by the loop. (b) How is the work
done during the process in part (a) related to the work
done if the loop is traversed in the opposite direction,
$1 \rightarrow 4 \rightarrow 3 \rightarrow 2 \rightarrow 1$. Explain. (c) How much work is done in
the cycle $3 \rightarrow 4 \rightarrow 2 \rightarrow 3$?

42. •• **Work done by the lungs.** The graph in Figure 15.33 shows
BIO a pV diagram of the air in a human lung when a person is inhal-
ing and then exhaling a deep breath. Such graphs, obtained in
clinical practice, are normally somewhat curved, but we have
modeled one as a set of straight lines of the same general shape.
(*Important:* The pressure shown is the *gauge* pressure, *not* the
absolute pressure.) (a) How many joules of *net* work does this
person's lung do during one complete breath? (b) The process
illustrated here is somewhat different from those we have been
studying, because the pressure change is due to changes in the
amount of gas in the lung, not to temperature changes. (Think
of your own breathing. Your lungs do not expand because
they've gotten hot.) If the temperature of the air in the lungs
remains a reasonable 20°C, what is the maximum number of
moles in this person's lungs during a breath?

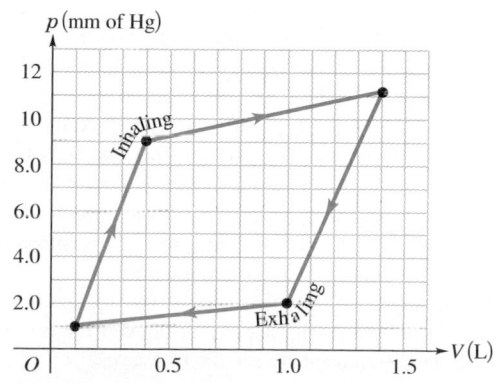

▲ **FIGURE 15.33** Problem 42.

43. • In a certain chemical process, a lab technician supplies
254 J of heat to a system. At the same time, 73 J of work
are done on the system by its surroundings. What is the
increase in the internal energy of the system?

44. • A gas in a cylinder expands from a volume of $0.110 \, m^3$ to $0.320 \, m^3$. Heat flows into the gas just rapidly enough to keep the pressure constant at $1.80 \times 10^5 \, Pa$ during the expansion. The total heat added is $1.15 \times 10^5 \, J$. (a) Find the work done by the gas. (b) Find the change in internal energy of the gas.

45. • A gas in a cylinder is held at a constant pressure of $2.30 \times 10^5 \, Pa$ and is cooled and compressed from $1.70 \, m^3$ to $1.20 \, m^3$. The internal energy of the gas decreases by $1.40 \times 10^5 \, J$. (a) Find the work done by the gas. (b) Find the amount of the heat that flowed into or out of the gas, and state the direction (inward or outward) of the flow.

46. • Five moles of an ideal monatomic gas with an initial temperature of $127°C$ expand and, in the process, absorb 1200 J of heat and do 2100 J of work. What is the final temperature of the gas?

47. •• When a system is taken from state a to state b in Figure 15.34 along the path acb, 90.0 J of heat flows into the system and 60.0 J of work is done by the system. (a) How much heat flows into the system along path adb if the work done by the system is 15.0 J? (b) When the system is returned from b to a along the curved path, the absolute value of the work done by the system is 35.0 J. Does the system absorb or liberate heat? How much heat? (c) If $U_a = 0$ and $U_d = 8.0 \, J$, find the heat absorbed in the processes ad and db.

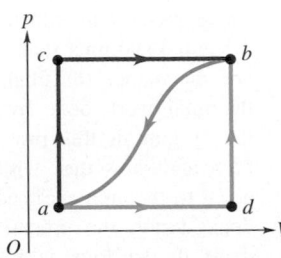

▲ **FIGURE 15.34** Problem 47.

15.6 Thermodynamic Processes

15.7 Properties of an Ideal Gas

48. • An ideal gas expands while the pressure is kept constant. During this process, does heat flow into the gas or out of the gas? Justify your answer.

49. • You are keeping 1.75 moles of an ideal gas in a container surrounded by a large ice-water bath that maintains the temperature of the gas at $0.00°C$. (a) How many joules of work would have to be done on this gas to compress its volume from 4.20 L to 1.35 L? (b) How much heat came into (or out of) the gas during this process? Was it into or out of?

50. • Suppose you do 457 J of work on 1.18 moles of ideal He gas in a perfectly insulated container. By how much does the internal energy of this gas change? Does it increase or decrease?

51. •• A cylinder with a movable piston contains 3.00 mol of N_2 gas (assumed to behave like an ideal gas). (a) The N_2 is heated at constant volume until 1557 J of heat have been added. Calculate the change in temperature. (b) Suppose the same amount of heat is added to the N_2, but this time the gas is allowed to expand while remaining at constant pressure. Calculate the temperature change. (c) In which case, (a) or (b), is the final internal energy of the N_2 higher? How do you know? What accounts for the difference between the two cases?

52. •• Figure 15.35 shows a pV diagram for an ideal gas in which its pressure tripled from a to b when 534 J of heat was put into the gas. (a) How much work was done on or by the gas between a and b? (b) Without doing any calculations, decide whether the temperature of this gas increased, decreased, or remained the same between

a and b. Explain your reasoning. (c) By how much did the internal energy of the gas change between a and b? Did it increase or decrease? (d) What is the temperature of the gas at point b in terms of its temperature at a, T_a?

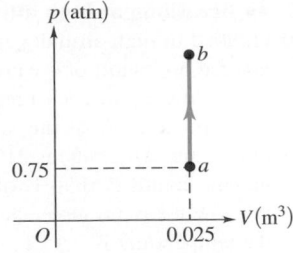

▲ **FIGURE 15.35** Problem 52.

53. •• Figure 15.36 shows a pV diagram for an ideal gas in which its absolute temperature at b is one-fourth of its absolute temperature at a. (a) What volume does this gas occupy at point b? (b) How many joules of work was done by or on the gas in this process? Was it done by or on the gas? (c) Did the internal energy of the gas increase or decrease from a to b? How do you know? (d) Did heat enter or leave the gas from a to b? How do you know?

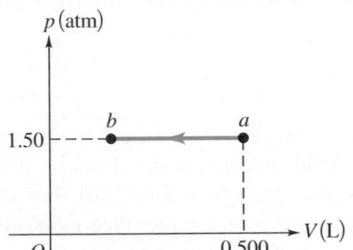

▲ **FIGURE 15.36** Problem 53.

54. •• The pV diagram in Figure 15.37 shows a process abc involving 0.450 mole of an ideal gas. (a) What was the temperature of this gas at points a, b, and c? (b) How much work was done by or on the gas in this process? (c) How much heat had to be put in during the process to increase the internal energy of the gas by 15,000 J?

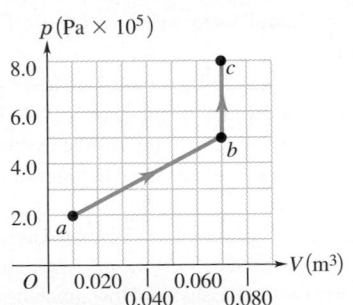

▲ **FIGURE 15.37** Problem 54.

55. •• A volume of air (assumed to be an ideal gas) is first cooled without changing its volume and then expanded without changing its pressure, as shown by the path abc in Fig. 15.38. (a) How does the final temperature of the gas compare with its initial temperature? (b) How much heat does the air exchange with its surroundings during the process abc? Does

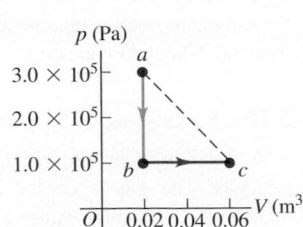

▲ **FIGURE 15.38** Problem 55.

the air absorb heat or release heat during this process? Explain. (c) If the air instead expands from state *a* to state *c* by the straight-line path shown, how much heat does it exchange with its surroundings?

56. •• In the process illustrated by the *pV* diagram in Figure 15.39, the temperature of the ideal gas remains constant at 85°C. (a) How many moles of gas are involved? (b) What volume does this gas occupy at *a*? (c) How much work was done by or on the gas from *a* to *b*? (d) By how much did the internal energy of the gas change during this process?

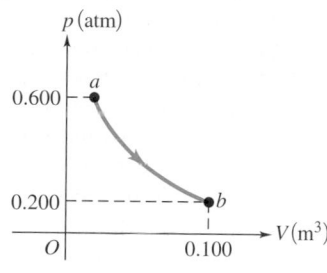

▲ **FIGURE 15.39** Problem 56.

57. • A cylinder contains 0.250 mol of carbon dioxide (CO_2) gas at a temperature of 27.0°C. The cylinder is provided with a frictionless piston, which maintains a constant pressure of 1.00 atm on the gas. The gas is heated until its temperature increases to 127.0°C. Assume that the CO_2 may be treated as an ideal gas. (a) Draw a *pV* diagram of this process. (b) How much work is done by the gas in the process? (c) On what is this work done? (d) What is the change in internal energy of the gas? (e) How much heat was supplied to the gas? (f) How much work would have been done if the pressure had been 0.50 atm?

58. •• **Heating air in the lungs.** Human lung capacity varies
BIO from about 4 L to 6 L, so we shall use an average of 5.0 L. The air enters at the ambient temperature of the atmosphere and must be heated to internal body temperature at an approximately constant pressure of 1.0 atm in our model. Suppose you are outside on a winter day when the temperature is −10°F. (a) How many moles of air does your lung hold if the 5.0 L is at the internal body temperature of 37°C? (b) How much heat must your body have supplied to get the 5.0 L of air up to internal body temperature, assuming that the atmosphere is all N_2? (See Table 15.4.) (c) Suppose instead that you manage to inhale the full 5.0 L of air in one breath and hold it in your lungs without expanding (or contracting) them. How much heat would your body have had to supply in that case to raise the air up to internal body temperature?

59. •• The graph in Figure 15.40 shows a *pV* diagram for 1.10 moles of *ideal* oxygen, O_2. (a) Find the temperature at points *a*, *b*, *c*, and *d*. (b) How many joules of heat enters (or leaves) the oxygen in segment (i) *ab*, (ii) *bc*, (iii) *cd*, (iv) *da*? (c) In each of the preceding segments, does the heat enter or leave the gas? How do you know?

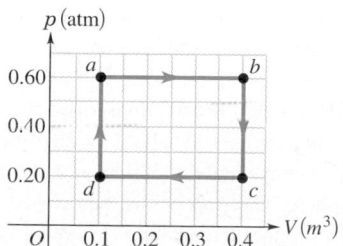

▲ **FIGURE 15.40** Problem 59.

60. • An ideal gas at 4.00 atm and 350 K is permitted to expand adiabatically to 1.50 times its initial volume. Find the final pressure and temperature if the gas is (a) monatomic with $C_p/C_V = \frac{5}{3}$, (b) diatomic with $C_p/C_V = \frac{7}{5}$.

61. •• An experimenter adds 970 J of heat to 1.75 mol of an ideal gas to heat it from 10.0°C to 25.0°C at constant pressure. The gas does +223 J of work during the expansion. (a) Calculate the change in internal energy of the gas. (b) Calculate γ for the gas.

62. •• Heat Q flows into a monatomic ideal gas, and the volume increases while the pressure is kept constant. What fraction of the heat energy is used to do the expansion work of the gas?

63. •• A player bounces a basketball on the floor, compressing it to 80.0% of its original volume. The air (assume it is essentially N_2 gas) inside the ball is originally at a temperature of 20.2°C and a pressure of 2.00 atm. The ball's diameter is 23.9 cm. (a) What temperature does the air in the ball reach at its maximum compression? (b) By how much does the internal energy of the air change between the ball's original state and its maximum compression?

64. •• In the *pV* diagram shown in Figure 15.41, 85.0 J of work was done by 0.0650 mole of ideal gas during an adiabatic process. (a) How much heat entered or left this gas from *a* to *b*? (b) By how many joules did the internal energy of the gas change? (c) What is the temperature of the gas at *b*?

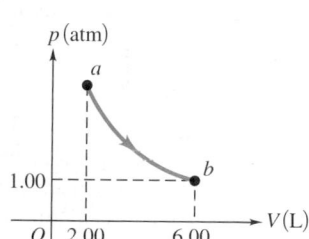

▲ **FIGURE 15.41** Problem 64.

General Problems

65. • Modern vacuum pumps make it easy to attain pressures on the order of 10^{-13} atm in the laboratory. At a pressure of 9.00×10^{-14} atm and an ordinary temperature of 300 K, how many molecules are present in 1.00 cm^3 of gas?

66. •• **How many atoms are you?** Estimate the number of atoms
BIO in the body of a 65 kg physics student. Note that the human body is mostly water, which has molar mass 18.0 g/mol, and that each water molecule contains three atoms.

67. •• **The effect of altitude on the lungs.** (a) Calculate the *change*
BIO in air pressure you will experience if you climb a 1000 m mountain, assuming that the temperature and air density do not change over this distance and that they were 22°C and 1.2 kg/m^3, respectively, at the bottom of the mountain. (b) If you took a 0.50 L breath at the foot of the mountain and managed to hold it until you reached the top, what would be the volume of this breath when you exhaled it there?

68. •• (a) Calculate the mass of nitrogen present in a volume of 3000 cm^3 if the temperature of the gas is 22.0°C and the absolute pressure is 2.00×10^{-13} atm, a partial vacuum easily obtained in laboratories. The molar mass of nitrogen (N_2) is 28.0 g/mol. (b) What is the density (in kg/m^3) of the N_2?

69. • An automobile tire has a volume of 0.0150 m^3 on a cold day when the temperature of the air in the tire is 5.0°C and atmospheric pressure is 1.02 atm. Under these conditions, the gauge pressure is measured to be 1.70 atm (about 25 lb/in^2). After the car is driven on the highway for 30 min, the temperature of the air in the tires has risen to 45.0°C and the volume to 0.0159 m^3. What is the gauge pressure at that time?

70. •• A cylinder 1.00 m tall with inside diameter 0.120 m is used to hold propane gas (molar mass 44.1 g/mol) for use in a barbecue. It is initially filled with gas until the gauge pressure is 1.30×10^6 Pa and the temperature is 22.0°C. The temperature of the gas remains constant as it is partially emptied out of the tank, until the gauge pressure is 2.50×10^5 Pa. Calculate the mass of propane that has been used.

71. •• **The surface of the sun.** The surface of the sun has a temperature of about 5800 K and consists largely of hydrogen atoms. (a) Find the rms speed of a hydrogen atom at this temperature. (The mass of a single hydrogen atom is 1.67×10^{-27} kg.) (b) What would be the mass of an atom that had half the rms speed of hydrogen?

72. •• **Atmosphere of Titan.** Titan, the largest satellite of Saturn, has a thick nitrogen atmosphere. At its surface, the pressure is 1.5 Earth-atmospheres and the temperature is 94 K. (a) What is the surface temperature in °C? (b) Calculate the surface density in Titan's atmosphere in molecules per cubic meter. (c) Compare the density of Titan's surface atmosphere to the density of Earth's atmosphere at 22°C. Which body has a denser atmosphere?

73. •• Helium gas expands slowly to twice its original volume, doing 300 J of work in the process. Find the heat added to the gas and the change in internal energy of the gas if the process is (a) isothermal, (b) adiabatic, (c) isobaric.

74. •• A cylinder with a piston contains 0.250 mol of ideal oxygen at a pressure of 2.40×10^5 Pa and a temperature of 355 K. The gas first expands isobarically to twice its original volume. It is then compressed isothermally back to its original volume, and finally it is cooled isochorically to its original pressure. (a) Show the series of processes on a pV diagram. (b) Compute the temperature during the isothermal compression. (c) Compute the maximum pressure. (d) Compute the total work done by the piston on the gas during the series of processes.

75. •• You blow up a spherical balloon to a diameter of 50.0 cm until the absolute pressure inside is 1.25 atm and the temperature is 22.0°C. Assume that all the gas is N_2, of molar mass 28.0 g/mol. (a) Find the mass of a single N_2 molecule. (b) How much translational kinetic energy does an average N_2 molecule have? (c) How many N_2 molecules are in this balloon? (d) What is the *total* translational kinetic energy of all the molecules in the balloon?

76. •• (a) One-third of a mole of He gas is taken along the path abc shown as the solid line in Figure 15.42. Assume that the gas may be treated as ideal. How much heat is transferred into or out of the gas? (b) If the gas instead went from state a to state c along the horizontal dashed line in Fig. 15.42, how much heat would be transferred into or out of the gas? (c) How does Q in part (b) compare against Q in part (a)? Explain.

77. •• A bicyclist uses a tire pump whose cylinder is initially full of air at an absolute pressure of 1.01×10^5 Pa. The length of stroke of the pump (the length of the cylinder) is 36.0 cm. At what part of the stroke (i.e., what length of the air column) does air begin to enter a tire in which the gauge pressure is 2.76×10^5 Pa? Assume that the temperature remains constant during the compression.

78. •• **The bends.** If deep-sea divers rise to the surface too quickly,
BIO nitrogen bubbles in their blood can expand and prove fatal. This phenomenon is known as the *bends*. If a scuba diver rises quickly from a depth of 25 m in Lake Michigan (which is fresh water), what will be the volume at the surface of an N_2 bubble that occupied 1.0 mm^3 in his blood at the lower depth? Does it seem that this difference is large enough to be a problem? (Assume that the pressure difference is due only to the changing water pressure, not to any temperature difference, an assumption that is reasonable, since we are warm-blooded creatures.)

79. •• Figure 15.43 shows a pV diagram for 0.0040 mole of *ideal* H_2 gas. The temperature of the gas does not change during segment bc. (a) What volume does this gas occupy at point c? (b) Find the temperature of the gas at points a, b, and c. (c) How much heat went into or out of the gas during segments ab, ca, and bc? Indicate whether the heat has gone into or out of the gas. (d) Find the change in the internal energy of this hydrogen during segments ab, bc, and ca. Indicate whether the internal energy increased or decreased during each of these segments.

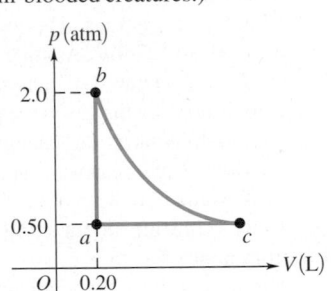

▲ **FIGURE 15.43** Problem 79.

80. •• The graph in Figure 15.44 shows a pV diagram for 3.25 moles of *ideal* helium (He) gas. Part ca of this process is isothermal. (a) Find the pressure of the He at point a. (b) Find the temperature of the He at points a, b, and c. (c) How much heat entered or left the He during segments ab, bc, and ca? In each segment, did the heat enter or leave? (d) By how much did the internal energy of the He change from a to b, from b to c, and from c to a? Indicate whether this energy increased or decreased.

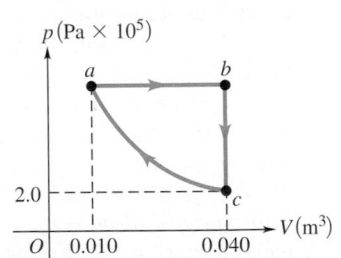

▲ **FIGURE 15.44** Problem 80.

81. •• A flask with a volume of 1.50 L, provided with a stopcock, contains ethane gas (C_2H_6) at 300 K and atmospheric pressure $(1.013 \times 10^5$ Pa$)$. The molar mass of ethane is 30.1 g/mol. The system is warmed to a temperature of 380 K, with the stopcock open to the atmosphere. The stopcock is then closed, and the flask is cooled to its original temperature. (a) What is the final pressure of the ethane in the flask? (b) How many grams of ethane remain in the flask?

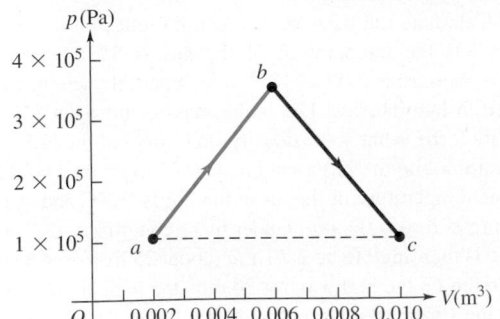

▲ **FIGURE 15.42** Problem 76.

82. •• **Diesel ignition.** Conventional engines ignite their fuel by using the spark from a spark plug. But in a diesel engine, the air enters the chamber at the temperature of the atmosphere and is compressed by the piston until it reaches 550°C, at which time the fuel is injected into the chamber and ignited by the hot air. There is no spark plug and no heat is put into the air. (One of the drawbacks of diesel engines is that they are hard to start in cold weather, as we shall shortly see.) Suppose that a certain chamber has a maximum volume of 0.50 L and uses 0.050 mole of air. We can model the air as all ideal N_2 and use the appropriate values from Table 15.4. (a) If the air temperature is 20°C, what is the volume of the air (which started at 0.50 L) when it has been compressed enough so that its temperature has risen to 550°C? (b) What is the change in the internal energy of the air during this compression? (c) How much work did the piston do on this gas while compressing it? (d) Suppose it is a cold winter morning, with the air temperature 10° F. If the piston compressed the air by the same amount as before, what will be the highest temperature the gas will reach in this case? (e) Do you now see why a diesel engine is hard to start in cold weather? Can you suggest any reasonable technological solutions to help start a diesel engine on a cold day?

83. •• Initially at a temperature of 80.0°C, 0.28 m^3 of air expands at a constant gauge pressure of 1.38×10^5 Pa to a volume of 1.42 m^3 and then expands further adiabatically to a final volume of 2.27 m^3 and a final gauge pressure of 2.29×10^4 Pa. Draw a pV diagram for this sequence of processes, and compute the total work done by the air. C_V for air is 20.8 J/(mol · K).

84. •• In a cylinder, 4.00 mol of helium initially at 1.00×10^6 Pa and 300 K expands until its volume doubles. Compute the work done by the gas if the expansion is (a) isobaric, (b) adiabatic. (c) Show each process on a pV diagram. In which case is the magnitude of the work done by the gas the greatest? (d) In which case is the magnitude of the heat transfer greatest? (e) In which case is the magnitude of the change in internal energy greatest?

85. •• Starting with 2.50 mol of N_2 gas (assumed to be ideal) in a cylinder at 1.00 atm and 20.0°C, a chemist first heats the gas at constant volume, adding 1.52×10^4 J of heat, then continues heating and allows the gas to expand at constant pressure to twice its original volume. (a) Calculate the final temperature of the gas. (b) Calculate the amount of work done by the gas. (c) Calculate the amount of heat added to the gas while it was expanding. (d) Calculate the change in internal energy of the gas for the whole process.

86. ••• A hot-air balloon stays aloft because hot air at atmospheric pressure is less dense than cooler air at the same pressure. (The calculation of the buoyant force is discussed in Chapter 13.) If the volume of the balloon is 500.0 m^3 and the surrounding air is at 15.0°C, what must the temperature of the air in the balloon be for it to lift a total load of 290 kg (in addition to the mass of the hot air)? The density of air at 15.0°C and atmospheric pressure is 1.23 kg/m^3.

Passage Problems

BIO Temperature and degrees of freedom. The internal energy of an ideal monatomic gas is simply the kinetic energy associated with the translational motion of its atoms as they move randomly in each of the three independent spatial dimensions. However, for a diatomic ideal gas we must also take into account the kinetic and potential energies associated with molecular vibration, and the kinetic energy associated with molecular rotation. Roughly speaking, each independent way that energy can be stored is known as a *degree of freedom*.

Although translational motion can occur at any temperature, rotational and vibrational motions typically cannot occur at lower temperatures—thus, the number of available degrees of freedom can change as the temperature changes. For example, an ideal *monatomic* gas has three degrees of freedom (one for each of its independent directions of translational motion) at all temperatures. In contrast, *diatomic* hydrogen (H_2) has five degrees of freedom near room temperature (3 translational and 2 rotational). However, at higher temperatures, where molecular vibrations can occur, diatomic hydrogen has seven degrees of freedom (3 translational, 2 rotational, and 2 vibrational).

The equipartition theorem states that at equilibrium each degree of freedom contributes $\frac{1}{2}nRT$ to the internal energy of the gas. For example, the internal energy of a monatomic gas, which has 3 degrees of freedom, would be $\frac{3}{2}nRT$.

87. Near room temperature, how does the internal energy of one mole of a diatomic ideal gas compare to that of one mole of a monatomic ideal gas?
    A. They have the same internal energy.
    B. The diatomic gas has 2 times as much internal energy as the monatomic gas.
    C. The diatomic gas has 2/3 times as much internal energy as the monatomic gas.
    D. The diatomic gas has 3/2 times as much internal energy as the monatomic gas.
    E. The diatomic gas has 5/3 times as much internal energy as the monatomic gas.

88. For an ideal gas with 9 degrees of freedom, the *molar heat capacity* at constant volume (J/(mol · K)) would be
    A. $\frac{3}{2}RT$
    B. $\frac{3}{2}nR$
    C. $\frac{9}{2}nR$
    D. $\frac{9}{2}R$
    E. $\frac{11}{2}R$

89. As the temperature of a monatomic gas increases, we expect its specific heat to
    A. remain the same
    B. increase
    C. decrease

90. As the temperature of a diatomic gas increases, we expect its specific heat to
    A. remain the same
    B. increase
    C. decrease

16 The Second Law of Thermodynamics

When you put 0.1 kg of boiling water and 0.1 kg of ice in an insulated cup, you end up with 0.2 kg of water at about 10°C. That's not surprising. But you'd be very surprised if you came back later and found that the water had turned back to 0.1 kg of ice and 0.1 kg of boiling water. That wouldn't violate the first law of thermodynamics; energy would be conserved. But it doesn't happen in nature. Why not? Why does a power plant convert less than half of the heat from burning coal into electrical energy, discarding the remainder of the heat? Why does heat always flow spontaneously from hotter places to cooler places, never the reverse? When you drop ink into water, it mixes spontaneously, coloring the water, but it never spontaneously unmixes. Why not? What do all these things have in common?

A study of inherently one-way processes, such as the flow of heat from hotter to colder regions and the conversion of work into heat by friction, leads to the *second law of thermodynamics*. This law places fundamental limitations on the efficiency of an engine or a power plant and on the minimum energy input needed to operate a refrigerator. So the second law is directly relevant to many important practical problems. We can also state the second law in terms of the concept of *entropy*, a quantitative measure of the degree of disorder, or randomness, of a system.

In this chapter we'll encounter the concept of entropy, which represents the fact that order tends to disintegrate spontaneously into disorder. To fend off entropy, our bodies require a constant input of energy—which we take in the form of food.

16.1 Directions of Thermodynamic Processes

Heat flows spontaneously from a hotter object to a cooler object, never the reverse. Spontaneous heat flow from a cool object to a hot object would not violate the first law, but it doesn't happen in nature. Or suppose all the air in a box could rush to one side, leaving vacuum in the other side, the reverse of the free expansion we described in Section 15.5. This phenomenon doesn't occur in nature either, although the first law doesn't forbid it. It's easy to convert mechanical energy completely into heat; we do this every time we use a car's brakes to stop it. It is not so easy to convert heat into mechanical energy. Many would-be inventors have proposed cooling some air to extract heat from it and then converting that heat to mechanical energy to propel a car or an airplane. None has ever succeeded; no one has ever built a machine that converts heat *completely* into mechanical energy.

What all these examples have in common is a preferred *direction*. In each case, a process proceeds spontaneously in one direction, but not in the other. Still, despite the fact that there is a preferred direction for every natural process, we can think of a class of idealized processes that are *reversible*. We say that a system undergoes a **reversible process** if the system is always very close to being in thermodynamic equilibrium, within itself and with its surroundings. When this is the case, any change of state that takes place can be reversed (i.e., made to go the other way) by making only an infinitesimal change in the conditions of the system. For example, heat flow between two objects whose temperatures differ only infinitesimally can be reversed by making only a very small change in one temperature or the other. Thus, a gas expanding slowly and adiabatically can be compressed slowly and adiabatically by an infinitesimal increase in pressure.

Reversible processes are thus **equilibrium processes.** In contrast, heat flow with a finite (as opposed to an infinitesimal) temperature difference, the free expansion of a gas, and the conversion of work to heat by friction are all **irreversible processes:** No small change in conditions could make any of them go to the other way. They are also all *non-equilibrium* processes.

Reversible process

A reversible process is a transition from one state of a thermodynamic system to another, during which the system is always very close to a state of mechanical and thermal equilibrium (including uniform temperature and pressure).

A reversible process is an idealization that can never be precisely attained in the real world. But by making the temperature gradients and the pressure differences in the substance very small, we can keep the system very close to equilibrium states.

Finally, we'll find that there is a relation between the direction of a process and the *disorder,* or *randomness,* of the resulting state. For example, imagine a tedious sorting job, such as alphabetizing a thousand book titles written on file cards. Throw the alphabetized stack of cards into the air. Do they come down in alphabetical order? No, their tendency is to come down in a random, or disordered, state. In the free expansion of a gas that we saw in Figure 15.25, the air is more disordered after it has expanded into the entire container than when it was confined in one side, because the molecules are scattered over more space.

Similarly, macroscopic kinetic energy is energy associated with organized, coordinated motions of many molecules in a moving macroscopic object, but heat transfer involves changes in energy of random, disordered molecular motion. Therefore the conversion of mechanical energy into heat involves an increase in randomness, or disorder.

In the sections that follow, we'll introduce the second law of thermodynamics by considering two broad classes of devices: *heat engines,* which are partly successful in converting heat into work, and *refrigerators,* which are partly successful in transporting heat from cooler to hotter objects.

▲ **Application** **Near-reversible control.** The energy released by the sun comes from nuclear fusion reactions in its core. The rate of these reactions depends sensitively on temperature: A slight increase in temperature would cause a large increase in energy output. For life on earth, it is essential that the sun's brightness be steady. Luckily, the rate of fusion is controlled by a "thermostat." Energy escapes from the core rather slowly. Suppose the rate of fusion increased slightly. Then the extra heat output would cause the core to swell and thus to cool down, reducing the fusion rate. (The core consists of a plasma rather than a gas per se, but thermodynamically, it behaves like a gas.) Similarly, a drop in the rate of fusion would cause the core to shrink and heat up, increasing the rate again. Inferno though it is, the core is always close to equilibrium, so that a slight change in conditions can reverse the direction of a thermodynamic process.

Conceptual Analysis 16.1

Reversible versus irreversible processes

Which of the following mechanical processes are not reversible, in the sense that energy converted from one form to another cannot be completely converted from the second form back to the first form?

A. A ball is tossed upward, gaining height.
B. A moving cart is caught by a spring, stretching the spring.
C. A moving block slides to a stop because of friction.

SOLUTION Processes A and B involve only mechanical energy. In each, the system is given some kinetic energy, which is then transformed into potential energy. In A, the kinetic energy is recovered as the ball falls back to its original elevation; in B, the kinetic energy is recovered when the spring is permitted to relax back to its original length. If friction can be neglected, processes A and B are reversible. Process C is not reversible, because mechanical energy is converted to internal energy and there is no way to convert it back to mechanical energy without supplying some work from outside the system.

16.2 Heat Engines

ActivPhysics 8.12: Cyclic Process—Strategies
ActivPhysics 8.13: Cyclic Process—Problems

▲ **Application Earth power.** One type of heat engine uses the earth itself as a source of heat. At the Svartsengi geothermal power plant in Iceland, the hot reservoir is a subsurface body of hot volcanic rock. (Iceland is one of the most active volcanic regions in the world.) The rock heats the local briny groundwater to temperatures of 240°C. Drill holes bring this water to the surface as high-pressure steam. The steam is used both to power turbines for electrical generation and to heat fresh water, which is distributed by the surrounding towns for home heating and hot water. The condensed steam is then released into the Blue Lagoon next to the power plant, where bathers can relax in 40°C waters even in the middle of an Icelandic winter. The power plant pumps about 150 MJ/s (150 MW) of heat from below ground and produces about 50 MW of electricity.

The essence of a technological society is its ability to use sources of mechanical energy other than muscle power. Sometimes mechanical energy is directly available; water power is an example. But most of our energy comes from the burning of fossil fuels (coal, oil, and gas) and from nuclear reactions. These processes supply energy that is transferred as *heat.* This transformation of energy is directly useful for heating buildings, for cooking, and for chemical and metallurgical processing. But to run an electrical generator or to operate a machine or propel a vehicle, we need *mechanical* energy.

Thus, it's important to know how to take heat from a source and convert as much of it as possible into mechanical energy or work. This is what happens in gasoline and Diesel engines in automobiles and trucks, jet engines in airplanes, steam turbines in electric power plants, and many other systems. A device that transforms heat partly into work or mechanical energy is called a **heat engine.** In a heat engine, a quantity of matter inside the engine undergoes addition and subtraction of heat, expansion and compression, and, sometimes, a phase change. We call this matter the **working substance** of the engine. In internal combustion engines (such as automobile engines), the working substance is a mixture of air and fuel; in a steam engine or steam turbine, it is water.

The simplest kind of engine to analyze is one in which the working substance undergoes a **cyclic process—a sequence of processes that eventually leaves the substance in the same state as that in which it started.** In a steam turbine, the water is recycled and used over and over. Internal combustion engines do not use the same air over and over, but we can still analyze them in terms of cyclic processes that approximate their actual operation.

All the heat engines just mentioned absorb heat from a source at a relatively high temperature, perform some mechanical work, and discard some heat at a lower temperature. As far as the engine is concerned, the discarded heat is wasted. In internal combustion engines, the waste heat is discarded in the hot exhaust gases and the cooling system; in a steam engine or steam turbine, the waste heat is the heat that must be taken out of the used steam in order to condense and recycle the water.

When a system is carried through a cyclic process, its initial and final internal energies are equal, so the change ΔU in internal energy of the system is zero. Thus, for any cyclic process, the first law of thermodynamics requires that

$$\Delta U = Q - W = 0 \quad \text{and} \quad Q = W.$$

That is, **in a cyclic process, the net heat Q flowing into the engine equals the net work W done by the engine.**

When we analyze heat engines, it helps to think of two objects with which the working substance of the engine can interact. One of these, called the *hot reservoir,* can give the working substance large amounts of heat at a constant temperature T_H without appreciably changing its own temperature. The other object, called the *cold reservoir,* can absorb large amounts of discarded heat from the engine at a constant lower temperature T_C. In a steam-turbine system, the flames and hot gases in the boiler are the hot reservoir, and the cold water and air used to condense and cool the used steam are the cold reservoir.

We denote the quantities of heat transferred from the hot and cold reservoirs as Q_H and Q_C, respectively. A quantity of heat Q is positive when heat is transferred *from* a reservoir *into* the working substance. When heat leaves the working substance, Q is negative. Thus, in a heat engine, Q_H is positive, but Q_C is negative, representing heat *leaving* the working substance. When the working substance does positive work on its surroundings, W is positive. These sign conventions are consistent with the rules that we stated in Chapter 15; we'll continue to use those rules here. Sometimes it clarifies the relationships to state them in terms of the absolute values of the Q's and W's, because absolute values are always positive. When we do this, our notation will show it explicitly.

We can represent the energy transformations in a heat engine by the *energy-flow diagram* of Figure 16.1. The engine itself is represented by the circle. The amount of heat Q_H supplied to the engine by the hot reservoir is proportional to the width of the incoming "pipeline" at the top of the diagram. The width of the outgoing pipeline at the bottom is proportional to the magnitude $|Q_C|$ of the heat discarded. The branch line to the right represents that portion of the heat supplied that the engine converts to mechanical work W.

When an engine repeats the same cycle over and over, Q_H and Q_C represent the quantities of heat absorbed and rejected, respectively, by the engine *during one cycle*; Q_H is positive, Q_C negative. The *net* heat Q absorbed per cycle is

$$Q = Q_H + Q_C = |Q_H| - |Q_C|. \tag{16.1}$$

The useful output of the engine is the net work W done by the working substance; from the first law,

$$W = Q = Q_H + Q_C = |Q_H| - |Q_C|. \tag{16.2}$$

Ideally, we would like to convert *all* the heat Q_H into work; in that case, we would have $Q_H = W$ and $Q_C = 0$. Experience shows that this is impossible; there is always some heat wasted, and Q_C can never be zero. We define the **thermal efficiency** of an engine, denoted by e, as the quotient of W and Q_H:

Thermal efficiency of a heat engine

The thermal efficiency of a heat engine is the useful work W done by the engine, divided by the heat input Q_H:

$$e = \frac{W}{Q_H}. \tag{16.3}$$

The thermal efficiency e represents the fraction of Q_H that *is* converted to work. To put it another way, e is what you get, divided by what you pay for. This is always less than 1, an all-too-familiar experience! In terms of the flow diagram of Figure 16.1, the most efficient engine is the one for which the branch pipeline representing the work output is as wide as possible and the exhaust pipeline representing the heat thrown away is as narrow as possible.

When we substitute the two expressions for W given by Equation 16.2 into Equation 16.3, we get the following equivalent expressions for e:

$$e = \frac{W}{Q_H} = 1 + \frac{Q_C}{Q_H} = 1 - \frac{|Q_C|}{|Q_H|}. \tag{16.4}$$

Note that e is a quotient of two energy quantities and thus is a pure number, without units. Of course, we must always express W, Q_H, and Q_C in the same units.

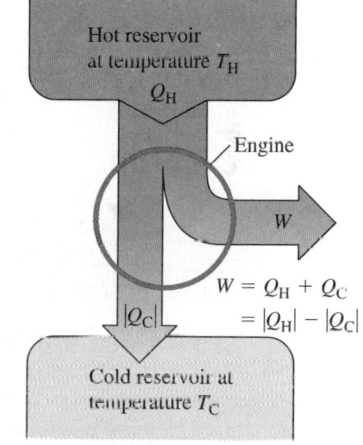

Hot reservoir at temperature T_H

Q_H

Engine

W

$|Q_C|$

$W = Q_H + Q_C$
$= |Q_H| - |Q_C|$

Cold reservoir at temperature T_C

▲ **FIGURE 16.1** Schematic energy-flow diagram for a heat engine.

PROBLEM-SOLVING STRATEGY 16.1 **Heat engines** (MP)

We suggest that you reread the strategy in Section 15.5; those suggestions are equally useful throughout the present chapter. The following points may need additional emphasis:

1. Be very careful with the sign conventions for W and the various Q's. W is positive when the system expands and does work, negative when it is compressed. Each Q is positive if it represents heat entering the working substance of the engine or other system, negative when heat leaves the system. When in doubt, use the first law, if possible, to check consistency. When you know that a quantity is negative, such as Q_C in the preceding discussion, it sometimes helps to write it as $Q_C = -|Q_C|$.

2. As always, unit consistency is essential. Be careful not to mix thermal units such as calories or Btu with mechanical quantities such as joules. We suggest you convert everything to joules whenever possible. Efficiency ratings of air conditioners are usually expressed in a peculiar mix of British and SI units; be careful!

3. Some problems deal with power rather than energy quantities. Power is work done per unit time $(P = W/t)$, and the rate of heat transfer (heat current) H is heat transferred per unit time $(H = Q/t)$. Sometimes it helps to ask, "What is W or Q in 1 second (or 1 hour)?"

EXAMPLE 16.1 **Fuel consumption in a truck**

A gasoline engine in a large truck takes in 2500 J of heat and delivers 500 J of mechanical work per cycle. The heat is obtained by burning gasoline with heat of combustion $L_c = 5.0 \times 10^4$ J/g. **(a)** What is the thermal efficiency of this engine? **(b)** How much heat is discarded in each cycle? **(c)** How much gasoline is burned during each cycle? **(d)** If the engine goes through 100 cycles per second, what is its power output in watts? In horsepower? **(e)** How much gasoline is burned per second? Per hour?

SOLUTION

SET UP It's often useful to make an energy-flow sketch for heat-engine problems; Figure 16.2 shows what we draw. We are given that $Q_H = 2500$ J and $W = 500$ J.

SOLVE Part (a): The thermal efficiency is found from Equation 16.3:

$$e = \frac{W}{Q_H} = \frac{500 \text{ J}}{2500 \text{ J}} = 0.20 = 20\%.$$

Part (b): From Equation 16.2, the heat Q_C discarded per cycle is the difference between the heat absorbed (Q_H) and the work W done by the engine:

$$W = Q_H + Q_C,$$
$$500 \text{ J} = 2500 \text{ J} + Q_C,$$
$$Q_C = -2000 \text{ J}.$$

Thus, 2000 J of heat leaves the engine during each cycle.

Part (c): Let m be the mass of gasoline burned during each cycle; then Q_H is m times the heat of combustion: $Q_H = mL_c$. Thus,

$$m = Q_H/L_c = (2500 \text{ J})/(5.0 \times 10^4 \text{ J/g}) = 0.050 \text{ g}.$$

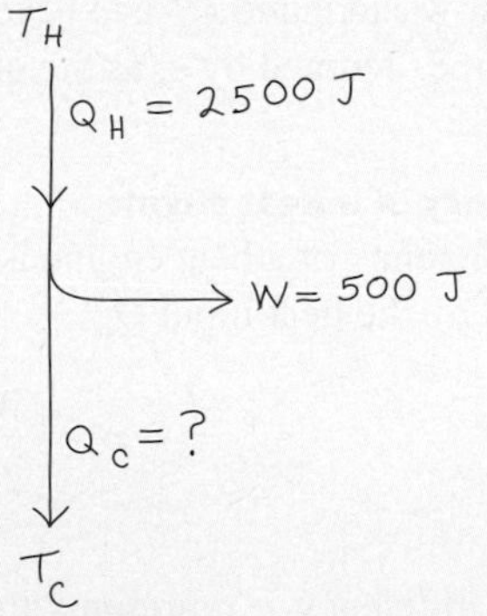

▲ **FIGURE 16.2** Our sketch for this problem.

Part (d): The power P (rate of doing work) is the work per cycle multiplied by the number of cycles per second.

$$P = (500 \text{ J/cycle})(100 \text{ cycles/s}) = 50,000 \text{ W} = 50 \text{ kW}, \quad \text{or}$$
$$P = (50,000 \text{ W})(1 \text{ hp}/746 \text{ W}) = 67 \text{ hp}.$$

Part (e): The mass of gasoline burned per second is the mass burned per cycle multiplied by the number of cycles per second:

$$(0.050 \text{ g/cycle})(100 \text{ cycles/s}) = 5.0 \text{ g/s}.$$

Continued

The mass burned per hour is

$$(5.0 \text{ g/s})\left(\frac{3600 \text{ s}}{1 \text{ h}}\right) = 18,000 \text{ g/h} = 18 \text{ kg/h}.$$

REFLECT The efficiency from part (a) is a fairly typical figure for cars and trucks if W includes only the work actually delivered to the wheels. The density of gasoline is about 0.70 g/cm³; the volume of fuel burned per hour is about 25,700 cm³, 25.7 L, or 6.8 gallons of gasoline per hour. If the truck is traveling at

55 mi/h (88 km/h), that volume represents fuel consumption of about 8.1 miles per gallon (mpg) (3.4 km/L).

Each cylinder in the engine goes through one cycle for every two revolutions of the crankshaft. For a four-cylinder engine, 100 cycles per second corresponds to 50 crankshaft revolutions per second, or 3000 rev/min (typical for highway speeds).

Practice Problem: A gasoline engine with a thermal efficiency of 28% has a power output of 50 hp. How much heat must be supplied per second to the engine? *Answer:* $Q_H = 133$ kJ.

16.3 Internal Combustion Engines

The gasoline engine, used in automobiles and many other types of machinery, is a familiar example of a heat engine. Figure 16.3 shows the sequence of processes. First, a mixture of air and gasoline vapor flows into a cylinder through an open intake valve while the piston descends, increasing the volume of the cylinder from a minimum of V (when the piston is all the way up) to a maximum of rV (when it is all the way down). The quantity r is called the **compression ratio;** for present-day automobile engines, it is typically about 8.

At the end of this *intake stroke,* the intake valve closes and the air–gasoline mixture is compressed, quickly and approximately adiabatically, to volume V during the *compression stroke.* The mixture is then ignited by the spark plug, and the heated gas expands, again approximately adiabatically, back to volume rV. As it does so, it pushes on the piston and does work on it; this is the *power stroke.* Finally, the exhaust valve opens and the combustion products are pushed out (during the *exhaust stroke*), leaving the cylinder ready for the next intake stroke.

The Otto Cycle

Figure 16.4 is a pV diagram showing an idealized model of the thermodynamic processes in a gasoline engine. This model is called the **Otto cycle.** We'll assume that the gasoline–air mixture behaves as an ideal gas. At point a, the mixture has entered the cylinder. The mixture is compressed adiabatically (line ab) and is then ignited. Heat Q_H (heat of combustion) is added to the system by the burning

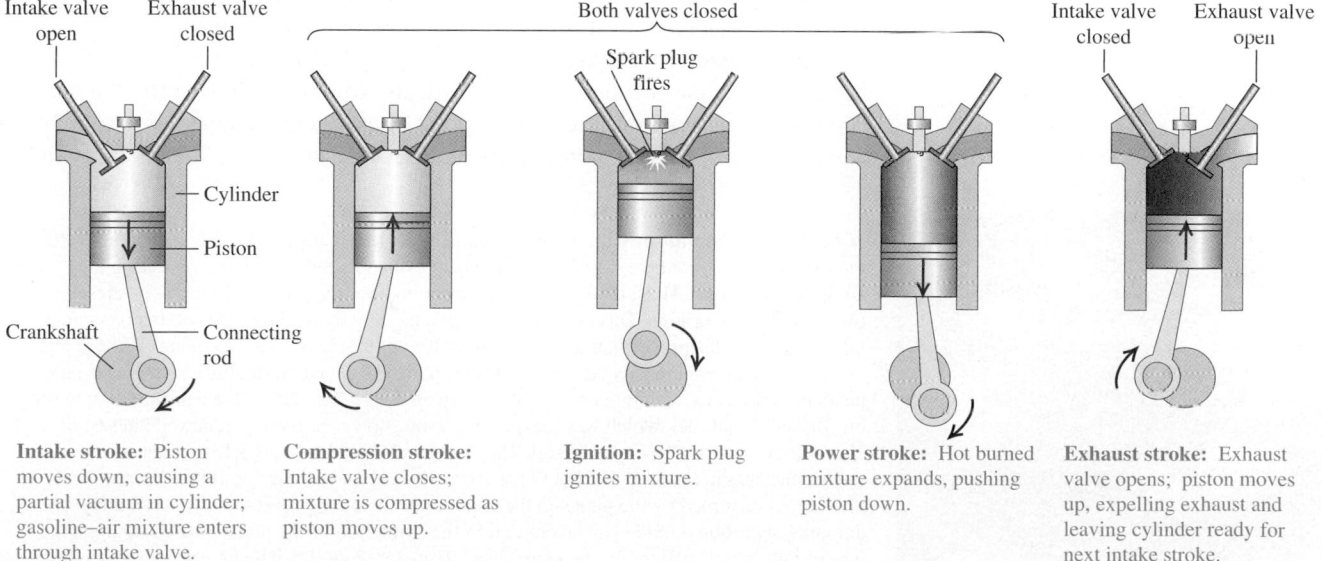

Intake valve open   Exhaust valve closed   Both valves closed   Spark plug fires   Intake valve closed   Exhaust valve open

Cylinder — Piston — Crankshaft — Connecting rod

Intake stroke: Piston moves down, causing a partial vacuum in cylinder; gasoline–air mixture enters through intake valve.

Compression stroke: Intake valve closes; mixture is compressed as piston moves up.

Ignition: Spark plug ignites mixture.

Power stroke: Hot burned mixture expands, pushing piston down.

Exhaust stroke: Exhaust valve opens; piston moves up, expelling exhaust and leaving cylinder ready for next intake stroke.

▲ **FIGURE 16.3** Cycle of a four-stroke internal combustion engine.

Otto cycle

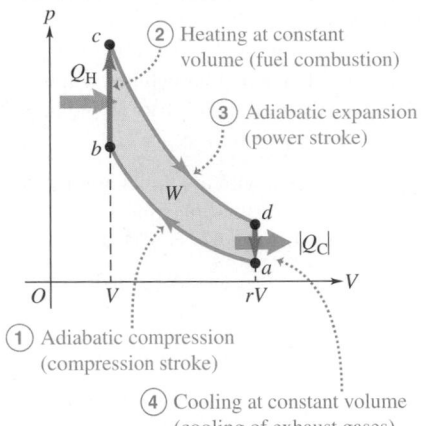

② Heating at constant volume (fuel combustion)

③ Adiabatic expansion (power stroke)

① Adiabatic compression (compression stroke)

④ Cooling at constant volume (cooling of exhaust gases)

▲ **FIGURE 16.4** The pV diagram for the Otto cycle, an idealized model of the thermodynamic processes that take place in a gasoline engine.

Diesel cycle

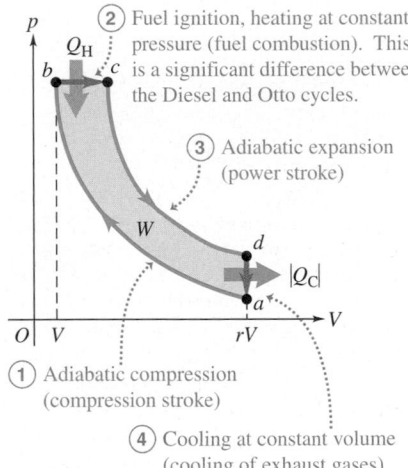

② Fuel ignition, heating at constant pressure (fuel combustion). This is a significant difference between the Diesel and Otto cycles.

③ Adiabatic expansion (power stroke)

① Adiabatic compression (compression stroke)

④ Cooling at constant volume (cooling of exhaust gases)

▲ **FIGURE 16.5** The pV diagram for the Diesel cycle.

gasoline (line bc), and the power stroke is the adiabatic expansion cd. The gas is cooled to the temperature of the outside air (da); during process (da), heat Q_C is released. In practice, this same air does not enter the engine again, but since an equivalent amount does enter, we may consider the process to be cyclic.

It's possible, although a bit complicated, to calculate the efficiency e of this idealized cycle; we'll omit the details. The result is surprisingly simple; it depends only on the compression ratio r and the ratio $\gamma = C_p/C_V$ of the molar heat capacities at constant volume and at constant pressure, which we introduced in Section 15.7. It turns out that the efficiency for the idealized Otto cycle is given by

$$e = 1 - \frac{1}{r^{\gamma-1}}. \tag{16.5}$$

The quantities r and γ are pure numbers, and they are always greater than 1. As we would expect, the thermal efficiency given by Equation 16.5 is always less than 1, even for this idealized model. With $r = 8$ and $\gamma = 1.4$ (the value for air), the theoretical efficiency is $e = 0.56$, or 56%. The efficiency can be increased by increasing r. However, doing this also increases the temperature reached by the air–fuel mixture at the end of the adiabatic compression. If the temperature is too high, the mixture explodes spontaneously during compression instead of burning evenly after the spark plug ignites it. This is called *pre-ignition,* or *detonation;* it causes a knocking sound and can damage the engine. The octane rating of a gasoline is a measure of its antiknock qualities. The maximum practical compression ratio for high-octane, or "premium," gasoline is about 10. Higher ratios can be used with more exotic fuels.

The Otto cycle is a highly idealized model. It assumes that the air–fuel mixture behaves as an ideal gas; it neglects friction, turbulence, loss of heat to cylinder walls, incomplete combustion, and many other effects that combine to reduce the efficiency of a real engine. Thermal efficiencies of real gasoline engines are typically around 20%.

The Diesel Cycle

The Diesel engine is similar in operation to the gasoline engine. The most important difference is that, at the beginning of the compression stroke, the cylinder contains air, but no fuel. A little before the beginning of the power stroke, the injectors start to inject fuel directly into the cylinder, just fast enough to keep the pressure approximately constant during the first part of the power stroke. Because of the high temperature developed during the adiabatic compression, the fuel ignites spontaneously as it is injected; no spark plugs are needed.

The idealized **Diesel cycle** is shown in the pV diagram of Figure 16.5. Starting at point a, air is compressed adiabatically to point b, heated (by the heat of combustion) at constant pressure to point c, expanded adiabatically to point d, and cooled at constant volume to point a. Because there is no fuel in the cylinder

◄ **Application Peanut power.** As we've learned, the key feature of the Diesel engine is that the fuel is ignited, not by a spark, but by the heat from adiabatic compression during the compression stroke. Thus, there's no real connection between Diesel engines and "diesel fuel." In fact, the original Diesel engine, invented in 1892 by Rudolph Diesel, ran on peanut oil! Diesel saw this property as a key feature of his engine: In addition to being far more efficient than steam engines, it could run on fuel produced by local agriculture, which was independent of the fossil fuel monopolies of the day. In the 1920s, the engine was modified to run on "diesel" fossil fuel, which was inexpensive. Now, however, there is renewed interest in "bio-diesel"—basically, vegetable oil. This fuel can be produced anywhere plants grow and contributes less net carbon dioxide to the atmosphere than burning fossil fuel because carbon dioxide is consumed by the plants in the creation of the oils. Unfortunately, substantial amounts of carbon dioxide can be emitted in the extraction of the oils and in the manufacture of nitrogen fertilizers for the plants used to create biodiesel. The use of filtered, cast-off cooking oil is a better solution that makes use of wasted oil and produces no sulphur emissions (unlike diesel and biodiesel fuel).

during most of the compression stroke, pre-ignition cannot occur, and the compression ratio r can be much higher than for a gasoline engine. This feature improves the efficiency of the engine and ensures reliable ignition when the fuel is injected (because of the high temperature reached during the adiabatic compression). Values of r of 15 to 20 are typical; with these values and $\gamma = 1.4$, the theoretical efficiency of the idealized Diesel cycle is about 0.65 to 0.70. As with the Otto cycle, the efficiency of any actual engine is substantially less than this. Diesel engines are usually more efficient than gasoline engines. They are also heavier (per unit power output) and often harder to start.

16.4 Refrigerators

We can think of a **refrigerator** as a heat engine operating in reverse. A heat engine takes heat from a hot place and gives off heat to a colder place. A refrigerator does the opposite; it takes heat from a cold place (the inside of the refrigerator) and gives it off to a warmer place (usually the air in the room where the refrigerator is located). A heat engine has a net *output* of mechanical work; the refrigerator requires a net *input* of mechanical work, as shown in Figure 16.6, an energy-flow diagram for a refrigerator. With the symbols we used in Section 16.2, Q_C is positive for a refrigerator, but both W and Q_H are negative, so $|W| = -W$ and $|Q_H| = -Q_H$.

For a cyclic process, the first law gives $\Delta U = 0$ and

$$Q_H + Q_C - W = 0, \quad \text{or} \quad -Q_H = Q_C - W.$$

Or, because both Q_H and W are negative,

$$|Q_H| = |Q_C| + |W|. \tag{16.6}$$

Thus, as the diagram shows, the heat $|Q_H|$ leaving the working substance and given to the hot reservoir is always *greater* than the heat Q_C taken from the cold reservoir. Note that the absolute-value relation given by Equation 16.6 is valid for both heat engines and refrigerators.

From an economic point of view, the best refrigeration cycle is one that removes the greatest amount of heat Q_C from the refrigerator for the least expenditure of mechanical work W. The relevant ratio is therefore $Q_C/|W|$. We call this the **performance coefficient,** denoted by K. Also, from Equation 16.6, we have $|W| = |Q_H| - |Q_C|$, so K can be expressed as follows:

Performance coefficient of a refrigerator

$$K = \frac{Q_C}{|W|} = \frac{|Q_C|}{|Q_H| - |Q_C|}. \tag{16.7}$$

As always, we measure Q_H, Q_C and W all in the same energy units; K is then a dimensionless number.

The principles of the common refrigeration cycle are shown schematically in Figure 16.7. The fluid "circuit" contains a refrigerant (the working substance). The left side of the circuit (including the cooling coils inside the refrigerator) is at low temperature and low pressure; the right side (including the condenser coils outside the refrigerator) is at high temperature and high pressure. Ordinarily, both sides contain liquid and vapor in phase equilibrium.

The compressor takes in fluid, compresses it adiabatically, and delivers it to the condenser coil at high pressure. The temperature of the fluid is then higher than that of the air surrounding the condenser, so the refrigerant gives off heat $|Q_H|$ and partially condenses to liquid. The fluid then expands adiabatically, at a rate controlled by the expansion valve, into the evaporator. As it does so, it cools

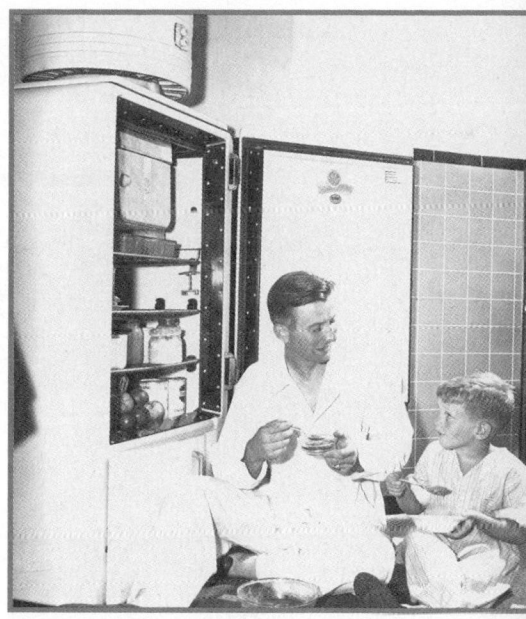

▲ **Application Cooling power.** The modern refrigerator is based on the physical principle that, as a gas expands or a liquid evaporates, it absorbs heat from its surroundings. The ancient Egyptians used this principle to make ice by setting water out in clay pots on cold, windy nights and keeping the pots wet on the outside. Mechanical refrigerators from the early 20th century, like the one shown here, used alternate compression and expansion of a volatile gas, such as ether or ammonia, to cool a closed container. As the gas is compressed on the outside of the container, it heats up, and this heat is released into the room. The compressed gas is allowed to expand in cooling coils inside the container, where it absorbs heat from the container, lowering the container's temperature. The net effect is that heat is transferred from a cold area to a warm area, but not without an energy cost: the energy required to compress the gas in each cycle.

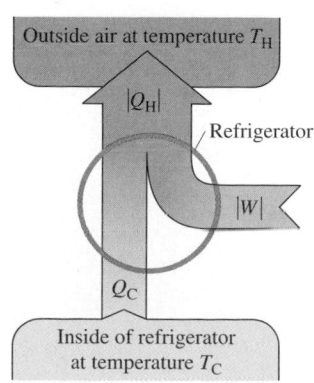

▲ **FIGURE 16.6** Schematic energy-flow diagram for a refrigerator.

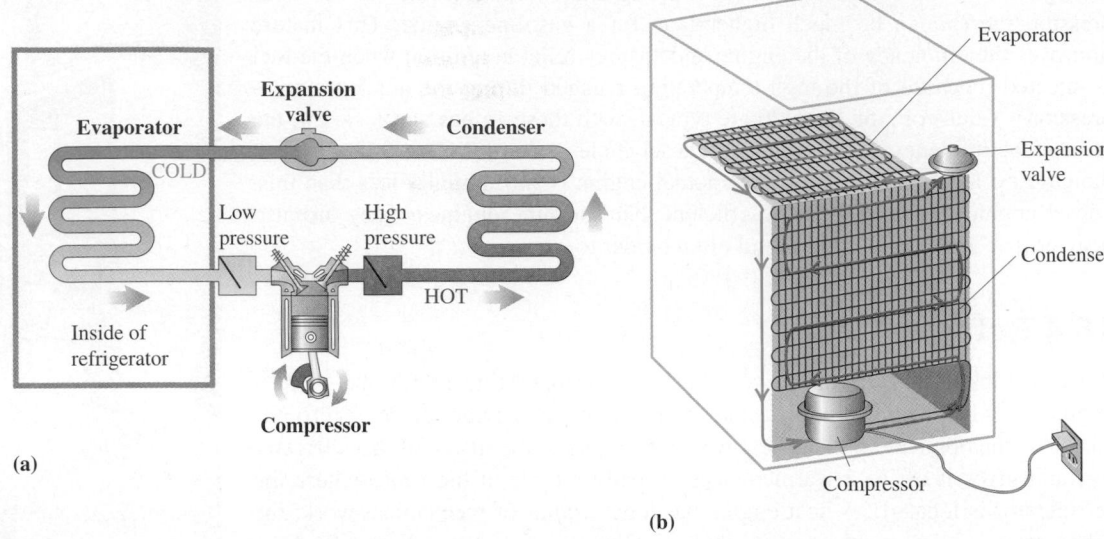

(a)

(b)

▲ **FIGURE 16.7** Principle of the mechanical refrigeration cycle.

considerably, enough so that the fluid in the evaporator coil is colder than its surroundings. The fluid absorbs heat Q_C from its surroundings, cooling them while partially vaporizing. It then enters the compressor to begin another cycle. The compressor, usually driven by an electric motor, requires energy input and does work $|W|$ *on* the working substance during each cycle.

An air conditioner operates on exactly the same principle. In this case, the refrigerator box becomes a room or an entire building. The evaporator coils are inside, the condenser is outside, and fans circulate air through these components (Figure 16.8). In large installations, the condenser coils are often cooled by water. For air conditioners, the quantities of greatest practical importance are the *rate* of heat removal (the heat current H from the region being cooled) and the *power* input $P = W/t$ to the compressor. If heat Q_C is removed in time t, then $H = Q_C/t$. We can then express the performance coefficient as

$$K = \frac{Q_C}{|W|} = \frac{Ht}{Pt} = \frac{H}{P}.$$

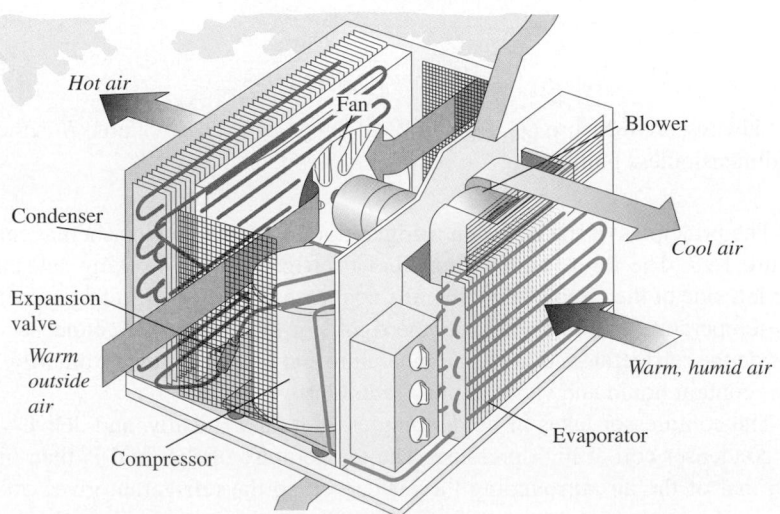

▲ **FIGURE 16.8** An air conditioner works on the same principle as a refrigerator.

Typical room air conditioners have heat removal rates H of 5000 to 10,000 Btu/h, or about 1500 to 3000 W, and require a power input of about 500 to 1000 W. Performance coefficients are typically about 3, with somewhat larger values for larger-capacity units. Unfortunately, K is usually expressed commercially in mixed units, with H in Btu per hour and P in watts. In these units, H/P is called the **energy efficiency rating (EER).** Because 1 W = 3.413 Btu/h, the EER is numerically 3.413 times as large as the dimensionless K. Room air conditioners typically have an EER of about 10. The units, customarily omitted, are $(\text{Btu/h})/\text{W}$.

A variation on this theme is the **heat pump,** used to heat buildings by cooling the outside. It functions like a refrigerator turned inside out. The evaporator coils are outside, where they take heat from cold air or water, and the condenser coils are inside, where they give off heat to the inside region. With proper design, the heat $|Q_H|$ delivered to the inside during each cycle can be considerably greater than the work $|W|$ required to get it there. Heat pumps are typically designed so that they can be run as air conditioners in the summer (by reversing the pumps so that the condenser becomes the evaporator and vice versa). In this mode, a heat pump has no extra efficiency over a standard air conditioner.

For example, suppose you have an air conditioner with $K = 3$. In normal use, the unit removes 300 J of heat from the inside of the house while expending 100 J of electrical energy and discarding 400 J to the outside. But if you mount the air conditioner backwards, so that 400 J goes *into* the house and 300 J is taken from the outside environment, you are getting a total of 400 J of heat into the house while paying the electric company for only 100 J of electrical energy.

Nevertheless, some work is *always* needed to transfer heat from a colder to a hotter object. Heat flows spontaneously from hotter to colder, and to reverse this flow requires the addition of work from the outside. Experience shows that it is impossible to make a refrigerator that transports heat from a cold object to a hotter object without the addition of work. If no work were needed, the performance coefficient would be infinite. We call such a device a *workless refrigerator;* it is a mythical beast, like the unicorn, the frictionless plane, and the free lunch.

Conceptual Analysis 16.2

Heating up or cooling down?

One method sometimes used in an attempt to cool a kitchen on a hot day is to leave the refrigerator door open. Does this actually cool the kitchen?

A. Yes, because refrigerators function on the same principle as air conditioners.
B. No, it actually warms the kitchen.
C. There is no net effect. The heat that is absorbed in the front of the refrigerator is offset by heat dissipated from the back of the refrigerator.

SOLUTION It is true that refrigerators and air conditioners function in the same way, but air conditioners move heat from the inside to the outside of the building, whereas refrigerators move heat from the refrigerator box to the rest of the kitchen. We've learned that, to move heat from the refrigerator box to the warmer kitchen, the refrigerator must convert work to heat. This additional heat is also released to the kitchen. Thus, the refrigerator *always* adds net heat to the kitchen, whether its door is open or shut. Opening the door makes the refrigerator work harder and thus heats the kitchen *more* than if you kept the door shut. Answer B is correct.

Conceptual Analysis 16.3

Efficient heating

Which of the following key advantages does a heat pump have over an ordinary electric (hot-wire) heater for heating a house?

A. The heat pump can be used as an air conditioner in the summer.
B. The heat pump pulls in warm air from the outside to add to the heat it produces.
C. The heat pump uses work to bring additional heat from the outside to the inside of the house.

SOLUTION As we've discussed, a heat pump is basically an air conditioner installed backward; indeed, most heat pumps are designed to switch direction and work as air conditioners when needed. Thus, A is correct. C is also correct; a standard hot-wire heater converts electrical energy to heat on a joule-for-joule basis, but a heat pump uses work (obtained from electric energy) to move heat from the outside to the inside *and also* converts this work to additional heat released inside the house. B is not correct: A heat pump moves heat, not hot air, from the outside to the inside.

16.5 The Second Law of Thermodynamics

Experimental evidence strongly suggests that it is impossible to build a heat engine that converts heat completely to work—that is, an engine with 100% thermal efficiency. This impossibility is the basis of one form of the **second law of thermodynamics:**

Second law of thermodynamics (engine statement)

It is impossible for any system to undergo a process in which it absorbs heat from a reservoir at a single temperature and converts the heat completely into mechanical work with the system ending in the same state in which it began.

We'll call this the "engine" statement of the second law.

The basis of the second law of thermodynamics lies in the difference between the nature of internal energy and that of macroscopic mechanical energy. In a moving object, the molecules have random motion, but superimposed on this is a coordinated motion of every molecule in the direction of the object's velocity. The kinetic energy associated with this *coordinated* macroscopic motion is what we call the kinetic energy of the moving object. The kinetic and potential energies associated with the *random* motion of the object's molecules constitute the internal energy. When a moving object comes to rest as result of friction or an inelastic collision, the organized part of the motion is converted to random motion. **Since we cannot control the motions of individual molecules, we cannot convert this random motion completely back to organized motion.** We can convert *part* of it, and that is what a heat engine does.

If the second law were *not* true, we could power an automobile or run a power plant simply by cooling the surrounding air and thus taking energy from it. Neither of these impossible scenarios violates the *first* law of thermodynamics. The second law is not a deduction from the first, but stands by itself as a separate law of nature. The first law denies the possibility of creating or destroying energy; the second law limits the *availability* of energy and the ways in which it can be used and converted.

Our analysis of refrigerators in Section 16.4 forms the basis for an alternative statement of the second law of thermodynamics. Heat flows spontaneously from hotter to colder objects, never the reverse. A refrigerator does take heat from a colder to a hotter object, but its operation depends on an input of mechanical energy or work. Generalizing this observation, we state the following principle:

Second law of thermodynamics (refrigerator statement)

It is impossible for any process to have as its sole result the transfer of heat from a cooler to a hotter object.

We'll call this the "refrigerator" statement of the second law. It may not seem to be very closely related to the "engine" statement. In fact, though, the two statements are completely equivalent. For example, if we could build a workless refrigerator, violating the "refrigerator" statement of the second law, we could use it in conjunction with a heat engine to pump the heat rejected by the engine back into the hot reservoir, to be reused. This composite machine (Figure 16.9a) would violate the "engine" statement of the second law because its net effect would be to take a net quantity of heat $Q_H - |Q_C|$ from the hot reservoir and convert it completely to work W.

Alternatively, if we could make an engine with 100% thermal efficiency, in violation of the first statement, we could run it by taking heat from the hot reservoir. Then we could use the work output to drive a refrigerator that pumps heat

If a workless refrigerator were possible, it could be used in conjunction with an ordinary heat engine to form a 100%-efficient engine, converting heat $Q_H - |Q_C|$ completely to work.

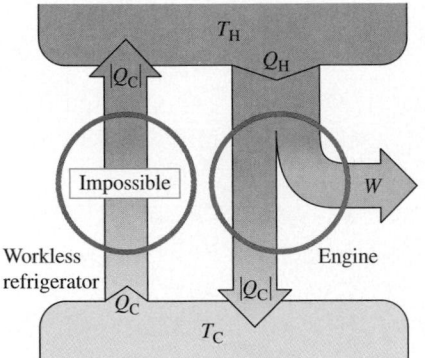

(a) The "engine" statement of the second law

If a 100%-efficient engine were possible, it could be used in conjunction with an ordinary refrigerator to form a workless refrigerator, transferring heat Q_C from the cold to the hot reservoir with no net input of work.

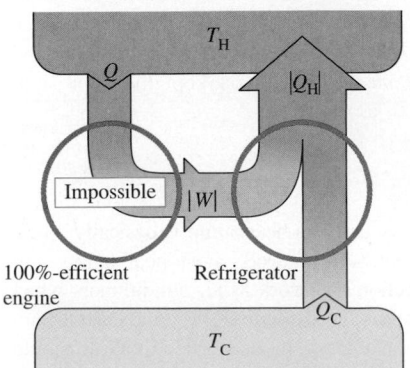

(b) The "refrigerator" statement of the second law

▲ **FIGURE 16.9** Energy-flow diagrams for equivalent forms of the second law. Since either of these machines is impossible, the other must be also.

from the cold reservoir to the hot (Figure 16.9b). This composite device would violate the "refrigerator" statement because its net effect would be to take heat Q_C from the cold reservoir and deliver it to the hot reservoir without requiring any input of work. Thus, any device that violates one form of the second law can also be used to make a device that violates the other form. In other words, if violations of the first form are impossible, so are violations of the second!

The conversion of work to heat, as in friction or viscous fluid flow, and heat flow from hot to cold across a finite temperature gradient are *irreversible* processes. The "engine" and "refrigerator" statements of the second law state that these processes can be only partially reversed. Here are two other examples: Gases always seep through an opening spontaneously from a region of high pressure to a region of low pressure; and gases left by themselves always tend to mix, never to unmix. Figure 16.10 shows two liquids mixing spontaneously.

The second law of thermodynamics is an expression of the inherent one-way aspect of these and many other irreversible processes. Energy conversion is an essential aspect of all plant and animal life and of many mechanical devices, so the second law of thermodynamics is of the utmost fundamental importance in the world we live in.

▲ **FIGURE 16.10** These liquids mix spontaneously—and they never spontaneously unmix.

16.6 The Carnot Engine: The Most Efficient Heat Engine

According to the second law, no heat engine can have 100% efficiency. But how great an efficiency *can* an engine have, given two heat reservoirs at temperatures T_H and T_C? This question was answered in 1824 by the French engineer Sadi Carnot (1796–1832), who developed a hypothetical, idealized heat engine that has the *maximum possible* efficiency consistent with the second law. In his honor, the cycle of this engine is called the **Carnot cycle.**

To understand the rationale of the Carnot cycle, we return to a recurrent theme in this chapter: *reversibility* and its relation to directions of thermodynamic processes. The conversion of work to heat is an irreversible process; the purpose of a heat engine is a *partial* reversal of this process: the conversion of heat to work with as great efficiency as possible. For maximum heat-engine efficiency, therefore, *we must avoid all irreversible processes.* This requirement turns out to be enough to determine the basic sequence of steps in the Carnot cycle, as we'll show next.

Heat flow through a finite temperature drop is an irreversible process. Therefore, during heat transfer in the Carnot cycle, there must be *no* finite temperature differences. When the engine takes heat from the hot reservoir at temperature T_H, the working substance of the engine must also be at T_H; otherwise, irreversible heat flow would occur. Similarly, when the engine discards heat to the cold reservoir at T_C, the engine itself must be at T_C. That is, every process that involves heat transfer must be *isothermal* at either T_H or T_C.

Conversely, in any process in which the temperature of the working substance of the engine is intermediate between T_H and T_C, there must be *no* heat transfer between the engine and either reservoir, because such heat transfer cannot be reversible. Therefore, any process in which the temperature T of the working substance changes must be adiabatic. The bottom line is that *every process* in our idealized cycle must be either *isothermal* or *adiabatic.* In addition, thermal and mechanical equilibrium must be maintained at all times, so that each process is completely reversible.

MasteringPHYSICS

ActivPhysics 8.14: Carnot Cycle

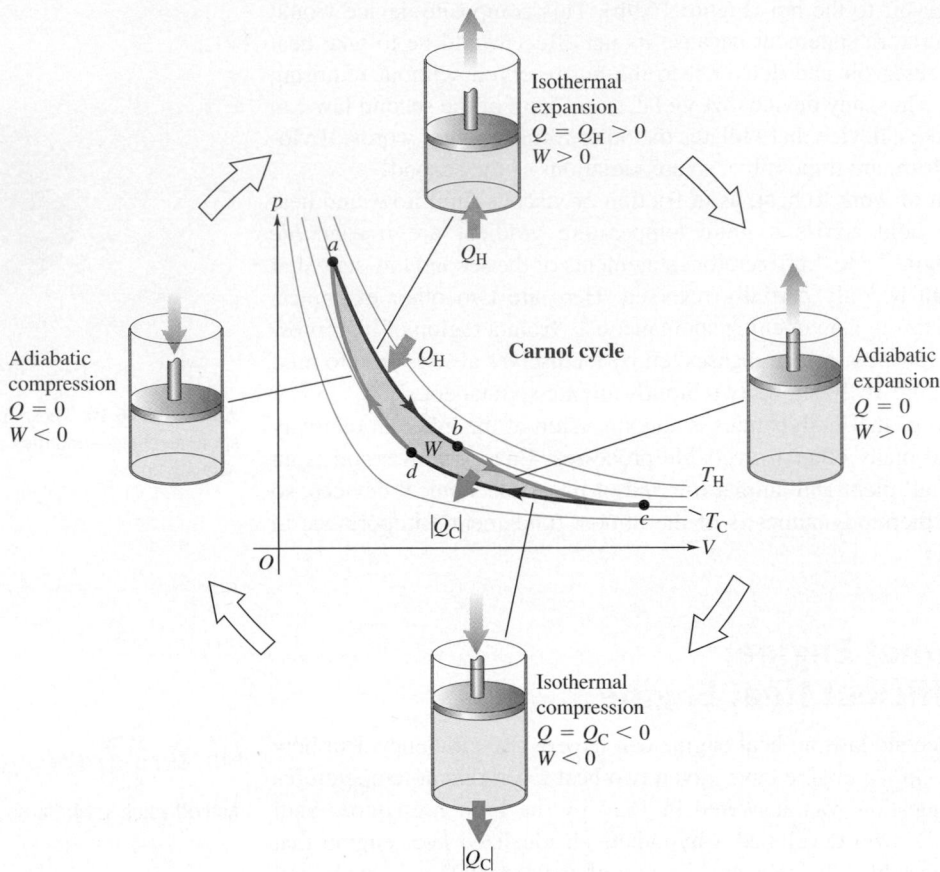

▲ **FIGURE 16.11** The Carnot cycle for an ideal gas.

The Carnot Cycle

The Carnot cycle consists of two isothermal and two adiabatic processes. Figure 16.11 shows a Carnot cycle using an ideal gas in a cylinder with a piston as its working substance. It consists of the following steps:

1. The gas expands isothermally at temperature T_H, absorbing heat Q_H (ab).
2. It expands adiabatically until its temperature drops to T_C (bc).
3. It is compressed isothermally at T_C, rejecting heat $|Q_C|$ (cd).
4. It is compressed adiabatically back to its initial state at temperature T_H (da).

When the working substance in a Carnot engine is an ideal gas, it's a straightforward calculation to find the heat and work for each step. Then from these quantities we can obtain the thermal efficiency, using the definition of e (Equation 16.4). The details of the calculations are a little involved, so we'll simply state the results here, without any derivation.

The ratio Q_C/Q_H of the two quantities of heat turns out to be equal (apart from a negative sign) to the ratio of the absolute temperatures of the reservoirs:

$$\frac{Q_C}{Q_H} = -\frac{T_C}{T_H}. \tag{16.8}$$

Combining this equation with the general expression for the thermal efficiency of *any* heat engine, $e = 1 + Q_C/Q_H$ (Equation 16.4), we find that the thermal efficiency of the Carnot engine depends only on the temperatures T_H and T_C:

Thermal efficiency of Carnot engine

$$e_{\text{Carnot}} = 1 - \frac{T_C}{T_H} = \frac{T_H - T_C}{T_H}. \qquad (16.9)$$

This surprisingly simple result says that the thermal efficiency of the Carnot engine depends only on the ratio T_C/T_H of the absolute temperatures of the two heat reservoirs. Note that the temperatures must be *absolute* (Kelvin). The efficiency is large when the temperature *difference* is large, and it is very small when the temperatures are nearly equal. The efficiency can never be exactly 1 unless $T_C = 0$. As we shall see later, this, too, is impossible.

EXAMPLE 16.2 A Carnot engine

A Carnot engine takes 2000 J of heat from a reservoir at 500 K, does some work, and discards some heat to a reservoir at 350 K. How much work does it do, how much heat is discarded, and what is its efficiency?

SOLUTION

SET UP AND SOLVE From Equation 16.8, the heat Q_C discarded by the engine is

$$Q_C = -Q_H \frac{T_C}{T_H} = -(2000 \text{ J}) \frac{350 \text{ K}}{500 \text{ K}}$$
$$= -1400 \text{ J}.$$

Then, from the first law, the work W done by the engine is

$$W = Q_H + Q_C = 2000 \text{ J} + (-1400 \text{ J})$$
$$- 600 \text{ J}.$$

From Equation 16.9, the thermal efficiency is

$$e_{\text{Carnot}} = 1 - \frac{T_C}{T_H} = 1 - \frac{350 \text{ K}}{500 \text{ K}} = 0.30 = 30\%.$$

Alternatively, from the basic definition of thermal efficiency,

$$e = \frac{W}{Q_H} = \frac{600 \text{ J}}{2000 \text{ J}} = 0.30 = 30\%.$$

REFLECT The efficiency of any engine is always less than 1. For the Carnot engine, the greater the ratio of T_H to T_C, the greater is the efficiency.

Practice Problem: What temperature T_H would the high-temperature reservoir need to have in order to increase the efficiency to $e = 0.50$? *Answer:* 700 K.

Carnot Refrigerator

Because each step in the Carnot cycle is reversible, the *entire cycle* may be reversed, converting the engine into a refrigerator. The performance coefficient of the Carnot refrigerator is obtained by combining the general definition of K, Equation 16.7, with Equation 16.8 for the Carnot cycle. We first rewrite Equation 16.7 as

$$K = \frac{|Q_C|}{|Q_H| - |Q_C|} = \frac{|Q_C|/|Q_H|}{1 - |Q_C|/|Q_H|}.$$

From Equation 16.8, $|Q_C|/|Q_H| = T_C/T_H$. Substituting this into the previous equation, we find that

$$K_{\text{Carnot}} = \frac{T_C}{T_H - T_C}. \qquad (16.10)$$

When the temperature difference $T_H - T_C$ is small, K is much larger than 1; in this case, a lot of heat can be "pumped" from the lower to the higher temperature with only a little expenditure of work. But the greater the temperature difference, the smaller K is and the more work is required to transfer a given quantity of heat.

EXAMPLE 16.3 A Carnot refrigerator

If the Carnot engine described in Example 16.2 is run backwards as a refrigerator, what is its performance coefficient?

Continued

SOLUTION

SET UP AND SOLVE We use the values from Example 16.2, with all the signs reversed because the cycle is run backwards:

$$Q_C = +1400 \text{ J}, \quad Q_H = -2000 \text{ J}, \quad W = -600 \text{ J}.$$

From Equation 16.7, the performance coefficient K is

$$K = \frac{Q_C}{|W|} = \frac{1400 \text{ J}}{600 \text{ J}} = 2.33.$$

Because the cycle is a Carnot cycle, we may also use Equation 16.10:

$$K = \frac{T_C}{T_H - T_C} = \frac{350 \text{ K}}{500 \text{ K} - 350 \text{ K}} = 2.33.$$

REFLECT For a Carnot cycle, e and K depend only on the temperatures, as shown by Equations 16.9 and 16.10, and we don't need to calculate Q and W. For cycles containing irreversible processes, however, these equations are not valid, and more detailed calculations are necessary.

Practice Problem: Suppose you want to increase the performance coefficient K in Example 16.3. Would it be better to decrease T_H by 50 K or increase T_C by 50 K? *Answer:* Increase T_C.

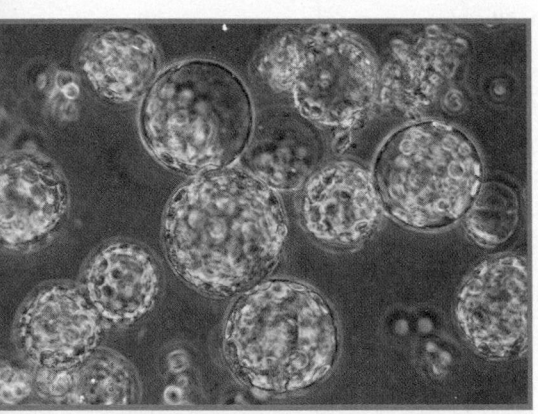

▲ **BIO Application Is photosynthesis efficient?** In photosynthesis, the chloroplasts of plants and algae are the subcellular organelles that capture bits of energy emitted by the sun and convert it to useful cellular energy in the form of ATP. If the sun and the plant are viewed as thermodynamic systems, the chloroplasts can be considered machines that work between two reservoirs of different temperatures to extract useful energy. We know that the efficiency of an engine depends on the temperature difference between the reservoirs, and in this case, the difference is very large indeed. The temperature of the Sun is 5600 K whereas the temperature of a plant might be 293 K. Thus the laws of thermodynamics tell us that photosynthesis is a perfectly practical process.

The Carnot Cycle and the Second Law

We can prove that **no engine can be more efficient than a Carnot engine operating between the same two temperatures.** The key to the proof is the preceding observation that, since each step in the Carnot cycle is reversible, the *entire cycle* may be reversed. Run backward, the engine becomes a refrigerator. Suppose we have a super-efficient engine that is more efficient than a Carnot engine (Figure 16.12). The Carnot engine, run backward as a refrigerator with negative work $-W$, takes in heat Q_C from the cold reservoir and expels heat $|Q_H|$ to the hot reservoir. The super-efficient engine expels heat $|Q_C|$, but to do this, it takes in a greater amount of heat $Q_H + Q$. Its work output is then $W + Q$, and the net effect of the two machines together is to take a quantity of heat Q and convert it completely into work. Doing this, however, violates the "engine" statement of the second law. We could construct a similar argument that a super-efficient engine could be used to violate the "refrigerator" statement of the second law. (Note that we don't have to assume that the super-efficient engine is reversible.)

Thus, the statement that no engine can be more efficient than a Carnot engine is yet another equivalent statement of the second law of thermodynamics. It also follows directly that **all Carnot engines operating between the same two temperatures have the same efficiency, irrespective of the nature of the working substance.** Although we discussed Equation 16.9 in the context of a Carnot engine using an ideal gas as its working substance, that equation is in fact valid for *any* Carnot engine, no matter what its working substance.

Equation 16.9 also shows how to maximize the efficiency of a real engine, such as a steam turbine in an electric generating plant. The intake temperature T_H must be made as high as possible and the exhaust temperature T_C as low as possi-

If a superefficient engine were possible, it could be used in conjunction with a Carnot refrigerator to convert the heat Q completely to work, with no net transfer to the cold reservoir.

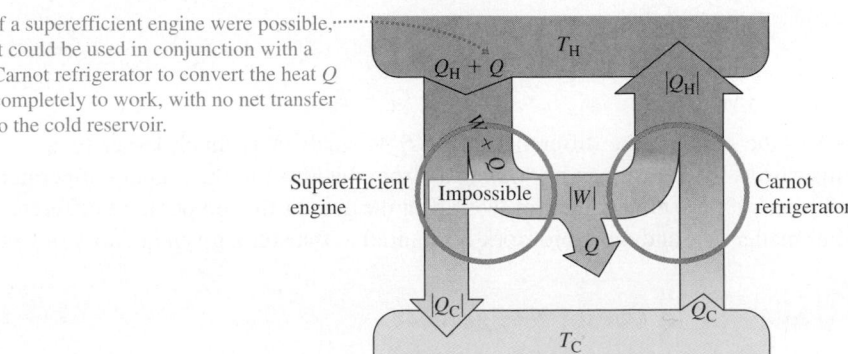

▲ **FIGURE 16.12** If there were a more efficient engine than a Carnot engine, it could be used in conjunction with a Carnot refrigerator to construct the impossible device shown here.

ble. The exhaust temperature can't be lower than the lowest temperature available for cooling the exhaust—usually the temperature of the air, or perhaps of river or lake water if this is available at the plant. The vapor pressures of all liquids increase rapidly with increasing temperature, so T_H is limited by the mechanical strength of the boiler. At 540°C (1000°F), the vapor pressure of water is about 125×10^5 Pa (120 atm, or 1800 psi), and this is about the maximum practical pressure in large present-day steam boilers.

The unavoidable exhaust heat loss in electric power plants creates a serious environmental problem. When a lake or river is used for cooling, the temperature of the body of water may be raised several degrees. Such a temperature change has a severely disruptive effect on the overall ecological balance, since relatively small temperature changes can have significant effects on metabolic rates in plants and animals. Because **thermal pollution,** as this effect is called, is an inevitable consequence of the second law of thermodynamics, careful planning is essential to minimize the ecological impact of new power plants.

16.7 Entropy

The second law of thermodynamics, as we have stated it, is a rather strange beast, compared with many familiar physical laws. It is not an equation or a quantitative relationship, but rather a statement of *impossibility.* However, the second law *can* be stated as a quantitative relation by use of the concept of *entropy,* the subject of this section.

We've talked about several processes that proceed naturally in the direction of increasing disorder. Irreversible heat flow increases disorder because the molecules are initially sorted into hotter and cooler regions, and this sorting is lost when the system comes to thermal equilibrium. Adding heat to an object increases its disorder because it increases average molecular speeds and therefore the randomness of molecular motion. Free expansion of a gas increases its disorder because the molecules have greater randomness of *position* after the expansion than before. **Entropy** provides a *quantitative* measure of disorder.

Entropy and Disorder

To introduce the concept of entropy, let's consider an infinitesimal isothermal expansion of an ideal gas. We add heat Q and let the gas expand just enough so that the temperature remains constant. Because the internal energy of an ideal gas depends only on its temperature, the internal energy is also constant. Thus, from the first law, the work W done by the gas is equal to the heat Q added. That is,

$$Q = W = p\,\Delta V = \frac{nRT}{V}\,\Delta V \qquad \text{and} \qquad \frac{Q}{T} = nR\frac{\Delta V}{V}.$$

The gas is in a more disordered state after the expansion than before because the molecules are moving in a larger volume and have more randomness of position. The fractional change in volume $\Delta V/V$ is a measure of the increase in disorder, and it is proportional to the quantity Q/T. We introduce the symbol S for the entropy of the system in a particular state, and we *define* the entropy change $\Delta S = S_2 - S_1$ during a reversible isothermal process at absolute temperature T as follows:

Definition of entropy change in a reversible isothermal process
In a reversible isothermal process from state 1 to state 2, the change in entropy $\Delta S = S_2 - S_1$ of a system equals the heat Q added to or removed from the system, divided by the absolute temperature T:

$$\Delta S = S_2 - S_1 = \frac{Q}{T}. \tag{16.11}$$

Unit: Energy divided by temperature: J/K.

Conceptual Analysis 16.4

Increasing entropy

An amount of heat Q is added to each of two quantities of an ideal gas. Gas A is at absolute temperature T, and gas B is at absolute temperature $2T$. Each gas is allowed to expand isothermally. Which gas undergoes the greater entropy change?

A. gas A.
B. gas B.
C. Both have the same entropy change.

SOLUTION For each gas, the change in entropy for the isothermal process is the amount of heat added, divided by the absolute temperature. The absolute temperature for gas B is twice as great as that for gas A, and the two amounts of heat are the same. So the entropy change of B is half that of A. The correct answer is A.

EXAMPLE 16.4 Entropy change in melting ice

Compute the change in entropy of 1.00 kilogram of ice at 0°C when it is melted and converted to water at 0°C.

SOLUTION

SET UP The temperature T is constant at 273 K, and the heat of fusion Q needed to melt the ice is $Q = mL_f = (1.00 \text{ kg})(334 \times 10^3 \text{ J/kg}) = 334 \times 10^3$ J.

SOLVE From Equation 16.11, the increase in entropy of the system is

$$\Delta S = \frac{Q}{T} = \frac{3.34 \times 10^5 \text{ J}}{273 \text{ K}} = 1220 \text{ J/K}.$$

REFLECT The entropy of the system increases as heat is added to it. The disorder increases because the water molecules in the liq-uid state have more freedom to move than when they were locked in a crystal structure. In any *isothermal* reversible process, the change of entropy equals the heat transferred divided by the absolute temperature. When we refreeze the water, its entropy change is $\Delta S = -1220$ J/K.

Practice Problem: When 50 g of liquid water at 0°C freezes into a 50 g ice cube at 0°C, what is the change in entropy of the material? *Answer:* $\Delta S = -61$ J/K.

We can generalize the definition of entropy change to include *any* reversible process leading from one state to another, whether it is isothermal or not. We represent the process as a series of infinitesimal reversible steps. During a typical step, an infinitesimal quantity of heat ΔQ is added to the system at absolute temperature T. Then we sum the quotients $\Delta Q/T$ for the entire process. If T varies during the process, then carrying out this sum requires calculus.

Because entropy is a measure of the disorder of a system in any specific state, it must depend only on the current state of the system, not on its past history. We'll show later, by using the second law of thermodynamics, that this is indeed the case. When a system proceeds from an initial state with entropy S_1 to a final state with entropy S_2, the change in entropy $\Delta S = S_2 - S_1$, defined by Equation 16.11, does not depend on the path leading from the initial to the final state, but is the same for *all possible* processes leading from state 1 to state 2. Thus, if the entropy of one state of a system is assigned a certain value, the entropies of all other states of that system also have definite values, and those values depend only on the state of the system. We recall that *internal energy,* which was introduced in Chapter 15, also has this property, although entropy and internal energy are very different quantities.

As with internal energy, the preceding discussion does not define entropy itself, but only the *change* in entropy in any given process. To complete the definition, we may arbitrarily assign a value to the entropy of a system in one specified reference state and then calculate the entropy of any other state with reference to this assigned value.

EXAMPLE 16.5 **An adiabatic expansion**

A gas expands adiabatically and reversibly. What is its change in entropy?

SOLUTION

SET UP AND SOLVE In an adiabatic process, no heat enters or leaves the system; $Q = 0$, and there is *no* change in entropy. $\Delta S = 0$. Every *reversible* adiabatic process is a constant-entropy process.

REFLECT The increase in disorder resulting from the gas occupying a greater volume after the expansion is exactly balanced by the decrease in disorder associated with the lowered temperature of the gas and its decreased molecular speeds.

EXAMPLE 16.6 **Entropy and the Carnot cycle**

The Carnot engine in Example 16.2 (Section 16.6) takes 2000 J of heat from a reservoir at 500 K, does some work, and discards some heat to a reservoir at 350 K. Find the total entropy change in the engine during one cycle.

SOLUTION

SET UP AND SOLVE During the isothermal expansion at 500 K, the engine takes in 2000 J and its entropy change is

$$\Delta S = \frac{Q}{T} = \frac{2000 \text{ J}}{500 \text{ K}} = 4.0 \text{ J/K}.$$

During the isothermal compression at 350 K, the engine gives off 1400 J of heat and its entropy change is

$$\Delta S = \frac{-1400 \text{ J}}{350 \text{ K}} = -4.0 \text{ J/K}.$$

The entropy change of the engine during each of the two adiabatic processes is zero. Thus, the total entropy change in the engine during one cycle is 4.0 J/K + 0 − 4.0 J/K + 0 = 0.

REFLECT The total entropy change of the two heat reservoirs is also zero, although each individual reservoir has a nonzero entropy change. This cycle contains no irreversible processes, and the total entropy change of the system and its surroundings is zero.

Entropy in Irreversible Processes

In an idealized reversible process involving only equilibrium states, the total entropy change of the system and its surroundings is zero. But all *irreversible* processes involve an increase in entropy. Unlike energy, entropy is *not* a conserved quantity. The entropy of an isolated system can change, but, as we shall see, it can never decrease. **An entropy increase occurs in every natural (irreversible) process if all systems taking part in the process are included.** Example 16.7 (below) illustrates an increase in entropy during an irreversible process in an isolated system.

The fact that entropy is a function only of the *state* of a system shows us how to compute entropy changes in *irreversible* (non-equilibrium) processes, to which Equation 16.11 is not applicable. We simply invent a path connecting the given initial and final states that *does* consist entirely of reversible equilibrium processes, and we compute the total entropy change for that alternative path. It is not the actual path, but the entropy change must be the same as that for the actual path.

EXAMPLE 16.7 **A free expansion**

A thermally insulated box is divided by a partition into two compartments, each having volume V. (Figure 16.13). Initially, one compartment contains n moles of an ideal gas at temperature T, and the other compartment is evacuated. We then break the partition, and the gas expands to fill both compartments. What is the entropy change in this free-expansion process?

Continued

SOLUTION

SET UP For the given process, $Q = 0$, $W = 0$, $\Delta U = 0$, and therefore (because the system is an ideal gas), $\Delta T = 0$. We might think that the entropy change is zero, because there is no heat exchange. But Equation 16.11 is valid only for *reversible* processes; the free expansion described here is *not* reversible, and there is an entropy change. To calculate ΔS, we use the fact that the entropy change depends only on the initial and final states. We can devise a *reversible* process having the same end points, use Equation 16.11 to calculate its entropy change, and thus determine the entropy change in the original process.

SOLVE The appropriate reversible process in this case is an isothermal expansion from V to $2V$ at temperature T, shown on the pV diagram of Figure 16.13c. The gas does work during this substitute expansion, so heat must be supplied to keep the internal energy constant. The total heat equals the total work, which is given by Equation 15.17:

$$W = Q = nRT \ln \frac{V_2}{V_1}.$$

Thus, the entropy change is

$$\Delta S = \frac{Q}{T} = \frac{nRT \ln \frac{2V}{V}}{T} = nR \ln 2,$$

and this is also the entropy change for the free expansion. For 1 mole,

$$\Delta S = (1 \text{ mol})[8.314 \text{ J}/(\text{mol} \cdot \text{K})](0.693)$$
$$= 5.76 \text{ J/K}.$$

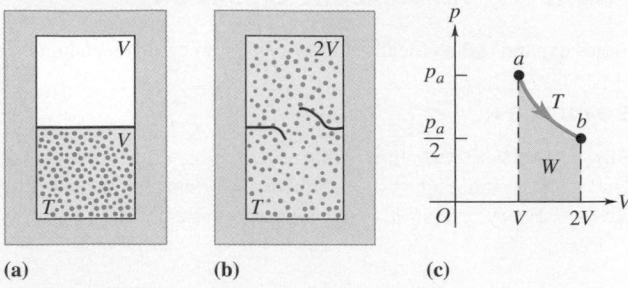

▲ **FIGURE 16.13** (a, b) Free expansion of an insulated ideal gas, an irreversible process. (c) An example of a reversible process that has the same initial and final states as the free expansion and thus has the same change in entropy.

REFLECT The entropy change ΔS is proportional to the amount of substance (n moles) and is independent of the temperature T (provided that it is constant). The increase in disorder is due to the increased randomness of position resulting from the increase in volume.

Practice Problem: In a free expansion, 2.00 mol of an ideal gas expands from an initial volume of 1.50 L to a final volume of 6.00 L. What is the increase in entropy? *Answer:* 23.1 J/K.

EXAMPLE 16.8   Mixing hot and cold water

Suppose 1.00 kg of water at 100°C is placed in thermal contact with 1.00 kg of water at 0°C. Find the approximate total change in entropy when the hot water cools to 99°C and the cold water warms to 1°C. Assume that the specific heat capacity of water is constant at 4190 J/(kg · K) over the given temperature range.

SOLUTION

SET UP AND SOLVE This process involves irreversible heat flow because of the temperature differences. The first 4190 J of heat transferred cools the hot water to 99°C (372 K) and warms the cold water from 0°C to 1°C (274 K). The net change of entropy of the hot water is approximately

$$\Delta S_{\text{hot}} = \frac{-4190 \text{ J}}{372 \text{ K}} = -11.3 \text{ J/K}.$$

(During this process, the temperature actually varies from 373 K to 372 K, but the resulting error in the calculation is very small.) The change of entropy of the cold water is approximately

$$\Delta S_{\text{cold}} = \frac{+4190 \text{ J}}{274 \text{ K}} = +15.3 \text{ J/K}.$$

The total change of entropy of the system during this process is

$$\Delta S_{\text{hot}} + \Delta S_{\text{cold}} = -11.3 \text{ J/K} + 15.3 \text{ J/K} = +4.0 \text{ J/K}.$$

Further increases in entropy occur as the system approaches thermal equilibrium at 50°C (323 K). Because the two temperatures vary continuously during this heat exchange, we would need methods of calculus to calculate the entropy changes precisely. It turns out that the total entropy change of the hot water is $\Delta S_{\text{hot}} = -603 \text{ J/K}$, the total entropy change of the cold water is $\Delta S_{\text{cold}} = 705 \text{ J/K}$, and the total entropy change of the system is $705 \text{ J/K} - 603 \text{ J/K} = 102 \text{ J/K}$.

REFLECT The total entropy of the final state is greater than that of the initial state, corresponding to increasing disorder as the distinction between the hotter and colder molecules gradually disappears. We could have reached the same end state by simply mixing the two quantities of water, also an irreversible process. The entropy depends only on the state of the system, and the total entropy change would be the same, 102 J/K.

Practice Problem: Repeat the approximate calculation for the total entropy change when the hot water cools from 76°C to 75°C and the cool water warms from 24°C to 25°C. *Answer:* 2.0 J/K.

Entropy and the Second Law

The mixing of substances at different temperatures and the flow of heat from a higher to a lower temperature are characteristic of *all* natural (that is, irreversible) processes. When all the entropy changes in the process are included, the increases in entropy are always found to be greater than the decreases. In the special case of a *reversible* process, the increases and decreases are equal. Thus, we can state the following general principle: **When all systems taking part in a process are included, the total entropy either remains constant or increases.** Put another way, **no process is possible in which the total entropy decreases when all systems taking part in the process are included.** This alternative statement of the second law of thermodynamics, in terms of entropy, is equivalent to the "engine" and "refrigerator" statements discussed earlier.

The increase of entropy in every natural (irreversible) process measures the increase in disorder, or randomness, in the universe associated with that process. Consider again the example of mixing hot and cold water. We *might* have used the hot and cold water as the high- and low-temperature reservoirs of a heat engine. While removing heat from the hot water and giving heat to the cold water, we could have obtained some mechanical work. But once the hot and cold water have been mixed and have come to a uniform temperature, this opportunity to convert heat to mechanical work is lost irretrievably. The lukewarm water will never *unmix* itself and separate into hotter and colder portions. No decrease in *energy* occurs when the hot and cold water are mixed. What has been lost is not *energy*, but *opportunity*— the opportunity to convert part of the heat from the hot water into mechanical work. When entropy increases, energy becomes less *available,* and the universe has become more random, or "run down."

16.8 The Kelvin Temperature Scale

When we studied temperature scales in Chapter 14, we expressed the need for a temperature scale that doesn't depend on the properties or behavior of any particular material. We can now use the Carnot cycle to define such a scale. We've learned that the efficiency of a Carnot engine operating between two heat reservoirs at absolute temperatures T_H and T_C is independent of the nature of the working substance and that it depends only on the temperatures. If several Carnot engines with different working substances operate between the same two heat reservoirs, their thermal efficiencies are all the same:

$$e = \frac{Q_H + Q_C}{Q_H} = 1 + \frac{Q_C}{Q_H}.$$

Therefore, the ratio Q_C/Q_H is the same for *all* Carnot engines operating between two given temperatures T_H and T_C.

Kelvin proposed that we *define* the ratio of the temperatures, T_C/T_H, to be equal to the magnitude of the ratio Q_C/Q_H of the quantities of heat absorbed and rejected:

$$\frac{T_C}{T_H} = \frac{|Q_C|}{|Q_H|} = -\frac{Q_C}{Q_H}. \tag{16.12}$$

Equation 16.12 looks identical to Equation 16.8, but there is a subtle and crucial difference. The temperatures in Equation 16.8 are based on an ideal-gas thermometer, as defined in Section 14.2. But Equation 16.12 *defines* a temperature scale that is based on the Carnot cycle and the second law of thermodynamics and is independent of the behavior of any particular substance. Thus, the **Kelvin temperature scale** is truly *absolute.*

▲ Application **Flower power.** Sometimes the second law of thermodynamics is invoked to make the process of evolution seem impossible. If the universe tends toward disorder (increasing entropy), how could complex life forms have evolved from simpler, primitive forms? Similarly, it would seem impossible that a complex plant or animal could develop from a single cell. The flaw in this reasoning is that neither the earth nor an organism is a closed system. The earth constantly receives energy from the sun (and radiates "waste" heat to space). Organisms also constantly consume energy and transform it to power their activities. You eat food; plants use sunlight; the embryo in a bird's egg uses chemical energy stored in the yolk and white. If you were sealed off from a suitable source of energy, you would quickly die—and then, indeed, entropy would take its course.

▲ **Application Zero power.** Using knowledge of the principles of physics and energy conservation, architects are designing homes of the future that will consume no net power. These homes, called "zero-power homes," are designed to generate as much power as they consume, allowing their owners to have an annual energy bill of $0. They can do this by using a combination of rooftop photovoltaic cells, solar water heaters, state-of-the-art insulation, energy-efficient appliances, and computer-controlled power distribution systems. Some can even generate excess power, which can be sold back to "the grid" to generate income. Of course, this net negative energy use does not violate the laws of thermodynamics: The house is not a closed system. The negative energy use by these homes results from the efficient collection and use of the vast amounts of solar energy reaching the homes each year.

To complete the definition of the Kelvin scale, we assign the arbitrary value of 273.16 K to the temperature of the triple point of water. As we discussed in Section 15.2, this is the unique temperature at which gas, liquid, and solid phases of water can coexist in phase equilibrium. When a substance is taken around a Carnot cycle, the ratio of the heat absorbed to the heat rejected, $|Q_H|/|Q_C|$, is equal to the ratio of the temperatures of the reservoirs, *as expressed on the gas scale,* defined in Section 14.2. Since, on both scales, the triple point of water is chosen to be 273.16 K, it follows that *the Kelvin and the ideal-gas scales are identical.*

The zero point on the Kelvin scale is called **absolute zero.** There are theoretical reasons for believing that absolute zero cannot be attained experimentally, although temperatures below 10^{-7} K have been achieved. The more closely we approach absolute zero, the more difficult it is to get yet closer. Absolute zero can also be interpreted on a molecular level, although this must be done with some care. Because of quantum effects, it would *not* be correct to say that at $T = 0$, all molecular motion ceases. At absolute zero, the system has its *minimum* possible total energy (kinetic plus potential). One statement of the *third law of thermodynamics* is that it is impossible to reach absolute zero in a finite number of thermodynamic steps.

16.9 Energy Resources: A Case Study in Thermodynamics

The laws of thermodynamics place very general limitations on the conversion of energy from one form to another. In these times of increasing energy demand and diminishing resources, such matters are of the utmost practical importance. We conclude this chapter with a brief discussion of a few energy-conversion systems, present and proposed.

About half of the electric power generated in the United States is obtained from coal-fired steam-turbine generating plants. Modern boilers can transfer about 80% to 90% of the heat of combustion of coal into steam. The theoretical maximum thermal efficiency of the turbine, given by Equation 16.9, is usually limited to about 0.55, and the actual efficiency is typically 90% of this value, or approximately 0.50. The efficiency of large generators in converting mechanical power to electrical power is very high, typically 99%. Thus, the overall thermal efficiency of such a plant is roughly (0.85) (0.50) (0.99), or at most about 40%.

The Tennessee Valley Authority's Kingston power plant has a generating capacity of about 1 gigawatt (1000 MW, or 10^9 W) of electrical power. The steam is heated to about 540°C (1000°F) at a pressure of more than 120 atm (1.24×10^7 Pa, or 1800 psi). The plant burns 14,000 tons of coal per day (13 million kg, the equivalent of 140 railroad cars), and the overall thermal efficiency is about 35%.

In a nuclear power plant, the heat that generates steam is supplied by a nuclear reaction rather than the chemical reaction of burning coal. The steam turbines in nuclear power plants have the same theoretical efficiency limit as those in coal-fired plants. At present, it is not practical to run nuclear reactors at temperatures and pressures as high as those in coal boilers, so the thermal efficiency of a nuclear plant is somewhat lower, typically 30%.

In the year 2005, about 20% of the world's electric power supply came from nuclear power plants. High construction costs, questions of public safety, and the problem of disposal of radioactive waste have slowed the development of additional nuclear power. However, there is also increasing concern about the environmental impact of coal-burning power plants. The health hazards of coal smoke are serious and well documented, and the burning of coal contributes to global warming through the greenhouse effect. It is important to manage our development of energy resources so as to minimize the risks to human life and health, and nuclear power continues to receive attention as a future energy resource. We noted at the end of Section 16.6 that thermal pollution from both coal-fired and nuclear plants poses serious environmental problems.

Solar energy is an inviting possibility. The power per unit surface area in the sun's radiation (before it passes through the earth's atmosphere) is approximately $1.4 \, \text{kW/m}^2$. A maximum of about $1.0 \, \text{kW/m}^2$ reaches the surface of the earth on a clear day, and the average over a 24-hour period is about $0.2 \, \text{kW/m}^2$. This radiation can be collected and used to heat water for home use (Figure 16.14a).

A different solar-energy scheme is to use banks of photovoltaic (PV) cells (Figure 16.14b) for the direct conversion of solar energy to electricity. Such a system is not a heat engine in the usual sense and is not limited by the Carnot efficiency. There are other fundamental limitations on photocell efficiency, but 50% seems attainable in multilayer semiconductor photocells. Because of high capital costs, the price of PV-produced energy is about 25 cents per kilowatt-hour (compared with about 12 cents for conventional coal-fired steam-turbine plants). But such a system offers many advantages, such as lack of noise, lack of moving parts, minimal maintenance requirements, and freedom from pollution. Capital costs are likely to drop with improved technology, and PV systems continue to be studied and developed.

The energy of wind, fundamentally solar in origin, can be gathered and converted by "wind farms" of windmills (Figure 16.14c). The largest wind farms are located in California, where the state government encourages their use. However, the Great Plains states alone have enough wind energy to theoretically meet the electrical power needs of the entire United States. As with PV systems, the capital costs of such systems are higher than for coal-fired generating plants of equal capacity.

Biomass is yet another mechanism for using solar energy. Plants absorb solar energy and store it; the energy can be utilized by burning the plant matter or by using fermentation and distillation to make alcohols. Ethanol made from fermented corn is used in the United States as a gasoline additive, although its cost is not quite competitive at present crude-oil price levels. An important feature of biomass materials is that they take as much carbon dioxide out of the atmosphere when they grow as is released when they burn; thus, they don't contribute to the greenhouse effect.

An indirect scheme for the collection and conversion of solar energy would use the temperature gradient in the oceans. In the Caribbean, for example, the water temperature near the surface is about 25°C, and at a depth of a few hundred meters it is about 10°C. The second law of thermodynamics forbids cooling the ocean and converting the extracted heat completely into work, but there is nothing to forbid running a heat engine between those two temperatures. The thermodynamic efficiency would be very low (about 5%), but a vast supply of energy would be available.

The present level of activity in energy-conversion research in the United States is rather low in comparison to the period of the 1970s, when we experienced a serious energy shortage. The disadvantages of excessive reliance on non-renewable resources such as oil—especially imported oil—are not hard to see. Critics have accused the U.S government of not having a coherent and effective energy policy compared with other countries. As we humans use up our available fossil-fuel resources, and as developing countries require more of the world's supply, we will certainly have to become more active in developing alternative energy sources. The principles of thermodynamics will play a central role in this development.

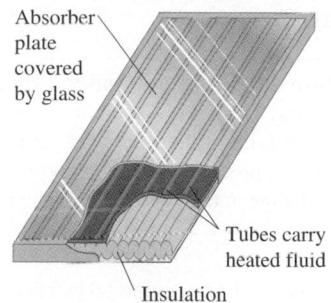

(a) Passive solar collecting panel for home water heating

(b) A telephone powered by photovoltaic cells

(c) Windmills generating electricity

▲ **FIGURE 16.14** Three sources of energy other than the burning of fossil fuels.

SUMMARY

Directions of Thermodynamic Processes
(Section 16.1) Heat flows spontaneously from a hotter object to a cooler object, never the reverse. A **reversible process (equilibrium process)** is a process whose direction can be reversed by an infinitesimal change in the conditions of the process.

Heat Engines
(Section 16.2) A **heat engine** takes heat Q_H from a source, converts part of it to work W, and discards the remainder, $|Q_C|$, at a lower temperature. The **thermal efficiency** e of a heat engine is

$$e = \frac{W}{Q_H} = 1 + \frac{Q_C}{Q_H} = 1 - \frac{|Q_C|}{|Q_H|}. \quad (16.4)$$

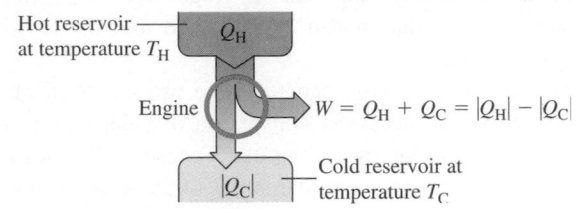

Internal Combustion Engines and Refrigerators
(Sections 16.3–16.4) An idealized model of the thermodynamic processes that take place in a gasoline engine is called the **Otto cycle.** The theoretical efficiency of the Otto cycle is about 50%; those of real gasoline engines are typically around 20%.

A **refrigerator** takes heat Q_C from a cold place, has an input W of work, and discards heat $|Q_H|$ at a warmer place. The **performance coefficient** K of a refrigerator is defined as

$$K = \frac{Q_C}{|W|} = \frac{|Q_C|}{|Q_H| - |Q_C|}. \quad (16.7)$$

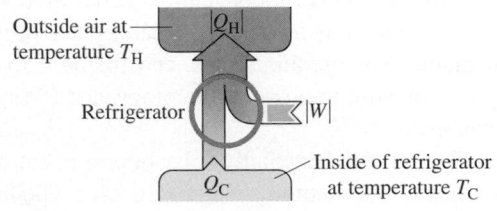

The Second Law of Thermodynamics
(Section 16.5) The **second law of thermodynamics** describes the directionality of natural thermodynamic processes. It can be stated in several equivalent forms. The "engine" statement is "No cyclic process can convert heat completely into work." The "refrigerator" statement is "No cyclic process can transfer heat from a cold place to a hotter place with no input of mechanical work."

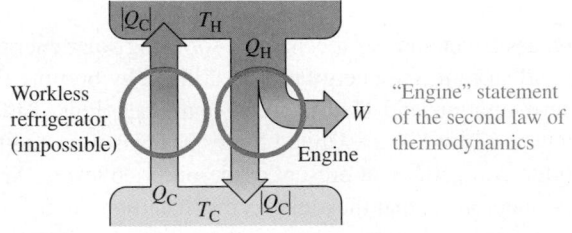

The Carnot Engine: The Most Efficient Heat Engine
(Section 16.6) The **Carnot cycle** operates between two heat reservoirs at temperatures T_H and T_C and uses only reversible processes. Its thermal efficiency is

$$e_{Carnot} = 1 - \frac{T_C}{T_H} = \frac{T_H - T_C}{T_H}. \quad (16.9)$$

No engine operating between the same two temperatures can be more efficient than a Carnot engine. All Carnot engines operating between the same two temperatures have the same efficiency.

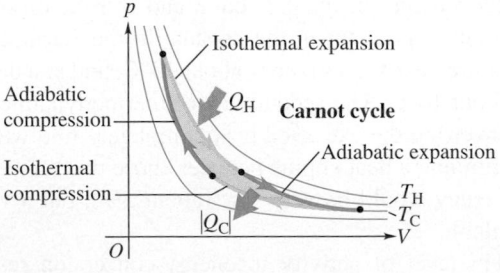

Entropy
(Section 16.7) **Entropy** is a quantitative measure of the disorder of a system. The entropy change in a reversible isothermal process is $\Delta S = Q/T$ (Equation 16.11), where T is absolute temperature. The second law can be stated in terms of entropy: The entropy of an isolated system may increase, but can never decrease. When a system interacts with its surroundings, the total entropy of system and surroundings can never decrease.

Continued

The Kelvin Temperature Scale

(Section 16.8) The Kelvin temperature scale is based on the efficiency of the Carnot cycle and is independent of the properties of any specific material. The zero point on the Kelvin scale is absolute zero.

Energy Resources: A Case Study in Thermodynamics

(Section 16.9) About half of the electric power generated in the United States is obtained from coal-fired steam-turbine generating plants. The theoretical maximum thermal efficiency of the turbine, given by Equation 16.9, is usually limited to about 0.55, and the actual overall efficiency is about 40%. In a nuclear power plant, the heat that generates steam is supplied by a nuclear reaction rather than the chemical reaction of burning coal; the thermal efficiency of a nuclear plant is typically 30%. Other sources of energy include the sun, wind, ethanol from fermented corn, and thermal gradients in the ocean.

 For instructor-assigned homework, go to www.masteringphysics.com

Conceptual Questions

1. Household refrigerators have arrays or coils of tubing on the outside, usually at the back or bottom. When the refrigerator is running, the tubing becomes quite hot. Where does the heat come from?

2. A chef tries to cool his kitchen on a hot day by leaving the door of the walk-in refrigerator open. Will this strategy work? Would the result be different if the chef used an old-fashioned icebox (which is simply an insulated box with a compartment holding ice)?

3. Imagine a special air filter placed in a window of a house. The tiny holes in the filter allow only air molecules moving faster than a certain speed to exit the house, and allow only air molecules moving slower than that speed to enter the house from outside. Explain why such an air filter would cool the house, and why the second law of thermodynamics makes building such a filter an impossible task.

4. A growing plant creates a highly complex and organized structure out of simple materials, such as air, water, and minerals. Does this violate the second law of thermodynamics? What is the plant's ultimate source of energy?

5. Why must a room air conditioner be placed in a window rather than just set on the floor and plugged in? Why can a refrigerator be set on the floor and plugged in?

6. If you pour a cup of hot water into a cup of cold water, you wind up with two cups of lukewarm water. (a) Is this process reversible or irreversible? Why? (b) Does the entropy increase or decrease? How can this be, since no heat was added to (or removed from) the system?

7. How can the thermal conduction of heat from a hot object to a cold object increase entropy when the same amount of heat that flows out of the hot object flows into the cold one?

8. How can the free expansion of a gas into a vacuum increase entropy when no heat is added to (or subtracted from) this gas?

9. Does the second law of thermodynamics say that heat can *never* go from a cold object to a hot object? Explain.

10. When a wet cloth is hung up in a hot wind in the desert, it is cooled by evaporation to a temperature that may be 20°C or so below that of the air. Discuss this process in light of the second law of thermodynamics.

11. In what ways is a heat pump different from (or similar to) a refrigerator?

12. Does the second law of thermodynamics say that entropy (disorder) can never decrease? Explain and give a few examples to illustrate your answer.

13. *Why* does heat *not* flow from a cold to a hot object? How does heat "know" it should obey the second law of thermodynamics? Explain by thinking about collisions at the molecular level.

14. Some people have suggested that the idea that life originated from nonliving materials violates the second law of thermodynamics because the molecules would have to go from a less ordered to a more ordered state, decreasing entropy. What is wrong with this argument?

15. What would be the efficiency of a Carnot engine operating with $T_H = T_C$? What would be the efficiency if $T_C - 0$ K and T_H were any temperature above 0 K? Interpret your answers.

16. The first law of thermodynamics is sometimes paraphrased as "You can't get something for nothing." The second law has been paraphrased as "You can't even break even." Explain why these statements are good summaries of those two laws.

17. Would it be more economical to run a refrigerator with a high or a low coefficient of performance? Why?

Multiple-Choice Problems

1. An insulated box has a barrier that confines a gas to only one side of the box. The barrier springs a leak, allowing the gas to

flow and occupy both sides of the box. Which statement best describes the entropy of this system?

A. The entropy is greater in the first state, with all the gas on one side of the box.

B. The entropy is greater in the second state, with the gas on both sides of the box.

C. The entropy is the same in both states, since no heat was added to the gas and its temperature did not change.

2. Suppose you put a hot object in contact with a cold object and observe (with some wonder) that heat flows from the cold object to the hot object, making the cold one colder and the hot one hotter. This process would violate

A. only the first law of thermodynamics.

B. only the second law of thermodynamics.

C. both the first and the second laws of thermodynamics.

3. Carnot engine A operates between temperatures of 500°C and 300°C; Carnot engine B operates between temperatures of 900°C and 700°C. How do the efficiencies of the two engines compare?

A. Engine A has higher efficiency.

B. Engine B has higher efficiency.

C. The efficiencies of the two engines are the same.

4. The thermal efficiency formula $e = W/Q_H$ is valid for which of the following heat engines? (There may be more than one correct choice.)

A. Carnot engine.

B. Otto engine.

C. Diesel engine.

D. Any other type of heat engine.

5. The thermal efficiency formula $e = 1 - T_C/T_H$ is valid for which of the following heat engines? (There may be more than one correct choice.)

A. Carnot engine.   B. Otto engine.

C. Diesel engine.   D. Any other type of heat engine.

6. A refrigerator consumes a certain amount of electrical energy in order to operate. In terms of the thermodynamic cycle of this refrigerator, this electrical energy is equal to

A. the work done on the gas.

B. the heat extracted from the food in the refrigerator.

C. the heat expelled into the room.

D. the sum of all three quantities in choices A, B, and C.

7. Which of the following statements is a consequence of the *second* law of thermodynamics? (There may be more than one correct choice.)

A. The efficiency of a heat engine can never be greater than unity.

B. Heat can never flow from a cold object to a hot object unless work is done.

C. Energy is conserved in all thermodynamics processes.

D. Even a perfect heat engine cannot convert all of the heat put into it to work.

8. You want to increase the efficiency of a Carnot heat engine by changing the temperature of either its hot or its cold reservoir by the same number of degrees ΔT. Which of the following choices will produce the *greater* increase in its thermodynamic efficiency?

A. Increase the temperature of the hot reservoir by ΔT.

B. Decrease the temperature of the cold reservoir by ΔT.

C. It would not matter. Both choices would produce the same increase in efficiency, since ΔT is the same for both.

9. Two Carnot heat engines operate between the same hot and cold reservoirs, but one has a greater compression ratio than the other. Which statement is true about their thermal efficiencies?

A. The one with the greater compression ratio will have the greater efficiency.

B. The one with the greater compression ratio will have the lesser efficiency.

C. Both will have the same efficiency.

10. You put 250 g of water into the freezer and make ice cubes. Which of the following statements is true about this process? (There may be more than one correct choice.)

A. The entropy of the water increases.

B. The entropy of the room increases.

C. The amount of heat removed from the water is equal to the amount of heat ejected into the room.

D. More heat is ejected into the room than was removed from the water.

E. The work done by the refrigerator is equal to the heat removed from the water.

11. If you mix cold milk with hot coffee in an insulated Styrofoam™ cup, which of the following things happen? (There may be more than one correct choice.)

A. The entropy of the milk increases.

B. The entropy of the coffee decreases by the same amount that the entropy of the milk increased.

C. The net entropy of the coffee–milk mixture does not change, because no heat was added to this system.

D. The entropy of the coffee–milk mixture increases.

12. Suppose a Carnot refrigerator is used to cool the interior of a food storage locker from room temperature to 5°C. As the temperature of the locker drops, the performance coefficient of the refrigerator

A. increases.

B. decreases.

C. is constant.

13. A glass of water left outside on a cold night freezes into solid ice, and its entropy decreases. Which statement is true of this process:

A. The entropy of the surrounding air does not change.

B. The entropy of the water changes more than the entropy of the surrounding air.

C. The entropy of the surrounding air changes more than the entropy of the water.

D. The entropy of the water and the entropy of the surrounding air change by equal amounts.

14. If we run an ideal Carnot heat engine in *reverse*, which of the following statements about it must be true? (There may be more than one correct choice.)

A. Heat enters the gas at the cold reservoir and goes out of the gas at the hot reservoir.

B. The amount of heat transferred at the hot reservoir is equal to the amount of heat transferred at the cold reservoir.

C. It is able to perform a net amount of useful work, such as pumping water from a well, during each cycle.

D. It can transfer heat from a cold object to a hot object.

15. Which of the following processes would be a *violation* of the second law of thermodynamics? (There may be more than one correct choice.)

A. All the kinetic energy of an object is transformed into heat.

B. All the heat put into the operating gas of a heat engine during one cycle is transformed into work.

C. A refrigerator removes 100 cal of heat from milk while using only 75 cal of electrical energy to operate.

D. A heat engine does 25 J of work while expelling only 10 J of heat to the cold reservoir.

Problems

16.2 Heat Engines

1. • A coal-fired power plant that operates at an efficiency of 38% generates 750 MW of electric power. How much heat does the plant discharge to the environment in one day?

2. • Each cycle, a certain heat engine expels 250 J of heat when you put in 325 J of heat. Find the efficiency of this engine and the amount of work you get out of the 325 J heat input.

3. • A diesel engine performs 2200 J of mechanical work and discards 4300 J of heat each cycle. (a) How much heat must be supplied to the engine in each cycle? (b) What is the thermal efficiency of the engine?

4. • An aircraft engine takes in 9000 J of heat and discards 6400 J each cycle. (a) What is the mechanical work output of the engine during one cycle? (b) What is the thermal efficiency of the engine?

5. • **A gasoline engine.** A gasoline engine takes in 1.61×10^4 J of heat and delivers 3700 J of work per cycle. The heat is obtained by burning gasoline with a heat of combustion of 4.60×10^4 J/g. (a) What is the thermal efficiency of the engine? (b) How much heat is discarded in each cycle? (c) What mass of fuel is burned in each cycle? (d) If the engine goes through 60.0 cycles per second, what is its power output in kilowatts? In horsepower?

6. • A gasoline engine has a power output of 180 kW (about 241 hp). Its thermal efficiency is 28.0%. (a) How much heat must be supplied to the engine per second? (b) How much heat is discarded by the engine per second?

7. • A certain nuclear power plant has a mechanical power output (used to drive an electric generator) of 330 MW. Its rate of heat input from the nuclear reactor is 1300 MW. (a) What is the thermal efficiency of the system? (b) At what rate is heat discarded by the system?

8. •• Figure 16.15 shows a pV diagram for a heat engine that uses 1.40 moles of an ideal diatomic gas. (a) How much heat goes into this gas per cycle, and where in the cycle does it occur? (b) How much heat is ejected by the gas per cycle, and where does it occur? (c) How much work does this engine do each cycle? (d) What is the thermal efficiency of the engine?

9. •• The pV diagram in Figure 16.16 shows a cycle of a heat engine that uses 0.250 mole of an ideal gas having $\gamma = 1.40$. The curved part ab of the cycle is adiabatic. (a) Find the pressure of the gas at point a. (b) How much heat enters this gas per cycle, and where does it happen? (c) How much heat leaves this gas in a cycle, and where does it occur? (d) How much work does this engine do in a cycle? (e) What is the thermal efficiency of the engine?

16.3 Internal Combustion Engines

10. • An Otto engine uses a gas having $\gamma = 1.40$ and has a compression ratio of 8.50. (a) Out of every gallon of fuel this engine burns, what fraction is wasted (i.e., produces energy that is wasted)? (b) If you could change the compression ratio to 10.0 instead, what percentage of the fuel would be wasted?

11. • What compression ratio r must an Otto cycle have to achieve an ideal efficiency of 65.0% if the gas used in the chamber has $\gamma = 1.40$?

12. • For an Otto engine with a compression ratio of 7.50, you have your choice of using an ideal monatomic or ideal diatomic gas. Which one would give you greater efficiency? Calculate the efficiency in both cases to find out.

13. •• (a) Calculate the theoretical efficiency for an Otto cycle engine with $\gamma = 1.40$ and $r = 9.50$. (b) If this engine takes in 10,000 J of heat from burning its fuel, how much heat does it discard to the outside air?

16.4 Refrigerators

14 • In one cycle, a freezer uses 785 J of electrical energy in order to remove 1750 J of heat from its freezer compartment at 10°F. (a) What is the coefficient of performance of this freezer? (b) How much heat does it expel into the room during this cycle?

15. • A refrigerator has a coefficient of performance of 2.10. Each cycle, it absorbs 3.40×10^4 J of heat from the cold reservoir. (a) How much mechanical energy is required each cycle to operate the refrigerator? (b) During each cycle, how much heat is discarded to the high-temperature reservoir?

16. • A window air-conditioner unit absorbs 9.80×10^4 J of heat per minute from the room being cooled and in the same period deposits 1.44×10^5 J of heat into the outside air. What is the power consumption of the unit in watts?

17. •• A freezer has a coefficient of performance of 2.40. The freezer is to convert 1.80 kg of water at 25.0°C to 1.80 kg of ice at −5.0°C in 1 hour. (a) What amount of heat must be removed from the water at 25.0°C to convert it to ice at −5.0°C? (b) How much electrical energy is consumed by the freezer during this hour? c) How much wasted heat is rejected to the room in which the freezer sits?

18. •• A cooling unit for chilling the water of an aquarium gives specifications of 1/10 hp and 1270 Btu/h. Assuming the unit produces its 1/10 hp at 70.0% efficiency, calculate its performance coefficient.

16.6 The Carnot Engine: The Most Efficient Heat Engine

19. • A Carnot engine whose high-temperature reservoir is at 620 K takes in 550 J of heat at this temperature in each cycle and gives up 335 J to the low-temperature reservoir. (a) How

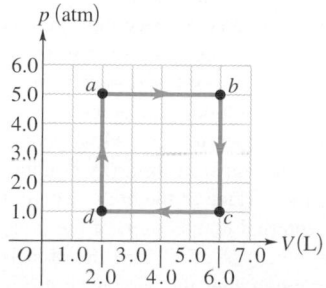

▲ **FIGURE 16.15** Problem 8.　　▲ **FIGURE 16.16** Problem 9.

much mechanical work does the engine perform during each cycle? (b) What is the temperature of the low-temperature reservoir? (c) What is the thermal efficiency of the cycle?

20. • A heat engine is to be built to extract energy from the temperature gradient in the ocean. If the surface and deepwater temperatures are 25°C and 8°C, respectively, what is the maximum efficiency such an engine can have?

21. • A Carnot engine is operated between two heat reservoirs at temperatures of 520 K and 300 K. (a) If the engine receives 6.45 kJ of heat energy from the reservoir at 520 K in each cycle, how many joules per cycle does it reject to the reservoir at 300 K? (b) How much mechanical work is performed by the engine during each cycle? (c) What is the thermal efficiency of the engine?

22. •• A Carnot engine has an efficiency of 59% and performs 2.5×10^4 J of work in each cycle. (a) How much heat does the engine extract from its heat source in each cycle? (b) Suppose the engine exhausts heat at room temperature $(20.0°C)$. What is the temperature of its heat source?

23. •• An ice-making machine operates in a Carnot cycle. It takes heat from water at 0.0°C and rejects heat to a room at 24.0°C. Suppose that 85.0 kg of water at 0.0°C are converted to ice at 0.0°C. (a) How much heat is rejected to the room? (b) How much energy must be supplied to the device?

24. •• A Carnot freezer that runs on electricity removes heat from the freezer compartment, which is at −10°C, and expels it into the room at 20°C. You put an ice-cube tray containing 375 g of water at 18°C into the freezer. (a) What is the coefficient of performance of this freezer? (b) How much energy is needed to freeze this water? (c) How much electrical energy must be supplied to the freezer to freeze the water? (d) How much heat does the freezer expel into the room while freezing the ice?

25. •• The *pV* diagram in Figure 16.17 shows a general Carnot cycle for an engine, with the hot and cold thermal reservoir segments identified. Segments *ab* and *cd* are isothermal, while the other two are adiabatic. (a) If this engine is used as a heat engine, what is the direction of the cycle, clockwise or counterclockwise? In which segments does heat enter the gas, and in which ones does it leave the gas? (b) If the engine is used as a refrigerator, what is the direction of the cycle? In which segments does heat enter the gas, and in which ones does it leave the gas? Also, which segments take place at the inside of the refrigerator, and which ones occur in the air of the room in which the refrigerator is operating? (c) If the engine is used as a heat pump–air conditioner, what is the direction of the cycle, and in which segments does heat enter the gas and in which ones does it leave the gas? Which segments take place inside of the house and which ones occur outside? (d) If the engine is used as a heat pump–house heater, what is the direction of the cycle? In which segments does heat enter the gas and in which ones does it leave the gas? Which segments take place inside of the house and which ones occur outside?

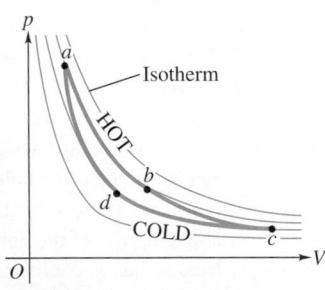

▲ **FIGURE 16.17** Problem 27.

16.7 Entropy

26. • A sophomore with nothing better to do adds heat to 0.350 kg of ice at 0.00°C until it is all melted. (a) What is the change in entropy of the water? (b) The source of the heat is a very massive body at a temperature of 25.0°C. What is the change in entropy of this body? (c) What is the total change in entropy of the water and the heat source?

27. • A 4.50 kg block of ice at 0.00°C falls into the ocean and melts. The average temperature of the ocean is 3.50°C, including all the deep water. By how much does the melting of this ice change the entropy of the world? Does it make it larger or smaller? (*Hint:* Do you think that the ocean will change temperature appreciably as the ice melts?)

28. •• A large factory furnace maintained at 175°C at its outer surface is wrapped in an insulating blanket of thermal conductivity 0.055 W/(m·K) which is thick enough that the outer surface of the insulation is at 42°C while heat escapes from the furnace at a steady rate of 125 W for each square meter of surface area. By how much does each square meter of the furnace change the entropy of the factory every second?

29. •• You decide to take a nice hot bath but discover that your thoughtless roommate has used up most of the hot water. You fill the tub with 270 kg of 30.0°C water and attempt to warm it further by pouring in 5.00 kg of boiling water from the stove. (a) Is this a reversible or an irreversible process? Use physical reasoning to explain. (b) Calculate the final temperature of the bath water. (c) Calculate the net change in entropy of the system (bath water + boiling water), assuming no heat exchange with the air or the tub itself.

30. •• If 25.0 g of the metal gallium melts in your hand (see Fig. 14.14), what is the change in entropy of the gallium in this process? What about the change in entropy of your hand? Is it positive or negative? Is its magnitude greater or less than that of the change in entropy of the gallium? The melting temperature of gallium is 29.8°C, and its heat of fusion is 8.04×10^4 J/kg.

31. •• Three moles of an ideal gas undergo a reversible isothermal compression at 20.0°C. During this compression, 1850 J of work is done on the gas. What is the change in entropy of the gas?

32. •• **Entropy change due to driving.** Premium gasoline produces 1.23×10^8 J of heat per gallon when it is burned at a temperature of approximately 400°C (although the amount can vary with the fuel mixture). If the car's engine is 25% efficient, three-fourths of that heat is expelled into the air, typically at 20°C. If your car gets 35 miles per gallon of gas, by how much does the car's engine change the entropy of the world when you drive 1.0 mile? Does it decrease or increase it?

33. •• **Entropy of metabolism.** An average sleeping person **BIO** metabolizes at a rate of about 80 W by digesting food or burning fat. Typically, 20% of this energy goes into bodily functions, such as cell repair, pumping blood, and other uses of mechanical energy, while the rest goes to heat. Most people get rid of all of this excess heat by transferring it (by conduction and the flow of blood) to the surface of the body, where it is radiated away. The normal internal temperature of the body (where the metabolism takes place) is 37°C, and the skin is typically 7 C° cooler. By how much does the person's entropy change per second due to this heat transfer?

34. •• **Entropy change from digesting fat.** Digesting fat pro-
BIO duces 9.3 food calories per gram of fat, and typically 80% of
this energy goes to heat when metabolized. The body then
moves all this heat to the surface by a combination of thermal
conductivity and motion of the blood. The internal temperature
of the body (where digestion occurs) is normally 37°C, and the
surface is usually about 30°C. By how much does the digestion
and metabolism of a 2.50 g pat of butter change your body's
entropy? Does it increase or decrease?

16.9 Energy Resources: A Case Study in Thermodynamics

35. • **Solar collectors.** A well-insulated house of moderate size in
a temperate climate requires an average heat input rate of
20.0 kW. If this heat is to be supplied by a solar collector with
an average (night and day) energy input of 300 W/m² and a
collection efficiency of 60.0%, what area of solar collector is
required?

36 • **Solar power.** A solar power plant is to be built with an average power output capacity of 2500 MW in a location where
the average power from the sun's radiation is 200 W/m² at
the earth's surface. What land area (in km² and mi²) must
the solar collectors occupy if they are (a) photocells with 42%
efficiency, (b) mirrors that generate steam for a turbine-
generator unit with an overall efficiency of 21%?

37. •• An experimental power plant at the Natural Energy Laboratory of Hawaii generates electricity from the temperature
gradient of the ocean. The surface and deep-water temperatures are 27°C and 6°C, respectively. (a) What is the maximum
theoretical efficiency of this power plant? (b) If the power
plant is to produce 210 kW of power, at what rate must heat be
extracted from the warm water? At what rate must heat be
absorbed by the cold water? Assume the maximum theoretical
efficiency. (c) The cold water that enters the plant leaves it at a
temperature of 10°C. What must be the flow rate of cold water
through the system? Give your answer in kg/h and L/h.

38. •• **Solar water heater.** A solar water heater for domestic hot-
water supply uses solar collecting panels with a collection efficiency of 50% in a location where the average solar-energy
input is 200 W/m². If the water comes into the house at 15.0°C
and is to be heated to 60.0°C, what volume of water can be
heated per hour if the area of the collector is 30.0 m²?

General Problems

39. •• You are designing a Carnot engine that has 2 mol of CO_2
as its working substance; the gas may be treated as ideal. The
gas is to have a maximum temperature of 527°C and a maximum pressure of 5.00 atm. With a heat input of 400 J per
cycle, you want 300 J of useful work. (a) Find the temperature
of the cold reservoir. (b) For how many cycles must this engine
run to melt completely a 10.0-kg block of ice originally at
0.0°C, using only the heat rejected by the engine?

40. •• A heat engine takes 0.350 mol of an ideal diatomic gas
around the cycle shown in the pV diagram of Figure 16.18.
Process $1 \rightarrow 2$ is at constant volume, process $2 \rightarrow 3$ is adiabatic, and process $3 \rightarrow 1$ is at a constant pressure of 1.00 atm.
The value of γ for this gas is 1.40. (a) Find the pressure and
volume at points 1, 2, and 3. (b) Calculate Q, W, and ΔU for
each of the three processes. (c) Find the net work done by the

gas in the cycle. (d) Find the
net heat flow into the engine
in one cycle. (e) What is the
thermal efficiency of the
engine? How does this efficiency compare with that
of a Carnot-cycle engine
operating between the same
minimum and maximum
temperatures T_1 and T_2?

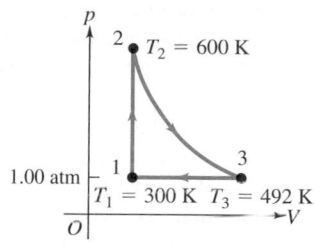

▲ **FIGURE 16.18** Problem 40.

41. •• As a budding mechanical engineer, you are called upon to
design a Carnot engine that has 2.00 moles of He gas (see
Table 15.4) as its working substance and that operates from a
high-temperature reservoir at 500°C. The engine is to lift a
15.0 kg weight 2.00 m per cycle, using 500 J of heat input. The
gas in the engine chamber can have a minimum volume of 5.00 L
during the cycle. (a) Draw a pV diagram for this cycle. In your
diagram, show where heat enters and leaves the gas. (b) What
must be the temperature of the cold reservoir? (c) What is the
thermal efficiency of the engine? (d) How much heat energy does
this engine waste per cycle? (e) What is the maximum pressure
that the gas chamber will have to withstand?

42. •• The Kwik-Freez Appliance Co. wants you to design a food
freezer that will keep the freezing compartment at −5.0°C and
will operate in a room at 20.0°C. The freezer is to make 5.00 kg
of ice at 0.0°C, starting with water at 20.0°C. Find the least possible amount of electrical energy needed to make this ice and the
smallest possible amount of heat expelled into the room.

43. •• A Carnot engine operates between two heat reservoirs at
temperatures T_H and T_C. An inventor proposes to increase the
efficiency by running one engine between T_H and an intermediate temperature T' and a second engine between T' and T_C,
using as input the heat expelled by the first engine. Compute
the efficiency of this composite system, and compare it to that
of the original engine.

44. •• A cylinder contains oxygen gas (O_2) at a pressure of
2.00 atm. The volume is 4.00 L, and the temperature is 300 K.
Assume that the oxygen may be treated as an ideal gas. The
oxygen is carried through the following processes:
  (i)   Heated at constant pressure from the initial state (state 1)
        to state 2, which has $T = 450$ K.
  (ii) Cooled at constant volume to 250 K (state 3).
  (iii) Compressed at constant temperature to a volume of
        4.00 L (state 4).
  (iv) Heated at constant volume to 300 K, which takes the sys-
        tem back to state 1.
  (a) Show these four processes in a pV diagram, giving the
numerical values of p and V in each of the four states. (b) Calculate Q and W for each of the four processes. (c) Calculate the
net work done by the oxygen. (d) What is the efficiency of this
device as a heat engine? How does this efficiency compare
with that of a Carnot-cycle engine operating between the same
minimum and maximum temperatures of 250 K and 450 K?

45. •• **Human entropy.** A person having skin of surface area
BIO 1.85 m² and temperature 30.0°C is resting in an insulated room
where the ambient air temperature is 20.0°C. In this state, a
person gets rid of excess heat by radiation. By how much does
the person change the entropy of the air in this room each second? (Recall that the room radiates back into the person and
that the emissivity of the skin is 1.00.)

46. •• A typical coal-fired power plant generates 1000 MW of usable power at an overall thermal efficiency of 40%. (a) What is the rate of heat input to the plant? (b) The plant burns anthracite coal, which has a heat of combustion of 2.65×10^7 J/kg. How much coal does the plant use per day, if it operates continuously? (c) At what rate is heat ejected into the cool reservoir, which is the nearby river? (d) The river's temperature is 18.0°C before it reaches the power plant and 18.5°C after it has received the plant's waste heat. Calculate the river's flow rate, in cubic meters per second. (e) By how much does the river's entropy increase each second?

47. •• **A human engine.** You decide to use your body as a Carnot
BIO heat engine. The operating gas is in a tube with one end in your mouth (where the temperature is 37.0°C) and the other end at the surface of your skin, at 30.0°C. (a) What is the maximum efficiency of such a heat engine? Would it be a very useful engine? (b) Suppose you want to use this human engine to lift a 2.50 kg box from the floor to a tabletop 1.20 m above the floor. How much must you increase the gravitational potential energy and how much heat input is needed to accomplish this? (c) How many 350 calorie (those are *food* calories, remember) candy bars must you eat to lift the box in this way? Recall that 80% of the food energy goes into heat.

48. •• What is the thermal efficiency of an engine that operates by taking n moles of diatomic ideal gas through the cycle $1 \rightarrow 2 \rightarrow 3 \rightarrow 4 \rightarrow 1$ shown in Fig. 16.19?

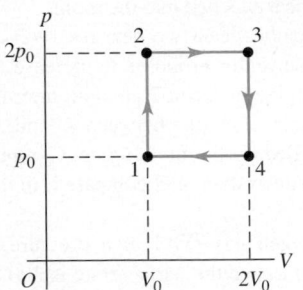

▲ **FIGURE 16.19** Problem 48.

49. •• One end of a copper rod is immersed in boiling water at 100°C and the other end in an ice–water mixture at 0°C. The sides of the rod are insulated. After steady-state conditions have been achieved in the rod, 0.160 kg of ice melts in a certain time interval. For this time interval, find (a) the entropy change of the boiling water, (b) the entropy change of the ice–water mixture, (c) the entropy change of the copper rod, (d) the total entropy change of the entire system.

50. •• The pV diagram in Figure 16.20 shows a heat engine operating on 0.850 mol of H_2. (See Table 15.4.) Segment ca is isothermal. (a) Without doing any calculations, identify the segments during which heat enters the gas and those during which it leaves the gas.

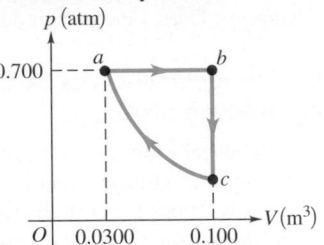

▲ **FIGURE 16.20** Problem 50.

Explain your reasoning. (b) Find the thermal efficiency of this heat engine, treating the gas as ideal.

51. •• Calculate the coefficient of performance of the engine in the previous problem if it is run in reverse, as a refrigerator.

17 Electric Charge and Electric Field

When you scuff your shoes across a carpet, you can get zapped by an annoying spark of static electricity That same spark could totally destroy the function of a computer chip. Lightning, the same phenomenon on a vast scale, can destroy a lot more than chips. The clinging of newly laundered synthetic fabrics is related to such sparks. All these phenomena involve electric charge and electrical interactions, one of nature's fundamental classes of interactions.

In this chapter, we'll study the interactions among electric charges that are at rest in our frame of reference; we call these **electrostatic interactions.** They are governed by Coulomb's law and are most conveniently described using the concept of *electric field.* We'll find that charge is *quantized;* it can have only certain values. The total electric charge in a system must be an integer multiple of the charge of a single electron. Electric charge obeys a *conservation* law: The total electric charge in a closed system must be constant. Electrostatic interactions hold atoms, molecules, and our bodies together, but they also are constantly trying to tear apart the nuclei of atoms. We'll explore all these concepts in this chapter and the ones that follow.

In a thundercloud, collisions between ice and slush particles give the ice particles a slight positive charge, and powerful updrafts carry the lighter ice particles toward the cloud top. When the resulting charge separation becomes strong enough, a lightning bolt results.

17.1 Electric Charge

The ancient Greeks discovered as early as 600 B.C. that when they rubbed amber with wool, the amber could then attract other objects. Today we say that the amber has acquired a net **electric charge,** or has become *charged.* The word *electric* is derived from the Greek word *elektron,* meaning "amber." When you scuff your shoes across a nylon carpet, you become electrically charged, and you can charge a comb by passing it through dry hair.

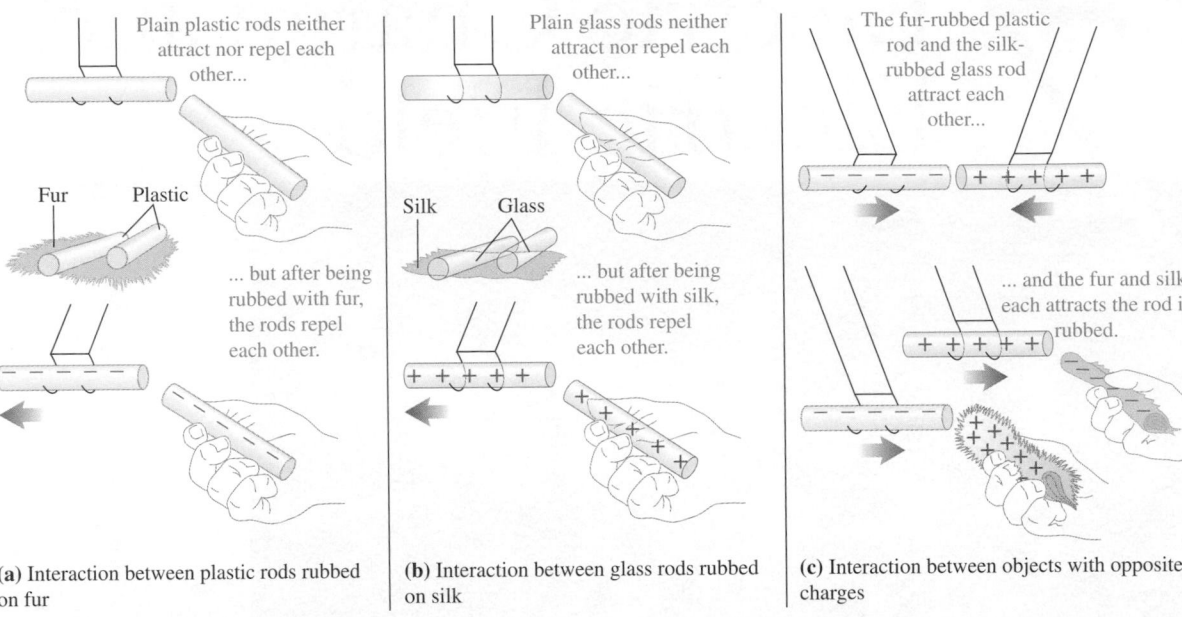

(a) Interaction between plastic rods rubbed on fur

(b) Interaction between glass rods rubbed on silk

(c) Interaction between objects with opposite charges

▲ **FIGURE 17.1** Experiments illustrating the nature of electric charge.

Plain plastic rods neither attract nor repel each other...

... but after being rubbed with fur, the rods repel each other.

Fur   Plastic

Plain glass rods neither attract nor repel each other...

... but after being rubbed with silk, the rods repel each other.

Silk   Glass

The fur-rubbed plastic rod and the silk-rubbed glass rod attract each other...

... and the fur and silk each attracts the rod it rubbed.

▲ **Application Run!** The person in this vacation snapshot, taken at a scenic over-look in Sequoia National Park, was amused to find her hair standing on end. Luckily, she and her companion left the overlook after taking the photo—and before it was hit by lightning. Just before lightning strikes, strong charges build up in the ground and in the clouds overhead. If you're standing on charged ground, the charge will spread onto your body. Because like charges repel, all your hairs tend to get as far from each other as they can. But the key thing is for *you* to get as far from that spot as *you* can!

Plastic rods and fur (real or fake) are particularly good for demonstrating electric-charge interactions. In Figure 17.1a, we charge two plastic rods by rubbing them on a piece of fur. We find that the rods repel each other. When we rub glass rods with silk (Figure 17.1b), the glass rods also become charged and repel each other. But a charged plastic rod *attracts* a charged glass rod (Figure 17.1c, top). Furthermore, the plastic rod and the fur attract each other, and the glass rod and the silk attract each other (Figure 17.1c, bottom).

These experiments and many others like them have shown that there are exactly two (no more) kinds of electric charge: the kind on the plastic rod rubbed with fur and the kind on the glass rod rubbed with silk. Benjamin Franklin (1706–1790) suggested calling these two kinds of charge *negative* and *positive,* respectively, and these names are still used.

Like and unlike charges

Two positive charges or two negative charges repel each other; a positive and a negative charge attract each other.

In the preceding discussion, the plastic rod and the silk have negative charge; the glass rod and the fur have positive charge.

When we rub a plastic rod with fur (or a glass rod with silk), *both* objects acquire net charges, and the net charges of the two objects are always equal in magnitude and opposite in sign. These experiments show that in the charging process we are not *creating* electric charge, but *transferring* it from one object to another. We now know that the plastic rod acquires extra electrons, which have negative charge. These electrons are taken from the fur, which is left with a deficiency of electrons (that is, fewer electrons than positively charged protons) and thus a net positive charge. The *total* electric charge on *both* objects does not change. This is an example of *conservation of charge;* we'll come back to this important principle later.

Conceptual Analysis 17.1

The sign of the charge

Three balls made of different materials are rubbed against different types of fabric—silk, polyester, etc. It is found that balls 1 and 2 repel each other and that balls 2 and 3 repel each other. From this result we can conclude that

A. Balls 1 and 3 carry charges of opposite sign.
B. Balls 1 and 3 carry charges of the same sign; ball 2 carries a charge of the opposite sign.
C. All three balls carry charges of the same sign.

SOLUTION Since balls 1 and 2 repel, they must be of the same sign, either both positive or both negative. Since 2 and 3 repel, they also must be of the same sign. This means that 1 and 3 both have the same sign as 2, so all three balls have the same sign.

The Physical Basis of Electric Charge

When all is said and done, we can't say what electric charge *is;* we can only describe its properties and its behavior. We *can* say with certainty that electric charge is one of the fundamental attributes of the particles of which matter is made. The interactions responsible for the structure and properties of atoms and molecules—and, indeed, of all ordinary matter—are primarily *electrical* interactions between electrically charged particles.

The structure of ordinary matter can be described in terms of three particles: the negatively charged **electron,** the positively charged **proton,** and the uncharged **neutron.** The protons and neutrons in an atom make up a small, very dense core called the **nucleus,** with dimensions on the order of 10^{-15} m (Figure 17.2). Surrounding the nucleus are the electrons, extending out to distances on the order of 10^{-10} m from the nucleus. If an atom were a few miles across, its nucleus would be the size of a tennis ball.

The negative charge of the electron has (within experimental error) *exactly* the same magnitude as the positive charge of the proton. In a neutral atom, the number of electrons equals the number of protons in the nucleus, and the net electrical charge (the algebraic sum of all the charges) is exactly zero (Figure 17.3a). The number of protons or electrons in neutral atoms of any element is called the **atomic number** of the element. If one or more electrons are removed, the remaining positively charged structure is called a **positive ion** (Figure 17.3b). A **negative ion** is an atom that has *gained* one or more electrons (Figure 17.3c). This gaining or losing of electrons is called **ionization.**

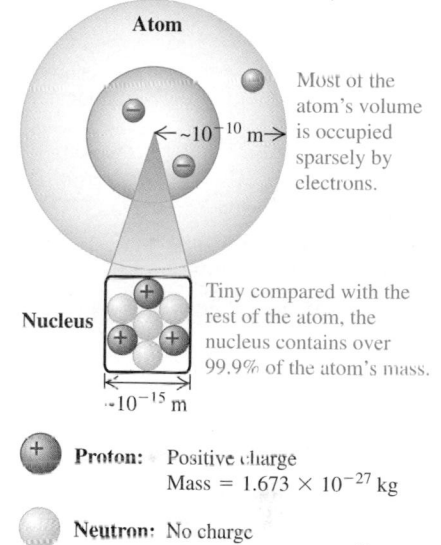

Atom

Most of the atom's volume is occupied sparsely by electrons.

$\leftarrow \sim 10^{-10}$ m $\rightarrow$

Nucleus

Tiny compared with the rest of the atom, the nucleus contains over 99.9% of the atom's mass.

$\sim 10^{-15}$ m

+ **Proton:** Positive charge
Mass = 1.673×10^{-27} kg

Neutron: No charge
Mass = 1.675×10^{-27} kg

− **Electron:** Negative charge
Mass = 9.109×10^{-31} kg

The charges of the electron and proton are equal in magnitude.

▲ **FIGURE 17.2** Schematic depiction of the structure and components of an atom.

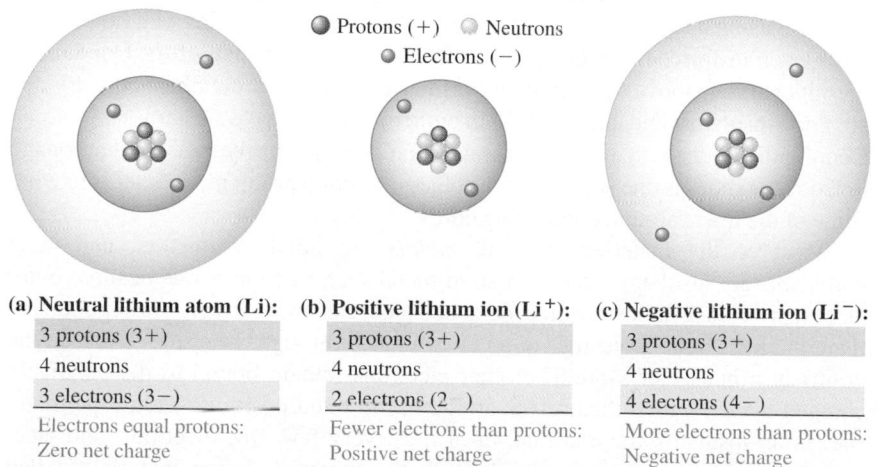

● Protons (+) ◉ Neutrons
◉ Electrons (−)

(a) Neutral lithium atom (Li):

| 3 protons (3+) |
| 4 neutrons |
| 3 electrons (3−) |

Electrons equal protons:
Zero net charge

(b) Positive lithium ion (Li⁺):

| 3 protons (3+) |
| 4 neutrons |
| 2 electrons (2) |

Fewer electrons than protons:
Positive net charge

(c) Negative lithium ion (Li⁻):

| 3 protons (3+) |
| 4 neutrons |
| 4 electrons (4−) |

More electrons than protons:
Negative net charge

▲ **FIGURE 17.3** The neutral lithium (Li) atom and positive and negative lithium ions.

The masses of the individual particles, to the precision that they are currently known, are as follows:

$$\text{Mass of electron} = m_e = 9.1093826(16) \times 10^{-31} \text{ kg};$$

$$\text{Mass of proton} = m_p = 1.67262171(29) \times 10^{-27} \text{ kg};$$

$$\text{Mass of neutron} = m_n = 1.67492728(29) \times 10^{-27} \text{ kg}.$$

The numbers in parentheses are the uncertainties in the last two digits. Note that the masses of the proton and neutron are nearly equal (within about 0.1%) and that the mass of the proton is roughly 2000 times that of the electron. Over 99.9% of the mass of any atom is concentrated in its nucleus.

When the number of protons in an object equals the number of electrons in the object, the total charge is zero, and the object as a whole is electrically neutral. To give a neutral object an excess negative charge, we may either *add negative* charges to it or *remove positive* charges from it. Similarly, we can give an excess positive charge to a neutral body by either *adding positive* charge or *removing negative* charge. When we speak of the charge on an object, we always mean its *net* charge.

An **ion** is an atom that has lost or gained one or more electrons. Ordinarily, when an ion is formed, the structure of the nucleus is unchanged. In a solid object such as a carpet or a copper wire, the nuclei of the atoms are not free to move about, so a net charge is due to an excess or deficit of electrons. However, in a liquid or a gas, net electrical charge may be due to movements of ions. Thus, a positively charged region in a fluid could represent an excess of positive ions, a deficit of negative ions, or both.

17.2 Conductors and Insulators

Some materials permit electric charge to move from one region of the material to another; others do not. For example, Figure 17.4 shows a copper wire supported by a nylon thread. Suppose you touch one end of the wire to a charged plastic rod and touch the other end to a metal ball that is initially uncharged. When you bring another charged object up close to the ball, the ball is attracted or repelled, showing that it has become electrically charged. Electric charge has been transferred through the copper wire between the ball and the surface of the plastic rod.

The wire is called a **conductor** of electricity. If you repeat the experiment, but this time using a rubber band or nylon thread in place of the wire, you find that *no* charge is transferred to the ball. These materials are called **insulators.** Conductors permit charge to move through them; insulators do not. Carpet fibers on a dry day are good insulators and allow charge to build up on us as we walk across the carpet. Coating the fibers with an antistatic layer that does not easily transfer electrons to or from our shoes is one solution to the charge-buildup problem; another is to wind some of the fibers around conducting cores.

Most of the materials we call *metals* are good conductors, and most *nonmetals* are insulators. Within a solid metal such as copper, one or more outer electrons in each atom become detached and can move freely throughout the material, just as the molecules of a gas can move through the spaces between the grains in a bucket of sand. The other electrons remain bound to the positively charged nuclei, which themselves are bound in fixed positions within the material. In an insulator, there are no, or at most very few, free electrons, and electric charge cannot move freely through the material. Some materials called

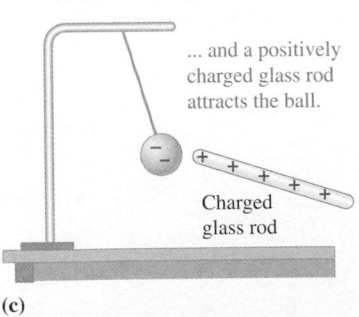

The wire conducts charge from the negatively charged plastic rod to the metal ball.

(a)

(b)

(c)

▲ **FIGURE 17.4** Charging by conduction. A copper wire is a good conductor. (a) The wire conducts charge between the plastic rod and the metal ball, giving the ball a negative charge. The charged ball is then (b) repelled by a like charge and (c) attracted by an unlike charge.

semiconductors are intermediate in their properties between good conductors and good insulators.

We've noted that, in a liquid or gas, charge can move in the form of positive or negative ions. Ionic solutions are usually good conductors. For example, when ordinary table salt (NaCl) dissolves in water, each sodium (Na) atom loses an electron to become a positively charged sodium ion (Na^+), and each chlorine (Cl) atom gains an electron to become a negatively charged chloride ion (Cl^-). These charged particles can move freely in the solution and thus conduct charge from one region of the fluid to another, providing a mechanism for conductivity. Ionic solutions are the dominant conductivity mechanism in many biological processes.

Induction

When we charge a metal ball by touching it with an electrically charged plastic rod, some of the excess electrons on the rod move from it to the ball, leaving the rod with a smaller negative charge. There is also a technique in which the plastic rod can give another object a charge of *opposite* sign without losing any of its own charge. This process is called charging by **induction.**

Figure 17.5 shows an example of charging by induction. A metal sphere is supported on an insulating stand (step 1). When you bring a negatively charged rod near the sphere, without actually touching it (step 2), the free electrons on the surface of the sphere are repelled by the excess electrons on the rod, and they shift toward the right, away from the rod. They cannot escape from the sphere because the supporting stand and the surrounding air are insulators. So we get excess negative charge at the right side of the surface of the sphere and excess positive charge (or a deficiency of negative charge) at the left side. These excess charges are called **induced charges.**

Not all of the free electrons move to the right side of the surface of the sphere. As soon as any induced charge develops, it exerts forces toward the *left* on the other free electrons. These electrons are repelled from the negative induced charge on the right and attracted toward the positive induced charge on the left. The system reaches an equilibrium state in which the force toward the right on an electron, due to the charged rod, is just balanced by the force toward the left due to the induced charge. If we remove the charged rod, the free electrons shift back to the left, and the original neutral condition is restored.

What happens if, while the plastic rod is nearby, you touch one end of a conducting wire to the right surface of the sphere and the other end to the earth (step 3 in Figure 17.5)? The earth is a conductor, and it is so large that it can act as a practically infinite source of extra electrons or sink of unwanted electrons. Some of the negative

▲ **Application Good conductor, bad conductor.** Salt water is salty because it contains an abundance of dissolved ions. These ions are charged and can move freely, so salt water is an excellent conductor of electricity. Ordinary tap water contains enough ions to conduct electricity reasonably well—which is why you should never, ever, use an electrical device in a bath. However, absolutely pure distilled water is an insulator, because it consists only of neutral water molecules.

PhET: Balloons and Static Electricity
PhET: John Travoltage

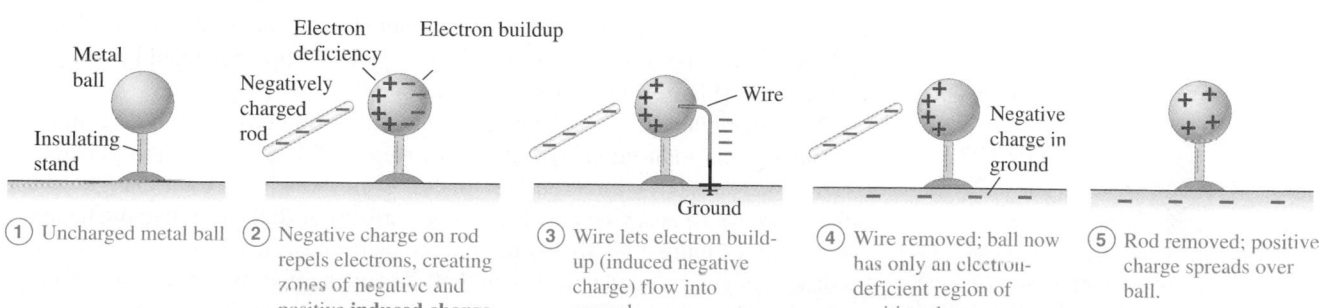

① Uncharged metal ball   ② Negative charge on rod repels electrons, creating zones of negative and positive **induced charge**.   ③ Wire lets electron buildup (induced negative charge) flow into ground.   ④ Wire removed; ball now has only an electron-deficient region of positive charge.   ⑤ Rod removed; positive charge spreads over ball.

▲ **FIGURE 17.5** Charging a metal ball by induction.

▲ **FIGURE 17.6** A charged plastic comb picks up *uncharged* bits of paper.

charge flows through the wire to the earth. Now suppose you disconnect the wire (step 4) and then remove the rod (step 5); a net positive charge is left on the sphere. The charge on the negatively charged rod has not changed during this process. The earth acquires a negative charge that is equal in magnitude to the induced positive charge remaining on the sphere.

Charging by induction would work just as well if the mobile charges in the sphere were positive charges instead of (negatively charged) electrons or even if both positive and negative mobile charges were present (as would be the case if we replaced the sphere with a flask of salt water). In this book, we'll talk mostly about metallic conductors, in which the mobile charges are negative electrons. However, even in a metal, we can describe conduction *as though* the moving charges were positive. In terms of transfer of charge in a conductor, a movement of electrons to the left is exactly equivalent to a movement of imaginary positive particles to the right. In fact, when we study electrical currents, we will find that, for historical reasons, currents in wires are described as though the moving charges were positive.

Conceptual Analysis 17.2

Charge and induction

If you charge a metal ball on an insulating stand *by induction,* using a charged glass or plastic rod, what happens to the charge on the rod?

A. It increases.
B. It decreases.
C. It stays the same.
D. The answer depends on whether the charge is positive or negative.

SOLUTION Look at Figure 17.5 to remind yourself how induction works. The rod *never touches the ball;* instead, its charge pushes or pulls the electrons in the ball so that they crowd (just slightly) to one or the other side of the ball, creating induced charges. A conducting wire lets the local excess of electrons drain into the ground and is then removed. No charges move from the rod to the ball (or to the ground), so the rod's charge is unchanged; thus, C is correct. It makes no difference whether the rod's charge is positive or negative.

When excess charge is placed on a solid conductor and is at rest (i.e., an electrostatic situation), the excess charge rests entirely on the surface of the conductor. If there were excess charge in the interior of the conductor, there would be electrical forces among the excess charges that would cause them to move, and the situation couldn't be electrostatic.

Polarization

A charged object can exert forces even on objects that are *not* charged themselves. If you rub a balloon on a rug and then hold it against the ceiling, the balloon sticks, even though the ceiling has no net electric charge. After you electrify a comb by running it through your hair, you can pick up uncharged bits of paper on the comb (Figure 17.6). How is this possible?

The interaction between the balloon and the ceiling or between the comb and the paper is an induced-charge effect. In step 2 of Figure 17.5, the plastic rod exerts a net attractive force on the sphere, even though the total charge on the sphere is zero, because the positive charges are closer to the rod than the negative charges are. Figure 17.7 shows this effect more clearly. The large ball *A* has a positive charge; the conducting metal ball *B* is uncharged. When we bring *B* close to *A*, the positive charge on *A* pulls on the electrons in *B*, setting up induced charges. Because the negative induced charge on the surface of *B* is closer to *A* than the positive induced charge is, *A* exerts a net attraction on *B*. (We'll study the dependence of electric forces on distance in Section 17.4.) Even in an insulator, the electric charges can shift back and forth a little when there is charge nearby. Figure 17.8 shows how a static charge enables a charged plastic comb to pick up

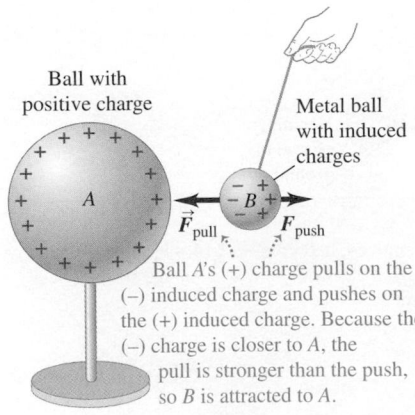

Ball with positive charge

Metal ball with induced charges

$\vec{F}_{pull}$ $\vec{F}_{push}$

Ball *A*'s (+) charge pulls on the (−) induced charge and pushes on the (+) induced charge. Because the (−) charge is closer to *A*, the pull is stronger than the push, so *B* is attracted to *A*.

▲ **FIGURE 17.7** The charge on ball *A* induces charges in ball *B*, resulting in a net attractive force between the balls.

uncharged bits of paper. Although the electrons in the paper are bound to their molecules and cannot move freely through the paper, they can still shift slightly to produce a net charge on one side and the opposite charge on the other. Thus, the comb causes each molecule in the paper to develop induced charges (an effect called **polarization).** The net result is that the scrap of paper shows a slight induced charge—enough to enable the comb to pick it up.

17.3 Conservation and Quantization of Charge

As we've discussed, an electrically neutral object is an object that has equal numbers of electrons and protons. The object can be given a charge by adding or removing either positive or negative charges. Implicit in this discussion are two very important principles. First is the principle of **conservation of charge:**

> ### Conservation of charge
> The algebraic sum of all the electric charges in any closed system is constant. Charge can be transferred from one object to another, and that is the only way in which an object can acquire a net charge.

Conservation of charge is believed to be a *universal* conservation law; there is no experimental evidence for any violation of this principle. Even in high-energy interactions in which charged particles are created and destroyed, the total charge of any closed system is exactly constant.

Second, the magnitude of the charge of the electron or proton is a natural unit of charge. Every amount of observable electric charge is always an integer multiple of this basic unit. Hence we say that charge is *quantized.* A more familiar example of quantization is money. When you pay cash for an item in a store, you have to do it in 1 cent increments. If grapefruits are selling three for a dollar, you can't buy one for $33\frac{1}{3}$ cents; you have to pay 34 cents. Cash can't be divided into smaller amounts than 1 cent, and electric charge can't be divided into smaller amounts than the charge of one electron or proton.

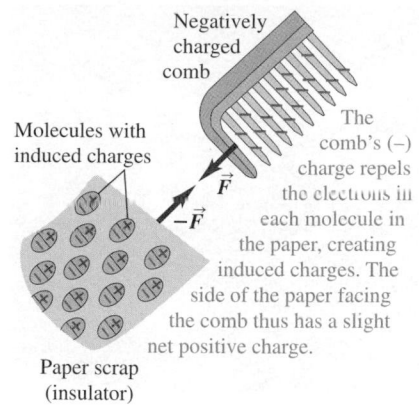

The comb's (−) charge repels the electrons in each molecule in the paper, creating induced charges. The side of the paper facing the comb thus has a slight net positive charge.

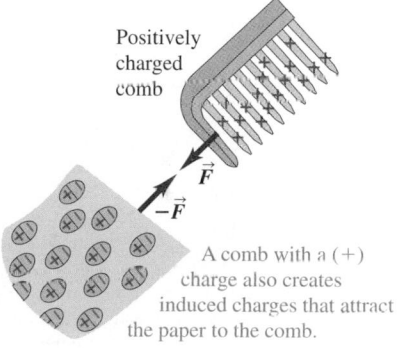

A comb with a (+) charge also creates induced charges that attract the paper to the comb.

▲ **FIGURE 17.8** A charged comb picks up uncharged paper by polarizing the paper's molecules.

Quantitative Analysis 17.3

Determine the charge

Three identical metal balls *A*, *B*, and *C* are mounted on insulating rods. Ball *A* has a charge $+q$, and balls *B* and *C* are initially uncharged (*q* is the usual symbol for electric charge). Ball *A* is touched first to ball *B* and then separately to ball *C*. At the end of this experiment, the charge on ball *A* is

A. $+q/2$.
B. $+q/3$.
C. $+q/4$.

SOLUTION When identical metal objects come in contact, any net charge they carry is shared equally between them. Thus, when *A* touches *B*, each ends up with a charge $+q/2$. When *A* then touches *C*, this charge is shared equally, leaving *A* and *C* each with a charge of $+q/4$.

Electric forces have many important practical applications. Electrostatic dust precipitators first create a charge on dust particles in the air and then use it to catch the dust. When a car is painted, electrostatic charges attract the droplets of sprayed paint to the body of the car. In a photocopy machine or a laser printer, charged areas on the paper attract the toner particles to the correct spots on the paper.

The forces that hold atoms and molecules together are fundamentally electrical in nature. The attraction between electrons and protons holds the electrons in atoms, holds atoms together to form polyatomic molecules, holds molecules together to form solids or liquids, and accounts for phenomena such as surface tension and the stickiness of glue. Within the atom, the electrons repel each other,

◄ **BIO Application Static cling.** As you know, the information that tells your body how to build itself is carried by the "double helix" of DNA, which consists of two DNA strands wound around each other. The two strands stick together by what is essentially static cling. Along each strand, specific molecular groups form dipoles, with a positive or negative end projecting outward. The positive charges on one strand interact precisely with the negative charges on the other, "zipping" the two strands together. Crucially, these interactions are strong enough to keep the strands from coming apart on their own, but weak enough that the cellular machinery can "unzip" the strands for copying.

but they are held in the atom by the attractive force of the protons in the nucleus. But what keeps the positively charged protons together in the tiny nucleus despite their mutual repulsion? As we will learn in Chapter 30, they are held by another, even stronger interaction called the *nuclear force.*

17.4 Coulomb's Law

Charles Augustin de Coulomb (1736–1806) studied the interaction forces of charged particles in detail in 1784. He used a torsion balance (Figure 17.9a) similar to the one used 13 years later by Cavendish to study the (much weaker) gravitational interaction, as we discussed in Section 6.3. For *point charges* (charged bodies that are very small in comparison with the distance r between them), Coulomb found that the electric force is proportional to $1/r^2$. That is, when the distance doubles, the force decreases to one fourth of its initial value.

The force also depends on the quantity of charge on each object, which we'll denote by q or Q. To explore this dependence, Coulomb divided a charge into two equal parts by placing a small charged spherical conductor into contact with an identical but uncharged sphere; by symmetry, the charge is shared equally between the two spheres. (Note the essential role of the principle of conservation of charge in this procedure.) Thus, he could obtain one-half, one-quarter, and so on, of any initial charge. He found that the forces that two point charges q_1 and q_2 exert on each other are proportional to each charge and therefore are proportional to the *product* $q_1 q_2$ of the two charges.

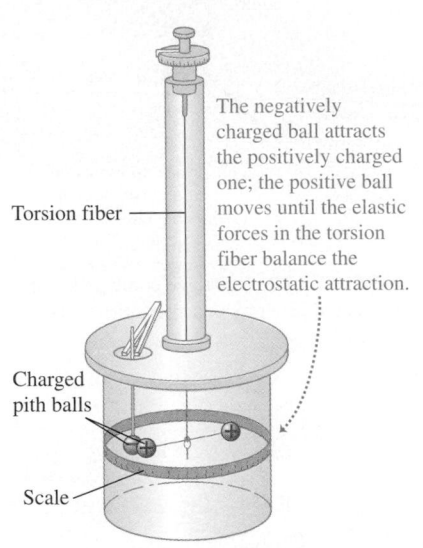

The negatively charged ball attracts the positively charged one; the positive ball moves until the elastic forces in the torsion fiber balance the electrostatic attraction.

Torsion fiber

Charged pith balls

Scale

(a) A torsion balance of the type used by Coulomb to measure the electric force

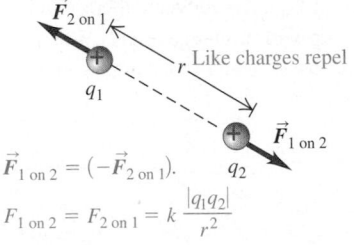

$\vec{F}_{2\ \text{on}\ 1}$

r Like charges repel

q_1

$\vec{F}_{1\ \text{on}\ 2}$

q_2

$\vec{F}_{1\ \text{on}\ 2} = (-\vec{F}_{2\ \text{on}\ 1}).$

$F_{1\ \text{on}\ 2} = F_{2\ \text{on}\ 1} = k\dfrac{|q_1 q_2|}{r^2}$

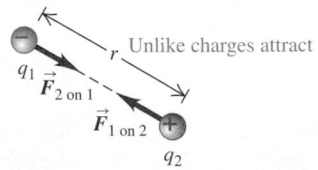

r Unlike charges attract

q_1 $\vec{F}_{2\ \text{on}\ 1}$

$\vec{F}_{1\ \text{on}\ 2}$

q_2

(b) Interaction of like and unlike charges

▲ **FIGURE 17.9** Schematic depiction of the apparatus Coulomb used to determine the forces between charged objects that can be treated as point charges.

Coulomb's law

The magnitude F of the force that each of two point charges q_1 and q_2 a distance r apart exerts on the other (Figure 17.9b) is directly proportional to the product of the charges $(q_1 q_2)$ and inversely proportional to the square of the distance between them (r^2). The relationship is expressed symbolically as

$$F = k\frac{|q_1 q_2|}{r^2}. \tag{17.1}$$

This relationship is called **Coulomb's law.**

The value of the proportionality constant k in Equation 17.1 depends on the system of units used. In the chapters on electricity and magnetism in this book, we'll use SI units exclusively. The SI electrical units include most of the familiar units, such as the volt, the ampere, the ohm, and the watt. (There is *no* British system of electrical units.) The SI unit of electric charge is called one **coulomb** (1 C). In this system, the constant k in Equation 17.1 is

$$k = 8.987551789 \times 10^9 \ \text{N} \cdot \text{m}^2/\text{C}^2.$$

In numerical calculations in problems, we'll often use the approximate value

$$k \simeq 8.99 \times 10^9 \, \text{N} \cdot \text{m}^2/\text{C}^2,$$

which is in error by about 0.03%.

The forces that two charges exert on each other always act along the line joining the charges. The two forces are always equal in magnitude and opposite in direction, even when the charges are not equal. *The forces obey Newton's third law.*

As we've seen, q_1 and q_2 can be either positive or negative quantities. When the charges have the same sign (both positive or both negative), the forces are repulsive; when they are unlike, the forces are attractive. We need the absolute value bars in Equation 17.1 because F is the magnitude of a vector quantity. By definition, F is always positive, but the product $q_1 q_2$ is negative whenever the two charges have opposite signs.

The proportionality of the electrical force to $1/r^2$ has been verified with great precision. There is no experimental evidence that the exponent is anything different from precisely 2. The form of Equation 17.1 is the same as that of the law of gravitation, but electrical and gravitational interactions are two distinct classes of phenomena. The electrical interaction depends on electric charges and can be either attractive or repulsive; the gravitational interaction depends on mass and is always attractive (because there is no such thing as negative mass).

Strictly speaking, Coulomb's law, as we have stated it, should be used only for point charges *in vacuum.* If matter is present in the space between the charges, the net force acting on each charge is altered because charges are induced in the molecules of the intervening material. We'll describe this effect later. As a practical matter, though, we can use Coulomb's law unaltered for point charges in air; at normal atmospheric pressure, the presence of air changes the electrical force from its vacuum value by only about 1 part in 2000.

In SI units, the constant k in Equation 17.1 is often written as

$$k = \frac{1}{4\pi\epsilon_0},$$

where $\epsilon_0 = 8.854 \times 10^{-12} \, \text{C}^2/(\text{N} \cdot \text{m}^2)$ is another constant. This alternative form may appear to complicate matters, but it actually simplifies some of the formulas that we'll encounter later. When we study electromagnetic radiation in Chapter 23, we'll show that the numerical value of ϵ_0 is closely related to the speed of light.

The most fundamental unit of charge is the magnitude of the charge of an electron or a proton, denoted by e. The most precise value available, as of 2005, is

$$e = 1.60217653(14) \times 10^{-19} \, \text{C}.$$

The number (14) in parentheses represents the uncertainty in the last two digits.

One coulomb represents the total charge carried by about 6×10^{18} protons, or the negative of the total charge of about 6×10^{18} electrons. For comparison, the population of the earth is about 6×10^9 persons, and a cube of copper 1 cm on a side contains about 2.4×10^{24} electrons.

In electrostatics problems, charges as large as 1 coulomb are very unusual. Two charges with magnitude 1 C, at a distance 1 m apart, would exert forces of magnitude 9×10^9 N (about a million tons) on each other! A more typical range of magnitudes is 10^{-9} to 10^{-6} C. The *microcoulomb* ($1 \, \mu\text{C} = 10^{-6} \, \text{C}$) and the *nanocoulomb* ($1 \, \text{nC} = 10^{-9} \, \text{C}$) are often used as practical units of charge. The total charge of all the electrons in a penny is about 1.4×10^5 C. This number shows that we can't disturb electrical neutrality very much without using enormous forces.

▲ **Application Great balls of fire?** Before the invention of the cyclotron, which uses both electric and magnetic fields to accelerate subatomic particles, physicists used electric field generators in atom-smashing experiments. These generators, like the huge Van de Graaff generators shown here, can accumulate either positive or negative charges on the surface of a metal sphere, thus generating immense electric fields. Charged particles in such an electric field are acted upon by a large electrical force, which can be used to accelerate the particles to very high velocities. When all excess charge is located on the outer surface of a conductor in an electrostatic situation, the electric field inside is zero. Thus, scientists actually set up small laboratories *inside* each of the spheres of the generator to study subatomic particles subjected to millions of volts in a tube on the outside connecting the two spheres.

MasteringPHYSICS

ActivPhysics 11.1: Electric Force: Coulomb's Law

ActivPhysics 11.2: Electric Force: Superposition Principle

ActivPhysics 11.3: Electric Force: Superposition Principle (Quantitative)

Conceptual Analysis 17.4

Charged spheres in motion

Two small identical balls A and B are held a distance r apart on a frictionless surface; r is large compared with the size of the balls. Ball A has a net charge q; ball B has a net charge $4q$. The balls are released at the same instant and begin to move apart. The magnitudes of their accelerations are

A. constant.
B. equal and decreasing.
C. unequal and decreasing.

SOLUTION Coulomb's law states that the magnitude of the force between two charged objects that can be treated as particles is

$F = (k|q_1 q_2|)/r^2$. Is this force somehow divided between the two objects? Does the object with the larger charge exert a stronger force? Should the force on each object be calculated separately? No; Newton's third law gives the answer. Whenever two objects interact, the forces that the two objects exert on each other are equal in magnitude (and opposite in direction). Since the balls experience the same magnitude of force and have the same mass, by Newton's second law they have the same magnitude of acceleration at any instant. As they move apart and r increases, the magnitude of acceleration decreases. The correct answer is B.

Superposition

When two charges exert forces simultaneously on a third charge, the total force acting on that charge is the *vector sum* of the forces that the two charges would exert individually. This important property, called the **principle of superposition,** holds for any number of charges. Coulomb's law, as we have stated it, describes only the interaction between two *point* charges, but by using the superposition principle, we can apply it to *any* collection of charges. Several of the examples that follow illustrate the superposition principle.

PROBLEM-SOLVING STRATEGY 17.1 **Coulomb's law**

SET UP

1. As always, consistent units are essential. With the value of k given earlier, distances *must* be in meters, charges in coulombs, and forces in newtons. If you are given distances in centimeters, inches, or furlongs, don't forget to convert! When a charge is given in microcoulombs, remember that $1\,\mu C = 10^{-6}\,C$.

SOLVE

2. When the forces acting on a charge are caused by two or more other charges, the total force on the charge is the *vector sum* of the individual forces. If you're not sure you remember how to do vector addition, you may want to review Sections 1.7 and 1.8. It's often useful to use components in an *x-y* coordinate system. As always, it's essential to distinguish between vectors, their magnitudes, and their components (using correct notation!) and to treat vectors properly as such.

3. Some situations involve a continuous distribution of charge along a line or over a surface. In this book, we'll consider only situations for which the vector sum described in Step 2 can be evaluated by using vector addition and symmetry considerations. In other cases, methods of integral calculus would be needed.

REFLECT

4. Try to think of particular cases where you can guess what the result should be, and compare your intuitive expectations with the results of your calculations.

EXAMPLE 17.1 Charge imbalance

(a) A large plastic block has a net charge of $-1.0\ \mu C = -1.0 \times 10^{-6}$ C. How many more electrons than protons are in the block? (b) When rubbed with a silk cloth, a glass rod acquires a net positive charge of 1.0 nC. If the rod contains 1.0 mole of molecules, what fraction of the molecules have been stripped of an electron? Assume that at most one electron is removed from any molecule.

SOLUTION

SET UP AND SOLVE Part (a): We want to find the number of electrons N_e needed for a net charge of -1.0×10^{-6} C on the object. Each electron has charge $-e$. We divide the total charge by $-e$:

$$N_e = \frac{-1.0 \times 10^{-6}\ \text{C}}{-1.60 \times 10^{-19}\ \text{C}} = 6.2 \times 10^{12}\ \text{electrons}.$$

Part (b): First we find the number N_{ion} of positive ions needed for a total charge of 1.0 nC if each ion has charge $+e$. The number of molecules in a mole is Avogadro's number, 6.02×10^{23}, so the rod contains 6.02×10^{23} molecules. As in part (a), we divide the total charge on the rod by the charge of one ion. Remember that $1\ \text{nC} = 10^{-9}$ C. Thus,

$$N_{ion} = \frac{1.0 \times 10^{-9}\ \text{C}}{1.6 \times 10^{-19}\ \text{C}} = 6.25 \times 10^{9}.$$

The fraction of all the molecules that are ionized is

$$\frac{N_{ion}}{6.02 \times 10^{23}} = \frac{6.25 \times 10^{9}}{6.02 \times 10^{23}} = 1.0 \times 10^{-14}.$$

REFLECT A charge imbalance of about 10^{-14} is typical for charged objects. Common objects contain a huge amount of charge, but they have very nearly equal amounts of positive and negative charge.

Practice Problem: A tiny object contains 5.26×10^{12} protons and 4.82×10^{12} electrons. What is the net charge on the object? *Answer:* 7.0×10^{-8} C.

EXAMPLE 17.2 Gravity in the hydrogen atom

A hydrogen atom consists of one electron and one proton. In an early, simple model of the hydrogen atom called the *Bohr model*, the electron is pictured as moving around the proton in a circular orbit with radius $r = 5.29 \times 10^{-11}$ m. (In Chapter 29, we'll study the Bohr model and also more sophisticated models of atomic structure.) What is the ratio of the magnitude of the electric force between the electron and proton to the magnitude of the gravitational attraction between them? The electron has mass $m_e = 9.11 \times 10^{-31}$ kg, and the proton has mass $m_p = 1.67 \times 10^{-27}$ kg.

SOLUTION

SET UP Figure 17.10 shows our sketch. The distance between the proton and electron is the radius r. Each particle has charge of magnitude e. The electric force is given by Coulomb's law and the gravitational force by Newton's law of gravitation.

SOLVE Coulomb's law gives the magnitude F_e of the electric force between the electron and proton as

$$F_e = k\frac{|q_1 q_2|}{r^2} = k\frac{e^2}{r^2},$$

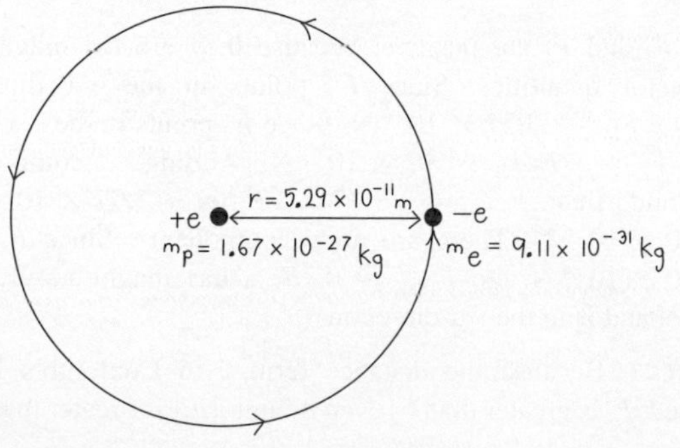

where $k = 8.99 \times 10^9\ \text{N} \cdot \text{m}^2/\text{C}^2$. The gravitational force $\vec{F}_g$ has magnitude F_g:

$$F_g = G\frac{m_1 m_2}{r^2} = G\frac{m_e m_p}{r^2},$$

where $G = 6.67 \times 10^{-11}\ \text{N} \cdot \text{m}^2/\text{kg}^2$. The ratio of the two forces is

$$\frac{F_e}{F_g} = \left(\frac{ke^2}{r^2}\right)\left(\frac{r^2}{Gm_e m_p}\right) = \frac{ke^2}{Gm_e m_p}$$

$$= \left(\frac{8.99 \times 10^9\ \text{N} \cdot \text{m}^2/\text{C}^2}{6.67 \times 10^{-11}\ \text{N} \cdot \text{m}^2/\text{kg}^2}\right)$$

$$\times \frac{(1.60 \times 10^{-19}\ \text{C})^2}{(9.11 \times 10^{-31}\ \text{kg})(1.67 \times 10^{-27}\ \text{kg})},$$

$$\frac{F_e}{F_g} = 2.27 \times 10^{39}.$$

REFLECT In our expression for the ratio, all the units cancel and the ratio is dimensionless. The astonishingly large value of F_e/F_g—about 10^{39}—shows that, in atomic structure, the gravitational force is completely negligible compared with the electrostatic force. The reason gravitational forces dominate in our daily experience

▲ FIGURE 17.10 Our sketch for this problem.

Continued

is that positive and negative electric charges are always nearly equal in number and thus cancel nearly completely. Since there is no "negative" gravitation, gravitational forces always add. Note also that because both F_e and F_g are proportional to $1/r^2$, the ratio F_e/F_g does not depend on the distance between the two particles.

Practice Problem: A hydrogen atom is at the earth's surface. The electron and proton in the atom are separated by a distance of 5.29×10^{-11} m. What is the ratio of the magnitude of the electrical force exerted by the proton on the electron to the weight of the electron? *Answer:* 9.2×10^{21}.

For all fundamental particles, the gravitational attraction is always much, much weaker than the electrical interaction. But suppose that the electric force were a million (10^6) times weaker than it really is. In that case, the ratio of electric to gravitational forces between an electron and a proton would be about $10^{39} \times 10^{-6} = 10^{33}$ and the universe would be a very different place. Materials would be a million times weaker than we are used to, because they are held together by electrostatic forces. Insects would need to have much thicker legs to support the same mass. In fact, animals couldn't get much larger than an insect unless they were made of steel, and even a hypothetical animal made of steel could be only a few centimeters in size before collapsing under its own weight. More significantly, if the electric force were a million times weaker, the lifetime of a typical star would decrease from 10 billion years down to 10 thousand years! This is hardly enough time for *any* living organisms—much less such complicated ones as insects or humans—to evolve.

EXAMPLE 17.3 Adding forces

Two point charges are located on the positive x axis of a coordinate system. Charge $q_1 = 3.0$ nC is 2.0 cm from the origin, and charge $q_2 = -7.0$ nC is 4.0 cm from the origin. What is the total force (magnitude and direction) exerted by these two charges on a third point charge $q_3 = 5.0$ nC located at the origin?

SOLUTION

SET UP We sketch the situation and draw a free-body diagram for charge q_3, using $\vec{F}_1$ to denote the force exerted by q_1 on q_3 and $\vec{F}_2$ for the force exerted by q_2 on q_3 (Figure 17.11). The directions of these forces are determined by the rule that like charges repel and unlike charges attract, so $\vec{F}_1$ points in the $-x$ direction and $\vec{F}_2$ points in the $+x$ direction. We don't yet know their relative magnitudes, so we draw them to arbitrary length.

SOLVE We use Coulomb's law to find the magnitudes of the forces $\vec{F}_1$ and $\vec{F}_2$; then we add these two forces (as vectors) to find the resultant force on q_3:

$$\vec{F}_{total} = \vec{F}_1 + \vec{F}_2,$$
so $\quad F_{total,x} = F_{1x} + F_{2x}$
and $\quad F_{total,y} = F_{1y} + F_{2y}.$

$$F_1 = k\frac{|q_1 q_3|}{r_{12}^2}$$
$$= (8.99 \times 10^9 \text{ N} \cdot \text{m}^2/\text{C}^2)\frac{(3.0 \times 10^{-9} \text{ C})(5.0 \times 10^{-9} \text{ C})}{(0.020 \text{ m})^2}$$
$$= 3.37 \times 10^{-4} \text{ N},$$

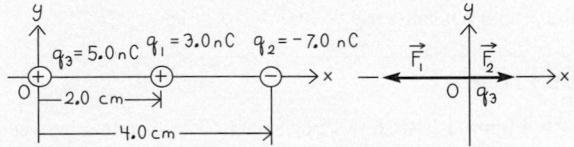

(a) Our diagram of the situation   **(b)** Free-body diagram for q_3

▲ **FIGURE 17.11** Our sketches for this problem.

$$F_2 = k\frac{|q_2 q_3|}{r_{23}^2}$$
$$= (8.99 \times 10^9 \text{ N} \cdot \text{m}^2/\text{C}^2)\frac{(7.0 \times 10^{-9} \text{ C})(5.0 \times 10^{-9} \text{ C})}{(0.040 \text{ m})^2}$$
$$= 1.97 \times 10^{-4} \text{ N}.$$

Both F_1 and F_2 are positive, because they are the magnitudes of vector quantities. Since $\vec{F}_1$ points in the $-x$ direction, $F_{1x} = -F_1 = -3.37 \times 10^{-4}$ N. Since $\vec{F}_2$ points in the $+x$ direction, $F_{2x} = +F_2 = +1.97 \times 10^{-4}$ N. Adding x components, we find that $F_{total,x} = -3.37 \times 10^{-4}$ N $+ 1.97 \times 10^{-4}$ N $= -1.40 \times 10^{-4}$ N. There are no y components. Since $F_{total,x} = -1.40 \times 10^{-4}$ N and $F_{total,y} = 0$, $\vec{F}_{total}$ has magnitude 1.40×10^{-4} N and is in the $-x$ direction.

REFLECT Because the distance term r in Coulomb's law is squared, F_1 is greater than F_2 even though $|q_2|$ is greater than $|q_1|$.

Practice Problem: In Example 17.3, what is the total force (magnitude and direction) exerted on q_1 by q_2 and q_3? *Answer:* 8.1×10^{-4} N, in the $+x$ direction.

EXAMPLE 17.4 **Vector addition of forces**

A point charge $q_1 = 2.0 \ \mu C$ is located on the positive y axis at $y = 0.30$ m, and an identical charge q_2 is at the origin. Find the magnitude and direction of the total force that these two charges exert on a third charge $q_3 = 4.0 \ \mu C$ that is on the positive x axis at $x = 0.40$ m.

SOLUTION

SET UP As in the previous example, we sketch the situation and draw a free-body diagram for q_3 (Figure 17.12), using $\vec{F}_1$ and $\vec{F}_2$ for the forces exerted on q_3 by q_1 and q_2, respectively. The directions of $\vec{F}_1$ and $\vec{F}_2$ are determined by the fact that like charges repel.

SOLVE As in Example 17.3, the net force acting on q_3 is the vector sum of $\vec{F}_1$ and $\vec{F}_2$. We use Coulomb's law to find the magnitudes F_1 and F_2 of the forces:

$$F_1 = k \frac{|q_1 q_3|}{r_{13}^2}$$

$$= (8.99 \times 10^9 \ \text{N} \cdot \text{m}^2/\text{C}^2) \frac{(2.0 \times 10^{-6} \ \text{C})(4.0 \times 10^{-6} \ \text{C})}{(0.50 \ \text{m})^2}$$

$$= 0.288 \ \text{N},$$

$$F_2 = k \frac{|q_2 q_3|}{r_{23}^2}$$

$$= (8.99 \times 10^9 \ \text{N} \cdot \text{m}^2/\text{C}^2) \frac{(2.0 \times 10^{-6} \ \text{C})(4.0 \times 10^{-6} \ \text{C})}{(0.40 \ \text{m})^2}$$

$$= 0.450 \ \text{N}.$$

We now calculate the x and y components of $\vec{F}_1$ and add them to the x and y components of $\vec{F}_2$, respectively, to obtain the components of the total force $\vec{F}_{total}$ on q_3. From Figure 17.12a, $\sin \theta = (0.30 \ \text{m})/(0.50 \ \text{m}) = 0.60$ and $\cos \theta = (0.40 \ \text{m})/(0.50 \ \text{m}) = 0.80$. Since the y component of $\vec{F}_2$ is zero and its x

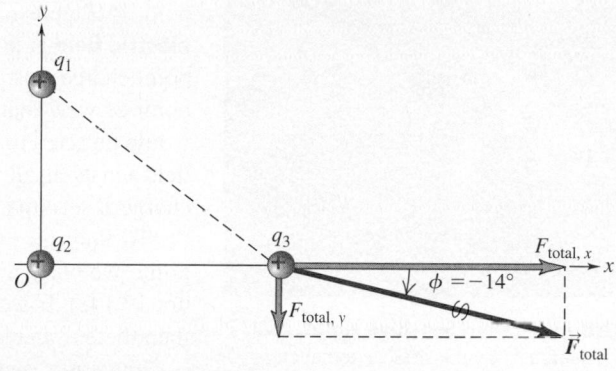

▲ FIGURE 17.13 The total force on q_3.

component is positive, the x component of $\vec{F}_2$ is $F_{2x} = F_2 = 0.450$ N. The total x and y components are

$$F_{total,x} = F_{1x} + F_{2x}$$
$$= (0.288 \ \text{N}) \cos \theta + 0.450 \ \text{N}$$
$$= (0.288 \ \text{N})(0.80) + 0.450 \ \text{N} = 0.680 \ \text{N},$$
$$F_{total,y} = F_{1y} + F_{2y} = -(0.288 \ \text{N}) \sin \theta + 0$$
$$= -(0.288 \ \text{N})(0.60) = -0.173 \ \text{N}.$$

These components combine to form $\vec{F}_{total}$, as shown in Figure 17.13:

$$F_{total} = \sqrt{F_{total,x}^2 + F_{total,y}^2}$$
$$= \sqrt{(0.680 \ \text{N})^2 + (-0.173 \ \text{N})^2} = 0.70 \ \text{N},$$
$$\tan \phi = \frac{F_{total,y}}{F_{total,x}} = \frac{-0.173 \ \text{N}}{0.680 \ \text{N}} \quad \text{and} \quad \phi = -14°.$$

The resultant force has magnitude 0.70 N and is directed at 14° below the $+x$ axis.

REFLECT The forces exerted by q_1 and q_2 both have components in the $+x$ direction, so these components add. The force exerted by q_1 also has a component in the $-y$ direction, so the net force is in the fourth quadrant. Even though q_1 and q_2 are identical, the force exerted by q_2 is larger than the force exerted by q_1 because q_3 is closer to q_2.

Practice Problem: In Example 17.4, what is the net force on q_3 if $q_1 = 2.0 \ \mu C$, as in the example, but $q_2 = -2.0 \ \mu C$? *Answer:* 0.28 N, 38° below the $-x$ axis.

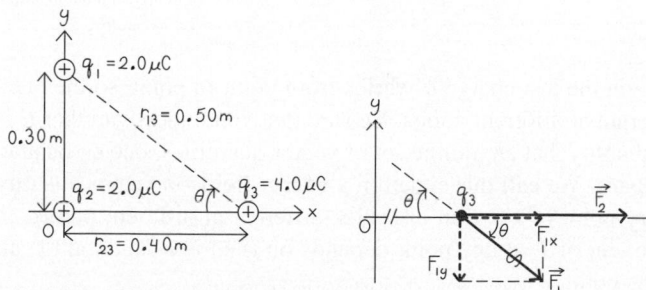

(a) Our sketch of the situation

(b) Free-body diagram for q_3

▲ FIGURE 17.12 Our sketches for this problem.

17.5 Electric Field and Electric Forces

When two electrically charged particles in empty space interact, how does each one "know" that the other is there? What goes on in the space between them to transmit the effect of each one to the other? We can begin to answer these questions, and at the same time reformulate Coulomb's law in a very useful way, by using the concept of *electric field*. To introduce this concept, let's look at the mutual repulsion of two positively charged objects A and B (Figure 17.14a).

MasteringPHYSICS

ActivPhysics 11.9: Motion of a Charge in an Electric Field: Introduction
ActivPhysics 11.10: Motion in an Electric Field: Problems

▲ **BIO Application Sensitive snout.** As a rule, mammals cannot sense external electric fields, but the platypus is an exception. It feeds on small underwater creatures, which it finds by nosing around the bottom of streams and ponds. It hunts with its eyes shut, and usually at night, so it cannot see its prey. Instead, its rubbery bill detects the tiny electric fields created by the nerves and muscles of the prey. (The bill is also highly sensitive to touch.) Because water is a good conductor but air is not, the ability to sense electric fields is found almost exclusively among water-dwelling creatures (mainly fishes).

A and B exert electric forces on each other.

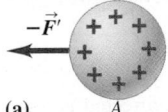 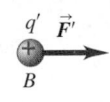

(a)　　　A

B is removed; point P marks its position.

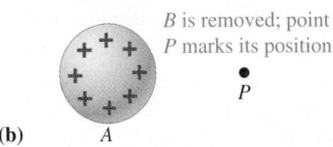

(b)　　A

A test charge placed at P is acted upon by a force $\vec{F'}$ due to the electric field $\vec{E}$ of charge A. $\vec{E}$ is the force per unit charge exerted on the test charge.

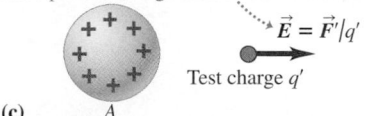

(c)　　A

▲ **FIGURE 17.14** A charged object creates an electric field in the space around it.

Suppose B is a point charge q', and let $\vec{F'}$ be the force on B, as shown in the figure. One way to think about this force is as an "action-at-a-distance" force—that is, as a force that acts across empty space without needing any matter (such as a pushrod or a rope) to transmit it through the intervening space.

Now think of object A as having the effect of somehow modifying the properties of the space around it. We remove object B and label its former position as point P (Figure 17.14b). We say that the charged object A produces or causes an **electric field** at point P (and at all other points in the neighborhood). Then, when point charge B is placed at point P and is acted upon by the force $\vec{F'}$, we take the point of view that the force is exerted on B *by the electric field* at P. Because B would be acted upon by a force at *any* point in the neighborhood of A, the electric field exists at all points in the region around A. (We could also say that point charge B sets up an electric field, which in turn exerts a force on object A.)

To find out experimentally whether there is an electric field at a particular point, we place a charged object, which we call a **test charge,** at the point (Figure 17.14c). If we find that the test charge experiences a non-zero electric force, then there is an electric field at that point.

Force is a vector quantity, so electric field is also a vector quantity. (Note the use of boldface letters with arrows on top of them in the discussion that follows.) To define the *electric field* $\vec{E}$ at any point, we place a test charge q' at the point and measure the electric force $\vec{F'}$ on it (Figure 17.14c). We define $\vec{E}$ at this point to be equal to $\vec{F'}$ divided by q':

Definition of electric field

When a charged particle with charge q' at a point P is acted upon by an electric force $\vec{F'}$, the electric field $\vec{E}$ at that point is defined as

$$\vec{E} = \frac{\vec{F'}}{q'}. \tag{17.2}$$

The test charge q' can be either positive or negative. If it is positive, the directions of $\vec{E}$ and $\vec{F'}$ are the same; if it is *negative,* they are opposite (Figure 17.15).

Unit: In SI units, in which the unit of force is the newton and the unit of charge is the coulomb, the unit of electric-field magnitude is 1 newton per coulomb (1 N/C).

The force acting on the test charge q' varies from point to point, so the electric field is also different at different points. Be sure that you understand that $\vec{E}$ is not a single vector quantity, but an infinite set of vector quantities, one associated with each point in space. We call this situation a **vector field**—a vector quantity associated with every point in a region of space, different at different points. In general, each component of $\vec{E}$ at any point depends on (i.e., is a function of) all the coordinates of the point.

NOTE ▶ There's a slight difficulty with our definition of electric field: In Figure 17.14, the force exerted by the test charge q' on the charge distribution A may cause the distribution to shift around, especially if object A is a conductor, in which charge is free to move. So the electric field around A when q' is present may not be the same as when q' is absent. But if q' is very small, the redistribution of charge on object A is also very small. So we refine our definition of electric field by taking the limit of Equation 17.2 as the test charge q' becomes very small and its disturbing effect on the charge distribution becomes negligible:

$$\vec{E} = \lim_{q' \to 0} \frac{\vec{F'}}{q'}. \tag{17.3} ◀$$

▶ **FIGURE 17.15** The direction of the electric force on a positive and negative test charge relative to the direction of the electric field.

The force on a positive test charge points in the direction of the electric field.

 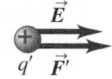

If an electric field exists within a *conductor,* the field exerts a force on every charge in the conductor, causing the free charges to move. By definition, an *electrostatic* situation is a situation in which the charges *do not* move. We conclude that **in an electrostatic situation, the electric field at every point within the material of a conductor must be zero.** (In Section 17.9, we'll consider the special case of a conductor that has a central cavity.)

The force on a negative test charge points opposite to the electric field.

 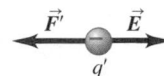

In general, the magnitude and direction of an electric field can vary from point to point. If, in a particular situation, the magnitude and direction of the field are *constant* throughout a certain region, we say that the field is *uniform* in that region.

Conceptual Analysis 17.5

A moving electron

A vacuum chamber contains a uniform electric field directed downward. If an electron is shot horizontally into this region, its acceleration is:

A. downward and constant.
B. upward and constant.
C. upward and changing.
D. downward and changing.

SOLUTION The electron has a negative charge, so it is acted upon by a force directed *opposite* to the electric field—that is, an upward force giving an upward acceleration. We're told that the electric field is *uniform,* meaning that its magnitude and direction are constant. Therefore, the force exerted on the electron by the electric field is constant $(\vec{F}' = \vec{E}/q')$. Because the force is constant, so is the acceleration (by Newton's second law). The correct answer is B.

EXAMPLE 17.5 **Accelerating an electron**

When the terminals of a 100 V battery are connected to two large parallel horizontal plates 1.0 cm apart, the resulting charges on the plates produce an electric field $\vec{E}$ in the region between the plates that is very nearly uniform and has magnitude $E = 1.0 \times 10^4$ N/C. Suppose the lower plate has positive charge, so that the electric field is vertically upward, as shown in Figure 17.16. (The thin pink arrows represent the field.) If an electron is released from rest at the upper plate, what is its speed just before it reaches the lower plate? How much time is required for it to reach the lower plate? The mass of an electron is $m_e = 9.11 \times 10^{-31}$ kg.

The thin arrows represent the uniform electric field.

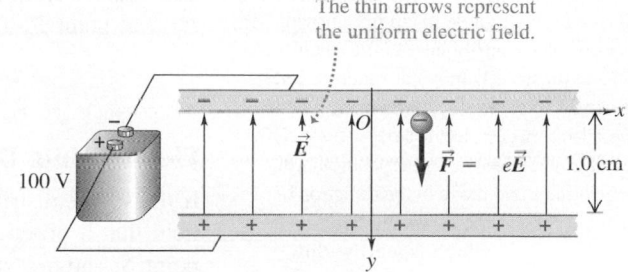

▲ **FIGURE 17.16**

SOLUTION

SET UP We place the origin of coordinates at the upper plate and take the $+y$ direction to be downward, toward the lower plate. The electron has negative charge, $q = -e$, so the direction of the force on the electron is downward, opposite to the electric field. The field is uniform, so the force on the electron is constant. Thus the electron has constant acceleration, and we can use the constant-acceleration equation $v_y^2 = v_{0y}^2 + 2a_y y$. The electron's initial velocity v_{0y} is zero, so

$$v_y^2 = 2a_y y.$$

SOLVE The force on the electron has only a y component, which is positive, and we can solve Equation 17.2 to find this component:

$$F_y = |q|E = (1.60 \times 10^{-19} \text{ C})(1.0 \times 10^4 \text{ N/C})$$
$$= 1.60 \times 10^{-15} \text{ N}.$$

Newton's second law then gives the electron's acceleration:

$$a_y = \frac{F_y}{m_e} = \frac{1.60 \times 10^{-15} \text{ N}}{9.11 \times 10^{-31} \text{ kg}} = +1.76 \times 10^{15} \text{ m/s}^2.$$

We want to find v_y when $y = 0.010$ m. The equation for v_y gives

$$v_y = \sqrt{2a_y y} = \sqrt{2(1.76 \times 10^{15} \text{ m/s}^2)(0.010 \text{ m})}$$
$$= 5.9 \times 10^6 \text{ m/s}.$$

Finally, $v_y = v_{0y} + a_y t$ gives the total travel time t:

$$t = \frac{v_y - v_{0y}}{a_y} = \frac{5.9 \times 10^6 \text{ m/s} - 0}{1.76 \times 10^{15} \text{ m/s}^2} = 3.4 \times 10^{-9} \text{ s}.$$

REFLECT The acceleration produced by the electric field is enormous; to give a 1000 kg car this acceleration, we would need a force of about 2×10^{18} N, or about 2×10^{14} tons. The effect of

Continued

gravity is completely negligible. The electron's final speed is only 2% of the speed of light, so we don't have to include relativistic effects. Note again that negative charges gain speed when they move in a direction opposite to the direction of the electric field.

Practice Problem: In this example, suppose a proton ($m_p = 1.67 \times 10^{-27}$ kg) is released from rest at the positive plate. What is its speed just before it reaches the negative plate? *Answer:* 1.38×10^5 m/s.

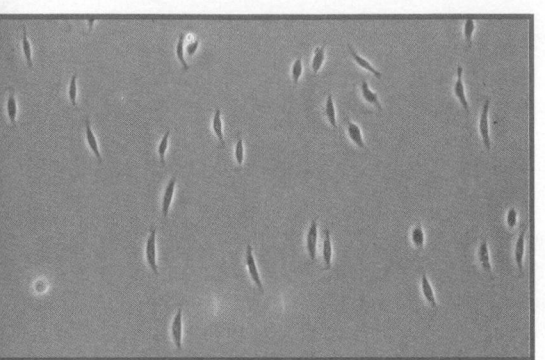

▲ BIO Application They got their electrical marching orders. Many cells, including nerve cells and skin cells, are remarkably sensitive to electrical fields. The photograph shows cultured skin cells of the zebrafish (an important experimental animal for biology and medicine). These cells are highly mobile in culture, moving at speeds of 10 μm/minute. Left to their own devices, these cells move at random, independently of each other; however, when exposed to a modest electrical field of 100 N/C, they align their long axes perpendicular to the field lines and move in the direction of the field. These cells respond to fields as small as 7 N/C, which is well within the range of electrical fields that have been measured near skin wounds in vertebrates. It may be that the wound healing response is controlled in part by natural electrical fields.

17.6 Calculating Electric Fields

In this section, we'll discuss several situations in which electric fields produced by specific charge distributions can be determined with fairly simple calculations. The key to these calculations is the principle of superposition, which we mentioned in Section 17.4 Restated in terms of electric fields, the principle is as follows:

Principle of superposition
The total electric field at any point due to two or more charges is the vector sum of the fields that would be produced at that point by the individual charges.

To find the field caused by several charges or an extended distribution of charge, we imagine the source to be made up of many point charges. We call the location of one of these points a **source point** (denoted by S, possibly with a subscript), and the point where we want to find the field is called the **field point** (denoted by P). We calculate the fields $\vec{E}_1$, $\vec{E}_2$, $\vec{E}_3, \cdots$ at point P caused by the individual point charges q_1, q_2, $q_3, \cdots$ located at points S_1, S_2, $S_3, \cdots$ and take their vector sum (using the superposition principle) to find the total field $\vec{E}_{total}$ at point P. That is,

$$\vec{E}_{total} = \vec{E}_1 + \vec{E}_2 + \vec{E}_3 + \cdots.$$

Electric Field Due to a Point Charge

If the source distribution is a single point charge q, it is easy to find the electric field that it produces. As before, we call the location of the charge the source point S, and we call the point P where we are determining the field the field point. If we place a small test charge q' at the field point P, at a distance r from the source point, the magnitude of the force $\vec{F}'$ is given by Coulomb's law, Equation 17.1:

$$F' = k\frac{|qq'|}{r^2}.$$

From Equation 17.3, we find the magnitude E of the electric field at P:

Electric field due to a point charge
The magnitude E of the electric field $\vec{E}$ at point P due to a point charge q at point S, a distance r from P, is given by

$$E = k\frac{|q|}{r^2}. \tag{17.4}$$

By definition, the electric field produced by a positive point charge always points *away from* it, but the electric field produced by a negative point charge points *toward* it.

Quantitative Analysis 17.6

Magnitude and direction

A small object S with a charge of magnitude q creates an electric field. At a point P located 0.36 m west of S, the field has a value of 40 N/C directed to the west. At a point 0.36 m east of S, the field is

A. 40 N/C, directed westward.
B. 40 N/C, directed eastward.
C. 80 N/C, directed westward.
D. 80 N/C, directed eastward.

SOLUTION As we just saw, the electric field of a positive point charge is directed radially away from the charge; that of a negative point charge is directed radially toward the charge. The fact that the field at P is directed to the west (away from S) means that S has a positive charge. Thus, at a point east of S, the field will point east. Equation 17.4, $E = k(|q|/r^2)$, tells us that the field has the same magnitude at all points that are the same radial distance r from S. Therefore, the correct answer is B.

Spherical Charge Distributions

In applications of electrostatics, we often encounter charge distributions that have spherical symmetry. Familiar examples include electric charge distributed uniformly over the surface of a conducting sphere and charge distributed uniformly throughout the volume of an insulating sphere. It turns out that the electric field produced by *any* spherically symmetric charge distribution, at all points outside this distribution, is the same as though all the charge were concentrated at a point at the center of the sphere. In field calculations, the field outside any spherical charge distribution can be obtained by replacing the distribution with a single point charge at the center of the sphere and equal to the total charge of the sphere.

EXAMPLE 17.6 **Electric field in a hydrogen atom**

(a) In the Bohr model of the hydrogen atom (described in Example 17.2), when the atom is in its lowest-energy state, the distance from the proton to the electron is 5.29×10^{-11} m. Find the electric field due to the proton at this distance. (b) A device called a Van de Graaff generator (a staple in science museums) can build up a large static charge on a metal sphere. Suppose the sphere of a Van de Graaff generator has a radius of 0.50 m and a net charge of 1.0 μC. What is the magnitude of the electric field 1.0 m from the center of the sphere? Compare this electric field with the field calculated in part (a).

SOLUTION

SET UP Figure 17.17 shows our diagrams for these cases.

SOLVE Part (a): We are asked to calculate the electric-field magnitude E at a distance of 5.29×10^{-11} m from a point charge (the proton). We use Equation 17.4; a proton has charge $q = +e = 1.60 \times 10^{-19}$ C, so

$$E = k\frac{|q|}{r^2}$$

$$= (8.99 \times 10^9 \text{ N} \cdot \text{m}^2/\text{C}^2)\frac{(1.60 \times 10^{-19} \text{ C})}{(5.29 \times 10^{-11} \text{ m})^2}$$

$$= 5.14 \times 10^{11} \text{ N/C}.$$

Part (b) To calculate the field of the van de Graaff sphere, we use the principle discussed above: a uniform spherical distribution of charge creates the same field as an equal point charge located at the center of the sphere. Thus, we can again use Equation 17.4:

$$E = k\frac{|q|}{r^2} = (8.99 \times 10^9 \text{ N} \cdot \text{m}^2/\text{C}^2)\frac{(1.0 \times 10^{-6} \text{ C})}{(1.0 \text{ m})^2}$$

$$= 9.0 \times 10^3 \text{ N/C}.$$

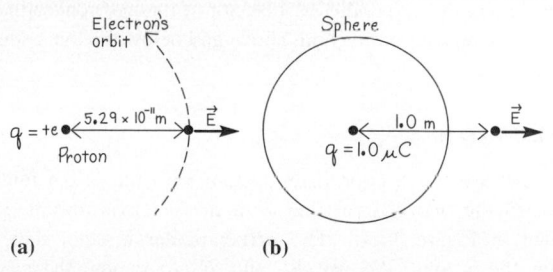

▲ **FIGURE 17.17** Our sketches for this problem.

The electric field in part (a) is larger than that in part (b) by a factor of 5.7×10^7.

REFLECT The electric field in an atom is extremely large compared with the electric fields of macroscopic objects with easily obtainable electric charges.

Practice Problem: At what distance from a proton does the electric field of the proton have magnitude 9.0×10^3 N/C? How does this distance compare with the Bohr orbit radius ($r = 5.29 \times 10^{-11}$ m) of the electron in the lowest-energy state of the hydrogen atom? *Answers:* 4.0×10^{-7} m; 7.6×10^3 times larger.

PROBLEM-SOLVING STRATEGY 17.2 **Electric field calculations**

SET UP

1. Be sure to use a consistent set of units. Distances must be in meters, charges in coulombs. If you are given cm or nC, don't forget to convert.
2. Usually, you will use components to compute vector sums. As we suggested for problems involving Coulomb's law, it may be helpful to review Sections 1.7 and 1.8. Use proper vector notation; distinguish carefully between scalars, vectors, and components of vectors. Indicate your coordinate axes clearly on your diagram, and be certain that the components are consistent with your choice of axes.

SOLVE

3. In working out directions of $\vec{E}$ vectors, be careful to distinguish between the *source point S* and the *field point P*. The field produced by a positive point charge always points in the direction from source point to field point; the opposite is true for a negative point charge.

REFLECT

4. If your result is a symbolic expression, check to see whether it depends on the variables in the way you expect. If it is numeric, estimate what you expect the result to be and check for consistency with the result of your calculations.

EXAMPLE 17.7 **Electric field of an electric dipole**

Point charges q_1 and q_2 of $+12$ nC and -12 nC, respectively, are placed 10.0 cm apart (Figure 17.18). This combination of two charges with equal magnitude and opposite sign is called an **electric dipole.** Compute the resultant electric field (magnitude and direction) at **(a)** point *a*, midway between the charges, and **(b)** point *b*, 4.0 cm to the left of q_1. **(c)** What is the direction of the resultant electric field produced by these two charges at points along the perpendicular bisector of the line connecting the charges? Consider points both above and below the line connecting the charges.

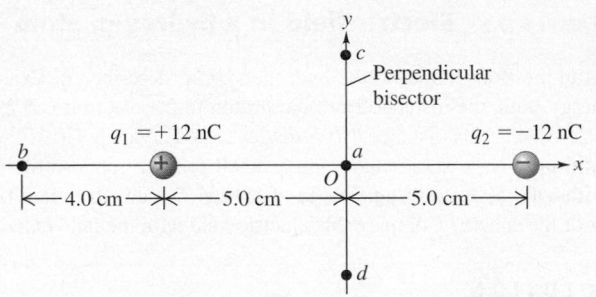

▲ **FIGURE 17.18**

SOLUTION

SET UP We use a coordinate system with the origin midway between the two charges and with the $+x$ axis toward q_2, as shown in Figure 17.18. The perpendicular bisector then lies along the y axis. We use $\vec{E}_1$ and $\vec{E}_2$ to denote the electric fields due to q_1 and q_2, respectively; the resultant electric field is the vector sum of these fields. The point charges are the source points, and points *a*, *b*, *c*, and *d* are the field points.

SOLVE For a point charge, the magnitude of the electric field is given by $E = k\dfrac{|q|}{r^2}$.

Part (a): The electric fields at point *a* are shown in Figure 17.19. $\vec{E}_1$ points away from q_1 (because q_1 is positive), and $\vec{E}_2$ points toward q_2 (because q_2 is negative). Thus,

$$E_1 = E_2 = k\frac{|q_1|}{r_1^2} = (8.99 \times 10^9 \text{ N}\cdot\text{m}^2/\text{C}^2)\frac{(12 \times 10^{-9} \text{ C})}{(0.050 \text{ m})^2}$$

$$= 4.32 \times 10^4 \text{ N/C}.$$

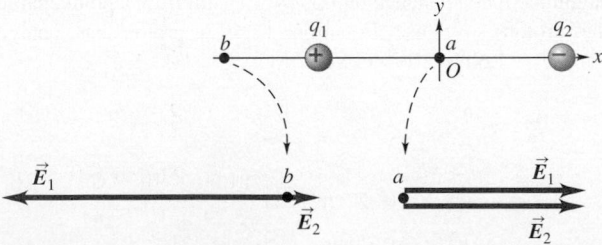

▲ **FIGURE 17.19** The electric fields due to the two charges at points *a* and *b*.

Since $\vec{E}_1$ and $\vec{E}_2$ point in the same direction, $E_{\text{total}} = E_1 + E_2 = 8.6 \times 10^4$ N/C and E_{total} points in the $+x$ direction, from the positive charge toward the negative charge.

Continued

Part (b): The electric fields at point b are shown in Figure 17.19. Again, $\vec{E}_1$ points away from q_1 and $\vec{E}_2$ points toward q_2. Hence,

$$E_1 = k\frac{|q_1|}{r_1^2} = (8.99 \times 10^9 \text{ N} \cdot \text{m}^2/\text{C}^2)\frac{(12 \times 10^{-9} \text{ C})}{(0.040 \text{ m})^2}$$
$$= 6.74 \times 10^4 \text{ N/C},$$

$$E_2 = k\frac{|q_2|}{r_2^2} = (8.99 \times 10^9 \text{ N} \cdot \text{m}^2/\text{C}^2)\frac{(12 \times 10^{-9} \text{ C})}{(0.140 \text{ m})^2}$$
$$= 5.50 \times 10^3 \text{ N/C}.$$

E_1 is larger than E_2 because point b is closer to q_1 than to q_2.

Since $\vec{E}_1$ and $\vec{E}_2$ point in opposite directions, $E_{\text{total}} = E_1 - E_2 = 6.2 \times 10^4 \text{ N/C}$. $\vec{E}_{\text{total}}$ points to the left, in the direction of the stronger field.

Part (c): At point c in Figure 17.18, the two electric fields are directed as shown in Figure 17.20a. In Figure 17.20b, each electric field is replaced by its x and y components. Point c is equidistant from the two charges, and $|q_1| = |q_2|$, so $E_1 = E_2$. The y

components of $\vec{E}_1$ and $\vec{E}_2$ are equal in magnitude and opposite in direction, and their sum is zero. The x components are equal in magnitude and are both in the $+x$ direction, so the resultant field is in the $+x$ direction.

At point d in Figure 17.18, the two electric fields are directed as shown in Figure 17.20c. The resultant field is again in the $+x$ direction. At all points along the perpendicular bisector of the line connecting the two charges, the resultant field is in the $+x$ direction, parallel to the direction from the positive charge toward the negative charge.

REFLECT Our general result in part (c) is consistent with the direction of the electric field calculated at point a. The resultant electric field has the same direction at every point along the perpendicular bisector, but the magnitude decreases at points farther from the charges.

Practice Problem: Repeat the calculations of this example, using the same value of q_1 as previously, but with $q_2 = +12$ nC (so that both charges are positive). *Answers:* (a) 0; (b) 7.3×10^4 N/C, in the $-x$ direction; (c) along the y axis and away from the charges.

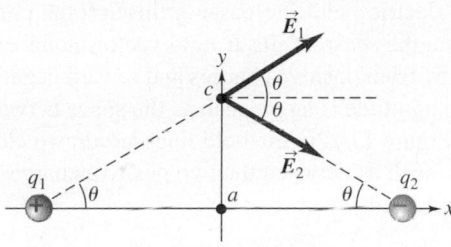

(a) The electric fields at point c

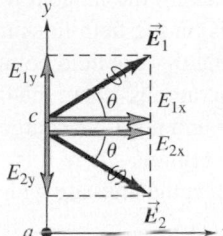

(b) The electric fields at point c and their components

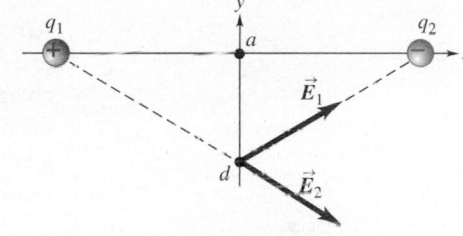

(c) The electric fields at point d

▲ **FIGURE 17.20**

17.7 Electric Field Lines

The concept of an electric field may seem rather abstract; you can't see or feel one (although some animals can). It's often useful to draw a diagram that helps to visualize electric fields at various points in space. A central element in such a diagram is the concept of **electric field lines.** An electric field line is an imaginary line drawn through a region of space so that, at every point, it is tangent to the direction of the electric field vector at that point. The basic idea is shown in Figure 17.21. Michael Faraday (1791–1867) first introduced the concept of field lines. He called them "lines of force," but the term "field lines" is preferable.

Electric field lines show the direction of $\vec{E}$ at each point, and their spacing gives a general idea of the *magnitude* of $\vec{E}$ at each point. Where $\vec{E}$ is strong, we draw lines bunched closely together; where $\vec{E}$ is weaker, they are farther apart. At any particular point, the electric field has a unique direction, so only one field line can pass through each point of the field. In other words, *field lines never intersect.*

Electric field lines always have these characteristics:

1. At every point in space, the electric field vector $\vec{E}$ at that point is tangent to the electric field line through that point.
2. Electric field lines are close together in regions where the magnitude of $\vec{E}$ is large, farther apart where it is small.
3. Field lines point away from positive charges and toward negative charges.

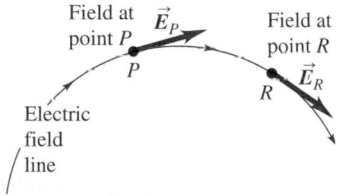

▲ **FIGURE 17.21** The direction of the electric field at any point is tangent to the field line through that point.

MasteringPHYSICS

PhET: Charges and Fields
PhET: Electric Field Hockey
PhET: Electric Field of Dreams
ActivPhysics 11.4: Electric Field: Point Charge
ActivPhysics 11.5: Electric Field Due to a Dipole
ActivPhysics 11.6: Electric Field: Problems

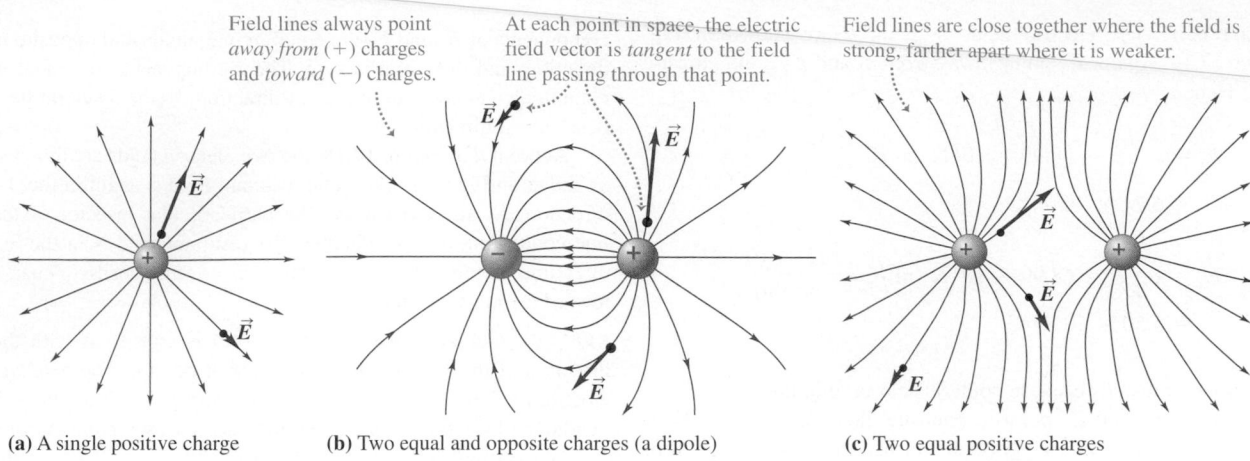

Field lines always point *away from* (+) charges and *toward* (−) charges.

At each point in space, the electric field vector is *tangent* to the field line passing through that point.

Field lines are close together where the field is strong, farther apart where it is weaker.

(a) A single positive charge   **(b)** Two equal and opposite charges (a dipole)   **(c)** Two equal positive charges

▲ **FIGURE 17.22** Electric field lines for several charge distributions.

Figure 17.22 shows some of the electric field lines in a plane containing (a) a single positive charge, (b) two equal charges, one positive and one negative (a dipole), and (c) two equal positive charges. These are cross sections of the actual three-dimensional patterns. The direction of the total electric field at every point in each diagram is along the tangent to the electric field line passing through the point. Arrowheads on the field lines indicate the sense of the $\vec{E}$ field vector along each line (showing that the field points away from positive charges and toward negative charges). In regions where the field magnitude is large, such as the space between the positive and negative charges in Figure 17.22b, the field lines are drawn close together. In regions where it is small, such as between the two positive charges in Figure 17.22c, the lines are widely separated.

NOTE ▶ There may be a temptation to think that when a charged particle moves in an electric field, its path always follows a field line. Resist that temptation; the thought is erroneous. The direction of a field line at a given point determines the direction of the particle's *acceleration,* not its velocity. We've seen several examples of motion in which the velocity and acceleration vectors have different directions. ◀

Parallel-Plate Capacitor

In a *uniform* electric field, the field lines are straight, parallel, and uniformly spaced, as in Figure 17.23. When two conducting sheets carry opposite charges and are close together compared with their size, the electric field in the region between them is approximately uniform. This arrangement is often used when a uniform electric field is needed, as in setups to deflect electron beams. A similar configuration of conductors, consisting of two sheets separated by a thin insulating layer, forms a device called a **parallel-plate capacitor,** which is widely used in electronic circuits and which we'll study in the next chapter.

17.8 Gauss's Law and Field Calculations

Gauss's law is an alternative formulation of the principles of electrostatics. It is logically equivalent to Coulomb's law, but for some problems it provides a useful alternative approach to calculating electric fields. Coulomb's law enables us to find the field at a *point P* caused by a single *point* charge *q*. To calculate fields produced by an *extended* charge distribution, we have to represent that distribution as an assembly of point charges and use the superposition

Between the plates of the capacitor, the electric field is nearly uniform, pointing from the positive plate toward the negative one.

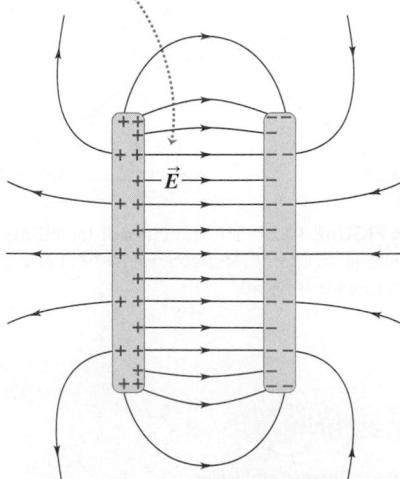

▲ **FIGURE 17.23** The electric field produced by a parallel-plate capacitor (seen in cross section). Between the plates, the field is nearly uniform.

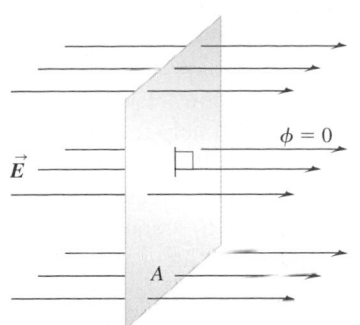

$\phi = 0$

$\vec{E}$

A

Electric field $\vec{E}$ is perpendicular to area A; the angle between $\vec{E}$ and a line perpendicular to the surface is zero.
The flux is $\Phi_E = EA$.

Area A is tilted at an angle ϕ from the perpendicular to $\vec{E}$.
The flux is $\Phi_E = EA \cos \phi$.

Area A is parallel to $\vec{E}$ (tilted at 90° from the perpendicular to $\vec{E}$).
The flux is $\Phi_E = EA \cos 90° = 0$.

▲ **FIGURE 17.24** The electric flux through a flat surface at various orientations relative to a uniform electric field.

principle. Gauss's law takes a more global view. Given any general distribution of charge, we surround it with an imaginary closed surface (often called a *Gaussian surface*) that encloses the charge. Then we look at the electric field at various points on this imaginary surface. Gauss's law is a relation between the field at *all* the points on the surface and the total charge enclosed within the surface.

Gauss's law is part of the key to using symmetry considerations in electric-field calculations. Calculations with a system that has symmetry properties can nearly always be simplified if we can make use of the symmetry, and Gauss's law helps us do just that.

Electric Flux

In formulating Gauss's law, we'll use the concept of **electric flux,** also called *flux of the electric field.* We'll define this concept first, and then we'll discuss an analogy with fluid flow that will help you to develop intuition about it.

The definition of electric flux involves an area A and the electric field at various points in the area. The area needn't be the surface of a real object; in fact, it will usually be an imaginary area in space. Consider first a small, flat area A perpendicular to a uniform electric field $\vec{E}$ (Figure 17.24a). We denote electric flux by Φ_E; we define the electric flux Φ_E through the area A to be the product of the magnitude E of the electric field and the area A:

$$\Phi_E = EA.$$

Roughly speaking, we can picture Φ_E in terms of the number of field lines that pass through A. More area means more lines through the area, and a stronger field means more closely spaced lines and therefore more lines per unit area.

If the area element A isn't perpendicular to the field $\vec{E}$, then fewer field lines pass through it. In this case, what counts is the area of the silhouette of A that we see as we look along the direction of $\vec{E}$; this is the area $A_{\perp}$ in Figure 17.24b, the *projection* of the area A onto a surface perpendicular to $\vec{E}$. Two sides of the projected rectangle have the same length as the original one, but the other two are foreshortened by a factor $\cos\phi$; so the projected area $A_{\perp}$ is equal to $A\cos\phi$. We generalize our definition of electric flux for a uniform electric field to

$$\Phi_E = EA\cos\phi. \tag{17.5}$$

Thus, $E\cos\phi$ is the component of the vector $\vec{E}$ perpendicular to the area. Calling this component $E_{\perp}$, we can rewrite Equation 17.5 as

$$\Phi_E = E_{\perp}A. \tag{17.6}$$

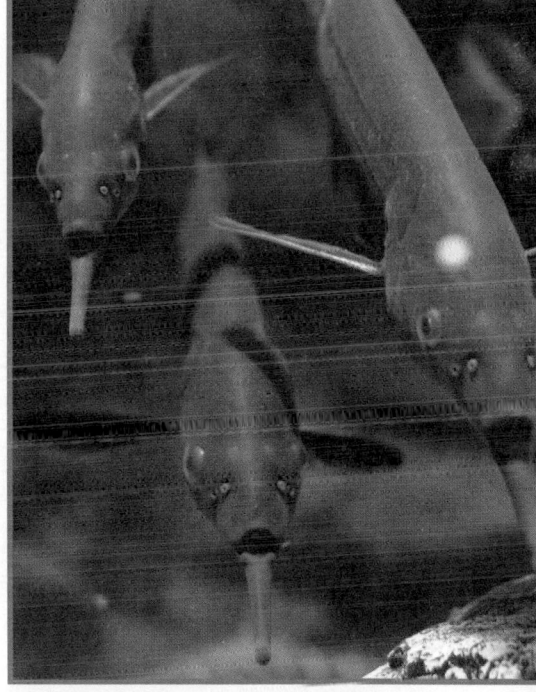

▲ **BIO Application Feeling my way.** These African elephant-nose fish "feel" their way through their murky freshwater environment by producing an electric field and sensing how objects distort the field. The field is produced in pulses by an electric organ near each fish's tail; it is detected by receptors covering portions of the fish's skin. An object that conducts electricity better than fresh water, such as an animal or a plant, causes the nearby field lines to bunch together, creating a spot of stronger field on the fish's skin. An object that conducts less well than water, such as a rock, causes the field lines to spread apart, which the fish sense as a spot of weaker field. By integrating the information from their receptors, the fish can perceive their surroundings. Several groups of fish generate and use electric fields in this way.

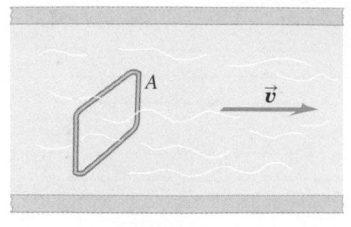

(a)

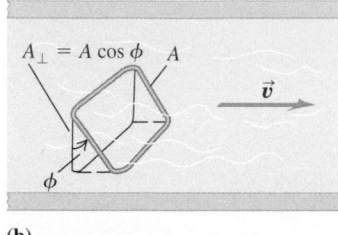

(b)

◀ **FIGURE 17.25** The volume flow rate of water through the wire rectangle is $vA\cos\phi$, just as the electric flux through an area A is $EA\cos\phi$.

The word *flux* comes from a Latin word meaning "flow." Even though an electric field is *not* a flow, an analogy with fluid flow will help to develop your intuition about electric flux. Imagine that $\vec{E}$ is analogous to the velocity of flow $\vec{v}$ of water through the imaginary area bounded by the wire rectangle in Figure 17.25. The flow rate through the area A is proportional to A and v, and it also depends on the angle between $\vec{v}$ and a line perpendicular to the plane of the rectangle. When the area is perpendicular to the flow velocity $\vec{v}$ (Figure 17.25a), the volume flow rate is just vA. When the rectangle is tilted at an angle ϕ (Figure 17.25b), the area that counts is the silhouette area that we see when looking in the direction of $\vec{v}$. That area is $A\cos\phi$, as shown, and the volume flow rate through A is $vA\cos\phi$. This quantity is called the *flux* of $\vec{v}$ through the area A; *flux* is a natural term because it represents the volume rate of flow of fluid through the area. In the electric-field situation, *nothing is flowing,* but the analogy to fluid flow may help you to visualize the concept.

EXAMPLE 17.8 Electric flux through a disk

A disk with radius 0.10 m is oriented with its axis (the line through the center, perpendicular to the disk's surface) at an angle of 30° to a uniform electric field $\vec{E}$ with magnitude 2.0×10^3 N/C (Figure 17.26). **(a)** What is the total electric flux through the disk? **(b)** What is the total flux through the disk if it is turned so that its plane is parallel to $\vec{E}$? **(c)** What is the total flux through the disk if it is turned so that its axis (marked by the dashed line perpendicular to the disk in the figure) is parallel to $\vec{E}$?

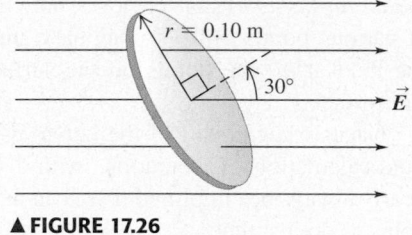

▲ **FIGURE 17.26**

SOLUTION

SET UP AND SOLVE Part (a): The area is $A = \pi(0.10 \text{ m})^2 = 0.0314 \text{ m}^2$. From Equation 17.5,

$$\Phi_E = EA\cos\phi = (2.0 \times 10^3 \text{ N/C})(0.0314 \text{ m}^2)(\cos 30°)$$
$$= 54 \text{ N} \cdot \text{m}^2/\text{C}.$$

Part (b): The axis of the disk is now perpendicular to $\vec{E}$, so $\phi = 90°$, $\cos\phi = 0$, and $\Phi_E = 0$.

Part (c): The axis of the disk is parallel to $\vec{E}$, so $\phi = 0$, $\cos\phi = 1$, and, from Equation 17.5,

$$\Phi_E = EA\cos\phi = (2.0 \times 10^3 \text{ N/C})(0.0314 \text{ m}^2)(1)$$
$$= 63 \text{ N} \cdot \text{m}^2/\text{C}.$$

REFLECT The flux through the disk is greatest when its axis is parallel to $\vec{E}$, and it is zero when $\vec{E}$ lies in the plane of the disk. That is, it is greatest when the most electric field lines pass through the disk, and it is zero when no lines pass through it.

Practice Problem: What is the flux through the disk if its axis makes an angle of 45° with $\vec{E}$? *Answer:* 44 N · m²/C.

EXAMPLE 17.9 Electric flux through a sphere

A positive point charge with magnitude 3.0 μC is placed at the center of a sphere with radius 0.20 m (Figure 17.27). Find the electric flux through the sphere due to this charge.

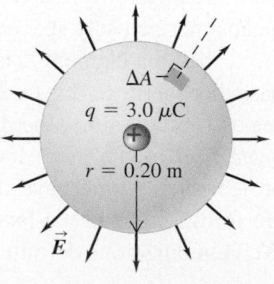

▶ **FIGURE 17.27**

Continued

SOLUTION

SET UP AND SOLVE At any point on the sphere, the magnitude of $\vec{E}$ is

$$E = \frac{kq}{r^2} = \frac{(8.99 \times 10^9 \text{ N} \cdot \text{m}^2/\text{C}^2)(3.0 \times 10^{-6} \text{ C})}{(0.20 \text{ m})^2}$$
$$= 6.75 \times 10^5 \text{ N/C}.$$

From symmetry, the field is perpendicular to the spherical surface at every point (so that $E_\perp = E$), and it has the same magnitude at every point. The flux through any area element ΔA on the sphere is just $E \, \Delta A$, and the flux through the entire surface is E times the total surface area $A = 4\pi r^2$. Thus, the total flux coming out of the sphere is

$$\Phi_E = EA = (6.75 \times 10^5 \text{ N/C})(4\pi)(0.20 \text{ m})^2$$
$$= 3.4 \times 10^5 \text{ N} \cdot \text{m}^2/\text{C}.$$

REFLECT The symmetry of the sphere plays an essential role in this calculation. We made use of the facts that E has the same value at every point on the surface and that at every point $\vec{E}$ is perpendicular to the surface.

Practice Problem: Repeat this calculation for the same charge, but a radius of 0.10 m. You should find that the result is the same as the one you obtained previously. We would have obtained the same result with a sphere of radius 2.0 m or 200 m. There's a good physical reason for this, as we'll soon see. *Answer:* $3.4 \times 10^5 \text{ N} \cdot \text{m}^2/\text{C}$

Gauss's Law

Gauss's law is an alternative to Coulomb's law for expressing the relationship between electric charge and electric field. It was formulated by Karl Friedrich Gauss (1777–1855), one of the greatest mathematicians of all time. Many areas of mathematics, from number theory and geometry to the theory of differential equations, bear the mark of his influence, and he made equally significant contributions to theoretical physics.

Gauss's law

The total electric flux Φ_E coming out of any closed surface (that is, a surface enclosing a definite volume) is proportional to the total (net) electric charge Q_{encl} inside the surface, according to the relation

$$\sum E_\perp \, \Delta A = 4\pi k Q_{encl}. \qquad (17.7)$$

The sum on the left side of this equation represents the operations of dividing the enclosing surface into small elements of area ΔA, computing $E_\perp \, \Delta A$ for each one, and adding all these products.

To develop Gauss's law, we'll start with the field due to a single positive point charge q. The field lines radiate out equally in all directions. We place this charge at the center of an imaginary spherical surface with radius R. The magnitude E of the electric field at every point on the surface is given by

$$E = k\frac{q}{R^2}.$$

At each point on the surface, $\vec{E}$ is perpendicular to the surface, and its magnitude is the same at every point, just as in Example 17.9. The total electric flux is just the product of the field magnitude E and the total area $A = 4\pi R^2$ of the sphere:

$$\Phi_E = EA = k\frac{q}{R^2}(4\pi R^2) = 4\pi kq. \qquad \text{(spherical surface)} \qquad (17.8)$$

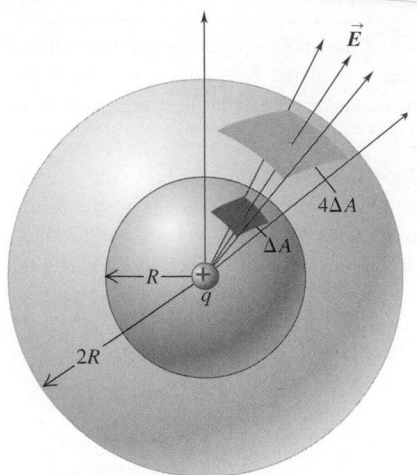

▲ FIGURE 17.28 Projection of an element of area ΔA, on a spherical surface of radius R, onto a sphere of radius $2R$. The projection multiplies each linear dimension by two, so the area element on the larger sphere is $4\Delta A$. The same number of field lines and the same flux pass through each area element.

We see that *the flux is independent of the radius R of the sphere.* It depends only on the charge q enclosed by the sphere.

We can also interpret this result in terms of field lines. We consider two spheres with radii R and $2R$, respectively (Figure 17.28). According to Coulomb's law, the field magnitude is one-fourth as great on the larger sphere as on the smaller, so the number of field lines per unit area should be one-fourth as great. But the area of the larger sphere is four times as great, so the *total* number of field lines passing through is the same for both spheres.

We've derived Equation 17.8 only for spherical surfaces, but we can generalize it to *any* closed surface surrounding an electric charge. We imagine the surface as being divided into small elements of area ΔA. If the electric field $\vec{E}$ is perpendicular to a particular element of area, then the number of field lines passing through that area is proportional to $E \Delta A$—that is, to the flux through A. If $\vec{E}$ is not perpendicular to the given element of area, we take the component of $\vec{E}$ perpendicular to ΔA; we call this component $E_\perp$, as before. Then the number of lines passing through ΔA is proportional to $E_\perp \Delta A$. (We don't consider the component of $\vec{E}$ *parallel* to the surface, because it doesn't correspond to any lines passing *through* the surface.)

To get the *total* number of field lines passing outward through the surface, we add up all the products $E_\perp \Delta A$ for all the surface elements that together make up the whole surface. This sum is the total flux through the entire surface. The total number of field lines passing through the surface is the same as that for the spherical surfaces we have discussed. Therefore, this sum is again equal to $4\pi kq$, just as in Equation 17.8, and our generalized relation is

$$\Phi_E = \sum E_\perp \Delta A = 4\pi kq. \quad \text{(for any closed surface)} \quad (17.9)$$

There is one further detail: We have to keep track of which lines point *into* the surface and which ones point *out;* we may have both types in some problems. Let's agree that $E_\perp$ and Φ_E are positive when the vector $\vec{E}$ has a component pointing *out of* the surface and negative when the component points *into* the surface.

Here's a further generalization: Suppose the surface encloses not just one point charge q, but several charges $q_1, q_2, q_3, \cdots$. Then the total (resultant) electric field $\vec{E}$ at any point is the vector sum of the $\vec{E}$ fields of the individual charges. Let $Q_{encl} = q_1 + q_2 + q_3 + \cdots$ be the *total* charge enclosed by the surface, and let $E_\perp$ be the component of the *total* field perpendicular to ΔA. Then the general statement of Gauss's law is

$$\sum E_\perp \Delta A = 4\pi k Q_{encl}. \quad (17.10)$$

Gauss's law is usually written in terms of the constant ϵ_0 we introduced in Section 17.4, defined by the relation $k = 1/4\pi\epsilon_0$. In terms of ϵ_0,

$$\sum E_\perp \Delta A = \frac{Q_{encl}}{\epsilon_0}. \quad \text{(for any closed surface)} \quad (17.11)$$

In Equations 17.7, 17.10, and 17.11, Q_{encl} is always the algebraic sum of all the (positive and negative) charges enclosed by the surface, and $\vec{E}$ is the *total* field at each point on the surface. Also, note that this field is in general caused partly by charges inside the surface and partly by charges outside. The outside charges don't contribute to the total (net) flux through the surface, so Equation 17.11 is still correct even when there are additional charges outside the surface that contribute to the electric field at the surface. When $Q_{encl} = 0$, the total flux through the surface must be zero, even though some areas may have positive flux and others negative.

NOTE ▶ The surface used for applications of Gauss's law need not be a real physical surface; in fact, it is usually an imaginary surface, enclosing a definite volume and a definite quantity of electric charge. ◀

EXAMPLE 17.10 **Field due to a spherical shell of charge**

A positive charge q is spread uniformly over a thin spherical shell of radius R (Figure 17.29) Find the electric field at points inside and outside the shell.

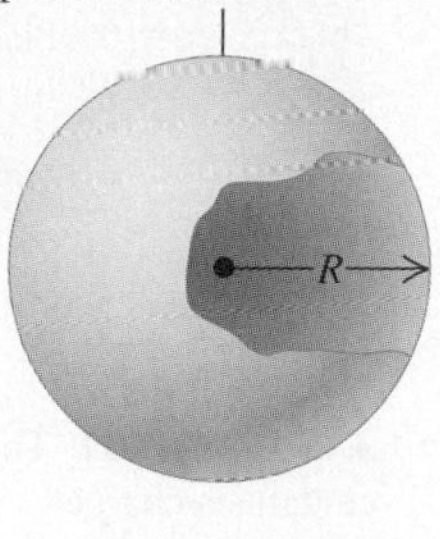

Thin spherical shell with total charge q

▶ **FIGURE 17.29**

SOLUTION

SET UP The system is spherically symmetric. This means that it is unchanged if we rotate it through any angle about an axis through its center. The field pattern of the rotated system must be identical to that of the original system. If the field had a component at some point that was perpendicular to the radial direction, that component would have to be different after at least some rotations. Thus, there can't be such a component, and the field must be radial.

We conclude that at every point outside the shell, the electric field due to the charge on the shell must be along a radial line—that is, along a line from the center of the shell to the field point. For the same reason, the magnitude E of the field depends only on the distance r from the center. Thus, the magnitude E is the same at all points on a spherical surface with radius r, concentric with the conductor.

SOLVE Because of the spherical symmetry, we take our Gaussian surface to be an imaginary sphere with radius r and concentric with the shell. We'll locate this surface first inside and then outside the shell of charge.

Inside the shell $(r < R)$**:** The Gaussian surface has area $4\pi r^2$. Since, by symmetry, the electric field is uniform over the Gaussian sphere and perpendicular to it at each point, the electric flux is $\Phi_E = EA = E(4\pi r^2)$. The Gaussian surface is inside the shell and encloses none of the charge on the shell, so $Q_{encl} = 0$.

Gauss's law $\Phi_E = Q_{encl}/\epsilon_0$ then says that $\Phi_E = E(4\pi r^2) = 0$, so $E = 0$. The electric field is zero at all points inside the shell.

Outside the shell $(r > R)$**:** Again, $\Phi_E = E(4\pi r^2)$. But now all of the shell is inside the Gaussian surface, so $Q_{encl} = q$. Gauss's law $\Phi_E = Q_{encl}/\epsilon_0$ then gives $E(4\pi r^2) = q/\epsilon_0$, and it follows that

$$E = \frac{q}{4\pi\epsilon_0 r^2} = k\frac{q}{r^2}.$$

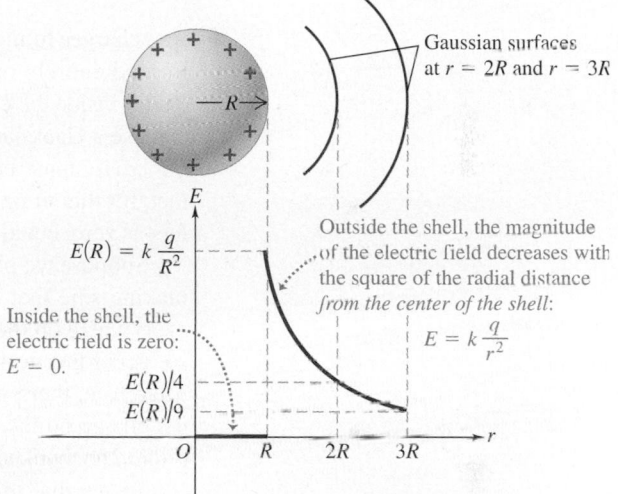

▲ FIGURE 17.30 The electric field of a charged spherical shell as a function of distance from the center. Outside the sphere, the field is the same as though the sphere's charge were all located at the center of the sphere.

REFLECT Figure 17.30 shows a graph of the field magnitude E as a function of r. The electric field is zero at all points inside the shell. At points outside the shell, the field drops off as $1/r^2$. Note that the magnitude of the electric field due to a point charge q is $E = kq/r^2$, so at points outside the shell the field is the same as if all the charge were concentrated at the center of the shell.

Practice Problem: What total charge q must be distributed uniformly over a spherical shell of radius $R = 0.50$ m to produce an electric field with magnitude 680 N/C at a point just outside the surface of the shell? *Answer:* ±19 nC.

17.9 Charges on Conductors

Early in our discussion of electric fields, we made the point that in an electrostatic situation (where there is no net motion of charge) the electric field at every point within a conductor is zero. (If it were not, the field would cause the conductor's

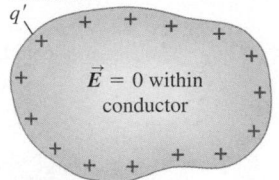

The charge q' is distributed over the surface of the conductor. The situation is electrostatic, so $\vec{E} = 0$ within the conductor.

(a) Solid conductor with charge q'

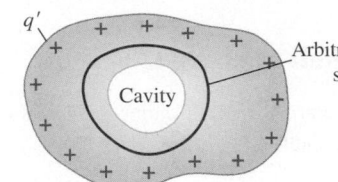

Because $\vec{E} = 0$ at all points within the conductor, the electric field at all points on the Gaussian surface must be zero.

Arbitrary Gaussian surface A

(b) The same conductor with an internal cavity

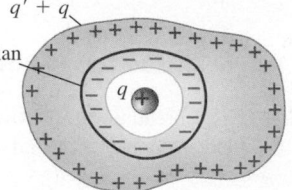

For $\vec{E}$ to be zero at all points on the Gaussian surface, the surface of the cavity must have a total charge $-q$.

(c) An isolated charge q is placed in the cavity

▲ **FIGURE 17.31** The charge on a solid conductor, on a conductor with a cavity, and on a conductor with a cavity that contains a charge.

free charges to move.) We've also learned that the charge on a solid conductor is located entirely on its surface, as shown in Figure 17.31a. But what if there is a cavity inside the conductor (Figure 17.31b)? If there is no charge in the cavity, we can use a Gaussian surface such as A to show that the net charge on the surface *of the cavity* must be zero because $\vec{E} = 0$ everywhere on the Gaussian surface. In fact, for this situation, we can prove not only that the *total* charge on the cavity surface is zero, but also that there can't be any charge *anywhere* on the cavity surface.

Suppose we place a small object with a charge q inside a cavity in a conductor, making sure that it does not touch the conductor (Figure 17.31c). Again, $\vec{E} = 0$ everywhere on the Gaussian surface A (because the situation is still electrostatic), so, according to Gauss's law, the *total* charge inside this surface must be zero. Therefore, there must be a total charge $-q$ on the cavity surface. Of course, the *net* charge on the conductor (counting both the inner and the outer surface) must remain unchanged, so a charge $+q$ must appear on its outer surface.

To see that this charge must be on the outer surface and not in the material, imagine first shrinking surface A so that it's just barely bigger than the cavity. The field everywhere on A is still zero, so, according to Gauss's law, the total charge inside A is zero. Now let surface A expand until it is just inside the outer surface of the conductor. The field is still zero everywhere on surface A, so the total charge enclosed is still zero. We have not enclosed any additional charge by expanding surface A; therefore, there must be no charge in the interior of the material. We conclude that the charge $+q$ must appear on the outer surface. By the same reasoning, if the conductor originally had a charge q', then the total charge on the outer surface after the charge q is inserted into the cavity must be $q + q'$.

EXAMPLE 17.11 Location of net charge on conductors

A hollow conductor carries a net charge of $+7$ nC. In its cavity, insulated from the conductor, is a small, isolated object with a net charge of -5 nC. How much charge is on the outer surface of the hollow conductor? How much is on the wall of the cavity?

SOLUTION

SET UP Figure 17.32 shows our sketch. We know that in this electrostatic situation the electric field in the conducting material must be zero. We draw a Gaussian surface within the material of the conductor. and apply Gauss's law.

SOLVE We apply Gauss's law $\Phi_E = Q_{encl}/\epsilon_0$ to the Gaussian surface shown in Figure 17.32. The Gaussian surface lies within the conducting material, so $E = 0$ everywhere on that surface. By Gauss's law, $\Phi_E = Q_{encl}/\epsilon_0$. Thus, $\Phi_E = 0$, so $Q_{encl} = 0$. But then, in order to have $Q_{encl} = 0$, there must be a charge of $+5$ nC

Continued

on the inner surface of the cavity, to cancel the charge in the cavity. The conductor carries a total charge of $+7$ nC, and all of its net charge is on its surfaces. So, if there is $+5$ nC on the inner surface, the remaining $+2$ nC must be on the outer surface, as shown in our sketch.

REFLECT Field lines pass between the $+5$ nC on the inner surface of the cavity and the -5 nC on the object in the cavity. Each field line going to the -5 nC charge originated on the $+5$ nC charge; the field lines don't continue into the conducting material, since $E = 0$ there. There is an electric field outside the conductor, due to the $+2$ nC on its surface.

Practice Problem: Repeat this example for the case where the conductor has a net charge of $+3$ nC. *Answers:* inner surface: $+5$ nC; outer surface: -2 nC.

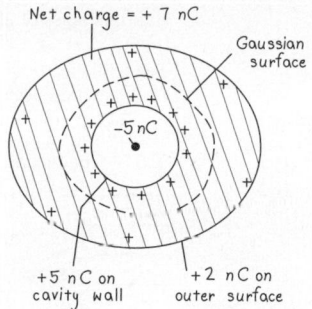

▲ **FIGURE 17.32** Our sketch for this problem.

Faraday Ice Pail

We can now consider a historic experiment, shown in Figure 17.33a. We mount a conducting container, such as a metal pail with a lid, on an insulating stand. The container is initially uncharged. Then we hang a charged metal ball from an insulating thread, lower it into the pail, and put the lid on (Figure 17.33b). Charges are induced on the walls of the container as shown. But now we let the ball *touch* the inner wall (Figure 17.33c). The surface of the ball becomes, in effect, part of the cavity surface. The situation is now the same as Figure 17.31b; if Gauss's law is correct, the net charge on this surface must be zero. Thus, the ball must lose all its charge. Finally, we pull the ball out, to find that it has indeed lost all its charge.

This experiment was performed by Michael Faraday, using a metal ice pail with a lid, and it is called **Faraday's ice-pail experiment.** (Similar experiments had been carried out earlier by Benjamin Franklin and Joseph Priestley, although with much less precision.) The result confirms the validity of Gauss's law and therefore of Coulomb's law. Faraday's result was significant because Coulomb's experimental method, using a torsion balance and dividing the charges, was not very precise. It is quite difficult to confirm the $1/r^2$ dependence of the electrostatic force with great precision by direct force measurements. Faraday's experiment tests the validity of Gauss's law, and therefore of Coulomb's law, with potentially much greater precision.

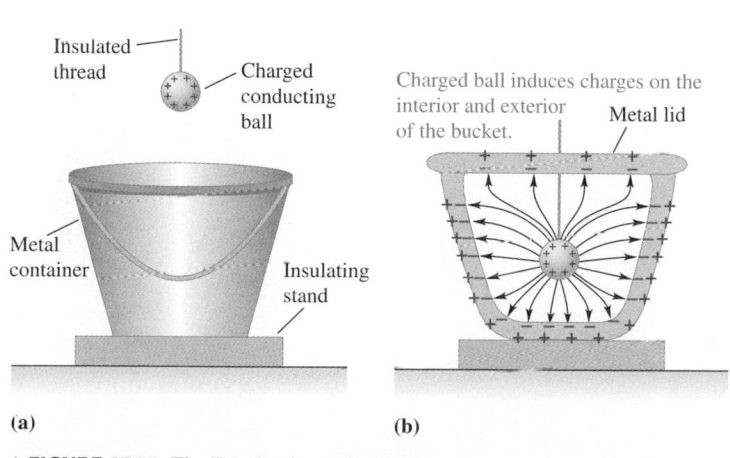

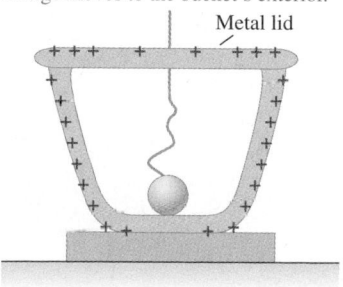

▲ **FIGURE 17.33** The Faraday ice-pail experiment.

The field induces charges on the left
and right sides of the conducting box.

The total electric field inside the box is
zero; the presence of the box distorts
the field in adjacent regions.

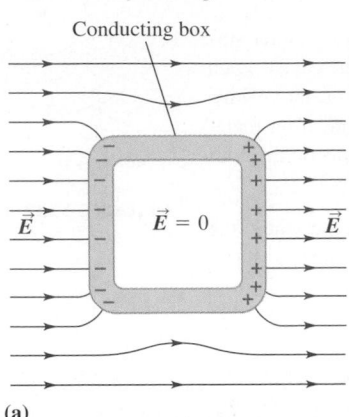

Conducting box

$\vec{E}$ $\vec{E} = 0$ $\vec{E}$

(a)

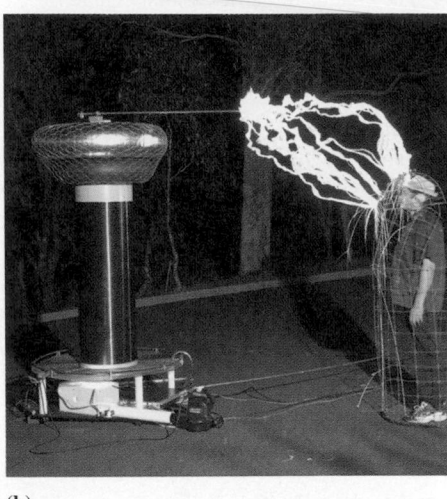

(b)

▲ **FIGURE 17.34** (a) The effect of putting a conducting box (an electrostatic shield) in a uniform electric field. (b) The conducting cage keeps the operator of this exhibit perfectly safe.

▲ **Application A Faraday cage when you need one.** If you find yourself in a thunderstorm while driving, *stay in your car*. If it gets hit by lightning, it will act as a Faraday cage and keep you safe.

Electrostatic Shielding

Suppose we have a highly sensitive electronic instrument that we want to protect from stray electric fields that might give erroneous measurements. We surround the instrument with a conducting box, or we line the walls, floor, and ceiling of the room with a conducting material such as sheet copper. The external electric field redistributes the free electrons in the conductor, leaving a net positive charge on the outer surface in some regions and a net negative charge in others (Figure 17.34). This charge distribution causes an additional electric field such that the *total* field at every point inside the box is zero, as Gauss's law says it must be. The charge distribution on the box also alters the shapes of the field lines near the box, as the figure shows. Such a setup is often called a *Faraday cage.*

SUMMARY

Electric Charge; Conductors and Insulators

(Sections 17.1–17.3) The fundamental entity in electrostatics is electric charge. There are two kinds of charge: positive and negative. Like charges repel each other; unlike charges attract. **Conductors** are materials that permit electric charge to move within them. **Insulators** permit charge to move much less readily. Most metals are good conductors; most nonmetals are insulators.

All ordinary matter is made of atoms consisting of protons, neutrons, and electrons. The protons and neutrons form the nucleus of the atom; the electrons surround the nucleus at distances much greater than its size. Electrical interactions are chiefly responsible for the structure of atoms, molecules, and solids.

Electric charge is conserved: It can be transferred between objects, but isolated charges cannot be created or destroyed. Electric charge is quantized: Every amount of observable charge is an integer multiple of the charge of an electron or proton.

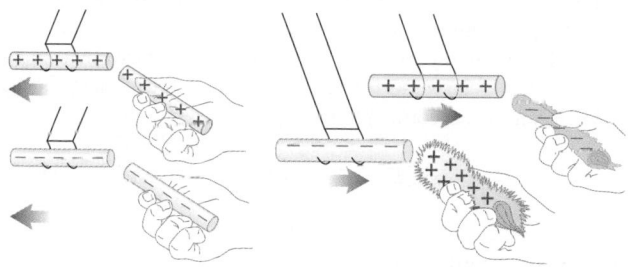

Like charges repel.   Unlike charges attract.

Coulomb's Law

(Section 17.4) Coulomb's law is the basic law of interaction for point electric charges. For point charges q_1 and q_2 separated by a distance r, the magnitude F of the force each charge exerts on the other is

$$F = k\frac{|q_1 q_2|}{r^2}. \qquad (17.1)$$

The force on each charge acts along the line joining the two charges. It is repulsive if q_1 and q_2 have the same sign, attractive if they have opposite signs. The forces form an action–reaction pair and obey Newton's third law.

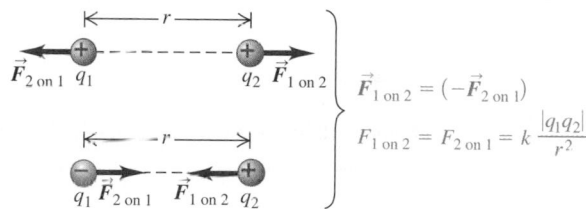

$$\vec{F}_{1 \text{ on } 2} = (-\vec{F}_{2 \text{ on } 1})$$

$$F_{1 \text{ on } 2} = F_{2 \text{ on } 1} = k\,\frac{|q_1 q_2|}{r^2}$$

Electric Field and Electric Forces

(Sections 17.5 and 17.6) Electric field, a vector quantity, is the force per unit charge exerted on a test charge at any point, provided that the test charge is small enough that it does not disturb the charges that cause the field. The principle of superposition states that the electric field due to any combination of charges is the vector sum of the fields caused by the individual charges. From Coulomb's law, the magnitude of the electric field produced by a point charge is

$$E = k\frac{|q|}{r^2}. \qquad (17.4)$$

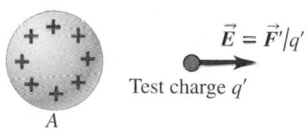

$$\vec{E} = \vec{F}'/q'$$

Test charge q'

Electric Field Lines

(Section 17.7) Field lines provide a graphical representation of electric fields. A field line at any point in space is tangent to the direction of $\vec{E}$ at that point, and the number of lines per unit area (perpendicular to their direction) is proportional to the magnitude of $\vec{E}$ at the point. Field lines point away from positive charges and toward negative charges.

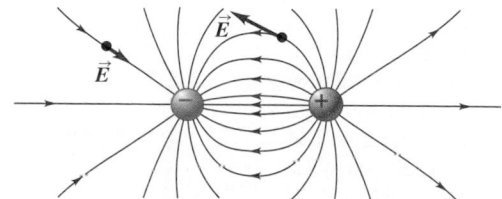

Continued

Gauss's Law

(Section 17.8) For a uniform electric field with component $E_\perp$ perpendicular to area A, the **electric flux** through the area is $\Phi_E = E_\perp A$ (Equation 17.6). **Gauss's law** states that the total electric flux Φ_E out of any closed surface (that is, a surface enclosing a definite volume) is proportional to the total electric charge Q_{encl} inside the surface, according to the relation

$$\sum E_\perp \, \Delta A = 4\pi k Q_{encl}. \qquad (17.7)$$

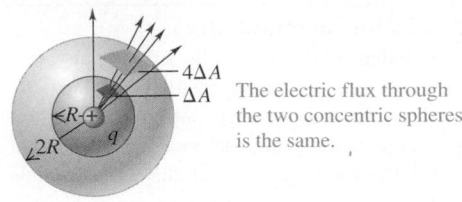

The electric flux through the two concentric spheres is the same.

Charges on Conductors

(Section 17.9) In a static configuration with no net motion of charge, the electric field is always zero within a conductor. The charge on a solid conductor is located entirely on its outer surface. If there is a cavity containing a charge $+q$ within the conductor, the surface of the cavity has a total induced charge $-q$.

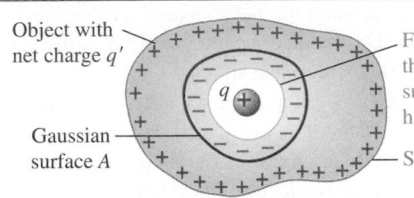

Object with net charge q'

Gaussian surface A

For $\vec{E}$ to remain zero across the Gaussian surface, the surface of the cavity must have a charge $-q$.

Surface charge $= q' + q$.

 For instructor-assigned homework, go to www.masteringphysics.com

Conceptual Questions

1. Bits of paper are attracted to an electrified comb or rod even though they have no net charge. How is this possible?

2. When you walk across a nylon rug and then touch a large metal object, you may get a spark and a shock. What causes this to happen?

3. What similarities does the electrical force have to the gravitational force? What are the most significant differences?

4. In a common physics demonstration, a rubber rod is first rubbed vigorously on silk or fur. It is then brought close to a small Styrofoam™ ball, which it attracts. If you then touch the ball with the rod, it suddenly repels the ball. Why does it first attract the ball, and why does it then repel the same ball?

5. How do we know that protons have positive charge and electrons have negative charge, rather than the reverse? Is there anything *inherently* positive about the proton's charge or inherently negative about the electron's charge?

6. Gasoline transport trucks sometimes have chains that hang down and drag on the road at the rear end of the truck. What are the chains for and how do they work?

7. A gold leaf electroscope, which is often used in physics demonstrations, consists of a metal tube with a metal ball at the top and a sheet of extremely thin gold leaf fastened at the other end. (See Fig. 17.35.) The gold leaf is attached in such a way that it can pivot about its upper edge. (a) If a charged rod is brought close to (but does not touch) the ball at the top, the gold leaf pivots outward, away from the

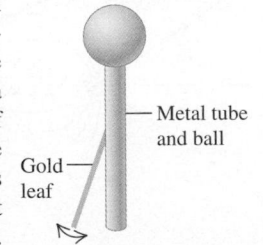

Metal tube and ball

Gold leaf

▲ **FIGURE 17.35** Question 7.

tube. Why? (b) What will the gold leaf do when the charged rod is removed? Why? (c) Suppose that the charged rod touches the metal ball for a second or so. What will the gold leaf do when the rod is removed in this case? Why?

8. Show how it is possible for *neutral* objects to attract each other electrically.

9. Suppose the disk in Example 17.8, instead of having its normal vector oriented at just two or three particular angles to the electric field, began to rotate continuously, so that its normal vector was first parallel to the field, then perpendicular to it, then opposite to it, and so on. Sketch a graph of the resulting electric flux versus time, for an entire rotation of 360°.

10. Atomic nuclei are made of protons and neutrons, a fact that, by itself, shows that there must be another kind of force in addition to the electrical and gravitational forces. Explain how we know this.

11. If an electric dipole is placed in a uniform electric field, what is the net force on it? Will the same thing necessarily be true if the field is *not* uniform?

12. *Why* do electric field lines point away from positive charges and toward negative charges?

13. A lightning rod is a pointed copper rod mounted on top of a building and welded to a heavy copper cable running down into the ground. Lightning rods are used in prairie country to protect houses and barns from lightning; the lightning current runs through the copper rather than through the building. Why does it do this?

14. A rubber balloon has a single point charge in its interior. Does the electric flux through the balloon depend on whether or not it is fully inflated? Explain your reasoning.

15. Explain how the electrical force plays an important role in understanding each of the following: (a) the friction force between two objects, (b) the hardness of steel, and (c) the bonding of amino acids to form proteins.

Multiple-Choice Problems

1. Just after two identical point charges are released when they are a distance D apart in outer space, they have an acceleration a. If you release them from a distance $D/2$ instead, their acceleration will be
   A. $a/4$. B. $a/2$. C. $2a$. D. $4a$.

2. If the electric field is E at a distance d from a point charge, its magnitude will be $2E$ at a distance
   A. $d/4$. B. $d/2$. C. $d/\sqrt{2}$.
   D. $d\sqrt{2}$. E. $2d$.

3. Two *unequal* point charges are separated as shown in Figure 17.36. The electric field due to this combination of charges can be zero
   A. only in region 1.
   B. only in region 2.
   C. only in region 3.
   D. in both regions 1 and 3.

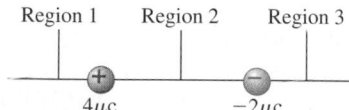

▲ **FIGURE 17.36** Multiple-choice problem 3.

4. Two protons close to each other are released from rest and are completely free to move. After being released (there may be more than one correct choice),
   A. their speeds gradually decrease to zero as they move apart.
   B. their speeds gradually increase as they move apart.
   C. their accelerations gradually decrease to zero as they move apart.
   D. their accelerations gradually increase as they move apart.

5. A spherical balloon contains a charge $+Q$ uniformly distributed over its surface. When it has a diameter D, the electric field at its surface has magnitude E. If the balloon is now blown up to twice this diameter without changing the charge, the electric field at its surface is
   A. $4E$. B. $2E$. C. E.
   D. $E/2$. E. $E/4$.

6. Two microscopic bags each contain two protons. When they are separated by a distance d, the electrical force on each bag due to the other bag is F. You now transfer a proton from one bag to another without changing anything else. The electrical force on each bag is now
   A. F. B. $\frac{3}{4}F$. C. $\frac{1}{2}F$. D. $\frac{1}{4}F$.

7. An electron is moving horizontally in a laboratory when a uniform electric field is suddenly turned on. This field points vertically downward. Which of the paths shown will the electron follow, assuming that gravity can be neglected?

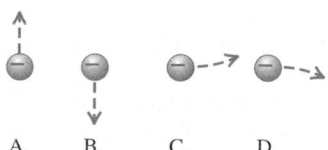

   A. B. C. D.

8. Point P in Figure 17.37 is equidistant from two point charges $\pm Q$ of equal magnitude. If a negative point charge is placed at

P without moving the original charges, the net electrical force the charges $\pm Q$ will exert on it is
   A. directly upward.
   B. directly downward.
   C. zero.
   D. directly to the right.
   E. directly to the left.

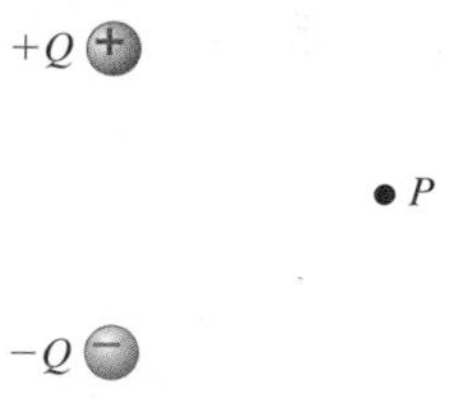

▲ **FIGURE 17.37** Multiple-choice problem 8.

9. A charge $+Q$ is suspended by a silk thread inside of a neutral metal box without touching the metal. What is true about the charge on the inner and outer surfaces of the box?
   A. The charge on both the inner and the outer surfaces is zero.
   B. The charge is $-Q$ on the inner surface and $+Q$ on the outer surface.
   C. The charge is $+Q$ on the inner surface and $-Q$ on the outer surface.
   D. The charge on both the inner and the outer surfaces is $+Q$.

10. A charge Q and a charge $3Q$ are released in a uniform electric field. If the force this field exerts on $3Q$ is F, the force it will exert on Q is
    A. F. B. $F/3$. C. $F/9$.

11. Three equal point charges are held in place as shown in Figure 17.38. If F_1 is the force on q due to Q_1 and F_2 is the force on q due to Q_2, how do F_1 and F_2 compare?
    A. $F_1 = 2F_2$. B. $F_1 = 3F_2$.
    C. $F_1 = 4F_2$. D. $F_1 = 9F_2$.

▲ **FIGURE 17.38** Multiple-choice problem 11.

12. An electric field of magnitude E is measured at a distance R from a point charge Q. If the charge is doubled to $2Q$ and the electric field is now measured at a distance of $2R$ from the charge, the new measured value of the field will be:
    A. $2E$ B. E
    C. $E/2$ D. $E/4$

13. A very small ball containing a charge $-Q$ hangs from a light string between two vertical charged plates, as shown in Figure 17.39. When released from rest, the ball will
    A. swing to the right.
    B. swing to the left.
    C. remain hanging vertically.

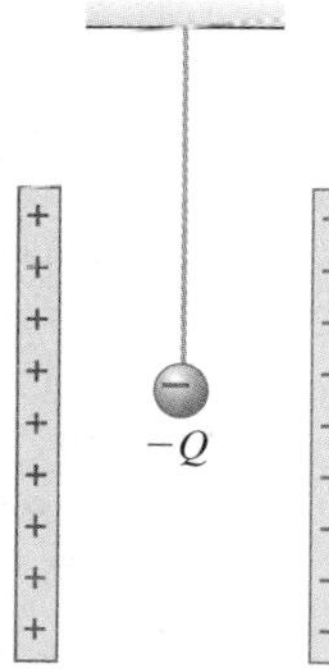

14. A point charge Q at the center of a sphere of radius R produces an electric flux of Φ_E coming out of the sphere. If the charge remains the same but the radius of the sphere is doubled, the electric flux coming out of it will be:
    A. $\Phi_E/2$ B. $\Phi_E/4$ C. $2\Phi_E$
    D. $4\Phi_E$ E. Φ_E

▲ **FIGURE 17.39** Multiple-choice problem 13.

15. Two charged small spheres are a distance R apart and exert an electrostatic force F on each other. If the distance is halved to $R/2$, the force exerted on each sphere will be
    A. $4F$. B. $2F$. C. $F/2$. D. $F/4$.

Problems

17.1 Electric Charge
17.2 Conductors and Insulators

1. • A positively charged glass rod is brought close to a *neutral* sphere that is supported on a nonconducting plastic stand as shown in Figure 17.40. Sketch the distribution of charges on the sphere if it is made of (a) aluminum, (b) nonconducting plastic.

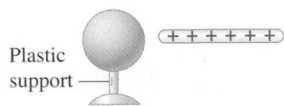
Plastic support
▲ **FIGURE 17.40** Problem 1.

2. • A positively charged rubber rod is moved close to a *neutral* copper ball that is resting on a nonconducting sheet of plastic. (a) Sketch the distribution of charges on the ball. (b) With the rod still close to the ball, a metal wire is briefly connected from the ball to the earth and then removed. After the rubber rod is also removed, sketch the distribution of charges (if any) on the copper ball.

3. • Two iron spheres contain excess charge, one positive and the other negative. (a) Show how the charges are arranged on these spheres if they are *very* far from each other. (b) If the spheres are now brought close to each other, but do not touch, sketch how the charges will be distributed on their surfaces. (c) In part (b), show how the charges would be distributed if both spheres were negative.

4. • **Electrical storms.** During an electrical storm, clouds can build up very large amounts of charge, and this charge can induce charges on the earth's surface. Sketch the distribution of charges at the earth's surface in the vicinity of a cloud if the cloud is positively charged and the earth behaves like a conductor.

17.3 Conservation and Quantization of Charge
17.4 Coulomb's Law

5. • In ordinary laboratory circuits, charges in the μC and nC range are common. How many excess electrons must you add to an object to give it a charge of (a) -2.50 μC, (b) -2.50 nC?

6. • **Signal propagation in neurons.** *Neurons* are components **BIO** of the nervous system of the body that transmit signals as electrical impulses travel along their length. These impulses propagate when charge suddenly rushes into and then out of a part of the neutron called an *axon*. Measurements have shown that, during the inflow part of this cycle, approximately 5.6×10^{11} Na$^+$ (sodium ions) per meter, each with charge $+e$, enter the axon. How many coulombs of charge enter a 1.5 cm length of the axon during this process?

7. •• **Particles in a gold ring.** You have a pure (24-karat) gold ring with mass 17.7 g. Gold has an atomic mass of 197 g/mol and an atomic number of 79. (a) How many protons are in the ring, and what is their total positive charge? (b) If the ring carries no net charge, how many electrons are in it?

8. • Two equal point charges of $+3.00 \times 10^{-6}$ C are placed 0.200 m apart. What are the magnitude and direction of the force each charge exerts on the other?

9. • At what distance would the repulsive force between two electrons have a magnitude of 2.00 N? Between two protons?

10. • A negative charge of -0.550 μC exerts an upward 0.200 N force on an unknown charge 0.300 m directly below it. (a) What is the unknown charge (magnitude and sign)? (b) What are the magnitude and direction of the force that the unknown charge exerts on the -0.550 μC charge?

11. • **Forces in an atom.** The particles in the nucleus of an atom are approximately 10^{-15} m apart, while the electrons in an atom are about 10^{-10} m from the nucleus. (a) Calculate the electrical repulsion between two protons in a nucleus if they are 1.00×10^{-15} m apart. If you were holding these protons, do you think you could feel the effect of this force? How many pounds would the force be? (b) Calculate the electrical attraction that a proton in a nucleus exerts on an orbiting electron if the two particles are 1.00×10^{-10} m apart. If you were holding the electron, do you think you could feel the effect of this force?

12. •• (a) What is the total negative charge, in coulombs, of all the electrons in a small 1.00 g sphere of carbon? One mole of C is 12.0 g, and each atom contains 6 protons and 6 electrons. (b) Suppose you could take out all the electrons and hold them in one hand, while in the other hand you hold what is left of the original sphere. If you hold your hands 1.50 m apart at arms length, what force will each of them feel? Will it be attractive or repulsive?

13. • As you walk across a synthetic-fiber rug on a cold, dry winter day, you pick up an excess charge of -55 μC. (a) How many excess electrons did you pick up? (b) What is the charge on the rug as a result of your walking across it?

14. •• Two small plastic spheres are given positive electrical charges. When they are 15.0 cm apart, the repulsive force between them has magnitude 0.220 N. What is the charge on each sphere (a) if the two charges are equal? (b) if one sphere has four times the charge of the other?

15. •• Two small aluminum spheres, each having mass 0.0250 kg, are separated by 80.0 cm. (a) How many electrons does each sphere contain? (The atomic mass of aluminum is 26.982 g/mol, and its atomic number is 13.) (b) How many electrons would have to be removed from one sphere and added to the other to cause an attractive force between the spheres of magnitude 1.00×10^{4} N (roughly 1 ton)? Assume that the spheres may be treated as point charges. (c) What fraction of all the electrons in each sphere does this represent?

16. •• Two small spheres spaced 20.0 cm apart have equal charge. How many excess electrons must be present on each sphere if the magnitude of the force of repulsion between them is 4.57×10^{-21} N?

17. •• An average human weighs about 650 N. If two such generic humans each carried 1.0 coulomb of excess charge, one positive and one negative, how far apart would they have to be for the electric attraction between them to equal their 650-N weight?

18. •• If a proton and an electron are released when they are 2.0×10^{-10} m apart (typical atomic distances), find the initial acceleration of each of them.

19. •• Three point charges are arranged on a line. Charge $q_3 = +5.00$ nC and is at the origin. Charge $q_2 = -3.00$ nC and is at $x = +4.00$ cm. Charge q_1 is at $x = +2.00$ cm. What is q_1 (magnitude and sign) if the net force on q_3 is zero?

20. •• If two electrons are each 1.50×10^{-10} m from a proton, as shown in Figure 17.41, find the magnitude and direction of the net electrical force they will exert on the proton.

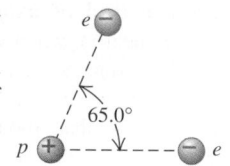

▲ **FIGURE 17.41** Problem 20.

21. •• Two point charges are located on the y axis as follows: charge $q_1 = -1.50$ nC at $y = -0.600$ m, and charge $q_2 = +3.20$ nC at the origin $(y = 0)$. What is the net force (magnitude and direction) exerted by these two charges on a third charge $q_3 = +5.00$ nC located at $y = -0.400$ m?

22. •• Two point charges are placed on the x axis as follows: Charge $q_1 = +4.00$ nC is located at $x = 0.200$ m, and charge $q_2 = +5.00$ nC is at $x = -0.300$ m. What are the magnitude and direction of the net force exerted by these two charges on a negative point charge $q_3 = -0.600$ nC placed at the origin?

23. •• Three charges are at the corners of an isosceles triangle as shown in Figure 17.42. The $+5.00\ \mu C$ charges form a dipole. (a) Find the magnitude and direction of the net force that the $-10.0\ \mu C$ charge exerts on the dipole. (b) For an axis perpendicular to the line connecting the two charges of the dipole at its midpoint and perpendicular to the plane of the paper, find the magnitude and direction of the torque exerted on the dipole by the $10.0\ \mu C$ charge.

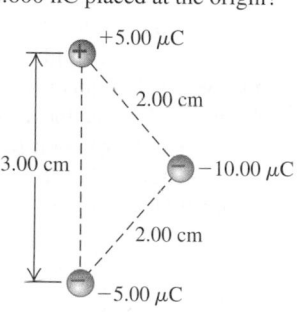

▲ **FIGURE 17.42** Problem 23.

24. •• **Base pairing in DNA, I.** The two sides of the DNA double helix are connected by pairs of bases (adenine, thymine, cytosine, and guanine). Because of the geometric shape of these molecules, adenine bonds with thymine and cytosine bonds with guanine. Figure 17.43 shows the thymine–adenine
BIO

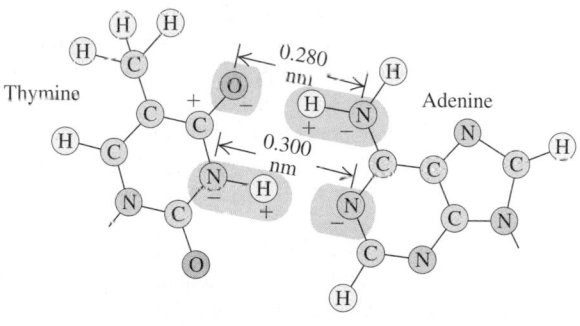

▲ **FIGURE 17.43** Problem 24.

bond. Each charge shown is $\pm e$, and the H—N distance is 0.110 nm. (a) Calculate the *net* force that thymine exerts on adenine. Is it attractive or repulsive? To keep the calculations fairly simple, yet reasonable, consider only the forces due to the O—H—N and the N—H—N combinations, assuming that these two combinations are parallel to each other. Remember, however, that in the O—H—N set, the O⁻ exerts a force on both the H⁺ and the N⁻, and likewise along

the N—H—N set. (b) Calculate the force on the electron in the hydrogen atom, which is 0.0529 nm from the proton. Then compare the strength of the bonding force of the electron in hydrogen with the bonding force of the adenine–thymine molecules.

25. •• **Base pairing in DNA, II.** Refer to the previous problem. Figure 17.44 shows the bonding of the cytosine and guanine molecules. The O—H and H—N distances are each 0.110 nm. In this case, assume that the bonding is due only to the forces along the O—H—O, N—H—N, and O—H—N combinations, and assume also that these three combinations are parallel to each other. Calculate the *net* force that cytosine exerts on guanine due to the preceding three combinations. Is this force attractive or repulsive?
BIO

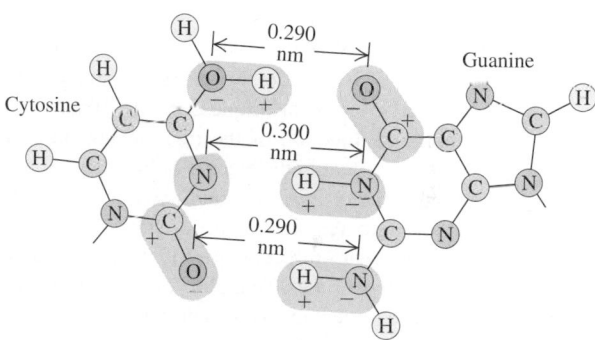

▲ **FIGURE 17.44** Problem 25.

26. •• **Surface tension.** Surface tension is the force that causes the surface of water (and other liquids) to form a "skin" that resists penetration. Because of this force, water forms into beads, and insects such as water spiders can walk on water. As we shall see, the force is electrical in nature. The surface of a polar liquid, such as water, can be viewed as a series of dipoles strung together in the stable arrangement in which the dipole moment vectors are parallel to the surface, all pointing in the same direction. Suppose now that something presses inward on the surface, distorting the dipoles as shown in Figure 17.45. Show that the two slanted dipoles exert a net upward force on the dipole between them and hence oppose the downward external force. Show also that the dipoles attract each other and thus resist being separated. Notice that the force between dipoles opposes penetration of the liquid's surface and is a simple model for surface tension.

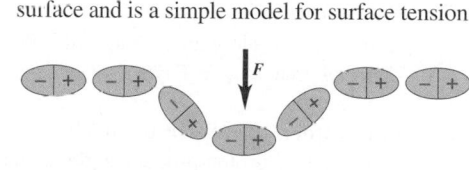

▲ **FIGURE 17.45** Problem 26.

27. •• If the central charge shown in Figure 17.46 is displaced 0.350 nm to the right while the other charges are held in place, find the magnitude and direction of the net force that the other two charges exert on it.

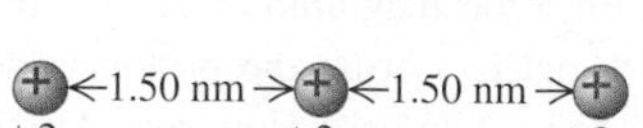

$\overset{+}{\underset{+2e}{\bigoplus}}\ \overset{1.50\ nm}{\longleftarrow}\ \overset{+}{\underset{+2e}{\bigoplus}}\ \overset{1.50\ nm}{\longleftarrow}\ \overset{+}{\underset{+2e}{\bigoplus}}$

▲ **FIGURE 17.46** Problem 27.

28. •• Two unequal charges repel each other with a force F. If both charges are doubled in magnitude, what will be the new force in terms of F?

29. •• In an experiment in space, one proton is held fixed and another proton is released from rest a distance of 2.50 mm away. (a) What is the initial acceleration of the proton after it is released? (b) Sketch qualitative (no numbers!) acceleration–time and velocity–time graphs of the released proton's motion.

30. ••• A charge $+Q$ is located at the origin and a second charge, $+4Q$, is at distance d on the x-axis. Where should a third charge, q, be placed, and what should be its sign and magnitude, so that all three charges will be in equilibrium?

17.5 Electric Field and Electric Forces

31. • A small object carrying a charge of -8.00 nC is acted upon by a downward force of 20.0 nN when placed at a certain point in an electric field. (a) What are the magnitude and direction of the electric field at the point in question? (b) What would be the magnitude and direction of the force acting on a proton placed at this same point in the electric field?

32. • (a) What must the charge (sign and magnitude) of a 1.45 g particle be for it to remain balanced against gravity when placed in a downward-directed electric field of magnitude 650 N/C? (b) What is the magnitude of an electric field in which the electric force it exerts on a proton is equal in magnitude to the proton's weight?

33. •• A uniform electric field exists in the region between two oppositely charged plane parallel plates. An electron is released from rest at the surface of the negatively charged plate and strikes the surface of the opposite plate, 3.20 cm distant from the first, in a time interval of 1.5×10^{-8} s. (a) Find the magnitude of this electric field. (b) Find the speed of the electron when it strikes the second plate.

17.6 Calculating Electric Fields

34. • A particle has a charge of -3.00 nC. (a) Find the magnitude and direction of the electric field due to this particle at a point 0.250 m directly above it. (b) At what distance from the particle does its electric field have a magnitude of 12.0 N/C?

35. • The electric field caused by a certain point charge has a magnitude of 6.50×10^3 N/C at a distance of 0.100 m from the charge. What is the magnitude of the charge?

36. • At what distance from a particle with a charge of 5.00 nC does the electric field of that charge have a magnitude of 4.00 N/C?

37. • **Electric fields in the atom.** (a) **Within the nucleus.** What strength of electric field does a proton produce at the distance of another proton, about 5.0×10^{-15} m away? (b) **At the electrons.** What strength of electric field does this proton produce at the distance of the electrons, approximately 5.0×10^{-10} m away?

38. •• A proton is traveling horizontally to the right at 4.50×10^6 m/s. (a) Find the magnitude and direction of the weakest electric field that can bring the proton uniformly to rest over a distance of 3.20 cm. (b) How much time does it take the proton to stop after entering the field? (c) What minimum field (magnitude and direction) would be needed to stop an electron under the conditions of part (a)?

39. •• **Electric field of axons.** A nerve signal is transmitted **BIO** through a neuron when an excess of Na^+ ions suddenly enters the axon, a long cylindrical part of the neuron. Axons are approximately 10.0 μm in diameter, and measurements show that about 5.6×10^{11} Na^+ ions per meter (each of charge $+e$) enter during this process. Although the axon is a long cylinder, the charge does not all enter everywhere at the same time. A plausible model would be a series of nearly point charges moving along the axon. Let us look at a 0.10 mm length of the axon and model it as a point charge. (a) If the charge that enters each meter of the axon gets distributed uniformly along it, how many coulombs of charge enter a 0.10 mm length of the axon? (b) What electric field (magnitude and direction) does the sudden influx of charge produce at the surface of the body if the axon is 5.00 cm below the skin? (c) Certain sharks can respond to electric fields as weak as 1.0 μN/C. How far from this segment of axon could a shark be and still detect its electric field?

40. •• Two point charges are separated by 25.0 cm (see Figure 17.47). Find the net electric field these charges produce at (a) point A, (b) point B. (c) What would be the magnitude and direction of the electric force this combination of charges would produce on a proton at A?

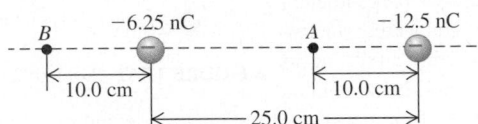

▲ **FIGURE 17.47** Problem 40.

41. •• A point charge of -4.00 nC is at the origin, and a second point charge of $+6.00$ nC is on the x axis at $x = 0.800$ m. Find the magnitude and direction of the electric field at each of the following points on the x axis: (a) $x = 20.0$ cm, (b) $x = 1.20$ m, (c) $x = -20.0$ cm.

42. •• In a rectangular coordinate system, a positive point charge $q = 6.00$ nC is placed at the point $x = +0.150$ m, $y = 0$, and an identical point charge is placed at $x = -0.150$ m, $y = 0$. Find the x and y components and the magnitude and direction of the electric field at the following points: (a) the origin; (b) $x = 0.300$ m, $y = 0$; (c) $x = 0.150$ m, $y = -0.400$ m, (d) $x = 0$, $y = 0.200$ m.

43. •• Two particles having charges of $+0.500$ nC and $+8.00$ nC are separated by a distance of 1.20 m. (a) At what point along the line connecting the two charges is the net electric field due to the two charges equal to zero? (b) Where would the net electric field be zero if one of the charges were negative?

44. •• Three negative point charges lie along a line as shown in Figure 17.48. Find the magnitude and direction of the electric field this combination of charges produces at point P, which lies 6.00 cm from the -2.00 μC charge measured perpendicular to the line connecting the three charges.

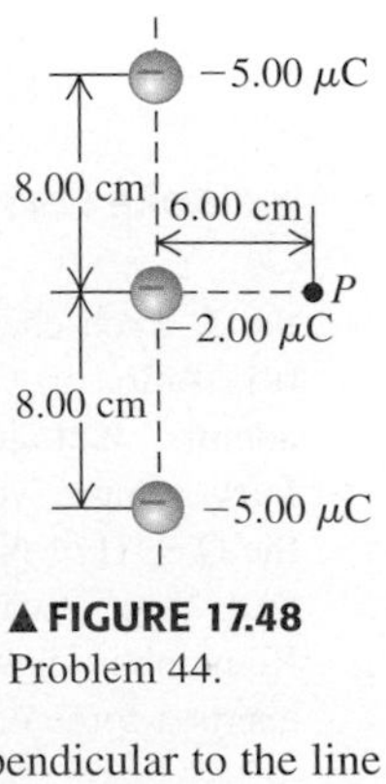

▲ **FIGURE 17.48** Problem 44.

45. •• **Torque and force on a dipole.** An electric dipole is in a uniform external electric field $\vec{E}$ as shown in Figure 17.49. (a) What is the net force this field exerts on the dipole? (b) Find the orientations of the dipole for which the torque on it about an axis through its center perpendicular to the plane of the figure is zero. (c) Which of the orientations in part (b) is stable, and which is unstable? (*Hint:* Consider a small displacement away from the equilibrium position, and see what happens.) (d) Show that, for the stable orientation in part (c), the dipole's own electric field *opposes* the external field for points between the charges.

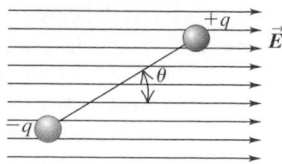

▲ **FIGURE 17.49** Problem 45.

46. •• (a) An electron is moving east in a uniform electric field of 1.50 N/C directed to the west. At point A, the velocity of the electron is 4.50×10^5 m/s toward the east. What is the speed of the electron when it reaches point B, 0.375 m east of point A? (b) A proton is moving in the uniform electric field of part (a). At point A, the velocity of the proton is 1.90×10^4 m/s, east. What is the speed of the proton at point B?

47. •• The electric field due to a certain point charge has a magnitude E at a distance of 1.0 cm from the charge. (a) What will be the magnitude of this field (in terms of E) if we move 1.0 cm farther away from the charge? (b) What will be the magnitude of the field (in terms of E) if we move an *additional* 1.0 cm farther away than in part (a)?

48. ••• For the dipole shown in Figure 17.50, show that the electric field at points on the x axis points vertically downward and has magnitude $kq(2a)/(a^2 + x^2)^{3/2}$. What does this expression reduce to when the distance between the two charges is much less than x?

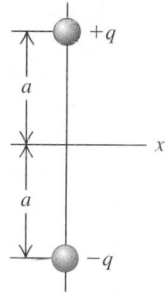

▲ **FIGURE 17.50** Problem 48.

17.7 Electric Field Lines

49. • Figure 17.51 shows some of the electric field lines due to three point charges arranged along the vertical axis. All three charges have the same magnitude. (a) What are the signs of the three charges? Explain your reasoning. (b) At what point(s) is the magnitude of the electric field the smallest? Explain your reasoning. Explain how the fields produced by each individual point charge combine to give a small net field at this point or points.

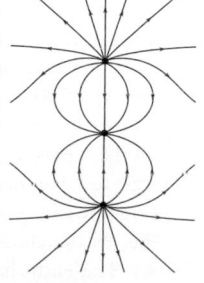

▲ **FIGURE 17.51** Problem 49.

50. • A proton and an electron are separated as shown in Figure 17.52. Points A, B, and C lie on the perpendicular bisector of the line connecting these two charges. Sketch the direction of the net electric field due to the two charges at (a) point A, (b) point B, and (c) point C.

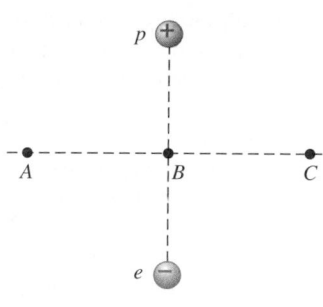

▲ **FIGURE 17.52** Problem 50.

51. •• Sketch electric field lines in the vicinity of two charges, Q and $-4Q$, located a small distance apart on the x-axis.

52. • Two point charges Q and $+q$ (where q is positive) produce the net electric field shown at point P in Figure 17.53. The field points parallel to the line connecting the two charges. (a) What can you conclude about the sign and magnitude of Q? Explain your reasoning. (b) If the lower charge were negative instead, would it be possible for the field to have the direction shown in the figure? Explain your reasoning.

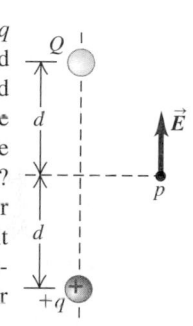

▲ **FIGURE 17.53** Problem 52.

53. •• Two very large parallel sheets of the same size carry equal magnitudes of charge spread uniformly over them, as shown in Figure 17.54. In each of the cases that follow, sketch the net pattern of electric-field lines in the region between the sheets, but far from their edges. (*Hint:* First sketch the field lines due to each sheet, and then add these fields to get the net field.) (a) The top sheet is positive and the bottom sheet is negative, as shown, (b) both sheets are positive, (c) both sheets are negative.

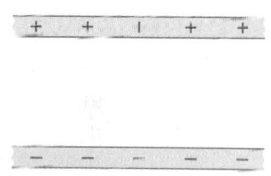

▲ **FIGURE 17.54** Problem 53.

17.8 Gauss's Law and Field Calculations

54. • (a) A closed surface encloses a net charge of 2.50 μC. What is the net electric flux through the surface? (b) If the electric flux through a closed surface is determined to be 1.40 N · m²/C, how much charge is enclosed by the surface?

55. • Figure 17.55 shows cross sections of five *closed* surfaces S_1, S_2, etc. Find the net electric flux passing through each of these surfaces.

56. •• A point charge 8.00 nC is at the center of a cube with sides of length 0.200 m. What is the electric flux through (a) the surface of the cube, (b) one of the six faces of the cube?

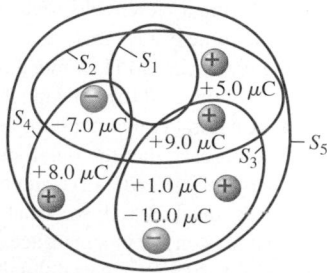

▲ **FIGURE 17.55** Problem 55.

57. • A charged paint is spread in a very thin uniform layer over the surface of a plastic sphere of diameter 12.0 cm, giving it a charge of -15.0 μC. Find the electric field (a) just inside the

paint layer, (b) just outside the paint layer, and (c) 5.00 cm out-side the surface of the paint layer.

58. •• On a humid day, an electric field of 2.00×10^4 N/C is enough to produce sparks about an inch long. Suppose that in your physics class, a van de Graaff generator (see Fig. 7.56) with a sphere radius of 15.0 cm is pro-ducing sparks 6 inches long. (a) Use Gauss's law to calculate the amount of charge stored on the surface of the sphere before you bravely discharge it with your hand. (b) Assume all the charge is concentrated at the center of the sphere, and use Coulomb's law to calculate the electric field at the surface of the sphere.

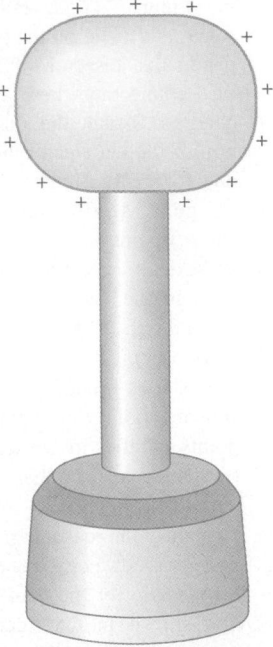

▲ FIGURE 17.56
Problem 58.

59. • (a) How many excess elec-trons must be distributed uni-formly within the volume of an isolated plastic sphere 30.0 cm in diameter to produce an elec-tric field of 1150 N/C just out-side the surface of the sphere? (b) What is the electric field at a point 10.0 cm outside the surface of the sphere?

60. •• In a certain region of space, the electric field $\vec{E}$ is uniform; i.e., neither its direction nor its magnitude changes in the region. (a) Use Gauss's law to prove that this region of space must be electrically neutral; that is, there must be no charge in this region. (b) Is the converse true? That is, in a region of space where there is no charge, must $\vec{E}$ be uniform? Explain.

61. •• A total charge of magnitude Q is distributed uniformly within a *thick* spherical shell of inner radius a and outer radius b. (a) Use Gauss's law to find the electric field within the cavity $(r \leq a)$. (b) Use Gauss's law to prove that the electric field outside the shell $(r \geq b)$ is exactly the same as if all the charge were concentrated as a point charge Q at the center of the sphere. (c) Explain why the result in part (a) for a *thick* shell is the same as that found in Example 17.10 for a *thin* shell. (*Hint:* A thick shell can be viewed as infinitely many thin shells.)

17.9 Charges on Conductors

62. • During a violent electrical storm, a car is struck by a falling high-voltage wire that puts an excess charge of -850 μC on the metal car. (a) How much of this charge is on the inner sur-face of the car? (b) How much is on the outer surface?

63. • A neutral conductor completely encloses a hole inside of it. You observe that the outer surface of this conductor carries a charge of -12 μC. (a) Can you conclude that there is a charge inside the hole? If so, what is this charge? (b) How much charge is on the inner surface of the conductor?

64. •• An irregular neutral conductor has a hollow cavity inside of it and is insulated from its surroundings. An excess charge of $+16$ nC is sprayed onto this conductor. (a) Find the charge on the inner and outer surfaces of the conductor. (b) Without touching the conductor, a charge of -11 nC is inserted into the cavity through a small hole in the conductor. Find the charge on the inner and outer surfaces of the conductor in this case.

General Problems

65. •• Three point charges are arranged along the x axis. Charge $q_1 = -4.50$ nC is located at $x = 0.200$ m, and charge $q_2 = +2.50$ nC is at $x = -0.300$ m. A positive point charge q_3 is located at the origin. (a) What must the value of q_3 be for the net force on this point charge to have magnitude 4.00 μN? (b) What is the direction of the net force on q_3? (c) Where along the x axis can q_3 be placed and the net force on it be zero, other than the trivial answers of $x = +\infty$ and $x = -\infty$?

66. •• An electron is released from rest in a uniform electric field. The electron accelerates vertically upward, traveling 4.50 m in the first 3.00 μs after it is released. (a) What are the magnitude and direction of the electric field? (b) Are we justified in ignor-ing the effects of gravity? Justify your answer quantitatively.

67. •• A charge $q_1 = +5.00$ nC is placed at the origin of an xy-coordinate system, and a charge $q_2 = -2.00$ nC is placed on the positive x axis at $x = 4.00$ cm. (a) If a third charge $q_3 = +6.00$ nC is now placed at the point $x = 4.00$ cm, $y = 3.00$ cm, find the x and y components of the total force exerted on this charge by the other two charges. (b) Find the magnitude and direction of this force.

68. •• A charge of -3.00 nC is placed at the origin of an xy-coordi-nate system, and a charge of 2.00 nC is placed on the y axis at $y = 4.00$ cm. (a) If a third charge, of 5.00 nC, is now placed at the point $x = 3.00$ cm, $y = 4.00$ cm, find the x and y com-ponents of the total force exerted on this charge by the other two charges. (b) Find the magnitude and direction of this force.

69. •• Point charges of 3.00 nC are situated at each of three cor-ners of a square whose side is 0.200 m. What are the magni-tude and direction of the resultant force on a point charge of -1.00 μC if it is placed (a) at the center of the square, (b) at the vacant corner of the square?

70. •• An electron is projected with an ini-tial speed $v_0 = 5.00 \times 10^6$ m/s into the uniform field between the parallel plates in Figure 17.57. The direction of the field is vertically downward, and the field is zero except in the space between the two plates. The electron enters the field at a point midway between the plates. If the electron just misses the upper plate as it emerges from the field, find the magnitude of the electric field.

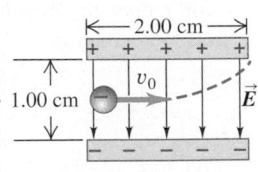

▲ FIGURE 17.57
Problem 70.

71. •• A small 12.3 g plastic ball is tied to a very light 28.6 cm string that is attached to the vertical wall of a room. (See Figure 17.58.) A uniform horizontal electric field exists in this room. When the ball has been given an excess charge of -1.11 μC, you observe that it remains suspended, with the string making an angle of 17.4° with the wall. Find the magnitude and direction of the electric field in the room.

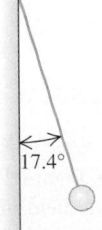

▲ FIGURE 17.58
Problem 71.

72. •• A -5.00 nC point charge is on the x axis at $x = 1.20$ m. A second point charge Q is on the x axis at -0.600 m. What must be the sign and magnitude of Q for the resultant electric field at the origin to be (a) 45.0 N/C in the $+x$ direction, (b) 45.0 N/C in the $-x$ direction?

73. •• The earth has a downward-directed electric field near its surface of about 150 N/m. If a raindrop with a diameter of 0.020 mm is suspended, motionless, in this field, how many excess electrons must it have on its surface?

74. •• A 9.60-μC point charge is at the center of a cube with sides of length 0.500 m. (a) What is the electric flux through one of the six faces of the cube? (b) How would your answer to part (a) change if the sides were 0.250 m long? Explain.

75. •• Two point charges q_1 and q_2 are held 4.00 cm apart. An electron released at a point that is equidistant from both charges (see Figure 17.59) undergoes an initial acceleration of 8.25×10^{18} m/s^2 directly upward in the figure, parallel to the line connecting q_1 and q_2. Find the magnitude and sign of q_1 and q_2.

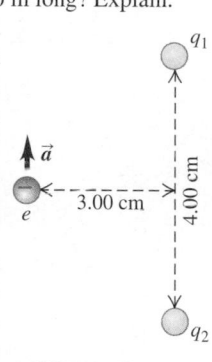

▲ **FIGURE 17.59** Problem 75.

76. •• **Electrophoresis.** Electrophoresis is a
BIO process used by biologists to separate different biological molecules (such as proteins) from each other according to their ratio of charge to size. The materials to be separated are in a viscous solution that produces a drag force F_D proportional to the size and speed of the molecule. We can express this relationship as $F_D = KRv$, where R is the radius of the molecule (modeled as being spherical), v is its speed, and K is a constant that depends on the viscosity of the solution. The solution is placed in an external electric field E so that the electric force on a particle of charge q is $F = qE$. (a) Show that when the

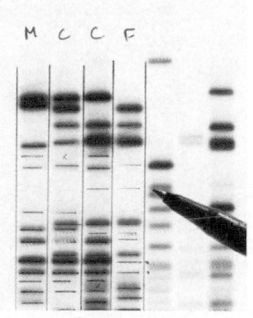

▲ **FIGURE 17.60** Problem 76.

electric field is adjusted so that the two forces (electrical and viscous drag) just balance, the ratio of q to R is Kv/E. (b) Show that if we leave the electric field on for a time T, the distance x that the molecule moves during that time is $x = (ET/k)(q/R)$. (c) Suppose you have a sample containing three different biological molecules for which the molecular ratio q/R for material 2 is twice that of material 1 and the ratio for material 3 is three times that of material 1. Show that the distances migrated by these molecules after the same amount of time are $x_2 = 2x_1$ and $x_3 = 3x_1$. In other words, material 2 travels twice as far as material 1, and material 3 travels three times as far as material 1. Therefore, we have separated these molecules according to their ratio of charge to size. In practice, this process can be carried out in a special gel or paper, along which the biological molecules migrate. (See Figure 17.60.) The process can be rather slow, requiring several hours for separations of just a centimeter or so.

77. •• An early model of the hydrogen atom viewed it as an electron orbiting a proton in a circular path with a radius of 5.29×10^{-11} m. What would be the speed of the electron in this model? You'll need some information from Appendix E, and may need to review Chapter 6 on circular motion.

Passage Problems

BIO How might cells respond to an electric field? Some cells can be observed to grow or migrate parallel to an applied electric field. (The field is applied by passing a current through the aqueous medium surrounding the cell. Because the cell membrane has a higher resistance than the medium, the current flows around the cell.) It has been hypothesized that this phenomenon may participate in the natural guidance of cells during embryonic development or wound repair. However, the mechanism by which the cells sense the electric field is not known.

In one proposed mechanism, the cell would use the distribution of cell-surface proteins to sense the electric field. The membrane surrounding the cell consists of a double layer of lipid molecules and has a viscosity similar to that of olive oil. The membrane is studded with protein molecules, which are free to move in the plane of the membrane. In the absence of a perturbing force, diffusion tends to distribute these molecules uniformly. However, for many membrane proteins, the portion of the protein that projects into the extracellular medium carries a net charge. An applied electric field will tend to move such charged proteins toward one or the other end of the cell. For a given type of charged protein, the resulting steady-state distribution depends on the applied field and on the concentration and net diffusion rate of the protein molecules. In theory, a cell could use an asymmetric protein distribution to sense and respond to an electric field. The graph in Figure 17.61 shows the steady-state distribution of a particular membrane protein in response to an electric field.

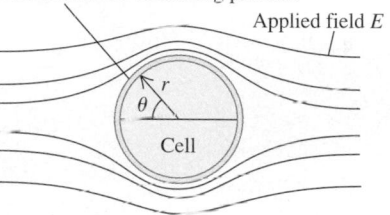

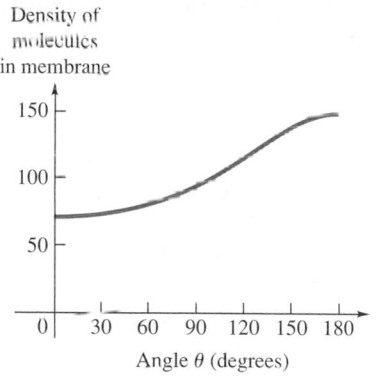

▲ **FIGURE 17.61** Problems 78–79.

78. What is the direction of the electrical field?
    A. It points from 0° to 180°.
    B. It points from 180° to 0°.
    C. Cannot tell without knowing the absolute value of the charge on the molecules.
    D. Cannot tell without knowing the sign of the charge on the molecules.

79. After the protein has reached the steady-state distribution shown by the graph, you turn the electric field off. Assuming diffusion acts unimpeded, the protein distribution will
    A. remain unchanged.
    B. become uniform, with the density at each location equal to that at 180° on the graph.
    C. become uniform with a density intermediate between those at 0° and 180° on the graph.
    D. become uniform, with the density at each location equal to that at 0° on the graph.

18 Electric Potential and Capacitance

This chapter is about energy associated with electrical interactions. Every time you turn on a light or an electric motor, you are making use of electrical energy, a familiar part of everyday life and an indispensable ingredient of our technological society. In Chapter 7, we introduced the concepts of *work* and *energy* in a mechanical context; now we combine these concepts with what we have learned about electric charge, Coulomb's law, and electric fields.

When a charged particle moves in an electric field created by charges at rest (i.e., an electrostatic field), the electric force does *work* on the particle. The force is *conservative;* the work can always be expressed in terms of a potential energy. This in turn is associated with a new concept called *electric potential,* or simply *potential.* In circuits, potential is often called *voltage.* The practical applications of this concept cover a wide range, including electric circuits, electron beams in TV picture tubes, high-energy particle accelerators, and many other devices and phenomena. The concept of potential is also essential for the analysis of a common circuit device called a *capacitor,* which we'll study later in the chapter.

The electrical potential between high-voltage wires and steel pylons is very high, so the wires are held away from the pylons by stacks of insulators.

18.1 Electric Potential Energy

The opening sections of this chapter are about work, potential energy, and conservation of energy. Let's begin by reviewing several essential points from Chapter 7. First, when a constant force $\vec{F}$ acts on a particle that moves in a straight line through a displacement $\vec{s}$ from point a to point b, the work $W_{a \to b}$ done by the force is

$$W_{a \to b} = Fs \cos\phi, \tag{18.1}$$

where ϕ is the angle between the force and displacement. We'll point the x axis in the direction of the particle's motion.

▶ **BIO** **Application** **Nanomachine.** This image shows the structure of a protein complex called a *voltage-gated potassium channel,* which participates in the functioning of nerve and muscle cells. The channel sits in the cell membrane and controls the flow of potassium ions out of the cell. Remarkably, the force that opens and closes the channel is provided by the electric field across the membrane. Each of the four subunits that make up the channel has an arm that carries positive charges. This arm is fairly mobile, so the electric field does work on it, pulling it toward one or the other side of the membrane, depending on the field's direction and strength. When the arm is pulled inward, the channel closes; when it is pulled outward, the channel opens.

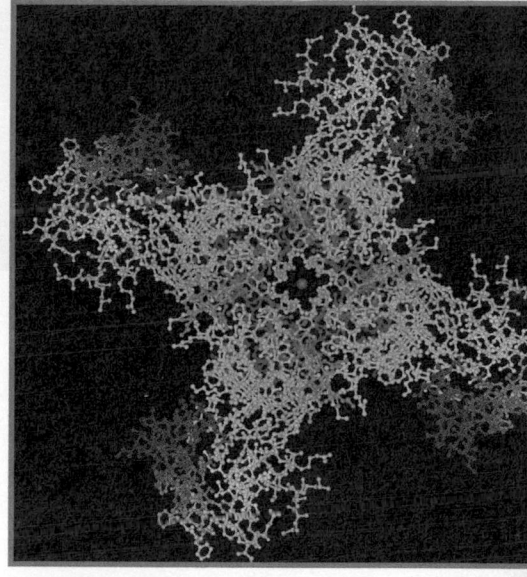

Second, because the force field is *conservative,* as we defined the term in Section 7.5, the work that is done can always be expressed in terms of a *potential energy U.* When the particle moves from a point where the potential energy is U_a to a point where it is U_b, the work $W_{a\to b}$ done by the force is

$$W_{a\to b} = U_a - U_b. \tag{18.2}$$

When $W_{a\to b}$ is positive, U_a is greater than U_b, and the potential energy *decreases.* That's what happens when a baseball falls from a high point (a) to a lower point (b) under the action of the earth's gravity (Figure 18.1a). The force of gravity does positive work, and the gravitational potential energy decreases. When a ball is thrown upward, the gravitational force does negative work during the ascent, and the potential energy increases. Figure 18.1b shows the analogous situation for electric fields; we'll discuss it in detail later in this section.

Third, the work–energy theorem says that the change in kinetic energy $\Delta K = K_b - K_a$ during any displacement is equal to the total work done on the particle. So if Equation 18.2 gives the *total* work, then $K_b - K_a = U_a - U_b$, which we usually write as

$$K_a + U_a = K_b + U_b. \tag{18.3}$$

Let's look at an electrical example of these basic concepts. In Figure 18.2, a pair of charged parallel metal plates sets up a uniform electric field with magnitude E. The field exerts a downward force with magnitude $F = q'E$ on a positive test charge q' as the charge moves a distance s from point a to point b. The force on the test charge is constant and independent of its location, so the work done by the electric field is

$$W_{a\to b} = Fs = q'Es. \tag{18.4}$$

We can represent this work in terms of a potential energy U, just as we did for gravitational potential energy in Section 7.5. The y component of force, $F_y = -q'E$, is constant, and there is no x or z component, so the work is independent of the path the particle takes from a to b. Just as the potential energy for the gravitational force $F_y = -mg$ was $U = mgy$, the potential energy for the electric-field force $F_y = -q'E$ is

$$U = q'Ey. \tag{18.5}$$

(We've chosen U to be zero at $y = 0$.) When the test charge moves from height y_a to height y_b, the work done on the charge by the field is given by

$$W_{a\to b} = U_a - U_b = q'Ey_a - q'Ey_b = q'E(y_a - y_b). \tag{18.6}$$

When y_a is greater than y_b (Figure 18.3a), the particle moves in the same direction as the $\vec{E}$ field, U decreases, and the field does positive work. When y_a is less than y_b (Figure 18.3b), the particle moves in the opposite direction to $\vec{E}$, U

Object moving in a uniform gravitational field:

$$W = -\Delta U_{\text{grav}} = mgh$$

Charge moving in a uniform electric field:

$$W = -\Delta U_E = qEs$$

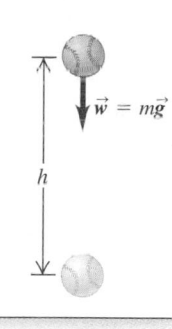

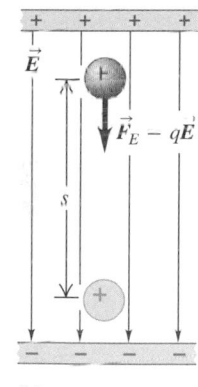

(a)   (b)

▲ **FIGURE 18.1** Because electric and gravitational forces are conservative, work done by either can be expressed in terms of a potential energy.

Work done on charge q' by the *constant* electric force between the plates: $W_{a\to b} = q'Es$

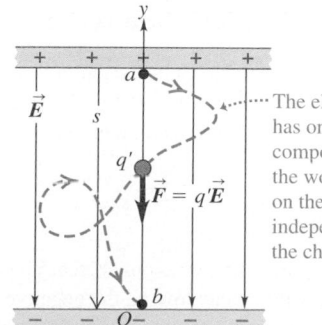

The electric force has only a y component, so the work it does on the charge is independent of the charge's path.

▲ **FIGURE 18.2** A test charge q' moves from point a to point b in a uniform electric field.

Positive charge moves in the direction of $\vec{E}$:
• Field does *positive* work on charge;
• *U decreases.*

Positive charge moves opposite to $\vec{E}$:
• Field does *negative* work on charge;
• *U increases.*

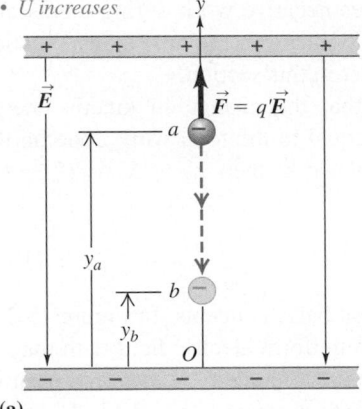

(a)

(b)

▲ **FIGURE 18.3** The work done by an electric field on a positive charge moving (a) in the direction of and (b) opposite to the electric field.

Negative charge moves in the direction of $\vec{E}$:
• Field does *negative* work on charge;
• *U increases.*

Negative charge moves opposite to $\vec{E}$:
• Field does *positive* work on charge;
• *U decreases.*

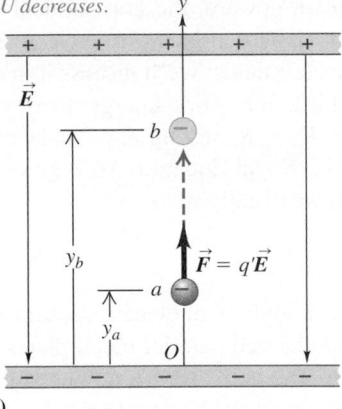

(a)

(b)

▲ **FIGURE 18.4** The work done by an electric field on a negative charge moving (a) in the direction of and (b) opposite to the electric field.

increases, and the field does negative work. In particular, if $y_a = s$ and $y_b = 0$, then Equation 18.6 gives $W_{a \to b} = q'Es$, in agreement with Equation 18.4.

If the test charge q' is negative, the potential energy increases when it moves with the field and decreases when it moves against the field (Figure 18.4).

EXAMPLE 18.1   **Work in a uniform electric field**

Two large conducting plates separated by 6.36 mm carry charges of equal magnitude and opposite sign, creating a uniform electric field with magnitude 2.80×10^3 N/C between the plates. An electron moves from the negatively charged plate to the positively charged plate. How much work does the electric field do on the electron?

SOLUTION

SET UP Figure 18.5 shows our sketch. The electric field is directed from the positive plate toward the negative plate. $\vec{F}_E = q\vec{E}$, so for an electron with negative charge $q = -e$, the electric force $\vec{F}_E$ points in the direction opposite to the electric field. Its magnitude is $F_E = eE$. The electric field is uniform, so the force it exerts on the electron is constant during the electron's motion.

Continued

SOLVE The force and displacement are parallel; the work W done by the electric-field force during a displacement of magnitude d is $W = F_e d \cos \phi$ with $\phi = 0$, so

$$W = F_e d = eEd$$
$$= (1.60 \times 10^{-19}\,\text{C})(2.80 \times 10^3\,\text{N/C})(6.36 \times 10^{-3}\,\text{m})$$
$$= 2.85 \times 10^{-18}\,\text{J}.$$

REFLECT The amount of work done is proportional to the electric field magnitude E and to the displacement magnitude d of the electron. The electric field does positive work on the electron. If there are no other forces, the electron's kinetic energy increases by the same amount as the work done on the electron by the electric field.

Practice Problem: In Example 18.1, how much work does the electric field do on the electron if the magnitude of the field is doubled, to $5.60 \times 10^3\,\text{N/C}$, and the separation between the plates is halved, to 3.18 mm? *Answer:* $2.85 \times 10^{-18}\,\text{J}$.

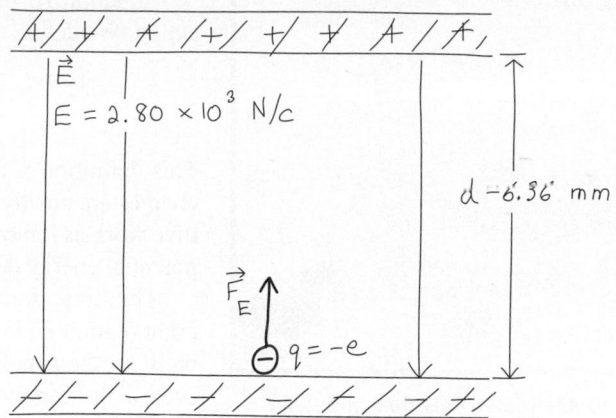

▲ **FIGURE 18.5** Our sketch for this problem.

Potential Energy of Point Charges

It's useful to calculate the work done on a test charge q' when it moves in the electric field caused by a single stationary point charge q. Suppose we place charge q at the origin of a coordinate system, and suppose the test charge q' moves along the x axis from point $x = a$ to point $x = b$ (Figure 18.6a). How much work does the force due to q do on q' during this displacement?

We can't simply multiply the force by the displacement, because the force isn't constant; it varies with the distance x according to the graph in Figure 18.6b. The work $W_{a \to b}$ done on the test charge is represented graphically by the area under the curve between $x = a$ and $x = b$. This area can be calculated with methods of integral calculus. The result is

$$W_{a \to b} = kqq'\left(\frac{1}{a} - \frac{1}{b}\right). \tag{18.7}$$

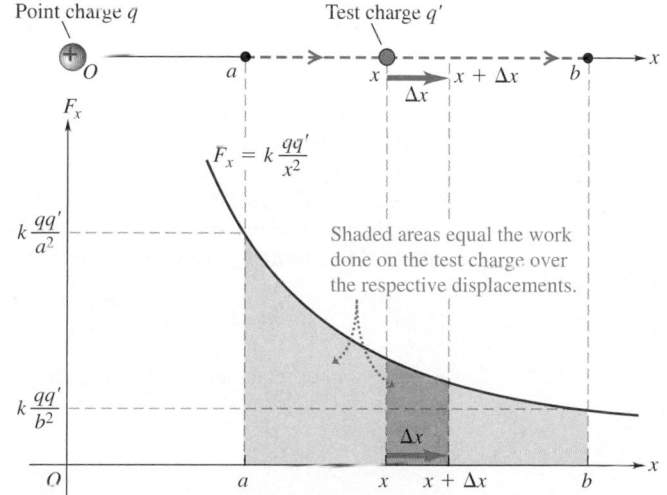

(a) Test charge moves from a to b.

(b) Force on the test charge as a function of position.

▲ **FIGURE 18.6** A test charge q' moves radially along a straight line extending from charge q. As it does so, the electric force on it decreases in magnitude.

▲ BIO Application Real-time molecular biology. The patch clamp technique is an ingenious way to investigate how cells work. To create a patch clamp, a polished microelectrode is carefully manipulated to the outer membrane of a cell to make a tight seal. Protein molecules called ion channels float in the oily membrane, and often a single channel is isolated within the patch. An experimenter can then study the electrophysiologic properties of the single ion channel and manipulate the voltage across the membrane (and thus the channel). In this way, we know that an electrical potential difference across the membrane controls the pore, and thus the movement of ions through the membrane. At the appropriate voltage, a channel flicks open for a time (perhaps 10 ms) and then closes. The opening and closing of voltage-gated channels underlie all electrical signaling in most cells to control many aspects of cellular function. The inventors of the patch clamp technique were awarded the Nobel Prize in Physiology or Medicine in 1991.

Equation 18.7 also shows how we can define a potential energy for this interaction. We define

$$U_a = \frac{kqq'}{a} \quad \text{and} \quad U_b = \frac{kqq'}{b}.$$

This definition is consistent with the requirement that $W_{a \to b} = U_a - U_b$. If both charges are positive and b is greater than a, the electric-field force on q' does positive work as it moves away from q. Correspondingly, U_a is greater than U_b, so the potential energy decreases as the force does positive work, just as we expect.

This important result can be generalized to the case where q, point a, and point b don't all lie along the same line. It can be shown that the work $W_{a \to b}$ done on q' by the $\vec{E}$ field produced by q is the same for *all possible paths* from a to b, even if these points don't lie on the same radial line from charge q. The work depends only on the distances a and b, not on the details of the path. Also, if q' returns to its starting point a by a different path, the total work done in the round-trip displacement is zero. These are the needed characteristics for a *conservative* force field, as we defined it in Chapter 7. Thus, we've verified that the force that q exerts on q' is a *conservative* force, and we've obtained an expression for the potential energy U when the test charge q' is at *any* distance r from charge q:

Potential energy of point charges

The potential energy U of a system consisting of a point charge q' located in the field produced by a stationary point charge q, at a distance r from the charge, is

$$U = k\frac{qq'}{r}. \tag{18.8}$$

We have *not* assumed anything about the signs of q and q'; Equation 18.8 is valid for any combination of signs.

Potential energy is always defined relative to some reference point at which $U = 0$. In Equation 18.8, U is zero when q and q' are infinitely far apart, or $r = \infty$. Therefore, U represents the work done on the test charge q' by the field of q when q' moves from an initial distance r to infinity. If q and q' have the same sign, the interaction is repulsive, the work is positive, and U is positive at any finite separation. If they have opposite signs, the interaction is attractive and U is negative.

Quantitative Analysis 18.1

Change in potential energy

Consider two positive point charges q_1 and q_2. Their potential energy is defined as zero when they are infinitely far apart, and it increases as they move closer. If q_2 starts at an initial distance r_i from q_1 and moves toward q_1 to a final distance $r_i - \Delta r$ (where Δr is positive), by how much does the system's potential energy change?

A. $kq_1q_2\left(\dfrac{1}{\Delta r} - \dfrac{1}{r_i - \Delta r}\right)$. B. $\Delta r\left(\dfrac{kq_1q_2}{r_i^2}\right)$.

C. $kq_1q_2\left(\dfrac{1}{r_i - \Delta r} - \dfrac{1}{r_i}\right)$.

SOLUTION The electric potential energy of the two charges depends on the distance r between them: $U = k(q_1q_2)/r$. Initially, the distance between them is r_i. After q_2 moves a distance Δr toward q_1, the distance is $r_i - \Delta r$. The change in potential energy depends on the reciprocal of these distances, so C must be the answer. More formally, the change in potential energy is

$$\Delta U = U_f - U_i = \frac{kq_1q_2}{r_i - \Delta r} - \frac{kq_1q_2}{r_i}.$$

Simplifying this equation yields expression C.

We can generalize Equation 18.8 to situations in which the $\vec{E}$ field is caused by *several* point charges q_1, q_2, q_3, ... at distances r_1, r_2, r_3, ..., respectively, from q'. The total electric field at each point is the *vector sum* of the fields due to the individual charges, and the total work done on q' during any displacement is the sum of the contributions from the individual charges. We conclude that the potential energy U associated with a test charge q' at point a in Figure 18.7, due to a collection of charges q_1, q_2, q_3, . . . at distances r_1, r_2, r_3, . . . , respectively, from q' is the *algebraic* sum (*not* a vector sum)

$$U = kq'\left(\frac{q_1}{r_1} + \frac{q_2}{r_2} + \frac{q_3}{r_3} + \cdots\right). \qquad (18.9)$$

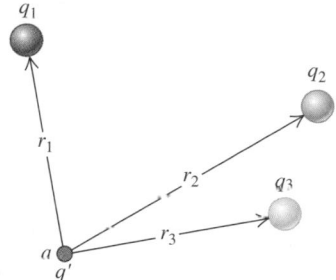

▲ **FIGURE 18.7** Potential energy associated with a charge q' at point a depends on charges q_1, q_2, and q_3 and on their respective distances r_1, r_2, and r_3 from point a.

When q' is at a different point b, the potential energy is given by the same expression, but r_1, r_2, . . . are the distances from q_1, q_2, . . . , respectively, to point b. The work $W_{a\rightarrow b}$ done on charge q' when it moves from a to b along *any* path is still equal to the difference $U_a - U_b$ between the potential energies when q' is at a and at b.

We can represent *any* charge distribution as a collection of point charges, so Equation 18.9 shows that we can always find a potential-energy function for *any* static electric field. It follows that **every electric field due to a static charge distribution is a conservative force field.**

Equations 18.8 and 18.9 define U to be zero when *all* the distances r_1, r_2, . . . are *infinite*—that is, when the test charge q' is very far away from all the charges that produce the field. As with any potential-energy function, the reference point is arbitrary: We can always add a constant to make U equal zero at any point we choose. Making $U = 0$ at infinity is a convenient reference level for electrostatic problems, but in circuit analysis other reference levels are often more convenient.

18.2 Potential

In Section 18.1, we looked at the potential energy U associated with a test charge q' in an electric field. Now we want to describe this potential energy on a "potential energy per unit charge" basis, just as the electric field describes the force on a charged particle in the field on a "force per unit charge" basis. Doing this leads us to the concept of **electric potential,** often called, simply, **potential.** The concept of electric potential is useful in calculations involving energies of charged particles. It also facilitates many electric-field calculations, because it is closely related to the concept of the $\vec{E}$ field. When we need to calculate an electric field, it is often easier to calculate the potential first and then find the field from it.

Definition of potential

The electric potential V at any point in an electric field is the electric potential energy U per unit charge associated with a test charge q' at that point:

$$V = \frac{U}{q'}, \qquad \text{or} \qquad U = q'V. \qquad (18.10)$$

Potential energy and charge are both scalars, so potential is a scalar quantity.

Unit: From Equation 18.10, the units of potential are energy divided by charge. The SI unit of potential, 1 J/C, is called one **volt** (1 V), in honor of the Italian scientist Alessandro Volta (1745–1827):

$$1\ \text{V} = 1\ \text{volt} = 1\ \text{J/C} = 1\ \text{joule/coulomb}.$$

In the context of electric circuits, potential is often called **voltage.** For instance, a 9 V battery has a difference in electric potential **(potential difference)** of 9 V between its two terminals. A 20,000 V power line has a potential difference of 20,000 V between itself and the ground.

MasteringPHYSICS

PhET: Charges and Fields
ActivPhysics 11.11: Electric Potential: Qualitative Introduction

▲ **Application** *Really* **high voltage.**
A lightning bolt occurs when the electric potential difference between cloud and ground becomes so great that the air between them ionizes and allows a current to flow. A typical bolt discharges about 10^9 J of energy across a potential difference of about 10^7 V. In a major electrical storm, the total potential energy accumulated and discharged is enormous.

To put Equation 18.2 on a "work per unit charge" basis, we divide both sides by q', obtaining

$$\frac{W_{a \to b}}{q'} = \frac{U_a}{q'} - \frac{U_b}{q'} = V_a - V_b, \qquad (18.11)$$

where $V_a = U_a/q'$ is the potential energy per unit charge at point a and V_b is that at b. We call V_a and V_b the *potential at point a* and *potential at point b*, respectively. The potential difference $V_a - V_b$ is called *the potential of a with respect to b*.

EXAMPLE 18.2 Parallel plates and conservation of energy

A 9.0 V battery is connected across two large parallel plates that are separated by 4.5 mm of air, creating a potential difference of 9.0 V between the plates. **(a)** What is the electric field in the region between the plates? **(b)** An electron is released from rest at the negative plate. If the only force on the electron is the electric force exerted by the electric field of the plates, what is the speed of the electron as it reaches the positive plate? The mass of an electron is $m_e = 9.11 \times 10^{-31}$ kg.

SOLUTION

SET UP Figure 18.8 shows our sketch. We use a to designate the electron's starting position at the negative plate and b for its final position at the positive plate. Then $V_b - V_a = +9.0$ V. The electric field is directed from the positive plate b toward the negative plate a (i.e., from higher potential toward lower potential), and it is uniform between the plates.

SOLVE Part (a): The expression $V_b - V_a$ is the potential at point b with respect to point a. This quantity (work per unit charge) is related to the electric field E (force per unit charge) between the plates by $V_b - V_a = Ed$, where d is the separation between the plates and Ed is the work per unit charge on a positively charged particle that moves from b to a. Thus,

$$E = \frac{V_b - V_a}{d} = \frac{9.0 \text{ V}}{4.5 \times 10^{-3} \text{ m}} = 2.0 \times 10^3 \text{ V/m}.$$

Part (b): Conservation of energy applied to points a and b at the corresponding plates gives

$$K_a + U_a = K_b + U_b.$$

Also, $U = q'V$, where $q' = -e$, the charge of an electron. Using this expression to replace U in the conservation-of-energy equation gives

$$K_a + q'V_a = K_b + q'V_b.$$

The electron is released from rest from point a, so $K_a = 0$. We next solve for K_b:

$$K_b = q'(V_a - V_b) = -e(V_a - V_b) = +e(V_b - V_a)$$
$$= (1.60 \times 10^{-19} \text{ C})(9.0 \text{ V})$$
$$= 1.44 \times 10^{-18} \text{ J}.$$

Then $K_b = \frac{1}{2}m_e v_b^2$ gives

$$v_b = \sqrt{\frac{2K_b}{m_e}} = \sqrt{\frac{2(1.44 \times 10^{-18} \text{ J})}{9.11 \times 10^{-31} \text{ kg}}} = 1.8 \times 10^6 \text{ m/s}.$$

ALTERNATIVE SOLUTION Part (b) could also be done by calculating the acceleration of the electron. We use a y coordinate with the origin at point b and the $+y$ axis pointing toward a. Newton's second law then gives

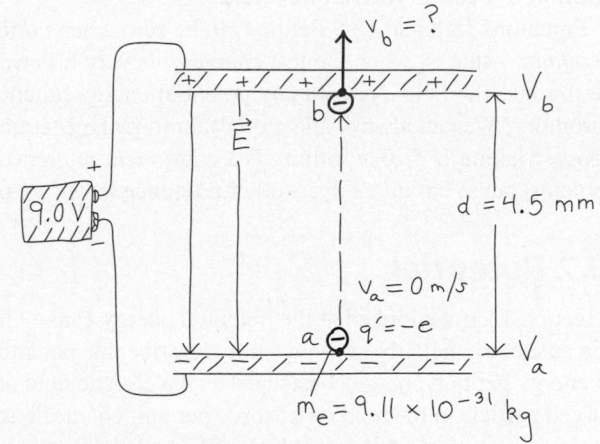

▲ **FIGURE 18.8** Our sketch for this problem.

$$a = \frac{F}{m} = \frac{eE}{m} = \frac{(1.60 \times 10^{-19} \text{ C})(2.0 \times 10^3 \text{ N/C})}{9.11 \times 10^{-31} \text{ kg}}$$
$$= 3.51 \times 10^{14} \text{ m/s}^2.$$

Also, $v_{0y} = 0$ and $y = 4.5 \times 10^{-3}$ m, so, from the relation $v^2 = v_{0y}^2 + 2a_y y$, we find

$$v_y = \sqrt{2a_y y} = \sqrt{2(3.51 \times 10^{14} \text{ m/s}^2)(4.5 \times 10^{-3} \text{ m})}$$
$$= 1.8 \times 10^6 \text{ m/s},$$

in agreement with the result from conservation of energy.

REFLECT Remember that electric fields point away from positive charges and toward negative charges and that they point in the direction of decreasing potential. Negative charges gain kinetic energy when they move to higher potential, because when V increases (becomes more positive), the electrical potential energy of a negative charge decreases (becomes less positive or more negative). When the electron moves from a to b, it loses potential energy and gains kinetic energy.

Practice Problem: Repeat the preceding problem, replacing the 9.0 V battery by an 18.0 V battery. *Answers:* (a) 4.0×10^3 V/m; (b) 2.5×10^6 m/s.

Potential of a Point Charge

The potential V (potential energy per unit charge) due to a point charge q, at any distance r from the charge, is obtained by dividing Equation 18.8 by the test charge q':

Potential of a point charge

When a test charge q' is a distance r from a point charge q, the potential V is

$$V = \frac{U}{q'} = k\frac{q}{r}, \qquad (18.12)$$

where k is the same constant as in Coulomb's law (Equation 17.1).

Similarly, to find the potential V at a point due to any collection of point charges $q_1, q_2, q_3, \ldots$ at distances $r_1, r_2, r_3, \ldots$, respectively, from q', we divide Equation 18.9 by q':

$$V = \frac{U}{q'} = k\left(\frac{q_1}{r_1} + \frac{q_2}{r_2} + \frac{q_3}{r_3} + \cdots\right). \qquad (18.13)$$

In deriving Equations 18.8 and 18.9, we assumed that the potential energy of a point charge is zero at an infinite distance from the charge; thus, the V defined by Equation 18.13 is zero at points infinitely far away from *all* the charges. We could add any constant to Equation 18.13 without changing the meaning, because only *differences* between potentials at two points are physically significant.

As noted earlier, the difference $V_a - V_b$ is called the *potential of a with respect to b;* we sometimes abbreviate this difference as $V_{ab} = V_a - V_b$. This is sometimes called the *potential difference between a and b,* but that's ambiguous unless we specify which is the reference point (that is, which point is at higher potential). Note that potential, like electric field, is independent of the test charge q' that we use to define it. When a positive test charge moves from a point of higher to a point of lower potential (that is, $V_a > V_b$), the electric field does positive work on it. A positive charge tends to "fall" from a high-potential region to a region with lower potential. The opposite is true for a negative charge.

▲ **BIO Application Are you all right up there?** Have you ever wondered why birds can perch safely on power lines? The answer is that no current flows through the bird because the bird does not offer a path to a point at lower potential—provided that it touches only the wire. Large hawks and other raptors are big enough that an outstretched wing can touch the power pole, a transformer, or another wire. If the bird bridges two structures at sufficiently different potentials, it can be electrocuted.

Quantitative Analysis 18.2 **Force and potential**

A positively charged particle is placed on the x axis in a region where the electrical potential due to other charges increases in the $+x$ direction, but does not change in the y or z direction. The particle

A. is acted upon by a force in the $+x$ direction.

B. is acted upon by a force in the $-x$ direction.

C. is not acted upon by any force.

SOLUTION Since potential is potential energy per unit charge, the direction of decreasing potential (the $-x$ direction) is the direction in which the particle's potential energy decreases. If the particle moves in this direction, the electric field does positive work on it, increasing its kinetic energy. Thus, that is the direction of the force exerted by the electric field on the particle, and B is correct. If this isn't clear, consider a gravitational analogy: When an object falls downward in a gravitational field, in the direction of the weight force, its potential energy U_{grav} decreases and its kinetic energy K increases.

EXAMPLE 18.3 Potential of two point charges

Two electrons are held in place 10.0 cm apart. Point a is midway between the two electrons, and point b is 12.0 cm directly above point a. **(a)** Calculate the electric potential at point a and at point b. **(b)** A third electron is released from rest at point b. What is the speed of this electron when it is far from the other two electrons? The mass of an electron is $m_e = 9.11 \times 10^{-31}$ kg.

Continued

SOLUTION

SET UP Figure 18.9 shows our sketch. Point b is a distance $r_b = \sqrt{(12.0 \text{ cm}^2) + (5.0 \text{ cm})^2} = 13.0 \text{ cm}$ from each electron.

SOLVE Part (a): The electric potential V at each point is the sum of the electric potentials of each electron: $V = V_1 + V_2 = k\dfrac{q_1}{r_1} + k\dfrac{q_2}{r_2}$, with $q_1 = q_2 = -e$. At point a, $r_1 = r_2 = r_a = 0.050$ m, so

$$V_a = -\frac{2ke}{r_a} = -\frac{2(8.99 \times 10^9 \text{ N} \cdot \text{m}^2/\text{C}^2)(1.60 \times 10^{-19} \text{ C})}{0.050 \text{ m}}$$
$$= -5.8 \times 10^{-8} \text{ V}.$$

At point b, $r_1 = r_2 = r_b = 0.130$ m, so

$$V_b = -\frac{2ke}{r_b} = -\frac{2(8.99 \times 10^9 \text{ N} \cdot \text{m}^2/\text{C}^2)(1.60 \times 10^{-19} \text{ C})}{0.130 \text{ m}}$$
$$= -2.2 \times 10^{-8} \text{ V}.$$

Part (b): Remember that our equation for potential assumes that U is zero when $r = \infty$. Thus, when the third electron is far from the other two (at a location we designate c), we can assume that $U = 0$. To find the electron's speed at point c, we use conservation of energy:

$$K_b + U_b = K_c + U_c.$$

We solve for K_c. First we use $U = q'V$ and $q' = -e$ to rewrite the preceding expression as

$$K_b - eV_b = K_c - eV_c.$$

We know that $V_c = 0$ because $V_c = \dfrac{kq}{r_c}$ and r_c is very large. Also, $K_b = 0$ because the electron is at rest before it is released. Then

$$K_c = -eV_b = -(1.60 \times 10^{-19} \text{ C})(-2.2 \times 10^{-8} \text{ V})$$
$$= +3.52 \times 10^{-27} \text{ J},$$
$$K_c = \tfrac{1}{2} m_e v_c^2,$$

so

$$v_c = \sqrt{\frac{2K_c}{m_e}} = \sqrt{\frac{2(3.52 \times 10^{-27} \text{ J})}{9.11 \times 10^{-31} \text{ kg}}} = 88 \text{ m/s}.$$

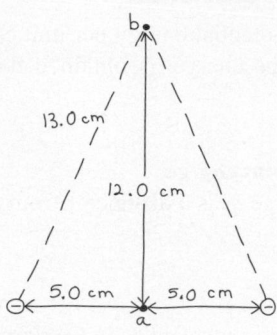

▲ **FIGURE 18.9** Our sketch for this problem.

REFLECT Remember that electric potential is a scalar quantity. We never talk about components of V; there is no such thing. When we add potentials caused by two or more point charges, the operation is simple scalar addition, not vector addition. But the sign of V, determined by the sign of the q that produces V, is important. Note that at point a the electric fields of the two electrons are equal in magnitude and opposite in direction and sum to zero. But the potentials for the electrons are both negative and *do not* add to zero. Make sure you understand this distinction.

When the third electron (the one that moves) is at point b, the electric potential energy is $U_b = -eV_b = +3.52 \times 10^{-27}$ J. At point c, the potential energy is zero. All of the initial electrical potential energy has been converted to kinetic energy because of the positive work done on it by the repulsive forces of the other two electrons. The negatively charged electron gains kinetic energy when it moves from a lower-potential point to a higher-potential point, in this case from $V_b = -2.2 \times 10^{-8}$ V to $V_c = 0$.

Note that the net force on the third electron decreases as that electron moves away from point b. Its acceleration is not constant, so constant-acceleration equations *cannot* be used to find its final speed. But conservation-of-energy principles are easy to apply.

Practice Problem: The electron at point b is replaced with a proton (mass $m_p = 1.67 \times 10^{-27}$ kg) that is released from rest and accelerates toward point a. What is the speed of the proton when it reaches point a? *Answer:* 2.6 m/s.

PROBLEM-SOLVING STRATEGY 18.1 **Calculation of potential**

SET UP AND SOLVE

1. Remember that potential is simply *potential energy per unit charge*. Understanding this simple statement can get you a long way.
2. To find the potential due to a collection of point charges, use Equation 18.13.
3. If you are given an electric field, or if you can find it by using any of the methods of Chapter 17, it may be easier to calculate the work done on a test charge during a displacement from point a to point b. When it's appropriate, make use of your freedom to define V to be zero at some convenient place. For point charges, this will usually be at infinity, but for other distributions of charge, it may be convenient or necessary to define V to be zero at some finite distance from the charge distribution—say, at point b. This is just like defining U to be zero at ground level in gravitational problems.

REFLECT

4. Remember that potential is a *scalar* quantity, not a *vector*. It doesn't have components. It would be seriously wrong to try to use components of potential.

EXAMPLE 18.4 **Parallel plates**

Find the potential at any height y between the two charged parallel plates discussed at the beginning of Section 18.1.

SOLUTION

SET UP Figure 18.10 shows our sketch. As before, we point the y axis upward. The electric field is uniform and directed vertically downward. We choose the potential V to be zero at $y = 0$ (point b in our sketch). The potential increases linearly as we move toward the upper plate.

SOLVE The potential energy U for a test charge q' at a distance y above the bottom plate is given by Equation 18.5, $U = q'Ey$. The potential V at point y is the potential energy per unit charge, $V = U/q'$, so

$$V = Ey.$$

Even if we had chosen a different reference level (at which $V = 0$), it would still be true that $V_y - V_b = Ey$. At point a, where $y = d$ and $V_y = V_a$, $V_a - V_b = Ed$ and

$$E = \frac{V_a - V_b}{d} = \frac{V_{ab}}{d}.$$

REFLECT The magnitude of the electric field equals the potential difference between the plates, divided by the distance between them. (*Caution!* This relation holds only for the parallel-plate

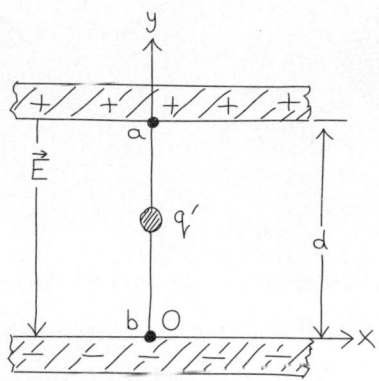

▲ **FIGURE 18.10** Our sketch for this problem.

arrangement described here, in which the electric field is uniform.)

Practice Problem: Suppose that in this problem we had chosen the potential to be zero at the upper plate, where $y = d$. Derive an expression for the potential at any value of y. *Answer:* $V = E(y - d)$.

We have defined potential as potential energy per unit charge, so

$$1\ V = \frac{1\ J}{1\ C} = \frac{(1\ N)(1\ m)}{1\ C}, \quad \text{and} \quad 1\ N/C = 1\ V/m.$$

Thus, the unit of electric field can be expressed as 1 *volt per meter* $(1\ V/m)$, as well as 1 N/C:

$$1\ V/m = 1\ N/C.$$

That is, we can think of electric field either as force per unit charge or as potential difference per unit distance. In practice, the latter (the volt per meter) is the usual unit of E.

18.3 Equipotential Surfaces

Field lines (Section 17.7) help us visualize electric fields. In a similar way, the potential at various points in an electric field can be represented graphically by **equipotential surfaces.** An equipotential surface is defined as a surface on which the potential is the same at every point. In a region where an electric field is present, we can construct an equipotential surface through any point. In diagrams, we usually show only a few representative equipotentials, often with equal potential differences between adjacent surfaces. No point can be at two different potentials, so equipotential surfaces for different potentials can never touch or intersect. We can't draw three-dimensional surfaces on a two-dimensional diagram, so we draw lines representing the intersections of equipotential surfaces with the plane of the diagram.

The potential energy for a test charge is the same at every point on a given equipotential surface, so the $\vec{E}$ field does no work on a test charge when it moves from point to point on such a surface. It follows that the $\vec{E}$ field can never have a

— Electric field lines

— Cross sections of equipotential
  surfaces at 20 V intervals

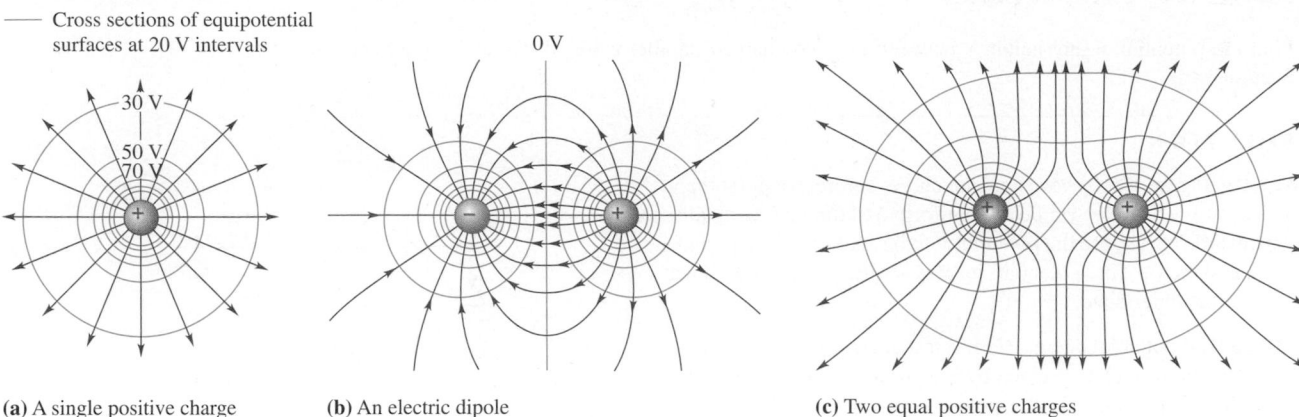

(a) A single positive charge   **(b)** An electric dipole   **(c)** Two equal positive charges

▲ **FIGURE 18.11** Equipotential surfaces and electric field lines for assemblies of point charges. How would the diagrams change if the charges were reversed?

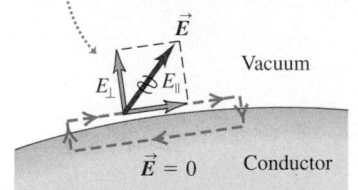

This doesn't happen!
If the electric field at the surface of a conductor had a tangential component $E_\parallel$, the electron could move in a loop with net work done.

▲ **FIGURE 18.12** At all points on the surface of a conductor, the electric field must be perpendicular to the surface. If it had a tangential component, conservation of energy could be violated, as shown here.

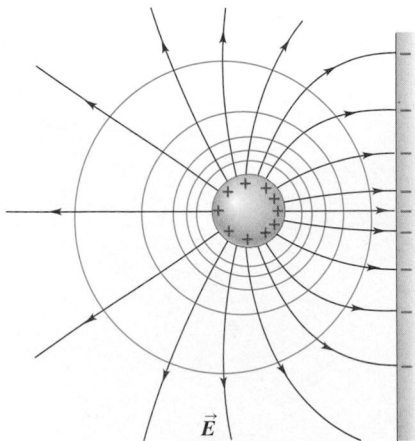

▲ **FIGURE 18.13** When charges are at rest, a conducting surface is always an equipotential surface. Field lines are perpendicular to a conducting surface.

component tangent to the surface; such a component would do work on a charge moving on the surface.

Therefore, $\vec{E}$ must be perpendicular to the surface at every point. **Field lines and equipotential surfaces are always mutually perpendicular.** In general, field lines are curves and equipotentials are curved surfaces. For the special case of a *uniform* field, in which the field lines are straight, parallel, and equally spaced, the equipotentials are parallel *planes* perpendicular to the field lines.

Figure 18.11 shows several arrangements of charges. The field lines in the plane of the charges are represented by red lines, and the intersections of the equipotential surfaces with that plane (that is, cross sections of those surfaces) are shown as blue lines. The actual field lines and equipotential surfaces are three dimensional. At each crossing of an equipotential and a field line, the two are perpendicular.

We can prove that when all charges are at rest, **the electric field just outside a conductor must be perpendicular to the surface at every point.** We know that $\vec{E} = 0$ at every point inside the conductor; otherwise, charges would move. In particular, the component of $\vec{E}$ tangent to the surface, just inside it at any point, is zero. It follows that the tangential component of $\vec{E}$ is also zero at every point just *outside* the surface. If it were not, a charge could move around a rectangular path partly inside and partly outside (Figure 18.12) and return to its starting point with a net amount of work having been done on it. This would violate the conservative nature of electrostatic fields. We conclude that the tangential component of $\vec{E}$ just outside the surface must be zero at every point on the surface. Thus, $\vec{E}$ is perpendicular to the surface at each point (Figure 18.13). It follows that, in an electrostatic situation, **a conducting surface is always an equipotential surface.**

Potential Gradient

We can draw equipotentials so that adjacent surfaces have equal potential differences. Then, in regions where the magnitude of $\vec{E}$ is large, the equipotential surfaces are close together because the field does a relatively large amount of work on a test charge in a relatively small displacement. Conversely, in regions where the field is weaker, the equipotential surfaces are farther apart.

To state this relationship more quantitatively, suppose we have two adjacent equipotential surfaces separated by a small distance Δs, with a potential difference ΔV between them. If Δs is very small, the electric field is approximately constant over that distance, so the work done by the electric field on a test charge q' that moves from one surface to the other in the direction of $\vec{E}$ is equal to $q'E\,\Delta s$. But from the definition of potential (potential energy per unit charge), this work is also equal to $-q' \Delta V$. Equating these two expressions, we find that $q'E \Delta s = -q' \Delta V$.

Electric field represented as potential gradient

The magnitude of the electric field at any point on an equipotential surface equals the rate of change of potential, ΔV, with distance Δs as the point moves perpendicularly from the surface to an adjacent one a distance Δs away:

$$E = -\frac{\Delta V}{\Delta s}. \qquad (18.14)$$

The negative sign shows that when a point moves in the direction of the electric field, the potential decreases. The quantity $\Delta V/\Delta s$, representing a rate of change of V with distance, is called the **potential gradient.** We see that this is an alternative name for electric field.

The relationship stated by Equation 18.14 can be expressed more precisely in terms of a coordinate system. If E_x represents the x component of electric field, then the correct relation is

$$E_x = -\frac{\Delta V}{\Delta x},$$

and similar equations hold for the y and z components of $\vec{E}$.

EXAMPLE 18.5 Equipotential surfaces within a capacitor

Suppose we have a parallel-plate capacitor like the one described at the beginning of this chapter. The plates are separated by 6.0 mm and carry charges of equal magnitude and opposite sign (Figure 18.14). The potential difference between the plates is 24.0 V. Let the potential of the negatively charged plate be zero; then the potential of the positive plate is $+24.0$ V. Draw an enlarged sketch of Figure 18.14; on it, sketch (1) the electric field lines between the plates and (2) the cross sections of the equipotential surfaces for which the potential is $+24.0$ V, $+18.0$ V, $+12.0$ V, $+6.0$ V, and 0.

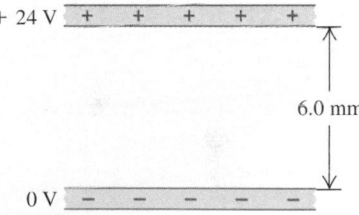

▲ FIGURE 18.14

SOLUTION

SET UP AND SOLVE Figure 18.15 shows our solution. Between the plates, the electric field is uniform and perpendicular to the plates, in the direction from the positive $(+)$ plate toward the negative $(-)$ plate. We draw the field lines evenly spaced to show that the field is uniform.

Conductors are equipotential surfaces, so the negative plate is an equipotential surface with $V = 0$, and the positive plate is an equipotential surface with $V = +24.0$ V. Also, $\Delta V = -E \Delta s$, so the equipotential surfaces between the plates are parallel to the plates. The potential increases linearly as we move from the negative to the positive plate; it changes by 24.0 V in the 6.0 mm between the plates. Thus, it changes by 4.0 V for every 1.0 mm, or by 6.0 V for every 1.5 mm. Thus, the $+6.0$ V, $+12.0$ V, and $+18.0$ V equipotential surfaces are located 1.5 mm, 3.0 mm, and 4.5 mm, respectively, from the negative plate.

REFLECT Our results demonstrate the general principle that electric field lines and equipotential surfaces are perpendicular to each other. Also, because the electric field is uniform between the plates, equipotential surfaces representing equal potential differences are equally spaced. If the electric field were not uniform,

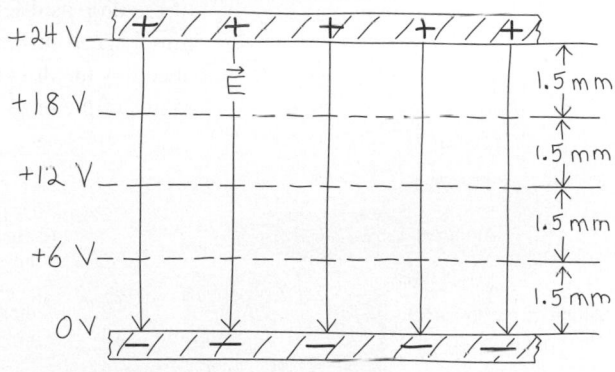

▲ FIGURE 18.15 Our solution to this problem.

the equipotential surfaces would get closer together with increasing electric field magnitude.

Practice Problem: For the capacitor in this example, find (a) the magnitude of the electric field and (b) the potential (again, assuming that $V = 0$ at the negative plate) at a point 4.0 mm from the negative plate and 2.0 mm from the positive plate. *Answers:* 4.0×10^3 V/m; $+16.0$ V.

18.4 The Millikan Oil-Drop Experiment

In Section 17.3, we talked a little about the *quantization* of charge. Have you ever wondered how the charge of an individual electron can be measured? The first solution to this formidable experimental problem was the **Millikan oil-drop experiment,** a brilliant piece of work carried out at the University of Chicago during the years 1909–1913 by Robert Andrews Millikan. In 1923 Millikan was awarded the Nobel prize for this and other related fundamental research.

Millikan's apparatus is shown schematically in Figure 18.16a. Two parallel horizontal metal plates *a* and *b* are insulated from each other and separated by a few millimeters. Oil is sprayed in very fine drops (with a diameter of around 10^{-4} mm) from an atomizer above the upper plate. A few drops are allowed to fall through a small hole in this plate and are observed with a telescope. A scale in the telescope permits precise measurements of the vertical positions of the drops, so their speeds can also be measured. In the process of atomization, some of the oil drops acquire a small electric charge—usually negative, but occasionally positive.

Here's how Millikan measured the charge on a drop. Suppose a drop has a negative charge with absolute value *q* and the plates are maintained at a potential difference such that there is a downward electric field with magnitude *E* between them. The forces on the drop are then its weight *mg* and the upward force *qE*. By adjusting the field *E*, Millikan could make *qE* equal to *mg* (Figure 18.16b). The drop was then in static equilibrium, and $q = mg/E$. The electric-field magnitude *E* is the potential difference V_{ab} divided by the distance *d* between the plates, as we found in Example 18.2. The drops are pulled into spherical shapes by surface tension; the mass *m* of a drop can be determined if its radius *r* is known, because the mass equals the product of the density ρ and the volume $4\pi r^3/3$. So the expression for *q* can be rewritten as

$$q = \frac{4\pi}{3} \frac{\rho r^3 g d}{V_{ab}}.$$

Everything on the right side of this equation is easy to measure, except for the radius *r* of the drop, which is much too small to measure directly with any degree of precision. But here comes an example of Millikan's genius: He determined the drop's radius by cutting off the electric field and measuring the *terminal speed* v_t of the drop as it fell (Figure 18.16c). (You may want to review the concept of terminal speed in Example 5.13). At the terminal speed, the weight *mg* is just balanced by the drag force F_D due to air resistance. This force in turn depends on the radius of the drop, so measuring the terminal speed enabled Millikan to calculate

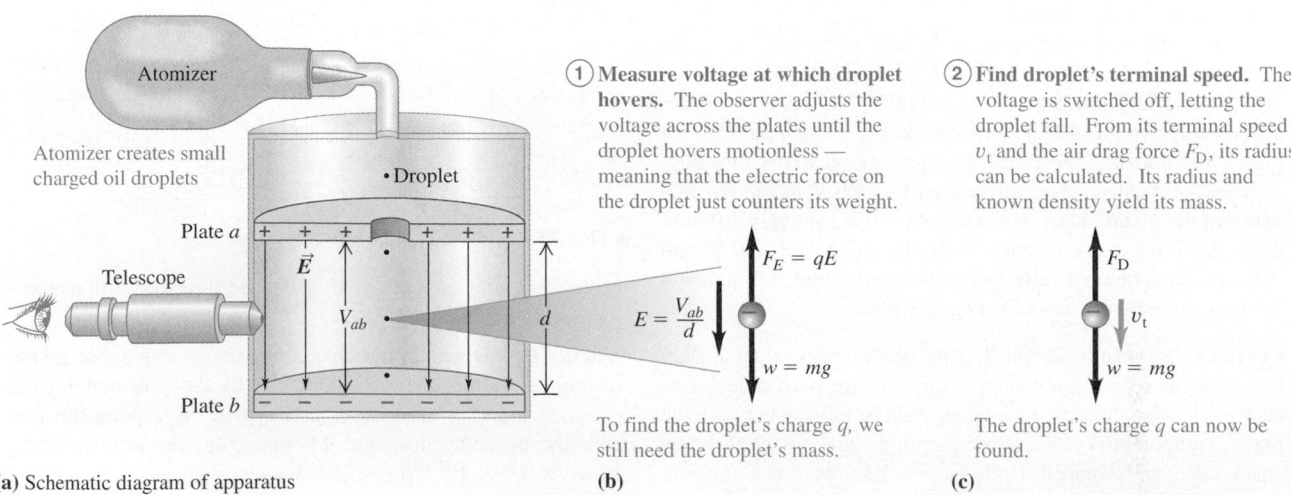

(a) Schematic diagram of apparatus

Atomizer

Atomizer creates small charged oil droplets

Plate *a*

Telescope

$\vec{E}$

•Droplet

V_{ab}

Plate *b*

d

(b)

①**Measure voltage at which droplet hovers.** The observer adjusts the voltage across the plates until the droplet hovers motionless — meaning that the electric force on the droplet just counters its weight.

$F_E = qE$

$E = \dfrac{V_{ab}}{d}$

$w = mg$

To find the droplet's charge *q*, we still need the droplet's mass.

(c)

②**Find droplet's terminal speed.** The voltage is switched off, letting the droplet fall. From its terminal speed v_t and the air drag force F_D, its radius can be calculated. Its radius and known density yield its mass.

F_D

v_t

$w = mg$

The droplet's charge *q* can now be found.

▲ **FIGURE 18.16** The Millikan oil-drop experiment, which demonstrated that charge is quantized and provided the first determination of *e*.

the radius of a drop. Thus, he was able to determine the electric charge q of a drop in terms of its terminal speed!

Millikan and his coworkers measured the charges of thousands of drops. Within the limits of their experimental error, every drop had a charge equal to some small integer multiple of a basic charge e. That is, they found drops with charges of $\pm 2e$, $\pm 5e$, and so on, but never with a value such as $0.76e$ or $2.49e$. A drop with charge $-e$ has acquired one extra electron; if its charge is $-2e$, it has acquired two extra electrons, and so on.

As we stated in Section 17.4, the present best experimental value of the absolute value of the charge of the electron is

$$|e| = 1.60217653(14) \times 10^{-19}\,\text{C},$$

where the (14) indicates the uncertainty in the last two digits, 53. This is far greater precision than Millikan was able to achieve.

The Electronvolt

The magnitude e of the charge of the electron can be used to define a unit of energy, the *electronvolt,* that is useful in many calculations with atomic and nuclear systems. When a particle with charge q moves from a point where the potential is V_a to a point where it is V_b, the change ΔU in the potential energy U is

$$\Delta U = q(V_b - V_a) = qV_{ba}.$$

If the charge q equals the magnitude e of the electron charge, namely, 1.602×10^{-19} C, and the potential difference is $V_{ba} = 1$ V, then the change in energy is

$$\Delta U = (1.602 \times 10^{-19}\,\text{C})(1\,\text{V}) = 1.602 \times 10^{-19}\,\text{J}.$$

This quantity of energy is defined to be 1 **electronvolt** (1 eV):

$$1\,\text{eV} = 1.602 \times 10^{-19}\,\text{J}.$$

The multiples meV, keV, MeV, GeV, and TeV are often used.

When a particle with charge e moves through a potential difference of 1 V, its change in potential energy has magnitude 1 eV. If the charge is an integer multiple of e, such as Ne, the change in potential energy in electronvolts is N times the potential difference in volts. For example, when an alpha particle, which has charge $2e$, moves between two points with a potential difference of 1000 V, the change in its potential energy is $2(1000\,\text{eV}) = 2000$ eV.

Although we've defined the electronvolt in terms of *potential* energy, we can use it for *any* form of energy, such as the kinetic energy of a moving particle. When we speak of a "1-million-volt electron," we mean an electron with a kinetic energy of 1 million electronvolts (1 MeV), equal to $(10^6)(1.602 \times 10^{-19}\,\text{J}) = 1.602 \times 10^{-13}$ J.

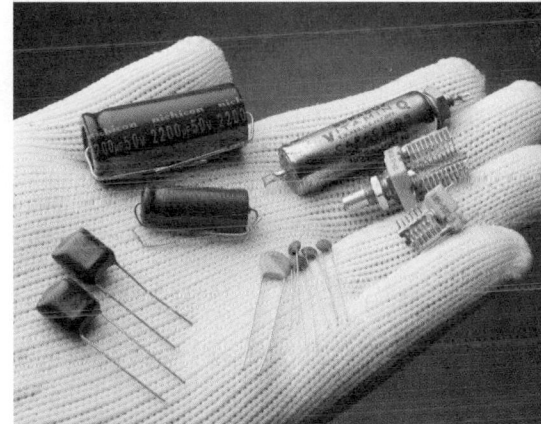

▲ **FIGURE 18.17** An assortment of practical capacitors.

18.5 Capacitors

A capacitor is a device that stores electric potential energy and electric charge. For instance, a camera flash requires a brief burst of power much greater than a camera battery can deliver. The flash is powered by a capacitor, which stores up energy from the battery and delivers it in a pulse when needed. Capacitors are also used in energy-storage units for pulsed lasers, in computer chips that store information, in circuits that improve the efficiency of power transmission lines, and in thousands of other devices. The study of capacitors will help us to develop insight into the behavior of electric fields and their interactions with matter. Figure 18.17 shows several commercial capacitors.

In principle, a capacitor consists of any two conductors separated by vacuum or an insulating material (Figure 18.18). When charges with equal magnitude and

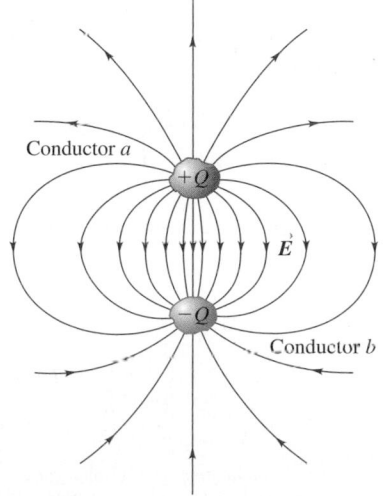

▲ **FIGURE 18.18** Any two conductors separated by vacuum or an insulating material form a capacitor.

opposite sign are placed on the conductors, an electric field is established in the region between the conductors, and there is a potential difference between them. In most practical applications, the conductors have charges with equal magnitude and opposite sign, and the *net* charge on the capacitor is zero. We'll assume throughout this section that that is the case. When we say that a capacitor has charge Q, we mean that the conductor at higher potential has charge Q and the conductor at lower potential has charge $-Q$ (assuming that Q is positive). Keep this in mind in the discussion and examples that follow.

For a capacitor with a given charged surface area, the electric field at any point in the region between the conductors is proportional to the magnitude Q of the charge on each conductor. It follows that the *potential difference V_{ab}* between the conductors is also proportional to Q. If we double the charge Q on the capacitor, the electric field at each point and the potential difference between the conductors both double, but the *ratio* of charge Q to potential difference V_{ab} does not change.

We define the **capacitance C** of a capacitor as follows:

Definition of capacitance

The capacitance C of a capacitor is the ratio of the magnitude of the charge Q on *either* conductor to the magnitude of the potential difference V_{ab} between the conductors:

$$C = \frac{Q}{V_{ab}}. \tag{18.15}$$

Unit: The SI unit of capacitance is called 1 **farad** (1 F), in honor of Michael Faraday. From Equation 18.15, 1 farad is equal to 1 *coulomb per volt* $(1\ \text{C/V})$: $1\ \text{F} = 1\ \text{C/V}$.

In circuit diagrams, a capacitor is represented by either of these symbols:

Parallel-Plate Capacitors

The most common form of capacitor consists of two parallel conducting plates, each with area A, separated by a distance d that is small in comparison with their dimensions (Figure 18.19). Nearly all the field of such a capacitor is localized in the region between the plates, as shown. There is some "fringing" of the field at the edges, also shown in the figure, but if the distance between the plates is small in comparison to their size, we can neglect this effect. The field between the plates is then *uniform,* and the charges on the plates are uniformly distributed over their opposing surfaces. We call this arrangement a **parallel-plate capacitor.**

The electric-field magnitude in the region between the plates is directly proportional to the electric charge *per unit area* of the plate, since the charge on each plate distributes itself evenly over the surface that faces the opposite plate. This quantity is called the **surface charge density** and is denoted by the small Greek letter sigma (σ). For a pair of plates with area A and total charges Q and $-Q$, the surface charge densities on the plates are $\sigma = Q/A$ and $\sigma = -Q/A$. Gauss's law (Section 17.8) can be used to prove that, for the parallel-plate situation, the field magnitude E is related very simply to Q and σ:

$$E = \frac{\sigma}{\epsilon_0} = \frac{Q}{\epsilon_0 A}.$$

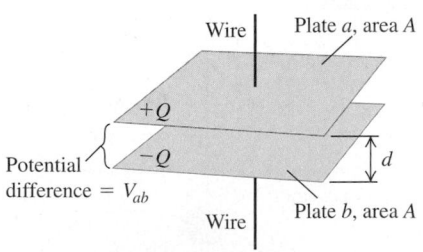

(a) A basic parallel-plate capacitor

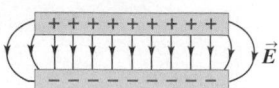

(b) Electric field due to a parallel-plate capacitor

▲ **FIGURE 18.19** The elements of a parallel-plate capacitor.

(We'll omit the details of this derivation.) We introduced the constant ϵ_0 in Section 17.4, where we said that the constant k in Coulomb's law can also be written as $k = 1/4\pi\epsilon_0$, where $\epsilon_0 = 8.854 \times 10^{-12}\,\text{C}^2/\text{N}\cdot\text{m}^2$. The field is uniform; if the distance between the plates is d, then the potential difference (voltage) between them is

$$V_{ab} = Ed = \frac{1}{\epsilon_0}\frac{Qd}{A}.$$

From this relation, we can derive a simple expression for the capacitance of a parallel-plate capacitor:

Capacitance of a parallel-plate capacitor

The capacitance C of a parallel-plate capacitor in vacuum is directly proportional to the area A of each plate and inversely proportional to their separation d:

$$C = \frac{Q}{V_{ab}} = \epsilon_0\frac{A}{d}. \qquad (18.16)$$

The quantities ϵ_0, A, and d are constants for a given capacitor. The capacitance C is therefore a constant, independent of the charge on the capacitor.

In Equation 18.16, if A is in square meters and d in meters, C is in farads. The units of ϵ_0 are $\text{C}^2/(\text{N}\cdot\text{m}^2)$; it follows that

$$1\,\text{F} = 1\,\text{C}^2/(\text{N}\cdot\text{m}) = 1\,\text{C}^2/\text{J}.$$

Because $1\,\text{V} = 1\,\text{J/C}$ (energy per unit charge), this relationship is consistent with our definition $1\,\text{F} = 1\,\text{C/V}$. Finally, the units of ϵ_0 can be expressed as

$$1\,\text{C}^2/(\text{N}\cdot\text{m}^2) = 1\,\text{F/m}.$$

This relation is useful in capacitance calculations, and it also helps us to verify that Equation 18.16 is dimensionally consistent.

The foregoing equations for parallel-plate capacitors are correct when there is only vacuum in the space between the plates. When matter is present, things are somewhat different. We'll return to this topic in Section 18.9.

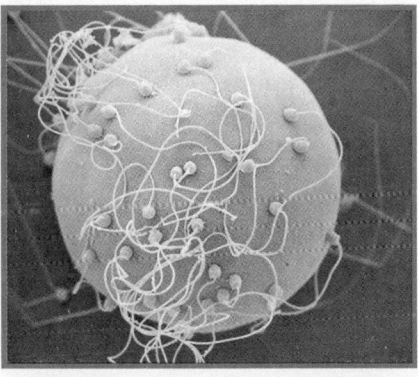

▲ **BIO Application Sea urchin contraception.** The sea urchin egg, when released into the ocean, is assaulted by thousands of sperm. If more than one sperm enters the egg, a condition known as polyspermy, the embryo will develop abnormally and die. In response to the entry of the first sperm, a protective extracellular envelope arises from the egg surface to prevent further sperm from entering. However, that process takes a minute or so, too slow to protect the egg from polyspermy. So a faster electrical response has evolved to block polyspermy. The cell membrane of the unfertilized egg has a potential difference of about 70 mV across it, with the inside of the egg negative with respect to the sea water. Since the membrane is quite thin, the electric field within the membrane is very high. When the first sperm enters the egg, ion channels in the egg membrane open to allow Na^+ to enter, rapidly reversing the potential inside the membrane to positive. A sperm can only fuse with the egg when the membrane's electric field is in the original orientation, so a second sperm trying to penetrate the egg is blocked. If an experimenter reverses the direction of the electrical field, no fertilization takes place.

Quantitative Analysis 18.3 The effect of plate spacing

Suppose two parallel-plate capacitors have the same area A and charge Q, but in capacitor 1 the spacing between the plates is d and in capacitor 2 it is $2d$. If the voltage between the plates in capacitor 1 is V, the voltage of capacitor 2 is

A. $\frac{1}{2}V$.
B. V.
C. $2V$.

SOLUTION There are two ways to approach this problem. First, since the surface charge density is the same for both capacitors, the electric field between their plates is the same. However, if you double the plate separation, a test charge gains or loses twice as much potential energy in moving from one plate to the other. Because voltage (potential) is potential energy divided by charge, doubling the plate separation doubles the voltage (answer C). Alternatively, we could use Equation 18.16, which tells us that $Q/V_{ab} = \epsilon_0 A/d$. Since Q and A are fixed (and ϵ_0 is a constant), doubling d also doubles V_{ab}.

EXAMPLE 18.6 Size of a 1.0 F capacitor

A parallel-plate capacitor has a capacitance of 1.0 F, and the plates are 1.0 mm apart. What is the area of the plates?

SOLUTION

SET UP AND SOLVE For a parallel-plate capacitor in air, $C = \epsilon_0 A/d$. Solving for A gives

$$A = \frac{Cd}{\epsilon_0} = \frac{(1.0\,\text{F})(1.0\times10^{-3}\,\text{m})}{8.85\times10^{-12}\,\text{F/m}} = 1.1\times10^8\,\text{m}^2.$$

Continued

REFLECT This area corresponds to a square about 10 km on a side, an area about a third larger than Manhattan Island! It used to be considered a good joke to send a newly graduated engineer to the stockroom for a 1.0 F capacitor. That is not as funny as it used to be; recently developed technology makes it possible to make 1.0 F capacitors that are a few *centimeters* on a side. One type uses activated carbon granules, of which 1 gram has a surface area of about 1000 m².

Practice Problem: What is the area of each plate of a parallel-plate air capacitor that has a capacitance of 1.0 pF = 1.0 × 10^{-12} F if the separation between the plates is 1.0 mm? If the plates are square, what is the length of each side? *Answers:* 1.1 × 10^{-4} m²; 11 mm.

In many applications, the most convenient units of capacitance are the *microfarad* $(1\ \mu F = 10^{-6}\ F)$ and the *picofarad* $(1\ pF = 10^{-12}\ F)$.

EXAMPLE 18.7 **Properties of a parallel-plate capacitor**

The plates of a parallel-plate capacitor are 5.00 mm apart and 2.00 m² in area. A potential difference of 10.0 kV is applied across the capacitor. Compute **(a)** the capacitance, **(b)** the charge on each plate, and **(c)** the magnitude of the electric field in the region between the plates.

SOLUTION

SET UP Figure 18.20 shows our sketch.

SOLVE Part (a): To find the capacitance, we use Equation 18.16, which expresses capacitance in terms of the area and separation distance of the plates:

$$C = \frac{\epsilon_0 A}{d} = \frac{(8.85 \times 10^{-12}\ F/m)(2.00\ m^2)}{5.00 \times 10^{-3}\ m}$$

$$= 3.54 \times 10^{-9}\ F = 0.00354\ \mu F.$$

Part (b): Now that we know the capacitance, we can use it and the voltage to find the charge on the plates. Remember that capacitance is charge divided by voltage: $C = Q/V_{ab}$. Solving for Q gives

$$Q = CV_{ab} = (3.54 \times 10^{-9}\ F)(1.00 \times 10^4\ V)$$

$$= 3.54 \times 10^{-5}\ C = 35.4\ \mu C.$$

Part (c): To find the electric field magnitude between the plates, we can use the relation $E = -\Delta V/\Delta s$; that is, the electric field magnitude equals the potential gradient. The minus sign in

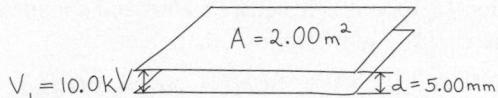

▲ **FIGURE 18.20** Our sketch for this problem.

this equation indicates that the electric field is in the direction of *decreasing* potential. Setting $\Delta V = V_{ab}$ and $\Delta s = d$, and considering only the magnitude of the electric field, we find that

$$E = \frac{V_{ab}}{d} = \frac{1.00 \times 10^4\ V}{5.00 \times 10^{-3}\ m} = 2.00 \times 10^6\ V/m.$$

REFLECT The physical dimensions of this capacitor are quite large. In practice, a capacitor with this C and large A can be constructed by rolling thin parallel conductors into a cylinder.

Practice Problem: Repeat the preceding problem for a capacitor with the same area and potential difference, but with a plate separation of 2.50 mm (half as large). Is the electric field magnitude between the plates larger or smaller than that for the original capacitor? *Answers:* (a) $C = 0.00708\ \mu F$; (b) 70.8 μC; (c) 4.00 × 10^6 V/m; larger.

18.6 Capacitors in Series and in Parallel

Capacitors are manufactured with certain standard capacitances and working voltages. However, these standard values may not be the ones you actually need in a particular circuit. You can obtain the values you need by combining capacitors; the simplest combinations are a series connection and a parallel connection.

Capacitors in Series

Figure 18.21a is a schematic diagram of a **series connection.** Two capacitors are connected in series (one after the other) between points a and b, and a constant potential difference V_{ab} is maintained. The capacitors are both initially uncharged. When a positive potential difference V_{ab} is applied between points a and b, the top plate of C_1 acquires a positive charge Q. The electric field of this positive charge pulls negative charge up to the bottom plate of C_1 until all of the field lines end on the bottom plate (Figure 18.21a) and the lower plate of C_1 has acquired charge $-Q$. These negative charges had to come from the top plate of C_2, which becomes positively charged with charge $+Q$. This positive charge then pulls negative charge $-Q$ from the connection at point b onto the bottom plate of C_2. The total charge on the lower plate of C_1 and the upper plate of C_2 must always be zero, because these plates aren't connected to anything except each other. **In a series connection, the magnitude of the charge on all of the plates is the same, because of conservation of charge. The total potential difference across all of the capacitors is the sum of the individual potential differences.**

Referring again to Figure 18.21a, we have

$$V_{ac} = V_1 = \frac{Q}{C_1}, \qquad V_{cb} = V_2 = \frac{Q}{C_2},$$

$$V_{ab} = V = V_1 + V_2 = Q\left(\frac{1}{C_1} + \frac{1}{C_2}\right),$$

$$\frac{V}{Q} = \frac{1}{C_1} + \frac{1}{C_2}.$$

The **equivalent capacitance** C_{eq} of the series combination is defined as the capacitance of a *single* capacitor for which the charge Q is the same as for the combination when the potential difference V is the same. For such a capacitor, shown in Figure 18.21b,

$$C_{eq} = \frac{Q}{V}, \qquad \text{or} \qquad \frac{1}{C_{eq}} = \frac{V}{Q}.$$

Combining the last two equations, we find that

$$\frac{1}{C_{eq}} = \frac{1}{C_1} + \frac{1}{C_2}.$$

We can extend this analysis to any number of capacitors in series:

Equivalent capacitance of capacitors in series

When capacitors are connected in series, **the reciprocal of the equivalent capacitance of a series combination equals the sum of the reciprocals of the individual capacitances:**

$$\frac{1}{C_{eq}} = \frac{1}{C_1} + \frac{1}{C_2} + \frac{1}{C_3} + \cdots. \qquad \text{(capacitors in series)} \qquad (18.17)$$

The magnitude of charge is the same on all of the plates of all of the capacitors, but the potential differences across individual capacitors are, in general, different.

Capacitors in Parallel

The arrangement shown in Figure 18.22a is called a **parallel connection.** Two capacitors are connected in parallel between points a and b. In this case, the upper plates of the two capacitors are connected together to form an equipotential surface, and the lower plates form another. **The potential difference is the same for both capacitors** and is equal to $V_{ab} = V$. The charges Q_1 and Q_2, which are not necessarily equal, are given by

$$Q_1 = C_1 V, \qquad Q_2 = C_2 V.$$

The *total* charge Q of the combination, and thus on the equivalent capacitor, is

$$Q = Q_1 + Q_2 = V(C_1 + C_2),$$

Capacitors in series:
- The capacitors have the same charge Q.
- Their potential differences add:
  $V_{ac} + V_{cb} = V_{ab}$.

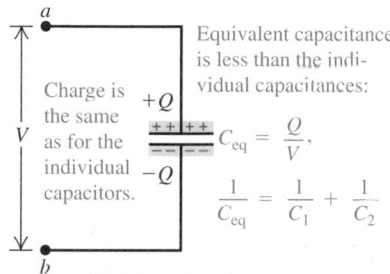

(a) Two capacitors in series

Charge is the same as for the individual capacitors.

Equivalent capacitance is less than the individual capacitances:

$$C_{eq} = \frac{Q}{V},$$

$$\frac{1}{C_{eq}} = \frac{1}{C_1} + \frac{1}{C_2}$$

(b) The equivalent single capacitor

▲ **FIGURE 18.21** The effect of connecting two capacitors in series.

Capacitors in parallel:
- The capacitors have the same potential V.
- The charge on each capacitor depends on its capacitance: $Q_1 = C_1 V$, $Q_2 = C_2 V$.

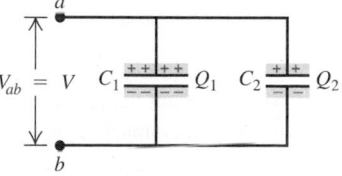

(a) Capacitors connected in parallel

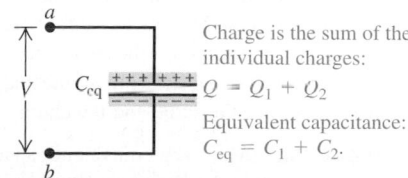

Charge is the sum of the individual charges:
$Q = Q_1 + Q_2$

Equivalent capacitance:
$C_{eq} = C_1 + C_2$.

(b) The equivalent single capacitor

▲ **FIGURE 18.22** The effect of connecting two capacitors in parallel.

so

$$\frac{Q}{V} = C_1 + C_2.$$

The *equivalent* capacitance C_{eq} of the parallel combination is defined as the capacitance of a single capacitor (Figure 18.22b) for which the total charge is the same as in Figure 18.22a. For this capacitor, $Q/V = C_{eq}$, so

$$C_{eq} = C_1 + C_2.$$

In the same way, we can derive an expression for the equivalent capacitance of any number of capacitors in parallel:

Equivalent capacitance of capacitors in parallel
When capacitors are connected in parallel, **the equivalent capacitance of the combination equals the** *sum* **of the individual capacitances:**

$$C_{eq} = C_1 + C_2 + C_3 + \cdots. \qquad \text{(capacitors in parallel)} \qquad (18.18)$$

In a parallel connection, the equivalent capacitance is always *greater than* any individual capacitance; in a series connection, it is always *less than* any individual capacitance.

Conceptual Analysis 18.4

Distribution of charge

The capacitors in Figure 18.23 have the same area and plate separation. The power source functions to separate charges, creating a buildup of positive charge on one side and an equal buildup of negative charge on the other. When the power source is switched on, which of the following choices describes the charge on the plates?

A. Plates A and D have equal and opposite charges; plates B and C are uncharged.
B. All plates have the same magnitude of charge; plates A and B have the same sign, plates C and D the opposite sign.
C. All plates have the same magnitude of charge; plates A and C have the same sign, plates B and D the opposite sign.

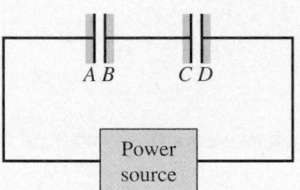

▲ **FIGURE 18.23**

SOLUTION The power source creates equal and opposite charges on plates A and D. The charge on plate A attracts an equal and opposite charge onto plate B, and the charge on plate D attracts an equal and opposite charge onto plate C. Thus, answer C is correct.

EXAMPLE 18.8 **Capacitors in series and in parallel**

Two capacitors, one with $C_1 = 6.0\ \mu$F and the other with $C_2 = 3.0\ \mu$F, are connected to a potential difference of $V_{ab} = 18$ V. Find the equivalent capacitance, and find the charge and potential difference for each capacitor when the two capacitors are connected **(a)** in series and **(b)** in parallel.

SOLUTION

SET UP Figure 18.24 shows our sketches of the two situations. We remember that when capacitors are connected in series, the charges are the same on the two capacitors and the potential differences add. When they are connected in parallel, the potential differences are the same and the charges add.

SOLVE Part (a): The equivalent capacitance for the capacitors in series is given by Equation 18.17:

$$\frac{1}{C_{eq}} = \frac{1}{C_1} + \frac{1}{C_2} = \frac{C_1 + C_2}{C_1 C_2}.$$

Thus,

$$C_{eq} = \frac{C_1 C_2}{C_1 + C_2} = \frac{(6.0\ \mu\text{F})(3.0\ \mu\text{F})}{6.0\ \mu\text{F} + 3.0\ \mu\text{F}} = 2.0\ \mu\text{F}.$$

The charge is $Q = C_{eq}V = (2.0\ \mu\text{F})(18\ \text{V}) = 36\ \mu$C, the same for both capacitors. The voltages are

$$V_1 = \frac{Q}{C_1} = \frac{36\ \mu\text{C}}{6.0\ \mu\text{F}} = 6.0\ \text{V} \quad \text{and} \quad V_2 = \frac{Q}{C_2} = \frac{36\ \mu\text{C}}{3.0\ \mu\text{F}} = 12.0\ \text{V}.$$

Note that $V_1 + V_2 = V_{ab}$ (i.e., 6.0 V + 12 V = 18 V).

Continued

Part (b): When capacitors are connected in parallel, the potential differences are the same and the charges add. The equivalent capacitance is given by Equation 18.18:

$$C_{eq} = C_1 + C_2 = 6.0\ \mu\text{F} + 3.0\ \mu\text{F} = 9.0\ \mu\text{F}.$$

The potential difference for the equivalent capacitor is equal to the potential difference for each capacitor:

$$V_1 = V_2 = V_{ab} = 18\ \text{V}.$$

The charges of the capacitors are

$$Q_1 = C_1V = (6.0\ \mu\text{F})(18\ \text{V}) = 108\ \mu\text{C},$$
$$Q_2 = C_2V = (3.0\ \mu\text{F})(18\ \text{V}) = 54\ \mu\text{C}.$$

The total charge is $Q_1 + Q_2 = Q$, so the charge on the equivalent capacitor is $Q = C_{eq}V = (9.0\ \mu\text{F})(18\ \text{V}) = 162\ \mu\text{C}$.

REFLECT For capacitors in series, the *larger* potential difference appears across the capacitor with the *smaller* capacitance. For capacitors in parallel, the *greater* charge is on the capacitor with the *larger* capacitance. For capacitors in series, the equivalent capacitance is *smaller* than the capacitance of any of the individual capacitors. (Don't try to memorize rules like these; instead, focus on understanding how capacitors work so that the "rules" become self-evident.) Also, for capacitors in parallel, the equivalent capacitance is *larger* than the capacitance of any of the individual capacitors, so, for a given total potential difference, the parallel combination stores more charge than the series combination.

Practice Problem: Repeat this problem for $V_{ab} = 18\ \text{V}$ and $C_1 = C_2 = 9.0\ \mu\text{F}$. Also, (c) find the ratio of total stored charge for the parallel combination to that for the series combination. *Answers:* (a) $C_{eq} = 4.5\ \mu\text{F}$, $Q_1 = Q_2 = 81\ \mu\text{C}$, $V_1 = V_2 = 9\ \text{V}$. (b) $C_{eq} = 18\ \mu\text{F}$, $V_1 = V_2 = 18\ \text{V}$, $Q_1 = Q_2 = 162\ \mu\text{C}$; (c) 4.0.

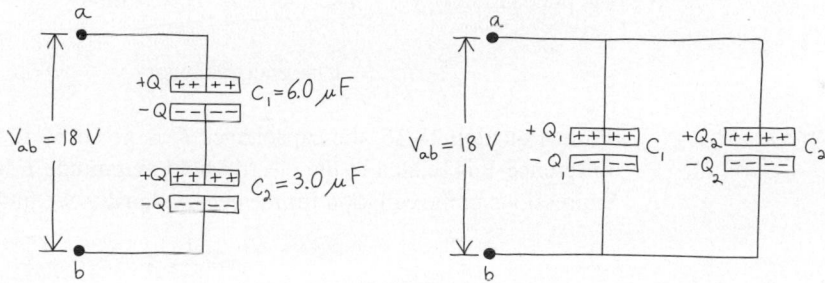

▲ **FIGURE 18.24** Our sketches for this problem.

18.7 Electric Field Energy

Many of the most important applications of capacitors depend on their ability to store energy. The capacitor plates, with opposite charges, separated and attracted toward each other, are analogous to a stretched spring or an object lifted in the earth's gravitational field. The potential energy corresponds to the energy input required to charge the capacitor and to the work done by the electrical forces when it discharges. This work is analogous to the work done by a spring or the earth's gravity when the system returns from its displaced position to the reference position.

One way to calculate the potential energy U of a charged capacitor is to calculate the work W required to charge it. The final charge Q and the final potential difference V are related by $Q = CV$. Let v and q be the varying potential difference and charge, respectively, during the charging process; then, at any instant, $v = q/C$. The work ΔW required to transfer an additional small element of charge Δq is

$$\Delta W = v\,\Delta q = \frac{q\,\Delta q}{C}.$$

Because the work needed to add Δq increases in direct proportion to the amount of charge q already present, and therefore to the potential difference v already created, the *total* work W needed to increase the charge q from zero to a final value Q is the *average* potential difference $V/2$ during the charging process, multiplied by the final charge Q:

$$U = W_{\text{total}} = \left(\frac{V}{2}\right)Q = \frac{Q^2}{2C} = \frac{1}{2}CV^2. \qquad (18.19)$$

If we define the potential energy of an *uncharged* capacitor to be zero, then W_{total} is equal to the potential energy U of the charged capacitor. When Q is in coulombs, C in farads (coulombs per volt), and V in volts (joules per coulomb), U is in joules.

A charged capacitor is the electrical analog of a stretched spring with elastic potential energy $U = \frac{1}{2}kx^2$. The charge Q is analogous to the elongation x, and the *reciprocal* of the capacitance, $1/C$, is analogous to the force constant k. The energy supplied to a capacitor in the charging process is analogous to the work we do on the spring when we stretch it.

Energy Density in an Electric Field

The energy stored in a capacitor is related directly to the electric field between the capacitor plates. In fact, we can think of the energy as stored *in the field* in the region between the plates. To develop this relation, let's find the energy *per unit volume* in the space between the plates of a parallel-plate capacitor with plate area A and separation d. We call this quantity the **energy density** and denote it by u. From Equation 18.19, the total energy is $U = \frac{1}{2}CV^2$, and the volume between the plates is simply Ad. The energy density is thus

$$u = \text{energy density} = \frac{\frac{1}{2}CV^2}{Ad}.$$

From Equation 18.16, the capacitance C is given by $C = \epsilon_0 A/d$. The potential difference V is related to the electric field magnitude E by $V = Ed$. Using these expressions in the equation for the energy density, we find that

$$u = \frac{1}{2}\epsilon_0 E^2. \tag{18.20}$$

We've derived Equation 18.20 only for one specific kind of capacitor, but it turns out to be valid for any capacitor in vacuum and, indeed, *for any electric field configuration in vacuum*. This result has an interesting implication: We think of vacuum as space with no matter in it, but vacuum can nevertheless have electric fields and therefore energy. In other words, it isn't necessarily just empty space. We'll use Equation 18.20 in Chapter 23 in connection with the energy transported through space by electromagnetic waves.

EXAMPLE 18.9   Stored energy

A capacitor with $C_1 = 8.0 \, \mu\text{F}$ is connected to a potential difference $V_0 = 120$ V, as shown in Figure 18.25a. **(a)** Find the magnitude of charge Q_0 and the total energy stored after the capacitor has become fully charged. **(b)** Without any charge being lost from the plates, the capacitor is disconnected from the source of potential difference and connected to a second capacitor $C_2 = 4.0 \, \mu\text{F}$ that is initially uncharged (Figure 18.25b). After the charge has finished redistributing between the two capacitors, find the charge and potential difference for each capacitor, and find the total stored energy.

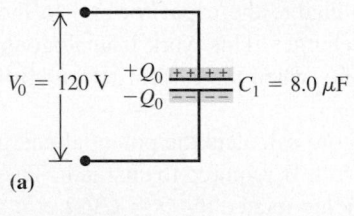

(a)

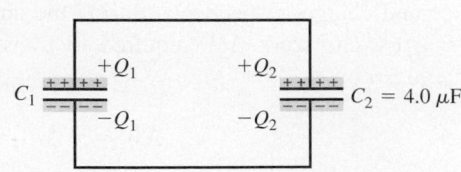

▶ **FIGURE 18.25**   **(b)**

SOLUTION

SET UP After the two capacitors are connected (Figure 18.25b), the two upper plates of the capacitors are connected by a conducting wire. Therefore, they become a single conductor and form a single equipotential surface. Both lower plates are also at the same potential, different from that of the upper plates. The final potential difference between the plates, V, is thus the same for both capacitors. The final charges are given by $Q_1 = C_1/V$ and $Q_2 = C_2/V$.

Continued

The positive charge Q_0 originally on one plate of C_1 becomes distributed over the upper plates of both capacitors, and the negative charge $-Q_0$ originally on the other plate of C_1 becomes distributed over the lower plates of both.

SOLVE Part (a): For the original capacitor, we use the potential difference and the capacitance to find the charge: $Q_0 = C_1 V_0 = (8.0 \,\mu\text{F})(120 \text{ V}) = 960 \,\mu\text{C}$. To find the stored energy, we use Equation 18.19:

$$U = \frac{1}{2} Q_0 V_0 = \frac{1}{2}(960 \times 10^{-6} \text{ C})(120 \text{ V}) = 0.058 \text{ J}.$$

Part (b): From conservation of charge, $Q_1 + Q_2 = Q_0$. Since V is the same for both capacitors, $Q_1 = C_1 V$ and $Q_2 = C_2 V$. When we substitute these equations into the conservation-of-charge equation, we find that $C_1 V + C_2 V = Q_0$ and

$$V = \frac{Q_0}{C_1 + C_2} = \frac{960 \,\mu\text{C}}{12 \,\mu\text{F}} = 80 \text{ V}.$$

Then $Q_1 = C_1 V = 640 \,\mu\text{C}$ and $Q_2 = C_2 V = 320 \,\mu\text{C}$.

The final total stored energy is the sum of the energies stored by each capacitor:

$$\frac{1}{2} Q_1 V + \frac{1}{2} Q_2 V = \frac{1}{2}(Q_1 + Q_2) V = \frac{1}{2} Q_0 V$$
$$= \frac{1}{2}(960 \times 10^{-6} \text{ C})(80 \text{ V}) = 0.038 \text{ J}.$$

REFLECT As we would expect, the potential difference across C_1 decreases when the capacitors are connected, because the first capacitor gives some energy to the second one. Since the two capacitors have the same V, they are in parallel, but the concept of an equivalent capacitance is not needed here.

We see that the final stored energy is only 65% of the initial value. As charge moves to the new capacitor, some of the stored energy is converted to other forms: The conductors become a little warmer, and some energy is radiated as electromagnetic waves. We'll study these phenomena in later chapters.

Practice Problem: In this example, it $C_2 = 8.0 \,\mu\text{F}$ (equal to C_1), what fraction of the original stored energy remains after C_1 is connected to C_2? *Answer:* 50%.

18.8 Dielectrics

Most capacitors have a nonconducting material, or **dielectric,** between their plates. A common type of capacitor uses two strips of metal foil for the plates and two strips of plastic sheet such as Mylar® for the dielectric. (Figure 18.26). The resulting sandwich is rolled up, forming a unit that can provide a capacitance of several microfarads in a compact package.

Placing a solid dielectric between the plates of a capacitor serves three functions. First, it solves the mechanical problem of maintaining two large metal sheets at a very small separation without actual contact.

Second, any dielectric material, when subjected to a sufficiently large electric field, undergoes *dielectric breakdown,* a partial ionization that permits conduction through a material that is supposed to be an insulator. Many insulating materials can tolerate stronger electric fields without breakdown than can air.

Third, the capacitance of a capacitor of given dimensions is *greater* when there is a dielectric material between the plates than when there is air or vacuum. We can demonstrate this effect with the aid of a sensitive electrometer, a device that measures the potential difference between two conductors without letting any appreciable charge flow from one to the other. Figure 18.27a shows an electrometer connected across a charged capacitor with charge of magnitude Q on each plate and potential difference V_0. When we insert a sheet of dielectric between the

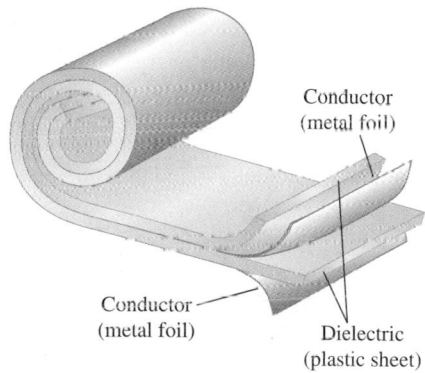

▲ **FIGURE 18.26** A common type of parallel-plate capacitor is made from a rolled-up sandwich of metal foil and plastic film.

Conductor (metal foil)

Conductor (metal foil)

Dielectric (plastic sheet)

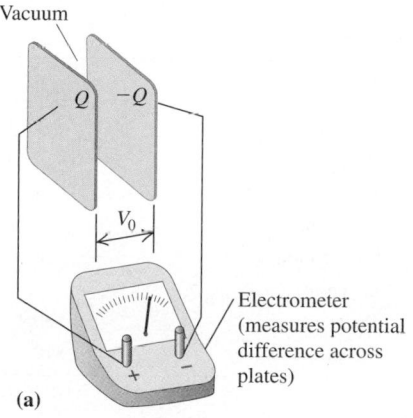

Vacuum

Electrometer (measures potential difference across plates)

(a)

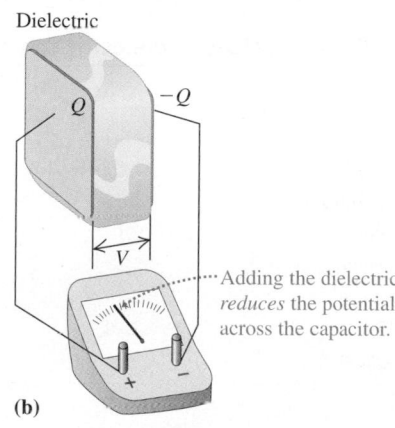

Dielectric

Adding the dielectric *reduces* the potential across the capacitor.

(b)

◄ **FIGURE 18.27** The effect of placing a dielectric between the plates of a parallel-plate capacitor.

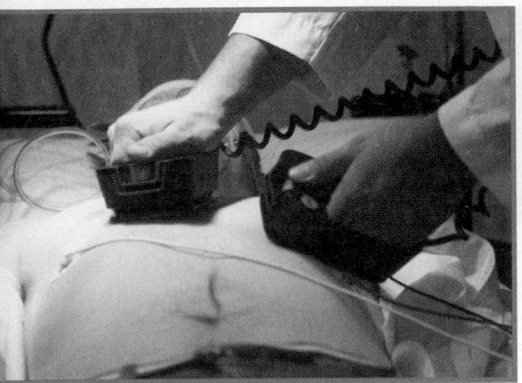

plates, the potential difference *decreases* to a smaller value V (Figure 18.27b). When we remove the dielectric, the potential difference returns to its original value V_0, showing that the original charges on the plates have not changed.

The original capacitance C_0 is given by $C_0 = Q/V_0$, and the capacitance C with the dielectric present is $C = Q/V$. The charge Q is the same in both cases, and V is less than V_0, so we conclude that the capacitance C with the dielectric present is *greater* than C_0. When the space between the plates is completely filled by the dielectric, the ratio of C to C_0 (equal to the ratio of V_0 to V) is called the **dielectric constant** of the material, K:

$$K = \frac{C}{C_0}. \tag{18.21}$$

When the charge is constant, the potential difference is *reduced* by a factor K:

$$V = \frac{V_0}{K}.$$

The dielectric constant K is a pure number. Because C is always greater than C_0, K is always greater than 1. A few representative values of K are given in Table 18.1. For vacuum, $K = 1$ by definition. For air at ordinary temperatures and pressures, K is about 1.0006; this is so nearly equal to 1 that, for most purposes, a capacitor in air is equivalent to one in vacuum.

When a dielectric material is inserted between the plates of a capacitor while the charge is kept constant, the potential difference between the plates decreases by a factor K. Therefore, the electric field between the plates must decrease by the same factor. (Refer to Equation 18.14 if you don't see why.) If E_0 is the vacuum value and E the value with the dielectric, then

$$E = \frac{E_0}{K}. \tag{18.22}$$

The fact that E is smaller when the dielectric is present means that the surface charge density is also smaller. The surface charge on the conducting plates does not change, but an *induced* charge of the opposite sign appears on each surface of the dielectric (Figure 18.28). This surface charge is a result of the redistribution of charge within the molecules of the dielectric material, a phenomenon called **polarization.** We'll discuss its molecular basis in Section 18.10.

For a given charge density σ_i, the induced charges on the dielectric's surfaces reduce the electric field between the plates.

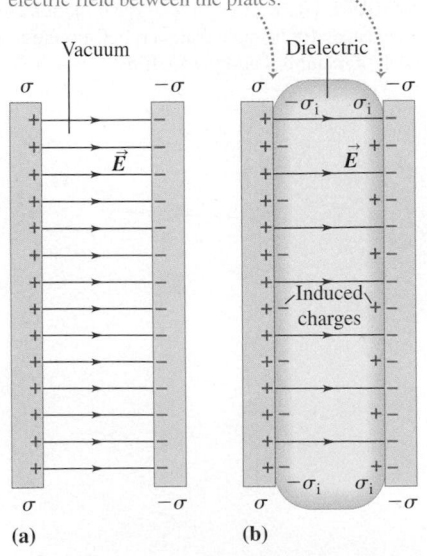

▲ **FIGURE 18.28** The effect of a dielectric on the electric field between the plates of a capacitor.

TABLE 18.1 **Values of dielectric constant K at 20°C**

Material	K	Material	K
Vacuum	1	Polyvinyl chloride	3.18
Air (1 atm)	1.00059	Plexiglas®	3.40
Air (100 atm)	1.0548	Glass	5–10
Teflon®	2.1	Neoprene	6.70
Polyethylene	2.25	Germanium	16
Benzene	2.28	Glycerin	42.5
Mica	3.–6	Water	80.4
Mylar®	3.1	Strontium titanate	310

EXAMPLE 18.10 Effect of a dielectric

The plates of an air-filled parallel-plate capacitor each have area 2.00×10^3 cm^2 and are 1.00 cm apart. The capacitor is connected to a power supply and charged to a potential difference of $V_0 = 3.00$ kV. The capacitor is then disconnected from the power supply without any charge being lost from its plates. After the capacitor has been disconnected, a sheet of insulating plastic material is inserted between the plates, completely filling the space between them. When this is done, the potential difference between the plates decreases to 1.00 kV, while the charge remains constant. (a) What is the capacitance of the capacitor before and after the dielectric is inserted? (b) What is the dielectric constant of the plastic? (c) What is the electric field between the plates before and after the dielectric is inserted?

SOLUTION

SET UP Figure 18.29 shows our before and after sketches.

SOLVE Part (a): The presence of the dielectric increases the capacitance. Without the dielectric, the capacitance is

$$C_0 = \frac{\epsilon_0 A}{d} = \frac{(8.85 \times 10^{-12}\text{ F/m})(0.200\text{ m}^2)}{0.010\text{ m}}$$
$$= 1.77 \times 10^{-10}\text{ F} = 177\text{ pF}.$$

The original charge on the capacitor is

$$Q = C_0 V_0 = (1.77 \times 10^{-10}\text{ F})(3.00 \times 10^3\text{ V})$$
$$= 5.31 \times 10^{-7}\text{ C} = 0.531\ \mu\text{C}.$$

After the dielectric is inserted, the charge is still $Q = 0.531\ \mu$C, but now $V = 1.00$ kV, so

$$C = \frac{Q}{V} = \frac{0.531 \times 10^{-6}\text{ C}}{1.00 \times 10^3\text{ V}} = 5.31 \times 10^{-10}\text{ F} = 531\text{ pF}.$$

Part (b): By definition, the dielectric constant is

$$K = C/C_0 = (531\text{ pF})/(177\text{ pF}) = 3.00.$$

Note that this is also

$$K = V_0/V = (3.00\text{ kV})/(1.00\text{ kV}) = 3.00.$$

Part (c): Since the electric field is uniform between the plates, $V = Ed$. Hence,

$$E_0 = \frac{V_0}{d} = \frac{3.00\text{ kV}}{0.0100\text{ m}} = 3.00 \times 10^5\text{ V/m}$$

before the dielectric is inserted, and

$$E = \frac{V}{d} = \frac{1.00\text{ kV}}{0.0100\text{ m}} = 1.00 \times 10^5\text{ V/m}$$

after it is inserted.

REFLECT Both the potential difference and the electric field decrease by a factor of K. The charge Q is the same with and without the dielectric; the induced surface charge on the dielectric causes a decrease in the net field between the plates. The electric field and the potential difference are proportional, so when E decreases, V decreases.

Practice Problem: Suppose we repeat the preceding experiment, but this time keep the capacitor connected to the power supply while inserting the dielectric, so that the potential difference across the capacitor remains at 3.00 kV. Find (a) the charge on the plates and (b) the electric field between the plates before and after the dielectric is inserted? *Answers:* (a) 0.531 μC; 1.59 μC. (b) 3.00×10^5 V/m; 3.00×10^5 V/m.

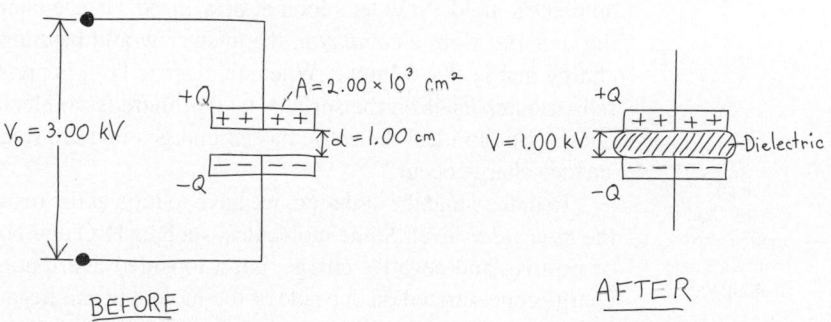

▲ **FIGURE 18.29** Our sketches for this problem.

Dielectric Breakdown

We mentioned earlier that when any dielectric material is subjected to a sufficiently strong electric field, it becomes a conductor. This phenomenon is called **dielectric breakdown.** Conduction occurs when the electric field is so strong that

▲ **FIGURE 18.30** Dielectric breakdown in the laboratory and in nature. The left-hand photo shows a block of Plexiglas® subjected to a very strong electric field; the pattern was etched by flowing charge.

electrons are ripped loose from their molecules and crash into other molecules, liberating even more electrons. This avalanche of moving charge, forming a spark or arc discharge, often starts quite suddenly. Figure 18.30a shows a beautiful laboratory example of dielectric breakdown. Lightning represents dielectric breakdown in air under the action of a sufficiently large potential difference between cloud and ground (Figure 18.30b).

Capacitors always have maximum voltage ratings. When a capacitor is subjected to excessive voltage, an arc may form through a layer of dielectric, burning or melting a hole in it. This hole then provides a conducting path between the conductors, creating a short circuit. If the path remains after the arc is extinguished, the device is rendered permanently useless as a capacitor. The maximum electric field a material can withstand without the occurrence of breakdown is called its **dielectric strength.** The dielectric strength of dry air is about $3 \times 10^6 \text{ V/m}$. Values of dielectric strength for common insulating materials are typically in the range from 1 to $6 \times 10^7 \text{ V/m}$, all substantially greater than that for air.

18.9 Molecular Model of Induced Charge

In the preceding section, we discussed induced surface charges on a dielectric in an electric field. Now let's look at how these surface charges can come about. If the material were a *conductor,* the answer would be simple: Conductors contain charge that is free to move. When an electric field is present, some of the charge redistributes itself on the surface, so that there is no electric field inside the conductor. But an ideal dielectric has *no* charges that are free to move, so how can a surface charge occur?

To understand the situation, we have to look at the rearrangement of charges at the *molecular* level. Some molecules, such as H_2O and N_2O, have equal amounts of positive and negative charge, but a lopsided distribution, with excess positive charge concentrated on one side of the molecule and negative charge on the other. This arrangement is called an *electric dipole,* and the molecule is called a *polar molecule.* When no electric field is present in a gas or liquid with polar molecules, the molecules are oriented randomly (Figure 18.31a). When they are placed in an electric field, however, they tend to orient themselves as in Figure 18.31b as a result of the electric-field forces.

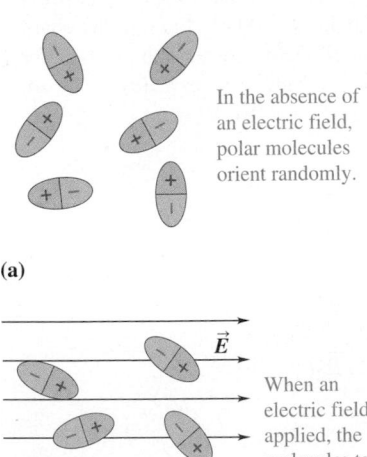

In the absence of an electric field, polar molecules orient randomly.

(a)

$\vec{E}$

When an electric field is applied, the molecules tend to align with it.

(b)

▲ **FIGURE 18.31** The effect of an electric field on a group of polar molecules.

Even a molecule that is *not* ordinarily polar *becomes* a dipole when it is placed in an electric field, because the field pushes the positive charges in the molecule in the direction of the field and pushes the negative charges in the opposite direction. This action causes a redistribution of charge within the molecules (Figure 18.32). Such dipoles are called *induced* dipoles.

With either polar or nonpolar molecules, the redistribution of charge caused by the field leads to the formation of a layer of charge on each surface of the dielectric material. The charges are not free to move indefinitely, as they would be in a conductor, because each charge is bound to a molecule. They are in fact called *bound charges,* to distinguish them from the *free charges* that are added to and removed from the conducting capacitor plates. In the interior of the material, the net charge per unit volume remains zero. This redistribution of charge is called **polarization,** and the material is said to be *polarized.*

The four parts of Figure 18.33 show the behavior of a slab of dielectric when it is inserted into the field between a pair of oppositely charged capacitor plates. Figure 18.33a shows the original field. Figure 18.33b is the situation after the dielectric has been inserted, but before any rearrangement of charges has occurred. The thinner arrows in Figure 18.33c show the additional field set up in the dielectric by the induced surface charges. This field is *opposite* to the original field, but it is not great enough to cancel it completely, because the charges in the dielectric are not free to move indefinitely. The field in the dielectric is therefore decreased in magnitude. The resultant field is shown in Figure 18.33d. In the field-line representation, some of the field lines leaving the positive plate go through the dielectric, and others terminate on the induced charges on the faces of the dielectric.

In the absence of an electric field, nonpolar molecules are not electric dipoles.

(a)

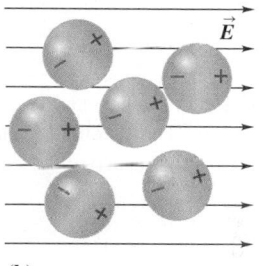

An electric field causes the molecules' positive and negative charges to separate slightly, making the molecule effectively polar.

(b)

▲ **FIGURE 18.32** The effect of an electric field on a group of nonpolar molecules.

MasteringPHYSICS®

PhET: Molecular Motors
PhET: Optical Tweezers and Applications
PhET: Stretching DNA

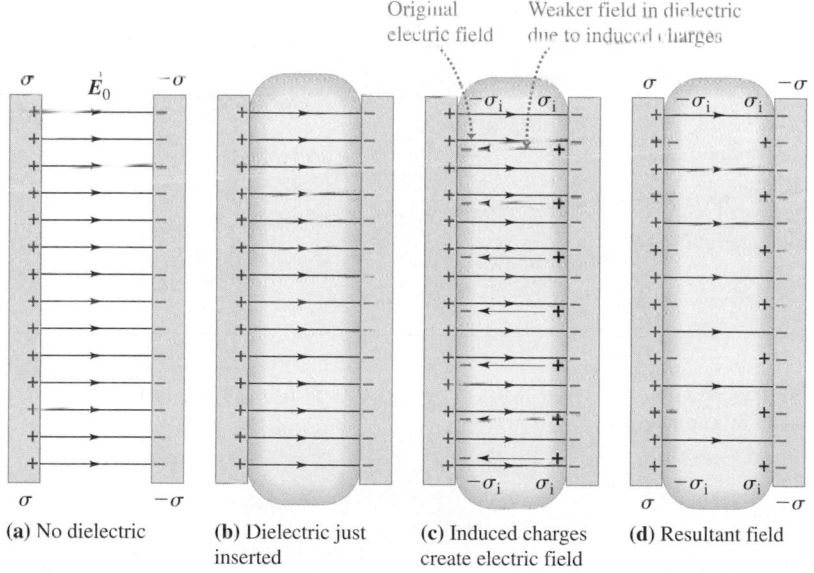

(a) No dielectric (b) Dielectric just inserted (c) Induced charges create electric field (d) Resultant field

▲ **FIGURE 18.33** How a dielectric reduces the electric field between capacitor plates.

SUMMARY

Electric Potential Energy

(Section 18.1) The work W done by the electric-field force on a charged particle moving in a field can be represented in terms of potential energy U: $W_{a \to b} = U_a - U_b$ (Equation 18.2). For a charge q' that undergoes a displacement $\vec{s}$ parallel to a uniform electric field, the change in potential energy is $U_a - U_b = q'Es$ (Equation 18.5). The potential energy for a point charge q' moving in the field produced by a point charge q at a distance r from q' is

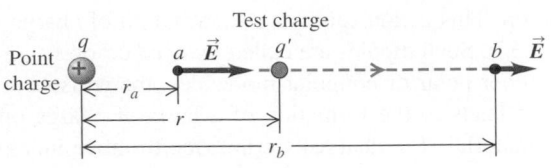

$$U = k\frac{qq'}{r}. \tag{18.8}$$

Potential

(Section 18.2) **Potential,** a scalar quantity denoted by V, is potential energy per unit charge. The potential at any point due to a point charge is

$$V = \frac{U}{q'} = k\frac{q}{r}. \tag{18.12}$$

A positive test charge tends to "fall" from a high-potential region to a low-potential region.

Equipotential Surfaces

(Section 18.3) An **equipotential surface** is a surface on which the potential has the same value at every point. At a point where a field line crosses an equipotential surface, the two are perpendicular. When all charges are at rest, the surface of a conductor is always an equipotential surface, and all points in the interior of a conductor are at the same potential.

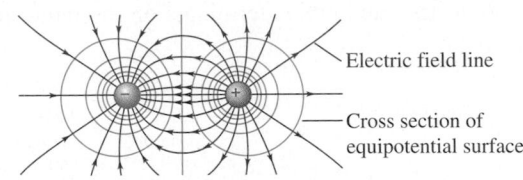

Electric field line

Cross section of equipotential surface

The Millikan Oil-Drop Experiment

(Section 18.4) The Millikan oil-drop experiment determined the electric charge of individual electrons by measuring the motion of electrically charged oil drops in an electric field. The size of a drop is determined by measuring its terminal speed of fall under gravity and the drag force of air.

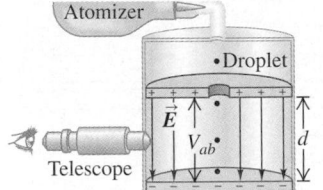

Capacitors

(Sections 18.5 and 18.6) A **capacitor** consists of any pair of conductors separated by vacuum or an insulating material. The **capacitance** C is defined as $C = Q/V_{ab}$ (Equation 18.14). A **parallel-plate capacitor** is made with two parallel plates, each with area A, separated by a distance d. If they are separated by vacuum, the capacitance is $C = \epsilon_0(A/d)$ (Equation 18.16).

When capacitors with capacitances C_1, C_2, C_3, . . . are connected in series, the equivalent capacitance C_{eq} is given by

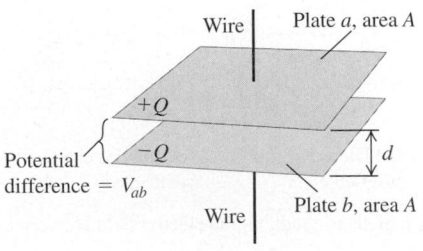

$$\frac{1}{C_{eq}} = \frac{1}{C_1} + \frac{1}{C_2} + \frac{1}{C_3} + \cdots. \tag{18.17}$$

When they are connected in parallel, the equivalent capacitance is

$$C_{eq} = C_1 + C_2 + C_3 + \cdots. \tag{18.18}$$

Continued

Electric Field Energy

(Section 18.7) The energy U required to charge a capacitor C to a potential difference V and a charge Q is equal to the energy stored in the capacitor and is given by

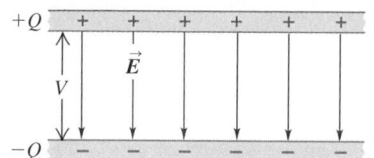

$$U = W_{\text{total}} = \left(\frac{V}{2}\right)Q = \frac{Q^2}{2C} = \frac{1}{2}CV^2. \qquad (18.19)$$

This energy can be thought of as residing in the electric field between the conductors; the energy density u (energy per unit volume) is $u = \frac{1}{2}\epsilon_0 E^2$ (Equation 18.20).

Dielectrics

(Section 18.8) When the space between the conductors is filled with a dielectric material, the capacitance *increases* by a factor K called the dielectric constant of the material. When the charges $\pm Q$ on the plates remain constant, charges induced on the surface of the dielectric *decrease* the electric field and potential difference between conductors by the same factor K. Under sufficiently strong fields, dielectrics become conductors, a phenomenon called dielectric breakdown. The maximum field that a material can withstand without breakdown is called its dielectric strength.

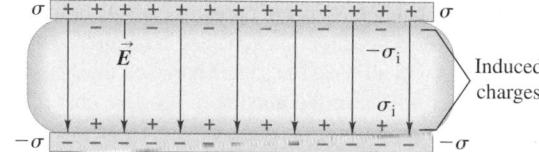

Molecular Model of Induced Charge

(Section 18.9) A *polar molecule* has equal amounts of positive and negative charge, but a lopsided distribution, with excess positive charge concentrated on one side of the molecule and negative charge on the other. When placed in an electric field, polar molecules tend to partially align with the field. For a material containing polar molecules, this microscopic alignment appears as an induced surface charge density. Even a molecule that is not ordinarily polar attains a lopsided charge distribution when it is placed in an electric field: The field pushes the positive charges in the molecule in the direction of the field and pushes the negative charges in the opposite direction.

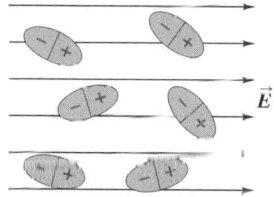

For instructor-assigned homework, go to www.masteringphysics.com

Conceptual Questions

1. *Why* must electric-field lines be perpendicular to equipotential surfaces?
2. Which way do electric-field lines point, from high to low potential or from low to high? Explain your reasoning.
3. If the electric field is zero throughout a certain region of space, is the potential necessarily zero in that region? If not, what *can* be said about the potential?
4. The potential (relative to a point at infinity) midway between two charges of equal magnitude and opposite sign is zero. Can you think of a way to bring a test charge from infinity to this midpoint in such a way that no work is done in any part of the displacement?
5. A high-voltage dc power line falls on a car, putting the entire metal body of the car at a potential of 10,000 V with respect to the ground. What happens to the occupants (a) when they are sitting in the car and (b) if they step out of the car?
6. Since potential can have any value you want depending on the choice of reference level of zero potential, how does a

voltmeter know what to read when you connect it between two points?
7. A capacitor is charged by being connected to a battery and is then disconnected from the battery. The plates are then pulled apart a little. How does each of the following quantities change as all this goes on? (a) the electric field between the plates, (b) the charge on the plates, (c) the potential difference across the plates, (d) the total energy stored in the capacitor.
8. A capacitor is charged by being connected to a battery of fixed potential and is kept connected to the battery. The plates are then pulled apart a little. How does each of the following quantities change as all this goes on? (a) the electric field between the plates, (b) the charge on the plates, (c) the potential difference across the plates, (d) the total energy stored in the capacitor?
9. Two parallel-plate capacitors, identical except that one has twice the plate separation of the other, are charged by the same voltage source. Which capacitor has a stronger electric field between the plates? Which capacitor has a greater charge? Which has greater energy density? Explain your reasoning.

10. The two plates of a capacitor are given charges $\pm Q$, and then they are immersed in a tank of benzene. Does the electric field between them increase, decrease, or remain the same?

11. Liquid dielectrics having polar molecules (such as water) have dielectric constants that *decrease* with increasing temperature. Why?

12. To store the maximum amount of energy in a parallel-plate capacitor with a given battery (voltage source), would it be better to have the plates far apart or closer together?

13. You have two capacitors and want to connect them across a voltage source (battery) to store the maximum amount of energy. Should they be connected in series or in parallel?

14. You have three capacitors, not necessarily equal, that you can connect across a battery of fixed potential. Show how you should connect these capacitors (in series or in parallel) so that (a) the capacitors will all have the same charge on their plates; (b) each capacitor will have the maximum possible charge on its plates; (c) you will store the most possible energy in the capacitor combination; (d) the capacitors will all have the same potential across them.

15. (a) If the potential (relative to infinity) is zero at a point, is the electric field necessarily zero at that point? (b) If the electric field is zero at a point, is the potential (relative to infinity) necessarily zero there? Prove your answers, using simple examples.

Multiple-Choice Problems

1. A surface will be an *equipotential* surface if (there may be more than one correct choice)
   A. the electric field is zero at all points on it.
   B. the electric field is tangent to the surface at all points.
   C. the electric field is perpendicular to the surface at all points.

2. In Figure 18.34, point P is equidistant from both point charges. At that point (there may be more than one correct choice),
   A. the electric field points directly to the right.
   B. the electric field is zero.
   C. the potential (relative to infinity) is zero.
   D. the potential (relative to infinity) points upward.

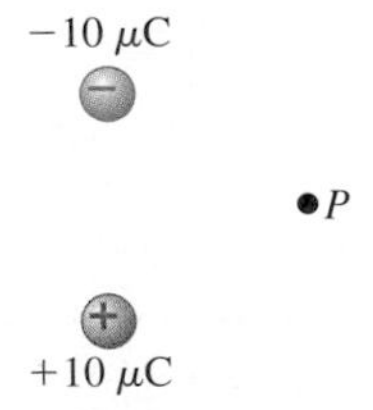

$-10\ \mu C$

$\bullet P$

$+10\ \mu C$

▲ **FIGURE 18.34** Multiple-choice problem 2.

3. For the capacitor network shown in Figure 18.35, a constant potential difference of 50 V is maintained across points a and b by a battery. Which of the following statements about this network is correct? (There may be more than one correct choice.)
   A. The 10 μF and 20 μF capacitors have equal charges.
   B. The charge on the 20 μF capacitor is twice the charge on the 10 μF capacitor.
   C. The potential difference across the 10 μF capacitor is the same as the potential difference across the 20 μF capacitor.
   D. The equivalent capacitance of the network is 60 μF.

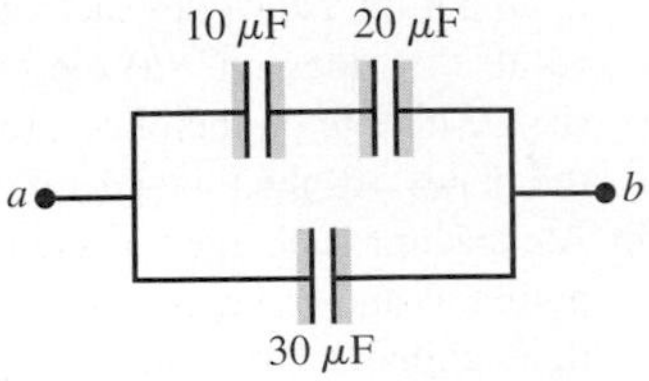

$10\ \mu F$ $20\ \mu F$

$a \bullet$ $\bullet b$

$30\ \mu F$

▲ **FIGURE 18.35** Multiple-choice problem 3.

4. A parallel-plate capacitor having circular plates of radius R and separation d is held at a fixed potential difference by a battery. If the plates are moved closer together while they are held at the same potential difference (there may be more than one correct choice),
   A. the amount of charge on each of them will increase.
   B. the amount of charge on each of them will decrease.
   C. the amount of charge on each of them will stay the same.
   D. the energy stored in the capacitor increases.

5. A parallel-plate capacitor having circular plates of radius R and separation d is charged to a potential difference by a battery. It is then *removed* from the battery. If the plates are moved closer together (there may be more than one correct choice),
   A. the amount of charge on each of them will increase.
   B. the amount of charge on each of them will decrease.
   C. the amount of charge on each of them will stay the same.
   D. the energy stored in the capacitor increases.

6. Two electrons close to each other are released from rest and are completely free to move. After being released (there may be more than one correct choice),
   A. their kinetic energies gradually decrease to zero as they move apart.
   B. their kinetic energies increase as they move apart.
   C. their electrical potential energy gradually decreases to zero as they move apart.
   D. their electrical potential energy increases as they move apart.
   E. their speeds gradually decrease to zero as they move apart.

7. The capacitor network shown in Figure 18.36 is connected across a fixed potential difference of 25 V. Which statements about this network must be true? (There may be more than one correct choice.)

$5\ \mu F$ $10\ \mu F$ $15\ \mu F$

$a \bullet$ ─┤├─┤├─┤├─ $\bullet b$

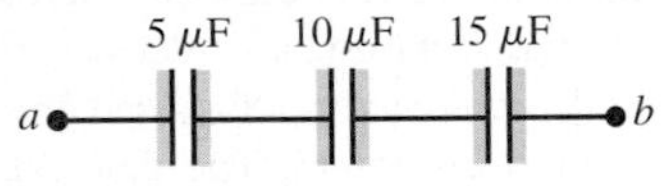

▲ **FIGURE 18.36** Multiple-choice problem 7.

   A. The potential difference is the same across each capacitor.
   B. The charge is the same on each capacitor.
   C. The equivalent capacitance of the network is 30 μF.
   D. The equivalent capacitance of the network is less than 30 μF.

8. If the potential (relative to infinity) due to a point charge is V at a distance R from this charge, the distance at which the potential (relative to infinity) is $2V$ is
   A. $4R$.   B. $2R$.   C. $R/2$.   D. $R/4$.

9. If the electrical potential energy of two point charges is U when they are a distance d apart, their potential energy when they are twice as far apart will be
   A. $U/4$.   B. $U/2$.   C. $2U$.   D. $4U$.

10. An electron is released between the plates of a charged parallel-plate capacitor very close to the right-hand plate. Just as it reaches the left-hand plate, its speed is v. If the distance between the plates were *halved* without changing the electric potential difference between them, then the speed of the electron when it reached the left-hand plate would be
    A. $2v$.   B. $v\sqrt{2}$.   C. v.
    D. $v/\sqrt{2}$.   E. $v/2$.

11. The plates of a parallel-plate capacitor are connected across a battery of fixed potential difference and that produces a

uniform electric field E between the plates. If the plates are pulled twice as far apart, but are kept connected to the battery, the electric field between the plates will be
A. $4E$.   B. $2E$.   C. E.
D. $E/2$.   E. $E/4$.

12. At a point P a distance d from a point charge, the potential relative to infinity is V and the electric-field magnitude is E. If you now move to a point S at which the potential is $V/2$, the electric-field magnitude at S will be
A. $E/4$.   B. $E/2$.   C. $2E$.   D. $4E$.

13. When a certain capacitor carries charge of magnitude Q on each of its plates, it stores energy U. In order to store twice as much energy, how much charge should it have on its plates?
A. $\sqrt{2}Q$.   B. $2Q$.   C. $4Q$.   D. $8Q$.

14. Two large metal plates carry equal and opposite charges spread over their surfaces, as shown in Figure 18.37. Which statements about these plates are correct? (There may be more than one correct choice.)

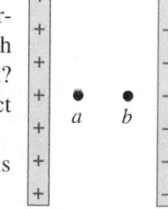

A. The electrical potential at point a is higher than the potential at point b.
B. The electrical potential at point a is equal to the potential at point b.
C. The electric-field strength at point a is equal to the field strength at point b.
D. If a positive point charge is released at point a, it will move with constant velocity toward point b.

▲ **FIGURE 18.37** Multiple-choice problem 14.

15. The electric potential (relative to infinity) due to a single point charge Q is $+400$ V at a point that is 0.90 m to the right of Q. The electric potential (relative to infinity) at a point 0.90 m to the left of Q is
A. -400 V.   B. $+200$ V.   C. $+400$ V.

Problems

18.1 Electrical Potential Energy

1. • A charge of 28.0 nC is placed in a uniform electric field that is directed vertically upward and that has a magnitude of 4.00×10^4 N/C. What work is done by the electric force when the charge moves (a) 0.450 m to the right; (b) 0.670 m upward; (c) 2.60 m at an angle of 45.0° downward from the horizontal?

2. • Two very large charged parallel metal plates are 10.0 cm apart and produce a uniform electric field of 2.80×10^6 N/C between them. A proton is fired perpendicular to these plates with an initial speed of 5.20 km/s, starting at the middle of the negative plate and going toward the positive plate. How much work has the electric field done on this proton by the time it reaches the positive plate?

3. • How far from a $-7.20\ \mu C$ point charge must a $+2.30\ \mu C$ point charge be placed in order for the electric potential energy of the pair of charges to be -0.400 J? (Take the energy to be zero when the charges are infinitely far apart.)

4. •• A point charge $q_1 = +2.40\ \mu C$ is held stationary at the origin. A second point charge $q_2 = -4.30\ \mu C$ moves from the point $x = 0.150$ m, $y = 0$, to the point $x = 0.250$ m, $y = 0.250$ m. How much work is done by the electric force on q_2?

5. •• Two stationary point charges of $+3.00$ nC and $+2.00$ nC are separated by a distance of 50.0 cm. An electron is released from rest at a point midway between the charges and moves along the line connecting them. What is the electric potential energy for the electron when it is (a) at the midpoint and (b) 10.0 cm from the $+3.00$ nC charge?

6. •• **Energy of DNA base pairing, I.** (See Problem 24 in **BIO** Chapter 17; see also Figure 17.43.) (a) Calculate the electric potential energy of the adenine–thymine bond, using the same combinations of molecules ($O\!-\!H\!-\!N$ and $N\!-\!H\!-\!N$) as in Problem 17.24. (b) Compare this energy with the potential energy of the proton–electron pair in the hydrogen atom.

7. •• **Energy of DNA base pairing, II.** (See Problem 25 in **BIO** Chapter 17; see also Figure 17.44.) Calculate the electric potential energy of the guanine–cytosine bond, using the same combinations of molecules ($O\!-\!H\!-\!O$, $N\!-\!H\!-\!N$, and $O\!-\!H\!-\!N$) as in Problem 17.25.

8. •• (a) A set of point charges is held in place at the vertices of an equilateral triangle of side 10.0 cm, as shown in Figure 18.38(a). Find the maximum amount of total kinetic energy that will be produced when the charges are released from rest in the frictionless void of outer space. (b) If the charges at the vertices of the right triangle in Figure 18.38(b) are released, how much total kinetic energy will they gain? When will this maximum kinetic energy be achieved, just following the release of the charges or after a very long time?

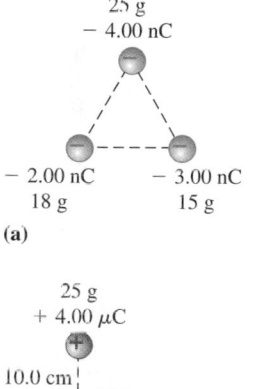

▲ **FIGURE 18.38** Problem 8.

9. •• Three equal 1.20-μC point charges are placed at the corners of an equilateral triangle whose sides are 0.500 m long. What is the potential energy of the system? (Take as zero the potential energy of the three charges when they are infinitely far apart.)

10. •• When two point charges are a distance R apart, their potential energy is -2.0 J. How far (in terms of R) should they be from each other so that their potential energy is -6.0 J?

18.2 Potential

11. • Two large metal parallel plates carry opposite charges of equal magnitude. They are separated by 45.0 mm, and the potential difference between them is 360 V. (a) What is the magnitude of the electric field (assumed to be uniform) in the region between the plates? (b) What is the magnitude of the force this field exerts on a particle with charge $+2.40$ nC?

12. • A potential difference of 4.75 kV is established between parallel plates in air. If the air becomes ionized (and hence electrically conducting) when the electric field exceeds 3.00×10^6 V/m, what is the minimum separation the plates can have without ionizing the air?

13. • **Oscilloscope.** Oscilloscopes are found in most science laboratories. Inside, they contain deflecting plates consisting of more-or-less square parallel metal sheets, typically about 2.5 cm on each side and 2.0 mm apart. In many experiments,

the maximum potential across these plates is about 25 V. For this maximum potential, (a) what is the strength of the electric field between the plates, and (b) what magnitude of acceleration would this field produce on an electron midway between the plates?

14. • **Axons.** Neurons are the
BIO basic units of the nervous system. They contain long tubular structures called *axons* that propagate electrical signals away from the ends of the neurons. The axon contains a solution of potassium ions K^+ and

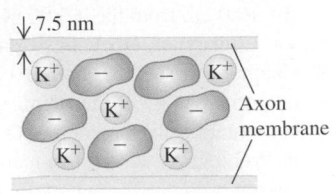

▲ **FIGURE 18.39** Problem 14.

large negative organic ions. The axon membrane prevents the large ions from leaking out, but the smaller K^+ ions are able to penetrate the membrane to some degree. (See Figure 18.39.) This leaves an excess negative charge on the inner surface of the axon membrane and an excess of positive charge on the outer surface, resulting in a potential difference across the membrane that prevents further K^+ ions from leaking out. Measurements show that this potential difference is typically about 70 mV. The thickness of the axon membrane itself varies from about 5 to 10 nm, so we'll use an average of 7.5 nm. We can model the membrane as a large sheet having equal and opposite charge densities on its faces. (a) Find the electric field inside the axon membrane, assuming (not too realistically) that it is filled with air. Which way does it point, into or out of the axon? (b) Which is at a higher potential, the inside surface or the outside surface of the axon membrane?

15. • **Electrical sensitivity of sharks.** Certain sharks can detect
BIO an electric field as weak as $1.0 \, \mu V/m$. To grasp how weak this field is, if you wanted to produce it between two parallel metal plates by connecting an ordinary 1.5 V AA battery across these plates, how far apart would the plates have to be?

16. •• A particle with a charge of $+4.20$ nC is in a uniform electric field $\vec{E}$ directed to the left. It is released from rest and moves to the left; after it has moved 6.00 cm, its kinetic energy is found to be $+1.50 \times 10^{-6}$ J. (a) What work was done by the electric force? (b) What is the potential of the starting point with respect to the endpoint? (c) What is the magnitude of $\vec{E}$?

17. •• Two very large metal parallel plates are 20.0 cm apart and carry equal, but opposite, surface charge densities. Figure 18.40 shows a graph of the potential, relative to the negative plate, as a function of x. For this case, x is the distance from the inner surface of the negative plate, measured perpendicular to the

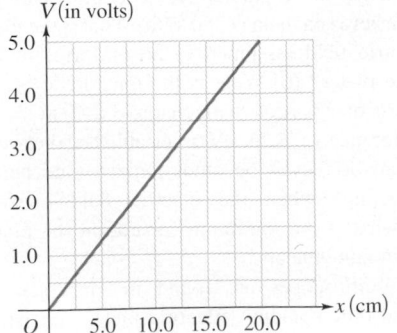

▲ **FIGURE 18.40** Problem 17.

plates, and points from the negative plate toward the positive plate. Find the electric field between the plates.

18. •• A uniform electric field has magnitude E and is directed in the negative x-direction. The potential difference between point a (at $x = 0.60$ m) and point b (at $x = 0.90$ m) is 240 V. (a) Which point, a or b, is at the higher potential? (b) Calculate the value of E. (c) A negative point charge $q = -0.200 \, \mu C$ is moved from b to a. Calculate the work done on the point charge by the electric field.

19. • A point charge has a charge of 2.50×10^{-11} C. At what distance from the point charge is the electric potential (a) 90.0 V? (b) 30.0 V? Take the potential to be zero at an infinite distance from the charge.

20. •• (a) An electron is to be accelerated from 3.00×10^6 m/s to 8.00×10^6 m/s. Through what potential difference must the electron pass to accomplish this? (b) Through what potential difference must the electron pass if it is to be slowed from 8.00×10^6 m/s to a halt?

21. •• A small particle has charge $-5.00 \, \mu C$ and mass 2.00×10^{-4} kg. It moves from point A, where the electric potential is $V_A = +200$ V, to point B, where the electric potential is $V_B = +800$ V. The electric force is the only force acting on the particle. The particle has speed 5.00 m/s at point A. What is its speed at point B? Is it moving faster or slower at B than at A? Explain.

22. •• Two point charges $q_1 = +2.40$ nC and $q_2 = -6.50$ nC are 0.100 m apart. Point A is midway between them; point B is 0.080 m from q_1 and 0.060 m from q_2. (See Figure 18.41.) Take the electric potential to be zero at infinity. Find (a) the potential

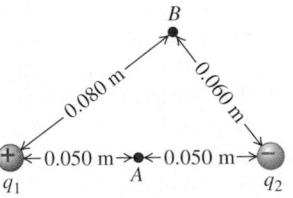

▲ **FIGURE 18.41** Problem 22.

at point A; (b) the potential at point B; (c) the work done by the electric field on a charge of 2.50 nC that travels from point B to point A.

23. •• A point charge $Q = +4.60 \, \mu C$ is held fixed at the origin. A second point charge $q = +1.20 \, \mu C$ with mass of 2.80×10^{-4} kg is placed on the x axis, 0.250 m from the origin. (a) What is the electric potential energy U of the pair of charges? (Take U to be zero when the charges have infinite separation.) (b) The second point charge is released from rest. What is its speed when its distance from the origin is (i) 0.500 m; (ii) 5.00 m; (iii) 50.0 m?

24. •• Two protons are released from rest when they are 0.750 nm apart. (a) What is the maximum speed they will reach? When does this speed occur? (b) What is the maximum acceleration they will achieve? When does this acceleration occur?

25. •• **Cathode-ray tube.** A *cathode-ray tube* (CRT) is an evacuated glass tube. Electrons are produced at one end, usually by the heating of a metal. After being focused electromagnetically into a beam, they are accelerated through a potential difference, called the *accelerating potential*. The electrons then strike a coated screen, where they transfer their energy to the coating through collisions, causing it to glow. CRTs are found in oscilloscopes and computer monitors, as well as in earlier versions of television screens. (a) If an electron of mass m and charge $-e$ is accelerated from rest through an accelerating

potential V, show that the speed it gains is $v = \sqrt{2eV/m}$. (We are assuming that V is small enough that the final speed is much less than the speed of light.) (b) If the accelerating potential is 95 V, how fast will the electrons be moving when they hit the screen?

26. •• **X-ray tube.** An X-ray tube is similar to a cathode-ray tube. (See previous problem.) Electrons are accelerated to high speeds at one end of the tube. If they are moving fast enough when they hit the target at the other end, they give up their energy as X-rays (a form of nonvisible light). (a) Through what potential difference should electrons be accelerated so that their speed is 1.0% of the speed of light when they hit the target? (b) What potential difference would be needed to give protons the same kinetic energy as the electrons? (c) What speed would this potential difference give to protons? Express your answer in m/s and as a percent of the speed of light.

27. •• A gold nucleus has a radius of 7.3×10^{-15} m and a charge of $+79e$. Through what voltage must an α-particle, with its charge of $+2e$, be accelerated so that it has just enough energy to reach a distance of 2.0×10^{-14} m from the surface of a gold nucleus? (Assume the gold nucleus remains stationary and can be treated as a point charge.)

18.3 Equipotential Surfaces

28. •• A parallel-plate capacitor having plates 6.0 cm apart is connected across the terminals of a 12 V battery. (a) Being as *quantitative* as you can, describe the location and shape of the equipotential surface that is at a potential of $+6.0$ V relative to the potential of the negative plate. Avoid the edges of the plates. (b) Do the same for the equipotential surface that is at $+2.0$ V relative to the negative plate. (c) What is the potential gradient between the plates?

29. •• Two very large metal parallel plates that are 25 cm apart, oriented perpendicular to a sheet of paper, are connected across the terminals of a 50.0 V battery. (a) Draw *to scale* the lines where the equipotential surfaces due to these plates intersect the paper. Limit your drawing to the region between the plates, avoiding their edges, and draw the lines for surfaces that are 10.0 V apart, starting at the low-potential plate. (b) These surfaces are separated equally in potential. Are they also separated equally in distance? (c) *In words*, describe the shape and orientation of the surfaces you just found.

30. •• (a) A $+5.00$ pC charge is located on a sheet of paper. (a) Draw *to scale* the curves where the equipotential surfaces due to these charges intersect the paper. Show only the surfaces that have a potential (relative to infinity) of 1.00 V, 2.00 V, 3.00 V, 4.00 V, and 5.00 V. (b) The surfaces are separated equally in potential. Are they also separated equally in distance? (c) *In words*, describe the shape and orientation of the surfaces you just found.

31. •• A metal sphere carrying an evenly distributed charge will have spherical equipotential surfaces surrounding it. Suppose the sphere's radius is 50.0 cm and it carries a total charge of $+1.50$ μC. (a) Calculate the potential of the sphere's surface. (b) You want to draw equipotential surfaces at intervals of 500 V outside the sphere's surface. Calculate the distance between the first and the second equipotential surfaces, and between the 20th and 21st equipotential surfaces. (c) What does the changing spacing of the surfaces tell you about the electric field?

32. •• Figure 18.42 shows a set of electric-field lines for a particular distribution of charges. Use these lines to draw a series of equipotential surfaces for this system. Limit yourself to the plane of the paper.

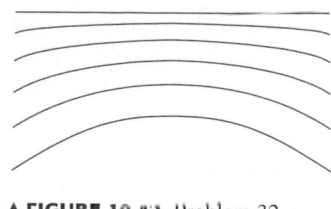

▲ **FIGURE 18.42** Problem 32.

33. •• **Dipole.** A dipole is located on a sheet of paper. (a) In the plane of that paper, carefully sketch the electric field lines for this dipole. (b) Use your field lines in part (a) to sketch the equipotential curves where the equipotential surfaces intersect the paper.

18.4 The Millikan Oil-Drop Experiment

34. •• In a particular Millikan oil-drop apparatus, the plates are 2.25 cm apart. The oil used has a density of 0.820 g/cm^3, and the atomizer that sprays the oil drops produces drops of diameter 1.00×10^{-3} mm. (a) What strength of electric field is needed to hold such a drop stationary against gravity if the drop contains five excess electrons? (b) What should be the potential difference across the plates to produce this electric field? (c) If another drop of the same oil requires a plate potential of 73.8 V to hold it stationary, how many excess electrons did it contain?

35. • (a) If an electron and a proton each have a kinetic energy of 1.00 eV, how fast is each one moving? (b) What would be their speeds if each had a kinetic energy of 1.00 keV? (c) If they were each traveling at 1.00% the speed of light, what would be their kinetic energies in keV?

18.5 Capacitors

36. • (a) You find that if you place charges of ± 1.25 μC on two separated metal objects, the potential difference between them is 11.3 V. What is their capacitance? (b) A capacitor has a capacitance of 7.28 μF. What amount of excess charge must be placed on each of its plates to make the potential difference between the plates equal to 25.0 V?

37. • The plates of a parallel-plate capacitor are 3.28 mm apart, and each has an area of 12.2 cm^2. Each plate carries a charge of magnitude 4.35×10^{-8} C. The plates are in vacuum. (a) What is the capacitance? (b) What is the potential difference between the plates? (c) What is the magnitude of the electric field between the plates?

38. • The plates of a parallel-plate capacitor are 2.50 mm apart, and each carries a charge of magnitude 80.0 nC. The plates are in vacuum. The electric field between the plates has a magnitude of 4.00×10^6 V/m. (a) What is the potential difference between the plates? (b) What is the area of each plate? (c) What is the capacitance?

39. • A parallel-plate air capacitor has a capacitance of 500.0 pF and a charge of magnitude 0.200 μC on each plate. The plates are 0.600 mm apart. (a) What is the potential difference between the plates? (b) What is the area of each plate? (c) What is the electric-field magnitude between the plates? (d) What is the surface charge density on each plate?

40. • **Capacitance of an oscilloscope.** Oscilloscopes have parallel metal plates inside them to deflect the electron beam. These plates are called the *deflecting plates.* Typically, they are

squares 3.0 cm on a side and separated by 5.0 mm, with vacuum in between. What is the capacitance of these deflecting plates and hence of the oscilloscope? (This capacitance can sometimes have an effect on the circuit you are trying to study and must be taken into consideration in your calculations.)

41. •• A 10.0 μF parallel-plate capacitor with circular plates is connected to a 12.0 V battery. (a) What is the charge on each plate? (b) How much charge would be on the plates if their separation were doubled while the capacitor remained connected to the battery? (c) How much charge would be on the plates if the capacitor were connected to the 12.0 V battery after the radius of each plate was doubled without changing their separation?

42. •• A 10.0 μF parallel-plate capacitor is connected to a 12.0 V battery. After the capacitor is fully charged, the battery is disconnected without loss of any of the charge on the plates. (a) A voltmeter is connected across the two plates without discharging them. What does it read? (b) What would the voltmeter read if (i) the plate separation were doubled; (ii) the radius of each plate was doubled, but the separation between the plates was unchanged?

43. •• You make a capacitor by cutting the 15.0-cm-diameter bottoms out of two aluminum pie plates, separating them by 3.50 mm, and connecting them across a 6.00-V battery. (a) What's the capacitance of your capacitor? (b) If you disconnect the battery and separate the plates to a distance of 3.50 cm without discharging them, what will be the potential difference between them?

44. •• A 5.00 pF parallel-plate air-filled capacitor with circular plates is to be used in a circuit in which it will be subjected to potentials of up to 1.00×10^2 V. The electric field between the plates is to be no greater than 1.00×10^4 N/C. As a budding electrical engineer for Live-Wire Electronics, your tasks are to (a) design the capacitor by finding what its physical dimensions and separation must be and (b) find the maximum charge these plates can hold.

45. •• How far apart would parallel pennies have to be to make a 1.00-pF capacitor? Does your answer suggest that you are justified in treating these pennies as infinite sheets? Explain.

46. •• A parallel-plate capacitor C is charged up to a potential V_0 with a charge of magnitude Q_0 on each plate. It is then disconnected from the battery, and the plates are pulled apart to twice their original separation. (a) What is the new capacitance in terms of C? (b) How much charge is now on the plates in terms of Q_0? (c) What is the potential difference across the plates in terms of V_0?

18.6 Capacitors in Series and in Parallel

47. • For the system of capacitors shown in Figure 18.43, find the equivalent capacitance (a) between b and c, (b) between a and c.

48. • **Electric eels.** Electric
BIO eels and electric fish generate large potential differences that are used to stun enemies and prey. These potentials are produced by cells that each can generate 0.10 V. We can plausibly model such cells as charged capacitors. (a) How should

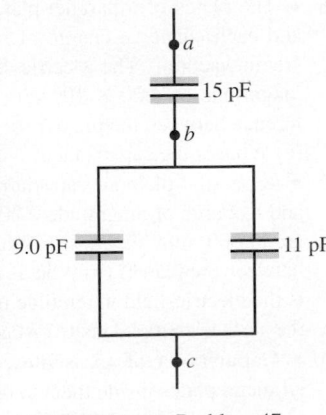
▲**FIGURE 18.43** Problem 47.

these cells be connected (in series or in parallel) to produce a total potential of more than 0.10 V? (b) Using the connection in part (a), how many cells must be connected together to produce the 500 V surge of the electric eel?

49. •• In Figure 18.44, $C_1 = 6.00\,\mu$F, $C_2 = 3.00\,\mu$F, and $C_3 = 5.00\,\mu$F. The capacitor network is connected to an applied potential V_{ab}. After the charges on the capacitors have reached their final values, the charge on C_2 is 40.0 μC. (a) What are the charges on capacitors C_1 and C_3? (b) What is the applied voltage V_{ab}?

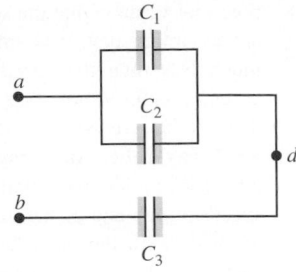
▲**FIGURE 18.44** Problem 49.

50. •• You are working on an electronics project requiring a variety of capacitors, but have only a large supply of 100 nF capacitors available. Show how you can connect these capacitors to produce each of the following equivalent capacitances: (a) 50 nF, (b) 450 nF, (c) 25 nF, (d) 75 nF.

51. •• In Figure 18.44, $C_1 = 3.00\,\mu$F and $V_{ab} = 120$ V. The charge on capacitor C_1 is 150 μC. Calculate the voltage across the other two capacitors.

52. •• A 4.00 μF and a 6.00 μF capacitor are connected in series, and this combination is connected across a 48.0 V potential difference. Calculate (a) the charge on each capacitor and (b) the potential difference across each of them.

53. •• In the circuit shown in Figure 18.45, the potential difference across ab is $+24.0$ V. Calculate (a) the charge on each capacitor and (b) the potential difference across each capacitor.

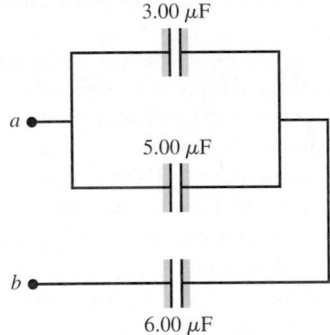
▲**FIGURE 18.45** Problem 53.

54. •• In Figure 18.46, each capacitor has $C = 4.00\,\mu$F and $V_{ab} = +28.0$ V. Calculate (a) the charge on each capacitor and (b) the potential difference across each capacitor.

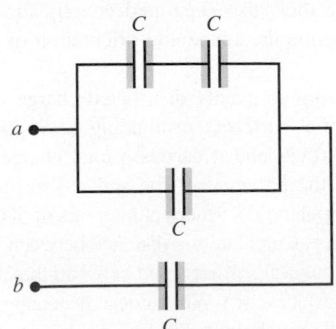
▲**FIGURE 18.46** Problem 54.

55. •• Figure 18.47 shows a system of four capacitors, where the potential difference across ab is 50.0 V. (a) Find the equivalent capacitance of this system between a and b. (b) How much charge is stored by this combination of capacitors? (c) How much charge is stored in each of the 10.0 μF and the 9.0 μF capacitors?

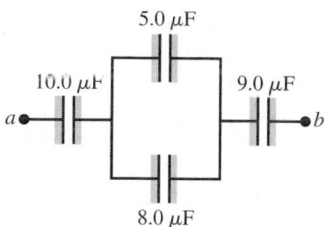

▲ **FIGURE 18.47** Problem 55.

56. •• For the system of capacitors shown in Figure 18.48, a potential difference of 25 V is maintained across ab. (a) What is the equivalent capacitance of this system between a and b? (b) How much charge is stored by this system? (c) How much charge does the 6.5 nF capacitor store? (d) What is the potential difference across the 7.5 nF capacitor?

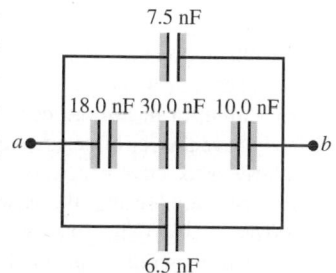

▲ **FIGURE 18.48** Problem 56.

18.7 Electric Field Energy

57. • How much charge does a 12 V battery have to supply to fully charge a 2.5 μF capacitor and a 5.0 μF capacitor when they're (a) in parallel, (b) in series? (c) How much energy does the battery have to supply in each case?

58. • A 5.80 μF parallel-plate air capacitor has a plate separation of 5.00 mm and is charged to a potential difference of 400 V. Calculate the energy density in the region between the plates, in units of J/m^3.

59. • (a) How much charge does a battery have to supply to a 5.0 μF capacitor to create a potential difference of 1.5 V across its plates? How much energy is stored in the capacitor in this case? (b) How much charge would the battery have to supply to store 1.0 J of energy in the capacitor? What would be the potential across the capacitor in that case?

60. • In the text, it was shown that the energy stored in a capacitor C charged to a potential V is $U = \frac{1}{2}QV$. Show that this energy can also be expressed as (a) $U = Q^2/2C$ and (b) $U = \frac{1}{2}CV^2$.

61. •• A parallel-plate vacuum capacitor has 8.38 J of energy stored in it. The separation between the plates is 2.30 mm. If the separation is decreased to 1.15 mm, what is the energy stored (a) if the capacitor is disconnected from the potential source so the charge on the plates remains constant, and (b) if the capacitor remains connected to the potential source so the potential difference between the plates remains constant?

62. •• (a) How many excess electrons must be added to one plate and removed from the other to give a 5.00 nF parallel-plate capacitor 25.0 μJ of stored energy? (b) How could you modify the geometry of this capacitor to get it to store 50.0 μJ of energy without changing the charge on its plates?

63. •• For the capacitor network shown in Figure 18.49, the potential difference across ab is 36 V. Find (a) the total charge stored in this network, (b) the charge on each capacitor, (c) the total energy stored in the network, (d) the energy stored in each capacitor, and (e) the potential difference across each capacitor.

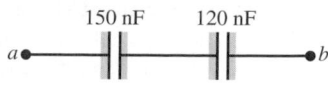

▲ **FIGURE 18.49** Problem 63.

64. •• For the capacitor network shown in Figure 18.50, the potential difference across ab is 220 V. Find (a) the total charge stored in this network, (b) the charge on each capacitor, (c) the total energy stored in the network, (d) the energy stored in each capacitor, and (e) the potential difference across each capacitor.

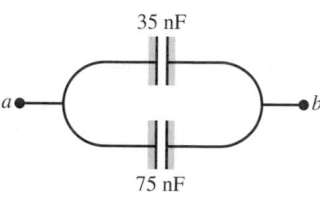

▲ **FIGURE 18.50** Problem 64.

65. •• A 20.0 μF capacitor is charged to a potential difference of 800 V. The terminals of the charged capacitor are then connected to those of an uncharged 10.0 μF capacitor. Compute (a) the original charge of the system, (b) the final potential difference across each capacitor, (c) the final energy of the system, and (d) the decrease in energy when the capacitors are connected.

66. •• For the capacitor network shown in Figure 18.51, the potential difference across ab is 12.0 V. Find (a) the total energy stored in this network and (b) the energy stored in the 4.80 μF capacitor.

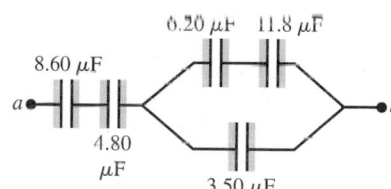

▲ **FIGURE 18.51** Problem 66.

67. •• A parallel plate air capacitor has a capacitance of 920 pF. The charge on each plate is 2.55 μC. (a) What is the potential difference between the plates? (b) If the charge is kept constant, what will be the potential difference between the plates if the separation is doubled? (c) How much work is required to double the separation?

18.8 Dielectrics

68. • A parallel-plate capacitor has capacitance $C_0 = 5.00$ pF when there is air between the plates. The separation between the plates is 1.50 mm. (a) What is the maximum magnitude of charge Q that can be placed on each plate if the electric field in the region between the plates is not to exceed 3.00×10^4 V/m? (b) A dielectric with $K = 2.70$ is inserted between the plates of the capacitor, completely filling the volume between the plates. Now what is the maximum magnitude of charge on each plate if the electric field between the plates is not to exceed 3.00×10^4 V/m?

69. • **Cell membranes.** Cell mem-
BIO branes (the walled enclosure around a cell) are typically about 7.5 nm thick. They are partially permeable to allow charged material to pass in and out, as needed. Equal but oppo-site charge densities build up on the inside and outside faces of such a membrane, and these charges prevent additional charges from passing through the cell wall. We can model a cell membrane as a parallel-plate capacitor, with the membrane itself containing proteins embedded in an organic material to give the membrane a dielectric constant of about 10. (See Figure 18.52.) (a) What is the capacitance per square centimeter of such a cell wall? (b) In its normal resting state, a cell has a potential difference of 85 mV across its membrane. What is the electric field inside this membrane?

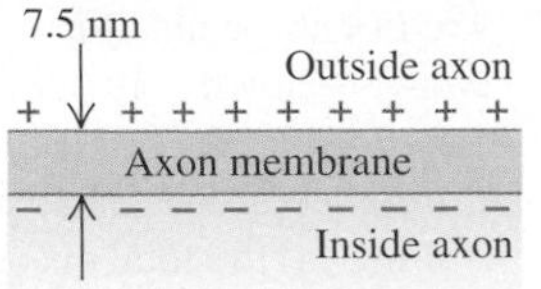

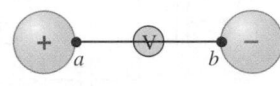

▲ **FIGURE 18.52** Problem 69.

70. •• A parallel-plate capacitor is to be constructed by using, as a dielectric, rubber with a dielectric constant of 3.20 and a dielectric strength of 20.0 MV/m. The capacitor is to have a capacitance of 1.50 nF and must be able to withstand a maxi-mum potential difference of 4.00 kV. What is the minimum area the plates of this capacitor can have?

71. •• A 12.5 μF capacitor is connected to a power supply that keeps a constant potential difference of 24.0 V across the plates. A piece of material having a dielectric constant of 3.75 is placed between the plates, completely filling the space between them. (a) How much energy is stored in the capacitor before and after the dielectric is inserted? (b) By how much did the energy change during the insertion? Did it increase or decrease?

72. •• The paper dielectric in a paper-and-foil capacitor is 0.0800 mm thick. Its dielectric constant is 2.50, and its dielectric strength is 50.0 MV/m. Assume that the geometry is that of a parallel-plate capacitor, with the metal foil serving as the plates. (a) What area of each plate is required for a 0.200 μF capacitor? (b) If the electric field in the paper is not to exceed one-half the dielectric strength, what is the maxi-mum potential difference that can be applied across the capacitor?

73. •• A constant potential difference of 12 V is maintained between the terminals of a 0.25-μF, parallel-plate, air capaci-tor. (a) A sheet of Mylar is inserted between the plates of the capacitor, completely filling the space between the plates. When this is done, how much additional charge flows onto the positive plate of the capacitor (see Table 18.1)? (b) What is the total induced charge on either face of the Mylar sheet? (c) What effect does the Mylar sheet have on the electric field between the plates? Explain how you can reconcile this with the increase in charge on the plates, which acts to *increase* the electric field.

General Problems

74. •• (a) If a spherical raindrop of radius 0.650 mm carries a charge of −1.20 pC uniformly distributed over its volume, what is the potential at its surface? (Take the potential to be zero at an infinite distance from the raindrop.) (b) Two identi-cal raindrops, each with radius and charge specified in part (a), collide and merge into one larger raindrop. What is the radius of this larger drop, and what is the potential at its surface, if its charge is uniformly distributed over its volume?

75. •• At a certain distance from a point charge, the potential and electric-field magnitude due to that charge are 4.98 V and 12.0 V/m, respectively. (Take the potential to be zero at infin-ity.) (a) What is the distance to the point charge? (b) What is the magnitude of the charge? (c) Is the electric field directed toward or away from the point charge?

76. •• Two oppositely charged identical insulating spheres, each 50.0 cm in diameter and carrying a uniform charge of magnitude 175 μC, are placed 1.00 m apart center to center (Fig. 18.53). (a) If a voltmeter is connected between the near-est points (a and b) on their surfaces, what will it read? (b) Which point, a or b, is at the higher potential? How can you know this without any calculations?

▲ **FIGURE 18.53** Problem 76.

77. •• **Potential in human cells.** Some cell walls in the human
BIO body have a layer of negative charge on the inside surface and a layer of positive charge of equal magnitude on the outside surface. Suppose that the charge density on either surface is ±0.50 × 10⁻³ C/m², the cell wall is 5.0 nm thick, and the cell-wall material is air. (a) Find the magnitude of $\vec{E}$ in the wall between the two layers of charge. (b) Find the potential difference between the inside and the outside of the cell. Which is at the higher potential? (c) A typical cell in the human body has a volume of 10^{-16} m³. Estimate the total electric-field energy stored in the wall of a cell of this size. (*Hint:* Assume that the cell is spherical, and calculate the vol-ume of the cell wall.) (d) In reality, the cell wall is made up, not of air, but of tissue with a dielectric constant of 5.4. Repeat parts (a) and (b) in this case.

78. •• An alpha particle with a kinetic energy of 10.0 MeV makes a head-on collision with a gold nucleus at rest. What is the distance of closest approach of the two particles? (Assume that the gold nucleus remains stationary and that it may be treated as a point charge. The atomic number of gold is 79, and an alpha particle is a helium nucleus consisting of two protons and two neutrons.)

79. •• In the *Bohr model* of the hydrogen atom, a single electron revolves around a single proton in a circle of radius r. Assume that the proton remains at rest. (a) By equating the electric force to the electron mass times its acceleration, derive an expression for the electron's speed. (b) Obtain an expression for the electron's kinetic energy, and show that its magnitude is just half that of the electric potential energy. (c) Obtain an expression for the total energy, and evaluate it using $r = 5.29 \times 10^{-11}$ m. Give your numerical result in joules and in electron volts.

80. •• A proton and an alpha particle are released from rest when they are 0.225 nm apart. The alpha particle (a helium nucleus) has essentially four times the mass and two times the charge of a proton. Find the maximum *speed* and maximum *acceleration* of each of these particles. When do these maxima occur, just following the release of the particles or after a very long time?

81. •• A parallel-plate air capacitor is made from two plates 0.200 m square, spaced 0.800 cm apart. It is connected to a 120-V battery. (a) What is the capacitance? (b) What is the

charge on each plate? (c) What is the electric field between the plates? (d) What is the energy stored in the capacitor? (e) If the battery is disconnected and then the plates are pulled apart to a separation of 1.60 cm, what are the answers to parts (a), (b), (c), and (d)?

82. •• In the previous problem, suppose the battery remains connected while the plates are pulled apart. What are the answers then to parts (a), (b), (c), and (d) after the plates have been pulled apart?

83. •• A capacitor consists of two parallel plates, each with an area of 16.0 cm², separated by a distance of 0.200 cm. The material that fills the volume between the plates has a dielectric constant of 5.00. The plates of the capacitor are connected to a 300-V battery. (a) What is the capacitance of the capacitor? (b) What is the charge on either plate? (c) How much energy is stored in the charged capacitor?

84. •• Electronic flash units for cameras contain a capacitor for storing the energy used to produce the flash. In one such unit, the flash lasts for $\frac{1}{675}$ s with an average light power output of 2.70×10^5 W. (a) If the conversion of electrical energy to light is 95% efficient (the rest of the energy goes to thermal energy), how much energy must be stored in the capacitor for one flash? (b) The capacitor has a potential difference between its plates of 125 V when the stored energy equals the value calculated in part (a). What is the capacitance?

85. •• In Figure 18.54, each capacitance C_1 is 6.9 μF and each capacitance C_2 is 4.6 μF. (a) Compute the equivalent capacitance of the network between points a and b. (b) Compute the charge on each of the three capacitors nearest a and b when $V_{ab} = 420$ V.

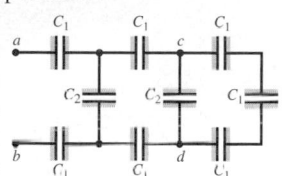

▲ **FIGURE 18.54** Problem 85.

86. •• A parallel-plate capacitor is made from two plates 12.0 cm on each side and 4.50 mm apart. Half of the space between these plates contains only air, but the other half is filled with Plexiglas® of dielectric constant 3.40. (See Figure 18.55.) An 18.0 V battery is connected across the plates. (a) What is the capacitance of this combination? (*Hint:* Can you think of this capacitor as equivalent to two capacitors in parallel?) (b) How much energy is stored in the capacitor? (c) If we remove the Plexiglas®, but change nothing else, how much energy will be stored in the capacitor?

▲ **FIGURE 18.55** Problem 86.

87. •• A parallel-plate capacitor with plate separation d has the space between the plates filled with two slabs of dielectric, one with constant K_1 and the other with constant K_2, and each having thickness $d/2$. (a) Show that the capacitance is given by
$$C = \frac{2\epsilon_o A}{d}\left(\frac{K_1 K_2}{K_1 + K_2}\right).$$ (*Hint:* Can you think of this combination as two capacitors in series?) (b) To see if your answer is reasonable, check it in the following cases: (i) There is only one dielectric, with constant K, and it completely fills the space between the plates. (ii) The plates have nothing but air, which we can treat as vacuum, between them.

Passage Problems

BIO The electric egg. The eggs of many species undergo a rapid change in the electrical potential difference across the outer membrane when they are fertilized. This change in potential difference affects the physiological development of the eggs. The potential difference across the membrane is called the membrane potential, V_m, defined as the inside potential minus the outside potential. The membrane potential V_m arises when protein enzymes use the energy available in ATP to actively expel sodium ions (Na⁺) and accumulate potassium ions (K⁺). Because the membrane of the unfertilized egg is selectively permeable to K⁺, the V_m of the resting sea urchin egg is about –70 mV; that is, the inside has a potential of 70 mV less than that of the outside. The egg membrane behaves as a capacitor with a specific capacitance of about 1 μF/cm². When a sea urchin egg is fertilized, Na⁺ channels in the membrane are opened, Na⁺ enters the egg, and V_m rapidly changes to +30 mV, where it remains for several minutes. The concentration of Na⁺ in the egg's interior is about 30 mmoles/liter (30 mM) and 450 mM in the surrounding sea water. The inside K⁺ concentration is about 200 mM and the outside K⁺ is 10 mM. A useful constant that connects electrical and chemical units is the Faraday number, which has a value of approximately 10^5 coulomb/mole. That is, an Avogadro number (a mole) of monovalent ions such as Na⁺ or K⁺ carries a charge of 10^5 C.

88. How many moles of Na⁺ must move per unit area of membrane to change V_m from –70 mV to +30 mV, making the assumption that the membrane behaves purely as a capacitor?
    A. 10^{-4} mole/cm²
    B. 10^{-9} mole/cm²
    C. 10^{-12} mole/cm²
    D. 10^{-14} mole/cm²

89. Suppose the egg has a diameter of 200 μm. What fractional change in internal Na⁺ concentration results from the fertilization-induced change in V_m? Assume that the Na⁺ ions are distributed throughout the cell volume.
    A. Increases by 1 part in 10^4
    B. Increases by 1 part in 10^5
    C. Increases by 1 part in 10^6
    D. Increases by 1 part in 10^7

90. Suppose the change in V_m was caused by the entry of Ca²⁺ instead of Na⁺. How many Ca²⁺ ions would have to enter the cell per unit membrane to produce the change?
    A. Half as many as for Na⁺
    B. The same as for Na⁺
    C. Twice as many as for Na⁺
    D. Cannot say without knowing the inside and outside concentrations of Ca²⁺

91. What is the minimum amount of work that must be done by the cell to restore V_m to its original value?
    A. 3 mJ
    B. 3 μJ
    C. 3 nJ
    D. 3 pJ

19 Current, Resistance, and Direct-Current Circuits

Electric circuits are at the heart of all radio and television transmitters and receivers, CD players, household and industrial power distribution systems, flashlights, computers, and the nervous systems of animals. In the past two chapters, we've studied the interactions of electric charges *at rest;* now we're ready to study charges *in motion.*

In this chapter, we'll study the basic properties of electrical conductors and how their behavior depends on temperature. We'll learn why a short, fat, cold copper wire is a better conductor than a long, skinny, hot steel wire. We'll study the properties of batteries and how they cause current and energy transfer in a circuit. In this analysis, we'll use the concepts of current, potential difference ("voltage"), resistance, and electromotive force.

This reliably sunny beachside roof is covered with electricity-generating photovoltaic panels.

19.1 Current

In the preceding two chapters, our primary emphasis was on electrostatics, the study of electrical interactions with charges at rest. We learned that, in electrostatic situations, there can be no electric field $\vec{E}$ in the interior of a conducting material; otherwise the mobile charges in the conductor would move, and the situation wouldn't be electrostatic.

In this chapter, we shift our emphasis to situations in which non-zero electric fields exist inside conductors, causing motion of the mobile charges within the conductors. A **current** (also called *electric current*) is any motion of charge from one region of a conductor to another. To maintain a steady flow of charge in a conductor, we have to maintain a steady force on the mobile charges, either with an electrostatic field or by other means that we'll consider later. For now, let's assume that there is an electric field $\vec{E}$ within the conductor, so a particle with charge q is acted upon by a force $\vec{F} = q\vec{E}$.

▶ **Application The secret of semiconductors.** Metals conduct electricity well because each atom gives up one or more outer electrons, which become free charge carriers. Insulators don't conduct electricity because all their electrons are firmly bound. *Semiconductors* such as silicon have loosely bound electrons that occasionally become free charge carriers. The positive "hole" left behind by a free electron can also migrate. A semiconductor can be modified so as to have mainly electrons (N-type) or mainly holes (P-type) as its free charge carriers. A junction between N and P semiconductors has the remarkable property of conducting electricity preferentially in one direction, from P to N. This property is central to the working of computer chips (like the one shown here) and many other electronic devices.

To help visualize current, let's think of conductors in the shape of wires. (Figure 19.1). Current is defined to be the amount of charge that moves through a given cross section of conductor per unit time. Thus, if a net charge ΔQ flows through a cross section during a time Δt, the current through that area, denoted by I, is defined as follows:

Definition of current

When a net charge ΔQ passes through a cross section of conductor during time Δt, the current is

$$I = \frac{\Delta Q}{\Delta t}. \tag{19.1}$$

Unit: 1 coulomb/second = 1 C/s = 1 ampere = 1 A.

Current is a *scalar* quantity. The SI unit of current is the **ampere;** one ampere is defined to be 1 *coulomb per second* ($1\ A = 1\ C/s$). This unit is named in honor of the French scientist André Marie Ampère (1775–1836). When an ordinary flashlight (D-cell size) is turned on, the current in the bulb is about 0.5 to 1 A; the current in the starter motor of a car engine is around 200 A. Currents in radio and television circuits are usually expressed in *milliamperes* ($1\ mA = 10^{-3}\ A$) or *microamperes* ($1\ \mu A = 10^{-6}\ A$), and currents in computer circuits are expressed in *nanoamperes* ($1\ nA = 10^{-9}\ A$) or *picoamperes* ($1\ pA = 10^{-12}\ A$).

In Figure 19.2a, the moving charges are positive and the current flows in the same direction as the moving charges. In Figure 19.2b, they are negative, and the current is opposite to the motion of the moving charges. But the *current* is still from left to right, because in both cases the result is a net transfer of positive charge from left to right. Particles flowing out at an end of the cylindrical section are continuously replaced by particles flowing *in* at the opposite end.

When there is a steady current in a closed loop (a "complete circuit"), the total charge in every segment of the conductor is constant. From the principle of

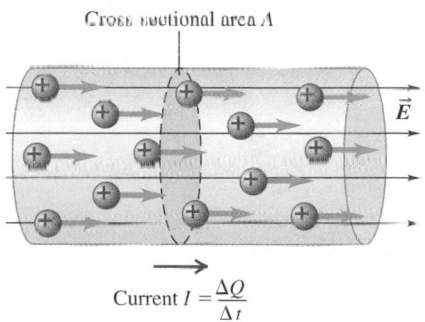

Current $I = \dfrac{\Delta Q}{\Delta t}$

▲ **FIGURE 19.1** The definition of current.

A **conventional current** is treated as a flow of positive charges, regardless of whether the free charges in the conductor are positive, negative, or both.

In a metallic conductor, the moving charges are electrons — but the *current* still points in the direction positive charges would flow.

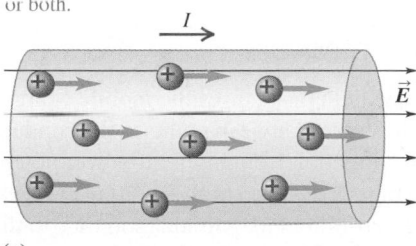

(a)

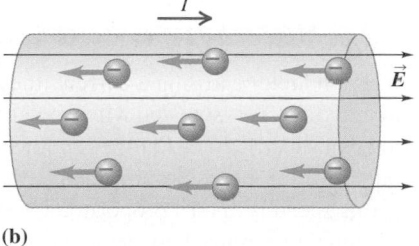

(b)

▲ **FIGURE 19.2** Conventional and electron currents.

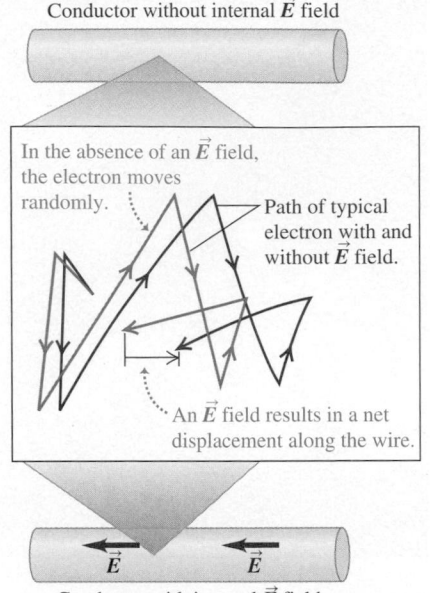

Conductor without internal $\vec{E}$ field

In the absence of an $\vec{E}$ field, the electron moves randomly.

Path of typical electron with and without $\vec{E}$ field.

An $\vec{E}$ field results in a net displacement along the wire.

$\vec{E}$ $\vec{E}$

Conductor with internal $\vec{E}$ field

▲ **FIGURE 19.3** The presence of an electric field imposes a small drift (greatly exaggerated here) on an electron's random motion.

conservation of charge (Section 17.3), the rate of flow of charge *out* at one end of a segment at any instant equals the rate of flow of charge *in* at the other end of the segment at that instant, and *the current at any instant is the same at all cross sections.* Current is *not* something that squirts out of the positive terminal of a battery and is consumed or used up by the time it reaches the negative terminal.

In circuit analysis, we'll always describe currents *as though* they consisted entirely of positive charge flow, even in cases in which we know the actual current is due to (negatively charged) electrons. In metals, the moving charges are always (negative) electrons, but in an ionic solution, both positive and negative ions are moving. We speak of "conventional current" to describe a flow of positive charge that is equivalent to the actual flow of charge of either sign.

In a metal, the free electrons have a lot of random motion, somewhat like the molecules in a gas. When an electric field is applied to the metal, the forces that it exerts on the electrons lead to a small net motion, or *drift,* in the direction of the force, in addition to the random motion (Figure 19.3). Thus, the motion consists of random motion with very large average speeds (on the order of 10^6 m/s) and a much slower drift speed (often on the order of 10^{-4} m/s) in the direction of the electric-field force. But when you turn on a light switch, the light comes on almost instantaneously because the electric fields in the conductors travel with a speed approaching the speed of light. You don't have to wait for individual electrons to travel from the switch to the bulb!

EXAMPLE 19.1 How many electrons?

One of the circuits in a small portable CD player operates on a current of 2.5 mA. How many electrons enter and leave this part of the player in 1.0 s?

SOLUTION

SET UP Conservation of charge tells us that when a steady current flows, the same amount of current enters and leaves the player per unit time.

SOLVE We use the current to find the total charge that flows in 1.0 s. We have

$$I = \frac{\Delta Q}{\Delta t}, \quad \text{so}$$
$$\Delta Q = I \, \Delta t = (2.5 \times 10^{-3}\,\text{A})(1.0\,\text{s}) = 2.5 \times 10^{-3}\,\text{C}.$$

Each electron has charge of magnitude $e = 1.60 \times 10^{-19}$ C. The number N of electrons is the total charge ΔQ, divided by the magnitude of the charge e of one electron:

$$N = \frac{\Delta Q}{e} = \frac{2.5 \times 10^{-3}\,\text{C}}{1.60 \times 10^{-19}\,\text{C}} = 1.6 \times 10^{16}.$$

REFLECT It's important to realize that charge is not "used up" in a CD player (or any other electrical device); the same amount that flows in also flows out. However, as we will see, the charge loses *potential energy* as it flows through the player; this is how it powers the player's operation.

Practice Problem: The current in a wire is 2.00 A. How much time is needed for 1 mole of electrons (6.02×10^{23} electrons) to pass a point in the wire? *Answer:* 13.4 h.

PhET: Conductivity
PhET: Ohm's Law
PhET: Resistance in a Wire

19.2 Resistance and Ohm's Law

In a conductor carrying a current, the electric field that causes the mobile charges to move also is associated with a potential difference between points in the conductor. The electric field $\vec{E}$ always points in the direction from higher to lower potential. The current I is proportional to the average drift speed of the moving charges. If this speed is in turn proportional to the electric-field magnitude (and thus to the potential difference between the ends of the conductor), then, for a given segment of conductor, the current I is approximately proportional to the potential difference

V between the ends. In this case, the ratio V/I is approximately constant. This ratio is called the **resistance** of the conductor; it is defined as follows:

Definition of resistance

When the potential difference V between the ends of a conductor is proportional to the current I in the conductor, the ratio V/I is called the resistance of the conductor:

$$R = \frac{V}{I}. \tag{19.2}$$

Unit: The SI unit of resistance is the **ohm,** equal to 1 volt per ampere. The ohm is abbreviated with a capital Greek omega, Ω. Thus, $1\ \Omega = 1\ \text{V/A}$. The *kilohm* $(1\ \text{k}\Omega = 10^3\ \Omega)$ and the *megohm* $(1\ \text{M}\Omega = 10^6\ \Omega)$ are also in common use.

The observation that for many conducting materials, current is proportional to potential difference (voltage) is called **Ohm's law:**

Ohm's law

The potential difference V between the ends of a conductor is proportional to the current I through the conductor; the proportionality factor is the resistance R.

We see that when Ohm's law is obeyed, the resistance defined by Equation 19.2 is a constant, independent of V and I. Like the ideal-gas equation or Hooke's law, Ohm's law represents an *idealized model;* it describes the behavior of some materials quite well, but it isn't a general description of all materials. We'll return to this point later.

A 100 W, 120 V light bulb has a resistance of 140 Ω at the bulb's operating temperature. A 100 m length of 12 gauge copper wire (the size usually used in household wiring) has a resistance of about 0.5 Ω at room temperature. Resistors with designated resistances are used in a wide variety of electric circuits; we'll study some of their applications later in this chapter. Resistors in the range from 0.01 Ω to 10^7 Ω can be bought off the shelf. Figure 19.4 illustrates a color code that is used to label the resistance of resistors.

Resistivity

For a conductor in the form of a uniform cylinder (such as a wire with uniform cross section), the resistance R is found to be proportional to the length L of the conductor and inversely proportional to its cross-sectional area A. These relationships are reasonable: For a given current, if we double the length of conductor, the total potential difference between the ends must double; and if the cross-sectional area is doubled, a given potential difference causes twice as great a current flow. This relationship can be expressed as follows:

Definition of resistivity

The resistance R is proportional to the length L and inversely proportional to the cross-sectional area A, with a proportionality factor ρ called the **resistivity** of the material. That is,

$$R = \rho\frac{L}{A}, \tag{19.3}$$

where ρ, in general different for different materials, characterizes the conduction properties of a material.

Unit: The SI unit of resistivity is 1 ohm · meter $= 1\ \Omega \cdot \text{m}$.

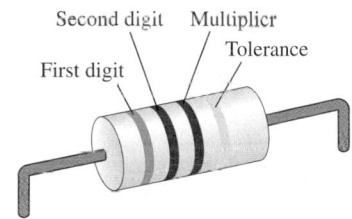

▲ **FIGURE 19.4** Commercial resistors use a code consisting of colored bands to indicate their resistance.

TABLE 19.1   Resistivities at room temperature

Substance	$\rho\ (\Omega \cdot m)$	Substance	$\rho\ (\Omega \cdot m)$
Conductors:		Mercury	95×10^{-8}
Silver	1.47×10^{-8}	Nichrome alloy	100×10^{-8}
Copper	1.72×10^{-8}	Insulators:	
Gold	2.44×10^{-8}	Glass	$10^{10} - 10^{14}$
Aluminum	2.63×10^{-8}	Lucite	$> 10^{13}$
Tungsten	5.51×10^{-8}	Quartz (fused)	75×10^{16}
Steel	20×10^{-8}	Teflon®	$> 10^{13}$
Lead	22×10^{-8}	Wood	$10^{8} - 10^{11}$

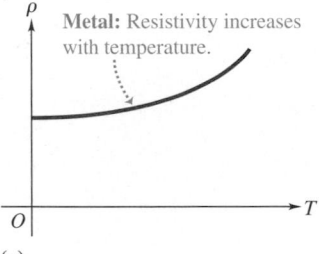

Metal: Resistivity increases with temperature.

(a)

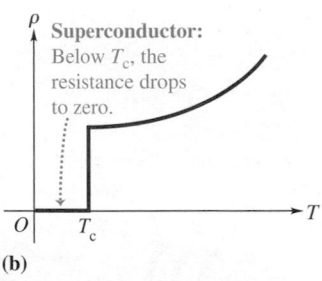

Superconductor: Below T_c, the resistance drops to zero.

(b)

▲ **FIGURE 19.5** The temperature dependence of resistance in (a) a typical metal and (b) a superconductor.

▲ **FIGURE 19.6** A maglev train in Shanghai. Maglev ("magnetic-levitation") trains use superconducting electromagnets to create magnetic fields strong enough to levitate a train off the tracks.

Equation 19.3 shows that the resistance R of a wire or other conductor with uniform cross section is directly proportional to the length of the wire and inversely proportional to its cross-sectional area. It is also proportional to the resistivity of the material of which the conductor is made. This behavior is analogous to water flowing through a hose. A thin hose offers more resistance to flow than a fat one. We can increase the resistance to flow by stuffing the hose with cotton or sand; this corresponds to increasing the resistivity. The flow rate is approximately proportional to the pressure difference between the ends. Flow rate is analogous to current, pressure difference to potential difference (voltage). Let's not stretch this analogy too far, though; the flow rate of water in a pipe is usually *not* proportional to the pipe's cross-sectional area.

NOTE ▶ It's important to distinguish between resistivity and resistance. Resistivity is a property of a material, independent of the shape and size of the specimen, while resistance depends on the size and shape of the specimen or device, as well as on its resistivity. ◀

A few representative values of resistivity are given in Table 19.1. A perfect conductor would have zero resistivity, and a perfect insulator would have infinite resistivity. Metals and alloys have the smallest resistivities and are the best conductors. Note that the resistivities of insulators are greater than those of metals by an enormous factor, on the order of 10^{18} to 10^{22}.

Temperature Dependence of Resistance

The resistance of every conductor varies somewhat with temperature. The resistivity of a *metallic* conductor nearly always increases with increasing temperature (Figure 19.5a). Over a small temperature range (up to 100 C° or so), the change in resistivity of a metal is approximately proportional to the temperature change. If R_0 is the resistance at a reference temperature T_0 (often taken as 0°C or 20°C) and R_T is the resistance at temperature T, then the variation of R with temperature is described approximately by the equation

$$R_T = R_0[1 + \alpha(T - T_0)]. \qquad (19.4)$$

The factor α is called the **temperature coefficient of resistivity.** For common metals, α typically has a value of 0.003 to 0.005 $(\text{C}°)^{-1}$. That is, an increase in temperature of 1 C° increases the resistance by 0.3% to 0.5%.

A small semiconductor crystal called a *thermistor* can be used to make a sensitive electronic thermometer. Its resistance is used as a thermometric property.

Superconductivity

Some materials, including several metallic alloys and oxides, show a phenomenon called *superconductivity*. In these materials, as the temperature decreases, the resistivity at first decreases smoothly, like that of any metal. But then, at a certain critical transition temperature T_c, a phase transition occurs, and the resistivity suddenly drops to zero, as shown in Figure 19.5b. Once a current has been established in a superconducting ring, it continues indefinitely without the presence of any driving field.

Superconductivity was discovered in 1911 in the laboratory of H. Kamerlingh-Onnes. He had just discovered how to liquefy helium, which has a boiling temperature of 4.2 K at atmospheric pressure. Measurements of the resistance of mercury at very low temperatures showed that below 4.2 K, its resistivity suddenly dropped to zero. In recent years, complex oxides of yttrium, copper, and barium have been found that have a much higher superconducting transition temperature. The current (2010) record for T_c is about 160 K, and materials that are superconductors at room temperature may well become a reality. The implications of these discoveries for power-distribution systems, computer design, and transportation are enormous. Meanwhile, superconducting electromagnets cooled by liquid helium are used in particle accelerators and some experimental magnetic-levitation (maglev) trains (Figure 19.6).

In 1913, Kamerlingh-Onnes was awarded a Nobel Prize for his research on properties of materials at very low temperatures. In 1987, J. Georg Bednorz and Karl Alexander Müller were awarded a Nobel Prize for their discovery of high-temperature superconductivity in ceramic materials.

Non-ohmic Conductors

If a conductor obeys Ohm's law, a graph of current versus voltage is a straight line with a slope of $1/R$ (Figure 19.7a). Such a material is said to be *ohmic*. By contrast, Figure 19.7b shows the graph for a semiconductor diode, a device that is decidedly *non* ohmic. Notice that the resistance of a diode depends on the *direction* of the current. Diodes act like one-way valves for current; they are used to perform a wide variety of logic functions in computer circuitry.

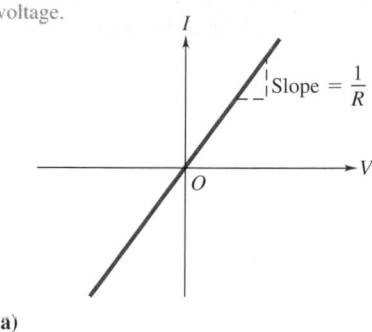

Ohmic resistor (e.g., typical metal wire): At a given temperature, current is proportional to voltage.

(a)

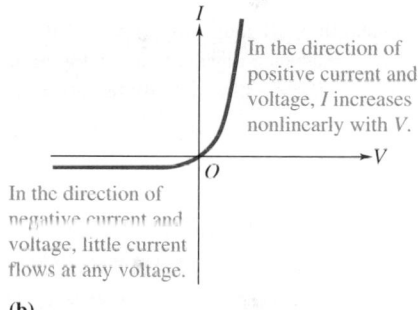

Semiconductor diode: a non-ohmic resistor

In the direction of positive current and voltage, I increases nonlinearly with V.

In the direction of negative current and voltage, little current flows at any voltage.

(b)

▲ **FIGURE 19.7** Graphs of current versus voltage for (a) a typical ohmic resistor and (b) a semiconductor diode.

Conceptual Analysis 19.1

Resistance and resistivity

Two wires made of pure copper have different resistances. These wires may differ in

A. length.
B. cross-sectional area.
C. resistivity.
D. temperature.

SOLUTION Since the two wires are made of the same material, their resistivities at a given temperature are the same. Any difference in resistance at a given temperature must be due to a difference in length or cross-sectional area (or both). If the wires are at significantly different temperatures however, their difference in resistance could be due to the resulting difference in resistivity, as well as to a difference in their length or cross-sectional area. Thus, *all* of the choices are correct.

EXAMPLE 19.2   Resistance in your stereo system

Suppose you're hooking up a pair of stereo speakers. **(a)** You happen to have on hand some 20-m-long pieces of 16 gauge copper wire (diameter 1.3 mm); you use them to connect the speakers to the amplifier. These wires are longer than needed, but you just coil up the excess length instead of cutting them. What is the resistance of one of these wires? **(b)** To improve the performance of the system, you purchase 3.0-m-long speaker cables that are made with 8 gauge copper wire (diameter 3.3 mm). What is the resistance of one of these cables?

Continued

SOLUTION

SET UP Figure 19.8 shows our sketch. The resistivity of copper at room temperature is $\rho = 1.72 \times 10^{-8}\,\Omega \cdot \text{m}$ (Table 19.1). The cross-sectional area A of a wire is related to its radius by $A = \pi r^2$.

SOLVE To find the resistances, we use Equation 19.3, $R = \rho L / A$.

Part (a): $R = \dfrac{(1.72 \times 10^{-8}\,\Omega \cdot \text{m})(20\,\text{m})}{\pi(6.5 \times 10^{-4}\,\text{m})^2} = 0.26\,\Omega.$

Part (b): $R = \dfrac{(1.72 \times 10^{-8}\,\Omega \cdot \text{m})(3.0\,\text{m})}{\pi(1.65 \times 10^{-3}\,\text{m})^2} = 6.0 \times 10^{-3}\,\Omega.$

REFLECT The shorter, fatter wires offer over forty times less resistance than the longer, skinnier ones.

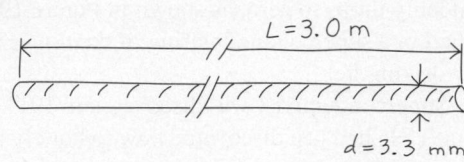

▲ **FIGURE 19.8** Our sketch for this problem.

Practice Problem: 14 gauge copper wire has a diameter of 1.6 mm. What length of this wire has a resistance of $1.0\,\Omega$? *Answer:* 120 m.

EXAMPLE 19.3 **Warm wires and cold wires**

A length of 18 gauge copper wire with a diameter of 1.02 mm and a cross-sectional area of $8.20 \times 10^{-7}\,\text{m}^2$ has a resistance of $1.02\,\Omega$ at a temperature of 20°C. Find the resistance at 0°C and at 100°C. The temperature coefficient of resistivity of copper is $0.0039\,(\text{C}°)^{-1}$.

SOLUTION

SET UP AND SOLVE We use Equation 19.4, with $T_0 = 20°\text{C}$ and $R_0 = 1.02\,\Omega$. At $T = 0°\text{C}$,

$$R = R_0[1 + \alpha(T - T_0)]$$
$$= (1.02\,\Omega)(1 + [0.0039(\text{C}°)^{-1}][0°\text{C} - 20°\text{C}])$$
$$= 0.94\,\Omega.$$

At $T = 100°\text{C}$,

$$R = (1.02\,\Omega)(1 + [0.0039(\text{C}°)^{-1}][100°\text{C} - 20°\text{C}])$$
$$= 1.34\,\Omega.$$

REFLECT Between the freezing and boiling temperatures of water, the resistance increases by about 40%.

Practice Problem: On a hot summer day in Death Valley, the resistance is $1.14\,\Omega$. What is the temperature? *Answer:* 51°C.

PhET: Battery Voltage

19.3 Electromotive Force and Circuits

For a conductor to have a steady current, it must be part of a path that forms a closed loop, or **complete circuit.** But the path cannot consist entirely of resistance. In a resistor, charge always moves in the direction of decreasing potential energy. There must be some part of the circuit where the potential energy *increases.*

The situation is analogous to the fountain in Figure 19.9a. The water emerges at the top, cascades down to the basin, and then is pumped back to the top for another trip. The pump does work on the water, raising it to a position of higher gravitational potential energy; without the pump, the water would simply fall to the basin and stay there. Similarly, an electric circuit must contain a battery or other device that does work on electric charges to bring them to a position of higher electric potential energy so that they can flow through a circuit to a lower potential energy. The situation of Figure 19.9b doesn't occur in the real world!

(a)   (b)

▲ **FIGURE 19.9** (a) A pump is needed to keep the water circulating in this fountain. (b) A current in an ordinary circuit with no source of emf would be analogous to this famous impossible scene by Escher.

Electromotive Force

The influence that moves charge from lower to higher potential (despite the electric-field forces in the opposite direction) is called **electromotive force** (abbreviated **emf** and pronounced "ee-em-eff"). Every complete circuit with a continuous current must include some device that provides emf. "Electromotive force" is a poor term because emf is *not* a force, but an "energy per unit charge" quantity, like potential. The SI unit of emf is the same as the unit for potential, the volt $(1 \text{ V} = 1 \text{ J/C})$. A battery with an emf of 1.5 V does 1.5 J of work on every coulomb of charge that passes through it. We'll use the symbol $\mathcal{E}$ for emf.

Batteries, electric generators, solar cells, thermocouples, and fuel cells are all sources of emf. Each such device converts energy of some form (mechanical, chemical, thermal, and so on) into electrical potential energy and transfers it into the circuit where the device is connected. An *ideal* source of emf maintains a constant potential difference between its terminals, independently of the current through it. We define emf quantitatively as the magnitude of this potential difference. As we will see, such an ideal source is a mythical beast, like the unicorn, the frictionless plane, and the free lunch. Nevertheless, it's a useful idealization.

Figure 19.10 is a schematic diagram of a source of emf that maintains a potential difference between points *a* and *b*, called the *terminals* of the device. Terminal *a*, marked +, is maintained at *higher* potential than terminal *b*, marked −. Associated with this potential difference is an electric field $\vec{E}$ in the region around the terminals, both inside and outside the source. The electric field inside the device is directed from *a* to *b* as shown. A charge *q* within the source is acted upon by an electric force $\vec{F}_E = q\vec{E}$. The source has to provide some additional influence, which we represent as a non-electrostatic force $\vec{F}_n$ that pushes charge from *b* to *a* inside the device (opposite to the direction of $\vec{F}_E$) and maintains the potential difference. The nature of this additional influence depends on the source. In a battery, it is due to chemical processes; in an electric generator, it results from magnetic forces. It could even be you, rubbing plastic on fur, as we saw in Chapter 17.

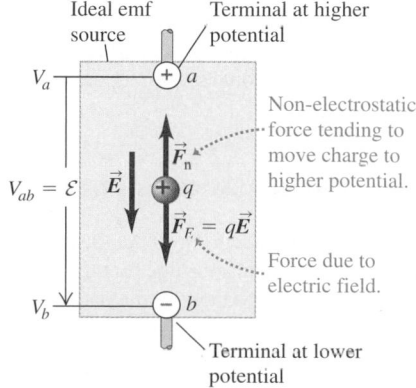

When the emf source is not part of a closed circuit, $F_n = F_E$ and there is no net motion of charge between the terminals.

▲ **FIGURE 19.10** Schematic diagram of an ideal emf source in an "open-circuit" condition.

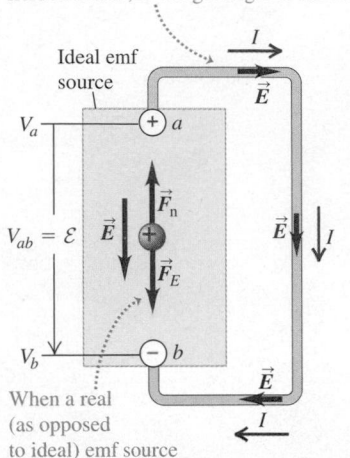

Potential across terminals creates electric field in circuit, causing charges to move.

When a real (as opposed to ideal) emf source is connected to a circuit, V_{ab} and thus F_E fall, so that $F_n > F_E$ and $\vec{F}_n$ does work on the charges.

▲ **FIGURE 19.11** Schematic diagram of an ideal emf source in a complete circuit.

The potential V_{ab} of point a with respect to point b is defined, as always, as the work per unit charge done by the electrostatic force $\vec{F}_E = q\vec{E}$ on a charge q that moves from a to b. The emf $\mathcal{E}$ of the source is the energy per unit charge supplied by the source during an "uphill" displacement from b to a. For the ideal source of emf that we have described, the potential difference V_{ab} is equal to the electromotive force $\mathcal{E}$:

$$V_{ab} = \mathcal{E}. \quad \text{(no complete circuit)} \quad (19.5)$$

Now let's make a complete circuit by connecting a wire with resistance R to the terminals of a source (Figure 19.11). The charged terminals a and b of the source set up an electric field in the wire, and this causes a current in the wire, directed from a toward b. From $V = IR$, the current I in the circuit is determined by

$$\mathcal{E} = V_{ab} = IR. \quad \text{(ideal source of emf)} \quad (19.6)$$

That is, when a charge q flows around the circuit, the potential rise $\mathcal{E}$ as it passes through the source is numerically equal to the potential drop $V_{ab} = IR$ as it passes through the resistor. Once $\mathcal{E}$ and R are known, this relation determines the current in the circuit. The current is the same at every point in the circuit.

Conceptual Analysis 19.2

Current in a circuit

The circuit in Figure 19.12 contains a battery and two identical lightbulbs, which are resistors. The battery maintains a constant potential difference between its terminals, independently of the current through it. The bulbs are equally bright when the switch is closed, meaning that they receive the same current. When the switch is opened, what happens to the brightness of bulb A?

A. It increases (because bulb A now receives the current that formerly went through bulb B).
B. It stays the same (meaning that the current through bulb A is unchanged).

SOLUTION The key to this problem is the fact that a battery or other emf source is a source of constant *voltage*, not constant *current*. A 9 V battery produces whatever current is determined by the resistance of the external circuit, but the emf is always 9 V. Be sure you understand this point; don't treat an emf source as though it produces constant current.

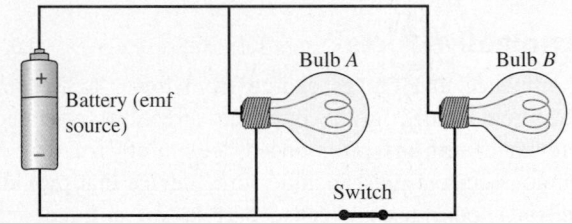

▲ **FIGURE 19.12**

The current drawn by each bulb depends on its resistance and the voltage across it. Since bulb B is connected in parallel with bulb A, both experience the full voltage produced by the battery. The presence or absence of B thus makes no difference to the current through bulb A, so opening the switch does not alter A's brightness. What *does* change is the amount of current supplied by the battery. When the switch is closed, the two bulbs receive equal current from the battery, draining it twice as fast as when only one bulb is lit.

EXAMPLE 19.4 **Electrical hazards in heart surgery**

A patient is undergoing open-heart surgery. A sustained current as small as 25 μA passing through the heart can be fatal. Assume that the heart has a constant resistance of 250 Ω; determine the minimum voltage that poses a danger to the patient.

SOLUTION

SET UP AND SOLVE Figure 19.13 shows our sketch. 1 μA = 1×10^{-6} A. We use Ohm's law to relate voltage, resistance, and current:

$$V = IR = (25 \times 10^{-6}\,\text{A})(250\,\Omega) = 6.25 \times 10^{-3}\,\text{V}$$
$$= 6.25\,\text{mV}.$$

▶ **FIGURE 19.13** Our sketch for this problem.

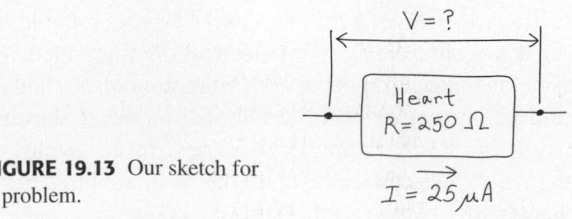

Continued

REFLECT Our result shows that even a small voltage can be dangerous if applied directly to the heart. As a safety measure, any electrical equipment near a patient during surgery must be "grounded." Grounding means that there is a low-resistance conducting path to the earth, so any undesirable current will pass into the ground instead of through the person. During some surgeries, even the patient is grounded to prevent electrical shock.

Practice Problem: A voltage of 12.0 V across the terminals of a device produces a current of 3.00 mA through the device. What is the resistance of the device? *Answer:* 4.00 kΩ.

Internal Resistance in a Source of emf

Real sources of emf don't behave exactly like the ideal sources we've described because charge that moves through the material of any real source encounters *resistance*. We call this the **internal resistance** of the source, denoted by r. If this resistance behaves according to Ohm's law, r is constant. The current through r has an associated drop in potential equal to Ir. The terminal potential difference V_{ab} is then

$$V_{ab} = \mathcal{E} - Ir. \quad \text{(source with internal resistance)} \quad (19.7)$$

The potential V_{ab}, called the **terminal voltage,** is less than the emf $\mathcal{E}$ because of the term Ir representing the potential drop across the internal resistance r.

The current in the external circuit is still determined by $V_{ab} = IR$. Combining this relationship with Equation 19.7, we find that $\mathcal{E} - Ir = IR$, and it follows that

$$I = \frac{\mathcal{E}}{R + r}. \quad \text{(source with internal resistance)} \quad (19.8)$$

That is, the current I equals the source emf $\mathcal{E}$, divided by the *total* circuit resistance $(R + r)$. Thus, we can describe the behavior of a source in terms of two properties: an emf $\mathcal{E}$, which supplies a constant potential difference that is independent of the current, and a series internal resistance r.

To summarize, a circuit is a closed conducting path containing resistors, sources of emf, and possibly other circuit elements. Equation 19.7 shows that the algebraic sum of the potential differences and emf's around the path is zero. Also, the current in a simple loop is the same at every point. Charge is conserved; if the current were different at different points, there would be a continuing accumulation of charge at some points, and the current couldn't be constant.

Table 19.2 shows the symbols usually used in schematic circuit diagrams. We'll use these symbols in most of the circuit analysis in the remainder of this chapter.

TABLE 19.2 Circuit symbols used in this chapter

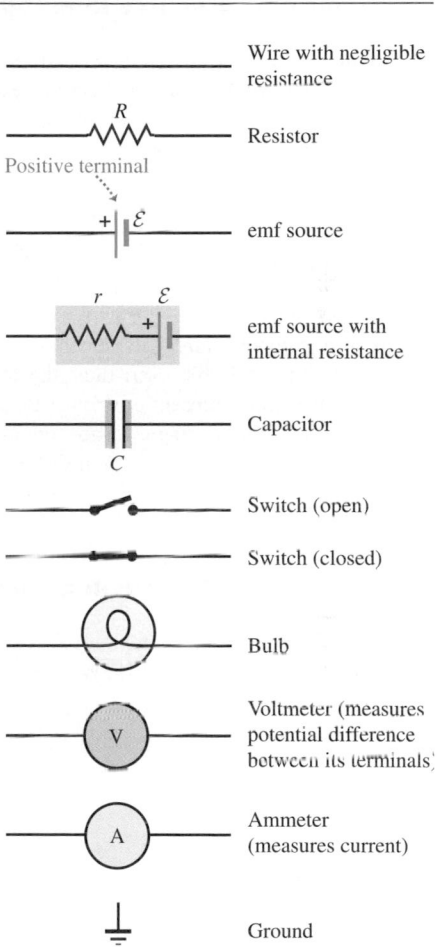

	Wire with negligible resistance
R (Resistor symbol)	Resistor
Positive terminal $+\| \mathcal{E}$	emf source
r $\mathcal{E}$ (symbol)	emf source with internal resistance
C (symbol)	Capacitor
(open switch)	Switch (open)
(closed switch)	Switch (closed)
(bulb symbol)	Bulb
V	Voltmeter (measures potential difference between its terminals)
A	Ammeter (measures current)
(ground symbol)	Ground

EXAMPLE 19.5 A dim flashlight

As a flashlight battery ages, its emf stays approximately constant, but its internal resistance increases. A fresh battery has an emf of 1.5 V and negligible internal resistance. When the battery needs replacement, its emf is still 1.5 V, but its internal resistance has increased to 1000 Ω. If this old battery is supplying 1.0 mA to a lightbulb, what is its terminal voltage?

SOLUTION

SET UP AND SOLVE The terminal voltage of a new battery is 1.5 V. The terminal voltage of the old, worn-out battery is given by $V_{ab} = \mathcal{E} - Ir$, so

$$V_{ab} = 1.5 \text{ V} - (1.0 \times 10^{-3} \text{ A})(1000 \ \Omega) = 0.5 \text{ V}.$$

REFLECT The terminal voltage is less than the emf because of the potential drop across the internal resistance of the battery.

When testing a battery, it is better to measure its terminal voltage when it is supplying current than just to measure its emf.

Practice Problem: An old battery with an emf of 9.0 V has a terminal voltage of 8.2 V when it is supplying a current of 2.0 mA. What is the internal resistance of the battery? *Answer:* 400 Ω.

NOTE ▶ In the examples that follow, it's important to understand how the meters in the circuit work. The symbol V in a circle represents an ideal voltmeter. It measures the potential difference between the two points in the circuit where it is connected, but *no current flows through the voltmeter*. The symbol A in a circle represents an ideal ammeter. It measures the current that flows through it, but *there is no potential difference between its terminals*. Thus, the behavior of a circuit doesn't change when an ideal ammeter or voltmeter is connected to it. ◀

EXAMPLE 19.6 A source in an open circuit

Figure 19.14 shows a source with an emf $\mathcal{E}$ of 12 V and an internal resistance r of 2.0 Ω. (For comparison, the internal resistance of a commercial 12 V lead storage battery is only a few thousandths of an ohm.) Determine the readings of the ideal voltmeter V and ammeter A.

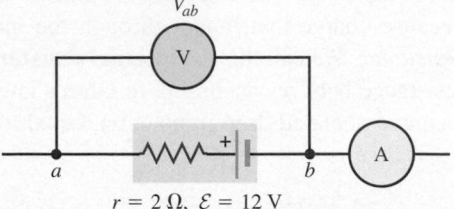

▶ **FIGURE 19.14** $r = 2\,\Omega,\ \mathcal{E} = 12$ V

SOLUTION

SET UP AND SOLVE First we see that the diagram does not show a complete circuit. (Remember that there is no current through an ideal voltmeter; thus, the loop containing the voltmeter does *not* represent a circuit.) Because there is no current through the battery, there is no potential difference across its internal resistance. The potential difference V_{ab} across its termi-nals is equal to the emf $(V_{ab} = \mathcal{E} = 12$ V$)$, and the voltmeter reads $V = 12$ V. Because there is no complete circuit, there is no current, so the ammeter reads zero.

REFLECT As soon as this battery is put into a complete circuit, its internal resistance causes its terminal voltage to be less than its emf.

EXAMPLE 19.7 A source in a complete circuit

Using the battery in Example 19.6, we add a 4.0 Ω resistor to form the complete circuit shown in Figure 19.15. What are the voltmeter and ammeter readings now?

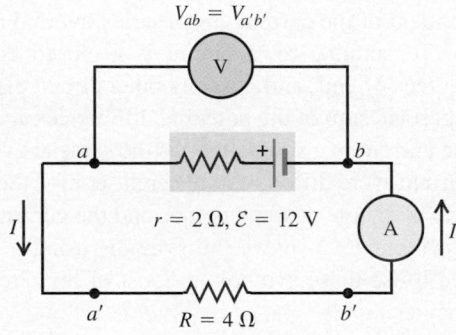

▶ **FIGURE 19.15** a' $R = 4\,\Omega$ b'

SOLUTION

SET UP AND SOLVE We now have a complete circuit and a current I through the resistor R, determined by Equation 19.8:

$$I = \frac{\mathcal{E}}{R + r} = \frac{12\ \text{V}}{4.0\ \Omega + 2.0\ \Omega} = 2.0\ \text{A}.$$

The ammeter A reads $I = 2.0$ A.

Our idealized conducting wires have zero resistance, so there is no potential difference between points a and a' or between b and b'. That is, $V_{ab} = V_{a'b'}$, and the voltmeter reading is this potential difference. We can find V_{ab} by considering a and b either as the terminals of the resistor or as the terminals of the source. Considering them as the terminals of the resistor, we use Ohm's law $(V = IR)$ to obtain

$$V_{a'b'} = IR = (2.0\ \text{A})(4.0\ \Omega) = 8.0\ \text{V}.$$

Considering them as the terminals of the source, we have

$$V_{ab} = \mathcal{E} - Ir = 12\ \text{V} - (2.0\ \text{A})(2.0\ \Omega) = 8.0\ \text{V}.$$

Either way, we conclude that the voltmeter reads $V_{ab} = 8.0$ V.

REFLECT The terminal voltage V_{ab} of the battery is less than the battery emf because of the potential drop across the battery's internal resistance r.

Practice Problem: If a different resistor R is used and the voltmeter reads 6.0 V, what is the ammeter reading? *Answer:* 3.0 A.

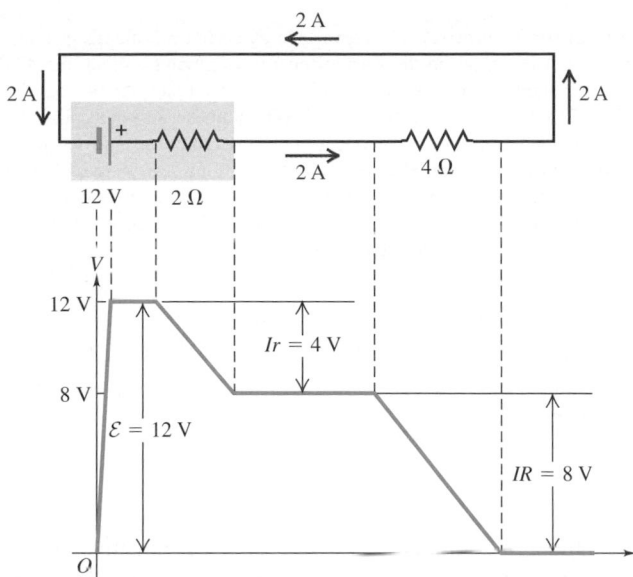

▲ **FIGURE 19.16** Potential rises and drops in the circuit.

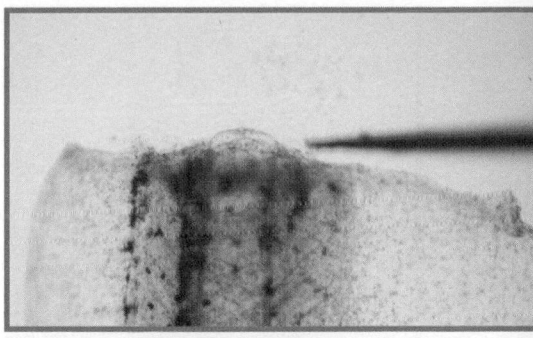

▲ **BIO Application The body electric.**
In addition to the impulses produced by excitable tissue such as nerve and muscle, the body has another kind of electrical signaling. The skin of vertebrates, including humans, typically generates an electrical potential across the epithelial layer of cells that is the active part of the skin. This potential is usually due to the net inward transport of Na⁺, and the interior spaces of the animal may be +50 mV with respect to the outside. If the integrity of the epithelial layer is compromised by a wound, the connected sheet of cells near the wound each act as tiny batteries in parallel to drive substantial current out through the wound, resulting in an electrical field in the tissues near the wound. Physiologists can measure these relatively steady currents using a vibrating electrode that samples a region near the wound and reports the potential along the excursion path of the electrode. An electrode (seen as a blur) placed on a frog tadpole tail where its tip was recently cut off measured current densities as large as 40 mA/m². The tadpole will regenerate the missing piece in a few days and electrical signaling appears to play a role in initiating this regeneration process.

Figure 19.16 is a graph showing how the potential varies around the complete circuit of Figure 19.15. The horizontal axis doesn't necessarily represent actual distances; rather, it shows various points in the loop. If we take the potential to be zero at the negative terminal of the battery, then we have a rise $\mathcal{E}$ and a drop Ir in the battery and an additional drop IR in the external resistor. As we finish our trip around the loop, the potential is back where it started.

The difference between a fresh flashlight battery and an old one is not so much in the emf, which decreases only slightly with use, but mostly in the internal resistance, which may increase from a few ohms when the battery is fresh to as much as 1000 Ω or more after long use. Similarly, a car battery can deliver less current to the starter motor on a cold morning than when the battery is warm, not because the emf is appreciably less, but because the internal resistance increases with decreasing temperature. Residents of northern Iowa have been known to soak their car batteries in warm water to provide greater starting power on very cold mornings!

19.4 Energy and Power in Electric Circuits

Let's now look at some energy and power relations in electric circuits. The box in Figure 19.17 represents a circuit element with potential difference $V_a - V_b = V_{ab}$ between its terminals and current I passing through it in the direction from a toward b. This element might be a resistor, a battery, or something else; the details don't matter. As charge passes through the circuit element, the electric field does work on the charge.

When a charge q passes through the circuit element, the work done on the charge is equal to the product of q and the potential difference V_{ab} (work per unit charge). When V_{ab} is positive, a positive amount of work qV_{ab} is done on the charge as it "falls" from potential V_a to the lower potential V_b. If the current is I, then in a time interval Δt an amount of charge $\Delta Q = I\,\Delta t$ passes through. The work ΔW done on this amount of charge is

$$\Delta W = V_{ab}\,\Delta Q = V_{ab}I\,\Delta t.$$

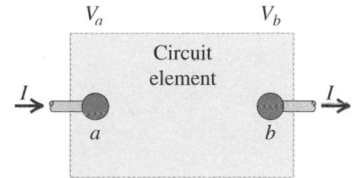

▲ **FIGURE 19.17** The power input P to the portion of the circuit between a and b is $P = V_{ab}I$.

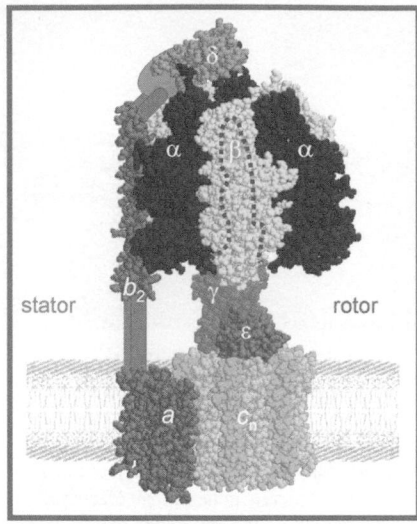

stator

rotor

◀ **BIO Application Life's currency.** The basic "energy currency" of cells is a molecule called ATP, which carries energy in the form of energy-rich bonds. When the ATP molecule is "discharged," the bonds break to power cellular reactions. Later, the bonds are re-formed. The cell's main ATP recharger is a protein called ATP synthase. In your cells, this protein spans the membrane of a cellular compartment, the mitochondrion. As food molecules are broken down, their energy is used to pump hydrogen ions (H^+)—which are simply protons—out of the mitochondrion, creating an electrical potential difference across the membrane (as well as a difference in H^+ concentration). ATP synthase allows the protons to flow back in, using their energy to recharge ATP.

This work represents electrical energy transferred *into* the circuit element. The time rate of energy transfer is *power,* denoted by P. Dividing the preceding equation by Δt, we obtain the *time rate* at which the rest of the circuit delivers electrical energy to the circuit element:

$$\frac{\Delta W}{\Delta t} = P = V_{ab}I. \tag{19.9}$$

If the potential at b is higher than at a, then V_{ab} is negative, and there is a net transfer of energy *out of* the circuit element. The element is then acting as a source, delivering electrical energy into the circuit to which it is connected. This is the usual situation for a battery, which converts chemical energy into electrical energy and delivers it to the external circuit.

The unit of V_{ab} is 1 volt, or 1 joule per coulomb, and the unit of I is 1 ampere, or 1 coulomb per second. We can now confirm that the SI unit of power is 1 watt:

$$(1 \text{ J/C})(1 \text{ C/s}) = 1 \text{ J/s} = 1 \text{ W}.$$

Conceptual Analysis 19.3

Buying electricity

What do you buy from the power company?

A. Only energy.
B. Electrons and energy.
C. Only electrons.

SOLUTION The wires to your house provide two things: a current and a potential difference (usually either 110 V or 220 V,

depending on where you live). Remember that a current is a flow of charge. Because charge is a conserved quantity, the rate at which charge enters a circuit element at any instant must equal the rate at which it leaves the element. Therefore, the electrons are more like a conveyor belt than they are like the material on the belt. You don't buy electrons and use them up. However, you *do* use the energy represented by the flow of charge down a potential difference. Thus, the power company sells you energy only.

PhET: Battery-Resistor Circuit
PhET: Signal Circuit

Pure Resistance

When current flows through a resistor, electrical energy is transformed into thermal energy. An electric toaster is an obvious example. We calculate the power dissipated through a resistor as follows: The potential difference across the resistor is $V_{ab} = IR$. From Equation 19.9, the electric power delivered to the resistor by the circuit is

$$P = V_{ab}I = I^2R = \frac{V_{ab}^2}{R}. \tag{19.10}$$

What becomes of this energy? The moving charges collide with atoms in the resistor and transfer some of their energy to these atoms, increasing the internal energy of the material. Either the temperature of the resistor increases, or there is a flow of heat out of it, or both. We say that energy is *dissipated* in the resistor at a rate I^2R. Too high a temperature can change the resistance unpredictably; the resistor may melt or even explode. Of course, some devices, such as electric heaters, are designed to get hot and transfer heat to their surroundings. But every resistor has a *power rating:* the maximum power that the device can dissipate without becoming overheated and damaged. In practical applications, the power rating of a resistor is often just as important a characteristic as its resistance.

- The emf source converts non-electrical to electrical energy at a rate $\mathcal{E}I$.
- Its internal resistance *dissipates* energy at a rate I^2r.
- The difference $\mathcal{E}I - I^2r$ is its power output.

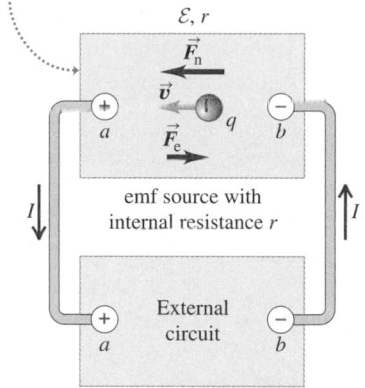

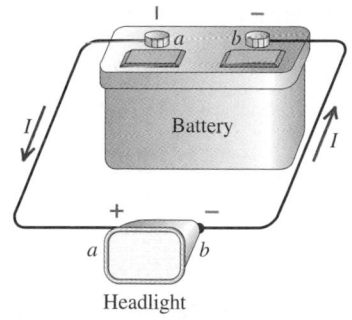

(a) Diagrammatic circuit

(b) A real circuit of the type shown in (a)

▲ **FIGURE 19.18** Power output of an emf source.

Power Output of a Source

The upper box in Figure 19.18a represents a source with emf $\mathcal{E}$ and internal resistance r, connected by ideal (resistanceless) conductors to an external circuit represented by the lower box. This arrangement could describe a car battery connected to the car's headlights (Figure 19.18b). Point a is at higher potential than point b; that is, $V_a > V_b$. But now the current I is *leaving* the device at the higher-potential terminal (rather than entering there). Energy is being delivered to the external circuit, and the rate of its delivery to the circuit is given by Equation 19.10: $P = V_{ab}I$. For a source that can be described by an emf $\mathcal{E}$ and an internal resistance r, we may use Equation 19.7: $V_{ab} = \mathcal{E} - Ir$.

Multiplying this equation by I, we find that

$$P = V_{ab}I = \mathcal{E}I - I^2r. \qquad (19.11)$$

The emf $\mathcal{E}$ is the work per unit charge performed on the charges as they are pushed "uphill" from b to a in the source. When a charge ΔQ flows through the source during a time interval Δt, the work done on it is $\mathcal{E} \Delta Q = \mathcal{E}I \Delta t$. Thus, $\mathcal{E}I$ is the *rate* at which work is done on the circulating charges. This term represents the rate of conversion of non-electrical energy to electrical energy within the source. The term I^2r is the rate at which electrical energy is *dissipated* in the internal resistance of the source. The difference, $\mathcal{E}I - I^2r$, is the *net* electrical power output of the source—that is, the rate at which the source delivers electrical energy to the remainder of the circuit.

▲ Application **Cheap light** If you've had incandescent flashlights or bicycle lights and changed to lights that use light-emitting diodes (LEDs), you know the large difference in energy consumption. A halogen bicycle headlight might go through a set of batteries in 3 hours, but an even brighter LED headlight will last 30 hours. Why the difference? The answer is that any incandescent bulb (including a halogen bulb) works by using the dissipation of electrical energy to heat a filament white hot. Some of the energy is converted to visible light, but most is lost as heat. In an LED, electrical energy is used to move semiconductor electrons to a region where they emit light. Most of the electrical energy, then, emerges as light; little is lost as heat.

Quantitative Analysis 19.4   **Power and current**

A 1200 W floor heater, a 360 W television, and a hand iron operating at 900 W are all plugged into the same 120 volt circuit in a house (that is, the same pair of wires that come from the basement fuse box). What is the total current flowing through this circuit?

A. 20.5 A.
B. 17.5 A.
C. 15 A.
D. 12.5 A.

SOLUTION The electric power input to an appliance is given by $P = VI$. Each device is attached to the same 120 volt source. That is, the devices are in parallel, and their currents add to give the total current in the pair of wires coming from the fuse box. The current I through each device is given by $I = P/120$ V. The currents through each, in the order listed, are 10 A, 3 A, and 7.5 A. The sum of these currents is 20.5 A. This would likely trip the circuit breaker for that circuit.

EXAMPLE 19.8 **Lightbulb resistance**

The bulb for an interior light in a car is rated at 15.0 W and operates from the car battery voltage of 12.6 V. What is the resistance of the bulb?

SOLUTION

SET UP AND SOLVE The electric power consumption of the bulb is 15.0 W. Thus,

$$ P = \frac{V^2}{R}, \quad \text{so} \quad R = \frac{V^2}{P} = \frac{(12.6 \text{ V})^2}{15.0 \text{ W}} = 10.6 \ \Omega. $$

REFLECT This is the resistance of the lightbulb at its operating temperature. When the light first comes on, the filament is cooler and the resistance is lower. Because $P = V^2/R$, the bulb has a greater rate of electrical energy consumption when it first comes on. Also, because $I = V/R$, the current drawn by the bulb is greater when it is cool and R is less.

Practice Problem: What is the resistance of a heating element when the current through it is 1.20 A and it is consuming electrical energy at a rate of 220 W? *Answer:* 153 Ω.

19.5 Resistors in Series and in Parallel

Resistors turn up in all kinds of circuits, ranging from hair dryers and space heaters to circuits that limit or divide current or that reduce or divide a voltage. Suppose we have three resistors with resistances R_1, R_2, and R_3. Figure 19.19 shows four different ways they might be connected between points a and b. In Figure 19.19a, the resistors provide only a single path between these points. When several circuit elements, such as resistors, batteries, and motors, are connected in sequence as in Figure 19.19a, we say that they are connected in **series.** Since no charge accumulates anywhere between point a and point b, the *current* is the same in all of the resistors when they are connected in series.

The resistors in Figure 19.19b are said to be connected in **parallel** between points a and b. Each resistor provides an alternative path between the points. For circuit elements that are connected in parallel, the *potential difference* is the same across each element. In Figure 19.19c, resistors R_2 and R_3 are in parallel, and this combination is in series with R_1. In Figure 19.19d, R_2 and R_3 are in series, and this combination is in parallel with R_1.

For any combination of resistors that obey Ohm's law, we can always find a single resistor that could replace the combination and result in the same total current and potential difference. The resistance of this single resistor is called the **equivalent resistance** of the combination. If any one of the networks in Figure 19.19 were replaced by its equivalent resistance R_{eq}, we could write

$$ V_{ab} = IR_{eq}, \quad \text{or} \quad R_{eq} = \frac{V_{ab}}{I}, $$

where V_{ab} is the potential difference between terminals a and b of the network and I is the current at point a or point b. To compute an equivalent resistance, we assume a potential difference V_{ab} across the actual network, compute the corresponding current I, and take the ratio V_{ab}/I.

Resistors in Series

For series and parallel combinations, we can derive general equations for the equivalent resistance. If the resistors are in *series,* as in Figure 19.19a, the current I must be the same in all of them. (Otherwise, charge would be piling up at the points where the terminals of two resistors are connected.) Applying $V = IR$ to each resistor, we have

$$ V_{ax} = IR_1, \quad V_{xy} = IR_2, \quad V_{yb} = IR_3. $$

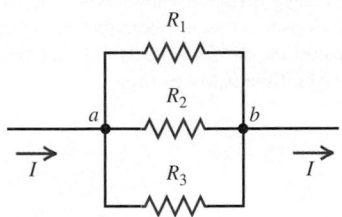

(a) Resistors in series

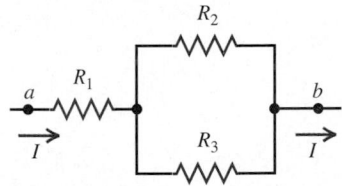

(b) Resistors in parallel

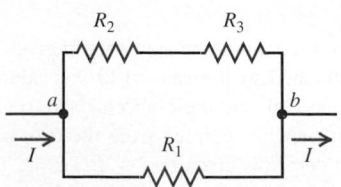

(c) R_1 in series with a parallel combination of R_2 and R_3

(d) R_1 in parallel with a series combination of R_2 and R_3

▲ **FIGURE 19.19** Four different ways of connecting three resistors.

The potential difference V_{ab} is the sum of these three quantities:

$$V_{ab} = V_{ax} + V_{xy} + V_{yb} = I(R_1 + R_2 + R_3),$$

or

$$\frac{V_{ab}}{I} = R_1 + R_2 + R_3.$$

But V_{ab}/I is, by definition, the equivalent resistance R_{eq}. Therefore,

$$R_{eq} = R_1 + R_2 + R_3.$$

It is easy to generalize this relationship to any number of resistors:

Equivalent resistance for resistors in series
The equivalent resistance of *any number* of resistors in series equals the sum of their individual resistances:

$$R_{eq} = R_1 + R_2 + R_3 + \cdots. \tag{19.12}$$

The equivalent resistance is always *greater than* any individual resistance.

Resistors in Parallel

For resistors in parallel, as in Figure 19.19b, the potential difference between the terminals of each resistor must be the same and equal to V_{ab}. Let's call the currents in the three resistors I_1, I_2, and I_3, respectively. Then, from $I = V/R$,

$$I_1 = \frac{V_{ab}}{R_1}, \qquad I_2 = \frac{V_{ab}}{R_2}, \qquad I_3 = \frac{V_{ab}}{R_3}.$$

Charge is neither accumulating at, nor draining out of, point a; all charge that enters point a also leaves that point. Thus, the total current I must equal the sum of the three currents in the resistors:

$$I = I_1 + I_2 + I_3 = V_{ab}\left(\frac{1}{R_1} + \frac{1}{R_2} + \frac{1}{R_3}\right),$$

or

$$\frac{I}{V_{ab}} = \frac{1}{R_1} + \frac{1}{R_2} + \frac{1}{R_3}.$$

But by the definition of the equivalent resistance R_{eq}, $I/V_{ab} = 1/R_{eq}$, so

$$\frac{1}{R_{eq}} = \frac{1}{R_1} + \frac{1}{R_2} + \frac{1}{R_3}.$$

Again, it is easy to generalize this relationship to *any number* of resistors in parallel:

Equivalent resistance for resistors in parallel
For *any number* of resistors in parallel, the *reciprocal* of the equivalent resistance equals the *sum of the reciprocals* of their individual resistances:

$$\frac{1}{R_{eq}} = \frac{1}{R_1} + \frac{1}{R_2} + \frac{1}{R_3} + \cdots. \tag{19.13}$$

The equivalent resistance is always *less than* any individual resistance.

MasteringPHYSICS

PhET: Circuit Construction Kit (DC Only)
ActivPhysics 12.1: DC Series Circuits (Qualitative)
ActivPhysics 12.2: DC Parallel Circuits
ActivPhysics 12.3: DC Circuit Puzzles

PROBLEM-SOLVING STRATEGY 19.1 **Resistors in series and in parallel**

SET UP

1. It helps to remember that when resistors are connected in series, the total potential difference across the combination is the sum of the individual potential differences. When resistors are connected in parallel, the potential difference is the same for every resistor and is equal to the potential difference across the parallel combination.

2. Also, keep in mind the analogous statements for current: When resistors are connected in series, the current is the same through every resistor and is equal to the current through the series combination. When resistors are connected in parallel, the total current through the combination is equal to the sum of currents through the individual resistors.

SOLVE

3. We can often consider networks such as those in Figures 19.19c and 19.19d as combinations of series and parallel arrangements. In Figure 19.19c, we first replace the parallel combination of R_2 and R_3 by its equivalent resistance; this then forms a series combination with R_1. In Figure 19.19d, the combination of R_2 and R_3 in series forms a parallel combination with R_1.

REFLECT

4. The rule for combining resistors in parallel follows directly from the principle of conservation of charge. The rule for combining resistors in series results from a fundamental principle about work: When a particle moves along a path, the total work done on it is the sum of the quantities of work done during the individual segments of the path.

EXAMPLE 19.9 **A resistor network**

Three identical resistors with resistances of 6.0 Ω are connected as shown in Figure 19.20 to a battery with an emf of 18.0 V and zero internal resistance. **(a)** Find the equivalent resistance of the resistor network. **(b)** Find the current in each resistor.

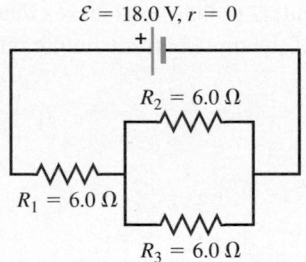

▶ **FIGURE 19.20**

SOLUTION

SET UP AND SOLVE Part (a): To find the equivalent resistance, we identify series or parallel combinations of resistors and replace them by their equivalent resistors, continuing this process until the circuit has just a single resistor that is the equivalent resistor for the network. Figure 19.21 shows the procedure.

In this network, R_2 and R_3 are in parallel, so their equivalent resistance R_{23} is given by

$$\frac{1}{R_{23}} = \frac{1}{R_2} + \frac{1}{R_3} = \frac{1}{6.0\ \Omega} + \frac{1}{6.0\ \Omega} = \frac{2}{6.0\ \Omega} = \frac{1}{3.0\ \Omega}.$$

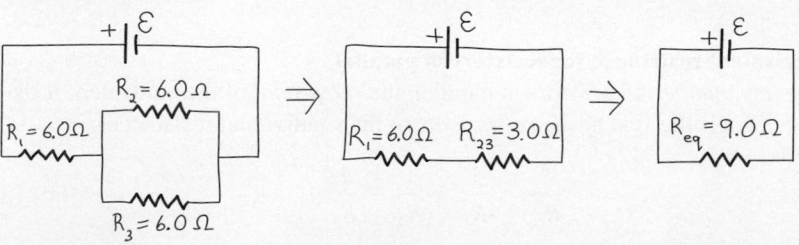

▶ **FIGURE 19.21** Our procedure for finding the equivalent resistance.

(a) Original circuit   **(b)** Parallel resistors combined   **(c)** Equivalent resistor

Continued

This gives $R_{23} = 3.0\ \Omega$. In Figure 19.21b, we've replaced the parallel combination of R_2 and R_3 with R_{23}. The circuit now has R_1 and R_{23} in series. Their equivalent resistance R_{eq} is given by

$$R_{eq} = R_1 + R_{23} = 6.0\ \Omega + 3.0\ \Omega = 9.0\ \Omega.$$

Thus, the equivalent resistance of the entire network is $9.0\ \Omega$ (Figure 19.21c).

Part (b): To find the currents and voltages, we work backward, as shown in Figure 19.22, starting with the single equivalent resistor. In Figure 19.22a, the voltage across R_{eq} is $\mathcal{E}$ (because the battery has no internal resistance). Also, $\mathcal{E} = IR_{eq}$, so $I = \mathcal{E}/R_{eq} = 18.0\ V/9.0\ \Omega = 2.0\ A$. This is the current through the battery and also the total current through the network.

In Figure 19.22b, R_1, R_{23}, and the battery are in series, so the same $2.0\ A$ current passes through each. The voltage V_1 across R_1 is

$$V_1 = IR_1 = (2.0\ A)(6.0\ \Omega) = 12.0\ V.$$

The voltage V_{23} across R_{23} is

$$V_{23} = IR_{23} = (2.0\ A)(3.0\ \Omega) = 6.0\ V.$$

(Note that $V_1 + V_{23} = \mathcal{E}$.)

The voltage across the parallel combination of R_2 and R_3 is the potential difference between points c and d, which is $V_{23} = 6.0\ V$. Since the voltage across R_2 is $6.0\ V$, we can calculate the current I_2 through R_2: $I_2 = \dfrac{6.0\ V}{6.0\ \Omega} = 1.0\ A$. A similar calculation gives the current I_3 through R_3: $I_3 = \dfrac{6.0\ V}{6.0\ \Omega} = 1.0\ A$.

In summary, $I_1 = 2.0\ A$ and $I_2 = I_3 = 1.0\ A$, as shown in Figure 19.22c. (Note that $I_2 + I_3 = I_1$.)

REFLECT The two identical resistors R_2 and R_3 form a parallel combination; the $2.0\ A$ current arriving at point c in the circuit divides equally: Half goes through R_2 and half through R_3. At point d, these two $1.0\ A$ currents recombine into a $2.0\ A$ current.

Practice Problem: What is the current through each resistor in the network in this example if the resistors aren't equal, but instead $R_1 = 4.0\ \Omega$, $R_2 = 6.0\ \Omega$, and $R_3 = 3.0\ \Omega$? *Answers:* $I_1 = 3.0\ A, I_2 = 1.0\ A, I_3 = 2.0\ A.$

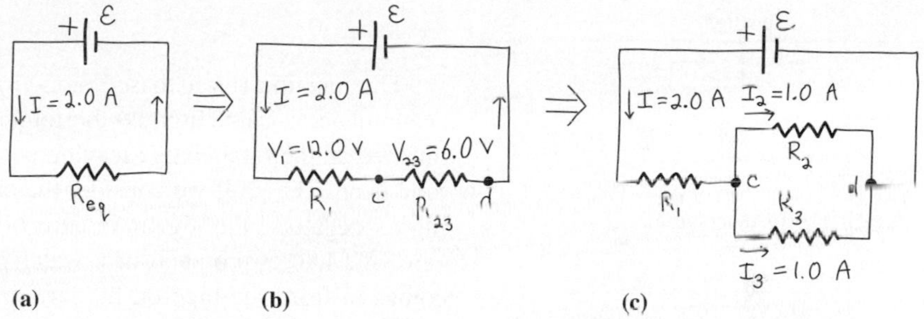

► FIGURE 19.22 Our procedure for finding the currents.

(a)   (b)   (c)

19.6 Kirchhoff's Rules

Many practical networks cannot be reduced to simple series–parallel combinations. Figure 19.23 shows two examples. Figure 19.23a is a circuit with two emf sources and a resistor. (This circuit might represent a battery feeding current to a lightbulb while being charged by a battery charger.) We don't need any new *principles* to compute the currents in networks such as these, but there are several techniques that help us to handle them systematically. We'll describe one of these, first developed by Gustav Robert Kirchhoff (1824–1887).

Mastering**PHYSICS**

ActivPhysics 12.5: Using Kirchhoff's Laws

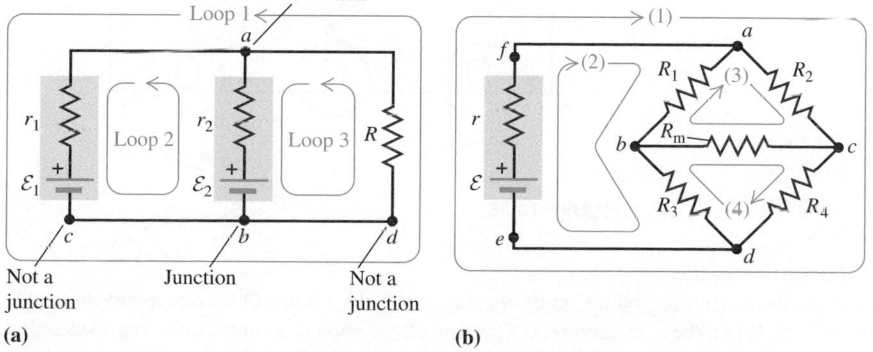

(a)   (b)

▲ FIGURE 19.23 Two networks that cannot be reduced to simple series–parallel combinations of resistors.

First, here are some terms that we'll use often: A **junction** in a circuit is a point where three or more conductors meet. Junctions are also called *nodes* or *branch points*. The circuit in Figure 19.23a has two junctions: *a* and *b*. Points *c* and *d* are *not* junctions. A **loop** is any closed conducting path. Figure 19.23 shows the possible loops for the layout of each circuit.

Kirchhoff's rules consist of the following two statements:

Kirchhoff's junction (or point) rule:
The algebraic sum of the currents into any junction is zero; that is,

$$\Sigma I = 0. \tag{19.14}$$

Currents *into* a junction are positive; currents *out of* a junction are negative.

Kirchhoff's loop rule:
The algebraic sum of the potential differences in any loop, including those associated with emf's and those of resistive elements, **must equal zero;** that is,

$$\sum_{\text{around loop}} V = 0. \tag{19.15}$$

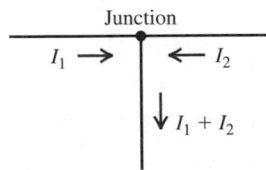

(a) Kirchhoff's junction rule

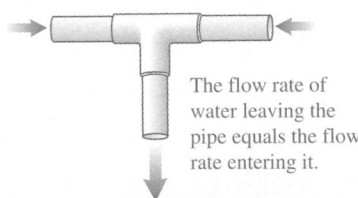

The flow rate of water leaving the pipe equals the flow rate entering it.

(b) Water-pipe analogy for Kirchhoff's junction rule

▲ **FIGURE 19.24** Kirchhoff's junction rule.

The junction rule is based on *conservation of electric charge*. No charge can accumulate at a junction, so the total charge entering the junction per unit time must equal the total charge leaving per unit time (Figure 19.24a). Charge per unit time is current, so if we consider the currents entering as positive and those leaving as negative, the algebraic sum of the currents entering a junction must be zero. It's like a T branch in a water pipe (Figure 19.24b); if 1 liter per minute comes in from the pipe on the left, and 1 liter per minute from the pipe on the right, you can't have 3 liters per minute going out the pipe at the bottom. We actually used the junction rule (without saying so) in Section 19.5, in the derivation of Equation 19.13 for resistors in parallel.

The loop rule is based on conservation of energy. The electrostatic field is a *conservative* force field. Suppose we go around a loop, measuring potential differences across successive circuit elements as we go. When we return to the starting point, we must find that the *algebraic sum* of these differences is zero. That is, when a charge travels in a loop and returns to its original location, the total change in the electric potential energy is zero; otherwise, the force wouldn't be conservative.

Quantitative Analysis 19.5 **Throw the switch!**

Bulbs *A*, *B*, and *C* in Figure 19.25 are identical. Closing the switch in the figure causes which of the following changes in the potential differences? More than one answer may be correct.

A. The potential differences across *A* and *B* are unchanged.
B. The potential difference across *C* drops by 50%.
C. The potential differences across *A* and *B* each increase by 50%.
D. The potential difference across *C* drops to zero.

SOLUTION When the switch is open, the same current passes through all the bulbs. Because the bulbs have the same resistance, the voltage $V = IR$ is also the same across all of them. When the switch is closed, the switch and bulb *C* form a loop. We assume that the switch is an ideal conductor, so the voltage drop across it is zero. Therefore, according to the loop rule, the voltage across

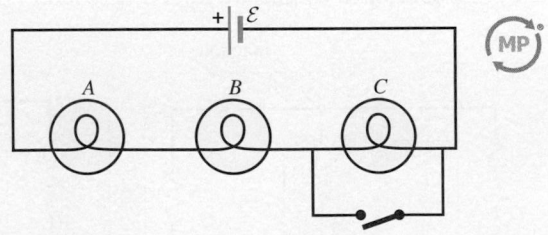

▲ **FIGURE 19.25**

bulb *C* must also be zero. Thus, choice D is correct and choice B is incorrect. The emf of the source is now "split" between only two bulbs: *A* and *B*. Hence, the potential difference across each of these bulbs increases by 50% (so choice C is correct and choice A is incorrect).

Using Kirchhoff's laws to find the currents and potentials in a circuit can be tricky. We suggest that you study the following problem-solving strategy carefully. Often, the hardest part of the solution is not in understanding the basic principles, but in keeping track of algebraic signs!

PROBLEM-SOLVING STRATEGY 19.2 **Kirchhoff's rules**

SET UP

1. Draw a *large* circuit diagram so that you have plenty of room for labels. Label all quantities, known and unknown, including an assumed direction for each unknown current and emf. Often you won't know in advance the actual direction of an unknown current or emf, but this doesn't matter. Carry out your solution, using the assumed direction. If the actual direction of a particular quantity is opposite to your assumption, the result will come out with a negative sign. If you use Kirchhoff's rules correctly, they give you the directions as well as the magnitudes of unknown currents and emf's. We'll illustrate this point in the examples that follow.

2. Usually, when you label currents, it is best to use the junction rule immediately, to express the currents in terms of as few quantities as possible. For example, Figure 19.26a shows a correctly labeled circuit; Figure 19.26b shows the same circuit, relabeled by applying the junction rule to point a to eliminate I_3.

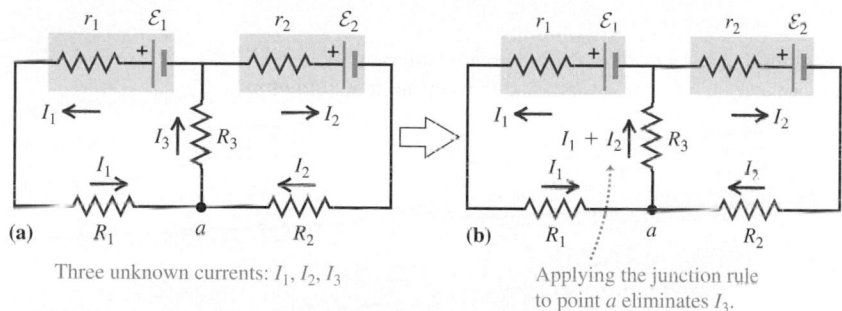

(a) Three unknown currents: I_1, I_2, I_3

(b) Applying the junction rule to point a eliminates I_3.

▲ **FIGURE 19.26** Using the junction rule to eliminate an unknown current.

SOLVE

3. Choose any closed loop in the network, and designate a direction (clockwise or counterclockwise) to go around the loop when applying the loop rule. The direction doesn't have to be the same as any assumed current's direction.

4. Go around the loop in the designated direction, adding potential differences as you cross them. An emf is counted as positive when you traverse it from − to + and negative when you traverse it from + to −. An IR product is negative if your path passes through the resistor in the *same* direction as the assumed current and positive if it passes through in the opposite direction. "Uphill" potential changes are always positive; "downhill" changes are always negative. Figure 19.27 summarizes these sign conventions. In each part of the figure, the direction of "travel" is the direction in which we are going around a loop, using Kirchhoff's loop rule, not necessarily the direction of the current.

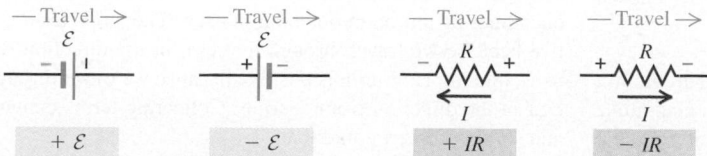

▲ **FIGURE 19.27** Sign conventions to use in traveling around a circuit loop when applying Kirchhoff's rules.

Continued

5. Apply Kirchhoff's loop rule to the potential differences obtained in Step 4: $\Sigma V = 0$.

6. If necessary, choose another loop to get a different relation among the unknowns, and continue until you have as many independent equations as unknowns or until every circuit element has been included in at least one of the chosen loops.

7. Finally, solve the equations, by substitution or some other means, to determine the unknowns. Be especially careful with algebraic manipulations; one sign error is fatal to the entire solution.

8. You can use this same bookkeeping system to find the potential V_{ab} of any point a with respect to any other point b. Start at b and add the potential changes that you encounter in going from b to a, using the same sign rules as in Step 4. The algebraic sum of these changes is $V_{ab} = V_a - V_b$.

REFLECT

9. Always remember that when you go around a loop, adding potential differences in accordance with Kirchhoff's loop rule, rises in potential are positive and drops in potential are negative. Follow carefully the procedure described; getting the signs right is absolutely essential.

EXAMPLE 19.10 Jump-start your car

The circuit shown in Figure 19.28 is used to start a car that has a dead (i.e., discharged) battery. It includes two batteries, each with an emf and an internal resistance, and two resistors. Find the current in the circuit and the potential difference V_{ab}.

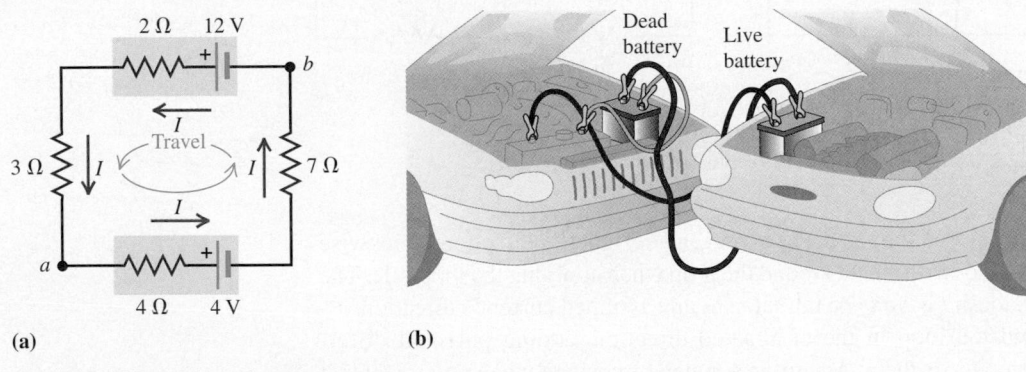

▲ FIGURE 19.28

SOLUTION

SET UP There is only one loop, so we don't need Kirchhoff's junction rule. To use the loop rule, we first assume a direction for the current. As shown in the figure, we assume that I is counterclockwise, coming from the + terminal of the battery with the larger emf. If this assumption is incorrect, then when we solve for I, we'll get a negative value. We choose to travel counterclockwise around the loop (but we could just as well have chosen to travel clockwise).

SOLVE To find the current in the circuit, we apply Kirchhoff's loop rule, starting at point a. The potential change has magnitude IR for a resistor and $\mathcal{E}$ for an emf. The sign of each potential

change is determined as described in step 4 of Problem Solving Strategy 19.2. The loop rule gives

$$-I(4.0\,\Omega) - 4.0\,\text{V} - I(7.0\,\Omega) + 12\,\text{V} - I(2.0\,\Omega) - I(3.0\,\Omega) = 0.$$

Be absolutely sure that you understand the signs in this equation. Each IR term in the sum has a minus sign because we travel through each resistor in the assumed direction of the current, encountering a drop in potential in each. The 4.0 V term is negative because we travel through that emf in the direction from + to −, and the 12 V term is positive because we travel through that emf in the direction from − to +. Collecting terms containing I and solving for I, we find that

Continued

$$8.0 \text{ V} = I(16 \, \Omega) \quad \text{and} \quad I = 0.5 \text{ A}.$$

Our positive result for I tells us that the current actually is in the counterclockwise direction we assumed. If we had assumed that I is clockwise, the loop rule would have given us $I = -0.5$ A. The minus sign in the result would tell us that the actual current is opposite to this assumed direction.

To find V_{ab}, we first take the upper path from b to a, through the 12 V battery. We get

$$V_b + 12 \text{ V} - (0.50 \text{ A})(2.0 \, \Omega) - (0.50 \text{ A})(3.0 \, \Omega) = V_a,$$
$$V_{ab} = V_a - V_b = 12 \text{ V} - 1.0 \text{ V} - 1.5 \text{ V} = 9.5 \text{ V}.$$

Point a is at higher potential than point b. The potential rise through the 12 V emf is greater than the total drop through the resistors.

Taking the lower path, through the 4.0 V emf, gives

$$V_b + (0.50 \text{ A})(7.0 \, \Omega) + 4.0 \text{ V} + (0.50 \text{ A})(4.0 \, \Omega) = V_a,$$
$$V_{ab} = V_a - V_b = 3.5 \text{ V} + 4.0 \text{ V} + 2.0 \text{ V} = 9.5 \text{ V}.$$

The result for V_{ab} is the same as before. In this equation, the IR terms are positive because, as we travel along this path, we pass through each resistor in a direction opposite to the current direction.

REFLECT To jump-start a car, we create a circuit by connecting the $+$ terminal of the live battery to the $+$ terminal of the dead one and connecting the two $-$ terminals together.

Practice Problem: In this example, if the emf of the 4 V battery is increased to 16 V and the rest of the circuit remains the same, what is the potential difference V_{ab}? *Answer:* 13.2 V.

EXAMPLE 19.11 **Recharging a battery**

In the circuit shown in Figure 19.29, a 12 V power supply with unknown resistance r (represented as a battery) is connected to a run-down rechargeable battery with unknown emf $\mathcal{E}$ and internal resistance 1 Ω and to a bulb with resistance 3 Ω. The current through the bulb is 2 A, and the current in the run-down battery is 1 A; the directions of both of these currents are shown in the figure. **(a)** Find the unknown current I in the 12 V battery, the internal resistance r of the power supply, and the emf $\mathcal{E}$ of the run-down battery. **(b)** Find the electrical power for each emf in the circuit.

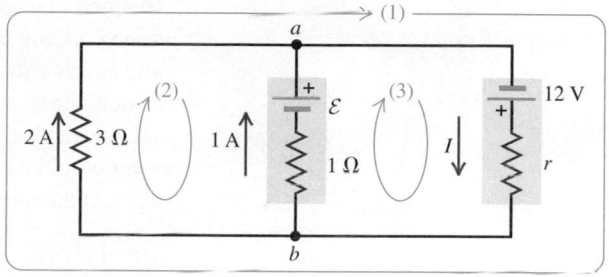

▲ **FIGURE 19.29**

SOLUTION

SET UP As shown in the figure, we assume that the current I is downward. We also identify the possible loops for applying the loop rule and choose a clockwise travel direction for each.

SOLVE Part (a): We first apply the junction rule to point a. When we sum the currents at a junction, currents into the junction are positive and currents out of the junction are negative. This gives

$$-I + 1 \text{ A} + 2 \text{ A} = 0 \quad \text{and} \quad I = 3 \text{ A}.$$

To find r, we use the loop rule for loop (1), since r is the only unknown quantity in that loop. Starting at point a and proceeding clockwise, we use the loop rule to obtain

$$+12 \text{ V} - (3 \text{ A})r - (2 \text{ A})(3 \, \Omega) = 0 \quad \text{and}$$
$$r = \frac{12 \text{ V} - 6 \text{ V}}{3 \text{ A}} = 2 \, \Omega.$$

The IR terms are negative because we pass through each resistance in the direction of the current. Note that the current may not be the same for all the resistors in a loop. To calculate $\mathcal{E}$, we use the loop rule with loop (2). Starting at point a and proceeding clockwise, we find that

$$-\mathcal{E} + (1 \text{ A})(1 \, \Omega) - (2 \text{ A})(3 \, \Omega) = 0 \quad \text{and}$$
$$\mathcal{E} = 1 \text{ V} - 6 \text{ V} = -5 \text{ V}.$$

The IR term for the 1 Ω resistor is positive because we travel through this resistor in the direction opposite to the current.

The fact that our result for $\mathcal{E}$ is negative shows that the actual polarity of this emf is opposite to what is shown in the figure.

As a check, we can use loop (3) in Figure 19.29. Starting at point a and again proceeding clockwise, we obtain the loop equation

$$+12 \text{ V} - Ir - (1 \text{ A})(1 \, \Omega) + \mathcal{E} = 0.$$

Since $Ir = (3 \text{ A})(2 \, \Omega) = 6 \text{ V}$, the equation becomes $+12 \text{ V} - 6 \text{ V} - 1 \text{ V} + \mathcal{E} = 0$. Solving gives $\mathcal{E} = -5 \text{ V}$, in agreement with the result from loop (2).

Part (b): The electrical power for an emf is $P = \mathcal{E}I$, where I is the current through the emf. When the current passes through the emf from $-$ to $+$, positive charges *gain* electrical potential energy, so the emf adds energy to the circuit. If the current is in the opposite direction, the emf removes electrical energy from the circuit. In our circuit, the emf of the 12 V battery adds energy to the circuit at a rate

$$P_{in} = \mathcal{E}I = (12 \text{ V})(3 \text{ A}) = 36 \text{ W}.$$

The 5 V emf of the battery being recharged removes energy from the circuit at a rate

$$P_{out} = \mathcal{E}I = (5 \text{ V})(1 \text{ A}) = 5 \text{ W}.$$

When a battery is recharged, electrical energy is converted to chemical energy.

REFLECT The total rate at which the emfs supply energy to the circuit is $P_{in} - P_{out} = 36 \text{ W} - 5 \text{ W} = 31 \text{ W}$. By energy

Continued

conservation, this rate must equal the total rate P_R at which electrical energy is dissipated in all the resistors. For a resistor, $P = IR^2$, so the total rate of electrical consumption in the resistors is

$$P_R = (2\,A)^2(3\,\Omega) + (1\,A)^2(1\,\Omega) + (3\,A)^2(2\,\Omega)$$
$$= 12\,W + 1\,W + 18\,W = 31\,W,$$

which agrees with the power supplied by the emfs, confirming energy conservation.

Practice Problem: In this example, find I, r, and $\mathcal{E}$ if the current through the bulb is 3 A rather than 2 A. *Answers:* 4 A, 0.75 Ω, −8 V.

19.7 Electrical Measuring Instruments

The concepts of current, potential difference, and resistance have played a central role in our analysis of circuits in this chapter and the preceding one. But how do we *measure* these quantities? Many common devices, including car instrument panels, battery chargers, and inexpensive electrical instruments, measure potential difference (voltage), current, or resistance using a device containing a pivoted coil of wire placed in the magnetic field of a permanent magnet (Figure 19.30). Attached to the coil is a coiled spring; in the equilibrium position, with no current in the coil, the pointer is at zero. When there is a current in the coil, the magnetic field exerts a torque on the coil that is proportional to the current. Thus, the angular deflection of the coil and pointer is directly proportional to the coil current, and the device can be calibrated to measure current. The ideal behavior for a meter would be for it to measure the circuit quantities of interest without disturbing or changing those quantities by its presence.

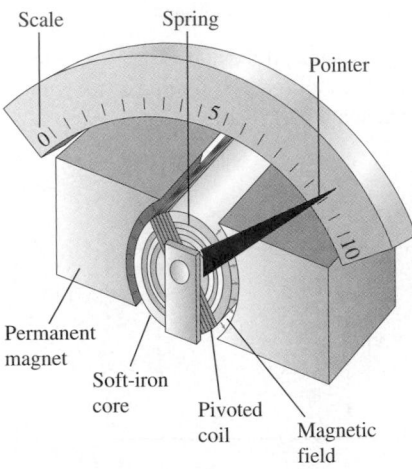

▲ FIGURE 19.30 One type of galvanometer.

Ammeters

An instrument that measures current is usually called an **ammeter.** The essential concept is that *an ammeter always measures the current passing through it.* An *ideal* ammeter would have *zero* resistance, so that including it in a branch of a circuit would not affect the current in that branch. All real ammeters have some finite resistance; it's always desirable for an ammeter to have as little resistance as possible, so that its presence disturbs the circuit behavior as little as possible.

Voltmeters

A **voltmeter** is a device that measures the potential difference (voltage) between two points. To make this measurement, a voltmeter must have its terminals connected to the two points in question. An ideal voltmeter would have *infinite* resistance, so that no current would flow through it. Then connecting it between two points in a circuit would not alter any of the circuit currents. Real voltmeters always have finite resistance, but a voltmeter should have a large enough resistance that connecting it in a circuit does not change the other currents appreciably.

For greater precision and mechanical ruggedness, pivoted-coil meters have been largely replaced by electronic instruments with direct digital readouts. These devices are more precise, stable, and mechanically rugged, and often more expensive, than the older devices. Digital voltmeters can be made with very high internal resistance, on the order of 100 MΩ. Figure 19.31 shows a *multimeter*—an instrument that can measure voltage, current, and resistance over a wide range.

19.8 Resistance–Capacitance Circuits

In our discussion of circuits, we've assumed that all the emf's and resistances are *constant.* As a result, all potentials and currents are constant and independent of time. Figure 19.32a shows a simple example of a circuit in which the current and voltages are *not* constant. We'll call this a resistance–capacitance $(R$–$C)$ circuit. The capacitor is initially uncharged; at some initial time $t = 0$, we close the switch, completing the circuit and permitting current around the loop to begin

ActivPhysics 12.4: Using Ammeters and Voltmeters

▲ FIGURE 19.31 A digital multimeter.

charging the capacitor (Figure 19.32b). The current begins at the same instant in every part of the circuit, and at each instant it is the same in every part.

We'll neglect the internal resistance of the battery, so its terminal voltage is constant and equal to the battery emf $\mathcal{E}$. The capacitor is initially uncharged; the potential difference across it is initially zero. At this time, from Kirchhoff's loop rule, the voltage across the resistor R is equal to the battery's terminal voltage $\mathcal{E}$. The initial current through the resistor, which we'll call I_0, is given by Ohm's law: $I_0 = \mathcal{E}/R$. The initial capacitor charge, which we'll call Q_0, is zero. We denote the time-varying current and charge as i and q, respectively.

As the capacitor charges, its voltage increases, so the potential difference across the resistor must decrease, corresponding to a decrease in current. From Kirchhoff's loop rule, the sum of these two is constant and equal to the battery emf $\mathcal{E}$:

$$\mathcal{E} = iR + \frac{q}{C}. \qquad (19.16)$$

After a long time, the capacitor becomes fully charged,. The current decreases to zero, the potential difference across the resistor becomes zero, and the entire battery voltage $\mathcal{E}$ appears across the capacitor. Thus, the capacitor charge and current vary with time as shown by the graphs in Figure 19.33.

To obtain a detailed description of the time variation of charge and current in the circuit, we would need to solve Equation 19.16, which is a differential equation for the charge q and its rate of change i. The solutions contain the exponential function $e^{-t/RC}$, where $e = 2.718$ is the base of natural logarithms. We'll omit the details of these calculations; it turns out that if the switch is closed at time $t = 0$, the current i and charge q vary with time t according to the equations

$$i = I_0 e^{-t/RC}, \qquad q = Q_{\text{final}}\left(1 - e^{-t/RC}\right). \qquad (19.17)$$

The graphs of Figure 19.33 are graphs of these equations. The graphs and the equations both show that as time goes on (and the exponential function $e^{-t/RC}$ approaches zero), the capacitor charge q approaches its final value, which we've called Q_{final}. The current decreases and eventually becomes zero. When $i = 0$, Equation 19.16 gives

$$\mathcal{E} = \frac{q}{C} \qquad \text{and} \qquad q = Q_{\text{final}} = C\mathcal{E}.$$

We note that the final charge Q_{final} doesn't depend on R, although the rate at which this final value is approached is slower when R is large than when it is small.

Returning to Figure 19.33, we note that at the instant the switch is closed $(t = 0)$ the current jumps from zero to its initial value $I_0 = \mathcal{E}/R$. After that, it decreases and gradually approaches zero. The capacitor charge starts at zero and gradually approaches the final value $Q_{\text{final}} = C\mathcal{E}$.

After a time $t = RC$, the exponential functions in Equations 19.17 have the value $e^{-1} = 1/e = 0.368$. At this time, the current has decreased from its initial value I_0 to $1/e$ of that value, and the capacitor charge has reached $(1 - 1/e) = 0.632$ of its final value $Q_{\text{final}} = C\mathcal{E}$. The product RC is therefore a measure of how quickly the capacitor charges; it is called the **time constant,** or the **relaxation time,** of the circuit, denoted by τ (the Greek letter tau):

$$\tau = RC. \qquad (19.18)$$

When τ is small, the capacitor charges quickly; when τ is larger, the charging takes more time. The horizontal axis (where $i = 0$) is an *asymptote* for the curve in Figure 19.33a. Strictly speaking, i never becomes precisely zero, but the longer

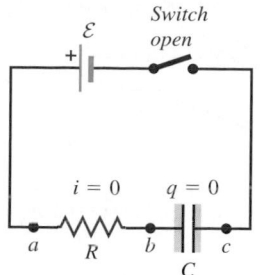

(a) Capacitor initially uncharged

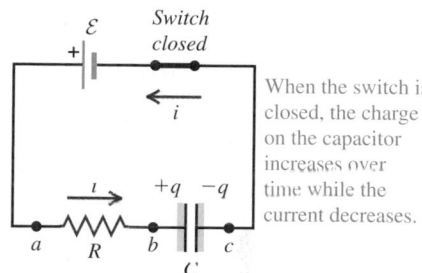

When the switch is closed, the charge on the capacitor increases over time while the current decreases.

(b) Charging the capacitor

▲ **FIGURE 19.32** Charging a capacitor in a resistance–capacitance circuit.

Mastering**PHYSICS**

ActivPhysics 12.6: Capacitance
ActivPhysics 12.7: Series and Parallel Capacitors
ActivPhysics 12.8: RC Circuit Time Constants

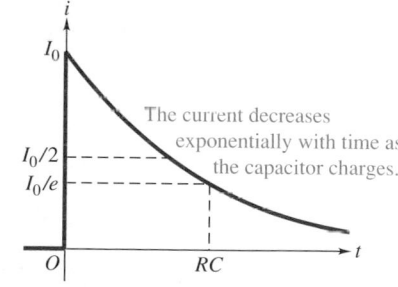

The current decreases exponentially with time as the capacitor charges.

(a) Graph of current versus time

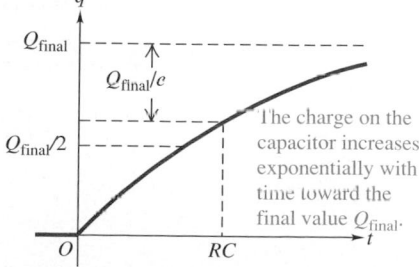

The charge on the capacitor increases exponentially with time toward the final value Q_{final}.

(b) Graph of capacitor charge versus time

▲ **FIGURE 19.33** Current i and capacitor charge q as functions of time for charging the capacitor in the circuit in Figure 19.32.

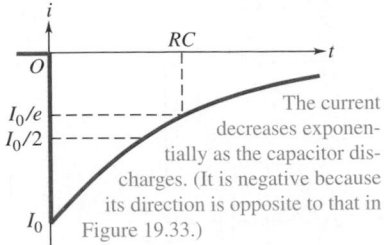

The current decreases exponentially as the capacitor discharges. (It is negative because its direction is opposite to that in Figure 19.33.)

(a) Graph of current versus time

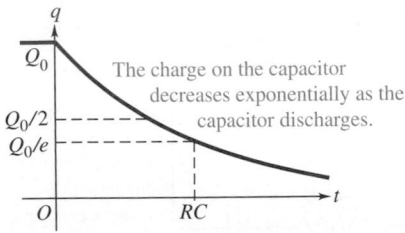

The charge on the capacitor decreases exponentially as the capacitor discharges.

(b) Graph of capacitor charge versus time

▲ **FIGURE 19.34** Current i and capacitor charge q as functions of time for discharging the capacitor in the circuit in Figure 19.32.

we wait, the closer it gets. For example, after a time equal to $10\tau = 10RC$, the current has decreased to about 0.00005 of its initial value. Similarly, the curve in Figure 19.33b approaches the horizontal broken line labeled Q_{final} as an asymptote. The charge q never attains precisely this value, but after a time equal to $10RC$, the difference between q and Q_{final} is about $0.00005Q_{final}$.

We could carry out a similar analysis for the *discharge* of a capacitor that is initially charged. We won't go into the details, but the results are similar to those obtained in the charging situation. The current i and the capacitor charge q vary with time according to the equations

$$i = I_0 e^{-t/RC} \quad \text{and} \quad q = Q_0 e^{-t/RC}, \quad (19.19)$$

where I_0 and Q_0 are the initial values at time $t = 0$. Both I and q decrease to $1/e = 0.368$ of their initial values after a time equal to $\tau = RC$, as shown in Figure 19.34. After a time equal to $10\tau = 10RC$, they have both decreased to about 0.00005 of their initial values. So after a time equal to a few time constants, the capacitor is, for all practical purposes, completely discharged.

19.9 Physiological Effects of Currents

Electrical potential differences and currents play a vital role in the nervous systems of animals. Conduction of nerve impulses is basically an electrical process, although the mechanism of conduction is much more complex than in simple materials such as metals.

A nerve fiber, or *axon*, along which an electrical impulse can travel, includes a cylindrical membrane with one conducting fluid (electrolyte) inside and another outside (Figure 19.35a). Chemical systems similar to those in batteries maintain a potential difference on the order of 0.1 V between these fluids. When an electrical pulse is initiated, the nerve membrane temporarily becomes more permeable to the ions in the fluids, leading to a local drop in potential (Figure 19.35b). As the pulse propagates, the membrane recovers and overshoots briefly before returning the potential to its initial value (Figure 19.35c).

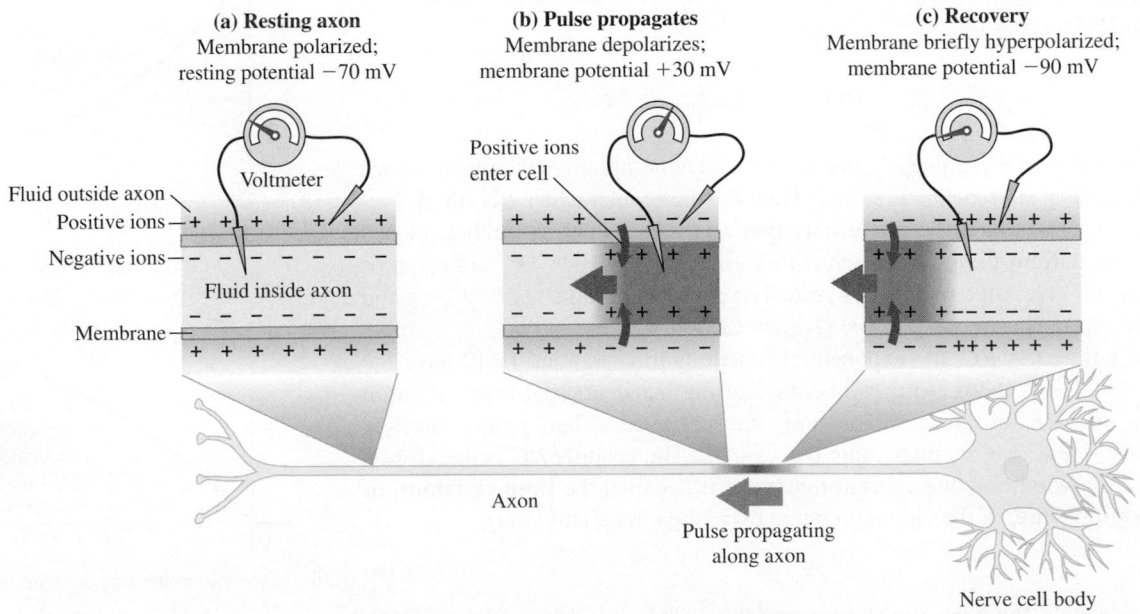

▲ **FIGURE 19.35** Propagation of a pulse along the axon of a nerve cell.

The electrical nature of nerve-impulse conduction is responsible for the great sensitivity of the body to externally supplied electric currents. (Impulses equivalent to those carried by nerve cells also occur in muscle cells, where they induce contraction. Thus, muscles as well as nerves are sensitive to currents.) Currents through the body as small as 0.1 A (much too small to produce significant heating) can be fatal because they interfere with nerve processes that are essential for vital functions, such as the heartbeat. The resistance of the human body is highly variable. Body fluids are usually quite good conductors because of their substantial ion concentrations, but the resistance of skin is relatively high, ranging from 500 kΩ for very dry skin to 1000 Ω or so for wet skin, depending also on the area of contact. If $R = 1000\ \Omega$, a current of 0.1 A requires a voltage of $V = IR = (0.1\ \text{A})(1000\ \Omega) = 100\ \text{V}$. This is within the range of household voltages, and it is one reason it is dangerous to receive a shock with wet skin.

Even smaller currents can be dangerous. A current of 0.01 A through an arm or leg causes strong, convulsive muscle action and considerable pain, and with a current of 0.02 A, a person who is holding the conductor that is inflicting the shock is typically unable to release it. Currents of this magnitude through the chest can cause ventricular fibrillation, a disorganized twitching of the heart muscles that pumps very little blood. Surprisingly, very large currents (over 0.1 A) are somewhat *less* likely to cause fatal fibrillation, because the heart muscle is "clamped" in one position. The heart actually stops beating and is more likely to resume normal beating when the current is removed. The electric defibrillators used for medical emergencies apply a large current pulse to stop the heart (and the fibrillation) and give it a chance to restore normal rhythm.

Thus, electric current poses three different kinds of hazards: interference with the nervous system, injury caused by convulsive muscle action, and burns from I^2R heating. The moral of this rather morbid story is that under certain conditions, voltages as small as 10 V can be dangerous. All electric circuits and equipment should always be approached with respect and caution.

On the positive side, rapidly alternating currents can have beneficial effects. Alternating currents with frequencies on the order of 10^6 Hz do not interfere appreciably with nerve processes and can be used for therapeutic heating for arthritic conditions, sinusitis, and a variety of other disorders. If one electrode is made very small, the resulting concentrated heating can be used for local destruction of tissue such as tumors or for cutting tissue in certain surgical procedures.

The study of particular nerve impulses is also an important *diagnostic* tool in medicine. The most familiar examples are electrocardiography (EKG) and electroencephalography (EEG). Electrocardiograms, obtained by attaching electrodes to the chest and back and recording the regularly varying potential differences, are used to study heart function. Similarly, electrodes attached to the scalp permit the study of potentials in the brain, and the resulting patterns can be helpful in diagnosing epilepsy, brain tumors, and other disorders.

19.10 Power Distribution Systems

We conclude this chapter with a brief discussion of practical household and automotive electric power distribution systems. Automobiles use direct-current (dc) systems, and nearly all household, commercial, and industrial systems use alternating current (ac) because of the ease of stepping voltage up and down with transformers. Most of the same basic wiring concepts apply to both systems. We'll talk about alternating-current circuits in greater detail in Chapter 22.

The various lamps, motors, and other appliances to be operated are always connected *in parallel* to the power source (the wires from the power company for houses or the battery and alternator for a car). The basic idea of house wiring is shown in Figure 19.36. One side of the "line," as the pair of conductors is called,

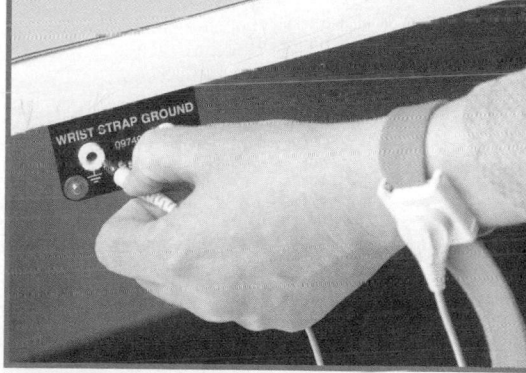

▲ **BIO Application Chained to your job.**
This electronics worker is wearing a grounding strap around the wrist. People easily acquire a static charge by rubbing against objects, and because we are conductors, we can deliver this charge as a current to any conducting object we touch. Such a current won't harm a doorknob, but it's quite enough to fry sensitive electronics. The wrist strap prevents such mishaps. It makes direct electrical contact with the skin and is connected to a grounded conductor, so it keeps static charge from accumulating. If you've ever installed memory cards in your own computer, you may have received and used a disposable wrist strap.

▶ **FIGURE 19.36** Schematic diagram of part of a house wiring system. Only two branch circuits are shown; an actual system might have four to thirty branch circuits. Lamps and appliances may be plugged into the outlets. The grounding conductors, which normally carry no current, are not shown. Modern household systems usually have two "hot" lines with opposite polarity with respect to neutral, and a voltage of 240 V between them. (Actual wires use a different color-coding system.)

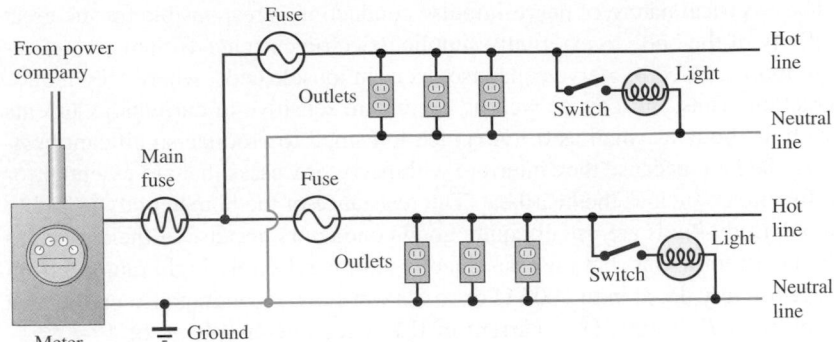

is called the *neutral* side; it is always connected to "ground" at the entrance panel. For houses, ground is an actual electrode driven into the earth (usually a good conductor) and also sometimes connected to the household cold-water pipes. Electricians speak of the "hot" side and the "neutral" side of the line. Most modern house wiring systems have *two* hot lines with opposite polarity with respect to the neutral; the voltage between them is twice the voltage between either hot line and the neutral line. (We'll return to this detail later.)

Household voltage is nominally 120 V in the United States and Canada, and often 240 V in Europe. (For alternating current, which varies sinusoidally with time, these numbers represent the *root-mean-square* voltage, which is $1/\sqrt{2}$ times the peak voltage. We'll discuss this further in Section 22.1.) The power input P to a device is given by Equation 19.9: $P = V_{ab}I$. For example, the current in a 100 W lightbulb is

$$I = \frac{P}{V} = \frac{100 \text{ W}}{120 \text{ V}} = 0.83 \text{ A}.$$

The resistance of this bulb at operating temperature is

$$R = \frac{V}{I} = \frac{120 \text{ V}}{0.83 \text{ A}} = 144 \text{ }\Omega, \quad \text{or} \quad R = \frac{V^2}{P} = \frac{(120 \text{ V})^2}{100 \text{ W}} = 144 \text{ }\Omega.$$

Similarly, a 1500 W waffle iron draws a current of $(1500 \text{ W})/(120 \text{ V}) = 12.5 \text{ A}$ and has a resistance of 9.6 Ω at its operating temperature. Because of the temperature dependence of resistivity, the resistances of these devices are considerably less when they are cold. If you measure the resistance of a lightbulb with an ohm-meter (whose small current causes very little temperature rise), you will probably get a value of about 10 Ω. When a light bulb is turned on, there is an initial surge of current as the filament heats up. That's why, when a light bulb gets ready to burn out, it nearly always happens when you turn it on.

The maximum current available from an individual circuit is limited by the resistance of the wires. The I^2R power loss in the wires causes them to become hot, and in extreme cases this can cause a fire or melt the wires. Ordinary lighting and outlet wiring in houses usually uses 12 gauge wire. This has a diameter of 2.05 mm and can carry a maximum current of 20 A safely (without overheating). Larger sizes, such as 8 gauge (3.26 mm) or 6 gauge (4.11 mm), are used for high-current appliances—for example, ranges and clothes dryers—and 2 gauge (6.54 mm) or larger is used for the main power lines entering a house.

Protection against overloading and overheating of circuits is provided by fuses or circuit breakers. A *fuse* contains a link of lead–tin alloy with a very low melting temperature; the link melts and breaks the circuit when its rated current is exceeded. A *circuit breaker* is an electromechanical device that performs the same function, using an electromagnet or a bimetallic strip to "trip" the breaker and interrupt the circuit when the current exceeds a specified value. Circuit breakers have the advantage that they can be reset after they are tripped. A blown fuse must be replaced, but fuses are somewhat more reliable in operation than circuit breakers are.

If your system has fuses and you plug too many high-current appliances into the same outlet, the fuse blows. *Do not* replace the fuse with one of larger rating; if you do, you risk overheating the wires and starting a fire. The only safe solution is to distribute the appliances among several circuits. Modern kitchens often have three or four separate 20 A circuits, each of which can carry a current of 20 A without overheating.

Contact between the hot and neutral sides of the line causes a *short circuit.* Such a situation, which can be caused by faulty insulation or by any of a variety of mechanical malfunctions, provides a very low resistance current path, permitting a large current that would quickly melt the wires and ignite their insulation if the current were not interrupted by a fuse or circuit breaker. An equally dangerous situation is a broken wire that interrupts the current path, creating an *open circuit.* This is hazardous because of the sparking that can occur at the point of intermittent contact.

In approved wiring practice, a fuse or breaker is placed *only* on the hot side of the line, never on the neutral side. Otherwise, if a short circuit should develop because of faulty insulation or other malfunction, the ground-side fuse could blow. The hot side would still be live and would pose a shock hazard if you touched the live conductor and a grounded object such as a water pipe. For similar reasons, the wall switch for a light fixture should always be on the hot side of the line, never the neutral side.

Further protection against shock hazard is provided by a third conductor called the *grounding wire,* included in all present-day wiring. This conductor corresponds to the long round or U-shaped prong of the three-prong connector plug on an appliance or power tool. It is connected to the neutral side of the line at the entrance panel, where the meter is. It normally carries no current, but it connects the metal case or frame of the device to ground. If a conductor on the hot side of the line accidentally contacts the frame or case, the grounding conductor provides a current path and the fuse blows. Without the ground wire, the frame could become "live"; that is, it could reach a potential 120 V above ground. Then, if you touched it and a water pipe (or even a damp basement floor) at the same time, you could get a dangerous shock (Figure 19.37). In some situations, especially for outlets located outdoors or near a sink or water pipes, a special kind of circuit breaker called a *ground-fault interrupter* (GFI or GFCI) is used. This device senses the difference in current between the hot and neutral conductors (which is normally zero) and trips when it exceeds some very small value, typically 5 mA.

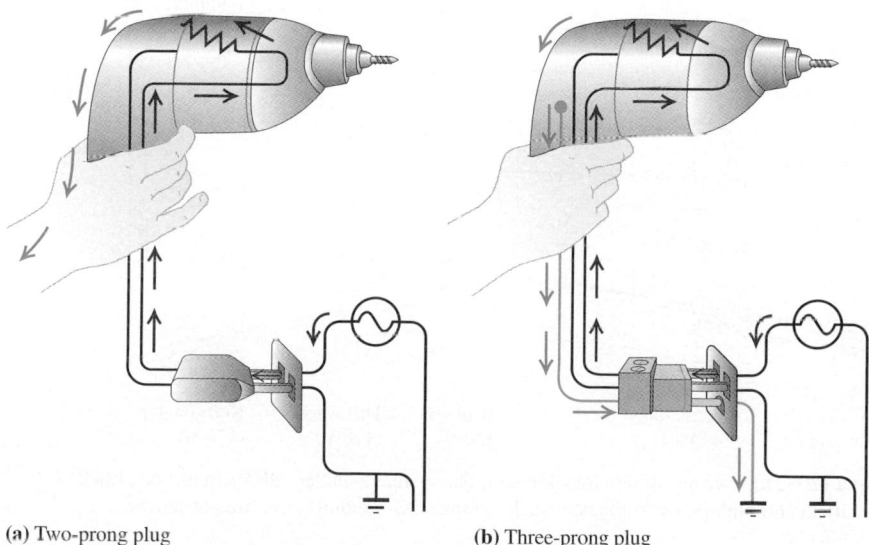

(a) Two-prong plug   **(b) Three-prong plug**

◀ **FIGURE 19.37** (a) If a malfunctioning electric drill is connected to a wall socket via a two-prong plug, a person may receive a shock. (b) When the drill malfunctions when connected via a three-prong plug, a person touching it receives no shock, since electric charge flows through the third prong and into the ground, rather than through the person's body. If the ground current is appreciable, the fuse blows.

Most household wiring systems actually use a slight elaboration of the system just described. Your power company provides *three* conductors (Figure 19.38). One is neutral; the other two are both at 120 V with respect to the neutral, but with opposite polarity, giving a voltage of 240 V between them. The power company calls this a *three-wire line,* in contrast to the 120 V two-wire (plus ground-wire) line described. With a three-wire line, 120 V lamps and appliances can be connected between the neutral conductor and either hot conductor, and high-power devices requiring 240 V are connected between the two hot lines. Ranges and dryers are usually designed for 240-V power input.

To help prevent wiring errors, household wiring uses a standardized color code in which the hot side of a line has black insulation (black and red for the two sides of a 240 V line), the neutral side has white insulation, and the grounding conductor is bare or has green insulation. In electronic devices and equipment, by contrast, the ground or neutral side of the line is usually black, so beware! (Our illustrations do not follow standard code, but use red for the hot line and blue for neutral.)

The preceding discussion can be applied directly to automobile wiring. The voltage is about 13 V (direct current); the power is supplied by the battery and the alternator, which charges the battery when the engine is running. The neutral side of the circuits is connected to the body and frame of the vehicle. For this low voltage, safety does not require a separate grounding conductor. The fuse or circuit breaker arrangement is the same, in principle, as in household wiring. Because of the lower voltage, more current is required for the same power; a 100 W headlight bulb requires a current of $(100\text{ W})/(13\text{ V}) = 7.7$ A.

Although we have spoken mostly of *power* in this section, what households really buy from their power company is *energy.* Power is energy transferred per unit time, so energy is average power multiplied by time. The usual unit of energy sold by the power company is the kilowatt-hour (1 kWh):

$$1\text{ kWh} = (10^3\text{ W})(3600\text{ s}) = 3.6 \times 10^6\text{ W}\cdot\text{s} = 3.6 \times 10^6\text{ J}.$$

One kilowatt-hour typically costs 2 to 10 cents, depending on one's location and the quantity of energy purchased. To operate a 1500 W (1.5 kW) waffle iron continuously for 1 hour requires 1.5 kWh of energy; at 10 cents per kWh, the cost is 15 cents (not including the cost of the flour and eggs in the waffle batter). The cost of operating any lamp or appliance for a specified time can be calculated in the same way if the power rating is known. However, many electric cooking utensils (including waffle irons) cycle on and off to maintain a constant temperature, and the *average* power may be less than the power rating marked on the device.

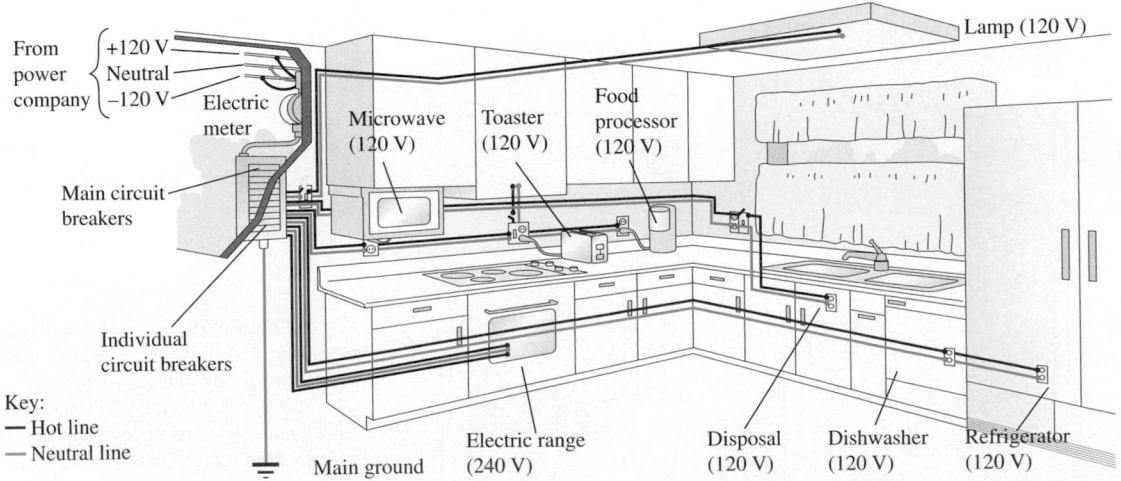

▲ **FIGURE 19.38** Schematic diagram of a 120–240 V wiring system for a kitchen. The system includes 120 V circuits on either side of the neutral line, as well as 240 V circuits for high-power appliances such as ranges. Grounding wires are not shown.

SUMMARY

Current

(Section 19.1) **Current** is the amount of charge flowing through a conductor per unit time. The SI unit of current is the ampere, equal to 1 coulomb per second $(1\ \text{A} = 1\ \text{C/s})$. If a net charge ΔQ flows through a wire in time Δt, the current through the wire is $I = \Delta Q/\Delta t$ (Equation 19.1).

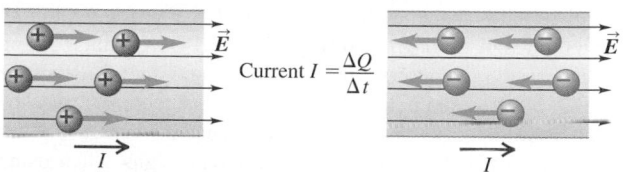

Current $I = \dfrac{\Delta Q}{\Delta t}$

Resistance and Ohm's Law

(Section 19.2) In a conductor, the **resistance** R is the ratio of voltage to current: $R = V/I$ (Equation 19.2). The SI unit of resistance is the **ohm** (Ω), equal to 1 volt per ampere. In materials that obey **Ohm's law,** the potential difference V between the ends of a conductor is proportional to the current I through the conductor; the proportionality factor is the resistance R.

For a given conducting material, resistance R is proportional to length and inversely proportional to cross-sectional area. For a specific material, this relationship can be expressed as $R = \rho(L/A)$ (Equation 19.3), where ρ is the **resistivity** of that material.

Resistance and resistivity vary with temperature; for metals, they usually increase with increasing temperature.

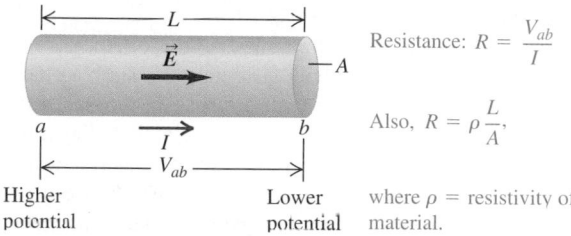

Resistance: $R = \dfrac{V_{ab}}{I}$

Also, $R = \rho\dfrac{L}{A}$,

where ρ = resistivity of material.

Electromotive Force and Circuits

(Section 19.3) A **complete circuit** is a conductor in the form of a loop providing a continuous current-carrying path. A complete circuit carrying a steady current must contain a source of electromotive force (emf), symbolized by $\mathcal{E}$. An ideal source of emf maintains a constant potential difference $V_{ab} = \mathcal{E}$ (Equation 19.5), but every real source of emf has some internal resistance r. The terminal potential difference V_{ab} then depends on current: $V_{ab} = \mathcal{E} - Ir$ (Equation 19.7).

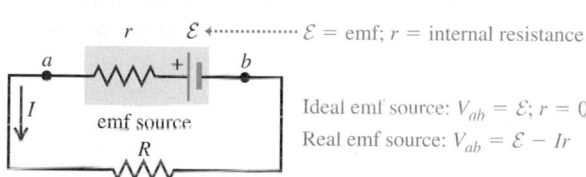

$\mathcal{E}$ = emf; r = internal resistance

Ideal emf source: $V_{ab} = \mathcal{E}$; $r = 0$
Real emf source: $V_{ab} = \mathcal{E} - Ir$

Energy and Power in Electric Circuits

(Section 19.4) A circuit element with a potential difference V and a current I puts energy into a circuit if the current direction is from lower to higher potential in the device and takes energy out of the circuit if the current is opposite. The power P (rate of energy transfer) is $P = VI$ (Equation 19.9). A resistor R always takes energy out of a circuit, converting it to thermal energy at a rate given by $P = V_{ab}I = I^2R = V_{ab}^2/R$ (Equation 19.10).

Resistors in Series and in Parallel

(Section 19.5) When several resistors $R_1, R_2, R_3, \cdots$ are connected in series, the **equivalent resistance** R_{eq} is the sum of the individual resistances: $R_{eq} = R_1 + R_2 + R_3 + \cdots$. (Equation 19.12). When several resistors are connected in parallel, the equivalent resistance R_{eq} is given by

$$\frac{1}{R_{eq}} = \frac{1}{R_1} + \frac{1}{R_2} + \frac{1}{R_3} + \cdots. \qquad (19.13)$$

Continued

Kirchhoff's Rules

(Section 19.6) Kirchhoff's junction rule is based on conservation of charge. It states that the algebraic sum of the currents into any junction must be zero: $\Sigma I = 0$ (Equation 19.14). Kirchhoff's loop rule is based on conservation of energy and the conservative nature of electrostatic fields. It states that the algebraic sum of the potential differences around any loop must be zero: $\Sigma V = 0$ (Equation 19.15). Be especially careful with signs when using Kirchhoff's rules.

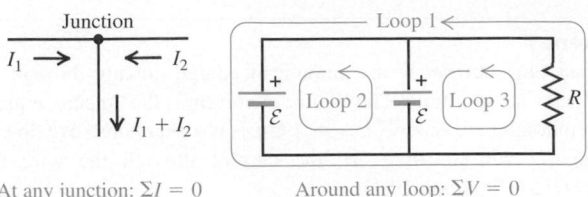

At any junction: $\Sigma I = 0$ Around any loop: $\Sigma V = 0$

Electrical Measuring Instruments

(Section 19.7) The ideal behavior for a meter is for it to measure the circuit quantities of interest without changing or disturbing them. An ammeter always measures the current passing through it. An *ideal* ammeter would have *zero* resistance, so that including it in a branch of a circuit would not affect the current in that branch. A voltmeter always measures the potential difference between two points. An ideal voltmeter would have *infinite* resistance, so that no current would flow through it.

Resistance–Capacitance Circuits

(Section 19.8) When a capacitor is charged by a battery in series with a resistor, the current and capacitor charge are not constant. The charge varies with time as $q = Q_{final}(1 - e^{-t/RC})$ (Equation 19.17). In a time $\tau = RC$, there is a significant change in the charge on the capacitor. This time is called the **time constant,** or **relaxation time,** and is the same for charging or discharging.

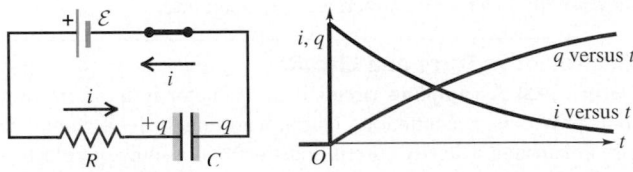

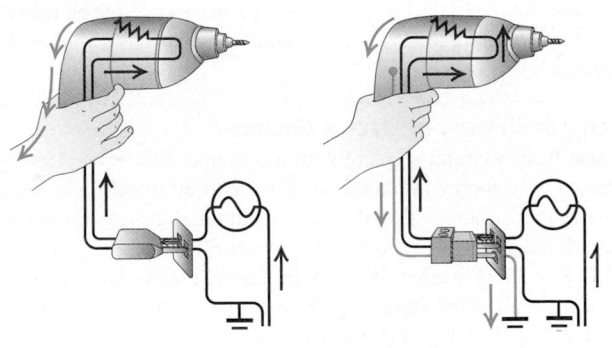

Applications of Currents

(Sections 19.9 and 19.10) The conduction of nerve impulses is basically an electrical process. Currents through the body as small as 0.1 A can be fatal because they interfere with this process (which also occurs in heart and other muscle cells).

In house wiring, one line entering the house is always *neutral,* or at the same voltage as the ground (to which it is connected). The other one or two wires are *hot.* The maximum current available from an individual circuit is limited by the resistance of the wires; if they carry too much current, I^2R power loss causes them to overheat. Protection against overloading of circuits is provided by fuses or circuit breakers.

 For instructor-assigned homework, go to www.masteringphysics.com

---

Conceptual Questions

1. A rule of thumb used to determine the internal resistance of a source is that it is the open-circuit voltage divided by the short-circuit current. Is this rule correct?

2. The energy that can be extracted from a storage battery is always less than the energy that goes into it while it is being charged. Why?

3. A cylindrical rod has resistivity ρ. If we triple its length and diameter, what is its resistivity, in terms of ρ?

4. A fuse is a device designed to break a circuit, usually by melting, when the current exceeds a certain value. Fuses are widely used in electronic equipment, but have been replaced by circuit breakers in household wiring. In the "old days," people would sometimes replace a blown fuse with a penny, which happened to be the same size as a fuse. Was this a safe practice? Why?

5. True or false? (a) Adding more resistance to a circuit increases its resistance. (b) Removing resistance from a circuit decreases its resistance. Justify your answers with simple examples.

6. Why does the resistance of an object increase with temperature? Explain by looking at the movement of charges at the atomic level.

7. *How* does a capacitor store energy? Can a resistor store energy?
8. High-voltage power supplies are sometimes designed to have a rather large internal resistance as a safety precaution. Why is such a power supply with a large internal resistance safer than one with the same voltage, but lower internal resistance?
9. If a 1.5 V AA battery produces *less than* 1.5 V across its terminals while operating in a circuit, does this *necessarily* mean that the battery is running down and near the end of its life? Why?
10. Can all combinations of resistors be reduced to series and parallel combinations? Illustrate your answer with some examples. (*Hint:* Check out some of the circuits in this chapter.)
11. In a two-cell flashlight, the batteries are normally connected in series. Why not connect them in parallel?
12. Old-time Christmas tree lights had the property that, when one bulb burned out, all the lights went out. How were these lights connected, in series or in parallel? How could you rewire them to prevent all the lights from going out when one of them burned out?
13. You connect a number of identical light bulbs to a flashlight battery. (a) What happens to the brightness of each bulb as more and more bulbs are added to the circuit if you connect them (i) in series and (ii) in parallel? Will the battery last longer if the bulbs are in series or in parallel? Explain your reasoning.
14. For very large resistances, it is easy to construct resistance–capacitance circuits having time constants of several seconds or minutes. How might this fact be used to measure such very large resistances, too large to measure by more conventional means?
15. When you scuff your shoes across a nylon carpet, you can easily produce a potential of several thousand volts between your body and the carpet. Yet contact with a power line of comparable voltage would probably be fatal. Why the difference?

Multiple-Choice Problems

1. A cylindrical metal rod has a resistance R. If both its length and its diameter are tripled, its new resistance will be:
   A. R B. $9R$ C. $R/3$ D. $3R$
2. A resistor R and another resistor $2R$ are connected in series across a battery. If heat is produced at a rate of 10 W in R, then in $2R$ it is produced at a rate of
   A. 40 W.   B. 20 W.   C. 10 W.   D. 5 W.
3. A resistor R and another resistor $2R$ are connected in parallel across a battery. If heat is produced at a rate of 10 W in R, in $2R$ it is produced at a rate of
   A. 40 W.   B. 20 W.   C. 10 W.   D. 5 W.
4. Which statements about the circuit shown in Figure 19.39 are correct? All meters are considered to be ideal, the connecting leads have no resistance, and the battery has no internal resistance. (There may be more than one correct choice.)
   A. The reading in ammeter A_1 is greater than the reading in A_2 because current is lost in the resistor.
   B. The two ammeters have exactly the same readings.

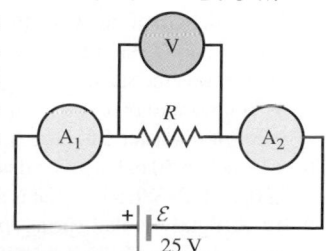

▲ **FIGURE 19.39** Multiple-choice problem 4.

C. The voltmeter reads less than 25 V because some voltage is lost in the resistor.
D. The voltmeter reads exactly 25 V.
5. When the switch S in Figure 19.40 is closed, the reading of the voltmeter V will
   A. increase.   B. decrease.
   C. stay the same.

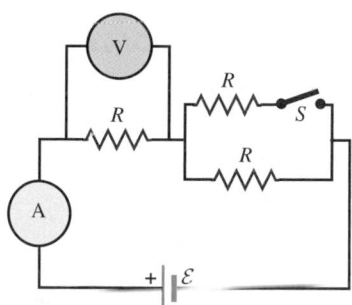

▲ **FIGURE 19.40** Multiple-choice problem 5.

6. When the switch S in the circuit in the previous question is closed, the reading of the ammeter A will
   A. increase.   B. decrease.
   C. stay the same.
7. Three identical lightbulbs are connected in the circuit shown in Figure 19.41. After the switch S is closed, what will be true about the brightness of these bulbs?
   A. B_1 will be brightest and B_3 will be dimmest.
   B. B_3 will be brightest and B_1 will be dimmest.
   C. All three bulbs will have the same brightness.

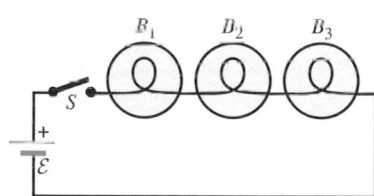

▲ **FIGURE 19.41** Multiple-choice problem 7.

8. A cylindrical metal rod of length L and diameter D is connected across a battery having no internal resistance. An ammeter in the circuit measures the current to be I. If we now double the diameter of the rod, but change nothing else, the ammeter will read
   A. $4I$.   B. $2I$.   C. $I/2$.   D. $I/4$.
9. Two identical metal rods are welded together end to end. If each rod has a length L and resistivity ρ, the resistivity of the combination will be
   A. 4ρ.   B. 2ρ.
   C. ρ.   D. $\rho/2$.
10. In the circuit shown in Figure 19.42, resistor A has three times the resistance of resistor B. Therefore,
   A. the current through A is three times the current through B.
   B. the current through B is three times the current through A.

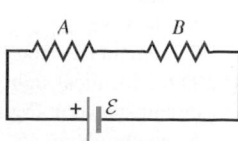

▲ **FIGURE 19.42** Multiple-choice problem 10.

C. the potential difference across *A* is three times the potential difference across *B*.

D. the potential difference is the same across both resistors.

11. In which of the two circuits shown in Figure 19.43 will the capacitors charge more rapidly when the switch is closed?

A. Circuit (a)

B. Circuit (b)

C. The capacitors will charge at the same rate in the two circuits.

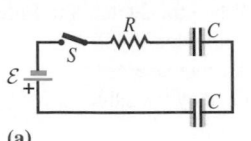

(a)

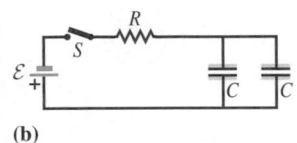

(b)

▲ **FIGURE 19.43** Multiple-choice problem 11.

12. The battery shown in the circuit in Figure 19.44 has some internal resistance. When we close *S*, the reading of the voltmeter *V* will

A. increase.

B. stay the same.

C. decrease.

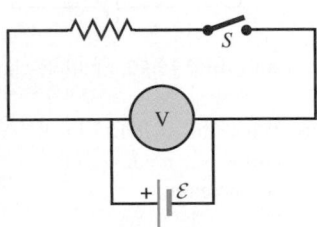

▲ **FIGURE 19.44** Multiple-choice problem 12.

13. A battery with no internal resistance is connected across identical lightbulbs as shown in Figure 19.45. When you close the switch *S*, bulbs *B₁* and *B₂* will

A. brighter than before.

B. dimmer than before.

C. just as bright as before.

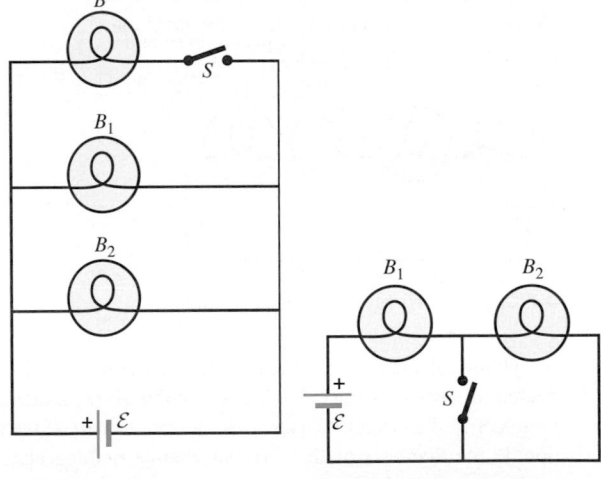

▲ **FIGURE 19.45** Multiple-choice problem 13.

▲ **FIGURE 19.46** Multiple-choice problem 14.

14. The battery shown in the circuit in Figure 19.46 has no internal resistance. After you close the switch *S*, the brightness of bulb *B₁* will

A. increase.   B. decrease.

C. remain the same.

15. Three identical light bulbs, *A*, *B*, and *C*, are connected in the circuit shown in Figure 19.47. When the switch *S* is closed,

A. the brightness of *A* and *B* remains the same as it was, but *C* goes out.

B. the brightness of *A* and *B* remains the same as it was, but *C* will be about half as bright as it was.

C. the brightness of *A* and *B* decreases, and *C* goes out.

D. the brightness of *A* and *B* increases, and *C* will be about half as bright as it was.

E. the brightness of *A* and *B* increases, but *C* goes out.

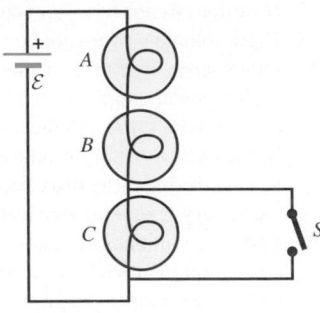

▲ **FIGURE 19.47** Multiple-choice problem 15.

Problems

19.1 Current

1. • Typical household currents are on the order of a few amperes. If a 1.50 A current flows through the leads of an electrical appliance, (a) how many electrons per second pass through it, (b) how many coulombs pass through it in 5.0 min, and (c) how long does it take for 7.50 C of charge to pass through?

2. • **Lightning strikes.** During lightning strikes from a cloud to the ground, currents as high as 25,000 A can occur and last for about 40 μs. How much charge is transferred from the cloud to the earth during such a strike?

3. • **Transmission of nerve impulses.** Nerve cells transmit electric signals through their long tubular axons. These signals propagate due to a sudden rush of Na^+ ions, each with charge $+e$, into the axon. Measurements have revealed that typically about 5.6×10^{11} Na^+ ions enter each meter of the axon during a time of 10 ms. What is the current during this inflow of charge in a meter of axon?

BIO

4. •• In an ionic solution, a current consists of Ca^{2+} ions (of charge $+2e$) and Cl^- ions (of charge $-e$) traveling in opposite directions. If 5.11×10^{18} Cl^- ions go from *A* to *B* every 0.50 min, while 3.24×10^{18} Ca^{2+} ions move from *B* to *A*, what is the current (in mA) through this solution, and in which direction (from *A* to *B* or from *B* to *A*) is it going?

5. •• Copper has 8.5×10^{28} electrons per cubic meter. (a) How many electrons are there in a 25.0 cm length of 12-gauge copper wire (diameter 2.05 mm)? (b) If a current of 1.55 A is flowing in the wire, what is the average drift speed of the electrons along the wire? (There are 6.24×10^{18} electrons in 1 coulomb of charge.)

19.2 Resistance and Ohm's Law

6. • A 14 gauge copper wire of diameter 1.628 mm carries a current of 12.5 mA. (a) What is the potential difference across a 2.00 m length of the wire? (b) What would the potential difference in part (a) be if the wire were silver instead of copper, but all else was the same?

7. • You want to precut a set of 1.00 Ω strips of 14 gauge copper wire (of diameter 1.628 mm). How long should each strip be?

8. • A wire 6.50 m long with diameter of 2.05 mm has a resistance of 0.0290 Ω. What material is the wire most likely made of?

9. •• A tightly coiled spring having 75 coils, each 3.50 cm in diameter, is made of insulated metal wire 3.25 mm in diameter. An ohmmeter connected across opposite ends of the spring reads 1.74 Ω. What is the resistivity of the metal?

10. •• What diameter must a copper wire have if its resistance is to be the same as that of an equal length of aluminum wire with diameter 3.26 mm?

11. •• An aluminum bar 3.80 m long has a rectangular cross section 1.00 cm by 5.00 cm. (a) What is its resistance? (b) What is the length of a copper wire 1.50 mm in diameter having the same resistance?

12. •• If you triple the length of a cable and at the same time double its diameter, what will be its resistance if its original resistance was R?

13. •• A ductile metal wire has resistance R. What will be the resistance of this wire in terms of R if it is stretched to three times its original length, assuming that the density and resistivity of the material do not change when the wire is stretched. (*Hint:* The amount of metal does not change, so stretching out the wire will affect its cross-sectional area.)

14. • What is the resistance of a Nichrome™ wire at 0.0°C if its resistance is 100.00 Ω at 11.5°C? The temperature coefficient of resistivity for Nichrome™ is 0.00040 $(C°)^{-1}$.

15. •• A 1.50-m cylindrical rod of diameter 0.500 cm is connected to a power supply that maintains a constant potential difference of 15.0 V across its ends, while an ammeter measures the current through it. You observe that at room temperature (20.0°C) the ammeter reads 18.5 A, while at 92.0°C it reads 17.2 A. You can ignore any thermal expansion of the rod. Find (a) the resistivity and (b) the temperature coefficient of resistivity at 20°C for the material of the rod.

16. • A carbon resistor having a temperature coefficient of resistivity of -0.00050 $(C°)^{-1}$, is to be used as a thermometer. On a winter day when the temperature is 4.0°C, the resistance of the carbon resistor is 217.3 Ω. What is the temperature on a spring day when the resistance is 215.8 Ω? (Take the reference temperature T_0 to be 4.0°C.)

17. •• In a laboratory experiment, you vary the current through an object and measure the resulting potential difference across it in each case. Figure 19.48 shows a graph of this potential V as a function of the current I (a) Does Ohm's law apply to this object? Why do you say so? (b) How is the resistance of the object related to the *slope* of the graph? Show why. (c) Use the slope of the graph to find the resistance of the object.

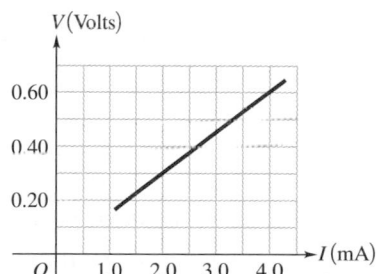

▲ **FIGURE 19.48** Problem 17.

18. •• The following measurements of current and potential difference were made on a resistor constructed of Nichrome™ wire, where V_{ab} is the potential difference across the wire and I is the current through it:

$I(A)$	0.50	1.00	2.00	4.00
$V_{ab}(V)$	1.94	3.88	7.76	15.52

(a) Graph V_{ab} as a function of I. (b) Does Ohm's law apply to Nichrome™? How can you tell? (c) What is the resistance of the resistor in ohms?

19. •• A battery-powered light bulb has a tungsten filament. When the switch connecting the bulb to the battery is first turned on and the temperature of the bulb is 20°C, the current in the bulb is 0.860 A. After the bulb has been on for 30 s, the current is 0.220 A. What is then the temperature of the filament?

19.3 Electromotive Force and Circuits

20. • When you connect an unknown resistor across the terminals of a 1.50 V AAA battery having negligible internal resistance, you measure a current of 18.0 mA flowing through it. (a) What is the resistance of this resistor? (b) If you now place the resistor across the terminals of a 12.6 V car battery having no internal resistance, how much current will flow? (c) You now put the resistor across the terminals of an unknown battery of negligible internal resistance and measure a current of 0.453 A flowing through it. What is the potential difference across the terminals of the battery?

21. • **Current in the body.** The resistance of the body varies
BIO from approximately 500 kΩ (when it is very dry) to about 1 kΩ (when it is wet). The maximum safe current is about 5.0 mA. At 10 mA or above, muscle contractions can occur that may be fatal. What is the largest potential difference that a person can safely touch if his body is wet? Is this result within the range of common household voltages?

22. • **"Current Baba."** According to a July 20, 2004, newspa-
BIO per article, a Hindu holy man known as "Current Baba" touches an electric wire three times daily to become "intoxicated." According to a doctor quoted in the article, "The human body can absorb currents up to 12 volts. In this case, however, repeated exposure to electricity seems to have built up ["Current Baba's"] body's tolerance levels to as much as 16 volts." (a) What is wrong with the doctor's statement? What do you think he really meant to say? (b) Since "Current Baba" was after the maximum "intoxication," he should have his body wet. In that case, how much current would he get with each jolt? Is this enough to be dangerous? (See the previous problem.)

23. • A copper transmission cable 100 km long and 10.0 cm in diameter carries a current of 125 A. What is the potential drop across the cable?

24. • A gold wire 6.40 m long and of diameter 0.840 mm carries a current of 1.15 A. Find (a) the resistance of this wire and (b) the potential difference between its ends.

25. •• When a solid cylindrical rod is connected across a fixed potential difference, a current I flows through the rod. What would be the current (in terms of I) if (a) the length were doubled, (b) the diameter were doubled, (c) both the length and the diameter were doubled?

26. • A 6.00 V lantern battery is connected to a 10.5 Ω lightbulb, and the resulting current in the circuit is 0.350 A. What is the internal resistance of the battery?

27. •• When switch S in Figure 19.49 is open, the voltmeter V across the battery reads 3.08 V. When the switch is closed, the voltmeter reading drops to 2.97 V and the ammeter A reads 1.65 A. Find the emf, the internal

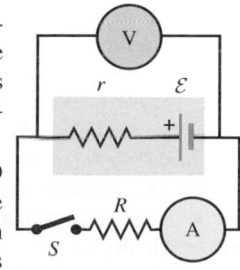

▲ **FIGURE 19.49** Problem 27.

resistance of the battery, and the circuit resistance R. Assume that the two meters are ideal, so that they don't affect the circuit.

28. •• A complete series circuit consists of a 12.0 V battery, a 4.70 Ω resistor, and a switch. The internal resistance of the battery is 0.30 Ω. The switch is open. What does an ideal voltmeter read when placed (a) across the terminals of the battery, (b) across the resistor, (c) across the switch? (d) Repeat parts (a), (b), and (c) for the case when the switch is closed.

29. •• With a 1500 MΩ resistor across its terminals, the terminal voltage of a certain battery is 2.50 V. With only a 5.00 Ω resistor across its terminals, the terminal voltage is 1.75 V. (a) Find the internal emf and the internal resistance of this battery. (b) What would be the terminal voltage if the 5.00 Ω resistor were replaced by a 7.00 Ω resistor?

30. • An automobile starter motor is connected to a 12.0 V battery. When the starter is activated it draws 150 A of current, and the battery voltage drops to 7.0 V. What is the battery's internal resistance?

31. •• Consider the circuit shown in Figure 19.50. The terminal voltage of the 24.0 V battery is 21.2 V. What is (a) the internal resistance r of the battery; (b) the resistance R of the circuit resistor?

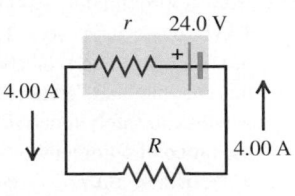
▲ FIGURE 19.50 Problem 31.

32. •• When switch S in Fig. 19.51 is open, the voltmeter V of the battery reads 3.08 V. When the switch is closed, the voltmeter reading drops to 2.97 V, and the ammeter A reads 1.65 A. Find the emf, the internal resistance of the battery, and the circuit resistance R. Assume that the two meters are ideal, so they don't affect the circuit.

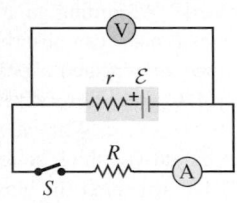
▲ FIGURE 19.51 Problem 32.

19.4 Energy and Power in Electric Circuits

33. • A resistor with a 15.0 V potential difference across its ends develops thermal energy at a rate of 327 W. (a) What is the current in the resistor? (b) What is its resistance?

34. • **Power rating of a resistor.** The *power rating* of a resistor is the maximum power it can safely dissipate without being damaged by overheating. (a) If the power rating of a certain 15 kΩ resistor is 5.0 W, what is the maximum current it can carry without damage? What is the greatest allowable potential difference across the terminals of this resistor? (b) If a 9.0 kΩ resistor is to be connected across a 120 V potential difference, what power rating is required for that resistor?

35. • An idealized voltmeter is connected across the terminals of a 15.0 V battery, and a 75.0 Ω appliance is also connected across its terminals. If the voltmeter reads 11.3 V: (a) how much power is being dissipated by the appliance, and (b) what is the internal resistance of the battery?

36. • **Treatment of heart failure.** A heart defibrillator is used to
BIO enable the heart to start beating if it has stopped. This is done by passing a large current of 12 A through the body at 25 V for a very short time, usually about 3.0 ms. (a) What power does the defibrillator deliver to the body, and (b) how much energy is transferred?

37. • **Lightbulbs.** The wattage rating of a lightbulb is the power it consumes *when it is connected across a 120 V potential difference.* For example, a 60 W lightbulb consumes 60.0 W of electrical power only when it is connected across a 120 V potential difference. (a) What is the resistance of a 60 W lightbulb? (b) Without doing any calculations, would you expect a 100 W bulb to have more or less resistance than a 60 W bulb? Calculate and find out.

38. • **Electrical safety.** This procedure is *not recommended!* You'll
BIO see why after you work the problem. You are on an aluminum ladder that is standing on the ground, trying to fix an electrical connection with a metal screwdriver having a metal handle. Your body is wet because you are sweating from the exertion; therefore, it has a resistance of 1.0 kΩ. (a) If you accidentally touch the "hot" wire connected to the 120 V line, how much current will pass through your body? Is this amount enough to be dangerous? (The maximum safe current is about 5 mA.) (b) How much electrical power is delivered to your body?

39. • **Electric eels.** Electric eels generate electric pulses along
BIO their skin that can be used to stun an enemy when they come into contact with it. Tests have shown that these pulses can be up to 500 V and produce currents of 80 mA (or even larger). A typical pulse lasts for 10 ms. What power and how much energy are delivered to the unfortunate enemy with a single pulse, assuming a steady current?

40. •• **Electric space heater.** A "540 W" electric heater is designed to operate from 120 V lines. (a) What is its resistance, and (b) what current does it draw? (c) At 7.4¢ per kWh, how much does it cost to operate this heater for an hour? (d) If the line voltage drops to 110 V, what power does the heater take, in watts? (Assume that the resistance is constant, although it actually will change because of the change in temperature.)

41. •• The battery for a certain cell phone is rated at 3.70 V. According to the manufacturer it can produce 3.15×10^4 J of electrical energy, enough for 5.25 h of operation, before needing to be recharged. Find the average current that this cell phone draws when turned on.

42. •• For the circuit in Fig. 19.52, find (a) the rate of conversion of internal (chemical) energy to electrical energy within the battery, (b) the rate of dissipation of electrical energy in the battery, (c) the rate of dissipation of electrical energy in the external resistor.

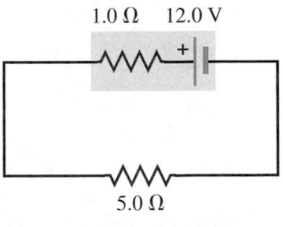
▲ FIGURE 19.52 Problem 42.

43. •• A 540-W electric heater is designed to operate from 120 V lines. (a) What is its resistance? (b) What current does it draw? (c) If the line voltage drops to 110 V, what power does the heater take? (Assume that the resistance is constant. Actually, it will change because of the change in temperature.) (d) The heater coils are metallic, so that the resistance of the heater decreases with decreasing temperature. If the change of resistance with temperature is taken into account, will the electrical power consumed by the heater be larger or smaller than what you calculated in part (c)? Explain.

44. •• **Electricity through the body, I.** A person with a body
BIO resistance of 10 kΩ between his hands accidentally grasps the terminals of a 14 kV power supply. (a) If the internal resistance of the power supply is 2000 Ω, what is the current through the person's body? (b) What is the power dissipated in

his body? (c) If the power supply is to be made safe by increasing its internal resistance, what should the internal resistance be for the maximum current in the situation just described to be 1.00 mA or less?

45. •• **Electricity through the body, II.** The average bulk resistivity of the human body (apart from surface resistance of the
BIO skin) is about 5.0 $\Omega \cdot$ m. The conducting path between the hands can be represented approximately as a cylinder 1.6 m long and 0.10 m in diameter. The skin resistance can be made negligible by soaking the hands in salt water. (a) What is the resistance between the hands if the skin resistance is negligible? (b) What potential difference between the hands is needed for a lethal shock current of 100 mA? (Note that your result shows that small potential differences produce dangerous currents when the skin is damp.) (c) With the current in part (b), what power is dissipated in the body?

19.5 Resistors in Series and in Parallel

46. • Find the equivalent resistance of each combination shown in Figure 19.53.

47. • Calculate the (a) maximum and (b) minimum values of resistance that can be obtained by combining resistors of 36 Ω, 47 Ω, and 51 Ω.

48. •• Each of two identical uniform metal bars has a resistance R. If they are welded together along one-third of their lengths (see Figure 19.54), what is the resistance of this combination in terms of R?

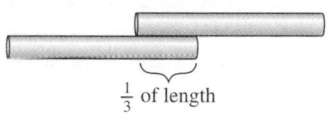

$\frac{1}{3}$ of length

▲ **FIGURE 19.54** Problem 48.

49. • A 40.0 Ω resistor and a 90.0 Ω resistor are connected in parallel, and the combination is connected across a 120-V dc line. (a) What is the resistance of the parallel combination? (b) What is the total current through the parallel combination? (c) What is the current through each resistor?

50. • Three resistors having resistances of 1.60 Ω, 2.40 Ω, and 4.80 Ω, respectively, are connected in parallel to a 28.0 V battery that has negligible internal resistance. Find (a) the equivalent resistance of the combination, (b) the current in each resistor, (c) the total current through the battery, (d) the voltage across each resistor, and (e) the power dissipated in each resistor. (f) Which resistor dissipates the most power, the one with the greatest resistance or the one with the least resistance? Explain why this should be.

51. • Now the three resistors of the previous problem are connected in series to the same battery. Answer the same questions for this situation.

52. •• Compute the equivalent resistance of the network in Figure 19.55, and find the current in each resistor. The battery has negligible internal resistance.

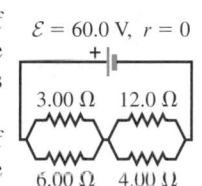

▲ **FIGURE 19.55** Problem 52.

53. •• Compute the equivalent resistance of the network in Figure 19.56, and find the current in each resistor. The battery has negligible internal resistance.

54. •• **Lightbulbs in series, I.** The power rating of a lightbulb is the power it consumes when connected across a 120 V outlet. (a) If you put two 100 W bulbs in series across a 120 V outlet, how much power would each consume if its resistance were constant? (b) How much power does each one consume if you connect them in parallel across a 120 V outlet?

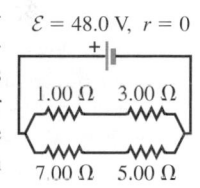

▲ **FIGURE 19.56** Problem 53.

55. •• You absentmindedly solder a 69.8 kΩ resistor into a circuit where a 36.5 kΩ should be. How can you get the proper resistance without replacing the bigger resistor or removing anything from the circuit?

56. •• You need to connect a 68 kΩ resistor and one other resistor to a 110 V power line. If you want the two resistors to use 4 times as much power when connected in parallel as they use when connected in series, what should be the value of the unknown resistor?

19.6 Kirchhoff's Rules

57. •• The batteries shown in the circuit in Figure 19.57 have negligibly small internal resistances. Find the current through (a) the 30.0 Ω resistor, (b) the 20.0 Ω resistor, and (c) the 10.0 V battery.

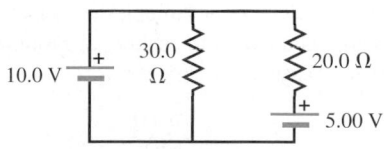

▲ **FIGURE 19.57** Problem 57.

58. •• Find the emf's $\mathcal{E}_1$ and $\mathcal{E}_2$ in the circuit shown in Figure 19.58.

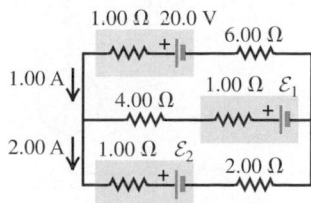

▲ **FIGURE 19.58** Problem 58.

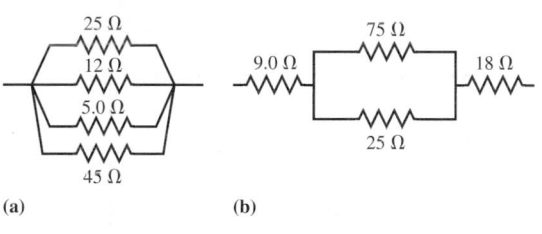

(a)   (b)

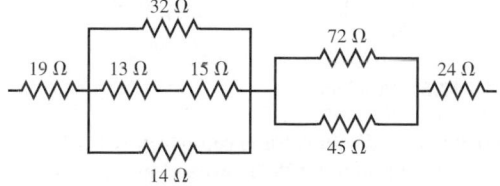

(c)

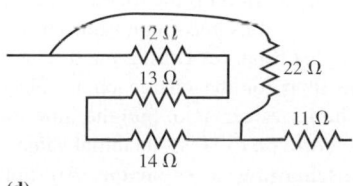

(d)

▲ **FIGURE 19.53** Problem 46.

59. •• In the circuit shown in Figure 19.59, ammeter A_1 reads 10.0 A and the batteries have no appreciable internal resistance. (a) What is the resistance of R? (b) Find the readings in the other ammeters.

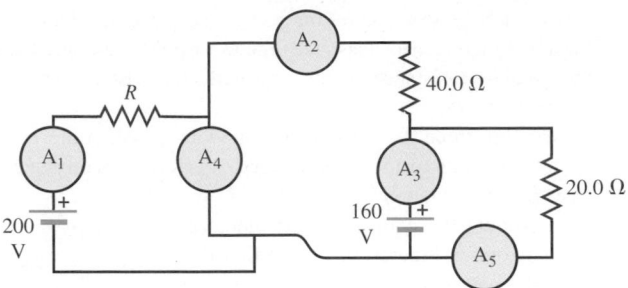

▲ **FIGURE 19.59** Problem 59.

60. •• In the circuit shown in Figure 19.60, find (a) the current in resistor R, (b) the value of the resistance R, and (c) the unknown emf $\mathcal{E}$.
61. •• In the circuit shown in Figure 19.61, current flows through the 5.00 Ω resistor in the direction shown, and this resistor is measured to be consuming energy at a rate of 20.0 W. The batteries have negligibly small internal resistance. What current does the ammeter A read?

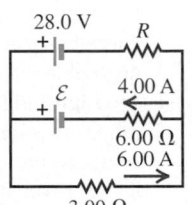

▲ **FIGURE 19.60** Problem 60.

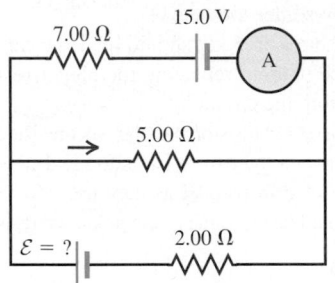

▲ **FIGURE 19.61** Problem 61.

62. •• In the circuit shown in Fig. 19.62, the 6.0 Ω resistor is consuming energy at a rate of 24 J/s when the current through it flows as shown. (a) Find the current through the ammeter A. (b) What are the polarity and emf of the battery $\mathcal{E}$, assuming it has negligible internal resistance?

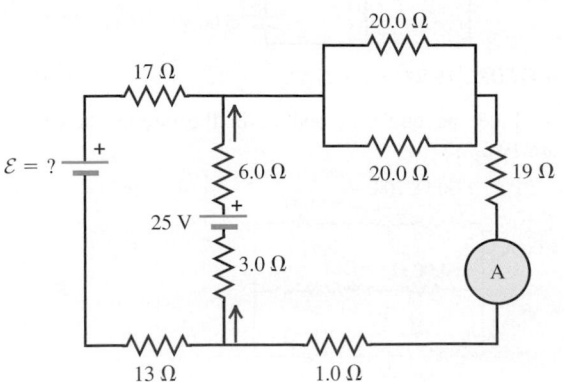

▲ **FIGURE 19.62** Problem 62.

19.8 Resistance-Capacitance Circuits

63. • A 500.0 Ω resistor is connected in series with a capacitor. What must be the capacitance of the capacitor to produce a time constant of 2.00 s?
64. • A fully charged 6.0 μF capacitor is connected in series with a 1.5×10^5 Ω resistor. What percentage of the original charge is left on the capacitor after 1.8 s of discharging?
65. •• A 12.4 μF capacitor is connected through a 0.895 MΩ resistor to a constant potential difference of 60.0 V. (a) Compute the charge on the capacitor at the following times after the connections are made: 0, 5.0 s, 10.0 s, 20.0 s, and 100.0 s. (b) Compute the charging currents at the same instants. (c) Graph the results of parts (a) and (b) for t between 0 and 20 s.
66. •• A 6.00 μF capacitor that is initially uncharged is connected in series with a 4500 Ω resistor and a 500 V emf source with negligible internal resistance. Just after the circuit is completed, what are (a) the voltage drop across the capacitor, (b) the voltage drop across the resistor, (c) the charge on the capacitor, and (d) the current through the resistor? (e) A long time after the circuit is completed (after many time constants), what are the values of the preceding four quantities?
67. •• A capacitor is charged to a potential of 12.0 V and is then connected to a voltmeter having an internal resistance of 3.40 MΩ. After a time of 4.00 s the voltmeter reads 3.0 V. What are (a) the capacitance and (b) the time constant of the circuit?
68. •• When a capacitor is being charged up, (a) how many time constants are required for it to receive 95% of its maximum charge, and (b) what is the current in the circuit at that time? (*Note:* You will have to solve an exponential equation.)
69. •• In the circuit shown in Figure 19.63, the capacitors are all initially uncharged and the battery has no appreciable internal resistance. After the switch S is closed, find (a) the maximum charge on each capacitor, (b) the maximum potential difference across each capacitor, (c) the maximum reading of the ammeter A, and (d) the time constant for the circuit.

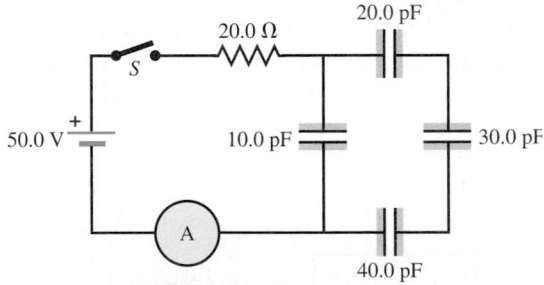

▲ **FIGURE 19.63** Problem 69.

70. •• **Charging and discharging a capacitor.** A 1.50 μF capacitor is charged through a 125 Ω resistor and then discharged through the same resistor by short-circuiting the battery. While the capacitor is being charged, find (a) the time for the charge on its plates to reach $1 - 1/e$ of its maximum value and (b) the current in the circuit at that time. (c) During the discharge of the capacitor, find the time for the charge on its plates to decrease to $1/e$ of its initial value. Also, find the time for the current in the circuit to decrease to $1/e$ of its initial value.
71. •• **Charging and discharging a capacitor.** An initially uncharged capacitor C charges through a resistor R for many time constants and then discharges through the same resistor.

Call Q_{max} the maximum charge on its plates and I_{max} the maximum current in the circuit. (a) Sketch clear graphs of the charge on the plates and the current in the circuit as functions of time for the charging process. (b) During the discharging process, the charge on the capacitor and the current both decrease exponentially from their maximum values. Use this fact to sketch graphs of the current in the circuit and the charge on the capacitor as functions of time.

General Problems

72. •• The circuit shown in Figure 19.64 contains two batteries, each with an emf and an internal resistance, and two resistors. Find (a) the current in the circuit (magnitude *and* direction) and (b) the terminal voltage V_{ab} of the 16.0 V battery.

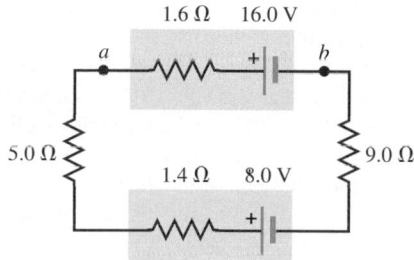

▲ **FIGURE 19.64** Problem 72.

73. •• If an ohmmeter is connected between points a and b in each of the circuits shown in Fig. 19.65, what will it read?

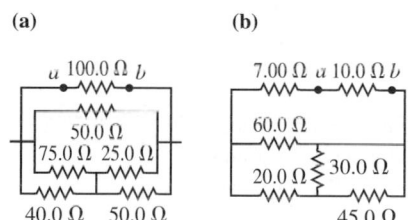

▲ **FIGURE 19.65** Problem 73.

74. •• A refrigerator draws 3.5 A of current while operating on a 120 V power line. If the refrigerator runs 50% of the time and electric power costs $0.12 per kWh, how much does it cost to run this refrigerator for a 30-day month?

75. •• A toaster using a Nichrome™ heating element operates on 120 V. When it is switched on at 20°C, the heating element carries an initial current of 1.35 A. A few seconds later, the current reaches the steady value of 1.23 A. (a) What is the final temperature of the element? The average value of the temperature coefficient of resistivity for Nichrome™ over the temperature range from 20°C to the final temperature of the element is $4.5 \times 10^{-4} \, (\text{C}°)^{-1}$. (b) What is the power dissipated in the heating element (i) initially; (ii) when the current reaches a steady value?

76. •• A piece of wire has a resistance R. It is cut into three pieces of equal length, and the pieces are twisted together parallel to each other. What is the resistance of the resulting wire in terms of R?

77. •• **Flashlight batteries.** A typical small flashlight contains two batteries, each having an emf of 1.5 V, connected in series with a bulb having resistance 17 Ω. (a) If the internal resistance of the batteries is negligible, what power is delivered to the bulb? (b) If the batteries last for 5.0 h, what is the total energy delivered to the bulb? (c) The resistance of real batteries increases as they run down. If the initial internal resistance is negligible, what is the combined internal resistance of both batteries when the power to the bulb has decreased to half its initial value? (Assume that the resistance of the bulb is constant. Actually, it will change somewhat when the current through the filament changes, because this changes the temperature of the filament and hence the resistivity of the filament wire.)

78. •• In the circuit of Figure 19.66, find (a) the current through the 8.0 Ω resistor and (b) the total rate of dissipation of electrical energy in the 8.0 Ω resistor and in the internal resistance of the batteries. (c) In one of the batteries, chemical energy is being converted into electrical energy. In which one is this happening and at what rate?

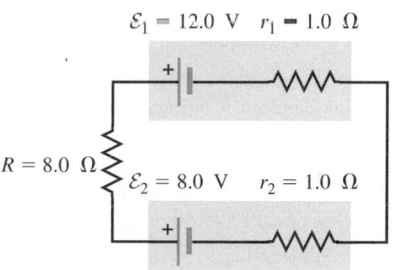

▲ **FIGURE 19.66** Problem 78.

79. •• **Struck by lightning.** Lightning strikes can involve currents as high as 25,000 A that last for about 40 μs. If a person is struck by a bolt of lightning with these properties, the current will pass through his body. We shall assume that his mass is 75 kg, that he is wet (after all, he is in a rainstorm) and therefore has a resistance of 1.0 kΩ, and that his body is all water (which is reasonable for a rough, but plausible, approximation). (a) By how many degrees Celsius would this lightning bolt increase the temperature of 75 kg of water? (b) Given that the internal body temperature is about 37°C, would the person's temperature actually increase that much? Why not? What would happen first?

80. •• **Navigation of electric fish.** Certain fish, such as the Nile fish (*Gnathonemus*), concentrate charges in their head and tail, thereby producing an electric field in the water around them. (See Figure 19.67.) This field creates a potential difference of a few volts between the head and tail, which in turn causes current to flow in the conducting seawater. As the fish swims, it passes near objects that have resistivities different from that of seawater, which in turn causes the current to vary. Cells in the skin of the fish are sensitive to this current and can detect changes in it. The changes in the current allow the fish to navigate. (In the next chapter, we shall investigate *how* the fish might detect this current.) Since the electric field is weak far from the fish, we shall consider only the field running directly from the head to the tail. We can model the seawater through which that field passes as a conducting tube of area

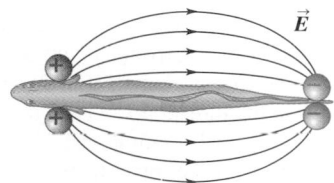

▲ **FIGURE 19.67** Problem 80.

1.0 cm² and having a potential difference of 3.0 V across its ends. The length of a Nile fish is about 20 cm, and the resistivity of seawater is 0.13 Ω·m. (a) How large is the current through the tube of seawater? (b) Suppose the fish swims next to an object that is 10 cm long and 1.0 cm² in cross-sectional area and has half the resistivity of seawater. This object replaces the seawater for half the length of the tube. What is the current through the tube now? How large is the *change* in the current that the fish must detect? (*Hint:* How are this object and the remaining water in the tube connected, in series or in parallel?)

81. •• Each of three resistors in Figure 19.68 has a resistance of 2.00 Ω and can dissipate a maximum of 32.0 W without becoming excessively heated. What is the maximum power the circuit can dissipate?

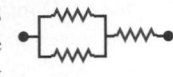

▲ **FIGURE 19.68** Problem 81.

82. •• **Leakage in a dielectric.** Two parallel plates of a capacitor have equal and opposite charges Q. The dielectric has a dielectric constant K and a resistivity ρ. Show that the "leakage" current I carried by the dielectric is given by $I = Q/K\epsilon_0\rho$. (*Note:* See Section 18.8 for a review of capacitors with dielectrics.)

83. •• **Energy use of home appliances.** An 1800 W toaster, a 1400 W electric frying pan, and a 75 W lamp are plugged into the same electrical outlet in a 20 A, 120 V circuit. (*Note:* When plugged into the same outlet, the three devices are in parallel with each other across the 120 V outlet.) (a) What current is drawn by each device? (b) Will this combination blow the circuit breaker?

84. •• Two identical 1.00 Ω wires are laid side by side and soldered together so that they touch each other for half of their lengths. (See Figure 19.69.) What is the equivalent resistance of this combination?

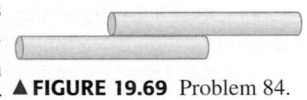

▲ **FIGURE 19.69** Problem 84.

85. •• Three identical resistors are connected in series. When a certain potential difference is applied across the combination, the total power dissipated is 27 W. What power would be dissipated if the three resistors were connected in parallel across the same potential difference?

86. •• (a) Calculate the equivalent resistance of the circuit of Figure 19.70 between x and y. (b) If a voltmeter is connected between points a and x when the current in the 8.0 Ω resistor is 2.4 A in the direction from left to right in the figure, what will it read?

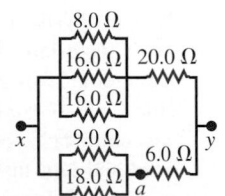

▲ **FIGURE 19.70** Problem 86.

87. •• A power plant transmits 150 kW of power to a nearby town, through wires that have total resistance of 0.25 Ω. What percentage of the power is dissipated as heat in the wire if the power is transmitted at (a) 220 V and (b) 22 kV?

88. •• What must the emf $\mathcal{E}$ in Figure 19.71 be in order for the current through the 7.00 Ω resistor to be 1.80 A? Each emf source has negligible internal resistance.

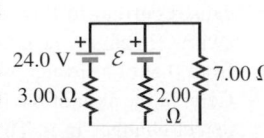

▲ **FIGURE 19.71** Problem 88.

89. •• For the circuit shown in Figure 19.72, if a voltmeter is connected across points a and b, (a) what will it read, and (b) which point is at a higher potential, a or b?

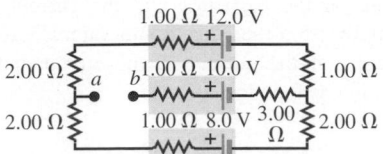

▲ **FIGURE 19.72** Problem 89.

90. •• A 4600 Ω resistor is connected across a charged 0.800 nF capacitor. The initial current through the resistor, just after the connection is made, is measured to be 0.250 A. (a) What magnitude of charge was initially on each plate of this capacitor? (b) How long after the connection is made will it take before the charge is reduced to $1/e$ of its maximum value?

91. •• A capacitor that is initially uncharged is connected in series with a resistor and a 400.0 V emf source with negligible internal resistance. Just after the circuit is completed, the current through the resistor is 0.800 mA and the time constant for the circuit is 6.00 s. What are (a) the resistance of the resistor and (b) the capacitance of the capacitor?

92. •• In the circuit shown in Fig. 19.73, R is a variable resistor whose value can range from 0 to ∞, and a and b are the terminals of a battery having an emf $\mathcal{E} = 15.0$ V and an internal resistance of 4.00 Ω. The ammeter and voltmeter are both idealized meters. As R varies over its full range of values, what will be the largest and smallest readings of (a) the voltmeter and (b) the ammeter? (c) Sketch qualitative graphs of the readings of both meters as functions of R, as R ranges from 0 to ∞.

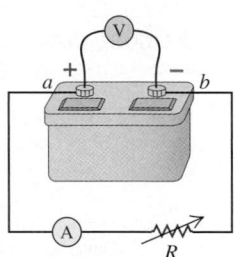

▲ **FIGURE 19.73** Problem 92.

Passage Problems

BIO The nerve membrane as an R–C circuit. We now know that the electrical properties of nerve cells are governed by ion channels, which are protein molecules that span the limiting membrane of the axon. Each ion channel has a water-filled pore that electrically connects the interior of the axon to the outside bathing medium. The lipid-rich membrane in which the channels reside has very little electrical conductivity. Long before the existence of ion channels was known, scientists inferred many of their properties from electrical measurements on the squid giant axon, which could be easily impaled by electrodes and its electrical properties manipulated.

93. If we model the pore of a hypothetical ion channel spanning the membrane of a nerve cell as a cylinder 0.3 nm long with a radius of 0.3 nm filled with a fluid of resistivity 100 Ω cm, what is the conductance of the channel? (Be careful with units).
    A. 1 GΩ   B. 1 nS   C. 100 GΩ   D. 100 nS

94. If the actual conductance of an axon's ion channel is 10 pS and the peak current during a squid axon's action potential (the electrical disturbance that propagates down the axon) is 5 mA/cm^2, what is the density of channels in the membrane? Assume that the voltage across the membrane is 50 mV.

    A. 1/m^2 B. 100/cm^2 C. 1/cm^2 D. 100/μm^2

95. Cell membranes across a wide variety of organisms have a specific capacitance of 1 μF/cm^2. In order for the electrical signal of the nerve to propagate down the axon, the charge on the membrane capacitor must be changed. What is the characteristic time (time constant) required to do this?

    A. 1 μs   B. 10 μs   C. 100 μs   D. 1 ms

20 Magnetic Field and Magnetic Forces

In industrial settings, electromagnets are often used to pick up and move iron-containing material, such as this shredded scrap. How can electric currents cause magnetic forces? We'll learn in this chapter.

Everybody uses magnetic forces. Without them, there would be no electric motors or generators, no microwave ovens, and no computer printers or disk drives. The most familiar aspects of magnetism are those associated with permanent magnets, which attract unmagnetized iron objects and can also attract or repel other magnets. A compass needle interacting with the earth's magnetism is an example of this interaction. But the fundamental nature of magnetism is that it is an interaction associated with moving *electric* charges.

A magnetic field is established by a permanent magnet, by an electric current in a conductor, or by other moving charges. This magnetic field, in turn, exerts forces on moving charges and current-carrying conductors. Magnetic forces are an essential aspect of the interactions among electrically charged particles. In the first several sections of this chapter, we study magnetic forces and torques; then we examine the ways in which magnetic fields are *produced* by moving charges and currents.

20.1 Magnetism

Magnetic phenomena were first observed at least 2500 years ago in fragments of magnetized iron ore found near the ancient city of Magnesia (now Manisa, in western Turkey). It was discovered that when an iron rod is brought into contact with a natural magnet, the rod also becomes magnetized. When such a rod is suspended by a string from its center, it tends to line itself up in a north–south direction, like a compass needle. Magnets have been used for navigation at least since the 11th century. Magnets of this sort are called **permanent magnets.** Figure 20.1 shows a more contemporary example of the magnetic forces associated with a permanent magnet.

▲ **FIGURE 20.1** This bar magnet picks up steel filings—but not the copper filings in the pile. Later in this chapter, we'll learn w some metals are strongly magnetic and oth are not.

Before the relation of magnetic interactions to moving charges was well understood, the interactions of bar magnets and of compass needles were described in terms of *magnetic poles*. The end of a bar magnet that points toward the earth's *geographic* north pole (which marks our planet's axis of rotation) is called a **north pole,** or **N pole,** and the other end is a **south pole,** or **S pole.** Two opposite poles attract each other, and two like poles repel each other (Figure 20.2). However, cutting a bar magnet in two does not give you two isolated poles. Instead, each half has N and S poles, as shown in Figure 20.3.

There is no evidence that a single isolated magnetic pole exists; poles always appear in pairs. The concept of magnetic poles is of limited usefulness and can be somewhat misleading. The existence of an isolated magnetic pole, or *magnetic monopole,* would have sweeping implications for theoretical physics. Extensive searches for magnetic monopoles have been carried out, so far without success.

A compass needle points north because the earth itself is a magnet; its geographical north pole is close to a magnetic *south* pole, as shown in Figure 20.4. The magnetic behavior of the earth is similar to that of a bar magnet with its axis not quite parallel to the geographic axis (the axis of rotation). Thus, a compass reading deviates somewhat from geographic north; this deviation, which varies with location, is called *magnetic declination.* Also, the earth's magnetic field is not horizontal at most points on the surface of our planet; its inclination up or down is described by the *angle of dip.*

Although permanent magnets are familiar to us, most of the magnets in our lives (and in the universe) are **electromagnets,** in which magnetic effects are produced by an electric current. Every appliance you own that has an electric motor contains an electromagnet. The phenomenon of electromagnetism was discovered

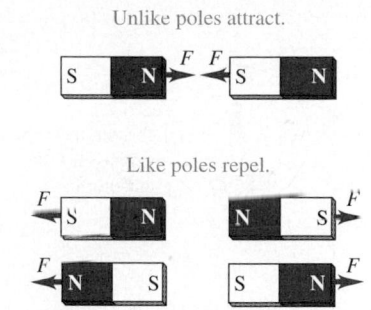

▲ **FIGURE 20.2** Unlike magnetic poles attract each other; like magnetic poles repel each other.

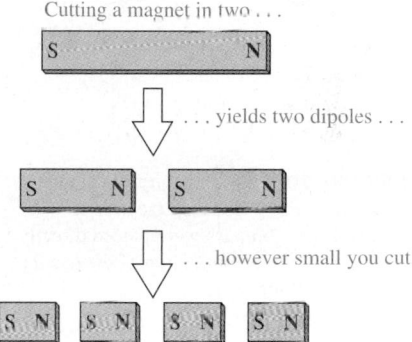

▲ **FIGURE 20.3** Magnets always have paired N and S poles.

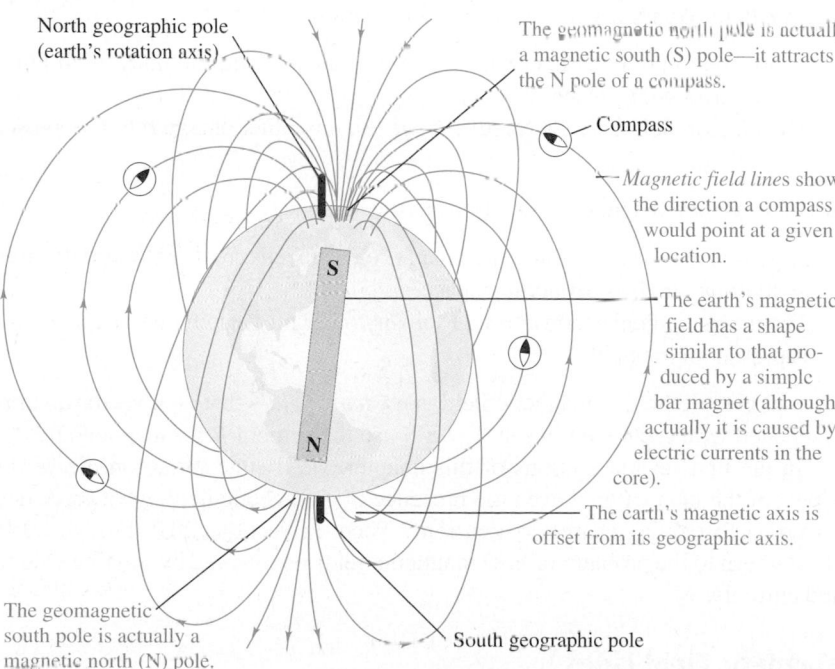

▲ **FIGURE 20.4** A compass placed at any point in the earth's magnetic field will point in the direction of the field line at that point. Representing the earth's field as that of a tilted bar magnet is only a crude approximation of its fairly complex configuration.

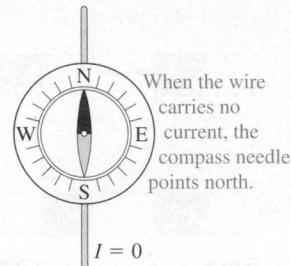

When the wire carries no current, the compass needle points north.

$I = 0$

When the wire carries a current, the compass needle deflects. The direction of deflection depends on the direction of the current.

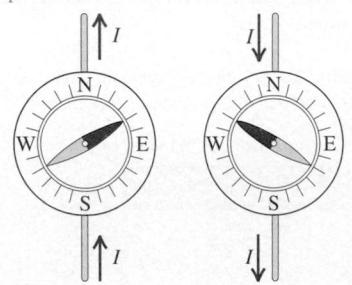

▲ **FIGURE 20.5** The behavior of a compass placed directly over a wire (seen from above). If the compass were placed directly *under* the wire, the deflection of the needle would be reversed.

Mastering**PHYSICS**

PhET: Magnet and Compass
PhET: Magnets and Electromagnets

At each point, the field line is tangent to the magnetic field vector $\vec{B}$.

The more densely the field lines are packed, the stronger the field is at that point.

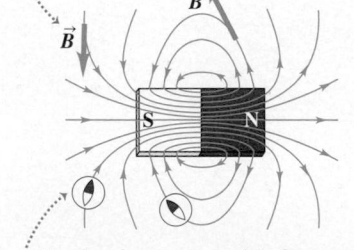

At each point, the field lines point in the same direction a compass would . . .

. . . therefore, magnetic field lines point *away from* N poles and *toward* S poles.

▲ **FIGURE 20.6** Magnetic field lines in a plane through the center of a permanent magnet.

in 1819 by the Danish scientist Hans Christian Oersted, who observed that a compass needle was deflected by a current-carrying wire, as shown in Figure 20.5. A few years later, it was found that moving a magnet near a conducting loop can cause a current in the loop and that a changing current in one conducting loop can cause a current in a separate loop.

We now know that electric and magnetic interactions are intimately intertwined. For example, the earth's magnetism results from electric currents circulating in its molten core. A compass needle points north because the earth itself is an electromagnet. The earth's outer core is made of conductive molten metal, which carries large electric currents while also circulating slowly. The currents in the core produce the earth's magnetic field.

This field changes through time, reflecting the complex dynamics of the circulating molten core. Geologic evidence shows that the magnetic poles wander relative to the geographic poles. Indeed, the earth's field periodically reverses, so that the former north magnetic pole becomes a south magnetic pole, and vice versa. Historically, reversals have occurred, on the average, a few times per million years. Currently, the earth's field is weakening; we may (or may not) be approaching a reversal.

Humans are hardly the first creatures to use earth's magnetic field for navigation. Many organisms, ranging from some bacteria to pigeons and perhaps even whales, can sense magnetic fields and use them for navigation or orientation. In many of these organisms, this sense appears to depend on tiny, intracellular crystals of an iron-containing mineral called magnetite—the same mineral that led to the human discovery of magnetism. These crystals presumably act as tiny bar magnets, orienting themselves in response to the local magnetic field.

20.2 Magnetic Field and Magnetic Force

To introduce the concept of a magnetic field, let's review our formulation of *electrical* interactions in Chapter 17, where we introduced the concept of an *electric* field. We represented electrical interactions in two steps:

1. A distribution of electric charge at rest creates an electric field $\vec{E}$ at all points in the surrounding space.
2. The electric field exerts a force $\vec{F} = q\vec{E}$ on any other charge q that is present in the field.

We can describe magnetic interactions in the same way:

1. A permanent magnet, a moving charge, or a current creates a **magnetic field** at all points in the surrounding space.
2. The magnetic field exerts a force $\vec{F}$ on any other moving charge or current that is present in the field.

Like an electric field, a magnetic field is a *vector field*—that is, a vector quantity associated with each point in space. We'll use the symbol $\vec{B}$ for magnetic field.

In the first several sections of this chapter, we'll concentrate on the *second* aspect of the interaction: Given the presence of a magnetic field, what force does it exert on a moving charge or a current? Then, in Sections 20.7 through 20.10, we'll return to the problem of how magnetic fields are *created* by moving charges and currents.

Magnetic Field Lines

We can represent any magnetic field by **magnetic field lines,** just as we did for the earth's magnetic field in Figure 20.4. As shown in Figure 20.6, we draw the lines so that the line through any point is tangent to the magnetic-field vector $\vec{B}$ at that point. Also, we draw the *number* of lines per unit area (perpendicular to the lines at a given point) to be proportional to the magnitude ("strength") of the field at that point. For a bar magnet, the field lines outside the magnet point from the N pole toward the S pole. As Figure 20.6 shows, a compass placed in a magnetic field always tends to orient its needle parallel to the magnetic field at each point; the *N* pole of the compass needle always tends to point in the direction of $\vec{B}$, and the *S* pole tends to point opposite to $\vec{B}$.

Magnetic field lines are sometimes called magnetic lines of force, but that's not a good name for them because, unlike electric field lines, they *do not* point in the direction of the force on a charge. At each point, the magnetic field line through the point *does* lie in the direction a compass needle would point when placed at that location; this may help you to visualize these lines. Just as with electric field lines, we draw only a few representative lines; otherwise, the lines would fill up all of space. Also, because the direction of $\vec{B}$ at each point is unique, field lines never intersect.

Figure 20.7 shows the magnetic field lines produced by several additional shapes of magnet. Notice especially the field of the C-shaped magnet in Figure 20.7a. In the space between its poles, the field lines are approximately straight, parallel, and equally spaced, showing that the magnetic field in this region is approximately *uniform* (constant in magnitude and direction). In Figure 20.7b, notice the graphical convention that we use to represent a magnetic field going into or out of the paper. Think of the dots as representing the points of vectors and the ×'s as the tail feathers of vectors. We'll use this scheme for diagrams later in the chapter. (The same convention can be used for electric fields.) Figure 20.7c shows that the magnetic field of a current-carrying loop resembles that of a bar magnet. Figure 20.4 shows that the magnetic field of the earth also resembles that of a bar magnet, with the field lines pointing from the N pole (near the geographic south pole) toward the S pole. Although the electric currents in the earth's core don't form simple loops or coils, the same principle is at work.

▲ **BIO Application Homeward bound.**
Homing pigeons are famous for being able to find their home roosts from thousands of kilometers away; they have been used since ancient times for carrying messages. But how do they do it? It appears that they use the earth's magnetic field, among other cues. Evidence indicates that they sense this field at least partly by means of small magnetite crystals located in their beaks. When the area of the beak containing the crystals is anaesthetized or the nerves to it are cut, the birds lose their ability to sense the magnetic field. The pigeon shown in this photo has a small magnet attached to its beak; this also interferes with the bird's ability to sense the earth's magnetic fields.

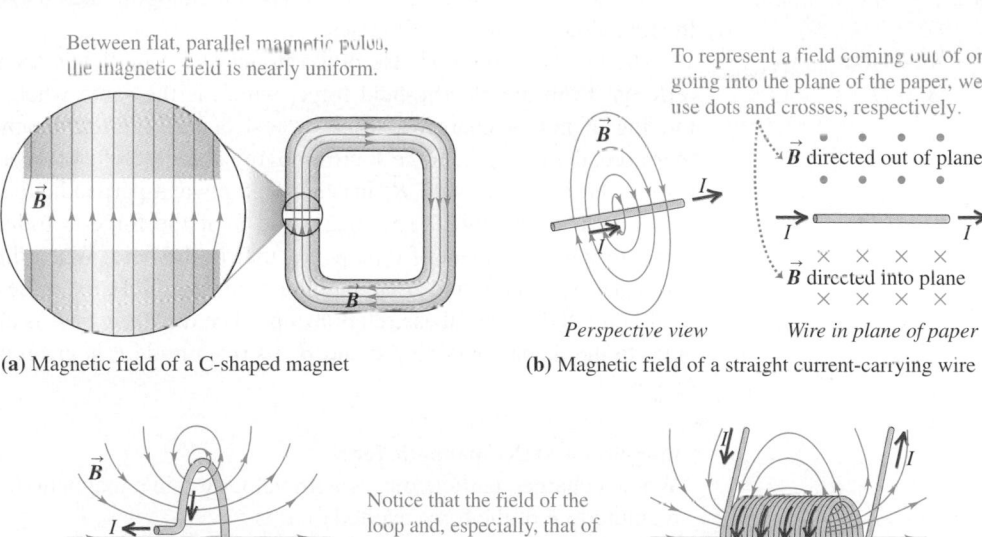

Between flat, parallel magnetic poles, the magnetic field is nearly uniform.

To represent a field coming out of or going into the plane of the paper, we use dots and crosses, respectively.

$\vec{B}$ directed out of plane

$\vec{B}$ directed into plane

Perspective view   *Wire in plane of paper*

(a) Magnetic field of a C-shaped magnet

(b) Magnetic field of a straight current-carrying wire

Notice that the field of the loop and, especially, that of the coil look like the field of a bar magnet (Figure 20.6).

(c) Magnetic fields of a current-carrying loop and a current-carrying coil (solenoid)

▲ **FIGURE 20.7** Some examples of magnetic fields.

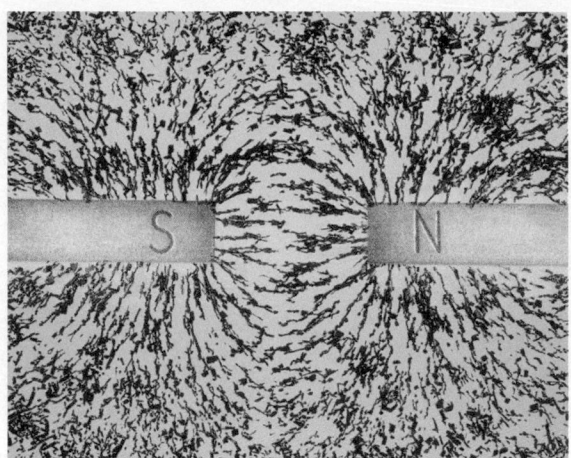

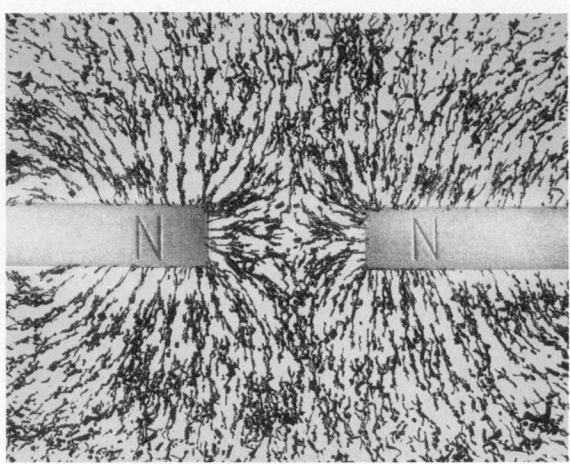

▲ **FIGURE 20.8** Magnetic field lines made visible by iron filings, which line up tangent to the field lines like little compass needles.

You may have had a chance to play with iron filings and a magnet. If you put the filings on a sheet of paper or plastic and hold the magnet beneath it, the filings line up tangent to magnetic field lines, as shown in Figure 20.8.

Magnetic Force

The most fundamental manifestation of a magnetic field is the force that it exerts on a moving charged particle. What are the characteristics of this force? First, its magnitude is proportional to the charge. If two particles, with charges q and $2q$, move in a given magnetic field with the same velocity, the force on the particle with charge $2q$ is twice as great in magnitude as that on charge q. The magnitude of the force is also proportional to the magnitude ("strength") of the field: If we double the magnitude of the field without changing the charge or its velocity, the magnitude of the force doubles.

The magnetic force is also proportional to the particle's speed. This is quite different from the electric-field force, which is the same whether the charge is moving or not. A charged particle at rest experiences no magnetic force. Furthermore, the magnetic force $\vec{F}$ on a moving charge does not have the same direction as the magnetic field $\vec{B}$; instead, it is always perpendicular to both $\vec{B}$ and the particle's velocity $\vec{v}$. The magnitude F of the force is found to be proportional to the component of $\vec{v}$ perpendicular to the field. When that component is zero (that is, when $\vec{v}$ and $\vec{B}$ are parallel or antiparallel), the force is zero.

Figure 20.9 shows these relationships. The direction of $\vec{F}$ is always perpendicular to the plane containing $\vec{v}$ and $\vec{B}$. Its magnitude F is given by the following expression:

Magnitude of the magnetic force
When a charged particle moves with velocity $\vec{v}$ in a magnetic field $\vec{B}$, the magnitude F of the force exerted on it is

$$F = |q|v_{\perp}B = |q|vB\sin\phi, \tag{20.1}$$

where $|q|$ is the magnitude of the charge and ϕ is the angle measured from the direction of $\vec{v}$ to the direction of $\vec{B}$, as shown in Figure 20.9.

A charge moving **parallel** to a magnetic field experiences **zero magnetic force.**

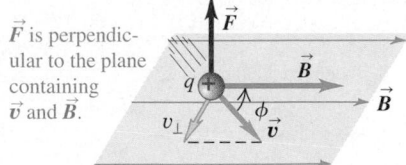

A charge moving at an angle ϕ to a magnetic field experiences a magnetic force with magnitude $F = |q|v_{\perp}B = |q|vB\sin\phi$.

$\vec{F}$ is perpendicular to the plane containing $\vec{v}$ and $\vec{B}$.

A charge moving **perpendicular** to a magnetic field experiences a maximal magnetic force with magnitude $F_{\max} = qvB$.

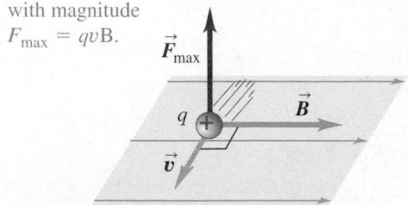

▲ **FIGURE 20.9** The magnetic force on a positive charge moving relative to a uniform magnetic field.

Right-hand rule for the direction of magnetic force on a **positive** charge moving in a magnetic field:

① Place the $\vec{v}$ and $\vec{B}$ vectors tail to tail.

② Imagine turning $\vec{v}$ toward $\vec{B}$ in the $\vec{v}$-$\vec{B}$ plane (through the smaller angle).

③ The force acts along a line perpendicular to the $\vec{v}$-$\vec{B}$ plane. Curl the fingers of your *right hand* around this line in the same direction you rotated $\vec{v}$. Your thumb now points in the direction the force acts.

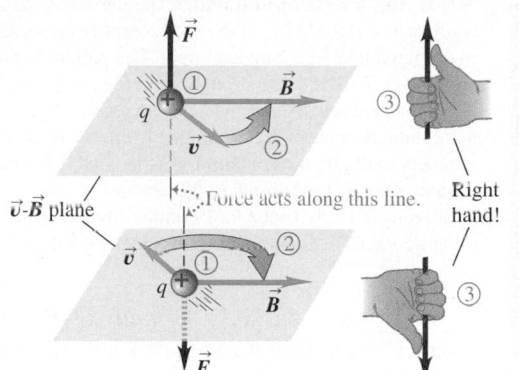

If the charge is negative, the direction of the force is *opposite* to that given by the right-hand rule.

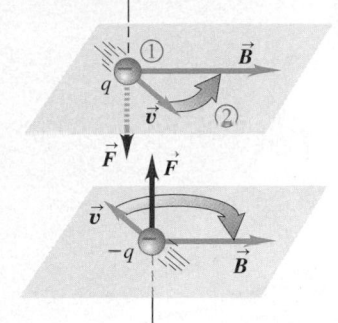

▲ **FIGURE 20.10** The right-hand rule for the direction of the magnetic force on a charge moving in a magnetic field.

Equation 20.1 can be interpreted in a different but equivalent way. Because ϕ is the angle between the directions of vectors $\vec{v}$ and $\vec{B}$, we may interpret $B\sin\phi$ as the component of $\vec{B}$ perpendicular to $\vec{v}$—that is, $B_\perp$. With this notation, the force expression (Equation 20.1) becomes

$$F = |q|vB_\perp. \tag{20.2}$$

This form is equivalent to Equation 20.1, but it's sometimes more convenient to use, especially in problems involving *currents* rather than individual particles. Later in the chapter, we'll discuss forces on conductors carrying currents.

This description of the magnetic-field force doesn't specify the direction of $\vec{F}$ completely; there are always two directions, opposite to each other and both perpendicular to the plane containing $\vec{v}$ and $\vec{B}$. To complete the description, we use a right-hand rule similar to the rule we used in connection with rotational motion in Chapter 10:

Right-hand rule for magnetic force

As shown in Figure 20.10, the right-hand rule for finding the direction of magnetic force on a *positive* charge is as follows:

1. Draw the vectors $\vec{v}$ and $\vec{B}$ with their tails together.
2. Imagine turning $\vec{v}$ in the plane containing $\vec{v}$ and $\vec{B}$ until it points in the direction of $\vec{B}$. Turn it through the smaller of the two possible angles.
3. The force then acts along a line perpendicular to the plane containing $\vec{v}$ and $\vec{B}$. Using your *right hand,* curl your fingers around this line in the same direction (clockwise or counterclockwise) that you turned $\vec{v}$. Your thumb now points in the direction of the force $\vec{F}$ on a *positive* charge.

If the charge is *negative,* the force on it points in the direction opposite to that given by the right-hand rule. In other words, for a negative charge, apply the right-hand rule and then reverse the direction of the force.

Figure 20.11 reinforces the fact that if two charges of opposite sign move with the same velocity in the same magnetic field, they experience magnetic forces in opposite directions. Figures 20.9 through 20.11 show several examples of the relationships of the directions of $\vec{F}$, $\vec{v}$, and $\vec{B}$ for both positive and negative charges. Be sure that you understand these relationships and can verify these figures for yourself.

Positive and negative charges moving in the same direction through a magnetic field experience magnetic forces in *opposite* directions.

▲ **FIGURE 20.11** The effect of the sign of a moving charge on the magnetic force exerted on it.

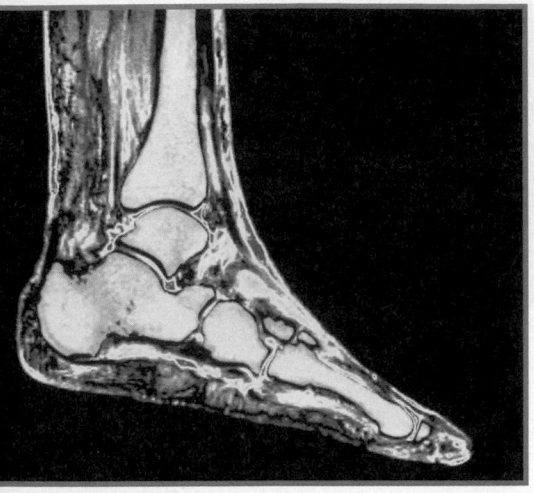

From Equation 20.1, the *units* of B must be the same as the units of the product F/qv. Therefore, the SI unit of B is equivalent to $1 \text{ N} \cdot \text{s}/(\text{C} \cdot \text{m})$, or, since 1 ampere is 1 coulomb per second $(1 \text{ A} = 1 \text{ C}/\text{s})$, it is equivalent to $1 \text{ N}/(\text{A} \cdot \text{m})$. This unit is called the **tesla** (abbreviated T), in honor of Nikola Tesla (1857–1943), the prominent Serbian-American scientist and inventor.

Definition of the tesla

$$1 \text{ tesla} = 1 \text{ T} = 1 \text{ N}/(\text{A} \cdot \text{m}).$$

The cgs unit of B, the **gauss** $(1 \text{ G} = 10^{-4} \text{ T})$, is also in common use. Instruments for measuring magnetic field are sometimes called gaussmeters or teslameters.

The magnetic field of the earth is about $0.5 \times 10^{-4} \text{ T}$ (0.5 G). Magnetic fields on the order of 10 T occur in the interior of atoms, as shown by analysis of atomic spectra. The electromagnets in MRI machines used in medical imaging generate fields from less than a tesla to a few teslas. The largest values of steady magnetic field that have been achieved in the laboratory are about 45 T. The magnetic field at the surface of a neutron star is believed to be on the order of 10^8 T.

Conceptual Analysis 20.1

Direction of magnetic force

Which of the three paths, 1, 2, or 3, does the electron in Figure 20.12 follow? (Remember that the blue ×'s represent a magnetic field pointing *into* the page, as explained in Figure 20.7b.)

A. Path 1.
B. Path 2, because the force on it is zero.
C. Path 2, because the force on it is perpendicular to the page. (We see the path projected onto the plane of the paper.)
D. Path 3.

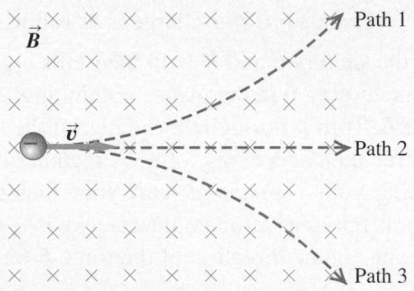

▲ **FIGURE 20.12**

SOLUTION The path depends on the direction of the force (if any) exerted by the magnetic field on the electron. The electron's velocity is not parallel to $\vec{B}$, so the electron experiences a force. To determine the force's direction, we use the right-hand rule. First, we identify the plane containing $\vec{v}$ and $\vec{B}$. (It is perpendicular to the page.) To turn $\vec{v}$ toward $\vec{B}$, we rotate it away from us into the page. Next, we hold our right hand so that the fingers can wrap around a line perpendicular to the plane of $\vec{v}$ and $\vec{B}$.

(This line is in the plane of the paper, parallel to the side of the page.) When we curl our fingers in the direction we turned $\vec{v}$, our thumb points toward the top margin of the page. But that is the direction of the force the magnetic field would exert on a *positive* charge. Since the electron is negative, the force exerted on it is in the plane of the paper, directed toward the bottom of the page. Thus, the electron follows path 3.

PROBLEM-SOLVING STRATEGY 20.1   **Magnetic forces**

SET UP

1. The biggest difficulty in determining magnetic forces is relating the directions of the vector quantities. In determining the direction of the magnetic-field force, draw the two vectors $\vec{v}$ and $\vec{B}$ with their tails together so that you can visualize and draw the plane in which they lie. This also helps you to identify the angle ϕ (always less than 180°) between the two vectors and to avoid getting its complement or some other erroneous angle. Then remember that $\vec{F}$ is always perpendicular to this plane, in a direction determined by the right-hand rule. Keep referring to Figures 20.9 and 20.10 until you're sure you understand this. If q is negative, the force is *opposite* to the direction given by the right-hand rule.

SOLVE AND REFLECT

2. Whenever you can, do the problem in two ways: using Equation 20.1 and then using Equation 20.2. Check that the results agree.

EXAMPLE 20.1   **A proton beam**

In Figure 20.13, a beam of protons moves through a uniform magnetic field with magnitude 2.0 T, directed along the positive z axis. The protons have a velocity of magnitude 3.0×10^5 m/s in the x-z plane at an angle of 30° to the positive z axis. Find the force on a proton. The charge of the proton is $q = +1.6 \times 10^{-19}$ C.

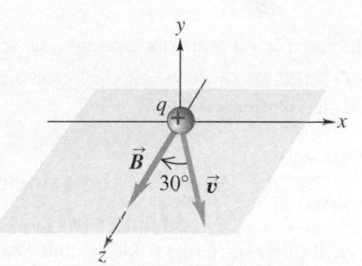

▶ FIGURE 20.13

SOLUTION

SET UP We use the right-hand rule to find the direction of the force. The force acts along the y axis, so we curl the fingers of our right hand around this axis in the direction from $\vec{v}$ toward $\vec{B}$. We find that the force acts in the $-y$ direction.

SOLVE To find the magnitude of the force, we use Equation 20.1:

$$F = qvB\sin\phi$$
$$= (1.6 \times 10^{-19}\,\text{C})(3.0 \times 10^5\,\text{m/s})(2.0\,\text{T})(\sin 30°)$$
$$= 4.8 \times 10^{-14}\,\text{N}.$$

REFLECT We could also obtain this result by finding $B_\perp$ and applying Equation 20.2: $F = |q|vB_\perp$. However, since we were given the angle ϕ, Equation 20.1 is more convenient. To check for consistency of units, we recall that $1\,\text{T} = 1\,\text{N}/(\text{A}\cdot\text{m})$.

Practice Problem: An electron beam moves through a uniform magnetic field with magnitude 3.8 T, directed in the $-z$ direction. The electrons have a velocity of 2.4×10^4 m/s in the y-z plane at an angle of 40° from the $-z$ axis toward the $+y$ axis. Find the force on an electron. *Answer:* $\vec{F} = 9.4 \times 10^{-15}$ N in the $+x$ direction.

Velocity Selector

To explore the principles we've learned, let's look at a device called a **velocity selector.** Many applications of present-day technology use a beam of charged particles that all have the same speed. Common sources of particle beams usually produce particles with a *range* of speeds. The velocity selector uses an arrangement of electric and magnetic fields that lets us select only particles with the desired speed.

ActivPhysics 13.8: Velocity Selector

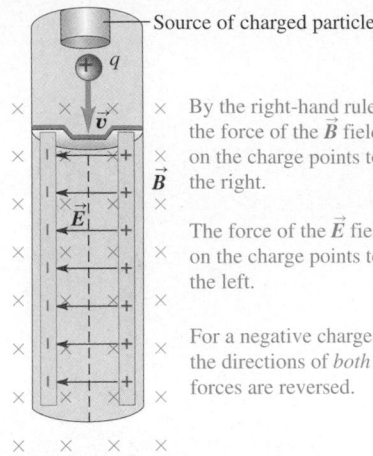

Source of charged particles

By the right-hand rule, the force of the $\vec{B}$ field on the charge points to the right.

The force of the $\vec{E}$ field on the charge points to the left.

For a negative charge, the directions of *both* forces are reversed.

(a) Schematic diagram of velocity selector

$F_E = qE$ $F_B = qvB$

Only if a charged particle has $v = E/B$ do the electric and magnetic forces cancel. All other particles are deflected.

(b) Free-body diagram for a positive particle

▲ **FIGURE 20.14** A velocity selector for charged particles.

The basic principle is shown in Figure 20.14. A charged particle with mass m, charge q, and speed v enters a region of space where the electric and magnetic fields are perpendicular to the particle's velocity and to each other, as shown in the figure. The electric field $\vec{E}$ points to the left, and the magnetic field $\vec{B}$ is into the plane of the page. If q is positive, the electric force is to the left, with magnitude qE, and the magnetic force is to the right, with magnitude qvB. By adjusting the magnitudes E and B, we can make these forces equal in magnitude. The *total* force is then zero, and the particle travels in a straight line with constant velocity. For zero total force, we need

$$\sum \vec{F} = 0, \qquad -qE + qvB = 0, \qquad \text{and} \qquad v = \frac{E}{B}. \qquad (20.3)$$

Only particles with speeds equal to E/B can pass through without being deflected by the fields. By adjusting E and B appropriately, we can select particles having a particular speed. Because q divides out, Equation 20.3 also works for electrons or other negatively charged particles. Do you understand why the electric and magnetic forces both have directions opposite to the preceding ones if q is negative?

Conceptual Analysis 20.2

Selecting the velocity of positive ions

A parallel beam of positive ions, differing from each other in mass, velocity, and charge, enters the magnetic and electric fields of the velocity selector shown in Figure 20.15. For the magnetic and electric forces on the particles to be in opposite directions, which of the charged plates has positive charge?

A. The upper plate.
B. The lower plate.

Charged plates

▲ **FIGURE 20.15**

SOLUTION The magnetic field is into the page, the velocity of the charges points to the right, and the charges are positive. Therefore, by the right-hand rule, the force on the charges points toward the top of the page. We want the electric force to point toward the bottom of the page, however; thus, since the charges are positive, the electric field must also point toward the bottom, so the top plate must be positive and the bottom plate negative.

Thomson's e/m Experiment

In one of the landmark experiments in physics at the turn of the 20th century, Sir J. J. Thomson used the idea just described to measure the ratio of charge to mass for the electron. The speeds of electrons in a beam were determined from the potential difference used to accelerate them, and measurements of the electric and magnetic fields enabled Thomson to determine the ratio of electric charge magnitude (e) to mass (m) of the electrons. (The electron charge and mass were not measured separately until 15 years later.)

The most significant aspect of Thomson's e/m measurements was that he found that all particles in the beam had the *same value* for this quantity and that the value was independent of the materials used for the experiment. This independence showed that the particles in the beam, which we now call electrons, are a common constituent of all matter. Thus, Thomson is credited with the first discovery of a subatomic particle, the electron.

Fifteen years after Thomson's experiments, Millikan succeeded in measuring the charge of the electron. This result, together with the value of e/m, enabled

▶ **Application Radarange®** This photo shows the first commercial microwave oven, called a Radarange®, produced in 1947. It was the size of a modern refrigerator and cost nearly $3000. Microwave ovens were developed as an accidental offshoot of radar research. During a radar testing session, one scientist reached into his pocket to find his candy bar melted! This occurred because radar uses a device called a magnetron to generate electromagnetic radiation of a defined frequency, and certain frequencies can cook food. A filament in the center of the magnetron emits electrons, which then travel toward a surrounding cylindrical conductor. Magnets cause the electrons to curve in their path to the cylinder, where they encounter specially engineered resonant cavities that force the electrons into defined current oscillations. This arrangement generates just the proper frequency of electromagnetic radiation to pop your popcorn.

him to determine, for the first time, the *mass* of the electron. The most precise value available at present is

$$m_e = 9.1093826(16) \times 10^{-31} \text{ kg}.$$

20.3 Motion of Charged Particles in a Magnetic Field

When a charged particle moves in a magnetic field, the motion is determined by Newton's laws, with the magnetic force given by Equation 20.2. Figure 20.16 shows a simple example: A particle with positive charge q is at point O, moving with velocity $\vec{v}$ in a uniform magnetic field $\vec{B}$ directed into the plane of the figure. The vectors $\vec{v}$ and $\vec{B}$ are perpendicular, so the magnetic force $\vec{F}$ has magnitude $F = qvB$, and its direction is as shown in the figure. The force is *always* perpendicular to $\vec{v}$ so it cannot change the *magnitude* of the velocity, only its direction. (To put it differently, the force can't do work on the particle, so the force can't change the particle's kinetic energy.) Thus, the magnitudes of both $\vec{F}$ and $\vec{v}$ are constant. At points such as P and S in Figure 20.16a, the *directions* of force and velocity have changed, but not their magnitudes.

The particle therefore moves in a plane under the influence of a constant-magnitude force that is always at right angles to the velocity of the particle. Comparing these conditions with the discussion of circular motion in Chapter 6, we see that the particle's path is a *circle*, traced out with constant speed v (Figure 20.16b). The radial acceleration is v^2/R, and, from Newton's second law,

$$F = |q|vB = m\frac{v^2}{R},$$

where m is the mass of the particle. The radius R of the circular path is

$$R = \frac{mv}{|q|B}. \tag{20.4}$$

This result agrees with our intuition that it's more difficult to bend the paths of fast and heavy particles into a circle, so the radius is larger for fast, massive particles than for slower, less massive particles. Likewise, for a given charge, a larger magnetic field increases the force and pulls the particle into a smaller radius. If the charge q is negative, the particle moves *clockwise* around the orbit in Figure 20.16a.

The angular velocity ω of the particle is given by Equation 9.13: $\omega = v/R$. Combining this relationship with Equation 20.4, we get

$$\omega = \frac{v}{R} = v\frac{|q|B}{mv} = \frac{|q|B}{m}. \tag{20.5}$$

A charge moving at right angles to a uniform $\vec{B}$ field moves in a circle at constant speed because $\vec{F}$ and $\vec{v}$ are always perpendicular to each other.

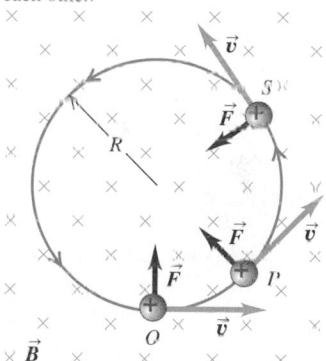

(a) The orbit of a charged particle in a uniform magnetic field

(b) An electron beam curving in a magnetic field

▲ **FIGURE 20.16** The circular orbit of a charged particle whose initial velocity is perpendicular to a magnetic field.

ActivPhysics 13.4: Magnetic Force on a Particle

The number of revolutions per unit time (the frequency of revolution) is $\omega/2\pi$. This frequency is independent of the radius R of the path. It is called the **cyclotron frequency;** in a particle accelerator called a *cyclotron,* particles moving in nearly circular paths are given a boost twice each revolution (by an electric field), increasing their energy and their orbital radii, but not their angular velocity. Similarly, a *magnetron*—a common source of microwave radiation for microwave ovens and radar systems—emits radiation with a frequency proportional to the frequency of the circular motion of electrons in a vacuum chamber between the poles of a magnet.

Quantitative Analysis 20.3

Period of cyclotron motion

A charged particle enters a uniform magnetic field with a velocity $\vec{v}$ at right angles to the field. It moves in a circle with period T. If an identical particle enters the field with a velocity $2\vec{v}$, its period is

A. $4T$.
B. $2T$.
C. T.
D. $T/4$.

SOLUTION As we just found in deriving Equation 20.4, the radius of the particle's path is directly proportional to the particle's speed: $R = mv/|q|B$. Doubling the speed doubles the radius R and hence the circumference $2\pi R$ of the circle. Thus, to complete a circle, the particle that is moving twice as fast must go twice as far. But because it's moving twice as fast, it covers the distance in the same amount of time as the slower particle. Since the period is the amount of time it takes to complete one circle, the correct answer is C. The period is independent of the particle's speed.

This particle's motion has components both parallel ($v_\parallel$) and perpendicular ($v_\perp$) to the magnetic field, so it moves in a spiral.

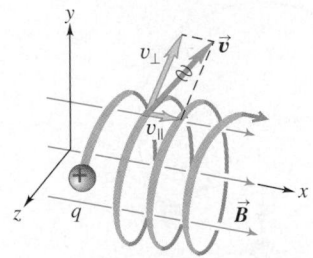

▲ **FIGURE 20.17** The spiral path of a charged particle whose initial velocity has components parallel and perpendicular to the magnetic field.

Helical Motion

If a particle moves in a uniform magnetic field and if the direction of its initial velocity is *not* perpendicular to the field, there is a component of velocity *parallel* to the field. This velocity component is constant, because there is no component of force parallel to the field. In this case, the particle moves in a helix (Figure 20.17). The radius R of the helix is given by Equation 20.4, where v is now the component of velocity perpendicular to the $\vec{B}$ field.

The motion we've just described is responsible for the earth's auroras (Figure 20.18). The sun emits a fast-moving "wind" of charged particles. When these particles encounter the earth's magnetic field far out in space, some are trapped and begin spiraling along magnetic field lines, which guide them toward the earth's poles. (See Figure 20.4.) In the upper atmosphere, these energetic charged particles collide with air molecules, causing the molecules to emit the light we call the auroras. The planets Jupiter and Saturn, which have strong magnetic fields, also have auroras.

▲ **FIGURE 20.18** Auroras on **(a)** earth and **(b)** Saturn.

Conceptual Analysis 20.4

The effects of magnetic force

A charged particle enters a uniform magnetic field. The field can change the particle's

A. velocity.
B. speed.
C. kinetic energy.

If the particle moves parallel to the magnetic field, the field has no effect on it. If the particle has a component of velocity $v_\perp$ perpendicular to the field, the field exerts a force on the particle. Because this force acts perpendicular to $v_\perp$, it changes the particle's direction but not its speed. Thus, a uniform magnetic field can change the velocity, but not the speed, of a charged particle. Because the kinetic energy of a given particle depends on speed $\left(K = \frac{1}{2}mv^2\right)$, the field also does not change the particle's kinetic energy. The correct answer is A.

EXAMPLE 20.2 Electron motion in a microwave oven

A magnetron in a microwave oven emits microwaves with frequency $f = 2450$ MHz. What magnetic field strength would be required for electrons to move in circular paths with this frequency?

SOLUTION

SET UP Figure 20.19 shows our diagram. Because the electron is negatively charged, the right-hand rule tells us that it circles clockwise. The frequency is $f = 2450$ MHz $= 2.45 \times 10^9$ s^{-1}. The corresponding angular velocity is $\omega = 2\pi f = (2\pi)(2.45 \times 10^9 \text{ s}^{-1}) = 1.54 \times 10^{10}$ rad/s.

SOLVE From Equation 20.5,

$$B = \frac{m\omega}{|q|} = \frac{(9.11 \times 10^{-31} \text{ kg})(1.54 \times 10^{10} \text{ rad/s})}{1.60 \times 10^{-19} \text{ C}} = 0.0877 \text{ T}.$$

REFLECT This is a moderate field strength, easily produced with a permanent magnet. Electromagnetic waves with this frequency can penetrate several centimeters into food with high water content.

Practice Problem: If the magnetron emits microwaves with frequency 2300 MHz, what magnetic field strength would be required

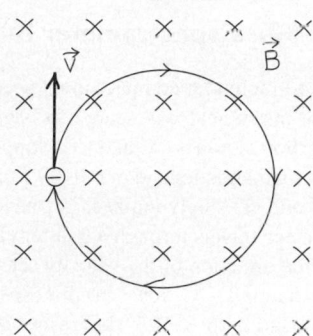

▲ **FIGURE 20.19** Our sketch for this problem.

for electrons to move in circular paths with that frequency? *Answer:* $B = 0.0823$ T.

20.4 Mass Spectrometers

Magnetic forces on charged particles play an important role in a family of instruments called **mass spectrometers.** These instruments are used to measure masses of positive ions and thus measure atomic and molecular masses. (Recall that a positive ion is an atom or molecule that has lost one or more electrons and hence has acquired a net positive charge.)

One type of mass spectrometer is shown in Figure 20.20. Ions from a source pass through the slits S_1 and S_2, forming a narrow beam. Then the ions pass through a *velocity selector* with crossed $\vec{E}$ and $\vec{B}$ fields, as we've described in Section 20.2, to block all ions except those with speeds v equal to E/B. Finally, the ions having this speed pass into a region with a magnetic field $\vec{B}'$ perpendicular to the figure, where they move in circular arcs with radius R determined by Equation 20.4, $R = mv/qB'$. Particles with different masses strike the detector at different points; a particle detector makes a scan to determine the number of particles collected at a given radius R. We assume that each ion has lost one electron, so the net charge of each ion is just e. With everything known in this equation except m, we can compute the mass m of the ion.

One of the earliest applications of the mass spectrometer was the discovery that the element neon has two species of atoms, with atomic masses 20 and 22 g/mol. We now call these species **isotopes** of the element. Later experiments

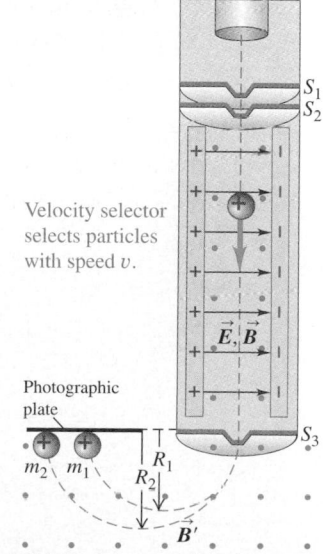

Velocity selector selects particles with speed v.

$\vec{E}, \vec{B}$

Photographic plate

Magnetic field separates particles by mass; the greater a particle's mass, the greater is the radius of its path.

▲ **FIGURE 20.20** One type of mass spectrometer.

have shown that many elements have several isotopes, with atoms that are very similar in chemical behavior, but different in mass. Each element has a definite number of electrons in its neutral, un-ionized atoms, and an equal number of protons in each nucleus. Mass differences between isotopes result from different numbers of *neutrons* in the nuclei.

For example, a neutral carbon atom always has six electrons, and there are always six protons in each carbon nucleus. But the nuclei can have a range of number of neutrons, most commonly six, but sometimes seven or eight. Using a prefixed superscript to indicate the total number of protons and neutrons (called the *mass number*), we denote these carbon isotopes as ^{12}C, ^{13}C, and ^{14}C.

An interesting application of radioactivity is the dating of archeological and geological specimens by measuring the concentration of radioactive isotopes of an element such as carbon (for once-living organisms) and potassium (for rock formations). We'll return to this topic in Chapter 30.

EXAMPLE 20.3   Mass spectrometer for a Mars rover

Scientists want to include a compact mass spectrometer on a future Mars rover. Among other things, the instrument will search for signs of life by measuring the relative abundances of the carbon isotopes ^{12}C and ^{13}C. Suppose the instrument is designed as shown in Figure 20.21, has a magnetic field of 0.0100 T, and selects carbon ions that have a speed of 5.00×10^3 m/s and are singly ionized (i.e., have a charge of $+e$). Each ion emerging from the velocity selector travels through a quarter circle before striking the detector plate (oriented at 45° to the direction of the velocity selector). (**a**) What are the radii R_{12} and R_{13} of the orbits of ^{12}C and ^{13}C ions in this spectrometer? These ions have masses of 1.99×10^{-26} kg and 2.16×10^{-26} kg, respectively. (**b**) How far apart are the spots these ions produce on the detector plate?

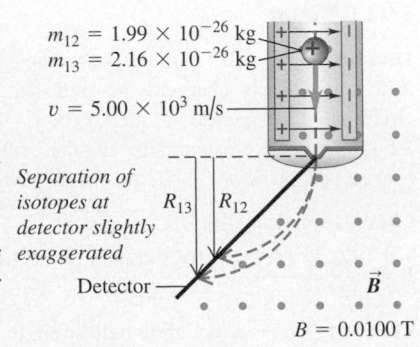

▲ **FIGURE 20.21**

SOLUTION

SET UP AND SOLVE Part (a):   To find the radii of the orbits, we use Equation 20.4. The charge $+e$ is 1.60×10^{-19} C. We have

$$R_{12} = \frac{m_{12}v}{eB} = \frac{(1.99 \times 10^{-26}\ \text{kg})(5.00 \times 10^3\ \text{m/s})}{(1.60 \times 10^{-19}\ \text{C})(0.0100\ \text{T})} = 0.0622\ \text{m};$$

$$R_{13} = \frac{m_{13}v}{eB} = \frac{(2.16 \times 10^{-26}\ \text{kg})(5.00 \times 10^3\ \text{m/s})}{(1.60 \times 10^{-19}\ \text{C})(0.0100\ \text{T})} = 0.0675\ \text{m}.$$

Part (b):   As the geometry of Figure 20.21 shows, the ions strike the detector plate at points separated by a distance of

$$\sqrt{2}\,(R_{13} - R_{12}) = \sqrt{2}\,(0.0675\ \text{m} - 0.0622\ \text{m})$$
$$= 7.50 \times 10^{-3}\ \text{m} = 7.50\ \text{mm}.$$

REFLECT Notice that the radius is quite small, about 6 cm. Even though the complete mass spectrometer will be larger than this, it will still fit easily on a robotic rover. The quarter-circle design also helps reduce size and weight.

Practice Problem: If the isotope ^{14}C is also present in the sample, what will its radius be at the detection plate? The mass of ^{14}C is 2.33×10^{-26} kg. *Answer: R = 7.28 cm.*

20.5 Magnetic Force on a Current-Carrying Conductor

What makes an electric motor work? The forces that make it turn are forces that a magnetic field exerts on a conductor carrying a current. The magnetic forces on the moving charges within the conductor are transmitted to the material of the conductor, and the resulting force is distributed along the conductor's length. The moving-coil galvanometer that we described in Section 19.7 also uses magnetic forces on conductors.

We can compute the force on a current-carrying conductor, starting with the magnetic-field force on a single moving charge. The magnitude F of that force is given by Equation 20.1: $F = |q|vB\sin\phi$. Figure 20.22 shows a straight segment of a conducting wire, with length l and cross-sectional area A; the direction of the current is from bottom to top. The wire is in a uniform magnetic field $\vec{B}$; the field is perpendicular to the plane of the diagram and directed *into* the plane. Let's assume first that the moving charges are positive. Later we'll see what happens when they are negative.

The drift velocity $\vec{v}_d$ (the average velocity of the moving charges, discussed in Section 19.1) is upward, perpendicular to $\vec{B}$. According to the right-hand rule, the force $\vec{F}$ on each charge is directed to the left, as shown in the figure. In this case, $\vec{v}_d$ and $\vec{B}$ are perpendicular, and the magnitude F_{av} of the average force on each charge is $F_{av} = qv_dB$.

We can derive an expression for the *total* force on all the moving charges in a segment of conductor with length l in terms of the *current* in the conductor. Let Q be the magnitude of the *total* moving charge in this segment of conductor. The time Δt needed for a charge to move from one end of the segment to the other is $\Delta t = l/v_d$, and the total amount of charge Q that flows through the wire during this time interval is related to the current I by $Q = I\,\Delta t$.

From Equation 20.1 ($F = |q|v_\perp B$), the total force on all the moving charges Q in this segment has magnitude $F = Qv_dB$. Using $Q = I\,\Delta t$ and $\Delta t = l/v_d$, we can rewrite Equation 20.1 as

$$F = Qv_dB = (I\Delta t)(l/\Delta t)B,$$
$$F = IlB. \qquad (20.6)$$

If the $\vec{B}$ field is not perpendicular to the wire, but makes an angle ϕ with it, we handle the situation the same way we did in Section 20.2 for a single charge. The component of $\vec{B}$ parallel to the wire (and to the drift velocities of the charges) exerts no force; the component perpendicular to the wire is $B_\perp = B\sin\phi$. Here is the general relation:

Magnetic force on a current-carrying conductor

The magnetic field force on a segment of conductor with length l, carrying a current I in a uniform magnetic field $\vec{B}$, is

$$F = IlB_\perp = IlB\sin\phi. \qquad (20.7)$$

The force is always perpendicular to both the conductor and the field, with the direction determined by the same right-hand rule that we used for a moving positive charge (Figure 20.23).

Figure 20.24 shows the relations among the various directions for several cases.

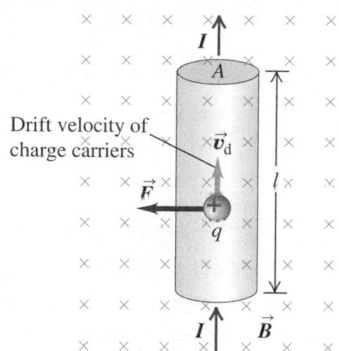

▲ **FIGURE 20.22** Magnetic force on a representative moving positive charge in a current-carrying conductor.

Force on a straight wire carrying a positive current and oriented at an angle ϕ to a magnetic field $\vec{B}$:

Magnitude is $F = IlB_\perp = IlB\sin\phi$

Direction is given by the right-hand rule.

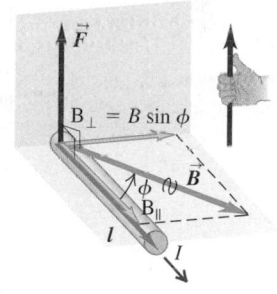

▲ **FIGURE 20.23** The magnetic force on a segment of current-carrying wire in a magnetic field.

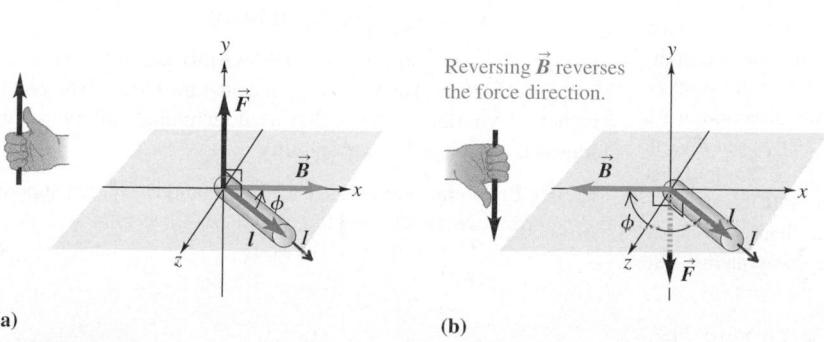

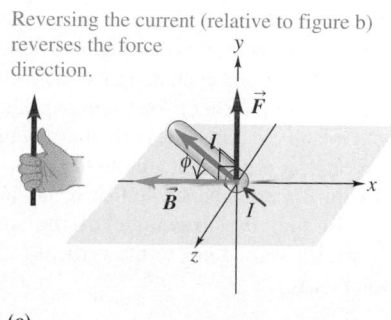

▲ **FIGURE 20.24** The relation of the direction of the magnetic force on a current-carrying conductor to the directions of the current and the magnetic field.

Finally, what happens when the moving charges are negative, such as electrons in a metal? Then, in Figure 20.22, an upward current corresponds to a *downward* drift velocity. But because q is negative in this case, the direction of the force $\vec{F}$ is the same as before. Thus, Equations 20.6 and 20.7 are valid for both positive and negative charges and even when *both* signs of charge are present at once (as sometimes occurs in semiconductor materials and in ionic solutions).

Conceptual Analysis 20.5

Magnetic rail gun

Figure 20.25 shows a magnetic rail gun—a device that uses magnetic forces to accelerate a conducting object (in this case, a bar). The bar is placed across a pair of stationary conducting rails that are in a plane perpendicular to a magnetic field. Closing the switch sends a current through the circuit formed by the rails and the bar, causing the magnetic-field force to accelerate the bar along the rails. In which orientation, A or B, must the battery be connected if the bar is to accelerate to the right?

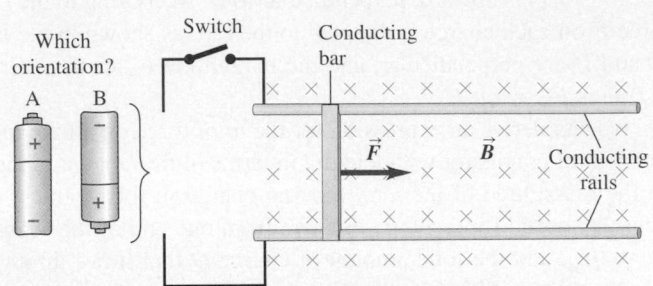

▲ **FIGURE 20.25**

SOLUTION To find the correct battery orientation, we use the right-hand rule in reverse. The magnetic force on the bar must point to the right, so we point the thumb of our right hand in that direction. The direction that our fingers curl is the direction in which the current direction in the bar would have to rotate to become parallel to $\vec{B}$. Thus, the current in the bar must point toward the bottom of the page. Since current flows from the positive to the negative terminal of a battery, we must connect the battery in orientation A.

EXAMPLE 20.4 Magnetic bird perch

The world's lightest bird is the bee hummingbird; its mass is 1.6 g (less than that of a penny). As shown in Figure 20.26, we design a perch for this bird that is supported by a magnetic force. The perch is 10 cm long, has a mass of 7.0 g, and contains a wire carrying a current $I = 4.0$ A. The perch is oriented at right angles to a uniform magnetic field that is just strong enough to support the perch and a bee hummingbird. What are the direction and magnitude of the magnetic field?

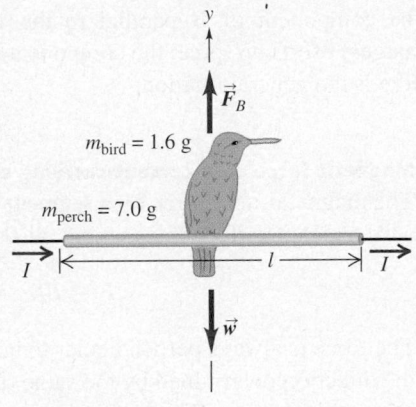

▶ **FIGURE 20.26**

SOLUTION

SET UP The magnetic field must generate an upward force on the perch equal to the combined weight of the perch and bird.

SOLVE To find the direction of the magnetic field, we use the right-hand rule in reverse. The magnetic field must be perpendicular to the plane containing the magnetic force $\vec{F}_B$ and I (that is, the plane of the page). We point our thumb in the direction of $\vec{F}_B$ and imagine rotating the current arrow in the direction our fingers curl until it is perpendicular to the plane of the page. We find that it points *into* the paper, so that is the direction of $\vec{B}$.

To find the magnitude of the magnetic field, we apply Newton's second law to the vertical (y) force components. The total mass is

$$M = m_{\text{bird}} + m_{\text{perch}} = 1.6\,\text{g} + 7.0\,\text{g} = 8.6\,\text{g} = 8.6 \times 10^{-3}\,\text{kg}.$$

The equilibrium condition $\sum F_y = 0$ gives

$$F_B - Mg = 0, \qquad IlB = Mg,$$

$$B = \frac{Mg}{Il} = \frac{(8.6 \times 10^{-3}\,\text{kg})(9.8\,\text{m/s}^2)}{(4.0\,\text{A})(0.10\,\text{m})} = 0.21\,\text{T}.$$

REFLECT Before you think of building this perch for your pet bird, consider what happens as soon as the bird leaves the perch. Magnetic levitation of this sort is used technologically, but with additional systems to ensure stability.

Practice Problem: How massive a bird could this perch support if the current were increased to $I = 5.0$ A? *Answer: m = 3.7 g.*

20.6 Force and Torque on a Current Loop

Current-carrying conductors usually form closed loops, so it's worthwhile to use the results of Section 20.4 to find the *total* magnetic force and torque on a conductor in the form of a loop. As an example, let's look at a rectangular current loop in a uniform magnetic field. We can represent the loop as a series of straight line segments. We'll find that the total *force* on the loop is zero, but that there is a net *torque* acting on the loop, with some interesting properties.

Figure 20.27 shows a rectangular loop of wire with side lengths *a* and *b*. A line perpendicular to the plane of the loop (that is, a *normal* to the plane) makes an angle ϕ with the direction of the magnetic field $\vec{B}$, and the loop carries a current *I*. The wires leading the current into and out of the loop and the source of emf are omitted to keep the diagram simple.

The force $\vec{F}$ on the right side of the loop (length *a*) is in the direction of the *x* axis, toward the right, as shown. On this side, $\vec{B}$ is perpendicular to the direction of the current, and the force on this side has magnitude $F = IaB$. A force $-\vec{F}$ with the same magnitude, but opposite direction, acts on the opposite side, as shown in the figure.

The sides with length *b* make an angle equal to $90° - \phi$ with the direction of $\vec{B}$. The forces on these sides are the vectors $\vec{F}'$ and $-\vec{F}'$; their magnitude F' is given by

$$F' = IbB \sin(90° - \phi) = IbB \cos\phi.$$

The lines of action of both forces lie along the *y* axis.

The *total* force on the loop is zero because the forces on opposite sides cancel out in pairs. The two forces $\vec{F}'$ and $-\vec{F}'$ lie along the same line, so they have no net torque with respect to any axis. The two forces $\vec{F}$ and $-\vec{F}$, equal in magnitude and opposite in direction, but not acting along the same line, form what is called a *couple*. It can be shown that the torque of a couple, with respect to any axis, is the magnitude of either force, multiplied by the perpendicular distance between

▲ **Application Hello?** No matter what their size, dc motors operate on the same fundamental principle to convert electrical energy to mechanical energy. A current is sent through a movable wire coil in a magnetic field. The magnetic force produced acts as a torque at right angles to both the wire and the magnetic field, causing the coil to rotate. To keep the torque from reversing in direction as the coil rotates, a device called a commutator changes the direction of the current through the coil each half revolution. This mechanism provides a convenient, and nowadays indispensable, means of converting electrical energy into mechanical energy. The tiny motor shown here has an off-center weight attached to the shaft, causing it to move back and forth when the shaft rotates. You may own such a motor; it produces the familiar buzz of a cell phone in vibrate mode.

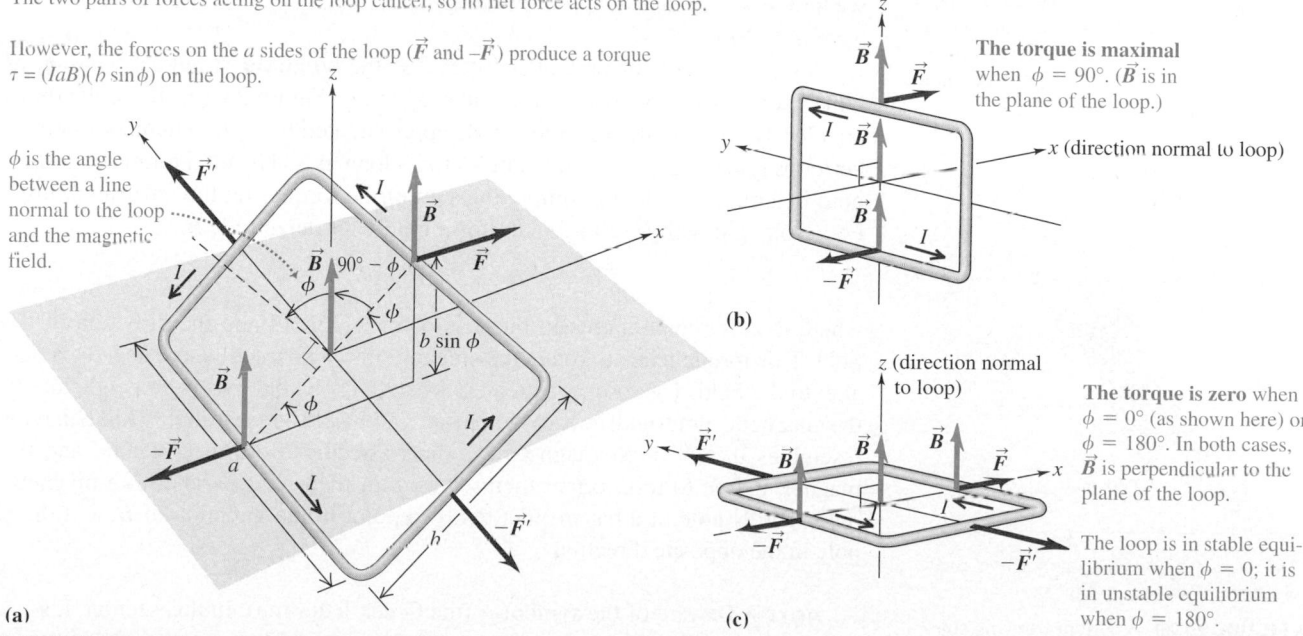

▲ **FIGURE 20.27** (a) Forces on the sides of a current-carrying loop rotating in a magnetic field. (b), (c) orientations at which the torque on the loop is maximal and zero, respectively.

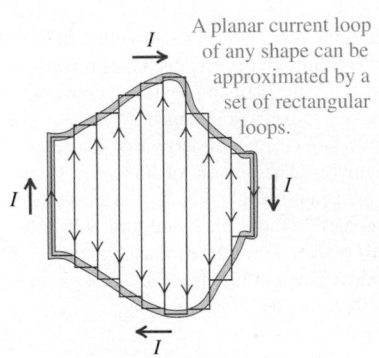

A planar current loop of any shape can be approximated by a set of rectangular loops.

▲ **FIGURE 20.28** An arbitrary planar shape can be approximated to any desired accuracy by a collection of rectangles. The narrower and more numerous the rectangles, the more closely they approximate the shape.

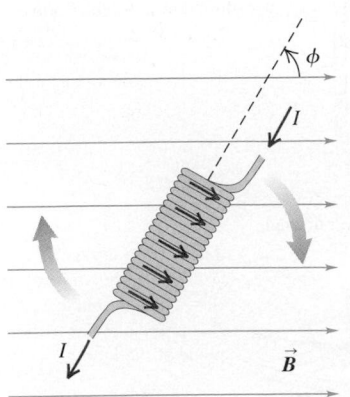

The torque tends to make the solenoid rotate clockwise in the plane of the page.

▲ **FIGURE 20.29** A current-carrying solenoid in a uniform magnetic field experiences a torque.

the lines of action of the two forces. From Figure 20.27a, this distance is $b \sin \phi$, so the torque τ is

$$\tau = (IaB)(b \sin \phi). \tag{20.8}$$

The torque is greatest when $\phi = 90°$, B is in the plane of the loop, and the normal to this plane is perpendicular to $\vec{B}$ (Figure 20.27b). The torque is zero when ϕ is 0° or 180° ($\sin \phi = 0$) and the normal to the loop is parallel or antiparallel to the field (Figure 20.27c). The value $\phi = 0$ is a stable equilibrium position because the torque is zero there, and when the coil is rotated slightly from this position, the resulting torque tends to rotate it back toward $\phi = 0$. The position $\phi = 180°$ is an *unstable* equilibrium position.

The area A of the loop is equal to ab, so we can rewrite Equation 20.8 as follows:

Torque on a current-carrying loop

When a conducting loop with area A carries a current I in a uniform magnetic field of magnitude B, the torque exerted on the loop by the field is

$$\tau = IAB \sin \phi, \tag{20.8}$$

where ϕ is the angle between the normal to the loop and $\vec{B}$.

The torque τ tends to rotate the loop in the direction of *decreasing* ϕ—that is, toward its stable equilibrium position, in which $\phi = 0$ and the loop lies in the x-y plane, perpendicular to the direction of the field $\vec{B}$ (Figure 20.27c). The product IA is called the **magnetic moment** of the loop, denoted by μ:

$$\mu = IA. \tag{20.9}$$

We've derived Equations 20.8 and 20.9 for a rectangular current loop, but it can be shown that these relations are valid for a plane loop of any shape. The proof rests on the fact that any planar loop may be approximated as closely as we wish by a large number of rectangular loops, as shown in Figure 20.28. Also, if we have N such loops wrapped close together, then, in place of Equation 20.9, we have $\mu = NIA$.

An arrangement of particular interest is the **solenoid,** a helical winding of wire, such as a coil wound on a circular cylinder (Figure 20.29). If the windings are closely spaced, the solenoid can be approximated by a large number of circular loops lying in planes at right angles to its long axis. The total torque on a solenoid in a magnetic field is simply the sum of the torques on the individual turns. For a solenoid with N turns in a uniform field with magnitude B,

$$\tau = NIAB \sin \phi,$$

where ϕ is the angle between the axis of the solenoid and the direction of the field. This torque tends to rotate the solenoid into a position where its axis is parallel to the field. The torque is greatest when the solenoid axis is perpendicular to the magnetic field and is zero when axis and field are parallel. This behavior resembles that of a bar magnet or compass needle: Both the solenoid and the magnet, if free to turn, orient themselves with their axes parallel to a magnetic field. The N pole of a bar magnet tends to point in the direction of $\vec{B}$, and the S pole in the opposite direction

NOTE ▶ Beware of the symbol μ (the Greek letter mu). In this section, it's the symbol for magnetic moment, but later μ_0 will represent a constant called the *permeability of vacuum.* Be careful! ◀

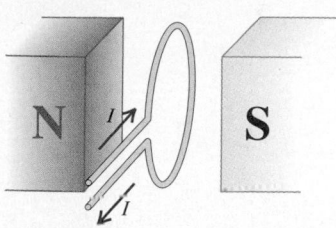

<table><tr><td>

Conceptual Analysis 20.6

Stability of a current loop

The loop in Figure 20.30 is oriented in the vertical plane at right angles to the magnetic field. Is this an equilibrium orientation? (That is, if you place the loop in exactly this orientation, will it stay that way?) If so, is the equilibrium stable? (That is, if you rotate the loop slightly away from this orientation, will it tend to rotate back?)

A. This is a stable equilibrium orientation.
B. This is an equilibrium orientation, but not a stable one.
C. This is not an equilibrium orientation.

SOLUTION The magnetic field points from the N to the S pole, to the right in the figure. The right-hand rule tells us that the force on the top part of the loop points toward the bottom of the page, and that the force on the bottom of the loop points toward the top

</td></tr></table>

▲ FIGURE 20.30

of the page. These forces create no torque, so the loop is in equilibrium. However, if you displace the loop slightly, the forces produce a torque that tends to rotate the loop farther from its original position—not back toward it. Thus, this orientation is not stable; B is correct. If you rotate the loop 180°, it is then in a stable equilibrium position.

EXAMPLE 20.5 **Torque on a circular coil**

A circular coil of wire with average radius 0.0500 m and 30 turns lies in a horizontal plane. It carries a current of 5.00 A in a counterclockwise sense when viewed from above. The coil is in a uniform magnetic field directed toward the right, with magnitude 1.20 T. Find the magnetic moment and the torque on the coil. Which way does the coil tend to rotate?

SOLUTION

SET UP Figure 20.31 shows our diagram. The area of the coil is $A = \pi r^2 = \pi(0.0500 \text{ m})^2 = 7.85 \times 10^{-3} \text{ m}^2$; the angle ϕ between the direction of $\vec{B}$ and the *axis* of the coil (perpendicular to its plane) is 90°.

SOLVE The magnetic moment for one turn of the coil is $\mu = IA$ (Equation 20.9). Therefore, the total magnetic moment for all 30 turns is

$$\mu_{\text{total}} = 30\mu = 30IA = 30(5.00 \text{ A})(7.85 \times 10^{-3} \text{ m}^2)$$
$$= 1.18 \text{ A} \cdot \text{m}^2.$$

From Equation 20.8, the torque on each turn of the coil is

$$\tau = IAB\sin\phi = (5.00 \text{ A})(7.85 \times 10^{-3} \text{ m}^2)(1.20 \text{ T})(\sin 90°)$$
$$= 0.0471 \text{ N} \cdot \text{m},$$

and the total torque on the coil of 30 turns is

$$\tau = (30)(0.0471 \text{ N} \cdot \text{m}) = 1.41 \text{ N} \cdot \text{m}.$$

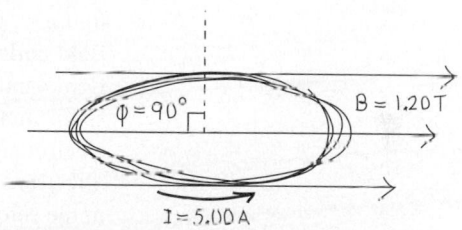

▲ FIGURE 20.31 Our sketch for this problem.

Using the right-hand rule on the two sides of the coil, we find that the torque tends to rotate the right side down and the left side up.

REFLECT The torque tends to rotate the coil toward the stable equilibrium orientation, in which the normal to the plane is parallel to $\vec{B}$.

Practice Problem: Calculate the torque on the coil when it is placed in a magnetic field along the direction of the axis of the coil. *Answer:* $\tau = 0 \text{ N} \cdot \text{m}$.

The Direct-Current Motor

No one needs to be reminded of the importance of electric motors in contemporary society. Their operation depends on magnetic forces on current-carrying conductors. As an example, let's look at a simple type of direct-current motor, shown in Figure 20.32. The center part is the *armature*, or *rotor;* it is a cylinder of soft steel that rotates about its axis.

Embedded in slots in the rotor surface (parallel to its axis) are insulated copper conductors. Current is led into and out of these conductors through stationary graphite blocks called *brushes* that make sliding contact with a segmented

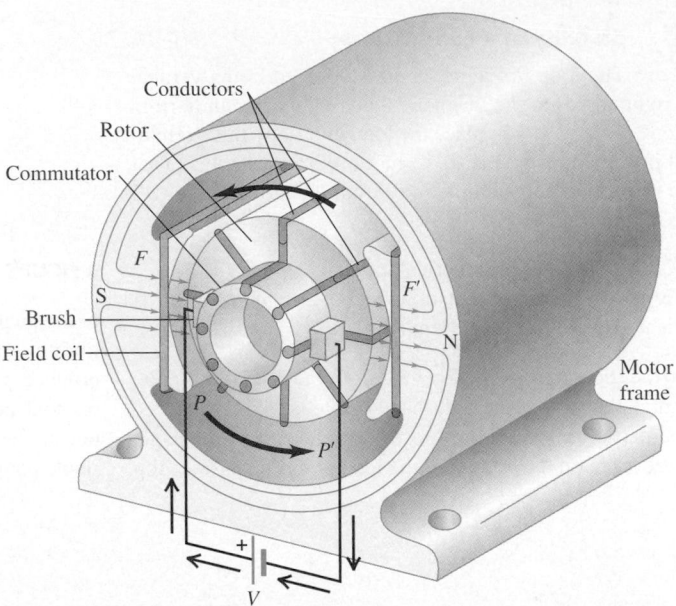

▲ **FIGURE 20.32** Schematic diagram of a dc motor. The rotor rotates on a shaft through its center, perpendicular to the plane of the figure.

cylinder, called the **commutator,** that turns with the rotor. The setup is shown in principle in Figure 20.33. The commutator is an automatic switching arrangement that maintains the currents in the conductors in the directions shown in the figure as the rotor turns. The coils F and F' shown in Figure 20.32 are called **field coils.** The steady current in these coils sets up a magnetic field in the motor frame and in the gap between the pole pieces P and P' and the rotor. (In some small motors, this magnetic field is supplied by permanent magnets instead of electromagnets.) Some of the magnetic field lines are shown in blue. With the directions of field and rotor currents as shown, the side thrust on each conductor in the rotor is such as to produce a *counterclockwise* torque on the rotor.

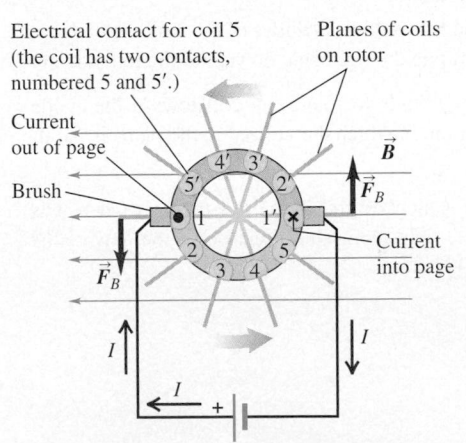

The brushes transmit current through contacts 1 and 1′ to coil 1, which is oriented to receive maximal torque from the magnetic field.

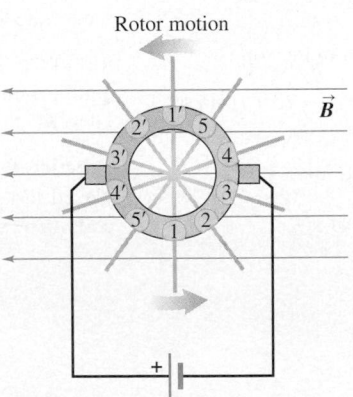

When the brushes are between contacts, inertia keeps the rotor turning.

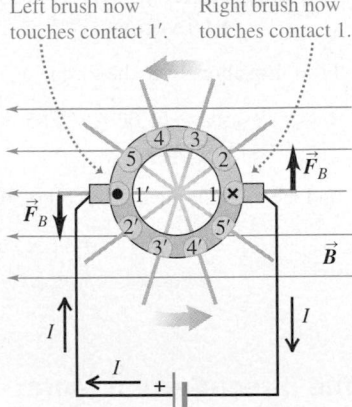

The brushes again contact coil 1, but with opposite polarity, so the torque on the coil is still counterclockwise.

▲ **FIGURE 20.33** By reversing the direction of the current in each coil once per half cycle, the commutator ensures that the torque on the coils always points in the same direction.

▶ Application **A clever solution.** As you know, a current traveling through a wire generates a magnetic field surrounding the wire. In a normal household extension cord, the two wires run side by side, and their fields add to produce a small net magnetic field. Because many types of sensitive electronic equipment cannot tolerate even these slight magnetic fields, the *coaxial* cable was developed. In a coaxial cable, one of the conductors has the form of a hollow tube, and the other runs through its center. (The cable is called "coaxial" because the conductors have the same axis.) As long as the currents in the two conductors are equal, the two magnetic fields cancel, so the cable produces no net magnetic field.

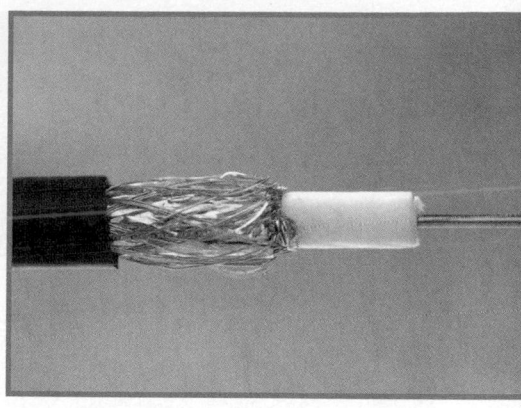

20.7 Magnetic Field of a Long, Straight Conductor

Thus far in this chapter, we've studied one aspect of the magnetic interaction of moving charges: the *forces* that a magnetic field exerts on moving charges and on currents in conductors. We didn't worry about what caused the magnetic field; we simply took its existence as a given fact. Now we're ready to return to the question of how magnetic fields are *produced* by moving charges and by currents. We begin by studying the magnetic fields produced by a few simple configurations, such as a long, straight wire and a circular loop. Then we look briefly at two more general methods for calculating the magnetic fields produced by more general configurations of currents in conductors.

The simplest current configuration to describe is a long, straight conductor carrying a steady current I. Experimentally, we find that the magnetic field produced by such a conductor has the general shape shown in Figure 20.34. The magnetic field lines are all *circles;* at each point, $\vec{B}$ is tangent to a circle centered on the conductor and lying in a plane perpendicular to it. Because of the axial symmetry, we know that $\vec{B}$ has the same *magnitude* at all points on a particular field line. The *magnitude* B at a distance r from the axis of the conductor is inversely proportional to r and is directly proportional to the current I. This I/r dependence can be derived from more general considerations, which we'll describe briefly in Section 20.10.

Magnetic field of a long, straight wire

The magnetic field $\vec{B}$ produced by a long, straight conductor carrying a current I, at a distance r from the axis of the conductor, has magnitude B given by

$$B = \frac{\mu_0 I}{2\pi r}. \tag{20.10}$$

In this equation, μ_0 is a constant called the *permeability of vacuum.* Its numerical value depends on the system of units we use. In SI units, the units of μ_0 are $(\text{T} \cdot \text{m}/\text{A})$. Its numerical value, which is related to the definition of the unit of current, is defined to be *exactly* $4\pi \times 10^{-7}$:

$$\mu_0 = 4\pi \times 10^{-7} \, \text{T} \cdot \text{m}/\text{A}.$$

A useful relationship for checks of unit consistency is $1 \, \text{T} \cdot \text{m}/\text{A} = 1 \, \text{N}/\text{A}^2$. We invite you to verify this equivalence.

NOTE ▶ The μ_0 just presented is a different quantity from the symbol μ for magnetic moment that we introduced in Section 20.6. This duplication in notation is unfortunately found everywhere in the literature. Beware! ◀

The shape of the magnetic field lines is completely different from that of the electric field lines in the analogous electrical situation. Electric field lines radiate outward from the charges that are their sources (or inward for negative charges).

MasteringPHYSICS

ActivPhysics 13.1: Magnetic Field of a Wire

Right-hand rule for the magnetic field around a current-carrying wire: Point the thumb of your right hand in the direction of the current. Your fingers now curl around the wire in the direction of the magnetic field lines.

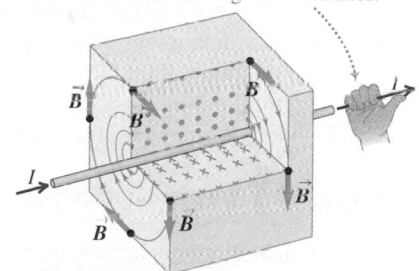

▲ **FIGURE 20.34** Right-hand rule for the direction of the magnetic field around a long, straight conductor carrying a current.

By contrast, magnetic field lines *encircle* the current that acts as their source. Electric field lines begin and end at charges, while experiments have shown that magnetic field lines *never* have endpoints, no matter what shape the conductor is. (If lines *did* begin or end at a point, this point would correspond to a "magnetic charge," or a single magnetic pole. As we mentioned in Section 20.1, there is no experimental evidence that such a pole exists.)

As Figure 20.34 shows, the direction of the $\vec{B}$ lines around a straight conductor is given by a new right-hand rule: Grasp the conductor with your right hand, with your thumb extended in the direction of the current. Your fingers then curl around the conductor in the direction of the $\vec{B}$ lines.

Conceptual Analysis 20.7

A wire and a compass

Suppose you lay a magnetic compass flat on a table and stretch a wire horizontally under it, parallel to and a short distance below the compass needle (Figure 20.35). The wire is connected to a battery; when you close the switch, the N end of the compass needle deflects to the left. In which orientation (A or B) is the battery connected to the circuit?

SOLUTION Recall that a compass needle or bar magnet in a magnetic field tends to align so that a line from its S pole to its N pole points in the same direction as the magnetic field lines. Since the N pole of the compass needle deflects to the left (and the S pole to the right), we know that the magnetic field above the wire points from right to left. Next, we use our new right-hand rule (Figure 20.34) to find the direction of the current in the wire.

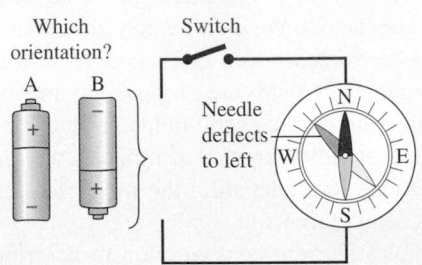

▲ **FIGURE 20.35**

We hold our right hand so that the fingers curl in the direction of the magnetic field around the wire and find that our thumb points toward the bottom of the page. Thus the current flows clockwise in the loop, and the battery is connected in orientation A.

EXAMPLE 20.6 ### Magnetic field from power lines

A long, straight dc power line carries a current of 100 A. A swarm of bees builds a hive next to it. It is hypothesized that bees use the earth's magnetic field as a reference direction when orienting their honeycombs. At what distance from the power line is the magnitude of the magnetic field from the current equal to the magnitude of Earth's magnetic field, about 5.0×10^{-5} T?

SOLUTION

SET UP AND SOLVE We need to find the distance r at which the magnitude of the field B_{power} from the current in the power line equals the magnitude of field B_{earth} from the earth. Setting $B_{power} = B_{earth}$ and using Equation 20.10 for B_{power}, we find that

$$\frac{\mu_0 I}{2\pi r} = B_{earth}.$$

We solve this equation for the distance r from the power line:

$$r = \frac{\mu_0 I}{2\pi B_{earth}} = \frac{(4\pi \times 10^{-7} \text{ T} \cdot \text{m/A})(100 \text{ A})}{2\pi(5.0 \times 10^{-5} \text{ T})} = 0.40 \text{ m}.$$

REFLECT The magnetic field of the power line at a distance of half a meter is comparable in magnitude to the earth's field. Depending on the orientation of the power line relative to the earth's field, the current could cause a significant disruption in the bees' perception of direction.

Practice Problem: At what distance from a power line carrying a current of 110 A would it create a magnetic field with the same magnitude as the earth's? *Answer: r = 0.44 m.*

20.8 Force between Parallel Conductors

Let's look next at the interaction force between two long, straight, parallel current-carrying conductors. This problem comes up in a variety of practical situations, and it also has fundamental significance in connection with the definition

The magnetic field of the lower wire exerts an attractive force on the upper wire. By the same token, the upper wire attracts the lower one.

If the wires had currents in *opposite* directions, they would *repel* each other.

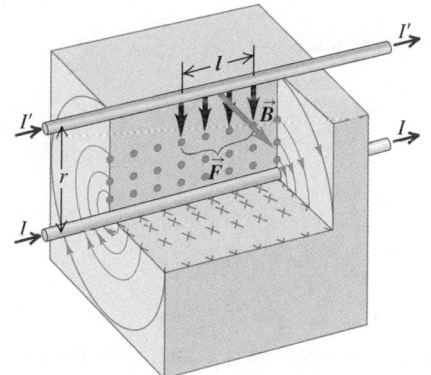

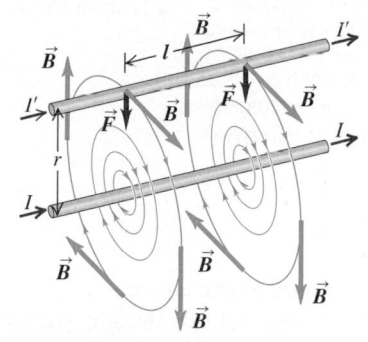

◄ **FIGURE 20.36** Attraction between parallel conductors carrying currents in the same direction.

of the ampere. Figure 20.36 shows segments of two long, straight, parallel conductors separated by a distance r and carrying currents I and I' in the same direction. Each conductor lies in the magnetic field set up by the other, so each is acted upon by a force. The diagram shows some of the field lines set up by the current in the *lower* conductor.

From Equation. 20.10, the magnitude of the $\vec{B}$ vector at points on the upper conductor is

$$B = \frac{\mu_0 I}{2\pi r}.$$

From Equation 20.6, the force on a length l of the upper conductor is

$$F = I'Bl = I'\left(\frac{\mu_0 I}{2\pi r}\right)l = \frac{\mu_0 l I I'}{2\pi r},$$

and the force *per unit length*, F/l, is

$$\frac{F}{l} = \frac{\mu_0 I I'}{2\pi r}. \tag{20.11}$$

The right-hand rule and Figure 20.36 show that the direction of the force on the upper conductor is *downward*. An equal (in magnitude) and opposite (in direction) upward force per unit length acts on the lower conductor; you can see that by looking at the field set up by the upper conductor. Therefore, the conductors *attract* each other. If the direction of either current is reversed, the forces reverse also. Parallel conductors carrying currents in *opposite* directions *repel* each other.

Mastering**PHYSICS**

ActivPhysics 13.5: Magnetic Force on a Wire

EXAMPLE 20.7 **A superconducting cable**

Two straight, parallel superconducting cables 4.5 mm apart (between centers) carry equal currents of 15,000 A in opposite directions. Find the magnitude and direction of the force per unit length exerted by one conductor on the other. Should we be concerned about the mechanical strength of these wires?

SOLUTION

SET UP Figure 20.37 shows our sketch. The two currents have opposite directions, so, from the preceding discussion, the forces are repulsive.

SOLVE To find the force per unit length, we use Equation 20.11:

$$\frac{F}{l} = \frac{\mu_0 I I'}{2\pi r} = \frac{(4\pi \times 10^{-7}\ \text{N/A}^2)(15{,}000\ \text{A})^2}{2\pi(4.5 \times 10^{-3}\ \text{m})}$$
$$= 1.0 \times 10^4\ \text{N/m}.$$

Continued

Therefore, the force exerted on a 1.0 m length of the conductor is 1.0×10^4 N.

REFLECT This is a large force, something over 1 ton per meter, so the mechanical strength of the conductors and insulating materials is certainly a significant consideration. Currents and separations of this magnitude are used in superconducting electromagnets in particle accelerators, and mechanical stress analysis is a crucial part of the design process.

Practice Problem: What is the maximum current in the conductors if the force per unit length is not to exceed 0.25×10^4 N/m? *Answer:* 7,500 A.

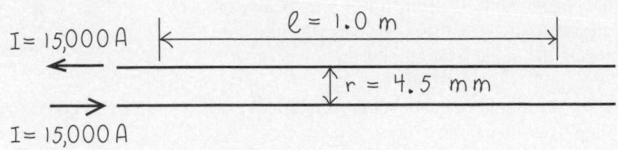

▲ **FIGURE 20.37** Our diagram for this problem.

Definition of the ampere

The forces that two straight, parallel conductors exert on one another form the basis for the official SI definition of the ampere, as follows:

One ampere is that unvarying current which, if present in each of two parallel conductors of infinite length and 1 meter apart in empty space, causes a force of exactly 2×10^{-7} newtons per meter of length on each conductor.

This definition is consistent with the definition of the constant μ_0 as *exactly* $4\pi \times 10^{-7}$ N/A^2, as we stated in Section 20.7.

20.9 Current Loops and Solenoids

In many practical devices, such as transformers and electromagnets, in which a current is used to establish a magnetic field, the wire carrying the current is wound into a *coil* consisting of many circular loops. So an expression for the magnetic field produced by a single circular conducting loop carrying a current is very useful. Figure 20.38 shows a circular conductor with radius R, carrying a current I. The current is led into and out of the loop through two long, straight wires side by side; the currents in these straight wires are in opposite directions, so their magnetic fields cancel each other.

Experimentally, we find that the magnetic field at the center of a circular loop has the direction shown in the figure and that its magnitude is directly proportional to the current I and inversely proportional to the radius R of the loop. Specifically, we have the following relationship:

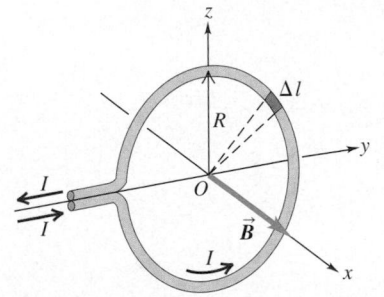

▲ **FIGURE 20.38** Magnetic field of a circular loop.

Magnetic field at center of circular loop

$$B = \frac{\mu_0 I}{2R} \qquad \text{(center of circular loop).} \qquad (20.12)$$

If we have a coil of N loops instead of a single loop, and if the loops are closely spaced and all have the same radius, then each loop contributes equally to the field, and the field at the center is just N times Equation 20.12:

$$B = \frac{\mu_0 N I}{2R} \qquad \text{(center of } N \text{ circular loops).} \qquad (20.13)$$

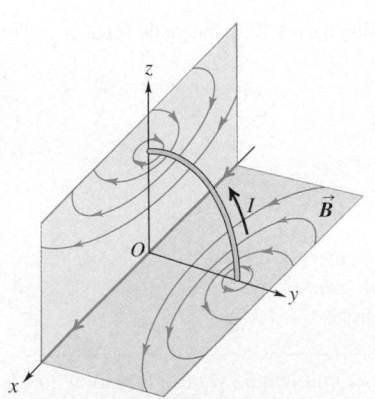

▲ **FIGURE 20.39** Magnetic field lines induced by the current in a circular loop. At points on the axis, the $\vec{B}$ field has the same direction as the magnetic moment of the loop.

Some of the magnetic field lines surrounding a circular loop and lying in planes through the axis are shown in Figure 20.39. The field lines encircle the conductor, and their directions are given by the same right-hand rule as that for a long, straight conductor. The magnetic field lines are *not* circles, but they are closed curves. At points along the axis of the loop, the $\vec{B}$ field is parallel to the axis; its magnitude is greatest at the center of the loop and decreases on both sides.

It's interesting to note the similarity of Equation 20.12 for a circular loop to Equation 20.10 for a long, straight conductor. The two expressions differ only by a factor of π: The field at the center of a circular loop with radius R is π times as great as the field at a distance R from a long, straight wire carrying the same current. Indeed, both equations can be derived from the same principles, which we'll mention briefly in Section 20.10.

EXAMPLE 20.8 A current loop for an electron beam experiment

A coil used to produce a magnetic field for an electron beam experiment has 200 turns and a radius of 12 cm. **(a)** What current is needed to produce a magnetic field with a magnitude of 5.0×10^{-3} T at the center of the coil? **(b)** Figure 20.40 shows an electron being deflected as it moves through the coil. What is the direction of current in the coil?

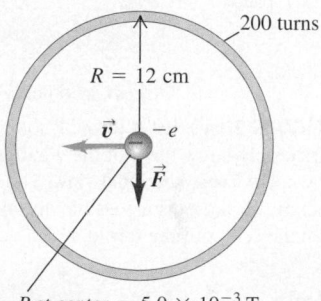

▶ **FIGURE 20.40** B at center $= 5.0 \times 10^{-3}$ T

SOLUTION

SET UP AND SOLVE To find the needed current, solve Equation 20.13 for I:

$$I = \frac{2RB}{\mu_0 N} = \frac{2(0.12\ \text{m})(5.0 \times 10^{-3}\ \text{T})}{(4\pi \times 10^{-7}\ \text{T} \cdot \text{m/A})(200)} = 4.8\ \text{A}.$$

From the right hand rule, with the velocity vector of the electron pointing to the left and the force vector toward the bottom of the page, the direction of the magnetic field must be out of the page. (Remember that we're dealing with a *negative* charge.) The direction of the current must be counterclockwise.

REFLECT The current required is directly proportional to the radius of the coil; the greater the distance from the center to the conductor, the greater the current required. And the current required varies inversely with the number of turns; the more turns, the smaller required current.

Practice Problem: A proton moving through the coil to the right is deflected toward the bottom of the page. What is the direction of the current in the coil? *Answer:* Counterclockwise.

Magnetic Field of a Solenoid

A **solenoid** is a helical winding of wire, usually wound around the surface of a cylindrical form. Ordinarily, the turns are so closely spaced that each one is very nearly a circular loop. There may be several layers of windings. The solenoid in Figure 20.41 is drawn with only a few turns so that the field lines can be shown. All turns carry the same current I, and the total $\vec{B}$ field at every point is the vector sum of the fields caused by the individual turns. The figure shows field lines in the x-y and x-z planes. The field is found to be most intense in the center, less intense near the ends.

If the length L of the solenoid is large in comparison with its cross-sectional radius R, the $\vec{B}$ field inside the solenoid near its center is very nearly uniform and parallel to the axis, and the field outside, adjacent to the center, is very small. The magnetic-field magnitude B at the center depends on the number of turns *per unit length* of the solenoid; we'll call that quantity n. If there are N turns distributed uniformly over a total length L, then $n = N/L$. The field magnitude at the center of the solenoid is given by

$$B = \mu_0 n I \qquad \text{(center of long solenoid).} \qquad (20.14)$$

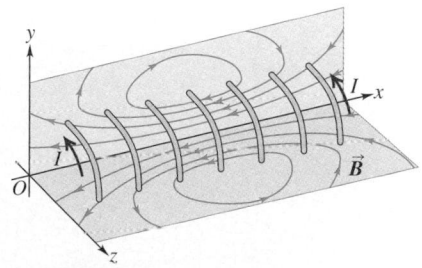

▲ **FIGURE 20.41** Magnetic field lines produced by the current in a solenoid. For clarity, only a few turns are shown.

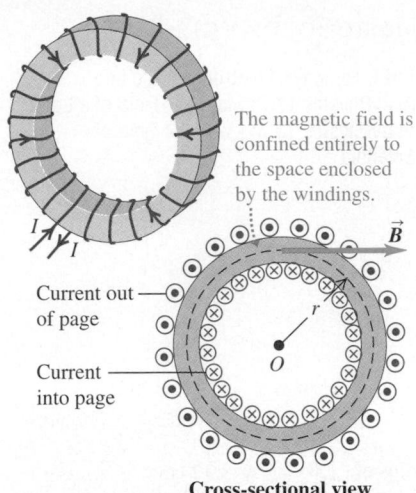

The magnetic field is confined entirely to the space enclosed by the windings.

Current out of page

Current into page

$\vec{B}$

r

O

Cross-sectional view

▲ **FIGURE 20.42** (a) A toroidal solenoid. For clarity, only a few turns of the winding are shown. (b) Cross-sectional view. The dashed black line represents a possible distance r from the center of the toroid.

A variation is the **toroidal** (doughnut-shaped) **solenoid,** more commonly called a *toroid,* shown in Figure 20.42. This shape has the interesting property that when there are many very closely spaced windings, the magnetic field is confined almost entirely to the space enclosed by the windings; there is almost no field at all outside this region. If there are N turns in all, then the field magnitude B at a distance r from the center of the toroid (*not* from the center of its cross section) is given by

$$B = \frac{\mu_0 NI}{2\pi r} \qquad \text{(toroidal solenoid).} \qquad (20.15)$$

The magnetic field is *not* uniform over a cross section of the core, but is inversely proportional to r. But if the radial thickness of the core is small in comparison with the overall radius of the toroid, the field is *nearly* uniform over a cross section. In that case, since $2\pi r$ is the circumference of the toroid and $N/2\pi r$ is the number of turns per unit length n, we can rewrite Equation 20.15 as

$$B = \mu_0 nI,$$

just as at the center of a long, *straight* solenoid.

Conceptual Analysis 20.8

Aligning a bar magnet

A bar magnet is suspended freely in line with the long axis of a solenoid, as shown in Figure 20.43. If the magnet aligns so that its N pole faces the solenoid, in which orientation (A or B) is the battery connected?

SOLUTION Because the magnet aligns with its N pole pointing left, the magnetic field along the axis of the solenoid also points left. (Remember that a magnet aligns so that a line from its S pole to its N pole points in the same direction as the local field.) To create this field, the current must enter the solenoid at its right end and exit at the left end. Thus, on the near side of each loop, the current is toward the top of the page. Therefore, the battery is connected in orientation A.

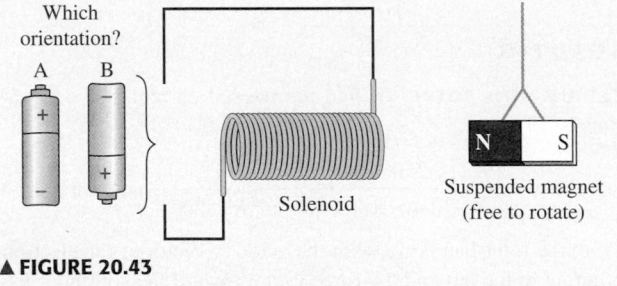

Which orientation?

A   B

Solenoid

Suspended magnet (free to rotate)

N   S

▲ **FIGURE 20.43**

20.10 Magnetic Field Calculations

In the previous three sections, we've stated without derivation several equations for the magnetic fields caused by currents in conductors. Deriving these relations can be quite complicated, so in this section we'll just sketch the principles on which the derivations are built. The basic relationships can be stated in two rather different forms. In the first, we begin with the $\vec{B}$ field produced by a current I in a short segment of conductor with length Δl. Then we add the contributions to $\vec{B}$ from all the segments of the conductor to find the *total* $\vec{B}$. This principle is called the *law of Biot and Savart.* In the second form, we consider the magnetic field at *all* points on a magnetic field line that encircles current-carrying conductors; this principle is called *Ampère's law.*

Law of Biot and Savart

The law of Biot and Savart gives the magnetic field $\Delta\vec{B}$ produced by a current I in a short segment of conductor with length Δl. The vector $\Delta\vec{l}$ is in the direction of the current. We call the location of the segment the **source point,** and the point P where we want to find the field is called the **field point.** The distance between the two points is r. The law is as follows:

Law of Biot and Savart

The magnitude ΔB of the magnetic field $\Delta \vec{B}$ due to a segment of conductor with length Δl, carrying a current I, is given by

$$\Delta B = \frac{\mu_0}{4\pi}\frac{I\,\Delta l\sin\theta}{r^2}. \tag{20.16}$$

This I/r^2 relationship reminds us of Coulomb's law. But the *direction* of $\Delta \vec{B}$ is *not* along the line from the source point to the field point. Instead, at each point it is perpendicular to the plane containing this line and the direction of the segment Δl, as shown in Figure 20.44. Furthermore, the field magnitude ΔB is proportional to the sine of the angle θ between these two directions.

Figure 20.44 shows the magnetic field $\vec{B}$ at several points in the vicinity of the segment Δl. At all points along the direction of Δl, the field is zero because, in Equation 20.16, $\theta = 0$ or $180°$ and $\sin\theta = 0$ at all such points. At any distance r from the segment, $\Delta \vec{B}$ has its greatest magnitude at points lying in the plane perpendicular to the segment; at all points in that plane, $\theta = 90°$ and $\sin\theta = 1$.

The magnetic field lines for the field $\Delta \vec{B}$ due to the current I in segment Δl are similar in character to those for a long, straight conductor; they are *circles* with centers along the line of Δl, lying in planes perpendicular to this line. We can use the same right-hand rule as for the long, straight conductor: Grasp the segment with your right hand so that your right thumb points in the direction of current flow; your fingers then curl around the segment in the same sense as the magnetic field lines.

To find the total magnetic field $\vec{B}$ at any point in space due to the current in a complete circuit, we represent the circuit as a large number of conducting segments Δl in series. We then use the *superposition principle*, which we've already encountered for *electric* fields, and find the *vector sum* of all the $\Delta \vec{B}$'s due to all the segments. This can become a sticky mathematical problem, but there are a few cases in which it is fairly simple.

For example, we can use the law of Biot and Savart to derive the expression for the magnetic field at the center of a circular conducting loop with radius R and current I (Equation 20.12). We represent the loop as a large number of segments with lengths Δl_1, Δl_2, and so on. A typical segment with length Δl is shown

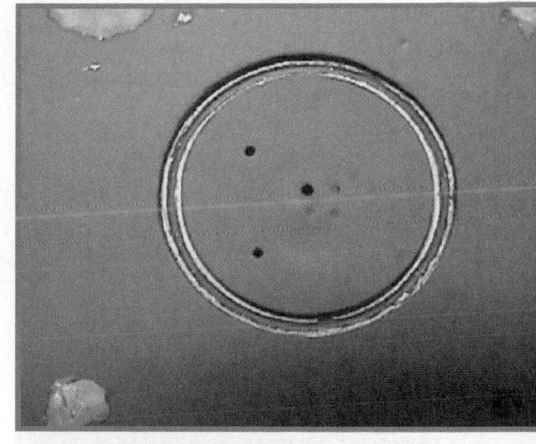

▲ **Application Micropump.** You may have heard of "lab on a chip" chemistry, in which complex procedures are performed on tiny samples as they move through microscale channels on a chip. For instance, the loop channel in this photo could be used for thermal cycling—you maintain different parts of the loop at different temperatures while pumping your sample around it. One challenge of microscale chemistry, however, is how to move fluids in tiny channels. *Magnetohydro-dynamics* provides an elegant answer. To create a magnetohydrodynamic (MHD) pump, you flank a channel with electrodes and magnets so that $\vec{E}$ and $\vec{B}$ fields cross the channel at right angles to each other. (In this photo, the gold rings are electrodes; the magnet poles are above and below the chip plane.) If the channel contains a conducting fluid, the effect of the two fields is to move the fluid *along* the conduit. There are no moving parts, and the pump is controlled by varying the voltage across the electrodes.

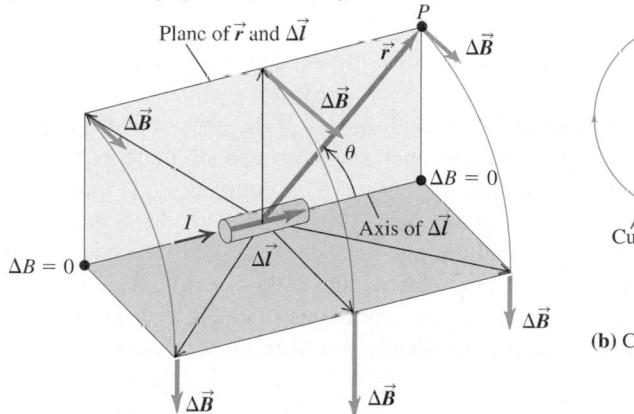

For these field points, $\vec{r}$ and $\Delta \vec{l}$ both lie in the tan-colored plane, and $\Delta \vec{B}$ is perpendicular to this plane.

Plane of $\vec{r}$ and $\Delta \vec{l}$

P

$\vec{r}$

$\Delta \vec{B}$

$\Delta \vec{B}$

θ

$\Delta B = 0$

I

Axis of $\Delta \vec{l}$

$\Delta \vec{B}$

$\Delta B = 0$

$\Delta \vec{l}$

$\Delta \vec{B}$

$\Delta \vec{B}$

$\Delta \vec{B}$

For these field points, $\vec{r}$ and $\Delta \vec{l}$ both lie in the orange-colored plane, and $\Delta \vec{B}$ is perpendicular to this plane.

(a) Perspective view

$\Delta \vec{B}$

Current into plane of paper

(b) Cross-sectional view

◀ **FIGURE 20.44** (a) Magnetic field vectors due to a short segment of current-carrying conductor, $\Delta \vec{l}$. At each point, $\Delta \vec{B}$ is perpendicular to the plane of $\vec{r}$ and $\Delta \vec{l}$ and its magnitude is proportional to the sine of the angle between them. (b) Magnetic field lines in a plane perpendicular to a short segment of current-carrying conductor with a current directed into the plane of the paper.

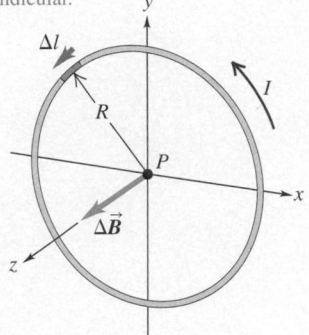

The segment Δl, the radial line R, and the magnetic field $\Delta \vec{B}$ are all mutually perpendicular.

▲ **FIGURE 20.45** Magnetic field $\Delta \vec{B}$ caused by a segment Δl of a circular conducting loop.

in Figure 20.45. All the segments are at the same distance R from the point P at the center, and each makes a right angle with the line joining it to P. The vectors $\Delta \vec{B}_1$, $\Delta \vec{B}_2$, and so on, due to the various segments, are all in the same direction, perpendicular to the plane of the loop, as shown. In Equation 20.16, $\theta = 90°$, $\sin \theta = 1$, and $r = R$, for every segment. The magnitude of the total $\vec{B}$ field is given by

$$B = \frac{\mu_0 I}{4 \pi R^2} (\Delta l_1 + \Delta l_2 + \cdots).$$

But $\Delta l_1 + \Delta l_2 + \cdots$ is the total distance around the loop—that is, the *circumference* of the loop, $2\pi R$—so

$$B = \frac{\mu_0 I}{4 \pi R^2} (2\pi R) = \frac{\mu_0 I}{2R},$$

in agreement with Equation 20.12.

This formulation is strictly valid only when the conductors are surrounded with vacuum. When air or any nonmagnetic material is present, the formulation is in error by only about 0.1% or less. In Section 20.11, we'll show how to modify the formulation to take account of the material around the conductors.

Ampère's Law

Ampère's law provides an alternative formulation of the relationship between a magnetic field and its sources. Ampère's law is analogous to Gauss's law, which offers an alternative to Coulomb's law for electric-field calculations. Like Gauss's law, Ampère's law is particularly useful when the problem at hand has symmetry properties that can be exploited.

To introduce the basic idea, let's consider the magnetic field caused by a long, straight conductor. We stated in Section 20.7, Equation 20.10, that the field at a distance r from a long, straight conductor carrying a current I has magnitude

$$B = \frac{\mu_0 I}{2 \pi r}$$

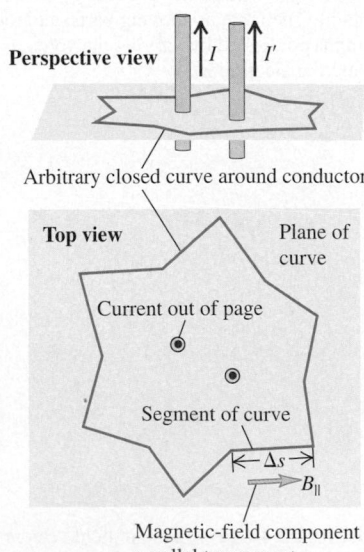

Ampère's law: If we take the products $B_{\parallel} \Delta s$ for all segments around the curve, their sum equals μ_0 times the total enclosed current:

$$\sum B_{\parallel} \Delta s = \mu_0 I_{\text{encl}}.$$

▲ **FIGURE 20.46** Ampère's law for an arbitrary closed curve of straight segments around a pair of conductors.

and that the magnetic field lines are circles centered on the conductor (Figure 20.34). The circumference of a circular field line with radius r is $2\pi r$. Now we note that, for any circular field line with radius r, centered on the conductor, the product of B and the circumference is

$$B(2 \pi r) = \frac{\mu_0 I}{2 \pi r} (2 \pi r) = \mu_0 I.$$

That is, this product is independent of r and depends only on the current I in the conductor.

This is a special case of Ampère's law. Here's the general statement: We construct an imaginary closed curve that encircles one or more conductors. We divide this curve into segments, calling a typical segment Δs (Figure 20.46). At each segment, we take the component of $\vec{B}$ parallel to the segment; we call this component $B_{\parallel}$. We take all the products $B_{\parallel} \Delta s$ and add them as we go completely around the closed curve. The result of this sum is always equal to μ_0 times the total current I_{encl} in all the conductors that are encircled by the curve. Expressing this relationship symbolically, we have Ampère's law:

Ampère's law
When a path is made up of a series of segments Δs, and when that path links conductors carrying total current I_{encl},

$$\sum B_{\parallel} \Delta s = \mu_0 I_{\text{encl}}. \tag{20.17}$$

Our initial example, the field of a long, straight conductor, used a circular path for which $B_\parallel$ was equal to B and was the same at each point of the path, so $\sum B_\parallel \Delta s$ was just equal to B multiplied by the circumference of the path. As with Gauss's law, the path that we use doesn't have to be the outline of any actual physical object; usually, it is a purely geometric curve that we construct to apply Ampère's law to a specific situation.

If several conductors pass through the surface bounded by the path, the total magnetic field at any point on the path is the vector sum of the fields produced by the individual conductors. Then we evaluate Equation 20.17, using the *total* $\vec{B}$ field at each point and the total current I_{encl} enclosed by the path. The result equals μ_0 times the *algebraic sum* of the currents. We need a sign rule for the currents; here it is: For the surface bounded by our Ampère's-law path, take a line perpendicular to the surface and wrap the fingers of your right hand around this line so that your fingers curl around in the same direction you plan to go around the path when you evaluate the $B_\parallel \Delta s$ sum. Then your thumb indicates the positive current direction. Currents that pass through the surface in that direction are positive; those in the opposite direction are negative.

20.11 Magnetic Materials

In all of the preceding discussion of magnetic fields caused by currents, we've assumed that the space surrounding the conductors contains only vacuum. If matter is present in the surrounding space, the magnetic field is changed. The atoms that make up all matter contain electrons in motion, and these electrons form microscopic current loops that produce magnetic fields of their own. In many materials, these currents are randomly oriented, causing no net magnetic field. But in some materials, the presence of an externally caused field can cause the loops to become oriented preferentially with the field so that their magnetic fields *add* to the external field. We then say that the material is *magnetized.*

Paramagnetism

A material showing the behavior we've just described is said to be **paramagnetic.** The magnetic field at any point in such a material is greater by a numerical factor K_m than it would be in vacuum. The value of K_m is different for different materials; it is called the **relative permeability** of a material. For a given material, K_m depends on temperature; values of K_m for common paramagnetic materials at room temperature are typically 1.000002 to 1.0004.

All the equations in this chapter that relate magnetic fields to their sources can be adapted to the situation in which the conductor is embedded in a paramagnetic material by replacing μ_0 everywhere with $K_m\mu_0$. This product is usually denoted as μ; it is called the **permeability** of the material:

$$\mu = K_m\mu_0. \tag{20.18}$$

NOTE ▶ Remember that in this context μ is magnetic permeability, *not* the magnetic moment we defined in Section 20.6. Be careful! ◀

Diamagnetism

In some materials, the total field due to the electrons in each atom sums to zero when there is no external field; such materials have *no* net atomic current loops. But even in these materials, magnetic effects are present, because an external field causes slight distortion of the electron current loops. The additional field caused by this distortion is always *opposite* in direction to that of the external field. Such materials are said to be **diamagnetic;** they always have relative permeabilities very slightly less than unity, typically on the order of 0.9998 to 0.99999. Thus, the **susceptibility,** K_m-1, of a diamagnetic material is very small and negative.

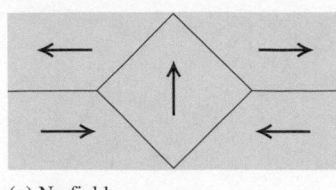

(a) No field

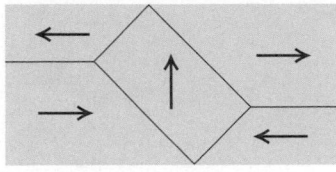

(b) Weak field $\vec{B}$

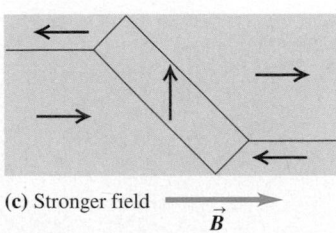

(c) Stronger field $\vec{B}$

▲ **FIGURE 20.47** In this drawing, adapted from a magnified photo, the arrows show the directions of magnetization in the domains of a single crystal of nickel. Domains magnetized in the direction of an applied magnetic field grow larger.

Ferromagnetism

In a third class of materials, called **ferromagnetic materials,** strong interactions between microscopic current loops cause the loops to line up parallel to each other in regions called **magnetic domains,** even when no external field is present. Figure 20.47 shows an example of magnetic domain structure. Within each domain, nearly all of the atomic current loops are parallel. When there is no externally applied field, the orientations of the domain magnetizations are random, and the net magnetization is zero. But when an externally applied field is present, it exerts torques on the atomic current loops, and they tend to orient themselves parallel to the field, like microscopic compass needles. The domain boundaries also shift; the domains magnetized in the direction of the field grow, and those magnetized in other directions shrink. This "cooperative" phenomenon leads to a relative permeability that is *much larger* than unity, typically on the order of 1000 to 10,000. Iron, cobalt, nickel, and many compounds and alloys containing these elements are ferromagnetic.

In ferromagnetic materials, as the external field increases, a point is reached at which nearly all the microscopic current loops have their axes parallel to that field. This condition is called *saturation magnetization;* after it is reached, a further increase in the external field causes no increase in magnetization. Some ferromagnetic materials retain their magnetization even after the external magnetic field is removed. These materials can thus become *permanent magnets.* Many kinds of steel and many alloys, such as alnico, are commonly used for permanent magnets. When such a material is magnetized to near saturation, the magnetic field in the material is typically on the order of 1 T. Magnetic tapes, strips, and computer disks use this same retention of magnetization.

Ferromagnetic materials are widely used in electromagnets, transformer cores, and motors and generators, where it is usually desirable to have as large a magnetic field as possible for a given current. In these applications, it is usually desirable for the material *not* to have permanent magnetization. Soft iron is often used because it has high permeability without appreciable permanent magnetization.

SUMMARY

Magnetism; Fields and Forces

(Sections 20.1 and 20.2) A bar magnet has a **north (N) pole** and a **south (S) pole.** Two opposite poles attract each other, and two like poles repel each other. A moving charge creates a **magnetic field** in the surrounding space. A moving charge, or current, experiences a force in the presence of a magnetic field. The direction of the force is given by the right-hand rule for magnetic forces, and the magnitude is given by Equation 20.1.

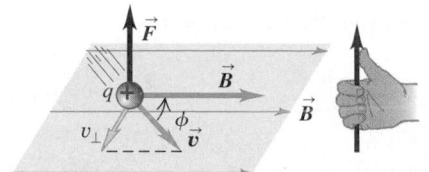

Motion of Charged Particles in Magnetic Fields

(Sections 20.3 and 20.4) The magnetic force is always perpendicular to $\vec{v}$; a particle moving under the action of a magnetic field alone moves with constant speed. In a uniform field, a particle with initial velocity perpendicular to the field moves in a circle with radius R given by $R = mv/|q|B$ (Equation 20.4). Mass spectrometers use this relationship to determine atomic masses. When a positive ion of known speed undergoes circular motion in a magnetic field of known strength, the mass can be determined by measuring the radius.

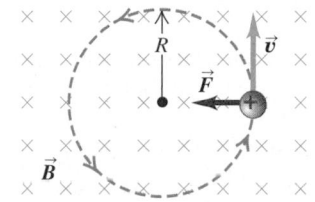

The orbit of a positive charge in a uniform magnetic field.

Magnetic Force on a Current-Carrying Conductor

(Section 20.5) When a current-carrying conductor is in the presence of a magnetic field, the field exerts a force on the conductor because each individual charge in the current is acted upon by a force given by $F = |q|vB\sin\phi$ (Equation 20.1). The direction of the force is determined by using the same right-hand rule that we used for a moving positive charge. The magnitude of the force is given by $F = IlB_{\perp} = IlB\sin\phi$ (Equation 20.7).

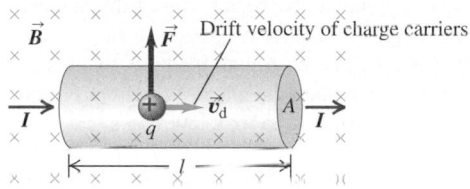

Drift velocity of charge carriers

Force and Torque on a Current Loop; Direct-Current Motors

(Section 20.6) A current loop can be represented as connected segments of current-carrying conductors. In the presence of a magnetic field, each segment is acted upon by a force due to the field. In a uniform magnetic field, the total force on a current loop is zero, regardless of its shape, but the magnetic forces create a torque τ given by $\tau = IAB\sin\phi$ (Equation 20.8). A direct-current motor is driven by torques on current-carrying conductors. The key component is the **commutator,** which is used to reverse the direction of the current in order to maintain the torque.

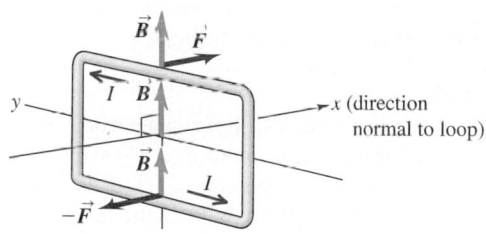

Magnetic Field of a Long, Straight Conductor

(Sections 20.7 and 20.8) Moving charges, and therefore currents, produce magnetic fields. When a current passes through a long, straight wire, the magnetic field lines are circles centered on the wire. Due to this symmetry of the field pattern, the magnitude of the magnetic field is the same at all points on a field line at radial distance r:

$$B = \frac{\mu_0 I}{2\pi r}. \qquad (20.10)$$

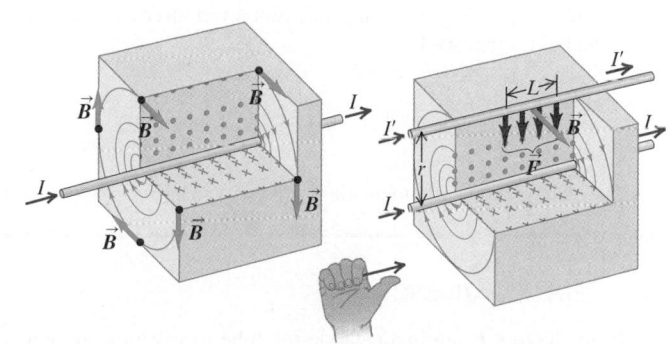

Current-carrying conductors can exert magnetic forces on each other. Thus, two parallel wires can attract or repel each other, depending on the direction of the currents they carry. For two long, straight parallel wires, the force per unit length is

$$\frac{F}{l} = \frac{\mu_0 II'}{2\pi r}. \qquad (20.11)$$

Continued

Current Loops and Solenoids

(Section 20.9) Many practical devices depend on the magnetic field produced at the center of a circular coil of wire. If a coil of radius R consists of N loops, the magnetic field at the center is

$$B = \frac{\mu_0 NI}{2R}. \quad (20.13)$$

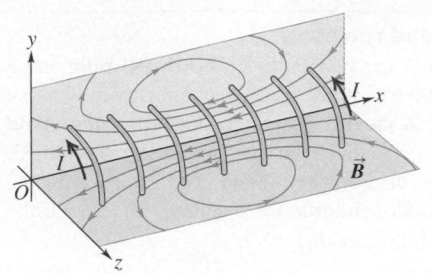

A long **solenoid** of many closely spaced windings produces a nearly uniform field in its interior, midway between its ends, having magnitude $B = \mu_0 nI$ (Equation 20.14).

Magnetic Field Calculations

(Section 20.10) The magnetic field magnitude ΔB created by a short current-carrying segment Δl is given by the law of Biot and Savart:

$$\Delta B = \frac{\mu_0}{4\pi} \frac{I \, \Delta l \sin\theta}{r^2}. \quad (20.16)$$

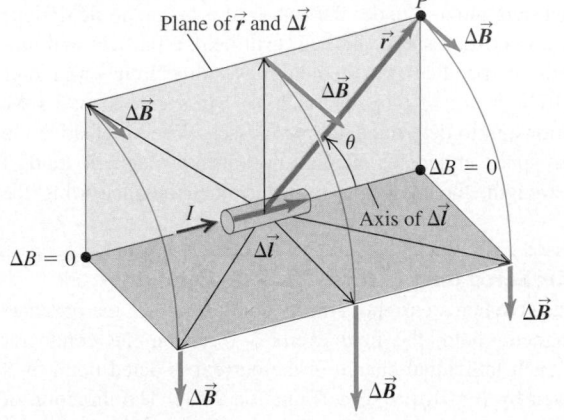

To determine the magnetic field from an extended current-carrying wire, first consider the extended wire as being made of many smaller segments. Then, using the principle of superposition at a particular point in space, compute the vector sum of the magnetic field due to each small segment of current.

When a current I is enclosed by a path made of many small segments Δs, and when at each point the magnetic field has a component $B_\parallel$ parallel to the segments, **Ampère's law** states that

$$\sum B_\parallel \, \Delta s = \mu_0 I_{\text{encl}}. \quad (20.17)$$

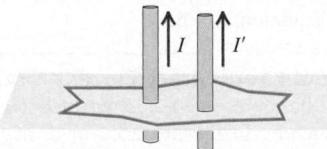

If the products of the magnetic field component $B_\parallel$ and each segment Δs (that is, the components $B_\parallel \, \Delta s$) are summed around *any* path enclosing a total current I_{encl}, the result is $\mu_0 I_{\text{encl}}$.

Magnetic Materials

(Section 20.11) For magnetic materials, the magnetization of the material causes an additional contribution to $\vec{B}$. For paramagnetic and diamagnetic materials, μ_0 is replaced in magnetic-field expressions by $\mu = K_m\mu_0$, where μ is the permeability of the material and K_m is its relative permeability. For **ferromagnetic** materials, K_m is much larger than unity. Some ferromagnetic materials are permanent magnets, retaining their magnetization even after the external magnetic field is removed.

 For instructor-assigned homework, go to www.masteringphysics.com

---

Conceptual Questions

1. If an electron beam in a cathode-ray tube travels in a straight line, can you be sure that no magnetic field is present?

2. Why is it *not* a good idea to call magnetic-field lines "magnetic lines of force"?

3. If the magnetic force does no work on a charged particle, how can it have any effect on the particle's motion? Can you think of other situations in which a force does no work, but has a significant effect on an object's motion?

4. A permanent magnet can be used to pick up a string of nails, tacks, or paper clips, even though these objects are not themselves magnets. How can this be?

5. Streams of charged particles emitted from the sun during unusual sunspot activity create a disturbance in the earth's magnetic field (called a *magnetic storm*). How can they cause such a disturbance?

6. A student once proposed to obtain an isolated magnetic pole by taking a bar magnet (N pole at one end, S at the other) and breaking it in half. Would this work? Why?

7. The magnetic force on a moving charged particle is always perpendicular to the magnetic field $\vec{B}$. Is the trajectory of a moving particle always perpendicular to the magnetic field lines? Explain your reasoning.

8. The text discusses the magnetic field of an infinitely long, straight conductor carrying a current. Of course, there is no such thing as an infinitely long *anything*. How would you decide whether a particular wire is long enough to be considered infinite?

9. Two parallel conductors carrying current in the same direction attract each other. If they are permitted to move toward each other, the forces of attraction do work. Where does the energy come from? Does this phenomenon contradict the assertion in this chapter that magnetic forces do no work on moving charges?

10. Household wires (such as lamp cords) often carry currents of several amps. Why do no magnetic effects show up near such wires?

11. You have a large bar magnet and want to identify its north and south poles. It is too heavy to show any effects from the Earth's weak magnetic field. You have an ordinary compass available, but alas, its poles are not marked. In spite of its lack of pole markings, how could you use this compass to determine the polarity of your bar magnet? (*Hint:* Can you first determine the north and south poles of the compass?)

12. When you bring a very strong permanent magnet close to an ordinary bar magnet, you often find that the north (or south) pole of the permanent magnet attracts *both* poles of the ordinary magnet. Why does this happen?

13. Students sometimes suggest that the reversals of the earth's magnetic field were caused when our planet's spin reversed its direction. Is this a plausible explanation? What are some good criticisms of it? (For example, would the energy required to cause this reversal be likely to leave some trace in the geological record?)

14. Can a charged particle move through a magnetic field that exerts no force on it? How? Could it move through an electric field that exerts no force on it?

15. If the magnitude of the magnetic field a distance R from a very long, straight, current-carrying wire is B, at what distance from the wire will the field have magnitude $3B$?

Multiple-Choice Problems

1. An electron traveling at a high speed enters a uniform magnetic field directed perpendicular to its path. Which of the following quantities will change while the electron travels through the field? (There may be more than one correct answer.)
   A. Its speed.
   B. Its velocity.
   C. Its acceleration.
   D. Its kinetic energy.
   E. Its potential energy.

2. A negatively charged particle shoots into a uniform magnetic field directed out of the paper, as shown in Figure 20.48. A possible path of this particle is
   A. 1.   B. 2.
   C. 3.   D. 4.

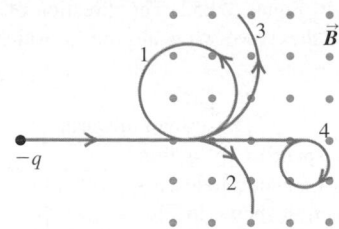

▲ **FIGURE 20.48** Multiple-choice problem 2.

3. A beam of protons is directed horizontally into the region between two bar magnets, as shown in Figure 20.49. The magnetic field in this region is horizontal. What is the effect of the magnetic field on the protons? (*Hint:* First refer to Figure 20.6 to find the direction of the magnetic field.)

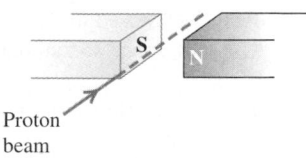

▲ **FIGURE 20.49** Multiple-choice problem 3.

   A. The protons are accelerated to the left, toward the S magnetic pole.
   B. The protons are accelerated to the right, toward the N magnetic pole.
   C. The protons are accelerated upward.
   D. The protons are accelerated downward.
   E. The protons are not accelerated, since the magnetic field does not change their speed.

4. A wire carrying a current in the direction shown in Fig. 20.50 passes between the poles of two bar magnets. What is the direction of the magnetic force on this wire due to the magnet? (*Hint:* Recall that magnetic-field lines point out of a north magnetic pole and into a south magnetic pole.)
   A. out of the paper.
   B. into the paper.
   C. toward the N pole of the magnet.
   D. toward the S pole of the magnet.

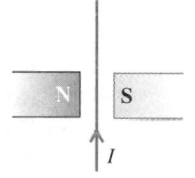

▲ **FIGURE 20.50** Multiple-choice problem 4.

5. A solenoid is connected to a battery as shown in Fig. 20.51, and a bar magnet is placed nearby. What is the direction of the magnetic force that this solenoid exerts on the bar magnet? (*Hint:* Think of the solenoid as a bar magnet, and identify what would be its north and south poles.)
   A. upward.
   B. downward.
   C. to the right, away from the solenoid.
   D. to the left, toward the solenoid

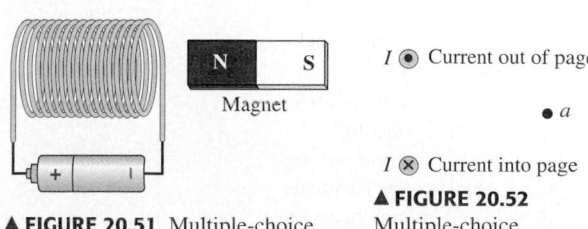

▲ **FIGURE 20.51** Multiple-choice problem 5.

▲ **FIGURE 20.52** Multiple-choice problem 6.

6. Two very long, straight parallel wires carry currents of equal magnitude, but opposite direction, perpendicular to the paper

in the directions shown in Figure 20.52. The direction of the net magnetic field due to these two wires at point a, which is equidistant from both wires, is

A. directly to the right.   B. directly to the left.
C. straight downward.   D. straight upward.

7. A light circular wire suspended by a thin silk thread in a uniform magnetic field carries a current in the direction shown in Figure 20.53. The magnetic field is perpendicular to the plane of the paper, and the wire is held at rest in that plane. If the wire is suddenly released so that it is free to rotate,
   A. It will rotate so that point a goes into the paper.
   B. It will rotate so that point a goes out of the paper.
   C. It will not rotate.

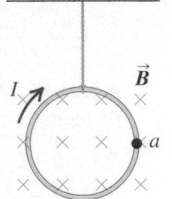

▲ FIGURE 20.53 Multiple-choice problem 7.

8. An electron is moving directly toward you in a horizontal path when it suddenly enters a uniform magnetic field that is either vertical or horizontal. If the electron begins to curve upward in its motion just after it enters the field, you can conclude that the direction of the magnetic field is
   A. upward.   B. downward.
   C. to your left.   D. to your right.

9. The two coils shown in Figure 20.54 are parallel to each other and are connected to batteries. Coil A is held in place, but coil C is free to move. After the switch S is closed, coil C will initially move
   A. toward coil A.
   B. away from coil A.
   C. upward.
   D. downward.

10. A loose, floppy coil of wire is carrying current I. The loop of wire is placed on a horizontal table in a uniform magnetic field $\vec{B}$ perpendicular to the plane of the table. This causes the loop to expand into a circular shape while still lying on the table. What orientation of the current and magnetic field could cause this to happen? (There may be more than one correct answer.)
   A. Current clockwise, $\vec{B}$ upward.
   B. Current clockwise, $\vec{B}$ downward.
   C. Current counterclockwise, $\vec{B}$ upward.
   D. Current counterclockwise, $\vec{B}$ downward.

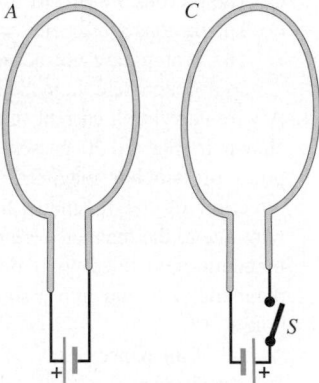

▲ FIGURE 20.54 Multiple-choice problem 9.

11. A metal bar connected by metal leads to the terminals of a battery hangs between the poles of a horseshoe magnet, as shown in Figure 20.55. Just after the switch S is closed, what will happen

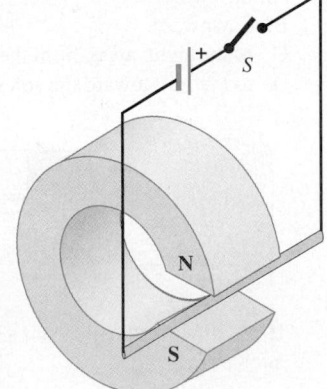

▲ FIGURE 20.55 Multiple-choice problem 11.

to the bar? (*Hint:* Recall that magnetic-field lines point out of a north magnetic pole and into a south magnetic pole.)
   A. It will be pushed upward, decreasing the tension in the leads.
   B. It will be pushed downward, increasing the tension in the leads.
   C. It will swing outward, away from the magnet.
   D. It will swing inward, into the magnet.

12. A certain current produces a magnetic field B near the center of a solenoid. If the current is doubled, the field near the center will be
   A. $4B$.   B. $2B$.   C. $B\sqrt{2}$.   D. B.

13. A coil is connected to a battery as shown in Figure 20.56. A bar magnet is suspended with its N pole just above the center of the coil. What will happen to the bar magnet just after the switch S is closed? (*Hint:* Think of the coil as a bar magnet, and identify what would be its north and south poles.)
   A. It will be pulled toward the coil.
   B. It will be pushed away from the coil.
   C. It will be pushed out of the paper.
   D. It will be pushed into the paper.

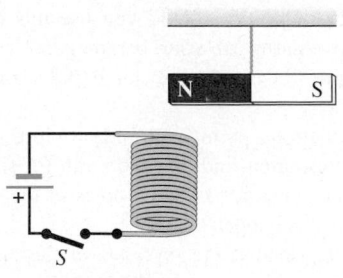

▲ FIGURE 20.56 Multiple-choice problem 13.

14. The force exerted by a constant uniform magnetic field on a current-carrying wire of length L produces a force per unit length of F_L on the wire. If a wire twice as long and carrying the same current is placed in the same field with the same orientation, the force per unit length would be
   A. $2F_L$.   B. F_L.   C. $F_L/2$.

15. A particle enters a uniform magnetic field initially traveling perpendicular to the field lines and is bent in a circular arc of radius R. If this particle were traveling twice as fast, the radius of its circular arc would be
   A. $2R$.   B. $R\sqrt{2}$.   C. $R/\sqrt{2}$.   D. $R/2$.

Problems

20.2 Magnetic Field and Magnetic Force

1. • In a 1.25 T magnetic field directed vertically upward, a particle having a charge of magnitude 8.50 μC and initially moving northward at 4.75 km/s is deflected toward the east. (a) What is the sign of the charge of this particle? Make a sketch to illustrate how you found your answer. (b) Find the magnetic force on the particle.

2. • An ion having charge $+6e$ is traveling horizontally to the left at 8.50 km/s when it enters a magnetic field that is perpendicular to its velocity and deflects it downward with an initial magnetic force of 6.94×10^{-15} N. What are the direction and magnitude of this field? Illustrate your method of solving this problem with a diagram.

3. • A proton traveling at 3.60 km/s suddenly enters a uniform magnetic field of 0.750 T, traveling at an angle of 55.0° with the field lines (Figure 20.57). (a) Find the magnitude and direction of the force this magnetic field exerts on the proton. (b) If you can vary the direction of the proton's velocity, find the magnitude of the maximum and minimum forces you could achieve, and show how the velocity should be oriented to achieve these forces. (c) What would the answers to part (a) be if the proton were replaced by an electron traveling in the same way as the proton?

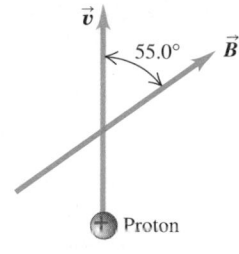

▲ **FIGURE 20.57** Problem 3.

4. • A particle having a mass of 0.195 g carries a charge of -2.50×10^{-8} C. The particle is given an initial horizontal northward velocity of 4.00×10^4 m/s. What are the magnitude and direction of the minimum magnetic field that will balance the earth's gravitational pull on the particle?

5. • At a given instant, a particle with a mass of 5.00×10^{-3} kg and a charge of 3.50×10^{-8} C has a velocity with a magnitude of 2.00×10^5 m/s in the $+y$ direction. It is moving in a uniform magnetic field that has magnitude 0.8 T and is in the $-x$ direction. What are (a) the magnitude and direction of the magnetic force on the particle and (b) its resulting acceleration?

6. •• If the magnitude of the magnetic force on a proton is F when it is moving at 15.0° with respect to the field, what is the magnitude of the force (in terms of F) when this charge is moving at 30.0° with respect to the field?

7. •• A ^{9}Be nucleus containing four protons and five neutrons has a mass of 1.50×10^{-26} kg and is traveling vertically upward at 1.35 km/s. If this particle suddenly enters a horizontal magnetic field of 1.12 T pointing from west to east, find the magnitude and direction of its acceleration vector the instant after it enters the field.

8. • A particle with a charge of -2.50×10^{-8} C is moving with an instantaneous velocity of magnitude 40.0 km/s in the xy-plane at an angle of 50° counterclockwise from the $+x$ axis. What are the magnitude and direction of the force exerted on this particle by a magnetic field with magnitude 2.00 T in the (a) $-x$ direction, and (b) $+z$ direction?

9. •• A particle with mass 1.81×10^{-3} kg and charge of $+1.22 \times 10^{-8}$ C has, at a given instant, a velocity of 3.00×10^4 m/s along the $+y$-axis, as shown in Figure 20.58. What are the magnitude and direction of the particle's acceleration produced by a magnetic field of magnitude 1.25 T in the xy-plane, directed at an angle of 45.0° counterclockwise from the $+x$-axis?

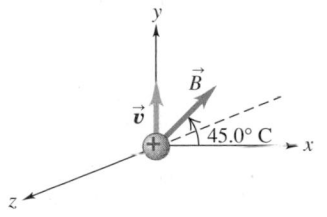

▲ **FIGURE 20.58** Problem 9.

10. •• A 150 V battery is connected across two parallel metal plates of area 28.5 cm² and separation 8.20 mm. A beam of alpha particles (charge $+2e$, mass 6.64×10^{-27} kg) is accelerated from rest through a potential difference of 1.75 kV and enters the region between the plates perpendicular to the electric field. What magnitude and direction of magnetic field are needed so that the alpha particles emerge undeflected from between the plates?

11. •• A velocity selector having uniform perpendicular electric and magnetic fields is shown in Figure 20.59. The electric field is provided by a 150 V DC battery connected across two large parallel metal plates that are 4.50 cm apart. (a) What must be the magnitude of the magnetic field so that charges having a velocity of 3.25 km/s perpendicular to the fields will pass through undeflected? (b) Show how the magnetic field should point in the region between the plates.

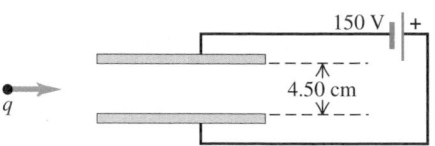

▲ **FIGURE 20.59** Problem 11.

12. •• An electron moves at 2.50×10^6 m/s through a region in which there is a magnetic field of unspecified direction and magnitude 7.40×10^{-2} T. (a) What are the largest and smallest possible magnitudes of the acceleration of the electron due to the magnetic field? (b) If the actual acceleration of the electron is one-fourth of the largest magnitude in part (a), what is the angle between the electron velocity and the magnetic field?

20.3 Motion of Charged Particles in a Magnetic Field

13. • In a cloud chamber experiment, a proton enters a uniform 0.250 T magnetic field directed perpendicular to its motion. You measure the proton's path on a photograph and find that it follows a circular arc of radius 6.13 cm. How fast was the proton moving?

14. • An alpha particle (a He nucleus, containing two protons and two neutrons and having a mass of 6.64×10^{-27} kg) traveling horizontally at 35.6 km/s enters a uniform, vertical, 1.10 T magnetic field. (a) What is the diameter of the path followed by this alpha particle? (b) What effect does the magnetic field have on the speed of the particle? (c) What are the magnitude and direction of the acceleration of the alpha particle while it is in the magnetic field? (d) Explain why the speed of the particle does not change even though an unbalanced external force acts on it.

15. • A deuteron particle (the nucleus of an isotope of hydrogen consisting of one proton and one neutron and having a mass of 3.34×10^{-27} kg) moving horizontally enters a uniform, vertical, 0.500 T magnetic field and follows a circular arc of radius 55.6 cm. (a) How fast was this deuteron moving just before it entered the magnetic field and just after it came out of the field? (b) What would be the radius of the arc followed by a proton that entered the field with the same velocity as the deuteron?

16. •• A beam of protons traveling at 1.20 km/s enters a uniform magnetic field, traveling perpendicular to the field. The beam exits the magnetic field in a direction perpendicular to its original direction (Fig. 20.60). The beam travels a distance of 1.18 cm *while in the field*. What is the magnitude of the magnetic field?

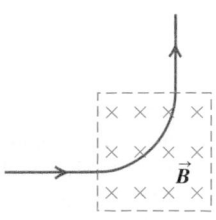

▲**FIGURE 20.60** Problem 16.

17. •• A uniform magnetic field bends an electron in a circular arc of radius R. What will be the radius of the arc (in terms of R) if the field is tripled?

18. •• An electron at point A in Figure 20.61 has a speed v_0 of 1.41×10^6 m/s. Find (a) the magnitude and direction of the magnetic field that will cause the electron to follow the semicircular path from A to B and (b) the time required for the electron to move from A to B. (c) What magnetic field would be needed if the particle were a proton instead of an electron?

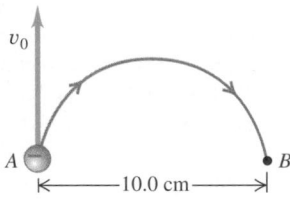

▲ **FIGURE 20.61** Problem 18.

19. •• A beam of protons is accelerated through a potential difference of 0.745 kV and then enters a uniform magnetic field traveling perpendicular to the field. (a) What magnitude of field is needed to bend these protons in a circular arc of diameter 1.75 m? (b) What magnetic field would be needed to produce a path with the same diameter if the particles were electrons having the same speed as the protons?

20. •• A 3.25 g bullet picks up an electric charge of 1.65 μC as it travels down the barrel of a rifle. It leaves the barrel at a speed of 425 m/s, traveling perpendicular to the earth's magnetic field, which has a magnitude of 5.50×10^{-4} T. Calculate (a) the magnitude of the magnetic force on the bullet and (b) the magnitude of the bullet's acceleration due to the magnetic force at the instant it leaves the rifle barrel.

21. •• An electron in the beam of a TV picture tube is accelerated through a potential difference of 2.00 kV. It then passes into a magnetic field perpendicular to its path, where it moves in a circular arc of diameter 0.360 m. What is the magnitude of this field?

20.4 Mass Spectrometers

22. •• (a) What is the speed of a beam of electrons when the simultaneous influence of an electric field of 1.56×10^4 V/m and a magnetic field of 4.62×10^{-3} T, with both fields normal to the beam and to each other, produces no deflection of the electrons? (b) In a diagram, show the relative orientation of the vectors $\vec{v}$, $\vec{E}$ and $\vec{B}$. (c) When the electric field is removed, what is the radius of the electron orbit? What is the period of the orbit?

23. • Singly ionized (one electron removed) atoms are accelerated and then passed through a velocity selector consisting of perpendicular electric and magnetic fields. The electric field is 155 V/m and the magnetic field is 0.0315 T. The ions next enter a uniform magnetic field of magnitude 0.0175 T that is oriented perpendicular to their velocity. (a) How fast are the ions moving when they emerge from the velocity selector? (b) If the radius of the path of the ions in the second magnetic field is 17.5 cm, what is their mass?

24. •• **Determining diet.** One method for determining the amount
BIO of corn in early Native American diets is the *stable isotope ratio analysis* (SIRA) technique. As corn photosynthesizes, it concentrates the isotope carbon-13, whereas most other plants concentrate carbon-12. Overreliance on corn consumption can then be correlated with certain diseases, because corn lacks the essential amino acid lysine. Archaeologists use a mass spectrometer to separate the ^{12}C and ^{13}C isotopes in samples of human

remains. Suppose you use a velocity selector to obtain singly ionized (missing one electron) atoms of speed 8.50 km/s and want to bend them within a uniform magnetic field in a semicircle of diameter 25.0 cm for the ^{12}C. The measured masses of these isotopes are 1.99×10^{-26} kg (^{12}C) and 2.16×10^{-26} kg (^{13}C). (a) What strength of magnetic field is required? (b) What is the diameter of the ^{13}C semicircle? (c) What is the separation of the ^{12}C and ^{13}C ions at the detector at the end of the semicircle? Is this distance large enough to be easily observed?

25. •• **Ancient meat eating.** The amount of meat in prehistoric
BIO diets can be determined by measuring the ratio of the isotopes nitrogen-15 to nitrogen-14 in bone from human remains. Carnivores concentrate ^{15}N, so this ratio tells archaeologists how much meat was consumed by ancient people. Use the spectrometer of the previous problem to find the separation of the ^{14}N and ^{15}N isotopes at the detector. The measured masses of these isotopes are 2.32×10^{-26} kg (^{14}N) and 2.49×10^{-26} kg (^{15}N).

20.5 Magnetic Force on a Current-Carrying Conductor

26. • A straight vertical wire carries a current of 1.20 A downward in a region between the poles of a large electromagnet where the field strength is 0.588 T and is horizontal. What are the magnitude and direction of the magnetic force on a 1.00 cm section of this wire if the magnetic-field direction is (a) toward the east, (b) toward the south, (c) 30.0° south of west?

27. • **Magnetic force on a lightning bolt.** Currents during lightning strikes can be up to 50,000 A (or more!). We can model such a strike as a 50,000 A vertical current perpendicular to the earth's magnetic field, which is about $\frac{1}{2}$ gauss. What is the force on each meter of this current due to the earth's magnetic field?

28. • A horizontal rod 0.200 m long carries a current through a uniform horizontal magnetic field of magnitude 0.067 T that points perpendicular to the rod. If the magnetic force on this rod is measured to be 0.13 N, what is the current flowing through the rod?

29. • A straight 2.5 m wire carries a typical household current of 1.5 A (in one direction) at a location where the earth's magnetic field is 0.55 gauss from south to north. Find the magnitude and direction of the force that our planet's magnetic field exerts on this wire if is oriented so that the current in it is running (a) from west to east, (b) vertically upward, (c) from north to south. (d) Is the magnetic force ever large enough to cause significant effects under normal household conditions?

30. •• Between the poles of a powerful magnet is a cylindrical uniform magnetic field with a diameter of 3.50 cm and a strength of 1.40 T. A wire carries a current through the center of the field at an angle of 65.0° to the magnetic field lines. If the wire experiences a magnetic force of 0.0514 N, what is the current flowing in it?

31. • A rectangular 10.0 cm by 20.0 cm circuit carrying an 8.00 A current is oriented with its plane parallel to a uniform 0.750 T magnetic field (Figure 20.62). (a) Find the magnitude and direction of the magnetic force on each segment

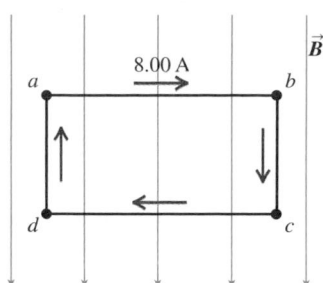

▲ FIGURE 20.62 Problem 31.

(*ab*, *bc*, etc.) of this circuit. Illustrate your answers with clear diagrams. (b) Find the magnitude of the net force on the entire circuit.

32. •• A long wire carrying a 6.00 A current reverses direction by means of two right-angle bends, as shown in Figure 20.63. The part of the wire where the bend occurs is in a magnetic field of 0.666 T confined to the circular region of diameter 75 cm, as shown. Find the magnitude and direction of the net force that the magnetic field exerts on this wire.

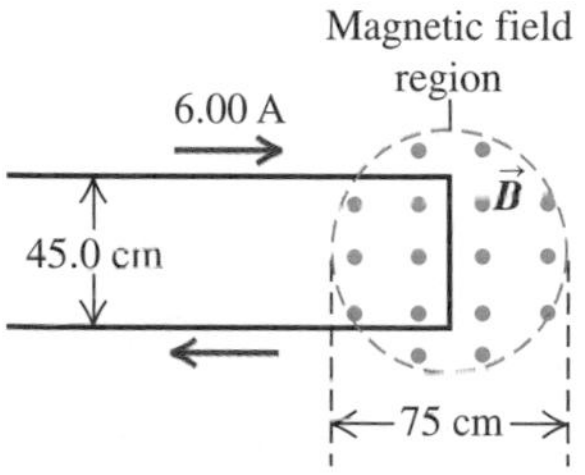

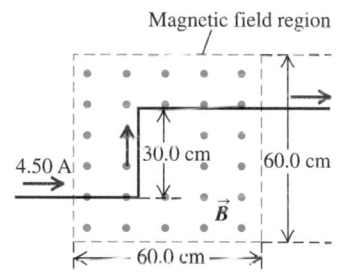

▲ FIGURE 20.63 Problem 32.

33. •• A long wire carrying 4.50 A of current makes two 90° bends, as shown in Figure 20.64. The bent part of the wire passes through a uniform 0.240 T magnetic field directed as shown in the figure and confined to a limited region of space. Find the magnitude and direction of the force that the magnetic field exerts on the wire.

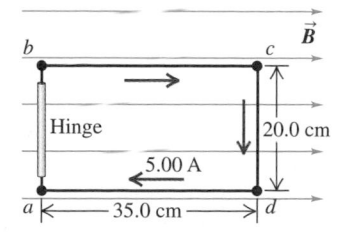

▲ FIGURE 20.64 Problem 33.

20.6 Force and Torque on a Current Loop

34. • The 20.0 cm by 35.0 cm rectangular circuit shown in Figure 20.65 is hinged along side *ab*. It carries a clockwise 5.00 A current and is located in a uniform 1.20 T magnetic field oriented perpendicular to two of its sides, as shown. (a) Make a clear diagram showing the direction of the force that the magnetic field exerts on each segment of the circuit (*ab*, *bc*, etc.). (b) Of the four forces you drew in part (a), decide which ones exert a torque about the hinge *ab*. Then calculate only those forces that exert this torque. (c) Use your results from part (b) to calculate the torque that the magnetic field exerts on the circuit about the hinge axis *ab*.

▲ FIGURE 20.65 Problem 34.

35. • The plane of a 5.0 cm by 8.0 cm rectangular loop of wire is parallel to a 0.19 T magnetic field, and the loop carries a current of 6.2 A. (a) What torque acts on the loop? (b) What is the magnetic moment of the loop?

36. •• A circular coil of wire 8.6 cm in diameter has 15 turns and carries a current of 2.7 A. The coil is in a region where the magnetic field is 0.56 T. (a) What orientation of the coil gives the maximum torque on the coil, and what is this maximum torque? (b) For what orientation of the coil is the magnitude of the torque 71% of the maximum found in part (a)?

37. •• A rectangular coil of wire 22.0 cm by 35.0 cm and carrying a current of 1.40 A is oriented with the plane of its loop perpendicular to a uniform 1.50 T magnetic field, as shown in Figure 20.66. (a) Calculate the net force and torque that the magnetic field exerts on this coil. (b) The coil is now rotated through a 30.0° angle about the axis shown, the left side coming out of the plane and the right side going into the plane. Calculate the net force and torque that the magnetic field exerts on the coil. (*Hint:* In order to help visualize this three-dimensional problem, make a careful drawing of the coil as viewed along its axis of rotation.)

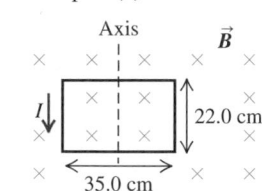

▲ FIGURE 20.66 Problem 37.

38. • A solenoid having 165 turns and a cross-sectional area of 6.75 cm² carries a current of 1.20 A. If it is placed in a uniform 1.12 T magnetic field, find the torque this field exerts on the solenoid if its axis is oriented (a) perpendicular to the field, (b) parallel to the field, (c) at 35.0° with the field.

39. •• A circular coil of 50 loops and diameter 20.0 cm is lying flat on a tabletop, and carries a clockwise current of 2.50 A. A magnetic field of 0.450 T, directed to the north and at an angle of 45.0° from the vertical down through the coil and into the tabletop is turned on. (a) What is the torque on the coil, and (b) which side of the coil (north or south) will tend to rise from the tabletop?

20.7 Magnetic Field of a Long, Straight Conductor

40. • You want to produce a magnetic field of magnitude 5.50×10^{-4} T at a distance of 0.040 m from a long, straight wire's center. (a) What current is required to produce this field? (b) With the current found in part (a), how strong is the magnetic field 8.00 cm from the wire's center?

41. • **Household magnetic fields.** Home circuit breakers typically have current capacities of around 10 A. How large a magnetic field would such a current produce 5.0 cm from a long wire's center? How does this field compare with the strength of the earth's magnetic field?

42. • (a) How large a current would a very long, straight wire have to carry so that the magnetic field 2.00 cm from the wire is equal to 1.00 G (comparable to the earth's northward-pointing magnetic field)? (b) If the wire is horizontal with the current running from east to west, at what locations would the magnetic field of the wire point in the same direction as the horizontal component of the earth's magnetic field? (c) Repeat part (b) except with the wire vertical and the current going upward.

43. • **Currents in the heart.** The body contains many small currents caused by the motion of ions in the organs and cells. Measurements of the magnetic field around the chest due to currents in the heart give values of about 1.0 μG. Although the
BIO

actual currents are rather complicated, we can gain a rough understanding of their magnitude if we model them as a long, straight wire. If the surface of the chest is 5.0 cm from this current, how large is the current in the heart?

44. • **Magnetic sensitivity of electric fish.** In a problem dealing
BIO with electric fish in Chapter 19, we saw that these fish navigate by responding to changes in the current in seawater. This current is due to a potential difference of around 3.0 V generated by the fish and is about 12 mA within a centimeter or so from the fish. Receptor cells in the fish are sensitive to the current. Since the current is at some distance from the fish, the sensitivity of these cells suggests that they might be responding to the magnetic field created by the current. To get some estimate of how sensitive the cells are, we can model the current as that of a long, straight wire with the receptor cells 2.0 cm away. What is the strength of the magnetic field at the receptor cells?

45. • In a conventional cheap flashlight, a straight copper strip runs along the tube of the flashlight to connect the bulb to the negative terminal of the battery at the bottom of the tube. If this strip carries a current of 0.65 A while you're holding the flashlight, what is the magnitude of the magnetic field at the surface of your hand, 0.30 cm from the strip? (You can treat the strip as a long, thin, straight wire.) How does your answer compare to the earth's magnetic field?

46. •• If the magnetic field due to a long, straight current-carrying wire has a magnitude B at a distance R from the wire's center, how far away must you be (in terms of R) for the magnetic field to decrease to $B/3$?

47. •• A current in a long, straight wire produces a magnetic field of 8.0 μT at 2.0 cm from the wire's center. Answer the following questions *without* finding the current: (a) What is the magnetic field strength 4.0 cm from the wire's center? (b) How far from the wire's center will the field be 1.0 μT? (c) If the current were doubled, what would the field be 2.0 cm from the wire's center?

48. •• **EMF.** Currents in DC transmission lines can be 100 A or
BIO more. Some people have expressed concern that the electromagnetic fields (EMFs) from such lines near their homes could cause health dangers. Using your own observations, estimate how high such lines are above the ground. Then use your estimate to calculate the strength of the magnetic field these lines produce at ground level. Express your answer in teslas and as a percent of the earth's magnetic field (which is 0.50 gauss). Does it seem that there is cause for worry?

49. • A long, straight telephone cable contains six wires, each carrying a current of 0.300 A. The distances between wires can be neglected. (a) If the currents in all six wires are in the same direction, what is the magnitude of the magnetic field 2.50 m from the cable? (b) If four wires carry currents in one direction and the other two carry currents in the opposite direction, what is the magnitude of the field 2.50 m from the cable?

50. •• Two insulated wires perpendicular to each other in the same plane carry currents as shown in Figure 20.67. Find the magnitude of the *net* magnetic field

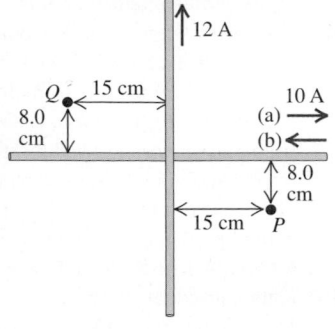

▲ **FIGURE 20.67** Problem 50.

these wires produce at points P and Q if the 10.0 A current is (a) to the right or (b) to the left.

51. •• Two long, straight parallel wires are 10.0 cm apart and carry 4.00 A currents in the same direction (Figure 20.68). Find the magnitude and direction of the magnetic field at (a) point P_1, midway between the wires, (b) point P_2, 25.0 cm to the right of P_1.

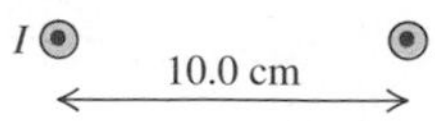

$I \odot$ 10.0 cm $\odot I$

▲ **FIGURE 20.68**
Problem 51.

52. •• Two long parallel transmission lines 40.0 cm apart carry 25.0 A and 75.0 A currents. Find all locations where the net magnetic field of the two wires is zero if these currents are in (a) the same direction, (b) opposite directions.

20.8 Force between Parallel Conductors

53. • Two high-current transmission lines carry currents of 25 A and 75 A in the same direction and are suspended parallel to each other 35 cm apart. If the vertical posts supporting these wires divide the lines into straight 15 m segments, what magnetic force does each segment exert on the other? Is this force attractive or repulsive?

54. •• Two long current-carrying wires run parallel to each other. Show that if the currents run in the same direction, these wires attract each other, whereas if they run in opposite directions, the wires repel.

55. •• A 2.0 m ordinary lamp extension cord carries a 5.0 A current. Such a cord typically consists of two parallel wires carrying equal currents in opposite directions. Find the magnitude and direction (attractive or repulsive) that the two segments of this cord exert on each other. (You will need to inspect an actual lamp cord at home and measure or reasonably estimate the quantities needed to do this calculation.)

56. •• An electric bus operates by drawing current from two parallel overhead cables, at a potential difference of 600 V, and spaced 55 cm apart. When the power input to the bus's motor is at its maximum power of 65 hp, (a) what current does it draw and (b) what is the attractive force per unit length between the cables?

20.9 Current Loops and Solenoids

57. • A circular metal loop is 22 cm in diameter. (a) How large a current must flow through this metal so that the magnetic field at its center is equal to the earth's magnetic field of 0.50×10^{-4} T? (b) Show how the loop should be oriented so that it can cancel the earth's magnetic field at its center.

58. • A closely wound circular coil with a diameter of 4.00 cm has 600 turns and carries a current of 0.500 A. What is the magnetic field at the center of the coil?

59. • A closely wound circular coil has a radius of 6.00 cm and carries a current of 2.50 A. How many turns must it have if the magnetic field at its center is 6.39×10^{-4} T?

60. • **Currents in the brain.** The magnetic field around the head
BIO has been measured to be approximately 3.0×10^{-8} gauss. Although the currents that cause this field are quite complicated, we can get a rough estimate of their size by modeling them as a single circular current loop 16 cm (the width of a typical head) in diameter. What is the current needed to produce such a field at the center of the loop?

61. • A closely wound, circular coil with radius 2.40 cm has 800 turns. What must the current in the coil be if the magnetic field at the center of the coil is 0.0580 T?

62. •• Two circular concentric loops of wire lie on a tabletop, one inside the other. The inner loop has a diameter of 20.0 cm and carries a clockwise current of 12.0 A, as viewed from above, and the outer wire has a diameter of 30.0 cm. What must be the magnitude and direction (as viewed from above) of the current in the outer loop so that the net magnetic field due to this combination of loops is zero at the common center of the loops?

63. •• Calculate the magnitude and direction of the magnetic field at point P due to the current in the semicircular section of wire shown in Figure 20.69. (*Hint:* The current in the long, straight section of wire produces no field at P. Can you relate the semicircle to a current loop?)

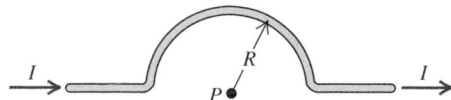

▲ **FIGURE 20.69** Problem 63.

64. • A solenoid contains 750 coils of very thin wire evenly wrapped over a length of 15.0 cm. Each coil is 0.800 cm in diameter. If this solenoid carries a current of 7.00 A, what is the magnetic field at its center?

65. • As a new electrical technician, you are designing a large solenoid to produce a uniform 0.150 T magnetic field near its center. You have enough wire for 4000 circular turns, and the solenoid must be 1.40 m long and 2.00 cm in diameter. What current will you need to produce the necessary field?

66. • A solenoid is designed to produce a 0.0279 T magnetic field near its center. It has a radius of 1.40 cm and a length of 40.0 cm, and the wire carries a current of 12.0 A. (a) How many turns must the solenoid have? (b) What total length of wire is required to make this solenoid?

67. •• A single circular current loop 10.0 cm in diameter carries a 2.00 A current. (a) What is the magnetic field at the center of this loop? (b) Suppose that we now connect 1000 of these loops in series within a 500 cm length to make a solenoid 500 cm long. What is the magnetic field at the center of this solenoid? Is it 1000 times the field at the center of the loop in part (a)? Why or why not?

68. •• A solenoid that is 35 cm long and contains 450 circular coils 2.0 cm in diameter carries a 1.75 A current. (a) What is the magnetic field at the center of the solenoid, 1.0 cm from the coils? (b) Suppose we now stretch out the coils to make a very long wire carrying the same current as before. What is the magnetic field 1.0 cm from the wire's center? Is it the same as you found in part (a)? Why or why not?

69. •• You have 25 m of wire, which you want to use to construct a 44 cm diameter coil whose magnetic field at its center will exactly cancel the earth's field of 0.55 gauss. What current will your coil require?

70. • A toroidal solenoid (see Figure 20.42) has inner radius $r_1 = 15.0$ cm and outer radius $r_2 = 18.0$ cm. The solenoid has 250 turns and carries a current of 8.50 A. What is the magnitude of the magnetic field at the following distances from the center of the torus: (a) 12.0 cm; (b) 16.0 cm; (c) 20.0 cm?

20.10 Magnetic Field Calculations

71. • A long, straight wire carries a current of 10.0 A, as shown in Figure 20.70. Use the law of Biot and Savart to find the magnitude and direction of the magnetic field at point P due to each of the following 2.00 mm segments of this wire: (a) segment A and (b) segment C.

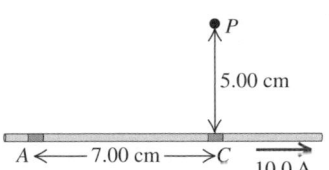

▲ **FIGURE 20.70** Problem 71.

72. • A long wire carrying a 5.00 A current makes an abrupt right-angle bend as shown in Figure 20.71. Use the law of Biot and Savart to determine the magnitude and direction of the magnetic field at point P due to the 1.50 cm bent segment if P is 15.0 cm from the midpoint of that segment.

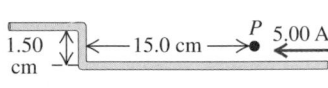

▲ **FIGURE 20.71** Problem 72.

73. • Three long, straight electrical cables, running north and south, are tightly enclosed in an insulating sheath. One of the cables carries a 23.0 A current southward; the other two carry currents of 17.5 A and 11.3 A northward. Use Ampere's law to calculate the magnitude of the magnetic field at a distance of 10.0 m from the cables.

74. •• A long, straight, cylindrical wire of radius R carries a current uniformly distributed over its cross section. At what location is the magnetic field produced by this current equal to half of its largest value? Use Ampere's law and consider points inside and outside the wire.

20.11 Magnetic Materials

75. • Platinum is a paramagnetic metal having a relative permeability of 1.00026. (a) What is the magnetic permeability of platinum? (b) If a thin rod of platinum is placed in an external magnetic field of 1.3500 T, with its axis parallel to that field, what will be the magnetic field inside the rod?

76. • When a certain paramagnetic material is placed in an external magnetic field of 1.5000 T, the field inside the material is measured to be 1.5023 T. Find (a) the relative permeability and (b) the magnetic permeability of this material.

General Problems

77. •• A 150 g ball containing 4.00×10^8 excess electrons is dropped into a 125 m vertical shaft. At the bottom of the shaft, the ball suddenly enters a uniform horizontal 0.250 T magnetic field directed from east to west. If air resistance is negligibly small, find the magnitude and direction of the force that this magnetic field exerts on the ball just as it enters the field.

78. •• **Magnetic balance.** The circuit shown in Figure 20.72 is used to make a magnetic balance to weigh objects. The mass m to be measured is hung from the center of the bar, which is in a uniform magnetic field of 1.50 T directed into the plane of the figure. The battery voltage can be adjusted to vary the current in the circuit. The

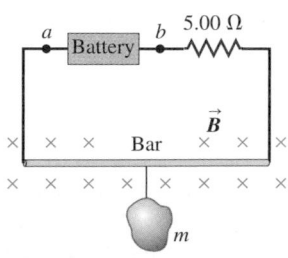

▲ **FIGURE 20.72** Problem 78.

horizontal bar is 60.0 cm long and is made of extremely light-weight material, so its mass can be neglected. It is connected to the battery by thin vertical wires that can support no appreciable tension; all the weight of the mass m is supported by the magnetic force on the bar. A 5.00 Ω resistor is in series with the bar, and the resistance of the rest of the circuit is negligibly small. (a) Which point, a or b, should be the positive terminal of the battery? (b) If the maximum terminal voltage of the battery is 175 V, what is the greatest mass m that this instrument can measure?

79. ●● A thin 50.0-cm-long metal bar with mass 750 g rests on, but is not attached to, two metal supports in a uniform 0.450 T magnetic field, as shown in Figure 20.73. A battery and a 25.0 Ω resistor in series are connected to the supports. What is the largest terminal voltage the battery can have without breaking the circuit at the supports?

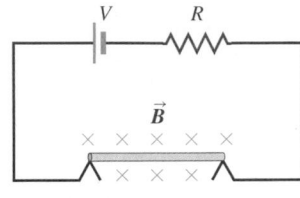
▲ **FIGURE 20.73** Problem 79.

80. ●● A long, straight wire containing a semicircular region of radius 0.95 m is placed in a uniform magnetic field of magnitude 2.20 T as shown in Figure 20.74. What is the net magnetic force acting on the wire when it carries a current of 3.40 A? (*Hint:* In Figure 20.74, what does symmetry tell you about the forces on the upper and lower halves of the semicircular region?)

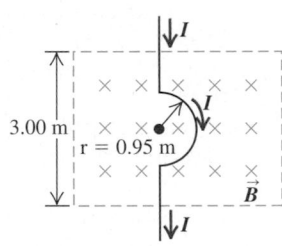
▲ **FIGURE 20.74** Problem 80.

81. ●● A singly charged ion of ^{7}Li (an isotope of lithium containing three protons and four neutrons) has a mass of 1.16×10^{-26} kg. It is accelerated through a potential difference of 220 V and then enters a 0.723 T magnetic field perpendicular to the ion's path. What is the radius of the path of this ion in the magnetic field?

82. ●● An insulated circular ring of diameter 6.50 cm carries a 12.0 A current and is tangent to a very long, straight insulated wire carrying 10.0 A of current, as shown in Figure 20.75. Find the magnitude and direction of the magnetic field at the center of the ring due to this combination of wires.

83. ●● **The effect of transmission lines.** Two hikers are reading a compass under an overhead transmission line that is 5.50 m above the ground and carries a current of 0.800 kA in a horizontal direction from north to south. (a) Find the magnitude and direction of the magnetic field at a point on the ground directly under the transmission line. (b) One hiker suggests that they walk 50 m away from the lines to avoid inaccurate compass readings due to the current. Considering that the earth's magnetic field is on the order of 0.5×10^{-4} T, is the current really a problem?

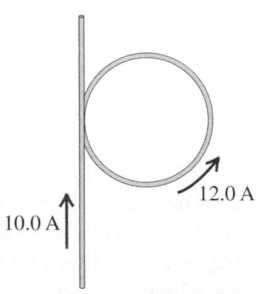
▲ **FIGURE 20.75** Problem 82.

84. ●● A long, straight horizontal wire carries a current of 2.50 A directed toward the right. An electron is traveling in the vicinity of this wire. (a) At the instant the electron is 4.50 cm above the wire's center and moving with a speed of 6.00×10^4 m/s directly toward it, what are the magnitude and direction of the force that the magnetic field of the current exerts on the electron? (b) What would be the magnitude and direction of the magnetic force if the electron were instead moving parallel to the wire in the same direction as the current?

85. ●● Two very long, straight wires carry currents as shown in Figure 20.76. For each case shown, find all locations where the net magnetic field due to these wires is zero.

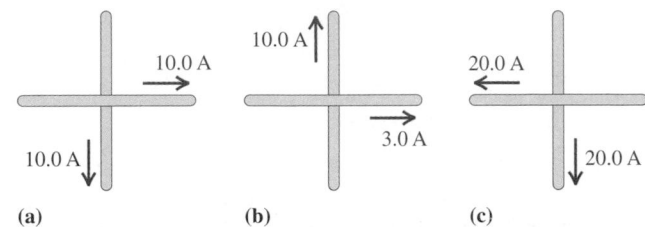

(a) (b) (c)

▲ **FIGURE 20.76** Problem 85.

86. ●● **Bubble chamber, I.** Certain types of bubble chambers are filled with liquid hydrogen. When a particle (such as an electron or a proton) passes through the liquid, it leaves a track of bubbles, which can be photographed to show the path of the particle. The apparatus is immersed in a known magnetic field, which causes the particle to curve. Figure 20.77 is a trace of a bubble chamber image showing the path of an electron. (a) How could you determine the *sign* of the charge of a particle from a photograph of its path? (b) How can physicists determine the *momentum* and the *speed* of this electron by using measurements made on the photograph, given that the magnetic field is known and is perpendicular to the plane of the figure? (c) The electron is obviously spiraling into smaller and smaller circles. What properties of the electron must be changing to cause this behavior? Why does this happen? (d) What would be the path of a neutron in a bubble chamber? Why?

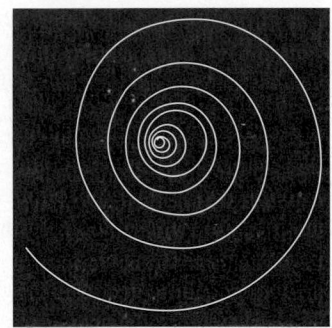
▲ **FIGURE 20.77** Problem 86.

87. ●● A 3.00 N metal bar, 1.50 m long and having a resistance of 10.0 Ω, rests horizontally on conducting wires connecting it to the circuit shown in Figure 20.78. The bar is in a uniform, horizontal, 1.60 T magnetic field and is not attached to the wires in the circuit. What is the acceleration of the bar just after the switch S is closed?

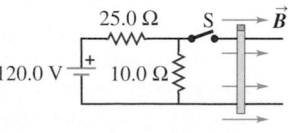
▲ **FIGURE 20.78** Problem 87.

88. ●● A pair of long, rigid metal rods, each of length L, lie parallel to each other on a perfectly smooth table. Their ends are connected by identical, very light conducting springs of force constant k (Figure 20.79) and negligible unstretched length. If a

current I runs through this circuit, the springs will stretch. At what separation will the rods remain at rest? Assume that k is large enough so that the separation of the rods will be much less than L.

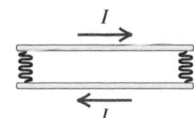

▲ **FIGURE 20.79** Problem 88.

89. •• **Atom smashers!** A *cyclotron particle accelerator* (sometimes called an "atom smasher" in the popular press) is a device for accelerating charged particles, such as electrons and protons, to speeds close to the speed of light. The basic design is quite simple. The particle is bent in a circular path by a uniform magnetic field. An electric field is pulsed periodically to increase the speed of the particle. The charged particle (or ion) of mass m and charge q is introduced into the cyclotron so that it is moving perpendicular to a uniform magnetic field $\vec{B}$. (a) Starting with the radius of the circular path of a charge moving in a uniform magnetic field, show that the time T for this particle to make one complete circle is $T = \dfrac{2\pi m}{|q|B}$. (*Hint:* You can express the speed v in terms of R and T because the particle travels through one circumference of the circle in time T.) (b) Which would take longer to complete one circle, an ion moving in a large circle or one moving in a small circle? Explain.

90. •• **Medical uses of cyclotrons.** The largest cyclotron (see the previous problem) in the United States is the *Tevatron* at Fermilab, near Chicago, Illinois. It is called a Tevatron because it can accelerate particles to energies in the TeV range $(1 \text{ tera-eV} = 10^{12} \text{ eV})$. Its circumference is 6.4 km, and it currently can produce a maximum energy of 2.0 TeV. In a certain medical experiment, protons will be accelerated to energies of 1.25 MeV and aimed at a tumor to destroy its cells. (a) How fast are these protons moving when they hit the tumor? (b) How strong must the magnetic field be to bend the protons in the circle indicated?

91. •• A plastic circular loop has radius R, and a positive charge q is distributed uniformly around the circumference of the loop. The loop is now rotated around its central axis, perpendicular to the plane of the loop, with angular speed ω. If the loop is in a region where there is a uniform magnetic field $\vec{B}$ directed parallel to the plane of the loop, calculate the magnitude of the magnetic torque on the loop.

92. •• A long wire carrying 6.50 A of current makes two bends, as shown in Figure 20.80. The bent part of the wire passes through a uniform 0.280 T magnetic field directed as shown in the figure and confined to a limited region of space. Find the magnitude and direction of the force that the magnetic field exerts on the wire.

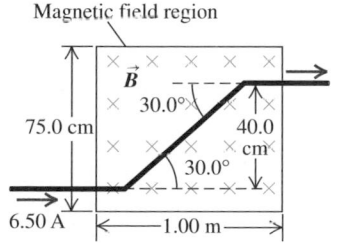

▲ **FIGURE 20.80** Problem 92.

Passage Problems

BIO **Magnetic fields and MRI.** Magnetic resonance imaging (MRI) is a powerful imaging method that, unlike x-ray imaging, allows sharp images of soft tissue to be made without exposure to potentially damaging radiation. While a full explanation of MRI is beyond the scope of an introductory physics textbook, some understanding can be achieved by the relatively simple application of the classical (that is, non-quantum) physics of magnetism. The starting point for MRI is nuclear magnetic resonance (NMR), a phenomenon that depends on the fact that protons in the atomic nucleus have a magnetic field, $\vec{B}$. The origin of the proton's magnetic field is the spin of the proton. Being charged, the spinning proton constitutes an electrical current analogous to a wire loop through which current flows. Like the wire loop, if the proton is subjected to an external magnetic field, $\vec{B}_0$, it experiences a torque and a magnetic moment, μ_p. The magnitude of the magnetic moment is about 1.4×10^{-26} J/T. The proton can be thought of as being in one of two states, with the magnetic moment oriented parallel or antiparallel to the applied magnetic field, and work must be done to flip the proton from the low energy state to the high energy state.

An important consideration is that the net magnetic field of any given nucleus, except for hydrogen, consists of protons and neutrons. The hydrogen atom, of course, has only a proton. If a nucleus has an even number of protons and neutrons, they will pair in a way such that half of the protons have spins in one orientation and half have spins in the other orientation, so net magnetic moment for the nucleus is zero. Only nuclei with a net magnetic moment are candidates for MRI. Hydrogen is the atom most commonly imaged.

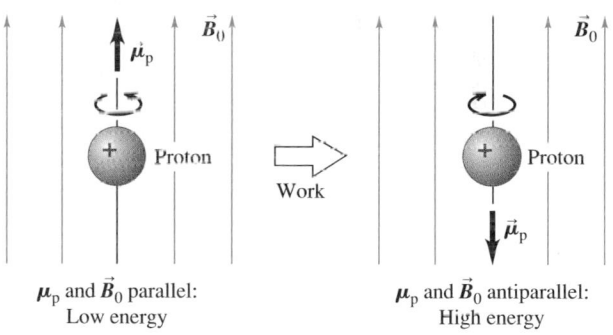

μ_p and $\vec{B}_0$ parallel:
Low energy

μ_p and $\vec{B}_0$ antiparallel:
High energy

▲ **FIGURE 20.81** Problems 93–95.

93. If a proton is exposed to an external magnetic field of 2T that has a direction perpendicular to the axis of the spin of the proton, what will be the torque on the proton?
    A. 0
    B. 1.4×10^{-26} N · m
    C. 2.8×10^{-26} N · m
    D. 0.7×10^{-26} N · m

94. Which of following elements is a candidate for MRI?
    A. $^{12}C_6$ B. $^{16}O_8$
    C. $^{40}Ca_{20}$ D. $^{31}P_{15}$

21 Electromagnetic Induction

Have you ever wondered how a card reader reads the magnetically encoded data on a credit card? The information is stored in tiny magnetized regions on the card (on the order of 10^{-8} m). In a card reader, the motion of the credit card moves these tiny magnets past the read/write head, inducing currents in the head that convey the data to a computer.

What goes on in the read/write head is an example of electromagnetic induction: A changing magnetic field causes a current in a circuit, even though there's no battery or other obvious source of electromotive force (emf).

In this chapter, we discuss the electromotive force (emf) that results from *magnetic* interactions. Many of the components of present-day electric power systems, including generators, transformers, and motors, depend directly on magnetically induced emfs. These systems would not be possible if we had to depend on chemical sources of emf, such as batteries.

The central principle in this chapter is *Faraday's law.* This law relates induced emf to changing *magnetic flux* through a loop, often a closed circuit. (We'll define magnetic flux in Section 21.2.) We also discuss *Lenz's law,* which helps us to predict the directions of induced emf's and currents. This chapter discusses the principles that we need to understand electrical energy conversion devices, including motors, generators, and transformers. It also paves the way for the analysis of electromagnetic waves in Chapter 23.

This engineer examines the turbine rotor of a power-plant generator. The rotor converts the kinetic energy of high-pressure steam to rotational energy, which in turn is converted to electrical energy.

21.1 Induction Experiments

We begin our study of magnetically induced electromotive force with a look at several pioneering experiments carried out in the 1830s by Michael Faraday in England and Joseph Henry (the first director of the Smithsonian Institution) in the

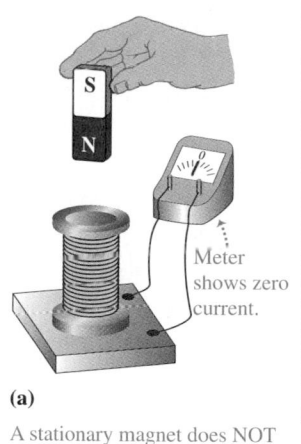
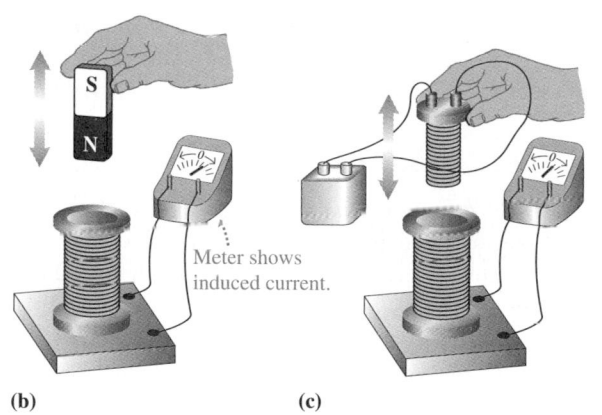
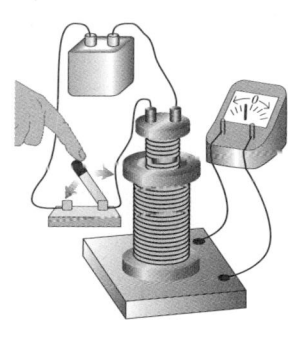

Meter shows zero current.

Meter shows induced current.

(a)

A stationary magnet does NOT induce a current in a coil.

(b)

Move the magnet toward or away from coil.

(c)

Move a second, current-carrying coil toward or away from the coil.

(d)

Vary the current in the second coil (by closing or opening a switch).

All these actions DO induce a current in the coil. What do they have in common?
(They cause the magnetic field through the coil to *change*.)

▲ **FIGURE 21.1** Examples of experiments in which a magnetic field does or does not induce a current in a coil.

United States. Figure 21.1 shows several examples. In Figure 21.1a, a coil of wire is connected to a current-measuring device such as the moving-coil galvanometer we described in Section 19.7. When the nearby magnet is stationary, the meter shows no current. This isn't surprising; there's no source of emf in the circuit. But when we *move* the magnet either toward or away from the coil (Figure 21.1b), the meter shows current in the circuit, but *only* while the magnet is moving. If we hold the magnet stationary and move the coil, we again detect a current during the motion. So something about the *changing* magnetic field through the coil is causing a current in the circuit. We call this an **induced current,** and the corresponding emf that has to be present to cause such a current is called an **induced emf.**

In Figure 21.1c, we replace the magnet with a second coil connected to a battery. We find that when the second coil is stationary, there is no current in the first coil. But when we move the second coil toward or away from the first, or the first coil toward or away from the second, there is a current in the first coil, but again only while one coil is moving relative to the other.

Finally, using the two-coil setup in Figure 21.1d, we keep both coils stationary and vary the current in the second coil, either by opening and closing the switch or by changing the resistance in its circuit (with the switch closed). We find that as we open or close the switch, there is a momentary current pulse in the first circuit. When we vary the current in the second coil, there is an induced current in the first circuit, but only while the current in the second circuit is changing. The setup shown in Figure 21.2 can be used for more detailed experiments.

▲ **FIGURE 21.2** An apparatus for investigating induced currents. A change in the $\vec{B}$ field or the shape or location of the coil induces a current.

▶ **Application Induction cooker.** An induction cooktop presents a smooth surface with just a few stencilled rings. You can set a "burner" on high and place your hand on it without feeling any heat—yet it cooks food efficiently. A standard electric range cooks by passing a current through a high-resistance heating element, which dissipates energy by getting red hot. In an induction cooktop, alternating current passes through a coil beneath the cooktop, inducing a strong, rapidly changing magnetic field. If you put a conducting pan on the cooktop, the currents induced in the pan cause the *pan* to heat up. Needless to say, an induction range won't work with nonconducting ceramic or Pyrex cookware.

What do all these phenomena have in common? The answer, which we'll explore in detail in this chapter, is that a *changing magnetic flux* through the coil, from whatever cause, induces a current in the circuit. This statement forms the basis of Faraday's law of induction, the main subject of this chapter.

21.2 Magnetic Flux

The concept of *magnetic flux* is analogous to that of electric flux, which we encountered in Section 17.8 in connection with Gauss's law. It is also closely related to magnetic field lines, which we studied in Section 20.2. Magnetic flux, denoted by Φ_B, is defined with reference to an area and a magnetic field. We can divide any surface into elements of area ΔA, as shown in Figure 21.3. For each element, we determine the component $B_\perp$ of the magnetic field $\vec{B}$ normal (perpendicular) to the surface at the position of that element, as shown in the figure. From the figure, $B_\perp = B\cos\phi$, where ϕ is the angle between the direction of $\vec{B}$ and a line perpendicular to the surface. In general, this component is different for different points on the surface.

We define the magnetic flux $\Delta\Phi_B$ through the element of area ΔA as

$$\Delta\Phi_B = B_\perp \, \Delta A = B\cos\phi \, \Delta A. \tag{21.1}$$

The *total* magnetic flux through the surface is the sum of the contributions from the individual area elements. In the special case in which $\vec{B}$ is uniform over a plane surface with total area A, we can simplify the preceding expression as follows:

Magnetic flux through a plane surface in a constant magnetic field
For a plane surface with area A in a uniform magnetic field $\vec{B}$, both $B_\perp$ and ϕ are the same at all points on the surface. The magnetic flux Φ_B through the surface is

$$\Phi_B = B_\perp A = BA\cos\phi. \tag{21.2}$$

If $\vec{B}$ happens to be perpendicular to the surface, then $\cos\phi = 1$ and Equation 21.2 reduces to $\Phi_B = BA$. Figure 21.4 shows three cases for a plane surface in a uniform magnetic field.

Unit: The SI unit of magnetic flux is the unit of magnetic field (1 T) times the unit of area (1 m^2):

$$(1\,\text{T})(1\,\text{m}^2) = [1\,\text{N}/(\text{A}\cdot\text{m})](1\,\text{m}^2) = 1\,\text{N}\cdot\text{m}/\text{A}.$$

This unit is called 1 **weber** (1 Wb), in honor of Wilhelm Weber (1804–1891):

$$1\,\text{weber} = 1\,\text{Wb} = 1\,\text{T}\cdot\text{m}^2 = 1\,\text{N}\cdot\text{m}/\text{A}.$$

Magnetic flux through element of area ΔA:
$\Delta\Phi_B = B\cos\phi\,\Delta A = B_\perp\Delta A$

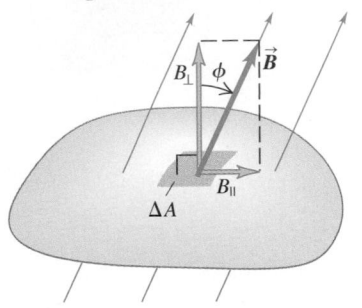

▲ FIGURE 21.3 The magnetic flux through an area element ΔA.

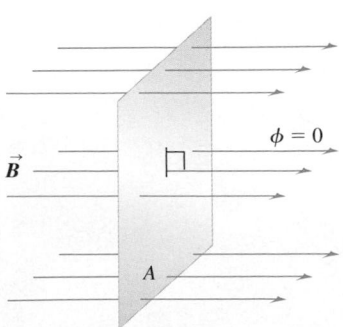

$\vec{B}$ perpendicular to A ($\phi = 0$): magnetic flux $\Phi_B = BA$.

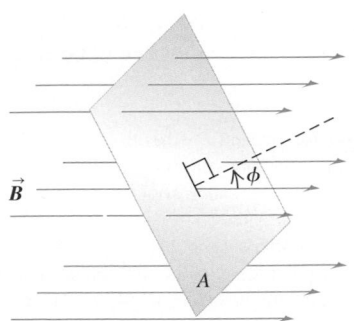

$\vec{B}$ at an angle ϕ to the perpendicular to A: magnetic flux $\Phi_B = BA\cos\phi$.

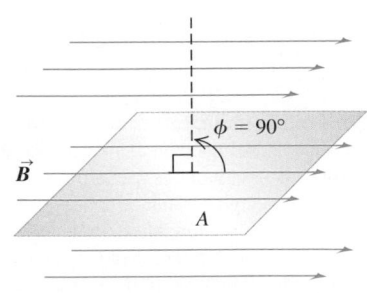

$\vec{B}$ parallel to A ($\phi = 90$): magnetic flux $\Phi_B = 0$.

▲ FIGURE 21.4 The magnetic flux through a flat surface at various orientations relative to a uniform magnetic field.

If the element of area ΔA in Equation 21.1 is at right angles to the field lines, then $B_\perp = B$, and

$$B = \frac{\Delta \Phi_B}{\Delta A}. \qquad (21.3)$$

That is, magnetic field is *magnetic flux per unit area* across an area at right angles to the magnetic field. For this reason, magnetic field $\vec{B}$ is sometimes called **magnetic flux density.** The unit of magnetic flux is 1 weber, so the unit of magnetic field, 1 tesla, is also equal to 1 *weber per square meter:*

$$1 \text{ T} = 1 \text{ Wb/m}^2.$$

We can picture the total flux through a surface as proportional to the number of field lines passing through the surface and the field (the flux density) as proportional to the number of lines *per unit area.*

Conceptual Analysis 21.1

Flux through a flexible coil

Suppose you hold a circular wire coil perpendicular to a constant magnetic field and then squeeze it into a narrow oval, as shown in Figure 21.5. How does the flux through the squeezed coil compare with that through the circular coil?

A. The flux is less, because the area of the coil is smaller.
B. The flux is the same, because the length of wire is the same.
C. The flux is the same, because the angle of the coil relative to the magnetic field stays the same.

SOLUTION As Equation 21.2 shows, the magnetic flux through a coil depends on (1) the magnetic field magnitude B, (2) the area A of the coil, and (3) the angle ϕ of the coil's axis with respect to the magnetic field: $\Phi_B = BA\cos\phi$. A change in any of these variables changes the flux through the coil. Squeezing the coil reduces its area, so A is correct. (Neither B nor ϕ changes.) The magnetic flux through the coil does not depend on the length of wire in the coil, so B isn't correct. And the change in A results in a change in Φ_B, even though ϕ doesn't change; so C is also incorrect.

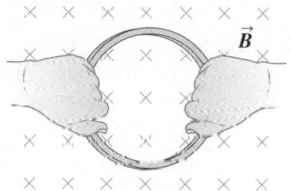

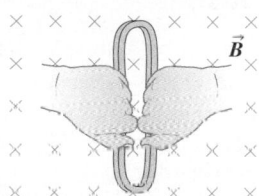

Circular wire coil   Coil squeezed into oval

▲ **FIGURE 21.5**

EXAMPLE 21.1 **Magnetic flux**

A plane surface with area 3.0 cm^2 is placed in a uniform magnetic field that is oriented at an angle of 30° to the surface. **(a)** What is the angle ϕ? **(b)** If the magnetic flux through this area is 0.90 mWb, what is the magnitude of the magnetic field?

SOLUTION

SET UP AND SOLVE Part (a): Figure 21.6 shows our sketch. As defined earlier, ϕ is the angle between the direction of $\vec{B}$ and a line *normal* to the surface, so $\phi = 60°$ (not 30°).

Part (b): Because B and ϕ are the same at all points on the surface, we can use Equation 21.2: $\Phi_B = BA\cos\phi$. We solve for B, remembering to convert the area to square meters:

$$B = \frac{\Phi_B}{A\cos\phi} = \frac{0.90 \times 10^{-3} \text{ Wb}}{(3.0 \times 10^{-4} \text{ m}^2)(\cos 60°)} = 6.0 \text{ T}.$$

REFLECT The flux through the surface depends on its angle relative to $\vec{B}$. Knowing this angle and the flux per unit area, we can find the magnitude of the magnetic field.

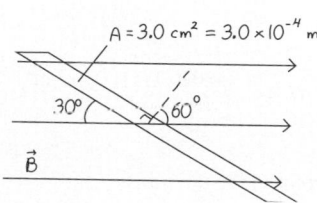

▲ **FIGURE 21.6** Our sketch for this problem.

Practice Problem: For the same B and A, find the angle ϕ at which the flux would have half the value given above (i.e., the angle ϕ at which $\Phi_B = 0.45 \times 10^{-3}$ Wb). *Answer:* $\phi = 76°$.

21.3 Faraday's Law

The common element in all induction effects is changing magnetic flux (defined in Section 21.2) through a circuit. *Faraday's law* states that the induced emf in a circuit is *directly proportional* to the time rate of change of the magnetic flux Φ_B through the circuit.

Faraday's law of induction:

The magnitude of the induced emf in a circuit equals the absolute value of the time rate of change of the magnetic flux through the circuit.

In symbols, Faraday's law is

$$\mathcal{E} = \left| \frac{\Delta \Phi_B}{\Delta t} \right|. \tag{21.4}$$

In this definition, $\mathcal{E}$ is the magnitude of the emf and is always positive.

Here's a simple example of this law in action:

EXAMPLE 21.2 Current induced in a single loop

In Figure 21.7, the magnetic field in the region between the poles of the electromagnet is uniform at any time, but is increasing at the rate of 0.020 T/s. The area of the conducting loop in the field is 120 cm², and the total circuit resistance, including the meter, is 5.0 Ω. Find the magnitudes of the induced emf and the induced current in the circuit.

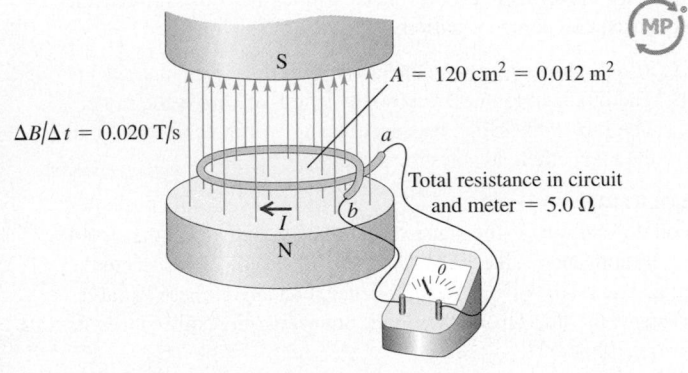

▶ **FIGURE 21.7**

SOLUTION

SET UP The area $A = 0.012 \ \text{m}^2$ is constant, so the absolute value of the rate of change of magnetic flux is A multiplied by the rate of change of the magnetic field magnitude B.

SOLVE The magnitude of the induced emf, $\mathcal{E}$, is equal to the absolute value of the rate of change of the magnetic flux, $|\Delta\Phi_B/\Delta t|$:

$$\mathcal{E} = \left| \frac{\Delta\Phi_B}{\Delta t} \right| = \frac{\Delta B}{\Delta t} A = (0.020 \ \text{T/s})(0.012 \ \text{m}^2) = 2.4 \times 10^{-4} \ \text{V}$$

$$= 0.24 \ \text{mV}.$$

The induced current in the circuit is

$$I = \frac{\mathcal{E}}{R} = \frac{2.4 \times 10^{-4} \ \text{V}}{5.0 \ \Omega} = 4.8 \times 10^{-5} \ \text{A} = 0.048 \ \text{mA}.$$

REFLECT It's worthwhile to verify unit consistency in this calculation. There are many ways to do this; one is to note that we can rewrite the magnetic force relation $(F = qvB)$ as $B = F/qv$, so

$$1 \ \text{T} = (1 \ \text{N})/(1 \ \text{C} \cdot \text{m/s}).$$

The units of magnetic flux can thus be expressed as

$$(1 \ \text{T})(1 \ \text{m}^2) = 1 \ \text{N} \cdot \text{s} \cdot \text{m/C}$$

and the rate of change of magnetic flux as

$$1 \ \text{N} \cdot \text{m/C} = 1 \ \text{J/C} = 1 \ \text{V}.$$

So the unit of $\Delta\Phi_B/\Delta t$ is the volt, as is required by Equation 21.4. Also, recall that the unit of magnetic flux is $1 \ \text{T} \cdot \text{m}^2 = 1 \ \text{Wb}$, so $1 \ \text{V} = 1 \ \text{Wb/s}$.

Practice Problem: Suppose we change the apparatus so that the magnetic field increases at a rate of 0.15 T/s, the area of the conducting loop in the field is 0.020 m², and the total circuit resistance is 7.5 Ω. Find the magnitude of the induced emf and the induced current in the circuit. *Answers:* $\mathcal{E} = 3.0 \ \text{mV}$; $I = 0.40 \ \text{mA}$.

If we have a coil with N identical turns, and the magnetic flux varies at the same rate through each turn, the induced emf's in the turns are all equal, are in *series*, and must be added. The total emf magnitude $\mathcal{E}$ is then

$$\mathcal{E} = N \left| \frac{\Delta \Phi_B}{\Delta t} \right|. \tag{21.5}$$

A magnet and a coil

Conceptual Analysis 21.2

When a magnet is plunged into a coil with two turns (Figure 21.8a), a voltage is induced in the coil and a current flows in the circuit. If the magnet is plunged with the same speed into a coil with *four* turns (Figure 21.8b), the induced voltage is

A. twice as great.
B. four times as great.
C. half as great.

SOLUTION As we just saw, the emf is the same for each turn of the coil, and the emf's for the turns add (because the turns are in series). Thus, doubling the number of turns doubles the total induced voltage. Answer A is correct.

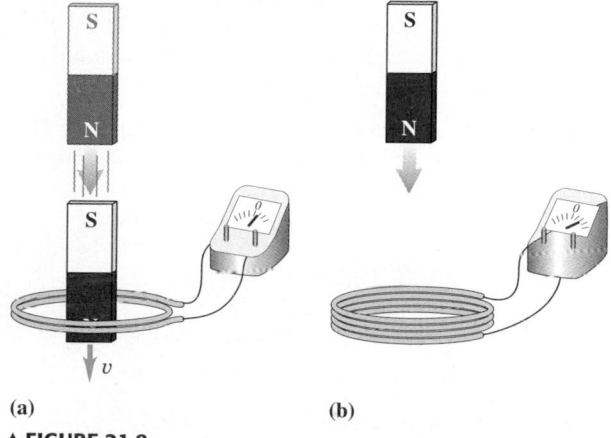

(a)   (b)

▲ **FIGURE 21.8**

PROBLEM-SOLVING STRATEGY 21.1   **Faraday's law**

SET UP

1. To calculate the rate of change of magnetic flux, you first have to understand what is making the flux change. Is the conductor moving? Is it changing orientation? Is the magnetic field changing? Remember that it's not the flux itself, but its *rate of change*, that counts.

SOLVE

2. If your conductor has N turns in a coil, don't forget to multiply by N.
3. In the next section, we'll learn a method for finding the directions of induced currents and emf's.

REFLECT

4. The shape of the coil doesn't matter; it can be circular, rectangular, or some other shape. Only the total rate of change of flux through the coil, and its number of turns, are significant.

NOTE ▶ In the statement of Faraday's law given by Equations 21.4 and 21.5, we've assumed that $\mathcal{E}$ and Φ_B are positive quantities. However, Faraday's law can also be stated in a more general way as $\mathcal{E} = -\Delta \Phi_B / \Delta t$, in which $\mathcal{E}$ and Φ_B can be either positive or negative. This form, combined with a fairly elaborate set of sign rules, gives the direction as well as the magnitude of an induced emf or current. For most of our applications, we won't need this generalization; instead, we'll rely on a principle called *Lenz's law,* to be discussed in Section 21.4, to determine the directions of the various quantities. ◄

EXAMPLE 21.3 **Induced emf in a coil of wire**

A circular coil of wire with 500 turns and an average radius of 4.00 cm is placed at a 60° angle to the uniform magnetic field between the poles of a large electromagnet. The field changes at a rate of −0.200 T/s. What is the magnitude of the resulting induced emf?

SOLUTION

SET UP Figure 21.9 shows our sketch. We remember that the angle ϕ is the angle between the magnetic field and the *normal* to the coil and thus is 30°, not 60°. The area of the coil is $A = \pi(0.0400 \text{ m})^2 = 0.00503 \text{ m}^2$.

SOLVE To find the magnitude of the emf, we use Equation 21.5: $\mathcal{E} = N|\Delta\Phi_B/\Delta t|$. We first need to find the rate of change of magnetic flux, $\Delta\Phi_B/\Delta t$. The flux at any time is given by $\Phi_B = BA\cos\phi$, so the magnitude of the rate of change of flux is

$$\left|\frac{\Delta\Phi_B}{\Delta t}\right| = \left|\frac{\Delta B}{\Delta t}\right|A\cos\phi$$
$$= (0.200 \text{ T/s})(0.00503 \text{ m}^2)(\cos 30°)$$
$$= 0.000871 \text{ T}\cdot\text{m}^2/\text{s} = 0.000871 \text{ Wb/s}.$$

Now we calculate the induced emf:

$$\mathcal{E} = N\left|\frac{\Delta\Phi_B}{\Delta t}\right| = (500)|(-0.000871 \text{ Wb/s})| = 0.436 \text{ V}.$$

REFLECT We need the absolute value signs on the right side of Faraday's law because we have defined $\mathcal{E}$ to be the *magnitude* of

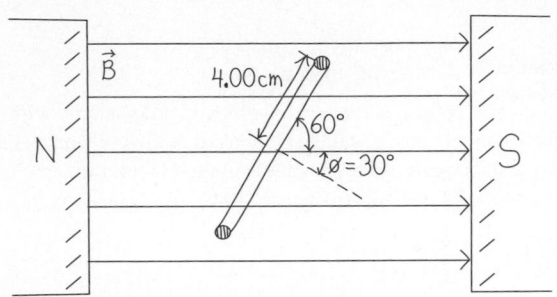

▲ **FIGURE 21.9** Our sketch for this problem.

the emf, always a positive quantity. As we've noted, Faraday's law can be stated so that the emf is positive or negative, depending on its direction. We avoid that complexity by stating the law in terms of the *magnitude* of the emf. In the next section, we'll learn how to determine the directions of induced currents.

Practice Problem: A 300-turn coil with an average radius of 0.03 m is placed in a uniform magnetic field so that $\phi = 40°$. The field increases at a rate of 0.250 T/s. What is the magnitude of the resulting emf? *Answer:* $\mathcal{E} = 0.162 \text{ V}$.

EXAMPLE 21.4 **A slide-wire generator**

The U-shaped conductor in Figure 21.10 lies perpendicular to a uniform magnetic field $\vec{B}$ with magnitude $B = 0.60$ T, directed into the page. We lay a metal rod with length $L = 0.10$ m across the two arms of the conductor, forming a conducting loop, and move the rod to the right with constant speed $v = 2.5$ m/s. What is the magnitude of the resulting emf?

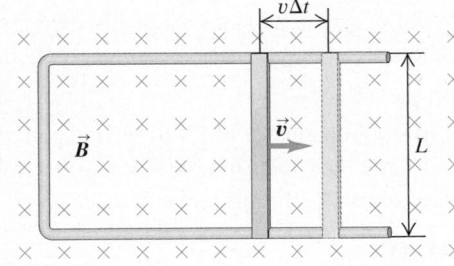

▶ **FIGURE 21.10**

SOLUTION

SET UP AND SOLVE The magnetic flux through the loop is changing because the area of the loop is increasing. During a time interval Δt, the sliding rod moves a distance $v\,\Delta t$ and the area increases by $\Delta A = Lv\,\Delta t$. Therefore, in time Δt, the magnetic flux through the loop increases by an amount $\Delta\Phi_B = B\,\Delta A = BLv\,\Delta t$. The magnitude of the induced emf is $\mathcal{E} = |\Delta\Phi_B/\Delta t| = BLv$. With the numerical values given, we find

$$\mathcal{E} = (0.60 \text{ T})(0.10 \text{ m})(2.5 \text{ m/s}) = 0.15 \text{ V}.$$

REFLECT As in Example 21.3, we've found the *magnitude* of the emf, but not the direction of the resulting induced current. Does the current flow clockwise or counterclockwise? We address this question in the next section.

Practice Problem: With the given magnetic field and rod length, what must the speed be if the induced emf has magnitude 0.75 V? *Answer:* 12.5 m/s.

Generators

A **generator** is a device that converts mechanical energy to electrical energy. A common design uses a conducting coil that rotates in a magnetic field, producing

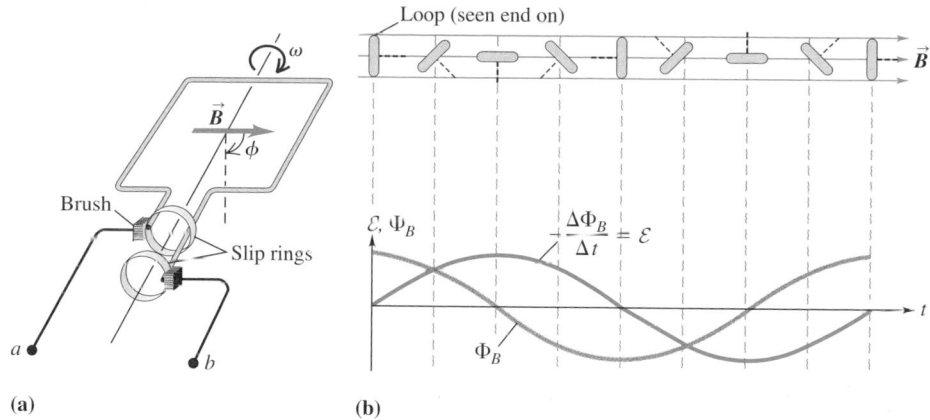

Loop (seen end on)

Brush

Slip rings

(a)   (b)

▲ **FIGURE 21.11** (a) Schematic diagram of a simple alternator, using a conducting loop rotating in a magnetic field. The loop connects to the external circuit via the slip rings. (b) Graphs of the flux through the loop (with loop position shown schematically) and of the resulting emf across terminals *ab*.

an induced emf according to Faraday's law. Figure 21.11a shows a simplified model of such a device. A rectangular loop with area A rotates with constant angular velocity ω about the axis shown. As the loop rotates, the magnetic flux through it changes sinusoidally with time, inducing a sinusoidally varying emf. If the shaft of the coil is connected, for instance, to a windmill, it converts wind power into electric power. The magnetic field $\vec{B}$ is uniform and constant. At time $t = 0$, $\phi = 0$; the figure shows the loop at the position $\phi = 90°$.

To analyze the behavior of this device, we need to use the generalized form of Faraday's law, mentioned above; we assume here that $\mathcal{E}$, Φ_B, and $\Delta\Phi_B/\Delta t$ can be either positive or negative. As the coil rotates, the signs of both $\mathcal{E}$ and Φ_B reverse twice during each revolution. The flux Φ_B through the loop equals its area A, multiplied by the component of B perpendicular to the plane of the loop—that is, $B\cos\phi$. Thus $\Phi_B = AB\cos\phi = AB\cos\omega t$, where now $\cos\phi$ and Φ_B can be either positive or negative.

Because Φ_B is a sinusoidal function of time, $\Delta\Phi_B/\Delta t$ isn't constant. To derive an expression for $\Delta\Phi_B/\Delta t$ would require calculus, but we can make an educated guess even without calculus. Figure 21.11b includes a graph of Φ_B as a function of time. At each point on the curve, the rate of change of Φ_B is equal to the *slope* of the curve. Where the curve is horizontal, Φ_B momentarily is not changing; where the slope is steepest, Φ_B is changing most rapidly with time. So, from the shape of the graph of Φ_B versus t, we can sketch the graph of $\Delta\Phi_B/\Delta t$ as a function of time. That sketch is shown in Figure 21.11b; it looks like a *sine* curve, and indeed it is. The rate of change of Φ_B turns out to equal $-\omega BA\sin\omega t$, so, from Faraday's law, the emf $\mathcal{E}$ is given by

$$\mathcal{E} = -\frac{\Delta\Phi_B}{\Delta t} = \omega AB\sin\omega t.$$

This equation shows that the induced emf $\mathcal{E}$ varies sinusoidally with time, as shown in Figure 21.11b. When the plane of the loop is perpendicular to $\vec{B}$ ($\phi = 0$ or $180°$), Φ_B reaches its maximum (AB) and minimum ($-AB$) values. At these times, its instantaneous rate of change is zero and $\mathcal{E}$ is zero. Also, $|\mathcal{E}|$ is greatest in magnitude when the plane of the loop is parallel to $\vec{B}$ ($\phi = 90°$ or $270°$) and Φ_B is changing most rapidly. Note that the induced emf depends, not on the *shape* of the loop, but only on its area. Finally, a real windmill generator of this type would use a coil with many turns, thus greatly increasing the induced emf.

The rotating loop is the prototype of a common kind of alternating-current (ac) generator, or *alternator;* it develops a sinusoidally varying emf. We can use

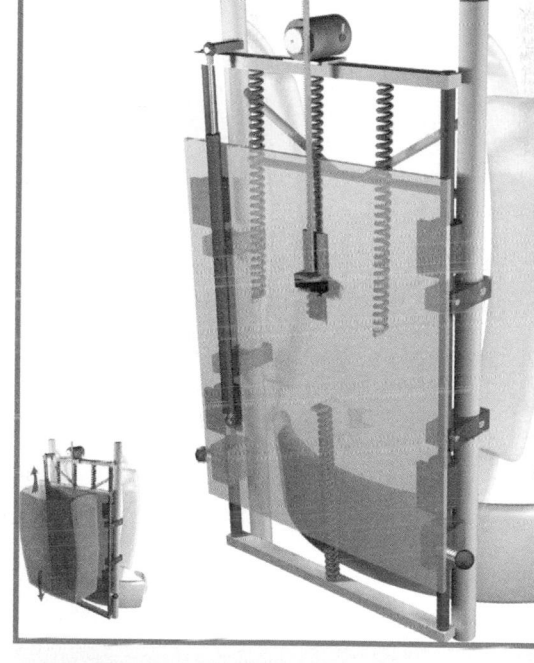

▲ **Application Geek alert!** Sure, you'd like to take your cell phone, GPS unit, PDA, and night-vision goggles on your two-month outback trek, but what about power? Who wants to crank a manual generator? The answer is this pack. Invented by Professor Larry Rome of the University of Pennsylvania, the device harnesses energy from the striding motion of walking that would otherwise be wasted. Instead of being fixed to the pack supports, the pack load is free to ride up and down as you stride. As it does so, a pinion uses the motion to turn a small generator. At a pace of 5.5 kilometers per hour, a 29 kilogram load can generate about 4 watts, plenty to power a number of small devices. Geek appeal aside, this device could be useful to field scientists, disaster-relief workers, and others operating in remote areas.

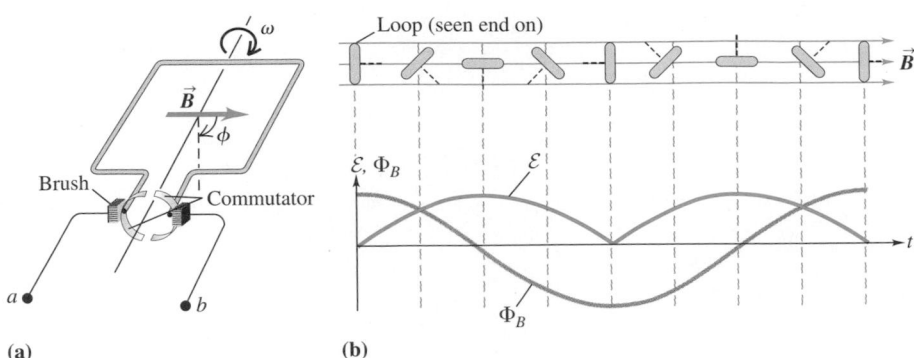

(a)

(b)

▲ **FIGURE 21.12** (a) Schematic diagram of a dc generator using a split-ring commutator. (b) Graph of the flux through the loop (with loop position shown schematically) and of the resulting emf across terminals *ab*.

it as a source of emf in an external circuit by adding two *slip rings,* which rotate with the loop, as shown in Figure 21.11a. Stationary contacts called *brushes* slide on the rings and are connected to the output terminals *a* and *b*.

We can use a similar scheme to obtain an emf that always has the same sign. The arrangement shown in Figure 21.12a is called a *commutator;* it reverses the connections to the external circuit at those angular positions where the emf reverses. The resulting emf is shown in Figure 21.12b. This device is the proto-type of a direct-current (dc) generator. Commercial dc generators have a large number of coils and commutator segments; this arrangement smooths out the bumps in the emf, so the terminal voltage is not only unidirectional, but also practically constant. This brush-and-commutator arrangement is also used in the direct-current motor that we discussed in Section 20.6.

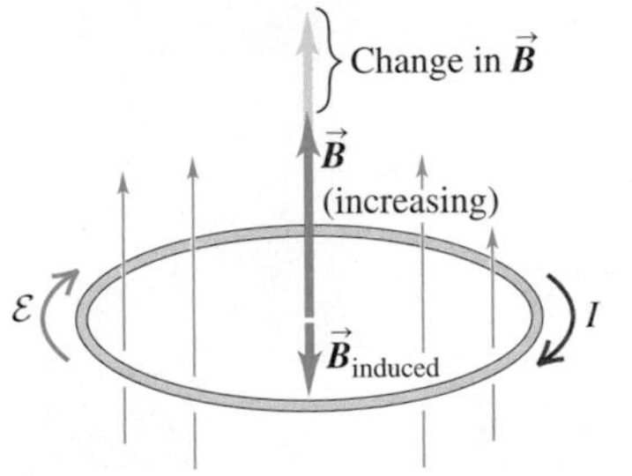

▲ **FIGURE 21.13** Relation among a changing external magnetic field, the resulting induced current, and the magnetic field induced by the current.

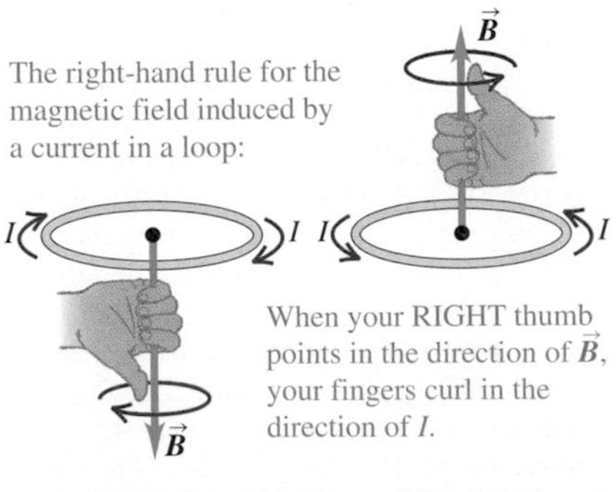

21.4 Lenz's Law

So far we've discussed the *magnitude,* but not the *direction,* of an induced emf and the associated current. To determine the direction of an induced current or emf, we use *Lenz's law.* This law is not an independent principle; it can be derived from Faraday's law with an appropriate set of sign conventions that we haven't discussed in detail. Lenz's law also helps us to gain an intuitive under-standing of various induction effects and of the role of energy conservation. H. F. E. Lenz (1804–1865) was a German scientist who duplicated independently many of the discoveries of Faraday and Henry. **Lenz's law** is as follows:

Lenz's law
The direction of any magnetically induced current or emf is such as to oppose the direction of the phenomenon causing it.

The right-hand rule for the magnetic field induced by a current in a loop:

When your RIGHT thumb points in the direction of $\vec{B}$, your fingers curl in the direction of *I*.

▲ **FIGURE 21.14** A review of the right-hand rule for the direction of the magnetic field induced by a current in a conducting loop.

The cause of a magnetically induced current or emf may be a changing flux through a stationary circuit, or the motion of a conductor in a magnetic field, or any combination of the two, as shown in Figure 21.1. A simple example of the first case is shown in Figure 21.13: A conducting loop is placed in an increasing magnetic field. According to Faraday's law, a current is induced in the loop; this current produces an additional magnetic field, as we discussed in Section 20.9 and as is shown in Figure 21.14. Lenz's law states that this additional field must

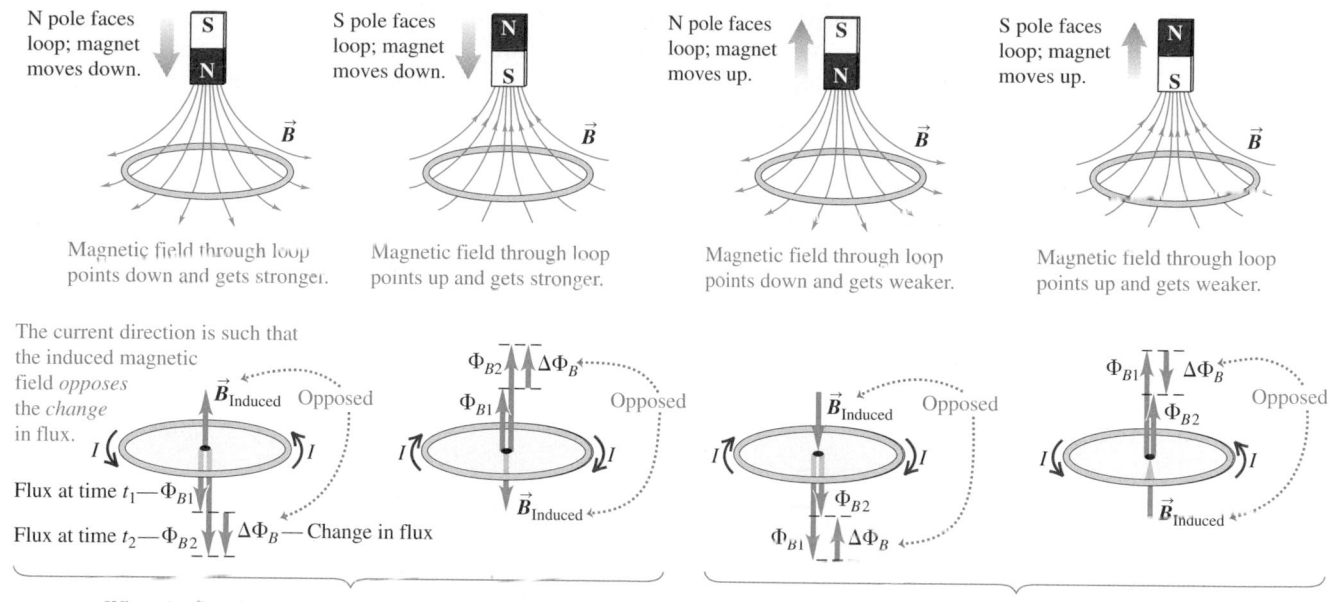

▲ FIGURE 21.15 Directions of induced currents as a bar magnet moves along the axis of a conducting loop.

oppose the direction in which the original field is increasing, so it must be downward in Figure 21.13. This in turn requires the induced current to be clockwise as seen from above.

Figure 21.15 shows how the foregoing analysis works out for four situations. Within the loop, the induced magnetic field points *opposite* to the original field if the original field or flux is *increasing.* Conversely, if the original field is *decreasing,* the additional field caused by the induced current points in the *same* direction as the original field. In each case, as Lenz's law requires, the induced magnetic field opposes the *change in flux* through the circuit (*not* the flux itself).

A similar analysis can be made for a conducting loop that moves in a magnetic field. If the magnetic flux through the loop changes, an induced current is produced. As in the preceding case, the current generates its own induced magnetic field. According to Lenz's law, this induced magnetic field *opposes* the *change* in flux that induced the current in the first place. Thus, if the flux is increasing, the induced magnetic field points opposite to the original magnetic field (opposing the increase in flux), and if the flux is decreasing, the induced magnetic field points in the same direction as the original magnetic field (opposing the decrease in flux).

There is another, equivalent, way to look at this second case. Remember from Chapter 20 that a current-carrying conductor is acted upon by a force when placed in a magnetic field. (See Figures 20.10 and 20.24 to review the relevant right-hand rule.) According to Lenz's law, this force is directed so as to *oppose* the *motion* that induces the current. The induced current adopts the direction that creates such a force.

NOTE ▶ In all these cases, the induced current tends to preserve the *status quo* by opposing the motion or the change of flux that originally induced it. ◀

To have an induced current, we need a complete circuit. If a conductor does *not* form a complete circuit, then we can mentally complete the circuit between the ends of the conductor and use Lenz's law to determine the direction of the current. We can then deduce the polarity of the ends of the open-circuit conductor. The direction from the − end to the + end within the conductor is the direction the current would have if the circuit were complete.

EXAMPLE 21.5 The force on a slide-wire rod

Figure 21.16a shows the slide-wire generator from Example 21.4. The conducting rod is moving to the right. Find the direction of the current induced in the loop and the direction of the force exerted on the rod by the resulting induced magnetic field.

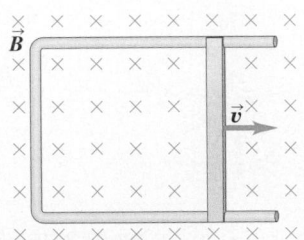

(a) Slide wire moving in magnetic field

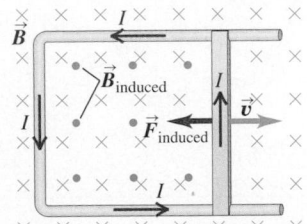

(b) Induced current, magnetic field, and magnetic force on slide wire

▲ FIGURE 21.16

SOLUTION

SET UP AND SOLVE The rod's motion increases the flux through the loop, so, by Lenz's law, the direction of the induced current must be such as to oppose this change—that is, to produce an induced magnetic field pointing out of the page (Figure 21.16b). By the right-hand rule for the magnetic field due to a current in a conductor (see Figure 21.14), the induced current is counterclockwise.

Lenz's law also tells us that the induced current creates a left-directed force on the moving bar, opposing the motion that causes the increase in flux. Does our counterclockwise current produce such a force? We apply the right-hand rule for the force on a current-carrying conductor in a magnetic field. (See Figure 20.24.) As predicted, the force points to the left. (Note that we use the direction of the *original* magnetic field in apply-

ing this right-hand rule, not the direction of the induced field. The original field is always stronger than the induced field and hence dominates.)

REFLECT We can apply Lenz's law to a moving conductor by looking at either the induced magnetic field (which opposes the change in magnetic flux) or the force on the conductor (which opposes the motion that causes the change in flux). The two perspectives give the same result and can be used to check each other.

The behavior described in this example is also directly related to energy conservation. If the induced current were in the opposite direction, the resulting magnetic force on the rod would accelerate it to an ever-increasing speed with no external energy source, despite the fact that electrical energy is being dissipated in the loop. This effect would be a clear violation of energy conservation.

EXAMPLE 21.6 Induced current in a resistance–capacitance circuit

Loop *A* in Figure 21.17 is part of a resistance–capacitance circuit. When the switch is closed, the capacitor discharges. While the capacitor is discharging, what are the directions of the induced emf, the induced current, and the induced magnetic field in loop *B*?

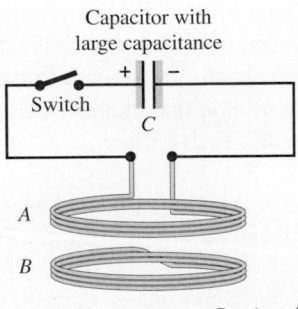

▶ FIGURE 21.17

Continued

SOLUTION

SET UP AND SOLVE When we close the switch, the current is from the positive to the negative plate of the capacitor (counterclockwise in loop *A* as viewed from above), creating an upward magnetic field through loop *B*. As we learned in Section 19.8, the current decreases with time, so the flux through loop *B* is decreas-

ing. To counter the decrease in flux, the induced magnetic field must point upward, meaning that the induced emf and the current in loop *B* are counterclockwise as viewed from above.

REFLECT As we've noted, when the flux is decreasing, the induced magnetic field points in the *same direction* as the original magnetic field.

21.5 Motional Electromotive Force

When a conductor moves in a magnetic field, we can gain added insight into the resulting induced emf by considering the magnetic forces on charges in the conductor. Figure 21.18a shows the same moving rod that we discussed in Example 21.5, separated for the moment from the U-shaped conductor. The magnetic field $\vec{B}$ is uniform and directed into the page, and we give the rod a constant velocity $\vec{v}$ to the right. A charged particle q in the rod then experiences a magnetic force with magnitude $F_B = |q|vB$. In the following discussion, we'll assume that q is positive; in that case, the direction of this force is *upward* along the rod, from b toward a. (The emf and the current are the same whether q is positive or negative.)

This magnetic force causes the free charges in the rod to move, creating an excess of positive charge at the upper end a and negative charge at the lower end b. This, in turn, creates an electric field $\vec{E}$ in the direction from a to b, which exerts a downward force $\vec{F}_E$ on the charges. The accumulation of charge at the ends of the rod continues until $\vec{E}$ is large enough for the downward electric force (with magnitude qE) to cancel exactly the upward magnetic force (magnitude qvB). Then $qE = qvB$ and the charges are in equilibrium, with point a at higher potential than point b.

What is the magnitude of the potential difference V_{ab}? It is equal to the electric field magnitude E, multiplied by the length L of the rod. From the preceding discussion, $E = vB$, so

$$V_{ab} = EL = vBL, \qquad (21.6)$$

with point a at higher potential than point b.

Now suppose the moving rod slides along a stationary U-shaped conductor, forming a complete circuit (Figure 21.18b). No *magnetic* force acts on the charges in the stationary conductor, but there is an *electric* field caused by the accumulations of charge at a and b. Under the action of this field, a current is

MasteringPHYSICS

ActivPhysics 13.10: Motional emf

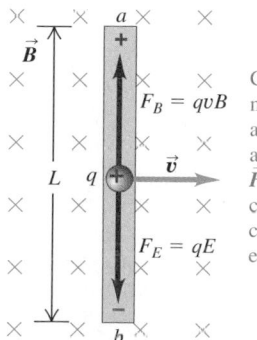

Charges in the moving rod are acted upon by a magnetic force $\vec{F}_B$; the resulting charge separation creates a canceling electric force $\vec{F}_E$.

(a) Isolated moving rod

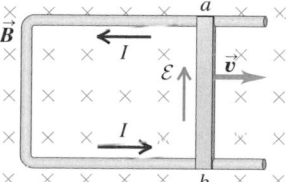

The motional emf $\mathcal{E}$ in the moving rod creates an electric field in the stationary conductor.

(b) Rod connected to stationary conductor

▲ **FIGURE 21.18** (a) A conducting rod moving in a uniform magnetic field. The rod, its velocity, and the magnetic field are mutually perpendicular. (b) The direction of the induced current when the rod is connected to a circuit.

◀ **Application Tether power** What to do about space debris? Every year, we put more satellites into low earth orbit. When their useful life is over, we should remove them to prevent collisions. One method would be to have the satellite use its thruster rocket to lower its orbit until it enters the atmosphere and burns up. But that requires lifting extra fuel into orbit and works only if the satellite is functional. The Terminator Tether offers a low-weight, reliable alternative. On a command from the ground, the tether module reels out a kilometers-long, very thin conducting cable. As the satellite orbits, the earth's magnetic field creates an upward emf in the cable. The cable can exchange electrons with the earth's ionosphere, so the emf drives an upward current in the cable. The resulting interaction with the earth's magnetic field produces a drag force tending to slow the satellite. As the satellite slows, its orbit decays until it enters the atmosphere and burns up.

established in the counterclockwise sense around the complete circuit. The moving rod has become a source of emf; within it, charge moves from lower to higher potential, and in the remainder of the circuit, charge moves from higher to lower potential. We call this emf a **motional emf;** and as with other emf's, we denote it as $\mathcal{E}$. When the velocity, magnetic field, and length are mutually perpendicular,

$$\mathcal{E} = vBL. \tag{21.7}$$

This is the same result that we obtained with Faraday's law in Section 21.3.

The emf associated with the moving rod in Figure 21.18 is analogous to that of a battery with its positive terminal at a and its negative terminal at b, although the origins of the two emf's are completely different. In each case, a non-electrostatic force acts on the charges in the device in the direction from b to a, and the emf is the work per unit charge done by this force when a charge moves from b to a in the device. When the device is connected to an external circuit, the direction of current is from b to a in the device and from a to b in the external circuit (in this case, the U-shaped conductor to the left of the rod).

If we express v in meters per second, B in teslas, and L in meters, $\mathcal{E}$ is in joules per coulomb, or volts. We invite you to verify this statement.

Conceptual Analysis 21.3

Forces on a square loop

The square loop in Figure 21.19 is pulled upward from between the poles of the magnet. How do the magnetic forces on the top and bottom wires compare?

A. The forces are directed downward on both wires and are equal in magnitude.
B. The forces are directed downward on both wires, but the force on the bottom wire is stronger.
C. The forces are directed upward on the top wire and downward on the bottom wire and are equal in magnitude.
D. The forces are directed upward on the top wire and downward on the bottom wire; the force on the bottom wire is stronger.

SOLUTION As the loop moves up from between the poles of the magnet, the flux through it decreases, inducing a current. Lenz's law tells us that a net downward force acts on the loop, opposing its upward motion. However, Lenz's law doesn't tell us about the relative magnitudes of the forces on the top and bottom wires. For that information, we must consider the magnetic forces acting on these two current-carrying wires.

First we determine the direction of the current. We could use Lenz's law, but instead let's use the right-hand rule for the force on a moving charge. The wires move upward and the magnetic

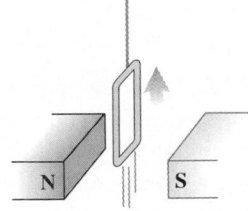

▲ **FIGURE 21.19**

field points to the right, so, by the right-hand rule, the free electrons in both wires experience a magnetic force out of the plane of the page. Because the bottom wire is in a stronger magnetic field, its electron forces are greater, so the overall conventional current is clockwise as viewed from the N pole of the magnet. The current on the top wire is therefore out of the plane of the page, and the force on this wire (by the right-hand rule) is upward. The current in the bottom wire is into the plane of the page, so the force on that wire is downward. Because the bottom wire is in a stronger magnetic field, the force exerted on it is stronger, so answer D is correct. As Lenz's law predicts, the net force on the loop is downward.

EXAMPLE 21.7   **Motional electromotive force**

Suppose the length L of the slide-wire rod in Figure 21.18b is 0.10 m, its speed v is 2.5 m/s, the total resistance of the loop is 0.030 Ω, and B is 0.60 T. Find the emf $\mathcal{E}$, the induced current, the force acting on the rod, and the mechanical power needed to keep the rod moving at constant speed.

SOLUTION

SET UP AND SOLVE To find the emf, we use Equation 21.7:

$$\mathcal{E} = vBL = (2.5 \text{ m/s})(0.60 \text{ T})(0.10 \text{ m}) = 0.15 \text{ V}.$$

From Ohm's law, the current in the loop is

$$I = \mathcal{E}/R = (0.15 \text{ V})/(0.030 \,\Omega) = 5.0 \text{ A}.$$

Lenz's law tells us that an induced force acts on the rod, opposite to the rod's direction of motion. The magnitude of this force is

$$F_{\text{induced}} = ILB = (5.0 \text{ A})(0.10 \text{ m})(0.60 \text{ T}) = 0.30 \text{ N}.$$

To keep the rod moving at constant speed, a force equal in magnitude and opposite in direction to $\vec{F}_{\text{induced}}$ must act on the rod. Therefore, the mechanical power P needed to keep the rod moving is

$$P = F_{\text{induced}}v = (0.30 \text{ N})(2.5 \text{ m/s}) = 0.75 \text{ W}.$$

REFLECT The expression for the emf is the same result we found in Example 21.4 from Faraday's law. The rate at which the induced emf delivers electrical energy to the circuit is

$$P = \mathcal{E}I = (0.15 \text{ V})(5.0 \text{ A}) = 0.75 \text{ W}.$$

This is equal to the mechanical power input, $F_{\text{induced}}v$, as we should expect. The system is converting mechanical energy (work) into electrical energy. Finally, the rate of *dissipation* of electrical energy in the circuit resistance is

$$P = I^2R = (5.0 \text{ A})^2(0.030 \,\Omega) = 0.75 \text{ W},$$

which is also to be expected.

Practice Problem: If $L = 0.15$ m, $v = 3.0$ m/s, the total resistance of the loop is 0.020 Ω, and $B = 0.5$ T, find the rate at which the induced emf delivers electrical energy to the circuit. *Answer:* $P = 2.5$ W.

21.6 Eddy Currents

In the examples of induction effects that we've studied, the induced currents have been confined to well-defined paths in conductors and other components, forming a *circuit*. However, many pieces of electrical equipment contain masses of metal moving in magnetic fields or located in changing magnetic fields. In situations like these, we can have induced currents that circulate throughout the volume of a conducting material. Because their flow patterns resemble swirling eddies in a river, we call these **eddy currents**.

As an example, consider a metallic disk rotating in a magnetic field perpendicular to the plane of the disk, but confined to a limited portion of the disk's area, as shown in Figure 21.20a. Sector Ob is moving across the field and has an emf induced in it. Sectors Oa and Oc are not in the field, but they provide conducting paths for charges displaced along Ob to return from b to O. The result is a circulation of eddy currents in the disk, somewhat as sketched in Figure 21.20b. The downward current in the neighborhood of sector Ob experiences a sideways magnetic-field force toward the right that *opposes* the rotation of the disk, as Lenz's law predicts. The return currents lie outside the field, so no such forces are exerted on them. The interaction between the eddy currents and the field causes a braking action on the disk and a conversion of electrical energy to heat because of the resistance of the material.

The shiny metal disk in the electric meter outside a house rotates as a result of eddy currents that are induced in it by magnetic fields caused by sinusoidally varying currents in a coil. Similar effects can be used to stop the rotation of a circular saw quickly when the power is turned off. Eddy current braking is also used on some electrically powered rapid-transit vehicles. Electromagnets mounted in the cars induce eddy currents in the rails; the resulting magnetic fields cause braking forces on the electromagnets and thus on the cars. Finally, the familiar metal detectors seen at security checkpoints in airports (Figure 21.21a) operate by detecting eddy currents induced in metallic objects. Similar devices (Figure 21.21b) are used to locate buried treasure such as bottle caps and lost pennies.

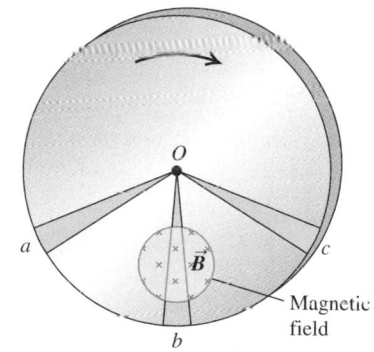

(a) Metal disk rotating through a magnetic field

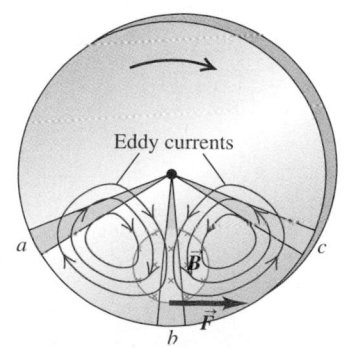

(b) Resulting eddy currents and braking force

▲ **FIGURE 21.20** The origin of eddy currents.

▲ **Application** **Plug-free power.** New technology may allow you to recharge one or more electronic devices simultaneously, while eliminating the need for specific plugs or holders for each device. A recharging pad plugs into a regular household outlet and generates a changing magnetic field just above its surface. When electronic devices with the appropriate circuitry are placed on this surface, the pad acts much like a transformer, inducing an emf in the devices. The resulting alternating current can be converted to direct current, which recharges the batteries of objects on the pad. A similar principle has been used for years in electric toothbrush rechargers, which operate without direct electrical contact between the toothbrush and the recharging stand, avoiding problems with moisture and possible short-circuiting of the charger.

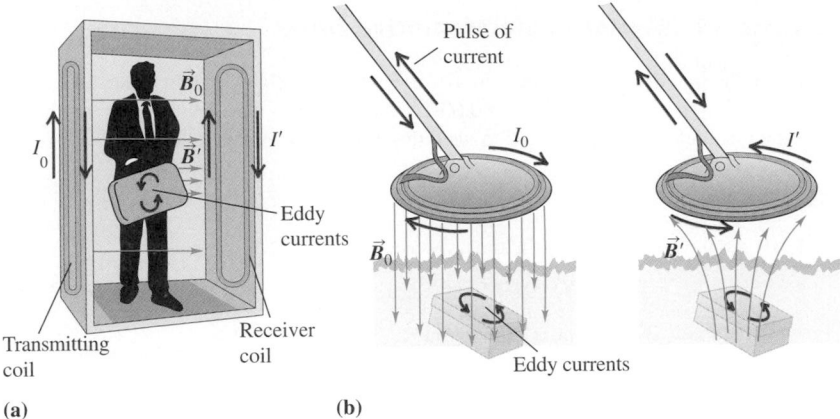

▲ **FIGURE 21.21** (a) A metal detector at an airport security checkpoint detects eddy currents induced in conducting objects by an alternating magnetic field. (b) Portable metal detectors work on the same principle.

21.7 Mutual Inductance

How can a 12 volt car battery provide the thousands of volts needed to produce sparks across the gaps of the spark plugs in the engine? Electrical transmission lines often operate at 500,000 volts or more. Applied directly to your household wiring, this voltage would incinerate everything in sight. How can it be reduced to the relatively tame 120 or 240 volts required by familiar electrical appliances?

The solutions to both of these problems, and to many others concerned with varying currents in circuits, involve the *induction* effects we've studied in this chapter. A changing current in a coil induces an emf in an adjacent coil. The coupling between the coils is described by their *mutual inductance*. This interaction is the operating principle of a *transformer*, which we'll study in Section 21.9.

Figure 21.22 is a cross-sectional view of two coils of wire. A current i_1 in coil 1 sets up a magnetic field, as indicated by the blue lines, and some of these field lines pass through coil 2. We denote the magnetic flux through each turn of coil 2, caused by the current i_1 in coil 1, as Φ_{B2}. (We're using the lowercase letter i, with a subscript, for time-varying currents.) The magnetic field is proportional to i_1, so Φ_{B2} is also proportional to i_1. When i_1 changes, Φ_{B2} changes, inducing an emf $\mathcal{E}_2$ in coil 2 with magnitude

$$\mathcal{E}_2 = N_2 \left| \frac{\Delta \Phi_{B2}}{\Delta t} \right|. \tag{21.8}$$

We could represent the proportionality of Φ_{B2} and i_1 in the form $\Phi_{B2} = (\text{constant}) \, i_1$, but instead it is more convenient to include the number of turns, N_2, in the relation. Introducing a proportionality constant M_{21}, we write

$$N_2 |\Phi_{B2}| = M_{21} |i_1|. \tag{21.9}$$

From this relation, it follows that

$$N_2 \left| \frac{\Delta \Phi_{B2}}{\Delta t} \right| = M_{21} \left| \frac{\Delta i_1}{\Delta t} \right|,$$

and we can rewrite Equation 21.8 as

$$\mathcal{E}_2 = M_{21} \left| \frac{\Delta i_1}{\Delta t} \right|. \tag{21.10}$$

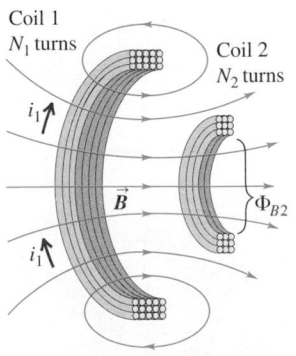

▲ **FIGURE 21.22** In mutual inductance, a portion of the magnetic flux set up by a current in coil 1 links with coil 2.

We can repeat this discussion for the opposite case, in which a changing current i_2 in coil 2 causes a changing flux Φ_{B1}, and an emf $\mathcal{E}_1$, in coil 1. The results, corresponding to Equations 21.8 and 21.10, are

$$\mathcal{E}_1 = N_1 \left| \frac{\Delta \Phi_{B1}}{\Delta t} \right| \quad \text{and} \quad \mathcal{E}_1 = M_{12} \left| \frac{\Delta i_2}{\Delta t} \right|.$$

We might expect that the corresponding constant M_{12} would be different from M_{21}, because, in general, the two coils are not identical and the flux through them is not the same. It turns out, however, that M_{12} is *always* equal to M_{21}, even when the two coils are not symmetric. We call this common value the **mutual inductance** M; it characterizes completely the induced-emf interaction of two coils.

Mutual inductance

The mutual inductance M of two coils is given by

$$M = M_{21} = M_{12} = \left| \frac{N_2 \Phi_{B2}}{i_1} \right| = \left| \frac{N_1 \Phi_{B1}}{i_2} \right|. \qquad (21.11)$$

From the preceding analysis, we can also write

$$\mathcal{E}_2 = M \left| \frac{\Delta i_1}{\Delta t} \right| \quad \text{and} \quad \mathcal{E}_1 = M \left| \frac{\Delta i_2}{\Delta t} \right|. \qquad (21.12)$$

Unit: The SI unit of mutual inductance is called the **henry** (1 H), in honor of Joseph Henry (1797–1878), one of the discoverers of electromagnetic induction. From Equation 21.11, one henry is equal to *one weber per ampere*. Other equivalent units, obtained by reference to Equation 21.10, are *one volt-second per ampere* and *one ohm-second:*

$$1\ \text{H} = 1\ \text{Wb/A} = 1\ \text{V} \cdot \text{s/A} = 1\ \Omega \cdot \text{s}.$$

If the coils are in vacuum, M depends only on their geometry. If a magnetic material is present, M also depends on the magnetic properties of the material.

EXAMPLE 21.8   The Tesla coil

A Tesla coil is a type of high-voltage generator; you may have seen one in a science museum. In one form of Tesla coil, shown in Figure 21.23, a long solenoid with length l and cross-sectional area A is closely wound with N_1 turns of wire. A coil with N_2 turns surrounds it at its center. Find the mutual inductance of the coils.

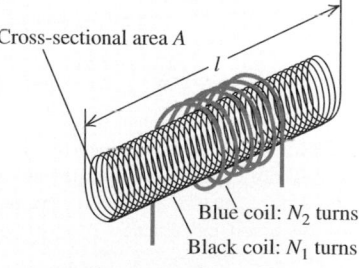

Cross-sectional area A

Blue coil: N_2 turns
Black coil: N_1 turns

▶ **FIGURE 21.23** One form of Tesla coil, a long solenoid with cross-section area A and N_1 turns, surrounded at its center by a small coil with N_2 turns.

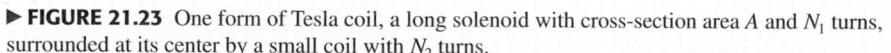

SOLUTION

SET UP A current i_1 in the solenoid sets up a magnetic field $\vec{B}_1$ at its center. According to Equation 20.14 (Section 20.9), the magnitude B of the magnetic field at the center of a solenoid is given by $B = \mu_0 nI$, where $n = N/l$ is the number of turns per unit length. Using this equation, we find that

$$B_1 = \mu_0 ni = \frac{\mu_0 N_1 i_1}{l}.$$

The flux through each turn at the center of coil 1, Φ_{B1}, equals $B_1 A$. All of this flux also passes through coil 2, so $\Phi_{B2} = \Phi_{B1} = B_1 A$.

SOLVE From Equation 21.11, the mutual inductance M is

$$M = \left| \frac{N_2 \Phi_{B2}}{i_1} \right| = \left| \frac{N_2}{i_1} B_1 A \right| = \left| \frac{N_2}{i_1} \frac{\mu_0 N_1 i_1}{l} A \right| = \frac{\mu_0 A N_1 N_2}{l}.$$

REFLECT Here's a numerical example to give you an idea of magnitudes. Suppose $l = 0.50\ \text{m}$, $A = 10\ \text{cm}^2 = 1.0 \times 10^{-3}\ \text{m}^2$, $N_1 = 1000$ turns, and $N_2 = 10$ turns. Then

$$M = \frac{[4\pi \times 10^{-7}\ \text{Wb/(A} \cdot \text{m)}](1.0 \times 10^{-3}\ \text{m}^2)(1000)(10)}{0.50\ \text{m}}$$

$$= 25 \times 10^{-6}\ \text{Wb/A} = 25 \times 10^{-6}\ \text{H} = 25\ \mu\text{H}.$$

EXAMPLE 21.9 **The average magnetic flux through a Tesla coil**

Suppose the current i_2 in the smaller coil in Example 21.8 is given by $i_2 = (2.0 \times 10^6 \text{ A/s})t$. At time $t = 3.0 \, \mu\text{s}$, what is the average magnetic flux through each turn of the solenoid caused by the current in the smaller coil? What is the induced emf in the solenoid?

SOLUTION

SET UP AND SOLVE At time $t = 3.0 \, \mu\text{s}$, the current in coil 2 is

$$i_2 = (2.0 \times 10^6 \text{ A/s})(3.0 \times 10^{-6} \text{ s}) = 6.0 \text{ A}.$$

To find the flux in the solenoid, we use Equation 21.9, with the roles of coils 1 and 2 interchanged, so that $N_1|\Phi_{B1}| = M|i_2|$. Solving for the average magnetic flux Φ_{B1}, we obtain

$$|(\Phi_{B1})_{\text{av}}| = \frac{M|i_2|}{N_1} = \frac{(25 \times 10^{-6} \text{ H})(6.0 \text{ A})}{1000} = 1.5 \times 10^{-7} \text{ Wb}.$$

This is an average value; the flux will vary considerably from the center to the ends.

The induced emf $\mathcal{E}_1$ is given by Equation 21.12, with the change in current per unit time, $\Delta i_2/\Delta t$, equal to 2.0×10^6 A/s:

$$\mathcal{E}_1 = \left| M\frac{\Delta i_2}{\Delta t} \right| = (25 \times 10^{-6} \text{ H})(2.0 \times 10^6 \text{ A/s}) = 50 \text{ V}.$$

REFLECT In an operating Tesla coil, $\Delta i_2/\Delta t$ would be alternating much more rapidly, and its magnitude would be much larger than in this example.

Practice Problem: For the given Tesla coil, how many turns (N_1) should coil 1 have in order to get an induced emf magnitude of 650 V? *Answer:* $N_1 = 13{,}000$ turns.

21.8 Self-Inductance

In our discussion of mutual inductance, we assumed that one circuit acted as the source of a magnetic field and that the emf under consideration was induced in a separate, independent circuit when some of the magnetic flux created by the first circuit passed through the second. However, when a current is present in *any* circuit, this current sets up a magnetic field that links with *the same* circuit and changes when the current changes. Any circuit that carries a varying current has an induced emf in it resulting from the variation in *its own* magnetic field. Such an emf is called a **self-induced emf.**

As an example, consider a coil with N turns of wire, carrying a current i (Figure 21.24). As a result of this current, a magnetic flux Φ_B passes through each turn. In analogy to Equation 21.11, we define the **inductance** L of the circuit (sometimes called **self-inductance**):

$$L = \left| \frac{N\Phi_B}{i} \right|, \qquad \text{or} \qquad N|\Phi_B| = L|i|. \tag{21.13}$$

If Φ_B and i change with time, then

$$N\left| \frac{\Delta \Phi_B}{\Delta t} \right| = L\left| \frac{\Delta i}{\Delta t} \right|.$$

From Faraday's law, Equation 21.4, the magnitude $\mathcal{E}$ of the self-induced emf is $\mathcal{E} = N|\Delta\Phi_B/\Delta t|$, so we can also state the definition of L as follows:

Definition of self-inductance

The self-inductance L of a circuit is the magnitude of the self-induced emf $\mathcal{E}$ per unit rate of change of current, so that:

$$\mathcal{E} = L\left| \frac{\Delta i}{\Delta t} \right|. \tag{21.14}$$

From the definition, the units of self-inductance are the same as those of mutual inductance; the SI unit of self-inductance is *one henry*.

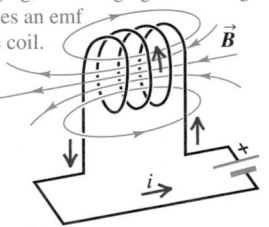

Self-inductance: If the current i in the coil is changing, the changing flux through the coil induces an emf in the coil.

$\vec{B}$

i

▲ **FIGURE 21.24** A changing current in a coil produces a changing magnetic field that in turn induces an emf in the coil—the phenomenon of self-inductance.

A circuit, or part of a circuit, that is designed to have a particular inductance is called an **inductor,** or a *choke.* Like resistors and capacitors, inductors are among the indispensable circuit elements of modern electronics. In later sections, we'll explore the circuit behavior of inductors. The usual circuit symbol for an inductor is

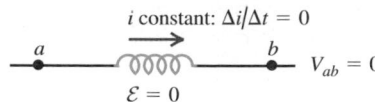

(a) When the current is constant, there is no self-induced emf, so the voltage across the inductor is zero.

(b) and (c) show that a changing current induces an emf and hence a voltage across the inductor.

We can find the direction of the self-induced emf from Lenz's law. The cause of the induced emf is the *changing current* in the conductor, and the emf always acts to oppose this change. Figure 21.25 shows several cases. In Figure 21.25a, the current is constant, Δi is always zero, and $V_{ab} = 0$. In Figure 21.25b, the current is increasing and $\Delta i/\Delta t$ is positive. According to Lenz's law, the induced emf must oppose the increasing current. The emf therefore must be in the sense from b to a; that is, a becomes the terminal with higher potential, and V_{ab} is *positive,* as shown in the figure. The emf opposes the increase in current caused by the external circuit. The direction of the emf is analogous to that of a battery with a as its + terminal.

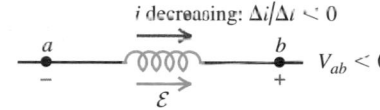

(b) If the current is *increasing,* then, by Lenz's law, the emf points opposite to i.

In Figure 21.25c, the situation is the opposite. The current is decreasing and $\Delta i/\Delta t$ is negative. The induced emf $\mathcal{E}$ opposes this decrease, and V_{ab} is negative. In both cases, the induced emf opposes, not the current itself, but the *rate of change* $\Delta i/\Delta t$ of the current. Thus, the circuit behavior of an inductor is quite different from that of a resistor.

(c) If the current is *decreasing,* then, by Lenz's law, the emf points in the same direction as i.

▲ **FIGURE 21.25** Relation of current and self-induced emf in an inductor with negligible resistance. The voltage V_{ab} depends on the time rate of change of current.

EXAMPLE 21.10 Inductance of a toroidal solenoid

An air-core toroidal solenoid with cross-sectional area A and mean radius r is closely wound with N turns of wire (Figure 21.26). Determine its self-inductance L. In calculating the flux, assume that B is uniform across a cross section; neglect the variation of B with distance from the toroidal axis.

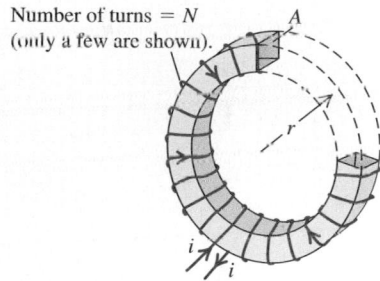

Number of turns = N (only a few are shown).

▲ **FIGURE 21.26**

SOLUTION

SET UP From Equation 21.13, which defines inductance,

$$L = \left| \frac{N\Phi_B}{i} \right|.$$

SOLVE To find Φ_B, we first find the field magnitude B. From Equation 20.15, the magnetic field magnitude B at a distance r from the toroidal axis is $B = \mu_0 Ni/2\pi r$. If we assume that the field has this magnitude over the entire cross-sectional area A, then the total flux Φ_B through the cross section is

$$\Phi_B = BA = \left| \frac{\mu_0 NiA}{2\pi r} \right|.$$

All the flux links with each turn, and the self-inductance L is

$$L = \left| \frac{N\Phi_B}{i} \right| = \frac{\mu_0 N^2 A}{2\pi r}.$$

REFLECT As an example, suppose $N = 200$ turns, $A = 5.0 \text{ cm}^2 = 5.0 \times 10^{-4} \text{ m}^2$, and $r = 0.10$ m; then

$$L = \frac{[4\pi \times 10^{-7} \text{ Wb}/(\text{A} \cdot \text{m})](200)^2(5.0 \times 10^{-4} \text{ m}^2)}{2\pi(0.10 \text{ m})}$$

$$= 40 \times 10^{-6} \text{ H} = 40 \ \mu\text{H}.$$

This inductor with 200 turns is somewhat bigger than an ordinary doughnut.

Practice Problem: For this toroidal solenoid, how many turns would be required for an inductance of $360 \ \mu\text{H}$? *Answer:* 600 turns.

The inductance of a circuit depends on its size, shape, and number of turns. For N turns close together, it is always proportional to N^2. The inductance also depends on the magnetic properties of the material enclosed by the circuit. In the preceding examples, we assumed that the conductor was surrounded by vacuum. If the flux is concentrated in a region containing a magnetic material with permeability μ, then, in the expression for B, we must replace μ_0 (the permeability of vacuum) by $\mu = K_m\mu_0$, as we discussed in Section 20.11. If the material is *ferromagnetic,* this difference is of crucial importance: An inductor wound on a soft iron core having $K_m = 5000$ has an inductance approximately 5000 times as great as the same coil with an air core. Iron-core and ferrite-core inductors are widely used in a variety of electric-power and electronic applications.

21.9 Transformers

One of the great advantages of alternating current (ac, usually varying sinusoidally with time) over direct current (dc, not varying with time), for electric-power distribution is that it is much easier to step voltage levels up and down with ac than with dc. For long-distance power transmission, it is desirable to use as high a voltage and as small a current as possible; this approach reduces I^2R losses in the transmission lines, so smaller wires can be used for a given power level, enabling us to save on material costs. Present-day transmission lines routinely operate at voltages on the order of 500 kV. However, safety considerations and insulation requirements dictate relatively low voltages in generating equipment and in household and industrial power distribution. The standard voltage for household wiring is 120 V in the United States and Canada and 240 V in most of Western Europe. The necessary voltage conversion is accomplished by the use of **transformers.**

In the discussion that follows, the symbols for voltage and current usually represent amplitudes—that is, *maximum* magnitudes of quantities that vary sinusoidally with time. In fact, alternating voltages and currents are usually described in terms of rms (root-mean-square) values, which, for sinusoidally varying quantities, are less than the maximum values by a factor of $1/\sqrt{2}$. For example, the standard 120 V household voltage has a *maximum* (peak) value of $(120\text{ V})\sqrt{2} = 170$ V. (We'll discuss this distinction in greater detail in Chapter 22.)

A transformer consists of two coils (usually called *windings*), electrically insulated from each other but wound on the same core. In Section 21.7, we discussed the mutual inductance of such a system. The winding to which power is supplied is called the **primary;** the winding from which power is delivered is called the **secondary.** Transformers used in power-distribution systems have soft iron cores. The circuit symbol for an iron-core transformer is

Here's how a transformer works: An alternating current in either winding sets up an alternating flux in the core, and according to Faraday's law, this induces an

▶ Application **The last step down.** You've probably seen "cans" like these attached to the tops of local utility poles. These are step-down transformers; they bring the voltage down from the 2400 V put out by local substations to the 240 V commonly distributed to houses. The substations have transformers that bring down the voltage from the much higher values—typically, 240,000 V or more—of long-distance power lines. Power stations use step-up transformers to create these high voltages.

emf in each winding. Energy is transferred from the primary winding to the secondary winding via the core flux and its associated induced emf's.

An idealized transformer is shown in Figure 21.27. We assume that all the flux is confined to the iron core, so at any instant, the magnetic flux Φ_B is the same in the primary and secondary coils. We also neglect the resistance of the windings. The primary winding has N_1 turns, and the secondary winding has N_2 turns. When the magnetic flux changes because of changing currents in the two coils, the resulting induced emf's are

$$\mathcal{E}_1 = N_1 \left| \frac{\Delta \Phi_B}{\Delta t} \right| \quad \text{and} \quad \mathcal{E}_2 = N_2 \left| \frac{\Delta \Phi_B}{\Delta t} \right|.$$

Because the same flux links both primary and secondary, these expressions show that the induced emf *per turn* is the same in each. The ratio of the secondary emf $\mathcal{E}_2$ to the primary emf $\mathcal{E}_1$ is therefore equal to the ratio of secondary to primary turns, often called the *turns ratio:*

$$\frac{\mathcal{E}_2}{\mathcal{E}_1} = \frac{N_2}{N_1}.$$

If the windings have zero resistance, the induced emf's $\mathcal{E}_1$ and $\mathcal{E}_2$ are respectively equal to the corresponding terminal voltages V_1 and V_2, and we find the following:

Relation of voltage to winding turns for a transformer

For an ideal transformer (with zero resistance),

$$\frac{V_2}{V_1} = \frac{N_2}{N_1}. \tag{21.15}$$

By choosing the appropriate turns ratio N_2/N_1, we may obtain any desired secondary voltage from a given primary voltage. If $V_2 > V_1$, we have a *step-up* transformer; if $V_2 < V_1$, we have a *step-down* transformer. The V's can be either both amplitudes or both rms values.

If the secondary circuit is completed by a resistance R, then $I_2 = V_2/R$. From energy considerations, the power delivered to the primary equals that taken out of the secondary, so

$$V_1 I_1 = V_2 I_2. \tag{21.16}$$

We can combine Equations 21.15 and 21.16 with the relation $I_2 = V_2/R$ to eliminate V_2 and I_2:

$$\frac{V_1}{I_1} = \frac{R}{(N_2/N_1)^2}. \tag{21.17}$$

This equation shows that when the secondary circuit is completed through a resistance R, the result is the same as if the *source* had been connected directly to

The induced emf *per turn* is the same in both coils, so we adjust the ratio of terminal voltages by adjusting the ratio of turns:

$$\frac{V_2}{V_1} = \frac{N_2}{N_1}$$

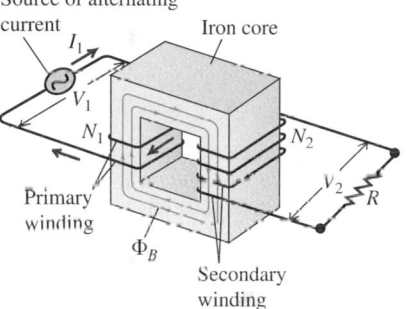

▲ **FIGURE 21.27** Schematic diagram of an idealized step-up transformer.

a resistance equal to R divided by the square of the turns ratio, $(N_2/N_1)^2$. In other words, the transformer "transforms" not only voltages and currents, but resistances as well.

EXAMPLE 21.11 A high-voltage coffee maker

A friend brings back from Europe a device that she claims to be the world's greatest coffee maker. Unfortunately, it was designed to operate from a 240 V line, standard in Europe. At this rms voltage, the coffee draws 960 W of power. (a) If your friend wants to operate the coffee maker in the United States, where the rms line voltage is 120 V, what does she need to do? (b) What current will the coffee maker draw from the 120 V line? (c) What is its resistance? (These voltages and currents are rms values, to be discussed in Chapter 22, so the power quantities are given by Equation 21.16.)

SOLUTION

SET UP Our friend needs to step up the 120 V that comes into her home to the 240 V required to operate the coffee maker, so she has to use a step-up transformer. From Equation 21.15, our friend needs a transformer with a turns ratio equal to the voltage ratio.

SOLVE Part (a): The input (primary) voltage is $V_1 = 120$ V, and we need an output (secondary) voltage $V_2 = 240$ V. From Equation 21.15, the required turns ratio is

$$\frac{N_2}{N_1} = \frac{V_2}{V_1} = \frac{240 \text{ V}}{120 \text{ V}} = 2.$$

Part (b): For a power of 960 W, the current on the 240 V side is 4.0 A. From Equation 21.16, the primary current (from the 120 V source) is 8.0 A. The *power* is the same on the primary and secondary sides:

$$P = (120 \text{ V})(8.0 \text{ A}) = (240 \text{ V})(4.0 \text{ A}) = 960 \text{ W}.$$

Part (c): For a current of 4.0 A on the 240 V side, the resistance R of the coffee maker must be

$$R = \frac{V_2}{I_2} = \frac{240 \text{ V}}{4.0 \text{ A}} = 60 \text{ } \Omega.$$

REFLECT The ratio of primary voltage to current is $R' = \dfrac{V_1}{I_1} = \dfrac{120 \text{ V}}{8.0 \text{ A}} = 15 \text{ } \Omega$. The primary voltage–current relation is the same as though we had connected a 15 Ω resistor directly to the 120 V source.

Practice Problem: Suppose we want to use the coffee maker on an airplane on which the voltage is 480 V. Find the turns ratio of the required transformer and the ratio R' of primary voltage to current. *Answer:* $N_2/N_1 = 1/2, R' = 240 \text{ } \Omega$.

Real transformers always have some energy losses. The windings have some resistance, leading to i^2R losses, although superconducting transformers may appear on the horizon in the next few years. There are also energy losses through eddy currents (Section 21.6) in the core, as shown in Figure 21.28. The alternating current in the primary winding sets up an alternating flux within the core and a current in the secondary winding. However, the iron core is also a conductor, and any section, such as AA, can be pictured as several conducting circuits, one within the other (Figure 21.28b). The flux through each of these

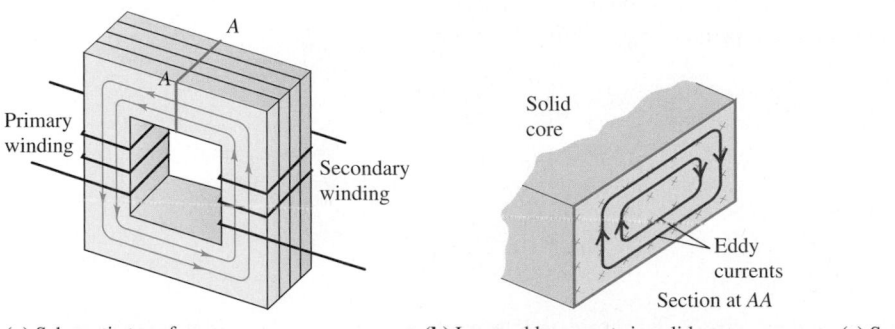
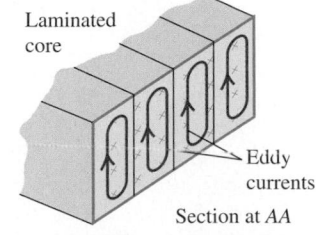

(a) Schematic transformer   **(b)** Large eddy currents in solid core   **(c)** Smaller eddy currents in laminated core

▲ **FIGURE 21.28** Eddy currents in a transformer can be minimized by use of a laminated core.

circuits is continually changing, so eddy currents circulate in the entire volume of the core, wasting energy through I^2R heating (due to the resistance of the core material) and setting up unwanted opposing flux.

In actual transformers, the eddy currents are greatly reduced by the use of a *laminated* core—that is, a core built up of thin sheets, or laminae. The large electrical surface resistance of each lamina, due to either a natural coating of oxide or an insulating varnish, effectively confines the eddy currents to individual laminae (Figure 21.28c). The possible eddy-current paths are narrower, the induced emf in each path is smaller, and the eddy currents are greatly reduced. Transformer efficiencies are usually well over 90%; in large installations, they may reach 99%.

In small transformers in which it is important to keep eddy-current losses to an absolute minimum, the cores are sometimes made of *ferrites,* which are complex oxides of iron and other metals. These materials are ferromagnetic, but their resistivity is much greater than that of pure iron.

21.10 Magnetic Field Energy

Establishing a current in an inductor requires an input of energy, so an inductor carrying a current has energy stored in it, associated with its magnetic field. Let's see how this comes about. A changing current in an inductor causes an emf $\mathcal{E}$ between the inductor's terminals. The source that supplies the current must maintain a corresponding potential difference V_{ab} between its terminals while the current is changing; therefore, the source must supply energy to the inductor. We can calculate the total energy input U needed to establish a final current I in an inductor with inductance L if the initial current is zero.

Let the current at some instant be i and its rate of change be $\Delta i/\Delta t$. Then the terminal voltage at that instant is $\mathcal{E} = V_{ab} = L\,\Delta i/\Delta t$, and the average power P supplied by the current source during the small time interval Δt is

$$P = V_{ab}i = Li\frac{\Delta i}{\Delta t}. \tag{21.18}$$

The energy ΔU supplied to the inductor during this time interval is approximately $\Delta U = P\,\Delta t$, so $\Delta U = Li\,\Delta i$.

To find the total energy U supplied while the current increases from zero to a final value I, we note that the *average* value of Li during the entire increase is equal to half of the final value—that is, to $LI/2$. The product of this quantity and the *total* increase in current I gives the total energy U supplied, and we find the following relationship:

Energy stored in an inductor

The energy stored by an inductor with inductance L carrying a current I is

$$U = \frac{1}{2}LI^2. \tag{21.19}$$

This result can also be derived with the use of integral calculus.

After the current has reached its final steady value I, $\Delta i/\Delta t = 0$ and the power input is zero. We can think of the energy U as analogous to a *kinetic energy* associated with the current. This energy is zero when there is no current; when the current is I, the energy is $\frac{1}{2}LI^2$.

When the current decreases from I to zero, the inductor acts as a source, supplying a total amount of energy $\frac{1}{2}LI^2$ to the external circuit. If we interrupt the circuit suddenly by opening a switch, the current decreases very rapidly, the induced

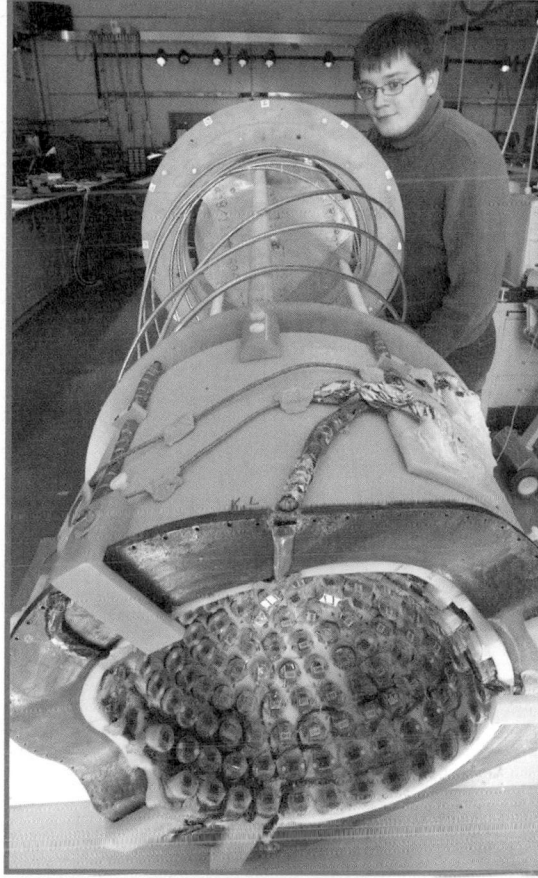

▲ **BIO Application SQUID tales.** The transient electrical signals from excitable animal tissues generate tiny magnetic fields, generally below the detection limit of the usual devices for measuring magnetic fields. However, a specialized instrument called SQUID (for Superconducting Quantum Interference Device) uses a detector coil that is superconducting; that is, it has zero electrical resistance. With careful shielding from stray magnetic fields and other tricks, magnetic fields smaller than 1 fT can be measured. These devices have been used to detect the fields generated by single nerve impulses; the contraction of abdominal smooth muscle in intact, living animals; and the fetal heartbeat. Arrays of SQUID detectors as shown in the figure can be used to pinpoint the electrical storm in the brain that accompanies epilepsy. An important difference between SQUID detection and magnetic resonance (MRI) imaging of brain tissue is that SQUID directly detects electrical activity by way of the tiny magnetic fields, while MRI detects increased blood flow to active areas of the brain—an indirect (and slower) measure of nervous activity.

emf is correspondingly very large, and the energy may be dissipated in an arc across the switch contacts. This large emf is the electrical analog of the large force exerted on a car that runs into a concrete bridge abutment and stops suddenly.

The energy in an inductor is actually stored in the magnetic field within the coil, just as the energy of a capacitor is stored in the electric field between its plates. We can develop a relation for magnetic-field energy analogous to the one we obtained for electric-field energy in Section 18.7 (Equation 18.20). We'll concentrate on one simple case: the ideal toroidal solenoid. This system has the advantage that its magnetic field is confined completely to a finite region of space within its core. As in Example 21.10, we assume that the cross-sectional area A is small enough that we can consider the magnetic field to be uniform over the area. The volume V enclosed by the toroid is approximately equal to the circumference $2\pi r$ multiplied by the area A, or $V = 2\pi rA$. From Example 21.10, the self-inductance of the toroidal solenoid (with vacuum in the core) is

$$L = \frac{\mu_0 N^2 A}{2\pi r},$$

and the stored energy U when the current is I is

$$U = \frac{1}{2}LI^2 = \frac{1}{2}\frac{\mu_0 N^2 A}{2\pi r}I^2. \tag{21.20}$$

The magnetic field, and therefore the energy U, are localized in the volume $V = 2\pi rA$ enclosed by the windings. The energy *per unit volume*, or **energy density,** $u = U/V$, is then

$$u = \frac{U}{2\pi rA} = \frac{1}{2}\mu_0\frac{N^2 I^2}{(2\pi r)^2}.$$

We can express this energy in terms of the magnetic field B inside the toroid. From Equation 20.15, B is given by

$$B = \frac{\mu_0 NI}{2\pi r}, \quad \text{and} \quad \frac{N^2 I^2}{(2\pi r)^2} = \frac{B^2}{\mu_0^2}.$$

When we substitute the last expression into the preceding equation for u, we finally find that

$$u = \frac{B^2}{2\mu_0}. \tag{21.21}$$

This is the magnetic analog of the energy per unit volume in the *electric* field of an air capacitor, $u = \frac{1}{2}\epsilon_0 E^2$, which we derived in Section 18.7.

EXAMPLE 21.12 Storing energy in an inductor

The electric-power industry would like to find efficient ways to store surplus energy generated during low-demand hours to help meet customer requirements during high-demand hours. Perhaps superconducting coils can be used. What inductance would be needed to store 1.00 kWh of energy in a coil carrying a current of 200 A?

SOLUTION

SET UP AND SOLVE The problem tells us that $U = 1.00\,\text{kWh} = (1.00 \times 10^3\,\text{W})(3600\,\text{s}) = 3.6 \times 10^6\,\text{J}$ and $I = 200\,\text{A}$. Solving Equation 21.20 for L, we obtain

$$L = \frac{2U}{I^2} = \frac{2(3.60 \times 10^6\,\text{J})}{(200\,\text{A})^2} = 180\,\text{H}.$$

REFLECT A 180 H inductor using conventional wire heavy enough to carry 200 A would be very large (the size of a room), but a superconducting inductor could be much smaller.

Practice Problem: How much energy, in kWh, could be stored in a 200 H inductor carrying a current of 350 A? *Answer:* $U = 3.40\,\text{kWh}$.

21.11 The *R–L* Circuit

An inductor is primarily a circuit device. A circuit containing a resistor and an inductor is called an *R–L* circuit. Let's look at the behavior of a simple *R–L* circuit. One thing is clear already: We aren't going to see any sudden changes in the current through an inductor. Equation 21.14 shows that the greater the rate of change of current, $\Delta i / \Delta t$, the greater must be the potential difference between the inductor terminals. This equation, together with Kirchhoff's rules, gives us the principles that we need to analyze circuits containing inductors.

PROBLEM-SOLVING STRATEGY 21.2 **Inductors in circuits** (MP)

SET UP

1. When an inductor is used as a *circuit* device, all the voltages, currents, and capacitor charges are, in general, functions of time, not constants as they have been in most of our previous circuit analysis. But Kirchhoff's rules, which we studied in Chapter 19, are still valid. When the voltages and currents vary with time, Kirchhoff's rules hold at each instant of time.

2. As in all circuit analysis, getting the signs right is sometimes more challenging than understanding the principles. We suggest that you review the strategy in Section 19.2. In addition, study carefully the sign rule described with Equation 21.14 and Figure 21.25. In Kirchhoff's loop rule, when we go through an inductor in the *same* direction as the assumed current, we encounter a voltage *drop* equal to $L \Delta i / \Delta t$, so the corresponding term in the loop equation is $-L \Delta i / \Delta t$.

Current Growth in an *R–L* Circuit

We can learn several basic things about inductor behavior by analyzing the circuit of Figure 21.29. The resistor *R* may be a separate circuit element, or it may be the resistance of the inductor windings; every real-life inductor has some resistance, unless it is made of superconducting wire. By closing the switch, we can connect the *R–L* combination to a source having a constant emf $\mathcal{E}$. (We assume that the source has zero internal resistance, so the terminal voltage equals the emf.) Suppose the switch is initially open, and then, at some initial time $t = 0$, we close it. As we have mentioned, the current cannot change suddenly from zero to some final value, because of the infinite induced emf that would result. Instead, it begins to grow at a definite rate that depends on the value of *L* in the circuit.

Let *i* be the current at some time *t* after the switch is closed, and let $\Delta i / \Delta t$ be the rate of change of *i* at that time. Then the potential difference v_{bc} across the inductor at that time is

$$v_{bc} = L \frac{\Delta i}{\Delta t},$$

and the potential difference v_{ab} across the resistor is

$$v_{ab} = iR.$$

We apply Kirchhoff's voltage rule, starting at the negative terminal of the source and proceeding counterclockwise around the loop:

$$\mathcal{E} - iR - L \frac{\Delta i}{\Delta t} = 0. \tag{21.22}$$

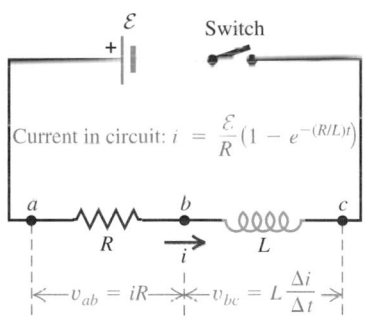

The voltage across the resistor depends on the current; the voltage across the inductor depends on the *rate of change of* the current.

▲ **FIGURE 21.29** An *R–L* circuit.

Solving this equation for $\Delta i / \Delta t$, we find that the rate of increase of current is

$$\frac{\Delta i}{\Delta t} = \frac{\mathcal{E} - iR}{L} = \frac{\mathcal{E}}{L} - \frac{R}{L} i. \tag{21.23}$$

At the instant the switch is first closed, $i = 0$ and the potential drop across R is zero. The initial rate of change of current is

$$\left(\frac{\Delta i}{\Delta t}\right)_{\text{initial}} = \frac{\mathcal{E}}{L}. \tag{21.24}$$

The greater the inductance L, the more slowly the current increases.

As the current increases, the term $(R/L)i$ in Equation 21.23 also increases, and the *rate* of increase of current becomes smaller and smaller. When the current reaches its final *steady-state* value I, its rate of increase is zero. Then Equation 21.23 becomes

$$\frac{\Delta i}{\Delta t} = 0 = \frac{\mathcal{E}}{L} - \frac{R}{L} I,$$

and the final current I is

$$I = \frac{\mathcal{E}}{R}.$$

That is, the *final* current I doesn't depend on the inductance L; it is the same as it would be if the resistance R alone were connected to the source with emf $\mathcal{E}$.

We can use Equation 21.23 to derive an expression for the current i as a function of time. The derivation requires calculus, and we'll omit the details. The final result is

$$i = \frac{\mathcal{E}}{R}\left(1 - e^{-(R/L)t}\right). \tag{21.25}$$

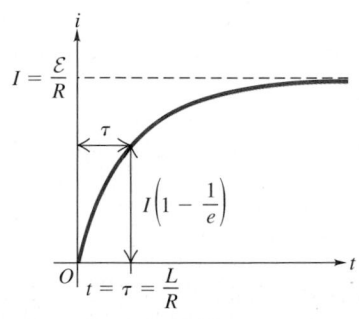

▲ **FIGURE 21.30** Graph of i versus t for the growth of current in an R–L circuit. The final current is $I = \mathcal{E}/R$; after one time constant, the current is $1 - (1/e)$ of this value.

Figure 21.30 is a graph of Equation 21.25, showing the variation of current with time. At time $t = 0$, $i = 0$, and the initial slope of the curve is $\Delta i / \Delta t = \mathcal{E}/L$. As $t \rightarrow \infty$, $i \rightarrow \mathcal{E}/R$ and the slope $\Delta i / \Delta t$ approaches zero, as we predicted.

As Figure 21.30 shows, the instantaneous current i first rises rapidly, then increases more slowly, and approaches the final value $I = \mathcal{E}/R$ asymptotically. At a time τ equal to L/R, the current has risen to $1 - (1/e)$, or about 0.63 of its final value. The quantity $\tau = L/R$ is called the **time constant** for the circuit:

$$\tau = \frac{L}{R}. \tag{21.26}$$

In a time equal to 2τ, the current reaches 86% of its final value; in 5τ, 99.3%; and in 10τ, 99.995%.

For a given value of R, the time constant τ is greater for greater values of L. When L is small, the current rises rapidly to its final value; when L is large, it rises more slowly. For example, if $R = 100 \, \Omega$ and $L = 10 \, \text{H}$, then

$$\tau = \frac{L}{R} = \frac{10 \, \text{H}}{100 \, \Omega} = 0.10 \, \text{s},$$

and the current increases to about 0.632 of its final value in 0.10 s. But if $R = 100 \, \Omega$ and $L = 0.010 \, \text{H}$, then $\tau = 1.0 \times 10^{-4} \, \text{s} = 0.10 \, \text{ms}$, and the rise is much more rapid.

This entire discussion should look familiar; the situation is similar to that of a charging and discharging capacitor, which we analyzed in Section 19.8. It would be a good idea to compare that section with our discussion of the L–R circuit here.

EXAMPLE 21.13 **Current in an *R–L* circuit**

In Figure 21.31, suppose $\mathcal{E} = 120$ V, $R = 200 \,\Omega$, and $L = 10.0$ H. Find (a) the final current after the switch has been closed a long time, (b) the initial time rate of change of the current, immediately after the switch has been closed, and (c) the time at which the current has reached 63.2% of its final value.

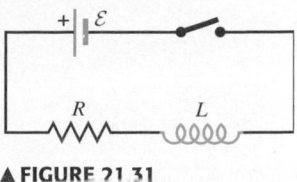

▲ FIGURE 21.31

SOLUTION

SET UP AND SOLVE Part (a): The final current I, which does not depend on the inductance L, is

$$I = \frac{\mathcal{E}}{R} = \frac{120 \text{ V}}{200 \,\Omega} = 0.60 \text{ A}.$$

Part (b): The current can't change discontinuously, so immediately after the switch is closed, $i = 0$ and there is no voltage drop across the resistor. At this time, Kirchhoff's loop rule (Equation 21.22) gives

$$\mathcal{E} - L\left(\frac{\Delta i}{\Delta t}\right)_{\text{initial}} = 0, \quad \text{and} \quad \left(\frac{\Delta i}{\Delta t}\right)_{\text{initial}} = \frac{\mathcal{E}}{L} = \frac{120 \text{ V}}{10.0 \text{ H}} = 12.0 \text{ A/s}.$$

Part (c): From Equation 21.25, the current reaches 0.632 $(=1 - 1/e)$ of its final value when

$$1 - \frac{1}{e} = 1 - e^{-(R/L)t}, \quad -1 = -\frac{R}{L}t, \quad \text{and}$$

$$t = \frac{L}{R} = \frac{10.0 \text{ H}}{200 \,\Omega} = 0.050 \text{ s}.$$

Note that this value of t is equal to the time constant τ defined by Equation 21.26.

REFLECT The final current is independent of the inductance because the current is no longer changing with time and there is no induced emf in the inductor. The initial current is independent of the resistance because the initial current is zero and there is no voltage drop across the resistor. The time constant is directly proportional to the inductance: The greater the inductance, the more slowly the current increases.

Practice Problem: Suppose we now want the final current to be 0.300 A, with $\mathcal{E}$ and τ keeping their previous values. What values of R and L should be used? *Answer:* $400 \,\Omega$, 20 H.

Energy considerations offer us additional insight into the behavior of an *R–L* circuit. The instantaneous rate at which the source delivers energy to the circuit is $P = \mathcal{E}i$. The instantaneous rate at which energy is dissipated in the resistor is i^2R, and the rate at which energy is stored in the inductor is $iv_{bc} = Li\,\Delta i/\Delta t$. When we multiply Equation 21.23 by Li and rearrange terms, we find that

$$\mathcal{E}i = Li\frac{\Delta i}{\Delta t} + i^2R. \tag{21.27}$$

This shows that part of the power $\mathcal{E}i$ supplied by the source is dissipated (i^2R) in the resistor and part is stored $(Li\,\Delta i/\Delta t)$ in the inductor.

Current Decay in an *R–L* Circuit

In Figure 21.32, we show an *R–L* circuit with two switches. Suppose that switch S_1 has been closed for a long time and that the current has reached a steady value I_0. Resetting our stopwatch to redefine the initial time, we close switch S_2 at time $t = 0$, bypassing the battery. (At the same time, we should open S_1 to save the battery from ruin.) The current through R and L does not instantaneously go to zero, but decays smoothly, as shown in Figure 21.33. The Kirchhoff's-rule loop equation is obtained from Equation 21.23 by simply omitting the $\mathcal{E}$ term. When we retrace the steps in the preceding analysis, we find that the current i varies with time according to the relationship

$$i = I_0 e^{-(R/L)t}, \tag{21.28}$$

where I_0 is the initial current at time $t = 0$. The time constant, $\tau = L/R$, is the time required for the current to decrease to $1/e$, or about 0.37, of its original value. In time 2τ, it has dropped to $1/e^2 = 0.135$, and so on.

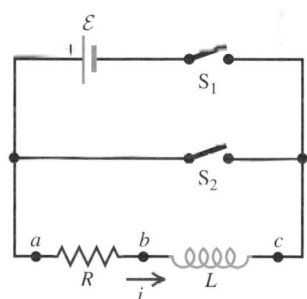

▲ FIGURE 21.32 An *R–L* circuit with a second switch that constitutes a short circuit, disconnecting the resistor and inductor from the emf source.

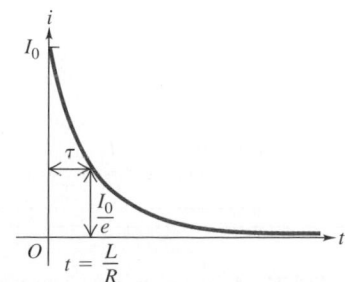

▲ FIGURE 21.33 Graph of i versus t for decay of current in an *R–L* circuit.

The energy needed to maintain the current during this decay is provided by the energy stored in the magnetic field of the inductor. The detailed energy analysis is simpler this time. In place of Equation 21.27, we have

$$0 = Li\frac{\Delta i}{\Delta t} + i^2 R. \qquad (21.29)$$

In this case, $Li\,\Delta i/\Delta t$ is negative, because Δi is negative. Equation 21.29 shows that the energy stored in the inductor *decreases* at a rate equal to the rate of dissipation of energy $i^2 R$ in the resistor.

21.12 The *L–C* Circuit

A circuit containing an inductor and a capacitor (which we'll call an **L–C circuit**) shows an entirely new mode of behavior, characterized by *oscillating* current and charge. This is in sharp contrast to the *exponential* approach to a steady-state situation that we have seen with both *R–C* and *R–L* circuits. In the *L–C* circuit in Figure 21.34, we charge the capacitor to a potential difference V_{max} and an initial charge $Q_{max} = CV_{max}$, as shown in Figure 21.34a, and then close the switch. What happens?

The capacitor begins to discharge through the inductor. Because of the induced emf in the inductor, the current cannot change instantaneously; rather, it starts at zero and eventually builds up to a maximum value I_{max}. During this buildup, the capacitor is discharging. At each instant, the capacitor potential difference equals the induced emf, so as the capacitor discharges, the *rate of change* of current decreases. When the capacitor potential difference becomes zero, the induced emf is also zero, and the current has leveled off at its maximum value I_{max}. Figure 21.34b shows this situation; the capacitor has completely discharged.

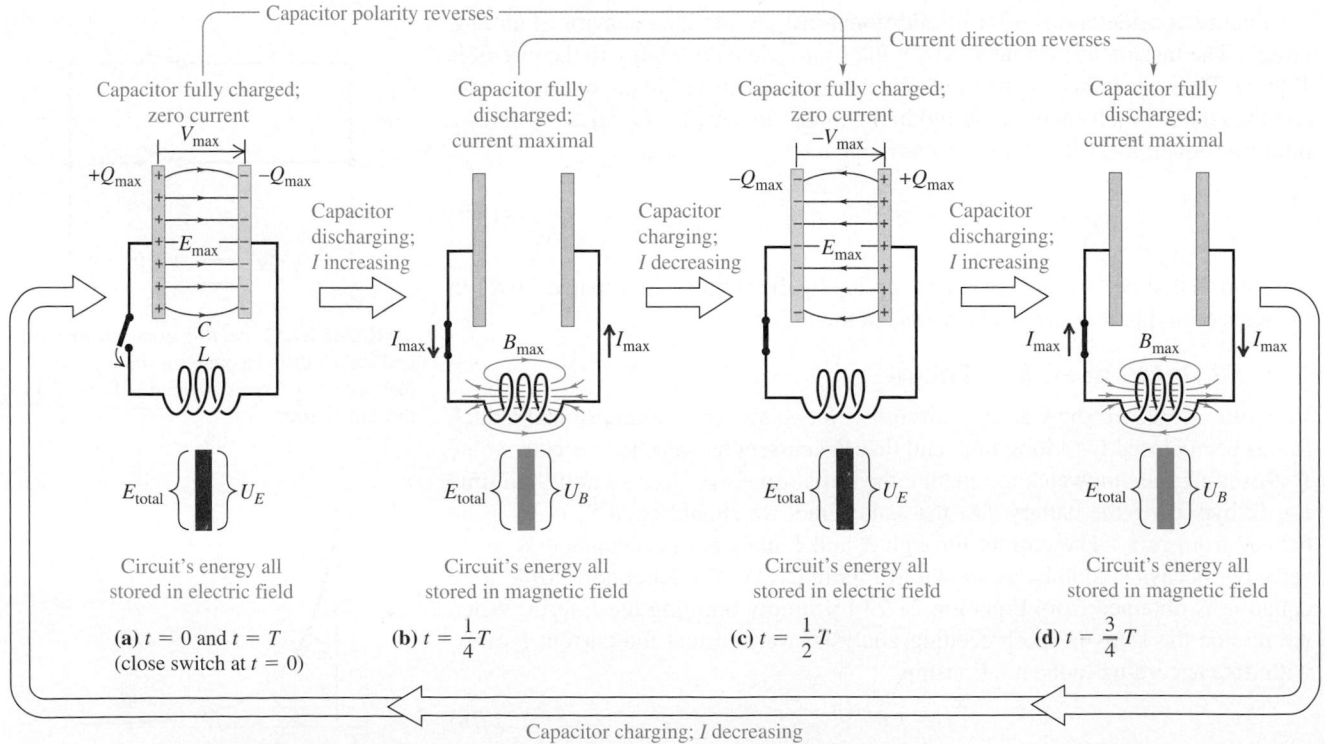

▲ **FIGURE 21.34** Energy transfer between electric and magnetic fields in an oscillating *L–C* circuit. The switch is closed at time $t = 0$, when the capacitor charge is Q_{max}. As in simple harmonic motion, the total energy E_{total} is constant.

The potential difference between its terminals (and those of the inductor) has decreased to zero, and the current has reached its maximum value I_{max}.

During the discharge of the capacitor, the increasing current in the inductor has established a magnetic field in the space around it, and the energy that was initially stored in the capacitor's electric field is now stored in the inductor's magnetic field.

The current cannot drop to zero instantaneously; as it persists, the capacitor begins to charge with polarity opposite to that in the initial state. As the current decreases, the magnetic field also decreases, inducing an emf in the inductor in the same direction as the current. Eventually, the current and the magnetic field reach zero, and the capacitor has been charged in the *opposite* sense to its initial polarity (Figure 21.34c), with a potential difference $-V_{max}$ and charge $-Q_{max}$.

The process now repeats in the reverse direction; a little later, the capacitor has again discharged, and there is a current in the inductor in the opposite direction (Figure 21.34d). The whole process then repeats once more. If there are no energy losses, the charges on the capacitor continue to oscillate back and forth indefinitely. This process is called an **electrical oscillation.**

From an energy standpoint, the oscillations of an electric circuit transfer energy from the capacitor's electric field to the inductor's magnetic field and back. The *total* energy associated with the circuit is constant. This phenomenon is analogous to the transfer of energy in an oscillating, frictionless mechanical system, from potential to kinetic and back, with constant total energy.

Detailed analysis shows that the angular frequency ω of the oscillations we've just described is given by

$$\omega = \sqrt{\frac{1}{LC}}. \tag{21.30}$$

Thus, the charge and current in the *L–C* circuit oscillate sinusoidally with time, with an angular frequency determined by the values of L and C. In many ways, this behavior is directly analogous to simple harmonic motion in mechanical systems.

▲ **Application** **Tuning in.** What happens when you turn the tuning knob on this radio? Radios often used *L–C* circuits to tune in on one desired frequency among all the frequencies bombarding the radio's antenna. Radio waves create small oscillating currents in the antenna. A given *L–C* circuit is most responsive to currents oscillating at its own resonance frequency; it tends to pass frequencies close to this and screen out others. When you twist the knob, you change the capacitance or inductance of the circuit and thus its resonant frequency.

SUMMARY

Electromagnetic Induction and Faraday's Law

(Sections 21.1–21.3) A changing magnetic flux through a circuit loop induces an emf in the circuit. **Faraday's law** states that the magnitude $\mathcal{E}$ of the induced emf in a circuit equals the absolute value of the time rate of change of magnetic flux through the circuit: $\mathcal{E} = |\Delta\Phi_B/\Delta t|$ (Equation 21.4). This relation is valid whether the change in flux is caused by a changing magnetic field, motion of the conductor, or both.

The magnet's motion causes a *changing* magnetic field through the coil, inducing a current in the coil.

Lenz's Law

(Section 21.4) **Lenz's law** states that an induced current or emf always acts to oppose the change that caused it. Lenz's law can be derived from Faraday's law and is a convenient way to determine the correct sign for any induced effect.

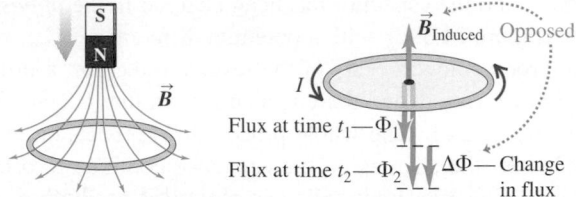

Motional Electromotive Force

(Section 21.5) When a conductor moves in a magnetic field, the charges in the conductor are acted upon by magnetic forces that create a current. When a conductor with length L moves with speed v perpendicular to a uniform magnetic field with magnitude B, the induced emf is $\mathcal{E} = vBL$ (Equation 21.7).

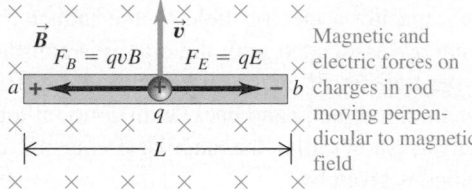

Magnetic and electric forces on charges in rod moving perpendicular to magnetic field

Eddy Currents

(Section 21.6) When a bulk piece of conducting material, such as a metal, is in a changing magnetic field or moves through a nonuniform field, **eddy currents** are induced in it.

Mutual Inductance and Self-Inductance

(Sections 21.7 and 21.8) When changing magnetic flux created by a changing current i_1 in one circuit links a second circuit, an emf with magnitude $\mathcal{E}_2$ is induced in the second circuit. A changing current i_2 in the second circuit induces an emf with magnitude $\mathcal{E}_1$ in the first circuit. The two emfs are given by

$$\mathcal{E}_2 = \left| M\frac{\Delta i_1}{\Delta t} \right| \quad \text{and} \quad \mathcal{E}_1 = \left| M\frac{\Delta i_2}{\Delta t} \right|, \quad (21.12)$$

where M is a constant called the **mutual inductance.** A changing current i in any circuit induces an emf $\mathcal{E}$ in that same circuit, called a self-induced emf, given by

$$\mathcal{E} = \left| L\frac{\Delta i}{\Delta t} \right|, \quad (21.14)$$

where L is a constant called **inductance** or self-inductance.

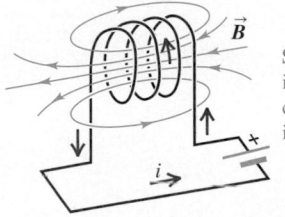

Self-inductance: If the current in the coil is changing, the changing flux through the coil induces an emf in the coil.

Continued

Transformers

(Section 21.9) A **transformer** is used to transform the voltage and current levels in an ac circuit. An alternating current in either winding results in an alternating flux in the other winding, inducing an emf. In an ideal transformer with no energy losses, if the primary winding has N_1 turns and the secondary winding has N_2 turns, the two voltages are related by

$$\frac{V_2}{V_1} = \frac{N_2}{N_1}. \qquad (21.15)$$

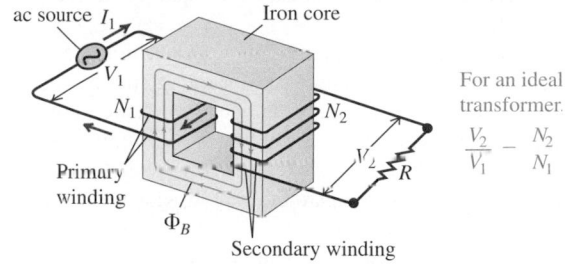

For an ideal transformer:

$$\frac{V_2}{V_1} = \frac{N_2}{N_1}$$

Magnetic Field Energy

(Section 21.10) An inductor with inductance L carrying current I has energy $U = \frac{1}{2}LI^2$ (Equation 21.19). This energy is stored in the magnetic field of the inductor. The energy density u (energy per unit volume) is given by $u = B^2/2\mu_0$ (Equation 21.21).

R–L and L–C Circuits

(Sections 21.11 and 21.12) In a circuit containing a resistor R, an inductor L, and a source of emf $\mathcal{E}$, the growth and decay of current are exponential, with a characteristic time τ called the **time constant,** given by $\tau = L/R$ (Equation 21.26). The time constant τ is the time required for the increasing current to approach within a fraction $1 - (1/e)$, or about 63%, of its final value.

A circuit containing an inductance L and a capacitance C undergoes electrical oscillations with angular frequency ω, where $\omega = \sqrt{1/LC}$ (Equation 21.30).

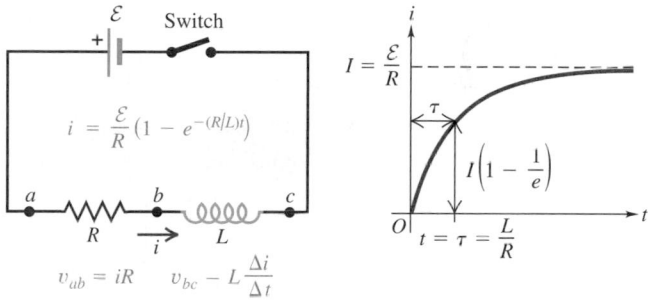

$$i = \frac{\mathcal{E}}{R}\left(1 - e^{-(R/L)t}\right)$$

$$v_{ab} = iR \qquad v_{bc} = L\frac{\Delta i}{\Delta t}$$

For instructor-assigned homework, go to www.masteringphysics.com

Conceptual Questions

1. Two circular loops lie adjacent to each other on a tabletop. One is connected to a source that supplies an increasing current; the other is a simple closed ring. Is the induced current in the ring in the same direction as that in the ring connected to the source, or opposite? What if the current in the first ring is decreasing?

2. Small one-cylinder gasoline engines sometimes use a device called a *magneto* to supply current to the spark plug. A permanent magnet is attached to the flywheel, and a stationary coil is mounted adjacent to it. Explain how this device is able to generate current. What happens when the magnet passes the coil?

3. A long, straight current-carrying wire passes through the center of a metal ring, perpendicular to its plane. If the current in the wire increases, is a current induced in the ring? Explain.

4. Two closely wound circular coils have the same number of turns, but one has twice the radius of the other. How are the self-inductances of the two coils related?

5. One of the great problems in the field of energy resources and utilization is the difficulty of storing electrical energy in large quantities economically. Discuss the feasibility of storing large amounts of energy by means of currents in large inductors.

6. Suppose there is a steady current in an inductor. If one attempts to reduce the current to zero instantaneously by opening a switch, a big fat spark appears at the switch contacts. Why? What happens to the induced emf in this situation? Is it physically possible to stop the current instantaneously?

7. Why does a transformer *not* work with dc current?

8. Does Lenz's law say that the induced current in a metal loop always flows to oppose the magnetic flux through that loop? Explain.

9. Does Faraday's law say that a large magnetic flux induces a large emf in a coil? Explain.

10. True or false? Inductors always oppose the current through them. Can you think of any situations in which they actually *help* the current through them?

11. An airplane is in level flight over Antarctica, where the magnetic field of the earth is mostly directed upward away from the ground. As viewed by a passenger facing toward the front of the plane, is the left or the right wingtip at higher potential? Does your answer depend on the direction the plane is flying?

12. Capacitors store energy by accumulating charges on their plates, but how do inductors store energy?

13. A metal ring can be moved in and out of the space between the poles of a horseshoe magnet. Show that you must do work *both* to push it in *and* to pull it out.

14. The ratio of the primary voltage to the secondary voltage in a certain ideal step-down transformer is 2:1. The ratio of the

secondary *current* to the primary *current* is also 2:1. Explain how conservation of energy makes the equality of the two ratios inevitable.

15. In an *R–C* circuit, a resistor, an uncharged capacitor, a dc battery, and an open switch are in series. In an *R–L* circuit, a resistor, an inductor, a dc battery, and an open switch are in series. Compare the behavior of the current in these circuits (a) just after the switch is closed and (b) long after the switch has been closed. In other words, compare the way in which a capacitor and an inductor affect a circuit.

16. You have a bar magnet with unidentified poles. Utilizing only simple laboratory equipment, such as conducting wire and an ammeter, show how you could use induction phenomena to determine which are the north and south poles of this magnet.

Multiple-Choice Problems

1. A square loop of wire is pulled upward out of the space between the poles of a magnet, as shown in Figure 21.35. As this is done, the current induced in this loop, as viewed from the N pole of the magnet, will be directed
   A. clockwise.   B. counterclockwise.   C. zero.

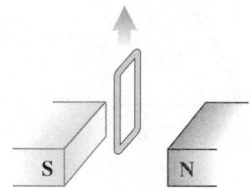

▲ **FIGURE 21.35** Multiple-choice problem 1.

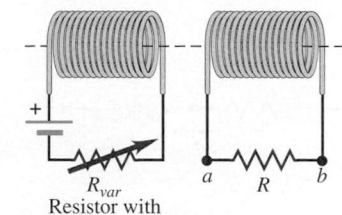

▲ **FIGURE 21.36** Multiple-choice problem 2.

2. The two solenoids in Figure 21.36 are coaxial and fairly close to each other. While the resistance of the variable resistor in the left-hand solenoid is *increased* at a constant rate, the induced current through the resistor *R* will
   A. flow from *a* to *b*.   B. flow from *b* to *a*.
   C. be zero because the rate is constant.

3. A metal ring is oriented with the plane of its area perpendicular to a spatially uniform magnetic field that increases at a steady rate. After the radius of the ring is doubled, while the rate of increase of the field is cut in half, the emf induced in the ring
   A. remains the same.
   B. increases by a factor of 2.
   C. increases by a factor of 4.
   D. decreases by a factor of 2.

4. The slide wire of the variable resistor in Figure 21.37 is moved steadily to the right, increasing the resistance in the circuit. While this is being done, the current induced in the small circuit *A* is directed
   A. clockwise.   B. counterclockwise.   C. zero.

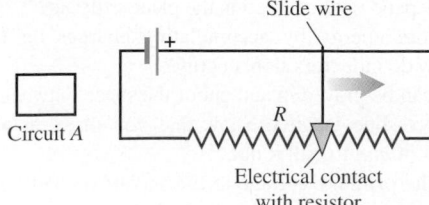

▲ **FIGURE 21.37** Multiple-choice problem 4.

5. The slide wire on the variable resistor in Figure 21.38 is moved steadily to the left. While this is being done, the current induced in the small circuit *A* is directed
   A. clockwise.
   B. counterclockwise.
   C. zero.

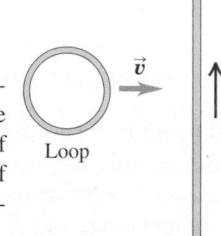

▲ **FIGURE 21.38** Multiple-choice problem 5.

6. A metal loop moves at constant velocity toward a long wire carrying a steady current *I*, as shown in Figure 21.39. The current induced in the loop is directed
   A. clockwise.
   B. counterclockwise.
   C. zero.

7. The primary coil of an ideal transformer carries a current of 2.5 A, while the secondary coil carries a current of 5.0 A. The ratio of number of turns of wire in the primary to that in the secondary is
   A. 1 : 1.
   B. 1 : 2.
   C. 2 : 1.

▲ **FIGURE 21.39** Multiple-choice problem 6.

8. A metal loop is held above the S pole of a bar magnet, as shown in Figure 21.40, when the magnet is suddenly dropped from rest. Just after the magnet is dropped, the induced current in the loop, as viewed from above it, is directed
   A. clockwise.
   B. counterclockwise.
   C. zero.

▲ **FIGURE 21.40** Multiple-choice problem 8.

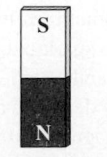

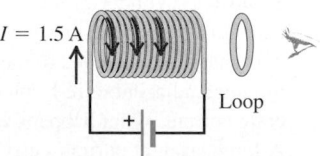

▲ **FIGURE 21.41** Multiple-choice problem 9.

9. A steady current of 1.5 A flows through the solenoid shown in Figure 21.41. The current induced in the loop, as viewed from the right, is directed
   A. clockwise.
   B. counterclockwise.
   C. zero.

10. A vertical bar moves horizontally at constant velocity through a uniform magnetic field, as shown in Figure 21.42. We observe that point *b* is at a higher potential than point *a*. We

can therefore conclude that the magnetic field must have a component that is directed
A. vertically downward.
B. vertically upward.
C. perpendicular to the plane of the paper, outward.
D. perpendicular to the plane of the paper, inward.

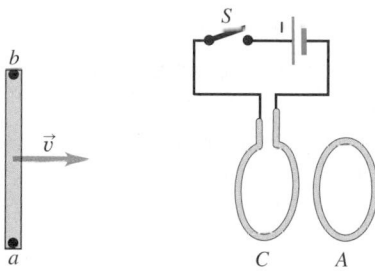

▲ **FIGURE 21.42**
Multiple-choice
problem 10.

▲ **FIGURE 21.43**
Multiple-choice
problem 11.

11. The vertical loops A and C in Figure 21.43 are parallel to each other and are centered on the same horizontal line that is perpendicular to both of them. Just after the switch S is closed, loop A will
A. not be affected by loop C.   B. be attracted by loop C.
C. be repelled by loop C.   D. move upward.
E. move downward.

12. After the switch S in Figure 21.43 has been closed for a very long time, loop A will
A. not be affected by loop C.
B. be attracted by loop C.
C. be repelled by loop C.
D. move upward.
E. move downward.

13. After the switch S in the circuit in Figure 21.44 is closed,
A. The current is zero 1.5 ms (one time constant) later.
B. The current is zero for a very long time afterward.
C. The largest current is 5.0 A and it occurs just after S is closed.
D. The largest current is 5.0 A and it occurs a very long time after S has been closed.

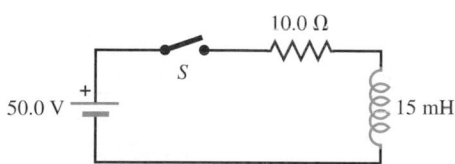

▲ **FIGURE 21.44** Multiple-choice problem 13.

14. A square metal loop is pulled to the right at a constant velocity perpendicular to a uniform magnetic field, as shown in Figure 21.45. The current induced in this loop is directed
A. clockwise.   B. counterclockwise.
C. zero.

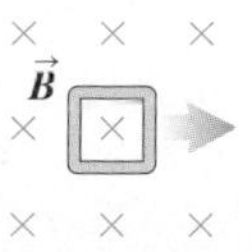

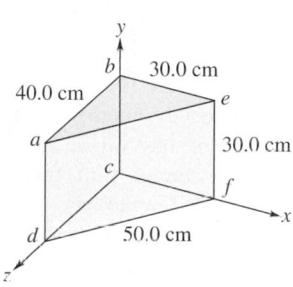

▲ **FIGURE 21.45**
Multiple-choice
problem 14.

15. A metal loop is being pushed at a constant velocity into a uniform magnetic field, as shown in Figure 21.46, but is only partway into the field. As a result of this motion,
A. End a of the resistor R is at a higher potential than end b.
B. End b of the resistor R is at a higher potential than end a.
C. Ends a and b are at the same potential.

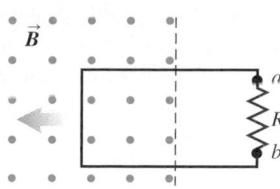

▲ **FIGURE 21.46** Multiple-choice problem 15.

Problems

21.2 Magnetic Flux

1. • A circular area with a radius of 6.50 cm lies in the x-y plane. What is the magnitude of the magnetic flux through this circle due to a uniform magnetic field $B = 0.230$ T that points (a) in the $+z$ direction? (b) at an angle of 53.1° from the $+z$ direction? (c) in the $+y$ direction?

2. • The magnetic field $\vec{B}$ in a certain region is 0.128 T, and its direction is that of the $+z$ axis in Figure 21.47. (a) What is the magnetic flux across the surface $abcd$ in the figure? (b) What is the magnetic flux across the surface $befc$? (c) What is the magnetic flux across the surface $aefd$? (d) What is the net flux through all five surfaces that enclose the shaded volume?

▲ **FIGURE 21.47** Problem 2.

3. •• An open plastic soda bottle with an opening diameter of 2.5 cm is placed on a table. A uniform 1.75 T magnetic field directed upward and oriented 25° from vertical encompasses the bottle. What is the total magnetic flux through the plastic of the soda bottle?

21.3 Faraday's Law

4. • A single loop of wire with an area of 0.0900 m² is in a uniform magnetic field that has an initial value of 3.80 T, is perpendicular to the plane of the loop, and is decreasing at a constant rate of 0.190 T/s. (a) What emf is induced in this loop? (b) If the loop has a resistance of 0.600 Ω, find the current induced in the loop.

5. • A coil of wire with 200 circular turns of radius 3.00 cm is in a uniform magnetic field along the axis of the coil. The coil has $R = 40.0$ Ω. At what rate, in teslas per second, must the magnetic field be changing to induce a current of 0.150 A in the coil?

6. • In a physics laboratory experiment, a coil with 200 turns enclosing an area of 12 cm² is rotated from a position where its plane is perpendicular to the earth's magnetic field to one where its plane is parallel to the field. The rotation takes 0.040 s. The earth's magnetic field at the location of the laboratory is 6.0×10^{-5} T. (a) What is the total magnetic flux through the coil before it is rotated? After it is rotated? (b) What is the average emf induced in the coil?

7. • A closely wound rectangular coil of 80 turns has dimensions of 25.0 cm by 40.0 cm. The plane of the coil is rotated from a position where it makes an angle of 37.0° with a magnetic field of 1.10 T to a position perpendicular to the field. The rotation takes 0.0600 s. What is the average emf induced in the coil?

8. •• A very long, straight solenoid with a cross-sectional area of 6.00 cm² is wound with 40 turns of wire per centimeter, and the windings carry a current of 0.250 A. A secondary winding of 2 turns encircles the solenoid at its center. When the primary circuit is opened, the magnetic field of the solenoid becomes zero in 0.0500 s. What is the average induced emf in the secondary coil?

9. •• A 30.0 cm × 60.0 cm rectangular circuit containing a 15 Ω resistor is perpendicular to a uniform magnetic field that starts out at 2.65 T and steadily decreases at 0.25 T/s. (See Figure 21.48.) While this field is changing, what does the ammeter read?

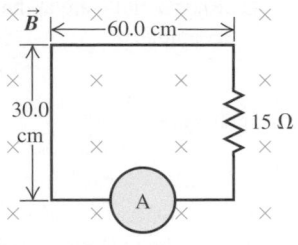

▲ **FIGURE 21.48** Problem 9.

10. •• A circular loop of wire with a radius of 12.0 cm is lying flat on a tabletop. A magnetic field of 1.5 T is directed vertically upward through the loop (Figure 21.49). (a) If the loop is removed from the field region in a time interval of 2.0 ms, find the average emf that will be induced in the wire loop during the extraction process. (b) If the loop is viewed looking down on it from above, is the induced current in the loop clockwise or counterclockwise?

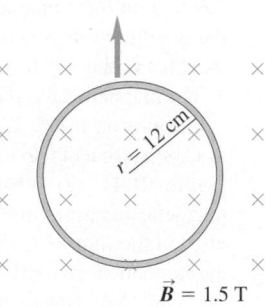

▲ **FIGURE 21.49** Problem 10.

11. •• A flat, square coil with 15 turns has sides of length 0.120 m. The coil rotates in a magnetic field of 0.0250 T. (a) What is the angular velocity of the coil if the maximum emf produced is 20.0 mV? (*Hint:* Look at the motional emf induced across the ends of the segments of the coil.) (b) What is the average emf at this angular velocity?

21.4 Lenz's Law

12. • A cardboard tube is wrapped with two windings of insulated wire, as shown in Figure 21.50. Is the induced current in the resistor R directed from left to right or from right to left in the following circumstances? The current in winding A is directed (a) from a to b and is increasing, (b) from b to a and is decreasing, (c) from b to a and is increasing, and (d) from b to a and is constant.

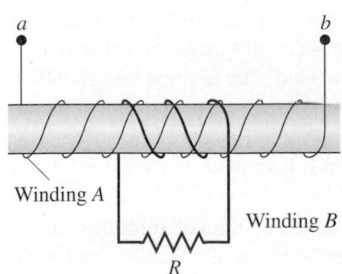

▲ **FIGURE 21.50** Problem 12.

13. • A circular loop of wire is in a spatially uniform magnetic field, as shown in Figure 21.51. The magnetic field is directed into the plane of the figure. Determine the direction (clockwise or counterclockwise) of the induced current in the loop when (a) B is increasing; (b) B is decreasing; (c) B is constant with a value of B_0. Explain your reasoning.

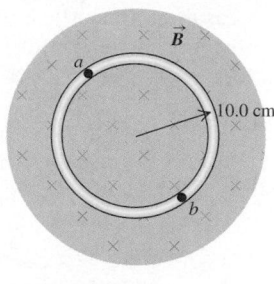

▲ **FIGURE 21.51** Problem 13.

14. • Using Lenz's law, determine the direction of the current in resistor ab of Figure 21.52 when (a) switch S is opened after having been closed for several minutes; (b) coil B is brought closer to coil A with the switch closed; (c) the resistance of R is decreased while the switch remains closed.

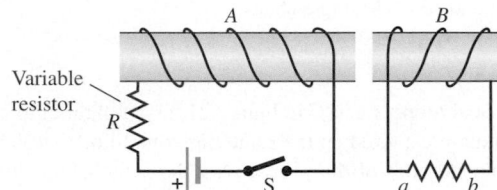

▲ **FIGURE 21.52** Problem 14.

15. • A solenoid carrying a current I is moving toward a metal ring, as shown in Figure 21.53. In what direction, clockwise or counterclockwise (as seen from the solenoid) is a current induced in the ring? In what direction will the induced current be if the solenoid now stops moving toward the ring, but the current in it begins to decrease?

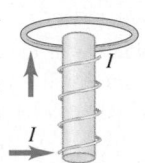

▲ **FIGURE 21.53** Problem 15.

16. • A metal bar is pulled to the right perpendicular to a uniform magnetic field. The bar rides on parallel metal rails connected through a resistor, as shown in Figure 21.54, so the apparatus makes a complete circuit. Find the direction of the current induced in the circuit in two ways: (a) by looking at the magnetic force on the charges in the moving bar and (b) using Lenz's law.

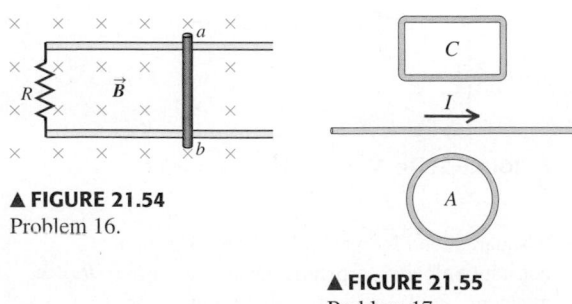

▲ **FIGURE 21.54** Problem 16.

▲ **FIGURE 21.55** Problem 17.

17. • Two closed loops A and C are close to a long wire carrying a current I. (See Figure 21.55.) Find the direction (clockwise or

counterclockwise) of the current induced in each of these loops if I is steadily *increasing*.

18. • A bar magnet is held above a circular loop of wire as shown in Figure 21.56. Find the direction (clockwise or counterclockwise, as viewed from *below* the loop) of the current induced in this loop in each of the following cases. (a) The loop is dropped. (b) The magnet is dropped. (c) Both the loop and magnet are dropped at the same instant.

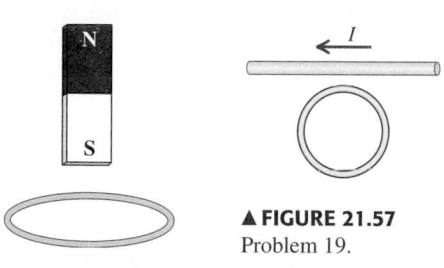

▲ **FIGURE 21.57**
Problem 19.

▲ **FIGURE 21.56**
Problem 18.

19. •• The current in Figure. 21.57 obeys the equation $Ie = I_0e^{-2bt}$, where $b > 0$. Find the direction (clockwise or counterclockwise) of the current induced in the round coil for $t > 0$.

20. • A bar magnet is close to a metal loop. When this magnet is suddenly moved to the left away from the loop, as shown in Figure 21.58, a counterclockwise current is induced in the coil, as viewed by an observer looking through the coil toward the magnet. Identify the north and south poles of the magnet.

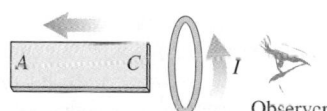

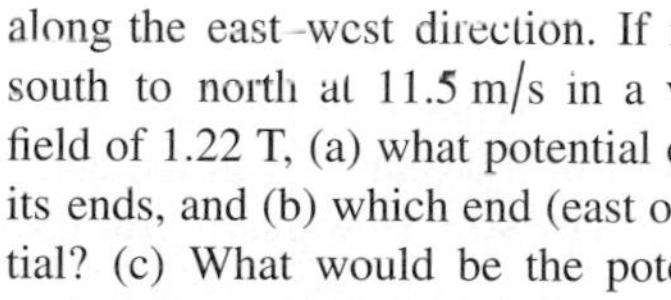

▲ **FIGURE 21.58** Problem 20.

21.5 Motional Electromotive Force

21. • A very thin 15.0 cm copper bar is aligned horizontally along the east–west direction. If it moves horizontally from south to north at 11.5 m/s in a vertically upward magnetic field of 1.22 T, (a) what potential difference is induced across its ends, and (b) which end (east or west) is at a higher potential? (c) What would be the potential difference if the bar moved from east to west instead?

22. • When a thin 12.0 cm iron rod moves with a constant velocity of 4.50 m/s perpendicular to the rod in the direction shown in Figure 21.59, the induced emf across its ends is measured to be 0.450 V. (a) What is the magnitude of the magnetic field? (b) Which point is at a higher potential, a or b? (c) If the bar is rotated clockwise by 90° in the plane of the paper, but keeps the same velocity, what is the potential difference induced across its ends?

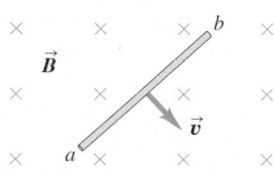

▲ **FIGURE 21.59**
Problem 22.

23. •• You're driving at 95 km/h in a direction 35° east of north, in a region where the earth's magnetic field of 5.5×10^{-5} T is horizontal and points due north. If your car measures 1.5 m from its underbody to its roof, calculate the induced emf between roof and underbody. (You can assume the sides of the car are straight and vertical.) Is the roof of the car at a higher or lower potential than the underbody?

24. •• A 1.41 m bar moves through a uniform, 1.20 T magnetic field with a speed of 2.50 m/s (Figure 21.60). In each case, find the emf induced between the ends of this bar and identify which, if any, end (a or b) is at the higher potential. The bar moves in the direction of (a) the $+x$-axis; (b) the $-y$-axis; (c) the $+z$-axis. (d) How should this bar move so that the emf across its ends has the greatest possible value with b at a higher potential than a, and what is this maximum emf?

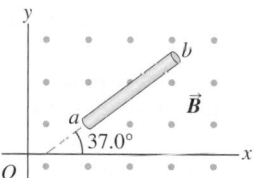

▲ **FIGURE 21.60** Problem 24.

25. •• The conducting rod ab shown in Figure 21.61 makes frictionless contact with metal rails ca and db. The apparatus is in a uniform magnetic field of 0.800 T, perpendicular to the plane of the figure. (a) Find the magnitude of the emf induced in the rod when it is moving toward the right with a speed 7.50 m/s. (b) In what direction does the current flow in the rod? (c) If the resistance of the circuit $abdc$ is a constant 1.50 Ω, find the magnitude and direction of the force required to keep the rod moving to the right with a constant speed of 7.50 m/s.

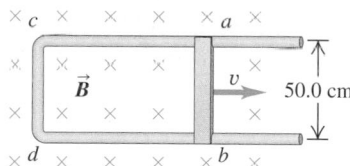

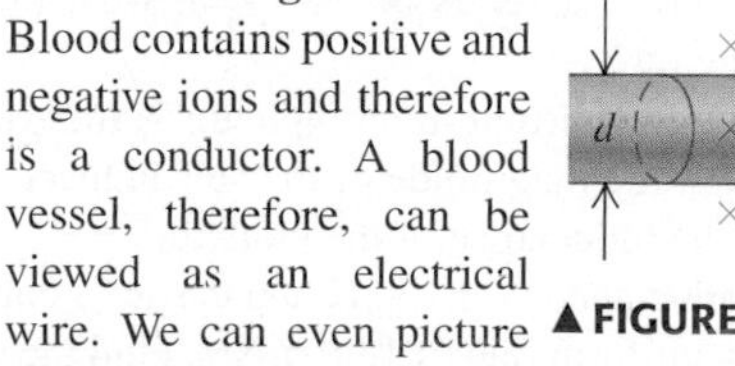

▲ **FIGURE 21.61** Problem 25.

26. •• **Measuring blood flow.**
BIO Blood contains positive and negative ions and therefore is a conductor. A blood vessel, therefore, can be viewed as an electrical wire. We can even picture

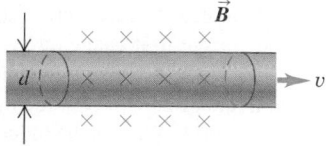

▲ **FIGURE 21.62** Problem 26.

the flowing blood as a series of parallel conducting slabs whose thickness is the diameter d of the vessel moving with speed v. (See Figure 21.62.) (a) If the blood vessel is placed in a magnetic field B perpendicular to the vessel, as in the figure, show that the motional potential difference induced across it is $\mathcal{E} = vBd$. (b) If you expect that the blood will be flowing at 15 cm/s for a vessel 5.0 mm in diameter, what strength of magnetic field will you need to produce a potential difference of 1.0 mV? (c) Show that the volume rate of flow (R) of the blood is equal to $R = \pi\mathcal{E}d/4B$. (*Note:* Although the method developed here is useful in measuring the rate of blood flow in a vessel, it is limited to use in surgery because measurement of the potential $\mathcal{E}$ must be made directly across the vessel.)

21.7 Mutual Inductance

27. • A toroidal solenoid has a mean radius of 10.0 cm and a cross-sectional area of 4.00 cm^2 and is wound uniformly with 100 turns. A second coil with 500 turns is wound uniformly on top of the first. What is the mutual inductance of these coils?

28. • A 10.0-cm-long solenoid of diameter 0.400 cm is wound uniformly with 800 turns. A second coil with 50 turns is wound around the solenoid at its center. What is the mutual inductance of the combination of the two coils?

29. • Two coils are wound around the same cylindrical form, like the coils in Example 21.8. When the current in the first coil is decreasing at a rate of 0.242 A/s, the induced emf in the second coil has magnitude 1.65 mV. (a) What is the mutual inductance of the pair of coils? (b) If the second coil has 25 turns, what is the average magnetic flux through each turn when the current in the first coil equals 1.20 A? (c) If the current in the second coil increases at a rate of 0.360 A/s, what is the magnitude of the induced emf in the first coil?

30. • One solenoid is centered inside another. The outer one has a length of 50.0 cm and contains 6750 coils, while the coaxial inner solenoid is 3.0 cm long and 0.120 cm in diameter and contains 15 coils. The current in the outer solenoid is changing at 37.5 A/s. (a) What is the mutual inductance of these solenoids? (b) Find the emf induced in the inner solenoid.

31. •• Two toroidal solenoids are wound around the same form so that the magnetic field of one passes through the turns of the other. Solenoid 1 has 700 turns, and solenoid 2 has 400 turns. When the current in solenoid 1 is 6.52 A, the average flux through each turn of solenoid 2 is 0.0320 Wb. (a) What is the mutual inductance of the pair of solenoids? (b) When the current in solenoid 2 is 2.54 A, what is the average flux through each turn of solenoid 1?

21.8 Self-Inductance

32. • A 4.5 mH toroidal inductor has 125 identical equally spaced coils. (a) If it carries an 11.5 A current, how much magnetic flux passes through each of its coils? (b) If the potential difference across its ends is 1.16 V, at what rate is the current in it changing?

33. • At the instant when the current in an inductor is increasing at a rate of 0.0640 A/s, the magnitude of the self-induced emf is 0.0160 V. What is the inductance of the inductor?

34. • An inductor has inductance of 0.260 H and carries a current that is decreasing at a uniform rate of 18.0 mA/s. Find the self-induced emf in this inductor.

35. • A 2.50 mH toroidal solenoid has an average radius of 6.00 cm and a cross-sectional area of 2.00 cm^2. (a) How many coils does it have? (Make the same assumption as in Example 21.10.) (b) At what rate must the current through it change so that a potential difference of 2.00 V is developed across its ends?

36. •• **Self-inductance of a solenoid.** A long, straight solenoid has N turns, a uniform cross-sectional area A, and length l. Use the definition of self-inductance expressed by Equation 21.13 to show that the inductance of this solenoid is given approximately by the equation $L = \mu_0 A N^2/l$. Assume that the magnetic field is uniform inside the solenoid and zero outside. (Your answer is approximate because B is actually smaller at the ends than at the center of the solenoid. For this reason, your answer is actually an upper limit on the inductance.)

37. •• When the current in a toroidal solenoid is changing at a rate of 0.0260 A/s, the magnitude of the induced emf is 12.6 mV. When the current equals 1.40 A, the average flux through each turn of the solenoid is 0.00285 Wb. How many turns does the solenoid have?

21.9 Transformers

38. • A transformer consists of 275 primary windings and 834 secondary windings. If the potential difference across the primary coil is 25.0 V, (a) what is the voltage across the secondary coil, and (b) what is the effective load resistance of the secondary coil if it is connected across a 125 Ω resistor?

39. • **Off to Europe!** You plan to take your hair blower to Europe, where the electrical outlets put out 240 V instead of the 120 V seen in the United States. The blower puts out 1600 W at 120 V. (a) What could you do to operate your blower via the 240 V line in Europe? (b) What current will your blower draw from a European outlet? (c) What resistance will your blower appear to have when operated at 240 V?

40. •• You need a transformer that will draw 15 W of power from a 220 V (rms) power line, stepping the voltage down to 6.0 V (rms). (a) What will be the current in the secondary coil? (b) What should be the resistance of the secondary circuit? (c) What will be the equivalent resistance of the input circuit?

41. • **A step-up transformer.** A transformer connected to a 120 V (rms) ac line is to supply 13,000 V (rms) for a neon sign. To reduce the shock hazard, a fuse is to be inserted in the primary circuit and is to blow when the rms current in the secondary circuit exceeds 8.50 mA. (a) What is the ratio of secondary to primary turns of the transformer? (b) What power must be supplied to the transformer when the rms secondary current is 8.50 mA? (c) What current rating should the fuse in the primary circuit have?

21.10 Magnetic Field Energy

42. • An air-filled toroidal solenoid has a mean radius of 15.0 cm and a cross-sectional area of 5.00 cm^2. When the current is 12.0 A, the energy stored is 0.390 J. How many turns does the winding have?

43. • **Energy in a typical inductor.** (a) How much energy is stored in a 10.2 mH inductor carrying a 1.15 A current? (b) How much current would such an inductor have to carry to store 1.0 J of energy? Is this a reasonable amount of current for ordinary laboratory circuit elements?

44. • (a) What would have to be the self-inductance of a solenoid for it to store 10.0 J of energy when a 1.50 A current runs through it? (b) If this solenoid's cross-sectional diameter is 4.00 cm, and if you could wrap its coils to a density of 10 coils/mm, how long would the solenoid be? (See problem 36.) Is this a realistic length for ordinary laboratory use?

45. •• A solenoid 25.0 cm long and with a cross-sectional area of 0.500 cm^2 contains 400 turns of wire and carries a current

of 80.0 A. Calculate: (a) the magnetic field in the solenoid; (b) the energy density in the magnetic field if the solenoid is filled with air; (c) the total energy contained in the coil's magnetic field (assume the field is uniform); (d) the inductance of the solenoid.

46. •• Large inductors have been proposed as energy-storage devices. (a) How much electrical energy is converted to light and thermal energy by a 200 W lightbulb in one day? (b) If the amount of energy calculated in part (a) is stored in an inductor in which the current is 80.0 A, what is the inductance?

47. •• When a certain inductor carries a current I, it stores 3.0 mJ of magnetic energy. How much current (in terms of I) would it have to carry to store 9.0 mJ of energy?

21.11 The *R–L* Circuit

48. • A 12.0 V dc battery having no appreciable internal resistance, a 150.0 Ω resistor, an 11.0 mH inductor, and an open switch are all connected in series. After the switch is closed, what are (a) the time constant for this circuit, (b) the maximum current that flows through it, (c) the current 73.3 μs after the switch is closed, and (d) the maximum energy stored in the inductor?

49. • An inductor with an inductance of 2.50 H and a resistor with a resistance of 8.00 Ω are connected to the terminals of a battery with an emf of 6.00 V and negligible internal resistance. Find (a) the initial rate of increase of the current in the circuit, (b) the initial potential difference across the inductor, (c) the current 0.313 s after the circuit is closed, and (d) the maximum current.

50. • In Figure 21.63, both switches S_1 and S_2 are initially open. S_1 is then closed and left closed until a constant current is established. Then S_2 is closed just as S_1 is opened, taking the battery out of the circuit. (a) What is the initial current in the resistor just after S_2 is closed and S_1 is opened? (b) What is the time constant of the circuit? (c) What is the current in the resistor after a large number of time constants have elapsed?

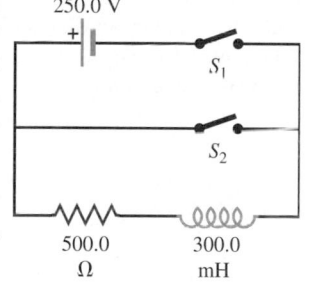

250.0 V

S_1

S_2

500.0 Ω   300.0 mH

▲ **FIGURE 21.63** Problem 50.

51. •• In the circuit shown in Figure 21.64, the battery and the inductor have no appreciable internal resistance and there is no current in the circuit. After the switch is closed, find the readings

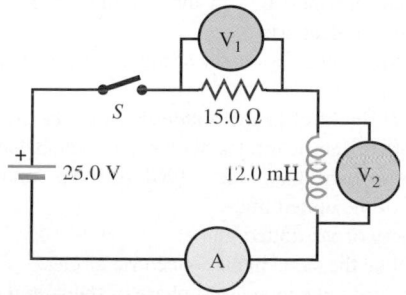

V_1

S

15.0 Ω

25.0 V   12.0 mH V_2

A

▲ **FIGURE 21.64** Problem 51.

of the ammeter (A) and voltmeters $(V_1 \text{ and } V_2)$ (a) the instant after the switch is closed; (b) after the switch has been closed for a very long time. (c) Which answers in parts (a) and (b) would change if the inductance were 24.0 mH instead?

52. •• A 35.0 V battery with negligible internal resistance, a 50.0 V resistor, and a 1.25 mH inductor with negligible resistance are all connected in series with an open switch. The switch is suddenly closed. (a) How long after closing the switch will the current through the inductor reach one-half of its maximum value? (b) How long after closing the switch will the energy stored in the inductor reach one-half of its maximum value?

53. •• A 1.50 mH inductor is connected in series with a dc battery of negligible internal resistance, a 0.750 kΩ resistor, and an open switch. How long after the switch is closed will it take for (a) the current in the circuit to reach half of its maximum value, (b) the energy stored in the inductor to reach half of its maximum value? (*Hint:* You will have to solve an exponential equation.)

21.12 The *L–C* Circuit

54. • A 12.0 μF capacitor and a 5.25 mH inductor are connected in series with an open switch. The capacitor is initially charged to 6.20 μC. What is the angular frequency of the charge oscillations in the capacitor after the switch is closed?

55. •• A 5.00 μF capacitor is initially charged to a potential of 16.0 V. It is then connected in series with a 3.75 mH inductor. (a) What is the total energy stored in this circuit? (b) What is the maximum current in the inductor? What is the charge on the capacitor plates at the instant the current in the inductor is maximal?

General Problems

56. •• A 15.0 μF capacitor is charged to 175 μC and then connected across the ends of a 5.00 mH inductor. (a) Find the maximum current in the inductor. At the instant the current in the inductor is maximal, how much charge is on the capacitor plates? (b) Find the maximum potential across the capacitor. At this instant, what is the current in the inductor? (c) Find the maximum energy stored in the inductor. At this instant, what is the current in the circuit?

57. •• An inductor is connected to the terminals of a battery that has an emf of 12.0 V and negligible internal resistance. The current is 4.86 mA at 0.725 ms after the connection is completed. After a long time the current is 6.45 mA. What are (a) the resistance R of the inductor and (b) the inductance L of the inductor?

58. •• A rectangular circuit is moved at a constant velocity of 3.0 m/s into, through, and then out of a uniform 1.25 T magnetic field, as shown in Figure 21.65. The magnetic field region is considerably wider than 50.0 cm. Find the magnitude and direction (clockwise or counterclockwise) of the current induced in the circuit as it is (a) going into the magnetic field, (b) totally

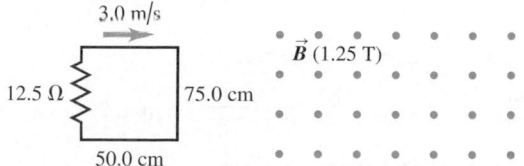

3.0 m/s

$\vec{B}$ (1.25 T)

12.5 Ω   75.0 cm

50.0 cm

▲ **FIGURE 21.65** Problem 58.

within the magnetic field, but still moving, and (c) moving out of the field. (d) Sketch a graph of the current in this circuit as a function of time, including the preceding three cases.

59. •• The rectangular loop in Figure 21.66, with area A and resistance R, rotates at uniform angular velocity ω about the y axis. The loop lies in a uniform magnetic field $\vec{B}$ in the direction of the x axis. Sketch graphs of the following quantities, as functions of time, letting $t = 0$ when the loop is in the position shown in the figure: (a) the magnetic flux through the loop, (b) the rate of change of flux with respect to time, (c) the induced emf in the loop, (d) the induced emf if the angular velocity is doubled.

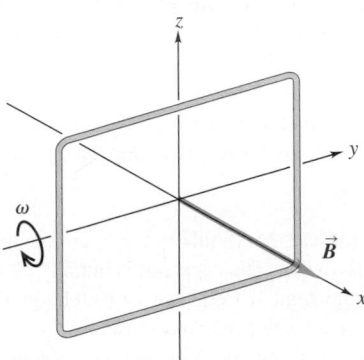

▲ **FIGURE 21.66** Problem 59.

60. •• A flexible circular loop 6.50 cm in diameter lies in a magnetic field with magnitude 0.950 T, directed into the plane of the page as shown in Figure 21.67. The loop is pulled at the points indicated by the arrows, forming a loop of zero area in 0.250 s. (a) Find the average induced emf in the circuit. (b) What is the direction of the current in R: from a to b or from b to a? Explain your reasoning.

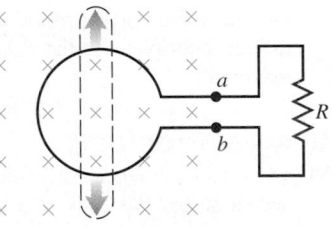

▲ **FIGURE 21.67** Problem 60.

61. •• **An electromagnetic car alarm.** Your latest invention is a car alarm that produces sound at a particularly annoying frequency of 3500 Hz. To do this, the car-alarm circuitry must produce an alternating electric current of the same frequency. That's why your design includes an inductor and a capacitor in series. The maximum voltage across the capacitor is to be 12.0 V (the same voltage as the car battery). To produce a sufficiently loud sound, the capacitor must store 0.0160 J of energy. What values of capacitance and inductance should you choose for your car-alarm circuit?

62. •• In the circuit shown in Figure 21.68, S_1 has been closed for a long enough time so that the current reads a steady 3.50 A. Suddenly, S_2 is closed and S_1 is opened at the same instant. (a) What is the maximum charge that the capacitor will receive? (b) What is the current in the inductor at this time?

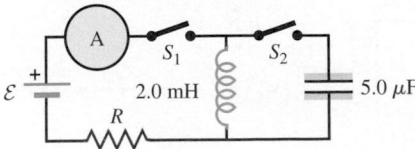

▲ **FIGURE 21.68** Problem 62.

63. •• Consider the circuit in Figure 21.69. (a) Just after the switch is closed, what is the current through each of the resistors? (b) After the switch has been closed a long time, what is the current through each resistor? (c) After S has been closed a long time, it is opened again. Just after it is opened, what is the current through the 20.0 Ω resistor?

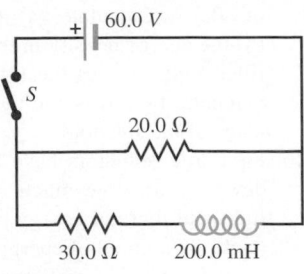

▲ **FIGURE 21.69** Problem 63.

Passage Problems

BIO Stimulating the brain. Communication in the nervous system is based on propagating electrical signals called action potentials that travel along the extended nerve cell processes, the axons. Action potentials are generated when the electrical potential difference across the membrane changes so that the inside of the cell becomes more positive. Researchers in clinical medicine and neurobiology want to stimulate nerves non-invasively at specific locations in conscious subjects. But using electrodes to apply current on the skin is painful and requires large currents.

Anthony Barker and colleagues at the University of Sheffield in England developed a technique that is now widely used called transcranial magnetic stimulation (TMS). In the TMS technique, a coil positioned near the skull produces a time-varying magnetic field, which induces electric currents in the conductive brain tissue sufficient to cause action potentials in nerve cells. For example, if the coil is placed near the motor cortex, the region of the brain that controls voluntary movement, scientists can monitor the contraction of muscles and assess the state of the connections between the brain and the muscle.

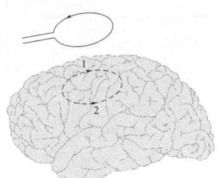

▲ **FIGURE 21.70** Problems 64–65.

64. In the diagram of TMS shown in Figure 21.70, a current pulse increases to a peak and then decreases to zero in the direction shown in the stimulating coil. What will be the direction (1 or 2) of the induced current (dotted line) in the brain tissue?
    A. 1   B. 2
    C. 1 while the current increases in the stimulating coil and 2 while the current decreases
    D. 2 while the current increases in the stimulating coil, 1 while the current decreases

65. The brain tissue at the level of the dotted line may be considered as a series of concentric circles, with each circle behaving independently. Where will the induced EMF be the greatest?
    A. At the center of the dotted line
    B. At the periphery of the dotted line
    C. The EMF will be the same in all concentric circles
    D. At the center during the increasing phase of the stimulating current and at the periphery during the decreasing phase

22 Alternating Current

During the 1880s, there was a heated and acrimonious debate over the best method of electric-power distribution. Thomas Edison favored direct current (dc)—that is, steady current that does not vary with time. George Westinghouse favored alternating current (ac)—current (and hence voltage) that varies sinusoidally with time. Westinghouse argued that transformers (which we studied in Section 21.9) could be used to step voltage up and down with ac, but not with dc. Edison claimed that dc was inherently safer.

Eventually, the arguments of Westinghouse prevailed, and most present-day household and industrial power-distribution systems operate with alternating current. Any appliance that you plug into a wall outlet uses ac, and many battery-powered devices, such as portable audio players and cell phones, make use of the dc supplied by the battery to create or amplify alternating currents. Circuits in modern communications equipment, including radios and televisions, make extensive use of ac.

In this chapter, we'll learn how resistors, inductors, and capacitors behave in circuits with sinusoidally varying voltages and currents. Many of the principles that we found useful in Chapters 18, 19, and 21 are applicable, and we'll explore several new concepts related to the circuit behavior of inductors and capacitors. A key idea in this discussion is the concept of *resonance*, which we studied in Chapter 13 in relation to mechanical systems.

As you likely know, good speakers have separate drivers for different parts of the frequency spectrum—small tweeters, larger midrange drivers, and still larger woofers for the low sounds. But the input to the amplifier is a single curve, not three. How can an amplifier sort out the frequencies and send them to the right drivers? You'll find out in this chapter.

22.1 Phasors and Alternating Currents

We've already studied a source of alternating emf (voltage) that can supply an **alternating current** to a circuit. A coil of wire rotating with constant angular

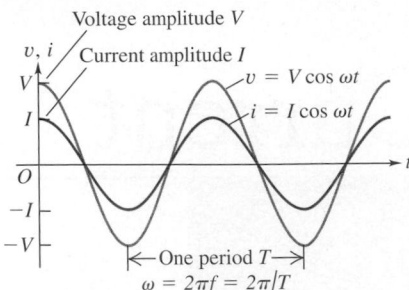

(a) Graphs of sinusoidal current and voltage versus time. The relative heights of the two curves are not significant.

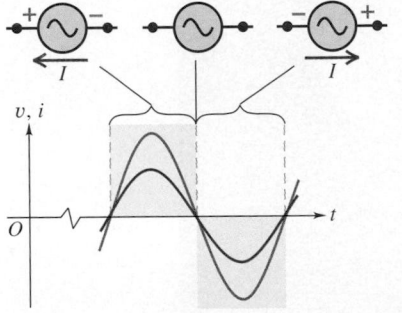

(b) The graphs related to a schematic ac source.

▲ **FIGURE 22.1** Sinusoidal current and voltage graphed as functions of time.

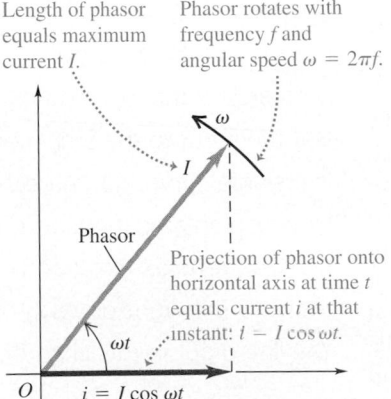

▲ **FIGURE 22.2** A phasor diagram.

velocity in a magnetic field (Section 21.3) develops a sinusoidally alternating emf and is the prototype of the commercial alternating-current generator, or *alternator.* An *L–C* circuit (Section 21.12) can be used to produce an alternating current with a frequency that may range from a few hertz to many millions.

We'll use the term **ac source** for any device that supplies a sinusoidally varying potential difference v or current i. A sinusoidal voltage might be described by a function such as

$$v = V\cos\omega t. \tag{22.1}$$

In this expression, V is the maximum potential difference, which we call the **voltage amplitude;** v is the *instantaneous* potential difference, and ω is the **angular frequency,** equal to 2π times the frequency f (the number of cycles per second). Figure 22.1 shows graphs of voltage and current that vary sinusoidally with time.

In North America, commercial electric-power distribution systems always use a frequency $f = 60$ Hz, corresponding to an angular frequency $\omega = (2\pi\ \text{rad})(60\ \text{s}^{-1}) = 377\ \text{rad/s}$. Similarly, a sinusoidal current i might be described by

$$i = I\cos\omega t, \tag{22.2}$$

where I is the maximum current, or **current amplitude** (Figure 22.1). In the next section, we'll look at the behavior of individual circuit elements when they carry a sinusoidal current. The usual circuit-diagram symbol for an ac source is

To represent sinusoidally varying voltages and currents, we'll use vector diagrams similar to the diagrams we used with the circle of reference in our study of harmonic motion in Section 11.4. In these diagrams, the instantaneous value of a quantity that varies sinusoidally with time is represented by the *projection* onto a horizontal axis of a vector with a length equal to the amplitude of the quantity. For example, the vector (from point O to point Q in Figure 11.20) has length A, representing the amplitude of the motion, and it rotates counterclockwise with constant angular velocity ω. These rotating vectors are called **phasors,** and diagrams containing them are called **phasor diagrams.** Figure 22.2 shows a phasor diagram for the sinusoidal current described by Equation 22.2. The length I of the phasor is the maximum value of I, and the projection of the phasor onto the horizontal axis at time t is $I\cos\omega t$. At time $t = 0$, $i = I$; this is why we chose to use the cosine function rather than the sine in Equation 22.2.

A phasor isn't a real physical quantity with a direction in space, as are velocity, momentum, and electric field. Rather, it is a *geometric* entity that provides a language for describing and analyzing physical quantities that vary sinusoidally with time. In Section 11.4, we used a single phasor to represent the position and motion of a point mass undergoing simple harmonic motion. In this chapter, we'll use phasors to *add* sinusoidal voltages and currents. Combining sinusoidal quantities with phase differences then becomes a matter of vector addition.

Conceptual Analysis 22.1

A current phasor

A sinusoidal current is described by $i = I\cos\omega t$, where $\omega = 1.57$ rad/s and I is the maximum magnitude of the current. At some time $t - t'$, where 1 s $< t' < 2$ s, the current is -5 amperes. Which of the phasors in Figure 22.3 (A–D) can represent the current at time t'?

SOLUTION Since the current is the phasor's projection onto the *horizontal* axis, the fact that the current at time t' is negative rules out phasors A and D. (They represent *positive* currents.) To decide between B and C, we need to know which quadrant the phasor is in at times between 1 s and 2 s. For that, we need to know the period T—the time taken for the phasor to make one complete revolution (or the time required for one cycle on a graph of i versus t). We get the period from the angular frequency ω: $T = 2\pi/\omega = 2\pi/(1.57$ rad/s$) = 4$ s. Thus, the phasor rotates through one rev-

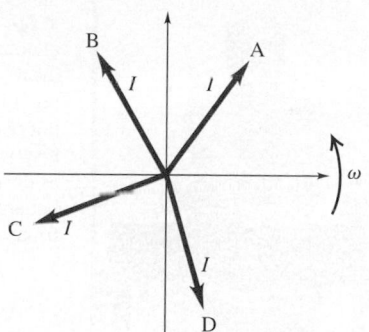

▲ **FIGURE 22.3**

olution in 4 s, or one quadrant each second; during the second second, it is in the second (upper left) quadrant, so B is correct.

We can describe and measure a sinusoidally varying current i in terms of its maximum value, which we can denote as I_{max} or simply I. An alternative notation that is often more useful is the concept of *root-mean-square (rms) value*. We encountered this concept in Section 15.3 in connection with the speeds of molecules in a gas. As shown graphically in Figure 22.4, we *square* the instantaneous current i, calculate the *average* value of i^2, and then take the *square root* of that average. This procedure defines the **root-mean-square (rms) current,** denoted as I_{rms}. Even when i is negative, i^2 is always positive, so I_{rms} is never zero (unless i is zero at every instant).

There's a simple relation between I_{rms} and the maximum current I. If the instantaneous current i is given by $I\cos\omega t$, then

$$i^2 = I^2\cos^2\omega t.$$

From trigonometry, we use the double-angle formula

$$\cos^2 A = \tfrac{1}{2}(1 + \cos 2A)$$

to find that

$$i^2 = \tfrac{1}{2}I^2(1 + \cos 2\omega t)$$
$$= \tfrac{1}{2}I^2 + \tfrac{1}{2}I^2\cos 2\omega t.$$

The average of $\cos 2\omega t$ is zero because it is positive half the time and negative half the time. Thus, the average of i^2 is simply $I^2/2$; the square root of this quantity is I_{rms}:

rms values of sinusoidally varying current and voltage
The rms current for a sinusoidal current with amplitude I is

$$I_{rms} = \frac{I}{\sqrt{2}}. \tag{22.3}$$

In the same way, the root-mean-square value of a sinusoidal voltage with amplitude (maximum value) V is

$$V_{rms} = \frac{V}{\sqrt{2}}. \tag{22.4}$$

Meaning of the rms value of a sinusoidal quantity (here, ac current with $I = 3$ A). We:

1. Graph current i versus time,
2. *square* the instantaneous current i,
3. take the *average* (mean) value of i^2,
4. take the *square root* of that average.

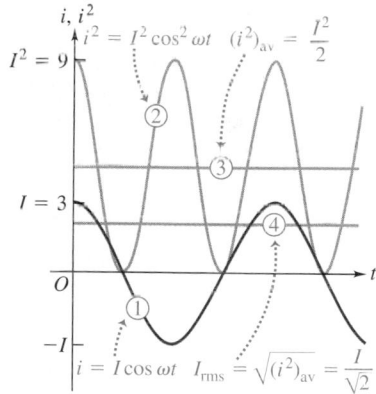

▲ **FIGURE 22.4** Root-mean-square value of a sinusoidally varying quantity.

◀ **Application Going nowhere.** You may wonder how we can power anything by use of alternating current, since the electrons simply jiggle back and forth, going nowhere, and the *average* current is zero. But direct current is not that different. In a dc circuit, there is no current unless the circuit is a closed loop, and then the electrons merely travel in closed loops, not really "going anywhere" either. In both cases, what matters is that the electrons move in response to a potential difference, so energy is available. In a lightbulb, ac is just as effective as dc at heating the filament. A motor that uses ac is constructed differently from a dc motor, but again, the current still interacts with the magnetic field to turn the rotor. Thus, this industrious runner is getting nowhere, thanks to electrons doing the same.

Voltages and currents in power-distribution systems are always described in terms of their rms values, not their amplitudes (maximum values). The usual household power supply in North America, designated 120 volt ac, has an rms voltage of $V_{rms} = 120$ V. The voltage *amplitude* V is

$$V = \sqrt{2}\,V_{rms} = \sqrt{2}\,(120\text{ V}) = 170\text{ V}.$$

Meters that are used for ac voltage and current measurements are nearly always calibrated to read rms values, not maximum values.

EXAMPLE 22.1 **Current in a personal computer**

The plate on the back of a personal computer says that the machine draws 2.7 A from a 120 V, 60 Hz line. For this PC, what are **(a)** the average current, **(b)** the average of the square of the current, and **(c)** the current amplitude?

SOLUTION

SET UP We're given the current, voltage, and frequency (60 Hz), but we actually need only the current. We remember that the stated current is an rms value. To clarify our thinking, we sketch the approximate graph in Figure 22.5.

SOLVE Part (a): The average of any sinusoidal alternating current over any whole number of cycles is zero.

Part (b): The rms current, 2.7 A, is the *square root* of the *mean* (average) of the *square* of the current. That is,

$$I_{rms} = \sqrt{(i^2)_{av}}, \quad \text{or}$$
$$(i^2)_{av} = (I_{rms})^2 = (2.7\text{ A})^2 = 7.3\text{ A}^2.$$

Part (c): From Equation 22.3, the current amplitude I is

$$I = \sqrt{2}\,I_{rms} = \sqrt{2}\,(2.7\text{ A}) = 3.8\text{ A}.$$

REFLECT We can always describe a sinusoidally varying quantity either in terms of its amplitude or in terms of its rms value. We'll see later that the rms value is usually more convenient for expressing energy and power relations.

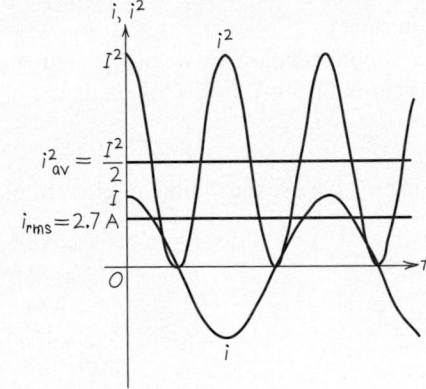

▲ **FIGURE 22.5** Our sketch for this problem.

Practice Problem: A lightbulb draws 0.50 A from a 120 V, 60 Hz line. For this bulb, find the current amplitude. *Answer: I =* 0.71 A.

22.2 Resistance and Reactance

In this section, we'll derive voltage–current relations for an individual resistor, inductor, or capacitor carrying a sinusoidal current.

Resistor in an ac Circuit

First, let's consider a resistor with resistance R, with a sinusoidal current given by Equation 22.2: $i = I\cos\omega t$, as in Figure 22.6a. The current amplitude (maximum current) is I. We denote the instantaneous potential of point a with respect to point b by v_R; from Ohm's law, v_R is given by

$$v_R = iR = IR\cos\omega t. \tag{22.5}$$

The maximum voltage V_R (the voltage amplitude) is the coefficient of the cosine function; that is,

$$V_R = IR. \tag{22.6}$$

So we can also write

$$v_R = V_R\cos\omega t. \tag{22.7}$$

The instantaneous voltage v_R and current i are both proportional to $\cos\omega t$, so the voltage is in step, or *in phase*, with the current. Thus, the voltage and current reach their peak values at the same time, go through zero at the same time, and so on. Equation 22.6 shows that the current and voltage amplitudes are related in the same way as in a dc circuit.

Figure 22.6b shows graphs of i and v_R as functions of time. The vertical scales for i and v_R are different and have different units, so the relative heights of the two curves are not significant. The corresponding phasor diagram is shown in Figure 22.6c. Because i and v_R are *in phase* and have the same frequency, the current and voltage phasors rotate together; they are parallel at each instant. Their projections on the horizontal axis represent the instantaneous current and voltage, respectively.

Inductor in an ac Circuit

Next, suppose we replace the resistor in Figure 22.6a with a pure inductor with self-inductance L and zero resistance (Figure 22.7a). Again, we assume that the current is $i = I\cos\omega t$. The induced emf in the direction of i is given by Equation 21.14, as shown in Figure 21.25: $\mathcal{E} = -L\,\Delta i/\Delta t$. This corresponds to the potential of point b with respect to point a; the potential v_L of point a with respect to point b is the negative of $\mathcal{E}$, and we have

$$v_L = L\,\Delta i/\Delta t.$$

The voltage across the inductor at any instant is proportional to the *rate of change* of the current. The points of maximum voltage on the graph correspond to maximum steepness of the current curve, and the points of zero voltage are the points where the current curve instantaneously levels off at its maximum and minimum values (Figure 22.7b). The voltage and current are "out of step," or *out of phase,*

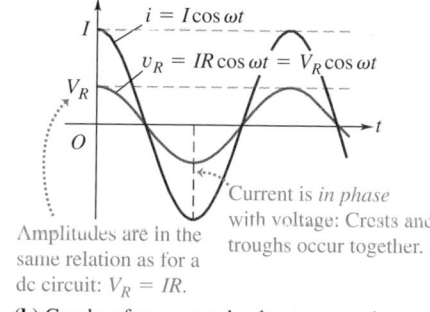

(a) Circuit with ac source and resistor

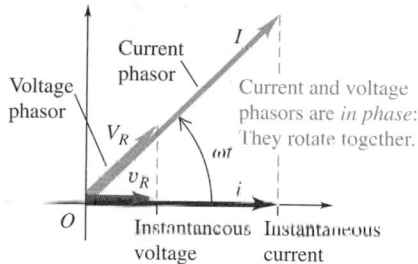

Amplitudes are in the same relation as for a dc circuit: $V_R = IR$.

Current is *in phase* with voltage: Crests and troughs occur together.

(b) Graphs of current and voltage versus time

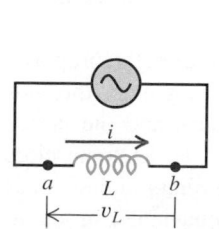

Current and voltage phasors are *in phase*: They rotate together.

(c) Phasor diagram

▲ **FIGURE 22.6** Current and voltage in a resistance R connected across an ac source.

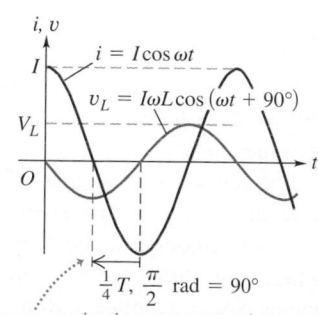

Voltage curve *leads* current curve by a quarter cycle (corresponding to $\phi = \pi/2$ rad $= 90°$).

(a) Circuit with ac source and inductor   **(b)** Graphs of current and voltage versus time

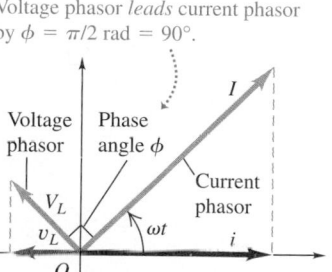

Voltage phasor *leads* current phasor by $\phi = \pi/2$ rad $= 90°$.

(c) Phasor diagram

▲ **FIGURE 22.7** Current and voltage in an inductance L connected across an ac source.

by a quarter cycle: The voltage peaks occur a quarter cycle *earlier* than the current peaks, and we say that the voltage *leads* the current by 90°. The phasor diagram in Figure 22.7c also shows this relationship; the voltage phasor is ahead of the current phasor by 90°.

The general shape of the graph of v_L as a function of t in Figure 22.7b is that of a *sine* function. Indeed, it can be shown by using calculus that in this case,

$$v_L = -I\omega L \sin\omega t. \qquad (22.8)$$

We can also obtain the phase relationship just described by rewriting Equation 22.8, using the identity $\cos(A + 90°) = -\sin A$:

$$v_L = I\omega L \cos(\omega t + 90°). \qquad (22.9)$$

This result and the phasor diagram show that the voltage can be viewed as a cosine function with a "head start" of 90°, $\pi/2$ radians, or $\frac{1}{4}$ cycle.

For consistency with later discussions, we'll usually follow this pattern, describing the phase of the *voltage* relative to the *current*, not the reverse. Thus, if the current i in a circuit is

$$i = I\cos\omega t$$

and the voltage v of one point with respect to another is

$$v = V\cos(\omega t + \phi),$$

we call ϕ the **phase angle,** understanding that it gives the phase of the *voltage* relative to the *current*. For a pure inductor, $\phi = 90°$, and the voltage *leads* the current by 90°, $\pi/2$ radians, or $\frac{1}{4}$ cycle. For a pure resistor, $\phi = 0$, and the voltage and current are *in phase*.

From Equation 22.8 or 22.9, the voltage amplitude V_L is

$$V_L = I\omega L. \qquad (22.10)$$

Inductive reactance

We define the **inductive reactance** X_L of an inductor as the product of the inductance L and the angular frequency ω:

$$X_L = \omega L. \qquad (22.11)$$

Using X_L, we can write Equation 22.10 in the same form as for a resistor $(V_R = IR)$:

$$V_L = IX_L. \qquad (22.12)$$

Because X_L is the ratio of a voltage to a current, its unit is the same as for resistance: the ohm.

The inductive reactance of an inductor is directly proportional both to its inductance L and to the angular frequency. The greater the inductance and the higher the frequency, the *larger* is the reactance and the larger is the induced voltage amplitude, for a given current amplitude. In some circuit applications, such as power supplies and radio-interference filters, inductors are used to block high frequencies while permitting lower frequencies or dc to pass through. Thus, inductors are also called *chokes*, because they choke off high-frequency currents.

EXAMPLE 22.2   Inductance in a radio

The current amplitude in an inductor in a radio receiver is to be 250 μA when the voltage amplitude is 3.60 V at a frequency of 1.60 MHz (corresponding to the upper end of the AM broadcast band). What inductive reactance is needed? What inductance is needed?

SOLUTION

SET UP AND SOLVE From Equation 22.12,

$$X_L = \frac{V_L}{I} = \frac{3.60 \text{ V}}{250 \times 10^{-6} \text{ A}} = 1.44 \times 10^4 \,\Omega = 14.4 \text{ k}\Omega.$$

From Equation 22.11, with $\omega = 2\pi f$, we find that

$$L = \frac{X_L}{2\pi f} = \frac{1.44 \times 10^4 \,\Omega}{2\pi(1.60 \times 10^6 \text{ Hz})} = 1.43 \times 10^{-3} \text{ H} = 1.43 \text{ mH}.$$

REFLECT The inductance L is constant, but the inductive reactance X_L is proportional to frequency. If the frequency approaches zero, the inductive reactance also approaches zero.

Practice Problem: A generator contains a 3.5 mH inductor. Find the inductive reactance in the circuit when the generator frequency is 2.5×10^3 Hz. *Answer:* $X_L = 55 \,\Omega$.

Capacitor in an ac Circuit

Finally, suppose we connect a capacitor with capacitance C to the source, as in Figure 22.8a, producing a current $i = I\cos\omega t$ through the capacitor. You may object that charge can't really move *through* the capacitor, because its two plates are insulated from each other. True enough, but at each instant, as the capacitor charges and discharges, we have a current i into one plate and an equal current out of the other plate, just as though the charge were being conducted through the capacitor. For this reason, we often speak about alternating current *through* a capacitor.

To find the voltage v_C of point a with respect to point b, we first note that v_C is the charge q on a capacitor plate, divided by the capacitance C: $v_C = q/C$ (as we learned in Section 18.7). Thus, the rate of change of v_C is equal to $1/C$ times the rate of change of q, where q is in turn equal to the current i in the circuit. That is,

$$\frac{\Delta v_C}{\Delta t} = \frac{1}{C}\frac{\Delta q}{\Delta t} = \frac{i}{C}.$$

If the current i is once again given by $i = I\cos\omega t$, then

$$\frac{\Delta v_C}{\Delta t} = \frac{1}{C}I\cos\omega t.$$

The current is proportional to the rate of change of the voltage. The current is greatest when v_C is increasing most rapidly, and it is zero when v_C instantaneously levels off at a maximum or minimum value. In short, the graph of v_C as a function of time must resemble a sine curve, as shown in Figure 22.8b.

In fact, it can be shown, using methods of calculus, that

$$v_C = \frac{I}{\omega C}\sin\omega t. \tag{22.13}$$

As Figure 22.8b shows, the current is greatest when the v_C curve is rising or falling most steeply (and the capacitor is charging or discharging most rapidly) and zero when the v_C curve instantaneously levels off at its maximum and minimum values.

The capacitor voltage and current are out of phase by a quarter cycle. The peaks of voltage occur a quarter cycle *after* the corresponding current peaks, and we say that the voltage *lags* the current by 90°. The phasor diagram in Figure 22.8c

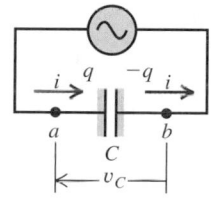

(a) Circuit with ac source and capacitor

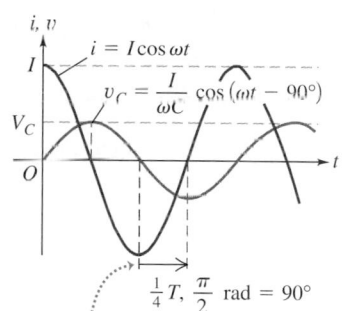

Voltage curve *lags* current curve by a quarter cycle (corresponding to $\phi = -\pi/2$ rad $= -90°$).

(b) Graphs of current and voltage versus time

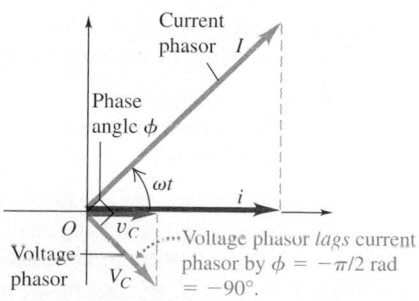

(c) Phasor diagram

▲ **FIGURE 22.8** Current and voltage in a capacitance C connected across an ac source.

shows this relationship; the voltage phasor is *behind* the current phasor by a quarter cycle, 90°, or $\pi/2$.

We can also derive this phase difference by rewriting Equation 22.13, using the identity $\cos(A - 90°) = \sin A$:

$$v_C = \frac{I}{\omega C}\cos(\omega t - 90°). \quad (22.14)$$

This cosine function has a "late start" of 90° compared with that for the current.

Equations 22.13 and 22.14 show that the *maximum* voltage V_C (the voltage amplitude) is

$$V_C = \frac{I}{\omega C}. \quad (22.15)$$

To put this expression in the same form as that for a resistor $(V_R = IR)$, we define a quantity X_C, called the **capacitive reactance** of the capacitor, as follows:

Capacitive reactance
We define the capacitive reactance of a capacitor as the inverse of the product of the angular frequency and the capacitance:

$$X_C = \frac{1}{\omega C}. \quad (22.16)$$

Then

$$V_C = IX_C. \quad (22.17)$$

Because X_C is the ratio of a voltage to a current, its unit is the same as that for resistance and inductive reactance: the ohm.

The capacitive reactance of a capacitor is inversely proportional both to the capacitance C and to the angular frequency ω. The greater the capacitance and the higher the frequency, the *smaller* is the reactance X_C. Capacitors tend to pass high-frequency current and to block low-frequency and dc currents, just the opposite of inductors.

Conceptual Analysis 22.2

Reactance and current

In Figure 22.9, the current in circuit 1, containing an inductor, has an angular frequency ω_1, and the current in circuit 2, containing a capacitor, has an angular frequency ω_2. What happens to the brightness of each lightbulb if we keep the source voltage constant while we increase ω_1 and decrease ω_2?

A. Both bulbs get brighter.
B. The brightness of each bulb remains constant.
C. Both bulbs get dimmer.

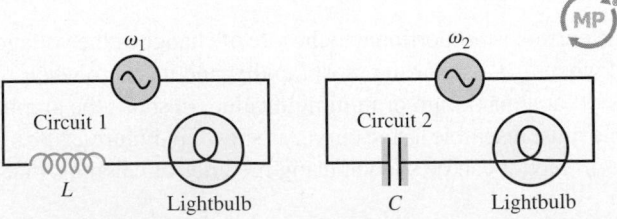

▲ FIGURE 22.9

SOLUTION For the inductor circuit, $X_L = \omega_1 L$, so the inductive reactance increases with increasing angular frequency ω_1. Since $X_C = 1/\omega_2 C$, the reactance of the capacitor circuit increases as its angular frequency ω_2 decreases. The current amplitude through each element for a given voltage amplitude V is given by $I = V/X_L$ and $I = V/X_C$, respectively. We see that the currents in both circuits decrease as their reactances increase, so both bulbs get dimmer.

EXAMPLE 22.3 Resistor and capacitor in series

A 300 Ω resistor is connected in series with a 5.0 μF capacitor, as shown in Figure 22.10. The voltage across the resistor is $v_R = (1.20\ \text{V})\cos(2500\ \text{rad/s})t$. **(a)** Derive an expression for the circuit current. **(b)** Determine the capacitive reactance of the capacitor. **(c)** Derive an expression for the voltage v_C across the capacitor.

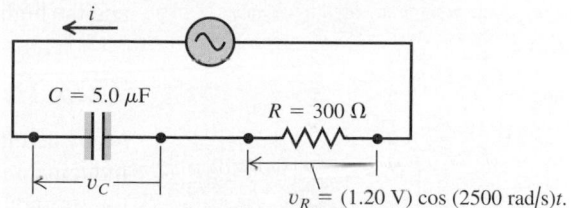

▲ **FIGURE 22.10**

SOLUTION

SET UP AND SOLVE Part (a): We recall that circuit elements in series have the same current. We can thus use $v_R = iR$ and solve for i:

$$i = \frac{v_R}{R} = \frac{(1.20\ \text{V})\cos(2500\ \text{rad/s})t}{300\ \Omega}$$

$$= (4.0 \times 10^{-3}\ \text{A})\cos(2500\ \text{rad/s})t.$$

Part (b): The expression for v_R tells us that $\omega = 2500\ \text{rad/s}$. Therefore, we can use Equation 22.16 to find the capacitive reactance of the capacitor:

$$X_C = \frac{1}{\omega C} = \frac{1}{(2500\ \text{rad/s})}\frac{1}{(5.0 \times 10^{-6}\ \text{F})} = 80\ \Omega.$$

Part (c): First we use Equation 22.17 to find the amplitude V_C of the voltage across the capacitor:

$$V_C = IX_C = (4.0 \times 10^{-3}\ \text{A})(80\ \Omega) = 0.32\ \text{V}.$$

The capacitor voltage lags the current by 90°. From Equation 22.14, the instantaneous capacitor voltage is

$$v_C = \frac{I}{\omega C}\cos(\omega t - 90°)$$

$$v_C = (0.32\ \text{V})\cos[(2500\ \text{rad/s})t - \pi/2\ \text{rad}].$$

REFLECT We have converted 90° to $\pi/2$ rad, so all the angular quantities have the same units. In ac circuit analysis, phase angles are often given in degrees, so be careful to convert to radians when necessary.

Practice Problem: A 200 Ω resistor is connected in series with a 5.8 μF capacitor. The voltage across the resistor is $v_R = (1.20\ \text{V})\cos(2500\ \text{rad/s})t$. Determine the capacitive reactance of the capacitor. *Answer:* $X_C = 69\ \Omega$.

TABLE 22.1 Circuit elements with alternating current

Circuit element	Circuit quantity	Amplitude relation	Phase of v
Resistor	R	$V_R = IR$	In phase with i
Inductor	$X_L = \omega L$	$V_L = IX_L$	Leads i by 90°
Capacitor	$X_C = 1/\omega C$	$V_C = IX_C$	Lags i by 90°

Table 22.1 summarizes the relations between voltage and current amplitudes for the three circuit elements that we've discussed.

The graphs in Figure 22.11 show how the resistance of a resistor and the reactances of an inductor and a capacitor vary with angular frequency. As the frequency approaches infinity, the reactance of the inductor approaches infinity and that of the capacitor approaches zero. As the frequency approaches zero, the inductive reactance approaches zero and the capacitive reactance approaches infinity. The limiting case of zero frequency corresponds to a dc circuit; in that case, there is *no* current through a capacitor, because $X_C \rightarrow \infty$, and there is no inductive effect, because $X_L \rightarrow 0$.

Figure 22.12 shows an application of the preceding discussion to a loudspeaker system. The woofer and tweeter are connected in parallel across the amplifier output. The capacitor in the tweeter branch blocks the low-frequency components of sound, but passes the higher frequencies; the inductor in the woofer

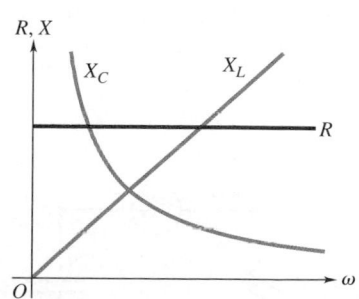

▲ **FIGURE 22.11** Graphs of R, X_L, and X_C as functions of angular frequency ω.

The inductor and capacitor feed low frequencies mainly to the woofer and high frequencies mainly to the tweeter.

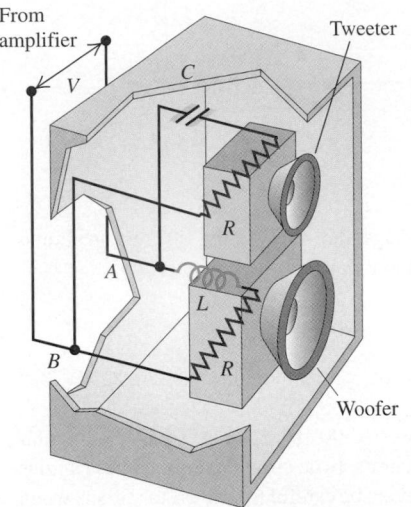

(a) A crossover network in a loudspeaker system

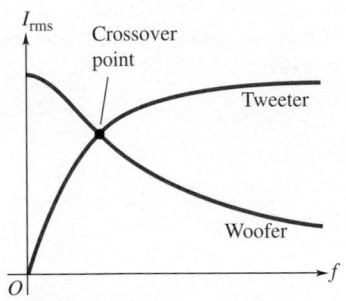

(b) Graphs of rms current as functions of frequency for a given amplifier voltage

▲ **FIGURE 22.12** How capacitance and inductance can be used to direct frequencies to the tweeter and woofer of a speaker.

PhET: Circuit Construction Kit (AC + DC)

branch does the opposite. Thus, the low-frequency sounds are routed to the woofer and the high-frequency sounds to the tweeter.

22.3 The Series *R–L–C* Circuit

Many ac circuits that are used in practical electronic systems involve resistance, inductive reactance, and capacitive reactance. A series circuit containing a resistor, an inductor, and a capacitor is shown in Figure 22.13a. To analyze this and similar circuits, we'll use a phasor diagram that includes the voltage and current phasors for each of the components. In this circuit, because of Kirchhoff's loop rule, the instantaneous *total* voltage v_{ad} across all three components is equal to the source voltage at that instant. We'll show that the phasor representing this total voltage is the *vector sum* of the phasors for the individual voltages. The complete phasor diagram for the circuit is shown in Figure 22.13b. This diagram may appear complicated; we'll explain it one step at a time.

Let's assume that the source supplies a current i given by $i = I\cos\omega t$. Because the circuit elements are connected in series, the current i at any instant is the same at every point in the circuit. Thus, a *single phasor I,* with length proportional to the current amplitude, represents the current in *all* circuit elements.

We use the symbols v_R, v_L, and v_C for the instantaneous voltages across R, L, and C, respectively, and we use V_R, V_L, and V_C for their respective maximum values (voltage amplitudes). We denote the instantaneous and maximum *source* voltages by v and V. Then $v = v_{ad}$, $v_R = v_{ab}$, $v_L = v_{bc}$, and $v_C = v_{cd}$.

We've shown that the potential difference between the terminals of a resistor is *in phase* with the current in the resistor and that its maximum value V_R is

$$V_R = IR. \tag{22.6}$$

The phasor V_R in Figure 22.13b, in phase with the current phasor I, represents the voltage across the resistor. Its projection onto the horizontal axis at any instant gives the instantaneous potential difference v_R.

The voltage across an inductor *leads* the current by 90°. Its voltage amplitude is

$$V_L = IX_L. \tag{22.12}$$

The phasor V_L in Figure 22.13b represents the voltage across the inductor, and its projection onto the horizontal axis at any instant equals v_L.

The voltage across a capacitor *lags* the current by 90°. Its voltage amplitude is

$$V_C = IX_C. \tag{22.17}$$

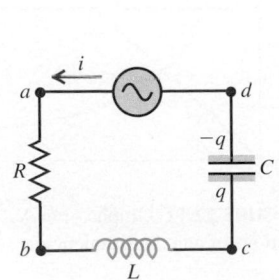

(a) Series *R-L-C* circuit

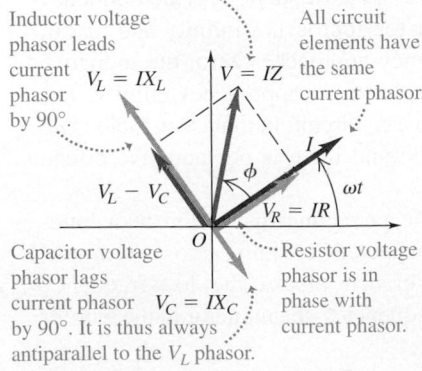

Source voltage phasor is the vector sum of the V_R, V_L, and V_C phasors.

Inductor voltage phasor leads current phasor by 90°.

All circuit elements have the same current phasor.

Capacitor voltage phasor lags current phasor by 90°. It is thus always antiparallel to the V_L phasor.

Resistor voltage phasor is in phase with current phasor.

(b) Phasor diagram for the case $X_L > X_C$

If $X_L < X_C$, the source voltage phasor lags the current phasor, $X < 0$, and ϕ is a negative angle between 0 and −90°.

(c) Phasor diagram for the case $X_L < X_C$

▲ **FIGURE 22.13** Current and voltage in a series *R–L–C* circuit.

The phasor V_C in Figure 22.13b represents the voltage across the capacitor, and its projection onto the horizontal axis at any instant equals v_C.

The instantaneous potential difference v between terminals a and d is equal at every instant to the (algebraic) sum of the potential differences v_R, v_L, and v_C. That is, it equals the sum of the *projections* of the phasors V_R, V_L, and V_C. But the sum of the projections of these phasors is equal to the *projection* of their *vector sum*. So the vector sum V must be the phasor that represents the source voltage v and the instantaneous total voltage v_{ad} across the series of elements.

To form this vector sum, we first subtract the phasor V_C from the phasor V_L. (These two phasors always lie along the same line and have opposite directions, because V_L is always 90° ahead of the I phasor, and V_C is always 90° behind it.) This gives the phasor $V_L - V_C$, which is always at right angles to the phasor V_R. So, from the Pythagorean theorem, the magnitude of the phasor V is

$$V = \sqrt{V_R^2 + (V_L - V_C)^2} = \sqrt{(IR)^2 + (IX_L - IX_C)^2}$$
$$= I\sqrt{R^2 + (X_L - X_C)^2}. \tag{22.18}$$

The quantity $(X_L - X_C)$ is called the **reactance** of the circuit, denoted by X:

$$X = X_L - X_C. \tag{22.19}$$

Finally, we define the **impedance** Z of the circuit as

$$Z = \sqrt{R^2 + (X_L - X_C)^2} = \sqrt{R^2 + X^2}. \tag{22.20}$$

We can now rewrite Equation 22.18 as

$$V = IZ. \tag{22.21}$$

Equation 22.21 again has the same form as $V_R = IR$, with the impedance Z playing the same role as the resistance R in a dc circuit. Note, however, that the impedance is actually a function of R, L, and C, as well as of the angular frequency ω. The complete expression for Z for a series circuit is as follows:

Impedance of a series *R–L–C* circuit

$$Z = \sqrt{R^2 + X^2}$$
$$= \sqrt{R^2 + (X_L - X_C)^2} \tag{22.22}$$
$$= \sqrt{R^2 + [\omega L - (1/\omega C)]^2}.$$

Impedance is always a ratio of a voltage to a current; the SI unit of impedance is the ohm.

Equation 22.22 gives the impedance Z only for a *series R–L–C* circuit, but, using Equation 22.21, we can *define* the impedance of *any* network as the ratio of the voltage amplitude to the current amplitude.

The angle ϕ shown in Figure 22.13b is the phase angle of the source voltage v with respect to the current i. From the diagram,

$$\tan\phi = \frac{V_L - V_C}{V_R} = \frac{I(X_L - X_C)}{IR} = \frac{X_L - X_C}{R} = \frac{X}{R},$$
$$\phi = \arctan\left(\frac{\omega L - 1/\omega C}{R}\right). \tag{22.23}$$

If $X_L > X_C$, as in Figure 22.13b, the source voltage leads the current by an angle ϕ between 0 and 90°. If the current is $i = I\cos\omega t$, then the source voltage v is

$$v = V\cos(\omega t + \phi).$$

If $X_L < X_C$, as in Figure 22.13c, then vector *V* lies on the opposite side of the current vector *I* and the voltage *lags* the current. In this case, $X = X_L - X_C$ is a *negative* quantity, $\tan \phi$ is negative, and ϕ is a negative angle between 0 and $-90°$.

NOTE ▶ All the relations that we have developed for a series *R–L–C* circuit are still valid even if one of the circuit elements is missing. If the resistor is missing, we set $R = 0$; if the inductor is missing, we set $L = 0$. But if the capacitor is missing, we set $C = \infty$, corresponding to zero potential difference $(v = q/C)$ or to zero capacitive reactance $(X_C = 1/\omega C)$. ◀

Conceptual Analysis 22.3

Limiting behavior of impedance

Figure 22.14 shows a series *R–L–C* circuit constructed by a student in a lab class. The resistance is due to the filament of a lightbulb. When the student closes the switch, the bulb does not light up. Which of the following is or are possible explanations?

A. The inductance is too small.
B. The frequency is too small.
C. The capacitance is too great.
D. The frequency is too great.

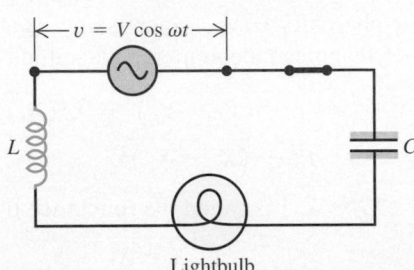

Lightbulb

▲ FIGURE 22.14

SOLUTION Since the bulb won't light and the current through the bulb is $I = V/Z$, we are looking for causes of high impedance. The impedance $Z = \sqrt{R^2 + [\omega L - (1/\omega C)]^2}$ approaches infinity in both the limits $\omega \to 0$ (from the $1/\omega C$ term) and $\omega \to \infty$ (from the ωL term). Thus, choices B and D are possible explanations.

PROBLEM-SOLVING STRATEGY 22.1 **Alternating-current circuits** (MP)

SET UP

1. In ac circuit problems, it is nearly always easiest to work with angular frequency ω. But you may be given the ordinary frequency *f*, expressed in hertz. Don't forget to convert, using $\omega = 2\pi f$.

2. Keep in mind a few basic facts about phase relationships: For a resistor, voltage and current are always *in phase*, and the two corresponding phasors in a phasor diagram always have the same direction. For an inductor, the voltage always *leads* the current by 90° (that is, $\phi = +90°$) and the voltage phasor is always turned 90° counterclockwise from the current phasor. For a capacitor, the voltage always *lags* the current by 90° (that is, $\phi = -90°$) and the voltage phasor is always turned 90° clockwise from the current phasor.

SOLVE

3. Remember that Kirchhoff's rules are applicable to ac circuits. All the voltages and currents are sinusoidal functions of time instead of being constant, but Kirchhoff's rules hold at each instant. Thus, in a series circuit, the instantaneous current is the same in all circuit elements; in a parallel circuit, the instantaneous potential difference is the same across all circuit elements.

REFLECT

4. Reactance and impedance are analogous to resistance: Each represents the ratio of voltage amplitude *V* to current amplitude *I* in a circuit element or combination of elements. But keep in mind that phase relations play an essential role: Resistance and reactance have to be combined by *vector* addition of the corresponding phasors. For example, when you have several circuit elements in series, you can't just *add* all the numerical values of resistance and reactance; doing that would ignore the phase relations.

EXAMPLE 22.4 **An *R–L–C* circuit**

In the series circuit of Figure 22.13a, suppose $R = 300\ \Omega$, $L = 60$ mH, $C = 0.50\ \mu$F, $V = 50$ V, and $\omega = 10{,}000$ rad/s. Find the reactances X_L, X_C, and X, the impedance Z, the current amplitude I, the phase angle ϕ, and the voltage amplitude across each circuit element.

SOLUTION

SET UP If the circuit and phasor diagrams were not already provided in Figure 22.13, we would sketch them as the first step in this problem.

SOLVE From Equations 22.11 and 22.16, the reactances are

$$X_L = \omega L = (10{,}000\ \text{rad/s})(60 \times 10^{-3}\ \text{H}) = 600\ \Omega,$$
$$X_C = \frac{1}{\omega C} = \frac{1}{(10{,}000\ \text{rad/s})(0.50 \times 10^{-6}\ \text{F})} = 200\ \Omega.$$

The reactance X of the circuit is

$$X = X_L - X_C = 600\ \Omega - 200\ \Omega = 400\ \Omega,$$

and the impedance Z is

$$Z = \sqrt{R^2 + X^2} = \sqrt{(300\ \Omega)^2 + (400\ \Omega)^2} = 500\ \Omega.$$

With source voltage amplitude $V = 50$ V, the current amplitude I is

$$I = \frac{V}{Z} = \frac{50\ \text{V}}{500\ \Omega} = 0.10\ \text{A}.$$

The phase angle ϕ is

$$\phi = \arctan\frac{X_L - X_C}{R} = \arctan\frac{400\ \Omega}{300\ \Omega} = 53°.$$

Because the phase angle ϕ is positive, the voltage *leads* the current by 53°. From Equation 22.6, the voltage amplitude V_R across the resistor is

$$V_R = IR = (0.10\ \text{A})(300\ \Omega) = 30\ \text{V}.$$

From Equation 22.12, the voltage amplitude V_L across the inductor is

$$V_L = IX_L = (0.10\ \text{A})(600\ \Omega) = 60\ \text{V}.$$

From Equation 22.17, the voltage amplitude V_C across the capacitor is

$$V_C = IX_C = (0.10\ \text{A})(200\ \Omega) = 20\ \text{V}.$$

REFLECT Note that because of the phase differences between voltages across the separate elements, the source voltage amplitude $V = 50$ V is *not* equal to the sum of the voltage amplitudes across the separate circuit elements. That is, 50 V $\neq$ 30 V + 60 V + 20 V. These voltages must be combined by *vector* addition of the corresponding phasors, *not* by simple numerical addition.

In this problem, the phase angle ϕ is positive, so the voltage *leads* the current by an angle (between 0 and 90°) equal to ϕ. If ϕ had turned out to be negative, then the voltage would *lag* the current by that angle.

Practice Problem: In a series circuit, suppose $R = 100\ \Omega$, $L = 200$ mH, $C = 0.60\ \mu$F, $V = 60$ V, and $\omega = 4000$ rad/s. Find the reactances X_L and X_C, the impedance Z, the current amplitude I, the phase angle ϕ, and the amplitude across each circuit element. *Answers:* $X_L = 800\ \Omega$; $X_C = 420\ \Omega$; $Z = 400\ \Omega$; $I = 0.15$ A, $\phi = 75°$; $V_R = 15$ V, $V_L = 120$ V, $V_C = 63$ V.

In this entire discussion, we've described magnitudes of voltages and currents in terms of their *maximum* values: the voltage and current *amplitudes*. But we remarked at the end of Section 22.1 that these quantities are usually described not in terms of their amplitudes but in terms of rms values. For any sinusoidally varying quantity, the rms value is always $1/\sqrt{2}$ times the amplitude. All the relations between voltage and current that we've derived in this and the preceding sections are still valid if we use rms quantities throughout instead of amplitudes. For example, if we divide Equation 22.21 by $\sqrt{2}$, we get

$$\frac{V}{\sqrt{2}} = \frac{I}{\sqrt{2}} Z,$$

which we can rewrite as

$$V_{\text{rms}} = I_{\text{rms}} Z. \qquad (22.24)$$

We can translate Equations 22.6, 22.12, 22.17, and 22.18 in exactly the same way.

Finally, we remark that what we have been describing throughout this section is the *steady-state* condition of a circuit: the state that exists after the circuit has been connected to the source for a long time. When the source is first connected, there may be additional voltages and currents, called *transients,* whose nature

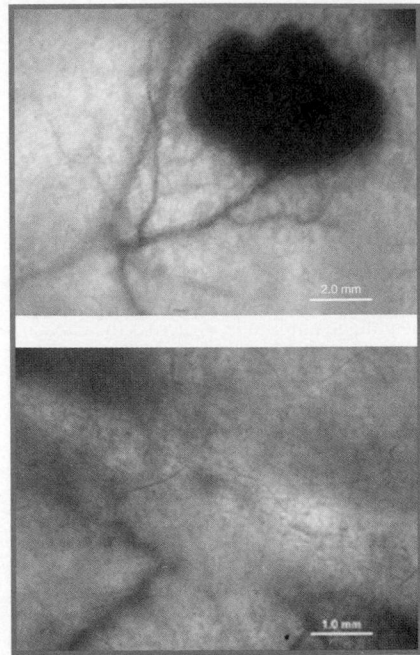

▲ **BIO Application Zapping skin cancer.**
Melanoma is an increasing human health problem and researchers are eagerly seeking effective new treatments. Recent work has shown that tumors in mice are brought into complete remission by the application of ultrashort (300 ns) high-electric-field (4 MV/m) pulses. The photographs above show a tumor before beginning electric pulse treatments (top) and after 28 days of treatment (bottom). The researchers designed the pulses based on an analysis of how high frequencies interact with cells. The membranes of cells can be modeled as a resistor in parallel with a capacitor each in series with the resistance of the fluid surrounding the cell. At high frequencies, the capacitive reactance, X_C, becomes negligibly small and the current flows into the cell, punching nanopores in the membrane that allow ions to redistribute, thus damaging the cell.

depends on the time in the cycle when the circuit is initially completed. A detailed analysis of transients is beyond our scope. They always die out after a sufficiently long time, and they do not affect the steady-state behavior of the circuit. But they can cause dangerous and damaging surges in power lines, and delicate electronic systems such as computers should always be provided with power-line surge protectors.

22.4 Power in Alternating-Current Circuits

Alternating currents play a central role in systems for distributing, converting, and using electrical energy, so it's important to look at power relationships in ac circuits. When a source with voltage amplitude V and instantaneous potential difference v supplies an instantaneous current i (with current amplitude I) to an ac circuit, the instantaneous power p that it supplies is $p = vi$. Let's first see what this means for individual circuit elements.

We'll assume in each case that $i = I\cos\omega t$. Suppose first that the circuit consists of a *pure resistance* R, as in Figure 22.6; then i and v are *in phase*. We obtain the graph representing p by multiplying the heights of the graphs of v (red curve) and i (purple curve) in Figure 22.6b at each instant. This graph is shown as the black curve in Figure 22.15a. The product vi is always positive, because v and i are always either both positive or both negative. Energy is supplied to the resistor at every instant for both directions of i, although the power is not constant.

The power curve is symmetrical about a value equal to one-half of its maximum value VI, so the *average power* P is

$$P = \tfrac{1}{2}VI. \tag{22.25}$$

An equivalent expression is

$$P = \frac{V}{\sqrt{2}}\frac{I}{\sqrt{2}} = V_{rms}I_{rms}. \tag{22.26}$$

Also, $V_{rms} = I_{rms}R$, so we can express P in any of these equivalent forms:

$$P = I_{rms}^2R = \frac{V_{rms}^2}{R} = V_{rms}I_{rms}. \tag{22.27}$$

Note that the preceding expressions have the same form as the corresponding relations for a dc circuit, Equation 19.10. Note also that they are valid only for pure resistors, not for more complicated combinations of circuit elements.

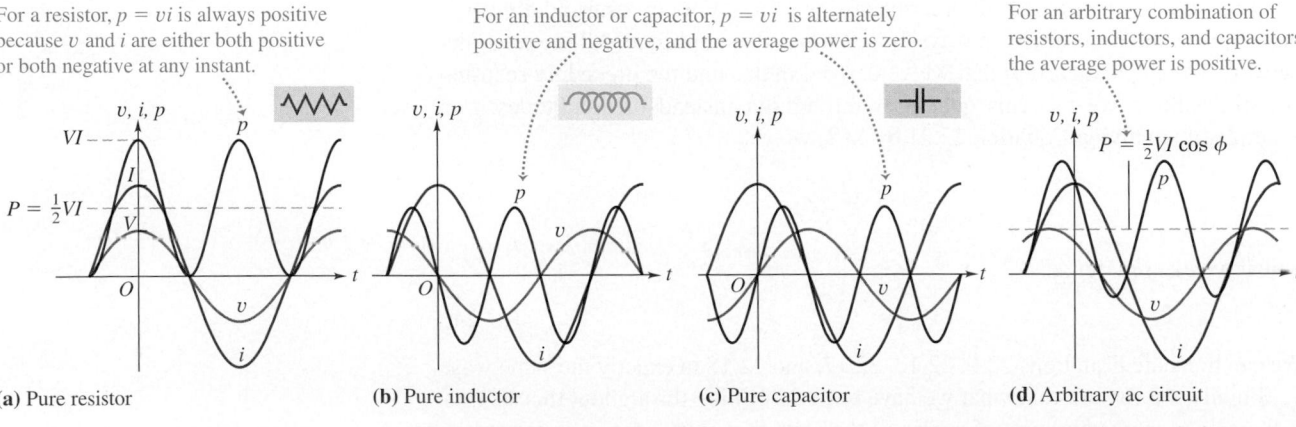

(a) Pure resistor (b) Pure inductor (c) Pure capacitor (d) Arbitrary ac circuit

KEY: Instantaneous current, i —— Instantaneous voltage across device, v —— Instantaneous power input to device, p ——

▲ **FIGURE 22.15** Graphs of voltage, current, and power as functions of time for various circuits.

Next, we connect the source to a pure inductor L, as in Figure 22.7. The voltage leads the current by $90°$. When we multiply the curves of v and i, the product vi is *negative* during the half of the cycle when v and i have *opposite* signs. We get the power curve in Figure 22.15b, which is symmetrical about the horizontal axis. It is positive half the time and negative the other half, and the *average* power is zero. When p is positive, energy is being supplied to set up the magnetic field in the inductor; when p is negative, the field is collapsing and the inductor is returning energy to the source. The net energy transfer over one cycle is zero.

Finally, we connect the source to a pure capacitor C, as in Figure 22.8. The voltage lags the current by $90°$. Figure 22.15c shows the power curve; the average power is again zero. Energy is supplied to charge the capacitor and is returned to the source when the capacitor discharges. The net energy transfer over one cycle is again zero.

In *any* ac circuit, with any combination of resistors, capacitors, and inductors, the voltage v has some phase angle ϕ with respect to the current i, and the instantaneous power p is given by

$$p = vi = [V\cos(\omega t + \phi)][I\cos\omega t]. \tag{22.28}$$

The instantaneous power curve has the form shown in Figure 22.15d. The area under the positive loops is greater than that under the negative loops, and the average power is positive.

To derive an expression for the *average* power, which we'll denote by capital P, we use the identity for the cosine of the sum of two angles:

$$\cos(a + b) = \cos a \cos b - \sin a \sin b$$

Applying this identity to Equation 22.28, we find:

$$p = [V(\cos\omega t \cos\phi - \sin\omega t \sin\phi)][I\cos\omega t]$$
$$= VI\cos\phi\cos^2\omega t - VI\sin\phi\cos\omega t\sin\omega t.$$

From the discussion leading to Equation 22.3 in Section 22.1, we see that in the first term the average value of $\cos^2\omega t$ (over one cycle) is $\frac{1}{2}$. In the second term, the average value of $\cos\omega t\sin\omega t$ is zero, because this product is equal to $\frac{1}{2}\sin 2\omega t$, whose average over a cycle is zero. So the average power P is

$$P = \frac{1}{2}VI\cos\phi = V_{rms}I_{rms}\cos\phi. \tag{22.29}$$

When v and i are *in phase*, $\phi = 0$, $\cos\phi = 1$, and the average power equals $V_{rms}I_{rms}$ (which also equals $\frac{1}{2}VI$). When v and i are $90°$ *out of phase*, the average power is zero. In the general case, when v has phase angle ϕ with respect to i, the average power equals $\frac{1}{2}I$ multiplied by $V\cos\phi$, the component of V that is *in phase* with I. The relationship of the current and voltage phasors for this case is shown in Figure 22.16. For the series R–L–C circuit, $V\cos\phi$ is the voltage amplitude for the resistor, and Equation 22.29 is the power dissipated in the resistor. The power dissipation in the inductor and capacitor is zero.

The factor $\cos\phi$ is called the **power factor** of the circuit. For a pure resistance, $\phi = 0$, $\cos\phi = 1$, and $P = V_{rms}I_{rms}$. For a pure (resistanceless) capacitor or inductor, $\phi = \pm90°$, $\cos\phi = 0$, and $P = 0$. For a series R–L–C circuit, the power factor is equal to R/Z. Can you prove this?

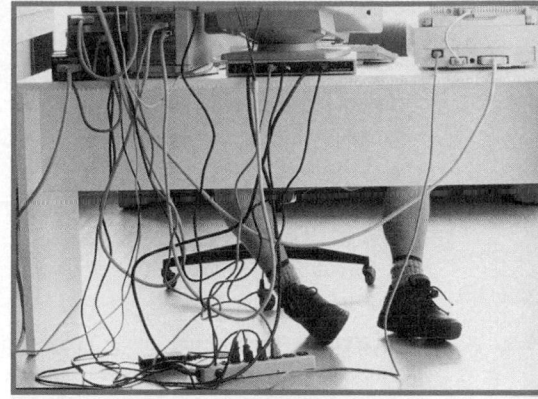

▲ Application **Too much power.** Many types of modern electronic equipment, such as computers and DVD players, can be damaged irreversibly by abrupt changes in electrical current known as power surges. Power surges can be caused by nearby lightning strikes, by malfunctioning transformers, or by switching on a large piece of electrical equipment such as a compressor or elevator that can momentarily disrupt the line voltage. A common type of surge protector has a component called a metal oxide varistor (a variable resistor containing semiconductors connecting the hot wire to the ground wire). At normal line voltage, the varistor resistance is very high and the current travels past it to the outlet. However, if the voltage exceeds a certain limit, the varistor resistance drops dramatically and the current flows through it directly to the ground wire, bypassing the outlet and protecting your computer from being "fried."

Average power $= \frac{1}{2}I(V\cos\phi)$, where $V\cos\phi$ is the component of V in phase with I.

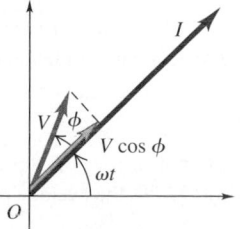

▲ **FIGURE 22.16** Average power in an ac circuit

Comparing power

Figure 22.17 shows phasors for two *R–L–C* circuits. The circuits have the same current amplitude *I*; circuit 1 has voltage amplitude V_1 and circuit 2 has voltage amplitude V_2. Which of the following statements is correct?

A. Circuit 1 has greater average power because it has a larger power factor: $\cos\phi_1 > \cos\phi_2$.
B. Circuit 2 has greater average power because its resistor dissipates more power.

SOLUTION The average power in a circuit is proportional to the component of the voltage that is in phase with the current: $P = \frac{1}{2}(V\cos\phi)I$. For a *single circuit* with a fixed *V* and *I*, the average power increases or decreases according to the power factor $\cos\phi$. Circuit 1 has a greater power factor $(\cos\phi_1 > \cos\phi_2)$. However, we're comparing the component of the voltage that is in phase

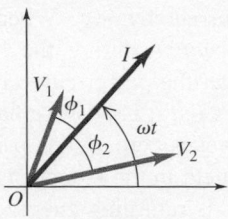

▲ FIGURE 22.17

with the current between *two circuits*. By inspection of Figure 22.17, $V_2\cos\phi_2 > V_1\cos\phi_1$, so choice B is correct. In addition, as stated in choice B, it's always true that the average power in an *R–L–C* circuit depends only on the energy dissipated through its resistance.

EXAMPLE 22.5 An electric hair dryer

An electric hair dryer is rated at 1500 W at 120 V. Calculate **(a)** the resistance, **(b)** the rms current, and **(c)** the maximum instantaneous power of the dryer. Assume pure resistance.

SOLUTION

SET UP AND SOLVE Part (a): We solve Equation 22.27 for *R* and substitute the given values:

$$R = \frac{V_{rms}^{2}}{P} = \frac{(120\ \text{V})^2}{1500\ \text{W}} = 9.6\ \Omega.$$

Part (b): From Equation 22.26,

$$I_{rms} = \frac{P}{V_{rms}} = \frac{1500\ \text{W}}{120\ \text{V}} = 12.5\ \text{A}.$$

Or, from Equation 22.6,

$$I_{rms} = \frac{V_{rms}}{R} = \frac{120\ \text{V}}{9.6\ \Omega} = 12.5\ \text{A}.$$

Part (c): The maximum instantaneous power is *VI*; from Equation 22.29,

$$VI = 2P = 2(1500\ \text{W}) = 3000\ \text{W}.$$

REFLECT To mislead the unwary consumer, some manufacturers of stereo amplifiers state power outputs in terms of the peak value rather than the lower average power. Caveat emptor!

Practice Problem: A toaster is rated at 900 W at 120 V. Calculate the resistance and the rms current. *Answers: R = 16 Ω; I_{rms} = 7.5 A.*

EXAMPLE 22.6 Power factor for an *R–L–C* circuit

For the series *R–L–C* circuit of Example 22.4, calculate the power factor and the average power to the entire circuit and to each circuit element.

SOLUTION

SET UP AND SOLVE The power factor is $\cos\phi = \cos 53° = 0.60$.
From Equation 22.29, the average power to the circuit is

$$P = \tfrac{1}{2}VI\cos\phi = \tfrac{1}{2}(50\ \text{V})(0.10\ \text{A})(0.60) = 1.5\ \text{W}.$$

REFLECT All of this power is dissipated in the resistor; the average power to a pure inductor or pure capacitor is always zero.

Practice Problem: If the inductance *L* in this circuit could be changed, what value of *L* would give a power factor of unity? *Answer:* 20 mH.

A low power factor (large angle of lag or lead) is usually undesirable in power circuits because, for a given potential difference, a large current is needed to supply a given amount of power. This results in large I^2R losses in the transmission lines. Your electric power company may charge a higher rate to a client with a low power factor. Many types of ac machinery draw a lagging current; the power factor can be corrected by connecting a capacitor in parallel with the load. The leading current drawn by the capacitor compensates for the lagging current in the other branch of the circuit. The capacitor itself absorbs no net power from the line.

22.5 Series Resonance

The impedance of a series R–L–C circuit depends on the frequency, as the following equation shows:

$$Z = \sqrt{R^2 + X^2}$$
$$= \sqrt{R^2 + (X_L - X_C)^2} \qquad (22.22)$$
$$= \sqrt{R^2 + [\omega L - (1/\omega C)]^2}.$$

Figure 22.18a shows graphs of R, X_L, X_C, and Z as functions of ω. We have used a logarithmic angular frequency scale so that we can cover a wide range of frequencies. Because X_L increases and X_C decreases with increasing frequency, there is always one particular frequency at which X_L and X_C are equal and $X = X_L - X_C$ is zero. At this frequency, the impedance Z has its *smallest* value, equal to just the resistance R.

Suppose we connect an ac voltage source with constant voltage amplitude V, but variable angular frequency ω, across a series R–L–C circuit. As we vary ω, the current amplitude I varies with frequency as shown in Figure 22.18b; its *maximum* value occurs at the frequency at which the impedance Z attains its *minimum* value. This peaking of the current amplitude at a certain frequency is called **resonance**. The angular frequency ω_0 at which the resonance peak occurs is called the **resonance angular frequency**. This is the angular frequency at which the inductive and capacitive reactances are equal, so

$$X_L = X_C, \qquad \omega_0 L = \frac{1}{\omega_0 C}, \qquad \omega_0 = \frac{1}{\sqrt{LC}}. \qquad (22.30)$$

The **resonance frequency** f_0 is $\omega_0/2\pi$.

Now let's look at what happens to the *voltages* in a series R–L–C circuit at resonance. The current at any instant is the same in L and C. The voltage across an inductor always *leads* the current by 90°, or a quarter cycle, and the voltage across a capacitor always *lags* the current by 90°. Therefore, the instantaneous voltages across L and C always differ in phase by 180°, or a half cycle; they have opposite signs at each instant. If the *amplitudes* of these two voltages are equal, then they add to zero at each instant and the *total* voltage v_{bd} across the L–C combination in Figure 22.13a is exactly zero! This occurs only at the resonance frequency f_0.

NOTE ▶ Depending on the numerical values of R, L, and C, the voltages across L and C individually can be larger than that across R. Indeed, at frequencies close to resonance, the voltages across L and C individually can be *much larger* than the source voltage! ◀

The *phase* of the voltage relative to the current is given by Equation 22.23. At frequencies below resonance, X_C is greater than X_L; the capacitive reactance dominates, the voltage *lags* the current, and the phase angle ϕ is between zero and −90°. Above resonance, the inductive reactance dominates; the voltage *leads* the current, and the phase angle is between zero and +90°. At resonance, $\phi = 0$ and the power factor is $\cos\phi = 1$. This variation of ϕ with angular frequency is shown in Figure 22.18b.

When we vary the inductance L or the capacitance C of a circuit, we can also vary the resonance frequency. That's how some older radio or television receiving

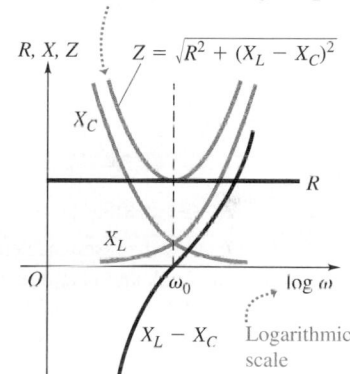

Impedance Z is least at the angular frequency at which $X_C = X_L$.

(a) Reactance, resistance, and impedance as functions of angular frequency

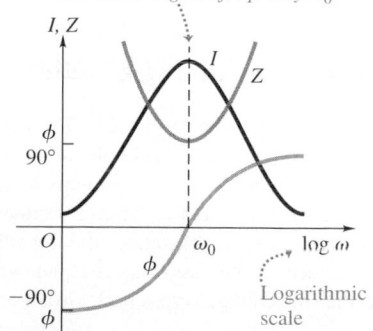

Current peaks at the angular frequency at which impedance is least. This is the *resonance angular frequency* ω_0.

(b) Impedance, current, and phase angle as functions of angular frequency

▲ **FIGURE 22.18** Graphs showing the impedance minimum in a series R–L–C circuit.

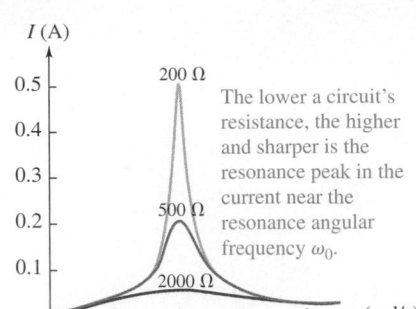

▲ **FIGURE 22.19** Resonance peaks for R–L–C circuits with different resistances.

The lower a circuit's resistance, the higher and sharper is the resonance peak in the current near the resonance angular frequency ω_0.

sets were "tuned" to receive particular stations. In the early days of radio, this was accomplished by the use of capacitors with movable metal plates whose overlap could be varied to change C. Alternatively, L could be varied with the use of a coil with a ferrite core that slid in or out.

In a series *R–L–C* circuit, the impedance reaches its minimum value and the current reaches its maximum value at the resonance frequency. Figure 22.19 shows a graph of rms current as a function of frequency for such a circuit, with $V = 100$ V, $R = 500$ Ω, $L = 2.0$ H, and $C = 0.50$ μF. This curve is called a *response curve*, or a *resonance curve*. The resonance angular frequency is $\omega_0 = 1/\sqrt{LC} = 1000$ rad/s. As we expect, the curve has a peak at this angular frequency.

The resonance frequency is determined by L and C; what happens when we change R? Figure 22.19 also shows graphs of I as a function of ω for $R = 200$ Ω and for $R = 2000$ Ω. The curves are all similar for frequencies far away from resonance, where the impedance is dominated by X_L or X_C. But near resonance, where X_L and X_C nearly cancel each other, the curve is higher and more sharply peaked for small values of R than for larger values. The maximum height of the curve is in fact inversely proportional to R: A small R gives a sharply peaked response curve, and a large value of R gives a broad, flat curve.

In the early days of radio and television, the shape of the response curve was of crucial importance. The sharply peaked curve made it possible to discriminate between two stations broadcasting on adjacent frequency bands. But if the peak was *too* sharp, some of the information in the received signal, such as the high-frequency sounds in music, was lost. A sharply peaked resonance curve corresponds to a small value of R and a lightly damped oscillating system; a broad flat curve goes with a large value of R and a heavily damped system.

Quantitative Analysis 22.5

The resonance peak

We draw the resonance curve for a circuit with electrical components L, R, and C and resonant frequency ω_0. If the values of L, R, and C are all doubled, how does the new resonance curve differ from the original one? Sketch the resonance curve for each case.

A. Peak twice as high, peak frequency twice as great.
B. Peak half as high, peak frequency twice as great.
C. Peak twice as high, peak frequency half as great.
D. Peak half as high, peak frequency half as great.

SOLUTION The greater the resistance, the flatter is the resonance peak. The maximum height is inversely proportional to R: When we double R, the peak height decreases to half its original value. This narrows the possibilities to choices B and D. Since $\omega_0 = 1/\sqrt{LC}$, doubling L and C halves the resonant frequency, so curve D is correct.

EXAMPLE 22.7 Tuning a radio

The series circuit in Figure 22.20 is similar to arrangements that are sometimes used in tuning circuits in simple radio receivers (often available in kit form). This circuit is connected to the terminals of an ac source with a constant rms terminal voltage of 1.0 V and a variable frequency. Find **(a)** the resonance frequency; **(b)** the inductive reactance, the capacitive reactance, the reactance, and the impedance at the resonance frequency; **(c)** the rms current at resonance; and **(d)** the rms voltage across each circuit element at resonance.

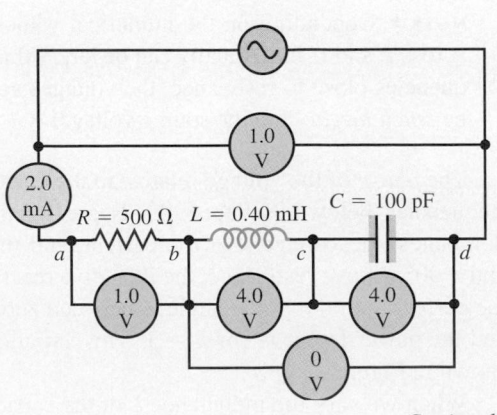

▶ **FIGURE 22.20**

Continued

SOLUTION

SET UP AND SOLVE Part (a): The resonance angular frequency is

$$\omega_0 = \frac{1}{\sqrt{LC}} = \frac{1}{\sqrt{(0.40 \times 10^{-3}\,\text{H})(100 \times 10^{-12}\,\text{F})}}$$
$$= 5.0 \times 10^6\,\text{rad/s}.$$

The corresponding frequency is $f = \omega/2\pi = 8.0 \times 10^5\,\text{Hz} = 800\,\text{kHz}$.

Part (b): At this frequency,

$$X_L = \omega L = (5.0 \times 10^6\,\text{rad/s})(0.40 \times 10^{-3}\,\text{H}) = 2000\,\Omega,$$

$$X_C = \frac{1}{\omega C} = \frac{1}{(5.0 \times 10^6\,\text{rad/s})(100 \times 10^{-12}\,\text{F})} = 2000\,\Omega,$$

$$X = X_L - X_C = 2000\,\Omega - 2000\,\Omega = 0.$$

From Equation 22.22, the impedance Z at resonance is equal to the resistance: $Z = R = 500\,\Omega$.

Part (c): At resonance, the rms current is

$$I = \frac{V}{Z} = \frac{V}{R} = \frac{1.0\,\text{V}}{500\,\Omega} = 0.0020\,\text{A} = 2.0\,\text{mA}.$$

Part (d): The rms potential difference across the resistor is

$$V_R = IR = (0.0020\,\text{A})(500\,\Omega) = 1.0\,\text{V}.$$

The rms potential differences across the inductor and capacitor are, respectively,

$$V_L = IX_L = (0.0020\,\text{A})(2000\,\Omega) = 4.0\,\text{V},$$
$$V_C = IX_C = (0.0020\,\text{A})(2000\,\Omega) = 4.0\,\text{V}.$$

The rms potential difference V_{bd} across the inductor–capacitor combination is

$$V_{bd} = IX = I(X_L - X_C) = 0.$$

REFLECT The frequency found in part (a) corresponds to the lower part of the AM radio band. At resonance, the instantaneous potential differences across the inductor and the capacitor have equal amplitudes; but they are 180° out of phase and so add to zero at each instant. Note also that, at resonance, V_R is equal to the source voltage V, but in this example, V_L and V_C are both considerably *larger* than V.

Practice Problem: In a radio tuning circuit, a 300 Ω resistor, a 0.50 mH inductor, and a 150 pF capacitor are connected in series with an rms terminal voltage of 1.5 V. Find the rms voltage across each circuit element at resonance. *Answers:* $V_R = 1.5\,\text{V}$; $V_L = 9.1\,\text{V}$; $V_C = 9.1\,\text{V}$.

Resonance phenomena occur in all areas of physics; we have already seen one example in the forced oscillation of the harmonic oscillator (Section 11.6). In that case, the amplitude of a mechanical oscillation peaked at a driving-force frequency close to the natural frequency of the system. The behavior of the R–L–C circuit is analogous to this behavior. We suggest that you review that discussion now, looking for the analogies. Other important examples of resonance occur in acoustics, in atomic and nuclear physics, and in the study of fundamental particles (high-energy physics).

22.6 Parallel Resonance

A different kind of resonance occurs when a resistor, an inductor, and a capacitor are connected in *parallel,* as shown in Figure 22.21a. This circuit has resonance behavior similar to that of the series R–L–C circuit we analyzed in Section 22.5, but the roles of voltage and current are reversed. This time, the instantaneous potential difference is the same for all three circuit elements and is equal to the source voltage $v = V\cos\omega t$, but the current is different in each of the three elements.

The instantaneous current i_R through the resistor is in phase with the source voltage. Its peak value is $I_R = V/R$. As we discussed in Section 22.2, the instantaneous current i_L through the inductor lags the source voltage by 90° and has peak value $I_L = V/X_L = V/\omega L$. The instantaneous current i_C through the capacitor leads the source voltage by 90° and has peak value $I_C = V/X_C = V\omega C$. The phasor diagram is shown in Figure 22.21b.

Thus, at any frequency, the inductor current and the capacitor current differ in phase by exactly a half cycle $(180°)$ and tend to cancel each other. At one particular frequency ω_0, the two reactances X_L and X_C are equal; at that frequency, the

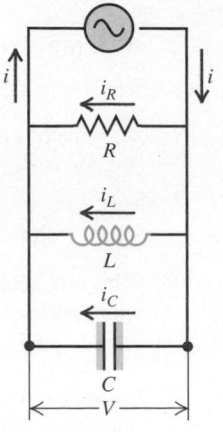

(a) A parallel *R-L-C* circuit

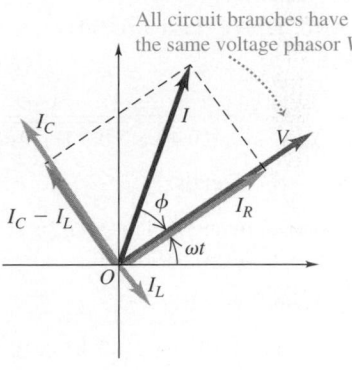

(b) Phasor diagram showing current phasors for the three branches

▲ **FIGURE 22.21** Voltage and current in a parallel *R–L–C* circuit.

inductor current and the capacitor current are equal in magnitude and opposite in direction at every instant and therefore add to zero. This occurs when

$$X_L = X_C, \qquad \omega_0 L = \frac{1}{\omega_0 C}, \qquad \text{and} \qquad \omega_0 = \frac{1}{\sqrt{LC}}.$$

We recognize this relation as the same as Equation 22.30, the condition for resonance in a *series R–L–C* circuit. The difference is that, at resonance, the total current through the parallel combination reaches a *minimum* because the total current through L and C is zero. Thus at resonance the impedance of the parallel circuit reaches a *maximum* value equal simply to $Z = R$.

A detailed analysis of the currents in the three branches shows that, at any frequency, the impedance Z of the parallel combination is given by

$$\frac{1}{Z} = \sqrt{\frac{1}{R^2} + \left(\omega C - \frac{1}{\omega L}\right)^2}. \tag{22.31}$$

This result confirms the earlier statement that at resonance $1/Z$ attains its minimum value and therefore Z attains its maximum value, $Z = R$.

NOTE ▶ The total current in a *parallel R–L–C* circuit is *minimum* at resonance. When $\omega C = 1/\omega L$, Equation 22.31 becomes simply $I = V/R$. This does *not* mean that there is *no* current in L or C at resonance, but only that the two currents cancel completely at every instant. If R is large, the impedance Z of the circuit near resonance is much *larger* than the individual reactances X_L and X_C, and the individual currents in L and C can be much larger than the total current. ◀

EXAMPLE 22.8   More current is less current

In the parallel circuit of Figure 22.21, suppose the circuit elements, the applied voltage, and the angular frequency have the same values as in Example 22.4, in which $R = 300 \ \Omega$, $X_L = 600 \ \Omega$, and $X_C = 200 \ \Omega$. Determine **(a)** the impedance of the parallel combination, **(b)** the current amplitude for each element, and **(c)** the total current amplitude.

Continued

SOLUTION

SET UP AND SOLVE Part (a): The impedance Z is given by Equation 22.31. Substituting the values from Example 22.4 into this equation, we get

$$\frac{1}{Z} = \sqrt{\frac{1}{(300\ \Omega)^2} + \left(\frac{1}{200\ \Omega} - \frac{1}{600\ \Omega}\right)^2},$$
$$Z = 212\ \Omega.$$

Part (b): The current through the resistor is given by Equation 22.6:

$$I_R = \frac{V}{R} = \frac{50\ \text{V}}{300\ \Omega} = 0.167\ \text{A}.$$

The current through the inductor is found from Equation 22.12:

$$I_L = \frac{V}{X_L} = \frac{50\ \text{V}}{600\ \Omega} = 0.083\ \text{A}.$$

The current through the capacitor is given by Equation 22.17:

$$I_C = \frac{V}{X_C} = \frac{50\ \text{V}}{200\ \Omega} = 0.25\ \text{A}.$$

Part (c): The amplitude of the total current is

$$I = \frac{V}{Z} = \frac{50\ \text{V}}{212\ \Omega} = 0.24\ \text{A},$$

or

$$I = \sqrt{I_R^2 + (I_L - I_C)^2}$$
$$= \sqrt{(0.167\ \text{A})^2 + (0.083\ \text{A} - 0.25\ \text{A})^2}$$
$$= 0.24\ \text{A}.$$

REFLECT The amplitude I of the total current is less than that of the current through the capacitor; the inductor and capacitor currents partially cancel each other because they are a half cycle out of phase. Make sure you understand that, because of the phase differences of the individual currents, you *cannot* simply add the individual current amplitudes to get the amplitude of the total current.

Practice Problem: Find the resonance angular frequency, and find the circuit impedance and the total current at resonance, if $V = 50$ V as before. *Answers:* 5.8×10^3 s^{-1}, 300 Ω, 0.17 A.

SUMMARY

Phasors and Alternating Currents

(**Section 22.1**) An ac source produces an emf that varies sinusoidally with time. A sinusoidal voltage or current can be represented by a **phasor**—a vector that rotates counterclockwise with constant angular velocity ω equal to the angular frequency of the sinusoidal quantity. Its projection on the horizontal axis at any instant represents the instantaneous value of the quantity. For a sinusoidal current i with maximum value I, the phasor is given by $i = I\cos\omega t$ (Equation 22.2). In power calculations, it is useful to use the **root-mean-square** (rms) value: $I_{rms} = I/\sqrt{2}$ (Equation 22.3).

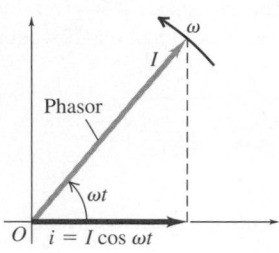

Resistance and Reactance

(**Section 22.2**) If the current is given by $i = I\cos\omega t$ (Equation 22.2) and the voltage v between two points is $v = V\cos(\omega t + \phi)$, then ϕ is called the **phase angle** of the voltage relative to the current.

The voltage across a resistor R is in phase with the current, and the voltage and current amplitudes are related by $V_R = IR$ (Equation 22.6). The voltage across an inductor L leads the current with a phase angle of $\phi = 90°$; the voltage and current amplitudes are related by $V_L = IX_L$ (Equation 22.12), where $X_L = \omega L$ (Equation 22.11) is the **inductive reactance** of the inductor. The voltage across a capacitor C lags the current with a phase angle $\phi = -90°$; the voltage and current amplitudes are related by $V_C = IX_C$ (Equation 22.17), where $X_C = 1/\omega C$ (Equation 22.16) is the **capacitive reactance** of the capacitor.

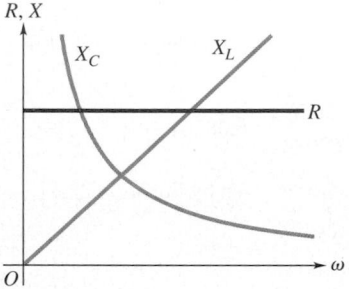

The Series R–L–C Circuit

(**Section 22.3**) In a series R–L–C circuit, the voltage and current amplitudes are related by $V = IZ$ (Equation 22.21), where Z is the **impedance** of the circuit: $Z = \sqrt{R^2 + [\omega L - (1/\omega C)]^2}$ (Equation 22.22). The phase angle ϕ of the voltage relative to the current is given by

$$\phi = \arctan\frac{\omega L - 1/\omega C}{R}. \qquad (22.23)$$

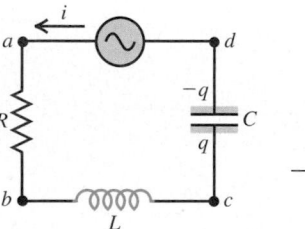
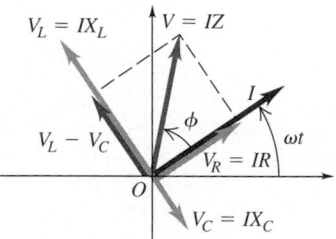

Power in Alternating-Current Circuits

(**Section 22.4**) The average power input P to an ac circuit is I times one-half the component of the voltage that is in phase with the current, or $P = \frac{1}{2}VI\cos\phi = V_{rms}I_{rms}\cos\phi$ (Equation 22.29), where ϕ is the phase angle of voltage with respect to current. Power is dissipated only through the resistor. For circuits containing only capacitors and inductors, $\phi = \pm 90°$ and the average power is zero. The quantity $\cos\phi$ is called the **power factor**.

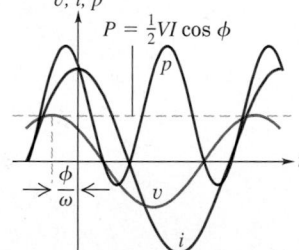

Graphs of p, v, and i versus time for an arbitrary combination of resistors, inductors, and capacitors. The average power is positive.

Series and Parallel Resonance

(**Sections 22.5 and 22.6**) The current in a series R–L–C circuit reaches a maximum, and the impedance reaches a minimum, at an angular frequency $\omega_0 = 1/(LC)^{1/2}$ known as the **resonance angular frequency**. This phenomenon is called **resonance**. At resonance, the voltage and current are in phase and the impedance Z is equal to the resistance R. The smaller the resistance, the sharper is the **resonance peak**. In an R–L–C parallel circuit, the total current attains a minimum, and the impedance attains a maximum, at the resonance angular frequency ω_0.

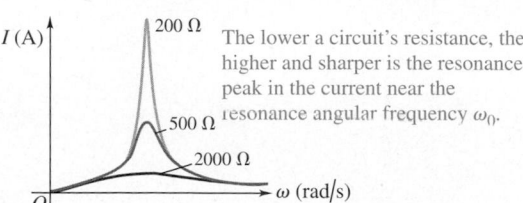

The lower a circuit's resistance, the higher and sharper is the resonance peak in the current near the resonance angular frequency ω_0.

Conceptual Questions

1. For a series ac R–L–C circuit, (a) why do the voltage amplitudes *not* obey the equation $V = V_R + V_L + V_C$; and (b) would the instantaneous voltages obey the equation $v = v_R + v_L + v_C$? Why or why not?
2. In Example 22.5, a hair dryer was treated as a pure resistor. But because there are coils in the heating element and in the motor that drives the blower fan, a hair dryer also has inductance. Qualitatively, does including an inductance increase or decrease the values of R, I_{rms}, and P?
3. Fluorescent lights often use an inductor, called a "ballast," to limit the current through the tubes. Why is it better to use an inductor than a resistor for this purpose?
4. At high frequencies, a capacitor becomes a short circuit. Discuss why this is so.
5. At high frequencies, an inductor becomes an open circuit. Discuss why this is so.
6. The current in an ac power line changes direction 120 times per second, and its average value is zero. So how is it possible for power to be transmitted in such a system?
7. Electric-power companies like to have their power factors as close to unity as possible. Why?
8. Some electrical appliances operate equally well on ac or dc, while others work only on ac or only on dc. Give examples of each, and explain the reasons for the differences.
9. When a series-resonant circuit is connected across a 120 V ac line, the voltage rating of the capacitor may be exceeded even if it is rated at 200 or 400 V. How can this be?
10. Is it possible for the power factor of an R–L–C ac series circuit to be zero? Justify your answer on *physical* grounds.
11. During the last quarter of the 19th century, there was great and acrimonious controversy over whether ac or dc should be used for power transmission. Edison favored dc, while George Westinghouse championed ac. What arguments might each have used to promote his scheme?
12. In an ac circuit, why is the average power delivered to a capacitor or an inductor equal to zero, and why is this not the case for a resistor?
13. dc voltage comes from batteries, but how is ac voltage generated?
14. In what ways is impedance similar to ordinary resistance, and in what ways is it different?
15. Transformers, such as those which plug into the wall socket for use with small electrical appliances, often feel warm to the touch. What is the source of this heat?

Multiple-Choice Problems

1. A piece of electrical equipment in an ac circuit draws a root-mean-square current of 5.00 A. The average current over each cycle is
   A. $5\sqrt{2}\,A = 7.07\,A$. B. $5.00\,A$.
   C. $5/\sqrt{2}\,A = 3.54\,A$. D. 0.
2. A sinusoidal current is described by $i = I\cos\omega t$, where $\omega = 1.57\,rad/s$. At some time t', where $2\,s < t' < 4\,s$, the current is $+3.0\,A$. Which phasor can represent the current at time t'?

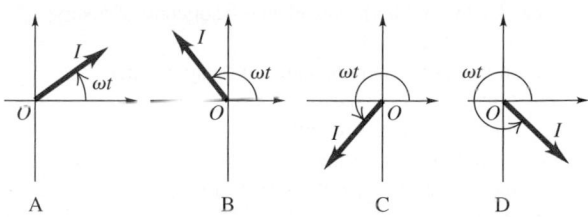

A   B   C   D

3. A lightbulb is the resistance in a series R–L–C circuit having an ac voltage source $v = V\cos\omega t$. As the frequency of the source is adjusted closer and closer to the value $1/\sqrt{LC}$, what happens to the brightness of the bulb?
   A. It increases. B. It decreases.
   C. It does not change.
4. A series R–L–C ac circuit with a sinusoidal voltage source of angular frequency ω has a total reactance X. If this frequency is doubled, the reactance becomes
   A. $4X$. B. $2X$. C. $X/2$. D. $X/4$.
   E. none of the above.
5. In a series R–L–C circuit powered by an ac sinusoidal voltage source, which phasor diagram best illustrates the relationship between the current i and the potential drop v_R across the *resistor*?

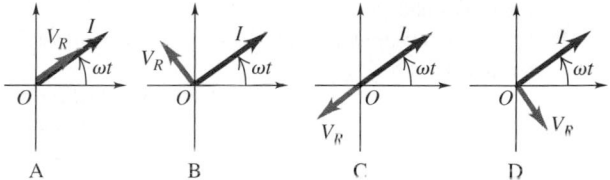

A   B   C   D

6. In a series R–L–C circuit powered by an ac sinusoidal voltage source, which phase diagram best illustrates the relationship between the current i and the potential drop v_C across the *capacitor*?

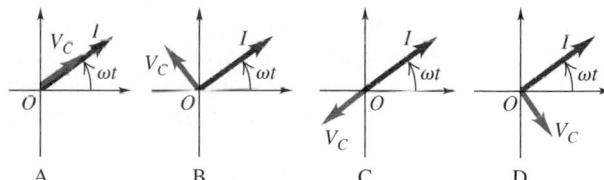

A   B   C   D

7. In a series R–L–C circuit powered by an ac sinusoidal voltage source, which phase diagram best illustrates the relationship between the current i and the potential drop v_L across the *inductor*?

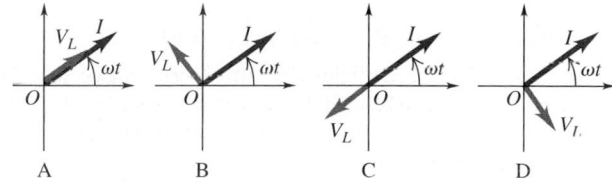

A   B   C   D

8. A series circuit contains an inductor, a resistor, a capacitor, and a sinusoidal voltage source of angular frequency ω. If we double this frequency (there may be more than one correct choice),
   A. the inductive reactance is doubled.
   B. the capacitive reactance is doubled.
   C. the total reactance is doubled.
   D. the impedance is doubled.

9. In order to double the resonance frequency of a series $R–L–C$ ac circuit, you could
   A. double both the inductance and capacitance.
   B. double the resistance.
   C. cut the resistance in half.
   D. cut both the inductance and capacitance in half.
10. For the current to have its maximum value in a series $R–L–C$ ac circuit,
   A. $\omega = 0$. B. $\omega \to \infty$. C. $\omega = 1/\sqrt{LC}$.
11. If the root-mean-square current in an ac circuit is 10.0 A, the current amplitude is approximately
   A. 7.07 A. B. 5.00 A. C. 14.1 A. D. 20.0 A.
12. In a series $R–L–C$ circuit, the voltage source produces an angular frequency ω. If this frequency is *increased* slightly (there may be more than one correct choice),
   A. The total impedance will *necessarily* increase.
   B. The reactance will *necessarily* increase.
   C. The impedance due to the resistor does not change.
   D. The total impedance increases if $\omega \geq 1/\sqrt{LC}$.
13. In an $R–L–C$ ac series circuit, if the resistance is much smaller than the inductive or capacitive reactance and the system is operated at a frequency much higher than the resonance frequency,
   A. the phase angle ϕ is close to zero and the power factor $\cos\phi$ is small.
   B. the phase angle ϕ is close to zero and the power factor $\cos\phi$ is close to 1.
   C. the phase angle is close to 90° and the power factor is small.
   D. the phase angle is close to 90° and the power factor is close to 1.
14. In a series $R–L–C$ ac circuit at resonance,
   A. The impedance is zero.
   B. The impedance has its maximum value.
   C. The reactance is equal to R.
   D. The total impedance has its minimum value, which is equal to R.
15. A circuit consists of a light bulb, a capacitor, and an inductor connected in series to an ac power source. The capacitor and the inductor have equal reactances. If both the capacitor and the inductor are removed from the circuit, what will happen to the brightness of the light bulb?
   A. It will increase. B. It will decrease.
   C. It will remain the same.

Problems

22.1 Phasors and Alternating Currents

1. • You have a special lightbulb with a *very* delicate wire filament. The wire will break if the current in it ever exceeds 1.50 A, even for an instant. What is the largest root-mean-square current you can run through this bulb?
2. • The plate on the back of a certain computer scanner says that the unit draws 0.34 A of current from a 120 V, 60 Hz line. Find (a) the root-mean-square current, (b) the current amplitude, (c) the average current, and (d) the average square of the current.

22.2 Resistance and Reactance

3. • A capacitance C and an inductance L are operated at the same angular frequency. (a) At what angular frequency will

they have the same reactance? (b) If $L = 5.00$ mH and $C = 3.50$ μF, what is the numerical value of the angular frequency in part (a), and what is the reactance of each element?
4. • (a) Compute the reactance of a 0.450 H inductor at frequencies of 60.0 Hz and 600 Hz. (b) Compute the reactance of a 2.50 μF capacitor at the same frequencies. (c) At what frequency is the reactance of a 0.450 H inductor equal to that of a 2.50 μF capacitor?
5. •• **A radio inductor.** You want the current amplitude through a 0.450-mH inductor (part of the circuitry for a radio receiver) to be 2.60 mA when a sinusoidal voltage with amplitude 12.0 V is applied across the inductor. What frequency is required?
6. • A 2.20 μF capacitor is connected across an ac source whose voltage amplitude is kept constant at 60.0 V, but whose frequency can be varied. Find the current amplitude when the angular frequency is (a) 100 rad/s; (b) 1000 rad/s; (c) 10,000 rad/s.
7. • The voltage amplitude of an ac source is 25.0 V, and its angular frequency is 1000 rad/s. Find the current amplitude if the capacitance of a capacitor connected across the source is (a) 0.0100 μF, (b) 1.00 μF, (c) 100 μF.
8. • Find the current amplitude if the self-inductance of a resistanceless inductor that is connected across the source of the previous problem is (a) 0.0100 H, (b) 1.00 H, (c) 100 H.

22.3 The Series $R–L–C$ Circuit

9. • A sinusoidal ac voltage source in a circuit produces a maximum voltage of 12.0 V and an rms current of 7.50 mA. Find (a) the voltage and current amplitudes and (b) the rms voltage of this source.
10. • A 65 Ω resistor, an 8.0 μF capacitor, and a 35 mH inductor are connected in series in an ac circuit. Calculate the impedance for a source frequency of (a) 300 Hz and (b) 30.0 kHz.
11. • In an $R–L–C$ series circuit, the rms voltage across the resistor is 30.0 V, across the capacitor it is 90.0 V, and across the inductor it is 50.0 V. What is the rms voltage of the source?
12. •• A 1500 Ω resistor is connected in series with a 350 mH inductor and an ac power supply. At what frequency will this combination have twice the impedance that it has at 120 Hz?
13. •• (a) Compute the impedance of a series $R–L–C$ circuit at angular frequencies of 1000, 750, and 500 rad/s. Take $R = 200$ Ω, $L = 0.900$ H, and $C = 2.00$ μF. (b) Describe how the current amplitude varies as the angular frequency of the source is slowly reduced from 1000 rad/s to 500 rad/s. (c) What is the phase angle of the source voltage with respect to the current when $\omega = 1000$ rad/s? (d) Construct a phasor diagram when $\omega = 1000$ rad/s.
14. •• A 200 Ω resistor is in series with a 0.100 H inductor and a 0.500 μF capacitor. Compute the impedance of the circuit and draw the phasor diagram (a) at a frequency of 500 Hz, (b) at a frequency of 1000 Hz. In each case, compute the phase angle of the source voltage with respect to the current and state whether the source voltage lags or leads the current.

22.4 Power in Alternating-Current Circuits

15. • The power of a certain CD player operating at 120 V rms is 20.0 W. Assuming that the CD player behaves like a pure resistance, find (a) the maximum instantaneous power, (b) the rms current, and (c) the resistance of this player.

16. • A series R–L–C circuit is connected to a 120 Hz ac source that has V_{rms} = 80.0 V. The circuit has a resistance of 75.0 Ω and an impedance of 105 Ω at this frequency. What average power is delivered to the circuit by the source?

17. • The circuit in Problem 13 carries an rms current of 0.250 A with a frequency of 100 Hz. (a) What is the average rate at which electrical energy is converted to heat in the resistor? (b) What average power is delivered by the source? (c) What is the average rate at which electrical energy is dissipated (converted to other forms) in the capacitor? in the inductor?

18. •• A series ac circuit contains a 250 Ω resistor, a 15 mH inductor, a 3.5 μF capacitor, and an ac power source of voltage amplitude 45 V operating at an angular frequency of 360 rad/s. (a) What is the power factor of this circuit? (b) Find the average power delivered to the entire circuit. (c) What is the average power delivered to the resistor, to the capacitor, and to the inductor?

22.5 Series Resonance

19. • An ac series R–L–C circuit contains a 120 Ω resistor, a 2.0 μF capacitor, and a 5.0 mH inductor. Find (a) the resonance angular frequency and (b) the length of time that each cycle lasts at the resonance angular frequency.

20. • (a) At what angular frequency will a 5.00 μF capacitor have the same reactance as a 10.0 mH inductor? (b) If the capacitor and inductor in part (a) are connected in an L–C circuit, what will be the resonance angular frequency of that circuit?

21. •• In an R-L-C series circuit, R = 150 Ω, L = 0.750 H, and C = 0.0180 μF. The source has voltage amplitude V = 150 V and a frequency equal to the resonance frequency of the circuit. (a) What is the power factor? (b) What is the average power delivered by the source? (c) The capacitor is replaced by one with C = 0.0360 μF and the source frequency is adjusted to the new resonance value. Then what is the average power delivered by the source?

22. • You need to make a series ac circuit having a resonance angular frequency of 1525 rad/s using a 138 Ω resistor, a 10.5 μF capacitor, and an inductor. (a) What should be the inductance of the inductor, and (b) what is the impedance of this circuit when you use it with an ac voltage source having an angular frequency of 1525 Hz?

23. •• A series circuit consists of an ac source of variable frequency, a 115 Ω resistor, a 1.25 μF capacitor, and a 4.50 mH inductor. Find the impedance of this circuit when the angular frequency of the ac source is adjusted to (a) the resonance angular frequency, (b) twice the resonance angular frequency, and (c) half the resonance angular frequency.

24. • In a series R–L–C circuit, R = 400 Ω, L = 0.350 H, and C = 0.0120 μF. (a) What is the resonance angular frequency of the circuit? (b) The capacitor can withstand a peak voltage of 550 V. If the voltage source operates at the resonance frequency, what maximum voltage amplitude can it have if the maximum capacitor voltage is not exceeded?

25. • In a series R–L–C circuit, I = 0.200 H, C = 80.0 μF, and the voltage amplitude of the source is 240 V. (a) What is the resonance angular frequency of the circuit? (b) When the source operates at the resonance angular frequency, the current amplitude in the circuit is 0.600 A. What is the resistance R of the resistor? (c) At the resonance frequency, what are the peak voltages across the inductor, the capacitor, and the resistor?

26. •• In an R–L–C series circuit, R = 300 Ω, L = 0.400 H, and C = 6.00 × 10^{-8} F. When the ac source operates at the resonance frequency of the circuit, the current amplitude is 0.500 A. (a) What is the voltage amplitude of the source? (b) What is the amplitude of the voltage across the resistor, across the inductor, and across the capacitor? (c) What is the average power supplied by the source?

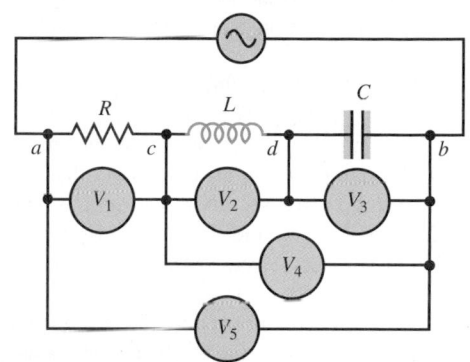

▲ **FIGURE 22.22** Problems 26, 35, and 36.

22.6 Parallel Resonance

27. • A 125 Ω resistor, an 8.50 μF capacitor, and an 11.2 mH inductor are all connected in parallel across an ac voltage source of variable frequency. (a) At what angular frequency will the impedance have its maximum value, and (b) what is that value?

28. • For the circuit in Figure 22.23, R = 300 Ω, L = 0.500 H, and C = 0.600 μF. The voltage amplitude of the source is 120 V. (a) What is the resonance frequency of the circuit? (b) Sketch the phasor diagram at the resonance frequency. (c) At the resonance frequency, what is the current amplitude through the source? (d) At the resonance frequency, what is the current amplitude through the resistor? Through the inductor? Through the branch containing the capacitor?

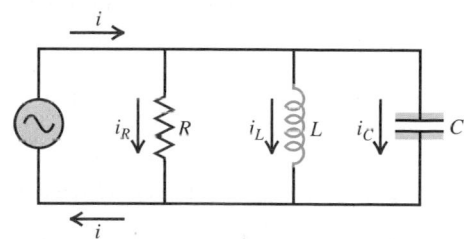

▲ **FIGURE 22.23** Problem 28.

29. • For the circuit in Figure 22.23, R = 200 Ω, L = 0.800 H, and C = 5.00 μF. When the source is operated at the resonance frequency, the current amplitude in the inductor is 0.400 A. Determine the current amplitude (a) in the branch containing the capacitor and (b) through the resistor.

30. •• (a) Use the phasor diagram for a parallel R–L–C circuit (see Figure 22.21) to show that the current amplitude I for the current i through the source is given by $I = \sqrt{I_R^2 + (I_C - I_L)^2}$. (b) Show that the result of part (a) can be written as $I = V/Z$, with $1/Z = \sqrt{1/R^2 + (\omega C - 1/\omega L)^2}$.

General Problems

31. •• A coil has a resistance of 48.0 Ω. At a frequency of 80.0 Hz, the voltage across the coil leads the current in it by 52.3°. Determine the inductance of the coil.

32. •• A large electromagnetic coil is connected to a 120 Hz ac source. The coil has resistance 400 Ω, and at this source frequency the coil has inductive reactance 250 Ω. (a) What is the inductance of the coil? (b) What must the rms voltage of the source be if the coil is to consume an average electrical power of 800 W?

33. •• A parallel-plate capacitor having square plates 4.50 cm on each side and 8.00 mm apart is placed in series with an ac source of angular frequency 650 rad/s and voltage amplitude 22.5 V, a 75.0 Ω resistor, and an ideal solenoid that is 9.00 cm long, has a circular cross section 0.500 cm in diameter, and carries 125 coils per centimeter. What is the resonance angular frequency of this circuit? (See problem 36 in Chapter 21.)

34. •• At a frequency ω_1, the reactance of a certain capacitor equals that of a certain inductor. (a) If the frequency is changed to $\omega_2 = 2\omega_1$, what is the ratio of the reactance of the inductor to that of the capacitor? Which reactance is larger? (b) If the frequency is changed to $\omega_3 = \omega_1/3$, what is the ratio of the reactance of the inductor to that of the capacitor? Which reactance is larger?

35. •• Five voltmeters, calibrated to read rms values, are connected as shown in Figure 22.22. Let $R = 200\,\Omega$, $L = 0.400$ H, and $C = 6.00\,\mu$F. The source voltage amplitude is $V = 30.0$ V. What is the reading of each voltmeter if (a) $\omega = 200$ rad/s; (b) $\omega = 1000$ rad/s?

36. •• Consider the circuit sketched in Figure 22.22. The source has a voltage amplitude of 240 V, $R = 150\,\Omega$, and the reactance of the capacitor is 600 Ω. The voltage amplitude across the capacitor is 720 V. (a) What is the current amplitude in the circuit? (b) What is the impedance? (c) What two values can the reactance of the inductor have?

37. •• In a series R–L–C circuit, the components have the following values: $L = 20.0$ mH, $C = 140$ nF, and $R = 350\,\Omega$. The generator has an rms voltage of 120 V and a frequency of 1.25 kHz. Determine (a) the power supplied by the generator; and (b) the power dissipated in the resistor.

38. •• (a) Show that for an R–L–C series circuit the power factor is equal to R/Z. (*Hint:* Use the phasor diagram; see Figure 22.13b.) (b) Show that for any ac circuit, not just one containing pure resistance only, the average power delivered by the voltage source is given by $P_{av} = I_{rms}^2 R$.

39. •• In an R–L–C series circuit the magnitude of the phase angle is 54.0°, with the source voltage lagging the current. The reactance of the capacitor is 350 Ω, and the resistor resistance is 180 Ω. The average power delivered by the source is 140 W. Find (a) the reactance of the inductor; (b) the rms current; (c) the rms voltage of the source.

40. •• In a series R–L–C circuit, $R - 300\,\Omega$, $X_C = 300\,\Omega$, and $X_L = 500\,\Omega$. The average power consumed in the resistor is 60.0 W. (a) What is the power factor of the circuit? (b) What is the rms voltage of the source?

41. •• In a series R–L–C circuit, the phase angle is 40.0°, with the source voltage leading the current. The reactance of the capacitor is 400 Ω, and the resistance of the resistor is 200 Ω. The average power delivered by the source is 150 W. Find (a) the reactance of the inductor, (b) the rms current, (c) the rms voltage of the source.

42. •• A 100.0 Ω resistor, a 0.100 μF capacitor, and a 300.0 mH inductor are connected in series to a voltage source with amplitude 240 V. (a) What is the resonance angular frequency? (b) What is the maximum current in the resistor at resonance? (c) What is the maximum voltage across the capacitor at resonance? (d) What is the maximum voltage across the inductor at resonance? (e) What is the maximum energy stored in the capacitor at resonance? in the inductor?

43. •• Consider the same circuit as in the previous problem, with the source operated at an angular frequency of 400 rad/s. (a) What is the maximum current in the resistor? (b) What is the maximum voltage across the capacitor? (c) What is the maximum voltage across the inductor? (d) What is the maximum energy stored in the capacitor? in the inductor?

Passage Problems

BIO Converting dc to ac. Individual cells such as eggs are often organized spatially, as manifested in part by asymmetries in the cell membrane. These asymmetries include non-uniform distributions of ion transport mechanisms that result in net electrical current entering one region of the membrane and leaving another. Because these steady cellular currents may regulate cell polarity, leading (in the case of eggs) to embryonic polarity, scientists are interested in measuring them.

The cellular currents move in loops through the extracellular fluid around the cells. Ohm's Law requires that there be voltage differences between any two points in the fluid near current-producing cells. While the currents may be significant, the extracellular voltage differences are tiny, on the order of nanovolts. If the voltage differences in the medium near a cell can be mapped, the current density can be calculated using Ohm's Law. One way to measure these voltage differences might be to use two electrodes spaced 10 or 20 μm apart, but that fails because the dc impedance (the resistance) of such electrodes is high and the inherent noise in these high-impedance electrodes swamps the cellular voltages.

One successful method uses a platinum ball electrode moved sinusoidally between two points near a cell. The electrical potential that the electrode measures (with respect to a distant reference electrode) also varies sinusoidally, so the dc potential difference between the two extremes of the electrode's excursion is converted to a sine-wave ac potential difference. The platinum electrode behaves as a capacitor in series with the resistance of the fluid, called the access resistance. The access resistance (R_A) has a value of about, $\rho/(10a)$, where ρ is the resistivity of the medium (usually expressed in $\Omega \cdot$ cm) and a is the radius of the electrode. The platinum ball typically has a diameter of 20 μm and a capacitance of 10 nF, and the resistivity of many biological fluids is 100 $\Omega \cdot$ cm.

44. What is the dc impedance of the electrode, assuming that it behaves as an ideal capacitor?
A. 0　　B. Infinite　　C. $\sqrt{2} \times 10^4\,\Omega$　　D. $\sqrt{2} \times 10^6\,\Omega$

45. If the electrode is oscillated between two points 20 μ apart with a frequency of $(5000/\pi)$ Hz, what is the impedance of the electrode?
A. 0　　B. Infinite　　C. $\sqrt{2} \times 10^4\,\Omega$　　D. $\sqrt{2} \times 10^6\,\Omega$

46. The signal from the oscillating electrode is fed into an amplifier, which reports the measured voltage as an rms value, V_{rms}. However, the number of interest for analyzing currents driven by the cell is the peak-to-peak voltage difference (V_{pp}), that is, the voltage difference between the two extremes of the electrode's excursion. What is the value of V_{pp} in terms of V_{rms}?
A. $V_{rms}/\sqrt{2}$　　B. $V_{rms}/2\sqrt{2}$　　C. $\sqrt{2}V_{rms}$　　D. $2\sqrt{2}V_{rms}$

23 Electromagnetic Waves

Energy from the sun, an essential requirement for life on earth, reaches us by means of electromagnetic waves (light) that travel through 93 million miles of (nearly) empty space. Electromagnetic waves occur in an astonishing variety of physical situations, including TV and radio transmission, cellular phones, microwave oscillators for ovens and radar, lightbulbs, x-ray machines, and radioactive nuclei. So it's important for us to make a careful study of their properties and behavior.

The telescope in the foreground is Gemini North, one of a pair of twin 8.1-m telescopes located on Mauna Kea in Hawaii (Gemini North) and in the Chilean Andes (Gemini South). Together, these two telescopes can see the entire visible universe.

The existence of electromagnetic waves depends on two facts. First, a time-varying magnetic field acts as a source of an electric field, as implied by Faraday's law of induction. Second (and less familiar), a time-varying electric field acts as a source of a magnetic field, as we'll discuss later. Thus, a time-varying field of either kind gives rise to a time-varying field of the other kind in neighboring regions of space. In this way, time-varying electric and magnetic fields can propagate through space as a *wave*. Such waves carry energy and momentum and have the property of polarization. A wave can be *sinusoidal,* in which case the $\vec{E}$ and $\vec{B}$ fields are sinusoidal functions of time. The spectrum of electromagnetic waves covers an extremely broad range of frequencies and wavelengths. In particular, light consists of electromagnetic waves, and our study of optics in later sections of this chapter will be based in part on the electromagnetic nature of light.

23.1 Introduction to Electromagnetic Waves

In the last several chapters, we've studied various aspects of electric and magnetic fields. When the fields don't vary with time, such as an electric field produced by charges at rest, or the magnetic field of a steady current, we can analyze

A rudimentary electromagnetic wave. The electric and magnetic fields are uniform behind the advancing wave front and zero in front of it

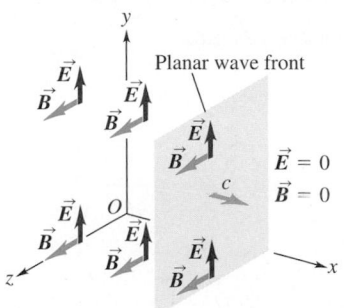

▲ **FIGURE 23.1** A planar electromagnetic wave front.

the electric and magnetic fields independently, without considering interactions between them. But when the fields vary with time, they're no longer independent.

Faraday's law (Section 21.3) tells us that a time-varying magnetic field acts as a source of an electric field, as shown by induced emf's in inductances and transformers. In 1865, the Scottish mathematician and physician James C. Maxwell proposed that a time-varying *electric* field could play the same role as a current, as an additional source of *magnetic* field. This two-way interaction between the two fields can be summarized elegantly in a set of four equations now called *Maxwell's equations.* We won't study these equations here because they require calculus, but you may encounter them in a later course.

Thus, when *either* an electric or a magnetic field is changing with time, a field of the other kind is induced in adjacent regions of space. In this way, time-varying electric and magnetic fields can propagate through space from one region to another, even when there is no matter in the intervening region. Such a propagating disturbance is called an **electromagnetic wave.**

One outcome of Maxwell's analysis was his prediction that an electromagnetic disturbance should propagate in free space with a speed equal to that of light and that light waves were therefore very likely to be electromagnetic in nature. In 1887, Heinrich Hertz used oscillating L–C circuits (discussed in Section 21.12) to produce, in his laboratory, electromagnetic waves with wavelengths of the order of a few meters. Hertz also produced electromagnetic *standing waves* and measured the distance between adjacent nodes (one half-wavelength) to determine their wavelength λ. Knowing the resonant frequency f of his circuits, he then determined the speed v of the waves from the wavelength–frequency relation $v = \lambda f$. In this way, Hertz confirmed that the wave speed was the same as that of light, thus verifying Maxwell's theoretical prediction directly. The SI unit of frequency, one cycle per second, is named the *hertz* in honor of the great German scientist.

Maxwell's synthesis, wrapping up the basic principles of electromagnetism neatly and elegantly in four equations, stands as a towering intellectual achievement. It is comparable to the Newtonian synthesis we described at the end of Section 6.5 and to the 20th-century development of relativity, quantum mechanics, and the understanding of DNA. All are beautiful, and all are monuments to the achievements of which the human intellect is capable!

23.2 Speed of an Electromagnetic Wave

We're now ready to introduce the basic ideas of electromagnetic waves and their relation to the principles of electromagnetism. To start, we'll postulate a simple field configuration that has wavelike behavior. We'll assume an electric field that has only a y component and a magnetic field with only a z component, and we'll assume that both fields move together in the $+x$ direction with a speed c that is initially unknown. Then we'll ask whether these fields are physically possible—that is, whether they are consistent with the laws of electromagnetism we've studied.

A Simple Plane Electromagnetic Wave

Using an x-y-z coordinate system (Figure 23.1), we imagine that all space is divided into two regions by a plane perpendicular to the x axis (parallel to the y-z plane). At every point to the left of this plane there exist a uniform electric field $\vec{E}$ in the $+y$ direction and a uniform magnetic field $\vec{B}$ in the $+z$ direction, as shown. Furthermore, we suppose that the boundary plane, which we call the *wave front*, moves to the right with a constant speed c, as yet unknown. Thus, the $\vec{E}$ and $\vec{B}$ fields travel to the right into previously field-free regions with a definite speed. The situation, in short, describes a rudimentary electromagnetic wave.

We won't concern ourselves with the problem of actually *producing* such a field configuration. Instead, we simply ask whether it is consistent with the laws of electromagnetism.

The analysis leading to the answer to this question is somewhat complicated, so we'll simply state the answer: It is: Yes, this primitive electromagnetic wave *is* consistent with the laws of electromagnetism, provided that two conditions are satisfied: (1) When an electromagnetic wave travels in vacuum, the magnitudes of $\vec{E}$ and $\vec{B}$ are in a definite, constant ratio:

$$E = cB. \tag{23.1}$$

(2) The wave front moves with a speed c given by

$$c = \frac{1}{\sqrt{\epsilon_0 \mu_0}}, \tag{23.2}$$

where ϵ_0 and μ_0 are the proportionality constants we encountered in Coulomb's law (Equation 17.1) and the law of Biot and Savart (Equation 20.16), respectively. Inserting the numerical values of these quantities, we find that

$$c = \frac{1}{\sqrt{[8.85 \times 10^{-12} \ \mathrm{C^2/(N \cdot m^2)}](4\pi \times 10^{-7} \ \mathrm{N/A^2})}}$$
$$= 3.00 \times 10^8 \ \mathrm{m/s}.$$

Our assumed wave is consistent with the principles of electromagnetism, provided that the wave front moves with the speed c just given. We recognize this as the speed of light in vacuum! We shouldn't be too surprised by this result; the constant ϵ_0 appears in the equation that relates an electric field to its sources, and μ_0 plays a similar role for magnetic fields. Since electromagnetic waves depend on the interaction between these two fields, we might expect that the two constants should play a central role in their propagation.

We've chosen a simple and primitive wave for our study in order to avoid mathematical complications, but this special case illustrates several important features of *all* electromagnetic waves:

Characteristics of electromagnetic waves in vacuum

1. The wave is **transverse:** Both $\vec{E}$ and $\vec{B}$ are perpendicular to the direction of propagation of the wave and to each other.
2. There is a definite ratio between the magnitudes of $\vec{E}$ and $\vec{B}$: $E = cB$.
3. The wave travels in vacuum with a definite and unchanging speed c.
4. Unlike mechanical waves, which need the oscillating particles of a medium such as water or air to be transmitted, electromagnetic waves require no medium. What's "waving" in an electromagnetic wave are the electric and magnetic fields.

NOTE ▶ The relation $E = cB$ is correct only in the SI unit system. It would be wrong to conclude that E is larger than B. Comparing E and B is like comparing apples and oranges: They are different physical quantities and have different units. ◀

Figure 23.2 shows a "right-hand rule" to determine the directions of $\vec{E}$ and $\vec{B}$: Point the thumb of your right hand in the direction the wave is traveling. Imagine rotating the $\vec{E}$ field vector 90° in the sense your fingers curl; that gives you the direction of the $\vec{B}$ field.

▲ **Application The eerie glow of faster-than-light travel** Nuclear reactors and their radioactive fuel rods are often submerged in pools of water that absorb their radiation. These pools glow with the gorgeous cerulean blue light you see in the photo. This glow, an example of a phenomenon called Cerenkov radiation, is caused by energetic electrons that travel *faster* than the speed of light in water. Nothing can travel faster than the speed of light in vacuum, but there's no law against exceeding the speed of light in a given medium. The fast electrons, which originate in nuclear reactions, interact with atoms in the water, which emit light. Because of the electrons' speed, these interactions are confined to a shock wave like that of a supersonic jet, and the emitted light is mainly blue.

Right-hand rule for an electromagnetic wave:

① Point the thumb of your right hand in the wave's direction of propagation.

② Imagine rotating the $\vec{E}$ field vector 90° in the sense your fingers curl.

That is the direction of the $\vec{B}$ field.

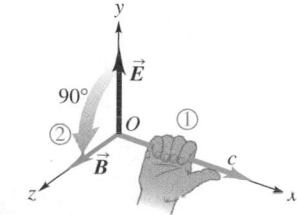

▲ **FIGURE 23.2** The right-hand rule for the directions of $\vec{E}$ and $\vec{B}$ in an electromagnetic wave.

Conceptual Analysis 23.1

The directions of $\vec{E}$ and $\vec{B}$

Consider a simple plane electromagnetic wave that is traveling vertically upward with its electric field pointing eastward. In which direction does the magnetic field point?

A. North
B. South
C. West

SOLUTION Using the right-hand rule just given, we point the thumb of our right hand in the direction of the wave's propagation (upward). We now imagine rotating the $\vec{E}$ vector 90° in the direction our fingers curl. It starts out pointing east and ends up pointing north, so northward is the direction of $\vec{B}$.

EXAMPLE 23.1 The speed of light in vacuum

(a) What distance does light travel in 1 second, expressed in units of the circumference of the earth? The equatorial circumference of the earth is 4.01×10^7 m. **(b)** How much time is required for light to travel 1 foot?

SOLUTION

SET UP AND SOLVE Part (a): In 1 second, light travels a distance

$$x = ct = (3.00 \times 10^8 \text{ m/s})(1 \text{ s}) = 3.00 \times 10^8 \text{ m}$$
$$= (3.00 \times 10^8 \text{ m})\left(\frac{1 \text{ circumference unit}}{4.01 \times 10^7 \text{ m}}\right)$$
$$= 7.48 \text{ circumferences.}$$

Part (b): With 1 ft = 0.3048 m, the time t for light to travel 1 foot is

$$t = \frac{x}{c} = \frac{0.3048 \text{ m}}{3.00 \times 10^8 \text{ m/s}} = 1.02 \times 10^{-9} \text{ s.}$$

REFLECT The straight-line distance light travels in 1 second is about 7.5 circular trips around the earth. The time to travel 1 ft is 1.02×10^{-9} s = 1.02 ns. A useful approximation is that light travels at about 1 ft/ns.

Practice Problem: How far does light travel in a year (a light year)? *Answer:* $x = 9.46 \times 10^{15}$ m.

23.3 The Electromagnetic Spectrum

Electromagnetic waves cover an extremely broad spectrum of wavelengths and frequencies. Radio and TV transmission, visible light, infrared and ultraviolet radiation, x rays, and gamma rays all form parts of the **electromagnetic spectrum.** The extent of this spectrum is shown in Figure 23.3, which gives approximate wavelength and frequency ranges for the various segments. Despite vast

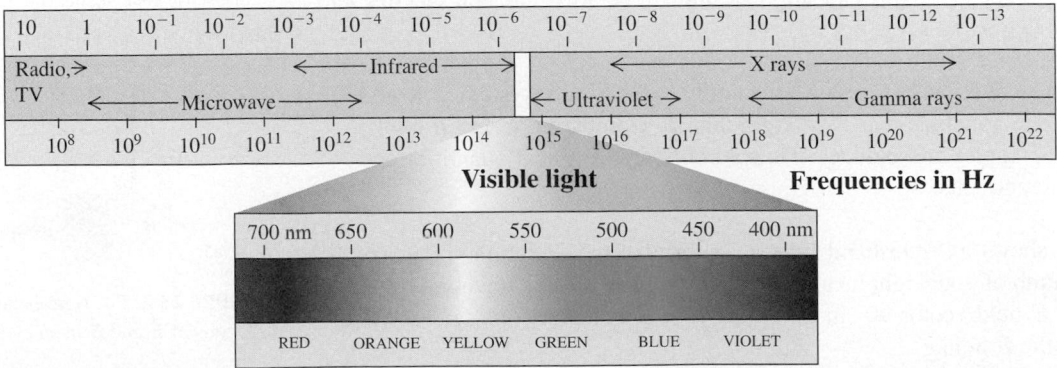

▲ **FIGURE 23.3** The electromagnetic spectrum. The frequencies and wavelengths found in nature extend over such a wide range that we must use a logarithmic scale to graph them. The boundaries between bands are somewhat arbitrary.

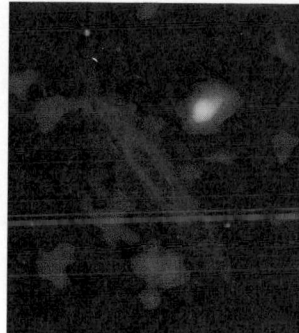 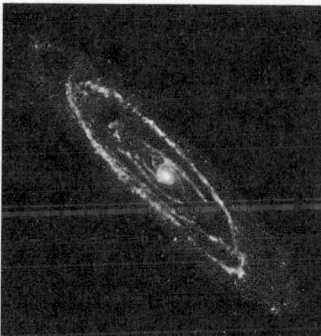  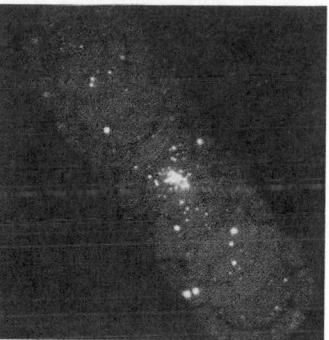

Radio This image combines a high-resolution radio image of the galaxy disk (blue) with a larger-scale image of the surrounding regions of space (red). Both images are sensitive to hydrogen gas. The clouds of hydrogen gas surrounding the galaxy are invisible in other spectral bands.

Infrared The wavelengths used for this image are particularly sensitive to the warm dust present in star-forming regions. Because these regions occur in galactic arms, the image shows the galaxy's arm structure especially clearly.

Visible In the visible we see mainly the light of stars, blocked in places by dark streamers of dust. Regions containing mostly old stars appear yellow white; star-forming regions containing young stars have a blue cast.

X-ray This x-ray image of the central part of the galaxy's disk shows many point sources of x rays, including a cluster near the galaxy's center. These points are mostly *x-ray binaries* containing a normal star orbiting a neutron star or black hole. The hot disk around the galaxy's central black hole also radiates x rays.

▲ **FIGURE 23.4** Images of the nearby Andromeda galaxy taken in several spectral regions. The Andromeda galaxy is a large spiral galaxy like our own Milky Way; the Andromeda and Milky Way galaxies dominate the local group of galaxies. Except for the visible-light image, these images are all in false color.

differences in their uses and means of production, all these electromagnetic waves have the general characteristics described in the preceding sections, including the common propagation speed $c = 3.00 \times 10^8$ m/s (in vacuum). All are the same in principle; they differ in frequency f and wavelength λ, but the relation $c = \lambda f$ holds for each.

As Figure 23.3 shows, we can detect only a very small segment of this spectrum directly through our sense of sight. Within the visually detectable range, we perceive wavelength (or frequency) in terms of color, from long-wavelength red to short-wavelength violet. Some animals (including bees and birds) can see into the ultraviolet; pit vipers use their pit organs to "see" infrared radiation.

Light from many familiar sources (including the sun) is a mixture of many different wavelengths. By using special sources or filters, we can select a narrow band of wavelengths with a range of, say, from 500 to 501 nm. Such light is approximately *monochromatic* (single-color) light. Absolutely monochromatic light with only a single wavelength is an unattainable idealization. When we use the expression "monochromatic light with wavelength 500 nm" with reference to a laboratory experiment, we really mean a small band of wavelengths *around* 500 nm. One distinguishing characteristic of light from a *laser* is that it is much more nearly monochromatic than light produced in any other way.

Despite the limitations of the human eye, science and technology use all parts of the electromagnetic spectrum. Figure 23.4 shows how wavelengths in the radio, infrared, visible, and x-ray parts of the spectrum are used to explore the Andromeda galaxy, a neighbor of our own Milky Way.

23.4 Sinusoidal Waves

Sinusoidal electromagnetic waves are analogous to sinusoidal transverse mechanical waves on a stretched string. We studied mechanical waves in Chapter 12; we suggest that you review that discussion. In a sinusoidal electromagnetic wave, the

▲ **BIO** Application **Ultraviolet vision.**
What we call "visible light" is just the part of the electromagnetic spectrum that human eyes see. Many other animals would define "visible" somewhat differently. For instance, many animals, including insects and birds, see into the UV, and the natural world is full of signals that they see and we don't. The left-hand photo shows how a black-eyed Susan looks to us; the right-hand photo (in false color) shows the same flower in UV light. The bees that pollinate these flowers see the prominent central spot that is invisible to us. Similarly, many birds—including bluebirds, budgies, parrots, and even peacocks—have ultraviolet patterns that make them even more vivid to each other than they are to us.

$\vec{E}$ and $\vec{B}$ fields at any point in space are sinusoidal functions of time, and at any instant of time the *spatial* variation of the fields is also sinusoidal.

Some sinusoidal electromagnetic waves share with the primitive wave we described in Section 23.2 the property that at any instant the fields are uniform over any plane perpendicular to the direction of propagation (as shown in Figure 23.1). Such a wave is called a **plane wave.** The entire pattern travels in the direction of propagation with speed c. The directions of $\vec{E}$ and $\vec{B}$ are perpendicular to the direction of propagation (and to each other), so the wave is *transverse.*

The frequency f, the wavelength λ, and the speed of propagation c of any periodic wave are related by the usual wavelength–frequency relation $c = \lambda f$. For visible light, a typical frequency is $f = 5 \times 10^{14}$ Hz; the corresponding wavelength is

$$\lambda = \frac{c}{f} = \frac{3 \times 10^8 \text{ m/s}}{5 \times 10^{14} \text{ Hz}} = 6 \times 10^{-7} \text{ m} = 600 \text{ nm},$$

which is similar in size to some bacteria and about one-hundredth the size of a human hair! If the frequency is 10^8 Hz (100 MHz), typical of commercial FM radio stations, the wavelength is

$$\lambda = \frac{3 \times 10^8 \text{ m/s}}{10^8 \text{ Hz}} = 3 \text{ m}.$$

Figure 23.5 shows a sinusoidal electromagnetic wave traveling in the $+x$ direction. The $\vec{E}$ and $\vec{B}$ vectors are shown only for a few points on the positive side of the x axis. Imagine a plane perpendicular to the x axis at a particular point and a particular time; the fields have the same values at all points in that plane. Of course, the values are different in different planes. In the planes where the $\vec{E}$ vector is in the $+y$ direction, $\vec{B}$ is in the $+z$ direction; where $\vec{E}$ is in the $-y$ direction, $\vec{B}$ is in the $-z$ direction. These directions illustrate the direction relations that we described in Section 23.2.

We can describe electromagnetic waves by means of *wave functions,* just as we did in Section 12.4 for waves on a string. One form of the wave function for a transverse wave traveling to the right along a stretched string is Equation 12.5,

$$y(x, t) = A \sin 2\pi \left(\frac{t}{T} - \frac{x}{\lambda} \right) = A \sin (\omega t - kx),$$

where y is the transverse displacement from its equilibrium position at time t of a point with coordinate x on the string. The quantity A is the maximum displacement, or *amplitude,* of the wave; ω is its *angular frequency,* equal to 2π times the frequency f; and k is the **wave number,** or *propagation constant,* equal to $2\pi/\lambda$, where λ is the wavelength.

In Figure 23.5, let E and B represent the instantaneous values of the electric and magnetic fields, respectively, and let E_{max} and B_{max} represent the maximum values, or *amplitudes,* of those fields. The wave functions for the wave are then

$$E = E_{max} \sin 2\pi \left(\frac{t}{T} - \frac{x}{\lambda} \right) = E_{max} \sin (\omega t - kx),$$

$$B = B_{max} \sin 2\pi \left(\frac{t}{T} - \frac{x}{\lambda} \right) = B_{max} \sin (\omega t - kx). \quad (23.3)$$

The sine curves in Figure 23.5 represent instantaneous values of E and B as functions of x at time $t = T/4$. The wave travels to the right with speed c.

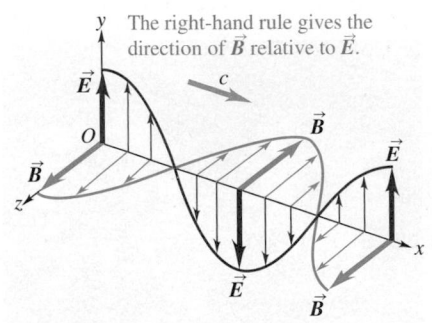

▲ **FIGURE 23.5** Representation of the electric and magnetic fields as functions of x at time $t = T/4$ for a sinusoidal electromagnetic wave traveling in the $+x$ direction. The fields are shown only for points on the positive side of the x axis.

Equations 23.3 show that, at any point, the sinusoidal oscillations of $\vec{E}$ and $\vec{B}$ are *in phase*. From Equation 23.1, the amplitudes must be related by

$$E_{max} = cB_{max}. \tag{23.4}$$

Figure 23.6 shows the electric and magnetic fields of a wave traveling in the *negative x* direction. At points where $\vec{E}$ is in the positive y direction, $\vec{B}$ is in the *negative z* direction; where $\vec{E}$ is in the negative y direction, $\vec{B}$ is in the *positive z* direction. The wave functions for this wave are

$$E = -E_{max}\sin 2\pi\left(\frac{t}{T} + \frac{x}{\lambda}\right) = -E_{max}\sin(\omega t + kx),$$

$$B = B_{max}\sin 2\pi\left(\frac{t}{T} + \frac{x}{\lambda}\right) = B_{max}\sin(\omega t + kx). \tag{23.5}$$

As with the wave traveling in the $+x$ direction, the sinusoidal oscillations of the $\vec{E}$ and $\vec{B}$ fields at any point are *in phase*.

A wave whose $\vec{E}$ field always lies along the same line is said to be *linearly polarized*. Both of these sinusoidal waves are linearly polarized in the y direction (because the $\vec{E}$ field always lies along the y axis).

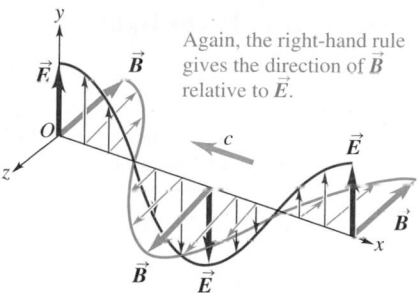

▲ **FIGURE 23.6** An electromagnetic wave like the one in Figure 23.5, but traveling in the $-x$ direction, shown at time $t = 3T/4$.

PROBLEM-SOLVING STRATEGY 23.1 **Electromagnetic waves**

SET UP

1. For the problems posed in this chapter, the most important advice that we can give is to concentrate on basic relationships, such as the relation of $\vec{E}$ to $\vec{B}$ (both magnitude and direction), how the wave speed is determined, the transverse nature of the waves, and so on. Don't get sidetracked by mathematical details.

SOLVE

2. In the discussions of sinusoidal waves, both traveling and standing, you need to use the language of sinusoidal waves from Chapter 12. Don't hesitate to go back and review that material, including the problem-solving strategies presented in those chapters. Keep in mind the basic relationships for periodic waves: $v = \lambda f$ and $\lambda = vT$. For electromagnetic waves in vacuum, $v = c$. Be careful to distinguish between ordinary frequency f, usually expressed in hertz, and angular frequency $\omega = 2\pi f$, expressed in rad/s. Remember that the wave number k is $k = 2\pi/\lambda$, and that $\omega = ck$.

EXAMPLE 23.2 **Remote control**

A remote-control unit for a stereo system emits radiation with a frequency of 1.0×10^{14} Hz. Calculate the wavelength of the radiation it emits. Identify the corresponding region of the electromagnetic spectrum by referring to Figure 23.3.

SOLUTION

SET UP AND SOLVE The wavelength of the radiation is

$$\lambda = c/f = (3.00 \times 10^8 \text{ m/s})/(1.0 \times 10^{14} \text{ Hz})$$
$$= 3.0 \times 10^{-6} \text{ m}.$$

REFLECT This is in the infrared (IR) band of the spectrum.

Practice Problem: Calculate the wavelength of a 92.9 MHz FM-station radio wave. *Answer:* $\lambda = 3.23$ m.

EXAMPLE 23.3 Laser light

A carbon dioxide laser emits a sinusoidal electromagnetic wave that travels in vacuum in the negative x direction, like the wave in Figure 23.6. The wavelength is 10.6 μm, and the $\vec{E}$ field is along the z axis, with a maximum magnitude of 1.5 MV/m. Find the equations for the magnitudes of vectors $\vec{E}$ and $\vec{B}$ as functions of time and position.

SOLUTION

SET UP From the right-hand rule, when $\vec{E}$ is in the positive z direction, $\vec{B}$ is in the positive y direction; and when $\vec{E}$ is in the negative z direction, $\vec{B}$ is in the negative y direction (consistent with the right-hand rule). Thus, we don't need the negative sign in the expression for E in Equation 23.5.

SOLVE Since the wave is traveling along the negative x axis, the general equations for the wave are

$$E = E_{max}\sin(\omega t + kx) \qquad B = B_{max}\sin(\omega t + kx).$$

To find B_{max}, we use Equation 23.4: $E_{max} = cB_{max}$. The wavelength is $\lambda = 10.6 \times 10^{-6}$ m, so

$$k = \frac{2\pi}{\lambda} = \frac{2\pi \text{ rad}}{10.6 \times 10^{-6} \text{ m}} = 5.93 \times 10^5 \text{ rad/m}.$$

Also,

$$\omega = ck = (3.00 \times 10^8 \text{ m/s})(5.93 \times 10^5 \text{ rad/m})$$
$$= 1.78 \times 10^{14} \text{ rad/s}.$$

Substituting into the above equations, with

$$B_{max} = \frac{E_{max}}{c} = \frac{1.50 \times 10^6 \text{ V/m}}{3.00 \times 10^8 \text{ m/s}} = 5.00 \times 10^{-3} \text{ T},$$

we get

$$E = E_{max}\sin(\omega t + kx)$$
$$= (1.5 \times 10^6 \text{ V/m})\sin[(1.78 \times 10^{14} \text{ rad/s})t$$
$$+ (5.93 \times 10^5 \text{ rad/m})x],$$
$$B = B_{max}\sin(\omega t + kx)$$
$$= (5.0 \times 10^{-3} \text{ T})\sin[(1.78 \times 10^{14} \text{ rad/s})t$$
$$+ (5.93 \times 10^5 \text{ rad/m})x].$$

REFLECT Note that no negative sign is needed in the expression for E; the right-hand rule is obeyed without it. At any point, the two fields are in phase.

PhET: Microwaves

23.5 Energy in Electromagnetic Waves

It is a familiar fact that energy is associated with electromagnetic waves. Think of the sun's radiation and the radiation in microwave ovens. To derive detailed relationships for the energy in an electromagnetic wave, we begin with the expressions derived in Sections 18.7 and 21.10 for the **energy densities** (energy per unit volume) in electric and magnetic fields; we suggest that you review those derivations now. Specifically, Equations 18.20 and 21.21 show that the total energy density u in a region of space where $\vec{E}$ and $\vec{B}$ fields are present is given by the following expressions:

Energy density in electric and magnetic fields
The energy density u (energy per unit volume) in a region of empty space where electric and magnetic fields are present is

$$u = \frac{1}{2}\epsilon_0 E^2 + \frac{1}{2\mu_0}B^2. \qquad (23.6)$$

The two field magnitudes are related by Equation 23.1:

$$B = \frac{E}{c} = \sqrt{\epsilon_0\mu_0}E.$$

Combining this equation with Equation 23.6, we can also express the energy density u as

$$u = \frac{1}{2}\epsilon_0 E^2 + \frac{1}{2\mu_0}(\sqrt{\epsilon_0\mu_0}E)^2 = \epsilon_0 E^2. \qquad (23.7)$$

This result shows that the energy density associated with the $\vec{E}$ field is equal to that of the $\vec{B}$ field.

In the simple wave described in Section 23.2, the $\vec{E}$ and $\vec{B}$ fields advance in the $+x$ direction into regions where originally no fields were present, so it is clear that the wave transports energy from one region to another. We can describe this energy transfer in terms of *energy transferred per unit time per unit cross-sectional area,* or *power per unit area,* for an area perpendicular to the direction of wave travel. The average value of this quantity, for any wave, is called the *intensity* of the wave.

To see how the energy flow is related to the fields, consider a stationary plane, perpendicular to the x axis, that coincides with the wave front at a certain time. In a time Δt after this, the wave front moves a distance $\Delta x = c\,\Delta t$ to the right of the plane. Considering an area A on the stationary plane (Figure 23.7), we note that the energy in the space to the right of this area must have passed through it to reach the new location. The volume ΔV of the relevant region is the base area A times the length $c\,\Delta t$. The energy ΔU in this region is the energy density u times this volume:

$$\Delta U = u\,\Delta V = (\epsilon_0 E^2)(Ac\,\Delta t).$$

This energy passes through the area A in time Δt. The energy flow per unit time per unit area, which we'll denote by S, is

$$S = \frac{1}{A}\frac{\Delta U}{\Delta t} = \epsilon_0 c E^2.$$

Using Equations 23.1 $(E = cB)$ and 23.2 $(c = 1/\sqrt{\epsilon_0\mu_0})$, we can derive the alternative forms

$$S = \epsilon_0 c E^2 = \frac{\epsilon_0}{\sqrt{\epsilon_0\mu_0}}E^2 = \sqrt{\frac{\epsilon_0}{\mu_0}}E^2 = \frac{EB}{\mu_0} = cu. \qquad (23.8)$$

The units of S are energy per unit time per unit area, or power per unit area. The SI unit of S is $1\ \mathrm{J/(s\cdot m^2)}$, or $1\ \mathrm{W/m^2}$. That is, S is power per unit area.

In all the preceding equations, E and B are the *instantaneous* values of the electric and magnetic field magnitudes, respectively. For a wave in which the fields vary with time, S also varies. The *average* value of S is the average energy transmitted across a given area perpendicular to the direction of propagation, per unit area and per unit of time—that is, the average power per unit area.

For a *sinusoidal* wave, the average value of E^2 is one-half the square of the amplitude $E_{\max}$. In this case, we can find the average value of S simply by replacing E^2 in Equation 23.8 with $E_{\max}^2/2$. Equation 23.8 then becomes

$$S_{\mathrm{av}} = \frac{1}{2}\sqrt{\frac{\epsilon_0}{\mu_0}}E_{\max}^2 = \frac{1}{2}\epsilon_0 c E_{\max}^2 = \frac{E_{\max}B_{\max}}{2\mu_0}. \qquad \text{(sinusoidal wave)} \qquad (23.9)$$

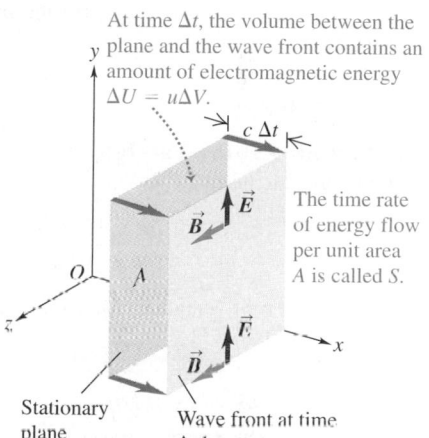

At time Δt, the volume between the y plane and the wave front contains an amount of electromagnetic energy $\Delta U = u\Delta V$.

$c\,\Delta t$

$\vec{E}$

$\vec{B}$

The time rate of energy flow per unit area A is called S.

O

A

z

$\vec{E}$

$\vec{B}$

x

Stationary plane

Wave front at time Δt later

▲ **FIGURE 23.7** A wave front at a time Δt after it passes through the stationary plane with area A.

Making the same substitution in Equation 23.7, we obtain $u_{av} = \frac{1}{2}\epsilon_0 E_{max}^2$. Comparing this equation with Equation 23.9, we find that S_{av} and u_{av} are simply related:

$$S_{av} = \frac{u_{av}}{\sqrt{\epsilon_0 \mu_0}} = u_{av}c. \quad \text{(sinusoidal wave)} \tag{23.10}$$

The average power per unit area in an electromagnetic wave is also called the **intensity** of the wave, denoted by I. That is, $I = S_{av}$, and we can write

$$I = S_{av} = \frac{1}{2}\sqrt{\frac{\epsilon_0}{\mu_0}}E_{max}^2 = \frac{1}{2}\epsilon_0 c E_{max}^2 = \frac{E_{max}B_{max}}{2\mu_0}. \tag{23.11}$$

EXAMPLE 23.4 Laser cutter

A laser cutter used for cutting thin sheets of material emits a beam with electric-field amplitude $E_{max} = 2.76 \times 10^5$ V/m over an area of 2.00 mm^2. Find **(a)** the maximum magnetic field B_{max}, **(b)** the maximum energy density u_{max}, **(c)** the intensity $S_{av} = I$ of the beam, and **(d)** the average power of the beam.

SOLUTION

SET UP AND SOLVE Part (a): From Equation 23.1, the maximum magnetic field B_{max} is

$$B_{max} = \frac{E_{max}}{c} = \frac{2.76 \times 10^5 \text{ V/m}}{3.00 \times 10^8 \text{ m/s}} = 9.20 \times 10^{-4} \text{ T}.$$

Part (b): From Equation 23.7, the maximum energy density u_{max} is

$$u_{max} = \epsilon_0 E_{max}^2$$
$$= [8.85 \times 10^{-12} \text{ C}^2/(\text{N}\cdot\text{m}^2)](2.76 \times 10^5 \text{ N/C})^2$$
$$= 0.674 \text{ N/m}^2 = 0.674 \text{ J/m}^3.$$

The *average* energy density is half of this:

$$u_{av} = \frac{1}{2}\epsilon_0 E_{max}^2 = 0.337 \text{ J/m}^3.$$

Part (c): The intensity $I = S_{av}$ is given by Equation 23.10:

$$I = S_{av} = u_{av}c = (0.337 \text{ J/m}^3)(3.00 \times 10^8 \text{ m/s})$$
$$= 1.01 \times 10^8 \text{ J/(m}^2\cdot\text{s)} = 1.01 \times 10^8 \text{ W/m}^2.$$

Alternatively, from Equation 23.9,

$$I = S_{av} = \frac{1}{2}\epsilon_0 c E_{max}^2$$
$$= \frac{1}{2}[8.85 \times 10^{-12} \text{ C}^2/(\text{N}\cdot\text{m}^2)](3.00 \times 10^8 \text{ m/s})$$
$$\cdot (2.76 \times 10^5 \text{ N/C})^2$$
$$= 1.01 \times 10^8 \text{ W/m}^2.$$

Part (d): The average total power P_{av} is the intensity S_{av} multiplied by the cross-sectional area A of the beam:

$$P_{av} = (1.01 \times 10^8 \text{ W/m}^2)(2.00 \times 10^{-6} \text{ m}^2)$$
$$= 202 \text{ W}.$$

REFLECT A power of 200 W is enough power to cut cardboard and thin wood. A typical laser pointer has an output power on the order of a few milliwatts.

Practice Problem: Find the maximum energy density for a 2000 W laser with an electric-field amplitude $E_{max} = 8.68 \times 10^5$ N/C. *Answer:* $u_{max} = 6.67$ J/m^3.

Radiation Pressure

The fact that electromagnetic waves transport energy follows directly from the fact that energy is required to establish electric and magnetic fields. It can also be shown that electromagnetic waves carry *momentum p*, with a corresponding momentum density (momentum p per volume V) of magnitude

$$\frac{p}{V} = \frac{\epsilon_0 E^2}{c} = \frac{EB}{\mu_0 c^2} = \frac{S}{c^2}.$$

For a sinusoidal wave, the average value of E^2 is $E_{max}^2/2$; from Equation 23.9, the average momentum density is

$$\frac{p_{av}}{V} = \frac{\epsilon_0 E_{max}^2}{2c} = \frac{S_{av}}{c^2}. \tag{23.12}$$

The momentum p of an electromagnetic wave is a property of the field; it is not associated with the mass of a moving particle in the usual sense.

There is a corresponding momentum *flow rate,* equal to the momentum per unit volume (Equation 23.12) multiplied by the wave speed c. For a sinusoidal wave, the average momentum flow Δp in a time interval Δt, per unit area A, is

$$\frac{1}{A}\frac{\Delta p}{\Delta t} = \frac{1}{2}\epsilon_0 E^2 = \left(\frac{S_{av}}{c^2}\right)c = \frac{S_{av}}{c} = \frac{I}{c}. \quad \text{(sinusoidal wave)} \quad (23.13)$$

This is the average momentum transferred per unit surface area per unit time.

This momentum transfer is responsible for the phenomenon of **radiation pressure.** When an electromagnetic wave is completely absorbed by a surface perpendicular to the direction of propagation of the wave, the rate of change of momentum with respect to time equals the *force* on the surface. Thus, the average force per unit area, or, simply, the pressure, is equal to I/c. If the wave is totally reflected, the change in momentum is twice as great, and the pressure is $2I/c$. For example, the value of I (or S_{av}) for direct sunlight before it passes through the earth's atmosphere is about 1.4 kW/m². The corresponding radiation pressure on a completely absorbing surface is

$$\frac{I}{c} = \frac{1.4 \times 10^3 \text{ W/m}^2}{3.0 \times 10^8 \text{ m/s}} = 4.7 \times 10^{-6} \text{ Pa}.$$

The average radiation pressure on a totally *reflecting* surface is twice this, $2I/c$, or 9.4×10^{-6} Pa. These are very small pressures, on the order of 10^{-10} atmosphere, but they can be measured with sensitive instruments.

EXAMPLE 23.5 Solar sails

Suppose a spacecraft with a mass of 2.50×10^4 kg has a solar sail made of perfectly reflective aluminized Kapton® film with an area of 2.59×10^6 m² (about 1 square mile). If the spacecraft is launched into earth orbit and then deploys its sail at right angles to the sunlight, what is the acceleration due to sunlight? (At the earth's distance from the sun, the pressure exerted by sunlight on an absorbing surface is 4.70×10^{-6} Pa.)

SOLUTION

SET UP We sketch the situation in Figure 23.8. Because our sail is perfectly reflective, the pressure exerted on it is twice the value given in the statement of the problem: $p = 9.40 \times 10^{-6}$ Pa. Before the sail is deployed, the spacecraft's net radial velocity relative to the sun is zero.

SOLVE The radiation pressure p is the magnitude of force per unit area, so we start by finding the magnitude F of force exerted on the sail:

$$F = pA = (9.40 \times 10^{-6} \text{ N/m}^2)(2.59 \times 10^6 \text{ m}^2) = 24.3 \text{ N}.$$

We now find the magnitude a of the spacecraft's acceleration due to the radiation pressure:

$$a = F/m = (24.3 \text{ N})/(2.50 \times 10^4 \text{ kg}) = 9.72 \times 10^{-4} \text{ m/s}^2.$$

REFLECT For this spacecraft, a square mile of sail provides an acceleration of only about $10^{-4}g$, small compared with the accelerations provided by chemical rockets. However, rockets burn out; sunlight keeps pushing. Even on its first day, the craft travels

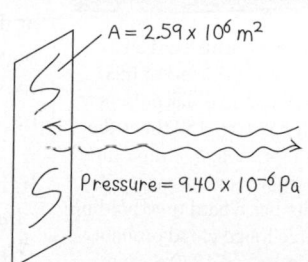

▲ FIGURE 23.8 Our diagram for this problem.

more than 3000 km in the radial direction, and on day 12 its radial speed passes 1 km/s.

Practice Problem: A communications satellite has solar-energy collecting panels with a total area of 4.0 m². What is the average magnitude of total force on these panels associated with radiation pressure, if the radiation is completely absorbed? *Answer:* $F = 1.9 \times 10^{-5}$ N.

Dust tail: Consists of fine dust accelerated by light pressure. It points away from the sun, but curves slightly because light accelerates the dust particles only gradually.

Ion tail: Consists of gas molecules ionized by the sun's ultraviolet light; it is quickly entrained by the sun's "wind" of charged particles and points straight away from the sun.

▲ **FIGURE 23.9** A comet actually has two tails. Both point away from the sun, but one is accelerated quickly by electric interactions and the other more slowly by light pressure.

The pressure of the sun's radiation is partially responsible for pushing the tail of a comet away from the sun (Figure 23.9). Also, while stars the size of our sun are supported against gravitational collapse mainly by the pressure of their hot gas, for some massive stars, radiation pressure dominates, and gravitational collapse of the star is prevented mainly by the light radiating outward from its core. Figure 23.10 shows an extreme example.

23.6 Nature of Light

The remainder of this chapter is devoted to optics. We'll lay some of the foundation needed for the study of many recent developments in this area of physics, including optical fibers, holograms, optical computers, and new techniques in medical imaging. We begin with a study of the laws of reflection and refraction and the concepts of dispersion, polarization, and scattering of light. Along the way, we'll compare the various possible descriptions of light in terms of *rays* and *waves,* and we'll look at Huygens's principle, an important connecting link between these two viewpoints.

Until the time of Isaac Newton (1642–1727), most scientists thought that light consisted of streams of particles (called *corpuscles*) emitted by visible objects. Galileo and others tried (unsuccessfully) to measure the speed of light. Around 1665, evidence of *wave* properties of light began to emerge. By the early 19th century, evidence that light is a wave had grown very persuasive. The picture of light as an electromagnetic wave isn't the whole story, however. Several effects associated with the emission and absorption of light reveal that it also has a particle aspect, in that the energy carried by light waves is packaged in discrete bundles called *photons* or *quanta.* These apparently contradictory wave and particle properties have been reconciled since 1930 with the development of quantum electrodynamics, a comprehensive theory that includes *both* wave and particle properties. The *propagation* of light is best described by a wave model, but understanding emission and absorption by atoms and nuclei requires a particle approach.

The fundamental sources of all electromagnetic radiation are electric charges in accelerated motion. All objects emit electromagnetic radiation as a result of thermal motion of their molecules; this radiation, called *thermal radiation,* is a

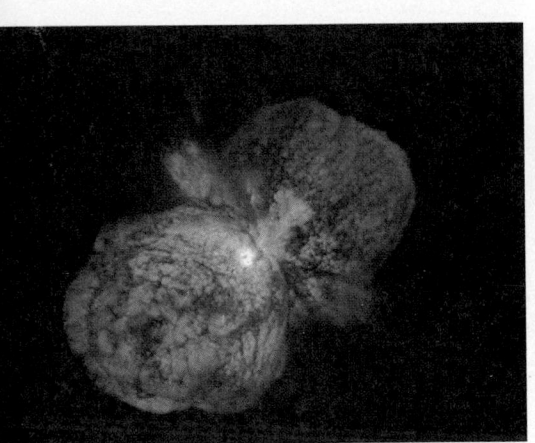

▲ **FIGURE 23.10 Massive star Eta Carinae.** The blue-white monster at the heart of this cloud may be the most massive star presently alive in our galaxy. It has 100–150 times the mass of our sun, but it is about 4 *million* times more luminous, and light pressure makes it very unstable. Gravity has a hard time holding it together. The strange lobed cloud probably dates from an episode around 1840 during which this star temporarily became the second brightest in earth's sky. During such flare-ups, light blows vast quantities of material off the star's surface. Even when relatively quiescent, this star loses matter at a high rate owing to light pressure. Indeed, light pressure limits how massive a star can be: A star's luminosity depends on its mass, but luminosity increases much faster than mass. Eta Carinae is probably close to the limit of stellar stability.

▶ **FIGURE 23.11** We use incandescent bulbs because of the visible light the hot filament emits. However, most of the filament's radiation is in the infrared. That is why incandescent bulbs are hot, and it is also why they are energy inefficient.

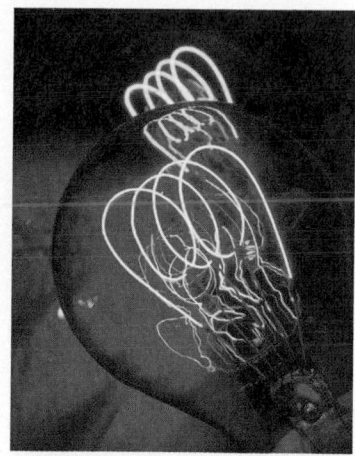

mixture of different wavelengths. At sufficiently high temperatures, all matter emits enough visible light to be self-luminous; a very hot body appears "red hot" or even "white hot." Thus, hot matter in any form is a source of light. Familiar examples are a candle flame (hot gas), hot coals in a campfire, the coils in an electric room heater, and an incandescent lamp filament (which usually operates at a temperature of about 3000°C) (Figure 23.11).

Light is also produced during electrical discharges through ionized gases. The bluish light of mercury-arc lamps, the orange-yellow of sodium-vapor lamps, and the various colors of "neon" signs are familiar. A variation of the mercury-arc lamp is the *fluorescent* lamp. This light source uses a material called a *phosphor* to convert the ultraviolet radiation from a mercury arc into visible light. This conversion makes fluorescent lamps more efficient than incandescent lamps in converting electrical energy into light.

A special light source that has attained prominence in the last 50 years is the *laser,* which can produce a very narrow beam of enormously intense radiation. High-intensity lasers are used to cut steel, fuse high-melting-point materials, carry out microsurgery, and in many other applications. A significant characteristic of laser light is that it is much more nearly *monochromatic,* or single frequency, than light from any other source. (Figure 23.12)

The speed of light in vacuum is a fundamental constant of nature. As we discussed in Section 1.3, the speed of light in vacuum is *defined* to be precisely 299,792,458 m/s, and 1 meter is defined to be the distance traveled by light in vacuum in a time of $1/299,792,458$ s. The second is defined by the cesium clock, which can measure time intervals with a precision of one part in 10^{13}. If future work results in greater precision in measuring the speed of light, the value just cited won't change, but a small adjustment will be made in the definition of the meter.

Wave Fronts

We often use the concept of a **wave front** to describe wave propagation. We define a wave front as *the locus of all adjacent points at which the phase of vibration of the wave is the same.* That is, at any instant, all points on a wave front are at the same part of the cycle of their periodic variation. During wave propagation, the wave fronts all move with the same speed in the direction of propagation of the wave.

A familiar example of a wave front is a crest of a water wave. When we drop a pebble in a calm pool, the expanding circles formed by the wave crests are wave fronts. Similarly, when sound waves spread out in still air from a pointlike source, any spherical surface concentric with the source is a wave front, as shown in Figure 23.13. The "pressure crests"—the surfaces over which the pressure is maximum—form sets of expanding spheres as the wave travels outward from the source. In diagrams of wave motion, we usually draw only parts of a few wave fronts, often choosing consecutive wave fronts that have the same phase, such as crests of a water wave. These consecutive wave crests are separated from each other by one wavelength.

For a light wave (or any other electromagnetic wave), the quantity that corresponds to the displacement of the surface in a water wave or the pressure in a sound wave is the electric or magnetic field. We'll often use diagrams that show the shapes of the wave fronts or their cross sections. For example, when electromagnetic waves are radiated by a small light source, we can represent the wave fronts as spherical surfaces concentric with the source or, as in Figure 23.14a, by

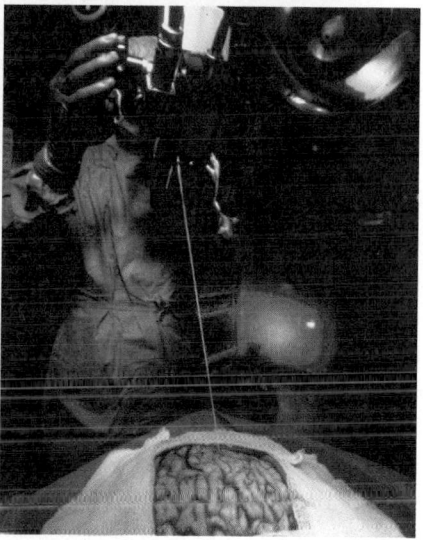

▲ **FIGURE 23.12** A laser being used for brain surgery. Lasers can be used as ultraprecise, bloodless "scalpels" to reach and remove tumors with minimal damage to neighboring healthy tissues.

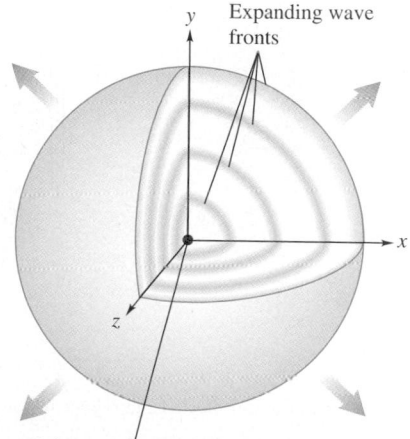

▲ **FIGURE 23.13** Spherical wave fronts, such as those from a point source of sound, spread out uniformly in all directions (provided that the medium is uniform and isotropic).

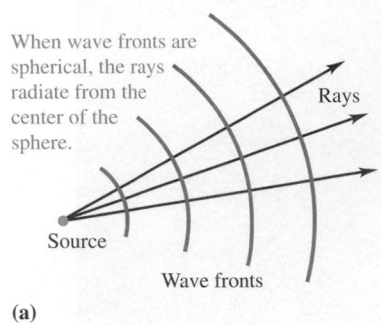

When wave fronts are spherical, the rays radiate from the center of the sphere.

Rays

Source

Wave fronts

(a)

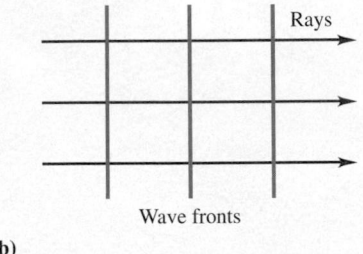

When wave fronts are planar, the rays are perpendicular to the wave fronts and parallel to each other.

Rays

Wave fronts

(b)

▲ **FIGURE 23.14** Spherical and planar wave fronts and rays.

the intersections of these surfaces with the plane of the diagram. Far away from the source, where the radii of the spheres have become very large, a section of a spherical surface can be considered as a plane, and we have a *plane* wave (Figure 23.14b).

It's often convenient to represent a light wave by **rays** rather than by wave fronts. Rays were used to describe light long before its wave nature was firmly established, and in a particle theory of light, rays are the paths of the particles. From the wave viewpoint, *a ray is an imaginary line along the direction of travel of the wave.* In Figure 23.14a, the rays are the radii of the spherical wave fronts; in Figure 23.14b, they are straight lines perpendicular to the wave fronts. When waves travel in a homogeneous, isotropic material (a material with the same properties in all of its regions and in all directions), the rays are always straight lines normal to the wave fronts. At a boundary surface between two materials, such as the surface of a glass plate in air, the wave speed and the direction of a ray usually change, but the ray segments in each material (the air and the glass) are straight lines.

In the remainder of this chapter and in the next three, we'll have many opportunities to see the interplay among the ray, wave, and particle descriptions of light. The branch of optics for which the ray description is adequate is called **geometric optics** (Chapters 24 and 25); the branch dealing specifically with wave behavior is called **physical optics** (Chapter 26).

23.7 Reflection and Refraction

In this section, we'll explore the basic elements of the *ray* model of light. When a light wave strikes a smooth interface (a surface separating two transparent materials, such as air and glass or water and glass), the wave is, in general, partly reflected and partly *refracted* (transmitted) into the second material, as shown in Figure 23.15a. For example, when you look into a store window from the street

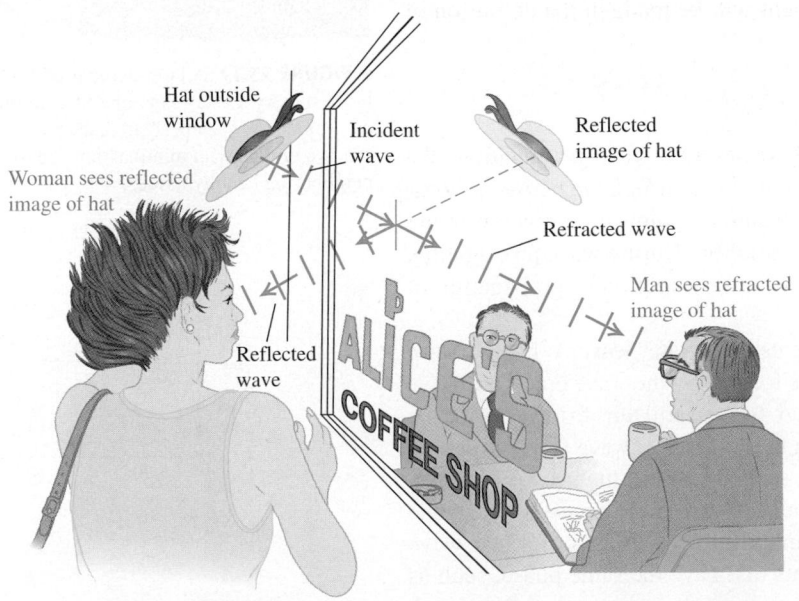

Hat outside window

Incident wave

Reflected image of hat

Woman sees reflected image of hat

Refracted wave

Man sees refracted image of hat

Reflected wave

ALICE'S COFFEE SHOP

(a) Plane waves reflected and refracted from a window.

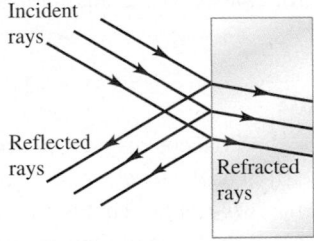

Incident rays

Reflected rays

Refracted rays

(b) The waves in the outside air and glass represented by rays.

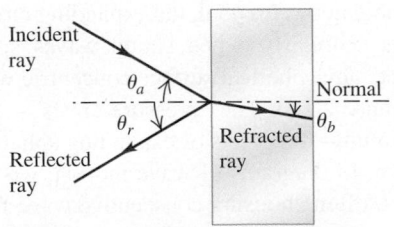

Incident ray

θ_a

θ_r

Normal

θ_b

Refracted ray

Reflected ray

(c) The representation simplified to show just one set of rays.

▲ **FIGURE 23.15** A plane wave is in part reflected and in part refracted at the boundary between two media (in this case, air and glass).

and see a reflection of the street scene, a person inside the store can look out *through* the window at the same scene, as light reaches him by refraction.

The segments of plane waves shown in Figure 23.15b can be represented by bundles of rays forming *beams* of light. For simplicity, we often draw only one ray in each beam (Figure 23.15c). Representing these waves in terms of rays is the basis of *geometric optics*. We begin our study with the behavior of an individual ray.

We describe the directions of the incident, reflected, and refracted rays at a smooth interface between two optical materials in terms of the angles they make with the *normal* to the surface at the point of incidence, as shown in Figure 23.15c. If the interface is rough, both the transmitted light and the reflected light are scattered in various directions, and there is no single angle of transmission or reflection. Reflection at a definite angle from a very smooth surface is called *specular reflection;* scattered reflection from a rough surface is called *diffuse reflection*. This distinction is illustrated in Figure 23.16. Specular reflection also occurs at a very smooth opaque surface, such as one made of highly polished metal or plastic.

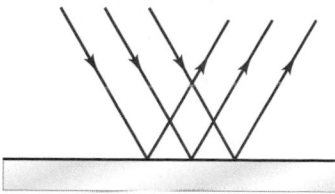

(a) Specular reflection

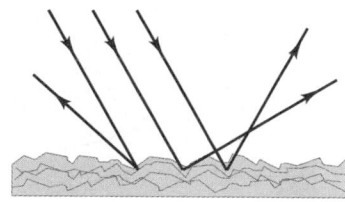

(b) Diffuse reflection

▲ **FIGURE 23.16** Two types of reflection.

Conceptual Analysis 23.2

Laser beam demonstration

When a laser beam is aimed at the wall of a lecture hall, every student in the class observes a red dot on the wall. What can you infer about the direction of the beam as it leaves the wall?

A. The beam is reflected in all directions, an example of diffuse reflection.
B. The beam is reflected at a definite angle, an example of specular reflection.
C. Nothing can be inferred from this experiment. Other results are needed.

SOLUTION The fact that *everyone* in the class can see the dot means that the beam is reflected diffusely, in all directions. If the beam were pointed at a mirror, it would reflect in only one direction and be seen by only one person (or nobody).

The **index of refraction** of an optical material (also called the *refractive index*), denoted by n, plays a central role in geometric optics.

Definition of index of refraction

The index of refraction of an optical material, denoted as n, is the ratio of the speed of light in vacuum (c) to the speed of light in the material (v):

$$n = \frac{c}{v}. \tag{23.14}$$

Light always travels *more slowly* in a material than in vacuum, so n for any material is always greater than one. For vacuum, $n - 1$ by definition.

Experimental studies of the directions of the incident, reflected, and refracted rays at an interface between two optical materials lead to the following conclusions (see Figure 23.17):

Principles of geometric optics

1. **The incident, reflected, and refracted rays, and the normal to the surface, all lie in the same plane.** If the incident ray is in the plane of the diagram and the boundary surface between the two materials is perpendicular to this plane, then the reflected and refracted rays are in the plane of the diagram.

1. The incident, reflected, and refracted rays and the normal to the surface all lie in the same plane.
 Angles θ_a, θ_b, and θ_r are measured *from the normal*.

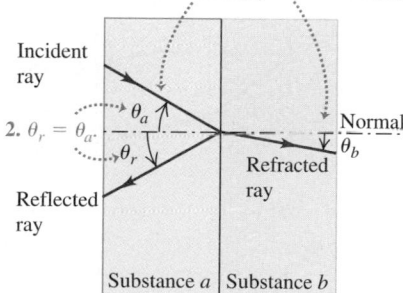

2. $\theta_r = \theta_a$

3. When a monochromatic light ray crosses the interface between two given substances a and b, the angles θ_a and θ_b are related to the indexes of refraction of a and b by

$$\frac{\sin\theta_a}{\sin\theta_b} = \frac{n_b}{n_a}$$

▲ **FIGURE 23.17** The principles of geometric optics.

A ray entering a material of *larger* index of refraction bends *toward* the normal.

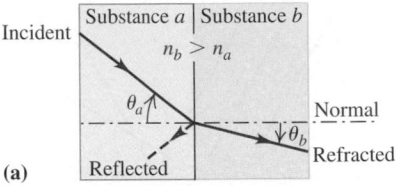

(a)

A ray oriented perpendicular to the surface does not bend, regardless of the materials.

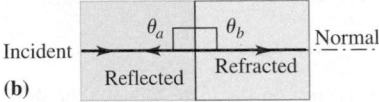

(b)

A ray entering a material of *smaller* index of refraction bends *away from* the normal.

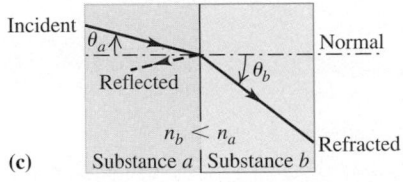

(c)

▲ **FIGURE 23.18** Refraction on crossing an interface to a material of larger or smaller index of refraction.

The path of a refracted ray is reversible (the same from either direction).

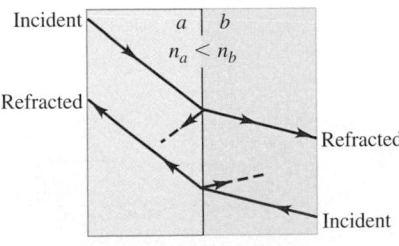

The same is true for a reflected ray.

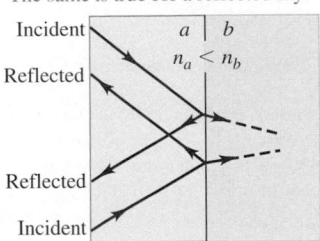

▲ **FIGURE 23.19** Reversibility of refraction and reflection.

2. **The angle of reflection θ_r is equal to the angle of incidence θ_a for all wavelengths and for any pair of substances.** That is, in Figure 23.17,

$$\theta_r = \theta_a. \tag{23.15}$$

This relationship, together with the fact that the incident and reflected rays and the normal all lie in the same plane, is called the **law of reflection.**

3. For monochromatic light and for a given pair of substances a and b on opposite sides of the interface, **the ratio of the sines of the angles θ_a and θ_b, where both angles are measured from the normal to the surface, is equal to the inverse ratio of the two indexes of refraction:**

$$\frac{\sin\theta_a}{\sin\theta_b} = \frac{n_b}{n_a}, \qquad \text{or} \qquad n_a\sin\theta_a = n_b\sin\theta_b. \tag{23.16}$$

These experimental results, together with the fact that the incident and refracted rays and the normal to the surface all lie in the same plane (with the incident and refracted rays always on opposite sides of the normal), is called the **law of refraction,** or **Snell's law,** after Willebrord Snell (1591–1626), Dutch mathematician and physicist. There is some doubt that Snell actually discovered the law named after him. The discovery that $n = c/v$ came much later.

Equation 23.16 shows that when a ray passes from one material (a) into another material (b) having a larger index of refraction and a smaller wave speed ($n_b > n_a$), the angle θ_b with the normal is *smaller* in the second material than the angle θ_a in the first, and the ray is bent *toward* the normal (Figure 23.18a). This is the case in Figure 23.15, where light passes from air into glass. When the second index is *less than* the first ($n_b < n_a$), the ray is bent *away from* the normal (Figure 23.18c). The index of refraction of vacuum is 1, by definition. When a ray passes from vacuum into a material, it is always bent *toward* the normal; when passing from a material into vacuum, it is always bent *away from* the normal. When the incident ray is perpendicular to the interface, $\theta_a = 0$ and $\sin\theta_a = 0$. In this special case, $\sin\theta_b = 0, \theta_b = 0$, and the transmitted ray is not bent at all (Figure 23.18b).

When a ray of light approaches the interface from the opposite side (Figure 23.19), there are again reflected and refracted rays; these two rays—the incident ray and the normal to the surface—again lie in the same plane. The laws of reflection and refraction apply regardless of whether the incident ray is in material a or material b in the figure. The path of a refracted ray is *reversible:* The ray follows the same path when going from b to a as when going from a to b. The path of a ray *reflected* from any surface is also reversible.

The *intensities* of the reflected and refracted rays depend on the angle of incidence, the two indexes of refraction, and the polarization of the incident ray. For unpolarized light, the fraction reflected is smallest at *normal* incidence ($0°$), where it is about 4% for an air–glass interface, and it increases with increasing angle of incidence up to 100% at grazing incidence, when $\theta_a = 90°$.

Most glasses used in optical instruments have indexes of refraction between about 1.5 and 2.0. A few substances have larger indexes; two examples are diamond, with 2.42, and rutile (a crystalline form of titanium dioxide), with 2.62. The index of refraction depends not only on the substance, but also on the wavelength of the light. The dependence on wavelength is called *dispersion;* we'll discuss it in Section 23.9. Indexes of refraction for several solids and liquids are given in Table 23.1.

The index of refraction of *air* at standard temperature and pressure is about 1.0003, and we will usually take it to be exactly 1. The index of refraction of a gas increases in proportion to its density.

TABLE 23.1 **Index of refraction for yellow sodium light** ($\lambda_0 = 589$ nm)

Substance	Index of refraction, n	Substance	Index of refraction, n
Solids		Medium flint	1.62
Ice (H_2O)	1.309	Dense flint	1.66
Fluorite (CaF_2)	1.434	Lanthanum flint	1.80
Polystyrene	1.49	*Liquids at 20°C*	
Rock salt (NaCl)	1.544	Methanol (CH_3OH)	1.329
Quartz (SiO_2)	1.544	Water (H_2O)	1.333
Zircon ($ZrO_2 \cdot SiO_2$)	1.923	Ethanol (C_2H_5OH)	1.36
Fabulite ($SrTiO_3$)	2.409	Carbon tetrachloride (CCl_4)	1.460
Diamond (C)	2.417	Turpentine	1.472
Rutile (TiO_2)	2.62	Glycerine	1.473
Glasses (typical values)		Benzene	1.501
Crown	1.52	Carbon disulfide (CS_2)	1.628
Light flint	1.58		

When light passes from one material into another, the frequency f of the wave doesn't change. The boundary surface cannot create or destroy waves; the number arriving per unit time must equal the number leaving per unit time; otherwise, incident and transmitted waves couldn't have a definite phase relationship. In any material, $v = \lambda f$. Because f is the same in any material as in vacuum and v is always less than the wave speed c in vacuum by the factor n, λ is also correspondingly reduced. Thus, the wavelength λ of the light in a material is *less than* its wavelength λ_0 in vacuum by the factor n. That is,

$$\lambda = \frac{\lambda_0}{n}. \qquad (23.17)$$

Conceptual Analysis 23.3

Properties of a refracted wave

A monochromatic beam of light passes from air into a block of clear plastic. Which of the following properties may differ between the part of the beam in the air and the part in the plastic?

A. Frequency.
B. Wavelength.
C. Speed.

SOLUTION As we've noted, the frequency cannot change when the wave crosses an interface, because that would necessitate creating or destroying waves at the interface. The speed of light is less in any material than that in vacuum and may differ between different materials. (We expect it to differ between air and plastic.) Because wavelength depends on frequency and speed, a change in speed requires a corresponding change in wavelength.

Conceptual Analysis 23.4

Propagating from air to glass

Figure 23.20 shows a light ray passing from air into glass. Which of the choices A–D represents the ray within the glass?

SOLUTION Glass has a greater index of refraction than air, so we expect the refracted ray in the glass to bend toward the normal. Ray A bends *away from* the normal; this is the result we would get if the index of refraction of glass were less than that of air. Ray B doesn't bend at all; this would be the result if glass and air had exactly the same index of refraction. Ray C is a correct choice. What about ray D? For this ray, $\sin\theta_{glass} = 0$, so it doesn't satisfy Snell's law (Equation 23.16) $n_{air}\sin\theta_{air} = n_{glass}\sin\theta_{glass}$.

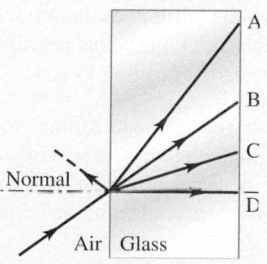

▲ **FIGURE 23.20**

PROBLEM-SOLVING STRATEGY 23.2 **Reflection and refraction**

SET UP

1. In geometric optics problems involving rays and angles, *always* start by drawing a large, neat diagram. Label all known angles and indexes of refraction.
2. Don't forget that, by convention, we always measure the angles of incidence, reflection, and refraction from the *normal* to the surface, *never* from the surface itself.

SOLVE

3. You'll often have to use some simple geometry or trigonometry in working out angular relations: The sum of the interior angles in a triangle is 180° and so on. (See Appendix A.6 for a review.) It often helps to think through the problem, asking yourself, "What information am I given?" "What do I need to know in order to find this angle?" or "What other angles or other quantities can I compute, using the information given in the problem?"

REFLECT

4. Refracted light is always bent toward the normal when the second index is greater than the first, away from the normal when the second index is less than the first. Check whether your results are consistent with this rule.

EXAMPLE 23.6 **A fishpond**

You kneel beside the fishpond in your backyard and look at one of the fish. You see it by sunlight that reflects off the fish and refracts at the water–air interface. If the light from the fish to your eye strikes the water–air interface at an angle of 60.0° to the interface, what is its angle of refraction of the ray in the air?

SOLUTION

SET UP Figure 23.21 shows our sketch. We take the water as medium a and the air as medium b; from Table 23.1, $n_a = 1.33$ and $n_b = 1.00$. Note that θ_a is 30.0°, not 60.0°! We use Snell's law to find θ_b.

SOLVE From Snell's law,

$$n_a \sin\theta_a = n_b \sin\theta_b,$$

$$\theta_b = \sin^{-1}\frac{n_a \sin\theta_a}{n_b} = \sin^{-1}\frac{1.33\sin 30.0°}{1.00} = 41.7°.$$

REFLECT Because the difference in refractive index between water and air is substantial, the actual position of the fish is quite different from its apparent position to you.

Practice Problem: You are spearfishing from a boat and eye a large bass swimming below. It is *apparently* at an angle of 40.0° from the normal. At what angle should you aim your spear? *Answer:* $\theta_b = 28.9°$, measured from the normal.

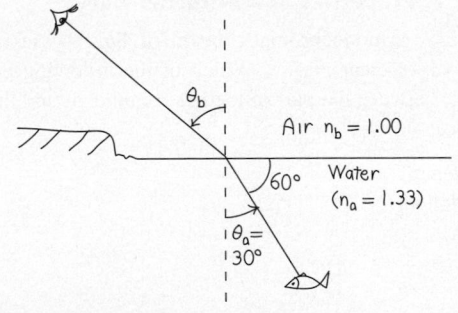

▲ **FIGURE 23.21** Our diagram for this problem.

Conceptual Analysis 23.5

Looking through the glass

Figure 23.22 shows a setup in which two pairs of pins are separated by a block of glass. You observe the pins so that the ray reaching your eye passes through all the pins (Figure 23.22a). Which of the choices in Figure 23.22b represents the *apparent* position of the pins on the far side of the glass block?

SOLUTION The actual position of the pins below the glass is B. But the eyes can see only the direction of the ray emerging from the top surface of the glass. Nothing about that light ray tells you that it refracted twice as it crossed the two air–glass interfaces. Thus, the apparent position of the pins is A. Notice that the light refracts through the same angle as it crosses from air into glass and then from glass into air. This illustrates the point made in Figure 23.19.

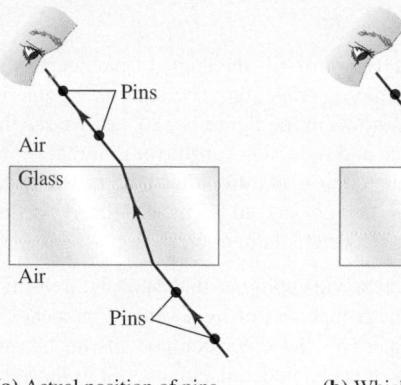

(a) Actual position of pins

(b) Which choice represents the apparent position of the pins?

▲ FIGURE 23.22

EXAMPLE 23.7 **Index of refraction in the eye**

The wavelength of the red light from a helium–neon laser is 633 nm in air, but 474 nm in the jellylike fluid inside your eyeball, called the *vitreous humor* (Figure 23.23). Calculate the index of refraction of the vitreous humor and the speed and frequency of light passing through it.

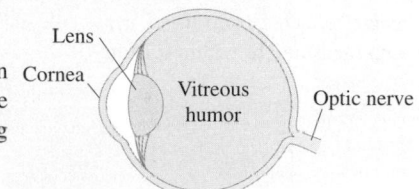

▲ FIGURE 23.23

SOLUTION

SET UP AND SOLVE We don't need to sketch this problem, because we're interested only in the index of refraction of a material and the speed of light in it, not in the path a particular ray follows. The index of refraction of air is very close to one, so we assume that the wavelength of the laser light is the same in vacuum as in air. Then the wavelength λ in the material is given by Equation 23.17:

$$n = \frac{\lambda_0}{\lambda} = \frac{633 \text{ nm}}{474 \text{ nm}} = 1.34,$$

which is about the same as for water. Then, to find the speed of light in the vitreous humor, we use Equation 23.14, $n = c/v$:

$$v = \frac{c}{n} = \frac{3.00 \times 10^8 \text{ m/s}}{1.34} = 2.24 \times 10^8 \text{ m/s}.$$

The frequency of the light is

$$f = \frac{v}{\lambda} = \frac{2.24 \times 10^8 \text{ m/s}}{474 \times 10^{-9} \text{ m}} = 4.73 \times 10^{14} \text{ Hz}.$$

REFLECT Note that while the speed and wavelength have different values in air and in the vitreous humor, the *frequency* f_0 in air is the same as the frequency f in the vitreous humor (and the frequency in vacuum):

$$f_0 = \frac{c}{\lambda_0} = \frac{3.00 \times 10^8 \text{ m/s}}{633 \times 10^{-9} \text{ m}} = 4.73 \times 10^{14} \text{ Hz}.$$

This result confirms the general rule that when a light wave passes from one material into another, the frequency doesn't change.

Practice Problem: The newest laser pointers emit green light with a wavelength of $\lambda_0 = 532$ nm in air. What is the wavelength of this light in the vitreous humor of the eyeball? *Answer:* $\lambda = 397$ nm.

EXAMPLE 23.8 **Reflected light rays**

Two mirrors are perpendicular to each other. A ray traveling in a plane perpendicular to both mirrors is reflected from one mirror and then the other, as shown in Figure 23.24. What is the ray's final direction relative to its original direction?

Continued

SOLUTION

SET UP AND SOLVE For mirror 1, the angle of incidence is θ_1, and this equals the angle of reflection. The sum of the interior angles in the triangle shown in the figure is $180°$, so we see that the angles of incidence and reflection for mirror 2 are $90° - \theta_1$. The total change in direction of the ray after both reflections is therefore $(180° - 2\theta_1) + 2\theta_1 = 180°$. That is, the ray's final direction is opposite to its original direction.

REFLECT An alternative viewpoint is that specular reflection reverses the sign of the component of light velocity perpendicular to the surface, but leaves the other components unchanged. We invite you to verify this fact in detail and to use it to show that when a ray of light is successively reflected by three mirrors forming a corner of a cube (a "corner reflector"), its final direction is again opposite to its original direction. The principle of a corner reflector is widely used in taillight lenses and highway signs to improve their nighttime visibility. Apollo astronauts placed arrays of corner reflectors on the moon. By use of laser beams reflected from these arrays, the earth–moon distance has been measured to within 0.15 m.

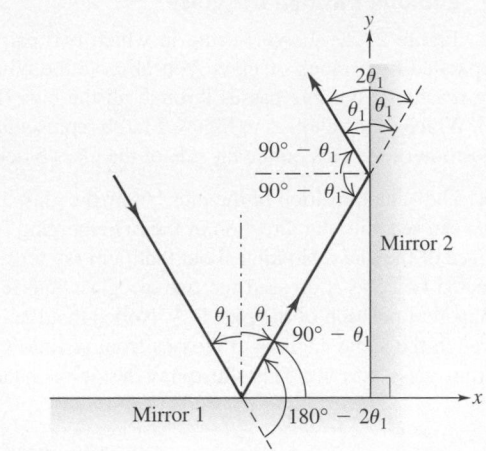

▲ **FIGURE 23.24**

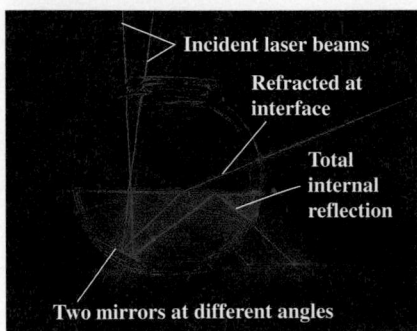

Total internal reflection occurs only if $n_b < n_a$.

At the critical angle of incidence, θ_{crit}, the angle of refraction, $\theta_b = 90°$.

Any ray with $\theta_a > \theta_{\text{crit}}$ shows total internal reflection.

(a) Total internal reflection

Incident laser beams

Refracted at interface

Total internal reflection

Two mirrors at different angles

(b) Total internal reflection demonstrated with a laser, mirrors, and water in a fishbowl

▲ **FIGURE 23.25** Total internal reflection.

23.8 Total Internal Reflection

Figure 23.25a shows several rays diverging from a point source P in a material a with index of refraction n_a. The rays strike the surface of a second material b with index n_b, where $n_b < n_a$. From Snell's law,

$$\sin\theta_b = \frac{n_a}{n_b}\sin\theta_a.$$

Because $n_a/n_b > 1$, $\sin\theta_b$ is larger than $\sin\theta_a$, so the ray is bent *away from* the normal. Thus, there must be some value of θ_a *less than* $90°$ for which Snell's law gives $\sin\theta_b = 1$ and $\theta_b = 90°$. This is shown by ray 3 in the diagram, which emerges just grazing the surface, at an angle of refraction of $90°$.

The angle of incidence for which the refracted ray emerges tangent to the surface is called the **critical angle,** denoted by θ_{crit}. If the angle of incidence is *greater than* the critical angle, then the sine of the angle of refraction, as computed by Snell's law, has to be greater than unity, which is impossible. Hence, for angles of incidence greater than the critical angle, the ray *cannot* pass into the upper material; it is trapped in the lower material and is completely reflected internally at the boundary surface, as shown in Figure 23.25. This situation, called **total internal reflection,** occurs only when a ray is incident on an interface with a second material whose index of refraction is *smaller* than that of the material in which the ray is traveling.

We can find the critical angle for two given materials by setting $\theta_b = 90°$ and $\sin\theta_b = 1$ in Snell's law. We then have the following result:

Total internal reflection

When a ray traveling in a material a with index of refraction n_a reaches an interface with a material b having index n_b, where $n_b < n_a$, it is totally reflected back into material a if the angle incidence is greater than the critical angle given by

$$\sin\theta_{\text{crit}} = \frac{n_b}{n_a}. \tag{23.18}$$

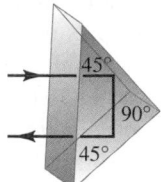

If the incident beam is oriented as shown, total internal reflection occurs on the 45° faces (because, for a glass–air interface, $\theta_{crit} = 41°$).

45°

90°

45°

(a) Total internal reflection in a Porro prism.

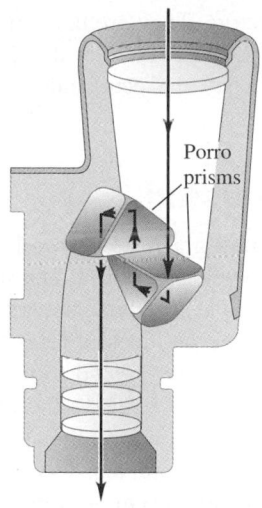

Porro prisms

(b) Binoculars use Porro prisms to reflect the light to each eyepiece.

▲ **FIGURE 23.26** Total internal reflection in Porro prisms.

For a glass–air surface, with $n = 1.52$ for the glass,

$$\sin\theta_{crit} = \frac{1}{1.52} = 0.658, \qquad \theta_{crit} = 41.1°.$$

The fact that this critical angle is slightly less than 45° makes it possible to use a triangular prism with angles of 45°, 45°, and 90° as a totally reflecting surface. As reflectors, totally reflecting prisms have some advantages over metallic surfaces such as ordinary coated-glass mirrors. Light is *totally* reflected by a prism, but no metallic surface reflects 100% of the light incident on it. Also, the reflecting properties are permanent and not affected by tarnishing.

A 45°–45°–90° prism, used as in Figure 23.26a, is called a *Porro prism.* Light enters and leaves at right angles to the hypotenuse and is totally reflected at each of the shorter faces. The total change in direction of the rays is 180°. Binoculars often use combinations of two Porro prisms, as shown in Figure 23.26b.

When a beam of light enters at one end of a transparent rod (Figure 23.27), the light is totally reflected internally and is "trapped" within the rod even if the rod is curved, provided that the curvature is not too great. Such a rod is sometimes called a *light pipe.* A bundle of fine plastic fibers behaves in the same way and has the advantage of being flexible. A bundle may consist of thousands of individual fibers, each on the order of 0.002–0.01 mm in diameter. If the fibers are assembled in the bundle so that the relative positions of the ends are the same (or mirror images) at both ends, the bundle can transmit an image, as shown in Figure 23.28.

Fiber-optic devices have found a wide range of medical applications in instruments called *endoscopes,* which can be inserted directly into the bronchial tubes, the knee joint, the colon, and so on, for direct visual examination. A bundle of fibers can be enclosed in a hypodermic needle for the study of tissues and blood vessels far beneath the skin.

Fiber optics are also widely used in communication systems, where they are used to transmit a modulated laser beam. The number of binary digits that can be transmitted per unit time is proportional to the frequency of the wave. Infrared and visible-light waves have much higher frequencies than radio waves, so a modulated laser beam can transmit an enormous amount of information through a single fiber-optic cable. Another advantage of fiber-optic cables is that they are

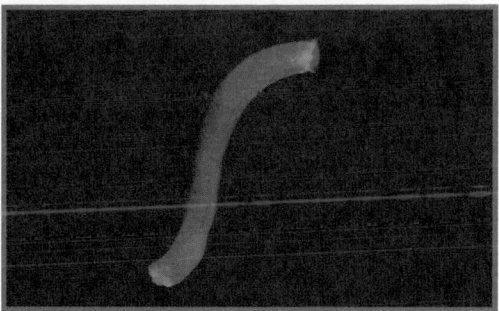

▲ **BIO Application Let the light shine in.** In addition to photosynthesis, plants and algae use light to regulate many aspects of their physiology, including seed germination, stem elongation, and growth direction. These responses are tuned to specific parts of the visible spectrum, and a variety of molecules are used as photoreceptors. Some of these responses take place in seedlings underground. How is the light transmitted to the responding portions? Scientists found that the columns of cells in seedlings act as fibers that can guide light from near the surface of the soil to portions of the seedling farther below. The photograph shows light piping in an isolated, curved stem of an oat seedling.

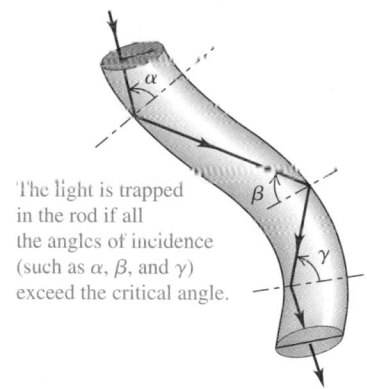

The light is trapped in the rod if all the angles of incidence (such as α, β, and γ) exceed the critical angle.

▲ **FIGURE 23.27** The principle of a light pipe, the basis for fiber optics.

▲ **FIGURE 23.28** An image transmitted by a bundle of optical fibers.

electrical insulators. They are thus immune to electrical interference from lightning and other sources, and they don't allow unwanted currents between a source and a receiver. They are secure and difficult to "bug," but they are also difficult to splice and tap into.

23.9 Dispersion

Ordinarily, white light is a superposition of waves with wavelengths extending throughout the visible spectrum. The speed of light *in vacuum* is the same for all wavelengths, but the speed in a material substance is different for different wavelengths. Therefore, the index of refraction of a material depends on wavelength. The dependence of wave speed and index of refraction on wavelength is called **dispersion.** Figure 23.29 shows the variation of index of refraction with wavelength for a few common optical materials. The value of *n* usually *decreases* with increasing wavelength and thus *increases* with increasing frequency. Hence, light of longer wavelength usually has greater speed in a given material than light of shorter wavelength.

Figure 23.30 shows a ray of white light incident on a prism. The deviation (change in direction) produced by the prism increases with increasing index of refraction and frequency and with decreasing wavelength. Violet light is deviated most and red least, and other colors show intermediate deviations. When it comes out of the prism, the light is spread out into a fan-shaped beam, as shown. The light is said to be *dispersed* into a spectrum. The amount of dispersion depends on the *difference* between the refractive indexes for violet light and for red light. From Figure 23.29, we can see that for a substance such as fluorite, whose refractive index for yellow light is small, the difference between the indexes for red and violet is also small. For silicate flint glass, both the index for yellow light and the difference between extreme indexes are larger. (The values of index of refraction in Table 23.1 are values for a wavelength of 589 nm, near the center of the visible range of wavelengths.)

The brilliance of diamond is due in part to its large dispersion and in part to its unusually large refractive index. Crystals of rutile and of strontium titanate, which can be produced synthetically, have about eight times the dispersion of diamond!

When you experience the beauty of a rainbow, as in Figure 23.31a, you are seeing the combined effects of dispersion and internal reflection. The light comes from behind you and is refracted into many small water droplets in the air. Each ray undergoes internal reflection from the back surface of the droplet and is reflected back to you (Figure 23.31b). Dispersion causes different colors to be refracted preferentially at different angles, so you see the various colors as coming

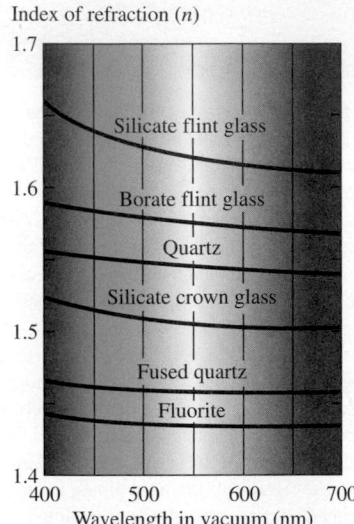

Index of refraction (*n*)

Silicate flint glass

Borate flint glass

Quartz

Silicate crown glass

Fused quartz

Fluorite

Wavelength in vacuum (nm)

▲ **FIGURE 23.29** Variation of index of refraction with wavelength for several materials.

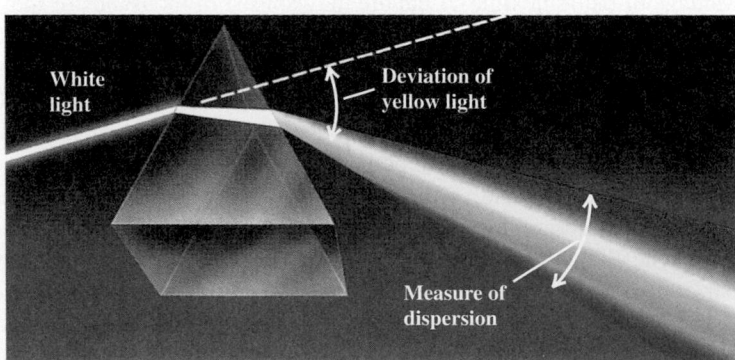

▲ **FIGURE 23.30** Schematic representation of dispersion by a prism.

(a) A double rainbow.

The rays of sunlight that form the primary rainbow refract into the droplets, undergo internal reflection, and refract out.

The two refractions disperse the colors.

Incident white light

Water droplets in cloud

42°

40°

O

Angles exaggerated for clarity. Only a primary rainbow is shown.

P

Observer at P

(b) How refraction and reflection in cloud droplets forms a rainbow. The x-y plane is horizontal, the z axis vertical.

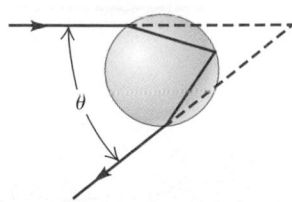

(c) An incoming ray undergoes two refractions and one internal reflection. The angle θ is greater for red light than for violet.

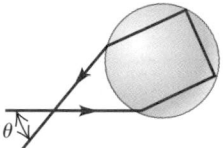

(d) A secondary rainbow is formed by rays that undergo two refractions and two internal reflections; the angle θ is greater for violet light than for red.

▲ **FIGURE 23.31** How refraction and internal reflection in water droplets create a double rainbow.

from different regions of the sky, forming concentric circular arcs (Figure 23.31b). When you see a second, slightly larger rainbow with its colors reversed, you are seeing the results of dispersion and *two* internal reflections (Figure 23.31d).

23.10 Polarization

Polarization occurs with all transverse waves. This chapter is mainly about light, but to introduce basic polarization concepts, let's go back to some of the ideas presented in Chapter 12 about transverse waves on a string. For a string whose equilibrium position is along the x axis, the displacements may be along the y direction, as in Figure 23.32a. In this case, the string always vibrates in the x-y plane. But the displacements might instead be along the z axis, as in Figure 23.32b; then the string vibrates in the x-z plane.

When a wave has only y displacements, we say that it is **linearly polarized** in the y direction; similarly, a wave with only z displacements is linearly polarized in the z direction. For mechanical waves, we can build a **polarizing filter** that permits only waves with a certain polarization direction to pass. In Figure 23.32c, the string can slide vertically in the slot without friction, but no horizontal motion is possible. This filter passes waves polarized in the y direction but blocks those polarized in the z direction.

This same language can be applied to electromagnetic waves, which also have polarization. As we learned in Section 23.2, an electromagnetic wave is a *transverse* wave: The fluctuating electric and magnetic fields are perpendicular to the direction of propagation and to each other. We always define the direction of polarization of an electromagnetic wave to be the direction of the *electric*-field vector, not the magnetic-field vector, because most common electromagnetic-wave detectors (including the human eye) respond to the *electric* forces on electrons in materials, not the magnetic forces.

Polarizing Filters

Polarizing filters can be made for electromagnetic waves; the details of their construction depend on the wavelength. For microwaves with a wavelength of a few centimeters, a grid of closely spaced, parallel conducting wires that are insulated

MasteringPHYSICS

ActivPhysics 16.9: Polarization

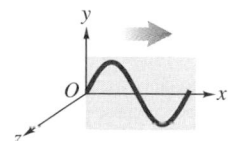

(a) Transverse wave linearly polarized in the y direction

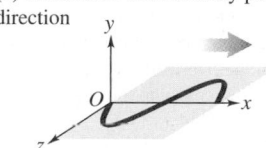

(b) Transverse wave linearly polarized in the z direction

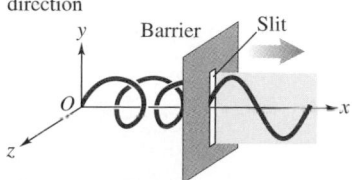

(c) The slit functions as a polarizing filter, passing only motion in the y direction

▲ **FIGURE 23.32** The concept of wave polarization applied to a transverse wave in a string.

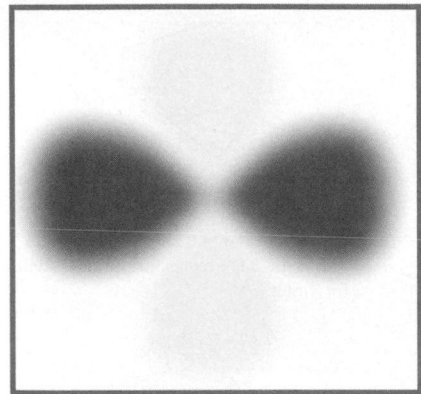

▲ **BIO Application The sky is polarized—and we can see it!** The scattering of sunlight by molecules of the atmosphere produces the intense blue of the clear sky. The shorter (bluish) wavelengths are scattered and thus seen by us while the longer (reddish) wavelengths are less affected and not directed to our eyes. Were it not for this scattering, the sky would appear black, even during the day. This scattering also polarizes the light in any direction perpendicular to the sun's rays. Light scattered back to our eyes from the horizon is only weakly polarized. Insects can detect the polarization of sky light, but this ability is almost completely missing in humans. Looking toward the zenith of the sky, you may with practice observe Haidinger's brush, a yellowish dumbbell-shaped figure flanked by bluer areas. The axis of the yellow figure is perpendicular to the direction of polarization of the light, so it points toward the sun. This phenomenon and many others are explored in an elegant 1954 book by the Belgian-born physicist Marcel Minnaert, *The Nature of Light and Colour in the Open Air*, in which he says of Haidinger's brush, "I can see it particularly clearly in the twilight if I stare at the zenith; the whole sky seems to be covered by a network, as it were, and everywhere I look I see this characteristic figure." He goes on to say, "It is very pleasing to be able to determine the direction of polarization without an instrument in this way and even to obtain an estimate of its degree." Haidinger's brush is an optical response of the human eye and thus cannot be photographed.

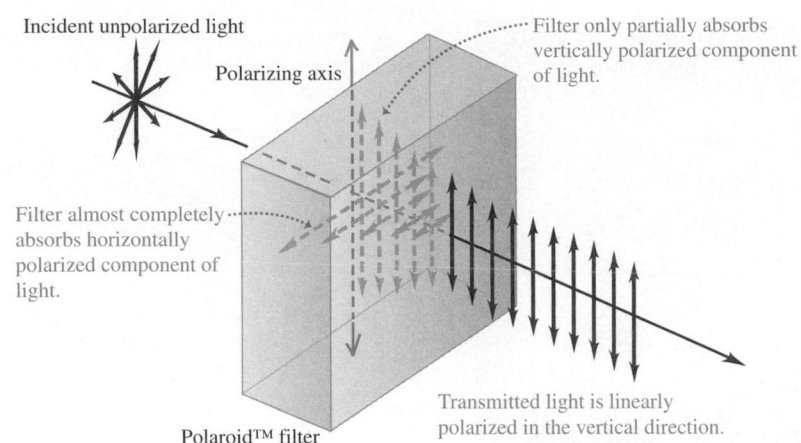

▲ **FIGURE 23.33** How a Polaroid™ filter produces polarized light.

from each other passes waves whose $\vec{E}$ fields are perpendicular to the wires, but not those with $\vec{E}$ fields parallel to the wires. The most common polarizing filter for light is a material known by the trade name Polaroid™, widely used for sunglasses and polarizing filters for camera lenses. This material, developed originally by Edwin H. Land, incorporates substances that exhibit **dichroism,** the selective absorption of one of the polarized components much more strongly than the other (Figure 23.33). A Polaroid™ filter transmits 80% or more of the intensity of waves polarized parallel to a certain axis in the material (called the **polarizing axis**), but only 1% or less of waves polarized perpendicular to this axis.

Waves emitted by radio transmitter antennas are usually linearly polarized. Vertical-rod antennas emit waves that, in a horizontal plane around the antenna, are polarized in the vertical direction (parallel to the antenna). Rooftop TV antennas have horizontal elements in the United States and vertical elements in Great Britain because the transmitted waves have different polarizations.

Light from ordinary sources doesn't have a definite polarization. The "antennas" that radiate light waves are the molecules that make up the sources. The waves emitted by any one molecule may be linearly polarized, like those from a radio antenna. But any actual light source contains a tremendous number of molecules with random orientations, so the light emitted is a random mixture of waves that are linearly polarized in all possible transverse directions.

An ideal polarizing filter, or **polarizer,** passes 100% of the incident light polarized in the direction of the filter's *polarizing axis,* but blocks all light polarized perpendicular to that axis. Such a device is an unattainable idealization, but the concept is useful in clarifying the basic ideas. In the discussion that follows, we'll assume that all polarizing filters are ideal. In Figure 23.34, unpolarized light (a random mixture of all polarization states) is incident on a polarizer in the form of a flat plate. The polarizing axis is represented by the blue line. The $\vec{E}$ vector of the incident wave can be represented in terms of components parallel and perpendicular to the polarizing axis. The polarizer transmits only the components of $\vec{E}$ parallel to that axis. The light emerging from the polarizer is linearly polarized parallel to the polarizing axis.

When we measure the intensity (power per unit area) of the light transmitted through an ideal polarizer, using the photocell in Figure 23.34, we find that it is exactly half that of the incident light, no matter how the polarizing axis is oriented. Here's why: We can resolve the $\vec{E}$ field of the incident wave into a component parallel to the polarizing axis and a component perpendicular to it. Because the incident light is a random mixture of all states of polarization, these two components are, on average, equal. The ideal polarizer transmits only the component parallel to the polarizing axis, so half the incident intensity is transmitted.

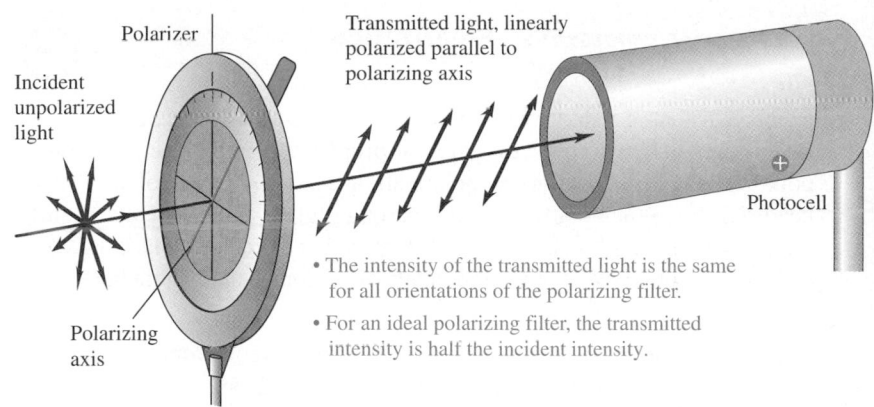

▲ FIGURE 23.34 The effect of a polarizing filter on unpolarized incident light.

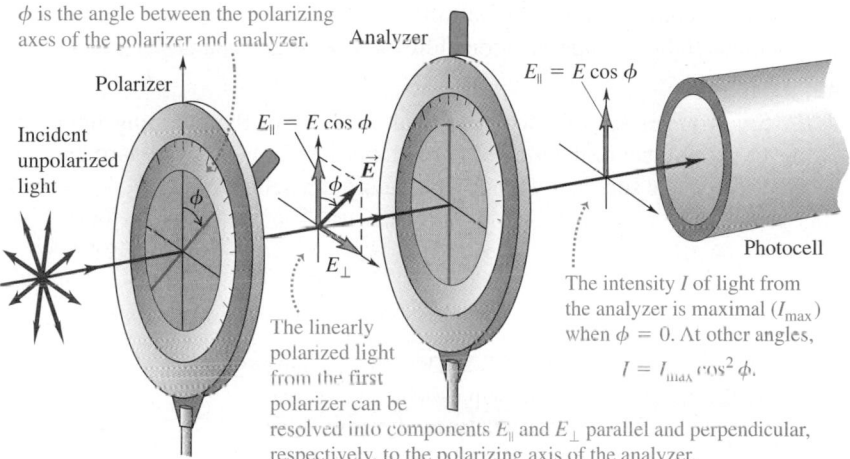

▲ FIGURE 23.35 The effect of passing unpolarized light through two polarizers.

▲ BIO Application Polarized communication. Numerous marine invertebrates have optical arrangements in their eyes that permit the perception of the orientation of polarized light. Some of these animals also can vary the degree of polarization of the light reflected from their bodies as a method of camouflage or communication. The accompanying photographs show a true color image (top) of a cuttlefish (not a fish at all, but an invertebrate related to squids and octopi) and an image in which the degree of polarization of the reflected light is represented by color. Horizontal polarization (in red) is seen in the vertical stripes on the animal. This pattern of polarization changes during mating and response to prey.

Now suppose we insert a second polarizer between the first polarizer and the photocell (Figure 23.35). The polarizing axis of the second polarizer, or *analyzer,* is vertical, and the axis of the first polarizer makes an angle ϕ with the vertical. That is, ϕ is the angle between the polarizing axes of the two polarizers. We can resolve the linearly polarized light transmitted by the first polarizer into two components, as shown in Figure 23.35, one parallel and the other perpendicular to the vertical axis of the analyzer. Only the parallel component, with amplitude $E \cos\phi$, is transmitted by the analyzer. The transmitted intensity is greatest when $\phi = 0$; it is zero when $\phi = 90°$—that is, when polarizer and analyzer are *crossed* (Figure 23.36).

To find the transmitted intensity at intermediate angles, we recall from our discussion in Section 23.5 that the intensity of an electromagnetic wave is proportional to the *square* of the amplitude of the wave. The ratio of the transmitted to the incident *amplitude* is $\cos\phi$, so the ratio of the transmitted to the incident *intensity* is $\cos^2\phi$. Thus, we obtain the following result:

Light transmitted by polarizing filter
When linearly polarized light strikes a polarizing filter with its axis at an angle ϕ to the direction of polarization, the intensity of the transmitted light is

$$I = I_{max}\cos^2\phi, \tag{23.19}$$

where I_{max} is the maximum intensity of light transmitted (at $\phi = 0$) and I is the amount transmitted at angle ϕ. This relationship, discovered experimentally by Etienne Louis Malus in 1809, is called **Malus's law.**

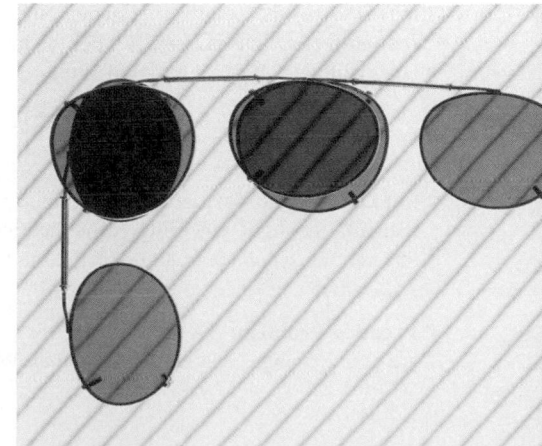

▲ FIGURE 23.36 Polarizing sunglasses oriented with their polarizing axes parallel (top center) and perpendicular (top left). The crossed polarizers in the top left transmit no light.

PROBLEM-SOLVING STRATEGY 23.3 **Polarization**

SET UP

1. Remember that in light waves (and all other electromagnetic waves), the $\vec{E}$ field is perpendicular to the direction of propagation and is the direction of polarization. The direction of polarization can be thought of as a two-headed arrow. When working with polarizing filters, you are really dealing with components of $\vec{E}$ parallel and perpendicular to the polarizing axis. Everything you know about components of vectors is applicable here.

SOLVE

2. The intensity (average power per unit area) of a wave is proportional to the *square* of its amplitude. If you find that two waves differ in amplitude by a certain factor, their intensities differ by the square of that factor.

3. Unpolarized light is a random mixture of all possible polarization states, so, on average, it has equal components in any two perpendicular directions, with each component having half the total light intensity. Partially linearly polarized light is a superposition of linearly polarized and unpolarized light.

REFLECT

4. In any arrangement of filters, the total intensity of the outgoing light can never exceed that of the incoming light, because of conservation of energy. Check to make sure that your results satisfy this requirement.

EXAMPLE 23.9 **Linear polarization**

You and a friend each have a pair of polarizing sunglasses and decide to test Malus's law by using the light sensors in your physics lab. You orient the sunglasses so that the angle between the polarizing axes of two of the lenses is 30°; then you direct a narrow beam of unpolarized light through both lenses. Relative to the intensity I_0 of the incident unpolarized beam, what intensity do you expect to find after the beam passes through the first lens? After it passes through the second?

SOLUTION

SET UP Figure 23.37 shows our sketch.

SOLVE The intensity of the light transmitted through an ideal polarizer is exactly half that of the unpolarized incident light I_0, regardless of the polarizer's orientation. Therefore, the intensity I_1 after the first lens is $I_1 = I_0/2$. For the intensity after the second lens, with $\phi = 30°$, Malus's law gives

$$I_2 = I_1 \cos^2 30° = \left(\frac{I_0}{2}\right)\cos^2 30° = \left(\frac{I_0}{2}\right)\left(\frac{\sqrt{3}}{2}\right)^2 = \frac{3}{8}I_0.$$

REFLECT If the polarizing axes of the two lenses are parallel, then *all* the light emerging from the first lens is transmitted through the second lens. In this case, the transmitted intensity is $I_0/2$, which is also equal to $(I_0/2)\cos^2 0°$.

Practice Problem: Now you rotate the polarizing axes so they are at a 45° angle relative to each other. What is the intensity of

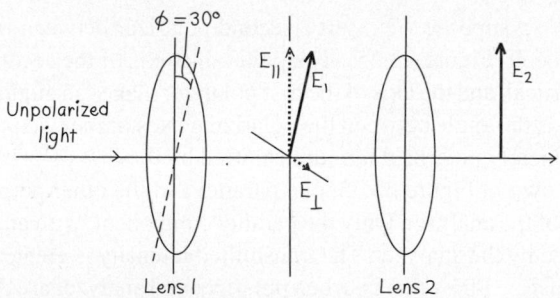

▲ **FIGURE 23.37** Our sketch for this problem.

the light emerging from the second polarizing lens? *Answer:* $I_2 = I_0/4$.

Polarization by Reflection

Unpolarized light can be partially polarized by *reflection*. When unpolarized light strikes a reflecting surface between two optical materials, preferential reflection occurs for those waves in which the electric-field vector is parallel to the reflect-

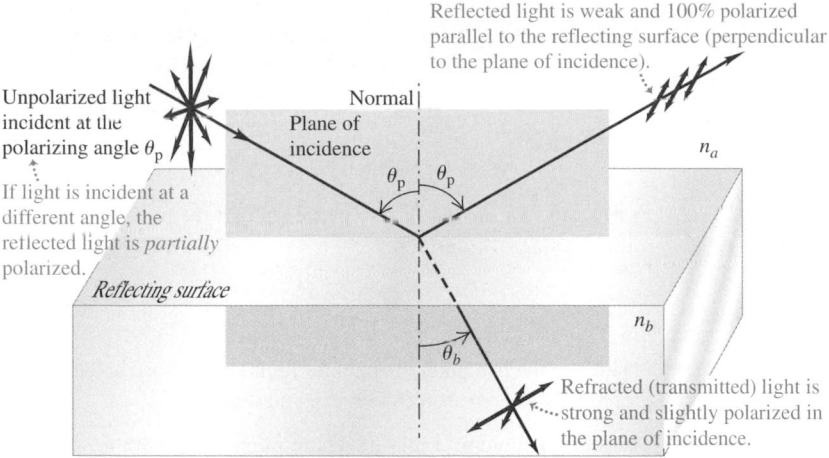

▲ **FIGURE 23.38** When light is incident at the polarizing angle, the reflected light is 100% linearly polarized.

ing surface. In Figure 23.38, the plane containing the incident and reflected rays and the normal to the surface is called the **plane of incidence.** At one particular angle of incidence, called the **polarizing angle** θ_p, only the light for which the $\vec{E}$ vector is perpendicular to the plane of incidence (and parallel to the reflecting surface) is reflected. The reflected light is therefore linearly polarized perpendicular to the plane of incidence (parallel to the reflecting surface), as shown in Figure 23.38.

When light is incident at the polarizing angle θ_p, *none* of the $\vec{E}$-field component *parallel* to the plane of incidence is reflected; this component is transmitted 100% in the *refracted* beam. So the *reflected* light is *completely* polarized. The *refracted* light is a mixture of the component parallel to the plane of incidence, all of which is refracted, and the remainder of the perpendicular component; it is therefore *partially* polarized.

In 1812, Sir David Brewster noticed that when the angle of incidence is equal to the polarizing angle θ_p, the reflected ray and the refracted ray are perpendicular to each other, as shown in Figure 23.39. In this case, the angle of refraction θ_b is equal to $90° - \theta_p$ so $\sin\theta_b = \cos\theta_p$. From the law of refraction, $n_a\sin\theta_p = n_b\sin\theta_b$, so we find that $n_a\sin\theta_p = n_b\cos\theta_p$ and

$$\frac{\sin\theta_p}{\cos\theta_p} = \tan\theta_p = \frac{n_b}{n_a}. \tag{23.20}$$

This relation is known as **Brewster's law.** It states that when the tangent of the angle of incidence equals the ratio of the two indexes of refraction, the reflected light is completely polarized.

Component perpendicular to plane of page

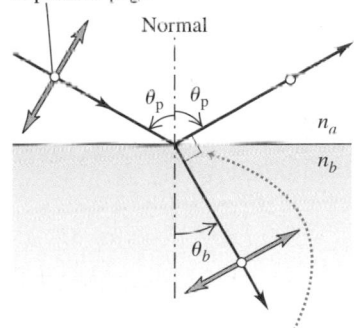

When light strikes a surface at the polarizing angle, the reflected and refracted rays are perpendicular to each other and

$$\tan\theta_p = \frac{n_b}{n_a}$$

▲ **FIGURE 23.39** Brewster's law. This diagram shows a side view of the scene in Figure 23.38.

EXAMPLE 23.10 **Reflection from a swimming pool's surface**

Sunlight reflects off the smooth surface of an unoccupied swimming pool. (a) At what angle of reflection is the light completely polarized? (b) What is the corresponding angle of refraction for the light that is transmitted (refracted) into the water? (c) At night, an underwater floodlight is turned on in the pool. Repeat parts (a) and (b) for rays from the floodlight that strike the smooth surface from below.

Continued

SOLUTION

SET UP Figure 23.40 shows the sketches we draw (one for each situation).

SOLVE Part (a): We're looking for the polarizing angle for light that passes from air into water, so $n_a = 1.00$ (air) and $n_b = 1.33$ (water). From Equation 23.20,

$$\theta_p = \tan^{-1}\frac{n_b}{n_a} = \tan^{-1}\frac{1.33}{1.00} = 53.1°.$$

Part (b): The incident light is at the polarizing angle, so the reflected and refracted rays are perpendicular; hence,

$$\theta_p + \theta_b = 90°,$$
$$\theta_b = 90° - 53.1° = 36.9°.$$

Part (c): Now the light goes from water to air, so $n_a = 1.33$ and $n_b = 1.00$. Again using Equation 23.20, we have

$$\theta_p = \tan^{-1}\frac{n_b}{n_a} = \tan^{-1}\frac{1.00}{1.33} = 36.9°,$$
$$\theta_b = 90° - 36.9° = 53.1°.$$

REFLECT We can check our answer in part (b) by using Snell's law, $n_a \sin\theta_p = n_b \sin\theta_b$, or

$$\sin\theta_b = \frac{n_a \sin\theta_p}{n_b} = \frac{1.00 \sin 53.1°}{1.33} = 0.600, \qquad \theta_b = 36.9°.$$

Note that the two polarizing angles found in parts (a) and (c) add to 90°. This is *not* an accident; can you see why?

Practice Problem: Light travels through water with $n_a = 1.33$ and reflects off a glass surface with $n_b = 1.52$. At what angle of reflection is the light completely polarized? *Answer:* $\theta_p = 48.8°$.

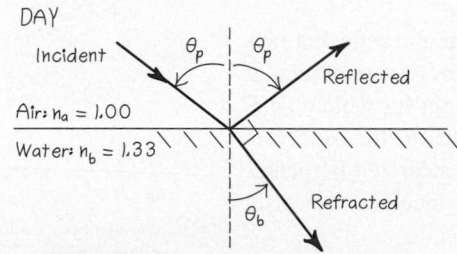 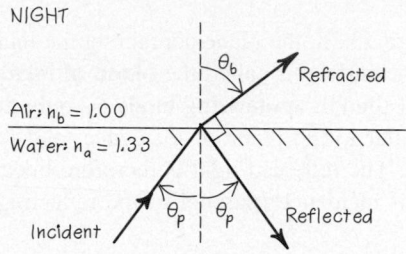

▲ **FIGURE 23.40** Our sketches for this problem.

Polarizing filters are widely used in sunglasses. When sunlight is reflected from a horizontal surface, the reflected light contains a preponderance of light polarized in the horizontal direction. When the reflection occurs at a smooth asphalt road surface or the surface of a lake, it causes unwanted glare. Vision can be improved by eliminating this glare. The manufacturer makes the polarizing axis of the lens material vertical, so very little of the horizontally polarized light is transmitted to the eyes. The glasses also reduce the overall intensity in the transmitted light to somewhat less than 50% of the intensity of the unpolarized incident light.

Photoelasticity

Some optical materials, when placed under mechanical stress, develop the property that their index of refraction is different for different planes of polarization. The result is that the plane of polarization of incident light can be rotated, by an amount that depends on the stress. This effect is the basis of the science of **photoelasticity.** Stresses in girders, boiler plates, gear teeth, and cathedral pillars can be analyzed by constructing a transparent model of the object, usually of a plastic material, subjecting it to stress, and examining it between a polarizer and an analyzer in the crossed position. Very complicated stress distributions can be studied by these optical methods. Figure 23.41 shows photographs of photoelastic models under stress.

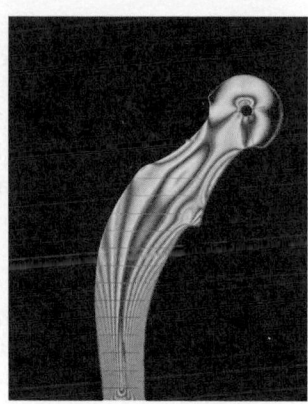

▲ **FIGURE 23.41** In photoelastic stress analysis, a structure such as a cathedral or an artificial hip is modeled in a transparent photoelastic material, subjected to stress, and analyzed in polarized light. The resulting images show where stress is concentrated in the object, allowing engineers to design it appropriately. In the case of cathedrals, such models have explained spectacular historical collapses by showing that inadequate buttressing and high winds can cause tensile stresses in masonry members. Masonry is strong in compression, but not in tension —which is one reason that modern large buildings use both steel and concrete in their frames.

23.11 Huygens's Principle

The laws of reflection and refraction of light rays that we introduced in Section 23.7 were discovered experimentally long before the wave nature of light was firmly established. However, we can *derive* these laws from wave considerations and show that they are consistent with the wave nature of light.

We begin with a principle called **Huygens's principle,** stated originally by Christiaan Huygens in 1678. Huygens's principle offers a geometrical method for finding, from the known shape of a wave front at some instant, the shape of the wave front at some later time. Huygens made the following hypothesis:

Huygens's principle:

> Every point of a wave front may be considered the source of secondary wavelets that spread out in all directions with a speed equal to the speed of propagation of the wave.

The new wave front at a later time is then found by constructing a surface *tangent* to the secondary wavelets, which is called the *envelope* of the wavelets.

All the results that we obtain from Huygens's principle can also be obtained from Maxwell's equations. Thus, it is not an independent principle, but it is very helpful in demonstrating the close relationship between the wave and ray models of light.

Huygens's principle is illustrated in Figure 23.42. The original wave front AA' is traveling outward from a source, as indicated by the small arrows. We want to find the shape of the wave front after a time interval t. Let v be the speed of propagation of the wave; then, in time t, it travels a distance vt. We construct several circles (traces of spherical wavelets) with radius $r = vt$, centered at points along AA'. The trace of the envelope of these wavelets, which is the new wave front, is the curve BB'. Throughout this discussion we're assuming that the speed v is the same at all points and in all directions.

The law of reflection can be derived from Huygens's principle; here's a brief sketch of the derivation: We consider a plane wave approaching a plane reflecting surface. In Figure 23.43a, the lines AA', BB', and CC' represent successive positions of a wave front approaching the reflecting surface MM'. As the points on

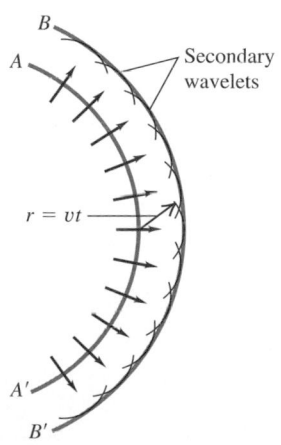

▲ **FIGURE 23.42** Applying Huygens's principle to wave front AA' to construct a new wave front BB'.

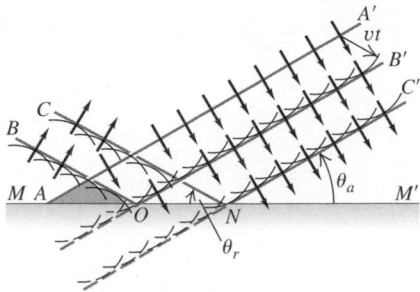

(a) Successive positions of a plane wave AA' as it is reflected from a plane surface

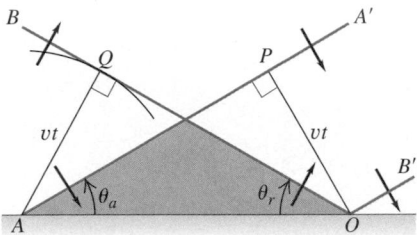

(b) Magnified portion of part (a)

▲ **FIGURE 23.43** Reflection of a plane wave from a surface.

this wave front successively reach points on surface MM', secondary wavelets are produced, as shown. The effect of the reflecting surface is to *change the direction* of travel of the wavelets that strike it, so part of a wavelet that would have penetrated the surface actually lies above it.

Figure 23.43b is an enlarged view of the colored area in Figure 23.43a. The right triangles OAP and AOQ are equal, so $\theta_a = \theta_r$. The angle the *wave front* makes with the surface equals the angle the *ray* makes with the *normal* to the surface, so the angle of reflection of the ray equals the angle of incidence, showing the relation between the law of reflection and Huygens's principle.

We can derive the law of *refraction* from Huygens's principle by a similar procedure. In Figure 23.44a, we consider a wave front, represented by line AA', for which point A has just arrived at the boundary surface SS' between two transparent materials a and b with wave speeds v_a and v_b, respectively. (The *reflected* waves are not shown in the figure; they proceed exactly as in Figure 23.43.) As successive points on the wavefront AA', arrive at the interface, secondary wavelets originate at these points.

Figure 23.44b is an enlarged view of a portion of the interface. Note that AO is the hypotenuse of triangle AOQ and also of triangle AOB. The angles θ_a and θ_b between the surface and the incident and refracted wave fronts are the angle of incidence and the angle of refraction, respectively. In time t, point A' on the wave front moves a distance $OQ = v_a t$, and point A moves a distance $AB = v_b t$. For the right triangles AOQ and AOB, we have

$$\sin\theta_a = \frac{v_a t}{AO} \quad \text{and} \quad \sin\theta_b = \frac{v_b t}{AO}.$$

Combining these two equations, we obtain

$$\frac{\sin\theta_a}{\sin\theta_b} = \frac{v_a}{v_b}. \tag{23.21}$$

We've defined the index of refraction n of a material as the ratio of the speed of light in vacuum (c) to its speed v in the material: $n_a = c/v_a$ and $n_b = c/v_b$. Thus,

$$\frac{n_b}{n_a} = \frac{c/v_b}{c/v_a} = \frac{v_a}{v_b},$$

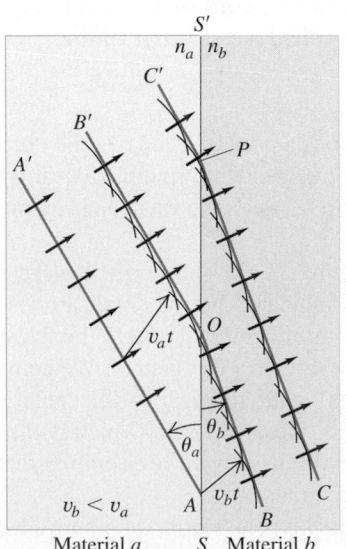

(a) Successive positions of a plane wave AA' as it is refracted by a plane surface

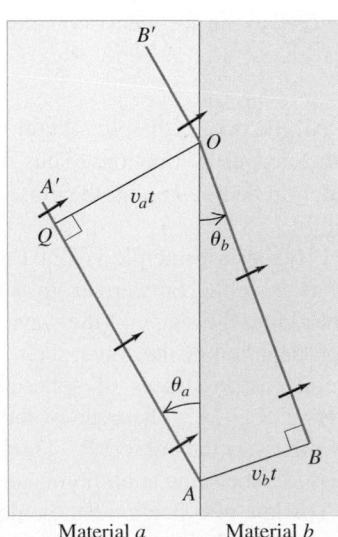

(b) Magnified portion of part (a)

▲ **FIGURE 23.44** Refraction of a plane wave from a surface for the case $v_b < v_a$.

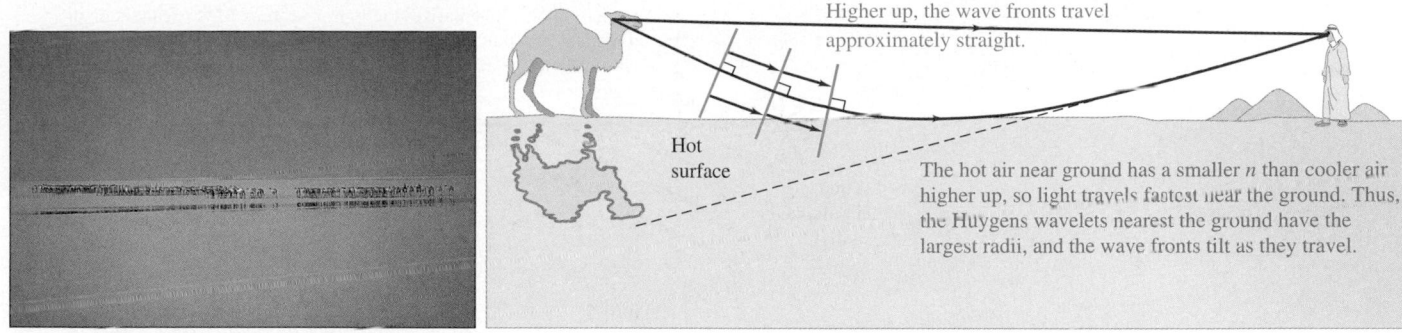

Higher up, the wave fronts travel approximately straight.

Hot surface

The hot air near ground has a smaller n than cooler air higher up, so light travels fastest near the ground. Thus, the Huygens wavelets nearest the ground have the largest radii, and the wave fronts tilt as they travel.

▲ **FIGURE 23.45** The cause of a mirage.

and we can rewrite Equation 23.21 as

$$\frac{\sin\theta_a}{\sin\theta_b} = \frac{n_b}{n_a},$$

or

$$n_a\sin\theta_a = n_b\sin\theta_b,$$

which we recognize as Snell's law, Equation 23.16. So we have derived Snell's law from a wave theory. Alternatively, we may choose to regard Snell's law as an experimental result that defines the index of refraction of a material; in that case, the preceding analysis helps to confirms the relationship $v = c/n$ for the speed of light in a material.

Mirages offer an interesting demonstration of Huygens's principle in action. When a surface of pavement or desert sand is heated intensely by the sun, a hot, less dense layer of air with smaller n forms near the surface. The speed of light is slightly greater in this hotter air near the ground, and the Huygens wavelets have slightly larger radii (because they move slightly faster and travel farther in a given time interval). As a result, the wave fronts tilt somewhat, and rays that were headed slightly toward the ground (with an incident angle near $90°$) can be bent upwards, as shown in Figure 23.45. Light farther from the ground is bent less and travels nearly in a straight line. The observer sees the object in its natural position, with an inverted image below it, as though seen in a horizontal reflecting surface. Even when the turbulence of the heated air prevents a clear inverted image from being formed, the mind of the thirsty traveler can interpret the apparent reflecting surface as a sheet of water.

It's important to keep in mind that Maxwell's equations are the fundamental relations for electromagnetic wave propagation. But it is a remarkable fact that Huygens's principle anticipated Maxwell's analysis by two centuries! Indeed, Maxwell provided the theoretical underpinning for Huygens's principle. Every point in an electromagnetic wave, with its time-varying electric and magnetic fields, acts as a source of the continuing wave, as predicted by Ampère's and Faraday's laws.

23.12 Scattering of Light

The sky is blue. Sunsets are red. Skylight is partially polarized; you can see this by looking at the sky directly overhead through a polarizing filter. It turns out that one phenomenon is responsible for all three of these effects.

In Figure 23.46, sunlight (unpolarized) comes from the left along the x axis and passes over an observer looking vertically upward along the y axis. (We are viewing the situation from the side.) Molecules of the earth's atmosphere are located at point O. The electric field in the beam of sunlight sets the electric

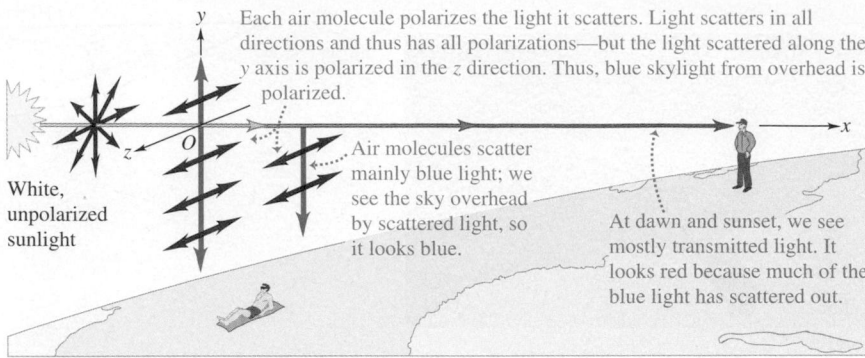

▲ **FIGURE 23.46** The sky is blue because light scattered by the atmosphere is primarily blue (and linearly polarized). At sunset and sunrise, we see primarily transmitted light, which is red because the blue scattered light has been subtracted from it.

charges in the molecules into vibration. Light is a transverse wave; the direction of the electric field in any component of the sunlight lies in the y-z plane, and the motion of the charges takes place in this plane. There is no field, and therefore no vibration, in the direction of the x axis.

An incident light wave whose $\vec{E}$ field has y and z components causes the electric charges in the molecules to vibrate along the line of $\vec{E}$. We can resolve this vibration into two components, one along the y axis (blue arrows), the other along the z axis (red arrows). The two components in the incident light thus produce the equivalent of two molecular "antennas," oscillating with the same frequency as the incident light and lying along the y and z axes, respectively.

Such an antenna doesn't radiate in the direction of its own length. The antenna along the y axis doesn't send any light to the observer directly below it, although it does emit light in other directions. Therefore, the only light reaching that observer comes from the other antenna, corresponding to the component of vibration along the z axis. This light is linearly polarized, with its electric field parallel to the antenna. The vectors on the y axis below point O show the direction of polarization of the light reaching the observer.

The process that we've just described is called **scattering.** The energy of the scattered light is removed from the original beam, reducing its intensity. Detailed analysis of the scattering process shows that the intensity of the light scattered from air molecules increases in proportion to the fourth power of the frequency (inversely to the fourth power of the wavelength). Thus, the intensity ratio for the two ends of the visible spectrum is $(700 \text{ nm}/400 \text{ nm})^4 = 9.4$. Roughly speaking, scattered light contains nine times as much blue light as red, and that's why the sky is blue.

Because skylight is partially polarized, polarizers are useful in photography. The sky in a photograph can be darkened by appropriate orientation of the polarizer axis. The effect of atmospheric haze can be reduced in exactly the same way, and unwanted reflections can be controlled just as with polarizing sunglasses, discussed in Section 23.10.

Toward evening, when sunlight has to travel a long distance through the earth's atmosphere, a substantial fraction of the blue light is removed by scattering. White light minus blue light appears yellow or red. Thus, when sunlight with the blue component removed is incident on a cloud, the light reflected from the cloud to the observer has the yellow or red hue so often observed at sunset. If the earth had no atmosphere, we would receive *no* skylight at the earth's surface, and the sky would appear as black in the daytime as it does at night. Thus, to an astronaut in a spacecraft or on the moon, the sky appears black, not blue.

SUMMARY

Electromagnetic Waves, the Speed of Light, and the Electromagnetic Spectrum

(Sections 23.1–23.3) When either an electric or a magnetic field changes with time, a field of the other kind is induced in the adjacent regions of space. This electromagnetic disturbance, or **electromagnetic wave,** can travel through space even when there is no matter in the intervening region. In vacuum, electromagnetic waves travel at the speed of light $c = 3.00 \times 10^8$ m/s.

The **electromagnetic spectrum** covers a range of frequencies from at least 1 to 10^{24} Hz and a correspondingly broad range of wavelengths. Radio waves have low frequencies (long wavelengths), and gamma rays have high frequencies (short wavelengths). Visible light is a very small part of this spectrum, with wavelengths in vacuum from 400 to 700 nm.

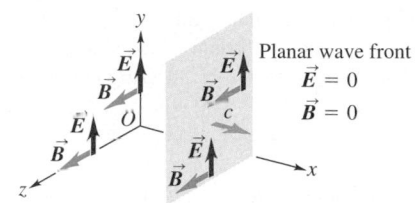

Sinusoidal Waves and Energy

(Sections 23.4 and 23.5) In a sinusoidal electromagnetic plane wave, the $\vec{E}$ and $\vec{B}$ fields vary sinusoidally in space and time. For a sinusoidal plane wave traveling in the $+x$ direction, $E = E_{max} \sin(\omega t - kx)$ and $B = B_{max} \sin(\omega t - kx)$ (Equations 23.3)

The electric and magnetic fields each contain energy. The maximum magnitudes of the electric and magnetic fields are related by the equation $E_{max} = cB_{max}$ (Equation 23.4); the intensity I can be expressed as

$$I = \frac{1}{2}\sqrt{\frac{\epsilon_0}{\mu_0}}E_{max}^2 = \frac{1}{2}\epsilon_0 c E_{max}^2. \qquad (23.11)$$

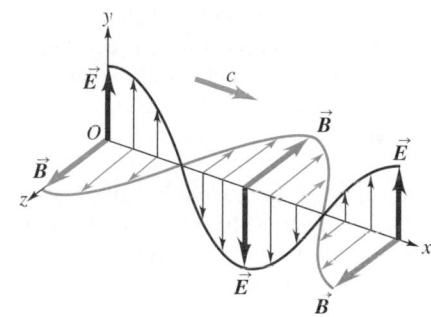

The Nature of Light

(Section 23.6) Light is an electromagnetic wave. When emitted or absorbed, it also shows particle properties. Light is emitted by accelerated electric charges that have been given excess energy by heat or electrical discharge. The speed of light, c, is a fundamental physical constant.

A **wave front** is a surface of constant phase; wave fronts move with a speed equal to the propagation speed of the wave. A ray is a line along the direction of propagation, perpendicular to the wave fronts.

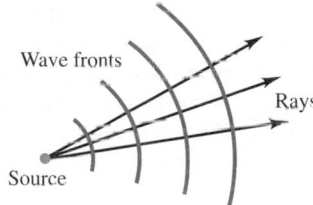

Reflection and Refraction; Total Internal Reflection

(Sections 23.7 and 23.8) The **index of refraction,** n, of a material is the ratio of the speed of light in vacuum, c, to the speed v in the material: $n = c/v$ (Equation 23.14). The incident, reflected, and refracted rays and the normal to the interface all lie in a single plane called the **plane of incidence.** Angles of incidence, reflection, and refraction are always measured from the normal to the interface. The law of reflection states that the angles of incidence and reflection are equal: $\theta_r = \theta_a$ (Equation 23.15). The law of refraction is $n_a \sin\theta_a = n_b \sin\theta_b$ (Equation 23.16).

When a ray travels within a material of greater index of refraction n_a, toward an interface with one of smaller index n_b, total internal reflection occurs when the angle of incidence exceeds a critical value θ_{crit} given by $\sin\theta_{crit} = n_b/n_a$ (Equation 23.18).

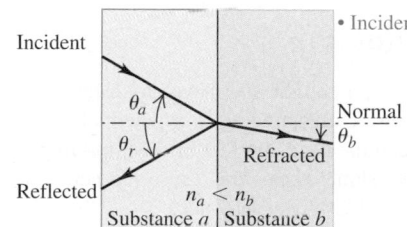

- Incident, reflected, and refracted rays and the normal lie in the same plane.
- $\theta_r = \theta_a$.
- $\dfrac{\sin\theta_a}{\sin\theta_b} = \dfrac{n_b}{n_a}$

Continued

Dispersion

(Section 23.9) The variation of index of refraction, n, with wavelength λ is called **dispersion.** Usually, n increases with decreasing λ. Thus, refraction is greater for light of shorter wavelength (higher frequency) in a given material than light of longer wavelength. When white light is incident on a prism, dispersion causes light with differing wavelengths, and therefore differing colors, to be refracted at different angles.

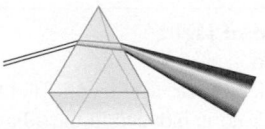

Polarization

(Section 23.10) The direction of polarization of a linearly polarized electromagnetic wave is the direction of the $\vec{E}$ field. A **polarizing filter** passes radiation that is linearly polarized along its polarizing axis and blocks radiation polarized perpendicular to that axis. **Malus's law** states that when polarized light of intensity I_{max} is incident on an analyzer and ϕ is the angle between the polarizing axes of the polarizer and analyzer, the transmitted intensity I is $I = I_{max}\cos^2\phi$ (Equation 23.19).

Brewster's law states that when unpolarized light strikes an interface between two materials, the reflected light is completely polarized perpendicular to the plane of incidence if the angle of incidence, θ_p, is given by $\tan\theta_p = n_b/n_a$ (Equation 23.20).

Huygens's Principle

(Section 23.11) If the position of a wave front at one instant is known, the position of the front at a later time can be constructed by using **Huygens's principle:** Every point of a wave front may be considered the source of secondary wavelets that spread out in all directions, and the new wave front is the surface that is tangent to the wavelets.

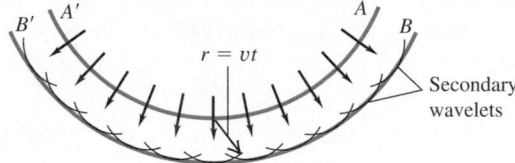

Scattering of Light

(Section 23.12) Light is scattered by air molecules, and the scattered light is partially polarized. Light with higher frequencies is scattered more than light with lower frequencies. The sky is blue because air molecules scatter into view more higher-frequency blue light than lower-frequency red light.

 For instructor-assigned homework, go to www.masteringphysics.com

Conceptual Questions

1. The light beam from a searchlight may have an electric-field magnitude of 1000 V/m, corresponding to a potential difference of 1500 V between the head and feet of a 1.5-m-tall person on whom the light shines. Does this cause the person to feel a strong electric shock? Why or why not?

2. If a light beam carries momentum, should a person holding a flashlight feel a recoil analogous to the recoil of a rifle when it is fired? Why is this recoil not actually observed?

3. Why is the average radiation pressure on a perfectly reflecting surface twice as great as on a perfectly absorbing surface?

4. When an electromagnetic wave is reflected from a moving reflector, the frequency of the reflected wave is different from that of the initial wave. Explain physically how this happens. Also, show why a higher-than-normal frequency results if the reflector is moving toward the observer and a lower frequency

if it is moving away. (Some radar systems used for highway-speed control operate on this principle.)

5. When hot air rises around a radiator or from a heating duct, objects behind it appear to shimmer or waver. What is happening?

6. How does the refraction of light account for the twinkling of starlight? (*Hint:* The earth's atmosphere consists of layers of varying density.)

7. Light requires about 8 min to travel from the sun to the earth. Is it delayed appreciably by the earth's atmosphere?

8. Sometimes when looking at a window, one sees two reflected images, slightly displaced from each other. What causes this phenomenon?

9. A student claimed that, because of atmospheric refraction, the sun can be seen after it has set and that the day is therefore longer than it would be if the earth had no atmosphere. First, what does the student mean by saying the sun can be seen after

it has set? Second, comment on the validity of the conclusion. Does the same effect also occur at sunrise?

10. If you look at your pet fish through the corner of your aquarium, you may see a double image of the fish, one image on each side of the corner. Explain how this could happen.

11. How could you determine the direction of the polarizing axis of a single polarizer? (*Hint:* Is there any naturally occurring polarized light you could use?)

12. In three-dimensional movies, two images are projected on the screen, and the viewers wear special glasses to sort them out. How does the polarization of light allow this effect to work?

13. Can sound waves be reflected? Refracted? Give examples to back up your answer.

14. Why should the wavelength, but not the frequency, of light change in passing from one material into another?

15. When light is incident on an interface between two materials, the angle of the refracted ray depends on the wavelength, but the angle of the reflected ray does not. Why should this be?

Multiple-Choice Problems

1. Light having a certain frequency, wavelength, and speed is traveling through empty space. If the frequency of this light were doubled, then
   A. its wavelength would remain the same, but its speed would double.
   B. its wavelength would remain the same, but its speed would be halved.
   C. its wavelength would be halved, but its speed would double.
   D. its wavelength would be halved, but its speed would remain the same.
   E. both its speed and its wavelength would be doubled.

2. Unpolarized light with an original intensity I_0 passes through two ideal polarizers having their polarizing axes turned at $120°$ to each other. After passing through both polarizers, the intensity of the light is
   A. $\dfrac{\sqrt{3}}{2}I_0$.   B. $\frac{1}{2}I_0$.   C. $\dfrac{\sqrt{3}}{4}I_0$.
   D. $\frac{1}{4}I_0$.   E. $\frac{1}{8}I_0$.

3. Light travels from water (with index of refraction 1.33) into air (index of refraction 1.00). Which of the following statements about this light is true? (There may be more than one correct choice.)
   A. The light has the same frequency in the air as it does in the water.
   B. The light travels faster in the air than in the water.
   C. The light has the same wavelength in the air as it does in the water.
   D. The light has the same speed in the air as in the water.
   E. The wavelength of the light in the air is greater than the wavelength in the water.

4. If a sinusoidal electromagnetic wave with intensity $10 \ W/m^2$ has an electric field of amplitude E, then a $20 \ W/m^2$ wave of the same wavelength will have an electric field of amplitude
   A. $4E$.   B. $2\sqrt{2}E$.   C. $2E$.   D. $\sqrt{2}E$.

5. A plane electromagnetic wave is traveling vertically downward with its magnetic field pointing northward. Its electric field must be pointing
   A. toward the south.   B. toward the east.
   C. toward the west.   D. vertically upward.
   E. vertically downward.

6. Suppose that a reflective solar sail (see Example 23.5) is deployed not perpendicular to the sun's rays but at some other angle. In what direction will the sail accelerate?
   A. In the direction the sun's rays are moving.
   B. Perpendicular to the surface of the sail.
   C. At an angle somewhere between that of the sun's rays and the perpendicular to the surface of the sail.
   D. The sail will not accelerate unless it is perpendicular to the sun's rays.

7. The index of refraction, n, has which of the following range of values?
   A. $n \geq 1$.   B. $0 \leq n \leq 1$.   C. $n \geq 0$.

8. A ray of light going from one material into another follows the path shown in Figure 23.47. What can you conclude about the relative indexes of refraction of these two materials?
   A. $n_a \geq n_b$.   B. $n_a > n_b$.
   C. $n_a < n_b$.   D. $n_a \leq n_b$.

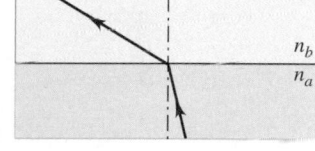
▲ FIGURE 23.47 Multiple-choice problem 8.

9. Which of the following statements about radio waves, infrared radiation, and x rays are correct? (There may be more than one correct choice.)
   A. They all have the same wavelength in vacuum.
   B. They all have the same frequency in vacuum.
   C. They all have exactly the same speed as visible light in vacuum.
   D. The short-wavelength x rays travel faster through vacuum than the long-wavelength radio waves.

10. Two lasers each produce 2 mW beams. The beam of laser B is wider, having twice the cross-sectional area as the beam of laser A. Which of the following statements about these two laser beams are correct? (There may be more than one correct choice.)
    A. Both of the beams have the same average power.
    B. Beam A has twice the intensity of beam B.
    C. Beam B has twice the intensity of beam A.
    D. Both beams have the same intensity.

11. A ray of light follows the path shown in Figure 23.48 as it reaches the boundary between two *transparent* materials. What can you conclude about the relative indexes of refraction of these two materials?
    A. $n_1 \geq n_2$.   B. $n_1 > n_2$.
    C. $n_1 < n_2$.   D. $n_1 \leq n_2$.

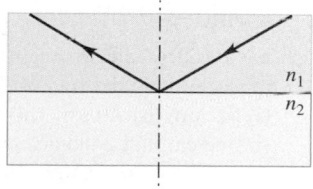
▲ FIGURE 23.48 Multiple-choice problem 11.

12. A light beam has a wavelength of 300 nm in a material of refractive index 1.5. In a material of refractive index 3.0, its wavelength will be
    A. 450 nm.   B. 300 nm.   C. 200 nm.
    D. 150 nm.   E. 100 nm.

13. A light beam has a frequency of 300 MHz in a material of refractive index 1.5. In a material of refractive index 3.0, its frequency will be
    A. 450 MHz. B. 300 MHz. C. 200 MHz.
    D. 150 MHz. E. 100 MHz.

14. You are sunbathing in the late afternoon when the sun is relatively low in the western sky. You are lying flat on your back, looking straight up through Polaroid sunglasses (see Figure 23.46 and the discussion of sunglasses in Section 23.10). To minimize the amount of light reaching your eyes, you should lie with your feet pointing in what direction? (There may be more than one correct answer.)
    A. north. B. east. C. south.
    D. west. E. some other direction.

15. A beam of light takes time t to travel a distance L in a certain liquid. If we now add water to the liquid to reduce its index of refraction by half, the time for the beam to travel the same distance will be
    A. $2t$. B. $\sqrt{2}t$. C. $t/\sqrt{2}$. D. $t/2$.

Problems

23.2 Speed of an Electromagnetic Wave

1. ● When a solar flare erupts on the surface of the sun, how many minutes after it occurs does its light show up in an astronomer's telescope on earth? (Consult Appendix E.)

2. ● **TV ghosting.** In a TV picture, faint, slightly offset ghost images are formed when the signal from the transmitter travels to the receiver both directly and indirectly after reflection from a building or some other large metallic mass. In a 25 inch set, the ghost is about 1.0 cm to the right of the principal image if the reflected signal arrives 0.60 μs after the principal signal. In this case, what is the difference in the distance traveled by the two signals?

3. ● (a) How much time does it take light to travel from the moon to the earth, a distance of 384,000 km? (b) Light from the star Sirius takes 8.61 years to reach the earth. What is the distance to Sirius in kilometers?

4. ●● A geostationary communications satellite orbits the earth directly above the equator at an altitude of 35,800 km. Calculate the time it takes for a signal to travel from a point on the equator to the satellite and back to the ground at another point on the equator exactly halfway around the earth. (See Appendix E.)

23.3 The Electromagnetic Spectrum
23.4 Sinusoidal Waves

5. ● Consider electromagnetic waves propagating in air. (a) Determine the frequency of a wave with a wavelength of (i) 5.0 km, (ii) 5.0 μm, (iii) 5.0 nm. (b) What is the wavelength (in meters and nanometers) of (i) gamma rays of frequency 6.50×10^{21} Hz, (ii) an AM station radio wave of frequency 590 kHz?

6. ● Most people perceive light having a wavelength between 630 nm and 700 nm as red and light with a wavelength between 400 nm and 440 nm as violet. Calculate the approximate frequency ranges for (a) violet light and (b) red light.

7. ● The electric field of a sinusoidal electromagnetic wave obeys the equation $E = -(375 \text{ V/m}) \sin[(5.97 \times 10^{15} \text{ rad/s})t + (1.99 \times 10^{7} \text{ rad/m})x]$. (a) What are the

amplitudes of the electric and magnetic fields of this wave? (b) What are the frequency, wavelength, and period of the wave? Is this light visible to humans? (c) What is the speed of the wave?

8. ● A sinusoidal electromagnetic wave having a magnetic field of amplitude 1.25 μT and a wavelength of 432 nm is traveling in the $+x$ direction through empty space. (a) What is the frequency of this wave? (b) What is the amplitude of the associated electric field? (c) Write the equations for the electric and magnetic fields as functions of x and t in the form of Equations (23.3).

9. ● **Visible light.** The wavelength of visible light ranges from 400 nm to 700 nm. Find the corresponding ranges of this light's (a) frequency, (b) angular frequency, (c) wave number.

10. ● **Ultraviolet radiation.** There are two categories of ultraviolet light. Ultraviolet A (UVA) has a wavelength ranging from 320 nm to 400 nm. It is not so harmful to the skin and is necessary for the production of vitamin D. UVB, with a wavelength between 280 nm and 320 nm, is much more dangerous, because it causes skin cancer. (a) Find the frequency ranges of UVA and UVB. (b) What are the ranges of the wave numbers for UVA and UVB?

11. ● **Medical x rays.** Medical x rays are taken with electromagnetic waves having a wavelength around 0.10 nm. What are the frequency, period, and wave number of such waves?

12. ● Radio station WCCO in Minneapolis broadcasts at a frequency of 830 kHz. At a point some distance from the transmitter, the magnetic-field amplitude of the electromagnetic wave from WCCO is 4.82×10^{-11} T. Calculate (a) the wavelength, (b) the wave number, (c) the angular frequency, and (d) the electric-field amplitude.

13. ●● A sinusoidal electromagnetic wave of frequency 6.10×10^{14} Hz travels in vacuum in the $+x$-direction. The magnetic field is parallel to the y-axis and has amplitude 5.80×10^{-4} T. (a) Find the magnitude and direction of the electric field. (b) Write the wave functions for the electric and magnetic fields in the form of Equations (23.3).

14. ● Consider each of the electric- and magnetic-field orientations given next. In each case, what is the direction of propagation of the wave? (a) $\vec{E}$ in the $+x$ direction, $\vec{B}$ in the $+y$ direction. (b) $\vec{E}$ in the $-y$ direction, $\vec{B}$ in the $+x$ direction. (c) $\vec{E}$ in the $+z$ direction, $\vec{B}$ in the $-x$ direction. (d) $\vec{E}$ in the $+y$ direction, $\vec{B}$ in the $-z$ direction.

15. ●● An electromagnetic wave has a magnetic field given by $B = (8.25 \times 10^{-9} \text{ T}) \sin[(\omega t + 1.38 \times 10^{4} \text{ rad/m})x]$, with the magnetic field in the $+y$ direction. (a) In which direction is the wave traveling? (b) What is the frequency f of the wave? (c) Write the wave function for the electric field.

23.5 Energy in Electromagnetic Waves

16. ● **Laboratory lasers.** He–Ne lasers are often used in physics demonstrations. They produce light of wavelength 633 nm and a power of 0.500 mW spread over a cylindrical beam 1.00 mm in diameter (although these quantities can vary). (a) What is the intensity of this laser beam? (b) What are the maximum values of the electric and magnetic fields? (c) What is the average energy density in the laser beam?

17. ● **Fields from a lightbulb.** We can reasonably model a 75 W incandescent lightbulb as a sphere 6.0 cm in diameter. Typically, only about 5% of the energy goes to visible light; the rest

goes largely to nonvisible infrared radiation. (a) What is the visible light intensity (in W/m²) at the surface of the bulb? (b) What are the amplitudes of the electric and magnetic fields at this surface, for a sinusoidal wave with this intensity?

18. • **Threshold of vision.** Under controlled darkened condi-
BIO tions in the laboratory, a light receptor cell on the retina of a person's eye can detect a single photon (more on photons in Chapter 28) of light of wavelength 505 nm and having an energy of 3.94×10^{-19} J. We shall assume that this energy is absorbed by a single cell during one period of the wave. Cells of this kind are called *rods* and have a diameter of approximately 0.0020 mm. What is the intensity (in W/m²) delivered to a rod?

19. • **High-energy cancer treatment.** Scientists are working on
BIO a new technique to kill cancer cells by zapping them with ultrahigh-energy (in the range of 10^{12} W) pulses of light that last for an extremely short time (a few nanoseconds). These short pulses scramble the interior of a cell without causing it to explode, as long pulses would do. We can model a typical such cell as a disk 5.0 μm in diameter, with the pulse lasting for 4.0 ns with an average power of 2.0×10^{12} W. We shall assume that the energy is spread uniformly over the faces of 100 cells for each pulse. (a) How much energy is given to the cell during this pulse? (b) What is the intensity (in W/m²) delivered to the cell? (c) What are the maximum values of the electric and magnetic fields in the pulse?

20. • At the floor of a room, the intensity of light from bright overhead lights is 8.00 W/m². Find the radiation pressure on a totally absorbing section of the floor.

21. • The intensity at a certain distance from a bright light source is 6.00 W/m². Find the radiation pressure (in pascals and in atmospheres) on (a) a totally absorbing surface and (b) a totally reflecting surface.

22. •• A sinusoidal electromagnetic wave from a radio station passes perpendicularly through an open window that has area 0.500 m². At the window, the electric field of the wave has rms value 0.0200 V/m. How much energy does this wave carry through the window during a 30.0 s commercial?

23. •• Two sources of sinusoidal electromagnetic waves have average powers of 75 W and 150 W and emit uniformly in all directions. At the same distance from each source, what is the ratio of the maximum electric field for the 150 W source to that of the 75 W source?

24. •• Radiation falling on a perfectly reflecting surface produces an average pressure p. If radiation of the same intensity falls on a perfectly absorbing surface and is spread over twice the area, what is the pressure at that surface in terms of p?

25. •• A sinusoidal electromagnetic wave emitted by a cellular phone has a wavelength of 35.4 cm and an electric field amplitude of 5.40×10^{-2} V/m at a distance of 250 m from the antenna. Calculate: (a) the frequency of the wave; (b) the magnetic-field amplitude; (c) the intensity of the wave.

23.7 The Reflection and Refraction of Light

26. • Two plane mirrors intersect at right angles. A laser beam strikes the first of them at a point 11.5 cm from their point of intersection, as shown in Figure 23.49. For what angle of incidence at the first mirror will this ray strike the midpoint of the second mirror (which is 28.0 cm long) after reflecting from the first mirror?

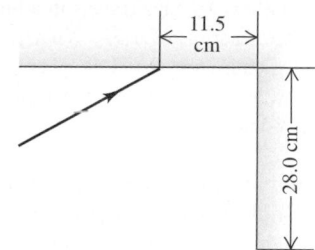

▲ **FIGURE 23.49** Problem 26.

27. • Three plane mirrors intersect at right angles. A beam of laser light strikes the first of them at an angle θ with respect to the normal. (See Figure 23.50.) (a) Show that when this ray is reflected off of the other two mirrors and crosses the original ray, the angle α between these two rays will be $\alpha = 180° - 2\theta$. (b) For what angle θ will the two rays be perpendicular when they cross?

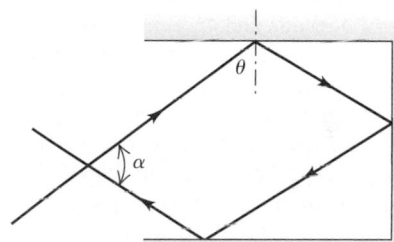

▲ **FIGURE 23.50** Problem 27.

28. •• Two plane mirrors A and B intersect at a 45° angle. Three rays of light leave point P (see Figure 23.51) and strike one of the mirrors. What is the subsequent path of each of the following rays until they no longer strike either of the mirrors? (a) Ray 1, which strikes A at 45° with respect to the normal. (b) Ray 2, which strikes B traveling perpendicular to mirror A. (c) Ray 3, which strikes B perpendicular to its surface.

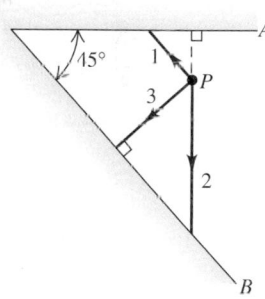

▲ **FIGURE 23.51** Problem 28.

29. •• Prove that when a ray of light travels at any angle into the corner formed by two mirrors placed at right angles to each other, the reflected ray emerges parallel to the original ray (see Figure 23.52).

30. •• A light beam travels at 1.94×10^8 m/s in quartz. The wavelength of the light in quartz is 355 nm. (a) What is the index of refraction of quartz at this wavelength? (b) If this same light travels through air, what is its wavelength there?

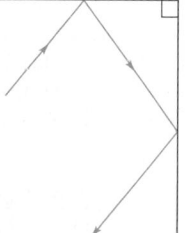

▲ **FIGURE 23.52** Problem 29.

31. •• Using a fast-pulsed laser and electronic timing circuitry, you find that light travels 2.50 m within a plastic rod in 11.5 ns. What is the refractive index of the plastic?

32. • Light with a frequency of 5.80×10^{14} Hz travels in a block of glass that has an index of refraction of 1.52. What is the wavelength of the light (a) in vacuum and (b) in the glass?

33. • The speed of light with a wavelength of 656 nm in heavy flint glass is 1.82×10^8 m/s. What is the index of refraction of the glass at this wavelength?

34. • **Light inside the eye.** The vitreous humor, a transparent,
BIO gelatinous fluid that fills most of the eyeball, has an index of refraction of 1.34. Visible light ranges in wavelength from 400 nm (violet) to 700 nm (red), as measured in air. This light travels through the vitreous humor and strikes the rods and cones at the surface of the retina. What are the ranges of (a) the wavelength, (b) the frequency, and (c) the speed of the light just as it approaches the retina within the vitreous humor?

35. •• Light of a certain frequency has a wavelength of 438 nm in water. What is the wavelength of this light (a) in benzene, (b) in air? (See Table 23.1.)

36. •• A 1.55-m-tall fisherman stands at the edge of a lake, being watched by a suspicious trout who is 3.50 m from the fisherman in the horizontal direction and 45.0 cm below the surface of the water. At what angle from the vertical does the fish see the top of the fisherman's head?

37. • Show that when a light ray travels from air through a sheet of glass with parallel surfaces and back into the air, it emerges traveling parallel to its original direction, although slightly displaced. (*Note:* The result of this problem is important, because it shows us that a sheet with parallel faces does not change the direction of a ray. It can also be shown that a very thin sheet does not displace the beam significantly. This is the case with a thin lens, since its opposite faces near its center are essentially parallel to each other. Rays that strike the center of such a lens go essentially straight through.)

38. • A glass plate having parallel faces and a refractive index of 1.58 lies at the bottom of a liquid of refractive index 1.70. A ray of light in the liquid strikes the top of the glass at an angle of incidence of 62.0°. Compute the angle of refraction of this light in the glass.

39. • A beam of light in air makes an angle of 47.5° with the *surface* (*not* the normal) of a glass plate having a refractive index of 1.66. (a) What is the angle between the reflected part of the beam and the *surface* of the glass? (b) What is the angle between the refracted beam and the *surface* (*not* the normal) of the glass?

40. •• A laser beam shines along the surface of a block of transparent material. (See Figure 23.53.) Half of the beam goes straight to a detector, while the other half travels through the block and then hits the detector. The time delay between the arrival of the two light beams at the detector is 6.25 ns. What is the index of refraction of this material?

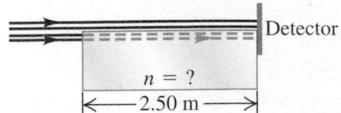

▲ **FIGURE 23.53** Problem 40.

41. •• **Reversibility of rays.** Ray 1 of light in medium A (see Figure 23.54) strikes the surface at 51.0° with respect to the normal. (a) Show that the angle of refraction of ray 1 with respect to the normal in medium B is 35.8°. (b) Now suppose that ray 2 is the reverse of ray 1, so that it strikes the surface at 35.8° with the normal in B. Show that ray 2 will come out in A at 51.0° with the normal. In other words, show that rays 1 and 2 follow the same path, except reversed from each other.

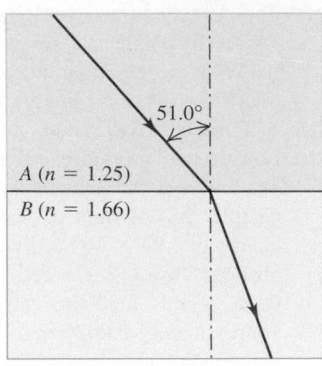

▲ **FIGURE 23.54** Problem 41.

42. •• You (height of your eyes above the water, 1.75 m) are standing 2.00 m from the edge of a 2.50-m-deep swimming pool. You notice that you can barely see your cell phone, which went missing a few minutes before, on the bottom of the pool. How far from the side of the pool is your cell phone?

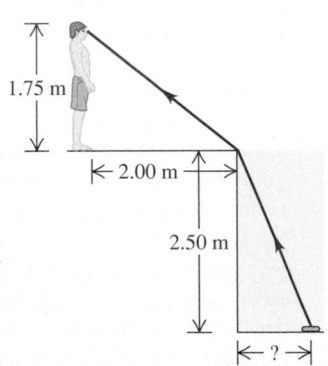

▲ **FIGURE 23.55** Problem 42.

43. •• A parallel-sided plate of glass having a refractive index of 1.60 is in contact with the surface of water in a tank. A ray coming from above makes an angle of incidence of 32.0° with the top surface of the glass. What angle does this ray make with the normal in the water?

44. •• As shown in Figure 23.56, a layer of water covers a slab of material X in a beaker. A ray of light traveling upwards follows the path indicated. Using the information on the figure, find (a) the index of refraction of material X and (b) the angle the light makes with the normal in the *air*.

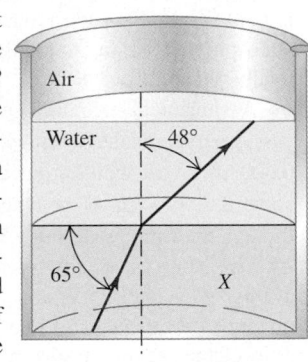

▲ **FIGURE 23.56** Problem 44.

23.8 Total Internal Reflection

45. • A ray of light in diamond (index of refraction 2.42) is incident on an interface with air. What is the *largest* angle the ray can make with the normal and not be totally reflected back into the diamond?

46. • The critical angle for total internal reflection at a liquid–air interface is 42.5°. (a) If a ray of light traveling in the liquid has an angle of incidence of 35.0° at the interface, what angle does the refracted ray in the air make with the normal? (b) If a ray of light traveling in air has an angle of incidence of 35.0° at the interface, what angle does the refracted ray in the liquid make with the normal?

47. •• At the very end of Wagner's series of operas *The Ring of the Nibelung,* Brünnhilde takes the golden ring from the finger of the dead Siegfried and throws it into the Rhine, where it

sinks to the bottom of the river. Assuming that the ring is small enough to be treated as a point compared with the depth of the river and that the Rhine is 10.0 m deep where the ring goes in, what is the *area* of the largest circle at the surface of the water over which light from the ring could escape from the water?

48. •• A ray of light is traveling in a glass cube that is totally immersed in water. You find that if the ray is incident on the glass–water interface at an angle to the normal greater than 48.7°, no light is refracted into the water. What is the refractive index of the glass?

49. •• Light is incident along the normal to face AB of a glass prism of refractive index 1.52, as shown in Figure 23.57. Find the largest value the angle α can have without any light refracted out of the prism at face AC if (a) the prism is immersed in air and (b) the prism is immersed in water.

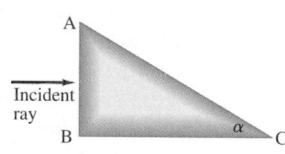

▲ **FIGURE 23.57** Problem 49.

50. •• **Light pipe.** Light enters a solid tube made of plastic having an index of refraction of 1.60. The light travels parallel to the upper part of the tube. (See Figure 23.58.) You want to cut the face *AB* so that all the light will reflect back into the tube after it first strikes that face.

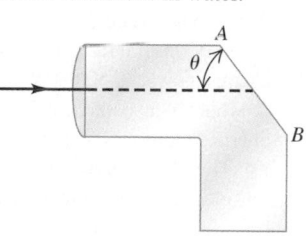

▲ **FIGURE 23.58** Problem 50.

(a) What is the largest that θ can be if the tube is in air? (b) If the tube is immersed in water of refractive index 1.33, what is the largest that θ can be?

51. •• An optical fiber consists of an outer "cladding" layer and an inner core with a slightly higher index of refraction. Light rays entering the core are trapped inside by total internal reflection and forced to travel along the fiber (see Figure 23.59). Suppose the cladding has an index of refraction of 1.46 and the core has an index of refraction of 1.48. Calculate the largest angle θ between a light ray and the longitudinal axis of the fiber (see the figure) for which the ray will be totally internally reflected at the core/cladding boundary.

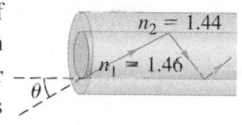

▲ **FIGURE 23.59** Problem 51.

23.9 Dispersion

52. • A beam of light strikes a sheet of glass at an angle of 57.0° with the normal in air. You observe that red light makes an angle of 38.1° with the normal in the glass, while violet light makes a 36.7° angle. (a) What are the indexes of refraction of this glass for these colors of light? (b) What are the speeds of red and violet light in the glass?

53. • Use the information from the graph in Figure 23.29 to construct a graph of the index of refraction of silicate flint glass as a function of the frequency of light.

54. •• A narrow beam of white light strikes one face of a slab of silicate flint glass. The light is traveling parallel to the two adjoining faces, as shown in Figure 23.60. For the transmitted light inside the glass, through what angle $\Delta\theta$ is the complete visible spectrum of light dispersed? (Consult the graph in Figure 23.29.)

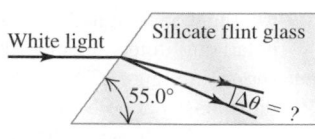

▲ **FIGURE 23.60** Problem 54.

55. •• Use the graph in Figure 23.29 for silicate flint glass. (a) What are the indexes of refraction of this glass for extreme violet light of wavelength 400 nm and for extreme red light of wavelength 700 nm? (b) What are the wavelengths of 400 nm violet light and 700 nm red light in this glass? (c) Calculate the *ratio* of the speed of extreme red light to that of extreme violet light in the glass. Which of these travels faster in the glass? (d) If a beam of white light in air strikes a sheet of this glass at 65.0° with the normal in air, what will be the angle of *dispersion* between the extremes of visible light in the glass? In other words, what will be the angle between extreme red and extreme violet light in the glass?

56. • The indices of refraction for violet light ($\lambda = 400$ nm) and red light ($\lambda = 700$ nm) in diamond are 2.46 and 2.41, respectively. A ray of light traveling through air strikes the diamond surface at an angle of 53.5° to the normal. Calculate the angular separation between these two colors of light in the refracted ray.

23.10 Polarization

57. • Unpolarized light with intensity I_0 is incident on an ideal polarizing filter. The emerging light strikes a second ideal polarizing filter whose axis is at 41.0° to that of the first. Determine (a) the intensity of the beam after it has passed through the second polarizer and (b) its state of polarization.

58. • Two ideal polarizing filters are oriented so that they transmit the *maximum* amount of light when unpolarized light is shone on them. To what fraction of its maximum value I_0 is the intensity of the transmitted light reduced when the second filter is rotated through (a) 22.5°, (b) 45.0°, and (c) 67.5°?

59. •• A beam of unpolarized light of intensity I_0 passes through a series of ideal polarizing filters with their polarizing directions turned to various angles as shown in Figure 23.61. (a) What is the light intensity (in terms of I_0) at points A, B, and C? (b) If we remove the middle filter, what will be the light intensity at point C?

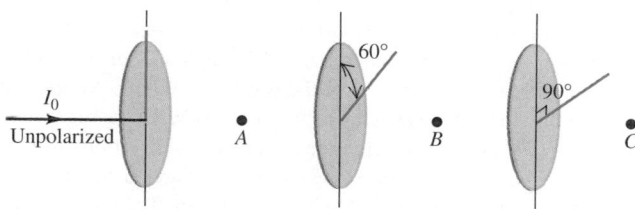

▲ **FIGURE 23.61** Problem 59.

60. •• Three ideal polarizing filters are stacked, with the polarizing axis of the second and third filters at 23.0° and 62.0°, respectively, to that of the first. If unpolarized light is incident on the stack, the light has intensity 75.0 W/cm² after it passes through the stack. If the incident intensity is kept constant, what is the intensity of the light after it has passed through the stack if the second polarizer is removed?

61. •• Light of original intensity I_0 passes through two ideal polarizing filters having their polarizing axes oriented as shown in Figure 23.62. You want to adjust the angle ϕ so that the intensity at point P is equal to $I_0/10$. (a) If the original light is unpolarized, what should ϕ be? (b) If the original light is linearly polarized in the same direction as the polarizing axis of the first polarizer the light reaches, what should ϕ be?

▲ **FIGURE 23.62** Problem 61.

62. • The polarizing angle for light in air incident on a glass plate is 57.6°. What is the index of refraction of the glass?

63. •• A beam of polarized light passes through a polarizing filter. When the angle between the polarizing axis of the filter and the direction of polarization of the light is θ, the intensity of the emerging beam is I. If you instead want the intensity to be $I/2$, what should be the angle (in terms of θ) between the polarizing angle of the filter and the original direction of polarization of the light?

64. •• A beam of unpolarized light in air is incident at an angle of 54.5° (with respect to the normal) on a plane glass surface. The reflected beam is completely linearly polarized. (a) What is the refractive index of the glass? (b) What is the angle of refraction of the transmitted beam?

65. •• Plane-polarized light passes through two polarizers whose axes are oriented at 35.0° to each other. If the intensity of the original beam is reduced to 15.0%, what was the polarization direction of the original beam, relative to the first polarizer?

General Problems

66. •• The energy flow to the earth from sunlight is about 1.4 kW/m². (a) Find the maximum values of the electric and magnetic fields for a sinusoidal wave of this intensity. (b) The distance from the earth to the sun is about 1.5×10^{11} m. Find the total power radiated by the sun.

67. •• A plane sinusoidal electromagnetic wave in air has a wavelength of 3.84 cm and an $\vec{E}$ field amplitude of 1.35 V/m. (a) What is the frequency of the wave? (b) What is the $\vec{B}$ field amplitude? (c) What is the intensity? (d) What average force does this radiation exert perpendicular to its direction of propagation on a totally absorbing surface with area 0.240 m²?

68. •• A powerful searchlight shines on a man. The man's cross-sectional area is 0.500 m² perpendicular to the light beam, and the intensity of the light at his location is 36.0 kW/m². He is wearing black clothing, so that the light incident on him is *totally* absorbed. What is the magnitude of the force the light beam exerts on the man? Do you think he could sense this force?

69. •• **Laser surgery.** Very short pulses of high-intensity laser
BIO beams are used to repair detached portions of the retina of the eye. The brief pulses of energy absorbed by the retina welds the detached portion back into place. In one such procedure, a laser beam has a wavelength of 810 nm and delivers 250 mW of power spread over a circular spot 510 μm in diameter. The vitreous humor (the transparent fluid that fills most of the eye) has an index of refraction of 1.34. (a) If the laser pulses are each 1.50 ms long, how much energy is delivered to the retina with each pulse? (b) What average pressure does the pulse of the laser beam exert on the retina as it is fully absorbed by the circular spot? (c) What are the wavelength and frequency of the laser light inside the vitreous humor of the eye? (d) What are the maximum values of the electric and magnetic fields in the laser beam?

70. •• A small helium-neon laser emits red visible light with a power of 3.20 mW in a beam that has a diameter of 2.50 mm. (a) What are the amplitudes of the electric and magnetic fields of the light? (b) What are the average energy densities associated with the electric field and with the magnetic field? (c) What is the total energy contained in a 1.00 m length of the beam?

71. •• Radio receivers can comfortably pick up a broadcasting station's signal when the electric field strength of the signal is about 10.0 mV/m. If a radio station broadcasts in all directions with an average power of 50.0 kW, what would be the maximum distance at which you could easily pick up its transmissions? (Atmospheric conditions can have major effects on this distance.)

72. •• The 19th-century inventor Nikola Tesla proposed to transmit electric power via sinusoidal electromagnetic waves. Suppose power is to be transmitted in a beam of cross-sectional area 100 m². What electric- and magnetic-field amplitudes are required to transmit an amount of power comparable to that handled by modern transmission lines (which carry voltages and currents of the order of 500 kV and 1000 A)?

73. •• **Solar sail.** NASA is doing research on the concept of *solar sailing*. A solar sailing craft uses a large, low-mass sail and the energy and momentum of sunlight for propulsion. (a) Should the sail be absorptive or reflective? Why? (b) The total power output of the sun is 3.9×10^{26} W. How large a sail is necessary to propel a 10,000 kg spacecraft against the gravitational force of the sun? Express your result in square kilometers. (c) Explain why your answer to part (b) is independent of the distance from the sun.

74. • A thick layer of oil is floating on the surface of water in a tank. A beam of light traveling in the oil is incident on the water interface at an angle of 30.0° from the normal. The refracted beam travels in the water at an angle of 45.0° from the normal. What is the refractive index of the oil?

75. •• A thin beam of light in air is incident on the surface of a lanthanum flint glass plate having a refractive index of 1.80. What is the angle of incidence, θ_a, of the beam with this plate, for which the angle of refraction is $\theta_a/2$? Both angles are measured relative to the normal.

76. •• You want to support a sheet of fireproof paper horizontally, using only a vertical upward beam of light spread uniformly over the sheet. There is no other light on this paper. The sheet measures 22.0 cm by 28.0 cm and has a mass of 1.50 g. (a) If the paper is black and hence absorbs all the light that hits it, what must be the intensity of the light beam? (b) For the light in part (a), what are the maximum values of its electric and magnetic fields? (c) If the paper is white and hence reflects all the light that hits it, what intensity of light beam is needed to support it? (d) To see if it is physically reasonable to expect

to support a sheet of paper this way, calculate the intensity in a typical 0.500 mW laser beam that is 1.00 mm in diameter and compare this value with your answer in part (a).

77. •• A light ray in air strikes the right-angle prism shown in Figure 23.63. This ray consists of two different wavelengths. When it emerges at face *AB,* it has been split into two different rays that diverge from each other by 8.50°. Find the index of refraction of the prism for each of the two wavelengths.

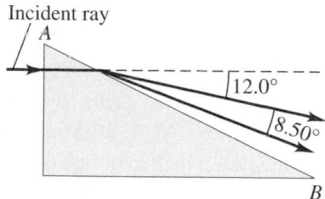

Incident ray

▲ **FIGURE 23.63** Problem 77.

78. •• A ray of light is incident in air on a block of a transparent solid whose index of refraction is *n*. If $n = 1.38$, what is the *largest* angle of incidence, θ_a, for which total internal reflection will occur at the vertical face (point *A* shown in Figure 23.64)?

79. •• A light beam is directed parallel to the axis of a hollow cylindrical tube. When the tube contains only air, it takes the light 8.72 ns to travel the length of the tube, but when the tube is filled with a transparent jelly, it takes the light 2.04 ns longer to travel its length. What is the refractive index of this jelly?

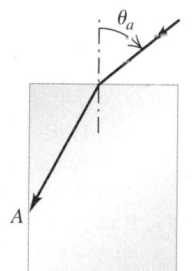

▲ **FIGURE 23.64** Problem 78.

80. •• **Heart sonogram.** Physicians use high-frequency ($f = $
BIO 1 MHz to 5 MHz) sound waves, called ultrasound, to image internal organs. The speed of these ultrasound waves is 1480 m/s in muscle and 344 m/s in air. We define the index of refraction of a material for sound waves to be the ratio of the speed of sound in air to the speed of sound in the material. Snell's law then applies to the refraction of sound waves. (a) At what angle from the normal does an ultrasound beam enter the heart if it leaves the lungs at an angle of 9.73° from the normal to the heart wall? (Assume that the speed of sound in the lungs is 344 m/s.) (b) What is the critical angle for sound waves in air incident on muscle?

81. •• The prism shown in Figure 23.65 has a refractive index of 1.66, and the angles *A* are 25.0°. Two light rays *m* and *n* are parallel as they enter the prism. What is the angle between them after they emerge?

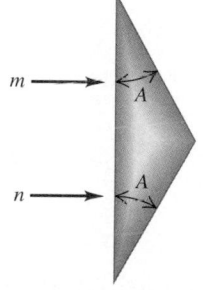

▲ **FIGURE 23.65** Problem 81.

82. •• A 45°–45°–90° prism is immersed in water. A ray of light is incident normally on one of the prism's shorter faces. What is the minimum index of refraction that the prism must have if this ray is to be totally reflected within the glass at the long face of the prism?

83. •• A beaker with a mirrored bottom is filled with a liquid whose index of refraction is 1.63. A light beam strikes the top surface of the liquid at an angle of 42.5° from the normal. At what angle from the normal will the beam exit from the liquid after traveling down through it, reflecting from the mirrored bottom, and returning to the surface?

84. •• A ray of light traveling *in* a block of glass ($n = 1.52$) is incident on the top surface at an angle of 57.2° with respect to the normal in the glass. If a layer of oil is placed on the top surface of the glass, the ray is totally reflected. What is the maximum possible index of refraction of the oil?

85. •• A block of glass has a polarizing angle of 60.0° for red light and 70.0° for blue light, for light traveling in air and reflecting from the glass. (a) What are the indexes of refraction for red light and for blue light? (b) For the same angle of incidence, which color is refracted more on entering the glass?

86. •• In a physics lab, light with wavelength 490 nm travels in air from a laser to a photocell in 17.0 ns. When a slab of glass 0.840 m thick is placed in the light beam, with the beam incident along the normal to the parallel faces of the slab, it takes the light 21.2 ns to travel from the laser to the photocell. What is the wavelength of the light in the glass?

87. •• (a) Light passes through three parallel slabs of different thicknesses and refractive indexes. The light is incident in the first slab and finally refracts into the third slab. Show that the middle slab has no effect on the final direction of the light. That is, show that the direction of the light in the third slab is the same as if the light had passed directly from the first slab into the third slab. (b) Generalize this result to a stack of *N* slabs. What determines the final direction of the light in the last slab?

88. • The refractive index of a certain glass is 1.66. For what angle of incidence is light that is reflected from the surface of this glass completely polarized if the glass is immersed in (a) air or (b) water?

89. •• A thin layer of ice ($n = 1.309$) floats on the surface of water ($n = 1.333$) in a bucket. A ray of light from the bottom of the bucket travels upward through the water. (a) What is the largest angle with respect to the normal that the ray can make at the ice–water interface and still pass out into the air above the ice? (b) What is this angle after the ice melts?

90. •• **Optical activity of biological molecules.** Many biologi-
BIO cally important molecules are optically active. When linearly polarized light traverses a solution of compounds containing these molecules, its plane of polarization is rotated. Some compounds rotate the polarization clockwise; others rotate the polarization counterclockwise. The amount of rotation depends on the amount of material in the path of the light. The following data give the amount of rotation through two amino acids over a path length of 100 cm:

Rotation (degrees) *l*-leucine	*d*-glutamic acid	Concentration (g/100 mL)
−0.11	0.124	1.0
−0.22	0.248	2.0
−0.55	0.620	5.0
−1.10	1.24	10.0
−2.20	2.48	20.0
−5.50	6.20	50.0
−11.0	12.4	100.0

From these data, find the relationship between the concentration *C* (in grams per 100 mL) and the rotation of the polarization (in degrees) of each amino acid. (*Hint:* Graph the concentration as a function of the rotation angle for each amino acid.)

91. •• A horizontal cylindrical tank 2.20 m in diameter is half full of water. The space above the water is filled with a pressurized gas of unknown refractive index. A small laser can move along the curved bottom of the water and aims a light beam toward the center of the water surface (Figure 23.66). You observe that when the laser has moved a distance $S = 1.09$ m or more (measured along the curved surface) from the lowest point in the water, no light enters the gas. (a) What is the index of refraction of the gas? (b) How long does it take the light beam to travel from the laser to the rim of the tank when (i) $S > 1.09$ m and (ii) $S < 1.09$ m?

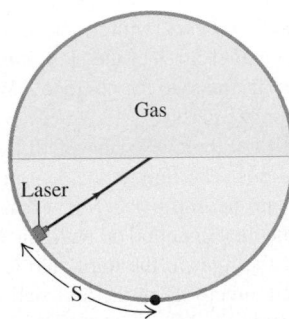

▲ **FIGURE 23.66** Problem 91.

Passage Problems

Reflection and refraction. We can see a transparent object that is illuminated because of the incident light that is reflected from its surfaces and the refracted light that is reflected by small imperfections within the object itself.

When a light ray is incident on a transparent surface we can easily predict the *direction* of the reflected and refracted rays by using the laws of reflection and refraction. However, the *amount* of light reflected from a surface is more difficult to determine since this depends on the direction and polarization of the incident ray, and the refractive indices for both surfaces. For example, when light strikes the boundary between two surfaces (with refractive indices of n_1 and n_2) at an angle of incidence that is near 90°, the fractional intensity of light reflected from the boundary is given by

$$\left(\frac{n_1 - n_2}{n_1 + n_2}\right)^2$$

According to this result, typical optical glass (with an index of refraction of 1.5) should reflect about 4% (0.04) of the light normally incident from the surrounding air, which is in fact the case.

92. If a light beam strikes a 10 cm thick slab of glass (which is immersed in air) at an angle of 30° from the normal to the surface, what will be its angle to the normal when it leaves the back side of the slab? Assume that the slab has parallel sides and an index of refraction of 1.5.
    A. 0°   B. 30°   C. 60°   D. 58.3°

93. If a layer of oil, with an index of refraction of 1.8, is placed on the top of the glass, what will be the new angle at which the light beam leaves the glass?
    A. 0°   B. 16°   C. 30°   D. 46°

94. If the entire slab (without the oil) is submerged in a fluid with an index of refraction of 1.5, what will be the effect?
    A. The slab will appear to change color.
    B. Light striking the slab could be totally reflected.
    C. The slab will be very difficult to see.
    D. Light exiting the slab could be totally reflected.

$\mathcal{24}$ Geometric Optics

When you look at yourself in a flat mirror, you appear your actual size and right side up. But in a curved mirror you may appear larger or smaller, or even upside down. Why is this? Mirrors, magnifying glasses, and telescopes are a familiar part of everyday life. We can use the ray model of light, introduced in the preceding chapter, to understand the behavior of these and similar instruments. All we need are the laws of reflection and refraction and some simple geometry and trigonometry.

In this chapter and the next, we'll make frequent use of the concept of *image*. When a politician worries about his image, he is thinking of how he looks to the voters. We'll use the term in a related, but more precise, way, having to do with the behavior of a collection of rays that converge toward or appear to diverge from a point called an *image point*. In this chapter, we analyze the formation of images by a single reflecting or refracting surface and by a thin lens. This discussion lays the foundation for analyses of many familiar optical instruments, including camera lenses, magnifiers, the human eye, microscopes, and telescopes. We'll study these instruments in Chapter 25.

24.1 Reflection at a Plane Surface

The concept of **image** is the most important new idea in this chapter and the next. Consider Figure 24.1. Several rays diverge from point P and are reflected at the plane mirror, according to the law of reflection. After they are reflected, their final directions are the same as though they had come from point P'. We call point P an *object point* and point P' the corresponding *image point,* and we say that the mirror forms an *image* of point P. The outgoing rays (those going away from the mirror) don't really come from point P', but their directions are the same as though they had come from that point.

The lens in this photo has the same fundamental shape as the lenses used for nearsightedness. The lens is held with the convex side facing us, and the objects seen through it look shrunken. You will see the same effect if you look at a nearsighted person's eyes through his or her eyeglasses.

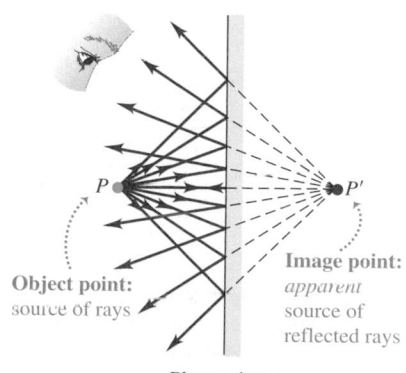

Object point: source of rays

Image point: *apparent* source of reflected rays

Plane mirror

▲ **FIGURE 24.1** The rays entering the eye after reflection from a plane mirror look as though they had come from point P', the image point for object point P.

803

Something similar happens at a plane refracting surface, as shown in Figure 24.2. Rays coming from object point P are refracted at the interface between two optical materials. When the angles of incidence are small, the final directions of the rays after refraction are the same as though they had come from point P', and again we call P' an *image point.*

In both of these cases, the rays spread out from point P, whether it is an actual point source of light or a point that scatters the light shining on it. An observer who can see only the rays spreading out from the surface (after they are reflected or refracted) *thinks* that they come from the image point P'. This image point is therefore a convenient way to describe the directions of the various reflected or refracted rays, just as the object point P describes the directions of the rays arriving at the surface *before* reflection or refraction.

To find the precise location of the image point P' that a plane mirror forms of an object point P, we use the construction shown in Figure 24.3. The figure shows two rays diverging from an object point P at a distance s to the left of a plane mirror. We call s the **object distance.** The ray PV is incident normally on the mirror (that is, it is perpendicular to the surface of the mirror), and it returns along its original path.

The ray PB makes an angle θ with PV. It strikes the mirror at an angle of incidence θ and is reflected at an equal angle with the normal. When we extend the two reflected rays backward, they intersect at point P', at a distance s' behind the mirror. We call s' the **image distance.** The line between P and P' is perpendicular to the mirror. The two triangles have equal angles, so P and P' are at equal distances from the mirror, and s and s' have equal magnitudes.

We can repeat the construction of Figure 24.3 for each ray diverging from P. The directions of *all* the outgoing rays are the same *as though* they had originated at point P', confirming that P' is the *image* of P. The rays do not actually pass through point P'. In fact, for an ordinary mirror, there is no light at all on the back side. In cases like this, and like that of Figure 24.2, the outgoing rays don't actually come from P', and we call the image a **virtual image.** Later we will see cases in which the outgoing rays really *do* pass through an image point, and we will call the resulting image a **real image.** The images formed on a projection screen or the photographic film (or digital sensor array) in a camera are real images.

Sign Rules

Before we go further, let's introduce some general sign rules. These may seem unnecessarily complicated for the simple situations discussed so far, but we want to state the rules in a form that will be applicable to *all* the situations that we'll encounter. These will include image formation by a plane or spherical reflecting or refracting surface and by a pair of refracting surfaces forming a lens. Here are the rules:

When $n_a > n_b$, P' is closer to the surface than P; for $n_a < n_b$, the reverse is true.

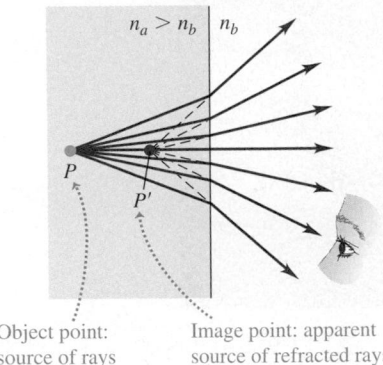

Object point: source of rays

Image point: apparent source of refracted rays

▲ **FIGURE 24.2** The rays entering the eye after refraction at the plane interface look as though they had come from point P', the image point for object P. The angles of incidence have been exaggerated for clarity.

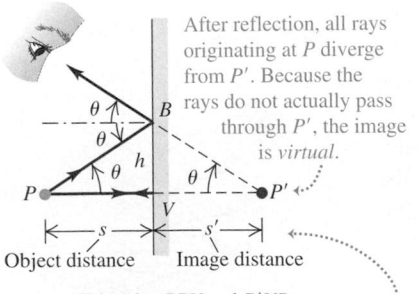

After reflection, all rays originating at P diverge from P'. Because the rays do not actually pass through P', the image is *virtual.*

Object distance   Image distance

Triangles PBV and $P'VB$ are congruent, so $|s| = |s'|$.

▲ **FIGURE 24.3** Reflected rays originating at the object point P appear to come from the virtual image point P'.

Sign rules for object and image distances

Object distance: When the object is on the same side of the reflecting or refracting surface as the incoming light, the object distance s is positive; otherwise, it is negative.

Image distance: When the image is on the same side of the reflecting or refracting surface as the outgoing light, the image distance s' is positive; otherwise, it is negative.

Figure 24.4 shows how these rules are applied to a plane mirror and to a plane refracting surface. For a mirror, the incoming and outgoing sides are always the same.

In Figure 24.4, the object distance s is *positive* in both cases because the object point P is on the incoming side of the reflecting or refracting surface. The image distance s' is *negative* because the image point P' is *not* on the outgoing side of the surface.

Let's now consider just the case of a plane mirror (Figure 24.4a). For a plane mirror, the object and image distances s and s' are related simply by

$$s = -s'. \quad \text{(plane mirror)} \quad (24.1)$$

Next we consider an object with finite size, parallel to the mirror. In Figure 24.5, this object is the blue arrow, which extends from P to Q and has height y. Two of the rays from Q are shown; *all* the rays from Q appear to diverge from its image point Q' after reflection. The image of the arrow, represented by the dashed magenta arrow, extends from P' to Q' and has height y'. Other points of the arrow PQ have image points between P' and Q'. The triangles PQV and $P'Q'V$ have equal angles, so the object and image have the same size and orientation, and $y = y'$.

The ratio of the image height to the object height, y'/y, in *any* image-forming situation is called the **lateral magnification** m; that is,

Definition of lateral magnification

For object height y and image height y', the lateral magnification m is

$$m = \frac{y'}{y}. \quad (24.2)$$

For a plane mirror, the lateral magnification m is unity. In other words, when you look at yourself in a plane mirror, your image is the same size as the real you.

We'll often represent an object by an arrow. Its image may be an arrow pointing in the *same* direction as the object or in the *opposite* direction. When the directions are the same, as in Figure 24.5, we say that the image is **erect;** when they are opposite, the image is **inverted.** The image formed by a plane mirror is always erect. A positive value of lateral magnification m corresponds to an erect image; a negative value corresponds to an inverted image. That is, for an erect image, y and y' always have the *same* sign, while for an inverted image, they always have *opposite* signs.

Figure 24.6 shows a three-dimensional virtual image of a three-dimensional object—a hand—formed by a plane mirror. The index and middle fingers of the

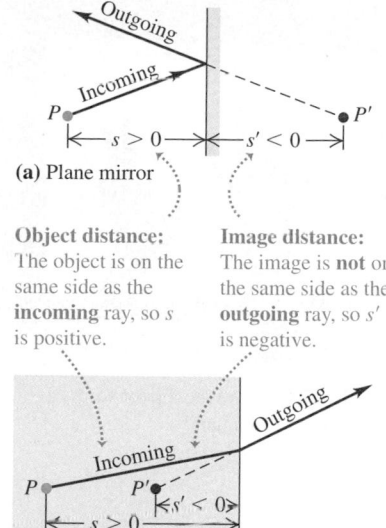

(a) Plane mirror

Object distance: The object is on the same side as the **incoming** ray, so s is positive.

Image distance: The image is **not** on the same side as the **outgoing** ray, so s' is negative.

(b) Plane refracting interface

▲ **FIGURE 24.4** The sign rules for the object and image distances as applied to a plane reflecting mirror and a plane refracting interface.

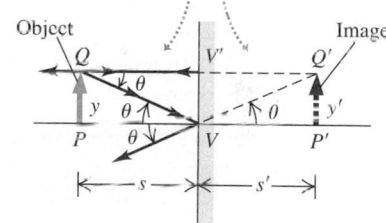

For a plane mirror, PQV and $P'Q'V$ are congruent, so $y = y'$ and the lateral magnification is 1 (the object and image are the same size).

▲ **FIGURE 24.5** Construction for determining the height of an image formed by reflection from a plane mirror.

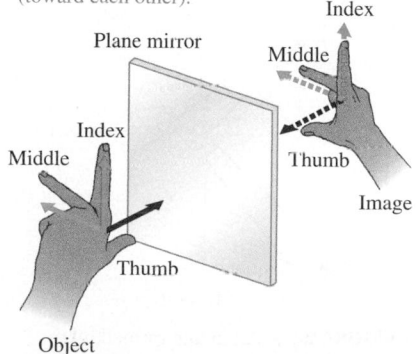

The object and image are *reversed*: The object and image thumbs point in opposite directions (toward each other).

▲ **FIGURE 24.6** The image formed by the mirror is virtual, erect, and reversed. It is the same size as the object.

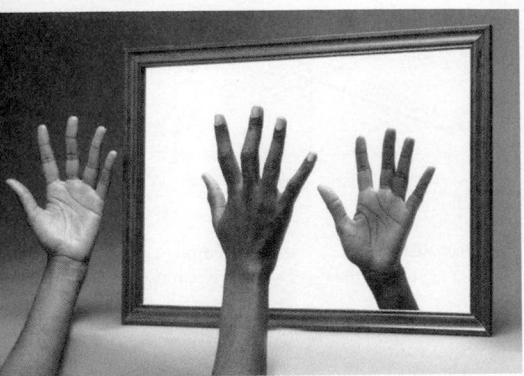

▲ **FIGURE 24.7** A plane mirror forms a reversed image of a hand.

image point in the same direction as those of the actual hand, but the object and image *thumbs* point in opposite directions: The actual thumb points toward the mirror, but the image thumb points "out of the mirror." The image of a three-dimensional object formed by a plane mirror is the same *size* as the object in all its dimensions, but the image and object are *not* identical. They are related in the same way as a right hand and a left hand, and indeed we speak of a pair of objects with this relationship as mirror-image objects. To verify this object–image relationship, try arranging your left and right hands to match the object and image hands in Figure 24.6. When an object and its image are related in this way, the image is said to be **reversed.** When the transverse dimensions of object and image are in the same direction, the image is erect. A plane mirror always forms an erect, but reversed, image. Figure 24.7 illustrates this point.

An important property of all images formed by reflecting or refracting surfaces is that an image formed by one surface or optical device can serve as the object for a second surface or device. Figure 24.8 shows a simple example. Mirror 1 forms an image P_1' of the object point P, and mirror 2 forms another image P_2', each in the way we have just discussed. But in addition, the image P_1' formed by mirror 1 can serve as an object for mirror 2, which then forms an image of this object at point P_3' as shown. Similarly, mirror 1 can use the image P_2' formed by mirror 2 as an object and form an image of it. We leave it to you to show that this image point is also at P_3'. Later in this chapter, we'll use this principle to locate the image formed by two successive curved-surface refractions in a lens.

Mirror, mirror on the wall

You want to install the shortest possible wall mirror that will let you see all of yourself when you stand straight. The bottom edge of the mirror will be

A. halfway between the floor and the level of your eyes.
B. nearer to the floor than to the level of your eyes.
C. nearer to the level of your eyes than to the floor.
D. at floor level.

SOLUTION Because you want the mirror to be as short as possible, the rays that go from your toes to your eyes should reflect from the bottom edge of the mirror. Because the angles of incidence and reflection are the same, the bottom of the mirror should be halfway between your toes and your eyes.

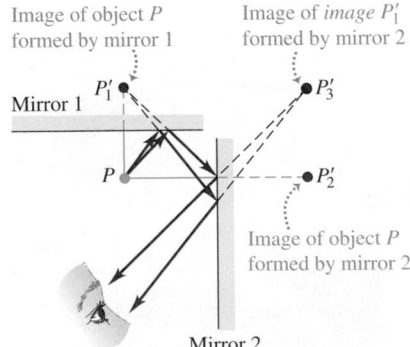

Image of object P
formed by mirror 1

Image of *image P_1'*
formed by mirror 2

Mirror 1

P_1' •

• P_3'

P •

• P_2'

Image of object P
formed by mirror 2

Mirror 2

▲ **FIGURE 24.8** An image formed by the reflection of an image, treated as an object. This figure shows mirror 2 reflecting the image from mirror 1 to form image P_3'. The same image is obtained when mirror 1 reflects the image P_2' from mirror 2.

24.2 Reflection at a Spherical Surface

Continuing our analysis of reflecting surfaces, we consider next the formation of an image by a *spherical* mirror. Figure 24.9a shows a spherical mirror with radius of curvature R, with its concave side facing the incident light. The **center of curvature** of the surface (the center of the sphere of which the surface is a part) is at C. Point P is an object point; for the moment, we assume that the distance from P to V is greater than R. The ray PV, passing through C, strikes the mirror normally and is reflected back on itself. Point V, at the center of the mirror surface, is called the **vertex** of the mirror, and the line PCV is called the **optic axis.**

Ray PB, at an angle α with the axis, strikes the mirror at B, where the angles of incidence and reflection are θ. The reflected ray intersects the axis at point P'. We'll show that *all* rays from P intersect the axis at the *same* point P', as in Figure 24.9b, no matter what α is, provided that it is a *small* angle. Point P' is therefore the *image* of object point P. The object distance, measured from the vertex V, is s, and the image distance is s'. The object point P is on the same side as the incident light, so, according to the sign rule in Section 24.1, the object distance s is positive. The image point P' is on the same side as the reflected light, so the image distance s' is also positive.

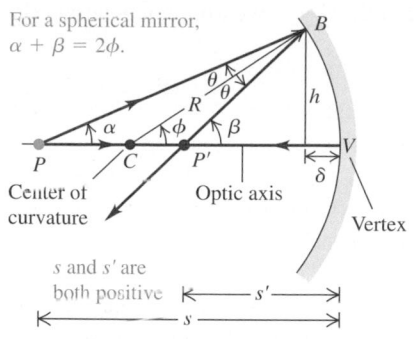

For a spherical mirror, $\alpha + \beta = 2\phi$.

Center of curvature

Optic axis

Vertex

s and s' are both positive

(a) Construction for finding the position P' of an image formed by a concave spherical mirror.

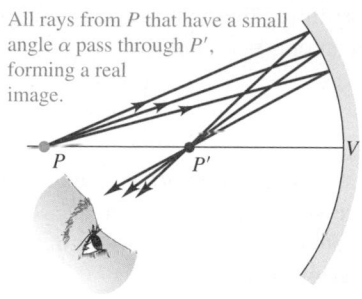

All rays from P that have a small angle α pass through P', forming a real image.

(b) The paraxial approximation, which holds for rays with small α.

▲ **FIGURE 24.9** Reflection from a concave spherical mirror.

Unlike the reflected rays in Figure 24.1, the reflected rays in Figure 24.9b actually do intersect at point P'; then they diverge from P' *as if* they had originated at that point. The image point P' is thus a *real* image point.

A theorem from plane geometry states that an exterior angle of a triangle equals the sum of the two opposite interior angles. Using this theorem with triangles PBC and $P'BC$ in Figure 24.9a, we have

$$\phi = \alpha + \theta, \qquad \beta = \phi + \theta.$$

Eliminating θ between these equations gives

$$\alpha + \beta = 2\phi. \tag{24.3}$$

Now we need a sign rule for the radii of curvature of spherical surfaces:

Sign rule for the radius of curvature

When the center of curvature C is on the same side as the outgoing (reflected) light, the radius of curvature R is positive; otherwise, it is negative.

In Figure 24.10a, R is positive because the center of curvature C is on the same side of the mirror as the reflected light. This is always the case when reflection occurs at the *concave* side of a surface. For a *convex* surface (Figure 24.10b), the center of curvature is on the opposite side from the reflected light, and R is negative.

We may now compute the image distance s'. In Figure 24.9a, let h represent the height of point B above the axis, and let δ denote the short distance from V to the foot of this vertical line. We now write expressions for the tangents of α, β, and ϕ, remembering that s, s', and R are all positive quantities:

$$\tan\alpha = \frac{h}{s - \delta}, \qquad \tan\beta = \frac{h}{s' - \delta}, \qquad \tan\phi = \frac{h}{R - \delta}.$$

These trigonometric equations cannot be solved as simply as the corresponding algebraic equations for a plane mirror. However, *if the angle α is small,* then the angles β and ϕ are also small. Now, the tangent of a small angle is nearly equal to the angle itself (in radians), so we can replace $\tan\alpha$ by α, and so on, in the previous equations. Also, if α is small, we can neglect the distance δ compared with s', s, and R. So, for small angles, we have the approximate relations

$$\alpha = \frac{h}{s}, \qquad \beta = \frac{h}{s'}, \qquad \phi = \frac{h}{R}.$$

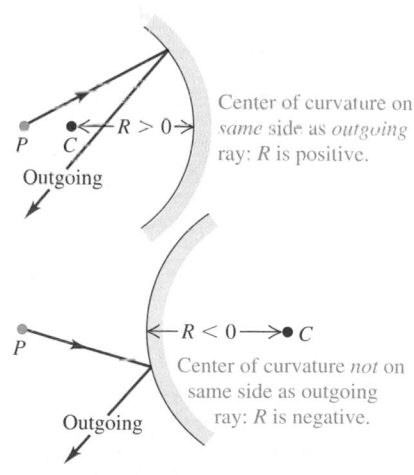

Center of curvature on *same* side as *outgoing* ray: R is positive.

Outgoing

Center of curvature *not* on same side as outgoing ray: R is negative.

Outgoing

▲ **FIGURE 24.10** Sign rule for the radius of curvature of a spherical surface.

Substituting these into Equation 24.3 and dividing by h, we obtain a general relation among s, s', and R:

$$\frac{1}{s} + \frac{1}{s'} = \frac{2}{R}. \qquad \text{(spherical mirror)} \qquad (24.4)$$

This equation does not contain the angle α; this means that *all* rays from P that make sufficiently small angles with the optic axis intersect at P' after they are reflected. Such rays, close to the axis and nearly parallel to it, are called **paraxial rays.**

Be sure you understand that Equation 24.4, as well as many similar relations that we will derive later in this chapter and the next, is only *approximately* correct. It results from a calculation containing approximations, and it is valid only for paraxial rays. The term **paraxial approximation** is often used for the approximations that we've just described. As the angle α increases, the point P' moves somewhat closer to the vertex; a spherical mirror, unlike a plane mirror, does not form a precise point image of a point object. This property of a spherical mirror is called *spherical aberration.*

If $R = \infty$, the mirror becomes *plane,* and Equation 24.4 reduces to Equation 24.1 $(s' = -s)$, which we derived earlier for a plane reflecting surface.

Focal Point

When the object point P is very far from the mirror $(s = \infty)$, the incoming rays are parallel. From Equation 24.4, the image distance s' is then given by

$$\frac{1}{\infty} + \frac{1}{s'} = \frac{2}{R}, \qquad s' = \frac{R}{2}.$$

The situation is shown in Figure 24.11a. A beam of incident rays parallel to the axis converges, after reflection, to a point F at a distance $R/2$ from the vertex of the mirror. Point F is called the **focal point,** or the *focus,* and its distance from the vertex, denoted by f, is called the **focal length.** We see that f is related to the radius of curvature R by

$$f = \frac{R}{2}. \qquad (24.5)$$

We can discuss the opposite situation, shown in Figure 24.11b. When the *image* distance s' is very large, the outgoing rays are parallel to the optic axis. The object distance s is then given by

$$\frac{1}{s} + \frac{1}{\infty} = \frac{2}{R}, \qquad s = \frac{R}{2}.$$

In Figure 24.11b, all the rays coming to the mirror from the focal point are reflected parallel to the optic axis. Again, we see that $f = R/2$.

Thus, the focal point F of a concave spherical mirror has the following properties:

Focal point of a concave spherical mirror
1. Any incoming ray parallel to the optic axis is reflected through the focal point.
2. Any incoming ray that passes through the focal point is reflected parallel to the optic axis.

For spherical mirrors, these statements are true only for paraxial rays; for parabolic mirrors, they are *exactly* true.

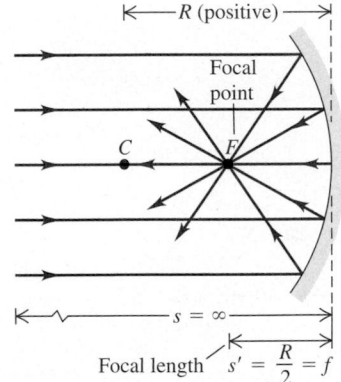

(a) All parallel rays incident on a spherical mirror reflect through the focal point.

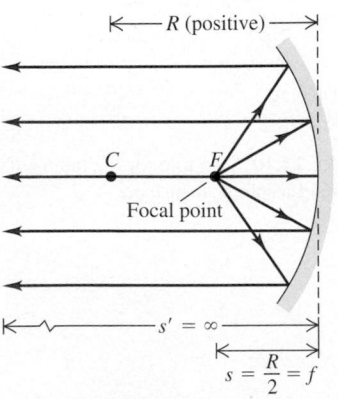

(b) Rays diverging from the focal point reflect to form parallel outgoing rays.

▲ **FIGURE 24.11** Reflection and production of parallel rays by a concave spherical mirror. The angles are exaggerated for clarity.

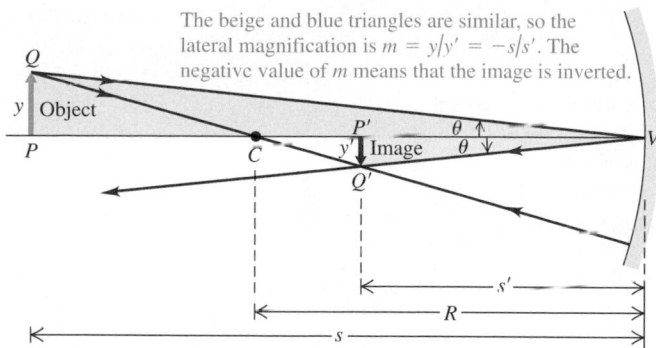

The beige and blue triangles are similar, so the lateral magnification is $m = y/y' = -s/s'$. The negative value of m means that the image is inverted.

▲ **FIGURE 24.12** Construction for determining the position, orientation, and height of an image formed by a concave spherical mirror.

We'll usually express the relationship between object and image distances for a mirror, Equation 24.4, in terms of the focal length f:

$$\frac{1}{s} + \frac{1}{s'} = \frac{1}{f}. \qquad \text{(spherical mirror)} \qquad (24.6)$$

Now suppose we have an object with finite height, represented by the blue arrow PQ in Figure 24.12, perpendicular to the axis PV. The image of P formed by paraxial rays is at P'. The object distance for point Q is very nearly equal to that for point P, so the image $P'Q'$ is nearly straight and is perpendicular to the axis. Note that object and image have different heights, y and y', respectively, and that they have opposite orientations. We've defined the *lateral magnification m* as the ratio of the image height y' to the object height y:

$$m = \frac{y'}{y}.$$

Because triangles PVQ and $P'VQ'$ in Figure 24.12 are *similar*, we also have the relation $y/s = -y'/s'$. The negative sign is needed because object and image are on opposite sides of the optic axis; if y is positive, y' is negative. Therefore, the lateral magnification m is given by

$$m = \frac{y'}{y} = -\frac{s'}{s}. \qquad \text{(spherical mirror)} \qquad (24.7)$$

A negative value of m indicates that the image is *inverted* relative to the object, as the figure shows. In cases that we'll consider later, in which m may be either positive or negative, a positive value always corresponds to an erect image and a negative value to an inverted one. For a *plane* mirror, $s = -s'$, so $y' = y$ and the image is erect, as we have already shown.

Although the ratio of the image height to the object height is called the *lateral magnification,* the image formed by a mirror or lens may be either larger or smaller than the object. If it is smaller, then the magnification is less than unity in absolute value. For instance, the image formed by an astronomical telescope mirror or a camera lens is usually *much* smaller than the object. For three-dimensional objects, the ratio of image distances to object distances measured *along* the optic axis is different from the ratio of *lateral* distances (the lateral magnification). In particular, if m is a small fraction, the three-dimensional image of a three-dimensional object is reduced *longitudinally* much more than it is reduced *transversely,* and the image appears squashed along the optic axis. Also, the image formed by a spherical mirror, like that of a plane mirror, is always reversed.

Conceptual Analysis 24.2

An image problem

Three students stand in front of a large concave mirror at a carnival (Figure 24.13). An apple hangs from a string as shown, at the same level as the center of curvature, but farther from the mirror than its center of curvature. Each student is asked to report to a bystander where the image of the apple (modeled as a point) is located. The bystander suspects that they aren't all telling the truth. Chandra claims to see the image of the apple at position (1). Joe says he sees it at position (2). Michi thinks they're both fibbing; she believes that all of them see it at position (3), where she sees it. Who is telling the truth?

A. Chandra and Joe are telling the truth; the position of the image depends on the observer's position.
B. Michi is correct; all observers see the image at the same position.

SOLUTION As shown in Figure 24.9b, all the reflected rays converge at the image point; it follows that all three observers must

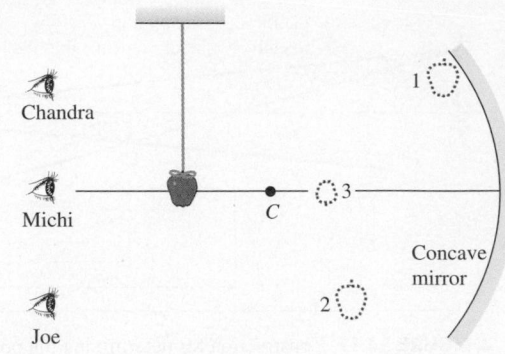

▲ FIGURE 24.13

see the image as located at the same point. The object and the center of curvature are on the optic axis, so the image is also on the optic axis. Michi is right; Chandra and Joe are fibbing.

EXAMPLE 24.1 **Image from a concave mirror**

A lamp is placed 10 cm in front of a concave spherical mirror that forms an image of the filament on a screen placed 3.0 m from the mirror. What is the radius of curvature of the mirror? If the lamp filament is 5.0 mm high, how tall is its image? What is the lateral magnification?

SOLUTION

SET UP Figure 24.14 shows our diagram.

SOLVE Both object distance and image distance are positive; we have $s = 10$ cm and $s' = 300$ cm. To find the radius of curvature, we use Equation 24.4:

$$\frac{1}{s} + \frac{1}{s'} = \frac{2}{R},$$
$$\frac{1}{10 \text{ cm}} + \frac{1}{300 \text{ cm}} = \frac{2}{R},$$

and $R = 19.4$ cm. To find the height of the image, we use Equation 24.7:

$$m = \frac{y'}{y} = -\frac{s'}{s},$$
$$\frac{y'}{5.0 \text{ mm}} = -\frac{300 \text{ cm}}{10 \text{ cm}},$$

and $y' = -150$ mm. The lateral magnification m is

$$m = \frac{y'}{y} = \frac{-150 \text{ mm}}{5 \text{ mm}} = -30.$$

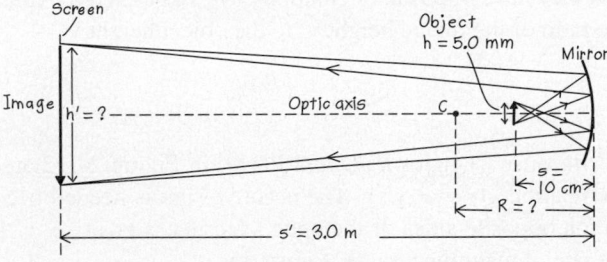

▲ FIGURE 24.14 Our diagram for this problem (not to scale).

REFLECT The image is inverted (as we know because $m = -30$ is negative) and is 30 times taller than the object. Notice that the filament is *not* located at the mirror's focal point; the image is not formed by rays parallel to the optic axis. (The focal length of this mirror is $f = R/2 = 9.7$ cm.)

Practice Problem: A concave mirror has a radius of curvature $R = 25$ cm. An object of height 2 cm is placed 15 cm in front of the mirror. What is the image distance? What is the height of the image? *Answers:* $s' = 75$ cm, $y' = -10$ cm.

Convex Mirrors

In Figure 24.15a, the *convex* side of a spherical mirror faces the incident light. The center of curvature is on the opposite side from the outgoing rays, so, according to our sign rule, R is negative. Ray PB is reflected, with the angles of incidence and reflection both equal to θ. The reflected ray, projected backward, intersects the axis at P'. *All* rays from P that are reflected by the mirror appear to

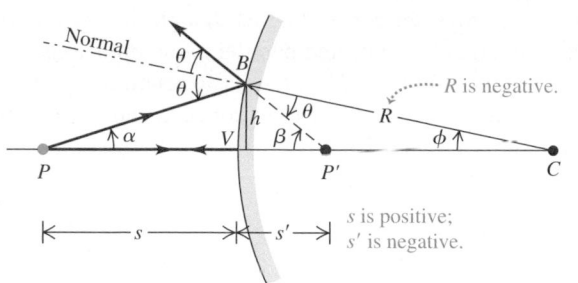

(a) Construction for finding the position of an image formed by a convex mirror.

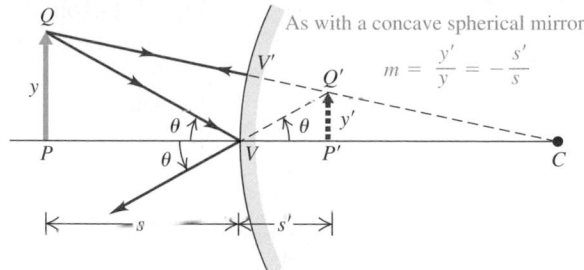

(b) Construction for finding the magnification of an image formed by a convex mirror.

▲ **FIGURE 24.15** Reflection from a convex spherical mirror. As with a concave spherical mirror, rays from P for which α is small form an image at P'.

diverge from the same point P', provided that the angle α is small. Therefore, P' is the image point of P. The object distance s is positive, the image distance s' is negative, and the radius of curvature R is negative.

Figure 24.15b shows two rays diverging from the head of the arrow PQ and the virtual image $P'Q'$ of this arrow. We can show, by the same procedure that we used for a concave mirror, that

$$\frac{1}{s} + \frac{1}{s'} = \frac{2}{R},$$

and that the lateral magnification m is

$$m = \frac{y'}{y} = -\frac{s'}{s}.$$

These expressions are exactly the same as those for a concave mirror.

When R is negative (as with a convex mirror), incoming rays parallel to the optic axis are not reflected through the focal point F. Instead, they diverge as though they had come from the point F at a distance f behind the mirror, as shown in Figure 24.16a. In this case, f is the focal length and F is called a **virtual focal point**. The corresponding image distance s' is negative, so both f and R are negative, and Equation 24.5 holds for convex as well as concave mirrors. In

▲ **Application A wide, wide world.** Professional truck drivers rely heavily on their mirrors, and you have probably seen the phrase "If you can't see my mirrors, I can't see you" on a truck or two rolling down the highway. In addition to regular plane mirrors to see the traffic behind them, truckers usually have a smaller, convex mirror that gives them a "wide-angle" view backwards. As we've learned, a convex mirror gives an upright virtual image with a lateral magnification less than one. Because the image size is reduced, a convex mirror has a larger field of view than a plane mirror, so the driver can see more lanes of traffic. Surveillance mirrors are convex for the same reason. *Concave* mirrors are not used for such purposes, because the image they produce is upside down.

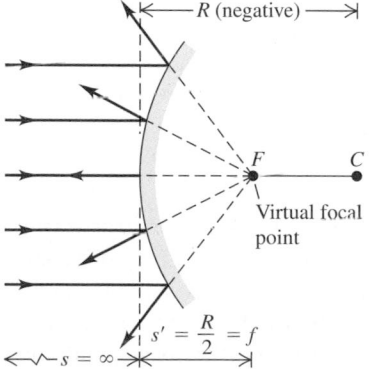

(a) Paraxial rays incident on a convex spherical mirror diverge from a virtual focal point.

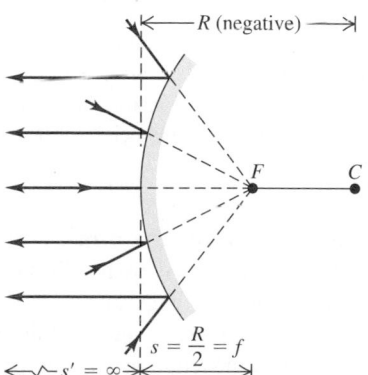

(b) Rays aimed at the virtual focal point are parallel to the axis after reflection.

▲ **FIGURE 24.16** Reflection and production of paraxial rays by a convex spherical mirror. The angles are exaggerated for clarity.

Figure 24.16b, the incoming rays are converging as though they would meet at the virtual focal point F, and they are reflected parallel to the optic axis.

In summary, Equations 24.4 through 24.7, the basic relationships for image formation by a spherical mirror, are valid for *both* concave and convex mirrors, provided that we use the sign rules consistently.

EXAMPLE 24.2 Santa's image problem

Santa checks himself for soot, using his reflection in a shiny, silvered Christmas tree ornament 0.750 m away (Figure 24.17). The diameter of the ornament is 7.20 cm. Standard reference works state that Santa is a "right jolly old elf"; we estimate his height as 1.6 m. Where and how tall is the image of Santa formed by the ornament? Is it erect or inverted?

▶ **FIGURE 24.17**

SOLUTION

SET UP Figure 24.18 shows our diagram. (To limit its size, we drew it not to scale; the angles are exaggerated, and Santa would actually be much taller and farther away.) The surface of the ornament closest to Santa acts as a convex mirror with radius $R = -(7.20 \text{ cm})/2 = -3.60$ cm and focal length $f = R/2 = -1.80$ cm. The object distance is $s = 0.750$ m $= 75.0$ cm.

SOLVE From Equation 24.6,

$$\frac{1}{s'} = \frac{1}{f} - \frac{1}{s} = \frac{1}{-1.80 \text{ cm}} - \frac{1}{75.0 \text{ cm}},$$
$$s' = -1.76 \text{ cm}.$$

The lateral magnification m is given by Equation 24.7:

$$m = \frac{y'}{y} = -\frac{s'}{s} = -\frac{-1.76 \text{ cm}}{75.0 \text{ cm}} = 2.35 \times 10^{-2}.$$

Because m is positive, the image is erect, and it is only about 0.0235 as tall as Santa himself. Thus, the image height y' is

$$y' = my = (0.0235)(1.6 \text{ m}) = 3.8 \times 10^{-2} \text{ m} = 3.8 \text{ cm}.$$

REFLECT The object is on the same side of the mirror as the incoming light, so the object distance s is positive. Because s' is negative, the image is behind the mirror—that is, in Figure 24.18 it is on the side opposite to that of the outgoing light—and it is

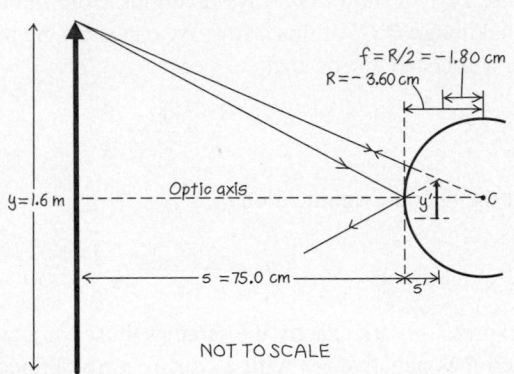

▲ **FIGURE 24.18** Our diagram for this problem.

virtual. The image is about halfway between the front surface of the ornament and its center. Thus, this convex mirror forms an erect, virtual, diminished, reversed image. In fact, when the object distance s is positive, a convex mirror *always* forms an erect, virtual, diminished, reversed image.

Practice Problem: One of Santa's elves scoots in halfway between Santa and the ornament to check and see that his hat is on straight. His image height in the ornament is 2.00 cm. What is the height of the elf? *Answer:* $y = 43.7$ cm.

Spherical mirrors have many important uses. A concave mirror forms the light from a flashlight or headlight bulb into a parallel beam. Convex mirrors are used to give a wide-angled view to car and truck drivers, for shoplifting surveillance in stores, and at "blind" intersections. A concave mirror with a focal length long enough so that your face is between the focal point and the mirror functions as a magnifier. Some solar-power plants use an array of plane mirrors to simulate an approximately spherical concave mirror. This array is used to collect and direct the sun's radiation to the focal point, where a steam boiler is placed. You can probably think of other examples from your everyday experience.

24.3 Graphical Methods for Mirrors

We can find the position and size of the image formed by a mirror by a simple graphical method. This method consists of finding the point of intersection of a few particular rays that diverge from a point of the object (such as point Q in Figure 24.19) and are reflected by the mirror. Then (neglecting aberrations) *all* rays from this point that strike the mirror will intersect at the same point. For this construction, we always choose an object point that is *not* on the optic axis. Four rays that we can usually draw easily are shown in Figure 24.19. These are called **principal rays.**

Definitions of principal rays for spherical mirrors
1. A ray parallel to the axis, after reflection, passes through the focal point F of a concave mirror or appears to come from the (virtual) focal point of a convex mirror.
2. A ray through, away from, or proceeding toward the focal point F is reflected parallel to the axis.
3. A ray along the radius through, away from, or proceeding toward the center of curvature C strikes the surface normally and is reflected back along its original path.
4. A ray to *the vertex V* is reflected forming equal angles with the optic axis.

Once we have found the position of the real or virtual image point by means of the (real or virtual) intersection of any two of these four principal rays, we can draw the path of any other ray from the object point to the same image point.

① Ray parallel to axis reflects through focal point.
② Ray through focal point reflects parallel to axis.
③ Ray through center of curvature intersects the surface normally and reflects along its original path.
④ Ray to vertex reflects symmetrically around optic axis.

① Reflected parallel ray appears to come from focal point.
② Ray toward focal point reflects parallel to axis.
③ As with concave mirror: Ray radial to center of curvature intersects the surface normally and reflects along its original path.
④ As with concave mirror: Ray to vertex reflects symmetrically around optic axis.

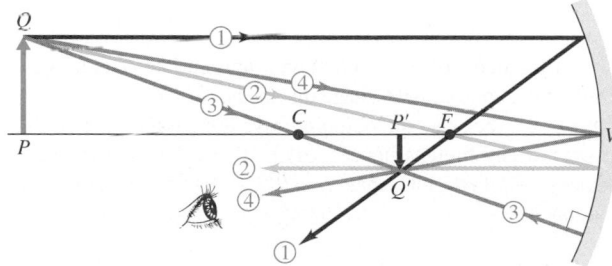

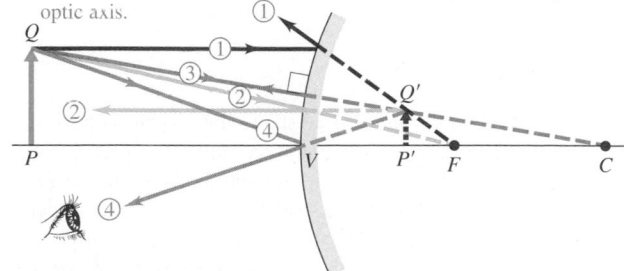

(a) Principal rays for concave mirror

(b) Principal rays for convex mirror

▲ **FIGURE 24.19** Principal-ray diagrams for concave and convex mirrors. To find the image point Q', we draw any two of these rays; the image point is located at their intersection.

PROBLEM-SOLVING STRATEGY 24.1 **Image formation by mirrors**

SET UP

1. The principal-ray diagram is to geometric optics what the free-body diagram is to mechanics! When you attack a problem involving image formation by a mirror, *always* draw a principal-ray diagram first if you have enough information. (And apply the same advice to lenses in the sections that follow.) It's usually best to orient your diagrams consistently, with the incoming rays traveling from left to right. Don't draw a lot of other rays at random; stick with the principal rays—the ones that you know something about.

2. If your principal rays don't converge at a real image point, you may have to extend them straight backward to locate a virtual image point. We recommend drawing the extensions with broken lines. Another useful aid is to color-code your principal rays consistently; for example, referring to the preceding definitions of principal rays, Figure 24.19 uses purple for 1, green for 2, orange for 3, and pink for 4.

SOLVE

3. Identify the known and unknown quantities, such as s, s', R, and f. Make lists of the known and unknown quantities, and identify the relationships among them; then substitute the known values and solve for the unknowns. Pay careful attention to *signs* of object and image distances, radii of curvature, focal lengths, lateral magnifications, and object and image heights.

REFLECT

4. Make certain you understand that the same sign rules work for all four cases in this chapter: reflection and refraction from both plane and spherical surfaces. A negative sign on any one of the quantities mentioned in item 3 *always* has significance; use the equations and the sign rules carefully and consistently, and they will tell you the truth!

EXAMPLE 24.3 **Graphical construction of an image from a mirror**

A concave mirror has a radius of curvature with absolute value 20 cm. Find graphically the image of a real object in the form of an arrow perpendicular to the axis of the mirror at each of the following object distances: **(a)** 30 cm; **(b)** 20 cm; **(c)** 10 cm; and **(d)** 5 cm. Check the construction by *computing* the image distance and lateral magnification of each image.

SOLUTION

SET UP AND SOLVE Figure 24.20 shows the graphical constructions. Study each of these diagrams carefully, comparing each numbered ray to the definitions set out earlier. Note that in Figure 24.20b the object and image distances are equal. Ray 3 cannot be drawn in this case because a ray from Q through (or proceeding from) the center of curvature C does not strike the mirror. For a similar reason, ray 2 cannot be drawn in Figure 24.20c. In this case, the outgoing rays are parallel, corresponding to an infinite image distance. In Figure 24.20d, the outgoing rays have no real intersection point; they must be extended backward to find the point from which they appear to diverge—that is, the virtual image point Q'.

Measurements of the figures, with appropriate scaling, give the following approximate image distances: (a) 15 cm, (b) 20 cm, (c) ∞ or −∞ (because the outgoing rays are parallel and do not converge at any finite distance), (d) −10 cm. To *compute* these distances, we first note that $f = R/2 = 10$ cm; then we use Equation 24.6:

$$\frac{1}{s} + \frac{1}{s'} = \frac{1}{f}.$$

The lateral magnifications measured from the figures are approximately (a) $-\frac{1}{2}$, (b) -1, (c) $\pm\infty$ (because the image distance is infinite), (d) $+2$. To *compute* the lateral magnifications, use Equation 24.7, $m = -s'/s$. Here are the computations:

Continued

Part (a): With $s = 30$ cm and $f = 10$ cm, we obtain

$$\frac{1}{30 \text{ cm}} + \frac{1}{s'} = \frac{1}{10 \text{ cm}}, \qquad s' = 15 \text{ cm},$$

$$m = -\frac{15 \text{ cm}}{30 \text{ cm}} = -\frac{1}{2}.$$

Part (b): With $s = 20$ cm and $f = 10$ cm, we get

$$\frac{1}{20 \text{ cm}} + \frac{1}{s'} = \frac{1}{10 \text{ cm}}, \qquad s' = 20 \text{ cm},$$

$$m = -\frac{20 \text{ cm}}{20 \text{ cm}} = -1.$$

Part (c): With $s = 10$ cm and $f = 10$ cm, we have

$$\frac{1}{10 \text{ cm}} + \frac{1}{s'} = \frac{1}{10 \text{ cm}}, \qquad s' = \pm\infty,$$

$$m = -\frac{\pm\infty}{10 \text{ cm}} = \mp\infty.$$

Part (d): With $s = 5$ cm and $f = 10$ cm, we find that

$$\frac{1}{5 \text{ cm}} + \frac{1}{s'} = \frac{1}{10 \text{ cm}}, \qquad s' = -10 \text{ cm},$$

$$m = -\frac{-10 \text{ cm}}{5 \text{ cm}} = +2.$$

REFLECT When the object is at the focal point, the outgoing rays are parallel and the image is at infinity. When the object distance is greater than the focal length, the image is inverted and real; when the object distance is less than the focal length, the image is erect and virtual.

Practice Problem: If an object is placed 15 cm in front of this mirror, where is the image? *Answer:* $s' = 30$ cm; light rays are reversible!

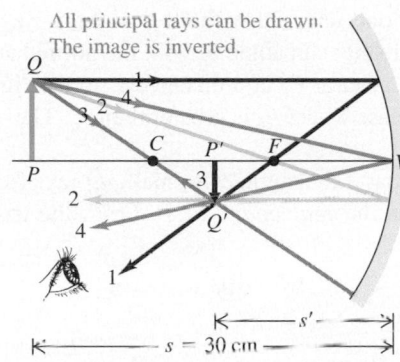

(a) Construction for $s = 30$ cm

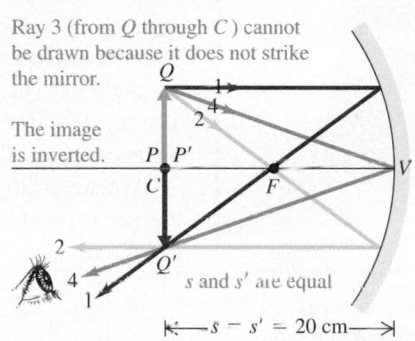

(b) Construction for $s = 20$ cm

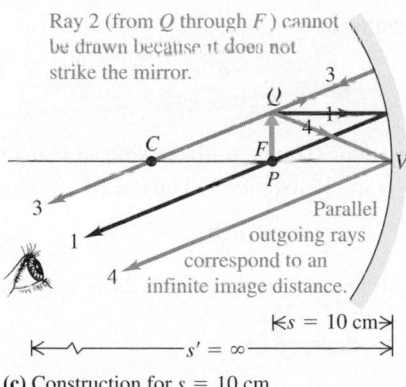

(c) Construction for $s = 10$ cm

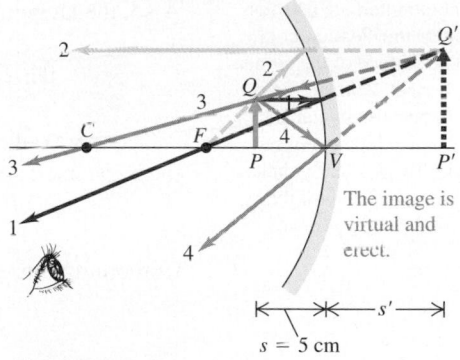

(d) Construction for $s = 5$ cm

▲ **FIGURE 24.20**

24.4 Refraction at a Spherical Surface

Our next topic is refraction at a spherical surface—that is, at a spherical interface between two optical materials with different indexes of refraction. This analysis is directly applicable to some real optical systems, such as the human eye. It also provides a stepping-stone for the analysis of lenses, which usually have *two* spherical surfaces.

In Figure 24.21, a spherical surface with radius R forms an interface between two materials with indexes of refraction n_a and n_b. The surface forms an image

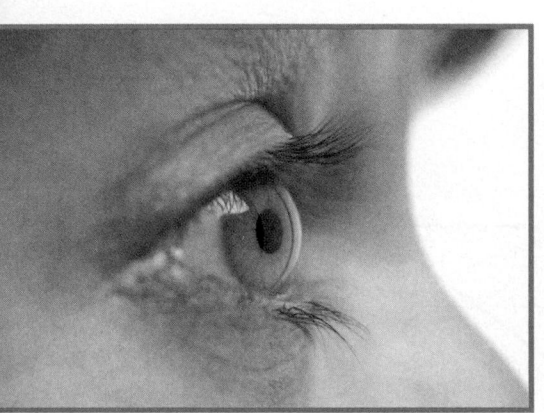

▲ FIGURE 24.21 Construction for finding the position of the image point P' of an object point P, formed by refraction at a spherical surface.

▲ BIO Application Seeing in focus. To see clearly, your eye must focus the light reaching it to form a crisp image on the retina (the light-receptive tissue at the back of the eye). Most of this focusing happens through refraction at the interface between the air and the cornea—the clear outer element of your eye. The surface of the cornea is approximately spherical. Corneal focusing works only in air, because it depends on the difference in refractive index between air and corneal tissue. To see clearly under water, you need to wear a face mask that creates a pocket of air around your eyes.

point P' of an object point P; we want to find how the object and image distances (s and s') are related. We'll use the same sign rules that we used for spherical mirrors. The center of curvature C is on the outgoing side of the surface, so R is positive. Ray PV strikes the vertex V and is perpendicular to the surface at that point. It passes into the second material without deviation. Ray PB, making an angle α with the axis, is incident at an angle θ_a with the normal and is refracted at an angle θ_b. These rays intersect at P' at a distance s' to the right of the vertex. The figure is drawn for the case where n_b is greater than n_a. The object and image distances are both positive.

We are going to prove that if the angle α is small, *all* rays from P intersect at the same point P', so P' is the *real image* of P. From the triangles PBC and $P'BC$, we have

$$\theta_a = \alpha + \phi, \qquad \phi = \beta + \theta_b. \tag{24.8}$$

From the law of refraction,

$$n_a \sin\theta_a = n_b \sin\theta_b.$$

Also, the tangents of α, β, and ϕ are

$$\tan\alpha = \frac{h}{s+\delta}, \qquad \tan\beta = \frac{h}{s'-\delta}, \qquad \tan\phi = \frac{h}{R-\delta}. \tag{24.9}$$

For paraxial rays, we may approximate both the sine and tangent of an angle by the angle itself and neglect the small distance δ. The law of refraction then becomes

$$n_a\theta_a = n_b\theta_b.$$

Combining this equation with the first of Equations 24.8, we obtain

$$\theta_b = \frac{n_a}{n_b}(\alpha + \phi).$$

When we substitute this expression into the second of Equations 24.8, we get

$$n_a\alpha + n_b\beta = (n_b - n_a)\phi. \tag{24.10}$$

Now we use the approximations $\tan\alpha = \alpha$, and so on, in Equations 24.9, and we neglect δ; those equations then become

$$\alpha = \frac{h}{s}, \qquad \beta = \frac{h}{s'}, \qquad \phi = \frac{h}{R}.$$

Finally, we substitute these equations into Equation 24.10 and divide out the common factor h; we finally obtain

$$\frac{n_a}{s} + \frac{n_b}{s'} = \frac{n_b - n_a}{R}. \tag{24.11}$$

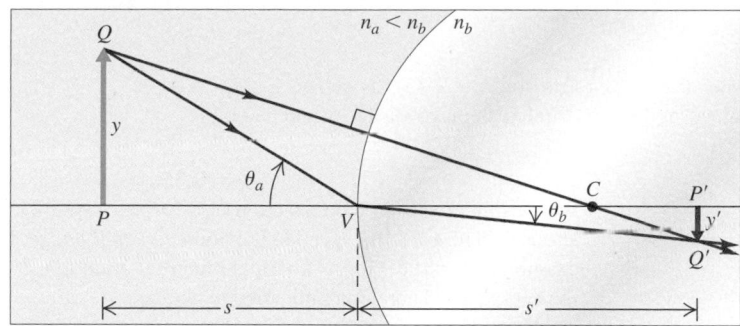

▲ **FIGURE 24.22** Construction for determining the height of an image formed by refraction at a spherical surface.

This equation does not contain the angle α, so the image distance is the same for *all* paraxial rays from P.

To obtain the lateral magnification for this situation, we use the construction in Figure 24.22. We draw two rays from point Q, one through the center of curvature C and the other through the vertex V. From the triangles PQV and $P'Q'V$,

$$\tan\theta_a = \frac{y}{s}, \qquad \tan\theta_b = \frac{-y'}{s'},$$

and from the law of refraction,

$$n_a \sin\theta_a = n_b \sin\theta_b.$$

For small angles,

$$\tan\theta_a = \sin\theta_a, \qquad \tan\theta_b = \sin\theta_b,$$

so, finally,

$$\frac{n_a y}{s} = -\frac{n_b y'}{s'},$$

or

$$m = \frac{y'}{y} = -\frac{n_a s'}{n_b s}. \tag{24.12}$$

▲ **Application Sometimes the sun is green.** As the sun approaches the horizon, a number of interesting phenomena can be observed, especially if the view of the horizon is unrestricted, such as at sea. One of these is the "green flash" in which the last portion of the setting sun appears to be green. While the observation and analysis of this phenomenon is complicated by atmospheric conditions, the phenomenon is essentially a result of refraction along the long path the light must take through the atmosphere at sunset. Under the right conditions, the atmosphere acts as a lens to refract the setting sun's rays into their component parts. The blue-green rays of the spectrum are refracted more than the red, and thus as the sun sinks below the horizon, we can see the shorter wavelengths for a bit longer than the longer wavelengths. As the blue light is scattered strongly by the atmosphere, we see the rim of the sinking sun as a green segment. This may be followed by a rarely observed vertical green ray. The green ray has been the subject of stories and myths, including an 1882 novel by Jules Verne, *The Green Ray*.

EXAMPLE 24.4 Glass rod in air

The cylindrical glass rod in Figure 24.23 has index of refraction 1.52. One end is ground to a hemispherical surface with radius $R = 2.00$ cm. The rod is surrounded by air. **(a)** Find the image distance of a small object on the axis of the rod and 8.00 cm to the left of the vertex. **(b)** Find the lateral magnification.

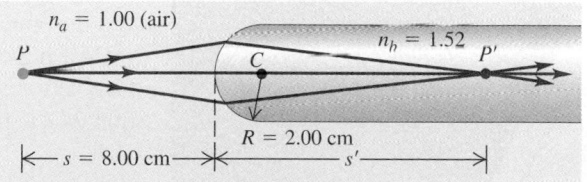

▲ **FIGURE 24.23**

SOLUTION

SET UP AND SOLVE We are given that

$$n_a = 1.00, \qquad n_b = 1.52,$$
$$R = +2.00 \text{ cm}, \qquad s = +8.00 \text{ cm}.$$

Part (a): From Equation 24.11,

$$\frac{1.00}{8.00 \text{ cm}} + \frac{1.52}{s'} = \frac{1.52 - 1.00}{+2.00 \text{ cm}},$$
$$s' = +11.3 \text{ cm}.$$

Part (b): From Equation 24.12, the lateral magnification is

$$m = -\frac{n_a s'}{n_b s} = -\frac{(1.00)(11.3 \text{ cm})}{(1.52)(8.00 \text{ cm})} = -0.929.$$

REFLECT Our result for part (a) tells us that the image is formed to the right of the vertex (because s' is positive) at a distance of 11.3 cm from it. From part (b), we know that the image is somewhat smaller than the object, and it is inverted. If the object is an arrow 1.00 mm high pointing upward, the image is an arrow 0.929 mm high pointing downward.

EXAMPLE 24.5 Glass rod in water

The glass rod of Example 24.4 is immersed in water (index of refraction $n = 1.33$), as shown in Figure 24.24. The other quantities have the same values as before. Find the image distance and lateral magnification.

SOLUTION

SET UP AND SOLVE From Equation 24.11,

$$\frac{1.33}{8.00 \text{ cm}} + \frac{1.52}{s'} = \frac{1.52 - 1.33}{+2.00 \text{ cm}},$$
$$s' = -21.3 \text{ cm}.$$

The fact that s' is negative means that after the rays are refracted by the surface, they are not converging, but *appear* to diverge from a point 21.3 cm to the *left* of the vertex. We have

seen a similar case in the refraction of diverging rays by a plane surface (Figure 24.2); we called the point a *virtual image*. In this example, the surface forms a virtual image 21.3 cm to the left of the vertex. The lateral magnification m is then

$$m = -\frac{(1.33)(-21.3 \text{ cm})}{(1.52)(8.00 \text{ cm})} = +2.33.$$

REFLECT In this case, the image is erect (because m is positive) and 2.33 times as large as the object.

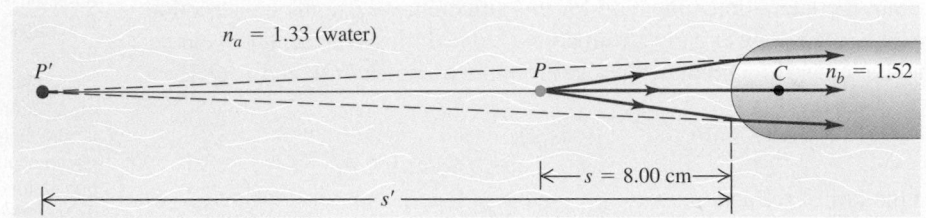

$n_a = 1.33$ (water)

P' P C $n_b = 1.52$

$\longleftarrow s = 8.00 \text{ cm} \longrightarrow$

► **FIGURE 24.24** s'

Equations 24.11 and 24.12 can be applied to both convex and concave refracting surfaces, provided that you use the sign rules consistently. It doesn't matter whether n_b is greater or less than n_a. We suggest that you construct diagrams like Figures 24.21 and 24.22 when R is negative and when n_b is less than n_a, and use them to derive Equations 24.11 and 24.12 for these cases.

Here's a final note on the sign rule for the radius of curvature R of a surface: For the convex reflecting surface in Figure 24.15 we considered R negative, but in Figure 24.21 the refracting surface with the same orientation has a *positive* value of R. This may seem inconsistent, but it isn't. Both cases are consistent with the rule that R is positive when the center of curvature is on the outgoing side of the surface and negative when it is *not* on the outgoing side. When both reflection and refraction occur at a spherical surface, R has one sign for the reflected light and the opposite sign for the refracted light.

An important special case of a spherical refracting surface is a *plane* surface between two optical materials. This corresponds to setting $R = \infty$ in Equation 24.11. In this case,

$$\frac{n_a}{s} + \frac{n_b}{s'} = 0. \qquad (24.13)$$

To find the lateral magnification m for this case, we combine Equation 24.13 with the general relation Equation 24.12, obtaining the simple result

$$m = 1. \qquad (24.14)$$

That is, the image formed by a *plane* refracting surface is always the same size as the object, and it is always erect. For a real object, the image is always virtual.

A familiar example of image formation by a plane refracting surface is the appearance of a partly submerged drinking straw or canoe paddle (Figure 24.25). When viewed from some angles, the object appears to have a sharp bend at the water surface because the submerged part appears to be only about three-quarters of its actual distance below the surface.

▲ **FIGURE 24.25** An object partially submerged in water appears to bend at the surface, owing to the refraction of the light rays.

EXAMPLE 24.6 How deep is the pool?

Swimming-pool owners know that the pool always looks shallower than it really is and that it is important to identify clearly the deep parts so that people who can't swim won't jump into water that's over their heads. If a nonswimmer looks down into water that is actually 2.00 m (about 6 ft, 7 in) deep, how deep does it appear to be?

SOLUTION

SET UP Figure 24.26 shows a diagram for the problem. We pick an arbitrary object point P on the bottom of the pool; the image distance s' from the surface of the pool to the image point P' is the apparent depth of the pool. (Note that the image is virtual and that s' is negative.) The rays we draw confirm that the pool does indeed appear shallower than it actually is. (P' is above the actual bottom.)

SOLVE To find s', we use Equation 24.13:

$$\frac{n_a}{s} + \frac{n_b}{s'} = 0,$$

$$\frac{1.33}{2.00 \text{ m}} + \frac{1.00}{s'} = 0,$$

$$s' = -1.50 \text{ m}.$$

REFLECT The apparent depth is only about three-quarters of the actual depth, or about 4 ft, 11 in. A 6 ft nonswimmer who didn't allow for this effect would be in trouble. The negative sign shows that the image is virtual and on the incoming side of the refracting surface.

Practice Problem: An ice fisherman would like to check the thickness of the ice to see whether it is safe to walk on. He knows

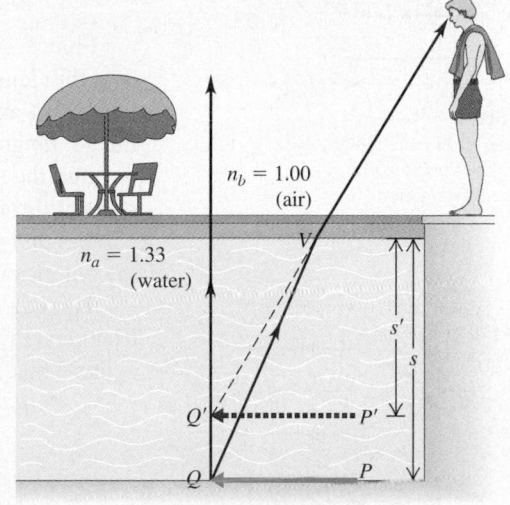

▲ **FIGURE 24.26**

that the ice ($n_{ice} = 1.309$) should be 15.0 cm thick to be safe. If he looks straight down through the ice, how thick will the ice appear to be if it is just thick enough for safe walking? *Answer:* $s' = -11.5$ cm.

24.5 Thin Lenses

PhET: Geometric Optics

The most familiar and widely used optical device (after the plane mirror) is the *lens.* A lens is an optical system with two refracting surfaces. The simplest lens has two *spherical* surfaces close enough together that we can neglect the distance between them (the thickness of the lens); we call this a **thin lens.** We can analyze such a system in detail by using the results of Section 24.4 for refraction by a single spherical surface. However, we postpone this analysis until later in this section so that we can first discuss the properties of thin lenses.

Focal Points of a Lens

Figure 24.27 shows a lens with two spherical surfaces; this type of lens is thickest at its center. The central horizontal line is called the *optic axis,* as with spherical mirrors. The centers of curvature of the two spherical surfaces lie on and define the optic axis.

When a beam of rays parallel to the optic axis passes through the lens, the rays converge to a point F_2 (Figure 24.27a). Similarly, rays passing through point F_1 emerge from the lens as a beam of rays parallel to the optic axis (Figure 24.27b). The points F_1 and F_2 are called the first and second *focal points,* and the distance f (measured from the center of the lens) is called the *focal length.* We've already used the concepts of focal point and focal length for spherical mirrors in Section 24.2. The two focal lengths in Figure 24.27, both labeled f, are *always equal* for a thin lens, even when the two sides have different curvatures. We will derive this somewhat surprising result later in this section, when we

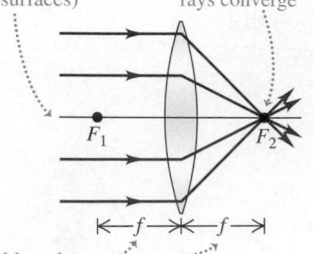

Optic axis (passes through centers of curvature of both lens surfaces)

Second focal point: the point to which incoming parallel rays converge

Focal length
• Measured from lens center.
• Always the same on both sides of the lens.
• For a converging thin lens, f is positive.

(a)

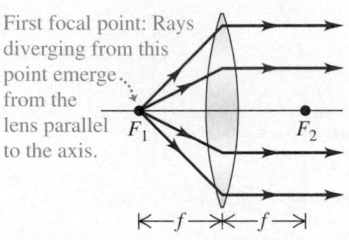

First focal point: Rays diverging from this point emerge from the lens parallel to the axis.

(b)

▲ **FIGURE 24.27** The focal points of a converging thin lens. The numerical value of f is positive in this case.

derive the relationship of f to the index of refraction of the lens and the radii of curvature of its surfaces.

NOTE ▶ A ray passing through a thin lens is refracted at both spherical surfaces, but we'll usually draw the rays as bent at the midplane of the lens rather than at the spherical surfaces. This scheme is consistent with the assumption that the lens is very thin. ◀

Figure 24.28 shows how we can determine the position of the image formed by a thin lens. Using the same notation and sign rules as before, we let s and s' be the object and image distances, respectively, and let y and y' be the object and image heights. Ray QA, parallel to the optic axis before refraction, passes through the second focal point F_2 after refraction. Ray QOQ' passes undeflected straight through the center of the lens, because at the center the two surfaces are parallel and (we have assumed) very close together. There is refraction where the ray enters and leaves the material, but no net change in direction.

The two angles labeled α in Figure 24.28 are equal. Therefore, the two right triangles PQO and $P'Q'O$ are *similar*, and ratios of corresponding sides are equal. Thus,

$$\frac{y}{s} = -\frac{y'}{s'}, \quad \text{or} \quad \frac{y'}{y} = -\frac{s'}{s}. \quad (24.15)$$

(The reason for the negative sign is that the image is below the optic axis and y' is negative.) Also, the two angles labeled β are equal, and the two right triangles OAF_2 and $P'Q'F_2$ are similar. Hence, we have

$$\frac{y}{f} = -\frac{y'}{s' - f},$$

or

$$\frac{y'}{y} = -\frac{s' - f}{f}. \quad (24.16)$$

We now equate Equations 24.15 and 24.16, divide by s', and rearrange the resulting equation to obtain

$$\frac{1}{s} + \frac{1}{s'} = \frac{1}{f}. \quad (24.17)$$

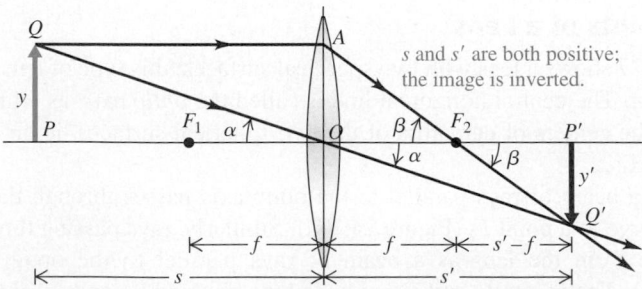

s and s' are both positive; the image is inverted.

▲ **FIGURE 24.28** Construction used to find the image position of a thin lens. To emphasize that the lens is assumed to be very thin, the upper ray is shown bent at the midplane and the lower ray is shown as a straight line.

▶ Application **Coming soon to a disc drive near you . . . .** Data stored on optical media such as DVDs is read with a laser. To increase the density of optical data storage, it was necessary to devise very small lenses that were capable of accurately focusing light of short wavelengths (in the blue and ultraviolet). Such lenses are often made out of diamond, which has an exceptionally high index of refraction (enabling the lens to be small) and is transparent to the desired wavelengths. The lens shown is about 1 mm across and is made from pure synthetic diamond.

This analysis also gives us the lateral magnification $m = y'/y$ of the system; from Equation 24.15,

$$m = \frac{y'}{y} = -\frac{s'}{s}. \tag{24.18}$$

The negative sign tells us that when s and s' are both positive, the image is *inverted,* and y and y' have opposite signs.

Equations 24.17 and 24.18 are the basic equations for thin lenses. It is pleasing to note that their *form* is exactly the same as that of the corresponding equations for spherical mirrors, Equations 24.6 and 24.7. As we'll see, the same sign rules that we used for spherical mirrors are also applicable to lenses.

Figure 24.29 shows how a lens forms a three-dimensional image of a three-dimensional object. The real thumb points toward the lens, and the thumb's image points away from the lens. Thus, in contrast to the image produced by a plane mirror (Figure 24.6), the image is *not* reversed: The image of a left hand is a left hand. However, the real index finger points up, while its image points down, so the image produced by the lens is inverted. To make a real hand match the image produced by this lens, you would simply rotate the hand by 180° around the lens axis. Inversion of an image is equivalent to a rotation of 180° about the lens axis.

A bundle of parallel rays incident on the lens shown in Figure 24.27 converges to a real image point after passing through the lens. This lens is called a **converging lens.** Its focal length is a positive quantity, so it is also called a *positive lens.*

A bundle of parallel rays incident on the lens in Figure 24.30 *diverges* after refraction, and this lens is called a **diverging lens.** Its focal length is a negative quantity, and the lens is also called a *negative lens.* The focal points of a negative lens are *virtual* focal points, and they are reversed relative to the focal points of a positive lens. The second focal point, F_2, of a negative lens is the point from which rays that are originally parallel to the axis *appear to diverge*

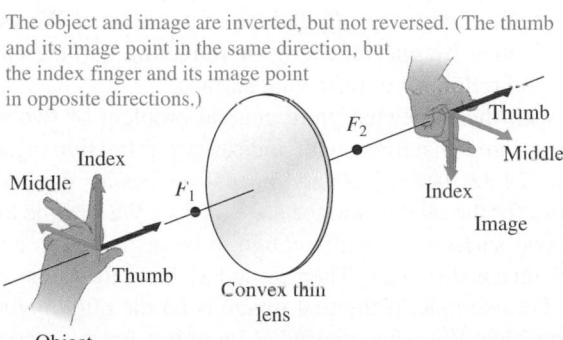

The object and image are inverted, but not reversed. (The thumb and its image point in the same direction, but the index finger and its image point in opposite directions.)

▲ **FIGURE 24.29** A lens does not produce a reversed image.

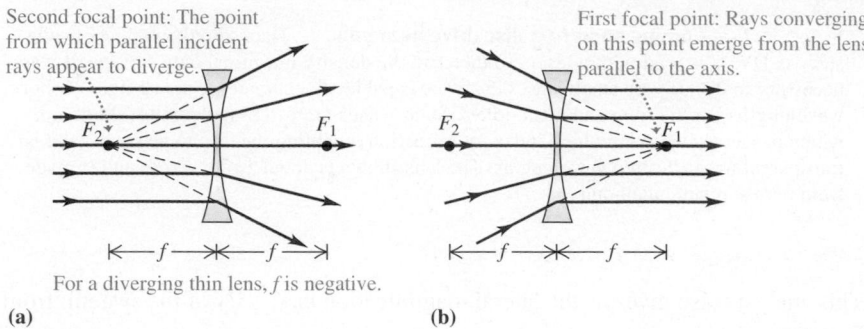

Second focal point: The point from which parallel incident rays appear to diverge.

First focal point: Rays converging on this point emerge from the lens parallel to the axis.

For a diverging thin lens, *f* is negative.

(a)   **(b)**

▲ **FIGURE 24.30** F_2 and F_1 are, respectively, the second and first focal points of a diverging thin lens.

after refraction, as in Figure 24.30a. Incident rays converging toward the first focal point F_1, as in Figure 24.30b, emerge from the lens parallel to its axis.

Equations 24.17 and 24.18 apply to *both* negative and positive lenses. Various types of lenses, both converging and diverging, are shown in Figure 24.31. Any lens that is thicker in the center than at the edges is a converging lens with positive f, and any lens that is thinner at the center than at the edges is a diverging lens with negative f (provided that the lens material has a greater index of refraction than the surrounding material has). We can prove this statement using Equation 24.20, which we haven't yet derived.

Conceptual Analysis 24.3

Half of a lens

A lens projects an image on a screen. If the left half of the lens is covered, then

A. the left half of the image disappears.
B. the right half of the image disappears.
C. the entire image disappears.
D. the image becomes fainter.

SOLUTION When you look at nearly any object, you can usually see it from many different positions of your eye. This means that

light rays are leaving the object and traveling out in many directions. If the object is in front of a lens, then the many rays from a single point on the object leave that point and hit every point of the lens. These rays are all refracted, and they all intersect at the position of the image. When you block half the lens, you block only half the rays leaving a given point on the object, so the resulting image is dimmer, but complete.

The Thin-Lens Equation

Now we proceed to derive Equation 24.17 in more detail. At the same time, we'll derive the relationship between the focal length f of the lens, its index of refraction, n, and the radii of curvature, R_1 and R_2, of its surfaces. We use the principle that an image formed by one reflecting or refracting surface can serve as the object for a second reflecting or refracting surface.

We begin with the somewhat more general problem of two spherical interfaces separating three materials with indexes of refraction n_a, n_b, and n_c, as shown in Figure 24.32. The object and image distances for the first surface are s_1 and s_1', and those for the second surface are s_2 and s_2'. We assume that the distance t between the two surfaces is small enough to be neglected in comparison with the object and image distances. Then s_2 and s_1' have the same magnitude, but opposite sign. For example, if the first image is on the outgoing side of the first surface, s_1' is positive. But when viewed as an object for the second surface, it is *not* on the incoming side of that surface. So we can say that $s_2 = -s_1'$.

We need to use the single-surface equation, Equation 24.11, twice, once for each surface. The two resulting equations are

$$\frac{n_a}{s_1} + \frac{n_b}{s_1'} = \frac{n_b - n_a}{R_1},$$

$$\frac{n_b}{s_2} + \frac{n_c}{s_2'} = \frac{n_c - n_b}{R_2}.$$

Ordinarily, the first and third materials are air or vacuum, so we set $n_a = n_c = 1$. The second index, n_b, is that of the lens material; we now call that index simply n. Substituting these values and the relation $s_2 = -s_1'$ into the preceding equations, we get

$$\frac{1}{s_1} + \frac{n}{s_1'} = \frac{n - 1}{R_1},$$

$$-\frac{n}{s_1'} + \frac{1}{s_2'} = \frac{1 - n}{R_2}.$$

To get a relation between the initial object position s_1 and the final image position s_2', we add these two equations. This eliminates the term n/s_1', and we obtain

$$\frac{1}{s_1} + \frac{1}{s_2'} = (n - 1)\left(\frac{1}{R_1} - \frac{1}{R_2}\right).$$

Finally, thinking of the lens as a single unit, we call the object distance simply s instead of s_1 and the final image distance s' instead of s_2'. Making these substitutions, we have

$$\frac{1}{s} + \frac{1}{s'} = (n - 1)\left(\frac{1}{R_1} - \frac{1}{R_2}\right). \tag{24.19}$$

Now we compare this equation with the other thin-lens equation, Equation 24.17. We see that the object and image distances s and s' appear in exactly the same places in both equations and that the focal length f is given by

$$\frac{1}{f} = (n - 1)\left(\frac{1}{R_1} - \frac{1}{R_2}\right). \tag{24.20}$$

This relation is called the **thin-lens equation,** or the *lensmaker's equation.* In the process of rederiving the thin-lens equation, we have also derived an expression for the focal length f of a lens in terms of its index of refraction n and the radii of curvature R_1 and R_2 of its surfaces. Combining Equations 24.19 and 24.20 gives the general form of the thin-lens equation:

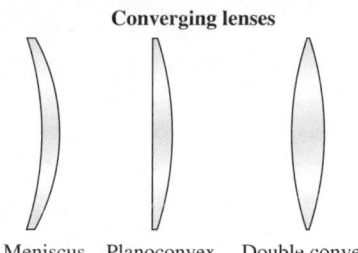

Converging lenses

Meniscus   Planoconvex   Double convex

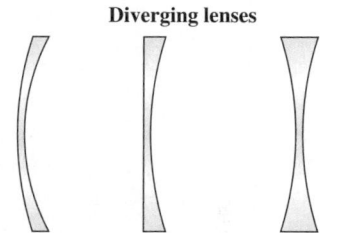

Diverging lenses

Meniscus   Planoconcave   Double concave

▲ **FIGURE 24.31** A selection of lenses.

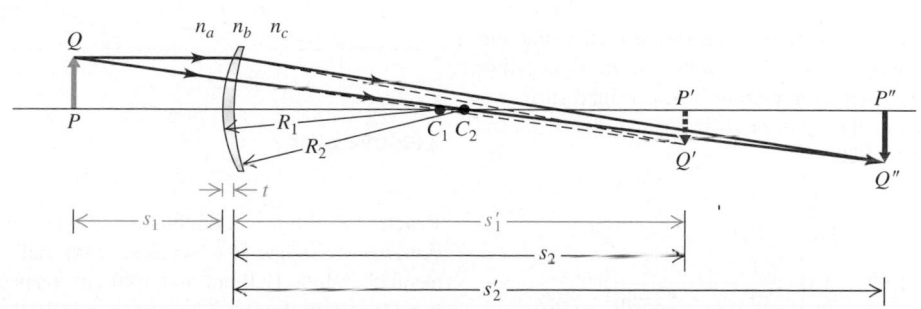

▲ **FIGURE 24.32** The image formed by the first surface of a lens serves as the object for the second surface.

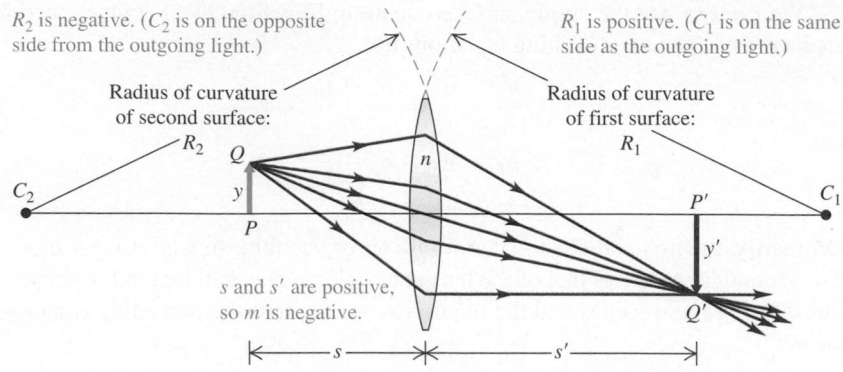

R_2 is negative. (C_2 is on the opposite side from the outgoing light.)

R_1 is positive. (C_1 is on the same side as the outgoing light.)

Radius of curvature of second surface: R_2

Radius of curvature of first surface: R_1

s and s' are positive, so m is negative.

▲ **FIGURE 24.33** A thin lens. This lens is converging and thus has a positive focal length f.

Thin-lens equation

$$\frac{1}{s} + \frac{1}{s'} = \frac{1}{f} = (n - 1)\left(\frac{1}{R_1} - \frac{1}{R_2}\right).$$ (24.21)

We use all our previous sign conventions with Equations 24.19, 24.20, and 24.21. For example, in Figure 24.33, s, s', and R_1 are positive, but R_2 is negative.

Quantitative Analysis 24.4 **Object distance and magnification**

A certain lens forms a real image of a very distant object on a screen that is 10 cm from the lens. As the object is moved closer to the lens, at what distance of the object from the lens are the object and image the same height?

A. only at 20 cm.
B. somewhere between 20 cm and 40 cm.
C. at 40 cm.
D. only at 10 cm.

SOLUTION If the object distance s is infinite, the image distance s' is equal to the focal length f. Thus, the focal length of our lens is 10 cm. For the object and image to be the same height, the object and image distances must be the same; hence, $s = s'$. The thin-lens equation (Equation 24.17 or 24.21) then gives us

$$\frac{1}{s} + \frac{1}{s} = \frac{1}{f} = \frac{1}{10\text{ cm}}, \qquad s = 20\text{ cm}.$$

The object and image distances (and therefore their heights) are equal only when $s = s' = 20$ cm.

EXAMPLE 24.7 **Double-concave diverging lens**

The two surfaces of the lens shown in Figure 24.34 have radii of curvature with absolute values of 20 cm and 5.0 cm, respectively. The index of refraction is 1.52. What is the focal length f of the lens?

SOLUTION

SET UP AND SOLVE The center of curvature of the first surface is on the incoming side of the light, so R_1 is negative: $R_1 = -20$ cm. The center of curvature of the second surface is on the outgoing side of the light, so R_2 is positive: $R_2 = 5.0$ cm. Then, from Equation 24.20,

$$\frac{1}{f} = (n - 1)\left(\frac{1}{R_1} - \frac{1}{R_2}\right)$$

$$= (1.52 - 1)\left(\frac{1}{-20\text{ cm}} - \frac{1}{5.0\text{ cm}}\right),$$

$$f = -7.7\text{ cm}.$$

REFLECT This lens is a diverging lens—a negative lens—with a negative focal length.

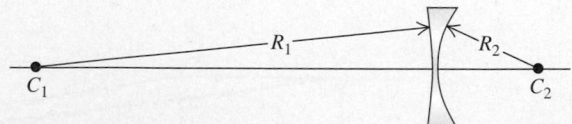

▲ **FIGURE 24.34**

Practice Problem: A double-concave lens with index of refraction $n = 1.50$ has two surfaces with radii of curvature with absolute values 12.0 cm and 10.0 cm, respectively. What is the focal length of the lens? *Answer:* $f = -10.9$ cm.

EXAMPLE 24.8   The eye's lens

When light enters your eye, most of the focusing happens at the interface between the air and the cornea (the outermost element of the eye). The eye also has a doubly-convex *lens,* lying behind the cornea, that completes the job of forming an image on the retina. (The lens also enables us to shift our distance of focus; it gets rounder for near vision and flatter for far vision.) The lens has an index of refraction of about 1.40. **(a)** For the lens in Figure 24.35, find the focal length. **(b)** If you could consider this lens in isolation from the rest of the eye, what would the image distance be for an object 0.20 m in front of this crystalline lens?

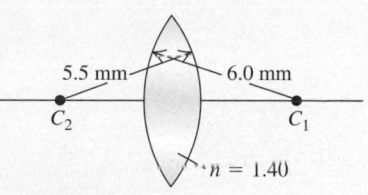

▲ **FIGURE 24.35**

SOLUTION

SET UP The center of curvature of the first surface of the lens is on the outgoing side, so $R_1 = +6.0$ mm. The center of curvature for the second surface is *not* on the outgoing side, so $R_2 = -5.5$ mm. We solve for f, then use the result in Equation 24.21.

SOLVE Part (a):   Using Equation 24.20, we find that

$$\frac{1}{f} = (1.40 - 1)\left(\frac{1}{+6.0 \text{ mm}} - \frac{1}{-5.5 \text{ mm}}\right),$$
$$f = 7.2 \text{ mm}.$$

Part (b):   The object distance is $s = 0.20$ m = 200 mm. Using Equation 24.21, we obtain

$$\frac{1}{200 \text{ mm}} + \frac{1}{s'} = \frac{1}{7.2 \text{ mm}}, \qquad s' = 7.5 \text{ mm}.$$

REFLECT   The image is slightly farther from the lens than it would be for an infinitely distant object. As expected for a converging lens, the focal length is positive.

It's not hard to generalize Equation 24.20 or 24.21 to the situation in which the lens is immersed in a material with an index of refraction greater than unity. We invite you to work out the thin-lens equation for this more general situation.

24.6 Graphical Methods for Lenses

We can determine the position and size of an image formed by a thin lens by using a graphical method that is similar to the one we used in Section 24.3 for spherical mirrors. Again, we draw a few special rays called *principal rays* that diverge from a point of the object that is *not* on the optic axis. The intersection of these rays, after they pass through the lens, determines the position and size of the image. In using this graphical method, as in Section 24.5, we consider the entire deviation of a ray as occurring at the midplane of the lens, as shown in Figure 24.36; this is consistent with the assumption that the distance between the lens surfaces is negligible.

① Parallel incident ray refracts to pass through second focal point F_2
② Ray through center of lens (does not deviate appreciably)
③ Ray through the first focal point F_1 that emerges parallel to the axis

① Parallel incident ray appears after refraction to have come from the second focal point F_2
② Ray through center of lens (does not deviate appreciably)
③ Ray aimed at the first focal point F_1 that emerges parallel to the axis

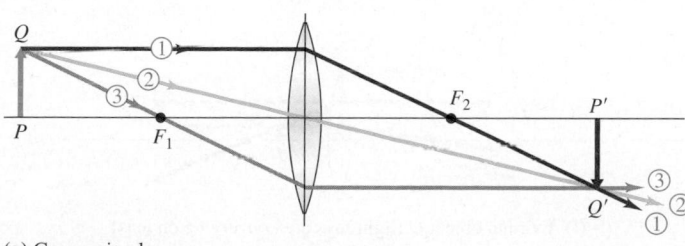

(a) Converging lens

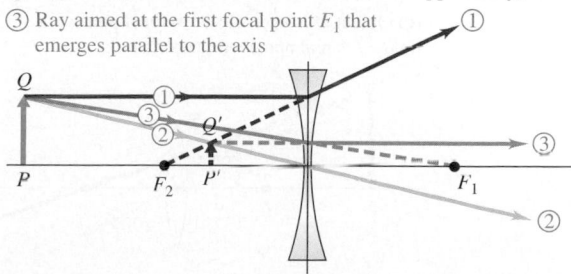

(b) Diverging lens

▲ **FIGURE 24.36** Principal-ray diagrams showing the graphical method for locating an image produced by a thin lens.

The three principal rays whose paths are usually easy to trace for lenses are shown in Figure 24.36. They are as follows:

Principal rays for thin lenses

1. *A ray parallel to the axis,* after refraction by the lens, passes through the second focal point F_2 of a converging lens or appears to come from the second focal point of a diverging lens (purple in diagrams).
2. *A ray through the center of the lens* is not appreciably deviated, because, at the center of the lens, the two surfaces are parallel and close together (green in diagrams).
3. *A ray through, away from, or proceeding toward the first focal point F_1* emerges parallel to the axis (orange in diagrams).

When the image is real, the position of the image point is determined by the intersection of any two of the three principal rays (Figure 24.36a). When the image is virtual, the outgoing rays diverge. In this case, we extend the diverging outgoing rays backward to their intersection point (Figure 24.36b). Once the image position is known, we can draw any other ray from the same point. Usually, nothing is gained by drawing a lot of additional rays.

Figure 24.37 shows several principal-ray diagrams for a converging lens for several object distances. We suggest that you study each of these diagrams very carefully, comparing each numbered ray with the preceding description. Several points are worth noting. In Figure 24.37d, the object is at the focal point, and ray 3 cannot be drawn because it does not pass through the lens. In Figure 24.37e, the object distance is less than the focal length. The outgoing rays are divergent, and the *virtual image* is located by extending the outgoing rays backward. In this case, the image distance s' is negative. Figure 24.37f corresponds to a *virtual*

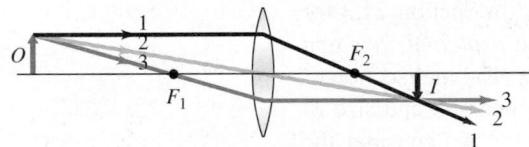

(a) Object O is outside focal point; image I is real.

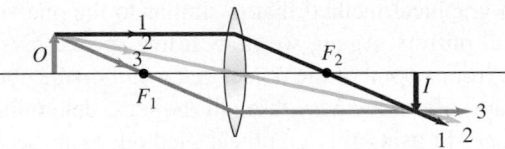

(b) Object O is closer to focal point; image I is real and farther away.

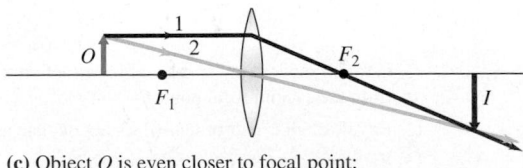

(c) Object O is even closer to focal point; image I is real and even farther away.

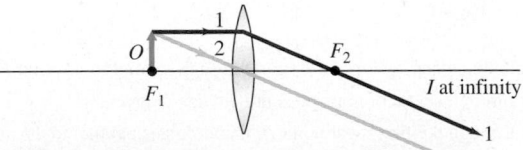

(d) Object O is at focal point; image I is at infinity.

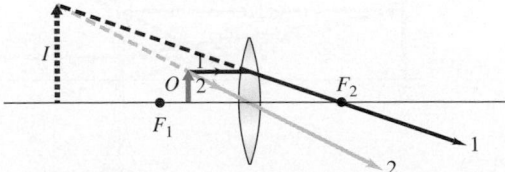

(e) Object O is inside focal point; image I is virtual and larger than object.

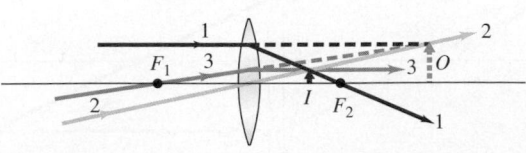

(f) A virtual object O (light rays are *converging* on lens)

▲ **FIGURE 24.37** Formation of images by a thin converging lens for various object distances. The principal rays are numbered.

object. The incoming rays do not diverge from a real object point *O*, but are *converging* as though they would meet at the virtual object point *O* on the right side. The object distance *s* is negative in this case. The image is real; the image distance *s'* is positive and less than *f*.

Conceptual Analysis 24.5

Possible rays

In Figure 28.38, which of the rays (A–D) *did not* originate from point *Q* at the top of the object?

SOLUTION Because the lens is a diverging lens, rays spread apart after passing through it. Ray C comes from point *Q* and passes in a straight line through the lens, crossing the optic axis at its center. When rays A and D are projected backward, they pass through the focal point F_2; thus, before they passed through the lens, they were parallel to the axis of the lens. Ray A therefore came from *Q*; ray D did not. Ray B is parallel to the axis after refraction, so it was directed toward focal point F_1 before being refracted. Thus, rays D and B do not come from *Q*.

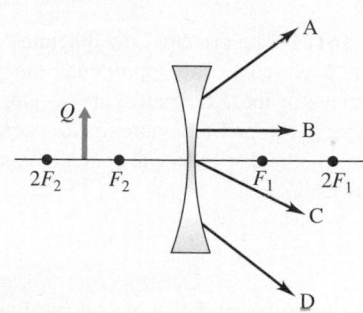

▲ **FIGURE 24.38**

PROBLEM-SOLVING STRATEGY 24.2   **Image formation by a thin lens**

SET UP

1. The strategy outlined in Section 24.3 is equally applicable here, and we suggest that you review it now. Always begin with a principal-ray diagram if you have enough information. Orient your diagrams consistently so that light travels from left to right. For a lens, there are only three principal rays, compared to four for a mirror. Don't just sketch these diagrams; draw the rays with a ruler and measure the distances carefully. Draw the rays so that they bend at the midplane of the lens, as shown in Figure 24.37. Be sure to draw *all three* principal rays whenever possible. The intersection of any two determines the image, but if the third doesn't pass through the same intersection point, you know that you've made a mistake. Redundancy can be useful in spotting errors.

SOLVE

2. If the outgoing principal rays don't converge at a real image point, the image is virtual. Then you have to extend the outgoing rays backward to find the virtual image point, which lies on the *incoming* side of the lens.

3. The same sign rules that we've used for mirrors and single refracting surfaces are also applicable to thin lenses. Be extremely careful to get your signs right and to interpret the signs of results correctly.

REFLECT

4. Always determine the image position and size *both* graphically and by calculating. Then compare the results. This gives an extremely valuable consistency check.

5. Remember that the *image* from one lens or mirror may serve as the *object* for another. In that case, be careful in finding the object and image *distances* for this intermediate image; be sure you include the distance between the two elements (lenses and/or mirrors) correctly.

EXAMPLE 24.9 Image formed by a converging lens

A converging lens has a focal length of 20 cm. Find graphically the image location for an object at each of the following distances from the lens: **(a)** 50 cm; **(b)** 20 cm; **(c)** 15 cm; **(d)** −40 cm. Determine the lateral magnification in each case. Check your results by calculating the image distance and lateral magnification.

SOLUTION

SET UP AND SOLVE The principal-ray diagrams are shown in Figures 24.37a, d, e, and f. The approximate image distances, from measurements of these diagrams, are 35 cm, ∞, −40 cm, and 15 cm, respectively, and the approximate lateral magnifications are $-\frac{2}{3}$, ∞, +3, and $+\frac{1}{3}$. To calculate the image positions, we use Equation 24.17:

$$\frac{1}{s} + \frac{1}{s'} = \frac{1}{f}.$$

To calculate the lateral magnifications, we use Equation 24.18, $m = -s'/s$.

Part (a): $s = 50$ cm and $f = 20$ cm:

$$\frac{1}{50 \text{ cm}} + \frac{1}{s'} = \frac{1}{20 \text{ cm}}, \qquad s' = 33.3 \text{ cm},$$

$$m = -\frac{33.3 \text{ cm}}{50 \text{ cm}} = -\frac{2}{3}.$$

Part (b): $s = 20$ cm and $f = 20$ cm:

$$\frac{1}{20 \text{ cm}} + \frac{1}{s'} = \frac{1}{20 \text{ cm}}, \qquad s' = \pm\infty,$$

$$m = -\frac{\pm\infty}{20 \text{ cm}} = \mp\infty.$$

Part (c): $s = 15$ cm and $f = 20$ cm:

$$\frac{1}{15 \text{ cm}} + \frac{1}{s'} = \frac{1}{20 \text{ cm}}, \qquad s' = -60 \text{ cm},$$

$$m = -\frac{-60 \text{ cm}}{15 \text{ cm}} = +4.$$

Part (d): $s = -40$ cm and $f = 20$ cm:

$$\frac{1}{-40 \text{ cm}} + \frac{1}{s'} = \frac{1}{20 \text{ cm}}, \qquad s' = 13.3 \text{ cm},$$

$$m = -\frac{13.3 \text{ cm}}{-40 \text{ cm}} = +\frac{1}{3}.$$

REFLECT The graphical results for the image distances are fairly close to the calculated results, except for those from Figure 24.37e, where the precision of the diagram is limited because the rays extended backward have nearly the same direction.

Practice Problem: If an object is placed at a point that is 5.0 cm to the left of the lens in this problem, where is the image and what is the lateral magnification of the image? *Answer:* 6.7 cm to the left of the lens; $s' = -6.7$ cm, $m = 1.3$.

EXAMPLE 24.10 Image formed by a diverging lens

You are given a thin diverging lens. You find that a beam of parallel rays spreads out after passing through the lens, as though all the rays came from a point 20.0 cm from the center of the lens. You want to use this lens to form an erect virtual image that is one-third the height of a real object. **(a)** Where should the object be placed? **(b)** Draw a principal-ray diagram.

SOLUTION

SET UP AND SOLVE Part (a): The behavior of the parallel incident rays indicates that the focal length is negative: $f = -20.0$ cm. We want the lateral magnification to be $m = +\frac{1}{3}$ (positive because the image is to be erect). From Equation 24.18, $m = \frac{1}{3} = -s'/s$. So we use $s' = -s/3$ in Equation 24.17:

$$\frac{1}{s} + \frac{1}{s'} = \frac{1}{f},$$

$$\frac{1}{s} + \frac{1}{-s/3} = \frac{1}{-20.0 \text{ cm}},$$

$$s = 40.0 \text{ cm},$$

$$s' = -\frac{40.0 \text{ cm}}{3} = -13.3 \text{ cm}.$$

Part (b): Figure 24.39 shows our principal-ray diagram for this problem, with the rays numbered the same way as in Figure 24.36b.

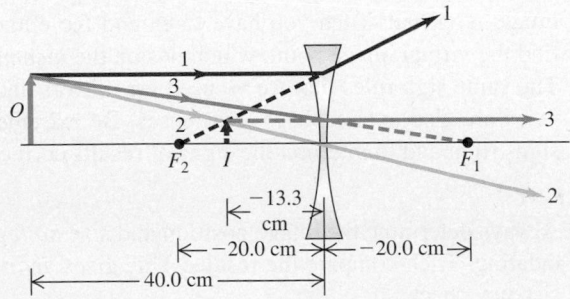

▲ FIGURE 24.39

REFLECT In part (a), the image distance is negative, so the real object and the virtual image are on the same side of the lens.

Practice Problem: For a diverging lens with a focal length of $f = -10$ cm, where should an object be placed to form an erect virtual image that is one-fourth the height of the object? *Answer:* $s = 30$ cm.

SUMMARY

Reflection at a Plane Surface

(Section 24.1) When rays diverge from an **object point** Q and are reflected at a plane surface, the directions of the outgoing rays are the same as though they had diverged from a point Q' called the **image point.** Since the rays don't actually converge at Q', the image formed is a **virtual image.** An image formed by a plane mirror is always **reversed;** for example, the image of a right hand is a left hand.

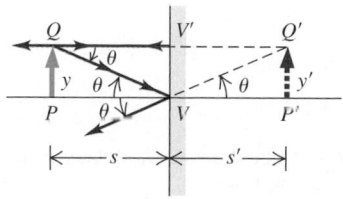

The **lateral magnification** m in any reflecting or refracting situation is defined as the ratio of the image height y' to the object height y:

$$m = \frac{y'}{y}. \qquad (24.2)$$

When m is positive, the image is erect; when m is negative, the image is inverted.

Reflection at a Spherical Surface

(Section 24.2) The **focal point** of a mirror is the point at which parallel rays converge after reflection from a concave mirror or the point from which they appear to diverge after reflection from a convex mirror. Rays diverging from the focal point of a concave mirror are parallel after reflection; rays converging toward the focal point of a convex mirror are parallel after reflection. The distance from the focal point to the vertex is called the focal length f, where $f = R/2$ (Equation 24.5). The object distance, image distance, and focal length are related by

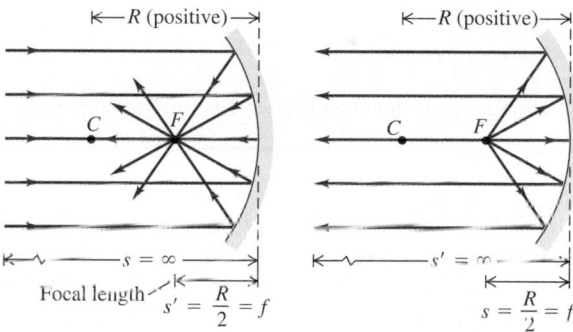

$$\frac{1}{s} + \frac{1}{s'} = \frac{1}{f}. \qquad (24.6)$$

Graphical Methods for Mirrors

(Section 24.3) Four principal rays can be drawn to find the size and position of the image formed by a mirror:

1. *A ray parallel to the axis,* after reflection, passes through the focal point F of a concave mirror or appears to come from the (virtual) focal point of a convex mirror.
2. *A ray through (or proceeding toward) the focal point F is* reflected parallel to the axis.
3. *A ray along the radius through or away from the center of curvature C* intersects the surface normally and is reflected back along its original path.
4. *A ray to the vertex V is* reflected and forms equal angles with the optic axis.

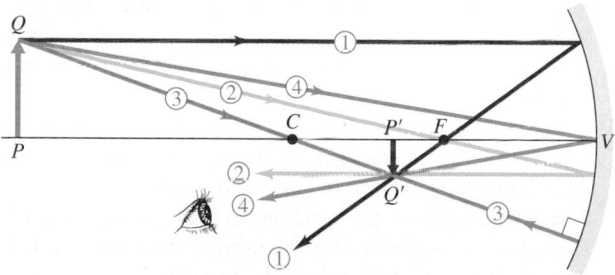

Refraction at a Spherical Surface and Thin Lenses

(Sections 24.4 and 24.5) When a spherical surface forms the interface between two materials, the magnification of the image is given by

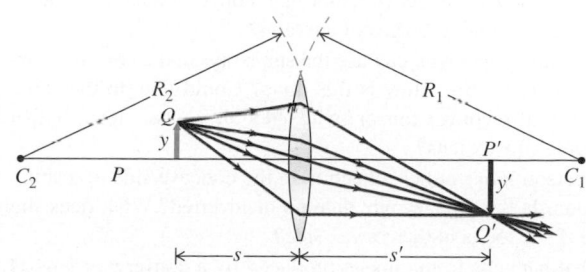

$$m = -\frac{n_a s'}{n_b s}. \qquad (24.12)$$

A **thin lens** has two spherical surfaces close enough together that we can ignore the distance between them. The magnification of a thin lens is given by $m = -s'/s$ (Equation 24.18). The focal length

Continued

is the same on both sides of a thin lens, even when the two sides have different curvatures. The behavior of a thin lens is described by the **thin-lens equation:**

$$\frac{1}{s} + \frac{1}{s'} = \frac{1}{f} = (n-1)\left(\frac{1}{R_1} - \frac{1}{R_2}\right). \qquad (24.19, 24.20, 24.21)$$

Graphical Methods for Lenses

(Section 24.6) Three principal rays can be drawn to find the size and position of the image formed by a thin lens:

1. *A ray parallel to the axis,* after refraction by the lens, passes through the second focal point F_2 of a converging lens or appears to come from the second focal point of a diverging lens.
2. *A ray through the center of the lens* is not appreciably deviated, because, at the center of the lens, the two surfaces are parallel and close together.
3. *A ray through (or proceeding toward) the first focal point F_1* emerges parallel to the axis.

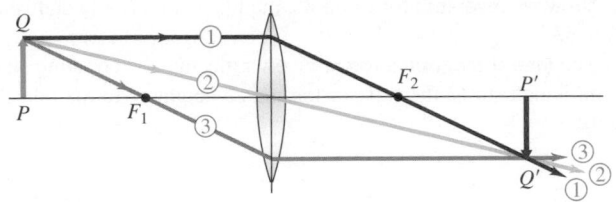

 For instructor-assigned homework, go to www.masteringphysics.com

Conceptual Questions

1. If a spherical mirror is immersed in water, does its focal length change? What if a lens is immersed in water?
2. For what range of object positions does a concave spherical mirror of focal length f form a real image? What about a convex spherical mirror?
3. If a screen is placed at the location of a real image, will the image appear on the screen? What happens if the image is a virtual one?
4. Is it possible to view a virtual image directly with your eye? How?
5. When a room has mirrors on two opposite walls, an infinite series of images can be seen. Show why this happens. Why do the distant images appear progressively darker?
6. A spherical mirror is cut in half horizontally. Will an image be formed by the bottom half of the mirror? If so, where will the image be formed, and how will its appearance compare to the image formed by the full mirror?
7. A concave mirror (sometimes surrounded by lights) is often used as an aid for applying cosmetics to the face. Why is such a mirror always concave rather than convex? What considerations determine its radius of curvature?
8. On a sunny day, you can use the sun's rays and a concave mirror to start a fire. How is this done? Could you do the same thing with a convex mirror? with a double concave lens? with a double convex lens?
9. A person looks at her reflection in the concave side of a shiny spoon. Is this image right side up or inverted? What does she see if she looks in the convex side?
10. What happens to the image produced by a converging lens as the object is slowly moved *through* the focal point? Does the same thing happen for a diverging lens?

11. Without measuring its radius of curvature (which is not so easy to do), explain how you can experimentally determine the focal length of (a) a concave mirror, (b) a convex mirror. Your apparatus consists of viewing screens and an optical bench on which to mount the mirrors and measure distances.
12. Without measuring its radii of curvature (which is not so easy to do), explain how you can experimentally determine the focal length of (a) a converging lens, (b) a diverging lens. Your apparatus consists of viewing screens and an optical bench on which to mount the lenses and measure distances.
13. What happens to the focal length of a thin lens if you turn it around? Which things change and which ones stay the same?
14. A spherical air bubble in water can function as a lens. Is it a converging or a diverging lens? How is its focal length related to its radius?
15. You have a curved spherical mirror about a foot across. You find that when your eye is very close to the mirror, you see an erect image of your face. But as you move farther and farther from the mirror, your face suddenly looks upside down. (a) What kind of mirror (convex or concave) is this? (b) Are the images you see real or virtual?

Multiple-Choice Problems

1. A ray from an object passes through a thin lens, as shown in Figure 24.40. What can we conclude about the lens from this ray?
   A. It is a converging lens.
   B. It is a diverging lens.
   C. It is not possible to tell which type of lens it is.

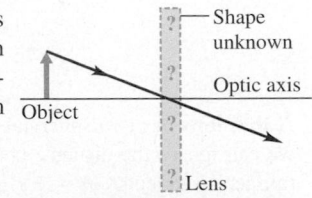

▲ **FIGURE 24.40** Multiple-choice problem 1.

2. If a single lens forms a real image, we can conclude that
   A. It is a converging lens.
   B. It is a diverging lens.
   C. It could be either type of lens.

3. If a single lens forms a virtual image, we can conclude that
   A. It is a converging lens.
   B. It is a diverging lens.
   C. It could be either type of lens.

4. An object lies outside the focal point of a converging lens. Which of the following statements about the image formed by this lens must be true? (There may be more than one correct choice.)
   A. The image is always real and inverted.
   B. The image could be real or virtual, depending on how far the object is past the focal point.
   C. The image could be erect or inverted, depending on how far the object is past the focal point.
   D. The image is always on the opposite side of the lens from the object.

5. An object lies outside the focal point of a diverging lens. Which of the following statements about the image formed by this lens must be true? (There may be more than one correct choice.)
   A. The image is always virtual and inverted.
   B. The image could be real or virtual, depending on how far the object is past the focal point.
   C. The image could be erect or inverted, depending on how far the object is past the focal point.
   D. The image is always virtual and on the same side of the lens as the object.

6. A spherical mirror is shown in Figure 24.41, and the object is placed in front of the curved front surface. Which of the following statements about this mirror must be true? (There may be more than one correct choice.)
   A. Its radius of curvature is negative.
   B. The image it produces is inverted.
   C. A ray parallel to the optic axis is reflected from the mirror so that it passes through the focal point of the mirror.
   D. It always produces a virtual image on the opposite side of the mirror from the object.

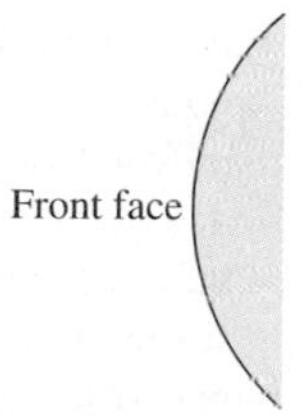

▲ **FIGURE 24.41** Multiple-choice problem 6.

7. An object is beyond the focal point of a converging lens. If you want to bring its image closer to the lens, you should move the object
   A. away from the lens.
   B. toward the lens.
   C. to the focal point of the lens.

8. A thin glass lens has a focal length f in air. If you now make a lens of identical shape, using glass having twice the refractive index of the original glass, the focal length of the new lens will be
   A. $f/2$.
   B. less than $f/2$.
   C. $2f$.
   D. greater than $2f$.

9. A ray from an object passes through a thin lens, as shown in Figure 24.42. What can we conclude about the lens from this ray?
   A. It must be a converging lens.
   B. It must be a diverging lens.
   C. It could be either a converging or a diverging lens.

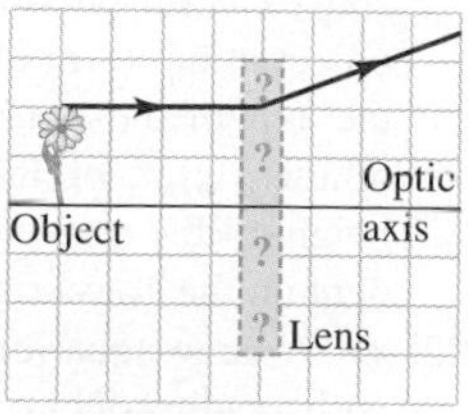

▲ **FIGURE 24.42** Multiple-choice problem 9.

10. As you move an object from just outside to just inside the focal point of a converging lens, its image
    A. goes from real to virtual and from inverted to erect.
    B. goes from inverted to erect, but remains real.
    C. goes from inverted to erect, but remains virtual.
    D. goes from real to virtual, but remains inverted.
    E. remains both erect and virtual.

11. As you move an object from just outside to just inside the focal point of a diverging lens, its image
    A. goes from real to virtual and from inverted to erect.
    B. goes from inverted to erect, but remains real.
    C. goes from inverted to erect, but remains virtual.
    D. goes from real to virtual, but remains inverted.
    E. remains both erect and virtual.

12. You have a shiny salad bowl with a spherical shape. If you hold the bowl at arm's length with the inside of the bowl facing you, the image of your face that you see will be:
    A. upside down and smaller than your face.
    B. right side up and smaller than your face.
    C. right side up and bigger than your face.
    D. upside down and bigger than your face.

13. A very long tube of transparent glass has a hemispherical end, as shown in Figure 24.43, and is surrounded by air. A small object is placed to the left of the curved glass surface. Which of the following statements must be true about the image of this object formed by the glass? (There may be more than one correct choice.)
    A. It will lie within the glass.
    B. It will lie in the air to the left of the glass.
    C. It will be real.   D. It will be virtual.

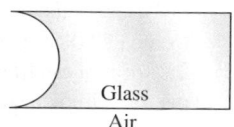

▲ **FIGURE 24.43** Multiple-choice problem 13.

14. A very long glass tube, with index of refraction 1.50, has a convex hemispherical end (the reverse of the tube in question 13). Suppose the tube is immersed in a fluid whose index of refraction is 1.60, and a small object is placed to the left of the curved glass surface. Which of the following statements must be true about the image of this object formed by the glass? (There may be more than one correct choice.)
    A. It will lie within the glass.
    B. It will lie in the fluid to the left of the glass.
    C. It will be real.   D. It will be virtual.

15. A certain thin lens has a focal length f. If you double both of its radii of curvature, but change nothing else, its focal length will now be
    A. $4f$.   B. $2f$.   C. $f/2$.   D. $f/4$.

Problems

24.1 Reflection at a Plane Surface

1. • A candle 4.85 cm tall is 39.2 cm to the left of a plane mirror. Where is the image formed by the mirror, and what is the height of this image?

2. • What is the size of the smallest vertical plane mirror in which a 10 ft tall giraffe standing erect can see her full-length image? (*Hint:* Locate the image by drawing a number of rays from the giraffe's body that reflect off the mirror and go to her eye. Then eliminate that part of the mirror for which the reflected rays do not reach her eye.)

3. •• An object is placed between two plane mirrors arranged at right angles to each other at a distance d_1 from the surface of one mirror and a distance d_2 from the surface of the other. (a) How many images are formed? Show the location of the images in a diagram. (b) Draw the paths of rays from the object to the eye of an observer.

4. •• If you run away from a plane mirror at 2.40 m/s, at what speed does your image move away from you?

24.2 Reflection at a Spherical Surface

5. • A concave spherical mirror has a radius of curvature of 10.0 cm. Calculate the location and size of the image formed of an 8.00-mm-tall object whose distance from the mirror is (a) 15.0 cm, (b) 10.0 cm, (c) 2.50 cm, and (d) 10.0 m.

6. • Repeat the previous problem, except use a convex mirror with the same magnitude of focal length.

7. • The diameter of Mars is 6794 km, and its minimum distance from the earth is 5.58×10^7 km. (a) When Mars is at this distance, find the diameter of the image of Mars formed by a spherical, concave telescope mirror with a focal length of 1.75 m. (b) Where is the image located?

8. • A concave mirror has a radius of curvature of 34.0 cm. (a) What is its focal length? (b) A ladybug 7.50 mm tall is located 22.0 cm from this mirror along the principal axis. Find the location and height of the image of the insect. (c) If the mirror is immersed in water (of refractive index 1.33), what is its focal length?

9. • **Rearview mirror.** A mirror on the passenger side of your car is convex and has a radius of curvature with magnitude 18.0 cm. (a) Another car is seen in this side mirror and is 13.0 m behind the mirror. If this car is 1.5 m tall, what is the height of its image? (b) The mirror has a warning attached that objects viewed in it are closer than they appear. Why is this so?

10. •• Examining your image in a convex mirror whose radius of curvature is 25.0 cm, you stand with the tip of your nose 10.0 cm from the surface of the mirror. (a) Where is the image of your nose located? What is its magnification? (b) Your ear is 10.0 cm behind the tip of your nose; where is the image of your ear located, and what is its magnification? Do your answers suggest reasons for your strange appearance in a convex mirror?

11. •• A coin is placed next to the convex side of a thin spherical glass shell having a radius of curvature of 18.0 cm. An image of the 1.5-cm-tall coin is formed 6.00 cm behind the glass shell. Where is the coin located? Determine the size, orientation, and nature (real or virtual) of the image.

12. •• (a) Show that when an object is *outside* the focal point of a concave mirror, its image is always *inverted* and *real*. Is there any limitation on the magnification? (b) Show that when an object is *inside* the focal point of a concave mirror, its image is always *erect* and *virtual*. Is there any limitation on the magnification?

24.3 Graphical Methods for Mirrors

13. • A spherical, concave shaving mirror has a radius of curvature of 32.0 cm. (a) What is the magnification of a person's face when it is 12.0 cm to the left of the vertex of the mirror? (b) Where is the image? Is the image real or virtual? (c) Draw a principal-ray diagram showing the formation of the image.

14. • An object 0.600 cm tall is placed 16.5 cm to the left of the vertex of a concave spherical mirror having a radius of curvature of 22.0 cm. (a) Draw a principal-ray diagram showing

the formation of the image. (b) Calculate the position, size, orientation (erect or inverted), and nature (real or virtual) of the image.

15. • Repeat the previous problem for the case in which the mirror is *convex*.

16. • The stainless steel rear end of a tanker truck is convex, shiny, and has a radius of curvature of 2.0 m. You're tailgating the truck, with the front end of your car only 5.0 m behind it. Making the not very realistic assumption that your car is on the axis of the mirror formed by the tank, (a) determine the position, orientation, magnification, and type (real or virtual) of the image of your car's front end that forms in this mirror; (b) draw a principal-ray diagram of the situation to check your answer.

17. • The thin glass shell shown in Figure 24.44 has a spherical shape with a radius of curvature of 12.0 cm, and both of its surfaces can act as mirrors. A seed 3.30 mm high is placed 15.0 cm from the center of the mirror along the optic axis, as shown in the figure. (a) Calculate the location and height of the image of this seed. (b) Suppose now that the shell is reversed. Find the location and height of the seed's image.

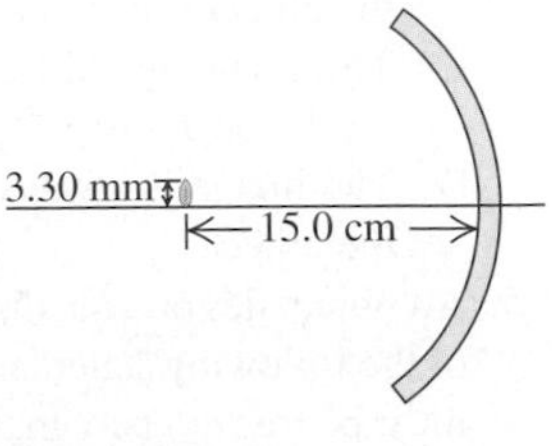

3.30 mm ↑ |← 15.0 cm →|

▲ **FIGURE 24.44** Problem 17.

18. •• **Dental mirror.** A dentist uses a curved mirror to view teeth on the upper side of the mouth. Suppose she wants an erect image with a magnification of 2.00 when the mirror is 1.25 cm from a tooth. (Treat this problem as though the object and image lie along a straight line.) (a) What kind of mirror (concave or convex) is needed? Use a ray diagram to decide, without performing any calculations. (b) What must be the focal length and radius of curvature of this mirror? (c) Draw a principal-ray diagram to check your answer in part (b).

24.4 Refraction at a Spherical Surface

19. • The left end of a long glass rod 6.00 cm in diameter has a convex hemispherical surface 3.00 cm in radius. The refractive index of the glass is 1.60. Determine the position of the image if an object is placed in air on the axis of the rod at the following distances to the left of the vertex of the curved end: (a) infinitely far, (b) 12.0 cm, and (c) 2.00 cm.

20. • The rod of the previous problem is immersed in a liquid. An object 90.0 cm from the vertex of the left end of the rod and on its axis is imaged at a point 1.60 m inside the rod. What is the refractive index of the liquid?

21. • The left end of a long glass rod 8.00 cm in diameter and with an index of refraction of 1.60 is ground and polished to a convex hemispherical surface with a radius of 4.00 cm. An object in the form of an arrow 1.50 mm tall, at right angles to the axis of the rod, is located on the axis 24.0 cm to the left of the vertex of the convex surface. Find the position and height of the image of the arrow formed by paraxial rays incident on the convex surface. Is the image erect or inverted?

22. •• A large aquarium has portholes of thin transparent plastic with a radius of curvature of 1.75 m and their convex sides facing into the water. A shark hovers in front of a porthole, sizing

up the dinner prospects outside the tank. (a) If one of the shark's teeth is exactly 45.0 cm from the plastic, how far from the plastic does it appear to be to observers outside the tank? (You can ignore refraction due to the plastic.) (b) Does the shark appear to be right side up or upside down? (c) If the tooth has an actual length of 5.00 cm, how long does it appear to the observers?

23. • **A spherical fishbowl.** A small tropical fish is at the center of a water-filled spherical fishbowl 28.0 cm in diameter. (a) Find the apparent position and magnification of the fish to an observer outside the bowl. The effect of the thin walls of the bowl may be ignored. (b) A friend advised the owner of the bowl to keep it out of direct sunlight to avoid blinding the fish, which might swim into the focal point of the parallel rays from the sun. Is the focal point actually within the bowl?

24. •• **Focus of the eye.** The cornea of the eye has a radius of
BIO curvature of approximately 0.50 cm, and the aqueous humor behind it has an index of refraction of 1.35. The thickness of the cornea itself is small enough that we shall neglect it. The depth of a typical human eye is around 25 mm. (a) What would have to be the radius of curvature of the cornea so that it alone would focus the image of a distant mountain on the retina, which is at the back of the eye opposite the cornea? (b) If the cornea focused the mountain correctly on the retina as described in part (a), would it also focus the text from a computer screen on the retina if that screen were 25 cm in front of the eye? If not, where would it focus that text, in front of or behind the retina? (c) Given that the cornea has a radius of curvature of about 5.0 mm, where does it actually focus the mountain? Is this in front of or behind the retina? Does this help you see why the eye needs help from a lens to complete the task of focusing?

25. • A speck of dirt is embedded 3.50 cm below the surface of a sheet of ice having a refractive index of 1.309. What is the apparent depth of the speck, when viewed from directly above?

26. • A skin diver is 2.0 m below the surface of a lake. A bird flies overhead 7.0 m above the surface of the lake. When the bird is directly overhead, how far above the diver does it appear to be?

27. •• A zoo aquarium has transparent walls, so that spectators on both sides of it can watch the fish. The aquarium is 5.50 m across, and the spectators on both sides of it are standing 1.20 m from the wall. How far away do spectators on one side of the aquarium appear to those on the other side? (Ignore any refraction in the walls of the aquarium.)

28. • To a person swimming 0.80 m beneath the surface of the water in a swimming pool, the diving board directly overhead appears to be a height of 5.20 m above the swimmer. What is the actual height of the diving board above the surface of the water?

24.5 The Thin Lens

29. • A converging lens with a focal length of 7.00 cm forms an image of a 4.00-mm-tall real object that is to the left of the lens. The image is 1.30 cm tall and erect. Where are the object and image located? Is the image real or virtual?

30. • A converging lens with a focal length of 90.0 cm forms an image of a 3.20-cm-tall real object that is to the left of the lens. The image is 4.50 cm tall and inverted. Where are the

object and image located in relation to the lens? Is the image real or virtual?

31. • You are standing in front of a lens that projects an image of you onto a wall 1.80 m on the other side of the lens. This image is three times your height. (a) How far are you from the lens? (b) Is your image erect or inverted? (c) What is the focal length of the lens? Is the lens converging or diverging?

32. •• Figure 24.45 shows an object and its image formed by a thin lens. (a) What is the focal length of the lens and what type of lens (converging or diverging) is it? (b) What is the height of the image? Is it real or virtual?

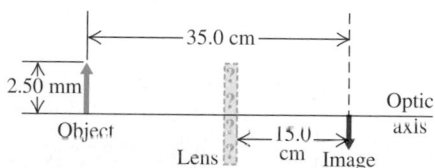

▲ **FIGURE 24.45** Problem 32.

33. •• Figure 24.46 shows an object and its image formed by a thin lens. (a) What is the focal length of the lens and what type of lens (converging or diverging) is it? (b) What is the height of the image? Is it real or virtual?

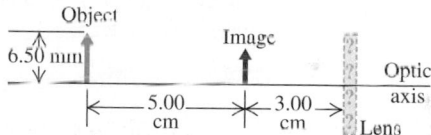

▲ **FIGURE 24.46** Problem 33.

34. •• Figure 24.47 shows an object and its image formed by a thin lens. (a) What is the focal length of the lens and what type of lens (converging or diverging) is it? (b) What is the height of the image? Is it real or virtual?

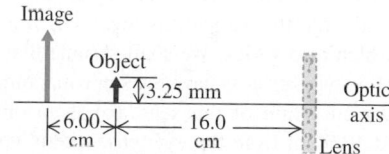

▲ **FIGURE 24.47** Problem 34.

35. • The two surfaces of a plastic converging lens have equal radii of curvature of 22.0 cm, and the lens has a focal length of 20.0 cm. Calculate the index of refraction of the plastic.

36. • The front, convex, surface of a lens made for eyeglasses has a radius of curvature of 11.8 cm, and the back, concave, surface has a radius of curvature of 6.80 cm. The index of refraction of the plastic lens material is 1.67. Calculate the focal length of the lens.

37. • For each of the thin lenses (L_1 and L_2) shown in Figure 24.48, the index of refraction of the lens' glass is 1.50, and the object is to the left of the lens. The radii of curvature indicated are just the magnitudes. Calculate the focal length of each lens.

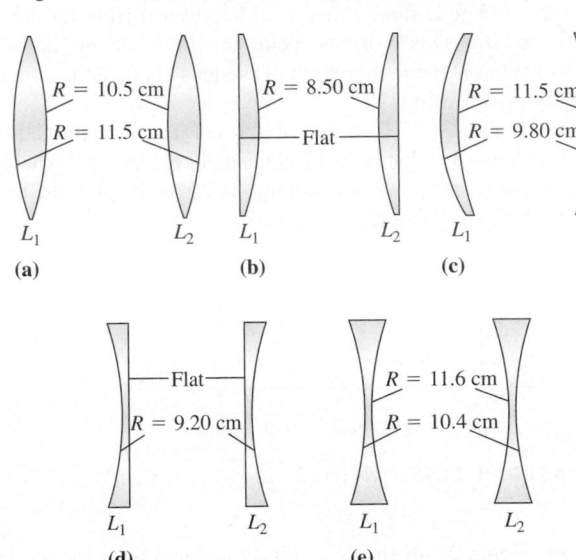

▲ **FIGURE 24.48** Problem 37.

38. • For each thin lens shown in Figure 24.49, calculate the location of the image of an object that is 18.0 cm to the left of the lens. The lens material has a refractive index of 1.50, and the radii of curvature shown are only the magnitudes.

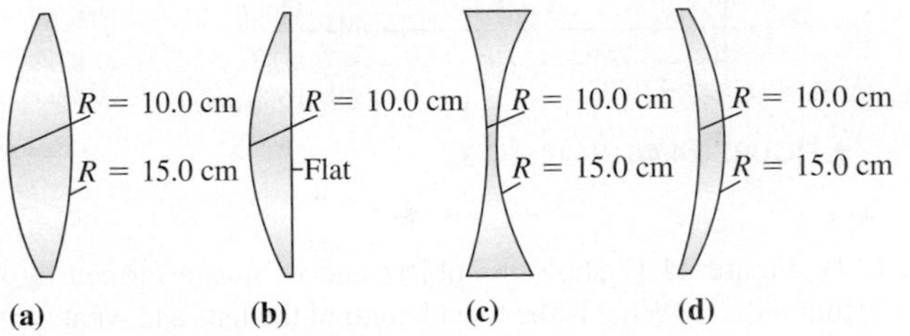

▲ **FIGURE 24.49** Problem 38.

39. • **The lens of the eye.** The crystalline lens of the human eye is
BIO a double-convex lens made of material having an index of refraction of 1.44 (although this varies). Its focal length in air is about 8.0 mm, which also varies. We shall assume that the radii of curvature of its two surfaces have the same magnitude. (a) Find the radii of curvature of this lens. (b) If an object 16 cm tall is placed 30.0 cm from the eye lens, where would the lens focus it and how tall would the image be? Is this image real or virtual? Is it erect or inverted? (*Note:* The results obtained here are not strictly accurate, because the lens is embedded in fluids having refractive indexes different from that of air.)

40. • **The cornea as a simple lens.** The cornea behaves as a
BIO thin lens of focal length approximately 1.8 cm, although this varies a bit. The material of which it is made has an index of refraction of 1.38, and its front surface is convex, with a radius of curvature of 5.0 mm. (a) If this focal length is in air, what is the radius of curvature of the back side of the cornea? (b) The closest distance at which a typical person can focus on an object (called the near point) is about 25 cm,

although this varies considerably with age. Where would the cornea focus the image of an 8.0-mm-tall object at the near point? (c) What is the height of the image in part (b)? Is this image real or virtual? Is it erect or inverted? (*Note:* The results obtained here are not strictly accurate, because, on one side, the cornea has a fluid with a refractive index different from that of air.)

41. • An insect 3.75 mm tall is placed 22.5 cm to the left of a thin planoconvex lens. The left surface of this lens is flat, the right surface has a radius of curvature of magnitude 13.0 cm, and the index of refraction of the lens material is 1.70. (a) Calculate the location and size of the image this lens forms of the insect. Is it real or virtual? erect or inverted? (b) Repeat part (a) if the lens is reversed.

42. •• A double-convex thin lens has surfaces with equal radii of curvature of magnitude 2.50 cm. Looking through this lens, you observe that it forms an image of a very distant tree, at a distance of 1.87 cm from the lens. What is the index of refraction of the lens?

43. •• A converging meniscus lens (see Fig. 24.31) with a refractive index of 1.52 has spherical surfaces whose radii are 7.00 cm and 4.00 cm. What is the position of the image if an object is placed 24.0 cm to the left of the lens? What is the magnification?

44. •• A converging lens with a focal length of 12.0 cm forms a virtual image 8.00 mm tall, 17.0 cm to the right of the lens. Determine the position and size of the object. Is the image erect or inverted? Are the object and image on the same side or opposite sides of the lens?

45. •• **Combination of lenses, I.** When two lenses are used in combination, the first one forms an image that then serves as the object for the second lens. The magnification of the combination is the ratio of the height of the final image to the height of the object. A 1.20-cm-tall object is 50.0 cm to the left of a converging lens of focal length 40.0 cm. A second converging lens, this one having a focal length of 60.0 cm, is located 300.0 cm to the right of the first lens along the same optic axis. (a) Find the location and height of the image (call it I_1) formed by the lens with a focal length of 40.0 cm. (b) I_1 is now the object for the second lens. Find the location and height of the image produced by the second lens. This is the final image produced by the combination of lenses.

46. •• (a) You want to use a lens with a focal length of 35.0 cm to produce a real image of an object, with the image twice as long as the object itself. What kind of lens do you need, and where should the object be placed? (b) Suppose you want a virtual image of the same object, with the same magnification—what kind of lens do you need, and where should the object be placed?

47. •• **Combination of lenses, II.** Two thin lenses with a focal length of magnitude 12.0 cm, the first diverging and the second converging, are located 9.00 cm apart. An object 2.50 mm tall is placed 20.0 cm to the left of the first (diverging) lens. (a) How far from this first lens is the final image formed? (b) Is the final image real or virtual? (c) What is the height of the final image? Is it erect or inverted? (*Hint:* See Problem 45.)

48. •• A lens forms a real image, which is 214 cm away from the object and 1⅔ times its length. What kind of lens is this, and what is its focal length?

24.6 Graphical Methods for Lenses

49. • A converging lens has a focal length of 14.0 cm. For each of two objects located to the left of the lens, one at a distance of 18.0 cm and the other at a distance of 7.00 cm, determine (a) the image position, (b) the magnification, (c) whether the image is real or virtual, and (d) whether the image is erect or inverted. Draw a principal-ray diagram in each case.

50. • A converging lens forms an image of an 8.00-mm-tall real object. The image is 12.0 cm to the left of the lens, 3.40 cm tall, and erect. (a) What is the focal length of the lens? (b) Where is the object located? (c) Draw a principal-ray diagram for this situation.

51. • A diverging lens with a focal length of −48.0 cm forms a virtual image 8.00 mm tall, 17.0 cm to the right of the lens. (a) Determine the position and size of the object. Is the image erect or inverted? Are the object and image on the same side or opposite sides of the lens? (b) Draw a principal-ray diagram for this situation.

52. • When an object is 16.0 cm from a lens, an image is formed 12.0 cm from the lens on the same side as the object. (a) What is the focal length of the lens? Is the lens converging or diverging? (b) If the object is 8.50 mm tall, how tall is the image? Is it erect or inverted? (c) Draw a principal-ray diagram.

53. •• Figure 24.50 shows a small plant near a thin lens. The ray shown is one of the principal rays for the lens. Each square is 2.0 cm along the horizontal direction, but the vertical direction is not to the same scale. Use information from the diagram to answer the following questions: (a) Using only the ray shown, decide what type of lens (converging or diverging) this is. (b) What is the focal length of the lens? (c) Locate the image by drawing the other two principal rays. (d) Calculate where the image should be, and compare this result with the graphical solution in part (c).

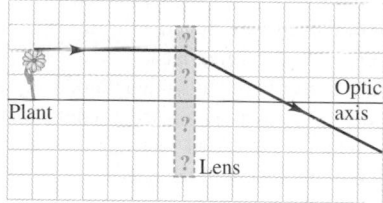

▲ **FIGURE 24.50** Problem 53.

54. •• Figure 24.51 shows a small plant near a thin lens. The ray shown is one of the principal rays for the lens. Each square is 2.0 cm along the horizontal direction, but the vertical direction is not to the same scale. Use information from the diagram to answer the following questions: (a) Using only the ray shown, decide what type of lens (converging or diverging) this is. (b) What is the focal length of the lens? (c) Locate the image by drawing the other two principal rays. (d) Calculate where the image should be, and compare this result with the graphical solution in part (c).

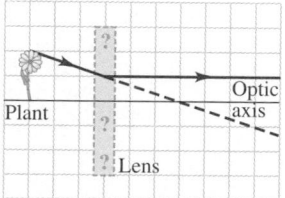

▲ **FIGURE 24.51** Problem 54.

55. •• Figure 24.52 shows a small plant near a thin lens. The ray shown is one of the principal rays for the lens. Each square is 2.0 cm along the horizontal direction, but the vertical direction is not to the same scale. Use information from the diagram to answer the following questions: (a) Using only the ray shown, decide what type of lens (converging or diverging) this is. (b) What is the focal length of the lens? (c) Locate the image by drawing the other two principal rays. (d) Calculate where the image should be, and compare this result with the graphical solution in part (c).

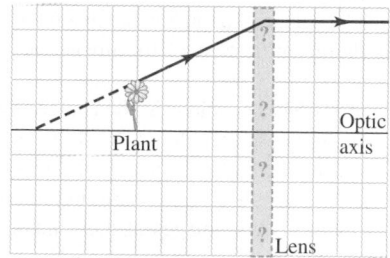

▲ **FIGURE 24.52** Problem 55.

56. •• Figure 24.53 shows a small plant near a thin lens. The ray shown is one of the principal rays for the lens. Each square is 2.0 cm along the horizontal direction, but the vertical direction is not to the same scale. Use information from the diagram to answer the following questions: (a) Using only the ray shown, decide what type of lens (converging or diverging) this is. (b) What is the focal length of the lens? (c) Locate the image by drawing the other two principal rays. (d) Calculate where the image should be, and compare this result with the graphical solution in part (c).

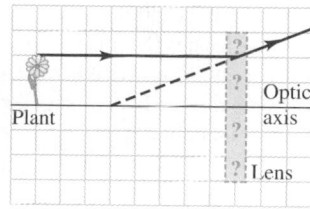

▲ **FIGURE 24.53** Problem 56.

General Problems

57. •• A layer of benzene $(n = 1.50)$ 2.60 cm deep floats on water $(n = 1.33)$ that is 6.50 cm deep. What is the apparent distance from the upper benzene surface to the bottom of the water layer when it is viewed at normal incidence?

58. •• Where must you place an object in front of a concave mirror with radius R so that the image is erect and $2\frac{1}{2}$ times the size of the object? Where is the image?

59. •• A luminous object is 4.00 m from a wall. You are to use a concave mirror to project an image of the object on the wall, with the image 2.25 times the size of the object. How far should the mirror be from the wall? What should its radius of curvature be?

60. •• A concave mirror is to form an image of the filament of a headlight lamp on a screen 8.00 m from the mirror. The filament is 6.00 mm tall, and the image is to be 36.0 cm tall. (a) How far in front of the vertex of the mirror should the filament be placed? (b) To what radius of curvature should you grind the mirror?

61. •• A plastic lens $(n = 1.67)$ has one convex surface of radius 12.2 cm and one concave surface of radius 15.4 cm. If an object is placed 35.0 cm from the lens, (a) where is the image located and (b) what is its magnification?

62. • A 3.80-mm-tall object is 24.0 cm from the center of a silvered spherical glass Christmas tree ornament 6.00 cm in diameter. What are the position and height of its image?

63. •• A lensmaker wants to make a magnifying glass from glass with $n = 1.55$ and with a focal length of 20.0 cm. If the two surfaces of the lens are to have equal radii, what should that radius be?

64. •• An object is placed 18.0 cm from a screen. (a) At what two points between object and screen may a converging lens with a 3.00 cm focal length be placed to obtain an image on the screen? (b) What is the magnification of the image for each position of the lens?

65. •• As shown in Figure 24.54, a candle is at the center of curvature of a concave mirror whose focal length is 10.0 cm. The converging lens has a focal length of 32.0 cm and is 85.0 cm to the right of the candle. The candle is viewed through the lens from the right. The lens forms two images of the candle. The first is formed by light passing directly through the lens. The second image is formed from the light that goes from the candle to the mirror, is reflected, and then passes through the lens. (a) For each of these two images, draw a principal-ray diagram that locates the image. (b) For each image, answer the following questions: (i) Where is the image? (ii) Is the image real or virtual? (iii) Is the image erect or inverted with respect to the original object?

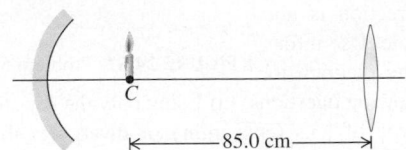

▲ **FIGURE 24.54** Problem 65.

66. •• In the text, Equations (24.4) and (24.7) were derived for the case of a concave mirror. Give a similar derivation for a convex mirror, and show that the same equations result if you use the sign convention established in the text.

67. •• **A lens in a liquid.** A lens obeys Snell's law, bending light rays at each surface an amount determined by the index of refraction of the lens and the index of the medium in which the lens is located. (a) Equation (24.20) assumes that the lens is surrounded by air. Consider instead a thin lens immersed in a liquid with refractive index n_{liq}. Prove that the focal length f' is then given by Eq. (24.20), with n replaced by n/n_{liq}. (b) A thin lens with index n has focal length f in vacuum. Use the result of part (a) to show that when this lens is immersed in a liquid of index n_{liq}, it will have a new focal length given by

$$f' = \left[\frac{n_{liq}(n-1)}{n - n_{liq}}\right]f.$$

Passage Problems

Refraction of liquids. The focal length of a mirror can be determined entirely from the shape of the mirror. In contrast, to determine the focal length of a lens we must know both the shape of the lens and its index of refraction—and the index of refraction of the surrounding medium. For instance, when a thin lens is immersed in a liquid we must modify the thin-lens equation to take into account the refractive properties of the surrounding liquid:

$$\frac{1}{f} = \left(\frac{n}{n_{liq}} - 1\right)\left(\frac{1}{R_1} - \frac{1}{R_2}\right)$$

where n_{liq} is the index of refraction of the liquid and n is the index of refraction of the glass.

68. If you place a glass lens ($n = 1.5$), which has a focal length of 0.5 meters in air, into a tank of water ($n = 1.33$), what will happen to its focal length?
    A. Nothing will happen.
    B. The focal length of the lens will be reduced.
    C. The focal length of the lens will be increased.
    D. There is not enough information to answer the question.

69. If you place a concave glass lens into a tank of a liquid that has an index of refraction that is greater than that of the lens, what will happen?
    A. The lens will no longer be able to create any images.
    B. The focal length of the lens will become longer.
    C. The focal length of the lens will become shorter.
    D. The lens will become a converging lens.

70. If you place a concave mirror with a focal length of 1 m into a liquid that has an index of refraction of 3, what will happen?
    A. The mirror will no longer be able to focus light.
    B. The focal length of the mirror will decrease.
    C. The focal length of the mirror will increase.
    D. Nothing will happen.

25 Optical Instruments

How does a camera resemble the human eye? What are the significant differences? What does a projector operator have to do to "focus" the picture on the screen? Why do large telescopes usually include curved mirrors as well as lenses? This chapter is concerned with these and other questions having to do with a number of familiar optical systems, some of which use several lenses or combinations of lenses and mirrors. In answering such questions, we can apply the basic principles of mirror and lens behavior that we studied in Chapter 24.

The concept of *image* plays an important role in the analysis of optical instruments. We continue to base our analysis on the *ray* model of light, so the content of this chapter comes under the general heading *geometric optics*.

The eyes and the camera viewing this landscape form similar images by similar means. Both have lenses that focus light on a recording surface (retina or CCD array). Both can adjust their focus for near or distant objects. In this chapter we'll study these and similar optical devices.

25.1 The Camera

The essential elements of a **camera** are a lens equipped with a shutter, a light-tight enclosure, and a light-sensitive film (or an electronic sensor, which we'll discuss later) that records an image (Figure 25.1). The lens forms an inverted, usually reduced, real image, in the plane of the film or sensor, of the object being photographed. To provide proper image distances for various object distances, the lens is moved closer to or farther from the film or sensor, often by being turned in a threaded mount. Most lenses have several elements, permitting partial correction of various *aberrations,* including the dependence of index of refraction on wavelength and the limitations imposed by the paraxial approximation. (We'll discuss lens aberrations in Section 25.7).

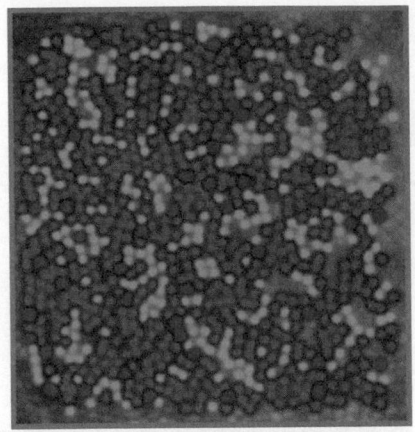

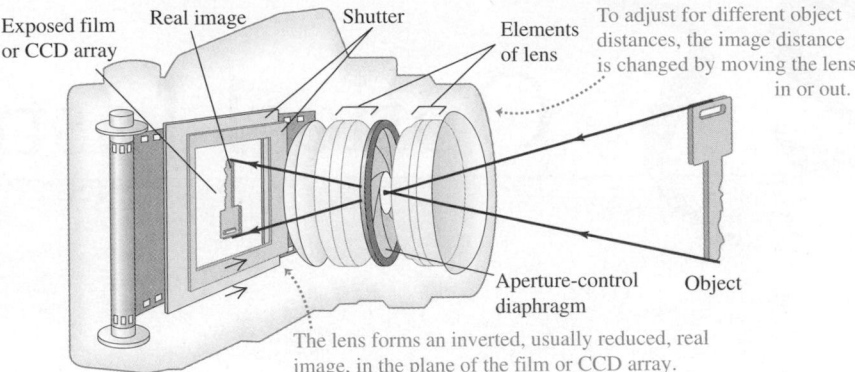

▲ FIGURE 25.1 Camera with aperture control.

▲ **BIO Application Seeing the light in living color.** As shown in Figure 25.7, the lens of the human eye focuses light on the retina, which comprises an array of photoreceptor cells, each of which generates an electrical signal when it absorbs light. The eyes of primates contain three types of cone cells responsible for color vision, each containing pigment molecules tuned to absorb different portions of the visible spectrum. The protein portion of the pigment is different in the three types and determines the peak wavelength response of each cell. The three molecular types have peak sensitivities in the red, green, and blue portions of the visible spectrum, but their sensitivities overlap considerably. Our perception of a color depends on the ratio of activation among the three, which explains why two pigments can be mixed to provide an impression of a color that is quite different from either color of the pigments. The photograph shows an actual image of a human retina in which the three cone types have been identified and artificially colored red, green, and blue to show the distribution of the three types.

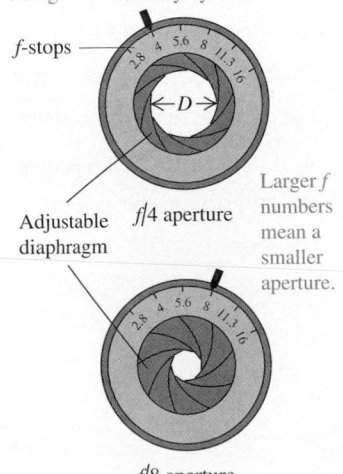

▲ FIGURE 25.2 The diaphragm that controls a camera's aperture.

In order for the film or sensor to record the image properly, the total light energy per unit area reaching the film (the exposure) must fall within certain limits. This is controlled by the *shutter* and the *lens aperture.* The shutter controls the time interval during which light enters the lens. The interval is usually adjustable in steps corresponding to factors of about two, often from 1 s to $\frac{1}{1000}$ s.

The intensity of light reaching the film is proportional to the effective area of the lens, which may be varied by means of an adjustable aperture, or *diaphragm*—a nearly circular hole with variable diameter D. The aperture size is usually described as a fraction of the focal length f of the lens. A lens with a focal length $f = 50$ mm and an aperture diameter of 25 mm has an aperture of $f/2$; many photographers would call this an *f*-number of 2, or simply "$f/2$." In general,

$$f\text{-number} = \frac{\text{Focal length}}{\text{Aperture diameter}} = \frac{f}{D}. \qquad (25.1)$$

Because the light intensity at the film or sensor is proportional to the area of the lens aperture and thus to the *square* of its diameter, changing the diameter by a factor of $\sqrt{2}$ changes the intensity by a factor of 2. Adjustable apertures (often called *f-stops*) usually have scales labeled with successive numbers related by factors of approximately $\sqrt{2}$, such as $f/2$, $f/2.8$, $f/4$, $f/5.6$, $f/8$, $f/11$, $f/16$, $f/22$, and so on. The larger numbers represent smaller apertures and intensities, and each step corresponds to a factor of 2 in intensity (Figure 25.2). The actual exposure is proportional to both the aperture area and the time of exposure. Thus, $\left(f/4 \text{ and } \frac{1}{500} \text{ s}\right)$, $\left(f/5.6 \text{ and } \frac{1}{250} \text{ s}\right)$, and $\left(f/8 \text{ and } \frac{1}{125} \text{ s}\right)$ all correspond to the same exposure.

The choice of focal length for a camera lens depends on the film size and the desired angle of view, or *field*. The popular 35 mm camera has an image size of 24×36 mm on the film. The normal lens for such a camera usually has a focal length of about 50 mm, permitting an angle of view of about 45°. A lens with a longer focal length, often 135 mm or 200 mm, provides a *smaller* angle of view and a larger image of *part* of the object, compared with a normal lens. This gives the impression that the camera is *closer* to the object than it really is, and such a lens is called a *telephoto lens*. A lens with a shorter focal length, such as 35 mm or 28 mm, permits a wider angle of view and is called a *wide-angle lens*. An extreme wide-angle, or "fish-eye," lens may have a focal length as small as 6 mm. Figure 25.3 shows a scene photographed from the same point with lenses of various focal lengths.

(a) $f = 28$ mm

(b) $f = 105$ mm

(c) $f = 300$ mm

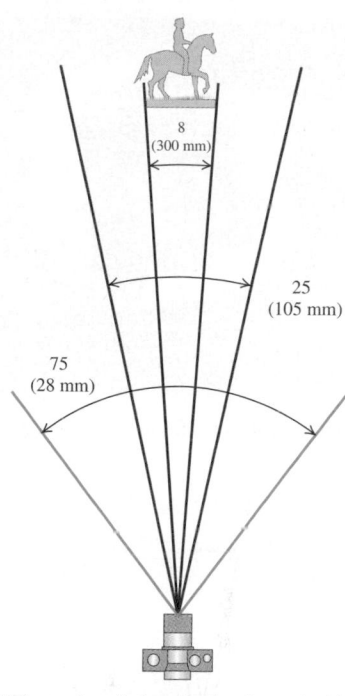

(d) The angles of view for the photos in (a)–(c)

▲ **FIGURE 25.3** As f increases, the image size increases proportionally.

EXAMPLE 25.1 **Photographic exposures**

A common telephoto lens for a 35 mm camera has a focal length of 200 mm and a range of f-stops from $f/5.6$ to $f/45$. **(a)** What is the corresponding range of aperture diameters? **(b)** What is the corresponding range of intensities of the film image?

SOLUTION

SET UP AND SOLVE Part (a): From Equation 25.1, the diameters range from

$$D = \frac{f}{f\text{-number}} = \frac{200 \text{ mm}}{5.6} = 36 \text{ mm}$$

to

$$D = \frac{200 \text{ mm}}{45} = 4.4 \text{ mm}.$$

Part (b): Because the intensity is proportional to the square of the diameter, the ratio of the intensity at $f/5.6$ to that at $f/45$ is approximately

$$\left(\frac{36 \text{ mm}}{4.4 \text{ mm}}\right)^2 \cong \left(\frac{45}{5.6}\right)^2 \cong 65 \qquad (\text{about } 2^6).$$

REFLECT If the correct exposure time at $f/5.6$ is $(1/1000 \text{ s})$, then at $f/45$ it is $(65)(1/1000 \text{ s}) = 1/15$ s.

Practice Problem: If the correct exposure at $f/5.6$ is $1/1000$ s, what is the correct exposure at $f/11$? At $f/16$? *Answers:* $1/250$ s, $1/125$ s.

▲ **FIGURE 25.4** A CCD chip on a circuit board.

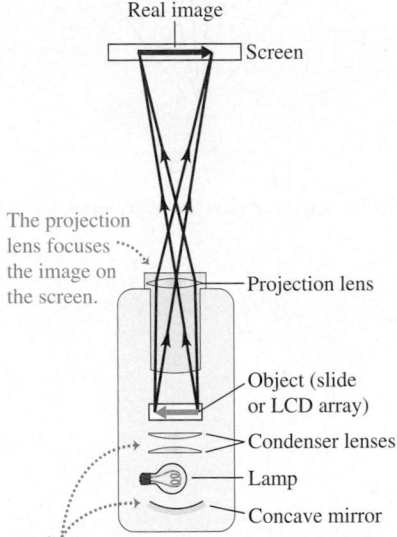

The projection lens focuses the image on the screen.

Real image
Screen
Projection lens
Object (slide or LCD array)
Condenser lenses
Lamp
Concave mirror

The mirror and condenser lenses direct the lamp's light onto the slide or LCD array.

▲ **FIGURE 25.5** The optical components of a slide projector.

The Digital Camera

In a digital camera, the light-sensitive film is replaced by an array of tiny photocells fabricated on a semiconductor chip called a charge-coupled device (CCD) chip. This device covers the image plane, dividing it into many rectangular areas called *pixels*. The total number of pixels is typically 2 to 10 million; a 2 megapixel camera has about 2 million pixels. The intensity of light of each primary color striking each pixel is recorded and stored digitally. Once stored, the image can be processed digitally to crop or enlarge it, change the color balance, and so on. Figure 25.4 shows a CCD chip.

The exposure time is controlled electronically, so a mechanical shutter is not needed. The dimensions of the CCD chip can be made smaller than the image size in a typical 35 mm camera by a factor on the order of $\frac{1}{5}$ in each dimension. The focal length of the lens can be correspondingly smaller, so the entire camera can be made much more compact than the usual 35 mm camera. This rapidly developing technology is used in applications such as cell-phone cameras.

25.2 The Projector

A **projector** for viewing slides or motion pictures operates very much like a camera in reverse. The essential elements are shown in Figure 25.5. Light from the source (an incandescent bulb or, in large motion-picture projectors, a carbon-arc lamp) shines through the film, and the projection lens forms a real, enlarged, inverted image of the film on the projection screen. Additional lenses called *condenser lenses* are placed between lamp and film. Their function is to direct the light from the source so that most of it enters the projection lens after passing through the film. A concave mirror behind the lamp also helps direct the light. The condenser lenses must be large enough to cover the entire area of the film. The image on the screen is always real and inverted; this is why slides have to be put into a projector upside down.

The position and size of the image projected on the screen are determined by the position and focal length of the projection lens.

EXAMPLE 25.2 **A slide projector**

An ordinary 35 mm color slide has a picture area of 24 × 36 mm. What focal length would a projection lens need in order to project a 1.2 m × 1.8 m image of this picture on a screen 5.0 m from the lens?

SOLUTION

SET UP Figure 25.6 diagrams the situation.

SOLVE We use the thin-lens analysis from Section 24.5. We need a lateral magnification (disregarding the sign) of

$$|m| = \frac{y'}{y} = \frac{1.2 \text{ m}}{24 \times 10^{-3} \text{ m}} = 50.$$

From Equation 24.18,

$$m = \frac{y'}{y} = -\frac{s'}{s},$$

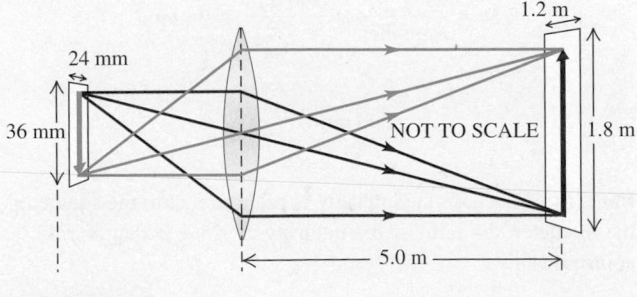

24 mm
36 mm
1.2 m
NOT TO SCALE
1.8 m
5.0 m

▲ **FIGURE 25.6**

Continued

so the ratio s'/s must also be 50. (The image is real, so s' is positive.) We are given that $s' = 5.0$ m. Thus,

$$s = \frac{5.0 \text{ m}}{50} = 0.10 \text{ m}.$$

The object and image distances are related by Equation 24.17, which we solve for the focal length f:

$$\frac{1}{f} = \frac{1}{s} + \frac{1}{s'} = \frac{1}{0.10 \text{ m}} + \frac{1}{5.0 \text{ m}},$$
$$f = 0.098 \text{ m} = 98 \text{ mm}.$$

REFLECT A popular focal length for home slide projectors is 100 mm; such lenses are readily available and would be the appropriate choice in this situation.

Practice Problem: For this same projector, what size image would appear on a screen 1.5 m away? *Answer:* 34 cm × 51 cm.

▶ **BIO Application The original digital image?** Insects view the world through compound eyes, which are arrays of individual light-sensing units packed together over the surface of the eye. (Honeybees, with fairly good vision, have about 7000 units; dragonflies have about 30,000.) Each unit records the intensity and color of the light entering it, producing a mosaic image. The modern digital camera uses a similar approach: A CCD is an array of individual photoreceptors (pixels) that record intensity at specific colors. However, with several million pixels, typical digital cameras now have better resolution than any insect. Unlike the honeybee photoreceptors, which respond to yellow, blue, and ultraviolet light, the pixels in a digital camera record red, green, and blue light, creating a color image that looks natural to our eyes.

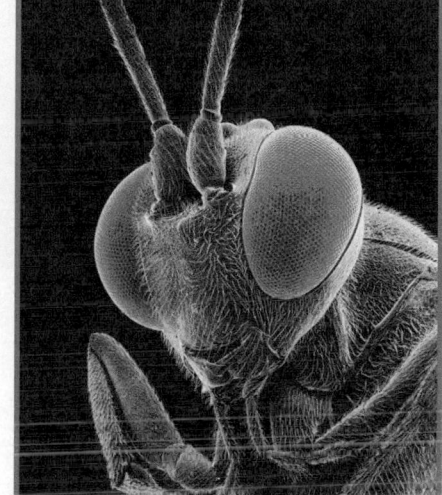

Digital Projectors

Digital projectors (also called *data projectors*) use the same optical principles as the projectors just described, but the film or slide is replaced by a rectangular array of pixels consisting of liquid-crystal diodes (LCDs). Each diode can be made transparent or opaque by imposing appropriate electrical signals on it, and the resulting image is projected onto a screen. This scheme is widely used in computer monitors and TV display systems.

25.3 The Eye

The optical behavior of the eye is similar to that of a camera. The essential parts of the human eye, considered as an optical system, are shown in Figure 25.7. The eye is nearly spherical in shape and about 2.5 cm in diameter. The front portion is somewhat more sharply curved and is covered by a tough, transparent membrane called the *cornea*. The region behind the cornea contains a liquid called the

PhET: Color Vision

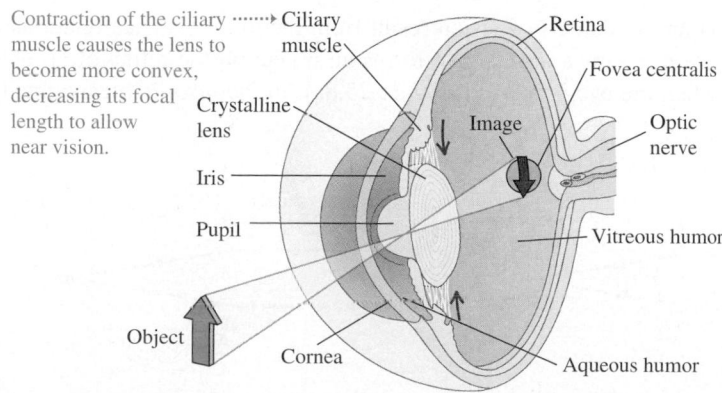

Contraction of the ciliary ·······▶ muscle causes the lens to become more convex, decreasing its focal length to allow near vision.

Ciliary muscle

Crystalline lens

Iris

Pupil

Object

Cornea

Retina

Fovea centralis

Optic nerve

Image

Vitreous humor

Aqueous humor

(a) Diagram of the eye

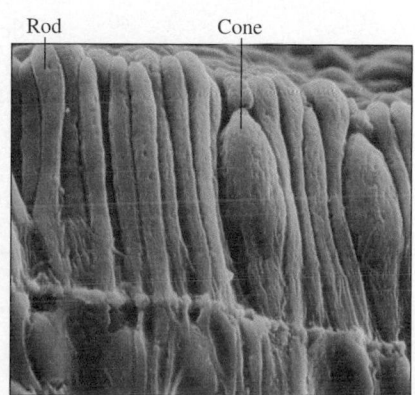

Rod   Cone

(b) Scanning electron micrograph showing retinal rods and cones in cross section

▲ **FIGURE 25.7** The human eye.

▲ BIO Application Focus-o-rama. The crystalline lens and ciliary muscle found in humans and other mammals is only one of a number of focusing mechanisms used by animals. Birds, for instance, can change the shape not only of their lens, but also of the corneal surface. In aquatic animals, the corneal surface is not very useful for focusing, because its refractive index is close to that of water. Thus, focusing is accomplished entirely by the lens, which is nearly spherical. Fishes change their focal length by using a muscle to move the lens either inward or outward. Whales and dolphins achieve the same effect, but hydraulically: They fill or empty a fluid chamber behind the lens to move the lens in or out. In the compound eyes of insects, each unit has its own lens, but adaptive focusing is not needed.

TABLE 25.1 Receding of near point with age

Age (years)	Near point (cm)
10	7
20	10
30	14
40	22
50	40
60	200

aqueous humor. Next comes the *crystalline lens,* a capsule containing a fibrous jelly, hard at the center and progressively softer at the outer portions. The crystalline lens is held in place by ligaments that attach it to the ciliary muscle, which encircles it. Behind the lens, the eye is filled with a thin, watery jelly called the *vitreous humor.* The indexes of refraction of both the aqueous humor and the vitreous humor are nearly equal to that of water, about 1.336. The crystalline lens, although not homogeneous, has an average index of refraction of 1.437. This is not very different from the indexes of the aqueous and vitreous humors; most of the bending of light rays entering the eye occurs at the outer surface of the cornea.

Refraction at the cornea and the surfaces of the lens produces a *real image* of the object being viewed; the image is formed on the light-sensitive *retina* that lines the rear inner surface of the eye. The retina plays the same role as the CCD in a digital camera. The *rods* and *cones* in the retina act like an array of photocells, sensing the image and transmitting it via the *optic nerve* to the brain. Vision is most acute in a small central region called the *fovea centralis,* about 0.25 mm in diameter.

In front of the lens is the *iris.* This structure contains an aperture with variable diameter called the *pupil,* which opens and closes to adapt to changing light intensity. The receptors of the retina also change their sensitivity in response to light intensity (as when your eyes adapt to darkness).

For an object to be seen sharply, the image must be formed exactly on the retina. The lens-to-retina distance, corresponding to s', does not change, but the eye accommodates to different object distances s by changing the focal length of its lens. When the ciliary muscle surrounding the lens contracts, the lens bulges and the radii of curvature of its surfaces *decrease* thereby decreasing the focal length. For the normal eye, an object at infinity is sharply focused when the ciliary muscle is relaxed. With increasing tension, the focal length decreases to permit sharp imaging on the retina of closer objects. This process is called *accommodation.*

The extremes of the range over which distinct vision is possible are known as the *far point* and the *near point* of the eye. The far point of a normal eye is at infinity. The position of the near point depends on the amount by which the ciliary muscle can increase the curvature of the crystalline lens. The range of accommodation gradually diminishes with age as the crystalline lens loses its flexibility. For this reason, the near point gradually recedes as one grows older. This recession of the near point is called *presbyopia;* it is the reason that people need reading glasses when they get older, even if their vision is good otherwise. Table 25.1 shows the approximate position of the near point for an average person at various ages. For example, an average person 50 years of age cannot focus on an object closer than about 40 cm.

Defects of Vision

Several common defects of vision result from incorrect distance relations in the eye. As we just saw, a normal eye forms an image on the retina of an object at infinity when the eye is relaxed (Figure 25.8a). In the *myopic* (nearsighted) eye,

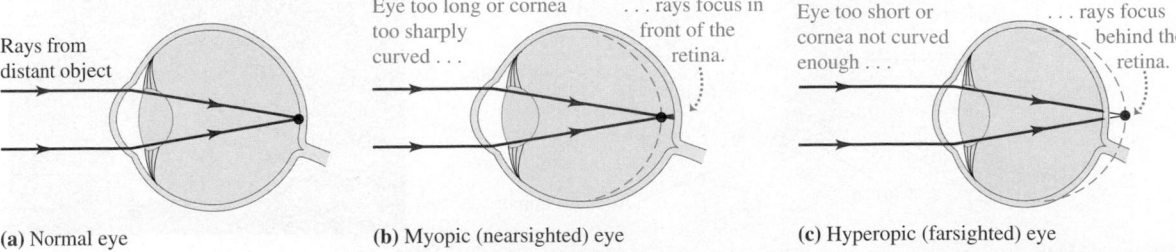

(a) Normal eye

(b) Myopic (nearsighted) eye

(c) Hyperopic (farsighted) eye

▲ FIGURE 25.8 Refractive errors for nearsighted and farsighted eyes.

the eyeball is too long from front to back in comparison to the radius of curvature of the cornea (or the cornea is too sharply curved), and rays from an object at infinity are focused in front of the retina (Figure 25.8b). The most distant object for which an image can be formed on the retina is then nearer than infinity. The *hyperopic* (farsighted) eye has the opposite problem: The eyeball is too short or the cornea is not curved enough, and the image of an infinitely distant object is behind the retina (Figure 25.8c). The myopic eye produces *too much* convergence in a parallel bundle of rays for an image to be formed on the retina; the hyperopic eye produces *not enough* convergence.

Astigmatism refers to a defect in which the surface of the cornea is not spherical, but is more sharply curved in one plane than another. As a result, horizontal lines may be imaged in a different plane from vertical lines (Figure 25.9a). Astigmatism may make it impossible, for example, to focus clearly on the horizontal and vertical bars of a window at the same time.

All these defects can be corrected by the use of corrective lenses (eyeglasses or contact lenses) or, in recent years, by refractive surgery in which the cornea itself is reshaped. The near point of either a presbyopic or a hyperopic eye is *farther* from the eye than normal. To see an object clearly at normal reading distance (often assumed to be 25 cm), such an eye needs an eyeglass lens that forms a virtual image of the object at or beyond the near point. This can be accomplished by a converging (positive) lens, as shown in Figure 25.10. In effect, the lens moves the object farther away from the eye, to a point where a sharp retinal image can be formed. Similarly, eyeglasses for myopic eyes use diverging (negative) lenses to move the image closer to the eye than the actual object, as shown in Figure 25.11.

Astigmatism is corrected by use of a lens with a *cylindrical* surface. For example, suppose the curvature of the cornea in a horizontal plane is correct for focusing rays from infinity on the retina, but the curvature in the vertical plane is not great enough to form a sharp retinal image. Then, when a cylindrical lens with its axis horizontal is placed before the eye, the rays in a horizontal plane are unaffected, but the additional divergence of the rays in a vertical plane causes these to be sharply imaged on the retina, as shown in Figure 25.9b.

Lenses for correcting vision are usually described in terms of the **power,** defined as the reciprocal of the focal length, expressed in meters. The unit of power is the **diopter.** Thus, a lens with $f = 0.50$ m has a power of 2.0 diopters, $f = -0.25$ m corresponds to -4.0 diopters, and so on. The numbers on a prescription for glasses are usually powers expressed in diopters. When the

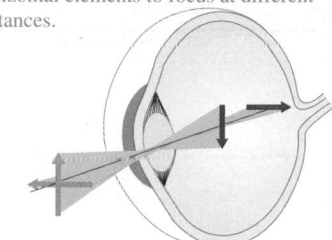

Shape of eyeball or lens causes vertical and horizontal elements to focus at different distances.

(a) Vertical lines are imaged in front of the retina.

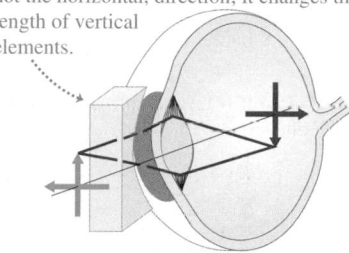

This cylindrical lens is curved in the vertical, but not the horizontal, direction; it changes the focal length of vertical elements.

(b) A cylindrical lens corrects for astigmatism.

▲ **FIGURE 25.9** (a) An uncorrected astigmatic eye. (b) Correction of the astigmatism by a cylindrical lens.

▲ Application **Near- or farsighted?** Can you tell by looking at these eyeglasses whether their owner is myopic (nearsighted) or hyperopic (farsighted)? Yes, easily: Just notice whether the glasses seem to magnify or shrink the person's face. Nearsighted people need eyeglasses with diverging lenses, which make their eyes look smaller from the outside; farsighted people wear glasses with converging lenses, which make their eyes look larger. This man is nearsighted.

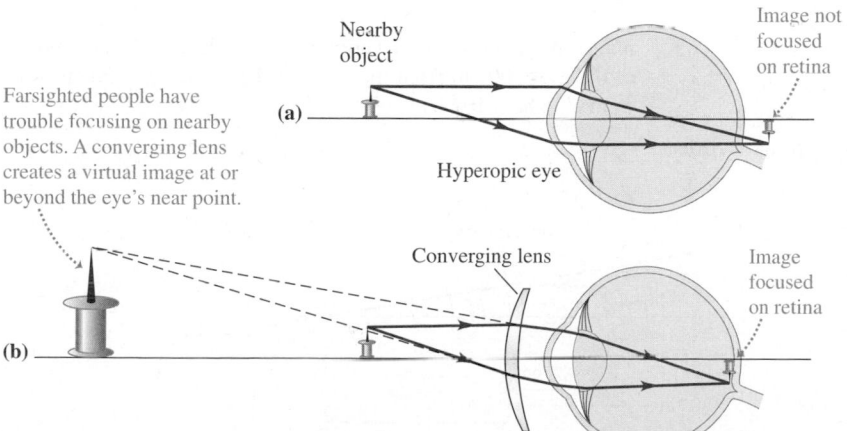

Farsighted people have trouble focusing on nearby objects. A converging lens creates a virtual image at or beyond the eye's near point.

Nearby object

Image not focused on retina

(a)

Hyperopic eye

Converging lens

Image focused on retina

(b)

▲ **FIGURE 25.10** (a) An uncorrected farsighted (hyperopic) eye. (b) A positive lens gives the extra convergence needed to focus the image on the retina.

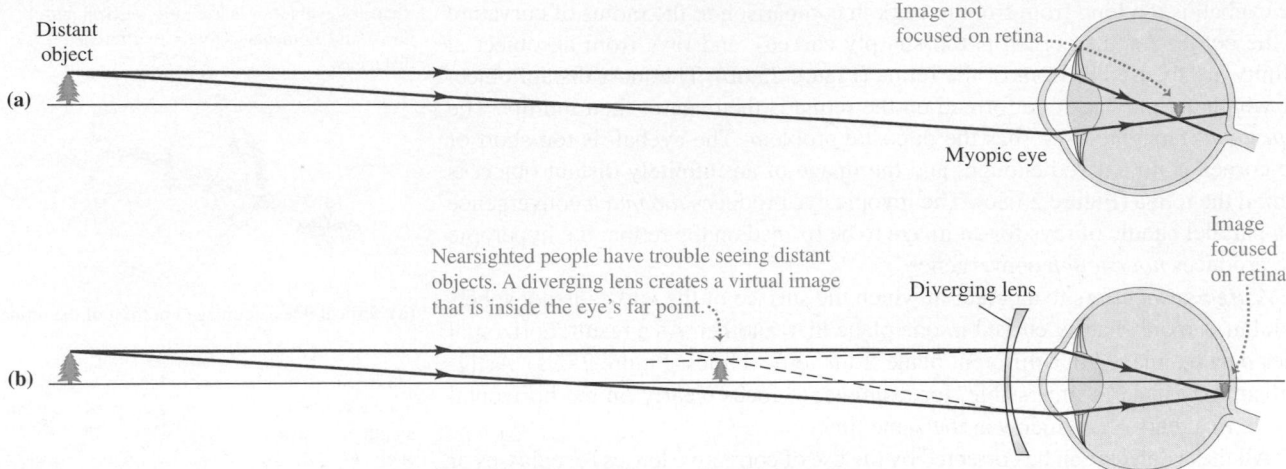

▲ **FIGURE 25.11** (a) An uncorrected nearsighted (myopic) eye. (b) A negative lens spreads the rays farther apart to compensate for the eye's excessive convergence.

correction involves both astigmatism and myopia or hyperopia, there are three numbers for each lens: one for the spherical power, one for the cylindrical power, and an angle to describe the orientation of the cylinder axis.

NOTE ► The use of the term *power* in reference to lenses is unfortunate; it has nothing to do with the familiar meaning of "energy per unit time" in various other areas of physics. The unit of power of a lens is $(1/\text{meter})$, *not* $(1\ \text{joule}/\text{second})$. ◄

With ordinary eyeglasses, the corrective lens is typically 1 to 2 cm from the eye. In the examples that follow, we'll ignore this small distance and assume that the corrective lens and the eye coincide.

EXAMPLE 25.3 **Correcting for farsightedness**

The near point of a certain hyperopic eye is 100 cm in front of the eye. What lens should be used to enable the eye to see clearly an object that is 25 cm in front of the eye? (Neglect the small distance from the lens to the eye.)

SOLUTION

SET UP Figure 25.12 shows the relevant diagram; it is essentially the same as Figure 25.10b, except that it emphasizes the role of the corrective lens. We want the lens to form a virtual image of the object at a location corresponding to the near point of the eye, 100 cm from the lens. That is, when $s = 25$ cm, we want s' to be -100 cm.

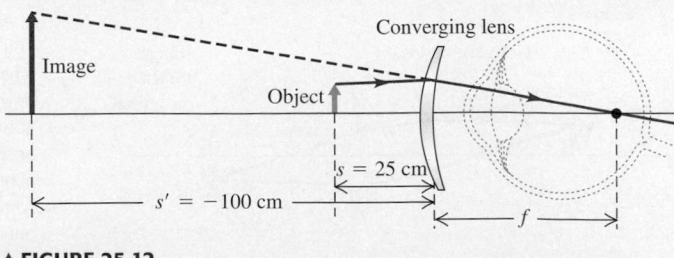

▲ **FIGURE 25.12**

Continued

SOLVE From the basic thin-lens equation (Equation 24.17),

$$\frac{1}{f} = \frac{1}{s} + \frac{1}{s'} = \frac{1}{+25 \text{ cm}} + \frac{1}{-100 \text{ cm}},$$
$$f = +33 \text{ cm}.$$

REFLECT We need a converging lens with focal length $f = 33$ cm. The corresponding power is $1/(0.33 \text{ m})$, or $+3.0$ diopters.

Practice Problem: A 60-year-old man has a near point of 200 cm. What lens should he have to see clearly an object that is 25 cm in front of the lens? *Answer:* $f = 29$ cm (about 3.5 diopters).

EXAMPLE 25.4 Correcting for nearsightedness

The far point of a certain myopic eye is 50 cm in front of the eye. What lens should be used to focus sharply an object at infinity? (Assume that the distance from the lens to the eye is negligible.)

SOLUTION

SET UP Figure 25.13 shows the relevant diagram. The far point of a *myopic* eye is nearer than infinity. To see clearly objects that are beyond the far point, such an eye needs a lens that forms a virtual image of the object no farther from the eye than the far point. We assume that the virtual image is formed at the far point. Then, when $s = \infty$, we want s' to be -50 cm.

SOLVE From the basic thin-lens equation,

$$\frac{1}{f} = \frac{1}{s} + \frac{1}{s'} = \frac{1}{\infty} + \frac{1}{-50 \text{ cm}},$$
$$f = -50 \text{ cm}.$$

REFLECT Because all rays originating at an object distance of infinity are parallel to the axis of the lens, the image distance s' equals the focal length f. We need a *diverging* lens with focal length $f = -50$ cm $= -0.50$ m. The power is $-1/(0.50 \text{ m}) = -2.0$ diopters.

Practice Problem: If the far point of an eye is 75 cm, what lens should be used to see clearly an object that is at infinity? *Answer:* $f = -75$ cm (about -1.3 diopters).

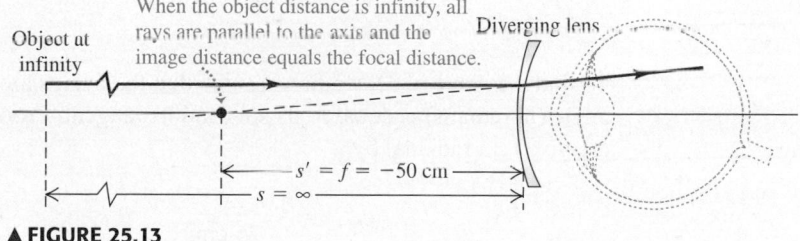

Object at infinity

When the object distance is infinity, all rays are parallel to the axis and the image distance equals the focal distance.

Diverging lens

$s' = f = -50$ cm

$s = \infty$

▲ **FIGURE 25.13**

25.4 The Magnifier

The apparent size of an object is determined by the size of its image on the retina. If the eye is unaided, this size depends upon the *angle* θ subtended by the object at the eye, called its **angular size** (Figure 25.14a).

To look closely at a small object, such as an insect or a crystal, you bring it close to your eye, making the subtended angle and the retinal image as large as possible. But your eye cannot focus sharply on objects closer than the near point, so the angular size of an object is greatest (subtends the largest possible viewing angle) when it is placed at the near point. In the discussion that follows, we'll assume an average viewer with a near point 25 cm from the eye.

A converging lens can be used to form a virtual image that is larger and farther from the eye than the object itself, as shown in Figure 25.14. Then the object can be moved closer to the eye, and the angular size of the image may be substantially larger than the angular size of the object at 25 cm without the lens. A

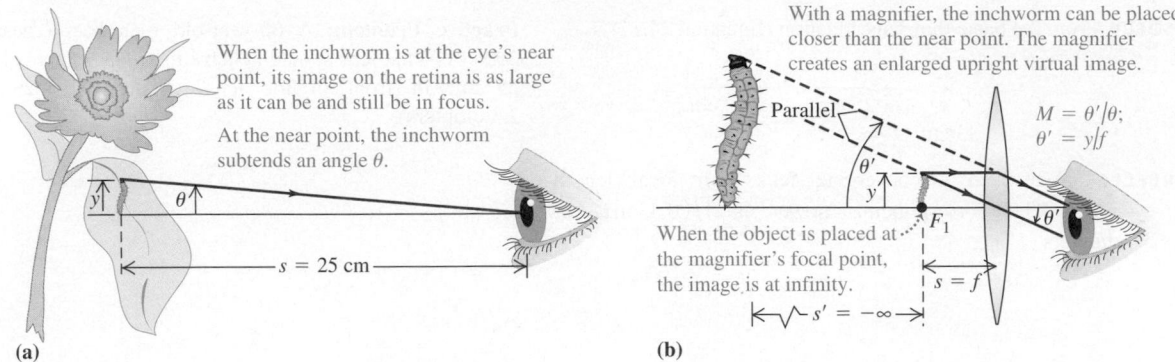

When the inchworm is at the eye's near point, its image on the retina is as large as it can be and still be in focus.

At the near point, the inchworm subtends an angle θ.

$s = 25$ cm

(a)

With a magnifier, the inchworm can be placed closer than the near point. The magnifier creates an enlarged upright virtual image.

Parallel

$M = \theta'/\theta;$
$\theta' = y/f$

When the object is placed at the magnifier's focal point, the image is at infinity.

$s = f$

$s' = -\infty$

(b)

▲ **FIGURE 25.14** (a) The subtended angle θ is largest when an object is at the near point. A simple magnifier produces a virtual image at infinity, which acts as a real object subtending a larger angle θ' for the eye.

lens used in this way is called a *magnifying glass,* or simply a **magnifier.** The virtual image is most comfortable to view when it is placed at infinity, and in the discussion that follows we'll assume that this is done.

In Figure 25.14a, the object is at the near point, where it subtends an angle θ at the eye. In Figure 25.14b, a magnifier in front of the eye forms an image at infinity, and the angle subtended at the magnifier is θ'. We define the **angular magnification** M as follows:

Definition of angular magnification

Angular magnification M is the ratio of the angle θ' subtended by an object at the eye when the magnifier is used to the angle θ subtended without the magnifier:

$$M = \frac{\theta'}{\theta}. \tag{25.2}$$

To find the value of M, we first assume that the angles are small enough that each angle (in radians) is equal to its sine and its tangent. From Figure 25.14, θ and θ' are given (in radians) by

$$\theta = \frac{y}{25 \text{ cm}}, \qquad \theta' = \frac{y}{f}.$$

Combining these expressions with Equation 25.2, we find that

$$M = \frac{\theta'}{\theta} = \frac{y/f}{y/25 \text{ cm}} = \frac{25 \text{ cm}}{f}. \tag{25.3}$$

NOTE ▶ Be careful not to confuse *angular magnification* M (a ratio of two angles) with *lateral magnification* m (the ratio of image to object height, which we defined in Chapter 24). In some of the examples we'll discuss later, M is the more relevant quantity. Be on the lookout for this distinction, and make sure you understand why one or the other is relevant in a specific situation. ◀

It may seem that we can make the angular magnification as large as we like by decreasing the focal length f. However, the aberrations of a simple double-convex lens (to be discussed in Section 25.7) set an upper limit on M of about $3\times$ to $4\times$. If these aberrations are corrected, the angular magnification may be made as great as $20\times$. When greater magnification than this is needed, we usually use a compound microscope, discussed in the next section.

Comparing magnifying lenses

A simple magnifier with a shorter focal length gives a greater angular magnification than one with a longer focal length primarily because

A. the image is the same angular size for both lenses, but the image can be seen clearly closer to the eye with the lens having smaller focal length;
B. The angular image size is larger for the lens having smaller focal length;
C. The object can be held farther from the eye and still be seen clearly;
D. the magnification depends, *not* on the focal length, but on the location of your eye relative to the lens.

SOLUTION The purpose of any magnifying device is to create an image with larger angular size than is possible with the unaided eye, in order to produce the largest possible image on the retina. The minimum eye-to-object distance for the normal unaided eye is about 25 cm, so an object with height y has maximum angular size $y/25$ cm. When a converging lens with focal length f is placed closer to the eye than 25 cm, the lens forms a virtual image. If this image is at infinity, its angular size is y/f, as discussed earlier. The angular magnification is the quotient of these two angular sizes, which, from Equation 25.3, is $(25\,\text{cm}/f)$. If the focal length is anything less than 25 cm, the angular size of the image is *greater than* the angular size of the object. The object must be at or inside the focal point to create a virtual image. Shortening the focal length means that the object must be closer, creating a larger image size and greater angular magnification. The correct answer is B.

EXAMPLE 25.5 A simple magnifier

You have two plastic lenses, one double convex and the other double concave, each with a focal length with absolute value 10.0 cm. **(a)** Which lens can you use as a simple magnifier? **(b)** What is the angular magnification?

SOLUTION

SET UP AND SOLVE Part (a): To act as a magnifier, a lens must produce an upright virtual image that is larger than the object. Figure 25.14 shows that a double-convex lens does this.

REFLECT A double-*concave* lens would produce an upright virtual image *smaller* than the object.

Part (b): From Equation 25.3, the angular magnification M is

$$M = \frac{25\,\text{cm}}{f} = \frac{25\,\text{cm}}{10\,\text{cm}} = 2.5.$$

Practice Problem: A simple magnifier has a focal length of 14 cm. What is the angular magnification of this magnifier? *Answer: M = 1.8.*

25.5 The Microscope

When we need greater magnification than we can get with a simple magnifier, the instrument we usually use is the **microscope,** sometimes called a *compound microscope.* The essential elements of a microscope are shown in Figure 25.15a. To analyze this system, we use the principle that an image formed by one optical element, such as a lens or mirror, can serve as the object for a second element. We used this principle in Section 24.5 when we derived the thin-lens equation by repeated application of the single-surface refraction equation.

The object O to be viewed is placed just beyond the first focal point F_1 of the **objective,** a lens that forms a real, enlarged, inverted image I (Figure 25.15b). In a properly designed instrument, this image lies just inside the first focal point F_2 of the **eyepiece** (also called the *ocular*), which forms a final virtual image of I at I'. The position of I' may be anywhere between the near and far points of the eye. Both the objective and the eyepiece of an actual microscope are highly corrected compound lenses, but for simplicity, we show them here as simple thin lenses.

The notation in Figure 25.15b merits careful study. The objective lens has focal length f_1 and focal points F_1 and F_1'; the eyepiece lens has focal length f_2 and focal points F_2 and F_2'. The object and image distances for the objective lens are s_1 and s_1', respectively, and the object and image distances for the eyepiece lens are s_2 and s_2', respectively (not shown in the figure).

Mastering**PHYSICS**

ActivPhysics 15.12: Two-Lens Optical Systems

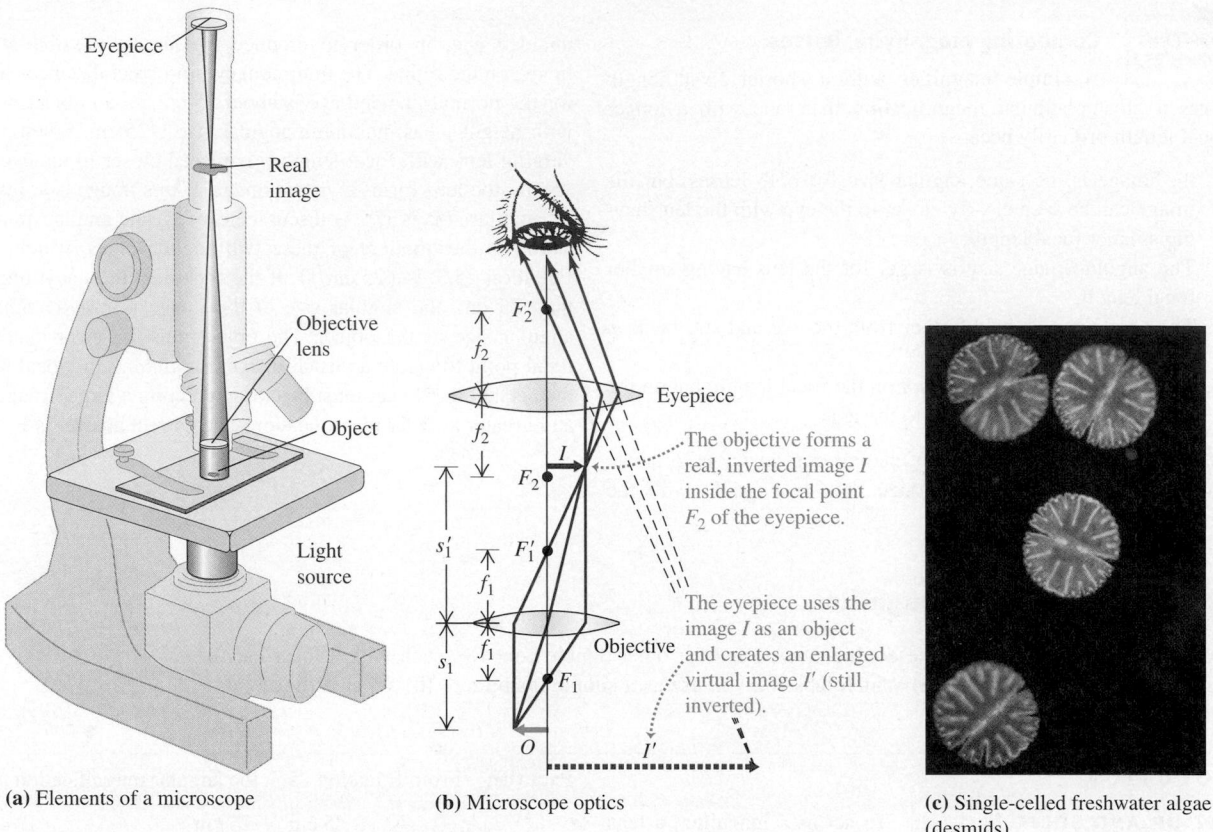

(a) Elements of a microscope

(b) Microscope optics

(c) Single-celled freshwater algae (desmids).

Eyepiece

Real image

Objective lens

Object

Light source

F_2'

f_2

f_2

F_2

I

Eyepiece

The objective forms a real, inverted image I inside the focal point F_2 of the eyepiece.

s_1'

F_1'

f_1

f_1

s_1

Objective

F_1

The eyepiece uses the image I as an object and creates an enlarged virtual image I' (still inverted).

O

I'

▲ **FIGURE 25.15** A light microscope. Typically, light microscopes can resolve details as small as 200 nm, comparable to the wavelength of light.

As with a simple magnifier, the most significant quantity for a microscope is the *angular* magnification M. The objective forms an enlarged, real, inverted image that is viewed through the eyepiece. Thus, the overall angular magnification M of the compound microscope is the product of the *lateral* magnification m_1 of the objective and the *angular* magnification M_2 of the eyepiece. The first is given by

$$m_1 = -\frac{s_1'}{s_1},$$

where s_1 and s_1' are the object and image distances, respectively, for the objective. Ordinarily, the object is very close to the focal point, and the resulting image distance s_1' is very great in comparison to the focal length f_1 of the objective. Thus, s_1 is approximately equal to f_1, and $m_1 \simeq -s_1'/f_1$.

The eyepiece functions as a simple magnifier, as discussed in Section 25.4. From Equation 25.3, its angular magnification is $M_2 = (25 \text{ cm})/f_2$, where f_2 is the focal length of the eyepiece, considered as a simple lens. The overall magnification M of the compound microscope (apart from a negative sign, which is customarily ignored) is the product of the two magnifications; that is,

$$M = m_1 M_2 = \frac{(25 \text{ cm}) s_1'}{f_1 f_2}, \tag{25.4}$$

where s_1', f_1, and f_2 are measured in centimeters. The final image is inverted with respect to the object (Figure 25.15b). Microscope manufacturers usually specify the values of m_1 and M_2 for microscope components, rather than the focal lengths of the objective and eyepiece.

Conceptual Analysis 25.2

Lenses in a compound microscope

The basic compound microscope consists of

A. two diverging lenses, both of short focal length;
B. two converging lenses, both of short focal length;
C. two converging lenses, one of short and one of long focal length;
D. one diverging lens and one converging lens.

SOLUTION The eyepiece of a compound microscope serves as a simple magnifier to view the image created by the objective lens. (The telescope, discussed in the next section, uses the same arrangement.) Simple magnifiers are always converging lenses. As we discussed in Conceptual Analysis 25.1, the magnifier (the eyepiece in this case) requires a focal length less than 25 cm. The total length of the microscope must be small enough so that the observer can reach the sample while looking into the eyepiece. The focal length of the objective must be small enough to get a large lateral magnification and still keep the objective's image inside the microscope. Thus, the focal length of the objective must be much less than the barrel length of the microscope; the correct answer is B.

EXAMPLE 25.6 Watching bacteria

The maximum magnification attainable with a compound microscope is about 2000×. The smallest bacteria are roughly 0.5 micron in size. **(a)** For an eyepiece with 15×, what magnification must the objective lens have to reach an overall magnification of 2000×? **(b)** How big does a half-micron bacterium appear? Is this larger or smaller than the period at the end of this sentence?

SOLUTION

SET UP We are given the angular magnification of the eyepiece: $M_2 = 15\times$. We want an overall magnification $M = 2000\times$. The size of the bacterium is $0.5\ \mu m = 0.5 \times 10^{-6}$ m. We want to find the lateral magnification m_1 of the objective.

SOLVE Part (a): From Equation 25.4, $m_1 = M/M_2 = 2000\times/15\times = 130\times$.

Part (b): The apparent size of the bacterium is

$$(2000)(0.5 \times 10^{-6}\ m) = 1.0 \times 10^{-3}\ m = 1.0\ mm.$$

REFLECT The apparent size is a few times larger than the period at the end of this sentence. Thus, the bacterium is visible with 2000× magnification, but without much detail.

Practice Problem: If the only available eyepiece has an angular magnification of 10×, what is the overall magnification of this microscope? *Answer:* 1300×.

25.6 Telescopes

The optical system of a refracting **telescope** is similar to that of a compound microscope. In both instruments the image formed by an objective is viewed through an eyepiece. The difference is that the telescope is used to view large objects at large distances, while the microscope is used to view small objects close at hand.

An *astronomical telescope* is shown in Figure 25.16. The objective forms a reduced real image *I* of the object, and the eyepiece forms an enlarged virtual image of *I*. As with the microscope, the image *I'* may be formed anywhere between the near and far points of the eye. Objects viewed with a telescope are usually so far away from the instrument that the first image *I* is formed very nearly at the second focal point of the objective. This image is the object for the eyepiece. If the final image *I'* formed by the eyepiece is at infinity (for the most comfortable viewing by a normal eye), the first image must be at the first focal point of the eyepiece. The distance between the objective and the eyepiece is the length of the telescope, which is therefore the *sum* $f_1 + f_2$ of the focal lengths of the objective and the eyepiece.

The angular magnification *M* of a telescope is defined as the ratio of the angle subtended at the eye by the final image *I'* to the angle subtended at the (unaided) eye by the object. We can express this ratio in terms of the focal lengths of the objective and the eyepiece. In Figure 25.16, the ray passing through F_1, the first

MasteringPHYSICS

ActivPhysics 15.12: Two-Lens Optical Systems

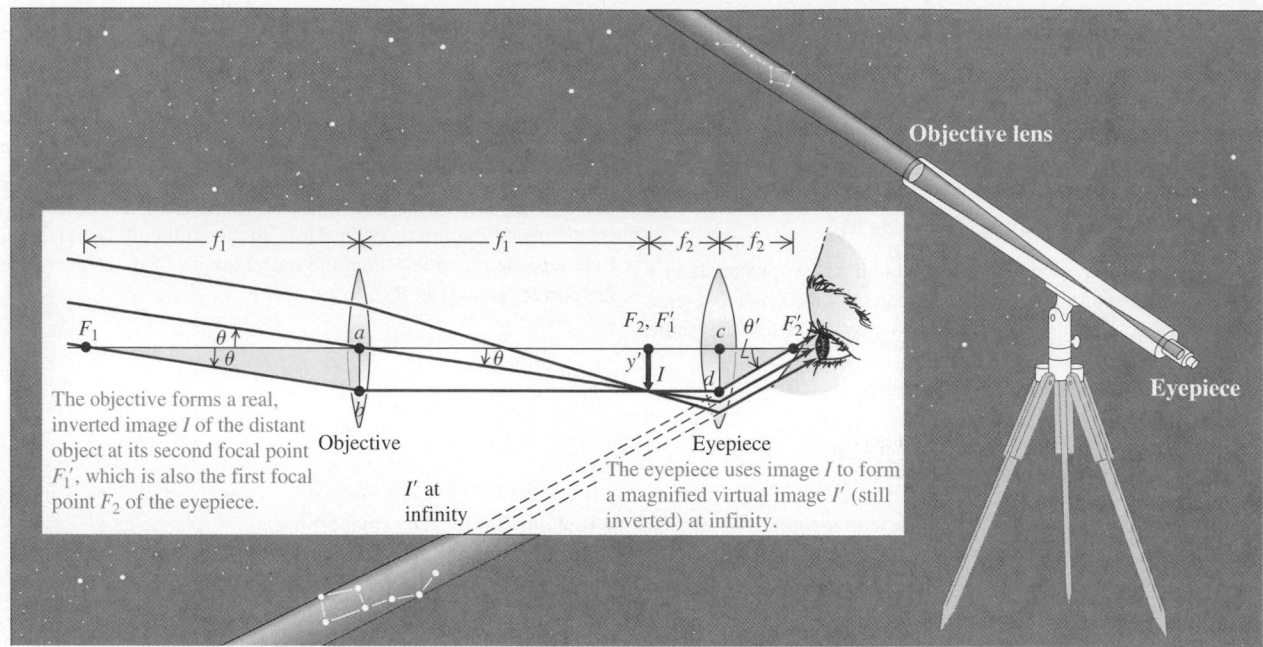

▲ **FIGURE 25.16** Optical system of an astronomical telescope.

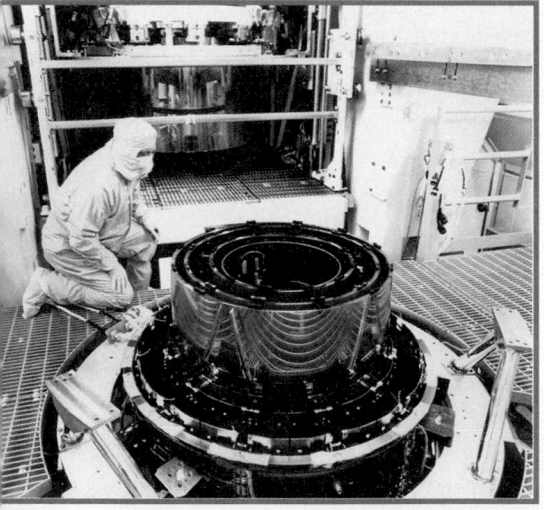

▲ **Application X-ray telescope.** It's not easy to make a telescope that works on x rays, because these rays simply pass through the surface of a normal mirror. They can be reflected only by a metal surface that they strike at a grazing angle of incidence. Therefore, an x-ray telescope consists of a sheaf of concentric, barrel-shaped mirrors, each shaped to reflect grazing-incidence x rays toward a focus. Because x rays do not penetrate earth's atmosphere, x-ray telescopes must be placed in space. This photo shows the mirrors of the Chandra x-ray telescope being assembled.

focal point of the objective, and through F_2', the second focal point of the eyepiece, has been emphasized. The object subtends an angle θ at the objective and would subtend essentially the same angle at the unaided eye. Also, since the observer's eye is placed just to the right of the focal point F_2', the angle subtended at the eye by the final image is very nearly equal to the angle θ'. Because bd is parallel to the optic axis, the distances ab and cd are equal to each other and also to the height y' of the image I. Because θ and θ' are small, they may be approximated by their tangents. From the right triangles F_1ab and $F_2'cd$, we obtain

$$\theta = \frac{-y'}{f_1}, \quad \theta' = \frac{y'}{f_2},$$

and the angular magnification M is

$$M = \frac{\theta'}{\theta} = -\frac{y'/f_2}{y'/f_1} = -\frac{f_1}{f_2}. \tag{25.5}$$

The angular magnification M of a telescope is equal to the ratio of the focal length of the objective to that of the eyepiece. The negative sign shows that the final image is inverted.

An inverted image is no particular disadvantage for astronomical observations. When we use a telescope or binoculars on earth, though, we want the image to be right side up. Inversion of the image is accomplished in *prism binoculars* by a pair of 45°–45°–90° totally reflecting prisms called *Porro prisms* (introduced in Section 23.8). Porro prisms are inserted between objective and eyepiece, as shown in Figure 25.17. The image is inverted by the four internal reflections from the faces of the prisms at 45°. The prisms also have the effect of folding the optical path and making the instrument shorter and more compact than it would otherwise be. (See Figure 23.26.) Binoculars are usually described by two numbers separated by a multiplication sign, such as 7 × 50. The first number is the angular magnification M, and the second is the diameter of the objective lenses (in millimeters). The diameter determines the light-gathering capacity of the objective lenses and thus the brightness of the image.

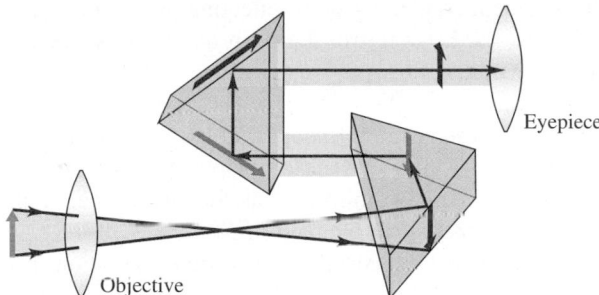

▲ **FIGURE 25.17** Inversion of an image in prism binoculars.

 **Conceptual Analysis 25.3**

The objective lens

The purpose of the objective lens of a telescope is to

A. produce an image whose angular size is greater than that of the object;
B. take parallel light rays and spread them apart;
C. produce an image that is larger than the object;
D. form an image that can be looked at closely with the aid of an eyepiece.

SOLUTION When we look at a distant object (for example, a person 100 m away), the angular size is the ratio of the height of the person to the distance to the person. The image created by the objective lens is much smaller than the person. However, using an eyepiece, the observer can get very close to the objective's image to view that image. Viewed through the eyepiece, the angular size of the objective's image is much larger than the angular size of the distant object. The correct answer is D.

Reflecting Telescopes

In the *reflecting telescope* (Figure 25.18a), the objective lens is replaced by a concave mirror. In large telescopes, this scheme has many advantages, both theoretical and practical. Mirrors are inherently free of chromatic aberrations (the dependence of focal length on wavelength), and spherical aberrations (associated with the paraxial approximation) are easier to correct than for a lens. The reflecting surface is sometimes parabolic rather than spherical. The material of the mirror need not be transparent, and the mirror can be made more rigid than a lens, which can be supported only at its edges.

Because the image in a reflecting telescope is formed in a region traversed by incoming rays, it can be observed directly with an eyepiece only by blocking off part of the incoming beam (Figure 25.18a), an arrangement that is practical only for the very largest telescopes. Alternative schemes use a mirror to reflect the image through a hole in the mirror or out the side, as shown in Figures 25.18b and 25.18c. This scheme is also used in some long-focal-length telephoto lenses for cameras. In the context of photography, such a system is called a *catadioptric*

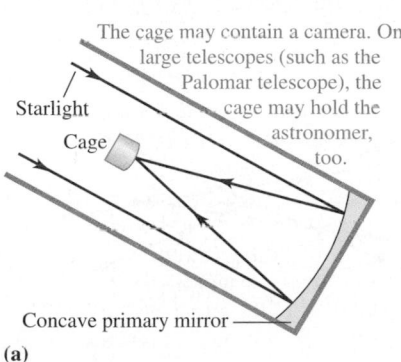
(a)

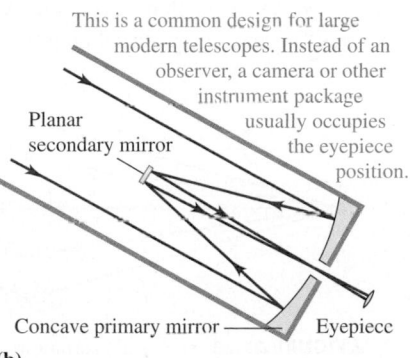
(b)

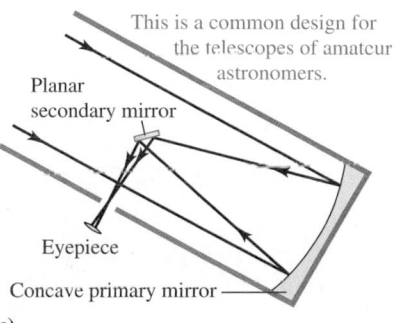
(c)

▲ **FIGURE 25.18** Three optical systems for reflecting telescopes.

(a) One of the 8.4 m single primary mirrors of the Large Binocular Telescope being shaped by a computer-controlled polishing machine. (The mirror is not yet silvered.)

(b) A worker in front of the last segment of the Southern African Large Telescope (SALT).

▲ **FIGURE 25.19** Examples of single and composite telescope mirrors.

lens—an optical system containing both reflecting and refracting elements. The Hubble Space Telescope has a mirror 2.4 m in diameter, with an optical system similar to that in Figure 25.18b.

It is very difficult and costly to fabricate large mirrors and lenses with the accuracy needed for astronomical telescopes. Lenses larger than 1 m in diameter are usually not practical. A few current telescopes have single mirrors over 8 m in diameter. The largest single-mirror reflecting telescope in the world, the Large Binocular Telescope on Mt Graham in Arizona, will have two 8.4 m mirrors on a common mount; the first mirror went into operation in October 2005 (Figure 25.19a).

Another technology for creating very large telescope mirrors is to construct a composite mirror of individual hexagonal segments and use computer control for precise alignment. The twin Keck telescopes, atop Mauna Kea in Hawaii, have mirrors with an overall diameter of 10 m, made up of 36 separate 1.8 m hexagonal mirrors. The orientation of each mirror is controlled by computer to within one millionth of an inch. In the Southern Hemisphere, the Southern African Large Telescope (SALT) has an 11 m mirror composed of 91 hexagonal segments (Figure 25.19b). It went into operation in September 2005.

25.7 Lens Aberrations

An **aberration** is any failure of a mirror or lens to behave precisely according to the simple formulas that we've derived. Aberrations can be classified as **chromatic aberrations,** which involve wavelength-dependent imaging behavior, and **monochromatic aberrations,** which occur even with monochromatic (single-wavelength) light. Lens aberrations are not caused by faulty construction of the lens, such as irregularities in its surfaces, but rather are inevitable consequences of the laws of refraction at spherical surfaces.

Monochromatic aberrations are all related to the *paraxial approximation.* Our derivations of equations for object and image distances, focal lengths, and magnification have all been based upon this approximation. We have assumed that all rays are *paraxial*—that is, that they are very close to the optic axis and make very small angles with it. Real lenses never conform exactly to this condition.

For any lens with an aperture of finite size, the cone of rays that forms the image of any point has a finite size. In general, nonparaxial rays that proceed from a given object point *do not* all intersect at precisely the same point after they are refracted by a lens. For this reason, the image formed by these rays is never perfectly sharp. *Spherical aberration* is the failure of rays from a point object on the optic axis to converge to a point image. Instead, the rays converge within a circle of minimum radius, called *the circle of least confusion,* and then diverge again, as shown in Figure 25.20. The corresponding effect for points off the optic axis produces images that are comet-shaped figures rather than circles; this effect is called *coma.* Note that decreasing the aperture size cuts off the larger-angle rays, thus decreasing spherical aberration.

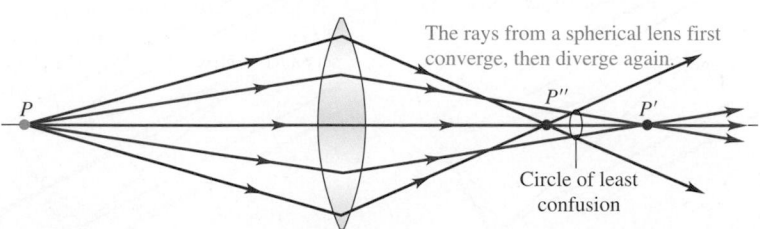

▲ **FIGURE 25.20** Spherical aberration.

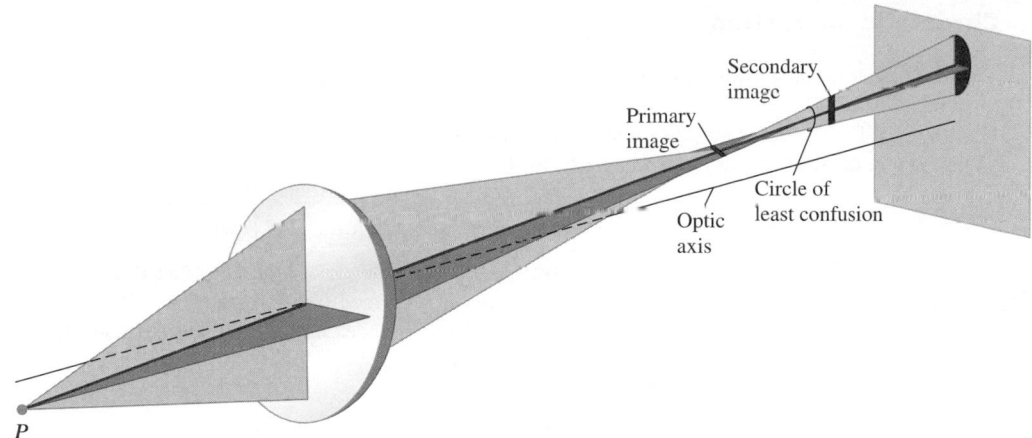

▲ **FIGURE 25.21** Astigmatism of a lens for a point below the optic axis. The lens forms two images of the point, in planes perpendicular to each other.

Spherical aberration is also present in spherical mirrors. Mirrors used in astronomical reflecting telescopes are usually paraboloidal rather than spherical; paraboloidal shapes eliminate spherical aberration for points on the axis, but are much more difficult to fabricate precisely than are spherical shapes.

Astigmatism is the imaging of a point off the axis as two perpendicular lines in different planes. In this aberration, the rays from a point object converge, at a certain distance from the lens, to a line called the *primary image,* perpendicular to the plane defined by the optic axis and the object point. At a somewhat different distance from the lens, they converge to a second line called the *secondary image,* which is *parallel* to that plane. This effect is shown in Figure 25.21. The circle of least confusion (greatest convergence) appears between these two positions.

The location of the circle of least confusion depends on the object point's *transverse* distance from the axis, as well as its *longitudinal* distance from the lens. As a result, object points lying in a plane are, in general, imaged not in a plane, but in some curved surface; this effect is called *curvature of field.*

Finally, the image of a straight line that does not pass through the optic axis may be curved. As a result, the image of a square with the axis through its center may resemble a barrel (sides bent outward) or a pincushion (sides bent inward). This effect, called *distortion,* is not related to lack of sharpness of the image, but results from a change in lateral magnification with distance from the axis.

Chromatic aberrations are a result of *dispersion*—the variation of index of refraction with wavelength that we discussed in Section 23.9. Dispersion causes the focal length of a lens to be somewhat different for different wavelengths. When an object is illuminated with light containing a mixture of wavelengths, different wavelengths are imaged at different points. The magnification of a lens also varies with wavelength; this effect is responsible for the rainbow-fringed images that are seen with inexpensive binoculars or telescopes. Reflectors are inherently free of chromatic aberrations; this is one of the reasons for their usefulness in large astronomical telescopes.

It is impossible to eliminate all these aberrations from a single lens, but in a compound lens with several optical elements, the aberrations of one element may partially cancel those of another. The design of such lenses is an extremely complex undertaking, aided greatly in recent years by the use of computers. It is still impossible to eliminate all aberrations, but it *is* possible to decide which ones are most troublesome for a particular application and to design accordingly.

EXAMPLE 25.7 **Chromatic aberration**

A glass planoconvex lens has its flat side toward the object, and the other side has a radius of curvature with magnitude 30.0 cm. The index of refraction of the glass for violet light (wavelength 400 nm) is 1.537, and for red light (700 nm) it is 1.517. The color purple is a mixture of red and violet. A purple object is placed 80.0 cm from this lens. Where are the red and violet images formed?

SOLUTION

SET UP AND SOLVE We use the thin-lens equation in the form given by Equation 24.19:

$$\frac{1}{s} + \frac{1}{s'} = (n-1)\left(\frac{1}{R_1} - \frac{1}{R_2}\right).$$

In this case, using the usual sign rules, we have $R_1 = \infty$ and $R_2 = -30.0$ cm. For violet light $(n = 1.537)$,

$$\frac{1}{80.0 \text{ cm}} + \frac{1}{s'} = (1.537 - 1)\left(\frac{1}{\infty} - \frac{1}{-30.0 \text{ cm}}\right)$$
$$s' = 185 \text{ cm}.$$

For red light $(n = 1.517)$, we find that $s' = 211$ cm.

REFLECT The violet light is refracted more than the red, and its image is formed closer to the lens. We see that a rather small variation in index of refraction causes a substantial displacement of the image.

SUMMARY

The Camera and the Projector

(Sections 25.1 and 25.2) A camera forms a real, inverted, usually reduced image of the object being photographed. For cameras using film, the amount of light striking the film is controlled by the shutter speed and the aperture. A projector is essentially a camera in reverse: A lens forms a real, inverted, enlarged image on a screen of the slide or motion-picture film.

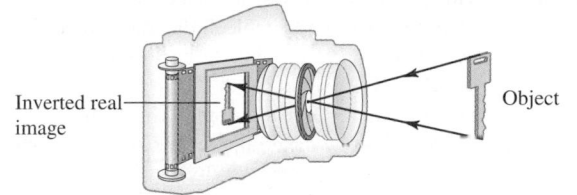

Inverted real image

Object

The Eye

(Section 25.3) In the eye, a real image is formed on the retina and transmitted to the optic nerve. Adjustment for various object distances is made by the ciliary muscle; for close vision, the lens becomes more convex, decreasing its focal length. For sharp vision, the image must form exactly on the retina. In a nearsighted (myopic) eye, the image is formed in front of the retina. In a farsighted (hyperopic) eye, the image is formed behind the retina. The power of a corrective lens, in diopters, is the reciprocal of the focal length, in meters.

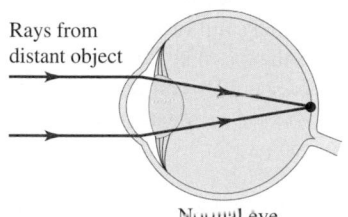

Rays from distant object

Normal eye

Magnifiers, Microscopes, and Telescopes

(Section 25.4–25.6) The apparent size of an object is determined by the size of its image on the retina. A simple magnifier creates a virtual image whose angular size is larger than that of the object itself. The angular magnification is the ratio of the angular size of the virtual image to that of the object.

In a compound microscope, the objective lens forms a first image in the barrel of the instrument, and this image becomes the object for the second lens, called the eyepiece. The eyepiece forms a final virtual image, often at infinity, of the first image. A telescope operates on the same principle, but the object is far away. In a reflecting telescope, the objective lens is replaced by a concave mirror, which eliminates chromatic aberrations.

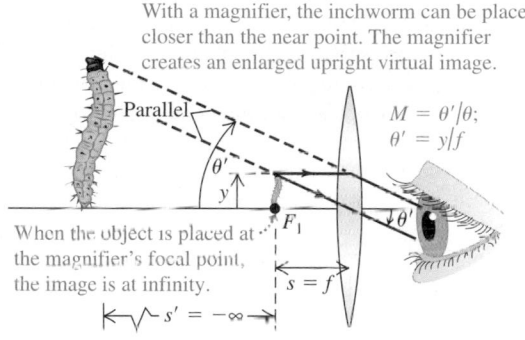

With a magnifier, the inchworm can be placed closer than the near point. The magnifier creates an enlarged upright virtual image.

Parallel

$M = \theta'/\theta;$
$\theta' = y/f$

When the object is placed at the magnifier's focal point, the image is at infinity.

F_1

$s = f$

$s' = -\infty$

Lens Aberrations

(Section 25.7) Lens aberrations account for the failure of a lens to form a perfectly sharp image of an object. Monochromatic aberrations occur because of limitations of the paraxial approximation. Chromatic aberrations result from the dependence of index of refraction on wavelength, causing the focal length of a lens to be somewhat different for different wavelengths.

Conceptual Questions

1. Sometimes a wine glass filled with white wine forms an image of an overhead light on a white tablecloth. Show how this image is formed. Would the same image be formed with an empty glass? With a glass of water?

2. How could one very quickly make an approximate measurement of the focal length of a converging lens? Could the same method be applied to a diverging lens? Why?

3. The human eye is often compared to a camera. In what ways is it similar to a camera? In what ways does it differ?

4. How could one make a lens for sound waves?

5. A diver proposed using a plastic bag full of air, immersed in water, as an underwater lens. Is this possible? If the lens is to be a converging lens, what shape should the air pocket have?

6. When a converging lens is immersed in water, does its focal length increase or decrease in comparison to its focal length in air?

7. You are marooned on a desert island and want to use your eyeglasses to start a fire. Can this be done if you are nearsighted? If you are farsighted?

8. If the sensor array of a digital camera is placed at the location of a real image, the sensor will record this image. Can this be done with a virtual image? How might one record a virtual image?

9. There have been reports of round fishbowls starting fires by focusing the sun's rays coming in a window. Is this possible?

10. Since a refracting telescope and a compound microscope have the same basic design, can they be used interchangeably? In other words, could you use a telescope as a microscope and vice versa? Why? In what ways are they different and in what ways similar?

11. You are selecting a converging lens for a magnifier. By simply feeling several lenses, you find that all are flat on one side, yet on the other side one lens is highly curved while the another one is almost flat. Which of these two lenses would give a greater angular magnification?

12. While choosing between two refracting astronomical telescopes, you notice that one is twice as long as the other one. Will the longer telescope necessarily produce a greater angular magnification than the shorter one? Why?

13. When choosing between two refracting astronomical telescopes, you notice that the tube of one is twice as wide as the tube of the other one because the objective lens is twice as wide. Will the telescope with the wider tube necessarily produce a greater angular magnification than the narrower one? Why? Would there be *any* advantages to the wider-tube telescope? If so, what would they be?

14. You've entered a survival contest that will include building a crude telescope. You are given a large box of lenses. Which two lenses do you pick? How do you quickly identify them?

15. If a person is severely nearsighted, can the optometrist prescribe a *single* lens that will correct this defect and allow him to see clearly at all the usual distances?

16. Ads for amateur telescopes sometimes contain statements such as "maximum magnification 600×." Are such statements meaningful? Why can they be misleading? Is there really a maximum magnification for a telescope?

17. The focal length of a simple lens depends on the color (wavelength) of light passing through it. Why? Is it possible for a lens to have a positive focal length for some colors and negative for others? Explain.

Multiple-Choice Problems

1. The focusing mechanism of the human eye most closely resembles that of a
   A. telescope.
   B. microscope.
   C. camera.
   D. magnifier.

2. Which of the following statements are true about the lenses used in eyeglasses to correct nearsightedness and farsightedness? (There may be more than one correct choice.)
   A. They produce a real image.
   B. They produce a virtual image.
   C. Both nearsightedness and farsightedness are corrected with a converging lens.
   D. Both nearsightedness and farsightedness are corrected with a diverging lens.

3. The discussion of simple magnifiers in Sec. 25.4 was based on the assumption of a viewer with a normal near point of 25 cm. *Compared to the angular size they can see with their own unaided eyes,* how does the angular magnification obtained by a myopic person compare with that obtained by a person with normal vision, if they look at the same object with the same magnifying glass?
   A. The myopic and the normal person obtain the same magnification.
   B. The myopic person obtains a greater magnification.
   C. The myopic person obtains less magnification.

4. If, without changing anything else, we double the focal lengths of both of the lenses in a refracting telescope that had an original angular magnification M, its new angular magnification will be
   A. $4M$.
   B. $2M$.
   C. M.
   D. $M/2$.
   E. $M/4$.

5. If a person's eyeball is 2.7 cm deep instead of the usual 2.5 cm, he most likely suffers from
   A. astigmatism.
   B. color blindness.
   C. nearsightedness.
   D. farsightedness.

6. Which of the following statements are true about the eye? (There may be more than one correct choice.)
   A. When focusing on a very distant object, the lens is curved the most.
   B. When focusing on a nearby object, the lens is curved the most.
   C. Most of the bending of light is accomplished by the lens.
   D. Most of the bending of light is accomplished by the cornea.
   E. Most of the focusing is due to variations in the diameter of the pupil.

7. If a camera lens gives the proper exposure for a photograph at a shutter speed of $1/200$ s at an *f*-stop of $f/2.8$, the proper shutter speed at $f/5.6$ is
   A. $1/800$ s.
   B. $1/400$ s.
   C. $1/100$ s.
   D. $1/50$ s.

8. A slide projector produces
   A. a real erect image of the slide.
   B. a virtual erect image of the slide.
   C. a real inverted image of the slide.
   D. a virtual inverted image of the slide.
9. Which of the following operations would increase the angular magnification of a refracting telescope? (There may be more than one correct choice.)
   A. Increase the focal length of the objective lens.
   B. Increase the focal length of the eyepiece.
   C. Decrease the focal length of the objective.
   D. Decrease the focal length of the eyepiece.
   E. Increase the focal length of both the objective lens and the eyepiece by the same factor.
10. Which of the following statements are true about a compound microscope? (There may be more than one correct choice.)
    A. The objective lens produces an inverted virtual image of the object.
    B. The image produced by the objective lens is just inside the focal point of the eyepiece.
    C. The final image is inverted compared with the original object.
    D. The object to be viewed is placed just inside the focal point of the objective lens.
    E. The final image is virtual.
11. Laser eye surgery can correct various vision defects by altering the shape of the cornea's outer surface. Suppose a surgeon performs an operation that decreases the radius of curvature of the cornea's surface; this would be an appropriate procedure for what kind of vision defect?
    A. Astigmatism
    B. Presbyopia
    C. Nearsightedness
    D. Farsightedness
12. A camera is focusing on an animal. As the creature moves *closer* to the lens, what must be done to keep the animal in focus?
    A. The lens must be moved closer to the film (or light sensors in a digital camera).
    B. The lens must be moved farther from the film (or light sensors in a digital camera).
    C. The *f*-number of the lens must be increased.
    D. The *f*-number of the lens must be decreased
13. Your eye is focusing on a person. As he walks toward you, what must be done to keep him in focus?
    A. The pupil of your eye must open up.
    B. The distance from the eye lens to the retina must decrease.
    C. The distance from the eye lens to the retina must increase.
    D. The focal length of your eye must decrease.
    E. The focal length of your eye must increase.
14. An astronomical telescope is made with an objective lens of focal length 150 cm and an eyepiece with focal length of 2.0 cm. The magnification of this instrument is
    A. 75×.   B. 300×.
    C. 152×.   D. 148×.
15. A simple magnifying glass produces a
    A. real inverted image.
    B. real erect image.
    C. virtual inverted image.
    D. virtual erect image.

Problems

25.1 The Camera

1. • The focal length of an $f/4$ camera lens is 300 mm. (a) What is the aperture diameter of the lens? (b) If the correct exposure of a certain scene is $\frac{1}{250}$ s at $f/4$, what is the correct exposure at $f/8$?
2. • Camera A has a lens with an aperture diameter of 8.00 mm. It photographs an object, using the correct exposure time of $\frac{1}{30}$ s. What exposure time should be used with camera B in photographing the same object with the same film if camera B has a lens with an aperture diameter of 23.1 mm?
3. • (a) A small refracting telescope designed for individual use has an objective lens with a diameter of 6.00 cm and a focal length of 1.325 m. What is the f-number of this instrument? (b) The 200-inch-diameter objective mirror of the Mt. Palomar telescope has an f-number of 3.3. Calculate its focal length. (c) The distance between lens and retina for a normal human eye is about 2.50 cm, and the pupil can vary in size from 2.0 mm to 8.0 mm. What is the range of f-numbers for the human eye?
4. •• A 135 mm telephoto lens for a 35 mm camera has f-stops that range from $f/2.8$ to $f/22$. (a) What are the smallest and largest aperture diameters for this lens? What is the diameter at $f/11$? (b) If a 50 mm lens had the same f-stops as the telephoto lens, what would be the smallest and largest aperture diameters for that lens? (c) At a given shutter speed, what is the ratio of the greatest to the smallest light intensity of the film image? (d) If the shutter speed for correct exposure at $f/22$ is 1/30 s, what shutter speed is needed at $f/2.8$?
5. • A camera lens has a focal length of 200 mm. How far from the lens should the subject for the photo be if the lens is 20.4 cm from the film?
6. •• A camera with a 90-mm-focal-length lens is focused on an object 1.30 m from the lens. To refocus on an object 6.50 m from the lens, by how much must the distance between the lens and the film be changed? To refocus on the more distant object, is the lens moved toward or away from the film?
7. • A certain digital camera having a lens with focal length 7.50 cm focuses on an object 1.85 m tall that is 4.25 m from the lens. (a) How far must the lens be from the sensor array? (b) How tall is the image on the sensor array? Is it erect or inverted? Real or virtual? (c) A SLR digital camera often has pixels measuring 8.0 μm × 8.0 μm. How many such pixels does the height of this image cover?
8. •• Your digital camera has a lens with a 50 mm focal length and a sensor array that measures 4.82 mm × 3.64 mm. Suppose you're at the zoo, and want to take a picture of a 4.50-m-tall giraffe. If you want the giraffe to exactly fit the longer dimension of your sensor array, how far away from the animal will you have to stand?
9. •• You want to take a full-length photo of your friend who is 2.00 m tall, using a 35 mm camera having a 50.0-mm-focal-length lens. The image dimensions of 35 mm film are 24 mm × 36 mm, and you want to make this a vertical photo in which your friend's image completely fills the image area. (a) How far should your friend stand from the lens? (b) How far is the lens from the film?
10. •• **Zoom lens, I.** A *zoom lens* is a lens that varies in focal length. The zoom lens on a certain digital camera varies in focal length from 6.50 mm to 19.5 mm. This camera is focused

on an object 2.00 m tall that is 1.50 m from the camera. Find the distance between the lens and the photo sensors and the height of the image (a) when the zoom is set to 6.50 mm focal length and (b) when it is at 19.5 mm. (c) Which is the telephoto focal length, 6.50 mm or 19.5 mm?

25.2 The Projector

11. • A slide projector uses a lens of focal length 115 mm to focus a 35 mm slide (having dimensions 24 mm × 36 mm) on a screen. The slide is placed 12.0 cm in front of the lens. (a) Where should you place the screen to view the image of this slide? (b) What are the dimensions of the slide's image on the screen?

12. •• An LCD projector (see Sec. 25.2) has a projection lens with *f*-number of 1.8 and a diameter of 46 mm. The LCD array measures 3.30 cm × 3.30 cm and will be projected on a screen 8.00 m from the lens. If the array is 800 × 600 pixels, what will be the dimensions of a single pixel on the screen?

13. • The dimensions of the picture on a 35 mm color slide are 24 mm × 36 mm. An image of the slide is projected onto a screen 9.00 m from the projector lens with a lens of focal length 150 mm. (a) How far is the slide from the lens? (b) What are the dimensions of the image on the screen?

14. •• You are designing a projection system for a hall having a screen measuring 4.00 m square. The lens of a 35 mm slide projector in the projection booth is 15.0 m from this screen. You want to focus the image of 35 mm slides (which are 24 mm × 36 mm) onto this screen so that you can fill as much of the screen as possible without any part of the image extending beyond the screen. (a) What focal-length lens should you use in the projector? (b) How far from the lens should the slide be placed? (c) What are the dimensions of the slide's image on the screen?

15. •• In a museum devoted to the history of photography, you are setting up a projection system to view some historical 4.0 inch × 5.0 inch color slides. Your screen is 6.0 m from the projector lens, and you want the image to be 4.0 ft × 5.0 ft on the screen. (a) What focal-length lens do you need? (b) How far from the lens should you put the slide?

25.3 The Eye

16. • **The cornea as a thin lens.**
BIO Measurements on the cornea of a person's eye reveal that the magnitude of the front surface radius of curvature is 7.80 mm, while the magnitude of the rear surface radius of curvature is 7.30 mm (see Figure 25.22), and that the index of refraction of the cornea is 1.38. If the cornea were simply a thin lens in air, what would be its focal length and its power in diopters? What type of lens would it be?

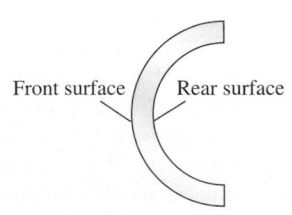

Front surface / Rear surface

▲ **FIGURE 25.22** Problem 16.

17. • **Range of the focal length of the eye.** We can model the eye
BIO as a sphere 2.50 cm in diameter with a thin lens in the front and the retina along the back surface. A 20-year-old person should be able to focus on objects from her near point (see Table 25.1) up to infinity. What is the range of the effective focal length of the lens of the eye for this person?

18. • A 40-year-old optometry patient focuses on a 6.50-cm-tall
BIO photograph at his near point. (See Table 25.1.) We can model his eye as a sphere 2.50 cm in diameter, with a thin lens at the front and the retina at the rear. (a) What is the effective focal length of his eye and its power in diopters when he focuses on the photo? (b) How tall is the image of the photo on his retina? Is it erect or inverted? Real or virtual? (c) If he views the photograph from a distance of 2.25 m, how tall is its image on his retina?

19. •• **Crystalline lens of the eye.** The crystalline lens of the eye
BIO is double convex and has a typical index of refraction of 1.43. At minimum power, the front surface has a radius of 10.0 mm and the back surface has a radius of 6.0 mm; at maximum power, these radii are 6.0 mm and 5.5 mm, respectively (although the values do vary from person to person). (a) Find the maximum and minimum power (in diopters) of the crystalline lens if it were in air. (b) What is the range of focal lengths the eye can achieve? (c) At minimum power, where does it focus the image of a very distant object? (d) At maximum power, where does it focus the image of an object at the near point of 25 cm?

20. • **Contact lenses.** Contact lenses are placed right on the eye-
BIO ball, so the distance from the eye to an object (or image) is the same as the distance from the lens to that object (or image). A certain person can see distant objects well, but his near point is 45.0 cm from his eyes instead of the usual 25.0 cm. (a) Is this person nearsighted or farsighted? (b) What type of lens (converging or diverging) is needed to correct his vision? (c) If the correcting lenses will be contact lenses, what focal length lens is needed and what is its power in diopters?

21. •• **Ordinary glasses.** Ordinary glasses are worn in front of
BIO the eye and usually 2.0 cm in front of the eyeball. Suppose that the person in the previous problem prefers ordinary glasses to contact lenses. What focal length lenses are needed to correct his vision, and what is their power in diopters?

22. • A person can see clearly up close, but cannot focus on
BIO objects beyond 75.0 cm. She opts for contact lenses to correct her vision. (a) Is she nearsighted or farsighted? (b) What type of lens (converging or diverging) is needed to correct her vision? (c) What focal-length contact lens is needed, and what is its power in diopters?

23. •• In one form of cataract surgery the person's natural lens,
BIO which has become cloudy, is replaced by an artificial lens. The refracting properties of the replacement lens can be chosen so that the person's eye focuses on distant objects. But there is no accommodation, and glasses or contact lenses are needed for close vision. What is the power, in diopters, of the corrective contact lenses that will enable a person who has had such surgery to focus on the page of a book at a distance of 24 cm?

24. •• **Bifocals, I.** A person can focus clearly only on objects
BIO between 35.0 cm and 50.0 cm from his eyes. Find the focal length and power of the correcting contact lenses needed to correct (a) his closeup vision, (b) his distant vision.

25. •• A student's far point is at 17.0 cm, and she needs glasses to
BIO view her computer screen comfortably at a distance of 45.0 cm. What should be the power of the lenses for her glasses?

26. •• (a) Where is the near point of an eye for which a contact
BIO lens with a power of +2.75 diopters is prescribed? (b) Where is the far point of an eye for which a contact lens with a power of −1.30 diopters is prescribed for distant vision?

27. •• **Corrective lenses.** Determine the power of the corrective
BIO contact lenses required by (a) a hyperopic eye whose near
point is at 60.0 cm, and (b) a myopic eye whose far point is at
60.0 cm.

25.4 The Magnifier

28. • You want to view an insect 2.00 mm in length through a
magnifier. If the insect is to be at the focal point of the magnifier, what focal length will give the image of the insect an
angular size of 0.025 radian?

29. • A simple magnifier for viewing postage stamps and other
pieces of paper consists of a
thin lens mounted on a stand.
When this device is placed on a
stamp, the stand holds the lens at
the proper distance from the
stamp for viewing. (See Figure 25.23.) If the lens has a focal
length of 11.5 cm, and the image
is to be at infinity, (a) what is the
angular magnification of the device, and (b) how high should
the stand hold the lens above the stamp?

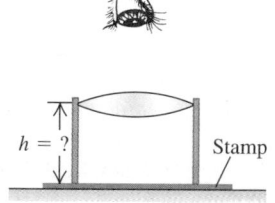
▲ **FIGURE 25.23** Problem 29.

30. • A thin lens with a focal length of 6.00 cm is used as a simple magnifier. (a) What angular magnification is obtainable
with the lens if the object is at the focal point? (b) When an
object is examined through the lens, how close can it be
brought to the lens? Assume that the image viewed by the eye
is at infinity and that the lens is very close to the eye.

31. •• The focal length of a simple magnifier is 8.00 cm. Assume
the magnifier to be a thin lens placed very close to the eye.
(a) How far in front of the magnifier should an object be
placed if the image is formed at the observer's near point,
25.0 cm in front of her eye? (b) If the object is 1.00 mm high,
what is the height of its image formed by the magnifier?

25.5 The Microscope

32. • A microscope has an objective lens with a focal length of
BIO 12.0 mm. A small object is placed 0.8 mm beyond the focal
point of the objective lens. (a) At what distance from the
objective lens does a real image of the object form? (b) What
is the magnification of the real image? (c) If an eyepiece with a
focal length of 2.5 cm is used, with a final image at infinity,
what will be the overall angular magnification of the object?

33. • A compound microscope has an objective lens of focal
BIO length 10.0 mm with an eyepiece of focal length 15.0 mm, and
it produces its final image at infinity. The object to be viewed
is placed 2.0 mm beyond the focal point of the objective lens.
(a) How far from the objective lens is the first image formed?
(b) What is the overall magnification of this microscope?

34. •• An insect 1.2 mm tall is placed 1.0 mm beyond the focal
BIO point of the objective lens of a compound microscope. The
objective lens has a focal length of 12 mm, the eyepiece a focal
length of 25 mm. (a) Where is the image formed by the objective lens and how tall is it? (b) If you want to place the eyepiece so that the image it produces is at infinity, how far should
this lens be from the image produced by the objective lens?
(c) Under the conditions of part (b), find the overall magnification of the microscope.

35. •• The objective lens and the eyepiece of a microscope are
BIO 16.5 cm apart. The objective lens has a magnification of 62×
and the eyepiece has a magnification of 10×. Calculate (a) the
overall magnification of the microscope, (b) the focal length of
each lens, and (c) where an object should be placed in order
for a normal eye to focus comfortably on the image.

36. •• The focal length of the eyepiece of a certain microscope is
BIO 18.0 mm. The focal length of the objective is 8.00 mm. The
distance between objective and eyepiece is 19.7 cm. The final
image formed by the eyepiece is at infinity. Treat all lenses as
thin. (a) What is the distance from the objective to the object
being viewed? (b) What is the magnitude of the linear magnification produced by the objective? (c) What is the overall angular magnification of the microscope?

37. •• A certain microscope is provided with objectives that have
BIO focal lengths of 16 mm, 4 mm, and 1.9 mm and with eyepieces that have angular magnifications of 5× and 10×. Each
objective forms an image 120 mm beyond its second focal
point. Determine (a) the largest overall angular magnification
obtainable and (b) the smallest overall angular magnification
obtainable.

38. •• **Resolution of a microscope.** The image formed by a
BIO microscope objective with a focal length of 5.00 mm is
160 mm from its second focal point. The eyepiece has a focal
length of 26.0 mm. (a) What is the angular magnification of
the microscope? (b) The unaided eye can distinguish two
points at its near point as separate if they are about 0.10 mm
apart. What is the minimum separation that can be resolved
with this microscope?

25.6 Telescopes

39. • A refracting telescope has an objective lens of focal length
16.0 in and eyepieces of focal lengths 15 mm, 22 mm, 35 mm,
and 85 mm. What are the largest and smallest angular magnifications you can achieve with this instrument?

40. • The eyepiece of a refracting astronomical telescope (see
Figure 25.16) has a focal length of 9.00 cm. The distance
between objective and eyepiece is 1.80 m, and the final image is
at infinity. What is the angular magnification of the telescope?

41. • **Galileo's telescopes, I.** While Galileo did not invent the
telescope, he was the first known person to use it astronomi-
cally, beginning around 1609. Five of his original lenses have
survived (although he did work with others). Two of these
have focal lengths of 1710 mm and 980 mm. (a) For greatest
magnification, which of these two lenses should be the eyepiece and which the objective? How long would this telescope be between the two lenses? (b) What is the greatest
angular magnification that Galileo could have obtained with
these lenses? (*Note:* Galileo actually obtained magnifications
up to about 30×, but by using a diverging lens as the eyepiece.) (c) The Moon subtends an angle of $\frac{1}{2}°$ when viewed
with the naked eye. What angle would it subtend when
viewed through this telescope (assuming that all of it could
be seen)?

42. •• The objective mirror of the Hubble Space telescope has
a focal length of 57.6 meters. The planet Mars's closest
approach to the earth is about 35 million miles. Use data from
Appendix E to help you calculate the size of the real image of
Mars formed by the Hubble's objective mirror when the planet
is closest to earth.

43. •• The largest refracting telescope in the world is at Yerkes Observatory in Wisconsin. The objective lens is 1.02 m in diameter and has a focal length of 19.4 m. Suppose you want to magnify Jupiter, which is 138,000 km in diameter, so that its image subtends an angle of $\frac{1}{2}°$ (about the same as the moon) when it is 6.28×10^8 km from earth. What focal-length eyepiece do you need?

25.7 Lens Aberrations

44. • A double-concave lens having radii of curvature of magnitudes 32.0 cm and 24.0 cm is made of a glass for which the refractive index for red light of wavelength 700 nm is 1.44 and for violet light of wavelength 400 nm is 1.57. A white object is placed 50.0 cm in front of this lens. Where will the red light and violet light be focused?

45. • A thin double-convex lens has radii of curvature of magnitudes 25.0 cm and 35.0 cm and is made of silicate flint glass. (See Figure 23.29.) (a) What is its focal length for (i) red light of wavelength 700 nm and (ii) violet light of wavelength 400 nm? (b) A color chart is placed 30.0 cm from the front of this lens. How far from the lens will the red and violet light be focused?

46. • A thin planoconvex lens has a radius of curvature of magnitude 22.5 cm on the curved side. When a color chart is placed 48.0 cm from the lens, green light of wavelength 550 nm is focused 277 cm from the lens and blue light of wavelength 450 nm is focused 171 cm from the lens. What are the indices of refraction for these two wavelengths of light?

47. •• A thin planoconvex lens with a radius of curvature of magnitude 28.0 cm on the curved face is made of silicate flint glass. (See Figure 23.29.) When a colorful object is placed 65.0 cm from this lens, the yellow light of wavelength 550 nm is perfectly focused on a screen. (a) How far is the screen from the lens? (b) Where will red light of wavelength 700 nm and violet light of wavelength 400 nm be focused?

General Problems

48. • A photographer takes a photograph of a Boeing 747 airliner (length 70.7 m) when it is flying directly overhead at an altitude of 9.50 km. The lens has a focal length of 5.00 m. How long is the image of the airliner on the film?

49. •• **Curvature of the cornea.** In a simplified model of the
BIO human eye, the aqueous and vitreous humors and the lens all have a refractive index of 1.40, and all the refraction occurs at the cornea, whose vertex is 2.60 cm from the retina. What should be the radius of curvature of the cornea such that the image of an object 40.0 cm from the cornea's vertex is focused on the retina? (See Sec. 24.4.)

50. •• **A nearsighted eye.** A certain very nearsighted person cannot
BIO focus on anything farther than 36.0 cm from the eye. Consider the simplified model of the eye described in problem 49. If the radius of curvature of the cornea is 0.75 cm when the eye is focusing on an object 36.0 cm from the cornea vertex and the indexes of refraction are as described in problem 49, what is the distance from the cornea vertex to the retina? What does this tell you about the shape of the nearsighted eye?

51. •• **What is the smallest thing we can see?** The smallest
BIO object we can resolve with our eye is limited by the size of the light receptor cells on the retina. In order to distinguish any

detail in an object, its image cannot be any smaller than a single retinal cell. Although the size depends on the type of cell (rod or cone), a diameter of a few microns (μm) is typical near the center of the eye. We shall model the eye as a sphere 2.50 cm in diameter with a single thin lens at the front and the retina at the rear, with light receptor cells 5.0 μm in diameter. (a) What is the smallest object you can resolve at a near point of 25 cm? (b) What angle is subtended by this object at the eye? Express your answer in units of minutes ($1° = 60$ min), and compare it with the typical experimental value of about 1.0 min. (*Note:* There are other limitations, such as the bending of light as it passes through the pupil, but we shall ignore them here.)

52. •• You are examining a flea with a converging lens that has a focal length of 4.00 cm. If the image of the flea is 6.50 times the size of the flea, how far is the flea from the lens? Where, relative to the lens, is the image?

53. •• **Physician, heal thyself!** (a) Experimentally determine the
BIO near and far points for both of your own eyes. Are these points the same for both eyes? (All you need is a tape measure or ruler and a cooperative friend.) (b) Design correcting lenses, as needed, for your closeup and distant vision in one of your eyes. If you prefer contact lenses, design that type of lens. Otherwise design lenses for ordinary glasses, assuming that they will be 2.0 cm from your eye. Specify the power (in diopters) of each correcting lens.

54. •• **Laser eye surgery.** The distance from the vertex of
BIO the cornea to the retina for a certain nearsighted person is 2.75 cm, and the radius of curvature of her cornea is 0.700 cm. She decides to get laser surgery to correct her vision. Using the simplified model of the eye described in problem 49, calculate the radius of curvature for her cornea that the surgeon should aim for, in order to allow her to view distant objects with a relaxed eye.

55. •• **It's all done with mirrors.** A photographer standing 0.750 m in front of a plane mirror is taking a photograph of her image in the mirror, using a digital camera having a lens with a focal length of 19.5 mm. (a) How far is the lens from the light sensors of the camera? (b) If the camera is 8.0 cm high, how high is its image on the sensors?

56. •• During a lunar eclipse, a picture of the moon (which has a diameter of 3.48×10^6 m and is 3.86×10^8 m from the earth) is taken with a camera whose lens has a focal length of 300 mm. (a) What is the diameter of the image on the film? (b) What percent is this of the width of a 24 mm $\times$ 36 mm color slide?

57. •• A person with a digital camera uses a lens of focal length 15.0 mm to take a photograph of the image of a 1.40-cm-tall seedling located 10.0 cm behind a double-convex lens of 17.0 cm focal length. (See Figure 25.24.) The camera's lens is 5.00 cm from the convex lens. (a) How far is the camera's lens from the photoreceptors? (b) What is the height of the seedling on the photoreceptors? (c) Will the camera actually record *two* images, one of the seedling directly and the other of the seedling as viewed through the lens?

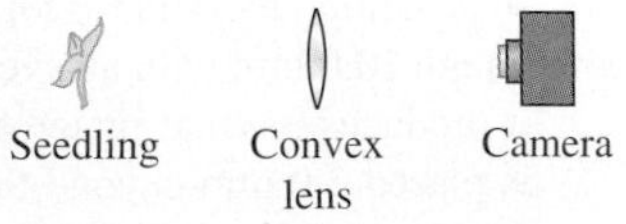

▲ FIGURE 25.24 Problem 57.

Seedling Convex lens Camera

58. •• A microscope with an objective of focal length 8.00 mm
BIO and an eyepiece of focal length 7.50 cm is used to project an image on a screen 2.00 m from the eyepiece. Let the image

distance of the objective be 18.0 cm. (a) What is the lateral magnification of the image? (b) What is the distance between the objective and the eyepiece?

59. •• A person with a near point of 85 cm, but excellent distant
BIO vision, normally wears corrective glasses. But he loses them while traveling. Fortunately, he has his old pair as a spare. (a) If the lenses of the old pair have a power of +2.25 diopters, what is his near point (measured from his eye) when he is wearing the old glasses if they rest 2.0 cm in front of his eye? (b) What would his near point be if his old glasses were contact lenses instead?

60. •• A telescope is constructed from two lenses with focal lengths of 95.0 cm and 15.0 cm, the 95.0-cm lens being used as the objective. Both the object being viewed and the final image are at infinity. (a) Find the angular magnification of the telescope. (b) Find the height of the image formed by the objective of a building 60.0 m tall and 3.00 km away. (c) What is the angular size of the final image as viewed by an eye very close to the eyepiece?

61. •• **Galileo's telescopes, II.** The characteristics that follow are characteristics of two of Galileo's surviving double-convex lenses. The numbers given are *magnitudes* only; you must supply the correct signs. L_1: front radius = 950 mm, rear radius = 2700 mm, refractive index = 1.528; L_2: front radius = 535 mm, rear radius = 50,500 mm, refractive index = 1.550. (a) What is the largest angular magnification that Galileo could have obtained with these two lenses? (b) How long would this telescope be between the two lenses?

62. •• **Water drop magnifier.** You can make a pretty good magnifying lens by putting a small drop of water on a piece of transparent kitchen wrap. Suppose your drop has an upper surface with a radius of curvature of 1.6 cm and the side on the kitchen wrap is essentially flat. (a) Calculate the focal length of your water lens. (b) What's the angular magnification of the lens? (c) Suppose you place this planoconvex water lens directly onto the surface of a table, so that the tabletop is in effect about half the thickness of the drop, or 1.0 mm, away from the lens. Where does the image of the tabletop form, what type is it, and what is its magnification? (Use the thin lens equation here, even though the small object distance relative to the thickness of the lens makes it a poor approximation in this case.) What does this result tell you about how a simple magnifier works?

26 Interference and Diffraction

I f you've ever blown soap bubbles, you know that part of the fun is watching the multicolored reflections from the bubbles. An ugly black oil spot on the pavement can become a thing of beauty after a rain, when it reflects a rainbow of colors. These familiar sights give us a hint that there are aspects of light that we haven't yet explored. In our discussion of lenses, mirrors, and optical instruments, we've used the model of *geometric optics,* which represents light as *rays*—straight lines that are bent at a reflecting or refracting surface.

But many aspects of the behavior of light (including colors in soap bubbles and oil films) *can't* be understood on the basis of rays. We have already learned that light is fundamentally a *wave,* and in some situations we have to consider its wave properties explicitly. In this chapter, we'll study *interference* and *diffraction* phenomena. When light passes through apertures or around obstacles, the patterns that are formed are a result of the *wave* nature of light; they cannot be understood on the basis of rays. Such effects are grouped under the heading *physical optics.* We'll look at several practical applications of physical optics, including diffraction gratings, x-ray diffraction, and holography.

A soap bubble is just a film of soapy water. Why is it iridescent? And how is this effect related to the iridescence on a CD or a hummingbird's throat? In this chapter we'll study these and related optical phenomena.

26.1 Interference and Coherent Sources

In our discussions of mechanical waves in Chapter 12 and electromagnetic waves in Chapter 23, we often talked about *sinusoidal* waves with a single frequency and a single wavelength. In optics, such a wave is characteristic of **monochromatic light** (light of a single color). Common sources of light, such as an incandescent lightbulb or a flame, *do not* emit monochromatic light; rather, they give off a continuous distribution of wavelengths.

A precisely monochromatic light wave is an idealization, but monochromatic light can be *approximated* in the laboratory. For example, some optical filters block all but a very narrow range of wavelengths. Gas discharge lamps, such as a mercury-vapor lamp, emit light with a discrete set of colors, each having a narrow band of wavelengths. The bright green line in the spectrum of a mercury-vapor lamp has a wavelength of about 546.1 nm, with a spread on the order of ± 0.001 nm. By far the most nearly monochromatic light source available at present is the *laser*. The familiar helium–neon laser, inexpensive and readily available, emits red light at 632.8 nm, with a line width (wavelength range) on the order of ± 0.000001 nm, or about 1 part in 10^9. As we analyze interference and diffraction effects in this chapter and the next, we'll often assume that we're working with monochromatic light.

The term **interference** refers to any situation in which two or more waves overlap in space. When this occurs, the total displacement at any point at any instant of time is governed by the **principle of linear superposition.** We introduced these ideas in Sections 12.5 and 12.8 in relation to mechanical waves; you may want to review those sections. Linear superposition is the most important principle in all of physical optics, so make sure you understand it well.

The principle of superposition

When two or more waves overlap, the resultant displacement at any point and at any instant may be found by adding the instantaneous displacements that would be produced at that point by the individual waves if each were present alone.

▲ **Application Waves are waves.** All waves exhibit the phenomena of diffraction and interference that we discuss in this chapter for light waves. This photo shows water waves diffracting and interfering as they pass between offshore barriers. Indeed, the enormous tsunami waves created by the Sumatra–Andaman earthquake of December 2004, which spread around the whole globe, exhibited interference and diffraction as they passed through the gaps between islands and continents.

We use the term *displacement* in a general sense. For waves on the surface of a liquid, we mean the actual displacement of the surface above or below its normal level. For sound waves, the term refers to the excess or deficiency of pressure. For electromagnetic waves, we usually mean a specific component of the electric or magnetic field.

Another term we'll often use is **phase,** introduced in Chapter 22. When we say that two periodic motions are *in phase,* we mean that they are in step; they reach their maximum values at the same time, their minimum values at the same time, and so on. When two periodic motions are *one-half cycle out of phase,* the positive peaks of one occur at the same times as the negative peaks of the other, and so on.

To introduce the essential ideas of interference, let's consider two identical sources of monochromatic waves separated in space by a certain distance. The two sources are permanently *in phase;* they vibrate in unison. They might be two agitators in a ripple tank, two loudspeakers driven by the same amplifier, two radio antennas powered by the same transmitter, or two small holes or slits in an opaque screen illuminated by the same monochromatic light source.

We position the sources at points S_1 and S_2 along the y axis, equidistant from the origin, as shown in Figure 26.1a. Let P_0 be any point on the x axis. By symmetry, the distance from S_1 to P_0 is *equal* to the distance from S_2 to P_0; waves from the two sources thus require equal times to travel to P_0. Accordingly, waves that leave S_1 and S_2 in phase arrive at P_0 in phase. The two waves add, and the total amplitude at P_0 is *twice* the amplitude of each individual wave.

Similarly, the distance from S_2 to point P_1 is exactly one wavelength *greater* than the distance from S_1 to P_1. Hence, a wave crest from S_1 arrives at P_1 exactly one cycle earlier than a crest emitted at the same time from S_2, and again the two waves arrive *in phase.* For point P_2, the path difference is *two* wavelengths, and again the two waves arrive in phase, and so on.

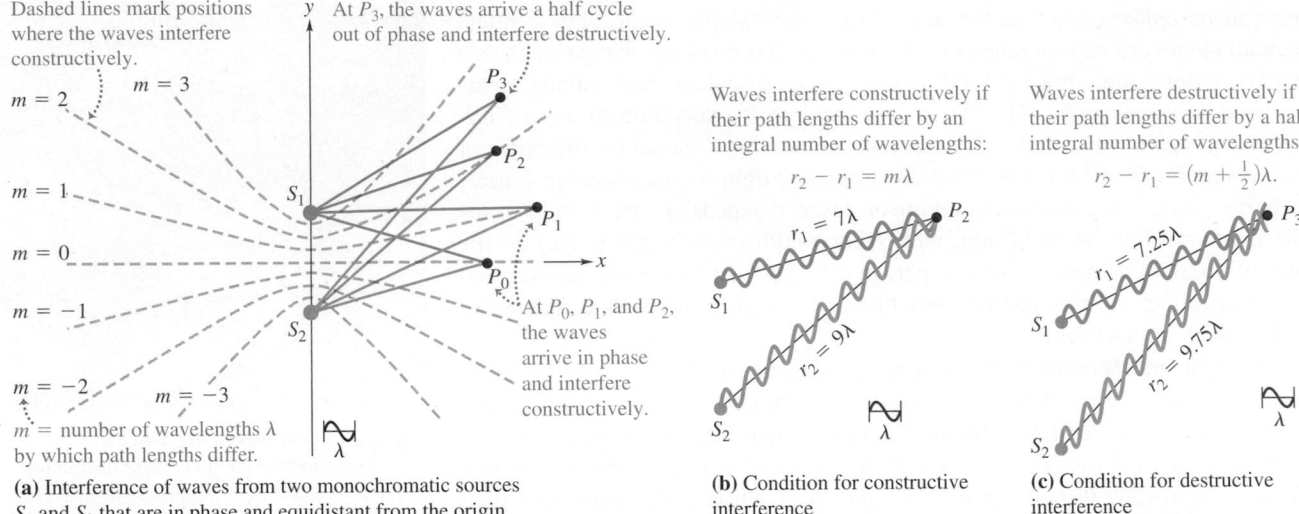

▲ FIGURE 26.1 Interference of monochromatic light waves from two sources that are in phase. In this example, the distance between the sources is 4λ.

The addition of amplitudes that results when waves from two or more sources arrive at a point *in phase* is called **constructive interference,** or *reinforcement* (Figure 26.1b). Let the distance from S_1 to any point P be r_1 and the distance from S_2 to P be r_2. Then the condition that must be satisfied for constructive interference to occur at P is that the path difference $r_2 - r_1$ for the two sources must be an integral multiple of the wavelength λ:

Constructive interference

Constructive interference of two waves arriving at a point occurs when the path difference from the two sources is an integer number of wavelengths:

$$r_2 - r_1 = m\lambda \qquad (m = 0, \pm 1, \pm 2, \pm 3, \cdots). \qquad (26.1)$$

For our example, all points satisfying this condition lie on the dashed curves in Figure 26.1a.

Intermediate between the dashed curves in Figure 26.1a is another set of curves (not shown) for which the path difference for the two sources is a *half-integral* number of wavelengths. Waves from the two sources arrive at a point on one of these lines (such as point P_3 in Figure 26.1a) exactly a half cycle out of phase. A crest of one wave arrives at the same time as a "trough" (a crest in the opposite direction) from the other wave (Figure 26.1c), and the resultant amplitude is the *difference* between the two individual amplitudes. If the individual amplitudes are equal, then the *total* amplitude is zero! This condition is called **destructive interference,** or *cancellation*. For our example, the condition for destructive interference is as follows:

Destructive interference

Destructive interference of two waves arriving at a point occurs when the path difference from the two sources is a half-integer number of wavelengths:

$$r_2 - r_1 = \left(m + \tfrac{1}{2}\right)\lambda \qquad (m = 0, \pm 1, \pm 2, \pm 3, \cdots). \qquad (26.2)$$

Figures 26.1b and 26.1c show the phase relationships for constructive and destructive interference of two waves.

An example of this interference pattern is the familiar ripple-tank pattern shown in Figure 26.2. The two wave sources are two agitators driven by the same vibrating mechanism. The regions of both maximum and zero amplitude are clearly visible.

For Equations 26.1 and 26.2 to hold, the two sources must *always* be in phase. With light waves, there is no practical way to achieve such a relationship with two *separate* sources because of the way light is emitted. In ordinary light sources, atoms gain excess energy by thermal agitation or by impact with accelerated electrons. An atom thus "excited" begins to radiate energy and continues until it has lost all the energy it can, typically in a time on the order of 10^{-8} s. The many atoms in a source ordinarily radiate in an unsynchronized and random phase relationship, and the separate beams of light emitted from *two* such sources have no definite phase relation to each other.

However, the light from a single source can be split so that parts of it emerge from two or more regions of space, forming two or more *secondary sources*. Then any random phase change in the source affects these secondary sources equally and does not change their *relative* phase. Light from two such sources, derived from a single primary source and with a definite, constant phase relation, is said to be **coherent.** We'll consider the interference of light from two secondary sources in the next section.

The distinguishing feature of light from a *laser* is that the emission of light from many atoms is *synchronized* in frequency and phase. As a result, the random phase changes mentioned in the paragraphs above occur *much* less frequently. Definite phase relations are preserved over correspondingly much greater lengths in the beam, and laser light is much more *coherent* than ordinary light.

▲ **FIGURE 26.2** Photograph of ripple pattern on water. The ripples are produced by two objects moving up and down in phase just under the water surface.

26.2 Two-Source Interference of Light

One of the earliest quantitative experiments involving the interference of light was performed in 1800 by the English scientist Thomas Young. His experiment involved interference of light from two sources, which we discussed in Section 26.1. Young's apparatus is shown in Figure 26.3a. Monochromatic light emerging from a narrow slit S_0 (1.0 μm or so wide) falls on a screen with two other narrow slits S_1 and S_2, each 1.0 μm or so wide and a few micrometers apart. According to Huygens's principle, cylindrical wave fronts spread out from slit S_0 and reach slits S_1 and S_2 *in phase* because they travel equal distances from S_0. The waves emerging from slits S_1 and S_2 are therefore always in phase, so S_1 and S_2 are *coherent* sources (Figure 26.3b). But the waves from these sources do not necessarily arrive at point P in phase, because of the path difference $(r_2 - r_1)$.

To simplify the analysis that follows, we assume that the distance R from the slits to the screen is so large in comparison to the distance d between the slits that the lines from S_1 and S_2 to P are very nearly parallel, as shown in Figure 26.3c. This is usually the case for experiments with light. The difference in path length is then given by

$$r_2 - r_1 = d\sin\theta. \tag{26.3}$$

We found in Section 26.1 that constructive interference (reinforcement) occurs at a point P (in a brightly illuminated region of the screen) when the path difference $d\sin\theta$ is an integral number of wavelengths, $m\lambda$, where $m = 0, \pm 1, \pm 2, \pm 3, \cdots$. So we have the following principles:

MasteringPHYSICS

PhET: Wave Interference
ActivPhysics 16.1: Two-Source Interference: Introduction
ActivPhysics 16.2: Two-Source Interference: Qualitative Questions
ActivPhysics 16.3: Two-Source Interference: Problems

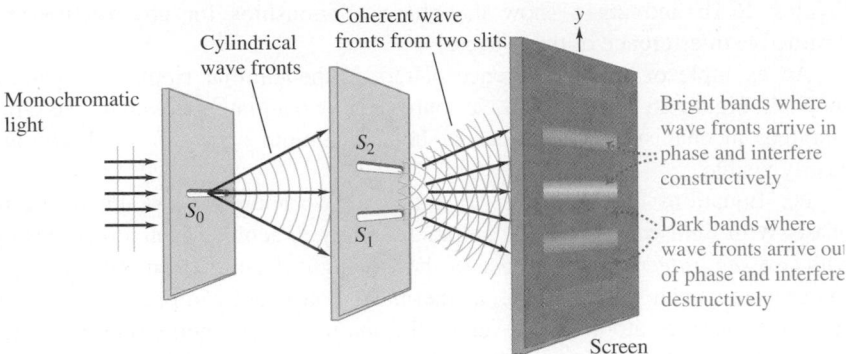

(a) Interference of light waves passing through two slits

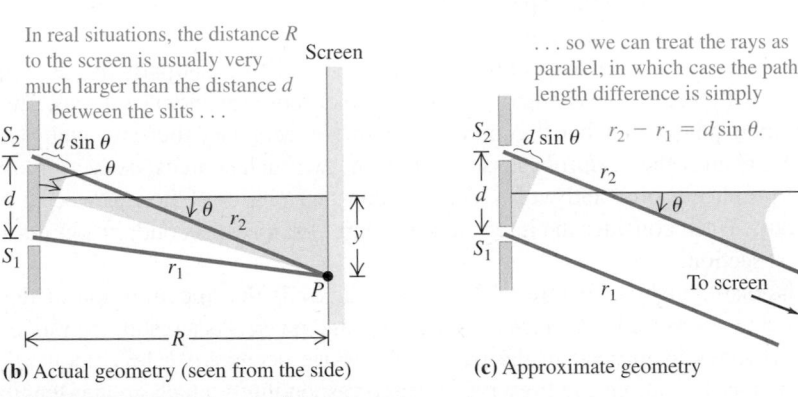

(b) Actual geometry (seen from the side) **(c)** Approximate geometry

▲ **FIGURE 26.3** Young's two-slit experiment.

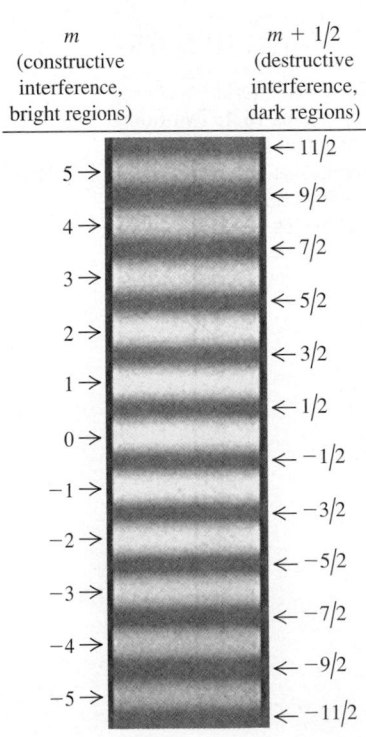

▲ **FIGURE 26.4** Interference fringes produced by Young's two-slit experiment.

Constructive and destructive interference, two slits

Constructive interference occurs at angles θ for which

$$d\sin\theta = m\lambda \qquad (m = 0, \pm 1, \pm 2, \cdots). \qquad (26.4)$$

Similarly, destructive interference (cancellation) occurs, forming dark regions on the screen, at points for which the path difference is a half-integral number of wavelengths, $(m + \frac{1}{2})\lambda$:

$$d\sin\theta = (m + \tfrac{1}{2})\lambda \qquad (m = 0, \pm 1, \pm 2, \cdots). \qquad (26.5)$$

Thus, the pattern on the screen of Figure 26.3a is a succession of bright and dark bands, often called *fringes*. A photograph of such a pattern is shown in Figure 26.4.

We can derive an expression for the positions of the centers of the bright bands on the screen. In Figure 26.3b, y is measured from the center of the pattern, corresponding to the distance from the center of Figure 26.4. Let y_m be the distance from the center of the pattern $(\theta = 0)$ to the center of the mth bright band. We denote the corresponding value of θ as θ_m; then, when $R \gg d$,

$$y_m = R\tan\theta_m.$$

In experiments such as this, the y_m's are often much smaller than R. This means that θ_m is very small, $\tan\theta_m$ is very nearly equal to $\sin\theta_m$, and

$$y_m = R\sin\theta_m.$$

Combining this relationship with Equation 26.4, we obtain the equation for constructive interference in Young's experiment:

Constructive interference, Young's experiment

$$y_m = R\frac{m\lambda}{d} \qquad (m = 0, \pm 1, \pm 2, \cdots). \qquad (26.6)$$

We can measure R and d, as well as the positions y_m of the bright fringes, so this experiment provides a direct measurement of the wavelength λ. Young's experiment was in fact the first direct measurement of wavelengths of light.

With this same approximation, the condition for formation of a *dark* fringe is

$$y_m = R\frac{(m + \frac{1}{2})\lambda}{d} \qquad (m = 0, \pm 1, \pm 2, \cdots).$$

NOTE ▶ The distance between adjacent bright bands in the pattern is *inversely* proportional to the distance d between the slits. The closer together the slits are, the more the pattern spreads out. When the slits are far apart, the bands in the pattern are closer together. ◀

Quantitative Analysis 26.1 **Constructive interference**

The experimental arrangement in Figure 26.5 is used to observe a two-slit interference pattern. Which of the following changes would increase the separation between the bright bands? (A lightbulb filament emits all the colors of the visible spectrum.) More than one answer may be correct.

A. Move the screen closer to the slits.
B. Replace the green filter by a red one.
C. Increase the width of each slit.
D. Decrease the separation of the slits.
E. Move the filament closer to the slits.

SOLUTION Equation 26.6, $y_m = R(m\lambda/d)$, predicts the results of this experiment. The separation of the bright lines (fringes) decreases with the distance R to the screen, so A won't work. Red light has a greater wavelength λ than green light, so B will work. The width of each slit has an effect on the pattern, but not on the

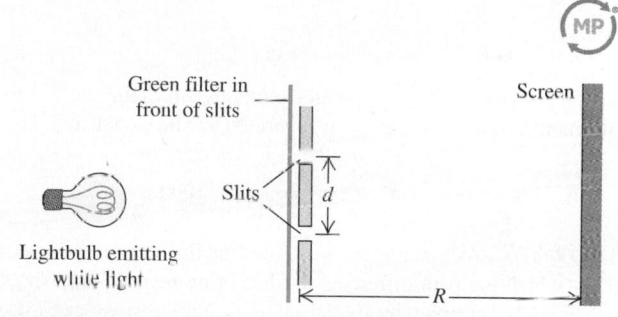

▲ **FIGURE 26.5**

separation, of the bright bands, so C won't work. The separation of the fringes is inversely related to the separation d of the slits, so D will work. The distance from the source to the slits has no effect on any part of the pattern, so E won't work.

EXAMPLE 26.1 **Determining wavelength**

A two-slit interference experiment is used to determine the unknown wavelength of a laser light source, as shown in Figure 26.6. With the slits 0.200 mm apart and a screen at a distance of 1.00 m, the third bright band out from the central bright band is found to be 9.49 mm from the center of the screen. **(a)** What is the wavelength of the light? **(b)** How far apart would the slits have to be so that the fourth minimum (dark band) would occur at 9.49 mm from the center of the screen?

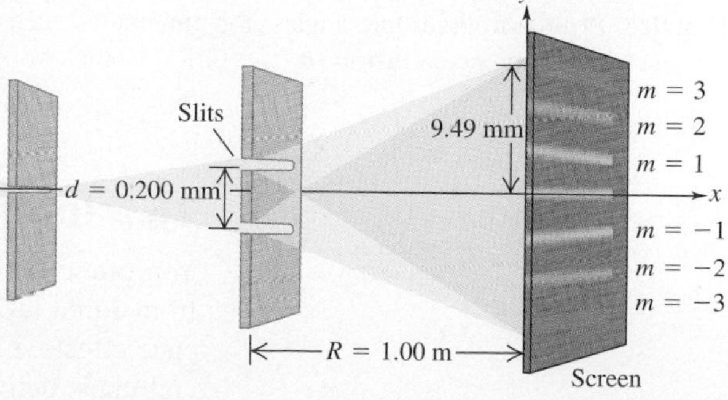

▶ **FIGURE 26.6**

Continued

SOLUTION

SET UP AND SOLVE Part (a): From Equation 26.6, with $m = 3$,

$$\lambda = \frac{y_m d}{mR} = \frac{(9.49 \times 10^{-3}\,\text{m})(0.200 \times 10^{-3}\,\text{m})}{(3)(1.00\,\text{m})}$$
$$= 633 \times 10^{-9}\,\text{m} = 633\,\text{nm}.$$

Part (b): For destructive interference, and therefore a minimum, $m = 0$ gives the first minimum, $m = 1$ the second, $m = 2$ the third, and $m = 3$ the fourth. Thus,

$$d = \frac{\left(m + \frac{1}{2}\right)\lambda R}{y_m} = \frac{\left(3 + \frac{1}{2}\right)(633 \times 10^{-9}\,\text{m})(1.00\,\text{m})}{9.49 \times 10^{-3}\,\text{m}}$$
$$= 2.33 \times 10^{-4}\,\text{m} = 0.233\,\text{mm}.$$

REFLECT This wavelength corresponds to the red light from a helium–neon laser.

Practice Problem: If a green laser pointer with $\lambda = 523$ nm is used, where would the screen have to be placed in order to find the third bright fringe 9.49 mm from the central fringe? *Answer: R = 1.21 m.*

EXAMPLE 26.2 Broadcast pattern of a radio station

A radio station operating at a frequency of 1500 kHz $= 1.5 \times 10^6$ Hz (near the top end of the AM broadcast band) has two identical vertical dipole antennas spaced 400 m apart, oscillating in phase. At distances much greater than 400 m, in what directions is the intensity greatest in the resulting radiation pattern?

SOLUTION

SET UP The two antennas, seen from above in Figure 26.7, correspond to sources S_1 and S_2 in Figure 26.3. The wavelength is

$$\lambda = \frac{c}{f} = \frac{3.0 \times 10^8\,\text{m/s}}{1.5 \times 10^6\,\text{Hz}} = 200\,\text{m}.$$

SOLVE The directions of the intensity maxima are the values of θ for which the path difference is zero or an integral number of wavelengths, as given by Equation 26.4. (Note that we can't use Equation 26.6, because the angles θ_m aren't necessarily small.) Inserting the numerical values, with $m = 0, \pm 1, \pm 2$, we find that

$$\sin\theta = \frac{m\lambda}{d} = \frac{m(200\,\text{m})}{400\,\text{m}} = \frac{m}{2}, \qquad \theta = 0, \pm 30°, \pm 90°.$$

REFLECT In this example, values of m greater than 2 or less than -2 give values of $\sin\theta$ greater than 1 or less than -1, which is impossible. There is no direction for which the path difference is three or more wavelengths. Thus, in this example, values of m of ± 3 and beyond have no physical meaning.

Practice Problem: Find the angles for minimum intensity (destructive interference). *Answer: $\theta = \pm 14.5°, \pm 48.6°$.*

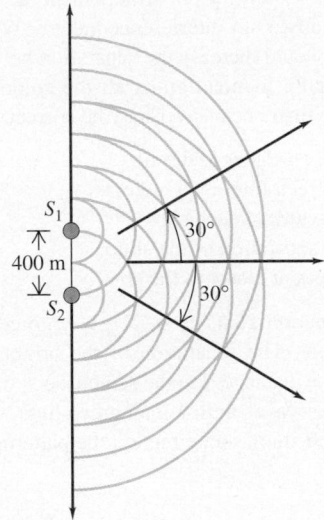

▲ **FIGURE 26.7**

26.3 Interference in Thin Films

You often see bright bands of color when light reflects from a soap bubble or from a thin layer of oil floating on water. These bands are the results of interference effects. Light waves are reflected from opposite surfaces of the thin films, and constructive interference between the two reflected waves (with different path lengths) occurs in different places for different wavelengths. The situation is shown schematically in Figure 26.8a. Light shining on the upper surface of a thin film with thickness t is partly reflected at the upper surface (path abc). Light *transmitted* at the upper surface is partly reflected at the lower surface (path $abdef$). The two reflected waves come together at point P on the retina of the eye. Depending on the phase relationship, they may interfere constructively or destructively. Different colors have different wavelengths, so the interference may be constructive for some colors and destructive for others. That's why you see rainbow-colored rings or fringes, like the ones in Figure 26.8b.

Here's an example involving *monochromatic* light reflected from two nearly parallel surfaces at nearly normal incidence. Figure 26.9 shows two plates of glass separated by a thin wedge of air. We want to consider interference between the two light waves reflected from the surfaces adjacent to the air wedge, as shown. (Reflections also occur at the top surface of the upper plate and the bottom surface of the lower plate; to keep our discussion simple, we won't include those reflections.) The situation is the same as in Figure 26.8, except that the thickness of the film isn't uniform. The path difference between the two waves is just twice the thickness t of the air wedge at each point. At points where $2t$ is an integral number of wavelengths, we expect to see constructive interference and a bright area; where $2t$ is a half-integral number of wavelengths, we expect to see destructive interference and a dark area. Along the line where the plates are in contact, there is practically *no* path difference, and we expect a bright area.

When we carry out the experiment, the bright and dark fringes appear, but they are interchanged! Along the line of contact, we find a *dark* fringe, not a bright one. This suggests that one or the other of the reflected waves has undergone a half-cycle phase shift during its reflection. In that case, the two waves reflected at the line of contact are a half cycle out of phase even though they have the same path length.

This unexpected phase shift illustrates a general principle that we can state as follows: When a wave traveling in medium a is reflected at an interface between this material and a different material b, there may or may not be an additional phase shift associated with the reflection, depending on the refractive indexes n_a and n_b of the two materials. If the second material (b) has *greater* refractive index than the first (a) (that is, when $n_b > n_a$), the reflected wave undergoes a half-cycle phase shift during reflection. When the second material (b) has *smaller* refractive index than the first (a) (that is, when $n_b < n_a$), there is *no* phase shift.

In fact, this phase shift behavior is predicted by Maxwell's equations from a detailed description of electromagnetic waves. The derivation is beyond the scope of this text, but let's check the prediction with the situation shown in Figure 26.9. For the wave reflected from the upper surface of the air wedge, n_b (air) is less than n_a (glass), so this wave has zero phase shift. For the wave reflected from the lower surface, n_b (glass) is greater than n_a (air), so this wave has a half-cycle phase shift. Waves reflected from the line of contact have no path difference to give additional phase shifts, and they interfere destructively; this is what we observe. We invite you to use the foregoing principle to show that, in Figure 26.8, the wave reflected at point b is shifted by a half cycle and the wave reflected at d is not (assuming the index of the bottom layer is less than n).

We can summarize this discussion symbolically: For a film with thickness t and light at normal (perpendicular) incidence, the reflected waves from the two surfaces interfere *constructively* if neither or both have a half-cycle reflection phase shift whenever the condition

$$2t = m\lambda \qquad (m = 0, 1, 2, \cdots), \qquad (26.7)$$

Light reflected from the upper and lower surfaces of the film comes together in the eye at P and undergoes interference.

Some colors interfere constructively and others destructively, creating the iridescent color bands we see.

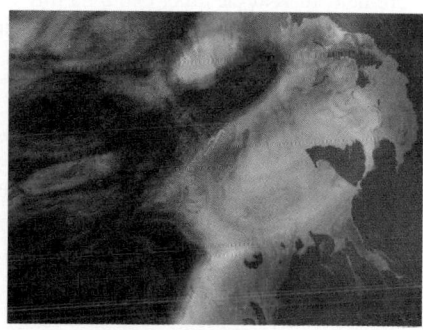

(a) Interference between rays reflected from the two surfaces of a thin film

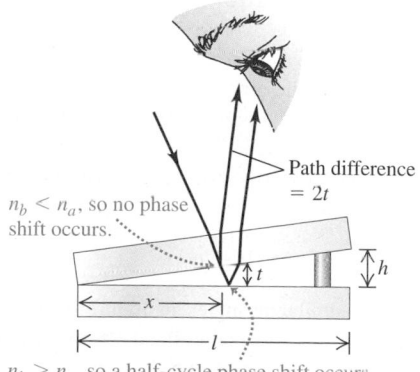

(b) The rainbow fringes of an oil slick on water

▲ **FIGURE 26.8** Interference patterns produced by reflection from a thin film.

$n_b < n_a$, so no phase shift occurs.

Path difference $= 2t$

$n_b > n_a$, so a half-cycle phase shift occurs.

▲ **FIGURE 26.9** Interference between two light waves reflected from the two sides of an air wedge separating two glass plates.

(where λ is the wavelength *in the film*), is satisfied. However, when *one* of the two waves has a half-cycle reflection phase shift, the same equation is the condition for *destructive* interference.

Similarly, if neither wave, or both, has a half-cycle phase shift, the condition for *destructive* interference in the reflected waves is

$$2t = \left(m + \tfrac{1}{2}\right)\lambda \qquad (m = 0, 1, 2, \cdots). \tag{26.8}$$

However, if one wave has a half-cycle phase shift, the same equation is the condition for *constructive* interference.

Quantitative Analysis 26.2

Destructive interference

Two very flat planes of glass touch at one end, but are held apart at the other end by a tiny filament of nylon placed between them. When the planes of glass are viewed by reflection from a sodium lamp, 23 evenly spaced dark bands are counted. The width of the filament

A. is 23 wavelengths of sodium light.
B. is 11 wavelengths of sodium light.
C. cannot be determined without a knowledge of the length of the glass plates.

SOLUTION Equation 26.7 gives the thickness of the film of air between the two glass plates at the locations of the dark bands. The first dark band occurs where the thickness is zero—that is, where the glass planes touch. The second occurs where the air thickness is $\lambda/2$, the third where it is $2\lambda/2$, and the fourth where it is $3\lambda/2$. Continuing this pattern, we see that, at the 23rd dark line, the thickness is $22\lambda/2$, or 11 wavelengths of light. If the filament is just beyond the 23rd dark line, it will actually be just a little thicker than 11 wavelengths of light.

EXAMPLE 26.3 Interference fringes

Suppose the two glass plates in Figure 26.10 are two microscope slides 10 cm long. At one end, they are in contact; at the other end, they are separated by a piece of paper 0.020 mm thick. What is the spacing of the interference fringes seen by reflection? Is the fringe at the line of contact bright or dark? Assume monochromatic light with $\lambda_0 = 500$ nm.

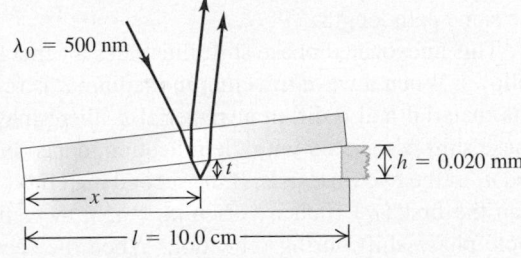

▶ **FIGURE 26.10**

SOLUTION

SET UP AND SOLVE For the wave reflected from the upper surface of the air wedge, n_b (air) is less than n_a (glass), so this wave has zero phase shift. For the wave reflected from the lower surface of the air wedge, n_b (glass) is greater than n_a (air), so this wave has a half-cycle wave shift. Thus, the fringe at the line of contact is dark.

The condition for *destructive* interference (a dark fringe) is Equation 26.7:

$$2t = m\lambda \qquad (m = 0, 1, 2, \cdots).$$

From similar triangles in Figure 26.10, the thickness t at each point is proportional to the distance x from the line of contact:

$$\frac{t}{x} = \frac{h}{l}.$$

Combining this equation with Equation 26.7, we find that

$$\frac{2xh}{l} = m\lambda;$$

$$x = m\frac{l\lambda}{2h} = m\frac{(0.10 \text{ m})(500 \times 10^{-9} \text{ m})}{(2)(0.020 \times 10^{-3} \text{ m})} = m(1.25 \text{ mm}).$$

Successive dark fringes, corresponding to successive integer values of m, are spaced 1.25 mm apart.

REFLECT Both glass slides have refractive indexes greater than unity. The indexes don't need to be equal; the results would be the same if the bottom slide had $n = 1.50$ and the top slide had $n = 1.60$.

Practice Problem: Suppose we use two layers of paper, so that $h = 0.040 \times 10^{-3}$ m. What is the fringe spacing? *Answer:* 0.62 mm.

EXAMPLE 26.4 A thin film of water

Suppose the glass plates in Example 26.3 have $n = 1.52$ and the space between the plates contains water $(n = 1.33)$ instead of air. What happens now?

SOLUTION

SET UP AND SOLVE The glass has a greater refractive index than either water or air, so the phase shifts are the same as in Example 26.3. Thus, the fringe at the line of contact is still dark.
The wavelength in water is

$$\lambda = \frac{\lambda_0}{n} = \frac{500 \text{ nm}}{1.33} = 376 \text{ nm}.$$

The fringe spacing is reduced by a factor of 1.33 and is therefore equal to 0.94 mm.

REFLECT If the space between the glass plates were filled with a liquid, such as carbon disulfide, that had a *greater* index of refraction than glass, a phase shift would occur at the top surface, and not at the bottom surface, of the liquid wedge. However, the waves from the contact point would still be out of phase by a half cycle, and the fringe at the line of contact would still be dark.

EXAMPLE 26.5 Grease as a thin film

Suppose the upper of the two plates in Example 26.3 is made of a plastic material with $n = 1.40$, the wedge is filled with a silicone grease having $n = 1.50$, and the bottom plate is of a dense flint glass with $n = 1.60$ (Figure 26.11). What is the spacing of the interference fringes seen by reflection? Is the fringe at the line of contact bright or dark?

▶ **FIGURE 26.11**

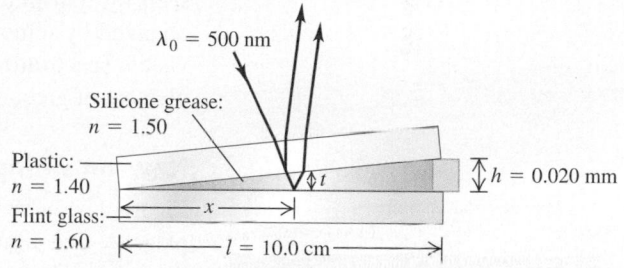

SOLUTION

SET UP For the wave reflected from the upper surface of the silicone grease wedge, n_b (silicone grease) is greater than n_a (plastic), so this wave has a half-cycle phase shift. For the wave reflected from the lower surface of the silicone grease wedge, n_b (flint glass), is greater than n_a (silicone grease), so this wave also has a half-cycle wave shift. The fringe spacing is again determined by the wavelength in the film (that is, in the silicone grease), which is

$$\lambda = \frac{\lambda_0}{n} = \frac{500 \text{ nm}}{1.50} = 333 \text{ nm}.$$

With half-cycle phase shifts at both surfaces, the condition for constructive interference (a bright fringe) is $2t = m\lambda$ ($m = 0, 1, 2, \cdots$).

SOLVE With half-cycle phase shifts at both surfaces, the fringe at the line of contact is bright. From the similar triangles in Figure 26.11,

$$\frac{t}{x} = \frac{h}{l}.$$

Combining this equation with Equation 26.7, we find that

$$\frac{2xh}{l} = m\lambda,$$

$$x = m\frac{l\lambda}{2h} = m\frac{(0.10 \text{ m})(333 \times 10^{-9} \text{ m})}{(2)(0.020 \times 10^{-3} \text{ m})} = m(0.833 \text{ mm}).$$

Successive bright fringes, corresponding to successive integer values of m, are spaced 0.833 mm apart.

REFLECT In this case, both reflected waves undergo half-cycle phase shifts on reflection, so the *relative* phase doesn't change. Thus, the line of contact is a bright fringe.

Practice Problem: Suppose we repeat this experiment with light having a wavelength in vacuum of $\lambda_0 = 600$ nm. What is the fringe spacing? *Answer:* 1.0 mm.

The results of Examples 26.4 and 26.5 show that the fringe spacing is proportional to the wavelength of the light used. Thus, for a given choice of materials, the fringes are farther apart with red light (larger λ) than with blue light (smaller λ). If we use white light, the reflected light at any point is a mixture of wavelengths for which constructive interference occurs; the wavelengths that interfere

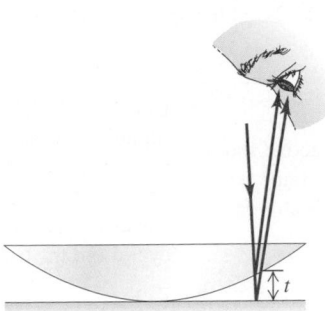

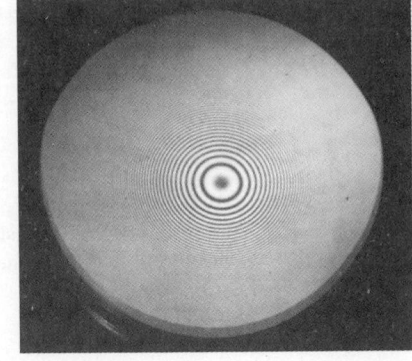

(a) A convex lens in contact with a glass plane

(b) Newton's rings: circular interference fringes

▲ **FIGURE 26.12** Newton's rings.

destructively are weak or absent in the reflected light. But the colors that are weak in the *reflected* light are strong in the *transmitted* light. At any point, the color of the wedge as viewed by reflected light is the *complement* of its color as seen by transmitted light! Roughly speaking, the complement of any color is obtained by removing that color from white light (a mixture of all colors in the visible spectrum). For example, the complement of blue is yellow, and the complement of green is magenta.

Newton's Rings

Figure 26.12a shows the convex surface of a lens in contact with a plane glass plate. A thin film of air is formed between the two surfaces. When you view the setup with monochromatic light, you see circular interference fringes (Figure 26.12b). These fringes were studied by Newton and are called *Newton's rings*. When viewed by reflected light, the center of the pattern is black. Can you see why a black center should be expected?

We can compare the surfaces of two optical parts by placing the two in contact and observing the interference fringes that are formed. Figure 26.13 is a photograph made during the grinding of a telescope objective lens. The lower, larger-diameter, thicker disk is the correctly shaped master, and the smaller upper disk is the lens under test. The "contour lines" are Newton's interference fringes; each one indicates an additional distance of a half-wavelength between the specimen and the master. At 10 lines from the center spot, the distance between the two surfaces is 5 wavelengths, or about 0.003 mm. This isn't very good; high-quality lenses are routinely ground with a precision of less than 1 wavelength. The surface of the primary mirror of the Hubble Space Telescope was ground to a precision of better than $\frac{1}{50}$ wavelength.

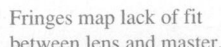

Fringes map lack of fit between lens and master.

Master   Lens being tested

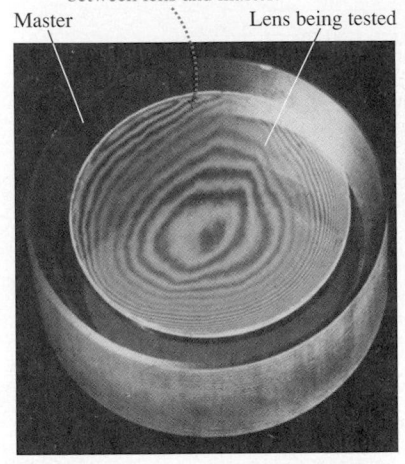

▲ **FIGURE 26.13** The surface of a telescope objective lens under inspection.

◄ **Application** **See my smile.** The Mona Lisa by Leonardo da Vinci may be the most famous painting of all time; millions of visitors come to the Louvre in Paris to see her every year. For security, the painting is protected by a bulletproof glass case. If this case did not have a nonreflective coating, visitors would perpetually be bobbing their heads trying to see the painting without the glare of obscuring reflections.

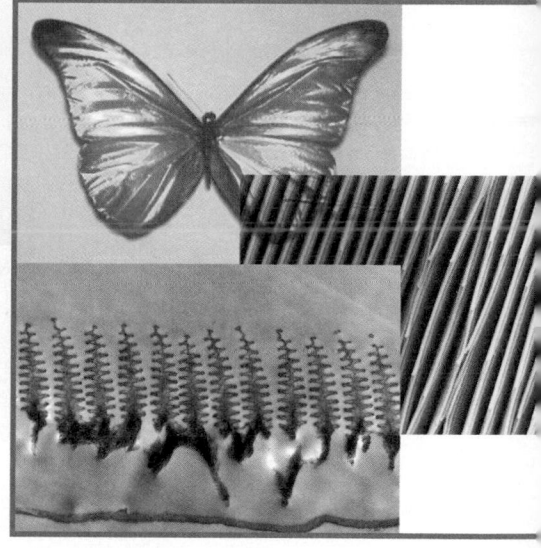

Nonreflective Coatings

Nonreflective coatings for lens surfaces make use of thin-film interference. A thin layer or film of hard transparent material with an index of refraction smaller than that of the lens material is deposited on the lens surface, as shown in Figure 26.14. Light is reflected from both surfaces of the layer. In both reflections, the light is reflected from a medium of greater index than that in which it is traveling, so the same phase change occurs. If the film thickness is a quarter (one-fourth) of the wavelength *in the film* (assuming normal incidence), the total path difference is a half-wavelength. Light reflected from the first surface is then a half cycle out of phase with light reflected from the second surface, and there is destructive interference.

The thickness of the nonreflective coating can be a quarter-wavelength for only one particular wavelength, usually chosen in the central yellow-green portion of the spectrum (550 nm), where the eye is most sensitive. Then there is somewhat more reflection at both longer and shorter wavelengths, and the reflected light has a purple hue. The overall reflection from a lens or prism surface can be reduced in this way from 4–5% to less than 1%. This treatment is particularly important in eliminating stray reflected light in highly corrected lenses with many air–glass surfaces. The same principle is used to minimize reflection from silicon photovoltaic solar cells by means of a thin surface layer of silicon monoxide (SiO).

Reflective Coatings

If a material with an index of refraction *greater* than that of glass is deposited on glass, forming a film with a thickness of a quarter-wavelength, then the reflectivity of the glass is *increased*. In this case, there is a half-cycle phase shift at the air–film interface, but none at the film–glass interface, and reflections from the two sides of the film interfere constructively. For example, a coating with an index of refraction of 2.5 allows 38% of the incident energy to be reflected, compared with 4% or so with no coating. By the use of multiple-layer coatings, it is possible to achieve nearly 100% transmission or reflection for particular wavelengths. These coatings are used for infrared "heat reflectors" in motion-picture projectors, solar cells, and astronauts' visors and for color separation in color television cameras, to mention only a few applications.

26.4 Diffraction

According to *geometric* optics, when an opaque object is placed between a point light source and a screen, as in Figure 26.15, the shadow of the object should form a perfectly sharp line. No light at all should strike the screen at points within the shadow. But we've already seen in this chapter that the *wave* nature of light causes things to happen that can't be understood with the simple model of geometric optics. The edge of the shadow is never perfectly sharp. Some light appears in the area that we expect to be in the shadow, and we find alternating bright and dark fringes in the illuminated area. More generally, light emerging from apertures doesn't behave precisely according to the predictions of the

Destructive interference occurs when
• the film is about $\frac{1}{4}\lambda$ thick and
• the light undergoes a phase change at both reflecting surfaces,
so that the two reflected waves emerge from the film about $\frac{1}{2}$ cycle out of phase.

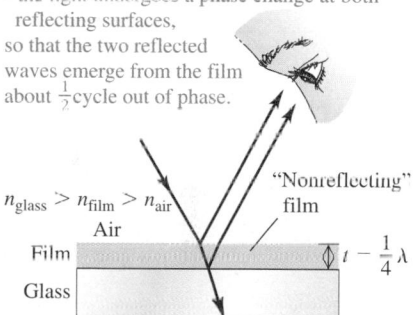

▲ **FIGURE 26.14** How a nonreflective coating works.

Geometric optics predicts that this situation should produce a sharp boundary between illumination and solid shadow. That's NOT what really happens!

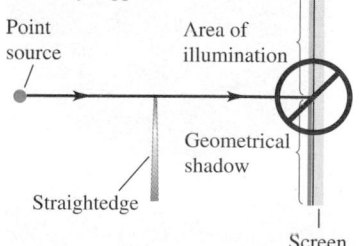

▲ **FIGURE 26.15** Setup for the shadow produced when a sharp edge obstructs monochromatic light from a point source.

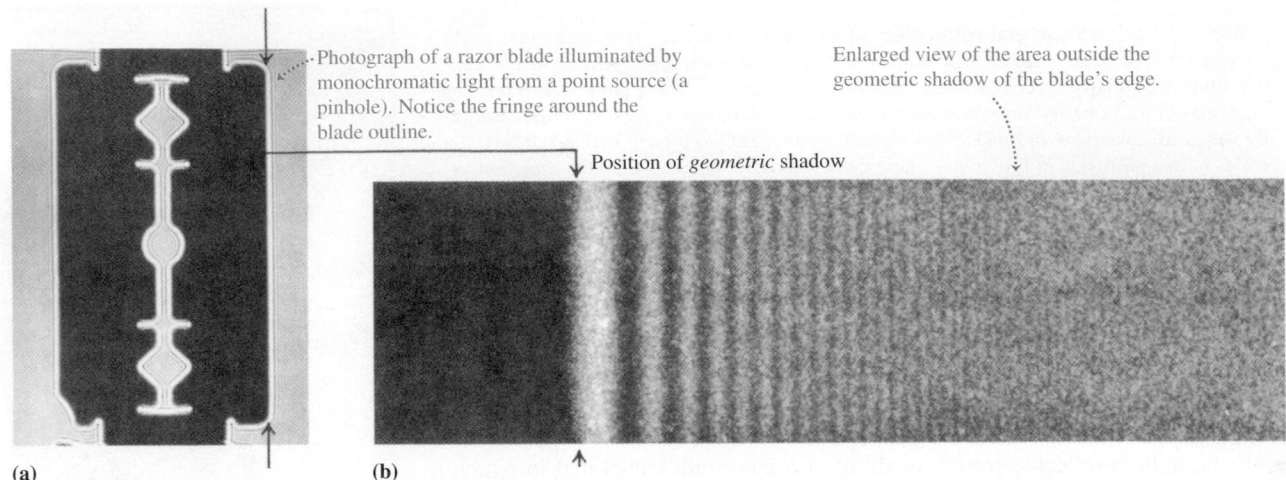

Photograph of a razor blade illuminated by monochromatic light from a point source (a pinhole). Notice the fringe around the blade outline.

Position of *geometric* shadow

Enlarged view of the area outside the geometric shadow of the blade's edge.

(a)　　　　(b)

▲ **FIGURE 26.16** Actual shadow of a razor blade produced in a situation like that of Figure 26.15.

straight-line ray model of geometric optics. An important class of such effects occurs when light strikes a barrier with an aperture or an edge. The interference patterns formed in such a situation are grouped under the heading **diffraction.**

Here's an example: The photograph in Figure 26.16a was made by placing a razor blade halfway between a pinhole, illuminated by monochromatic light, and a photographic film. The film recorded the shadow cast by the blade. Figure 26.16b is an enlargement of a region near the shadow of the right edge of the blade. The position of the *geometric* shadow line is indicated by arrows. The area outside the geometric shadow is bordered by alternating bright and dark bands. There is some light in the shadow region, although this is not very visible in the photograph. The first bright band, just outside the geometric shadow, is actually *brighter* than in the region of uniform illumination to the extreme right. This simple experiment gives us some idea of the richness and complexity of a seemingly simple phenomenon: the casting of a shadow by an opaque object.

We don't often observe diffraction patterns such as Figure 26.16 in everyday life, because most ordinary light sources are neither monochromatic nor point sources. If a frosted lightbulb were used instead of a point source in Figure 26.15, each wavelength of the light from every point of the bulb would form its own diffraction pattern, and the patterns would overlap to such an extent that we wouldn't see any individual pattern.

Diffraction is sometimes described as "the bending of light around an obstacle." But the process that causes diffraction effects is present in the propagation of *every* wave. When part of the wave is cut off by some obstacle, we observe diffraction effects that result from interference of the remaining parts of the wave fronts. Every optical instrument uses only a limited portion of a wave; for example, a telescope uses only the part of a wave admitted by its objective lens. Thus, diffraction plays a role in nearly all optical phenomena.

Figure 26.17 shows a diffraction pattern formed by a steel ball about 3 mm in diameter. Note the rings in the pattern, both outside and inside the geometric shadow area, and the bright spot at the very center of the shadow. The existence of this spot was predicted in 1818, on the basis of a wave theory of light, by the French mathematician Siméon-Denis Poisson, during an extended debate in the French Academy of Sciences concerning the nature of light. Ironically, Poisson himself was *not* a believer in the wave theory of light, and he published this *apparently* absurd prediction hoping to deal a death blow to the wave theory. But the prize committee of the Academy arranged for an experimental test, and soon the bright spot was actually observed. (It had in fact been seen as early as 1723, but those experiments had gone unnoticed.)

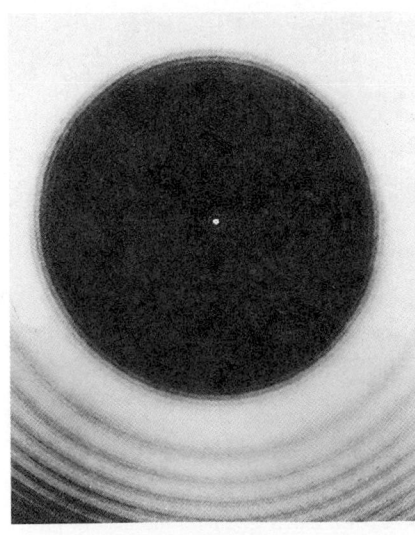

▲ **FIGURE 26.17** Fresnel diffraction pattern formed by a steel ball 3 mm in diameter. The Poisson bright spot is seen at the center of the area in shadow.

The fringe patterns formed by diffraction effects can be analyzed using Huygens's principle (Section 23.11). Let's review that principle briefly. Every point of a wave front can be considered the source of secondary wavelets that spread out in all directions with a speed equal to the speed of propagation of the wave. The position of the wave front at any later time is the *envelope* of the secondary wavelets at that time. To find the resultant displacement at any point, we combine all the individual displacements produced by these secondary wavelets, using the superposition principle and taking into account their amplitudes and relative phases.

In Figure 26.15, both the point source and the screen are at finite distances from the obstacle forming the diffraction pattern. This situation is described as *near-field diffraction,* or **Fresnel diffraction** (after Augustin Jean Fresnel, 1788–1827). If the source, obstacle, and screen are far enough away that all lines from the source to the obstacle can be considered parallel and all lines from the obstacle to a point in the pattern can be considered parallel, the phenomenon is called *far-field diffraction,* or **Fraunhofer diffraction** (after Joseph von Fraunhofer, 1787–1826). Fraunhofer diffraction is usually simpler to analyze in detail than is Fresnel diffraction, so we'll confine our discussion in this chapter to Fraunhofer diffraction.

Finally, we emphasize that there is no fundamental distinction between *interference* and *diffraction*. In preceding sections of this chapter, we used the term *interference* for effects involving waves from a small number of sources, usually two. *Diffraction* usually involves a *continuous* distribution of Huygens's wavelets across the area of an aperture or a very large number of sources or apertures. But the same physical principles—superposition and Huygens's principle—govern both categories.

26.5 Diffraction from a Single Slit

In this section, we'll discuss the diffraction pattern formed by plane-wave (parallel-ray) monochromatic light when it emerges from a long, narrow slit, as shown in Figure 26.18. (We call the narrow dimension the *width,* even though, in this figure, it is a vertical dimension. We'll assume that the slit width a is much greater than the wavelength λ of the light.) According to geometric optics, the transmitted beam should have the same cross section as the slit, as in Figure 26.18a. What is *actually* observed, however, is the pattern shown in Figure 26.18b. The beam spreads out vertically after passing through the slit. The diffraction pattern consists of a central bright band, which may be much broader than the width of the vertical slit, bordered by alternating dark and

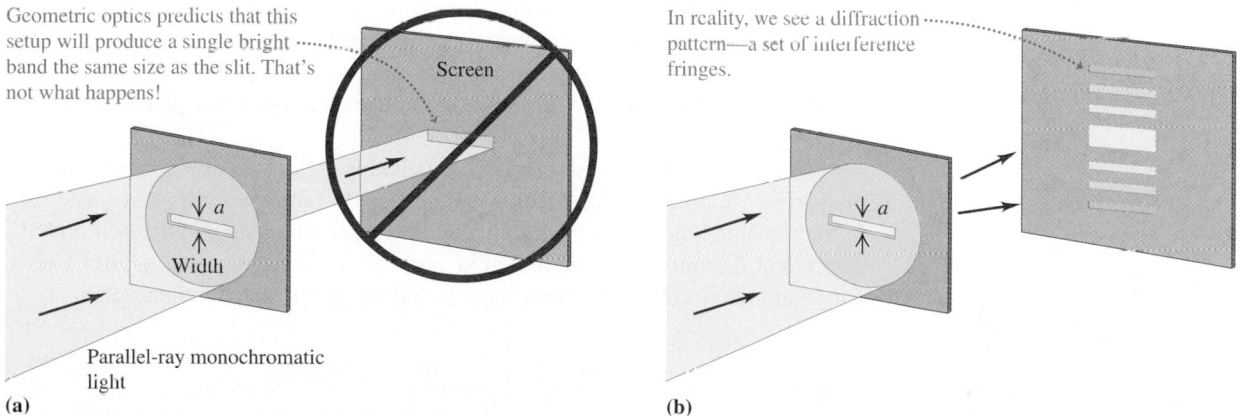

Geometric optics predicts that this setup will produce a single bright band the same size as the slit. That's not what happens!

Screen

$\downarrow a$
$\uparrow$
Width

Parallel-ray monochromatic light

(a)

In reality, we see a diffraction pattern—a set of interference fringes.

$\downarrow a$
$\uparrow$

(b)

▲ **FIGURE 26.18** Comparison of the result predicted by geometric optics and the actual result when monochromatic light passes through a narrow slit and illuminates a screen.

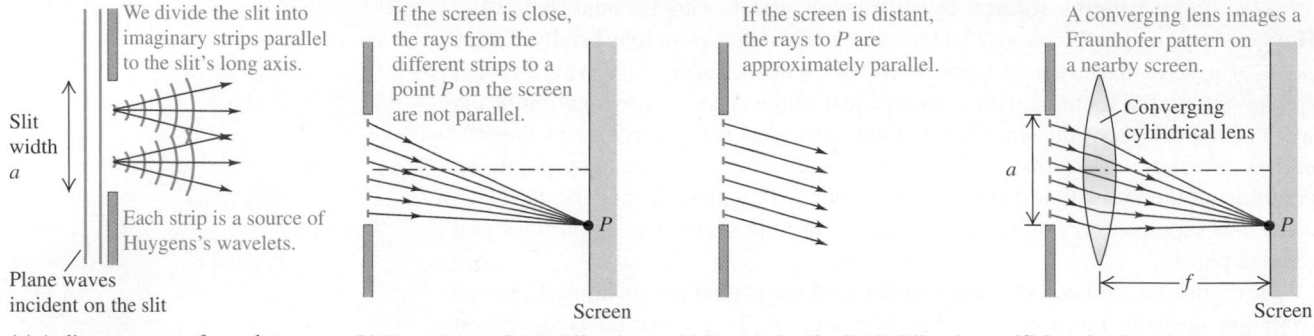

(a) A slit as a source of wavelets   **(b)** Fresnel (near-field) diffraction   **(c)** Fraunhofer (far-field) diffraction   **(d)** Imaging Fraunhofer diffraction

▲ **FIGURE 26.19** Analysis of diffraction by a single rectangular slit.

bright bands with rapidly decreasing intensity. When a is much greater than λ, about 85% of the total energy is in this central bright band, whose width is found to be *inversely* proportional to the width of the slit. You can easily observe a similar diffraction pattern by looking at a point source, such as a distant street light, through a narrow slit formed between two fingers in front of your eye. The retina of your eye then corresponds to the screen.

Figure 26.19 shows a side view of the same setup; the long sides of the slit are perpendicular to the figure. According to Huygens's principle, each element of area of the slit opening can be considered as a source of secondary wavelets. In particular, imagine dividing the slit into several narrow strips with equal width, parallel to the long edges and perpendicular to the page in Figure 26.19a. Cylindrical secondary wavelets spread out from each strip, as shown in cross section.

In Figure 26.19b, a screen is placed to the right of the slit. We can calculate the resultant intensity at a point P on the screen by adding the contributions from the individual wavelets, taking proper account of their various phases and amplitudes. We assume that the screen is far enough away that all the rays from various parts of the slit to a particular point P on the screen are parallel, as in Figure 26.19c. An equivalent situation is Figure 26.19d, in which the rays to the lens are parallel and the lens forms a reduced image of the same pattern that would be formed on an infinitely distant screen without the lens. We might expect that the various light paths through the lens would introduce additional phase shifts, but in fact it can be shown that all the paths have *equal* phase shifts, so this is not a problem.

The situation of Figure 26.19b is Fresnel diffraction; those of Figures 26.19c and 26.19d, in which the outgoing rays are considered parallel, are Fraunhofer diffraction. We can derive quite simply the most important characteristics of the diffraction pattern from a single slit. First consider two narrow strips, one just below the top edge of the drawing of the slit and one at its center, shown in end view in Figure 26.20. The difference in path length to point P is $(a/2)\sin\theta$, where a is the slit width. Suppose this path difference happens to be equal to $\lambda/2$; then light from the two strips arrives at point P with a half-cycle phase difference, and cancellation occurs.

Similarly, light from two strips immediately *below* those two in the figure also arrives at point P a half-cycle out of phase. In fact, the light from *every* strip in the top half cancels out the light from a corresponding strip in the bottom half. The result is complete cancellation at point P for the entire slit, giving a dark fringe in the interference pattern. That is, a dark fringe occurs whenever

$$\frac{a}{2}\sin\theta = \pm\frac{\lambda}{2}, \quad \text{or} \quad \sin\theta = \pm\frac{\lambda}{a}.$$

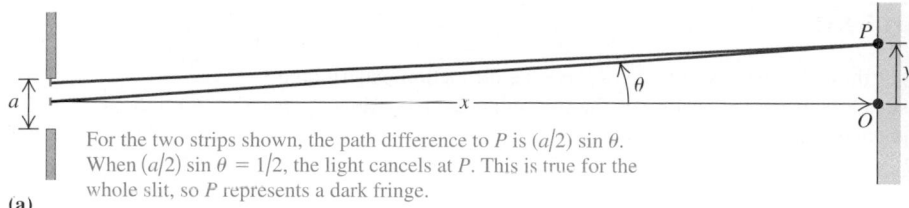

For the two strips shown, the path difference to P is $(a/2) \sin\theta$. When $(a/2) \sin\theta = 1/2$, the light cancels at P. This is true for the whole slit, so P represents a dark fringe.

(a)

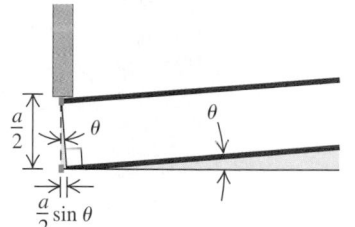

θ is usually very small, so we can use the approximations $\sin\theta = \theta$ and $\tan\theta = \theta$. Then the condition for a dark band is

$$y_m = R\frac{m\lambda}{a}.$$

(b) Enlarged view of the top half of the slit

▲ **FIGURE 26.20** Geometry of rays passing through a single slit.

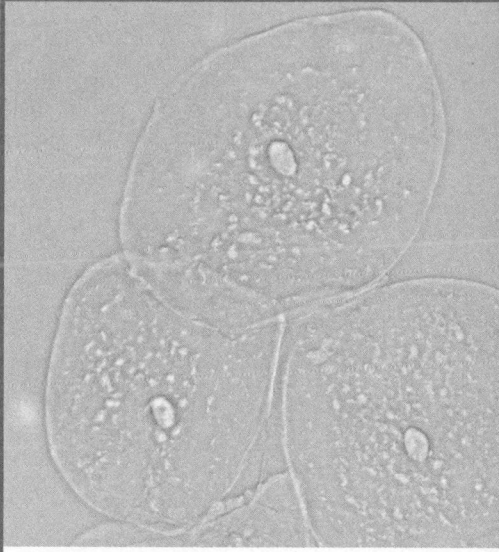

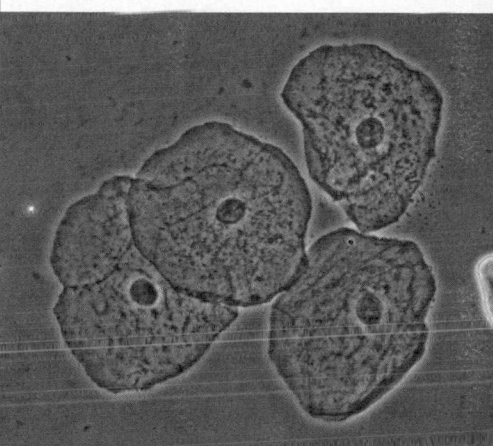

We may also divide the screen into quarters, sixths, and so on and use the preceding argument to show that a dark fringe occurs whenever $\sin\theta = \pm 2\lambda/a$, $\pm 3\lambda/a$, and so on. Thus, the condition for a *dark* fringe is

$$\sin\theta = \frac{m\lambda}{a} \qquad (m = \pm 1, \pm 2, \pm 3, \cdots). \qquad (26.9)$$

For example, if the slit width is equal to 10 wavelengths $(a = 10\lambda)$, dark fringes occur at $\sin\theta = \pm\frac{1}{10}$, $\pm\frac{2}{10}$, $\pm\frac{3}{10}, \cdots$. Between the dark fringes are bright fringes.

NOTE ▶ In the "straight ahead" direction $(\sin\theta = 0)$ is a *bright* band; in this case, light from the entire slit arrives at P in phase. It would be wrong to put $m = 0$ in Equation 26.9. The central bright fringe is wider than the others, as Figure 26.18 shows. In the small-angle approximation we'll use in the following discussion, it is exactly *twice* as wide. ◀

With light, the wavelength λ is on the order of 500 nm $= 5 \times 10^{-7}$ m. This is often much smaller than the slit width a; a typical slit width is 10^{-2} cm $= 10^{-4}$ m. Therefore, the values of θ in Equation 26.9 are often so small that the approximation $\sin\theta = \theta$ is very good. In that case, we can rewrite that equation as

$$\theta = \frac{m\lambda}{a} \qquad (m = \pm 1, \pm 2, \pm 3, \cdots).$$

Also, if the distance from slit to screen is R, and the vertical distance of the mth dark band from the center of the pattern is y_m, then $\tan\theta = y_m/R$. For small θ, we may approximate $\tan\theta$ by θ, and we then find that

$$y_m = R\frac{m\lambda}{a} \qquad (m = \pm 1, \pm 2, \pm 3, \cdots). \qquad (26.10)$$

NOTE ▶ This equation has the same form as the equation for the two-slit pattern, Equation 26.6, but here it gives the positions of the *dark* fringes in a *single-slit* pattern rather than the *bright* fringes in a *double-slit* pattern. Also, $m = 0$ is *not* a dark fringe. Be careful! ◀

▲ **BIO Application Enhancing the image.** A major problem facing the biological or medical microscopist is the generation of contrast in the object under study. If the object strongly absorbs light, then the outline of the object will be clearly seen; however, individual cells are often thin and transparent, and the cells are nearly invisible in the microscope. One solution is to stain the cells, but this generally kills them. A far better solution was achieved by Frits Zernike, who applied the principles of wave optics to create a microscope in which the small modifications in the phase of the illuminating light due to slight differences in the index of refraction between the cell and the surrounding medium were converted to differences in light amplitude. Unlike phase differences, amplitude differences can be detected by the eye. The upper photograph shows living human epithelial cells in ordinary bright-field view, and the lower one shows the same cells in phase-contrast microscopy. Zernike was awarded the Nobel Prize in Physics in 1953, and the phase-contrast technique is still used routinely in research laboratories.

EXAMPLE 26.6 A single-slit experiment

You pass 633 nm (helium–neon) laser light through a narrow slit and observe the diffraction pattern on a screen 6.0 m away. You find that the distance between the centers of the first minima (dark fringes) on either side of the central bright fringe in the pattern is 27 mm. How wide is the slit?

SOLUTION

SET UP The angle θ in this situation is very small, so we can use the approximate relation of Equation 26.10. Figure 26.21 shows the various distances. The distance y_1 from the central maximum to the first minimum on either side is half the distance between the two first minima, so $y_1 = (27 \text{ mm})/2$.

SOLVE Solving Equation 26.10 for the slit width a and substituting $m = 1$, we find that

$$a = \frac{R\lambda}{y_1} = \frac{(6.0 \text{ m})(633 \times 10^{-9} \text{ m})}{(27 \times 10^{-3} \text{ m})/2}$$

$$= 2.8 \times 10^{-4} \text{ m} = 0.28 \text{ mm}.$$

REFLECT It can also be shown that the distance between the *second* minima on the two sides is $2(27 \text{ mm})$, and so on.

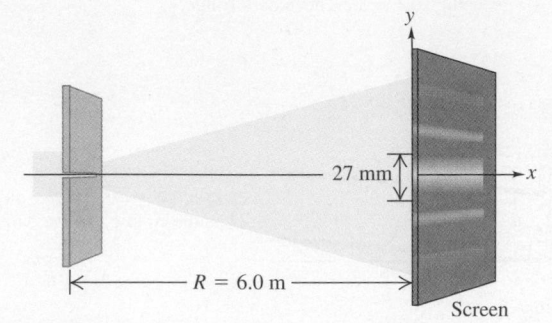

▲ **FIGURE 26.21**

Practice Problem: What is the distance between the fifth minima on the two sides of the central bright fringe? *Answer:* $2y_5 = 5(27 \text{ mm}) = 135 \text{ mm}$.

In the preceding analysis, we've located the maxima and minima in the diffraction pattern formed by a single slit. It's also possible to calculate the intensity at any point in the pattern, using Huygens's principle and some nontrivial mathematical analysis. We'll omit the details, but Figure 26.22a is a graph of intensity as a function of position for a single-slit pattern. Figure 26.22b is a photograph of the same pattern. Figure 26.23 shows how the width of the diffraction pattern varies inversely with that of the slit: The narrower the slit, the wider is the diffraction pattern.

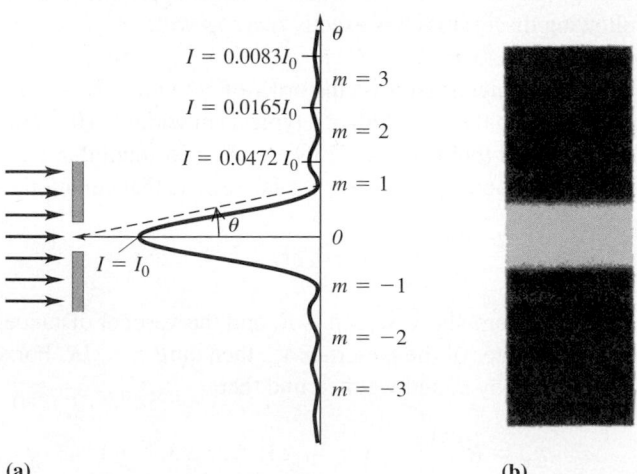

(a)   (b)

▲ **FIGURE 26.22** (a) Intensity distribution for diffraction from a single slit. (b) Photograph of the Fraunhofer diffraction pattern of a single slit.

If the slit width is equal to or narrower than the wavelength, only one broad maximum forms.

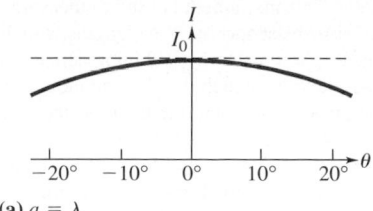

(a) $a = \lambda$

The wider the slit (or the shorter the wavelength), the narrower and sharper is the central peak.

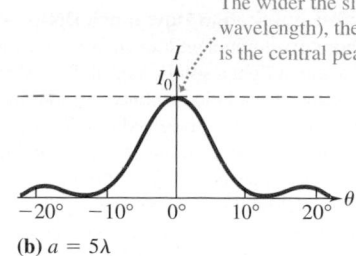

(b) $a = 5\lambda$

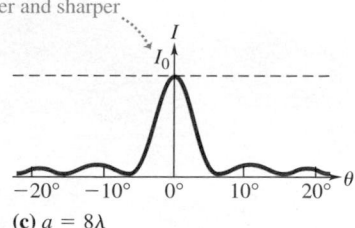

(c) $a = 8\lambda$

▲ **FIGURE 26.23** Effect of slit width or wavelength on the diffraction pattern formed by a single slit.

26.6 Multiple Slits and Diffraction Gratings

In Sections 26.1 and 26.2, we analyzed interference from two point sources or from two very thin slits. In Section 26.5, we carried out a similar analysis for a single slit with finite width. Now let's consider patterns produced by *several* very narrow slits. Assume that each slit is narrow in comparison to the wavelength, so its diffraction pattern spreads out nearly uniformly. Figure 26.24 shows an array of several narrow slits, with distance d between adjacent slits. Constructive interference occurs for those rays that are at an angle θ to the normal and that arrive at point P when the path difference between adjacent slits is an integral number of wavelengths:

$$d\sin\theta = m\lambda \qquad (m = 0, \pm 1, \pm 2, \cdots). \qquad (26.11)$$

That is, the maxima in the pattern occur at the same positions as for a two-slit pattern with the same spacing (Equation 26.4). In this respect, the pattern resembles the two-slit pattern.

But what happens *between* the maxima? In the two-slit pattern, there is exactly one intensity minimum between each pair of maxima, corresponding to angles for which

$$d\sin\theta = \left(m + \tfrac{1}{2}\right)\lambda \qquad (m = 0, 1, 2, \cdots),$$

or for which the phase difference between waves from the two sources is $\tfrac{1}{2}$ cycle, $\tfrac{3}{2}$ cycle, $\tfrac{5}{2}$ cycle, and so on. (See Figure 26.25a.) In the eight-slit pattern, these are also minima, because the light from adjacent slits cancels out in pairs. Detailed

Maxima occur where the path difference for adjacent slits is a whole number of wavelengths: $d\sin\theta = m\lambda$.

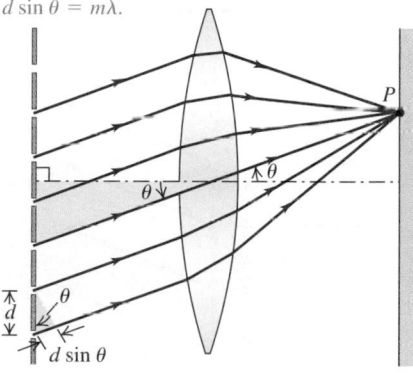

▲ **FIGURE 26.24** In multiple-slit diffraction, rays from every slit arrive in phase to give a sharp maximum if the path difference between adjacent slits is a whole number of wavelengths.

Two slits produce one minimum between adjacent maxima.

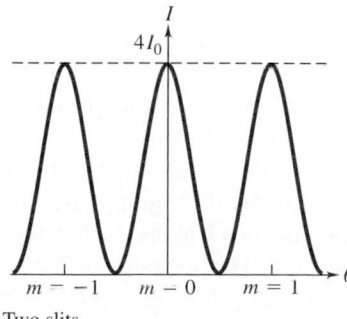

(a) Two slits

Eight slits produce larger, narrower maxima in the same locations, separated by seven minima.

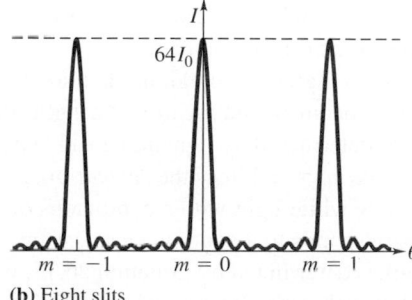

(b) Eight slits

With sixteen slits, the maxima are still taller and narrower, with more intervening minima.

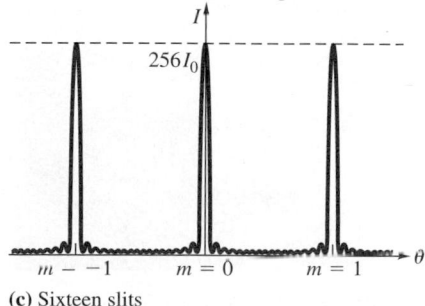

(c) Sixteen slits

▲ **FIGURE 26.25** Effect of the number of slits on a diffraction pattern, for a given slit width and spacing.

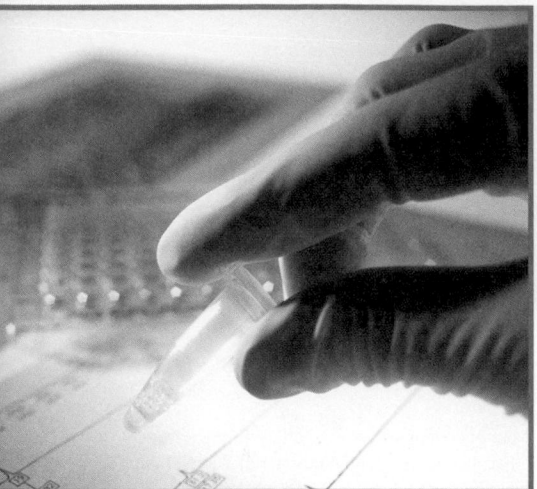

calculation shows that the interference pattern is as shown in Figure 26.25b. The maxima are in the same positions as for the two-slit pattern of Figure 26.25a, but they are much narrower. Figure 26.25c shows the corresponding pattern with 16 slits. Increasing the number of slits in an interference experiment while keeping the spacing of adjacent slits constant gives interference patterns with the maxima in the same positions as with two slits, but progressively sharper and narrower.

Diffraction Gratings

An array of a large number of parallel slits, all with the same width a and spaced equal distances d between centers, is called a **diffraction grating.** The first one was constructed out of fine wires by Fraunhofer. Gratings can be made by using a diamond point to scratch many equally spaced grooves on a glass or metal surface or by photographically reducing a pattern of black and white stripes drawn on paper with a pen or a computer. For a grating, what we have been calling *slits* are often called *rulings* or *lines.*

In Figure 26.26, GG' is a cross section of a grating; the slits are perpendicular to the plane of the page. Only six slits are shown in the diagram, but an actual grating may contain several thousand. The spacing d between centers of adjacent slits is called the *grating spacing;* typically, it is about 0.002 mm. A plane wave of monochromatic light is incident normally on the grating from the left side. We assume that the pattern is formed on a screen far enough away that all rays emerging from the grating in a particular direction are parallel.

We noted earlier that the principal intensity maxima occur in the same directions as for the two-slit pattern, directions for which the path difference for adjacent slits is an integer number of wavelengths. So the positions of the maxima are once again given by Equation 26.11:

$$d\sin\theta = m\lambda \qquad (m = 0, \pm1, \pm2, \pm3, \cdots).$$

As we saw in Figure 26.25, the larger the number of slits, the sharper are the peaks of the diffraction pattern.

When a grating containing hundreds or thousands of slits is illuminated by a parallel beam of monochromatic light, the pattern is a series of sharp lines at angles determined by Equation 26.11. The $m = \pm1$ lines are called the *first-order lines,* the $m = \pm2$ lines the *second-order* lines, and so on. If the grating is illuminated by white light with a continuous distribution of wavelengths, each value of m corresponds to a continuous spectrum in the pattern. The angle for each wavelength is determined by Equation 26.11, which shows that, for a given value of m, long wavelengths (at the red end of the spectrum) lie at larger angles (i.e., are deviated more from the straight-ahead direction) than the shorter wavelengths (at the violet end of the spectrum). Figure 26.27 shows a familiar example.

As Equation 26.11 shows, the sines of the deviation angles of the maxima are proportional to the ratio λ/d. For substantial deviation to occur, the grating

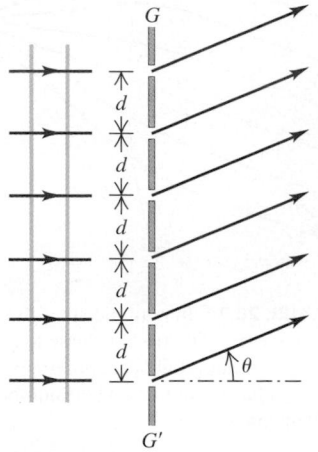

▲ **FIGURE 26.26** A portion of a transmission diffraction grating.

▲ **FIGURE 26.27** Unless you download all your music electronically, you probably have a few diffraction gratings around the house. The tracks in a commercial CD act as a reflection diffraction grating for visible light, producing rainbow patterns.

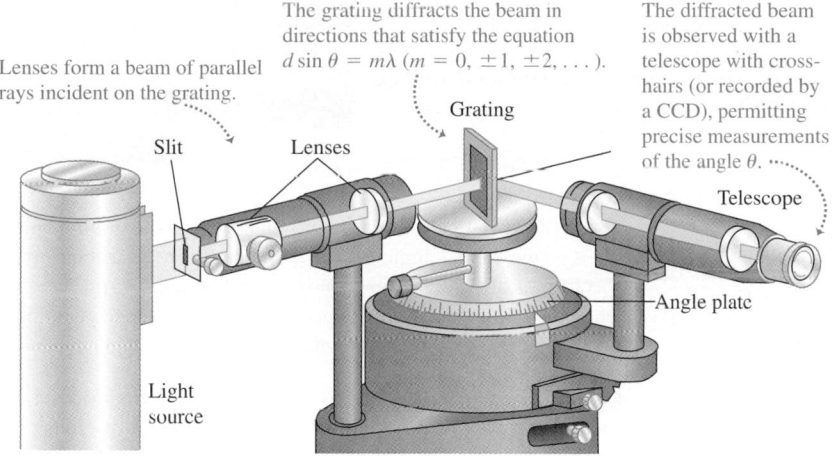

The grating diffracts the beam in directions that satisfy the equation $d \sin \theta = m\lambda$ ($m = 0, \pm 1, \pm 2, \ldots$).

Lenses form a beam of parallel rays incident on the grating.

The diffracted beam is observed with a telescope with cross-hairs (or recorded by a CCD), permitting precise measurements of the angle θ. ⋯⋯

Grating

Slit

Lenses

Telescope

Angle plate

Light source

▲ **FIGURE 26.28** A diffraction grating spectrometer.

spacing d should be of the same order of magnitude as the wavelength λ. Gratings for use in the visible spectrum usually have about 500 to 1500 slits per millimeter, so d (which equals the reciprocal of the number of slits per unit width) is on the order of 1000 nm.

Grating Spectrometers

Diffraction gratings are widely used in spectrometry as a means of dispersing a light beam into spectra. If the grating spacing is known, then we can measure the angles of deviation and use Equation 26.11 to compute the wavelength. A typical setup is shown in Figure 26.28. A prism can also be used to disperse the various wavelengths through different angles, because the index of refraction always varies with wavelength. But there is no simple relationship that describes this variation, so a spectrometer using a prism has to be calibrated with known wavelengths that are determined in some other way. Another difference is that a prism deviates red light the least and violet the most, while a grating does the opposite.

ActivPhysics 16.4: The Grating: Introduction and Qualitative Questions
ActivPhysics 16.5: The Grating: Problems

EXAMPLE 26.7   Width of a grating spectrum

The wavelengths of the visible spectrum are approximately 400 nm (violet) through 700 nm (red). Find the angular width of the first-order visible spectrum produced by a plane grating with 600 lines per millimeter when white light falls normally onto the grating.

SOLUTION

SET UP The first-order spectrum corresponds to $m = 1$. The grating has 600 lines per millimeter, so the grating spacing (the distance d between adjacent lines is

$$d = \frac{1}{600} \text{ mm} = 1.67 \times 10^{-6} \text{ m}.$$

SOLVE From Equation 26.11, with $m = 1$, the angular deviation θ_v of the violet light is given by

$$\sin\theta_v = \frac{m\lambda}{d} = \frac{(1)400 \times 10^{-9} \text{ m}}{1.67 \times 10^{-6} \text{ m}} = 0.240,$$

so that

$$\theta_v = 13.9°.$$

The angular deviation of the red light is given by

$$\sin\theta_r = \frac{700 \times 10^{-9} \text{ m}}{1.67 \times 10^{-6} \text{ m}} = 0.419;$$

thus,

$$\theta_r = 24.8°.$$

Therefore, the angular width of the first-order visible spectrum is

$$24.8° - 13.9° = 10.9°.$$

REFLECT Increasing the number of lines per millimeter in the grating increases the angular width of the spectrum, but the grating spacing d can't be less than the wavelengths of the spectrum being observed. Do you see why?

Practice Problem: If the angular width of the first-order visible spectrum is instead $20.5° - 11.5° = 9.0°$, how many lines per millimeter are there in the grating? *Answer:* 500 lines/mm.

EXAMPLE 26.8 **Spectra that overlap**

Show that, in the situation of Example 26.7, the violet end of the third-order spectrum overlaps the red end of the second-order spectrum (Figure 26.29).

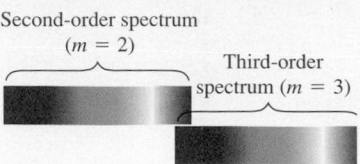

▶ **FIGURE 26.29**

SOLUTION

SET UP AND SOLVE From Equation 26.11, with $m = 3$, the angular deviation of the third-order violet end is given by

$$\sin \theta_{\text{v}} = \frac{(3)(400 \times 10^{-9}\,\text{m})}{d} = \frac{1.20 \times 10^{-6}\,\text{m}}{d}.$$

The deviation of the second-order red end, with $m = 2$, is given by

$$\sin \theta_{\text{r}} = \frac{(2)(700 \times 10^{-9}\,\text{m})}{d} = \frac{1.40 \times 10^{-6}\,\text{m}}{d}.$$

These two equations show that no matter what the grating spacing d is, the largest angle θ_{r} (at the red end) for the second-order spectrum is always greater than the smallest angle θ_{v} (at the violet end) for the third-order spectrum, so the second and third orders *always* overlap, as shown in Figure 26.29.

REFLECT If we could separate the third-order and second-order spectra, then we *would* be able to distinguish the third-order violet from the second-order red. In actuality, they overlap, so the separate spectra can't be viewed clearly.

26.7 X-Ray Diffraction

X rays were discovered by Wilhelm Röntgen (1845–1923) in 1895, and early experiments suggested that they were electromagnetic waves with wavelengths on the order of 10^{-10} m. At about the same time, the idea began to emerge that the atoms in a crystalline solid are arranged in a lattice in a regular repeating pattern, with spacing between adjacent atoms also on the order of 10^{-10} m. Putting these two ideas together, Max von Laue (1879–1960) proposed in 1912 that a crystal might serve as a kind of three-dimensional diffraction grating for x rays. That is, a beam of x rays might be scattered (absorbed and re-emitted) by the individual atoms in a crystal, and the scattered waves might interfere just like waves from a diffraction grating.

The first experiments in **x-ray diffraction** were performed in 1912 by Walther Friederich, Paul Knipping, and von Laue, using the experimental setup sketched in Figure 26.30a. The scattered x rays *did* form an interference pattern, which the three scientists recorded on photographic film. Figure 26.30b is a photograph of such a pattern. These experiments verified that x rays *are* waves, or at least have wavelike properties, and also that the atoms in a crystal *are*

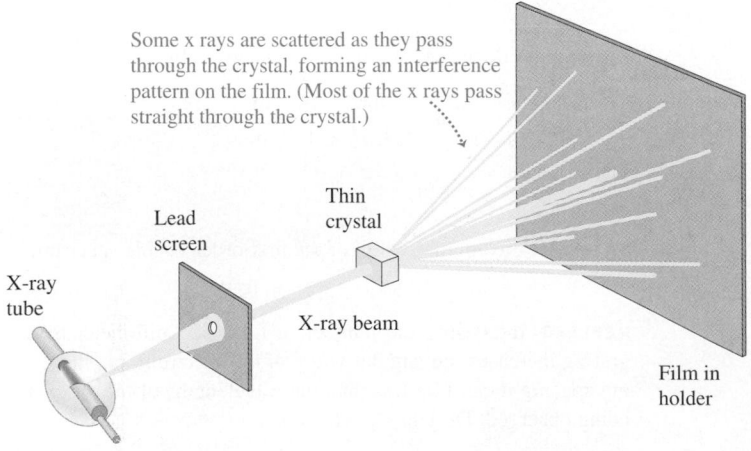

Some x rays are scattered as they pass through the crystal, forming an interference pattern on the film. (Most of the x rays pass straight through the crystal.)

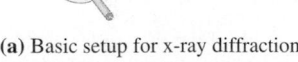

(a) Basic setup for x-ray diffraction

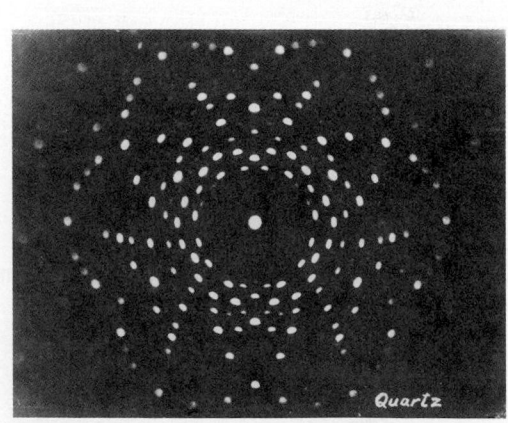

(b) Laue diffraction pattern for a thin section of quartz crystal

▲ **FIGURE 26.30** X-ray diffraction.

arranged in a regular pattern (Figure 26.31). Since that time, x-ray diffraction has proved an invaluable research tool, both for measuring wavelengths of x rays and for studying crystal structure.

To introduce the basic ideas, we consider first a two-dimensional scattering situation, as shown in Figure 26.32a, in which a plane wave is incident on a rectangular array of scattering centers. The situation might involve a ripple tank with an array of small posts, or 3 cm microwaves striking an array of small conducting spheres, or x rays incident on an array of atoms. In the case of x rays, the wave induces vibrations in the individual atoms, which then act like little antennas, emitting scattered waves. The resulting interference pattern is the superposition of all these scattered waves. The situation is different from that obtained with a diffraction grating, in which the waves from all the slits are emitted *in phase* (for a plane wave at normal incidence). With x rays, the scattered waves are *not* all in phase, because their distances from the *source* are different. To compute the interference pattern, we have to consider the *total* path differences for the scattered waves, including the distances both from source to scatterer and from scatterer to observer.

As Figure 26.32b shows, the path length from source to observer is the same for all the scatterers in a single row if the angles θ_a and θ_r are equal. Scattered radiation from *adjacent* rows is *also* in phase if the path difference for adjacent rows is an integral number of wavelengths. Figure 26.32c shows that this path difference is $2d\sin\theta$. Therefore, the conditions for radiation from the *entire array* to reach the observer in phase are that (1) the angle of incidence must equal the angle of scattering and (2) the path difference for adjacent rows must equal $m\lambda$, where m is an integer. We can express the second condition as

$$2d\sin\theta = m\lambda \qquad (m = 1, 2, 3, \cdots). \qquad (26.12)$$

In directions for which this condition is satisfied, we see a strong maximum in the interference pattern. We can describe this interference in terms of *reflections* of the wave from the horizontal rows of scatterers in Figure 26.32a. Strong reflection (constructive interference) occurs at angles such that the angles of the incident and scattered x rays are equal and Equation 26.12 is satisfied.

NOTE ▶ The angle θ is customarily measured with respect to the surface of the crystal. This approach is different from our usual one, in which we measure θ with respect to the normal to the plane of an array of slits or a grating. Also, Equation 26.12 is *not* the same as Equation 26.11. Be careful! ◀

We can extend this discussion to a three-dimensional array by considering *planes* of scatterers instead of *rows*. Figure 26.33 shows two different sets of parallel planes that pass through all the scatterers. Waves from all the scatterers in a

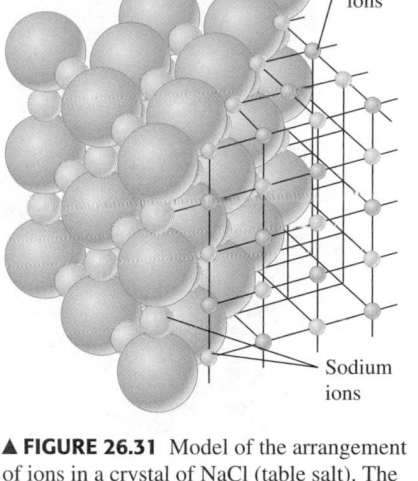

▲ **FIGURE 26.31** Model of the arrangement of ions in a crystal of NaCl (table salt). The spacing of adjacent ions is 0.282 nm. For convenience, the ions are represented as spheres, although their electron clouds actually overlap somewhat.

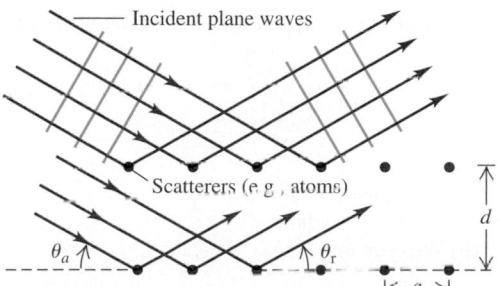

(a) Scattering of waves from a rectangular array

Interference from adjacent atoms in a row is constructive when the path lengths $a\cos\theta_a$ and $a\cos\theta_r$ are equal.

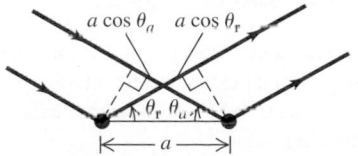

(b) Scattering from adjacent atoms in row

Interference from atoms in adjacent rows is constructive when the path difference $2d\sin\theta$ is an integral number of wavelengths.

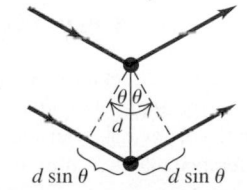

(c) Scattering from atoms in adjacent rows

▲ **FIGURE 26.32** Scattering of waves from rows of atoms (or other scatterers) in a two-dimensional rectangular array.

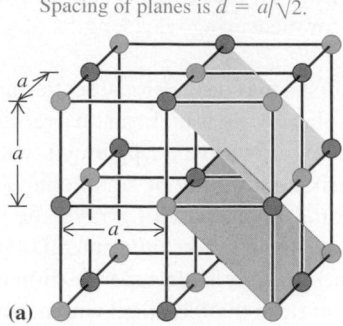

Spacing of planes is $d = a/\sqrt{2}$.

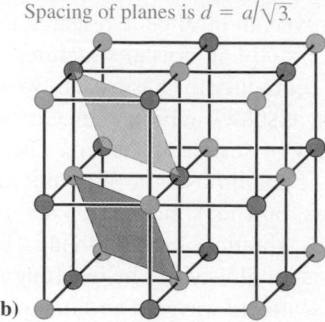

Spacing of planes is $d = a/\sqrt{3}$.

(a)

(b)

▲ **FIGURE 26.33** Cubic crystal lattice showing two different families of crystal planes. There are also three sets of planes parallel to the cube faces, with spacing a.

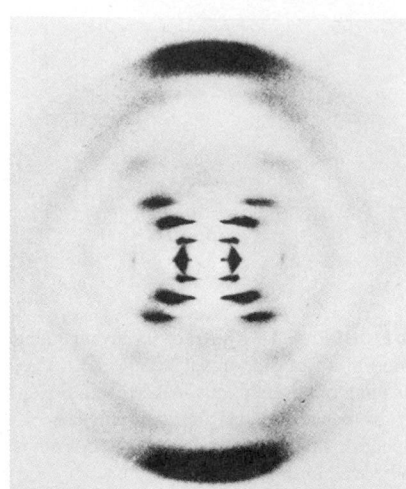

The historic x-ray diffraction image of DNA fibers obtained by Rosalind Franklin in 1953. This image was central to the elucidation of the double-helix structure of DNA.

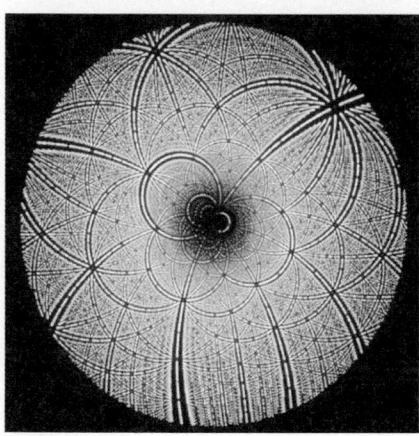

The x-ray diffraction pattern of the enzyme rubisco, which plants use to "fix" atmospheric carbon dioxide into carbohydrate.

▲ **FIGURE 26.34** Historic and modern examples of x-ray diffraction spectra.

given plane interfere constructively if the angles of incidence and scattering are equal. There is also constructive interference between planes when Equation 26.12 is satisfied, where d is now the distance between adjacent planes. Because there are many different sets of parallel planes, there are many values of d and many sets of angles that give constructive interference for the whole crystal lattice. This phenomenon is called **Bragg reflection,** and Equation 26.12 is called the **Bragg condition,** in honor of Sir William Bragg and his son Laurence Bragg, two pioneers in x-ray analysis.

NOTE ▶ Don't let the term *reflection* obscure the fact that we are dealing with an *interference* effect. In fact, the reflections from various planes are closely analogous to interference effects in thin films (Section 26.3). ◀

As Figure 26.30b shows, nearly complete cancellation occurs for all but certain very specific directions, where constructive interference occurs and forms bright spots. Such a pattern is usually called an x-ray *diffraction* pattern, although *interference* pattern might be more appropriate. This particular type of pattern is called a *Laue pattern.*

If the crystal lattice spacing is known, we can determine the wavelength of the x rays (just as we determined wavelengths of visible light in Section 26.6 by measuring diffraction patterns from slits or gratings). For example, we can determine the crystal lattice spacing for sodium chloride from its density and Avogadro's number. Then, once we know the x-ray wavelength, we can use x-ray diffraction to explore the structure and lattice spacing of crystals with unknown structure.

X-ray diffraction is by far the most important experimental tool in the investigation of the crystal structure of solids. Atomic spacings in crystals can be measured precisely, and the lattice structure of complex crystals can be determined. X-ray diffraction also plays an important role in studies of the structures of liquids and of organic molecules. Indeed, it was one of the chief experimental techniques used in working out the double-helix structure of DNA and the structures of proteins (Figure 26.34).

EXAMPLE 26.9 **X-ray diffraction with silicon crystal**

Suppose you direct an x-ray beam with wavelength 0.154 nm at certain planes of a silicon crystal. As you increase the angle of incidence from zero, you find the first strong interference maximum from these planes when the beam makes an angle of 34.5° with them. **(a)** How far apart are the planes? **(b)** Will you find other interference maxima from these planes at larger angles?

Continued

SOLUTION

SET UP AND SOLVE Part (a): To find the plane spacing d, we solve the Bragg equation (Equation 26.12) for d and set $m = 1$:

$$d = \frac{m\lambda}{2\sin\theta} = \frac{(1)(0.154 \text{ nm})}{2\sin 34.5°} = 0.136 \text{ nm}.$$

This is the distance between adjacent planes.

Part (b): To calculate other angles, we solve Equation 26.12 for $\sin\theta$:

$$\sin\theta = \frac{m\lambda}{2d} = m\frac{0.154 \text{ nm}}{2(0.136 \text{ nm})} = m(0.566).$$

REFLECT Values of m of 2 or greater give values of $\sin\theta$ greater than unity, which is impossible. Therefore, there are no other angles for interference maxima for this particular set of crystal planes.

26.8 Circular Apertures and Resolving Power

We've studied in detail the diffraction patterns formed by long, thin slits or arrays of such slits. But an aperture of *any* shape forms a diffraction pattern. The diffraction pattern formed by a *circular* aperture is of special interest because of its role in limiting the resolving power of optical instruments. In principle, we could compute the intensity at any point P in the diffraction pattern by dividing the area of the aperture into small elements, finding the resulting wave amplitude and phase at P, and then summing all these elements to find the resultant amplitude and intensity at P. In practice, this calculation requires the use of integral calculus with numerical approximations, and we won't pursue it further here. We'll simply *describe* the pattern and quote a few relevant numbers.

The diffraction pattern formed by a circular aperture consists of a central bright spot surrounded by a series of bright and dark rings, as shown in Figure 26.35. We can describe the pattern in terms of the angle θ, representing the angular size of each ring. If the aperture diameter is D and the wavelength is λ, then the angular size θ_1 of the first *dark* ring is given by the following expression:

First dark ring from a circular aperture

$$\sin\theta_1 = 1.22\frac{\lambda}{D}. \qquad (26.13)$$

MasteringPHYSICS®

ActivPhysics 16.7: Circular Hole Diffraction
ActivPhysics 16.8: Resolving Power

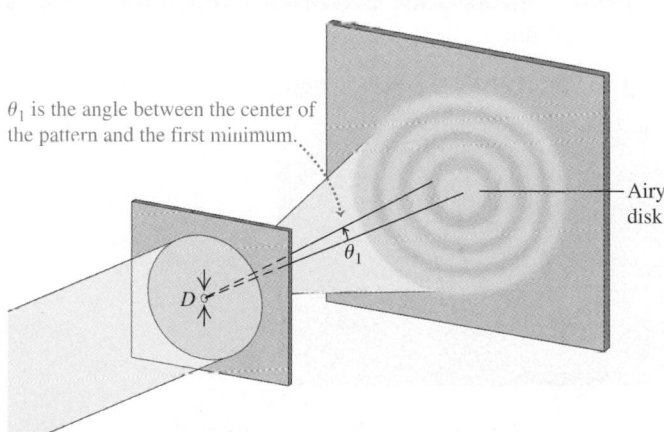

θ_1 is the angle between the center of the pattern and the first minimum.

Airy disk

D

▲ **FIGURE 26.35** Diffraction pattern formed by a circular aperture of diameter D. The pattern consists of a central bright spot and alternating bright and dark rings. (Not to scale.)

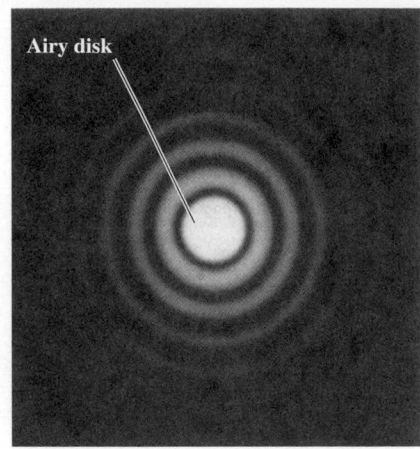

▲ FIGURE 26.36 Diffraction pattern formed by a circular aperture. The Airy disk is over-exposed so that the other rings may be seen.

The angular sizes $(\theta_2 \text{ and } \theta_3)$ of the next two dark rings are given, respectively, by

$$\sin\theta_2 = 2.23\frac{\lambda}{D}, \qquad \sin\theta_3 = 3.24\frac{\lambda}{D}.$$

The central bright spot is called the **Airy disk,** in honor of Sir George Airy (1801–1892), Astronomer Royal of England, who first derived the expression for the intensity in the pattern. The angular size of the Airy disk is that of the first dark ring, given by Equation 26.13.

The *intensities* in the bright rings drop off very quickly. When D is much larger than the wavelength λ, which is the usual case for optical instruments, the peak intensity in the center of the first bright ring is only 1.7% of the value at the center of the Airy disk, and at the center of the second bright ring it is only 0.4%. Most (85%) of the total energy falls within the Airy disk. Figure 26.36 shows a diffraction pattern from a circular aperture 1.0 mm in diameter.

This analysis has far-reaching implications for image formation by lenses and mirrors. In our study of optical instruments in Chapter 25, we assumed that a lens with focal length f focuses a parallel beam (plane wave) to a *point* at a distance f from the lens. This assumption ignored diffraction effects. We now see that what we get is not a point, but the diffraction pattern just described. If we have two point objects, their images are not two points, but two diffraction patterns. When the objects are close together, their diffraction patterns overlap; if they are close enough, their patterns overlap almost completely and cannot be distinguished. The effect is shown in Figure 26.37, which shows the patterns for four (very small) "point" objects. In Figure 26.37a, the images of objects 1 and 2 are well separated, but the images of objects 3 and 4 on the right have merged together. In

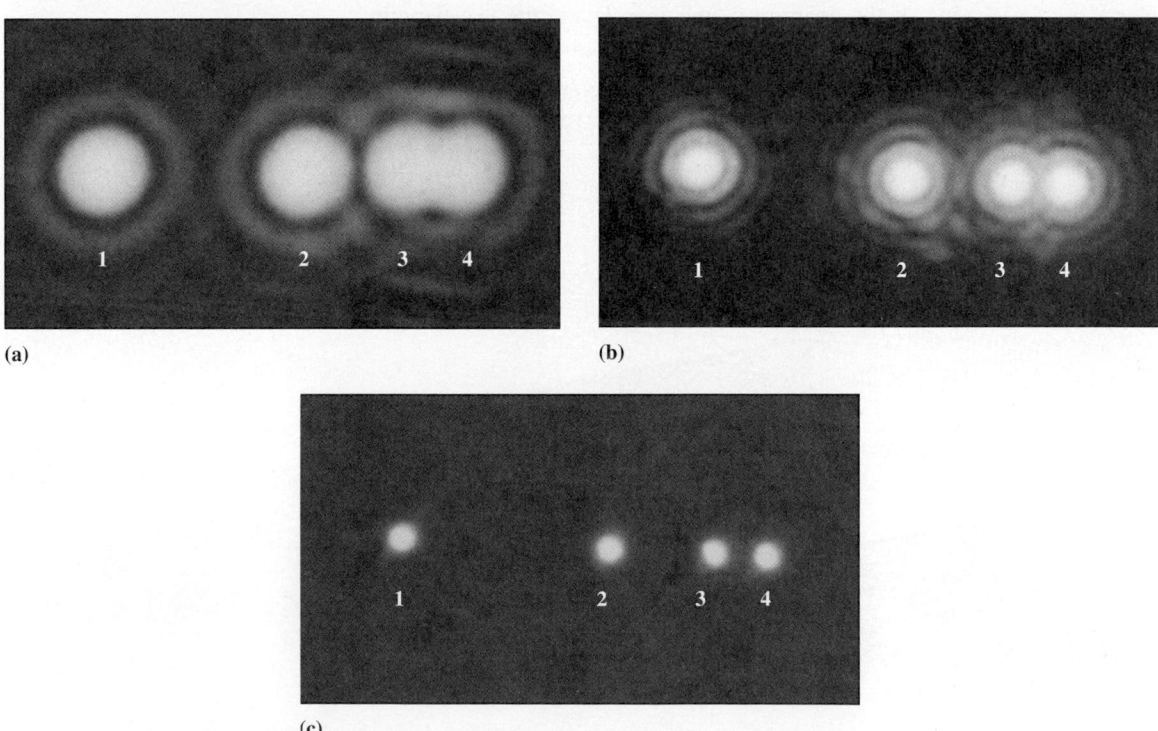

(a)

(b)

(c)

▲ FIGURE 26.37 Diffraction pattern of four "point" sources as seen through a circular opening whose size increases from (a) to (c). In (a), the aperture is so small that the patterns of sources 3 and 4 overlap and are barely resolved by Rayleigh's criterion (discussed later in text). Increasing the aperture size decreases the size of the diffraction patterns, as shown in (b) and (c).

Figure 26.37b, with a larger aperture diameter and resulting smaller Airy disks, images 3 and 4 are better resolved. In Figure 26.37c, with a still larger aperture, they are well resolved.

A widely used criterion for the resolution of two point objects, proposed by Lord Rayleigh (1887–1905) and called **Rayleigh's criterion,** is that the objects are just barely resolved (that is, distinguishable) if the center of one diffraction pattern coincides with the first minimum of the other. In that case, the angular separation of the image centers is given by Equation 26.13. The angular separation of the *objects* is the same as that of the *images,* so two point objects are barely resolved, according to Rayleigh's criterion, when their angular separation is given by Equation 26.13. The angles are usually very small, so $\sin\theta \simeq \theta$, and we have

Rayleigh's criterion

The minimum angular separation of two objects that can barely be resolved by an optical instrument is called the **limit of resolution,** θ_{res}, of the instrument:

$$\theta_{res} = 1.22\frac{\lambda}{D}. \qquad (26.14)$$

The smaller the limit of resolution, the greater the *resolution,* or **resolving power,** of the instrument. Diffraction sets the ultimate limits on the resolution of lenses and mirrors. *Geometric* optics may make it seem that we can make images as large as we like. Eventually, though, we always reach a point at which the image becomes larger, but does not gain in detail. The images in Figure 26.37 would not become sharper with further enlargement.

EXAMPLE 26.10 **Resolving power of the human eye**

The iris of the eye is a circular aperture that allows light to pass into the eye. **(a)** For an iris with radius 0.25 cm, and for visible light with a wavelength of 550 nm, what is the resolving angle, or limiting resolution, of the eye, based on Rayleigh's criterion? **(b)** In fact, the actual limiting resolution of the human eye is about four times poorer: $\theta_{res} = 4\theta_{Rayleigh}$. What is the farthest distance s from a tree that you could stand and resolve two birds sitting on a limb, separated transversely by a distance $y = 10$ cm?

SOLUTION

SET UP AND SOLVE Part (a): Figure 26.38 shows the situation. We use Equation 26.14, with the diameter $D = 2 \times$ radius $= 2(0.25\text{ cm}) = 0.50\text{ cm} = 5.0 \times 10^{-3}$ m:

$$\theta_{Rayleigh} = 1.22\frac{\lambda}{D} = (1.22)\frac{(550 \times 10^{-9}\text{ m})}{(5.0 \times 10^{-3}\text{ m})}$$
$$= 1.34 \times 10^{-4}\text{ rad.}$$

Part (b): Using $\theta_{res} = 4\theta_{Rayleigh}$, we find that the actual limiting resolution of the human eye is

$$\theta_{res} = 4\theta_{Rayleigh} = 4(1.34 \times 10^{-4}\text{ rad}) = 5.4 \times 10^{-4}\text{ rad.}$$

To find the farthest distance s from the tree that you could stand and still resolve two birds separated by 10 cm, we use the small-angle approximation

$$\sin\theta \simeq \tan\theta \simeq \theta \simeq \frac{y}{s}, \text{ and}$$
$$s = \frac{y}{\theta} = \frac{10\text{ cm}}{5.4 \times 10^{-4}} = 19{,}000\text{ cm} = 190\text{ m.}$$

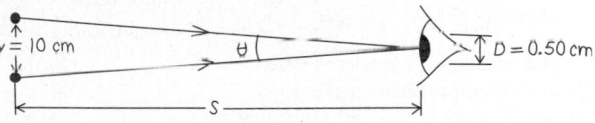

▲ **FIGURE 26.38**

REFLECT This result is a rather optimistic estimate of the resolving power of the human eye. Many other factors, including the illumination level and small defects of vision, also act to limit the actual resolution.

Practice Problem: For a person whose vision has angular resolution eight times that given by the Rayleigh criterion, what is the maximum distance at which the person can distinguish two birds sitting 10 cm apart on a tree limb? *Answer:* 93 m.

Radio interferometry. The Very Large Array in Soccoro, New Mexico, consists of 27 radio dishes that can be moved on tracks; at their greatest separation, their resolution equals that of a single dish 36 km across.

Optical interferometry. The four 8.2 m telescopes of the European Southern Observatory's Very Large Telescope in Cerro Paranal, Chile, can be combined optically in pairs. Functioning together, the outer two telescopes have the resolution of a single much larger telescope.

▲ **FIGURE 26.39** Some modern telescopes use *interferometry* to combine the waves from several source telescopes, producing an image that has the resolving power of a telescope with the diameter of the whole array.

An important lesson to be learned from this analysis is that resolution improves with shorter wavelengths. Thus, ultraviolet microscopes have higher resolution than visible-light microscopes. In electron microscopes, the resolution is limited by the wavelengths associated with the wavelike behavior of electrons. These wavelengths can be made 100,000 times smaller than wavelengths of visible light, with a corresponding gain in resolution. Finally, one reason for building very large reflecting telescopes is to increase the aperture diameter and thus minimize diffraction effects. Such telescopes also provide a greater light-gathering area for viewing very faint stars.

The Hubble Space Telescope, launched April 25, 1990, from the space shuttle *Discovery,* has a mirror diameter of 2.4 m. The telescope was designed to resolve objects 2.8×10^{-7} rad apart with 550 nm light. This was at least a factor of six better than the much larger earth-based telescopes of the time, whose resolving power was limited primarily by the distorting effects of atmospheric turbulence. Many of the current generation of earth-based telescopes use *adaptive optics* to counteract atmospheric distortion. Adaptive optics is a technique in which computer-controlled actuators distort the telescope mirror in real time to compensate for the effects of the atmosphere, allowing such telescope to achieve resolutions closer to their native diffraction limits.

Diffraction is an important consideration for satellite "dishes"—parabolic reflectors designed to receive satellite transmission. Satellite dishes have to be able to pick up transmissions from two satellites that are only a few degrees apart and are transmitting at the same frequency. The need to resolve two such transmissions determines the minimum diameter of the dish. As higher frequencies are used, the diameter needed decreases. For example, when two satellites 5.0° apart broadcast 7.5 cm microwaves, the minimum dish diameter required to resolve them (by Rayleigh's criterion) is about 1.0 m.

The effective diameter of a telescope can be increased in some cases by using arrays of smaller telescopes. The Very Long Baseline Array (VLBA), a group of 10 radio telescopes scattered at locations from Mauna Kea to the Virgin Islands, has a maximum separation of about 8000 km and can resolve radio signals to 10^{-8} rad. This astonishing resolution is comparable, in the optical realm, to seeing a parked car on the moon. The same technique is harder to apply to optical telescopes because of the shorter wavelengths involved. Nevertheless, large optical telescopes are now achieving this goal (Figure 26.39). The use of widely spaced satellite arrays will increase resolution even more in the future.

26.9 Holography

Holography is a technique for recording and reproducing an image of an object without the use of lenses. Unlike the two-dimensional images recorded by an ordinary photograph or television system, a holographic image is truly three dimensional. Such an image can be viewed from different directions to reveal different sides and from various distances to reveal changing perspective. If you had never seen a hologram, you wouldn't believe it was possible!

The basic procedure for making a hologram is shown in Figure 26.40a. We illuminate the object to be holographed with monochromatic light, and we place a photographic film so that it is struck by scattered light from the object (the object beam) and also by direct light from the source (the reference beam). In practice, the source must be a laser, for reasons that we'll discuss later. Interference between the direct and scattered light leads to the formation and recording of a complex interference pattern on the film.

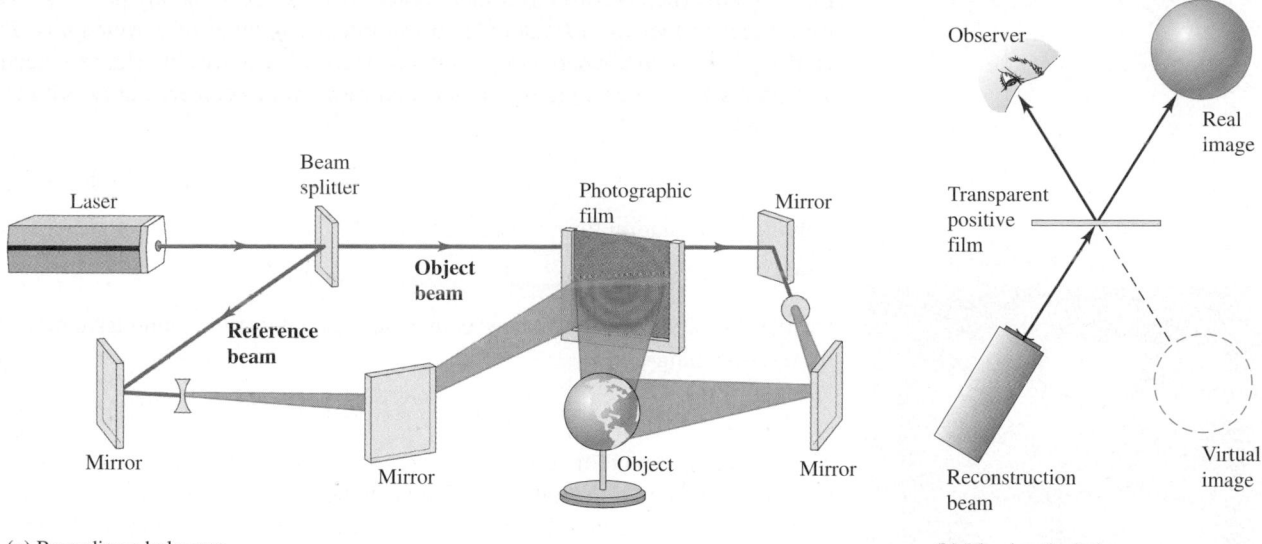

(a) Recording a hologram

(b) Viewing the hologram

▲ **FIGURE 26.40** (a) A hologram is the record on film of the interference pattern formed with light from a coherent source and light scattered from an object. (b) Images are formed when light is projected through the hologram.

To form the images, we simply project light (the reconstruction beam) through the developed film, as shown in Figure 26.40b. Two images are formed: a virtual image on the side of the film nearer the source and a real image on the opposite side.

A complete analysis of holography is beyond our scope, but we can gain some insight into the process by looking at how a single point is imaged to form a hologram. Consider the interference pattern formed on a photographic film by the superposition of an incident plane wave and a spherical wave, as shown in Figure 26.41a. The spherical wave originates at a point source P at a distance b_0 from the film. P may in fact be a small object that scatters part of the incident plane wave. We assume that the two waves are monochromatic and coherent and that the phase relation is such that constructive interference occurs at point O on

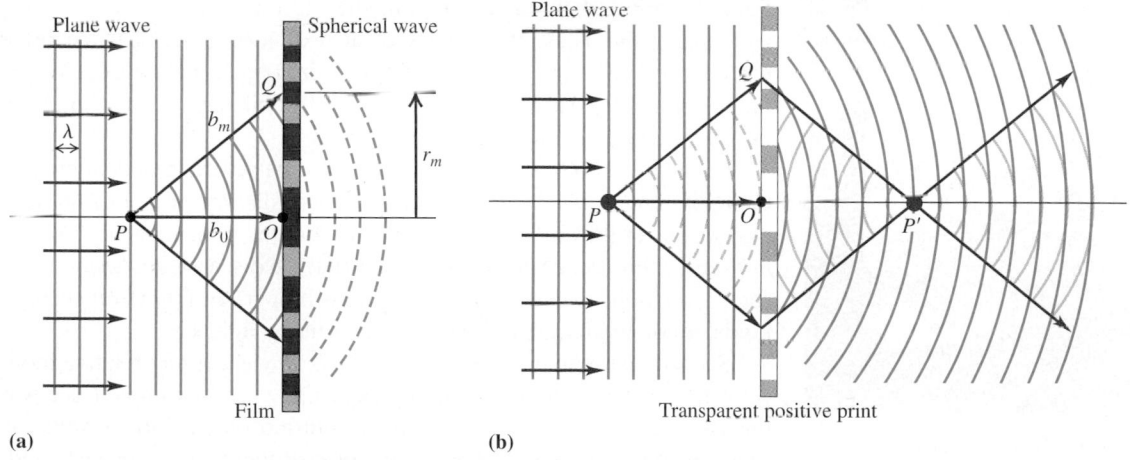

(a)   **(b)**

▲ **FIGURE 26.41** (a) Constructive interference of plane and spherical waves occurs in the plane of the film at every point Q for which the distance b_m from P is greater than the distance b_0 from P to O by an integral number of wavelengths $m\lambda$. For the point shown, $m = 2$. (b) When a plane wave strikes the developed film, the diffracted wave consists of a wave converging to P' and then diverging again and a diverging wave that appears to originate at P. These waves form real and virtual images, respectively.

the diagram. Then constructive interference will *also* occur at any point Q on the film that is farther from P than O is by an integral number of wavelengths. That is, if $b_m - b_0 = m\lambda$, where m is an integer, then constructive interference occurs. The points where this condition is satisfied form circles centered at O, with radii r_m given by

$$b_m - b_0 = \sqrt{b_0^2 + r_m^2} - b_0 = m\lambda \qquad (m = 1, 2, 3, \cdots). \quad (26.15)$$

Solving this equation for r_m^2, we find that

$$r_m^2 = \lambda(2mb_0 + m^2\lambda).$$

Ordinarily, b_0 is very much larger than λ, so we neglect the second term in parentheses, obtaining

$$r_m = \sqrt{2m\lambda b_0} \qquad (m = 1, 2, 3, \cdots). \quad (26.16)$$

The interference pattern consists of a series of concentric bright circular fringes with radii given by Equation 26.16. Between these bright fringes are dark fringes.

Now we develop the film and make a transparent positive print, so the bright-fringe areas have the greatest transparency on the film. Then we illuminate the print with monochromatic plane-wave light of the same wavelength λ that we used initially. In Figure 26.41b, consider a point P' at a distance b_0 along the axis from the film. The centers of successive bright fringes differ in their distances from P' by an integral number of wavelengths; therefore, a strong *maximum* in the diffracted wave occurs at P'. That is, light converges to P' and then diverges from it on the opposite side. Hence, P' is a *real image* of point P.

This is not the entire diffracted wave, however. The interference of the wavelets that spread out from all the transparent areas form a second spherical wave that is diverging rather than converging. When traced back behind the film, this wave appears to be spreading out from point P. Thus, the total diffracted wave from the hologram is a superposition of a spherical wave converging to form a real image at P' and a spherical wave that diverges as though it had come from the virtual image point P.

Because of the principle of linear superposition, what is true for the imaging of a single point is also true for the imaging of any number of points. The film records the superposed interference pattern from the various points, and when light is projected through the film, the various image points are reproduced simultaneously. Thus, the images of an extended object can be recorded and reproduced just as they can for a single point object. Figure 26.42 shows photographs of a holographic image from two different angles, revealing the changing perspective in this three-dimensional image.

In making a hologram, we have to overcome several practical problems. First, the light used must be *coherent* over distances that are large in comparison to the dimensions of the object and its distance from the film. Ordinary light sources *do not* satisfy this requirement, for reasons that we discussed in Section 26.1. Therefore, laser light is essential for making a hologram. (However, many common kinds of holograms can be *viewed* with ordinary light.)

Second, extreme mechanical stability is needed. If any relative motion of the source, object, or film occurs during exposure, even by as much as a wavelength, the interference pattern on the film is blurred enough to prevent satisfactory image formation. These obstacles are not insurmountable, however, and holography promises to become increasingly important in research, entertainment, and a wide variety of technological applications.

▲ **FIGURE 26.42** Two views of a hologram, showing how the perspective of the image changes with the angle from which it is viewed.

SUMMARY

Interference and Coherent Sources

(Section 26.1) The overlap of waves from two sources of monochromatic light forms an **interference pattern.** The principle of linear superposition states that the total wave disturbance at any point is the sum of the disturbances from the separate waves. **Constructive interference** results when two waves arrive at a point in phase; **destructive interference** results when two waves arrive at a point exactly a half cycle out of phase.

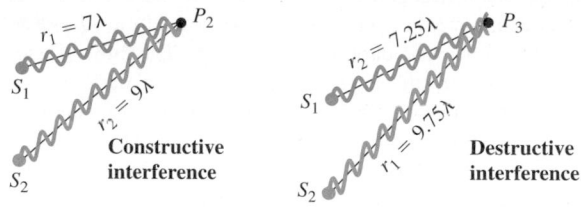

Two-Source Interference of Light

(Section 26.2) When two light sources are in phase, constructive interference occurs at a point when the difference in path length from the two sources is zero or an integral number of wavelengths; destructive interference occurs when the path difference is a half-integral number of wavelengths. When the lines from the sources to a distant point P make an angle θ with the horizontal line in the figure, the condition for constructive interference is $d\sin\theta = m\lambda$ (Equation 26.4) and the condition for destructive interference is $d\sin\theta = \left(m + \frac{1}{2}\right)\lambda$ (Equation 26.5), where $m = 0$, ± 1, ± 2, $\cdots$ for both cases.

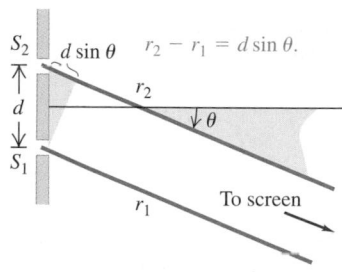

Interference in Thin Films

(Section 26.3) When light is reflected from both sides of a thin film with thickness t, constructive interference between the reflected waves occurs when $2t = m\lambda$ $(m = 0, 1, 2, \cdots)$ (Equation 26.7), unless a half-cycle phase shift occurs at only one surface; then this is the condition for destructive interference. A half-cycle phase shift occurs during reflection whenever the index of refraction of the second material is greater than that of the first.

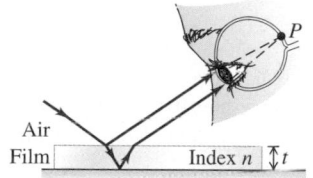

Diffraction; Single and Multiple Slits

(Sections 26.4–26.6) Diffraction occurs when light passes through an aperture or around an edge. For a single narrow slit with width a, the condition for destructive interference (a dark fringe) at a point P at an angle θ from the direction of the incident light is $\sin\theta = m\lambda/a$, where $m = \pm 1, \pm 2, \pm 3, \cdots$ (Equation 26.9).

A **diffraction grating** consists of a large number of thin parallel slits spaced a distance d apart. The condition for maximum intensity in the interference pattern is $d\sin\theta = m\lambda$, where $m = 0$, ± 1, ± 2, ± 3, $\cdots$ (Equation 26.11). This is the same condition as for the two-source pattern, but for the grating, the maxima are very sharp and narrow.

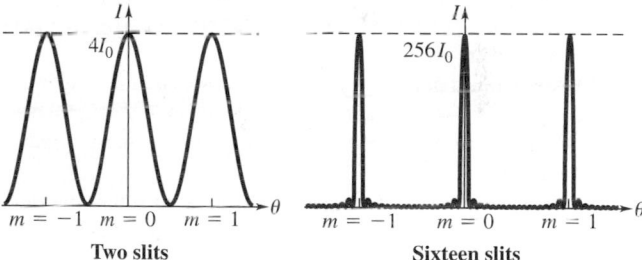

X-Ray Diffraction

(Section 26.7) A crystal having its atoms arranged in a lattice in a regularly repeating pattern can serve as a three-dimensional diffraction grating for x rays with wavelengths of the same order of magnitude as the lattice spacing, typically 10^{-10} m. The scattered x rays emerging from the crystal form an interference pattern.

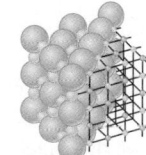

Continued

Circular Apertures and Resolving Power

(Section 26.8) The diffraction pattern from a circular aperture with diameter D consists of a central bright spot, called the **Airy disk,** and a series of concentric dark and bright rings. The limit of resolution, determined by the angular size of the first dark ring, is $\theta_{res} = 1.22\lambda/D$ (Equation 26.14). Diffraction sets the ultimate limit on the resolution (sharpness of the image) of optical instruments. According to Rayleigh's criterion, two point objects are just barely resolved when their angular separation θ is given by Equation 26.14.

Holography

(Section 26.9) A hologram is a photographic record of an interference pattern formed by light scattered from an object and light coming directly from the source. When properly illuminated, a hologram forms a true three-dimensional image of the object.

 For instructor-assigned homework, go to www.masteringphysics.com

Conceptual Questions

1. Could an experiment similar to Young's two-slit experiment be performed with sound? How might it be carried out? Does it matter that sound waves are longitudinal and electromagnetic waves are transverse?

2. One refracting astronomical telescope has a tube twice as wide as another one because the objective lens is twice the diameter. Are there any advantages to the wide telescope over the narrow one? What are they?

3. A two-slit interference experiment is set up, and the fringes are displayed on a screen. Then the whole apparatus is immersed in the nearest swimming pool. How does the fringe pattern change?

4. Would the headlights of a distant car form a two-source interference pattern? If so, how might it be observed? If not, why not?

5. Coherent red light illuminates two narrow slits that are 25 cm apart. Will a two-slit interference pattern be observed when the light from the slits falls on a screen? Explain.

6. If a two-slit interference experiment were done with white light, what would be seen?

7. Could x ray diffraction effects with crystals be observed by using visible light instead of x rays? Why or why not?

8. Does a microscope have better resolution with red light or blue light? Why?

9. Around harbors, where oil from boat engines is on the water, you often see patterns of closed colored fringes, like the ones in Figure 26.8b. Why do the patterns make *closed* fringes? Why are the fringes different colors?

10. What happens to the width of the central bright *region* (not just the central point) of a single-slit diffraction pattern as you make the slit thinner and thinner?

11. What are some advantages to making a telescope with a large-diameter objective lens (or mirror)? Does the large diameter contribute to the magnification of the telescope?

12. A *very* thin soap film ($n = 1.33$), whose thickness is much less than a wavelength of visible light, looks black; it appears to reflect no light at all. Why? By contrast, an equally thin layer of soapy water ($n = 1.33$) on glass ($n = 1.50$) appears quite shiny. Why is there a difference?

13. When monochromatic light passes through two thin slits, what characteristics of the light and of the slits limit the number of bright spots (interference maxima) that will occur, or are there an infinite number of them? Assume that you can detect all the bright spots, even if they are very dim.

14. If we view a double-slit interference pattern on a fairly distant screen, are the bright spots necessarily equally spaced on the screen? Consider the cases of small and large angles from the central spot.

15. Optical telescopes having a principal mirror only a few meters in diameter can produce extremely sharp images, yet radiotelescopes need to be hundreds (or even thousands) of meters in diameter to make sharp images. Why do they need to be so much larger than optical telescopes?

Multiple-Choice Problems

1. Two sources of waves are at A and B in Figure 26.43. At point P, the path difference for waves from these two sources is
   A. $x + y$.
   B. $\dfrac{x + y}{2}$.
   C. $x - y$.
   D. $\dfrac{x - y}{2}$.

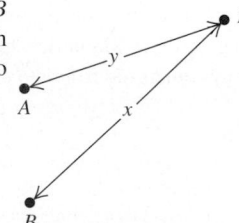

▲ **FIGURE 26.43** Multiple-choice problem 1.

2. If the sources A and B in the previous problem are emitting waves of wavelength λ that are in

phase with each other, *constructive* interference will occur at point *P* if (there may be more than one correct choice):

A. $x = y$.
B. $x + y = \lambda$.
C. $x - y = 2\lambda$.
D. $x - y = 5\lambda$.

3. To obtain the greatest resolution from a microscope,
   A. You should view the object in long-wavelength visible light.
   B. You should view the object in short-wavelength visible light.
   C. You can use any wavelength of visible light, since the resolution is determined only by the diameter of the microscope lenses.

4. A monochromatic beam of laser light falls on a thin slit and produces a series of bright and dark spots on a screen. If this apparatus (slit and screen) is submerged in water, the dark spots
   A. will not change their location on the screen.
   B. will move away from the center spot.
   C. will move toward the center spot.

5. A person is standing at a distance *x* from a stereo speaker that is emitting a continuous tone. She hears the sound directly from the speaker, as well as the sound reflected from a wall a distance *y* ($y > x$) away. (See Figure 26.44.) The path difference between these two sound waves as they reach the listener is
   A. 2*y*.   B. 2*x*.   C. *y* − *x*.
   D. 2*y* − *x*.   E. *x* + *y*.

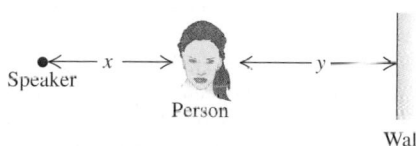

▲ FIGURE 26.44 Multiple-choice problem 5.

6. When a thin oil film spreads out on a puddle of water, the thinnest part of the film looks dark in the resulting interference pattern. This implies that
   A. the oil has a higher index of refraction than water.
   B. water has a higher index of refraction than the oil.
   C. the oil and water have identical indexes of refraction.

7. A laser beam of wavelength 500 nm is shone through two different diffraction gratings *A* and *B*, and the pattern is viewed on a distant screen. The pattern from grating *A* consists of many closely spaced bright dots, while that from *B* contains few dots spaced widely apart. What can you conclude about the line densities N_A and N_B (the number of lines/cm) of these two gratings?
   A. $N_A > N_B$.   B. $N_A < N_B$.
   C. You cannot conclude anything about the line densities without knowing the order of the bright spots.

8. A film contains a single thin slit of width *a*. When monochromatic light passes through this slit, the first two dark fringes on either side of the center on a distant screen are a distance *x* apart. If you increase the width of the slit, these two dark fringes will
   A. move closer together.
   B. move farther apart.
   C. remain the same distance apart.

9. Light of wavelength λ strikes a pane of glass of thickness *T* and refractive index *n*, as shown in Figure 26.45. Part of the beam is reflected off the upper surface of the glass, and part is transmitted and then reflected off the lower surface of the glass. *Destructive* interference between these two beams will occur if

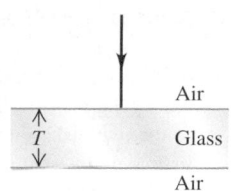

▲ FIGURE 26.45 Multiple-choice problem 9.

A. $T = \dfrac{\lambda}{2}$.   B. $2T = \dfrac{\lambda}{2}$.

C. $T = \dfrac{\lambda}{2n}$.   D. $2T = \dfrac{\lambda}{2n}$.

10. Two thin parallel slits are a distance *d* apart. Monochromatic light passing through them produces a series of interference bright spots on a distant screen. If you *decrease* the distance between these slits, the bright spots will
    A. move closer to the center spot.
    B. move farther from the center spot.
    C. not change position.

11. Laser light of wavelength λ passes through a thin slit of thickness *a* and produces its first dark fringes at angles of ±45° with the original direction of the beam. The slit is then reduced in size to a *circle* of diameter *a*. When the same laser light is passed through the circle, the first dark fringe occurs at
    A. ±59.6°.   B. ±54.9°.   C. ±36.9°.   D. ±35.4°.

12. The formula $y_m = R\dfrac{m\lambda}{d}$ for the location of the points of constructive interference from two slits is valid
    A. only for large angles θ.
    B. only for small angles θ.
    C. for all angles θ, because it is a general formula.

13. A light beam strikes a pane of glass as shown in Figure 26.46. Part of it is reflected off the air–glass surface, and part is transmitted and then reflected off the glass–water surface. Which statements are true about this light? (There may be more than one correct choice.)

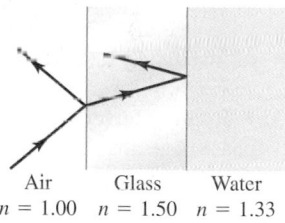

▲ FIGURE 26.46 Multiple-choice problem 13.

   A. Half-cycle phase shifts occur at both the air–glass and glass–water surfaces.
   B. Half-cycle phase shifts occur at neither of the surfaces.
   C. A half-cycle phase shift occurs at the glass–water surface, but not at the air–glass surface.
   D. A half-cycle phase shift occurs at the air–glass surface, but not at the glass–water surface.
   E. The phase shifts at both surfaces cancel each other out.

14. Light of wavelength λ and frequency *f* passes through a single slit of width *a*. The diffraction pattern is observed on a screen a distance *x* from the slit. Which of the following will decrease the width of the central maximum? (There may be more than one correct answer.)
   A. Decrease the slit width.
   B. Decrease the frequency of the light.
   C. Decrease the wavelength of the light.
   D. Decrease the distance *x* of the screen from the slit.

15. Both CDs and DVDs will flash a rainbow spectrum when viewed from certain angles. The "track pitch," or distance between rows of pits, is 1600 nm on a CD but only 740 nm on a DVD. This is part of the reason why a DVD can store much more data. Which of the two disks should produce a wider separation between red light and violet light in the reflected spectrum?
    A. The CD.   B. The DVD.
    C. The same for both.

Problems

26.1 Interference and Coherent Sources

1. • Two small stereo speakers A and B that are 1.40 m apart are sending out sound of wavelength 34 cm in all directions and all in phase. A person at point P starts out equidistant from both speakers and walks (see Figure 26.47) so that he is always 1.50 m from speaker B. For what values of x will the sound this person hears be (a) maximally reinforced, (b) cancelled? Limit your solution to the cases where $x \le 1.50$ m.

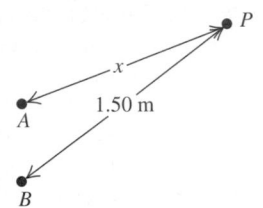
▲ **FIGURE 26.47** Problem 1.

2. • A person with a radio-wave receiver starts out equidistant from two FM radio transmitters A and B that are 11.0 m apart, each one emitting in-phase radio waves at 92.0 MHz. She then walks so that she always remains 50.0 m from transmitter B. (See Figure 26.48.) For what values of x will she find the radio signal to be (a) maximally enhanced, (b) cancelled? Limit your solution to the cases where $x \ge 50.0$ m.

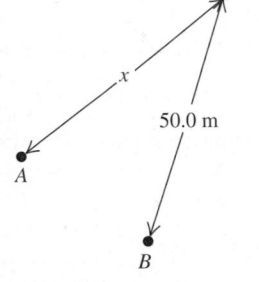
▲ **FIGURE 26.48** Problem 2.

3. • **Radio interference.** Two radio antennas A and B radiate in phase. Antenna B is 120 m to the right of antenna A. Consider point Q along the extension of the line connecting the antennas, a horizontal distance of 40 m to the right of antenna B. The frequency, and hence the wavelength, of the emitted waves can be varied. (a) What is the longest wavelength for which there will be destructive interference at point Q? (b) What is the longest wavelength for which there will be constructive interference at point Q?

4. •• Two speakers that are 15.0 m apart produce in-phase sound waves of frequency 250.0 Hz in a room where the speed of sound is 340.0 m/s. A woman starts out at the midpoint between the two speakers. The room's walls and ceiling are covered with absorbers to eliminate reflections, and she listens with only one ear for best precision. (a) What does she hear, constructive or destructive interference? Why? (b) She now walks slowly toward one of the speakers. How far from the center must she walk before she first hears the sound reach a minimum intensity? (c) How far from the center must she walk before she first hears the sound maximally enhanced?

5. •• Suppose that the situation is the same as in the previous problem, except that the two speakers are 180° out of phase. Repeat parts (a), (b), and (c) of that problem.

26.2 Two-Source Interference of Light

6. • Coherent light of wavelength 525 nm passes through two thin slits that are 0.0415 mm apart and then falls on a screen 75.0 cm away. How far away from the central bright fringe on the screen is (a) the fifth bright fringe (not counting the central bright fringe); (b) the eighth dark fringe?

7. • Coherent light from a sodium-vapor lamp is passed through a filter that blocks everything except for light of a single wavelength. It then falls on two slits separated by 0.460 mm. In the resulting interference pattern on a screen 2.20 m away, adjacent bright fringes are separated by 2.82 mm. What is the wavelength of the light that falls on the slits?

8. • Young's experiment is performed with light of wavelength 502 nm from excited helium atoms. Fringes are measured carefully on a screen 1.20 m away from the double slit, and the center of the 20th fringe (not counting the central bright fringe) is found to be 10.6 mm from the center of the central bright fringe. What is the separation of the two slits?

9. • Coherent light of frequency 6.32×10^{14} Hz passes through two thin slits and falls on a screen 85.0 cm away. You observe that the third bright fringe occurs at ± 3.11 cm on either side of the central bright fringe. (a) How far apart are the two slits? (b) At what distance from the central bright fringe will the third dark fringe occur?

10. •• Coherent light with wavelength 600 nm passes through two very narrow slits and the interference pattern is observed on a screen 3.00 m from the slits. The first-order bright fringe is at 4.84 mm from the center of the central bright fringe. For what wavelength of light will the first-order dark fringe be observed at this same point on the screen?

11. • Two slits spaced 0.450 mm apart are placed 75.0 cm from a screen. What is the distance between the second and third dark lines of the interference pattern on the screen when the slits are illuminated with coherent light with a wavelength of 500 nm?

12. •• Coherent light that contains two wavelengths, 660 nm (red) and 470 nm (blue), passes through two narrow slits separated by 0.300 mm, and the interference pattern is observed on a screen 5.00 m from the slits. What is the distance on the screen between the first-order bright fringes for the two wavelengths?

13. •• Two thin parallel slits that are 0.0116 mm apart are illuminated by a laser beam of wavelength 585 nm. (a) On a very large distant screen, what is the *total* number of bright fringes (those indicating complete constructive interference), including the central fringe and those on both sides of it? Solve this problem without calculating all the angles! (*Hint:* What is the largest that $\sin\theta$ can be? What does this tell you is the largest value of m?) (b) At what angle, relative to the original direction of the beam, will the fringe that is most distant from the central bright fringe occur?

14. •• Two small loudspeakers that are 5.50 m apart are emitting sound in phase. From both of them, you hear a singer singing C# (frequency 277 Hz), while the speed of sound in the room is 340 m/s. Assuming that you are rather far from these speakers, if you start out at point P equidistant from both of them and walk around the room in front of them, at what angles (measured relative to the line from P to the midpoint between the speakers) will you hear the sound (a) maximally enhanced, (b) cancelled? Neglect any reflections from the walls.

26.3 Interference in Thin Films

15. • The walls of a soap bubble have about the same index of refraction as that of plain water, $n = 1.33$. There is air both inside and outside the bubble. (a) What wavelength (in air) of visible light is most strongly reflected from a point on a soap bubble where its wall is 290 nm thick? To what color does this correspond (see Figure 23.3)? (b) Repeat part (a) for a wall thickness of 340 nm.

16. • What is the thinnest soap film (excluding the case of zero thickness) that appears black when viewed by reflected light with a wavelength of 480 nm? The index of refraction of the film is 1.33, and there is air on both sides of the film.

17. • A thin film of polystyrene of refractive index 1.49 is used as a nonreflecting coating for Fabulite (strontium titanate) of refractive index 2.409. What is the minimum thickness of the film required? Assume that the wavelength of the light in air is 480 nm.

18. • **Conserving energy.** You want to coat the *inner* surfaces of your windows (which have refractive index of 1.51) with a film in order to enhance the reflection of light back into the room so that you can use bulbs of lower wattage than usual. You find that MgF_2, with $n = 1.38$, is not too expensive, so you decide to use it. Since incandescent home lightbulbs emit reddish light with a peak wavelength of approximately 650 nm, you decide that this wavelength is the one to enhance in the light reflected back into the room. (a) What is the minimum thickness of film that you will need? (b) If this layer seems too thin to be able to put on accurately, what other thicknesses would also work? Give only the three thinnest ones.

19. • **Nonglare glass.** When viewing a piece of art that is behind glass, one often is affected by the light that is reflected off the front of the glass (called *glare*), which can make it difficult to see the art clearly. One solution is to coat the outer surface of the glass with a thin film to cancel part of the glare. (a) If the glass has a refractive index of 1.62 and you use TiO_2, which has an index of refraction of 2.62, as the coating, what is the minimum film thickness that will cancel light of wavelength 505 nm? (b) If this coating is too thin to stand up to wear, what other thicknesses would also work? Find only the three thinnest ones.

20. • The lenses of a particular set of binoculars have a coating with index of refraction $n = 1.38$, and the glass itself has $n = 1.52$. If the lenses reflect a wavelength of 525 nm the most strongly, what is the minimum thickness of the coating?

21. • A plate of glass 9.00 cm long is placed in contact with a second plate and is held at a small angle with it by a metal strip 0.0800 mm thick placed under one end. The space between the plates is filled with air. The glass is illuminated from above with light having a wavelength in air of 656 nm. How many interference fringes are observed per centimeter in the reflected light?

22. • Two rectangular pieces of plane glass are laid one upon the other on a table. A thin strip of paper is placed between them at one edge, so that a very thin wedge of air is formed. The plates are illuminated at normal incidence by 546 nm light from a mercury-vapor lamp. Interference fringes are formed, with 15.0 fringes per centimeter. Find the angle of the wedge.

23. •• A researcher measures the thickness of a layer of benzene $(n = 1.50)$ floating on water by shining monochromatic light onto the film and varying the wavelength of the light. She finds that light of wavelength 575 nm is reflected most strongly from the film. What does she calculate for the minimum thickness of the film?

24. •• **Compact disc player.** A compact disc (CD) is read from the bottom by a semiconductor laser beam with a wavelength of 790 nm that passes through a plastic substrate of refractive index 1.8. When the beam encounters a pit, part of the beam is reflected from the pit and part from the flat region between the pits, so these two beams interfere with each other. (See Figure 26.49.) What must the minimum pit depth be so that the part of the beam reflected from a pit cancels the part of the beam reflected from the flat region? (It is this cancellation that allows the player to recognize the beginning and end of a pit. For a fuller explanation of the physics behind CD technology, see the article "The Compact Disc Digital Audio System," by Thomas D. Rossing, in the December 1987 issue of *The Physics Teacher*.)

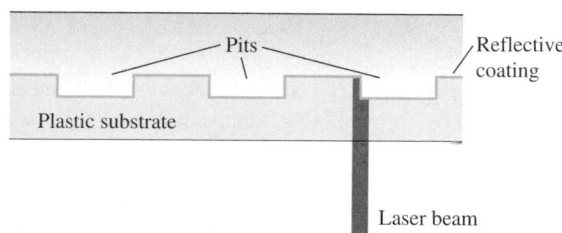

▲ **FIGURE 26.49** Problem 24.

26.5 Diffraction from a Single Slit

25. • A beam of laser light of wavelength 632.8 nm falls on a thin slit 0.00375 mm wide. After the light passes through the slit, at what angles relative to the original direction of the beam is it completely cancelled when viewed far from the slit?

26. • Parallel rays of green mercury light with a wavelength of 546 nm pass through a slit with a width of 0.437 mm. What is the distance from the central maximum to the first minimum on a screen 1.75 m away from the slit?

27. • Parallel light rays with a wavelength of 600 nm fall on a single slit. On a screen 3.00 m away, the distance between the first dark fringes on either side of the central maximum is 4.50 mm. What is the width of the slit?

28. • Monochromatic light from a distant source is incident on a slit 0.750 mm wide. On a screen 2.00 m away, the distance from the central maximum of the diffraction pattern to the first minimum is measured to be 1.35 mm. Calculate the wavelength of the light.

29. •• Red light of wavelength 633 nm from a helium–neon laser passes through a slit 0.350 mm wide. The diffraction pattern is observed on a screen 3.00 m away. Define the width of a bright fringe as the distance between the minima on either side. (a) What is the width of the central bright fringe? (b) What is the width of the first bright fringe on either side of the central one?

30. • Light of wavelength 633 nm from a distant source is incident on a slit 0.750 mm wide, and the resulting diffraction pattern is observed on a screen 3.50 m away. What is the distance between the two dark fringes on either side of the central bright fringe?

31. •• **Doorway diffraction.** Diffraction occurs for all types of waves, including sound waves. Suppose sound of frequency 1250 Hz leaves a room through a 1.00-m-wide doorway. At which angles relative to the centerline perpendicular to the doorway will someone outside the room hear no sound? Use

344 m/s for the speed of sound in air and assume that the source and listener are both far enough from the doorway for Fraunhofer diffraction to apply. You can ignore effects of reflections.

32. •• Light of wavelength 585 nm falls on a slit 0.0666 mm wide. (a) On a very large distant screen, how many *totally* dark fringes (indicating complete cancellation) will there be, including both sides of the central bright spot? Solve this problem *without* calculating all the angles! (*Hint:* What is the largest that $\sin\theta$ can be? What does this tell you is the largest that *m* can be?) (b) At what angle will the dark fringe that is most distant from the central bright fringe occur?

33. •• A glass sheet measuring 10.0 cm × 25.0 cm is covered by a very thin opaque coating. In the middle of this sheet is a thin, straight scratch 0.00125 mm thick, as shown in Figure 26.50. The

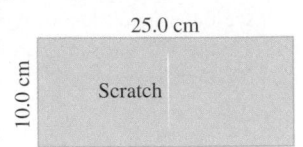

▲ **FIGURE 26.50** Problem 33.

sheet is totally immersed beneath the surface of a liquid having an index of refraction of 1.45. Monochromatic light strikes the sheet perpendicular to its surface and passes through the scratch. A screen is placed under water a distance 30.0 cm away from the sheet and parallel to it. You observe that the first dark fringes on either side of the central bright fringe on this screen are 22.4 cm apart. What is the wavelength of the light in air?

26.6 Multiple Slits and Diffraction Gratings

34. • A laser beam of unknown wavelength passes through a diffraction grating having 5510 lines/cm after striking it perpendicularly. Taking measurements, you find that the first pair of bright spots away from the central maximum occurs at ±15.4° with respect to the original direction of the beam. (a) What is the wavelength of the light? (b) At what angle will the next pair of bright spots occur?

35. • A laser beam of wavelength 600.0 nm is incident normally on a transmission grating having 400.0 lines/mm. Find the angles of deviation in the first, second, and third orders of bright spots.

36. • When laser light of wavelength 632.8 nm passes through a diffraction grating, the first bright spots occur at ±17.8° from the central maximum. (a) What is the line density (in lines/cm) of this grating? (b) How many additional bright spots are there beyond the first bright spots, and at what angles do they occur?

37. • A diffraction grating has 5580 lines/cm. When a beam of monochromatic light goes through it, the *second* pair of bright spots occurs at ±26.3 cm from the central spot on a screen 42.5 cm past the grating. (a) What is the wavelength of this light? (b) How far from the central spot does the next pair of bright spots occur on the screen?

38. • Monochromatic light is at normal incidence on a plane transmission grating. The first-order maximum in the interference pattern is at an angle of 8.94°. What is the angular position of the fourth-order maximum?

39. •• **CDs and DVDs as diffraction gratings.** A laser beam of wavelength $\lambda = 632.8$ nm shines at normal incidence on the reflective side of a compact disc. (a) The tracks of tiny pits in which information is coded onto the CD are 1.60 μm apart. For what angles of reflection (measured from the normal) will the intensity of light be maximum? (b) On a DVD, the tracks are only 0.740 μm apart. Repeat the calculation of part (a) for the DVD.

40. •• Light of wavelength 631 nm passes through a diffraction grating having 485 lines/mm. (a) What is the *total* number of bright spots (indicating complete constructive interference) that will occur on a large distant screen? Solve this problem *without* finding the angles. (*Hint:* What is the largest that $\sin\theta$ can be? What does this imply for the largest value of *m*?) (b) What is the angle of the bright spot farthest from the center?

41. •• If a diffraction grating produces a third-order bright spot for red light (of wavelength 700 nm) at 65.0° from the central maximum, at what angle will the second-order bright spot be for violet light (of wavelength 400 nm)?

26.7 X-Ray Diffraction

42. • X-rays of wavelength 0.0850 nm are scattered from the atoms of a crystal. The second-order maximum in the Bragg reflection occurs when the angle θ in Figure 26.32 is 21.5°. What is the spacing between adjacent atomic planes in the crystal?

43. • Monochromatic x rays are incident on a crystal for which the spacing of the atomic planes is 0.440 nm. The first-order maximum in the Bragg reflection occurs when the incident and reflected x rays make an angle of 39.4° with the crystal planes. What is the wavelength of the x rays?

44. • Electromagnetic waves of wavelength 0.173 nm fall on a crystal surface. As the angle from the plane is gradually increased, starting at 0°, you find that the first strong interference maximum occurs when the beam makes an angle of 22.4° with the surface of the crystal planes in the Bragg reflection. (a) What is the distance between the crystal planes? (b) At what other angles will interference maxima occur?

26.8 Circular Apertures and Resolving Power

45. • A converging lens 7.20 cm in diameter has a focal length of 300 mm. If the resolution is diffraction limited, how far away can an object be if points on it transversely 4.00 mm apart are to be resolved (according to Rayleigh's criterion) by means of light of wavelength 550 nm?

46. • A telescope is used to observe two distant point sources transversely 2.50 m apart with light of wavelength 600 nm. The objective of the telescope is covered with a slit of width 0.350 mm. What is the maximum distance in meters at which the two sources may be distinguished if the resolution is diffraction limited and Rayleigh's criterion is used?

47. • Two satellites at an altitude of 1200 km are separated by 28 km. If they broadcast 3.6-cm microwaves, what minimum receiving-dish diameter is needed to resolve (by Rayleigh's criterion) the two transmissions?

48. •• **Resolution of telescopes.** Due to blurring caused by atmospheric distortion, the best resolution that can be obtained by a normal, earth-based, visible-light telescope is about 0.3 arcsecond (there are 60 arcminutes in a degree and 60 arcseconds in an arcminute). (a) Using Rayleigh's criterion, calculate the diameter of an earth-based telescope that gives this resolution with 550-nm light. (b) Increasing the telescope diameter beyond the value found in part (a) will increase the light-gathering power of the telescope, allowing more distant and dimmer astronomical objects to be studied, but it will not improve the resolution. In what ways are the Keck telescopes (each of 10-m diameter) atop Mauna Kea in Hawaii superior to the Hale Telescope (5-m diameter) on Palomar Mountain in California? In what ways are they *not* superior? Explain.

49. ●● **Resolution of the eye, I.** Even if the lenses of our eyes
BIO functioned perfectly, our vision would still be limited due to
diffraction of light at the pupil. Using Rayleigh's criterion,
what is the smallest object a person can see clearly at his near
point of 25.0 cm with a pupil 2.00 mm in diameter and light of
wavelength 550 nm? (To get a reasonable estimate without
having to go through complicated calculations, we'll ignore
the effect of the fluid in the eye.) Based upon your answer,
does it seem that diffraction plays a significant role in limiting
our visual acuity?

50. ●● **Resolution of the eye, II.** The maximum resolution of the
BIO eye depends on the diameter of the opening of the pupil (a diffraction effect) and the size of the retinal cells, as illustrated in
problem 51 in Chapter 25. In that problem, we saw that the
size of the retinal cells (about 5.0 μm in diameter) limits the
size of an object at the near point (25 cm) of the eye to a height
of about 50 μm. (To get a reasonable estimate without having
to go through complicated calculations, we shall ignore the
effect of the fluid in the eye.) (a) Given that the diameter of the
human pupil is about 2.0 mm, does the Rayleigh criterion
allow us to resolve a 50-μm-tall object at 25 cm from the eye
with light of wavelength 550 nm? (b) According to the
Rayleigh criterion, what is the shortest object we could resolve
at the 25 cm near point with light of wavelength 550 nm?
(c) What angle would the object in part (b) subtend at the eye?
Express your answer in minutes ($60 \text{ min} = 1°$), and compare
it with the experimental value of about 1 min. (d) Which effect
is more important in limiting the resolution of our eyes, diffraction or the size of the retinal cells?

51. ●● **Spy satellites?** Rumor has it that the U.S. military has spy
satellites in orbit carrying telescopes that can resolve objects
on the ground as small as the width of a car's license plate. If
we assume that such satellites orbit about 400 km above the
ground (which is typical for orbiting telescopes) and that they
focus light of wavelength 500 nm, what would have to be the
minimum diameter of the mirror (or objective lens) of this
kind of a telescope in order to resolve such small objects?
(Measure the width of a car's license plate.)

General Problems

52. ●● Two identical audio speakers connected to the same
amplifier produce in-phase sound waves with a single frequency that can be varied between 300 and 600 Hz. The speed
of sound is 340 m/s. You find that where you are standing,
you hear minimum-intensity sound. (a) Explain why you hear
minimum-intensity sound. (b) If one of the speakers is moved
39.8 cm toward you, the sound you hear has maximum intensity. What is the frequency of the sound? (c) How much closer
to you from the position in part (b) must the speaker be moved
to the next position where you hear maximum intensity?

53. ●● Suppose you illuminate two thin slits by monochromatic
coherent light in air and find that they produce their first interference *minima* at $\pm 35.20°$ on either side of the central bright
spot. You then immerse these slits in a transparent liquid and
illuminate them with the same light. Now you find that the first
minima occur at $\pm 19.46°$ instead. What is the index of refraction of this liquid?

54. ● **Coating eyeglass lenses.** Eyeglass lenses can be coated on
BIO the *inner* surfaces to reduce the reflection of stray light to the
eye. If the lenses are medium flint glass of refractive index

1.62 and the coating is fluorite of refractive index 1.432,
(a) what minimum thickness of film is needed on the lenses to
cancel light of wavelength 550 nm reflected toward the eye at
normal incidence, and (b) will any other wavelengths of visible light be cancelled or enhanced in the reflected light?

55. ●● **Sensitive eyes.** After an eye examination, you put some
BIO eyedrops on your sensitive eyes. The cornea (the front part of
the eye) has an index of refraction of 1.38, while the eyedrops
have a refractive index of 1.45. After you put in the drops, your
friends notice that your eyes look red, because red light of
wavelength 600 nm has been reinforced in the reflected light.
(a) What is the minimum thickness of the film of eyedrops on
your cornea? (b) Will any other wavelengths of visible light be
reinforced in the reflected light? Will any be cancelled?
(c) Suppose you had contact lenses, so that the eyedrops went
on them instead of on your corneas. If the refractive index of
the lens material is 1.50 and the layer of eyedrops has the same
thickness as in part (a), what wavelengths of visible light will
be reinforced? What wavelengths will be cancelled?

56. ●● A wildlife photographer uses a moderate telephoto lens of
focal length 135 mm and maximum aperture $f/4.00$ to photograph a bear that is 11.5 m away. Assume the wavelength is
550 nm. (a) What is the width of the smallest feature on the
bear that this lens can resolve if it is opened to its maximum
aperture? (b) If, to gain depth of field, the photographer stops
the lens down to $f/22.0$, what would be the width of the smallest resolvable feature on the bear?

57. ●● **Thickness of human hair.** Although we have discussed
BIO single-slit diffraction only for a slit, a similar result holds
when light bends around a straight, thin object, such as a
strand of hair. In that case, a is the width of the strand. From
actual laboratory measurements on a human hair, it was found
that when a beam of light of wavelength 632.8 nm was shone
on a single strand of hair, and the diffracted light was viewed
on a screen 1.25 m away, the first dark fringes on either side of
the central bright spot were 5.22 cm apart. How thick was this
strand of hair?

58. ●● An oil tanker spills a large amount of oil ($n = 1.45$) into
the sea ($n = 1.33$). (a) If you look down onto the oil spill
from overhead, what predominant wavelength of light do you
see at a point where the oil is 380 nm thick? (b) In the water
under the slick, what visible wavelength (as measured in air) is
predominant in the transmitted light at the same place in the
slick as in part (a)?

59. ●● A glass plate ($n = 1.53$) that is 0.485 μm thick and surrounded by air is illuminated by a beam of white light normal
to the plate. What wavelengths (in air) within the limits of the
visible spectrum ($\lambda = 400$ nm to 700 nm) (a) are intensified in
the reflected beam, (b) are cancelled in the reflected light?

60. ●● The radius of curvature of the convex surface of a
planoconvex lens is 95.2 cm. The lens is placed convex side
down on a perfectly flat glass plate that is illuminated from
above with red light having a wavelength of 580 nm. Find
the diameter of the second bright ring in the interference
pattern.

61. ● **X-ray diffraction of salt.** X rays with a wavelength of
0.125 nm are scattered from a cubic array (of a sodium chloride crystal), for which the spacing of adjacent atoms is
$a = 0.282$ nm. (a) If diffraction from planes parallel to a cube
face is considered, at what angles θ of the incoming beam relative

to the crystal planes will maxima be observed? (b) Repeat part (a) for diffraction produced by the planes shown in Fig. 26.33a, which are separated by $a/\sqrt{2}$.

62. •• **Searching for planets around other stars.** If an optical telescope focusing light of wavelength 550 nm had a perfectly ground mirror, what would have to be the minimum diameter of its mirror so that it could resolve a Jupiter-size planet around our nearest star, Alpha Centauri, which is about 4.3 light years from earth? (Consult Appendix E.)

63. •• You need a diffraction grating that will disperse the visible spectrum (400 nm to 700 nm) through 30.0° in the first-order pattern. What must be the line density of this grating? (*Hint:* Use the fact that $\sin(A + B) = \sin A \cos B + \cos A \sin B$.)

64. •• A uniform thin film of material of refractive index 1.40 coats a glass plate of refractive index 1.55. This film has the proper thickness to cancel normally incident light of wavelength 525 nm that strikes the film surface from air, but it is somewhat greater than the minimum thickness to achieve this cancellation. As time goes by, the film wears away at a steady rate of 4.20 nm per year. What is the minimum number of years before the reflected light of this wavelength is now enhanced instead of cancelled?

65. •• A diffraction grating has 650 slits/mm. What is the highest order that contains the entire visible spectrum? (The wavelength range of the visible spectrum is approximately 400–700 nm.)

Passage Problems

Interference and sound waves. The phenomenon of interference occurs not only with light waves, but also with all frequencies of electromagnetic waves and all other types of waves, such as sound and water waves. Suppose that your professor sets up two sound speakers in the front of your classroom and uses an electronic oscillator to produce sound waves of a single frequency. When the professor begins, you and many other students hear a loud tone, while other students hear nothing. The speed of sound in air is 340 m/s.

66. The professor then does something to the apparatus. The frequency that you hear does not change, but the loudness decreases and now all your fellow student can also hear the tone. What did the professor do?
   A. She did nothing.
   B. She turned down the volume of the speakers.
   C. She changed the phase relationship of the speakers.
   D. She disconnected one speaker.

67. The professor now returns the apparatus to the same condition that it was in at the start of lecture. She then does something else to the speakers and all the students who heard nothing before now hear a loud tone, while you and those who heard the tone hear nothing. What did the professor do?
   A. She did nothing.
   B. She turned down the volume of the speakers.
   C. She changed the phase relationship of the speakers.
   D. She disconnected one speaker.

68. The professor now returns the apparatus to its original configuration (the one it had at the start of the lecture) so you again hear the original loud tone. She now slowly moves one of the speakers away from you until it first reaches a point where you can no longer hear the tone. If the speaker was moved 0.34 meters further from you, what is the frequency of the tone?
   A. 1000 Hz
   B. 2000 Hz
   C. 500 Hz
   D. 250 Hz

27 Relativity

When the year 1905 began, 25-year-old Albert Einstein was an unknown clerk in the Swiss patent office. By the end of that year, he had published three papers of extraordinary importance. The first was an analysis of Brownian motion; the second (for which he was later awarded the Nobel Prize) was on the photoelectric effect. In the third, Einstein introduced his **special theory of relativity,** in which he proposed drastic revisions in the classical concepts of space and time.

The concept of an **inertial frame of reference** plays a central role in our discussion. This concept was introduced in Section 4.2, and we strongly recommend that you review that section now. Briefly, an inertial frame of reference is a frame of reference in which Newton's laws are valid. Any frame of reference that moves with constant velocity with respect to an inertial frame is also an inertial frame.

The special theory of relativity is based on the simple statement that all the laws of physics are the same in every inertial frame of reference. This innocent-sounding proposition has far-reaching implications. Here are three that we'll explore in this chapter:

1. When two observers who are moving relative to each other measure a time interval or a length, they may not get the same results.
2. Two events that appear to one observer to occur at the same time may appear to another observer to occur at different times.
3. For the conservation principles of momentum and energy to be valid in all inertial systems, Newton's second law and the definitions of momentum and kinetic energy have to be revised.

A recurring theme in our discussions will be the role of the observer in the formulation of physical laws.

Einstein in 1932. At this point the former patent clerk is Director of the great Kaiser Wilhelm Physical Institute in Berlin. In the following year he will leave Nazi Germany for the United States, where he will settle at Princeton.

In studying this material, you'll confront some ideas that at first sight may seem too strange to be believed. You'll find that your intuition is often unreliable when you consider phenomena far removed from everyday experience, especially when objects are moving with speeds much faster than those we encounter in everyday life. But the theory rests on a solid foundation of experimental evidence. It has far-reaching consequences in *all* areas of physics (and beyond), including thermodynamics, electromagnetism, optics, atomic and nuclear physics, and high-energy physics. It doesn't *refute* Newton's laws, but it generalizes them to a vastly expanded realm of physical phenomena.

27.1 Invariance of Physical Laws

Although the implications of Einstein's theory were revolutionary, they are based on simple principles. What's more, we can work out several important results using only straightforward algebra.

Principle of Relativity

Einstein's **principle of relativity** is as follows:

> **Fundamental Postulate of Special Relativity**
> All the laws of physics, including mechanics and electromagnetism, are the same in all inertial frames of reference.

In other words, all inertial frames are equivalent; any one inertial frame is as good as any other.

Here are two examples: Suppose you watch two children playing catch with a ball while the three of you are aboard a train that moves with constant velocity with respect to the earth, assumed to be an inertial frame of reference. No matter how carefully you study the motion of the ball, you can't tell how fast (or, in fact, whether) the train is moving. This is because the laws of mechanics (Newton's laws) are the same in every inertial system (in this case, a system stationary on the ground and a system moving with the train).

Another example is the electromotive force (emf) induced in a coil of wire by a nearby moving permanent magnet (Figure 27.1a). In the frame of reference in which the coil is stationary, the moving magnet causes a change in magnetic flux through the coil, and this change induces an emf. In a frame of reference where the magnet is stationary (Figure 27.1b), the motion of the coil through a magnetic field causes magnetic forces to act upon the mobile charges in the conductor, inducing an emf. According to the principle of relativity, both of these points of view have equal validity and both must predict the same induced emf. As we saw in Chapter 21, Faraday's law of electromagnetic induction can be applied to either description, and it does indeed satisfy this requirement. If the moving-magnet and moving-coil situations *did not* give the same results, we could use this experiment to distinguish one inertial frame from another, and that would contradict the principle of relativity.

Equally significant is the fact that Maxwell's equations predicted the speed of electromagnetic radiation, as discussed in Chapter 23. The laws of electromagnetism show that light and all other electromagnetic waves travel with a constant speed $c = 299{,}792{,}458$ m/s. (We'll often use the approximate value $c = 3.00 \times 10^8$ m/s, which is within one part in 1000 of the exact value.) If these laws are the same in all inertial frames, then the principle of relativity requires that this speed must be the same in all inertial frames.

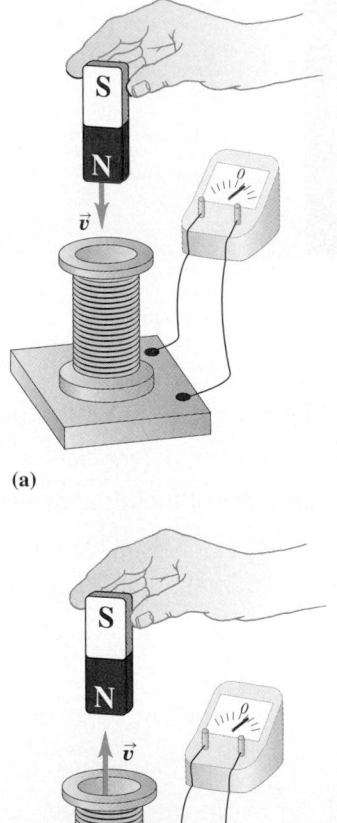

(a)

(b)

▲ **FIGURE 27.1** Induction experiments using (a) a magnet that is moving relative to the coil and (b) a stationary magnet.

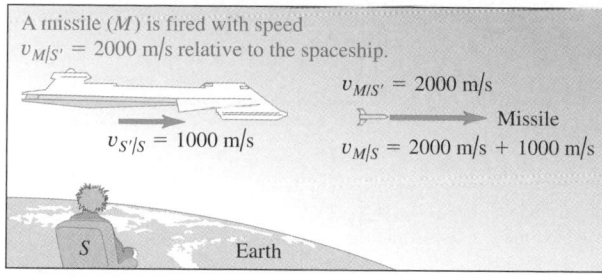

A missile (M) is fired with speed
$v_{M/S'}$ = 2000 m/s relative to the spaceship.

$v_{M/S'}$ = 2000 m/s

$v_{S'/S}$ = 1000 m/s

Missile

$v_{M/S}$ = 2000 m/s + 1000 m/s

S Earth

▲ **FIGURE 27.2** A spaceship (S') moves with speed $v_{S'/S}$ = 1000 m/s relative to an observer on earth (S).

Speed of Light

During the nineteenth century, most physicists believed that light traveled through a hypothetical medium called the *ether*, just as sound waves travel through air. If so, the speed of light would depend on the motion of the observer relative to the ether and would therefore be different in different directions. The Michelson–Morley experiment, an interference experiment similar to those described in Chapter 26, was an effort to detect the motion of the earth relative to the ether. Einstein's conceptual leap was to recognize that if Maxwell's equations for the electromagnetic field are to be valid in all inertial frames, then the speed of light in vacuum should also be the same in all inertial frames and in all directions. In fact, Michelson and Morley detected *no* motion of their laboratory relative to the supposed ether. Indeed, there is no experimental evidence for the existence of an ether, and the concept has been discarded.

Thus, a direct consequence of Einstein's principle of relativity, sometimes called the second postulate of relativity, is this statement:

Invariance of speed of light
The speed of light in vacuum is the same in all inertial frames of reference and is independent of the motion of the source.

Let's think about what this statement means. Suppose two observers measure the speed of light. One is at rest with respect to the light source, and the other is moving away from it. Both are in inertial frames of reference. According to the principle of relativity, the two observers must obtain the same result for the speed of light, despite the fact that one is moving with respect to the other.

To explore this statement further, let's consider the following situation: A spaceship moving past earth at 1000 m/s fires a missile straight ahead with a speed of 2000 m/s (relative to the spacecraft) (Figure 27.2). What is the missile's speed relative to earth? This looks like an elementary problem in relative velocity. The correct answer, according to Newtonian mechanics, is 3000 m/s. But now suppose there is a searchlight in the spaceship, pointing in the same direction that the missile was fired. An observer on the spaceship measures the speed of light emitted by the searchlight and obtains the value c. According to our previous discussion, the observer on earth who measures the speed of this same light must *also* obtain the value c, *not* c + 1000 m/s. This contradicts our elementary notion of relative velocities; it may not appear to agree with common sense, and it certainly represents a break from our Newtonian notions of relative velocity. But "common sense" is intuition based on everyday experience, and such experience does not usually include measurements of the speed of light.

Let's restate this argument symbolically, using the two inertial frames of reference, labeled S for the observer on earth and S' for the moving spaceship,

▲ **Application It's all relative.** If this child throws the ball upward, it lands back in his hands, regardless of whether he is standing still or running at constant velocity. In *his* inertial frame of reference, the two events are the same. To a person standing still, they aren't—the ball moves along a line in one case and along a parabolic path in the other. The ball's path depends on the frame in which it is observed. Similarly, in this chapter we'll see that measurements of distances and time intervals also vary depending on the relative motion of the observer and the observed, especially at speeds approaching the speed of light.

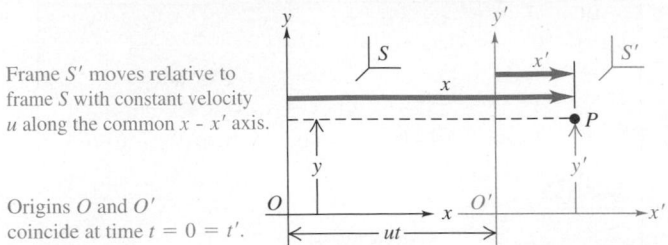

Frame S' moves relative to frame S with constant velocity u along the common x - x' axis.

Origins O and O' coincide at time $t = 0 = t'$.

▲ **FIGURE 27.3** A moving coordinate system, showing that the position of particle P can be described by the earth coordinates x and y in S or by the spaceship coordinates x' and y' in S'.

shown in Figure 27.3. To keep things as simple as possible, we have omitted the z axes. The x axes of the two frames lie along the same line, but the origin O' of frame S' moves relative to the origin O of frame S with constant velocity u along the common x-x' axis. We set our clocks so that the two origins coincide at time $t = 0$. Then their separation at a later time t is ut.

Now think about how we describe the motion of a particle P. This particle might represent an exploratory vehicle launched from the spaceship or a flash of light from a searchlight. We can describe the *position* of point P by using the earth coordinates (x, y) in S or the spaceship coordinates (x', y') in S'. The figure shows that these coordinates are related by the transformation equations $x = x' + ut,\quad y = y'$. For completeness, we add the assumption that all time measurements are the same in the two frames of reference, so $t = t'$. These three equations, based on the familiar Newtonian notions of space and time, are called the **Galilean coordinate transformation.**

Galilean coordinate transformation:
$$x = x' + ut, \qquad y = y', \qquad t = t'. \tag{27.1}$$

As we'll see in the sections that follow, these equations aren't consistent with the principle of relativity. Developing the needed modifications will yield the fundamental equations of relativity.

Relative Velocity

If point P moves in the x direction, its velocity v as measured by an observer who is stationary in S is given by $v = \Delta x / \Delta t$. Its velocity v' measured by an observer at rest in S' is $v' = \Delta x' / \Delta t$. From our discussion of relative velocities in Sections 2.7 and 3.5, we know that these velocities are related by

$$v = v' + u, \tag{27.2}$$

which agrees with the relative velocity equation we derived at the end of Chapter 2. We can also derive this relation from Equations 27.1: Suppose that the particle is at a point described by coordinate x_1 or x_1' at time t_1 and by x_2 or x_2' at time t_2. Then $\Delta t = t_2 - t_1$. From Equation 27.1,

$$\Delta x = x_2 - x_1 = (x_2' - x_1') + u(t_2 - t_1) = \Delta x' + u\,\Delta t,$$
$$\frac{\Delta x}{\Delta t} = \frac{\Delta x'}{\Delta t} + u,$$

and

$$v = v' + u,$$

in agreement with Equation 27.2.

Now here's the fundamental problem: If we apply Equation 27.2 to the speed of light, using c for the speed seen by the observer in S and c' for the speed seen by the observer in S', then Equation 27.2 says that $c = c' + u$. Einstein's principle of relativity, supported by experimental evidence, says that $c = c'$. This is a genuine inconsistency, not an illusion, and it demands resolution. If we accept the principle of relativity, we are forced to conclude that Equations 27.1 and 27.2, intuitively appealing as they are, *cannot* be correct for the speed of light and hence aren't consistent with the principle of relativity. They have to be modified to bring them into harmony with that principle.

The resolution involves some fundamental modifications in our kinematic concepts. First is the seemingly obvious assumption that the observers in frames S and S' use the same time scale. We stated this assumption formally by including in Equations 27.1 the equation $t = t'$. But we are about to show that the assumption $t = t'$ *cannot* be correct; the two observers must have different time scales. The difficulty lies in the concept of simultaneity, which is our next topic. A careful analysis of simultaneity will help us to develop the appropriate modifications of our notions about space and time. In this analysis, we'll make frequent use of hypothetical experiments, often called thought experiments or, in Einstein's original German, *gedanken* experiments, to help clarify concepts and relationships.

Conceptual Analysis 27.1

Two spaceships and a star

Two spaceships can travel in any direction. Each ship carries equipment to measure the speed of light coming from any targeted star. In one experiment, each ship travels at a constant speed and they measure the speed of light from the same star. Which of the following statements is correct?

A. If the spaceships travel in a direction perpendicular to the direction of the incoming light, they both measure a light speed of zero.
B. No matter in what direction each spaceship travels, they measure a light speed of $c = 3.00 \times 10^8$ m/s.

C. For each ship to measure a light speed $c = 3.00 \times 10^8$ m/s, both ships must travel in the same direction, but they can have unequal speeds.
D. If the spaceships travel parallel to the incoming light, but in opposite directions, the ship traveling towards the star measures a greater speed of light.

SOLUTION In all of the choices, each ship travels at a constant speed in a constant direction. Therefore, each ship is an inertial frame of reference. The invariance of the speed of light means that the speed of light is the same in all inertial frames of reference, regardless of the direction of travel. So choice B is correct. Note that choice D would be correct if we were measuring the speed of an object that has mass (such as a proton emitted by the star) instead of measuring light.

27.2 Relative Nature of Simultaneity

When two events occur at the same time, we say that they are *simultaneous*. If you awoke at seven o'clock this morning, then two events (your awakening and the arrival of the hour hand of your clock at the number seven) occurred simultaneously. The fundamental problem in measuring time intervals is that, in general, two events that are simultaneous when seen by an observer in one frame of reference are *not* simultaneous as seen by an observer in a second frame if it is moving in relation to the first, *even if both are inertial frames.* This is the relative nature of **simultaneity.**

This idea may seem to be contrary to common sense. But here is a thought experiment, devised by Einstein, that illustrates the point: Consider a long train moving with uniform velocity, as shown in Figure 27.4. Two lightning bolts strike the train, one at each end. Each bolt leaves a mark on the train and one on the ground at the same instant. The points on the ground are labeled A and B in the figure, and the corresponding points on the train are A' and B'. An observer is standing on the ground at O, midway between A and B. Another observer is at O' at the middle of the train, midway between A' and B', moving with the train.

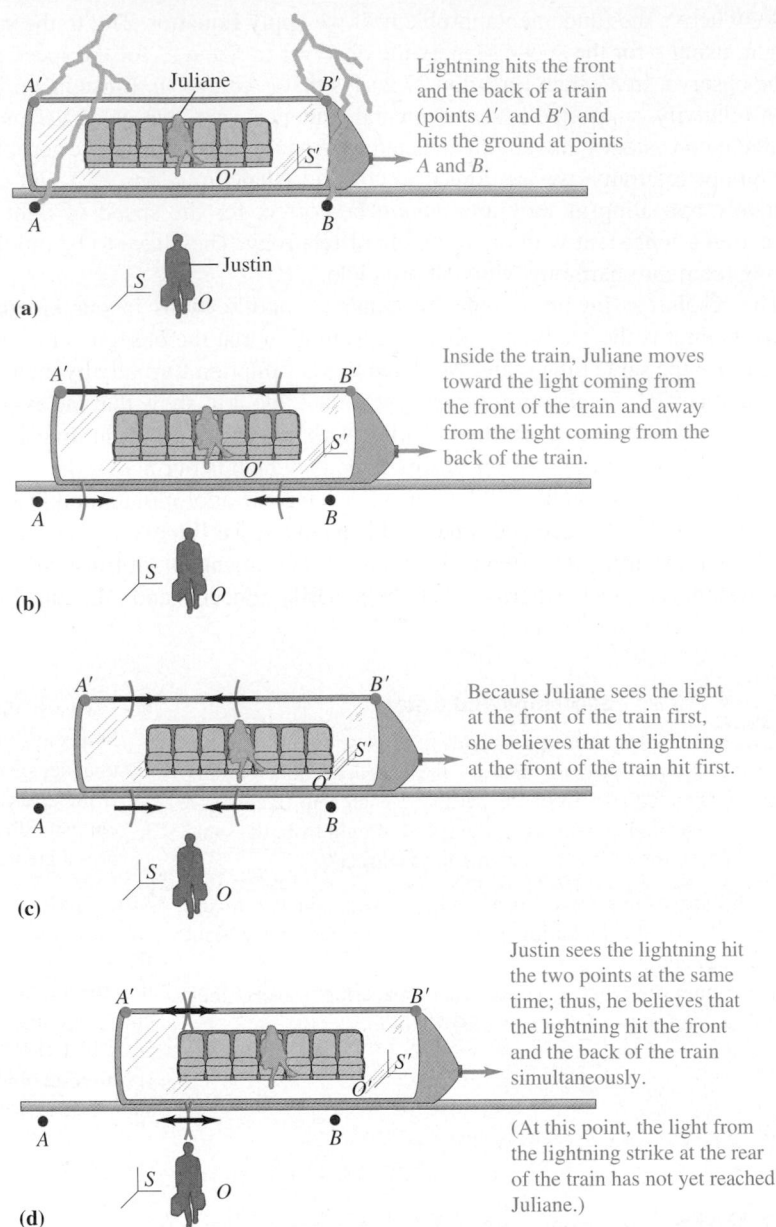

Lightning hits the front and the back of a train (points A' and B') and hits the ground at points A and B.

(a)

Inside the train, Juliane moves toward the light coming from the front of the train and away from the light coming from the back of the train.

(b)

Because Juliane sees the light at the front of the train first, she believes that the lightning at the front of the train hit first.

(c)

Justin sees the lightning hit the two points at the same time; thus, he believes that the lightning hit the front and the back of the train simultaneously.

(At this point, the light from the lightning strike at the rear of the train has not yet reached Juliane.)

(d)

▲ **FIGURE 27.4** A thought experiment illustrating simultaneity.

Both observers see both light flashes emitted from the points where the lightning strikes.

Suppose the two light flashes reach the observer at O simultaneously. He knows he is the same distance from A as from B, so he concludes that the two bolts struck A and B simultaneously. But the observer at O' is moving to the right, with the train, with respect to the observer at O, and the light flash from B' reaches her before the light flash from A' does. Because the observer at O' is the same distance from A' as from B', she concludes that the lightning bolt at the front of the train struck *earlier* than the one at the rear. The two events appear simultaneous to observer O, but not to observer O'.

Relative nature of simultaneity

Whether two events at different space points are simultaneous depends on the state of motion of the observer.

Furthermore, there is no basis for saying that O is right and O' is wrong, or the reverse. According to the principle of relativity, no inertial frame of reference is preferred over any other in the formulation of physical laws. Each observer is correct in his or her own frame of reference. In other words, simultaneity is not an absolute concept. Whether two events are simultaneous depends on the frame of reference. Because of the essential role of simultaneity in measuring time intervals, it also follows that *the time interval between two events depends on the frame of reference in which it is measured.* So our next task is to learn how to compare time intervals in different frames of reference.

27.3 Relativity of Time

We need to learn how to relate time intervals between events that are observed in two different frames of reference. If a train passes through a certain station at 12:00 and another at 12:05, the stationmaster sees a time interval of 5 minutes between trains. How would the conductor on the front train measure this time interval, and what result would he get? To answer these questions, let's consider another thought experiment. As before, a frame of reference S' (perhaps the moving train) moves along the common x-x' axis with constant speed u relative to a frame S (the station platform). For reasons that will become clear later, we assume that u is always less than the speed of light c. A passenger Juliane at O', moving with frame S', directs a flash of light at a mirror a distance d away, as shown in Figure 27.5a. She measures the time interval Δt_0 for light to make the round-trip to the mirror and back. (We use the subscript zero as a reminder that the timing apparatus is at rest in frame S'.) The total distance is $2d$, and the light flash moves with constant speed c, so the time interval Δt_0 is the distance $2d$ that the light flash travels, divided by the speed c:

$$\Delta t_0 = \frac{2d}{c}. \qquad (27.3)$$

We call S' Juliane's **rest frame,** because she is not moving with respect to S'.

The time for the round-trip, as measured by Justin in frame S, is a different interval Δt. During this time, the light source moves at constant speed u in relation to S. In time Δt, it moves a distance $u\,\Delta t$. The total round-trip distance, as seen in S, is not just $2d$, but $2l$. Figure 27.5 shows that d and $u\,\Delta t/2$ are the two legs of a right triangle, with l as the hypotenuse, so

$$l = \sqrt{d^2 + \left(\frac{u\,\Delta t}{2}\right)^2}.$$

ActivPhysics 17.1: Relativity of Time

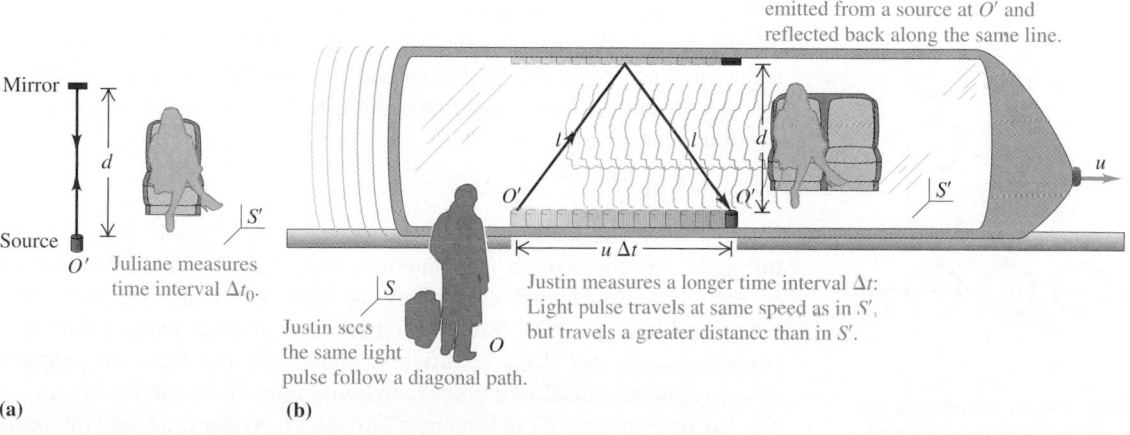

▲ **FIGURE 27.5** A thought experiment illustrating the relativity of time.

(In writing this expression, we've used the fact that the distance d is the same for both observers. We'll discuss this point later.) The speed of light is the same for both observers, so the relation in S analogous to Equation 27.3 is

$$\Delta t = \frac{2l}{c} = \frac{2}{c}\sqrt{d^2 + \left(\frac{u\,\Delta t}{2}\right)^2}. \tag{27.4}$$

We'd like to have a relation between the two time intervals Δt and Δt_0 that doesn't contain d. (We don't want the time intervals to depend on the dimensions of our clock.) To get this relation, we solve Equation 27.3 for d and substitute the result into Equation 27.4, obtaining

$$\Delta t = \frac{2}{c}\sqrt{\left(\frac{c\,\Delta t_0}{2}\right)^2 + \left(\frac{u\,\Delta t}{2}\right)^2}. \tag{27.5}$$

Now we square both sides of this equation and solve for Δt; the result is

$$\Delta t = \frac{\Delta t_0}{\sqrt{1 - u^2/c^2}}.$$

We may generalize this important result: If two events occur at the same point in space in a particular frame of reference, and if the time interval between them, as measured by an observer at rest in this frame (which we call the **rest frame** of this observer) is Δt_0, then an observer in a second frame moving with constant velocity u relative to the first frame measures the time interval to be Δt:

Time dilation:

$$\Delta t = \frac{\Delta t_0}{\sqrt{1 - u^2/c^2}}. \tag{27.6}$$

The denominator is always smaller than unity, so Δt is always larger than Δt_0. Think of an old-fashioned pendulum clock that ticks once a second, as observed in its rest frame; this is Δt_0. When the time between ticks is measured by an observer in a frame that is moving with velocity u with respect to the clock, the time interval Δt that he observes is *longer* than 1 second, and he thinks the clock is running slow. This effect is called **time dilation.** A clock moving with respect to an observer appears to run more slowly than a clock that is at rest in the observer's frame. Note that this conclusion is a direct result of the fact that the speed of light is the same in both frames of reference.

Two special cases are worth mentioning. First, if the velocity u of frame S' relative to S is much less than c, then u^2/c^2 is much smaller than one; in that case, the denominator in Equation 27.6 is nearly equal to one, and that equation reduces to the Newtonian relation $\Delta t = \Delta t_0$. That's why we don't see relativistic effects in everyday life. Second, if u is larger than c, the denominator is the square root of a negative number and is thus an imaginary number. This suggests that speeds greater than c are impossible. We'll examine this conclusion in more detail in later sections.

Proper Time

There is only one frame of reference in which the clock is at rest, but there are infinitely many in which it is moving. Therefore, the time interval measured between two events (such as two ticks of the clock) that occur at the same point (as viewed in a particular frame) is a more fundamental quantity than the interval between events that occur at different points. We use the term **proper time** to describe the time interval Δt_0 between two events that occur *at the same point.*

The time interval Δt in Equation 27.6 isn't a proper time interval, because it is the interval between two events that occur *at different spatial points* in the frame

▲ **Application Not so fast.** Is the measured time for this runner to complete the race absolute or does it depend on who has the clock? A clock in the pocket of a runner crossing the finish line at constant velocity would be running at a very slightly slower rate from the viewpoint of an official with a stationary clock at the finish line. This would make the official measured time infinitesimally longer than that measured by the runner's clock. Fortunately, this time dilation effect would be significant only if this runner were approaching the speed of light, in which case we probably wouldn't need a clock to tell who won the race.

The grid is three dimensional; there are identical planes of clocks that are parallel to the page, in front of and behind it, connected by grid lines perpendicular to the page.

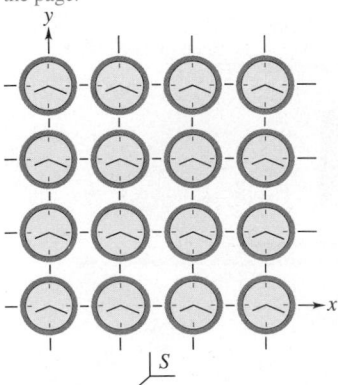

▲ **FIGURE 27.6** A frame of reference pictured as a coordinate system with a grid of synchronized clocks.

of reference S. Therefore, it can't really be measured by a single observer at rest in S. But we can use *two* observers, one stationary at the location of the first event and the other at the second, each with his or her own clock. We can synchronize these two clocks without difficulty, as long as they are at rest in the same frame of reference. For example, we could send a light pulse simultaneously to the two clocks from a point midway between them. When the pulses arrive, the observers set their clocks to a prearranged time. (But note that clocks synchronized in one frame of reference are, in general, *not* synchronized in any other frame.)

In thought experiments, it's often helpful to imagine *many observers* with synchronized clocks at rest at various points in a particular frame of reference. We can picture a frame of reference as a coordinate grid with lots of synchronized clocks distributed around it, as suggested by Figure 27.6. Only when a clock is *moving* relative to a given frame of reference do we have to watch for ambiguities of synchronization or simultaneity.

Conceptual Analysis 27.2

Pendulums in relative motion

Two identical pendulums have a period of τ_0 when measured in the factory. While one pendulum swings on earth, the other is taken by astronauts on a spaceship that travels at 99% the speed of light. Which statement is correct about observations of the pendulums? (There may be more than one correct choice.)

A. When observed from the earth, the pendulum on the ship has a period less than τ_0.
B. When observed from the earth, the pendulum on the ship has a period greater than τ_0.

C. The astronauts observe the pendulum on earth with a period greater than τ_0.
D. The astronauts observe the pendulum on the ship with a period greater than τ_0.

SOLUTION Recall that a clock moving with respect to an observer appears to run more slowly than a clock that is at rest in the observer's frame. Therefore, people on earth will observe the pendulum on the ship with a period greater than τ_0 (choice B). In addition, the pendulum on earth is moving with respect to the astronauts, so the astronauts observe the pendulum on earth with a period greater than τ_0 (choice C).

EXAMPLE 27.1 **Time dilation at 0.99c**

A spaceship flies past earth with a speed of $0.990c$ (about 2.97×10^8 m/s) relative to earth (Figure 27.7). A high-intensity signal light (perhaps a pulsed laser) blinks on and off; each pulse lasts $2.20\ \mu s$ ($=2.20 \times 10^{-6}$ s), as measured on the spaceship. At a certain instant, the ship is directly above an observer on earth and is traveling perpendicular to the line of sight. **(a)** What is the duration of each light pulse, as measured by the observer on earth? **(b)** How far does the ship travel in relation to the earth during each pulse?

SOLUTION

SET UP First we label the frames of reference. Let S be the earth's frame of reference and S' that of the spaceship. Next, we note that the duration of a laser pulse, measured by an observer on the spaceship, is a proper time in S', because the laser is at rest in S' and the two events (the starting and stopping of the pulse) occur at the same point relative to S'. In Equation 27.6, we denote this interval by Δt_0. The time interval Δt between these events, as seen in any other frame of reference, is greater than Δt_0. The speed of the spaceship (frame S') relative to earth (frame S) is $u = 0.990c = 2.97 \times 10^8$ m/s.

SOLVE Part (a): In the preceding discussion, the proper time is $\Delta t_0 = 2.20\ \mu s$. According to Equation 27.6, the corresponding interval Δt measured by an observer on the earth (S) is

$$\Delta t = \frac{\Delta t_0}{\sqrt{1 - u^2/c^2}} = \frac{2.20\ \mu s}{\sqrt{1 - (0.990)^2}} = 15.6\ \mu s.$$

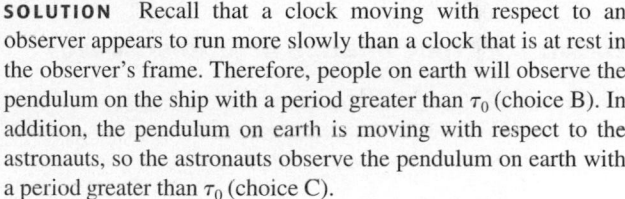

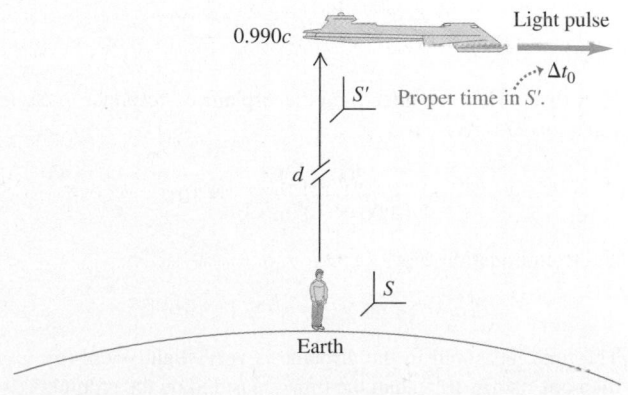

▲ **FIGURE 27.7** A spaceship traveling perpendicular to earth transmits a high-intensity light signal.

Continued

Part (b): During this interval, the spaceship travels a distance d relative to earth, given by

$$d = u \, \Delta t = (0.990)(3.00 \times 10^8 \text{ m/s})(15.6 \times 10^{-6} \text{ s})$$
$$= 4600 \text{ m} = 4.6 \text{ km}.$$

REFLECT The time dilation in S is about a factor of 7. To an observer on earth, each pulse appears to last about seven times as long as it does to an observer on the spaceship.

Practice Problem: Suppose the spaceship changes velocity relative to earth, so that the duration of the laser pulses as seen on earth is 4.40×10^{-6} s. What is the spaceship's velocity relative to earth? *Answer:* 2.60×10^8 m/s.

An experiment that is similar in principle to this spaceship example, although different in detail, provided the first direct experimental confirmation of Equation 27.6. μ leptons, or *muons,* are unstable particles, first observed in cosmic rays. They decay with a mean lifetime of 2.20 μs, as measured in a frame of reference in which the particles are at rest. But in cosmic-ray showers, the particles are moving very fast. The mean lifetime of a muon with speed of $0.99c$ has been measured to be 15.6 μs. Note that the numbers are identical to those in the spaceship example; the duration of the light pulse is replaced by the mean lifetime of the muon. These measurements provide a direct experimental confirmation of Equation 27.6.

EXAMPLE 27.2 Time dilation at jetliner speeds

An airplane flies from San Francisco to New York (about 4800 km, or 4.80×10^6 m) at a steady speed of 300 m/s (about 670 mi/h) (Figure 27.8). How much time does the trip take, as measured **(a)** by observers on the ground; **(b)** by an observer in the plane?

SOLUTION

SET UP We denote the airplane's speed relative to the ground as u. The time between the plane's departure and arrival, as measured by an observer on the plane, is a proper time, measured between two events at the same point in the plane's frame of reference, and we denote it as Δt_0. The corresponding time interval as seen by observers who are stationary on the ground is Δt, and the distance d traveled during this interval is $d = u \, \Delta t$.

SOLVE Part (a): The time Δt measured by the ground observers is simply the distance divided by the speed:

$$\Delta t = \frac{4.80 \times 10^6 \text{ m}}{300 \text{ m/s}} = 1.60 \times 10^4 \text{ s},$$

or about $4\frac{1}{2}$ hours.

Part (b): The time interval in the airplane corresponds to Δt_0 in Equation 27.6. We have

$$\frac{u^2}{c^2} = \frac{(300 \text{ m/s})^2}{(3.00 \times 10^8 \text{ m/s})^2} = 10^{-12},$$

and from Equation 27.6,

$$\Delta t_0 = (1.60 \times 10^4 \text{ s}) \sqrt{1 - 10^{-12}}.$$

The time measured in the airplane is very slightly less (by less than one part in 10^{12}) than the time measured on the ground.

REFLECT To measure the time interval Δt seen by the ground observers, we need *two* observers with synchronized clocks, one in San Francisco and one in New York, because the two events

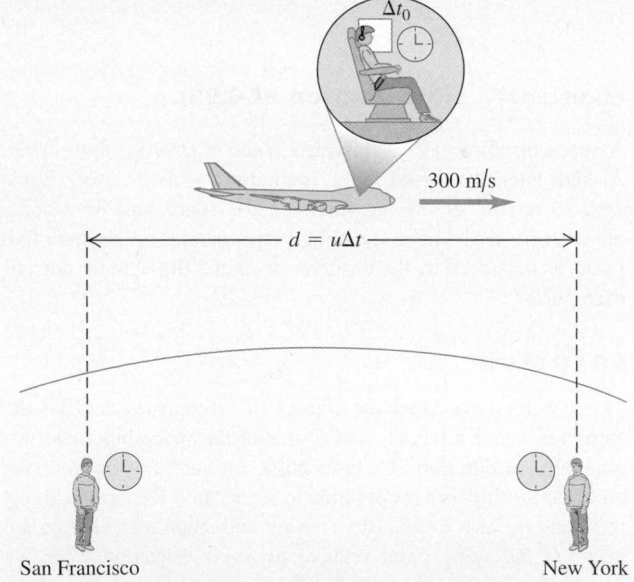

▲ FIGURE 27.8 An airplane flying from San Francisco to New York, observed from the ground and from the plane.

(takeoff and landing) occur at different spatial points in the ground frame of reference. In the airplane's frame, they occur at the *same* point, so Δt_0 can be measured by a single observer. (But note that if the plane develops engine trouble and has to return to San Francisco, it must be in an accelerated, non-inertial frame of reference during its turnaround. In this case, the preceding

Continued

analysis isn't valid, because we have used only inertial frames in our derivations.)

We don't notice such tiny effects in everyday life. But as we mentioned in Section 1.3, present-day atomic clocks can attain a precision of about one part in 10^{13}, and measurements similar to the one in this example have been carried out, verifying Equation 27.6 directly.

Practice Problem: The π^+ meson (or positive pion) is an unstable particle that has a lifetime (measured in its rest frame) of about 2.60×10^{-8} s. If the particle is moving with a speed of $0.992c$ with respect to your laboratory, what is its lifetime as measured in the laboratory? What distance does it travel in the laboratory during this time? *Answer:* 2.06×10^{-7} s; 61.3 m.

Space Travel and the Twin Paradox

Time dilation has some interesting consequences for space travel. Suppose we want to explore a planet in a distant solar system 200 light years away. Could a person possibly live long enough to survive the trip? One light year is the distance light travels in 1 year. In ordinary units, this distance is $(3.00 \times 10^8 \text{ m/s})(3.15 \times 10^7 \text{ s}) = 9.45 \times 10^{15}$ m. Even if the spaceship can attain a speed of $0.999c$ relative to earth, the trip requires nearly 200 years, as observed in earth's reference frame.

But the aging of the astronauts is determined by the time measured in the spaceship's frame of reference. With reference to Equation 27.6, Δt is the time interval measured on earth, and Δt_0 is the corresponding proper time interval measured in the moving spaceship. If the ship's speed relative to earth is $0.999c$, then Equation 27.6 gives

$$\sqrt{1 - (0.999c)^2/c^2} = \sqrt{1 - (0.999)^2} = 0.0447 \quad \text{and}$$
$$\Delta t_0 = \Delta t \sqrt{1 - (0.999)^2} = (200 \text{ y})(0.0447) = 8.94 \text{ years.}$$

Observers on earth think the trip lasts 200 years, but to the astronauts on the spaceship, the interval is only 8.94 years.

But there's another problem. Suppose the astronauts want to come back home to earth. The time-dilation equation (Equation 27.6) suggests an apparent paradox called the **twin paradox.** Consider two identical-twin astronauts named Eartha and Astrid. Eartha remains on earth, while Astrid takes off on a high speed trip through the galaxy. Because of time dilation, Eartha sees all of Astrid's life processes, such as her heartbeat, proceeding more slowly than her own. Thus, Eartha thinks that Astrid ages more slowly, so when Astrid returns to earth, she is younger than Eartha.

Now here's the paradox: All inertial frames are equivalent. Can't Astrid make exactly the same arguments to conclude that Eartha is in fact the younger? Then each twin thinks that the other is younger, and that's a paradox.

To resolve the paradox, we recognize that the twins are *not* identical in all respects. If Eartha remains in an inertial frame at all times, Astrid must have an acceleration with respect to inertial frames during part of her trip in order to turn around and come back. Eartha remains always at rest in the same inertial frame; Astrid does not. Thus, there is a real physical difference between the circumstances of the twins. Careful analysis shows that Eartha is correct: When Astrid returns, she *is* younger than Eartha.

▲ **Application Who's the grandmother?**
The answer to this question may seem obvious, but to the student of relativity, it could depend on which person had traveled to a distant planet at 99.9% of the speed of light. Imagine that a 20-year-old woman had given birth to a child and then immediately left on a 100-light year trip at nearly the speed of light. Because of time dilation for the traveler, only about 10 years would pass, and she would appear to be about 30 years old when she returned, even though 100 years had passed by for people on Earth. Meanwhile, the child she left behind at home could have had a baby 20 years after her departure, and this grandchild would now be 80 years old!

27.4 Relativity of Length

Just as the time interval between two events depends on the observer's frame of reference, the distance between two points also depends on the observer's frame of reference. The concept of simultaneity is involved. Suppose you want to measure the length of a moving car. One way is to have two friends make marks on the pavement *at the same time* at the positions of the front and rear bumpers. Then you measure the distance between the marks. If you mark the position of the front bumper at one time and that of the rear bumper half a second later, you won't get the car's true length. But we've learned that simultaneity isn't an absolute concept,

MasteringPHYSICS

ActivPhysics 17.2: Relativity of Length

so we have to be very careful to treat times and time intervals correctly in various reference frames.

To develop a relationship between lengths in various coordinate systems, we consider another thought experiment. We attach a source of light pulses to one end of a ruler and a mirror to the other end, as shown in Figure 27.9. The ruler is at rest in reference frame S', and its length in that frame is l_0. Then the time Δt_0 required for a light pulse to make the round-trip from source to mirror and back (a distance $2l_0$) is given by

$$\Delta t_0 = \frac{2l_0}{c}. \tag{27.7}$$

This is a proper time interval in S', because departure and return occur at the same point in S'.

In reference frame S, the ruler is moving to the right with speed u during the time the light pulse is traveling. The length of the ruler in S is l, and the time of travel from source to mirror, as measured in S, is Δt_1. During this interval, the ruler, with source and mirror attached, moves a distance $u\,\Delta t_1$. The total length of path d from source to mirror is not l, but

$$d = l + u\,\Delta t_1.$$

The light pulse travels with speed c, so it is also true that

$$d = c\,\Delta t_1.$$

Combining these two equations to eliminate d, we find that

$$c\,\Delta t_1 = l + u\,\Delta t_1 \qquad \text{or} \qquad \Delta t_1 = \frac{l}{c - u}.$$

In the same way, we can show that the time Δt_2 for the return trip from mirror to source, as measured in S, is

$$\Delta t_2 = \frac{l}{c + u}.$$

The *total* time $\Delta t = \Delta t_1 + \Delta t_2$ for the round-trip, as measured in S, is the sum of these two time intervals:

$$\Delta t = \frac{l}{c - u} + \frac{l}{c + u} = \frac{2l}{c(1 - u^2/c^2)}. \tag{27.8}$$

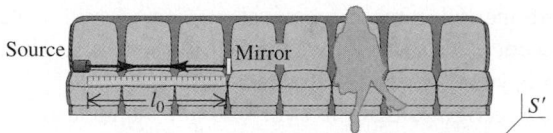

Source — Mirror

The ruler is stationary in Juliane's frame of reference S'. The light pulse travels a distance l_0 from the light source to the mirror.

(a)

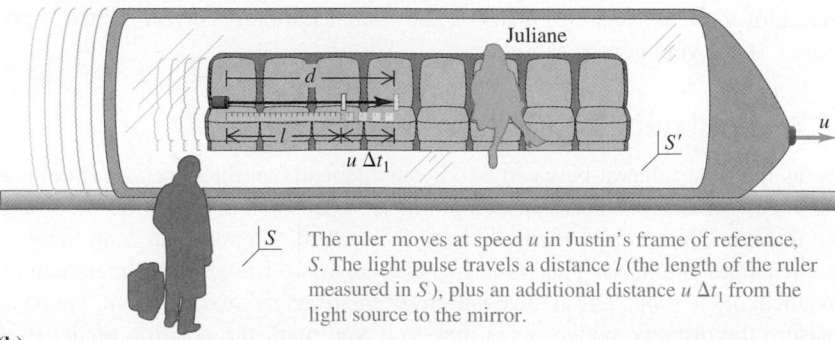

Juliane

The ruler moves at speed u in Justin's frame of reference, S. The light pulse travels a distance l (the length of the ruler measured in S), plus an additional distance $u\,\Delta t_1$ from the light source to the mirror.

(b)

▶ **FIGURE 27.9** A thought experiment illustrating the relativity of length.

We also know that Δt and Δt_0 are related by Equation 27.6, because Δt_0 is a proper time in S'. When we solve Equation 27.6 for Δt_0 and substitute the result into Equation 27.7, we get

$$\Delta t\sqrt{1 - \frac{u^2}{c^2}} = \frac{2l_0}{c}.$$

Finally, combining this equation with Equation 27.8 to eliminate Δt and simplifying, we obtain the following relationship:

Length contraction parallel to motion:

$$l = l_0\sqrt{1 - \frac{u^2}{c^2}}. \qquad (27.9)$$

Thus, the length l measured in S, in which the ruler is moving, is *shorter* than the length l_0 measured in S', where the ruler is at rest. A length measured in the rest frame of the body is called a **proper length;** thus, in Equation 27.9, l_0 is a proper length in S', and the length measured in any other frame is *less than* l_0. This effect is called **length contraction.**

EXAMPLE 27.3 **The shrinking spaceship**

In Example 27.1 (Section 27.3), suppose a crew member on the spaceship measures its length, obtaining the value 400 m. What is the length measured by observers on earth?

SOLUTION

SET UP The length of the spaceship in the frame in which it is at rest (400 m) is a proper length in this frame, corresponding to l_0 in Equation 27.9. We want to find the length l measured by observers on earth, who see the ship moving with speed u.

SOLVE From Equation 27.9,

$$l = l_0\sqrt{1 - \frac{u^2}{c^2}} = (400 \text{ m})\sqrt{1 - (0.990)^2} = 56.4 \text{ m}.$$

REFLECT To measure l, we need two observers, because we have to observe the positions of the two ends of the spaceship simultaneously in the earth's reference frame. We could use two observers with synchronized clocks (Figure 27.10). These two observations are simultaneous in the earth's reference frame, but they are *not* simultaneous as seen by an observer in the spaceship.

Practice Problem: Suppose your car is 5.00 m long when it is sitting at rest in your driveway. How fast would it have to move for its length (as seen by someone at rest in the driveway) to be 4.00 m? *Answer:* 1.80×10^8 m/s.

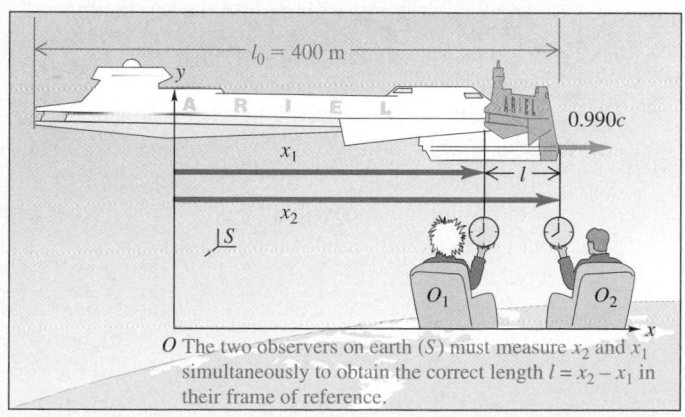

▲ **FIGURE 27.10**

When u is very small compared with c, the square-root factor in Equation 27.9 approaches unity, and in the limit of small speeds we recover the Newtonian

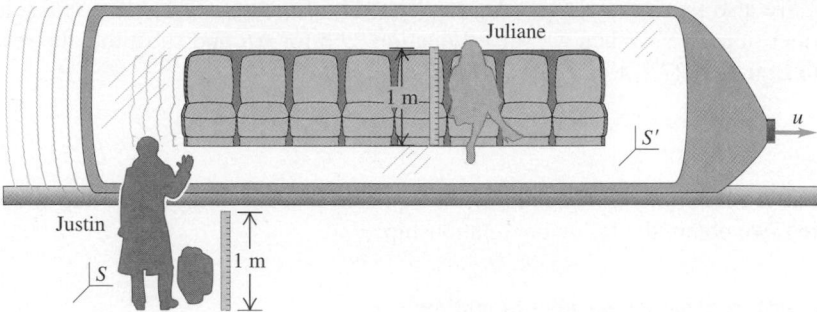

▲ **FIGURE 27.11** The rulers are perpendicular to the relative velocity, so, for any value of u, the observers measure each ruler to have a length of 1 meter.

relation $l = l_0$. This and the corresponding result for time dilation show that Equations 27.1 retain their validity when all speeds are much smaller than c; only at speeds comparable to c are modifications needed.

Lengths Perpendicular to Relative Motion

We have derived Equation 27.13 for lengths measured in the direction *parallel* to the relative motion of the two frames of reference. Lengths measured *perpendicular* to the direction of motion are *not* contracted. To prove this, let's consider two identical rulers (Figure 27.11). One ruler is at rest in frame S and lies along the y axis with one end at O, the origin of S. The other ruler is at rest in frame S' and lies along the y' axis with one end at O', the origin of S'. At the instant the two origins coincide, observers in the two frames of reference S and S' simultaneously make marks at the positions of the upper ends of the rulers. Because the observations occur at the same space point for both observers, they agree that they are simultaneous.

If the marks made by the two observers *do not* coincide, then one must be higher than the other. This would imply that the two frames aren't equivalent. But any such asymmetry would contradict the basic postulate of relativity that all inertial frames of reference are equivalent. Consistency with the postulates of relativity requires that both observers see the two rulers as having the same length, even though, to each observer, one of them is stationary and one is moving.

Length contraction perpendicular to motion
There is no length contraction perpendicular to the direction of relative motion.

(a) Array at rest

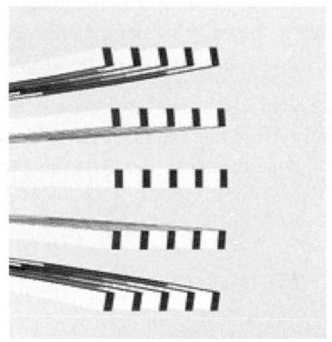

(b) Array moving to the right at 0.2c

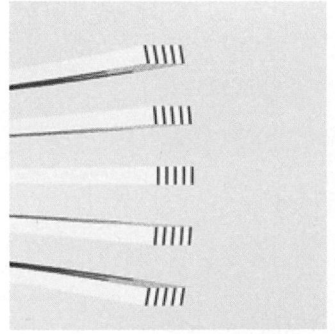

(c) Array moving to the right at 0.9c

▲ **FIGURE 27.12** Computer-generated images of 25 rods with square cross sections.

Finally, let's consider the visual appearance of a moving three-dimensional body. If we could see the positions of all points of the body simultaneously, it would appear just to shrink in the direction of motion. But we *don't* see all the points simultaneously; light from points farther from us takes longer to get to us than light from points near to us, so we see the positions of farther points as they were at earlier times.

Suppose we have a cube with its faces parallel to the coordinate planes. When we look straight-on at the closest face of such a cube at rest, we see only that face. But when the cube is moving past us toward the right, we can also see the left side because of the effect just described. More generally, we can see some points that we couldn't see when the body was at rest, because it moves out of the way of the light rays. Conversely, some light that can get to us when the body is at rest is blocked by the moving body. Because of all this, the cube appears rotated and distorted. Figure 27.12 shows a computer-generated image of a more complicated body, moving at a relativistic speed relative to the observer, that illustrates this effect.

Quantitative Analysis 27.3   **A moving square**

An engineer in space measures the area of a square of girders with sides of length l_0. If she is moving with constant velocity parallel to one side of the square, which of the following is true of the area $A_{observer}$ that she measures?

A. $A_{observer} > l_0^2$.
B. $A_{observer} = l_0^2$.
C. $A_{observer} < l_0^2$.

SOLUTION Because the engineer moves parallel to one side of the square, she sees the square shortened by length contraction in her direction of motion—to her, it looks like a rectangle that is narrower than it is tall, rather than a square. Since the measured *height* of the square is not affected by her motion, the area she measures is less than the area she would measure if she were at rest relative to the square (so $A_{observer} < l_0^2$). Being a space engineer, she naturally takes this length contraction into account and computes the correct area.

27.5 The Lorentz Transformation

In Section 27.1, we discussed the Galilean coordinate transformation equations (Equations 27.1), which relate the coordinates (x, y) of a point in frame of reference S to the coordinates (x', y') of the same point in a second frame S' when S' moves with constant velocity u relative to S along the common x-x' axis (Figure 27.13).

As we have seen, this Galilean transformation is valid only when u is much smaller than c. The relativistic equations for time dilation and length contraction (Equations 27.6 and 27.9) can be combined to form a more general set of transformation equations that are consistent with the principle of relativity. The more general relations are called the **Lorentz transformation.** In the limit of very small u, they reduce to the Galilean transformation, but they may also be used when u is comparable to c.

The basic question is this: When an event occurs at point (x, y) at time t, as observed in a frame of reference S, what are the coordinates (x', y') and time t' of the event as observed in a second frame S' moving relative to S with constant velocity u along the x direction?

As before, we assume that the origins of the two frames S and S' coincide at the initial time $t = t' = 0$. Deriving the transformation equations involves several intermediate steps. We'll omit the details; here are the results:

Lorentz transformation equations

$$x' = \frac{x - ut}{\sqrt{1 - u^2/c^2}}, \qquad y' = y, \qquad t' = \frac{t - ux/c^2}{\sqrt{1 - u^2/c^2}}. \qquad (27.10)$$

The Lorentz transformation equations are the relativistic generalization of the Galilean transformation, Equations 27.1. When u is much smaller than c, the square roots in the denominators approach one, and the second term in the numerator of the t' equation approaches zero. In this limit, Equations 27.10 become

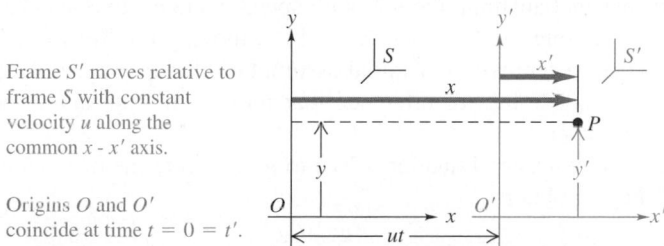

Frame S' moves relative to frame S with constant velocity u along the common x - x' axis.

Origins O and O' coincide at time $t = 0 = t'$.

▲ **FIGURE 27.13** As measured in frame of reference S, x' is contracted to $x'\sqrt{1 - u^2/c^2}$ so $x = ut + x'\sqrt{1 - u^2/c^2}$ and $x' = (x - ut)/\sqrt{1 - u^2/c^2}$.

▲ Application Space and time are what??
This may be how Newton would feel after talking to Einstein and finding out that neither distances in space nor intervals of time have absolute meanings, independent of the frame of reference. Both space and time, according to relativity theory, have meaning only when considering the relative motion of the observer in relation to what is being observed and measured. Much like Maxwell's equations of electromagnetism, which described the interdependence of electricity and magnetism, Einstein's theory of relativity demonstrated the intimate association between space and time. For example, length contraction can be compensated for by time dilation. In fact, time is often considered to be a fourth dimension, and relativistic events may be described in terms of the four-dimensional space-time continuum.

identical to Equations 27.1: $x' = x - ut$ and $t' = t$. In general, though, both the coordinates and the time of an event in one frame depend on both its coordinates and time in another frame. *Space and time have become intertwined; we can no longer say that length and time have absolute meanings, independent of a frame of reference.*

Relativistic Velocity Transformation

We can use Equations 27.10 to derive the relativistic generalization of the Galilean velocity transformation, Equation 27.2. First we use Equations 27.10 to obtain relations for distance and time *intervals* between two events in the two coordinate systems. Suppose a body is moving along the x axis. As observed in S, it arrives at point x_1 at time t_1 and at point x_2 at time t_2. The distance interval as observed in S is $\Delta x = x_2 - x_1$, the time interval is $\Delta t = t_2 - t_1$, and the velocity as seen in S (strictly speaking, the x component of velocity) is $v = \Delta x / \Delta t$. The corresponding intervals as observed in the moving frame S' (moving with x component of velocity u relative to S) are $\Delta x' = x_2' - x_1'$ and $\Delta t = t_2' - t_1'$. The x component of velocity as seen in S' is $v' = \Delta x' / \Delta t'$. Using the first and third of Equations 27.10 for each event and subtracting, we find that

$$\Delta x' = \frac{\Delta x - u\,\Delta t}{\sqrt{1 - u^2/c^2}}, \qquad \Delta t' = \frac{\Delta t - u\,\Delta x/c^2}{\sqrt{1 - u^2/c^2}}. \qquad (27.11)$$

Finally, we divide the first of Equations 27.11 by the second:

$$v' = \frac{\Delta x'}{\Delta t'} = \frac{\Delta x - u\,\Delta t}{\Delta t - u\,\Delta x/c^2} = \frac{(\Delta x/\Delta t) - u}{1 - (u/c^2)(\Delta x/\Delta t)}.$$

But $\Delta x / \Delta t$ equals v, the x component of velocity as measured in S, so we finally obtain the relation between v and v':

Relativistic velocity transformation

$$v' = \frac{v - u}{1 - uv/c^2}. \qquad (27.12)$$

This surprisingly simple result says that if a body is moving with x component of velocity v' relative to a frame S', and if S' is moving with x component of velocity u relative to another frame S, then the x components of velocity are related by Equation 27.12.

When u and v are much smaller than c, the second term in the denominator in Equation 27.12 is very small, and we obtain the nonrelativistic result $v' = v - u$. The opposite extreme is the case $v = c$; then we find that

$$v' = \frac{c - u}{1 - uc/c^2} = c.$$

This result says that anything moving with speed c relative to S also has speed c relative to S', despite the fact that frame S' is moving relative to S. So Equation 27.12 is consistent with our initial assumption that the speed of light is the same in all inertial frames of reference that move with constant velocity with respect to each other.

We can also rearrange Equation 27.12 to give v in terms of v'. Solving that equation for v, we obtain

$$v = \frac{v' + u}{1 + uv'/c^2}. \qquad (27.13)$$

Note that this equation has the same form as Equation 27.12, with v and v' interchanged and the sign of u reversed. Indeed, this resemblance is in fact *required* by the principle of relativity, which insists that there is no basic distinction between the two inertial frames S and S'.

PROBLEM-SOLVING STRATEGY 27.1 **Lorentz transformation**

SET UP

1. Observe carefully the roles of the concepts of proper time and proper length. A time interval between two events that happen at the same space point in a particular frame of reference is a proper time in that frame, and the time interval between the same two events is greater in any other frame. The length of a body measured in a frame in which it is at rest is a proper length in that frame, and the length is less in any other frame.

SOLVE

2. The Lorentz transformation equations tell you how to relate measurements made in different inertial frames of reference. When you use them, it helps to make a list of coordinates and times of events in the two frames, such as x_1, x_1', t_2, and so on. List and label carefully what you know and don't know. Do you know the coordinates in one frame? The time of an event in one frame? What are your knowns and unknowns?

3. In velocity-transformation problems, if you have two observers measuring the motion of a body, decide which you want to call S and which S', identify the velocities v and v' clearly, and make sure you know the velocity u of S' relative to S. Use either form of the velocity transformation equation, Equation 27.12 or 27.13, whichever is more convenient.

REFLECT

4. Don't be surprised if some of your results don't seem to make sense or if they disagree with your intuition. Reliable intuition about relativity takes time to develop; keep trying!

EXAMPLE 27.4 **Relative velocities in space travel**

A spaceship moving away from the earth with speed $0.90c$ fires a robot space probe in the same direction as its motion, with speed $0.70c$ relative to the spaceship (Figure 27.14). What is the probe's speed relative to the earth?

▶ **FIGURE 27.14**

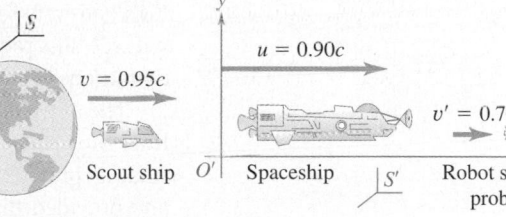

SOLUTION

SET UP We resist the Newtonian temptation to simply add the two speeds, obtaining the impossible result $1.60c$. Instead, we use the relativistic formulation given by Equations 27.12 and 27.13, which you may want to review before proceeding. Let the earth's frame of reference be S, and let the spaceship's be S'. Then u is the spaceship's velocity relative to earth, and v' is the velocity of the probe relative to the spaceship. We want to find v, the velocity of the probe relative to earth.

SOLVE From the statement of the problem and the discussion in the preceding paragraph, $u = 0.90c$, $v' = 0.70c$, and v is to be determined. We use Equation 27.13:

$$v = \frac{v' + u}{1 + uv'/c^2} = \frac{0.70c + 0.90c}{1 + (0.90c)(0.70c)/c^2} = 0.98c.$$

REFLECT We can check that the *incorrect* value obtained from the Galilean velocity-addition formula $(v = v' + u)$ is a velocity of $1.60c$ relative to earth; this velocity is larger than the speed of light, and larger than the correct relativistic value by a factor of about 1.63.

Practice Problem: Suppose the robot space probe is aimed directly toward the earth instead of away from it. What is its velocity relative to earth? *Answer:* $0.54c$.

EXAMPLE 27.5 **Space travel again**

A scout ship from the earth tries to catch up with the spaceship of Example 27.4 by traveling at $0.95c$ relative to the earth. What is its speed relative to the spaceship?

SOLUTION

SET UP Again we let the earth's frame of reference be S and the spaceship's frame be S'. Again we have $u = 0.90c$, but now $v = 0.95c$.

SOLVE According to nonrelativistic velocity addition, the scout ship's velocity relative to the spaceship would be $0.05c$. We get the correct result from Equation 27.12:

$$v' = \frac{v - u}{1 - uv/c^2} = \frac{0.95 - 0.90c}{1 - (0.90c)(0.95c)} = 0.34c.$$

REFLECT Are you surprised that v' is greater than $0.05c$? Try to think of an argument as to why this might have been expected.

Practice Problem: Referring to Equation 27.12, verify that when both v and u are much smaller than c, the expression for v' reduces to the nonrelativistic expression.

Equations 27.12 and 27.13 can be used to show that when the relative velocity u of two frames is less than c, an object moving with a speed less than c in one frame of reference also has a speed less than c in *every other* frame of reference. This is one reason for thinking that no material object may travel with a speed greater than that of light relative to *any* frame of reference. The relativistic generalizations of energy and momentum, which we'll consider in the sections that follow, give further support to this hypothesis.

27.6 Relativistic Momentum

Newton's laws of motion have the same form in all inertial frames of reference. When we transform coordinates from one inertial frame to another using the Galilean coordinate transformation, the laws should be *invariant* (unchanging). But we have just learned that the principle of relativity forces us to replace the Galilean transformation with the more general Lorentz transformation. As we will see, this requires corresponding generalizations in the laws of motion and the definitions of momentum and energy.

The principle of conservation of momentum states that **when two objects collide, the total momentum is constant,** provided that they are an isolated system (that is, provided that they interact only with each other, not with anything else) and provided that the velocities of the objects are measured in the same inertial reference frame. If conservation of momentum is a valid physical law, then, according to the principle of relativity, it must be valid in *all* inertial frames of reference.

But suppose we look at a collision of two particles in one inertial coordinate system S and find that the total momentum is conserved. Then we use the Lorentz transformation to obtain the velocities of the particles in a second inertial system S'. It turns out that if we use the Newtonian definition of momentum, $\vec{p} = m\vec{v}$, momentum is *not* conserved in the second system. If we're confident that the principle of relativity and the Lorentz transformation are correct, the only way to save momentum conservation is to generalize the Newtonian definition of momentum.

We'll omit the detailed derivation of the correct relativistic generalization of momentum, but here is the result: If a particle has mass m when it is at rest (or moving with a speed much smaller than c), then the **relativistic momentum $\vec{p}$** of the particle when it is moving with velocity $\vec{v}$ is given by the following expression:

Relativistic momentum

$$\vec{p} = \frac{m\vec{v}}{\sqrt{1 - v^2/c^2}}. \qquad (27.14)$$

When the particle's speed v is much less than c, Equation 27.14 is approximately equal to the Newtonian expression $\vec{p} = m\vec{v}$, but in general, the momentum is greater in magnitude than mv (Figure 27.15). In fact, as v approaches c, the magnitude of the momentum approaches infinity.

It's often convenient to use the abbreviation

$$\gamma = \frac{1}{\sqrt{1 - v^2/c^2}}. \qquad (27.15)$$

Using this abbreviation, we can rewrite Equation 27.14 as

$$\vec{p} = \gamma m\vec{v}. \qquad (27.16)$$

When v is much smaller than c, γ is approximately equal to one; when v is almost as large as c, γ approaches infinity. Thus, γ provides a measure of the amount by which Equation 27.14 and similar relations differ from their Newtonian counterparts. Figure 27.16 shows how γ increases as v approaches c.

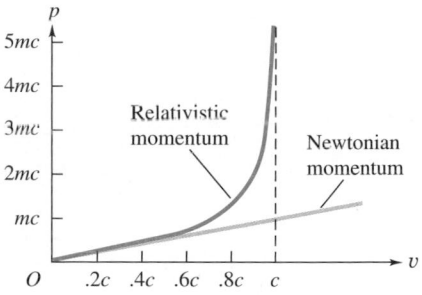

▲ **FIGURE 27.15** The magnitude of momentum as a function of speed.

Quantitative Analysis 27.4

The possible range of γ

Two objects, 1 and 2, are shot away from an observer with equal speeds and in opposite directions. Which of the following statements are correct about their respective values of γ (that is, γ_1 and γ_2)?

A. If $\gamma_1 > 0$, then $\gamma_2 < 0$.
B. If $\gamma_1 > 1$, then $\gamma_2 < 1$.
C. $1 \leq \gamma_1 < \infty$ and $1 \leq \gamma_2 < \infty$.
D. $0 \leq \gamma_1 < 1$ and $0 \leq \gamma_2 < 1$.

SOLUTION In Equation 27.15, the function γ depends on v^2, so objects with $+v$ and $-v$ have the same γ, since $(+v)^2 = (-v)^2 = v^2$. Therefore, $\gamma_1 = \gamma_2$, which is one way to rule out choices A and B. In addition, the speed of each object must be somewhere in the range $0 \leq v < c$. If we consider γ in the limit as $v \to 0$, then $\gamma \to 1$; likewise, in the limit as $v \to c$, $\gamma \to \infty$. Thus, for both objects, $1 \leq \gamma < \infty$.

In Equation 27.14, m is a *constant* that describes the inertial properties of a particle. Because $\vec{p} = m\vec{v}$ is still valid in the limit of very small velocities, m must be the same quantity we used (and learned to measure) in our study of Newtonian mechanics. In relativistic mechanics, m is often called the **rest mass** of a particle.

What about the relativistic generalization of Newton's second law? In Newtonian mechanics, one form of the second law, as we discussed in Section 8.1, is

$$\vec{F} = \frac{\Delta \vec{p}}{\Delta t}.$$

That is, force equals rate of change of momentum with respect to time. Experiments show that this result is still valid in relativistic mechanics, provided that we use the *relativistic* momentum given by Equation 27.14. An interesting aspect of this relation is that, because momentum is no longer directly proportional to velocity, the rate of change of momentum is no longer directly proportional to acceleration. As a result, **a constant force does not cause constant acceleration.** Furthermore, if the force has components both parallel and perpendicular to the object's instantaneous velocity, the relation between force and acceleration turns out to be different for the two components. In this case, the force and the acceleration are in different directions!

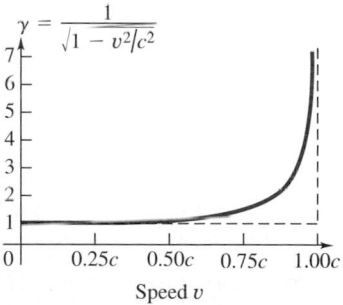

▲ **FIGURE 27.16** As v approaches c, $\gamma = 1/\sqrt{1 - v^2/c^2}$ grows larger without bound.

If the force and velocity have the same direction, it turns out that the acceleration a is given by

$$F = \frac{ma}{(1 - v^2/c^2)^{\frac{3}{2}}} = \gamma^3 ma. \tag{27.17}$$

When v is much smaller than c, γ is approximately one, and this expression reduces to the familiar Newtonian $F = ma$.

As Figure 27.16 shows, the factor $\gamma = 1/\sqrt{1 - v^2/c^2}$ in the relativistic momentum expression approaches infinity as the object's speed increases. Therefore, the acceleration caused by a given constant force continuously *decreases*. As the speed approaches c, the magnitude of the momentum approaches infinity and the acceleration approaches zero, no matter how great a force is applied. **It is impossible to accelerate a particle from a state of rest to a speed equal to or greater than c.** The speed of light is sometimes called **the ultimate speed;** no material object can travel faster than light.

EXAMPLE 27.6 Relativistic dynamics of a proton

A proton (rest mass 1.67×10^{-27} kg, charge 1.60×10^{-19} C) is moving parallel to an electric field that has magnitude $E = 5.00 \times 10^5$ N/C. Find the magnitudes of momentum and acceleration at the instants when $v = 0.010c$, $0.90c$, and $0.99c$.

SOLUTION

SET UP The relativistic definition of momentum, Equation 27.14, differs from its Newtonian counterpart by the factor γ. As Equation 27.15 shows, γ is only slightly greater than unity when v is much smaller than c, but γ becomes very large when v approaches c. Thus, our first step is to compute the value of γ for each speed given in the problem. Then we can find the corresponding momentum values from Equation 27.16. We'll also need to know that the force (magnitude F) exerted by the electric field on the proton is

$$F = qE = (1.60 \times 10^{-19}\,\text{C})(5.00 \times 10^5\,\text{N/C})$$
$$= 8.00 \times 10^{-14}\,\text{N}.$$

SOLVE Using Equation 27.15, we find that $\gamma = 1.00$, 2.29, and 7.09. From Equation 27.16, the corresponding values of p are

$$p_1 = \gamma mv = (1.00)(1.67 \times 10^{-27}\,\text{kg})(0.010)(3.00 \times 10^8\,\text{m/s})$$
$$= 5.01 \times 10^{-21}\,\text{kg}\cdot\text{m/s},$$

$$p_2 = (2.29)(1.67 \times 10^{-27}\,\text{kg})(0.90)(3.00 \times 10^8\,\text{m/s})$$
$$= 1.03 \times 10^{-18}\,\text{kg}\cdot\text{m/s},$$

$$p_3 = (7.09)(1.67 \times 10^{-27}\,\text{kg})(0.99)(3.00 \times 10^8\,\text{m/s})$$
$$= 3.52 \times 10^{-18}\,\text{kg}\cdot\text{m/s}.$$

From Equation 27.17,

$$a = \frac{F}{\gamma^3 m}.$$

When $v = 0.01c$ and $\gamma = 1.00$,

$$a_1 = \frac{8.00 \times 10^{-14}\,\text{N}}{(1)^3(1.67 \times 10^{-27}\,\text{kg})} = 4.79 \times 10^{13}\,\text{m/s}^2.$$

The other accelerations are smaller by factors of γ^3:

$$a_2 = 3.99 \times 10^{12}\,\text{m/s}^2, \qquad a_3 = 1.34 \times 10^{11}\,\text{m/s}^2.$$

▲ **FIGURE 27.17** The Stanford Linear Accelerator (SLAC).

These are only 8.33% and 0.280% of the values predicted by nonrelativistic mechanics.

REFLECT As the proton's speed increases, the relativistic values of momentum differ more and more from the nonrelativisitic values computed from $p = mv$. The momentum at $0.99c$ is more than three times as great as that at $0.90c$ because of the increase in the factor γ. We also note that as v approaches c, the acceleration drops off very quickly. In the Stanford Linear Accelerator (Figure 27.17), a path length of 3 km is needed to give electrons the speed that, according to classical physics, they could acquire in 1.5 cm.

Practice Problem: At what speed is the momentum of a proton twice as great as the nonrelativistic result (from $p = mv$)? Ten times as great? Would your results be different for an electron? *Answers:* $0.866c$; $0.995c$; no.

Equation 27.14 is sometimes interpreted to mean that a rapidly moving particle undergoes an increase in mass. If the mass at zero velocity (the rest mass) is denoted by m, then the "relativistic mass" m_{rel} is given by

$$m_{rel} = \frac{m}{\sqrt{1 - v^2/c^2}} = \gamma m. \qquad (27.18)$$

Indeed, when we consider the motion of a system of particles (such as gas molecules in a moving container), the total mass of the system is the sum of the relativistic masses of the particles, not the sum of their rest masses.

The concept of relativistic mass also has its pitfalls, however. As Equation 27.17 shows, it is *not* correct to say that the relativistic generalization of Newton's second law is $\vec{F} = m_{rel}\vec{a}$, and it is *not* correct that the relativistic kinetic energy of an object is $K = \frac{1}{2}m_{rel}v^2$. Thus, this concept must be approached with great caution. We prefer to think of m as a constant, unvarying quantity for any given object and to incorporate the correct relativistic relationships into the definitions of momentum (as we have done) and kinetic energy (the subject of the next section).

27.7 Relativistic Work and Energy

When we developed the relationship between work and kinetic energy in Chapter 7, we used Newton's laws of motion. But these laws have to be generalized to bring them into harmony with the principle of relativity, so we also need to generalize the definition of kinetic energy. It turns out that the classical definition of work $(W = F\,\Delta x)$ can be retained. Einstein showed that when a force does work W on a moving body, causing acceleration and changing the object's speed from v_1 to v_2, these quantities are related by the equation

$$W = \frac{mc^2}{\sqrt{1 - v_2^2/c^2}} - \frac{mc^2}{\sqrt{1 - v_1^2/c^2}}. \qquad (27.19)$$

This result suggests that we might define kinetic energy as

$$K \stackrel{?}{=} \frac{mc^2}{\sqrt{1 - v^2/c^2}}. \qquad (27.20)$$

But this expression isn't zero when the particle is at rest. Instead, when $v = 0$, K becomes equal to mc^2. Thus, a more reasonable relativistic definition of kinetic energy K is

Relativistic kinetic energy

$$K = \frac{mc^2}{\sqrt{1 - v^2/c^2}} - mc^2. \qquad (27.21)$$

This looks rather different from the Newtonian expression $K = \frac{1}{2}mv^2$, but the two must agree whenever v is much smaller than c. Indeed, it can be shown that as v approaches zero, the Newtonian and relativistic expressions give the same result, and we obtain the classical $\frac{1}{2}mv^2$. Figure 27.18 shows a graph of relativistic kinetic energy as a function of v, compared with the Newtonian expression.

But what is the significance of the term mc^2 that we had to subtract in Equation 27.21? Although Equation 27.20 does not give the *kinetic energy* of the particle, perhaps it represents some kind of *total* energy, including both the kinetic energy and an additional energy mc^2 that the particle possesses even when it is not moving. We'll call this additional energy the **rest energy** of the particle,

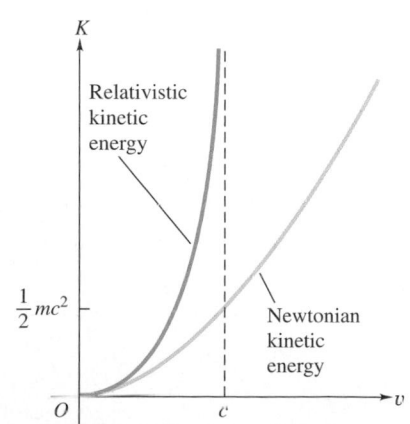

▲ **FIGURE 27.18** Comparison of relativistic and classical expressions for kinetic energy, as functions of speed v.

▲ Application Watts happening?
Einstein's theory of relativity forced us to revise our Newtonian conceptions of force, momentum, and energy. We now know that, in relativistic terms, a constant force does not necessarily produce a constant acceleration and that kinetic energy can be defined in terms of the speed of a particle and the speed of light. In one of the most famous equations in history, Einstein showed that a particle at rest contains a quantity of "rest energy" represented by $E = mc^2$ and that matter and energy are interconvertible. Much like conceptual breakthroughs in understanding electromagnetism, space-time, and the dual wave/particle nature of light, this astounding relationship showed that matter and energy are merely different forms of each other. This has practical applications in the nuclear power plant shown above, in which a tiny amount of the mass of a uranium nucleus being split is converted to a tremendous amount of energy.

denoted by $E_{rest} = mc^2$. Then when we denote the *total* energy by E and use Equation 27.21, we obtain this result:

Total energy of an object

$$E = K + mc^2 = K + E_{rest} = \frac{mc^2}{\sqrt{1 - v^2/c^2}}. \quad (27.22)$$

Rest energy of an object

$$E_{rest} = mc^2. \quad (27.23)$$

There is, in fact, direct experimental evidence that rest energy really does exist. The simplest example is the decay of the π^0 meson (also called the *neutral pion*). This is an unstable particle, produced in high-energy collisions of other particles. When it decays, it disappears and electromagnetic radiation appears. When the particle, with mass m, is at rest (and therefore has no kinetic energy) before its decay, the total energy of the radiation produced is found to be exactly equal to mc^2. There are many other examples of fundamental particle transformations in which the sum of the rest masses of the particles in the system changes. In every case, a corresponding energy change occurs, consistent with the assumption of a rest energy mc^2 associated with a rest mass m. We'll study such particle transformations in greater detail in Chapter 30.

Historically, the principles of conservation of mass and of energy developed quite independently. The theory of relativity now shows that they are actually two special cases of a single broader conservation principle: the **principle of conservation of mass and energy.** In some physical phenomena, neither the sum of the rest masses of the particles nor the total energy (other than rest energy) is separately conserved, but there is a more general conservation principle: **In an isolated system, when the sum of the rest masses changes, there is always an equal and opposite change in the total energy other than the rest energy.**

The conversion of mass into energy is the fundamental principle involved in the generation of power through nuclear reactions, a subject we will discuss in Chapter 30. When a uranium nucleus undergoes fission in a nuclear reactor, the total mass of the resulting fragments is less than that of the parent nucleus, and the total kinetic energy of the fragments is equal to this mass deficit multiplied by c^2. This kinetic energy can be used in a variety of ways, such as producing steam to operate turbines for electric-power generators.

Energy–Momentum Relations

In Chapter 8, we learned that in Newtonian mechanics, the kinetic energy K and the magnitude of momentum p for a particle with mass m are related by $K = p^2/2m$. In relativistic mechanics, there is an analogous (and equally simple) relation between the total energy E (kinetic energy plus rest energy) of an object and its magnitude of momentum p.

Relativistic energy–momentum relation

$$E^2 = (mc^2)^2 + (pc)^2. \quad (27.24)$$

This equation shows again the existence of rest energy; for a particle at rest $(p = 0)$, the total energy is $E = mc^2$.

Equation 27.24 also suggests the possible existence of particles that have energy and momentum even when they have zero rest mass. In such a case, $m = 0$ and the relation between E and p is even simpler:

Energy and momentum of massless particles

$$E = pc. \qquad (27.25)$$

In fact, massless particles *do* exist. One example is the *photon*—the quantum of electromagnetic radiation. Photons always travel with the speed of light; they are emitted and absorbed during changes of state of an atomic or nuclear system when the energy and momentum of the system change.

EXAMPLE 27.7 **Rest energy of the electron**

(a) Find the rest energy of an electron ($m = 9.09 \times 10^{-31}$ kg, $e = 1.60 \times 10^{-19}$ C) in joules and in electronvolts. (b) Find the speed of an electron that has been accelerated by an electric field from rest through a potential difference of 20.0 kV (typical of TV picture tubes in the pre-flat-panel era) and through 5.00 MV (a high-voltage x-ray machine).

SOLUTION

SET UP We need the expression for rest energy, $E_{\text{rest}} = mc^2$, the definition of the electronvolt (which we suggest you review in Section 18.3), and the conversion factor from electronvolts to joules. We find the total energy E in Equation 27.22 by adding the kinetic energy (from the accelerating potential difference) to the rest energy, and then we solve Equation 27.22 for v.

SOLVE Part (a): The electron's rest energy is

$$mc^2 = (9.09 \times 10^{-31} \text{ kg})(2.998 \times 10^8 \text{ m/s})^2$$
$$= 8.187 \times 10^{-14} \text{ J}.$$

From Section 18.3, 1 eV $= 1.602 \times 10^{-19}$ J, so

$$mc^2 = (8.187 \times 10^{-14} \text{ J})(1 \text{ eV}/1.602 \times 10^{-19} \text{ J})$$
$$= 5.11 \times 10^5 \text{ eV} = 0.511 \text{ MeV}.$$

Part (b): An electron accelerated through a potential difference of 20.0 kV gains a kinetic energy of $K = 20.0$ keV, equal to about 4% of the rest energy found in part (a). Combining these values with Equation 27.22, we get

$$20.0 \times 10^3 \text{ eV} + 5.11 \times 10^5 \text{ eV} = \frac{5.11 \times 10^5 \text{ eV}}{\sqrt{1 - v^2/c^2}}.$$

The rest is arithmetic; we first simplify this equation by dividing both sides by 5.11×10^5 eV and inverting both sides, obtaining

$$\sqrt{1 - v^2/c^2} = 0.962 \quad \text{and} \quad v = 0.272c.$$

REFLECT When $K = 20$ keV, the kinetic energy is a small fraction of the rest energy $E_{\text{rest}} = 511$ keV, and the speed is about one-fourth the speed of light. We invite you to use Equation 27.15 to show that in this case $\gamma = 1.04$. For this energy, the electron's behavior is not very different from what Newtonian mechanics would predict. But when we repeat the calculation with a potential difference of 5.00 MV and kinetic energy $K = 5.00$ MeV $= 5.00 \times 10^6$ eV, we find that $v = 0.996c$ and $\gamma = 10.8$. In this case, the kinetic energy is much *larger* than the rest energy, the speed is very close to c, and the particle's behavior is emphatically non-Newtonian. Such a speed is said to be in the *extreme relativistic range*.

Practice Problem: In Example 27.7, suppose the particle is a proton accelerated through the same potential differences as was the electron in that example. Find the speed in each case. Could nonrelativistic approximations be used? *Answers:* 1.96×10^6 m/s, 3.08×10^7 m/s; yes (to 0.4%).

EXAMPLE 27.8 **Colliding protons and meson production**

Two protons, each with rest mass $M = 1.67 \times 10^{-27}$ kg, have equal speeds in opposite directions. They collide head-on, producing a π^0 meson (also called a *neutral pion*) with mass $m = 2.40 \times 10^{-28}$ kg (Figure 27.19). If all three particles are at rest after the collision, find the initial speed of each proton and find its kinetic energy, expressed as a fraction of its rest energy.

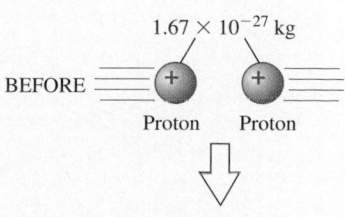

1.67×10^{-27} kg

BEFORE

Proton Proton

AFTER

Meson (2.40×10^{-28} kg)

▲ **FIGURE 27.19**

Continued

SOLUTION

SET UP The key is to understand that the total energy of the system (including the rest energy) is conserved, so we can write the "before-and-after" equation:

$$\frac{2Mc^2}{\sqrt{1 - v^2/c^2}} = 2Mc^2 + mc^2.$$

SOLVE Substituting numerical values, taking care to use consistent units, we find that

$$\frac{2(1.67 \times 10^{-27}\,\text{kg})}{\sqrt{1 - v^2/c^2}} = 2(1.67 \times 10^{-27}\,\text{kg})$$
$$+ (2.40 \times 10^{-28}\,\text{kg}).$$

We divide by the proton mass M and invert, obtaining

$$\sqrt{1 - v^2/c^2} = 0.933 \quad \text{and} \quad v = 0.360c.$$

The initial kinetic energy K of each proton is half the pion rest energy, $K = mc^2/2$, so the ratio of K to the proton rest energy Mc^2 is

$$\frac{mc^2/2}{Mc^2} = \frac{m}{2M} = \frac{2.40 \times 10^{-28}\,\text{kg}}{2(1.67 \times 10^{-27}\,\text{kg})} = 0.0719.$$

The rest energy of a proton is 938 MeV, so its initial kinetic energy is $(0.0719)(938\,\text{MeV}) = 67.4\,\text{MeV}$.

REFLECT We invite you to verify that the rest energy of the π^0 meson is twice this result, or 135 MeV. Also, note that the initial kinetic energies of the protons are not much different from the values they would have according to Newtonian mechanics. Thus, the speeds might be called nonrelativistic, but the direct conversion of kinetic energy into mass is definitely a relativistic process.

Practice Problem: In this example, suppose the colliding particles are electrons instead of protons. Find the initial energy of each electron, and show that the energies and speeds are in the extreme relativistic range. *Answer:* 67.4 MeV (electron rest energy is 0.511 MeV).

27.8 Relativity and Newtonian Mechanics

The sweeping changes required by the principle of relativity go to the very roots of Newtonian mechanics, including the concepts of length and time, the equations of motion, and the conservation principles. Thus it may appear that we have destroyed the foundations on which Newtonian mechanics is built. In one sense this is true, and yet the Newtonian formulation is still valid whenever speeds are small in comparison to the speed of light. In such cases, time dilation, length contraction, and the modifications of the laws of motion are so small that they are unobservable. In fact, every one of the principles of Newtonian mechanics survives as a special case of the more general relativistic formulation.

So the laws of Newtonian mechanics are not *wrong;* they are *incomplete.* They are a limiting case of relativistic mechanics; they are approximately correct when all speeds are small in comparison to c and they become exactly correct in the limit when all speeds approach zero. Thus, relativity does not destroy the laws of Newtonian mechanics, but rather generalizes them. Newton's laws rest on a solid base of experimental evidence, and it would be strange to advance a new theory that is inconsistent with this evidence. There are many situations for which Newtonian mechanics is clearly inadequate, including all phenomena in which particle speeds are comparable to that of light or in which the direct conversion of mass to energy occurs. But there is still a large area, including nearly all the behavior of macroscopic bodies in mechanical systems, in which Newtonian mechanics is perfectly adequate. It's interesting to speculate how different the experiences of everyday life would be if the speed of light were 10 m/s; we would be living in a very different world indeed!

At this point, we may ask whether relativistic mechanics is the final word on this subject or whether *further* generalizations are possible or necessary. For example, inertial frames of reference have occupied a privileged position in our

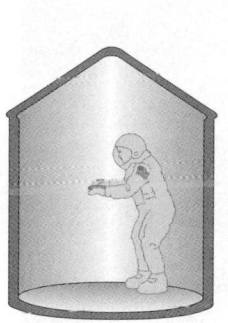

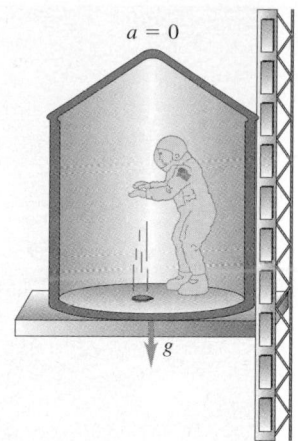

An astronaut is about to drop her watch in a spaceship.

(a)

In gravity-free space, the floor accelerates upward at $a = g$ and hits the watch.

(b)

On the earth's surface, the watch accelerates downward at $a = g$ and hits the floor.

(c)

▲ **FIGURE 27.20** Without information from outside the spaceship, the astronaut cannot distinguish situation (b) from situation (c).

discussion. Should the principle of relativity be extended to noninertial frames as well?

Here's an example that illustrates some implications of this question. A student is standing in an elevator. The elevator cables have all broken, and the safety devices have all failed at once, so the elevator is in free fall with an acceleration of 9.8 m/s^2 relative to earth. Temporarily losing her composure, the student loses her grasp of her physics textbook. The book doesn't fall to the floor, because it, the student, and the elevator all have the same free-fall acceleration. But the student can view the situation in two ways: Is the elevator really in free fall, with her in it, or is it possible that somehow the earth's gravitational attraction has suddenly vanished? As long as she remains in the elevator and it remains in free fall, she can't tell whether she is indeed in free fall or whether the gravitational interaction has vanished.

A more realistic illustration is an astronaut in a space station in orbit around the earth. Objects in the spaceship *seem* to be weightless because both they and the station are in circular or elliptical orbits and are constantly accelerating toward the earth. But without looking outside the ship, the astronauts have no way to determine whether gravitational interactions have disappeared or whether the spaceship is in a noninertial frame of reference accelerating toward the center of the earth (Figure 27.20).

These considerations form the basis of Einstein's **general theory of relativity.** If we can't distinguish experimentally between a gravitational field at a particular location and an accelerated reference system, then there can't be any real distinction between the two. Pursuing this concept, we may try to represent *any* gravitational field in terms of special characteristics of the coordinate system. This turns out to require even more sweeping revisions of our space–time concepts than did the special theory of relativity, and we find that, in general, the geometric properties of space are non-Euclidean (Figure 27.21) and gravitational fields are closely related to the geometry of space.

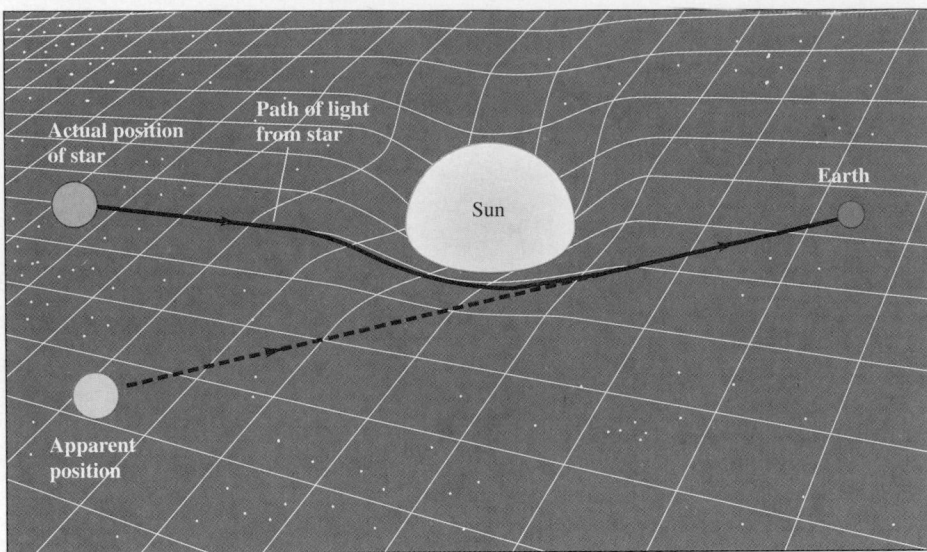

▲ **FIGURE 27.21** Curved space. The change of position is greatly exaggerated.

The general theory of relativity has passed several experimental tests, including three proposed by Einstein. One test has to do with understanding the rotation of the axes of the planet Mercury's elliptical orbit, called the *precession of the perihelion* of Mercury. (The perihelion is the point of closest approach to the sun.) Another test concerns the apparent bending of light rays from distant stars when they pass near the sun, and the third test is the *gravitational red shift,* the increase in wavelength of light proceeding outward from a massive source. Some details of the general theory are more difficult to test and remain speculative in nature, but this theory has played a central role in cosmological investigations of the structure of the universe, the formation and evolution of stars, and related matters.

SUMMARY

Relativity and Simultaneity

(Sections 27.1 and 27.2) All the fundamental laws of physics have the same form in all inertial frames of reference. The speed of light in vacuum is the same in all inertial frames of reference and is independent of the motion of the source. Simultaneity is not an absolute concept: two events that appear to be simultaneous in one inertial frame may not appear simultaneous in a second frame moving relative to the first.

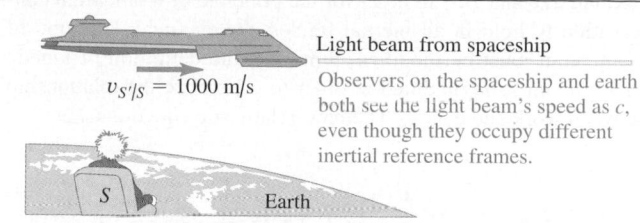

Light beam from spaceship

$v_{S'/S} = 1000$ m/s

Observers on the spaceship and earth both see the light beam's speed as c, even though they occupy different inertial reference frames.

S Earth

Relativity of Time

(Section 27.3) If Δt_0 is the time interval between two events that occur at the same space point in a particular frame of reference, it is called a **proper time.** If the first frame moves with a constant velocity u relative to a second frame, the time interval Δt between the events, as observed in the second frame, is longer than Δt_0:

$$\Delta t = \frac{\Delta t_0}{\sqrt{1 - u^2/c^2}}. \tag{27.6}$$

This effect is called **time dilation:** A clock moving with respect to an observer appears to run more slowly (Δt) than a clock that is at rest in the observer's frame (Δt_0).

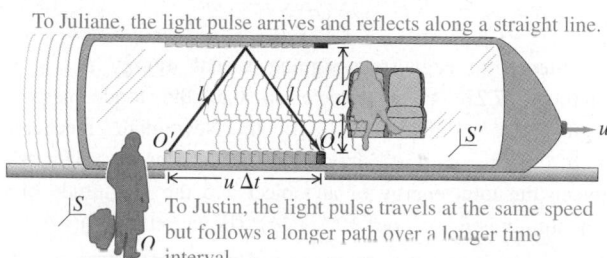

To Juliane, the light pulse arrives and reflects along a straight line.

l l d

O' O' S' u

$u\,\Delta t$

To Justin, the light pulse travels at the same speed but follows a longer path over a longer time interval.

Relativity of Length

(Section 27.4) If l_0 is the distance between two points that are at rest in a particular frame of reference, it is called a **proper length.** If this first frame is moving with a constant velocity u relative to a second frame, and distances in each frame are measured parallel to that frame's velocity, then the distance l between the points, as observed in the second frame, is shorter than l_0, in accordance with the formula.

$$l = l_0\sqrt{1 - \frac{u^2}{c^2}}. \tag{27.9}$$

This effect is called **length contraction:** A ruler oriented parallel to the frame's velocity and moving with respect to an observer appears shorter (l) than when measured in a frame at rest (l_0) in the observer's frame.

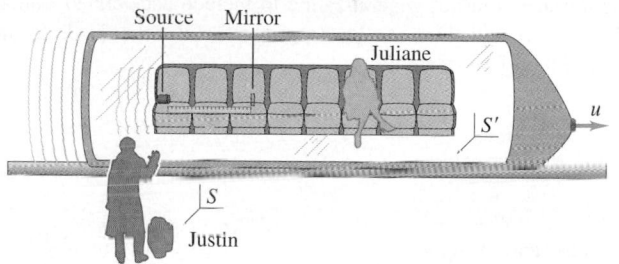

Source   Mirror

Juliane

S' u

S

Justin

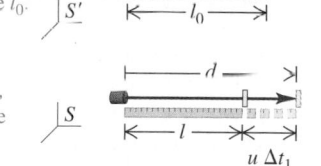

Source   Mirror

In Juliane's frame of reference S', the light pulse travels a distance l_0.

S' l_0

In Justin's frame of reference S, the light pulse travels a distance $l + u\,\Delta t_1$.

S

d

l $u\,\Delta t_1$

The Lorentz Transformation

(Section 27.5) The **Lorentz transformation** relates the coordinates and time of an event in an inertial coordinate system S to the coordinates and time of the same event as observed in a second inertial frame S' moving with constant velocity u relative to the first. The Lorentz transformation equations are

$$x' = \frac{x - ut}{\sqrt{1 - u^2/c^2}}, \qquad y' = y, \qquad t' = \frac{t - ux/c^2}{\sqrt{1 - u^2/c^2}}. \tag{27.10}$$

For one-dimensional motion, the velocity v' in S' is related to the velocity v in S by

$$v' = \frac{v - u}{1 - uv/c^2}. \tag{27.12}$$

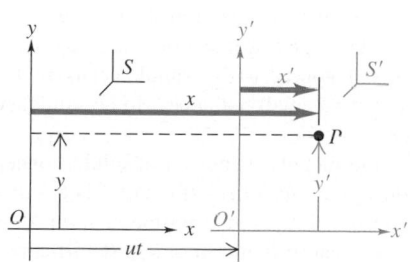

Continued

Relativistic Momentum and Energy

(Sections 27.6 and 27.7) In order for the principle of momentum conservation to hold in all inertial frames, the classical definition of momentum must be modified. Similarly, the definition of kinetic energy K must be modified in order to generalize the relationship between work and energy. The new, relativistic equations are

$$\vec{p} = \frac{m\vec{v}}{\sqrt{1 - v^2/c^2}} \qquad (27.14)$$

and

$$K = \frac{mc^2}{\sqrt{1 - v^2/c^2}} - mc^2 \qquad (27.21)$$

The latter form suggests assigning a **rest energy** $E_{rest} = mc^2$ (Equation 27.23) to a particle so that the total energy is $E = K + mc^2$ (Equation 27.22). In Newtonian mechanics, $K = p^2/2m$. In relativistic mechanics, the analogous relationship between the total energy of an object and the magnitude of its momentum p is $E^2 = (mc^2)^2 + (pc)^2$ (Equation 27.24).

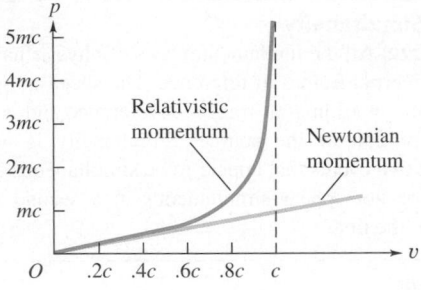

Relativity and Newtonian Mechanics

(Section 27.8) The special theory of relativity is a generalization of Newtonian mechanics. All the principles of Newtonian mechanics are present as limiting cases when all the speeds are small in comparison to c. Further generalization to include accelerated frames of reference and their relation to gravitational fields leads to the general theory of relativity.

 For instructor-assigned homework, go to www.masteringphysics.com

Conceptual Questions

1. Suppose the speed of light were 30 m/s instead of its actual value. Describe various ways in which the behavior of your everyday world would be different than it is now.

2. The average life span in the United States is about 70 years. Does this mean that it is impossible for an average person to travel a distance greater than 70 light years away from the earth? (A light year is the distance light travels in a year.)

3. Two events occur at the same spatial point in a particular frame of reference and appear to be simultaneous in that frame. Is it possible that they will not appear to be simultaneous in another frame?

4. Does the fact that simultaneity is not an absolute concept also destroy the concept of *causality?* If event A is to *cause* event B, A must occur first. Is it possible that in some frames A will appear to cause B and in others B will appear to cause A?

5. You are standing on a train platform watching a high-speed train pass by. A light inside one of the train cars is turned on and then a little later it is turned off. (a) Who can measure the proper time interval for the duration of the light: you or a passenger on the train? (b) Who can measure the proper length of the train car, you or a passenger on the train? (c) Who can measure the proper length of a sign attached to a post on the train platform, you or a passenger on the train? In each case, explain your answer.

6. According to the twin paradox mentioned in Section 27.3, if one twin stays on earth while the other takes off in a spaceship at a relativistic speed and then returns, one will be older than the other. Can you think of a practical experiment, perhaps using two very precise atomic clocks, that would test this conclusion?

7. Photons are considered to be "massless" particles. Yet since they have energy, you might expect that they also have mass. So what does it mean to call a particle "massless"?

8. A student asserted that a massless particle must always travel at exactly the speed of light. Why do you think this is correct? If so, how do massless particles such as photons acquire that speed? Can they start from rest and accelerate?

9. The theory of relativity sets an upper limit on the speed a particle can have. Does this mean that there are also upper limits on its energy and momentum?

10. Why do you think the development of Newtonian mechanics preceded the more refined relativistic mechanics by so many years?

11. You're approaching the star Betelgeuse in your spaceship at a speed of 0.2*c*. At what speed does the light from the star reach you?

12. Does relativity say that we can travel fast enough to actually get younger than when we started? Explain.

13. Discuss several good reasons for believing that no matter can reach the speed of light. (*Hint:* Look at what happens to several physical quantities, such as kinetic energy and momentum, as $v \rightarrow c$.)

14. People sometimes interpret the theory of relativity as saying that "everything is relative." Is this really what the theory says? Can you think of any physical quantities that are *not* relative in this theory? (In fact, Einstein thought of his theory as a theory of *absolutes* rather than *relativity*.)

15. Some people have expressed doubt about the theory of relativity, dismissing it as little more than a *theory*. What *experimental evidence* can you cite that this theory is actually correct?

Multiple-Choice Problems

1. A rocket flies toward the earth at $\frac{1}{2}c$, and the captain shines a laser light beam in the forward direction. Which of the following statements about the speed of this light are correct? (There may be more than one correct answer.)
   A. The captain measures speed *c* for the light.
   B. An observer on earth measures speed *c* for the light.
   C. An observer on earth measures speed $\frac{3}{2}c$ for the light.
   D. The captain measures speed $\frac{1}{2}c$ for the light.

2. A rocket is traveling at $\frac{1}{3}c$ relative to earth when a lightbulb in the center of a cubical room is suddenly turned on. An astronaut at rest in the rocket and a person at rest on earth both observe the light hit the opposite walls of the room. (See Figure 27.22.) We shall call these events *A* and *B*. Which statements about the two events are true? (There may be more than one true statement.)

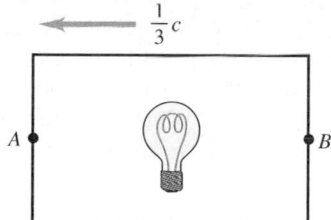

▲ FIGURE 27.22 Multiple-choice problem 2.

   A. To the observer on earth, both events happen at the same time.
   B. To the astronaut, both events happen at the same time.
   C. To the observer on earth, event *B* happens before event *A*.
   D. To the astronaut, event *A* happens after event *B*.

3. According to the principles of special relativity (there may be more than one correct choice),
   A. The speed of light is the same for all observers in inertial reference frames, no matter how fast those frames are moving.
   B. All physical quantities are relative, their values depending on the motion of the observer.

   C. Two events that occur at the same time for one observer must occur at the same time for all other observers.
   D. The laws of physics are the same in all inertial reference frames, no matter how fast they are moving.

4. A square measuring 1 m by 1 m is moving away from observer *A* along a direction parallel to one of its sides at a speed such that γ is equal to 2. The *area* of this square, as measured by observer *A*, is
   A. 4 m^2.   B. 2 m^2.   C. 1 m^2.   D. $\frac{1}{2}$ m^2.   E. $\frac{1}{4}$ m^2.

5. To an observer moving along with the square in the previous question, the area of the square is
   A. 4 m^2.   B. 2 m^2.   C. 1 m^2.   D. $\frac{1}{2}$ m^2.   E. $\frac{1}{4}$ m^2.

6. To the observer moving along with the square in Question 4, the time interval between consecutive blinks of her eyes is 1 s. To the stationary observer *A*, this time interval is
   A. 2 s.   B. 1 s.   C. $\frac{1}{2}$ s.   D. $\frac{1}{4}$ s.

7. A high-speed train passes a train platform. Anthony is a passenger on the train, Miguel is standing on the train platform, and Carolyn is riding a bicycle toward the platform in the same direction as the train is traveling. Choose the proper order of how long each of these observers measures the train to be, from longest to shortest.
   A. Carolyn, Miguel, Anthony.
   B. Miguel, Carolyn, Anthony.
   C. Miguel, Anthony, Carolyn.
   D. Anthony, Carolyn, Miguel.
   E. Anthony, Miguel, Carolyn.

8. For a material object, such as a rocket ship, the possible range of γ is
   A. $0 \leq \gamma \leq 1$.   B. $\gamma \geq 1$.   C. $0 \leq \gamma < \infty$.

9. If it requires energy *U* to accelerate a rocket from rest to $\frac{1}{2}c$, the energy needed to accelerate that rocket from $\frac{1}{2}c$ to *c* would be
   A. $\frac{1}{2}U$.   B. *U*.   C. 2*U*.   D. infinite.

10. The reason we do not observe relativistic effects (such as time dilation or length contraction) at ordinary speeds on earth is that
    A. Special relativity is valid at all speeds, but the effects are normally too small to observe at ordinary speeds on earth.
    B. Special relativity is valid only when the speed of an object approaches that of light.
    C. We do readily observe relativistic effects for objects such as jet planes.

11. A rocket is traveling toward the earth at $\frac{1}{2}c$ when it ejects a missile forward at $\frac{1}{2}c$ relative to the rocket. According to *Galilean* velocity addition, the speed of this missile as measured by an observer on earth would be
    A. 0.   B. $\frac{4}{5}c$.   C. *c*.

12. For the missile in the previous question, the *correct* value for its speed measured by an observer on earth would be
    A. 0.   B. $\frac{4}{5}c$.   C. *c*.

13. Suppose a rocket traveling at 99.99% of the speed of light measured relative to the earth makes a trip to a star 100 light years from earth (meaning that it would take light 100 years to make the trip). During this rocket trip (there may be more than one correct choice),
    A. People on earth would age essentially 100 years.
    B. The astronauts in the rocket would age more than 100 years.
    C. The astronauts in the rocket would age less than 100 years.
    D. The astronauts in the rocket would age the same as the people on earth.

14. A rocket ship is moving toward earth at $\frac{2}{3}c$. The crew is using a telescope to watch a Cubs baseball game in Chicago. The

batter hits the ball (event *A*), which is soon caught (event *B*) by a player 175 ft away, as measured in the ball park. Which one of the following is the proper length of the distance the ball traveled?
A. the 175 ft measured in the ball park.
B. the distance measured by the rocket's crew.
C. Both distances are equal, and hence both are the proper length.

15. A large constant force is used to accelerate an object from rest to a high speed. In which form of Newton's second law—relativistic or classical nonrelativistic—does the object take a longer time to reach a speed of 0.9*c*?
A. Relativistic.   B. Nonrelativistic.   C. Same for both.

Problems

27.1 Invariance of Physical Laws
27.2 Relative Nature of Simultaneity

1. •• A spaceship is traveling toward earth from the space colony on Asteroid 1040A. The ship is at the halfway point of the trip, passing Mars at a speed of 0.9*c* relative to Mars's frame of reference. At the same instant, a passenger on the spaceship receives a radio message from her boyfriend on 1040A and another from her hairdresser on earth. According to the passenger on the ship, were these messages sent simultaneously or at different times. If at different times, which one was sent first? Explain your reasoning.

2. •• A rocket is moving to the right at half the speed of light relative to the earth. A lightbulb in the center of a room inside the rocket suddenly turns on. Call the light hitting the front end of the room event *A* and the light hitting the back of the room event *B*. (See Figure 27.23.) Which event occurs first, *A* or *B*, or are they simultaneous, as viewed by (a) an astronaut riding in the rocket and (b) a person at rest on the earth?

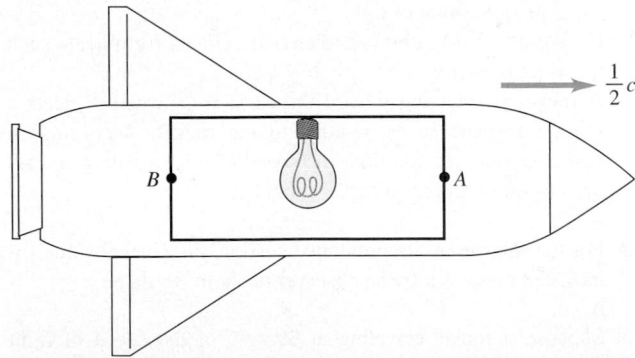

▲ **FIGURE 27.23** Problem 2.

27.3 Relativity of Time

3. • A futuristic spaceship flies past Pluto with a speed of 0.964c relative to the surface of the planet. When the spaceship is directly overhead at an altitude of 1500 km, a very bright signal light on the surface of Pluto blinks on and then off. An observer on Pluto measures the signal light to be on for 80.0 μs. What is the duration of the light pulse as measured by the pilot of the spaceship?

4. • Inside a spaceship flying past the earth at three-fourths the speed of light, a pendulum is swinging. (a) If each swing takes 1.50 s as measured by an astronaut performing an experiment inside the spaceship, how long will the swing take as measured by a person at mission control on earth who is watching the experiment? (b) If each swing takes 1.50 s as measured by a person at mission control on earth, how long will it take as measured by the astronaut in the spaceship?

5. • You take a trip to Pluto and back (round trip 11.5 billion km), traveling at a constant speed (except for the turnaround at Pluto) of 45,000 km/h. (a) How long does the trip take, in hours, from the point of view of a friend on earth? About how many years is this? (b) When you return, what will be the difference between the time on your atomic wristwatch and the time on your friend's? (*Hint:* Assume the distance and speed are highly precise, and carry a lot of significant digits in your calculation!)

6. • The negative pion (π^-) is an unstable particle with an average lifetime of 2.60×10^{-8} s (measured in the rest frame of the pion). (a) If the pion is made to travel at very high speed relative to a laboratory, its average lifetime is measured in the laboratory to be 4.20×10^{-7} s. Calculate the speed of the pion expressed as a fraction of *c*. (b) What distance, as measured in the laboratory, does the pion travel during its average lifetime?

7. • An alien spacecraft is flying overhead at a great distance as you stand in your backyard. You see its searchlight blink on for 0.190 s. The first officer on the craft measures the searchlight to be on for 12.0 ms. (a) Which of these two measured times is the proper time? (b) What is the speed of the spacecraft relative to the earth, expressed as a fraction of the speed of light, *c*?

8. •• How fast must a rocket travel relative to the earth so that time in the rocket "slows down" to half its rate as measured by earth-based observers? Do present-day jet planes approach such speeds?

9. •• A spacecraft flies away from the earth with a speed of 4.80×10^6 m/s relative to the earth and then returns at the same speed. The spacecraft carries an atomic clock that has been carefully synchronized with an identical clock that remains at rest on earth. The spacecraft returns to its starting point 365 days (1 year) later, as measured by the clock that remained on earth. What is the difference in the elapsed times on the two clocks, measured in hours? Which clock, the one in the spacecraft or the one on earth, shows the smaller elapsed time?

27.4 Relativity of Length

10. • You measure the length of a futuristic car to be 3.60 m when the car is at rest relative to you. If you measure the length of the car as it zooms past you at a speed of 0.900c, what result do you get?

11. • A meterstick moves past you at great speed. Its motion relative to you is parallel to its long axis. If you measure the length of the moving meterstick to be 1.00 ft (1 ft = 0.3048 m)—for example, by comparing it with a 1-foot ruler that is at rest relative to you, at what speed is the meterstick moving relative to you?

12. • In the year 2084, a spacecraft flies over Moon Station III at a speed of 0.800c. A scientist on the moon measures the length of the moving spacecraft to be 140 m. The spacecraft later lands on the moon, and the same scientist measures the length of the now stationary spacecraft. What value does she get?

13. • A rocket ship flies past the earth at 85.0% of the speed of light. Inside, an astronaut who is undergoing a physical examination is having his height measured while he is lying down parallel to the direction the rocket ship is moving. (a) If his height is measured to be 2.00 m by his doctor inside the ship, what height would a person watching this from earth measure for his height? (b) If the earth-based person had measured 2.00 m, what would the doctor in the spaceship have measured for the astronaut's height? Is this a reasonable height? (c) Suppose the astronaut in part (a) gets up after the examination and stands with his body perpendicular to the direction of motion. What would the doctor in the rocket and the observer on earth measure for his height now?

14. •• A spaceship makes the long trip from earth to the nearest star system, Alpha Centauri, at a speed of $0.955c$. The star is about 4.37 light years from earth, as measured in earth's frame of reference (1 light year is the distance light travels in a year). (a) How many years does the trip take, according to an observer on earth? (b) How many years does the trip take according to a passenger on the spaceship? (c) How many light years distant is Alpha Centauri from earth, as measured by a passenger on the speeding spacecraft? (Note that, in the ship's frame of reference, the passengers are at rest, while the space between earth and Alpha Centauri goes rushing past at $0.955c$.) (d) Use your answer from part (c) along with the speed of the spacecraft to calculate another answer for part (b). Do your two answers for that part agree? Should they?

15. •• A muon is created 55.0 km above the surface of the earth (as measured in the earth's frame). The average lifetime of a muon, measured in its own rest frame, is 2.20 μs, and the muon we are considering has this lifetime. In the frame of the muon, the earth is moving toward the muon with a speed of $0.9860c$. (a) In the muon's frame, what is its initial height above the surface of the earth? (b) In the muon's frame, how much closer does the earth get during the lifetime of the muon? What fraction is this of the muon's original height, as measured in the muon's frame? (c) In the earth's frame, what is the lifetime of the muon? In the earth's frame, how far does the muon travel during its lifetime? What fraction is this of the muon's original height in the earth's frame?

27.5 The Lorentz Transformation

16. • An enemy spaceship is moving toward your starfighter with a speed of $0.400c$, as measured in your reference frame. The enemy ship fires a missile toward you at a speed of $0.700c$ relative to the enemy ship. (See Figure 27.24.) (a) What is the speed of the missile relative to you? Express your answer in terms of the speed of light. (b) If you measure the enemy ship to be 8.00×10^6 km away from you when the missile is fired, how much time, measured in your frame, will it take the missile to reach you?

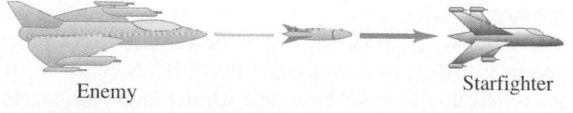

Enemy   Starfighter

▲ **FIGURE 27.24** Problem 16.

17. •• An imperial spaceship, moving at high speed relative to the planet Arrakis, fires a rocket toward the planet with a speed of $0.920c$ relative to the spaceship. An observer on Arrakis measures the rocket to be approaching with a speed of $0.360c$. What is the speed of the spaceship relative to Arrakis? Is the spaceship moving toward or away from Arrakis?

18. • Two particles in a high-energy accelerator experiment are approaching each other head-on, each with a speed of $0.9520c$ as measured in the laboratory. What is the magnitude of the velocity of one particle relative to the other?

19. • A pursuit spacecraft from the planet Tatooine is attempting to catch up with a Trade Federation cruiser. As measured by an observer on Tatooine, the cruiser is traveling away from the planet with a speed of $0.600c$. The pursuit ship is traveling at a speed of $0.800c$ relative to Tatooine, in the same direction as the cruiser. What is the speed of the cruiser relative to the pursuit ship?

20. • Two particles are created in a high-energy accelerator and move off in opposite directions. The speed of one particle, as measured in the laboratory, is $0.650c$, and the speed of each particle relative to the other is $0.950c$. What is the speed of the second particle, as measured in the laboratory?

21. •• Neutron stars are the remains of exploded stars, and they rotate at very high rates of speed. Suppose a certain neutron star has a radius of 10.0 km and rotates with a period of 1.80 ms. (a) Calculate the surface rotational speed at the equator of the star as a fraction of c. (b) Assuming the star's surface is an inertial frame of reference (which it isn't, because of its rotation), use the Lorentz velocity transformation to calculate the speed of a point on the equator with respect to a point directly opposite it on the star's surface.

27.6 Relativistic Momentum

22. • At what speed is the momentum of a particle three times as great as the result obtained from the nonrelativistic expression mv?

23. •• (a) At what speed does the momentum of a particle differ by 1.0% from the value obtained with the nonrelativistic expression mv? (b) Is the correct relativistic value greater or less than that obtained from the nonrelativistic expression?

24. • **Relativistic baseball.** Calculate the magnitude of the force required to give a 0.145 kg baseball an acceleration of $a = 1.00 \text{ m/s}^2$ in the direction of the baseball's initial velocity, when this velocity has a magnitude of (a) 10.0 m/s; (b) $0.900c$; (c) $0.990c$.

25. •• Sketch a graph of (a) the nonrelativistic Newtonian momentum as a function of speed v and (b) the relativistic momentum as a function of v. In both cases, start from $v = 0$ and include the region where $v \to c$. Does either of these graphs extend beyond $v = c$?

26. • An electron is acted upon by a force of 5.00×10^{-15} N due to an electric field. Find the acceleration this force produces in each case: (a) The electron's speed is 1.00 km/s. (b) The electron's speed is 2.50×10^8 m/s and the force is parallel to the velocity.

27.7 Relativistic Work and Energy

27. • Using both the nonrelativistic and relativistic expressions, compute the kinetic energy of an electron and the ratio of the two results (relativistic divided by nonrelativistic), for speeds of (a) 5.00×10^7 m/s, (b) 2.60×10^8 m/s.

28. • What is the speed of a particle whose kinetic energy is equal to (a) its rest energy, (b) five times its rest energy?

29. • **Particle annihilation.** In proton–antiproton annihilation, a proton and an antiproton (a negatively charged particle with the mass of a proton) collide and disappear, producing electromagnetic radiation. If each particle has a mass of 1.67×10^{-27} kg and they are at rest just before the annihilation, find the total energy of the radiation. Give your answers in joules and in electron volts.

30. • The sun produces energy by nuclear fusion reactions, in which matter is converted into energy. By measuring the amount of energy we receive from the sun, we know that it is producing energy at a rate of 3.8×10^{26} W. (a) How many kilograms of matter does the sun lose each second? Approximately how many tons of matter is this? (b) At this rate, how long would it take the sun to use up all its mass? (See Appendix E.)

31. •• A proton (rest mass 1.67×10^{-27} kg) has total energy that is 4.00 times its rest energy. What are (a) the kinetic energy of the proton; (b) the magnitude of the momentum of the proton; (c) the speed of the proton?

32. • In a hypothetical nuclear-fusion reactor, two deuterium nuclei combine or "fuse" to form one helium nucleus. The mass of a deuterium nucleus, expressed in atomic mass units (u), is 2.0136 u; that of a helium nucleus is 4.0015 u. $(1 \text{ u} = 1.661 \times 10^{-27} \text{ kg.})$ (a) How much energy is released when 1.0 kg of deuterium undergoes fusion? (b) The annual consumption of electrical energy in the United States is on the order of 1.0×10^{19} J. How much deuterium must react to produce this much energy?

33. •• **An antimatter reactor.** When a particle meets its antiparticle (more about this in Chapter 30), they annihilate each other and their mass is converted to light energy. The United States uses approximately 1.0×10^{20} J of energy per year. (a) If all this energy came from a futuristic antimatter reactor, how much mass would be consumed yearly? (b) If this antimatter fuel had the density of Fe (7.86 g/cm^3) and were stacked in bricks to form a cubical pile, how high would it be? (Before you get your hopes up, antimatter reactors are a *long* way in the future—if they ever will be feasible.)

34. • A particle has a rest mass of 6.64×10^{-27} kg and a momentum of 2.10×10^{-18} kg·m/s. (a) What is the total energy (kinetic plus rest energy) of the particle? (b) What is the kinetic energy of the particle? (c) What is the ratio of the kinetic energy to the rest energy of the particle?

35. •• (a) Through what potential difference does an electron have to be accelerated, starting from rest, to achieve a speed of $0.980c$? (b) What is the kinetic energy of the electron at this speed? Express your answer in joules and in electronvolts.

36. •• Sketch a graph of (a) the nonrelativistic Newtonian kinetic energy as a function of speed v, (b) the relativistic kinetic energy as a function of speed v. In both cases, start from $v = 0$ and include the region where $v \rightarrow c$. Does either of these graphs extend beyond $v = c$?

General Problems

37. •• The starships of the Solar Federation are marked with the symbol of the Federation, a circle, while starships of the Denebian Empire are marked with the Empire's symbol, an ellipse whose major axis is 1.40 times its minor axis

($a = 1.40b$ in Figure 27.25). How fast, relative to an observer, does an Empire ship have to travel for its markings to be confused with those of a Federation ship?

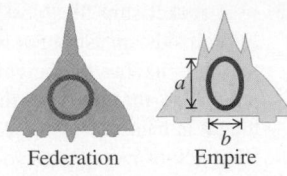
▲ **FIGURE 27.25** Problem 37.

38. • A space probe is sent to the vicinity of the star Capella, which is 42.2 light years from the earth. (A light year is the distance light travels in a year.) The probe travels with a speed of $0.9910c$ relative to the earth. An astronaut recruit on board is 19 years old when the probe leaves the earth. What is her biological age when the probe reaches Capella, as measured by (a) the astronaut and (b) someone on earth?

39. •• Two events are observed in a frame of reference S to occur at the same space point, the second occurring 1.80 s after the first. In a second frame S' moving relative to S, the second event is observed to occur 2.35 s after the first. What is the difference between the positions of the two events as measured in S'?

40. •• **Why are we bombarded by muons?** Muons are unstable subatomic particles (more on them in Chapter 30) that decay to electrons with a mean lifetime of 2.2 μs. They are produced when cosmic rays bombard the upper atmosphere about 10 km above the earth's surface, and they travel very close to the speed of light. The problem we want to address is why we see any of them at the earth's surface. (a) What is the greatest distance a muon could travel during its 2.2 μs lifetime? (b) According to your answer in part (a), it would seem that muons could never make it to the ground. But the 2.2 μs lifetime is measured in the frame of the muon, and they are moving very fast. At a speed of $0.999c$, what is the mean lifetime of a muon as measured by an observer at rest on the earth? How far could the muon travel in this time? Does this result explain why we find muons in cosmic rays? (c) From the point of view of the muon, it still lives for only 2.2 μs, so how does it make it to the ground? What is the thickness of the 10 km of atmosphere through which the muon must travel, as measured by the muon? Is it now clear how the muon is able to reach the ground?

41. •• How fast does a muon (see the previous problem) have to move (according to an outside observer) in order to travel 1.0 km during its brief lifetime of 2.2 μs?

42. • A cube of metal with sides of length a sits at rest in the laboratory with one edge parallel to the x axis. Therefore, in the laboratory frame, its volume is a^3. A rocket ship flies past the laboratory parallel to the x axis with a velocity v. To an observer in the rocket, what is the volume of the metal cube?

43. •• In an experiment, two protons are shot directly toward each other, each moving at half the speed of light relative to the laboratory. (a) What speed does one proton measure for the other proton? (b) What would be the answer to part (a) if we used only nonrelativistic Newtonian mechanics? (c) What is the kinetic energy of each proton as measured by (i) an observer at rest in the laboratory and (ii) an observer riding along with one of the protons? (d) What would be the answers to part (c) if we used only nonrelativistic Newtonian mechanics?

44. •• A 0.100 μg speck of dust is accelerated from rest to a speed of $0.900c$ by a constant 1.00×10^6 N force. (a) If the nonrelativistic form of Newton's second law $(\Sigma F = ma)$ is used, how far does the object travel to reach its final speed?

(b) Using the correct relativistic form of Equation 27.19, how far does the object travel to reach its final speed? (c) Which distance is greater? Why?

45. •• By what minimum amount does the mass of 4.00 kg of ice increase when the ice melts at 0.0°C to form water at that same temperature? (The heat of fusion of water is 3.34×10^5 J/kg.)

46. •• In certain radioactive beta decay processes (more about these in Chapter 30), the beta particle (an electron) leaves the atomic nucleus with a speed of 99.95% the speed of light relative to the decaying nucleus. If this nucleus is moving at 75.00% the speed of light, find the speed of the emitted electron relative to the laboratory reference frame if the electron is emitted (a) in the same direction that the nucleus is moving, (b) in the opposite direction from the nucleus's velocity. (c) In each case in parts (a) and (b), find the kinetic energy of the electron as measured in (i) the laboratory frame and (ii) the reference frame of the decaying nucleus.

47. •• Starting from Equation 27.24, show that in the classical limit $(pc \ll mc^2)$ the energy approaches the classical kinetic energy plus the rest mass energy. (*Hint:* If $x \ll 1$, $\sqrt{1+x} \approx 1 + x/2$.)

48. •• **Space travel?** Travel to the stars requires hundreds or thousands of years, even at the speed of light. Some people have suggested that we can get around this difficulty by accelerating the rocket (and its astronauts) to very high speeds so that they will age less due to time dilation. The fly in this ointment is that it takes a *great deal* of energy to do this. Suppose you want to go to the immense red giant Betelgeuse, which is about 500 light years away. (A light year is the distance that light travels in one year.) You plan to travel at constant speed in a 1000 kg rocket ship (a little over a ton), which, in reality, is far too small for this purpose. In each case that follows, calculate the time for the trip, as measured by people on earth and by astronauts in the rocket ship, the energy needed in joules, and the energy needed as a percent of U.S. yearly use (which is 1.0×10^{20} J). For comparison, arrange your results in a table showing v_{Rocket}, t_{Earth}, t_{Rocket}, E (in J), and E (as % of U.S. use). The rocket ship's speed is (a) 0.50c, (b) 0.99c, and (c) 0.9999c. On the basis of your results, does it seem likely that any government will invest in such high-speed space travel any time soon?

49. •• A nuclear device containing 8.00 kg of plutonium explodes. The rest mass of the products of the explosion is less than the original rest mass by one part in 10^4. (a) How much energy is released in the explosion? (b) If the explosion takes place in 4.00 μs, what is the average power developed by the bomb? (c) What mass of water could the released energy lift to a height of 1.00 km?

50. •• Electrons are accelerated through a potential difference of 750 kV, so that their kinetic energy is 7.50×10^5 eV. (a) What is the ratio of the speed v of an electron having this energy to the speed of light, c? (b) What would the speed be if it were computed from the principles of classical mechanics?

51. •• The distance to a particular star, as measured in the earth's frame of reference, is 7.11 light years (1 light year is the distance light travels in 1 year). A spaceship leaves earth headed for the star, and takes 3.35 years to arrive, as measured by

passengers on the ship. (a) How long does the trip take, according to observers on earth? (b) What distance for the trip do passengers on the spacecraft measure? (*Hint:* What is the speed of light in units of ly/y?)

52. •• **Čerenkov radiation.** The Russian physicist P. A. Čerenkov discovered that a charged particle traveling in a solid with a speed exceeding the speed of light *in that material* radiates electromagnetic radiation. (This phenomenon is analogous to the sonic boom produced by an aircraft moving faster than the speed of sound in air.) Čerenkov shared the 1958 Nobel Prize for this discovery. What is the minimum kinetic energy (in electronvolts) that an electron must have while traveling inside a slab of crown glass $(n = 1.52)$ in order to create Čerenkov radiation?

53. •• Scientists working with a particle accelerator determine that an unknown particle has a speed of 1.35×10^8 m/s and a momentum of 2.52×10^{-19} kg·m/s. From the curvature of its path in a magnetic field they also deduce that it has a positive charge. Using this information, identify the particle.

Passage Problems

Speed of light. Our universe has properties that are determined by the values of the fundamental physical constants, and it would be a much different place if the charge of the electron, the mass of the proton, or the speed of light were substantially different from what they actually are. For instance, the speed of light is so large that the effects of relativity usually go unnoticed in everyday events. Let's imagine an alternate universe where the speed of light is 1,000,000 times smaller than it is in our universe to see what would happen.

54. What is the speed of light in the alternate universe?
A. 3×10^8 m/s
B. 3×10^6 m/s
C. 3000 m/s
D. 300 m/s

55. An airplane has a length of 60 m when measured at rest. When the airplane is moving at 180 m/s (400 mph) in the alternate universe, how long would it appear to be to a stationary observer?
A. 24 m   B. 36 m
C. 48 m   D. 60 m
E. 75 m

56. If the airplane has a rest mass of 20,000 kg, what is its relativistic mass when moving at 180 m/s?
A. 8000 kg
B. 12,000 kg
C. 16,000 kg
D. 25,000 kg
E. 33,300 kg

57. In our universe the rest energy of an electron is approximately 8.2×10^{-14} J. What would it be in the alternate universe?
A. 8.2×10^{-8} J
B. 8.2×10^{-26} J
C. 8.2×10^{-2} J
D. 0.82 J

28 Photons, Electrons, and Atoms

W hat is light? The work of Maxwell, Hertz, and others established firmly that light is an electromagnetic wave. Interference, diffraction, and polarization phenomena show convincingly the wave nature of light and other electromagnetic radiation.

But there are also many phenomena, particularly those involving the emission and absorption of electromagnetic radiation, that show a completely different aspect of the nature of light, in which it seems to behave as a stream of *particles*. In such phenomena, the energy of light is emitted and absorbed in packages with a definite size, called *photons* or *quanta*. The energy of a single photon is proportional to the frequency of the radiation, and we say that the energy is *quantized*.

The energy associated with the internal motion within atoms is also quantized. Each kind of atom has a set of possible energy values called *energy levels*. The internal energy of an atom must be equal to one of these values; an atom cannot have an energy between two values. Understanding the internal structure of atoms requires a new language in which electrons sometimes behave like waves rather than particles.

The three common threads woven through this chapter are the quantization of electromagnetic radiation, the existence of discrete energy levels in atoms, and the dual wave–particle nature of both particles and electromagnetic radiation. These three basic concepts take us a long way toward understanding a wide variety of otherwise puzzling phenomena, including the photoelectric effect (the emission of electrons from a surface when light strikes it), the line spectra emitted by gaseous elements, the operation of lasers, and the production and scattering of x rays. Analysis of these phenomena and their relation to atomic structure takes us to the threshold of quantum mechanics, which involves some radical changes in our views of the nature of radiation and of matter itself.

When you take a snapshot, you use light. However, this false-color scanning electron micrograph of a fruit fly was made using a beam of electrons instead. Up to now, we've treated electrons as particles and light as a wave. Now we'll see that both of them can actually be treated in both ways.

28.1 The Photoelectric Effect

The **photoelectric effect,** first observed by Heinrich Hertz in 1887, is the emission of electrons from the surface of a conductor when light strikes the surface. The liberated electrons absorb energy from the incident radiation and are thus able to overcome the potential-energy barrier that normally confines them inside the material. The process is analogous to **thermionic emission,** discovered in 1883 by Edison, in which the escape energy is supplied by heating the material to a high temperature, liberating electrons by a process analogous to boiling a liquid. The minimum amount of energy an individual electron has to gain in order to escape from a particular surface is called the **work function** for that surface; it is denoted by ϕ.

The photoelectric effect was investigated in detail by Wilhelm Hallwachs and Philipp Lenard during the years 1886–1900. These two researchers used an apparatus shown schematically in Figure 28.1a. Two conducting electrodes are enclosed in an evacuated glass tube. The negatively charged electrode is called the **cathode,** and the positively charged electrode is called the **anode.** The battery, or other source of potential difference, creates an electric field (red arrows) in the direction from anode to cathode. Monochromatic (single-frequency) light (purple arrows) falls on the surface of the cathode, causing electrons to be emitted from the cathode. A high vacuum, with residual pressure of 0.01 Pa $(10^{-7}$ atm$)$ or less, is needed to avoid collisions of electrons with gas molecules.

Once emitted, the electrons are pushed toward the anode by the electric field, causing a current i in the external circuit; the current is measured by the galvanometer G. Hallwachs and Lenard studied how this current varies with voltage and with the frequency and intensity of the light.

Hallwachs and Lenard found that when monochromatic light falls on the cathode, *no* electrons are emitted unless the frequency of the light is greater than some minimum value f, called the **threshold frequency,** that depends on the material of the cathode. For most metals, this frequency corresponds to that of light in the ultraviolet range (wavelengths of 200 to 300 nm), but for potassium and cesium oxides, it is in the visible spectrum (400 to 700 nm). This experimental result is consistent with the idea that each liberated electron absorbs an amount of energy E proportional to the frequency f of the light. When the frequency f is not great enough, the energy E is not great enough for the electron to surmount the potential-energy barrier ϕ at the surface.

When f is *greater than* the threshold value, it is found that some electrons are emitted from the cathode with substantial initial speeds. The evidence for this conclusion is the fact that, even with *no* potential difference between anode and cathode, a few electrons reach the anode, causing a small current in the external circuit. Indeed, even when the polarity of the potential difference V is reversed (Figure 28.1b) and the associated electric-field force on the electrons is back toward the cathode, some electrons still reach the anode. The electron flow stops completely only when the reversed potential difference V is made large enough that the corresponding potential energy eV is greater than the maximum kinetic energy $\frac{1}{2}mv_{max}^2$ of the emitted electrons.

Stopping potential

The reversed potential difference required to stop the electron flow completely is called the stopping potential, denoted by V_0. From the preceding discussion,

$$\frac{1}{2}mv_{max}^2 = eV_0. \qquad (28.1)$$

Light causes the cathode to emit electrons, which are pushed toward the anode by the electric field force.

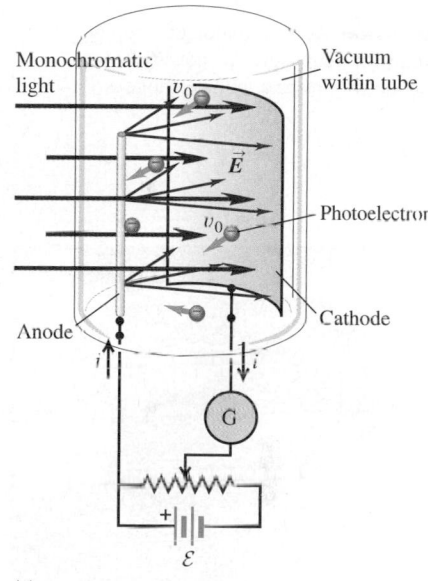

(a)

Even when the direction of $\vec{E}$ field is reversed, some electrons still reach the anode.

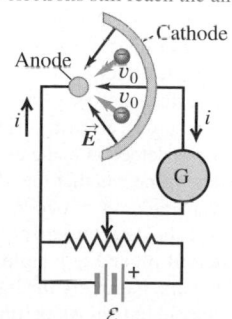

(b) Overhead view with $\vec{E}$ field reversed

▲ **FIGURE 28.1** A phototube demonstrating the photoelectric effect. The stopping potential V_0 is the minimum absolute value of the reverse potential difference that gives zero current.

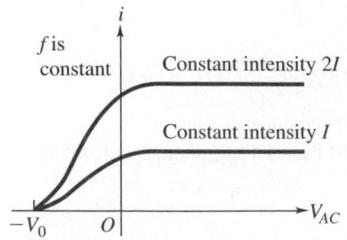

▲ **FIGURE 28.2** Photocurrent I as a function of V_{AC}.

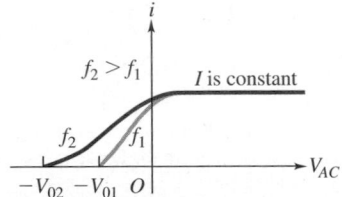

▲ **FIGURE 28.3** Variation of stopping potential with frequency. The potential of the anode with respect to the cathode is V_{AC}.

▲ **Application Fire!** Countless lives have been saved from tragic fires by smoke detectors, thanks to our understanding and use of the photoelectric effect, in which light shining on a surface causes ejection of electrons and a resulting electric current. Some types of smoke detectors make use of this effect to determine whether the air in a room contains particles of smoke. A tiny beam of light inside the detector passes by a photosensitive cell placed at a right angle to the light beam. Normally the light bypasses the photocell, but smoke particles scatter the light, causing some of it to hit the photocell and produce a small current via the photoelectric effect. This current results in activation of a very loud alarm, giving any unwary occupants of a burning building advance warning of a potentially life-threatening fire.

Measuring the stopping potential V_0 therefore gives us a direct measurement of the maximum kinetic energy electrons have when they leave the cathode.

A classical wave theory of light would predict that when we increase the *intensity* of the light striking the cathode, the electrons should come off with greater energy and the stopping potential should be greater. But that isn't what actually happens. Figure 28.2 shows how the photocurrent varies with voltage for two different intensity levels. When the potential V_{AC} of the anode with respect to the cathode is sufficiently large and positive, the curve levels off, showing that *all* the emitted electrons are being collected by the anode. When the light intensity is increased (say, from I to $2I$), the maximum current increases, but the stopping potential is found to be the same. However, when the frequency f of the light is increased, the stopping potential V_0 increases linearly (Figure 28.3). These results suggest that the maximum kinetic energy of an emitted electron *does not* depend on the *intensity* of the incident light, but it *does* depend on the *frequency* or *wavelength*.

The correct analysis of the photoelectric effect was developed by Albert Einstein in 1905. Extending a proposal made five years earlier by Max Planck, Einstein postulated that a beam of light consists of small bundles of energy called **quanta** or **photons**.

Energy of a photon

The energy E of an individual photon is equal to a constant times the frequency f of the photon; that is,

$$E = hf, \tag{28.2}$$

where h is a universal constant called **Planck's constant.** The numerical value of h, to the accuracy presently known, is

$$h = 6.6260693(11) \times 10^{-34} \text{ J} \cdot \text{s}.$$

In Einstein's analysis, a photon striking the surface of a conductor is absorbed by an electron. The energy transfer is an "all-or-nothing" process: The electron gets either *all* the photon's energy or *none* of it. If this energy is greater than the surface potential-energy barrier (the work function ϕ), the electron can escape from the surface.

It follows that the maximum kinetic energy $\frac{1}{2}mv_{max}^2$ for an emitted electron is the energy hf it gains by absorbing a photon, minus the work function ϕ:

$$\frac{1}{2}mv_{max}^2 = hf - \phi. \tag{28.3}$$

Combining this relationship with Equation 28.1, we find that

$$eV_0 = hf - \phi, \quad \text{or} \quad V_0 = \frac{h}{e}f - \frac{\phi}{e}. \tag{28.4}$$

That is, if we measure the stopping potential V_0 for each of several values of frequency f, we expect V_0 to be a linear function of f. A graph of V_0 (on the vertical axis) as a function of f (on the horizontal axis) should then be a straight line. Measurements of V_0 and f confirm this prediction. Furthermore, the slope of the line is h/e, and the intercept on the vertical axis (corresponding to $f = 0$) is at $V_0 = -\phi/e$. Thus, we can determine both the work function ϕ (in electronvolts) for the material and the value of the quantity h/e. (Example 28.3 shows in detail how this can be done.) After the electron charge e was measured directly by Robert Millikan in 1909, Planck's constant h could also be determined from these measurements.

Electron energies and work functions are usually expressed in electronvolts (Section 18.4):

$$1 \text{ eV} = 1.602 \times 10^{-19} \text{ J}.$$

In terms of electronvolts, Planck's constant is

$$h = 4.136 \times 10^{-15} \text{ eV} \cdot \text{s}.$$

Table 28.1 lists a few typical work functions of elements. The values are approximate because they are sensitive to surface impurities. For example, a thin layer of cesium oxide can reduce the work function of a metal to about 1 eV.

TABLE 28.1   Work functions of elements

Element	Work function (eV)
Aluminum	4.3
Carbon	5.0
Copper	4.7
Gold	5.1
Nickel	5.1
Silicon	4.8
Silver	4.3
Sodium	2.7

Conceptual Analysis 28.1

Vary the frequency

Light falling on a metal surface causes electrons to be emitted from the metal by the photoelectric effect. In a particular experiment, the intensity of the incident light and the temperature of the metal are held constant. As we decrease the frequency of the incident light,

A. the work function of the metal increases.
B. the number of electrons emitted from the metal decreases steadily to zero.
C. the maximum speed of the emitted electrons decreases steadily until no electrons are emitted.
D. the stopping potential increases.

SOLUTION In the photoelectric effect, some of an incident photon's energy is used to remove an electron from the metal; the remainder becomes the electron's kinetic energy. The energy to remove the electron from the metal, called the work function ϕ, is constant for a given material. The smaller the frequency of light, the smaller the energy of each photon $(E = hf)$. Therefore, as we decrease the frequency of the incident light, less and less of each photon's energy is available to become kinetic energy of the emitted electrons, and the speed of the emitted electrons decreases steadily to zero. The correct answer is C.

EXAMPLE 28.1   Conductivity enhanced by photons

Silicon films become better electrical conductors when illuminated by photons with energies of 1.14 eV or greater. This behavior is called **photoconductivity.** What is the corresponding photon wavelength? In what portion of the electromagnetic spectrum does it lie?

SOLUTION

SET UP The energy E of a photon is related to its frequency f by $E = hf$. The wavelength λ is related to the frequency by the general wave relation $\lambda = c/f$. Combining these relations and solving for λ, we get $\lambda = hc/E$.

SOLVE It's easiest to perform our calculations using units of $(\text{eV} \cdot \text{s})$ for h. That is, $h = 4.136 \times 10^{-15} \text{ eV} \cdot \text{s}$. Using the preceding expression for λ, we get

$$\lambda = \frac{hc}{E} = \frac{(4.136 \times 10^{-15} \text{ eV} \cdot \text{s})(3.00 \times 10^8 \text{ m/s})}{1.14 \text{ eV}}$$
$$= 1.09 \times 10^{-6} \text{ m} = 1090 \text{ nm}.$$

The wavelengths of the visible spectrum are about 400 to 700 nm, so the wavelength we have found is in the near-infrared region of the spectrum.

REFLECT The frequency of a photon is inversely proportional to its wavelength, so the minimum energy of 1.14 eV corresponds to the maximum wavelength that causes photoconduction for silicon, in this case about 1090 nm. Thus, for silicon, all light with a wavelength less than 1090 nm, including all light in the visible spectrum, contributes to photoconductivity.

Practice Problem: If we need a material that is photoconductive for any wavelength in the visible spectrum (i.e., 400 to 700 nm), to what range of photon energies must it respond? *Answer:* 1.77 to 3.10 eV.

EXAMPLE 28.2 **Reverse potential needed to stop photoelectric current**

In a photoelectric-effect experiment with light of a certain frequency, a reverse potential difference of 1.25 V is required to reduce the current to zero. Using this value, together with the mass and charge of the electron, find the maximum kinetic energy and the maximum speed of the photoelectrons emitted.

SOLUTION

SET UP If the maximum electron kinetic energy doesn't exceed 1.25 eV, *all* the electrons will be stopped by a reverse potential difference of 1.25 V or greater. For consistency of units, we need to convert this energy to joules and then use the definition of kinetic energy to find the maximum speed.

SOLVE The maximum kinetic energy is

$$K_{max} = eV_0 = (1.60 \times 10^{-19}\,\text{C})(1.25\,\text{V}) = 2.00 \times 10^{-19}\,\text{J}.$$

From the definition of kinetic energy, $K_{max} = \frac{1}{2}mv_{max}^2$, we get

$$v_{max} = \sqrt{\frac{2K_{max}}{m}} = \sqrt{\frac{2(2.00 \times 19^{-19}\,\text{J})}{9.11 \times 10^{-31}\,\text{kg}}}$$
$$= 6.63 \times 10^5\,\text{m/s}.$$

Alternatively, from Equation 28.1,

$$\frac{1}{2}mv_{max}^2 = eV_0,$$

$$v_{max} = \sqrt{\frac{2eV_0}{m}} = \sqrt{\frac{2(1.60 \times 10^{-19}\,\text{C})(1.25\,\text{V})}{9.11 \times 10^{-31}\,\text{kg}}}$$
$$= 6.63 \times 10^5\,\text{m/s}.$$

REFLECT This speed is much *smaller* than c, the speed of light, meaning that we are justified in using the nonrelativistic kinetic-energy expression. An equivalent statement is that the kinetic energy $eV_0 = 1.25$ eV is much smaller than the electron's rest energy $mc^2 = 0.511$ MeV.

Practice Problem: What reverse potential is required to reduce the current to zero if the maximum electron speed is 9.38×10^5 m/s? *Answer:* 2.50 V.

EXAMPLE 28.3 **How to measure Planck's constant**

For a certain cathode material used in a photoelectric-effect experiment, a stopping potential of 3.0 V was required for light of wavelength 300 nm, 2.0 V for 400 nm, and 1.0 V for 600 nm. Determine the work function for this material and the value of Planck's constant, as obtained from these data.

SOLUTION

SET UP From the preceding discussion, a graph of V_0 (on the vertical axis) as a function of f (on the horizontal axis) should be a straight line, as given by Equation 28.4:

$$V_0 = \frac{h}{e}f - \frac{\phi}{e}.$$

Our graph is shown in Figure 28.4. We see that the slope of the line is h/e and the intercept on the vertical axis (corresponding to $f = 0$) is at $V_0 = -\phi/e$. (But note that the portion of the graph below the horizontal axis corresponds to negative values of V_0, for which f is less than the threshold value and no photoelectrons are actually emitted.)

SOLVE The frequencies, obtained from $f = c/\lambda$ and $c = 3.00 \times 10^8$ m/s, are 1.0, 0.75, and 0.5×10^{15} Hz. From the slope of the graph, we find that

$$\frac{h}{e} = \frac{1.0\,\text{V}}{0.25 \times 10^{15}\,\text{s}^{-1}} = 4.0 \times 10^{-15}\,\text{J} \cdot \text{s/C},$$
$$h = (4.0 \times 10^{-15}\,\text{J} \cdot \text{s/C})(1.60 \times 10^{-19}\,\text{C})$$
$$= 6.4 \times 10^{-34}\,\text{J} \cdot \text{s}.$$

The intercept on the vertical axis occurs at

$$-\frac{\phi}{e} = -1.0\,\text{V}, \qquad \phi = 1.0\,\text{eV} = 1.60 \times 10^{-19}\,\text{J}.$$

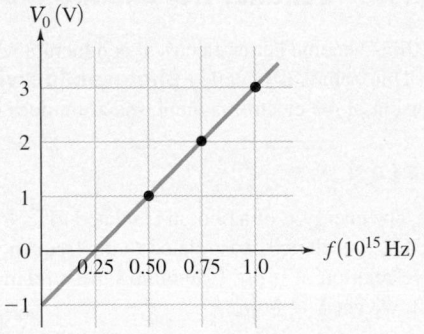

▲ **FIGURE 28.4** Our graph for this problem.

The work function ϕ is 1.0 eV. The graph shows that the minimum frequency for emission of electrons is 0.25×10^{15} Hz. The corresponding *maximum* wavelength for which electrons are emitted is

$$\lambda = \frac{c}{f} = \frac{3.0 \times 10^8\,\text{m/s}}{0.25 \times 10^{15}\,\text{s}^{-1}} = 1.2 \times 10^{-6}\,\text{m} = 1200\,\text{nm}.$$

REFLECT This experimental value of h differs by about 3% from the correct value.

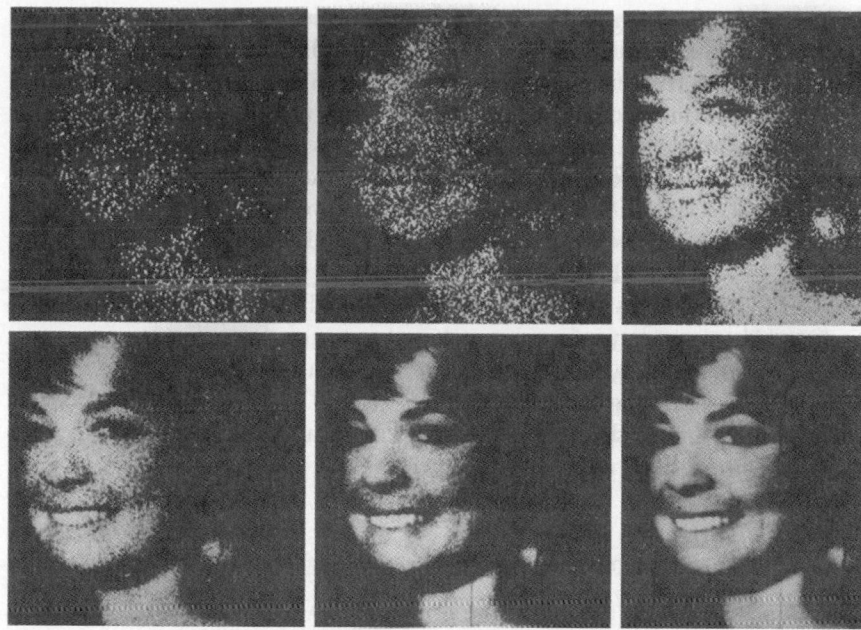

▲ **FIGURE 28.5** Comparison of photographs made with few and many photons.

Figure 28.5 shows a direct illustration of the particle aspects of light. When relatively few photons strike the photographic film, the impacts of individual photons are clearly visible. Only when the number of photons is much larger does a distinct picture emerge.

Photon frequency and wavelength

The concept of photons is applicable to *all* regions of the electromagnetic spectrum, including radio waves, x rays, and gamma rays. Photons always travel with the speed of light, c. A photon of any frequency f and wavelength $\lambda = c/f$ has energy E given by

$$E = hf = \frac{hc}{\lambda}. \tag{28.5}$$

Furthermore, according to relativity theory, every particle that has energy must also have momentum, even if it has no rest mass. Photons have zero rest mass. According to Equation 27.25, a photon with energy E has momentum with magnitude p given by $E = pc$. Thus, the wavelength λ of a photon, its momentum p, and its frequency f are related simply by

$$p = \frac{E}{c} = \frac{hf}{c} = \frac{h}{\lambda}. \tag{28.6}$$

EXAMPLE 28.4 **Photons from a radio station**

Radio station WQED in Pittsburgh broadcasts at 89.3 MHz with a radiated power of 43.0 kW. How many photons does it emit each second?

Continued

SOLUTION

SET UP First we need to find the energy of one photon. From that result, we can find the number of photons needed per second for a total power output of 43.0 kW. We also need the conversion factor 1 W = 1 J/s. (Remember that the watt is a unit of power, not energy; power is energy per unit time. Remember also that 1 Hz = 1 s^{-1}.)

SOLVE The station sends out 43.0×10^3 joules each second. The energy of each photon emitted is

$$E = hf = (6.626 \times 10^{-34} \text{ J} \cdot \text{s})(89.3 \times 10^6 \text{ Hz})$$
$$= 5.92 \times 10^{-26} \text{ J}.$$

The number of photons per second is therefore

$$\frac{43.0 \times 10^3 \text{ J/s}}{5.92 \times 10^{-26} \text{ J/photon}} = 7.26 \times 10^{29} \text{ photons/s}.$$

REFLECT With this huge number of photons leaving the station each second, the discreteness of the tiny individual bundles of energy isn't noticed; the radiated energy appears to be a continuous flow.

Practice Problem: During a diagnostic x ray, a total energy of 2.5×10^{-3} J is absorbed by about 5.0 kg of tissue. If the x-ray photons have an energy of 50 keV and the tissue has the same density as water, how many photons are absorbed by 1.0 cm^3 of tissue? *Answer:* 6.2×10^7 photons.

28.2 Line Spectra and Energy Levels

The existence of line spectra has been known for more than 200 years. A prism or a diffraction grating can be used to separate the various wavelengths in a beam of light into a *spectrum*. If the light source is a very hot solid or liquid (such as the filament in a lightbulb), the spectrum is *continuous;* light of all wavelengths is present (Figure 28.6a). But if the source is a gas carrying an electrical discharge (as in a neon sign), or if it is a volatile salt heated in a flame (as when table salt is thrown into a campfire), only a few colors appear, in the form of isolated sharp parallel lines (Figure 28.6b). (Each "line" is an image of the spectrograph slit, deviated through an angle that depends on the wavelength of the light forming the image. A spectrum of this sort is called a *line spectrum;* each line corresponds to a definite wavelength and frequency.

Early in the 19th century, it was discovered that each chemical element has a definite, unchanging set of wavelengths in its line spectrum, and the identification of elements by their spectra became a useful analytical technique. The characteristic spectrum for each element was assumed to be related to the internal structure

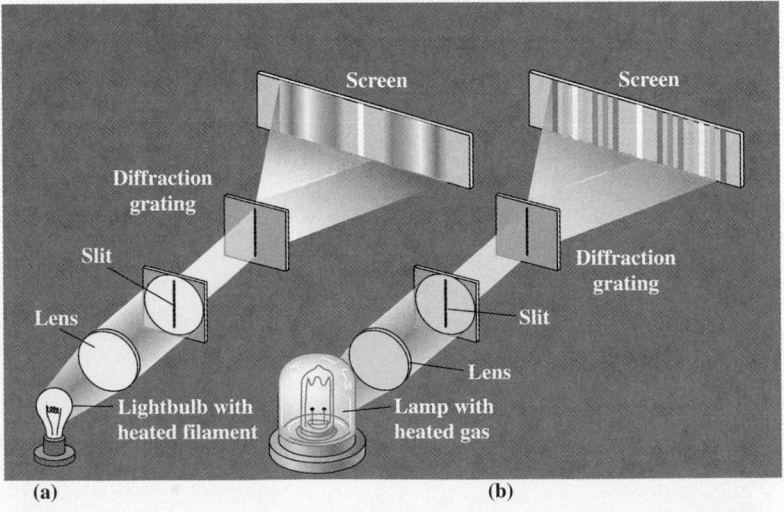

▲ **FIGURE 28.6** (a) Continuous spectrum from a very hot light source (the lightbulb filament). (b) Line spectrum emitted by a lamp containing a heated gas.

of its atoms, but attempts to understand this relation on the basis of classical mechanics and electrodynamics were not successful, even as recently as 1900. Two key pieces of the puzzle were missing: the photon concept and a new and revolutionary picture of the structure of the atom.

The key idea was finally found in 1913 by Niels Bohr, with an insight that, from a historical vantage point more than 90 years later, seems almost obvious; yet in its time, it represented a bold and brilliant stroke. Here's Bohr's reasoning: Every atom has some internal structure and internal motion and therefore some internal energy. But each atom has a set of possible energy levels. An atom can have an amount of internal energy corresponding to any one of these levels, but it cannot have an energy intermediate between two levels. If the atoms of a particular element can emit and absorb photons with only certain particular energies (corresponding to the line spectrum of that element), then *the atoms themselves must be able to possess only certain particular quantities of energy.* Bohr's hypothesis is shown schematically in Figure 28.7. Bohr assumed that while an atom is in one of these "permitted" energy states, it does not radiate. However, an atom can make a transition from one energy level to a lower level by emitting a photon with energy equal to the energy difference between the initial and final levels.

For the simplest atom, hydrogen, Bohr pictured these levels in terms of the electron revolving in various circular orbits around the proton. Only certain orbit radii were permitted. We'll return to this picture in the next section.

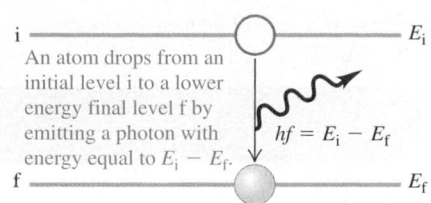

▲ **FIGURE 28.7** Energy levels and photon emission.

Bohr's hypothesis

If E_i is the initial energy of an atom before a transition from one energy level to another, E_f is the atom's (smaller) final energy after the transition, and the energy of the emitted photon is hf, then

$$hf = E_i - E_f. \tag{28.7}$$

For example, a photon of orange light with wavelength $\lambda = 600$ nm has a frequency f given by

$$f = \frac{c}{\lambda} = \frac{3.00 \times 10^8 \text{ m/s}}{600 \times 10^{-9} \text{ m}} = 5.00 \times 10^{14} \text{ s}^{-1}.$$

The corresponding photon energy is

$$E = hf = (6.63 \times 10^{-34} \text{ J} \cdot \text{s})(5.00 \times 10^{14} \text{ s}^{-1})$$
$$= 3.31 \times 10^{-19} \text{ J} = 2.07 \text{ eV}.$$

This photon must be emitted in a transition between two states of the atom (each with a definite energy level) that differ in energy by 2.07 eV.

The same principle applies when a photon is *absorbed* by an atom. In this case, the atom's final energy is greater than its initial energy, and instead of Equation 28.7, we have

$$hf = E_f - E_i.$$

Hydrogen Spectrum

Let's see how Bohr's hypothesis fits in with what was known about spectra in 1913. The spectrum of hydrogen, the least massive atom, had been studied intensively. Under proper conditions, atomic hydrogen emits a series of four lines in the visible region of the spectrum, shown in Figure 28.8. The line with longest wavelength, or lowest frequency, in the red, is called H_α; the next line, in the blue-green, is H_β, and so on.

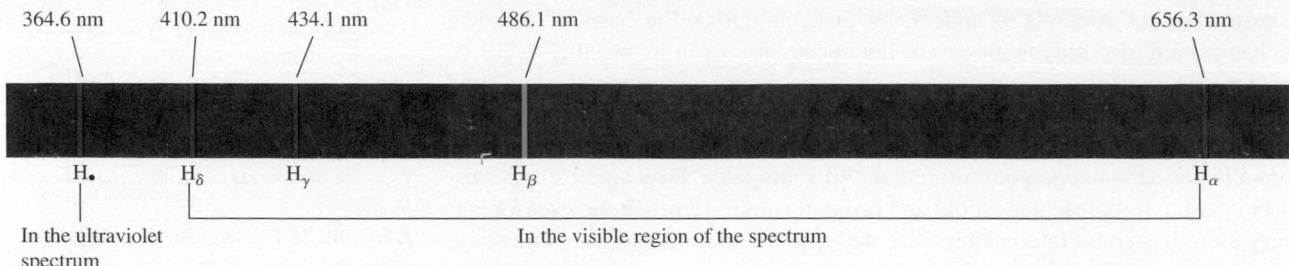

▲ **FIGURE 28.8** The Balmer series of spectral lines for atomic hydrogen.

In 1885, Johann Balmer (1825–1898) found (by trial and error) a formula that gives the wavelengths of these lines, which are now called the Balmer series.

Balmer's formula for the hydrogen spectrum

Balmer's formula is

$$\frac{1}{\lambda} = R\left(\frac{1}{2^2} - \frac{1}{n^2}\right), \tag{28.8}$$

where λ is the wavelength, R is a constant called the **Rydberg constant,** and n may have the integer values $3, 4, 5, \cdots$. If λ is in meters, the numerical value of R (determined from measurements of wavelengths) is

$$R = 1.097 \times 10^7 \, \text{m}^{-1}.$$

Substituting $n = 3$ in Equation 28.8, we obtain the wavelength of the H_α line:

$$\frac{1}{\lambda} = \left(1.097 \times 10^7 \, \text{m}^{-1}\right)\left(\frac{1}{2^2} - \frac{1}{3^2}\right), \quad \lambda = 656.3 \, \text{nm}.$$

For $n = 4$, we obtain the wavelength of the H_β line, and so on.

Balmer's formula has a direct relation to Bohr's hypothesis about energy levels. Using the relations $f = c/\lambda$ and $E = hf$, we can find the photon energies E corresponding to the wavelengths of the Balmer series. Multiplying Equation 28.8 by hc, we find that

$$\frac{hc}{\lambda} = hf = E = hcR\left(\frac{1}{2^2} - \frac{1}{n^2}\right) = \frac{hcR}{2^2} - \frac{hcR}{n^2}. \tag{28.9}$$

Comparing Equations 28.7 and 28.9, we see that the two agree if we identify $-hcR/n^2$ as the initial energy E_i of the atom and $-hcR/2^2$ as its final energy E_f in a transition in which a photon with energy $hf = E_i - E_f$ is emitted.

The Balmer series (as well as others that we'll mention shortly) therefore suggests that the hydrogen atom has a series of energy levels, which we'll denote as E_n, given by

$$E_n = -\frac{hcR}{n^2}, \qquad n = 2, 3, 4, \cdots. \tag{28.10}$$

(These energies are negative because we have arbitrarily chosen the potential energy to be zero at very large values of n. As we'll see, the state with $n = \infty$ corresponds to the state where the electron is completely separated from the nucleus—a single proton—and is at rest.) Each wavelength in the Balmer series corresponds to a transition from a state having n equal to 3 or greater to the state where $n = 2$.

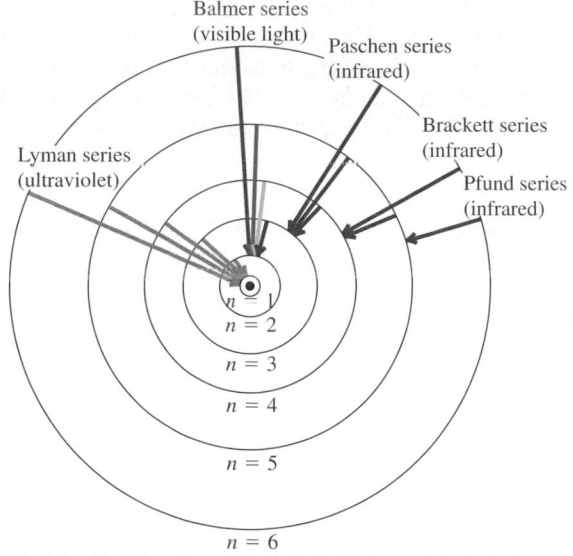

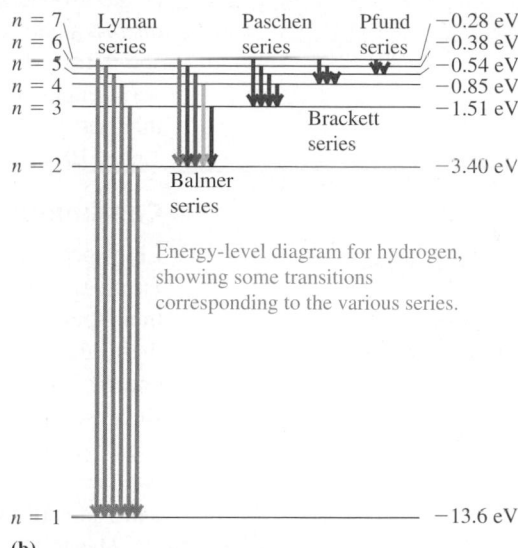

Energy-level diagram for hydrogen, showing some transitions corresponding to the various series.

"Permitted" orbits of an electron in the Bohr model of a hydrogen atom (not to scale). Arrows indicate the transitions responsible for some of the lines of various series.

(a)

(b)

▲ **FIGURE 28.9** Bohr orbits and spectral series.

The numerical value of the product hcR is

$$hcR = (6.626 \times 10^{-34}\,\text{J}\cdot\text{s})(2.998 \times 10^8\,\text{m/s})(1.097 \times 10^7\,\text{m}^{-1})$$
$$= 2.179 \times 10^{-18}\,\text{J} = 13.6\,\text{eV}.$$

Thus, the magnitudes of the energy levels given by Equation 28.10 are approximately -13.6 eV, -3.40 eV, -1.51 eV,···.

Other series of spectrum lines for hydrogen have since been discovered experimentally. They are named for their discoverers; their wavelengths are represented by formulas similar to the Balmer formula. One series is in the ultraviolet region, the others in the infrared. *All* the spectral series of hydrogen can be understood on the basis of transitions from one energy level (and corresponding electron orbit) to another, with the energy levels given by Equation 28.10. For the Lyman series, the final state is always $n = 1$; for the Paschen series, it is $n - 3$; and so on. Taken together, these spectral series give strong support to Bohr's picture of energy levels in atoms. The relation of the various spectral series to the energy levels and electron orbits is shown in Figure 28.9.

Energy Levels

Only atoms or ions with a single electron can be represented by a simple formula of the Balmer type. But it is *always* possible to analyze the more complicated spectra of other elements in terms of transitions among various energy levels and to deduce the numerical values of these levels from the measured spectrum wavelengths. Every atom has a lowest energy level (or energy state), representing the minimum energy the atom can have. This is called the **ground state,** and all states with energy greater than that of the ground state are called **excited states.** A photon corresponding to a particular spectrum line is emitted when an atom makes a transition from an excited state to a lower excited state or to the ground state.

In some cases, an atom can have two or more states with different electron configurations, but with the same energy. Thus, we'll sometimes need to distinguish between energy *states* and energy *levels;* one level can sometimes correspond to several states. We'll return to this distinction later; for now, we'll use the terms *state* and *level* interchangeably.

▲ **Application** **The physics of fireworks.**
Fireworks were invented by the ancient Chinese, long before we understood the physics of how the various colors are produced by atomic emission. As you have learned, when atoms absorb energy, their electrons can be temporarily boosted to higher energy levels within the atom. When these electrons drop back to lower energy levels, they emit photons of characteristic wavelengths that correspond to energy differences between pairs of energy levels. When these wavelengths are in the visible region of the electromagnetic spectrum, we see different colors depending on which chemical element or compound has been excited. In practice, fireworks makers often use compounds of sodium to produce bright yellow, calcium for orange, strontium for red, barium for green, and copper for blue.

The sodium atom has two closely spaced levels called *resonance levels* at about 2.1 eV above the ground state. The characteristic yellow light with wavelengths 589.0 and 589.6 nm is emitted during transitions from one of these levels to the ground state. Conversely, a sodium atom initially in the ground state can *absorb* a photon with wavelength 589.0 or 589.6 nm. After a short time, the atom spontaneously returns to the ground state by emitting a photon with the same wavelength. The average time spent in the excited state is called the **lifetime** of the state; for the resonance levels of the sodium atom, the lifetime is about 1.6×10^{-8} s = 16 ns.

Continuous Spectra

Line spectra are produced by matter in the gaseous state, in which the atoms are far apart and interactions between them are negligible. If the atoms are identical, their spectra are also identical. But in condensed states of matter (liquid or solid), there are strong interactions between atoms. These interactions cause shifts in energy levels, and levels are shifted by different amounts for different atoms. Because of the very large numbers of atoms, practically any photon energy is possible. Therefore, hot condensed matter always emits a spectrum with a *continuous* distribution of wavelengths, not a line spectrum. Such a spectrum is called a **continuous emission spectrum.**

The total rate of radiation of energy from a hot liquid or solid is proportional to the fourth power of the absolute temperature T, as we learned in Section 14.7. The radiation consists of a continuous distribution of wavelengths; it is called **blackbody** radiation because of its relation to the radiation from a material that is an ideal absorber of radiation. Blackbody radiation from a hot material is most intense in the vicinity of a certain wavelength that is *inversely* proportional to the absolute temperature. As the material's temperature increases, the intensity peak shifts to shorter wavelengths, and the total intensity increases. When a body that is glowing dull red is heated further, it gets brighter and more orange or yellow. Historically, Planck's analysis of blackbody radiation initially led him to the photon concept five years before Einstein's analysis of the photoelectric effect.

PROBLEM-SOLVING STRATEGY 28.1 **Photons and energy levels**

SET UP

1. Remember that with photons, as with any other periodic wave, the wavelength λ and frequency f are related by $f = c/\lambda$. The energy E of a photon can be expressed as hf or hc/λ, whichever is more convenient for the problem at hand. Be careful with units: If E is in joules, h must be in joule-seconds, λ in meters, and f in s^{-1}, or hertz. The magnitudes are in such unfamiliar ranges that common sense may not help if your calculation is wrong by a factor of 10^{10}, so be careful when you add and subtract exponents with powers of 10.

SOLVE

2. It's often convenient to measure energy in electronvolts. The conversion $1 \text{ eV} = 1.602 \times 10^{-19}$ J is often useful. When energies are in eV, you may want to express h in electronvolt-seconds; in those units, $h = 4.136 \times 10^{-15}$ eV · s. We invite you to verify this value.

3. Keep in mind that an electron moving through a potential difference of 1 V gains or loses an amount of energy equal to 1 eV. You will use the electronvolt a lot in this chapter and the next two, so it's important that you get familiar with it now.

EXAMPLE 28.5 **A mythical atom**

A hypothetical atom has three energy levels: the ground-state level and levels 1.00 eV and 3.00 eV above the ground state. **(a)** Find the frequencies and wavelengths of the spectrum lines for this atom. **(b)** What wavelengths can be *absorbed* by the atom if it is initially in the ground state?

SOLUTION

SET UP First we draw an energy-level diagram (Figure 28.10a). **(a)** Draw arrows to show the possible transitions from higher to lower energy levels, and label the energy for each transition leading from a higher to a lower energy level. These are the photon energies; from these, we can find the frequencies and wavelengths. **(b)** Draw arrows on the diagram showing the possible transitions from lower to higher energy levels, starting from the lowest energy state. These are the energies of photons that can be absorbed, and again, we can find the associated frequencies and wavelengths.

SOLVE Part (a): The possible photon energies E, corresponding to the transitions shown, are 1.00 eV, 2.00 eV, and 3.00 eV. Each photon frequency f is given by $f = E/h$. For 1.00 eV, we get

$$f = \frac{E}{h} = \frac{1.00 \text{ eV}}{4.136 \times 10^{-15} \text{ eV} \cdot \text{s}} = 2.42 \times 10^{14} \text{ Hz}.$$

For 2.00 eV and 3.00 eV, $f = 4.84 \times 10^{14}$ Hz and 7.25×10^{14} Hz, respectively. We find the wavelengths from $\lambda = c/f$. For 1.00 eV,

$$\lambda = \frac{c}{f} = \frac{3.00 \times 10^8 \text{ m/s}}{2.42 \times 10^{14} \text{ Hz}} = 1.24 \times 10^{-6} \text{ m} = 1240 \text{ nm}.$$

For 2.00 eV and 3.00 eV, the wavelengths are 620 nm and 414 nm, respectively.

Part (b): For an atom that is initially in the ground state, only a 1.00 eV or 3.00 eV photon can be absorbed; a 2.00 eV photon cannot, because there is no energy level 2.00 eV above the

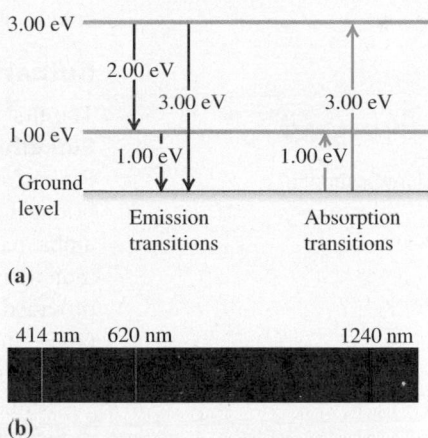

(a)

414 nm   620 nm   1240 nm

(b)

▲ **FIGURE 28.10**

ground state. From the preceding calculations, the corresponding wavelengths are 1240 nm and 414 nm, respectively; these two lines appear in the absorption spectrum for this atom.

REFLECT The lines in the emitted spectrum are 1240 nm (in the near infrared), 620 nm (red), and 414 nm (violet). Those in the absorption spectrum are 1240 nm and 414 nm. If the atom is initially in the state that is 1.00 eV above the ground state, then it can absorb a photon with an energy of 2.00 eV.

Practice Problem: How would your answers differ from those given above if the middle level were 2.00 eV above the ground state instead of 1.00 eV? *Answer:* (a) Same as before; (b) 620 nm, 1240 nm.

The Bohr hypothesis was successful in relating the wavelengths in line spectra to energy levels of atoms, but it provided no basis for predicting what the energy levels should be for any particular kind of atom. Bohr did provide a partial solution for this problem for the simplest atom, hydrogen; we'll discuss this solution in the next section. Then we'll introduce some general principles of quantum mechanics that are needed for a more general understanding of the structure and energy levels of atoms.

28.3 The Nuclear Atom and the Bohr Model

What does the inside of an atom look like? In one sense, this is a silly question: We know that atoms are much smaller than wavelengths of visible light, so there is no hope of actually *seeing* an atom. But we can still describe the interior of an atom by describing how the mass and electric charge are distributed through its volume.

Here's where things stood in 1910: J. J. Thomson had discovered the electron and measured its charge-to-mass ratio (e/m) in 1897; by 1910, Millikan had completed his first measurements of the electron charge e. These and other experiments showed that most of the mass of an atom had to be associated with the

ActivPhysics 18.1: The Bohr Model
ActivPhysics 19.1: Particle Scattering

positive charge, not with the electrons. It was also known that the overall size of atoms is on the order of 10^{-10} m and that all atoms except hydrogen contain more than one electron. What was *not* known was how the mass and charge were distributed within the atom. Thomson had proposed a model of the atom that included a sphere of positive charge on the order of 10^{-10} m in diameter, with the electrons embedded in it like chocolate chips in a more or less spherical cookie.

Rutherford Scattering

The first experiments designed to probe the inner structure of the atom were the **Rutherford scattering** experiments, carried out in 1910–1911 by Sir Ernest Rutherford and two of his students, Hans Geiger and Ernest Marsden, at Cambridge University in England. Rutherford's experiment consisted of projecting alpha particles (emitted with speeds on the order of 10^7 m/s from naturally radioactive elements) at the atoms under study. The deflections of these particles provided information about the internal structure and charge distribution of the target atoms. (Alpha particles are now known to be identical to the nuclei of helium atoms: two protons and two neutrons bound together.)

Rutherford's experimental setup is shown schematically in Figure 28.11. A radioactive material at the left emits alpha particles. Thick lead screens stop all particles except those in a narrow beam defined by small holes. The beam then passes through a target consisting of a thin gold foil. (Gold was used because it can be beaten into extremely thin sheets.) After passing through the foil, the beam strikes a screen coated with zinc sulfide, similar in principle to the screen of an older model TV picture tube. A momentary flash or **scintillation** can be seen on the screen whenever it is struck by an alpha particle (just as spots on the TV screen glow when the electron beam strikes them). Rutherford's students were assigned the tedious task of counting the numbers of particles deflected through various angles. This unpleasant duty led Geiger to the invention of what is now called the Geiger counter, a commonly used radiation detector.

The fact that an alpha particle can pass through a thin metal foil shows that alpha particles must be able actually to penetrate into the interiors of atoms. The mass of an alpha particle is about 7300 times that of an electron. Momentum considerations show that the alpha particle can be scattered only a small amount by its interaction with the much lighter electrons. It's like driving a car through a hailstorm; the hailstones don't deflect the car much. Only interactions with the positive charge, which is tied to most of the mass of the atom, can deflect the alpha particle appreciably, perhaps comparable to colliding with an oncoming truck.

MasteringPHYSICS

PhET: Rutherford Scattering

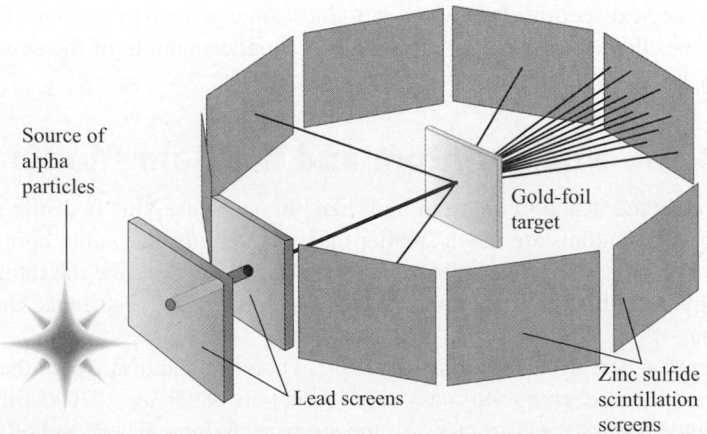

Source of alpha particles

Gold-foil target

Lead screens

Zinc sulfide scintillation screens

▲ **FIGURE 28.11** The scattering of alpha particles by a thin metal foil.

In the Thomson model, the positive charge is distributed through the whole atom. Rutherford calculated that the maximum deflection angle an alpha particle can have in this situation is only a few degrees, as Figure 28.12a suggests. The experimental results, however, were very different from this and were totally unexpected. Some alpha particles were scattered by nearly 180°—that is, almost straight backward! Rutherford wrote later:

> **It was quite the most incredible event that ever happened to me in my life. It was almost as incredible as if you had fired a 15-inch shell at a piece of tissue paper and it came back and hit you.**

Back to the drawing board! Suppose the positive charge, instead of being distributed through a sphere with atomic dimensions (on the order of 10^{-10} m), is all concentrated in a much *smaller* space. Rutherford called this concentration of positive charge the **nucleus** of the atom. Then the atom would act like a point charge down to much smaller distances. The maximum repulsive force on the alpha particle would be much larger, and large-angle scattering would be possible, as in Figure 28.12b. Rutherford again computed the numbers of particles expected to be scattered through various angles. Within the precision of his experiments, the computed and measured results agreed, down to distances on the order of 10^{-14} m.

Rutherford's experiments therefore established that the atom *does* have a nucleus—a very small, very dense structure no larger than 10^{-14} m in diameter. The nucleus occupies only about 10^{-12} of the total volume of the atom, but it contains all the positive charge and at least 99.95% of the total mass of the atom!

Figure 28.13 shows a computer simulation of the paths of 5.0 MeV alpha particles scattered by gold nuclei with a radius of 7.0×10^{-15} m (the actual value) and by nuclei with a hypothetical radius ten times this great. In the second case, there is no large-angle scattering.

The Bohr Model

At the same time (1913) that Bohr established the relationship between spectral wavelengths and energy levels, he also proposed a mechanical model of the hydrogen atom. Using this model, now known as the **Bohr model,** he was able to *calculate* the energy levels and obtain agreement with values determined from spectra.

In the light of Rutherford's discovery of the nucleus, the question arose as to what kept the negatively charged electrons at relatively large distances $(\sim10^{-10}$ m$)$ from the extremely small $(\sim10^{-15}$ m$)$, positively charged nucleus despite the electrostatic attraction that tended to pull the electrons into the

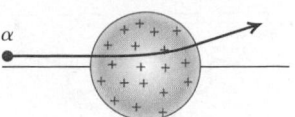

Thomson's model of the atom: An alpha particle is scattered through only a small angle.

(a)

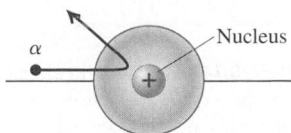

Rutherford's model of the atom: An alpha particle can be scattered through a large angle by the compact, positively charged nucleus.

(b)

▲ **FIGURE 28.12** Rutherford scattering for the Thomson version of the atom vs. the nuclear atom. (Not drawn to scale.)

Mastering**PHYSICS**

PhET: Models of the Hydrogen Atom

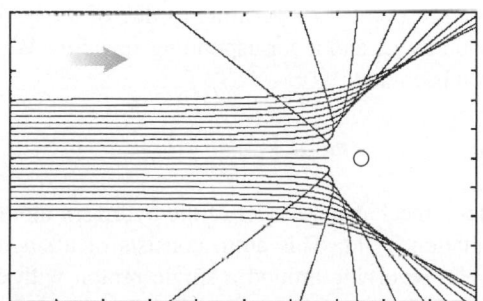

(a) A gold nucleus with radius 7.0×10^{-15} m gives large angle scattering.

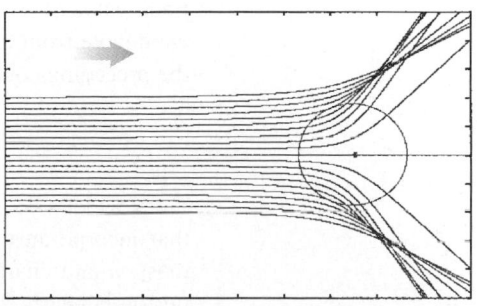

(b) A nucleus with 10 times the radius of the nucleus in (a) shows no large-scale scattering.

▲ **FIGURE 28.13** Computer simulation of scattering of 5.0-MeV alpha particles from a gold nucleus. Each curve shows a possible alpha-particle trajectory.

▲ **FIGURE 28.14** A paradox in classical models of the atom: An electron, spiraling in toward the nucleus, emits a continuous spectrum of electromagnetic radiation.

nucleus. Rutherford suggested that perhaps the electrons *revolve* in orbits about the nucleus, just as the planets revolve around the sun, with the electrical attraction of the nucleus providing the necessary centripetal force.

But according to classical electromagnetic theory, any accelerating electric charge, either oscillating or revolving, radiates electromagnetic waves, just as a radio or TV transmitting antenna does. The total energy of a revolving electron should therefore decrease continuously, its orbit should become smaller and smaller, and it should rapidly spiral into the nucleus (Figure 28.14). Furthermore, according to classical theory, the *frequency* of the electromagnetic waves emitted should equal the frequency of revolution of the electrons in their orbits. As the electrons radiated energy, their angular velocities would change continuously, and they would emit a *continuous* spectrum (a mixture of all frequencies), not the *line* spectrum actually observed.

To solve this problem, Bohr made a revolutionary proposal. In effect, he postulated that an electron in an atom can revolve in certain **stable orbits,** with a definite energy associated with each orbit, without emitting radiation, contrary to the predictions of classical electromagnetic theory. According to Bohr, an atom emits or absorbs radiation only when it makes a transition from one of these stable orbits to another, at the same time emitting (or absorbing) a photon with appropriate energy and frequency given by Equation 28.7.

To determine the radii of the "permitted" orbits, Bohr introduced what we have to regard in hindsight as a brilliant intuitive guess. He postulated that the angular momentum L of the orbiting electron is **quantized;** that is, it can have only certain specific values. Specifically, Bohr proposed that the angular momentum must be an integer multiple of $h/2\pi$. (Note that the units of Planck's constant h, usually written as J · s, are the same as the units of angular momentum, usually written as kg · m^2/s.) Bohr postulated that in the hydrogen atom, only those orbits are permitted for which the angular momentum is an integral multiple of $h/2\pi$.

Recall (Equation 10.11) that the angular momentum L of a particle with mass m, moving with speed v in a circle of radius r, is $L = mvr$. So Bohr's assumption may be stated as follows:

Quantization of angular momentum in the Bohr model

In Bohr's hypothesis, the allowed values of angular momentum L for an electron in a circular orbit are given by

$$L = mvr = n\frac{h}{2\pi}, \qquad n = 1, 2, 3, \cdots.$$

Each value of n also corresponds to a permitted value of the orbit radius, which we denote from now on by r_n, and a corresponding speed v_n. With this notation, the preceding equation becomes

$$mv_n r_n = n\frac{h}{2\pi}. \qquad (28.11)$$

Now let's consider a mechanical model of the hydrogen atom (Figure 28.15) that incorporates Equation 28.11. This atom consists of a single electron with mass m and charge $-e$, revolving around a single proton with charge $+e$. The proton is nearly 2000 times as massive as the electron, so we'll assume that the proton doesn't move. We learned in Chapter 6 that when a particle with mass m moves with speed v_n in a circular orbit with radius r_n, its acceleration has magnitude v_n^2/r_n. According to Newton's second law, a force with magnitude

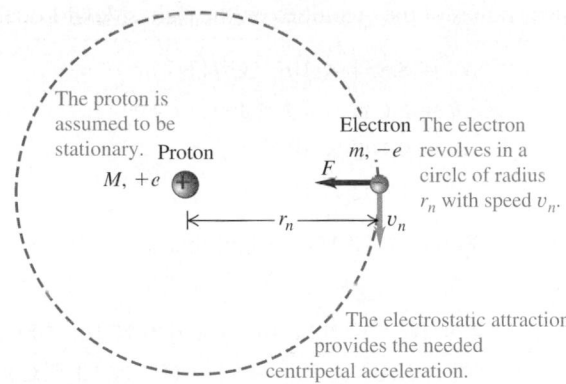

▲ FIGURE 28.15 Bohr model of the hydrogen atom.

$F = mv_n^2/r_n$ is needed to cause this acceleration. The force F is provided by the electrical attraction between the two charges, each with magnitude e, according to Coulomb's law, so

$$F = k\frac{e^2}{r_n^2} = \frac{1}{4\pi\epsilon_0}\frac{e^2}{r_n^2},$$

where $k = 1/4\pi\epsilon_0$ is the constant in Coulomb's law (Section 17.4).

The $F = ma$ equation is

$$\frac{1}{4\pi\epsilon_0}\frac{e^2}{r_n^2} = m\frac{v_n^2}{r_n}. \tag{28.12}$$

We can combine Equations 28.11 and 28.12 and rearrange terms to obtain separate expressions for r_n and v_n. When we do this, we find the following relationships.

Radius and speed of Bohr orbits

In the Bohr model, an electron in a state with quantum number n has orbit radius r_n and speed v_n given respectively by

$$r_n = \epsilon_0\frac{n^2h^2}{\pi me^2}, \tag{28.13}$$

$$v_n = \frac{1}{\epsilon_0}\frac{e^2}{2nh}. \tag{28.14}$$

Equation 28.13 shows that the orbit radius r_n is proportional to n^2; the smallest orbit r_1 corresponds to $n = 1$. This minimum radius r_1 is called the **Bohr radius:**

$$r_1 = \epsilon_0\frac{h^2}{\pi me^2}. \tag{28.15}$$

With this notation, we can rewrite Equation 28.13 as

$$r_n = n^2r_1. \tag{28.16}$$

The permitted, nonradiating orbits have radii r_1, $4r_1$, $9r_1$, and so on. The value of n for each orbit is called the **principal quantum number** for that orbit.

The numerical values of the quantities on the right side of Equation 28.15 are

$$\epsilon_0 = 8.854 \times 10^{-12}\, C^2/(N \cdot m^2),$$
$$h = 6.626 \times 10^{-34}\, J \cdot s,$$
$$m = 9.109 \times 10^{-31}\, kg,$$
$$e = 1.602 \times 10^{-19}\, C.$$

Using these values in Equation 28.15, we find that the radius r_1 of the smallest Bohr orbit is

$$r_1 = \frac{(8.854 \times 10^{-12}\, C^2/(N \cdot m^2))(6.626 \times 10^{-34}\, J \cdot s)^2}{(3.142)(9.109 \times 10^{-31}\, kg)(1.602 \times 10^{-19}\, C)^2}$$
$$= 0.5293 \times 10^{-10}\, m.$$

This result gives us an estimate of the size of the hydrogen atom that is consistent with atomic dimensions estimated by various other methods.

We can also use the preceding result with Equation 28.14 to find the orbital speed of the electron for any value of n. We leave this calculation as an exercise. The result is that, for the $n = 1$ state, $v_1 = 2.19 \times 10^6$ m/s. This speed is less than 1% of the speed of light, so we don't need to be concerned about relativistic effects.

Energy Levels

We can use Equations 28.13 and 28.14 to find the kinetic and potential energies, K_n and U_n, for an electron in the orbit with quantum number n. For K_n, we use Equation 28.14:

$$K_n = \frac{1}{2}mv_n^2 = \frac{1}{\epsilon_0^2}\frac{me^4}{8n^2h^2}.$$

We use Equation 28.13 to express the potential energy U_n in terms of n:

$$U_n = -\frac{1}{4\pi\epsilon_0}\frac{e^2}{r_n} = -\frac{1}{\epsilon_0^2}\frac{me^4}{4n^2h^2}.$$

The total energy E_n for any n is the sum of the kinetic and potential energies:

$$E_n = K_n + U_n = -\frac{1}{\epsilon_0^2}\frac{me^4}{8n^2h^2}. \tag{28.17}$$

This expression has a negative sign because we have taken the reference level of potential energy to be zero when the electron is at rest at an infinite distance from the nucleus. We are interested only in energy *differences* between levels.

The preceding discussion shows that the possible energy levels and electron orbits for the atom are labeled by values of the quantum number n. For each value of n, there are corresponding values of the orbit radius r_n, speed v_n, angular momentum $L_n = nh/2\pi$, and total energy E_n. The energy of the atom is least when $n = 1$ and E_n has its largest negative value. This is the *ground state* of the atom—the state with the smallest orbit, with radius r_1. For $n = 2, 3, \cdots$, the absolute value of E_n is smaller and the energy is progressively larger (less negative). The orbit radius increases as n^2, as shown by Equation 28.13.

Comparing the expression for E_n in Equation 28.17 with Equation 28.10 ($E_n = -hcR/n^2$, the energy-level equation deduced from spectrum analysis), we see that they agree only if the coefficients are equal:

Relation of the Rydberg constant to fundamental constants:

$$hcR = \frac{1}{\epsilon_0^2}\frac{me^4}{8h^2}, \qquad \text{or} \qquad R = \frac{me^4}{8\epsilon_0^2 h^3 c}. \qquad (28.18)$$

This equation shows us how to *calculate* the value of the Rydberg constant R from the fundamental physical constants m, c, e, h, and ϵ_0, all of which can be determined quite independently of the Bohr theory.

When we substitute the numerical values of these quantities into Equation 28.18, we obtain the value $R = 1.097 \times 10^7 \text{ m}^{-1}$. This value is within 0.01% of the value determined from wavelength measurements, a major triumph for Bohr's theory! We invite you to substitute the numerical values into Equation 28.18 and compute the value of R to confirm these statements.

We can also calculate the energy required to remove the electron completely from a hydrogen atom. This energy is called the **ionization energy.** Ionization corresponds to a transition from the ground state $(n = 1)$ to an infinitely large orbit radius $(n = \infty)$. The predicted energy is 13.6 eV. The ionization energy can also be measured directly, and the two values agree within 0.1%.

Quantitative Analysis 28.2

Transitions and frequency

According to the Bohr model of the hydrogen atom, as we look at higher and higher values of n, the wavelength of the photon emitted due to transitions between adjacent electron orbits

A. gets progressively smaller.
B. gets progressively larger.
C. remains constant.

SOLUTION In the Bohr model, the magnitude of the energy of a hydrogen atom in state n is proportional to $1/n^2$ (Equation 28.17). The energy of the photon emitted during a transition between two adjacent states is simply the difference of the energies $hf = E_i - E_f$ of the two states. As n gets larger, the magnitude of E_n and the difference $(E_i - E_f)$ get smaller. Therefore, the energy and frequency of the emitted photon are smaller for larger values of n, and, from $\lambda = c/f$, the wavelength increases. The correct answer is B.

EXAMPLE 28.6 The energies depend on the quantum number

Find the kinetic, potential, and total energies of the hydrogen atom in the $n = 2$ state, and find the wavelength of the photon emitted in the transition $n = 2 \rightarrow n = 1$.

SOLUTION

SET UP We'll need to use the expressions for kinetic and potential energy that we derived earlier, leading to Equation 28.17. We can simplify the calculations a lot by noting that the product of constants that appears in this equation is equal to hcR, where R is the Rydberg constant we encountered in Equation 28.18. The numerical value is 13.6 eV:

$$\frac{me^4}{8\epsilon_0^2 h^2} = hcR = 13.6 \text{ eV}.$$

SOLVE Using this equation, we can express the kinetic, potential, and total energies for the state with quantum number n respectively as

$$K_n = \frac{13.6 \text{ eV}}{n^2}, \qquad U_n = \frac{-27.2 \text{ eV}}{n^2}, \qquad E_n = \frac{-13.6 \text{ eV}}{n^2}.$$

For the $n = 2$ state, we find that $K_2 = 3.40$ eV, $U_2 = -6.80$ eV, and $E_2 = -3.40$ eV.

The energy of the emitted photon is

$$E_i - E_f = -3.40 \text{ eV} - (-13.6 \text{ eV}) = 10.2 \text{ eV}$$
$$= 1.63 \times 10^{-18} \text{ J}.$$

This is equal to hc/λ, so we obtain

$$\lambda = \frac{hc}{E_i - E_f} = \frac{(6.626 \times 10^{-34} \text{ J} \cdot \text{s})(3.00 \times 10^8 \text{ m/s})}{1.63 \times 10^{-18} \text{ J}}$$
$$= 122 \times 10^{-9} \text{ m} = 122 \text{ nm}.$$

REFLECT The wavelength we just found is that of the Lyman alpha line, the longest-wavelength line in the Lyman series of ultraviolet lines in the hydrogen spectrum.

Practice Problem: Find the wavelength for the transition $n = 3 \rightarrow n = 1$ for singly ionized helium, which has one electron and a nuclear charge of $2e$. (Note that the value of the Rydberg constant is four times as great as for hydrogen because it is proportional to the square of the product of the nuclear charge and the electron charge. (See the next paragraph.) *Answer:* 25.6 nm.

The Bohr model can be extended to other one-electron atoms, such as the singly ionized helium atom, the doubly ionized lithium atom, and so on. If the nuclear charge is Ze (where Z is the atomic number) instead of just e, the effect in the preceding analysis is to replace e^2 everywhere by Ze^2. In particular, the orbit radii r_n given by Equation 28.13 become smaller by a factor of Z, and the energy levels E_n given by Equation 28.17 are multiplied by Z^2. We invite you to verify these statements.

Limitations of the Bohr Model

Although the Bohr model predicted the energy levels of the hydrogen atom correctly, it raised as many questions as it answered. It combined elements of classical physics with new postulates that were inconsistent with classical ideas. There was no clear justification for restricting the angular momentum to multiples of $h/2\pi$, except that it led to the right answer. The model provided no insight into what happens *during* a transition from one orbit to another. The stability of certain orbits was achieved at the expense of discarding the only picture available in Bohr's time of the electromagnetic mechanism by which the atom radiated energy. The angular velocities of the electron motion were not the same as the angular frequencies of the emitted radiation, a result that is contrary to classical electrodynamics. Attempts to extend the model to atoms with two or more electrons were not successful. In Chapter 29, we'll find that even more radical departures from classical concepts were needed before the understanding of atomic structure could progress further.

28.4 The Laser

A **laser** is a light source that produces a beam of highly coherent and very nearly monochromatic light as a result of "cooperative" emission from many atoms. The name *laser* is an acronym for "light amplification by stimulated emission of radiation." We can understand the principles of laser operation on the basis of photons and atomic energy levels.

If an atom has an excited state with energy level E above the ground state, then an atom in the ground state can absorb a photon whose frequency f is given by the Planck equation $E = hf$. This process is shown schematically in Figure 28.16a, which shows a gas in a transparent container. An atom A absorbs a photon, reaching an excited state, labeled A* in the figure. Some time later, the excited atom returns to the ground state by emitting a photon with the same frequency and energy as the one originally absorbed. This process is called **spontaneous emission;** the direction of the emitted photon is random (Figure 28.16b), and its vibrations aren't synchronized with those of the incoming photon that started the process.

In **stimulated emission** (Figure 28.16c), an incident photon encounters an atom that's already in an excited state with the same energy above the ground state as the photon's energy. By a kind of resonance effect, the incoming photon causes the atom to emit another photon with the same frequency and direction as the incoming photon and with its vibrations synchronized with those of the incoming photon. One photon goes in, and two come out, with their vibrations synchronized—hence the term *light amplification*. Under proper conditions, the laser uses excitation and stimulated emission to produce a beam consisting of a large number of such synchronized photons. Such a beam of synchronized photons is said to be *coherent.*

Here's how one kind of laser works: Suppose we have a large number of identical atoms in a gas or vapor in a container with transparent walls, as in Figure 28.16a. At moderate temperatures, if no radiation is incident on the

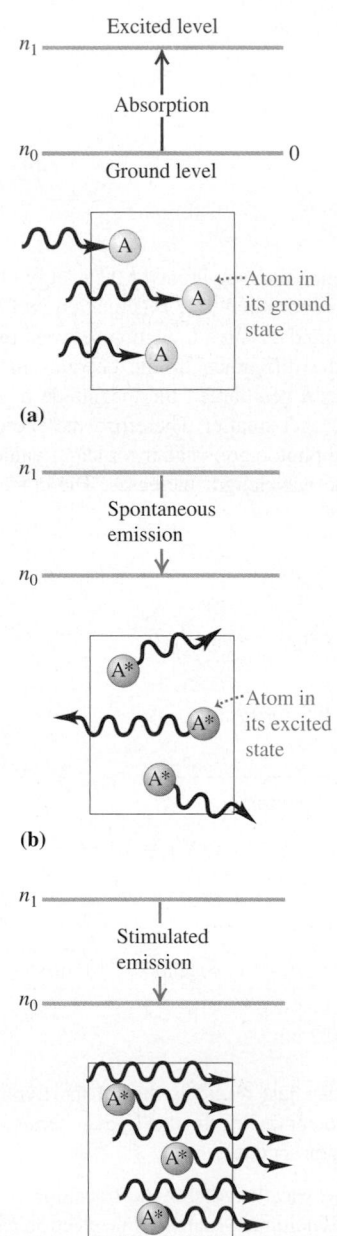

▲ **FIGURE 28.16** Three interaction processes for an atom and electromagnetic waves.

container, most of the atoms are in the ground state, and only a few are in excited states. If there is a state with energy level E above the ground state, the ratio of the number n_E of atoms in this state (the *population* of the state) to the number of atoms n_0 in the ground state is small.

Suppose we send through the container a beam of radiation with frequency f corresponding to the energy difference E. Some of the atoms absorb photons of energy E and are raised to the excited state, and the population ratio n_E/n_0 increases. Because n_0 is originally so much larger than n_E, an enormously intense beam of light would be required to increase n_E to a value comparable to n_0.

But now suppose that we can create a situation in which n_E is substantially increased compared with its normal equilibrium value; this condition is called a **population inversion.** If n_E can be increased enough, the rate of energy radiation by stimulated emission can actually exceed the rate of absorption. The system then acts as a source of radiation with photon energy E. Furthermore, because the photons are the result of stimulated emission, they all have the same frequency, phase, polarization, and direction, and they are all synchronized. The resulting emitted radiation is therefore much more *coherent* than light from ordinary sources, in which the emissions of individual atoms are *not* coordinated. This coherent emission is exactly what happens in a laser.

Lasers can be made to operate over a broad range of the electromagnetic spectrum, from microwaves to x rays. In recent years, lasers have found a wide variety of practical applications. The high intensity of a laser beam makes it a convenient drill. For example, a laser beam can drill a very small hole in a diamond for use as a die in making very small diameter wire. Because the photons in a laser beam are strongly correlated in their directions, the light output from a laser can be focused to a very narrow beam, and it can travel long distances without appreciable spreading. Surveyors often use lasers, especially in situations in which great precision is required, such as a long tunnel drilled from both ends.

Lasers are widely used in medical science. A laser can produce a narrow beam with extremely high intensity, high enough to vaporize anything in its path. This property is used in the treatment of a detached retina. A short burst of radiation damages a small area of the retina, and the resulting scar tissue "welds" the retina back to the choroid from which it has become detached. Laser beams have many surgical applications; blood vessels cut by the beam tend to seal themselves off, making it easier to control bleeding. Lasers are also used for selective destruction of tissue, as in the removal of tumors.

▲ **Application As red as rubies.** The beautiful red color of the ruby has been enjoyed for centuries, and in 1960 this gemstone was used in the first laser. A ruby laser uses a high-quality cylindrical ruby crystal surrounded by a high-intensity lamp that provides the energy to excite electrons of chromium atoms in the ruby to higher energy levels. As these excited electrons return to their ground state, they emit photons of red light at 694 nm, and photons from one atom stimulate emission of photons from other atoms, rapidly amplifying the light intensity. The result is a coherent beam of synchronized photons that can travel very long distances with little spreading, even to the moon and back. After Apollo astronauts placed mirrors on the moon in 1969, a ruby laser beam shot from Texas was used to accurately measure the earth-to-moon distance to within 10 cm, a precision of nine significant figures.

PhET: Lasers
ActivPhysics 18.3: The Laser

Electrons are emitted thermionically from the heated cathode and are accelerated toward the anode; when they strike it, x rays are produced.

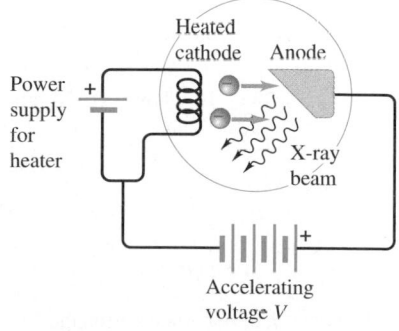

▲ **FIGURE 28.17** An apparatus used to produce x rays, similar to Röntgen's 1895 apparatus.

28.5 X-Ray Production and Scattering

The production and scattering of **x rays** provides additional examples of the quantum view of electromagnetic radiation. **X rays** are produced when rapidly moving electrons that have been accelerated through a potential difference on the order of 10^3 to 10^6 V strike a metal target. X rays were first produced in 1895 by Wilhelm Röntgen (1845–1923), using an apparatus similar in principle to that shown in Figure 28.17. Electrons are "boiled off" from the heated cathode by thermionic emission and are accelerated toward the anode (the target) by a large potential difference V. The bulb is evacuated (to a residual gas pressure of 10^{-7} atm or less) so that the electrons can travel from cathode to anode without colliding with air molecules. When V is a few thousand volts or more, a penetrating radiation is emitted from the anode surface.

X-Ray Photons

Because of their electromagnetic origin, it is clear that x rays are electromagnetic waves. Like light, they are governed by quantum relations in their interaction with matter. Thus, we can talk about x-ray photons or quanta, and the energy of an x-ray photon is related to its frequency and wavelength in the same way as for photons of light: $E = hf = hc/\lambda$. Typical x-ray wavelengths are 0.001 to 1 nm $(10^{-12}$ to 10^{-9} m$)$. X-ray wavelengths can be measured quite precisely by crystal diffraction techniques, which we've discussed previously.

X-ray emission is the inverse of the photoelectric effect. In photoelectric emission, the energy of a photon is transformed into kinetic energy of an electron; in x-ray production, the kinetic energy of an electron is transformed into energy of a photon. The energy relation is exactly the same in both cases. In x-ray production, we can neglect the work function of the target and the initial kinetic energies of the electrons because they are ordinarily only a few electronvolts, very small in comparison to the energies of the accelerated electrons when they strike the target.

Each element has a set of atomic energy levels associated with x-ray photons and therefore also has a characteristic x-ray spectrum. These energy levels, called **x-ray energy levels,** are rather different in character from those associated with visible spectra. They are associated with vacancies in the inner electron configurations of complex atoms. The energy levels can be hundreds or thousands of electronvolts above the ground state, rather than a few electronvolts, as is typical with optical spectra.

EXAMPLE 28.7   Wavelength of an x-ray photon

Electrons are accelerated by a potential difference of 10.0 kV. If an electron produces a photon on impact with the target, what is the minimum wavelength of the resulting x rays?

SOLUTION

SET UP We simply note that the *maximum* photon energy $hf_{max} = hc/\lambda_{min}$ and *minimum* wavelength occur when *all* the initial kinetic energy eV of the electron is used to produce a single photon:

$$eV = hf = \frac{hc}{\lambda}.$$

All that's left is to solve for λ and substitute the numbers.

SOLVE Following the program just outlined, we get

$$\lambda = \frac{hc}{eV} = \frac{(6.626 \times 10^{-34} \text{ J} \cdot \text{s})(3.00 \times 10^8 \text{ m/s})}{(1.602 \times 10^{-19} \text{ C})(1.00 \times 10^4 \text{ V})}$$
$$= 1.24 \times 10^{-10} \text{ m} = 0.124 \text{ nm}.$$

REFLECT We can measure the x-ray wavelength by crystal diffraction and confirm this prediction directly. The wavelength is *smaller* than that for typical visible-light photons by a factor on the order of 5000, and the photon energy is greater by the same factor.

Practice Problem: What accelerating voltage is needed to produce x rays with a wavelength of 0.050 nm? *Answer:* 24.8 kV.

ActivPhysics 17.6: Uncertainty Principle

Compton Scattering

A phenomenon called **Compton scattering,** first observed in 1924 by A. H. Compton, provides additional direct confirmation of the quantum nature of x rays. When x rays strike matter, some of the radiation is scattered, just as visible light falling on a rough surface undergoes diffuse reflection. Compton discovered that some of the scattered radiation has a smaller frequency and longer wavelength than the incident radiation and that the increase in wavelength, $\Delta\lambda$, depends on the angle through which the radiation is scattered.

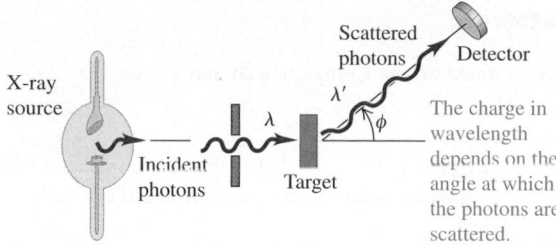

▲ **FIGURE 28.18** A Compton-effect experiment.

If the scattered radiation emerges at an angle ϕ with respect to the incident direction, as shown in Figure 28.18, and if λ and λ' are the wavelengths of the incident and scattered radiation, respectively, then the wavelength increase $\Delta\lambda$ is given by a simple expression:

Wavelength increase in Compton scattering

$$\Delta\lambda = \lambda' - \lambda = \frac{h}{mc}(1 - \cos\phi), \qquad (28.19)$$

where m is the electron mass.

Compton scattering cannot be understood on the basis of classical electromagnetic theory, which predicts that the scattered wave has the *same* wavelength as the incident wave. In contrast, the quantum theory provides a beautifully clear explanation. We imagine the scattering process as a collision of two *particles:* the incident photon and an electron initially at rest, as in Figure 28.19a. The photon gives up some of its energy and momentum to the electron, which recoils as a result of the impact. The x-ray photon carries both energy and momentum; the *total* energy of the photon and the electron is conserved during the collision, and the same is true of the total momentum. The final scattered photon has less energy, smaller frequency, and longer wavelength than the initial one (Figure 28.19b).

Equation 28.19 can be derived from the principles of conservation of energy and momentum, treating the electron and the photons as particles, using the relativistic energy–momentum relations, and using the wavelength–momentum relation $p = h/\lambda$ for the incident and scattered photons. The quantity h/mc that appears in this equation has units of length, as it must for dimensional consistency. Its numerical value is

$$\frac{h}{mc} = \frac{6.626 \times 10^{-34}\,\text{J}\cdot\text{s}}{(9.109 \times 10^{-31}\,\text{kg})(2.998 \times 10^8\,\text{m/s})}$$
$$= 2.426 \times 10^{-12}\,\text{m} = 2.426\,\text{pm}.$$

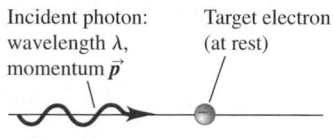

(a) Before collision: The target electron is at rest.

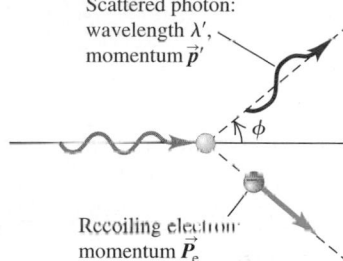

(b) After collision. The angle between the directions of the scattered photon and the incident photon is ϕ.

(c) Vector diagram showing the conservation of momentum in the collision: $\vec{p} - \vec{p}' + \vec{P}_\text{e}$.

▲ **FIGURE 28.19** Schematic diagram of Compton scattering.

Conceptual Analysis 28.3

Photon strikes electron

A photon of light scatters off of an electron which is initially at rest. After the collision

A. The wavelength and momentum of the photon are less than before the collision.
B. The wavelength of the photon and the momentum of the electron are greater than before the collision.
C. The energy of the photon is less than before the collision, while its momentum is unchanged.

SOLUTION The scattering process can be thought of as a collision of two particles—the photon and the electron. Therefore, the energy and momentum lost by the photon are gained by the electron. When the photon loses energy, its wavelength increases, according to Equation 28.5 ($E = hc/\lambda$), and the electron gains momentum. The correct answer is B.

EXAMPLE 28.8 **Wavelength shift in Compton scattering**

For the x-ray photons in Example 28.7 $(\lambda = 0.124 \text{ nm})$, at what angle do the Compton-scattered x rays have a wavelength that is 1.0% longer than that of the incident x rays?

SOLUTION

SET UP The scattered x-ray photons have less energy, lower frequency, and longer wavelength than the incident photons. Conservation of energy and momentum led us to the relation

$$\Delta\lambda = \lambda' - \lambda = \frac{h}{mc}(1 - \cos\phi)$$

for the wavelengths λ and λ' of the incident and scattered photons.

SOLVE In the preceding equation, we want $\Delta\lambda = \lambda' - \lambda$ to be 1.0% of 0.124 nm, or 1.24 pm. Using the value $h/mc = 2.426$ pm, we find that

$$\Delta\lambda = \frac{h}{mc}(1 - \cos\phi),$$
$$1.24 \text{ pm} = (2.426 \text{ pm})(1 - \cos\phi),$$
$$\phi = 60.7°.$$

REFLECT We note that if the scattering angle is less than this value, the interaction is weaker and the wavelength shift is also less.

Practice Problem: What is the scattering angle if the scattered wavelength is 0.050% longer than the incident wavelength? *Answer:* 13.0°.

Applications of X Rays

Because many materials that are opaque to ordinary light are transparent to x rays, this radiation can be used to visualize the interiors of materials such as broken bones or defects in structural steel. In the past 20 to 30 years, several highly sophisticated x-ray imaging techniques have been developed. One widely used system is *computerized axial tomography;* the corresponding instrument is called a CAT scanner. The x-ray source produces a thin, fan-shaped beam that is detected by an array of several hundred detectors in a line. The entire apparatus is rotated around the subject, and the changing photon-counting rates of the detectors are recorded digitally. A computer processes this information and reconstructs a picture of density over an entire cross section of the subject. Density differences as small as 1% can be detected with CAT scans, and tumors and other anomalies much too small to be seen with older x-ray techniques can be detected.

X rays cause damage to living tissues. As x-ray photons are absorbed in tissues, they break molecular bonds and create highly reactive free radicals (such as neutral H and OH), which in turn can disturb the molecular structure of proteins and especially genetic material. Young and rapidly growing cells are particularly susceptible; thus, x rays are useful for the selective destruction of cancer cells. However, a cell may be damaged by the radiation, but survive, continue dividing, and produce generations of defective cells; hence, paradoxically, x rays can *cause* cancer as well as treat it. The medical use of x rays requires careful assessment of the balance between risks and benefits of radiation exposure in each individual case.

28.6 The Wave Nature of Particles

Some aspects of emission and absorption of light can be understood on the basis of photons and atomic energy levels. But a complete theory should also offer some means of *predicting* the values of these energy levels for any particular atom. The Bohr model of the hydrogen atom was a start, but it could not be generalized to many-electron atoms. For more general problems, even more sweeping revisions of 19th-century ideas were needed. The wave–particle duality of electromagnetic radiation had to be extended to include *particles* as well as radiation. As a consequence, a particle can no longer be described as a single point moving in space. Instead, it is an inherently spread-out entity. As the particle moves, the spread-out character has some of the properties of a *wave*. The new theory, called *quantum mechanics,* is the key to understanding the structure of atoms and molecules, including their spectra, chemical behavior, and many other properties. It has the happy effect of restoring unity to our description of both particles and radiation, and wave concepts are a central theme.

De Broglie Waves

A major advance in the understanding of atomic structure began in 1924 (about 10 years after Bohr published his analysis of the hydrogen atom) with a proposition made by a young French physicist, Louis de Broglie. His reasoning, freely paraphrased, went like this:

Nature loves symmetry. Light is dualistic in nature; in some situations, it behaves like a wave, in others like a particle. If nature is symmetric, this duality should also hold for matter. Electrons and protons, which we usually think of as *particles,* may in some situations behave like *waves*.

If an electron acts like a wave, it should have a wavelength. Using the nonrelativistic definitions of momentum and energy, de Broglie postulated that a free electron with mass m, moving with speed v should have a wavelength λ related to the magnitude of its momentum, $p = mv$, in exactly the same way as for a photon, as expressed by Equation 28.6, $\lambda = h/p$.

De Broglie wavelength of an electron

$$\lambda = \frac{h}{p} = \frac{h}{mv},\qquad(28.20)$$

where h is Planck's constant.

De Broglie asserted that the relation $\lambda = h/p$ should hold for *both* electrons and photons, and, indeed, for all particles. (But we shouldn't expect to see the wavelengths of *macroscopic* objects: A pitched baseball might have a momentum with magnitude $6 \text{ kg} \cdot \text{m/s}$; the corresponding wavelength would be about 10^{-34} m, much too small to be seen with even the most powerful microscope.)

De Broglie's proposal was radical, but it was clear that a radical idea was needed. The dual nature of electromagnetic radiation had led to adoption of the photon concept, also a radical idea. A similar revolution was needed in the mechanics of particles. An essential concept of quantum mechanics is that a particle is no longer described as located at a single point, but instead is an inherently spread-out entity described in terms of a function that has various values at various points in space. The spatial distribution describing a *free* electron has a recurring pattern characteristic of a *wave* that propagates through space. Electrons within atoms can be visualized as diffuse clouds surrounding the nucleus.

Conceptual Analysis 28.4 **de Broglie wavelength**

If the de Broglie wavelength of a particle is too small, its wavelike properties are difficult to observe. For which particle would the wavelike properties be easier to observe?

A. A low-mass, slow particle
B. A high-mass particle moving at nearly the speed of light
C. A low-mass particle moving at nearly the speed of light.

SOLUTION The larger a particle's de Broglie wavelength, the easier it is to observe its wavelike properties. Since $\lambda = h/mv$ (Equation 28.20), slow, low-mass particles have the largest wavelengths and thus have the most easily observed wavelike properties. The correct answer is A.

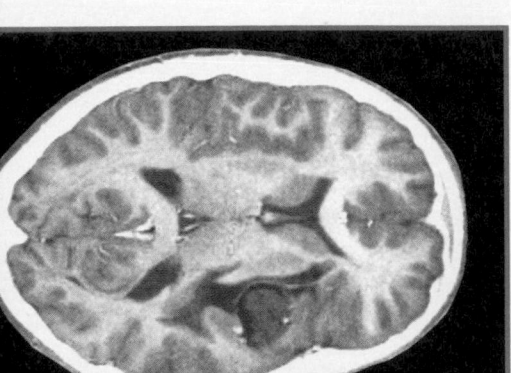

▲ BIO Application Boron neutron capture therapy. We have learned that particles exhibit wave-like behavior and that the wavelength and energy of a particle depend upon its velocity and mass. An experimental form of brain cancer therapy uses beams of slow-moving neutrons called thermal neutrons that are the proper wavelength to have just the right amount of energy to be absorbed by boron atoms, which are then converted into an unstable state. These unstable boron atoms then split into high-energy lithium nuclei and alpha particles that are deadly to nearby tissue. But because they can only travel a few micrometers, less than the width of a human cell, they will kill only the cell that had contained the boron atom. Medical researchers have developed boron-containing compounds that are selectively taken up by tumor cells, ensuring that only tumor cells will be destroyed by the beam of thermal neutrons.

PROBLEM-SOLVING STRATEGY 28.2 Atomic physics (MP)

1. In atomic physics, the orders of magnitude of physical quantities are so unfamiliar that common sense often isn't much help in judging the reasonableness of a result. It helps to remind yourself of some typical magnitudes of various quantities:

   Size of an atom: 10^{-10} m
   Mass of an atom: 10^{-26} kg
   Mass of an electron: 10^{-30} kg
   Energy of an atomic energy level: 1 to 10 eV, or 10^{-18} J (but some interaction energies in many-electron atoms are much larger).
   Speed of an electron in the Bohr atom: 10^6 m/s
   Electron charge: 10^{-19} C
   kT at room temperature: $1/40$ eV

   You may want to add items to this list. These values will also help you in Chapter 30 when we have to deal with magnitudes characteristic of *nuclear* rather than atomic structure, often different by factors of 10^4 to 10^6. In working out problems, be careful to handle powers of ten properly. A gross error might not be obvious.

2. As in preceding sections, energies may be expressed either in joules or in electronvolts. Be sure to use consistent units. Lengths, such as wavelengths, are always in meters if you use the other quantities consistently in SI units, such as $h = 6.626 \times 10^{-34}$ J · s. If you want nanometers or something else, don't forget to convert. In some problems, it's useful to express h in eV: $h = 4.136 \times 10^{-15}$ eV · s.

3. Kinetic energy can be expressed either as $K = \frac{1}{2}mv^2$ or (because $p = mv$) as $K = p^2/2m$. The latter form is often useful in calculations involving the de Broglie wavelength.

EXAMPLE 28.9 Energy of a thermal neutron

Neutrons have a wavelength, obeying the same relation as for electrons and protons: Equation 28.6 $(\lambda = h/p)$. Find the speed and kinetic energy of a neutron $(m = 1.675 \times 10^{-27}$ kg$)$ that has a de Broglie wavelength $\lambda = 0.200$ nm, typical of atomic spacing in crystals. Compare the energy with the average kinetic energy of a gas molecule at room temperature $(T = 20°C)$.

SOLUTION

SET UP We know the wavelength λ, the mass m, and Planck's constant h, so we can use the de Broglie relationship in the form $\lambda = h/mv$ to find the neutron's speed. From that result, we can find its kinetic energy.

SOLVE From the de Broglie relationship,

$$v = \frac{h}{\lambda m} = \frac{6.626 \times 10^{-34} \text{ J} \cdot \text{s}}{(0.200 \times 10^{-9} \text{ m})(1.675 \times 10^{-27} \text{ kg})}$$
$$= 1.98 \times 10^3 \text{ m/s};$$
$$K = \frac{1}{2}mv^2 = \frac{1}{2}(1.675 \times 10^{-27} \text{ kg})(1.98 \times 10^3 \text{ m/s})^2$$
$$= 3.28 \times 10^{-21} \text{ J} = 0.0205 \text{ eV}.$$

The average translational kinetic energy of a molecule of an ideal gas at absolute temperature $T = 293$ K (equal to 20°C or 68°F) is given by Equation 15.8, where k is Boltzmann's constant:

$$K = \frac{3}{2}kT = \frac{3}{2}(1.38 \times 10^{-23} \text{ J/K})(293 \text{ K})$$
$$= 6.07 \times 10^{-21} \text{ J} = 0.0379 \text{ eV}.$$

REFLECT The two energies are comparable in magnitude. In fact, a neutron with kinetic energy in this range is called a *thermal neutron*. Diffraction of thermal neutrons can be used to study crystal and molecular structure in the same way as x-ray diffraction is used. Neutron diffraction has proved especially useful in the study of large organic molecules.

Practice Problem: Find the kinetic energy of an electron with a de Broglie wavelength of 0.100 nm. *Answer:* 150 eV.

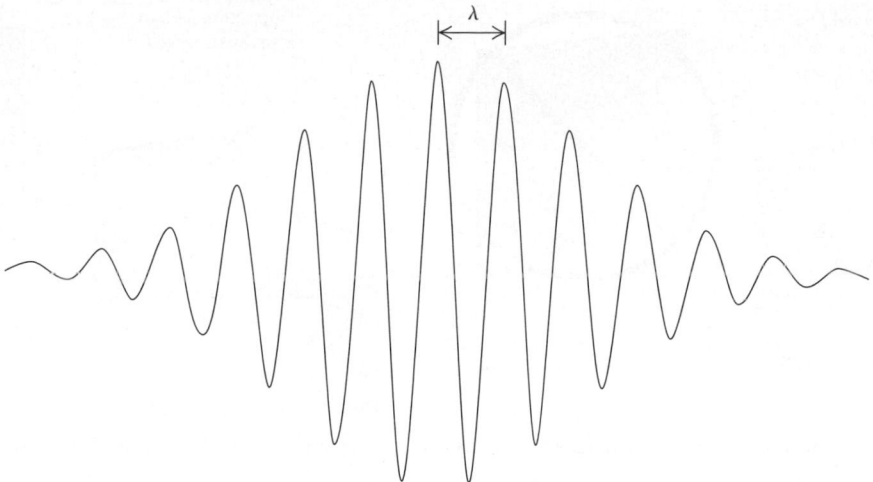

▲ **FIGURE 28.20** A wave that has both wave and particle aspects.

Davisson–Germer Experiment

De Broglie's wave hypothesis, unorthodox though it seemed at the time, received almost immediate direct experimental confirmation. We described in Section 26.7 how the atoms in a crystal can function as a three-dimensional diffraction grating for x rays. An x-ray beam is strongly reflected when it strikes a crystal at an angle that gives constructive interference in the various scattered waves. The existence of these strong reflections is evidence for the wave nature of x rays, and it provides a means for measuring the wavelengths of x rays.

In 1927, C. J. Davisson and L. H. Germer, working at Bell Telephone Laboratories, discovered that a beam of electrons was reflected in almost the same way that x rays would be reflected from the same crystal; that is, they were being diffracted. Davisson and Germer could determine the speeds of the electrons from the accelerating voltage, so they could compute the de Broglie wavelength $\lambda = h/mv$ from Equation 28.20. They found that the angles at which strong reflection took place were the same as those at which x rays with the same wavelength would be reflected. This phenomenon is called **electron diffraction,** and its discovery gave strong support to de Broglie's hypothesis.

Additional experiments were carried out in many laboratories. In Germany, Estermann and Stern demonstrated diffraction of alpha particles. More recently, diffraction experiments have been performed with low-energy neutrons and large molecules. Thus, the wave nature of particles, so strange in 1924, became firmly established in the ensuing years.

Figure 28.20 suggests how a possible wave function for a free electron might look. The wave is more or less localized in space; if we don't look too closely, it looks like a point particle. Yet it clearly has wave properties, including a definite wavelength. In short, it has both wave and particle properties.

28.7 Wave–Particle Duality

The discovery of the dual wave–particle nature of matter has forced us to reevaluate the kinematic language we use to describe the behavior of a particle. In classical Newtonian mechanics, we think of a particle as a point. At any instant in time, it has a definite position and a definite velocity. But in general, such a specific description isn't possible. When we look on a small enough scale, there are fundamental limitations on the precision with which we can describe the position and velocity of a particle. Many aspects of a particle's behavior can be stated only in terms of probabilities.

MasteringPHYSICS

PhET: Davisson–Germer: Electron Diffraction

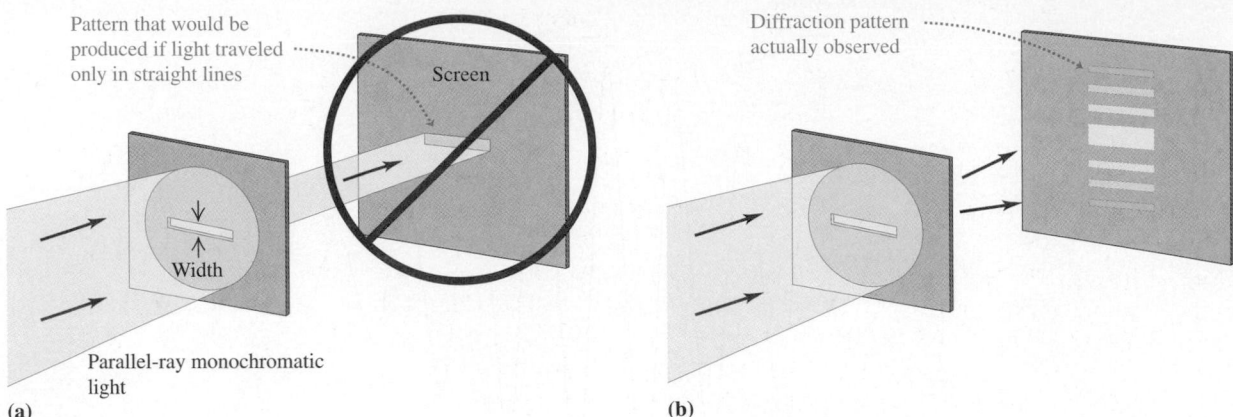

Pattern that would be produced if light traveled only in straight lines

Screen

Width

Parallel-ray monochromatic light

(a)

Diffraction pattern actually observed

(b)

▲ **FIGURE 28.21** (a) Geometric "shadow" of a horizontal slit. (b) Diffraction pattern of a horizontal slit. The slit width has been greatly exaggerated.

Single-Slit Diffraction

To get some insight into the nature of the problem, let's review an optical single-slit diffraction experiment. Figure 28.21 shows a light beam incident on a narrow horizontal slit. If light traveled only in straight lines, the pattern on the screen would be as shown in Figure 28.21a. The actual pattern (Figure 28.21b), however, is a diffraction pattern, showing the wave nature of light. Detailed analysis shows that the width of the central maximum in the pattern is inversely proportional to the height of the aperture. Most (about 85%) of the total energy is concentrated in the central maximum of the pattern.

Now we perform the same experiment again, but using a beam of electrons instead of a beam of monochromatic light (Figure 28.22). We have to do the experiment in vacuum (10^{-7} atm or less) so that the electrons don't bump into air molecules. We can produce the electron beam with a setup similar in principle to the electron gun in a cathode-ray tube. This produces a narrow beam of electrons that all have the same direction and speed and therefore also the same de Broglie wavelength.

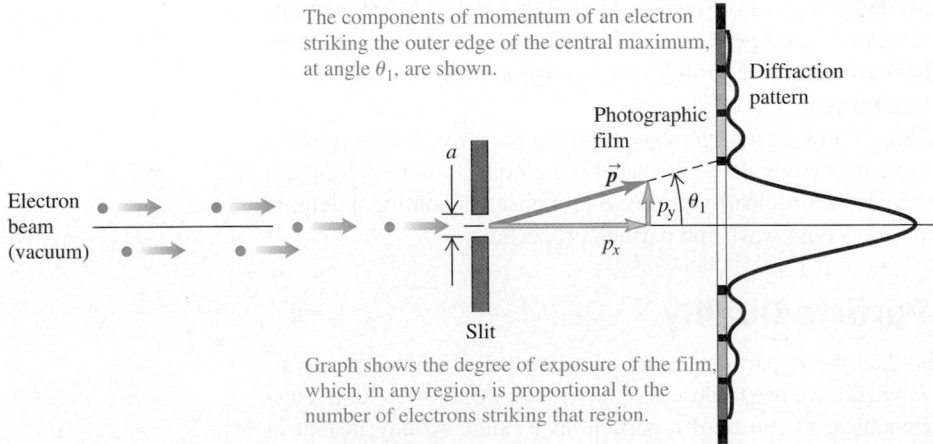

The components of momentum of an electron striking the outer edge of the central maximum, at angle θ_1, are shown.

Diffraction pattern

Photographic film

a

$\vec{p}$

p_y θ_1

p_x

Electron beam (vacuum)

Slit

Graph shows the degree of exposure of the film, which, in any region, is proportional to the number of electrons striking that region.

▲ **FIGURE 28.22** An electron diffraction experiment.

The result of this experiment, recorded on photographic film or with the use of more sophisticated detectors, is a diffraction pattern (Figure 28.22) identical to the one shown in Figure 28.21b. Most (85%) of the electrons strike the film in the region of the central maximum, but a few strike farther from the center, in the subsidiary maxima on both sides.

The electrons don't all follow the same path, even though they all have the same initial state of motion. In fact, we can't predict the trajectory of an individual electron from knowledge of its initial state. The best we can do is to say that most of the electrons go to a certain region, fewer go to other regions, and so on. That is, we can describe only the *probability* that an individual electron will strike each of various areas on the film. This fundamental indeterminacy has no counterpart in Newtonian mechanics, in which the motion of a particle or a system can always be predicted if we know the initial position and motion with great enough precision.

There are fundamental uncertainties in both the position and the momentum of an individual particle, and these two uncertainties are related inseparably and in a reciprocal way. The uncertainty Δy in the y position of a particle is determined by the slit width a, and the uncertainty Δp_y in the y component of momentum, p_y, is directly related to the width of the central maximum in the diffraction pattern, as shown in Figure 28.22. If we reduce the uncertainty in y by decreasing the slit width, the diffraction pattern becomes broader, corresponding to an *increase* in the uncertainty of p_y.

More detailed analysis shows that the *product* of the two uncertainties must be at least as great as h:

$$\Delta p_y a \geq h. \tag{28.21}$$

The slit width a represents the uncertainty in the *position* of an electron as it passes through the slit, and the width of the diffraction pattern determines the uncertainty Δp_y. We can reduce Δp_y only by increasing the slit width, which increases the position uncertainty. Conversely, when we make the slit narrower, the diffraction pattern broadens, and the corresponding uncertainty in momentum increases.

The Uncertainty Principle

In more general discussions of uncertainty relations, the uncertainty of a quantity is usually described in terms of the statistical concept of *standard deviation,* a measure of the spread or dispersion of a set of numbers around their average value. If a coordinate x has an uncertainty Δx defined in this way, and if the corresponding momentum component p_x has an uncertainty Δp_x, then the two uncertainties are found to be related in general by an inequality named for its discoverer, Werner Heisenberg:

Heisenberg uncertainty principle for position and momentum

$$\Delta x \, \Delta p_x \geq \frac{h}{2\pi}. \tag{28.22}$$

(This relationship differs from Equation 28.21 by a numerical factor of 2π because of the way "standard deviation" is defined.)

Equation 28.22 is one form of the Heisenberg uncertainty principle. It states that, in general, neither the momentum nor the position of a particle can be measured simultaneously with arbitrarily great precision, as classical physics would predict. Instead, the uncertainties in the two quantities play complementary roles, as we have described. There are corresponding uncertainty relations for the y and z coordinates and momentum components.

PhET: Fourier: Making Waves
PhET: Quantum Wave Interference
ActivPhysics 17.6: Uncertainty Principle

Uncertainty in Energy

There is also an uncertainty principle involving *energy*. It turns out that the energy of a system has an inherent uncertainty.

Heisenberg uncertainty principle: energy and time

The uncertainty ΔE in the energy of a system in a particular state depends on the *time interval* Δt during which the system remains in that state. The relation is

$$\Delta E \, \Delta t \geq \frac{h}{2\pi}. \tag{28.23}$$

A system that remains in a certain state for a long time (large Δt) can have a well-defined energy (small ΔE), but if it remains in that state for only a short time (small Δt), the uncertainty in energy must be correspondingly greater (large ΔE).

EXAMPLE 28.10 Electron in a box

An electron is confined inside a cubical box 1.0×10^{-10} m on a side. **(a)** Estimate the minimum uncertainty in the x component of the electron's momentum. **(b)** If the electron has momentum with magnitude equal to the uncertainty found in part (a), what is its kinetic energy? Express the result in joules and in electronvolts.

SOLUTION

SET UP Let's concentrate on x components. The uncertainty Δx is the size of the box: $\Delta x = 1.0 \times 10^{-10}$ m. (a) We get the uncertainty Δp_x in momentum from the uncertainty principle, Equation 28.22. (b) Knowing Δp_x, we can estimate the uncertainty in the kinetic energy K from the relation $K = p_x^2/2m$.

SOLVE Part (a): From Equation 28.22, the minimum uncertainty in p_x is

$$\Delta p_x = \frac{h}{2\pi \, \Delta x} = \frac{6.63 \times 10^{-34} \, \text{J} \cdot \text{s}}{2\pi (1.0 \times 10^{-10} \, \text{m})} = 1.1 \times 10^{-24} \, \text{kg} \cdot \text{m/s}.$$

Part (b): Assuming that the energy is in the nonrelativistic range, an electron with this magnitude of momentum has kinetic energy

$$K = \frac{p_x^2}{2m} = \frac{(1.1 \times 10^{-24} \, \text{kg} \cdot \text{m/s})^2}{2(9.11 \times 10^{-31} \, \text{kg})}$$
$$= 6.1 \times 10^{-19} \, \text{J} = 3.8 \, \text{eV}.$$

REFLECT The box is roughly the same size as an atom, so it isn't surprising that the energy is of the same order of magnitude as typical electron energies in atoms. If our result had differed from typical atomic energy levels by a factor of 10^6, we might be suspicious.

Practice Problem: Suppose the box is the size of a nucleus (on the order of 10^{-15} m) and the particle is a proton. Make a rough estimate of the uncertainty in energy of the particle. *Answer:* 21 MeV.

EXAMPLE 28.11 Energy uncertainty in the sodium atom

A sodium atom in one of the "resonance levels" remains in that state for an average time of 1.6×10^{-8} s before it makes a transition back to the ground state by emitting a photon with wavelength 589 nm and energy 2.109 eV. Find the uncertainty in energy of the resonance level and the wavelength spread of the corresponding spectrum line.

Continued

SOLUTION

SET UP We need to use the uncertainty principle for energy given by Equation 28.23; we are given that $\Delta t = 1.6 \times 10^{-8}$ s, so we can find ΔE. Then the fractional uncertainty $\Delta \lambda / \lambda$ in wavelength should be proportional to the fractional uncertainty $\Delta E / E$ in energy.

SOLVE From Equation 28.23,

$$\Delta E = \frac{h}{2\pi \, \Delta t} = \frac{6.626 \times 10^{-34} \, \text{J} \cdot \text{s}}{(2\pi)(1.6 \times 10^{-8} \, \text{s})}$$
$$= 6.6 \times 10^{-27} \, \text{J} = 4.1 \times 10^{-8} \, \text{eV}.$$

The uncertainty in the resonance-level energy and in the corresponding photon energy amounts to about 2 parts in 10^8. (The atom remains an indefinitely long time in the ground state, so there is *no* uncertainty there.) Assuming that the corresponding spread in wavelength, or "width," of the spectrum line is also approximately two parts in 10^8, we find that

$$\Delta \lambda = (2 \times 10^{-8})(589 \, \text{nm}) = 0.000012 \, \text{nm}.$$

REFLECT The irreducible uncertainty $\Delta \lambda$ is called the *natural line width* of this particular spectrum line. Although small, it is within the limits of resolution of present-day spectrometers. Ordinarily, the natural line width is much smaller than line broadening from other causes, such as collisions among atoms.

28.8 The Electron Microscope

The **electron microscope** offers us an important and interesting example of the interplay between wave and particle properties of electrons. An electron beam can be used to form an image of an object in exactly the same way as a light beam. A ray of light is bent by reflection or refraction, and an electron trajectory is bent by an electric or magnetic field. Rays of light diverging from a point on an object can be brought to convergence by a converging lens or concave mirror, and electrons diverging from a small region can be brought to convergence by an electrostatic or magnetic lens. Figures 28.23a and 28.23b show the behavior of a simple type of electrostatic lens, and Figure 28.23c shows the analogous optical system. In each case, the image can be made larger than the object, so both devices can act as magnifiers.

The analogy between light rays and electrons goes deeper. The *ray* model of geometrical optics (ray optics) is an approximate representation of the more general *wave* model. Geometrical optics is valid whenever interference and diffraction effects can be neglected. Similarly, the model of an electron as a point

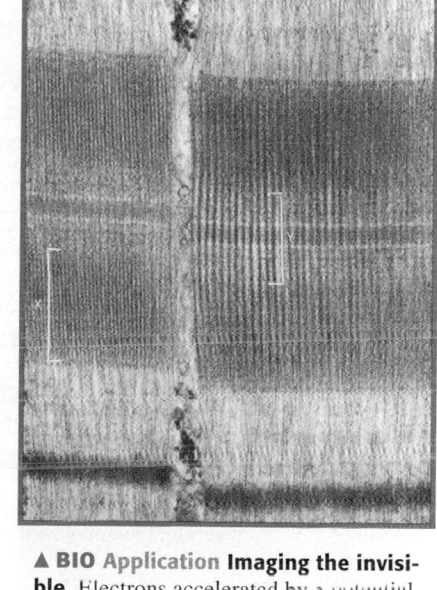

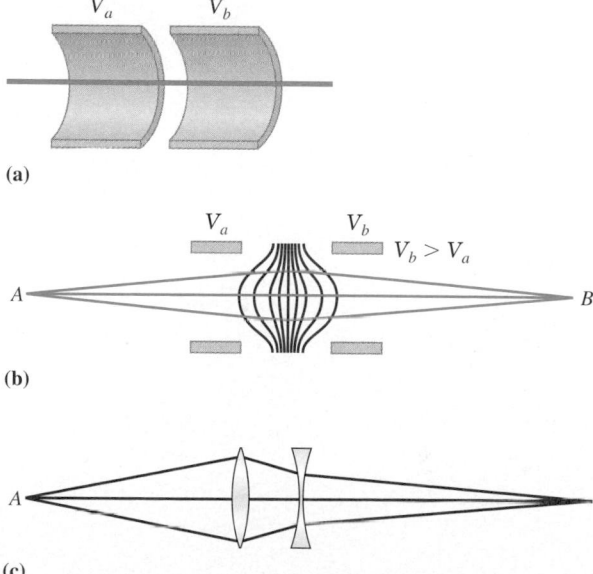

(a)

(b)

(c)

▲ **FIGURE 28.23** Electrostatic lens for an electron microscope.

▲ **BIO Application Imaging the invisible.** Electrons accelerated by a potential of 200 kV have de Broglie wavelengths several orders of magnitude smaller than photons of visible light. The electron microscope creates such energetic electrons and then focuses them with magnetic and electrostatic lenses onto specimens. Biological materials sliced into thin sections can be examined in the electron microscope, producing images with far better resolution than is possible in the light microscope. Individual macromolecules can also be imaged by these means. The photograph shows the beautifully arranged alternating thin (actin) and thick (myosin) filaments of muscle. The diameter of the thin filaments is about 7 nm, and the myosin filaments are 10 to 11 nm thick. The bracket on the left (X) indicates the region of overlap between the actin and myosin molecules; the right bracket (Y) shows the region containing only myosin. The major uncertainty in using this technique is the degree to which the sample is altered by necessarily elaborate preparation for examination in a high-vacuum system.

particle following a line trajectory is an approximate description of the actual behavior of the electron; this model is useful when we can neglect effects associated with the wave nature of electrons.

How is an electron microscope superior to an optical microscope? The *resolution* of an optical microscope is limited by diffraction effects, as we discussed in Section 26.8. Using wavelengths in the visible spectrum, around 500 nm, an optical microscope can't resolve objects smaller than a few hundred nanometers, no matter how carefully its lenses are made. The resolution of an electron microscope is similarly limited by the wavelengths of the electrons, but these may be many thousands of times *smaller* than wavelengths of visible light. As a result, the useful magnification of an electron microscope can be thousands of times as great as that of an optical microscope.

Note that the ability of the electron microscope to form an image *does not* depend on the wave properties of electrons. We can compute the trajectories of electrons by treating them as classical charged particles under the action of electric- and magnetic-field forces. Only when we talk about *resolution* do the wave properties become important.

EXAMPLE 28.12 Electron wavelengths

An electron beam is formed by a setup similar to the electron gun in an older-model TV set, accelerating electrons through a potential difference (voltage) of several thousand volts. What accelerating voltage is needed to produce electrons with wavelength 10 pm = 0.010 nm (roughly 50,000 times smaller than typical visible-light wavelengths)?

SOLUTION

SET UP The wavelength λ is determined by the de Broglie relation, $\lambda = h/p = h/mv$. The momentum $p = mv$ is related to the kinetic energy $K = p^2/2m$, which in turn is determined by the accelerating voltage V: $K = eV$.

SOLVE When we put the preceding pieces together, we get

$$K = eV = \frac{p^2}{2m} = \frac{(h/\lambda)^2}{2m}.$$

Solving for V and inserting the appropriate numbers, we find that

$$V = \frac{h^2}{2me\lambda^2}$$

$$= \frac{(6.626 \times 10^{-34}\,\text{J} \cdot \text{s})^2}{2(9.109 \times 10^{-31}\,\text{kg})(1.602 \times 10^{-19}\,\text{C})(10 \times 10^{-12}\,\text{m})^2}$$

$$= 1.5 \times 10^4\,\text{V} = 15,000\,\text{V}.$$

REFLECT The voltage we just obtained is approximately equal to the accelerating voltage for the electron beams formerly used in older TV picture tubes and computer monitors. This example shows, incidentally, that the sharpness of a TV picture is *not* limited by electron diffraction effects. Also, the kinetic energy of an electron is 15,000 eV, a small fraction of the electron's *rest* energy of about 511,000 eV; this shows that relativistic effects need not be considered in solving this problem.

SUMMARY

The Photoelectric Effect

(Section 28.1) The energy E of an individual photon is proportional to its frequency f, $E = hf$ (Equation 28.2), where h is Planck's constant. The wavelength of a photon is given by $\lambda = c/f$, and the momentum p is inversely proportional to the wavelength:

$$p = \frac{E}{c} = \frac{hf}{c} = \frac{h}{\lambda} \qquad (28.6)$$

In the **photoelectric effect,** a photon striking a conductor is absorbed by an electron. If the photon energy is greater than the work function ϕ of the material, the electron can escape from the surface.

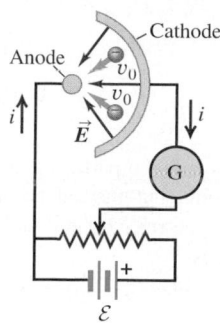

The photoelectric effect

Line Spectra and Energy Levels

(Section 28.2) If an atom makes a transition from an initial energy E_i to a lower final energy E_f, the energy of the emitted photon is hf, where $hf = E_i - E_f$ (Equation 28.7). The photon energies of the hydrogen atom's spectral lines are all differences between its energy levels. The energy levels are given by

$$E_n = -\frac{hcR}{n^2} = -\frac{13.6 \text{ eV}}{n^2}, \qquad n = 1, 2, 3, \cdots. \quad (28.10)$$

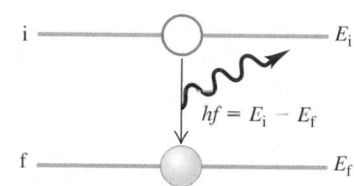

The Nuclear Atom and the Bohr Model

(Section 28.3) In Bohr's model of the hydrogen atom, the allowed values of angular momentum L for an electron in a circular orbit are integer multiples of the constant $h/2\pi$:

$$L = mv_n r_n - n\frac{h}{2\pi}, \qquad n = 1, 2, 3, \cdots. \quad (28.11)$$

The integer n is called the **principal quantum number;** it also determines the electron's radius and orbital speed for each level:

$$r_n = \epsilon_0 \frac{n^2 h^2}{\pi m e^2}, \qquad v_n = \frac{1}{\epsilon_0}\frac{e^2}{2nh} \quad (28.13, 28.14)$$

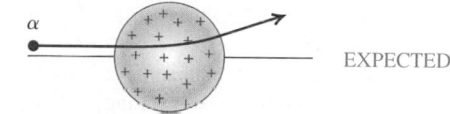

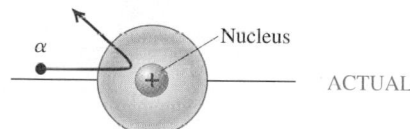

The Laser

(Section 28.4) A laser requires a population inversion in which the population of atoms in higher energy states is enhanced. Through stimulated emission, many photons are radiated with the same frequency, phase, and polarization—they are all synchronized.

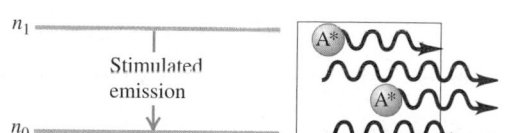

X-Ray Production and Scattering

(Section 28.5) When rapidly moving electrons that have been accelerated through a potential difference of the order of 10^3 to 10^6 V strike a metal target, x rays are produced. In Compton scattering, x rays strike matter, some of the radiation is scattered with a larger wavelength than the incident radiation, and the increase in wavelength $\Delta\lambda$ depends on the angle through which the radiation is scattered.

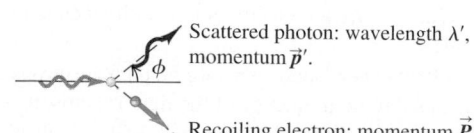

Continued

The Wave Nature of Particles

(Section 28.6) Just as photons have a wavelength $\lambda = h/p$ (Equation 28.6), using $p = mv$ in this equation gives the de Broglie wavelength λ for a particle of mass m:

$$\lambda = \frac{h}{p} = \frac{h}{mv}. \qquad (28.20)$$

An essential concept of quantum mechanics is that a particle is no longer described as located at a single point, but instead is an inherently spread-out entity. Electrons within atoms can be visualized as diffuse clouds surrounding the nucleus.

Wave–Particle Duality

(Section 28.7) Neither the momentum nor the position of a particle can be measured simultaneously with arbitrarily great precision, as classical physics would predict. Instead, the uncertainties in the two quantities play complementary roles: The less the uncertainty about a particle's position, the more uncertain is its momentum, and vice versa. This fundamental property of nature is described by Heisenberg's uncertainty principle:

$$\Delta x \, \Delta p_x \geq \frac{h}{2\pi}. \qquad (28.22)$$

The Electron Microscope

(Section 28.8) An electron beam can be used to form an image of an object in exactly the same way as a light beam. While reflection and refraction are used to bend light rays, electric and magnetic fields are used to bend an electron's trajectory. Just as the resolution of an optical microscope is limited by the wavelength of light, the resolution of an electron microscope is limited by the wavelength of the electrons. Since the wavelengths of the electrons are much smaller than those for light used in an optical microscope, the resolution of an electron microscope is typically thousands of times better.

 For instructor-assigned homework, go to www.masteringphysics.com

Conceptual Questions

1. In the photoelectric effect, electrons are knocked out of metals by photons of light. So why are electrons left in the metals in an ordinary room?

2. In photoelectric-effect experiments, why are no photoelectrons ejected from the metal if the frequency of the light is below the threshold frequency, even though the intensity of the light is very strong?

3. In photoelectric-effect experiments, why does the stopping potential increase with the frequency of the light, but *not* with the intensity of the light?

4. In what ways do photons resemble familiar particles such as electrons? In what ways do they differ? For example, do they have mass? Electric charge? Can they be accelerated? What mechanical properties do they have?

5. Is the frequency of a light beam the same thing as the number of photons per second that it carries? If not, what is the difference between the light frequency and the number of photons per second? Could you change the number of photons per second in a beam of light without changing the frequency of the light? How?

6. Would you expect quantum effects to be generally more important at the low-frequency end of the electromagnetic spectrum (radio waves) or at the high-frequency end (x rays and gamma rays)? Why?

7. Except for some special-purpose films, most black-and-white photographic film is less sensitive at the far-red end of the

visible spectrum than at the blue end and has almost no sensitivity to infrared, even at high intensities. How can these properties be understood on the basis of photons?

8. Human skin is relatively insensitive to visible light, but ultraviolet radiation can be quite destructive to skin cells. Why do you think this is so?

9. Figure 28.3 shows that in a photoelectric-effect experiment the photocurrent i for large values of V_{AC} has the same value no matter what the light frequency f (provided that f is higher than the threshold frequency f_0). Explain why.

10. Particles such as electrons are sometimes whimsically called "wavicles." Explain why this is actually an appropriate name for them.

11. The phosphorescent materials that coat the inside of a fluorescent tube convert ultraviolet radiation (from the mercury-vapor discharge inside the tube) to visible light. Could one also make a phosphor that converts visible light to ultraviolet? Why or why not?

12. As an object is heated to a very high temperature and becomes self-luminous, the apparent color of the emitted radiation shifts from red to yellow and finally to blue with the increasing temperature. What causes the color shift?

13. If a proton and an electron have the same speed, which has the longer de Broglie wavelength? Explain.

14. Why go through the expense of building an electron microscope for studying very small objects such as organic molecules? Why not just use extremely short electromagnetic waves, which are much cheaper to generate?

15. Do the planets of the solar system obey a distance law $(r_n = n^2 r_1)$ like the electrons of the Bohr atom? Should they? Why (or why not)? (Consult Appendix E for the appropriate distances.)

16. Which has more total energy, a hydrogen atom with an electron in a high shell (large n) or in a low shell (small n)? Which is moving faster, the high-shell electron or the low-shell electron? Is there a contradiction here? Explain.

Multiple-Choice Problems

1. Light falling on a metal surface causes electrons to be emitted from the metal by the photoelectric effect. As we increase the *intensity* of this light, but keep its wavelength the same (there may be more than one correct answer),
   A. The number of electrons emitted from the metal increases.
   B. The number of electrons emitted from the metal does not change.
   C. The maximum speed of the emitted electrons does not change.
   D. The maximum speed of the emitted electrons increases.

2. Light falling on a metal surface causes electrons to be emitted from the metal by the photoelectric effect. As we increase the *frequency* of this light, but do not vary anything else (there may be more than one correct answer),
   A. The number of electrons emitted from the metal increases.
   B. The maximum speed of the emitted electrons increases.
   C. The maximum speed of the emitted electrons does not change.
   D. The work function of the metal increases.

3. According to the Bohr model of the hydrogen atom, as we look at higher and higher values of n, the distance between adjacent electron orbits
   A. gets progressively smaller and smaller.
   B. gets progressively larger and larger.
   C. remains constant.

4. According to the Bohr model of the hydrogen atom, as we look at higher and higher values of n, the difference in energy between adjacent electron orbits
   A. gets progressively smaller and smaller.
   B. gets progressively larger and larger.
   C. remains constant.

5. A photon of wavelength λ has energy E. If its wavelength were doubled, its energy would be
   A. $4E$.   B. $2E$.   C. $\frac{1}{2}E$.   D. $\frac{1}{4}E$.

6. Consider a hypothetical single-electron Bohr atom for which an electron in the $n = 1$ shell has a total energy of -81.0 eV while one in the $n = 3$ shell has a total energy of -9.00 eV. Which statements about this atom are correct? (There may be more than one correct choice.)
   A. The energy needed to ionize the atom is 90.0 eV.
   B. It takes 90.0 eV to move an electron from the $n = 1$ to the $n = 3$ shell.
   C. If an electron makes a transition from the $n = 3$ to the $n = 1$ shell, it will give up 72.0 eV of energy.
   D. If an electron makes a transition from the $n = 3$ to the $n = 1$ shell, it must absorb 72.0 eV of energy.

7. Which of the following energies would be a physically *reasonable* photoelectric work function for a typical metal?
   A. 5.0 J.   B. 5.0 GeV.   C. 5.0 MeV.   D. 5.0 eV.

8. If the Bohr radius of the $n = 3$ state of a hydrogen atom is R, then the radius of the ground state is
   A. $9R$.   B. $3R$.   C. $R/3$.   D. $R/9$.

9. If the energy of the $n = 3$ state of a Bohr-model hydrogen atom is E, the energy of the ground state is
   A. $9E$.   B. $3E$.   C. $E/3$.   D. $E/9$.

10. Electrons are accelerated from rest through a potential difference V. As V is increased, the de Broglie wavelength of these electrons
    A. increases.   B. decreases.   C. does not change.

11. Proton A has a de Broglie wavelength λ. If proton B has twice the speed (which is much less than the speed of light) of proton A, the de Broglie wavelength of proton B is
    A. 2λ.   B. $\lambda\sqrt{2}$.   C. $\lambda/\sqrt{2}$.   D. $\lambda/2$.

12. Electron A has a de Broglie wavelength λ. If electron B has twice the kinetic energy (but a speed much less than that of light) of electron A, the de Broglie wavelength of electron B is
    A. 2λ.   B. $\lambda\sqrt{2}$.   C. $\lambda/\sqrt{2}$.   D. $\lambda/2$.

13. When a photon of light scatters off of a free stationary electron, the wavelength of the photon
    A. decreases.   B. increases.   C. remains the same.
    D. could either increase or decrease, depending on the initial energy of the photon.

14. Which of the following is NOT a characteristic of laser light:
    A. All the photons in laser light are in phase with each other.
    B. Laser light contains a very broad spectrum of wavelengths.
    C. The photons produced by a laser all travel in the same direction.
    D. The photons in laser light all have the same frequency.

15. Some lasers emit light in pulses that are only 10^{-12} s in duration. Compared to a laser that emits a steady, continuous beam of light, the Heisenberg uncertainty principle leads us to expect the photons from a pulsed laser to have
    A. higher energy.
    B. lower energy.
    C. greater momentum.
    D. a broader range of frequencies.
    E. a narrower range of frequencies.

16. Light of frequency f falls on a metal surface and ejects electrons of maximum kinetic energy K by the photoelectric effect. If the frequency of this light is doubled, the maximum kinetic energy of the emitted electrons will be
    A. $K/2$.   B. K.
    C. $2K$.   D. greater than $2K$.

Problems

28.1 Photoelectric Effect

1. • **Response of the eye.** The human eye is most sensitive to
   BIO green light of wavelength 505 nm. Experiments have found that when people are kept in a dark room until their eyes adapt to the darkness, a *single* photon of green light will trigger receptor cells in the rods of the retina. (a) What is the frequency of this photon? (b) How much energy (in joules and eV) does it deliver to the receptor cells? (c) To appreciate what a small amount of energy this is, calculate how fast a typical bacterium of mass 9.5×10^{-12} g would move if it had that much energy.

2. • An excited nucleus emits a gamma-ray photon with an energy of 2.45 MeV. (a) What is the photon's frequency? (b) What is the photon's wavelength? (c) How does this wavelength compare with a typical nuclear diameter of 10^{-14} m?

3. • A laser used to weld detached retinas emits light with a
   BIO wavelength of 652 nm in pulses that are 20.0 ms in duration. The average power expended during each pulse is 0.600 W. (a) How much energy is in each pulse, in joules? In electron volts? (b) What is the energy of one photon in joules? In electron volts? (c) How many photons are in each pulse?

4. • A radio station broadcasts at a frequency of 92.0 MHz with a power output of 50.0 kW. (a) What is the energy of each emitted photon, in joules and electron volts? (b) How many photons are emitted per second?

5. • The predominant wavelength emitted by an ultraviolet lamp is 248 nm. If the total power emitted at this wavelength is 12.0 W, how many photons are emitted per second?

6. • A photon has momentum of magnitude 8.24×10^{-28} kg·m/s. (a) What is the energy of this photon? Give your answer in joules and in electron volts. (b) What is the wavelength of this photon? In what region of the electromagnetic spectrum does it lie?

7. • In the photoelectric effect, what is the relationship between the threshold frequency f_0 and the work function ϕ?

8. • A clean nickel surface is exposed to light of wavelength 235 nm. What is the maximum speed of the photoelectrons emitted from this surface? Use Table 28.1.

9. • The photoelectric threshold wavelength of a tungsten surface is 272 nm. (a) What are the threshold frequency and work function (in eV) of this tungsten? (b) Calculate the maximum kinetic energy (in eV) of the electrons ejected from this tungsten surface by ultraviolet radiation of frequency 1.45×10^{15} Hz.

10. • What would the minimum work function for a metal have to be for visible light (having wavelengths between 400 nm and 700 nm) to eject photoelectrons?

11. •• When ultraviolet light with a wavelength of 400.0 nm falls on a certain metal surface, the maximum kinetic energy of the emitted photoelectrons is measured to be 1.10 eV. What is the maximum kinetic energy of the photoelectrons when light of wavelength 300.0 nm falls on the same surface?

12. •• When ultraviolet light with a wavelength of 254 nm falls upon a clean metal surface, the stopping potential necessary to terminate the emission of photoelectrons is 0.181 V. (a) What is the photoelectric threshold wavelength for this metal? (b) What is the work function for the metal?

13. •• The photoelectric work function of potassium is 2.3 eV. If light having a wavelength of 250 nm falls on potassium, find (a) the stopping potential in volts; (b) the kinetic energy, in electron volts, of the most energetic electrons ejected; (c) the speeds of these electrons.

14. •• In a photoelectric effect experiment it is found that no current flows unless the incident light has a wavelength shorter than 289 nm. (a) What is the work function of the metal surface? (b) What stopping potential will be needed to halt the current if light of 225 nm falls on the surface?

15. •• Light with a wavelength range of 145–295 nm shines on a silicon surface in a photoelectric effect apparatus, and a reversing potential of 3.50 V is applied to the resulting photoelectrons. (a) What is the longest wavelength of the light that will eject electrons from the silicon surface? (b) With what maximum kinetic energy will electrons reach the anode?

28.2 Line Spectra and Energy Levels

16. • (a) How much energy is needed to ionize a hydrogen atom that is in the $n = 4$ state? (b) What would be the wavelength of a photon emitted by a hydrogen atom in a transition from the $n = 4$ state to the $n = 2$ state?

17. • Use Balmer's formula to calculate (a) the wavelength, (b) the frequency, and (c) the photon energy for the H_γ line of the Balmer series for hydrogen.

18. • Find the longest and shortest wavelengths in the Lyman and Paschen series for hydrogen. In what region of the electromagnetic spectrum does each series lie?

19. • (a) Calculate the longest and shortest wavelengths for light in the Balmer, Lyman, and Brackett series. (b) Use your results from part (a) to decide in which part of the electromagnetic spectrum each of these series lies.

20. •• The energy-level scheme for the hypothetical one-electron element searsium is shown in Fig. 28.24. The potential energy is taken to be zero for an electron at an infinite distance from the nucleus. (a) How much energy (in electron volts) does it take to ionize an electron from the ground level? (b) An 18 eV

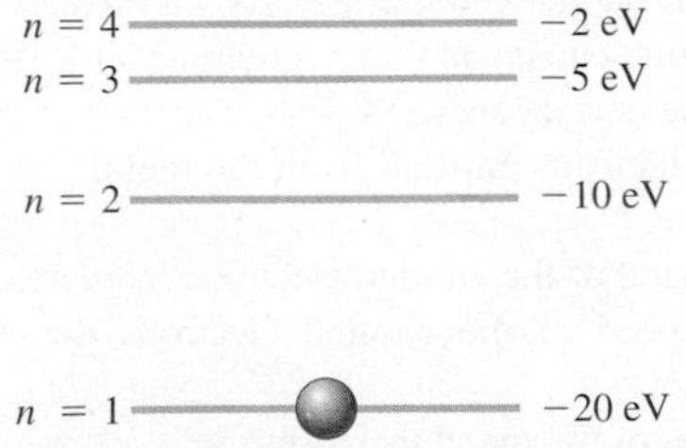

▲ FIGURE 28.24 Problem 20.

photon is absorbed by a searsium atom in its ground level. As the atom returns to its ground level, what possible energies can the emitted photons have? Assume that there can be transitions between all pairs of levels. (c) What will happen if a photon with an energy of 8 eV strikes a searsium atom in its ground level? Why? (d) Photons emitted in the searsium transitions $n = 3 \rightarrow n - 2$ and $n = 3 \rightarrow n = 1$ will eject photoelectrons from an unknown metal, but the photon emitted from the transition $n = 4 \rightarrow n = 3$ will not. What are the limits (maximum and minimum possible values) of the work function of the metal?

21. •• In a set of experiments on a hypothetical one-electron atom, you measure the wavelengths of the photons emitted from transitions ending in the ground state $(n = 1)$, as shown in the energy-level diagram in Fig. 28.25. You also observe that it takes 17.50 eV to ionize this atom. (a) What is the energy of the atom in each of the levels $(n = 1, n = 2,$ etc.$)$ shown in the figure? (b) If an electron made a transition from the $n = 4$ to the $n = 2$ level, what wavelength of light would it emit?

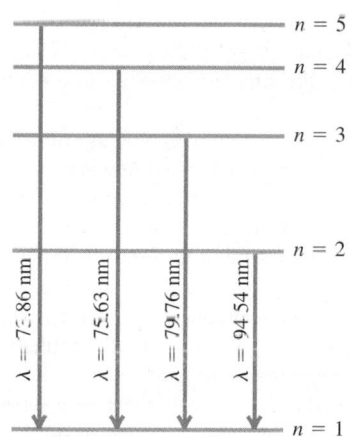

▲ **FIGURE 28.25** Problem 21.

28.3 The Nuclear Atom and the Bohr Model

22. • For a hydrogen atom in the ground state, determine, in electron volts, (a) the kinetic energy of the electron, (b) the potential energy, (c) the total energy, (d) the minimum energy required to remove the electron completely from the atom. (e) What wavelength does a photon with the energy calculated in part (d) have? In what region of the electromagnetic spectrum does it lie?

23. • Use the Bohr model for the following calculations: (a) What is the speed of the electron in a hydrogen atom in the $n = 1, 2,$ and 3 levels? (b) Calculate the radii of each of these levels. (c) Find the total energy (in eV) of the atom in each of these levels.

24. •• An electron in an excited state of hydrogen makes a transition from the $n = 5$ level to the $n = 2$ level. (a) Does the atom emit or absorb a photon during this process? How do you know? (b) Calculate the wavelength of the photon involved in the transition.

25. •• A hydrogen atom initially in the ground state absorbs a photon, which excites it to the $n - 4$ state. Determine the wavelength and frequency of the photon.

26. •• Light of wavelength 59 nm ionizes a hydrogen atom that was originally in its ground state. What is the kinetic energy of the ejected electron?

27. •• A triply ionized beryllium ion, Be^{3+} (a beryllium atom with three electrons removed), behaves very much like a hydrogen atom, except that the nuclear charge is four times as great. (a) What is the ground-level energy of Be^{3+}? How does this compare with the ground-level energy of the hydrogen atom? (b) What is the ionization energy of Be^{3+}? How does this compare with the ionization energy of the hydrogen atom? (c) For the hydrogen atom, the wavelength of the photon emitted in the transition $n = 2$ to $n = 1$ is 122 nm. (See Example 28.6.) What is the wavelength of the photon emitted when a Be^{3+} ion undergoes this transition? (d) For a given value of n, how does the radius of an orbit in Be^{3+} compare with that for hydrogen?

28.4 The Laser

28. • (a) Use the information for neon shown in Fig. 28.26 to compute the energy difference for the $5s$-to-$3p$ transition in neon. Express your result in electron volts and in joules. (b) Calculate the wavelength of a photon having this energy, and compare your result with the observed wavelength of the laser light. (c) What is the wavelength of the light from the $3p$-to-$3s$ transition in neon?

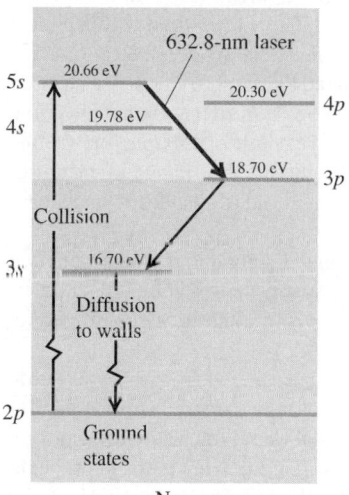

◄ **FIGURE 28.26** Problem 28.

29. • The diode laser keychain you use to entertain your cat has a wavelength of 645 nm. If the laser emits 4.50×10^{17} photons during a 30.0 s feline play session, what is its average power output?

30. • **Laser surgery.** Using a mixture of CO_2, N_2, and sometimes
BIO He, CO_2 lasers emit a wavelength of 10.6 μm. At power outputs of 0.100 kW, such lasers are used for surgery. How many photons per second does a CO_2 laser deliver to the tissue during its use in an operation?

31. • **PRK surgery.** Photorefractive keratectomy (PRK) is a
BIO laser-based surgery process that corrects near- and farsightedness by removing part of the lens of the eye to change its curvature and hence focal length. This procedure can remove layers 0.25 μm thick in pulses lasting 12.0 ns with a laser beam of wavelength 193 nm. Low-intensity beams can be used because each individual photon has enough energy to break the covalent bonds of the tissue. (a) In what part of the electromagnetic spectrum does this light lie? (b) What is the energy of a single photon? (c) If a 1.50 mW beam is used, how many photons are delivered to the lens in each pulse?

32. • **Removing birthmarks.** Pulsed dye lasers emit light of
BIO wavelength 585 nm in 0.45 ms pulses to remove skin blemishes such as birthmarks. The beam is usually focused onto a circular spot 5.0 mm in diameter. Suppose that the output of one such laser is 20.0 W. (a) What is the energy of each photon, in eV? (b) How many photons per square millimeter are delivered to the blemish during each pulse?

28.5 X-Ray Production and Scattering

33. • (a) What is the minimum potential difference between the filament and the target of an x-ray tube if the tube is to accelerate electrons to produce x rays with a wavelength of 0.150 nm? (b) What is the shortest wavelength produced in an x-ray tube operated at 30.0 kV? (c) Would the answers to parts (a) and (b) be different if the tube accelerated protons instead of electrons? Why or why not?

34. • The cathode-ray tubes that generated the picture in early color televisions were sources of x rays. If the acceleration voltage in a television tube is 15.0 kV, what are the shortest-wavelength x rays produced by the television? (Modern televisions contain shielding to stop these x rays.)

35. • An x ray with a wavelength of 0.100 nm collides with an electron that is initially at rest. The x ray's final wavelength is 0.110 nm. What is the final kinetic energy of the electron?

36. • If a photon of wavelength 0.04250 nm strikes a free electron and is scattered at an angle of 35.0° from its original direction, find (a) the change in the wavelength of this photon, (b) the wavelength of the scattered light, (c) the change in energy of the photon (is it a loss or a gain?), and (d) the energy gained by the electron.

37. • X rays with initial wavelength 0.0665 nm undergo Compton scattering. What is the longest wavelength found in the scattered x rays? At which scattering angle is this wavelength observed?

38. •• An incident x-ray photon is scattered from a free electron that is initially at rest. The photon is scattered straight back at an angle of 180° from its initial direction. The wavelength of the scattered photon is 0.0830 nm. (a) What is the wavelength of the incident photon? (b) What is the magnitude of the momentum of the electron after the collision? (c) What is the kinetic energy of the electron after the collision?

39. •• Protons are accelerated from rest by a potential difference of 4.00 kV and strike a metal target. If a proton produces one photon on impact, what is the minimum wavelength of the resulting x rays? How does your answer compare to the minimum wavelength if 4.00 keV electrons are used instead? Why do x-ray tubes use electrons rather than protons to produce x rays?

28.6 The Wave Nature of Particles

40. • (a) An electron moves with a speed of 4.70×10^6 m/s. What is its de Broglie wavelength? (b) A proton moves with the same speed. Determine its de Broglie wavelength.

41. • How fast would an electron have to move so that its de Broglie wavelength would be 1.00 mm?

42. • (a) Approximately what range of photon energies (in eV) corresponds to the visible spectrum? (b) Approximately what range of *wavelengths* and *kinetic energies* would electrons in this energy range have?

43. •• In the Bohr model of the hydrogen atom, what is the de Broglie wavelength for the electron when it is in (a) the $n = 1$ level and (b) the $n = 4$ level? In each case, compare the de Broglie wavelength to the circumference $2\pi r_n$ of the orbit.

44. • (a) What is the de Broglie wavelength of an electron accelerated through 800 V? (b) What is the de Broglie wavelength of a proton accelerated through the same potential difference?

45. •• Find the wavelengths of a photon and an electron that have the same energy of 25 eV. (The energy of the electron is its kinetic energy.)

28.7 Wave–Particle Duality

46. • (a) The uncertainty in the x component of the position of a proton is 2.0×10^{-12} m. What is the minimum uncertainty in the x component of the velocity of the proton? (b) The uncertainty in the x component of the velocity of an electron is 0.250 m/s. What is the minimum uncertainty in the x coordinate of the electron?

47. •• A certain atom has an energy level 3.50 eV above the ground state. When excited to this state, it remains 4.0 μs, on the average, before emitting a photon and returning to the ground state. (a) What is the energy of the photon? What is its wavelength? (b) What is the smallest possible uncertainty in energy of the photon?

48. •• A pesky 1.5 mg mosquito is annoying you as you attempt to study physics in your room, which is 5.0 m wide and 2.5 m high. You decide to swat the bothersome insect as it flies toward you, but you need to estimate its speed to make a successful hit. (a) What is the maximum uncertainty in the horizontal position of the mosquito? (b) What limit does the Heisenberg uncertainty principle place on your ability to know the horizontal velocity of this mosquito? Is this limitation a serious impediment to your attempt to swat it?

49. •• Suppose that the uncertainty in position of an electron is equal to the radius of the $n = 1$ Bohr orbit, about 0.5×10^{-10} m. Calculate the minimum uncertainty in the corresponding momentum component, and compare this with the magnitude of the momentum of the electron in the $n = 1$ Bohr orbit.

28.8 The Electron Microscope

50. •• (a) What accelerating potential is needed to produce electrons of wavelength 5.00 nm? (b) What would be the energy of photons having the same wavelength as these electrons? (c) What would be the wavelength of photons having the same energy as the electrons in part (a)?

51. •• (a) In an electron microscope, what accelerating voltage is needed to produce electrons with wavelength 0.0600 nm? (b) If protons are used instead of electrons, what accelerating voltage is needed to produce protons with wavelength 0.0600 nm? (*Hint:* In each case the initial kinetic energy is negligible.)

52. •• **Structure of a virus.** To investigate the structure of
BIO extremely small objects, such as viruses, the wavelength of the probing wave should be about one-tenth the size of the object for sharp images. But as the wavelength gets shorter, the energy of a photon of light gets greater and could damage or destroy the object being studied. One alternative is to use electron matter waves instead of light. Viruses vary considerably in size, but 50 nm is not unusual. Suppose you want to study such

a virus, using a wave of wavelength 5.00 nm. (a) If you use light of this wavelength, what would be the energy (in eV) of a single photon? (b) If you use an electron of this wavelength, what would be its kinetic energy (in eV)? Is it now clear why matter waves (such as in the electron microscope) are often preferable to electromagnetic waves for studying microscopic objects?

General Problems

53. •• **Exposing photographic film.** The light-sensitive compound on most photographic films is silver bromide (AgBr). A film is "exposed" when the light energy absorbed dissociates this molecule into its atoms. (The actual process is more complex, but the quantitative result does not differ greatly.) The energy of dissociation of AgBr is 1.00×10^5 J/mol. For a photon that is just able to dissociate a molecule of silver bromide, find (a) the photon's energy in electron volts, (b) the wavelength of the photon, and (c) the frequency of the photon. (d) Light from a firefly can expose photographic film, but the radiation from an FM station broadcasting 50,000 W at 100 MHz cannot. Explain why this is so, basing your answer on the energy of the photons involved.

54. •• A 2.50 W beam of light of wavelength 124 nm falls on a metal surface. You observe that the maximum kinetic energy of the ejected electrons is 4.16 eV. Assume that each photon in the beam ejects an electron. (a) What is the work function (in electron volts) of this metal? (b) How many photoelectrons are ejected each second from this metal? (c) If the power of the light beam, but not its wavelength, were reduced by half, what would be the answer to part (b)? (d) If the wavelength of the beam, but not its power, were reduced by half, what would be the answer to part (b)?

55. •• A sample of hydrogen atoms is irradiated with light with a wavelength of 85.5 nm, and electrons are observed leaving the gas. If each hydrogen atom were initially in its ground level, what would be the maximum kinetic energy, in electron volts, of these photoelectrons?

56. •• An unknown element has a spectrum for absorption from its ground level with lines at 2.0, 5.0, and 9.0 eV. Its ionization energy is 10.0 eV. (a) Draw an energy-level diagram for this element. (b) If a 9.0 eV photon is absorbed, what energies can the subsequently emitted photons have?

57. •• (a) What is the least amount of energy, in electron volts, that must be given to a hydrogen atom which is initially in its ground level so that it can emit the H_α line in the Balmer series? (b) How many different possibilities of spectral line emissions are there for this atom when the electron starts in the $n = 3$ level and eventually ends up in the ground level? Calculate the wavelength of the emitted photon in each case.

58. •• A specimen of the microorganism *Gastropus hyptopus* **BIO** measures 0.0020 cm in length and can swim at a speed of 2.9 times its body length per second. The tiny animal has a mass of roughly 8.0×10^{-12} kg. (a) Calculate the de Broglie wavelength of this organism when it is swimming at top speed. (b) Calculate the kinetic energy of the organism (in eV) when it is swimming at top speed.

59. •• A photon with a wavelength of 0.1800 nm is Compton scattered through an angle of 180°. (a) What is the wavelength of the scattered photon? (b) How much energy is given to the electron? (c) What is the recoil speed of the electron? Is it necessary to use the relativistic kinetic-energy relationship?

60. •• (a) Calculate the maximum increase in photon wavelength that can occur during Compton scattering. (b) What is the energy (in electron volts) of the smallest-energy x-ray photon for which Compton scattering could result in doubling the original wavelength?

61. •• An incident x-ray photon of wavelength 0.0900 nm is scattered in the backward direction from a free electron that is initially at rest. (a) What is the magnitude of the momentum of the scattered photon? (b) What is the kinetic energy of the electron after the photon is scattered?

62. •• A photon with wavelength of 0.1100 nm collides with a free electron that is initially at rest. After the collision, the photon's wavelength is 0.1132 nm. (a) What is the kinetic energy of the electron after the collision? What is its speed? (b) If the electron is suddenly stopped (for example, in a solid target), all of its kinetic energy is used to create a photon. What is the wavelength of this photon?

63. • From the kinetic-molecular theory of an ideal gas (Chapter 15), we know that the average kinetic energy of an atom is $\frac{3}{2}kT$. What is the wavelength of a photon that has this energy for a temperature of 27°C?

64. •• **Doorway diffraction.** If your wavelength were 1.0 m, you would undergo considerable diffraction in moving through a doorway. (a) What must your speed be for you to have this wavelength? (Assume that your mass is 60.0 kg.) (b) At the speed calculated in part (a), how many years would it take you to move 0.80 m (one step)? Will you notice diffraction effects as you walk through doorways?

65. • What is the de Broglie wavelength of a red blood cell with a **BIO** mass of 1.00×10^{-11} g that is moving with a speed of 0.400 cm/s? Do we need to be concerned with the wave nature of the blood cells when we describe the flow of blood in the body?

66. •• **Removing vascular lesions.** A pulsed dye laser emits light **BIO** of wavelength 585 nm in 450 μs pulses. Because this wavelength is strongly absorbed by the hemoglobin in the blood, the method is especially effective for removing various types of blemishes due to blood, such as port-wine-colored birthmarks. To get a reasonable estimate of the power required for such laser surgery, we can model the blood as having the same specific heat and heat of vaporization as water (4190 J/kg·K, 2.256×10^6 J/kg). Suppose that each pulse must remove 2.0 μg of blood by evaporating it, starting at 33°C. (a) How much energy must each pulse deliver to the blemish? (b) What must be the power output of this laser? (c) How many photons does each pulse deliver to the blemish?

67. •• (a) What is the energy of a photon that has wavelength 0.10 μm? (b) Through approximately what potential difference must electrons be accelerated so that they will exhibit wave nature in passing through a pinhole 0.10 μm in diameter? What is the speed of these electrons? (c) If protons rather than electrons were used, through what potential difference would protons have to be accelerated so they would exhibit wave nature in passing through this pinhole? What would be the speed of these protons?

68. •• In a parallel universe, the value of Planck's constant is 0.0663 J·s. Assume that the physical laws and all other physical constants are the same as in our universe. In this other universe, two physics students are playing catch with a baseball. They are 50 m apart, and one throws a 0.10 kg ball with a

speed of 5.0 m/s. (a) What is the uncertainty in the ball's horizontal momentum in a direction perpendicular to that in which it is being thrown if the student throwing the ball knows that it is located within a cube with volume 1000 cm^3 at the time she throws it? (b) By what horizontal distance could the ball miss the second student?

69. •• The neutral π° meson is an unstable particle produced in high-energy particle collisions. Its mass is about 264 times that of the electron, and it exists for an average lifetime of 8.4×10^{-17} s before decaying into two gamma-ray photons. Assuming that the mass and energy of the particle are related by the Einstein relation $E = mc^2$, find the uncertainty in the mass of the particle and express it as a fraction of the particle's mass.

70. •• A beam of electrons is accelerated from rest and then passes through a pair of identical thin slits that are 1.25 nm apart. You observe that the first double-slit interference dark fringe occurs at ± 18.0° from the original direction of the beam when viewed on a distant screen. (a) Are these electrons relativistic? How do you know? (b) Through what potential difference were the electrons accelerated?

Passage Problems

BIO Radiation therapy of tumors. Malignant tumors are often treated with targeted x-ray radiation therapy. To generate these medical x rays, a linear accelerator directs a high-energy beam of electrons toward a metal target (typically tungsten). As the electrons pass near the heavy metal nuclei, they are deflected and accelerated, emitting high-energy photons in a process known as Bremsstrahlung (from the German for "braking radiation"). The resultant x rays are collimated into a beam that is focused on the tumor. The tissue absorbs the energy predominately via Compton interactions, so it is important to know how many Compton interactions occur and how many ionizations a single Compton electron produces. A linear accelerator used in radiation therapy produces x ray photons with an average energy of about 2 MeV, each of which impart 1 MeV to the Compton electrons. A typical tumor has 10^8 cells/cm^3 and a full treatment may involve a dose of 70 Gy in 35 fractional exposures on different days. The gray (Gy) is a measure of the absorbed energy dose of radiation per unit mass of tissue, expressed in the units of J/kg.

71. How much energy is imparted to a cell during one day's treatment? Assume that the specific gravity of the tumor is 1 and that 1J = 6×10^{18}eV.
    A. 12 MeV/cell
    B. 120 MeV/cell
    C. 120 MeV/cell
    D. 120×10^3 MeV/cell

72. Suppose the answer to problem 71 is 12 MeV/cell. How many Compton interactions will occur per cell in a single day's treatment?
    A. 120×10^6
    B. 120×10^4
    C. 120×10^2
    D. 12

73. Each Compton electron causes a series of ionization in the tissue as it interacts with molecules (mainly water), and each ionization takes about 40 eV. How many ionizations occur in a single cell as a result of a day's treatment?
    A. 3
    B. 3×10^2
    C. 3×10^4
    D. 3×10^6

29 Atoms, Molecules, and Solids

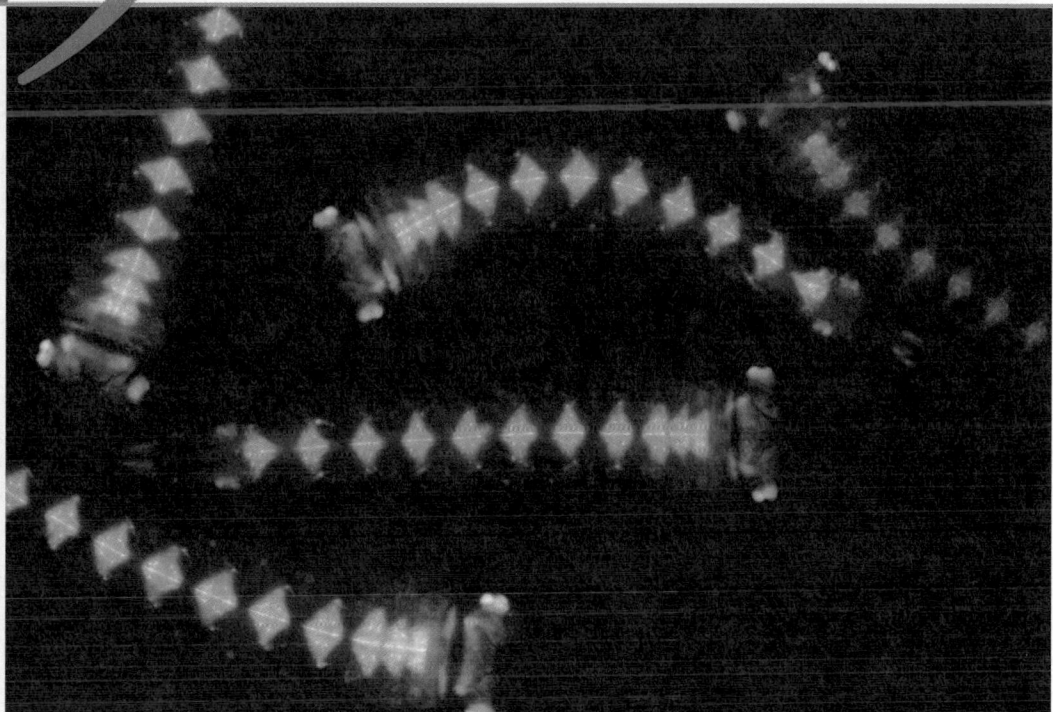

The basic concepts of quantum mechanics that we studied in Chapter 28 provide the key to understanding many aspects of the structure of atoms, molecules, and solid materials. We'll use the concept of a wave function as the basic language for describing electron configurations and motions. An additional property of electrons, called electron spin, plays a central role in describing these configurations. And we need one more crucial principle: the exclusion principle, which states that two electrons may not occupy the same quantum-mechanical state.

Using these three ideas, we can apply the principles of quantum mechanics to the analysis of a variety of atomic systems, such as the structure and chemical behavior of multielectron atoms (including the periodic table of the elements), molecular bonds (the binding of two or more atoms in a stable structure), and the large-scale binding of many atoms into solid structures. Many properties of solid materials can be understood at least qualitatively on the basis of electron configurations. We'll discuss semiconductors, a particular class of solid materials, in some detail because of their inherent interest and their great practical importance in present-day technology. Finally, we'll look briefly at the phenomenon of superconductivity: the complete disappearance of electrical resistance at low temperatures.

These mosquito larvae glow green because they are labeled with green fluorescent protein, a protein isolated from jellyfish that fluoresces (glows) green when illuminated with blue or UV light. In this chapter, you'll learn how such phenomena work.

29.1 Electrons in Atoms

We've learned that the classical description of the motion of a particle, in terms of three coordinates and three velocity components, can't be applied to motion on an atomic scale. In some situations, electrons behave like waves rather than particles, and their wavelength can be measured by diffraction experiments, as we discussed in Section 28.6. So we need to develop a description of electron motion that incorporates the idea of a spatially spread-out entity. In developing this description, we can take a cue from the language of classical wave motion.

971

Wave Functions

In Chapter 12, we described the motion of transverse waves on a string by specifying the position of each point in the string at each instant of time. To do this, we used a *wave function,* a concept we introduced in Section 12.4. If a point on the string, at some distance x from the origin, is displaced transversely a distance y from its equilibrium position at some time t, then there is a function $y = f(x, t)$, or $y(x, t)$, that represents the displacement of that point at *any* x at *any* time t. Once we know the wave function that describes a particular wave motion, we can find the position, velocity, and acceleration of any point on the string at any time, and so on.

In some situations, electrons behave like waves, so it's natural to adapt the language of waves to describe the motion of electrons. We'll use a **wave function** as the central element of our generalized language for describing the possible dynamic states (analogous to position and motion) of electrons (and, indeed, particles in general). The symbol usually used for such a wave function is Ψ. For a single electron, Ψ is a function of the space coordinates (x, y, z) and time. For a system with many electrons, it is a function of all the space coordinates of all the particles, and time. Just as the wave function $y(x, t)$ for mechanical waves on a string provides a complete description of the motion, the wave function $\Psi(x, y, z, t)$ for a particle contains all the information that can be known about the dynamic state of that particle.

Two questions arise. First, how do we relate the wave function to observable behavior of the particle? Second, how do we determine what Ψ is for any particular dynamic state of the system? With reference to the first question, the wave function describes the distribution of the particle in space. We can think of it as a cloudlike entity, more dense in some regions and less dense in others. The wave function is related to the *probability* of finding the particle in each of various regions: The particle is most likely to be found in regions where Ψ is large. We've already used this interpretation in our discussion of electron diffraction experiments. If the particle has a charge, the wave function can be used to find the charge density at any point in space.

In addition, from Ψ we can calculate the average position of the particle, its average velocity, and dynamic quantities such as momentum, energy, and angular momentum. The mathematical techniques required are far beyond the scope of this discussion, but they are well established and well supported by experimental results. Figure 29.1 shows cross sections of wave functions for three possible

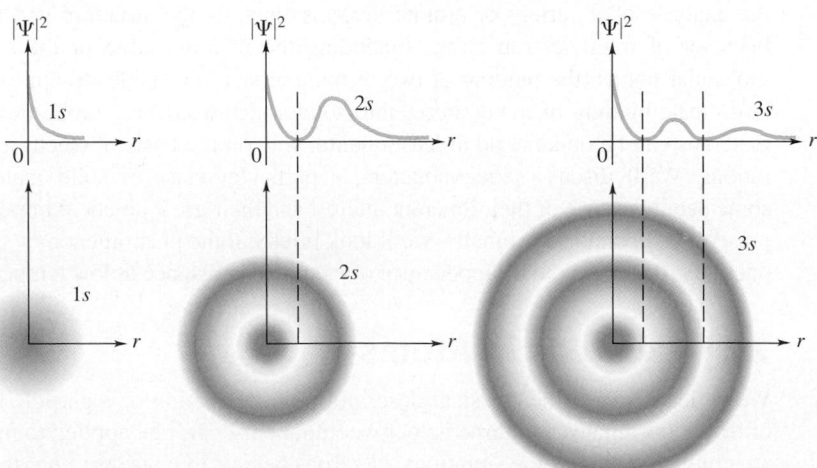

▲ **FIGURE 29.1** Three-dimensional probability distribution functions $|\Psi|^2$ for the spherically symmetric $1s$, $2s$, and $3s$ wave functions.

spherically symmetric states of the hydrogen atom. Of course, there is no possibility of actually seeing this distribution, since the atom is much smaller than wavelengths of visible light.

The answer to the second question is that the wave function Ψ must be one of a set of solutions of a certain differential equation. The **Schrödinger equation,** as it is called, was developed by Erwin Schrödinger (1887–1961) in 1926. In principle, we can set up a Schrödinger equation for any given physical situation. The wave functions that are solutions of this equation represent various possible physical states of the system, the analog of the orbits in the Bohr model.

Furthermore, it turns out that, for many systems, it is not even *possible* to find acceptable solutions of the Schrödinger equation unless some physical quantity, such as the energy of the system, has certain special values. Thus, the solutions of the Schrödinger equation for any particular system also yield a set of allowed energy levels. This discovery is of the utmost importance. Before the development of the Schrödinger equation, there was no way to predict energy levels from any fundamental theory, except for the very limited success of the Bohr model for hydrogen.

Soon after the Schrödinger equation was developed, it was applied to the problem of the hydrogen atom. The predicted energy levels E_n for this, the simplest of all atoms, turned out to be identical to those from the Bohr model, Equation 28.17. Therefore, these results also agreed with experimental values obtained from spectrum analysis.

In addition, the solutions have quantized values of angular momentum. As we learned in Section 10.7, angular momentum is fundamentally a vector quantity. In the Bohr model, the angular momentum is a vector perpendicular to the plane of the electron's orbit. The quantization of angular momentum was put into the Bohr model as an *ad hoc* assumption with no fundamental justification. With the Schrödinger equation, it appears naturally as a condition for the existence of acceptable solutions. Specifically, it is found that, for an electron in a hydrogen atom in a state with energy E_n and quantum number n, acceptable solutions of the Schrödinger equation exist only when the magnitude L of the vector angular momentum $\vec{L}$ is given by the following expression:

Permitted values of angular momentum
The possible values of angular momentum of the electron in a hydrogen atom are

$$L = \sqrt{l(l + 1)}\,\frac{h}{2\pi} \qquad (l = 0, 1, 2, \cdots, n - 1). \qquad (29.1)$$

The *component* of $\vec{L}$ in a given direction—say, the z component L_z, can have only the set of values

$$L_z = m_l\frac{h}{2\pi} \qquad (|m_l| = 0, 1, 2, \cdots, l). \qquad (29.2)$$

That is, m_l is a positive or negative integer or zero, with magnitude no greater than l.

Thus, instead of a single quantum number n as in the Bohr model, the possible wave functions for the hydrogen atom (corresponding to the various solutions of the Schrödinger equation) are labeled according to the values of *three* integers (quantum numbers) n, l, and m_l. These are called the **principal quantum number** (n), the **angular momentum quantum number** (l), and the **magnetic quantum number** (m_l). (There is also a fourth quantum number associated with

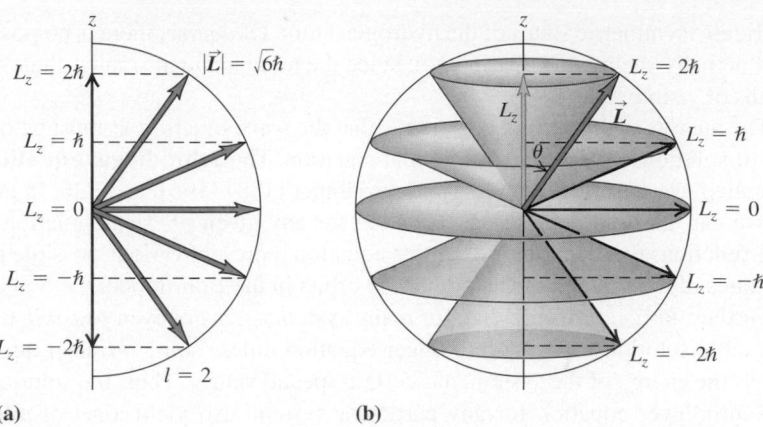

▲ **FIGURE 29.2** Magnitude and components of angular momentum, L and L_z.

electron spin; we'll discuss it later in this section.) For each energy level E_n, there are several distinct states having the same energy (and the same value of n), but different values of l and m_l. The only exception is the ground state, for which $n = 1$ and the only possibility for the other two quantum numbers is $(l = 0, m_l = 0)$. For any value of n, l can be no greater than $n - 1$. The term *magnetic quantum number* refers to the slight shifts in energy levels (and thus in spectral-line wavelengths) that occur when the atom is placed in a magnetic field. These shifts are different for states having the same n and l, but different values of m_l.

The quantity $h/2\pi$ appears so often in quantum mechanics that we give it a special symbol, $\hbar$, pronounced "h-bar." That is,

$$\hbar = \frac{h}{2\pi} = 1.054 \times 10^{-34} \, \text{J} \cdot \text{s}.$$

In terms of $\hbar$, Equations 29.1 and 29.2 respectively become

$$L = \sqrt{l(l + 1)} \, \hbar \qquad (l = 0, 1, 2, \cdots, n - 1) \tag{29.3}$$

and

$$L_z = m_l \hbar \qquad (m_l = 0, \pm 1, \pm 2, \cdots, \pm l). \tag{29.4}$$

Note that the component L_z can never be quite as large as L. For example, when $l = 2$ and $m_l = 2$, we find that

$$L = \sqrt{2(2 + 1)} \, \hbar = 2.45\hbar, \qquad L_z = 2\hbar.$$

Figure 29.2 shows the situation. For $l = 2$ and $m_l = 2$, the angle between the vector $\vec{L}$ and the z axis is given by

$$\cos\theta = \frac{L_z}{L} = \frac{2}{2.45}, \qquad \theta = 35°.$$

We can't know the precise direction of $\vec{L}$; it can have any direction on the cone in the figure at an angle of $\theta = 35°$ with the z axis. This result also follows from the uncertainty principle, which makes it impossible to predict the *direction* of the angular momentum vector with complete certainty.

Figure 29.2 shows the angular momentum relations for the particular states with $l = 2$. In each state, L and L_z have definite values, but the component of $\vec{L}$ perpendicular to the z axis does not; we can say only that this component is confined to a cone of directions, as shown in the figure.

EXAMPLE 29.1 **Angular momentum states**

Consider a hydrogen atom in a state with $n = 4$. Find expressions for the largest magnitude L of angular momentum, the largest positive value of L_z, and the corresponding values of the quantum numbers l and m_l. For the corresponding quantum state, find the smallest angle that the angular momentum vector can make with the $+z$ axis.

SOLUTION

SET UP If $n = 4$, the largest possible value of l is 3, and the maximum positive value of m_l is 3. The possible values of L and L_z are given by Equations 29.3 and 29.4.

SOLVE From Equation 29.3, when $l = 3$, $L = \sqrt{l(l+1)}\,\hbar = \sqrt{3(3+1)}\,\hbar = 3.46\hbar$. From Equation 29.4, $L_z = m_l\hbar = 3\hbar$. The geometry is similar to Figure 29.1; the minimum value of θ is given by

$$\cos\theta_{\min} = \frac{L_z}{L} = \frac{m_l\hbar}{\sqrt{l(l+1)}\,\hbar} = \frac{3}{\sqrt{3(3+1)}}, \qquad \theta_{\min} = 30.0°.$$

REFLECT For the same value of l, but other values of the quantum number l_z, the angle is greater. For $m_l = -3$, it is 150°.

Practice Problem: In this problem, L_x and L_y don't have definite values, but the quantity $L_x^2 + L_y^2$ does. What is it? *Answer:* $[l(l+1) - m_l^2]\hbar^2$.

The new quantum mechanics we have just described is much more complex, both conceptually and mathematically, than Newtonian mechanics. It deals with probabilities rather than certainties, and it predicts discrete rather than continuous behavior. However, quantum mechanics enables us to understand physical phenomena and to analyze physical problems for which classical mechanics is completely powerless.

Electron Spin

All atoms show shifts or splitting of their energy levels and spectral wave lengths, when placed in a magnetic field. This isn't surprising; electric charges moving in a magnetic field are acted upon by forces, as we discussed in Chapter 20. But some atoms show unexpected shifts even when *no* magnetic field is present. It was suggested in 1925 (initially in the context of the Bohr model) that perhaps the electron should be pictured as a spinning sphere of charge rather than an orbiting point charge, a concept called **electron spin.** If so, the electron has not only angular momentum associated with its orbital motion, but also additional angular momentum associated with rotation on its axis. The resulting additional magnetic-field interactions might cause the observed energy-level anomalies.

Here's an analogy: The earth travels in a nearly circular orbit around the sun, and at the same time it rotates on its axis. Each motion has its associated angular momentum, which we call the orbital and spin angular momentum, respectively. The total angular momentum of the system is the vector sum of the two.

Precise spectroscopic analysis, as well as a variety of other experimental evidence, has shown conclusively that the electron *does* have angular momentum that doesn't depend on its orbital motion, but is intrinsic to the particle itself. Like orbital angular momentum, spin angular momentum (usually denoted by $\vec{S}$) is found to be quantized. Suppose we have an apparatus that measures the component of $\vec{S}$ in a particular direction, such as the z axis. Denoting the z component of $\vec{S}$ by S_z, we find that the only possible values are

$$S_z = \pm\tfrac{1}{2}\hbar.$$

This relation is similar to Equation 29.4 for L_z, the z component of orbital angular momentum, but the magnitude of S_z is one-half of $\hbar$ instead of an integral

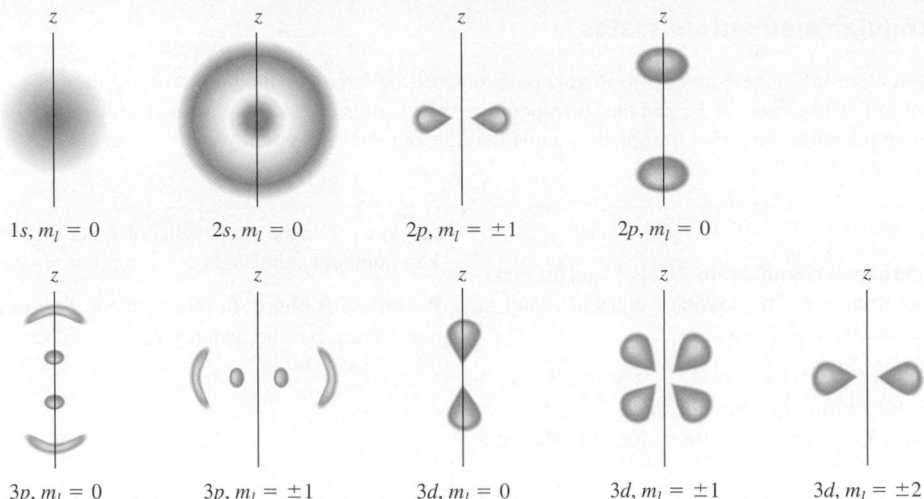

▲ **FIGURE 29.3** Cross sections of three-dimensional probability distributions for a few quantum states of the hydrogen atom.

multiple. Thus, an electron is often called a "spin-$\frac{1}{2}$ particle." The magnitude $S = \left| \vec{S} \right|$ of the spin angular momentum (in analogy to Equation 29.3) is

$$S = \sqrt{\tfrac{1}{2}\left(\tfrac{1}{2} + 1\right)}\hbar = \frac{\sqrt{3}}{2}\hbar.$$

In quantum mechanics, in which the Bohr orbits are superseded by wave functions, we can't really picture an electron as a solid object that spins. But if we visualize the electron wave functions as "clouds" surrounding the nucleus, then we can imagine many tiny arrows distributed throughout the cloud, all pointing in the same direction, either all $+z$ or all $-z$. But don't take this picture too seriously!

In any event, the concept of electron spin is well established by a variety of experimental evidence. To label completely the state of the electron in a hydrogen atom, we now need a *fourth* quantum number *s,* which we'll call the *spin quantum number,* to specify the electron spin orientation. If *s* can take the value $\frac{1}{2}$ or $-\frac{1}{2}$, then the *z* component of spin angular momentum is given by

$$S_z = s\hbar \quad \left(s = \pm\tfrac{1}{2}\right). \tag{29.5}$$

Thus, the possible states of the electron in the hydrogen atom are labeled with a set of four quantum numbers (n, l, m_l, s). The principal quantum number *n* determines the energy, and the other three quantum numbers specify the angular momentum and its component in a specified direction, usually taken to be the *z* axis.

Figure 29.3 shows cross-sectional representations of several hydrogen-atom wave functions.

Many-Electron Atoms

The hydrogen atom, with one electron and a nucleus consisting of a single proton, is the simplest of all atoms. All other electrically neutral atoms contain more than one electron. The analysis of atoms with more than one electron increases in complexity very rapidly. Each electron interacts not only with the positively charged nucleus, but also with all the other electrons. In principle, the motion of the electrons is still governed by the Schrödinger equation, but the mathematical problem of finding appropriate solutions of that equation is so complex that it has not been solved exactly even for the helium atom, which has a mere two electrons.

Various approximation schemes can be used to apply the Schrödinger equation to multielectron atoms. The most drastic is to ignore the interactions of the electrons with each other and assume that each electron moves under the influence only of the electric field of the nucleus, which is considered to be a point charge. A less drastic and more useful model is to think of all the electrons together as making up a charge cloud that is, on the average, spherically symmetric. In this model, each individual electron moves under the action of the total electric field due to the nucleus and this averaged-out electron cloud. It turns out that when an electron moves in an electric field produced by a spherically symmetric charge distribution, the angular momentum states given by the quantum numbers l, m_l, and s are exactly the same as those of the hydrogen atom.

This model is called the **central-field approximation;** it provides a useful starting point for the understanding of the structure of complex atoms. The energy levels now depend on both n and l; usually, for a given value of n, the levels increase in energy with increasing l. But individual levels can still be labeled by the set of four quantum numbers (n, l, m_l, s), with the same restrictions as before:

$$0 \le l \le n - 1, \qquad |m_l| \le l, \qquad \text{and} \qquad s = \pm\tfrac{1}{2}. \qquad (29.6)$$

The Exclusion Principle

To implement the central-field approximation, we need one additional principle: the **exclusion principle.** To understand the essential role this principle plays, we need to consider the lowest energy state, or **ground state,** of a multielectron atom. In the central-field model, which ignores interactions between electrons, there is a lowest energy state (corresponding roughly to the $n = 1$ state for the single electron in the hydrogen atom). We might expect that, in the ground state of a complex atom, *all* the electrons should be in this lowest state. If so, then the behavior of atoms with increasing numbers of electrons should show gradual changes in physical and chemical properties as the number of electrons in the atoms increases.

A variety of evidence shows conclusively that this is *not* what happens at all. For example, the elements fluorine, neon, and sodium respectively have 9, 10, and 11 electrons per atom. Fluorine is a halogen; it tends strongly to form compounds in which each atom acquires an extra electron. Sodium, an alkali metal, forms compounds in which it *loses* an electron, and neon is an inert gas (or noble gas) that ordinarily forms no compounds at all. These observations, and many others, show that, in the ground state of a complex atom, the electrons *cannot* all be in the lowest energy states.

The key to resolving this puzzle was discovered by the Swiss physicist Wolfgang Pauli in 1925, and it is accordingly named for its discoverer:

The Pauli exclusion principle

No two electrons in an atom can occupy the same quantum-mechanical state. Alternatively, no two electrons in an atom can have the same values of all four of their quantum numbers.

Different quantum states correspond to different spatial distributions, including different distances from the nucleus, and to different electron-spin orientations. Roughly speaking, in a complex atom there isn't enough room for all the electrons in the states nearest the nucleus. Some electrons are forced by the Pauli principle into states farther away, with higher energies.

TABLE 29.1   Quantum states of electrons in the first four shells

n	l	m_l	Spectroscopic notation	Number of states		Shell
1	0	0	$1s$	2		K
2	0	0	$2s$	2	} 8	L
2	1	$-1, 0, 1$	$2p$	6		
3	0	0	$3s$	2	} 18	M
3	1	$-1, 0, 1$	$3p$	6		
3	2	$-2, -1, 0, 1, 2$	$3d$	10		
4	0	0	$4s$	2	} 32	N
4	1	$-1, 0, 1$	$4p$	6		
4	2	$-2, -1, 0, 1, 2$	$4d$	10		
4	3	$-3, -2, -1, 0, 1, 2, 3$	$4f$	14		

We can now make a list of all the possible sets of quantum numbers, and thus of the possible states of electrons, in an atom. Such a list is given in Table 29.1, which also indicates two alternative notations. It is customary to designate the value of l by a letter, according to this scheme:

$$l = 0: s \text{ state}$$
$$l = 1: p \text{ state}$$
$$l = 2: d \text{ state}$$
$$l = 3: f \text{ state}$$
$$l = 4: g \text{ state}$$

This peculiar choice of letters originated in the early days of spectroscopy and has no fundamental significance. A state for which $n = 2$ and $l = 1$ is called a $2p$ state, and so on, as shown in Table 29.1, which also shows the relationship between values of n and the x-ray levels $(K, L, M, \cdots)$ that we'll describe in the next section. The $n = 1$ levels are designated as K, $n = 2$ levels as L, and so on.

EXAMPLE 29.2   Quantum states for the hydrogen atom

How many different $4f$ states does the hydrogen atom have? Make a list.

SOLUTION

SET UP AND SOLVE The "4" in $4f$ means $n = 4$, and the f means $l = 3$. The maximum magnitude of m_l is 3, so there are seven possible values of m_l: $m_l = -3, -2, -1, 0, 1, 2, 3$. For each of these, there are two possible values of the spin quantum number s: $\pm\frac{1}{2}$. The total number of states is 14.

REFLECT For any value of l, the number of possible values of m_l is $2l + 1$.

Practice Problem: How many $4d$ states does the hydrogen atom have? How many $4g$ states? *Answer:* 10; 0.

With some exceptions, the average distance of an electron from the nucleus increases with n. Thus, each value of n corresponds roughly to a region of space around the nucleus in the form of a spherical **shell.** Hence we speak of the K shell as the region occupied by the electrons in the $n = 1$ states (closest to the nucleus), the L shell as the region of the $n = 2$ states, and so on. States with the same n but different l, form *subshells,* such as the $3p$ subshell.

Each quantum state corresponds to a certain distribution of the electron cloud in space. Therefore, the exclusion principle says, in effect, "No more than two electrons (with opposite values of the spin quantum number s) can occupy the same region of space." The wave functions that describe electron distributions don't have sharp, definite boundaries, but the exclusion principle limits the degree of overlap of electron wave functions that is permitted. The maximum numbers of electrons in each shell and subshell are shown in Table 29.1.

29.2 Atomic Structure

We're now ready to use the exclusion principle, along with the electron energy states described in Section 29.1, to derive the most important features of the structure and chemical behavior of multielectron atoms, including the periodic table of the elements. The number of electrons in an atom in its normal (electrically neutral) state is called the **atomic number,** denoted by Z. The nucleus contains Z protons and some number of neutrons. The neutron has *no* charge; the proton and electron charges have the same magnitude, but opposite sign, so in the normal atom the total electric charge is zero. Because the electrons are attracted to the nucleus, the quantum states corresponding to regions nearest the nucleus have the lowest energies.

We can imagine constructing an atom by starting with a bare nucleus with Z protons and adding Z electrons, one by one. To obtain the ground state, we fill the lowest-energy states (those closest to the nucleus, with the smallest values of n and l) first, and we use successively higher states until all the electrons are in place. In filling up these states, we must give careful attention to the limitations imposed by the exclusion principle, as shown in Table 29.1. The chemical properties of an atom are determined principally by interactions involving the outermost electrons, so we particularly want to learn how those electrons are arranged.

Let's look at the ground-state electron configurations for the first few atoms (in order of increasing Z). For hydrogen, the ground state is $1s$; the single electron is in the state $n = 1$, $l = 0$, $m_l = 0$, and $s = \pm\frac{1}{2}$. In the helium atom ($Z = 2$), both electrons are in $1s$ states, with opposite spin components; we denote this configuration as $1s^2$. For helium, the K shell ($n = 1$) is completely filled, and all others are empty. Helium is an inert gas; it has no tendency to gain or lose an electron, and ordinarily it forms no compounds.

Lithium ($Z = 3$) has three electrons; in the ground state, two are in $1s$ states (filling the $1s$ level), and one is in a $2s$ state. We denote this configuration as $1s^2 2s$. (The exclusion principle forbids all three electrons being in $1s$ states.) On the average, the $2s$ electron is considerably farther from the nucleus than the $1s$ electrons, as shown schematically in Figure 29.4. Thus, the net charge influencing the $2s$ electron is approximately $+e$, rather than $+3e$, as it would be without the two $1s$ electrons present. The $2s$ electron is loosely bound; only 5.4 eV is required to remove it, compared with 13.6 eV needed to ionize the hydrogen atom. In chemical behavior, lithium is an *alkali metal*. It forms ionic compounds in which each atom *loses* an electron, corresponding to a valence of $+1$.

Next comes beryllium ($Z = 4$); its ground-state configuration is $1s^2 2s^2$, with a filled K shell and two electrons in the L shell. Beryllium is the first of the *alkaline-earth* elements, forming ionic compounds in which the valence of the atoms is $+2$.

Table 29.2 on the next page shows the ground-state electron configurations of the first 30 elements. The L shell can hold a total of eight electrons. At $Z = 10$, both the K and L shells are filled, and there are no electrons in the M shell. We expect this to be a particularly stable configuration with little tendency either to gain or to lose electrons. This element is neon, a noble gas with no known compounds. The next element after neon is sodium ($Z = 11$), with filled K and L shells and one electron in the M shell. Its "filled-shell-plus-one-electron" configuration resembles that of lithium, and indeed, both are alkali metals (a group that also includes potassium, rubidium, cesium, and francium). The element before neon is fluorine, with $Z = 9$. It has a vacancy in the L shell and has an affinity for

On average, $2s$ electron is considerably farther from the nucleus than the $1s$ electrons. Therefore, it experiences a net nuclear charge of approximately $+e$ (rather than $+3e$).

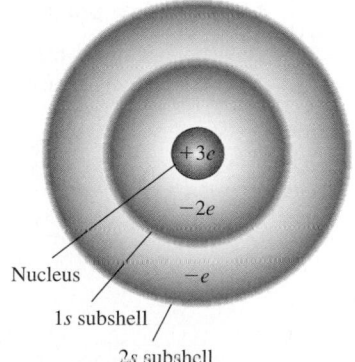

▲ **FIGURE 29.4** Schematic representation of the charge distribution in a lithium atom.

an extra electron to fill that shell. Fluorine forms ionic compounds in which it has a valence of -1. This behavior is characteristic of the halogens (fluorine, chlorine, bromine, iodine, and astatine), all of which have similar configurations.

TABLE 29.2 Ground-state electron configurations of atoms

Element	Symbol	Atomic number (Z)	Electron configuration
Hydrogen	H	1	$1s$
Helium	He	2	$1s^2$
Lithium	Li	3	$1s^2 2s$
Beryllium	Be	4	$1s^2 2s^2$
Boron	B	5	$1s^2 2s^2 2p$
Carbon	C	6	$1s^2 2s^2 2p^2$
Nitrogen	N	7	$1s^2 2s^2 2p^3$
Oxygen	O	8	$1s^2 2s^2 2p^4$
Fluorine	F	9	$1s^2 2s^2 2p^5$
Neon	Ne	10	$1s^2 2s^2 2p^6$
Sodium	Na	11	$1s^2 2s^2 2p^6 3s$
Magnesium	Mg	12	$1s^2 2s^2 2p^6 3s^2$
Aluminum	Al	13	$1s^2 2s^2 2p^6 3s^2 3p$
Silicon	Si	14	$1s^2 2s^2 2p^6 3s^2 3p^2$
Phosphorus	P	15	$1s^2 2s^2 2p^6 3s^2 3p^3$
Sulfur	S	16	$1s^2 2s^2 2p^6 3s^2 3p^4$
Chlorine	Cl	17	$1s^2 2s^2 2p^6 3s^2 3p^5$
Argon	Ar	18	$1s^2 2s^2 2p^6 3s^2 3p^6$
Potassium	K	19	$1s^2 2s^2 2p^6 3s^2 3p^6 4s$
Calcium	Ca	20	$1s^2 2s^2 2p^6 3s^2 3p^6 4s^2$
Scandium	Sc	21	$1s^2 2s^2 2p^6 3s^2 3p^6 3d 4s^2$
Titanium	Ti	22	$1s^2 2s^2 2p^6 3s^2 3p^6 3d^2 4s^2$
Vanadium	V	23	$1s^2 2s^2 2p^6 3s^2 3p^6 3d^3 4s^2$
Chromium	Cr	24	$1s^2 2s^2 2p^6 3s^2 3p^6 3d^5 4s$
Manganese	Mn	25	$1s^2 2s^2 2p^6 3s^2 3p^6 3d^5 4s^2$
Iron	Fe	26	$1s^2 2s^2 2p^6 3s^2 3p^6 3d^6 4s^2$
Cobalt	Co	27	$1s^2 2s^2 2p^6 3s^2 3p^6 3d^7 4s^2$
Nickel	Ni	28	$1s^2 2s^2 2p^6 3s^2 3p^6 3d^8 4s^2$
Copper	Cu	29	$1s^2 2s^2 2p^6 3s^2 3p^6 3d^{10} 4s$
Zinc	Zn	30	$1s^2 2s^2 2p^6 3s^2 3p^6 3d^{10} 4s^2$

▲ **Application It's a gas.** An interesting and sometimes colorful group of elements are the noble gases, located in the extreme right column of the periodic table. As with any group of elements in a column of the periodic table, they share similar chemical characteristics. The noble gases, including helium, neon, argon, and krypton, are unusual because all have completely filled electron shells or subshells. They have almost no tendency to form molecular bonds with other atoms and therefore exist as monatomic gases. When noble gases are confined in a thin glass tube and subjected to high voltage, electrons in the atoms are excited to higher energy levels. As the electrons return to the ground state, the atoms emit photons of visible light. Each element produces a characteristic color and, in various combinations, they can produce the wide variety of eye-catching colors that we see in "neon" lights today.

Conceptual Analysis 29.1

Spectroscopic notation

The atom having the electron configuration $1s^2$, $2s^2$, $2p^6$ has (there may be more than one correct choice)

A. electrons in states $n = 1, 2$.
B. electrons in states $n = 2, 6$.
C. 6 orbital electrons.
D. 10 orbital electrons.

SOLUTION In spectroscopic notation, the numbers preceding s, p, d, f, etc., indicate the quantum number n. Thus, $1s^2$ corresponds to electrons in state $n = 1$, and $2s^2$ and $2p^6$ correspond to $n = 2$ (choice A). Furthermore, the *exponent* in the notation $1s^2$ indicates that there are 2 orbital electrons in an s state, $2s^2$ indicates another 2 electrons in an s state, and $2p^6$ indicates 6 more electrons in $2l$ states. Therefore, there are $2 + 2 + 6 = 10$ orbital electrons, and choice D is also correct.

The Periodic Table of the Elements

Proceeding down the list, we can understand many of the regularities in chemical behavior displayed by the **periodic table of the elements** (Appendix D) on the basis of electron configurations. A slight complication occurs with the M and N

shells because the 3*d* and 4*s* subshells ($n = 3$, $l = 2$ and $n = 4$, $l = 0$, respectively) overlap in energy. Argon $(Z = 18)$ has all the 1*s*, 2*s*, 2*p*, 3*s*, and 3*p* states filled, but in potassium $(Z = 19)$ the additional electron goes into a 4*s* level rather than a 3*d* level (because the 4*s* level has slightly lower energy).

The next several elements have one or two electrons in the 4*s* states and increasing numbers in the 3*d* states. These elements are all metals with rather similar chemical and physical properties; they form the first *transition series*, starting with scandium $(Z = 21)$ and ending with zinc $(Z = 30)$, for which all the 3*d* and 4*s* levels are filled. Something similar happens with $Z = 57$ through $Z = 71$, which have two electrons in the 6*s* levels but only partially filled 4*f* and 5*d* levels. These are the rare-earth elements, with similar physical and chemical properties. Yet another such series, called the *actinide series*, starts with $Z = 91$.

The similarity of the elements in each group (vertical column) of the periodic table reflects a corresponding similarity in outer electron configuration. All the noble gases (helium, neon, argon, krypton, xenon, and radon) have filled-shell or filled-subshell configurations. All the alkali metals (lithium, sodium, potassium, rubidium, cesium, and francium) have "filled-shell-or-subshell-plus-one" configurations. All the alkaline-earth metals (beryllium, magnesium, calcium, strontium, barium, and radium) have "filled-shell-or-subshell-plus-two" configurations, and all the halogens (fluorine, chlorine, bromine, iodine, and astatine) have "filled-shell-or-subshell-minus-one" configurations.

PROBLEM-SOLVING STRATEGY 29.1 **Atomic structure**

1. Be sure you know how to count the energy states for electrons in the central-field approximation. There are four quantum numbers: n, l, m_l, and s; n is always positive, l can be zero or positive, m_l can be zero, positive, or negative, and $s = \pm\frac{1}{2}$. Be sure you know how to count the number of states in each shell and subshell; study Tables 29.1 and 29.2 carefully.
2. As in Chapter 28, familiarizing yourself with some numerical magnitudes is useful. Here are two examples to work out: The magnitude of the electrical potential energy of a proton and an electron (or two electrons) 0.10 nm apart (typical of atomic dimensions) is 14.4 eV, or 2.31×10^{-18} J. Think of other examples like this, and work them out to help you know what kinds of magnitudes to expect in atomic physics.
3. As in Chapter 28, you'll need to use both electronvolts and joules. The conversion $1 \text{ eV} = 1.602 \times 10^{-19}$ J and Planck's constant in eV, $h = 4.136 \times 10^{-15}$ eV · s, are useful quantities. Nanometers are convenient for atomic and molecular dimensions, but don't forget to convert to meters in calculations.

EXAMPLE 29.3 **Periodic table of the elements**

The element fluorine has atomic number $Z = 9$ and ground-state electron configuration $1s^2 2s^2 2p^5$. What element of next-larger Z has chemical properties that are similar to those of fluorine? Explain the reasoning behind your answer by giving the ground-state electron configuration for that element.

SOLUTION

SET UP Each *s* subshell ($l = 0$ levels) can accommodate two electrons, and each *p* subshell ($l = 1$ levels) can hold six elec-

trons. Subshells fill in the order 1*s*, 2*s*, 2*p*, 3*s*, 3*p*, 4*s*, etc. Fluorine is one electron short of having a filled 2*p* subshell.

Continued

SOLVE The next element with similar properties is one electron short of having a filled $3p$ subshell. The $1s$, $2s$, $2p$, $3s$, and $3p$ levels together can accommodate $2 + 2 + 6 + 2 + 6 = 18$ electrons. With one $3p$ vacancy, the number of electrons in the atom in question is 17. The element we seek is chlorine $(Z = 17)$.

REFLECT Fluorine and chlorine are in adjacent rows of the same column of the periodic table, and both are halogens.

Practice Problem: The element lithium has atomic number $Z = 3$ and ground-state electron configuration $1s^2 2s$. What element of next-larger Z has chemical properties that are similar to those of lithium? *Answer:* Sodium $(Z = 11)$.

The periodic table of the elements was first formulated by the Russian chemist Dmitri Ivanovich Mendeleev in about 1870 as a purely *empirical* structure based on observations of the similarities in behavior of groups of chemical elements. The understanding of this behavior on the basis of atomic structure came nearly 50 years later.

X-Ray Energy Levels

The outer electrons of an atom are responsible for optical spectra. Their excited states are usually only a few electronvolts above the energy of the ground state. In transitions from excited states to the ground state, they usually emit photons in or near the visible region, with photon energies of about 2 to 3 eV. There are also **x-ray energy levels,** corresponding to vacancies in the *inner* shells of a complex atom. We mentioned these levels briefly in Section 28.5. In an x-ray tube, an electron may strike the target with enough energy to knock an electron out of an inner shell of the target atom. These inner electrons are much closer to the nucleus than the electrons in the outer shells; they are therefore much more tightly bound, and hundreds or thousands of electronvolts may be required to remove them.

Suppose an electron is knocked out of the K shell. The atom is then left with a vacancy; we'll call this state a K *x-ray energy level.* This vacancy can subsequently be filled by an electron falling in from one of the outer shells, such as the L, M, N, $\cdots$ shell. The transition is accompanied by a *decrease* in the energy of the atom (because *less* energy would be needed to remove an electron from the L, M, N, $\cdots$, shell), and an x-ray photon is emitted with energy equal to this decrease. Each state (for any specific element) has a definite energy, so the emitted x rays have definite wavelengths, and the spectrum is a *line spectrum.*

If the outermost electrons are in the N shell, the x-ray spectrum has three lines, resulting from transitions in which the vacancy in the K shell is filled by an L, M, or N electron. This series of lines is called a K series. Figure 29.5 shows the K series for tungsten, molybdenum, and copper.

There are other series of x-ray lines, called the L, M, and N series, produced by the ejection of an electron from the L, M, or N shell rather than the K shell. Electrons in these outer shells are farther away from the nucleus and are not held as tightly as those in the K shell. Their removal thus requires less energy, and the x-ray photons emitted when these vacancies are filled have lower energy than those in the K series have.

The three lines in each series are called the K_α, K_β, and K_γ lines. The K_α line is produced by the transition of an L electron to the vacancy in the K shell, the K_β line by an M electron, and the K_γ line by an N electron.

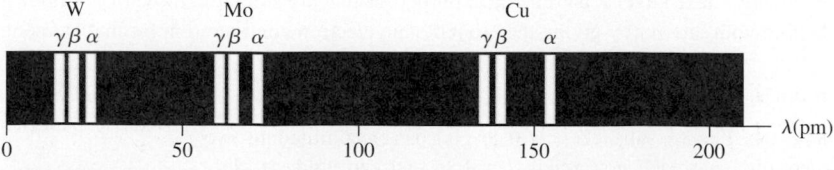

▲ **FIGURE 29.5** K series for tungsten (W), molybdenum (Mo), and copper (Cu).

X-ray line emission

Consider x-ray line emission from an iron (Fe) atom. Which of the following statements is correct?

A. The energy of x-ray photons is less than the energy of visible-light photons.

B. X rays are emitted when electrons make a transition from higher energy shells to the K shell.

C. A vacancy or hole is left behind in the K shell after the x ray is emitted.

SOLUTION For line emission of x rays, an electron must first be knocked out from one of the lowest-energy shells, such as the K shell. Then an electron from a higher energy shell drops down to fill this vacancy, emitting an x ray in the process. The correct answer is B.

EXAMPLE 29.4 Counting electrons in shells

(a) For silicon (Si, $Z = 14$) and germanium (Ge, $Z = 32$), make lists of the numbers of electrons in each subshell $(1s, 2s, 2p, \cdots)$. Use the allowed values of the quantum numbers along with the exclusion principle; do not refer to Table 29.2. Explain why these two elements have similar chemical behavior.

SOLUTION

SET UP Each s subshell ($l - 0$ levels) can accommodate two electrons $(l = 0,\ s = \pm\frac{1}{2})$. Each p subshell ($l = 1$ levels) can hold six) $(l = 1,\ m = -1, 0, +1,\ s = \pm\frac{1}{2})$, and so on.

SOLVE (a) For silicon, the $(1s)$, $(2s)$, and $(2p)$ subshells together contain $2 + 2 + 6 = 10$ electrons. The remaining 4 of the total of 14 must be in the $(3s)$ and $(3p)$ subshells, with two electrons in each. For germanium, the $(1s)$, $(2s)$, $(2p)$, $(3s)$, $(3p)$, and $(3d)$ subshells together contain $2 + 2 + 6 + 2 + $ $6 + 10 = 28$ electrons. The remaining 4 electrons go in the $(4s)$ and $(4p)$ subshells, with 2 electrons in each. Both elements have filled shells plus 4 extra electrons in outer, unfilled shells.

REFLECT The element carbon $(Z = 6)$ also has two s and two p electrons in its outermost shell. Its chemical behavior is similar to that of silicon and germanium.

Practice Problem: Carry out a similar analysis for the elements having $Z = 12$ and $Z = 20$, and predict the similarity in chemical behavior of these two elements.

EXAMPLE 29.5 Potential energy of two particles

Verify the statement in the Problem-Solving Strategy that the magnitude of the potential energy of an electron and a proton separated by 0.100 nm is 14.4 eV.

SOLUTION

SET UP The potential energy of two point charges q_1 and q_2 separated by a distance r is

$$U = \frac{1}{4\pi\epsilon_0} \frac{q_1 q_2}{r},$$

a relationship we derived in Section 18.1 (Equation 18.8).

SOLVE The magnitude of the charge of each particle is

$$q_1 = q_2 = 1.60 \times 10^{-19}\,\text{C}.$$

When the distance r between them is

$$r = 0.100\,\text{nm} = 1.00 \times 10^{-10}\,\text{m},$$

the potential energy is

$$U = \frac{1}{4\pi\epsilon_0} \frac{|q_1 q_2|}{r} = (8.99 \times 10^9\,\text{N}\cdot\text{m}^2/\text{C}^2)\frac{(1.60 \times 10^{-19}\,\text{C})^2}{1.00 \times 10^{-10}\,\text{m}}$$
$$= 2.30 \times 10^{-18}\,\text{J} = 14.4\,\text{eV}.$$

REFLECT This result is the same order of magnitude as the ionization energy of hydrogen (the energy required to remove the electron completely—i.e., 13.6 eV).

Practice Problem: If one of the two electrons in a helium atom has been removed, what is the potential energy of the remaining electron at a distance of 0.100 nm from the nucleus? Express your result both in joules and in eV. *Answer:* -28.8 eV $= -4.60 \times 10^{-18}$ J.

29.3 Diatomic Molecules

The study of electron configurations in atoms provides valuable insight into the nature of chemical bonds—the interactions that hold atoms together to form stable structures such as molecules and solids. There are several types of chemical bonds, including ionic, covalent, van der Waals, and hydrogen bonds.

PhET: Double Wells and Covalent Bonds
PhET: The Greenhouse Effect

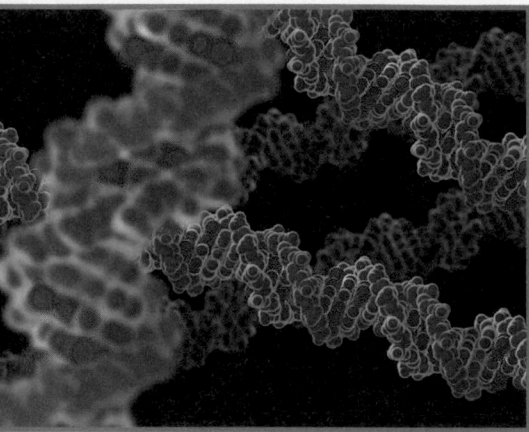

▲ **BIO Application Bonding for life.** The structure of the DNA double helix is based on several of the types of chemical bonds we have discussed. The atoms in each of the two strands are held together by covalent bonds resulting from the sharing of electron pairs between adjacent atoms. Each strand also contains numerous negatively charged phosphate groups that are involved in ionic bonding via electrostatic attractions to positive ions that help stabilize the structure. Weaker hydrogen bonds hold the two DNA strands together, and can be easily broken to allow the DNA strands to be copied during cellular reproduction. And, in processes involving DNA–protein interactions, the DNA typically fits into a precisely shaped pocket of the protein; it is held there in part by close-range interactions involving van der Waals bonds.

Ionic Bond

The **ionic bond,** also called the electrovalent or heteropolar bond, is an interaction between two ionized atoms.. The most familiar example is sodium chloride ($NaCl$), in which the sodium atom gives its one $3s$ electron to the chlorine atom, filling the vacancy in the $3p$ subshell of chlorine.

Let's look at the energy balance in this transaction. Removing the $3s$ electron from the sodium atom requires 5.1 eV of energy; this is called the **ionization energy,** or ionization potential, of sodium. Chlorine has an **electron affinity** of 3.6 eV. That is, the neutral chlorine atom can attract an extra electron into the vacancy in its $2p$ subshell, where the electron is attracted to the nucleus, with an attractive potential energy of magnitude 3.6 eV. Thus, creating the separated Na^+ and Cl^- ions requires a net expenditure of only 5.1 eV − 3.6 eV = 1.5 eV. When the two mutually attracting ions come together, the magnitude of their negative potential energy is determined by the closeness to which they can approach each other. This in turn is limited by the exclusion principle, which forbids extensive overlap of the electron clouds of the two ions.

The minimum potential energy for NaCl turns out to be −5.7 eV at a separation (between centers) of 0.24 nm. At distances less than this, the interaction becomes repulsive. The net energy given up by the system in creating the ions and letting them come together to the equilibrium separation of 0.24 nm is −5.7 eV + 1.5 eV = −4.2 eV, and this is the binding energy of the molecule. That is, 4.2 eV of energy is needed to dissociate the molecule into separate neutral atoms.

Ionic bonds are interactions between charge distributions that are nearly spherically symmetric. Their electrical interaction is similar to that of two point charges, so they are not highly directional. They can involve more than one electron per atom. The alkaline-earth elements form ionic compounds in which an atom loses two electrons; an example is $Mg^{++}(Cl^-)_2$. Loss of more than two electrons is relatively rare; instead, a different kind of bond comes into operation.

Covalent Bond

The **covalent,** or homopolar, bond is characterized by a more nearly symmetric participation of the two atoms, in contrast to the asymmetry of the electron-transfer process of the ionic bond. The simplest example of a covalent bond is that in the hydrogen molecule, a structure containing two protons and two electrons. This bond is shown schematically in Figure 29.6. As the separate atoms come together, the electron wave functions are distorted from the configurations of isolated atoms and become more concentrated in the region between the two protons. The net attraction of the electrons for each proton more than balances the repulsion of the protons and that of the two electrons. The energy of the covalent bond in the hydrogen molecule H_2 is −4.48 eV.

The exclusion principle permits two electrons to occupy the same region of space only when they have opposite spins. When the spins are parallel, the state that would be most favorable from energy considerations (i.e., both electrons in the region between atoms) is forbidden by the exclusion principle. Thus, opposite spins are an essential requirement for an electron-pair bond, and no more than two electrons can participate in such a bond.

However, an atom with several electrons in its outermost shell can form several electron-pair bonds. The bonding of carbon and hydrogen atoms, of central importance in organic chemistry, is an example. In the methane molecule (CH_4), the carbon atom is at the center of a regular tetrahedron, with a hydrogen atom at

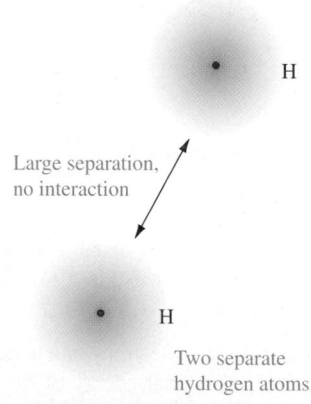

Large separation, no interaction

H

H

Two separate hydrogen atoms.

The covalent bond; the charge clouds for the two electrons with opposite spin s are concentrated in the region between the nuclei (protons).

H_2

▲ **FIGURE 29.6** Formation of covalent bond in hydrogen.

each corner. The carbon atom has four electrons in its L shell $(n = 2)$, and one of these electrons forms a covalent bond with each of the four hydrogen atoms, as shown in Figure 29.7. Similar patterns occur in more complex organic molecules.

Ionic and covalent bonds represent two extremes in the nature of **molecular bonds,** but there is no sharp division between the two types. Often, there is a partial transfer of one or more electrons (corresponding to a greater or smaller distortion of the electron wave functions) from one atom to another. As a result, many molecules having dissimilar atoms have electric dipole moments (a preponderance of positive charge at one end and of negative charge at the other). Such molecules are called **polar molecules.** Water molecules have exceptionally large electric dipole moments that are responsible for the exceptionally large dielectric constant of liquid water.

Weak Bonds

Ionic and covalent bonds, with typical bond energies of 1 to 5 eV, are considered strong bonds. There are also two types of much *weaker* bonds with typical energies of 0.5 eV or less. One of these, the **van der Waals bond,** is an interaction between the electric dipole moments of two atoms or molecules. The bonding of water molecules in the liquid and solid states results partly from dipole–dipole interactions. The interaction potential energy drops off very quickly with distance r between molecules, usually in proportion to $1/r^6$.

Even when an atom or molecule has no permanent dipole moment, fluctuating charge distributions can lead to fluctuating dipole moments that, in turn, can induce dipole moments in neighboring structures. The resulting dipole–dipole interaction can be attractive and can lead to weak bonding of atoms or molecules. The low-temperature liquefaction and solidification of the inert gases and of such molecules as H_2, O_2, and N_2 is due to induced-dipole van der Waals interactions. Not much thermal agitation energy is needed to break these weak bonds, so such substances usually exist in the liquid and solid states only at very low temperatures.

Another type of weak bond, the **hydrogen bond,** is analogous to the covalent bond, in which an electron pair binds two positively charged structures. In the hydrogen bond, a proton (H^+ ion) gets between two atoms, polarizing them and attracting them by means of the induced dipoles. This bond is unique to hydrogen-containing compounds, because only hydrogen has a singly ionized state with no remaining electron cloud. (The hydrogen ion is a bare proton, much smaller than any other singly ionized atom.) The energy required to break a hydrogen bond is small, usually less than 0.5 eV. Hydrogen bonding plays an essential role in many organic molecules. For example, it provides the cross-linking of polymer chains such as polyethylene and the cross-link bonding between the two strands of the famous double-helix DNA molecule.

All these types of bonds play roles in the structure of solids as well as of molecules. Indeed, a solid is in many respects a giant molecule. Still another type of bonding, the *metallic bond,* comes into play in the structure of metallic solids. We'll return to this subject in the next section.

Molecular Spectra

All molecules have quantized energy levels associated with the internal motion of their electrons. In addition, an entire molecule can rotate. The simplest example, a diatomic molecule, can be thought of as a rigid dumbbell (Figure 29.8a) that can rotate about an axis through its center of mass. The angular momentum

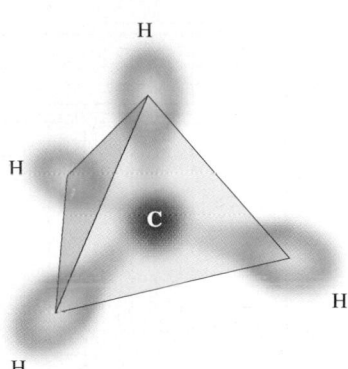

▲ **FIGURE 29.7** Schematic diagram of the methane molecule. The carbon atom is at the center of a regular tetrahedron and forms four covalent bonds with the hydrogen atoms at the corners. Each covalent bond includes two electrons with opposite spins, forming a charge cloud that is concentrated between the carbon atom and a hydrogen atom.

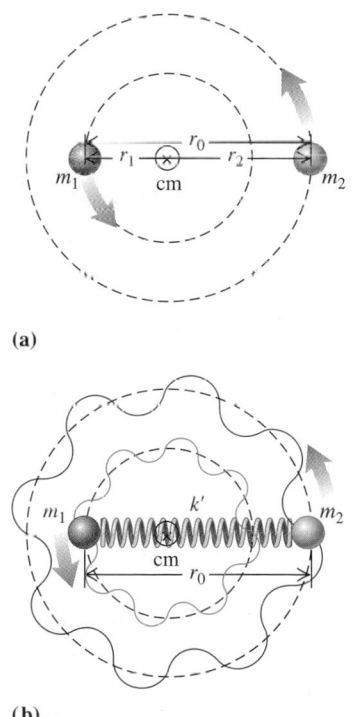

▲ **FIGURE 29.8** A model of (a) a rotating diatomic molecule and (b) vibration and rotation. Vibrational energies are typically on the order of 0.1 eV, much smaller than those of atomic spectra, but usually much larger than energies of the rotational levels.

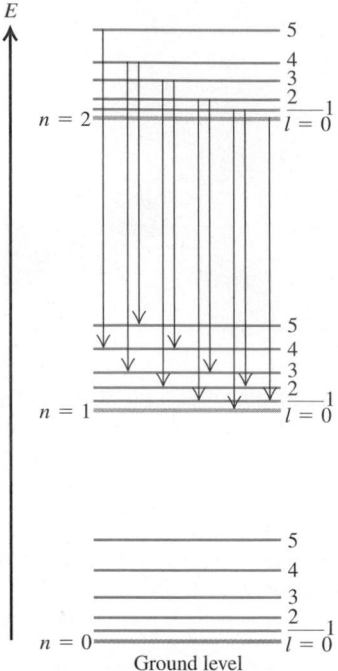

▲ FIGURE 29.9 Energy-level diagram for vibrational and rotational energy levels of a diatomic molecule. For each vibrational level (n), there is a series of more closely spaced rotational levels (l). Several transitions corresponding to a single band in a band spectrum are shown.

associated with this rotation is quantized, just as the angular momentum of an electron in an atom is quantized. Thus, the kinetic energy associated with the rotational motion of a molecule is also quantized: The molecule has a series of rotational energy levels. Because a molecule is always much more massive than an individual electron, these levels are much more closely spaced (typically on the order of 10^{-3} eV) than are the usual atomic energy levels (typically a few eV). The corresponding photon energies associated with transitions among these levels are in the far-infrared region of the spectrum.

In addition, a molecule is never completely rigid; the bonds can stretch and flex. Hence, a more realistic model of a diatomic molecule (Figure 29.8b) consists of two masses connected by a spring rather than a rigid rod. Then, in addition to rotating, the molecule can vibrate, with the atoms moving along the line joining them. The additional energy (both kinetic and potential) associated with vibrational motion is quantized. A typical scheme of vibrational and rotational levels is shown in Figure 29.9. Transitions among these levels lead to complex infrared spectra with *bands* of spectral lines (Figure 29.10). All molecules can also have excited states of their *electrons,* in addition to the rotational and vibrational states we have described. In general, these lie at much higher energies than the rotational and vibrational states.

Infrared spectroscopy has proved to be an extremely valuable analytical tool. It provides information about the strength and rigidity of molecular bonds and the structure of complex molecules. Also, because every molecule (like every atom) has its unique characteristic spectrum, infrared spectroscopy can be used to identify unknown compounds.

EXAMPLE 29.6 **Wavelengths for atomic and molecular transitions**

The separation between adjacent energy levels is typically a few eV for atomic energy levels, on the order of 0.1 eV for vibrational levels, and on the order of 10^{-3} eV for rotational levels. Find the wavelength of the photon emitted during a transition in which the energy of the molecule decreases by 5.00 eV, by 0.500 eV, and by 5.00×10^{-3} eV. In each case, in what region of the electromagnetic spectrum does the photon lie?

SOLUTION

SET UP The transition energy for the molecule equals the energy of the photon. The photon wavelength λ is related to its energy E by Equation 28.2:

$$E = hf = \frac{hc}{\lambda}, \quad \text{or} \quad \lambda = \frac{hc}{E}.$$

The energies are given in electronvolts, so we use the value of Planck's constant h in electronvolt seconds: $h = 4.136 \times 10^{-15}$ eV · s.

SOLVE For a transition energy of 5.0 eV,

$$\lambda = \frac{hc}{E} = \frac{(4.136 \times 10^{-15} \text{ eV} \cdot \text{s})(3.00 \times 10^8 \text{ m/s})}{5.00 \text{ eV}}$$
$$= 2.48 \times 10^{-7} \text{ m} = 248 \text{ nm}.$$

This wavelength is in the ultraviolet region of the spectrum. (The visible spectrum is from about 400 to 700 nm.)

For a photon energy of 0.500 eV (smaller by a factor of 10), the wavelength is 10 times as great, 2480 nm = 2.48 μm, in the infrared region. For $E = 5.00 \times 10^{-3}$ eV, $\lambda = 0.248$ mm, in the microwave region.

REFLECT These photon energies (and corresponding wavelengths) range over a factor of a thousand, yet they represent only a small segment of the entire electromagnetic spectrum. Compared with photons of visible light, x-ray energies are typically higher by a factor of 10^4 to 10^6, gamma-ray energies are higher still, and radio and TV photon energies are usually smaller by a factor on the order of 10^6 to 10^8.

Practice Problem: What is the smallest energy of a transition that produces a photon in the visible region of the spectrum? *Answer:* 1.77 eV.

▲ **FIGURE 29.10** A typical molecular band spectrum.

29.4 Structure and Properties of Solids

The term *condensed matter* includes both solids and liquids. In both states, the interactions between atoms or molecules are strong enough to give the material a definite volume that changes relatively little with applied stress. The distances between centers of adjacent atoms or molecules in condensed matter are of the same order of magnitude as the sizes of the atoms or molecules themselves, typically 0.1 to 0.5 nm.

A crystalline solid is characterized by **long-range order,** a recurring pattern of atomic positions that extends over many atoms. This pattern is called the **crystal structure** or the **lattice structure** of the solid. Four examples of crystal lattices are shown in Figure 29.11. Most liquids have only **short-range order** (correlations between neighboring atoms or molecules) but not long-range order. There are also amorphous (noncrystalline) solids, such as glass. Conversely, some liquids show long-range order, such as the organic compounds used in liquid crystal digital display devices.

Ionic and Covalent Crystals

In some cases, the forces responsible for the regular arrangement of atoms in a crystal are the same as those involved in molecular bonds. Ionic and covalent molecular bonds are found in ionic and covalent crystals, respectively. The most familiar **ionic crystals** are the alkali halides, such as ordinary salt (NaCl). The positive sodium ions and the negative chlorine ions occupy alternate positions in a cubic crystal lattice, as shown in Figure 29.12. The forces are the familiar Coulomb's-law forces between charged particles. These forces have no preferred direction, and the arrangement in which the material crystallizes is determined by the relative size of the two ions.

An example of a **covalent crystal** is the diamond structure, found in the diamond form of carbon and also in silicon, germanium, and tin (Figure 29.13). All these elements are in Group IV of the periodic table, with four electrons in the outermost shell. In the diamond structure, each atom is situated at the center of a regular tetrahedron, with four nearest-neighbor atoms at the corners. The centrally located atom forms a covalent bond with each of these nearest neighbor atoms. These bonds are strongly directional because of the asymmetrical electron distributions, and the result is a tetrahedral structure.

A third type of crystal, less directly related to the chemical bond than are ionic or covalent crystals, is the **metallic crystal.** In this structure, the outermost electrons are not localized at individual atomic lattice sites, but are detached from their parent atoms and are free to move through the crystal. The corresponding charge clouds (and their associated wave functions) extend over many atoms. Thus, we can picture a metallic crystal roughly as an array of positive ions (atoms from which one or more electrons have been removed) immersed in a sea of electrons whose attraction for the positive ions holds the crystal together (Figure 29.14). This "sea" has many of the properties of a gas, and indeed we speak of the electron-gas model of metallic solids.

In a metallic crystal, the atoms would form shared-electron bonds if they had enough valence electrons, but they don't. Instead, electrons are shared among

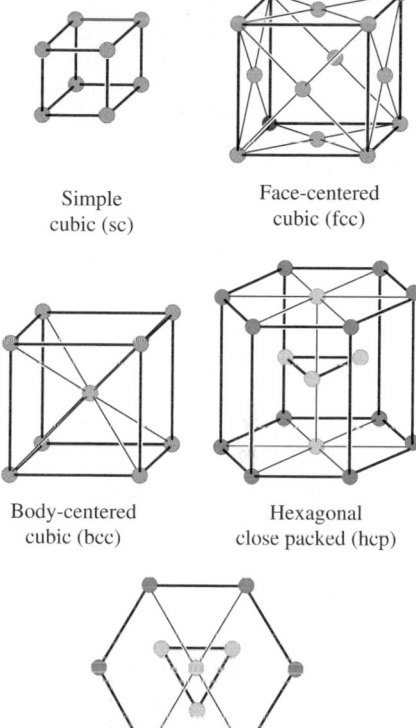

Simple cubic (sc) Face-centered cubic (fcc)

Body-centered cubic (bcc) Hexagonal close packed (hcp)

Top view, hexagonal close packed

▲ **FIGURE 29.11** Portions of some common types of crystal lattices.

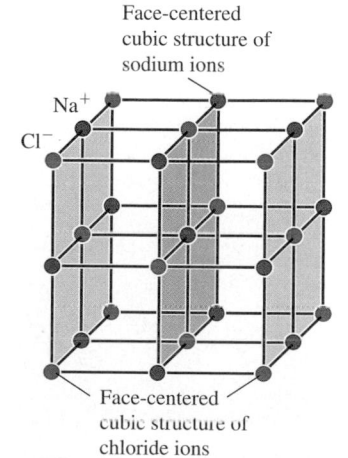

Face-centered cubic structure of sodium ions

Na⁺
Cl⁻

Face-centered cubic structure of chloride ions

▲ **FIGURE 29.12** Representation of part of the sodium chloride crystal structure. The distances between ions are exaggerated.

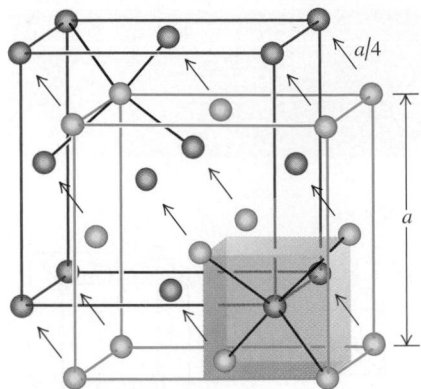

▲ **FIGURE 29.13** The diamond structure, shown as two interpenetrating face-centered cubic structures, with distances between atoms exaggerated.

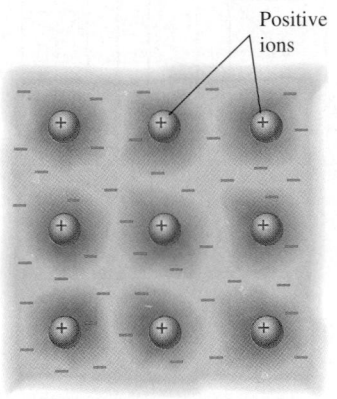

Positive ions

▲ **FIGURE 29.14** In a metallic solid, one or more electrons are detached from each atom and are free to wander around the crystal, forming an "electron gas." The wave functions for these electrons extend over many atoms. The positive ions vibrate around fixed locations in the crystal.

The irregularity is seen most easily by viewing the figure from various directions at a grazing angle with the page.

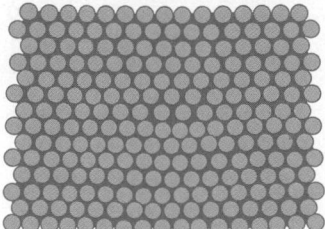

▲ **FIGURE 29.15** An edge dislocation in two dimensions. In three dimensions, an edge dislocation would look like an extra plane of atoms slipped partway into the crystal.

many atoms. This bonding is not strongly directional. The shape of the crystal structure is determined primarily by considerations of **close packing**—that is, the maximum number of atoms that can fit into a given volume. The two most common metallic crystal structures—the face-centered cubic structure and the hexagonal close-packed structure—are shown in Figure 29.11. In each of these structures, each atom has 12 nearest neighbors.

In a perfect crystal, the repeating crystal structure extends uninterrupted throughout the entire material. Real crystals show a variety of departures from this idealized structure. Materials are often polycrystalline, composed of many small perfect single crystals bonded together at grain boundaries. Within a single crystal, interstitial atoms may occur in places where they do not belong, and there may be vacancies—lattice sites that should be occupied by an atom but aren't. An imperfection of particular interest in semiconductors, which we'll discuss in Section 29.6, is the *impurity atom,* a foreign atom (e.g., arsenic in a silicon crystal) substituting for the usual occupant of a lattice site.

A more complex kind of imperfection is the *dislocation,* shown schematically in Figure 29.15, in which one plane of atoms slips in relation to another. The mechanical properties of metallic crystals are influenced strongly by the presence of dislocations. The ductility and malleability of some metals depend on the presence of dislocations that move through the crystal during plastic deformations.

Resistivity

We can understand many macroscopic properties of solids on the basis of the microscopic structure of materials. Electrical properties are of particular interest because of their dominant role in present-day technology. The electrical **resistivity** of a crystalline solid material is determined by the amount of freedom the electrons have to move within the crystal lattice. In a metallic crystal, the valence electrons are not bound to individual lattice sites, but are free to move through the crystal. Metals are usually good conductors.

In a covalent crystal, the valence electrons are tied up in the bonds responsible for the crystal structure and are therefore not free to move. There are no mobile charges available for conduction, and such materials are usually insulators. Similarly, an ionic crystal such as NaCl has no charges that are free to move, and solid NaCl is an insulator. However, when salt is melted, the ions are no longer locked to their individual lattice sites, but are free to move, and molten NaCl is a good conductor. There are, of course, no perfect conductors (except for superconductors at low temperatures) or perfect insulators, but the resistivities of good insulators are greater than those of good conductors by an enormous factor, on the order of at least 10^{15}.

In addition, the resistivities of all materials depend on temperature; in general, the large resistivity of an insulator decreases with temperature, but that of a good conductor usually increases at increased temperatures. Two competing effects are responsible for this difference. In metals, the number of electrons available for conduction is nearly independent of temperature, and the resistivity is determined by the frequency of collisions between electrons and the positively charged ion cores in the lattice. As the temperature increases, the amplitude of vibration of the ion cores increases, and the electrons collide more frequently with the ion cores. This effect causes the resistivities of most metals to increase with temperature.

In insulators, the small amount of conduction that does take place is due to electrons that have gained enough energy from thermal motion of the lattice to break away from their "home" atoms and wander through the lattice. The number of electrons able to gain this much energy is very strongly dependent on temperature; a twofold increase in the number of mobile electrons for a 10 C° rise in temperature is typical. Partially offsetting this energy gain is the increased frequency of collisions at higher temperatures, as with metals, but the increased number of carriers is a far larger effect. Resistivities of insulators invariably decrease rapidly (i.e., insulators become better conductors) as the temperature increases.

Electrical resistivity (or conductivity) is closely related to thermal conductivity, which involves the transport of microscopic mechanical energy rather than electric charge. Wave motion associated with vibrations of the crystal lattice is one mechanism for energy transfer. In metals, the mobile electrons also carry kinetic energy from one region to another. This effect turns out to be much larger than that of the lattice vibrations. As a result, metals are usually much better thermal conductors than are nonmetals, which have few free electrons. Good electrical conductors are nearly always also good thermal conductors.

29.5 Energy Bands

The concept of **energy bands** in solids offers us additional insight into several properties of solids. To introduce the idea, suppose we have a large number N of identical atoms, far enough apart that their interactions are negligible. Every atom has the same energy-level diagram. We can draw an energy-level diagram for the entire system. It looks just like the diagram for a single atom, but the exclusion principle, applied to the entire system, permits each state to be occupied by N electrons instead of just one.

Now we begin to push the atoms closer together. Because of the electrical interactions and the exclusion principle, the wave functions—especially those of the outer, or *valence,* electrons—begin to distort. The energy levels also shift somewhat; some move upward and some downward, depending on the environment of each individual atom. Thus, each valence electron state for the system, formerly a sharp energy level that could accommodate N electrons, splits into a band containing N closely spaced levels, as shown in Figure 29.16. Ordinarily, N is very large, on the order of Avogadro's number (10^{23}), so we can think of the levels as forming a continuous distribution of energies within a band. Between adjacent energy bands are gaps, or forbidden regions, where there are *no* allowed energy levels. The width of a gap is called a *band gap,* denoted by E_g. The inner electrons in an atom are affected much less by nearby atoms than the valence electrons are, and their energy levels remain relatively sharp.

What does all this have to do with electrical conductivity? In insulators and semiconductors, the valence electrons fill completely the highest occupied band, called the **valence band.** The next-higher band, called the **conduction band,** is completely empty. The **energy gap** E_g separating the two may be on the order of 1 to 5 eV. This situation is shown in Figure 29.17a. The electrons in the valence

Mastering**PHYSICS**

PhET: Band Structure
PhET: Conductivity

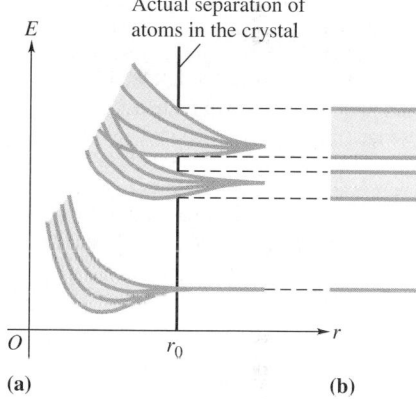

▲ **FIGURE 29.16** Origin of energy bands in a solid. (a) As the distance r between atoms decreases, the energy levels spread into bands. The vertical line at r_0 shows the actual atomic spacing in the crystal. (b) Symbolic representation of energy bands.

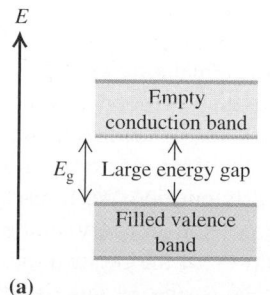
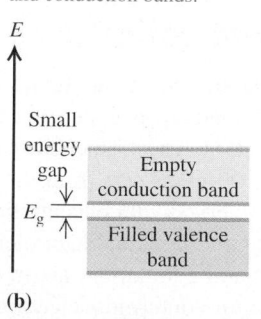
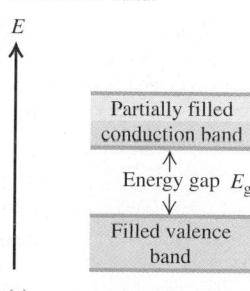

▲ **FIGURE 29.17** Three types of energy-band structures.

▲ **Application Packing them in.** This beautiful diamond is made of pure carbon and is the hardest natural substance known. The pile of graphite, a soft mineral used in powdered form as a dry lubricant, is also pure carbon. How is this possible? It's all in the packing—how the individual carbon atoms are bonded together in space. In a diamond, each atom is covalently bonded to four other atoms in a regular tetrahedral arrangement, giving diamond its extreme toughness. But in graphite, each carbon atom is tightly bound only to three others, forming a series of planar hexagonal sheets. Although the bonds within the sheets are stronger than those in diamond, the sheets themselves are held together only weakly. Therefore, individual sheets can easily slide past each other, accounting for the slippery feel of pure graphite and its use in pencils and lubricants.

At absolute zero, a completely filled valence band is separated by a narrow energy gap E_g of 1 eV or less from a completely empty conduction band. At ordinary temperatures, a number of electrons are excited to the conduction band.

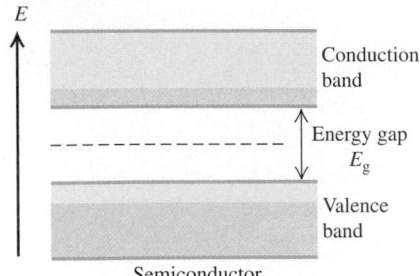

▲ **FIGURE 29.18** Band structure of a semiconductor.

band are not free to move in response to an applied electric field; to move, an electron would have to go to a different quantum-mechanical state with slightly different energy, but all the nearby states are already occupied. The only way an electron can move is to jump into the conduction band. This would require an additional energy of a few electronvolts (at least as great as the band-gap energy E_g), and that much energy is ordinarily not available. (The average energy available from thermal motion is typically on the order of kT, which, at room temperature, is about 1/40 eV.) The situation is like a completely filled parking lot: None of the cars can move because there is no place to go. If a car could jump over the others, it could move.

However, at any temperature above absolute zero, the atoms of the crystal have some vibrational motion, and there is some probability that an electron can gain enough energy (at least as great as E_g) from thermal motion to jump to the conduction band. Once in the conduction band, an electron is free to move in response to an applied electric field because there are plenty of nearby empty states available. There are always a few electrons in the conduction band, so no material is a perfect insulator. Furthermore, as the temperature increases, the population in the conduction band increases very rapidly. A doubling of the number of conduction electrons for a temperature rise of 10 C° is typical.

With metals, the situation is different, because the valence band is only partly filled. The metal sodium is an example. For an isolated sodium atom in its ground state, the valence electron is in a $3s$ state. There are $3p$ excited states at about 2.1 eV above the $3s$ ground state. But in the crystal lattice of *solid* sodium, the $3s$ and $3p$ *bands* spread out enough that they actually overlap, forming a single band that is only one-quarter filled. The situation is similar to the one shown in Figure 29.17c. Electrons in states near the top of the filled portion of the band have many adjacent unoccupied states available, and they can easily gain or lose small amounts of energy in response to an applied electric field. Therefore, these electrons are mobile and can contribute to electrical and thermal conductivity. Metallic crystals always have partly filled bands. In the *ionic* NaCl crystal, by contrast, there is no overlapping of bands; the valence band is completely filled, and solid sodium chloride is an insulator.

The band picture also adds insight into the phenomenon of dielectric breakdown, in which materials that are normally insulators become conductors when they are subjected to a large enough electric field. If the electric field in a material is so large that there is a potential difference of a few volts over a distance comparable to atomic sizes (i.e., a field on the order of 10^{10} V/m), then the field can do enough work on a valence electron to boost it over the forbidden region and into the conduction band. Real insulators usually have dielectric strengths much less than this because of structural imperfections that provide some energy states in the forbidden region.

The concept of energy bands is useful in understanding the properties of semiconductors, which we'll study in the next section.

29.6 Semiconductors

As the name implies, a **semiconductor** is a material with an electrical resistivity that is intermediate between the resistivities of good conductors and those of good insulators. The tremendous importance of semiconductors in present-day electronics stems in part from the fact that their electrical properties are highly sensitive to very small concentrations of impurities. We'll discuss the basic concepts, using the semiconductor elements silicon and germanium as examples.

Silicon and germanium are in Group IV of the periodic table. Each has four electrons in the outermost electron subshells (the $3s$ and $3p$ levels for Si, the $4s$ and $4p$ levels for Ge). Both crystallize in the diamond structure (Section 29.4) with covalent bonding. Each atom lies at the center of a regular tetrahedron, forming a covalent bond with each of four nearest neighbors at the corners of the tetrahedron. All the valence electrons are involved in the bonding. The band structure (Section 29.5) has a valence band (which, at a temperature of absolute zero, would be completely filled), separated by an energy gap E_g from a nearly empty conduction band (Figure 29.17b).

At low temperatures, these materials are insulators. Electrons in the valence band have no nearby levels available into which they can move in response to an applied electric field. However, the energy gap E_g between the valence and conduction bands is unusually small: 1.14 eV for silicon and only 0.67 eV for germanium, compared with 5 eV or more for many insulators. Thus, even at room temperature, a substantial number of electrons are dissociated from their parent atoms and jump the gap into the conduction band (Figure 29.18). This number increases rapidly with increasing temperature.

When an electron is removed from a covalent bond, it leaves a vacancy where there would ordinarily be an electron. An electron from a neighboring atom can easily drop into this vacancy, leaving the neighbor with the vacancy. In this way, the vacancy, usually called a **hole,** can travel through the crystal and serve as an additional current carrier. A hole behaves like a positively charged particle, even though the moving charges are electrons. It's like describing the motion of a bubble in a liquid. In a pure semiconductor, holes and electrons are always present in equal numbers. When an electric field is applied, they move in opposite directions (Figure 29.19). This conductivity is called **intrinsic conductivity,** to distinguish it from conductivity due to impurities, which we'll discuss later.

The parking-lot analogy we mentioned earlier helps to clarify the mechanisms of conduction in a semiconductor. A covalent crystal with no bonds broken is like a filled floor of a parking garage. No cars (electrons) can move because there is nowhere for them to go (i.e., no unoccupied energy states with nearly the same energy as the occupied states). But if one car is moved to the empty floor above, it can move freely, and the empty space it leaves also permits other cars to move on the nearly filled floor. The motion of the vacant space corresponds to a hole in the normally filled valence band.

Impurities

Suppose we mix into melted germanium ($Z = 32$) a small amount of arsenic ($Z = 33$), the next element after germanium in the periodic table. (The deliberate addition of impurity elements is often called *doping.*) Arsenic is in Group V of the periodic table (Appendix D); it has five valence electrons. When one of these electrons is removed, the remaining electron structure is essentially that of germanium. The only difference is that it is scaled down in size by the insignificant factor 32/33, because the arsenic nucleus has a charge of $+33e$ rather than $+32e$. An arsenic atom can comfortably take the place of a germanium atom in the lattice (Figure 29.20a). Four of its five valence electrons form the necessary covalent bonds with the nearest neighbors. The fifth valence electron is very loosely bound, with a binding energy of only about 0.01 eV. In the band picture, this valence electron corresponds to an isolated energy level lying in the gap, 0.01 eV below the bottom of the conduction band (Figure 29.20b). This level is called a *donor level,* and the impurity atom that is responsible for it is called a *donor impurity.*

All Group V elements, including nitrogen, phosphorus, arsenic, antimony, and bismuth, can serve as donor impurities. Even at ordinary temperatures, substantial numbers of electrons in donor levels can gain enough energy to climb into the conduction band, where they are free to wander through the lattice. The corresponding positive charge is associated with the nuclear charge of the arsenic atom ($+33e$ instead of $+32e$). It is *not* free to move; in contrast to the situation

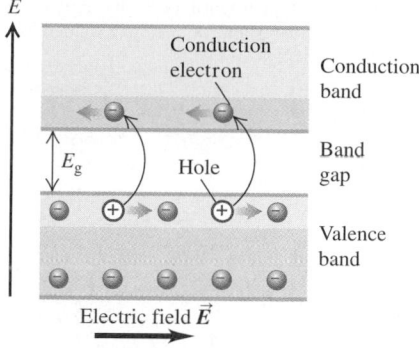

▲ **FIGURE 29.19** Motion of electrons in the conduction band, and of holes in the valence band of a semiconductor under the action of an applied electric field $\vec{E}$.

MasteringPHYSICS

PhET: Semiconductors
PhET: Conductivity

A donor (*n*-type) impurity atom has a fifth valence electron that does not participate in the covalent bonding and is very loosely bound.

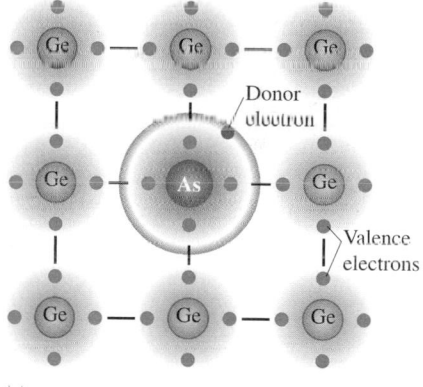

(a)

Energy-band diagram for an *n*-type semiconductor at a low temperature. One donor electron has been excited from the donor levels into the conduction band.

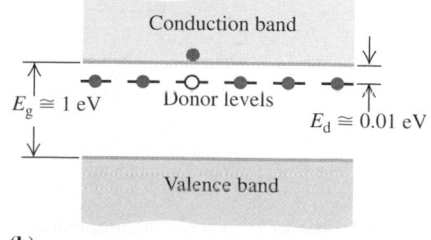

(b)

▲ **FIGURE 29.20** *n*-type impurity conductivity.

An acceptor *p*-type impurity atom has only three valence electrons, so it can borrow an electron from a neighboring atom.

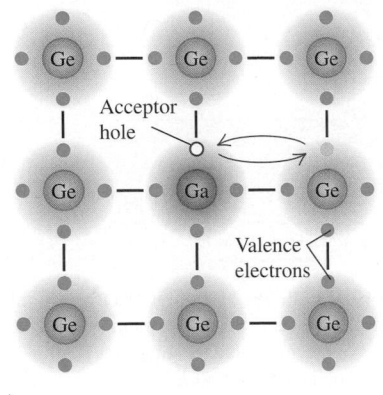

(a)

Energy-band diagram for a *p*-type semiconductor at a low temperature. One acceptor level has accepted an electron from the valence band, leaving a hole behind.

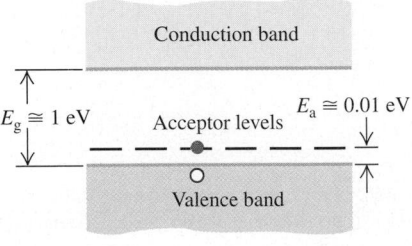

(b)

▲ **FIGURE 29.21** *p*-type semiconductor.

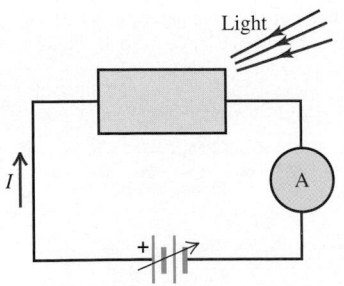

▲ **FIGURE 29.22** A semiconductor photocell in a circuit.

with electrons and holes in pure germanium, this positive charge does not participate in conduction.

At ordinary temperatures, only a very small fraction of the valence electrons in pure germanium are able to escape their sites and participate in conduction. A concentration of arsenic atoms as small as one part in 10^{10} can increase the conductivity so drastically that conduction due to impurities becomes by far the dominant mechanism. In this case, the conductivity is due almost entirely to negative-charge (electron) motion. We call the material an ***n*-type semiconductor,** with *n*-type impurities.

Adding atoms of an element in Group III, with only three valence electrons, has an analogous effect. An example is gallium $(Z = 31)$; placed in the germanium lattice, the gallium atom would like to form four covalent bonds, but it has only three outer electrons. It can, however, "steal" an electron from a neighboring germanium atom to complete the required four covalent bonds. The resulting atom has the same electron configuration as Ge, but is larger by a factor of 32/31 because the nuclear charge is $+31e$ instead of $+32e$. This "theft" leaves the neighboring atom with a hole, or missing electron, which can then move through the lattice just as in intrinsic conductivity.

The "stolen" electron is bound to the gallium atom in a level called an *acceptor level,* about 0.01 eV above the top of the valence band (Figure 29.21). The gallium atom thus completes the needed four covalent bonds, but it has a net charge of $-e$ (because there are 32 electrons and a nuclear charge of $+31e$). In this case, the corresponding negative charge is associated with the deficiency of positive charge of the gallium nucleus $(+31e$ instead of $+32e)$, so that negative charge is not free to move. A semiconductor with Group III impurities is called a ***p*-type semiconductor**—a material with *p*-type impurities. The two types of impurities, *n* and *p*, are also called *donors* and *acceptors,* respectively. The deliberate addition of these impurity elements is called *doping.*

29.7 Semiconductor Devices

The first transistor was invented in 1947. Since then, semiconductor devices have revolutionized the electronics industry, with applications in communications, computer systems, control systems, and many other areas. Semiconductor devices play an indispensable role in contemporary electronics. Semiconductor-based large-scale integrated circuits can incorporate the equivalent of many thousands of transistors, capacitors, resistors, and diodes on a silicon chip less than 1 cm square. Such chips form the heart of every pocket calculator, personal computer, and mainframe computer.

One simple semiconductor device is the *photocell.* In Figure 29.22, a thin slab of pure silicon or germanium (with small intrinsic conductivity) is connected to the terminals of a battery. The resulting current is small because there are very few mobile charges. But now we irradiate the slab with electromagnetic waves whose photons have at least as much energy as the band gap E_g between the valence and conduction bands. (This radiation corresponds to photon energies in the visible or near-infrared region of the electromagnetic spectrum.) Now an electron in the valence band can absorb a photon and jump to the conduction band, where it and the hole it left behind contribute to the conductivity of the slab. Therefore, the conductivity and the circuit current increase with the intensity of the radiation.

Detectors for charged particles operate on the same principle. When a high-energy charged particle passes through the semiconductor material, it collides inelastically with valence electrons, exciting them from the valence to the conduction

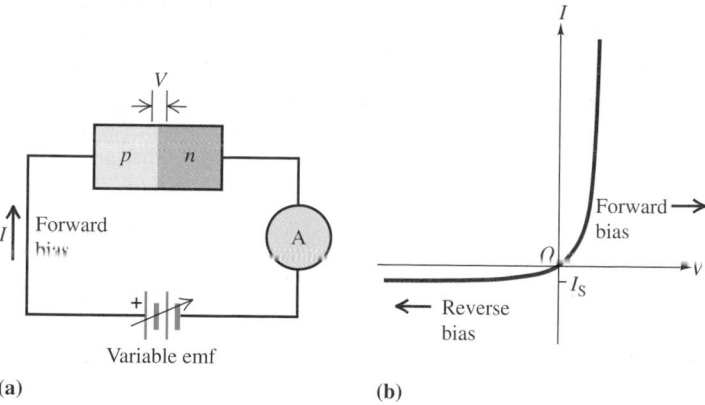

▲ FIGURE 29.23 (a) Schematic diagram of a *p–n* junction in a circuit. (b) Graph of the current–voltage relationship.

band and creating pairs of holes and conduction electrons. The conductivity increases momentarily, causing a pulse of current in an external circuit. Semiconductor detectors are widely used in nuclear and high-energy physics research.

The *p-n* Junction

In many semiconductor devices, the essential principle is that the conductivity of the material is controlled by impurity concentrations, which can be varied within wide limits from one region of a device to another. An example is the **p–n junction**—the boundary layer between two regions of a semiconductor, one with *p*-type impurities and the other with *n*-type impurities. This boundary region is called a *junction;* such a region can be fabricated by the deposition of *n*-type material on a clean surface of some *p*-type material, or in various other ways.

When a *p–n* junction is connected in an external circuit, as in Figure 29.23a, and the potential difference *V* across the junction is varied, the current varies as shown in Figure 29.23b. The device conducts much more readily in the direction from *p* to *n* than the reverse, and the current *I* is not proportional to the potential difference *V*. This behavior is in sharp contrast to the symmetric and linear behavior of a resistor that obeys Ohm's law. Such a one-way device is called a **diode.**

We can understand the behavior of a *p–n* junction diode qualitatively on the basis of the conductivity mechanisms in the two regions. In Figure 29.23, we connect the *p* region to the positive terminal of the battery, so it is at higher potential than the *n* region. The resulting electric field is in the direction from *p* to *n*. This is called the *forward direction,* and the positive potential difference is called **forward bias.** Holes in the *p* region flow into the *n* region, and electrons in the *n* region move into the *p* region; this flow constitutes a forward current.

When the polarity is reversed, the condition is called **reverse bias.** In this case, the field tends to push electrons from *p* to *n* and holes from *n* to *p*. But there are very few mobile electrons in the *p* region, only those associated with intrinsic conductivity and some that diffuse over from the *n* region. Similarly, there are very few holes in the *n* region. As a result, the current in the reverse direction is much smaller than that obtained with the same potential in the forward direction.

A **light-emitting diode** (LED) is, as the name implies, a *p–n* junction that emits light. When the junction is forward biased, many holes are injected across the junction into the *n* region, and electrons are injected into the *p* region. When an electron falls into a hole, the pair can emit a photon with energy approximately

▲ Application Is there a room available?
The electronic behavior of semiconductors involves not only the movement of electrons but also the movement of electron vacancies known as "holes" where electrons are absent. If a Group IV element such as germanium contains a very tiny impurity of a group V element such as arsenic, the arsenic fits into the tetrahedral lattice of the germanium crystal, but has an extra electron. This extra electron is free to move when an external voltage is applied, creating an *n*-type semiconductor. If, on the other hand, the germanium contains a tiny impurity of a Group III element such as gallium, there is a deficiency of one electron, creating a hole. Under an applied voltage, a hole in this *p*-type semiconductor behaves like a mobile positive charge, moving as an electron would, but in the opposite direction.

When $V_e = 0$, the current is very small. When a potential V_e is applied between emitter and base, holes travel from the emitter to the base. When V_c is sufficiently large, most of the holes continue into the collector.

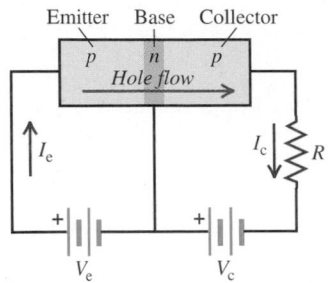

▲ FIGURE 29.24 Schematic diagram of a p–n–p transistor and circuit.

When $V_b = 0$, I_c is very small, and most of the voltage V_c appears across the base–collector junction. As V_b increases, the base–collector potential decreases, and more holes can diffuse into the collector; thus, I_c increases. Ordinarily, I_c is much larger than I_b.

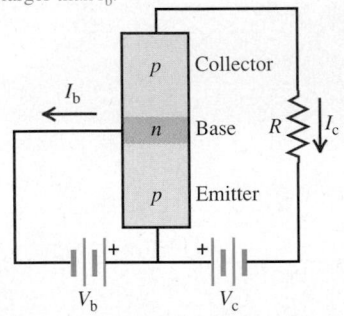

▲ FIGURE 29.25 A common-emitter circuit.

equal to the band-gap energy E_g. This energy (and therefore the photon wavelength and the color of the light) can be varied by using materials with different band-gap energies. Light-emitting diodes are widely used for digital displays in clocks, electronic equipment, automobile instrument panels, and many other applications.

The reverse process is called the *photovoltaic effect*. Here the material absorbs photons, and electron–hole pairs are created. Pairs are created close enough to a p–n junction so that they can migrate to it. The electrons are swept to the n side and the holes to the p side. When we connect this device to an external circuit, it becomes a source of emf and power. Such a device is often called a solar cell, although it can function with any light with photon energies greater than the band-gap energy. The production of low-cost photovoltaic cells for large-scale solar energy conversion is an active field of research.

Transistors

One type of **transistor** includes two p–n junctions in a "sandwich" configuration, which may be either p–n–p or n–p–n. A p–n–p transistor is shown in Figure 29.24. The outer regions are called the **emitter** and **collector,** and the center region is called the **base.** When there is no current in the left loop of the circuit, there is only a small current through the resistor R, because the voltage across the base–collector junction is in the reverse direction. But when a forward-bias voltage is applied between emitter and base, as shown in the figure, the holes traveling from emitter to base can travel through the base to the second junction, where they come under the influence of the collector-to-base potential difference and flow on through the resistor.

In this way, the current in the collector circuit is controlled by the current in the emitter circuit. Furthermore, V_c may be considerably larger than V_e, so the power dissipated in R may be much larger than the power supplied to the emitter circuit by the battery. Thus, the device functions as a **power amplifier.** If the potential drop across R is greater than V_e, it may also be a voltage amplifier.

In this configuration, the base is the common element between the "input" and "output" sides of the circuit. Another widely used arrangement is the common-emitter circuit, shown in Figure 29.25. In this circuit, the current in the collector side of the circuit is much larger than that in the base side, and the result is current amplification.

An important variation is the field-effect transistor (Figure 29.26). A slab of p-type silicon is made with two n-type regions on the top, called the *source* and the *drain;* a metallic conductor is fastened to each. Separated from the slab by an insulating layer of silicon dioxide (SiO_2) is a third electrode called the *gate.* When there is no charge on the gate and a potential difference of either polarity is

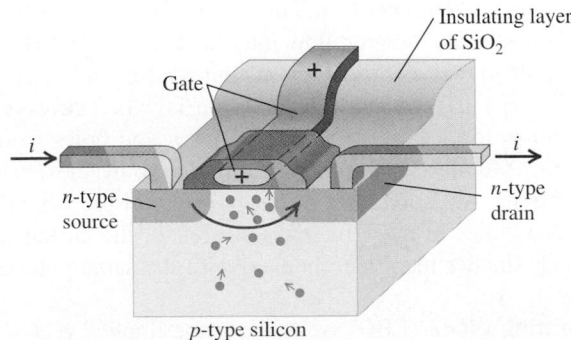

▲ FIGURE 29.26 A field-effect transistor. The current from the source to the drain is controlled by the potential difference between the source and the drain and by the charge on the gate; no current flows through the gate.

applied between the source and the drain, there is very little current, because one of the *p–n* junctions is reverse biased.

Now we place a positive charge on the gate. There aren't many free electrons in the *p*-type material, but there are some, and they are attracted toward the positively charged gate. The concentration of electrons near the gate (and between the two junctions) is greatly enhanced, and the electron current between source and drain increases accordingly. The conductivity of the region between source and drain depends critically on the electron concentration. Thus, the current is extremely sensitive to the gate charge and potential, and the device functions as an amplifier. Note that there is very little current into or out of the gate. The device just described is called a metal-oxide-semiconductor field-effect transistor (MOSFET).

A further refinement in semiconductor technology is the **integrated circuit,** commonly referred to as a **chip.** By successively depositing layers of material and etching patterns to define current paths, we can combine the functions of several MOSFET transistors, capacitors, and resistors on a single square of semiconductor material that may be only a few millimeters on a side. An elaboration of this idea leads to large-scale integrated circuits and very-large-scale integration (VLSI). Starting with a silicon chip base, various layers are built up, including evaporated metal layers for conducting paths and silicon dioxide layers for insulators and for dielectric layers in capacitors. Appropriate patterns are etched into each layer by the use of photosensitive etch-resistant materials, onto which optically reduced patterns are projected.

A circuit containing the functional equivalent of many thousands of transistors, diodes, resistors, and capacitors can be built up on a single chip. These metal-oxide-semiconductor (MOS) chips are the heart of pocket calculators and nearly all present-day computers, large and small. An example is shown in Figure 29.27.

▲ **FIGURE 29.27** An integrated circuit chip the size of your thumb can contain millions of transistors.

29.8 Superconductivity

The electrical resistivity of all conducting materials varies with temperature. Some materials, including several metallic alloys and oxides, show a phenomenon called **superconductivity,** or the complete disappearance of electrical resistance at low temperatures. As the temperature decreases, the resistivity at first decreases smoothly. But then, at a certain temperature T_c, called the **critical temperature,** a phase transition occurs and the resistivity suddenly drops to exactly zero, or at most to a value so small that it cannot be measured. When a current is established in a superconducting ring, it continues indefinitely with no need for an external energy source.

Superconductivity was discovered in 1911 by the Dutch physicist H. Kamerlingh Onnes. He had just discovered how to liquefy helium, which has a boiling temperature of 4.2 K at atmospheric pressure (lower at reduced pressure). Measurements of the resistivity of mercury at very low temperatures showed that when very pure solid mercury is cooled to 4.16 K, it undergoes a phase transition in which its resistivity suddenly drops to zero.

There are two types of superconducting materials. A magnetic field can never exist inside a type-I superconductor. When we place such a material in a magnetic field, eddy currents are induced that exactly cancel the applied field everywhere inside the material (except for a surface layer a hundred or so atoms thick). The critical temperature always *decreases* when the material is placed in a magnetic field. When the field is sufficiently strong, the superconducting phase transition is eliminated.

Each superconducting material has a critical temperature T_c above which it is no longer a superconductor. Until 1986, the highest T_c attained was about 23 K, with a niobium–germanium alloy. This meant that superconductivy occurred only when the material was cooled by means of (expensive and scarce) liquid helium or (explosive) liquid hydrogen. But in January 1986, Karl Muller and Johannes

Bednorz discovered an alloy of barium, lanthanum, and copper with a T_c of nearly 40 K. By 1987, a ceramic material consisting of a complex oxide of yttrium, copper, and barium had been found that has $T_c = 93$ K. This temperature is significant, because it is above the 77 K boiling temperature of (inexpensive and safe) liquid nitrogen. The current (2005) record for T_c for these ceramic high-temperature superconductors is about 160 K. Materials that are superconductors at room temperatures may well become a reality in the future.

Conceptual Analysis 29.3

The critical temperature

A superconducting material is cooled below its critical temperature. Which statement is correct?

A. The critical temperature is the temperature at which the resistivity of a material remains constant with temperature.
B. The critical temperature is always close to absolute zero.
C. The critical temperature is the temperature at which the resistivity of a material becomes zero.

SOLUTION The electrical resistivity of a superconductor makes an abrupt drop to zero at temperatures below the critical temperature. This transition to superconductivity doesn't need to be near absolute zero; indeed, some materials have a critical temperature near 160 K. The correct answer is C.

▲ Application **We've come a long way, baby.** The invention of the transistor in 1947 sparked a revolution in many areas of electronics, including the ways we are able to enjoy our music. Transmitted signals in the earliest radios were amplified using bulky vacuum tubes, making these radios large and relatively immobile. The development of *n*-type and *p*-type semiconductors and their combined use to make a transistor allowed amplification of radio signals in a pocket-sized set, aptly named the "transistor radio." But the one-inch-long transistors in these portable units were enormous by today's standards. Now, using photoetching and metal vapor deposition on a microscopic scale, thousands of individual transistors and circuit elements can be produced on a single 1-mm-square integrated circuit. These "chips" are used in computers, cell phones, and nearly all modern electronic devices that we now take for granted.

Superconductivity was not well understood on a theoretical basis until 1957. In that year, Bardeen, Cooper, and Schrieffer published the theory, now called the BCS theory, that was to earn them the Nobel Prize in 1972. The key to the BCS theory is an interaction between *pairs* of conduction electrons, called *Cooper pairs,* caused by an interaction with the positive ion cores of the crystal structure. A free electron exerts attractive forces on nearby positive ion cores, pulling them slightly closer together and distorting the structure. The resulting slight concentration of positive charge then exerts an attractive force on another electron with momentum opposite that of the first electron. At ordinary temperatures this electron-pair interaction is small in comparison to energies of thermal motion, but at very low temperatures it becomes significant.

Thus bound together, the pairs of electrons cannot *individually* gain or lose small amounts of energy, as they could ordinarily do in a partly filled conduction band. The result is an energy gap in the allowed states of the pairs, and at low temperatures there is not enough energy for a pair to jump this gap. Therefore, the electrons can move freely through the lattice without any exchange of energy through collisions.

When a type-II superconductor is placed in a magnetic field, the bulk of the material is superconducting, but there are thin filaments of material in the normal state, aligned parallel to the field. Currents circulate around the boundaries of these filaments, and there is magnetic flux inside the filaments. Type-II superconductors are used for electromagnets because much larger magnetic fields can usually be present without destroying the superconducting state than is possible with ordinary superconductors.

Superconducting electromagnets are widely used in research laboratories. Once a current is established in the coil of such a magnet, no additional power input is required because there is no resistive energy loss. The coils can also be made more compact because there is no need to provide channels for cooling fluids. Thus, superconducting magnets can attain large fields much more easily and economically than conventional magnets can; fields on the order of 10 T are fairly routine. These considerations also make superconductors attractive for long-distance electric-power transmission, an active area of development.

One of the most glamorous applications of superconductors is in the field of magnetic levitation. Imagine a superconducting ring mounted on a railroad car that runs on a conducting guideway. The current induced in the guideway leads to a repulsive interaction with the rail, and levitation is possible. A magnetically levitated train is now in regular service on a 30 km guideway to and from the Pudong International Airport, outside Shanghai, China. The usual maximum speed is 267 mi/h.

The search for room-temperature superconductors continues. The implications of such materials, if they can be found, are breathtaking. They would have important applications for long-distance power transmission, magnetic levitation, computer design, and many other areas. The high-temperature superconductors discovered thus far are mechanically weak and brittle, like many ceramics, and are often chemically unstable. Fabricating conductors from them will pose difficult technological problems.

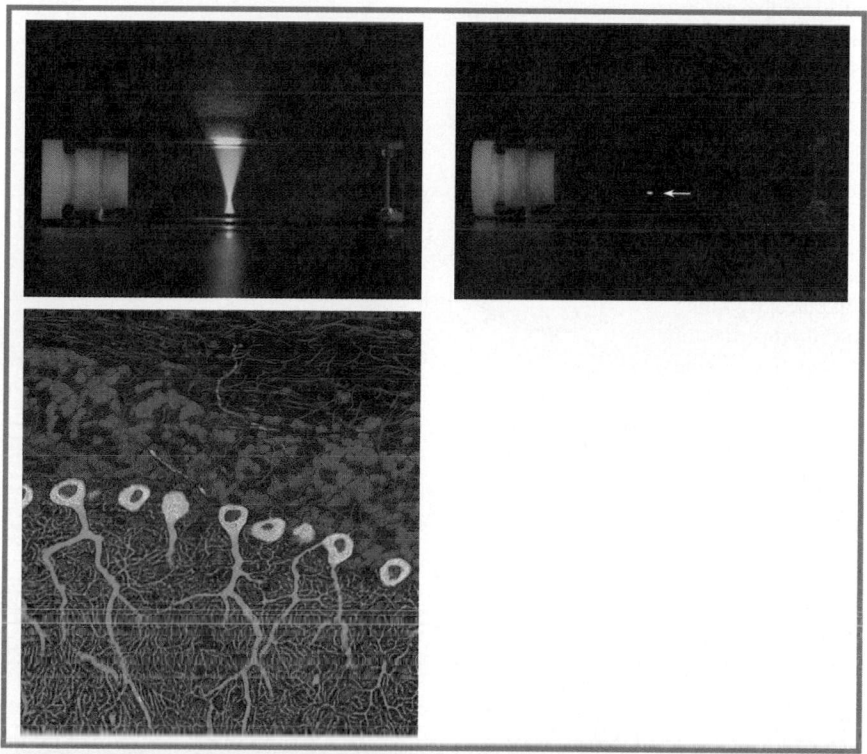

▲ **BIO Application Looking into living tissue via quantum physics.** Molecules have electrons with quantized excited states; in some cases, the energy gap between states matches the energies of photons from the near-UV and the visible regions of the spectrum. An electron excited by a photon rapidly loses some energy through vibrational relaxation (see Fig. 29.9) and then returns to the ground state with the emission of a photon that has somewhat less energy (longer wavelength) than the exciting photon. This process of fluorescence has important applications in biology and medicine. Fluorescent molecules can be introduced into cells by a number of means, including engineering the genes so that the cells themselves produce the molecules. These fluorophores then give information about many processes and structures in the living cells.

One problem with fluorescence microscopy is the absorption of energy by molecules that are not in the plane of focus of the microscope objective. The fluorescent emission from the out-of-focus region degrades the image, and the absorbed energy can be damaging to cells. One solution to this problem is an application of the two-photon effect, a quantum mechanical phenomenon described in the 1930s. If two photons each of approximately half the energy required to excite an electron are absorbed within a very short time, the electron moves to the excited state. A fluorophore that is efficiently excited by absorbing single photons of 450 nm can be excited by absorbing two photons at about 900 nm. However, the photon density must be very high; sufficient photon density is achieved only in the focal plane of the objective. For the remainder of the optical path, the 900 nm photons are not absorbed, so no out-of-focus light is generated and no energy is imparted to the cell.

The upper figure shows two side views (perpendicular to the optical axis of the microscope) of the fluorescent emissions from beams of focused exciting light passing through a medium containing a fluorophore. The right image is of single photon fluorescence and the left image is of two-photon fluorescence; the difference in out-of-focus fluorescence is obvious. The lower photograph is of two-photon imaging of brain tissue showing Purkinje cells in yellow.

SUMMARY

Electrons in Atoms

(Section 29.1) The Bohr model fails to fully describe atoms because it combines elements of classical physics with some principles of quantum mechanics. To explain the observed properties of an atom, electrons must be described with a **wave function** that is determined by solving the Schrödinger equation. As a result, some properties of the atom are quantized, including orbital and spin angular momentum. The **quantum numbers** (n, l, m_l, s) specify the quantum-mechanical states of the atom, and according to the **Pauli exclusion principle,** no two electrons in an atom can occupy the same state.

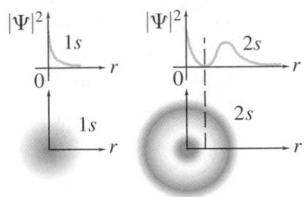

Atomic Structure

(Section 29.2) We can imagine constructing a neutral atom by starting with a bare nucleus with Z protons and adding Z electrons, one by one, respecting the Pauli exclusion principle. To obtain the ground state, we fill the lowest-energy states (those closest to the nucleus, with the smallest values of n and l) first, and we use successively higher states until all the electrons are in place. By filling the atomic states we begin to learn the chemical properties of atoms, which are determined principally by interactions involving the *outermost* electrons.

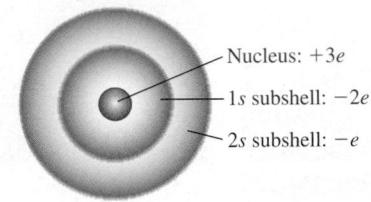

Diatomic Molecules

(Section 29.3) An **ionic bond** is a bond between two ionized atoms—one atom gives at least one electron to fill a vacancy in a shell of the other. In a **covalent bond,** the electron cloud tends to concentrate between the atoms—the positive nucleus of each atom is attracted to the somewhat centralized electron cloud. There are also weaker bonds such as the **van der Waals** and **hydrogen bonds.**

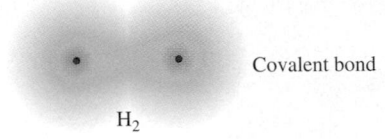

Structure and Properties of Solids

(Section 29.4) A crystalline solid is characterized by long-range order, a recurring pattern of atomic positions that extends over many atoms. Liquids have short-range order. We can understand many macroscopic properties of solids, including mechanical, thermal, electrical, magnetic, and optical properties, by considering their relation to the microscopic structure of the material.

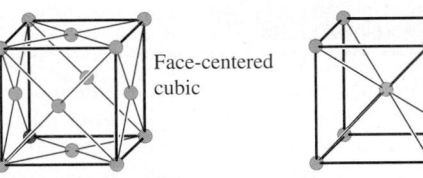
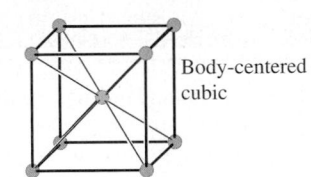

Semiconductors

(Sections 29.5–29.7) A **semiconductor** is a material with an electrical resistivity that is intermediate between those of good conductors and those of good insulators. A **hole** is a vacancy in a bond where there would normally be an electron—the hole acts like a positive charge. In **n-type** semiconductors, the conductivity is due to the motion of electrons; in **p-type,** holes act as the moving charges. A **transistor** can be made by layering two *p-n* junctions—the resulting devices can act as power, current, or voltage amplifiers.

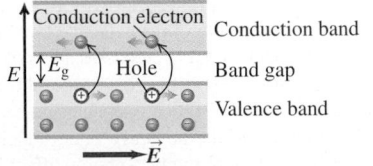

Superconductivity

(Section 29.8) As the temperature decreases in a superconductor, the resistivity at first decreases smoothly. But then at the **critical temperature,** a phase transition occurs and the resistivity suddenly drops to zero. One recently discovered superconductor has a critical temperature of 160 K. When we place a superconductor in a magnetic field, eddy currents are induced that exactly cancel the applied field everywhere inside the material—a magnetic field can never exist inside a type-I superconducting material.

Conceptual Questions

1. In the ground state of the helium atom, the electrons must have opposite spins. Why?
2. What does it mean for a quantity to be *quantized?* Give several examples of quantized quantities.
3. The exclusion principle is sometimes described as the quantum version of the principle that no two objects can be in the same place at the same time. Is this description appropriate?
4. In describing the atom, we often refer to the "electron cloud." Does this mean that the electron is spread out in space in the form of a cloud? Just what is that cloud?
5. In what ratio would you expect Mg and Cl to combine? Why?
6. A student asserted that any filled shell (i.e., all the levels for a given n occupied by electrons) must have zero total angular momentum and hence must be spherically symmetric. Do you believe this? What about a filled *subshell* (all values of m_l for given values of n and l)?
7. The central-field approximation is more accurate for alkali metals than for transition metals such as iron, nickel, or copper. Why do you think this is? (Figure 29.3 may be helpful in answering this question.)
8. The nucleus of a gold atom contains 79 protons. How would you expect the energy required to remove a $1s$ electron from a gold atom to compare with the energy required to remove a $1s$ electron from a hydrogen atom? In what region of the electromagnetic spectrum would a photon of the appropriate energy lie?
9. Elements can be identified by their visible spectra. Could analogous techniques be used to identify compounds from their molecular spectra? In what region of the electromagnetic spectrum would the appropriate radiation lie?
10. The ionization energies of the alkali metals (i.e., the energies required to remove an outer electron from the metal) are in the range from 4 to 6 eV, while those of the inert gases are in the range from 15 to 25 eV. Why the difference?
11. The energy required to remove the $3s$ electron from a sodium atom in its ground state is about 5 eV. Would you expect the energy required to remove an additional electron to be about the same, more, or less? Why?
12. Individual atoms have discrete energy levels, but certain solids (which are made up of only individual atoms) show energy bands and gaps. What causes the solids to behave so differently from the atoms of which they are composed?
13. Increasing the temperature of an ordinary conductor normally makes it a poorer conductor, whereas increasing the temperature of an ordinary nonconductor normally makes it a *better* conductor. Why this difference in behavior? Explain at the atomic level.
14. Why does a diode conduct current in one direction better than in the opposite direction? Why do ordinary resistors *not* behave this way?
15. Why is it advantageous to add impurities to semiconductors? That is, why are they doped?
16. What is the difference between an n-type and a p-type semiconductor?

Multiple-Choice Problems

1. Which statement about an electron in the $5f$ state is correct? (There may be more than one correct choice.)
   A. The magnitude of its orbital angular momentum is $2\sqrt{3}\hbar$.
   B. The largest orbital angular momentum it could have in any given direction is $5\hbar$.
   C. Its magnetic quantum number could have the value ± 3, ± 2, ± 1, or 0.
   D. Its spin angular momentum is $2\sqrt{3}\hbar$.
2. The atom having the electron configuration $1s^2\, 2s^2\, 2p^6\, 3s^2\, 3p^5$ has (there may be more than one correct choice)
   A. 17 orbital electrons.
   B. 11 orbital electrons.
   C. electrons with $l = 0, 1, 2$.
   D. electrons with $m_l = 0$ and ± 1.
3. Which statement about x rays is correct? (There may be more than one correct choice.)
   A. Their wavelengths are greater than those of visible light.
   B. They are emitted when electrons in complex atoms (those having a large atomic number) make transitions from higher shells to the K shell.
   C. They are emitted when electrons in hydrogen make transitions from higher shells to the K shell.
   D. The energy of x-ray photons is greater than the energy of visible-light photons.
4. Table 29.2 shows that for the ground state of the potassium atom the outermost electron is in a $4s$ state. This tells you that, compared to the $3d$ level for this atom, the $4s$ level
   A. has a lower energy.
   B. has a higher energy.
   C. has approximately the same energy.
5. Which statement about semiconductors is correct?
   A. Their resistivity is normally somewhat greater than that of most metals.
   B. A "hole" in a semiconductor is due to the removal of a proton.
   C. Electrical conduction in an n-type semiconductor is due to the transfer of neutrons.
   D. Electrical conduction in a p-type semiconductor is due to the transfer of electrons.
6. An electron in the N shell can have which of the following quantum numbers? (There may be more than one correct choice.)
   A. $s = -\frac{1}{2}$. B. $l = 1$. C. $m_l = 4$.
   D. $n = 4$. E. $m_l = -3$.
7. Bonds that involve the essentially complete transfer of electrons from one atom to another are
   A. van der Waals bonds.
   B. hydrogen bonds.
   C. ionic bonds.
   D. covalent bonds.
8. Consider an atom having electron configuration $1s^2\, 2s^2\, 2p^4$. The atom with the next higher Z and having similar chemical properties would have Z equal to
   A. 9. B. 15. C. 16. D. 34.
9. The number of $5f$ states for hydrogen is
   A. 22. B. 18. C. 14. D. 7.
10. If an electron has the quantum numbers $n = 4$, $l = 2$, $m_l = -1$, $s = \frac{1}{2}$, it is in the state
   A. $2p$.
   B. $2g$.
   C. $4p$.
   D. $4d$.
   E. $4f$.

Problems

29.1 Electrons in Atoms

1. • The orbital angular momentum of an electron has a magnitude of 4.716×10^{-34} kg · m^2/s. What is the angular-momentum quantum number l for this electron?

2. • Consider states with $l = 3$. (a) In units of $\hbar$, what is the largest possible value of L_z? (b) In units of $\hbar$, what is the value of L? Which is larger, L or the maximum possible L_z? (c) Assume a model in which $\vec{L}$ is described as a classical vector. For each allowed value of L_z, what angle does the vector $\vec{L}$ make with the $+z$ axis?

3. •• An electron is in the hydrogen atom with $n = 3$. (a) Find the possible values of L and L_z for this electron, in units of $\hbar$. (b) For each value of L, find all the possible angles between L and the z axis.

4. •• An electron is in the hydrogen atom with $n = 5$. (a) Find the possible values of L and L_z for this electron, in units of $\hbar$. (b) For each value of L, find all the possible angles between L and the z axis. (d) What are the maximum and minimum values of the magnitude of the angle between L and the z axis?

5. •• Consider an electron in the N shell. (a) What is the smallest orbital angular momentum it could have? (b) What is the largest orbital angular momentum it could have? Express your answers in terms of $\hbar$ and in SI units. (c) What is the largest orbital angular momentum this electron could have in any chosen direction? Express your answers in terms of $\hbar$ and in SI units. (d) What is the largest spin angular momentum this electron could have in any chosen direction? Express your answers in terms of $\hbar$ and in SI units. (e) For the electron in part (c), what is the ratio of its spin angular momentum in the z direction to its orbital angular momentum in the z direction?

6. • (a) How many different $3d$ states does hydrogen have? Make a list showing all of them. (b) How many different $3f$ states does it have?

7. •• (a) How many different $5g$ states does hydrogen have? (b) Which of the states in part (a) has the largest angle between $\vec{L}$ and the z axis and what is that angle? (c) Which of the states in part (a) has the smallest angle between $\vec{L}$ and the z axis, and what is that angle?

8. •• In a particular state of the hydrogen atom, the angle between the angular momentum vector $\vec{L}$ and the z axis is $\theta = 26.6°$. (See Figure 29.2.) If this is the smallest angle for this particular value of the angular momentum quantum number l, what is l?

29.2 Atomic Structure

9. • Make a list of the four quantum numbers n, l, m_l, and s for each of the 12 electrons in the ground state of the magnesium atom.

10. • (a) List the different possible combinations of quantum numbers l and m_l for the $n = 5$ shell. (b) How many electrons can be placed in the $n = 5$ shell?

11. • For bromine $(Z = 35)$, make a list of the number of electrons in each subshell $(1s, 2s, 2p, \text{etc.})$.

12. • (a) Write out the electron configuration $(1s^2\, 2s^2, \text{etc.})$ for Li and Na. (b) How many electrons does each of these atoms have in its outer shell?

13. •• (a) Write out the ground-state electron configuration $(1s^2, 2s^2, \text{etc.})$ for the carbon atom. (b) What element of next-larger Z has chemical properties similar to those of carbon? (See Example 29.3.) Give the ground-state electron configuration for this element.

14. •• (a) Write out the ground-state electron configuration $(1s^2\, 2s^2, \text{etc.})$ for the beryllium atom. (b) What element of next-larger Z has chemical properties similar to those of beryllium? (See Example 29.3.) Give the ground-state electron configuration of this element. (c) Use the procedure of part (b) to predict what element of next-larger Z than in (b) will have chemical properties similar to those of the element you found in part (b), and give its ground-state electron configuration.

15. •• Write out the electron configuration $(1s^2\, 2s^2, \text{etc.})$ for Ne, Ar, and Kr. (b) How many electrons does each of these atoms have in its outer shell? (c) Predict the chemical behavior of these three atoms. Explain your reasoning.

16. •• Calculate, in units of $\hbar$, the magnitude of the maximum orbital angular momentum for an electron in a hydrogen atom for states with a principal quantum number of 2, 20, and 200. Compare each with the value of $n\hbar$ postulated in the Bohr model. What trend do you see?

17. • (a) What is the orbital angular momentum of any s-subshell electron? (b) If we try to model the atom classically as a scaled-down version of a solar system, with the electrons orbiting the nucleus the way the planets orbit the sun, what does the result in part (a) tell us would be the speed of an s-subshell electron? Is this result physically possible? What would happen to an electron with that speed?

18. •• The energies for an electron in the K, L, and M shells of the tungsten atom are $-69,500$ eV, $-12,000$ eV, and -2200 eV, respectively. Calculate the wavelengths of the K_α and K_β x rays of tungsten.

29.3 Diatomic Molecules

19. • If the energy of the H_2 covalent bond is -4.48 eV, what wavelength of light is needed to break that molecule apart? In what part of the electromagnetic spectrum does this light lie?

20. • (a) A molecule decreases its vibrational energy by 0.250 eV by giving up a photon of light. What wavelength of light does it give up during this process, and in what part of the electromagnetic spectrum does that wavelength of light lie? (b) An atom decreases its energy by 8.50 eV by giving up a photon of light. What wavelength of light does it give up during this process, and in what part of the electromagnetic spectrum does that wavelength of light lie? (c) A molecule decreases its rotational energy by 3.20×10^{-3} eV by giving up a photon of light. What wavelength of light does it give up during this process, and in what part of the electromagnetic spectrum does that wavelength of light lie?

21. •• **An ionic bond.** (a) Calculate the electric potential energy for a K^+ ion and a Br^- ion separated by a distance of 0.29 nm, the equilibrium separation in the KBr molecule. Treat the ions as point charges. (b) The ionization energy of the potassium atom is 4.3 eV. Atomic bromine has an electron affinity of 3.5 eV. Use these data and the results of part (a) to estimate the binding energy of the KBr molecule. Do you expect the actual binding energy to be higher or lower than your estimate? Explain your reasoning.

29.4 Structure and Properties of Solids

22. •• The spacing of adjacent atoms in a NaCl crystal is 0.282 nm, and the masses of the atoms are 3.82×10^{-26} kg (Na) and

5.89×10^{-26} kg (Cl). Use this information to calculate the density of sodium chloride.

23. •• Potassium bromide (KBr) has a density of 2.75×10^3 kg/m^3 and the same crystal structure as NaCl. The mass of potassium is 6.49×10^{-26} kg, and that of bromine is 1.33×10^{-25} kg. (a) Calculate the average spacing between adjacent atoms in a KBr crystal. (b) Compare the spacing for KBr with the spacing for NaCl. (See previous problem.) Is the relation between these two values qualitatively what you would expect? Explain your reasoning.

29.5 Energy Bands
29.6 Semiconductors
29.7 Semiconductor Devices

24. • Look at the graph for a diode in Figure 29.23(b) in the text. Why does this graph go below the horizontal axis to the left of the origin? Would it do this if the diode were replaced by an ordinary resistor?

25. •• The gap between valence and conduction bands in diamond is 5.47 eV. (a) What is the maximum wavelength of a photon that can excite an electron from the top of the valence band into the conduction band? In what region of the electromagnetic spectrum does this photon lie? (b) Explain why pure diamond is transparent and colorless. (*Hint:* Will photons of visible light that strike a diamond be absorbed or transmitted?) (c) Most gem diamonds have a yellow color. Explain how impurities in the diamond can cause this color.

26. •• A variable dc power supply having reversible polarity is connected to a diode having a p–n junction as shown in Figure 29.28. Starting with the power supply's polarity as shown in the figure, its potential is gradually decreased to zero and then gradually increased in the reverse direction. (a) Sketch a graph of the reading in the ammeter as a function of the potential difference V across the power supply. Make sign differences clear. (b) Suppose now that the terminals of the diode are reversed and the same procedure is followed. Sketch a graph of the reading in the ammeter as a function of the potential difference V across the power supply. Make sign differences clear. (c) Suppose now that the diode is replaced by an ordinary resistor and the same procedure is followed. Sketch a graph of the reading in the ammeter as a function of the potential difference V across the power supply. Make sign differences clear. (d) Explain the reasons for the differences between the graphs in (i) parts (a) and (b), (ii) parts (a) and (c).

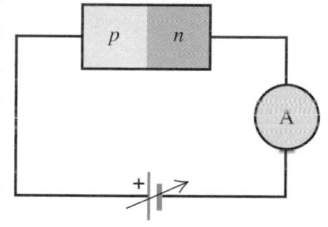

▲ **FIGURE 29.28** Problem 26.

27. •• The gap between valence and conduction bands in silicon is 1.12 eV. A nickel nucleus in an excited state emits a gamma-ray photon with wavelength 9.31×10^{-4} nm. How many electrons can be excited from the top of the valence band to the bottom of the conduction band by the absorption of this gamma ray?

29.8 Superconductivity

28. • Sketch a qualitative (no numbers) graph of the resistance as a function of temperature for (a) an ordinary conductor, such as Cu, including temperatures approaching 0 K; (b) a superconductor; include temperatures above and below the critical temperature, and let the temperature approach 0 K.

General Problems

29. •• For magnesium, the first ionization potential is 7.6 eV; the second (the additional energy required to remove a second electron) is almost twice this, 15 eV, and the third ionization potential is much larger, about 80 eV. Why do these numbers keep increasing?

30. •• **Failure of the classical model.** (a) What is the spin angular momentum of an electron along the z direction, in SI units? (b) In the rest of this problem, we shall try to understand the behavior of an electron by modeling it as a classical ball of matter. We shall also use the nonrelativistic formulas, although a thorough treatment would require use of the more complicated relativistic ones. If we model this electron classically as a solid uniform sphere of diameter 1.0×10^{-15} m, what would be its angular velocity in rad/s? (c) What would be the speed of the surface of this classical sphere? Is there anything suspicious about this result? (d) In part (c), the result cannot be correct because the surface of the electron cannot move faster than the speed of light. Suppose we try to "correct" this model by assuming that the surface is moving at the speed of light— the upper limit of its speed. What would have to be the diameter of an electron in that case? Given that atomic nuclei are about 10×10^{-15} m in diameter, does this result seem plausible for the size of an electron? (e) Would the bizarre results in parts (c) and (d) be affected appreciably if we instead modeled the electron as a hollow spherical shell instead of a solid? Why? Notice that we encounter results that contradict observation when we try to think of the electron simply as a spinning sphere. We simply cannot understand the behavior of an electron (and other subatomic particles) by using our ordinary classical models.

31. •• An electron has spin angular momentum and orbital angular momentum. For the 3d electron in scandium, what percent of its total orbital angular momentum is its spin angular momentum in the z direction?

32. •• The dissociation energy of the hydrogen molecule (i.e., the energy required to separate the two atoms) is 4.48 eV. In the gas phase (treated as an ideal gas), at what temperature is the average translational kinetic energy of a molecule equal to this energy?

33. •• The maximum wavelength of light that a certain silicon photocell can detect is 1.11 μm. (a) What is the energy gap (in electron volts) between the valence and conduction bands for this photocell? (b) Explain why pure silicon is opaque. (*Hint:* Will visible light that strikes silicon be transmitted or absorbed?)

34. •• Use the electron configurations of He, Ne, and Ar to explain why these atoms normally do not combine chemically with other atoms.

35. •• Use the electron configurations of H and O to explain why these atoms combine chemically in a two-to-one ratio to form water.

36. •• Use the electron configurations of Si and O to explain why these atoms combine chemically in a one-to-two ratio to form sand.

37. •• Consider an electron in hydrogen having total energy −0.5440 eV. (a) What are the possible values of its orbital angular momentum (in terms of $\hbar$)? (b) What wavelength of light would it take to excite this electron to the next higher shell? Is this photon visible to humans?

38. ●● The energy of the van der Waals bond, which is responsible for a number of the characteristics of water, is about 0.50 eV. (a) At what temperature would the average translational kinetic energy of water molecules be equal to this energy? (b) At that temperature, would water be liquid or gas? Under ordinary everyday conditions, do van der Waals forces play a role in the behavior of water?

39. ●● (a) What is the lowest possible energy (in electron volts) of an electron in hydrogen if its orbital angular momentum is $\sqrt{12}\hbar$? (b) What are the largest and smallest values of the z component of the orbital angular momentum (in terms of $\hbar$) for the electron in part (a)? (c) What are the largest and smallest values of the spin angular momentum (in terms of $\hbar$) for the electron in part (a)? (d) What are the largest and smallest values of the orbital angular momentum (in terms of $\hbar$) for an electron in the M shell of hydrogen?

40. ●● An electron in hydrogen is in the $5f$ state. (a) Find the largest possible value of the z component of its angular momentum. (b) Show that for the electron in part (a), the corresponding x and y components of its angular momentum satisfy the equation $\sqrt{L_x^2 + L_y^2} = \hbar\sqrt{3}$.

30 Nuclear and High-Energy Physics

During the 20th century, nuclear physics has had enormous effects on humankind, many beneficial, some catastrophic. Many people have strong opinions about the uses of nuclear physics. Ideally, these opinions should be based on understanding rather than on prejudice or emotion, and we hope that this chapter will help you reach that understanding.

Every atom contains at its center an extremely dense, positively charged *nucleus*, much smaller than the overall size of the atom, but containing most of its mass. In this chapter, we'll look first at several important properties of nuclei and of the interactions that hold them together. The stability or instability of a particular nucleus is determined by the competition between the attractive nuclear forces among the protons and neutrons and the repulsive electrical interactions among the protons. Unstable nuclei decay by a variety of processes, transforming themselves spontaneously into other structures. Structure-altering nuclear reactions can also be induced by collision of a nucleus with a particle or another nucleus. Two classes of reactions of special interest are fission and fusion. Research into the nature of, and interactions among, fundamental particles has required the construction of large experimental facilities, such as particle accelerators, as scientists probe more deeply into this most fundamental aspect of the nature of our physical universe.

This bog mummy was found in a peat deposit in Denmark. Radiocarbon dating, performed on a sample of his liver, indicates that he died in the late third century BCE. His body was preserved by burial in a peat bog, which also turned his dark hair red.

30.1 Properties of Nuclei

Nuclei are composed of protons and neutrons (collectively referred to as *nucleons*). These two particles have nearly equal masses (within about 0.1%), about 1.67×10^{-27} kg. They aren't fundamental particles, but are composed of more

▲ Application **Where's the nucleus?** If a hydrogen atom were the size of this baseball stadium, how large do you think the nucleus would be? The answer: Only the size of a pencil eraser in the center. Therefore, textbook drawings of atoms that show the nucleus are far out of scale. If the nucleus were the size of a baseball, the electron cloud would be about 20 km across. In fact, because so much of the atom is empty space, it has been estimated that, without this space, all the solid matter in the human body would fit on the head of a pin!

basic entities called *quarks*. We'll begin our discussion by describing the behavior of nuclei in terms of protons and neutrons; later we'll return to their relation to quarks.

The total number of protons in a nucleus is the **atomic number** Z. In an electrically neutral atom, this is equal to the number of electrons. Hence, Z determines the chemical properties of each element. The number of neutrons, denoted by N, is called the **neutron number.** The total number of protons and neutrons is called the **nucleon number,** denoted by A:

$$A = Z + N. \tag{30.1}$$

The mass of a nucleus is approximately proportional to the total number of nucleons, so A is also (and more commonly) called the **mass number.**

The most obvious feature of the atomic nucleus is its size: The nucleus is 20,000 to 200,000 times smaller than the atom itself. We can picture a nucleus as roughly spherical in shape. The radius depends on the mass, which in turn depends primarily on the total number A of nucleons. The radii of most nuclei are represented fairly well by an empirical equation:

Radii of nuclei

The radii of nuclei are given approximately by the formula

$$R = R_0 A^{1/3}, \tag{30.2}$$

where R_0 is an empirical constant:

$$R_0 = 1.2 \times 10^{-15} \text{ m} = 1.2 \text{ fm}.$$

The volume V of a sphere is equal to $4\pi R^3/3$, so Equation 30.2 shows that the *volume* of a nucleus is proportional to its mass number A and thus to its total mass. It follows that the *density* (mass per unit volume, proportional to A/R^3) is the same for all nuclei. That is, **all nuclei have approximately the same density.** This fact is of crucial importance in understanding nuclear structure.

EXAMPLE 30.1   Nuclear density

The most common variety of iron nucleus has $A = 56$. Find the radius, approximate mass, and approximate density of the iron nucleus.

SOLUTION

SET UP The mass number A and the radius R are related by Equation 30.2. The proton and neutron masses are approximately equal, about 1.67×10^{-27} kg each. The total mass M of the nucleus is A times the mass m of a single proton or neutron.

SOLVE From Equation 30.2, the radius R of the nucleus is given by

$$R = R_0 A^{1/3} = (1.2 \times 10^{-15} \text{ m})(56^{1/3}) = 4.6 \times 10^{-15} \text{ m}.$$

The total mass M (neglecting the small difference in mass between protons and neutrons) is

$$M = (56)(1.67 \times 10^{-27} \text{ kg}) = 9.4 \times 10^{-26} \text{ kg}.$$

The volume V is

$$V = \tfrac{4}{3}\pi R^3 = \tfrac{4}{3}\pi (4.6 \times 10^{-15} \text{ m})^3 = 4.1 \times 10^{-43} \text{ m}^3,$$

and the density ρ is

$$\rho = \frac{M}{V} = \frac{9.4 \times 10^{-26} \text{ kg}}{4.1 \times 10^{-43} \text{ m}^3} = 2.3 \times 10^{17} \text{ kg/m}^3.$$

REFLECT The radius of the iron nucleus is about 4.6×10^{-15} m, on the order of 10^{-5} of the overall radius of an atom of iron. The density of the element iron (in the solid state) is about 8000 kg/m³ (8 g/cm³), so we see that the nucleus is on the order of 10^{13} times as dense as the bulk material. Densities of this magnitude are also found in white dwarf stars, which are similar to gigantic nuclei. A 1 cm cube of material with this density would have a mass of 2.3×10^{11} kg, or 230 million metric tons!

Practice Problem: Find the approximate mass, radius, and density of the uranium nucleus with $A = 238$, and compare the results with the values for iron. Assume that the total mass is 238 times the mass of a single neutron, neglecting binding-energy corrections. *Answer:* 4.0×10^{-25} kg, 7.4×10^{-15} m, 2.3×10^{17} kg/m³.

Nuclear Masses

A single nuclear species having specific values of both Z and N is called a **nuclide.** Table 30.1 lists values of A, Z, and N for several nuclides. An atom's electron structure, which determines its chemical properties, is determined by the charge Ze of the nucleus. The table shows some nuclei with the same Z but different N. These are nuclei of the same element, but they have different masses. Nuclei of a given element with different mass numbers are called **isotopes** of the element. An example that is abundant in nature is chlorine $(Cl, Z = 17)$. About 76% of chlorine nuclei have $N = 18$; the other 24% have $N = 20$. Because of the differing nuclear masses, different isotopes of an element have slightly different physical properties, such as melting and boiling temperatures and diffusion rates. The two common isotopes of uranium, with $A = 235$ and $A = 238$, are separated industrially by means of centrifuges or by taking advantage of the different diffusion rates of gaseous uranium hexafluoride (UF_6) containing the two isotopes.

Table 30.1 also shows the usual notation for individual nuclides: the symbol of the element, with a pre-subscript equal to the atomic number Z and a pre-superscript equal to the mass number A. For example, $^{13}_{6}C$ denotes the isotope of carbon with $Z = 6$, $A = 13$, and $N = 7$. The general format for an element El is $^{A}_{Z}El$. The isotopes of chlorine, with $A = 35$ and $A = 37$, are written as $^{35}_{17}Cl$ and $^{37}_{17}Cl$, respectively. This notation is redundant, because the name of the element determines the atomic number Z, but it's a useful aid to memory. The pre-subscript (the value of Z) is sometimes omitted, as in ^{35}Cl.

The proton and neutron masses are, respectively,

$$m_p = 1.67262171(29) \times 10^{-27} \text{ kg},$$
$$m_n = 1.67492728(29) \times 10^{-27} \text{ kg}.$$

(The two digits in parentheses indicate the uncertainty in the last two digits of each value.) These values are nearly equal, so it is not surprising that many nuclear masses are approximately integer multiples of the proton or neutron mass. For precise measurements, it's useful to define a new mass unit equal to

TABLE 30.1   Compositions of some common nuclei

Nucleus	Mass number (total number of nuclear particles), A	Atomic number (number of protons), Z	Neutron number, $N = Z - A$
$^{1}_{1}H$	1	1	0
$^{2}_{1}D$	2	1	1
$^{4}_{2}He$	4	2	2
$^{6}_{3}Li$	6	3	3
$^{7}_{3}Li$	7	3	4
$^{9}_{4}Be$	9	4	5
$^{10}_{5}B$	10	5	5
$^{11}_{5}B$	11	5	6
$^{12}_{6}C$	12	6	6
$^{13}_{6}C$	13	6	7
$^{14}_{7}N$	14	7	7
$^{16}_{8}O$	16	8	8
$^{23}_{11}Na$	23	11	12
$^{65}_{29}Cu$	65	29	36
$^{200}_{80}Hg$	200	80	120
$^{235}_{92}U$	235	92	143
$^{238}_{92}U$	238	92	146

one-twelfth of the mass of the neutral carbon atom with mass number $A = 12$. This mass is called one **unified atomic mass unit** $(1\,\text{u})$:

$$1\,\text{u} = 1.66053886(28) \times 10^{-27}\,\text{kg}.$$

In unified atomic mass units, the masses of the proton, neutron, and electron are

$$m_\text{p} = 1.00727646688(13)\,\text{u},$$
$$m_\text{n} = 1.00866491560(55)\,\text{u},$$
$$m_\text{e} = 0.00054857990945(24)\,\text{u},$$
$$m_\text{H} = m_\text{p} + m_\text{e} = 1.0078250\,\text{u}.$$

These values are more precise than the u-to-kg conversion factor because atomic masses can be compared with each other with greater precision than they can be compared with the standard kilogram. Also note that 1 mole (see Section 15.1) of particles with a mass of 1 u each would have a total mass of exactly 1 gram.

We can find the energy equivalent of 1 u from the relation $E = mc^2$:

$$E = (1.66054 \times 10^{-27}\,\text{kg})(2.99792 \times 10^8\,\text{m/s})^2$$
$$= 1.49242 \times 10^{-10}\,\text{J} = 931.494\,\text{MeV}.$$

The masses of some common atoms, including their electrons, are shown in Table 30.2. Such tables always give masses of *neutral* atoms (including Z electrons), rather than masses of bare nuclei (atoms with all their electrons removed), because it is much more difficult to measure masses of bare nuclei with high precision. To obtain the mass of a bare nucleus, subtract Z times the electron mass from the atomic mass.

The total mass of a nucleus is always *less* than the total mass of its constituent parts because of the mass equivalent $(E = mc^2)$ of the internal kinetic energy and of the (negative) potential energy associated with the attractive forces that hold the nucleus together. This mass difference is called the **mass defect,** denoted by ΔM. The magnitude of the total potential energy is called the **binding energy,** denoted by E_B. Thus, $E_\text{B} = (\Delta M)c^2$.

PhET: Simplified MRI
ActivPhysics 19.2: Nuclear Binding Energy

TABLE 30.2 Atomic masses of light elements

Element	Atomic number, Z	Neutron number, N	Atomic mass, u	Mass number, A
Hydrogen, H	1	0	1.007825	1
Deuterium, H	1	1	2.014101	2
Helium, He	2	1	3.016029	3
Helium, He	2	2	4.002603	4
Lithium, Li	3	3	6.015123	6
Lithium, Li	3	4	7.016005	7
Beryllium, Be	4	5	9.012182	9
Boron, B	5	5	10.012937	10
Boron, B	5	6	11.009305	11
Carbon, C	6	6	12.000000	12
Carbon, C	6	7	13.003355	13
Nitrogen, N	7	7	14.003074	14
Nitrogen, N	7	8	15.000109	15
Oxygen, O	8	8	15.994915	16
Oxygen, O	8	9	16.999132	17
Oxygen, O	8	10	17.999161	18

Source: Atomic Mass Evaluation 2003, Nuclear Physics A 729 (2003).

Mass defect

The mass defect ΔM for a nucleus with mass M containing Z protons and N neutrons is defined as

$$\Delta M = Zm_p + Nm_n - M. \qquad (30.3)$$

(The term *mass defect* is somewhat misleading. This quantity doesn't represent a defect or discrepancy in our calculations; it is simply the mass difference defined in Equation 30.3.)

The simplest nucleus, that of hydrogen, is a single proton. Next comes the nucleus of ^2_1H, the isotope of hydrogen with mass number 2, usually called *deuterium*. Its nucleus consists of a proton and a neutron bound together to form a particle called the *deuteron*. To find the binding energy of the deuteron, we calculate the mass defect ΔM, using the values in Table 30.2:

$$\Delta M = m_H + m_n - m_D$$
$$= 1.007825 \text{ u} + 1.008665 \text{ u} - 2.014101 \text{ u} = 0.00239 \text{ u}.$$

(Note that m_H and m_D each include one electron, so in the calculations of ΔM, these atomic masses can be used in place of nuclear masses; the electron masses cancel out.) The energy equivalent is

$$E_B = (0.00239 \text{ u})(931.5 \text{ MeV/u}) = 2.23 \text{ MeV}.$$

This is the amount of energy required to pull the deuteron apart into a proton and a neutron; we call it the *binding energy* of the deuteron. This value is unusually small; binding energies for most nuclei are about 8 MeV per nucleon.

PROBLEM-SOLVING STRATEGY 30.1   Nuclear structure

1. As in Chapters 28 and 29, familiarity with numerical magnitudes is helpful. The scale of things in nuclear structures is very different from that of atomic structures. The size of a nucleus is on the order of 10^{-15} m; the potential energy of interaction between two protons at this distance is 2.31×10^{-13} J, or 1.44 MeV. Typical nuclear interaction energies are on the order of a few million electronvolts (MeV), rather than a few electronvolts (eV) as with atoms. Protons and neutrons are about 1840 times as massive as electrons. The binding energy per nucleon is roughly 1% of the rest energy of a nucleon; compare this percentage with that of the ionization energy of the hydrogen atom, which is only 0.003% of the electron rest energy. By contrast, angular momentum is of the same order of magnitude in both atoms and nuclei, because it is determined by the value of Planck's constant. Compare the results of your calculations with these ranges of magnitude.

2. When you're doing energy calculations that involve the mass defect, binding energies, and so on, note that mass tables nearly always list the masses of *neutral* atoms, with their full complements of electrons. If you need the mass of a bare nucleus, you have to subtract the masses of these electrons. The binding energies of the electrons are much smaller than nuclear binding energies, and we won't worry about them. Calculations of binding energies often involve subtracting two nearly equal quantities. To get enough precision in the difference, you often have to carry five or six significant figures, if that many are available. If not, you may have to be content with an approximate result.

3. Nuclear masses are usually measured in atomic mass units (u). A useful conversion factor is the energy equivalence 1 u = 931.5 MeV.

EXAMPLE 30.2 **Mass defect for carbon**

Find the mass defect, the total binding energy, and the binding energy per nucleon for the common isotope of carbon, $^{12}_{6}C$.

SOLUTION

SET UP The neutral carbon atom consists of six protons, six neutrons, and six electrons. The total mass of six protons and six electrons is equal to the mass of six hydrogen atoms. Therefore, the total mass of the separate parts is $6m_H + 6m_n$, so the mass defect ΔM is given by $\Delta M = 6m_H + 6m_n - m_C$. We find the masses from Table 30.2 and the preceding discussion.

SOLVE We substitute numerical values in this expression for ΔM, obtaining

$$\Delta M = 6(1.007825 \text{ u}) + 6(1.0086649 \text{ u}) - 12.000000 \text{ u}$$
$$= 0.09894 \text{ u}.$$

The energy equivalent of this mass is

$$(0.09894 \text{ u})(931.5 \text{ MeV/u}) = 92.16 \text{ MeV}.$$

Thus, the total binding energy for the 12 nucleons is 92.16 MeV, and the binding energy per nucleon is 92.16 MeV/12 = 7.68 MeV per nucleon.

Alternate Solution: We could first find the mass of the *bare* carbon nucleus by subtracting the mass of six electrons from the mass of the neutral atom (including six electrons), given in Table 30.2 as 12.00000 u. Then we find the total mass of six protons and six neutrons. The mass defect is that total minus the mass of the bare nucleus. This method is more complicated than the first solution because the electron masses have to be subtracted explicitly. We don't recommend this alternate solution method.

REFLECT To pull the carbon nucleus completely apart into 12 separate nucleons would require a minimum of 92.16 MeV. The binding energy *per nucleon* is one-twelfth of this, or 7.68 MeV per nucleon. Nearly all stable nuclei, from the lightest to the most massive, have binding energies in the range from 6 to 9 MeV per nucleon, for reasons we'll discuss in the next section.

Practice Problem: Find the mass defect, the total binding energy, and the binding energy per nucleon for the common isotope of helium, ^{4}He. *Answer:* 0.03038 u, 28.30 MeV, 7.075 MeV.

In our discussion of masses, mass defects, and binding energies of nuclei, we have always assumed (without actually saying so) that the nuclei are in their lowest energy state, or ground state. But just as with atoms and molecules, nuclei have internal motion and sets of allowed energy levels, including a ground state (state of lowest energy) and excited states. Because of the great strength of nuclear interactions, excitation energies of nuclei are typically on the order of 1 MeV, compared with a few eV for atomic energy levels. In ordinary physical and chemical transformations, the nucleus always remains in its ground state. When a nucleus is placed in an excited state, either by bombardment with high-energy particles or by a radioactive transformation, it can decay to the ground state by the emission of one or more photons, called in this case gamma rays or gamma-ray photons, with typical energies of a few tenths MeV to a few MeV.

In Section 29.1, we discussed the orbital and spin angular momentum of electrons in atoms. The proton and neutron are also spin-$1/2$ particles; each has a spin angular momentum that can have a component $\pm\frac{1}{2}\hbar$ in any given axis direction. In addition, each particle can have orbital angular momentum due to motion within the nucleus, with possible components in a given axis direction that are integer multiples of $\hbar$. The total angular momentum of a nucleus is usually called **nuclear spin,** although in general it is associated with both orbital and spin angular momentum of the protons and neutrons.

30.2 Nuclear Stability

Nearly all stable nuclei, from the lightest to the most massive, have binding energies in the range of 6 to 9 MeV per nucleon (Figure 30.1). The forces that hold protons and neutrons together in the nucleus, despite the electrical repulsion of the protons, are an example of the strong interactions we mentioned in Section 5.5.

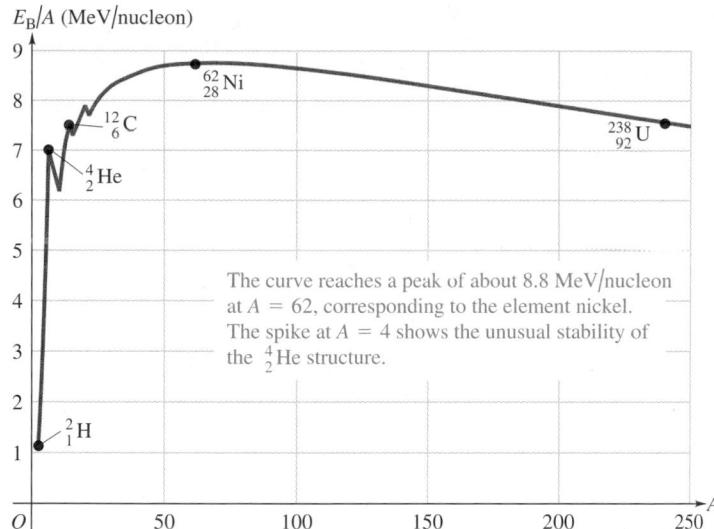

▲ FIGURE 30.1 Approximate binding energy per nucleon as a function of mass number A (the total number of nucleons) for stable nuclides.

In the present context, this kind of interaction is called the **nuclear force.** Here are its most important characteristics:

1. The nuclear force does not depend on charge; both neutrons and protons are bound, and the binding is the same for both.
2. The nuclear force has a short range, on the order of nuclear dimensions, 10^{-15} m. (Otherwise, the nucleus would pull in additional protons and neutrons.) But within its range, the nuclear force is much stronger than electrical forces; otherwise, the nucleus could never be stable because of the mutual repulsion of the protons.
3. The nearly constant density of nuclear matter and the nearly constant binding energy per nucleon show that a particular nucleon cannot interact simultaneously with *all* the other nucleons in a nucleus, but only with those few in its immediate vicinity. This property is different from its counterpart in electrical forces: Every proton in the nucleus repels every other one. The limited number of interactions of which nucleons are capable is called *saturation;* it is analogous in some respects to covalent bonding in molecules. (For example, a carbon atom can form at most four covalent bonds with other atoms.)
4. The nuclear force favors the binding of *pairs* of protons or neutrons with opposite spins and of pairs of pairs (i.e., a pair of protons and a pair of neutrons), with each pair having a total spin of zero. Thus, the alpha particle (two protons and two neutrons) is an exceptionally stable nucleus.

These qualitative features of the nuclear force are very helpful in understanding why some nuclear structures are stable and others are unstable. Of about 2500 different nuclides now known, only about 300 are stable. The others are **radioactive**—unstable structures formed during the early history of the universe or from the decay of other unstable nuclei. They decay to form other nuclides by emitting particles and electromagnetic radiation. The time scale of these decay processes ranges from a small fraction of a microsecond to billions of years.

The stable nuclides are plotted on the graph in Figure 30.2, where the neutron number N and proton number (or atomic number) Z for each nuclide are shown. (This chart is called a *Segrè chart* after its inventor, Emilio Segrè.) Each light blue line perpendicular to the line $N = Z$ represents a specific value of the mass number $A = Z + N$. Most lines of constant A pass through only one or two stable nuclides. This means that there are usually only one or two stable nuclides

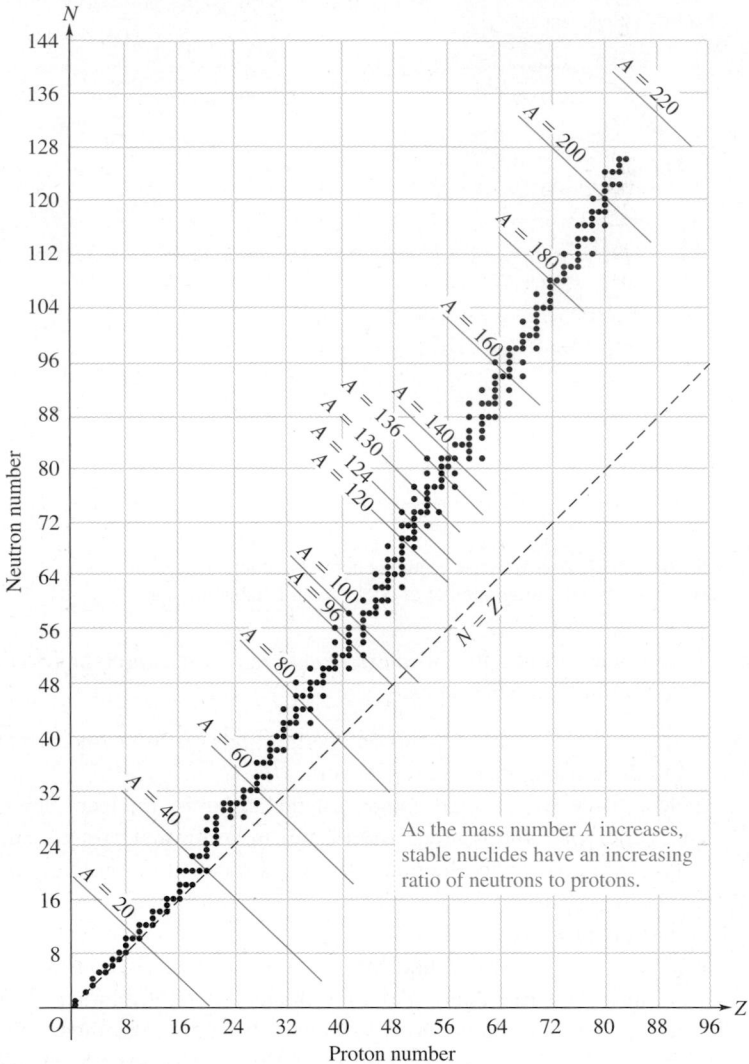

▲ **FIGURE 30.2** Segrè chart showing neutron number and proton number for stable nuclides.

with a given mass number. The lines at $A = 20$, $A = 40$, $A = 60$, and $A = 80$ are examples. In four cases, these lines pass through *three* stable nuclides, namely, at $A = 96$, 124, 130, and 136. Only four stable nuclides have both odd Z and odd N:

$$^2_1\text{H}, \qquad ^6_3\text{Li}, \qquad ^{10}_5\text{B}, \qquad ^{14}_7\text{N};$$

these are called odd–odd nuclides. Also, there is no stable nuclide with $A = 5$ or $A = 8$. The ^4He nucleus (alpha particle), with a pair of protons and a pair of neutrons, does not readily accept a fifth particle into its structure, and a collection of eight nucleons splits rapidly into two ^4He nuclei.

The points representing stable nuclides define a rather narrow region of stability. For small mass numbers, the numbers of protons and neutrons are approximately equal: $N = Z$. The ratio N/Z increases gradually with A, up to about 1.6 at large mass numbers. The graph shows that no nuclide with A greater than 209 or Z greater than 83 is stable. Note also that there is *no* stable nuclide with $Z = 43$ (technetium) or 61 (promethium).

The stability or instability of a nucleus is determined primarily by the competition between the attractive nuclear force and the repulsive electrical force. As we've mentioned, the nuclear force favors pairs of nucleons and pairs of pairs. If there were no electrical interactions, the most stable nuclei would be those with equal numbers of neutrons and protons, $N = Z$. The electrical repulsion shifts the balance to favor a greater numbers of neutrons, but a nucleus with too many neutrons is unstable because not enough of them are paired with protons. A nucleus with too many protons has too much repulsive electrical interaction, compared with the attractive nuclear interaction, to be stable.

As A increases, the positive electric energy per nucleon grows faster than the negative nuclear energy per nucleon, because of the saturation effects with the (attractive) nuclear force that aren't present with the (repulsive) electrical force. At sufficiently large A, a point is reached at which stability is impossible. Thus, the competition between electric and nuclear forces accounts for the fact that the neutron–proton ratio in stable nuclei increases with Z and also for the fact that a nucleus cannot be stable if either A or Z is too large.

Conceptual Analysis 30.1

The force between nucleons

Which, if any, of the following statements concerning the nuclear force are false?

A. The nuclear force is a very short-range force.
B. The nuclear force is attractive.
C. The nuclear force acts on both protons and neutrons.

SOLUTION All of the preceding statements about the nuclear force are correct.

30.3 Radioactivity

As mentioned earlier, of the 2500 known nuclides, only about 300 are stable. The remaining ones decay by emitting particles and transforming themselves into other nuclides, which also may be unstable. The spontaneous disintegration of nuclides that don't meet the stability requirements we discussed in the preceding section is called **radioactivity.** A nucleus is unstable if it is too big $(Z > 83$ or $A > 209)$ or if its neutron–proton ratio is wrong.

Alpha Decay

The two most common decay modes are the emission of alpha (α) particles and the emission of beta (β) particles. When a nucleus disintegrates and emits an alpha particle, it is said to undergo **alpha decay.** An **alpha particle** is identical to the ^{4}He nucleus: two protons and two neutrons bound together, with zero total spin. Alpha emission occurs principally with nuclei that are too large to be stable. When a nucleus emits an α particle, its N and Z values each decrease by two, and A decreases by four. Figure 30.2 shows that the resulting nucleus is closer to stable territory on the Segrè chart than the original nucleus is.

A familiar example of an alpha emitter is radium, ^{226}Ra (Figure 30.3). The speed of the emitted α particle can be determined from the curvature of its trajectory when it passes through a magnetic field. This speed turns out to be about 1.5×10^7 m/s. Although large, it is only 5% of the speed of light, so we can use the nonrelativistic kinetic-energy expression. The α particle mass is 6.64×10^{-27} kg, so the kinetic energy K is

$$K = \tfrac{1}{2}Mv^2 = \tfrac{1}{2}(6.64 \times 10^{-27} \text{ kg})(1.5 \times 10^7 \text{ m/s})^2$$
$$= 7.5 \times 10^{-13} \text{ J} = 4.7 \times 10^6 \text{ eV} = 4.7 \text{ MeV}.$$

Alpha particles are always emitted with a definite kinetic energy, determined by conservation of momentum and energy. They can travel several centimeters in air or a few tenths or hundredths of a millimeter through solids before they are brought to rest by collisions.

MasteringPHYSICS

PhET: Alpha Decay
ActivPhysics 19.4: Radioactivity

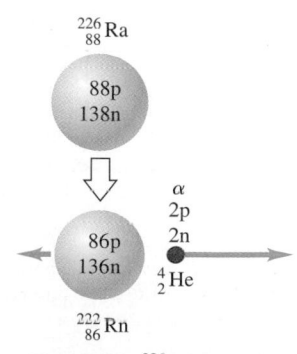

The nuclide $^{226}_{88}$Ra decays by alpha emission to $^{222}_{86}$Rn.

▲ **FIGURE 30.3** Radium decay.

▲ BIO Application Radiation protection.
Depending on their charge and kinetic energy, different types of radiation penetrate matter to different degrees. Alpha particles are electrically charged, and because of their mass they are ejected from radioactive nuclei at relatively low speeds. Therefore, it takes only a little matter to stop them—a sheet of paper, ordinary clothing, or the outer layer of your skin. These clothes are ample protection against an external alpha emitter. However, if an alpha emitter (such as radon gas) decays within the body, it can do serious damage to nearby cells. Beta particles (electrons) leave the nucleus at high speeds and thus have more penetrating power than alpha particles. They can be stopped by heavy clothing and penetrate only a few millimeters through body tissues. ^{131}I, a beta emitter, is used to treat thyroid cancer. The isotope is taken up by the cancerous tissue and kills it. Gamma rays—being uncharged, energetic photons— require heavy shielding, such as a lead shield or concrete wall.

Beta Decay

Early in the 20th century, it was established that the β particle is an *electron*. You might well ask how a nucleus can emit an electron if there aren't any electrons in the nucleus. The emission of a β particle involves the transformation of a neutron in the nucleus into a proton and an electron. We'll discuss such transformations in more detail later in this chapter. Even a free neutron decays into a proton and an electron, with an average lifetime of about 15 minutes. Today, this emitted particle is called a **beta-minus (β^-) particle,** to distinguish it from another particle, the positron (β^+), that is identical, except that it has a positive rather than negative charge. We'll talk more about the whole menagerie of fundamental particles in Section 30.8.

Beta-minus decay usually occurs with nuclei for which the neutron-to-proton ratio N/Z is too great for stability. In β^- decay, N decreases by one, Z increases by one, and A doesn't change. Note that both α and β^- emission, by changing the Z value of a nucleus, have the effect of changing one element into another—the dream of medieval alchemists. When a nucleus has a neutron-to-proton ratio that is too *small* for stability, it may emit a positron (β^+), increasing N by one and decreasing Z by one.

Beta particles can be identified and their speeds measured by measuring the curvature of their paths in a magnetic field. The speeds range up to 0.9995 of the speed of light, in the extreme-relativistic range. If the only two particles involved were the β^- particle and the recoiling nucleus, conservation of momentum and energy would require each to have a definite energy. But in fact, β^- particles are found to have a continuous spectrum of energies. Thus, in 1930, the Swiss theoretical physicist Wolfgang Pauli proposed that a third particle, electrically neutral and unseen in the original experiments, must be involved.

Enrico Fermi christened this particle the **neutrino.** It wasn't observed directly until 1953, although by then its existence had been firmly established by indirect evidence. Today, this particle is called the **antineutrino,** symbolized $\bar{\nu}_e$. The basic process of β^- decay can be represented as

$$n \rightarrow p + \beta^- + \bar{\nu}_e. \qquad (30.4)$$

After α or β^- emission, the remaining nucleus is sometimes left in an excited state. It can then decay to its ground state by emitting a photon, often called a gamma-ray photon or simply a **gamma** (γ). Typical γ energies are 10 keV to 5 MeV. For example, α particles emitted from radium have two possible kinetic energies, either 4.784 MeV or 4.602 MeV, corresponding to a *total* released energy of 4.871 MeV or 4.685 MeV. When an α with the smaller energy is emitted, the resulting nucleus (which corresponds to the element *radon*), is left in an excited state. It then decays to its ground state by emitting a γ photon with energy

$$4.871 \text{ MeV} - 4.685 \text{ MeV} = 0.186 \text{ MeV}.$$

A photon with that amount of energy is in fact observed during this decay. These processes are shown in Figure 30.4.

Decay Series

When a radioactive nucleus decays, the resulting nucleus may also be unstable. In this case, there is a series of successive decays until a stable configuration is reached. The most abundant radioactive nucleus found on earth is uranium ^{238}U, which undergoes a series of 14 decays, including eight α emissions and six β^- emissions, terminating at the stable isotope of lead, ^{206}Pb.

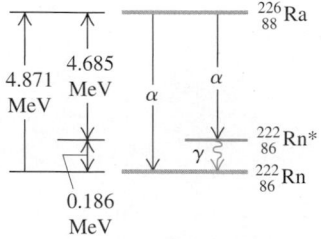

▲ FIGURE 30.4 Energy-level diagram of γ photon with energy 0.186 MeV observed during decay.

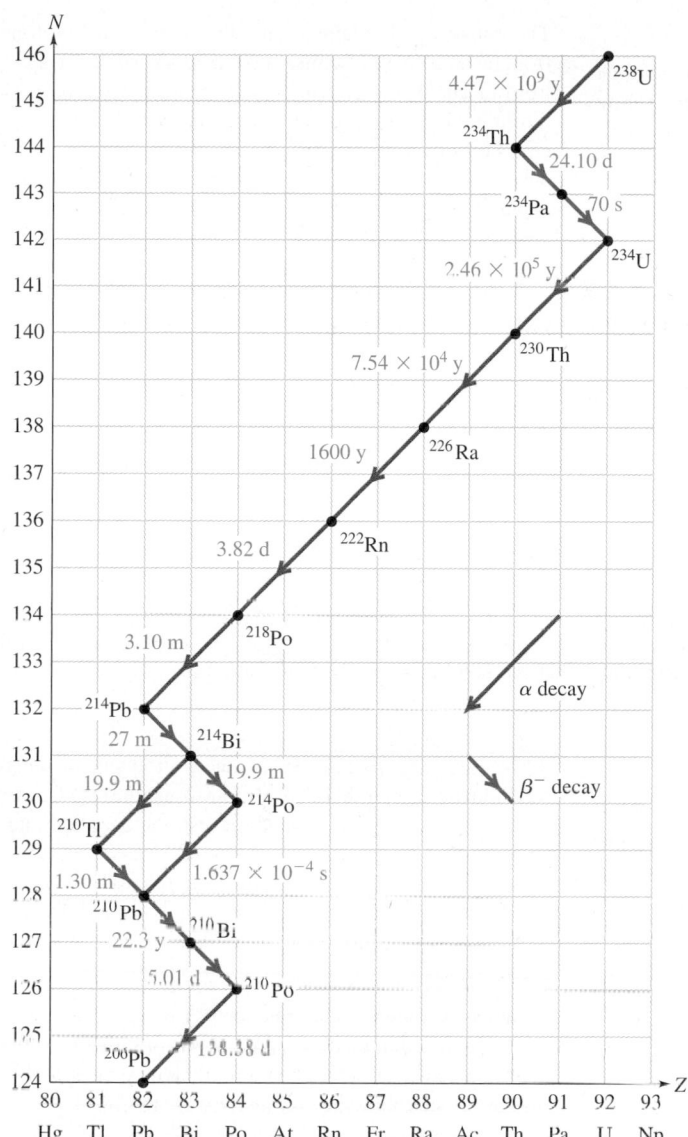

▲ **FIGURE 30.5** Segrè chart showing the uranium ^{238}U decay series, terminating with the stable nuclide ^{206}Pb.

Radioactive decay series can be represented on a Segrè chart, as in Figure 30.5. In alpha emission, N and Z each decrease by two; in beta emission, N decreases by one and Z increases by one. The half-lives (discussed later in this section) of the individual decays are given in years (y), days (d), hours (h), minutes (m), or seconds (s). Figure 30.5 shows the uranium decay series, which begins with the common uranium isotope ^{238}U and ends with an isotope of lead, ^{206}Pb. We note that the series includes unstable isotopes of several elements that also have *stable* isotopes, including thallium (Tl), lead (Pb), and bismuth (Bi). The unstable isotopes of these elements that occur in the ^{238}U series all have too many neutrons to be stable, as we discussed in Section 30.1.

Another decay series starts with the uncommon isotope ^{235}U and ends with ^{207}Pb; a third starts with thorium (^{232}Th) and ends with ^{208}Pb. A fourth series, produced in nuclear reactors, starts with neptunium (^{237}Np) and ends with ^{209}Bi.

Quantitative Analysis 30.2

A change in charge

An element of atomic number 84 decays radioactively to an element of atomic number 83. Which of the following emissions achieves this result?

A. alpha particle only.
B. beta particle only.
C. alpha particle plus beta particle.

SOLUTION A change in atomic number from 84 to 83 means that the number of protons in the nucleus has decreased by one.

The emission of an alpha particle removes two protons from the nucleus. The emission of a beta particle (an electron) means that a neutron has changed to a proton plus the emitted electron, so the atomic number increases by one. The emission of an alpha particle and a beta particle will give the desired result. (So would the emission of a positron; during positron emission, a proton changes to a neutron plus the emitted positron.)

EXAMPLE 30.3   Radium decay

Suppose you are given the following atomic masses:

$$^{226}_{88}\text{Ra:} \quad 226.025410 \text{ u,}$$
$$^{222}_{86}\text{Rn:} \quad 222.017578 \text{ u,}$$
$$^{4}_{2}\text{He:} \quad 4.002603 \text{ u.}$$

Show that α emission is energetically possible for radium, and find the kinetic energy of the emitted α particle.

SOLUTION

SET UP First note that the masses given are those of the neutral atoms, including 88, 86, and 2 electrons, respectively, so we have the same numbers of electrons in the initial and final states. This is essential. Next, we have to compare the mass of ^{226}Ra with the sum of the masses of ^{222}Rn and ^{4}He. If the former is greater, then α decay is possible.

SOLVE The difference between the mass of ^{226}Ra and the sum of the masses of ^{222}Rn and ^{4}He is

$$226.025410 \text{ u} - (222.017578 \text{ u} + 4.002603 \text{ u}) = 0.005229 \text{ u.}$$

The fact that this quantity is positive shows that the decay is energetically possible. The energy equivalent of 0.005229 u is

$$E = (0.005229 \text{ u})(931.5 \text{ MeV/u}) = 4.871 \text{ MeV.}$$

REFLECT The total mass of the system has decreased by 0.005229 u, and the corresponding increase in kinetic energy is

4.871 MeV. Thus, we expect the decay products to emerge with a total kinetic energy of 4.871 MeV. Momentum is also conserved: If the parent nucleus is initially at rest, the daughter nucleus and the α particle have momenta with equal magnitude and opposite direction. The kinetic energy is $K = p^2/2m$, so the kinetic energy divides inversely as the masses of the two particles. The α gets $222/(222 + 4)$ of the total, or 4.78 MeV, equal to the observed α energy.

Practice Problem: The nuclide ^{60}Co, an odd–odd unstable nucleus, is used in medical applications of radiation. Show that it is unstable against β decay, and find the total energy of the decay products. The following atomic masses are given:

$$^{60}_{27}\text{Co:} \quad 59.933817 \text{ u,}$$
$$^{60}_{28}\text{Ni:} \quad 59.930786 \text{ u.}$$

Answer: 2.82 MeV.

Decay Rates

In any sample of a radioactive element, the number of radioactive nuclei decreases continuously as the nuclei decay. This is a statistical process; there is no way to predict when any individual nucleus will decay. The rates of decay of various nuclei cover a very wide range, from microseconds to billions of years.

Let N be the number of radioactive nuclei in a sample at time t, and let ΔN be the (negative) change in that number during a short time interval Δt. (*Note:* We've previously used N to denote the number of neutrons in a nucleus. The present N is something completely different, so be careful.) The rate of change of N is $\Delta N/\Delta t$. The larger the number of nuclei in the sample, the larger is the number that decay during any interval, so the rate of change of N is proportional to N. That is, it is equal to a constant λ multiplied by N:

Rate of decay of radioactive nuclei

$$\frac{\Delta N}{\Delta t} = -\lambda N. \tag{30.5}$$

The constant λ is called the **decay constant.** It has different values for different nuclides.

A large value of λ corresponds to rapid decay, a small value of λ to slower decay. The situation is analogous to a discharging capacitor, which we studied in Section 19.8. Figure 30.6 shows the number N of nuclei as a function of time for the decay of polonium, ^{214}Po.

It's possible to derive an expression for the number N of nuclei remaining after any time t if the number at time $t = 0$ is N_0. The derivation requires calculus, and the result is an exponential function:

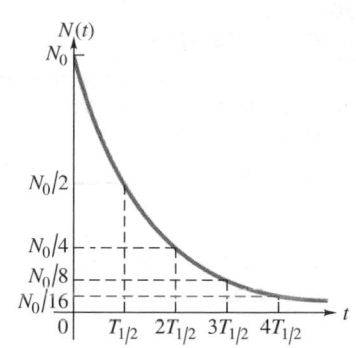

▲ **FIGURE 30.6** Graph of exponential decay.

Number of radioactive nuclei remaining after time *t*

If there are N_0 nuclei with decay constant λ at time $t = 0$, the number N remaining after time t is

$$N = N_0 e^{-\lambda t}. \tag{30.6}$$

The **half-life** $T_{1/2}$ of a radioactive substance is the time required for the number of radioactive nuclei to decrease to half the original number N_0. Then half of those remaining nuclei decay during a second interval $T_{1/2}$, and so on. The numbers remaining after successive intervals of $T_{1/2}$ are $N_0/2$, $N_0/4$, $N_0/8$, ...

The half-life $T_{1/2}$ and the decay constant λ are related by the following equation:

Decay constant and half-life

$$T_{1/2} = \frac{\ln 2}{\lambda} - \frac{0.693}{\lambda}. \tag{30.7}$$

In particle physics, the life of an unstable nucleus or particle is usually described by the lifetime, not the half-life. The **lifetime** (or mean lifetime) T_{mean} is the average time for a nucleus or particle to decay:

$$T_{\text{mean}} - \frac{1}{\lambda} - \frac{T_{1/2}}{\ln 2} = \frac{T_{1/2}}{0.693}. \tag{30.8}$$

The **activity** of a specimen is the number of decays per unit time. A common unit is the **curie,** abbreviated Ci, defined to be 3.70×10^{10} decays per second:

$$1 \text{ Ci} = 3.70 \times 10^{10} \text{ decays/s}.$$

This is approximately equal to the activity of 1 g of radium. The number of decays is proportional to the number of radioactive nuclei, so the activity decreases with time. Thus, Figure 30.6 is also a graph of the *activity* of polonium, ^{210}Po, which has a half-life of 140 days.

The SI unit of activity is the **becquerel,** abbreviated Bq. One becquerel is one decay per second:

$$1 \text{ Bq} = 1 \text{ decay/s}, \qquad 1 \text{ Ci} = 3.70 \times 10^{10} \text{ Bq}.$$

Conceptual Analysis 30.3

Half-life

A pure sample of a radioactive substance in a container has a half-life of 1 h. Which of the following statements is or are true?

A. After 1 h, half of the atoms will have decayed, and the remainder will decay in the next 1 h.
B. After 1 h, half of the atoms will have decayed, and half of the remainder decay in the next 1 h.
C. After 1 h, the mass in the container is half its original value.

SOLUTION During a half-life, half of the sample of the element is converted, usually to a different element. Half of the remainder of the sample is converted during the next half-life. Thus, B is correct. Because the atoms do not disappear, the mass of the sample remains the same, except for the tiny loss representing the conversion of mass to energy.

Radiocarbon Dating

An interesting application of radioactivity is the dating of archeological and geological specimens by measuring the concentration of radioactive isotopes. The most familiar example is *radiocarbon dating.* The unstable isotope ^{14}C is produced by nuclear reactions in the atmosphere caused by cosmic-ray bombardment; thus, there is a small proportion of ^{14}C in the carbon dioxide in the atmosphere. Plants that obtain their carbon from this source contain the same proportion of ^{14}C as the atmosphere. When a plant dies, it stops taking in carbon, and the ^{14}C it has already absorbed decays, with a half-life of 5730 years. By measuring the proportion of ^{14}C in the remains, we can determine how long ago the organism died.

One difficulty with radiocarbon dating is that the concentration of ^{14}C in the atmosphere has varied by a few percent during the last 100 years because of the burning of fossil fuels and atmospheric testing of nuclear weapons. Corrections can be made on the basis of other data, such as measurements on annual-growth rings of trees. Similar techniques are used with other isotopes for dating geological specimens. Some rocks, for example, contain the unstable potassium isotope ^{40}K, which decays to the stable nuclide ^{40}Ar, with a half-life of 1.3×10^9 y. The age of the rock can be determined by comparing the concentrations of ^{40}K and ^{40}Ar.

EXAMPLE 30.4 **Radiocarbon dating**

The activity of atmospheric carbon due to the presence of ^{14}C is about 0.255 Bq per gram of carbon. What fraction of carbon atoms are ^{14}C? If the activity of an archeological specimen is 0.0637 Bq per gram, what is its approximate age?

SOLUTION

SET UP We know that the half-life of ^{14}C is $T_{1/2} = 5730$ y. We convert years to seconds and then find the decay constant λ from Equation 30.7, $T_{1/2} = 0.693/\lambda$. The number of decays per second in the atmosphere is 0.255, and we can find the *total* number of ^{14}C atoms from Equation 30.5:

$$\frac{\Delta N}{\Delta t} = -\lambda N.$$

Finally, we compare this N with the total number of all carbon atoms, using the fact that one gram of the common isotope ^{12}C is $\frac{1}{12}$ mol.

SOLVE First, the half-life is

$$T_{1/2} = (5730 \text{ y})(3.156 \times 10^7 \text{ s/y}) = 1.81 \times 10^{11} \text{ s}.$$

The decay constant is $\lambda = 0.693/T_{1/2} = 3.38 \times 10^{-12} \text{ s}^{-1}$.

Next, with $\Delta t = 1$ s, Equation 30.5 gives

$$N = -\frac{1}{\lambda}\frac{\Delta N}{\Delta t} = -\left(\frac{1}{3.83 \times 10^{-12} \text{ s}^{-1}}\right)\left(\frac{-0.255}{1 \text{ s}}\right)$$
$$= 6.66 \times 10^{10}.$$

The total number of C atoms in 1 g $(=1/12 \text{ mol})$ is $(1/12 \text{ mol})(6.023 \times 10^{23} \text{ atoms/mol})$, or 5.02×10^{22}. The ratio of ^{14}C atoms to all carbon atoms is

$$\frac{6.66 \times 10^{10}}{5.02 \times 10^{22}} = 1.33 \times 10^{-12}.$$

The activity of the archeological specimen, per unit mass, is one-fourth the activity of ^{14}C in the present atmosphere. Thus, the ^{14}C in the specimen has been decaying for a time equal to twice the half-life, or about 11,400 y.

Continued

30.4 Radiation and the Life Sciences

The interaction of radiation with living organisms is a vitally important topic. In this discussion, we define **radiation** to include radioactivity (alpha, beta, gamma, and neutrons) and electromagnetic radiation such as x rays and gamma rays. As these particles pass through matter, they lose energy, breaking molecular bonds and creating ions—hence the term *ionizing radiation.* Charged particles interact through direct electrical forces on the electrons in the material. X rays and gamma rays interact by the photoelectric effect, in which an electron absorbs a photon and is freed from its molecular bond, or by Compton scattering (Section 28.5). Neutrons cause ionization indirectly by colliding with nuclei or by being absorbed by nuclei, with subsequent alpha or beta decay of the resulting unstable nucleus.

The interactions of radiation with living tissue are extremely complex. It is well known that excessive exposure to radiation, including sunlight, x rays, and all the nuclear radiations, can destroy tissues. In mild cases this results in a burn, as with common sunburn. Greater exposure can cause severe illness or death by a variety of mechanisms, including massive destruction of tissue cells, the alteration of genetic material, and destruction of the components of bone marrow that produce red blood cells.

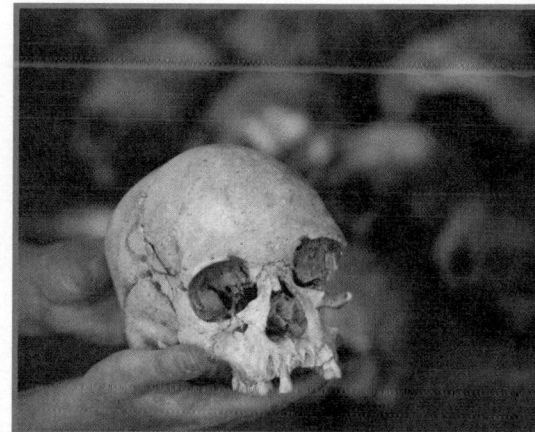

▲ **BIO Application The oldest American**
This young woman's skull, found in Brazil, is the oldest human skull known from the Americas. Radiocarbon dating puts its age at about 11,500 years old—more than 2000 years older than the celebrated Kennewick Man. Some scientists argue that her skull measurements and teeth resemble those of Africans or South Sea islanders, raising the possibility that the Americas were peopled from the sea as well as by migration from northeast Asia across the Bering land bridge during the last Ice Age.

Radiation Doses

The quantitative description of the effect of radiation on living tissue is called **radiation dosimetry.** The **absorbed dose** of radiation is defined as the energy delivered to the tissue, per unit mass of tissue. The SI unit of absorbed dose, the joule per kilogram, is called the **gray** (Gy); that is, 1 Gy = 1 J/kg. Another unit, in more common use at present, is the **rad,** defined as 0.01 J/kg = 1 cGy:

Definition of unit of absorbed radiation dose

$$1 \text{ Gy} = 1 \text{ J/kg}, \qquad 1 \text{ rad} = 0.01 \text{ J/kg} = 0.01 \text{ Gy} = 1 \text{ cGy}. \quad (30.9)$$

By itself, the absorbed dose isn't an adequate measure of biological effect, because equal energies of different kinds of radiation cause different degrees of biological effect. This variation is described by a numerical factor called the **relative biological effectiveness (RBE)** of each specific radiation. X rays with 200 keV energy are defined to have an RBE of unity, and the effects of other kinds of radiation can be compared experimentally. Table 30.3 shows approximate values of RBE for several types of radiation. All these values depend somewhat on the energy of the radiation and on the kind of tissue in which the radiation is absorbed.

The biological effect of radiation is measured by the product of the absorbed dose and the RBE of the radiation; this product is called the *biologically equivalent dose,* or simply, equivalent dose. The SI unit of equivalent dose for humans is the **Sievert** (Sv):

TABLE 30.3 Relative biological effectiveness (RBE) for several types of radiation

Radiation	RBE
X rays and gamma rays	1
Electrons	1
Protons	5
Alpha particles	20
Heavy ions	20
Slow neutrons	5–20
	(energy dependent)

Units of equivalent dose

$$\text{equivalent dose (Sv)} = \text{RBE} \times \text{absorbed dose (Gy)}. \quad (30.10)$$

A more common unit, corresponding to the rad, for equivalent dose for humans is the **rem:**

$$\text{equivalent dose (rem)} = \text{RBE} \times \text{absorbed dose (rad)}. \quad (30.11)$$

Thus, 1 rem = 0.01 Sv.

Conceptual Analysis 30.4

Absorbing radiation

Suppose a person who has one hand in a wide beam of x rays receives an equivalent dose of 90 millirem in 30 sec. If the person puts both hands in this beam, side by side, for 30 sec, then the equivalent dose the person will receive is

A. 180 millirem, because twice the number of ions will be created.
B. 90 millirem, because the number of ions created per gram is unchanged.
C. 45 millirem, because the radiation is spread over twice as much of the body.

SOLUTION The millirem is related to the number of molecules ionized by radiation in one gram of tissue. So whether your body receives an equivalent dose of 90 millirem to the tip of your finger or 90 millirem to the whole body, the dose is the same. The correct answer is B. However, the number of ions created in the body and the possible destructive effects, and the potential for cancer, are vastly different for the two cases. Any time you receive ionizing radiation (such as that from diagnostic x rays), the area exposed should be kept as small as possible.

EXAMPLE 30.5 Radiation exposure during an x ray

During a diagnostic x ray, a broken leg with a mass of 5 kg receives an equivalent dose of 50 mrem. Determine the total energy absorbed and the number of x-ray photons absorbed if the x-ray energy is 50 keV.

SOLUTION

SET UP We note that for x rays, RBE = 1, so the absorbed dose is equal to the equivalent dose. We express the absorbed dose in J/kg, find the energy of one x-ray photon, and then find the number of x-ray photons needed to supply the specified total energy.

SOLVE For x rays, RBE = 1, so the absorbed dose is

$$50 \text{ mrad} = 0.050 \text{ rad} = (0.050)(0.01 \text{ J/kg})$$
$$= 5.0 \times 10^{-4} \text{ J/kg}.$$

The total energy absorbed is

$$(5.0 \times 10^{-4} \text{ J/kg})(5 \text{ kg}) = 2.5 \times 10^{-3} \text{ J}$$
$$= 1.56 \times 10^{16} \text{ eV}.$$

The energy of one x-ray photon is 50 keV = 5.0×10^4 eV.

The number of x-ray photons is

$$\frac{1.56 \times 10^{16} \text{ eV}}{5.0 \times 10^4 \text{ eV}} = 3.1 \times 10^{11} \text{ photons}.$$

REFLECT If the ionizing radiation had been a beam of alpha particles, for which RBE = 20, the absorbed dose needed for an equivalent dose of 50 mrem would be 2.5 mrad, corresponding to a total absorbed energy of 1.25×10^{-4} J.

Practice Problem: In a diagnostic x-ray procedure, 5.00×10^{10} photons are absorbed by tissue with a mass of 0.600 kg. The x-ray wavelength is 0.0200 nm. Find the total energy absorbed by the tissue and the equivalent dose in rem. *Answer:* 4.97×10^{-4} J, 0.0828 rem.

Radiation Hazards

Ionizing radiation is a double-edged sword; it poses serious health hazards, yet it also provides many benefits to humanity, including the diagnosis and treatment of disease and a wide variety of analytical applications. Here are a few numbers for perspective: An ordinary chest x ray delivers about 0.20 to 0.40 mSv to about 5 kg of tissue. (To convert from Sv to rem, simply multiply by 100.) Radiation exposure from cosmic rays and natural radioactivity in soil, building materials, and so on is on the order of 1.0 mSv (0.10 rem) per year at sea level and twice that at an elevation of 1500 m (about 5000 ft). It's been esti-mated that the average total radiation exposure for the entire U.S. population is roughly 360 mrem or 3.6 mSv per year, with about 80% of this from natural sources and 20% from human-made sources.

A whole-body dose of up to about 0.2 Sv causes no immediately detectable effect. A short-term whole-body dose of 5 Sv or more usually causes death within a few days or weeks. A localized dose of 100 Sv causes complete destruction of the exposed tissue.

The long-term hazards of radiation exposure in causing various cancers and genetic defects have been widely publicized, and the question of whether there is any "safe" level of radiation exposure has been hotly debated. U.S. government regulations are based on a predetermined maximum yearly exposure, from all except natural sources, of 2 to 5 mSv. Workers with occupational exposure to radiation are permitted 50 mSv per year. Recent studies suggest that these limits are too high and that even extremely small exposures carry hazards, but it is very difficult to gather reliable statistics on the effects of low dosages. It has become clear that any use of x rays for medical diagnosis should be preceded by a very careful consideration of the relation of risk to possible benefit.

Another sharply debated question is that of radiation hazards from nuclear power plants. The radiation level from these plants is *not* negligible. However, to make a meaningful evaluation of hazards, we must compare these levels with the alternatives, such as levels from coal-powered plants. The health hazards of coal smoke are serious and well documented, and the natural radioactivity in the smoke from a coal-fired power plant is believed to be roughly 100 times as great as that from a properly operating nuclear plant with equal capacity.

The comparison is complicated by the possibility of a catastrophic nuclear accident and the very serious problem of disposing of radioactive waste from nuclear plants. It is clearly impossible to eliminate *all* hazards to health. Our goal should be rather to try to take a rational approach to the problem of *minimizing* the total hazard from all sources. Figure 30.7 shows one estimate of the various sources of radiation exposure for the U.S. population.

Radon Hazards

A serious health hazard in some regions is the accumulation in houses of radon, ^{222}Rn, an inert, colorless, odorless radioactive gas. Looking at the ^{238}U decay chain (Figure 30.5), we see that radon (^{222}Rn) is produced by the decay of ^{226}Ra, which occurs in minute quantities in the rocks and soil on which houses are built. In turn, ^{222}Rn decays to polonium, ^{218}Po, with a half-life of 3.82 days. It's a dynamic equilibrium situation in which the rate of production equals the rate of decay. The hazard from ^{222}Rn is greater than for the other elements in the ^{238}U decay series because ^{222}Rn is a *gas*. During its short half life, it can migrate from the soil into your house, especially via basements. If a ^{222}Rn nucleus decays in your lungs, it emits a damaging α particle. And the radioactive daughter nucleus, ^{218}Po, which is *not* chemically inert, is likely to stay there until *it* decays, emits another damaging α particle, and so on through the ^{238}U decay series.

So how great a hazard *is* radon? The average activity in the air inside Amer-ican homes due to ^{222}Ra is about 1.5 pCi per liter of air, although in some local areas it is as high as 3500 pCi/L. (Recall that 1 pCi $= 10^{-12}$ Ci and 1 Ci $= 3.7 \times 10^{10}$ Bq.) It has been estimated that lifetime exposure to a concentration of 1.5 pCi per liter of air reduces average life expectancy by about 40 days. For com-parison, life-long smoking of one pack of cigarettes per day reduces life expectancy by about 6 years, and lifelong exposure to the average emission from all the nuclear power plants in the world reduces life expectancy by something in the range from 0.01 day to 5 days. These figures include catastrophes such as the 1986 nuclear reactor disaster at Chornobyl in Ukraine, for which the *local* effect on life expectancy is much greater and more difficult to determine.

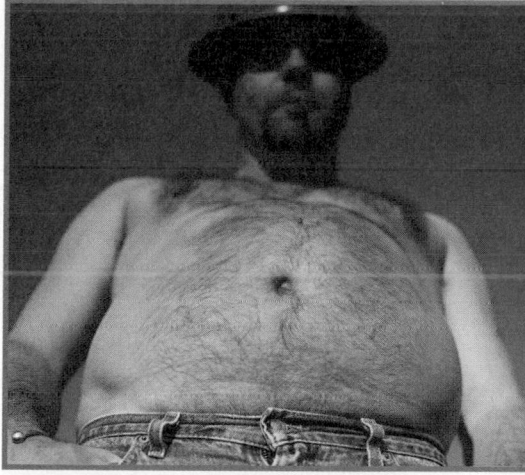

▲ **BIO Application How long is it in here?** When studying human exposure to radioactive isotopes, it is important to con-sider not only an element's physical half-life but also its biological half-life, a measure of how long an element is retained in the human body. As you know, the physi-cal half-life is the time it takes for half of the atoms in a radioactive isotope to decay. The biological half-life is the time it takes for half of an element to be excreted from the body. Although the physical half-lives of radioactive ^3H (12 years) and ^{14}C (5700 years) are relatively long, both are cleared from the body quickly, having biological half-lives of only 12 days. On the other hand, radioactive strontium (^{90}Sr), with a physical half-life of 29 years, is a more serious problem because it is a chemical cousin to calcium and is deposited in bones, giving it a biological half-life of 50 years.

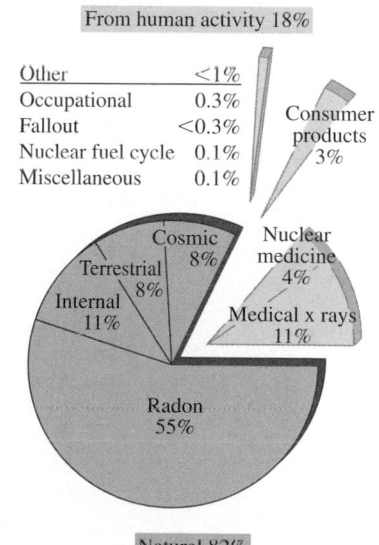

From human activity 18%

Other	<1%
Occupational	0.3%
Fallout	<0.3%
Nuclear fuel cycle	0.1%
Miscellaneous	0.1%

Consumer products 3%

Cosmic 8%

Nuclear medicine 4%

Terrestrial 8%

Internal 11%

Medical x rays 11%

Radon 55%

Natural 82%

▲ **FIGURE 30.7** Contributions of various sources to the total average radiation exposure in the U.S. population, expressed as percent-ages of the total.

EXAMPLE 30.6 Radioactivity in the air

What is the average activity of a liter of air, in Bq? In decays per hour?

SOLUTION

SET UP We use the average value just cited, 1.5 pCi/L, along with the conversion factor $1 \text{ Ci} = 3.7 \times 10^{10}$ Bq.

SOLVE $1.5 \text{pCi/L} = (1.5 \times 10^{-12} \text{ Ci/L})\left(\dfrac{3.70 \times 10^{10} \text{ Bq}}{1 \text{ Ci}}\right)$

$= 0.056 \text{ Bq}.$

Since $1 \text{ Bq} = 1 \text{ decay/s}$, the number of decays in one hour is

$(0.056 \text{ decays/s})(3600 \text{ s/h}) = 200 \text{ decays/h}.$

REFLECT A level of activity of air of 1.5 pCi/L represents, on average, the decay of one Rn nucleus every 18 s in each liter of air. If the basement of a medium-size house has a volume of 10^6 L, the total activity of the air in this basement is about 1.5×10^{-6} Ci, or about 5×10^4 disintegrations per second.

Practice Problem: If your lungs contain an average of 5.0 L of air, how many decays occur each minute? *Answer: 17.*

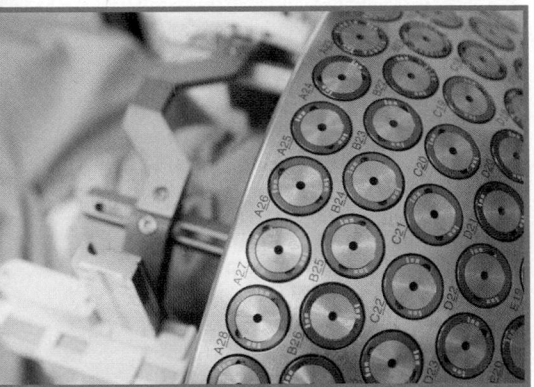

▲ **BIO Application Not science fiction.**
This woman is wearing a gamma knife radiosurgery helmet—a device that allows gamma rays to kill a tumor deep in the brain without harming the intervening tissue. Gamma beams aimed through the 101 holes in the helmet converge on the precise site of the tumor. Each beam alone is too weak to do significant harm to the tissue it passes through, but where the beams add up they are lethal. Radiation is used in many other medical techniques, both diagnostic and therapeutic, including CT, PET, and MRI scans, plus others mentioned in the text.

Benefits of Radiation

Radiation is widely used in medicine for intentional selective destruction of tissue such as tumors. The hazards are not negligible, but if the disease would be fatal without treatment, any hazard may be preferable. Artificially produced isotopes are often used as sources of radiation treatment. Such artificial sources have several advantages over naturally radioactive isotopes. They can have shorter half-lives and correspondingly greater activity. Isotopes can be chosen that do not emit alpha particles, which are usually not wanted, and the electrons they emit are easily stopped by thin metal sheets without blocking the desired gamma radiation. Photon and electron beams from linear accelerators have also been used as sources of radiation.

Here are several examples of the expanding field called *nuclear medicine.* Radioactive isotopes have the same electron configurations and chemical behavior as stable isotopes of the same element. But the location and concentration of radioactive isotopes can be detected easily by measurements of the radiation they emit. A familiar example is the use of an unstable isotope of iodine for thyroid studies. A minute quantity of ^{131}I is fed or injected into the patient, and the speed with which it becomes concentrated in the thyroid provides a measure of thyroid function. The half-life is about 8 days, so there are no long-lasting radiation hazards. Nearly all the iodine taken in is either eliminated or stored in the thyroid, and the body does not discriminate between the unstable isotope ^{131}I and the stable isotope ^{127}I.

With the use of more sophisticated scanning detectors, one can also obtain a "picture" of the thyroid, which shows enlargement and other abnormalities. This procedure, a type of autoradiography, is comparable to photographing the glowing filament of an incandescent lightbulb, using the light emitted by the filament itself. If cancerous thyroid nodules are detected, they can be destroyed by much larger doses of ^{131}I.

Similar techniques are used to visualize coronary arteries. A thin tube or catheter is threaded through a vein into the heart, and a radioactive material is injected. Narrowed or blocked arteries can actually be photographed by use of a scanning detector; such a picture is called an *angiogram* or *arteriogram.* A useful isotope for such purposes is technetium (^{99}Tc), formed in the beta decay of molybdenum-99 (^{99}Mo). Technetium-99 is formed in an excited state, from which it decays to the ground state by γ emission with a half-life of about 6 hours, unusually long for γ emission. The chemistry of technetium is such that it can readily be attached to organic molecules that are taken up by various organs of the body. A small quantity of such technetium-bearing molecules is injected into a patient, and a scanning detector or gamma camera is used to produce an image that reveals which parts of the body take up these γ-emitting molecules. This technique, in which ^{99}Tc acts as a radioactive *tracer,* plays an important role in locating cancers, embolisms, and other pathologies (Figure 30.8).

Tracer techniques have many other applications. Tritium (^{3}H), a radioactive isotope of hydrogen, is used to tag molecules in complex organic reactions. In the world of machinery, radioactive iron can be used to study piston ring wear; radioactive tags on pesticide molecules can be used to trace their passage through food chains. Laundry detergent manufacturers test the effectiveness of their products by using radioactive dirt.

Many direct effects of radiation are also useful, such as strengthening polymers by cross-linking, sterilizing surgical tools, dispersing unwanted static electricity in the air, and ionizing the air in smoke detectors. Gamma rays are used to sterilize and preserve some food products.

30.5 Nuclear Reactions

In Section 30.3, we studied the decay of unstable nuclei by the spontaneous emission of an α or β particle, sometimes followed by γ emission. Nothing was done to initiate this emission, and nothing could be done to control it. Now let's consider some processes in which nuclear particles are rearranged as a result of the bombardment of a nucleus by a particle, rather than through a spontaneous natural process.

Rutherford suggested in 1919 that a massive particle with sufficient kinetic energy might be able to penetrate a nucleus. The result would be either a new nucleus with greater atomic number and mass number or a decay of the original nucleus. Rutherford bombarded nitrogen with alpha particles from naturally radioactive sources and obtained an oxygen nucleus and a proton, according to the equation

$$^{4}_{2}\text{He} + ^{14}_{7}\text{N} \rightarrow ^{17}_{8}\text{O} + ^{1}_{1}\text{H}. \tag{30.12}$$

Such a process is called a **nuclear reaction.**

Nuclear reactions obey the classical conservation principles for electric charge, momentum, angular momentum, and energy (including kinetic and rest energies). Another conservation law, not anticipated by classical physics, is conservation of the total number of nucleons. The numbers of protons and neutrons need not be conserved separately. (We've already seen that in β decay a neutron changes into a proton.) We'll look at the basis for the principle of conservation of nucleon number in Section 30.8.

When two nuclei interact, charge conservation requires the sum of the initial atomic numbers (Z) to be equal to the sum of the final atomic numbers. From conservation of the total number of nucleons, the sum of the initial mass numbers (A) is also equal to the sum of the final mass numbers. But the initial rest mass is, in general, *not* equal to the final rest mass, reflecting the fact that the collisions may be inelastic.

Reaction Energy

The difference between the rest masses before and after the reaction corresponds to the **reaction energy,** according to the mass–energy relation $E = mc^2$. If initial particles A and B interact to produce final particles C and D, the reaction energy Q is defined as

$$Q = (M_A + M_B - M_C - M_D)c^2. \tag{30.13}$$

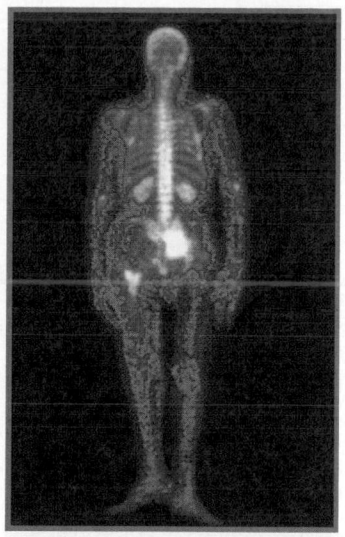

▲ **BIO Application A high "tech" helper.** The most commonly used radioisotope in medicine is technetium-99 $(^{99}$Tc$)$, which is quite useful in diagnostic procedures for a number of reasons. It decays to produce a low-energy beta particle, which is not very damaging to tissue, plus a non-destructive gamma ray that is energetic enough to leave the body to provide an image, such as the one shown in the bone scan above. ^{99}Tc has a half-life of only 6 hours, which is long enough to monitor metabolic processes in the body but short enough to disappear quickly afterward. Due to its versatility in chemical bonding, it can be incorporated into a wide variety of chemical compounds that can be targeted to specific tissues or organs such as red blood cells, heart muscle, or the spleen. One promising derivative of ^{99}Tc binds selectively to tumor tissues, allowing imaging of cancers that might otherwise be difficult to detect.

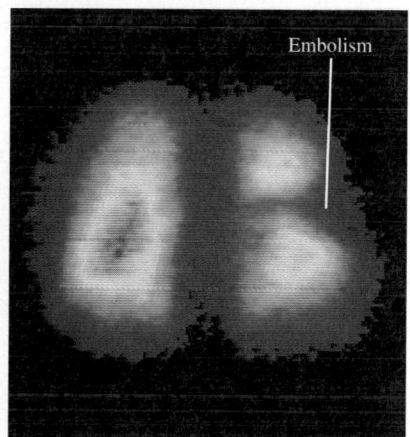

▲ **FIGURE 30.8** Lung scan with radioactive tracer ^{99}Tc. The orange glow in the lung on the left indicates strong γ-ray emission by the ^{99}Tc, which shows that the chemical was able to pass into this lung through the bloodstream. The lung on the right shows weaker emission, indicating an embolism that is restricting blood flow.

To balance the electrons, we use the neutral atomic masses in Equation 30.13 ($^1_1$H for a proton, $^4_2$He, and so on). When Q is positive, the total mass decreases and the total kinetic energy increases. Such a reaction is called an **exoergic reaction** or (borrowing a term from chemistry) an **exothermic reaction**. When Q is negative, the mass increases and the kinetic energy decreases, and the reaction is said to be **endoergic** or **endothermic**. Note that an endoergic reaction cannot occur at all unless the initial kinetic energy is at least as great as $|Q|$; this energy is called the **threshold energy** for the reaction. (1 u is equivalent to 931.5 MeV.)

EXAMPLE 30.7 **An exoergic reaction**

When lithium is bombarded by a proton, two alpha particles are produced. Find the reaction energy Q.

SOLUTION

SET UP The reaction can be represented as

$$^1_1\text{H} + ^7_3\text{Li} \rightarrow ^4_2\text{He} + ^4_2\text{He}.$$

We need to use the definition of reaction energy, Equation 30.13. The masses we need are given in Table 30.2

SOLVE From Table 30.2, we find the initial and final masses:

A:	$^1_1$H	1.007825 u	C:	$^4_2$He	4.002603 u
B:	$^7_3$Li	7.016005 u	D:	$^4_2$He	4.002603 u
		8.023830 u			8.005206 u

Four electron masses are included on each side. We see that

$$M_A + M_B - M_C - M_D = 0.018624 \text{ u}.$$

The mass decreases by 0.018624 u; from Equation 30.13, the reaction energy is

$$Q = (0.018624 \text{ u})(931.5 \text{ MeV/u}) = 17.35 \text{ MeV}.$$

REFLECT The total mass decreases, so the total kinetic energy increases. The final total kinetic energy of the two separating α particles is 17.35 MeV greater than the initial total kinetic energy of the proton and the lithium nucleus.

Practice Problem: Find the Q value for the reaction $^1_0$n $+ ^{10}_5$B $\rightarrow ^7_3$Li $+ ^4_2$He. Is this reaction exoergic or endoergic? *Answer:* 2.79 MeV; exoergic.

EXAMPLE 30.8 **An endoergic reaction**

Find the reaction energy for Rutherford's experiment, described by Equation 30.12.

SOLUTION

SET UP As in Example 30.7, we use the rest masses found in Table 30.2 to evaluate the reaction energy Q, given by Equation 30.12. The initial and total masses each include nine electron masses.

SOLVE The mass calculation, in tabular form, is

A:	$^4_2$He	4.002603 u	C:	$^{17}_8$O	16.999132 u
B:	$^{14}_7$N	14.003074 u	D:	$^1_1$H	1.007825 u
		18.005677 u			18.006957 u

We see that the total rest mass increases by 0.001280 u, and the corresponding reaction energy is

$$Q = (-0.001280 \text{ u})(931.5 \text{ MeV/u}) = -1.192 \text{ MeV}.$$

REFLECT This amount of energy is absorbed in the reaction. In a head-on collision with zero total momentum, the minimum total initial kinetic energy for this reaction to occur is 1.192 MeV. Ordinarily, though, this reaction would be produced by bombarding stationary ^{14}N nuclei with α particles. In this case, the α energy must be *greater* than 1.192 MeV. The α can't give up *all* of its kinetic energy, because then the final total kinetic energy would be zero and momentum would not be conserved. It turns out that, to conserve momentum, the initial α energy must be at least 1.533 MeV.

Practice Problem: Consider the reaction $^6_3$Li $+ ^4_2$He $\rightarrow ^9_4$Be $+ ^1_1$H, produced by bombarding a solid lithium target with α particles. Show that this reaction is endoergic, and find the amount by which the total initial kinetic energy exceeds the total final value. *Answer:* 2.125 MeV.

For a charged particle such as a proton or an α particle to penetrate the nucleus of another atom, it usually must have enough initial kinetic energy to overcome the potential-energy barrier caused by the repulsive electrostatic forces. For example, in the reaction of Example 30.7, suppose we treat the proton and the ^{7}Li nucleus as spherically symmetric charges with radius given by

$R = R_0 A^{1/3}$ (Equation 30.2) and with a distance of about 3.5×10^{-15} m between their centers. Then the repulsive potential energy U of the proton (charge $+e$) and the lithium nucleus (charge $+3e$) at this distance is

$$U = \frac{1}{4\pi\epsilon_0} \frac{(e)(3e)}{r} = \frac{(9.0 \times 10^9 \, \text{N} \cdot \text{m}^2 \cdot \text{C}^{-2})(3)(1.6 \times 10^{-19} \, \text{C})^2}{3.5 \times 10^{-15} \, \text{m}}$$
$$= 1.98 \times 10^{-13} \, \text{J} = 1.24 \times 10^6 \, \text{eV} = 1.24 \, \text{MeV}.$$

Even though energy is liberated in this reaction, the proton must have a minimum, or *threshold,* energy of about 1.2 MeV for the reaction to occur.

The absorption of neutrons by nuclei forms an important class of nuclear reactions. Heavy nuclei bombarded by neutrons in a nuclear reactor can undergo a series of neutron absorptions alternating with beta decays, in which the mass number A increases by as much as 25. Some of the transuranic elements (elements having Z larger than 92, which don't occur in nature) are produced in this way. Many transuranic elements, having Z as high as 116, have been identified.

The analytical technique of neutron activation analysis uses similar reactions. When stable nuclei are bombarded by neutrons, some nuclei absorb neutrons and then undergo β^- decay. The energies of the β^- and γ emissions depend on the unstable parent nuclide and provide a means of identifying it. The presence of quantities of elements far too small for conventional chemical analysis can be detected in this way.

30.6 Nuclear Fission

Nuclear fission is a decay process in which an unstable nucleus splits into two fragments of comparable mass instead of emitting an α or β particle. Fission was discovered in 1939, when Otto Hahn and Fritz Strassman bombarded uranium $(Z = 92)$ with neutrons. The resulting radiation did not coincide with that of any known radioactive nuclide. Meticulous chemical analysis led to the astonishing conclusion that they had found radioactive isotopes of barium $(Z = 56)$ and krypton $(Z = 36)$. They concluded, correctly, that the uranium nuclei were splitting into two massive fragments, which they called **fission fragments.** A few free neutrons usually appeared along with the fission fragments. The energy released during fission, almost 200 MeV per nucleus, emerged as kinetic energy of the fission fragments.

Both the common (99.3%) isotope ^{238}U and the uncommon (0.7%) isotope ^{235}U (as well as several other nuclides) can be split by neutron bombardment, ^{235}U by slow neutrons, but ^{238}U only by neutrons with at least 1.2 MeV of energy. Fission resulting from neutron absorption is called **induced fission.** Some nuclei can also undergo **spontaneous fission,** which occurs without initial neutron absorption. More than 100 different nuclides representing more than 20 different elements have been found among the fission products. Figure 30.9 shows the distribution of mass numbers for fission fragments from the nuclide ^{235}U.

Fission reactions are always exoergic, because the fission fragments have greater binding energy per nucleon than the original nucleus. The total kinetic energy of the fission fragments is enormous, about 200 MeV (compared with typical α and β energies of a few MeV).

Fission fragments always have too many neutrons to be stable. The N/Z value for stable nuclides is about 1.3 at $A = 100$ and 1.4 at $A = 150$. The fragments have approximately the same N/Z as ^{235}U, about 1.55. They respond to this surplus of neutrons by emitting two or three free neutrons and by undergoing a series of β^- decays (each of which increases Z by one and decreases N by one), until a stable value of N/Z is reached.

Mastering**PHYSICS**

PhET: Nuclear Fission

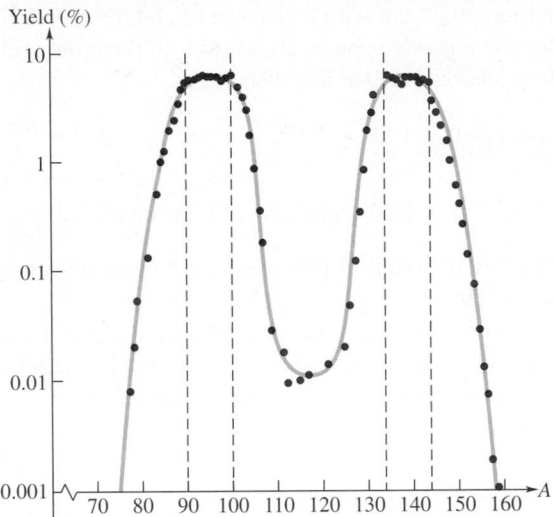

▲ **FIGURE 30.9** Mass distribution of fission fragments.

Chain Reactions

Fission of a uranium nucleus, triggered by neutron bombardment, releases other neutrons that can trigger more fissions; thus, a **chain reaction** is possible (Figure 30.10). The reaction can be made to proceed slowly and in a controlled manner, as in a nuclear reactor, or explosively, as in a bomb. The energy released in a chain reaction is enormous, far greater than in any chemical reaction. For example, when uranium is oxidized, or "burned," to uranium dioxide (UO_2), the heat of combustion is about 4500 J/g. Expressed as energy per atom, this is about 11 eV per uranium atom. By contrast, fission liberates about 200 MeV per atom, 20 million times as much energy.

Nuclear Reactors

A **nuclear reactor** is a system in which a controlled nuclear chain reaction is used to liberate energy. In a nuclear power plant, this energy is used to generate steam, which operates a turbine and turns an electrical generator. On the average, each fission of a ^{235}U nucleus produces about 2.5 free neutrons, so 40% of the neutrons are needed to sustain a chain reaction. The probability of neutron absorption by a nucleus is much greater for low-energy (less than 1 eV) neutrons than for the higher-energy (1 MeV) neutrons liberated during fission. The emitted neutrons are slowed down by collisions with nuclei in the surrounding material, called the *moderator,* so that they can cause further fissions. In nuclear power plants the moderator is often water, occasionally graphite.

The rate of the reaction is controlled by inserting or withdrawing control rods made of elements (often cadmium) whose nuclei absorb neutrons without undergoing any additional reaction. The isotope ^{238}U can also absorb neutrons, leading to ^{239}U, but not with high enough probability for it to sustain a chain reaction by itself. Thus, uranium used in reactors is "enriched" by increasing the proportion of ^{235}U from the natural value of 0.7%, typically to 3% or so, by isotope-separation processing.

In a nuclear power plant, the fission energy appears as kinetic energy of the fission fragments, and its immediate result is to heat the fuel elements and the surrounding water. This heat generates steam to drive turbines, which in turn drive the electrical generators (Figure 30.11). The steam generator is a heat exchanger that takes heat from this highly radioactive water and generates non-radioactive steam to run the turbines.

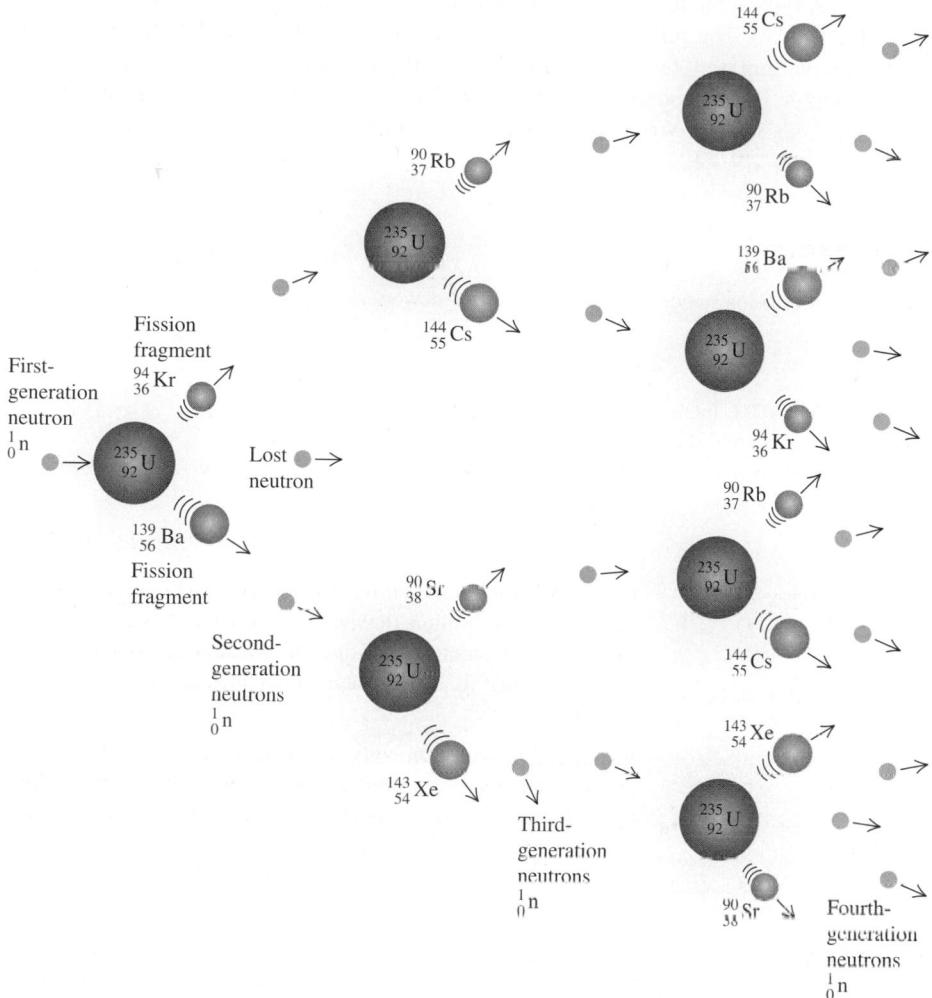

▲ **FIGURE 30.10** Schematic diagram of a nuclear fission chain reaction.

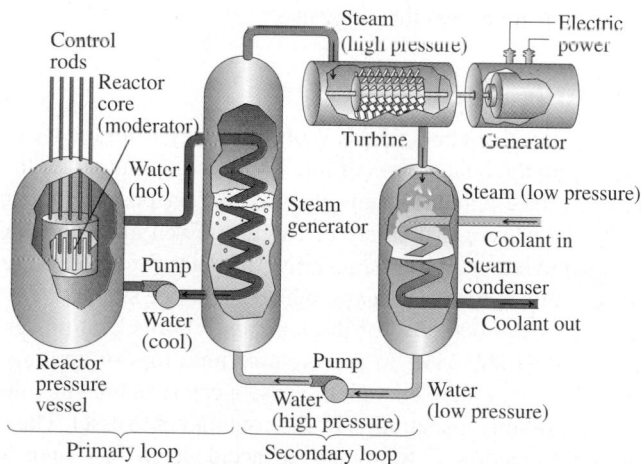

▲ **FIGURE 30.11** Schematic diagram of a nuclear power plant.

A typical nuclear power plant has an electric generating capacity of 1000 MW (or 10^9 W). The turbines are heat engines and are subject to the efficiency limitations imposed by the second law of thermodynamics, as we discussed in Chapter 16. In modern nuclear power plants, the overall efficiency is about one-third, so 3000 MW of thermal power from the fission reaction is needed to generate 1000 MW of electrical power.

EXAMPLE 30.9 Fuel consumption in a reactor

How much uranium has to undergo fission per unit time to provide 3000 MW of thermal power?

SOLUTION

SET UP Each second, we need 3000 MJ, or 3000×10^6 J. Each fission provides 200 MeV. From these numbers, we can find the number of fissions per second. Then, using the mass of a uranium atom, we can find the total mass of U needed per unit time.

SOLVE Each fission provides an amount of energy

$$200 \text{ MeV} = (200 \text{ MeV})(1.6 \times 10^{-13} \text{ J/MeV})$$
$$= 3.2 \times 10^{-11} \text{ J}.$$

The number of fissions needed per second is

$$\frac{3000 \times 10^6 \text{ J}}{3.2 \times 10^{-11} \text{ J}} = 9.4 \times 10^{19}.$$

Each uranium atom has a mass of about

$$(235)(1.67 \times 10^{-27} \text{ kg}) = 3.9 \times 10^{-25} \text{ kg,}$$

so the mass of uranium needed per second is

$$(9.4 \times 10^{19})(3.9 \times 10^{-25} \text{ kg}) = 3.7 \times 10^{-5} \text{ kg} = 37 \text{ mg.}$$

REFLECT The total consumption of uranium per day (86,400 seconds) is

$$(3.7 \times 10^{-5} \text{ kg/s})(86,400 \text{ s/d}) = 3.2 \text{ kg/d.}$$

For comparison, note that the 1000 MW coal-fired power plant we described in Section 16.9 burns 14,000 tons (about 10^7 kg) of coal per day. Combustion of one carbon atom yields about 2 eV of energy, while fission of one uranium nucleus yields 200 MeV, 10^8 times as much.

Practice Problem: In this example, the calculated mass includes only the mass of the fissionable nuclide $^{235}_{92}$U. If the reactor fuel has been enriched to 3% by isotope separation of the natural mix of uranium nuclides, what total mass of uranium is required per day? *Answer:* About 100 kg/d.

Nuclear fission reactors have many other practical uses. Among these are the production of artificial radioactive isotopes for medical and other research, producing high-intensity neutron beams for research in nuclear structure, and producing fissionable transuranic elements such as plutonium ^{239}Pu from the common uranium isotope ^{238}U. The last is the function of *breeder reactors,* which can actually produce more fuel than they use.

Hazards of Nuclear Reactors

We mentioned earlier that about 15 MeV of the energy from the fission of a ^{235}U nucleus comes from the beta decays of the fission fragments rather than from the kinetic energy of the fragments themselves. This fact poses a serious problem with respect to the control and safety of reactors. Even after the chain reaction has been completely stopped by the insertion of control rods into the core, heat continues to be evolved by the β decays, which cannot be stopped. For the reactor in Example 30.9 (with 3000 MW of thermal power), this heat power is initially very large, more than 200 MW. In the event of total loss of cooling water, this amount of power is more than enough to cause a catastrophic "meltdown" of the reactor core and, possibly, penetration of the containment vessel. The difficulty in achieving a "cold shutdown" following an accident at the Three Mile Island nuclear power plant in Pennsylvania in March 1979 was a result of the continued evolution of heat due to β decays.

The Chornobyl catastrophe of April 26, 1986 resulted from a combination of an inherently unstable design and several human errors committed during a test of the emergency core cooling system. Too many control rods were withdrawn to compensate for a decrease in power caused by a buildup of neutron absorbers such as ^{135}Xe. The power level rose from 1% of normal to 100 times normal in 4 seconds; a steam explosion ruptured pipes in the core cooling system and blew the heavy concrete cover off the reactor vessel. The graphite moderator caught fire and burned for several days. The total activity of the radioactive material released into the atmosphere has been estimated as about 10^8 Ci.

30.7 Nuclear Fusion

In a **nuclear fusion** reaction, two or more small light nuclei combine to form a larger nucleus. Fusion reactions release energy for the same reason as fission reactions: The binding energy per nucleon after the reaction is greater than before. Referring to Figure 30.1, we see that the binding energy per nucleon increases with A up to about $A = 60$, so the fusion of nearly any two light nuclei to make a nucleus with A less than 60 is likely to be an exoergic reaction.

Here are three examples of energy-liberating fusion reactions:

ActivPhysics 19.3: Fusion

$$\,^1_1\text{H} + \,^1_1\text{H} \rightarrow \,^2_1\text{H} + \beta^+ + \nu_e,$$

$$\,^1_1\text{H} + \,^2_1\text{H} \rightarrow \,^3_2\text{He} + \gamma,$$

$$\,^3_2\text{He} + \,^3_2\text{He} \rightarrow \,^4_2\text{He} + \,^1_1\text{H} + \,^1_1\text{H}.$$

In the first reaction, two protons combine to form a deuteron, a β^+ or positron (a positively charged electron that we'll talk about in Section 30.8), and an electron neutrino. In the second, a proton and a deuteron (^2_1H, a proton and a neutron bound together) combine to form the light isotope of helium, ^3_2He. In the third, two ^3_2He nuclei unite to form ordinary helium (^4_2He) and two protons.

The positrons produced during the first step of this sequence collide with electrons; mutual annihilation takes place, and their energy is converted into γ radiation. The net effect of the sequence is therefore the combination of four hydrogen nuclei into a helium nucleus and gamma radiation. The net energy release, which can be calculated from the mass balance, turns out to be 26.7 MeV.

These fusion reactions, collectively known as the proton–proton chain, take place in the interior of the sun and other stars. Each gram of the sun's mass contains about 4.5×10^{23} protons. If all of these protons were fused into helium, the energy released would be about 130,000 kWh. If the sun were to continue to radiate at its present rate, it would take about 75 billion years to exhaust its supply of protons.

For two nuclei to fuse, they must come together to within the range of the nuclear force, typically on the order of 2×10^{-15} m. To do this, they must overcome the electrical repulsion of their positive charges. For two protons a distance of 2×10^{-15} m apart, the corresponding potential energy is on the order of 1.1×10^{-13} J, or 0.7 MeV; this amount represents the initial kinetic energy the fusing nuclei must have.

Such energies are available at extremely high temperatures. According to Section 15.3 (Equation 15.8), the average translational kinetic energy of a gas molecule at temperature T is $\frac{3}{2}kT$, where k is Boltzmann's constant. For this energy to be equal to 1.1×10^{-13} J, the temperature must be on the order of 5×10^9 K. Not all the nuclei have to have this energy, but the temperature must be on the order of millions of kelvins if any appreciable fraction of the nuclei are to have enough kinetic energy to surmount the electrical repulsion and achieve fusion. Fusion chain reactions with such high temperatures occur in the interiors of stars. Because of these extreme temperature requirements, such reactions are called **thermonuclear reactions.** An uncontrolled thermonuclear reaction occurs in a hydrogen bomb, with enormous destructive power.

Intensive efforts are underway in many laboratories to achieve controlled fusion reactions, which potentially represent an enormous new energy resource. In one kind of experiment, a plasma is heated to an extremely high temperature by an electrical discharge, while being contained by appropriately shaped magnetic fields. In another experiment, pellets of the material to be fused are heated by a high-intensity laser beam. As yet, no one has succeeded in producing fusion reactions under controlled conditions to yield a net surplus of usable energy.

Methods of achieving fusion that don't require high temperatures are also being studied; these are called **cold fusion.** A few researchers have claimed to have achieved cold fusion in an electrolytic process, but their results have not been confirmed by other investigators.

30.8 Fundamental Particles

It can be argued that the study of fundamental particles began in about 400 B.C., when the Greek philosophers Democritus and Leucippus suggested that matter is made of indivisible particles that they called *atoms*. This idea lay dormant until about 1804, when John Dalton (1766–1844), often called the father of modern chemistry, discovered that many chemical phenomena could be explained on the basis of atoms of each element as the fundamental, indivisible building blocks of matter. But toward the end of the 19th century, it became clear that atoms are *not* indivisible. The electron, the photon, and the proton were all discovered within a span of 20 years at the beginning of the 20th century. By 1925, quantum mechanics was in full flower as the key to understanding atomic structure, although many details remained to be worked out. At that time, it appeared that the proton and the electron were the basic building blocks of all matter.

The Neutron

The discovery of the neutron in 1930 was an important milestone. In that year, two German physicists, Bothe and Becker, observed that when beryllium, boron, or lithium was bombarded by high-energy (several MeV) α particles from the radioactive element polonium, the bombarded material emitted a radiation that had much greater penetrating ability than the original α particles. Experiments by James Chadwick the following year showed that this emanation consisted of uncharged (electrically neutral) particles with mass approximately equal to that of the proton. Chadwick christened these particles *neutrons*. A typical reaction, using a beryllium target, is

$$\,^4_2\text{He} + \,^9_4\text{Be} \rightarrow \,^{12}_6\text{C} + \,^1_0\text{n}, \tag{30.14}$$

where $\,^1_0\text{n}$ denotes a neutron.

Because neutrons have no charge, they produce no ionization when they pass through gases, and they are not deflected by electric or magnetic fields. They interact only with nuclei; they can be slowed down during elastic scattering with a nucleus, and they can penetrate the nucleus. Slow neutrons can be detected by means of another nuclear reaction—the ejection of an α particle from a boron nucleus, according to the reaction

$$\,^1_0\text{n} + \,^{10}_5\text{B} \rightarrow \,^7_3\text{Li} + \,^4_2\text{He}. \tag{30.15}$$

Because of its electric charge, the ejected α particle is easy to detect with a Geiger counter or other particle detector

The discovery of the neutron cleared up a mystery about the composition of the nucleus. Before 1930, it had been thought that the total mass of a nucleus was due only to its protons, but no one understood why the charge-to-mass ratio wasn't the same for all nuclei. It soon became clear that all nuclei (except hydrogen) contain both protons and neutrons. In fact, the proton, the neutron, and the electron are the basic building blocks of atoms. One might think that that would be the end of the story. On the contrary, it is barely the beginning!

The Positron

The positive electron, or **positron** (denoted by β^+ or e^+), was first observed in 1932 by Carl D. Anderson during an investigation of cosmic rays. Anderson used a cloud chamber, a common experimental tool in the early days of particle physics. In the cloud chamber, supercooled vapor condenses around a line of ions created by the passage of a charged particle; the result is a visible track whose density depends on the particle's speed. When the cloud chamber is placed in a magnetic field, the paths of charged particles are curved; measurements of the curvature of their trajectories can be used to determine the charge and momentum, and thus the mass, of the particles.

In one historic cloud-chamber photograph (Figure 30.12), the density and curvature of a certain track suggested a mass equal to that of the electron. But the track curved the wrong way, showing that the particle had positive charge. Anderson concluded correctly that the track had been made by a positive electron, or **positron.** The mass of the positron is equal to that of an ordinary (negative) electron; its charge is equal in magnitude, but opposite in sign, to the electron charge. Pairs of particles related to each other in this way are said to be **antiparticles** of each other.

Positrons do not form a part of ordinary matter. They are produced in high-energy collisions of charged particles or gamma rays with matter in a process called **pair production,** in which an ordinary electron (e^- or β^-) and a positron (e^+ or β^+) are produced simultaneously (Figure 30.13). Electric charge is conserved in this process, but enough energy E must be available to account for the energy equivalent of the rest masses m of the two particles. The minimum energy for pair production is

$$E = 2mc^2 = 2(9.11 \times 10^{-31}\,\text{kg})(3.00 \times 10^8\,\text{m/s})^2$$
$$= 1.64 \times 10^{-13}\,\text{J} = 1.02\,\text{MeV}.$$

The inverse process, e^+e^- annihilation, occurs when a positron and an electron collide. Both particles disappear, and two or three gamma-ray photons appear, with total energy $2mc^2$. Decay into a single photon is impossible because such a process cannot conserve both energy and momentum.

Positrons also occur in the decay of some unstable nuclei. Recall that nuclei with too many neutrons for stability often emit a β^- particle (an electron), decreasing N by one and increasing Z by one. A nucleus with too few neutrons for stability may respond by converting a proton to a neutron and emitting a positron (β^+), thereby increasing N by one and decreasing Z by one. Such nuclides don't occur in nature, but they can be produced artificially by neutron bombardment of stable nuclides in nuclear reactors. An example is the unstable odd–odd nuclide ^{22}Na, which has one less neutron than the stable ^{23}Na. The nuclide ^{22}Na decays with a half-life of 2.6 y by emitting a positron, leaving the stable even–even nuclide ^{22}Ne, with the same mass number $A = 22$.

Mesons as Force Mediators

In classical physics, we describe the interaction of charged particles in terms of Coulomb's-law forces. In quantum mechanics, we can describe this interaction in terms of the emission and absorption of photons. Two electrons repel each

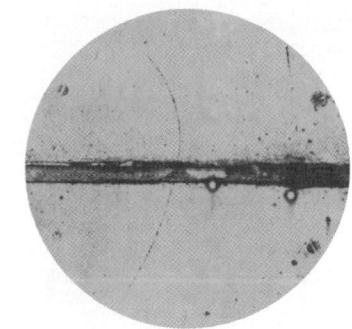

▲ **FIGURE 30.12** The cloud-chamber track made by the first positron ever identified. The lead plate in this photograph is 6 mm thick.

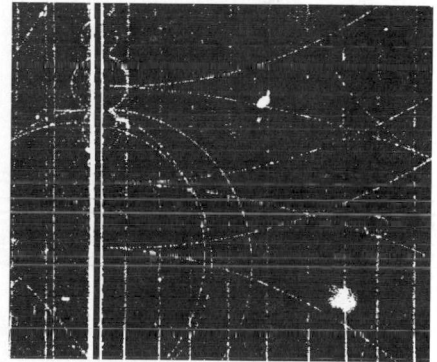

(a)

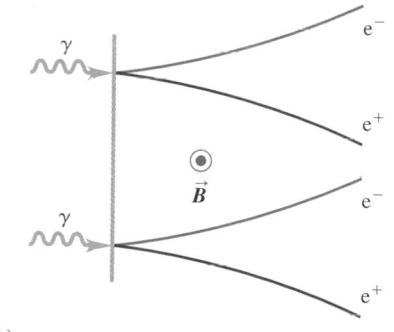

(b)

▲ **FIGURE 30.13** (a) Photograph of bubble-chamber tracks of electron–positron pairs that are produced when 300 MeV photons strike a lead sheet. A magnetic field directed out of the photograph made the electrons and positrons curve in opposite directions. (b) Diagram showing the pair-production process for two of the photons.

Two skaters exert repulsive forces on each other by tossing a ball back and forth.

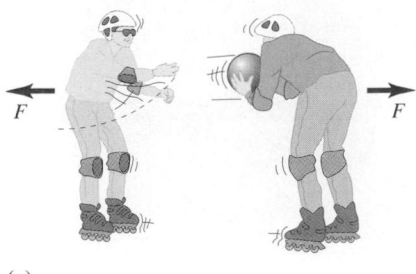

(a)

Two skaters exert attractive forces on each other when one tries to grab the ball out of the other's hands.

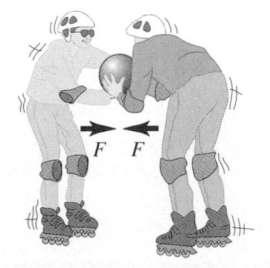

(b)

▲ **FIGURE 30.14** Exchange of a ball as an example of a force mediator.

ActivPhysics 19.5: Particle Physics

other as one emits a photon and the other absorbs it, just as two skaters can push each other apart by tossing a ball back and forth between them (Figure 30.14a). If the charges are opposite and the force is attractive, we imagine the skaters grabbing the ball away from each other (Figure 30.14b). We say that the electromagnetic interaction between two charged particles is *mediated,* or transmitted, by photons.

In 1935, the nature of the nuclear force, which binds protons and neutrons together despite the electrical repulsions of the protons, was a complete mystery. In that year, the Japanese physicist Hideki Yukawa (1907–1981) proposed a hypothetical particle he called a **meson,** with mass intermediate between those of the electron and the proton. Yukawa suggested that nucleons could interact by emitting and absorbing mesons in a process analogous to the electrical interaction between charged particles by the exchange of photons. At the time, there was not the slightest shred of evidence that mesons actually existed. But it was clear that a radical idea was needed to understand nuclear forces, and the scientific world was receptive to Yukawa's proposal.

There were some blind alleys encountered in developing Yukawa's idea. In 1937, two new particles were discovered. These particles are now called **muons.** The μ^- has a charge equal to that of the electron, and its antiparticle, the μ^+, has a positive charge of equal magnitude. The two particles have equal mass, about 207 times the electron mass (that is, $207m_e$). The energy equivalent is about 106 MeV, so the rest mass can also be expressed as $106\ \text{MeV}/c^2$. Muons, like electrons, have spin $\frac{1}{2}$. They are unstable; each decays, usually into an electron with the same sign, plus two neutrinos, with a lifetime of about 2.2×10^{-6} s. In short, muons act very much like massive electrons.

It soon became clear that these particles had no strong interactions with nuclei and therefore could not be Yukawa's mesons. But in 1947 yet another family of unstable particles was discovered. These were christened π mesons or **pions;** they have charges $+e$, $-e$, and zero. The charged pions $(\pi^+$ and $\pi^-)$ have masses of about 273 times the electron mass $(273m_e)$, with rest energy 140 MeV. Each usually decays into a muon with the same sign, plus a neutrino, with a lifetime of about 2.6×10^{-8} s. The neutral pion (π^0) has a smaller mass, about 264 electron masses $(264m_e)$, with a corresponding rest energy of 135 MeV, and decays into two gamma-ray photons with an extremely short lifetime of about 8.4×10^{-17} s. The pions *do* interact strongly with nuclei, and they *are* the particles predicted by Yukawa as mediators of the nuclear force.

The discovery of pions in 1947 resolved some questions about the nuclear force. But more surprises loomed on the horizon, and an entire new branch of physics, now called *high-energy physics,* began to unfold. In the next section, we'll sketch briefly some aspects of this new field.

30.9 High-Energy Physics

From the security of 1931, when it was thought that there were three permanent, unchanging fundamental particles, we enter the partly mapped territory of present-day particle physics, in the midst of a veritable deluge of new particles. The entities that we're accustomed to calling particles are not permanent; they can be created and destroyed during interactions with other particles. Each particle has an associated antiparticle; for a few, the particle and its antiparticle are identical. Many particles are unstable, decaying spontaneously into other particles. Some particles serve as mediators, or transmitters, of the various interactions.

The Four Forces

Here's where things stood in 1950: Four categories of interactions (often called *forces*) had been identified at that time. They are, in order of decreasing strength,

1. the strong interaction,
2. the electromagnetic interaction,
3. the weak interaction, and
4. the gravitational interaction.

The electromagnetic and gravitational forces are familiar from classical physics. Both have a $1/r^2$ dependence on distance, but the gravitational force is very much weaker than the electromagnetic force. For example, the gravitational attraction of two protons is smaller than their electrical repulsion by a factor of about 10^{36}. The gravitational force is of primary importance in the structure of stars and the large-scale behavior of the universe, but it is not believed to play a significant role in particle interactions at currently attainable energies.

The other two forces are less familiar. The strong interaction is responsible for the nuclear force and also for the production of pions and several other particles in high-energy collisions. Within its range, the strong interaction is roughly 100 times as strong as the electromagnetic interaction, but it drops off with distance more quickly than $1/r^2$. The fourth interaction, called the weak interaction, is responsible for beta decay, such as the conversion of a neutron into a proton, an electron, and an antineutrino. It is also responsible for the decay of many unstable particles, such as pions into muons, muons into electrons, Σ particles into protons, and so on. Like the strong interaction, the weak interaction is a short-range interaction, but it is weaker by a factor of about 10^9.

More Particles

We've outlined the most important properties of muons and pions. The next major event was the discovery of the **antiproton** (denoted by $\bar{p}$). The existence of this particle had been suspected ever since the discovery of the positron in 1932. The $\bar{p}$ was finally found in 1955, when proton–antiproton (p–$\bar{p}$) pairs were created with the use of a beam of 6 GeV protons. The antineutron was found soon afterward. Especially after 1960, as higher-energy accelerators and more sophisticated detectors were developed, a veritable blizzard of new unstable particles were identified. To describe and classify them, we need a small blizzard of new terms.

It became clear that it's useful to classify particles in terms of their interactions. The two principal categories are **hadrons,** which have strong interactions, and **leptons,** which do not. Hadrons include baryons and mesons; baryons are nucleons and more massive particles that resemble nucleons. We'll also distinguish between **bosons,** which always have zero or integer spins, and **fermions,** which have half-integer spins. Fermions obey the exclusion principle; bosons do not.

Leptons

Leptons include the electrons ($e^{\pm}$), the muons ($\mu^{\pm}$), the tau particles ($\tau^{\pm}$), and three kinds of neutrinos. The taus, discovered in 1975, have spin $\frac{1}{2}$ and mass 1784 MeV/c^2. In all, there are six leptons and six antileptons. All leptons have spin $\frac{1}{2}$. Taus and muons are unstable; a tau can decay into a muon plus two neutrinos, and a muon decays into an electron plus two neutrinos. Originally, the neutrinos were believed to have zero rest mass, but recent experimental evidence shows that they have small nonzero masses. Note that the τ particles are more massive than nucleons; they are classified as leptons rather than baryons because they have no strong interactions. Leptons obey a *conservation principle.* For the three kinds of

TABLE 30.4 The leptons

Particle name	Symbol	Anti-particle	Mass (MeV/c^2)	L_e	L_μ	L_τ	Lifetime (s)	Principal decay modes
Electron	e^-	e^+	0.511	+1	0	0	Stable	
Electron neutrino	ν_e	$\bar{\nu}_e$	$<3 \times 10^{-6}$	+1	0	0	Stable	
Muon	μ^-	μ^+	105.7	0	+1	0	2.20×10^{-6}	$e^- \bar{\nu}_e \nu_\mu$
Muon neutrino	ν_μ	$\bar{\nu}_\mu$	<0.19	0	+1	0	Stable	
Tau	τ^-	τ^+	1777	0	0	+1	2.9×10^{-13}	$\mu^- \bar{\nu}_\mu \nu_\tau, e^- \bar{\nu}_e \nu_\tau$
Tau neutrino	ν_τ	$\bar{\nu}_\tau$	<18.2	0	0	+1	Stable	

leptons there are three lepton numbers L_e, L_μ, and L_τ. In all interactions, each lepton number is separately conserved. Table 30.4 lists all the known leptons.

Hadrons

Hadrons, the strongly interacting particles, are a much more complex family than leptons. There are two subclasses: mesons and baryons. **Mesons** have spin 0 or 1, and baryons have spin $\frac{1}{2}$ or $\frac{3}{2}$. Therefore, all mesons are bosons, and all baryons are fermions. Mesons include the pions, already mentioned, and several other particles, including K mesons (or *kaons*), η mesons, and other particles that we'll mention later. **Baryons** include the nucleons and several particles called *hyperons,* including the Λ, Σ, Ξ, and Ω. These resemble the nucleons but are more massive. All the hyperons are unstable, decaying by various processes to other hyperons or to nucleons. All the mesons are unstable because all can decay to less massive particles, in accordance with all the conservation laws governing such decays. Each hadron has an associated antiparticle, denoted with an overbar, like the antiproton $\bar{p}$. Table 30.5 is a sampling of the many known hadrons (including mesons and baryons). We'll discuss the terms *strangeness, isospin,* and *quark content* later.

TABLE 30.5 Some known hadrons and their properties

Symbol	Mass, MeV/c^2	Charge	Spin	Isospin	Strangeness	Mean lifetime, s	Typical decay modes	Quark content
Mesons								
π^0	135.0	0	0	1	0	8.4×10^{-17}	$\gamma\gamma$	$u\bar{u}, d\bar{d}$
π^+	139.6	+1	0	1	0	2.60×10^{-8}	$\mu^+ \nu_\mu$	$u\bar{d}$
π^-	139.6	−1	0	1	0	2.60×10^{-8}	$\mu^- \bar{\nu}_\mu$	$\bar{u}d$
K^+	493.7	+1	0	1/2	+1	1.24×10^{-8}	$\mu^+ \nu_\mu$	$u\bar{s}$
K^-	493.7	−1	0	1/2	−1	1.24×10^{-8}	$\mu^- \bar{\nu}_\mu$	$\bar{u}s$
η^0	547.3	0	0	0	0	$\sim 10^{-18}$	$\gamma\gamma$	$u\bar{u}, d\bar{d}, s\bar{s}$
Baryons								
p	938.3	+1	1/2	1/2	0	stable	—	uud
n	939.6	0	1/2	1/2	0	886	$pe^- \bar{\nu}_e$	udd
Λ^0	1115	0	1/2	0	−1	2.63×10^{-10}	$p\pi^-$ or $n\pi^0$	uds
Σ^+	1189	+1	1/2	1	−1	8.02×10^{-11}	$p\pi^0$ or $n\pi^+$	uus
Δ^{++}	1232	+2	3/2	3/2	0	$\sim 10^{-23}$	$p\pi^+$	uuu
Ξ^-	1321	−1	1/2	1/2	−2	1.64×10^{-10}	$\Lambda^0 \pi^-$	dss
Ω^-	1672	−1	3/2	0	−3	8.2×10^{-11}	$\Lambda^0 K^-$	sss
Λ_c^+	2285	1	1/2	0	0	2.0×10^{-13}	$pK^- \pi^+$	udc

Baryons obey the principle of **conservation of baryon number.** We assign a baryon number $B = 1$ to each baryon and $B = -1$ to each antibaryon. **In all interactions, the total baryon number is conserved.** This principle is the reason that in all nuclear reactions the total mass number A must be conserved. Protons and neutrons are baryons, and A is the total number of nucleons in a nucleus.

EXAMPLE 30.10 Conservation of baryon number

Which of the following reactions obey the law of conservation of baryon number?

A. $n + p \rightarrow n + p + p + \bar{p}$;
B. $n + p \rightarrow n + p + \bar{p}$

SOLUTION

SET UP AND SOLVE In each case, the initial baryon number is $1 + 1 = 2$. (A) The final baryon number for this reaction is $1 + 1 + 1 + (-1) = 2$, so baryon number is conserved. (B) The final baryon number for this reaction is $1 + 1 + (-1) = 1$.

REFLECT Baryon number is *not* conserved, and reaction B does not occur in nature.

Strangeness

The K mesons and the Λ and Σ hyperons appeared on the scene during the late 1950s. Because of their unusual behavior, they were called *strange particles.* They were produced in high-energy collisions such as $\pi^+ + p$, and a K and a hyperon were always produced together. The frequency of occurrence of the production process suggested that it was a strong-interaction process, but the relatively long lifetimes of these particles suggested that their decay was a weak-interaction processes. Even stranger, the K^0 appeared to have two lifetimes, one (about 9×10^{-11} s) characteristic of strong-interaction decays, the other nearly 600 times longer. So were the K mesons hadrons or not?

This question led physicists to introduce a new quantity called **strangeness.** The hyperons Λ^0 and $\Sigma^{\pm,0}$ were assigned a strangeness value $S = -1$, and the associated K^0 and K^+ mesons were assigned a value of $S = 1$ (as shown in Table 30.5). The corresponding antiparticles have opposite strangeness, $S = +1$ for $\overline{\Lambda}^0$ and $\overline{\Sigma}^{\pm,0}$ and $S = -1$ for $\overline{K}^0$ and K^-. Strangeness is *conserved* in production processes such as

$$p + \pi^- \rightarrow \Sigma^- + K^+ \quad \text{and} \quad p + \pi^- \rightarrow \Lambda^0 + K^0.$$

The process $p + \pi^- \rightarrow p + K^-$ is forbidden by conservation of strangeness, and it doesn't occur in nature.

When strange particles decay individually, strangeness is usually not conserved. Typical processes include

$$\Sigma^+ \rightarrow n + \pi^+,$$
$$\Lambda^0 \rightarrow p + \pi^-,$$
$$K^- \rightarrow \pi^+ + \pi^- + \pi^-.$$

In each of these processes, the initial strangeness is 1 or -1, and the final strangeness is zero. All observations of these particles are consistent with the conclusion that strangeness is conserved in strong interactions, but that it can change by zero or one unit in weak interactions.

Conservation Laws

The classical conservation laws of energy, momentum, angular momentum, and electric charge are believed to be obeyed in all interactions. These laws are called

absolute conservation laws. Baryon number is also conserved in all interactions. The decay of strange particles provides our first example of a **conditional conservation law,** one that is obeyed in some interactions but not in others. Strangeness is conserved in strong and electromagnetic interactions, but not in all weak interactions.

Conceptual Analysis 30.5

Particle creation

One reason a photon could not create an odd number of electrons plus positrons is that such a process would

A. not conserve charge.
B. not conserve energy.
C. require photon energies that are not attainable.
D. result in the creation of mass.

SOLUTION One of the observed rules of nature is that during any type of particle interaction or change, the net charge (the number of positive charges minus the number of negative charges) does not change, a rule called conservation of charge. If a photon, a zero-charge particle, converts into one positron and one electron, the net charge still adds to zero. An odd number of particles must include either an extra electron or an extra positron, violating conservation of charge.

Quarks

The leptons form a fairly neat package: three mass particles and three neutrinos, each with its antiparticle, together with a conservation law relating their numbers. Physicists now believe that leptons are genuinely fundamental particles. The hadron family, by contrast, is a mess. Table 30.5 shows only a small sample of more than 250 hadrons that have been discovered since 1960. It has become clear that these particles *do not* represent the most fundamental level of the structure of matter; instead, there is at least one additional level of structure.

Our present understanding of the nature of this level is built on a proposal made initially in 1964 by Murray Gell-Mann and his collaborators. In this proposal, hadrons are *not* fundamental particles, but are composite structures whose constituents are spin-$\frac{1}{2}$ fermions called **quarks.** Each baryon is composed of three quarks (qqq), and each meson is a quark–antiquark pair $(q\bar{q})$. No other combinations seem to be necessary. This scheme requires that quarks have electric charges with magnitudes one-third and two-thirds of the electron charge e, which was previously thought to be the smallest unit of charge. Quarks also have fractional values of the baryon number B. Two quarks can combine with their spins parallel to form a particle with spin 1 or with their spins antiparallel to form a particle with spin 0. Similarly, three quarks can combine to form a particle with spin $\frac{1}{2}$ or $\frac{3}{2}$.

The first quark theory included three types (flavors) of quarks, labeled u (up), d (down), and s (strange), as shown in Table 30.6. The corresponding antiquarks $\bar{u}$, $\bar{d}$, and $\bar{s}$ have opposite values of Q, B, and S. Protons, neutrons, π and K mesons, and several hyperons can be constructed from these three quarks. (We describe the charge Q of a particle as a multiple of the magnitude e of the electron charge.) For example, the proton quark content is uud; a proton has $Q/e = +1$, baryon number $B = +1$, and strangeness $S = 0$. From Table 30.6, the u quark has $Q/e = \frac{2}{3}$ and $B = \frac{1}{3}$, and the d quark has $Q/e = -\frac{1}{3}$ and $B = \frac{1}{3}$. So the values of Q/e add to 1, and the values of the baryon number B also add to 1, as we would expect.

TABLE 30.6 Properties of the three original quarks

Particle	Q/e	Spin	Baryon number	Strangeness	Charm	Bottomness	Topness
u	$\frac{2}{3}$	$\frac{1}{2}$	$\frac{1}{3}$	0	0	0	0
d	$-\frac{1}{3}$	$\frac{1}{2}$	$\frac{1}{3}$	0	0	0	0
s	$-\frac{1}{3}$	$\frac{1}{2}$	$\frac{1}{3}$	-1	0	0	0

The neutron is **udd**, with total $Q = 0$ and $B = 1$. The π^+ meson is **u$\bar{\text{d}}$**, with $Q/e = 1$ and $B = 0$, and the K^+ meson is **u$\bar{\text{s}}$**. The antiproton is $\bar{p} = \overline{\text{uud}}$, the negative pion is $\pi^- = \overline{\text{ud}}$, and so on. Table 30.5 lists the quark content of some of the hadrons (mesons and baryons).

In the standard model, the attractive interactions that hold quarks together are assumed to be mediated by massless spin-1 bosons called **gluons,** just as photons mediate the electromagnetic interaction and pions mediate the nucleon–nucleon interaction in the old Yukawa theory. Quarks, having spin $\frac{1}{2}$, are fermions and so are subject to the exclusion principle. They come in three colors, and the exclusion principle is assumed to apply separately to each color. A baryon always contains one red, one green, and one blue quark, so the baryon itself has no color. Each gluon has a color and an anticolor, and color is conserved during the emission and absorption of a gluon by a quark. The color of an individual quark changes continually because of gluon exchange.

The theory of strong interactions is known as **quantum chromodynamics** (QCD). Individual free quarks have not been observed. In most QCD theories, there are phenomena associated with the creation of quark–antiquark pairs that make it impossible to observe a single free, isolated quark. Nevertheless, an impressive body of experimental evidence supports the correctness of the quark structure of hadrons and the belief that quantum chromodynamics is the key to understanding the strong interactions.

More Quarks

Although the three quarks we've described seem to fit the particles known in 1970, they are not by any means the end of the story. Additional particles discovered since 1970, as well as theoretical considerations based on symmetry properties, point to the existence of three additional quarks: **c**, **b**, and **t**. Experimental confirmation of this conjecture came in 1995 with the discovery of the **t** quark. Table 30.7 lists some properties of the six quarks.

The Standard Model

The particles and interactions we've described, including six quarks (from which all the hadrons are made), six leptons, and the particles that mediate those interactions, form a reasonably comprehensive picture of the fundamental building blocks of nature; this scheme has come to be called the **standard model.** The strong interaction among quarks is mediated by gluons, and the electromagnetic interaction among charged particles is mediated by photons, both spin-1 bosons. The weak interaction is mediated by exchange of the weak bosons $\left(W^\pm \text{ and } Z^0 \right)$, spin-1 particles with enormous masses, 80 GeV/c^2 and 91 GeV/c^2. The gravitational interaction is thought to be mediated by a massless spin-2 boson called the graviton, which has not yet been observed experimentally.

It has long been a dream of particle theorists to be able to combine all four interactions of nature into a single unified theory. In 1967, Steven Weinberg and

TABLE 30.7 Properties of quarks

Particle	Q/e	Spin	Baryon number	Strangeness	Charm	Bottomness	Topness
u	$\frac{2}{3}$	$\frac{1}{2}$	$\frac{1}{3}$	0	0	0	0
d	$-\frac{1}{3}$	$\frac{1}{2}$	$\frac{1}{3}$	0	0	0	0
s	$-\frac{1}{3}$	$\frac{1}{2}$	$\frac{1}{3}$	1	0	0	0
c	$\frac{2}{3}$	$\frac{1}{2}$	$\frac{1}{3}$	0	+1	0	0
b	$-\frac{1}{3}$	$\frac{1}{2}$	$\frac{1}{3}$	0	0	+1	0
t	$\frac{2}{3}$	$\frac{1}{2}$	$\frac{1}{3}$	0	0	0	+1

Abdus Salam proposed a theory that treats the weak and electromagnetic forces as two aspects of a single interaction at sufficiently high energies. This **electroweak theory** was successfully verified in 1983 with the discovery of the weak-force intermediary particles, the Z^0 and $W^{\pm}$ bosons by two experimental groups working at the $p\bar{p}$ collider at the CERN Laboratory in Geneva, Switzerland. Weinberg, Salam, and Sheldon Glashow, who also contributed to the theory, received the 1979 Nobel Prize in Physics for their work.

Grand Unified Theories

Can the theory of strong interaction and the electroweak theory be unified to give a comprehensive theory of strong, weak, and electromagnetic interactions? Such schemes, called **grand unified theories** (GUTs), are still speculative in nature. One interesting feature of some grand unified theories is that they predict the decay of the proton (violating the conservation of baryon number), with an estimated lifetime on the order of 10^{30} years. Experiments are under way that, theoretically, should have detected the decay of the proton if its lifetime is 10^{30} years or less. Such decays have not been observed, but experimental work continues.

In the standard model, the neutrinos have zero mass, although experiments designed to determine neutrino masses are extremely difficult to perform. In most GUTs, neutrinos must have nonzero masses. If they do have mass, transitions called neutrino oscillations can occur in which one type of neutrino $(\nu_e, \nu_\mu,$ or $\nu_\tau)$ changes into another type. Recent (1998) experiments have indeed confirmed the existence of neutrino oscillations, showing that neutrinos do have nonzero masses. The present upper limits on neutrino masses are shown in Table 30.4. This discovery has cleared up a long-standing mystery about neutrinos coming from the sun. The observed flux of solar electron neutrinos is only one-third of the predicted value. It now appears that the sun is indeed producing electron neutrinos at the predicted rate, but that two-thirds of them are transformed into muon or tau neutrinos as they pass through and interact with the material of the sun.

The ultimate dream of theorists is to unify all four fundamental interactions, including gravitation as well as the strong and electroweak interactions included in GUTs. Such a unified theory is whimsically called a **theory of everything** (TOE). Such theories range from the speculative to the fantastic. One popular ingredient is a space-time continuum with more than four dimensions, containing structures called *strings*. Another element is supersymmetry, which gives every boson a fermion "superpartner." Such concepts lead to the prediction of whole new families of particles, including sleptons, photinos, squarks bound together by gluinos, and even winos and wimps, none of which has been observed experimentally. Theorists are still very far away from a satisfactory TOE.

30.10 Cosmology

In this final section of our book, we'll explore briefly the connections between the early history of the universe and the interactions among fundamental particles. It is surprising and remarkable that at the dawn of the 21st century physicists have found such close ties between the smallest things we know about (the range of the weak interaction, on the order of 10^{-18} m) and the largest (the universe itself, on the order of at least 10^{26} m).

The Expanding Universe

Until early in the 20th century, it was usually assumed that the universe was *static;* stars might move relative to each other, but without any general expansion or contraction. But if everything is initially sitting still in the universe, why doesn't gravity just pull it all together into one big glob? Newton himself recognized the seriousness of this troubling question.

About 1930, astronomers began to find evidence that the universe is *not* static. The motions of distant galaxies relative to earth can be measured by observing the shifts in the wavelengths of their spectra due to the Doppler effect. (This is the same effect for *light* that we studied with *sound* in Section 12.12.) The shifts are called **red shifts,** because they are always toward longer wavelengths, showing that distant galaxies appear to be receding from us and from each other. The astronomer Edwin Hubble measured red shifts from many distant galaxies and came to the astonishing conclusion that the speed of recession, v, of a galaxy is approximately proportional to its distance r from us. This relation is now called **Hubble's law.**

Another aspect of Hubble's observations was that distant galaxies appeared to be receding in all directions. There is no particular reason to think that our galaxy is at the center of the universe; if we were moving along with some other galaxy, everything else would still seem to be receding from us. Thus, the universe looks the same from all locations, and we believe that the laws of physics are the same everywhere.

The Big Bang

An appealing hypothesis suggested by Hubble's law is that at some time in the past, all the matter in the universe was concentrated in a very small space and was blown apart in an immense explosion labeled the **Big Bang,** giving all observable matter more or less the velocities we observe today. By correlating distances with speeds of recession, it has been established that the Big Bang occurred about 14 billion years ago. In this calculation, we've neglected any slowing down due to gravitational attraction. Whether or not this omission is justified depends on the average density of matter; we'll return to this point later.

Critical Density

We've mentioned that the law of gravitation isn't consistent with a static universe. But what is its role in an expanding universe? Gravitational attractions should slow the initial expansion, but by how much? If they are strong enough, the universe should expand more and more slowly, eventually stop, and then begin to contract, perhaps all the way down to what's been called a "Big Crunch." But if the gravitational forces are much weaker, they slow the expansion only a little, and the universe continues to expand forever.

The situation is analogous to the problem of the escape velocity of a projectile launched from earth. A similar analysis can be carried out for the universe. Whether or not the universe continues to expand indefinitely depends on the average density of matter. If matter is relatively dense, there is a lot of gravitational attraction to slow, and eventually stop, the expansion and make the universe contract again. If not, the expansion continues indefinitely. The critical density, denoted by ρ_c, turns out to be

$$\rho_c = 5.8 \times 10^{-27} \text{ kg/m}^3.$$

If the average density of the universe is less than this, the universe continues to expand indefinitely; if it is greater, it eventually stops expanding and begins to contract, possibly leading to the Big Crunch and then another Big Bang. Note that ρ_c is a very small number by terrestrial standards; the mass of a hydrogen atom is 1.67×10^{-27} kg, so ρ_c corresponds to about three hydrogen atoms per cubic meter.

Recent research shows that the average density of all matter in the universe is about 27% of the critical density ρ_c, but that the average density of *luminous* matter (i.e., matter that emits some sort of radiation) is only 4% of ρ_c. In other words, most of the matter in the universe does not emit electromagnetic radiation of any kind. At present, the nature of this **dark matter** remains an outstanding mystery.

We mentioned some candidates for dark matter at the end of Section 30.9, including WIMPs (weakly interactive massive particles) and other subatomic particles far more massive than any that can be produced in accelerator experiments. But whatever the true nature of dark matter, it is by far the dominant form of matter in the universe.

Because the average density of matter in the universe is less than the critical density, we might expect that the universe will continue to expand indefinitely. Gravitational attractions should slow the expansion down, but not enough to stop it. One way to test this prediction is to look at redshifts of extremely distant objects. The surprising results from such measurements show that distant galaxies actually have *smaller* redshifts than values predicted by Hubble's law. The implication is that the expansion has been speeding up rather than slowing down.

Why is the expansion speeding up? The explanation generally accepted by astronomers and physicists is that space is permeated with a kind of energy that has no gravitational effect and emits no electromagnetic radiation, but rather acts as a kind of "antigravity" that produces a universal repulsion. This invisible, immaterial energy is called **dark energy.** The nature of dark energy is poorly understood, but it is the subject of very active research. Present estimates indicate that the energy *density* of dark energy is about three times greater than that of matter, a result which suggests that the total energy density of the universe is greater than the critical density ρ_c and that the universe will continue to expand forever.

The Beginning of Time

The evolution of the universe has been characterized by the continuous growth of a scale factor R, which we can think of, very roughly, as characterizing the size of the universe. In Section 30.9, we mentioned the unification of the electromagnetic and weak interactions at energies of several hundred GeV, constituting the electroweak interaction. Most theorists believe that the strong and electroweak interactions become unified at energies on the order of 10^{14} GeV (as in the GUT models) and that at energies on the order of 10^{19} GeV all four of the fundamental forces (strong, electromagnetic, weak, and gravitational) become unified. Average particle energies were in the latter range at a time on the order of 10^{-43} s after the Big Bang. If we mentally go backward in time, we have to stop at $t = 10^{-43}$ s because we have no adequate theory that unifies all four interactions.

The Standard Cosmological Model

The brief chronology of the Big Bang that follows is called the *standard model.* The figure on pages 1040–1041 presents a graphical description of this history, with the characteristic sizes, particle energies, and temperatures at various times. Referring to this chart frequently will help you understand the discussion that follows. In this model, the temperature at time $t = 10^{-43}$ s was about 10^{32} K, and the average energy per particle was on the order of 10^{19} GeV. In the unified theories, this is about the energy below which gravity begins to behave as a separate interaction. The time $t = 10^{-43}$ s therefore marks the end of any proposed TOE and the beginning of the GUT period.

During the GUT period, from $t = 10^{-43}$ s to $t = 10^{-35}$ s, the strong and electroweak forces were still unified, and the universe consisted of a soup of quarks and leptons transforming into each other so freely that there was no distinction between the two families of particles. One important characteristic of GUTs is that at sufficiently high energies, baryon number is not conserved. Thus, by the end of the GUT period, the numbers of quarks and antiquarks may have been unequal. This point has important implications; we'll return to it shortly.

By $t = 10^{-35}$ s, the temperature had decreased to about 10^{27} K and the average energy to about 10^{14} GeV. At this energy, the strong force separated from the electroweak force, and baryon number and lepton number began to be separately conserved. Some models postulate an enormous and very rapid expansion at this time, a factor on the order of 10^{50} in 10^{-32} s. As the expansion continued and the average particle energy continued to decrease, quarks began to bind together to form nucleons and antinucleons. After the average energy fell below the threshold for nucleon–antinucleon pair production, many of the nucleons annihilated nearly all the less numerous antinucleons. Later still, the average energy dropped below the threshold for electron–positron pair production; most of the remaining positrons were annihilated, leaving the universe with many more protons and electrons than the antiparticles of each.

By time $t = 225$ s, the average energy dropped below the binding energy of the deuteron, about 2.23 MeV, and the age of *nucleosynthesis* had begun. At first, only the simplest bound states—^{2}H, ^{3}He, and ^{4}He—appeared. Further nucleosynthesis began very much later, at about $t = 2 \times 10^{13}$ s (approximately 700,000 y). By this time, the average energy was a few eV, and electrically neutral H and He atoms could form. With the electrical repulsions of the nuclei canceled out, gravitational attraction could pull the neutral atoms together to form galaxies and, eventually, stars. Thermonuclear reactions in stars are believed to have produced all of the more massive nuclei and, ultimately, the nuclides and chemical elements we know today.

Matter and Antimatter

One of the most remarkable features of our universe is the asymmetry between matter and antimatter. One might think that the universe should have equal numbers of protons and antiprotons, and of electrons and positrons, but this doesn't appear to be the case. There is no evidence for the existence of substantial amounts of antimatter (matter composed of antiprotons, antineutrons, and positrons) anywhere in the universe. Theories of the early universe must confront this imbalance.

We've mentioned that most GUTs allow for the violation of conservation of baryon number at energies at which the strong and electroweak interactions have converged. If particle–antiparticle symmetry is also violated, we have a mechanism for making more quarks than antiquarks, more leptons than antileptons, and, eventually, more nucleons than antinucleons. But any asymmetry created in this way during the GUT era might be wiped out by the electroweak interaction after the end of the GUT era. The problem of the matter–antimatter asymmetry is still very much an open one.

We hope that this qualitative discussion has conveyed at least a hint as to the close connections between particle theory and cosmology. There are many unanswered questions. We don't know what happened during the first 10^{-43} s after the Big Bang because we have no quantum theory of gravity; there are many versions of GUTs from which to choose. We don't know what dark matter and dark energy are. We are still very far from having a suitable theory that unifies all four interactions in nature. But this search for understanding of the physical world we live in continues to be one of the most exciting adventures of the human mind.

1040

AGE OF QUARKS AND GLUONS (GUT Period)
Dense concentration of matter and antimatter; gravity a separate force; more quarks than antiquarks. Inflationary period (10^{-35} s): rapid expansion, strong force separates from electroweak force.

AGE OF NUCLEONS AND ANTINUCLEONS
Quarks bind together to form nucleons and antinucleons; energy too low for nucleon-antinucleon pair production at 10^{-2} s

AGE OF NUCLEOSYNTHESIS
Stable deuterons; matter 74% H, 25% He, 1% heavier nuclei

AGE OF LEPTONS
Leptons distinct from quarks; $W^{\pm}$ and Z^0 bosons mediate weak force (10^{-12} s)

BIG BANG 10^{-43} s 10^{-32} s 10^{-6} s   225 s

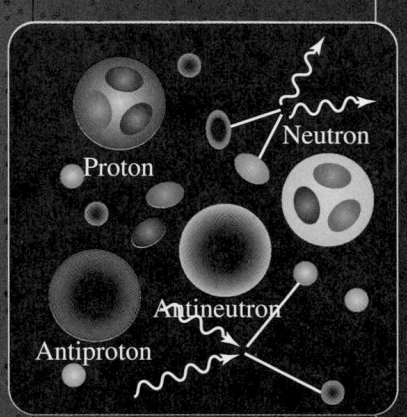

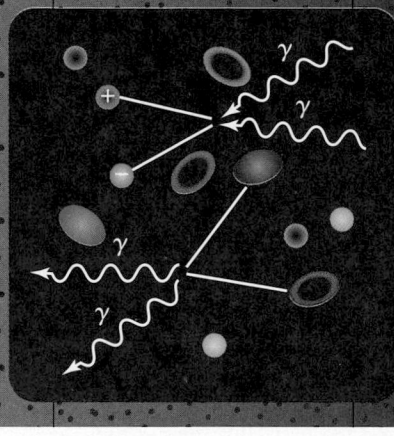

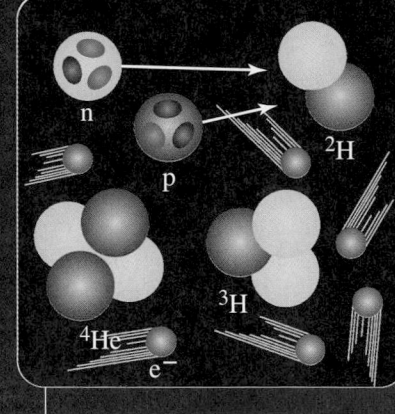

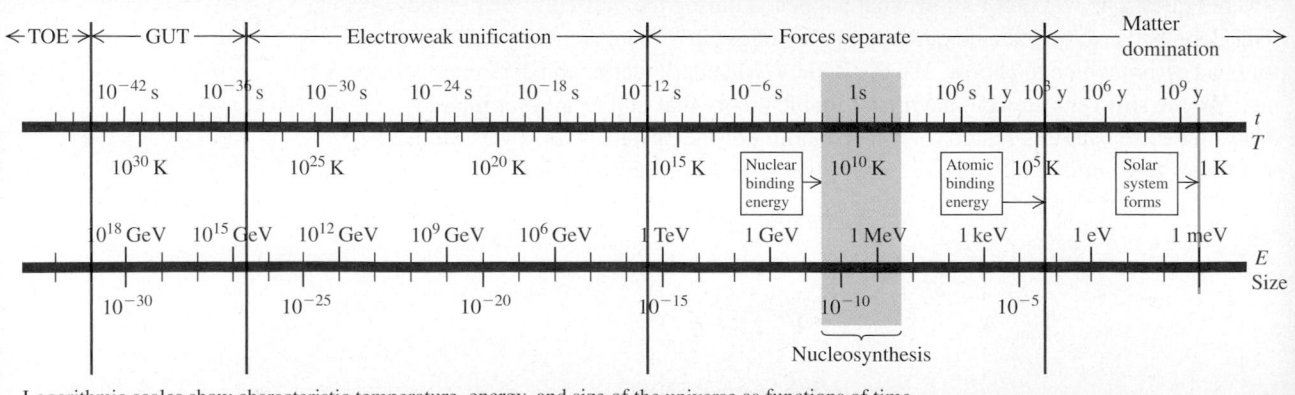

Logarithmic scales show characteristic temperature, energy, and size of the universe as functions of time.

A Brief History of the Universe

AGE OF IONS
...panding, cooling
gas of ionized
H and He

AGE OF ATOMS
Neutral atoms form, pulled
together by gravity; universe becomes
transparent to most light

**AGE OF STARS
AND GALAXIES**
Thermonclear fusion
begins in stars, forming
heavier nuclei

10^{13} s 10^{15} s   **NOW**

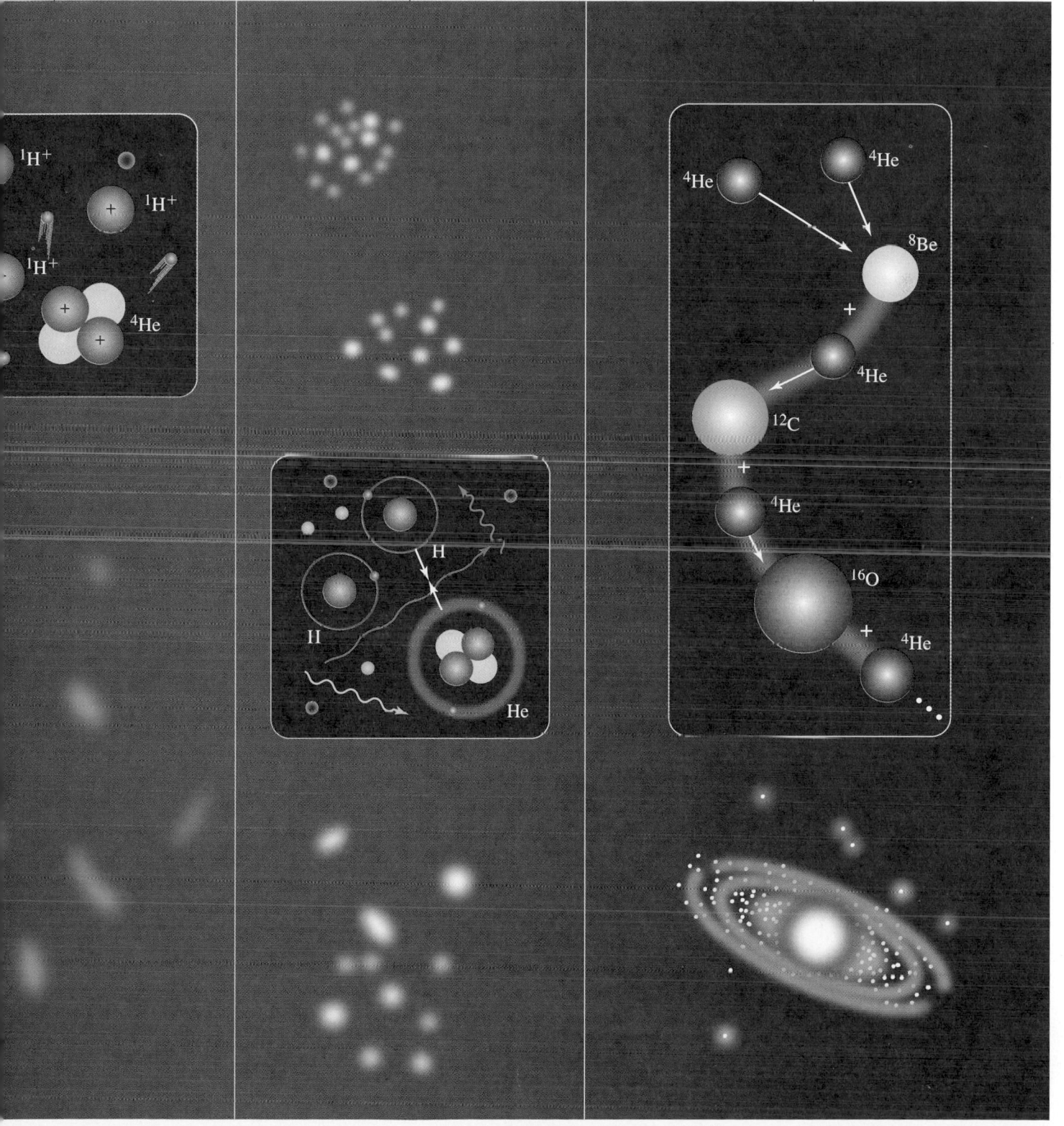

SUMMARY

Properties of Nuclei

(Section 30.1) An atomic nucleus is roughly spherical in shape, with a radius that increases with the number of nucleons (protons and neutrons) A in the nucleus: $R = R_0 A^{1/3}$ (Equation 30.1). The total mass of a nucleus is always *less* than the total mass of its constituent parts because of the mass equivalent $(E = mc^2)$ of the binding energy that hold the nucleus together.

Nuclear Stability

(Section 30.2) The most important reason that some nuclei are stable and others are not is the competition between the attractive nuclear force and the repulsive electrical force. The nuclear force favors *pairs* of nucleons with opposite spin, and *pairs of pairs*. Electrical repulsion tends to favor greater numbers of neutrons, but a nucleus with too many neutrons is unstable because not enough of them are paired with protons.

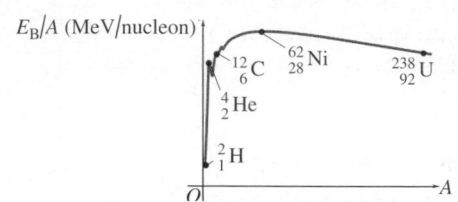

Radioactivity, Radiation, and the Life Sciences

(Sections 30.3 and 30.4) The two most common decay modes are the emission of alpha (α) and beta (β) particles. The **alpha particle** has two protons and two neutrons bound together, with zero total spin. Alpha emission occurs principally with nuclei that are too large to be stable. The **beta particle** (or more specifically, the beta-minus β^- particle) is an electron. The antineutrino $\bar{\nu}_e$, is also involved in the process of β^- decay represented as $n \to p + \beta^- + \bar{\nu}_e$ (Equation 30.4).

Radiation exposure can cause cancers and genetic defects, yet it can be effective in medicine for intentional selective destruction of tissue such as tumors. Radiation can also help diagnose a problem: By injecting low doses of radioactive isotopes into the body, "images" can be made of potentially problematic regions where the radioactive material builds up.

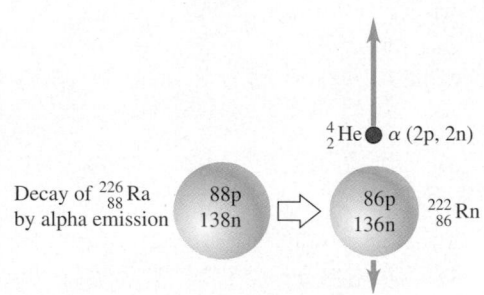

Nuclear Reactions, Fission, and Fusion

(Sections 30.5–30.7) A **nuclear reaction** is the result of bombardment of a nucleus by a particle, rather than a spontaneous natural process as in radioactive decay. For a charged particle such as a proton or α particle to penetrate the nucleus of another atom, it must usually have enough initial kinetic energy to overcome the potential-energy barrier caused by the repulsive electrostatic forces.

Nuclear fission is a decay process in which an unstable nucleus splits into two fragments. The total mass of the resulting fragments is less than the original atom, and this released mass-energy appears mostly as kinetic energy of the fragments. In a **nuclear fusion** reaction, two or more small light nuclei come together or *fuse*, to form a larger nucleus. Fusion reactions release energy for the same reason as fission reactions; the binding energy per nucleon after the reaction is greater than before.

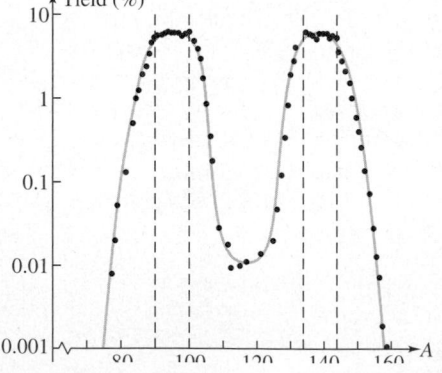

Continued

Fundamental Particles and High-Energy Physics

(Sections 30.8 and 30.9) For each type of massive particle (electron, proton, neutron, etc.) there exists an antiparticle. When a particle and antiparticle collide, they can annihilate. Protons and neutrons, as well as other hadrons, are made of quarks of fractional charge. The proton consists of three quarks, **uud**, while the neutron is made of **udd**. Mesons are also made of quarks, combining a quark and antiquark, as in the π^+ meson, $u\bar{d}$. The world we encounter every day has numerous substances with varying properties. While the properties are many, the fundamental building blocks are few—**u** quarks, **d** quarks, electrons, and photons.

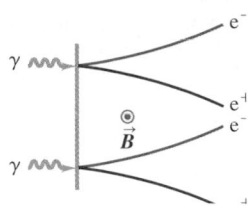

Cosmology

(Section 30.10) The universe has been expanding since the Big Bang, about 14 billion years ago. As expressed in **Hubble's law,** the more distant the galaxy, the greater its speed of recession. The energy content of the universe falls into three categories—regular luminous matter, dark matter, and dark energy. Only a few percent of the universe is regular matter made of protons and neutrons (forming stars and galaxies). About 25% of energy in the universe is **dark matter,** and we still don't know what it's made of. More recently, there's evidence that about 75% of the energy in the universe is **dark energy**—we don't know how this energy comes about, but we know it isn't mass. Taken together, the total energy appears to exceed critical density; if so, we will expand forever.

 For instructor-assigned homework, go to www.masteringphysics.com

Conceptual Questions

1. Since different isotopes of the same element have the same chemical behavior, how can they be separated from each other?

2. In beta decay, a neutron becomes a proton, an electron, and an antineutrino. This type of decay also occurs with free neutrons, with a half-life of about 15 min. Could a free *proton* undergo a similar decay to become a neutron, a positron (positive electron), and a neutrino? Why or why not?

3. In the ^{238}U decay chain shown in Figure 30.5, some nuclides in the chain are found much more abundantly in nature than others, despite the fact that every ^{238}U nucleus goes through every step in the chain before finally becoming ^{206}Pb. Why are the abundances of the intermediate nuclides not all the same?

4. Why aren't the masses of all nuclei integer multiples of the mass of a single nucleon?

5. True or false? During one half-life, the *mass* of a radioisotope is reduced by half. Explain.

6. Why is the decay of an unstable nucleus not affected by the *chemical* situation of the atom, such as the nature of the molecule in which it is bound?

7. Changing the temperature of atoms affects their chemical reaction rate, but has no effect on their rate of radioactive decay. Why is one rate affected, but not the other?

8. The only two stable nuclides with more protons than neutrons are $^1_1$H and $^3_2$He. Why is $Z > N$ so uncommon?

9. Why do high-Z nuclei require more neutrons than protons to be stable, as indicated in the Segrè diagram in Figure 30.2?

10. The binding energy curve in Figure 30.1 slopes *downward* for heavy nuclei. Why does this happen? Why does the binding energy per nucleon *not* keep increasing?

11. Nuclear power plants use nuclear fission reactions to generate steam to run steam-turbine generators. How does the nuclear reaction produce *heat?*

12. **Electron capture.** There are cases in which a nucleus having too few neutrons for stability can capture one of the electrons in the K shell of the atom. This process is known as *electron capture* (or, sometimes, *K*-capture). What is the effect of electron capture on N, A, and Z? The atom ^{103}Pd decays this way; identify the daughter nucleus that is produced.

13. **Positron emission.** One form of beta decay (known as *positron emission*) is due to the decay of a proton to a positron (a positive electron, β^+), a neutron, and a neutrino. What is the effect of positron emission on N, A, and Z? The atom ^{40}K decays this way; identify the daughter nucleus that is produced. (Note that a *free* proton cannot decay this way, only a proton in the nucleus.)

14. Is it possible that some parts of the universe contain antimatter whose atoms have nuclei made of antiprotons and antineutrons, surrounded by positrons? How might we detect this condition without actually going there? What problems could arise if we actually *did* go there?
15. Why are so many health hazards associated with fission fragments that are produced during the fission of heavy nuclei?
16. According to the standard model of the fundamental particles, what are the similarities between baryons and leptons? What are the most important differences?

Multiple-Choice Problems

1. Which of the following statements about the atomic nucleus are correct? (There may be more than one correct choice.)
   A. It is held together by the strong force that affects protons and neutrons, but not electrons.
   B. It has a typical radius on the order of 10^{-10} m.
   C. Its volume is nearly the same for all atoms.
   D. The density of heavy nuclei is considerably greater than the density of light nuclei.

2. A nucleus containing 60 nucleons has a radius R. The radius of a nucleus containing 180 nucleons would be closest to
   A. $3R$.
   B. $3^{2/3}R = 2.08R$.
   C. $3^{1/3}R = 1.44R$.
   D. $\sqrt{3}R = 1.73R$.

3. The hypothetical atom $^{37}_{17}X$ contains (there may be more than one correct choice)
   A. 17 orbital electrons.
   B. 37 protons.
   C. 17 protons.
   D. 20 neutrons.
   E. 54 nucleons.

4. Which of the following statements about a typical atomic nucleus are correct? (There may be more than one correct choice.)
   A. The number of neutrons is equal to the number of protons.
   B. It contains protons, neutrons, and neutrinos.
   C. It usually contains more neutrons than protons.
   D. It has a typical radius on the order of 10^{-15} m.

5. Which nuclide X would properly complete the following reaction:
   $$^1_0n + ^{235}_{92}U \rightarrow ^{88}_{38}Sr + X + 12^1_0n$$
   A. $^{146}_{54}Xe$
   B. $^{148}_{52}Te$
   C. $^{136}_{54}Xe$
   D. $^{136}_{42}Mo$

6. One problem in radiocarbon dating of biological samples, especially very old ones, is that they can easily be contaminated with modern biological material during the measurement process. What effect would this contamination have on the measurements?
   A. Contaminated samples would be measured to be older than they really are.
   B. The half-life of the radioactive carbon would appear to be shorter.
   C. Contaminated samples would be measured to be younger than they really are.
   D. The decay rate of the radioactive carbon would appear to be increased.

7. In a nuclear accident, a lab worker receives a dose of D rads of radiation of x rays having an RBE of 1. If instead he had been exposed to the same amount of energy from alpha particles having an RBE of 20, his exposure would have been (there may be more than one correct choice)
   A. D rads.
   B. $20D$ rads.
   C. D rem.
   D. $20D$ rem.

8. Atom A has twice as long a half-life as atom B. If both of them start out with an amount N_0, then in the time that A has been reduced to $N_0/2$, B has been reduced to
   A. $N_0/2$.
   B. $N_0/4$.
   C. $N_0/8$.
   D. $N_0/16$.

9. If the binding energy per nucleon for hypothetical element $^{13}_7X$ is 5 MeV/nucleon, the total binding energy of its nucleus is
   A. 5 MeV.
   B. 30 MeV.
   C. 35 MeV.
   D. 65 MeV.
   E. 100 MeV.

10. A certain atomic nucleus containing 50 nucleons has a density ρ. If another nucleus contains 150 nucleons, its density will be closest to
   A. 3ρ.
   B. $3^{2/3}\rho = 2.08\rho$.
   C. $3^{1/3}\rho = 1.44\rho$.
   D. $\sqrt{3}\rho = 1.73\rho$.
   E. ρ.

11. A beaker contains a pure sample of a radioactive substance having a half-life of 1 h. Which statement about this substance is correct? (There may be more than one correct choice.)
   A. After 1 h, half of the atoms will have decayed, and the remainder will decay in the next 1 h.
   B. After 1 h, half of the atoms will have decayed, and half of the remainder will decay in the next 1 h.
   C. After 1 h, the mass in the beaker will be only half its original value.

12. After a particle was emitted from the nucleus of an atom, it was found that the atomic number of the atom had *increased*. The emitted particle could have been (there may be more than one correct choice)
   A. a β^- particle.
   B. an alpha particle.
   C. a positron (β^+).
   D. a gamma ray.

13. In a nuclear reactor accident, one worker receives a dose of 20 rad of x rays, while a second worker receives a dose of 10 rad of alpha particles. How do the two workers' biologically equivalent doses compare?
   A. The first worker receives twice the equivalent dose of the second worker.
   B. The first worker receives 40 times the equivalent dose of the second worker.
   C. The first worker receives $1/200$ the equivalent dose of the second worker.
   D. The first worker receives $1/10$ the equivalent dose of the second worker.

14. When copper-64, $^{64}_{29}Cu$, undergoes β^- decay, the daughter nucleus contains
   A. 34 protons and 30 neutrons.
   B. 30 protons and 34 neutrons.
   C. 28 protons and 36 neutrons.
   D. 27 protons and 33 neutrons.

15. The radiation from a radioactive sample of a single isotope decreases to one-eighth of the original intensity I_0 in 30 years. What would the intensity be after 10 *more* years?
   A. $I_0/16$.
   B. $I_0/32$.
   C. $I_0/64$.

Problems

30.1 Properties of Nuclei

1. • How many protons and how many neutrons are there in a nucleus of (a) neon, $^{21}_{10}$Ne (b) zinc, $^{65}_{30}$Zn (c) silver, $^{108}_{47}$Ag.

2. • Calculate the approximate (a) radius, (b) volume, and (c) density (in kg/m^3) of a nucleus of gold, $^{197}_{79}$Au. (You can ignore the binding energy of the nucleus for this calculation.)

3. •• **Density of the nucleus.** (a) Using the empirical formula for the radius of a nucleus, show that the volume of a nucleus is directly proportional to its nucleon number A. (b) Give a reasonable argument concluding that the mass m of a nucleus of nucleon number A is approximately $m = m_p A$, where m_p is the mass of a proton. (c) Use the results of parts (a) and (b) to show that all nuclei should have about the same *density*. Then calculate this density in kg/m^3, and compare it with the density of lead (which is 11.4 g/cm^3) and a neutron star (about 10^{17} kg/m^3).

4. • For the common isotope of nitrogen, ^{14}N, calculate (a) the mass defect, (b) the binding energy, and (c) the binding energy per nucleon.

5. • Calculate the binding energy (in MeV) and the binding energy per nucleon of (a) the deuterium nucleus, $^2_1$H, and (b) the helium nucleus, $^4_2$He. (c) How do the results of parts (a) and (b) compare?

6. •• (a) Calculate the total binding energy (in MeV) of the nuclei of ^{56}Fe (of atomic mass 55.934937 u) and of ^{207}Pb (of atomic mass 206.975897 u). (b) Calculate the binding energy per nucleon for each of these atoms. (c) How much energy would be needed to totally take apart each of these nuclei? (d) For which of these atoms are the individual nucleons more tightly bound? Explain your reasoning.

7. • What is the maximum wavelength of a γ ray that could break a deuteron into a proton and a neutron? (This process is called photodisintegration.)

8. •• A photon with a wavelength of 3.50×10^{-13} m strikes a deuteron, splitting it into a proton and a neutron. (a) Calculate the kinetic energy released in this interaction. (b) Assuming the two particles share the energy equally, and taking their masses to be 1.00 u, calculate their speeds after the photodisintegration.

30.2 Nuclear Stability

30.3 Radioactivity

9. • The isotope ^{218}Po decays via α decay. The measured atomic mass of ^{218}Po is 218.008973 u, and the daughter nucleus atomic mass is 213.999805 u. (See also Table 30.2.) (a) Identify the daughter nucleus by name, nucleon number, atomic number, and neutron number. (b) Calculate the kinetic energy (in MeV) of the α particle if we can ignore the recoil of the daughter nucleus.

10. • Tritium ($^3_1$H) is an unstable isotope of hydrogen; its mass, including one electron, is 3.016049 u. (a) Show that tritium must be unstable with respect to beta decay because the decay products ($^3_2$He plus an emitted electron) have less total mass than the tritium. (b) Determine the total kinetic energy (in MeV) of the decay products, taking care to account for the electron masses correctly.

11. •• **Thorium series.** The following decays make up the thorium decay series (the X's are unknowns for you to identify):

$$^{232}\text{Th} \xrightarrow{\alpha} X_1, \qquad ^{228}\text{Ra} \xrightarrow{\beta^-} {}^{228}\text{Ac}, \qquad X_2 \xrightarrow{\beta^-} {}^{228}\text{Th},$$

$$^{228}\text{Th} \xrightarrow{X_3} {}^{224}\text{Ra}, \qquad ^{224}\text{Ra} \xrightarrow{\alpha} {}^{220}\text{Rn}, \qquad ^{220}\text{Rn} \xrightarrow{\alpha} X_4,$$

$X_5 \xrightarrow{\alpha} {}^{212}\text{Pb}$, and $^{212}\text{Pb} \xrightarrow{X_6} {}^{212}\text{Bi}$. The ^{212}Bi then decays by an α decay and a β^- decay, which can occur in either order (α followed by β or β followed by α). (a) Identify each of the six unknowns (X_1, X_2, etc.) by nucleon number, atomic number, neutron number, and name. (b) Write out the decays of ^{212}Bi and indicate the end product of this series. (For some guidance, see the discussion under "Decay Series" in Section 30.3.) (c) Draw a Segrè chart for the thorium series, similar to the one shown in Figure 30.5.

12. •• Suppose that 8.50 g of a nuclide of mass number 105 decays at a rate of 6.24×10^{11} Bq. What is its half-life? (*Hint:* Use the fact that $\Delta N/\Delta t = -\lambda N$. You are given $\Delta N/\Delta t$ and can figure out N knowing the mass number and mass of your sample.)

13. • A radioisotope has a half-life of 5.00 min and an initial decay rate of 6.00×10^3 Bq. (a) What is the decay constant? (b) What will be the decay rate at the end of (i) 5.00 min, (ii) 10.0 min, (iii) 25.0 min?

14. •• A radioactive isotope has a half-life of 76.0 min. A sample is prepared that has an initial activity of 16.0×10^{10} Bq. (a) How many radioactive nuclei are initially present in the sample? (b) How many are present after 76.0 min? What is the activity at this time? (c) Repeat part (b) for a time of 152 min after the sample is first prepared.

15. •• Calcium-47 is a β^- emitter with a half-life of 4.5 days. If a bone sample contains 2.24 g of this isotope, at what rate will it decay?

16. •• A 12.0-g sample of carbon from living matter decays at the rate of 180.0 decays/min due to the radioactive ^{14}C in it. What will be the decay rate of this sample in (a) 1000 years and (b) 50,000 years? (*Hint:* The decay rate is proportional to the number of radioactive carbon atoms remaining; you can therefore replace N and N_0 in Eq. 30.6 with decay rates once you have a value for λ.)

BIO

17. •• **Radioactive tracers.** Radioactive isotopes are often introduced into the body through the bloodstream. Their spread through the body can then be monitored by detecting the appearance of radiation in different organs. ^{131}I, a β^- emitter with a half-life of 8.0 d, is one such tracer. Suppose a scientist introduces a sample with an activity of 375 Bq and watches it spread to the organs. (a) Assuming that the sample all went to the thyroid gland, what will be the decay rate in that gland 24 d (about $2\frac{1}{2}$ weeks) later? (b) If the decay rate in the thyroid 24 d later is actually measured to be 17.0 Bq, what percent of the tracer went to that gland? (c) What isotope remains after the I-131 decays?

BIO

18. •• The common isotope of uranium, ^{238}U, has a half-life of 4.47×10^9 years, decaying to ^{234}Th by alpha emission. (a) What is the decay constant? (b) What mass of uranium is required for an activity of 1.00 curie? (c) How many alpha particles are emitted per second by 10.0 g of uranium?

19. •• A sample of the radioactive nuclide ^{199}Pt is prepared that has an initial activity of 7.56×10^{11} Bq. (a) 92.4 min after the sample is prepared, the activity has fallen to 9.45×10^{10} Bq. What is the half-life of this nuclide? (b) How many radioactive nuclei were initially present in the sample?

20. •• A sample of charcoal from an archaeological site contains 55.0 g of carbon and decays at a rate of 0.877 Bq. How old is it?

21. •• **We are stardust.** In 1952, spectral lines of the element technetium-99 (^{99}Tc) were discovered in a red-giant star. Red giants are very old stars, often around 10 billion years old, and near the end of their lives. Technetium has *no* stable isotopes, and the half-life of ^{99}Tc is 200,000 years. (a) For how many half-lives has the ^{99}Tc been in the red-giant star if its age is 10 billion years? (b) What fraction of the original ^{99}Tc would be left at the end of that time? This discovery was extremely important because it provided convincing evidence for the theory (now essentially known to be true) that most of the atoms heavier than hydrogen and helium were made inside of stars by thermonuclear fusion and other nuclear processes. If the Tc had been part of the star since it was born, the amount remaining after 10 billion years would have been so minute that it would not have been detectable. This knowledge is what led the late astronomer Carl Sagan to proclaim that "we are stardust."

22. •• Radioactive isotopes used in cancer therapy have a "shelf-life," like pharmaceuticals used in chemotherapy. Just after it has been manufactured in a nuclear reactor, the activity of a sample of ^{60}Co is 5000 Ci. When its activity falls below 3500 Ci, it is considered too weak a source to use in treatment. You work in the radiology department of a large hospital. One of these ^{60}Co sources in your inventory was manufactured on October 6, 2008. It is now April 6, 2011. Is the source still usable? The half-life of ^{60}Co is 5.271 years.

30.4 Radiation and the Life Sciences

23. • **Radiation overdose.** If a person's entire body is exposed to
BIO 5.0 J/kg of x rays, death usually follows within a few days. (Consult Table 30.3.) (a) Express this lethal radiation dose in Gy, rad, Sv, and rem. (b) How much total energy does a 70.0 kg person absorb from such a dose? (c) If the 5.0 J/kg came from a beam of protons instead of x rays, what would be the answers to parts (a) and (b)?

24. • A radiation specialist prescribes a dose of 125 rem for a
BIO patient, using an apparatus that emits alpha particles. (Consult Table 30.3.) (a) How many rads does this dose provide to the patient? (b) How much energy does a 5.0 g sample of irradiated tissue receive? (c) Suppose your hospital has only an electron source available. How many rads should you administer to this patient to achieve the same 125 rem dose?

25. •• A nuclear chemist receives an accidental radiation dose
BIO of 5.0 Gy from slow neutrons (RBE = 4.0). What does she receive in rad, rem, and J/kg?

26. •• (a) If a chest x ray delivers 0.25 mSv to 5.0 kg of tissue, how
BIO many *total* joules of energy does this tissue receive? (b) Natural radiation and cosmic rays deliver about 0.10 mSv per year at sea level. Assuming an RBE of 1, how many rem and rads is this dose, and how many joules of energy does a 75 kg person

receive in a year? (c) How many chest x rays like the one in part (a) would it take to deliver the same *total* amount of energy to a 75 kg person as she receives from natural radiation in a year at sea level, as described in part (b)?

27. •• **To scan or not to scan?** It has become popular for some
BIO people to have yearly whole-body scans (CT scans, formerly called CAT scans), using x rays, just to see if they detect anything suspicious. A number of medical people have recently questioned the advisability of such scans, due in part to the radiation they impart. Typically, one such scan gives a dose of 12 mSv, applied to the *whole body*. By contrast, a chest x ray typically administers 0.20 mSv to only 5.0 kg of tissue. How many chest x rays would deliver the same *total* amount of energy to the body of a 75 kg person as one whole-body scan?

28. •• In an industrial accident a 65-kg person receives a lethal
BIO whole-body equivalent dose of 5.4 Sv from x rays. (a) What is the equivalent dose in rem? (b) What is the absorbed dose in rad? (c) What is the total energy absorbed by the person's body? How does this amount of energy compare to the amount of energy required to raise the temperature of 65 kg of water 0.010°C?

29. •• **Food irradiation.** Food is often irradiated with the use of
BIO either x rays or electron beams, to help prevent spoilage. A low dose of 5–75 kilorads (krad) helps to reduce and kill inactive parasites, a medium dose of 100–400 krad kills microorganisms and pathogens such as salmonella, and a high dose of 2300–5700 krad sterilizes food so that it can be stored without refrigeration. (a) A dose of 175 krad kills spoilage microorganisms in fish. If x rays are used, what would be the dose in Gy, Sv, and rem, and how much energy would a 150 g portion of fish absorb? (See Table 30.3.) (b) Repeat part (a) if electrons of RBE 1.50 are used instead of x rays.

30. •• In a diagnostic x-ray procedure, 5.00×10^{10} photons are
BIO absorbed by tissue with a mass of 0.600 kg. The x-ray wavelength is 0.0200 nm. (a) What is the total energy absorbed by the tissue? (b) What is the equivalent dose in rem?

31. •• A person ingests an amount of a radioactive source with a
BIO very long lifetime and activity 0.72 μCi. The radioactive material lodges in the lungs, where all of the 4.0-MeV α particles emitted are absorbed within a 0.50-kg mass of tissue. Calculate the absorbed dose and the equivalent dose for one year.

32. •• **Irradiating ourselves!** The radiocarbon in our bodies is
BIO one of the naturally occurring sources of radiation. Let's see how large a dose we receive. ^{14}C decays via β^- emission, and 18% of our body's mass is carbon. (a) Write out the decay scheme of carbon-14 and show the end product. (A neutrino is also produced.) (b) Neglecting the effects of the neutrino, how much kinetic energy (in MeV) is released per decay? The atomic mass of C-14 is 14.003242 u. (See Table 30.2.) (c) How many grams of carbon are there in a 75 kg person? How many decays per second does this carbon produce? (d) Assuming that all the energy released in these decays is absorbed by the body, how many MeV/s and J/s does the C-14 release in this person's body? (e) Consult Table 30.3 and use the largest appropriate RBE for the particles involved. What radiation dose does the person give himself in a year, in Gy, rad, Sv, and rem?

30.5 Nuclear Reactions

33. • Consider the nuclear reaction $^4_2\text{He} + ^7_3\text{Li} \rightarrow ^{10}_5\text{B} + ^1_0\text{n}$. Is energy absorbed or liberated? How much energy?

34. •• Consider the nuclear reaction

$$^2_1\text{H} + ^{14}_7\text{N} \rightarrow \text{X} + ^{10}_5\text{B}$$

where X is a nuclide. (a) What are Z and A for the nuclide X? (b) Calculate the reaction energy Q (in MeV).

35. •• How much energy (in J and MeV) would a proton need for its surface to reach the surface of the nucleus of the ^{17}O atom? Assume that oxygen and the proton obey the empirical radius formula of Section 30.1.

30.6 Nuclear Fission

36. • Assuming that 200 MeV is released per fission, how many fissions per second take place in a 100 MW reactor?

37. •• The United States uses 1.0×10^{19} J of electrical energy per year. If all this energy came from the fission of ^{235}U, which releases 200 MeV per fission event, (a) how many kilograms of ^{235}U would be used per year; (b) how many kilograms of uranium would have to be mined per year to provide that much ^{235}U? (Recall that only 0.70% of naturally occurring uranium is ^{235}U.)

38. •• At the beginning of Section 30.6, a fission process is illustrated in which ^{235}U is struck by a neutron and undergoes fission to produce ^{144}Ba, ^{89}Kr, and three neutrons. The measured masses of these isotopes are 235.043930 u (^{235}U), 143.922953 u (^{144}Ba), 88.917630 u (^{89}Kr), and 1.0086649 u (neutron). (a) Calculate the energy (in MeV) released by each fission reaction. (b) Calculate the energy released per gram of ^{235}U, in MeV/g.

30.7 Nuclear Fusion

39. • Calculate the energy released in the fusion reaction $^2_1\text{H} + ^3_1\text{He} \rightarrow ^4_2\text{H} + ^1_0\text{n}$. The atomic mass of ^3_1H (tritium) is 3.016049 u.

40. • Consider the fusion reaction $^2_1\text{H} + ^2_1\text{H} \rightarrow ^3_2\text{He} + ^1_0\text{n}$. (a) Compute the energy liberated in this reaction in MeV and in joules. (b) Compute the energy *per mole* of deuterium, remembering that the gas is diatomic, and compare it with the heat of combustion of diatomic molecular hydrogen, which is about 2.9×10^5 J/mol.

41. •• **Comparison of energy released per gram of fuel.** (a) When gasoline is burned, it releases 1.3×10^8 J per gallon (3.788 L) of energy. Given that the density of gasoline is 737 kg/m^3, express the quantity of energy released in J/g of fuel. (b) During fission, when a neutron is absorbed by a ^{235}U nucleus, about 200 MeV of energy is released for each nucleus that undergoes fission. Express this quantity in J/g of fuel. (c) In the proton–proton chain that takes place in stars like our sun, the overall fusion reaction can be summarized as six protons fusing to form one ^4He nucleus with two leftover protons and the liberation of 26.7 MeV of energy. The fuel is the six protons. Express the energy produced here in units of J/g of fuel. Notice the huge difference between the two forms of nuclear energy, on the one hand, and the chemical energy from gasoline, on the other. (d) Our sun produces energy at a measured rate of 3.92×10^{26} W. If its mass of 1.99×10^{30} kg were all gasoline, how long could it last before consuming all its fuel? (*Historical note:* Before the discovery of nuclear fusion

and the vast amounts of energy it releases, scientists were confused. They knew that the earth was at least many millions of years old, but could not explain how the sun could survive that long if its energy came from chemical burning.)

42. • Show that the net result of the proton–proton fusion chain that occurs inside our sun can be summarized as

$$6\,\text{p}^+ \rightarrow ^4_2\text{He} + 2\,\text{p}^+ + 2\,\beta^+ + 2\,\gamma + 2\,\nu_e.$$

30.8 Fundamental Particles
30.9 High-Energy Physics

43. •• **Pair annihilation.** Consider the case where an electron e$^-$ and a positron e$^+$ annihilate each other and produce photons. Assume that these two particles collide head-on with equal, but small, speeds. (a) Show that it is not possible for only *one* photon to be produced. (*Hint:* Consider the conservation law that must be true in any collision.) (b) Show that if only two photons are produced, they must travel in opposite directions and have equal energy. (c) Calculate the wavelength of each of the photons in part (b). In what part of the electromagnetic spectrum do they lie?

44. •• **Radiation therapy with π^- mesons.** Beams of π^-
BIO mesons are used in radiation therapy for certain cancers. The energy comes from the complete decay of the π^- to *stable* particles. (a) Write out the complete decay of a π^- meson to stable particles. What are these particles? (*Hint:* See the end of Section 30.8.) (b) How much energy is released from the complete decay of a single π^- meson to stable particles? (You can ignore the very small masses of the neutrinos.) (c) How many π^- mesons need to decay to give a dose of 50.0 Gy to 10.0 g of tissue? (d) What would be the equivalent dose in part (c) in Sv and in rem? Consult Table 30.3 and use the largest appropriate RBE for the particles involved in this decay.

45. • If a Σ^+ at rest decays into a proton and a π^0, what is the total kinetic energy of the decay products?

46. • A positive pion at rest decays into a positive muon and a neutrino. (a) Approximately how much energy is released in the decay? (Assume the neutrino has zero rest mass. Use the muon and pion masses given in terms of the electron mass near the end of Section 30.8.) (b) Why can't a positive muon decay into a positive pion?

47. •• A proton and an antiproton annihilate, producing two photons. Find the energy, frequency, and wavelength of each photon emitted (a) if the initial kinetic energies of the proton and antiproton are negligible and (b) if each particle has an initial kinetic energy of 830 MeV.

48. • Which of the following reactions obey the conservation of baryon number? (a) $\text{p} + \text{p} \rightarrow \text{p} + \text{e}^+$, (b) $\text{p} + \text{n} \rightarrow 2\,\text{e}^+ + \text{e}^-$, (c) $\text{p} \rightarrow \text{n} + \text{e}^- + \bar{\nu}_e$, (d) $\text{p} + \bar{\text{p}} \rightarrow 2\,\gamma$.

49. •• **Comparing the strengths of the four forces.** Both the strong and the weak interactions are short-range forces having a range of around 1 fm. If you have two protons separated by this distance, they will be influenced by all four of the fundamental forces (or interactions). (a) Calculate the strengths of the electrical and gravitational interactions that will influence each of these protons. (b) Use the information given under "The Four Forces" at the beginning of Section 30.9 to estimate the approximate strengths of the strong interaction and the

weak interaction on each of these protons. (c) Arrange the four forces in order of strength, starting with the strongest. (d) Express each force as a multiple of the weakest of the four.

50. ●● Determine the electric charge, baryon number, strangeness quantum number, and charm quantum number for the following quark combinations: (a) uus, (b) $c\bar{s}$, (c) $\overline{ddu}$, and (d) $\bar{c}b$.

30.10 Cosmology

51. ●● The critical density of the universe is $5.8 \times 10^{-27} \text{ kg/m}^3$. (a) Assuming that the universe is all hydrogen, express the critical density in the number of H atoms per cubic meter. (b) If the density of the universe is equal to the critical density, how many atoms, on the average, would you expect to find in a room of dimensions $4 \text{ m} \times 7 \text{ m} \times 3 \text{ m}$? (c) Compare your answer in part (b) with the number of atoms you would find in this room under normal conditions on the earth.

General Problems

52. ●● The results of activity measurements on a radioactive sample are given in the table. (a) Estimate the half-life of the sample. (b) Find the sample's decay constant. (c) How many radioactive nuclei were present in the sample at $t = 0$? (d) How many were present after 7.0 h?

Time (h)	Decays/s
0	20,000
0.5	15,900
1.0	12,600
1.5	9,980
2.0	7,940
2.5	6,300
3.0	4,970
4.0	3,150
5.0	1,980
6.0	1,250
7.0	790

53. ● The starship *Enterprise*, of television and movie fame, is powered by the controlled combination of matter and antimatter. If the entire 400 kg antimatter fuel supply of the *Enterprise* combines with matter, how much energy is released?

54. ●● A 70.0 kg person experiences a whole-body exposure to **BIO** alpha radiation with energy of 1.50 MeV. A total of 5.00×10^{12} alpha particles is absorbed. (a) What is the absorbed dose in rad? (b) What is the equivalent dose in rem? (c) If the source is 0.0100 g of ^{226}Ra (half-life 1600 years) somewhere in the body, what is the activity of the source? (d) If all the alpha particles produced are absorbed, what time is required for this dose to be delivered?

55. ●● A ^{60}Co source with activity 15.0 Ci is imbedded in a tumor **BIO** that has a mass of 0.500 kg. The Co source emits gamma-ray photons with average energy of 1.25 MeV. Half the photons are absorbed in the tumor, and half escape. (a) What energy is delivered to the tumor per second? (b) What absorbed dose (in rad) is delivered per second? (c) What equivalent dose (in rem) is delivered per second if the RBE for these gamma rays is 0.70? (d) What exposure time is required for an equivalent dose of 200 rem?

56. ●● The nucleus $^{15}_{8}\text{O}$ has a half-life of 2.0 min. $^{19}_{8}\text{O}$ has a half-life of about 0.5 min. (a) If, at some time, a sample contains equal amounts of $^{15}_{8}\text{O}$ and $^{19}_{8}\text{O}$, what is the ratio of $^{15}_{8}\text{O}$ to $^{19}_{8}\text{O}$ after 2.0 min? (b) After 10.0 min?

57. ●● The unstable isotope ^{40}K is used to date rock samples. Its half-life is 1.28×10^8 years. (a) How many decays occur per second in a sample containing 6.00×10^{-6} g of ^{40}K? (b) What is the activity of the sample in curies?

58. ●● **Radiation treatment of prostate cancer.** In many cases, **BIO** prostate cancer is treated by implanting 60 to 100 small seeds of radioactive material into the tumor. The energy released from the decays kills the tumor. One isotope that is used (there are others) is palladium (^{103}Pd), with a half-life of 17 days. If a typical grain contains 0.250 g of ^{103}Pd, (a) what is its initial activity rate in Bq, and (b) what is the rate 68 days later?

59. ●● An unstable isotope of cobalt, ^{60}Co, has one more neutron in its nucleus than the stable ^{59}Co and is a beta emitter with a half-life of 5.3 years. This isotope is widely used in medicine. A certain radiation source in a hospital contains 0.0400 g of ^{60}Co. (a) What is the decay constant for that isotope? (b) How many atoms are in the source? (c) How many decays occur per second? (d) What is the activity of the source, in curies?

60. ●● **An oceanographic tracer.** Nuclear weapons tests in the 1950s and 1960s released significant amounts of radioactive tritium ($^{3}_{1}\text{H}$, half-life 12.3 years) into the atmosphere. The tritium atoms were quickly bound into water molecules and rained out of the air, most of them ending up in the ocean. For any of this tritium-tagged water that sinks below the surface, the amount of time during which it has been isolated from the surface can be calculated by measuring the ratio of the decay product, $^{3}_{2}\text{He}$, to the remaining tritium in the water. For example, if the ratio of $^{3}_{2}\text{He}$ to $^{3}_{1}\text{H}$ in a sample of water is 1:1, the water has been below the surface for one half-life, or approximately 12 years. This method has provided oceanographers with a convenient way to trace the movements of subsurface currents in parts of the ocean. Suppose that in a particular sample of water, the ratio of $^{3}_{2}\text{He}$ to $^{3}_{1}\text{H}$ is 4.3 to 1.0. How many years ago did this water sink below the surface?

61. ●● A bone fragment found in a cave believed to have been **BIO** inhabited by early humans contains 0.21 times as much ^{14}C as an equal amount of carbon in the atmosphere when the organism containing the bone died. (See Example 30.4 in Section 30.3.) Find the approximate age of the fragment.

62. ● **Radioactive fallout.** One of the problems of in-air testing **BIO** of nuclear weapons (or, even worse, the *use* of such weapons!) is the danger of radioactive fallout. One of the most problematic nuclides in such fallout is strontium-90 (^{90}Sr), which breaks down by β^- decay with a half-life of 28 years. It is chemically similar to calcium and therefore can be incorporated into bones and teeth, where, due to its rather long half-life, it remains for years as an internal source of radiation. (a) What is the daughter nucleus of the ^{90}Sr decay? (b) What percent of the original level of ^{90}Sr is left after 56 years? (c) How long would you have to wait for the original level to be reduced to 6.25% of its original value?

63. ● Consider the nuclear reaction $^{2}_{1}\text{H} + ^{14}_{7}\text{N} \rightarrow ^{6}_{3}\text{Li} + ^{10}_{5}\text{B}$. Is energy absorbed or liberated? How much?

64. ●● The atomic mass of $^{56}_{26}\text{Fe}$ is 55.934939 u, and the atomic mass of $^{56}_{27}\text{Co}$ is 55.939847 u. (a) Which of these nuclei will decay into the other? (b) What type of decay will occur? (c) How much kinetic energy will the products of the decay have?

65. ● A K^+ meson at rest decays into two π mesons. (a) What are the allowed combinations of π^0, π^+, and π^- as decay products? (b) Find the total kinetic energy of the π mesons.

66. •• The measured energy width of the ϕ meson is 4.0 MeV, and its mass is 1020 MeV/c^2. Using the uncertainty principle (in the form $\Delta E \, \Delta t \geq h/2\pi$), estimate the lifetime of the ϕ meson.

67. •• Given that each particle contains only combinations of u, d, s, $\bar{u}$, $\bar{d}$, and $\bar{s}$, deduce the quark content of (a) a particle with charge $+e$, baryon number 0, and strangeness $+1$; (b) a particle with charge $+e$, baryon number -1, and strangeness $+1$; (c) a particle with charge 0, baryon number $+1$, and strangeness -2.

Passage Problems

BIO Looking under the hood of PET. Positron Emission Tomography (PET), a kind of imaging, involves injecting a patient with artificially produced atoms that have nuclei containing an excess of neutrons. As these neutrons decay into protons, they emit positrons at fairly slow, non-relativistic speeds. When a positron encounters an electron, they annihilate each other and emit two x ray photons in opposite directions. The patient is enclosed in a circular array of photodetectors, with the tissue to be imaged centered in the detector array. If two photons strike detectors simultaneously (within 10 ns), we can conclude that they are produced by an e^+e^- annihilation event somewhere along a line connecting the two photodetectors. By observing many such simultaneous events, we can create a map of the distribution of the positron-emitting atoms in the tissue. The index of refraction of biological tissue for x rays is 1.

68. What is the energy of each of the photons resulting from an annihilation event?
    A. $\frac{1}{2} m_e v^2$, where v is the speed of the positron.
    B. $m_e v^2$
    C. $\frac{1}{2} m_e c^2$
    D. $m_e c^2$

69. What is the wavelength of each photon produced in an annihilation event?
    A. $\dfrac{m_e v^2}{hc}$
    B. $\dfrac{hc}{m_e v^2}$
    C. $\dfrac{h}{m_e c}$
    D. $\dfrac{m_e c^2}{h}$

70. Suppose that an annihilation event occurs on the line 3 cm from the center of the line connecting the two photodetectors that receive the resultant x rays. Often, a section of the whole brain is being imaged, so the x rays do not come from a point source. Will those photons be counted as having arrived simultaneously?
    A. No, because the time difference will be 100 ms.
    B. No, because the time difference will be 200 ms.
    C. Yes, because the time difference will be 0.1 ns.
    D. Yes because the time difference will be 0.2 ns.

The International System of Units

The Système International d'Unités, abbreviated SI, is the system developed by the General Conference on Weights and Measures and adopted by nearly all the industrial nations of the world. The following material is adapted from B. N. Taylor, ed., National Institute of Standards andTechnology Spec. Pub. 811 (U.S. Govt. Printing Office, Washington, DC, 1995). See also **http://physics.nist.gov/cuu**

Quantity	Name of unit	Symbol	
	SI base units		
length	meter	m	
mass	kilogram	kg	
time	second	s	
electric current	ampere	A	
thermodynamic temperature	kelvin	K	
amount of substance	mole	mol	
luminous intensity	candela	cd	
	SI derived units		**Equivalent units**
area	square meter	m^2	
volume	cubic meter	m^3	
frequency	hertz	Hz	s^{-1}
mass density (density)	kilogram per cubic meter	kg/m^3	
speed, velocity	meter per second	m/s	
angular velocity	radian per second	rad/s	
acceleration	meter per second squared	m/s^2	
angular acceleration	radian per second squared	rad/s^2	
force	newton	N	$kg \cdot m/s^2$
pressure (mechanical stress)	pascal	Pa	N/m^2
kinematic viscosity	square meter per second	m^2/s	
dynamic viscosity	newton-second per square meter	$N \cdot s/m^2$	
work, energy, quantity of heat	joule	J	$N \cdot m$
power	watt	W	J/s
quantity of electricity	coulomb	C	$A \cdot s$
potential difference, electromotive force	volt	V	$J/C, W/A$
electric field strength	volt per meter	V/m	N/C
electric resistance	ohm	Ω	V/A
capacitance	farad	F	$A \cdot s/V$
magnetic flux	weber	Wb	$V \cdot s$
inductance	henry	H	$V \cdot s/A$
magnetic flux density	tesla	T	Wb/m^2
magnetic field strength	ampere per meter	A/m	
magnetomotive force	ampere	A	
luminous flux	lumen	lm	$cd \cdot sr$
luminance	candela per square meter	cd/m^2	
illuminance	lux	lx	lm/m^2
wave number	1 per meter	m^{-1}	
entropy	joule per kelvin	J/K	
specific heat capacity	joule per kilogram-kelvin	$J/(kg \cdot K)$	
thermal conductivity	watt per meter-kelvin	$W/(m \cdot K)$	

Quantity	Name of unit	Symbol	Equivalent units
radiant intensity	watt per steradian	W/sr	
activity (of a radioactive source)	becquerel	Bq	s^{-1}
radiation dose	gray	Gy	J/kg
radiation dose equivalent	sievert	Sv	J/kg
SI supplementary units			
plane angle	radian	rad	
solid angle	steradian	sr	

Definitions of SI Units

meter (m) The *meter* is the length equal to the distance traveled by light, in vacuum, in a time of $1/299,792,458$ second.

kilogram (kg) The *kilogram* is the unit of mass; it is equal to the mass of the international prototype of the kilogram. (The international prototype of the kilogram is a particular cylinder of platinum-iridium alloy that is preserved in a vault at Sèvres, France, by the International Bureau of Weights and Measures.)

second (s) The *second* is the duration of 9,192,631,770 periods of the radiation corresponding to the transition between the two hyperfine levels of the ground state of the cesium-133 atom.

ampere (A) The *ampere* is that constant current that, if maintained in two straight parallel conductors of infinite length, of negligible circular cross section, and placed 1 meter apart in vacuum, would produce between these conductors a force equal to 2×10^{-7} newton per meter of length.

kelvin (K) The *kelvin,* unit of thermodynamic temperature, is the fraction $1/273.16$ of the thermodynamic temperature of the triple point of water.

ohm (V) The *ohm* is the electric resistance between two points of a conductor when a constant difference of potential of 1 volt, applied between these two points, produces in this conductor a current of 1 ampere, this conductor not being the source of any electromotive force.

coulomb (C) The *coulomb* is the quantity of electricity transported in 1 second by a current of 1 ampere.

candela (cd) The *candela* is the luminous intensity, in a given direction, of a source that emits monochromatic radiation of frequency 540×10^{12} hertz and that has a radiant intensity in that direction of $1/683$ watt per steradian.

mole (mol) The *mole* is the amount of substance of a system that contains as many elementary entities as there are carbon atoms in 0.012 kg of carbon 12. The elementary entities must be specified and may be atoms, molecules, ions, electrons, other particles, or specified groups of such particles.

newton (N) The *newton* is that force that gives to a mass of 1 kilogram an acceleration of 1 meter per second per second.

joule (J) The *joule* is the work done when the point of application of a constant force of 1 newton is displaced a distance of 1 meter in the direction of the force.

watt (W) The *watt* is the power that gives rise to the production of energy at the rate of 1 joule per second.

volt (V) The *volt* is the difference of electric potential between two points of a conducting wire carrying a constant current of 1 ampere, when the power dissipated between these points is equal to 1 watt.

weber (Wb) The *weber* is the magnetic flux that, linking a circuit of one turn, produces in it an electromotive force of 1 volt as it is reduced to zero at a uniform rate in 1 second.

lumen (lm) The *lumen* is the luminous flux emitted in a solid angle of 1 steradian by a uniform point source having an intensity of 1 candela.

farad (F) The *farad* is the capacitance of a capacitor between the plates of which there appears a difference of potential of 1 volt when it is charged by a quantity of electricity equal to 1 coulomb.

henry (H) The *henry* is the inductance of a closed circuit in which an electromotive force of 1 volt is produced when the electric current in the circuit varies uniformly at a rate of 1 ampere per second.

radian (rad) The *radian* is the plane angle between two radii of a circle that cut off on the circumference an arc equal in length to the radius.

steradian (sr) The *steradian* is the solid angle that, having its vertex in the center of a sphere, cuts off an area of the surface of the sphere equal to that of a square with sides of length equal to the radius of the sphere.

SI Prefixes The names of multiples and submultiples of SI units may be formed by application of the prefixes listed in Appendix F.

The Greek Alphabet

Name	Capital	Lowercase	Name	Capital	Lowercase
Alpha	A	α	Nu	N	ν
Beta	B	β	Xi	Ξ	ξ
Gamma	Γ	γ	Omicron	O	o
Delta	Δ	δ	Pi	Π	π
Epsilon	E	ϵ	Rho	P	ρ
Zeta	Z	ζ	Sigma	Σ	σ
Eta	H	η	Tau	T	τ
Theta	Θ	θ	Upsilon	Y	υ
Iota	I	ι	Phi	Φ	ϕ
Kappa	K	κ	Chi	X	χ
Lambda	Λ	λ	Psi	Ψ	ψ
Mu	M	μ	Omega	Ω	ω

Periodic Table of the Elements

Group: 1 2 3 4 5 6 7 8 9 10 11 12 13 14 15 16 17 18
Period

Period	1	2	3	4	5	6	7	8	9	10	11	12	13	14	15	16	17	18
1	1 **H** 1.008																	2 **He** 4.003
2	3 **Li** 6.941	4 **Be** 9.012											5 **B** 10.811	6 **C** 12.011	7 **N** 14.007	8 **O** 15.999	9 **F** 18.998	10 **Ne** 20.180
3	11 **Na** 22.990	12 **Mg** 24.305											13 **Al** 26.982	14 **Si** 28.086	15 **P** 30.974	16 **S** 32.065	17 **Cl** 35.453	18 **Ar** 39.948
4	19 **K** 39.098	20 **Ca** 40.078	21 **Sc** 44.956	22 **Ti** 47.867	23 **V** 50.942	24 **Cr** 51.996	25 **Mn** 54.938	26 **Fe** 55.845	27 **Co** 58.933	28 **Ni** 58.693	29 **Cu** 63.546	30 **Zn** 65.409	31 **Ga** 69.723	32 **Ge** 72.64	33 **As** 74.922	34 **Se** 78.96	35 **Br** 79.904	36 **Kr** 83.798
5	37 **Rb** 85.468	38 **Sr** 87.62	39 **Y** 88.906	40 **Zr** 91.224	41 **Nb** 92.906	42 **Mo** 95.94	43 **Tc** (98)	44 **Ru** 101.07	45 **Rh** 102.906	46 **Pd** 106.42	47 **Ag** 107.868	48 **Cd** 112.411	49 **In** 114.818	50 **Sn** 118.710	51 **Sb** 121.760	52 **Te** 127.60	53 **I** 126.904	54 **Xe** 131.293
6	55 **Cs** 132.905	56 **Ba** 137.327	71 **Lu** 174.967	72 **Hf** 178.49	73 **Ta** 180.948	74 **W** 183.84	75 **Re** 186.207	76 **Os** 190.23	77 **Ir** 192.217	78 **Pt** 195.078	79 **Au** 196.967	80 **Hg** 200.59	81 **Tl** 204.383	82 **Pb** 207.2	83 **Bi** 208.980	84 **Po** (209)	85 **At** (210)	86 **Rn** (222)
7	87 **Fr** (223)	88 **Ra** (226)	103 **Lr** (262)	104 **Rf** (261)	105 **Db** (262)	106 **Sg** (266)	107 **Bh** (264)	108 **Hs** (269)	109 **Mt** (268)	110 **Ds** (271)	111 **Rg** (272)	112 **Uub** (285)	113 **Uut** (284)	114 **Uuq** (289)	115 **Uup** (288)	116 **Uuh** (292)	117 **Uus** (294)	118 **Uuo**

Lanthanoids

57 **La** 138.905	58 **Ce** 140.116	59 **Pr** 140.908	60 **Nd** 144.24	61 **Pm** (145)	62 **Sm** 150.36	63 **Eu** 151.964	64 **Gd** 157.25	65 **Tb** 158.925	66 **Dy** 162.500	67 **Ho** 164.930	68 **Er** 167.259	69 **Tm** 168.934	70 **Yb** 173.04

Actinoids

89 **Ac** (227)	90 **TH** (232)	91 **Pa** (231)	92 **U** (238)	93 **Np** (237)	94 **Pu** (244)	95 **Am** (243)	96 **Cm** (247)	97 **Bk** (247)	98 **Cf** (251)	99 **Es** (252)	100 **Fm** (257)	101 **Md** (258)	102 **No** (259)

For each element the average atomic mass of the mixture of isotopes occurring in nature is shown. For elements having no stable isotope, the approximate atomic mass of the longest-lived isotope is shown in parentheses. For elements that have been predicted but not yet confirmed, no atomic mass is given. All atomic masses are expressed in atomic mass units ($1\ u = 1.660538728(83) \times 10^{-27}\ kg$), equivalent to grams per mole (g/mol).

Unit Conversion Factors

LENGTH

$1\text{ m} = 100\text{ cm} = 1000\text{ mm} = 10^6\text{ }\mu\text{m} = 10^9\text{ nm}$

$1\text{ km} = 1000\text{ m} = 0.6214\text{ mi}$

$1\text{ m} = 3.281\text{ ft} = 39.37\text{ in.}$

$1\text{ cm} = 0.3937\text{ in.}$

$1\text{ in.} = 2.540\text{ cm}$

$1\text{ ft} = 30.48\text{ cm}$

$1\text{ yd} = 91.44\text{ cm}$

$1\text{ mi} = 5280\text{ ft} = 1.609\text{ km}$

$1\text{ Å} = 10^{-10}\text{ m} = 10^{-8}\text{ cm} = 10^{-1}\text{ nm}$

$1\text{ nautical mile} = 6080\text{ ft}$

$1\text{ light year} = 9.461 \times 10^{15}\text{ m}$

AREA

$1\text{ cm}^2 = 0.155\text{ in.}^2$

$1\text{ m}^2 = 10^4\text{ cm}^2 = 10.76\text{ ft}^2$

$1\text{ in.}^2 = 6.452\text{ cm}^2$

$1\text{ ft}^2 = 144\text{ in.}^2 = 0.0929\text{ m}^2$

VOLUME

$1\text{ liter} = 1000\text{ cm}^3 = 10^{-3}\text{ m}^3$
$\quad = 0.03531\text{ ft}^3 = 61.02\text{ in.}^3$

$1\text{ ft}^3 = 0.02832\text{ m}^3 = 28.32\text{ liters} = 7.477\text{ gallons}$

$1\text{ gallon} = 3.788\text{ liters}$

TIME

$1\text{ min} = 60\text{ s}$

$1\text{ h} = 3600\text{ s}$

$1\text{ d} = 86{,}400\text{ s}$

$1\text{ y} = 365.24\text{ d} = 3.156 \times 10^7\text{ s}$

ANGLE

$1\text{ rad} = 57.30° = 180°/\pi$

$1° = 0.01745\text{ rad} = \pi/180\text{ rad}$

$1\text{ revolution} = 360° = 2\pi\text{ rad}$

$1\text{ rev/min (rpm)} = 0.1047\text{ rad/s}$

SPEED

$1\text{ m/s} = 3.281\text{ ft/s}$

$1\text{ ft/s} = 0.3048\text{ m/s}$

$1\text{ mi/min} = 60\text{ mi/h} = 88\text{ ft/s}$

$1\text{ km/h} = 0.2778\text{ m/s} = 0.6214\text{ mi/h}$

$1\text{ mi/h} = 1.466\text{ ft/s} = 0.4470\text{ m/s} = 1.609\text{ km/h}$

$1\text{ furlong/fortnight} = 1.662 \times 10^{-4}\text{ m/s}$

ACCELERATION

$1\text{ m/s}^2 = 100\text{ cm/s}^2 = 3.281\text{ ft/s}^2$

$1\text{ cm/s}^2 = 0.01\text{ m/s}^2 = 0.03281\text{ ft/s}^2$

$1\text{ ft/s}^2 = 0.3048\text{ m/s}^2 = 30.48\text{ cm/s}^2$

$1\text{ mi/h}\cdot\text{s} = 1.467\text{ ft/s}^2$

MASS

$1\text{ kg} = 10^3\text{ g} = 0.0685\text{ slug}$

$1\text{ g} = 6.85 \times 10^{-5}\text{ slug}$

$1\text{ slug} = 14.59\text{ kg}$

$1\text{ u} = 1.661 \times 10^{-27}\text{ kg}$

$1\text{ kg has a weight of }2.205\text{ lb when }g = 9.80\text{ m/s}^2$

FORCE

$1\text{ N} = 10^5\text{ dyn} = 0.2248\text{ lb}$

$1\text{ lb} = 4.448\text{ N} = 4.448 \times 10^5\text{ dyn}$

PRESSURE

$1\text{ Pa} = 1\text{ N/m}^2 = 1.450 \times 10^{-4}\text{ lb/in.}^2 = 0.209\text{ lb/ft}^2$

$1\text{ bar} = 10^5\text{ Pa}$

$1\text{ lb/in.}^2 = 6895\text{ Pa}$

$1\text{ lb/ft}^2 = 47.88\text{ Pa}$

$1\text{ atm} = 1.013 \times 10^5\text{ Pa} = 1.013\text{ bar}$
$\quad = 14.7\text{ lb/in.}^2 = 2117\text{ lb/ft}^2$

$1\text{ mm Hg} = 1\text{ torr} = 133.3\text{ Pa}$

ENERGY

$1\text{ J} = 10^7\text{ ergs} = 0.239\text{ cal}$

$1\text{ cal} = 4.186\text{ J (based on }15°\text{ calorie)}$

$1\text{ ft}\cdot\text{lb} = 1.356\text{ J}$

$1\text{ Btu} = 1055\text{ J} = 252\text{ cal} = 778\text{ ft}\cdot\text{lb}$

$1\text{ eV} = 1.602 \times 10^{-19}\text{ J}$

$1\text{ kWh} = 3.60 \times 10^6\text{ J}$

MASS–ENERGY EQUIVALENCE

$1\text{ kg} \leftrightarrow 8.988 \times 10^{16}\text{ J}$

$1\text{ u} \leftrightarrow 931.5\text{ MeV}$

$1\text{ eV} \leftrightarrow 1.074 - 10^{-9}\text{ u}$

POWER

$1\text{ W} = 1\text{ J/s}$

$1\text{ hp} = 746\text{ W} = 550\text{ ft}\cdot\text{lb/s}$

$1\text{ Btu/h} = 0.293\text{ W}$

Numerical Constants

Fundamental Physical Constants*

Name	Symbol	Value
Speed of light in vacuum	c	2.99792458×10^8 m/s
Magnitude of charge of electron	e	$1.60217653(14) \times 10^{-19}$ C
Gravitational constant	G	$6.6742(10) \times 10^{-11}$ N $\cdot$ m^2/kg^2
Planck's constant	h	$6.6260693(11) \times 10^{-34}$ J $\cdot$ s
Boltzmann constant	k	$1.3806505(24) \times 10^{-23}$ J/K
Avogadro's number	N_A	$6.0221415(10) \times 10^{23}$ molecules/mol
Gas constant	R	$8.314472(15)$ J/(mol $\cdot$ K)
Mass of electron	m_e	$9.1093826(16) \times 10^{-31}$ kg
Mass of proton	m_p	$1.67262171(29) \times 10^{-27}$ kg
Mass of neutron	m_n	$1.67492728(29) \times 10^{-27}$ kg
Permeability of vacuum	μ_0	$4\pi \times 10^{-7}$ Wb/T $\cdot$ m/A
Permittivity of vacuum	$\epsilon_0 = 1/\mu_0 c^2$	$8.854187817 \cdots \times 10^{-12}$ C^2/(N $\cdot$ m^2)
	$1/4\pi\epsilon_0$	$8.987551787 \cdots \times 10^9$ N $\cdot$ m^2/C^2

Other Useful Constants*

Name	Symbol	Value
Mechanical equivalent of heat		4.186 J/cal ($15°$ calorie)
Standard atmospheric pressure	1 atm	1.01325×10^5 Pa
Absolute zero	0 K	$-273.15°$C
Electron volt	1 eV	$1.60217653(14) \times 10^{-19}$ J
Unified atomic mass unit	1 u	$1.66053886(28) \times 10^{-27}$ kg
Electron rest energy	$m_e c^2$	$0.510998918(44)$ MeV
Volume of ideal gas (0°C and 1 atm)		$22.413996(39)$ liter/mol
Acceleration due to gravity (standard)	g	9.80 m/s^2

*Source: National Institute of Standards and Technology (**http://physics.nist.gov/cuu**). Numbers in parentheses show the uncertainty in the final digits of the main number; for example, the number 1.6454(21) means 1.6454 $\pm$ 0.0021. Values shown without uncertainties are exact.

Astronomical Data[†]

Body	Mass (kg)	Radius (m)	Orbit radius (m)	Orbit period
Sun	1.99×10^{30}	6.96×10^{8}	—	—
Moon	7.35×10^{22}	1.74×10^{6}	3.84×10^{8}	27.3 d
Mercury	3.30×10^{23}	2.44×10^{6}	5.79×10^{10}	88.0 d
Venus	4.87×10^{24}	6.05×10^{6}	1.08×10^{11}	224.7 d
Earth	5.97×10^{24}	6.38×10^{6}	1.50×10^{11}	365.3 d
Mars	6.42×10^{23}	3.40×10^{6}	2.28×10^{11}	687.0 d
Jupiter	1.90×10^{27}	6.91×10^{7}	7.78×10^{11}	11.86 y
Saturn	5.68×10^{26}	6.03×10^{7}	1.43×10^{12}	29.45 y
Uranus	8.68×10^{25}	2.56×10^{7}	2.87×10^{12}	84.02 y
Neptune	1.02×10^{26}	2.48×10^{7}	4.50×10^{12}	164.8 y
Pluto	1.31×10^{22}	1.15×10^{6}	5.91×10^{12}	247.9 y

[†]Source: NASA Jet Propulsion LaboratorySolar System Dynamics Group (**http://ssd.jpl.nasa.gov**), and P. Kenneth Seidelmann, ed., ***Explanatory Supplement to the Astronomical Almanac*** (University Science Books, Mill Valley, CA, 1992), pp. 704–706. For each body, "radius" is its radius at its equator and "orbit radius" is its average distance from the sun (for the planets) or from the earth (for the moon).

Prefixes for Powers of 10

Power of ten	Prefix	Abbreviation	Pronunciation
10^{-24}	yocto-	y	*yoc*-toe
10^{-21}	zepto-	z	*zep*-toe
10^{-18}	atto-	a	*at*-toe
10^{-15}	femto-	f	*fem*-toe
10^{-12}	pico-	p	*pee*-koe
10^{-9}	nano-	n	*nan*-oe
10^{-6}	micro-	μ	*my*-crow
10^{-3}	milli-	m	*mil*-i
10^{-2}	centi-	c	*cen*-ti
10^{3}	kilo-	k	*kil*-oe
10^{6}	mega-	M	*meg*-a
10^{9}	giga-	G	*jig*-a or *gig*-a
10^{12}	tera-	T	*ter*-a
10^{15}	peta-	P	*pet*-a
10^{18}	exa-	E	*ex*-a
10^{21}	zetta-	Z	*zet*-a
10^{24}	yotta-	Y	*yot*-a

Examples:

1 femtometer = 1 fm = 10^{-15} m

1 picosecond = 1 ps = 10^{-12} s

1 nanocoulomb = 1 nC = 10^{-9} C

1 microkelvin = 1 μK = 10^{-6} K

1 millivolt = 1 mV = 10^{-3} V

1 kilopascal = 1 kPa = 10^{3} Pa

1 megawatt = 1 MW = 10^{6} W

1 gigahertz = 1 GHz = 10^{9} Hz

Answers to Selected Odd-Numbered Problems

Chapter 0

Problems

1. $9x^8y^4$
3. $4x^4y^{-6}$
5. 4.75×10^5
7. 1.23×10^{-4}
9. $x = 4$
11. $x = \pm 2\sqrt{3}$
13. $x = 2, 3$
15. $t = 1.8$ s, and $t = -2.25$
17. $x = 2, y = 3$
19. 200 N
21. 1.2×10^{-6} m
23. 186 N
25. (a) $\log\left(\dfrac{x^4y}{(x+y)^3}\right)$ (b) $\log(xz)$
27. 10^{-6}
29. (a) 0.75 m, 0.045 m² (b) 0.55 m², 0.039 m³
    (c) 0.6 m², 0.02 m³ (d) 0.58 m², 0.034 m³
31. 22 cm

Chapter 1

Conceptual Questions

5. Numerous possibilities exist, such as the time of swing of a pendulum of a certain length; the time required for an object to fall from a specific height; the pulse rate of a healthy adult; the flow rate of a specific volume of sand (e.g., an hourglass); and the constant-frequency electric signal generated by the vibrations of a quartz crystal. Each of these potential standards, however, has deficiencies and is thus less accurate than the atomic standard.

7. No, the formula cannot be correct. V has units of length cubed (e.g., m³), r and h each have the units of length and π is dimensionless; therefore, $\pi r^3 h$ has units of length raised to the fourth power (e.g., m⁴), which is dimensionally incorrect.

9. Measure the height of a stack of 100 pages, and divide that distance by 100 to obtain the thickness of a single page.

11. Consider the SI System of units: A has units of m²; V has units of m³; C has units of m; R has units of m. If we compare the equivalency of only the units for each formula, we obtain (a) $\{m^2\} \neq \{m\}$;

(b) $\{m^3\} \neq \{m^2\}$; (c) $\{m^2\} = \{m^2\}$;
(d) $\{m^3\} = \{m^3\}$; (e) $\{m\} \neq \{m^2\}$;
(f) $\{m^3\} = \{m^3\}$. Therefore, formulas
(a), (b), and (e) cannot be correct.

13. (a) Yes, since scalar addition only involves adding numbers. (b) No, because vector addition is not done simply by adding magnitudes. The magnitude of the vector sum depends on the direction, as well as the magnitude, of each vector being added. For example, if two vectors have equal magnitude, but lie in opposite directions, their vector sum is zero.

Multiple-Choice Problems

1. B
3. C
5. B
7. B
9. C
11. B
13. D

Problems

1. (a) 2.40 grams/day
   (b) 1.20×10^{-4} grams/day
   (c) 0.500 grams/day (d) 1.50 kilohms
   (e) 20 milliamps
3. (a) 4.10×10^5 μg/day (b) 0.900 g/day
   (c) 2 tablets/day (d) 0.070 mg/day
5. (a) 1.00×10^3 kg/m³
   (b) 1.05 g/cm³ (c) 1.00 kg, 2.20 lb
7. 1.02 ns
9. 15.9 km/L
11. $6.39 per gallon
13. (a) 3600 s
    (b) 8.64×10^4 s (c) 3.15×10^7 s
15. (a) 3.14; 3.1416; 3.1415927
    (b) 2.72; 2.7183; 2.7182818
    (c) 3.61; 3.6056; 3.6055513
17. (a) 0.9969 to 0.9981 (b) 0.061 to 0.078
    (c) 0.061 to 0.079
19. (a) 5.50 g/cm³
    (b) 1.1×10^6 g/cm³
    (c) 4.7×10^{14} g/cm³
21. 9.0 cm
23. 18.9 g
25. 10^6
27. 4×10^8
29. 10^{12}; no
31. 3×10^9 beats; 4×10^7 gallons

33. 9×10^{14}; 3×10^6 per person
39. 1190 N; 13.4° above forward direction
41. (a) 6.4 m; 51° (b) 6.7 km; 243°
    (c) 19 m/s; 298° (d) 14 N; 124°
43. (a) 10.0 lb; 37° (b) 39 m/s; 232°
    (c) 2500 km; 127° (d) 89.9 N; 323°
45. (a) 5.40 cm, −1.50 cm (b) 5.60 cm,
    344° counterclockwise from +x axis
    (c) 2.80 cm, −6.00 cm (d) 6.62 cm,
    295° counterclockwise from +x axis
47. (b) 44° N
49. (a) 5.06 km; 20.2° north of west
51. 142 g to 149 g
53. (a) 2200 g (b) 2.1 m
55. 6.7×10^{-2} L = 67 cm³
57. (a) 5.7×10^6 steps (b) 120 days
59. (a) 0.5 kg (b) 6×10^{-14} g (c) 0.3 g
61. 143 m; 41° south of west
63. 3.30 N
65. 160 N, 13° below the horizontal

Passage Problems

67. C
69. C

Chapter 2

Conceptual Questions

1. (a) The magnitude of the displacement equals the distance traveled only when the motion is always in the same direction and lies along a straight line in space. An example is a car traveling due east on a straight, level road. (b) A car travels due east for 0.4 mi and then due west for 0.1 mi. The displacement has magnitude 0.3 mi, while the distance traveled is 0.5 mi. (c) No.

3. The average velocity equals the instantaneous velocity if the velocity is constant.

5. (a) True. The average speed can be zero only if the object is not moving. (b) False. If the object returns to its starting point, then its displacement is zero and its average velocity is zero; however, the distance it traveled is not zero, and thus its average speed is nonzero. An example is a car that travels 50 m north for 5 s and then 50 m south for 5 s. For the 10-second interval, the average velocity is zero, but the average speed is $(100\text{ m})/(10\text{ s}) = 10$ m/s.

9. In each of these cases, the slope of velocity versus time is the acceleration. The sign of the slope gives the sign of the acceleration, in terms of the direction taken to be positive for plotting the velocity. Consequently, a positive slope indicates that the object is either speeding up in the positive x direction or slowing down in the negative x direction. Similarly, a negative slope indicates an increasing speed in the negative x direction or a decreasing speed in the positive x direction. A zero slope implies constant speed and thus no acceleration.

11. (a) The velocity must be positive, since x is increasing in the positive x direction. The intervals between time values, Δt, and between displacements, Δx, are both constants for all five intervals; thus, the velocity is constant and the acceleration is zero. (b) For each time interval, the average velocity is $v_{av,x} = \Delta x/\Delta t = (0.25\ \text{m})/(2.00\ \text{s}) = 0.125\ \text{m/s}$. Since the velocity does not change, $a_{av,x} = \Delta v_x/\Delta t = (0.0\ \text{m/s})/(2.00\ \text{s}) = 0.0\ \text{m/s}^2$ for each time interval. (c) The displacement is $7.75\ \text{m} - 6.50\ \text{m} = 1.25\ \text{m}$.

15. (a) Since all three objects have the same v at A and at B, Δv is the same for all three. And given that $a_{av} = \Delta v/\Delta t$, a_{av} is the same for all three objects. (b) At A, they all have the same velocity. By comparing the slope of a tangent line drawn to the v–t graph for each object at point A, we can conclude that the acceleration is different for all three objects.

17. (a) Since the object's velocity is always positive, the object does not reverse its direction of motion. (b) No, the object is always moving away from its starting point; the velocity is always in the same direction. (c) Yes, $v = 0$ at the beginning and at the end of the motion. (d) Yes, initially the object has a positive acceleration; however, the acceleration subsequently decreases in magnitude, passes through zero, and then becomes increasingly negative (increasing in magnitude).

Multiple-Choice Problems

1. C, D
3. C, D
5. D
7. A, D
9. D
11. A
13. B, D
15. C

Problems

1. (a) $x_A = -5\ \text{cm}, x_B = +45\ \text{cm},$ $x_C = +15\ \text{cm}, x_D = -5\ \text{cm}$
   (b) (i) +50 cm; 50 cm (ii) −30 cm; 30 cm (iii) −20 cm; 20 cm (iv) 0; 100 cm

3. 1.3 m/s; 2.0 m/s
5. (a) 3.0 m/s; 1.5 m/s; 1.8 m/s (b) same
7. (a) 7.0 m (b) (i) 1.0 m/s (ii) 3.0 m/s (iii) 5.0 m/s (iv) 7.0 m/s (v) 4.0 m/s
9. (a) 15.0 mi/h (b) 22.0 ft/s (c) 6.70 m/s
11. (a) 5.0 m (b) 1.1×10^7 yr
13. 1.6 mi
15. (a) 26 m/s (b) 140 m
17. 40 min
19. (a) IV (b) I (c) V (d) II (e) III
23. (a) (in m/s^2) 0, 1.0, 1.5, 2.5, 2.5, 2.5, 1.0, 0; no; yes (b) 2.5 m/s^2; 1.0 m/s^2; 0
27. (a) 6.3 ft/s^2 (b) 347 ft
29. (a) 4.9 m/s (b) 0.049 s (c) 100 m/s^2, 10 g's
31. (a) 675 m/s^2 (b) 0.067 s
33. 38 cm
35. (a) 4.42×10^5 m/s$^2 = 4.51 \times 10^4 g$ (b) 0.758 ms
37. (a) 1.67 m/s^2 (b) 12 s (c) 240 m
39. (a) braking: -26.5 ft/s^2; speeding up: 6.67 ft/s^2 (b) 90.4 mph, (c) 3.32 s
41. $4A$; $2C$
43. $R_{\text{large}}/R_{\text{small}} = 1.10$
45. (a) 62.5 km (b) 175 m/s
47. (a) 2000 m (b) $2T$
49. (a) 30.6 m (b) 24.5 m/s
51. (a) 33.5 m (b) 15.8 m/s
53. (a) 0.250 s: 40.9 m aboveground; 2.55 m/s upward. 1.00 s: 40.1 m aboveground; 4.80 m/s downward (b) 3.41 s (c) 28.4 m/s (d) 41.2 m
55. (a) 70.6 m (b) 15.3 s
57. (a) 4.93 s (b) 36.3 m/s
59. (a) 3.3 s (b) $9H$
61. (a) 107 mph (b) 107 mph
63. (a) 9.9 m/s, downward (b) 9.9 m/s, upward
65. 1300 yr
67. (a) 2.7 mi/h (b) 24 mi/h (c) no
69. (a) $H/6$ (b) 24.0 s
71. (a) 3.1×10^6 m/s$^2 = 3.2 \times 10^5 g$ (b) 1.6 ms (c) no
73. 3.60 m from point directly below the drop
75. (a) 2.04 s (b) 6.12 s (c) 8.16 s (d) 4.08 s (e) 9.8 m/s^2 downward at all points
77. 510 cm/s^2
79. (a) 628 m (b) 0; 9.8 m/s^2 downward (c) 36.4 s; 111 m/s
81. (a) 250 km (b) 49 s
83. (a) 383 m (b) overestimated

Passage Problems

84. D
85. B
86. C

Chapter 3

Conceptual Questions

1. The acceleration is downward at all points in the motion. The football's velocity has a constant nonzero horizontal component, so the velocity is never vertical and therefore never parallel to the acceleration. At the highest point in the trajectory, the vertical component of the velocity is zero; thus, the velocity is horizontal, and the acceleration and velocity are perpendicular at this point.

3. The constant speed of the elevator does not affect the rate at which the object falls relative to the elevator. She obtains $a = 9.8$ m/s^2 downward for either direction of the elevator's motion.

5. As it falls, the package retains the same horizontal component of velocity that it had in the airplane. The package and airplane travel horizontally at the same speed; thus, relative to the pilot, the package falls straight downward. A person at rest on the ground sees the package move with constant horizontal velocity and with a vertical velocity of increasing magnitude. Relative to this person, the path of the package is a parabola.

7. The time in the air is governed solely by the vertical component of the motion: $t = 2v_{0y}/g = 2v_0\sin\theta_0/g$. To achieve the maximum time in the air, maximize $\sin\theta_0$ by launching the projectile straight upward at a launch angle of 90° with respect to the horizontal. Since the horizontal component of velocity would be zero, the range would be zero ($x = (v_0\cos 90°)t$).

9. (a) Yes, at the maximum height of the projectile's motion, the vertical component of its velocity is zero and thus its velocity is purely horizontal. (b) No, since the projectile's velocity has a constant nonzero horizontal component, the velocity is never vertical.

11. Two aspects of this path violate the physical principles of projectile motion: (1) The horizontal component of the velocity must be constant; this path provides for a varying horizontal velocity component that equals zero at the path's endpoints. (2) The shape of the path must be parabolic; this path is a semicircle.

15. The trajectory does not allow for a constant horizontal velocity component; the horizontal component varies from zero, to a maximum nonzero value at the peak of travel, and back to zero. Also, this path does not have the requisite parabolic shape that results from motion having a constant downward acceleration and a constant horizontal velocity component.

Multiple-Choice Problems

1. B
3. C
5. A
7. B
9. C
11. A
13. D
15. D
17. B

Problems

1. (a) 1.09 km/s; −0.24 km/s
   (b) 1.12 km/s; 12.4° below +x axis
3. (a) 2.0 m, 1.0 m (b) 2.2 m, 26.6° counterclockwise from +x axis (c) 5.3 m/s; 3.3 m/s (d) 6.2 m/s; 32° counterclockwise from +x axis
5. (a) $(v_x)_{av} = 3.8$ m/s; $(v_y)_{av} = 3.8$ m/s; $(a_x)_{av} = 0.46$ m/s^2; $(a_y)_{av} = -0.46$ m/s^2
   (b) $(v_x)_{av} = 3.8$ m/s; $(v_y)_{av} = 0$; $(a_x)_{av} = 0$; $(a_y)_{av} = -0.46$ m/s^2
   (c) $(v_x)_{av} = -3.8$ m/s; $(v_y)_{av} = -3.8$ m/s; $(a_x)_{av} = -0.46$ m/s^2; $(a_y)_{av} = 0.46$ m/s^2
   (d) $(v_x)_{av} = (v_y)_{av} = 0$; $(a_x)_{av} = (a_y)_{av} = 0$ (e) 5.4 m/s; no; distance traveled is larger than displacement
7. (a) 0.945 m
9. (a) 0.391 s (b) 3.58 m/s (c) 5.24 m/s at 46.9° below horizontal
11. 1.28 m/s^2
13. (a) 30.6 m/s (b) 36.3 m/s
15. (a) 13 m/s, 20 m/s (b) 20 m (c) 2.0 s (d) 13 m/s, horizontal; 9.80 m/s^2, downward (e) 4.0 s (f) 52 m
17. (a) 0.682 s, 2.99 s (b) 24.0 m/s, 11.3 m/s; 24.0 m/s, −11.3 m/s (c) 30.0 m/s at 36.9° below the horizontal
19. (a) 118 ft/s (b) 54.4 ft
21. (a) 4.00 m/s (b) 1.47 m
23. (a) 53.1° (b) 15.0 m/s; 9.80 m/s^2, downward (c) 15.9 m; 17.7 m/s
25. (a) 45° (b) 63.8 m (c) 3.61 s
27. 6.69 m/s
29. 45°; 50.0 m
31. 2.7 m/s
33. (a) 2.97×10^4 m/s (b) 5.91×10^{-3} m/s^2 (c) 4.78×10^4 m/s; 3.97×10^{-2} m/s^2
35. 1.64×10^{-3} m/s^2
37. 25 m/s^2 = 2.5g
39. 0.36 m/s, 38° west of south
41. (a) 28° north of east
    (b) 3.7 m/s, east
    (c) 217 s
43. (a) 24° west of south (b) 5.5 hr
45. 9.87 m/s
47. 275 m
49. (a) 81.6 m (b) 8.16 s (c) in the cart
51. (a) 9.16 m/s^2 (b) 2.72 s (c) 8.45 m
53. (a) 8.50 m/s (b) 13.3 m
55. 8.8 m; 52.4 m
57. 6.91 m
59. $3\sqrt{gD}$ min; $3.13\sqrt{gD}$ max
61. (a) 3.50 m/s^2, upward
    (b) 3.50 m/s^2, downward
    (c) 12.6 s
63. (a) 49 m/s (b) 50 m
65. (a) 863 m (b) 4360 m

Passage Problems

67. B
69. B

Chapter 4

Conceptual Questions

1. The force is applied at the wheels by the surface of the road.
3. Use Newton's second law. Apply a known force that imparts rectilinear motion, and measure the acceleration that is produced. If the person is initially at rest and the force is applied for time t, the acceleration is $a = v/t$, where v is the final speed of the person. The mass is then calculated as $m = F/a$.
5. The bus must be slowing down (braking). The friction force on the ball is small; consequently, the ball continues to move with its initial velocity: the velocity of the bus prior to braking.
7. By Newton's third law, the magnitude of the force the compact exerts on the SUV equals the magnitude of the force the SUV exerts on the compact. The effect of the force on the motion—the acceleration that is produced—is greater for the compact, and thus for its passengers, since the compact has less mass.
9. The bowling ball weighs less on the moon than on the earth, but it has the same mass. It would therefore be just as hard to stop the thrown bowling ball as it is on earth.
11. The impact force gives the car a large forward acceleration. Consequently, the car seat imparts a forward force, proportional to the forward acceleration, on the passenger's torso. However, if the car does not have a headrest, or if the headrest is improperly adjusted, the motion of the passenger's unsupported head follows Newton's first law. The head, which was not acted upon by an outside force, initially maintains the speed of the car and thus moves backward *relative* to the passenger's torso. Next, the head is rapidly thrust forward. This back-and-forth or "whipping" motion can cause injury to the neck.
13. While the ball is in the air, there is no horizontal force on it, and it travels in a straight line in accordance with Newton's first law. If the plane turns, the ball no longer travels in a straight line relative to the passenger to whom the ball was thrown. There is now a net horizontal force on the passenger, causing him to accelerate in the new direction of the plane; but there is no corresponding force on the ball. The passenger thus cannot catch the ball.
15. With no air resistance, all objects fall with the same acceleration and thus have identical speeds when they reach the ground. The pillow is softer, so it travels through a greater distance as it comes to rest. The pillow therefore has a smaller magnitude of acceleration than the rock, and the stopping

force on the pillow is less. By Newton's third law, since your head exerts less force on the pillow than on the rock, the pillow exerts less force on your head.
17. (a) You feel this because the seat exerts a forward force on you, while your body exerts a backward reaction force on the seat. You feel as though you are being pushed back, but you are actually being pushed forward. (b) The free-body diagram is sketched in the figure below. $\vec{F}_1$ and $\vec{F}_2$ are the horizontal and vertical components, respectively, of the force the car seat exerts on you. F_2 equals the weight mg of the person, while F_1 produces the horizontal acceleration.

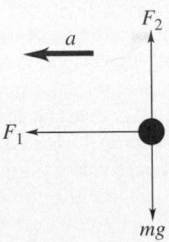

Multiple-Choice Problems

1. C
3. B
5. D
7. C
9. A
11. A, D
13. B
15. B

Problems

1. 7.1 N to the right; 7.1 N downward
3. 241 N; 287 N
5. (a) $F_{1x} = 844$ N, $F_{1y} = 507$ N; $F_{2x} = -418$ N, $F_{2y} = 668$ N; $F_{3x} = -248$ N, $F_{3y} = -328$ N (b) 866 N; 78.1°
7. (a) 4.31 m/s^2 (b) 215 m
9. (a) 680 N (b) the ground
11. 90.9 kg
13. 2940 N
15. (a) 2.06×10^{-6} N (b) 1.21×10^{-4} N (c) 10.1 lb; 4.6 kg
17. (a) 1.81 m/s^2 (b) 1.79 kg
19. (a) 138 N (b) 138 N (c) 1350 N for both (a) and (b) (d) the same, 138 N
21. (a) 3.71 m/s^2 (b) 8.69 kg
23. (a) 2.45 m/s^2 (b) 191 m
25. (a) earth (b) 4 N; the book (c) no (d) 4 N; the earth; the book; upward (e) 4 N; the hand; the book; downward (f) second (g) third
27. 0.452 m/s^2, downward
31. (b) yes
37. (a) 490 N (b) 735 N (c) 612 N
39. (a) 502 N (b) 345 N (c) 424 N
41. (b) 142 N
43. (a) 784 N (b) 47.4 N
45. (a) 21.2 m/s^2 (b) 1700 N (c) 1700 N; 173 kg

47. 3.7×10^6 N
49. (b) upward (c) 1.5 m/s^2, upward
51. (a) 4.8 m/s (b) 16 m/s^2 (d) 2360 N
53. (a) 140 N
57. (a) 539 N (b) 81 N, upward
    (c) 1.47 m/s^2, upward

Passage Problems

59. D
61. A

Chapter 5

Conceptual Questions

1. No, the object is in equilibrium only if the vector sum of all of the forces on it is zero. With only one applied force, the net force cannot be zero.
5. The coefficient of static friction is much less for icy pavement.
7. The vertical component of the vector $\vec{T}$ must equal the weight of the object. Since the magnitude of any vector is greater than either of its components, T must be greater than the weight of the object.
9. With no friction, the weight W will accelerate downward when the system is released. This downward acceleration is possible only if a net downward force exists. The downward force W on the weight must therefore be greater than the upward tensile force in the wire. The tension will be less than W.
11. (a) No, without friction, there would be no horizontal force on you. (b) Yes, you could push down on the rung of a ladder, and the ladder would exert an upward normal force on you. (c) No, the pole could not exert a vertical force, parallel to the pole, without friction. (d) Yes, you could push down on the ground, and the ground would push up on you. (e) No, without friction, the ground could not exert the horizontal force on the wheels necessary for traction. (f) No, without friction, there would be no horizontal force on the car; therefore, the car could travel only in a straight line.

Multiple-Choice Problems

1. A
3. B
5. A
7. B
9. A
11. C
13. D

Problems

1. (a) 3.6 kg (b) 410 kg (c) 6×10^3 kg
3. (a) 25.0 N (b) 50.0 N
5. (a) 2.54×10^3 N (b) 1.01°
7. (a) N/m^2; kg/(m · s^2) (b) 25 cm^2
9. (a) 47.0 N (b) 85.2 N
11. (b) 649 N; 314 N; 813 N (c) no
13. (b) 2.65 N (c) 0.193 N
15. (a) $w\sin\alpha$ (b) $2w\sin\alpha$ (c) $w\cos\alpha$ for each block
17. (a) 4.41×10^4 N (b) 60 times the weight

19. 250W
21. (a) 3.4 m/s (c) 2.2W
23. (a) 48.0 N (b) $a = 2.65$ m/s^2, downward (c) 0
25. (b) 2.08 m/s^2; 3.07 N
27. (a) 4.0 m/s^2, down the slope
    (b) 4.0 m/s^2, down the slope
    (c) 4.0 m/s^2, down the slope
29. (b) up the ramp (c) 50.0 kg box: 0.612 m/s^2 up the ramp; 30.0 kg box: 0.612 m/s^2 downward
31. (a) 0 (b) 6.0 N (c) 16.0 N (d) 8.0 N
33. (a) 266 lb (b) 0
35. 0.16
37. 0.25
39. (a) 22.0 N (b) 1.96 m/s^2, opposite to the motion
41. $\mu_s = \tan\theta_1$; $\mu_k = \tan\theta_2$
43. 50°
45. (a) 283 N (b) 7.39 m/s^2
47. (a) 19.3° (b) 0.92 m/s^2 (c) 3 m/s
49. (a) 0.710, 0.472 (b) 258 N
    (c) (i) 51.8 N (ii) 1.97 m/s^2
55. (a) 357 N/m (b) 2.08 kg
57. (a) 5100 N (b) 85 m/s^2, 8.70,
59. (a) yes; graph of m versus x is a straight line (b) 24.5 N/m (c) 94 cm
61. (a) 15.2 cm (b) 17.6 cm
63. (a) 4.4 m/s (b) 440 m/s^2, upward; 2.5×10^4 N (c) 7.0 ms
65. 32 N/m
67. (a) 621 N (b) 161 N
69. (a) 1.10×10^8 N (b) 5W (c) 8.4 s
71. (a) 12.0 N (b) 15.0 N
73. (a) yes; the graph of W versus L is a straight line (b) 50.0 N/m (c) 4.0 N
75. (a) 2.0×10^4 N (b) 6.19×10^3 N (c) 3.26×10^4 N
77. (a) 250 N/m (b) 5.23 m/s^2; in the backward direction
79. 4.71 s
81. (a) 16.9 N (b) 10.1 N
83. (b) 20 N (c) 100 N
85. 5.9 m/s
87. g/μ_s

Passage Problems

89. B
91. D

Chapter 6

Conceptual Questions

3. This is incorrect. The magnitude of the force of the earth on the apple equals the magnitude of the force of the apple on the earth. Equal forces have different effects on the two objects because the mass of the earth is much greater than the mass of the apple. The same magnitude of force produces a much larger acceleration for the apple.
5. No, there is no way to tell the difference without making observations outside the laboratory.
7. (a) No. The gravitational pull on the 1.0 kg object is $F_g = G\dfrac{(1.0\,\text{kg})m_p}{R_p^2}$,

where m_p is the mass of the planet and R_p is its radius. If the planets have different radii, then the gravitational pulls will be different; the planet of smaller radius exerts a greater pull. (b) Yes. The gravitational pull is $F_g = G\dfrac{(1.0\,\text{kg})m_p}{r^2}$, where r is the distance of the object from the center of the planet. At the same r and for planets of the same m_p, the force F_g will be the same.
9. False. F_g is proportional to $1/r^2$, where r is the distance of the object from the center of the earth. Doubling the height above the surface does not double the distance from the center of the earth.
11. At one particular speed, the size of the centripetal force needed to keep you moving in a circle of the Ferris wheel's radius is exactly equal to your weight. At that speed, gravity provides all the centripetal force at the top, but with nothing "left over" to hold you to the seat, which therefore exerts no upward force on you. At the same speed, at the bottom, the seat must not only support your normal weight, but must also provide a centripetal force equal to your weight. The seat therefore pushes upward on you with a force equal to twice your weight at that point in the ride.
13. (a) The car turns while the person tends to continue traveling in a straight line. There is no outward force on the person. (b) The free-body diagram is given in the figure below. The acceleration $\vec{a}_{\text{rad}}$ is inward toward the center of the curve. $\vec{F}$ is the force exerted by the car on the person.

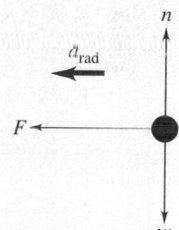

15. First, there is no outward force on the satellite. Second, the forces on the satellite do not balance. There is, however, a net inward force (gravity) that produces the inward acceleration associated with the circular motion. The free-body diagram is given in the figure below.

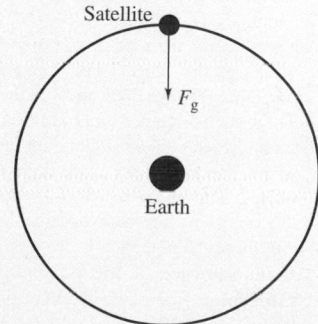

Multiple-Choice Problems

1. D
3. D
5. C
7. C
9. D
11. B
13. A
15. C

Problems

3. (b) 77 N (c) 12 times weight of hand
5. (b) 6.19 s (c) no
7. (a) 35.0 km/s (b) 5.50×10^{22} N
9. 20.0 m/s = 44.7 mph
11. (a) 5.24 m/s (b) highest: 833 N; lowest: 931 N (c) 14.2 s (d) 1760 N
13. (a) 1.73 m/s^2 (c) 0.0115 N, upward (d) 9.8×10^{-3} N
15. 230 m
17. 1.87×10^{-34} N
19. 2.18
21. (a) 2.59×10^8 m (b) no
23. 0.0261 mm; no
25. 2.2×10^{-9} m/s^2, toward the 8.00 kg mass
27. (a) 2.25×10^{25} kg (b) $3.76 m_E$
29. 433 N
31. 9.16×10^{13} N
33. (a) 175 min (b) 3.71 m/s^2
35. 7.46×10^3 m/s
37. (a) 86,400 s = 24 h (c) 3.58×10^7 m
39. 37.7 days
41. 4.7 m/s
43. 2.55 rpm
45. 92 km
47. (b) 0.28 (c) no
49. (a) 2.9×10^{15} kg; 7.7×10^{-3} m/s^2 (b) 6.2 m/s; yes
51. 1.1 m/s
53. (a) 6.28×10^{-5} N (b) 3.77×10^3 N
55. 16.8 m/s

Passage Problems

57. D
59. B

Chapter 7

Conceptual Questions

1. The force one surface exerts on the other displaces the atoms of the surface from their equilibrium positions and imparts oscillatory motion to them. The energy associated with this motion is thermal energy.
5. The force of the rope on the object does work on the object. The force of the object on the rope does work on the rope. Thus, each force performs work on a different object; the work done on the rope is not part of the net work done on the object. The nonzero net work done on the object by the rope provides kinetic energy to the object.
7. In the absence of friction in the jack mechanism, the same work is done in both cases. When the jack is used, a smaller force is exerted over a larger distance to do the same amount of work as lifting the car directly.
9. The elastic potential energy of the compressed spring is converted into thermal energy, which is dispersed within the solution.
11. The kinetic energy comes from potential energy stored in your tensed muscles. When you jump, you push downward on the floor with a force greater than your weight. By Newton's third law, the floor pushes up on you with an upward force of the same magnitude. The work this force does on you increases your kinetic energy and lifts you into the air. You could not jump if you were floating in outer space, since there is nothing to exert a force on you.
13. The change in kinetic energy is proportional to the change in v^2. For 15 mph to 20 mph, the change in v^2 is 175 mph^2. For 10 mph to 15 mph, the change in v^2 is 125 mph^2. Thus, the change is greater for 15 mph to 20 mph.
15. Energy is transported from the sun to the earth by electromagnetic waves, such as visible light, that are radiated by the sun. Part of this energy is converted to chemical energy by plants, using photosynthesis. This chemical energy is ingested when we eat the plants or eat an animal that has eaten the plants. Our bodies convert the chemical energy (food energy) into mechanical energy through the operation of our muscles.

Multiple-Choice Problems

1. B
3. A
5. C
7. A, D
9. A, D, E
11. B, D
13. C
15. C

Problems

1. 300 J
3. 5.22×10^4 J
5. (a) 3.60 J (b) −0.900 J (c) 0 (d) 0 (e) 2.70 J
7. (a) 74 N (b) 330 J (c) −330 J (d) 0; 0 (e) 0
9. 2.62×10^9 J
11. (a) 324 J (b) −324 J (c) 0
13. 4.21 g
15. (a) 7.07 m/s (b) 3/4
17. 74.2 J
19. (a) 38.6 m (b) 154 m (c) converted to thermal energy
21. (a) 30.3 m/s (b) 46.9 m
23. (a) 0.0399 (b) 0.160
25. (a) 2.6×10^4 N/m (b) 650 N
27. 0.290 m, 0.375 J

29. (a) 6.2 J (b) 11.2 J
31. 420 J; from potential energy stored in his tensed muscles
33. (a) 3.03 m/s; as mass leaves spring (b) 95.9 m/s^2; just after mass is released
35. (a) 199 N/m (b) 69.3 cm, 47.8 J
37. (a) 2.08×10^5 N/m (b) no
39. (a) 920 (b) 180 m
41. 1.1×10^3 jumps; 37 min.; no
43. (a) 7.24×10^3 J (b) 7.24×10^3 J, 120 m/s; 14 m/s, 32 mph (c) 4850 ft; 9700 ft
45. (a)–(d) $\sqrt{v_0^2 + 2gh}$ (e) unchanged
47. (a) 2.0 m/s (b) 9.8×10^{-7} J, 2.0 J/kg (c) 200 m, 63 m/s (d) 5.9 J/kg (e) in its tensed legs
49. 7.9 m/s
51. $4K$
53. 7.38 cm
55. 127 J, converted to thermal energy
57. 0.233 J
59. 5.2 m/s
61. (a) 300 N (b) (i) 56% (ii) 33%
63. (a) 3.6×10^5 J (b) 100 m/s; no (c) 520 m; no
65. (a) $4.39 (b) $0.13
67. 28
69. (a) 1.3×10^3 km^2 (b) 36 km, 22 mi; yes
71. (a) 0.134 hp (b) 56 kW (c) 2.68×10^{-3} hp (d) 5.25×10^{23} hp (e) 28 kWh
73. 7.10×10^4 W
75. (a) 1.3×10^5 J (b) 1.5 W (c) kinetic energy, thermal energy
77. (a) 1.0×10^5 N/m, 250 J (b) 0.46 m/s
79. 1.83 m/s
81. 3.00 m/s
83. 390 W
85. 2.1 m/s, 1.9 N
87. (a) 16.0 m/s (b) 1.15×10^4 N
89. (a) 180 lb (b) 0.55 hp
91. 10 m/s
93. 0.602 m
95. (a) 20 m (b) −78 J
97. (a) 1.6×10^3 food calories/day (b) thermal energy (c) 2.48×10^3 food calories
99. (a) 4.43 m/s (b) less
101. (a) 1.68×10^3 N (b) 13.8 hp (c) 154 hp
103. 3.6×10^{13} J; 8.6×10^{-3} megaton bombs
105. (a) yes (b) 250 N/m (c) 28.3 cm (d) 70.8 N

Chapter 8

Conceptual Questions

1. The free-body diagrams are given in the figure below. n_{BA} is the normal force that B exerts on A, n_{AB} is the normal force that A exerts on B, and n is the normal force that the table exerts on B. f_{BA} is the friction force that B exerts on A, f_{AB} is the friction force that A exerts on B, and f is the friction force that the table exerts on B. (a) For block A, n_{BA}, f_{BA}, and w_A are all external forces. (b) For

block B, n, n_{AB}, f_{AB}, f, P, and w_B are all external forces. (c) P, f, n, w_A, and w_B are external forces; n_{BA}, n_{AB}, f_{BA}, and f_{AB} are internal forces.

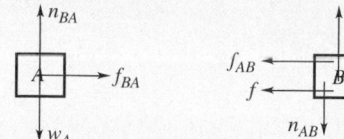

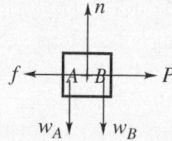

3. Since the same force gives the lighter fragment a greater acceleration, the lighter fragment gains more velocity in the same time.

5. The follow-through increases the time the force acts and therefore increases the impulse applied to the ball. This in turn increases the final momentum and speed of the struck object.

9. (a) Yes. $K = p^2/2m$, so if $p = 0$, then $K = 0$. (b) No. The system can have zero total momentum because the objects have equal nonzero momenta in opposite directions. But the kinetic energy of each object does not depend on the direction of the object's velocity and thus is never negative. Therefore, the kinetic energies do not add to zero. (c) Yes. The only way the total kinetic energy can be zero is for each object to be at rest.

11. If a drop is considered to be the system, the system is not isolated. The drop gains momentum as it falls, because of the external force of gravity. When the drop strikes the ground, the momentum of the drop is removed by the net upward impulse due to the force exerted by the ground. If the drop and earth are considered to be an isolated system, the total momentum is zero before the drop starts to fall; remains zero as the drop falls, because the earth moves toward the drop; and is zero after the drop lands and both drop and earth come to rest.

13. Since the change in momentum is greater when the bullets bounce off, the impulse applied by the plate is greater in that case. By Newton's third law, the impulse applied to the plate is also greater.

15. No. The horizontal component of momentum must be conserved. It is nonzero just before the explosion, so it cannot be zero immediately after the explosion.

Multiple-Choice Problems

1. A, D
3. A, B, C
5. C
7. A

9. B
11. B, D
13. A, C, D
15. D

Problems

1. (a) 44 kg·m/s (b) 0 (c) -12 kg·m/s (d) -2 kg·m/s
3. (a) -30 kg·m/s; -55 kg·m/s (b) 0; 52 kg·m/s (c) 0; -3 kg·m/s
5. (a) 6.5 kg·m/s; 150 J (b) (i) 73 m/s (ii) 110 m/s
7. (a) $3p$ (b) $\sqrt{3}p$
9. 2.5×10^{-23} m/s
11. 9.9 cm/s
13. (a) 0.790 m/s (b) -0.0023 J
15. (a) 7.20 m/s at $38.0°$ (b) -680 J
17. 0.960 m/s
19. 1.8 m
21. (a) 15.0 m/s (b) 25.0%; converted to other forms
23. (a) 81.8 kg (b) 216 J
25. 229 m/s
27. 0.763 m/s at $27.6°$ south of east
29. (a) 7.1 m/s (b) 5.2 m/s
31. (a) 0.300 kg: 0.27 m/s to the right; 0.150 kg: 1.07 m/s to the right (b) 0.300 kg: 0.011 J; 0.150 kg: 0.086 J
33. (a) proton: 154 km/s to the left; alpha particle: 104 km/s to the right (b) 3.58×10^{-17} J (c) total kinetic energy of the system is conserved
35. (a) 0.333 m/s, 3.33 J (b) -1.33 m/s (A), $+0.67$ m/s (B)
37. (a) 5.22 kg·m/s (b) 261 N
39. (a) 4.2 kg·m/s; 140 N (b) -7.3 kg·m/s; -240 N
41. (a) 120.0 N/s, to the left (b) (i) 80 m/s, to the left (ii) 50 m/s, to the left
43. 2.1 m/s
45. 4670 km
47. (a) 3.70×10^9 m (b) 4.48×10^5 m (c) star moves in orbit of larger radius
49. (a) 0.700 m upward and 0.700 m to the right
51. (a) 16.0 m behind leading car (b) 5.04×10^4 kg·m/s (c) 16.8 m/s (d) 5.04×10^4 kg·m/s; same
53. (a) 80.0 N (b) Yes. Eject gas in the appropriate direction.
55. 2.4 km/s
57. (a) 53 g (b) 5.19 N
59. 91.3 m south
61. (a) 14.6 m/s, at $31.0°$ from the original direction of the bullet (b) no; $\Delta K = -189$ J
63. (a) 5.28 m/s (b) 5.7 m
65. $R/4$
67. (a) 2.60 m/s (b) 325 m/s
69. 0.105 m/s
73. 0.400 m/s
75. 65.5 m/s
77. 4.98×10^5 m/s
79. (a) A: v_0 to the left; B: v_0 to the right; elastic (b) both carts are at rest; inelastic
81. 250 J

83. $v_{\text{suv}} = 12$ m/s
    $v_{\text{sedan}} = 21$ m/s

Passage Problems

85. E
87. C

Chapter 9

Conceptual Questions

1. The tangential acceleration is parallel to the instantaneous velocity at each point and is equal to the rate of change of the speed of the point. The radial acceleration is associated with the rate of change of the direction of the point's velocity.

3. Yes, the point has both tangential and radial acceleration. The tangential acceleration has a constant magnitude $r\alpha$. It is tangential to the circular path and changes direction as the flywheel rotates. The radial acceleration is $r\omega^2$ and is not constant in magnitude, since α changes ω. The radial acceleration is directly radially inward and changes direction as the flywheel rotates.

5. (a) Yes, the ring is a collection of point masses, all the same distance R from the axis. (b) No, the point mass has $I = MR^2$, and the solid uniform disk has $I = \frac{1}{2}MR^2$. The mass of the disk is spread between the axis and a distance R from the axis, so I for the disk is less than MR^2.

7. Some of the energy supplied to the bike goes into rotational kinetic energy of the wheels, rather than translational kinetic energy of the bike. The amount of translational kinetic energy lost to rotation of the wheels increases as the mass of the wheels increases.

9. When the raw egg rolls, only the liquid near the shell rotates, whereas all the contents of the hard-boiled egg rotate. The raw egg therefore has a smaller moment of inertia and will reach the bottom of the slope first.

13. His legs and arms move in circular arcs relative to his hips and shoulders, respectively.

15. The acceleration depends on the slope of the hill. If the slope changes along the path, then the translational acceleration of the ball changes and constant-acceleration equations cannot be used.

Multiple-Choice Problems

1. B
3. D
5. A
7. B
9. C
11. C
13. B
15. A

Problems

1. 17.2°
3. (a) 0.105 rad/s; 1.75×10^{-3} rad/s;
   1.45×10^{-4} rad/s (b) 1 min; 1 hr; 12 hr
5. (a) 2.25 s (b) 2.79 rad/s
7. (a) 5.76×10^5 m/s (b) 1.56×10^{-5} rad
9. (a) 200 rad/s (b) 3.1×10^{-3} s
   (c) 170 rpm (d) 0.032 s
11. (a) 2.33 rad/s² (b) 818°
13. 0.0135 rad/s²
15. (a) 12.5 s (b) 7.96 rev
17. (a) 300 rpm (b) 75.0 s; 312 rev
19. 10.5 rad/s
21. 92 rad/s
23. (a) 18.0 m/s² (b) 3.00 m/s, 18.0 m/s²
25. 28 rpm
27. (a) 0.831 m/s (b) 109 m/s²
29. (a) 2.29 (b) 1.51 (c) 15.7 m/s, 108g
31. (a) 2.33 kg·m² (b) 7.33 kg·m² (c) 0
33. (a) 0.0640 kg·m² (b) 0.0320 kg·m²
   (c) 0.0320 kg·m²
35. (a) 3.16×10^{23} J (b) 126 years
37. (a) 0.0225 kg·m² (b) 0.500 kg
39. 0.600 kg·m²
41. (a) 2.25 m (b) 5.63 m
43. (a) 0.664 m (b) 7.69 rad/s
45. (a) 140 J (b) 33%
47. (a) 1.80 m/s (b) 7.18 J
49. (a) 67.9 rad/s (b) 8.35 J
51. (a) 1/3 (b) 2/7 (c) 2/5 (d) 5/13
53. 6.15 m/s
55. 18.4 rad/s²
57. (a) 84.0 rev (b) 5.95 mi (c) 583 mi
59. (a) 9.72 m/s (b) 8.50 J (c) 652 rad/s
61. (a) 2.66×10^{33} J (b) 2.57×10^{29} J
   (c) 9.66×10^{-3}%
63. (a) 1.70 m/s (b) 84.8 rad/s
65. (a) 3.3 kg·m² (b) 93 J
67. 6.67×10^{-3} kg·m²
69. (a) H (b) $5H/7$ (c) in (b), stone ends up
   with rotational kinetic energy
71. (a) 1.05 rad/s (b) 5.0 J (c) 78.5 J
   (d) 6.4%
73. (a) 23.0 m

Passage Problems

75. D
77. D

Chapter 10

Conceptual Questions

1. The extra length increases the torque applied to the bolt; the force exerted on the wrench by the mechanic acts with a longer moment arm. The increased torque can shear the bolt.
3. To prevent the bricks from falling, two criteria must be met. First, the center of gravity of the top brick cannot extend past the right-hand edge of the brick beneath it, so the minimum overlap is 6 inches. Second, the center of gravity of the two-brick combination cannot lie beyond the edge of the table. To determine

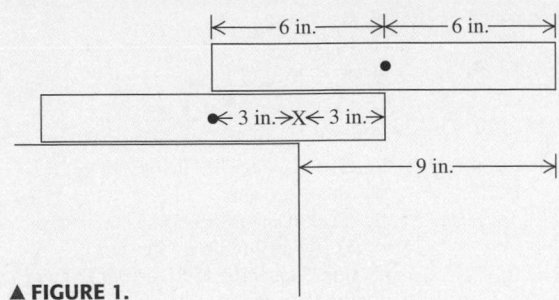

▲ **FIGURE 1.**

the maximum x, assume the first criteria is just met—the top brick extends 6 in. beyond the bottom brick. As Figure 1 shows, the center of gravity of the two-brick combination is then 9 in. to the left of the right-hand edge of the top brick. Therefore, the greatest x can be is 9.0 in.

5. (a) Yes, you can change the location of your center of mass by changing the shape of your body. This changes the way that your mass is distributed. (b) Yes, you can change your moment of inertia with respect to some axis by moving your arms or legs to change how your mass is distributed with respect to the axis.

7. The refrigerator starts to slide when the horizontal applied force exceeds the maximum possible static friction force. The size of the static friction force is proportional to the weight of the refrigerator and to the coefficient of static friction between the refrigerator and the floor. Consider an axis at the forward, lower corner of the refrigerator. The refrigerator starts to tip when the torque due to the applied force exceeds the torque due to the weight of the refrigerator. The torque due to the applied force depends on the applied force and on where the force is applied. The gravity torque depends on the weight of the refrigerator and the width of the refrigerator. Therefore, whether the refrigerator starts to slide or starts to tip first depends on: μ_s, the width of the refrigerator and the height at which the force is applied. The higher up you push, the more likely the refrigerator is to tip.

9. A glass tips when its center of gravity is pushed beyond the edge of its base. The answer to both questions depends on the location of the center of gravity. For a given base, the glass is easier to tip the higher its center of gravity. For a center of gravity at a given height, the more narrow the base the easier it is for the glass to tip. If the water lowers the center of gravity, it makes the glass harder to tip. If the water raises the center of gravity, it makes the glass easier to tip. The water also increases the weight of the object so more force is required to tilt it.

11. Torques tending to close the door are opposed by torques applied by the

object. The moment arm for the force exerted by the object is very small, so the force is very large and pulls horizontally against the hinges.

13. A free-body diagram for the hammer is sketched in Figure 2. The person using the hammer to pull the nail exerts a force $\vec{P}$ at the end of the handle. The nail exerts a downward force with magnitude F on the claw of the hammer and by Newton's third law the hammer exerts an upward force with magnitude $F' = F$ on the nail. The magnitude of the torque due to $\vec{P}$ equals the magnitude of the torque due to $\vec{F}$. But the force $\vec{P}$ has a much larger moment arm than $\vec{F}$, so $F \gg P$ and F' is large.

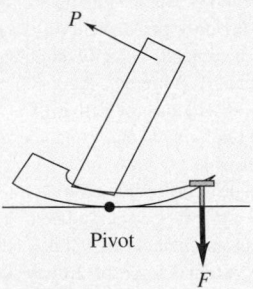

▲ **FIGURE 2.**

15. Since angular momentum is given by $L = I\omega$ and rotational kinetic energy is given by $K_{rot} = \frac{1}{2}I\omega^2$, $K_{rot} = \dfrac{L^2}{2I}$. If they have the same L they will have the same K_{rot} only if they have the same I. And since $L = \sqrt{2I(K_{rot})}$, if they have the same rotational kinetic energy they will have the same angular momentum only if they have the same I. And rotational kinetic energy is a scalar whereas angular momentum is a vector. Even if they have the same magnitude of angular momentum, the angular momenta of the two objects could be different if they are in different directions.

Multiple-Choice Problems

1. C
3. E
5. B
7. A
9. D

11. C
13. C
15. D

Problems

1. (a) 40.0 N · m, counterclockwise
   (b) 34.6 N · m, counterclockwise
   (c) 20.0 N · m, counterclockwise
   (d) 17.3 N · m, clockwise (e) 0 (f) 0
3. −0.31 N · m, clockwise
5. 2.50 N · m, counterclockwise
7. 3.22 rad/s^2
9. 0.482
11. (a) 2.23 m/s^2 (b) 15.2 N (c) 4.46 rad/s^2
13. (a) 3.16×10^4 N (b) 2.60×10^4 N; 2.67×10^4 N
15. −0.0524 N · m
17. 2.20 s
19. (a) 8.98 rad/s (b) 9.88×10^3 J
   (c) 494 W
21. (a) 358 N · m (b) 1.79×10^3 N
   (c) 83.8 m/s
23. (a) 2.67×10^{40} kg · m^2/s
   (b) 7.07×10^{33} kg · m^2/s
25. (a) 0 (b) 44.8 kg · m^2/s, counterclockwise
27. (a) 45,000 N · m (b) 45,000 kg · m^2/s^2
29. 1.14 rev/s
31. (a) 1.71 rad/s (b) external force but no external torque
33. 0.924 rad/s
35. (a) only rotational equilibrium (b) neither (c) only translational equilibrium (d) only translational equilibrium (e) both
37. 49.9 cm
39. hand: 272 N; foot: 130 N
41. left: 25.0 N; right: 50.0 N
43. (a) 1.41×10^3 N (b) 7000 N
45. (b) 300 N (c) 1200 N
47. (b) 6880 N (c) 2560 N
49. (a) 1920 N (b) 1140 N
51. (b) 569 N (c) 474 N (d) same
55. (a) perpendicular to face of clock and directed inward (b) perpendicular to hubcap of wheel and pointed outward (c) perpendicular to ice and pointed downward (d) aligned with earth's axis and pointing upward from north pole
57. (b) $W_A = 3.0$ N; $W_B = 15.0$ N; $W_C = 8.0$ N; $S_1 = 32.0$ N; $S_2 = 24.0$ N; $S_3 = 9.0$ N
59. (a) 3.7 kN, 2.0 kN vertically upward
61. (a) $0.012W$ (b) less (c) 25.0°; tips
63. 0.730 m/s^2; 6.09 rad/s^2; 36.3 N; 21.1 N
65. (a) 69.6 N (b) 85.1° (c) 139 N
67. (a) 24.0 cm to the right; 6.7 cm above the shoulder joint
   (b) 49.4 N (c) $F_v = 3.4$ N; $F_h = 28.3$ N
69. (b) $(Mg/2) \sin \theta$
71. (a) $mg \dfrac{\sqrt{2Rh - h^2}}{R - h}$
   (b) $(mg) \sqrt{2Rh - hv}/(2R - h)$
   (c) at top of wheel

73. 4.6×10^3 rad/s
75. 0.675 s
77. 7600 N

Chapter 11

Conceptual Questions

1. Swinging of a playground swing; this motion is simple harmonic only for small amplitudes and the motion also loses mechanical energy due to dissipative forces. Pistons in a car engine; force doesn't precisely obey Hooke's law.
3. The negative work done by friction removes kinetic energy from the object. The object slows down and takes more time to complete one cycle, for example more time to move from the initial maximum displacement to equilibrium, than when friction is absent. The period is increased by friction.
5. The period decreases. The effect is the same as increasing g by taking pendulum to another planet; the restoring force is increased.
7. Molecules in a gas have space in between them and can be pushed closer together. In a liquid or solid the molecules are more closely packed together. An attempt to push them together is opposed by a large repulsive force.
9. In a satellite the pendulum is in free-fall. There is no restoring force and the pendulum does not swing back and forth when released with an initial displacement. For a mass on a spring the restoring force is applied by the spring, and this force is not affected by being in orbit. The period of a mass on a spring could be used as a time keeping device but a pendulum could not.
11. The circular frequency $\omega = \dfrac{2\pi}{T} = 2\pi f$ is proportional to the number of complete swings the pendulum makes each second. It is inversely proportional to the time for one complete swing. The angular velocity is the rate at which the angular position of the pendulum is changing and is proportional to the linear speed of the pendulum bob along its path. The circular frequency describes a full cycle of the motion and is constant during the motion. The angular velocity changes during the motion. It is zero at the maximum displacement and is a maximum as the pendulum bob passes through its lowest point.

Multiple-Choice Problems

1. A
3. B
5. C
7. A, C, D
9. C
11. B
13. B
15. C

Problems

1. (a) 1.03×10^9 Pa (b) 1.47×10^{-3}
   (c) 7.01×10^{11} Pa
3. (a) steel: 1.1×10^{-4}; copper: 2.1×10^{-4}
   (b) steel: 8.3×10^{-5} m; copper: 1.6×10^{-4} m
5. relaxed: 3.3×10^4 Pa; maximum tension: 6.7×10^5 Pa
7. 1.4 mm
9. (a) 2.2 mm (b) smaller
11. (a) 24% (b) 12%
13. (a) -0.0531 m^3 (b) 1.09×10^3 kg/m^3
15. 1/9
17. 3.4×10^6 N
19. 8.6°
21. 10.2 m/s^2
23. (a) 2.15×10^{-3} s; 2.93×10^3 rad/s
   (b) 2.00×10^4 Hz; 1.26×10^5 rad/s
   (c) 1.3×10^{-15} s to 2.3×10^{-15} s; 4.3×10^{14} Hz to 7.5×10^{14} Hz
   (d) 2.0×10^{-7} s; 3.1×10^7 rad/s
25. (a) 0.120 m (b) 1.60 s (c) 0.625 Hz
27. (a) 0.067 s (b) 15 Hz (c) 94 rad/s
29. (a) 0.0284 J (b) 0.0138 m (c) 0.615 m/s
31. (a) 0.376 m (b) 59.2 m/s^2 (c) 118 N
33. $A/\sqrt{2}$
35. (a) 0.80 s (b) 1.25 Hz (c) 7.85 rad/s
   (d) 3.0 cm (e) 148 N/m
37. (a) 3.03 N/m (b) 1.08 m/s (c) 3.16 m/s^2
39. 1.05×10^3 N/m
41. (a) 1.60 s (b) 0.625 Hz (c) 3.93 rad/s
   (d) 5.1 cm; 0.4 s, 1.2 s, 1.8 s
   (e) 79 cm/s^2; 0.4 s, 1.2 s, 1.8 s (f) 4.9 kg
43. (b) 9.99×10^{-15} g = 9.99 fg
45. 1.88 s
47. 2.60 s
49. (a) $\sqrt{2} T$ (b) $L/9$ (c) $L_{new} = 10L$
   (d) $T/\sqrt{10}$ (e) $T_{new} = T$
51. (a) 0.543 J (b) converted to other forms by dissipative forces
53. (a) 0, 0.80 s, 1.60 s, 2.40 s, 3.20 s
   (b) 0.0533 J (c) 0.0218 J; converted to other forms by nonconservative forces
55. (a) 6.72×10^3 m/s^2 (b) 3.02×10^3 N
   (c) 18.3 m/s; 75.6 J
57. 29.4 m/s^2
59. 2.00 m
61. 0.705 Hz; 14.5°
63. (a) 4.2×10^4 N (b) 64 m
65. (b) 2000 N; 2.72 times his weight
   (c) 4.4 mm
67. (a) 1.63 m (b) brass: 2.00×10^8 Pa; nickel: 4.00×10^8 Pa (c) brass: 2.2×10^{-3}; nickel: 1.9×10^{-3}
69. 35,900 barrels

Chapter 12

Conceptual Questions

1. There is kinetic energy of the moving string and the elastic potential energy of the string when it is displaced from equilibrium. The kinetic energy could be transferred to an object attached to the end of the string.

3. Sound waves require a medium and cannot propagate through empty space.

5. The harmonic content (the extent to which frequencies higher than the fundamental are present) is different for the same note produced by different instruments.

7. This would be an example of resonance. When the sung note has the same frequency as one of the normal-mode frequencies of the glass, large amplitude standing waves can be set up in the glass.

9. The observed frequency of the rotating fork is shifted by the Doppler effect. Its pitch is increased when it is moving toward the listener and is decreased when it is moving away from the listener. The listener therefore hears two different frequencies from the two forks and beats occur.

11. Changing the tension in the string changes the speed of transverse waves on the string and this alters the standing wave frequencies for the string. The notes of a piano are produced by mallets striking stretched wires and the piano is tuned by adjusting the tension in these wires.

Multiple-Choice Problems

1. A
3. A
5. B
7. B, C
9. A
11. D
13. C
15. C

Problems

1. (a) 1.7 cm to 17 m (b) 4.3×10^{14} Hz to 7.5×10^{14} Hz (c) 1.5 cm (d) 6.4 cm
3. 2.2 km/s
5. 258 m/s
7. (a) 16.3 m/s (b) 0.136 m (c) 23.1 m/s; 0.192 m (both increase by a factor of $\sqrt{2}$)
9. (a) $\sqrt{2}V$ (b) V (c) $V/\sqrt{3}$ (d) $\sqrt{2}V$ (e) $2\sqrt{5}V$
11. (a) 25.0 Hz, 0.0400 s, 19.6 rad/m (b) $y(x, t) = -(0.0700 \text{ m}) \sin 2\pi \cdot \left(\dfrac{t}{0.0400 \text{ s}} - \dfrac{x}{0.320 \text{ m}} \right)$ (c) 0.0495 m
13. 0.469 g/cm
17. (a) 408 Hz (b) 24th
19. (a) 45.0 cm (b) no; 440 Hz is lowest frequency possible
23. (a) fundamental: 0.60 m; 1st overtone: 0.30 m, 0.90 m; 2nd overtone: 0.20 m, 0.60 m, 1.00 m (b) fundamental: 0; 1st overtone: 0, 0.80 m; 2nd overtone: 0, 0.48 m, 0.96 m
25. (a) 35.2 Hz (b) 17.6 Hz
27. (a) 16.0 cm, 8.0 cm, 5.3 cm, 4.0 cm; 2210 Hz, 4420 Hz, 6630 Hz, 8840 Hz (b) 32.0 cm, 10.7 cm, 6.4 cm, 4.6 cm; 1110 Hz, 3330 Hz, 5550 Hz, 7770 Hz
29. 39.1 cm; yes

31. (a) 70 Hz (b) 93 Hz
33. (a) 0.237 m (b) 0.711 m (c) 0.474 m
35. (a) $n(820 \text{ Hz})$, $n = 1, 2, 3, \ldots$ (b) $(2n + 1)(410 \text{ Hz})$, $n = 0, 1, 2, \ldots$
37. (a) 90 m (b) 102 kHz (c) 1.4 cm (d) 4.4 mm to 8.8 mm (e) 6.2 MHz
39. (a) 511 W (b) 0.829 W/m² (c) 511 J
41. (a) 81.0 dB (b) 28.1 dB (c) -18.2 dB
43. 15.0 cm
47. (a) 57 dB (b) 2.00×10^{-2} W/m²
49. 14 beats
51. (a) 433 Hz (b) loosen
53. (a) 375 Hz (b) 371 Hz (c) 4 Hz
55. 19.8 m/s
57. (a) 2120 Hz (b) 0.172 m
59. (a) 1.86 s (b) 3.48 s
61. (a) 2000 Hz (b) 1500 Hz
63. 0.390 s
65. (a) 12.6 m (b) 1.26 km
67. flute harmonic $3N$ ($N = 1, 3, \ldots$) resonates with string harmonic $4N$
69. (a) 1.88 m/s (b) 6.26 Hz
71. 10 octaves
73. (a) $f_{\text{L}-\max} = f_{\text{siren}} v/(v - 2\pi f_{\text{P}} A_{\text{P}})$; $f_{\text{L}-\min} = f_{\text{siren}} v/(v + 2\pi f_{\text{P}} A_{\text{P}})$
75. (a) 38 Hz (b) no
77. 1.3 Hz

Chapter 13

Conceptual Questions

1. The pressure in a fluid depends only on height, not the shape of the container. The height of both columns is the same because the pressure at the bottom of each column of water must be the same, if the water is not flowing in either direction. The water in the funnel is supported partly by the rest of the water and partly by the sides of the funnel.

3. Hot air has a smaller density than the surrounding air so the buoyancy force is larger than the weight when the balloon is filled with hot air. The density of the air in the balloon depends on the temperature of this air; the ascent or descent of the balloon is controlled by varying the temperature of the air in the balloon.

5. The pressure difference would be $(0.01)(1.0 \times 10^5 \text{ Pa}) = 1.0 \times 10^3$ Pa. For a door with area 2 m² the net force keeping the door closed would be 2×10^3 N. This is about 450 pounds. It would take a very strong person to pull the door open. But all that would be required would be to open it slightly, so air can move from the high pressure to the low pressure side. Once the pressure is equalized the door will easily open the rest of the way.

9. When the marble is sitting in the floating box it must cause an amount of water to be displaced that is equal to the weight of the marble. When the marble is sitting

at the bottom of the beaker the buoyant force on it is less than its weight, since it sinks, so it is displacing less than its weight of water. The marble causes less water to be displaced when it has been dropped into the water, so the water level in the beaker falls.

11. The pressure inside the rocket ship pushes the air through the hole. The vacuum doesn't pull the air out; the vacuum exerts no force.

13. (a) The average density of the submarine must equal that of the surrounding water when the submarine floats at equilibrium at some depth. The density of water increases with depth below the surface, so the average density of the submarine is different when it is floating at different depths. (b) The submarine could rise by decreasing its average density and descend by increasing its average density. It could do this either by pushing some water out of tanks within the submarine and replacing that water by air, or by letting more water into these tanks.

15. The balloons move forward toward the front of the car. The air in the car has the same acceleration as the car. The force that produces this acceleration for a volume of the air is exerted by the surrounding air. When this volume is replaced by a helium-filled balloon, the force on the balloon exerted by the surrounding air is the same as when just air was in the volume. But the helium-filled balloon has less mass than the same volume of air (the balloon rises if released), so the same force produces a greater acceleration when applied to the balloon and the balloon has a greater forward acceleration than does the air inside the car.

Multiple-Choice Problems

1. B
3. A, D
5. A
7. C
9. D
11. C
13. E
15. D

Problems

1. 7.02×10^3 kg/m³; yes
3. 860 N; 193 lb
5. (a) 19.3 cm (b) 2.29 million dollars
7. (a) 8ρ (b) $L/3^{1/3}$
9. 1.6
11. (a) 5.25 kg (b) 1.3 N (c) 1.1×10^{-9} g
13. (a) 2.52×10^6 Pa (b) 1.78×10^5 N
15. (a) From 4×10^6 Pa (40 atm) to 3×10^7 Pa (300 atm) (b) 7×10^9 N
17. 93.5 mm of Hg
19. 2.27×10^5 N
21. (a) 864 kg/m³ (b) 0.0475 atm
23. (a) 706 Pa (b) 3160 Pa
25. 117/77

27. (a) 4.64 m (b) 1.45 mm
29. (a) 3.4 L (b) 56 N
31. (a) 5.75 m^3 (b) 1150 kg
33. 0.562 m^3
35. (a) 6370 N (b) 558 kg (c) 85.9%
37. (a) 1.02×10^3 kg/m^3 (b) 8.0%
39. 2.86 Pa
41. 20.0
43. (a) 2.33 m/s (b) 5.21 m/s
45. 9.6 m/s
47. 16.6 m/s
49. 1.47×10^5 Pa
51. 500 N, downward
53. 0.41 cm
55. 2.03×10^4 Pa
57. 9.59 cm/s
59. $(2^{1/4})D$
61. (a) 1.47×10^3 Pa (b) 13.9 cm
63. 4.66×10^{-4} m^3; 5.27 kg
65. 732 kg/m^3
67. 33.5 N
69. 2.05 m
71. (a) 1.30×10^5 Pa (b) 13.3 m
73. (c) Use $v_1 A_1 = v_2 A_2$.

Chapter 14

Conceptual Questions

1. When the material is heated the average distance between the atoms or molecules in the material increases uniformly in all directions. The piece of material with the hole expands just like the piece without the hole, and the hole expands just like a solid piece of the material.
3. The two metals have different coefficients of linear expansion. The strips are in good thermal contact, so when the temperature changes both strips have the same ΔT. The strip with the larger α has a greater change of length. If ΔT is positive the composite strip bends such that the material with the larger α is on the outside of the curve. If ΔT is negative the composite strip bends such that the material with the larger α is on the inside of the curve.
5. The spheres have an average density that is slightly less than the density of the liquid at low temperatures. And the spheres differ a little from one another in their average densities. As the temperature of the liquid increases, it expands and its density decreases. When the average density of a particular sphere exceeds that of the liquid, that sphere sinks to the bottom of the cylinder. This occurs at different temperatures for different spheres, with the ones with the greater average density descending first. The volume of each sphere also increases as the temperature increases, so its average density decreases, but the coefficient of volume expansion for the liquid is much larger than for the solid spheres and the expansion of the liquid dominates.

9. False. Conservation of energy requires that the amount of heat that flows into the cooler object equals the amount of heat energy that flows out of the warmer object. But the temperature changes associated with these heat flows depend on the mass and specific heat capacities of the objects. Equal heat flows need not mean equal temperature changes.
11. Phase transitions happen at fixed temperatures. As long as two phases are present the temperature of the system remains at the phase transition temperature. The heat added during the phase change is required for the phase change and doesn't produce a temperature change.
13. There will be small effects due to different emissivity of the coffee when its color changes. But the predominant effect on the rate of heat transfer is the temperature difference between the coffee and its surroundings; the greater the temperature difference the greater the rate of heat loss to the environment. The heat loss during the five minutes is less if the cream is added at the beginning of the period, so the coffee is cooler as it sits during the five-minute period.
15. No. The heat your furnace must provide equals the heat that passes out of the house into the environment. The rate of heat transfer decreases when the temperature difference between the house and its surroundings is less. Less heat is lost while you are away if the house is at a lower temperature during that time.

Multiple-Choice Problems

1. C
3. D
5. C
7. C
9. A
11. A
13. B
15. A

Problems

1. (a) Yes. 104.4°F (b) Yes. 54°F (c) No. 77°C, 171°F
3. (a) 27.2 C° 27.2 K (b) −55.6 C°, −55.6 K
5. (a) −40.0° (b) No
7. (a) 6.8 in. (b) 6.7×10^{-6} (F°)$^{-1}$
9. 2.3×10^{-5} (C°)$^{-1}$
11. 11 L
13. 160 m^3
15. 1.7×10^{-5} (C°)$^{-1}$
17. 0.008 cm
19. 3.03×10^5 J
21. (a) 38 J (b) 4.5×10^4 J
23. 240 J/kg · K
25. (a) 113 C° (b) 6.35 C°
27. 0.0613 C°
29. (a) 2.51×10^3 J/(kg · K)
    (b) overestimate

31. (a) 3680 J (b) 1.31×10^5 J
    (c) 0.404 kg
33. 3.7 kg
35. 5.05×10^{15} kg
37. 1.43×10^5 J, 34.2 kcal
39. (a) 6.33×10^4 J (b) 6.91×10^3 J
    (c) steam burns are more severe
41. (a) 5.9 C° (b) 109°F; yes
43. (a) 215 J/(kg · K) (b) water (c) too small
45. 35.1°C
47. 2.10 kg
49. 3.08 g
51. 8.0 cm
53. 105.5 C°
55. (a) 2.1×10^4 W (b) 6.4×10^3 W
57. (a) 29.6°C (b) 9.1 cm
59. (a) 1.96 m^2 (b) 937 W (c) 118 W
61. (a) 140 W (b) 180 W
63. 2.1 cm^2
65. (a) 0.793 g (b) 1.43 g
67. 2.9 m
69. 26 kW
71. 456 s
73. 37.5°C
75. 341 m
77. 3.45 L
79. 2.7 h
81. (a) 8.6 C° (b) 114° F
83. 5.0×10^{-2} W/m · k
85. 5.2 %
87. 5.82 g

Chapter 15

Conceptual Questions

1. No. The conversion from kelvins to Celsius degrees involves an additive rather than a multiplicative factor, so the conversion factor cannot be absorbed into the gas constant. For example, there are negative temperatures on the Celsius scale and negative values for T don't make sense in the ideal gas equation, no matter what the value is for the gas constant.
5. In a hot object the random motions of the atoms (or molecules) are more energetic than in a cold object. When the two objects are in contact their atoms collide and in the collision kinetic energy is transferred from the atoms in the hot object to the atoms in the cold object.
7. As the temperature is decreased the average kinetic energy of the molecules in the material decreases. Collisions between molecules in the material become less frequent and are less energetic.
9. False. At thermal equilibrium the two kinds of atoms have the same temperature and the same average kinetic energy. To have the same average kinetic energy, the lighter atoms have a greater speed.
11. For any process of an ideal gas, $\Delta U = nC_V \Delta T$. The constants C_V and C_p are defined by the equations for the heat flow for constant volume ($Q = nC_V \Delta T$) and

constant pressure $(Q = nC_p \Delta T)$ processes. For a constant volume process, $W = 0$ and $Q = nC_V \Delta T$ so by $\Delta U = Q - W$, $\Delta U = nC_V \Delta T$. For a constant pressure process, $Q = nC_p \Delta T$, but $W \neq 0$ and ΔU is not equal to Q.

13. For an ideal gas U depends only on T. Since $\Delta U = nC_V \Delta T$ for constant volume, ΔU is also given by this expression for any process.

15. No. To derive these equations we used the ideal gas law. The equation of state for liquids and solids differs greatly from the ideal gas law, so equations 15.22 and 15.23 don't apply.

Multiple-Choice Problems

1. C
3. D
5. C
7. B
9. D
11. B
13. B, C, D
15. A, B, D

Problems

1. 0.959 atm
3. (a) $-175°C$ (b) 1.00 L
5. (a) 5.78×10^3 mol (b) 185 kg
7. 3.36×10^5 Pa
9. 0.159 L
11. 503°C
13. 2.28×10^4 Pa
15. 18 cm³; 56 cm³
17. no; no
19. 3.35×10^{25} molecules
21. (a) could be true (b) could be true (c) not true (d) must be true (e) could be true
23. (a) 4.84×10^2 m/s (b) 6.21×10^{-21} J
25. 6.1×10^{-6} m
27. (a) 1.90 km/s (b) no (c) O_2: 0.477 km/s; N_2: 0.510 km/s (d) some molecules move faster than v_{rms}
29. (a) 22.4 L (b) 0.243 L
31. (a) 10.5 breaths/min (b) 2.62×10^{21} molecules
33. (a) 49.6°C (b) 1110 J
35. (a) 4.09 C°
37. (b) 0
39. (b) 1.33×10^3 J
41. (a) 1.2×10^3 J (b) negative of work done in (a) (c) 600 J
43. 327 J
45. (a) -1.15×10^5 J (b) 2.55×10^5 J; heat flows out of the gas
47. (a) 45.0 J (b) liberates 65.0 J (c) ad: 23.0 J db: 22.0 J
49. (a) -4510 J (b) 4510 J out of gas
51. (a) 25.0 K (b) 17.9 K (c) a
53. (a) 0.125 L (b) 57.0 J done on gas (c) decreased (d) left the gas
55. (a) same (b) absorbs 4000 J (c) absorbs 8000 J
57. (b) 208 J (c) on the piston (d) 712 J (e) 920 J (f) 208 J

59. (a) $T_a = 665$ K; $T_b = 2660$ K; $T_c = 887$ K; $T_d = 222$ K
(b) (i) 6.39×10^4 J (ii) 4.05×10^4 J (iii) 2.13×10^4 J (iv) 1.01×10^4 J
(c) (i) enters (ii) leaves (iii) leaves (iv) enters
61. (a) 747 J (b) 1.30
63. (a) 47.4°C (b) 345 J
65. 2.20×10^6 molecules
67. (a) 1.18×10^4 Pa (b) 0.5667
69. 1.92 atm
71. (a) 1.2×10^4 m/s (b) 6.68×10^{-27} kg
73. (a) $Q = 300$ J; $\Delta U = 0$
(b) $Q = 0$; $\Delta U = -300$ J
(c) $Q = 750$ J; $\Delta U = 450$ J
75. (a) 4.65×10^{-26} kg (b) 6.11×10^{-21} J
(c) 2.04×10^{24} molecules
(d) 1.25×10^4 J
77. 9.6 cm
79. (a) 0.80 L (b) $T_a = 305$ K; $T_b = T_c = 1220$ K (c) ab: 76 J into gas; ca: 107 J out of gas; bc: 56 J into gas (d) ab: $\Delta U = 76$ J; bc: $\Delta U = 0$; ca: $\Delta U = -76$ J
81. (a) 8.00×10^4 Pa (b) 1.45 g
83. 4.18×10^5 J
85. (a) 1172 K (b) 1.22×10^4 J (c) 4.26×10^4 J (d) 4.56×10^4 J

Passage Problems

87. E
89. A

Chapter 16

Conceptual Questions

1. For a refrigerator device, $|Q_H| = |Q_C| + |W|$. This says that the heat energy $|Q_H|$ deposited into the room is the same as the heat energy $|Q_C|$ removed from the food compartment plus the amount of mechanical energy required to operate the refrigerator (inputed to the device as electrical energy). The tubing is where heat flows from the device into the room.

5. For an air conditioner or refrigerator the amount of heat $|Q_H|$ that flows out of the device is greater than the heat $|Q_C|$ that flows into the device. In fact, $|Q_H| = |Q_C| + |W|$, where $|W|$ is the energy supplied to operate the device. If an air conditioner is set on the floor and plugged in, $|Q_C|$ is removed from the room but $|Q_H|$ is ejected into the room. Since $|Q_H|$ is larger than $|Q_C|$, the room is heated rather than cooled. A refrigerator also adds heat to the room in which it is placed. That heat must be removed, for example by a window air conditioner, or else the room will be heated.

7. The change in entropy depends not only on the heat flow but is also inversely proportional to the Kelvin temperature at which that heat flow occurs. The negative heat flow out of the hot object occurs at a higher temperature than the

positive heat flow into the cold object, so the positive entropy change is greater than the negative entropy change and the net entropy change is positive.

9. No. The second law says this can never be the sole result of a process for an isolated system. But a device can cause this to happen if mechanical energy is inputed to the device. A refrigerator is an example of a device that moves heat from a cold object (the food) to a warm object (the room).

11. Both devices move heat energy from a cooler object to a warmer object. For a refrigerator the purpose of the device is to remove heat from the cooler object (the food compartment). The heat deposited into the warmer object (the room) is waste heat. For a heat pump operating in the winter, the purpose of the device is to deposit heat into the warmer object (the interior of the building).

13. In a hot object the random motions of the atoms (or molecules) are more energetic than in a cold object. When the two objects are in contact their atoms collide and in the collision kinetic energy is transferred from the atoms in the hot object to the atoms in the cold object. In the collisions energy is transferred in this direction and not in the reverse direction.

17. The coefficient of performance of a refrigerator is the ratio of the magnitude of the heat $|Q_C|$ removed at the low temperature reservoir to the magnitude of the work $|W|$ required to operate the device. Removal of $|Q_C|$ is the useful action of the device and $|W|$ is what you have to pay for, so it is most economical to operate a refrigerator with a large coefficient of performance.

Multiple-Choice Problems

1. B
3. A
5. A
7. B, D
9. C
11. A, D
13. C
15. B

Problems

1. 1.1×10^{14} J
3. (a) 6500 J (b) 34%
5. (a) 23% (b) 12,400 J (c) 0.350 g (d) 222 kW, 298 hp
7. (a) 25% (b) 970 MW
9. (a) 12.3 atm (b) 5470 J; process ca (c) 3723 J; process bc (d) 1747 J (e) 31.9%
11. 13.8
13. (a) 59.4% (b) 4060 J
15. (a) 1.62×10^4 J (b) 5.02×10^4 J

17. (a) 8.09×10^5 J (b) 3.37×10^5 J
    (c) 1.14×10^6 J
19. (a) 215 J (b) 378 K (c) 39.1%
21. (a) -3.72×10^3 J (b) 2.73×10^3 J
    (c) 42.3%
23. (a) -3.09×10^7 J (b) 2.5×10^6 J
25. (a) clockwise; heat enters in ab and
    leaves in cd (b) counterclockwise; heat
    enters in dc and leaves in ba; dc inside
    refrigerator and ba in room (c) counter-
    clockwise; heat enters in dc and leaves
    in ba; dc inside house and ba outside
    (d) counterclockwise; heat enters in dc
    and leaves in ba; dc outside house and
    ba inside
27. 70 J/K; increases
29. (a) irreversible (b) 304.42 K (c) 470 J/K
31. -6.31 J/K
33. -4.8×10^{-3} J/K
35. 111 m^2
37. (a) 7.0% (b) 3.0 MW; 2.8 MW
    (c) 6×10^5 kg/h $= 6 \times 10^5$ L/h
39. (a) $-73°$ C (b) 3.34×10^4 cycles
41. (b) 45° C (c) 58.9% (d) 206 J
    (e) 2.57×10^6 Pa
43. $1 - T_C/T_H$; same
45. 0.379 J/K
47. (a) 2.26%; no (b) 29.4 J; 1.30×10^3 J
    (c) 1.11×10^{-3} candy bars
49. (a) -143 J/K (b) 196 J/K (c) 0
    (d) 53 J/K
51. 6.17

Chapter 17

Conceptual Questions

1. The electrified object induces a charge
   separation in the neutral object. For example, a positively charged rod displaces
   negative charge in the bit of paper
   toward the rod. Since the negative
   charge in the paper is then closer to the
   rod than the positive charge in the paper
   and since the electrical force is inversely
   proportional to the distance squared, the
   attractive force between positive and negative charges is larger in magnitude than the
   repulsive force between positive and positive charges and the net force is attractive.
3. *Similarities:* Both forces are proportional
   to the square of the distance between the
   objects. Both forces obey Newton's third
   law. *Differences:* The gravitational force
   is always attractive. The electrical force
   can be either attractive or repulsive,
   depending on the signs of the charges.
   The electrical force is stronger than the
   gravitational force. For example, the
   electrical force between two 1 Coulomb
   charges separated by 1 meter is immense
   whereas the gravitational force between
   two 1 kilogram masses separated by
   1 meter is very small.
5. Positive and negative are just names
   given to the sign of charge that protons
   and electrons have. There is no inherent
   significance.

7. (a) The charged rod pulls charge of the
   opposite sign onto the metal ball. This
   leaves the gold leaf and metal tube with a
   net charge of the same sign as the rod.
   Since the gold leaf and the tube have
   charge of the same sign, they repel.
   (b) When the rod is removed the charge
   on the ball spreads back over the tube and
   leaf and they become neutral again.
   There is no net charge and no electrical force so the leaf hangs vertically.
   (c) When the charged rod touches the ball
   it transfers charge to it. This net charge
   spreads over the ball, tube and leaf. When
   the rod is removed this net charge stays
   on the leaf and tube and they continue to
   repel. The gold leaf continues to hang at
   an angle away from the tube.
11. The dipole consists of equal amounts of
    positive and negative charge. In a uni-
    form electric field the force on the posi-
    tive charge is equal in magnitude and
    opposite in direction to the force on the
    negative charge, and the net force is zero.
    If the field is not uniform and is stronger
    at one of the charges in the dipole, the
    force on that charge is greater in magni-
    tude than the force on the other charge
    and the net force is not zero.
13. The lightning is attracted to the sharp
    point of the rod. Copper is an excellent
    conductor and the lightning current travels
    through it rather than through the wood of
    the barn, since wood is a poor conductor
    of electricity.
15. (a) In a solid material the molecules are
    held to each other by electrical forces.
    When one surface slides over another,
    molecules of the surface are pulled off.
    The resistance to this happening, due to
    the electrical bonding, gives rise to the
    friction force. (b) The steel atoms are held
    together by strong electrical forces. In
    order to penetrate the surface, the surface
    atoms must be displaced and the electrical
    bonding forces oppose this. (c) Chemical
    bonding arises from electrical forces
    within the molecules, due to charge dis-
    placement within the molecules.

Multiple-Choice Problems

1. D
3. C
5. E
7. C
9. B
11. D
13. B
15. A

Problems

5. (a) 1.56×10^{13} electrons
   (b) 1.56×10^{10} electrons
7. (a) 4.27×10^{24} protons; 6.83×10^5 C
   (b) 4.27×10^{24} electrons
9. 1.07×10^{-14} m; 1.07×10^{-14} m

11. (a) 230 N; yes; 52 lbs
    (b) 2.30×10^{-8} N; no
13. (a) 3.4×10^{14}
    (b) $+55$ μC
15. (a) 7.25×10^{24} electrons
    (b) 5.27×10^{15} electrons
    (c) 7.27×10^{-10}
17. 3.7 km
19. 0.750 nC
21. 2.59×10^{-6} N; $-y$ direction
23. (a) 1680 N; from the $+5.00$ μC charge
    toward the -5.00 μC charge
    (b) 22.3 N $\cdot$ m, clockwise
25. 1.26×10^{-8} N; attractive
27. 4.27×10^{-10} N, to the left
29. 2.21×10^4 m/s^2
31. (a) 2.50 N/C, upward
    (b) 4.00×10^{-19} N, upward
33. (a) 1.62×10^3 N/C (b) 4.27×10^6 m/s
35. 7.23×10^{-9} C
37. (a) 5.8×10^{19} N/C (b) 5.8×10^9 N/C
39. (a) 9.0×10^{-12} C (b) 32 N/C; away
    from axon (c) 280 m
41. (a) 1050 N/C; $-x$ direction (b) 312 N/C;
    $+x$ direction (c) 845 N/C; $+x$ direction
43. (a) between the charges, 0.24 m from
    the $+0.500$ nC charge
    (b) 0.40 m from ± 0.500 nC charge,
    1.60 m from ∓ 8.00 nC charge
45. (a) zero (b) vector from $-q$ to $+q$ paral-
    lel to $\vec{E}$; vector from $-q$ to $+q$ antiparal-
    lel to $\vec{E}$ (c) stable: vector from $-q$ to $+q$
    is parallel to $\vec{E}$; unstable: vector from $-q$
    to $+q$ is antiparallel to $\vec{E}$
47. (a) $E/4$ (b) $E/9$
49. (a) top, positive; middle, negative; bottom,
    positive; (b) on horizontal line through
    middle charge
55. S_1: 0; S_2: 7.9×10^5 N $\cdot$ m^2/C;
    S_3: 0; S_4: 1.1×10^5 N $\cdot$ m^2/C;
    S_5: 6.8×10^5 N $\cdot$ m^2/C
57. (a) zero
    (b) 3.75×10^7 N/C; radially inward
    (c) 1.11×10^7 N/C; radially inward
59. (a) 1.80×10^{10} (b) 414 N/C
61. (a) zero
63. (a) yes, -12 μC (b) $+12$ μC
65. (a) 3.2 nC (b) to the right
    (c) $x = -1.76$ m
67. (a) $F_x = 8.64 \times 10^{-5}$ N; $F_y = -5.52 \times 10^{-5}$ N
    (b) 1.03×10^{-4} N; 32.6° below $+x$ axis
69. (a) 1.35×10^{-3} N; away from vacant
    corner (b) 1.29×10^{-3} N; toward center
    of square
71. 3.41×10^4 N/C, to the left
73. 1.7×10^6 electrons
75. $q_1 = 6.14 \times 10^{-6}$ N/C;
    $q_2 = -6.14 \times 10^{-6}$ N/C
77. 2.2×10^6 m/s

Passage Problems

79. C

Chapter 18

Conceptual Questions

1. The electrical force at a point is tangent to the electric field lines at that point. Therefore, there is no component of electrical force in the direction perpendicular to the field line. No work is done on a charge when it moves perpendicular to a field line so the potential doesn't change in this direction and this direction is along an equipotential surface.

3. Zero electric field means zero force which in turn means no electrical work done on a test charge when it moves from point A to point B within the region. Therefore, any two points A and B in the region are at the same potential. In this region the potential is constant but not necessarily zero.

5. Current flows between points of different potential. (a) When they are in the car and touch two points on the car body no current flows through them because the two points are at the same potential. (b) The body of the car and the ground have a large potential difference so when an occupant touches both the car and the ground a dangerously large current flows through him.

7. Since the capacitor is disconnected from the battery before the plates are pulled apart, the charge on the plates stays constant while the plates are moved apart. (a) The electric field between the plates depends only on the charge per unit area on the plates. This doesn't change and the electric field stays constant. (b) The charge is trapped on the plates and doesn't change. (c) Since the distance between the plates increases while the electric field stays the same, more work is done by the electric force on a test charge when it moves from one plate to the other and the potential difference increases. (d) The plates attract each other so work must be done to pull them apart and this increases the stored energy. We can also see this from the equation $U = QV/2$. V increases while Q stays the same, so U increases.

11. As the temperature of the liquid increases the random motion of the molecules increases and this decreases the alignment of the molecular dipoles along the electric field direction. This reduces the induced electric field within the dielectric and therefore reduces the effect of the dielectric on the net field between the plates. The dielectric constant is the ratio of the net field without the dielectric to the field with the dielectric. Therefore, the dielectric constant becomes closer to 1.0 as the temperature increases.

13. The stored energy for a capacitor is given by $U = \frac{1}{2}CV^2$. In parallel the potential difference V across each capacitor equals the battery voltage whereas in series the potentials add to give the battery voltage. Therefore, the voltage for each capacitor is greater in parallel and the stored energy is greater when they are connected in parallel.

15. (a) No. At a point midway between two point charges of equal magnitude and opposite sign the potential (relative to infinity) is zero but the electric field is not zero. The electric field is directed toward the negative charge. (b) No. At a point midway between two equal point charges (equal in magnitude and sign) the electric field is zero and the potential (relative to infinity) is not zero.

Multiple-Choice Problems

1. A, C
3. A
5. C
7. B, D
9. B
11. D
13. A
15. C

Problems

1. (a) 0 (b) $+7.50 \times 10^{-4}$ J (c) -2.06×10^{-3} J
3. 0.373 m
5. (a) -2.88×10^{-17} J (b) -5.04×10^{-17} J
7. -1.42×10^{-18} J
9. 0.078 J
11. (a) 8000 N/C (b) 1.92×10^{-5} N
13. (a) 1.25×10^4 V/m (b) 2.20×10^{15} m/s^2
15. 1.5×10^3 km
17. 25 V/m
19. (a) 2.5 mm (b) 7.5 mm
21. 7.42 m/s; faster
23. (a) 0.199 J (b) (i) 26.6 m/s (ii) 36.7 m/s (iii) 37.6 m/s
25. (b) 5.8×10^6 m/s
27. 4.2×10^6 V
29. (b) yes (c) flat sheets parallel to the plates
31. (a) 27.0 kV (b) 0.98 cm; 2.40 m (c) the increasing spacing shows that the field is weaker at greater distances from the charged sphere
35. (a) electron: 5.93×10^5 m/s; proton: 1.38×10^4 m/s (b) electron: 1.87×10^7 m/s; proton: 4.38×10^5 m/s (c) electron: 0.0256 keV; proton: 47.0 keV
37. (a) 3.29 pF (b) 13.2 kV (c) 4.02×10^6 V/m
39. (a) 400 V (b) 3.39×10^{-2} m^2 (c) 6.67×10^5 V/m (d) 5.90×10^{-6} C/m^2
41. (a) 120 μC (b) 60 μC (c) 480 μC
43. (a) 0.447 μF (b) 60.0 V
45. 2.8 mm
47. (a) 20 pF (b) 8.6 pF
49. (a) $Q_1 = 80.0 \times 10^{-6}$ C; $Q_3 = 120.0 \times 10^{-6}$ C (b) 37.4 V

51. $V_2 = 50$ V; $V_3 = 70$ V
53. (a) 3.00 μF: 30.8 μC; 5.00 μF: 51.3 μC; 6.00 μF: 82.1 μC (b) 3.00 μF: 10.3 V; 5.00 μF: 10.3 V; 6.00 μF: 13.7 V
55. (a) 3.47 μF (b) 174 μC (c) 174 μC on each
57. (a) 90 μC (b) 20 μC (c) 5.4×10^{-4} J (parallel); 1.2×10^{-4} J (series)
59. (a) 7.5 μC; 5.6 μJ (b) 3.2×10^{-3} C; 630 V
61. (a) 4.19 J (b) 16.8 J
63. (a) 2.4 μC (b) 2.4 μC (c) 43.2 μJ (d) 150 nF: 19.2 μJ; 120 nF: 24.6 μJ (e) 150 nF: 16 V; 120 nF: 20 V
65. (a) 0.0160 C (b) 533 V for each (c) 4.27 J (d) 2.13 J
67. (a) 2770 V (b) 5540 V (c) 3.53×10^{-3} J
69. (a) 1.18 μF per cm^2 (b) 1.13×10^6 V/m
71. (a) 3.6 mJ; 13.5 mJ (b) increased by 9.9 mJ
73. (a) 6.3×10^{-6} C (b) 6.3×10^{-6} C (c) no effect
75. (a) 0.415 m (b) 2.30×10^{-10} C (c) away from the point charge
77. (a) 5.67×10^7 V/m (b) 0.28 V; outside (c) 7×10^{-15} J (d) 1.0×10^7 V/m; 0.052 V
79. (a) $\dfrac{mv^2}{r} = \dfrac{ke^2}{r^2}$ and $v = \sqrt{\dfrac{ke^2}{mr}}$

(b) $K = \dfrac{1}{2}mv^2 = \dfrac{1}{2}\dfrac{ke^2}{r} = -\dfrac{1}{2}U$

(c) $E = K + U = \dfrac{1}{2}U = -\dfrac{1}{2}\dfrac{ke^2}{r} = -\dfrac{1}{2}\dfrac{k(1.60 \times 10^{-19}\ \text{C})^2}{5.29 \times 10^{-11}\ \text{m}} = -2.17 \times 10^{-18}$ J $= -13.6$ eV
81. (a) 4.43×10^{-11} F (b) 5.31×10^{-9} C (c) 1.50×10^4 V/m (d) 3.19×10^{-7} J (e) (a) 2.21×10^{-11} F; (b) 5.31×10^{-9} C; (c) 1.50×10^4 V/m; (d) 6.37×10^{-7} J
83. (a) 3.54×10^{-11} F (b) 1.06×10^{-8} C (c) 1.59×10^{-6} J
85. (a) 2.3 μF (b) $Q_1 = 9.7 \times 10^{-4}$ C; $Q_2 = 6.4 \times 10^{-4}$ C

Passage Problems

89. B
91. D

Chapter 19

Conceptual Questions

1. Yes. The open circuit voltage V_{open} is the emf $\mathcal{E}$ of the source. When the source is short-circuited, $\mathcal{E} - I_{short}r = 0$, where r is the internal resistance of the source. $I_{short} = \dfrac{\mathcal{E}}{r} \cdot \dfrac{V_{open}}{I_{short}} = \dfrac{\mathcal{E}}{(\mathcal{E}/r)} = r.$

5. (a) False. Consider a circuit containing a single resistor R_1. Adding a second resistor R_2 in parallel with R_1 lowers the net resistance of the circuit, because the second resistor adds an alternative current path. (b) False. A simple example is two resistors in parallel. If one is removed, the net resistance doubles.

7. A capacitor stores energy in the electric field that is created between its plates when the capacitor is charged. A resistor cannot store energy. A resistor dissipates electrical energy by converting it to thermal energy and the thermal energy cannot be converted back to electrical energy in the resistor at some later time.

9. No. The terminal voltage V_{ab} is equal to $\mathcal{E} - Ir$. The terminal voltage can be noticeably less than the emf when a large current passes through the battery.

11. In series, the emf of the combination is twice the emf of a single battery and the bulb is much brighter than if a single battery were used. In parallel the emf applied to the bulb is the emf of a single battery. In parallel the current through each battery is half the current through the bulb. The batteries last longer, but the brightness of the bulb is the same as if only a single battery is used.

15. Energy is charge times potential and it is the energy dissipated in your body that is harmful. When you receive a net charge by scuffing your shoes across the carpet, the net charge you receive is very small. A large current can flow when you discharge, but only for a very short time. The total charge that flows is very small and the energy generated is also very small. A power line is a continuous source of current and the current can flow long enough for a fatal amount of energy to be transferred to your body.

Multiple-Choice Problems

1. C
3. D
5. A
7. C
9. C
11. A
13. C
15. E

Problems

1. (a) 9.38×10^{18} electrons/s (b) 450 C (c) 5.00 s
3. 9.0 μA
5. (a) 7.0×10^{22} (b) 0.035 mm/s
7. 121 m
9. $1.75 \times 10^{-6}\ \Omega \cdot m$
11. (a) $2.00 \times 10^{-4}\ \Omega$ (b) 2.05 cm
13. $9R$
15. (a) $1.06 \times 10^{-5}\ \Omega \cdot m$ (b) $1.05 \times 10^{-3}\ (^\circ C)^{-1}$
17. (a) yes, graph of V versus I is a straight line (b) R is the slope (c) 150 Ω
19. 666°C
21. 5.0 V; yes
23. 27.4 V
25. (a) $I/2$ (b) $4I$ (c) $2I$
27. 3.08 V; 0.067 Ω; 1.8 Ω
29. (a) 2.50 V; 2.14 Ω (b) 1.91 V
31. (a) 0.700 Ω (b) 5.30 Ω

33. (a) 21.8 A (b) 0.688 Ω
35. (a) 1.70 W (b) 24.5 Ω
37. (a) 240 Ω (b) less; 144 Ω
39. 40 W; 0.40 J
41. 0.450 A
43. (a) 26.7 Ω (b) 4.50 A (c) 453 W (d) larger
45. (a) 1000 Ω (b) 100 V (c) 10 W
47. (a) 134 Ω (b) 15 Ω
49. (a) 27.7 Ω (b) 4.33 A (c) 40 Ω: 3.00 A; 90 Ω: 1.33 A
51. (a) 8.80 Ω (b) 3.18 A in each (c) 3.18 A (d) 1.60 Ω: 5.09 V; 2.40 Ω: 7.64 V; 4.80 Ω: 15.3 V (e) 1.60 Ω: 16.2 W; 2.40 Ω: 24.3 W; 4.80 Ω: 48.5 W (f) greatest resistance; $P = I^2 R$ and I is the same for each
53. 3.00 Ω; 1.00 Ω and 3.00 Ω: 12.0 A; 7.00 Ω and 5.00 Ω: 4.00 A
55. Solder a 76.5 kΩ resistor in parallel with the 69.8 kΩ resistor.
57. (a) 0.333 A (b) 0.250 A (c) 0.583 A
59. (a) 20.0 Ω (b) A_2: 4.00 A; A_3: 12.0 A; A_4: 14.0 A; A_5: 8.00 A
61. 0.714 A
63. 4.00×10^{-3} F
65. (a) 0 s, 0 C; 5 s, 2.7×10^{-4} C; 10 s, 4.42×10^{-4} C; 20 s, 6.21×10^{-4} C; 100 s, 7.44×10^{-4} C (b) 0 s, 6.70×10^{-5} A; 5 s, 4.27×10^{-5} A; 10 s, 2.72×10^{-5} A; 20 s; 1.11×10^{-5} A; 100 s, 8.20×10^{-9} A
67. (a) 8.49×10^{-7} F (b) 2.89 s
69. (a) $Q_{10} = 500$ pC; $Q_{20} = Q_{30} = Q_{40} = 461$ pC (b) $V_{10} = 50.0$ V; $V_{20} = 23.1$ V; $V_{30} = 15.4$ V; $V_{40} = 11.5$ V (c) 2.50 A (d) 384 ps
73. (a) 18.7 Ω (b) 7.5 Ω
75. (a) 237°C (b) (i) 162 W (ii) 148 W
77. (a) 0.529 W (b) 9530 J (c) 6.8 Ω
79. (a) 80 C° (b) no; person would vaporize
81. 48.0 W
83. (a) toaster: 15.0 A; frypan: 11.7 A; lamp: 0.625 A (b) yes
85. 243 W
87. (a) 77% (b) 0.0077%
89. (a) 0.22 V (b) point a
91. (a) $5.00 \times 10^5\ \Omega$ (b) 1.20×10^{-5} F

Passage Problems

93. B
95. B

Chapter 20

Conceptual Questions

1. No. There could be magnetic field along the direction of motion. Such a field would produce no force on the electrons.

3. The magnetic force is always perpendicular to the velocity and hence does no work and doesn't change the speed of the particle. But the force perpendicular to the velocity changes the direction of the velocity and alters the path of the particle. Another example is a rock swinging on the end of a string. The tension in the string does no work but is necessary in order to keep the rock moving in an arc of a circle.

5. A stream of charged particles produces a magnetic field, just as current flow in a wire does. The field of the particles adds vectorially to the earth's field and alters the net field.

9. The energy comes from the source that is producing the current in the wires. The magnetic forces do work because the current-carrying charges (electrons) are confined to the wires. Forces perpendicular to the motion of the electrons produce bulk motion of the wires. There is no contradiction, it is just a different context.

11. You could use either a current carrying wire or a bar magnet that does have its poles marked to first determine which ends of the compass needle is a north pole and which is a south pole. Then the north pole of the large bar magnet attracts the south pole of the compass needle.

13. No, this is not plausible. In such a reversal the direction of the earth's angular momentum would reverse. A huge external torque would have to be applied to the earth to change the direction of its angular momentum. Conservation of angular momentum prohibits a spin reversal.

15. No. The force on a moving charge is perpendicular to the magnetic field and hence to the tangent to the field line at a point. An example is a charge moving in a circular path whose plane is perpendicular to the magnetic field. In this case the charge moves perpendicular to the field lines.

Multiple-Choice Problems

1. B, C
3. C
5. D
7. C
9. B
11. C
13. A
15. A

Problems

1. (a) positive (b) 0.0505 N
3. (a) 3.54×10^{-16} N, into page (b) maximum: $F = 4.32 \times 10^{-16}$ N, $\vec{v}$ perpendicular to $\vec{B}$; minimum: $F = 0$, $\vec{v}$ parallel or antiparallel to $\vec{B}$ (c) 3.45×10^{-16} N, out of page
5. (a) 5.60×10^{-3} N, $+z$ direction (b) 1.12 m/s²
7. 6.45×10^{10} m/s², north
9. 1.79×10^{-1} m/s², in the $-z$ direction
11. (a) 1.02 T (b) out of page
13. 1.47×10^6 m/s
15. (a) 1.33×10^7 m/s; 1.33×10^7 m/s (b) 27.7 cm
17. $R/3$

19. (a) 4.51×10^{-3} T (b) 2.46×10^{-6} T
21. 8.38×10^{-4} T
23. (a) 4920 m/s (b) 9.95×10^{-26} kg
25. 2.0 cm
27. 2.5 N
29. (a) 2.1×10^{-4} N, upward
    (b) 2.1×10^{-4} N, west (c) zero (d) no
31. (a) *da*: zero; *ab*: 1.2 N, into page;
    bc: zero; *cd*: 1.2 N, out of page (b) zero
33. 0.724 N, 63.4° below the direction that
    is horizontal and to the right
35. (a) 4.71×10^{-3} N · m (b) 0.025 A · m^2
37. (a) zero, zero
    (b) $F = 0, \tau = 0.0809$ N · m
39. (a) 1.25 N · m (b) the north side of the
    coil will tend to rise
41. 4.0×10^{-5} T; same order as the earth's
    field
43. 25 μA
45. 4.3×10^{-5} T
47. (a) 4.0 μT (b) 16.0 cm (c) 16.0 μT
49. (a) 1.44×10^{-7} T (b) 4.80×10^{-8} T
51. (a) zero (b) 6.67 μT; toward top of page
53. 0.0161 N, attractive
55. 2×10^{-3} N; repulsive
57. (a) 8.8 A
59. 24
61. 2.77 A
63. $\frac{\mu_0 I}{4R}$; into page
65. 41.8 A
67. (a) 2.51×10^{-5} T (b) 5.03×10^{-4} T;
    no
69. 1.1 A
71. (a) 1.57×10^{-7} T; out of page
    (b) 8.00×10^{-7} T; out of page
73. 1.16×10^{-7} T
75. (a) 1.257×10^{-6} T · m/A (b) 1.3504 T
77. 7.93×10^{-10} N, south
79. 817 V
81. 7.81 mm
83. (a) 2.90×10^{-5} T, east (b) yes
85. (a) along line passing through intersec-
    tion of wires and with slope -1.00
    (b) along line passing through intersec-
    tion of wires and with slope $+0.333$
    (c) along line passing through intersec-
    tion of wires and with slope $+1.00$
87. 5.88 m/s^2
89. (b) same for both (c) the ion in the
    large circle
91. $\frac{1}{2} q\omega r^2 B$

Passage Problems

93. B
95. D

Chapter 21

Conceptual Questions

1. Let *A* be the loop that is connected to the
   current source. View the two loops from
   above and assume the current in *A* is
   clockwise. Inside *A* its magnetic field is
   directed away from you and outside *A*
   and at points in the plane of the loops its

field is directed toward you. Therefore,
the flux through the second loop is
toward you and is increasing. Inside the
second loop the field of the induced current is away from you, to oppose the
increase in flux in the opposite direction.
To produce a field in this direction, the
induced current in the second loop is
clockwise. Therefore, when the current
in the first loop is increasing, the currents
in the two loops are in the same direction. On the other hand, when the current
is decreasing the flux through the second
loop is decreasing and is toward you, so
the field of the induced current is toward
you inside the second loop. This means
the induced current is counterclockwise.
Therefore, when the current in the first
loop is decreasing the currents in the two
loops are in opposite directions.

3. The field lines for the magnetic field of
   the wire are sets of concentric circles
   whose planes are perpendicular to the
   wire. The magnetic field of the wire is
   parallel to the plane of the ring and produces no flux through the ring. There is
   no current induced in the ring.

5. To store useful amounts of energy, large
   currents and large values of inductance *L*
   are needed. Large *L* for a solenoid means
   a large number of turns. This requires a
   long length of wire and this means the
   inductor will have resistance that isn't
   small. Large I^2R heating will dissipate
   electrical energy. To avoid this, superconducting solenoids would have to be used.

7. Transformer operation depends on induced emf. For dc, the current is constant
   and the current in the primary doesn't
   cause a changing flux in the secondary.
   There is no induced emf and no current
   and voltage in the secondary.

9. The induced emf doesn't depend on the
   size of the magnetic flux. It instead
   depends on the rate of change of the flux.

13. Let the magnetic field of the magnet be
    directed away from you, perpendicular
    to the plane of the ring. When the ring is

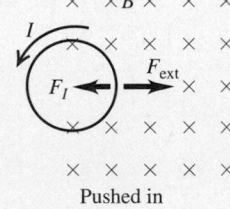
Pushed in

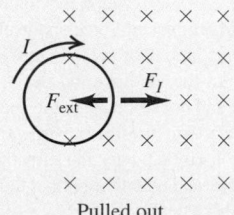
Pulled out

pushed into the field, the induced current
is counterclockwise in the figure and the
force on the ring due to the induced
current is to the left. To move the ring
into the field an external force directed
into the field region must be applied.
When the ring is pulled from the field,
the induced current is clockwise in the
figure and the force on the ring due to the
induced current is to the right. To move the
ring out of the field region an external
force directed away from the field region
must be applied. These results agree with
Lenz's law.

15. The voltage across the resistor is propor-
    tional to the current through it, the volt-
    age across the inductor depends on the
    rate of change of the current, and the
    voltage across the capacitor depends on
    the charge on its plates. (a) In the *R-C*
    circuit, just after the switch is closed the
    charge on the capacitor is zero and
    the full battery voltage appears across
    the resistor. The current has its maxi-
    mum value. In the *R-L* circuit, just after
    the switch is closed the current is still
    zero but is increasing at its maximum
    rate. The full battery voltage is across
    the inductor. (b) In the *R-C* circuit, the
    full battery voltage is across the capaci-
    tor and the current is zero. In the *R-L* cir-
    cuit, the current is no longer changing,
    the full battery voltage is across the
    resistor and the current is a maximum.
    The capacitor has no effect initially but
    causes the current to go to zero as time
    progresses. The inductor initially limits
    the rate at which the current can increase
    but after a long time it has no effect on
    the circuit.

Multiple-Choice Problems

1. A
3. B
5. B
7. C
9. C
11. C
13. D
15. A

Problems

1. (a) 3.05×10^{-3} Wb
   (b) 1.83×10^{-3} Wb (c) 0
3. -7.8×10^{-4} Wb
5. 10.6 T/s
7. 58.4 V
9. 3.0 mA
11. (a) 3.70 rad/s (b) 0
13. (a) counterclockwise (b) clockwise
    (c) induced current is zero
15. counterclockwise; clockwise
17. *A*: counterclockwise; *C*: clockwise
19. clockwise
21. (a) 2.10 V (b) east (c) zero
23. 1.25 mV; higher

25. (a) 3.00 V (b) from b to a (c) 0.800 N
27. 4.00×10^{-5} H
29. (a) 6.82 mH (b) 3.27×10^{-4} Wb
    (c) 2.45 mV
31. (a) 1.96 H (b) 7.11×10^{-3} Wb
33. 0.250 H
35. (a) 1940 (b) 800 A/s
37. 238 turns
39. (a) use step-down transformer
    (b) 6.67 A (c) 36.0 Ω
41. (a) 108 (b) 110 W (c) 0.918 A
43. (a) 6.74 mJ (b) 14 A; no
45. (a) 0.161 T (b) 1.03×10^4 J/m^3
    (c) 0.129 J (d) 4.02×10^{-5} H
47. $I\sqrt{3}$
49. (a) 2.40 A/s (b) 6.00 V (c) 0.475 A
    (d) 0.750 A
51. (a) $V_1 = 0$, $V_2 = 25.0$ V, $A = 0$
    (b) $V_1 = 25.0$ V, $V_2 = 0$ V, $A = 1.67$ A
    (c) none would change
53. (a) 1.39 μs (b) 2.46 μs
55. (a) 0.640 mJ (b) 0.584 A
57. (a) 1860 Ω (b) 0.963 H
61. 222 μF; 9.31 μH

Passage Problems

65. B

Chapter 22

Conceptual Questions

1. (a) The individual voltages achieve their maximum values at different times. (b) Yes, Kirchhoff's loop rule requires that at any instant of time the instantaneous voltages obey this equation.
3. The inductor doesn't consume electrical energy. It stores energy but later releases that energy back to the circuit. A resistor dissipates electrical energy by converting it to unwanted thermal energy
5. At high frequencies the current is changing rapidly and the induced emf across the inductor is very large. This large induced emf opposes the current and current doesn't flow, just as if there were a break (an infinite resistance) in the circuit.
7. When the power factor is small, a large current is needed to supply a given amount of power. This results in large i^2R rates of electrical energy losses in the transmission lines.
9. The voltages across the capacitor and inductor are 180° out of phase so the voltage across the combination is zero at all times. The current is limited only by the resistance in the circuit, and the source voltage amplitude is equal to IR. If $X_C > R$, then the voltage amplitude across the capacitor exceeds the voltage amplitude across the resistor and hence exceeds the source voltage amplitude.
11. An advantage of ac is that transformers allow ac voltage to be stepped up or down. Power can be transmitted at high

voltage and low current to minimize I^2R losses in the transmission lines, and the voltage can be stepped down to safer values for distribution to consumers. A disadvantage of ac is that the power delivered depends on the rms voltage and peak voltages are larger by $\sqrt{2}$.
13. ac voltage is generated by application of Faraday's law, by rotating coils in a magnetic field.
15. There is I^2R heating in the wires of the primary and secondary windings. There is generation of thermal energy through hysteresis in the transformer core. And eddy currents induced in the core leads to I^2R heating.

Multiple-Choice Problems

1. D
3. A
5. A
7. B
9. D
11. C
13. C
15. C

Problems

1. 1.06 A
3. (a) $\dfrac{1}{\sqrt{LC}}$ (b) 7560 rad/s; 37.8 Ω
5. 1.63×10^6 Hz
7. (a) 0.250 mA (b) 25.0 mA (c) 2.50 A
9. (a) 10.6 mA (b) 8.48 V
11. 50.0 V
15. (a) 40.0 W (b) 0.167 A (c) 720 Ω
17. (a) 12.5 W (b) 12.5 W (c) 0
19. (a) 1.0×10^4 rad/s (b) 6.28×10^{-4} s
21. (a) 1.00 (b) 75.0 W (c) 75.0 W
23. (a) 115 Ω (b) 146 Ω (c) 146 Ω
25. (a) 250 rad/s (b) 400 Ω
    (c) $V_C = V_L = 30.0$ V; $V_R = 240$ V
27. (a) 3.24×10^3 rad/s (b) 125 Ω
29. (a) 0.400 A (b) 0.800 A
31. 0.124 H
33. 3.59×10^7 rad/s
35. (a) 20.5 V (b) 21.2 V
37. (a) 7.32 W (b) 7.32 W
39. (a) 102 Ω (b) 0.882 W (c) 270 V
41. (a) 568 Ω (b) 0.866 A (c) 226 V
43. 9.64 mA (b) 241 V (c) 1.16 V
    (d) 1.39×10^{-5} J

Passage Problems

45. C

Chapter 23

Conceptual Questions

3. The momentum change of the light, and hence the momentum transferred to the surface, is greater when the light reflects. For absorption, the final momentum is zero. For reflection, the final momentum is equal in magnitude and opposite in direction to the initial momentum, so the change in momentum is twice as great as for absorption.

5. The hot air has a slightly smaller index of refraction than room temperature air. As light passes into and out of the hot air its rays are bent by the change in refractive index.
7. The refractive index of air is only slightly greater than unity and the distance the light travels through the atmosphere is a very, very small fraction of the distance from the sun to the earth, so the delay introduced by the earth's atmosphere is much, much less than 8 minutes.
9. Atmospheric refraction causes the sun to still be seen after the sun's disk has passed below the horizon. The same effect occurs at sun rise, and allows the sun to be seen before it rises above the horizon. This effect does increase the time between when the sun is seen to rise until it is seen to set.
11. As described in Figure 23.38, light reflected at an angle θ_p is polarized parallel to the reflecting surface. And, as described in Figure 23.46, skylight from overhead is polarized. Either of these sources of polarized light could be used to determine the direction of the axis of a polarizer, by observing what orientation of the polarizer gave maximum transmitted intensity.
13. Any wave exhibits both reflection and refraction. An echo is one example of reflection of sound. When we stand in a boat we hear sounds that are generated underwater because sound waves refract from the water into the air.
15. The incident and reflected rays travel in the same material. The reflection occurs because of the abrupt change in refractive index but in reflection the light stays in the same material. The refracted ray leaves one material and enters the other, and there is a change in wavelength.

Multiple-Choice Problems

1. D
3. A, B, E
5. C
7. A
9. C
11. B
13. B
15. D

Problems

1. 8.33 min
3. (a) 1.28 s (b) 8.16×10^{13} km
5. (a) (i) 6.0×10^4 Hz (ii) 6.0×10^{13} Hz
    (iii) 6.0×10^{16} Hz (b) (i) 4.62×10^{-14} m = 4.62×10^{-5} nm
    (ii) 508 m = 5.08×10^{11} nm
7. (a) 375 V/m; 1.25 μT
    (b) 9.52×10^{14} Hz; 316 nm; 1.05×10^{-15} s; not visible (c) 3.00×10^8 m/s

9. (a) 4.92×10^{14} Hz to 7.50×10^{14} Hz
    (b) 2.70×10^{15} rad/s
    to 4.71×10^{15} rad/s
    (c) 8.98×10^{6} rad/m to 1.57×10^{7} rad/m

11. 3.0×10^{18} Hz; 3.3×10^{-19} s; 6.3×10^{10} rad/m

13. (a) 1.74×10^{5} V/m, parallel to the $-z$-axis (b) $E = -(1.74 \times 10^{5}$ V/m$) \cdot \sin\left[(3.83 \times 10^{15}$ rad/s$)t - (1.28 \times 10^{7}$ rad/m$)x\right]$; $B = (5.80 \times 10^{-4}$ T$)\sin\left[(3.83 \times 10^{15}$ rad/s$)t - (1.28 \times 10^{7}$ rad/m$)x\right]$

15. (a) $-x$ direction (b) 6.59×10^{11}
    (c) $E = (2.48$ V/m$)\sin\left[(4.14 \times 10^{12}$ rad/s$)t + (1.38 \times 10^{4}$ rad/m$)x\right]$

17. (a) 330 W/m^2 (b) 500 V/m; 1.7 μT

19. (a) 80 J (b) 1.0×10^{21} W/m^2
    (c) 8.7×10^{11} V/m; 2.9×10^{3} T

21. (a) 2.00×10^{-8} Pa (b) 4.00×10^{-8} Pa

23. $\sqrt{2}$

25. (a) 8.47×10^{8} Hz (b) 1.80×10^{-10} T
    (c) 3.87×10^{-6} W/m^2

27. (b) 45°

31. 1.38

33. 1.65

35. (a) 389 nm (b) 584 nm

39. (a) 47.5° (b) 66.0°

43. 23.4°

45. 24.4°

47. 401 m^2

49. (a) 48.9° (b) 28.7°

51. 9.43°

55. (a) violet: 1.67; red: 1.62 (b) violet: 240 nm; red: 432 nm (c) 1.03; red (d) 1.1°

57. (a) $0.285I_0$ (b) linearly polarized

59. (a) A: $I_0/2$; B: $0.125I_0$; C : $0.0938I_0$
    (b) zero

61. (a) 63.4° (b) 71.6°

63. $\alpha = \arccos\left(\dfrac{\cos\theta}{\sqrt{2}}\right) = \cos^{-1}\left(\dfrac{\cos\theta}{\sqrt{2}}\right)$

65. (a) 61.8°

67. (a) 7.81×10^{9} Hz (b) 4.50×10^{-9} T
    (c) 2.42×10^{-3} W/m^2
    (d) 1.93×10^{-12} N

69. (a) 0.375 mJ (b) 4.08×10^{-3} Pa
    (c) 604 nm; 3.70×10^{14} Hz
    (d) 3.03×10^{4} V/m; 1.01×10^{-4} T

71. 173 km

73. (a) reflective (b) 6.42 km^2 (c) both gravitational force and radiation pressure due to sun are inversely proportional to the square of the distance from the sun

75. 51.7°

77. $n_1 = 1.10$; $n_2 = 1.14$

79. 1.23

81. 39.2°

83. 42.5°

85. (a) red: 1.73; blue: 2.75 (b) blue

87. (b) angle of incidence for first slab and index of refraction of first and last slabs

89. (a) 48.6° (b) 48.6°

91. (a) $n = 1.11$ (b) (i) 9.78 ns (ii) 4.09 ns; total $= 8.98$ ns

Passage Problems

93. C

Chapter 24

Conceptual Questions

1. The image formation for a spherical mirror is based on the law of reflection. This law is the same whether the light is propagating in air or water. So, no the focal length of the spherical mirror does not change. The image formation for a lens depends on the law of refraction. The bending of light by the lens depends on the change in refractive index as the light enters and leaves the lens. When a lens is immersed in water its focal length changes.

3. Real images can be projected onto a screen but virtual images cannot.

5. The image formed by the first mirror serves as the object for the second mirror. Each successive image is farther behind the mirror surface than the previous one. Each successive image arises from one additional reflection. Some intensity is lost at each reflection, so the images are progressively dimmer.

7. It is desired to have an enlarged, upright image. The image formed by a convex mirror is always smaller than the object. The image formed by a concave mirror is enlarged and upright when the object distance is smaller than the focal length. The radius of curvature of the mirror is related to the focal length f by $f = R/2$. So, the radius of curvature of the mirror is chosen so that person's face is inside the focal point.

9. The spoon has a small radius of curvature and hence a small focal length. When she looks at the concave side of the spoon, her face is outside the focal point and the image is real and inverted. When she looks at the convex side of the spoon the image is virtual and upright.

11. (a) For a concave mirror the image of a distant object is formed at the focal point in front of the mirror. (b) For a convex mirror the image of a distant object is formed at the focal point behind the mirror.

13. If light passes through a lens in the opposite direction, the focal length of the lens is unchanged. This can be seen as follows: $\dfrac{1}{f} = (n-1)\left(\dfrac{1}{R_1} - \dfrac{1}{R_2}\right)$.
    When the lens is turned around, $R_1 \to -R_2$ and $R_2 \to -R_1$, so $\dfrac{1}{f}$ is unchanged.

15. (a) A convex mirror always forms a virtual, erect image, for all object distances, so this mirror must be concave. (b) The erect image is virtual and the upside down image is real.

Multiple-Choice Problems

1. C
3. C
5. D
7. A
9. B
11. E
13. B, D
15. B

Problems

1. 39.2 cm to the right of the mirror; 4.85 cm

3. (a) 3

5. (a) 7.5 cm in front of mirror; 4.00 mm tall
    (b) 10.0 cm in front of mirror; 8.00 mm tall
    (c) 5.00 cm behind mirror; 16.0 mm tall
    (d) 5.00 cm in front of mirror; 0.040 mm tall

7. (a) 0.213 mm (b) 1.75 m in front of mirror

9. (a) 0.0103 m (b) height of image is much less than height of car so car appears to be farther away than its actual distance

11. 18.0 cm from the vertex; 50.0 cm tall, erect, virtual

13. (a) 4.00 (b) 48.0 cm behind mirror; virtual

15. (b) 6.60 cm to right of mirror; 0.240 cm tall; erect; virtual

17. (a) 10.0 cm to left of the mirror; 2.20 mm tall (b) 4.29 cm to right of the mirror; 0.944 mm tall

19. (a) 8.00 cm to right of vertex
    (b) 13.7 cm to right of vertex
    (c) 5.33 cm to left of vertex

21. 14.8 cm to right of vertex; 0.578 mm tall; inverted

23. (a) at center of bowl; $+1.33$ (b) no

25. 2.67 cm

27. 6.53 m

29. object: 4.85 cm from lens; image: 15.75 cm to left of lens; virtual

31. (a) 0.600 m (b) inverted (c) $+0.450$ m; converging

33. (a) -4.80 cm; diverging (b) 2.44 mm; virtual

35. $n = 1.55$

37. (a) $+11.0$ cm; $+11.0$ cm (b) $+17.0$ cm; $+17.0$ cm (c) $+133$ cm; -133 cm
    (d) -18.4 cm; -18.4 cm (e) -11.0 cm; -11.0 cm

39. (a) ± 7.0 mm (b) 8.2 mm on other side of lens; 4.4 mm; real; inverted

41. (a) 107 cm to right of lens; 17.8 mm tall; real; inverted (b) same as in part (a)

43. 71.2 cm to right of lens, $m = -2.97$

45. (a) 200 cm to right of first lens; 4.80 cm tall (b) 150 cm to right of second lens; 7.20 cm tall

47. (a) 53.0 cm to right (b) real (c) 2.50 mm; inverted

49. $s = 18.0$ cm: (a) 63.0 cm to right of lens (b) -3.50 (c) real (d) inverted
    $s = 7.00$ cm: (a) 14.0 cm to left of lens (b) $+2.00$ (c) virtual (d) erect

51. (a) 26.3 cm from lens with height 12.4 mm; image is erect; same side
53. (a) converging (b) 8.0 cm (c) 24.0 cm to right of lens (d) 24.0 cm to right of lens
55. (a) converging (b) 18.0 cm (c) 22.5 cm to left of lens (d) 22.5 cm to left of lens
57. −6.62 cm
59. 7.20 m; 4.43 m
61. (a) image is 58 cm from lens, on same side as object; (b) $m = 1.7$
63. 220 cm
65. light directly through lens:
(b) (i) 51.3 cm to right of lens (ii) real (iii) inverted
light reflecting off mirror:
(b) (i) 51.3 cm to right of lens (ii) real (iii) erect

Passage Problems

69. D

Chapter 25

Conceptual Questions

1. The filled wine glass acts as a thick, converging lens. The light is bent when it passes from air to glass, from glass to wine, from wine to glass and from glass to air. The refracting properties of the glass are changed when it is empty. If the glass is thick, an image can still be formed but with a different image distance. Water has a different refractive index from wine. When the glass is filled with water an image can be formed, but with a different image distance.
3. The eye is similar to a camera since a lens forms an image on a screen, either the retina or the film. For a camera, the lens has fixed focal length and the distance from the lens to the film is changed in order to focus the image on the film. For the eye, the lens to retina distance is fixed and the focal length can be changed.
5. Compared to a glass lens in air, the refraction air → glass is replaced by water → air and the refraction glass → air is replaced by air → water. In each case the light is bent in the opposite direction relative to the normal to the lens surface. A shape that is a diverging lens for glass in air is a converging lens for air surrounded by water. The air pocket must be thinner in the center than at its edges in order to serve as a converging lens.
7. To start a fire you need a converging lens, that forms a real image at the focal point of the lens. A converging lens corrects farsightedness and a diverging lens corrects nearsightedness. Your eyeglasses work if you are farsighted.
9. Yes, the fishbowl acts as a converging lens and can form a bright image at its focus.

11. The angular magnification M is proportional to $1/f$, so the lens with a smaller focal length gives the larger M. For a lens with one flat side, the lensmaker's equation gives $f = \left(\dfrac{1}{n-1}\right)R$, where R is the radius of curvature of the curved side. Small R gives small f, so the lens with one highly curved side gives the greater angular magnification.
13. No. The angular magnification depends on the focal lengths of the lenses and is independent of the diameter of the lenses. The larger diameter lenses will gather more light and will therefore produce brighter images.
15. No. To correct the severe nearsightedness, a diverging lens with power of large magnitude is required. For an object at the normal near point, this lens will produce an image closer to the eye than the near point. The lens that corrects far vision impairs close vision. Bifocals can be used to correct far vision without impairing, or even correcting, near vision, but they essentially have two lenses.

Multiple-Choice Problems

1. C
3. C
5. C
7. D
9. A, D
11. D
13. D
15. D

Problems

1. (a) 75 mm (b) $\dfrac{1}{62.5}$ s
3. (a) $f/22.1$ (b) 660 inches (c) $f/3.1$ to $f/13$
5. 10.2 m
7. (a) 7.63 cm (b) 3.33 cm; inverted; real (c) 4.2×10^3 pixels
9. (a) 2.83 m (b) 5.09 cm
11. (a) 2.76 m behind lens (b) 0.552 m by 0.828 m
13. (a) 15.3 cm (b) 1.41 m by 2.12 m
15. (a) 462 mm (b) 50 cm
17. 2.00 cm to 2.50 cm
19. (a) 150 diopters; 115 diopters (b) 6.67 mm to 8.70 mm (c) 8.70 mm (d) 6.85 mm
21. +49.4 cm; +2.02 diopters
23. 4.17 diopters
25. −3.66 diopters
27. (a) +2.33 diopters (b) −1.67 diopters
29. (a) 2.17 (b) 11.5 cm
31. (a) 6.06 cm (b) 4.13 mm
33. (a) 60.0 mm (b) 100
35. (a) 620× (b) $f_e = 2.50$ cm; $f_o = 0.226$ cm (c) $s = 0.230$ cm
37. (a) 640 (b) 43
39. 27, 4.8

41. (a) eyepiece: 980 mm; objective: 1710 mm; 2.69 m (b) 1.74 (c) 0.87°
43. 0.504 m
45. (a) red: 23.9 cm; violet: 22.1 cm (b) 118 cm; violet: 83.9 cm
47. (a) 140 cm (b) red: 16 cm beyond screen; violet: 18 cm in front of screen
49. 2.60 cm
51. (a) 50 μm (b) 0.70 min
55. (a) 19.8 mm (b) 0.106 cm
57. (a) 1.58 cm (b) 0.183 cm (c) no
59. (a) 30.9 cm (b) 29.2 cm
61. (a) 1.38 (b) 2.29 m

Chapter 26

Conceptual Questions

1. Yes. Use two coherent sources of sound, such as two speakers connected to the same amplifier. The distance from the speakers should be on the order of the wavelength of the sound. Observe the sound intensity at distances from the speakers that are much larger than the separation between the speakers. All that is required is that the waves obey the principle of superposition and exhibit interference; this is the case for both longitudinal and transverse waves. To actually observe interference effects reflections of the sound waves must be suppressed.
3. The water alters the wavelength of the light. $y_m = R\dfrac{m\lambda}{d}$, so the shorter wavelength in water means the bright fringes are closer together.
7. No. The Bragg condition is $2d \sin \theta = m\lambda$. For visible light $\lambda \gg d$ and the first maximum is not observed; for $m = 1$ the equation says $\sin \theta > 1$, which doesn't occur.
9. Each thickness of oil produces constructive interference in the reflected light for a particular wavelength (color). Contours of constant thickness of oil are closed lines so the fringes of each color are closed lines.
11. A large-diameter objective lens has more light gathering power and produces brighter images. The magnification depends on the focal length of the lens, which in turn depends on the radius of curvature of the lens surface and not on its diameter.
13. The number of bright spots is related to the maximum m in $d \sin \theta = m\lambda$. The largest value $\sin \theta$ can have is $\sin 90° = 1$, so the maximum m is less than d/λ. The number of bright spots depends on the wavelength of the light and the separation of the two slits.
15. The diffraction limit on resolution is given by $\sin \theta = 1.22\dfrac{\lambda}{D}$, where θ is the

smallest angular separation of objects whose images can be resolved. When λ is larger, as it is for radio waves as compared to visible light, a larger aperture diameter D is needed to achieve the same resolution.

Multiple-Choice Problems

1. B
3. B
5. A
7. B
9. C
11. A
13. D
15. C

Problems

1. (a) 150 cm, 116 cm, 82 cm, 48 cm, 14 cm
   (b) 133 cm, 99 cm, 65 cm, 31 cm
3. (a) 240 m (b) 120 m
5. (a) destructive (b) 68.0 cm (c) 34.0 cm
7. 590 nm
9. (a) 0.0389 mm (b) 2.60 cm
11. 0.833 mm
13. (a) 39 (b) ±73.4°
15. (a) 514 nm; green (b) 603 nm; orange
17. 80.5 nm
19. (a) 96.4 nm (b) 192 nm, 289 nm, 386 nm
21. 27.1 fringes/cm
23. 95.8 nm
25. ±9.71°, ±19.7°, ±30.4°, ±42.4°, ±57.5°
27. 0.800 mm
29. (a) 10.9 mm (b) 5.4 mm
31. ±16.0°, ±33.4°, ±55.6°
33. 634 nm
35. 13.9°, 28.7°, 46.1°
37. (a) 472 nm (b) 54.7 cm
39. (a) 23.3°, 52.3° (b) 58.8°
41. 20.2°
43. 0.559 nm
45. 429 m
47. 1.88 m
49. 0.084 mm; no
51. 81 cm
53. 1.73
55. (a) 103 nm (b) no; no (c) destructive: 600 nm; constructive: none
57. 30.2 μm
59. (a) 594 nm, 424 nm (b) 495 nm
61. 1.82 mm
63. 198 m
65. 22.3 yr

Passage Problems

67. C

Chapter 27

Conceptual Questions

1. Relativistic effects, such as time dilation and length contraction, would be part of our everyday experience.
3. No. If two events occurring at different points in a particular frame appear to be simultaneous in that frame, they need not be simultaneous in other frames. But if they occur at the same space point in one frame and are simultaneous in that frame, then they must be simultaneous in all frames. If they are both at the same space point and same time point in a frame then there is nothing to distinguish one from the other with respect to time and space.
7. They are massless in the sense of the equation $E^2 = (mc^2)^2 + (pc)^2$. For a photon there is no mc^2 term and energy and momentum are related by $E = pc$.
9. No. In the relativistic expressions for energy and momentum, both these quantities approach infinity as the speed of the particle approaches the speed of light.
11. c
13. A force in the direction of $\vec{v}$ increases the speed of an object. For a force in this direction, $a = F/\gamma^3 m$. As $v \to c$, $\gamma \to \infty$ and the acceleration produced by the force approaches zero. The closer the speed approaches c, the more difficult it is to increase the speed still further. Or, since $K \to \infty$ as $v \to c$, an infinite amount of work would be required to bring an object to the speed of light.
15. There are many experimental verifications of the theory of relativity. For example, experiments with atomic clocks and decay of unstable elementary particles have verified time dilation and length contraction. The Doppler effect for electromagnetic waves is based on relativity and is commonly used in radar guns. And the expression $E = mc^2$ is used to accurately predict energy release in the decay of unstable nuclei and elementary particles.

Multiple-Choice Problems

1. A, B
3. A, D
5. C
7. D
9. D
11. C
13. A, C
15. A

Problems

1. From the passenger's point of view, the signal from earth was sent before the one from the asteroid.
3. 0.301 ms
5. (a) 2.555555556×10^5 h = 29.2 y; (b) 0.8 s
7. (a) 12.0 ms (b) 0.998c
9. 1.12 h; clock in spacecraft
11. 2.86×10^8 m/s
13. (a) 1.05 m (b) 3.80 m; no (c) 2.00 m
15. (a) 9.17 km (b) 0.65 km; 7.1% (c) 1.32×10^{-5} s; 3.90 km; 7.1%
17. 0.837c; away from
19. 0.385c
21. (a) 0.116c (b) 0.229c
23. (a) 4.21×10^7 m/s (b) greater
25. yes, for nonrelativistic momentum
27. (a) 1.14×10^{-15} J; 1.16×10^{-15} J; 1.02
    (b) 3.08×10^{-14} J; 8.26×10^{-14} J; 2.68
29. 3.01×10^{-10} J = 1.88×10^9 eV
31. (a) 4.5×10^{-10} J (b) 1.94×10^{-18} kg·m/s
    (c) 0.968c
33. (a) 1.11×10^3 kg (b) 52.1 cm
35. (a) 2.06 MeV
    (b) 3.30×10^{-13} J = 2.06 MeV
37. 2.10×10^8 m/s
39. 4.53×10^2 m
41. 0.83c
43. (a) $4c/5$ (b) c (c) (i) 145 MeV
    (ii) 625 MeV (d) (i) 117 MeV
    (ii) 469 MeV
45. 1.49×10^{-11} kg
49. (a) 7.20×10^{13} J (b) 1.80×10^{19} W
    (c) 7.35×10^9 kg
51. (a) 8.34 y (b) 3.02 ly
53. m = 1.67×10^{-27} kg, proton

Passage Problems

55. C
57. B

Chapter 28

Conceptual Questions

1. The threshold wavelength is generally shorter than the wavelength for visible light, even more so if the surface is not clean and not free of coatings.
3. Increasing the frequency of the light increases the energy of each photon in the light and thereby increases the energy given to each photoelectron. Increasing the intensity means more photons but doesn't change their energy. Increasing the intensity of the light therefore means more photoelectrons are produced but doesn't change the maximum energy each can have.
5. The frequency of a light beam is totally different from the number of photons per second. The frequency is related to the wavelength and to the energy of each photon. The number of photons per second depends on the intensity of the light. To change the intensity of the light without changing the frequency you could for example pass a monochromatic laser beam though an absorbing medium. The emerging light has lower intensity but the same frequency.
7. Photons of different wavelength have different energy, with longer wavelength photons having less energy. Infrared photons individually have too little energy to produce the chemical change that exposes the film.
11. No. An ultraviolet photon has more energy than a visible-light photon. The phosphor reduces the energy of the photons by taking energy away. There is no way to increase the energy of the photons

with a phosphor. Converting visible light to ultraviolet would violate conservation of energy.

15. If we apply Bohr's angular momentum quantization to a planet, the quantum numbers n are huge and orbits for successive n are infinitesimally close in radius. No discrete nature of the orbit radii of the planets is observable.

Multiple-Choice Problems

1. A, C
3. B
5. C
7. D
9. A
11. D
13. B
15. D

Problems

1. (a) 5.94×10^{14} Hz
   (b) 3.94×10^{-19} J = 2.46 eV
   (c) 9.1 mm/s
3. (a) 0.0120 J = 7.49×10^{16} eV
   (b) 3.04×10^{-19} J = 1.90 eV
   (c) 3.94×10^{16} photons
5. 1.50×10^{19} photons/s
7. $\phi = hf_0$
9. (a) 1.10×10^{15} Hz; 4.55 eV
   (b) 1.44 eV
11. 2.13 eV
13. (a) 2.7 V (b) 2.7 eV (c) 9.7×10^5 m/s
15. (a) 259 nm (b) 0.27 eV
17. (a) 433 nm (b) 6.93×10^{14} Hz
   (c) 2.87 eV
19. (a) Balmer: 656 nm, 365 nm; Lyman: 122 nm, 91.2 nm; Brackett: 4051 nm, 1459 nm (b) Balmer: mostly visible; Lyman: ultraviolet; Brackett: infrared
21. (a) -17.50 eV; -4.38 eV; -1.95 eV; -1.10 eV; -0.71 eV (b) 378 nm
23. (a) 2.16×10^6 m/s; 1.09×10^6 m/s; 7.29×10^5 m/s (b) 0.529×10^{-10} m; 2.12×10^{-10} m; 4.76×10^{-10} m
   (c) -13.6 eV; -3.40 eV; -1.51 eV
25. 97.2 nm; 3.09×10^{15} Hz
27. (a) -218 eV; 16 times hydrogen value
   (b) 218 eV; 16 times hydrogen value
   (c) 7.63 nm
   (d) 1/4 times hydrogen value
29. 4.62 mW
31. (a) ultraviolet (b) 6.42 eV
   (c) 1.75×10^7 photons
33. (a) 8.29 kV (b) 0.0414 nm (c) no
35. 1.13 keV
37. 0.0714 nm, 180°
39. 3.11×10^{-10} m, the same
41. 0.727 m/s
43. (a) 3.32×10^{-10} m (b) 1.33×10^{-9} m
45. photon: 49.6 nm; electron: 0.245 nm
47. (a) 3.50 eV; 354 nm
   (b) 1.65×10^{-10} eV
49. 2.0×10^{-24} kg · m/s; comparable

51. (a) 419 V (b) 0.229 V
53. (a) 1.04 eV (b) 1.20×10^{-6} m
   (c) 2.51×10^{14} Hz (d) FM radio photons individually have too little energy
55. 14.5 eV
57. (a) 12.1 eV (b) 3; 658 nm, 103 nm, 122 nm
59. (a) 0.1849 nm (b) 183 eV
   (c) 8.02×10^6 m/s; no
61. (a) 6.99×10^{-24} kg · m/s (b) 705 eV
63. 32.0 μm
65. 1.66×10^{-17} m; no
67. (a) 12 eV (b) 1.5×10^{-4} V; 7.3×10^3 m/s
   (c) 8.2×10^{-8} V; 4.0 m/s
69. 1.4×10^{-35} kg; 5.8×10^{-8}

Passage Problems

71. B
73. D

Chapter 29

Conceptual Questions

1. The n, l and m_l quantum numbers are all zero for both electrons, so the Pauli exclusion principle requires that their spin quantum number s be different. The spin quantum number determines the z component of the spin angular momentum and for different s this component has opposite sign.

3. Somewhat. The Exclusion Principle refers to quantum numbers not position and the quantum description is in terms of position probabilities and not orbits of precise radii. But the quantum numbers of an electron determine the spatial distribution of the position probability.

5. Mg has two electrons outside a filled shell and Cl is one electron short of a filled shell, so we expect them to combine in a one to two ratio and form $MgCl_2$.

9. Yes, and in fact this is routinely done. The wavelengths of light emitted or absorbed in molecular transitions between vibrational or rotational levels are characteristic of each molecule. The radiation is in the infrared and micro-wave regions.

11. More. Due to screening by the other electrons, the $3s$ electron experiences an effective nuclear charge of about $+e$. There is less screening for the outer electron in Mg^+ and this electron is more tightly bound.

13. In a p-n junction diode a forward bias causes the plentiful holes in the p region to flow into the n region and the plentiful free electrons in the n region to flow into the p region. Both these flows correspond to conventional current in the p to n direction. But a reverse bias pushes the charges to flow in the opposite directions, and the p region has few electrons to flow into the n region and the n has few holes

to flow into the p region. Current easily flows in the forward direction but very little current flows with reverse bias. The conduction by a resistor has no directionality and current flows with equal ease (and difficulty) in either direction.

15. An n-type semiconductor has impurities that add free electrons that can conduct current. A p-type semiconductor has impurities that create holes (missing electrons) that are free to move through the material and thereby conduct electricity.

Multiple-Choice Problems

1. A, C
3. B, D
5. A, D
7. C
9. C

Problems

1. $l = 4$
3. (a) $L = 0$, $L_z = 0$. $L = \sqrt{2}\hbar$, $L_z = 0$, $\pm\hbar$. $L = \sqrt{6}\hbar$, $L_z = 0$, $\pm\hbar$, $\pm 2\hbar$
   (b) $L = 0$: θ not defined. $L = \sqrt{2}\hbar$: 45.0°, 90.0°, 135.0°.
   $L = \sqrt{6}\hbar$: 35.3°, 65.9°, 90.0°, 114.1°, 144.7°
5. (a) zero
   (b) $2\sqrt{3}\hbar = 3.65 \times 10^{-34}$ kg · m²/s
   (c) $3\hbar = 3.16 \times 10^{-34}$ kg · m²/s
   (d) $\hbar/2 = 5.27 \times 10^{-35}$ kg · m²/s
   (e) 1/6
7. (a) 18 (b) $m_l = -4$; 153.4° (c) $m_l = 4$; 26.6°
9. $n = 1, l = 0, m_l = 0, s = \frac{1}{2}$;
   $n = 1, l = 0, m_l = 0, s = -\frac{1}{2}$;
   $n = 2, l = 0, m_l = 0, s = \frac{1}{2}$;
   $n = 2, l = 0, m_l = 0, s = -\frac{1}{2}$;
   $n = 2, l = 1, m_l = 1, s = \frac{1}{2}$;
   $n = 2, l = 1, m_l = 1, s = -\frac{1}{2}$;
   $n = 2, l = 1, m_l = 0, s = \frac{1}{2}$;
   $n = 2, l = 1, m_l = 0, s = -\frac{1}{2}$;
   $n = 2, l = 1, m_l = -1, s = \frac{1}{2}$;
   $n = 2, l = 1, m_l = -1, s = -\frac{1}{2}$;
   $n = 3, l = 0, m_l = 0, s = \frac{1}{2}$;
   $n = 3, l = 0, m_l = 0, s = -\frac{1}{2}$
11. $1s^2 2s^2 2p^6 3s^2 3p^6 3d^{10} 4s^2 4p^5$
13. (a) $1s^2 2s^2 2p^2$
   (b) silicon; $1s^2 2s^2 2p^6 3s^2 3p^2$
15. (a) Ne: $1s^2 2s^2 2p^6$;
   Ar: $1s^2 2s^2 2p^6 3s^2 3p^6$;
   Kr: $1s^2 2s^2 2p^6 3s^2 3p^6 3d^{10} 4s^2 4p^6$
   (b) six (c) chemically inert
17. (a) zero (b) zero; no; would fall into nucleus
19. 277 nm; ultraviolet
21. (a) 5.0 eV (b) -4.2 eV
23. (a) 0.330 nm (b) larger for KBr
25. (a) 227 nm; ultraviolet
27. 1.20×10^6 electrons
31. $\pm 20.4\%$

33. (a) 1.12 eV

35. H has one electron in an unfilled shell and O is two electrons short of a filled shell

37. (a) 0, $\sqrt{2}\,\hbar$, $\sqrt{6}\,\hbar$, $\sqrt{12}\,\hbar$, $\sqrt{20}\,\hbar$,
    (b) 7470 nm, infrared, not visible

39. (a) -0.8500 eV (b) largest: $3\hbar$, smallest: $-3\hbar$ (c) $S = \sqrt{3/4}\,\hbar$ for all electrons
    (d) largest: $\sqrt{6}\,\hbar$, smallest: 0

Chapter 30

Conceptual Questions

1. They have slightly different masses and this mass difference can be used to separate them. Techniques include diffusion and ultracentrifuging of gaseous compounds.

3. The abundance is greater for the intermediate nuclides that have a longer half-life.

5. False. The number of radioactive nuclei is reduced by half, but they have decayed to daughter nuclei that retain most of the mass.

7. Chemical reactions are between atoms and molecules, and the average speeds with which they collide and the frequency of the collisions depend on the temperature. Radioactivity involves forces within the nucleus and the nucleus is unaffected by the temperature of the atoms. Also, typical kinetic energies of atoms near room temperature are similar to energy changes in chemical reactions whereas energies in radioactive decay correspond to kinetic energies at extremely high temperatures.

9. Every proton exerts an electrical force on every other proton and the number of pairs of protons and hence the total electrical energy increases more than linearly as the number of protons increases. More neutrons are needed, to increase the average distance between protons and to add nuclear force without adding electrical repulsion.

11. The kinetic energy of the reaction products is converted to thermal energy when they are stopped in a material.

13. A is unchanged. Z decreases by one, so N increases by one. β^+ decay of $^{40}_{19}$K produces the nucleus $^{40}_{18}$Ar.

15. Many fission fragments have a long half-life so stay radioactive for many years after they are produced. Also, many of them are taken up into the biosphere. For example, radioactive strontium has a half-life of 30 years and is deposited in bones because it is a chemical cousin to calcium.

Multiple-Choice Problems

1. A

3. A, C, D

5. C

7. A, D

9. D

11. B

13. D

15. A

Problems

1. (a) 10, 11 (b) 30, 35 (c) 47, 61

3. (c) 2.3×10^{17} kg/m³

5. (a) 2.23 MeV; 1.11 MeV/nucleon
   (b) 28.3 MeV; 7.07 MeV/nucleon
   (c) binding energy per nucleon is much larger for $^4_2$He

7. 5.575×10^{-13} m

9. (a) lead, $A = 214$, $Z = 82$, $N = 132$
   (b) 6.12 MeV

11. (a) X_1: $A = 228$, $Z = 88$, $N = 140$, radon;
    X_2: $A = 228$, $Z = 89$, $N = 139$, actinium;
    X_3: $A = 4$, $Z = 2$, $N = 2$, α particle;
    X_4: $A = 216$, $Z = 84$, $N = 132$, polonium;
    X_5: $A = 216$, $Z = 84$, $N = 132$, polonium;
    X_6: $A = 0$, $Z = -1$, $N = 0$, electron (β^-)
    (b) $^{212}_{83}$Bi $\xrightarrow{\alpha}$ $^{208}_{81}$Tl $\xrightarrow{\beta^-}$ $^{208}_{82}$Pb or
    $^{212}_{83}$Bi $\xrightarrow{\beta^-}$ $^{212}_{84}$Po $\xrightarrow{\alpha}$ $^{208}_{82}$Pb; end product is $^{208}_{82}$Pb

13. (a) 2.31×10^{-3} s^{-1}
    (b) (i) 3.00×10^3 Bq (ii) 1.50×10^3 Bq (iii) 188 Bq

15. 5.30×10^{16} Bq

17. (a) 46.9 Bq (b) 36.2% (c) $^{131}_{54}$Xe

19. (a) 30.8 min (b) 2.02×10^{15}

21. (a) 5.0×10^4 (b) $\times 10^{-15,000}$

23. (a) 5.0 Gy = 500 rad; 5.0 Sv = 500 rem
    (b) 350 J (c) 5.0 Gy, 500 rad, 50 Sv, 5000 rem, 350 J

25. 500 rad, 2000 rem, 5.0 J/kg

27. 900

29. (a) 1.75 kGy, 1.75 kSv, 175 krem; 262 J
    (b) 1.75 kGy, 2.62 kSv, 262 krem; 262 J

31. absorbed dose: 108 rad, equivalent dose: 2160 rem

33. 2.80 MeV absorbed

35. 4.3×10^{-13} J = 2.7 MeV

37. (a) 1.23×10^5 kg (b) 1.76×10^7 kg

39. 17.6 MeV

41. (a) 4.7×10^4 J/g (b) 8.2×10^{10} J/g
    (c) 4.26×10^{11} J/g (d) 7600 yr

43. (c) 2.42 pm; gamma rays

45. 116 MeV

47. 938.3 MeV, 2.27×10^{23} Hz, 1.32×10^{-15} m (b) 1768 MeV, 42.8×10^{22} Hz, 7.02×10^{-16} m

49. (a) electrical: 200 N; gravitational: 2×10^{-34} N (b) strong: 2×10^4 N; weak: 2×10^{-5} N (c) strong, electrical, weak, gravitational
    (d) $F_{strong} \approx 1 \times 10^{38}F_g$; $F_e \approx 1 \times 10^{36}F_g$; $F_{weak} \approx 1 \times 10^{29}F_g$

51. (a) 3.5 atoms/m³ (b) 294 atoms
    (c) 2×10^{27} atoms

53. 7.19×10^{19} J

55. (a) 0.0556 J/s (b) 11.1 rad/s
    (c) 7.78 rem/s (d) 25.7 s

57. (a) 15.4 decays/s (b) 4.16×10^{-10} Ci

59. (a) 4.14×10^{-9} s^{-1}
    (b) 4.01×10^{20} atoms
    (c) 1.7×10^{12} decays/s (d) 45 Ci

61. 1.287×10^4 y

63. 10.1 MeV absorbed

65. (a) $\pi^0 + \pi^+$ (b) 219 MeV

67. (a) $u\bar{s}$ (b) $\overline{dd}\,\bar{s}$ (c) uss

Passage Problems

69. C

CREDITS

Chapter 0

Page **0–1**: Shutterstock.

Chapter 1

Page **1**: Scott Cunningham/Getty Images. Page **2** TL: Shutterstock. Page **2** TR: Space Telescope Science Institute. Page **2** B: Bruce Ayres/Getty Images. Page **3**: Gerard Lacz/Peter Arnold/Photolibrary. Page **4** T: National Institute of Standards and Technology. Page **4** B: Bureau International des Poids et Mesures. Page **6** TL: Anglo-Australian Observatory/David Malin. Page **6** TR: Janice Carr/CDC. Page **6** ML: NASA. Page **6** MM: Getty Images Inc.—PhotoDisc. Page **6** MR: National Institute of Standards and Technology. Page **6** BL: NASA. Page **7**: NASA. Page **10**: Roger Viollet/Getty Images. Page **14**: Tom Walker/Getty Images.

Chapter 2

Page **29**: iStockphoto. Page **32**: Shutterstock. Page **37**: AP Photo/Dave Parker. Page **39**: AP Photo/Alastair Grant. Page **42**: Getty Images—BC. Page **49**: Dr. Kevin Eggan. Page **51**: James Sugar/Stock Photo/Black Star. Page **52**: Richard Megna/Fundamental Photographs.

Chapter 3

Page **68**: Daisy Gilardini/The Image Bank/Getty Images. Page **69**: AP Photo/Toby Talbot. Page **72**: Cindy Lewis/Carphotos/Alamy. Page **73**: Mark Boulton/Alamy. Page **75**: Richard Megna/Fundamental Photographs. Page **77**: Ken Davies/Masterfile. Page **78** T: Richard Megna/Fundamental Photographs. Page **78** B: Joe Raedle/Getty Images. Page **79**: Pascal Ribollet. Page **82**: Elsa/Getty Images. Page **87**: National Executive Committee for Space-Based PNT. Page **88**: Shutterstock.

Chapter 4

Page **99**: Ulrich Doering/Alamy. Page **102**: Sami Sarkis/Getty Images—Photodisc. Page **103**: Loomis Dean/Time & Life Pictures/Getty Images. Page **107**: CERN/Geneva. Page **110**: NASA. Page **111**: Shutterstock. Page **113**: Agence Zoom/Getty Images. Page **115**: NASA. Page **116** T: Rensselaer County Historical Society. Page **116** (a): John McDonough. Page **116** (b): John McDonough/Getty Images. Page **116** (c): Mark M. Lawrence/Corbis.

Chapter 5

Page **128**: Eric Draper/Getty Images. Page **129**: European Southern Observatory. Page **134**: Richard Megna/Fundamental Photographs. Page **138**: iStock-photo. Page **139**: Dean Conger/Corbis. Page **140**: Dr. Paul Selvin. Page **148** (a): NASA. Page **148** (b): Helen Hansma. Page **148** (c): Shutterstock. Page **148** (d): Anglo-Australian Observatory/David Malin.

Chapter 6

Page **161**: Clive Mason/Getty Images. Page **162**: Cornell University Press. Page **163**: Jed Jacobsohn/Getty Images. Page **170**: David P. Hall/Masterfile. Page **173**: NASA. Page **174**: NASA. Page **176**: Corbis. Page **178**: NASA. Page **179** TL: NASA. Page **179** TR: NASA. Page **179** B: NASA. Page **186**: NASA. Page **187**: NASA.

Chapter 7

Page **188**: Audio-kinetic ball machine by George Rhoads. Page **190**: Karlene V. Schwartz. Page **191**: Martin Harvey/Photo Researchers, Inc. Page **192**: NASA. Page **193**: Johannes Simon/AFP/Getty Images. Page **201**: Biology Media/Photo Researchers, Inc. Page **203**: Michael Yamashita/Corbis. Page **207** T: Adamsmith/Getty Images. Page **207** B: Bechara Kachar/NIH. Page **208**: Stephen Dalton/Photo Researchers, Inc. Page **216**: Thomas, D. D., D. Kast, and V. Korman. 2009. Site-Directed Spectroscopic Probes of Actomyosin Structural Dynamics. Annu Rev Biophys. 38:347–369. Page **218**: Shutterstock. Page **225**: Getty Images Inc.—PhotoDisc.

Chapter 8

Page **231**: Steve Dunwell/Getty Images. Page **232**: Archives of Ontario. Page **235**: Jean Louis Batt/Getty Images. Page **241**: Franck Seguin/Corbis. Page **244**: Mark Garlick/Photo Researchers, Inc. Page **248** T: Getty Images. Page **248** B: Stephen Dalton/Photo Researchers, Inc. Page **253**: John Slater/Corbis. Page **254** T: Richard Megna/Fundamental Photographs. Page **254** B: Shutterstock. Page **255**: Berit Myrekrok/Getty Images/Royalty Free. Page **256**: NASA.

Chapter 9

Page **267**: EFE/Javier Lizón/NewsCom. Page **268**: Mike Powell/Getty Images. Page **269**: Image Source/Getty Images/Royalty Free. Page **274**: SSPL/The Image Works. Page **277**: f8 Imaging/Hulton Archive/Getty Images. Page **282**: JPL/NASA.

Chapter 10

Page **294**: NASA. Page **297**: David Cumming/Corbis. Page **303**: Timothy Ryan. Page **304**: Chris Pelkie and Daniel Ripoll. Page **305**: Mark Dadswell/Getty Images. Page **306**: NASA. Page **308** T: NASA. Page **308** BL: Chris Trotman/Corbis. Page **308** BR: Duomo/Corbis. Page **313**: Corbis. Page **318**: Walter Sanders/Time & Life Pictures/Getty Images.

Chapter 11

Page **333**: Ahmad Masood/Reuters. Page **335**: Eric Cabanis/AFP/Getty Images. Page **336**: Lowell Georgia/Corbis. Page **337**: Ken Robinson. Page **340**: Hans Pfletschinger/Peter Arnold/Photolibrary. Page **342**: Gabe Palmer/Corbis. Page **348**: Wald De Heer. Page **352**: Frank Herholdt/Getty Images. Page **353**: Javier Larrea/AGE Fotostock. Page **354**: Pixtal/Age Fotostock. Page **356**: AP Photo.

Chapter 12

Page **365**: Shutterstock. Page **366**: Shutterstock. Page **367**: Photolibrary. Page **368**: Super Stock/AGE Fotostock. Page **373**: Education Development Center, Inc. Page **375** T: Richard Megna/Fundamental Photographs. Page **375** B: Andrew Davidhazy. Page **384**: AP Photo. Page **385**: Roger Ressmeyer/Corbis. Page **386**: Susumu Nishinaga/Photo Researchers, Inc. Page **390**: Shutterstock. Page **391**: NOAA. Page **395**: Jim Reed/Corbis. Page **396**: Kretztechnik/Photo Researchers, Inc. Page **403**: Dorling Kindersley Media Library.

Chapter 13

Page **407**: Shutterstock. Page **408**: Shutterstock. Page **411**: Sargent-Welch/VWR International. Page **412**: Richard T. Nowitz/Photo Researchers, Inc. Page **413**: Alan Becker/Getty Images. Page **418**: Dallas and John Heaton/Stock Connection. Page **420** (a): Mario Beauregard/Stock Connection. Page **420** (b): Adam Hart-Davis/Photo Researchers, Inc. Page **420** (c): Shaun Lowe/iStockphoto. Page **421**: Adam Hart-Davis/Photo Researchers, Inc.

Page **422**: Creatas/Thinkstock. Page **423** T: Pearson Education. Page **423** B: Alix/Photo Researchers, Inc. Page **424**: Peter/Georgina Bowater/Creative Eye/Mira. Page **425**: Mark Wilson/Getty Images. Page **430**: Harold E. Edgerton/Palm Press, Inc. Page **431** L: Getty Images, Inc.—Photodisc./Royalty Free. Page **431** R: Colin Barker/Getty Images.

Chapter 14

Page **441**: Rick & Nora Bowers/Alamy. Page **444** T: NOAA. Page **444** B: Sargent-Welch/VWR International. Page **447**: Jeff Daly/Fundamental Photographs. Page **450**: Shutterstock. Page **452**: Hugh D. Young. Page **455** T: Ted Kinsman/Photo Researchers, Inc. Page **455** B: Richard Megna/Fundamental Photographs. Page **457**: Image Source/Photolibrary. Page **458**: USDA/ARS. Page **461** T: Nature Picture Library. Page **461** B: Russ Underwood/Lockheed Martin. Page **464**: Getty Images Inc.—Punchstock. Page **465**: Philip Rosenberg/Pacific Stock/Photolibrary. Page **466**: OSF/Photolibrary. Page **467**: David Mauriuz/Corbis.

Chapter 15

Page **477**: Shutterstock. Page **478** T: Richard Megna/Fundamental Photographs. Page **478** B: NASA. Page **479**: Colin Garratt/Corbis. Page **480**: NASA. Page **481**: Shutterstock. Page **482**: Shutterstock. Page **483**: Miguel Angelo Silva/iStockphoto. Page **485**: National Researcher Council of Canada. Page **486**: ThermoMicroscopes. Page **489**: NASA. Page **490**: Calvin Hamilton. Page **494**: USGS. Page **495**: U.S. Navy photo by Mass Communication Specialist Seaman Ryan Steinhour (Released).

Chapter 16

Page **516**: Shutterstock. Page **517**: Shutterstock. Page **518**: iStockphoto. Page **522**: Mike Kepka/Corbis. Page **523**: Lambert/Hulton Archive/Getty Images. Page **527**: Shutterstock. Page **530**: Sinclair Stammers/Photo Researchers, Inc. Page **535**: NASA. Page **536**: Samantha Brown/AFP/Getty Images. Page **537** (b): Amos Zezmer/Omni-Photo Communications, Inc. Page **537** (c): Shutterstock.

Chapter 17

Page **545**: Shutterstock. Page **546**: NOAA. Page **549**: Shutterstock. Page **550**: Richard Megna/Fundamental Photographs. Page **552**: Getty Images, Inc.—Photodisc./Royalty Free. Page **553**: Massachusetts Institute of Technology. Page **558**: Dave Watts/Nature Picture Library. Page **560**: Kenneth Robinson. Page **565**: Gary Retherford/Photo Researchers, Inc. Page **572** T: Peter Terren/Tesladownunder.com. Page **572** B: Tom Bean/Corbis. Page **581**: David Parker/Photo Researchers, Inc.

Chapter 18

Page **582**: Shutterstock. Page **583**: American Association for the Advancement of Science. Page **587**: Gandee Vasan/Getty Images. Page **589**: Shutterstock. Page **595**: Eric Schrader—Pearson Education. Page **596**: Sandia National Laboratories. Page **597**: David M. Phillips/Photo Researchers, Inc. Page **604**: Thinkstock. Page **606** L: Stanford Linear Accelerator/SPL/Photo Researchers, Inc. Page **606** R: C. Mooney/Getty Images.

Chapter 19

Page **618**: Shutterstock. Page **619**: Alfred Pasteka/SPL/Photo Researchers, Inc. Page **622**: Shutterstock. Page **625** (a): Getty Images—Digital Vision. Page **625** (b): The M. C. Escher Company BV. Page **625** R: George Grall/Getty Images. Page **629**: Kenneth Robinson. Page **630**: Alan Senior. Page **631**: Reuters/Corbis. Page **640**: Richard Megna/Fundamental Photographs. Page **643**: Hemera/AGE Fotostock.

Chapter 20

Page **658**: Alex Bartel/Photo Researchers, Inc. Page **659**: Charles D. Winters/Photo Researchers, Inc. Page **661**: Iain MacDonald.

Page **662**: Richard Megna/Fundamental Photographs. Page **664**: Simon Fraser/Photo Researchers, Inc. Page **667** T: Bettmann/Corbis. Page **667** B: Sargent-Welch/VWR International. Page **668** L: NASA. Page **668** R: JPL/NASA. Page **673**: Shutterstock. Page **677**: Eric Schrader, Pearson Education. Page **683**: Haim Bau.

Chapter 21

Page **698**: ITAR-TASS/Nikolai Kuznetsov/Newscom. Page **699**: Peter Anderson-Dorling Kindersley Media Library. Page **705**: American Association for the Advancement of Science. Page **710**: Tethers Unlimited, Inc. Page **712**: Splashpower Ltd. Page **716**: Martyn Goddard/Corbis. Page **717**: Shutterstock. Page **719**: Los Alamos National Laboratory. Page **725**: H. Armstrong Roberts/Corbis.

Chapter 22

Page **735**: Casey Fleser/CC-By-2.0—http://creativecommons.org/licenses/by/2.0/deed.en. Page **738**: The Image Bank/Getty Images. Page **748**: Richard Nuccitelli et al. Int. J. Cancer: 125, 438–445 (2009). Page **749**: Dynamic Graphics/Photis/Alamy.

Chapter 23

Page **761**: Shutterstock. Page **762**: Shutterstock. Page **763**: Oakridge National Laboratory. Page **765** (a): Image courtesy of NRAO/AUI and David Thilker (JHU), Robert Braun (Astron), WSRT. Page **765** (b): NASA. Page **765** (c): NASA. Page **765** (d): Max-Planck-Institut für extraterrestrische Physik (MPE)/NASA. Page **765** B: Andrew Davidhazy. Page **772** T: Jerry Lodriguss/Photo Researchers, Inc. Page **772** B: NASA. Page **773** T: Thinkstock. Page **773** B: Alxander Tsiaras. Page **780**: Susan Schwartzenberg. Page **781** T: Dr. Dina Mandoli. Page **781** B: Barry Blanchard. Page **783**: Randy O' Rourke. Page **785** T: Nadav Shashar. Page **785** B: Kristen Brochmann/Fundamental Photographs. Page **789** L: Sepp Seitz/Woodfin Camp & Associates, Inc. Page **789** R: Peter Aprahamian/Sharples Stress Engineers Ltd./Photo Researchers, Inc. Page **791**: K. Nomachi/Fundamental Photographs. Page **794**: Kristen Brochmann.

Chapter 24

Page **803**: Derrick Alderman/Alamy. Page **804**: Ed Kashi/Corbis. Page **806**: Martin Bough/Fundamental Photographs. Page **811**: Pearson Education. Page **816**: Thinkstock. Page **817**: Tom Fleming/Photo Researchers, Inc. Page **818**: Richard Megna/Fundamental Photographs. Page **821**: Technodiamant USA.

Chapter 25

Page **837**: Stefan Schuetz/Getty Images. Page **838**: Heidi Hofer. Page **839** (a–c): Marshall Henrichs. Page **840**: Fotolia. Page **841** T: Susumu Nishinaga/Photo Researchers, Inc. Page **841** B: Lennart Nilsson/Scanpix Sweden AB. Page **842**: Shutterstock. Page **843**: Photodisc Red/Getty Images. Page **848**: Jan Hinsch/Photo Researchers, Inc. Page **850**: Eastman Kodak. Page **852** (a): Large Binocular Telescope Corporation. Page **852** (b): Southern African Large Telescope.

Chapter 26

Page **862**: Fotolia. Page **863**: Shutterstock. Page **865**: Roger Freedman. Page **866**: Pearson Education. Page **869**: Shutterstock. Page **872** T: Bausch & Lomb Inc. Page **872** M: Bausch & Lomb Inc. Page **872** B: Graeme Harris/Getty Images. Page **873**: Dr. Peter Vukusic. Page **874** T: Pearson Education. Page **874** M: Pearson Education. Page **874** B: Pearson Education/PH College. Page **877** L: Photographed for Pearson Science by Elisabeth Pierson, Radboud University, Nijmegen, Netherlands. Page **877** R: Michael W. Davidson, National High Magnetic Field Laboratory, The Florida State University. Page **880** T: TEK Image/Photo Researchers, Inc. Page **880** B: Photodisc Green/Getty Images. Page **882**: Estate of Bertram Eugene Warren. Page **884** T: SPL/Photo Researchers, Inc. Page **884** B: SPL/Photo Researchers, Inc. Page **886** T: Springer-Verlag GmbH & Co. KG. Page **886** B: Pearson Education.

888 T: South African Astronomical Observatory. Page **888** B: European Southern Observatory. Page **890**: Paul Silverman/ Fundamental Photographs.

Chapter 27

Page **899**: Bettmann/Corbis. Page **901**: A. Inden/Zefa/Corbis. Page **906**: Kimimasa/Corbis. Page **909**: iStockphoto. Page **914**: Henrick Sorensen/Getty Images—BC. Page **918**: David Parker/Photo Researchers, Inc. Page **920**: SuperStock, Inc.

Chapter 28

Page **932**: Peter Arnold, Inc./Photolibrary. Page **934**: Shutterstock. Page **937**: Corning Corporation. Page **941**: BrandX. Page **951**: Hank Morgan/Photo Researchers, Inc. Page **956**: Scott Camazine/Photo Researchers, Inc. Page **961**: V. Brockhaus/zefa/Corbis.

Chapter 29

Page **971**: Sinclair Stammers/Photo Researchers, Inc. Page **980**: Shutterstock. Page **984**: iStockphoto. Page **987**: Roger Freedman. Page **990**: Paul Silverman/Fundamental Photographs. Page **993**: Shutterstock. Page **995**: Getty Images—Photodisc-Royalty Free. Page **996**: Lebrecht Music & Arts Photo Library. Page **997**: Tom Deerinck.

Chapter 30

Page **1003**: C Gascoigne/Robert Harding. Page **1004**: Buddy Mays/ Corbis. Page **1012**: Shutterstock. Page **1017**: Kenneth Garrett/National Geographic Image Collection. Page **1019**: David Stuart/Masterfile. Page **1020**: Stockbroker/Photolibrary. Page **1021** T: GJLP/Photo Researchers, Inc. Page **1021** B: Dept. of Nuclear Medicine, Charing Cross Hospital/Photo Researchers, Inc. Page **1029** T: Ernest Orlando Lawrence Berkeley National Laboratory. Page **1029** B: Carl D. Anderson/Ernest Orlando Lawrence Berkeley National Laboratory.

INDEX